DISCARD

Construction Glossary

Construction Glossary
An Encyclopedic Reference and Manual
Second Edition

J. Stewart Stein, AIA, FCSI

John Wiley & Sons, Inc.

New York ● Chichester ● Brisbane ● Toronto ● Singapore

This text is printed on acid-free paper.

Copyright © 1993 by John Wiley & Sons, Inc.

All rights reserved. Published simultaneously in Canada.

Reproduction or translation of any part of this work beyond
that permitted by Section 107 or 108 of the 1976 United
States Copyright Act without the permission of the copyright
owner is unlawful. Requests for permission or further
information should be addressed to the Permissions Department,
John Wiley & Sons, Inc., 605 Third Avenue, New York, NY
10158-0012.

This publication is designed to provide accurate and
authoritative information in regard to the subject
matter covered. It is sold with the understanding that
the publisher is not engaged in rendering legal, accounting,
or other professional services. If legal advice or other
expert assistance is required, the services of a competent
professional person should be sought. *From a Declaration
of Principles jointly adopted by a Committee of the
American Bar Association and a Committee of Publishers.*

Library of Congress Cataloging in Publication Data:

Stein, J. Stewart.
    Construction glossary : an encyclopedic reference and manual /
  J. Stewart Stein.—2nd ed.
      p. cm.
    Includes index.
    ISBN 0-471-56933-X (Cloth)
      1. Building—Dictionaries.   2. Construction industry—Dictionaries.   I. Title.
    TH9.S78   1993
    690′.03—dc20
                                                                                        92-443

Printed in the United States of America

10 9 8 7 6 5 4 3 2 1

To my grandchildren, Lauren, Steven, Tracey and Leah, and my great-grandchildren, Whitney, Brian, Gregory, Kiley, Casey, Hillary and Samantha, whose existence creates the will and the motivation to pursue further accomplishments.

Preface

Construction industry "jargon" remains unique in its variety of meanings and phraseology. Over the last century, the same words and phrases have been interpreted to mean different things to different people, groups, and officials. New technological advancements and discoveries in science and engineering add to the confusion. The proof of this statement has been confirmed by continuous research and review of several hundred building codes and zoning ordinances, U.S. Government Printing office documents, and publications from technical and professional organizations and trade associations.

Having been licensed in 48 states of the continental United States and the District of Columbia and having practiced in 42 of these states, I have firsthand experience in the variations of established interpretations by local and state authorities. As new materials, products, and equipment are manufactured and new types of systems are designed, new words and meanings will be established and recorded. Over a period of several years they will become commonplace language, interpreted to satisfy the area and the location. The definitions included reflect multiple meanings, historical references, specification language, code interpretations, reference standards, manufacturers' recommended descriptions, and scientific and engineering analysis. Readers may observe some repetitions, but only because the similarity in definitions may have different meanings in its own classified subject matters.

In assembling and compiling this glossary, I endeavored to select available subject information for use by the construction industry and its related and allied constituents and to acquaint readers with the past and present meanings of words, terms, phraseology, and terminology unique to this great industry.

The aim of the Second Edition of the *Construction Glossary* is to increase its original coverage, while staying within the confines of the original format, and to assemble an identifiable product wholly related to the construction industry. The book is organized into 16 divisions, including three author's informational divisions. The division and section format is reproduced by permission of the Construction Specifications Institute from the MASTERFORMAT©1988 Edition.

J. STEWART STEIN, AIA, FCSI

Scottsdale, Arizona,
January, 1992

Exceptions

1. Definitions that include formulas and scientific data have been taken from published reports; the author assumes no responsibility for their authenticity.
2. Definitions that include reference numbers for standards published by technical and scientific organizations, federal agencies and trade associations must be verified, since standards are revised and updated, some become obsolete.
3. Definitions that make statements regarding specific code requirements will vary with each local, state, and federal jurisdiction and should be verified based on the location of the project.
4. Definitions that quote manufacturers' descriptions of products, materials, or equipment should be verified. The product, material, or piece of equipment may no longer be on the market.
5. Definitions that list manufacturers' addresses or their distributor may no longer apply as of the date of this publication and should be verified.
6. Words importing the masculine gender shall include the feminine and neuter.

Introduction

DEFINITION is the explanation of the meaning or meanings of a word. The question is, whose explanation of the word is correct, the originator's, the user of the word's, or the legal interpretation by the courts. DEFINITIONS usually are written by the producers for their materials, products, or equipment or by the originators of systems designed around materials, products, and equipment produced by others.

Before DEFINITIONS are accepted, they may be rewritten by technical and professional organizations, trade associations, and governmental agencies, based on their judgements as they affect their particular view. The use of words based on a DEFINITION is hazardous for the user. The word must represent the true meaning of its understanding by the originator, the user, the interpreter, the legal profession, and the general public. The DEFINITION of a word can fulfill a contract obligation or destroy a relationship between parties to a contract, creating a controversy that may finally lead to legal action.

In many instances, "job slang language" has been used in documents to the detriment of the established meaning of the word, which may represent the local common definition of the meaning. Many DEFINITIONS, unless they are of a highly technical nature containing descriptions that relate to scientific or engineering formulas, suddenly change their character and meaning due to job-related slang language. This creates a situation where the same materials, products, or methods of installation are described differently depending on location. This may obliterate the acceptable standard meaning of the word.

DEFINITION confusion also arises in the writing of building codes and zoning ordinances, where the DEFINITION is definitely related to the area, jurisdiction, and the inhabitants, and reflects a particular view. The word "standard" needs explanation; according to Webster a standard is "that which is established as a model or example by authority, custom, or general consent." The question is whose authority, what custom, and by whose consent? How confusing. Does it not really depend on interpretation in a specific area or location where their standard has been established and not the generally accepted standard? Having had experience with lawyers and the courts during my many years of practice, I beseech readers to be careful in signing contracts that are not explicit and in the selection of words to describe anything, especially in the writing of specifications.

J. STEWART STEIN, AIA, FCSI

Acknowledgments

Special acknowledgment is due to the following organizations for their cooperation and their permission to use selected and adapted information and data from their publications.

American Arbitration Association
140 West 51st Street, New York, New York 10020-1203
Permission to reprint "Construction Industry Arbitration Rules" dated January 1, 1991

American Concrete Institute
P.O. Box 19159, 22400 West Seven Mile Road, Detroit, Michigan 48219
Permission to quote selected representative terms from "Cement and Concrete Terminology" current edition of SP-19, 1990

American Society for Metals
Materials Park, Ohio 44073
Permission to quote selected information and data from the ASM *Metals Reference Book*, Second Edition, 1983.

American Society for Testing and Materials
1916 Race Street, Philadelphia, Pennsylvania 19103-1187
Permission to adapt and reprint selection of construction definitions from the Seventh Edition of the *Compilation of ASTM Definitions*, 1990.

American Society of Heating, Refrigeration and Air Conditioning Engineers, Inc.
1791 Tullie Circle NE, Atlanta, Georgia 30329
Reprinted with permission from the 1989 ASHRAE Fundamentals Volume.

Construction Specifications Institute
601 Madison Street, Alexandria, Virginia 22314
CSI format reproduced courtesy of CSI from the MASTERFORMAT, copyright October 1988. Further reproduction not authorized.

Illuminating Engineering Society of North America
345 East 47th Street, New York, New York 10017-2377
Permission to quote selected lighting terms from the *IES Lighting Handbook*, Current Edition 1984 Reference Volume.

National Fire Protection Association
1 Batterymarch Park, P.O. Box 9101, Quincy, Massachusetts 02269-9101
Definitions selected and adapted from the *National Electrical Code*—1990 Handbook, and *National Fire Codes*, 1991 Edition. Latest editions can be obtained from NFPA.

Contents

Division 1 General Requirements 1

1

General Requirements

BONDS FOR CONSTRUCTION

Bid bond

1. The function of a bid bond is to guarantee the good faith of the bidder, so that if awarded the contract within the time stipulated, he will enter into the contract and furnish the prescribed performance and payment bonds. If he fails to do so without justification, there will be paid to the owner the difference, not to exceed the penal sum of the bond, between his bid and such larger amount for which the owner may in good faith contract with another party to perform the work covered by said bid.

2. Form of bid security executed by the bidder as principal, and by a surety to guarantee that the bidder will enter into the contract within a specified time and furnish any required performance bond and labor and material payment bond.

3. Bond given by the contractor to the owner guaranteeing that if awarded the contract he will accept it and furnish final performance or payment bond as required.

4. A guarantee that the contractor enters into a contract, if it is awarded to him, and furnishes such contract bond as is required by the terms of the contract.

Blanket fidelity bond. Covers losses to an employer by dishonest acts of employees.

Bond. An obligation of the insurance company to protect the insured against financial loss caused by the acts or omissions of another person or persons known as the principal. In noninsurance sense a bond is a debt instrument consisting of a promise to pay depending on the terms of the bond.

Bond form. The bonding company, or surety, in most situations will provide their own form; the contractor in most cases agrees to accept the bonding company's bond form. However, architects have the choice of suggesting to the owner AIA Document A312, which consists of two separate bonds, the Performance Bond and the Payment Bond.

Bonding. Involves a three-way contractual relationship between the principal, surety, and obligee, related to financial stability.

Bond obligation. The obligation under the bond usually applies only to labor, material, equipment, and other services provided directly to the project.

Claimant. An individual or entity having a direct contract with the contractor or with a subcontractor of the contractor to furnish labor, materials, or equipment for use in the performance of the contract.

Completion bond. Form of surety or guaranty agreement that contains the promise of a third party, usually a bonding company, to complete or pay for the cost of completion of a construction contract if the construction contractor defaults.

Consent of surety. Written consent of the surety on a performance bond and/or labor and material payment bond to such contract changes as change orders or reduction in the contractor's retainage or to final payments or waiving notification of contract changes. The term is also used with respect to an extension of time in a bid bond.

Construction contract. The agreement between the owner and the contractor, including all contract documents such as bidding documents, plans, specifications, General Conditions of the Contract, and all related documents listed in the agreement.

Contract bond. Guarantee of the faithful performance of a contract. The bond provides indemnity against failure of a contractor to comply with the terms of his contract.

Contractor. Person or entity identified as such in the agreement and referred to throughout the contract documents as if singular in number. The term "contractor" means the contractor or the contractor's representative.

Contractor default. Contractor's failure, which has neither been remedied nor waived, to perform or otherwise to comply with the terms of the construction contract.

Fidelity bond. Surety bond that reimburses an employer named in the bond for loss sustained by reason of dishonest acts of an employee covered by the bond.

Filing claim. Bonds usually provide that any lawsuit must be brought within a specified period of time after the last of the labor, materials, or services were provided, or the right to action will be nullified.

Guarantee bond. Agreements under seal to assure that a project is completed in accordance with contract documents and/or that all obligations incurred on account thereof have been discharged.

Guaranty bond. Guarantee by a surety company that the contractor will either complete the work or pay all obligations, or both, depending on how the bond is worded. It is sometimes called a "surety bond." If the bond guarantees payment of financial obligation, it is called a "payment bond." If the bond guarantees the completion of the work, it is called a "performance bond" or a "completion bond."

Indemnity agreement. Agreement between the principal and the surety whereby the principal guarantees that the surety will incur no loss by providing the bond.

Labor and material payment bond

1. The purpose and function of the labor and material payment bond are that contractor's bills for such items are guaranteed to be paid by this bond.

2. Bond of the contractor in which a surety guarantees to the owner that the contractor will pay for labor and materials used in the performance of the contract. It is sometimes referred to as a payment bond. Claimants under the bond are defined as those who are employed by or have direct contracts with the contractor or any subcontractors. Usually issued in conjunction with a performance bond.

3. Bond given by the contractor to the owner guaranteeing that he will pay all labor and material bills arising out of the contract.

Lease bond. The type of lease bond of interest to architects is not the usual one guaranteeing that the lessee will pay the stipulated rental, but rather comes about in connection with leases under which the lessee is obligated to erect a building in accordance with the provisions of the lease. It is similar to a completion bond.

Liability bond. A bond that is intended to protect the assured from liability for damages or to protect the persons damaged by injuries occasioned by the assured when such liability should accrue and be imposed by law, as by a court.

License bond. Term used interchangeably with "permit bond" to describe bonds required by state law, municipal ordinance, or regulation as a condition precedent to the granting of a license to engage in a specified business or the granting of a permit to exercise a certain privilege. Such bonds provide payment to the obligee for loss or damage resulting from the operations permitted by law, ordinance, or regulation under which the bond is required and for violations by the licensee of the duties and obligations imposed upon him.

Lien bond. Bond generally given by a contractor indemnifying the owner of realty against loss due to filing of liens against his property for unpaid bills for the labor or material for work done on that property, etc.

Maintenance bond

1. Since the maintenance obligation contained in the contractor's agreement with the owner is usually covered by the performance bond, there are relatively few bonds written covering maintenance only. Exceptions are roofing guarantees, which may be furnished separately by a manufacturer or applicator. Usually maintenance guarantees in excess of 12 months involve an additional premium.

2. Bond given by the contractor to the owner guaranteeing to rectify defects in workmanship or materials for a specified time following completion. A 1-year maintenance bond is normally included in the performance bond without additional charge.

Mechanic's lien. Claim created by state statute for the purpose of securing priority of payment for the price or value of work performed and for the price of materials furnished in erecting or repairing a building or other structure.

No lien bond. Type of performance bond in some states, where in accordance with statute the agreement has been filed as a "no lien" contract, the effect of which is to deny the right to file a lien in connection with that contract.

Notice of claim. Any claimant other than a subcontractor must give notice as prescribed by statute. The purpose of the notice is to place the contractor and the surety on notice, so they may take the proper steps to protect their own interests.

Obligee. Person, firm, or corporation protected by the surety bond. The obligee under a bond is similar to the insured under an insurance policy.

Obligor. One bound by an obligation, commonly called "principal." Under a bond both the "principal" and "surety" are "obligors."

Owner. The person or entity identified as such in the agreement and referred to throughout the contract documents as if singular in number. The term "owner" means the owner or the owner's representative.

Owner default. The owner's failure, which has neither been remedied nor waived, to pay the contractor as required by the construction contract or to perform and complete or comply with the other terms thereof.

Penal amount. Bonds are usually written to limit the amount of the guarantee. The limiting amount, frequently 100% of the contract amount, is called the "penal amount" of the bond.

Penalty. Limit of an insurer's liability under a surety bond.

Performance bond

1. Bond of the contractor in which a surety guarantees to the owner that the work will be performed in accordance with the contract documents. Except where prohibited by statute, the performance bond is usually combined with the labor and material payment bond.

2. Bond given by the contractor to the owner guaranteeing that he will complete the contract as specified.

3. The basic function of the performance bond is to guarantee to the owner that the contractor will perform all the terms and conditions of the contract between him and the owner, and in default thereof to protect the owner against loss up to the bond penalty.

Principal. Party named in a surety bond for whose obligations the surety agrees to be equally liable. In the performance bond and labor and material payment bond the contractor is the principal.

Proper claimant. Obligations of the construction payment bond are available to the subcontractors of the prime contractor and the parties that have a direct contractual relationship with a subcontractor or to suppliers who have a direct relationship with the prime contractor.

Purpose of bonds. Bonds generally provide the owner, or funding agency, subcontractors, and suppliers, with assurance that should the contractor be financially incapable of making contractual payments, fulfilling other contractual obligations, or completing the contract, there will be a financially solvent party, the surety, available to fulfill the obligations.

Release of retained percentage bond. Under some statutes retained percentages may not be released until the contract has been completed and accepted, unless a contractor gives bond to indemnify the owner against loss by reason of the premature release.

Roofing bond. Limited type of guarantee offered by manufacturers of roofing materials. In many instances these bonds are written for a long term of years and although referred to as "bonds," they may be merely the unsupported warranty of the manufacturer without surety.

Self-insurers' workmen's compensation bonds. Bonds given by a self-insured employer to the state guaranteeing payment of statutory benefits to injured employees.

Statutory bond. General term describing a bond given to comply with the terms of a statute. Such a bond must cover whatever liability the statute imposes on the principal and the surety.

Subcontract bonds. Performance and payment bonds, required by general contractors from their subcontractors, guarantee that the subcontractor will faithfully perform the subcontract in accordance with its terms and pay the bills for labor and material incurred in the prosecution of the subcontract work.

Subdivision bonds. Bonds given by the developer to a public body, guaranteeing construction of all necessary improvements and utilities, similar to completion bonds.

Supply bonds—materials and equipment. Bonds given by the manufacturer or supply distributor to the owner guaranteeing that the materials or equipment contracted for will be delivered as specified in the contract.

Surety. Person or organization who, for a consideration, promises in writing to make good the debt or default of another.

Surety bond. Legal instrument under which one party agrees to answer to another party for the debt, default, or failure to perform of a third party.

Surety bond restrictions. Some bonding companies may restrict the bidding list to those contractors who have been approved and are financially qualified.

Surety company. A surety company is one that engages in the business of issuing surety bonds or becoming surety for trustees, guardians, executors, administrators, employees, etc. If any of these listed should be unfaithful to his trust and make away with property, the company makes good the loss.

Suretyship. Agreement on the part of a person or an organization (the surety) to be responsible for the debt, default, or wrongdoing of another (the principal). Surety is similar to guaranty; if the principal does not perform a certain obligation, his surety is bound to.

Termite bond. Form of bond given by manufacturers or applicators of substances intended to prevent the damage caused by termites.

Union wage bonds. Bonds given by the contractor to a union, guaranteeing that the contractor will pay union scale wages to employees and remit to the union any welfare funds withheld.

Work. Term includes the construction and services required by the contract documents, whether completed or partially completed, and includes all other labor, materials, equipment, and services provided or to be provided by the contractor to fulfill the contractual agreement and other obligations. The "work" may constitute the whole or a part of the project.

CONSTRUCTION EQUIPMENT

Absolute Pressure. Sum of the atmospheric pressure and gauge pressure (psig).

A-frame. Open structure tapering from a wide base to a load-bearing top.

A-frame derrick. Derrick in which the boom is hinged from a cross member between the bottom ends of two upright members spread apart at the lower ends and joined at the top; the boom point is secured to the junction of the side members, and the side members are braced or guyed from this junction point.

Aftercooler. Device that cools compressed air after it is fully compressed.

Air compressor. Permanent installation or mobile for many uses at the job site, usually electrically operated with various-size tanks and controls.

Air receiver. Air storage tank on a compressor.

Angle indicator (boom). Accessory that measures the angle of the boom to the horizontal.

Angle of loading. Inclination of a leg or branch of a sling measured from the horizontal or vertical plane provided that an angle of loading of 5° or less from the vertical may be considered a vertical angle of loading.

Angle of repose. Greatest angle above the horizontal plane at which a material will lie without sliding.

Angling dozer. Bulldozer with a blade that can be pivoted on a vertical center pin so as to cast its load to either side.

Angulated roping. System of platform suspension in which the upper wire rope sheaves or suspension points are closer to the plane of the building face than the corresponding attachment points on the platform, thus causing the platform to press against the face of the building during its vertical travel.

Articulating boom platform. Aerial device with two or more hinged boom sections.

Axis of rotation. Vertical axis around which the crane superstructure rotates.

Axle. Shaft or spindle with which or about which a wheel rotates. On truck and wheel-mounted cranes it refers to an automotive type of axle assembly, including housings, gearing, differential, bearings, and mounting appurtenances.

Axle (bogie). Two or more automotive type axles mounted in tandem in a frame so as to divide the load between the axles and permit vertical oscillation of the wheels.

Backhoe-pullshovel. Shovel that digs by pulling a boom-and-stick-mounted bucket toward itself.

Base mounting. Traveling base or carrier on which the rotating superstructure is mounted, such as a car, truck, crawlers, or wheel platform.

Basket derrick. Derrick without a boom, similar to a gin pole, with its base supported by ropes attached to corner posts or other parts of the structure. Base is at a lower elevation than its supports. Location of the base of a basket derrick can be changed by varying the length of the rope supports. Top of the pole is secured with multiple reeved guys to position the top of the pole to the desired location by varying the length of the upper guy lines. Load is raised and lowered by ropes through a sheave or block secured to the top of the pole.

Basket hitch. Sling configuration whereby the sling is passed under the load and has both ends, end attachments, eyes, or handles on the hook or a single master link.

Batter boards. Horizontal boards placed with vertical stakes to mark line and grade of a proposed building.

Blade. Part of an excavator that digs and pushes dirt but does not carry it.

Boatswain's chair. Seat supported by slings attached to a suspended rope, designed to accommodate one worker in a sitting position.

Boom

1. Timber or metal section or strut, pivoted or hinged at the heel (lower end), at a location fixed in height on a frame, mast, or vertical member, and with its point (upper end) supported by chains, ropes, or rods to the upper end of the frame, mast, or vertical member.

2. In a revolving shovel, a beam hinged to the deck front, supported by cables; also, any heavy beam that is hinged at one end and carries a weight-lifting device at the other.

Boom angle. Angle between the longitudinal centerline of the boom and the horizontal. Boom longitudinal centerline is a straight line between the boom foot pin (heel pin) centerline and boom point sheave pin centerline.

Boom harness. Block and sheave arrangement on the boom point to which the topping lift cable is reeved for lowering and raising the boom.

Boom hoist. Hoist drum and rope reeving system used to raise and lower the boom. Rope system may be all live reeving or a combination of live reeving and pendants.

Boom point. Outward end of the top section of the boom.

Boom stop. Device used to limit the angle of the boom at the highest position.

Booster pump. Pump that operates in the discharge line of another pump, either to increase pressure or to restore pressure lost by friction in the line or by lift.

Braided nylon rope. Exceptionally high-strength, abrasion-, rot-, mildew-, or swell-resistant rope.

Braided wire rope. Wire rope formed by plaiting component wire ropes.

Brake. Device used for retarding or stopping motion by friction or power means.

Breast derrick. Derrick without boom. Mast consists of two side members spread farther apart at the base than at the top and tied together at top and bottom by rigid members. Mast is prevented from tipping forward by guys connected to its top. Load is raised and lowered by ropes through a sheave or block secured to the top crosspiece.

Bridle wire rope sling. Sling composed of multiple wire rope legs with the top ends gathered in a fitting that goes over the lifting hook.

Bucket loader. Usually a chain bucket loader, tractor loader, or shovel dozer.

Bulldozer. Tractor equipped with a front pusher blade.

Bull float. Tool used to spread out and smooth the concrete.

Cab. Housing that covers the rotating superstructure machinery and/or operator's station. On truckcrane trucks a separate cab usually covers the driver's station.

Cable. Rope made of steel wire strands.

Cable laid endless sling-mechanical joint. Wire rope sling made endless by joining the ends of a single length of cable laid rope with one or more metallic fittings.

Cable-laid grommet—hand tucked. Endless wire rope sling made from one length of rope wrapped six times around a core formed by hand tucking the ends of the rope inside the six wraps.

Cable-laid rope. Wire rope composed of six wire ropes wrapped around a fiber or wire rope core.

Cable-laid rope sling—mechanical joint. Wire rope sling made from a cable-laid rope with eyes fabricated by pressing or swaging one or more metal sleeves over the rope junction.

Carbide. Tungsten carbide, a very hard and abrasion-resistant compound used in drill bits and other tools.

Cat. Trademark designation for any machine made by the Caterpillar Tractor Company. Widely used to indicate a crawler tractor or mounting of any make.

Chain bucket loader. Mobile loader that uses a series of small buckets on a roller chain to elevate spoil to the dumping point.

Chain hoist. Hoist for use in lifting light loads, triple spur geared with ball bearing on load sheaves, reinforced steel for gears, and twofold cover on brake housing; hooks made to prevent slip-off.

Cherry picker. Small derrick made up of a sheave on an A-frame, a winch and winch line, and a hook. Usually mounted on a truck.

Chicago boom derrick. Boom attached to a structure, an outside upright member of the structure serving as the mast and the boom being stepped in a fixed socket clamped to the upright. Derrick is complete with load, boom, and boom point swing line falls.

Choker. Chain or cable so fastened that it tightens on its load as it is pulled.

Choker hitch. Sling configuration with one end of the sling passing under the load and through an end attachment, handle, or eye on the other end of the sling.

Clamshell. Shovel bucket with two jaws that clamp together by their own weight when it is lifted by the closing line.

Clutch. Friction, electromagnetic, hydraulic, pneumatic, or positive mechanical device for engagement or disengagement of power.

Coating. Elastomer or other suitable material applied to a sling or to a sling component to impart desirable properties.

Counterweight. Weight used to supplement the weight of the machine in providing stability for lifting working loads.

Crane. Mobile machine used for lifting and moving loads without use of a bucket.

Crane and hoist signals. Regulations for crane and hoist signaling in accordance with applicable regulations of the American National Standards Institute.

Crane boom

1. Member hinged to the front of the rotating superstructure with the outer end supported by ropes leading to a gantry or A-frame and used for supporting the hoisting tackle.

2. Long, light metal boom, usually of lattice construction.

Crawler crane. Crane consisting of a rotating superstructure with power plant, operating machinery, and boom mounted on a base, equipped with crawler treads for travel. Function is to hoist and swing loads at various radii.

Derrick

1. Apparatus consisting of a mast or equivalent member held at the head by guys or braces, with or without a boom, for use with a hoisting mechanism and operating ropes.

2. Non-mobile tower equipped with a hoist. Used as a synonym for crane.

Derrick bullwheel. Horizontal ring or wheel, fastened to the foot of a derrick for the purpose of turning the derrick by means of ropes leading from this wheel to a powered drum.

Digging line. On a shovel; the cable that forces the bucket into the soil. Called "crowd" in a dipper shovel, "drag" in a pull shovel, and "dragline" and "closing line" in a clamshell.

Donkey. Winch with two drums that are controlled separately by clutches and brakes.

Dozer. Abbreviation for bulldozer or shovel dozer.

Dozer shovel. Tractor equipped with a front-mounted bucket that can be used for pushing, digging, and truck loading.

Dragline. Revolving shovel that carries a bucket attached only by cables and digs by pulling the bucket toward itself.

Dragshovel. Shovel equipped with a jack boom, live boom, hinged stick, and rigidly attached bucket that digs by pulling toward itself.

Drop hammer. Pile driving hammer that is lifted by a cable and obtains striking power by falling freely.

Drum. Cylindrical members around which ropes are wound for raising and lowering the load or boom.

Dynamic loading. Loads introduced into the machine or its components by forces in motion.

Explosive-actuated fastening tool (high velocity). Tool or machine that when used with a load, propels or discharges a stud, pin, or fastener at velocities in excess of 300 ft/second when measured 6.5 ft from the muzzle end of the barrel, for the purpose of impinging it upon, affixing it to, or penetrating another object or material.

Explosive-actuated fastening tool (low velocity). Tool with a heavy mass hammer supplemented by a load that moves a piston designed to be captive to drive a stud, pin, or fastener into a work surface, always starting the fastener at rest and in contact with the work surface.

Explosive-actuated fastening tool (low-velocity pistol tool). Tool that utilizes a piston designed to be captive to drive a stud, pin, or fastener into a work surface. Designed so that when used with any load that accurately chambers it, it will not cause stud, pin, or fastener to have a mean velocity in excess of 300 ft/second when measured 6.5 ft from the muzzle end of the barrel.

Extensible boom platform. Aerial device (except ladders) with a telescopic or extensible boom. Telescopic derricks with personnel platform attachments are considered to be extensible boom platforms when used with a personnel platform.

Eye. Loop formed at the end of a rope by securing the dead end to the live end at the base of the loop.

Fabric (metal mesh). Flexible portion of a metal mesh sling consisting of a series of transverse coils and cross rods.

Factor of safety. Ratio of the ultimate strength of the material to the allowable or working stress.

Female handle (choker). Handle with a handle eye and a slot of such dimension as to permit passage of a male handle, thereby allowing the use of a metal mesh sling in a choker hitch.

Fiddle block. Block consisting of two sheaves in the same plane held in place by the same cheek plates.

Filament nylon rope. Rope that is highly resistant to flexing, chafing, and abrasion. High stretch and working elasticity provide energy absorption. Resistant to common solvents and alkalies.

Foot. In tampering rollers, one of a number of projections, from a cylindrical drum.

Foot bearing or foot block. Lower support on which the mast rotates.

Gantry
1. Overhead structure that supports machines or operating parts.
2. Upward extension of a shovel revolving frame that holds the boom line sheaves.

Gantry (A-frame). Structural frame, extending above the superstructure, to which the boom support ropes are reeved.

Gauge (gage). Thickness of wire or sheet metal.

Gauge pressure (psig). Pressure measured by a gauge and indicating the pressure exceeding atmospheric.

Gin pole derrick. Derrick without a boom. Guys are so arranged from its top as to permit leaning the mast in any direction. Load is raised and lowered by ropes reeved through sheaves or blocks at the top of the mast.

Gin poles. Single, verticle, guyed pole for supporting lifting tackle.

Grader. Machine with a centrally located blade that can be angled to cast to either side, with independent hoist control on each side.

Grinding wheel flanges. Collars, discs, or plates between which wheels are mounted, referred to as adaptor, sleeve, or back-up type.

Ground Pressure. Weight of a machine or other floor-supported equipment divided by the area in square feet of the ground directly supporting it.

Gudgeon pin. Pin connecting the mast cap to the mast, allowing rotation of the mast.

Guy
1. Line that steadies a high piece of structure by pulling against an off-center load.
2. Line used to steady or secure the mast or other member in the desired position.

Guy derrick. Fixed derrick consisting of a mast capable of being rotated, supported in a vertical position by guys, and a boom, whose bottom end is hinged or pivoted to move in a vertical plane with a reeved rope between the head of the mast and the boom point for raising and lowering the boom, and a reeved rope from the boom point for raising and lowering the load.

Half-track. Heavy truck with high-speed crawler track drive in the rear and driving wheels in front.

Handle. Terminal fitting to which metal mesh fabric is attached.

Handle eye. Opening in a handle of a metal mesh sling shaped to accept a hook, shackle, or other lifting device.

Hardwood dragline mats. Mats made preferably from oak or metal, used as a mat or base for crawler cranes or other such equipment for support on ground surfaces too soft or muddy for proper operation of the equipment.

Hitch. Sling configuration whereby the sling is fastened to an object or load, either directly to it or around it.

Hoist. Mechanism by which a bucket or blade is lifted; the process of lifting it.

Hydraulic dredge. Floating pump that sucks up a mixture of water and soil, and usually discharges it on land through pipes.

Industrial tractor. Class of wheeled type tractor, of more than 20 engine horsepower; other than rubber-tired loaders and dozers used in operations such as landscaping, construction services, loading, digging, grounds keeping, and highway maintenance.

Insulated aerial device. Device designed for work on energized power lines and apparatus.

Jack boom

1. Boom that supports sheaves between the hoist drum and the main boom in a pull shovel or a dredge.
2. Boom whose function is to support sheaves that carry lines to a working boom.

Jib. Extension attached to the boom point to provide added boom length for lifting specified loads. Jib may be in line with the boom or offset to various angles.

Jib boom. Extension piece hinged to the upper end of a crane boom.

Ladder ditcher. Machine that digs ditches by means of buckets in a chain that travels around a boom.

Lattice boom. Long, light shovel boom fabricated of crisscrossed steel or aluminum angles or tubing.

Live boom. Shovel boom that can be lifted and lowered without interrupting the digging cycle.

Load hoist. Hoist drum and rope reeving system used for hoisting and lowering loads.

Load ratings. Crane ratings in pounds established by the manufacturers.

Locomotive crane. Consists of a rotating superstructure with power plant, operating machinery, and boom mounted on a base or car equipped for travel on railroad track. May be self-propelled or propelled by an outside source. Function is to hoist and swing loads at various radii.

Lower load block. Assembly of hook or shackle, swivel, sheaves, pins, and frame suspended by the hoisting ropes.

Manilla rope. Three-strand manilla rope, usually made from selected fibers and cured with water-pliant treatment to keep rope flexible in all weather conditions.

Mast cap (spider). Fitting at the top of the mast to which the guys are connected.

Master coupling link. Alloy-steel-welded coupling link used as an intermediate link to join alloy steel chain to master links.

Master link or gathering ring. Forged or welded steel link used to support all members (legs) of an alloy steel chain sling or wire rope sling.

Maximum rated load. Total of all loads, including the working load, the weight of the scaffold, and such other loads as may be reasonably anticipated.

Mechanical coupling link. Nonwelded, mechanically closed steel link used to attach master links, hooks, etc., to alloy steel chain.

Monkey. Common expression for a light pole derrick.

Mud. In rotary drilling, a mixture of water with fine drill cuttings and added material, which is pumped through the drill string to clean the hole and cool the bit.

Mobile. Manually propelled.

Mobile job crane. Crane for lifting light loads in maintenance shops, on job sites, and by light machinery, powered by ordinary hydraulic jack and equipped with two-speed retractable boom with several settings.

Outrigger. Outward extension of a frame that is supported by a jack or block. Used to increase stability.

Outriggers. Extendable or fixed metal arms that are attached to the mounting base and rest on supports at the outer ends.

Overwinding. Rope or cable wound and attached so that it stretches from the top of a drum to the load.

Pad (shoe or plate). Ground contact part of a crawler-type track.

Pneumatic. Powered or inflated by compressed air.

Pole derrick. Small portable derrick, guyed by ropes and equipped with a hand winch on which a cable is mounted, for raising moderately heavy objects. Sometimes referred to as a monkey.

Power take-off. Place in a transmission or engine to which a shaft can be so attached as to drive an outside mechanism.

Proof load. Load applied in performance of a proof test.

Proof test

1. Nondestructive tension test performed by the sling manufacturer or an equivalent entity to verify construction and workmanship of a sling.
2. Nondestructive tests performed by professional engineers to verify construction and workmanship on structural components of a structure.

Pull shovel. Shovel with a hinge-and-stick mounted bucket that digs while being pulled inward.

Pusher. Tractor that pushes a scraper to help it pick up a load.

Rail clamp. Tonglike metal device, mounted on a locomotive crane car, that can be connected to the track.

Reach. Effective length of an alloy steel chain sling measured from the top bearing surface of the upper terminal component to the bottom bearing surface of the lower terminal component.

Reeving. Rope system in which the rope travels around drums and sheaves.

Rig. General term denoting any machine. More specifically, the front or attachment of a revolving shovel.

Rigger. Mechanic whose function is to brace, guy, and arrange for hoisting materials.

Ripper. Towed machine equipped with teeth, used primarily for loosening hard soil and soft rock.

Rope. In crane operations refers to a wire rope.

Rops. Rollover protection structures used to protect equipment operators when machines overturn.

Runner. Lengthwise horizontal bracing or bearing members, or both.

Sandblaster. Equipment used for cleaning or removing surfaces by a controlled stream of pressure-ejected material consisting of sand or manufactured grit.

Scarifier. Accessory on a grader, roller, or other machine, used chiefly for shallow loosening of road surfaces.

Scraper. Digging, hauling, and grading machine having a cutting edge, carrying bowl, movable front wall (apron), and dumping or ejecting mechanism.

Selvage edge. Finished edge of synthetic webbing designed to prevent unraveling.

Semitrailer. Towed vehicle whose front rests on the towing unit.

Shaking screen

1. Screen moved with a back-and-forth or rotary motion to move material along it and through it.
2. Suspended screen moved with a back-and-forth or rotary motion with a throw of several inches or more.

Shearleg derrick. Derrick without a boom and similar to a breast derrick; mast is wide at the bottom and narrow at the top, hinged at the bottom and with top secured by a multiple reeved guy to permit handling loads at various radii by means of load tackle suspended from the mast top.

Sheepsfoot. Tamping roller with feet expanded at their outer tips.

Sheeting driver. Air hammer attachment fitted on plank ends allowing driving without splintering.

Shore. Supporting member that resists a compressive force imposed by a load.

Shovel dozer. Tractor equipped with a front-mounted bucket that can be used for pushing, digging, and truck loading.

Side loading. Load applied at an angle to the vertical plane of the boom.

Sill. Member connecting the foot block and stiff leg or a member connecting the lower ends of a double-member mast.

Skullcracker. Steel ball swung from a crane boom. Used for demolishing buildings and for breaking boulders.

Slat bucket. Digging bucket of basket construction used in handling sticky, chunky mud.

Sling. Assembly that connects the load to the material-handling equipment.

Smoother bar. Drag that breaks up lumps behind a leveling machine.

Standby crane. Crane not in regular service but used occasionally or intermittently as required.

Standby derrick. Derrick not in regular service but used occasionally or intermittently as required.

Standing (guy) rope. Supporting rope that maintains a constant distance between the points of attachment to the two components connected by the rope.

Static balance. Condition of rest created by inertia (dead weight) sufficient to oppose outside forces.

Static load. Load that is at rest and exerts downward pressure only.

Stiff leg. Rigid member supporting the mast at the head.

Stiff-leg derrick. Derrick similar to a guy derrick except that the mast is supported or held in place by two or more stiff members, called stiff legs, that are capable of resisting either tensile or compressive forces. Sills are generally provided to connect the lower ends of the stiff legs to the foot of the mast.

Strand-laid endless sling—mechanical joint. Wire rope sling made endless from one length of rope with the ends joined by one or more metallic fittings.

Strand-laid grommet—hand tucked. Endless wire rope sling made from one length of strand wrapped six times around a core formed by hand tucking the ends of the strand inside the six wraps.

Strand-laid rope. Wire rope made with strands (usually six or eight) wrapped around a fiber core, wire strand core, or independent wire rope core.

Stress. Force per unit area. When the force is one of compression it is known as "pressure." An internal force that resists an external force.

Stripping shovel. Shovel with an especially long boom and stick that enable it to reach farther and pile higher.

Structural competance. Ability of the machine or equipment and its components to withstand the stresses imposed by applied loads.

Superstructure

1. Rotating upper frame structure of the machine and the operating machinery mounted thereon.
2. Upper portion of a building above grounded level.

Swing

1. Rotation of the superstructure of a machine for movement of loads in a horizontal direction about the axis of rotation.
2. Relation of the mast and/or boom for movement of loads in a horizontal direction about the axis of rotation.

Swing mechanism. Machinery involved in providing rotation of the superstructure.

Tackle. Assembly of ropes and sheaves arranged for hoisting and pulling.

Tag line. Line from a crane boom to a clamshell bucket that holds the bucket from spinning out of position.

Ten wheeler. Truck with tandem rear axles.

Three-part line. Single strand of rope or cable doubled back around two sheaves so that three parts of it pull a load together.

Tilting dozer. Bulldozer whose blade can be pivoted on a horizontal center pin to cut low on either side.

Track. One of a pair of roller chains used to support and propel a machine. Has an upper surface that provides a track to carry the wheels of the machine and a lower surface providing continuous ground contact.

Tractor. Motor vehicle on tracks or wheels used for towing or operating vehicles or equipment.

Trailer (full trailer). Towed carrier that rests on its own wheels, both front and rear.

Trench jack. Screw or hydraulic type jacks used as cross bracing in a trench shoring system.

Truck crane. Crane consisting of a rotating superstructure with power plant, operating machinery, and boom mounted on an automotive truck equipped with a power plant for travel. Function is to hoist and swing loads at various radii.

Two-part line. Single strand of rope or cable doubled back around a sheave so that two parts of it pull a load together.

Universal joint. Connection between two shafts that allows them to turn or swivel at an angle.

Upper load block. Assembly of hook or shackle, swivel, sheaves, pins, and frame suspended from the boom point.

Vapor pressure. Pressure, measured in pounds per square inch (absolute), exerted by a volatile liquid as determined by the "Standard Method of Test for Vapor Pressure of Petroleum Products (Reid Method)" (ASTM D 323).

Vertical hitch. Method of supporting a load by a single vertical part or leg of the sling.

Vertical tower. Aerial device designed to elevate a platform in a substantially vertical axis.

Vibrating screen

1. Screen that is vibrated to move material along it and through it.
2. Screen that is vibrated to separate and move pieces resting on it.

Viscosity. Resistance of a fluid to flow. Liquid with a high-viscosity rating will resist flow more readily than will liquid with a low viscosity. The Society of Automotive Engineers (S.A.E.) has developed a series of viscosity numbers for indicating viscosities of lubricating oils.

Vitrified bonded grinding wheels. Wheels bonded with clay, glass, porcelain, or related ceramic materials.

Walking dragline. Dragline shovel that drags itself along the ground by means of side-mounted shoes.

Welded alloy steel chain slings. Identified by permanently affixed durable identification stating size, grade, rated capacity, and sling manufacturer.

Well point. Pipe, fitted with a driving point and a fine mesh screen, used to remove underground water; also, a complete set of equipment for drying up ground, including well points, connecting pipes, and a pump.

Wheelbase. Distance between centers of front and rear axles. For a multiple-axle assembly the axle center for wheelbase measurement is taken as the midpoint of the assembly.

Wheel-mounted crane. Crane that consists of a rotating superstructure with power plant, operating machinery, and boom mounted on a base or platform equipped with axles and rubber-tired wheels for travel. Base is usually propelled by the engine in the superstructure, but it may be equipped with a separate engine controlled from the superstructure. Function is to hoist and swing loads at various radii.

Whip line (auxiliary hoist). Separate hoist rope system of lighter load capacity and higher speed than provided by the main hoist.

Winch. Mechanical lifting device attached to derricks on which cable is wound up by means of a crank and locked in position by a ratchet.

Winch head. Power-driven spool for handling of loads by means of friction between fiber or wire rope and spool.

Working load

1. Load imposed by personnel, materials, and equipment.
2. External load, in pounds, applied to a crane, including the weight of load-attaching equipment, such as load blocks, shackles, and slings.

CONSTRUCTION INSURANCE

Accident

1. Sudden unexpected event, identifiable as to time and place, which may result in bodily injury or damage to property.
2. Sudden unexpected event identifiable as to time and place of occurrence.

Accidental means. Unintended, unexpected, and unforeseeable cause of an injury.

Accident insurance. Insurance that provides indemnity for loss of time and for medical expenses due to accident.

Accident prevention. Reduction of accident frequency through engineering and inspection or by changing or removing conditions that would be likely to cause accidents.

Accord and satisfaction. Adjustment of a disagreement about what is due and the payment of the agreed amount.

Act of God. An accident or event that is the result of natural causes, without human intervention or agency, that could not have been prevented by reasonable foresight or care.

Actual cash value. Actual value of property as of the time of its loss or damage. With reference to property insurance, it is the basis for determining the amount of the claim. In most states it is the current replacement cost less actual physical depreciation.

Actual damage. Damage that really exists in fact, as distinguished from potential and possible damage.

Adequate limits of liability. Theoretically, the maximum requirement is limited only by the solvency of the insured, and to be properly covered would require limits equal to at least the total amount of his assets. Certainly, this is not often practical, but most firms of standing decide in favor of conservatism, that is, high enough limits to protect against a catastrophe loss, often with a deductible or self-insured layer below to reduce overall premiums.

Adjustment. The process of determining the cause and amount of loss.

Admirality or maritime law. Extension to seamen of the rights and remedies under the Federal Employer's Liability Act of 1908. It affords seamen a right of recovery for bodily injury or death caused by the negligence of the owner or master of a vessel. Coverage is provided by an endorsement to a workmen's compensation policy.

Admitted company. Insurance company licensed to operate in the insurance business in a state under the laws governing insurance writings.

Advisory groups. The function of rating bureaus and similar groups that advise on premium costs and the possible reduction of premiums by revisions in construction material uses, but do not require member organizations to follow recommendations.

Affidavit of claim. Form required when a claim is filed. In general, it should contain the facts on which a claim is filed.

Affiliated companies. Companies or corporations related through common ownership of their securities or interlocking directorates.

Agent. One who is delegated by an insurance company certain authority and powers, specified by contract, and who within this authority can act for and bind the company. The agent acts as a representative of the insurance company.

Aggregate operations liability. The total amount of money that the insurance company will pay under the terms of the liability policy for claims for damages caused by the insured or his agents in the operation of a business.

Aggregate protective liability. The total amount of money that the insurance company will pay under the terms of the liability policy for claims that may arise from acts of independent contractors.

Agreed amount clause. The policyholder agrees to carry a specified amount of insurance in lieu of being subject to the policy's coinsurance clause.

All-risk policy. Policy that protects the insured from loss arising from any cause other than those perils or causes specifically excluded by name. This contrasts with the ordinary or "named peril" policy, which names the peril(s) insured against.

All-states endorsement. An insurance carrier can endorse the policy without charge to provide coverage A automatically in any state except those monopolistic fund states which do not permit private insurance.

Amount subject. The amount or value that may be reasonably lost in any single fire or other casualty.

Apportionment. Determination of the amount to be contributed by each policy when there is more than one in force, in the event of loss.

Appraisal. Survey of property made for determining its insurable value or the amount of loss sustained. A fire insurance policy provides for an appraisal after a loss on demand of either insured or company.

Appreciation. An increased conversion value of property or mediums of exchange due to economic or related causes, which may be temporary or permanent.

Appreciation rate. The published index figure used against the actual estimated cost of a property and computing its new replacement cost as of the date of the estimate.

Approved. The indication of an underwriting company to an insurance company that a particular risk, individual, product, or building meet the requirements for insurance.

Arbitration. If a dispute arises between the insured and the company or their representatives regarding the amount of the loss, the loss can be referred to arbitration.

Arbitration clause. Clause in a policy providing that if the policyholder and the company cannot agree on a settlement amount on a claim, both parties select a neutral umpire who has the authority to bind them to the settlement.

Arson. The willful and malicious burning of property.

Assigned risk. Any risk that the underwriters refuse to insure, but are required by state law to insure. The risk insurance is then assigned to a selected pool of insurance companies to divide the coverage responsibilities.

Assignment. Transfer of interest in a policy to another party, generally following the sale of property covered by the policy. For such assignment to be valid, it must be assented to by the company.

Assumed liability. Liability that arises from agreement between people as opposed to liability that arises from common or statutory law.

Automatic cover. In policies covering certain kinds of property clause stating that coverage will be provided for a limited period of time on property newly acquired by the insured.

Automatic reinstatement. In a policy provision that after a loss has been paid or the property restored, the amount of insurance will automatically return to its original amount, the amount of the loss being automatically reinstated. When a policy does not contain this provision, the original amount is usually reduced by the amount of loss paid.

Automobile foreign coverage. The use of an automobile outside the United States, its territories, possessions, or Canada requires special foreign coverage.

Average risk. Risk that is in accordance with the conditions called for in the established basic rate.

Aviation insurance. Insurance covering aircraft or their contents or the liability of the owner or operator; accident insurance on passengers; in general, insurance pertaining to aircraft.

Awareness protection. At any time during the policy period, the insured may report a circumstance to the company that may result in a claim, even though no claim has, in fact, been made. Should a claim subsequently be made, coverage would be afforded by the policy in force when the circumstance was reported. This feature provides an extra measure of protection, since many policies define the term "claim" restrictively and will not cover the insured for those matters falling within the restrictions.

Basic premium. Percentage of the standard premium retained by the company to cover acquisition costs, general administration, safety engineering, audit expenses, company profit, loss limitation charges, etc.

Basic rate. The manual or experience rate, from which discounts or added charges are made to compensate for the individual circumstance of the risk.

Binder. A written binder is necessary as a preliminary agreement to provide immediate insurance coverage until a policy can be written. It contains an effective date and definite time limit, and designates the company in which the risk is bound, as well as amount, perils insured against, and type of insurance.

Blanket crime policy. Policy that insures against employee dishonesty, losses of money and securities inside and outside the premises, money orders, counterfeit currency, and depositor's forgery. It covers money, securities, and other properties on a single-limit basis applying to all coverages, none of which may be eliminated.

Blanket policy. Insurance policy with broad coverage frequently used in burglary and fire insurance.

Blanket rate. Insurance rate applied to more than one property or subject of insurance.

Bodily injury. Injury pertaining to the body; generally, a physical injury.

Bodily injury liability insurance. Protection against loss arising out of the liability imposed upon the insured by law for damages because of bodily injury, sickness, or disease, including death resulting therefrom, sustained by any person or persons other than employees.

Broad form hold harmless. To indemnify and save harmless the owner with respect to the liability imposed upon said owner by law, but only to the extent that such liability arises out of the operations of the named insured under its contract with the owner.

Broad form indemnification—sole negligence of indemnitee. Broad form indemnifies the other party even where he is solely responsible for a loss.

Broad form property damage. Endorsement to the general liability policy which modifies the exclusion for property in the care, custody, and control of the insured and, in effect, provides broader coverage.

Broad insured coverage. Coverage provided for the present principals and employees of the named insured, but includes former principals and employees, who are also insured with respect to professional services performed prior to retirement or the termination of employment.

Broker. Licensed person who acts as a representative of the insured in the procurement and placement of insurance contracts.

Builders' risk and installation floater policies. Perils covered by builders' risk and installation floater policies vary. Policies may insure against fire, lightning, and the perils of extended coverage, such as windstorm, hail, explosion, riot, civil commotion, aircraft, vehicles, smoke, vandalism, and malicious mischief.

Builders' risk insurance

1. Specialized form of property insurance to cover work in the course of construction.

2. Form of insurance policy, to cover a building in the course of construction against the perils of fire and lightning.

Building rate. Fire insurance term that refers to the rates on buildings rather than on the contents of the building.

Business interruption insurance. Protection against loss of earnings and profits of a business during the time required to rebuild or repair property damaged or destroyed by fire or another insured peril.

Capacity. Maximum amount of insurance that a company will write on a single risk.

Care, custody and control. Standard exclusion in liability insurance policies. Under this exclusion, the liability insurance does not apply to damage to property in the care or custody of the insured, or to damage to property over which the insured is for any purpose exercising physical control.

Casualty insurance. Insurance primarily concerned with losses caused by injuries to persons and legal liability imposed upon the insured for such injury or for damage to property of others.

Catastrophe. Sudden and severe calamity or disaster. An event, insurancewise, that causes a loss of extraordinarily large amounts.

Certificate of insurance

1. Statement issued by an insurance company evidencing that an insurance policy has been issued. It states the policy number, coverage amounts, limits, effective and expiration dates, etc.

2. Certificate issued by an authorized insurance company representative stating the types, amounts, and effective dates of insurance in force for a designated insured.

3. Evidence of adequate protection.

Civil authority clause. Provision in a fire insurance policy that agrees to pay the policyholder for loss suffered should his property be destroyed by civil authority in the effort to prevent spread of fire.

Civil commotion. Disturbance involving large numbers of people which may result in damage to public and private property; uprising of people creating a prolonged disturbance.

Claim. Demand for payment under an insurance policy for loss that may come within the terms of that policy; demand against an insured for damages covered by a policy held by the insured. In the latter, the claim is referred to the insurer acting on behalf of the insured in accordance with the terms of the policy. After the amount of the claim has been determined, it becomes a loss.

Class rate. The premium rate applicable to a specified class of risk.

Clause. Section of a policy or endorsement attached to the policy dealing with a particular subject in the contract, for example, the coinsurance clause.

Coinsurance. Policy provision that requires the insured to carry insurance equal to a named percentage of the value of the property covered, or share in the loss as a coinsurer in event of a claim.

Coinsurer. One who shares a loss sustained under an insurance policy or policies as a result of failure to comply with the coinsurance provision. Also, when two or more companies share a risk they are coinsurers of it.

Collapse insurance. Insurance that covers building and contents loss or damage caused by collapse.

Collision insurance. Insurance against loss or damage caused by collision or upset of the insured vehicle.

Combination policy. Insurance policy made up of the contracts of two or more companies, where at least two kinds of insurance, each provided by a different insurer, are involved.

Commitments. Risk acceptable by an insurance company.

Commutation. Clause in contract which provides for estimation, payment, and complete discharge of all future obligations for reinsurance loss or losses incurred regardless of the continuing nature of certain losses.

Completed operations and products liability. Optional under the comprehensive general liability policy form, this division covers the insured's legal liability for bodily injury or property damage arising out of several causes covered by the policy.

Completed operations insurance

1. Liability insurance covers injuries to persons or damage to property occurring after an operation is completed, but attributed to that operation. An operation is completed when all operations under the contract have been completed or abandoned, when all operations at one job site are completed, or when the portion of the work out of which the injury or damage arises has been put to its intended use by the person for whom that portion of the work was done. Completed operations insurance does not apply to damage to the completed work itself.

2. Insurance that covers accidents that occur after the operations have been either completed and turned over to the owner or abandoned.

Completed value property insurance. Policy written at the start of a project in a predetermined amount, usually derived from the contract sum, less the cost of specified exclusions and adjusted to the final insurable cost on completion of the work.

Composite rating. Instead of paying general liability premiums geared to payroll, number of elevators, amount of sublet work, gross receipts, and contract cost, some contractors (and others) are billed a composite rate for all five divisions of coverage combined.

Comprehensive general liability policy

1. Policy that insures the contractor against his legal liability for bodily injury and property damage to third persons caused by accident and arising out of many situations and conditions listed in the policy.

2. Liability policy that automatically includes all forms of general (as differentiated from specific forms such as aircraft, marine, or auto) liability, the source of which may be either tort or contract. The policy may exclude products or completed operations, property damage, and most forms of contractural liabilities.

Comprehensive material damage insurance. Insurance against loss or damage resulting from numerous miscellaneous causes such as fire, theft, windstorm, flood, or vandalism, but normally not including loss by collision or upset.

Comprehensive dishonesty, disappearance, and destruction policy. Policy providing on an optional basis coverage against employee dishonesty, loss of money and securities inside and outside the premises, money orders and counterfeit paper currency, forgery, burglary of merchandise, and theft of merchandise and equipment.

Comprehensive general liability insurance. Broad form insurance covers claims for bodily injury and property damage and combines, under one policy, coverage for all liability exposures (except those specifically excluded) on a blanket basis. It automatically covers new and unknown hazards that may develop.

Concealment. Application for insurance, and failure to disclose known material facts with intent to defraud.

Condition. Provision in the policy setting forth the obligation of both insured and insurer in dealing with each other, the breach of which may suspend or void the policy coverage; for example, the requirement for prompt notice by the insured of an accident or loss or prior notice by insurer of cancellation.

Conflagration. A highly destructive fire. Fire that extends beyond a single risk and over a wide area.

Consequential loss. Loss not directly caused by damage to property but that may arise as a result of such damage, for example, spoilage of refrigerated food due to fire damage to refrigeration equipment.

Construction management. Coverage is available by endorsement to the basic policy form.

Constructive total loss. Situation where the cost of repair of damaged property exceeds its actual cash value; it is thus considered a total loss even though it may be repairable.

Contents rate. Fire insurance term that refers to the insurance rate on the contents of the building rather than the building itself.

Contingent liability. Liability imposed upon an individual, corporation, or partnership because of accidents caused by those, other than employees, for whose acts or omissions the first party may be legally responsible.

Contractor's equipment floater. Floater that insures against loss or damage to the equipment and tools of a contractor that are usually used away from the contractor's premises.

Contractor's liability insurance. Insurance purchased and maintained by the contractor to protect him from specified claims that may arise out of or result from his operations, whether such operations be by himself or by anyone directly employed by him.

Contractual liability

1. Liability for bodily injury or property damage assumed under written contract with another party. This is someone else's legal liability which the insured contractually agrees to pay for.
2. Liability of another assumed by a party under a contract or agreement, for example, indemnification or hold harmless clauses.

Contribution. Insurance company's payment of, or obligation to pay, all or part of a loss.

Cost of reproduction new. The normal cost of exact duplication of a property with the same or closely similar materials, as of a certain date or period.

Coverage. "Coverage" is used synonymously with "insurance" or "protection."

Debris removal. Policy that may include a clause or that is included in the policy form under which the insurer assumes liability for the cost of removal of debris resulting from damage to property covered by the insured.

Declaration. Statement made by a policyholder to a company or its agent upon which the company may reply in undertaking the insurance.

Deductible clause. Clause that specifies an amount or percentage to be deducted from any loss. The company is made liable only for the excess over a stated amount.

Demolition. The standard fire policy excludes liability for loss occasioned by the demolition of an undamaged part of a building because of the enforcement of an ordinance or law regulating construction or repair. Insurance against such loss may be provided by a demolition endorsement or by a "contingent liability from operations of building laws" endorsement.

Depreciation. Loss in value. Also the difference between the values as of two different dates.

Design-build coverage. Coverage is available where one has been acting in the capacity as contractor and professional, by endorsement under the basic policy form.

Disability benefit law. Statute that imposes upon an employer the legal liability to provide weekly benefits to employees who are disabled due to accidents or sickness outside of their employment.

Discovery form. Form that covers claims made during the policy period even though the circumstances giving rise to the claim occurred prior to the policy period, for example, completed operations coverage.

Drive-other-car insurance. Insurance that covers an insured for liability arising out of his use of an automobile not owned by or regularly furnished to the insured, usually subject to the policy condition that the automobile coverage, if any, of the owner of the car is primary coverage and the operator's insurance is excess coverage.

Effective date. The date on which an insurance policy goes into effect and from which time protection is provided.

Elevator liability. Legal liability for bodily injury or property damage caused by an occurrence arising from ownership, maintenance, and use of elevators owned, controlled, or operated by the insured.

Employee benefit liability. The employer can be held legally responsible for incomplete or questionable advice given to an employee regarding its pension plan.

Employee injuries. Injury to employees of the insured and obligations under workmen's compensation, unemployment compensation, or disability benefit laws are excluded to avoid conflict with such policies.

Employers' liability insurance

1. Protection for the employer against common lawsuits by employees for damages that may arise out of injuries or diseases in the course of their work. The limit of liability is per accident, not per employee.
2. Protection for the employer against claims by employees or employees' dependents for damages that arise out of injuries or disease sustained in the course of their work and that are based on common law negligence, rather than on liability under workmen's compensation acts.

Endorsement. Amendment added to and made part of the insurance contract.

Endorsements defining the status of executive officers or partners. Many states require an endorsement to clarify or define the scope of coverage provided under the applicable workmen's compensation law with respect to injury to corporate officers or partners.

Equipment floater. Inland marine insurance covering the equipment used by the contractor, including mobile equipment, excepting that intended for vehicular use on streets or highways.

Errors and omissions insurance. Insurance that indemnifies the insured for any loss sustained because of an error or oversight on his part.

Excess insurance. Policy or bond that covers the insured against certain hazards; it applies only to loss or damage in excess of a stated amount.

Exclusion C. Exclusion of property damage arising out of collapse of or structural injury to any building or structure due to grading of land, excavating, borrowing, filling, backfilling, tunneling, pile driving, or coffer-dam or caisson work or to the moving, shoring, underpinning, raising, or demolition of any building or structure, or the removal or rebuilding of any structural support thereof.

Exclusion U. Exclusion of property damage to wires, conduits, pipes, mains, sewers, tanks, tunnels, and any similar property, and any apparatus in connection therewith, beneath the surface of the ground or water, caused by and occurring during the use of mechanical equipment for the purpose of grading land, paving, excavating, drilling, borrowing, filling, backfilling or pile driving.

Exclusion X. Exclusion of property damage arising out of blasting or explosion other than the explosion of air or steam vessels, piping under pressure, prime movers, machinery, or power-transmitting equipment.

Expense of withdrawal or recall of insured's products or work. The exclusion simply clarifies the original intent not to cover expense of withdrawal, recall, or replacement of the named insured's products or work completed because of some known or suspected defect or deficiency.

Experience rating. When the premium level at normal manual rates reaches a practical level, the insured becomes eligible for experience rating; that is, a credit or debit is applied to manual rates, depending on the ratio of premiums to losses over a given number of years. Rating plans vary a good deal with the kind of coverage involved. In principle, they are similar; however, all are closely scrutinized by state insurance departments, interstate rating authorities, and independent rating organizations.

Expiration. Date on which a policy ceases to cover unless previously canceled.

Explosion, collapse, or underground damage. The basic policy excludes property damage arising out of explosion, collapse, or underground damage, but the exclusions (any one, two, or all three of them) apply only to certain classifications of work identified in the policy by the letters X, C, and/or U after the four digit classification code number.

Exposure. Estimate of the probability of loss from some hazard, contingency, or circumstance, such as proximity of insured property to adjoining property or lack of proximity to fire hydrants; also, the estimate of an insurer's liability under a policy from any one loss or accident.

Extended coverage endorsement. Endorsement extending the fire insurance policy to cover additional hazards, such as windstorm, hail, explosion, riot, riot attending a strike, civil commotion, aircraft, vehicles, and smoke.

Extra expense insurance. The policyholder is reimbursed for the additional money (beyond his property loss) he may be forced to spend to continue his normal operations because he has had a fire loss or other insured loss.

Factory Insurance Association. Association of stock insurance companies providing property damage and business interruption insurance for highly protected risks in the United States, Puerto Rico, and, to some extent, in Canada.

Factory Mutual Association. Association made up of mutual insurance companies.

Failure to perform as intended. Exclusion designed to spell out that the policy is no guarantee that the products or work performed will function properly. This, of course, is a business risk. If the failure to perform as intended is caused by a mistake or deficiency in design, formula, plan, specification, advertising material, or printed instructions, there is no coverage. If, on the other hand, there is an active malfunctioning that causes bodily injury or damage to other property, there is coverage.

Fallen building clause. Provision included in some fire insurance policies stipulating that if any material part of the building which is insured falls or collapses from causes other than fire or explosion, the fire insurance becomes void.

Faulty design, maps, plans, or specifications. Policies will usually contain an exclusion of liability arising out of faulty design, maps, plans, or specifications. Rather than

being "accidents," these concern professional competence and experience of the architect or engineer, and exposure better insured under professional liability policies available for that purpose.

Fire. Combustion sufficient to produce a spark, flame, or glow, but not an explosion. The result must be "hostile" as opposed to "friendly."

Fire legal liability. Liability imposed by law for loss of or damage to property of others in the care, custody, or control of the insured and caused by fire.

Following-form excess liability insurance. Excess insurance that provides coverage subject to the same terms and conditions as the underlying primary policy.

Foundation exclusion. Clause in a policy stating it does not insure foundations below ground level. As a result, their value is not used to determine the proper amount of insurance under a coinsurance clause.

Friendly fire. Fire confined to an intended place, for example, a furnace or fireplace.

Hazard. Condition that may create or increase the probability of a loss from a given peril.

Hazard insurance. Insurance that covers conditions that create the possibility of a loss, for example, ownership or use of premises, elevators, conduct of operations. The term is also used to designate the divisions of a liability insurance policy.

Health insurance. Insurance that provides indemnity for loss of time and for medical expenses due to sickness.

Hired automobiles. Insurance covers the name insured and, in this case, not the lessor, for bodily injury or property damage caused by an occurrence and arising from use of a hired vehicle.

Hired car. Automobile whose exclusive use and control has been temporarily given to the insured for a consideration.

Hold harmless or contractual liability. Liability assumed by a party under a written contract. The degree of liability assumed is generally described as limited, intermediate, or broad form.

Hostile fires. Fires that occur where they are not intended to be.

Incurred losses. Amounts paid and held in reserve by the insurance company, sometimes limited to maximum amounts chargeable, because of any one claim or accident.

Indemnification. Contractual obligation by which one person or organization agrees to secure another against loss or damage from specified liabilities.

Indemnity insurance. Insurance that provides indemnity against loss, in contrast to contracts that provide for indemnity against liability.

Independent contractor's insurance. Insurance that covers claims caused by occurrence and based on a contractor's contingent liability, usually resulting from the operation of his subcontractors.

Inherent explosion. Explosion resulting from the normal processes of a risk on the premises in which the explosion occurred, and not brought on by external causes—for example, a dust explosion in a grain mill.

Inland marine insurance

1. Insurance applicable to a contractor's equipment, other than vehicles designed for use on public highways, is called an equipment floater, because it applies to things of a mobile or "floating" nature. Almost anything movable can be insured, whether a large

crane, power shovel, caterpillar tractor, lift truck, or small tool. Coverage can be made to apply automatically to new or replacement equipment.

2. Insurance developed by marine underwriters to cover goods while in transit by other than ocean vessels. This includes any goods in transit, except transocean, the essential condition being that the insured property be movable. Bridges and tunnels are also considered an inland marine because they act as instruments of transportation.

Inspection bureau. Organization that reviews drawings for proposed buildings, investigates location and conditions, establishes risks, and sets rates for risks.

Installation floater. Floater that covers machinery and other prefabricated or preassembled equipment while it is being transported to the job site. This coverage continues until the items covered have been installed, tested, and accepted.

Insurable interest. Any interest in property or relation thereto of such a nature that damage to the property will cause pecuniary loss to the insured.

Insurance. Contractual relationship that exists when one party (the insurer), for a consideration (the premium), agrees to reimburse another (the insured) or pay on his behalf for loss on a specified subject caused by designated contingencies, the contingencies being the hazard or peril.

Insurance carrier. Insurance company that assumes financial responsibility for the risks of the policyholder.

Insurance coverage. The total amount of insurance carried by the policyholder.

Insurance policy. Written contract of an agreement to insure.

Insurance risk. General or relative term denoting the hazard involved in the insuring of property. The cost or premium is predicated on the relative risk or hazard.

Insure. To make or secure, to guarantee (as to insure safety to anyone); to engage to indemnify a person against pecuniary loss from specified perils or possible liability.

Insurer. The underwriter or insurance company with whom a contract of insurance is made; the one who assumes risk or underwrites the policy.

Intermediate form hold harmless. Policy issued to indemnify and save harmless the owner with respect to liability imposed upon said owner by law, but only to the extent that such liability arises out of the sole negligence of the named insured or the joint or concurring negligence of the named insured and the owner, and out of the operations of the named insured under the contract with the owner.

Intermediate form indemnification—joint negligence. Intermediate form adds agreement to defend and pay where both parties to the contract may be negligent and therefore legally liable for a loss.

Joint venture coverage. Coverage available for participation in a temporary partnership, for professionals or others who join together as an entity to provide services for a particular project.

Joint ventures. Ventures wherein two or more contractors team to perform a specific job; these call for specialized treatment in establishing a suitable insurance program.

Landlord's protective liability. Coverage for owners of property who lease entire premises to others who assume full control of the property.

Larceny. Theft of personal property, an offense usually defined by statute in most states.

Leasehold insurance. Protection against loss sustained by the termination of a lease due to hazards specifically insured against.

Legal defense provisions. Defense costs in excess of the deductible will be paid, in addition to the liability limit of the policy. Defense costs will not be charged against the liability coverage and thus will not reduce overall liability protection.

Legal liability insurance. Type of business insurance coverage that protects the insured against injury or damage claims from other parties.

Liability insurance. Insurance that covers suits against the insured for such damages as injury or death to other drivers or passengers, property damage, and the like.

Liability insurance (contractual). Insurance against loss under a contractual liability agreement.

Liability insurance (property damage). Insurance against loss due to claims for damages to other's property.

Liability limits. The sum or sums beyond which a liability insurance company does not protect the insured on a particular policy.

Limited form hold harmless. Policy issued to indemnify and save harmless the owner with respect to liability imposed upon said owner by law, but only to the extent that such liability arises out of the sole negligence of the named insured and out of work performed by the named insured under its contract with the owner.

Limited form indemnification—contractor's negligence. Limited form holds someone harmless against claims due to contractors' operations or negligence, or that of subcontractors.

Longshoremen's and Harbor Workers' Act coverage. Special coverage under the workmen's compensation policy applied when employees might come under the federal longshoremen's and Harbor Workers' Act if the work is maritime in nature and is not covered under a state's workmen's compensation act.

Longshoremen's and Harbor Workers' Act coverage— Maritime and Federal Employers' Liability Act. If any work involving maritime or railroad exposures is contemplated, the policy should provide coverage under these acts, which may prove applicable in lieu of a state's workmen's compensation act.

Loss. The basis for a claim for indemnity or damages under the terms of the policy.

Loss conversion factor. Percentage loading added to incurred losses to cover claim investigation and adjustment expenses.

Loss limitation. Maximum amount chargeable under the plan because of any one claim or accident, where such limitation applies. The charge for limiting losses is sometimes indicated separately, sometimes included in the basic premium factor.

Loss of use insurance. Protection against financial loss during the time required to repair or replace property damaged or destroyed by an insured peril.

Malpractice. Alleged professional misconduct or lack of skill in the performance of a professional act; term used to denote professional liability insurance coverage for physicians, surgeons, dentists, druggists, morticians, and beauty parlor operators.

Manual. Book published by an agency, insurance company, or rating bureau for the use of its policyholders. The book includes rates, classifications, specifications, and all rules governing the subject of insurance.

Manual premium. Premium developed by manual rates applied to units of exposure, such as payroll, contract cost, gross receipts, or number of vehicles.

Manufacturers' and contractors' liability insurance. Insurance that covers liability, excluding automobile liability, arising from business operations primarily manufacturing, construction, and installation work, and including ownership and maintenance of premises.

Marine insurance. Coverage primarily concerned with the protection of goods in transit and the means of transportation. The term is applied commonly to risks involving ocean transit.

Maritime and Federal Employers' Liability Act coverage. Special employees' liability statute applying to maritime workers, harbor workers, and federal employees.

Maximum premium. Highest premium possible under retrospective rating, regardless of losses, expressed as a percentage of the standard premium.

Minimum premium. Percentage of the standard premium designed to fix the lowest possible premium under retrospective rating.

Mortgage clause—mortgagee clause. Provision in a fire or other direct-damage policy covering mortgaged property that states that in the event of loss the mortgagee shall be paid to the extent of his interest in the property. No violation of the policy conditions by the insured voids the policy as to the mortgagee. The clause also gives the mortgagee other rights and privileges.

Multiperil policy. Single policy providing property, casualty, and inland marine coverage. Other types of coverage are provided on an optional basis according to the rules of the various states.

Mutual insurance company. Insurance company that is owned by its policyholders.

Named insured. Person(s), firm, or corporation designated in the policy by name as the insured(s) contrasted to one who may have an interest as an insured in a policy but is not designated as a "named insured."

Named perils. Method of writing a contract of insurance that specifies the perils that are covered as opposed to "all-risk" coverage, which covers all perils except those specifically excluded.

Negligence

1. Failure to exercise the degree of care that a reasonable and prudent person would exercise under the same circumstances. The legal liability for the consequences of an act or omission frequently turns to whether or not there has been negligence.

2. Failure to use the degree of care that a reasonable and prudent person would ordinarily use under the circumstances in a particular case. Negligence may involve acts of commission or omission, or both.

Nonconcurrency. Situation where two or more policies insure the same property under different descriptions, terms, or conditions; for example, one policy may be written with a coinsurance requirement and another policy without.

Noninsurable risk. Risk for which no insurance can be written, because the chance of loss is very high or cannot be accurately assessed.

Nonownership automobile liability. Insurance against legal liability arising out of the use of an automobile on behalf of the insured which is neither owned nor hired by an insured. Owner of the automobile is not covered.

Nuclear hazards. Special risk, such as work involving nuclear facilities, is normally insurable through pools of insurance carriers and by underwriters in cooperation with the construction contractors. Information is available through the Nuclear Energy Liability–Property Insurance Association, Farmington, CT.

Numerical rating systems. System is based on the principle that a large number of factors are involved in determining the composition of a risk.

Occupancy. Term that refers to the type and character of the use of a risk property.

Occupancy permit. An indorsement on an insurance policy permitting occupancy that might otherwise suspend the contract or make it invalid.

Occupational accident. Accident occurring in the course of one's employment and caused by inherent or related hazards.

Occupational disease. Impairment of health caused by continued exposure to conditions inherent in a person's occupation.

Occurrence

1. Accident or continuous exposure to conditions that result in injury or damage, provided the injury or damage is neither expected nor intended.

2. Happening or a continuous or repeated exposure to conditions that result in injury or damage, provided the injury or damage is accidentally caused.

Occurrence form. Insuring agreement that covers claims arising from both accidents and occurrences as opposed to the more limited accident-only form.

Office contents special form "all risk." Coverage designed to cover office contents, available to all office occupancies except the medical professions and those incidental to mercantile and manufacturing risks.

Open-stock burglary policy. Policy that covers merchandise, furniture, fixtures, and equipment against loss by burglary or the robbery of a watchman while the premises are not open for business.

Operations—premises. Insurance that covers legal liability for bodily injury or property damage caused by an occurrence.

Ordinance violation insurance. Property insurance used to cover additional costs resulting from violations of and compliance with ordinances.

Other insurance clause. Clause stating the course of action to be followed in case any other insurance embraces the same property or hazard.

Owned automobiles, automobiles under long-term lease. Insurance that covers legal liability for bodily injury or property damage caused by an occurrence (as defined) and arising from the ownership, maintenance, or use of owned automobiles anywhere in the United States, its territories, possessions, or Canada. This type of policy usually permits coverage for vehicles operated by the named insured under long-term lease, where the lessee provides primary insurance, and the lessor is named as additional insured.

Owner's and contractor's protective liability. Third-party legal liability insurance coverage protecting a contractor or owner from claims arising from the activities of their subcontractors.

Owner's, landlord's and tenant's liability insurance. Coverage for liability arising from the ownership or occupancy of premises and operations incidental thereto. It may be included in the owner's liability insurance.

Owner's liability insurance. Insurance that protects the owner against claims arising from the ownership of property and that may be extended to cover claims that may arise from operations of others under the construction contract.

Owner's protective liability insurance. General liability insurance to protect against liability arising out of the conduct of any business.

Package policy. Single policy that includes several different forms of insurance coverages.

Partial loss. Loss under an insurance policy that does not either completely destroy or render worthless the property of the insured or exhaust the insurance applied to the property.

Peril. Cause of loss insured against by a policy such as fire, windstorm, explosion, or burglary.

Perpetual insurance. A form of fire insurance policy that is uncommon because it has no expiration date.

Per risk excess insurance. Reinsurance on a per risk basis as contrasted from a per accident or aggregate basis.

Personal injury. Injury or damage to the character of a person as opposed to the body. Personal injury may arise from false arrest, slander, libel, malicious prosecution, etc.

Personal injury liability insurance. Insurance that protects against loss arising out of liability imposed on the insured by law for injury or damage to the character or reputation of a person, for example, false arrest, malicious prosecution, willful detention or imprisonment, libel, invasion of privacy, or wrongful entry.

Physical hazard. The material, structural, or operational features of risk itself, apart from the persons owning or managing it.

Plate glass insurance. Insurance that indemnifies the insured against loss caused by breakage of glass (or by chemicals accidentally or maliciously applied), excluding only loss caused by fire, war, and nuclear hazards.

Policy. A document on a printed form issued to the insured by the insurance company, stating the terms of the insurance contract.

Policyholder. Usually synonymous with "insured"; one who holds or owns an insurance contract on property or life not his own.

Premises. Particular location or portion thereof as stated on the policy.

Products liability insurance. Insurance that covers liability imposed for damages caused by an occurrence arising out of goods or products manufactured, sold, handled or distributed by the insured or others trading under his name. The occurrence must be after the product has been relinquished to others and away from premises of insured.

Professional liability insurance

1. Insurance that covers legal liability of the insured architect or engineer for damages caused by an error, omission, or negligent act in the insured's professional capacity.

2. Insurance that covers alleged professional misconduct or lack of ordinary skills in the performance of professional services.

Project coverage. Subject to underwriting considerations, a separate policy can be purchased to cover a specific project.

Property damage insurance. Part of general liability insurance covering injury to or destruction of tangible property, including loss of use resulting therefrom but usually not including property that is in the care, custody, or control of the insurance.

Property insurance. Collective term for fire, extended, coverage, vandalism and malicious mischief, and steam boiler and machinery insurance. It is referred to in the insurance industry as "builder's risk insurance."

Pro rata liability clause. Clause providing that the insurance company is liable for not more than the proportion of loss which the amount insured under the policy bears to all insurance policies covering the loss.

Protected. In fire insurance, risk located in an area protected by a fire department; in burglary insurance, risk equipped with a burglar alarm.

Protection. Safeguarding against loss provided under the terms of the insurance policy, known as "coverage."

Protective liability insurance. Provides against claims which may arise out of secondary or indirect responsibilities, such as association with independent contractors.

Proximate cause. Cause that directly produces the effect; dominating cause of loss or damage; unbroken chain of cause and effect. For example, fire is the proximate cause of damage done by water used in extinguishing a fire.

Public and institutional property plan. Insurance against fire and allied perils designed for properties such as hospitals, churches, schools.

Public liability insurance

1. Insurance covering liability of the insured for negligent acts resulting in bodily injury, disease, or death of others. Coverage excludes employees of the insured and/or damage to the property of others, other than property in the care, custody, or control of the insured.

2. Insurance that protects only the insured. It protects the contractor against legal liability for injuries to the person and property of members of the public, who are also referred to as "third persons."

Rating bureau. An organization that classifies and promulgates rates, compiles data on fires, safety provisions, and measures hazards of individual risks in the terms of rates for a given territory.

Release. Relinquishment of a right or claim for a stated consideration.

Rental value insurance. Insurance that reimburses the owner-occupant of a building for the expense of renting another location should the insured property become unusable as a result of damage by an insured peril.

Rents insurance. Time-element coverage providing for payment to an owner of a building for loss of rents due to damage of property by an insured peril.

Replacement cost insurance. Insurance that pays the replacement value of the damaged property without deduction for depreciation. It may be required that the property actually be replaced before the insured can collect the replacement cost of the damaged property.

Reporting form property insurance. Policy used when property values fluctuate during the policy term. In construction insurance coverage, the "property insurance" may be written in this form, which requires monthly statements showing the increase in value over the previous month of work in place.

Reservation of rights. Because of certain facts of a claim or the allegations of a complaint against an insured, the insurance company may initially feel that there may not be coverage under the policy. The company therefore asks that a "reservation of rights" document be executed. The document generally states that while the company will defend the insured and conduct investigations, it reserves its rights to deny coverage if and when its investigation indicates the claim is not covered under the policy. The insured in accepting the "reservation of rights" maintains his right to bring with action against the carrier, should he disagree with the carrier's possible subsequent denial of coverage.

Retirement protection. Professionals retiring from practice should be protected with coverage that will provide the insured protection against claims made at any time subsequent to the last policy, if claims arise from alleged negligent acts or negligent omissions that occurred during the term of practice.

Retrospective rating. Guaranteed cost experience rating plans take into account the insured's premium-loss record over the past several years to arrive at a fixed renewal rate. "Retrospective rating" does exactly the same thing and then goes one step further to determine the final premium for policies subject to the plan only after they expire. By adjusting the standard premium in direct relation to losses reported under those very policies, "retrospective rating" reflects more promptly and more closely the effect of such loss experience, good or bad. In short, it is cost-plus insurance.

Riot. Violent and tumultuous actions by a number of people (in most jurisdictions, such action must involve three or more people to be a riot).

Risk. Chance of loss; the insured or the property to which the policy relates.

Risk and hazard. Risk is often confused with hazard. Hazard is a condition that increases the likelihood of a loss. Risk is the uncertainty as to the outcome of an event when two or more possibilities exist.

Risk and peril. Risk is sometimes confused with peril, and the two terms are often used interchangeably. Peril is the source of loss as distinct from the uncertainty of loss. A fire, earthquake, and a liability judgment are sources from which losses emerge. They are perils that give rise to risk, but are not risks in themselves.

Robbery. Felonious taking, either by force or threat of force, of another's personal property. Robbery is commonly known as "hold up."

Safe burglary insurance. Insurance that insures against loss of property caused by forcible entry into a safe or vault. Damage to safes, vaults, and other property on the premises resulting from burglary is also covered, unless caused by fire.

Salvage. The recovery of materials, products or equipment that would reduce the amount of the loss. This includes the repair of damaged items.

Salvage scrap. Materials, products, or equipment beyond repair that have to be sold as scrap.

Settlement consent. The company must have one's consent to settle a claim. The settlement must be in the best interest of the insured.

Smoke damage insurance. Insurance against damage done by smoke from the sudden, unusual, and faulty operation of a heating or cooking unit, but only when such a unit is connected to a chimney by a smoke pipe and is on the premises described in the policy. Smoke damage from fireplaces and industrial apparatus is usually excluded. Damage caused by smoke from a hostile fire is covered by a fire insurance policy.

Special hazards insurance. Additional perils insurance to be included in property insurance, covering such perils as sprinkler leakage, collapse, water damage, and all physical loss; insurance on materials and supplies at other locations and/or in transit to the site.

Specific insurance. Policy coverage that describes in detail the property covered by the policy, rather than a blanket coverage policy.

Sprinkler leakage insurance. Insurance that provides indemnity against direct property loss or damage caused by the accidental (not caused by fire) discharge or leakage of water or other substances from an automatic sprinkler system.

Standard form. Insurance policy that has been adopted by many insurance companies and approved by respective state insurance departments.

Standard premiums. Manual premiums plus credit or debit factors produced by experience rating plans. In theory, the same as premiums developed under guaranteed cost policies.

Standard provisions. Those clauses that certain state codes require to be included in contracts of insurance.

Statutory law. Law established by a legislative act. In connection with workmen's compensation acts, conditions of coverage and benefits are determined by the governing bodies of each state and thus vary from state to state.

Statutory law liability. Liability arising from written legislative acts and imposed by statutory law.

Steam boiler and machinery insurance. Insurance that insures against loss arising from the operation of boilers, or other pressure vessels, and machinery. The insurance will cover loss to designated boilers, pressure vessels, and machinery and includes damage to other property. It may be extended to cover bodily injury liability and business interruption loss.

Stock insurance company. Insurance company owned and controlled by stockholders and conducted for profit.

Subrogation. By terms of the policy or by law, assignment to an insurer, after payment of a loss, of the rights of the insured to recover the amount of the loss from one legally liable for it.

Tangible property damage. Property damage or injury to or destruction of tangible property.

Tax multiplier. Fixed percentage factor designed to cover premium taxes, regulated by the individual states involved.

Temporary total disability benefits. Weekly benefit payable to an employee under the "workmen's compensation law" because he is temporarily unable to perform any duties for his employer as a result of an accident or sickness sustained within and arising out of his employment.

Termite insurance. Insurance concerned with the exposure to infestation and protection provided against infestation.

Theft. Illegal taking of property from premises or person without use of a threat of force or evidence of forcible entry.

Third-party insurance. Insurance protecting the insured against liability arising out of bodily injury to others or damage to the property of others.

Time element. Type of insurance that reimburses the insured for loss of the use of property, the amount of loss being dependent on the period of time required to rebuild, repair or recover. Examples are business interruption insurance and rent insurance.

Total loss. The complete destruction of the property. The claim of sufficient amount to require maximum settlement.

Umbrella excess liability coverage. Excess liability coverage, usually for considerably higher limits, over existing liability policies such as employer's liability, public liability, or automobile liability, providing direct coverage for most losses uninsured under existing liability policies after a self-retention limit (deductible) is exceeded.

Unoccupied. Furnished but not lived in. The standard fire insurance policy prohibits vacancy and unoccupancy beyond a 60-day period. Permission for unlimited vacancy and unoccupancy is usually given in protected territory without charge. In unprotected territory, permission is given as necessary and a charge is made.

USL&HW Compensation Act. United States Longshoreman's and Harbor Worker's Compensation Act. This legislation covers maritime employment on navigable waters but excludes the master or members of the crew. Coverage is provided by an endorsement to the workmen's compensation policy.

Vacant. Not lived in or regularly used and unfurnished (containing no property of value).

Valued policy. A valued policy provides that a stipulated insured amount will be paid in the event of a total loss. In some states there are "valued policy" laws which require that fire insurance policies on buildings be treated as "valued policies."

Vandalism and malicious mischief coverage. Coverage against loss or damage to the insured's property, caused by willful and malicious damage or destruction.

Vendor's liability. Insurance to protect against bodily injury or property damage claims resulting from use of materials or products after sale to the public. Policy may also endorse policy to protect those who sell the manufactured product, either individually or on a blanket basis, from contingent liability claims arising from such sale. Although final responsibility may accrue to the manufacturer, this optional vendor's endorsement assures the seller of legal representation and defense.

Waiver. Intentional relinquishment or abandonment of some known right, claim, or privilege.

Warranty. Absolute statement or stipulation in a policy as to the existence of a fact or condition of the subject of insurance, the truth of which may be necessary for the contractual validity of the policy.

War risk. War, civil war, insurrection, rebellion, or revolution risks and expenses are standard exclusions.

Watercraft liability coverage. Liability coverage for bodily injury or property damage arising out of the ownership and/or operation of watercraft.

Water damage legal liability. Liability imposed by law for loss of or damage to property of others caused by the accidental discharge, leakage, or overflow of water on or from premises owned by or rented to the insured. The usual care, custody, and control exclusions apply.

Water damage on or from insured's premises. If the damage occurs on or from premises owned or rented by the insured, the policy may exclude damage to building or inside property caused by the discharge, leakage, or overflow of water or steam from plumbing, heating, refrigerating, or air-conditioning systems, standpipes for fire hose or industrial or domestic appliances, or any substance from automatic sprinkler systems; or the collapse or fall of tanks or component parts of supports thereof which are part of automatic sprinkler systems; or rain or snow admitted directly into the building through defective doors, windows, skylights, transoms, or ventilators; excepting loss due to fire or operations performed by independent contractors.

Workmen's compensation and employers' liability insurance. Insurance bought by the employer in his own name. Neither the owner nor the architect can be added as additional insureds, nor should they be.

Workmen's compensation insurance. Insurance that covers liability of an employer to his employees for compensation and other benefits required by workmen's compensation laws with respect to injury, sickness, disease, or death arising from their employment.

Workmen's compensation law. "Workmen's compensation" benefits vary under the various state laws, but insurance is bought usually without any reference to limits of liability. The insurance company must provide the insurance up to the total liability of the employer under workmen's compensation laws.

Wrap-up insurance policy. Program combining all the interests involved in a construction project for insurance purposes with one insurer chosen by either the owner or contractor. It is usually appropriate only for very large projects. The policy must be specifically designed and negotiated for the individual project. It is not permitted in all states.

Wrap-up insurance programs. The owner, architect, engineer, general contractor, and subcontractors are protected under a single insurance package applicable to all.

XCU

1. Letters that refer to "exclusions from coverage" for property damage liability arising out of explosion or blasting, collapse of or structural damage to any building or structure, and underground damage caused by and occurring during the use of mechanical equipment.

2. Coverage under which the contractor is protected against property damage caused by explosion X, collapse of structures C, and damage to underground pipes and utilities caused by digging with mechanical equipment U.

CONSTRUCTION SCHEDULING

Activity. Task, event or item that can be identified and scheduled, so that its start and completion can be recognized.

Arrow diagram. Method of drawing network schedules using arrows to represent activities and their sequence relation to each other.

Backward pass. Process to latest start times and latest finish times for all activities.

Bar chart. Non-network scheduling technique used by contractors to originate data and information, which can then be processed to a network schedule.

Calendar. Normal monthly calendar indicating workdays and non-workdays for a project.

Cash flow. During an ongoing project, the inflow and outflow of cash required to make payments for materials and office and field operations and to subcontractors, and the receipt of payments due from the owner.

Central Limit Theorem. Mathematical theorem that allows the user to combine PERT activities with continuous probability distributions to determine the probability distribution of a critical path.

Complex compression. Compression, when direct costs of activities vary nonlinearly with their durations or the overhead costs vary nonlinearly with time.

Compression. Shortening a project schedule.

Cost control. Accounting for the resources expended on the project activities, comparing them to the estimated or budgeted costs, noting variances, and taking action to prevent future overruns.

Crash. Shortening project activities in order to reduce costs and shorten project length.

Crash cost. Cost of an activity at each crash duration.

Crash duration. Minimum practicable duration of any activity.

Critical path. The longest path or paths from project start to completion.

Critical Path Method (CPM). Time analysis of a network of activities that outlines the steps needed to complete a project. The Critical Path is the path that takes the longest amount of time to perform all of the activities in their logical sequence between the first event and the last event.

Database. Computer file containing data on a particular subject which can be automatically searched and from which desired data can be retrieved.

Dependency. Particular work activity, contingent upon the completion of one or more work activities that precede it.

Deterministic process. Process used to find the total duration of a project based on the sum of known activity durations.

Direct costs. All costs, usually nonlinear, directly attributable to the project work items; to accelerate or to delay the work to reduce work costs.

Dummy. Activity with zero time duration used to express logic and to provide a unique numbering for each activity, in order to start or complete a network schedule.

Duration. Time alloted in work days to complete an activity, and where indicated on the schedule.

Early event time. The earliest that an event or activity can start.

Early finish time. The earliest that an event or activity can be completed as scheduled.

Event. Connection between two or more related activities, as scheduled.

Float. Flexibility in the starting and/or completion time of an activity.

Forward pass. Process to find earliest start and earliest completion for all activities.

Fragnet. Detailed breakdown of an activity.

Free float. Maximum time an activity can be delayed without delaying the start of a succeeding activity.

Gnatts. Bar chart indicating non-network scheduling.

Independent float. Float that can neither interfere with a succeeding activity nor be interfered with by a preceding activity.

Interfering float. Float that can interfere with succeeding activities but will not delay the project.

Labor allocation. Assigning workers to a project in order to increase work progress or to stabilize the present work force.

Late-event time. Latest event that is scheduled to occur if the project is to be completed on time.

Late finish. Latest an activity can finish without delaying scheduled project completion.

Late start. Latest an event or activity can start without delaying scheduled project completion.

Lead-lag factor. Factor indicating the amount of time one activity or event leads or lags another.

Least cost. The anticipated estimated least cost, based on original budgeted cost, for a project.

Link. Line connecting two activities on a node diagram.

Logic. Relationship between work activities in a network diagram. The relation is determined by whether an activity must precede another, can be done concurrently with others, or must follow others.

Matrix method. Tabular method of solving Critical Path Method calculations.

Milestone. Event that has special significance in the scheduling, such as a special start or completion event.

Monitor. Continuous comparison of actual project timing with that indicated on the final approved schedule.

Narrative. Written report indicating the status of activities: manpower, network, billings, cash expenditures, deliveries, submittals, and approvals.

Network diagram. Graphic display indicating the logical order and sequence of events and work activities necessary to complete a project.

Node diagram. Method of drawing network schedules using circles or squares, called "nodes," to represent activities.

Normal cost. Cost of an activity at normal duration.

Normal duration. Duration of an activity in comparison to schedule.

Optimistic time. Shortest time that a project will take to complete, based on weather conditions, labor supply, and shipment of materials to the project as scheduled.

Optimum time. The scheduled project duration at least cost.

Overhead costs. Costs which are allocated over the entire project, based on operational costs and properly prorated.

Overlapping activities. Activities that are interdependent but can proceed simultaneously with other activities during some period of the scheduling.

Owner's checklist. Time established in the schedule for the owner or his representative to provide a checklist of items which are questionable and need answers.

Permit. Scheduled time allotment for the process of having the plans and specifications reviewed by the authorities and receiving the final approval and permit.

Pessimistic time. The longest time possible that a project can be completed, based on weather delays, strikes, stopping work at the site, delays in shipping materials to the project, and other calamities.

Planning. Process for development of a sequence for the project from the preliminary phases through the construction period and final cleanup to owner's occupancy.

Precedence diagramming. Network scheduling method, similar to CPM, which allows overlapping activities.

Probability. The probability that the project will be completed in accordance with the schedule and the contract agreement.

Program evaluation and review technique (PERT). Network scheduling technique. Activity durations are described by a probability distribution.

Photographic reports. Scheduled appearance of the photographer at the project to take photographs of selected views of the work in place as directed by the project superintendent.

Progress payments. Contractor's application for payment in accordance with contract stipulations and architect's review and acceptance.

Project. Well-defined scope of the construction work and as further described in the contract documents.

Project control. Continuous systematic review of the schedule and comparison with the actual progress at the site.

Resource. Requirements to proceed with a project, such as financing, labor, materials, equipment, and time.

Resource allocation. Allocating resources to a specific project as estimated and scheduled.

Restraints. Management decisions that affect the sequence of the work activities and the overall construction time.

Schedule. Time-based arrangement of all project activities, including all preliminary preparations.

Schedule analysis. Analysis of a category of activities to anticipate a change in duration that would affect the project completion date.

Scheduling. Determination of the time of the operations comprising the project and of the assembling of personnel and preliminary work required to set the construction in motion in accordance with a prescribed schedule.

Sequence. Logical order of work activities, depending upon restraints, availability of information, resources, and authorization.

Software. Information the computer uses; computer programs.

Standard deviation. Measure of the expected variation of an activity or a project's duration.

Superintendent's reports. Review and analysis of the reports of actual project progress with special notations on items that could affect schedules.

Target schedule. The approved schedule for the completion of the project, as stipulated in the contract.

Total float. Maximum time an activity can be delayed without delaying completion of the project.

Trend analysis. Analysis of the activities to review if certain events are habitually early or late.

Updating. Review, critique, and revision of the network diagram and computer print-outs, based on the timely and adequate input of information received from the construction team assigned to the project.

Workday. One calendar day of work on the project.

PLANS, SPECIFICATIONS, SURVEYS, PLATS, AND MAPS

Addendum. Document issued during the bidding period to clarify, revise, or add to the original construction documents or previously issued addenda.

Approval of changes on drawings. When it is necessary to change approved drawings or specifications, revisions should be approved before installation is commenced.

Approval of drawings and specifications. Where heating, ventilating and air conditioning equipment is required, complete drawings, specifications, and data sheets should be submitted to the building department for approval. Approval should be obtained before affected work is commenced, and all work should be executed according to the approved drawings and specifications.

Approved drawings kept at building. A complete set of approved drawings should be kept available at the job site.

Architectural drawings. Drawings should contain a plot plan, building floor plans, elevations, sections, and details. Schedules of finishes, doors, windows, decorating, and other schedules to coordinate with the specifications.

Base bid specifications. Specifications that define quality, function, and performance of products or equipment. In addition, products are identified by name of manufacturer, model or type, size, finish, and similar characteristics. Bidder has no selection of products. No substitutions are possible.

Base bid with alternates—specifications. Base bid specifications that allow the contractor to submit alternates or substitutions for the products specified.

Checklists. List established for each trade division as a check on preliminary specifications, so that no item that could affect the cost estimate is omitted.

Commercial specifications. Specifications recommended by manufacturers.

Complete plans. Plans consisting of floor plans, elevations, sections, structural plans and details, building service equipment plans and details, and such graphic representations, diagrams, and delineations as are necessary in the design of a building to show its size, construction, location, design, design live loads, unit stresses used in its structural design, foundation data, and other pertinent information as is necessary to determine that the building, when constructed in compliance with said plans, will comply with the code of the city pertaining to location, safety and health.

Compliance approval. Approval in compliance with the provisions of the building code.

Compliance required. No building, structure or land should be used, and no building, structure or part thereof should be erected, converted, altered, moved, demolished or removed, except as in conformity with the ordinance.

Comprehensive development plan. The allocation of land areas to the several varieties of physical development, present and future, of the regulated area, the same having been prepared in accordance with the principles of comprehensive planning or having been developed through the approval of subdivisions previously submitted, wherever such plan exists and has been officially adopted by the planning commission and recorded; wherever the term "development plan" is used, it should have the same meaning as the term "master plan" or "comprehensive plan."

Comprehensive plan. Includes charts, maps, descriptive matter officially adopted by a planning commission or governing body showing, among other things, recommendations for the most appropriate use of land, for the most desirable density of population, for a system of thoroughfares, parkways and recreation areas, for the general location and extent of facilities for water, sewer, light and power, for the general location, character and extent of community facilities.

Controlled construction. The construction of a building or structure or a specific part thereof which has been designated and erected under the supervision of a licensed or registered engineer or architect using controlled materials as defined in compliance with accepted engineering practice under the procedure of the code.

Coordination drawings. Drawings prepared during the construction phase to supplement working drawings and shop drawings, as necessary, to correlate the work of various trades.

CSI format. Standard table of contents for construction specifications, consisting of 16 major groups called "divisions." A division is then subdivided into units of work, called "sections." (Format developed by the Construction Specifications Institute, Washington, DC).

Descriptive specifications. A detailed written document of required properties of a product, material or equipment, describing their installation and the results to be obtained.

Design plan. Plan prepared by the city for implementing components of the general plan and may include, but is not limited to, the design, bulk, use, height, location and arrangements of buildings in respect to streets, open spaces, other structures, and natural features.

Detailed mechanical plans. Plans showing each feature of the sewer or sewage works design and including elevations, grades, profiles, sections, all special designs and structures, typical and special manholes, flush tanks, intersections and stream crossings, pumping stations, catch basins, storm water inlets, and all like appurtenances.

Development plan. Dimensioned presentation of the proposed development of a specified parcel of land which reflects on the location of buildings, easements, parking arrangement, public access and street pattern, and other similar features.

Drainage plan. Plan showing all proposed and existing facilities to collect and convey surface drainage, described by grades, contours and other topographical data.

Drawing preparation. Drawings should be made to scale, upon substantial quality paper, plastic, or cloth, and the drawings and specifications should be complete and complementary. The work proposed should indicate in detail that the building will conform to provisions of the code.

Drawings. The graphic means of indicating work to be done, which includes shapes, dimensions, locations, and relations between components or materials, including a generally accepted list of symbols to identify materials and a descriptive list of all abbreviations shown on the drawings.

Electrical engineering drawings. Drawings should contain, proposed electrical wiring for each floor, working spaces for switchgear and other switching and control panels, ampere capacity, service entrance section, location, voltage and horsepower, line diagrams for all wiring systems, and schedules and descriptions of all other miscellaneous wiring for intercommunications, fixture schedules, and schedule for amperage for each circuit and panel.

Engineering plans. Plans which include layouts, profiles, cross-sections, and other required details for the construction of public improvements.

Engineer's report. A comprehensive report describing the project, the basis of the design, together with design data and all other pertinent data necessary to present an accurate understanding of the work to be undertaken and the reasons for the same.

Equal to or approved by. This, or other synonymous terms, when used in the specifications, is expressly understood to mean that the approval of such submissions is vested in the architect/engineer whose decision is final, unless otherwise stated in the contract with the owner.

Federal and military specifications. Specifications listed in the index of *Federal Specifications*, *Standards and Handbooks*, which is issued yearly by the Superintendent of Documents, U.S. Government Printing Office, Washington, DC 20402.

Final development plan. Final plan prepared by a developer based on the approved preliminary plan of a proposed development or development area that consists of detailed drawings, specifications, cost estimates and agreements for the construction of the site improvements and buildings for the proposed development or development area.

Fire protection plans. Drawings indicating a complete system or systems of fire protection for the building, which should include water supply, pumps if required, riser mains, approved automatic fire extinguishing system, fire alarms, standpipes, smoke control, smoke detectors, Siamese connection for use by fire apparatus and fire pumps, and all other requirements of the local Fire Department. Plans must be submitted to the Fire Department Engineers before a permit will be issued.

General Conditions. Document relative to the responsibilities of the parties engaged in a construction project.

General development plan. Description of the development proposed within a particular Planned Community District includes map showing the location and arrangement of all proposed uses, and a written statement of the general regulations proposed to govern them.

Generally accepted standard. Specification, code, rule, guide, or procedure in the field of construction, or related thereto, recognized and accepted as authoritative.

General plan. Plan and statement of the objectives and recommendations for the general location and extent of desirable future land development, community facilities and street plans for the city, duly adopted or officially accepted.

Grid survey (entire site). Elevations on a rectangular grid section and at points where definite breaks in grade occur. Grid lines 50 ft or 100 ft unless otherwise specified and extended 10 ft beyond all property lines.

Guide specifications. Generally used as tools for use in preparing specification sections; usually an outline of information to be filled in to complete a section.

Landscaping plans. Drawings should contain information relative to the grading of the site, indicating grade elevations and contours of the soil, the location and description of all trees, shrubs, vines, plants, and ground cover, underground sprinkler systems and controls, and proposed or suggested low voltage lighting system.

Manufacturer's specifications. Information published by product manufacturers. Specifications list properties of materials, ratings of equipment, recommended use, installation, and workmanship requirements.

Map. A drawing showing geographic, topographic, or other physical features of the land.

Master Plan. Comprehensive plan including graphic and written proposals indicating the general location for streets, alleys, parks, schools, public buildings, and all physical development of the city and including any unit or part of such plan and any amendment to such plan or parts thereof. Such plan or part thereof may or may not be adopted by the planning board.

Master specifications assembly

1. Specifications prepared as a standard for a particular type of building or structure. Office production of master specifications vary in procedures: 1. Master paragraph cards are assembled for copy typing and revisions are made on duplicate cards, so as not to destroy the original. 2. Computer reproduction requires punch cards which are available and new punch cards are made with revisions. 3. Special typewriter tapes run automatically until a revision has to be made, and the operator stops the machine to manually type in the revision.

2. Automated text-handling systems providing a modern and efficient means of utilizing the text of an established master.

Mechanical drawings, specifications and data. All drawings and specifications and associated calculations and data for the installation of heating, ventilating and air conditioning systems should be designed and prepared to satisfy the requirements of the code. All drawings, specifications, calculations and associated design data should be submitted for review and approval under the provisions of the code and should be sealed or stamped by an engineer registered in accordance with the laws of the state.

Metes and bounds survey. System of land survey and description based on starting at a known reference point and tracing the boundary lines around an area.

Minimum building plans. Plans should include drawings of floor plans of all habitable floors and the basement or foundation plan, and such drawings should clearly indicate sizes and spacings of all supporting members, sizes of rooms, glass areas, sizes of all footings and reinforcing, thickness of basement or foundation walls and the reinforcing, thickness of all floor slabs with or without reinforcing, exterior and interior wall construction, sizes and spacing of all framing members, ceiling heights, and parapet wall heights.

Murder clause. An inclusion that does not define quality, but merely attempts to afford an unreasonable advantage to the writer of the specifications.

Or-equal specification

1. Specification permitting bidder or contractor to propose substitutions.

2. Specification that usually names one, two, or several products, followed by the term "or equal," or "or-approved equal."

Or-equal with prequalifications. Specification permitting bidder or contractor to propose substitutions to the architect/engineer or the owner for approval.

Outline specification. Limited, brief, simple material and equipment listing.

Performance specifications. Specifications that usually utilize Federal Specifications, ASTM Standards, AASHO Specifications, Commercial Standards, and other recognized reliable standards. Where no satisfactory standards exist, a complete description of the system and its guaranteed results are stated with descriptions of the materials, products, designs, functions, and performance to be obtained.

Plans and specifications. One or more drawings or documents indicating and describing the amount, arrangement, kind, and quality of the materials to be used for the construction of a building or structure.

Plat. Map, plan or chart of a city, town, section or sub-division, indicating the location and boundaries of individual properties.

Plat required. All applications for building permits should be accompanied by a plat in duplicate of a dimensioned sketch or to-scale plan signed by the owner or his authorized agent, showing the actual dimensions of the lot to be built upon, the location and size of the building or structure to be erected, the location of adjoining or surrounding buildings or structures, and such other information necessary to provide for the enforcement of the ordinance as may be required by the building inspector.

Preliminary development plan. Drawing prepared by a developer, which may include explanatory exhibits and text, submitted to the designated authority for the purpose of study of a proposed development of land, or a preliminary plan of land use of a development area, which, if approved by the designated authority, provides the basis for proceeding with the preparation of the final plan of a development or development area.

Preliminary specifications. Specification that briefly lists the materials and finishes to be used. Should accompany preliminary drawings. Serves as a basis for a preliminary cost estimate.

Preliminary subdivision plan. Complete and exact subdivision plan to define property rights and proposed streets and other improvements, presented for purposes of securing preliminary approval.

Proposal form. Many professional offices include a proposal form in the specification document, which the bidder may copy, or may be instructed to use own form or write the proposal on letterhead; public agencies usually provide forms to submit.

Proprietary specifications. Specifications permitting the professional and the owner to exercise close control over product selection. Names of products and manufacturers are given.

Record plat. Exact copy of the approved final plat, reproducible in standard size, prepared for necessary signatures and recording with the county recorder of deeds.

Reference standard. Standard specification that may be incorporated into a job specification by reference to a number or title. Reference becomes a part of the specification as though it were included word for word.

Seal required. Plans and specifications must bear the seal and signature of a registered architect or engineer in conformity with and when required by the statutes of the state.

Site map for project. Drawings should include a site map drawn to scale, adequately dimensioned, clearly showing the exact location of all structures existing or to be constructed. When a private water supply or sewage disposal system is necessary, the site map should show the location of proposed well, septic tank and disposal field in addition to existing wells, septic tanks, sewer lines, drains, sewage disposal fields, seepage pits, privies, and cesspools within 100 feet of the dwelling.

Specifications. Consists of three (3) important phases, and each section should include the following: (a) administrative and procedural requirements, (b) quality of products that are required to produce the specified results, and (c) how the products are to be incorporated into the construction to produce these results.

Specification work sheets. Work sheets prepared as the working drawings are being completed to keep an accurate take off of every item shown, to prepare for the start of the writing of the final specifications.

Stamped plans on job. The stamped set of plans and specifications issued to the applicant should be kept at the site of the construction or work and should be available to the authorized representative of the building department. There should be no deviation from the stamped or approved application, plans, or specifications without official approval.

Standard. That which is established by authority, custom, or general consent, as a model or example.

Standard specification. Specification that has been established as a model or example by authority, custom, or general consent.

Structural engineering drawings. Drawings should contain foundation plans, floor plans, framing plans, elevations, and sections indicating all of the structural requirements.

Surveys. Absolute requirements in order to prepare any other plans. The survey should include all the following information: soil or grade contours with given elevations prepared on a grid system, location and identification of all utilities entering the property, location of all existing trees, plants, shrubs, identification of private or public poles, existing structures or buildings, establishment of a datum elevation, and a bench mark for future use, identification of property lines, and, most important a legal description of the property. Include all public walks, roads, streets and alleys as indicated on plats and maps on record.

Supplementary general conditions. Document containing changes and additions to the general conditions in the specifications document.

System of rectangular surveys. Land survey system based on geographical coordinates of longitude and latitude, originally established by acts of Congress to survey the lands of public domain.

Trade association specifications. Specifications that are standards for the industry.

Working drawings. Graphic means of indicating work to be done. In particular, shows where materials are located and includes size, dimensions, and placement.

POLLUTION AND ENVIRONMENTAL CONTROLS

Abate. Put an end to or to reduce in degree or intensity; to reduce in value or amount.

Acid rainfall. Rainfall that is increasingly more acid, and thus more toxic to the environment, under conditions of increasing air pollution. Rainfall in the eastern United States has become 32 times more acid than it was in the 1950s.

Activated carbon. Form of amorphous carbon having an enormous number of pores per surface area, whose characteristics enable it to be highly absorbent to gases, vapors, and other noxious fumes.

Activated sludge. Sludge associated with sewage that is treated by subjecting it to aeration and bacteria action.

Activated sludge plants. Waste water treatment facilities in which the waste water passes through an aerated tank containing a suspension of aerobic bacteria that feed on the nutrients in the waste water.

Aeration. The introduction of air into a liquid, causing the air to bubble through the liquid.

Afterburner. Device added to any combustion system to burn up the smoke and flue gases that have penetrated the combustion chamber of the equipment.

Air contaminant. Term that includes products and by-products of smoke, soot, fly ash, dust, cinders, dirt, noxious or obnoxious acids, fumes, oxides, gases, vapors, odors, toxic or radioactive substances, waste, particulate, solid, liquid, or gaseous matter, or any other materials in such place, manner or concentration that may cause injury, detriment, nuisance, or annoyance to the public, or to endanger the health, safety or welfare of the public, or to cause or have a tendency to cause injury or damage to business or property.

Air furnace. Horizontal furnace, externally fired with a natural draught stack that is used to melt or treat ferrous materials for production of castings.

Air jets. Apparatus operated by steam or compressed air or a mechanically-driven blower for the purpose of causing high velocity air to be introduced into a furnace and/or to cause a more complete mixture of oxygen with the gases of combustion above the fuel bed.

Air pollutant. Dust, fumes, gas, mist, smoke, vapor, odor, particulate matter, or any combination thereof, except that such term should not include uncombined water in the atmosphere unless it presents a safety hazard.

Air pollutants. Matter in the air capable of creating or causing air pollution. Such matter may originate from any kind of combustion process or industrial or laboratory processes, both chemical and physical, and may appear as, but is not limited to, smoke, dusts, fumes, droplets, mists, vapors, gases, odors, or a combination of these.

Air pollution. Condition occurring in the outdoor atmosphere due to the expelling of contaminents to the air consisting of dust, fumes, gas, smoke, odors, vapors, and industrial mists.

Air quality standard. Ambient air quality goal established for the purpose of protecting the public health and welfare.

Ambient. General condition surrounding a given point.

Amine. Any of various basic compounds derived from ammonia by replacement of hydrogen by one or more univalent hydrocarbon radicals; compound containing one or more halogen atoms attached to nitrogen.

Anaerobic bacteria. Bacteria that can exist with partial or no oxygen.

Architectural coating. Coating used for residential, commercial, or industrial buildings and their appurtenances that is on-site applied.

Asbestos. Fibrous, rock-forming mineral including, but not limited to, such amphibole varieties as tremolite, actinolite, anthophyllite, grunerite, richxerite, edenite, amosite, crocidolite, and such serpentine varieties as ami-anthus and chrysolite, as well as synthetic asbestos fibers including, but not limited to, fluor-tremolite, fluorrichterite, and fluor-edenite.

Ashes. Includes cinders, fly ash, or any other solid material resulting from combustion, and may include unburned combustibles.

Atmospheric pollution. Discharge from stacks, chimneys, exhausts, vents, ducts, openings, buildings, structures, premises, open fires, portable boilers, vehicles, processes, or any other source of smoke, soot, fly ash, dust, cinders, dirt, noxious or obnoxious acids, fumes, oxides, gases, vapors, odors, toxic or radioactive substances, waste, particulate, solid, liquid, or gaseous matter, or any other materials in such place, manner, or concentration as to cause injury, detriment, nuisance, or annoyance to the public, or to endanger the health, comfort, repose, safety, or welfare of the public, or to cause or have a natural tendency to cause injury or damage to business or property.

Atmospheric pollution sources. Privately or publicly owned and operated equipment and devices that provide the pollution to the atmosphere.

Automobile and/or truck sales lot. Land area used or intended to be used for the display and/or sale of passenger automobiles and/or commercial vehicles.

Auxiliary fuel-firing equipment. Equipment for supplying additional heat by the combustion of an auxiliary fuel, for the purpose of attaining temperatures sufficiently high to dry and ignite waste material, to maintain ignition thereof, and to promote complete combustion of combustible solids, vapors, and gases.

Baffling. Any row(s) or plane(s) of refractory or other material that causes the gases in a steam boiler or other vessel, duct, or device to assume a definite and predetermined path of travel before reaching the chimney or smokestack.

Bessemer converters and pneumatic steel-making processes. Processes by which steel is made directly from molten iron or scrap metal by forcing gases through or over the molten metal to oxidize and carry off the carbon and other impurities in the metal.

Best management practice (BMP). Alternatives designed to reduce or prevent runoff of pollution discharges or emissions that adversely affect air, land, or water, including cost-effective determinations and evaluation of social and economic impacts.

Bioassay. Particular technique for testing a substance against a living tissue for toxic, mutagenic, carcinogenic, and teratogenic agents.

Biodegradable. Capable of being broken down into innocuous products by the action of living beings (microorganisms).

Biomass. Amount of living matter (e.g., in a unit area or volume of habitat).

Blast furnace and auxiliary equipment. Furnace and equipment used in connection with the smelting process of reducing metallic ores to molten metal in order to remove, primarily, the oxygen from the ore and producing gas as a by-product. The furnace and equipment consist of, but are not limited to, the furnace proper, charging equipment, stoves, bleeders, gas dust catcher, gas cleaning devices, and other auxiliaries pertinent to the process.

Blasting agent. Any material or mixture consisting of a fuel and oxidizer intended for blasting, not otherwise classified as an explosive, in which none of the ingredients are classified as explosives, provided that the finished product, as mixed and packaged for use or shipment, cannot be detonated by means of a No. 8 test blasting cap when unconfined. Materials or mixtures classified as nitrocarbonitrates by Department of Transportation regulations should be included in this definition.

BOD. Biochemical oxygen demand; biological oxygen demand.

Boiler burning fuel in suspension. Fuel-burning device in which fuel is conditioned or pulverized previous to admitting the fuel into the furnace for combustion. The combustion process is completed with the fuel in suspension.

Breeching. Conduit for the transport of products of combustion or processes to the atmosphere or to any intermediate device discharging them into the atmosphere. It does not include the chimney or stack.

Bridge wall. Wall at the rear of the grate or stoker that acts as a deflector or radiant heat reflector for the furnace gases and as a stop to the fuel bed; rear wall of the ash pit.

Brine. Water that is saturated or strongly impregnated with common salt; strong saline solution (e.g., calcium chloride).

BROX System. System that treats organically contaminated brines generated in glycol production.

Building fires. The expression "a new fire being built," indicates the period during which a fresh fire is being started and does not mean the process of replenishing an existing fuel bed with additional fuel.

By-product coke plant. Plant used in connection with the distillation process to produce coke. In it the volatile matter in coal is expelled, collected, and recovered. Such a plant consists of, but is not limited to, coal- and coke-handling equipment and by-product chemical plant and other equipment associated with and attendant on the coking chambers or ovens making up a single battery operated and controlled as a unit.

Carbon dioxide (CO_2). The colorless, odorless, electrically nonconductive, inert gas.

Carbon monoxide. Gaseous compound consisting of the chemical formula CO.

Carcinogen. Substance or agent which produces or incites cancer.

Catalyst. Substance (e.g., an enzyme) that initiates a chemical reaction and enables it to proceed under milder conditions (e.g., at a lower temperature) than otherwise possible.

Catalytic combustion system. An oven heater or any construction that employs catalysts to accelerate oxidization or combustion of fuel–air or fume–air mixtures for eventual release of heat to an oven process.

Chemosterilant. Substance that produces sterility (e.g., of an insect) without marked alteration of mating habits or life expectancy.

CHESS (Community Health Effects Surveillance Studies). The study gives significant evidence of the adverse effects of atmospheric sulfates in the continental United States. From such studies there may be enough evidence to show a cause/effect relationship between ambient sulfur oxides and pulmonary disease such as emphysema.

Chimney or stack. Conduit, duct, vent, flue, or opening of any kind whatsoever arranged to conduct any products of combustion to the atmosphere vertically.

Cinders. Particles not ordinarily considered as fly ash or dust because of their greater size. They consist essentially of fused ash and/or unburned matter.

Cleaning fires. Removing ashes from the fuel bed or furnace.

Climatology. Science that deals with climates and their phenomena.

Closed-cycle technology. Processes designed to prevent pollutants from escaping into the environment.

Coke. The solid fuel obtained by the carbonization of coal or the solid residue of petroleum product manufacture.

Cold boiler or furnace. A boiler or furnace in which fuel has not been consumed for a period of 24 hours or more.

Combined sewer. Sewer receiving both surface runoff and sewage.

Combined sources research. Development of technology to treat industrial wastes from several plants in a region with a single facility or in combination with municipal waste management.

Combustible refuse. Combustible waste material containing carbon in a free or combined state other than a liquid or gas.

Combustible rubbish. Rags, old clothes, leather, rubber, carpets, wood, excelsior, sawdust, tree branches, yard trimmings, wood furniture, and other combustible solids not considered by the city to be of a highly volatile or explosive nature.

Combustible trade waste. Paper, rags, leather, rubber, cartons, boxes free of wire and metal scraps, wood excelsior, sawdust, garbage, and other combustible solids except manure, and not considered by the city to be of a highly volatile or explosive nature.

Combustible waste matter. Magazines, books, trimmings from lawns, trees, or flower gardens, leaves, pasteboard boxes, rags, paper, straw, sawdust, packing material, shavings, boxes, and all rubbish and refuse that will ignite through contact with flames or ordinary temperatures.

Combustion. A chemical process that involves oxidation sufficient to produce light or heat.

Compliance monitoring. Monitoring that is undertaken to gather specific evidence from a point source or discharge for use in possible litigation.

Condensed fumes. Fumes that have cooled and returned to a liquid or solid.

Conservation vent valve. Weight-loaded valve designed and used to reduce evaporation losses of volatile organic substances by limiting the amount of air admitted to or vapors released from the vapor space of a closed storage vessel.

Construction. Installation or erection of fuel-burning combustion or process equipment or device.

Control apparatus. Device that prevents, eliminates, or controls the emission of an air contaminant.

Control device. Any device which has as its primary function the control of emissions from fuel burning, refuse burning, or from a process, and thus reduces the creation of, or the emission of, or both, air pollutants into the atmosphere.

Control technology. Combination of hardware, operating procedures, and process changes used to reduce the harmfulness of gaseous, liquid, or solid effluents from a pollution source. The technology is normally based on contaminant removal and isolation, transformation of the contaminants chemically and/or physically to a less harmful form, or dispersion of the contaminants to prevent localized high levels.

Corrosive liquids. Includes those acids, alkaline caustic liquids, and other corrosive liquids which, when in contact with living tissue, will cause damage to such tissue by chemical action or are liable to cause fire when in contact with organic matter or with certain chemicals.

Criteria. Information used as guidelines for decisions when establishing air quality goals, air quality standards, and the various air quality alert levels. In no case are criteria to be confused with air quality standards or goals.

Criteria pollutants. Pollutants for which an ambient air quality standard has been set. Currently, standards have been set for six pollutants, sulfur dioxide (SO_2), carbon monoxide (CO), total suspended particles (TSP), hydrocarbons (HC), oxidants (OX) and nitrogen oxides (NO_x).

Cupola. Vertical furnace in which alternate layers of basic material and coke are charged to produce molten ferrous and nonferrous metal for the production of castings. Auxiliary equipment consists of, but is not limited to, blowers, charging mechanism, collection equipment, heat exchangers, and slagging equipment.

Cytogenetics. Branch of biology that deals with the study of heredity and variation by the methods of both cytology (history of cells) and genetics.

Dairy beverage and food-processing equipment. Equipment used in the production of milk and dairy products, foods, and beverages, including their processing, preparation, or packaging for consumption.

Damper—automatic or manual. Device for regulating the volumetric flow of gas or air.

Decibel (dB). Unit for measuring the volume of a sound, equal to the logarithm of the ratio of the intensity of the sound to the intensity of an arbitrarily chosen standard sound.

Defect in vehicle or noisy load. The use of any automobile, motorcycle, street car, or other vehicle so out of repair or loaded in such a manner as to create loud or unnecessary grating, grinding, rattling, or other noise.

Diesel-powered motor vehicle. Vehicle that is selfpropelled by a compression-ignition type of internal combustion engine.

Discrete pulses. Pluses which do not exceed one hundred (100) impulses per minute.

Discrete tone. Sound wave whose instantaneous sound pressure varies essentially as a simple sinusoidal function of the time.

Domestic heating plant. Plant generating heat usually for a single-family residence, for two residences in either duplex or double-house form, or for multiple-dwelling units in which such a plant serves fewer than three apartments.

Domestic refuse-burning equipment. Refuse-burning equipment or incinerator used for a single-family residence, for two residences in either duplex or double-house form, or for multiple-dwelling units in which such equipment or incinerator serves fewer than three apartments.

Downdraft furnace. In this furnace there are two separate grates, one above the other. The top grate consists of water tubes, the bottom grate consists of common grate bars and is fed by half-consumed fuel falling from the upper grate. Air for combustion enters the upper fire door and passes through the bed of green fuel on the upper grate and then over the incandescent fuel on the lower grate.

Dryer. Device for drying by heat or forced ventilation, or both; apparatus such as a furnace, oven, or revolving kiln for expelling moisture or volatiles by evaporation or volatization.

Dual-use fallout shelter. A dual-use fallout shelter is a space having a normal, routine use and occupancy as well as having an emergency use as a fallout shelter.

Dump. Lot or land or part thereof used primarily for disposal by abandonment, dumping, burial, burning, or any other means and for whatever purpose, of garbage, offal, sewage, trash, refuse, junk, discarded machinery, vehicles or parts thereof, or waste material of any kind.

Dust. Particulate matter released into the air by natural forces; by any fuel-burning, combustion, or process equipment or device; by construction work; or by mechanical or industrial processes, such as crushing, grinding, milling, drilling, demolishing, shoveling, bagging, sweeping, covering, conveying, transferring, transporting, and the like.

Dust-free. Property which is maintained dust-free by paving with one of the following methods: asphaltic concrete, cement concrete, penetration treatment of bituminous material and a seal coat of bituminous binder and mineral aggregate, or the equivalent of these methods.

Dust separating equipment. Any device for separating the solid products of any combustion process, i.e., dust, solids, particulate matter, fly-ash, or any combination thereof, from the gases in which they are carried.

Ecological criteria development. Work that includes laboratory studies (such as bioassays) to establish tolerable pollutant levels. Under ecological criteria work is performed in direct response to legislative mandates to define numerical standards for pollutants.

Ecological processes and effects subprogram. Research subprogram that provides EPA with the knowledge and theoretical structure on which to base environmental criteria, standards, and regulations.

Ecosystem. Living community and the physical environment associated with it functioning as a unit in nature.

Effluent. Outflow from a pollution source.

Electric furnace. Furnace in which the melting and refining of metals is accomplished by means of electrical energy.

Emission. The act of releasing or discharging air pollutants from any source into the outdoor atmosphere.

Energy Conservation; Utilization; and Technology Assessment Subprogram. Subprogram that focuses on identification, characterization, assessment, and development of control technology for pollutants associated with utility and industrial combustion sources.

Energy Extraction and Processing Technology Subprogram. The objective of this subprogram is to permit a rapid increase in the extraction and processing of domestic energy resources and to enable these energy sources to be used effectively in an environmentally compatible manner.

Environment. Totality of natural and induced conditions occurring or encountered at any one time and place.

Environmental control. Method by which the severity of a damaging environmental stress is reduced to a level tolerable by equipment or personnel.

Environmental design criteria. Environmental parameters that represent a given degree of severity of conditions existing in nature, in equipment operation, or in storage, which are to be incorporated in the design of equipment.

Environmental engineering. Branch of engineering concerned with the designing, developing, and testing of equipment or materials to function reliably under all environmental conditions expected during their intended operational transportation or storage life.

Environmental factor. One of the components of an environment; environmental element. Environmental factors may be either induced, including those conditions resulting from the operation of a structure or item of equipment, or natural, including those conditions generated by the forces of nature and whose effects are experienced when the equipment or structure is at rest, as well as when it is in operation. The distinction between natural and induced environmental factors cannot always be clearly discerned or precisely defined.

Environmental geology. Applying geological data, information, and principles to the solutions of conditions created by human occupancy and activity that are detrimental to land and water resources.

Environmental impact analysis. Predetermination of the extent of pollution or environmental degradation that will be involved in a mining or processing project.

Environmental impact report. Detailed report of the existing conditions of a site or location and the environmental impact on the site or location that occurs due to its occupancy and specified use.

Environmental impact statement. Report of the potential effect of plans for land use in terms of the environmental, engineering, esthetic, and economic aspects of the proposed objective.

Environmental Management Subprogram. The objective of this subprogram is to provide regional environmental planners and managers with methods to determine feasible alternative solutions to specific environmental problems and to provide techniques for selecting the lowest cost solutions.

Environmental operating conditions. Factors of the environment that singly or in combination have a significant effect on industrial operations and must, therefore, be considered in the design and testing of materials.

Environmental protection. Research and its application designed to maintain or improve the degree of effective performance of man and equipment under all types of environmental stress.

Environmental Protection Agency. Agency of the U.S. Government, established in 1970, and given the responsibilities for controlling pollution of water and air, solid wastes, noise nuisance, radiation, and all other functions that create pollution and nuisances that are detrimental to the public.

Environmental research. Systematic study and investigation of any environmental factor or combination of factors for the purpose of discovering basic rules or principles governing their cause and behavior, extending knowledge of their occurrence and distribution, or ascertaining the relation between them and other aspects of the environment, both natural and induced.

Environmental resistance features. Characteristics or properties of a product that protect the product against the effects of an environmental exposure and prevent internal conditions that might lead to deterioration.

Environmental suitability of a product. Suitability of a product for serving its intended purpose under prescribed environmental conditions.

Environmental test. Laboratory test conducted to determine the functional performance of a component or system under conditions that simulate the real environment in which the component or system is expected to operate.

Environment of sedimentation. More or less destructive geomorphologic setting where sediments are deposited as beach environment.

Epideminology. Branch of medical science that deals with the incidence, distribution, and control of disease in a population; sum of factors controlling the presence or absence of a disease or pathogen.

Episode stage. A level of air pollution in excess of the ambient air quality standard that may result in an imminent and substantial danger to public health or welfare. This term should include alert, warning, and emergency stages.

Equipment causing pollution. Device capable of causing the emission of an air contaminant into the open air and any stack, chimney, conduit, flue, duct, vent or similar device connected, attached to or serving the equipment. Included is equipment in which the preponderance of the air contaminants emitted is caused by a manufacturing process.

Excess air. Air supplied in addition to the theoretical quantity necessary for complete combustion of all fuel and/or combustible waste material present.

Exhaust emissions. Substances emitted into the atmosphere from any opening downstream from the exhaust ports of a motor vehicle engine.

Existing source. Equipment, machines, devices, articles, contrivances, or installations which are under construction or in operation on the effective date of the regulation, except that any existing equipment, machine, device, article, contrivance, or installation which is altered, replaced, or rebuilt after the effective date of the regulation shall be defined as a new source.

Explosive. A chemical compound or mechanical mixture that is commonly used or intended for the purpose of producing an explosion, that contains any oxidizing and combustible units, or other ingredients, in such proportions, quantities, or packing, that an ignition by fire, by friction, by concussion, by percussion, or by detonator of any part of the compound or mixture may cause such a sudden generation of highly heated gases that the resultant gaseous pressures are capable of producing destructive effects on contiguous objects or of destroying life and limb.

Extension furnace (Dutch oven). Masonry structure or combination of masonry and metal built on the front of a boiler or other combustion device for the purpose of obtaining additional furnace volume.

Fallout shelter. A fallout shelter is any room, structure or space designated as such and providing its occupants with protection from fallout at a minimum protection factor of forty (40) from gamma radiation, as determined by an architect or engineer certified by the Office of Civil Defense as a Qualified Fallout Shelter Analyst.

Fate. Final destination for a substance that has traveled through the biosystem.

Floc. Loose, fluffy mass formed by the aggregation of a number of fine particles suspended in a liquid medium, usually water.

Flocculate. Cause to a aggregate into a fluocculent (loose, fluffy organization) mass.

Fluctuating noise. Noise whose sound pressure level varies significantly but does not equal the ambient environmental level more than once during the period of observation.

Flue-gas desulfurization (FGD). Process by which flue gas from coal-fired utility and industrial boilers is cleaned by passage over or through a bed of chemically active minerals, such as lime or limestone. The process is one of the few coal pollution control techniques available in the 1970s that meets Clean Air Act requirements.

Fluidized-bed combustion (FBC). Technique for burning coal on a suspended bed of mineral matter.

Fly ash. Particulate matter capable of being gas or airborne and consisting essentially of fused ash and/or burned or unburned material.

Fossil fuel. Natural gas, petroleum, coal, and any form of solid, liquid, or gaseous fuel derived from such materials.

Fossil-fuel-fired steam generating unit. A furnace or boiler, or combination of furnaces or boilers, connected to a common stack, used in the process of burning fossil fuel for the primary purpose of producing steam by heat transfer.

Freon. Generic term, originally a trade name, applied to various nonflammable fluorinated hydrocarbons used as refrigerants and/or propellants for aerosols.

Foundries—ferrous and nonferrous. Processes, devices and equipment used for producing castings, other than die castings, from basic material. Such processes, devices, and equipment consist of, but are not limited to, charging equipment, furnaces, collection equipment, and cleaning operations. Basic materials used include, but are not limited to, iron, brass, aluminum, and magnesium.

Fuel. Any form of combustible matter—solid, liquid, vapor, or gas.

Fuel-burning, combustion or process equipment or device. Any furnace, incinerator, fuel-burning equipment, refuse-burning equipment, boiler, apparatus, device, mechanism, fly ash collector, electrostatic precipitator, smoke-arresting or prevention equipment, stack, chimney, breeching or structure, used for the burning of fuel or other combustible material, or for the emission of products of combustion, or used in connection with any process that generates heat and may emit products of combustion. Included are process furnaces, such as heat-treating furnaces; by-product coke plants, core-baking ovens, mixing kettles, cupolas, blast furnaces, open-hearth furnaces, heating and reheating furnaces, puddling furnaces, sintering plants, Bessemer converters, electric steel furnaces, ferrous foundries, nonferrous foundries, kilns, stills, dryers, roasters, and appliances used in connection with any of the above equipment or devices and all other methods or forms of manufacturing, chemical, metallurgical or mechanical processing that may emit smoke, or particulate, liquid, gaseous, or other contaminated matter.

Fuel-burning equipment. Any furnace, boiler, apparatus, stack, and all appurtenances thereto, used in the process of burning fuel for the primary purpose of producing heat or power by indirect heat transfer.

Fuel-burning equipment—hand-fired type. Any fuel-burning, combustion, or process equipment or device, other than process equipment, or downdraft furnaces in which fuel is manually introduced directly into the furnace.

Fuel-burning equipment—mechanical. Fuel-burning, combustion, or process equipment or device incorporating a device by means of which fuel is mechanically introduced from outside the furnace into the zone of combustion.

Fuel oil. Oil commonly used as a fuel.

Fugitive dust. Solid, airborne, particulate matter emitted from any source other than through a stack.

Fumes—gases. Vapors or particulate matter that are of such character as to cause atmospheric pollution.

Fumigant. Includes any substance that by itself or in combination with any other substance emits or liberates a gas, fume, or vapor used for the destruction or control of insects, fungi, vermin, germs, rodents, or other pests, and should be distinguished from insecticides and disinfectants, which are essentially effective in the solid or liquid phases. Examples are methyl bromide, ethylene dibromide, hydrogen cyanide, carbon disulphide, and sulfuryl fluoride.

Furnace. Enclosed space provided for the ignition and/or combustion of fuel.

Furnace volume. Volume of the chamber or enclosure in which the combustion process takes place.

Garbage

1. Animal and vegetable matter such as that originating in houses, kitchens, restaurants and hotels, produce markets, food service and processing establishments, and greenhouses.
2. Solid wastes from the preparation, cooking, and dispensing of food; from the handling, storage, or sale of meat, fish, fowl, fruit, or vegetables; and from condemned food.

Geothermal. Of or relating to the heat of the earth's interior.

Glare. Excessive illumination. Flickering or intense sources of light should be controlled or shielded so as not to cross lot lines.

Glaring light near streets. Use of spotlights and other directional types of lights to illuminate buildings, displays, or signs are not usually permitted unless the lights are directed or shielded in such a manner that they do not cause glare or other annoyance to pedestrains and drivers of vehicles on near-by streets and roads.

Glycol. Ethylene glycol-alcohol containing two hydroxyl groups.

Goal. Expected level of air quality; air quality level to be obtained.

Ground or comminuted garbage. Wastes from the preparation, cooking, and dispensing of foods that has been comminuted to such a degree that all particles will be carried freely in suspension under conditions normally prevailing in public sewers, with no particle greater than $\frac{1}{2}$ in. in any dimension.

Groundwater. Water within the earth that supplies wells and springs.

Halogen. Any of five elements—fluorine, chlorine, bromine, iodine, and astatine—that form part of group VII A of the period table and normally exist in the free state as diatomic molecules.

Hazardous air pollutant. Air pollutant to which no ambient air quality standard is applicable and which in the judgment of the EPA administrator may cause or contribute to an increase in mortality or increase in serious irreversible or incapacitating illness.

Health and Ecological Effects Program. Program that is fundamental to EPA's responsibility to set criteria standards and guidelines to enhance environmental quality. It provides information for the establishment of water quality criteria, air quality criteria, ocean disposal criteria, pesticide registration guidelines, effluent standards for toxic and hazardous materials, and radiation standards.

Health and Ecological Effects / Energy Subprogram. Subprogram that identifies all adverse environmental aspects (essential for criteria development and control technology requirements) associated with energy extraction, conversion, and use.

Heating and reheating furnace. Furnace in which metal is heated to permit shaping or forming, or to achieve specific physical properties.

Heating or low-pressure boilers. All boilers designed for operating at a steam pressure of 15 psig or less.

Heating plant—other than domestic. Fuel-burning equipment used for the space heating of multiple-dwelling units containing more than two apartments, hotels, rooming houses, boarding houses, garages, schools, hospitals, churches, office buildings, stores, institutions, and for all commercial, industrial, or other establishments.

Heating surface. Any surface having steam, water, or other fluid on one side and hot gases on the other side, as found in a boiler or a warm-air heating furnace, not excepting any surface covered by arches or refractory.

Heating value. Heat released by combustion of 1 lb. of waste or fuel measured in British thermal units on an as-received basis. For solid fuels the heating value shall be determined using the latest revision of ASTM D 2015.

Horns—signaling devices and the like. The sounding of any horn or signal device on any automobile, motorcycle, bus, street car or other vehicle while not in motion is prohibited, except as a danger signal that another vehicle is approaching apparently out of control; or if in motion, is allowed only as a danger signal or where the motor vehicle statutes requires the sounding of such horn or signal device.

Hurricane requirements. Lateral support securely anchored to all walls provides the best, and only, sound structural stability against horizontal thrusts, such as winds of exceptional velocity.

Hydrocarbons. Class of chemical compounds consisting of hydrogen and carbon.

Idle speed. Revolutions per minute of a spark-ignition engine with a closed throttle operating at the manufacturer's recommended speed.

Impact noise. A short-duration sound which is incapable of being accurately measured on a sound level meter.

Impact vibration. Vibrations occurring in discrete pulses separated by an interval of at least one (1) minute and numbering no more than eight (8) per twenty-four (24) hour period.

Impulse. Discrete vibration pulsation occurring no more often than one (1) per second.

Impulsive noise. Noise characterized by brief excursions of sound pressure (acoustic impulses) that significantly exceed the ambient environmental sound pressure. The duration of a single impulse is usually less than 1 second.

Incinerator

1. Device intended or used for the destruction of garbage or other combustible refuse or waste materials by burning.
2. Combustible apparatus designed for high-temperature operation in which solid, semisolid, liquid, or gaseous combustible wastes are ignited and burned efficiently and from which the solid residues contain little or no combustible material.

Indirect heat exchanger. Equipment in which fuel is burned for the primary purpose of producing steam, hot water, or hot air or for the other indirect heating of liquids, gases, or solids, in which the products of combustion do not come into direct contact with process materials. Fuels may include, but are not limited to, coal, coke, lignite, coke breeze, gas, fuel oil and wood, but do not include refuse. When any products or by-products of a manufacturing process are burned for the same purpose, or in conjunction with any fuel, the same maximum limitation shall be governed by the most stringent limitation when refuse burning and indirect heat exchanger emissions are both considered.

Industrial cleaning equipment. Machinery and other tools used in cleaning processes during the course of industrial manufacturing, production, and assembly.

Industrial wastes. Solid, liquid, or gaseous wastes resulting from any industrial, manufacturing, trade, or business process or from the development, recovery, or processing of natural resources, including decayed wood, sawdust, shavings, bark, lime, refuse, ashes, garbage, offal, oil, tar, chemicals, and all other substances except sewage and industrial wastes.

Integrated Assessment Subprogram (part of the energy environmental program). The subprogram's objectives are to integrate the complex, environmental, social, and economic issues of various technologies under alternative environmental management systems.

Intercepting sewer. Sewer installed for the purpose of receiving sewage or combined sewage and storm flow from one or more sewers.

Intermittent controls. Controls that are used at intervals, that is, put into use when pollution is heavy and then later turned off or discontinued.

Intermittent noise. Noise whose sound-pressure level equals the ambient environmental level at least twice during the period of observation. The period of time during which the level of the noise remains at an essentially constant value different from that of the ambient is on the order of 1 second or more.

Internal combustion engine. Engine in which the combustion of gaseous, liquid, or pulverized solid fuel takes place within one or more cylinders.

Kiln. Furnace or heated chamber used for hardening, burning, or drying in the manufacture of such products as clay, brick, cement, pottery, ceramics, or limestone.

Lichen. Any of numerous complex thallophytic plants made up of an alga and a fungus growing in symbiotic association on a solid surface (as a rock).

Low-volatile coal. Coal that consists of 78 to 86% fixed carbon, and 14 to 22% volatile matter. A coal that is nonagglomerating.

Lugging. Reduction of speed from the maximum governed speed due to increased load.

Machine dishwasher. Equipment manufactured for the purpose of cleaning dishes, glassware, and other utensils involved in food preparation, consumption, or use, by means of a combination of water agitation and high temperatures.

Malathion. Thiophosphate insecticide with a lower mammalian toxicity than parathion.

Manufacturing process. Any action, operation or treatment embracing chemical, industrial, manufacturing, or processing factors, methods, or forms, including, but not limited to, furnaces, kettles, ovens, converters, cupolas, kilns, crucibles, stills, dryers, roasters, crushers, grinders, mixers, reactors, regenerators, separators, filters, reboilers, columns, classifiers, screens, quenchers, cookers, digesters, towers, washers, scrubbers, mills, condensers or absorbers.

Marginal land. Land that is barely productive agriculturally owing to its nonproductive capacity or its limited water supply. Attempts made to use such marginal land often result in accelerated land erosion, resource degradation, and the impairment of wildlife habitat and aquatic environments.

Materials processing research. Research that covers many industrial activities that mechanically or chemically change a material from one form to another, such as metal working or electroplating.

Materials production research. Research that includes problems of industries concerned with exploration for and production of raw materials such as iron, aluminum, and limestone.

Maximum allowable emission rate. Maximum amount of an air contaminant that may be emitted into the outdoor air during any prescribed interval of time.

Mechanical combustion equipment or mechanically fired apparatus. Fuel-burning, combustions, or process equipment or device in which the fresh fuel or combustible material is mechanically introduced from outside the furnace into the zone of combustion, the same being actuated by controls.

Measurement Techniques and Equipment Development Subprogram. Subprogram that involves development, evaluation, and demonstration of field and laboratory measurement and monitoring methods and instrumentation.

Microcosm. Community or other unity that is a typical or ideal example of a larger unity.

Minerals Processing and Manufacturing Industries Subprogram. Subprogram that considers point sources of water, air, and residue pollution produced by industry.

Minority Insitutions Research Support (MIRS). EPA program conducted to direct research grants to minority institutions in the area of environmental research.

Mobile source pollutants. Pollutants resulting from a source that moves, such as automobile emissions.

Modification. Any physical change in, or change in the method of operation of, a stationary source which increases or decreases the amount of any air pollutant emitted by such facility, or which results in the emission of any air pollutant not previously emitted, except that such term should not include the following: (a) routine maintenance, repair, replacement; (b) an increase in the production rate, if such increase does not exceed the operating design capacity of the affected facility; (c) an increase in hours of operation, if such increase does not exceed the operating design capacity of the facility; (d) use of an alternative fuel or raw material if, prior to the date any standard under the code becomes applicable to such facility, the affected facility is designed to accommodate such alternative use.

Monitoring and Technical Support Program. Program that includes research, development, and demonstration activities and direct assistance and support to all of EPA. The program includes three subprograms: Measurement Techniques and Equipment Development, Quality Assurance, and Technical Support.

Motor vehicle

1. Every self-propelled over-the-road vehicle, except electrically powered vehicles.

2. Passenger vehicle, truck, truck-tractor, trailer, or semitrailer propelled or drawn by mechanical power.

Multiple chamber incinerator. Any incinerator consisting of three or more refractory lined combustion chambers in series, physically separated by refractory walls, interconnected by gas passage ports or ducts, and employing adequate design parameters necessary for maximum combustion of the material to be burned. The combustion chamber should include, as a minimum, one chamber principally for ignition, one chamber principally for mixing, and one chamber for combustion.

Mutagenesis. Occurrence or induction of a relatively permanent change in hereditary material involving either a physical change in chromosone relations or a biochemical change in the codons (a triplet of nucleotides that is part of the genetic code and that specifies a particular amino acid in a protein or starts or stops protein synthesis) that make up genes.

Natural outlet. Outlet into a water course, pond, ditch, lake, or other body of surface water not man-made.

New fire being built. Period during which a fresh fire is being started. The expression does not include the process of replenishing an existing fuel bed with additional fuel.

New source. Equipment, machines, devices, articles, contrivances, or installations built or installed on or after the effective date of a regulation, or existing at such time, that are later altered, repaired, or rebuilt. Any such equipment, machines, devices, articles, contrivances, or installations moved to a new address or operated by a new owner or a new lessee after the effective date of the regulation should be considered a new source.

Noise. A disturbing sound of any kind caused by any circumstance.

Noises to attract attention. The use of any drum, loudspeaker, bell, or other instrument or device for the purpose of attracting, by creation of such noise, to any performance, show or sale, or display of merchandise is prohibited, except where the director of police has given permission for the use of same on a given occasion, at a given place, which permission should limit such uses to a period of not exceeding 48 hours.

Noncombustible rubbish. Metals, metal shavings, tin cans, glass, crockery, and other similar materials, but not the wastes resulting from building construction or alteration work. It should also include any small accumulation of yard dirt if stored as required in the code.

Noncombustible trade waste. Metals, metal shavings, wire, tin cans, cinders, earth, and other materials, but not the wastes resulting from building construction or alteration work.

Noncriteria pollutant. Hazardous pollutant (such as mercury, fluorides, vinyl chloride), for which no ambient air quality standard has been established. Insufficient health effects data have been developed to establish a "safe" exposure level for these pollutants.

Nonpoint source pollutants. Pollutants arising from certain management practices in the area of renewable resources, such as application of fertilizers or pesticides to productive land.

NO_x control technology. Research and development seeks to identify, assess, and promote development of cost-effective commercial methods for control of oxides of nitrogen (NO_x) from both existing and new stationary combustion sources.

Noxious acids. Anhydrous or hydrous acid forms in concentrations high enough to be toxic, to cause atmospheric pollution, or to constitute a nuisance.

Noxious matter. Material that is capable of causing injury or malaise to living organisms by chemical reaction or is capable of causing detrimental effects upon the health

or the psychological, social or economic well-being of human beings.

Octave band. A prescribed interval of sound frequencies that classifies sound according to pitch.

Odor. Emission of noxious, odorous matter is prohibited in such quantities as to be readily detectable at any point along lot lines, when diluted in the ratio of one volume of odorous air to four or more volumes of clean air, or as to produce a public nuisance or hazard beyond lot lines.

Odor nuisance. Any noxious odor in sufficient quantities and of such characteristics and duration as to be injurious to human, plant, or animal life, to health, or to property, or to unreasonably interfere with the enjoyment of life or property.

Odorous matter. Any material that produces an olfactory response among human beings.

Odor threshold. The concentration of odorous matter in the atmosphere necessary to be perceptible to the olfactory nerve of normal persons.

Oil burner. Device for the introduction of vaporized or atomized fuel oil into a furnace.

Oil-effluent water separator. Tank, box, sump, or other container or group of such containers in which an organic material floating on, entrained in, or contained in the water entering such containers is physically separated and removed from water prior to the exit from the container.

Opacity

1. Fraction of light transmitted from a source that is prevented from reaching the observer or instrument by reason of smoke and exhaust emissions.

2. Light-obscuring ability of an emission plume other than black.

Open air. All spaces outside of buildings, stacks, or exterior ducts.

Open fire. Fire from which the products of combustion are emitted directly into the open air without passing through a stack or chimney.

Open-hearth furnace. Furnace in which the melting and refining of metal is accomplished by the application of heat to a saucer-type or shallow hearth in an enclosed chamber. Such furnace consists of, but is not limited to, the furnace proper, checkers, flues, and stack and may include a waste heat boiler and other auxiliaries pertinent to the process.

Organic solvents. Volatile organic compounds that are liquids at standard conditions and are used as dissolvers, viscosity reducers, or cleaning agents.

Organic substances. Chemical compounds of carbon, including diluents and thinners that are liquids at standard conditions and are used as dissolvers, viscosity reducers, or cleaning agents, but excluding methane, carbon monoxide, carbon dioxide, carbonic acid, metallic carbonic acid, metallic carbide, metallic carbonates, and ammonium carbonate.

Organic vapor. Gaseous phase of an organic substance or a mixture of organic substances present in the atmosphere.

OSWMP (Office of Solid Waste Management Programs). Office established by EPA to deal with the national solid waste problem.

Outfall. Outlet of a body of water; mouth of a drain or sewer.

Ozone. Triatomic form of oxygen formed naturally in the upper atmosphere by a photochemical reaction with solar ultraviolet radiation. It is also generated commercially by an electric discharge in ordinary oxygen or air. It is a major agent in the formation of smogs and is used especially in disinfection and deodorization and in oxidation and bleaching. Its natural role in the upper atmosphere is to shield the earth from excess ultraviolet radiation.

Paper. Newspapers, periodicals, cardboard, and all other wastepaper.

Particulate matter

1. Solid or liquid material, other than water, that exists in finely divided form.

2. Material, other than water, that is suspended in or discharged into the atmosphere in a finely divided form as a liquid or solid.

Pathogen. Specific disease-causing agent such as a bacterium or virus.

Performance standard. A criterion established to control noise, vibration, smoke and particulate matter, toxic matter, odorous matter, fire and explosive hazards, and glare and radiation hazards generated by or inherent in uses of land or structures.

Period of observation. Time interval during which accoustical data are obtained. The period of observation is determined by the characteristics of the noise being measured and should also be at least ten times as long as the response time of the instrumentation. The greater the variance in indicated sound level, the longer must be the observation time for a given expected accuracy of the measurement.

Pesticide registration. EPA process by which a pesticide is approved for use.

Petrochemical. Chemical isolated or derived from petroleum or natural gas.

Petroleum coke. The residue of various petroleum processes which may be handled and burned as a solid fuel.

Pheromone. Chemical substance that is produced by an animal and serves as a stimulus to others of the same species for one or more behavioral responses.

Physical and chemical coal cleaning. Process involving methods to physically or chemically remove sulfur from coal with a moderate sulfur content (1 to 2%). It allows coal to be burned in conformity with Clean Air Act standards.

Poisonous gas. Any noxious gas of such nature that a small amount of the gas in air is dangerous to life. Examples are chlorine, cyanogen, fluorine, hydrogen cyanide, nitric oxide, nitrogen tetraoxide, and phosgene.

Poisonous gases or liquids. Includes gases that are highly poisonous even when present in the air in very small proportions; also liquids which give off highly poisonous vapors at ordinary temperatures.

Pollutant-of-the-month syndrome. Crisis atmosphere produced by a continuing series of revelations that show new substances to be harmful.

Pollution to waterways. Includes but is not limited to the discharge of deposits in or on public waterways of sewage, industrial wastes, or other wastes containing soluble or insoluble solids of organic or inorganic nature that may deplete the dissolved oxygen content of the waterways, contribute settleable solids that may form sludge deposits, contain oil, grease, or floating solids that may cause unsightly appearance on the surface of the waterways, or contains soluble materials detrimental to aquatic life, all beyond the content of such like substances present in an equal volume of the effluent discharge from the sewage treatment works into similar receiving waterways.

Polyphosphate builder of phosphorus. Water softening and soil-suspending agent made from condensed phosphates, including pyrophosphates, triphosphates, tripolyphosphates, metaphosphates and glassy phosphates, used as a detergent ingredient, but not including "polyphosphate builders" of "phosphorus" which are essential for medical, scientific, or special engineering use.

Portable boiler. Boiler used separately or in connection with a power shovel, road roller, hoist, derrick, pile driver, steam locomotive, diesel locomotive, steamboat, tugboat, tar kettle, asphalt kettle, or any other portable equipment capable of emitting smoke, particulate, or other matter.

Portable equipment. Equipment designed for the purpose of being readily transferred from one location to another.

Potable water. Water free from impurities present in amounts sufficient to cause disease or harmful physiological effects. Its bacteriological and chemical quality should conform to the requirements of the department of health.

Power boiler. A boiler carrying more than 15 psig steam and of more than 10 boiler horse power.

Power or high-pressure boilers. All boilers designed for operating at a steam pressure greater than 15 psig.

PPM. Parts per million.

PPM (vol)—parts per million (volume). Volume over volume ratio that expresses the volumetric concentration of a gaseous air contaminant in million-unit volumes of gas —for example, the number of microliters of sulfur dioxide per million microliters of air would be expressed in ppm (vol).

Pressure tank. Tank in which fluids are stored at a pressure greater than atmospheric pressure.

Primary air quality standards. Primary standards are defined as "allowing an adequate margin of safety" in protecting the public health.

Process furnace. Any furnace, kiln, still, or combustion device, other than a boiler furnace, used for the generation of heat or power.

Processes or process equipment. Action, operation, or treatment embracing chemical, industrial, or manufacturing factors, such as heat-treating furnaces, by-product coke plants, core-baking ovens, mixing kettles, cupolas, blast furnaces, open-hearth furnaces, heating and reheating furnaces, puddling furnaces, sintering plants, Bessemer converters, electric steel furnaces, ferrous and nonferrous foundries, kilns, stills, dryers, roasters, equipment used in connection therewith, and all other methods or forms of manufacturing or processing that may emit smoke, particulate matter, or other matter.

Processing. Any operation changing the nature of material or materials, such as the chemical composition or physical qualities. Does not include operations described as fabrication.

Processing of fuel. The washing, cleaning, screening, drying and pulverizing, flotation, coking, carbonization, quenching, briquetting, bagging, and packaging of solid fuels; the refining of liquid fuels; the manufacture of gaseous fuels.

Process weight. The total weight in pounds of all materials introduced into any specific process.

Public Sector Program. Program that includes three research subprograms: Waste Management, Water Supply, and Environmental Management.

Pyrolysis. Breaking up of large organic molecules brought about by the action of heat.

Pyrophoric dust. A dust in a finely-divided state that is spontaneously combustible in air.

Quality Assurance Subprogram. Subprogram that focuses on standardizing measurement methods providing standard reference materials and samples, and developing quality control guidelines and manuals, on-site evaluation of analytical laboratories, etc.

Radiation hazards. The deleterious and harmful effects of all ionizing radiation, which should include all radiation capable of producing ions in their passage through matter. Such radiations should include, but are not limited to, electromagnetic radiations, such as x-rays and gamma rays, and particulate radiation, such as electrons or beta particles, protons, neutrons, and alpha particles.

Radios, TVs, phonographs, etc. The playing or permitting the playing of any radio, television set, phonograph, musical instrument, or machine or device for the production or reproducing of sound is not allowed in such a manner or with such volume as to unreasonably annoy or disturb the quiet, comfort, or repose of persons in any dwelling, hotel, or any other type of residence, particularly between the hours of midnight and 8 a.m.

Radon. Colorless, odorless gas that results from the decay of radium, a common element found in soil, rock, and groundwater. Radon as it decays produces radon progeny, particles that emit high levels of alpha radiation. Radon progeny frequently attach themselves to other particulate matter and, when inhaled, bombard lung tissue with radiation levels associated with cancer.

Recommended-use level. Amount of synthetic detergent or other detergent recommended by the manufacturer for use per wash load; level at which detergent will effectively perform its intended function.

Reconstruction. Material change or alteration of existing fuel-burning, combustion, or process equipment or device from the physical or operating condition for which approval was last obtained; addition, removal, or replacement of appurtenances or devices that materially affect the method of preventing the discharge of pollutants into the atmosphere or its efficiency.

Refuse. Garbage, rubbish, and trade wastes.

Refuse burning equipment. Any destructor, incinerator, furnace, oven, or other apparatus and appurtenances thereto used primarily for the purpose of destroying, reducing, or consuming refuse as herein defined or any other material by combustion. This should also include crematories.

Regenerable. Term describing substances that can be reconstituted and used again.

Rendering. Heating process, including cooking, dehydrating, digesting, or evaporating, that leaves protein concentrations of animal or marine matter.

Renewable Resources Subprogram. Subprogram that includes food, fiber, and wood production and related activities ranging from agricultural production through harvesting.

Repair. Any work that requires the equipment to be wholly or partially dismantled and that results in the restoration of the equipment to its original state.

Residual oil. Fuel oil having a viscosity heavier than 125 seconds Saybolt Universal at 100°F, referred to as grades numbered 5 and 6 in Commercial Standard CS-12 U.S. Department of Commerce.

Retrofit. Furnish with new parts or equipment not available or installed at the time of manufacture.

Ringlemann Chart. Chart published and described in the U.S. Bureau of Mines Information Circular 8333, and on which are illustrated graduated shades of gray to black for use in estimating the light-obscuring capacity of smoke.

Ringlemann Number. The number of the area on the Ringlemann Chart that most nearly coincides with the visual density of emission or light-obscuring capacity of smoke.

Roaster. Device used to effect the expelling of volatile matter or oxidation as required in the manufacture of such products as prepared meats, grain, coffee beans, or nuts.

RPM. Engine crankshaft revolutions per minute.

Rubbish material. All miscellaneous matter, such as bottles, rags, mattresses, worn-out furniture, old clothes, old shoes, broken glass, leather, carpets, crockery, metal, rubber, cut grass, leaves, tree branches, lumber, or any materials that may accumulate as the result of building operations.

Safe Drinking Water Act (SDWA). The EPA administrator may conduct research, studies, and demonstrations relating to the causes, diagnosis, treatment, control, and prevention of physical and mental diseases and other impairments of man resulting directly or indirectly from contaminants in water, or to the provision of a dependably safe supply of drinking water.

Saline. Consisting of or containing salt.

Salmonid. Genus name (Salmonidae) of any of a family of elongate soft-finned fishes (as a salmon or trout) that have the last vertebrae upturned.

Salvage operation. Any business, trade, or industry engaged in whole or in part in salvaging or reclaiming any product or material, including, but not limited to, metals, chemicals, shipping containers, or drums.

Science Advisory Board (SAB). Board established to provide a strong, direct link between EPA's administrator and the scientific community.

Scrubber. Large-scale and relatively expensive device for accomplishing flue-gas desulfurization. In addition, some scrubbers accomplish partial removal of NO_x particles.

Sealed source. A quantity of radiation so enclosed as to prevent the escape of any radioactive material but at the same time permitting radiation to come out for use.

Secondary. Term characterizing a backup system or program.

Secondary air quality standards. These standards protect the public "from any known or anticipated adverse effects," not necessarily health effects.

Secondary treatment plants. Examples of secondary treatment plants are: waste water lagoons, trickling filters, or activated sludge plants. These plants alleviate the need for installation of entirely new treatment systems.

Second-generation flue-gas desulfurization process. Any of several processes that yield usable sulfur compounds as a by-product and/or permit reuse of the chemicals required in the desulfurization process.

Septic tank. A watertight settling tank in which solid sewage is decomposed by natural bacterial action.

Serious hazard. Hazard of considerable consequence to safety or health through the design, location, construction, or equipment of a building, or the condition thereof, which hazard has been established through experience to be of certain or probable consequence, or which can be determined to be, or which is obviously such a hazard.

Sewage. Water-carried wastes from residences, business buildings, institutional, and industrial establishments, together with such ground, surface, and storm waters as may be present.

Sewage system. Network of sewers and appurtenances for collection, transportation, and pumping of sewage and industrial wastes.

Sewage treatment works. Arrangement of devices and structures for treating sewage and industrial wastes.

Sewage works. All facilities for collecting, pumping, treating, and disposing of sewage and industrial wastes.

Sewer. Pipe or conduit for carrying sewage or other waste liquids.

Sintering plant. Plant used in connection with the fusing of fine particles of metallic ores causing agglomeration of such particles. Such a plant consists of, but is not limited to, sintering machines, handling facilities, wind boxes, stacks, and other auxiliaries pertinent to the process.

Sludge. Muddy deposit (as on a river bed); muddy or slushy mass, deposit, or sediment, for example, as a precipitated solid matter produced by water and sewage treatment process or muddy sediment in a steam boiler.

Small particle control technology. Control technology to reduce fine particle emissions (less than 3 μm in diameter).

Smoke

1. Small gas-borne particles resulting from incomplete combustion, consisting predominantly of carbon and other combustible material, and present in sufficient quantity to be observable independently of the presence of other solids.

2. Small gas-borne particles, other than water, that form a visible plume in the air from a source of atmospheric pollution.

Smoke (as applied to mobile sources). Matter in exhaust emissions that obscures the transmission of light.

Smokemeter. Device constructed in such a manner as to measure smoke opacity by obstructing light between a light source and photoelectric cell which will indicate the percent opacity of smoke. The design of this device is in substantial conformity with Technical Report J255 of the Society of Automotive Engineers.

Smoke monitor. Device using a light source and a detector to automatically measure and record the light-obscuring power of smoke at a specific location in the flue or stack of a source. The measuring and recording is at intervals of not less than 15 seconds.

Smoke oven. Any piece of equipment which is used for smoking food products.

Smoke unit. The number obtained when the smoke density in the Ringlemann Number is multiplied by the time of emission in minutes.

Solid fuel. Any material in its solid state capable of being consumed by a combustion process.

Solid Waste Disposal Act (SWDA). Act that directs the EPA administrator to conduct and cooperate in research efforts relating to any adverse health and welfare effects of the release into the environment of materials present in solid waste and methods to eliminate such effects; operation and financing of solid waste disposal programs; reduction of the amount of such waste and unsalvageable materials; development and application of new and improved methods of collecting and disposing of solid waste; and processing and recovering materials and energy from solid wastes.

Soot. Agglomerated particles consisting essentially of carbonaceous material.

Sound level (noise level). For airborne sound, sound level (noise level) is a weighted sound-pressure level, obtained by the use of metering characteristics.

Sound level meter. An instrument standardized by the American National Standards Institute for the measurement of the intensity of sound.

Sound pressure level. The sound pressure level, in decibels, of a sound is 20 times the logarithm to the base ten of the ratio of the pressure of the sound to the reference sound pressure. Unless otherwise specified, the effective (rms) pressure understood. The reference sound pressure is 20 μN/m^2.

Spark-ignition-powered motor vehicle. Vehicle that is self-propelled by a spark-ignition type of internal combustion engine, which includes, but is not limited to, engines fueled by gasoline, propane, butane, and methane compounds.

Stack or chimney

1. Conduit, duct, vent, flue, or opening of any kind arranged to conduct products of combustion to the atmosphere. The term does not include breeching.

2. Flue, conduit, or opening designed and constructed to emit air contaminants into the outdoor air.

Stack spray. Nozzle or series of nozzles installed in a stack above the breeching and used to inject wetting agents at high pressure to suppress the discharge of particulate matter from the stack.

Standard conditions. A dry gas temperature of 70°F and a gas pressure of 14.7 psia.

Standard cubic foot. Standard cubic foot is a measure of the volume of gas under standard conditions.

Standard of performance. Standard for the emission of air pollutants that reflects the degree of emission limitation achievable through the application of the best system of emission reduction that (taking into account the cost of achieving such reduction) the EPA Administrator determines has been adequately demonstrated.

Stationary source pollutants. Pollutants caused by sources that do not move, such as factories and power plants.

Steady noise. Noise whose level remains essentially constant (fluctuations are negligibly small) during the period of observation is a steady noise.

Steam whistles. The blowing of any steam whistle attached to any stationary boiler is not permitted except to give notice of the time to begin, stop work, or as a danger warning.

Stokers. Mechanical device that feeds solid fuel uniformly onto a grate or heart within a furnace.

Storm sewer. Pipe or conduit that carries storm and surface water and drainage, but excludes sewage and industrial wastes. It may, however, carry cooling waters and unpolluted waters.

Strategic Environmental Assessment System (SEAS). Operational tool for environmental forecasting and policy analysis. The EPA planned to have SEAS support an impact assessment of energy, environment, and recovery tradeoffs and alternatives.

Synergism. Cooperative action of discrete agencies (such as chemicals or muscles) so that the total effect is greater than the sum of two or more effects taken independently.

Synthetic detergent or detergent. Cleaning compound that is available for household use, laundry use, personal use, or industrial use, and is composed of organic and inorganic compounds. Included are soaps, water softeners, surface-active agents, dispersing agents, foaming agents, buffering agents, builders, fillers, dyes, enzymes, and fabric softeners, in the form of crystals, powders, flakes, bars, liquids, sprays, etc.

Tertiary. Term applied to a system that follows both a main and secondary effort.

Teratology. Study of malformations, monstrosities, or serious derivations from the normal type in organisms.

Thermal insecticidal fogging. The use of insecticidal liquids which are passed through thermal fog-generating units where they are, by means of heat, pressure, and turbulence, transformed and discharged in the form of a fog or mist that is blown into an area to be treated.

Toxic matter or material. Those matters or materials which are capable of causing injury to living organisms by chemical means.

Toxicology. Science that deals with poisons and their effects and the problems involved (clinical, industrial, or legal).

Toxic substance. Substance, either gaseous, liquid, or solid that when discharged into the sewer system in sufficient quantities will interfere with any sewage-treatment process, constitute a hazard to human beings or animals, inhibit aquatic life, or create a hazard to recreation in the receiving waters of the effluent from the sewage-treatment works.

Trace metals. Possibly toxic metals that move through the environment and humans in very small quantities.

Trade waste. All solid or liquid material or rubbish resulting from construction, building operations, or the prosecution of any business, trade, or industry. Such waste includes, but is not limited to, plastic products, chemicals, cinders, and other forms of solid or liquid waste materials.

Transport. Movement of a substance through the ecosystem.

Transport and Fate of Pollutants Subprogram. Subprogram responsible for the development of empirical and analytical techniques that relate air and water pollution source emissions and discharges to ambient exposures.

Triazine. Any of three compounds containing a ring composed of three nitrogen atoms; also, any of various derivatives of these including several used as herbicides.

Trickling filters. Waste water treatment equipment in which waste water is sprayed on and trickles down through an aerated bed of rocks, the surfaces of which are coated with bacterial populations that feed on the nutrient in the waste water.

Tritium. Radioactive isotope of hydrogen of a mass three times the mass of ordinary light hydrogen.

Trophic. Of, relating to, or characterized by nutrition.

Trophic level. One of the hierarchical strata of a food web characterized by organisms that are the same number of steps removed from the primary producers.

True vapor pressure. Equilibrium partial pressure exerted by a petroleum liquid as determined in accordance with methods described in American Petroleum Institute Bulletin 2517.

Unit operations. Methods whereby raw materials undergo physical change; methods by which raw materials may be altered into different states, such as vapor, liquid, or solid, without changing into a new substance with different properties and by composition.

Urea. Soluble basic nitrogenous compound that is the chief solid component of mammalian urine and an end product of protein decomposition. It is synthesized from carbon dioxide and ammonia and is used especially in synthesis (of resins and plastics) and in fertilizers and animal rations.

Vapor. Any material in a gaseous state that is formed from a substance, usually a liquid, by increase in temperature or release of pressure.

Vehicle. Self-propelled mechanism, such as a truck, machine, tractor, roller, derrick, crane, trencher, portable hoisting engine, or automobile; conveyance used for carrying persons or things; trailer, semitrailer, boat, tug, or other apparatus that is not ordinarily permanently installed in one location, but is used in various places over a wide area.

Vibration. All machinery mounted and operated to prevent transmission of ground vibration exceeding a displacement of three thousandths (0.003) of one (1) inch measured anywhere outside the lot line of its source or ground vibration which can be readily perceived by a person standing anywhere outside the lot lines of its source.

Volatile matter. Gaseous constituents of solid fuels as determined by ASTM D.3286.

Volatile organic material. Organic substance that has a vapor pressure of 2.5 psia or greater at 70°F.

Waste Management Subprogram. Subprogram that focuses on prevention, control, treatment, and management of the pollution produced by community, residential, or other nonindustrial activities. Research concerns municipal and domestic waste water and collection/transport systems, urban land surface runoff, municipal solid wastes, and associated air pollutants.

Waste water lagoons. Shallow earthen ponds, usually lined, in which liquid wastes are stored for an extended period to promote the natural setting of suspended solids and decomposition of organic compounds in the stored fluid.

Water. All water of any river, stream, watercourse, pond, or lake wholly or partly within territorial boundaries.

Watercourse. Any channel, natural or artificial, either lined or unlined, for drainage of storm water, groundwater, or clear water.

Watershed. Water parting; region or area bounded peripherally by a water parting and draining ultimately to a particular watercourse or body of water; crucial dividing point or line.

Water Supply Subprogram. Subprogram that focuses on three areas of concentration: health effects, water treatment, and systems and groundwater management.

PROJECT PREPARATION AND SAFETY CONTROL

Accepted engineering requirements. Requirements or practices that are compatible with standards required by a registered architect, a registered professional engineer, or other duly licensed or recognized authority.

Accident prevention tags. Temporary means of warning workers of an existing hazard, such as defective tools or equipment. Are not used as a substitute for accident prevention signs.

Acid. A corrosive and combustible material when used in conjunction with other ingredients.

Approved. Sanctioned, endorsed, accredited, certified, or accepted as satisfactory by a duly constituted and nationally recognized authority or agency.

Approved material, equipment and methods. Approved by the building official or by a recognized authoritative agency.

Approved rules. The legally adopted rules of the building official or of a recognized authoritative agency.

Approved storage facility. Facility for the storage of explosive materials conforming to the requirements covered by a license or permit issued under authority of a government agency.

Authorized person. A person approved or assigned by the employer to perform a specific type of duty or duties or to be at a specific location or locations at the jobsite.

Barricade

1. Obstruction to deter the passage of persons or vehicles.
2. Obstruction usually made of steel angle legs with high-impact plastic panel and reflectorized top panel.

Barricade Lantern. Lantern encased in plastic consisting of a 6-V-operated, 7-in. amber lensed reflector and battery case.

Batching plant. Permanent or mobile plant for the mixing and dispensing of weighed concrete materials to transit-mix trucks for job delivery.

Beginning of construction. Incorporation of labor and materials on the site, tract, or lot where a building or structure is proposed to be constructed; the incorporation of labor and material within the walls of a building where repairs or remodeling are proposed to be made; the incorporation of labor and materials at the site, lot, or parcel where land is to be used for purposes other than construction of a building.

Bench mark. Point of known or assumed elevation used as a reference in determining and recording other elevations.

Blasting cap. A cap or detonator with wires attached for exploding the same by means of electricity.

Blasting powder. An explosive substance composed of sulphur, charcoal, and sodium nitrate, specially prepared for the purpose of blasting.

Carbon dioxide (CO_2) fire extinguishers. Carbon dioxide is retained under its own pressure in a liquid condition at room temperature. The agent is self-expelling and is discharged by operation of a valve, which causes the carbon dioxide to be expelled through a horn in its vapor and solid phase.

Caution signs. Signs used only to warn against potential hazards or to caution against unsafe practices. With yellow as the predominating color, black upper panel and borders, yellow lettering of "caution" on the black panel, and the lower yellow panel for additional sign wording, with black lettering.

Class I flammable liquid. Liquid having a flash point at or below 20°F.

Class II flammable liquid. Liquid having a flash point above 20°F and below 70°F.

Class III flammable liquid. Liquid having a flash point above 70°F.

Closed container. Container so sealed by means of a lid or other device that neither liquid nor vapor will escape from it at ordinary temperatures.

Combustible dust. Fine particles of matter, such as lint, shavings, sawdust, flour, starch, sulphur, metal powders, and powdered plastics, liable to spontaneous ignition or explosion or constituting a dust hazard, except when handled, stored, or confined to eliminate the hazard.

Combustible liquid. Liquid having a flash point at or above 40°F (60°C) and below 200°F (93.4°C).

Combustion. Chemical process that involves oxidation sufficient to produce light or heat.

Competent person. One who is capable of identifying existing and predictable hazards in the surroundings, or working conditions which are unsanitary, hazardous, or dangerous to employees, and who has authorization to take prompt corrective measures to eliminate them.

Construction. Putting together and assembling of materials in order to erect or build a structure.

Construction and demolition wastes. The waste building materials and rubble resulting from construction, remodeling, repair, and demolition operation on houses, commercial buildings, pavements, and other structures.

Construction equipment. Construction machinery, tools, derricks, hoists, scaffolds, platforms, runways, ladders, and all materials-handling-equipment safeguards and protective devices used in construction operations.

Construction equipment and protection. Constructed, installed, and maintained in a safe manner and operated so as to ensure protection to the workers engaged thereon and to the general public. Usually unlawful to remove or render inoperative any structural, fire-protective, or sanitary safeguard or device, except when necessary for the actual installation and prosecution of the work.

Construction loads. Provisions provided to ensure that stresses due to wind loads, dead loads, and loads due to material storage and erection equipment occurring during the erection of any structure does not exceed the allowable stresses for materials as limited by the provisions of the applicable code.

Construction operation. Erection, alteration, repair, renovation, demolition, or removal of any building or structure.

Construction shed. A temporary building incidental to the construction of another building for which a permit has been issued and which is removed within 30 days after the completion of the building for which the permit was issued.

Construction site security. The owner, contractor, or responsible party constructing a new building should maintain security measures as deemed necessary or as required by the building department safety superintendent to control vandalism, fires, and other mischievous and deliberate acts of destruction.

Constructor. Person who contracts with the owner of a project for the work thereon and includes an owner who contracts with more than one person for the work on a project or undertakes the work on a project or any part thereof.

Container assembly. Assembly consisting essentially of the container and fittings for all container openings, including shutoff values, excess flow valves, liquid-level gauging devices, safety relief devices, and protective housing.

Contaminant. A material that by reason of its action upon, within, or to a person is likely to cause physical harm.

Controlled materials. Materials which are certified by an accredited authoritative agency as meeting accepted engineering standards for quality and a provided in the code.

Conveyance of hazardous materials. Unit for transporting explosives or blasting agents, including, but not limited to, trucks, trailers, rail cars, barges, and vessels.

Corrosive liquids. Includes hydrochloric, nitric, sulphuric, hydrofluoric, perchuloric, and other corrosive acids; alkaline caustic liquids and other corrosive liquids.

Crawling board. Plank with cleats spaced and secured at equal intervals for use by a worker on roofs; not designed to carry any material.

Danger signs. Sign used only where an immediate hazard exist. Usually with red as the predominating color for the upper panel, black outline on the borders, and a white lower panel for additional sign wording.

Datum. Any level surface taken as a plane of reference from which to measure elevations.

Decanting. Method used for decompressing under emergency circumstances; e.g., when workers are brought to atmospheric pressure with a very high gas tension in the tissues and then immediately recompressed in a second and separate chamber or lock.

Decay. Disintegration of wood substance due to action of wood-destroying fungi. Also known as dote and rot.

Decompression. Process of reducing high air pressure gradually enough so as not to injure personnel who have been working in it.

Defect. Any characteristic or condition which tends to weaken or reduce the strength of the tool, object, or structure of which it is a part.

Demolition. The dismantling or razing of all or part of a building, including all operations incidental thereto.

Directional signs. Signs other than automotive traffic signs. Usually white with a black panel and a white directional symbol. Any additional wording on the sign usually consists of black letters on the white background.

Dry chemical fire extinguisher. Extinguisher containing specially treated potassium bicarbonate dry chemical agent, providing protection against Class B and Class C fires—$2\frac{1}{2}$ lb capacity, U.L. rated 10-BC.

Dust mask. Lightweight aluminum mask with removable filter pads.

Exit signs. Signs usually having legible red letters not less than 6 in. high on a white field, with the principal stroke of the letters at least $\frac{3}{4}$ in. wide.

Explosion hazard gases. Includes, but is not limited to, the following gases: acetylene, ether, ethyl chloride, ethylene, liquified petroleum gases, hydrogen, illuminating gas, methyl chloride gas, and similar gases susceptible to explosion.

Explosion hazards. Every structure, room, or space occupied for uses involving explosion hazards should be equipped and vented with explosion relief systems and devices arranged for automatic release under predetermined increase in pressure for specific uses or in accordance with accepted engineering standards and practices.

Explosive. Any chemical compound or mechanical mixture that is intended for the purpose of producing an explosion; that contains any oxidizing and combustible units or other ingredients in such proportions, quantities, or packing that an ignition by fire, by friction, by concussion, by percussion, or by detonator of any part of the compound or mixture may cause such a sudden generation of highly heated gases that the resultant gaseous pressures are capable of producing destructive effects on contiguous objects, or of destroying life or limb.

Explosive gases. Acetylene, ether, ethyl chloride, ethylene, hydrogen illumination gas, petroleum gases, methyl chloride gas, and oxygen.

Explosive material. Any dangerous chemical classified as an explosive material in the code.

Exposed. Term applied to live part that can be inadvertently touched or approached nearer than a safe distance by a person. Applies to parts not suitably guarded, isolated, or insulated.

Fire brigade. Organized group of employees knowledgeable, trained, and skilled in the safe evacuation of employees during emergency situations and in assisting in firefighting operations.

Fire control portable equipment. Portable fire extinguishers and control equipment, provided in such quantities and types as are needed for the special hazards of operation and storage.

Fire extinguishers in operable condition. Fire extinguishers that are fully charged in operable condition and kept in designated places at all times when not being used.

Fire hazard. Any building, device, appliance, apparatus, equipment, tank, vehicle, combustible waste, fence, or vegetation that, in the opinion of the fire department, is in such a condition as to cause a fire or explosion.

Flaggers. When operations are such that signs, signals, and barricades do not provide the necessary protection on or adjacent to a highway or street, flaggers usually are used.

Flame gun. Large blowtorch using kerosene for fuel.

Flammable. Capable of being easily ignited, burning intensely, or having a rapid rate of flame spread.

Flammable liquids. Any liquid having a flash point below 140°F and a vapor pressure not exceeding 40 psia at 100°F.

Flash point of a liquid. Temperature at which liquid gives off vapor sufficient to form an ignitable mixture with the air near the surface of the liquid or within the vessel used, as determined by appropriate test procedure and apparatus.

Floor hole. Opening usually measuring less than 12 but more than 1 in. in its least dimension in any floor, platform, pavement, or yard through which materials, but not persons, may fall, such as a belt hole, pipe opening, or slot opening.

Floor opening. Opening usually measuring 12 in. or more in its least dimension in any floor, platform, pavement, or yard through which persons may fall, such as a hatchway, stair or ladder opening, pit, or large manhole. Floor openings occupied by elevators, dumb waiters, conveyors, machinery, or containers are excluded from this definition.

General signs and symbols. Regulatory signs and symbols visible at all times when work is being performed, and removed or covered when the hazards no longer exist.

Hard hats

1. Lightweight, high-impact-resistant fiberglass shells and accessories consisting of plastic head cradle and adjusting band and plastic forehead section, with available chin straps.
2. Aircraft-type aluminum shell, heat treated for toughness and impact absorption, anodized or in bright colors, with plastic head cradle and adjusting band and plastic forehead section, with available chin steps.

Hazard. Danger or injury to life or limb including casualty, fire, and shock when applicable.

Hazardous chemicals. Chemicals having serious flame or explosion hazards when coming into contact with water or moisture, such as metallic sodium, metallic potassium, sodium peroxide, calcium phosphide, yellow phosphorous, metallic magnesium powder, aluminum powder, calcium carbide, red phosphorous, and similar chemicals and solutions.

Hazardous gases. Includes, but is not limited to, such gases as ammonia, chlorine, phosgene, carbon bisulphide, and other toxic irritant, corrosive, or fume hazard gases, such as acetylene, ether, ethyl chloride, ethylene, liquified hydocarbons, and methyl chloride gas.

Hazardous material. Any material included under the definitions of flammable dust, flammable fiber, combustible liquid, dangerous chemical, flammable gas, liquified flammable gas, and flammable liquid.

Hazardous substance. Substance likely to cause death or injury by reason of being explosive, flammable, poisonous, corrosive, oxidizing, irritating, or otherwise harmful.

High air. Air pressure used to supply power to pneumatic tools and devices.

Horse. Sawhorse or other simple frame or support.

Hot work. Riveting, welding, burning, or other fire- or spark-producing operations.

Housekeeping. Cleaning of every floor, working place, and passageway to maintain such areas free of protruding nails, splinters, loose boards, unnecessary holes and openings, debris, and broken materials.

Inflammable fibrous materials. Includes such materials as hay, straw, broomcorn, hemp, tow, jute, sisal, kapok, hair, excelsior, oakum, and the like. ("Inflammable" and "flammable" are identical in meaning.)

Inorganic grinding wheels. Wheels that are bonded by means of inorganic material, such as clay, glass, porcelain, sodium silicate, magnesium oxychloride, or metal.

Insurer. Includes the State Compensation Insurance Fund and any private company, corporation, mutual association, reciprocal or interinsurance exchange authorized under the laws of the state to insure employers against liability for compensation and any employer to whom a certificate of consent to self-insure has been issued.

Jack. Mechanical or hydraulic lifting device. Hydraulic ram or cylinder.

Jackhammer. Air drill that hammers and rotates a hollow steel sleeve and a bit, and can be operated by one person.

Jetting. Drilling with high-pressure water or air jets.

Leveling rod (surveying). Telescoping rod marked in feet and fractions of feet, and fitted with a movable target or sighting disc.

Lifeline. Rope, suitable for supporting one person, to which a lanyard, safety belt, or harness is attached.

Load factor. Average load carried by an engine, machine, or plant, expressed as a percentage of its maximum capacity.

Loading. It is unlawful in most codes to load any structure, temporary support, scaffolding, sidewalk bridge, sidewalk shed, or any other device or construction equipment during the construction or demolition of any building or structure in excess of its safe working capacity.

Loading capacity. Safe working capacity determined in accordance with the applicable code for allowable loads and working stresses or by subjecting the structure device or equipment to a load $2\frac{1}{2}$ times the maximum live load to which it may be subjected during the construction without failure of any part thereof.

Local order. Any ordinance, order, rule, or determination of the governing body of any county, city, district, or other public or quasi-public corporation, or an order or direction of any public official, board, or department upon any matter over which the division has jurisdiction.

Low air. Air supplied to pressurize working chambers and locks.

Low density wood. Wood that is exceptionally light in weight and usually deficient in strength properties for the species.

Man lock. Chamber through which workers pass from one air pressure environment into another.

Mat. Heavy, flexible fabric of woven wire rope or chain used to confine blasts; a wood or metal platform used in sets to support machinery on soft ground.

Material. An established size, quality, composition, or strength, or with respect to an established size, quality, composition, or strength.

Material debris chutes. Chutes designed and constructed of materials of such strength as to eliminate failure due to impact of waste materials or debris loaded therein.

Materials lock. Chamber through which materials and equipment pass from one air pressure environment into another.

Maximum safe load. Load limits of floors within buildings and structures, in psf, conspicuously posted in all storage areas, except for floor or slab on grade.

Mobile construction office. Vehicle which can be installed at the site. Electrical service and telephone service installed after the vehicle is in place. The construction office usually contains space for a private office, contractor's room for display of plans, and toilet facilities.

Mobile construction storage. Vehicle for the storing of small tools, lamp supplies, finished hardware, spare parts for job equipment, and miscellaneous other small items.

Mobile space heaters. Gas, oil-fired, or electric type blowers used as temporary space heaters during cold weather conditions to provide ventilated heat for the installation of materials requiring space temperatures over 50°F.

Mobile unit. Combination of an aerial device, its vehicle, and related equipment.

Moisture hazard substance. Magnesium powder, calcium carbide, metallic sodium, and sodium peroxide.

Octane number. Perfect of isooctane by volume in a mixture of isooctane and normal heptane that has the same antiknock character in a standard variable-compression Cooperative Fuel Research test engine as the fuel under test. Octane has antiknock characteristics. Mixture having 75% octane and 25% heptane has an octane rating of 75.

Off-hand grinding. Grinding of any material or part that is held in the operator's hand.

One-part line. Single strand of rope or cable.

Operating device. Pushbutton, lever, or other manual device used to actuate a control.

Organic bonded grinding wheels. Wheels that are bonded by means of an organic material such as resin, rubber, shellac, or other similar bonding agent.

OSHA. Occupational Safety and Health Administration. Federal agency that regulates health and safety procedures.

Pioneering. First working over of rough or overgrown areas.

Pioneer road. Temporary road built along the route of a job to provide means for moving equipment and personnel.

Plumb bob. Pointed weight hung from a string. Used for vertical alignment.

Portable fire extinguisher locations. Fire extinguishers should be located so as to be readily accessible and immediately available in the event of fire, should be located along normal paths of travel.

Portable fire extinguishers. Fire extinguishers classified for use on certain classes of fires and rated for relative extinguishing effectiveness at a temperature of $+70°F$ by nationally recognized testing laboratories. Usually based on the types and classification of fires and the fire extinguisher potentials as determined by fire tests.

Portable grinder. Power-driven rotatable grinding, polishing, or buffing wheel mounted in such manner that it may be manually manipulated.

Portable grinding. Grinding operation where the grinding machine is designed to be hand held and may be easily moved from one location to another.

Portable tank. Closed container having a liquid capacity of more than 60 U.S. gal and not intended for fixed installation.

Portable truck scale. Aboveground truck scale for weighing material trucks arriving at and leaving large construction jobs.

Portable welder. Welding equipment used at job site for emergency uses.

Potable water. Water that meets the quality standards approved for drinking purposes by the state or local authority having jurisdiction.

Powder-actuated fastening tool. Tool or machine that drives a stud, pin, or fastener by means of an explosive charge.

Pressure. Force acting on a unit area. Usually shown as pounds per square inch (psi).

Pressurized water fire extinguisher. Fire extinguisher used for protection against small fires in Class A materials, such as wood, paper, rags, plastics, cloth, and rubber. Extinguisher shell made of stainless steel, requires compressed air with a pressure capacity of 100 psi, has $2\frac{1}{2}$-gal capacity; U.L. rated 2A.

Protective shield or guard. Device or guard that is attached to the muzzle end of the tool or equipment and designed to confine flying particles and dust.

Quality of materials. All materials, assemblies, construction, and equipment should conform to the regulations of the code, and should conform to generally accepted standards with respect to strength, durability, corrosion resistance, fire resistance, and other qualities recognized under those standards. All test specimens and construction should be truly representative of the material, workmanship, and details to be used in actual practice.

Radiation. Includes alpha rays, beta rays, gamma rays, x-rays, neutrons, high-speed electrons, high-speed protons, and other atomic particles.

Radiation restricted area. Any area whose access is controlled by the contractor for the purposes of protecting workers from exposure to radiation or radioactive materials.

Radioactive material. Material that emits by spontaneous-nuclear disintegration corpuscular or electromagnetic emanations.

Railings. One or a combination of railings constructed in accordance with OSHA requirements. Standard railing is a vertical barrier erected along exposed edges of floor openings, wall openings, ramps, platforms, and runways to prevent accidents.

Rated capacity or working load limit. Maximum working load permitted by the provisions of the applicable code, or as specified and identified on the equipment by the manufacturer.

Reinforced grinding wheels. Wheels that contain strengthening fabric or filament. Term "reinforced" does not cover wheels using such mechanical additions as steel rings, steel cup backs, or wire or tape winding.

Respirator masks. Made in single- or double-cartridge type, automatically designed to fit the facial contour.

Runway

1. Any aisle or walkway constructed or maintained as a temporary passageway for pedestrians or vehicles.

2. Passageway for persons elevated above the surrounding floor or ground level, such as a footwalk along shafting or a walkway between buildings.

Safety belt. Device, usually worn around the waist, attached to a lanyard and lifeline, of a structure.

Safety can. Approved closed container, with not more than 5-gal capacity, having a flash-arresting screen, spring-closing lid, and spout cover and designed so that it will safely relieve internal pressure when subjected to fire exposure.

Safety during construction. Construction, within the scope of this code, should be performed in such a manner that the workmen and public should be protected from injury and adjoining property should be protected from damage by the use of scaffolding, underpinning or other approved methods.

Safety during demolition. Safe and sanitary conditions should be provided where demolition and wrecking operations are being carried on. Work should be done in such a manner that hazard from fire, possibility of injury, danger to health, and conditions which may constitute a public nuisance will be minimized, in conformity with generally accepted standards.

Safety factor. Ratio of the ultimate breaking strength of a member or piece of material or equipment to the actual working stress or safe load when in use.

Safety goggles

1. Frames of molded plastic with tinted nonglare vinyl plastic lenses, including facial top and bottom cushion and adjustable elastic headband.

2. Molded plastic safety frames with folding side shields perforated for ventilation and clear heat-treated safety lenses.

Safety guard. Enclosure designed to furnish all possible protection to the person operating the grinding wheel and people in the immediate vicinity and restrain the travel of pieces of the grinding wheel in the event that the wheel is broken in operation.

Safety hook. Hook with a latch to prevent slings or load from accidentally slipping off the hook.

Safety instruction signs. Signs are usually white with green upper panel with white letters to convey the principal message. Additional wording on the sign consists of black letters on a white background.

Safety screen. Air- and water-tight diaphragm placed across the upper part of a compressed air tunnel between the face and bulkhead in order to prevent the flooding of the crown of the tunnel between the safety screen and the bulkhead, thus providing a safe place of refuge and means of exit from a flooding or flooded tunnel.

Schematic. Graphic illustration showing principles of construction or operation without accurate mechanical representation.

Screen. Mesh or bar surface used for separating pieces or particles of different sizes; also a filter.

Sidewalk shed. A construction over a public sidewalk, used to protect pedestrains from falling objects.

Signals. Moving signs provided by workers, such as flaggers, or by devices, such as flashing lights, to warn of possible or existing hazards.

Signs. Warnings of hazard temporarily or permanently affixed or placed at locations where hazards exist.

Slag. Refuse from steel making, used for fill for temporary roads on job sites and crushed to be used for many purposes.

Snagging. Grinding wheel that removes relatively large amounts of material without regard to close tolerances or surface finish requirements.

Special decompression chamber. Chamber that provides greater comfort for workers when the total decompression time exceeds 75 minutes.

Special permits. All special licenses and permits for the storage of materials on sidewalks and highways, for the use of water or other public facilities, and for the storage and handling of explosives should be secured from the proper authorities.

Spotter. In truck use, the person who directs the operator into loading or dumping position or, in a pile driver, the horizontal connection between the machinery deck and the lead (pile guide).

Steel landing mats. For use on roads having soft or muddy soil conditions, such as swamps and beach areas.

Temporary construction. Before any construction operation is started, plans and specifications are usually submitted for approval to the building official. Plans and specifications indicating the design, construction, and location of all sidewalk sheds, truck runways, trestles, foot bridges, guard fences, canopies, and other similar devices required in the operation in such detail as may be necessary to permit determination of conformance with the requirements of the applicable code.

Temporary encroachments. Subject to the approval of the building official, sidewalk sheds, underpinning, and other temporary protective guards and devices may project beyond the interior and street lot lines as may be required to ensure the safety of the adjoining property and the public. When necessary, the consent of the adjoining property owner is obtained to prevent possible legal action.

Temporary heaters. Heaters used while permanent heating equipment is being installed. When in use, set horizon-

tally level, unless otherwise permitted by the manufacturer's markings.

Temporary oil-fired heaters. Flammable liquid-fired heaters equipped with a primary safety control to stop the flow of fuel in the event of flame failure.

Toilet facility. Fixture connected to a sewer system, maintained within a toilet room.

Toilet room. Room maintained within or on the premises of any building containing toilet facilities.

Toxic material. Material in concentration or amount that exceeds the applicable limit of established standards. Material that is of such toxicity as to constitute a recognized hazard, causing or likely to cause death or serious physical harm.

Traffic cone. One piece of polyvinyl chloride material colored Blaze Orange.

Traffic signs. Legible traffic signs posted at points of hazard at all construction areas. All traffic control signs or devices used for protection of construction workmen are to conform to American National Standard Manual on Uniform Traffic Control Devices for Streets and Highways, ANSI D6.1.

Vehicle. Any carrier that is not manually propelled.

Vermin control. Enclosed storage area or workplace constructed, equipped, and maintained, to prevent the entrance or harborage of rodents, insects, and other vermin.

Water closet. Toilet fixture that is connected to a sewer system, maintained within a toilet room, and flushed with water.

Water supply. Available in volume and at adequate pressure to supply water hose streams as required by special hazards of operation.

Welding goggles. Cup-type goggle with one-piece, durable, injection-molded eyecups, for use in acetylene welding and cutting, burning, brazing and open-hearth furnace work.

Welding helmets. Made from fiberglass, in various types and protective shapes.

Working chamber. Space or compartment under air pressure in which the work is being done.

Working drawing. Any drawing showing sufficient detail so that whatever is shown can be built without other drawings or instructions except proper specifications.

SCAFFOLDING—PLATFORMS AND LADDERS

Aerial ladder. Aerial device consisting of a single- or multiple-section extensible ladder.

Approved

1. Term applied to equipment listed or approved by a nationally recognized testing laboratory such as Factory Mutual Engineering Corp., Underwriters' Laboratories Inc. or the appropriate federal agencies.

2. Sanctioned, endorsed, accredited, certified, or accepted as satisfactory by a duly constituted and nationally recognized authority or agency.

Bricklayer's square scaffold. Scaffold composed of framed wood squares that support a platform, limited to light and medium duty.

Cage. Guard that may be referred to as a cage or basket guard. Enclosure fastened to the side rails of the fixed ladder or to the structure to encircle the climbing space of the ladder for the safety of the person who must climb the ladder.

Carpenters' bracket scaffold. Scaffold consisting of wood or metal brackets supporting a platform.

Catch platform. A platform or other construction projecting from the face of a building, supported therefrom, and used to intercept the fall of objects and to protect individuals and property from falling debris.

Cleats. Ladder crosspieces of rectangular cross section placed on an edge on which a person may step in ascending or descending.

Coupler. Device for locking together the component parts of a tubular metal scaffold.

Double-cleat ladder. Similar to a single-cleat ladder, but wider, with an additional center rail that allows for two-way traffic for workers in ascending and descending.

Double-pole or independent pole scaffold. Scaffold supported from the base by a double row of uprights, independent of support from the walls and constructed of uprights, ledgers, horizontal platform bearers, and diagonal bracing.

Extension ladder. Non-self-supporting portable ladder adjustable in length. Consists of two or more sections traveling in guides or brackets so arranged as to permit length adjustment. Size is designated by the sum of the lengths of the sections measured along the side rails.

Extension trestle ladder. Self-supporting portable ladder, adjustable in length, consisting of a trestle ladder base and a vertically adjustable single ladder, with suitable means for locking the ladders together. Size is designated by the length of the trestle ladder base.

Fixed ladder. Ladder permanently attached to a structure, building, or equipment.

Float or ship scaffold. Scaffold hung from overhead supports by means of ropes and consisting of a substantial platform having diagonal bracing underneath, resting upon and securely fastened to two parallel plank bearers at right angles to the span.

Grab bars. Individual handholds placed adjacent to or as an extension above ladders for the purpose of providing access beyond the limits of the ladder.

Guardrail. Rail secured to uprights and erected along the exposed sides and ends of platforms.

Guide roller. Rotating, bearing-mounted, generally cylindrical member, operating separately or as part of a guide shoe assembly, attached to the platform, and providing rolling contact with building guideways or other building contact members.

Guide shoe. Assembly of rollers, slide members, or the equivalent attached as a unit to the operators' platform and designed to engage with the building members provided for the vertical guidance of the operators' platform.

Handrail

1. Single bar or pipe supported on brackets from a wall or partition, as on stairway or ramp, to furnish persons with a handhold.
2. Rail connected to a ladder stand, running parallel to the slope and/or top step.

Heavy-duty scaffold. Scaffold designed and constructed to carry a working load not to exceed 100 lb/ft^2.

Horse scaffold. Scaffold for light or medium duty, composed of wood or metal horses supporting a work platform.

Individual-rung ladder. Fixed ladder, each rung of which is individually attached to a structure, building, or equipment.

Interior hung scaffold. Scaffold suspended from the ceiling or roof structure.

Jacob's ladder. Marine ladder consisting of rope or chain side rails, with wood or metal step rungs.

Ladder

1. Appliance usually consisting of two side rails joined at regular intervals by crosspieces called steps, rungs, or cleats, on which a person may step in ascending or descending.
2. Digging boom assembly in a hydraulic dredge or chain-and-bucket ditcher.

Ladder cleats. Ladder crosspieces of rectangular cross section placed on edge, on which a person may step in ascending or descending.

Ladder jack scaffold. Light-duty scaffold supported by brackets attached to ladders, constructed to support not less than 40 lb/ft^2.

Ladder safety device. Device, other than a cage or well, designed to eliminate or reduce the possibility of accidental falls from a ladder.

Ladder stand. Mobile, fixed-size, self-supporting ladder consisting of a wide, flat thread ladder in the form of stairs. Assembly may include handrails.

Ledger stringer. Horizontal scaffold member that extends from post to post and supports the putlogs or bearers forming a tie between the posts.

Manually propelled mobile scaffold. Portable rolling scaffold supported by casters, capable of being moved to areas as required by workers at the job.

Masons' adjustable multiple-point suspension scaffold. Scaffold having a continuous platform suported by bearers suspended from wire rope from overhead supports, arranged and operated so as to permit the raising or lowering of the platform to desired working postions.

Medium-duty scaffold. Scaffold designed and constructed to carry a working load not to exceed 75 lb/sq ft.

Midrail. Rail approximately midway between the guard rail and platform, secured to the uprights erected along the exposed side and ends of platforms. Height established by OSHA requirements.

Mobile scaffold (tower). Light-, medium-, or heavy-duty scaffold mounted on casters or wheels.

Mobile work platform. Fixed work level one frame high on casters or wheels, with bracing diagonally from platform to vertical frame.

Needle beam scaffold. Light-duty scaffold consisting of needle beams supporting a platform.

Outrigger scaffold. Scaffold supported by outriggers projecting beyond the wall or face of the building or structure, the inboard ends of which are secured inside the building or structure.

Overhead protection for scaffolding. Approved overhead protection not more than 10 ft above the scaffold platform.

Platform

1. Working space for persons that is elevated above the surrounding floor or ground, such as a balcony or platform for the operation of machinery and equipment.

2. Any personnel-carrying device (basket or bucket) that is a component of an aerial device.

3. Wood mat used in sets to support machinery on soft ground. Also called a pontoon and an operator's station on a large machine, particularly when on rollers.

Platform ladder. Self-supporting ladder of fixed size with a platform provided at the working level. Size is determined by the distance along the front rail from the platform to the base of the ladder.

Putlog. Scaffold member on which the platform rests.

Rail ladder. Fixed ladder consisting of side rails joined at regular intervals by rungs or cleats and fastened in full length or in sections to a building, structure, or equipment.

Ramp. Incline connecting two levels.

Rungs. Ladder crosspieces of circular or oval cross section on which a person may step in ascending or descending.

Scaffold. Any temporary elevated platform and its necessary vertical, diagonal, and horizontal members used for supporting workers and materials. Also known as a scaffold tower.

Scaffolding. Scaffolds used in the erection, repair, alteration, or removal of buildings, constructed to ensure the safety of persons working on or passing under the scaffold.

Scaffolding bearer. Horizontal member of a scaffold on which the platform rests and that may be supported by ledgers.

Sectional ladder. Non-self-supporting portable ladder, nonadjustable in length, consisting of two or more sections so constructed that the sections may be combined to function as a single ladder. Size is designated by the overall length of the assembled sections.

Side-rolling ladder. Semifixed ladder, nonadjustable in length, supported by attachments to a guide rail, which is generally fastened to shelving, the plane of the ladder being also its plane of motion.

Side-step ladder. Ladder from which a person getting off at the top must step sideways in order to reach the landing.

Single-cleat ladder. Consists of a pair of side rails, usually parallel, but with flared side rails permissible, connected together with cleats that are joined to the side rails at regular intervals.

Single ladder. Non-self-supporting portable ladder, nonadjustable in length, consisting of but one section. Size is designated by the overall length of the side rail.

Single-point adjustable suspension scaffold. Manually or power-operated unit designed for light-duty use, supported by a single wire rope from an overhead support so arranged and operated as to permit the raising or lowering of the platform to desired working positions.

Single-pole scaffold. Platforms resting on putlogs or cross beams, the outside ends of which are supported on ledgers secured to a single row of posts or uprights, and the inner ends of which are supported on or in a wall.

Special-purpose ladder. Portable ladder that represents either a modification or a combination of design or construction features in one of the general-purpose ladders previously defined, in order to adapt the ladder to special or specific uses.

Standard railing. Vertical barrier erected along exposed edges of a floor opening, wall opening, ramp, platform, or runway to prevent falls of person, to be erected in accordance with OSHA requirements.

Stepladder. Self-supporting portable ladder, nonadjustable in length, having flat steps and a hinged back. Size is designated by the overall length of the ladder measured along the front edge of the side rails.

Stringers (wales). Horizontal members of a shoring system whose sides bear against the uprights or earth.

Stone setters' adjustable multiple-point suspension scaffold. Swinging scaffold having a platform supported by hangers suspended at four points so as to permit the raising or lowering of the platform to the desired working position by the use of hoisting machines.

Through ladder. Ladder on which a person getting off at the top must step through in order to reach the landing.

Trestle ladder. Self-supporting portable ladder, nonadjustable in length, consisting of two sections hinged at the top to form equal angles with the base. Size is designated by the length of the side rails measured along the front edge.

Trolley ladder. Semifixed ladder, nonadjustable in length, supported by attachments to an overhead track, the plane of the ladder being at right angles to the plane of motion.

Tube and coupler scaffold. Assembly consisting of tubing that serves as posts, bearers, braces, ties, and runners, a base supporting the posts, and special couplers which serve to connect the uprights and to join the other various members; usually used in fixed locations.

Tubular, welded, frame scaffold. Sectional, panel, or frame metal scaffold substantially built of prefabricated welded sections that consist of posts and bearers with intermediate connecting members and braced with diagonal or cross braces.

Tubular, welded, sectional, folding scaffold. Sectional, folding, metal scaffold, either of ladder frame or inside stairway design, substantially built of prefabricated welded sections that consist of end frames, platform frame, inside inclined stairway frame and braces, or hinged connected diagonal and horizontal braces, capable of being folded into a flat package when the scaffold is not in use.

Two-point suspension scaffold (swinging scaffold). Scaffold, the platform of which is supported by hangers (stirrups) at two points, suspended from overhead supports so as to permit the raising or lowering of the platform to the desired working position by tackle or hoisting machines.

Wall jack scaffold. Scaffold, the platform of which is supported by a bracket or jack that projects through a wall or window opening.

Work level. Elevated platform used for supporting workers and their materials, comprising the necessary vertical, horizontal, and diagonal braces, guard rails, and ladder for access to the work platform.

TEMPORARY SERVICES AND INSTALLATIONS

Access to site. Necessary arrangements with the local authorities to provide the location of a temporary driveway to the site.

Electric service. Temporary electric service provided as soon as a permit has been issued, to assure service to the site office, to the necessary lighting outlets where required during the progress of the building, and to workshops requiring power service.

Employee identification. Each employee or tradesman who is employed on the site is required to wear a badge that is color-coded in accordance with the trade on the job. Identification cards also are issued in case badges are lost or misplaced.

Heating and ventilation. Installed and maintained in areas requiring heating during the cold weather and ventilation during the hot weather.

Hoists. Temporary hoists and chutes installed at locations required by the building progress.

Insects. The site needs continuous insecticide spraying because it will be invaded, during certain periods of the year, depending on location, by all types of insects in the area to prepare hiding and housing places for their nests.

Noise level controls. Noise on the site is to be kept at a required level as stated in the code.

Office, tool sheds, and storage sheds. Temporary housing of office employees and provisions for work areas, storage facilities, and open storage bins to be provided and located on a site plan before the project begins in the field.

Off-site maintenance. In accordance with local ordinances the contractor is responsible for cleaning and removal of all debris on walks and streets in front of the project.

Portable fire protection. Fire extinguishers, attached to panels painted red, are placed at several locations on the site, especially at locations where welding is to be done. Portable fire extinguishers on wheels, which contain more chemicals or water, are used on large projects. Temporary fire protection equipment and their locations should be approved by the local fire department.

Portable toilet facilities. Chemical toilet facilities and ventilated enclosures installed at appropriate locations to reduce travel time for tradesmen.

Project sign. Contractor is to provide a project sign as indicated on the drawings, located as mutually agreed on by the owner and his architect.

Pumping equipment. Installation of pumping equipment to remove all standing water at the site, to prevent erosion of road bases and flooding of building site.

Rodent control. Due to the garbage created by food brought to the site by the tradesmen, a service should be provided, monthly if necessary, to prevent the lodging and growth of rodent families.

Safety precautions. Installation of barricades and signs which will direct traffic at the site and warn visitors not to stray into hazardous areas. Hard hats assigned to all visitors and others entering the work areas. OSHA and local inspectors should be consulted.

Security services. For large projects, arrangements are made to provide guard services during working hours and watchman services after working hours. If the site is large, dogs may be added for further security.

Site enclosure. Installation of a metal fence, at the property lines or as located by the contractor, to enclose the entire site. Sliding gates included at driveway entrance.

Telephone service. Service provided for the site office and coin phones that can be used by the tradesmen installed at several locations on the site.

Temporary roads on site. Temporary roads installed on safe and secure bases to provide access to all parts of the site. Due to heavy loads, constant maintenance is required. Roads should be as close as possible to the permanent road installations.

Tree and plant protection. Immediate protection for all trees and plants to remain on the site.

Visitor registration. All visitors, which includes local officials, OSHA inspectors, union officials, material salesmen, contractor officials, and utility representatives, must register.

Waste removal. Periodic removal from the site of all waste materials, which have been stored in waste disposal units located in appropriate areas.

Water service. Arrangements made with the utility company or the local water department to provide a water service to the site. Water outlets to be located as deemed necessary by the contractor.

Watering and dust control. Required by most local authorities; the site and adjacent areas must be kept dust free. A water-spraying tank truck is used in most situations, unless a permit is issued for the contractor to use the water from a fire hydrant.

Weather protection. Provisions made for weather protection as the building progress demands.

2

Site Work

DEMOLITION AND WRECKING

Approved storage facility. Facility for the storage of explosive materials conforming to permit issued by the proper authorities.

Barricade. Painted barriers to prevent vehicular or pedestrian traffic from entering the area.

Barricade lanterns. Warning lights placed on barricade. 6-V, 7-in. amber lens commonly used. Batteries placed in a combination plastic case supporting flashing light beam.

Blast area. Area in which explosives loading and blasting operations are conducted.

Blaster. Person or persons authorized to use explosives for blasting purposes and meeting the qualifications of the permit.

Blasting agent. Any material or mixture consisting of a fuel and oxidizer used for blasting, but not classified as an explosive and in which none of the ingredients is classified as an explosive, provided the finished (mixed) product cannot be detonated with a No. 8 test blasting cap when confined.

Blasting cap. Metallic tube closed at one end, containing a charge of one or more detonating compounds, and designed for and capable of detonation from the sparks or flame from a safety fuse inserted and crimped into the open end.

Block holing. Breaking of boulders by firing a charge of explosives that has been loaded in a drill hole.

Burning of debris. Allowable destruction of debris caused by demolition. Permits issued for burning where allowed. Environmental laws prevent burning in most localities.

Bus wire. Expendable wire, used in parallel or series in circuits, to which are connected the leg wires of electric blasting caps.

Class A explosives. Explosives possessing detonating hazard, such as dynamite, nitroglycerin, picric acid, lead azide, fulminate of mercury, black powder, as well as blasting caps, and detonating primers.

Class B explosives. Explosives possessing flammable hazard such as propellant explosives, including some smokeless propellants.

Class C explosives. Included are certain types of manufactured articles that contain Class A or Class B explosives, or both, as components, but in restricted quantities.

Cleaning up. Removal of all debris from public areas to allow normal traffic conditions.

Connecting wire. Insulated expandable wire used between electric blasting caps and the leading wires or between the bus wire and the leading wires.

Detonating cord. Flexible cord containing a center core of high explosives that when detonated will have sufficient strength to detonate other cap-sensitive explosives with which it is in contact.

Detonator. Blasting caps, electric blasting caps, delay electric blasting caps, and nonelectric delay blasting caps.

Dismantling of structure. Removal of all salvage materials that are usable or salable, such as ornamental items, doors, hardware, plumbing fixtures, heating equipment, and salvageable electrical fixtures and equipment.

Electric blasting cap. Cap designed for and capable of detonation by means of an electric current.

Electric delay blasting cap. Cap designed to detonate at a predetermined period of time after energy is applied to the ignition system.

Emergency first-aid kits. Medical first-aid equipment and material for emergency uses at the job.

Engineering survey. Survey to determine the condition of the framing, floors, and walls and the possibility of collapse. A safe plan to proceed with the demolition location by location.

Explosive. Chemical compound, mixture, or device, the primary or common purpose of which is to function by explosion, with substantially instantaneous release of gas and heat.

Extermination of rodents. It is a prerequisite of some local laws that rodents and vermin be exterminated before demolition takes place.

Fire extinguishers. For emergency uses, of several types: pressurized water, dry chemical, and carbon dioxide.

Flagger vests. For protection of directional and supervisory personnel, in fluorescent orange color.

Flasher bulbs. Long-life bulbs used in barricade lanterns; used in combination with flasher lantern batteries.

Folding litter (stretcher). Stretcher of wood rails and duck covering, for emergency uses.

Fuse lighters. Special devices for the purpose of igniting safety fuse.

Hearing protectors. Headgear offering a noise reduction level to protect personnel in and about the blasting area.

Initial electrical inspection. Tests or inspections of existing conditions before work is started. Conditions include, but are not limited to, energized lines and communication lines.

Leading wire. Insulated wire used between the electric power source and the electric blasting cap circuit.

License. Statutory or local laws require all wrecking or demolition contractors to be registered and secure a license to operate.

Magazine. Any building or structure used for the storage of explosive.

Material chute. Chutes designed and constructed of materials to provide the strength to eliminate failure due to impact loads caused by materials or debris entering the chute.

Mechanism of an explosion. Objective of a blasting operation is to shatter material so that it may be removed. Beyond the region of material actually shattered and displaced, there is generally a relatively small region of plastic deformation and cracking; beyond this the remaining energy is propagated as an elastic wave in the ground. If the charge is near the surface, there may also be propagation through the air.

Monitoring procedures. Measurement of vibration velocity is usually done by means of an electromagnetic pickup, whose resonance frequency is below the range of interest (2 or 3 Hz), connected to a suitable indicator. Indicator may take the form of a recording system or a simple meter. Recorder provides information about wave form, as well as amplitude. Portable seismograph is used in many instances.

Multiple delay blasting. Common practice in blasting is to use not a single charge but a series of charges distributed throughout the batch of material to be removed. It is also useful to arrange for charges to be fired not simultaneously but at intervals of a few milliseconds. This technique provides better results in terms of the blasting operation by improving the fragmentation of the material. It also results in greatly reduced vibration levels for a given total weight of explosive because the individual delays spread the energy over a longer time period.

Night work. It is usually required that permission be granted by local authorities to perform any work after sundown. When working at night, spotlights or portable lights for emergency lighting should be provided as needed to perform the work safely.

Nonelectric delay blasting cap. Blasting cap with an integral delay element in conjunction with and capable of being detonated by a detonation impulse or signal from miniaturized detonating cord.

Notice to adjacent property owners. Legal notice required by statutes or local ordinance to notify adjacent property owners in advance of schedule of demolition process.

Notice to fire department. Legal notice or safety precautionary notice to fire department to have water and hoses available in case of fire and to wash down dust clouds.

Notice to utilities. Legal or safety precautionary notice to all utility companies to shut off all power, gas, and water and for the telephone company to take steps to disconnect lines at poles that enter the site.

Ownership of materials. Agreement in advance between owner and contractor as to the ownership of usable salvage material that will become available before complete demolition.

Pedestrian protection. Canopy or overhead protection over walkways and driveways.

Permanent blasting wire. Permanently mounted insulated wire used between the electric power source and the electric blasting cap circuit.

Permit. Local requirements for licensed contractors to secure a permit and describe method and schedule of demolitions; list of all personnel and number and description of equipment and materials to be used.

Primer. Cartridge or container of explosives into which a detonator or detonating cord is inserted or attached.

Protective screening. Suspended metal screening of substantial construction to catch falling objects of considerable weight.

Safety fuse. Flexible cord containing an internal burning medium by which fire is conveyed at a continuous and uniform rate for the purpose of firing blasting caps.

Signs. Posting of painted signs required by statutory or local requirements to notify public of demolition.

Springing. Creation of a pocket in the bottom of a drill hole by the use of a moderate quantity of explosives in order that larger quantities of explosives may be inserted therein.

Temporary power. Power that needs to be maintained for electric tools, equipment, and lighting.

Temporary water. Water required for hosing down dust and in case of fire.

Traffic cones. One-piece polyvinyl chloride material painted in "Blaze Orange" used for traffic direction.

Traffic control. Detailed traffic control requirements issued by local police department for operation of demolition and the entry and departure of vehicles.

Utilities. Electric, gas, water, steam, sewer, and all other service lines disconnected and shut off safely to prevent fire, flooding, or explosions.

Watchers. Security guards or watchers to enforce safety requirements to those entering demolition area during and after working hours.

Water gels or slurry explosives. Wide variety of materials used for blasting. They all contain substantial proportions of water and high proportions of ammonium nitrate, some of which is in solution in the water. Two broad classes of water gels are those that are sensitized by a material classed as an explosive, such as TNT or smokeless powder, and those that contain no ingredient classified as an explosive, and are sensitized with metals such as aluminum or with other fuels. Water gels may be premixed at an explosives plant or mixed at the site immediately before delivery into the bore hole.

Water trucks. Truck body with water tank and spreading spray arms for dusting down.

Wrecking balls. For smashing masonry or concrete floors and walls; pear shaped, ball shaped or spherical shaped; made of cast iron or semisteel.

EARTHWORK

Adjusted net fill. Net fill after making allowance for shrinkage during compaction.

Angle of repose. Greatest angle above the horizontal plane at which a material will lie without sliding.

Backfill. Material used in refilling an excavation, such as for a foundation or subterranean pipe.

Backhaul. Line that pulls a drag scraper bucket backward from the dump point to the digging.

Backhaul cable. In a cable excavator, the line that pulls the bucket from the dumping point back to the digging.

Backhoe. Fully or partly revolving excavator, whose bucket is attached through a boom and stick, that digs by pulling the bucket toward itself.

Ballast. Heavy material, such as water, sand, or iron, that has no function in a machine except increase of weight.

Bank. Mass of soil rising above a digging or trucking level. Any soil that is to be dug from its natural position.

Bank gravel. Natural mixture of cobbles, gravel, sand, and fines.

Bank measure. Volume of soil or rock in its original place in the ground.

Bank yard. A yard of soil or rock measured in its original position before digging.

Bearing materials. Soil or compacted fill on which the foundations are supported.

Bedrock. The solid undisturbed rock in place either at the ground surface or beneath surficial deposits of natural soil or fill.

Belled excavation. Part of a shaft or footing excavation, usually near the bottom and bell shaped.

Belt loader. Machine whose forward motion cuts soil with a plowshare or disk and pushes it to a conveyor belt that elevates it to a dumping point.

Bench. A relatively level step excavated into each material on which fill is to be placed.

Berm. Artificial ridge of earth.

Blade. Part of an excavating equipment that digs and pushes dirt but does not carry it.

Blinding. Compacting soil immediately over a tile drain to reduce its tendency to move into the tile.

Boom. In a revolving shovel, a heavy beam hinged to the deck front, supported by cables. Heavy beam that is hinged at one end and carries a weight-lifting device at the other.

Borrow pit. Excavation from which material is taken to a nearby job.

Bucket. Part of an excavator that digs, lifts, and carries dirt.

Bucket loader. Machine having a digging and gathering rotor and a set of chain-mounted buckets to elevate the material to a dumping point.

Bulldozer. Tractor equipped with a front pusher blade.

Cave-in. Collapse of an unstable bank.

Cherry picker. Small derrick made up of a sheave on an A-frame, a winch and winch line, and a hook. Usually mounted on a truck.

Clamshell. Shovel bucket with two jaws that clamp together by their own weight when the bucket is lifted by the closing line.

Clay

1. Fine-grained inorganic soil possessing sufficient cohesion when dry to form hard lumps that cannot readily be pulverized by the fingers.

2. "Heavy" soil composed of particles less than $\frac{1}{256}$ mm in diameter.

Clay mineral. Naturally occurring inorganic material (usually crystalline) found in soils and other earthy deposits, the particles being of clay size, that is, not greater than 0.002 mm in diameter.

Claypan. A horizon of accumulation or a stratum of dense compact and relatively impervious clay. Claypan is not cemented, but is hard when dry and plastic or stiff when wet. Its presence, like that of a true hardpan, may interfere with water movement.

Clean material. Material that is free of foreign material; in reference to sand or gravel, means lack of binder.

Coarse sand. Sand consisting chiefly of grains that will be retained on a 65-mesh sieve.

Cobble. Rounded stone with diameter sizes ranging from 4 to 12 in.

Compacted fill. Usually mixtures of sand, gravel and predominately granular materials, crushed stone or non-corrosive slag compacted in layers in accordance with accepted engineering practice. Other materials having the same characteristics as sand, gravel, crushed stone or noncorrosive slag may be approved by the authority as compacted fills.

Compacted fill density. The density of the soil as it is compacted into a man-made fill.

Compacted yard. Soil or rock measured after it has been placed and compacted in a fill.

Compact gravel, compact sand. Deposits requiring picking for removal and offering high resistance to penetration by excavating tools.

Compaction. Reduction in bulk of fill by rolling, tamping, or soaking.

Compression. The volume change produced by application of a static external load.

Consolidation. Volume change that is achieved with the passage of time.

Crawler. One of a pair of roller chain tracks used to support and propel a machine, or any machine mounted on such tracks.

Crumber. Bulldozer blade that follows the wheel or ladder of a ditching machine to clean and shape the bottom.

Cutting. Part of the excavating process; also, lowering a grade.

Deep excavations. Whenever an excavation is made to a depth of more than four (4) feet below adjacent ground surfaces, the person who causes such excavation to be made, if afforded the necessary license to enter the adjoining premises, shall preserve and protect from injury at all times and at his own expense such adjoining structure or premises as may be affected by the excavation. If the necessary license is not afforded, it shall then be the duty of the owner of the adjoining premises to make his building or structure safe by installing proper underpinning or foundations or otherwise, and such owner, if it be necessary for the prosecution of his work, should be granted the necessary license to enter the premises where the excavation or demolition is contemplated.

Density. Ratio of the weight of a substance to its volume.

Depth of excavation. Depth is measured from the elevation of the street grade nearest to the point of excavation.

Dewatering. Removing water by pumping, drainage, or evaporation.

Diesel. Engine that uses nonvolatile fuel injected into the cylinders and ignites it by heat of compression.

Digging line. On a shovel, the cable that forces the bucket into the soil. Called "crowd" in a dipper shovel, "drag" in a pull shovel, and "dragline" and "closing line" in a clamshell.

Dip. Slope of layers of soil or rock.

Ditcher. Machine that digs ditches by continuous picking with cutters or small buckets. Generally, machine that digs a ditch.

Dozer. Word used in the industry for bulldozer or shovel dozer.

Drag. Pulling a bucket into the digging; the mechanism by which the pulling is done or controlled.

Dragline. Revolving shovel that carries a bucket attached only by cables and digs by pulling the bucket toward itself.

Drag scraper. Digging and transporting device consisting of a bottomless bucket working between a mast and an anchor. Also, towed bottomless scraper used for land leveling, called "leveling drag scraper" to distinguish it from cable type.

Drag trencher. Ditcher whose cutters drag dirt to the surface instead of lifting it in buckets.

Dredge. To dig underwater; also, a machine that digs under water.

Dry density. Term normally used for expressing the unit weight of soil. The dry density is computed from the wet density and the water content data.

Earth materials. Natural soil, bedrock and fill.

Elevating scraper. Scraper in which the apron is replaced by a chain-driven elevator that lifts soil into the bowl.

Elevation. Distance above or below a prescribed datum established by the city, county, state, or a federal agency from an established reference marker or monument.

Embankment. Fill whose top is higher than the adjoining surface.

Excavating cycle. Complete set of operations an excavating machine performs before repeating them.

Excavation

1. Any man-made cavity or depression in the earth's surface, including its sides, walls, or faces, formed by earth removal and producing unsupported earth conditions by reason of the excavation. If installed forms or similar structures reduce the depth-to-width relationship, an excavation may become a trench.

2. Cavity formed in the earth by excavating.

Existing grade. The grade prior to grading.

Expansion. Opposite of consolidation. The amount of expansion depends on the type of clay mineral and the availability of water and is a function of time, confining load, initial density, and initial water content.

Exploit. Excavate in such a manner as to utilize material in a particular vein or layer and waste or avoid surrounding material.

Feather. Blend the edge of new fill materials smoothly into the old materials.

Fill. Soil and other materials deposited at a site, or secured from the site, to be used to change the grades on the site.

Filling. Depositing or dumping of any matter including septic tank effluent onto, or into, the ground, except common household gardening and ground care.

Fines

1. Clay or silt particles in soil.

2. Clay or silt particles in the excavated materials.

Fine sand. Sand consisting chiefly of grains that will pass a 65-mesh sieve.

Finish grade. Final grade required by plans and specifications.

Front end loader. Tractor loader that both digs and dumps in front.

Front loader. Tractor carrying a front bucket for digging and loading. May be called just loader, or if crawler mounted, a dozer shovel.

Frost. Frozen soil that has been permeated with surface water.

Frost action. Phenomenon occurring in wet soils that results in volume increase or the build-up of stresses when subjected to freezing.

Frostline. Depth of frost penetration in soil. This depth varies in different parts of the country. Footings should be placed below this depth to prevent movement.

Full trailer. Towed vehicle whose weight rests entirely on its own wheels or crawlers.

Gradation. A descriptive term which refers to the distribution and size of grains in a soil.

Grade. Change the level of a part of the earth's surface, other than cause an excavation.

Grade stake. Stake indicating the amount of cut or fill required to bring the ground to a specified level.

Grading. The excavating and filling of land for altering the contours of the ground to prescribed grades for the purpose of erecting buildings or structures thereon or other use thereof, and does not include the processing of material for use on another site.

Grapple. Clamshell bucket having three or more jaws.

Gravel

1. Uncemented mixture of mineral grains $\frac{1}{4}$ in. or more in diameter.

2. Rock fragments from 2 to 64 mm (0.08 to 2.5 in.) in diameter, or a mixture of such gravel with sand, cobbles, boulders, and not over 15% fines.

Grizzly. Coarse screen used to remove oversized pieces from earth or blasted rock.

Gross cut. Total amount of excavation in a road or road section without regard to fill requirements.

Grubbing. Digging out roots.

Hard clay. Clay requiring picking for removal, a fresh sample of which cannot be molded in the fingers.

Hard compact soil. All earth materials not classified as running or unstable.

Hardpan

1. Thoroughly compact mixture of clay, sand, gravel, and boulders, for example, boulder clay; or a cemented mixture of sand or sand and gravel, with or without boulders, and difficult to remove by picking.

2. Hard, tight soil; also, a hard layer that may form just below plow depth on cultivated land.

Hard rock. Rock which requires drilling and blasting for removal.

Heave. Volume change produced by frost action or expansive soils.

Heavy soil. Fine-grained soil, made up largely of clay or silt.

Hoe. Backhoe.

Humus. Decayed organic matter; also, a dark fluffy swamp soil composed chiefly of decayed vegetation that is also called peat.

Hydraulicking. Excavating on dry land by means of water jets.

Jacknife. Tractor and trailer assuming such an angle to each other that the tractor cannot move forward.

Jib boom. Extension piece hinged to the upper end of a crane boom.

Kickout. Accidental release or failure of a shore or brace.

Ladder ditcher. Machine that digs trenches by means of buckets mounted on a pair of chains traveling on the exterior of a boom.

Liquid limit. That water content expressed as a percentage of the dry weight of soil at which the soil first shows a small but definite shearing strength as the water content is reduced.

Live boom. Shovel boom that can be lifted and lowered without interrupting the digging cycle.

Loam. Soft, easily worked soil containing sand, silt, and clay.

Loose gravel, loose sand. Deposits readily removable by shoveling only.

Loosening or scarifying. The operation opposite to that of compaction.

Loose yard. Soil or rock measured after it has been loosened by digging or blasting.

Mass diagram. Plotting of cumulative cuts and fills used for engineering computation of highway jobs.

Mass profile. Road profile showing cut and fill.

Medium clay. Clay that can be removed by spading, a fresh sample of which can be molded by a substantial pressure of the fingers.

Mica. A small flake of metamorphic rock, gives the appearance of a glistening fish scale.

Mica schist. A metamorphic rock consisting of naturally cemented, closely spaced, approximately parallel layers of scale-like flakes. The rock, if hard, sound, and massive, requires large explosive charges for removal.

Muck. Mud rich in humus; also, finely blasted rock, particularly from tunnels.

Mud. Soil containing enough water to make it soft.

Mudcapping. Blasting boulders or other rock by means of explosive laid on the surface and covered with mud.

Natural density (in place). the unit weight of a soil, expressed in lb/cu ft as it exists in a natural deposit at any particular time.

Natural soil. Naturally occurring surficial deposits overlying bedrock.

Net cut

1. In sidehill work, the cut required less the fill required at a particular station or part of a road.

2. Amount of excavated material to be removed from a road section, after completing fills in that section.

Net fill

1. Fill required less the cut required at a particular station or part of a road.

2. In sidehill work, the yardage of fill required at any station less the yardage of material obtained from the cut at that station.

One on two (one to two). Slope in which the elevation rises one foot in two horizontal feet.

Open cut. Excavation in which the working area is kept open to the sky, as opposed to cut-and-cover and underground work.

Ordinary fill. Usually comprised of clay, sand, gravel, and rocks artificially deposited at a site. Some specifications allow 5% of the total weight of the fill to be noncorrosive slag.

Ordinary gravel. Gravel containing particles up to $1\frac{1}{2}$ in. in size.

Organic soil. A soil composed mainly of organic matter on a volume basis. (Twenty per cent or more organic matter by weight.)

Overburden. Soil or rock lying on top of a pay formation.

Overhead shovel. Tractor that digs at one end, swings the bucket overhead, and dumps at the other end.

Paddle loader. Belt loader equipped with chain-driven paddles that move loose material to the belt.

Pan. Carrying scraper.

Part-swing shovel. Shovel in which the upper works can rotate through only part of a circle.

Peat (humus). Soft light swamp soil consisting mostly of decayed vegetation.

Perlite. Acid, igneous, glassy rock of the composition of obsidian, expanded by heating and divided into small spherical bodies by the tension developed by its contraction on cooling.

Permanent water level. "Sea level" unless special conditions exist. If special conditions exist, the term "permanent water level" shall mean such lower level as the building official in his opinion may deem to represent the permanent water level.

Permeability. The state of water movement in soil is called percolation; the measure of it is called "permeability"; and the factor relating permeability to unit conditions of control is called "coefficient of permeability."

Pit. Mine, quarry, or excavation area worked by the open-cut method to obtain material of value.

Plastic limit. That water content expressed as a percentage of the dry weight of soil at which the soil mass ceases to be plastic and becomes brittle, as determined by a procedure for rolling the soil mass into threads $\frac{1}{8}$ in. in diameter. The plastic limit is determined by reducing the water content of the soil mass.

Plasticity index. The difference between the liquid and plastic limits; represents the range of moisture within which soil is plastic.

Platforms. Movable wood or metal platforms used in sets to support machines on soft ground.

Pond. Area in which water is held long enough to allow fine soil particles to settle.

Porosity. The percentage of space in the soil mass not occupied by the solid, with respect to the total volume of the mass.

Porosity and void ratio. The measures of the state or condition of a soil structure.

Positive drainage. Sufficient slope to drain surface water away from buildings without ponding.

Primary excavation. Excavation made by digging in undisturbed soil, as distinguished from rehandling stockpiles.

Profile. Charted line indicating grades and distances, and usually depth of cut and height of fill for excavation and grading work. It is commonly taken along the centerline.

Pusher. Tractor or other machine that pushes a scraper to help it pick up a load.

Quagmire. Saturated area with a surface of soft mud, or, at best, a surface providing a shaky and precarious footing.

Quarrying. Removal of rock that has value because of its physical characteristics.

Quicksand. Fine sand or silt that is prevented from settling firmly together by upward movement of groundwater; also, any wet inorganic soil so unsubstantial that it will not support any load.

Rebound. Opposite of compression. The normal rebound phenomenon that occurs on release of a compressive load.

Reversed loader. Front end loader mounted on a wheel tractor having the driving wheels in front and steering at the rear.

Revolving shovel. Digging machine in which the upper works can revolve independently of its supporting unit.

Rig. Any machine; more specifically, the front or attachment of a revolving shovel.

Rock. Hard, firm, and stable parts of earth's crust; also, any material that requires blasting before it can be dug by available equipment.

Rock flour (inorganic silt). Fine-grained soil consisting chiefly of grains that will pass a 200-mesh sieve, and possessing sufficient cohesion when dry to form lumps that can readily be pulverized with the fingers.

Rotary tiller. Machine that loosens and mixes soil and vegetation by means of a high-speed rotor equipped with tines.

Sand

1. Loose soil composed of particles between $\frac{1}{16}$ and 2 mm in diameter.

2. Type of soil possessing practically no cohesion when dry, and consisting of mineral grains smaller than $\frac{1}{4}$ in. in diameter.

Scarifier. Accessory on a grader, roller, or other machine, used chiefly for shallow loosening of road surfaces.

Schist. Fine-textured laminated rock with a more or less wavy cleavage, containing mica or other flaky minerals.

Scraper (pan). Machine that has a cutting edge, bowl, and apron and that digs, transports, and spreads soil.

Seam. Layer of rock, coal, ore, granite, marble, etc.

Sedimentary rock. Rock formed by deposit of small particles of soil and consolidated over eons of time, which, when sound, cannot be removed by a pick and usually requires blasting for removal.

Sensitive clay. A type of clay that, when tested in an undisturbed condition, has substantial strength, but when disturbed and remolded loses its strength.

Selective digging. Separating two or more types of soil while digging.

Shale

1. Laminated, fine-textured, soft rock composed of consolidated clay or silt, which cannot be molded without the addition of water but can be reduced to a plastic condition by moderate grinding and mixing with water.

2. Rock formed of consolidated mud.

Shallow excavations. Wherever an excavation is made to a depth less than four (4) feet below the curb, the owner of a neighboring building or structure should preserve and protect from injury and shall support his building or structure by the necessary underpinning or foundation. If necessary for the purpose, he shall be afforded a license to enter the premises where the excavation is contemplated.

Sheet pile. Pile or sheeting that may form one of a continuous interlocking line, or a row of timber, concrete, or steel piles driven in close contact to provide a tight wall to resist the lateral pressure of water, adjacent earth, or other materials.

Shoring. Temporary bracing to hold the sides of an excavation from caving.

Shrinkage. Loss of bulk of soil when compacted in a fill, usually computed on the basis of bank measure.

Side casting. Piling spoil alongside the excavation from which it is taken.

Side hill. Slope that crosses the line of work.

Side hill cut. Long excavation in a slope that has a bank on one side and is near the original grade on the other.

Side stake. On a road job, a stake on the line of the outer edge of the proposed pavement; also, any stake not on the centerline.

Sides, walls, or faces. Vertical or inclined earth surfaces formed as a result of excavation work.

Silt

1. Soil composed of particles between $\frac{1}{256}$ and $\frac{1}{16}$ mm in diameter.

2. Heavy soil intermediate between clay and sand.

Silting. Filling with soil or mud deposited by water.

Skiving. Dig in thin layers.

Slat bucket. Open work bucket made of bars instead of plates, used in digging sticky soil.

Slope. Angle with the horizontal at which a particular earth material will stand indefinitely without movement.

Slope stake. Stake marking the line where a cut or fill meets the original grade.

Soft clay. Clay that when freshly sampled can be molded under relatively slight pressure of the fingers.

Soft rock. Characteristic that allows rock to be easily penetrated by using a pick.

Soil. Loose surface material of the earth's crust.

Soil consistency. The physical properties of soils affected by water content.

Soil consistency tests. The four stages, or states, recognized for describing the consistency of a soil. These are: (1) the liquid state, (2) the plastic state, (3) the semi-solid state, and (4) the solid state.

Soil cover (ground cover). Light roll roofing or plastic used on the ground of crawl spaces to minimize moisture permeation of the area.

Soil engineering. The application of the principles of soil mechanics in the investigation, evaluation, and design of civil works involving the use of earth materials and the inspection and testing of the construction thereof by a soil engineer.

Soil mechanics.   The application of the laws of solid and fluid mechanics to soils and similar granular materials as a basis for design, construction, and maintenance of stable foundations and earth structures.

Soil or soils.   Earth or earth stratum condition found at and below the point of bearing of building, footing, and foundations.

Soil shear strength.   The maximum resistance of a soil to shearing stresses.

Soil stabilizer.   Chemical that alters the engineering property of natural soil, used to stabilize soil slopes, to prepare for building foundations, and to prevent erosion.

Soil structure.   Arrangement of soil into various aggregates, each differing in the characteristics of its particles.

Soil survey.   The systematic examination of soils, their description and classification, mapping of soil types, and the assessment of soils for various agricultural and engineering uses.

Sound rock.   Rock having no visible cracks or seams.

Specific gravity.   The ratio between the unit weight of a substance and the unit weight of water at 4°C.

Spoil.   Dirt or rock that has been removed from its original location.

Spur.   Rock ridge projecting from a side wall after inadequate blasting.

Stabilize.   Make soil firm and prevent it from moving.

Stockpile.   Material dug and piled for future use.

Stope.   Underground excavation that is made in a series of steps or benches.

Stringers.   Horizontal members of a shoring system whose sides bear against the uprights or earth.

Strip.   Remove overburden or thin layers of pay material.

Stripping.   Removal of a surface layer or deposit, usually for the purpose of excavating other material under it.

Stripping shovel.   Shovel with an especially long boom and stick that enables it to reach farther and pile higher.

Stumper.   Narrow heavy dozer attachment used in pushing out tree stumps.

Subgrade.   Surface produced by grading native earth, or cheap imported materials, which serves as a base for a more expensive paving.

Swell.   Increase of bulk in soil or rock when it is dug or blasted.

Swing loader.   Tractor loader that digs in front and can swing the bucket to dump to the side of the tractor.

Talus.   Loose rock or gravel formed by disintegration of a steep rock slope.

Tamp.   Pound or press soil to compact it, either manually or by pulsating, vibrating equipment.

Tamper.   Tool for compacting soil in spots not accessible to rollers.

Tamping roller.   One or more steel drums filled with sand or water, fitted with projecting feet, and towed by means of a box frame.

Taproot.   Bit root that grows downward from the base of a tree.

Through cut.   Excavation between parallel banks that begins and ends at original grade.

Tight.   Soil or rock formations lacking veins of weakness.

Tilth.   Soil condition in relation to lump or particle size.

Tilting dozer.   Bulldozer whose blade can be pivoted on a horizontal center pin to cut low on either side.

Topsoil.   Topmost layer of soil. Usually contains humus and is capable of supporting good plant growth.

Tower loader.   Front end loader whose bucket is lifted along tracks on a more or less vertical tower.

Tractor.   Motor vehicle on tracks or wheels used for towing or operating vehicles or excavating equipment.

Tractor loading.   Tractor equipped with a bucket that can dig and be elevated to dump at truck height.

Trench

1.   Narrow excavation made below the surface of the ground. In general, the depth is greater than the width, but the width is not usually greater than 15 ft.
2.   Continuous ditch.

Trench braces.   Horizontal members of the shoring system whose ends bear against the uprights or stringers.

Trencher.   A ditcher.

Trench jack.   Screw-type or hydraulic jack used as cross bracing in a trench shoring system.

Trench shield.   Shoring system composed of steel plates and bracing, welded or bolted together, which support the walls of a trench from the ground level to the trench bottom and which can be moved along as work progresses.

Truck roller.   In a crawler machine, the small wheels that are under the track frame and that rest on the track.

Unclassified excavation.   Excavation paid for at a fixed price per yard, regardless of the material to be excavated.

Unstable soil.   Earth material, other than running, that because of its nature or the influence of related conditions, cannot be depended on to remain in place without extra support, such as that furnished by a system of shoring.

Uprights.   Vertical members of a shoring system.

Ultimate bearing capacity.   Average load per unit of area required to produce failure by rupture of a supporting soil mass.

Vein.   Layer, seam, or narrow irregular body of material different from surrounding formations.

Void ratio.   The ratio of the space not occupied by the solid particles to the volume of the solid particles.

Wagon.   Full trailer with a dump body.

Walking dragline.   Dragline shovel that drags itself along the ground by means of side-mounted shoes.

Waste removal.   Digging, hauling, and dumping of valueless material to get it off the premises.

Water table.   The upper surface of free ground water in a zone of saturating, except when separated from an underlying body of ground water by unsaturated material.

Well point.   Pipe having a fine mesh screen and a drive point at the bottom. Used for pumping out ground water.

Well point pump.   Centrifugal pump that can handle considerable quantities of air and is used for removing underground water to dry up an excavation.

Wet density.   The unit weight of the solid particles and the contained moisture expressed in lb/cu ft.

Wheel ditcher.   Machine that digs trenches by rotation of a wheel fitted with toothed buckets. Wheel equipped with digging buckets, supported and controlled by a tractor unit.

Windrow. Ridge of loose dirt.

Working cycle. Complete set of operations. For an excavator, it usually includes loading, moving, dumping, and returning to the loading point.

FENCES AND GATES

Aluminum barbed wire. Two strands, 0.110-in. diameter; four-point barbs, 0.080-in. diameter; alloy 5052-H38.

Aluminum brace bands and tension bands. Fabricated of extruded aluminum alloy 6063 T42, with beveled edges.

Aluminum cantilever slide gates. Gate frames of 2-in. sq aluminum tubing, alloy 6063-T6, weighting .94 lbs per lin ft, and welded at the corners so as to form a rigid one-piece unit. Fabric securely stretched and held on all four sides of the 2-in. sq tubing by use of hook bolts and tension rods. Two side-rolling wheels installed to insure alignment of truck in track for each gate leaf. All accessories galvanized after fabrication from malleable iron or steel.

Aluminum castings (ASTM B 26). Sand castings, alloy ZG61A, ZG61B, or ZC81B.

Aluminum-coated fabric. Coated steel wire, corrosion resistant; 0.40-oz. minimum deposit.

Aluminum colored fabric. Aluminum fabric colored to a colorfast green by a phosphate-chromate conversion treatment.

Aluminum extension arms for barbed wire. Formed from 0.090-in.-thick sheet, alloy 5052-H34.

Aluminum fabric. Chain link fabric, 2-in. sq mesh, woven from alloy 6061 wire.

Aluminum posts. Schedule 40 pipe, square tubing, or H-beams, alloy 6063-T6.

Aluminum rail couplings. Outside type, 0.080-in. wall, self-centering, alloy 6063.

Aluminum tension bars. Oval extruded bar, alloy 6063-T6.

Aluminum tension wire. Wire, 0.192 in., alloy 6061.

Aluminum top rail. Schedule 40 pipe weighing 0.786 lb/ft (linear), alloy 6063-T6.

Aluminum truss braces. Used for fences 6 ft. or over and for all fences where no top rail is used. Truss brace alloy 6063-T6, schedule 40 pipe, with a aluminum truss rod installed between terminal post and each adjacent line post.

Bands. Bands made from flat or beveled steel for attaching the fabric and stretcher bars to all terminal posts at intervals usually not exceeding 15 in.

Barbed steel wire. Galvanized high-tensile-strength steel wire with four-point round barbs spaced approximately 5 in. on center; wire can be aluminized finished.

Barbed wire. Wire with four-point barbs spaced 4 to 6 in. on center. Fabricated of aluminum-coated steel, aluminum-clad steel, vinyl-coated steel, aluminum alloy, or zinc-coated steel wire.

Barbed wire extension arms. Usually used for supporting three strands of barbed wire. Available in vertical or 45° angle types. Fabricated in steel or aluminum.

Barbed wire supporting arm. Fabricated of heavy pressed steel, with provisions for attachment to corner and line posts. Arm usually supports three rows of barbed wire.

Bethamized steel fabric. Trade name of the Bethlehem Steel Co. Steel is passed through a high-current-density electrolytic cell containing a highly purified solution of zinc sulfate.

Bottom tension wire. Used without bottom rail; usually 14-gauge galvanized twisted wire for residential installations and 7-gauge galvanized spiral wire for industrial installations.

Bottom tension wire to fabric. Fabric attached to bottom tension wire with hog rings or wire ties, usually not over 24 in. on center.

Brace bands. Bands for securing top rail and brace rails to end, gate, pull, or corner posts.

Brace rails and truss rods. Brace rails and adjustable truss rods at each gate post, end post, pull post, and both sides of corner posts, to align and make the installation rigid.

Center gate stop. Metal stop set and anchored in concrete, for all double swing gates.

Chain link fabric. Woven mesh available in various widths, gauges, wire breaking strengths, and protective coatings.

Clips and tie wires. Used for attaching fabric to all line posts at intervals usually not exceeding 15 in.

Concrete footings. The alternative to drive anchors is concrete footing. Footings should be a minimum of 12 in. in width or diameter and as deep as 36 in., depending on soil and frost conditions. The post is grouted into the footing, in a space provided in the center of the footing.

Diagonal truss rods. Standard steel sizes, adjustable type.

Diagonal truss rods for gates. Usually adjustable.

Double-gate stops. Consist of mushroom type or flush plate with anchors. Set in concrete to engage the center drop rod or plunger bar. Locking device and padlock eyes are an integral part of the latch.

Double-hinged gates. Electrically welded steel frames and fittings, using fence fabric. Installed with stretcher bars at vertical edges and tie wires at top and bottom edges, with diagonal cross bracing. Hinges made of pressed steel or malleable iron. Barbed wire supporting arms and barbed wire strands over the top of gates are available.

Drive anchors. Post footing consisting of two (2) angle blades driven on 45° angles through a special clamp (shoe) creating root-like stability, depending on soil conditions.

Expansion couplings. Minimum 6-in.-long top rail expansion couplings for continuous top rail.

Fabric colors. Medium green, dark green, and white for vinyl- and PVC-coated wire only.

Fabric metals. Usually of zinc-coated steel, aluminum-coated steel, aluminum-clad steel, aluminum alloy, and vinyl-coated steel.

Fabric widths. One piece for fence heights up to 12 ft.

Fabric wire gauges. Usually available: nos. 6, 9, and 11. Wire gauge includes coating or cladding.

Fasteners. Consist of nuts for tension band and hardware bolts on side of fence opposite fabric side.

Finish hardware for gates. Padlocks and locking devices are not usually an integral part of gate latch system.

Framing. All steel parts hot-dipped galvanized in accordance with ASTM standards, prior to vinyl coating. Thickness of vinyl coating to be 10–14 mils applied to fusion bonding.

Galvanized fabric. Galvanized steel fabric with hot-dipped coating of zinc.

Gate frames. Electrically welded or corner fitting type.

Gate framework. Consists of ends, corners, and pull posts made of the following materials: zinc-coated steel pipe, formed steel, aluminum pipe, steel H sections, steel square sections, aluminum H sections, and aluminum square sections.

Gate hinges. Non-lift-off or offset type for 180° gate swing. Hinge sizes to suit gate weight and sizes. Usually $1\frac{1}{2}$ pair for each gate leaf of 6-ft nominal height or over.

Gate latches. Fork or plunger-bar type. Operable from either side of gate. Padlock eye is an integral part of gate latch.

Gate posts. Available shapes for posts are round, square, and H section.

Gates. Swing or sliding gates include latches, stops, keepers, hinger or rollers, and roller tracks. Available with provisions for barbed wire above the fabric.

Gate types. Single swing, double swing, double sliding, vertical lift, single cantilever slide, single overhead slide, and electrically operated.

Groundings. Groundings are as follows: direct grounding system adjacent to each terminal post, ground cable attached to fence system, or 6-ft-long ground conductor rod driven to maximum depth for each ground cable and connected to ground cable. Ground conductor systems depend on local soil conditions.

Hinged gates. Electrically welded steel frames and fittings, using fence fabric. Installed with stretcher bars at vertical edges and tie wires at top and bottom edges, with diagonal cross bracing. Hinges made of pressed steel or malleable iron. Latch forked or plunger-bar type with padlock eye as part of latch. Barbed wire supporting arms and barbed wire strands over the top of the gate are available.

Hog rings. Metal rings for attaching fence fabric to tension wires.

Horizontal members. Zinc-coated steel pipe or tubing, vinyl-coated steel pipe or tubing, PVC-coated steel pipe or tubing, aluminum alloy pipe or tubing.

Horizontal members for gates. Available in round or square zinc-coated steel, round or square vinyl-coated steel, round or square PVC-coated steel, and round or square aluminum alloy.

Intermediate gate posts. Available shapes are round, square, and H section.

Keeper. Automatically engages the gate leaf and holds it in the open position until manually released.

Kennel system fencing. Portable, welded, modular fabricated; no concrete is needed for post holes.

Knuckling. Term used to describe the type of selvage obtained by interlocking adjacent pairs of wire ends and then bending the wire ends back into a closed loop.

Lacing. Wire mesh laced or fastened to frame through each link at points of strain.

Line post. Intermediate posts spaced a maximum of ten (10) feet apart.

Line post barbed wire arm. Post extensions which provide for three (3) strands of barbed wire on an inward or outward 45° angle. V-type barbed wire arms are available to accommodate six (6) strands of barbed wire.

Line post foundation depths. Recommended minimum depth of 3 ft., but usually governed by soil conditions at the site.

Line post hole diameters for footings. Recommended minimum diameter of 10 in. and at least 12 in. for larger size posts, at gates in particular.

Line posts other than pipe columns. I-beam shape, used as an H column.

Line posts to fabric. Fabric attached to line posts with tie bands or wire ties usually not more than 15 in. on center.

Line post ties or clips. Fasteners are installed approximately every 15 in. and are used to attach the fabric to the terminal posts.

Line post top. Cap that attaches top rail to line post. When top rail is not required, a moisture-excluding cap is used on tubular posts.

Maximum security. V-shaped barbed wire supporting arms carrying total of six strands of barbed wire. Barbs located on 3-in. centers.

Plain post caps. Standard watertight plain post caps, for use when neither top rail nor barbed wire is required.

Post caps for use with top rail and barbed wire. Standard watertight post caps with aperture for passage of top rail and including 45° angle barbed wire extension arm.

Privacy fence. Combination of steel wire fabric or mesh with interwoven seasoned wood pickets. Top rail of pipe or tubing. Pipe or tubing corner on end posts (not a chain link fence).

Rail end fittings. Standard closed rail end fittings for ends of top rail and all brace rails.

Rail tie wire. Double wrap of 13-gauge or single wrap of 9-gauge wire attaches the chain link fabric to the rail at approximately 24 in. intervals.

Selvage edges. Top and bottom of fabric with twisted and barbed selvages or with knuckled and knuckled or custom selvages.

Sliding gate operator. Electrically controlled operating device with heavy gauge steel frame for pad or post mounting. Features include all-weather protection, adjustable safety function clutch, limit switches, magnetic solenoid locking brake, emergency disconnect for manual operation, heavy-duty roller chain drive, permanently lubricated heavy-duty bearings, and high-starting torque motor with electric locking device.

Sliding gates. Electrically welded steel frames and fittings using fence fabric. Installed with stretcher bars at vertical edges and tie wires at top and bottom edges, with diagonal cross bracing. Installed with heavy-duty track ball bearing hanger sheaves and guides. Closure hardware includes padlock eyes.

Stretcher bar. Vertical metal bar next to corner or end posts, slipped in between mesh openings, held by hooks to post, and fastened to top and bottom rails.

Swing gate frames. Two (2) in. sq galvanized members weighing 2.60 lbs/lin ft or two (2) in. sq aluminum members alloy 6063-T6, weighing .94 lbs/lin ft. Galvanized or aluminum members may be vinyl coated. Gates equipped with galvanized steel hinges and latch. Fabric attached on all four sides of the gate by means of hook bolts and tension rods.

Swing gate operator. Electrically controlled operating device with all-weather housing, friction safety clutch, limit switches, positive locking of the gate arm in closed position, precision worm gear, speed reducer in constant-viscosity-lubricant, contactor-type reversing starter with overload protection, permanently lubricated heavy-duty bearings, and high-starting torque motor with electric locking device.

Tension bands. Used for securing tension bars to posts.

Tension bars. Usual length is 2 in. less than the height of the fence fabric; used for each corner post, pole post, and the gate post. Usually not used when roll-formed posts are installed.

Tension bars for gates. Used at vertical edges and also for top and bottom edges. Attached to frame with tension bands usually spaced not more than 15 in. on center.

Tension clip. Clip attaching tension bars to square terminal posts. Inserted into the post, they eliminate need for wrap-around bands.

Tension wire. Galvanized steel wire tied between each line post with ties or clips. Installed before fabric installation.

Terminal and line posts. Includes all end, corner, gate, and pull posts. Line posts at equidistant spaces usually not exceeding 10-ft centers between terminal posts. Line posts or terminal posts usually occur at major changes in ground contours.

Terminal post. Generic name for an end post, corner post, pull post, or gate post, depending on its function.

Terminal post barbed wire. Corner, end, and gate posts extend one (1) foot above fabric to allow the barbed wire to be connected directly to the terminal post. This allows the barbed wire to be stretched taut and maintains the highest possible degree of security.

Terminal post cap. Watertight ornamental or plain metal cap on top of post.

Tie bands. Metal tie bands for attaching fence fabric to posts and top rail.

Ties. Wire ties for attaching fence fabric to line posts and top rail.

Tie wires. U-shaped galvanized steel or aluminum alloy clips or wire used to fasten fabric to posts or rails.

Top rail. Pipe or roll-formed section. Top rail forms continuous brace from end to end of each run of fence.

Top rail and post braces for gates. Usually consists of steel pipe, formed steel, or aluminum pipe.

Top rail to fabric. Fabric attached to top rail with tie bands or wire ties, usually not over 24 in. on center.

Top tension wire. Used when top rail is not required. Usually 14-gauge galvanized twisted wire for residential installations and 7-gauge galvanized spiral wire for industrial installations.

Top tension wire to fabric. Fabric is attached to top tension wire with hog rings or wire ties usually not more than 24 in. on center.

Truss braces. Tubing or pipe used from end or corner post to adjacent line post, to stabilize the installation.

Truss rod system. Standard adjustable truss rod systems at each gate post, end post, pull post, and both sides of corner posts.

Twisting. Term used to describe the type of selvage obtained by twisting adjacent pairs of wire ends together in a close helix of one and one-half machine turns, which is equivalent to three full twists.

Vertical lift gates. Gate frames made of two (2) in. sq aluminum members, alloy 6063-T6, weighing .94 lbs/lin ft

and welded at all corners to form a rigid frame. Two (2) 12-in. flanged 53 lb steel H-beam posts of specified height as required for clearance, flanged and bolted to anchor bolt assemblies. Leveling cables and counterweights installed to operate the gate satisfactorily. Safety device provided to stop the gate in case of a malfunction. Electric operator capable of automatic operation.

Vertical members. Zinc-coated steel pipe or tubing, vinyl-coated steel pipe or tubing, PVC-coated steel pipe or tubing, aluminum alloy or tubing.

Vertical members for gates. Usually consist of round or square zinc-coated steel; round or square vinyl-coated steel; round or square PVC-coated steel; round or square aluminum alloy.

Vinyl-coated fabric. Chain link fabric usually made of 6- or 9-gauge vinyl-coated steel wire in accordance with Federal Specification RR-F-191. Fabric woven in accordance with approved standards from vinyl-coated strand having a bright, smooth core of carbon steel wire, hot-dipped galvanized to Class 1 coating; minimum, 30 oz/sq ft of zinc, coated by the thermal extrusion process with PVC compound.

Wire gauge. Thickness of wire gauge includes coating or cladding.

Zinc-coated barbed wire. Consists of two strands of $12\frac{1}{2}$-gauge wire with 14-gauge, four-point barbs, spaced approximately 5 in. apart. Wire 0.80 oz/sq ft zinc coated with a minimum coating of 0.80 oz/sq ft of surface area on $12\frac{1}{2}$ gauge and 0.60 oz/ft sq of surface area on 14-gauge wire.

FLOOD CONTROL AND SITE DRAINAGE

Artificial lake. A man-made basin other than a swimming pool, designed or intended to permanently contain water, which should have a depth of not less than two (2) feet or a minimum area of two hundred and fifty (250) sq ft at the low-water stage.

Artificial obstruction. Any obstruction that is not a natural obstruction, including any that, while not a significant obstruction in itself, is capable of accumulating debris and thereby reducing the flood-carrying capacity of the stream.

Base flood. A flood having a one (1) percent chance of being equalled or exceeded in any given year.

Base flood elevation. An elevation equal to that which reflects the height of the base flood as defined in the code.

Bypass channel. Channel built to carry excess water from a stream. Also known as a "flood relief channel."

Channel. A natural or artificial water-course with definite bed and banks to confine and conduct the normal flow of water.

Drainage. Removal of ground water or surface water or of water from structures, by gravity or pumping.

Drainage basin. An area in which surface runoff is collected and is carried by a drainage system such as a creek or river and its tributaries.

Drainage canal. Artificial canal built to drain water from an area having no natural outlet for precipitation accumulation.

Drainage channels. Channels used for interrupted flow of surface water.

Drainage facility. Any ditch, gutter, pipe, culvert, storm sewer, or other structure designed, intended, or constructed for the purpose of carrying surface waters off streets, public right-of-ways, parks, recreational areas, or any part of any subdivision or contiguous land areas.

Drainage pattern. The configuration of a natural or artificial drainage system.

Drainage right-of-way. The lands required for the installation of storm water sewers or drainage ditches or required along a natural stream or watercourse, for preserving the channel and providing for the flow of water therein to safeguard the public against flood damage, in accordance with the code.

Drywell. Covered pit with open-jointed lining or covered pit filled with coarse aggregate through which drainage from roofs, basement floors, foundation drain tile, or areaways may seep or leach into the surrounding soil.

Encroachment. Any fill, structure, building use, accessory use, or development in the floodway.

Encroachment floodway lines. Lines that are limits of obstruction to flood flows. These lines are on both sides of, and generally parallel to, the stream. The lines are established by assuming that the area landward (outside) of the encroachment lines will be ultimately developed in such a way that it will not be available to convey flood flows.

Federal flood insurance. Insurance available through the National Flood Insurance Program to property owners in limited amounts from private insurance carriers against casualty losses due to floods.

Fifty-year flood. A flood that has a two (2) percent chance of occurring in any one (1) year based upon criteria established by the state water commission.

Flood. Condition that occurs when water overflows the natural or artificial confines of a stream or other body of water, or accumulates by drainage over low-lying areas.

Flood base elevation. That elevation of the highest flood on record, determined by the city engineer's record of the elevations of the highest flood locations as indicated on the flood plain map of the city on file in the office of the city engineer. Flood base elevations at intermediate locations should be interpolated along the water course between the two (2) nearest flood base elevations, one (1) each upstream and downstream. The controlling flood base elevation for any building site should be the same as the flood base elevation at the nearest point of the water course as measured on a line perpendicular to the direction of the water course.

Flood control. Use of levees, walls, reservoirs, floodways, and other means to protect land from water overflow.

Flood current. Tidal current associated with increase in the height of a tide.

Flood dam. Dam for storing floodwater or for supplying water by mechanical means when needed.

Flood fringe. Portion of the floodplain outside of the floodway that is covered by floodwaters during the regional flood; it is generally associated with standing water rather than rapidly flowing water.

Floodgate. Gate used to restrain a flow or, when opened, to allow a flood flow to pass where channeled.

Flood hazard. A hazard to land or improvements due to overflow water having sufficient velocity to transport or deposit debris, scour the surface soil, dislodge or damage buildings, or erode the banks of water courses.

Flooding. Land subject to flooding or other hazards to life, health, or property deemed to be topographically unsuitable should not be plotted for permanent residential occupancy or for such other uses as may increase danger to health, life, or property, or aggravate erosion or flood hazard until all such hazards have been eliminated or unless adequate safeguards against such hazards are provided by the development plans. Such land within the subdivision should be set aside on the plot for uses and should not be endangered by periodic or occasional inundation or should not produce unsatisfactory living conditions.

Floodplain

1. Relatively flat areas of lowlands adjoining the channel of a watercourse or areas where drainage is or may be restricted by man-made structures, which areas have been or may be covered partially or wholly by floodwater resulting from a 100-year flood.

2. Land which has been or may be covered by floodwater during the regional flood. The floodplain includes the floodway and the flood fringe.

3. Continuous land area adjacent to a water course, the elevation of which is equal to or below the flood base elevation.

Floodplain regulations. Includes the codes, ordinances, and other regulations relating to the use of land and construction within the channel and floodplain areas, including zoning ordinances, subdivision regulations, building codes, setback requirements, open area regulations, and similar methods of control affecting the use and development of the areas.

Flood profile. A graph or a longitudinal profile showing the relationship of the water surface elevation of a flood event to locations along a stream or river.

Floodproofing. Involves any combination of structural provisions, changes, or adjustments to properties and structures subject to flooding, primarily for the purpose of reducing or eliminating flood damage to properties, water and sanitary facilities, structures, and contents of buildings in flood hazard areas.

Flood reservoir. A ponding area created for the purpose of impounding flood waters and alleviating flood damage which might result from man-made fills.

Floodway. The channel of a river or stream and those portions of the floodplain adjoining the channel required to carry and discharge the floodwater or flood flows associated with the regional flood.

Fully-developed watershed. Planned or estimated intensity of development in the watershed or drainage area.

Hydraulic reach. Portion of a river or stream extending from one significant change in the hydraulic character of a river or stream to the next significant change. These changes are usually associated with breaks in the slope of the water surface profile and may be caused by bridges, dams, expansion and contraction of the water flow, and changes in stream-bed slope or vegetation.

Inundation. Ponded water or water in motion of sufficient depth to damage property due to the presence of the water or to deposits of silt.

Irrigation. The controlled application of water to arable lands to supply water requirements not satisfied by rainfall.

Modified floodway. Any land area subject to inundation by slow-moving water caused by natural forces.

Natural outlet. Any outlet from a watercourse, pond, ditch, lake, or other body of surface or ground water.

One-hundred-year flood. A flood that has a one (1) % chance of occurring in any one year based on criteria established by the state water commission.

Open floodway. Any land lying in the path of velocity flood waters caused by natural forces.

Permanent water level. Sea level unless special conditions exist. If special conditions exist, such level as the project engineer in his opinion may deem to represent the permanent water level.

Regional flood. A flood determined to be representative of large floods known to have generally occurred in the state and which may be expected to occur on a particular stream because of like physical characteristics. The flood frequency of the regional flood is once in every one-hundred (100) years; this means that in any given year there is a one (1) percent chance that the regional flood may occur or be exceeded.

Retention basin. A holding pond or reservoir for surface water drained at a site to prevent flooding and fast accumulation of water.

Site grading. The site should be graded to provide for natural drainage to the area or section of land, which drainage will receive the rain or melted snow and direct it to the proper channel or ditch without causing any overflow to parcels of land occupied by others, including public streets and roads.

Storage capacity of floodplain. The volume of space above an area of floodplain land that can be occupied by floodwater of a given stage at a given time, regardless of whether the water is moving.

Storm sewer. Sewer used for conveying rain water, melted snow, and other surface precipitation.

Storm water. Water that is discharged from a surface as a result of rainfall or snowfall.

Subject to flooding. Where any area within the proposed subdivision is known to be subject to flooding such area should be clearly marked "Subject to Periodic Flooding" on the lot plan and should not be plotted in the streets and lots. Land which normally will be inundated less frequently than once in five years may be used for recreational residential lots or other recreational uses. In any event, easements must be reserved from normal flow line to the annual high water flow line of any water course or lake.

Subsurface drain. Drain that receives only subsurface or seepage water and conveys it to a place of disposal.

Surface drainage system. Designed to collect and dispose of rainfall runoff. There are two basic types: the open ditch/swale and culvert or an entire open system.

Trench drains. Located in low areas to intercept runoff.

Watercourse. Any channel, creek, arroyo, lake, river, stream, or other body of water having natural banks and bed through which waters flow at least periodically. The term may include specifically designated areas in which substantial flood damage may occur.

Waterway. A channel of water not less than fifty (50) feet wide and navigable by small boats.

Waterway lane. Line marking the normal division between land and a waterway as established by the appropriate governmental agency having jurisdiction.

IRRIGATION SYSTEMS

ABS (plastic pipe). Acrylonitrile-butadiene-styrene.

Antisiphon device. Device installed to prevent foreign matter from backsiphoning into the household water supply.

Antisiphon valve. Specially designed valve to prevent back flow of contaminated water returning or being sucked into domestic supply lines.

Approved backflow prevention device. Device that has been investigated and approved by an appropriate regulatory agency.

Asbestos cement pipe. Pipe usually used for installation of mains and other large-sized pipes in underground systems.

Automatic control valves. Valves which are activated by a 24-v impulse, separate each zone of the system, and feeds through lateral piping to individual irrigation heads.

Back flow. Reverse flow of water into the distributing pipes of a potable supply of water from the sprinkler system or any source or sources other than its original intake. Backsiphonage is one type of backflow.

Backflow preventor. Device or means to check or stop water from flowing into a potable supply of water from the sprinkler system.

Backpressure. Pressure in the nonpotable piping system greater than that in the potable water system. Pressure caused by gravity due to elevation differences is one common cause. Pumps can be another cause.

Backsiphonage. Form of backflow due to a negative or sub-atmospheric pressure within a potable water system.

Ball drive rotary head. Sprinkler head utilizing a spinning metal ball to provide the powers to rotate the nozzle.

Bubbler head. Head that discharges water with the same open effect as water flowing from a faucet; amount of flow is adjustable.

Cam drive rotary heads. Rotary sprinkler employing cams mounted directly on a rotor.

Check valve device. Device used to eliminate water hammer when the pump is stopped.

Check valves. Valves should be installed under all low heads to prevent low-head drainage when water is turned off.

Clean-water supply. A clean water supply is of primary importance for good sprinkler system operation with a minimum of maintenance. Poor performance of system and malfunction of sprinkler equipment are often caused by foreign matter in the water supply.

Connection to water supply. Connection to the potable water supply may be made either to the service line or to the water main line. Suitable provisions should be made to drain the sprinkler supply line.

Control valves. Valves controlling distribution of water from the sprinkler supply line to sprinkler distribution pipe. The valves may be installed singly or in a manifold.

Copper water tube. Rigid, hard-drawn form with soldered fittings. Types K, L, and M available for underground use.

Cross connections. Connections should not be made between the potable supply of water and any well or other source of water.

Diameter. Nominal size for pipe and tubing as designated commercially.

Dirty-water supplies. Constant source of system malfunction requiring frequent cleaning and flushing.

Drain valve. Automatic spring-loaded valve that allows water to drain out of sprinkler system pipes before freezing weather sets in. Freezing does not damage the heads or the pipe and fittings if the water is removed.

Electrolysis. Electrolytic corrosion caused by stray electrical current affecting steel pipe when attached directly to piping or appurtenances of dissimilar metals.

Fittings and connections. Fittings and connections made from noncorrosive materials and installed according to manufacturer's instructions.

Flow adjustor. Device used at the base of the nipple riser to adjust the flow of water.

Flow pressure. Pressure at head when water is flowing.

Flushing lines. Sprinkler head units are removed from installed castings and then water is turned on to flush lines.

Flush lawn hydrant. Device installed flush with the turf and connected on the sprinkler system main pressure supply line or any constantly pressured water supply.

Friction loss. Loss of water pressure resulting from drag or friction of water flowing through pipe, waters valves, and fittings. Available water pressure is used up before water reaches sprinkler nozzles.

Gear drive heads. Device that provides a steady powerful rotation to the sprinkling streams.

GPM. Gallons per minute.

Hazard. Any condition in a water supply system that creates or may create a danger to the health and well-being of the occupant.

Identification marking of materials. Each length of pipe and each pipe fitting, sprinkler head, valve, or device used in a lawn sprinkler system has cast, stamped, or indelibly marked on it the manufacturer's name and type. Continuous marking on length of pipe showing manufacturer's mark or name and type or classification.

Impact drive rotary head. Device employing a weighted spring-loaded drive arm to provide the force to rotate the nozzle assembly.

Installation of vacuum breaker. Approved vacuum breaker or backflow preventor should be installed in all lawn sprinkler systems. The vacuum breaker or backflow preventor is installed at least 6 in. above the highest sprinkler head in the lawn and is located at the top of the loop in the main-water supply line. Device is installed to prevent back siphonage into the potable water supply.

Lateral lines. Lines running from each remote control valve into each zone or irrigated area from the rest of the water distribution system. Lateral lines should be class 160 PVC or 80 lb PE pipe (for cold climates) with an 8 in. minimum cover.

Lawn sprinkler system. System that includes apparatus and equipment normally used for purposes of irrigation affixed permanently to the property in the lawn, ground, flower bed, or fence and connected to the watersupply. "Connection to the water supply" is constructed to mean connections to the hose bibs, as well as permanent connections to the water supply line.

Low-angle spray. Normal spray trajectory, about 30 to 40° above horizontal. Part-circle sprays for both pop-up and shrub heads are available with a low-angle trajectory of about 10°.

Main-line piping. Main lines should be class 200 PVC pipe with a minimum of 18 in. of cover. In freezing climates, install drain valves at all low points in the main line.

Manifold. Group of valves closely spaced off one supply line.

Materials for pressure lines. Pipe consisting of copper tubing, galvanized iron, lead, cast iron, or other approved materials, used for all pressure lines within the building and leading to all control valves outside the building. In addition, tubing under pressure leading to remote control hydraulic valves should be included as pressure lines. Asbestos-cement pipe or other material approved by the applicable code may be used in lines 3 in. or more in diameter outside the building.

Microsprays. Drip, low-pressure microsprays, and/or compensating bubbler heads should be used for shrub or ground cover.

Minimum depth of pipe. Minimum depth of all sprinkler lines is based on local frost conditions. When draining is necessary, the pipes should slope to drain, and drain valves should be installed at all low points in such lines. Drains are not required for nonrigid plastic lines. Drain valves usually drain into a sump or gravel pocket of proper size.

Part-circle heads. Used to control water spray along borders and in the direction of lawn or plantings.

Plastic pipe. Simple, lightweight pipe with fast installation and an unlimited life expectancy. Two types are available: polyvinyl chloride (PVC), classified as rigid pipe, and polyethylene (PE), classified as flexible pipe.

Plastic pipe creep. When the pressure of the water being conveyed exceeds the pressure rating of thermoplastic pipe, creep occurs. Creep in this case means stretching of the pipe, causing a thinning of the pipe walls. This may cause pipe failure.

Point of cross connection. Specific point or location in a public or a consumer's potable water system where a cross connection exists.

Pollution. Presence in water of any foreign substance that constitutes a consumer hazard or impairs its potability or usefulness.

Pop-up heads. Usually referred to as "lawn heads." Heads installed flush with the turf. Nozzle pops up to deliver the water spray during operation and recedes within the body when inoperative.

Potable water. Water that is safe for drinking or for personal or culinary use.

Precipitation rate. Individual sprinkler heads should have a matched precipitation rate within each zone.

Pressure for sprinkler distribution pipe. All sprinkler distribution pipe, capable of withstanding a continuous working pressure of 75 psi.

Pressure lines. Water line designed or intended to contain water under continuous working pressure.

Pressure-reducing valves. Valves should be installed to control pressure in the system, after a review of the pressure in the water supply system.

PSI. Pounds or pressure per square inch.

PVC (plastic pipe). Polyvinyl chloride.

Rain. Rain shut-off devices should be provided for all automatic controllers.

Rotary head. Device used in large turf grass areas. It is the same as the pop-up device, with the head being completely concealed in the ground except for the cover plate, which is exposed at ground level.

Sandy water. Sandy water causes fouled and eroded orifices, sticking sprinklers, sticking valves, frozen bearings, and accelerated wear.

Service line. Pipe or pipes conveying water from water main into the property.

Short radii nozzle. Head is a part-circle spray with a small radius and a relative flat trajectory.

Shrub heads. Spray heads designed for installation above shrubbery or flower beds. Heads are made to a much smaller configuration.

Size of pipe. Sprinkler systems should not be connected to any water supply line of less than $\frac{1}{2}$-in. diameter.

Spray head nozzle arcs. These are full-circle, half-circle, third-circle and quarter-circle nozzles. Operating pressure should be $15-35$ psi at the base of the head.

Sprinkler distribution pipe. Water line not under continuous pressure, conveying water from the control valves to the sprinkler heads.

Static pressure. Water pressure in line with no water flowing.

Strainer. Strainer should be used with any raw water supply.

Stream sprays. Spray nozzles that dispense water over the rated coverage areas with tiny streams.

Strip sprays. Spray nozzles for watering long, narrow strips of turf or plantings, available in several types. Sprays are also referred to as line sprays.

Trenching. Trench or cut that receives the pipe. Heads are positioned flush with the turf and are supplied by water from the buried pipes.

Vacuum breaker. Same as back flow preventor.

Valve key. Key on a rod for operating manual control and drain valves and other manual valves from a standing position.

Water distribution. Distribution is a function of the pipe used to build the irrigation system. The main-line pipe, also called the pressure line, follows the path from the water source to the remote control valves. It contains the highest water pressure in the system.

Water velocity. The irrigation system should be designed and its components selected so that the velocity in the service line from tap to meter does not exceed 12 fps and so that the flow in all system piping does not exceed 80° of the maximum safe capacity through the meter. In mainline piping, velocity should not exceed 6 fps to prevent water hammer in the system.

PAVING AND SURFACING

Aggregate. Hard, inert, mineral material used for mixing in graduated fragments. Includes sand, gravel, crushed stone, and slag.

Aggregate base. Materials that are selected to be placed and compacted or prepared subgrade supporting the surface. May consist of crushed limestone, crushed or uncrushed gravel, pit run gravel, cinders, oyster shell, caliche (limestone conglomerate), and other locally acceptable materials.

Aggregate storage bins. Bins that store the necessary aggregate sizes and feed them to the dryer in substantially the same proportions required in the finished mix.

Air-blown asphalt. Process of forcing air through asphalt cement under specified conditions produces an asphalt with characteristics that make it better adapted for specific purposes. Product is known as air-blown or oxidized asphalt. Asphalt cement so treated becomes partially oxidized and has a tendency not to soften under heat or become brittle when cold.

Anionic emulsion. Type of emulsion, that a particular emulsifying agent establishes a predominance of negative charges on the discontinuous phase.

Asphalt

1. Dark brown to black cementitious material, solid, semisolid, or liquid in consistency, in which the predominating constituents are bitumens that either occur in nature or are obtained in the refining of petroleum. Asphalt is a constituent in varying proportions, of most crude petroleums.

2. All asphalts result directly from the distillation of petroleum crude oil, whether the distillation is accomplished by natural or artificial means. Natural asphalts are formed when crude petroleum oils work their way through cracks to the earth's surface. Action of the sun and wind drives off the lighter oils and gases, leaving a dark brown or black plastic residue called asphalt. Most natural asphalts are impregnated with a fairly high percentage of very fine sand, picked up in the passage through the earth's crust. Asphalts are generally so hard that it is necessary to flux them with a petroleum oil to make them suitable for road construction use. Natural asphalts used in the United States for road construction come chiefly from Trinidad and Venezuela.

Asphalt alligator cracks. Interconnected cracks forming a series of small blocks resembling an alligator's skin or chicken wire.

Asphalt base course. Foundation course consisting of mineral aggregate bound together with asphaltic material.

Asphalt binder course. Intermediate course between a base course and an asphalt surface course. Binder course is usually a coarse-graded aggregate asphaltic concrete containing little or no mineral matter, passing the No. 200 sieve.

Asphalt block. Asphalt concrete molded under high pressure. Type of aggregate mixture composition, amount and type of asphalt, and size and thickness of the block may be varied to suit usage requirements.

Asphalt block pavement. Pavement in which the surface course is constructed of asphalt blocks. Blocks are laid in regular courses similar to brick pavements.

Asphalt cements. Heavier asphaltic residues resulting from the refining process are sometimes used as binders in base and surface courses without being liquefied or softened by a fluxing or emulsifying agent, being reduced instead to a workable condition by heating only. Asphalt cements are designated according to penetration range or degree of hardness.

Asphalt concrete. High-quality, thoroughly controlled hot mixture of asphalt cement and well-graded, high-quality aggregate, thoroughly compacted into a uniform dense mass.

Asphalt concrete base. Mixed in a central plant using heated aggregate or sand and paving asphalt. Prepared mix is laid with paving machines or spreaders and compacted.

Asphalt curbs. Machine-molded asphalt applied as curbs to pavements. No forms are needed.

Asphalt emulsion. Asphalt cement in water containing a small amount of emulsifying agent.

Asphalt emulsion slurry seal. Mixture of slow-setting emulsified asphalt, fine aggregate, and mineral filler, with water added to produce slurry consistency.

Asphaltenes. Components of the bitumen in petroleums, petroleum products, malthas, asphalt cements, and solid native bitumens, that are soluble in carbon disulfide, but insoluble in paraffin napthas.

Asphalt flux. An oil used to reduce the consistency or viscosity of hard asphalt to the point required for use.

Asphalt fog seal. Light application of liquid asphalt without mineral aggregate cover. Slow-setting asphalt emulsion diluted with water is preferred type.

Asphalt gutter. Gutters placed by machine and requiring no forms.

Asphaltic road oil. Thick, fluid solution of asphalt.

Asphaltic sand. Deposits of sand grains cemented together with soft, natural asphalt.

Asphalt intermediate course. Course between base and asphalt surface course.

Asphaltite. Any of the dark-colored, solid, naturally occurring bitumens that are insoluble in water, but more or less completely soluble in carbon disulfide, benzol and other types of solutions compatible with bitumens.

Asphalt joint filler. Asphaltic product used for filling cracks and joints in pavement and other structures.

Asphalt leveling course. Course of variable thickness used to eliminate irregularities in the contour of an existing surface prior to superimposed treatment or construction.

Asphalt macadam. Type of pavement construction using a coarse, open-graded aggregate that is usually produced by crushing and screening stone, slag, or gravel. Such aggregate is called "macadam aggregate." Asphalt may be incorporated into macadam construction by either penetration or mixing.

Asphalt mastic. Mixture of asphalt and fine mineral material in such proportions that it may be poured hot or cold into place and compacted by troweling to a smooth surface.

Asphalt mastic seal coat. Dense, impervious, voidless mixture of asphalt, mineral aggregate, and mineral dust. May be installed either cold or hot.

Asphalt overlay. One or more courses of asphalt construction on an existing pavement. Overlay generally includes a leveling course to correct the contour of the old pavement, followed by uniform course or courses to provide needed thickness.

Asphalt paint. Liquid asphaltic product usually containing small amounts of other materials such as lampblack, aluminum flakes, and mineral pigments.

Asphalt pavement. Pavement consisting of a surface of course mineral aggregate coated and cemented together with asphalt cement on supporting courses such as asphalt bases or crushed stone, slag, or gravel or on portland cement concrete, brick, or block pavement.

Asphalt pavement structure. Courses of asphalt-aggregate mixtures plus any nonrigid courses between the asphalt construction and the foundation or subgrade.

Asphalt planks. Premolded mixtures of asphalt fiber and mineral filler, sometimes reinforced with steel or glass fiber mesh. Usually made in 3- to 8-ft lengths, 6 to 12 in. wide. Asphalt planks may also contain mineral grits that maintain a sandpaper texture throughout their life.

Asphalt prime coat. Application of low-viscosity liquid asphalt to an absorbent surface. Used to prepare an untreated base for an asphalt surface. Prime penetrates into the base and plugs the voids, hardens the top, and helps bind it to the overlying asphalt course. Reduces the necessity of maintaining an untreated base course prior to placing the asphalt pavement.

Asphalt primer. Liquid asphalt of low viscosity that penetrates into a nonbituminous surface on application.

Asphalt rubber. Blend of asphalt cement and reclaimed tire rubber pieces, in which the total blend reacts to the hot asphalt cement sufficiently to cause the swelling of the rubber particles.

Asphalt seal coat. Thin asphalt surface treatment used to waterproof and improve the texture of an asphalt-wearing surface. Depending on the purpose, seal coats may or may not be covered with aggregate. Main types of seal coats are aggregate seals, fog seals, emulsion slurry seals, and sand seals.

Asphalt soil stabilization. Treatment of naturally occurring nonplastic or moderately plastic soil with liquid asphalt at normal temperatures. After mixing, aeration and compaction provides water-resistant base and subbase courses of improved load-bearing qualities.

Asphalt surface course. Top course of an asphalt pavement, sometimes called "asphalt-wearing course".

Asphalt surface treatments. Applications of asphaltic materials to any type of road or pavement surface, with or without a cover of mineral aggregate, that produce an increase in thickness of less than 1 in.

Asphalt tack coat. Light application of liquid asphalt to an existing asphalt or portland cement concrete surface. Asphalt emulsion diluted with water is the preferred type. Used to ensure a bond between the surface being paved and the overlying course.

Asphalt-treated base. Base material prepared in central mixing plants. Mix is spread with paving machines or spreaders and compacted with suitable rollers.

Automatic cycling control. Control system in which the opening and closing of the weigh hopper discharge gate, the bituminous discharge valve, and the pugmill discharge gate are actuated by means of self-acting mechanical or electronic machinery without any intermediate manual control. System includes preset timing devices to control the desired periods of dry and wet mixing cycles.

Automatic drier control. System that automatically maintains the temperature of aggregates discharged from the drier within a preset range.

Automatic proportioning control. System in which proportions of the aggregate and bituminous fractions are controlled by gates or valves that are opened and closed by means of self-acting mechanical or electronic machinery without any intermediate manual control.

Bank gravel. Gravel found in natural deposits, usually more or less intermixed with fine material such as sand or clay, or combinations thereof; gravelly clay, gravelly sand, clayey gravel, and sandy gravel indicate the varying materials in the mixture.

Base course. Layer of material immediately beneath the surface or intermediate course. May be composed of crushed stone, crushed slag, crushed or uncrushed gravel and sand, or combinations of these materials; may be bound with asphalt.

Binders.   Binder materials include water, clay, asphalt, tar, portland cement, crusher dust, lime rock, various chemical substances, and other materials having binding properties. Portland cement and water combine chemically to form a mortar that when coated over the aggregate particles will set them up and bind them together. Bituminous materials also form a coating over the particles and thus bind them together.

Bitumen.   Class of black or dark-colored solid, semisolid, or viscous cementitious substances, natural or manufactured, composed principally of high-molecular-weight hydrocarbons, of which asphalts, tars, pitches, and asphalties are typical.

Bituminated filler or hydrofiller (patented process).   Material prepared by a special process in which atomized bitumen is introduced under air pressure into a chamber containing a measured quantity of filler dust. Dust particles are coated or partially coated with fine globules of bitumen and deposited in containers. Water is added to prevent a conglomeration of the mixture (approximately 10% wet mix). Material is called "hydrofiller" and used for seal-coating road surfaces. Material may be spread by hand or by a finishing machine, after which it is rolled and immediately ready for traffic. Rolling process squeezes out the bulk of the water. In the hydrofiller stage the material may be stored for an extended period and used at atmospheric temperatures.

Bituminous.   Containing or treated with bitumen, for example, bituminous concrete, bituminized felts and fabrics, bituminous pavement.

Bituminous cement.   Bituminous material suitable for use as a binder, having cementing qualities which are dependent mainly on its bituminous character.

Bituminous coating.   Coating made principally of bituminous material and used as a surfacing for roads and as a water repellent barrier in buildings.

Bituminous concrete.   Concrete made with bituminous material as a binder for sand and gravel.

Bituminous emulsion.   Suspension of minute globules of bituminous material in water or in an aqueous solution, or suspension of minute globules of water or of an aqueous solution in a liquid bituminous material.

Bituminous grout.   Mixture of bituminous material and fine sand that, when heated, will flow into place at joints without mechanical manipulation.

Blast-furnace slag.   Nonmetallic product, consisting essentially of silicates and alluminosilicates of lime and other bases, that is developed simultaneously with iron in a blast furnace.

Bleeding or flushing asphalt.   Upward movement of asphalt in an asphalt pavement resulting in the formation of a film of asphalt on the surface.

Blown or oxidized asphalt.   Asphalt that is treated by blowing air through it at elevated temperature to give it characteristics desired for certain special uses, such as roofing, pipe coating, undersealing portland cement concrete pavements, membrane envelopes, and hydraulic applications.

Blowup.   Localized buckling or shattering of a rigid pavement, occurring usually at a transverse crack or joint.

Carbenes.   Components of the bitumen in petroleums, petroleum products, malthas, asphalt cements, and solid native bitumens that are soluble in carbon disulfide, but insoluble in carbon tetrachloride.

Catalytically blown asphalt.   Air-blown asphalt produced by using a catalyst during the blowing process.

Cationic emulsion.   Type of emulsion such that a particular emulsifying agent establishes a predominance of positive charges on the discontinuous phase.

Channels (ruts).   Channelized depressions that may develop in the wheel tracks of an asphalt pavement.

Chip seal.   Spray application of asphalt on an existing surface, followed by a cover of rock chips or screenings to function as a seal coat. Asphalt used for chip seals must be fluid enough to wet and adhere to the chips and yet develop sufficient strength to retain the chips under fast traffic.

Clinker.   Fused or partly fused by-product of the combustion of coal, but also including lava, portland cement clinker, and partly vitrified slag and brick.

Coal tar.   Dark brown to black cementious material produced by the destructive distillation of bituminous coal.

Coarse aggregate.   Mineral aggregate of limestone, gravel, slag, or other suitable material, most of which is retained on a No. 8 sieve.

Coarse-graded aggregate.   Aggregate having a continuous grading of particle size from coarse through fine with a predominance of coarse sizes.

Coke-oven tar.   Coal tar produced in by-product coke ovens in the manufacture of coke from bituminous coal.

Cold-laid asphalt surface.   Mix is the wearing surface of an asphalt pavement that can be spread and compacted at ambient temperature. Cold mix surfacings are used on light- to moderately-used roads and for maintenance patching. Mix can be prepared in a central or travel plant, or mixed in place.

Cold-laid plant mixture.   Plant mixes that may be spread and compacted at atmospheric temperature.

Combination or composite-type pavement structure.   When an asphalt pavement is laid on an old portland cement concrete base or other rigid base layer, the pavement structure is referred to as a combination or composite type.

Continuous-mix plant.   Bituminous concrete mixing plant that proportions both aggregate and bituminous constituents into the mix by a continuous volumetric proportioning system without definite batch intervals.

Corrugations (washboarding).   Form of plastic movement typified by ripples across the pavement surface.

Crack.   Approximately vertical random cleavage of the pavement due to failure, natural causes or traffic action.

Crusher-run.   Total unscreened product of a stone crusher.

Cutback asphalt.   Asphalt cement that has been liquefied by blending with petroleum solvents. On exposure to atmospheric conditions the diluents evaporate, leaving the asphalt cement to perform its function.

Cutback products.   Petroleum or tar residua that have been blended with distillates.

Deep-lift asphalt pavement.   Pavement in which the asphalt base course is placed in one or more lifts of 4-in. or more compacted thickness.

Deflection.   Amount of downward vertical movement of a surface due to the application of a load to the surface.

Delivery tolerances.   Permissible variations from the exact desired proportions of aggregate and bituminous material as delivered into the pugmill.

Dense-graded aggregate. Mineral aggregate uniformly graded from the maximum size down to and including sufficient mineral dust to reduce the void space in the compacted aggregate to exceedingly small dimensions approximating the size of voids in the dust itself.

Design lane. Lane on which the greatest number of equivalent 18,000-lb single-axle loads is expected.

Design subgrade strength value. Subgrade strength value less twice the standard deviation.

Design traffic number. Average daily number of equivalent 18,000-lb single-axle-load applications expected for the design lane during the design period.

Digital recorder. Instrument that prints the weight of materials in digital form on a tape or ticket.

Disintegration. Breaking up of a pavement into small, loose fragments due to traffic or weathering.

Distortion. Any change of a pavement surface from its original shape.

Dryer. Apparatus that will dry the aggregates and heat them to the specified temperatures.

Dry mixing period. Interval of time between the initial charge of dry aggregates into the pugmill and the initial application of bituminous material.

Dust palliative. Light application of asphalt to knit fine surface particles together to minimize their displacement by wind or traffic.

Effective thickness. Thickness that a pavement would be if it could be converted to full-depth asphalt concrete.

Emulsified asphalt

1. Emulsion of asphalt cement and water that contains a small amount of an emulsifying agent; a heterogeneous system containing two normally immiscible phases (asphalt and water) in which the water forms the continuous phase of the emulsion and minute globules of asphalt form the discontinuous phase.

2. Asphalts produced by combining heated asphalt with water by intense agitation in the presence of an emulsifier that prevents the recombination or coalescence of the asphalt particles. Emulsifier is usually soap, but may be starch, glue, gum, colloidal clay, or some other material with similar properties. In straight emulsions the asphalt particles of colloidal size are dispersed in the water, and the emulsion can be further diluted with water.

Emulsion. Brown liquids that turn black when applied to mineral aggregate. In contact with mineral aggregate, the asphalt in the emulsion tends to coagulate on the stone or soil particles, and the water is lost by flowing away, by evaporation, and by capillary attraction, leaving an asphalt film on each stone or soil particle.

Emulsion slurry. Mixture of slow-setting emulsified asphalt, fine aggregate, and mineral filler, with water added to produce slurry consistency.

Fault. Difference in elevation of two slabs at a joint or crack.

Fine aggregate. Mineral aggregate passing through a No. 8 sieve.

Fine graded aggregate. Aggregate having a continuous grading in particle size from coarse through fine with a predominance of fine sizes.

Flux or flux oil. Thick, relatively nonvolatile fraction of petroleum, which may be used to soften asphalt to a desired consistency; often used as base stock for manufacture of roofing asphalts.

Fog seal. Light spray application of emulsified asphalt, without cover aggregate, to an existing pavement as a seal to inhibit raveling or enrich the surface.

Free carbon in tars. Hydrocarbon fraction that is precipitated from a tar by dilution with carbon disulfide or benzene.

Full-depth asphalt pavement. Pavement in which asphalt mixtures are employed for all courses above the subgrade or improved subgrade. Full-depth asphalt pavement is laid directly on the prepared subgrade.

Gashouse coal tar. Coal tar produced in gashouse retorts in the manufacture of illuminating gas from bituminous coal.

Gilsonite. Form of natural asphalt, hard and brittle, occurring in rock crevices or veins, from which it is mined.

Graded aggregate. Aggregate having a continuous grading in particle size from coarse to fine.

Grade depressions. Localized low areas of limited size that may or may not be accompanied by cracking.

Graphic recorder. Instrument that scribes a line or lines on a chart simultaneously with the indication of the scale or meter as the materials are being weighed.

Hot aggregate storage bins. Bins that store heated and separated aggregates prior to their final proportioning into the mixer.

Hot-laid mixtures. Mixes that must be spread and compacted while in a heated condition.

Hot-mix asphalt base course. Foundation course consisting of mineral aggregate bound together with asphaltic cement.

Hot-mix asphalt intermediate course. Intermediate or leveling course (also termed binder course) of mineral aggregate, bound together with asphaltic cement, between a base course, and an asphalt surface course.

Hot-mix asphalt surface course. Top course of an asphalt pavement, sometimes called asphalt wearing course.

Hot-mix plant mixture. Plant mixes that must be spread and compacted while at elevated temperature. To dry the aggregate and obtain sufficient fluidity of the asphalt (usually asphalt cement), both must be heated prior to mixing—giving origin to the term "hot mix."

Instability. Lack of resistance to forces tending to cause movement or distortion of a pavement structure.

Inverted asphalt emulsion. Emulsified asphalt in which the continuous phase is asphalt, usually a liquid asphalt, and the discontinuous phase is minute globules of water in relatively small quantities.

Job-mix formula. Formulation of aggregates and asphalt to comply with the requirements of the specifications.

Liquid asphalt. Asphaltic material having a soft or fluid consistency that is beyond the range of measurement by the normal penetration test, the limit of which is 300 maximum.

Liquid bituminous materials. Bituminous materials having a penetration at 77°F under a load of 50 g applied for 1 second, of more than 350.

Load-bearing aggregates. Load-bearing class includes crushed or uncrushed gravel and stone, sand, slag, cinders, caliche, shell, iron ore, and other materials, usually granular, that have sufficient load-bearing and wear-resisting characteristics but may lack binder for stability. Aggregates should be free from vegetable matter, clay, or other deleterious matter and should not contain an excess of flat, thin, or elongated pieces of material such as shale, sandstone, slate, or coal. Should also be uniform in quality and composed of sound, hard, tough, durable pebbles or fragments of rock, slag, or other material.

Longitudinal crack. Crack that follows a course approximately parallel to the centerline.

Los Angeles abrasion test. Test to determine the degradation and wearing characteristics of aggregates under laboratory conditions.

Macadam aggregate. Coarse aggregate of uniform size, usually of crushed stone, slag, or gravel.

Maintenance patching materials. Powdered asphalt mixtures are useful maintenance patching materials. Since the asphalt contains no volatile oils, it will remain workable for a considerable period and makes good stockpile material. The mixtures can be made rich or lean for various types of patching and feather out to a thin edge, leaving a smooth surface. Powdered asphalt is produced by pulverizing hard asphalt. To produce the mix, the mineral aggregate is usually dried to a low heat and treated with liquid fluxing agent; the powdered asphalt is introduced and mixing continued until a homogeneous material is formed.

Manual cycling control. Control system in which the opening and closing of the weigh hopper gate, bituminous discharge valve, and pugmill discharge gate are controlled by manual means, with or without assist devices or time-locking devices, or both, to prevent premature opening of the gates or valves.

Manual proportioning control. Control system in which proportions of the aggregate and bituminous fractions are controlled by gates or valves that are opened and closed by manual means.

Medium curing asphalt. Liquid asphalt composed of asphalt cement and a kerosene-type diluent of medium volatility.

Mesh. Square opening of a sieve used in testing.

Mineral dust. Portion of the fine aggregate passing a No. 200 sieve.

Mineral-filled asphalt. Asphalt containing finely divided mineral matter passing a No. 200 sieve.

Mineral filler. Finely divided mineral product of which at least 65% pass a No. 200 sieve. Pulverized limestone is the most commonly manufactured filler, although other stone dust, hydrated lime, portland cement, and certain natural deposits of finely divided mineral matter are also used.

Mixed in place (road mix). Asphalt course produced by mixing mineral aggregate and liquid asphalt at the road site by means of travel plants, motor graders, drags, or special road-mixing equipment.

Mixed-in-place surface treatments. Consists of aggregate into which liquid asphalt is mixed by any of several methods.

Multiple surface treatments. Commonly two or three successive applications of asphaltic material and mineral aggregate.

Natural (native) asphalt. Asphalt occurring in nature, which has been derived from petroleum by natural processes of evaporation of volatile fractions leaving the asphalt fractions.

Normal temperature. As applied to laboratory observations of the physical characteristics of bituminous materials at 77°F.

Oil-gas tars. Tars produced by cracking oil vapors at high temperatures in the manufacture of oil gas.

Open-graded aggregate. Aggregate containing little or no mineral filler or in which the void spaces in the compacted aggregate are relatively large.

Oxidized or blown asphalt. Asphalt treated by blowing air through it at elevated temperature to give it characteristics desired for certain special uses such as roofing, pipe coating, undersealing portland cement concrete pavements, membrane envelopes, and hydraulic applications.

Patching mix. Mixes prepared for maintenance patching, usually in a central plant and occasionally by road-mixing methods.

Pavement structure. All courses of selected material placed on the foundation or subgrade soil, other than layers or courses constructed in grading operations.

Pavement structure—combination or composite type. When the asphalt pavement is on an old portland cement concrete pavement, a portland cement concrete base, or other rigid base, the pavement structure is referred to as a combination or composite-type pavement structure.

Penetration treatment. Surface treatment consisting of a spray application of liquid asphalt applied to an untreated granular roadway to form a light-duty wearing course by penetrating the roadway material. Type and grade of asphalt are selected to fit climatic conditions and gradations of the material in the roadway.

Petroleum asphalt. Asphalt refined from crude petroleum.

Pitches. Black or dark brown solid cementitious materials that gradually liquefy when heated and that are obtained as residua in the partial evaporation or fractional distillation of tar.

Planned-stage construction. Construction of roads and streets by applying successive layers of asphalt concrete according to design and a predetermined time schedule.

Plant mix. Mixture produced in an asphalt mixing plant that consists of mineral aggregate uniformly coated with asphalt cement or liquid asphalt.

Plant-mix base. Foundation course, produced in an asphalt mixing plant, that consists of mineral aggregate uniformly coated with asphalt cement, liquid asphalt, or emulsified asphalt.

Plant-mixed surface treatment. Consists of asphalt and aggregate mixed in an asphalt plant and spread on the surface of a primed base, an asphalt base course, or existing asphalt pavement.

Plant screens. Screens located between the drier and hot bins that separate the heated aggregates into the proper hot bin sizes.

Plasticity index. Indicates measure of the cohesion and swell characteristics of soil.

Polished aggregate. Aggregate particles in a pavement surface with edges that have been rounded and surfaces polished smooth by traffic.

Polymer-modified bitumen. Dark colored cementitious, high-molecular weight hydrocarbon such as asphalt, tar, or pitch in which polymer has been dispersed to modify the properties of the bitumen.

Potholes. Bowl-shaped holes in the pavement, of varying sizes, resulting from localized disintegration.

Powdered asphalt. Hard, solid asphalts that have been crushed or pulverized to a fine state of subdivision, with 100% passage through a No. 10 sieve and at least 50% passage through a No. 100 sieve. Generally limited to use in the lower types of bases and surfaces and usually mixed with liquid asphalts.

Preformed asphalt joint fillers. Premolded strips of asphalt mixed with fine mineral substances, fibrous materials, cork, sawdust, etc., manufactured in dimensions suitable for construction joints.

Premolded asphalt panels. Made with a core of asphalt, minerals, and fibers, covered on each side by a layer of asphalt-impregnated felt or fabric coated on the outside with hot-applied asphalt. Panels are made under pressure and heat to a width of 3 to 4 ft, a thickness of $\frac{1}{8}$ to 1 in., and any desired length.

Prime coat. Application of low-viscosity liquid asphalt to an absorbent surface. Used to prepare an untreated base for an asphalt surface. Prime penetrates into the base and plugs the voids, hardens the top, and helps to bind it to the overlying asphalt course. Also reduces the necessity of maintaining an untreated base course prior to placing the asphalt pavement.

Proportioning tolerance interlock. Device that prevents continuance of the proportioning cycle when a component quantity varies outside a preset range.

Pugmill. Device for mixing the separate hot aggregate and bituminous components into a homogeneous bituminous concrete ready for discharge into a delivery truck.

Pumping or blowing. Slab movement under passing loads resulting in the ejection of mixtures of water, sand, clay, and/or silt along transverse or longitudinal joints and cracks, and along pavement edges.

Rapid-curing asphalt. Liquid asphalt composed of asphalt cement and a naptha or gasoline-type diluent of high volatility.

Raveling. Progressive separation of aggregate particles in a pavement from the surface downward or from the edges inward.

Rebound deflection. Amount of vertical rebound of a surface that occurs when a load is removed from the surface.

Recording device. Device that presents a record of the distribution of materials that is unquestionably legible and permanent.

Refined tar. Tar freed from water by evaporation or distillation that is continued until the residue is of desired consistency; a product produced by fluxing tar residuum with tar distillate.

Reflection cracks. Cracks in asphalt overlays that reflect the crack pattern in the pavement structure underneath.

Residual deflection. Difference between original and final elevations of a surface resulting from the application or removal of one or more loads from the surface.

Road oil. Heavy petroleum oil, usually one of the slow-curing grades of liquid asphalt.

Rock asphalt. Porous rock, such as sandstone or limestone, that has become impregnated with natural asphalt through geologic processes.

Rock asphalt pavements. Pavements constructed of rock asphalt, natural or processed, and treated with asphalt or flux as may be required for construction.

Roughometer. Single-wheeled trailer instrumented to measure the roughness of a pavement surface in accumulated inches per mile.

Rubberized bitumen. Special form of polymer-modified bitumen where asphalt, tar, or pitch has been combined with natural or synthetic rubber to modify its properties.

Rubble. Rough stones of irregular shapes and sizes, broken from larger masses either naturally or artificially, for example, by geologic action, in quarrying, or in stone cutting or blasting.

Ruts. Channelized depressions.

Sampling of asphalt. In order to test asphalt properly for compliance with specifications, it is necessary to take a representative sample. Samples taken in the field should be forwarded immediately to laboratory for testing.

Sand asphalt. Mixture of sand and asphalt cement or liquid asphalt prepared with or without special control of aggregate grading, with or without mineral filler.

Sand seal. Spray application of asphalt on an existing surface, followed by a cover of sand to function as a seal coat.

Screen. Apparatus with circular apertures for separating sizes of material.

Seal coat. Surface treatment applied on any type of existing pavement surface or on the surface of an asphalt base course.

Semisolid asphalt. Asphalt that is intermediate in consistency between liquid and solid or hard asphalt, that is, normally has a penetration of between 10 and 300 (ASTM D 3279).

Semisolid bituminous materials. Materials having a penetration at 77°F under a load of 100 g applied for 5 seconds, of more than 10 and a penetration at 77°F under a load of 50 g applied for 1 second, of not more than 350.

Sheet asphalt. Hot mix of asphalt cement with clean, angular, graded sand and mineral filler. Its use is ordinarily confined to the surface course, usually laid on an intermediate or leveling course.

Shoving. Form of plastic movement resulting in localized bulging of the pavement.

Shrinkage cracks. Interconnected cracks forming a series of large blocks, usually with sharp corners and angles.

Sieve. Apparatus with square apertures for separating sizes of material.

Skid hazard. Any condition that might contribute to making a pavement slippery when wet.

Slippage cracks. Cracks, sometimes crescent shaped, that point in the direction of the thrust of wheels on the pavement surface.

Slow-curing asphalt. Liquid asphalt composed of asphalt cement and relatively low-volatile oils.

Slurry seal. Slurry mixture of emulsified asphalt, sand, and water applied to an existing surface and squeegeed into place to serve as a seal coat. Only slow-setting emulsions are used.

Solid (hard) asphalt. Asphalt having a normal penetration of less than 10.

Spalling. Breaking or chipping of the pavement at joints, cracks, or edges, usually resulting in fragments with feather edges.

Special coal tar pitch. Pitches having a tendency to remain stable in warm weather and plastic in cold weather and used for this reason for subsurface sealing of concrete pavements.

Steam dispersion mix (Steamix, patented process). Process disperses bitumen into finely divided particles by introducing steam into a conditioning chamber and usually simultaneously introducing a "workability" compound. After undergoing the conditioning process, the bitumen is discharged with force upon the aggregate by means of steam-atomizing nozzles. Aggregate and processed bitumen are mixed in a twin pugmill mixer, after which mixture is ready for immediate use or to be placed in storage for future use.

Stone chips. Small angular fragments of stone containing no dust.

Stone-filled sheet asphalt. Sheet asphalt containing up to 25% coarse aggregate.

Straight-run pitch. Pitch run to the consistency desired in the initial process of distillation without subsequent fluxing.

Subbase. Course in the asphalt pavement structure immediately below the base course is called the subbase course. If the subgrade soil is of adequate quality, it may serve as the subbase.

Subgrade. Uppermost material placed in embankments or unmoved from cuts in the normal grading of the roadbed. Serves as foundation for the asphalt pavement structure. Subgrade soil is usually called basement soil or foundation soil.

Subgrade improved. Any course or courses of select or improved material between the foundation soil and the sbuabase is usually referred to as the improved subgrade. Improved subgrade can be made up of two or more courses of different quality materials.

Sulphate soundness test. Test to determine the relative resistance of aggregates to deterioration resulting from alternate freezing and thawing.

Surface treatment. Generic term covering the entire range of applications of asphalt, with or without aggregate, to any type of road or pavement surface.

Tack coat. Light application of liquid asphalt to an existing asphalt or portland cement concrete surface. Asphalt emulsion diluted with water is the preferred type. Used to ensure a bond between the surface being paved and the overlying course.

Tailings. Stones that after going through the crusher do not pass through the largest openings of the screen.

Tars. Tars used in road construction are of three types: coal tar, which is produced by the destructive distillation of bituminous coal; coke-oven tar, which is produced in the manufacture of coke from bituminous coal; and water-gas tar, which is produced by cracking oil vapors at high temperatures in the manufacture of carburetted water gas.

Timelocking device. Interlocking system that automatically locks the weigh box gate after the mixer is charged, locks out the bituminous material throughout the dry-mix cycle; and locks the mixer throughout the dry and wet mixing cycles.

Transverse crack. Crack that follows a course approximately at right angles to the centerline.

Upheaval. Localized upward displacement of a pavement due to swelling of the subgrade or some portion of the pavement structure.

Void-filling aggregates. Materials used to reduce the voids in coarse load-bearing aggregates may consist of limestone dust, portland cement, ground silica, hydrated lime, slate dust, soapstone dust, clay, silt, or other similar fine material. Some of these may also have some binding properties, but when they are used as filler, some other material, usually bitumen, is the principal binder.

Volumetric batch plant. Bituminous concrete mixing plant that proportions aggregate and bituminous constituents into the mix by volumetrically measured batches.

Waves. Transverse undulations at regular intervals in the surface of a pavement consisting of alternate valleys and crests 2 ft or more apart.

Weigh batch plant. Bituminous concrete mixing plant that proportions the aggregate constituents into the mix by weighed batches. Additions of bituminous material are made by either weight or volume.

Weigh box. Weigh box or hopper in weigh batch plants connected with the scales that weigh each aggregate fraction before dropping the aggregates into the pugmill.

Well-graded aggregate. Aggregate that is graded from the maximum size down to filler for the purpose of obtaining an asphalt mix with a controlled void content and high stability.

Wet mixing period. Interval of time between initial application of bituminous material and the opening of the mixer gate.

PILING AND CAISSONS

Air / steam hammer. Impact pile driver powered by compressed air or steam.

Allowable pile-bearing load. Maximum load that can be permitted on a pile, with adequate safety against movement of such magnitude that the structure is endangered.

Anvil. Cushion or block placed on the head of the pile during the driving operation to prevent undue damage to the pile head.

Banding. Strapped timber piles with high-tensile steel bands to prevent splitting while driving.

Batter pile. Pile driven a specified or required number of degrees from the vertical, as called for by the design loads.

Batter piles. Also called "spur" piles. Piles that are often driven in the foundations of abutments for arch bridges to resist the horizontal components of the reaction.

Bearing capacity of a pile. Load per pile required to produce a condition of failure.

Bearing pile

1. General term applying to any pile that carries a super-imposed load.

2. Driven piles occasionally penetrate only a thin layer of soft earth before bearing up on an unyielding stratum, such as rock, that can support heavy concentrated loads. Under these conditions, the pile is a column and is designed as such. However, lateral support can be assumed for the portion of the pile in the ground in all but the worst types of soils.

3. Member driven or jetted into the ground and deriving its support from the underlying strata and/or by the friction of the ground in its surface.

Belled. Having a butt or end shaped like a bell, often used at the bottom of concrete piles or caissons.

Bends (caisson disease). Cramping disease induced by too rapid decrease of air pressure after a stay in compressed atmosphere, as in a caisson.

Blow count. The observed blows of the pile hammer per increment of pile penetration; blows on soil sampler in standard penetration test.

Bored pile. Concrete pile, with or without a casing, cast in place in a hole previously bored in soil or rock.

Boring. Hole in the earth produced by various methods; method of exploring subsurface conditions by drilling or otherwise advancing a cased or uncased hole into the earth.

Box caisson

1. Caisson used where no excavating is required, and consisting merely of a box, open at the top and closed at the bottom, which is filled with concrete or stone-masonry to serve as a foundation for the pier or other structure to be built on the same location.

2. Caisson made of timber and of concrete. Except where placed on piles, the use of this type of caisson is limited, owing to the necessity of first excavating to the desired depth, to where firm bearing may be obtained, before placing the caisson. Depth to which it is possible to excavate is limited owing to the tendency of the wet material to flow into the hole. Box caissons are made to sink several feet by running pipes through the bottom and forcing water through the same, thus washing out the material from underneath and allowing sinking to take place.

Box pile. Pile made from two deep-arch sheet piles, channels, or other structural shapes, welded along their contact lines. The box pile may be driven open or closed ended and filled with concrete or left empty.

Brace pile. Batter pile connected to a structure in a way to resist lateral forces.

Breuchaud pile. Pile placed by driving an open steel pipe into the ground, washing or blowing it out clean by air pressure, and then filling it with concrete. If the steel is always under water, it will never rust out, and the pipe can be filled with concrete almost to the bottom.

Bruns pile. Trade name for a concrete pile precast with short pieces of pipe at the ends for adding lengths with a tapered drive splicer.

Bulb piles. Piles consisting of steel tube, steel shell, and concrete. By means of a drop hammer operating inside a heavy-wall steel drive tube, the tube, closed at the bottom with a gravel or no-slump concrete plug, is installed to the required depth. The plug is then driven out and additional no-slump concrete is placed in the drive tube and extruded with the drop hammer to form an enlarged concrete base or bulb. A compacted concrete base is formed in granular material under high dynamic driving energy, and thus relatively high design capacities can often be developed, up to about 150 tons per pile. After formation of the base, a steel shell is placed inside the drive tube and joined to the bulb. The outer drive tube is withdrawn and the inner steel shell filled with concrete.

Butt of a pile. The larger or head end of a tapered pile, usually the upper end of a pile as driven.

Button bottom (Western). Pile driven to resistance with steel pipe casing and steel point. Permanent casing is inserted and filled with concrete, and driving casing is then withdrawn.

Caisson

1. Concrete cylindrical foundation or tubular pier filled with concrete.

2. Shaft of concrete or other material formed in the ground by excavation and filling for the purpose of sustaining a load. Essential feature of a caisson is that the soil or rock at the bottom can be manually and visually inspected.

3. Watertight box or cylinder used in excavating for foundations or tunnel pits to hold out water so that concreting or other construction can be carried on.

Caisson chamber. Air and watertight chamber, fabricated of wood, steel, concrete, or reinforced concrete, in which it is possible for persons to work under air pressure greater than atmospheric pressure to excavate material below water level.

Caisson hammer. Gravity hammer that is mounted on top of a caisson containing a cushioning system and a rudimentary lead.

Caisson or pier foundations. Foundation system where the building structure is supported on a system of holes (usually lined) bored in the earth to a stratum that will provide adequate support of design loads and filled with concrete. Borings 2 ft or larger in diameter to permit bottom inspection are usually considered caissons, while borings less than 2 ft in diameter are considered piers. Where ground water is a problem, the pneumatic-caisson method is employed.

Caisson pile

1. Cast-in-place pile made by driving a tube, excavating it, and filling the cavity with concrete.

2. Excavated shaft with a minimum horizontal dimension of 2 ft that is filled with concrete to transmit loads from a structure to bearing materials.

Caisson rings. Reinforcing bars, circular in shape, placed inside a caisson to which vertical bars are tied.

Caisson types. Caisson is a box; box caisson is open at the top and closed at the bottom, open caisson if open at the top and bottom, and pneumatic caisson if closed at the top, open at the bottom, and uses compressed air. In all cases the caisson is merely a shell that must be filled with concrete or other masonry. Most caissons are surmounted with cofferdams because, from standpoints of appearance and durability, it is usually undesirable to extend the caisson above low-water level. All caissons have one characteristic in common: They form a permanent shell for, and are an integral part of, bridge and building foundations, being used simply as a convenient means of placing the foundation in position.

Cased concrete pile (MacArthur). Concrete pile that is driven to resistance with steel pipe casing and steel core. The core is then removed, the permanent casing is inserted and filled with concrete, and the driving casing is withdrawn.

Casing. Casings used in connection with the installation of cast-in-place concrete piles are made of steel, circular in cross section, tapered or straight sided, and ranging in thickness from light-gauge corrugated casing to heavy-wall pipe. Light-gauge corrugated casing is driven closed end with an integral mandrel that is removed after driving, leaving the casing in place to act as a form for the concrete. Heavier wall pipe may be installed closed end by driving from the top or bottom, or open end by alternate driving and removal of material from within the pipe. Pipe driven with a permanently closed end (boot plate welded on) is left in place to act as a form for the concrete. Pipe or casing driven with a temporary closure or open end may be left in place or, as in the case of some cast-in-situ shafts, may be removed as the concrete is placed, in which case the casing is of extra heavy wall thickness and is a part of the equipment and not the pile.

Casing blows. The blows usually of a 300-lb hammer falling 18 in. onto a soil sampler casing while making a soil boring.

Cast-in-place cased concrete piles. Cased piles driven without a mandrel that have steel casings heavy enough to stand driving. A widely used type, the Monotube, has a tapered steel shell of gauges varying from No. 3 to No. 11. Shells are fluted to give them added stiffness. They are driven, inspected internally, filled with concrete, and finally any excess shell at the top is cut off with a torch.

Cast-in-place concrete piles

1. Concrete pile constructed by forming an unlined hole in the ground, as by auger boring, and filling it with concrete. Operation does not involve pile driving, a feature that may be an advantage in working in the immediate vicinity of an existing structure. Unsuitable when the auger hole must be extended below groundwater or in caving soil. When a means of keeping the ground open is required, a metal tube or shell may be driven into the ground and then filled with concrete. Shells are available with such strength that they do not require internal support during driving. When a lighter material is used, the shell is first drawn over a heavy mandrel and then driven into the ground. The mandrel is withdrawn after the driving operation and the shell is filled with concrete. On small jobs especially there is usually an advantage in using shells that do not require support during driving or in using other methods that eliminate the need for short-term use of heavy equipment.

2. Pile made by driving a metal form into the ground and filling it with concrete. May be subclassified as mandrel-driven pile with thin metal shell; the Raymond pile is the most widely known of this type. A shell of the standard-type Raymond pile tapers at the rate of 0.4-in. diam/ft of length and is shipped in 8-ft sections that fit together with a tight overlap in telescopic fashion. Shell is of corrugated sheet steel, which is usually of 14 to 20 gauge, but metal as heavy as 10 and as light as 24 gauge has been used. The shell is driven with a heavy-steel mandrel inside it; after driving, the mandrel is removed, and the shell is inspected by lowering a light into it and then filled with concrete.

3. This type of pile is divided into two kinds: those in which a steel shell is left in the ground and those in which no shell is left in the ground. The purpose of the shell is to prevent mud and water from mixing with the fresh concrete and to provide a form to protect the concrete while it is setting and to provide a restraint to the cross section. With the shell-less pile, the fresh concrete is in direct contact with the surrounding soil.

Cast-in-place pile. Concrete pile concreted either with or without a casing in its permanent location, as distinguished from a precast pile.

Cast-in-place uncased piles. Uncased pile of the Simplex and MacArthur types. The pile is placed by driving a heavy untapered pipe casing and filling it with concrete; the casing is then withdrawn.

Caudill drive point pile. Trade name for a thin shell pipe driven with a mandrel striking on the point.

Champion splicer. Trade name for a structural connector for splicing H piles.

Chenoweth pile

1. One of the earliest piles manufactured in America, the invention of A. C. Chenoweth of Brooklyn, NY. Rolled reinforced concrete pile is made in a special machine. No forms are used. Reinforcement is arranged in a spiral form in completed cross section.

2. Pile made by spreading concrete mortar over a wire mesh and then rolling the wet mass onto the shape of a pile, which, after setting, is placed in an ordinary pile driver.

Chicago caisson. Open circular pit lined with vertical plank staves that are braced internally by steel rings wedged against the staves. Excavation is carried down to a depth equal to the length of the staves—4 to 6 ft, and usually two bracing rings are inserted for each section.

Type is suitable chiefly for easily cut clays that will stand unbraced for several hours and where free-flowing water or running silt is not encountered. Enlargements are made at the bottom into the bearing stratum. Where softer soils do not permit the use of box sheeting or lagging, a steel cylinder is sunk as a cutoff of soil and water until it reaches required soil-bearing layers and can be excavated. If the bottom cannot be sealed off, excavation is underwater and the concrete is poured by tremie method.

Chock. Member, usually of timber, used as a separator between piles or timbers.

Cobi mandrel. A proprietary, segmented steel mandrel designed to be pneumatically expanded inside a thin corrugated shell that holds it during driving.

Cofferdams. Structure constructed by driving steel or wood sheeting in advance of the excavation, or simultaneously with it, and inserting sufficient bracing to keep the sheeting in place. Ordinary construction is to have double walls and packing mixtures of clay, gravel, etc., between the walls.

Combination open and pneumatic caissons. Caissons that are partially sunk by open methods and partially under compressed air. Open and pneumatic caisson methods are used where the groundwater level is considerably below sinking level, but where excavation must be carried well below groundwater through water-bearing granular soils, such as quicksand; where the upper stages of sinking are through cohesive soils, such as stiff clays, but where such soils are underlaid by water-bearing quicksands or other water-bearing noncohesive soils; where the water level within a noncohesive soil may be lowered by artificial means, such as by the use of well points, when the caissons may be sunk by open methods to the lowered groundwater table; or where open dredging through deep water-bearing soils is permissible, but where manual preparation and inspection of the finished bottom is desired.

Compacted concrete pile. A cast-in-place pile formed with an enlarged base. Dry concrete in the base is placed in small batches, which are compacted by heavy blows while plastic.

Composite piles

1. Pile that has a concrete upper section and timber lower section or tip. It is essential that the junction between the concrete and timber portions of the pile be below the water table or the timber will rot. There is also the composite pile with cast-in-place upper section or composite pile with precast upper section. If used in a waterfront structure, the joint between the sections must be below the mud line.

2. Pile made with precast concrete pile superimposed on wood pile. Precast section is cast with a recess in the lower end to receive the tenon of the wood pile, and various devices are used to fasten the two sections of the pile together.

3. Wood pile driven below water level to resistance. Upper section spliced to wood pile and filled with concrete or fitted with precast concrete section.

4. Principal materials of which piles are made are wood, steel, and concrete. Under certain conditions a single pile may be constructed in two sections of different composition—one of wood, for example, the other of concrete. Controlling factors in specifying pile composition are the nature of the soil and the nature and elevation of the groundwater.

5. Pile made from a combination of different types of materials.

6. Pile consisting of an upper and lower section of dissimilar materials of varying lengths.

7. Pile with a timber lower section and concrete upper section. Upper section may be precast, connected by either a pipe sleeve or dowel pin to the previously driven timber section, and then used as a follower to drive the timber down until the entire pile is to grade. In other types of composite pile the concrete section is cast in place. Composite timber piles may be made by splicing a treated upper section onto an untreated lower section.

Compound batter pile. Pile driven at an angle in two directions from the principal line of the piles, normally expressed as a ratio of horizontal to vertical of the piles' centerline.

Compresol pile. Pile formed by making a hole in the ground with a pear-shaped weight operated by a pile driver and tamping concrete in the hole.

Concrete fill of drilled caissons. Controlled concrete, with a compressive strength of not less than 3500 lb/in. (or as specified) at 28 days, deposited with a slump of not more than 6 in. When deposited in water, the concrete is placed with an approved bottom dump bucket or tremie to eliminate segregation.

Concrete pile. Precast reinforced or prestressed concrete pile driven into the ground by a pile driver or otherwise placed.

Concrete pile foundations. Concrete piles may be divided roughly into two classes: "precast" and "cast in place"; cast-in-place piles may or may not be reinforced; precast piles are always reinforced.

Convertible diesel pile hammer. Diesel hammer that can be operated either open-end (single-acting) or closed-end (double-acting).

Core

1. Soil material enclosed within a tubular pile after driving (it may be replaced with concrete).

2. Mandrel used for driving casings for cast-in-place piles.

3. Structural shape used to reinforce a drilled-in-caisson internally.

Cored pile. Pile formed by removing earth prior to pile installation.

Creosoted pile. Timber pile impregnated with coal-tar creosote to minimize deterioration.

Cuneiform pile. A tapered or step-tapered pile.

Cushion block. Single-piece assembly of cushioning material located between the hammer and driving head for the purpose of protecting the pile, as well as the hammer, during driving. Cushion block may consist of a single piece of wood, masonite, micarta, etc.; multiple pieces or layers of these or other cushioning materials; or alternating layers of these materials combined with layers of more durable material, such as thin steel or aluminum plates. Occasionally wire rope, manilla rope, and various compressed paper products are also used as a cushion block. In the driving of precast or prestressed concrete piles, a thinner cushion, usually of wood, is used between the driving head and the pile to protect the head of the pile during driving.

Cutoff. Process of cutting off a pile at the proper elevation (pile cutoff elevation) to provide a plane surface at right angles to the axis of the pile.

Cutting off piles. Wood, when wholly under water, will remain perfectly sound, but if wet and dry alternately, it will soon be destroyed. Consequently, piles should be cut off so that they will always be under water. If wood caps are used, the caps should also be permanently under water.

Cylinder caisson. Caisson consisting of a cylindrical shell of masonry, wood, metal, or reinforced concrete, shod with some form of cutting edge, and sunk by excavating the material within and at the same time weighting it or using the waterjet around the sides to decrease the friction. Other methods sometimes used for sinking the caisson include jacking, driving, and drilling. Where the cylinder is of large diameter, there may be two shells, an outer and an inner, the space between the two being filled with concrete as the caisson sinks. Where the cylinder caisson is used, it is customary to construct the pier as an upward extension of the caisson.

Diesel hammers

1. Hammer, which is a self-contained unit, made up of a cylinder, piston or ram, fuel tank, fuel pump and injectors, mechanical lubricator, and lube-oil tank. Cylinder and piston each have two diameters, with the small diameter the firing chamber and the large diameter, at the top, the recoil compression chamber. To start the hammer, the ram is lifted and dropped, with a fuel injection in the air compressed by the ram causing ignition that drives the pile down and the ram up. The rising ram clears the exhaust ports and is cushioned by air in the recoil chamber that then acts as a spring to start the next downstroke. The amount of fuel injected is controlled by the operator, thus varying the energy of the blow from idling to maximum.

2. Mechanical hammer where the ram is raised by internal combustion.

Diesel pile hammer. Hammer that is a self-contained unit similar in action to the semiautomatic single-acting steam hammer, inasmuch as the blow is delivered by the falling cylinder, after it is raised to the top of its stroke by an explosion of compressed air and injected fuel. Hammer is started by lifting the cylinder 3 ft by means of the trigger gear and then tripping it so that it falls freely. The trip gear also allows the hammer to be used as a drop hammer to give the pile a few short blows to cause it to enter the ground before starting automatic operation at full stroke. Stroke is regulated by the amount of fuel delivered by the pump. Hammer is provided with a flat anvil block and some means for guiding the head of the pile in the leads.

Differential-acting steam hammers. Hammer that employs steam or air to raise the striking part and impart additional energy during the downstroke. Nonexpansive use of steam in the steam cycle obviates a drop from the entering steam pressure to mean effective pressure.

Displacement pile. Solid pile or hollow pile, driven with the lower end closed, that displaces an equivalent soil volume by compaction or by lateral or vertical displacement of the soil, such as timber, closed-end pipe, or precast concrete piles (a Franki pile).

Dog leg pile. Pile curved or bent in driving.

Double-acting hammer. Steam or air hammer in which both the upward and downward strokes are powered by steam or air pressure in the same manner as a two-cycle engine. Diesel hammer in which the ram is lifted by diesel combustion and forced downward by the alternate compression and expansion of air in the upper cylinder. Steam or air hammer will impart the same energy for each stroke of the ram at constant steam or air pressure and volume.

Double-acting steam hammers. Hammer that employs steam or air to raise the striking part and also to impart additional energy during the downstroke. Downward acceleration of the ram owing to gravity is increased by the acceleration due to steam pressure. These hammers run at greater speeds than do single-acting hammers.

Dowel. Short length of reinforcing steel embedded in the pile for eventual splicing with the main pier or pile cap above.

Drilled caisson. Shaft section of concrete filled pipe or other approved steel shell extending to bedrock; with an uncased socket drilled into the bedrock that is filled with concrete thoroughly bonded to the rock wall. The caisson may be provided with a structural steel core or other suitable reinforcement, installed so as to deliver its load to the rock through the socket filling. When such steel core is provided, it is bedded in cement grout at the base of the rock socket before initial set.

Drilled-in caisson

1. Shells for drilled-in caissons are of welded steel plate; the caissons derive their name from the fact that they are sunk to rock, excavated, and sealed against the inflow of water and soil by a tremie concrete plug, with the bedrock then core-drilled to receive a structural steel H bearing pile or heavy I-beam section to permit loading the finished caisson as a column to develop higher bearing intensities than would otherwise be obtainable.

2. For very large loads and where bad filled ground must be penetrated, the drilled-in caisson is employed. Heavy casing pipe is driven to rock or heavy-bearing material. Soil and obstructions encountered are removed. A well drill bit is used to provide a socket in the rock into which a steel H or pipe section can be driven to increase the normal column strength of the concrete core, which is placed within a thin steel shell lining of the casing pipe. The casing pipe is then removed.

3. Caisson whose shell is driven to rock and its casing cleaned out. Steel H bearing section is encased in 24-in.-diameter, $\frac{1}{2}$-in.-thick steel casing spliced with $12\frac{1}{2}$-in. steel plate, with cutting shoe at the bottom and a socket cut in the rock for an anchor.

Drilled-in caisson (Western, Spencer, White, and Prentis). Caisson whose shell is driven to rock and cleaned out. The rock socket is drilled, the core inserted, and shell filled with concrete.

Driving. Operation whereby a pile is inserted into the soil, usually by means of a hammer.

Driving foundation piles for bridge piers. When driving foundation piles for bridge piers inside of cofferdams, the work is often done from the top of the cofferdam. It is considered better for operational purposes to drive the foundation piles before the dam is closed in. Often the back line of the dam, consisting of piles and sheeting, is driven and also the sides of the dam are brought out partly or the entire distance; then the foundation piles are driven with a floating driver, if possible, but otherwise by a land driver, and then the front or remaining sides of the cofferdam are driven and the bracing placed.

Driving wood piles with butt downward. Although it is the general practice to drive piles with the tip downward, special conditions occasionally make it advisable to drive them with the butt downward. In very soft ground, the larger area afforded by the butt of the pile will often be found to carry the load.

Drop hammer

1. Heavy mass of steel that descends vertically down the pile driver frame in the axis of the pile under gravity or power.

2. A drop hammer is simply a heavy weight that can be raised to the top of the leads and then allowed to fall freely on the driving cap or head of the pile. Sometimes provision is made to trip the hammer at the top of its travel so that in falling it does not lose energy by overhauling the hoisting line, the winch, and the hoisting sheave.

3. Type of pile hammer that is raised by a rope or steel cable and then allowed to drop. Its essential features are a solid casting with jaws on each side that fit into the guides of the pile driver leads; a pin near the top for attaching the hoisting rope or nippers, as the case may be; and a broad base on which it strikes the pile, the idea being to keep the center of gravity of the hammer as low as possible.

4. Weight that is raised by a rope running over the top of a framework and extending back to a drum or geared shaft. It is released by tripping it to drop free of the rope or by releasing the drum to allow the rope to unwind. The drag of the rope and drum reduces efficiency.

Earthquake uplift. Uplift may occur as a result of rocking action of the structure, caused by an earthquake.

Electrolysis. Process that becomes a factor with a steel piles in salt water, particularly where some other metal is in the water adjacent or where direct electrical currents may be connected to the piles or induce current in them. Electrolysis is not commonly a factor, however, particularly where the tops of the piles are insulated by concrete caps.

End-bearing piles. Pile that is used when a pile or caisson is driven to firm bearing on a relatively hard soil or rock formation through material with little or no penetration resistance. Under these conditions the pile functions for all practical purposes as a column. It is assumed to derive no vertical support from the material in which it is embedded and to deliver its full load, undiminished, to the material at the tip. It is also often assumed that the material surrounding the pile provides significant lateral support and that the column strength of the pile is thereby increased. End-bearing piles are usually characterized by a fairly constant cross-section, rather than being tapered like friction piles.

Expanding pile. Pile with a mechanical device for expanding the bottom for greater bearing or resistance uplift.

Explosives. Explosives are used to drill and blast ahead of the pile tip for the purpose of removing obstructions to open-end piles under very severe conditions.

False pile. Piles used for temporary bridges, cofferdams, and pile templates. Reusable wood or steel piles are the usual materials.

Fender pile. Treated wood piles or timbers used in a fender system, except that untreated piles are often used for ferry slips. Also called "guard pile."

Final blow count. The number of blows per inch, foot, or other unit length of measure at which the driving of the pile or soil sampling device was stopped.

First-class wood piles. Piles cut from sound trees, above the ground swell, when the sap is down. Wood should be free from ring shakes, unsound spots or knots, and short bends. Knots should be trimmed close to the body and the piles peeled soon after cutting. Piles should have a uniform taper. Line drawn from the center of the tip to the center of the butt should lie within the body of the pile.

Floating pile drivers. Drop or steam hammers mounted on a pontoon or barge which may be towed, but which is moored or anchored during driving.

Follower

1. Temporary filler section between the hammer and pile top, preferably of the same material as the pile, used to drive pile top to elevation below reach of hammer or below water.

2. Follower is usually employed to drive a pile below the leads or below the ground or water surface. It is a member interposed between the hammer and the pile to transmit blows to the pile when the latter is out of the leads.

Foot of a pile. The lower end of a driven pile or the pile tip.

Foot-pound blow. The energy in foot-pounds delivered per blow by the ram of a pile hammer.

Foundation piles (classified). Friction piles that by skin friction distribute the load throughout their length to the soil into which they are driven; end-bearing piles that carry a superimposed load through soft soil to firmer underlying material on which their tips rest or into which they are driven a short distance.

Franki displacement caisson. Steel pipe is driven to desired depth, the bulb is formed by ramming concrete through the die. The pipe is withdrawn several inches at a time, concrete is rammed into the void, and the steel casing is left in place.

Friction piles

1. Term indicating that at least a part of the loading on the pile is transmitted to the soil through skin friction at the pile surface. This designation, although adequate as a means of describing the initial stress transfer from the pile to the soil, may be somewhat misleading in its implications as to the ultimate load-carrying capacity of the pile. It is possible during or after driving to create so much skin friction that the full strength of the pile material can be developed or exceeded, at least temporarily, and yet to find that the pile has a very limited safe working load. The safe working load depends on the characteristics of all the soil within the zone to which the pile transmits stress, not merely the soil in immediate contact with the pile.

2. Driven-in material that, though it may be relatively weak, is either relatively uniform in strength with depth or that increases somewhat in strength with depth, rather than being driven to a point where the soil has little if any supporting capacity. Thus a friction pile, if properly utilized, transmits part of its load to the soil through skin friction, and, the tip necessarily being driven to reasonable firm material, a part through end bearing. Skin friction alone will not suffice.

3. Pile that develops load-bearing capacity through friction with the surrounding soil mass that it penetrates. Loading tests are the best means of determining the load-carrying capacity of friction piles, and such tests are required as the basis for design.

4. Piles that are not driven to a bearing on hard ground or rock are known as friction piles and are assumed to have no end bearing. When the piles stop in soils underlaid by compressible layers, a careful analysis of the imposed loading on such layers is necessary to determine what settlements are to be expected from consolidation.

5. Load-bearing pile that receives its principal vertical support from skin friction between the surface of the buried pile and the surrounding soil.

Frictional resistance. Wood piles generally depend on frictional resistance of the ground, since a pile would not have very much strength as a long column, even if resting on rock. Piles are simply long straight trees driven with the small end down.

Gow caisson

1. Term refers to Gow's method of sinking a shaft to firm material underlying clay. Excavation of the circular shaft is started without sheeting. As excavation proceeds, welded or riveted circular liners are slipped down into place, each being 4 to 8 ft in length and smaller than the one previously placed to pass through it. Liner sections therefore fit together by telescoping, and the diameter of the shaft decreases as its depth increases. Any inflow of water is dealt with by pumping. Bottoms of the shaft of Gow caissons are frequently belled out in the clay. Lengths of caisson lining are usually, but not always, removed as concreting proceeds.

2. Caisson consisting of a nest of steel cylinders 8 to 16 ft high. Each cylinder is 2 in. smaller in diameter than the preceding cylinder. The caisson is driven down by impact or jacking as excavation proceeds. Since no internal bracing is required, the excavation is easily performed, and the cylinders can be withdrawn as concreting progresses.

Gow caisson pile

1. Pile that functions like an end-bearing pile, but necessarily of considerably larger diameter, since instead of being driven it is formed to an excavation that is extended to a firm bearing stratum. This type of construction makes it possible, when soil conditions permit, to widen or bell the bottom of the caisson to provide an increased bearing area. When these caissons are hand excavated, the surface of the bearing material can be thoroughly cleaned and inspected. This has obvious advantages. Foundation elements of this type are fully efficient only for large, concentrated loads capable of stressing the caisson material to its full capacity.

2. Bell bottom of caisson shaped by hand or cutting machine and concrete poured; the casing is removed as caisson is poured, being removed entirely when concrete pour is completed.

3. Shallow pit excavated by hand and top cylinder placed in pit. Second cylinder is placed inside first, with process repeated until caisson reaches its full depth.

Gravity hammer. A weight configured to slide in pile hammer leads or within a hollow pile which has a formed bale or swivel at its top by which it can be mechanically lifted and dropped to drive pile.

Group settlement. Total downward movement of a pile of group of piles caused by the application of a test load.

Guard pile. Loosely driven, usually treated, timber piles.

Guide pile. Pile used as a guide for driving other piles or serving as a support for wales for sheet piling.

Hammer classification. Pile-driving hammers may be classified as to the source of the energy with which they strike the pile. There are two main groups: hammers in which the striking part, after being raised in one manner or another, falls relatively freely on the pile, and hammers in which the downward motion of the striking part is accelerated in some manner, as by steam or air pressure. The first group includes drop hammers, single-acting steam hammers, and diesel hammers. The second group consists of double-acting and differential-acting steam hammers.

Hammer speed. The number of complete strokes of a pile hammer achieved by the ram per minute.

H bearing piles

1. Pile occasionally used as long friction piles in cohesive soils.
2. Because of their high strength, steel piles are desirable for use as end-bearing piles. H bearing piles were restricted to such use because it was believed that their straight sides and relatively smooth surfaces made them unsuitable for developing skin friction. It has been found, however, that H bearing piles function as well under friction as any other type of pile, even in soft clay, where, presumably, the skin friction that develops through adhesion between the clay and the steel is equal to or greater than the shearing strength attributable to cohesion within the soil itself.

Heave

1. Upward movement of soil caused by expansion of displacement resulting from such phenomena as moisture absorption, removal of overburden, driving of piles, and frost action.
2. In driving timber, precast concrete, or cast-in-place piles, it is necessary to keep a close check on the elevation of the heads of piles already driven lest they heave owing to the effect of soil displacement when piles are driven near them. Piles that heave can be redriven to their original elevations.

Heaved pile. Fully driven pile that has been lifted from its original seat, usually by the driving of adjacent piles and ground or water pressure. Usually occurs in wood pile work.

Heaving. Heaving or uplift caused when piles are driven into incompressible clays.

Hollow cylindrical cast-in-place piles. Piles manufactured by the centrifugal casting process in short sections with longitudinal holes cored in the walls through which steel cables are subsequently strung and used to prestress the sections by the method of post-tensioning and assemble them into a pile of any desired length. Prestressed cables are finally grouted into place. These types of piles are generally patented.

H-section steel bearing. Pile driven to refusal or resistance. Used where penetration is in hard material. Encased with concrete if H bearing pile is placed in 18-gauge shell.

Hydrostatic uplift. Uplift that results if the weight of the structure is not sufficient to resist flotation. Empty graving dry docks, basements of heavy structures subjected to floodwater levels before superstructure weight has been added, and floor construction between column reactions are examples.

Ice uplift. Uplift that results from the grip of ice or frozen ground on the pile. Ice grip may or may not exceed the holding-down value of the pile. In tidal waters constant change in elevation does not permit a thick ice sheet to grip a pile. Storms may raise the water level on a weather shore, drawing piles out of the ground. Rivers above tidal influence, unless confined by dams, are affected by rising water caused by melting snow or rains, which lifts the ice. When the temperature is below freezing, adhesion of ice to timber equals its cohesion.

Indicator pile. Piles, often of prestressed concrete, driven in advance of major work to determine needed length before ordering.

Intermittent static uplift. Uplift that results from such conditions as certain positions of revolving cranes, the added weight of ice on wires, and any other movement of loads.

Interpile. Proprietary cast-in-place concrete pile.

Jacking. Process whereby hydraulic or mechanical screw jacks are used to advance the pile. The pile is built up in short, convenient lengths. This method is used instead of a pile hammer where access is difficult and to eliminate vibrations.

Jacking piles. Means of avoiding effects of vibration from pile driving.

Jetting

1. Water, air, or a mixture of both forced through a pipe at high pressure and velocity. Jets are sometimes built into piles, used in practically all soils to reduce friction in strata, and are unsuitable for bearing during the driving of piles.
2. High-pressure water jets emitted through a conduit embedded in the soil to aid in pile driving.
3. Method for installing piles by which streams of water under pressure are used to wash away the soil at the place of installation.

Jetting piles into place. Methods of pile insertion that include the Franki drop weight for pulling a pipe casing into the ground by acting on a compressed gravel or concrete plug in the bottom of the casing, or water-jet loosening of the soil allowing the pile to drop by gravity or under the weight of a hammer resting on the butt.

King pile. The center pile in a cluster of piles, usually with the top higher than the adjacent piles to hold lines of a ship; added support for a precast concrete or steel sheet pile wall.

Lagged pile. To increase the area of contact and the cross section of the pile timber, timber may be bolted around the circumference of the pile. Settlement usually occurs less where lagged piles are used.

Lagging

1. Short timber of steel sections connected by bolting or welding to timber or steel pipes to increase frictional resistance along sides of pile and end-bearing resistance when mounted near tip.
2. Lagging consists of timbers bolted to the sides of a pile and is occasionally used on wood or steel; H piles when driven into very soft ground. Lagging is usually much shorter than the pile and consists of two, three, or four timbers symmetrically located. Lagging has about the same effect as using a larger diameter pile and gives an increased bearing area and friction surface. Friction surface should probably be considered as based on the length of the bounding perimeter.

Large diameter pile. Pile with a nominal diameter exceeding 24 in.

Lateral springing of pile. In driving piles into some types of clay, the lateral springing of the pile under the hammer blows makes a hole slightly larger than the diameter of the pile; in a foundation this allows surface water to find its way to the foot of the pile and thus reduces both the skin friction and the bearing value of the clay under the foot of the pile. May cause settlement of piles under heavy loads, particularly moving loads.

Leader. In a pile driver, the vertical hanging beam that guides the hammer and the pile.

Leads. High tower frame of wood or structural steel that holds and guides the hammer in its vertical descent.

Load test. Testing capacity and relation of load to movement by placing a load on the bearing element before actually building upon the foundation.

MacArthur pedestal pile. Shell-less, cast-in-place concrete pile. Driving tube is a piece of heavy steel pipe that is open at both ends. Steel core is used inside the driving tube, and the two are driven together into the ground. When sufficient driving resistance has been obtained, the core is removed and a pedestal on the bottom of the pile is formed in the following manner: After driving is finished and the core has been removed, enough concrete is dumped into the tube to fill its bottom to a depth of 5 or 6 ft. The cap is then placed back in the tube on top of this concrete, and the tube is lifted about 3 ft. The hammer is then operated on top of the core in order to drive the concrete out of the bottom of the tube to form the pedestal. Sometimes several such charges of concrete are used in this manner to make the pedestal.

MacArthur pile

1. Pile driven with a core that fills the casing and is withdrawn before the concrete is placed. If an enlarged foot or "pedestal" is desired it may be formed on the pile by replacing the core after the casing is partially filled and has been pulled up to 3 ft or so. Driving the core lightly then forces concrete out of the bottom of the casing to form the enlarged foot or pedestal. The pile is completed by filling the casing with concrete and withdrawing the casing. Uncased concrete piles are objectionable in soils of such character that driving a pile will deform a nearby pile, the concrete of which is still plastic. To offset this difficulty, a light casing may be slipped inside the heavy one before the concrete is poured. This light casing is left in place, and the pile then becomes a special type of cased pile.

2. Cast-in-place concrete pile in which a steel shell remains in the ground. Driving tube is a piece of heavy steel pipe open at both ends. Steel driving core is fitted inside this pipe, and the two are driven together into the ground. When suitable driving resistance has been reached, driving is stopped and the core is removed from the tube. A thin metal shell several inches smaller in diameter than the driving tube is then dropped down inside the tube. The thin shell is then filled with concrete and the driving tube is withdrawn. With a pile of this kind the driving resistance is determined by means of the outer driving tube, but the pile that remains in the ground is several inches smaller in diameter than this tube. Carrying capacity of such a pile depends largely on the point resistance, since the surrounding soil may or may not come back against the shell to provide frictional resistance along the side of the pile.

Mandrel. Piece of pile driving equipment consisting of a heavy-wall circular steel pipe used in the installation of cast-in-place concrete piles. Thin steel shell is threaded over the mandrel. The mandrel, which acts as a pile, is driven into the soil carrying the casing along with it. The mandrel is then withdrawn, leaving the casing ready to receive the concrete.

Metal pile shoes. Shoes attached so that they act as an integral part of the pile. They are sometimes used on piles that are driven in soils containing boulders, riprap, deposits of coarse gravel, or very hard clay and are advantageous in furnishing grips for falsework piles on rock bottom.

Micro piles. Small diameter piles, most often used in underpinning.

Montee Caisson. Caisson used for deep foundations, consisting of a reinforced steel shell, with cutting edge of serrated hardened plate, that is connected to a large motor for continuous rotation. Its action is similar to that of the circular saw used in some large granite quarries to remove 8- and 10-ft diameter cores, to serve as weakening planes for freeing the rock sheets. To maintain completely fluid conditions, a slurry is fed into the cylinder to float the interior soil outward under the cutting edge and up along the outer surface of the shell.

Mooring pile. Usually a treated or greenheart wood pile driven to anchor marine vessels from wind or wave action.

Needle piles. Very small diameter. Slender driven steel tubular or rail section piles used in underpinning operations.

Net settlement. Depth of penetration that a pile retains after removal of a test load.

Non-displacement piles. Piles formed by boring or other methods of excavation. H, open-end pipe, and sheet piles are considered low-displacement piles.

Open caissons

1. Open caissons comprise three general types because they are sunk through soil to an adequate supporting medium by open methods, without the use of compressed air, and are open top and bottom. Open caissons include cylindrical caissons, single-wall box-type caissons, and multiwall caisson piers with dredging wells.

2. Caissons constructed on the surface like pneumatic caissons and sunk into position, where they may be held down by weights if necessary.

Overdriving of wood piles. Overdriving causes bending of the pile, bouncing of the hammer, cutting of the driving plate into the head of the pile, separation of the wood along annual growth rings near the surface, and brooming of the head. However, damage may still occur, in the case of tapered piles, before any of these symptoms appear, since in tapered piles the stress at the tip is usually greater than at the head, owing to the much smaller cross-sectional area of the pile at the point.

Pedestal concrete pile. Pile driven in the same manner as a straight shaft, except that when shell is partly removed, concrete is poured for pedestal and rammed with core. Care must be used not to disturb adjacent piles with pedestal.

Pedestal piles

1. Pile designed to retain permanently the good resistance effects of the temporary bulbs of pressure formed during driving by forming a concrete bulb that hardens. This may be advantageous in transferring footing loads through soft material to a firmer distributing stratum capable of bearing the load thus spread over it. Particularly useful in case the bearing stratum is hard to penetrate to sufficient depth to obtain adequate frictional area, or in case it is so thin that a

friction pile might be in danger of punching through to softer material beneath. Piles with large pedestals, such as Franki piles, are often capable of sustaining very large loads.

2. Cast-in-place concrete pile constructed so that concrete is forced out into a widened bulb or pedestal shape at the foot of the pipe that forms the pile.

3. Spread footing is obtained by driving the concrete out at the bottom of the shaft, while at the same time compressing the surrounding soil.

Penetration. Downward axial movement of a pile into soil under installation pressure.

Permanent piling. Permanent piling requires treating or armoring. Creosote treatments for wood piles are usually efficient protection for piles in the ground. In tropical waters, sheathing or armoring provides the longest protection for wood piles, and this may be done with sheet metal, precast concrete, gunite, vitrified pipe, cast-iron pipe, etc. Steel or concrete piles are also used when permanent piling is necessary, using waterline protection where advisable. Concrete piles in seawater usually require protection, such as asphalt coatings.

Pile

1. Element of construction composed of timber, concrete, or metal, or a combination of these, that is either set, driven, or screwed into the ground vertically for the purpose of sustaining the weight that is to rest upon it, or resisting a lateral force.

2. Structural unit introduced into the ground to transmit loads to lower soil strata or to alter the physical properties of the ground, of such shape, size, and length that the supporting material immediately underlying the base of the unit cannot be manually inspected.

3. Column of wood, steel and/or reinforced concrete, or precast concrete, driven, jacked, or jetted into the ground to support superimposed loads.

4. Slender structural element that is driven or otherwise introduced into the soil, usually for the purpose of providing vertical or lateral support.

5. Slender wood, concrete, or steel structural element, driven, jetted, or otherwise embedded on end in the ground for the purpose of supporting a load or compacting the soil.

6. Timber, steel, or concrete shafts sunk into soft ground, upon which foundations are built.

Pile bent. Two or more piles driven in a row transverse to the long dimension of the structure and fastened together by capping and (sometimes) bracing.

Pile cap

1. Structural member placed on and usually fastened to the top of a pile or a group of piles and used to transmit loads into the pile or group of piles and, in the case of a group, to connect them into a bent.

2. Metal cap or helmet temporarily fitted over the head of a precast pile to protect it during driving. Some form of shock-absorbing material is often incorporated.

3. Slab or mat, usually of reinforced concrete, covering the tops of piles for the purpose of tying them together and transmitting to them as a group the load of the structure that they are to carry; metal plate often placed on top of steel piles to distribute the load to the concrete from the pile.

4. Building loads are transferred to the heads of the piles through a structural element known as a pile cap.

Although pile caps resemble spread footings and are often constructed in contact with the ground, it is not common practice to consider that they derive any direct support from the soil. It is not uncommon to find that after construction the soil settles away from the underside of the pile cap, leaving a gap between the soil and the concrete.

5. Cast-in-place reinforced concrete used to distribute a column load to a group of piles.

Pile cap for wood piles. Cap interposed between the hammer and the pile head. Cap is composed of a toggling element and a cushion against which the hammer strikes. Toggling element is an iron casting having jaws corresponding to those on the hammer so that it may engage the same leaders. In the bottom face of the casting is a deep conical recess for the tapered head of the pile, and on the top surface is a shallow recess into which is set a short, round, wood cushion block. A steel band is pressed over the top end of the block to prevent it from splitting. After driving, the cap is attached by rope slings to hooks on the hammer so that it can be lifted.

Pile driver

1. Equipment used for the support of operation of a pile hammer.

2. Equipment or apparatus for driving piles. Characterized by the leads, sometimes called "leaders," which are upright parallel members supporting the pulleys or sheaves used to hoist the hammer and piles and to guide the movements of the hammer. Leaders may be of steel or wood.

Pile driving. Forcing a pile into a definite position in the ground without previous excavation, accomplished by the use of either drop hammers or steam hammers.

Pile driving rigs. Equipment used for driving piles and pile shells. Usually consists of a crane with a set of guides, known as leads, for the pile and for the hammer. When a steam hammer is used, a boiler or steam generator is also required, and this is often mounted on the equipment and becomes a part of the rig.

Pile footing. Concrete, steel, or wood piles driven into the ground, on which a reinforced concrete cap is used to distribute the load and support the column, strut, or pier.

Pile foundations. Foundation system composed of timber, concrete, or steel members driven vertically to a hard soil or rock stratum that will support the design loads.

Pile head. Top of the pile.

Pile load capacity. Maximum allowable capacity of a pile or group of piles to support loads.

Piles driven out of plumb. Where piles are driven out of plumb, a tolerance of 2% (or as allowed by applicable code) of the pile length, measured by plumb bob inside a casing pile or by extending the exposed section of solid piles, is permissible. Greater variation requires special provision to balance the resultant horizontal and vertical forces, using ties to adjacent column caps where necessary. For hollow-pipe piles, which are driven out of plumb or where obstructions have caused deviations from the intended line, the exact shape of the pile can be determined by the use of a magnetic shape-measuring instrument or an electronic plumb bob. The deflection curve of pile is plotted, and consecutive differentiation gives the slope, moment, and shears acting on the pile, from which the reduction in allowable load can be determined.

Pile splices. Piles are often made up of special sections of similar or different structural elements. Such composite piles are usually evaluated as equal to the weakest section. The splice must be designed to hold the sections together during driving, as well as to resist later effects from the driving of adjacent piles or from lateral loads. Timber pile sleeves consist of a tight-fitting steel pipe section into which the pile ends are driven; sometimes drift pins or through bolts are used to hold the assembly together. Steel rolled sections and pipe piles are often extended by a full-strength weld.

Pile tip. Lower extremity of the pile.

Pilings. Columns extending below the ground to bear the loads of a structure when the surface soil cannot. They may extend down to bearing soil or support the load by skin friction. Sheet piling is used to form bulkheads or retaining walls.

Pipe column. Column made of steel pipe, often filled with concrete.

Pipe pile. Steel cylinder, usually between 10 and 24 in. in diameter, generally driven with open ends to firm bearing and then excavated and filled with concrete; this pile may consist of several sections from 5 to 40 ft long, joined by special fittings, such as cast-steel sleeves, and is sometimes used with its lower end closed by a conical steel shoe.

Pipe step-taper pile. Combination of steel pipe and step-taper shells providing a pile of practically unlimited length. The pile is driven with a rigid steel mandrel extending through the step-taper shells and, under special conditions, through the pipe. The pile tip is closed with a flat steel plate welded to the pipe. After being driven to the required penetration, the pile is internally inspected and both the pipe and shell portions filled with concrete.

Pneumatic caisson

1. Solid concrete or masonry box whose sidewalls are extended downward as "cutting edges" to form one or more rooms, called "working chambers," extending over the full bottom area. Rooms are floorless and of sufficient height to permit workmen to excavate the soil materials from the floor of the caisson. From the roof of each such rooms one or more vertical shaft-ways (usually steel cylinders of elliptical cross sections) extend up through the concrete filled box to connect to an air lock. Air is supplied to the caisson working chamber at desired pressure, and conduits for light, telephone, etc., in the working chamber are built into the concrete body of the caisson.

2. Pneumatic caisson has four sides (or it may be circular) and a roof, but no bottom. The roof has one or more holes for shafts, usually about 3 ft in diameter, for the passage of men or material from the outer air into the working chamber. An air lock prevents the air pressure in the working chamber from being seriously reduced while men or material are passing in or out. Air pressure in the working chamber is kept just high enough to balance the water pressure.

3. Caisson used for excavations carried to depths at which it would be impossible with ordinary caissons to control the flow of water or soil through the bottom. Caisson is capped to provide a top seal, and work is done under artificially applied air pressure, sufficient to counterbalance hydrostatic pressure. Entrance to the working chamber is through double-gated locks.

4. Pneumatic caisson may be defined as a structure, open at the bottom and closed at the top; in other words, an inverted box in which compressed air is utilized to keep the water and mud from coming into the box and that forms an integral part of the foundation.

Precast concrete pile

1. Concrete pile tapered or straight, of lengths to 150 ft, driven to resistance or refusal.

2. Concrete pile cast in one piece and designed to develop an ultimate compressive strength of at least 5000/psi in 28 days after pouring.

Precast pile

1. Pile that is cast in a regular mold above the ground. After proper curing it is driven or jetted into place similarly to a wood pile. Precast pile is divided into two general groups: tapered pile and parallel-sided pile. In order to withstand handling and driving stresses, precast piles are always reinforced with longitudinal rods in combination with lateral reinforcement of wire hoops or spiral wrapping.

2. Pile manufactured in several shapes: square, round, or six-sided, driven to refusal or resistance. May be tapered. Steel point at bottom used where rock occurs.

3. Reinforced concrete pile manufactured in a casting plant or at the site, but not in its final position.

4. Pile reinforced with steel rods and high-value concrete, then driven like wood piles.

Precast segmental pile. Precast pile manufactured in lengths which enable the pile to be extended on site relatively quickly with a mechanical splice.

Precoring. By continuous flight auger or churn drill, a hole is formed into which the pile is lowered. The pile is then driven to bearing below the cored hole. Precoring is used for driving through thick stratum of stiff hard clay, to avoid displacement and heave of surrounding soil, to avoid injury to timber and thin shell pipes, and to eliminate driving resistance in strata unsuitable for bearing.

Preexcavation. Hand or machine excavation, used for removal of obstruction close to ground surface.

Pressure-injected footing. Foundation formed by applying heavy pressure and impact to a plug of concrete at the bottom of a casing for the purpose of causing the shaft and concrete to penetrate bearing materials; then ejecting the plug at a designated depth to create an expanded base or bulb, and then filling the casing shaft with concrete.

Pressure-treated pile. Round or sawn timber used as a pile, which has been pressure impregnated with a chemical preservative.

Prestressed pile. Precast concrete pile that is prestressed or post-tensioned to reduce or eliminate cracking caused by tensile stresses during transportation, driving, and in service.

Production pile. Pile that is a part of a specified pile foundation, as opposed to a preliminary test pile.

Protecting heads of wood creosoted foundation piles at cutoff. Where creosoted piles are used to support a foundation, it is usually because of the difficulty or impracticability of providing cutoff of untreated timber piles below the existing or probable future water table. In such cases great care must be used to prevent the infection of the untreated heart wood of the pile with the spores of dry rot. A heavy brush coat of hot creosote is thus given immediately after the pile has been cut off. It is then advisable to follow up the brush coat with the "soaking in" procedure.

Protecting wood pile head. To prevent splitting and reduce brooming, the head of a pile may be hooped with a pile ring and the pile neatly chamfered down, so that the first blow of the pile hammer puts the ring in place.

Railway pile drivers. Railway pile drivers consist of drop or steam hammers mounted on special flatcars. Steel leads fold inside railway clearances and are usually mounted on a turntable.

Range pile. Pile serving as a guide for locating piles or other structures or for marine surveying or dredging.

Raymond pile. Pile formed by driving a steel shell into the ground on a mandrel that can be collapsed and withdrawn. The hole is filled with concrete and reinforced, if required. Permanent steel shell used outside the mandrel provides the advantage of preventing any sand from flowing in as the mandrel is withdrawn.

Raymond step-tapered pile. Trade name for a spirally corrugated light gauge shell made in 4 to 16 ft sections of increasing diameter to form a step-tapered pile.

Rectangular open caissons with dredging wells. Caisson based on the same principle as the double-wall, open-cylinder caisson, differing from the latter only in the matter of shape and size. Most notable difference is that the open-cylinder caisson has but one dredging well.

Refusal. Depth beyond which a pile cannot be driven.

Rock socket. Socket drilled in sound rock and thoroughly cleaned of all foreign matter and loose rock. After examination and approval of the rock surface, the concrete fill is deposited in the dry or by an approved method under a water seal. Depth of socket should develop the full load-bearing capacity of the caisson on the approved spread area of distribution and without overlapping of stress cones.

Rotary drilling methods of sinking caissons. Methods of sinking small-diameter cylindrical caissons, based on some form of rotary drilling. In less stable soils caving of the sides of deep-bored shafts has been prevented by keeping the holes filled with a thick, semifluid "drilling mud" as the work proceeds. After the hole reaches full depth, a steel plate liner is installed and the bottom sealed with sand and cement placed outside the casing, or with tremie concrete placed inside, before the mud is pumped out and the caisson filled with concrete.

Screw pile. Pile with a spiral blade fixed on a shaft or; a shaped precast pile, screwed into the ground by a rotating force.

Sealing the caisson. When the caisson has reached its final resting place, either on rock, hardpan, or in a few places on sand or clay, it is necessary to fill the working chamber with concrete.

Second-class wood piles. Class includes any suitable timber that will stand driving. Piles are limited in use to foundations where they will be continuously submerged or to cofferdams and other temporary structures.

Selective wood for piles. Class includes white oak, the cedars, cypress, redwood, longleaf pine, and Douglas fir.

Semipermanent piling. Wood piling having somewhat longer life, obtained by leaving the bark on, provided abrasions and knots are covered with metal of a life expectancy equal to that of the wood. Inner skin of the bark may decompose and become slippery. Other and better methods of obtaining somewhat longer life (for one or two seasons) are to paint the pile with proved preservatives or to sheath it with creosoted battens nailed in place over the danger zone.

Settlement. Vertical downward movement of pile under superimposed load.

Sheet piles. Piles that may be used as bearing piles. When they are driven into rock, the allowable load on sheet piles is determined in the same manner required for other piles. When they are used as a friction wall, the allowable load is determined in accordance with approved engineering practice on the basis of either the embedded area of one side or double the projected area of the embedded wall, whichever is greater.

Shell. Thin-walled steel tube that remains in place for poured-in-place piles that act only as a form for the concrete.

Shoes and reinforced tips. Metal reinforcing, such as bands and shoes, for all types of piles. Provides protection against damage of tip, and additional cutting power.

Shoring. Placing of temporary or permanent inclined supports to existing structures.

Simplex pile

1. Pile driven without a core, the metal shoe closing the end of the casing being left in the ground.

2. Pile driven to resistance with steel pipe casing and steel point. Casing is filled with concrete and then withdrawn.

3. Made by driving down a closed steel pipe and withdrawing it while concrete is forced out at the bottom.

Single-acting hammer. Steam, air, or diesel-operated hammer in which the ram is lifted by steam or air pressure or by diesel combustion. At the top of the upward stroke the steam, air, or spent gases are exhausted, allowing the hammer ram to fall freely on the downward stroke. In the case of the steam or air hammer, the height of face fall is constant, imparting the same energy for each stroke. The stroke or height of free fall for a diesel hammer varies, depending on the resistance offered to the pile in driving, increasing as the resistance to driving increases.

Single-acting steam hammer. Steam or air raises the movable mass of the hammer, which then drops by gravity alone. Hammers of this type, in which the ram is the movable part, are always employed in leads. Some single-acting steam hammers employ the casing as the movable part, the piston being stationary and resting on the pile by means of a hammer rail and moving down with the pile. Characteristics of the blow are a low-striking velocity, owing to the low fall, and a heavy striking weight. Blows are much more rapidly delivered than with a drop hammer.

Single-wall open caissons

1. Caissons of this type are built on the surface of the ground and sunk by excavating or dredging through the open well with clamshell or orange-peel buckets. When difficulty caused by striking an obstruction arises, an air lock can be fitted and sinking continued by the pneumatic process. Open-well caissons are used more widely for bridge foundations than building foundations.

2. Open caisson constructed of timber or concrete, box-like self-braced structure, partly or entirely open both top and bottom, sunk through soil, or water and soil, to an adequate supporting medium by excavating soil materials below its open floor as it is sunk, after which it is filled with masonry or concrete. This type of caisson is used occasionally on land but is employed more extensively for the construction of bridge piers.

3. Open caisson that is a boxlike self-contained structure, either partly or entirely open at both top and bottom, and forming an integral and permanent part of the pier. Open caissons have the distinction of being employed for the deepest foundations and may be divided into three types: rectangular single-wall caisson, consisting of a frame with solid walls and without top, bottom, interior chambers, or cutting edges; cylinder caisson, consisting of open cylinders of iron or masonry; and rectangular caisson with dredging wells,

consisting of a structure partly closed at both the top and bottom, with open wells running vertically through it.

Slide pile. Pile driven into the earth to consolidate the soil and help to prevent it from sliding down a slope.

Soldier pile. A steel H section, normally driven vertically at intervals of several feet to hold horizontal lagging, installed to retain soil as excavation proceeds.

Spiral-formed steel shell. Shell driven with mandrel to resistance; pile poured after spandrel is removed.

Splicing piles. When it is necessary to use longer piles than can be obtained in the single length, then two piles may be spliced together, end to end, by timber fishplates bolted on four sides of the piles or by a metal sleeve used in the form of a heavy pipe. Half-lap joints fastened with bolts prove unsatisfactory because of the lack of lateral strength and stiffness. In swampy ground, one pile can be driven on top of another with only an iron dowel pin connecting the two. Pile splices are usually used where piles can be driven only in short sections owing to limited headroom, but usually this condition occurs with steel sheeting inside buildings, for pits, wells, shafts, etc., rather than for foundation piles.

Spudding. Heavy structural sections or closed-end pipes are alternately raised and dropped to form a hole into which a pile is lowered. The pile is then driven to bearing below the spudded hole. Used for driving past individual obstruction and through strata of fill with large boulders or rock fragments.

Static uplift. Static uplift may be present in designs where it is felt to be safe to rely on tension on piles, without the expectation of undue differentials in levels. Retaining walls on piles and fixed cranes are examples.

Steam hammer. Pile hammer that is automatically raised and then dropped a short distance by means of a steam cylinder and piston held in a frame in the leaders of the pile driver so that it follows the pile down in driving. The first steam hammer was invented about 1850. Steam hammers are of two general classes: single acting and double acting.

Steel H bearing piles

1. Piles that consist of heavy rolled H sections, the webs as well as the flanges of which are thick enough to prevent moderate corrosion from seriously weakening the pile. H-beams drive easily, make piles of great length, and therefore can be used as end-bearing piles to transmit loads to deep-seated firm strata that are impracticable to reach with timber or concrete piles. H bearing piles are occasionally spliced by riveting, bolting, or welding. A characteristic of H bearing pile is its ability to penetrate deep beds of soft soil containing thin layers of firm sand or sand and gravel. Small cross-sectional area eliminates any tendency of H bearing piles to leave the soil and with it nearby previously driven piles.

2. Profile usually used for this purpose is that known as the H-beam, or column. Has proved especially useful for structures in which the pile extends above the ground line and serves not only as pile but also as a column.

3. Where heavy resistances exist in layers that overlie the bearing depths, steel H bearing piles take heavy driving much better than any other type. After insertion in the ground, it has been found that the soil gripped between the web and inside faces of the flanges becomes an integral part of the pile. Frictional resistance

is measured along the surface of the enclosing rectangle and not along the metal surface of the section.

Steel piles. Usually either structural H-section bearing piles or steel pipe, the latter being filled with concrete after being driven. Because of the relatively slight section of pile material as compared with wood or concrete piles, steel H-section bearing piles are sometimes called "non-displacement piles."

Steel piles cutoff. Steel H bearing piles, steel pipes, or steel sheet piles can be cut under water by a diver using an oxyacetylene or oxyhydrogen flame and special burner, or a fully insulated arc-oxygen underwater cutting torch and tubular steel electrodes.

Steel pipe piles. Pile formed by driving ordinary heavy-wall steel pipe into the ground and filling it with concrete. Usually a steel shoe or boot plate is used at the bottom of the pipe, and the pile is then called a closed-end pipe pile.

Steel pipe piles, open end or point. Pile formed by blowing earth out with an air jet as it is being driven to refusal and loaded as a column (open end), or driven to refusal or to resistance with cast steel point.

Steel pipe pile with cast steel point. Pile driven to refusal or to resistance and loaded as a friction pile.

Steel pipe pile with open end. Pile formed by blowing earth out with an air jet as it is being driven. Driven to refusal and loaded as a column. Diameters vary, based on superimposed load.

Steel-sheet piling. Piers constructed as cellular shells of steel sheet piling filled with rock, soil or concrete, designed and constructed in accordance with accepted engineering practice as approved by the local building official.

Steel shell. Seamless or welded pipe of steel, fitted with an approved cutting shoe and structural cap or with other approved means of transmitting the superstructure load.

Step-taper piles. Pile installed by driving a closed-end steel shell (using a full-length steel mandrel) to required penetration and filling the shell with concrete.

Straight shaft MacArthur. Steel core and casing driven to resistance. Core removed and casing filled with concrete. Core removed with pressure on concrete.

Swage pile. Thin-wall pipe pile with bottom closed by a slightly tapered precast point.

Tapered fluted pile (Union). Steel shell with steel point driven to resistance and filled with concrete.

Tapered pile standard step (Raymond). Steel shell with inserted mandrel driven to resistance. Core withdrawn and casing filled with concrete.

Temporary casing. Open-end pipe casing driven and cleaned out. May be pulled later. Accomplishes the following: drives through minor obstructions, minimizes displacement, prevents caving or squeezing of holes, and permits concreting of pile prior to excavation to subgrade of foundation.

Temporary wood piling. Piling used in harbors where borer attack is known to be light, or in which the seasonal period of heavy attack has just passed, or in locations subject to decay or insect attack. Temporary piles may be of untreated wood. If practicable to pull the piling after it has served its purpose, unattacked creosoted piling may be reused if sound.

Test pile. Pile driven to ascertain driving conditions, probable required lengths, its capacity, and the carrying capacity of the soils.

Timber pile.   A tree trunk, usually debarked, driven into the ground. It may be pressure impregnated with preservatives.

TPT pile.   Proprietary precast concrete tip pile with thin-shell shaft, driven with the aid of a mandrel.

Treated pile.   Timber pile that has been chemically treated to withstand decay, rot, or fungus.

Tube pile.   A steel pipe pile.

Types of piles.   Piles may be divided into two general classifications, according to the manner in which they develop their capacity to support loads: end-bearing piles and friction piles. End-bearing piles transmit practically all of their load through their points to a firm substratum. Piles that receive only nominal support from the penetrated soils and that depend for their capacity on point bearing are friction piles.

Uncased concrete pile.   Column of concrete placed and left in the ground without encasement.

Uncased straight shaft (MacArthur and Western).   Shaft driven to resistance with steel pipe casing and steel core. Core is then removed and casing filled with concrete and then withdrawn.

Underpinning.   Adding new permanent support below existing foundations.

Uplift.   Piles are frequently required to provide anchorage against uplift. Many designs have been based on the assumption that a friction pile will safely resist an uplift equal to one-half its safe bearing value or an uplift producing a shearing stress of not over 250 lb/ft^2 on its imbedded surface, whichever is the lesser. Piles that are to resist uplift must have sufficient tensile strength to withstand the upward vertical forces to which they will be subjected. This may require the reinforcement of cast-in-place piles and special consideration of the design of the splices in spliced or composite piles.

Uplift due to lateral forces.   Uplift occurring from such causes as ice thrust behind retaining walls, impact from ships or floating objects striking a piled structure, or pressure from high-water-level and current velocity.

Uplift piles.   Piles that are sometimes required to resist uplift as part of the static design or for occasional or emergency conditions.

Upside down piles.   Tapered piles, specifically timber, driven with large butt downward.

Vibration.   High-amplitude vibrators are advantageous for driving in waterlogged sands and gravel and for driving sheet-piling.

Vibrator pile hammer.   Hammer that operates by vibration and not by impact forces as conventional pile hammers.

Wakefield pile.   Trade name for a timber sheet pile consisting of three planks bolted or spiked together, with the middle plank offset so as to form a tongue along one edge and a corresponding groove on the other.

Water jet.   In some soils, like quicksand, there is an advantage to water jetting the site of each pile and to working a jet pipe (pipe through which water is forced under pressure) up and down as the pile is being driven. In such soil conditions the driving is greatly facilitated. The pile is easily forced down and soil or sand flows back and binds or sticks to the wood, increasing the frictional resistance. In solids the water jet would make a hole that would not fill itself up again, in which case jetting is not desirable.

Water jet in driving piles.   Use of water jet in pile-driving operations differs radically in principle from driving with a hammer. The process consists of displacing the material at the proposed location of the pile by means of one or more water jets. In using a water jet the quantity of water is more important than the velocity. Velocity should be enough to excavate the sand, and when mixed with water, the volume of water should be sufficient to force the water to escape by rising to the surface and bringing material with it.

Wet caisson.   Steel cylinder sunk to rock as earth is removed. The bottom is sealed, water removed, and the cylinder filled with concrete.

Wind uplift.   Uplift occurring from pressure against such structures as chimneys, tall buildings, raised bascule bridges, and large surfaced areas.

Wing pile.   Bearing pile, usually of concrete, widened in the upper portion to form part of a sheet pile wall.

Wood for piles.   The kind of wood used for piles is determined by what is most easily obtained. Pine, hemlock, spruce, and many soft woods make usable piles. Cedar, hickory, oak, etc., are much tougher and more durable and therefore desirable when they can be obtained in proper lengths.

Wood piles

1.   Wood piles that are driven to acceptable resistance based on test pile.

2.   Southern pine and Douglas fir are common types, the former being procurable in quantity in lengths up to 65 ft and the latter in lengths up to 90 ft and over. Locally, other types, such as cedar (which is known for its resistance to decay), oak, and other hardwoods are used. Wood piles as a rule are used in permanent work only when it is reasonably certain that the entire pile will remain completely submerged, since the wood deteriorates rapidly above water unless treated piles are used.

RAILROAD WORK

Abrasion plate.   Special plate, generally steel fabricated, used in special situations (in special track work and under continuous insulated joints) where a standard tie plate cannot be used. It is used to support the rail and prevent abrasion to the tie.

Absorption.   Quantity of preservative solution absorbed by, or forced into, a timber or tie during treatment. Volumetric absorption is the ratio of the absorption to the total volume of the timber.

Acre.   As measured along railroad right-of-way, 8 ft 3 in. wide by 1 mile long.

Adzing machine.   Specially designed portable power-operated machine used to dress the rail seat on ties, thereby providing proper bearing for rail or tie plates.

Alignment (civil engineering).   A railroad's horizontal location as described by curves and tangents.

Alley.   Clear track in switching yard.

Anti-checking iron.   Piece of steel strip sharpened on one edge and bent in various shapes that is driven into the end of a tie to prevent checking and splitting.

Apron.   Waterfront railroad track on a bridge structure that connects the deck of a car ferry with the tracks on land, hinged at the shore end so that it is free to move vertically at the outboard end to accommodate varying elevations of the ferry.

Assigned siding. Side or team track owned by a carrier and assigned to one or several industrial concerns for loading or unloading.

Auxilliary fastenings. In track work, small parts such as tie plates, rail anchors, lock washers, rail braces, etc.

Ballast. Material selected for placement on the roadbed for the purpose of holding the track in line and at surface.

Ballast cleaning. The process of separating dirt from the ballast by shaking and depositing stone back onto the track.

Ballast curb. A length of timber placed longitudinally along the outer edge of the floor on ballast deck bridges to retain the ballast.

Ballast fork. Device used for the movement of rock ballast.

Ballast plow. Machine used to level ballast or move ballast from the track or to the shoulder of the track.

Ballast regulator. Device that distributes equal amounts of ballast.

Ballast section. Vertical cross section of the track from the subballast up, including ballast.

Ballast spreader. Device used to distribute ballast materials evenly on both sides of the track.

Ballast sweeper. Also called a "broom." Cleans ballast off the track.

Batter. A surface deformation of the head of the rail in the immediate vicinity of the end.

Branding. Markings hot rolled in raised figures and letters in the rail web to identify the weight of rail and section number, type of rail, kind of steel, name of manufacturer and mill, and year and month rolled.

Bumping post. A post at the end of a track used to stop moving equipment.

Burnett process. Full-cell pressure treatment application of salt preservative to ties.

Cambering machine. Device for giving hot rails emerging from the rolls the curvature required to compensate for the unequal cooling of head and base, so they will be as nearly straight as possible when cold.

Cant (of a rail). Rail's inward inclination effected by using inclined surface tie plates.

Claw bar. Special steel tool about 5 ft long with a claw end and long shank lever, designed to draw track spikes from railroad crossties by leverage.

Closure rails. The rails located between the parts of any special trackwork layout, as the rails between the switch and the frog in a turnout (sometimes called the lead rails or connecting rails); also the rails connecting the frogs of a crossing or of adjacent crossings, but not forming parts thereof.

Compound fissure (in a rail). Progressive fracture starting from a horizontal split head that turns up or down in the head of the rail, continuing until substantially at right angles to the length of the rail. Compound fissures require inspection of both faces of the fracture to locate the horizontal split head from which they originate.

Compromise joint. Rail joint between rails of different height and section; or rails of the same section but of different joint drillings.

Compromise rail. Rail of relatively short length, where the ends are of different section, corresponding with the sections of the rails to which they are to be joined. A compromise rail eliminates the need for direct joining of rails of different section.

Connecting track. Two turnouts with the track between the frogs arranged to form a continuous passage between one track and other intersecting or oblique track or another remote parallel track.

Construction station. A 100-ft distance of track as measured along the center line and designated by a stake bearing its number.

Continuous welded rail (CWR). Rails welded together in lengths of 400 or more ft.

Corrugated rail. Condition of roughness on the rail head seen as alternate ridges and grooves that develops in service.

Crib. The lateral space between two railroad ties. A retaining structure.

Cribbing machine. Removes rock between ties where the rail lies.

Crossing end frogs. The two frogs located at the opposite ends of the long diagonal of a crossing.

Crossing frogs. The two frogs located at the opposite ends of the short diagonal of a crossing.

Crossing plates. Plates that are interposed between a crossing and the ties or other timbers for the purpose of protecting the ties and better supporting the crossing by distributing the loads over larger areas.

Cross level. The distance one rail is above or below another. This should not be confused with superelevation on curves.

Crossover. Two turnouts where the track between the frogs is arranged to form a continuous passage between two nearby and generally parallel tracks.

Cross tie. The transverse member of the track structure to which the rails are spiked or otherwise fastened to provide proper gage and to cushion, distribute, and transmit the stresses of traffic through the ballast to the roadbed.

Curve (of a railroad line). In the United States, it is customary to express track curvature in degrees noted by the deflection from the tangent measured at stations 100 ft apart. In other words, the number of degrees of central angle subtended by a chord of 100 ft represents the "degree curve." One degree of curvature is equal to a radius of 5,750 ft.

Dead section. Section of track, either within a track circuit or between two track circuits, where the rails are not part of a track circuit.

Deck span bridge. Bridge where the track is carried on top of the stringers (girders) or trusses.

Departure track. One of the tracks in a departure yard where outgoing cars are placed.

Depth (of ballast). The depth measurement from the bottom of the tie to the top of the subgrade.

Detail fracture. Progressive fracture starting at or near the surface of the rail heads. Such fractures are entirely different from transverse fissures, compound fissures, or other defects that have internal origins.

DOT. The Department of Transportation. An agency of the U.S. government having jurisdiction over matters pertaining to all modes of transportation. The Federal Railroad Administration is the branch of the DOT that promulgates safety standards for rail equipment used in interchange.

Drift bolt. Piece of round or square steel with or without head or point and of specified length, driven as a spike.

Drift pin. Special railroad tool of round steel tapered for insertion to align holes by striking the large end.

Electric flash butt weld. Weld made by electrically heating two abutting rail ends. When the steel reaches the proper temperature, the rail ends are compressed together, eliminating the need for joint bars.

Elevation. In general, the vertical distance that some location is above some reference location, usually sea level. In railroad trackwork, the vertical distance that the outer rail is above the inner rail on a curve is called "the elevation" (or "superelevation") "of the curve."

Expansion shim. Spacer inserted between ends of abutting rails while track is being laid to provide allowance for expansion of steel when temperature changes.

Fastenings. Joint bars, bolts and spikes or other alternatives to spikes used in track construction.

Fatigue. Type of failure brought about by repeated cycles of load on a rail or other structural component.

Filler block. Steel block molded and designed to keep uniform the angle spread between lead and turnout rails and frogs, etc.

Flangeway. The open way through a track structure which provides a passageway for wheel flanges.

Floor height. The distance measured vertically from the top of the rail to the top surface of the floor in a railroad car.

Frog. Track structure used at the intersection of two running rails to provide support for wheels and passageways for their flanges, thus permitting wheels on either rail to cross the other.

Frog angle. The angle formed by the intersecting gage lines of a frog.

Gage line. Line $\frac{5}{8}$ in. below the top of the center line of the head of running rail along that side which is nearer the center of the track.

Gage plate. Metal plate, extending from rail to rail, used to maintain gage of track.

Gage rod. Device for holding track to correct gage, generally consisting of $1\frac{1}{4}$ in. rod with a forged jaw on one end and a malleable jaw on the other end, adjustable through a locknut.

Gagging. The work done on a rail at the straightening press with a steel "gag" or tool for the purpose of taking out a bend.

Grade (degree of). As used in connection with railway line, the rise or fall in a track expressed as a ratio to 100 ft of horizontal track.

Grooved tie. Cross tie that has had grooves machined across its top depressions into which ribs on the bottom of the time plate may fit.

Guard brace. Metal shape designed to fit the contour of the side of the guard rail and extend over the tie with provision for fastening thereto; the moving or tilting of the guard rail away from the running rail.

Guard check gage. The distance between guard line and gage line, measured across the track at right angles to the gage lines.

Guard clamp. Device consisting of a yoke and fastenings designed to engage the running rail and the guard rail and hold them in correct relation to each other.

Guard face gage. The distance between guard lines, measured across the track at right angles to the gage lines.

Guard line. Line along the side of the flangeway that is nearer the center of the track and at the same elevation as the gage line.

Guard rail. Rail or other structure laid parallel with the running rails of a track to prevent wheels from being derailed, or to hold wheels in correct alignment to prevent their flanges from striking the points of turnout, crossing frogs, or the points of switches.

Guard timber. Longitudinal timber placed outside the track rail to maintain the spacing of ties.

Incline. Inclined track or tracks and their supporting structure.

Industrial track. Switching track serving industries such as mines, mills, smelters, and factories.

Insulated rail joint. Joint where electrical insulation is provided between adjoining rails.

Interchange track. Track where cars are delivered or received, as between railways.

Joint bar. Steel bar, commonly used in pairs, for joining rail ends in railroad track.

Joint gap. Distance between the ends of contiguous rails in track, measured at a point $\frac{5}{8}$ in. below the top of the rail on the outside of the head.

Joint tie. Crosstie used under a rail joint.

Lead curve. Curve in a turnout, interposed between the switch and the frog.

Lead track. Extended track connecting either end of a yard with the main track.

Level. Condition of the track where the elevation of the two rails transversely is the same.

Line. Condition of the track with regard to uniformity in direction over short distances on tangents or uniformity of curvature over short distances on curves.

Line rail. Rail on which alignment is based.

Lining bar. Straight steel bar designed to be used for shifting and moving track and rails.

Location (trackwork). The established position of the center line and grade line of a railroad preparatory to its construction.

Main track. Track extending through yards and between stations, where trains are operated by timetable or train order, or both, or by the use of block signals.

Master track scale. Track scale especially designed for the calibration of railway test weight cars or for other special weighing where extreme accuracy is required.

Mate. Track structure having a fixed or immovable point used as a companion piece to a tongue switch on the opposite side of the track. A mate is termed "outside" or "inside" depending upon whether it is placed on the outside or inside of the curve; the "inside mate" is comparatively little used.

Middle ordinate. On curves, the distance from the gage line of a rail to the center of the curve to the midpoint of a string drawn taut and held in contact with the gage line of a rail at the ends of the curve. It forms a convenient means of measuring curvature.

Movable centerpoint. One of the two movable tapered rails of a movable point crossing or slip switch.

Movable point frog. Frog equipped with points which are movable in the same manner as the points on a switch.

Pad. Cushioning device located between the rail and the tie, usually employed where rails are installed on bridges.

Penetration (wood treatment). The depth to which preservative enters ties through both lateral and end surfaces.

Pipe carrier. Device to guide and support signal pipe across tracks.

Plate. Steel plate interposed between the rail, or other track structure, and the tie.

Plug. Rectangular pieces of wood, shaped like steel spikes, for driving into holes from which spikes have been withdrawn.

Private siding (track). Sidetrack, owned or leased by an industry, where cars are placed for loading or unloading.

Radius of curvature. A measure of the severity of a curve in a track structure based on the length of the radius of a circle that would be formed if the curve were continued. Freight cars designed to Association of American Railroads standards must negotiate curves of stated minimum radii without wheel or truck interference with brake rigging or structural underframe members.

Rail. Rolled steel shape, commonly a T-section, designed to be laid end-to-end in two parallel lines on cross ties or other suitable supports to form a track for railway rolling stock.

Rail anchor. Device attached to the base of a rail bearing against a crosstie to prevent the rail from moving longitudinally under traffic.

Rail bond. Short metal cable attached to adjacent rails at the joints to insure proper electrical continuity across the joint.

Rail brace. Bracing device for holding the rail in place, used at switches, movable point frogs, etc., in combination with switch, tie, or gage plates.

Rail clip. Device bolted or clamped to a rail, for supporting and guiding a detector bar.

Rail creep. Longitudinal sliding of rails in track under traffic or because of temperature changes.

Railroad tie. The transverse member of the track structure to which the rails are spiked or otherwise fastened to provide proper gage and to cushion, distribute, and transmit the stresses of traffic through the ballast to the roadbed.

Rail saw. A power machine provided with a saw of either tooth or friction type, used to cut steel rails.

Reinforcing rail. Bent rail placed with its head along the outside of and close to the head of a knuckle rail to strengthen it and to act as an easer rail; or a piece of rail similarly applied to a movable center point.

Roadbed. The foundation for the rails and ties of a railroad track.

Running rail. The rail or surface on which the tread of the wheel bears.

Running track. Track reserved for movement through a yard.

Screw spike. Cylindrical threaded spike, designed to be turned with a special wrench into holes bored in ties, to secure rails or to act as a tie plate holder in tie plates.

Self-guarded frog (flange frog). Frog provided with guides or flanges, above its running surface, that contact the tread rims of wheels to guide their flanges safely past the frog.

Separator guard. Metal block of two or more parts acting as a filler between the running rail and the guard rail and designed to provide varying widths of flangeways.

Shoulder (track). Portion of the ballast between the end of the tie and the toe of the ballast slope.

Side track. Track auxiliary to the main track.

Skate. Sliding device placed on a rail to engage with a car wheel so as to provide continuous braking by sliding friction.

Skate machine. Electrically controlled and electrically or pneumatically operated mechanism for placing a skate on, or removing it from, the rail.

Skeletonized track. Track where the ballast is removed from the cribs between the ties.

Special trackwork. All rails, track structures, and fittings, other than plain unguarded track, that is neither curved nor fabricated before laying.

Spike. Long steel square nail with a cutting edge, used to secure rail in place.

Spike puller. Tool with a claw end and two or three pairs of knobs on a straight bar, used to withdraw spikes.

Spiral. When used with respect to track, a form of easement curve where the change of degree of curve is uniform throughout its length.

Splice bar. A steel bar used to fasten together the ends of rails. Used in pairs, one on each side of the rail, and designed to fit the space between head and base closely. They are held in place by track bolts and suitable accessory equipment. Also called "joint bars," or "fishplates."

Splice drilling. The spacing of holes in the ends of rails or other track structures to receive the bolts for the fastening of joint bars.

Splice plate. Steel plate, usually of some specified dimension, spanning a break or a joint in some structural member and securely fastened (usually welded) to both parts so as to form a continuous member.

Split switch. Track structure consisting of two movable point rails and necessary fixtures, used to divert rolling stock from one track to another.

Split web. Longitudinal or diagonal transverse crack in the web of a rail.

Spot board (trackwork). Sighting board placed above and across the track at the proposed height to indicate the new surface and ensure its uniformity.

Spreader. Piece of maintenance-of-way equipment having plow-type blades for distributing ballast on the roadbed.

Spring washer. Member designed to prevent backward movement of the nut and looseness in the bolted members of a rail joint, due to wear, stretch, rust, or other deterioration.

Spur track. As distinguished from a side track, a spur track is of indefinite length, extending out from main line.

Standard gage. The standard distance between rails of North American railroads, 4 ft $8\frac{1}{2}$ in. measured between the inside faces of the rail heads.

Standard level. A term referring to the deck height of many intermodal and automobile rack flatcars, to differentiate them from so-called low level cars. Standard level cars have a deck height of $41\frac{1}{2}$ in.; low level cars have a $31\frac{1}{2}$ in. deck height.

Standing yard capacity. The sum of the capacities of all the track in a yard on which cars may be permitted to stand.

Stock rail. The rail against which the point of a switch, derail, or movable point frog rests.

Storage track. Track where cars are placed when held awaiting disposition or when not in service.

Stub track. Form of side track connected to a running track at one end only and usually protected at the end by some form of bumping post or other solid obstruction.

Subballast. Any material that is spread on the finished subgrade of the roadbed below the top-ballast to provide better drainage, prevent upheaval by frost, and better distribute the load over the roadbed.

Subgrade. Finished surface of the roadbed below the ballast and track.

Surface. The condition of the track as to vertical evenness or smoothness.

Switch heel. End of a switch rail that is farther from one point and nearer the frog.

Switch plate. Special metal tie plate for use on switch ties.

Switch point. Movable tapered track rail, the point of which is designed to fit against the stock rail.

Tamper. Power-driven machine for compacting ballast under ties.

Tamping bar. Steel bar with a blade on each end, used to drive ballast beneath the ties.

Tangent. Any straight portion of a railway alignment. Tangent track means a straight track with no curves.

Team track. Track where rail cars are placed for the use of the public in loading or unloading freight.

Templet or template. Gage for checking the shape and size of a rail section.

Throat. The point just ahead of the frog point where the converging wings of a frog are closest together.

Tipple. Track beneath or beside conveyor belts or bins that load ore, rock, coal, or other material into open-top cars.

Top ballast. Any material spread over a subballast to support the track structure, distribute the load to the subballast, and provide a good initial drainage.

Track. Assembly of rails, ties, and fastenings over which cars, locomotives, and trains are moved.

Track bolt. Bolt with a button head, oval or elliptical neck, and a threaded nut, designed to fasten together rails and joint bars.

Track brace. Fastening designed to brace the rail on the high side of a curve.

Track capacity. The number of cars that can stand in the clear on any given track.

Track centers. The distance, measured at right angles, between center lines of parallel tracks.

Track chart. Map-like representation of the grade and alignment of a section of a railroad.

Track circuit. Electrical circuit of which the rails of the track form a part.

Track circuit connector. Device used for connecting one or more wires to a rail.

Track crossing. Structure consisting of four connected frogs that permits one track to cross another at grade.

Track gage. Device that establishes or measures the gage of a track.

Track layout. Diagram of the physical location of tracks in a yard or terminal.

Track scale. Scale especially designed for weighing railway equipment.

Track shims. Flat wood boards of length and width similar to tie plates. They are placed between the ties and the tie plates to correct surface irregularities when the ballast is frozen.

Undercutter. Device used to remove ballast from under tracks.

Vertical split head. Rail defect appearing as a split along or near the middle of the head of a rail and extending into or through the head.

Welded rail. Two or more rails welded together at their ends to form a length less than 400 ft. See continuous-welded rail (CWR).

Wye. Track arrangement shaped like the letter "Y," but with a connecting segment between the two upper legs. This track layout is often used in small yards and at some rip tracks to enable equipment to be turned without a turntable.

SOILS: TREATMENT AND LANDSCAPING

Acid soil

1. Soil that is acid throughout most of all of its parts that plant roots occupy. Term is commonly applied to only the surface plowed layer or to some other specific layer or horizon of a soil. Soil having a preponderance of hydrogen over hydroxyl ions in the soil solution.

2. Soil with a large quantity of hydrogen and aluminum ions in proportion to hydroxyl ions. Term is usually applied to the surface layer or to the root zone.

Additive. Material added to fertilizer to improve its chemical or physical condition. An additive to liquid fertilizer might prevent crystals from forming in the liquid at temperatures at which crystallization would normally take place.

Adjuvant. Subsidiary or modifying ingredient, usually of a chemical mixture.

Adsorption complex. Group of substances in soil capable of adsorbing other materials. Colloidal particles account for most of this adsorption.

Agric horizon. Horizon immediately below the plow layer of cultivated soils containing accumulated clay and humus to the extent of at least 15% of the horizon volume.

Agricultural lime. Soil amendment consisting principally of calcium carbonate, but including magnesium carbonate and other materials and used to furnish calcium and magnesium as elements for the growth of plants and to neutralize soil acidity.

Agronomy. Specialization of agriculture concerned with the theory and practice of field crop production and soil management. Scientific management of land.

Alkaline soil

1. Precisely, any soil that has a pH value greater than 7. Practically, soil with a pH of greater than 7.3. Term is usually applied to surface layer or root zone and may be used to characterize a horizon or sample thereof.

2. Soil that is alkaline throughout most of all of its parts occupied by plant roots. Term is commonly applied to only a specific layer or horizon of a soil.

Alkali soil. Soil with a high degree of alkalinity (pH of 8.5 or higher) or more with a high exchangeable sodium content (15% or more of the exchange capacity), or both. Soil that contains sufficient alkali (sodium) to interfere with the growth of most crop plants.

Alkalization. Process whereby the exchangeable sodium content of a soil is increased.

All-aged. Applied to a stand in which, theoretically, trees of all ages up to and including those of the felling age are found.

Aluminosilicates. Compounds containing aluminum, silicon, and oxygen as main constituents, for example microcline.

Amendment. Material, such as lime, gypsum, sawdust, or synthetic conditioners, that is worked into the soil to make it more productive. Strictly, a fertilizer is also an amendment, but the term "amendment" is used most commonly for added materials other than fertilizer.

Ammonia. Colorless gas composed of one atom of nitrogen and three atoms of hydrogen. Ammonia liquefied under pressure is used as a fertilizer.

Ammonium nitrate (33.5-0-0). Mineral fertilizer containing 33.5% nitrogen.

Ammonium phosphate (16-20-0). Mineral fertilizer containing 16% nitrogen and 20% phosphorous.

Ammonium sulfate (21-0-0). Mineral fertilizer containing 21% nitrogen.

Annual. Lasting only one year or one growing season; refers to a plant that completes its growth in a single year.

Anthropic epipedon. Thick, dark, surface horizon, that is more than 50% saturated with bases and has a narrow C/N ratio formed under long-continued cultivation where large amounts of organic matter and fertilizers have been added.

Ap. Surface layer of soil disturbed by cultivation or pasturing.

Approved sod. Sod grown from certified seed of high quality and known origin and inspected by a certification agency for overall high quality, freedom from noxious weeds, and excessive amounts of other crop or weedy plants at time of harvest. Approved sod may consist of either a single variety or a mixture of varieties or species, provided that all seed in the mixture is certified. Sod must meet published standards to qualify for approval.

Arid climate. Dry climate like that of desert or semidesert regions. Limits of precipitation vary widely according to temperature, with an upper limit for cool regions of less than 10 in. and for tropical regions of as much as 20 in.

Aridisols. Soils characteristic of dry places, including soils such as desert, red desert, sierozem and solonchak.

Arid region. Areas where the potential water losses by evaporation and transpiration are greater than the amount of water supplied by precipitation. In the United States this area is broadly considered to be the dry parts of the 17 western states.

Artificial manure. Commercial fertilizers.

Association. Assemblage of plant, usually over a wide area, that has one or more dominant species from which it derives a definite aspect.

Available nutrient

1. That portion of any element or compound in the soil that can be readily assimilated by growing plants.

2. Nutrient in the soil that can be readily absorbed by growing plants. Nutrient in soluble or exchangeable form.

Available water. Portion of water in a soil that can be readily absorbed by plant roots. Considered to be that water held in the soil against a pressure of up to approximately 15 bars.

Backfill. Soil put back into the excavation around the plant. Backfill usually contains specified additives.

Balled and burlapped (B&B). Term describing field-grown plants tied in a burlap cover when removed from the nursery or original planting area.

Basin irrigation. Application of irrigation water to level areas that are surrounded by border ridges or levees. Usually irrigation water is applied at rates greater than the water intake rate of the soil.

Bed. Defined area for plant life, other than a lawn.

Bedding (soil). Arranging the surface of fields by plowing and grading into a series of elevated beds separated by shallow depressions or ditches for drainage.

Bedrock. Solid underlying soil, rock, and other earthly surface formations.

Bench terrace. Embankment constructed across sloping soils with a steep drop on the downslope side.

Blown-out land. Areas from which all or almost all of the soil and soil material has been removed by wind erosion, usually unfit for crop production; miscellaneous land type.

Bog soil

1. Intrazonal group of soils with muck or peaty surface soils underlaid by peat. Bog soils usually have swamp or marsh vegetation and are common in humid regions.

2. Great soil group of the intrazonal order and hydromorphic suborder. Includes muck and peat.

Bole. Lump of earth formed of any of several varieties of friable clay, usually colored red by iron oxide.

Buffer strips. Established strips of perennial grass or other erosion-resisting vegetation, usually on the contour in cultivated fields, to reduce runoff and erosion.

Bur. Any rough or prickly envelope of a seed or fruit.

Calcareous soil. Soil containing sufficient calcium carbonate, often with magnesium carbonate, to effervesce visibly when treated with cold hydrochloric acid.

Caliche

1. Broad term for the more or less cemented deposits of calcium carbonate in many soils of warm-temperature areas, like the southwestern states. When it is very near the surface or exposed by erosion, the material hardens.

2. Soft earthy limestone.

Certified sod. Turf grass sod of superior quality grown from high-quality certified seed of known origins or from plantings of certified grass seedlings or stolons. Certification agency inspects for assurance of satisfactory genetic identity and purity, overall high quality, and freedom from noxious weeds or crop plants at time of harvest; sod must meet published standards to qualify for certification.

Chaparral. Low, dense scrub vegetation, principally drought-resistant shrubs and bushes, characteristic of regions having a subtropical dry summer climate. In some places it is almost impenetrable, consisting of thickets of stiff or theory shrubs or dwarf trees. Term is also applied to the part of California that has the climatic conditions in which such vegetation flourishes.

Chelate. Type of chemical compound in which a metallic atom is firmly combined with a molecule by multiple chemical bonds.

Chemical fixation. Process or processes in a soil by which certain chemical elements essential for plant growth are converted from a soluble or exchangeable form to a much less soluble or nonexchangeable form. Contrast with nitrogen fixation.

Chisel. Tillage machine with one or more soil-penetrating points that can be drawn through the soil to loosen the subsoil, usually to a depth of 12 to 18 in.

Chlordane. Insecticide, nonharmful to plants.

Chlorosis

1. Condition in plants relating to the failure of chlorophyll (the green coloring matter) to develop. Chlorotic leaves range from light green through yellow to almost white.
2. Abnormal yellowish or whitish leaf color signifying a nutritional or other physiological disorder in plants.

Clay. Mineral soil particles less than 0.002 mm in diameter. As a soil textural class, soil material that contains 40% or more clay, less than 45% sand, and less than 40% silt.

Clay loam. Soil material that contains 27 to 40% clay and 20 to 45% sand.

Claypan. Compact, slowly permeable soil horizon rich in clay and separated more or less abruptly from the overlying soil. Claypans are commonly hard when dry and plastic or stiff when wet.

Clearing. Removal of trees, shrubs, bushes, and other organic matter found at or above original ground level.

Clod. Compact, coherent mass of soil ranging in size from 5 to 10 mm to as much as 8 or 10 in.; produced artificially, usually by the activity of plowing, digging, etc., especially when these operations are performed on soils that are either too wet or too dry for normal tillage operations.

Commercial raw bone meal. Bone meal that is finely ground and has a minimum analysis of 1% nitrogen and 18% phosphoric acid.

Common name. Plant name used by the public as distinguished from the botanical scientific name.

Compost. Mass of rotted organic matter made from waste plant residues. In organic fertilizers, nitrogen and a little soil are usually added to it. Organic residues usually are piled in layers, to which the fertilizers are added. Layers are separated by thin layers of soil, and the whole pile is kept moist and allowed to decompose. Pile is usually turned once or twice. Principal purpose in making compost is to permit the organic materials to become crumbly and to reduce the carbon-nitrogen ratio of the material.

Conifer. Tree or shrub belonging to the order Coniferales. Most conifers are needle leaf plants, but some have scalelike leaves ranging from small, as in many junipers and cedars, to more than an inch wide, as in the monkeypuzzle tree. Most species of conifers are evergreen, but the larches and some others are deciduous.

Consumptive use. Use of water by plants in transpiration and growth, plus water vapor loss from adjacent soil or snow or from intercepted precipitation in any specified time. Usually expressed as equivalent depth of free water per unit of time.

Contour. Imaginary line connecting points of equal elevation on the surface of the soil. Contour terrace is laid out on a sloping soil at right angles to the direction of the slope and level throughout its course. In contour plowing, the plowman keeps to a level line at right angles to the direction of the slope, which usually results in a curving furrow.

Crust. Thin, brittle layer of hard soil that forms on the surface of many soils when they are dry; exposed hard layer of materials cemented by calcium carbonate, gypsum, or other binding agents. Most desert crusts are formed by the exposure of such layers through removal of the upper soil by wind or running water and their subsequent hardening.

Cultipacking. Method by which a soil is firmly packed after plowing and seeding into V-shaped rows.

Cultivate. Prepare and use the land to raise crops by plowing it, planting seed, and taking care of the growing plants; to loosen the ground around growing plants to kill weeds and permit water penetration.

Damping-off. Sudden wilting and death of seedling resulting from attack by microorganisms.

DDT (dichloro-diphenyl-trichloroethane). Insecticide, not harmful to plants.

Dealkalization. Removal of exchangeable sodium (or alkali) from the soil usually by chemical treatment and leaching.

Deciduous. Losing leaves during a certain period of the year, generally either the cold or dry season.

Deep percolation. Downward movement of water beyond the reach of plant roots.

Deep soil. Soil deeper than 40 in. to rock or other strongly contrasting material. Also, a soil with a deep black surface layer; a soil deeper than about 40 in. to the parent material or to other unconsolidated rock material not modified by soil-forming processes; a soil in which the total depth of unconsolidated material, whether true soil or not, is 40 in. or more.

Deflocculate (soil conditioning). Separate or break up soil aggregates into the individual particles; to disperse the particles of a granulated clay to form a clay that runs together or puddles.

Defoliant. Chemical used to remove prematurely the leaves from plants.

Defoliate. Cause a tree or other plant to lose its leaves.

Denitrification. Biochemical reduction of nitrate or nitrite to gaseous nitrogen either as molecular nitrogen or as an oxide of nitrogen.

Desalinization. Removal of salts from saline soil, usually by leaching.

Desert soil. Zonal group of soils that have light-colored surface soils and usually are underlaid by calcaeous material and frequently by hard layers. Developed under extremely scanty scrub vegetation in warm to cool arid climates.

Drainage. Removal of excess surface water or excess water from within the soil by means of surface or subsurface drains.

Drain pipe. Porous concrete pipe, corrugated plastic tubing, or clay tile pipe used to conduct water from the soil.

Drought. Period of dryness of long duration. Usually considered to be any period of soil moisture deficiency within the plant root zone. Period of dryness of sufficient length to deplete soil moisture to the extent that plant growth is seriously retarded.

Drumlin. Smooth, streamlined hill composed of till.

Dryland farming. Practice of crop production in low-rainfall areas without irrigation.

Dune. Mount or ridge of loose sand piled up by the wind. Occasionally, during periods of extreme drought, granulated soil material of fine texture may be piled into low dunes.

Dust mulch. Loose, finely granular, or powdery condition on the surface of the soil, usually produced by shallow cultivation.

Edaphology. Science that deals with the influence of soils on living things, particularly plants, including man's use of land for plant growth.

Effective precipitation. Portion of the total precipitation that becomes available for such processes as plant growth and soil development.

Effluent. Outflowing of water from a subterranean storage space.

Eluviation. Removal of soil material in suspension from a layer or layers of a soil. Usually, the loss of material in solution is described by the term "leaching."

Environment. All external conditions that may act on an organism or soil to influence its development, including sunlight, temperature, moisture, and other organisms.

Enzymes. Substances produced by living cells that can bring about or speed up chemical reaction. They are organic catalysts.

Erosion. Wearing away of the land surface by detachment and transport of soil and rock materials through the action of moving water, wind, or other geological agents.

Erosion control. Any one of a number of different methods of preventing the washing away of soil (erosion) by excessive runoff or wind. Erosion control may consist of plantings to form a ground cover by rubble or crushed stone; may include terracing or any other method that will accomplish the desired results.

Eutrophic. Term applied to water that has a concentration of nutrients that is optimal, or nearly so, for plant or animal growth.

Evapotranspiration. Combined loss of water from a given area, and during a specified period of time, by evaporation from the soil surface and by transpiration from plants.

Evergreen. Any of the various needled and broad-leafed trees, shrubs, and herbs that retain their green foliage throughout the year and do not shed this foliage until new leaves have developed.

Exchangeable-cation percentage. Extent to which the adsorption complex of a soil is occupied by a particular cation.

Exchangeable-sodium percentage. Extent to which the adsorption complex of a soil is occupied by sodium.

Fallow

1. Cropland left idle in order to restore productivity, mainly through accumulation of water, nutrients, or both. Summer fallow is a common stage before cereal grain is planted in regions of limited rainfall. Soil is kept free of weeds and other vegetation, thereby conserving nutrients and water for the next year's crop.

2. Remove land from active crop production, commonly to allow its natural regeneration.

Fertilize

1. Enrich; to make (soil) fertile, for example, through the application of natural or manufactured commercial products.

2. Apply compost, manure, or commercial fertilizer to (a growing medium) in order to supply nutriments or make available nutriments already present.

Fertilizer

1. Complete product, part of the elements of which are derived from organic sources. Percentages by weight determined by the soil analysis furnished by the distributor.

2. Commercial product, uniform in composition, free flowing and suitable for application with approved equipment, delivered to the site in bags or other convenient containers, each fully labeled, and bearing the name, trade name, or trademark and warranty of the producer.

Fertilizer analysis. Expression that indicates the percentage of plant nutrients in a fertilizer. Thus a 10–20–10 grade contains 10% nitrogen, 20% phosphoric oxide, and 10% potash.

Fertilizer grade. Guaranteed minimum analysis, in percent, of the major plant nutrient elements contained in a fertilizer material or in a mixed fertilizer.

Fertilizer requirement. Quantity of certain plant nutrient elements needed, in addition to the amount supplied by the soil, to increase plant growth to a designated optimum.

Fertilizers. Substances added to soils to provide elements that are essential to plant growth, but that are not present in adequate quantities in the natural soil. Such substances may be manure or commercial fertilizers.

Field capacity

1. Percentage of water remaining in a soil two or three days after having been saturated and after free drainage has practically ceased.

2. Water remaining in a field soil that has been thoroughly wetted and drained.

Field sod. Sod obtained from grazing areas of pastures and meadows.

Fine-textured soil

1. Soil consisting of or containing large quantities of the fine fractions, particularly of silt and clay. (Includes all clay loams and clays; that is, clay loam, sandy clay loam, silty clay loam, sandy clay, silty clay, and clay textural classes.)

2. Texture exhibited by soils having clay as a part of their textural class name.

Fixation. Process or processes in a soil by which certain chemical elements essential for plant growth are converted from a soluble or exchangeable form to a much less soluble or to a nonexchangeable form; for example, phosphate "fixation," in contrast with nitrogen fixation.

Fixed phosphorus. Phosphorus that has been changed to a less-soluble form as a result of reaction with the soil; moderately available phosphorus.

Flood irrigation. Irrigation by running water over nearly level soil in a shallow flood.

Flood plain

1. Land bordering a stream, built up of sediments from overflow of the stream, and subject to inundation when the stream is at flood stage.

2. Flat or nearly flat land on the floor of a river valley that is covered by water during floods.

Flora. Vegetation of a region, zone, or environment. A list, catalog, or systematic report with keys or descriptions pertaining to the plants of a specific region, hence alpine flora, bog flora.

Foliar fertilization. Fertilization of plants by applying chemical fertilizers to their foliage.

Forb. Nongrasslike herbaceous plant.

Forest cover. All trees and other woody plants in a forest.

Forest land. Land bearing a stand of trees of any age or stature, including seedlings, and of species attaining a minimum of 6-ft average height at maturity; land from which such a stand has been removed, but which is not being used for other purposes. Term is commonly limited to land not on farms; forests on farms are commonly called "woodland" or "farm forests."

Friable

1. Easily crumbled.
2. Pertaining to the ease of crumbling of soils.

Fungi. Forms of plant life lacking chlorophyll and unable to make their own food.

Fungicide. Material that kills or checks growth of fungi.

Furrow

1. Water applied to row crops in ditches made by farm implements.
2. Channel worked into the surface of the soil by an implement such as plow or hoe.

Garden. Plot of land cultivated for the growth of selected ornamental planting.

Germinate. Start growing or developing.

Girdle. Encircle the stem of a living tree with cuts that completely sever bark and cambium and often are carried well into the outer sapwood, for the purpose of killing the tree by preventing passage of nutrients.

Granular fertilizer. Fertilizer composed of particles of roughly the same composition, about $\frac{1}{10}$ in. in diameter. This type of fertilizer contrasts with the normally fine or powdery fertilizer.

Granular structure. Soil structure in which the individual grains are grouped into spherical aggregates with indistinct sides. Highly porous granules are commonly called "crumbs." Well-granulated soil has the best structure for most ordinary crop plants.

Grass seed. Fresh, clean, new crop seed composed of the varieties mixed in the proportions by weight recommended for the locality and tested for purity and germination.

Green manure. Plant material incorporated with the soil while green, or soon after maturity, for improving the soil.

Ground cover. All herbaceous plants and low-growing shrubs on a specific area and the organic materials in various stages of decay.

Groundwater. Water that fills all the unblocked pores of underlying material below the water table, which is the upper limit of saturation.

Grubbing. Removal of stumps, roots, boards, logs, and other organic matter found at or below original ground level.

Gully. Channel resulting from water erosion that is deep enough to interfere with and not be obliterated by normal tillage operations.

Gypsiferous soil. Soil containing free gypsum.

Halophytic vegetation. Salt-loving or salt-tolerant vegetation, usually having fleshy leaves or thorns and resembling desert vegetation.

Hardening off. Roots that have reached the container edge and have become calloused.

Hardpan. Hardened or cement soil layer. Soil material may be sandy or clayey and may be cemented by iron oxide, silica, calcium carbonate, and other substances.

Hardwood. One of the botanical group of trees that have broad leaves, in contrast to the conifers.

Header. Material used to define the outline of a plant bed or to separate adjacent ground cover areas.

Headers and stakes. Usually made of construction-grade heart redwood, used to contain other materials from mixing with the growing areas.

Hedge. Plantings of bushes or woody plants in a row, usually to form a fence, divider, or windbreak.

Hedgerow. Linear thicket of bushes, commonly with some trees, left between two fields of cleared land or planted in order to separate fields.

Herb. Flowering plant without a persistent or woody stem above the ground. Herbs include all small seed plants, including grasses.

Humid climate. Climate with enough precipitation to support a forest vegetation, although there are exceptions where the plant cover includes no trees, as in high mountains. Precipitation may have a lower limit of as little as 15 in. in cool regions and as much as 60 in. in hot regions and an effectiveness index range of between 64 and 128. Climate having a high average relative humidity.

Humification. Processes involved in the decomposition of organic matter and leading to the formation of humus.

Humus. Semistable fraction of the soil organic matter remaining after the major portion of added plant and animal residues have decomposed. Usually it is dark colored.

Hydrophyte. Plant that typically grows in water or in saturated soil. Hydrophytes may be rooted or free floating, submerged, with floating leaves, or with leaves emergent above the water level.

Hydrous. Containing water.

Imbibition. Process by which plants absorb water from the soil.

Immature soil. Soil with indistinct or only slightly developed horizons because of the relatively short time it has been subjected to the various soil-forming processes. Soil that has not reached equilibrium with its environment.

Immobilization. Conversion of an element from the inorganic to the organic form in microbial tissues or in plant tissues, thus rendering the element not readily available to other organisms or to plants.

Impervious. Resistant to penetration by fluids or roots.

Impervious soil. Soil through which water, air, or roots penetrate slowly or not at all. No soil is absolutely impervious to water and air all the time.

Infiltration. Downward entry and absorption of water into the soil.

Inorganic

1. Substances occurring as minerals in nature or obtainable from them by chemical means. Refers to all matter except the compounds of carbon, but includes carbonates.
2. Not organic; designating or composed of matter other than animal or vegetable; occurring as minerals in nature or obtainable as minerals by chemical means. Not forming or not characteristic of an organism; designating matter other than compounds of carbon (but including carbonates).

Insecticide. Material or substance used to kill insects.

Irrigation

1. Natural, artificial, or mechanical watering of land.
2. Artificial application of water to the soil for the benefit of growing crops.

3. Controlled application of water to arable lands to supply water requirements not satisfied by rainfall.

Irrigation efficiency

1. Ratio of the water actually consumed by crops on an irrigated area to the amount of water diverted from the source onto the area.

2. Ratio of water actually consumed by crops on an irrigated area to total amount of water applied to the area.

Irrigation method. Manner in which water is artificially applied to an area. Methods and the manner of applying the water vary depending on water source.

Jungle. Dense, tangled second-growth vegetation of grasses, shrubs, small trees, and vines, generally associated with equatorial areas.

Land. Broader term than soil. In addition to soil, its attributes include other physical conditions such as mineral deposits and water supply; location in relation to centers of commerce, populations, and other land; the size of the individual tracts or holdings; and existing plant cover, works of improvement, and the like.

Landscape. Sum total of the characteristics that distinguish a certain kind of area on the earth's surface and give it a distinguishing pattern in contrast to other kinds of areas. Any one kind of soil is said to have a characteristic natural landscape, and under different uses it has one or more characteristic cultural landscapes.

Lawn. Land covered with grass or other ground cover or plant life.

Lawn soil. Friable soil of the region, free from lumps, toxic substances, debris, vegetation, and stones more than 1 in. in diameter and containing no caliche, high salts, or heavy alkali.

Lawn top mulch. Well-composted, screened, steer manure, uniform in composition, reasonably free from shavings, sawdust, or refuse, and not containing any harmful materials or more than 10% straw.

Leaching. Removal of materials in solution by the passage of water through soil.

Legume. Cloverlike plant having nitrogen nodule on its root, grown for forage or nitrogen enrichment soil.

Leveling (of land). Reshaping or modification of the land surface to a planned grade to provide a more suitable surface for the efficient application of irrigation water and to provide good surface drainage.

Liana. Woody climbing plant with ground roots.

Lichen. Any plant of the class Lichenes varying in size, form, and color, but always having a compound structure consisting of an alga and fungus. Lichen is an air plant that lacks roots, stems, branches, leaves, and flowers, reproduces by spores or through fragmentation, and is usually found attached to rocks, soil, wood, or bark.

Light soil. Coarse-textured soil; with a low drawbar pull and hence easy to cultivate.

Lime

1. Term applied to ground limestone (calcium carbonate) and hydrated lime (calcium hydroxide), which are used as amendments to reduce the acidity of acid soils.

2. In chemical terms, calcium oxide. In practical terms, material containing the carbonates, oxides, and/or hydroxides of calcium and/or magnesium used to neutralize soil acidity.

Lime requirement. Mass of agricultural limestone, or the equivalent of other specified liming material, required per acre to a soil depth of 6 in. (or on two million pounds of soil) to raise the pH of the soil to a desired value under field conditions.

Lister. Double moldboard plow that throws a furrow slice in opposite directions.

Loam. Mixture of sand, silt, or clay, or a combination of any of these, with organic matter. Sometimes called topsoil in contrast to the subsoils that contain little or no organic matter.

Loamy sand. The U.S. Department of Agriculture textural class name for soil containing more than 70% sand and less than 15% clay, less than 85% sand and 0% clay, and more than 10% clay and 0% silt.

Macronutrient. Chemical element necessary in large amounts (usually is greater than 1 ppm in the plant) for the growth of plants and usually applied artificially in fertilizer or liming materials. "Macro" refers to quantity and not to the essentiality of the element.

Maintenance. Includes all watering, weeding, cultivation, spraying, growing, and trimming necessary to establish the planted areas in a healthy growing condition, as well as work necessary to keep the areas neat, edged, and attractive.

Mangrove. One of a group of halophytic evergreen broadleaf trees and shrubs of tropical and subtropical coasts, typically growing in muddy areas, such as lagoons and estuaries, that are submerged perennially or at high tide by brackish or salty water. Many species of mangrove have prop roots or root knees above the ground.

Manure

1. Animal excreta used in manufactured organic fertilizers.

2. Dehydrated manure.

Marl. Soft and unconsolidated calcium carbonate, usually mixed with varying amounts of clay or other impurities.

Mature soil. Soil with well-developed soil horizons produced by the natural processes of soil formation and essentially in equilibrium with its present environment.

Medium-texture. Term applied to soil whose texture is intermediate between that of fine-textured and coarse-textured soils. Includes the following textural classes: very fine sandy loam, loam, silt loam, and silt.

Membrane. Usually recommended, 6-mil black polyethelene film, laid with at least 3-in. overlap at each joint, to prevent virgin soil growth from penetrating new soil fill.

Micronutrient. Chemical element necessary in only extremely small amounts (less than 1 ppm in the plant) for the growth of plants. "Micro" refers to quantity and not to the essentiality of the element.

Mildew. Any fungus growth that may be unsightly but not usually accompanied by severe degradation of the substrate (material) on which it is growing.

Mineral fertilizer. Fertilizer manufactured from chemicals, as opposed to those made from organic fertilizer.

Mineral soil. Soil consisting predominantly of, and having its properties determined predominantly by, mineral matter. Usually contains greater than 20% organic matter, but may contain an organic surface layer up to 30 cm thick.

Mites. Minute sucking insects.

Mixed forest. Forest composed of trees of two or more species. In practice, usually a forest in which at least 20% of the trees are of other than the principal species.

Mor. Raw humus; type of forest humus layer of unincorporated organic material, usually matted or compacted or both, distinct from the mineral soil, unless the latter has been blackened by washing in organic matter.

Muck

1. Highly decomposed organic material in which the original plant parts are not recognizable. Contains more mineral matter and is usually darker in color than peat.

2. Organic soil of very soft consistency.

Muck soil

1. Organic soil in which the organic matter is well decomposed.

2. Soil composed of thoroughly decomposed black organic material, with a considerable amount of mineral soil material, finely divided, and with few fibrous remains.

Mud. Mixture of soil and water in a fluid or weakly solid state.

Mulch. Natural or artificially applied layer of plant residues or other materials on the surface of the soil. Mulches are generally used to help conserve moisture, control temperature, prevent surface compaction or crusting, reduce runoff and erosion, improve soil structure, or control weeds. Common mulching materials include compost, sawdust, wood chips, and straw. Sometimes paper, fine brush, or small stones are used.

Mulch farming. System of farming in which the organic residues are not plowed into or otherwise mixed with the soil, but are left on the surface as a mulch.

Mull. Humus-rich layer of forested soils consisting of mixed organic and mineral matter. Mull blends into the upper mineral layers without an abrupt change in soil characteristics.

Muskeg. Terrain that is made up of a living organic mat of mosses, sedges, or grasses (with or without tree growth) underlaid by an extremely compressible mixture of partly disintegrated and decomposed organic material.

Nematocide. Substance that can be used to kill nematodes.

Nematodes. Very small worms abundant in many soils and important because many of them attack and destroy plant roots.

Nettle. Any of a genus of plants (Urtica), chiefly coarse herbs, armed with stinging hairs. Loosely, any prickly or stinging plants.

Neutral soil

1. Soil in which the surface layer, at least to normal plow depth, is neither acid nor alkaline in reaction. Practically, a soil in which the pH of the surface soil ranges between 6.6 and 7.4.

2. Soil that is neither significantly acid nor alkaline.

Nitrogen. One of the chemical elements regulating the ability of plants to make proteins vital to the formation of new protoplasm.

Normal soil. Soil having a profile in near equilibrium with its environment; soil developed under good, but not excessive, drainage from parent material of mixed mineral, physical, and chemical composition. In its characteristics it expresses the full effects of the forces of climate and living matter.

Nursery sod. Cultivated sod planted on agricultural land and grown specifically for sod purposes. Sod is carefully mowed and otherwise maintained from planting to harvest so as to maintain reasonable quality and uniformity.

Organic clay. Clay with a high organic content.

Organic fertilizer. Fertilizer composed of organic material.

Organic silt. Silt with a high organic content.

Organic soil

1. Soil that contains a high percentage (greater than 15 or 20%) of organic matter throughout the solum.

2. Term applied to a soil or soil horizon that consists primarily of organic matter, such as peat soils, muck soils, and peaty soil layers.

3. Soil with a high organic content. In general, organic soils are very compressible and have poor load-sustaining properties.

Ornamental plants. Plants grown for show and to satisfy the eye, rather than for food.

Palliative. Agent that serves to reduce an undesirable condition.

Park forest. Forest in which the trees stand apart from one another or in detached groups; very open forest in which the characteristic forest floor is usually replaced by grasses.

Pea gravel or washed river run rock. Clean, screened river rock, not crushed.

Peat

1. Unconsolidated soil material consisting largely of undecomposed organic matter accumulated under conditions of excessive moisture.

2. Fibrous mass of organic matter in various stages of decomposition, generally dark brown to black in color and of spongy consistency.

Peat moss. Commercial form of humus.

Perched water table. Upper water table that is separated from a lower water table by an impervious layer.

Perennial. Plants that live for more than two years.

Permanent wilting point. Moisture content at which the soil is incapable of maintaining succulent plants in an unwilted condition even though transpirational loss is negligibly low.

Permeability of soil. Quality of a soil horizon that enables water or air to move through it. Can be measured quantitatively in terms of rate of flow of water through a unit cross section in unit time under specified temperature and hydraulic conditions. Values for saturated soils usually are called "hydraulic conductivity." Permeability of a soil may be limited by the presence of one nearly impermeable horizon even though the others are permeable.

Pesticide. Substance for controlling insects, rodents, fungi, weeds, and other forms of plant or animal life that are considered to be pests.

pH. Symbol denoting the negative logarithm of the concentration of the hydrogen ion in gram atoms per liter, used in expressing both acidity and alkalinity. pH values run from 0 to 14, with 7 indicating neutrality, numbers less than 7 increasing acidity, and numbers greater than 7 increasing alkalinity.

Phosphate. Salt or ester of an acid containing phosphorous.

Phosphorous. One of the chemical elements vital to processes in all living cells.

Plant

1. Put in the ground and cover with soil to enable to grow.

2. Young tree, vine, shrub or herb planted or suitable for planting; vegetable, flower, fruit, or ornamental grown for or ready for transplanting.

Plant growth regulator. Chemical antiplant agent that regulates or inhibits plant growth.

Planting soil. Mixture of various ingredients to aid in plant growth.

Plant nutrient. Element taken in by a plant, essential to its growth, and used by it in elaboration of its food and tissue.

Plant pits. Holes dug to receive plants.

Plants. Trees, vines, shrubs, herbs, vegetable, flower or fruit. Sound, healthy, vigorous and free from insects, pests, plant diseases and injuries.

Plant yield. Measured production of plants or specific plant parts.

Plow pan. Compacted layer beneath the plow layer produced by pressure exerted on the soil during plowing.

Plugging. Method of planting by which shoots or seedlings are set or plugged into holes prepared in the soil.

Polyethelene (black), 6-mil thickness. Plastic sheet used for many purposes, for example, to foster plant growth and to prevent undergrowth.

Pore space. Space in the soil not occupied by solid particles.

Potash. Chemical essential to plant growth.

Potassium fixation. Conversion of exchangeable or water-soluble potassium to a form not easily exchanged from the adsorption complex with a cation of a neutral salt solution.

Potential acidity. Amount of acidity that must be neutralized to bring an acid soil to neutrality or to some predetermined higher pH value. It is approximated by the sum of the absorbed hydrogen and aluminum. Usually expressed in milliequivalents per unit mass of soil.

Prairie soils. Zonal great soil group consisting of soils formed under temperate to cool-temperate humid regions under tall grass vegetation.

Puddled soil

1. Dense, massive soil artificially compacted when wet and having no regular structure. Condition commonly results from the tillage of a clayey soil when it is wet.

2. Soil in which structure has been mechanically destroyed thereby allowing the soil to run together when saturated with water. Soil that has been puddled occurs in a massive nonstructural state.

Pulverize. Grind to powder or dust. In landscape work, usually means to break the soil into very fine pieces.

Pure forest. Forest composed principally of trees of one species, usually to the extent of 80% or more by number, but in special cases a percent of basal area or volume.

Purity and germination of grass seed. Bags must contain dealers guaranteed statement of composition of mixture and percentages of purity and germination. Seed should be delivered in unopened bags with label attached.

Red desert soil. Zonal great soil group consisting of soils formed in warm-temperate to hot dry regions under desert-type vegetation, mostly shrubs.

Red earth. Highly leached, red clayey soils of the humid tropics, usually with very deep profiles that are high in sesquioxides.

Relief. Difference in elevation between the high and low points on a landscape.

Respiration. Chemical reaction by which carbohydrates are oxidized and by which all animals and plants convert their food into energy. Carbon dioxide is released and oxygen used up.

Rhizobia. Bacteria capable of living symbiotically with higher plants, usually legumes, from which they receive their energy, and capable of using atmospheric nitrogen, hence, the term symbiotic nitrogen-fixing bacteria. Derived from the generic name Rhizobium.

Rhizosphere

1. Portion of the soil in the immediate vicinity of plant roots in which the abundance and composition of the microbial population are influenced by the presence of roots.

2. That portion of the soil directly affected by plant roots.

Rill

1. Small intermittent water course with steep sides, usually only a few inches deep, and hence no obstacle to tillage operations.

2. Small channel, such as a furrow prepared to conduct irrigation water across a field or that caused by erosive runoff. Where the cause is erosion, the term "rill" is limited to small channels that are only a few inches deep.

Rill erosion. Erosion process in which numerous small channels of only several inches in depth are formed; occurs mainly on recently cultivated soils.

Root bound. Term describing state of a plant when the roots have reached the container edge and start to encircle the root ball so that the plant is strangling itself. This is due to the plant's being in the container too long.

Root zone. Part of soil that is invaded by plant roots.

Runoff. Surface flow of water from an area; the total volume of surface flow during a specified time.

Saline-alkali soil. Soil having a combination of a harmful quantity of salts and either a high degree of alkalinity or a high amount of exchangeable sodium, or both, so distributed in the soil profile that the growth of most crop plants is less than normal.

Saline soil. Soil containing enough soluble salts to impair its productivity for plants, but not containing an excess of exchangeable sodium.

Saline-sodic soil. Soil containing both sufficient exchangeable sodium to interfere with the growth of most crop plants and appreciable quantities of soluble salts.

Saline soil

1. Term was formerly applied to any soil containing sufficient soluble salts to interfere with plant growth.

2. Soil containing sufficient soluble salts to impair its productivity. Specifically, a soil providing a saturation-paste extract having an electrical conductivity greater than 4 mmho/cm (at 25°C). The term "saline," when used alone, implies a low exchangeable sodium percentage of less than 15%.

Salt-affected soil. Soil that has been adversely modified for growth of most crop plants by the presence of certain types of exchangeable ions or soluble salts. Includes soil having an excess of salts or an excess of exchangeable sodium, or both.

Salts. Products, other than water, of the reaction of an acid with a base. Salts commonly found in soils break up into cations (sodium, calcium, etc.) and anions (chloride, sulfate, etc.), when dissolved in water.

Sand. Individual rock or mineral fragments having diameters ranging from 0.5 to 2.0 mm. Usually sand consists of quartz grains, but it may have any mineral composition. Soil that contains 85% or more sand particles and not more than 10% clay particles.

Sandy clay. Soil of a textural class containing 35% or more clay and 45% or more sand.

Sandy clay loam. Soil of a textural class containing 20 to 35% clay, less than 28% silt, and 45% or more sand.

Sandy loam. Soil of the sand loam textural class containing 50% sand and less than 20% clay.

Sapling. Young tree 3 ft or over in height and less than 4 in. in diameter.

Scrub forest. Area partially or completely covered with crowded bushes or stunted trees. More specifically, marginal forest on wind-exposed locations and on high elevations, composed of short, wind-trained trees or low shrubs forming a transitional zone along the upper timberline and the tropical or subtropical forests associated with subhumid or semiarid climatic conditions.

Seed

1. Source from which a flower or other plant grows. Small grainlike fruit from which a new plant will grow.
2. Sow with seed.

Seepage. Slow flow of water into or from a soil. Usual concept of seepage involves the lateral flow of water, as from an open body of water into neighboring soil or the reverse.

Self-mulching soil. Soil in which the surface layer becomes so well aggregated that it does not crust and seal under the impact of rain but instead serves as a surface mulch on drying.

Shear. Force, as of a tillage implement, acting at right angles to the direction of movement.

Shelterbelt. Belt of trees or shrubs arranged as protection against strong winds; a type of windbreak. Trees may be specially planted or left standing when the original forest is cut.

Shrub. Perennial plant that differs from a perennial herb in having persistent and woody stems; it differs from a tree in having low stature and a habit of branching from the base.

Silt. Individual mineral particles ranging in diameter between the upper size of clays (0.002 mm) and the lower size of very fine sand (0.05 mm); soil containing 80% or more particles of silt size and less than 12% clay; sediments deposited by water.

Site. In ecology, an area described or defined by its biotic, climatic, and soil conditions as related to its capacity to produce vegetation. Area sufficiently uniform in biotic, climatic, and soil conditions to produce a particular climate vegetation.

Slick spots. Small areas in a field that are slick when wet owing to a high content of exchangeable sodium.

Slope. Deviation of a plane surface from the horizontal. Slope is conventionally expressed in degrees, which are units of vertical distance for each 100 units of horizontal distance.

Snag. Standing dead tree from which the leaves and most of the branches have fallen, or a standing section of the stem of a tree broken off at a height of at least 20 ft; if less than 20 ft, properly termed a stub, sunken log, or a submerged stump.

Sod

1. Piece or layer of soil containing grass and its roots, grown elsewhere.

2. Should be freshly cultivated and weed, disease, and insect free. Sod is cut in minimum thickness of $\frac{3}{4}$ in., widths of 12 to 18 in., and minimum lengths of 6 ft. Installed with staggered joints, dried or damaged sod to be discarded. Upon installation sod should be tightly butted, tamped, or rolled firm and receive immediate watering.

3. Cover with sod.

Sodding. Removal of sections or pieces of an established lawn and resetting it elsewhere to establish a new lawn.

Sodic soil. Soil that contains sufficient sodium to interfere with the growth of most crop plants, and in which the exchangeable sodium percentage is 15% or more.

Sod removal. Commercial removal of soil filled with the roots of grass or herbs.

Softwood. Generally, one of the botanical group of trees that in most cases have needle or scalelike leaves; the conifers.

Soil

1. Natural medium for growth of land plants.
2. Dynamic natural body on the surface of the earth in which plants grow, composed of mineral and organic materials and living forms. Collection of natural bodies occupying parts of the earth's surface that support plants and that have properties due to the integrated effect of climate and living matter acting on parent material, as conditioned by relief, over periods of time.

Soil aeration. Process by which air in the soil is replaced by air from the atmosphere. In a well-aerated soil, the soil air is very similar in composition to the atmosphere above the soil. Poorly aerated soils usually contain a much higher percentage of carbon dioxide and a correspondingly lower percentage of oxygen than the atmosphere above the soil. Rate of aeration depends largely on the volume and continuity of pores within the soil.

Soil alkalinity. Degree or intensity of alkalinity of a soil, expressed by a value greater than 7.0 on the pH scale.

Soil amendment. Substance such as lime, sulfur, gypsum, and sawdust used to alter the properties of a soil, generally to make it more productive. Fertilizers are soil amendments, but the term is used most commonly for materials other than fertilizers.

Soil chemistry. The study and analysis of the inorganic and organic components and the life cycles within soils.

Soil class. Group of soils having a definite range in one or more properties such as acidity, degree of slope, texture, structure, land-use capability, degree of erosion, or drainage.

Soil colloid. Colloidal complex of soils composed principally of clay and humus.

Soil complex. Consists of two or more recognized classifications.

Soil conditioner. Material used for the improvement of soil.

Soil conservation. Combination of all management and land use methods that safeguard the soil against depletion or deterioration by natural or by man-induced factors.

Soil creep. Downward mass movement of sloping soil. Movement is usually slow and irregular and occurs most commonly when the lower soil is nearly saturated with water.

Soil erosion. The detachment and movement of topsoil by the action of wind and flowing water.

Soil fertility. Status of a soil with respect to the amount and availability to plants of elements necessary for plant growth.

Soil management. Sum total of all tillage operations, cropping practices, and fertilizer, lime, and other treatments conducted on or applied to a soil for the production of plants.

Soil management groups. Groups of soil-mapping units with similar adaptations or management requirements for one or more specific purposes, such as crop adaptations, crop rotations, drainage practices, fertilization, forestry, or highway engineering.

Soil moisture. Water contained in the soil.

Soil organic matter. Organic fraction of the soil; includes plant and animal residues at various stages of decomposition, cells and tissues of soil organisms, and substances synthesized by the soil population.

Soil phase. Subdivision of a soil type or other unit of classification having characteristics that affect the use and management of the soil but that do not vary sufficiently to differentiate it as a separate type. Variation in a property or characteristic such as degree of slope, degree of erosion, or content of stones.

Soil physics. The study of the physical character of soils.

Soil porosity. Degree to which the soil mass is permeated with pores or cavities.

Soil productivity. Capacity of a soil for producing a specified plant or sequence of plants under a specified system of management. Productivity emphasizes the capacity of soil to produce crops and should be expressed in terms of yields.

Soil profile. Vertical cross section of a soil, showing horizons and parent material.

Soil removal. Removal of any kind of soil or earth matter, including top soil, sand or other type of soil matter or combination thereof, except common household gardening or ground care.

Soil salinity. Amount of soluble salts in a soil.

Soil science. The study of the formation, properties, and classification of soil.

Soil series. Family of soils having similar profiles and developing from similar original materials under the influence of similar climate and vegetation.

Soil water percolation. Downward movement of excess water through soil.

Sow. Scatter or plant seed.

Spray mulch. Newly seeded area sprayed with specially prepared products for the prevention of erosion long enough to allow seed to germinate and sprout without being washed away.

Sprigging. Method of planting by which shoots and root portions of plants are impressed into the soil.

Sprinkling. Method of spraying system; spraying water over the soil surface through nozzles.

Stakes for guying trees. Stakes should be made of sound wood of uniform size, reasonably free of knots, and capable of standing in the ground at least two years. Wires or ropes ties to stakes to keep trees or plants in vertical position.

Steppe. Level or rolling treeless land, where temperature ranges usually are extreme.

Stolon. Stem growing horizontally on or just below the surface of the ground and capable of producing new roots, leaves, and stems at nodes.

Strip cropping. Practice of growing crops that require different types of tillage, such as row and sod, in alternate strips along contours or across the prevailing direction of wind.

Stubble mulch. Stubble of crops or crop residues left essentially in place on the land as a surface cover before and during the preparation of the seed bed and at least partly during the growing of a succeeding crop, for the purpose of erosion control.

Subsoil

1. The part of the soil below the depth of plowing.
2. B horizons of soils with distinct profiles. In soils with weak profile development, the subsoil can be defined as the soil below the plowed soil (or its equivalent of surface soil) in which roots normally grow.

Subsoil chisel

1. Tillage implement with one or more cultivator-type feet to which are attached strong knifelike units used to shatter or loosen hard, compact layers, usually in the subsoil, to depths below normal plow depth.
2. Tillage machine with one or more soil-penetrating points that can be drawn through the soil to loosen the subsoil, usually to a depth of 12 to 18 in.

Subsoiling. Breaking of compact subsoils, without inverting them, with a special knifelike instrument (chisel) that is pulled through the soil at depths usually of 12 to 24 in. and at spacings usually of 2 to 5 ft.

Subsurface tillage. Tillage with a special sweeplike plow or blade that is drawn beneath the surface at depths of several inches and cuts plant roots and loosens the soil without inverting it or without incorporating the surface cover.

Sulfur. Chemical used as fungicide and/or soil conditioner.

Super phosphate. Substance composed of finely ground phosphate rock; as commonly used for agricultural purposes contains not less than 18% available phosphoric acid.

Supplemental irrigation. Additional irrigation supplied by mechanical means during dry periods in regions where normal precipitation supplies most of the moisture for crops.

Surface soil. Uppermost part of the soil, ordinarily moved in tillage, or its equivalent in uncultivated soils, ranging in depth from 3 to 4 in. to 8 to 10 in. Frequently designated as the plow layer of the Ap horizon.

Swamp

1. Area of continuously saturated or spongy ground having poor drainage; therefore synonymous with marsh. Area of continuously saturated ground, supporting large aquatic plants having submerged or floating leafy shoots, often dominated by shrubs and trees.
2. Area saturated with water throughout much of the year, but with the surface of the soil usually not deeply submerged. Generally characterized by tree or shrub vegetation.

Terrace. Purpose of a terrace is to intercept surplus runoff and to retard it so that it will infiltrate into the soil. Terraces build up a reserve of water, materially assisting plant growth; they also slow down the rate of runoff, diverting excess water to prepared outlets. Terraces provide erosion control.

Thatch. Accumulation of an undecomposed layer of dead and dying stems, roots, rhizomes, and leaves on the soil surface below the green top growth of the turf.

Thin natural screen. Natural growth left in front of entrenchments and emplacements to aid in concealing them.

Till

1. Plow and prepare the soil for seeding; to seed or cultivate the soil.

2. Unconsolidated sediment containing all sizes of fragments from clay to boulders deposited by glacial action, usually unbedded.

Tillage. Operation of implements through the soil to prepare seed beds and root beds.

Tillage of subgrade. Loosening of subgrade by disking or by scarifying to a depth of at least 2 in. to permit bonding of the topsoil to the subgrade where compaction has occurred; removal of unsuitable subgrade material from the site.

Tilth. Physical condition of soil as related to its ease of tillage, fitness as a seed bed, and impedance to seedling emergence and root penetration.

Timberline. Upper limit of erect trees in mountainous regions. Northern limit of erect trees in the Arctic.

Top mulch

1. Peat moss applied as a mulch to plants and planting beds to a depth of 2 in. after all plants are set.
2. Applied layer of plant residues or other materials on the surface of the soil. Mulches are generally used to help conserve moisture, control temperature, prevent compaction or crusting, reduce runoff and erosion, improve soil structure, and control weeds. Common mulching materials include sawdust, wood chips and straw, paper, fine brush, and, occasionally, small stones.

Topography. Shape of the ground surface, such as hills, mountains, or plains. Steep topography indicates steep slopes or hilly land; flat topography indicates flat land with minor undulations and gentle slopes.

Topping. Removal of those portions of trees, bushes, and shrubs projecting above an elevation or plane shown or indicated on landscape drawings.

Topsoil

1. Presumed fertile soil or soil material, usually rich in organic matter, used to top dress road banks, lawns, and gardens; the surface plow layer of a soil and thus a synonym for surface soil; the original or present dark-colored upper soil, which ranges from a mere fraction of an inch to 2 or 3 ft on different kinds of soil, applied to soils in the field.
2. Friable loam, typical of cultivated topsoils of the locality, containing at least 2% decayed organic matter (humus). Fill taken from a well-drained arable site, reasonably free of subsoil, stones, earth clods, sticks, roots, or other objectionable extraneous matter or debris, and containing no toxic materials.

Trace elements. Elements found in plants in only small amounts, including several that are essential to plant growth.

Tree. Woody perennial plant usually having a single main stem that has generally few or no branches on its lower part and is crowned with a head of branches and foliage, or (as in palms) foliage only.

Tundra. Mixture of several types of vegetation occurring in arctic and alpine districts beyond and above the timberline; the plant cover includes low shrubs, herbs, sedges, grasses, lichens, and mosses.

Tussock. Tuft of grasses or grasslike plants.

Undergrowth. Growth of small trees and shrubby plants under a forest canopy.

Understory. Layer of young or stunted trees in a forest that is below the level of the main canopy.

Undue compaction. Term describing compaction of subsurface layers in soil by the application of weight, for example, by machines or tractors, so that the penetration of water and roots is interfered with. Because the traffic of machines is not the only cause of these pans, some persons call them "pressure pans" or "traffic pans."

Viable. Capable of germination or growth.

Vine. Plant having a woody or herbaceous stem that is too slender, flexible, or weak to hold itself erect.

Virgin soil. Soil that has not been significantly disturbed in its natural environment.

Wasteland. Land not suitable for or capable of producing materials or services of value. Miscellaneous land type.

Water content

1. Ratio, expressed as a percentage, of the weight of water in a given soil mass to the weight of solid particles.
2. As applied to soils work, the amount of water held in a soil expressed on a weight or volume basis. Conventionally, water content is expressed relative to the oven-dry weight or volume of soil.

Waterlogged. Saturated with water.

Water requirements. Water requirements vary with climatic conditions, soil moisture, and soil characteristics. Factors unfavorable to plant growth, for example, low fertility, disease, and drought, increase water requirements.

Water table. Upper limit of the part of the soil or underlying rock material that is wholly saturated with water. In some places an upper, or perched, water table may be separated from a lower one by a dry zone.

Weed. Common, unsightly growth that interferes with the healthy growth of other plants.

Weed killer. Chemical substance applied at the proper time in accordance with manufacturer's recommendations. Extreme care should be taken that work is done on a windless day so that no injury occurs to surrounding plant materials from drift.

Wilting. Loss of turgidity in plant tissue where the intake of water is insufficient to replace that lost by transpiration of other means, thus causing a deflation of plant cells.

Windbreak. Planting of trees, shrubs, or other vegetation, usually perpendicular or nearly so to the principal wind direction, to protect soil, crops, roads, etc., against the effects of winds, such as wind erosion, and the drifting of soil and snow.

Windrow. Row or line formed above ground by furrowing and placing the excavated earth adjacent to the plowed trench or furrow.

Xerophytes. Plants that grow in or on extremely dry soils or soil materials.

TERMITES AND TERMITE CONTROL

Aldrin

1. Chlorinated hydrocarbon that is not soluble in water. Usually formulated as a 25% emulsifiable concentrate by weight. Dilution ratio is 1 part Aldrin to 37 parts water by volume.
2. Chlorinated hydrocarbon not soluble in water, an emulsifiable concentrate containing 2 to 4 lb/gal.

Benzene hexachloride. Chemical commonly called "benhex." Product effectiveness depends on its gamma isomer.

Commonly available with a range of gamma contents from 8 to 36% by weight. Persistent characteristic odor of benzene hexachloride does not make it an advisable material for use around buildings.

Bethanaphithol. Toxicant once widely used for soil poisoning; has proved to be less effective than newly developed products.

Chlordane

1. Odorless liquid insecticide, chemically an indene derivative.
2. Most commonly used soil poison. Insoluble in water but readily emulsifiable. Usually available in concentrates of 4, 6, 7 or 8 lb/gal.

Cypermethrin. Mixture consisting of 0.3% emulsion, 1.25 gal of Prevail FT, and 98.75 gal of water.

Dichloro-diphenyl-trichloroethane (DDT). Synthetic insecticide remarkable for high toxicity to insects at low rates of application. Functions as stomach insecticide and also as contact insecticide, having the advantage of persistance of activity from residual products compared with older natural insecticides.

Diedrin. Chlorinated hydrocarbon. Recommended concentration of applied solution is 0.5%.

Dieldrin. Chlorinated hydrocarbon available in concentrate usually containing $1\frac{1}{2}$ lb of Diedrin per gallon.

Dragnet FT termiticide. (A registered trademark. Manufactured by the FMC Corporation, Princeton, New Jersey) Containing permethrin at the rate of 3.2 lbs/gal.

Formulator. Agency that combines basic toxicants into oil-soluble or water-emulsifiable concentrates.

Foundation applications. Recommended rate of 4 gal per each 10 lic on each side of the foundation wall and around masonry piers, at a minimum of 2 ft deep. Applicators will recommend type of toxicant for the area.

General application. Toxicant must not be applied when the soil or fill is excessively wet or after heavy rainfall. Area must be protected after treatment from disturbance by animals or humans.

Guarantees. Usual guarantees accepted by financing agencies from applicators; must be in writing and with a minimum of 5 years.

Heptachlor. Chemical toxicant manufactured by one company reported to withstand the standard 5-year test required by federal agencies. Recommended strength of working solution (by weight) 0.5%.

Isomers. Substances whose molecules are made of the same number of the same atoms but whose atoms are differently arranged, such as ethyl alcohol and methyl ether; both are isomers.

King and queen. This pair occupies a special place in the termite colony. The queen lays thousands of eggs annually and may live for 25 years.

Lindane. Gamma isomer of benzene hexachloride.

Nesting. Termite infestations usually occur in poorly vented areas where the soil is continuously warm and attracts moisture.

Nonsubterranean or dry-wood termites. Insects that live in the wood and do not require much moisture. Usually occur in the dry southern United States and the southeastern coastal plain and account for only a small portion of termite damage in the United States; they produce galleries and tunnels in the timber, much like the subterranean varieties. Because they do not require as much moisture, they often operate undetected and enter wood when it is exposed and unpainted. Termite nests can be poisoned by dry powder or liquid treatments of various kinds and by fumigation.

Nymph. Eggs hatch into nymphs and then mature into one of the termite castes.

Permethrin. Mixture consisting of 0.5% emulsion, 1.25 gal of Dragnet FT, and 98.75 gal of water.

Prevail FT termiticide. (A registered trademark. Manufactured by the FMC Corporation, Princeton, New Jersey). Containing cypermethrin at the rate of 2.0 lbs/gal.

Shield materials. Recommended minimum gauges; (1) galvanized iron (sheet metal), 26 gauge, (2) tempered sheet copper, 16 oz, (3) aluminum, 0.019 in., (4) stainless steel, 26 gauge.

Signs of termites. Piles of fine sawdust, piles of fecal pellets the size of Bermuda grass seed, little mud tubes against the foundation from the ground into the building.

Soldier. A termite with armored head and powerful jaws used to protect the colony from enemies.

Termitaphididae. Termite bugs, a small family of Hemiptera in the superfamily Aradoidea.

Termitarium. Termite's nest.

Termite. Soft-bodied insect of the order Isoptera, which contains the higher termites, representing 80% of the species.

Termites. Insects that superficially resemble ants in size, general appearance, and habit of living in colonies; hence, they are frequently called "white ants." Subterranean termites do not establish themselves in buildings by being carried in with lumber, but by entering from ground nests after the building has been constructed. If unmolested, they eat out the woodwork, leaving a shell of sound wood to conceal their activities, and damage may proceed so far as to cause collapse of parts of a structure before discovery. About 56 species of termites are known in the United States. Two major species, classified from the manner in which they attack wood, are ground-inhabiting or subterranean termites, the most common, and dry-wood termites, found in this country almost only along the extreme southern border and the Gulf of Mexico.

Termites and carpenter ants. Insects that consume wood substance. Subterranean termites cause the most termite damage to wood in the United States, particularly in the South, but are not unknown practically to the Canadian border, except at higher elevations. They feed on moist wood and will enter a structure and devour it internally, leaving a thin shelf of sound wood on the outside. Access can be impeded with metal termite shields. Untreated wood should be kept a minimum of 18 in. away from the ground. Soil poisoning near foundations is effective but may need to be renewed in areas of severe termite activity. Treatments effective against termites are usually also effective against decay.

Termite shield. Shield, usually of noncorrodible metal, placed in or on a foundation wall or other mass of masonry, or around pipes, to prevent passage of termites.

Winged reproductive. Termite, that, after shedding wings, pairs off with the supplementary reproductive to form a new colony.

Worker. Blind and sterile termite that forages for food to feed the colony. Most wood damage is done by workers.

WALKS, ROADS, AND PARKING

Accessory parking area. Open surfaced area on private property in a residential zone where permitted, immediately adjoining a commercial or industrial district, both fronting upon the same street surface to eliminate the dust nuisance, fenced, free of advertising signs, moderately illuminated, used for the parking of customers' and personnel automobiles.

Alley. A public right-of-way primarily designed to serve as secondary access to the side or rear of those properties whose principal frontage is on a street.

Alley or service drive. Minor right-of-way, privately or publicly owned, primarily for service access to the back or sides of property.

Arterial street. A street used for fast-moving traffic, located between developed areas in the city.

Business street. Street which services or is designed to serve as an access to abutting business properties.

Cartway. Graded portion of a street or alley, including travelway or shoulders.

Collector street. Street supplementary to and connecting the major street system to local streets, designated as a minor street.

Commercial parking area. Open surfaced area on private property in a commercial or industrial district other than a street or alley, used for parking of automobiles and trucks and available for public use at rental rates.

Commercial parking lot. Lot where cars are parked subject to renumeration.

Controlled intersection. Any intersection at which traffic control signals are in operation or a traffic officer is directing traffic.

Cross walk. Part of a roadway at an intersection included within the connections of the lateral lines of the sidewalks on opposite sides of the highway measured from the curbs or, in the absence of curbs, from the edges of the traversable roadway. Any portion of a roadway at an intersection or elsewhere distinctly indicated for pedestrian crossing by lines or other markings on the surface.

Cul-de-sac. A minor street intersecting another street at one end and terminating at the other by a vehicular turnaround.

Customer parking area. Off-street parking, designed and arranged to an approved standard established by the city engineer, that provides parking for all standard passenger vehicles with a minimum of maneuvering and made available as an accommodation to occupants and patrons of the property.

Dead end street. Street or portion of a street with only one vehicular outlet but which has a turnaround and which is designed to provide ample maneuvering area.

Expressway. Multi-lane divided highway for through traffic with full or partial control of access and with grade separations at some intersections and at all major railroad crossings.

Feeder road. Street or road intersecting with a limited access highway and having traffic interchange facilities with such limited access highway.

Freeway. Divided arterial highway for through traffic to which access from the abutting properties is prohibited and where all street crossings are made by grade separated intersections.

Frontage road. A street lying adjacent and approximately parallel to and separated from a freeway, and which affords access to abutting property.

Highway. The entire width between the boundary lines of every way publicly maintained when any part thereof is open to the use of the public for purposes of vehicular travel.

Industrial street. A street designed and constructed to serve both truck and bus movements within an industrial area. Abutting property will have free access. On-street parking and loading is prohibited.

Intersecting street. Any street which adjoins another street at an angle whether or not it crosses the other.

Limited access highway. A trafficway, including toll roads, for through traffic, in respect to which owners or occupants of abutting property or lands and other persons have no legal right of access to or from same, except at such points only and in such manner as may be determined by the public authority having jurisdiction over such trafficway.

Local street. Any street other than a collector street, major or secondary highway, or freeway, providing access to abutting property and serving local, as distinguished from through, traffic.

Loop street. A street used primarily for access to interior lots in a block, beginning and terminating at different points on the same abutting street.

Main street. Street upon which the majority of lots within a block frontage are fronted; or any street officially so designated; or the commercial or business street with the most traffic.

Major traffic streets. Those serving large volumes or comparatively high-speed and long distance traffic; includes facilities classified as main and secondary highways by the state highway department.

Marginal access road. Service road parallel to a feeder road; provides access to abutting properties and protection from through traffic.

Marginal access street. A local street providing access to lots which abut or are adjacent to a limited-access highway or major street.

Median. Portion of a divided highway separating the traveled ways of traffic proceeding in opposite directions.

Minor street. Any street not designated as a major or collector street and intended to serve or provide access exclusively to the properties abutting thereon.

Official traffic-control devices. All signs, barricades, signals, markings, and devices not inconsistant with the ordinance placed or erected by authority of a public body or official having jurisdiction, for the purpose of regulating, warning, or guiding traffic.

Off-street bicycle parking space. A paved and properly drained area, enclosed or unenclosed which is permanently reserved for parking one (1) bicycle, and which measures not less than two (2) ft in width and which should provide minimum security and support structures permanently fastened to the paved area.

Off-street parking. Space adequate for parking standard passenger vehicles together with properly related access to a public street or alley.

Parking. Temporary, transient storage of private passenger motor vehicles used for personal transportation while the operators of such vehicles are engaged in other activities.

Parking area. Open area, other than street or other public way, used for the parking of motor vehicles and available for public or private use whether for a fee or as a service or privilege for clients, customers, suppliers, or residents.

Parking lot. A parcel of land where members of the general public may park their motor vehicles for the purpose of utilizing an adjacent use or facility.

Parking space. A space within a public or private parking area or in a building, exclusive of driveways, ramps, columns, office, and work area, for the parking use.

Paved area. An area that has been drained, graded, compacted, provided with adequate base, and surfaced with asphaltic concrete at least two (2) inches thick, or equivalent, so as to provide sufficient durable surface to render the area usable for the purpose specified under normal weather conditions.

Pedestrian way. A public or private right-of-way solely for pedestrian circulation.

Primary residential street. A street that serves the prime function of collecting or distributing intracommunity residential traffic.

Private road. A private thoroughfare, other than a public street or alley, permanently reserved in order to provide a means of access to more than one zoning lot.

Private street. The area lying within the described limits of an easement or right-of-way, created by virtue of a recorded or registered instrument, for ingress and egress by one person, or any number of persons less than the public at large, over the land of another.

Public parking area. Open area other than a street or alley used for the parking of automobiles and available for public use whether free, for compensation, or as an accommodation for clients or customers.

Public street. Any thoroughfare or public way not less than 60 ft in width that has been dedicated to the public or deeded to the city for street purposes.

Public thoroughfare. Any legally established street or alley as defined in the code.

Public walkway. Any space designed or maintained solely for pedestrian use, without regard to ownership.

Public way. Any parcel of land more than 12 ft wide and appropriated to the free passage of the general public.

Residence street. Portion of a street between its intersections with two other streets where the majority of the frontage is within a residence district.

Roadway. Portion of a street that is used or intended to be used for the movement of motor vehicles.

Secondary residential street. A street used primarily for residential access.

Service road. Part of a major or secondary highway, containing a roadway which affords access to abutting property and is adjacent and approximately parallel to and separated from the principal roadway.

Side street. Street bounding a corner lot, extending in the same general direction as the line determining the depth of the lot.

Sidewalk. Portion of a street between the curb lines, or the lateral lines of a roadway and the adjacent property lines, intended for the use of pedestrians.

Street. A dedicated public passageway that affords a principal means of access to abutting property.

Thoroughfare. Public street, road, way, or other space customarily used for travel.

Through highway. Every highway or portion thereof where vehicle traffic is given preferential right-of-way, and at the entrances to which vehicular traffic from intersecting highways is required by law to yield right-of-way to vehicles on such through highway in obedience to either a stop sign or a yield sign, when such signs are erected as provided in the code.

Traffic. Pedestrians, ridden or herded animals, vehicles, and other conveyances either singly or together while using any highway for purposes of travel.

Traffic-control signal. Any device, whether manually, electrically, or mechanically operated, by which traffic is alternately directed to stop and to proceed.

Traffic lane. Strip of roadway intended to accommodate a single line of moving vehicles.

Traffic signaling device. A sign, device, or mechanical contrivance, used for the control of motor vehicular and pedestrian movement.

Traveled way. The portion of the roadway for the movement of vehicles, exclusive of shoulders.

Travelway. Portion of a public street or road which is intended for vehicular movement.

Way. Street, alley or other thoroughfare or easement permanently established for passage of persons or vehicles.

WASTE, SEWAGE, AND DISPOSAL SYSTEMS

Absorption bed

1. Trench exceeding 36 in. in width, containing a minimum of 12 in. of clean, coarse aggregate and a system of two or more distribution pipes through which treated sewage may seep into the surrounding soil.

2. Pit of relatively large dimensions filled with coarse aggregate, containing a distribution pipe system in which the septic tank effluent is absorbed through the exposed soil.

Absorption field

1. Arrangement of absorption trenches through which treated sewage is absorbed into the soil.

2. System of trenches containing coarse aggregate and a distribution pipe through which septic tank effluent may seep or leach into the surrounding soil.

Absorption lines. Tile construction laid with open joints. Bell-and-spigot tile laid with $\frac{1}{2}$-in. open joints at 2-ft intervals, with sufficient cement mortar at the bottom of the joint to ensure an even flow line. Agricultural tile sections are spaced not more than $\frac{1}{4}$ in., and the upper half of the joint is protected by asphalt-treated paper while the tile is being covered, unless the pipe is covered by at least 2 in. of gravel. Perforated clay tile or perforated bituminized-fiber pipe or asbestos cement pipe may be used, provided that sufficient openings are available for distribution of the effluent into the trench area.

Absorption trench. Trench not over 36 in. in width, containing a minimum of 12 in. of clean, coarse aggregate and a distribution pipe through which treated sewage is allowed to seep into the soil.

Anaerobic. Living without free oxygen. Anaerobic bacteria are found in septic tanks and are beneficial in digesting organic matter.

Baffles. When inlet and outlet baffles are used in a septic tank, they should extend the full width of the tank and be located 12 in. from the end walls. Baffles should extend at least 6 in. above the flow line. Inlet baffles should extend 12 in. and outlet baffles 15 to 18 in. below the flow line.

B.O.D. (Biochemical oxygen demand). The quantity of oxygen utilized in the biochemical oxidation of organic matter under standard laboratory procedure in 5 days at 20°C, expressed in parts per million by weight.

Bottom of seepage pit. The bottom of the pit is filled with coarse gravel to a depth of 1 ft.

Building drain. Lowest part of a house drainage system which receives the discharge from soil, waste, and other drainage pipes inside the building and conveys it to the building sewer beginning 3 ft outside the building wall. (The distance is dictated by contract arrangements.)

Building sewer. Horizontal piping of a drainage system which extends from the end of the building drain and which receives the discharge of the building drain and conveys it to a public sewer, private sewer, individual sewage disposal system, or other point of disposal.

Catch basin. An enlarged and trapped inlet to a sewer designed to capture debris and heavy solids carried by storm or surface water.

Cesspool

1. Covered pit with open-jointed lining in its bottom portions into which raw sewage is discharged, the liquid portion of the sewage being disposed of by seeping or leaching into the surrounding porous soil, and the solids or sludge being retained in the pit to undergo partial decomposition before occasional or intermittent removal.
2. Covered and lined underground pit used as a holding tank for domestic sewage and designed to retain the organic matter and solids but to permit the liquids to seep through the bottom and sides. Cesspools are prohibited and are not acceptable as a means of sewage disposal.
3. Receptacle in the ground that receives crude sewage and is so constructed that the organic portion of such sewage is retained while the liquid portion seeps through its walls or bottom.

Chlorine contact chamber. Septic tank with wood baffles or weirs, supplied with chlorine from a chlorinator at the intake point.

Collection and disposal of sewage. All sewage should be collected and disposed of in properly constructed and managed sewers, treatment facilities, septic tanks, chemical toilets, privies, or by other methods approved by the authority.

Combined sewer. A sewer receiving both surface runoff and sewage.

Combined sewer system. System which carries both sanitary sewage and/or industrial wastes and storm water or drainage.

Covers for seepage pits. Reinforced-concrete covers should be provided over seepage pits, preferably to finished grade.

Design. Design of an individual sewage disposal system must take into consideration location with respect to wells or other sources of water supply to prevent contamination, as well as topography, water table, soil characteristics, available area and location, and maximum occupancy of the building.

Discharge. System consisting of a septic tank discharging into either a subsurface disposal field or one or more seepage pits, or a combination of both.

Discharge of sewer systems. A sewer or septic tank outlet is not allowed to discharge directly into an open ditch, drain, or stream nor into a closed drain that eventually empties into an open ditch, drain, or stream.

Disposal drains. Cement or agricultural tile laid with loose abutting joints. Joints are covered with sections of roofing paper previous to refilling the pipe trench.

Disposal trenches. Disposal trenches are designed and constructed on the basis of the required effective percolation area.

Distribution box. Box provided to receive the effluent from the septic tank to ensure equal distribution to each individual line of the disposal field.

Distribution box connection. The distribution box is connected to the septic tank by a tight sewer line and located at the upper end of the disposal field.

Distribution box invert level. The invert of the inlet pipe is located 2 in. above the bottom of the box. The invert of the outlets to each distribution line are level with the bottom of the box and set at the same elevation.

Distribution line. Open joint or perforated pipe intended to permit soil absorption of effluent.

Domestic sewage. Liquid and waterborne wastes derived from the ordinary living processes, free from industrial wastes, and of such character as to permit satisfactory disposal, without special treatment, into the public sewer or by means of a private sewage disposal system.

Dosing chambers. Dosing chambers are not usually required in the case of individual septic tank disposal systems.

Dosing tank and siphon. For filter areas greater than 1800 sq ft and for distribution pipes longer than 300 ft, a closing tank is recommended. The capacity of the tank equals about 75% of the volume of the distribution pipes.

Drainage well. Drilled, driven, or natural cavity that taps the underground water and that receives surface waters, waste waters, domestic waste, and sewage.

Drywells. Used to provide an underground means of disposal for surface runoff. Effectiveness is dependent on surrounding soils for absorption.

Effluent. Partially treated liquid sewage flowing from any part of the disposal system, septic tank, or absorption system to place of final disposition.

Filter material for trenches. Material should cover the absorption tile, extend the full width of the trench, and not be less than 6 in. deep beneath the bottom of the tile. Filter material may be washed gravel, crushed stone, slag, or clean bank-run gravel ranging in size from $\frac{1}{2}$ to $2\frac{1}{2}$ in. Filter material should be covered by untreated paper or roofing felt strips as the laying of the pipe drain proceeds in the trench.

Food waste (garbage). Animal and vegetable waste resulting from the storage, handling, preparation, cooking, or serving of foods.

Garbage disposal. Where domestic garbage disposal units are installed or contemplated, it is recommended that septic tanks be at least 50% greater in capacity.

General information. Most local or state health departments will provide assistance in the design of a satisfactory private disposal system. Local contractors in the business of installing individual disposal systems can provide valuable assistance in respect to local soil conditions and the overall construction of a satisfactory system. In swampy or flooded areas, an absorption system is not satisfactory. The size of the property must be considered, inasmuch as an individual disposal system for a large number of homes will require a larger lot than is ordinarily required when connection is made to a sewer.

G.P.D. Gallons per day per person.

Grease trap. Constructed of concrete or a large tile pipe in which grease and oils are trapped before discharging to the disposal system.

Ground garbage. The residue from the preparation, cooking or dispensing of food that has been shredded to such degree that all particles will be carried freely in suspension under the flow conditions normally prevailing in public sewers, with no particle greater than $\frac{1}{2}$ in. in any dimension.

House sewer. Extension from the building drain to the public sewer or other place of disposal.

Impervious soil. Tight cohesive soil such as clay which does not allow the ready passage of water.

Increasing capacities of septic tank system. When it is found necessary to have increased septic tank capacity because of additional bedrooms or the installation of more bathroom fixtures or new plumbing-connected appliances, a new tank may be added to the existing installation. A new tank or compartment may be placed on either side of the existing tank. The liquid capacity of the new tank or compartment should be 50% greater but in no case less than one-third of the existing capacity.

Indirect waste pipe. Pipe that does not connect directly with the drainage system but conveys the liquid wastes by discharging into a plumbing fixture, interceptor, or other type of receptacle which then is directly connected to the drainage system.

Individual sewage disposal system. Combination of a sewage treatment plant (package plant or septic tank) and method of effluent disposal (soil absorption system or stream discharge) serving a single dwelling.

Industrial wastes. Particular liquid or other wastes resulting from any process of industry, agriculture, manufacture, trade, or business or the development of any natural resource.

Industrial wastewater. Liquid wastes resulting from the processes employed in industrial, manufacturing, trade, or business establishments, as distinct from domestic wastes.

Interceptor. Device designed and installed to separate and retain deleterious, hazardous, or undesirable matter from normal wastes and permit normal sewage or liquid wastes to discharge by gravity into a disposal terminal.

Invert. The invert of the inlet pipe in a septic tank should be located at least 3 in. above the invert of the outlet.

Junction box. Used in lieu of manhole for joining larger sewers, 48 in. and larger. Customized design is required for each juncture.

Lateral. Branch of the absorption field consisting of either the length from the distribution box or the length of distribution from the tee or cross fitting at the main line to the farthest point in a closed field.

Leaching cesspool. A leaching cesspool may be used in place of or to supplement tile fields. It is preferable where the soil below a depth of 2 or 3 ft is more porous than that above this depth. Minimum spacing 10 ft. When cesspool is located in fine sands, surround walls for entire height with layer of graded gravel.

Leaching well or cesspool. Pit or receptacle having porous walls that permit seepage into the ground.

Leaching well or pit. Individual sewage disposal system.

Leach line. In sewage disposal, a loose tile or perforated pipeline used to distribute sewage effluent through the soil.

Legality of septic tanks. Septic tanks or other private means of disposal age usually not approved where a public sewer is available, and such means of disposal are required to be discontinued when public sewers are made available when directed by the legal authorities.

Length of tanks. Septic tanks should be at least twice as long as they are wide.

Manholes. The inlet compartment of a septic tank must be provided with one manhole. Other compartments may also be provided with a manhole. Manholes should be at least 20 sq in. or 24 in. in diameter and provided with covers that can be sealed watertight and extended to grade. Where removable slab covers are provided, manholes are not usually required.

Multiple compartments. In a septic tank of more than one compartment, the inlet compartment should have a capacity of not less than two-thirds of the total tank capacity.

On-site (septic tank) sanitary sewage disposal. Covered watertight settling tank in which raw sewage is biochemically changed into solid, liquid, and gaseous states to facilitate further treatment and final disposal.

Parts per million. A weight to weight ratio as the parts per million value multiplied by the factor 8.345 are equivalent to pounds per million gallons of water.

Percolation. Ability of a liquid to be absorbed through a soil or pass through that soil.

Percolation test

1. Determination of the suitability of an area for subsoil effluent disposal by testing for the rate at which the undisturbed soil in an excavated pit of standard size will absorb water per unit of surface area.

2. Percolation test holes should be made over the entire area where the distribution field is going to be located, so as to get a good average condition. Holes should be 6 to 12 in. round and drilled vertical to the bottom of the proposed trench. Round holes may be bored with a post-type auger. Placing about 2 in. of coarse sand or gravel at the bottom of the hole will prevent scouring. It is desirable to saturate the soil of the hole thoroughly and then proceed with the test.

3. Subsurface exploration generally necessary to determine soil conditions, including depth of pervious ma-

terial, presence of rock formations, and level of groundwater. Augers with extension handles often are used for making such investigations.

pH. The logarithm of the reciprocal of the weight of hydrogen ions in grams per liter of solution.

Pipe inlet and outlet. In lieu of baffles in a septic tank, submerged pipe inlets and outlets consisting of a cast-iron sanitary tee with a short section of pipe may be installed to the depth required by local plumbing code.

Pit linings for seepage pits. Pits lined with stone, brick, or concrete blocks laid up dry with open joints and backed up with at least 3 in. of coarse gravel. Joints above the inlet may be sealed with cement mortar.

Private sewage disposal system. Privately owned plant for the treatment and disposal of sewage, such as a septic tank with an absorption field.

Private sewer. Sewer privately owned and not directly controlled by public authority.

Public sewage disposal. All plumbing fixtures installed in buildings intended for human habitation, occupancy, or use in premises abutting on a street, alley or easement in which there is a public sewer must be connected to such sewer.

Public sewer. A sewer in which all owners of abutting properties have equal rights, and which is controlled by the city.

Putrescible wastes. Wastes that are capable of being decomposed by microorganisms with sufficient rapidity as to cause nuisances from odors, gases, and similar objectionable conditions. Kitchen wastes, offal, and dead animals are examples of putrescible components of solid waste.

Sanitary building drain. Building drain that may conduct sewage and clear-water waste, but not storm water.

Sanitary building sewer. Building sewer that may conduct sewage and clear-water waste, but not storm water.

Sanitary drainage system. Drainage system that may conduct sewage or clear-water waste, and includes a combined building drain and combined building sewer.

Sanitary sewage. System designed to receive all sanitary sewage, including laundry waste from the building where allowed by code. Drainage from basement floor, footings, or roofs should not enter the system.

Sanitary sewer. Sewer that may conduct sewage or clear-water waste but not storm water.

Sanitary sewer facility. Public sanitary sewer facility, or a comparable common or package sanitary sewer facility approved by the authorities.

Scum clear space. In a septic tank, the distance between the bottom of the scum mat and the bottom of the outlet device (tee, baffle).

Seepage pit

1. A seepage pit is used either to supplement the subsurface disposal field or in lieu of such field where conditions are favorable, as may be found necessary and acceptable under local codes.

2. Covered underground pit with concrete or masonry lining designed to permit partially treated sewage to seep into the surrounding soil. It is also called a dry well.

Septic disposal system. System for the treatment and disposal of domestic sewage by means of a septic tank and soil absorption system.

Septic tank

1. Watertight reservoir or tank that receives sewage and by sedimentation and bacterial action effects a process of clarification and partial purification.

2. Watertight receptacle that receives the discharge of a drainage system or part thereof, and is designed and constructed so as to separate solids from the liquid, digest organic matter through a period of detention, and allow the liquids to discharge into the soil outside the tank through a system of open-joint or perforated piping, or a disposal pit.

Septic tank capacities. In computing septic tank capacity, the factor most often used is the number of bedrooms. When a food waste disposal unit is installed, there should be a slight increase in tank capacity to take care of the additional solids and extend the period of pumping a septic tank.

Septic tank construction

1. Septic tanks constructed of corrosion-resistant materials and of permanent construction. Cover of the tank, designed for a dead load of not less than 150 pounds per square foot and, if of concrete, reinforced and not less than 4 in. thick. Precast reinforced concrete septic tanks are available in many sizes.

2. Principal points in the construction of a septic tank are proper size for the intended liquid capacity, hermetically sealed construction, material selected for its ability to resist the corrosive effects of the soil surrounding the tank and the liquid contents of the tank, and sufficient strength to withstand earth loads on sides, top, and bottom.

Septic tank design. A tank is rectangular in shape with its length two to four times its width and a minimum depth of 4 ft. Tanks less than 8 ft long usually have one compartment; larger tanks usually have two compartments. The inlet compartment is about 75% of total tank capacity.

Serial distribution. Combination of several absorption trenches, seepage pits, or absorption beds arranged in sequence so that each is forced to utilize the total effective absorption area before liquid flows into the succeeding component.

Sewage. Liquid waste containing animal or vegetable matter in suspension or solution. It may include liquids containing minerals in solution from laboratories or industrial institutions.

Sewage disposal plant. Land, tanks, buildings, and apparatus used in the treatment of sewage by chemical precipitation, filtration, bacterial action, or some other developed method.

Sewage disposal system. A system for the disposal of sewage by means of a septic tank, cesspool, or mechanical treatment, all designed for use apart from a public sewer, to serve a single establishment, building, or development.

Sewage ejector. A mechanical device used to pump or eject sewage.

Sewage facility. Any sewer, sewage system, sewage treatment works or part thereof designed, intended, or constructed for the collection, treatment, or disposal of liquid waste, including industrial waste.

Sewage system. All service mains and intercepting sewers and structures by which sewage or industrial waste is collected, transported, treated, and disposed of. "Sewage system" does not include plumbing inside or in connection with buildings served or sewer laterals from a building or structure to city-owned mains.

Sewage treatment. Process for the purification of mixtures of human and other domestic and industrial wastes by aerobic or anaerobic methods.

Sewage treatment tank. Watertight receptacle so constructed as to promote the separation and decomposition of sewage.

Sewer. An underground pipe or open channel in a sewage system for carrying water or sewage to a disposal area.

Sewer gas. Gas evolved from decomposition of sewage; has a high content of methane and hydrogen sulfide.

Sludge. Accumulated solids that settle out of the sewage, forming a semiliquid mass on the bottom of the septic tank.

Sludge clear space. In a septic tank, the distance between the top of the sludge and the bottom of the outlet device.

Soil absorption system. System that utilizes the soil for subsurface absorption of treated sewage, such as an absorption trench, absorption bed, or seepage pit.

Soil structure. Heavy, tight clay, hardpan, rock, or other impervious soil formations are not suitable for seepage pit construction.

Special wastes. Special wastes require special methods of processing, such as the use of indirect waste piping and receptors, corrosion resistant piping, sand, oil, or grease interceptors, and other pretreatment facilities and devices.

S.S. Suspended solids.

Storm sewer. Sewer used for conveying groundwater, rainwater, surface water, or similar nonpollutional wastes.

Storm water drainage system. The piping system used for conveying rainwater or other precipitation to the storm sewer or other place of disposal.

Suitability of soil. The first step in the design of a septic sewage disposal system is to determine whether the soil is suitable for the absorption of septic tank effluent and, if it is, how much absorption area is required.

Sump pit. A tank or pit that receives clear liquid wastes that do not contain organic materials or compounds subject to decomposition, located below the normal grade of the gravity system and that must be emptied by mechanical means.

Sump pump. A mechanical device used to pump the liquid waste from a sump pit into the gravity drainage system.

Suspended solids. The total suspended matter that floats on the surface of, or is suspended in, water, wastewater, or other liquids, a high percentage of which is removable by laboratory filtering. Measurement of quantities of suspended solids should be made in accordance with procedures set forth in the code.

Trees. When seepage pits are located in close proximity to trees, it need not be lined; instead it may be filled with loose rock.

Type of system. The type of system is determined on the basis of location, soil permeability, and groundwater elevation.

Typical septic sewage disposal system. System consisting of house sewer, septic tank, and absorption field. Solids are separated and retained in the septic tank, liquid is passed to the absorption field, and gases are vented through the sewer and vent stack.

Wastes. Waste from all plumbing in the house, including kitchen sinks, food waste disposal units, laundry trays, and washing machines, discharge into the sewage system piping and then to the septic tank.

Wastewater. The liquid and water-carried industrial or domestic wastes from dwellings, commercial buildings, industrial facilities, and institutions, together with any groundwater, surface water, and stormwater that may be present, whether treated or untreated, which is discharged into or permitted to enter the system.

Water table. Avoid extending seepage pits into the groundwater table. Where the seepage pit is used to receive the septic tank effluent, the same limitations must be placed on the location of the pit as on the use of a cesspool.

WATERWAYS AND SHORELINES

Accretion. May be either natural or artificial. Natural accretion is the gradual buildup of land over a long period of time solely by the action of the forces of nature—on a beach by deposition of water- or airborne material. Artificial accretion is a similar buildup of land by reason of an act of man, such as the accretion formed by a groin, breakwater, or mechanically deposited beach fill.

Advance (of a beach). Continuing seaward movement of the shore line, or a net seaward movement of the shore line over a specified time.

Alluvium. Soil, sand, mud, or similar detrital material deposited by flowing water, or the deposits formed thereby.

Artificial nourishment. Process of replenishing a beach by artificial means, such as by the deposition of dredged materials.

Atoll. Ringlike "coral" island or islands encircling or nearly encircling a lagoon. The term "coral" island for most tropical islands is incorrect, as calcareous algae (Lithothamnion) often form much more than 50% of them.

Atoll reef. Ring-shaped coral reef, often carrying low sand islands, enclosing a body of water.

Awash. Nautical: Condition of an object that is nearly flush with the water level. Common usage: Condition of being tossed about or washed by waves or tide.

Backrush. Seaward return of the water following the uprush of the waves. For any given tide stage the point of farthest return seaward of the backrush is known as the limit of backrush or limit of backwash.

Backshore. That zone of the shore or beach lying between the foreshore and the coastline and acted on by waves only during severe storms, especially when combined with exceptionally high water.

Backwash. Water or waves thrown back by an obstruction, such as a ship, breakwater, or cliff.

Bank. Rising ground bordering a lake, river, or sea; on a river designated as right or left as it would appear facing downstream. Elevation of the sea floor of large area, surrounded by deeper water, but safe for surface navigation; a submerged plateau or shelf; a shoal or shallow.

Bar. Offshore ridge or mound of sand, gravel, or other unconsolidated material submerged at least at high tide, especially at the mouth of a river or estuary or lying a short distance from and usually parallel to the beach.

Barrier beach. Bar essentially parallel to the shore whose crest is above high water.

Barrier reef. Reef that is roughly parallel to the land but is some distance offshore, with deeper water intervening.

Bay. Recess in the shore or an inlet of a sea or lake between two capes or headlands, not so large as a gulf, but larger than a cove.

Baymouth bar. Bar extending partially or entirely across the mouth of a bay.

Bayou. Minor sluggish waterway or estuarial creek, tributary to or connecting other streams or bodies of water. Course is usually through lowlands or swamps.

Beach. Zone of unconsolidated material that extends landward from the low-water line to the place where there is marked change in material or physiographic form or to the line of permanent vegetation (usually the effective limit of storm waves). Seaward limit of the beach, unless otherwise specified, is the mean low-water line. Beach includes foreshore and backshore; sometimes refers to the material that is more or less in active transport alongshore or on- and offshore, rather than to the zone.

Beach barrier. Bar essentially parallel to the shore whose crest is above the high-water level.

Beach berm. Nearly horizontal portion of the beach or backshore formed by the deposit of material through wave action. Some beaches have no berms; others have one or several.

Beach cusp. One of a series of low mounds of beach material separated by crescent-shaped troughs spaced at more or less regular intervals along the beach face.

Beach erosion. Carrying away of beach materials by wave action, tidal currents, littoral currents, or wind.

Beach face. Section of the beach normally exposed to the action of the wave uprush. Foreshore zone of a beach.

Beach ridge. Essentially continuous mound of beach material behind the beach that has been heaped up by wave or other action. Ridges may occur singly or as a series of approximately parallel deposits.

Beach scarp. Almost vertical slope along the beach caused by erosion through wave action. May vary in height from a few inches to several feet depending on wave action and the nature and composition of the beach.

Beach width. Horizontal dimension of the beach as measured normal to the shore line.

Bell buoy. Buoy having a bell operated mechanically or by the action of waves, usually marking shoals or rocks.

Bench. Level or gently sloping erosion plane inclined seaward; nearly horizontal area at about the level of maximum high water on the sea side of a dike.

Bench mark. Fixed point used as a reference for elevations.

Berm crest. Seaward limit of a berm.

Bight. Slight indentation in the shore line of an open coast or bay, usually crescent shaped.

Blind rollers. Long, high swells that increase in height, almost to the breaking point, as they pass over shoals or run in shoaling water.

Bluff. High, steep bank or cliff.

Boat basin. Naturally or artificially enclosed or nearly enclosed body of water where small craft may anchor.

Bold coast. Prominent land mass that rises steeply from the sea.

Bore. Tidal flood with a high, abrupt front.

Bottom. Ground or bed under any body of water; the bottom of the sea.

Bottom (nature of). Composition or character of the bed of an ocean or other body of water, such as clay, coral, gravel, mud, ooze, pebbles, rock, shell, shingle, hard, or soft.

Boulder. Rounded rock more than 12 in. in diameter; larger than a cobblestone.

Breaker. Wave breaking on the shore, over a reef, etc. Breakers may be roughly classified into three kinds, although there is much overlapping: Spilling breakers break gradually over quite a distance; plunging breakers tend to curl over and break with a crash; and surging breakers peak up, but then instead of spilling or pumping they surge up the beach face.

Breaker depth. Still-water depth at the point where the wave breaks.

Breakwater. Structure protecting a shore area, harbor, anchorage, or basin from waves.

Bulkhead. Structure separating land and water areas, primarily designed to resist earth pressures.

Buoy. Float; especially, a floating object moored to the bottom to mark a channel, anchor, shoal rock, etc.

Buoyancy. Resultant of upward forces exerted by the water on a submerged or floating body equal to the weight of the water displaced by this body.

Canal. Artificial watercourse cut through a land area for use in navigation, irrigation, etc.

Can buoy. Squat and cylindrical or nearly cylindrical above water and conical below water.

Canyon. Oceanographical: deep submarine depression of valley form with relatively steep sides. Geographical: deep gorge or ravine with steep sides, often with a river flowing at its bottom.

Cape. Relatively extensive land area jutting seaward from a continent or large island that prominently marks a change in, or interrupts notably, the coastal trend; a prominent feature.

Capillary wave. Wave whose velocity of propagation is controlled primarily by the surface tension of the liquid in which the wave is traveling. Water waves of length less than 1 in. are considered to be capillary waves.

Causeway. Raised road across wet or marshy ground or across water.

Caustic. In refraction of waves, the name given to the curve to which adjacent orthogonals of waves, refracted by a bottom whose contour lines are curved, are tangents. Occurrence of a caustic always marks a region of crossed orthogonals and high wave convergence.

Channel. Natural or artificial waterway of perceptible extent that either periodically or continuously contains moving water, or that forms a connecting link between two bodies of water; part of a body of water deep enough to be used for navigation through an area otherwise too shallow for navigation; large strait, as the English Channel; deepest portion of a stream, bay, or strait through which the main volume or current of water flows.

Chart datum. Plane or level to which soundings on a chart are referred, usually taken to correspond to a low-water stage of the tide.

Chop. Short-crested waves that may spring up quickly in a fairly moderate breeze and break easily at the crest.

Clapotis. French equivalent for a type of standing wave; in American usage it is usually associated with the standing-wave phenomenon caused by the reflection of a wave train from a breakwater, bulkhead, or steep beach.

Cliff. High, steep face of rock; a precipice.

Coast. Strip of land of indefinite width (may be several miles) that extends from the seashore inland to the first major change in terrain features.

Coastal area. Land and sea area bordering the shore line.

Coastal current. One of the offshore currents flowing generally parallel to the shore line with a relatively uniform velocity (as compared to the littoral currents). Such currents are not related genetically to waves and resulting surf but may be composed of currents related to distribution of mass in ocean water (or local eddies), wind-driven currents, and/or tidal currents.

Coastal plain. Plain composed of horizontal or gently sloping strata of elastic materials fronting the coast and generally representing a strip of recently emerged sea bottom.

Coastline. Technically, the line that forms the boundary between the coast and the shore. Commonly, the line that forms the boundary between the land and the water.

Comber. Deep-water wave whose crest is pushed forward by a strong wind; much larger than a whitecap. Long-period spilling breaker.

Continental shelf. Zone bordering a continent and extending from the line of permanent immersion to the depth (usually about 100 fathoms) at which there is a marked or rather steep descent toward the great depths.

Contour. Line connecting the points on a land or submarine surface that have the same elevation; in topographic or hydrographic work, a line connecting all points of equal elevation above or below a datum plane.

Controlling depth. Least depth of water in the navigable parts of a waterway that limits the allowable draft of vessels.

Convergence. In refraction phenomena, the decreasing of the distance between orthogonals in the direction of wave travel. This denotes an area of increasing wave height and energy concentration. In wind setup phenomena, the increase in setup observed over that which would occur in an equivalent rectangular basin of uniform depth, caused by changes in planform or depth; also the decrease in basin width or depth causing such increase in setup.

Coral. Calcareous skeletons of various anthozoans and a few hydrozoans; also these skeletons when solidified into a stony mass. Many tropical islands, reefs, and atolls are formed of coral.

Cove. Small sheltered recess in a shore or coast, often inside a larger embayment.

Crest of berm. Seaward limit of a berm.

Crest of wave. Highest part of a wave; that part of the wave above still-water level.

Current. Flow of water.

Cusp. One of a series of naturally formed low mounds of beach material separated by crescent-shaped troughs spaced at more or less regular intervals along the beach face.

Cuspate bar. Crescent-shaped bar uniting with shore at each end. May be formed by a single spit growing from shore turning back to meet the shore again, or by two spits growing from shore uniting to form a bar of sharply cuspate form.

Cyloidal wave. Very steep, symmetrical wave whose crest forms an angle of 120°. Wave form is that of a cycloid. Trochoidal wave of maximum steepness.

Daily retardation of tides. Amount of time by which corresponding tidal phases grow later day by day (averages approximately 50 minutes).

Dan buoy. Buoy that carries a pole with a flag or light on it.

Drift current. Broad, shallow, slow-moving ocean or lake current.

Datum plane. Horizontal plane to which soundings, ground elevations, or water surface elevations are referred. Also called reference plane. Plane is called a tidal datum when defined by a certain phase of the tide. Common datum used on topographic maps is based on mean sea level.

Debris line. Line near the limit of storm wave uprush marking the landward limit of debris deposits.

Decay distance. Distance through which waves travel after leaving the generating area.

Decay of waves. Change that waves undergo after they leave a generating area (fetch) and pass through a calm or a region of lighter winds. In the process of decay, the significant wave height decreases and the significant wavelength increases.

Deep water. Water of depth such that surface waves are little affected by conditions on the ocean bottom. It is customary to consider water deeper than one-half the surface wavelength as deep water.

Deflation. Removal of material from a beach or other land surface by wind action.

Delta. Alluvial deposit, usually triangular, at the mouth of a river.

Depth. Vertical distance from the still-water level (or datum, as specified) to the bottom.

Depth of breaking. Still-water depth at the point where the wave breaks; also called breaker depth.

Derrick stone. Stone of such size as to require handling in individual pieces by mechanical means; generally 1 ton and up.

Diffraction of water waves. Phenomenon by which energy is transmitted laterally along a wave crest. When a portion of a train of waves is interrupted by a barrier, such as a breakwater, the effect of diffraction is manifested by propagation of waves into the sheltered region within the barrier's geometric shadow.

Dike (dyke). Wall or mound built around a low-lying area to prevent flooding.

Diurnal. Daily; recurring once each day, such as lunar day or solar day.

Diurnal tide. Tide with one high water and one low water in a tidal day.

Divergence. In refraction phenomena, the spreading of orthogonals in the direction of wave travel. This denotes an area of decreasing wave height and energy concentration. In wind setup phenomena, the decrease in setup observed under that which would occur in an equivalent rectangular basin of uniform depth, caused by changes in planform or depth. Also, the increase in basin width or depth causing such decrease.

Downcoast. In United States usage, the coastal direction generally trending toward the south.

Downdrift. Direction of predominant movement of littoral materials.

Drift. Speed at which a current runs; also, floating material deposited on a beach (driftwood) or deposit of a continental ice sheet, as a drumlin; sometimes used as an abbreviation of littoral drift.

Drift current. Broad, shallow, slow-moving ocean or lake current.

Dunes. Ridges or mounds of loose, wind-blown material, usually sand.

Duration. In wave forecasting, the length of time the wind blows in essentially the same direction over the fetch (generating area).

Ebb current. Movement of the tidal current away from shore or down a tidal stream.

Ebb tide. Nontechnical term referring to that period of tide between a high water and the succeeding low water; falling tide.

Echo sounder. Survey instrument that determines the depth of water by measuring the time required for a sound signal to travel to the bottom and return. May be either "sonic" or "supersonic" depending on the frequency of sound wave, the "sonic" being generally within the audible ranges (under 15,000 cycles per second).

Eddy. Circular movement of water formed on the side of a main current. Eddies may be created at points where the mainstream passes projecting obstructions.

Eddy current. Circular movement of water of comparatively limited area formed on the side of a main current. Eddies may be created at points where the mainstream passes projecting obstructions.

Eelgrass. Submerged marine plant with very long, narrow leaves, abundant along the North Atlantic coast.

Embankment. Artificial bank, mound, dike or the like, built to hold back water, carry a roadway, etc.

Embayed. Formed into a bay or bays, as an embayed shore.

Embayment. Indentation in a shore line forming an open bay.

Energy coefficient. Ratio of the energy in a wave per unit crest length transmitted forward with the wave at a point in shallow water to the energy in a wave per unit crest length transmitted forward with the wave in deep water. On refraction diagrams this is equal to the ratio of the distance between a pair of orthogonals at a selected point to the distance between the same pair of orthogonals in deep water. Also the square of the refraction coefficient.

Entrance. Avenue of access or opening to a navigable channel.

Erosion. Wearing away of land by the action of natural forces: on a beach, by the carrying away of beach material through wave action, tidal currents, littoral currents, or the action of the wind.

Escarpment. More or less continuous line of cliffs or steep slopes facing in one general direction, caused by erosion or faulting; also called "scarp."

Estuary. Portion of a stream influenced by the tide of the body of water into which it flows; a bay, as the mouth of a river, where the tide meets the river current.

Fairway. Parts of a waterway kept open and unobstructed for navigation.

Fathom. Unit of measurement used for soundings. Equal to 6 ft (1.83 m).

Fathometer. Copyrighted trade name for a type of echo sounder.

Feeder beach. Artificially widened beach serving to nourish downdrift beaches by natural littoral currents or forces.

Feeder current. Current that flows parallel to shore before converging and forming the neck of a rip current.

Fetch. In wave forecasting, the continuous area of water over which the wind blows in essentially a constant direction, sometimes used synonymously with fetch length; also generating area. In wind setup phenomena, for enclosed bodies of water, the distance between the points of maximum and minimum water surface elevations. This would usually coincide with the longest axis in the general wind direction.

Fetch length. In wave forecasting, the horizontal distance, in the direction of the wind, over which the wind blows.

Firth. Narrow arm of the sea; also, the opening of a river into the sea.

Fjord (fiord). Long, narrow arm of the sea between highlands.

Flood current. Movement of the tidal current toward the shore or up a tidal stream.

Flood tide. Nontechnical term referring to that period of tide between low water and the succeeding high water; a rising tide.

Foam line. Front of a wave as it advances shoreward after it has broken.

Following wind. In wave forecasting, wind blowing in the same direction in which waves are traveling.

Foreshore. That part of the shore lying between the crest of the seaward berm or the upper limit of wave wash at high tide and the ordinary low-water mark that is ordinarily traversed by the uprush and backrush of the waves as the tides rise and fall.

Freeboard. Additional height of a structure above design high-water level to prevent overflow; also, at a given time, the vertical distance between the water level at the top of the structure.

Freshet. Rapidly rising flood in a stream resulting from snow melt or rainfall.

Fringing reef. Reef attached to an insular or continental shore.

Front of the fetch. In wave forecasting, that end of the generating area toward which the wind is blowing.

Generating area. In wave forecasting, the continuous area of water surface over which the wind blows in essentially a constant direction. Sometimes used synonymously with fetch length.

Generation of waves. Creation of waves by natural or mechanical means. In wave forecasting, the creation and growth of waves caused by a wind blowing over a water surface for a certain period of time. Area involved is called "generating area" or "fetch."

Geometric mean diameter. Diameter equivalent of the arithmetic mean of the logarithmic frequency distribution. In the analysis of beach sands it is taken as that grain diameter determined graphically by the intersection of a straight line through selected boundary sizes (generally points on the distribution curve where 16 and 84% of the sample by weight are coarser) and a vertical line through the median diameter of the sample.

Geometric shadow. In wave diffraction theory, the area outlined by drawing straight lines paralleling the direction of wave approach through the extremities of the protective structure. It differs from the actual protected area to the extent that the diffraction and refraction effects modify the wave pattern.

Geomorphology. Branch of both physiography and geology that deals with the form of the earth, the general configuration of its surface, and the changes that take place in the evolution of land forms.

Gradient (grade). With reference to winds or currents, the rate of increase or decrease in speed, usually in the vertical, or the curve that represents this rate.

Gravity wave. Wave whose velocity of propagation is controlled primarily by gravity. Water waves of a length greater than 2 in. are considered gravity waves.

Groin. Shore protective structure (usually built perpendicular to the shore line) to trap littoral drift or retard erosion of the shore. It is built narrow in width (measured parallel to the shore line) with length that may vary from less than 100 to several hundred feet (extending from a point landward of the shore line out into the water). Groins may be classified as permeable or impermeable, impermeable groins having a solid or nearly solid structure and permeable groins having openings through them of sufficient size to permit passage of appreciable quantities of littoral drift.

Ground swell. Long, high ocean swell; also, this swell as it rises to prominent height in shallow water; however, usually does not become so high or dangerous as blind rollers.

Ground water. Subsurface water occupying the zone of saturation. Term is applied only to water below the water table.

Group velocity. Velocity at which a wave group travels. In deep water it is equal to one-half the velocity of the individual waves within the group.

Gulf. Relatively large portion of sea partly enclosed by land.

Gut. Narrow passage, such as a strait or inlet or a channel in otherwise shallower water, generally formed by water in motion.

Harbor. Protected part of a sea, lake, or other body of water used by vessels as a place of safety.

Head. Point or portion of land jutting out into the sea, lake, or other body of water; a cape or promontory; now, usually specifically, a promontory especially bold and cliff-like.

Head of rip. Section of a rip current that has widened out seaward of the breakers.

Height of wave. Vertical distance between a crest and the preceding trough.

Higher high water. Higher of the two high waters of any tidal day. Single high water occurring daily during periods when the tide is diurnal is considered to be a higher high water.

Higher low water. Higher of two low waters of any tidal day.

High-tide high water. Maximum height reached by each rising tide.

High-water line. In strictness, the intersection of the plane of mean high water with the shore. Shore line delineated on the nautical charts of the Coast and Geodetic Survey is an approximation of the mean high-water line.

High water of ordinary spring tides. Tidal datum appearing in some British publications, based on high water of ordinary spring tides.

Hindcasting—wave. Calculation from historic synoptic wind charts of the wave characteristics that probably occurred at some past time.

Hinterland. Region inland from the coast.

Hook. Spit or narrow cape turned landward at the outer end, resembling a hook in form.

Hydraulic jump. In fluid flow, a change in flow conditions accompanied by a stationary, abrupt turbulent rise in water level in the direction of flow. Type of stationary wave.

Hydrography. Configuration of an underwater surface including its relief, bottom materials, coastal structures, etc., and descriptions and study of sea, lakes, rivers, and other waters.

Inlet. Short, narrow waterway connecting a bay, lagoon, or similar body of water with a large parent body of water; arm of the sea (or other body of water) that is long compared to its width and that may extend a considerable distance inland.

Inshore current. Any current in or landward of the breaker zone.

Inshore (zone). In beach terminology, zone of variable width extending from the shore face through the breaker zone.

Insular shelf. Zone surrounding an island extending from the line of permanent immersion to the depth (usually about 100 fathoms) at which there is a marked or rather steep descent toward the great depths.

Internal waves. Waves that occur within a fluid whose density changes with depth, either abruptly at a sharp surface of discontinuity (an interface) or gradually. Their amplitude is greatest at the density discontinuity or, in the case of a gradual density change, somewhere in the interior of the fluid and not at the free upper surface where the surface waves have their maximum amplitude.

Isthmus. Narrow strip of land, bordered on both sides by water, that connects two larger bodies of land.

Jetty. On open seacoasts, a structure extending into a body of water, designed to prevent shoaling of a channel by littoral materials and to direct and confine the stream or tidal flow. Jetties are built at the mouth of a river or tidal inlet to help deepen and stabilize a channel.

Kelp. General name for several species of large seaweeds. Mass or growth of large seaweed or any of various large brown seaweeds.

Key. Low insular bank of sand, coral, etc., for example, one of the islets off the southern coast of Florida; also called "cay."

Kinetic energy of waves. In a progressive oscillatory wave, a summation of the energy of motion of the particles within the wave. This energy does not advance with the wave form.

Knoll. Submerged elevation of rounded shape rising from the ocean floor, but less prominent than a seamount or small rounded hill.

Knot. Unit of speed used in navigation. Equal to 1 nautical mile (6080.20 ft)/hour.

Lagoon. Shallow body of water, as a pond or lake, which usually has a shallow, restricted outlet to the sea.

Land breeze. Light wind blowing from the land caused by unequal cooling of land and water masses.

Landlocked. Area of water enclosed or nearly enclosed by land, as a bay or a harbor, thus protected from the sea.

Landmark. Conspicuous object—natural or artificial—located near or on land that aids in fixing the position of an observer.

Land-sea breeze. Combination of a land and a sea breeze as a diurnal phenomenon.

Lead line. Line, wire, or cord used in sounding. Weighted at one end with a plummet (sounding lead).

Lee. Shelter, or the part or side sheltered or turned away from the wind or waves. Chiefly nautical: the quarter or region toward which the wind blows.

Leeward. Direction toward which the wind is blowing; the direction toward which waves are traveling.

Length of wave. Horizontal distance between similar points on two successive waves measured perpendicularly to the crest.

Levee. Dike or embankment for the protection of land from inundation.

Littoral. Of or pertaining to a shore, especially of the sea; coastal region.

Littoral current. Nearshore currents primarily due to wave action, such as longshore currents and rip currents.

Littoral deposits. Deposits of littoral drift.

Littoral drift. Material moved in the littoral zone under the influence of waves and currents.

Littoral transport. Movement of materials along the shore in the littoral zone by waves and currents.

Longshore current. Current in the surf zone moving essentially parallel to the shore, usually generated by waves breaking at an angle to the shore line.

Lower high water. Lower of the two high waters of any tidal day.

Lower low water. Lower of the two low waters of any tidal day. Single low water occurring daily during periods when the tide is diurnal is considered to be a lower low water.

Low tide (low water). Minimum height reached by each falling tide.

Low-water datum. Approximation plane of mean low water that has been adopted as a standard reference plane.

Low-water line. Intersection of any standard low-tide datum plane with the shore.

Low water of ordinary spring tides. Tidal datum appearing in some British publications, based on low water of ordinary spring tides.

Mangrove. Particular kind of tropical tree or shrub with thickly matted roots, confined to low-lying brackish areas.

Marigram. Graphic record of the rise and fall of the tide.

Marsh. Tract of soft, wet, or periodically inundated land, generally treeless and usually characterized by grasses and other low growth.

Mass transport. Net transfer of water by wave action in the direction of wave travel.

Mean higher high water. Average height of the higher high water over a 19-year period. For shorter periods of observation, corrections are applied to eliminate known variations and reduce the result to the equivalent of a mean 19-year value.

Mean high water. Average height of the high waters over a 19-year period. For shorter periods of observations, corrections are applied to eliminate known variations and reduce the result to the equivalent of a mean 19-year value. All high-water heights are included in the average when the type of tide is either semidiurnal or mixed. Only the higher high-water heights are included in the average when the type of tide is diurnal. So determined, mean high water in the latter case is the same as mean higher high water.

Mean-high-water springs. Average height of the high waters occurring at the time of spring tide.

Mean lower low water. Average height of the lower low waters over a 19-year period. For shorter periods of observations, corrections are applied to eliminate known variations and reduce the result to the equivalent of a mean 19-year value.

Mean low water. Average height of the low waters over a 19-year period. For shorter periods of observations, corrections are applied to eliminate known variations and reduce the result to the equivalent of a mean 19-year value. All low-water heights are included in the average when the type of tide is either semidiurnal or mixed. Only the lower low-water heights are included in the average when the type of tide is diurnal. So determined, mean low water in the latter case is the same as mean lower low water.

Mean-low-water springs. Average height of low water occurring at the time of the spring tides. It is usually derived by taking a plane depressed below the half-tide level by an amount equal to one-half the spring range of tide, necessary corrections being applied to reduce the result to a mean value. This plane is used to a considerable extent for hydrographic work outside the United States and is the plane of reference for the Pacific approaches to the Panama Canal.

Mean sea level. Average height of the surface of the sea for all stages of the tide over a 19-year period, usually determined from hourly height readings.

Mean tide level (half-tide level). Plane midway between high water and mean low water.

Median diameter. Diameter that marks the division of a given sample into two equal parts by weight, one part containing all grains larger than that diameter and the other part containing all grains smaller.

Minimum duration. Time necessary for steady-state wave conditions to develop for a given wind velocity over a given fetch length.

Mixed tide. Type of tide in which the presence of a diurnal wave is conspicuous by a large inequality in either the high- or low-water heights with two high waters and two low waters usually occurring each tidal day. In strictness, all tides are mixed, but the name is usually applied without definite limits to the tide intermediate to those predominantly semidiurnal and those predominantly diurnal.

Mole. In coastal terminology, a massive solid-fill structure of earth (generally revetted), masonry, or large stone. May serve as a breakwater or pier.

Monolithic. Like a single stone or block. Therefore in breakwaters, the type of construction in which the structure's component parts are bound together to act as one.

Mud. Fluid-to-plastic mixture of finely divided particles of solid material and water.

Nautical mile. Length of a minute of arc, 1/21,600 of an average great circle of the earth. Generally 1 minute of latitude is considered equal to 1 nautical mile. Accepted United States value is 6080.20 ft—approximately 1.15 times as long as the statute mile of 5280 ft.

Neap tide. Tide occurring near the time of quadrature of the moon. Neap-tidal range is usually 10 to 30% less than the mean tidal range.

Nearshore (zone). In beach terminology, an indefinite zone extending seaward from the shore line somewhat beyond the breaker zone. It defines the area of nearshore currents.

Nearshore circulation. Ocean circulation pattern composed of nearshore and coastal currents.

Nearshore current system. Current system caused primarily by wave action in and near the breaker zone and consisting of four parts: shoreward mass transport of water; longshore currents; seaward return flow, including rip currents; and the longshore movement of the expanding heads of rip currents.

Neck. Narrow band of water flowing seaward through the surf.

Nip. Cut made by waves in a shore line of emergence.

Nodal zone. Area at which the predominant direction of the littoral transport changes.

Nourishment. Process of replenishing a beach. May be brought about by natural means, such as littoral drift, or by artificial means, such as by the deposition of dredged materials.

Nun or nut buoy. Buoy conical in shape.

Oceanography. Science treating of the oceans, their forms, physical features, and phenomena.

Offshore. In beach terminology, the comparatively flat zone of variable width extending from the breaker zone to the seaward edge of the continental shelf; direction seaward from the shore.

Offshore current. Any current in the offshore zone. Any current flowing away from shore.

Offshore wind. Wind blowing seaward from the land in the coastal area.

Onshore. Direction landward from the sea.

Onshore wind. Wind blowing landward from the sea in the coastal area.

Opposing wind. In wave forecasting, a wind blowing in the opposite direction to that in which the waves are traveling.

Orbit. In water waves, path of a water particle affected by the wave motion. In deep-water waves the orbit is nearly circular and in shallow-water waves the orbit is nearly elliptical. In general, orbits are slightly open in the direction of wave motion, giving rise to mass transport.

Orbital current. Flow of water accompanying the orbital movement of the water particles in a wave. Not to be confused with wave-generated littoral currents.

Orthogonal. On a refraction diagram, a line drawn perpendicular to the wave crests.

Oscillation. Periodic motion to and fro or up and down.

Oscillatory wave. Wave in which each individual particle oscillates about a point with little or no permanent change in position. Term is commonly applied to progressive oscillatory waves in which only the form advances, the individual particles moving in closed orbits. Distinguished from a wave of translation.

Outfall. Vent of a river, drain, etc. Structure extending into a body of water for the purpose of discharging sewage, storm runoff, or cooling water.

Overwash. Portion of the uprush that carries over the crest of a berm or a structure.

Parapet. Low wall built along the edge of a structure, as on a seawall or quay.

Particle velocity. For waves, the velocity induced by wave motion with which a specific water particle moves.

Pass. In hydrographic usage, a navigable channel through a bar, reef, or shoal or between closely adjacent islands.

Peninsula. Elongated portion of land nearly surrounded by water and connected to a larger body of land.

Periodic current. Current caused by the tide-producing forces of the moon and the sun; part of the same general movement of the sea that is manifested in the vertical rise and fall of the tides.

Permafrost. Permanently frozen subsoil.

Permanent currents. Current that runs continuously, independent of the tides and temporary causes. Permanent currents include the freshwater discharge of a river and the currents that form the general circulatory systems of the oceans.

Petrography. Description and systematic classification of rocks.

Pier. Structure extending out into the water from the shore to serve as a landing place, recreational facility, etc., rather than to afford coastal protection.

Pile. Long, slender piece of wood, concrete, or metal driven or jetted into the earth or seabed to serve as a support or protection.

Piling. Group of piles.

Pinnacle. Tall, slender, pointed, rocky mass.

Plain. Extent of level or nearly level land.

Planform. Outline or shape of a body of water as determined by the still-water line.

Plateau. Elevated plain, tableland, or flat-topped region of considerable extent.

Plunge point. For a plunging wave, the point at which the wave curls over and falls; final breaking point of the waves just before they rush up on the beach.

Point. Extreme end of a cape; the outer end of any land area protruding into the water, usually less prominent than a cape.

Port. Place where vessels may discharge or receive cargo. May be the entire harbor, including its approaches and anchorages, or may be the commercial part of a harbor, where the quays, wharves, and facilities for transfer of cargo and docks, repair shops, etc. are situated.

Potential energy of waves. In a progressive oscillatory wave, the energy resulting from the elevation or depression of the water surface from the undisturbed level. This energy advances with the wave form.

Profile beach. Intersection of the ground surface with a vertical plane. May extend from the top of the dune line to the seaward limit of sand movement.

Progressive wave. Wave manifested by the progressive movement of the wave form.

Promontory. High point of land projecting into a body of water; a headland.

Propagation of waves. Transmission of waves through water.

Prototype. In laboratory usage, original structure, concept, or phenomenon used as a basis for constructing a scale model or copy.

Quay. Stretch of paved bank or a solid artificial landing place parallel to the navigable waterway for use in loading and unloading vessels.

Quicksand. Loose, yielding, wet sand that offers no support to heavy objects. Upward flow of the water has a velocity that eliminates contact pressures between the sand grains and causes the sand-water mass to behave like a fluid.

Recession (of a beach). Continuing landward movement of the shore line; net landward movement of the shore line over a specified time.

Reef. Chain or range of rock or coral elevated above the surrounding bottom of the sea, generally submerged and dangerous to surface navigation.

Reference point. Specified location in plan and/or elevation to which measurements are referred.

Reference station. Station for which tidal constants have previously been determined and that is used as a standard for the comparison of simultaneous observations at a second station; also, a station for which independent daily predictions are given in the tide or current tables, from which corresponding predictions are obtained for other stations by means of differences or factors.

Reflected wave. Wave that is returned seaward when a wave impinges on a very steep beach, barrier, or other reflecting surface.

Refraction coefficient. Square root of the ratio of the spacing between adjacent orthogonals in deep water and in shallow water at a selected point. When multiplied by the shoaling factor, this becomes the wave-height coefficient or the ratio of the refracted wave height at any point to the deep-water wave height. Also, the square root of the energy coefficient.

Refraction diagram. Drawing showing positions of wave crests and/or orthogonals in a given area for a specific deep-water wave period and direction.

Refraction of water waves. Process by which the direction of a wave moving in shallow water at an angle to the contours is changed. Part of the wave advancing in shallower water moves more slowly than that part still advancing in deeper water, causing the wave crest to bend toward alignment with the underwater contours; bending of wave crests by currents.

Retardation. Amount of time by which corresponding tidal phase grow later day by day (averages approximately 50 minutes).

Retrogression (of a beach). Continuing landward movement of the shore line; net landward movement of the shore line over a specified time.

Revetment. Facing of stone, concrete, etc., built to protect a scarp, embankment, or shore structure against erosion by wave action or currents.

Ria. Long, narrow inlet, with depth gradually diminishing inward.

Rill marks. Tiny drainage channels in a beach caused by the flow seaward of water left in the sands of the upper part of the beach after the retreat of the tide or after the dying down of storm waves.

Rip. Body of water made rough by waves meeting an opposing current, particularly a tidal current; often found where tidal currents are converging and sinking.

Riparian. Pertaining to the banks of a body of water.

Riparian rights. Rights of a person owning land containing or bordering on a watercourse or other body of water in or to its banks, bed, or waters.

Rip current. Strong surface current of short duration flowing seaward from the shore. Usually appears as a visible band of agitated water and is the return movement of water piled up on the shore by incoming waves and wind. With the seaward movement concentrated in a limited band, its velocity is somewhat accentuated. Rip current consists of three parts: The "feeder current" flowing parallel to the shore inside the breakers; "neck," where the feeder currents converge and flow through the breakers in a narrow band or "rip"; and "head," where the current widens and slackens outside the breaker line. Rip current is often miscalled a "rip tide."

Ripple. Ruffling of the surface of water, hence a little curling wave or undulation. Wave controlled to a significant degree by both surface tension and gravity.

Ripple marks. Small, fairly regular ridges in the bed of a waterway or on a land surface caused by water currents or wind. As their form is approximately normal to the direction of current or wind, they indicate both the presence and the direction of currents or winds.

Riprap. Layer, facing, or protective mound of stones randomly placed to prevent erosion, scour, or sloughing of a structure or embankment; also, the stone so used.

Roadstead (nautical). Sheltered area of water near shore where vessels may anchor in relative safety.

Rock. Engineering: Natural aggregate of mineral particles connected by strong and permanent cohesive forces. In igneous and metamorphic rocks, it consists of interlocking crystals; in sedimentary rocks, or closely packed mineral grains, often bound together by a natural cement. Since the terms "strong" and "permanent" are subject to different interpretations, the boundary between rock and soil is necessarily an arbitrary one. Geological: Material that forms the essential part of the earth's solid crust and includes loose incoherent masses, such as beds of sand, gravel, clay, or volcanic ash, as well as the very firm, hard, and solid masses of granite, sandstone, and limestone. Most rocks are aggregates of one or more minerals, but some are composed entirely of glassy matter or of a mixture of glass and minerals.

Roller. Indefinite term. Roller sometimes considered to be one of a series of long-crested, large waves that roll in on a coast, as after a storm.

Rubble. Loose, angular, water-worn stones along a beach. Rough, irregular fragments of broken rock.

Runnel. Corrugation (trough) of the foreshore (or the bottom just offshore) formed by wave and/or tidal action. Larger than the trough between ripple marks.

Run-up. Rush of water up a structure on the breaking of a wave; also, uprush. Amount of run-up is the vertical height above still-water level that the rush of water reaches.

Saltation. Method of sand movement in a fluid in which individual particles leave the bed by bounding nearly vertically and, because the motion of the fluid is not strong or turbulent enough to retain them in suspension, return to the bed at some distance downstream. Travel path of the particles is a series of hops and bounds.

Salt marsh. Marsh periodically flooded by salt water.

Sand bar. In a river, ridge of sand built up to or near the surface by river currents.

Sand reef. Synonymous with bar.

Scarp. More or less continuous line of cliffs or steep slopes facing in one general direction, caused by erosion or faulting; also, escarpment.

Scour. Erosion, especially by moving water.

Sea. Ocean, or, alternatively, a large body of (usually) salt water smaller than an ocean; waves caused by wind at the place and time of observation.

Sea (state of). Description of the sea surface with regard to wave action.

Sea breeze. Breeze blowing from the sea toward the land; light wind blowing toward the land caused by unequal heating of land and water masses.

Sea cliff. Cliff situated at the seaward edge of the coast.

Sea mount. Submarine mountain rising more than 500 fathoms above the ocean floor.

Sea puss. Dangerous longshore current; a rip current, caused by return flow; loosely, the submerged channel or inlet through a bar caused by those currents.

Seashore. Shore of a sea or ocean.

Sea valley. Submarine depression of broad valley form without the steep side slopes that characterize a canyon.

Seawall. Structure separating land and water areas primarily designed to prevent erosion and other damage due to wave action.

Seiche. Periodic oscillation of a body of water whose period is determined by the resonant characteristics of the containing basin as controlled by its physical dimensions. These periods generally range from a few minutes to an hour or more. Originally the term was applied only to lakes, but now also to harbors, bays, oceans, etc.

Seismic sea wave (tsunami). Generally a long-period wave caused by an underwater seismic disturbance or volcanic eruption. Commonly misnamed "tidal wave."

Semidiurnal tides. Tide with two high and two low waters in a tidal day, with comparatively little diurnal inequality.

Set of current. Direction toward which a current flows.

Shallow water. Water of such depth that surface waves are noticeably affected by bottom topography. It is customary to consider water of depths less than half the surface wave length as shallow water. More strictly, in hydrodynamics with regard to progressive gravity waves, water in which the depth is less than one-twenty-fifth the wave length.

Sheet pile. Pile with a generally flat cross section to be driven into the ground or seabed and meshed or interlocked with like members to form a diaphragm, wall, or bulkhead.

Shingle. Loosely and commonly, beach material coarser than ordinary gravel, especially any having flat or flattish pebbles; strictly and accurately, beach material of smooth, well-rounded pebbles that are roughly the same size. Spaces between pebbles are not filled with finer materials.

Shoal. Detached elevation of the sea bottom comprised of any material except rock or coral, which may endanger surface navigation. To become shallow gradually; to cause to become shallow; to proceed from a greater to a lesser depth of water.

Shoaling coefficient. Ratio of the height of a wave in water of any depth to its height in deep water with the effect of refraction eliminated. Sometimes called "shoaling factor" or "depth factor."

Shore. Strip of ground bordering any body of water. Shore of unconsolidated material is usually called a beach.

Shore face. Narrow zone seaward from the low-tide shore line permanently covered by water, over which the beach sands and gravels actively oscillate with changing wave conditions.

Shore line. Intersection of a specified plane of water with the shore or beach; for example, high-water shore line would be the intersection of the plane of mean high water with the shore or beach. Line delineating the shore line on U.S. Coast and Geodetic Survey nautical charts and surveys approximates the mean-high-water line.

Significant wave. Statistical term denoting waves with the average height and period of the one-third highest waves of a given wave group. Composition of the higher waves depends on the extent to which the lower waves are considered. Wave of significant wave period and significant wave height.

Significant wave height. Average height of the one-third highest waves of a given wave group. Compositions of the highest waves depends on the extent to which the lower waves are considered. In wave record analysis, the average height of the highest one-third of a selected number of waves, this number being determined by dividing the time of record by the significant period.

Significant wave period. Arbitrary period generally taken as the period of the one-third highest waves within a given group. Composition of the highest waves depends on the extent to which the lower waves are considered. In wave record analysis this is determined as the average period of the most frequently recurring of the larger well-defined waves in the record under study.

Slack tide (slack water). State of a tidal current when its velocity is near zero, especially at the moment when a reversing current changes direction and its velocity is zero. Sometimes considered the intermediate period between ebb and flood during which the velocity of the currents is less than 0.1 knot.

Slip. Space between two piers, wharves, etc., for the berthing of vessels.

Slope. Degree of inclination to the horizontal, usually expressed as a ratio, such as 1 : 25 or 1 on 25, indicating 1 unit rise in 25 units of horizontal distance. Sometimes described by such adjectives as steep, moderate, gentle, mild, or flat.

Slough. Small muddy marshland or tidal waterway that usually connects other tidal areas; a tideland or bottomland creek.

Soil classification (size). Arbitrary division of a continuous scale of sizes such that each scale unit or grade may serve as a convenient class interval for conducting the analysis or for expressing the results of an analysis.

Solitary wave. Wave consisting of a single elevation (above the water surface), of height not necessarily small compared to the depth, and neither followed nor preceded by another elevation or depression of the water surfaces.

Sorting coefficient. Coefficient used in describing the distribution of grain sizes in a sample of unconsolidated material.

Sound

1. Wide waterway between the mainland and an island or connecting two sea areas; relatively long arm of the sea or ocean forming a channel between an island and a mainland or connecting two larger bodies, as a sea and the ocean, or two parts of the same body. Usually wider and more extensive than a strait.

2. To measure or ascertain the depth of water as with sounding lines.

Sounding. Measured depth of water. On hydrographic charts the soundings are adjusted to a specific plane of reference.

Sounding datum. Plane to which soundings are referred.

Sounding line. Line, wire, or cord used in sounding. It is weighted at one end with a plummet (sounding lead).

Spar buoy. Vertical, slender spar anchored at one end.

Spit. Small point of land or submerged ridge running into a body of water from the shore.

Spring tide. Tide that occurs at or near the time of new and full moon and that rises highest and falls lowest from the mean level.

Standing wave. Type of wave in which the surface of the water oscillates vertically between fixed points, called "nodes," without progression. Points of maximum vertical rise and fall are called "antinodes" or "loops." At the nodes the underlying water particles exhibit no vertical motion, but maximum horizontal motion. At the antinodes the underlying water particles have no horizontal motion and maximum vertical motion. They may be the result of two equal progressive wave trains traveling through each other in opposite directions. Sometimes called "stationary wave."

Stand of tide. Interval at high or low water when there is no sensible change in the height of the tide. Water level is stationary at high and low water for only an instant, but the change in level near these times is so slow that it is not usually perceptible.

Stationary wave. Wave of essentially stable form that does not move with respect to a selected reference point; a fixed swelling. Sometimes called "standing wave."

Still-water level. Elevation of the surface of the water if all wave action were to cease.

Stone. Rock or rocklike matter used as a building material; small or specific piece of rock.

Storm tide. Rise of water accompanying a storm caused by wind stresses on the water surface.

Strait. Relatively narrow waterway between two larger bodies of water.

Stream. Course of water flowing along a bed in the earth; a current in the sea formed by wind action, water density differences, etc.

Stream current. Narrow, deep, and fast-moving ocean current.

Submarine valley. Prolongation of a land valley into or across the continental or insular shelf, which generally gives evidence of having been formed by stream erosion.

Surf. Wave activity in the area between the shore line and the outermost limit of breakers.

Surf beat. Irregular oscillations of the nearshore water level, with periods of the order of several minutes.

Surf zone. Area between the outermost breaker and the limit of wave uprush.

Surge. Name applied to wave motion with a period intermediate between that of the ordinary wind wave and that of the tide, from $\frac{1}{2}$ to 60 minutes. It is of low height, usually less than 0.3 ft; in fluid flow, long interval variations in velocity and pressure, not necessarily periodic, perhaps even transient in nature.

Swamp. Tract of wet spongy land, frequently inundated by fresh- or saltwater, and characteristically dominated by trees and shrubs.

Swash. Rush of water up onto the beach following the breaking of a wave.

Swash channel. On the open shore, a channel cut by flowing water in its return to the parent body, such as a rip channel; secondary channel passing through or shoreward of an inlet or river bar.

Swash mark. Thin wavy line of fine sand, mica scales, bits of seaweed, etc., left by the uprush when it recedes from its upward limit of movement on the beach face.

Swell. Wind-generated waves that have advanced into regions of weaker winds or calm.

Terrace. Horizontal or nearly horizontal natural or artificial topographic feature interrupting a steeper slope, sometimes occurring in a series.

Tidal current. Current caused by the tide-producing forces of the moon and the sun; a part of the same general movement of the sea that is manifested in the vertical rise and fall of the tides.

Tidal day. Time of the rotation of the earth with respect to the moon, or the interval between two successive upper transits of the moon over the meridian of a place—about 24.84 solar hours (24 hours and 50 minutes) in length or 1.035 times as great as the mean solar day.

Tidal flats. Marshy or muddy land areas that are covered and uncovered by the rise and fall of the tide.

Tidal inlet. Natural inlet maintained by tidal flow; loosely, any inlet in which the tide ebbs and flows.

Tidal period. Interval of time between two consecutive like phases of the tide.

Tidal pool. Pool of water remaining on a beach or reef after recession of the tide.

Tidal prism. Total amount of water that flows into the harbor or out again with movement of the tide, excluding any freshwater flow.

Tidal range. Difference in height between consecutive high and low waters.

Tidal rise. Height of tide as referred to the datum of a chart.

Tide. Periodic rising and falling of the water that results from gravitational attraction of the moon and sun acting upon the rotating earth. Although the accompanying horizontal movement of the water resulting from the same cause is also sometimes called the "tide," it is preferable to designate the latter as tidal current, reserving the name "tide" for the vertical movement.

Tombolo. Area of unconsolidated material, deposited by wave action or currents, that connects a rock or island, etc., to the mainland or to another island.

Topography. Configuration of a surface, including its relief, the position of its streams, roads, buildings, etc.

Training wall. Wall or jetty to direct current flow.

Transitional water. In regard to progressive gravity waves, water whose depth is less than one-half but more than one-twenty-fifth the wavelength. Often called "shallow water."

Trochoidal wave. Progressive oscillatory wave whose form is that of a prolate cycloid or trochoid. It is approximated by waves of small amplitude.

Trough of wave. Lowest part of a wave form between successive crests; also, that part of a wave below still-water level.

Tsunami. Generally long-period wave caused by underwater seismic disturbance or volcanic eruption. Commonly misnamed "tidal wave."

Undertow. Current below water surface flowing seaward; also, the receding water below the surface from waves breaking on a shelving beach. Actually "undertow" is largely mythical. As the backwash of each wave flows down the beach, a current is formed that flows seaward; however, it is a periodic phenomenon. Most common phenomena referred to as "undertow" are actually the rip currents in the surf. Often uniform return flows seaward or lakeward are termed "undertow," though these flows will not be as strong as rip currents.

Underwater gradient. Slope of the sea bottom.

Undulation. Continuously propagated motion to and fro, in any fluid or elastic medium, with no permanent translation of the particles themselves.

Upcoast. In United States usage, the coastal direction generally trending toward the north.

Updrift. Direction opposite that of the predominant movement of littoral materials.

Uplift. Upward water pressure on the base of a structure or pavement.

Uprush. Rush of water up onto the beach following the breaking of a wave.

Variability of waves. Variation of heights and periods between individual waves within a wave train. Wave trains are not composed of waves of equal height and period, but rather of heights and periods that vary in a statistical manner; variation in direction of propagation of waves leaving the generating area. The variation in height along the crest. Usually called "variation along the wave."

Velocity of waves. Speed with which an individual wave advances.

Viscosity. Internal friction due to molecular cohesion in fluids. Internal properties of a fluid that offer resistance to flow.

Water line. Juncture of land and sea. This line migrates, changing with the tide or other fluctuation in the water level. Where waves are present on the beach, this line is also known as the limit of backrush. Approximately the intersection of the land with the still-water level.

Wave. Ridge, deformation, or undulation of the surface of a liquid.

Wave age. Ratio of wave velocity to wind velocity (in wave forecasting theory).

Wave amplitude. In hydrodynamics, one-half the wave height; in engineering usage, loosely, the wave height from crest to trough.

Wave crest. Highest part of a wave; also, that part of the wave above still-water level.

Wave crest length. Length of a wave along its crest. Sometimes called crest width.

Wave decay. Change that waves undergo after they leave a generating area (fetch) and pass through a calm or a region of lighter or opposing winds. In the process of decay, the significant wave height decreases and the significant wave length increases.

Wave direction. Direction from which a wave approaches.

Wave forecasting. Theoretical determination of future wave characteristics, usually from observed or predicted meteorological phenomena.

Wave generation. Creation of waves by natural or mechanical means. In wave forecasting, the growth of waves caused by a wind blowing over a water surface for a certain period of time. Area involved is called the generating area or fetch.

Wave group. Series of waves in which the wave direction, wavelength, and wave height vary only slightly.

Wave height. Vertical distance between a crest and the preceding trough.

Wave height coefficient. Ratio of the wave height at a selected point to the deep-water wave height. Refraction coefficient multiplied by the shoaling factor.

Wave hindcasting. Calculation from historic synoptic wind charts of the wave characteristics that probably occurred at some past time.

Wavelength. Horizontal distance between similar points on two successive waves measured perpendicularly to the crest.

Wave of translation. Wave in which the water particles are permanently displaced to a significant degree in the direction of wave travel. Distinguished from an oscillatory wave.

Wave period. Time for a wave crest to traverse a distance equal to one wavelength. Time for two successive wave crests to pass a fixed point.

Wave propagation. Transmission of waves through water.

Wave refraction. Process by which the direction of a train of waves moving in shallow water at an angle to the contours is changed. The part of the wave train advancing in shallower water moves more slowly than that part still advancing in deeper water, causing the wave crests to bend toward alignment with the underwater contours.

Wave steepness. Ratio of a wave's height to its length.

Wave train. Series of waves from the same direction.

Wave trough. Lowest part of a wave form between successive crests; also, that part of a wave below still-water level.

Wave variability. Variation of heights and periods between individual waves within a wave train. Wave trains are not composed of waves of equal height and period, but rather of heights and periods that vary in a statistical manner; variation in direction of propagation of waves leaving the generating area. The variation in height along the crest. Usually called "variation along the wave."

Wave velocity. Speed with which an individual wave advances.

Wharf. Structure built on the shore of a harbor, river, canal, etc., so that vessels may lie alongside to receive and discharge cargo and passengers.

Whistling buoy. Buoy operates by marking shoals or channel entrances.

Whitecap. On the crest of a wave, the white froth caused by wind.

Wind. Horizontal natural movement of air; air naturally in motion with any degree of velocity.

Windchop. Short-crested waves that may spring up quickly in a fairly moderate breeze and break easily at the crest.

Windward. Direction from which the wind is blowing.

Wind waves. Waves formed and built up by the wind; loosely, any wave generated by wind.

Wind setup. Vertical rise in the still-water level on the leeward side of a body of water caused by wind stresses on the surface of the water; difference in still-water levels on the windward and the leeward sides of a body of water caused by wind stresses on the surface of the water. Synonymous with wind tide. "Wind tide" is usually used in reference to the ocean and large bodies of water, whereas "wind setup" is usually used in reference to reservoirs and smaller bodies of water.

WELLS AND WATER DISTRIBUTION

Abandoned well. Well whose use has been discontinued or that is in such a state of disrepair that continued use for obtaining groundwater or for other useful purposes is impracticable.

Access port. Water supply wells are equipped with a usable access port or airline and an inside diameter opening of 0.5 in. minimum so that the position of the water level can be determined at any time. The port is installed and maintained to prevent entrance of water or foreign material.

Approved water supply. Any water supply approved by, or under the public health supervision of, a public health agency of the state.

Aquifer. Geologic formation, a group of such formations, or a part of such a formation that is water bearing.

Artesian well. Well tapping a confined or artesian aquifer.

Auxiliary supply. Any water supply on or available to the premises, other than the public water supply.

Backflow. Flow of water or other liquids, mixtures, or substances into the distributing pipes of a supply of potable water from any source other than its intended source and may be produced by the differential pressure existing between two systems either or both of which are at pressures greater than atmospheric.

Back-siphonage. Flow-back of water from a plumbing fixture or vessel or other source into a water supply pipe due to negative pressure in such pipe.

Branch distributing pipe. A pipe that is connected to a distributing pipe or riser pipe and conveys the water therefrom to the plumbing fixtures.

Branch supply pipe. A pipe that is connected to a principal supply pipe and conveys the water therefrom to the riser pipe or distributing pipe.

Bored well. Well excavated by means of a soil auger (hand or power), as distinguished from one that is dug or drilled.

Capacity of well test. Test capacity of water flow for a period of 4 hours after drilling has been completed to ascertain continuous yield.

Casing. Pipe inserted in water wells to prevent the sides from collapsing.

Chlorination of the well. Upon completion of the well construction and pump installation, water supply wells installed for the purpose of obtaining groundwater for domestic consumption should be sterilized in accordance with standards for sterilization of drinking water wells established by the U.S. Public Health Service.

Construction of wells. All acts necessary to construct wells for any intended purpose or use, including the location and excavation of the well, placement of casings, screens, and fittings, and development and testing.

Critical level. Highest level to which a back-siphonage preventer, when subjected to a specified test, can be submerged before backflow begins.

Cross-connection. Any unprotected connection between any part of a water system used or intended to supply water for drinking purposes and any source or system containing water or substance that is not or cannot be approved as safe, wholesome, and potable for human consumption.

Cross-connection (potable water system). A physical connection or arrangement between two otherwise separate piping systems, one of which contains potable water, and the other of which contains water of questionable safety, or steam, gases, or chemicals whereby there can be a flow from one system to another.

Disinfect. Recommended by code, well and piping should be disinfected as required before water is used for consumption.

Distributing pipe. A pipe that is connected to a riser pipe or a branch supply pipe and conveys the water therefrom to the branch distributing pipe.

Drill (drilling). All acts necessary to the construction of water well with power equipment, including the sealing of unused water well holes.

Drilled well. Well excavated wholly or in part by means of a drill (either percussion or rotary) that operates by cutting or abrasion or by use of a water jet.

Drinking water. Water provided or used for human consumption or for lavatory or culinary purposes.

Driven well. Well constructed by driving a casing, at the end of which there is a drive point and screen, without the use of any drilling, boring, or jetting device.

Dug well. Well excavated by means of picks, shovels, or other hand tools or by means of a power shovel or other dredging or trenching machinery, as distinguished from one put down by drill or auger.

Extension. The extension of a water main along a street, avenue or highway. An extension shall not include the water-service connection as defined in the code.

Flood level rim. Top edge at which water can overflow from a fixture or device.

Flushing and service water supply. Buildings in which there are waterclosets and other plumbing fixtures shall be provided with a supply of water adequate in volume and pressure for flushing and other building service purposes.

Frost conditions. In regions where frost occurs, safety precautions should be taken to install the water feed to at least 4 ft below ground.

Governmental regulations. The Environmental Protection Agency, Water Supply Division reprint of *Manual of Individual Water Supply Systems*, should be used as a guide.

Groundwater. Water of underground streams, channels, artesian basins, reservoirs, lakes, and other water under the surface of the ground, whether percolating or otherwise.

Ground water resources. The state finds that improperly constructed, operated, maintained, or abandoned wells can adversely affect the public health and the ground-water resources of the state. Consistent with the duty to safeguard the public welfare, safety, health, and to protect and beneficially develop the ground-water resources of the state, it is declared to be the policy of the state to require that the location, construction, repair and abandonment of wells, and the installation of pumps and pumping equipment conform to such reasonable requirements as may be necessary to protect the public, welfare, safety, health, and ground-water resources.

Individual water supply. A supply other than an approved public water supply that serves one or more families.

Installation of pumps and pumping equipment. Procedure employed in the placement and preparation for operation of pumps and pumping equipment, including all construction involved in making entrances to the well and establishing seals.

Main supply pipe. A pipe that is connected to the service pipe of any building, structure, or premises and conveys the water therefrom to the principal supply pipe.

Mineralized water. Whenever a water-bearing stratum or aquifer that contains nonpotable mineralized water is encountered in well construction, the stratum should be adequately cased or cemented off as conditions may require so that contamination of the overlying or underlying groundwater zones will not occur.

Minimum acceptable well capacity. Depending on the number of fixtures to be supplied, recommended gallons per minute (gpm) should be expected from the well.

Nonpotable mineralized water. Brackish, saline, or other water containing minerals of such quantity or type as to render the water unsafe, harmful, or generally unsuitable for human consumption and general use.

Non-potable water. Water not safe for drinking, personal, or culinary use.

Pitless adapter. A threaded or welded device that provides underground connection between the well casing and the buried piping, and that provides ready access to the drop pipe and any working parts within the well casing in a manner to protect the well from contamination.

Polluted water

1. Water containing organic or other contaminants of such type and quantity as to render it unsafe, harmful, or unsuitable for human consumption and general use.

2. In constructing any well, all water-bearing zones that are known to contain polluted water should be adequately cased or cemented off so that pollution of the overlying and underlying groundwater zones will not occur.

Potable water. Water that is satisfactory for drinking, culinary, and domestic purposes, and meets the requirements of the director of public health.

Potable water supply. All premises intended for human habitation or occupancy shall be provided with an adequate supply of pure and wholesome potable water, neither connected with unsafe water supplies nor subject to the hazards of backflow or back-siphonage.

Pressure tanks. Closed water storage containers constructed so as to operate under normal water system pressures.

Prevention of contamination. Well constructed and maintained in a condition whereby it is not a source or channel of contamination of the groundwater supply or any aquifer. "Contamination" as used in this definition means the act of introducing into water foreign materials of such a nature, quality, and quantity as to cause degradation of the quality of the water.

Principal supply pipes. Pipes that are the water supply arteries in buildings and structures. They are connected to the water supply main and convey the water therefrom to the pumps, tanks, filters, heaters, and other equipment together with all their appurtenances and to the branch supply pipes.

Private well. Permitted as a source of water when a public water facility is not available to the premise.

Pump. Any manufactured device designed to either raise the water from the well, or to discharge the water through a distribution system, or both.

Pump room or well room. Any enclosed structure, either above or below grade, that houses the pump, top of the well, any suction line, or any combination thereof.

Pumps and pumping equipment. Equipment or materials utilized or intended for use in withdrawing or obtaining groundwater, including well seals.

Quality of water. All premises intended for human habitation or occupancy shall be provided with potable water that has been declared safe for domestic use by the proper authorities.

Repair. Work involved in deepening, reaming, sealing, installing, or changing casing depths, perforating, screening, cleaning, acidizing or redeveloping a well excavation, or any other work that results in breaking or opening the well seal.

Rig permit and permit. A permit to operate a water well drilling rig required by the code.

Riser pipe. A pipe that is installed perpendicular to the horizontal through the floors, stories, and other open spaces of buildings and structures and conveys the water from the main or branch supply pipes to the distributing pipes or branch distributing pipes.

Shallow wells. Usually obtain their water from normal ground water and have no constant demanded flow, but are used for well points and water supply at construction jobs.

Testing. Well water should be tested for bacteria count and for chemical content before well is placed in operation, and should be tested thereafter as often as codes require.

Valves and casing in flowing artesian wells. Valves and casing on all flowing artesian wells must be maintained in such condition that the flow of water can be completely stopped when the well is not being put to a beneficial use.

Water pollution. Contamination of water by materials such as sewage effluent, chemicals, detergents, pesticides, and fertilizer runoff.

Water purification. Any of the several processes where undesirable impurities in water are removed or neutralized, such as chlorination, filtration, ion exchange, distillation, or other treatments.

Water-service connection. The pipe serving a premises from the main into the premises to a point 3 ft beyond the meter, including the meter.

Water service pipe. The pipe from the water supply to the building being served.

Water supply well. Wells usable as a source of water supply but that do not include a well constructed by an individual on land that is owned or leased by him, appurtenant to a single family dwelling, and intended for domestic use (including household purposes, farm livestock, or gardens).

Water-supply system. The "water-supply system" of a building or premises consists of the water-service pipe, the water-distributing pipes, and the necessary connecting pipes, fittings, control valves, and all appurtenances in or adjacent to the building or premises.

Well

1. Excavation that is cored, bored, drilled, jetted, dug, or otherwise constructed for the purpose of locating, testing, or withdrawing groundwater; for evaluating, testing, developing, draining, or recharging groundwater reservoirs or aquifers; or for controlling, diverting, or otherwise causing the movement of water from or into an aquifer.

2. Artificial excavation that derives water from the interstices of the rocks or soil which it penetrates. Wells are referred to as "shallow" or "deep" depending on whether they derive water from "free" or "confined" groundwater, respectively.

Well location. No well should be drilled closer than 100 ft from any septic tank or sewage disposal systems. Verify with local and state regulations regarding specified locations.

Well seal. Approved arrangement or device used to cap a well or to establish and maintain a junction between the casing or curbing of a well and the piping or equipment installed therein, whose purpose or function it is to prevent pollutants from entering the well at the upper terminal.

Well test. Water supply wells are tested for capacity by a method and for a period of time acceptable to the state health department and depending on the intended use of the well.

3

Concrete

BATCHING AND MIXING

Agitating speed. Rate of rotation of the drum or blades of a truck mixer or other device when used for agitating mixed concrete.

Agitating truck. Vehicle carrying a drum in which freshly mixed concrete can be conveyed from the point of mixing to that of placing, the drum being rotated continuously so as to agitate the contents.

Amount of mixing. Designation of the extent of mixer action employed in combining the ingredients for concrete. The mixing time of truck mixers is the number of revolutions of the drum or blades at mixing speed after the intermingling of the cement with water and aggregates.

Automatic batcher. Batcher equipped with gates or valves that, when actuated by a single starter switch, open automatically at the start of the weighing operation of each material and close automatically when the designated weight of each material has been reached.

Batch. Quantity of concrete mixed at one time.

Batch box. Container of known volume used to measure constituents of a batch of concrete in proper proportions.

Batched water. Mixing water added by a batcher to a concrete mixture before or during the initial stages of mixing.

Batcher. Device for measuring ingredients for a batch of concrete.

Batching. Weighing or volumetrically measuring and introducing into the mixer the ingredients for a batch of concrete.

Batching plant. Structure containing separate compartmented bins, each holding aggregates and cement until the materials can be discharged into a weighing hopper. Admixture are added (as determined by specifications) through a mechanical dispenser. Cement is weighed separately from the aggregates.

Batch mixer. Machine that mixes batches of concrete in contrast to a continuous mixer.

Batch weights. Weights of the various materials, such as cement, water, the several sizes of aggregate, and admixtures if used, of which a batch of concrete is composed.

Bulk cement. Cement that is transported and delivered in bulk (usually in specially constructed vehicles) instead of in bags.

Bulk loading. Loading of unbagged cement in containers, specially designed trucks, railroad cars, or ships.

Central-mixed concrete. Concrete that is mixed completely in a stationary mixer and then transported to the point of delivery in a truck agitator, or truck mixer operating at agitating speeds, or in nonagitating equipment.

Central mixer. Stationary concrete mixer from which the freshly mixed concrete is transported to the work.

Concrete batch plant. Includes such plants as portland cement concrete plant, transit concrete mixing plant, sand, gravel, and cement mixing plants, and soil cement mixing plants.

Continuous mixer. Mixer into which the ingredients of the mixture are fed without stopping, and from which the mixed product is discharged in a continuous stream.

Cumulative batching. Measuring more than one ingredient of a batch in the same container by bringing the batcher scale into balance at successive total weights as each ingredient is accumulated in the container.

Dry batch weight. Weight of the materials, excluding water, used to make a batch of concrete.

Gross vehicle load. Weight of a vehicle plus the weight of any load thereon.

Horizontal-axis mixer. Concrete mixer of the revolving-drum type in which the drum rotates about a horizontal axis.

Horizontal-shaft mixer. Mixer having a stationary cylindrical mixing compartment, with the axis of the cylinder horizontal, and one or more rotating horizontal shafts to which mixing blades or paddles are attached.

Inclined-axis mixer. Truck with revolving drum that rotates about an axis inclined to the bed of the truck chassis.

Manual batcher. Batcher equipped with gates or valves that are operated manually, with or without supplementary power from pneumatic, hydraulic, or electrical machinery, the accuracy of the weighing operation being dependent on the operator's observation of the scale.

Mixer. Machine used for blending the constituents of concrete, grout, mortar, cement paste, or other mixture.

Mixer efficiency. Adequacy of a mixer in rendering a homogenous product within a stated period. Homogeneity is determinable by testing for relative differences in physical properties of samples extracted from different portions of a freshly mixed batch.

Mixing cycle. Time taken for a complete cycle in a batch mixer, that is, the time elapsing between successive repetitions of the same operation (such as successive discharges of the mixer).

Mixing speed. Rotation rate of a mixer drum or the paddles in an open-top pan or trough mixer when mixing a batch; expressed in revolutions per minute (rpm) or in peripheral feet per minute of a point on the circumference at maximum diameter.

Mixing time. Period during which the constituents of a batch of concrete are mixed by a mixer; for a stationary mixer, time is given in minutes from the completion of mixer charging until the beginning of discharge; for a truck mixer, time is given in total minutes at a specified mixing speed or expressed in terms of total revolutions at a specified mixing speed.

Mixture. Assembled, blended, commingled ingredients of concrete or the like, or the proportions for their assembly.

Nonagitating unit. Truck-mounted container for transporting central-mixed concrete not equipped to provide agitation (slow mixing) during delivery.

Nontilting mixer. Horizontally rotating drum mixer that charges, mixes, and discharges without tilting.

Open-top mixer. Truck-mounted mixer consisting of a trough or a segment of a cylindrical mixing compartment within which paddles or blades rotate about the horizontal axis of the trough.

Ready-mixed concrete

1. Concrete manufactured for delivery in a plastic and unhardened state.
2. Concrete mixed and delivered to point designated.

Semiautomatic batcher. Batcher equipped with gates or valves that are separately opened manually to allow the material to be weighed, but that are closed automatically when the designated weight of each material has been reached.

Shrink-mixed concrete. Ready-mixed concrete mixed partially in a stationary mixer and then in a truck mixer.

Tilting mixer. Rotating drum mixer that discharges by tilting the drum about a fixed or movable horizontal axis at right angles to the drum axis. Drum axis may be horizontal or inclined while charging and mixing.

Transit-mixed concrete

1. Concrete, the mixing of which is wholly or principally accomplished in a truck mixer.
2. Dry materials placed in a truck at a batching plant. Water is added by the operator and the concrete is completely mixed in a truck mixer.

Trial batch. Batch of concrete prepared to establish or check proportions of the constituents.

Truck mixer. Concrete mixer suitable for mounting on a truck chassis and capable of mixing concrete in transit.

Vertical-shaft mixer. Cylindrical or annular mixing compartment having an essentially level floor and containing one or more vertical rotating shafts to which blades or paddles are attached. The mixing compartment may be stationary or rotate about a vertical axis.

Volumetric batch plant. Bituminous concrete mixing plant that proportions aggregate and bituminous constituents into the mix by volumetrically measured batches.

CAST-IN-PLACE CONCRETE AND RELATED CONSTRUCTION

Accredited laboratory test report. Test report including a statement by the testing laboratory that it is accredited for the test reported and that the test has been performed in accordance with the conditions prescribed by the accrediting body.

Architectural concrete. Concrete that will be permanently exposed to view and that therefore requires special care in selection of concrete materials, forming, placing, and finishing to obtain the desired architectural appearance.

Average concrete. Concrete mixed in accordance with the provisions of the specifications and any other applicable section of the code.

Backfill concrete. Non-structural concrete used to correct over-excavation, or fill excavated pockets in rock, or to prepare a surface to receive structural concrete.

Beam. Horizontal support or member carrying transverse loads.

Beam-and-slab floor construction. Ordinary type of construction in which the solid slab is supported by beams or girders.

Beam molds for testing. Mold that is rectangular in shape and of dimensions required to produce the test specimen desired.

Bleeding

1. Appearance of excess water rising to the surface shortly after placing of concrete.
2. Emergence of water from newly placed concrete, caused by the settlement of the solid materials within the mass; also called "water gain."
3. Autogenous flow of mixing water from within freshly placed concrete primarily due to settling of aggregates. Water usually rises to and collects on the upper surface of the concrete.

Bond. Adhesion and grip of concrete to reinforcement or to other surfaces against which it is placed, including friction due to shrinkage and longitudinal shear in the concrete engaged by the bar deformations.

Bond beam. Continuous structural member having the same thickness as the wall of which it is a part and which is designed and constructed to provide lateral stability to the wall.

Broom finish. Horizontal surface texture obtained by brooming plastic concrete to provide a nonslip surface.

Brushed surface. Sandy texture obtained by brushing the surface of freshly placed or slightly hardened concrete with a stiff brush for architectural effect or, in pavements, to increase skid resistance.

Bush-hammer finish. A finish on concrete obtained by means of a bush-hammer.

Camber. Deflection that is intentionally built into a structural element or form to improve appearance or to nullify the deflection of the element under the effects of loads, shrinkage, and creep.

Cantilever beam. Beam that extends beyond the support in an overhanging position.

Cast-in-place concrete. Concrete that is deposited in the place where it is required to harden as part of the structure, as opposed to precast concrete.

Cathead. Light frame and sheave at the top of a material tower through which the lifting cable is operated.

Cellular concrete. Lightweight product consisting of portland cement, cement-silica, cement-pozzolan, lime-pozzolan, or lime-silica pastes, or pastes containing blends of these ingredients and having a homogeneous void or cell structure, obtained by means of gas-forming chemicals or foaming agents. For cellular concretes containing binder ingredients other than or in addition to portland cement, autoclave curing is usually employed.

Cement. Binding agent capable of uniting dissimilar materials into a composite whole.

Chamfer. Beveled corner or edge on a beam or column.

Channel extender. Short piece of structural steel channel attached to the end of a steel beam or truss to provide for an adjustment in length of the beam or truss. Used in formwork.

Clay tile joist floor. Clay tile fillers between poured-in-place concrete joists.

Cold joint. Construction joint in concrete occurring at a place where the continuous pouring has been interrupted.

Column

1. Upright compression member, whose length exceeds three times its least lateral dimension.
2. Post or vertical member supporting a floor beam, girder, or other member and carrying axial load.

Column capital

1. Upper flared conical section of circular column; sometimes pyramidal capitals are made on square columns.
2. Enlargement of the upper end of a reinforced concrete column designed and built to act as a unit with the column and flat slab.

Column guard. Structural steel protection for the base of a column to protect against excessive traffic wear.

Column strip. Portion of a flat slab panel, one-half panel in width, occupying the two quarter-panel areas outside the middle strip and extending through the panel in the direction in which bending moments are being considered.

Combination column. Column in which a structural steel section is designed to carry the principal part of the load, wrapped with wire and encased in concrete of such quality that some additional load may be allowed.

Composite beams. Any rolled or fabricated steel floor beam entirely encased in a poured concrete haunch at least 4 in. wider, at its narrowest point than the flange of the beam, supporting a concrete slab on each side without openings adjacent to the beam.

Composite column. A column in which a steel or cast-iron section is completely encased in concrete containing spiral and longitudinal bar reinforcement.

Compression test. Test made on a cylinder of concrete, 6 in. in diameter and 12 in. in height, to determine the compressive strength of the concrete.

Compressive strength. Maximum compressive stress that a material such as portland cement, concrete, or grout is capable of sustaining. Compressive strength is calculated from the maximum load during a compression test, and the original cross-sectional area of the specimen is normally expressed in pounds per square inch.

Concrete

1. Mixture of portland cement, fine aggregate, coarse aggregate, and water, and, in some instances, an admixture.

2. Composite material that consists essentially of a binding medium within which are embedded particles or fragments of aggregate; in portland cement concrete, the binder is a mixture of portland cement and water.

Concrete strength. Concrete strength depends on many factors, such as mix design, placement, curing, material, and control.

Consistency

1. Uniformity of mixes or batches as measured by the slump test.
2. Relative mobility or ability of freshly mixed concrete or mortar to flow; the usual measurements are slump for concrete, flow for mortar or grout, and penetration resistance for neat cement paste.

Construction joint

1. Joint placed in concrete to permit practical placement of the work section by section.
2. Surface where two successive placements of concrete meet, across which it is desirable to develop and maintain the bond between the two concrete placements and through which any reinforcement that may be present is not interrupted.

Continuous beam (both ends). Beam that runs continuously through both supports.

Continuous beam (one end). Beam that is supported at one end by column, well, or other beam and extends into another beam at the opposite point of support.

Contraction joint. Formed, sawed, or tooled groove in a concrete structure to create a weakened plane and regulate the location of cracking resulting from the dimensional change of different parts of the structure.

Control joint. Joint placed in concrete to form a plane of weakness to prevent random cracks from forming due to shrinkage.

Controlled concrete. Concrete mixed in accordance with the requirements of the code.

Core drilling. Core specimen taken perpendicular to a horizontal surface (ASTM C 24).

Core test. Test similar to compression test, except that the concrete tested is cut from hardened concrete usually by means of a hollow drilling device.

Crazing

1. Occurrence of numerous fine hair cracks in the surface of newly hardened concrete due to surface shrinkage. Crack pattern is similar to that of a crushed eggshell.
2. Development of craze cracks; the pattern of craze cracks existing in a surface.

Creep. Time-dependent deformation due to sustained load.

Curing

1. Process consists of maintaining conditions of moisture and/or temperature sufficient to attain the desired degree of hydration of the cement.
2. Maintenance of humidity and temperature of freshly placed concrete during some definite period following placing, casting, or finishing to assure satisfactory hydration of the cementitious materials and proper hardening of the concrete.

Cylinder test. Laboratory test for compressive stress of a field sample of concrete 6 in. in diameter by 12 in. in length.

Drop-in-beam. A simple beam, usually supported by cantilever arms, with joints so arranged that it is installed by lowering into position.

Dropped panel

1. Structural portion of a flat slab that is thickened throughout an area surrounding the column capital.

2. In flat slab construction, plinth at the top of the column to thicken the slab.

Dusting

1. Appearance of a powdery material at the surface of hardened concrete.

2. Concrete floors tend to dust if not treated. Liquid surface treatments, such as solutions of magnesium fluorosilicate, sodium silicate, aluminum sulfate, or zinc sulfate, are usually used to harden concrete floors and decrease their tendency to dust.

Early strength. Strength of concrete usually as developed at various times during the first 72 hours after placement.

Eccentrically loaded columns. Columns with loads greater on one side than on the other.

Edge supported. Slab rests atop the perimeter foundation wall, or masonry walls supporting one or more slabs.

Edging. Finishing operation of rounding off the edge of a slab to prevent chipping or damage.

Efflorescence

1. Deposit of salts, usually white, formed on a surface, the substance having emerged in solution from within the concrete or masonry and been deposited by evaporation.

2. Growth of crystals on the surface and in the pores of the concrete, where a salt solution has evaporated and leached out of the concrete.

Elongated piece. Aggregate piece in which the ratio of the length to width of its circumscribing rectangular prism is greater than a specified value.

Expansion joint. Separation between adjoining parts of a concrete structure to allow small relative movements, such as those caused by thermal changes, to occur independently.

False set

1. Apparent set or stiffening of wet concrete prior to the time of the initial set.

2. Rapid development of rigidity in a freshly mixed portland cement paste, mortar, or concrete without the evolution of much heat; this rigidity can be dispelled and plasticity regained by further mixing without addition of water.

Fat concrete. Concrete containing a relatively large amount of plastic and cohesive mortar.

Field-cured cylinders. Test cylinders cured as nearly as practicable in the same manner as the concrete in the structure to indicate when supporting forms may be removed, additional construction loads imposed, or the structure placed in service.

Fillet. Beveled corner formed by a piece of wood, steel, or other material, triangular in shape and placed on the inside corner of a form to avoid a sharp 90° change in direction.

Final set. Degree of stiffening of the cement and water mixture. This is a degree greater than initial set and is generally stated as an emperical value, indicating the time in hours and minutes required for a cement paste to stiffen sufficiently to resist to an established degree the penetration of a weighted test needle (ASTM C 191–C 266).

Finishing. Leveling, smoothing, compacting, and otherwise treating surfaces of fresh or recently placed concrete or mortar to produce desired appearance and service.

Fins. Narrow linear concrete projections on a formed concrete surface, such as those resulting from grout flowing out between spaces in the form boards.

Flat beam. Wide, flat beam whose width is several times its depth; also, a special type of construction.

Flat piece. Aggregate piece in which the ratio of the width to thickness of its circumscribing rectangular prism is greater than a specified value.

Flat plate. Flat slab without column capitals or drop panels.

Flat slab. Concrete slab reinforced in two (2) or more directions, generally without beams or girders to transfer the loads to supporting columns.

Float finish. Concrete surface texture obtained by smoothing and leveling with a smooth wooden finishing tool.

Floating

1. Slab finishing operation that embeds aggregate, removes slight imperfections, humps, and voids to produce a level surface, and consolidates material rising to the surface.

2. Finishing a fresh concrete surface by use of a float; precedes troweling when that is the final finish.

Floating slab. Slab terminates at the inside face of the perimeter foundation wall and is said to "float" independently of the foundation wall.

Footing

1. Supporting section of a building that usually rests directly on the soil or supporting earth.

2. Structural unit used to distribute wall or column loads to the foundation materials.

3. Base of a foundation or column wall used to distribute the load over the subgrade.

Foundation. Substructure that distributes the load of the main structure to the earth.

Girder. Principal beam supporting other beams.

Granolithic concrete. Concrete suitable for use as a wearing surface finish to floors, made with specially selected aggregate of suitable hardness, surface texture, and particle shape.

Green concrete

1. Freshly placed concrete.

2. Concrete that has set but not appreciably hardened.

Grout

1. Liquid mixture of cement, water, and sand of pouring consistency.

2. Mixture of cementitious material and water, with or without aggregate, proportioned to produce a pourable consistency without segregation of the constituents; also, a mixture of other composition, but of similar consistency.

3. Mixture of sand, portland cement, and water; in some instances an admixture.

Grouted-aggregate concrete. Concrete that is formed by injecting grout into previously placed coarse aggregate.

Grouting. Process of filling with grout.

Gunite. Pneumatically placed concrete.

Haunch. Portion of a beam that increases in depth toward the support.

Heat of hydration

1. Heat liberated during chemical reactions that take place between the cement compounds and water during the periods of setting and hardening.

2. Heat evolved by chemical reactions with water, such as that evolved during the setting and hardening of portland cement.

Heat-resistant concrete. Any concrete which will not disintegrate when exposed to constant or cyclic heating at any temperature below which a ceramic bond is formed.

High-density concrete. Concrete of exceptionally high density, usually obtained by use of heavyweight aggregates, used especially for radiation shielding.

Hollow slab. Concrete slab ideally suited for bridge construction and deck structures. Slabs are placed side by side.

Honeycomb

1. Voids left in concrete due to failure of the mortar to fill the spaces among coarse aggregate particles effectively.

2. Voids around coarse aggregates caused by loss of matrix and/or by inadequate compaction of the concrete.

Hydration

1. Chemical reaction of water and cement that produces a hardened concrete.

2. Formation of a compound by the combination of water with some other substance; in concrete, the chemical reaction between hydraulic cement and water.

Imposed load. All loads, exclusive of dead load, that a structure is to sustain.

Inert. Having inactive chemical properties.

Initial set. Degree of stiffening of the cement and water mixture. This is a degree less than the final set and is generally stated as an empirical value indicating the time in hours and minutes required for a cement paste to stiffen sufficiently to resist, to an established degree, the penetration of a weighted test needle.

Inserts. Devices buried in the concrete to receive bolts or screws to support shelf angles, machinery, etc.

Insulating concrete. Concrete having low thermal conductivity; used as thermal insulation.

Integral. Elements that act together as a unit, as for example, concrete joists and a top slab. Concrete members may be made integral by bond, dowels, or being cast in one piece.

Integrally cast. Term describing elements that are cast in one piece, as, for example, concrete joists and a top slab.

Integral coloring. Coloring of concrete by coloring agents, chiefly metallic oxides which produce permanent coloring throughout the concrete.

Isolation joint. Separation between adjoining parts of a concrete structure, usually a vertical plane, at a designed location such as to interfere least with performance of the structure and yet allow relative movement and avoid formation of cracks elsewhere in the concrete, through which all or part of the bonded reinforcement is interrupted.

Joint filler. Compressible material used to fill a joint to prevent the infiltration of debris and provide support for sealants.

Joint sealant. Compressible material used to exclude water and solid foreign materials from joints.

Joint sealer. Mixture of materials that form a resilient and adhesive compound capable of effectively sealing joints in concrete against the infiltration of moisture and foreign material throughout repeated cycles of expansion and contraction with temperature change; will not flow from the joints in pavements or be picked up by vehicle tires at summer temperatures. Material capable of being brought to a uniform pouring consistency suitable for completely filling the joints without inclusion of large air holes or discontinuities and without damage to the material.

Joint sealing compound. Impervious material used to fill joints in pavements or structures.

Keys. Slotted joints in concrete, such as tongue and groove.

Lack of homogeneity. Causes a watery paste that can be rectified by adding finely divided minerals to stiffen the paste by increasing the ratio of surface area of solids to volume of water.

Laitance

1. Soft, weak layer of mortar appearing on the top of a horizontal surface of concrete as a result of segregation.

2. Layer of weak and nondurable material containing cement and fines from aggregates brought by bleeding water to the top of overwet concrete. The amount of laitance is generally increased by overworking or overmanipulating concrete at the surface by improper finishing.

3. Residue consisting of fine inert particles of cement, sand, or impurities that is brought to the surface of concrete by water rising to that surface.

Lightweight concrete. Concrete of substantially lower unit weight than that containing gravel or crushed stone aggregates.

Lightweight concrete (structural). Concrete having a 28-day compressive strength in excess of 2000 psi and an air dry unit weight less than 115 lb/cu ft.

Lintel. Structural beam over and above a window or door opening to support the wall above.

Live load. Loads and forces other than the dead load.

Load-sharing system. Construction composed of three or more essentially parallel members spaced at 24-in. centers or less, so arranged or connected that they mutually support the load.

Load test. Member or portion of the structure under consideration is subjected to a superimposed load equal to one and one-half times the live load plus one-half times the dead load. Load is left in position for a period of 24 hours before removal.

Low-density concrete. Concrete having an oven-dry unit weight of less than 50 pcf (800 kg/m^3).

Mass concrete. Volume of concrete with dimensions large enough to require that measures be taken to cope with the generation of heat from hydration of the cement and attendant volume change to minimize cracking.

Mass concrete pour. Use of finely divided minerals such as pozzolans (siliceous and aluminous materials), or cementitious materials, such as hydraulic lines, to lower cement content in mixes where liberated heat of hydration is of large proportion, as in dams or retaining walls.

Matrix. Cement, sand, and water portion of a concrete mix that bonds the coarse aggregates together.

Mattress or mat. Reinforced slab like the large footing used for structures where problems with soil conditions exist.

Middle strip. Portion of a flat-slab panel one-half the panel in width, symmetrical, to the panel in the direction in which bending moments are being considered.

Mix designs. Design for converting the cement, water and aggregates to batch weights on the basis of absolute volumes to produce 1 cu yd of concrete with the proper slump, specified compressive strength, and maximum density.

Monolithic. Term describing concrete cast in one operation. Monolithic concrete elements may be designed to act integrally or as separate elements if articulated by weakened plane contraction joints.

Monolithic concrete

1. Concrete placed in one continuous pour without joints.
2. Concrete cast with no joints other than construction joints.

Monolithic slab and foundation wall. Slab and foundation poured integrally, also called "slab-thickened edge."

Mortar. Mixture of portland cement, fine aggregate, and water.

No-fines concrete. A concrete mixture containing little or no fine aggregate.

Normal-weight concrete. Concrete having a unit weight of approximately 150 lb/cu ft made with aggregates of normal weight.

No-slump concrete. Concrete with a slump of $\frac{1}{4}$ in. (6 mm) or less.

Packaged dry concrete materials. Commercial production of dry, combined materials for use in concrete mix to which water is to be added.

Paneled ceiling. Ceiling of a flat slab in which approximately the portion of the area enclosed within the intersection of the two middle strips is reduced in thickness.

Panel length

1. Distance along a panel side from center of columns to a flat slab.
2. Distance in either rectangular direction between the center of two columns of a panel.

Pan floor. Series of concrete joists or small beams, joined together at the top with a thin slab, using prefabricated steel or fiberglass forms (pans).

Paste. Cement and water portion of a concrete mixture.

Pedestal

1. Upright compression member whose height does not exceed three times its least lateral dimension.
2. Short pier or plinth used as a base for a column.

Pedestal footing. Column footing projecting less than one-half its depth from the faces of the column on all sides and having a depth not more than three times its least width.

Pier. Isolated foundation member of plain or reinforced concrete.

Pilaster. Column embedded in a wall.

Placing. Act of putting concrete in position (sometimes incorrectly referred to as "pouring").

Plain concrete

1. Concrete without reinforcement; reinforced concrete that does not conform to the definition of reinforced concrete; also, used loosely to designate concrete containing no admixture and prepared without special treatment.
2. Concrete without metal reinforcement, or reinforced only for shrinkages or temperature changes.

Plastic concrete. Easily molded concrete that will change its form slowly only if the mold is removed.

Plastic cracking. Cracking that occurs in the surface of fresh concrete soon after it has been placed and while it is still plastic.

Ponding. Curing method for flat surfaces whereby a small earth dam or other water-retaining material is placed around the perimeter of the surface and the enclosed area is flooded with water.

Pop-out. Breaking away of small portions of a concrete surface due to internal pressure that leaves a shallow, typically conical depression. This may be either chemical or mechanical in nature.

Pneumatically placed concrete

1. Mixture of fine aggregate and cement pneumatically applied by suitable mechanism, to which water is added immediately prior to discharge from the applicator.
2. Mixture or portland cement and fine aggregate, mixed dry, passed through a cement gun or other similar device, hydrated at the nozzle, and deposited under pressure in its final position.

Puddling. Compacting or consolidating concrete with a rod or other tool.

Ready-mixed concrete. Concrete produced by a commercial establishment and delivered to the purchaser in the plastic state (ASTM C 94).

Rectangular beam. Beam of square or rectangular cross section, of which the top and bottom dimensions are equal and the two sides are the same.

Rectangular direction. Direction parallel to a side of the panel of a flat slab.

Reglet. Long, narrow, formed slot in concrete to receive flashing or serve as an anchorage.

Reinforced concrete. Concrete in which reinforcement other than that provided for shrinkage or temperature changes is embedded in such a manner that the two materials act together in resisting forces.

Retaining wall. Wall reinforced to hold or retain soil and other loads applied vertically or horizontally.

Retempering. Remixing of concrete that has started to stiffen by the addition of water, when permitted.

Sacking. Operation of removing or alleviating surface defects on a concrete surface by applying a mixture of sand and cement to the moistened surface and rubbing with a coarse material such as a burlap sack.

Salamander

1. Portable heating unit used to heat surrounding air.
2. Portable source of heat, customarily oil burning, used to heat an enclosure around or over newly placed concrete to prevent the concrete from freezing.

Sand. Fine granular material, usually less than $\frac{1}{2}$ in. in diameter, resulting from the natural disintegration of rock or from the crushing of friable sandstone rocks.

Sand plate. Bar support for slabs on the ground. Consists of steel wire or bar legs fastened to a steel plate that rests on compacted soil.

Scaffolding. A temporary structure for the support of deck forms, cartways, or workmen, or a combination of these such as an elevated platform for supporting workmen, tools, and materials; adjustable metal scaffolding is frequently adapted for shoring in concrete work.

Scale or mill scale. Thin blue covering of oxidized metal that clings to the surface of hot-rolled steel.

Scaling. Breaking away from the surface of hardened concrete, usually to a depth of from $\frac{1}{16}$ to $\frac{3}{16}$ in.; occurring at an early age of the concrete.

Scoring. Partial cutting of concrete flat work for the control of shrinkage cracking. Term also denotes the roughening of a slab to develop mechanical bond.

Screed. Template to guide finishers in leveling off the top of fresh concrete. Screeding is the operation of rough leveling.

Screed chairs. Supports to fix the depth of the slab and hold the guides for leveling off concrete.

Screed finish. Finish resulting from the leveling of plastic concrete with a screed or strike-off rod. This finish is very rough and is usually employed only when a topping is to be placed over the concrete.

Screeding. Striking off excess concrete in the finishing of concrete slab work.

Screen. Plate, sheet, woven cloth, or other device, with regularly spaced apertures of uniform size, mounted in a suitable frame or holder for use in separating material according to size.

Sealing compound. Impervious material applied as a coating or to fill joints or cracks in concrete.

Segregation

1. Separation of the heavier coarse aggregate from the mortar, or of water from the other ingredients of a concrete mix, during handling or placing.

2. Separation of the coarse and fine portions of the mixture in such a way as to make the aggregates or concrete unfit for the design use unless reblended or remixed.

Set. Condition reached by a cement paste, mortar, or concrete when it has lost plasticity to an arbitrary degree, usually measured in terms of resistance to penetration or deformation; initial set refers to first stiffening and final set to attainment of significant rigidity. Also, strain remaining after removal of stress.

Shear head. In the top of the columns, assembled unit of flat-slab or flat-plate construction to transmit loads from slab to column.

Shelf angles. Structural angles with holes or slots in one leg for bolting to the concrete to support brick work, stone, or other masonry facing.

Shoulders. Unintentional offset in a formed concrete surface occurring when the pressure of wet concrete forces a poorly anchored form out of alignment with another form.

Shrinkage

1. Decrease in initial volume due to the removal of moisture from fresh concrete. May also defer to decrease in volume due to subsequent decreases in temperature or moisture content.

2. Volume decrease caused by drying and chemical changes; a function of time, but not of temperature or stress due to external load.

Sieve. Metallic plate or sheet, woven wire cloth, or other similar device, with regularly spaced apertures of uniform size, mounted in a suitable frame or holder for use in separating material according to size; in mechanical analysis, an apparatus with square openings in a sieve, one with circular apertures is a screen.

Simple beam. Beam supported at two points and not continuous.

Sintering. Make or become cohesive by the combined action of heat and pressure.

Skewed. Placed at any angle except 90°.

Slab. Flat section of floor or roof either resting on the ground or supported by beams.

Slab-on grade. Nonsuspended, ground-supported concrete slab, often reinforced.

Slag cement

1. Hydraulic cement consisting essentially of an intimate and uniform blend of granulated blast-furnace slag and hydrated lime in which the slag constituent is more than a specified minimum percentage.

2. Finely divided material consisting essentially of an intimate and uniform blend of granulated blast-furnace slag and hydrated lime in which the slag constituent is at least 60% of the weight of the slag cement.

Slenderness ratio. Ratio of the unsupported length of a compression member to its least actual dimension.

Slump. Measure of consistency of plastic concrete. In standard testing methods the range of slump is from 0 to 12 in., and the lower the number, the stiffer the mix.

Slump test. Test used to determine the consistency of plastic concrete. The most commonly used test basically consists of placing a tapered, open-ended cylindrical mold on end on a flat surface, filling the mold in a prescribed manner with plastic concrete, immediately removing the mold vertically, and then measuring the amount of vertical subsidence or "slump" down from the top of the mold.

Spall. Disintegration of concrete surfaces or corners.

Spandrel beam. Marginal or edge beam; beam in an exterior wall.

Spread footings

1. Footings that distribute the building loads over a sufficient area of soil to secure adequate bearing capacity. They may be classified according to the manner in which they receive the loads.

2. Footings that support one or more columns or piers by bearing on earth or rock. A simple mat footing is usually called a "spread footing."

Steam curing. Subjecting the concrete members to hot, moist air by confining the members in an area into which steam is injected. Process is more accurately referred to as "low-pressure steam curing," which differentiates this process from steam curing with live steam at pressures above atmospheric pressure.

Story pole. Piece of wood or bar marked with the story height and vertical distances for spacing horizontal bars.

Strips. Technical name for the bands in flat-slab construction. The column strip is a quarter-panel wide each side of the column centerline and runs either way of the building, from column to column. The middle strip is half a panel in width, filling in between, and runs parallel to the column strips to fill in the center part of a panel.

Structural lightweight concrete. Concrete having a 28-day compressive strength in excess of 2000 psi and an air dry unit weight less than 115 lb/cu ft.

Strut

1. Short column.

2. Compression member other than a column or pedestal.

Subgrade. Fill or earth base on which concrete is placed.

Superstructure. Frame of the building above the foundations.

Surface water. Water carried by the aggregate, excluding that absorbed by the aggregate particles.

Tamping. The operation of compacting freshly-placed concrete by repeated blows or penetrations with a tamping device.

Template. Device for locating the pattern of an element under construction. Guide for finishing. Device for locating and holding dowels, laying out bolt holes and inserts, etc.

Tolerance. Tolerances for finished concrete are permissible deviations from established lines, grades, and dimensions.

Transverse. At right angles to the longitudinal axis (crosswise).

Tremie concrete. Subaqueous concrete placed by means of a tremie.

Troweling. Slab finishing operation that produces a smooth, hard surface.

Typical floors. Floors with the same general dimensions as to beam layout, slab layout, column spacing, etc. Dimensions of the columns may vary with the height of the structure.

Upturned beam. Concrete beam that extends above the slab or structure it is supporting.

Vacuum concrete. Concrete from which water and entrapped air are extracted by a vacuum process before hardening occurs.

Vibrated concrete. Concrete compacted by vibration during and after placing.

Vibrating. Mechanical method of compacting concrete.

Vibration. Energetic agitation of freshly mixed concrete, during placement, by mechanical devices, either pneumatic or electric, that create vibratory impulses of moderately high frequency that assist in consolidating the concrete in the form or mold.

Voids. Air spaces between pieces of aggregate within a cement paste.

Wall-bearing structure. Structure with the slabs (that is, the floors or roofs) supported on walls, generally of masonry, eliminating columns and beams.

Water-cement ratio. Total quantity of water entering the mixture, including the surface water carried by the aggregate, expressed in terms of the quantity of cement. Water-cement ratio, expressed in U.S. gal per sack (94 lb) of cement.

Water for concrete. Water for concrete should be clean and free from injurious amounts of oil, acid, alkali, organic matter, or other harmful substances.

Weep hole. Drainage opening in retaining wall.

Workability of concrete

1. Relative ease or difficulty with which concrete can be placed and worked into its final position within forms and around reinforcing.
2. Property determining the effort required to manipulate a freshly mixed quantity of concrete with a minimum loss of homogeneity.

Zero-slump concrete. Concrete of stiff or extremely dry consistency showing no measurable slump after removal of the slump cone.

CHEMICAL AND MATERIAL ADMIXTURES

Accelerated aging. Set of laboratory conditions designed to produce, in a short time, the results of normal aging.

Accelerating admixture

1. Admixture usually used in colder weather to accelerate the setting time of the concrete. Use of an accelerator also results in a higher initial strength of the concrete than without the admixture.
2. Admixture that accelerates the setting and early strength development of concrete.

Acceleration. Increase in velocity or rate of change; especially, the quickening of the natural progress of a process, such as hardening, setting, or strength development of concrete.

Accelerator. Substance that when added to concrete, mortar, or grout increases the rate of hydration of the hydraulic cement, shortens the time of setting, or increases the rate of hardening or strength development, or both.

Activator. Material that, when added to a chemical compound, speeds up the curing mechanism.

Addition. Material that is interground or blended in limited amounts into a hydraulic cement during manufacture either as a "processing addition" to aid in manufacturing and handling the cement or as a "functional addition" to modify the use properties of the finished product.

Additive. Term frequently used as a synonym for addition or admixture.

Admixture

1. Material other than water, aggregates, and hydraulic cement used as an ingredient of concrete or mortar and added to the concrete immediately before or during its mixing.
2. Material other than portland cement, water, or aggregate added to concrete to alter is properties, such as accelerators, retarders, air-entraining agents.

Admixture results. Admixtures result in changing the characteristics of the mix before set and the physical properties of the concrete after set.

Admixtures to architectural concrete. Water-reducing agents, plasticizers, and shrinkage reducers, used in various mixes to assist placing, improve surfaces, and stabilize volume.

Aggregate segregation. Water-reducing admixtures produce stiffer mixes, less subject to segregation. Air-entraining agents produce some buoyancy in fine aggregates, reducing segregation; gas-forming agents, such as aluminum, magnesium, or zinc powders, act in a similar manner.

Air content. Volume of air voids in concrete, exclusive of pore space in aggregate particles, usually expressed as a percentage of total volume of the concrete.

Air entraining. Capability of a material or process of developing a system of minute bubbles of air in cement paste, mortar, or concrete during mixing.

Air-entraining admixture. Purposeful addition of minute air bubbles, uniformly distributed throughout the concrete. Air-entrainment usually increases the durability of portland cement concrete exposed to the elements and somewhat decreases the compressive strength of cement mixes.

Air-entraining agent. Addition to hydraulic cement or admixture for concrete or mortar that causes entrained air to be incorporated into the concrete or mortar during mixing, usually to increase its workability and frost resistance (ASTM C 260).

Air-entraining hydraulic cement. Cement containing an air-entraining addition in such amount as to cause the product to entrain air in mortar within specified limits.

Air-entraining portland blast-furnace-slag cement. Sufficient air-entraining agent is added so that the resulting concrete complies with the air content of specified requirements (ASTM C 595-Type IS-A).

Air-entraining portland cements. Such cements are classified as Types IA, IIA, and IIIA. Their use is similar to Types I, II, and III portland cement respectively, except that small quantities of air-entraining materials, which make the concrete more resistant to frost action, have been interground during manufacture. (ASTM C 150).

Air-entraining portland-pozzolan cement. Portland-pozzolan cement with sufficient air-entraining addition used so that the resulting product complies with the air content of specified requirement (ASTM C 595-Type 1P-A).

Air-entraining slag cement. Slag cement with sufficient air-entrainment addition used so that the resulting concrete complies with the air content of specified requirements (ASTM C 595-Type SA).

Air entrainment

1. Occulsion of air in the form of minute bubbles (generally smaller than 1 mm) during the mixing of concrete or mortar.
2. Inclusion of minute bubbles of air within the cement paste during mixing.
3. Material used as an ingredient of concrete, added immediately before or during the mixing for the purpose of entraining air.

Air meter. Device for measuring the air content of concrete and mortar.

Alkali-aggregate expansion. Excessive expansion caused by reactions between aggregates and alkalis in the cement, sometimes aggravated by the addition of a calcium chloride accelerator.

Aluminate concrete. Concrete made with calcium-aluminate cement; used primarily where high-early-strength or refractory or corrosion-resistant concrete is required.

Binders. Cementing materials, either hydrated cements or products of cement or lime and reactive siliceous materials; the kinds of cement and curing conditions govern the general kind of binder formed. Also, materials such as asphalt, resins, and other materials forming the matrix of concretes, mortars, and sanded grouts.

Bleeding. Migration of excessive water to the surface of the concrete between the time of placement and the time of set.

Blended hydraulic cement. Cement consisting of two or more inorganic constitutents, which separately or in combination contribute to the strength gaining properties of the cement.

Bonding. For increased adhesion, patching, topping, etc., bonding admixtures, usually organic water emulsions of rubbers or polymers (polyvinyl chloride, polyvinyl acetates, acrylics, butadiene styrene), or epoxies increase bond between plastic mix and adjacent surfaces. Use of non-reemulsifiable types for high humidity or exterior applications is recommended.

Boron-loaded concrete. High-density concrete including a boron-containing admixture or aggregate, such as mineral colemanite, boron frits, or boron metal alloys, to act as a neutron attenuator.

Calcium. Silver white metallic element of the alkaline-earth group occurring only in combination with other elements.

Calcium chloride. Chemical composition admixture that is used to accelerate the setting of concrete (ASTM D 98).

Cellular concrete. Lightweight product consisting of portland cement, cement-silica, cement-pozzolan, lime-pozzolan, or lime-silica pastes, or pastes containing blends of these ingredients and having a homogeneous void or cell structure, obtained by means of gas-forming chemicals or foaming agents; for cellular concretes containing binder ingredients other than or in addition to portland cement, autoclave curing is usually employed.

Chemical-resistant polymer concrete. Material composed of a continuous phase (binder) of a polymer and a discontinuous phase (aggregate), generally used in applications where chemical resistance is required.

Corrosion-inhibiting admixtures. Sodium nitrate or sodium benzoate is used where reinforcing is subjected to saline or chloride exposure, where chloride accelerators are used in the mix, or where the reinforcing is pre- or poststressed and corrosion is critical.

Critical time schedules. Selected accelerators and water reducers promote faster strength development for earlier form removal and load acceptability to expedite construction schedules.

Curing agent. Additive incorporated in the finishing of a concrete slab or applied after the slab finish has attained the required setting period.

Diluent. Substance, liquid or solid, mixed with the active constituents of a formulation to increase the bulk or lower the concentration.

Dispersant. Material that deflocculates or disperses finely ground materials by satisfying the surface energy requirements of the particles; used as a slurry thinner or grinding aid.

Dispersing agent. Agent capable of increasing the fluidity of pastes, mortars, or concretes by reduction of inter-particle attraction.

Dispersion. Deflocculation or separation of cement particles by the reduction of attraction forces.

Entrained air. Microscopic air bubbles intentionally incorporated into mortar or concrete during mixing, usually by use of a surface-active agent; typically between 10 and 1000 mm in diameter and spherical or nearly so (ASTM C 173).

Epoxy concrete. Mixture of epoxy resin, catalyst, fine aggregate, and coarse aggregate.

Excessive air. Air-detraining agents such as tributyl phosphate eliminate excessive gas or air in the mix when ultimate strength of the concrete is in question.

Excessive stiffness. Selected water-reducing admixtures increase the slump and workability without changing the water-cement ratio. Air-entraining agents are also used to improve placeability.

Fibrous concrete. Concrete containing dispersed, randomly oriented fibers, made from many materials, such as, glass, asbestos, wood, steel, and plastic.

Flocculation. Coalescence of cement particles into larger groups or flocs.

Fly ash. Finely divided residue resulting from the combustion of ground or powdered coal, which is transported from the firebox through the boiler by flue gases (ASTM C 618).

Foamed concrete. Concrete made very light and cellular by the addition of a prepared foam or generation of gas within the unhardened mixture.

Foam-entraining admixture. Concentrated chemical polymer that creates ultramiscroscopic foam, producing a lightweight cellular concrete (ASTM C 869).

Formwork deflection. Selected retarders are used to maintain monolithic pours over long spans where formwork deflection at the last of the pour could cause cracks in the earlier placed concrete.

Form retarder. A deactivator that provides controlled retardation to temporarily and uniformly delay curing of a selected surface layer of concrete while the remainder of the concrete cures normally. The retarder can be used in any type of form. When the forms are stripped, the retarder paste is removed by washing, scouring, or sandblasting, which then reveals the aggregate surface.

Grouting-including admixtures. Expansion-producing (or shrinkage-preventing) admixtures used to provide crack-free solid bearing for column base plate or machinery grouting. Usual materials are finely divided iron with an oxidizer or anhydrous sulfoaluminates.

Hardener. Substance or mixture of substances added to an adhesive to promote or control the curing reaction by becoming part of the surface.

Heat of hydration. Temperature increase due to energy liberated during the reaction between the cement and water in the concrete.

High-early-strength concrete. Concrete that through the use of high-early-strength cement or admixtures is capable of attaining specified strength at an earlier age than normal concrete.

High-temperature pouring conditions (80°F and above). Selected retarders delay setting time and permit proper finishing by countering the accelerating effect of high temperatures without additional water. Water-reducing retarders (type D admixtures), used with caution on flat slabs in hot weather, will slow set but will not prevent rapid surface drying, causing finishing difficulties and possible shrinkage cracks (ASTM C 494).

Low-humidity or high-wind pouring conditions. Selected water-reducing agents (type A admixtures) are used without changing the water content to help compensate for excessive loss of water in slab surfaces. Admixtures cannot prevent loss of water by evaporation (ASTM C 494).

Low-temperature pouring conditions. Selected accelerators decrease setting time and increase strength development to compensate for low temperatures. Calcium chloride is not used where steel is prestressed, aluminum is imbedded, lightweight concrete is placed on a metal deck, corrosion is a problem, or concrete will be subjected to sulfate exposure.

Metallic floor hardener. Ready-mixed, iron/cement based; dry shake surface treatment, prepared to produce high-impact, abrasion-resistant floors. The material is distributed over freshly poured concrete floor surfaces.

Perlite. A volcanic glass having a perlitic structure, usually having a higher water content than obsidian; when expanded by heating, used as an insulating material and as a lightweight aggregate in concretes, mortars, and plasters.

Placing concrete against hydraulic pressure. Selected accelerating admixtures are used to reduce drastically the setting time of the mix down to less than 1 minute where necessary for emergency repair work.

Polymer. Product of polymerization; more commonly, a rubber or resin consisting of large molecules formed by polymerization.

Polymer-cement concrete. Mixture of water, hydraulic cement, aggregate, and a monomer or polymer; polymerized in place when a monomer is used.

Polymer concrete. Concrete in which an organic polymer serves as the binder; also known as resin concrete. Term is sometimes erroneously employed to designate hydraulic cement mortars or concretes in which part or all of the mixing water is replaced by an adqueous dispersion of a thermosplastic copolymer.

Portland-pozzolan cement

1. Intimate and uniform blend of portland cement or portland blast-furnace-slag cement and fine pozzolan produced either by intergrinding portland cement clinker and pozzolan or by blending portland cement or portland blast-furnace-slag cement and finely divided pozzolan, or by a combination of intergrinding and blending, in which the pozzolan constituent is between 15 and 40% of the portland-pozzolan cement (ASTM C 595-Type IP).

2. Cement made by blending not more than 50% pozzolan (a material consisting of siliceous or siliceous and aluminous material) with at least 50% portland cement.

3. Hydraulic cement consisting essentially of an intimate and uniform blend of portland cement of portland blast-furnace-slag cement and fine pozzolan produced by intergrinding portland cement clinker and pozzolan or by blending portland cement or portland blast-furnace-slag cement and finely divided pozzolan, or by a combination of intergrinding and blending, in which the pozzolan constituent is within specified limits.

Pozzolan. Siliceous or siliceous and aluminous material that in itself possesses little or no cementitious value but that in finely divided form and in the presence of moisture will chemically react with calcium hydroxide at ordinary temperatures to form compounds possessing cementitious properties (ASTM C 618).

Pumped concrete. Air-entraining admixtures and water reducers reduce segregation and increase the fluidity desirable for pumping.

Reactive aggregates. Crushed Pyrex Glass No. 7740 is graded into various sieve sizes and used to prevent excessive expansion of concrete due to the alkali aggregate reaction (ASTM C 441).

Reduction of bleeding. Air-entraining and gas-forming agents minimize bleeding by restricting the flow of water within the paste. Accelerators reduce bleeding by shortening the period of plasticity when bleeding occurs, finely divided minerals reduce bleeding by increasing the surface area of solids in respect to water content, and deflocculating or dispersive admixtures reduce bleeding by more uniform distribution of the cement particles within the paste. Water-reducing admixtures reduce bleeding indirectly by permitting mixes with lower water contents.

Restricted narrow pours. Both water-reducing and air-entraining admixtures are used to increase slump and placeability, without the addition of water, for confined pours or heavily reinforced sections, or where pumps or tremies are used in placing.

Retardation. Reduction in the rate of hardening or setting, that is, an increase in the time required to reach initial and final set or to develop early strength of fresh concrete, mortar, or grout.

Retarder

1. Admixture that delays the setting of cement paste and hence of mixtures, such as mortar or concrete, containing cement.
2. Admixture added to concrete to retard its set.

Segregation. Condition of nonuniform distribution of the concrete materials within the batch.

Self-desiccation. Removal of free water by chemical reaction so as to leave insufficient water to cover the solid surfaces and to cause a decrease in the relative humidity of the system. Term is applied to an effect occurring in sealed concretes.

Set. Change from fluid to solid form. The time at which concrete sets.

Set-retarding admixture. Admixture that will retard the setting action of concrete to allow proper finishing in hot weather. In addition, set-retarding admixtures may also be water-reducing admixtures.

Shop-poured precast concrete. Selected accelerating admixtures are employed by precasting yards where rapid stripping and reuse of forms is required.

Shrinkage cracks. Cracks caused by a reduction in the volume of concrete due to water loss by temperature, wind, or evaporation before strength is sufficient to hold shape. Air-entraining admixtures help prevent shrinkage cracks by reducing the volume of water lost before set. Selected retarders help prevent shrinkage cracks by postponing stiffening of the concrete until excessive water volume is lost.

Silicone. Resin, characterized by water-repellent properties, in which the main polymer chain consists of alternating silicon and oxygen atoms, with carbon-containing side groups; silicones may be used in caulking or coating compounds of admixtures for concrete.

Strength. Selected water-reducing admixtures increase compressive strength by improving water-cement ratio of mix. Finely divided minerals add to compressive strength by pozzolanic or cementitious action.

Surface retarder. Retarder applied to a form or to the surface of freshly mixed concrete to delay setting of the cement to facilitate construction joint cleanup or production of exposed aggregate finish.

Tall-form concrete pour. Accelerators provide faster set and strength development to relieve excessive form pressures.

Tall-structure concrete pour. Accelerators are often used to permit faster form removal and allow pour to support progressive overhead construction where time is important.

Time of final setting. Elapsed time, after initial contact of cement and water, required for the mortar sieved from the concrete to reach a penetration resistance of 4000 psi (ASTM C 403).

Time of initial setting. Elapsed time, after initial contact of cement and water, required for the mortar sieved from the concrete to reach a penetration resistance of 500 psi (ASTM C 403).

Vermicultie. Group name for certain platy minerals—hydrous silicates of aluminum, magnesium, and iron—characterized by marked exfoliation on heating; also, a constituent of clays.

Vermiculite concrete. Concrete in which the aggregate consists of exfoliated vermiculite.

Waterproofing compound. Material used to impart water repellency to a structure or a constructional unit.

Water-reducing admixtures

1. Admixture that reduces the quantity of mixing water required to produce concrete of a given consistency (ASTM C 494-Type A).
2. Agent that reduces the total quantity of mixing water required for a given workability, thereby increasing the compressive strength and density (ASTM C 494-Type A).

Water-reducing agent. Material that either increases slump of freshly mixed mortar or concrete without increasing water content or maintains workability with a reduced amount of water, the effect being due to factors other than air entrainment.

Water-reducing and accelerating admixture. Admixture that reduces the quantity of mixing water required to produce concrete of a given consistency and accelerates the setting and early-strength development of concrete (ASTM C 494-Type E).

Water-reducing and retarding admixture. Admixture that reduces the quantity of mixing water required to produce concrete of a given consistency and retards the setting of concrete (ASTM C 494-Type D).

Water-repellent admixture. Admixture that produces an internal barrier that restricts water from being absorbed by the concrete.

CONCRETE AGGREGATES

Aggregate

1. An inert material that when bound together into a conglomerate mass by cement paste forms concrete. Aggregates may be either naturally occurring or manufactured. Usually available both in fine and coarse grading, but occasionally these may be combined and only the combination is available.
2. Granular material, such as sand, gravel, crushed stone, and iron blast-furnace slag, used with a cementing medium to form concrete.
3. Hard inert material mixed with portland cement and water to form concrete. Fine aggregate has pieces less than and including $\frac{1}{4}$ in. in diameter; coarse aggregate has pieces larger than $\frac{1}{4}$ in. in diameter.

Aggregate gradation. Mixture lacking fines: Finely divided minerals decrease bleeding and increase strength and workability by providing a greater volume of solids per volume of water.

Alkali–silica reaction. Reaction between the alkalies (sodium and potassium) in portland cement and certain siliceous rocks or minerals, such as opaline chert and acidic volcanic glass, present in some aggregates; products of the reaction may cause abnormal expansion and cracking of concrete in service.

Air-cooled blast-furnace slag. Material resulting from solidification of molten blast-furnace slag under atmospheric conditions. Subsequent cooling may be accelerated by application of water to the solidified surface.

Bank gravel. Gravel found in natural deposits, such as sand or clay, or combinations there of: gravelly clay, gravelly sand, clayey gravel, and sandy gravel, with varying proportions of other materials in the mixtures.

Barrel (of cement). A quantity of portland cement; 376 lb (4 bags).

Blast-furnace slag. Nonmetrallic product, consisting essentially of silicates and aluminosilicates of lime and of other bases, that is developed in molten condition simultaneously with iron in a blast furnace.

Boulder gravel. Gravel containing particles up to 8 in. in size.

Cement factor. Quantity of cement contained in a unit volume of concrete, expressed as weight or volume.

Cementitious factor. Quantity of cement and other cementitious materials contained in a unit volume of concrete expressed as weight or volume.

Coarse aggregate. Includes, crushed stone, gravel, blast furnace slag, or other approved inert materials of similar characteristics, or combinations there of, having hard, strong, durable pieces, free from adherent coatings.

Coarse gravel. Gravel containing particles up to 3 in. in size.

Coarse sand. Sand of which more than 50% by weight is retained on a number 20 mesh sieve.

Concrete

1. Mixture of portland cement, fine aggregate, coarse aggregate, and water, and, in some instances, an admixture.
2. Composite material that consists essentially of a binding medium within which are embedded particles or fragments of aggregate; in portland cement concrete, the binder is a mixture of portland cement and water.

Coral aggregate. Sharply angular, dense rock pieces of hard reef coral, resembling pieces of limestone.

Crushed gravel. Product resulting from the artificial crushing of gravel, with substantially all fragments having at least one face resulting from fracture.

Crushed stone. Product resulting from the artificial crushing of rocks, boulders, or large cobblestones, substantially all faces of which have resulted from the crushing operation.

Dense-graded aggregate. Aggregates graded to produce low void content and maximum weight when compacted.

Dry-mix concrete. Mixture commonly sold in bags, containing sand, cement, and gravel without water; also a concrete of near zero slump or less.

Dry pack

1. Concrete mixture deposited and consolidated by dry packing.
2. Dry mixture of water, sand, and cement used where shrinkage must be kept to a minimum and generally compacted into place; also, the operation of placing such a mixture.

Dry-shake. A dry mixture of cement and fine aggregate distributed evenly on an unformed surface after water has largely disappeared following the strike-off, and then worked in by floating.

Expanded blast-furnace slag. Lightweight cellular material obtained by controlled processing of molten blast-furnace slag with water or with water and other agents, such as steam or compressed air, or both.

Expanded shale (clay or shale). Lightweight vesicular aggregate obtained by firing suitable raw materials in a kiln or on a sintering grate under controlled conditions.

Expansive cement (general). A cement that when mixed with water forms a paste that, after setting, tends to increase in volume to a significantly greater degree than portland cement paste; used to compensate for volume decrease due to shrinkage or to induce tensile stress in reinforcement.

Fine aggregate. Aggregate that will pass through a $\frac{3}{8}$-in sieve and almost entirely through a No. 4 sieve, most of which will be retained on a No. 200 sieve. May either occur naturally or be processed by crushing.

Fineness modulus

1. Empirical factor obtained by adding the total percentages of a sample of aggregate retained on each of a specified series of sieves and dividing the sum by 100. The higher the number, the coarser the aggregate.
2. Measure of the average size of aggregate calculated by passing aggregate through a series of screens of decreasing size.

Fine sand. Sand of which at least 50% by weight passes a Number 60 mesh sieve.

Flash coat. A light coat of shotcrete used to cover minor blemishes on a concrete surface.

Graded aggregate. Aggregate containing uniformly graduated particle sizes from the finest fine aggregate size to the maximum size of coarse aggregate.

Graded sand. Sand containing uniformly graduated particle sizes from very fine up to $\frac{1}{4}$ in.

Graded standard sand. Ottawa sand accurately graded between the U.S. Standard No. 30 (600 μm) and No. 100 (150 μm) sieves for use in the testing of cements.

Granulated blast-furnace slag. Nonmetallic product consisting essentially of silicates and aluminosilicantes of calcium that is developed simultaneously with iron in a blast furnace and granulated by quenching the molten material in water, or steam, and air.

Gravel. Granular material, predominantly retained on a No. 4 sieve, resulting from natural disintegration and abrasion of rock. Usually has smooth round surfaces.

Heavyweight aggregates. Aggregate of high specific gravity, such as barite, magnetite, limonite, ilmenite, iron, or steel, used to produce heavy concrete.

High-early-strength cement. Cement used to produce a concrete that develops strength more rapidly than normal.

Hot cement

1. Cement that is physically hot, usually due to inadequate or insufficient cooling after manufacture. Its use without sufficient cooling time will affect finishes on concrete slabs.
2. Newly manufactured cement that has not had an opportunity to cool after the burning and grinding of the component materials.

Hydraulic cement. A cement that sets and hardens by chemical interaction with water and that is capable of doing so under water.

Lightweight aggregate

1. Aggregate of low specific gravity, such as expanded or sintered clay, shale, slate, diatomaceous shale, perlite, vermiculite, or slag; natural pumice, scoria, volcanic

cinders, tuff, or diatomite; sintered fly ash or industrial cinders, used to produce lightweight concrete.

2. Prepared by expanding, calcining, or sintering products such as blast-furnace slag, clay, diatomite, fly ash, shale, or slate; aggregate prepared by processing natural materials, such as pumice, scoria, or tuff.

Lightweight concrete. Concrete of substantially lower unit weight than that containing gravel or crushed sonte aggregates.

Lightweight concrete (structural). Concrete having a 28-day compressive strength in excess of 2000 psi and an air-dry unit weight less than 115 lb/cu ft.

Manufactured sand. Fine aggregate produced by crushing rock, gravel, iron blast-furnace slag, or hydraulic cement concrete.

Matrix. Cement, sand, and water portion of a concrete mix that bonds the coarse aggregates together.

Medium sand. Sand of which at least 50% by weight passes a No. 20 mesh sieve and more than 50% by weight is retained on a No. 60 mesh sieve.

Natural cement. Hydraulic cement produced by calcining a naturally occurring argillaceous limestone at a temperature below the sintering point and then grinding to a fine powder.

Natural mineral aggregates. Silica materials, feldspars, ferromagnesian minerals, micaceous materials, clay minerals, iron oxide minerals, igneous rocks, sedimentary rocks, zeolites, carbonate materials, sulfate minerals, iron sulfide minerals, conglomerates, sandstones, quartizites, shales, argillites, siltstones, chert, claystones, and metamorphic rocks.

Natural sand. Sand resulting from natural distintegration and abrasion of rock.

Neat cement

1. Mixture of cement and water (no aggregates).
2. Hydraulic cement in the unhydrated state.

Normal-weight concrete. Concrete having a unit weight of approximately 150 lb/cu ft made with aggregates of normal weight.

Ottawa sand. Silica sand produced by processing of material obtained by hydraulic mining of massive ortho-quartzite situated in deposits near Ottawa, Illinois; composed almost entirely of naturally rounded grains of nearly pure quartz; used in mortars for testing of hydraulic cement.

Packaged dry concrete materials. Commercial production of dry, combined materials for use in concrete mix to which water is to be added.

Portland blast-furnace-slag cement

1. Intimate and uniform blend of portland cement and fine granulated blast-furnace slag produced either by intergrinding portland cement clinker and granulated blast-furnace slag or by blending portland cement and finely ground granulated blast-furnace slag, in which the slag constituent is between 25 and 65% of the weight of the portland blast-furnace-slag cement.
2. Cement made by grinding not more than 65% of granulated blast-furnace slag with at least 35% of portland cement.
3. Hydraulic cement consisting essentially of an intimately interground mixture of portland cement clinker and granulated blast-furnace slag or an intimate and

uniform blend of portland cement and fine granulated blast-furnace slag; in which the amount of the slag constituent is within specified limits.

Portland cement

1. Product obtained by pulverizing clinkers consisting essentially of hydraulic calcium silicates to which no additions have been made subsequent to calcination other than water and/or untreated calcium sulfate, except that additions not exceeding 1.0% of other materials may be innerground with the clinker at the option of the manufacturer, provided such materials in the amounts indicated have been shown not to be harmful by standard tests.
2. Cement made by heating clay and crushed limestone to a clinker and grinding to a fine pulverized state.
3. Resulting product obtained when material, essentially hydraulic calcium silicates, is pulverized.
4. Hydraulic cement produced by pulverizing clinker consisting essentially of hydraulic calcium silicates and usually containing one or more of the forms of calcium sulfate as an interground addition.

Portland cement clinker. Clinker from which portland cement can be made.

Preplaced aggregate concrete. Concrete produced by placing coarse aggregate in a form and later injecting a portland cement-sand or resin grout to fill the interstices.

Reactive aggregate. Aggregate containing substances capable of reacting chemically with the products of solution or hydration of the portland cement in concrete or mortar under ordinary conditions of exposure, resulting in some cases in harmful expansion, cracking, or staining.

Ready-mixed concrete. Concrete produced by a commercial establishment and delivered to the purchaser in the plastic state (ASTM C 94).

Sand. Fine granular material usually less than $\frac{1}{2}$ in. in diameter resulting from the natural disintegration of rock or from the crushing of friable sandstone rocks.

Shotcrete. Mortar or concrete pneumatically projected at high velocity onto a surface; also known as "air-blown mortar;" also pneumatically applied mortar or concrete, sprayed mortar and gunned concrete.

Slag cement

1. Hydraulic cement consisting essentially of an intimate and uniform blend of granulated blast-furnace slag and hydrated lime in which the slag constitutent is more than a specified minimum percentage.
2. Finely divided material consisting essentially of an intimate and uniform blend of granulated blast-furnace slag and hydrated lime in which the slag constituent is at least 60% of the weight of the slag cement.

Structural lightweight concrete. Concrete having a 28-day compressive strength in excess of 2000 psi and an air dry unit weight less than 115 lb/cu ft.

CONCRETE REINFORCEMENT

Accessory. Supplementary material of any sort. Term is incorrectly used to describe bar supports and spacers.

Anchorage. Embedment in concrete; portion of the reinforcement bar, either straight or with hooks, designed to prevent pulling out or slipping of the bar when subjected to stress. Anchorage of tension reinforcement in beams includes only the embedded length beyond a point of contraflexure or of zero moment.

Area of steel. Cross-sectional area of bars required for a given concrete section.

Average bond stress. Force in a bar divided by the product of its perimeter and its embedded length.

Axle steel. Steel rerolled from reclaimed freight car axles.

Axle steel bars. Bars rolled from carbon steel axles for railroad cars and locomotive tenders.

Band

1. Small bar or wire encircling the main reinforcement in a member to form a peripheral tie.
2. Group of bars distributed in a slab or wall.

Bar. Steel bar, usually deformed, used to reinforce concrete construction; sometimes called rod.

Bar chair. Individual supporting device used to support or hold reinforcing bars in proper position to prevent displacement before or during concreting.

Bar list. Bill of materials and bending details, where all sizes, lengths, and bending dimensions are shown.

Bar spacing. Distance between parallel reinforcing bars measured from center to center of the bars perpendicular to their longitudinal axes.

Bar supports and spacers. Devices, usually of formed wire, to support, hold, and space reinforcing bars within formwork.

Bar types. There are two classes of billet-steel concrete reinforcement bars: plain and deformed, consisting of three grades: structural, intermediate, and hard.

Basket. Wire assembly to support and space dowel bars and expansion joints in concrete slabs on the ground.

Beam-and-slab floor. A reinforced concrete floor system in which the floor slab is supported by beams of reinforced concrete.

Beam bolster. Continuous spacer used crosswise to support the bars in beams.

Beam-column. A structural member that is subjected to forces producing significant amounts of both bending and compression simultaneously.

Beam schedule. Table on working drawings or shop drawings giving the code number for each beam, the size of the beam, the number of bars, and the spacing of the stirrups.

Beam steel. Reinforcement in beams. Includes longitudinal straight, top, or bottom bars; truss bars; and stirrups, open or closed.

Bent bar. Reinforcing bar bent to a prescribed shape, such as a truss bar, straight bar with hook, stirrup, or column tie.

Billet. Piece of semifinished steel, nearly square in section, formed by hot-rolling an ingot or bloom, reducing it to about 4 sq in. in size.

Billet steel. Bars rolled from billets, in contrast to rail or axle steel.

Bond. Holding or gripping force between reinforcing steel and concrete.

Bond beam. A continuous beam, usually of reinforced concrete, but sometimes of reinforced brick or concrete block, placed in masonry walls to tie them together and add lateral stability. It also distributes concentrated vertical loads along the wall.

Bond reinforcement. Reinforcement bonded throughout its length to the surrounding concrete.

Break (a spiral). Opening a spiral to the round shape and forcing it completely in the opposite direction so that it will remain circular for placing.

Bundled bars. Group of not more than four parallel reinforcing bars in contact with each other, usually tied together.

Burning reinforcement. Cutting reinforcing bars with an oxyacetylene torch.

Cage. Rigid assembly of reinforcement, shop or job fabricated, neady for placing in position.

Cold-drawn wire reinforcement. Steel wire made from rods that have been hot-rolled from billets.

Column schedule. Table on working drawings or shop drawings giving the code number for the column, size, column number and size of reinforcing verticals, ties, or spirals.

Column ties. Lighter formed horizontal bars in a column to hold the verticles in place.

Column verticals. Bars standing upright or vertical in a column.

Combination column. Column in which a structural steel section, designed to carry the principal part of the load, is wrapped with wire and encased in concrete of such quality that some additional load may be allowed.

Composite beam. Rolled or fabricated steel floor beam entirely encased in poured concrete at least 4 in. wider at its narrowest point than the flange of the beam, supporting a concrete slab on each side without openings adjacent to the beam, provided that the top of the beam is at least $1\frac{1}{2}$ in. below the top and at least 2 in. above the bottom of the slab, that a good grade of stone or gravel concrete with portland cement is used, and that the concrete has adequate mesh or other reinforcing steel throughout its whole depth and across the soffit.

Composite column

1. Column in which a steel section is completely encased in concrete containing spiral and longitudinal bar reinforcement.
2. Column in which a concrete core enclosed by spiral reinforcement and further reinforced by longitudinal bars encases a structural steel column designed to carry a portion of the load.

Compression bar. Steel used to increase the resistance of concrete in compression.

Contact splice. Means of connecting reinforcing bars so that the bars are lapped and in direct contact.

Continuous high chairs. Support for steel at high points in a slab.

Crimped wire. Wire deformed into a curve that approximates a sine curve as a means of increasing the capacity of the wire to bond to concrete; also, welded wire fabric crimped to provide an integral chair.

Deformed bar

1. Bar with deformations (humps, diamonds, circles, or other ridges) to develop bond between steel and concrete.
2. Reinforcement bars with closely spaced shoulders, lugs, or projections formed integrally with the bar during rolling so as to firmly engage the surrounding concrete. Wire mesh with welded intersections not farther apart than 12 in. in the direction of the principal reinforcing and with cross wires not smaller than No. 10 may be rated as a deformed bar.

Designing. Preparation of engineering drawings to show general arrangement of structure, size, and reinforcement of member and other information for proper interpretation of the engineer's design.

Detailers. Drafting personnel who prepare shop fabrication and placing drawings from the engineering drawings prepared by the architect or engineer.

Diagonal band

1. Group of bars covering a width approximately four-tenths the average span, symmetrical with respect to the diagonal running from corner to corner of the panel of a flat slab.

2. In a four-way flat slab system, a group of bars covering a width approximately four-tenths the average span, symmetrical with the diagonal running from corner to corner of the panel.

Diagonal direction. Direction parallel or approximately parallel to the diagonal of the panel of a flat slab.

Diamond mesh. Reinforcing metallic fabric having rhomboidal openings in a geometric pattern.

Direct band. Group of bars, covering a width approximately 0.41, symmetrical with the centerlines of the supporting columns of a flat slab.

Distribution bar reinforcement. Small-diameter bars, usually at right angles to the main reinforcement, intended to spread a concentrated load on a slab and prevent cracking.

Doweled. Joined by dowels. Steel bars inserted in two separately cast concrete elements so as to cause them to function integrally.

Dowels. Short bars connecting two sections of concrete.

Dowel sleeve. Cap of light metal or shaped cardboard on one end of a dowel bar to allow free movement.

Dowel template. Frame or template that outlines the dimensions for setting dowel bars, usually on footings.

Edge-bar reinforcement. Tension steel sometimes used to strengthen otherwise inadequate edges in a slab, without resorting to edge thickening.

Effective area of concrete

1. Area of a section that lies between the centroid of the tensile reinforcement and the compression face of a flexural member.

2. Area of a section that lies between the centroid of the tension refinforcement and the compression surface in a beam or slab and has a width equal to the width of the rectangular beam or slab, or the effective width of the flange of a T-beam.

Effective area of reinforcement. Area obtained by multiplying the right cross-sectional area of the metal reinforcement by the cosine of the angle between its direction and that for which the effectiveness of the reinforcement is to be determined.

End-anchored reinforcement. Reinforcement, in concrete, provided at its ends with anchorage capable of transmitting the tensioning forces to the concrete.

Expanded metal lath. A metal network, often used as reinforcement in concrete or mortar construction, formed by suitably stamping or cutting sheet metal and stretching it to form open meshes, usually of diamond shape.

Fabricating. Actual shop work on the reinforcing steel, such as cutting, bending, bundling, and tagging.

Fireproofing. Encasement of steel with concrete to make it fire resistant; distance between face of concrete member and the reinforcing bars.

Flat slab

1. Concrete slab reinforced in two or more directions, generally without beams or girders to transfer the loads to supporting columns.

2. Reinforced concrete slab generally without beams or girders to transfer the loads to supporting members.

Flat-slab floor construction. Floor construction with few, if any, beams, the slab being of constant depth throughout, except for lintels or drop panels. In two-way flat-slab construction, the reinforcement has only column strips between columns and middle strips between them; in four-way construction, reinforcement runs not only directly between the columns, but also diagonally between them, reinforcement being laid in four different directions.

Hairpin. Light hairpin-shaped reinforcing bar used for shear reinforcement in beams, tie reinforcement in columns, or prefabricated column shear heads.

Header bars. Bars around the head of a column to support the steel in the upper portion of a slab; bars in short header joists that support other joists.

Heavy bending. Bending of bars (other than No. 2 and No. 3) at not more than 6 points; radius bending of one radius; other types of bending not defined as light bending.

Heavy-edge reinforcement. Wire fabric reinforcement, for highway pavement slabs, having one to four edge wires heavier than the other longitudinal wires.

Helical reinforcement (spiral reinforcement). Steel reinforcement of hot-rolled bar of cold-drawn wire fabricated into a helix.

Hickey. Hand tool with side opening jaw used in developing leverage for making bends on bars or pipes in place.

High chair. Wire support for steel in the upper portion of a slab or beam. This support can be either continuous or individual, the latter supporting a reinforcing bar to carry steel at right angles to it.

High-strength steel reinforcing. Steel with a high yield point, in the case of reinforcing bars, 60,000 psi and greater.

Hook. Complete semicircular turn with a radius of bend on the axis of the bar of not less than 3 and not more than 6 bar diameters plus an extension of at least 4 bar diameters at the free end of the bar; 90° bend having a radius of not less than 4 bar diameters plus an extension of 12 bar diameters. For stirrup anchorage only, a 135° turn with a radius on the axis of the bar of 3 diameters plus an extension of at least 6 bar diameters at the free end of the bar.

Horizontal bars. Bars that extend in a horizontal or level direction.

Horizontals. Bars running horizontally, that is, at right angles to the vertical bars.

Jack rod. Plain rod, usually $\frac{7}{8}$ or 1 in. in diameter, with square-cut or milled ends to support sliding forms in connection with a jacking device.

Jack rod sleeve. Piece of pipe that joins two jack rods for end-to-end butt splicing.

Joist bars. Bars used in reinforcing a joist (generally one straight bar and one bent bar are used).

Joist chairs. Supports that usually hold and space the two bars in a joist.

Joist schedule. Table in the placing plan giving the size and reinforcement of all joists in the portion of a building covered by that plan.

Lap

1. Length by which one bar or sheet of fabric reinforcement overlaps another.

2. Joining of two reinforcing bars by lapping them side by side for some 20 or more bar diameters, or as required by applicable code.

Light bending. Term describing all No. 2 and No. 3 bars, all stirrups and column ties, all bars of any size bent at more than 6 points, all bars bent in more than one plane, and all radius bending with more than one radius in any bar or a combination of radius and other types of bending, radius bending being defined as all bends having a radius of 10 in. or more to the outside of the bar.

Lock-woven steel fabric. Rectangular mesh woven together by the transverse wires. Longitudinal wires have an ultimate strength of 180,000 psi with an elastic limit of 125,000 psi.

Longitudinal bars. Run lengthwise of a parallel to the long member, such as in a beam or joist.

Longitudinal reinforcement. Reinforcement essentially parallel to the long axis of a concrete member.

Main reinforcement. Steel reinforcement designed to resist stresses resulting from design loads and moments, as opposed to reinforcement intended to resist secondary stresses.

Milled end. Square-cut or machined end of reinforcing bar making it perfectly square.

Mucking. Adjusting the reinforcing steel during the concreting operation.

Negative reinforcement. Reinforcement so placed as to take the tensile stress due to the negative bending moment.

Nominal diameter. For a deformed bar, the diameter of a plain round bar of the same weight per linear foot.

One-way system. Within a slab, arrangement of steel reinforcement that presumably bends in only one direction.

On the hook. Bar setter, who follows the pouring of concrete with a hooked bar to raise displaced reinforcing steel.

Pencil rod. Small plain round bars, usually No. 2, used in the top of a slab as temperature reinforcement; plain round wires (about No. 6, No. 5, or even $\frac{1}{4}$-in. diameter) used as hangers for suspended ceilings.

Plain bar. Reinforcement that does not conform to the definition of a deformed bar is classified as a plain bar.

Positive reinforcement. Reinforcement so placed as to take the tensile stress due to positive bending moment.

Principal reinforcement. Elements or configurations or reinforcement that provide the main resistance of reinforced concrete to loads borne by structures.

Rail steel. Reinforcing steel rolled from selected used railroad rails.

Rail steel reinforcement. Reinforcing bars hot-rolled from standard T-section rails.

Ratio of reinforcement. Ratio of the effective area of the reinforcement cut by a section of a beam or slab to the effective area of the concrete at that section.

Re-bar. Colloquial term for deformed steel reinforcing bar.

Reed clips. Light wire sections fastened to structural steel to attach concrete fireproofing to the structural steel.

Reinforced concrete

1. Concrete containing adequate reinforcement (either prestressed or not) and designed with the assumption that the two materials act together in resisting forces.

2. Concrete in which metal is embedded in such a manner that the two materials act together in resisting forces.

3. Concrete in which metal other than that for shrinkage or temperature changes is embedded so that the two materials act together for increased strength and ductility.

Reinforcement. Bars, wires, strands, and other slender members that are embedded in concrete in such a manner that the reinforcement and the concrete act together in resisting forces.

Reinforcing. Steel placed in concrete to take tensile stresses.

Rerolled steel. Steel rolled from reclaimed steel.

Schedule. Table shown on placing plans giving size, shape, and arrangement of a number of variations of reinforcements or similar items.

Shear bars. Bars in joists, beams, girders, or slabs, laid alongside the straight bars and bent up for shear.

Simple tie. One wrap of wire about two reinforcing bars twisted so that the bars are held in place.

Slab bolster. Bar support and spacer with corrugated top wire and supporting legs.

Slab spacer. Bar support and spacer for slab bars.

Sleeve. Tube that encloses a bar, dowel, anchor bolt, or used for passing pipes and conduits through slabs.

Spiraled columns. Columns that are usually circular and reinforced with longitudinal bars and a continuous, closely coiled spiral encircling them.

Spiral reinforcement. Continuously wound reinforcement in the form of a cylindrical helix.

Spirals. Coiled wire continuous hooping held to a definite pitch or spacing.

Spiral spacer bars. Bars, usually made of channel or angle irons, punched to form hooks, which are bent over the coiled spiral to maintain it at a definite pitch.

Splice. Development of one bar into another when it is impossible to ship the entire bar in one length. Sometimes a splice may be a butt weld or an overlap weld, but frequently it is an overlap of 20 or more bar diameters buried in the concrete, without the necessity of any mechanical connection of the two pieces.

Standard hook. Hook at the end of a reinforcing bar made in accordance with a standard.

Stem bars. Bars such as those used in the wall section of a cantilevered retaining wall. When a cantilever retaining wall and its footing are considered an integral unit, the wall is often referred to as the stem of the unit.

Stirrups. Loop-shaped (U or W) reinforcing bars ordinarily used in beams to limit diagonal cracking.

Stools. Common expression for chairs and other bar supports.

Structural slab. Suspended, self-supporting, reinforced concrete slab.

Tie bar. Bar added at right angles to and tied to minimum reinforcement to keep it in place.

Tied column

1. Column of almost any prismatic shape, reinforced with longitudinal bars and individual lateral ties a foot or more apart.

2. Column with vertical bars spaced and held together by ties, often assembled on the job.

Ties. Closed loops of small-sized reinforcing bars that encircle the longitudinal steel in columns (or beams or walls). Loops of wire twisted around two reinforcing bars to hold them in place. They add nothing to the strength of the structure.

Tie wire. 16- or 14-gauge black annealed wire in small coils used by bar setters.

Transverse reinforcement. Reinforcement at right angles to the principal axis of a member.

Triangle mesh wire fabric. Longitudinal wires, single or stranded, that are connected by transverse wires woven to them to form triangles.

Truss bars. Bent bars usually placed in the bottom of beams or joists through the middle of a span and bent up over the supports at each end.

Trussed bars. Bars bent up to act as both top and bottom reinforcement.

Two-way slab. Concrete slab reinforced in two directions and supported on four sides.

Two-way system. System of reinforcement: bars, rods, or wires are placed at right angles to each other in a slab to resist stresses due to the bending of the slab in two directions.

Unbonded reinforcement. Reinforcement not bonded throughout its length to the surrounding concrete.

Unit wire fabric. Rectangular mesh secured at the intersections by wires.

Upper beam bolster. Support for the upper layer of steel in beams.

Vertical bar

1. Bar used in an upright or vertical position.

2. Bar that extends in a vertical or plumb direction.

Wall beam. Reinforced concrete beam that extends from column to column along the outer edge of a wall panel.

Welded wire fabric

1. Heavy longitudinal wires held together by transverse wires that cross the longitudinal wires at right angles and are welded to them at the intersections. Longitudinal wires are spaced from 2 to 16 in. in $\frac{1}{2}$-in. increments. Transverse wires are spaced from 1 to 18 in. in 1-in. intervals.

2. Wire mesh put together or fabricated by welding the crossing joints. It comes usually in rolls (sometimes sheets) and is often used for temperature reinforcement.

Wire fabric reinforcement. Material used for floors, roofs, walls, vaults, etc. Wire fabric is made of cold-drawn steel wires crossing generally at right angles and secured at the intersections. Heavy wires run lengthwise and are called "carrying wires"; transverse wires are called "distributing" or "tie wires."

Woven wire fabric. Prefabricated steel reinforcement composed of cold-drawn steel wires mechanically twisted together to form hexagonally shaped openings.

EQUIPMENT AND TOOLS

Air lift. Equipment slurry or dry powder through pipes by means of compressed air.

Angle float. Finishing tool having a surface bent to form a right angle; used to finish reentrant angles.

Arrissing tool. Tool similar to a float but having a form suitable for rounding an edge of freshly placed concrete.

Bar bender. Machine for bending reinforcement.

Buggy. Two-wheeled or motor-driven cart, usually with rubber ties, for transporting small quantities of concrete from hoppers or mixers to forms; sometimes called a "concrete cart."

Bull float. Tool comprised of a large, flat, rectangular piece of wood, aluminum, or magnesium, usually 8 in. wide and 42 to 60 in. long, with a handle 4 to 16 ft in length, used to smooth unformed surfaces of freshly placed concrete.

Bush hammer

1. Hammer having a serrated face, as rows of pyramidal pints, used to roughen or dress a surface.

2. Finish a concrete surface by application of a bush hammer.

Chute. Sloping through or tube for conducting concrete, cement, aggregate, or other free-flowing materials from a higher to a lower point.

Concrete breaker. Compressed-air tool especially designed and constructed to break up concrete.

Concrete finishing machine. Machine mounted on flanged wheels that rides on the forms or on specially set tracks, used to finish surfaces such as those of pavements; a portable power-driven machine for floating and finishing floors and other slabs.

Concrete pave. Concrete mixer, usually mounted on crawler tracks, that mixes and places concrete pavement on the subgrade.

Concrete pump. Apparatus that forces concrete to the placing position through a pipeline or hose.

Concrete spreader. Machine, usually carried on side forms or on rails parallel thereto, designed to spread concrete from heaps already dumped in front of it, or to receive and spread concrete in a uniform layer.

Concrete vibrating machine. Machine that compacts a layer of freshly mixed concrete by vibration.

Conveyor. Device for moving materials; usually a continuous belt, an articulated system of buckets, a confined screw, or a pipe through which material is moved by air or water.

Curb tool. Tool used to give the desired finish and shape to the exposed surfaces of a concrete curb.

Cutting screed. Sharp-edged tool used to trim shotcrete to finished outline.

Darby

1. Tool used to level freshly placed concrete.
2. Hand-manipulated straight edge, usually 3 to 8 ft long, used in the early-stage leveling operations of concrete, preceding supplemental floating and finishing.

Darbying. Smoothing the surface of freshly placed concrete with a darby to level any raised spots and fill depressions.

Delivery hose. Hose through which shotcrete, grout, or pumped concrete or mortar passes; also known as "material hose" or "conveying hose."

Drop chute. Device used to confine or direct the flow of a falling stream of fresh concrete.

Edger. Finishing tool used on the edges of fresh concrete to provide a rounded corner.

Feather edge. Wood or metal tool having a beveled edge; used to straighten reentrant angles in finished plaster coat. Also, edge of a concrete or mortar placement, such as a path or topping that is beveled at an acute angle.

Finishing machine. Power-operated machine used to give the desired surface texture to a concrete slab.

Flat jerk. Hydraulic jack, consisting of light-gauge metal bent and welded to a flat shape, that expands under internal pressure.

Float. Tool (not a darby), usually of wood, aluminum, or magnesium, used in finishing operations to impact a relatively even but still open texture to an unformed fresh concrete surface.

Flow through. Sloping trough used to convey concrete by gravity flow from a transit mix truck or receiving hopper to the point of placement.

Groover. Tool used to form grooves or weakened plane joints in a concrete slab before hardening to control crack location or provide pattern.

Gutter tool. Tool used to give the desired shape and finish to concrete gutters.

Jitterbug. Grate tamper for pushing coarse aggregate slightly below the surface of a slab to facilitate finishing.

Jointer (concrete). Metal tool about 6 in. long and from 2 to $4\frac{1}{2}$ in. wide and having shallow, medium, or deep bits (cutting edges) ranging from $\frac{3}{16}$ to $\frac{3}{4}$ in. or deeper, used to cut a joint partly through fresh concrete.

Loading hopper. Hopper in which concrete or other free-flowing material is placed for loading by gravity into buggies or other conveyances for transport to the forms or other place of processing, use, or storage.

Power float. Motor-driven revolving disk that smooths, flattens, and compacts the surface of concrete floors or floor toppings.

Ramming. Heavy tamping of concrete, grout, or the like by means of a blunt tool forcibly applied.

Screed. Tool for striking off the concrete surface, sometimes referred to as a "strike off."

Stationary hopper. Container used to receive and temporarily store freshly mixed concrete.

Straightedge. Rigid, straight piece of wood or metal used to strike off or screed a concrete surface to proper grade or to check the planeness of a finished grade.

Surface vibrator. Vibrator used for consolidating concrete by application to the top surface of a mass of freshly mixed concrete. Four principal types exist: vibrating screeds, pan vibrators, plate or grid vibratory tampers, and vibratory roller screeds.

Tamper. Hand-operated device for compacting floor topping or other unformed concrete by impact from the dropped device in preparation for strike off and finishing; contact surface often consists of a screen or a grid of bars to force coarse aggregates below the surface to prevent interference with floating or troweling.

Tamping rod. Round, straight, steel rod having one or both ends rounded to a hemispherical tip.

Tremie

1. Pipe or tube through which concrete is deposited under water, having at its upper end a hopper for filling and a bail for moving the assemblage.
2. Pipe having a funnel-shaped upper end into which concrete is fed for depositing. Bottom end is normally kept continuously buried in the newly placed concrete and is raised with the rise of the concrete in the structure. Generally used when placing concrete under water, or when the location of depositing is too difficult or too far to reach by normal placing methods.

Trowel. Flat, broad-blade, steel hand tool used in the final stages of finishing operations to impart a relatively smooth surface to concrete floors and other unformed concrete surfaces; also, a flat, triangular-blade tool used for applying mortar to masonry.

Troweling machine. Motor-driven device that operates orbiting steel trowels on radial arms from a vertical shaft.

Truck mixer. Concrete mixer suitable for mounting on a truck chassis and capable of mixing concrete in transit.

Vertical-shaft mixer. Cylindrical or annular mixing compartment having an essentially level floor and containing one or more vertical rotating shafts to which blades or paddles are attached. The mixing compartment may be stationary or rotate about a vertical axis.

Vibrator

1. Mechanical tool that vibrates at a speed of 3000 to 10,000 rpm and is inserted into wet concrete or applied to formwork to compact concrete.
2. Oscillating machine used to agitate fresh concrete so as to eliminate gross voids, including entrapped air but not entrained air, and produce intimate contact with form surfaces and embedded materials.

Vibrators. Electrically powered vibrators with rigid or flexible shafts.

FORMWORK AND ACCESSORIES

Accessories. Items other than frames, braces, or post shores used to facilitate the construction of scaffold and shoring.

Adjustible wood shore. Two-piece shore consisting of overlapping wood members that are secured in place with post clamps.

Adjustment screw. Leveling device or jack composed of a threaded screw and adjusting handle, used for the vertical adjustment of shoring and formwork.

Aluminum forms. Prefabricated, lightweight, smooth-faced forms or forms prepared for patterned walls.

Anchor

1. Embedded fastening device.

2. Device used to secure formwork to previously placed concrete of adequate strength. Device is normally embedded in the concrete during placement.

Architectural concrete. Concrete that will be permanently exposed to view and that therefore requires special care in selection of concrete materials, forming, placing, and finishing to obtain the desired architectural appearance.

Architectural-grade concrete form panels. Panels consisting of exterior quality overlaid plywood, with overlay permanently fused to the plywood under heat and pressure. Properties of the overlay provide a nonglossy matte finish to the concrete surface, without surface discoloration from the overlay.

Architectural surfaces. Formed concrete surfaces where finished appearance is required by the design criteria.

Band iron. Thin metal strap used as form tie, hanger, etc.

Band-iron tightener. Hardened steel device with malleable iron jaw casting, used for pulling and tightening band iron.

Beam bottom. Soffit or bottom form for a beam.

Beam clamp

1. Various types of tying or fastening units used to hold the sides of beam forms.

2. Prefabricated, shaped punched clamps that hook over the top of steel beams and girders; used in combination with soffit spacers for hanging forms for fireproofing steel.

Beam forms

1. Retainer or mold, erected as to give the necessary shape, support, and finish to a concrete beam.

2. Prefabricated, reusable, wood, metal, or fiberglass forms.

Beam hangers. Wire, strap, and other hardware devices that support formwork from structural members.

Beam sides. Vertical side panels or parts of a beam form.

Beveled edge. Sloped edge or surface (chamfer), strip built into forms to eliminate sharp corners on columns and beams.

Bolt sleeve. Tube surrounding a bolt in a concrete wall to prevent concrete from adhering to the bolt.

Box out. Form an opening or pocket in concrete by a boxlike form.

Brace. member designed and located for the purpose of stiffening or restraining movement.

Bracing. Member used to support, strengthen, or position another piece or portion of a framework.

Break back ties

1. Weakened joint in tie designed for snap-off of external portion of tie so that tie ends terminate within formed surface.

2. Tie so designed as to allow it to be snapped off below the concrete surface.

Bulkhead. Partition in the forms blocking fresh concrete from a section of the forms or closing the end of a form, as, for example, a construction joint or change in foundation height.

Butt joint. Joint made by fastening two parts together end to end without overlapping.

Camber. Slight arching of a given beam or form to improve the appearance or compensate for design deflection.

Cathead. Notched wedge placed between two formwork members meeting at an oblique angle; spindle on a hoist; large, round retention nut used on the bolts.

Centering. Framework used in the construction of structures or any continuous structure where the entire falsework is lowered as a unit.

Chamfer. Oblique surface made by cutting away an edge or corner.

Chamfer strips. Strips that provides smooth, uninterrupted beveled edge to piers, beams, and all outside corners of poured concrete. Strips are easily applied, come equipped with or without nailing flange to facilitate installation, and are reusable.

Clay tile filler. Filler used to form concrete joists and slab.

Clean-out. Opening in the forms for the removal of debris, etc., closed before concrete is placed.

Cleat. Small board used to connect formwork members or used as a brace.

Climbing form. Form that is raised vertically for succeeding lifts of concrete in a given structure, usually supported on anchor bolts or rods embedded in the top of the previous lift.

Coating. Liquid applied to interior formwork surfaces, usually to promote easy release from the concrete.

Cold joints. Nonmoving joints developed in concrete elements to facilitate workable limits for placement.

Column capital form. Prefabricated galvanized steel capital form for round columns.

Column clamp. Fastening unit used to hold column form sides together.

Column side. One of the vertical panel components of a column form.

Concrete ceiling inserts. Inserts placed in concrete to carry hanger rods for suspending ceilings, ductwork, plumbing pipes, electrical conduits, and lighting fixtures.

Cone bolt. Form of tie rod for wall forms with cones at each end inside the forms so that a bolt can act as a spreader, as well as a tie.

Continuous high chair. Prefabricated galvanized steel wire or stainless steel chair, set on formwork to support steel reinforcing rods.

Continuous slotted inserts. Prefabricated heavy rolled steel insert with anchoring lugs for square-head bolts.

Control joints. Joints designed for use in solid wall construction. Control joints allow for expansion and contraction.

Coupling pin. Insert device used to connect lifts or tiers or formwork scaffolding vertically.

Cross bracing. System of members that connect frames or panels of scaffolding laterally to make a tower or continuous structure.

Cross joint. Joint at the end of individual formboards between subpurlins.

Crush plate. Expendable strip of wool attached to the edge of a form or intersection of fitted forms to protect the form from damage during prying, pulling, or other stripping operations.

Deck. Form upon which concrete for a slab is placed.

Dome form

1. Prefabricated wood, metal, or fiberglass domes, used in waffle design slabs.

2. Square prefabricated pan form used in two-way (waffle) concrete joist floor construction.

Dovetail anchor slots. Anchor slots prefabricated in galvanized steel, copper, stainless steel, and zinc alloy and used for permanent anchoring to concrete. Anchors are installed in slots for anchoring brick, block, tile, or stone.

Dowel cap. Prefabricated galvanized steel or plastic cap that fits over dowel and allows for movement.

Drip strip. Material nailed to the form and easily removed from the concrete. It can be reused several times.

Edge form. Formwork used to limit the horizontal spread of fresh concrete on flat surfaces such as floors.

Extension device. Any device, other than an adjustment screw, used to obtain vertical adjustment of shoring towers.

Face of form. Surface of form in contact with the concrete.

Falsework. Temporary structure erected to support work in the process of construction; composed of shoring or vertical posting, formwork for beams and slabs, and lateral bracing.

Fiber ducts. Prefabricated, laminated, tubular forms used in cast-in-place and precast concrete slab construction for air ducts or wiring ducts.

Fiber pans. Prefabricated, reusable, molded fiber pans for cast-in-place concrete joists and slabs.

Fiber tubes. Prefabricated, laminated, tubular forms used in cast-in-place and precast concrete slab construction.

Fillet. Narrow molding applied to the junction of two surfaces forming a corner.

Finnish birch panel forms. Face veneers of birch, with inner plies of alternating birch and spruce and/or pine. Grain of the inner birch plies runs perpendicular to the grain of the face plies. Plies are bonded with exterior phenolic resin adhesive (weather- and boilproof). Both faces have smooth phenolic-resin-impregnated paper surfaces applied under heat and pressure to produce smooth, even, tough surfaces. Highly resistant to alkalies in cement.

Fishtail. Wedge-shaped piece of wood used as part of the support form between tapered pans in concrete joist construction.

Flashing reglets. Prefabricated galvanized steel, copper, or stainless steel inserts, with foam filter or taped face to prevent seepage of concrete grout, removed when forms are removed. Used to receive flashing materials and roofing felt.

Flush-type slab handle. Handle for removable concrete slabs, manufactured in malleable iron or galvanized, painted, or cast bronze.

Flying forms. Large mechanically handled sections of formwork; frequently includes supporting truss, beam, or scaffolding units completely unitized. Term usually applies to floor framing system.

Form. Temporary structure or removable mold for the support of concrete while it is setting and gaining sufficient strength to be self-supporting.

Form anchors. form anchors are devices used in the securing of formwork to previously placed concrete of adequate strength. The devices normally are embedded in the concrete during placement. Actual load carrying capacity of the anchors depends on their shape and material, the strength and type of concrete in which they are embedded, the area of contact between concrete and anchor, and the depth of embedment and location in the member.

Form hangers. Devices used to support formwork loads from a structural steel or precast concrete framework.

Formers. Items attached to form facings and used to effect special profiles on concrete surfaces.

Formwork. Total system of support for freshly placed concrete, including the mold or sheathing that contacts the plastic concrete, as well as all supporting members, hardware, and necessary bracing.

Ganged forms. Prefabricated panels joined to make a much larger unit for convenience in erecting, stripping, and reusing.

Grade strip. Strip of wood tacked to the inside of a form at the line to which the top of the concrete lift is to come, either at a construction joint or the top of the structure.

Hanger

1. Device used to support formwork form a structural framework.

2. Device used to suspend one object from another.

Insulation. Insulating material or blankets applied to the outsides of forms in sufficient thickness and air tightness to conserve heat of hydration to maintain concrete at required temperatures in cold weather.

Jack. Mechanical device used for adjusting elevation of forms or form support. "Jack" is sometimes used instead of "Jack-shore."

Jack-shore. Telescoping, or otherwise adjustable, single-post metal shore used for supporting formwork.

Key or keyway. Groove made in a concrete joint giving shear strength to the joint.

Kicker. Board attached to a formwork member in a building frame or formwork to make the structure more stable.

Kickouts. Accidental release or failure of a shore or brace.

Kiln-dry lumber. Lumber form which moisture has been removed by placing material into a closed kiln or chamber, subjecting it to a high temperature, and providing it with proper ventilation during the heating process.

Knee brace. Brace between horizontal and vertical members in a building formwork to make the structure more stable.

Lacing. Horizontal bracing between shoring members.

Lifting inserts. Inserts placed in precast concrete panels to facilitate lifting procedures in the yard and at the job site.

Liner. Nonstructural form-facing element in contact with both form sheathing and concrete.

Lining

1. Selected materials used to line the concreting face of formwork to create a smooth or patterned finish to the concrete surface, to absorb moisture from the concrete, or to apply a set-retarding chemical to the formed surface.

2. Any sheet, plate, or layer of material attached directly to the inside face of the forms to improve or alter the surface texture of the finished concrete.

Manhole step rings. Prefabricated open-ended steel rods formed into U shapes with ends turned up for anchoring into concrete.

Mold. Cavity or surface against which concrete is cast to give it a desired shape or design pattern.

Moving forms. Large, prefabricated units of formwork incorporating supports, and designed to be moved horizontally on rollers or similar devices.

Mudsill. Wood plate or bulk timer used to distribute the load from props or shores to stable ground conditions.

Oil. Special oil is applied to interior surface of formwork previous to installation to promote easy release from the concrete when forms are moved.

Pan

1. Prefabricated form unit in concrete joist floor construction.
2. Preshaped metal, fiber, or fiberglass form used in the construction of ribbed floors or waffles.

Panel. Section of form sheating, constructed from boards, plywood, metal, or other materials, that can be erected and stripped as a unit.

Panel straps. Prefabricated galvanized steel straps nailed to forms to control racking and twisting of panels.

Paper form. Special heavy waxed paper mold used for casting concrete columns and other structural shapes.

Permanent shores. The original shores that remain in place without being disturbed during and after formwork removal; prevents the new concrete from supporting its own weight and additional construction loads above.

Plate. Horizontal timber acting as a sill or support for other members.

Plate saddle hanger. Hanger usually hung over the top of steel beams or masonry walls to carry supporting formwork for cast-in-place slab over steel supports.

Plywood concrete forms. Each panel bears the imprint of an authorizing agency certifying that the panel meets the grade requirements of Class 1, Exterior, with veneer face quality of A, as defined in U.S. Department of Commerce PS 1-66. All edge and core gaps exceeding $\frac{1}{4}$ in. must be filled with plastic filler; all mill cut edges sealed with two coats of polyurethane sealer, and all panels mill oiled on all exposed faces.

Polystyrene form liners. Used to add design accents that imitate stone, brick, or rough grain wood. Custom designs and creative patterns are available. The form liners are made for one-time use and are easy to apply and strip.

Post. Vertical formwork member used as a brace; also, shore, prop, or jack.

Pressed steel inserts. Inserts that are prefabricated in plain pressed or formed steel or galvanized finishes for inserting heads of bolts.

Pressure. Lateral pressure acting on vertical or inclined formed surfaces, resulting from the plastic behavior of the unhardened concrete confined by the forms.

PVC form liners. Non-porous form liners have a smooth alkali resistant plastic surface and can be coated for quick release. Liners are reusable up to at least ten times, depending on maintenance.

Radius formers. Formers made from a plastic material that provides a smooth, uninterrupted, rounded radius corner to poured concrete. Material strips cleanly from finished concrete and is easily removed from forms for reuse.

Raker. A sloping brace of metal or wood used in formwork as a shore head.

Rate of pour. Speed at which concrete is placed in the form.

Reshoring. Temporary vertical support for forms or completed structure placed after original shoring has been removed.

Rock anchor. Anchor for formwork, used where formwork occurs on one surface only with the rock used as the formwork for the inside face. The anchors are drilled into the rock, and a threaded coil rod and coil tie are used as a spacer.

Round column fiber tubes. Prefabricated, seamless, laminated tubes with a nonwater-sensitive adhesive. Exterior surface is uniformly wax impregnated for weather and moisture, interior surface is treated with vapor barrier to prevent penetration of moisture.

Round column steel forms. Prefabricated, removable galvanized steel forms, bolted angle reinforced with devices for steel wedge installation.

Rustication strip. Strip of wood or other material attached to a form surface to produce a groove or rustication in the concrete.

Scabbing. Inadvertent removal of the surface of the concrete due to adhesion to the form.

Screeding devices. Galvanized preformed wire shapes and devices used as screed chairs to establish finished height of concrete slabs.

Sheathing

1. Material forming the contact face of the forms; also called "lagging."
2. Form-facing element with structural capacity, in contact with form-framing members and with concrete when liner is not used.

She bolt. Type of form tie and spreader bolt in which the end fastenings are threaded into the end of the bolt, thus eliminating cones and reducing the size of holes left in the concrete surface.

Shore

1. Temporary support for formwork and fresh concrete or for recently built structures that have not developed full design strength.
2. Vertical or inclined members used as a prop.

Shore head. Wood or metal horizontal member placed on and fastened to vertical shoring member.

Shoring

1. Temporary system of supports for forms composed of wood, metal and metal posts, scaffolding frames, etc.
2. Props or posts of timber of other material in compression, used for the temporary support of excavations, formwork, or unsafe structures.
3. Process of erecting shore.
4. Temporary vertical supporting system for a nonvertical concrete element.

Side form spacers. A side form spacer is a device which maintains the desired distance between a vertical form and reinforcing bars.

Slab bolster. Fabricated, galvanized wire support placed in formwork for reinforcing bars.

Slip form

1. Form that moves, usually continuously, during the placing of the concrete; also called "sliding form." Form is made to move so that it leaves the formed

concrete only after concrete is strong enough to retain its shape and support its own weight. Use of this form is similar to an extrusion process.

2. Form that is raised vertically as the concrete is placed.

Slump of concrete. Number of inches that a mass of concrete settles after removal of a standard slump test cone.

Snap tie

1. Proprietary concrete wall form tie, the end of which can be twisted or snapped off after the forms have been removed.

2. Concrete wall form metal tie, with spreader washers, plastic cones, and a head that can be easily snapped off and the ends removed from the face of the concrete. Tapered cones are removed and the hole filled with grout.

Soffit. Underside or undersurface of a member.

Soffit spacer. Prefabricated steel round that serves as a separator and gauge between the soffit forms and structural steel. Two spacers are driven into the form at the beam hanger to ensure correct depth of fireproofing and assist in keeping the beam bottom rigid.

Soldiers. Vertical wales or stiffeners.

Spacer. Device that maintains reinforcement in proper position; wall forms at a given distance apart before and during pouring of concrete.

Spreader. Brace, usually of wood, used to separate two interior form faces until the concrete is poured.

Stake. Pointed piece of wood or other material to be driven into the ground.

Standard wall tube—plastic coated. Form for exposed columns. Tube is plastic coated inside to facilitate release and wax coated outside for moisture resistance.

Stone anchor inserts. Prefabricated, heavy-gauge stainless steel insert channel for vertical and horizontal adjustment in anchoring stone to concrete walls.

Strip. Job term that means to remove formwork.

Strong back. Frame attached to the back of a form or precast structural member to stiffen or reinforce it during concrete placing operations.

Structural surfaces

1. Formed surfaces where strength of the formed concrete is of major importance.

2. Formed concrete surfaces where the finished surface is free of honeycomb and structurally sound.

Stud. Vertical supporting member to which sheathing is attached.

Stud clamps. Malleable iron adjustable clamps used to joint together the outside studs of adjoining form panels.

Tape. Narrow band of fabric used to cover form joints.

Taper-rod-type tie. Generally a three-piece assembly, consisting of a completely removable tapered and treaded rod installed through the formwork with each end secured to horizontally walers on each side by means of nut washers or plates.

Telescopic metal joists. Adjustable metal joists used for supporting formwork.

Threaded inserts. Prefabricated malleable iron, gray iron, or cast bronze inserts with plate area for anchoring and setting in forms. Threaded for bolts from $\frac{1}{4}$ in. diameter to 1 in. in diameter.

Tie

1. Fabricated metal device used to secure and space both sides of formwork.

2. Tensile unit adapted to prevent concrete forms from spreading due to the fluid pressure of freshly placed, unhardened concrete.

3. Tensile unit for holding the forms secure against lateral pressures.

Tieback. Rod or cable fastened to a dead man, rigid foundation, or soil anchor to prevent lateral movement of formwork.

Tongue-and-groove joint. Joint consisting of a projecting rib on one edge of a board together with a corresponding groove in the edge of the adjacent board.

Top form. Form required on the upper or outer surface of certain types of poured concrete roof decks.

Tube-and-coupler shoring. Load-carrying assembly of tubing or pipe.

Utilitarian surface. Formed surfaces where appearance is not of major importance.

Vacuum system. Special type of form in which a partial vacuum is created to withdraw excess moisture from the concrete.

Variations from plumb. Such variations are usually specified as to level, grade, or vertical standards.

Waffle slab forms. Prefabricated steel, laminated fiber, or glass fiber forms.

Waler

1. Long horizontal member that holds vertical studs in position and forms in line; structural element in formwork framing systems.

2. Strong-back structural element in formwork framing systems, running at right angles to form studs, used to align studs and to tie them together into a structural entity.

Waler-rod-type tie. Multiple-piece (she-bolt) assembly, consisting of an inner fixed rod through the center of the framework with threaded connections at each end, secured to removable waler rods and extending through the waler and terminating with nut washers on threaded ends.

Wall form. Retainer or mold erected to give the necessary shape, support, and finish to a poured concrete wall.

Warping. Deformation from a plane surface.

Waterstop. Barrier embedded in concrete to prevent seepage of water through a joint.

Wedge. Piece of wood or metal tapering to a thin edge, used to adjust elevation or tighten formwork.

Wedge insert. Prefabricated malleable iron slip wedge insert set in a concrete form with a deep anchorage loop on the back side for reinforcing bar.

Wrecking strip. Small panel fitted into a formwork assembly in such a way that it can be easily removed ahead of main panels or forms, making it easier to release major form components.

Yoke. Tie or clamping device around column forms or over the top of wall or footing forms to keep them from spreading because of the lateral pressure of fresh concrete. Part of a structural assembly for slip forming, which keeps the forms from spreading and transfers form loads to the jacks.

PRECAST AND PRESTRESSED CONCRETE

Anchor. Device used to lock the stressed tendon in position so that it will retain its stressed condition.

Anchorage. Means by which the prestress force is permanently delivered to the concrete.

Anchorage deformation or slip. With certain types of post-tensioning systems, a small portion of the prestress is lost at the time the prestressing load is transferred from the prestressing jack to the anchorage device. This loss of stress is the result of the tendon's slipping in the anchorage device or the deformation of the anchorage device, or in some instances a combination of both.

Anchorage device. Device used in post-tensioning to anchor the tendon to the concrete member or in pretensioning to hold the tendon in the stressed condition during hardening of the concrete.

Auxiliary reinforcement. In a prestressed member, any reinforcement in addition to that participating in the prestressing function.

Bond. The bond serves a dual function in pretensioned-prestressed concrete. Its first function is to transfer the load from the steel to the concrete to accomplish the prestressing; thus it serves as a "prestress transfer bond." Its second function, is to distribute the steel stress to correspond to the magnitude of the change in moment at any cross section; thus it serves as a "flexural bond."

Bonded post-tensioning. Process whereby post-tensioned tendons are grouted after stressing in such a manner that the tendons become bonded to the concrete section.

Bonded tendons. Tendons that are bonded to the concrete either directly or through grouting. Unbonded tendons are free to move relative to the surrounding concrete.

Bond prevention. Stresses at the ends of pretensioned members are occasionally kept within the permissible values by preventing specific tendons from becoming bonded to the concrete at the ends by wrapping the tendons in plastic or paper tubes or by other means.

Broken bond. Term is occasionally used to describe bond prevention.

Cable. Post-tensioning tendon that is usually composed of a number of more or less parallel wires or strands.

Cap cable. Post-tensioned tendon that is placed through the ends of two precast flexural members after they have been erected and that, when post-tensioned, renders the members continuous.

Channel. Three-sided unit cast like a double T but lacking cantilevered edges Carries heavier loads than the double T and is considered excellent for floors and bridge decks. Comparatively large area can be covered by each unit.

Channel member. Member designed for heavy-duty, long-span use. Unit sees more use on bridge and heavy engineered construction than in building construction.

Circular prestressing. Prestressing principle applied to tanks, pipes, domes, and silos.

Compressive strength. Strength established in design so that the limiting values of working stresses may be set, and, to a lesser extent, so that the load-carrying capacity of prestressed units might be predicted.

Cracking load. Load that when applied to a prestressed member nullifies the prestressing and causes tensile stresses in the member that are equal in magnitude to the tensile strength of the concrete.

Creep

1. Increase in strain with time under constant load. At usual working stresses, creep is directly proportional to the applied unit stress. Proportionality ceases to exist under overload conditions, putting the stress–strain relationship at an unestablished high stress.

2. Permanent deformation of a material at a given temperature under sufficiently high sustained loading, continuing with time, but without increasing the load.

3. Time-dependent strain occurring under stress. Creep strain occurring at a diminishing rate is called "primary creep"; that occurring at a minimum and almost constant rate, "secondary creep," that occurring at an accelerating rate, "tertiary creep."

4. Slow deformation under stress.

5. Permanent deformation occurring over a period of time in a material subjected to constant stress at constant elevated temperatures.

Creep limit. Maximum stress that will cause less than a specified quantity of creep in a given time, the maximum nominal stress under which the creep strain rate decreases continuously with time under constant load and at constant temperature. Sometimes used synonymously with "creep strength."

Creep recovery. Time-dependent strain after release of load in a creep test.

Creep strength

1. Rate of continuous deformation under stress at a specific temperature. Commonly expressed as the stress, in pounds per square inch, required to produce a certain percentage of elongation in a specified number of hours.

2. Constant nominal stress that will cause a specified quantity of creep in a given time at constant temperature, constant nominal stress that will cause a specified creep rate at constant temperature.

3. Load that will produce only a specified size change when constantly applied for a given period of time at a specific temperature.

Creep test. Test for determining creep behavior of materials subjected to prolonged constant tension or compression loading at constant temperature.

Crimped wire. Wire that is deformed into a curve that approximates a sine curve as a means of increasing the capacity of the wire to bond to the concrete.

Curvature friction. Friction resulting from bends or curves in the specified cable profile.

Dead end. End of the tendon that is opposite the end from which the stress is applied in stressing a tendon from one end only.

Deferred deformation. Same as plastic flow.

Deflected or draped tendons. Tendons that have a curved trajectory through the member rather than being straight in either pre- or post-tensioned construction.

Deformation

1. Change in the shape or size of a body whenever it is subjected to a force.

2. Material when subjected to the action of a force changes shape. Change in shape is deformation or strain. Deformation per unit of length is the unit deformation.

Double T unit

1. Combining deck and joist in deck-framing member capable of spans 60 ft and over. The advantage of the unit is that large areas of deck can be simultaneously framed and formed in short periods of time. It is commonly employed where underside is to be left exposed.

2. Unit widely used for roof construction, containing beam and deck construction in one unit. Spans up to 60 ft are possible. Roofing and insultation are placed directly on smooth top surface.

Ducts. Holes in post-tensioned members that are provided for the tendons and that can be formed by a removable core or by a metallic or plastic tubing that remains in the member.

Effective prestress. Stress remaining in the tendons after all losses have occurred, excluding the effects of the dead load and superimposed loads.

Effective stress. Portion of the prestressing force that is retained in the structure permanently and is effective in resisting loads in the elastic range.

Elastic shortening. Elastic deformation of the concrete member that takes place as a result of the prestressing.

End anchorage. Anchorage device.

End block. Enlarged section at the ends of flexural members that is provided as a means of obtaining sufficient area in which to embed the end anchorages in post-tensioned construction.

Final stress. Effective stress.

Flat jack. Special hydraulic jack, formed of light-gauge metal, bent and welded into a flat shape that expands when hydraulic pressure is applied to it. Flat jacks are occasionally used in stressing special structures, such as pavements and dams, and the jacks are frequently left in the structure, where they can be used periodically to check or adjust the prestress.

Flat slab

1. Concrete slab that is prestressed or conventionally reinforced, mostly applicable to lightweight, short-span up to 8 ft. This type of slab requires joists or purlins for intermediate support.
2. Concrete slab resembling wood deck planking because of tongue and groove connection. Slab consists of a shallow section that takes little headroom.

Flexural bond. Bond stress between the concrete section and the tendon that results from the application of external load.

Friction loss. Stress loss in a tendon that results from the friction between the tendon and the duct or other device with which the tendon is in contact during stressing.

Gradual losses. Reduction from the initial force to the working force due to shrinkage, creep of the concrete, and stress relaxation of the steel.

Hoyer effect. Nature of bond stress along pretensioned tendons is based on the frictional forces that result from the tendons' attempting to regain the diameter they had before they were stressed.

I-Beam. Precast concrete beam employed principally in floor, roof, and bridge deck construction. Medium to long spans are possible where depth of member (head room) is not a factor. The unit combines well with a cast-in-place deck for composite construction. Spans are limited by loads and lateral support.

Indented wire. Wire used in pretensioning. The wire is put through a machine that causes small indentation on the surface intended to improve the bond characteristics of the wire.

Initial force. Force left in the tendons immediately after stressing and anchoring.

Initial stress or force. Stress in the prestressing tendons that is resisted by the concrete immediately after the completion of prestressing.

Inverted T-beam. Precast concrete beam commonly combined with cast-in-place or precast bridge deck to save head room. Lighter precast concrete joist used in building construction for support of masonry soffit blocks and concrete topping.

Jacking device or mechanism. Device used to stress the tendons.

Jacking force. Temporary force exerted by the device that introduces the tension into the tendons.

Jacking stress. Stress that occurs in the tendon prior to losses that may result from deformation of the anchorage.

Joist. Small precast concrete T-shaped beam used in parallel series in concrete joist floor construction.

Kern zone. Area within a particular cross section in which a force can be applied without the tensile stresses resulting in any of the extreme fibers.

Lateral support. Structural member or system of structural members that resists the horizontal component of loads.

Linear prestressing. Prestressing that is applied to elongated structural elements, such as beams, piles, columns, or girders.

Loss of prestress. Reduction in the prestressing force that results from the combined effects of creep, or relaxation of the steel, and plastic flow and shrinkage in the concrete. Term does not include frictional losses.

Modified I-beam. Precast concrete beam used mostly in bridge construction where poured deck acts as composite construction. Here stirrups are extended into slab to form shear connection.

Moment splice. Splice employed with conventional precast concrete members. The conventional practice with prestressed concrete is to use simple spans spliced over column points.

Multielement members. Post-tensioned members that are composed of several small precast elements held together by the post-tensioning.

Nonprestressed reinforcement. Reinforcement that is not prestressed and that may be either of high-quality or normal reinforcing steel.

Overstretching. Practice of stressing the tendons to a value higher than desired for the initial stress and holding this higher value for a specific time as a means of reducing the creep of the steel, which occurs after anchorage, or overcoming friction losses.

Partial prestressing. Prestressed construction in which the concrete is allowed to withstand nominal tensile stresses at design loads.

Plastic flow. Change of strain that takes place in concrete when subjected to constant stress.

Post-tensioned construction. Prestressed construction in which the prestressing operation takes place after the concrete has hardened.

Post-tensioning. Method of prestressing in which the tendons are tensioned after the concrete has hardened.

Precast concrete

1. Plain or reinforced concrete building element cast in other than its final position in the structure.

2. Concrete that is cast and allowed to cure into a desired shape prior to being placed in its final position.

Precast member. Concrete member that is cast and cured in other than its final position; the process of placing and finishing precast concrete.

Pre-post-tensioned construction. Prestressed concrete construction in which the prestressing is a combination of pre-tensioning and post-tensioning.

Prestressed concrete

1. Structural concept that combines concrete and steel; concrete is capable of resisting relatively high compressive stresses. However, its tensile strength is only 10 to 15% of its compressive strength.

2. Concrete in which there have been introduced internal stresses of such magnitude and distribution that the stresses resulting from service loads are adjusted to the desired degree.

3. Concrete shapes into which stress has been induced prior to their being placed in their final location.

Prestressed piles. Pile that is either octagonal or square, cast either hollow or solid. Prestressed piles are used as columns and exhibit little shrinkage or cracking. Units are available in long lengths for piling purposes.

Prestressed plank. Concrete plank commonly employed in continuous or clear span deck construction. Spans up to 60 ft are possible; joists also form deck surface. The voids between the planks receive dowel reinforcing and grout. Two- and three-core units in various plank widths, lengths, and thicknesses are available.

Pretensioned construction. Prestressed concrete construction that results from placing concrete around stressed tendons that are not released until the concrete has hardened.

Pretensioning. Method of prestressing in which the tendons are tensioned before the concrete is placed.

Proof stress. Specific stress to which some types of tendons are subjected in the manufacturing process as a means of reducting the deformation of anchorage, reducing the creep, or ensuring that the tendon is sufficiently strong.

Relaxation. Loss in stress that occurs in a material subjected to constant strain. Term is also used to mean the loss of prestress.

Single-wing double T unit. Unit that combines greater shear strength at joint owing to bearing ledge. When units are welded together, deck acts in diaphragm fashion. Construction joints are concea.ed by the stem of the unit, which makes for better exposed underside appearance. Long spans are available.

Sheath. Metal or plastic tube in which post-tensioned tendon may be encased in order to prevent its becoming bonded to the concrete during the placing of the plastic concrete.

Strand. Prestressing tendon that is composed of several wires, most of which are twisted.

Strength test. Average compressive strength of three companion compression test specimens tested at the same age.

Stress corrosion. Type of corrosion that occurs in many types of metals but that is particularly dangerous in cold-drawn wire. This type of corrosion is characterized by a disintegration of the cementitious portion of the steel and results in the steel's becoming brittle.

Stressing end. End of the tendon or member at which the prestress is applied in either pre- or post-tensioning, when the tendons are stressed from one end only.

T-concrete beam. Beam whose cross section resembles a "T." Several T-beams side-by-side, if acting as a unit, form a floor slab.

Tendon

1. Tensioned steel element used to impart prestress to the concrete.

2. Wire, strand, cable, or bar used to stress the concrete in prestressed construction.

Tilt-up construction. Method of construction where concrete wall sections are cast horizontally on the ground and tilted or lifted into position.

T-Joist

1. Unit used for moderately long spans where exposed ceilings are desired. There are several variations in section for this member.

2. Unit used in exposed ceilings because of design.

Transfer

1. Time or act of transferring the prestressing force from the jacking device to the concrete section.

2. Operation of transferring the tendon force to the concrete.

Transfer bond. Bond stress between the concrete and a pretensioned tendon at the end that transfers the stress in the tendon to the concrete.

Transmission length. Distance at the end of a pretensioned tendon necessary for the bond stress to develop the maximum tendon stress; sometimes called "transfer length."

Unbonded tendon. Post-tensioned tendon that is coated with grease or bituminous material and wrapped in paper or plastic tubing as a means of preventing the tendon from becoming bonded to the concrete.

Wide-flange beam. Precast concrete beam that is superior to the I-beam in design for long spans under heavy loading and where depths are restricted.

Wide-flange T-beam. Precast concrete beam that is superior to the I-beam design for wide spans under heavy loading. Wide upper flange permits longer unsupported span. Commonly employed on bridge and heavy engineered construction.

Wide precast, prestressed deck unit. Unit designed for rapid floor framing and decking use and made for spans up to and exceeding 30 ft.

Wobble friction. Friction caused by the unintended deviation of the prestressing steel from its specified profile.

Working force. Final force left in the tendons after all losses have occurred.

Masonry

BLUESTONE

Accent stone. Bluestone with unusual textures, available in a scabbled or bush hammered texture.

Anchors. Metal ties, cramps, dowels, and similar accessories used to hold bluestone units permanently in position in other than floors; usually made of galvanized or asphaltic coated steel, brass, bronze, or other corrosion-resistant metal.

Arris. Edge formed by the meeting of two surfaces.

Ashlar. Bluestone available in natural bed random-height rubble, sawed-bed multiple-height veneer, and mosaic shapes.

Back-up. Concealed construction of masonry or concrete acting as a mounting base.

Base. Lowest member of any structure or architectural feature occurring just above the finished floor line at the vertical wall surface.

Bluestone

1. Term applied to bluish gray sandstones, mica schist, and gray gneiss.
2. Dense, hard, fine-grained, commonly feldspathic sandstone or siltstone of medium to dark or bluish gray color that splits easily along original bedding planes to form thin slabs.
3. Hard sandstone of characteristic blue, gray, and buff colors, quarried in the states of New York and Pennsylvania.

Bluestone uses. The ease of working bluestone and the attractive edges, resulting from splitting or sawing across the strata, make its use available for treads, risers, coping, sills, mantels, hearths, and veneer.

Bond. Arrangement in which bluestones are laid so that vertical and horizontal joints occur in a particular pattern.

Border. Perimeter units enclosing a given area or space, usually differently shaped or laid in a pattern different from that of the enclosed units in the overall area of space.

Cleavage plane. Natural plane of weakness in a bluestone formation along which the material may be readily split or separated into segments; the angle of the cleavage plane may vary from horizontal to vertical, depending on the particular type of bluestone formation.

Cleft. Term "natural cleft surface" of a bluestone describes the natural texture resulting from splitting stone along its cleavage plane. Divided or split.

Coping. Top member of a wall, usually projecting beyond the vertical surfaces of the wall in order to protect the wall surfaces below.

Diamond-sawn. Cut with a saw that has commercial grade diamond particles embedded in the cutting edges.

Edges. Types of exposed edges for treads, coping, sills, etc., can be "Quarry Cut" or "Hand Tooled", with all front corners square. Where a less careful or less formal edge is desired, stone 2 in. or under in thickness usually can be snapped 90° to the surface and the ends cut full. Joints for paving (the field) are accurately hand cut and pitched to a straight line for $\frac{1}{2}$ in. joints.

Elk brook rubbed-finish stone. Stone used for interior flooring, base, treads, and stools and for exterior steps, paving, coping, sills, panel walls, benches; and sculpture bases. A product of Johnson & Rhodes Bluestone Co., East Branch, N.Y.

Exposed edges. Can be diamond-sawed, rubbed, snapped, rocked, or thermal.

Field. Overall horizontal area or space enclosed by borders, walls, partitions, or other construction components.

Finish. Texture and appearance resulting from a treatment of a surface.

Flagstone (flagging). Bluestone split into pieces suitable for floors and paving. May be of irregular broken shapes as taken from the quarry or of regular shapes having sawn edges.

Floor and paving flagging. Flagging in 1-, $1\frac{1}{2}$-, or 2-in. thickness, each piece having 1- to 12-sq ft surface area; larger sizes and heavier thicknesses are available. Standard sizes are cut to multiples of 6 in. less the allowance for the $\frac{1}{2}$ in. setting joint. Colors are variegated blue-gray, blue-green, buff, lilac and rust.

Gauge. Limit to uniform size, shape, etc., by selection, cutting, rubbing, etc.

Grooved back. Parallel grooves cut into the back or bed face of a bluestone unit to increase mortar bond.

Grout. Mortar used as a filler for joints and backing in stonework.

Gunny sack. Coarse jute sacking or a bag made of such material, used for cleaning mortar from bluestone.

Irregular flagstone. Random irregular bluestone shapes used for flooring and paving, available in several thicknesses.

Joint. Space between adjacent stones and other masonry materials.

Joint grouting and pointing. Joints are carefully packed with a mortar composed of one part portland cement, one-quarter part lime, and three parts clean sand, mixed to buttery consistency. In pointing, avoid deposits of excess mortar along the perimeter edges of the stones. Deposits should be completely removed at once with a cellulose sponge to avoid staining the material. Stones are finally finished by being rubbed in circular motion with dry sharp sand and small gunny sacks. The pointing of joints is done the same day the material is laid.

Mortar. Material consisting of cement and lime mixed with sand and water to form a hard-setting agent. Used for bedding and jointing stones.

Natural cleft finish. The face finish is usually the natural seam split, resulting in a texture that only nature can create. Colors of the natural cleft surface are variegated.

North Carolina bluestone. Hard, dense angillite, which, although a member of the slate family, is harder, stronger, and more stable than slate. Colors are almost uniformly dark blue-gray with a trace of green. Surface is smooth enough for installation upon splitting.

North River bluestone. Bluestone quarried from the old North River quarries of New York's Heldeberg and Catskill mountain range provides hard, fine-grained sandstone of the Devonian Age with a deep blue hue. A product of Heldberg Bluestone & Marble, Inc., East Berne, N.Y.

Panel wall. Panels available in natural cleft exposed face in pieces up to 8 sq ft and in rubbed or thermal finish in larger pieces.

Parge or parget. Cover a surface with a uniformly thick layer or coat of rich mortar. To apply a skim coat of portland cement mortar to the wet backs of the bluestones as they are laid.

Pattern. Design formed by the shape and arrangement of stones or other materials forming a floor or the facing of a wall. The four standard patterns are: European method, random rectangular, semirandom rectangular, and irregular.

Pennsylvania bluestone (not slate). Natural cleft, hard, durable quartzite sandstone that will not disintegrate in any climate. Hardness is approximately 25 by U.S. government standards. Porosity is approximately $3\frac{1}{2}\%$.

Plaster of paris. Highly calcined gypsum that, when mixed with water, forms a quick, hard-setting paste; used in some installations of bluestone facing units.

Quarry run. Bluestones taken directly from the quarry with no sawing, dressing, or other finishing performed on them are used in some types of construction.

Riser. Upright or vertical member used in closing up the back portion of a step in stair construction.

Rubbed finish. After wire sawing, surface is rubbed to a flat, even plane with coarse industrial diamonds.

Rubble. Roughly dressed stones of irregular shape, set dry, one on top of each other, as in a wall, or set in mortar.

Setting bed. Mortar used as a bed or undersurface of units when laid.

Sills. Bottom horizontal member of a window, door frame, or opening in an exterior wall.

Stool. Interior flat horizontal trim member at a window frame.

Thermal finish

1. The process of applying heat instantaneously to a sawn bluestone surface, producing a spalting effect resulting in a simulated, natural-appearing texture that is ideal for treads, wall caps, and spandrels. The color is constant throughout the pieces or panels.

2. Flame-textured finish accompanied by superheating the surface crystals so that they expand and "pop out" along their cleavage planes. Blue color and crystal pattern are retained in their muted state.

Tread. Horizontal non-slip bluestone member of a step in stair construction.

Underbed. Base material, such as a concrete slab or compacted sand, on which the mortar setting bed for bluestone for flooring or paving is placed.

BUILDING STONES

Agate. Variegated variety of quartz showing colored bands or bother markings (clouded, mosslike, etc.).

Alabaster. White, massive variety of gypsum. When polished often resembles some clouded stalagmites.

Anhydrite

1. Mineral that resembles dolomite in hand specimen but has three cleavages at right angles. Anhydrite is less soluble in hydrochloric acid than dolomite and is slightly soluble in water, it is harder than gypsum. Gypsum and anhydrite occurring in aggregates offer risks of sulfate attack on concrete and mortar.

2. Mineral, calcium sulfate in composition, that absorbs moisture from the atmosphere when used in base courses as a substitute for marble.

3. Mineral consisting primarily of anhydrous calcium sulfate.

Arch. Curved stone structure resting on supports at both extremities.

Architrave. Member of an entablature resting on the capitals of columns and supporting the frieze.

Argillite. Compact sedimentary rock composed mainly of clay and aluminum silicate minerals.

Arkose, arkosic sandstone, feldspathic sandstone. Sandstone containing 10% or more clastic grains of feldspar.

Arris. Sharp edge or exterior corner formed by the meeting of two surfaces.

Ashlar. Masonry wall having a face of square, rectangular, or irregular shaped stones, either smooth, textured, or quarry cut surfaces.

Back arch. Concealed arch carrying the backing of a wall where the exterior facing may be carried by a lintel.

Baluster. Miniature designed pillar or column supporting a rail. Used in balustrades.

Basalt. Dense-textured (asphanitic) igneous rock, relatively high in iron and magnesia minerals and relatively low in silica, generally dark gray to black, and feldspathic. General term in contradistinction to feldsite, a light-colored feldspathic and highly siliceous rock of similar texture and origin.

Bed

1. In granites and marbles, the layer or sheet of the rock mass that is horizontal, commonly curved, and lenticular, as developed by fractures. Term is sometimes also applied to the surface of parting between sheets. In stratified rocks, the unit layer formed by sedimentation, or variable thickness, and commonly tilted or distorted by subsequent deformation; it generally develops a rock cleavage, parting, or jointing along the planes of stratification.

2. Top or bottom of a joint. A natural bed is the surface of stone parallel to its stratification.

Belt course. Continuous horizontal course of flat stones placed in line marking a division in the wall plane.

Bevel. Shape created on the outer edge of a stone unit when the angle between two sides is greater or less than a right angle.

Bond stone. Stone used in varying percentages (usually defined by building codes) to anchor or bond the stone veneer to the backing material. Bond stones are generally cut to twice the bed thickness of the material being used.

Border stone. Stone used as an edging material. Generally used to retain the field of a platform.

Box. Tapered metal box wedged at the top of heavy stones during hoisting.

Briar Hill sandstone. Commonly known as the Massillon Sandstone. Stone is medium to coarse grained and composed mainly of quartz, with iron oxide, minerals and secondary silica comprising the cementing materials. A product of Briar Hill Stone Company, Gienmont, Ohio.

Broach. Drill or cut out material left between closely spaced drill holes. Also, a mason's sharp pointed chisel for dressing stone.

Brownstone

1. Brown sandstone used in the facing of buildings.

2. Sandstone of characteristic brown or reddish brown color that is due to a prominent amount of iron oxide as interstitial material.

Brushed finish. Finish obtained by brushing the stone with a coarse, rotary-type, wire brush.

Bull nose. Convex rounding of a stone member, such as usually found at the edge of a stair tread.

Calcite limestone. Limestone containing not more than 5% magnesium carbonate.

Calcite streaks. White or milkylike streak occurring in stone. It is a joint plane, usually wider than a glass seam, that has been recemented by deposition of calcite in the crack and is structurally sound.

Carve. Shape by cutting a design.

Cavity wall tie. Rigid, corrosion-resistant metal tie that bonds two wythes of a cavity wall. This is usually made of steel and formed in a Z shape or a rectangle.

Cement putty. Thick creamy mixture made with pure cement and water that is used to strengthen the bond between the stone and the setting bed.

Chamfer. Bevel the junction of an exterior angle.

Chat. Finish produced by using an abrasive when the gang saws are cutting a stone block either in a horizontal or vertical direction, cutting into the stone at a depth of at least $\frac{1}{8}$ in. Chat is the abrasive used. The abrasive leaves a finish that resembles the appearance of sandblasting.

Chat-sawed finish. Rough gang-sawed finish produced by sawing with coarse chat.

Cleavage. Breaking of a rock mass along natural surfaces; surface of natural parting.

Cleavage plane. Plane or planes along which a stone is likely to break or delaminate.

Composite wall. Wall in which the facing and backing are of different materials and bonded together with bond stones to act as a solid wall.

Coping. Sloped flat stone used as a cap on free-standing walls.

Corkstone. Stone weighing only about half as much as ordinary stone. Corkstone is light, very strong, and slightly flexible, and when a dry piece is dropped on a hard surface, it rings like steel. Its colors are charcoal brown, medium brown, tan, burnt orange. A product of the Georgia Marble Co., Atlanta, Georgia.

Cornerstone. Stone forming a part of a corner or angle in a wall; also, a stone laid at the formal inauguration of the erection of a building, usually incorporating a date or inscription with a hollow portion containing a metal box for documents.

Cornice. Molded projecting stone at the top of an entablature, or the top of a wall.

Corrugated anchors. Corrugated wall ties and dovetailed anchors.

Course. Single horizontal range of stone units the length of the wall.

Coursed veneer. Veneer achieved by using stones of the same or approximately the same heights. Horizontal joints run the entire length of the veneered area, vertical joints are constantly broken so that no two joints will be over one another.

Cross bedding. Arrangement of laminations of strata transverse or oblique to the main planes of stratification.

Crowfoot (styolite). Dark gray to black zigzag marking occurring in stone, usually structurally sound.

Curbing. Prefabricated shaped cut stone border at the edge of streets, roads and driveways at parkways or sidewalks.

Cut stone. Stone cut, machined, and finished to given sizes, dimensions, or shapes, and produced in accordance with the shop drawings that have been developed from the architect's working drawings.

Cutting stock. Slabs of varying sizes, finishes, and thicknesses that are used in fabricating treads, risers, copings, borders, sills, stools, hearths, mantels, and other special-purpose stones.

Dacite. Fine-grained, extrusive (volcanic) rock, intermediate in color and composition between basalt and rhyolite.

Dentil. Block projections on an entablature.

Dentil course. Lower part of the cornice with dentils. Cornice is usually joined to allow machine production of the dentils.

Dimension stone. Stone quarried, precut, and shaped to dimensions of specified sizes.

Dolomitic limestone. Limestone rich in magnesium carbonate, frequently somewhat crystalline in character, found in ledge formations in a wide variety of color tones and textures. Crushing and tensile strengths are greater than those of the oolitic limestones, and appearance shows greater variety in texture.

Dressed or hand dressed. Term applied to rough chunks of stone cut by hand. Stone that is sold as dressed stone generally is stone ready for installation.

Drip. Continuous recess cut under a sill or projecting stone to throw off water to prevent it from running down the face of the wall or any other vertical surface.

Dry. Open or unhealed joint plane condition not filled with calcite and not considered as structurally sound.

Dry wall. Stone wall that is constructed one stone upon the other without the use of mortar. Dry walls are generally used for retaining walls, to allow water to seep through dry joints, preventing unnecessary hydraulic pressures behind the backface of the wall.

Efflorescence. Crystalline deposit appearing on stone surfaces caused by soluble salts carried through or onto the stone by moisture, which has penetrated the brick, tile, concrete blocks, mortar, concrete and similar materials in the wall.

Entablature. Section consisting of an architrave, frieze, and cornice. The uppermost part of a classical order or columnar system.

Entasis. Shape of the upper two-thirds of a column, or extended throughout the whole length of a column depending on classical column order.

Expansion bolt. Socket that grips a drilled hole in stone by expanding its sides as the bolt is screwed into it.

Exposed aggregate. Large pieces of stone aggregate purposely exposed for their color and texture in a cast slab.

Face. Exposed portion of stone. Term is also used to refer to the edge treatment on various cutting stock materials.

Fascia. Horizontal belt of vertical face. Term is often used in combination with moldings.

Featherrock. Rock emerging from the depths of the Mono Craters located in California's remote Sierra Nevada region. Featherock is charcoal and silver gray in color. A product of Featherock, Inc., Burbank, California.

Field stone. Loose blocks separated from ledges by natural processes and scattered through or on the regolith (soil) cover. Term is also applied to similar transported materials, such as glacial boulders and cobbles.

Fines. Powder, dust, silt- and sand-sized material resulting from processing (usually crushing) rock.

Flat-stock anchors. Anchors fabricated from sheet steel or flat stock such as straps, cramps, dovetails, dowel straps and dowel, and two-way anchors.

Freestone. Stone that may be cut freely in any direction without fracture or splitting.

Frieze. Horizontal belt course, sometimes decorated with sculpture relief, occurring just under a cornice.

Gang sawed. Description of the granular surface of stone resulting from gang sawing alone.

Gauging. Grinding process for making all pieces of material that are to be used together the same thickness.

Glass seam. Description of a narrow glasslike streak occurring in stone; joint plane that has been recemented by deposition of translucent calcite in the crack and is structurally sound.

Grade course. Beginning course at the grade level, generally waterproofed with a damp check or damp course.

Grain. Easiest cleavage direction in a stone. Expression "with the grain" is the same as "natural bed." Also, particles (crystals, sand grains, etc.) of a rock.

Grout. Mortar of pouring consistency. Combination of cement, lime, sand, and water.

Hand-cut random rectangular ashlar. Pattern where all stone is hand cut into squares and rectangles. Joints are fairly consistent. Similar to sawed-bed ashlar in appearance.

Hand- or machine-pitch-faced (rock-faced) ashlar. Finish given to both veneer stone and cutting stock. Created by establishing a straight line back from the irregular face of the stone. Proper cutting tools are then used to cut along the line leaving a straight arris and the intended rock-faced finish on the exposed surface.

Head. End of a stone that has been tooled to match the face of the stone. Heads are used at outside corners, windows, door jambs, or any place where the veneering will be visible from the side.

Hearth. Stone part of the floor of a fireplace that also extends beyond the fire chamber.

Hearthstone. Single large stone (or stones) used for the hearth; the stone in front of the fire chamber, often extending to either or both sides.

Heldeberg fossil rock. Rock formed during the Silurian period of the Paleozoic age. A product of Heldeberg Bluestone & Marble, Inc., East Berne, N.Y.

Holes. Sinkages in the top beds of stones to engage Lewis pins for hoisting.

Honed finish. Superfine smooth finish.

Incise. Cut inwardly or engrave, as in an inscription.

Inscription. Lettering cut in stone.

Jack arch. Arch having horizontal or nearly horizontal upper and lower surfaces; also called "flat" or "straight arch."

Joint. Space between stone units, usually filled with mortar.

Joint filler tape. PVC foam joint tape that is adhesive on one or two sides and available in most sizes and hardnesses. Tape is weatherproof, may be used without additional sealants, and comes in black, gray, or white.

Jumper. In ashlar patterns, a piece of stone of higher rise than adjacent stones that is used to end a horizontal mortar joint at the point where it is set.

Keystone. Last wedge-shaped stone placed in the crown of an arch, regarded as binding the whole.

Lava. Term applied to igneous rocks such as basalt and rhyolite that erupted from the earth by volcanic action.

Lead buttons. Lead spacers in the solid horizontal joints to support the top stones until the mortar has set.

Lewis bolt. Galvanized steel bolt with tapered head, wedged in a tapered recess in the stone unit for hanging soffit stones.

Lewis holes. Holes in cut stones for use with the Lewis bolt, for lifting and support during setting of cut stones and sometimes for permanent support. Type of holes are selected for the particular Lewis (lifting device or hook) to be used.

Limestone

1. Sedimentary rock composed of calcium carbonate. It includes many varieties, such as, oolitic limestone, dolomitic limestone; and crystalline limestone.

2. Most widespread of carbonate rocks. It ranges from the pure limestones, consisting of the mineral calcite, to the pure dolomites, consisting of the mineral dolomite. Usually limestone contains both minerals in various proportions.

3. Rock mass consisting of calcium carbonate, which through metamorphism passes into a marble.

Limestone-faced precast panels. Technique of combining Indiana limestone of varying size and weight with precast concrete backing. Method welds the limestone veneer to the precast concrete by utilizing the bonding capabilities of high-strength epoxy.

Lipping. Term usually referring to flagging materials. Lipping occurs when two pieces of material to be joined together are slightly warped or twisted, causing one or more edges to be higher or lower than the adjoining material.

Lug sill. Stone sill set beyond the jambs on each side of the masonry opening.

Machine finish. Generally recognized standard machine finish produced by the planers.

Malpais. Refers to dark-colored rock, commonly lava, in rough terrain.

Masonry. Built-up construction, usually of a combination of masonry materials set in mortar.

Metamorphism. Change or alteration in a rock caused by exterior agencies, such as deep-seated heat and pressure or intrusion of rock materials.

Minnesota stone. Stone quarried from the Oneota dolomitic limestone high-ledge formation of Ordovician origin. Minnesota stone has a broad range of colors and is ideally suited to exterior and interior uses in a broad variety of machine- and hand-produced finishes. There are Minnesota stones that are classed as marble by the Marble Institute of America and subject to normal marble treatment and detailing in such applications as altars, wainscoting, and flooring. A product of Vetter Stone Company, Kasota, Minnesota.

Miter. Junction of two units at an angle, which the junction line usually bisects at 45°.

Modular, multiple cut, pattern cut. Standard patterns used throughout the stone industry. Patterns are usually based on multiples of a given height. Stone that is multiple cut or pattern cut is precut to allow typically for $\frac{1}{4}$- or $\frac{1}{2}$-in. joints or beds.

Moldings. Decorative stone deviating from a plane surface by projections, curved profiles, recesses, or any combination thereof.

Mortar. Plastic mixture of cement, lime, sand, and water, used to bond masonry units. Nonstaining waterproof mortar is preferable.

Mosaic. Veneering that is generally irregular with no definite pattern. Stone used in a mosaic pattern usually are irregular in shape.

Natural bed. Setting of the stone on the same plane as that on which it was formed in the ground. Term generally applies to all stratified materials.

Natural cleft. When stones that are formed in layers in the ground are cleaved or separated along a natural seam, the remaining surface is referred to as a "natural cleft surface."

Nicked-bit finish. Finish obtained by planing the stone with a planner tool in which irregular nicks have been made in the cutting edge.

Nonstaining mortar. Mortar composed of materials that individually or collectively do not contain material that will stain. Such mortars usually have a very low alkali content.

Obsidian. Glassy phase of lava. Highly siliceous natural glass.

Oolitic limestone. Calcite-cemented calcareous stone formed of shells and shell fragments, practically noncrystalline in character; this limestone is characteristically a freestone, without cleavage planes, possessing a uniformity of composition, texture, and structure. It also possesses a high internal elasticity, adapting itself without damage to extreme temperature changes.

Out of wind. Arris of the stone not in parallel or perpendicular lines.

Palletized. Term applied to stone stacked on wood pallets. Palletized stone is easily identified, moved, and transported.

Parging. Dampproofing by placing a coat of mortar to the backs of stones or the faces of the backup material.

Parquetry. Inlay of stone floors in geometrical or other patterns such as horizontal grain laid against vertical grain.

Perrons. Slabs of stone set on other stones serving as steps in gardens.

Pilaster. Engaged pier of shallow depth as part of a wall.

Pine log stone. True quartzite stone, a material that is harder than many granites; one of the most impervious of stone products and the only one recommended to be cleaned with acid. Available in three color ranges: regular —100% quarry yield; rustic—hand-selected, darkest, most "rustic" colors; sky blue—hand-selected, lightest colors. A product of Georgia Marble Company, Atlanta, Ga.

Pitched stone. Stone having the arris clearly defined. The face is roughly cut with a pitching chisel used along the line that becomes the arris.

Plinth. Lower square part of the base of a column; square base or a lower block, as of a pedestal; base block at the juncture of baseboard and trim around an opening.

Plucked finish. Finish obtained by rough-planing the surface of the stone; breaking or plucking out small particles to give a rough texture.

Pointing. Final filling and finishing of mortar joints that have been raked out.

Precast anchors. Anchors fit into the back of the stone for precast panels, produced of stainless steel or galvanized wire.

Pressure-relieving joint. Open horizontal joint below the supporting angle or hanger, located at approximately every floor line, not over 15 ft apart horizontally and every 20 to 30 ft vertically, to prevent the weight from being transmitted to the masonry below. These joints are caulked with a resilient nonstaining material to prevent moisture penetration.

Pressure-relieving lead joint fillers. Corrugated-lead-covered-type pad designed to relieve excessive loads on stonework. Lead pressure joint is placed in joints of buildings faced with stone.

Projections. Pulling out of stones from the normal plane of the wall to give an effect of ruggedness.

Pumice. Exceptionally cellular, glassy lava, resembling a solid froth, usually of silicic composition.

Quarry

1. Opening made in an outcrop of rock to obtain stone for commercial purposes.

2. Location of an operation where a natural deposit of stone is removed from the ground.

Quirt. Groove separating a head or other molding from the adjoining members.

Quoins. Stones at the corner of a wall emphasized by size, projection, rustication, or a different finish.

Recess. Sinkage in a wall plane.

Reglet. Narrow, flat molding of rectangular profile. Prepared recess cut into stone to receive flashing material.

Relief. Ornament in relief. Ornament or figure can be slightly, half, or greatly projected.

Relieving arch. Arch built over a lintel, to divert loads, thus relieving the lower member from excessive loading.

Return. Right angle turn of a molding or surface.

Return head. Stone facing with the finish appearing on both the face and the edge of the same stone, as on the corner of a building.

Reveal. Depth of stone between its outer face and a window or door set in an opening.

Rift

1. Microscopic cleavage in building stones that greatly aids in the process of quarrying.

2. Most pronounced direction of splitting or cleavage of a stone. Rift and grain may be obscure, as in some granites, but are important in both quarrying and processing stone.

Riprap. Irregular shaped stones. Stones thrown together without order to form a foundation or retaining walls.

Rock. Integral part of the earth's crust, composed of an aggregate of grains of one or more minerals. ("Stone" is the commercial term applied to quarry products.)

Rock (pitch) face. A rock or pitch face is similar to a split face, except that the face of the stone is pitched to a given line and plane, producing a bold appearance rather than the comparatively straight face obtained in split face.

Rose window. Circular stone window fitted with carved tracery.

Round-stock anchors. Rod cramp, rod anchor, eyebolt and dowel, flat-hook wall tie and dowel, dowel and wire toggle bolts.

Rubbed finish. Smoother finish obtained by mechanically rubbing.

Rubble. Product term applied to irregularly shaped pieces, partly trimmed or squared, generally with one split or finished face, and selected and specified within a size range.

Rustification or rustication. Recessing the margin of cut stone so that when placed together a channel is formed at each joint.

Saddle. Flat strip of stone projecting above the floor between the jambs of a door; a threshold.

Sand-sawed finish. Surface left as the stone comes from the gang saw. Moderately smooth, granular surface varying with the texture and grade of stone.

Sandstone (building). Material free of seams, cracks, or other imperfections that would impair its structural integrity. Sandstone containing minerals such as pyrite and marcasite that may upon exposure cause objectional stains should not be used for exterior uses (ASTM C 616).

Sawed face. Finish obtained from the process used in producing building stone. Varies in texture from smooth to rough and, coincident with the type of materials used in sawing, is characterized as diamond sawed, sand sawed, chat sawed, and shot sawed.

Sawed edge. Clean-cut edge generally achieved by cutting with a diamond blade, gang saw, or wire saw.

Scale. Thin lamina or paperlike sheets of rock, often loose, interrupting an otherwise smooth surface on stone.

Scoria. Irregular masses of lava resembling clinker of slag; may be cellular (vesicular), dark colored, and heavy.

Scotia. Concave molding.

Semi-rubbed. Finish achieved by hand or machine rubbing the rough or high spots off the surface to be used, leaving a certain amount of the natural surface along with the smoothed areas.

Shot sawed. Description of finish obtained by using steel shot in the gang-sawing process to produce random markings for a rough surface texture.

Shot-sawed finish. Rough gang-sawed finish produced by sawing with chilled steel shot.

Sill. Horizontal stone used under windows, doors, and other masonry openings.

Slab. Lengthwise cut of a large quarry block of stone.

Slip sill. Stone sill set between the jambs of a door or window.

Smooth finish. Finish produced by planer machines plus the removal of objectionable tool marks; also known as "smooth planer finish" and "smooth machine finish."

Snapped edge, quarry cut, or broken edge. Refers to a natural breaking of a stone either by hand or machine. Break usually is at right angles to the top and bottom surface.

Soapstone. Massive variety of talc with a soapy or greasy feel, used for hearths, table tops, carbed ornaments, chemical laboratories, counter tops, sinks, etc., known for its stainproof qualities.

Spall

1. Stone fragment that has split or broken off.

2. Sizes may vary from chip size to large size rocks. Spall pieces are primarily used for filling up large voids in rough rubble or mosaic patterns.

Splay. Beveled or slanted surface.

Split. Division of a rock by cleavage.

Split face (sawed bed). Usually split face is sawed on the beds and is split either by hand or machine so that the surface face of the stone exhibits the natural quarry texture.

Splitstone finish. Finish obtained by sawing to accurate heights, then breaking by machine to required bed widths. (Normal bed widths are $3\frac{1}{2}$ in.)

Stacked bond. Stone that is cut to one size and installed with unbroken vertical and horizontal joints running the entire length and height of the wall.

Stone. Term is generally accepted as synonymous with "rock," but is more properly applied to individual blocks, masses, or fragments taken from their original formation or considered for commercial use.

Stonehenge. Factory-produced cultured stone architectural panel used for exteriors and interiors as facing panels for walls, spandrels, columns, partitions, dividers, and floor covering. A product of Johns-Manville, New York, N.Y.

Storch stone anchors. Anchors that permit complete freedom of movement of stone veneer in a plane parallel to the wall face. They are comprised of a duoadjustable insert and an anchor that fits into the insert and into a hole in the stone. (Patented type of anchor available from Hohmann Barnard Inc., Hauppauge, N.Y.)

Stratification. Structure produced by deposition of sediments in beds or layers (strata), laminae, lenses, wedges, and other essentially tabular units.

Strip rubble. Rubble that comes from a ledge quarry. Beds of the stone, while uniformly straight, are of the natural cleft as the stone is removed from the ledge and then split by machine to approximately 4-in. widths.

Strips. Long pieces of stone, usually low-height ashlar courses, where length-to-height ratio is at maximum for the material used.

Tablet. Small flat slab or surface of stone, especially one bearing or intended to bear an inscription, carving, or the like.

Template. Pattern for repetitive marking or fabricating operation.

Texture. Three-dimensional surface enrichment independent of color.

Tolerance. Dimensional allowance made for the inability of men and machines to fabricate a product of exact dimensions.

Tooled finish. Finish customarily consisting of four, six, or eight parallel concave grooves to the inch.

Tracery. Ornamentation of panels, circular windows, door and window heads.

Tread. Horizontal flat stone with a nonslip finish used as the top walking surface on steps.

Trim. Stone used as decorative items only, such as sills, coping, or enframements, with various textures and finishes.

Trimmer arch. Stone arch, usually a low-rise arch, used for supporting a fireplace hearth.

Tuff. Cemented volcanic ash that includes many varieties.

Types of joints. Flush, rake, cove, weathered, bead, stripped and V joints are typical tool finished acceptable stone joints.

Typical standard anchors. Typical standard anchors are made of plain steel, hot-dipped galvanized steel, stainless steel, and other metals. They include strap anchors, coping anchors, plate and disk anchors, rod anchors, dove tail anchors, plug anchors, anchor buffs, and relieving anchors, as well as the use of solid or hollow metal dowels.

Undercut. Cut so as to present an overhanging part over a vertical surface.

Veneer or faced wall. Wall in which a facing and the backing are of different materials, but not so bonded as to exert a common reaction under load.

Veneer stone. Stone used as a decorative facing material and not meant to be load bearing.

Wall tie. Bonder or metal piece that connects wythes of masonry to each other or to other materials.

Walpole stone. Exceptionally hard stratified sandstone found in a range of natural colors from varied tones of violet to blue gray and rusty browns and used extensively in landscape work. Its flat surface makes it economical for retaining walls, riprap, stepping stones, and flagstone, and it is available in a range of sizes from small flat stones to large chunks weighing as much as 2 tons. (A product of Bates Bros., Seam-face Granite Co., East Weymouth, Mass.)

Wedging. Splitting of stone by driving wedges into planes of weakness.

Weep holes. Openings placed in mortar joints of facing material at the level of flashing to permit the escape of moisture.

Wire saw. Cut stone by passing a twisted, multistrand wire over the stone and immersing the wire in a slurry of abrasive material.

CHIMNEY CONSTRUCTION AND FIREPLACES

Anchorage. Chimneys in wood frame buildings are usually anchored laterally at each floor and ceiling line that is more than 6 ft above grade.

Appliance. Device that consumes fuel or electricity for heating, including incinerators.

Area of flues—throat and damper. Net cross-sectional area of the flue and of the throat between the firebox and the smoke chamber of a fireplace; usually must meet local code requirements. Where dampers are used, damper openings, when fully opened, should be equal to the required flue area.

Breeching
1. Casing at the end of a boiler to which a flue is connected as part of a system to convey products of combustion from the boiler to a chimney or vent.
2. Flue pipe or chamber for receiving flue gases from one or more flue connections and for discharging these gases through a single flue connection.

Chimney
1. Vertical shaft of masonry, reinforced concrete, or other approved noncombustible, heat-resisting material enclosing flues for the purpose of removing products of combustion.
2. That part of a building that contains a flue or flues for transmitting products of combustion from a furnace, fireplace, boiler, or any appliance to the outer air.
3. Vertical enclosure containing passageways for the discharge of products of combustion from fuel-burning devices.
4. Primarily vertical enclosure containing one or more flues.
5. Chimneys, stacks, or smoke flues intended for the purpose of removing the products of combustion from solid, gas, or liquid fuel.
6. Vertical masonry or reinforced concrete shaft enclosing flues, designed for the purpose of removing the products of combustion of solid, liquid, or liquefied petroleum gases or natural gas fumes to the outside atmosphere.
7. Primarily vertical shaft enclosing at least one flue for conducting flue gases to the outdoors.

Chimney cap. Masonry chimneys usually capped with a noncombustible weatherproof material, flashed to prevent moisture from entering around the flue lining. Masonry chimneys that conduct corrosive gases and vapors should be capped with special lead, metal-alloy, or ceramic caps.

Chimney connector. Pipe or breeching that connects the heating appliance to the chimney.

Chimney flue. Conduit for conveying the flue gases delivered into it by a breeching or vent connector to the outer air.

Chimney liner. Conduit containing a chimney flue used as a lining of a masonry or concrete chimney.

Chimneys of hollow clay tile. Chimneys of hollow clay tile units are considered an integral part of a wall of such units. Minimum of 8 in. of such wall may serve as one wall of the chimney.

Cleanouts. Accessible, approved cleanout opening with a tight-fitting cover provided at least 12 in. below the lowest vent inlet into any chimney, with a tag marked "clean-out," or cast into the door.

Clearance. Usual distance between the fireplace and combustibles, of at least 4 in. with the combustibles not closer than 6 in. to the fireplace opening. Wood facing or trim normally placed around the fireplace opening is permitted when the facing or trim is furred out from the fireplace wall at least 4 in. and attached to incombustible furring strips. Where the wall of the fireplace is 12 in. thick, facing or trim usually may be directly attached to the fireplace masonry.

Connector. Duct or pipe that connects an appliance to a chimney flue or gas vent.

Draft hood

1. Device in a ventilation system that regulates the flow of the products of combustion.
2. Device placed in and made part of the flue pipe from an appliance, or placed in the appliance itself, designed to ensure the ready escape of the products of combustion in the event of no draft, backdraft, or stoppage beyond the draft hood, prevent a backdraft from entering the appliance, and neutralize the effect of stack action of the chimney flue on the operation of the appliance.
3. Device placed in and made part of the gas vent from an appliance, or placed in the appliance itself.

Draft regulator. Device that functions to maintain a desired draft in the appliance by automatically reducing the draft to the desired value.

Draft regulator (barometric). Device that automatically reduces chimney draft to the desired value for an appliance. Double-acting regulator is designed to prevent back draft as well.

Exterior chimney. Chimney built outside the walls of a building, but receiving lateral support from the exterior walls of the building.

Factory-built chimney

1. Chimney that is shop fabricated, listed by an accredited authoritative testing agency, for venting gas appliances, gas incinerators, and solid or liquid fuel-burning appliances.
2. Listed metal chimney composed of factory-built components.
3. Chimney consisting entirely of factory-made parts, each designed to be assembled with the other without requiring fabrication on site.

Factory-built fireplace. Fireplace composed of listed factory-built components, assembled in accordance with the terms of listing to form the completed fireplace.

Fireplace. Lower portion of a chimney, usually a wide open recess in a wall designed for the burning, generally, of solid fuel.

Fireplace hearth. Usually masonry or other noncombustible materials, such as brick, concrete, or stone, at least 12 in. wider on each side than the fireplace opening and projecting at least 18 in.

Fireplace walls. Usually structural walls of fireplaces, must be at least 8 in. in thickness; back walls of fireboxes, at least 10 in. in thickness, except where a lining of firebrick is used, when they can be at least 8 in. in thickness.

Flue

1. Enclosed passage, primarily vertical, suitable for removing gaseous products of combustion to the outer air.
2. Non-combustible passageway for removing products of combustion from solid, liquid, or gas fuel.
3. Vertical passageway for products of combustion, suitable for devices or appliances using any type of fuel.
4. Space or passage in a chimney through which smoke, gas, or fumes ascend. Each passage is called a flue; all the flues, together with the surrounding masonry, make up the chimney.
5. Enclosed air tight passageway for conveying products of combustion to the outside air.
6. General term for the passages and conduits through which flue gases pass from the combustion chamber to the outer air.
7. Conduit or pipe, vertical or nearly so in direction, designed to convey all the products of combustion to the outside atmosphere.

Flue collar. Portion of a fuel-fired appliance designed for the attachment of the flue pipe or breeching.

Flue gas. Products of combustion gases including excess air.

Flue lining

1. Fire clay or terra-cotta pipe, round or square, used for the inner lining of chimneys with the concrete or masonry work around the outside. The flue lining usually runs from the concrete footing to several inches beyond the top of the chimney cap.
2. Masonry chimneys usually lined with fire-resistive clay tile flue lining, set with high-temperature moisture proof mortar. Lining extends from 8 in. below the lowest inlet to 4 in. above enclosing walls.

Flue pipe. Pipe connecting the flue collar of an appliance to a chimney.

Gas vent

1. Galvanized sheet metal, black iron, sheet aluminum, stainless steel, or copper fabricated flue for removing products of combustion from gas appliances, but not suitable for other fuels.
2. Opening in an appliance through which the products of combustion may escape; duct carrying or conveying smoke, gases, and vapors.
3. Portion of a venting system designed to convey vent gases vertically to the outside air from the vent connector of a gas-fired appliance, or directly from the appliance when a vent connector is not used and includes any offsets. All materials should be noncombustible, corrosion resistant.

High-heat-appliance-type chimney. Factory-built masonry or metal chimney suitable for removing the products of combustion from fuel-burning high-heat appliances producing combustion gases in excess of 2000°F, measured at the appliance flue outlet.

Hood. Canopy or similar device placed over a stove, range, or other heating installation connected to a ventilating duct or flue.

Interior chimney. Chimney built within the walls of a building and having lateral support from the building structure.

Isolated chimney. Chimney other than an interior or exterior chimney; also, that part of an interior or exterior chimney projecting above the building walls that give it lateral support.

Lining. Chimney flue lining, usually extends to a point at least 4 in. above the top of the enclosing masonry walls or the chimney cap.

Lintel. Steel member that supports masonry over the fireplace opening.

Low-heat-appliance-type chimney. Factory-built masonry or metal chimney suitable for removing the products of combustion from fuel-burning low-heat appliances producing combustion gases not in excess of 1000°F under normal operating conditions, but capable of producing combustion gases of 1400°F during intermittent forced firing for periods up to 1 hour. All temperatures are measured at the appliance flue outlet.

Masonry chimney. Field-constructed chimney, usually built in accordance with nationally recognized codes or standards.

Masonry fireplace. Fireplace built of masonry or reinforced concrete that includes a fire chamber, hearth, and chimney.

Masonry or concrete chimney. Chimney of brick, stone, concrete, or approved masonry units constructed on site.

Materials. Flue linings used in connection with solid or liquid fuel and bricks used in lieu of flue linings should have a softening point of not less than 1994°F.

Medium-heat-appliance-type chimney. Factory-built masonry or metal chimney suitable for removing the products of combustion from fuel-burning medium-heat appliances producing combustion gases not in excess of 2000°F, measured at the appliance flue outlet.

Metal chimney

1. Chimney constructed of metal with a minimum thickness not less than that of No. 10 manufacturers' standard gauge steel sheet.
2. Chimney made of metal of adequate thickness, galvanized or painted unless suitably corrosion resistant, properly welded or riveted, and built in accordance with nationally recognized codes or standards.
3. Single-wall chimney of metal constructed on site.

Metal heat circulators. Approved metal heat circulators installed in fireplaces.

Primary safety control. Device for shutting down the operation of an appliance when an unsafe condition occurs or for preventing start up when an unsafe condition exists.

Screens. Protection to stop flying embers from solid materials from traveling beyond the face of the fireplace, usually provided for fireplace openings.

Smoke chamber. Front and side walls are usually at least 8 in. in thickness; back walls are usually at least 6 in. in thickness. Chamber occurs immediately above the top of the fireplace opening to the underside of the damper.

Smoke pipe

1. Enclosed passage used to convey the products of combustion of any fuel to a flue.
2. Noncombustible, corrosive resistant pipe or breeching connecting a heat-producing device or appliance burning solid or liquid fuel, or a heat-producing device or appliance operating at more than 550°F with a flue.

3. Pipe or breeching that is primarily horizontal and that connects a heating appliance to a flue.

Smokestack

1. Vertical metal flue or chimney for removing products of combustion. Smokestacks can be located either inside or outside a building.
2. Enclosed passage, primarily vertical, used for removing the products of combustion of any fuel to the outer air.
3. Vertical flue constructed of metal (lined or unlined), uninsulated, to which one or more smoke pipes is connected.

Spark arrester. Corrosive resistant woven wire mesh placed over flue lining.

Stack. Structure or part thereof that contains a flue or flues for the discharge of gases.

Thimble. Noncombustible, corrosive resistant sleeve used where metal chimneys or flues pass through combustible roof construction.

Vent. Pipe or duct for conveying flue gases directly from an appliance to the outside atmosphere.

Vent connector

1. Passageway constructed of corrosive resistant metal to conduct the flue gases from the flue collar of the appliance to the chimney or gas vent.
2. Part of a venting system that conducts the flue or vent gases from the flue collar of a gas appliance to the chimney or gas vent; it may include a draft control device.

Vent pipe. Pipe for the removal of products of combustion from a gas-fired appliance.

Vent system. Gas vent or chimney and vent connector, if used, assembled to form a continuous open passageway from the gas appliance to the outside atmosphere for the purpose of removing vent gases.

Vent type B. Factory-made listed vent for use only on gas-fired appliances.

Vent type BW. Factory-made listed vent designed especially for use with wall-type gas heaters.

Vent type L. Factory-made listed vent designed for use with low-temperature venting systems.

Wythe. Partition between two chimney flues in the same stack; also, the inner and outer walls of a cavity wall.

EARTH MASONRY CONSTRUCTION

Adobe. Word used to describe earth-clay brick.

Adobe brick or block. Made from earth or clay mixture with straw interlaced, fashioned into bricks or blocks roughly molded and sun-dried.

Adobe crudo. Adobe block or brick that has been stabilized by either firing or asphalt emulsion.

Adobe file. File made from pieces of expanded metal lath that is nailed to a wood-shaped handle.

Adobe fireplaces. Traditional features of adobe homes include the use of fireplaces, such as the "Kiva" or the "Horno," or wall fireplaces such as the "Padercita."

Adobe floor. Slab of adobe mud, poured approximately 4 in.-thick on a well-compacted subbase.

Adobe form. Box form to cast adobes using the dry mud method or using a stiff mud that will keep its shape after the forms are removed. Forms are fabricated from wood or galvanized metal.

Adobe laying. The adobe is placed onto the wall, with each adobe gently shoved into its place. Adobes may be cut with a sharp axe or a skill saw using a masonry blade.

Adobe pinto. A type of adobe that has been fired in a hot fire, which brings out yellows, purples, and many other shades of color. Kiln temperatures must exceed 1200°F for a period of 24 hours.

Adobe quemado. Adobes that have been fired in a kiln to a ceramic-like condition.

Adobe soil. Heavy-textured clay soil found primarily in the southwestern areas of the U.S. and the northern regions of Mexico. Ideal soils are a gradation of particles from coarse sands down to very fine colloids or particles that can remain in suspension in water. Aggregates of $\frac{1}{4}$ in. in size are recommended for adobe brick or blocks. Larger pieces are acceptable for rammed earth construction.

Adobe jointer. A heavy duty spoon, is usually used to strike mortar joints.

Alisando. Process for smoothing the surfaces when mud plastering is applied.

Azada. Mortar mixing hoe.

Azaras. Reinforcing made from split palm trunks used with earth–clay mixture in the construction of floor slabs.

Bahareque. Form of earthen wall construction that has a proven record of antiseismic ability. Uses vertical posts about three feet or more apart with diagonal braces between them. To both sides of the verticals are nailed pieces of flexible, bamboo-like cane. The interior or space between is packed with mud and the wall cures in place.

Batter boards. Wood boards, 24 to 36 in. long, set up to hold the strings that determine the outer wall lines of the house.

Bond beam. Bond beams act as a "tie" or collar around the tops of all masonry walls.

Bond beam block. Type of hollow concrete block that is used with reinforcing rods and then filled with concrete grout to form the bond beam to be placed at the top of the adobe wall.

Burnt adobe masonry units. Units shall have an average ultimate compressive strength not less than 1000 psi. All units shall be laid in running bond with full head and bed joints.

Cajon. Type of earth-wall construction in which a clay-soil mix of appropriate consistency is used in the form of wall panels supported by a structural wall frame.

Clay. Smallest grain size division of soils, consisting usually of materials of flatish configuration and of 0.002 mm or less.

Climate zones. Adobe construction is not recommended in climate zones where high daily temperature changes can cause severe freeze-thaw cycle damage.

Corbel. Rectangular piece of wood, cut to an ornate and trapezoidal shape to transfer the load from a parapet or roof to the vertical support below.

Coursing. Referring to the courses of adobe in a wall. Often a reference to the quality of the workmanship.

Coving. Erosion along the base of an earthen wall, by salt, rain-splash, or a combination of both. This is a common condition in adobes that without adequate stem walls do not protect the earthen walls.

Crusher fine. Aggregate that passes a $\frac{3}{8}$ in. or $\frac{1}{4}$ in. screen. Usually regarded as the largest size rock allowable in an adobe mix.

Curing. Process of drying adobes in the sun. Adobes may be turned on edge to cure faster, as a greater amount of surface is exposed.

Dead earth. Earth that has lost its structural integrity through salt deterioration, freeze-thaw action, or in any manner that renders it nonstructural.

Double adobe. Walls of two adobes wide. Such walls may be interwoven for a solid wall or they may be laid to form two separate walls with a cavity in the center for insulation.

Dry mud method. Stiff or viscous mud that retains its shape when the form or frame is lifted off immediately in the adobe-making process.

Expanded metal lath. Metal lath is used for bullnosing around corners to window and door trim, to shape and hold wet plasters, as a tie between double adobe walls, and as a fastening device to attach earthen walls to roughbucks.

Exposed adobe. Adobe wall that is not plastered, with the coursing exposed. If the wall is exposed to the exterior, or weather side, it must be stabilized or in some way protected from the elements. If it is exposed to the inside, it need not be stabilized.

Fat. Adobe mix with too much clay.

FHA strap. A fastening device of galvanized steel, set in concrete or nailed to wood on one end, wrapped flat around the timber or beam at the other, then nailed down.

Fines. A term for the silts and clays in an adobe mix. There is a fairly wide range of the percentage of fines to the aggregates and sands that make up the rest of the mix. Most contractors agree that 15% in clay fines is nearly ideal.

Finish floor. The level of the top of the floor, regardless of the material used: brick, tile, wood, etc.

Flocculate. Condition in adobe manufacture in which upturned corners develop on adobes as the form is lifted off.

Fluff. Dead earth. Earth in an adobe wall that has deteriorated due to freeze-thaw, salts, etc. and has lost its structural integrity.

Form clips. Iron straps, made up in varying widths, used to hold wood forms to a certain dimension or width.

Foundation system. The foundation serves as a rigid collar on which the house sits. It must be rigid and unified so that the structure above will not settle.

Frame. In adobe block manufacture, constructions of wood or steel used to cast the block are known as "forms," and typically cast from 2 to 4 adobes per set. When the form is mounted onto a laydown machine and casts 30 or more block at a time, it becomes a "frame group."

Goat's milk plaster. Mixture of adobe fines and goat's milk, producing a very durable interior plaster, no doubt due to the latex-adhesive qualities in the milk.

Grade. The average pre-construction ground level around the building site. Most codes require that the top of the stem be from 6 to 10 in. above grade, so that back-splash or other water action cannot damage the adobe walls.

Gravel. Material that will not pass a USS No. 10 standard screen.

Gringo block. Wood nailer, set in the adobe wall as an attachment for door and window frames or interior frame walls. Made up of 2 × 4 in. lumber, they are hollow in the middle, then filled with the same mud the adobes are laid in.

Header. Horizontal structural member that supports the load over an opening such as a window or door. Similar to a lintel.

Hip roof. Roof assembly that rises from all four sides of the structure.

Illite. Clay type that usually does well in the manufacture of earthen materials.

Insulation. Effective values take into account the thermal mass, climate zones, wall compass orientation and color of the materials.

Jacal. Type of adobe-log structure, typical of higher elevations, where wood was more plentiful, and usually predating the time of sawmills. Jacales were made by setting up poles vertically, then packing mud into the spaces, from both outside and inside.

Jack rafter. Short rafter framing between the wall plate and a hip rafter or between the hip rafter and the ridge board.

Kaolinite. Clay type used in adobe and pressed adobe manufacture as well as rammed earth. It is considered a "stable" type of clay.

Ladrillera. Kiln for the firing of adobe bricks into burnt adobe or adobe quenado. The raw bricks are stacked in rows with passages between for firewood to be pushed in.

Laterite (or lateritic). Clay soils formed under tropical conditions by the weathering of igneous rock. They consist chiefly of stable clays and hydroxides of iron and aluminum.

Laydown machine. Adobe-making machine that is basically an improvement on the old hand-held form that uses a stiff mud. Laydown machines use the dry mud method. The mix is not "dry;" it is a fully saturated adobe mix that is stiff or viscous, so that when the laydown machine's form is lifted, all the adobes retain their shapes.

Lift. Layer of earth mix that is placed into rigid forms and then rammed. Lifts are about 7- to 8-in. thick on the average and compact down to about 5 in. Lifts may be placed by shoveling the material into the forms, then spreading out evenly. When a given number of lifts are rammed to the top of the form, the forms may be removed immediately and moved to another position.

Line. Mason's line, usually of colored nylon and stretched tightly between speed leads or story poles. The line determines the trueness of the wall face.

Liquid limit. Point at which the moisture content of a soil acts to dissolve the mechanical or chemical bond of the soil particles in adobe mortar or block. A mud flow is soil beyond the liquid limit.

Live and dead loads. Design loads as required by local building codes.

Location of manufacture. Bricks should be made near the point of use and will vary in size with the standards of the manufacturer.

Loggia wall. Perforated adobe wall, set to windward, so that air sweeping over and around the wall pulls air through the openings at a greater rate, increasing the cooling effect.

Lubricated soil. Condition in which too much asphalt emulsion is added to a soil mix, usually resulting in a weak, crumbly block that does not meet code.

Microfines. Very fine particles in an adobe or rammed earth mix. Composed of clays and some silts.

Monolithic element. Wall acts as one mass, as in an adobe or rammed earth wall.

Mortar wash. The practice of painting on a burnt adobe or adobe quemado wall; a thin slurry of portland cement and water.

Nailer. Wood, set into the adobe or rammed earth wall during construction, to which frame walls, cabinets, window, or door frames may be attached. Such nailers depend on the weight of the adobe mass above them to hold them in place.

Nicho. Niche or ledge cut into the adobe or rammed earth wall. Nichos serve as a place usually to display religious or art objects.

Nonbearing. Wall, adobe or frame, that carries only its own weight, not the roof load or other loads. Foundations will usually be lighter for nonbearing walls.

Overlap. Distance that one block overlaps the block below, usually 4 in. by code. Each course is laid so that the vertical or head joints are never over each other and so that corners and courses interlock.

Parapets. Buildings having parapets may use firebrick copings or may return the exterior stucco finish over the top of the wall, with proper metal flashing.

Parge. Smoothing over a masonry surface, usually with a trowled-on masonry cement. Parging helps to create a smooth draw in fireplaces, by skimming over protruding brick and other rough surfaces.

Party wall. Wall in adobe construction used jointly by two parties under easement agreement and different ownership; this also serves as a firewall for the protection of both parties.

Pilaster. Pilasters are no more than half again as thick as the wall they emerge from, whereas a buttress can project out from the wall it supports for several feet. Pilasters should be built into the wall.

Pitched tin roof. Roofing material, not tin, but galvanized steel. The style embraces the old "barn siding" material, with its corrugated pattern.

Plaster. Finish or coating applied manually or by machine. Cement plaster will vary from smooth, hard plaster to cement stucco finish applied directly to the adobe wall.

Plastic limit. Moisture content below which the soil is nonplastic in handling.

Plumb. Exactly perpendicular or vertical. The mason's line moving up the speed leads indicates a plumb adobe wall.

Portal. Covered porch, usually running the length or almost the length, of one side of a building.

Poured adobe. System in which wood forming moves along a wall, allowing large volumes of adobe to be poured in place. The mix is relatively liquid.

Pressed adobe. Different from regular adobe in its manufacture. Basically, the proper chosen soil mix is wetted to around 7 to 8% moisture content. This is less than rammed earth, which requires about 10%, and far less than regular adobe, which easily exceeds 10%. The mix is fed into the hopper of the machine and pressed either from the top or side into a chamber that determines the size of the block.

Pug mill. Mixing device for adobe mud. Pug mills are found at most larger adobe facilities. They receive the various ingredients going into the adobe-making process, such as soil, asphalt emulsion, water, and occasionally, straw. They consist of a long tank with a semicircular bottom. Through the center of the tank are usually two rotating shafts to which are welded the mixing paddles. Paddles rotate into each other, intermeshing so that the ingredients are thoroughly mixed. In some pug mills, a gate keeps the mix from being dumped into a storage pit until the viscosity of the mix is just right for adobe making, as in a laydown machine. In other operations, the mix is continuously being emptied into an adjacent pit, where a front-end loader scoops up the mud and dumps it into waiting forms.

Rafter. One of a series of structural members in a roofing system, designed to support loads.

Rammed earth wall construction. Walls are constructed by the use of slip forms, erected level and securely held in place. The forms are filled with damp earth in 4 in. lifts. Each lift is rammed with a tamper until full compaction is reached. When the form is filled and solidly compacted, it is moved to a new location. Walls should be a minimum of 12 in. thick and constructed in accordance with local codes.

Rammed earth wall materials. The basic materials are earth with allowable proportions of clay, silt, and small-sized aggregate.

Roll-lock. Course of red brick, set side-to-side, but lying down, so as to form the edge of a step or a change in floor level, usually in brick floor work.

Roof ridge. Horizontal line at the junction of the top edges of two-roof surfaces, where an external angle of 180° or more is formed.

Rough buck. Window or door frame of rough-cut lumber, usually of 2 in. thick stock.

Rough cut. Lumber that has not been surfaced to an exact dimension, but simply run through the mill saw to the full, or "rough" dimension.

Saddle. Small gable type of miniroof, placed where the roof slope meets the chimney, to divert rainwater, debris, etc. Constructed of galvanized sheet metal or copper as a flashing/saddle combination. Sometimes called a "cricket."

Salt erosion. Salts are an enemy to adobe or rammed earth construction. If moisture high in salt gets into an earth wall, the salt tends to crystalize at or near the surface of the earth wall, usually within about a 6-month period. As the salt crystals form, they expand and pop, turning the adobe into powder and eroding the wall.

Scoria. Crushed, volcanic aggregate used in landscaping, but also in adobe mixes to produce a hard, but relatively lightweight adobe block. Scoria is generally $\frac{1}{4}$ in. diameter or less, quite sharp edged and from brown to red and black in color.

Screening. Process of sifting clays or sands through screens of various sized openings to sort them for earth-building purposes. Soils going into adobe block typically go through a $\frac{1}{4}$ in. screen (hardware cloth). Soils used for adobe plasters or sands used for hard stuccos may be sifted through a $\frac{1}{8}$ screen, or finer.

Sharp sand. Sand, that, when examined under a microscope, shows sharp or very angular surfaces. Sharp sands are considered to be stronger or more binding in adobe mixes.

Side load. In pressed adobe machines, the action when the hydraulic ram presses the moistened earth with a side action into the chamber. The resulting blocks are always the same in depth, but may vary in width (side-to-side). Side load machines assure the mason of a block that is

always the same depth, allowing very thin mortar joints or slurries between the courses.

Sieve analysis. Particle size distribution in a particular earth or soil as determined by standard U.S. testing sieves.

Silt. Larger of the fine-grained soil particles in a mix, usually recognized as material passing through a U.S. standard No. 200 sieve, down to a minimum size of .005 mm.

Slush joint. In adobe work, another term for head joint, or the approximately 1 in. wide space between each adobe as placed in the wall.

Soil testing. General reference to equipment, tools, or assemblages that help the builder determine the suitability of a soil for adobe, rammed earth, or pressed block.

Solaradobe. The term used to describe adobe (or rammed earth) structures that use the sun, directly and indirectly, as a way to heat and cool inner spaces.

Spalling. A flaking-off of pieces of an earthen wall, caused by a number of factors. These include moisture getting into an earthen wall, then freezing, salts in the soil, or moisture coming from within a thick wall to the outside.

Speed leads. A set of vertical posts set up near the house corners, so that when a mason's line is stretched between them, that line defines the outside face of the adobe wall. The line moves up the face or one flat edge of the leads by means of line blocks (small blocks of wood with a groove for the line) and a right angle corner that "hooks" onto the lead. The tension of the line (which must be stretched very tight) keeps the blocks attached to the leads.

Stabilization. All earthen walls contain a percentage of clay; if moisture penetrates into them, they will swell when wetted and shrink as they dry. Some earths are almost unnoticeable in this action, while others with expansive clays will crack or spall. Whatever the case, unstabilized or unprotected earthen walls will slowly disintegrate over time.

Steel frame. Adobe construction in which the roof structure bears on the steel pipe frame rather than on the adobe wall.

Stucco. Three-coat, cement-based plaster that is applied to unstabilized or semistabilized earth walls. Either stucco or some other protective coating is required by code on earth walls that are not stabilized.

Top load. In pressed adobe manufacture, a machine that presses the moistened earth downward into the chamber. Resulting blocks are always consistent in width and length, but vary in depth because of the amount of moisture in the mix. To overcome slight differences in depth, such blocks are always laid in thick mortar joints.

Veneer adobe. Building consisting of adobe blocks backed up by 2 × 4 wood studs, which, if properly framed, can carry the roof load. Adobe wall is not considered "structural."

Wall anchor. In rammed earth, used to tie two massive walls together in situations where the entire corner or intersection of the walls was not rammed all at once.

Wall limitations. One-story walls should be 12 in.-thick for adobe brick or block construction (unless 10 in. thick is permitted by code) and 12 ft in height. Two-story walls should be 18 in. thick at first story and 12 in. thick at second story, but should not exceed 22 ft in height unless otherwise allowed by local codes.

Warp. Any variation from a true plane. Warping includes bow, crook, cup and twist in wood members.

Washed sand. Commercially sold plaster or masonry sand that has been washed to reduce the salt content. Unwashed sand with salts can cause streaking in masonry work.

Weathered wall. Earthen wall that has been eroded by the action of wind and rain over time. The wear can be slight or extensive, even threatening the structure.

Weephole. A drainhole, usually in a retaining wall, that allows excess moisture to the outside and relieves hydrostatic pressure against the wall.

Wet mud method. A block-making system in which a slurrylike adobe mud is prepared, then dumped into multi-ladder forms, usually of wood. The mud is then raked or pushed into the forms. Depending on weather, the forms may be cracked away from the blocks in 4 hours or so.

Whaler. In rammed earth, a horizontal member, often a 2×4 in. set on the outside of the rammed earth form work, so as to straighten and brace the forms.

Wind cracking. Surface cracking in adobe blocks that occurs during the curing process as a result of dry, windy conditions. The surface of the adobes is forced to dry out quickly, while the mass below remains wet. The resulting cracks usually do not render the adobes unusable.

Wood bond beam. Wood bond beams serve as a tie or bond around the top of the walls. They are leveled by first placing a course of adobe mud on top of the wall, then the wood bond beam, tapping it into position until level.

Wood lintel. Structural wood, timber, usually about the same width of the earth wall above or below it, that spans across a clear or unsupported space, such as a door or window.

GRANITE

Anchors-cramps, and dowels. All anchors, cramps, dowels, and other anchoring devices should be made of type 304 stainless steel or suitable nonferrous metal.

Arris. Sharp edge or exterior corner formed by the meeting of two surfaces, whether plane or curved.

Bed. Top or bottom horizontal surface of a piece, which is covered when the piece is set in place. Filled or open space extends horizontally between the adjacent pieces set in place.

Bed-and-joint width. Minimum recommended bed-and-joint width is $\frac{1}{4}$ in.

Boulder quarry. Term applied to a granite quarry confined to a group of large boulders. May be applied to any granite quarry where the stone is naturally broken up by joints into comparatively small blocks.

Building granite. Granite used either structurally or as a veneer for exterior or interior wall facings, steps, paving, curbing, copings, or other decorative or ornamental building features.

Cleaning. After being pointed, granite work is carefully cleaned, starting at the top and removing all dirt, excess mortar, stains, and other defacements. Stainless steel wire brushes or wool may be used, but the use of other wire brushes, acid, or other solutions that may cause discoloration is prohibited.

Coarse-stippled sandblasted finish. Coarse plane surface produced by blasting with an abrasive; coarseness varies with type of preparatory finish and grain structure of the granite.

Cobblestone. Natural rounded stone, large enough for use in paving. Commonly, paving blocks, usually granite, generally cut to rectangular shapes.

Commercial granite. Commercial granite includes gneiss, gneissic granite, granite gneiss, and other rock species known as seynite, monzonite, and granodiorite, as well as many species intermediate between them, and the gneissic varieties and gneisses of corresponding mineraogaic compositions and porphoritic textures.

Cramp. Anchoring device in the form of a metal bar bent at both ends in the shape of a flat U.

Cutoff. Direction along which granite must be channeled when it will not split in that direction.

Dimensions. Maximum variation in the dimensions of any piece may be one-fourth the specified bed-and-joint width.

Eight-cut finish. Fine, bush-hammered, interrupted parallel markings not over $\frac{3}{32}$ in. apart; corrugated finish, smoother near arris lines and on small surfaces.

Fine-rubbed finish. Finish that is smooth and free from scratches; finish with no sheen.

Fine-stippled sandblasted finish. Plane surface, slightly pebbled, with occasional slight trails or scratches.

Flatness tolerances. Variations from true plane or flat surfaces, determined by use of a 4-ft-long straight ledge applied in any direction on the surface.

Four-cut finish. Coarse, bush-hammered finish with same characteristics as six cut, but with markings not more than $\frac{7}{32}$ in. apart.

Granite

1. Crystalline igneous rock composed chiefly of feldspar, some quartz, and mica.
2. Igneous rock formed by plutonic intrusions, composed chiefly of quartz, feldspar, and mica. The term granite is sometimes applied to other high-strength igneous rocks.
3. Granular crystalline rock of predominantly interlocking feature, composed essentially of alkalis feldspars and quartz.
4. Hard rock composed of quartz and orthoclase or microline.
5. Fine-to-coarse-grained igneous rock, formed by volcanic action, consisting of quartz, feldspar, and mica, with accessory minerals. Granite-type rocks include those of similar texture and origin.

Honed finish. Dull sheen, without reflections.

Igneous rock. One of the three great classes of rock, igneous, sedimentary, and metamorphic, solidified from the molten state as granite and lavas.

Incidental cutting and drilling. Where thickness permits, all blocks weighing over 100 lb should have Lewis holes for lifting. Lewis holes extend no closer than 2 in. from the finished face and are not permitted on exposed surfaces. Pieces under 4 in. in thickness may have holes for C clamps.

Joint. End or side surface of a piece, which is covered when the piece is set in place. Filled or open space extends vertically between the adjacent pieces set in place.

Ledge. Term applied to a single bed or a group of beds occurring in a quarry. An outcrop of rock on the surface.

Lift. Term used by quarrymen to joint planes that are approximately in horizontal position.

Lifting (or Lewises). Lifting clamps used by the installing contractor for the handling and setting of pieces of granite weighing more than 100 lbs.

Masonry granite. Granite used in large blocks for retaining walls, piers for bridges, abutments, arch stones, and other similar purposes.

Minimum thickness. Suggested minimum thickness for exterior veneer: bush-hammered or pointed finish, 4 in. all other finishes 2 in. Tolerance of plus or minus $\frac{1}{2}$ in. allowed in the specified thickness. Wherever ashlar or veneer is used as a facing, a setting space of at least 1 in. measured from the nominal thickness of the piece is mandatory.

Mortar. Mortar for setting and pointing consists of one part portland cement and one part plastic lime hydrate to three parts of clean, nonstaining sand, mixed in small batches, using clean, nonalkaline water, until it is thoroughly homogeneous, stiff, and plastic.

Mount Airy granite. Perfectly solid granite deposit showing no natural bed planes or seams. The rift of easiest splitting way of the granite is approximately parallel to the top surface of the deposit. This fact enables the quarrymen to form sheets of granite several acres in area and of almost any required thickness. A product of The North Carolina Granite Corporation, Mount Airy, N.C.

Orbicular. Term describing igneous rock having its component minerals crystallized or segregated in spheroidal forms, applied to the granite of Craftsbury, Vt, and to the diorite of San Diego, Ca.

Orthoclase. Potassium aluminum silicate of the feldspar group, which crystalizes in the monoclinic system, a necessary mineral consistuent of all granites.

Phenocryst. In igneous rocks, the relatively large and conspicuous crystals in a finer grained matrix or ground mass.

Physical properties. Density of granite averages about 165 lbs/cu ft. Specific gravity or density in the metric system is 2.66.

Polished finish. Mirror gloss, with sharp reflections.

Porphyry. Igneous rock in which relatively large and conspicuous crystals (phenocrysts) are set in a matrix of finer crystals.

Quartzite. Compact granular rock composed of quartz crystals, usually so firmly cemented as to make the mass homogeneous. The stone is generally quarried in stratified layers, the surfaces of which are unusually smooth. Crushing and tensile strengths are extremely high, and the color range is wide.

Quartz porphyry. Igneous porphyritic rock of the same mineral composition as granite but with quartz occurring as phenocrysts.

Rubbed finish. Plane surface with occasional slight "trails" or scratches.

Sandstone. Sedimentary rock consisting usually of quartz cemented with silica, iron oxide, or calcium carbonate. Sandstone is durable and has a very high crushing and tensile strength and a wide range of colors and textures.

Sawed backs. Because of physical characteristics most granites cannot be split to a thickness of less than one-third the lesser face dimension. Consequently, sawed backs are specified for most veneers; they are also frequently specified for thicker ashlar because of design considerations.

Sawed finish. Relatively plane surface with texture ranging from wire sawed, a close approximation of rubbed finish, to shot sawed, with scorings $\frac{3}{32}$ in. in depth. Gang saws produce parallel scorings; rotary or circular saws make circular scorings. Shot-sawed surfaces are sandblasted to remove all rust stains and iron particles.

Scabbled. Roughly shaped or dressed to approximate shape and size.

Schist. Foliated metamorphic rock (recrystallized) characterized by thin foliae that are composed predominantly of minerals of thin platy or prismatic habits and whose long dimensions are oriented in approximately parallel positions along the planes of foliation. Because of this foliated structure schists split readily along these planes and so possess a pronounced rock cleavage. The more common schists are composed of the micas and other micalike minerals (such as chlorite) and generally contain subordinate quartz and/or feldspar of comparatively fine-grained texture; all graduations exist between schist and gneiss (coarsely foliated feldspathic rocks).

Seam. Crack of fissure in a rough quarry block.

Sheeted structure. Granite quarries that have well-defined horizontal joints, but few vertical joints.

Shot-ground finish. Plane surface with pronounced circular markings or trails having no regular pattern.

Six-cut finish. Medium bush-hammered finish, similar to but coarser than eight cut, with markings not more than $\frac{1}{8}$ in. apart.

Special finishes. Special finishes are available from manufacturers to meet special design requirements.

Staining. Materials especially likely to cause staining of granite are oak (when wet), knots in soft wood, oil, and asphalt-based compounds.

Start. Beginning of a crack, caused by quarrying, fabrication, or handing.

Structural granite. Includes all varieties of commercial granite that are sawed, cut, split, or otherwise shaped for building purposes. Classified into two grades: engineering and architectural grades (ASTM C 615).

Thermal finish. Plane surface with flame finish applied by mechanically controlled means to ensure uniformity. Surface coarseness varies depending on the grain structure of the granite.

Tolerance. Each piece of fabricated granite should have a maximum tolerance of plus or minus one-fourth of the joint width.

Wall surface finish. Various types of finishes available are: polished, honed, mello-toned, thermal, thermal-jet honed, split, and wire-sawed. Smooth finish surfaces such as polished, honed, rubbed, and mello-toned can be finished true and flat to within .0625 in.

GREENSTONE

Application. Greenstone is used for stairs, ramps, walkways, vestibules, and lobbies in stores, schools, hospitals, office buildings, hotels, and industrial plants. It is ideal for swimming pools, shower rooms, or dishwasher rooms.

Cleaning. Interior floors and treads should be thoroughly cleaned after installation and sealed to retain original finish.

Elk Brook greenstone. A product of Johnson & Rhodes Bluestone Co., East Branch, N.Y.

Exposed edges. Various types of exposed edges can be created by equipment, tools, and hand work depending on the use of the greenstone. Recommended edges are: diamond sawed, snapped, rubbed, thermal, and rocked. Sharp arrises are usually slightly rounded.

Exposed surfaces. Typical production produces wire sawed finishes, which then can be rubbed, or thermally treated to produce a special type finish. In any type of finish the greenstone surface is nonslip and remains that way even when wet.

Greenstone. Greenstone includes stones that have been metamorphosed or otherwise so altered that they have assumed a distinctive greenish color owing to the presence of one or more of the following minerals: chlorite, epidote, or actinolite.

Nonresonant qualities. The natural bond of greenstone provides nonresonant qualities that minimize noise and echoes of foot traffic.

Permanent nonslip protection. Surface remains nonslip even when wet, cannot wear smooth under heavy foot traffic, and is free from corrugations or grooves.

Rubbed finish. After wire sawing the surface is rubbed to a flat plane with coarse industrial diamonds.

Thermal finish. Flame-textured finish accomplished by superheating the surface crystals so that they expand and "pop out" along their cleavage planes. The soft green color and crystal pattern are retained in their muted state. Thermal-textured finish gives a natural look; its vertical use is for panel walls and its horizontal use is where the texture is desirable.

Wear resistance. Extremely hard, abrasive structure remains practically unaffected by severe service or concentrated traffic.

Wire sawed. Surface resulting from wire sawing, with wire marks occurring at random.

MARBLE

Abrasive hardness. Measured by a scuffing method that removes surface particles. A value of 10 has been established as the minimum acceptable value for marble chosen for general floor use.

Anchors. Mechanical devices for securing marble pieces to structural members, usually of nonstaining and noncorroding materials.

Architectural marble. Calcium carbonate with other components that give it color, pattern, and texture suitable for a desirable interior or exterior use building stone.

Arris. Edge of an external angle occurring at two surfaces.

Ashlar. Masonry having a face of square or rectangular stones laid in regular geometric patterns or irregular patterns depending on the type and shape of stones used.

Back-up wall. Masonry wall behind the exterior veneer or facing.

Bedding plane. Horizontal plane of a sedimentary stone in the position of its original formation.

Bleed. Staining action on marble caused by various oil-based putties, mastics, and other caulking materials.

Breccia. Fragmental rock of angular components cemented together into a coherent rock.

Buttering. Placing mortar or other joining material on a masonry unit with a trowel before laying it in place.

Calcite marble. Crystalline variety of limestone containing not more than 5% magnesium carbonate.

Clearance. Dimensional variation made to facilitate erection of units and provide for thermal and other estimated movements in the structure.

Commercial marble. Crystalline rock composed predominately of one or more of the following minerals: calcite, dolomite, or serpentine, and capable of taking high polish.

Conglomerate. Aggregate of rounded and water-worn pebbles and boulders cemented together into a coherent rock.

Cramp. U-shaped noncorrosive metal device for holding two adjacent pieces of marble together.

Crystalline limestone. Limestone, either calcitic or dolomitic, composed of interlocking crystalline grains of the constituent minerals and having pharneritic texture. Commonly used synonymously with marble, and thus representing a recrystallized limestone. Term is improperly applied to limestones that display some obviously crystalline grains in a fine-grained mass that are not of interlocking texture and do not compose the entire mass. All limestones are microscopically, or in part megascopically, crystalline; the term is confusing but should be restricted to stones that are completely crystalline and of megascopic and interlocking texture and that may be classed as marbles.

Curtain wall. Exterior wall that is not load bearing and that is supported by the structural framework of the building.

Cushion. Resilient pad intended to absorb or counteract severe stresses between adjoining marble slabs or marble slabs and other rigid materials.

Dolomite marble. Crystalline variety of limestone containing 40% magnesium carbonate as the dolomite molecule.

Dowel. Cylindrical noncorrosive metal pin used in aligning and strengthening joints of adjacent pieces of marble.

Exterior cubic marble. Term loosely applied in the stone industry to stones over 2 in. in thickness.

Exterior marble. Dimension marble for exterior building and structural applications. Crystalline rock composed predominantly of one or more of the following minerals: calcite, dolomite or serpentine, and capable of taking a polish (ASTM C 503).

Finisher. One who maintains and operates a plant and machinery for fabricating domestic and foreign marbles for building purposes.

Forest marble. Argillaceous limestone in which the coloring matter is so distributed as to resemble landscapes.

Grout. Concrete mortar with small aggregates and heavy liquid consistency, capable of being poured to fill small interstices.

Honed finish. Smooth surface with little or no gloss.

Liners. Structurally sound sections of marble that are cemented to the back of marble veneer slabs to give greater strength or additional bearing surface, or to increase joint depth.

Magnesian (dolomitic) marble. Crystalline variety of limestone containing not less than 5 nor more than 40% magnesium carbonate as the dolomite molecule.

Marble

1. Recrystallized medium-to-coarse-grained carbonate rock composed of calcite or dolomite, or calcite and dolomite. Original impurities are present in the form of new minerals, such as micas, amphiboles, pyroxenes, and graphite.

2. Metamorphosed limestone or dolomite. Massive variety of calcite that is capable of being polished and used for architectural and ornamental purposes.

3. Crystalline rock composed predominantly of one or more of the following minerals: calcite, dolomite, or serpentine, and capable of being polished as a finished product.

4. Metamorphic rock composed essentially of calcite and/or dolomite, generally a recrystallization of limestone to marble.

Marble-faced precast concrete panels. Combination of the two materials provides a structurally sound application to a building frame. Stainless steel anchors are used to fasten marble to precast reinforced concrete panels. Concrete panels are equipped with galvanized steel clip angle fasteners.

Mechanical properties. The compressive stress at which marble fails varies from 500 to 1500 psi. depending on variety. Tensile strength is less than 1000 psi. Flexural strength averages about 1500 psi, plus or minus one-third. Shear strength averages about 1500 psi.

Mullion. Nonbearing upright division member between windows or doors of a closed series.

Onyx. True onyx is a cryptocrystalline variety of quartz that closely resembles an agate. Onyx marble is a compact variety of limestone that is noted for its translucency and often for a delicate arrangement of colors more or less banded.

Onyx marble. Dense crystalline form of lime carbonate deposited usually from cold-water solutions. Onyx marble is generally translucent and shows a characteristic layering due to the mode of accumulation.

Opalized. Term describing rock of siliceous material in which a form of opal, a hydrous silicate, has been introduced.

Ophicalcite. Coarsely crystalline marble containing serpentine.

Panel wall. Non-load-bearing wall consisting of panels of materials, each panel being separately held in a frame. Frame may be structural itself or fastened to the structural framework of the building.

Pargeting

1. Plastering face of a backup wall or the back of the facing material with cement mortar to fill chance

voids. Referred to as "parging" in some parts of the country.

2. Monolithic coating over backup wall material and joints to provide waterproofing.

Permeability. Water vapor permeability of a homogeneous material is a property of the substance. Property may vary with conditions of exposure. Average permeability of a specimen is the product of its permeance and thickness. Accepted unit of permeability is a perm-inch, or 1 grain per square foot per hour per inch of mercury per inch of thickness.

Permeance. Water vapor permeance of a body between two specified surfaces is the ratio of its water vapor transmission to vapor pressure difference between the two surfaces. Accepted unit of permeance is a perm, or 1 grain per square foot per hour per inch of mercury per inch of thickness.

Phlogopite. Amber mica, which often occurs in marbles and serpentines.

Physical properties. Density of marble averages about 172 lbs/cu ft, or 0.1 lb/cu in. Specific gravity or density in the metric-system is 2.77.

Plinths. Lower square part of the base of a column. Square base or a lower block, as of a pedestal; base block at the juncture of baseboard and trim around an opening.

Polished finish

1. Smoothest finish available is stone, characterized by a gloss or reflective property and generally only possible on hard, dense materials.

2. Mirrorlike glossy surface that brings out the full color and character of the marble. Polished finish is usually not recommended for marbles to be used on building exteriors.

Quarry. A marble quarry is an excavation or pit, open to the air or underground, from which marble is obtained by cutting, drilling, or wire sawing.

Rate of transmission. The rate of water vapor transmission (WVT) of a body between two specified surfaces is the time rate of water vapor flow, under steady conditions, through unit area, under the conditions of the test. The accepted unit of WVT is 1 grain per square foot per hour.

Reglet. Narrow, flat, recessed molding of rectangular profile, also a prepared groove for flashing material.

Reveal. Depth of wall thickness between its outer face and a window or door jamb set in an opening.

Rodding. Reinforcement of a structurally unsound marble by cementing reinforcing rods into grooves or channels cut into the back of the slab.

Saddle. Usually a flat strip of marble projecting above the finished floor surface between jambs of a door; a threshold.

Sand and or abrasive finish. Flat, nonreflective surface. Finish is usually recommended for marbles intended for use on building exteriors.

Scientific definition of marble. Metamorphic (recrystallized) limestone composed predominately of crystalline grains of calcite or dolomite, or both, having interlocking or mosaic structure.

Serpentine. Hydrous magnesium silicate material of igneous origin, generally a very dark green color with markings of white, light green, or black. One of the hardest varieties of natural building stone.

Serpentine marble. Green marble characterized by a prominent amount of the mineral serpentine.

Setting space. Distance from the finished face of the marble to the face of the backup wall including space for spotting or grout.

Shear. Type of stress. A body is in shear when it is subjected to a pair of equal forces that are opposite in direction and that act along parallel planes.

Sill. Horizontal marble member, immediately supported by a foundation wall or piers.

Soffit. Underside of a beam, lintel, or reveal.

Soundness. Relative physical strength of marble. Sound marble is unbroken, relatively free from damage or defects, and of substantial and enduring character.

Spall. Chip or splinter removed from the main mass during quarrying or cutting operations.

Spandrel. Panel of wall between adjacent structural columns or between a windowsill and the window head immediately below it.

Spot or spotting. Adhesive contact, usually plaster of paris; applied between the back of marble veneer and the face of the backup wall to plumb or secure standing marble.

Sticking. Process of cementing together broken slabs or pieces of marble.

Stool

1. Inside vertical trim member under the sill of a window.

2. Flat marble piece, generally polished, used under the interior sill.

Styolite. Longitudinally streaked, columnar structure occurring in some marbles, of the same material as the marble in which it occurs.

Surface finishes. Appearance and texture of the surface of an area of finished or fabricated marble. Usual standard surface finishes are polished, honed, sand-rubbed and/or abrasive. Special finishes available are: axed, sand-blasted, sawn; tooled, bush-hammered, and rock-faced.

Template. Pattern used to fabricate many pieces of the same size or design.

Terrazzo. Type of concrete in which chips or pieces of stone, usually marble, are mixed with cement and ground to a flat surface, exposing the chips, which take a high polish.

Thermal properties. Marble has a medium thermal conductivity. With a "k" value of about 12 Btu/in./sq ft/hr/°F, it is both a good conductor as compared to lightweight insulation material and a good insulator as compared to metals.

Thin veneer. Term applied to stones 2 in. and less, when used on a building as a facing material.

Tolerance. Dimensional allowance made for the inability of men and machines to fabricate a product of exact dimensions.

Travertine

1. Terrestrial deposit of limestone formed in caves and around hot springs where cooling, carbonate-saturated groundwater is exposed to the air.

2. Calcareous rock deposited from solution. The compact, translucent variety is known as onyx marble.

3. Variety of limestone that has a partly crystalline or microcrystalline texture and porous or cellular layered structure, the cells being usually concentrated along certain layers and commonly displaying small stalactitic forms.

Travertine marble. Variety of limestone regarded as a product of chemical precipitation from hot springs. Travertine is cellular with the cells usually concentrated in thin layers that display a stalactitic structure. Some that take a polish are sold as marble and may be classified as travertine marble under the class of "commercial marble."

Veneer. Veneer layer of marble for a building wall is usually 2 in. or less in thickness.

Verdeantique. Marble composed chiefly of massive serpentine, capable of being polished, and commonly crossed by veinlets of other minerals, chiefly carbonates of calcium and magnesium.

Vug. Cavity in rock; sometimes lined or filled with either amorphous or crystalline material; common in calcereous rocks such as marble or limestone.

Waxing. Finishing by filling the natural voids in marble with color-blended materials.

WVT (water vapor transmission). Accepted unit of WVT is 1 grain per square foot per hour.

Zibell anchoring system. Arrangement of metal struts and special fastenings that provide anchoring for marble as thin as $\frac{7}{8}$ in. in both exterior and interior applications. The Zibell wall in weathertight and may be used with or without structural backup. Lightweight insulation may be installed between the marble and whatever interior finish is used. System developed by Georgia Marble Company, Atlanta, Ga.

MASONRY CONSTRUCTION

Abrasion of refractories. Wearing away of refractory surfaces by the scouring action of moving solids.

Absorption. Weight of water a brick or tile unit absorbs when immersed in either cold or boiling water for a stated length of time, expressed as a percentage of the weight of the dry unit.

Absorption rate. Weight of water absorbed when a clay brick is partially immersed for 1 minute, usually expressed in either grams or ounces per minute. Also called suction or initial rate of absorption.

Adhesion. Characteristic of mortar which enables it to cling to a masonry unit.

Adhesive strength. Quality of the bond that mortar has for holding two masonry units together.

Alignment. Construction units laid true to the line.

All-stretcher bond. Bond showing only stretchers on the face of the wall, each stretcher divided evenly over the stretchers under it.

American bond. Name sometimes given to common bond.

Angle iron. Structural piece of steel, used across openings and in other situations to support brickwork.

Anchor grooves. Grooves cut in construction units to hold anchors.

Approved masonry. Masonry constructed in accordance with local building codes.

Arch

1. Curved compressive structural member spanning openings or recesses; also built flat.

2. Form of construction in which a number of units span an opening by transferring vertical loads laterally to

adjacent units and thus to the supports. An arch is normally classified by the curve of its intrados.

3. An arch is essentially a beam in the plane of the loads.

Arch axis. Median line of the arch ring.

Arch spring line (or springing line). For minor arches, the line where the skewback cuts the soffit; for major parabolic arches, commonly the intersection of the arch axis with the skewback.

Arris. The sharp edge or salient angle formed by the meeting of two plane or curved surfaces.

Backboard. Temporary board on the outside of a scaffold.

Back filling. Rough masonry built behind a facing or between two faces; the filling over the extrados of an arch or in brickwork in spaces between structural timbers, sometimes called "brick nogging."

Backing. Portion of a masonry unit wall built in the rear of the face and bonded to the face; usually a more economical type of masonry unit.

Backing up. Part of all of the entire wall except the overhang face tier.

Backup. Part of a masonry wall behind the exterior facing.

Backset. Construction unit or course set back from the face of the balance of the masonry work.

Backwork. Mason lays up the back side of the wall, generally about 2 ft, so it can be parged or plastered before the front side is laid up.

Base course. Lowest course of masonry in a wall or pier.

Batter

1. Masonry recording or sloping back in successive courses; the opposite of corbel.

2. Face of a wall leaning from the vertical.

Batter stick. Tapering stick used in connection with a plumb rule for building battering surfaces.

Bearing blocks. Small blocks of cut stone built into a wall to provide bearing for wood beams.

Bed-depth. The thickness of a brick as opposed to height or length.

Bed (of mortar). Mortar on a wall ready to receive a brick upon it. Mortar on which a brick rests.

Bed joint

1. Horizontal layer of mortar on which a masonry unit is laid.

2. Horizontal joint between two courses.

Belt course. Narrow horizontal course of masonry, sometimes slightly projected, such as window sills that are made continuous; sometimes called "string course" or "sill course."

Beltstones or courses. Horizontal bands of stone encircling a building.

Bench mark. Established elevation that the mason can use for starting the first course of construction units.

Binders. Brick which extends only a part of the distance across a wall.

Blind bond. Masonry bond in which headers extend only half of the way through the tier of the face brick. The face bricks are all stretchers and some are split lengthwise to accommodate the headers.

Blind header. Concealed brick header in the interior of a wall not showing on the faces.

Block and bond. Laying a block bond on one side and a different unit and bond on the other side.

Block and cross bond. Combination of the two bonds. The face of the wall is in cross bond and the backing in block bond.

Blocking

1. Method of bonding two adjoining or intersecting walls not built at the same time by means of offsets whose vertical dimensions are not less than 8 in.

2. Method of building two adjoining or intersecting walls not built at the same time, by which the walls are tied together by offset and overhanging blocks of several courses of brick.

Blocking course. Course of stone placed on top of a cornice completing the wall.

Boasting. Cutting masonry with a broad-edged chisel.

Bond

1. Tie various parts of a masonry wall by lapping units one over another or by connecting with metal ties.

2. Patterns formed by exposed faces of units; adhesion between mortar or grout and masonry units or reinforcement.

3. Relative arrangement of vertical joints.

Bond beam. Horizontal reinforced masonry beam, resting on top of the last course of masonry, providing well-distributed loads from floor or roof joists.

Bond course. Course consisting of units that overlap more than one wythe of masonry.

Bonder. Bonding unit.

Bonder (header). Masonry unit; such as a header, that ties two or more wythes (leaves) of the wall together by overlapping.

Boning rods. Rods used to sight across to keep construction units level when laying them on a horizontal surface.

Box anchor. Square metal tie used to tie two masonry units together.

Breaking joints. Arrangement of masonry units that prevents continuous vertical joints from occurring in adjacent courses.

Brick and brick. Method of laying brick by which the brick are laid touching each other with only enough mortar to fill the irregularities of the surfaces.

Brick beam. A lintel constructed of reinforcing rods and grout in the void spaces of the bricks.

Brick chisel. Broad-edged chisel used to trim or cut brick.

Brick jointer. Hand tool (double ended) with $\frac{1}{2}$ to $\frac{3}{8}$ in. or $\frac{3}{4}$ to $\frac{7}{8}$ in. shaped in Z pattern, used to smooth out mortar joints.

Bricklayer. Person who has the knowledge and experience to place bricks in a wall, using mortar, bricks, concrete blocks, and other masonry materials.

Bricklayer's hammer. Hammer with a flat face and sharp peen used to break and dress bricks.

Brick masonry. Masonry of burned clay or shale brick, sand-lime brick, or concrete brick.

Brick nogging. Brick masonry installed between wood posts or beams.

Brock trowel. Flat triangular device, used in bricklaying to pick up and spread mortar on the joint or brick.

Brick veneer. A non-loadbearing single tier of brick applied to a wall of other materials.

Brickwork. Walls built of bricks laid in mortar.

Broken range. Masonry construction in which the continuity of the courses are broken at intervals.

Building. Combination of materials to form a construction that is safe and stable, and adapted to continuous occupancy for residence, business, industrial, assembly, or storage purposes.

Building brick in structure. All building brick should be free from cracks, laminations, and other defects of deficiencies which may interfere with proper laying of the brick or impair the strength or permanence of the structure.

Bull header. Brick laid on edge in the direction lengthwise of the wall; stretcher on edge.

Bull stretcher. Brick laid on its edge to show the broad side of the brick on the face of the wall.

Burning the joint. Black marks which appear in the mortar joints caused by a reaction between the mortar and the adjacent steel member.

Bush hammer. Hammer with a serated face for dressing masonry units.

Buttered joint. Very thin mortar joint made by placing a small quantity of mortar with the trowel on all edges of the brick and laying it without the usual mortar bed.

Camber

1. Relatively small rise of a jack arch.
2. Convexity of an upper surface.

Capacity insulation. Ability of masonry to store heat as a result of its mass, density, and specific heat.

Carve. To dress or cut a brick or stone.

C/B ratio. Ratio of the weight of water absorbed by a masonry unit during immersion in cold water to weight absorbed during immersion in boiling water. An indication of the probable resistance of brick to freezing and thawing. Also called "saturation coefficient."

Center. Temporary support to masonry, such as the support of an arch.

Centering. Temporary formwork for the support of masonry arches or lintels during construction. Also called "center."

Certificates. Documents from an approved testing agency attesting compliance with the applicable specifications for the grades, types, or classes of materials included in the specifications.

Chase. Continuous recess built into a wall to receive pipes, conduits, ducts, etc.

Chain bond. Bond formed by building in a metal tie or strap.

Chain course. Course of continuously fastened headers held by cramps.

Clip bond. Masonry wall bond formed by clipping off the inner corners of the block or face brick to obtain a diagonal bond with stretchers.

Clip course. The course of brick resting on a clip joint.

Clip joint. Joint of abnormal thickness to bring the course up to the required height. In no case should a clip joint exceed $\frac{1}{2}$ in. thick.

Clipped header. Bat placed to look like a header for the purpose of establishing a pattern. Also called "false header."

Closer

1. Last brick or tile laid in the course. May be whole or a portion of a unit.

2. Piece of brick laid to the line. Also, the last brick laid in any course of any tier.

Closure. Supplementary or short-length units used at corners or jambs to maintain bond patterns.

Closure brick. The last brick laid in a wall, usually located near the center of the wall and cut to fit.

Collar joint. Interior longitudinal vertical joint between two wythes (thickness) of masonry.

Common bond. Several courses of stretchers followed by one course of either Flemish or full headers.

Compass. An instrument or tool used to lay out work.

Consistency. Degree of density or viscosity of grout.

Control of expansion joint. Vertical joint built into a masonry unit wall, in places where it is presumed forces acting on the wall will cause certain movements which would create loosening of the mortar joints and the adjacent masonry units. The joint is filled with material which will provide flexibility, and some movement can occur without destroying the bonding.

Corbel. Shelf or ledge formed by projecting successive courses of masonry out from the face of the wall.

Corbel out. To build out one or more courses of brick or stone from the face of a wall to form a support shelf.

Corner. Place of directional change of walls or surfaces of a structure.

Corner pole. Simple tool that eliminates the building of corner leads in masonry work. Available in metal with adjustable coursing scales attached. Advantages: brick coursing is uniform and level; wall is plumb; joint thicknesses are uniform.

Course

1. One of the continuous horizontal layers of masonry units bonded with mortar. One course is equal to the thickness of the masonry unit plus the thickness of one mortar joint.

2. One horizontal layer of brick in a wall; radial layer of brick in an arch.

Course bed. The installation of stone, brick, or other masonry units in position upon which other materials are to be laid.

Coursed ashlar. Ashlar set to form continuous horizontal joints.

Coursed rubble. Masonry composed of roughly shaped stones fitting approximately on level beds, well bonded, and brought at vertical intervals to continuous level beds or courses.

Cross bond. Bond in which the joints of the second stretcher course comes in the middle of the first course; a course composed of headers and stretchers intervening.

Cross joint. Head joint between construction units in a course that connects the horizontal bed joints.

Crown. Apex of the arch ring. In symmetrical arches the crown is at midspan.

Cupping the mortar. Method used by the mason for cutting the mortar from the mortarboard and rolling the mortar onto a trowel in a cupping motion.

Curing. Natural curing of mortar joints by keeping the joints moist after set.

Cut joint. Protruding mortar of cross and bed joints that is cut off and removed with a trowel.

Damp course. Course or layer of impervious material that prevents capillary entrance of moisture from the ground or a lower course.

Dampproofing. One or more coatings of an asphaltic mastic compound or other material that is impervious to water. Some materials can be applied to back of stones or to the face of the back wall.

Dampproofing course. Layer of asphalt-impregnated felt roofing paper installed at the base course to intercept the infiltration of moisture.

Dead-burned magnesite. Granular product obtained by burning (firing) magnesite or other substances convertible to magnesia upon being heated above 1450°F long enough to form dense, weather-stable granules suitable for use as a refractory or in refractory products.

Deadman. Post or prop to which the line on a wall is attached.

Depth. In mason work, the distance from the face of the stone to the back of the stone.

Depth of arch. The depth of any arch is the dimension that is perpendicular to the tangent of the axis. The depth of a jack arch is taken to be its greatest vertical dimension.

Diagonal bond. Form of raking bond where the bricks are laid in an oblique direction in the center of a wall as a decorative feature.

Diaper. Any continuous pattern of brickwork where the various bonds are examples of decorative features.

Dog's tooth. Bricks laid with their corners projecting from the wall face.

Double jointing. Process of applying mortar to both ends of the closure brick as well as the ends of the bricks surrounding the area where the closure brick will be placed.

Double-screened ground refractory material. Refractory material that contains its original gradation of particle sizes resulting from crushing or grinding, or both, and from which particles coarser and finer than two specified sizes have been removed by screening.

Dovetail anchor. Galvanized tie in the shape of a dove's tail, used to tie masonry units to concrete walls or to other construction units.

Dowels. Straight galvanized metal bars used to connect two sections of masonry.

Draft. Margin on the surface of a stone cut approximately to the width of the chisel.

Dressing. Working the face of a stone to the required finish; also squaring a stone for ashlar.

Drip. Projecting piece of material shaped to throw off water and prevent its running down the face of a wall or other vertical surface.

Dry bonding. Laying of bricks without mortar to establish the bond for the wall.

Dry masonry. Masonry units that are not separated or held in place by mortar.

Dry stone walls. Walls in which natural or cut stones are installed tightly without mortar.

Dutch bond. Arrangement of bricks forming a modification of Old English bond, made by introducing a header at the second brick in every alternate stretching course, with a three-quarter brick beginning the other stretching courses.

Dutch corner. Method of starting off a corner by using a 6 in. piece of brick.

Edge set. Brick laid on edge showing the broad side of the brick.

Effective area of reinforcement. Area obtained by multiplying the right cross-sectional area of the metal reinforce-

ment by the cosine of the angle between its direction and the direction for which the effectiveness of the reinforcement is to be determined.

Effloresence

1. Powder or stain sometimes found on the surface of masonry, resulting from deposition of water-soluble salts.

2. Deposit of soluble salts, usually white in color, appearing on the exposed surface of masonry.

End-construction tile. Tile designed to be laid with axis of the cells vertical.

Engineered brick. Bricks whose nominal dimensions are 3.2 in. × 4 in. × 8 in.

English bond

1. Bond composed of alternate courses of headers and stretchers. Headers are centered on the stretchers, and joints between stretchers in all courses are aligned vertically. Snap headers are used in courses that are not structural bonding courses.

2. Alternate courses of full headers and stretchers, headers being plumb over each other, and the stretchers being plumb over each other. The headers are divided evenly over the stretchers and over the joints between the stretchers.

English corner. Method of starting a corner by laying a 2 in. piece of brick off the corner brick.

English cross bond. Modification of English bond in which stretcher courses break joints with each other.

English cross or Dutch bond

1. Variation of English bond that differs only in that vertical joints between the stretchers in alternate courses do not align vertically. These joints center on the stretchers themselves in the courses above and below.

2. Same as English bond, except that the alternating stretcher courses, instead of being plumb over each other, break evenly over each other.

Expansion anchor. Galvanized metal expandable unit, inserted into a drilled hole, that grips stone by expansion.

Expansion joint. Vertical joint or space to allow for expansion due to temperature changes.

Extrados. Convex curve that bounds the upper extremities of the arch.

Fat mortar

1. Mortar containing a high percentage of cementitious components. It is a sticky mortar that adheres to a trowel.

2. Mortar that tends to stick to the trowel, generally because of too little sand.

Field. Expanse of wall between openings, corners, etc., principally composed of stretchers.

Filling in. Laying brick on interior tiers after the face tiers in the same course have been laid.

Fire stop. Projection of brickwork on the walls between the joists or rafters to prevent spread of smoke or fire to adjacent areas.

Flare header. Header of darker color than the field of the wall.

Flashing

1. Thin, impervious sheet material, placed in mortar joints and across air spaces in masonry to collect water that may penetrate the wall and to direct it to the exterior.

2. Step during the burning process of clay masonry units that produces varying shades and colors in the units.

Flash set. The premature setting or hardening of mortar due to excessive heat or lack of water.

Flat stretcher course. Course of stretchers set on edge and exposing their flat sides on the surface of the wall.

Flemish bond

1. Each course of brick consists of alternate stretchers and headers, with the headers in alternate courses centered over the stretchers in the intervening courses. Where the headers are not used for the structural bonding, they may be obtained by using half brick called "clipped" or "snap" headers.
2. Courses of alternate stretchers and headers, with the center of the headers located plumb over the center of the stretchers beneath them.

Flemish bond (double). An arrangement of bricks which gives a Flemish bond look on both sides of the wall.

Flemish cross bond. Any bond having alternate courses of Flemish headers and stretchers, the Flemish headers being plumb over each other and the alternate stretcher courses being crossed over each other.

Flemish double-cross bond. Bond with odd-numbered stretcher courses divided evenly over each other, and even-numbered Flemish header courses in various locations with reference to the plumb of each other.

Flemish garden bond. Bricks laid so that each course has a header to every three or four stretchers.

Flemish header. One course of brick consisting of alternate stretchers and headers.

Flemish triple bond. Placing one header to three stretchers in a course.

Flush. Having the surface even with adjoining surfaces.

Flushing. Slushing mortar into the joints with a trowel.

Flush joint. Mortar joint which is even with the construction unit and in which no tooling is required.

Foot boards. A temporary arrangement of setting wide boards on blocks of concrete or bricks to allow the mason to reach a higher level of the wall without scaffolding.

Full header. Course consisting of all headers.

Furring. Method of finishing the interior face of a masonry wall to provide space for insulation, prevent moisture transmittance, or provide a level and plumb surface for finishing.

Furrowing. Slight indentation made in the mortar bed joint with the point of the trowel to prepare for laying of brick.

Galleting. Filling a masonry joint with rock chips to increase the strength of the joint.

Gang saw. Machine, electrically operated with multiple blades, used to saw rough quarry blocks into slabs.

Gantry. Form of scaffolding or staging constructed of timbers to provide a working platform.

Garden wall bond. Bond in which there are three stretchers alternating with a header on the same course.

Gauged putty. Successive batches of mortar mixed thin and in the same proportions to secure even setting time.

Gauge rod. Marked pole for measuring masonry coursing during construction.

Girth hitch. Form of knot used in masonry work to fasten a nail to a line.

Green staining. Reaction on masonry work caused by the salts contained in vanadium.

Grounds. Nailing strips placed in masonry walls as a means of attaching trim or furring.

Grouted brick masonry. Form of construction made with brick in which interior voids of the masonry are filled by pouring grout therein as the work progresses.

Grouted masonry. Solid masonry in which the outside and inside tiers or wythes are laid in full head and bed joints of mortar and all interior joints are filled with grout made of mortar of the same type as that used in laying the outer tiers or wythes.

Grout lift. An increment of grout height within the total pour; a pour may consist of one or more lifts.

Grout pour. The total height of masonry wall to be poured prior to the erection of additional masonry. A pour will consist of one of more lifts.

Gypsum mortar. Mortar usually composed by weight of one part of gypsum and not more than three parts of mortar aggregate.

Hacking. Stacking brick in a kiln or on a kiln car, or laying brick with the bottom edge set in from the plane surface of the wall.

Hardcore. Culled and broken bricks used to provide a base for concrete slabs, driveways, and paths.

Hawk. Small board with a handle on the bottom side, to hold small quantity of mortar for the mason's use.

Header

1. Brick laid so that only its end shows on the face of a wall.
2. Masonry unit that overlaps two or more adjacent wythes of masonry to tie them together. Often called "bonder."

Header bond. Bond showing only the headers on the face, with each header divided evenly on the header under it.

Header block. Concrete masonry units made with part of one side of the height removed to provide space for bonding with adjacent masonry units, such as brick.

Header joint. Joint between the ends of two bricks in the same course; also called "vertical joint."

Header tile. Tile containing recesses for brick headers in masonry-faced walls.

Heading course. Continuous bonding course of header brick. Also called "header course."

Head joint. Vertical mortar joint between ends of masonry units. Often called "cross joint."

Herringbone bond. Bricks laid in an angular or zigzag fashion.

High-lift grouting. Technique of grouting masonry in lifts up to 12 ft.

High-pressure steam curing. Method of curing concrete masonry units using saturated steam (365°F) under pressure, usually 125 to 150 psi. Also referred to as "autoclave curing."

Hod. Short wood or metal shaped trough (closed ends) on a pole, carried by hod carrier to deliver mixed mortar or bricks to the mason.

Inlet set. The drying of mortar which bonds it to the masonry unit.

Interlocking. Bonding masonry units by lapping them.

Joint. Space between masonry units, usually filled with mortar.

Jointer. Tool used for smoothing or indenting the surface of a mortar joint.

Joint grouting and pointing. Joints should be carefully packed with a mortar composed of one part portland cement, $\frac{1}{4}$ part lime, and three parts of clean sand, mixed to buttery consistency. In pointing, care should be exercised to avoid deposits of excess mortar along the perimeter edges of stones. Deposits should be completely removed at once with a cellulose sponge to avoid staining of the material. Stones should be finally finished by rubbing in circular motion with dry sharp sand and small gunny sacks. Pointing of joints should be done the same day the material is laid.

Jointing. The process of facing or tooling the mortar joints.

Key. The relative position of the headers of various courses with reference to vertical line.

Kiln. Specially constructed furnace in which brick, terra cotta, and tile are fired at controlled temperatures.

Lap. The distance one brick extends over another.

Laying overhand. Laying both faces of a wall when the scaffold is only on one side of a wall, necessitating reaching over to lay up the opposite side.

Laying to bond. Laying the brick of the entire course without a cut brick.

Lead

1. Section of a wall built up and racked back on successive courses. A line is attached to leads as a guide for constructing a wall between them.
2. Part of the wall built up ahead of the line to which to haul the line.

Line. String used by the bricklayer as a guide for laying the horizontal top edge of brick.

Line pin. Metal pin used to attach the line used for alignment of masonry units.

London trowel. Pattern of a trowel having a diamond-shaped heel.

Mason. Person who has knowledge and experience in laying masonry units.

Masonry. Architectural terra cotta, brick, and other solid masonry units of clay or shale; concrete masonry units, gypsum tile or block, plain concrete; stone; structural clay tile; structural glass block, or other similar building units of materials, or a combination of same, bonded together with mortar.

Masonry alkalinity. The alkalinity of the wall is a chemical property of masonry that may have a significant effect on paint durability and performance. Brick is normally neutral, but is set in mortars that are chemically basic.

Masonry bond. Bond obtained from the arrangement of the units and the use of mortar, epoxy, or other methods or materials.

Masonry filler unit. Masonry unit used to fill in between joists or beams to provide a platform for a cast-in-place concrete slab.

Masonry sealer. Waterproof varnish-like substance for sealing masonry against the absorption of water.

Masonry wall temporary clamp. Prefabricated noncorroding metal wire wall clamp used for holding protective coverings on the top of unfinished masonry walls during construction.

Material platform hoist. Manually or power operated suspended platform conveyance operating in guide rails for the exclusive raising or lowering of materials, operated and controlled from a point outside the conveyance.

Mechanical spalling of refractories. The spalling of a refractory unit is caused by stresses resulting from impact or pressure.

Mercury bob. Plumb bob filled with quicksilver to get the greatest weight in the smallest size.

Moist curing. Method of curing concrete masonry units using moisture at atmospheric pressure and a temperature of approximately 70°F.

Moisture content (concrete masonry units). Amount of water contained in a unit, expressed as a percentage of the total absorption; that is, a concrete masonry unit at 40% moisture content contains 40% of the water it could absorb.

Mortarboard. Flat, square, wood pallet for holding mortar.

Mortar box. Box in which mortar is mixed and softened by water for use by the mason.

Mud. Slang term used on the job to describe mortar or plaster.

Mullite refractories. Refractory products consisting predominantly of mullite crystals formed either by conversion of one or more of the sillimanite group of minerals or by synthesis from appropriate materials by means of either melting or sintering processes (ASTM C 467).

Muriatic acid. Commercial hydrochloric acid used for cleaning stonework diluted according to manufacturer's directions.

Nailing block. Wood block inserted into masonry wall with anchors attached to which the frame can be secured by nailing.

Nominal dimension–masonry. Dimension that may vary from actual masonry dimensions by the thickness of a mortar joint, but with variation not exceeding $\frac{1}{2}$ in.

Offset. Course that sets in from the course directly under it. Also called "setoff" or "setback," the opposite of corbel.

Outrigger. Beam projecting out of an opening in the wall and supporting an outside scaffold.

Oversail. Laying of brick so that each projects beyond the one below, same as a corbel.

Parging. Application of mortar to the back of the facing material or the face of the backing material. Also called "backplastering."

Pargeting. Process of applying a coat of cement mortar to masonry. Often spelled and/or pronounced "parging."

Pattern bond. Pattern formed by the masonry units and the mortar joints on the face of the wall. The pattern may result from the type of structural bond used or may be purely a decorative one unrelated to the structural bonding.

Peen. End of a hammer head that terminates in an edge.

Periodic kiln. Kiln that is loaded, fired, allowed to cool, and unloaded before reloading. Updraft kilns are not usually used for structural clay tile. Periodic downdraft kilns are commonly used in burning structural clay tile of all types, facing brick, and glazed ware—in fact, any product requiring close control of heat. The ware is stacked in such a way that the hot gasses travel up within "bag" walls, built inside the kiln walls, to the crown of the kiln and are then pulled down through the ware to flues under the floor and thence to the stack or the driers, if the waste heat is used for drying.

Perpend bond. Bond that signifies that a header extends through the whole thickness of the wall.

Philadelphia trowel. Pattern of a trowel having a square-shaped heel.

Pick and dip. Method of laying brick whereby the bricklayer simultaneously picks up a brick with one hand and enough mortar on a trowel to lay the brick with the other hand. Sometimes called the "eastern" or "New England method."

Plugging chisel. Chisel with a tapered blade, used for removing mortar from joints.

Plumb bob. Bob used by the mason to determine the direction of the absolute vertical. The instrument is a mason's line weighted with a metal shaped instrument (looks like a spinning top used by children).

Plumb bond. Another name for all-stretcher bond work built with particular effort to have corresponding joints exactly plumb with each other.

Plumb bond pole. Pole used for laying out the exact position of vertical joints.

Plumb glass. Slightly curved glass into which alcohol is sealed for use in a plumb rule.

Pointing. Process of removing deteriorated mortar from masonry and replacing it with new mortar; also the final patching, filling, or finishing of mortar joints in new masonry work.

Pointing trowel. Hand tool, triangularly shaped, $2\frac{1}{2}$ to 3 in. in width, used for immediate pointing of joints.

Primary member. Member of the structural frame of a building or structure used as a column, grillage beam, or to support masonry walls in which the facing and backing of the wall are bonded together with masonry units.

Prism. Assemblage of brick and mortar for the purpose of laboratory testing for design strength and quality control of materials and workmanship. Minimum dimensions for prisms are 12 in. in height with a slenderness ratio between 2 and 5.

Prism tests. When the ultimate compressive strength of brick masonry is to be established by tests, the tests are made well in advance of the proposed construction. Not less than five prisms should be built, stored in air at temperatures not less than 65°F, and tested after aging 28 days. Seven-day tests may be used provided the relation between 7- and 28-day strengths of the masonry is established by tests of the materials used; in the absence of an established 7- and 28-day strength relationship, 7-day strength may be assumed to attain 90% of the 28-day strength.

Puddling. Process of settling, distributing, or consolidating grout in a masonry reinforced wall, with a wood or steel rod to prevent the formation of voids in the reinforced sections of the wall.

Pug mill. Machine in which clay is ground, mixed, and tempered.

Queen closer. Cut brick having a nominal 2-in. horizontal face dimension.

Quoin

1. Projecting right-angle masonry corner.
2. Brickwork in a corner.

Racking

1. Method entailing stepping back successive courses of masonry.

2. Method of building the end of a wall so that it can be built on and against adjacent right-angle walls, without any toothers.

Raggle. Groove in a joint or special unit to receive roofing or flashing; also called "reglet."

Rake. In masonry, a method of laying brick courses in an angular designed fashion.

Raked joint. Joint made by removing the surface of the mortar while it is still soft. (While the joint may be compacted, it is difficult to make weathertight and is not recommended where heavy rain, high wind, or freezing is likely to occur.) Joint sets back about $\frac{1}{2}$ in. from the wall surface and creates a shadow that is very pronounced.

Raking bond. Brick laid in an angular or zigzag fashion.

Random ashlar. Ashlar set with stones of varying length and height so that neither vertical or horizontal joints are continuous.

Refractories. Materials, usually nonmetallic, used to withstand high temperature.

Refractory. Resistant to high temperature.

Refractory chrome ore. Refractory ore consisting essentially of chrome-bearing spinel, with only minor amounts of accessory minerals and with physical properties that are suitable for making refractory products.

Reinforced grouted masonry. Masonry construction made with solid masonry units in which interior joints of masonry are completely filled by pouring grout therein and in which reinforcement is embedded.

Reinforced hollow-unit masonry. Masonry construction made with hollow masonry units in which certain cells are continuously filled with concrete or grout and in which reinforcement is embedded.

Reinforced masonry

1. Masonry units, reinforcing steel, concrete grout and/or mortar combined to act together structurally.
2. Unit of masonry in which reinforcement is embedded in such a manner that the two materials act together in resisting forces.

Retempering of mortar. Restoring workability of mortar that has stiffened owing to evaporation by adding water and remixing.

Return. Surface turned back from the face of a principal surface.

Reveal

1. End of a wall, as at a jamb or return.
2. Portion of a jamb or recess that is visible from the face of a wall back to the frame placed between the jambs.

Rigid steel anchors. Anchors used for intersecting bearing or shear walls made of $1\frac{1}{2} \times \frac{1}{4}$ in. steel, not less than 2 ft. long, with ends turned up not less than 2 in. or with cross pins.

Riprap. Large, irregular blocks of stone quarried along natural lines of weakness. In some instances, broken concrete has been used as riprap.

Rolling scaffolding. Scaffolding mounted on wheels, used on small jobs and average height structures.

Roughly squared. Stone roughly squared with a hammer and laid as in rubble stonework.

Rough pointing. Smearing mortar joints with a trowel.

Round joint. Trade name for a half-circular indentation formed in a mortar joint by use of a convex jointer.

Rowlock

1. Brick laid on its face edge so that the normal bedding area is visible in the wall face.

2. Brick usually laid in the wall with its long dimension perpendicular to the wall face.

Rubbed joint. Flush mortar joint which has been rubbed flat with a burlap sack or a tool which has a rough surface.

Running bond

1. Simplest of the basic pattern bonds, the running bond consists of all stretchers. Since there are no headers in this bond, non-corroding metal ties are usually used. Running bond is used largely in cavity wall construction and veneered walls of brick, and often in facing tile walls where the bonding may be accomplished by extra-width stretcher tile.

2. Another name for all-stretcher bond.

Run of kiln. All brick in the kiln except brick that is too soft or misshapen to be laid even in the filling tiers.

Rustication. Emphasis of quoins and string courses in masonry, especially in ashlar work, by chamfering, rebating, and other methods.

Sailor brick. Brick that is laid in a vertical position with the largest side facing the front of the masonry wall.

Scaffold. Elevated platform that is used for supporting workmen or materials, or both.

Scutch. Tool resembling a pick on a small scale, with flat cutting edges for trimming bricks for particular uses.

Shot sawed. A finish obtained by using steel shot in a gang sawing process to produce random markings for a rough surface texture.

Shoved joints. Vertical joints filled by shoving a brick against the next brick when it is being laid in a bed of mortar.

Side construction tile. Tile intended for placement with axis of cells.

Silicon carbide defractories. Refractory products consisting predominantly of silicon carbide.

Silt. Stone particles ranging between 0.00015 and 0.0025 in. in diameter.

Slaking. Process of changing quickline into hydrated lime through chemical reaction by adding water.

Slushed joints. Vertical joints that are filled after units are laid by "throwing" mortar in with the edge of a trowel.

Snap header. Half brick laid in a wall to resemble a header.

Soffit

1. Underside of a beam, lintel, or arch.

2. Underside of a covering over an opening, such as the bottom of a cap or arch over a window or door.

Soldier

1. Stretcher set on end with narrow face showing on the wall surface.

2. Masonry unit set vertically on end with narrow face showing on the masonry surface.

Solid masonry

1. Masonry consisting wholly of solid masonry units laid contiguously in mortar, or consisting of plain concrete.

2. Masonry built without hollow spaces.

Spall. Small fragment removed from the face of a masonry unit by a blow, action of the elements, or chemical reaction of materials in the unit.

Spirit plumb rule. Plumb rule with a curved glass nearly full of alcohol or other thin, nonfreezing liquid. Location of the bubble with reference to a mark on the glass

indicates the plumb position of the edge face of the plumb rule.

Stack bond. A pattern bond where the masonry units are stacked over each other. There is no overlapping of units since all vertical joints are aligned. Usually this pattern is bonded to the backing with rigid steel ties, but when 8-in. bonder units are available, they may be used. In large wall areas and in load-bearing construction, it is advisable to reinforce the wall with steel reinforcement placed in the horizontal mortar joints. In stack bonding it is imperative that prematched or dimensionally accurate masonry units be used if the vertical alignment of the head joints is to be maintained.

Stacked ashlar. Ashlar set to form continuous vertical joints.

Straight edge. Board having one or two straight and parallel edges, used for leveling and plumbing longer surfaces than can be reached with an ordinary spirit level.

Stretcher

1. Unit laid with its length horizontal and parallel with the face of the wall or other masonry member.

2. Brick laid so that only its long side shows on the face of the wall.

String course. Course of stretcher masonry units.

Stringing mortar

1. Spreading enough mortar on a bed to lay several masonry units.

2. Method whereby a bricklayer picks up mortar for a large number of bricks and spreads it before laying the first brick.

Story-or-course pole. Use of a story-or-course pole, which is simply a board with markings 8 in. apart, provides an accurate method of finding the top of the concrete masonry unit for each course. Mortar joints for concrete masonry should be $\frac{3}{8}$ in. in thickness.

Story pole

1. Marked pole for measuring vertical masonry courses during construction.

2. Pole on which all measurements of courses, openings, projections, offsets, corbels, plates, and bottoms of beams of any one story are marked.

Strike. To cut off the projecting fresh mortar from a joint and to draw a trowel across the joint to densify and smooth the appearance at the same time.

Struck joint. Mortar joint that has been finished with a trowel. This is a common joint in ordinary brickwork. Mechanics often work from the inside of the wall, and it is easy to strike the joint with a trowel, since the joint starts flush with the top brick and slopes $\frac{3}{4}$ in. beyond the face of the lower brick.

Structural bond

1. Bond whereby individual masonry units are interlocked or tied together to cause the entire assembly to act as a single structural unit.

2. Bond made by tying wythes of a masonry wall together lapping units one over another or by connecting them with noncorroding metal ties.

Suction. Initial rate of water absorption by a clay masonry unit.

Swinging scaffolding. Scaffolding with heavy-duty suspended frame which supports the wood planks.

Tapping. Pounding a brick down into its bed of mortar with a trowel.

Temper. Moisten and mix clay, plaster, or mortar to a proper consistency.

Tempering mortar. Softening mortar by adding water.

Tensile bond. Adhesion between mortar and masonry units or reinforcement.

Termite shield. A sheet metal barrier placed in the first course of the masonry units to prevent passage of termites.

Test reports. Test reports submitted by an independent laboratory are usually required for each type of building and facing brick. Such reports indicate compressive strength, 24-hour cold-water absorption, 5-hour boil absorption, saturation coefficient, and initial rate of absorption (suction).

Through bonds. Bonds which extend entirely from back to front of the wall.

Tie. Unit of material that connects masonry to masonry or other materials.

Tier. Vertical layer of brick 4 in. or the width of one brick in thickness.

Tooled joint. Mortar joint whose face is compressed and shaped with a special concave or V-shaped tool.

Tooling

1. Compressing and shaping the face of a mortar joint with a special tool other than a trowel.

2. Brick mortar and joints that are exposed and have become "thumbprint" hard should be tooled with a round or other approved jointer. The jointer should be slightly larger than the width of the mortar joint so that complete contact is made along the edges of the units, compressing and sealing the surface of the joint. Exterior joints below grade should be trowel pointed or concave tooled, and all other joints not tooled should be flush cut.

Toother. Brick projecting from the end of a wall against which another wall will be built.

Toothing

1. Constructing the temporary end of a wall with the end stretcher of every alternate course projecting. Projecting units are toothers.

2. Projecting brick or block in alternate courses to provide for bond with adjoining masonry that will be laid later.

3. Temporary end of a wall built so that the end stretcher of every alternate course projects one-half its length.

Trig. Brick bedded to the proper height to hold a mason's line level in the center of a course.

Trowel. Triangular-blade hand tool made from steel, with a wooden handle, used by the mason for spreading mortar on masonry units. There are three styles of trowel: Philadelphia, London, and wide London.

Troweled joint. Mortar joint that has been finished with a trowel to form a struck joint or a weathered joint. In a raked joint the mortar is raked out to a specified depth while the mortar is still green.

Tuck pointing. In masonry, filling in cutout or defective mortar joints with fresh mortar.

Tuck point joint filler tool. Hand tool used by the mason to add mortar to joints. The tool is made from steel, has a wooden handle, and is long and narrow in shape. It is made in widths ranging from $\frac{1}{4}$ to 1 in. and averages about $6\frac{1}{2}$ in. in length.

Tunnel kiln. Kiln that is build as both a straight and circular tunnel, through which the ware passes while being burned. The ware is loaded on special cars, which then enter the tunnel and travel at the correct speed through the water-smoking, dehydration, oxidation, vitrifying, and cooling zones. Heat conditions in each zone are carefully controlled.

Unit masonry. Built-up construction or combination of masonry units set in mortar or grout.

Veneer

1. Single wythe of masonry for facing purposes, not structurally bonded.

2. Facing of brick, concrete, metal, stone, tile, or similar material attached to a wall for the purpose of providing ornamentation, protection, or insulation, but not counted as adding strength to the wall.

3. Wall facing attached to the wall, but not considered as contributing to the strength of the wall.

Vitrification. Condition resulting when kiln temperatures are sufficient to fuse grains and close pores of a clay product, making the mass impervious.

Wall plate. Horizontal member anchored to a masonry wall, to which other structural elements may be attached. Also called head plate.

Water retention (of mortar). Property of mortar that prevents the loss of water to masonry units having a high suction rate. Also prevents bleeding or water gain when mortar is in contact with units having a low suction rate.

Water retentivity. Property of a mortar that prevents the rapid loss of water to masonry units of high suction. Also prevents bleeding or water gain when mortar is in contact with relatively impervious units.

Water table. Projection of lower masonry on the outside of the wall slightly above the ground. Often a damp course is placed at the level of the water table to prevent upward penetration of groundwater.

Weathered joint. Joint requiring care as it must be worked from below. However, it is the best of the troweled joints, as it is compacted and sheds water readily. Joint is directly the opposite of a struck joint in that the joint starts flush with the bottom brick and slopes $\frac{1}{4}$ in. beyond the face at the top brick.

Weep holes. Openings placed in mortar joints of facing material at the level of flashing to permit the escape of moisture.

Wetting brick. Brick should be wetted sufficiently to ensure that each brick is nearly saturated, surface dry when laid. During freezing weather, units that require wetting should be sprinkled with warm or hot water just before laying. Brick is wetted to prevent water from being absorbed from the mortar, which would cause quick drying, and a faulty mortar would result.

Wythe (leaf)

1. Each continuous vertical section of a wall, 1 masonry unit in thickness and tied to its adjacent vertical section or sections (front or back) by bonders (headers), metal ties, or grout.

2. Thickness of masonry separating flues in a chimney.

3. Four-inch partition or tie between two walls, such as two walls of a chimney.

MASONRY MATERIALS

Accelerator. Material that speeds hardening of concrete or mortar.

Acid-resistant brick. Brick that is suitable for use where it will be in contact with chemicals, designed primarily for use in the chemical industry. Usually used with acid-resistant mortar.

Additive. Material added to an agglomerated mass for a specific purpose.

Adhesion-type ceramic veneer. Thin sections of ceramic veneer held in place by adhesion of mortar to unit and backing. No metal anchors are required.

Admixtures. Materials added to mortar as water-repellent or coloring agents or to retard or hasten setting.

Adobe. Soil of diatomaceous content mixed with sufficient water so that plasticity can be developed for molding into masonry units.

Adobe brick. Large clay brick, varying in size, roughly molded and sun dried.

Agglomerated mass. Matrix or binder and the aggregate of a mix such as plaster, concrete, and mortar.

Aggregate

1. Stone or pea gravel screened to adopted sizes for subsequent mixture with cement and sand in the production of concrete or masonry concrete grout (ASTM C 404).

2. Natural or manufactured sand for use in masonry mortar. Manufactured sand is the product obtained by crushing stone, gravel, or air-cooled iron blast-furnace slag (ASTM C 144).

Air-entraining agent. Material used to trap air in mortar to improve its workability and durability.

Air-setting refractory mortar. Composition of finely ground materials, marketed in either a wet or dry condition, which may require tempering with water to attain the desired consistency and which is suitable for laying refractory bricks and bonding them strongly upon drying and upon subsequent heating at furnace temperatures.

Alumina. Mineral contained in clay used for making bricks.

Amalgamation. Mixed blend or combination of materials such as lime, cement, sand, and water.

Anchor. Piece device or assemblage, usually noncorrosive metal, used to attach building parts to masonry or masonry materials.

Anchor bolt. Metal rod or bolt imbedded in masonry unit, surrounded by grout to hold or support other materials.

Anchored-type ceramic veneer. Thick sections of ceramic veneer held in place by grout and wire anchors connected to backing wall.

Angle brick. Brick shaped to an oblique angle to fit a salient corner.

Arch brick. Wedge-shaped brick for special use in an arch; extremely hard-burned brick from an arch of a scove kiln.

Architectural terra cotta

1. Custom-made, hard-burned, glazed or unglazed clay building units, plain or ornamental, machine extruded or hand molded, and generally larger in size than brick or facing tile.

2. Plain or ornamental (molded or extruded) hard-burned building units, usually larger in size than brick, consist-

ing of mixtures of plastic clays, fusible minerals, and grog, and having a glazed or unglazed ceramic finish.

Ashlar. Squared or cut block of stone, usually of rectangular dimensions. Also a flat-faced surface, generally square or rectangular, having sawed or dressed beds and joints.

Ashlar facing

1. Facing of a faced or veneered wall composed of solid rectangular units usually larger in size than brick, having sawed, dressed, or squared beds, and mortar joints.

2. Facing composed of solid rectangular units of burned clay or shale, or natural or cast stone.

Ashlar line. Main line of the surface of a wall of the superstructure.

Ashlar masonry

1. Masonry composed of rectangular units of burned clay or shale, or stone, generally larger in size than brick and properly bonded, having sawed, dressed, or squared beds, and joints laid in mortar. Often the unit size varies to provide a random pattern (random ashlar).

2. Masonry composed of cut stone with proper bond.

Autoclave. Special metal cylindrical chamber used to rapidly cure concrete block.

Autogenous healing. Ability of mortar with a high lime content to heal itself if the masonry cracks. Moisture is necessary for this healing to occur.

Bat

1. Piece of brick; the $4\frac{1}{2}$-in. end of a brick or broken brick.

2. Broken piece of brick, generally a half a brick.

Bearing plate. Steel plate laid in nonshrink mortar on a masonry wall to distribute the load placed on the wall.

Bearing stone. Stone that is capable of withstanding weight, thrust, or strain placed on it.

Benches. Bricks in the part of the kiln next to the fire that are generally baked to vitrification.

Block. Solid unit of masonry material formed in a uniform size.

Body brick. Best brick in the kiln. Brick that is baked hardest with the least distortion.

Bonding cement. White portland cement and glass fiber mixture that eliminates the use of mortar in the erection of concrete block walls. Material troweled to a thickness of $\frac{1}{8}$ in. on each side of the exposed surface of stacked block.

Brick

1. Structural unit of burned shale, clay sand-lime, or concrete, usually solid and about $8 \times 3\frac{3}{4} \times 2\frac{1}{2}$ in. in size.

2. Solid masonry unit having a shape approximately of a rectangular prism, usually not larger than $4 \times 4 \times 12$ in. Brick may be made of burned clay or shale, of fire clay or mixtures thereof, of lime and sand, of cement and suitable aggregates, or of other approved materials.

3. Small building unit, solid or cored not in excess of 25%, commonly in the form of a rectangular prism, formed from clay or shale, and hardened by heat.

4. Material of construction in small, regular units, formed from inorganic substances, and hardened in a shape approximating a rectangular prism; approximately $8 \times 3\frac{3}{4} \times 2\frac{1}{4}$ in. in size, the net cross-sectional area of

which in any plane parallel to the bearing surface is not less than 75% of its gross cross-sectional area measured in the same plane.

5. Solid masonry unit of clay or shale, formed into a rectangular prism while plastic, and burned or fired in a kiln.

Brick ashlar. Walls with ashlar facing backed with bricks.

Brick made from clay or shale. Building brick of clay or shale of a quality at least equal to that required by ASTM. Designation C 62 or C 216. When in contact with the ground, brick should be of at least Grade MW. Where severe frost action occurs in the presence of moisture, brick should be at least Grade SW.

Brick made from sand-lime. Building brick made from sand-lime should be of a quality at least equal to that required by ASTM. Designation C 73. When in contact with the ground, brick should be of at least Grade MW. Where severe frost action occurs in the presence of moisture, brick should be at least Grade SW.

Building brick. Brick made from clay or shale and fired to incipient fusion. Covers three grades of brick based on resistance to weathering. Intended for use in brick masonry (ASTM C 62).

Building (common) brick. Brick for building purposes, not especially treated for texture or color. Formerly called "common brick."

Building or structural unit. Unit that includes measures of durability, strength, and other structural properties, but not requirements affecting appearance.

Build-in items in masonry. Bolts, anchors, nailing blocks, inserts, flashing, lintels, etc.

Bullnose. Fabricated brick with a convex corner on one edge.

Burning (firing) of refractories. Final heat treatment in a kiln, to which refractory brick and shapes are subjected in the process of manufacture, for the purpose of developing bond and other necessary physical and chemical properties.

Calcining of refractory materials. Heat treatment to which raw refractory materials are subjected, preparatory to further processing or use, for the purpose of eliminating volatile chemically combined constituents and producing volume changes.

Calcium chloride. Chemical added to fresh mortar to retard freezing.

Calcium silicate face brick. Brick made principally from sand and lime and intended for use in brick masonry.

Cargon-ceramic refractory. Manufactured refractory comprised of carbon (including graphite) and one or more ceramic materials, such as fireclay and silicon carbide (ASTM C 73).

Casting. Fabricated, molded masonry units in a panel, ready for installation in a wall.

Cast stone. A quality at least equal to that required by A.C.I "Specifications for Cast Stone." All cast stone should be branded with a permanent identification mark of the manufacturer which should be registered.

Cell. An air space or void in a hollow clay tile unit or concrete block.

Cement. Chemically bonding material mixed with sand and/or gravel.

Ceramic color glaze. Opaque, colored glaze of satin or glossy finish obtained by spraying the clay body with a compound of metallic oxides, chemicals, and clays and

then burning them at high temperatures, fusing the glaze to body, making them inseparable.

Ceramic veneer. Architectural terra cotta, characterized by large face dimensions and thin sections.

Chase. Groove or shaft in a masonry wall provided for the accommodation of pipes, ducts, or conduits.

Chrome brick. Refractory brick manufactured substantially or entirely of chrome ore.

Chrome-magnesite brick. Refractory brick, which may be burned or unburned, manufactured substantially of a mixture of refractory chrome ore and dead-burned magnesite in which the chrome ore predominates by weight.

Cinder block. Concrete block, fabricated with cinders as an aggregate.

Cinder brick. Bricks fabricated in brick sizes, using cinders as an aggregate.

Clastic. Comglomerate stone made up of other particles of stones, as in granite.

Clay

1. Stone particles smaller than those that constitute silt.
2. Earthy or stony mineral aggregate consisting essentially of hydrous silicates of alumina, plastic when sufficiently pulverized and wetted, rigid when dry, and vitreous when fired to a sufficiently high temperature.

Clay building brick. All building brick made of burned clay or shale should conform to the requirements of ASTM C 62.

Clay masonry unit. Building unit, usually larger in size than a brick, composed of burned clay, shale, fireclay, or mixtures thereof.

Clay mortar mix. Finely ground clay used as a plasticizer for masonry mortars.

Clear ceramic glaze. Same as ceramic color glaze except that it is translucent or slightly tinted, with a glossy finish.

Clear ceramic glazed finish. Translucent or tinted glaze of lustrous finish compounded of metallic oxide, chemicals, and clays thoroughly ground together and sprayed on a previously formed body. Units are burned at high temperatures, fusing the glaze to the body and making them inseparable.

Cleavage plane. Cleavage that takes place in a masonry unit, or splitting in various directions.

Clinker brick. Very hard-burned brick whose shape is distorted or bloated due to nearly complete vitrification.

Clip. Portion of a brick cut to length.

CMU. Concrete masonry unit.

Coarse aggregate. Stones sized in diameter over $\frac{1}{4}$ in.

Cobble. Naturally rounded stone having a diameter ranging between 2.5 and 10 in.

Common brick. Natural brick made without texture or coloring. Used primarily in walls not exposed to the public, in commercial construction, but also used in the construction of economically built houses.

Concrete block. Hollow or solid block, made from portland cement and aggregates.

Concrete building brick. Solid concrete masonry unit, approximately a rectangular prism, usually not larger than $4 \times 4 \times 12$ in. (ASTM C 55).

Concrete masonry unit

1. Building unit made with such aggregate as sand, gravel, crushed stone, air-cooled slag, coal cinders, expanded shale or clay, expanded slag, volcanic cinders, pumice, and scoria. In some localities the term "concrete block" has been used to designate only those units made with

sand and gravel or crushed stone aggregates. Generally speaking, however, concrete block refers to a hollow concrete masonry unit, usually $8 \times 8 \times 16$ in. in dimension, made with aggregate.

2. Building unit or block larger in size than $12 \times 4 \times 4$ in. made of cement and suitable aggregates.

Concrete sand-lime building brick. Brick made from sand-lime should conform to requirements of the standard specification of ASTM C 73.

Conglomerate. Rounded fragments of sedimentary rock cemented into a mass.

Coping. Material or masonry units forming a cap or finish on top of a wall, pier, pilaster, chimney, etc., to protect masonry below from penetration of water from above.

Cored brick. Brick with three holes or voids in a single row near the center of the brick, or two rows of five holes. The amount of holes or voids varies with each manufacturer.

Corner block. Concrete masonry unit with a flat end for construction of the end or corner of a wall.

Corrosion of refractories. Destruction of refractory surfaces by the chemical action of external agencies.

Corrugated metal veneer ties. Ties not less than $\frac{7}{8}$ in. wide, of coated sheet steel not lighter than 0.0299 in. in thickness (22 gauge) nor 6 in. in length. Two 10-gauge coated wire ties may be used in lieu of corrugated metal ties.

Construction units. Units used in construction, such as brick, tile, blocks, and stone.

Cramps. Galvanized metal bars or straps bent at the ends to enter holes or voids in construction units to hold the units in place.

Cross-sectional area. Net cross-sectional area of a masonry unit is taken as the gross cross-sectional area minus the area of cores or cellular space. The gross cross-sectional area of scored units should be determined to the outside of the scoring, but the cross-sectional area of the grooves are not to be deducted from the gross cross-sectional area to obtain the net cross-sectional area.

Crushed stone. Stone resulting from compression of larger stone by means of mechanical devices.

Cubing. Assembling of concrete masonry units into cubes, after curing, for storage and delivery. Cube normally contains 6 layers of 15 to 18 blocks ($8 \times 8 \times 16$ in.) or an equivalent volume of other sized units.

Cull. Brick rejected as being below specified grade.

Curing (concrete masonry units). Atmospheric pressure steam curing; method of curing concrete masonry units, using steam at atmospheric pressure, usually at temperatures of 120°F to 180°F. Also called "low-pressure steam curing."

Cyclopean aggregate. Stones individually weighing more than 100 lbs.

Decorative blocks. Various types of concrete masonry units with beveled face shell recesses, which provide a special architectural appearance in wall construction.

Deformed bar. Reinforcing bar conforming to the Standard Specifications for Minimum Requirements for Deformations of Deformed Steel Bars for Concrete Reinforcement. Bars not conforming to these specifications are classed as "plain bars." (ASTM A 615, A 616, A 617).

Diaspore clay. Rock consisting essentially of diaspore bonded by fire clay.

Dry-press brick. Brick formed in molds under high pressures from relatively dry clay (5 to 7% moisture content).

Dry-press process bricks. Process in which a mixture of 90% clay and 10% water is mixed and formed into steel molds to form bricks under high pressure.

Economy brick. Brick whose nominal dimensions are $4 \times 4 \times 8$ in. Also called "jumbo brick."

Engineered brick. Brick whose nominal dimensions are $3.2 \times 4 \times 8$ in.

Exterior masonry. Building of exterior masonry construction if all enclosing walls are constructed of masonry or reinforced concrete with fire-resistive ratings as set forth in the code.

Face brick. Brick laid on the face of a wall. Face brick is also commonly known as "hard-burned brick."

Faced block. Concrete masonry units having a special ceramic, glazed, plastic, polished, or ground face.

Faced wall. A wall in which the masonry facing and backing are so bonded as to exert common action under load. A facing unit should be not less than nominal 4 in. in wall thickness.

Face shell. Side wall of a hollow masonry unit (also clay tile).

Facing. Any material forming a part of a wall used as a finished surface.

Facing brick. Brick made especially for facing purposes, often treated to produce surface texture. Such brick may be made of selected clays or treated to produce a desired color.

Facing brick (solid masonry unit). Brick made from clay, shale, fire clay, or mixtures of these and fired to incipient fusion (ASTM C 216).

Facing tile. Tile for exterior and interior masonry with exposed faces.

Fine aggregate. Stones sized less than $\frac{1}{4}$ in diameter.

Fire brick. Brick made of refractory ceramic material that will resist high temperatures.

Fireclay. Earthy or stony mineral aggregate that has as its essential constituent hydrous silicates of aluminum, with or without free silica, and is plastic when sufficiently pulverized and wetted, rigid when subsequently dried, and of suitable refractoriness for use in commercial refractory products.

Fireclay plastic refractory. Fireclay material tempered with water and suitable for ramming into place to form a monolithic furnace lining that will attain satisfactory physical properties when subjected to the heat of furnace operation.

Fireproofing. Material or combination of materials protecting structural members to increase their fire resistance.

Fireproofing tile. Tile designed for protecting structural members against fire.

Fire-resistive material. Noncombustible material.

Flat-bar or wire anchors with dovetails. For use with embedded slots or inserts, coated flat sheet steel not lighter and 0.0598 (16 gauge) $\times \frac{7}{8}$ in. wide, corrugated or turned up $\frac{1}{4}$ in. at the end or with a $\frac{1}{2}$-in. hole located within $\frac{1}{2}$ in. of the end; or No. 6 gauge coated steel wires looped and closed.

Flint fireclay. Hard or flint-like fireclay occurring as an unstratified massive rock, practically devoid of natural plasticity, and showing a conchoidal fracture.

Floor brick. Smooth dense brick, highly resistant to abrasion, used as finished floor surfaces.

Fluted block. Concrete block made with projected vertical ribs on the face of the block, used for obtaining textured walls.

Furring brick. Hollow brick large enough to bond. It is grooved to afford a key for plastering.

Furring tile. Tile used for lining the inside of walls, providing a plaster base and an air space between the plaster and the wall. Furring tile is non-load bearing and carries no superimposed load.

Gauged brick. Brick which has been ground or otherwise produced to accurate dimensions, or a tapered arch brick.

Glass block. Hollow, translucent blocks in various patterns and designs, made by fusing two halves of pressed glass together to create a partial vacuum and producing an R-factor of 1.79 in standard square units that equals the thermal efficiency of a 12 in. concrete wall.

Glazed building units. Should conform to the structural requirements for building brick of clay or shale, and glazed structural tile should conform to the structural requirements for structural clay tile.

Grade and types of brick. Brick subject to the action of weather or soil, but not to frost action when permeated with water, should be grade MW or SW, and brick subject to temperatures below freezing while in contact with the soil should be grade SW (MW—mild weather, SW—severe weather) ASTM C 216.

Grade LB (load bearing) structural tile. Tile suitable for general use in masonry where it is not exposed to frost action, or for use in exposed masonry where it is protected with a facing of 3 in. or more of stone, brick, terra cotta, or other masonry.

Grade LBX (load-bearing exposure) structural tile. Tile suitable for general use in masonry construction and adapted for use in masonry exposed to weathering, provided it is burned to the normal maturity of the clay. Such tile may also be considered suitable for the direct application of stucco.

Grade MW (mild weather) brick. Brick intended for use where it will be exposed to temperatures below freezing but unlikely to be permeated with water, or where a moderate and somewhat nonuniform degree of resistance to frost action is permissible.

Grade NW (normal weather) brick. Brick intended for use as backup or interior masonry. When water is in contact with a surface of a dry unit, the tendency is for the water to enter the unit by capillary suction. If there is enough water and the time of contact is sufficiently long, the water will strike through from face to face, giving a degree of saturation equaling or exceeding that resulting from a 24-hour submersion in water at room temperature.

Grade SW (severe weather) brick. Brick intended for use where a high degree of resistance to frost action is desired and the exposure is such that the brick may be frozen when permeated with water.

Granule. Stone having a diameter ranging between 0.07 and 0.15 in.

Gravel. Mixture of sand, granules, and pebbles, generally used for bedding purposes in building construction.

Green brick. Brick in its soft, pliable state before burning in the kiln.

Green brick work. Brickwork in which the mortar has not had time to set.

Grog fireclay mortar. Raw fireclay mixed with calcined fireclay or broken fireclay brick, or both, all ground to suitable fineness.

Ground fireclay. Fireclay or a mixture of fireclays that have been subjected to no treatment other than grinding or weathering, or both.

Ground stone. Stone pulverized from larger stone by means of mechanical devices.

Grout

1. Cementitious component of high water-cement ratio, permitting it to be poured into spaces within masonry walls. Grout consists of portland cement, lime, and aggregate and is often formed by adding water to mortar.

2. Mixture of cementitious material and aggregate to which sufficient water is added to produce pouring consistency without segregation of the constituents.

3. Mortar of a consistency that will flow or pour easily without segregation of the ingredients.

4. Thin, soupy mixture of cement, sand, and water.

Gypsum masonry. Form of construction made with gypsum blocks or tile in which the units are laid and set in gypsum mortar. No gypsum masonry should be used in any bearing wall or in any location where the gypsum will be directly exposed to weather or where subject to frequent or continuous wetting.

Gypsum units. Gypsum partition tile or block.

Hand-burned brick. Brick that receives the proper amount of controlled burning in the kiln.

Hard brick. Brick with a very dense composition and a very low absorption rate.

Hard burned. Term applied to nearly vitrified clay products that have been fired at high temperatures. These products have relatively low absorptions and high compressive strengths.

High-suction brick. Special kiln-burned brick that absorbs moisture rapidly.

Hollow clay tile. For exterior walls and bearing walls should meet the requirements for the "5-15" clay tile given in the ASTM "Standard Specifications and Tests for Structural Clay Load Bearing Wall Tile." The exterior shell of such tile should be not less than three-fourths of an inch ($\frac{3}{4}$ in.) thick, except that a tolerance of one-sixteenth of an inch ($\frac{1}{16}$ in.) will be permitted in such shell thickness. Hollow clay tile for non-bearing partitions, fire protection and furring, should meet the requirements of ASTM "Standard Specifications for Structural Clay Non-Load-Bearing Tile." Hollow Clay Tile for floor construction should meet the requirements of the "Standard Specifications for Structural Clay Floor Tile" (ASTM).

Hollow load-bearing concrete masonry units. Masonry units made from portland cement, mineral aggregates, and water with the possible inclusion of other materials (ASTM C 90).

Hollow masonry unit

1. Masonry unit whose net cross-sectional area in any plane parallel to the bearing surface is less than 75% of its gross cross-sectional area measured in the same plane.

2. Masonry unit having a core area greater than 25% of the total cross-sectional area of the unit.

Hydrated lime. Quicklime to which sufficient water has been added to convert the oxides to hydroxides.

Hydraulic mortar. Mortar used for masonry work under water.

Identification of brick. All building brick should be of distinctive design or appearance, or marked so that the manufacturer can be identified.

Insulating firebrick. Type of brick used primarily for lining certain kinds of industrial furnaces (ASTM C 155).

Jamb block. Concrete block especially formed with a slot for holding wood blocking or rough frames for the finished windows and door frames.

Jumbo brick. Generic term indicating a brick larger in size than the standard. Some producers use this term to describe oversized brick of specific dimensions manufactured by them.

Kiln run. Brick or tile from one kiln that has not been sorted or graded for size or color variation.

King closer. Brick cut diagonally to have one 2-in. end and one full-width end.

Lean lime. Lime that slakes slowly and does not yield much putty; a hydraulic lime.

Lean mortar

1. Mortar that is deficient in cementitious components, usually harsh and difficult to spread.
2. Mortar that does not adhere to the trowel, generally because of the presence of two much sand.

Light hard. Red brick, not the hardest in the kiln although suitable for supporting moderate loads; not able to withstand alternate freezing and thawing without damage.

Lime. The base of mortar and the result of limestone burned in a kiln until the carbon dioxide has been eliminated.

Lime putty

1. Hydrated lime in plastic form ready for addition to mortar.
2. Slaked lime without sand or cement.

Lintel

1. Horizontal structural member supporting part of a structure above an opening in a wall or partition.
2. Permanent horizontal support over an opening that may be curved or straight on the top.

Lintel block. Concrete block having the cell walls removed so that they can be used for reinforced, grout-filled lintels.

Load-bearing tile. Tile for use in masonry walls carrying superimposed loads.

Magnesite brick. Refractory brick manufactured substantially or entirely of dead-burned magnesite.

Magnesite chrome brick. Refractory brick, which may be burned or unburned, manufactured substantially of a mixture of dead-burned magnesite and refractory chrome ore in which the dead-burned magnesite predominates by weight.

Marble veneers. Class "A" marble should be hard, sound marble and free of any unsound lines. It should be of uniform thickness and sizes.

Masonry-bonded hollow wall. Hollow wall built of masonry units in which the inner and outer wythes of the wall are bonded together with masonry units.

Masonry cement

1. Hydraulic cement for use in mortars for masonry construction (ASTM C 219).

2. Mill-mixed mortar to which sand and water must be added.
3. Portland cement and other materials premixed and packaged, to which sand and water are added to make mortar (ASTM C 387).

Masonry mortar. Mortar used in masonry structures.

Masonry of hollow units

1. Masonry of hollow tile, concrete tile or blocks, or gypsum tile or blocks.
2. Masonry consisting wholly or in part of hollow masonry units laid contiguously in mortar.

Masonry reinforcing. Vertical steel reinforcing bars placed in the cells of concrete masonry units, and the cells filled with concrete grout.

Masonry rubble. Masonry composed of rough irregular shaped stones.

Masonry unit. Natural or manufactured building unit of burned clay, stone, glass, gypsum, etc.

Masonry units. Masonry unit whose net cross-sectional area in any plane parallel to the bearing surface is less than 75% of its gross cross-sectional area measured in the same plane.

Metal anchors and ties. Anchors and ties made of non-corrosive metal or coated with a corrosion-resistant metal plating.

Modular masonry unit. Unit whose nominal dimensions are based on the 4-in. module.

Mortar

1. Plastic mixture of one or more cementitious materials, sand, and water.
2. Plastic mixture of cementitious material, fine aggregates, and water used to bond masonry or other structural units. Mortar of pouring consistency is termed grout.
3. Plastic mixture used to fill the joints between bricks or masonry units.

Mortar colors. Mortar colors consist of inorganic compounds used in the proportions recommended by the manufacturer, but in no case exceed 15% of the weight of the cement, except that carbon black should not exceed 3% of the weight of the cement.

Mortar for unit masonry. Several types of mortars for use in the construction of unit masonry structures (ASTM C 270).

Multiwythe wall ties. Cavity wall ties or prefabricated welded joint reinforcements with No. 9 gauge, or larger, longitudinal and cross-tie wires. Cross ties are spaced not more than 16 in. apart.

Natural finish. Facing tile units having unglazed or uncoated surfaces burned to the natural color of the material used in forming the body.

Natural stone. Stone used in masonry that is sound, free from friable inclusions, and having sufficient strength, durability, and resistance to both impact and abrasion for the proposed use.

Neat cement. Cement and water mixed without any aggregates.

Nodular fireclay. Rock containing aluminous or ferruginous nodules, or both, bonded by fireclay.

Noncombustible. Term applied to material that will neither ignite nor actively support combustion in air at a temperature of 1200°F when exposed to fire.

Nonlustrous finish. Finish compounded of chemicals or clays, or both, thoroughly ground or mixed together and sprayed or blown on a previously formed body. Units are then burned at high temperatures, fusing the finish to the body and making them inseparable. This finish is used to cover the faces of facing tile units.

Norman brick. Brick whose nominal dimensions are $2.66 \times 4 \times 12$ in.

Open end block. Concrete block with an end web removed for placing the block around vertical steel reinforcing.

Packaged dry mortar. Mortar materials ready for mixing with water for general masonry uses, patching, and stucco work. (ASTM C 387).

Partition tile

1. Tile for use in the construction of non-load-bearing combination walls.
2. Tile designed for use in interior partitions.

Paving brick. Vitrified brick especially suitable for use in pavements where resistance to abrasion is important.

Pea gravel grout. Grout to which pea gravel is added.

Pebble. Naturally rounded stone having a diameter ranging between 0.15 and 2.15 in.

Place brick. Underburned, soft brick, used only for walls that are to be plastered and that will not be subjected to any loads.

Plain solid brick masonry. Form of construction made with brick without metal reinforcement.

Plastic or bond fire clay. Fireclay of sufficient natural plasticity to bond nonplastic materials.

Portland cement. Hydraulic cement produced by pulverizing clinker consisting of hydraulic calcium silicates usually containing one or more of the forms of calcium sulfate as an interground addition (ASTM C 150).

Precast concrete lintels. Lintels used over door and window openings. For modular window and door openings, they are designed with an offset on the underside.

Pressed brick. Brick pressed in the mold by mechanical means before it is baked.

Poultice. Paste made with a solvent and an inert material, used to remove objectional stains on brickwork.

Quicklime. White powder that remains after limestone has been burned at a high temperature in a kiln (ASTM C 5).

Random rubble

1. Masonry composed of irregular rough-shaped stones well-bonded and brought at irregular vertical intervals to discontinuous but approximately level beds or courses.
2. Masonry composed of roughly shaped stones laid without regularity of coursing, but well bonded and fitted together to form well-defined joints.

Reinforcement. Structural steel shape, such as a steel bar or rod, wire fabric, or expanded metal embedded or encased in masonry in such a manner that it works with the masonry in resisting forces.

Rich mix. Mortar with an excess of bonding material.

Rip block. Concrete block less than full size in height, generally used as a starting piece.

Rock. Compositional unit of solid mineral matter found in a natural state.

Rock face block. Concrete block made in a mold that resembles the design of a stone wall.

Roman brick. Brick whose nominal dimensions are $2 \times 4 \times 12$ in.

Rough or ordinary rubble

1. Masonry composed of unsquared field stones laid without regularity of coursing but well-bonded.
2. Masonry composed of rough nonshaped rock or field stones laid without regularity of coursing.

Rubble—aggregate. Stones individually weighing not more than 100 lb.

Rubble masonry. Masonry composed of roughly shaped stones.

Rough or ordinary rubble. Masonry composed of unsquared or field stones laid without regularity of coursing but well bonded.

Salmon brick

1. Relatively soft underburned brick, so named because of its color.
2. Brick that is softer than light hard and suitable for fire stopping.

Salt and pepper brick. Brick that has black marks on the finish, caused by iron spots in the clay.

Salt glaze. Gloss finish obtained by thermochemical reaction between silicates of clay and vapors of salt or chemicals.

Salt-glazed finish. Lustrous finish for facing tile units produced by salt and chemicals applied to a previously formed body at high temperatures.

Sand lime brick. Brick made from sand, lime, and cement.

Sand-struck brick. Bricks that have a sand finish because the mold is lubricated with sand to keep the brick from sticking to the mold.

Sash block. Special made concrete block that has a slot in the end to receive the flanges of a metal window frame.

SCR brick. Brick whose nominal dimensions are $2.66 \times 6 \times 12$ in. and that lays up three courses to 8 in. and produces a nominal 6-in.-thick wall. Developed by the Brick Institute of America.

SCR building panel. Prefabricated, structural ceramic panels, approximately $2\frac{1}{2}$ in. thick.

Secondhand masonry materials. May be used in masonry when such materials conform to the provisions of the code for corresponding new materials, provided that such secondhand materials are sound and free from defects that would impair their suitability for reuse and have been cleaned of old mortar and other adherent coatings that would prevent proper assembly or bond, and their use has been approved by the building department.

Selects. Bricks accepted as the best after culling.

Silica. Mineral contained in clay used for making bricks.

Silica fireclay. Refractory mortar consisting of a finely ground mixture of quartzite, silica brick, and fireclay of various proportions.

Single-block wall reinforcing (ladder type). Fabricated reinforcement designed to be embedded in the horizontal mortar joints of masonry walls, consisting of two parallel side rods with cross rods flush welded at 15 in. on center, thus forming a ladder configuration. Overall measurements (side rod to side rod) are approximately 2 in. less than the nominal thickness of the wall. Cross rods extend beyond the side rods approximately $\frac{1}{8}$ in., thereby forming two additional mortar locks.

Single-screened ground refractory material. Refractory material that contains its original gradation of particle sizes resulting from crushing, grinding, or both, and from which particles coarser than a specified size have been removed by screening.

Slump block. Concrete masonry unit produced so it "slumps," or sags before it hardens, for use in masonry wall construction.

Soap. Brick or tile of normal face dimensions having a nominal 2-in. thickness.

Soft burned. Term applied to clay products which have been fired at low-temperature ranges, producing relatively high absorptions and low compressive strengths.

Soft-mud brick. Brick produced by molding relatively wet clay (20 to 30% moisture), often a hand process. When the insides of molds are sanded to prevent the sticking of the clay, the product is sand-struck brick. When the molds are wetted to prevent sticking, the product is water-struck brick.

Solar screen tile. Tile manufactured for masonry screen construction, in various designs, finishes, and sizes.

Solid masonry unit

1. Masonry unit whose net cross-sectional area in every plane parallel to the bearing surface is 75% or more of its gross cross-sectional area measured in the same plane. Net cross-sectional area is taken as the gross cross-sectional area minus the area of the cores or cellular space. Gross cross-sectional area is determined to the outside of the scoring, but the cross-sectional area of the grooves are not considered part of the area of the coring and are not to be deducted from the gross cross-sectional area to obtain the net cross-sectional area.

2. Masonry unit having a core area less than 25% of its total cross-sectional area.

Split-face block. Solid or hollow concrete masonry unit that is machine fractured (split) lengthwise after hardening to produce a rough, varying surface texture.

Splitrock. Solid block of a coarse stone aggregate where the split face is used as a surface of a finished wall.

Standard brick. Brick with the nominal dimensions of 8 in. $\times$ 4 in. $\times$ 2-$\frac{2}{3}$ in.

Stiff-mud brick. Brick produced by extruding a stiff but plastic clay (12 to 15% moisture) through a die.

Structural clay backup tile—non-load bearing. Backup tile designed for use in the construction of load-bearing and non-load-bearing composite walls of brick or other masonry units and tile, in which the facing units are bonded to the backing and the superimposed load is supported by both the facing and backing. Backup tile includes header tile and stretcher units.

Structural clay facing tile. Load-bearing facing tile made from clay, shale, fire clay, or mixtures of these materials (ASTM C 212).

Structural clay fireproofing tile—non-load bearing. Fireproofing tile used for protecting structural members, particularly steel girders, beams, and columns, against fire.

Structural clay floor tile. Tile suitable for use in flat or segmented arches or in combination tile and concrete ribbed-slab construction.

Structural clay furring tile—Non-load bearing. Furring tile used for lining the inside of walls to provide a plaster base and an air space between plaster and wall.

Structural clay partition tile—non-load bearing. Partition tile designed for use in the construction of non-load-bearing interior partitions or for backing up non-load-bearing composite walls (ASTM C 56).

Structural clay tile. Hollow masonry unit composed of burned clay, shale, fireclay, or mixtures thereof and having parallel cells.

Structural clay wall tile—load bearing. Wall tile that is designed for use in the construction of exposed or faced load-bearing walls. Facing may consist of stucco, plaster, or other materials, but the tile is designed to carry the entire superimposed load, which may include the weight of the facing materials.

Structural facing tile. Tile units designed for use in interior and exterior unplastered walls and partitions and used for the construction of both load-bearing and non-load-bearing walls.

Structural facing unit. Structural or building unit designed for use where one or more faces will be exposed in the finished wall and for which specifications include requirements on color, finish, and other properties affecting appearance.

Structural glass block. Structural glass block should have unglazed surfaces to allow adhesion on all mortared faces.

Terra cotta. Hard-burned clay, usually molded into shapes for ornamentation of structural surfaces.

Texture. The surface appearance of the face of the brick.

Type FBA face brick (architectural). Brick manufactured and selected to produce characteristic architectural effects resulting from nonuniformity in size, color, and texture of the individual units.

Type FBS face brick (standard). Brick for general use in exposed exterior and interior masonry walls and partitions where wider color range and greater variation in size are permitted than are specified for type FBX.

Type FBX face brick (extra grade). Brick for general use in exposed exterior and interior masonry walls and partitions where a high degree of mechanical perfection, narrow color range, and minimum permissible variation in size are required.

Type FTS structural-grade facing tile (standard). Smooth-or rough-textured facing tile suitable for general use in exposed exterior and interior masonry walls and partitions, and adapted for use where tile of moderate absorption, moderate variation in face dimension, and medium color range may be used and minor defects in surface finish, including small handling chips, are not objectionable.

Type FTX structural clay facing tile (extra grade). Smooth facing tile suitable for general use in exposed exterior and interior masonry walls and partitions, and adapted for use where tile low in absorption, easily cleaned, and resistant to staining is required, and where a high degree of mechanical perfection, narrow color range, and minimum variation in face dimensions are desired.

Type I masonry cement. Cement used in masonry construction where high strength is not required (ASTM C 91).

Type II masonry cement. Cement for general use where mortars for masonry are required (ASTM C 91).

Unburned clay bricks. Unburned clay brick should conform to the requirements specified in the code.

Veneer wall tie. Strip or piece of noncorroding metal used to tie a facing veneer to the backing.

Wall faced. Substantial facing of a wall with a material that is bonded to the backing so as to become a part of and contribute strength to the wall as a whole.

Wall tie

1. Bonder or noncorroding metal piece that connects wythes of masonry to each other or to other materials.

2. Noncorroding metal band used to tie tiers of brick together or to tie the junction of two pieces of a wall, as, for example, at corners, angles, and toothing and backing.

Water-struck brick. Bricks that have a smooth finish because the molds are lubricated with water to keep the bricks from sticking to the molds.

Web. The cross walls in hollow concrete blocks.

Wire-cut brick. Brick having two of its surfaces formed by wires cutting the clay before it is baked.

Wire-mesh ties. Ties made from a minimum of 20-gauge, $\frac{1}{2}$-in. mesh, galvanized fabric, 4 in. wide.

Wire ties. Ties used for high-lift grouted reinforced brick masonry, made from 0.1483-in. nominal-diameter (9-gauge) steel wire, bent into a stirrup 4 in. wide and 2 in. shorter than the overall wall thickness. Two ends of such ties should meet in the center of one embedded end of the stirrup.

Wire track. Metal wire joint reinforcement used to strengthen or tie the masonry units together to form a composite wall. The reinforcement is usually placed at every second course.

Z-tie. Galvanized metal tie used to tie two tiers of masonry units together.

SLATE

Arris. Slight rounded edge to remove sharpness at exterior corners.

Buckingham slate. Slate of blue-black grade A unfading color. Large sizes are obtainable in natural cleft, sand-rubbed, or honed surfaces.

Buckingham or Virginia slate. Blue-black slate produced at several quarries in Buckingham County, Virginia. A product of Buckingham-Virginia Slate Corporation, Richmond, Va.

Caulking. Nonstaining polysulfide base type usually applied in accordance with manufacturer's instructions.

Cleaning. After slate is set (at least 14 days), it is thoroughly scrubbed with an approved detergent or cleaning agent and rinsed with clean water. On interior wall facings, especially of honed finish, the common practice is to rub down Pennsylvania slate until dry with a mixture of linseed oil and turpentine.

Clear. Term applied to Pennsylvania slate, meaning without ribbons or other markings.

Clear stock (Pennsylvania). Grade is characterized by a complete absence of ribbons. However, it may have small natural markings in the form of irregular spots, disks, or veinings.

Colors

1. Natural Buckingham unfading blue-black color variations occur only from natural markings in the slate or from the reflective sheen and shadow values of the graining of the natural cleft textures and cleavage planes.

2. Pennsylvania slate is available in one color only, gray, ranging in color value from a blue-gray to a gray-black.

Exposure. Shortest distance between exposed edges of overlapping slate shingles.

Finishes

1. The face finish of all exterior panels is natural cleft. Sand-rubbed and honed finishes are available. All exposed edges are honed to remove saw marks and darken the edge color. It is desirable to remove the arris of all exposed edges. The back surface of panels with exposed edges is gauge rubbed to give a better straight-line visual effect. Gauge rubbing the back surface of panels is also necessary for thin-set applications and other tight-fitting situations.

2. Pennsylvania structural slate is supplied in three standard finishes: natural cleft, sand-rubbed, and honed.

Fissility. Tendency shown in slate to separate into thin laminae.

Flagstone

1. Sandstone that splits readily into blocks suitable for flagging purposes.

2. Thin slabs of stone used for flagging or paving walks, driveways, patios, etc. Generally fine-grained sandstone, bluestone, quartzite, or slate, but thin slabs of other stones may be used.

Flagstone flooring. Flooring, usually of 1-in. thickness, from rectangularly shaped slate with sawed edges ranging from 1 to 4 sq ft per piece. Joints are at least $\frac{1}{2}$ in.

Gauging. Rubbing back of slate to obtain uniform thickness.

Grade A. Slate having unfading qualities, as recognized and defined by the National Bureau of Standards.

Graduated roof. Roof on which slates are arranged so that the thickest and longest occur at the eaves and gradually diminish in size and thickness toward the ridges.

Head lap. Shortest distance from the lower edge of an overlapping slate shingle to the upper edge of the unit in the second course below.

Honed

1. Finish similar to sand-rubbed finish, but using approximately 120 grit; it is very smooth and marks easily if used for flooring.

2. Finish is equivalent to approximately 120 grit in smoothness. It is semipolished without excessive sheen.

Installation. When laid on a mortar setting bed, stones could vary in thickness from $\frac{3}{4}$ to 1 in. When laid on a well-compacted sand or loam base, stones should vary from 1 to $1\frac{1}{2}$ in.

Kited. Cut on an angle; stair treads, for example.

Mechanical properties (Buckingham-Virginia slate). Compressive strength averages about 23000 psi. Modulus of rupture (shear) averages about 12000 psi.

Mechanical properties (Pennsylvania slate). Compressive strength averages from 9000 to 16000 psi. Flexural strength averages from 6000 to 10000 psi. Shear strength averages from 3000 to 3625 psi.

Natural cleft. Texture obtained by cleavage of slate, varying from $\frac{1}{16}$-in. deviation to 1 ft to $\frac{1}{4}$ in. per linear foot. Some texture variation can be expected, especially in lengths over 2 ft. In general, Pennsylvania has the most uniform natural cleft texture, Vermont next, and Virginia the most texture.

Natural cleft (Pennsylvania). Naturally split or cleaved face, moderately rough with some textural variation.

Nosing. Rounding edges, usually exposed edges of treads, etc.

Pennsylvania slate

1. Gray-black slate produced at several quarries in eastern Pennsylvania, near Bangor, Pen Argyl, and Slatington. Much of the Bangor area slate is used for roofing. Pen Argyl slate is used for structural purposes, blackboards, and flagging. Slatington area material is used for structural purposes and blackboards. Both ribbon slate and clear slate are quarried at most locations.

2. Dark gray slate only. Pennsylvania slate is either clear, that is, without ribbon, or with ribbon (a ribbonlike line across the face of the slate from $\frac{1}{2}$ to $1\frac{1}{2}$ wide). The clear is recommended for exterior use.

Physical properties (Buckingham-Virginia slate). Density averages about 180 pounds per cubic foot, impervious to moisture, acid and alkali resistivity .001%.

Planer. Knife-shaved surface. Customarily used on underside of treads, sills, and other slabs to facilitate setting and anchoring. The marks of the shaving knife usually are noticeable.

Pointing. Slate is pointed with a $1:2$ mixture of grout—one part portland cement, two parts No. 1 mason sand. If pointed with a tool, grout can be fairly dry. If pointing is by the pouring method, a wetter mix is required, and a tool is used for pointing after setting has commenced. Pointing is always done on the same day that the slate is installed to make joints and setting bed monolithic.

Quartzitic sandstone. Sandstone with a concentration of quartz grains and a siliceous cement.

Ribbon

1. Line of bedding or a thin bed appearing on the cleavage surface in slate. Sometimes it is of a different color from the slate.

2. Ribbonlike band of mica on the face of Pennsylvania slate.

Ribbon stock. Grade of slate distinguished by the presence of ornamental, integral bands or "ribbons" that usually are darker in color than the rest of the slate.

Roofing slate. Produced from natural slate and cut into square and rectangular shingles. Three grades of shingles are made, based on the length of service to be provided (ASTM C 406).

Sand rubbed

1. Surface is rubbed to remove all natural clefts, using the equivalent of a 60-grit abrasive.

2. Surface shows slight grain or stipple in an even plane, no natural cleft texture remains. Finish is equivalent to 60 grit and is obtained by wet sand placed on a rubbing bed.

Sculping. Fracturing the slate along the grain, that is, across the cleavage in the direction of the dip.

Semirubbed

1. Surface is obtained by rubbing to remove high spots, leaving approximately 50% natural cleft.

2. Surface is sand rubbed to remove high spots on the natural cleft. Approximately 50% natural cleft finish remains.

Sills. Sills are generally 2 in. in thickness. Slip sills of sand-rubbed slate are cut to detail.

Slate

1. Dense, very fine-textured soft rock, which is readily split along cleavage planes into thin sheets and which cannot be reduced to a plastic condition by moderate grinding and mixing with water.

2. Very fine-grained metamorphic rock derived from sedimentary rock shale. Slate is characterized by an excellent parallel cleavage entirely independent of the original bedding, by which cleavage, the rock may be split easily into relatively thin slabs.

3. Fine-grained metamorphic rock with a perfect slaty cleavage. Mineral composition usually cannot be determined by the unaided eye.

4. Metamorphic equivalent of shale; a hard gray, red, green, or black fine-grained rock with slaty cleavage.

5. Metamorphosed clay or shale, usually with a well-developed cleavage that is at right angles to the pressure that aided in the metamorphism.

6. Microcrystalline metamorphic rock composed of minerals of micaceous habit oriented in parallel positions so that the rock consists of very thin plates that impart a perfect rock cleavage to the mass and permit it to be split easily into thin, tough sheets.

Slater's nails. Nails designed for the purpose of attaching slate shingles.

Spandrel. Panel of wall between adjacent structural columns or between a windowsill and the window head immediately below the sill.

Standard roof. Roof covered with slate shingles of $\frac{3}{16}$-in. nominal thickness.

Structural slate. Slate primarily used for general building and structural purposes. Slate must be sound, free from spalls, pits, cracks, or other defects that would impair its strength or durability (ASTM C 629).

Textural roof. Sloping roof covered with slate shingles of various sizes, thicknesses, textures, and colors to achieve special architectural effects.

Texture. Textures range from smooth to rustic.

Treads and coping. Treads and coping are usually $1\frac{1}{4}$ in. thick with sand-rubbed top and face edges and $\frac{1}{8}$-in. arris on upper edge.

Unfading. Practically no weathering or fading.

Vein. A more or less regular mineral mass consisting of quartz, with or without calcite. Its presence is objectionable in a slate quarry.

Vermont slate. Colored slate quarried in Vermont and New York. Much of the unfading slate, that is, green, purple, and mottled green and purple, is quarried near Fair Haven, Vt. Weathering slates (used mostly for roofing) are quarried near Poultney, Vt., and Granville, N.Y.

Warping. Generally, a condition experienced only in thin flagging or flagstone materials; very common with flagstone materials that are taken from the ground and used in their natural slate. To eliminate warping in stones, it is necessary to finish the material further by, for example, machining, sand rubbing, honing, or polishing.

Weathering. The process of weathering effects some of our domestic slate. Causes are chemical changes in certain mineral impurities and the physical effect arising from the chemical weathering (ASTM C 406).

Wind rock. Slate unfit for commercial purposes.

WALLS, PARTITIONS, AND ARCHES

Abutment. Skewback and the masonry that it supports.

Airspace. Cavity or space in the wall, or between building materials or construction units.

Apron wall. Part of a panel wall between the windowsill and wall support.

Anta. Pier produced by thickening a wall at its termination.

Arcade. Series or range of arches with their supporting columns or piers.

Arch buttress. Called a "flying buttress"; an arch springing from a buttress or pier.

Area wall. Masonry surrounding or partly surrounding an area; the retaining wall around basement windows below grade.

Bearing partition. Interior bearing wall one story or less in height.

Bearing wall. Wall that supports any vertical load in addition to its own weight.

Brick cavity wall. A wall where a space is left between inner and outer tiers of brick. The space may be filled with insulation.

Brick masonry. Masonry of burned clay or shale brick, sand-lime brick, or concrete brick as required in the code.

Brick masonry-bearing walls. Should have a thickness of at least $\frac{1}{20}$ of their unsupported height or width, whichever is shorter, unless otherwise required in the code.

Brick veneer. Facing of brick laid against the frame or rough masonry wall construction.

Brick veneer wall. Usually used to describe a wall made up of brick veneer applied over wood framing.

Buttress. Projecting part of a masonry wall built integrally therewith, supported on proper foundations, and designed to furnish lateral stability.

Cavity wall (core wall)

1. Wall built of masonry units or of plain concrete, or of a combination of these materials, so arranged as to provide an air space within the wall (with or without insulating material), and in which the inner and outer wythes of the wall are tied together with noncorrosive metal ties.

2. Wall built of masonry units arranged to provide a continuous airspace 2 to 3 in. thick. Facing and backing wythes are connected with rigid noncorrosive metal ties.

3. Wall consisting of two walls separated by a continuous airspace and securely tied together with noncorroding metal ties embedded in the mortar joints. Ties, rectangular in shape, are made from No. 6 gauge wire and placed every 16 in. vertically and every 32 in. horizontally. When weepholes are required at the bottom of cavity walls, approved flashings are installed to keep any moisture that might collect in the cavity away from the inner wall. Weepholes can be formed by placing well-greased sash cord or rubber tubing in the hori-

zontal mortar joints and pulling them out after the mortar has hardened. To keep the cavity clean, a 1×2 in. board is laid across a level of wall ties to catch mortar droppings. The board can then be raised, cleaned, and laid in the wall at the next level.

Cavity wall reinforcing (ladder type). Fabricated ladder type masonry reinforcing designed for cavity-veneer walls. Consists of four parallel side rods with cross rods flush welded at 15 in. o.c. Two of the side rods act as reinforcing in the backup wythe and the other two side rods and cross rods act as a continuous tie and reinforce the other wythe.

Cavity wall tie

1. Rigid, corrosion-resistant metal tie that bonds two wythes of a cavity wall, made of steel $\frac{3}{16}$ in. in diameter and formed in a Z shape or rectangle.

2. Tie, $\frac{3}{16}$ in. diameter coated steel, rectangular, at least 2 in. wide with ends lapped or Z-shaped with 2 in. legs. Cavity wall ties are of such length as to provide 1 in. minimum mortar cover to ends or legs.

Common brickwork. Wall built of the most economical classes of brick, where appearance is not an important consideration.

Composite wall. Bonded wall with wythes constructed of different masonry units.

Composite wall reinforcing (parallel type). Fabricated reinforcement designed to be embedded in the horizontal mortar joints of composite and veneer masonry walls. Consists of three parallel side rods with cross rods flush welded at 15 in. o.c. Two of the side rods act as backup masonry reinforcement, and the third wire and the cross rods act as a continuous tie and reinforcement for the facing.

Composite wall reinforcing (truss type). Fabricated reinforcement designed to be embedded in the horizontal mortar joints of masonry walls. (Consists of two or more parallel side rods welded to a continuous, diagonally formed cross rod forming a truss design with alternating welds not exceeding 8 in. on center. Overall measurement (side rod to side rod) is approximately 2 in. less than the nominal thickness of the wall.

Constant cross-section arch. Arch whose depth and thickness remain constant throughout the span.

Contemporary bearing wall. Concept that requires floors and walls to work together as a system, each giving support to the other. Building of high strength, in which the structure provides finish, closure, partition, sound control, and fire resistance.

Curtain wall

1. Nonbearing wall. Built for the enclosure of a building, it is not supported at each story.

2. Nonbearing wall between columns or piers that is not supported by girders or beams but is usually supported at the ground level.

3. Exterior nonloading wall not wholly supported at each story. Such walls may be anchored to columns, spandrel beams, floors, or bearing walls but are not necessarily built between structural elements.

Dwarf wall. Wall or partition that does not extend to the ceiling. Also called "stub wall."

Elevation. The front view of a building; the building facade. Any of the other views of a building.

Enclosure wall. Enclosure wall may be exterior nonbearing wall in skeleton frame construction. Anchored to columns, piers, or floors but is not necessarily built between columns or piers nor wholly supported at each story.

Exterior wall

1. Outside wall or vertical enclosure of a building, other than a party wall.

2. Wall, bearing or nonbearing, that is used as an enclosing wall for a building, but that is not necessarily suitable for use as a party wall or fire wall.

Face. Front or visually exposed surface of a wall.

Faced wall. Wall in which the masonry facing and the backing are of different materials and are so bonded as to exert a common reaction under load.

Fire division wall. Wall that subdivides a building so as to resist the spread of fire. It is not necessarily continuous through all stories to and above the roof.

Fire wall. Wall of incombustible construction that subdivides a building or separates buildings to restrict the spread of fire, and that starts at the foundation and extends continuously through all stories to and above the roof, except where the roof is of fireproof or fire-resistive construction and the wall is carried up tightly against the underside of the roof slab.

Fixed arch. Arch whose skewback is fixed in position and inclination. Plain masonry arches are, by nature of their construction, fixed arches.

Flat arch. Arch whose soffit is in approximatley a level plane.

Flying buttress. Detached buttress or pier of masonry at some distance from a wall connected to the wall by an arch or part of an arch.

Gothic or pointed arch. Arch, with relatively high rise, whose sides consist of arcs of circles, the centers of which are at the level of the spring line. The Gothic arch is often referred to as a "drop," "equilateral" or "lancet" arch, depending on whether the spacings of the centers are, respectively, less than, equal to, or more than the clear span.

Hollow-bonded wall. Wall built of masonry units with or without any airspace within the wall, and in which the facing and backing of the wall are bonded together with required mortar.

Hollow wall

1. Wall of masonry so arranged as to provide an airspace within the wall between the inner and outer parts (wythes) of the wall.

2. Wall built of masonry units arranged to provide an airspace within the wall. The separated facing and backing are bonded together with masonry units.

3. Wall built of masonry units so arranged to provide an airspace within the wall, and in which the inner and outer parts of the wall are bonded together with masonry units or noncorrosive metal ties.

Interior wall or partition

1. Wall entirely surrounded by the exterior walls of a building.

2. Wall, either bearing or nonbearing, other than exterior, fire, or party walls.

Intrados. Concave curve that bounds the lower extremities of the arch. The distinction between soffit and intrados is that the intrados is linear, while the soffit is a surface.

Jack arch. Arch having horizontal or nearly horizontal upper and lower surfaces. Also called "flat" or "straight" arch.

Keystone. Last wedge-shaped stone placed in the crown of an arch; regarded as binding the whole.

Major arches. Arches with spans in excess of 6 ft and rise-to-span ratios greater than 0.15.

Masonry-bonded hollow wall. Wall built of masonry units installed as to provide an air space within the wall, and in which the inner and outer wythes of the wall are tied together with masonry units properly set.

Masonry crosswalls or pilasters. Construction which may be omitted on hollow concrete masonry bearing walls 12 in. or more in thickness where such walls are supported horizontally by floors or roofs at heights not exceeding 18 times the wall thickness, or as required by code.

Masonry wall. Bearing or nonbearing wall of hollow or solid masonry units.

Masonry walls below grade. Masonry walls that are in contact with the soil should be of sufficient strength and thickness to resist the lateral pressure from the adjacent earth and to support their vertical loads without exceeding the allowable stresses. The minimum thickness for masonry walls below grade should be 4 in. greater than the required thickness for the walls of the supported structures except that 12 in. walls will be accepted for buildings not more than 2 stories in height if substantial lateral support consisting of masonry walls, offsets or pilasters are provided at intervals of not to exceed 20 ft, or as required by code.

Masonry veneer. Masonry veneer on wood frame structures should be securely attached to the backing by corrosion-resistant corrugated metal ties, not less than No. 22 gauge in thickness and $\frac{7}{8}$ in. in width or equivalent. One tie should be used for at least each 2 sq ft of wall area and the distance between ties should not exceed 24 in. or by No. 13 gauge metal ties or equivalent located 36 in. horizontally and 18 in. vertically, or as required by code.

Minimum thickness of masonry-bearing walls. Walls may be decreased except for walls below grade, and the height or length to thickness ratio may be increased when data is submitted to the building department, which may take under consideration a justification in the reduction of the requirements in the code.

Minor arches. Arches with spans that do not exceed 6 ft and with maximum rise-to-open ratios of 0.15.

Multicentered arch. Arch whose curve consists of several arcs of circles that are normally tangent at their intersections.

Nonbearing masonry walls. All exterior nonbearing walls, if constructed with one wythe of brick to the weather may be backed with SW or MW classified clay or shale brick, concrete masonry units or clay tile conforming to the requirements of the code. If such walls are built of concrete masonry units or clay tile, such units should conform to the requirements of the code.

Nonbearing partition. Interior nonbearing wall 1 story or less in height.

Nonbearing wall. Wall that supports no vertical load other than its own weight.

Non-load-bearing tile. Tile designed for use in masonry walls carrying no superimposed loads.

Panel wall

1. Nonbearing wall in skeleton construction, built between columns or piers and wholly supported at each story.

2. Nonbearing wall, built between columns or piers, wholly supported on the structural frame at each floor elevation.

Parapet wall

1. That part of wall entirely above the roof line.

2. Vertical extension of a wall entirely above the roof line.

Partially reinforced masonry walls. Walls designed as plain masonry, except that reinforcement is provided in some portions to resist flexural tensile stresses.

Partition. Interior wall, 1 story or part of a story in height, that is not load bearing.

Party wall

1. Wall on an interior lot line used or adapted for joint service between two buildings.

2. Wall jointly owned and used by two parties under easement agreement or by right in law and erected at or on a line separating two parcels of land, each of which is, or is capable of being, a separate real-estate entity.

3. Wall used in common or adapted for joint use between two adjacent buildings, and which may be a fire wall or a fire division wall as determined by the limiting floor areas for the type of construction.

4. Wall used or adapted for joint service between two buildings and having a fire-resistance rating and structural stability at least equal to that of a fire wall.

5. Fire wall on an interior lot line used, or capable of being adapted for use, as part of two structures.

Perforated wall. Wall that contains a considerable number of relatively small openings. Often called "pierced wall" or "screen wall."

Pier

1. Isolated column of masonry. A bearing wall not bonded at the sides into associated masonry is considered a pier when its horizontal dimension measured at right angles to the thickness does not exceed four times its thickness, or as required by code.

2. Brickwork between two adjoining openings in the same story.

Pilaster

1. Part of a wall that projects not more than one-half of its own width beyond the outside or inside face of a wall, acting as an engaged pier.

2. Wall portion projecting from either or both wall faces and serving as a vertical column and/or beam.

3. Projection of masonry for the purpose of bearing concentrated loads, compensating for the reduction of a wall section by chases, openings, or recesses, or stiffening the wall against lateral forces.

4. Upright architectural member, rectangular in plan, structurally a pier or column, made part of a wall.

5. Portion of a wall that may serve as either a vertical beam or column, or both. In reinforced masonry the pilaster may or may not project beyond either face of the wall.

6. Thickened wall section or column built as an integral part of a wall.

7. Pier projecting from a wall.

Pocket. Chase, recess, or space, left in a wall to receive a beam or to allow for an adjoining wall.

Relieving arch. Arch built over a lintel, flat arch, or smaller arch to divert loads, thus relieving the lower member from excessive loading. Also known as "discharging" or "safety" arch.

Retaining wall. Wall used to resist the lateral displacement of any material.

Rise of an arch. The rise of a minor arch is the maximum height of the arch soffit above the level of its spring line. the rise of a major parabolic arch is the maximum height of the arch axis above its spring line.

Sample panel. Panel approximately 4 ft. long, approximately 3 ft. high, and of the proper thickness, showing proposed color range, texture, bond, mortar joint, and workmanship. The panel becomes the standard of comparison for all masonry work built of the same material and remains at job site until the masonry is completed.

Segmental arch

1. Arch whose curve is circular but less than a semicircle.

2. Arch whose bottom is the arc of a circle.

Semicircle arch. Arch whose curve is a semicircle.

Set. Wide bevel-edged chisel used for cutting brick.

Set in. Amount that the lower edge of a brick on the face tier is back from the line of the top edge of the brick directly below it.

Skewback. Inclined surface on which the arch joins the supporting wall. For jack arches the skewback is indicated by a horizontal dimension.

Solar screen. Perforated wall used as a sunshade.

Spandrel wall. That part of a curtain wall above the top of a window in one story and below the sill of the window in the story above.

Span of arch. Horizontal dimension between abutments. For minor arch calculations the clear span of the opening is used. For a major parabolic arch the span is the distance between the ends of the arch axis at the skewback.

Solid masonry wall. Wall built of masonry units laid with full mortar joints between them and with no type of framing present.

Stiffened masonry walls. Where solid masonry bearing walls are stiffened at distances not greater than 12 ft apart by masonry cross walls or by reinforced concrete floors, they may be of 12 in. thickness for the uppermost 50 ft, measured downward from the top of the wall, and should be increased 4 in. in thickness for each successive 50 ft or fraction thereof, or as required by code.

Stone walls. Rough or random or coursed rubbled-stone walls should be 4 in. thicker than is required by the code, but in no case less than 16 in. thick.

Straight arch. Arch whose soffit is in approximately a level plane.

Tudor arch. Pointed four-centered arch of medium rise-to-span ratio.

Veneered wall

1. Wall having a facing or veneering of masonry or other weather-resisting noncombustible materials securely attached to the backing, but not so bonded as to exert common action under load.

2. Wall having a facing that is not attached and bonded to the backing so as to form an integral part of the wall for purposes of load bearing and stability.

Voussoir. One of the wedge-shaped masonry units that forms an arch ring.

Wainscot. Lower portion of an interior wall, or the lower portion of the interior surface of an exterior wall, when finished differently from the remainder of the wall. Maximum height of a wainscot above the upper surface of the floor is usually 4 ft.

Wall

1. Vertical, platelike member, enclosing or dividing spaces and often used structurally.

2. Built-up structural element primarily vertical, designed to enclose or subdivide a building or structure.

Wall (common). Vertical separation completely dividing a portion of a building from the remainder of the building and creating in effect a building that from its roof to its lowest level is separate and complete into itself for its intended purpose, such wall being owned by one party, but jointly used by two parties, one or both of whom is entitled to such use by prior arrangement.

Walls below grade. Masonry walls that are in contact with the soil should be of sufficient strength and thickness to resist the lateral pressure from the adjacent earth and to support their vertical loads without exceeding the allowable stresses. The minimum thickness for masonry walls below grade should be 4 in. greater than the required thickness for the walls of the supported structures except that 12-in. walls will be accepted for buildings not more than 2 stories in height if substantial lateral support consisting of masonry walls, offsets or pilasters are provided at intervals of not to exceed 20 ft, or as required by code.

5

Metals

ALUMINUM

Age hardening. Increasing the hardness of an alloy by a relatively low-temperature heat treatment that causes precipitation of components or phases of the alloy from supersaturated solid solution.

Age softening. Loss of strength and hardness at room temperature, which takes place in certain alloys due to spontaneous reduction of residual stresses in the strain-hardened structure.

Aging. Process of allowing a heat-treatable aluminum alloy to remain at room temperature, over a period of time, after heat treatment (heating and quenching), to reach a stable state of increased strength.

Aging or precipitation hardening. Freshly quenched heat-treated material is relatively easy to form for several hours after quenching. Where maximum workability is required, as in deep drawing, material is fabricated immediately after quenching.

Aluminum alloy gains strength and becomes harder as the alloying constituents precipitate from the supersaturated solid solution at room temperature. For many alloys this natural aging results in a stable temper after about 4 days, but for others, the natural aging process would stretch out for years. Precipitation of these alloys is expedited by heating at moderate temperatures, called "artificial aging" or "thermal precipitation."

Alclad plate. Composite plate of an aluminum-alloy cored having on both surfaces (if on one side only, alclad one-side plate) a metallurgically bonded aluminum or aluminum alloy coating that is anodic to the core, thus electrolytically protecting the core against corrosion (ASTM B 209).

Alclad sheet

1. Clad product with an aluminum or aluminum alloy coating of high resistance to corrosion; the coating is anodic to the core alloy it covers, thus protecting it physically and electrolytically against corrosion.

2. Composite sheet produced by bonding either corrosion-resistant aluminum alloy or aluminum of high purity to base metal of structurally stronger aluminum alloy.

Alclad tube. Tube having on the inside surface a metallurgically bonded aluminum or aluminum-alloy coating that is anodic to the core alloy to which it is bonded, thus electrolitically protecting the core alloy against corrosion (ASTM B 221).

Alloy designation. Numerical system used in designating the various aluminum alloys.

Alloy sticks. Cast square bars of aluminum, magnesium, and copper that are added to high-purity zinc to provide the aluminum and magnesium constituents in certain type alloys.

Alumina. Hydrated form of aluminum oxide found in bauxite.

Aluminized steel. Steel coated with an aluminum-iron alloy coating; prepared by dip-coating and diffusing aluminum into steel; resists scaling and oxidation.

Aluminizing. Forming an aluminum or aluminum alloy coating on a metal by hot-dipping, hot-spraying, or diffusion.

Aluminum. Aluminum is not a single metal, but a whole family of metals, both cast and wrought. As the term is ordinarily employed, it includes all the aluminum alloys, as well as commercially pure and superpure aluminum itself. Alloys may contain as much as 20% of such other metals as copper, magnesium, silicon, and zinc.

Aluminum alloy extrusion ingot. Alloy ingot for extruding into rod, tubing, bar, or shapes. Chemical composition is closely controlled. Ingot size and shape are selected for the capacity of the extrusion press.

Aluminum alloy hardener ingot. Aluminum ingot containing elements that are difficult to put into solution in the pure form. The material is called "master alloy" or "hardener." The hardener ingot is used in casting operations to prepare alloys by melting mixtures of the hardener ingot, unalloyed ingot, and other alloying constituents.

Aluminum alloy hardener or master alloy ingot. Certain foundries prefer to buy pure metals and do their own alloying. Some elements are difficult to put into solution in the pure form; so the element is put into the melt by adding aluminum pig containing a high percentage of the element desired. This material is called "master alloy" or "hardened" and is usually produced by adding the alloying element to the virgin aluminum in the reduction pot.

Aluminum alloy ingot (casting or foundry ingot)

1. Aluminum ingot produced in numerous compositions to meet industry standards and provide the variety of characteristics required in aluminum castings. The ingot is a metallurgical combination of aluminum and various alloying constituents cast in the various sizes and shapes required by the foundry industry.

2. Aluminum alloy cast in one of the many ingot forms. The term is used commercially for the many casting alloys produced to specified composition for remelting and casting in a foundry.

Aluminum alloy rolling ingot. Ingot for rolling into sheet, plate, wire, rod, bar, or structural shapes. Chemical composition is closely controlled. Size and shape vary with their requirements of the end use.

Aluminum alloys. Aluminum alloys are divided into two general groups: binary alloys consisting of aluminum and a single controlled alloying element, and composite alloys consisting of aluminum and two or more controlled alloying elements.

Aluminum alloy sheet and plate. Materials used for applications requiring a ductile, light, nonrusting metal.

Aluminum and aluminum alloys. Aluminum and aluminum alloys are cast into convenient shapes for remelting or hot-forming. "Ingot" is the term used for all aluminum and aluminum alloy castings that are produced for remelting in a foundry or rolling or extruding in a mill.

Aluminum brass. Casting brass to which aluminum has been added as a flux to improve the casting qualities and, with the addition of lead, the machining qualities.

Aluminum bronze. Copper-aluminum alloy which may also contain iron, manganese, nickel, or zinc.

Aluminum coating. Film of aluminum applied to a metallic surface by spraying, electrolysis, or hot dipping.

Aluminum common nail. Plain-shank, aluminum alloy with flat head and medium diamond point.

Aluminum extrusion ingot (billet). Primary aluminum alloy ingot of shape (usually cylindrical) and size suitable for use in extrusion presses. An extrusion billet is a fast-chilled, metallurgically homogeneous form manufactured to specific standards. It is not remelted by users but is normally heated to extrusion temperatures and then undergoes severe mechanical work in a press to produce an extruded shape.

Aluminum foil. Foil made by rolling sheet in cold-rolling mills in thickness less than 0.006 in., usually supplied in coil form. Most common applications in the building industry are for thermal insulation and vapor retarders.

Aluminum ingots. Various types of ingot can be cast into any desired size. Ingots are normally supplied to the foundry industry as 50-lb unnotched castings but can be supplied in up to 1200-lb units. Ingots are also produced in 30-lb sizes or smaller, usually notched for ease in breaking. They are produced in various sizes and shapes up to 8000 lb for uses other than in foundries (rolling, extruding, and forging operations). Material is sometimes referred to as "billet."

Aluminum sheet ingot (cast). Primary aluminum alloy ingot of shape (usually rectangular) and size suited for use in rolling mills. Sheet ingot has fast-chilled structure, is cast to specific standards, and is normally reheated before rolling to produce sheet and plate.

Aluminum sheet ingot (rolled). Cast aluminum sheet ingot that has been rolled rather than cast to the size required. The product is then further rolled to produce sheet and plate.

Aluminum shot. Small round particles of 1/2-in. diameter or less, made by pouring molten aluminum through openings in a screen and allowing it to fall through the air into water. The material is used for deoxidizing steel and for alloying in the die casting industry.

Aluminum T-ingot. Direct-chilled primary ingot cast in long lengths. Dimensions and weights are uniform and there are no shrinkage depressions. The T-ingot is designed for ease of stacking and handling by forklift truck equipment and is available in pure grades of aluminum.

Aluminum tubing-bars and shapes. Manufactured products suitable for anodizing most color finishes.

Aluminum unalloyed ingot. Ingot cast from primary aluminum that has been reduced from alumina by the electrolytic process, refined, and cast in ingot form.

Angularity. Conformity to, or deviation from, specified angular dimensions in the cross section of a shape or bar.

Annealed foil. Foil completely softened by thermal treatment.

Annealing

1. Process in which an alloy is heated to a temperature of 600°F to 800°F and then slowly cooled to relieve internal stresses and return the alloy to its softest and most ductile condition.

2. Aluminum alloys that have been strain-hardend or strengthened by solution heat treatment can be returned to a dead-soft or fully annealed condition by heating to a specified temperature, with controlled cooling.

Anodes. Cast aluminum products designed for cathodic protection of ships, barges, pipelines, piers, and tanks, as well as other iron and steel structures subject to salt water exposure.

Anodic coating

1. Surface coating applied to an aluminum alloy by anodizing.

2. Electrochemical process that produces on an aluminum surface a film of hard aluminum oxide that increases resistance to atmospheric corrosion and abrasion. The film provides an excellent coating for aluminum, accurately reproducing the texture of the underlying surface, also provides an excellent base for paint.

Anodic color. Anodic color is produced from anodic oxide films that derive color from the aluminum alloys themselves. Color introduced into the anodic coating by dyes. Many coloring agents presently available are subject to noticeable fading under prolonged exposure to sunlight and are not recommended for exterior use.

Anodize. Apply an electrolytic oxide coating to an aluminum alloy by building up the natural surface film using an electric current (usually direct current) through an oxygen-yielding electrolyte with the alloy serving as the anode.

Anodizing. Characteristic rated on lightness of color, brightness, and uniformity of a clear anodized coating, applied in a sulfuric acid electrolyte.

Anodizing sheet. Sheet of suitable constitution and surface quality for the application of protective and decorative films by an anodic oxidation process.

Artificial aging. Heating of the aluminum alloy for a controlled time at an elevated temperature to accelerate and increase its strength gain after heat treatment.

Bar. Solid section that is long in relation to its cross-sectional dimensions, having a completely symmetrical cross section that is square or rectangular (excluding flattened wire) with sharp or rounded corners of edges; a regular hexagon or octagon, whose width or greatest distance between parallel faces is 3/8 in. or greater.

Bauxite. Raw ore of aluminum consisting of 45 to 60% aluminum oxide, 3 to 25% iron oxide, 2.5 to 18% silicon oxide, 2 to 5% titanium oxide, other impurities, and 12 to

30% water. The ore varies greatly in the proportions of its constituents, color, and consistency.

Bayer process

1. Process generally employed to refine alumina from bauxite.

2. Process for extracting alumina from bauxite ore before the electrolytic reduction. Bauxite is digested in a solution of sodium hydroxide, which converts the alumina to soluble aluminate. After the "red mud" residue has been filtered out, aluminum hydroxide is precipitated, filtered out, and calcined to alumina.

Bending. Temper and thickness greatly affect the minimum bend radius that can be used in forming. Certain aluminum alloys can be cold-formed to a sharp 90° bend; others must be formed on stipulated radii.

Beryllium. Beryllium has the important ability to reduce drossing during the pouring of the high-magnesium-content alloys (4% or more magnesium). This prevents loss of magnesium through burning out in melting operations. Only small amounts of beryllium are required, usually controlled to less than 0.01%.

Billet. Solid, semifinished, round or square product that has been hot-worked by forging, rolling, or extrusion.

Binary aluminum-copper alloys. These alloys are limited in number because the addition of nickel and/or magnesium to the aluminum and copper makes a more useful material.

Binary and aluminum-magnesium alloys. Alloys consisting of an excellent combination of mechanical and chemical properties. They are especially resistant to corrosion and tarnish, being superior to practically all other common aluminum casting alloys in this respect, and exceed the aluminum-silicon alloys in their resistance to marine atmosphere and mildly alkaline solutions.

Blank. Piece of metal cut or formed to regular or irregular shape for subsequent processing such as by forming, bending, or drawing.

Blanking. Frequently the first stage in a multistage die operation, using a punch press. In other operations a stack of blanks may be cut by a band saw or a router.

Blister. Raised spot on the surface of the metal caused by expansion of gas in a subsurface zone during thermal treatment.

Blocker-type forging. Die forging made in a single set of die impressions to the general contour of a finished part (ASTM B 247).

Block marks. Short, longitudinal scratches introduced during rolling, usually on the reeling equipment, by relative movement between adjacent wraps of the coil.

Bloom. Semifinished hot-rolled product, rectangular or square in cross section, produced on a blooming mill.

Bond blister. Blister that occurs at the interface between the coating and core in clad products. This type of blister is evident only on that surface of the sheet nearest the interface in question.

Boron. Boron provides grain-refining action when used with titanium. Without the benefits of boron, the grain-refining effects of titanium are reduced in remelting. Boron also improves tensile strength and ductility. While the amounts required are small, boron content is carefully controlled and seldom exceeds 0.01% in any of the common aluminum casting alloys.

Boss. Knoblike projection on the main body of a forging or casting.

Bottom draft. Taper or slope in the bottom of a forged depression to assist the flow of metal toward the sides of the depressed area.

Bow. Longitudinal curvature.

Brazed tube. Tube formed from sheet and fastened at the seams by brazing.

Brazing

1. Welding process where the filler metal is a nonferrous metal or alloy with a melting point higher than 1000°F, but lower than that of the metals joined.

2. Non-heat-treatable wrought alloys produced as sheet and plate are brazed either by torch, dip, or furnace methods. High-strength joints are obtainable even in very thin material. The process has little or no effect on the corrosion resistance of the material.

Brazing alloy. Alloy used as filler metal for brazing; certain types of aluminum-alloy materials.

Brazing aluminum. Method achieved by furnace, dip, or torch methods.

Brazing rod. Rod (rolled, extruded, or cast) for use in joining metals by brazing.

Brazing sheet

1. Nonclad or specially clad sheet for brazing purposes, with the surface of the specially clad sheet having a lower melting point than the core. Brazing sheet of the clad type may be clad on either one or two surfaces.

2. Special type of sheet that has a cladding that melts at a lower temperature than the core. Assemblies of parts made from this sheet can be brazed together by subjecting them to a temperature of about 1100°F, which does not melt the core alloy but only the cladding, joining the surfaces.

Brazing wire. Wire for use in joining metals by brazing.

Brinell hardness. Measure of resistance to indentation, obtained by applying a load through a ball indenter and measuring the permanent impression in the material. The hardness value is obtained by dividing the applied load in kilograms by the spherical area of the impression in square centimeters. In testing aluminum alloys, a load of 500 kg is applied to a ball of 10 mm in diameter for 30 seconds (ASTM E 10).

Broken edge. Edge containing cracks, splits, or tears.

Broken matte finish. An uneven finish on the matting surfaces of pack-rolled foil.

Broken surface

1. Surface of sheet running normal to the rolling direction and perpendicular to the sheet surface.

2. Surface showing surface fracturing, generally most pronounced at sharp corners of extrusions; the surface of a drawn product with minute cracks on normal to the direction of drawing.

Buckle. Distortion, such as a bulge, wave, or twist, causing the sheet to deviate from flat.

Buffing. Light polish produced by use of fine abrasives applied by cloth wheels running at high speed.

Burr. Thin edge or roughness left by sawing, shearing, or slitting.

Burred edge. Thin turned-down edge on a sheet or foil resulting from shearing.

Bus bar. Bus conductor of rectangular or square cross section of any dimension.

Bus conductor. Rigid electric conductor of any section, usually rectangular or square bar or tube, channel, or angles.

Butt-welded tube. Tube formed from sheet by placing one edge or end against the other and joining by welding.

Cast. Facsimile of a forging obtained by pouring plaster or a low-melting-point metal into a die cavity.

Casting. Pouring molten metal into molds to form desired shapes.

Cast products. Products manufactured in the foundry including sand castings, permanent-mold castings, and die castings.

Caustic stain. Superficial etching of the surface by caustic.

Center. There is a difference in thickness between the center and edges of the sheet.

Center buckle. Wavy condition in the center of a sheet in combination with flat edges.

Chatter. Uneven surface on extrusions or drawn products, usually formed by vibration on the metal during extrusion or drawing.

Chemical compositions. Chemical compositions depend on the alloy composition percentage limits considered standard by the industry, which the alloys are guaranteed to meet.

Chemical finishes

1. Chemical finishes produce surfaces not otherwise possible. Parts are usually submerged in a chemical solution to produce the desired reaction, and this limits the size of the castings that can be treated. Chemical finishes are decorative and corrosion resistant, but they do not have any resistance to abrasion and can be scratched or chipped off with careless handling.

2. Chemical finishes consist of three main types, used for decorative effect: caustic etching, acid etching, and chemical polishing. Etched surfaces are frosty or matte, while chemically polished surfaces are highly reflective. Chemically finished surfaces are subject to corrosion and must be protected by anodizing or lacquering.

Chemical properties. Properties of a material that describe its reactions with other substances.

Chop. Metal sheared from a vertical surface of a die forging that is spread by the die over an adjoining horizontal surface.

Chromium. Chromium is added to reduce stress cracks or stress corrosion in certain casting alloys, such as aluminum-zinc-magnesium alloy. In certain alloys it may be added to improve strength at elevated temperatures.

Chucking lug. Lug or boss added to a forging so that on center machining and forming may be performed with one setup or chucking. The lug is finally machined or cut away.

Circle. Circular blank fabricated from plate, sheet, or foil.

Clad alloys. Alloys having one or both surfaces of metallurgically bonded coating, the composition of which may or may not be the same as that of the core, and which is applied for such purposes as corrosion protection, surface appearance, or brazing.

Clad sheet. Sheet product having on one or both surfaces a metallurgically bonded coating, the composition of which may or may not be the same as that of the core, and which is applied for such purposes as corrosion protection, surface appearance, or brazing.

Class 1 hollow extruded shape. Hollow extruded shape whose void is round and 2 in. or more in diameter, and whose weight is equally distributed on opposite sides of two or more equally spaced axes.

Class 2 hollow extruded shape. Hollow extruded shape, other than Class 1, that does not exceed a 5-in.-diam circumscribing circle and has a single void of not less than 0.375-in. diam or 0.110-in. area.

Class 3 hollow extruded shape. Hollow extruded shape other than Class 1 or 2.

Close tolerance. Special tolerance that is closer than "standard."

Coating. Lacquer-type or wax films applied to one or both surfaces of foil for such purposes as heat-sealing, primer coat base for printing, or as protection against chemical attack.

Coating blister. Blister that occurs in the coating of clad products and is evident only on that side of the sheet containing the blistered cladding.

Coating steak. Streaked condition resulting from rolling on rolls containing bands of roll coating. In some instances when the coating is heavy, it may flake off onto the sheet surface and produce a speckled appearance known as "roll-coating pickup."

Coiled roofing sheet. Standardized coiled sheet material of specific temper, width, and thickness intended for the manufacture of corrugated or V-crimp roofing.

Coiled sheet. Sheet furnished in rolls and coils, with slit edges.

Coiled sheet circles. Coiled sheet cut into circular form.

Coiled sheet stock. Semifinished rough-rolled material, in coiled form, for further rolling into sheets.

Cold-drawing. Drawing products are formed by pulling (drawing) metal through a die. The purpose of cold-drawing is to obtain close dimensional tolerances and a better surface finish.

Cold-finished bar. Bar brought to final dimensions by cold-working to obtain improved surface finish and closer dimensional tolerances.

Cold-finished extruded bar. Cold-finished bar produced from extruded bar.

Cold-finished extruded rod. Cold-finished rod produced from extruded rod.

Cold-finished extruded shape. Cold-finished shape produced from an extruded shape.

Cold-finished products. Products finished to size at room temperature.

Cold-finished rod

1. Rod brought to final dimensions by cold-working to obtain improved surface finish and closer dimensional tolerances.

2. Cold-finished rod produced from hot-rolled rod.

Cold-finished rolled bar. Cold-finished bar produced from hot-rolled bar.

Cold-finished rolled shape. Cold-finished shape produced from a hot-rolled shape.

Cold-finished shape. Shape brought to final dimensions by cold-working to obtain improved surface finish and closer dimensional tolerances.

Cold-heading rod. Rod of a quality suitable for cold-heading applications such as rivets and bolts.

Cold-heading wire. Wire of a quality suitable for cold-heading applications such as rivets and bolts.

Cold-rolling. Forming sheet metal by rolling at room temperature metal previously hot-rolled to a thickness of approximately 0.125 in.

Cold shut

1. Linear discontinuity in a cast surface caused when meeting streams of metal fail to merge prior to solidification.

2. Forging defect developed by metal flowing into a section from two directions, resulting in a discontinuity at the junction.

Cold-working. Forming at room temperature by means of rolling, drawing, forging, stamping, or other mechanical methods of shaping.

Colored anodic coatings. Coatings obtained by dyeing and sealing the film. A wide variety of colors and shades are available for interior use. The number of colors suitable for exterior architectural use is limited.

Coloring. Final polishing process used to produce a high gloss, not to alter the color of the casting. Coloring is accomplished by cleaning the surface with benzene, drying with sawdust, and then using a muslin or flannel wheel and a compound of soft silica and grease to attain the desired gloss.

Combination process. Process used to retrieve additional alumina and soda from the red mud impurities of the Bayer process.

Commercially pure aluminum (99.00 to 99.9% pure). Aluminum as it comes direct from the reduction pot. High-purity aluminum (99.99% purity, minimum) is produced by a refining process. Aluminum alloys are made by adding controlled amounts of the various alloying elements.

Commercial tolerance. Term is sometimes used synonymously with "standard tolerance." In such cases the term "standard tolerance" is preferred.

Common alloy. Alloy that does not increase in strength when heat-treated (non-heat treatable). Common alloys may be strengthened by strain hardening.

Common binary alloys. Common binary alloys include aluminum-copper, aluminum-silicon, and aluminum-magnesium alloys. Typical composite alloys are the aluminum-copper-silicon, aluminum-copper-silicon-magnesium, and aluminum-silicon-magnesium alloys, as well as the aluminum-copper-nickel-magnesium alloy groups.

Concavity. Concave condition applicable to the width of any flat surface.

Concentricity. Adherence to a common center, as in the inner and outer walls of a tube.

Conduit. Tube used to protect low and high voltage electric wiring.

Contact with dissimilar material. Contact between aluminum and dissimilar metals or absorptive materials in the presence of moisture can cause corrosion of the aluminum. It is recommended that drainage of surface water from components of dissimilar metals onto aluminum surfaces be avoided.

Contour. Portion of the outline of a transverse cross section of an extruded shape that is represented by a curved line or curved lines.

Conveyor marks. Scratches and pits caused on one side of the sheet by contact with cables or other means of conveyance through the furnace during conveyor annealing.

Copper. Copper in an aluminum casting reduces its corrosion resistance severely owing to the galvanic reactions set up between the copper-rich particles of the constituent and the aluminum matrix when moisture is present. Thus in damp, salt atmospheres, pitting and corrosion products are the result. Corrosion rates increase rapidly from about 0.3% copper up to 1% copper content. They also increase further with added copper, but at a slower rate.

Core blister. Blister that occurs in the core of clad products.

Corrosion. Deleterious effect on an aluminum surface due to weathering, galvanic action, and direct chemical attack.

Coupon. Piece of metal from which a test specimen may be prepared. Usually produced as an integral extra piece on a casting or forging or as a separately cast or forged piece.

Covering area. Area in square inches covered by one pound of foil or laminated composite.

Cross hatch. Light, broken surface.

Cryolite. Sodium aluminum flouride used with alumina in the final electrolytic reduction of aluminum. Cryolite is found naturally in Greenland and is generally produced synthetically from alum, soda, and hydrofluoric acid.

Design seam. Line juncture resulting from the deliberate bonding of two or more edges by pressure, fusion, or mechanical interlocking.

Die casting. Metal object produced by the introduction of molten metal under substantial pressure into a metal die and characterized by a high degree of fidelity to the die cavity (ASTM B 85).

Die castings. Permanent-mold castings with the metal being forced into the mold under pressure. Die castings have smoother surfaces than permanent-mold castings and permit greater repetitive use of the molds. Die casting is generally used for the mass production of small items.

Die forging. Forging formed to the required shape and size by working in impression dies (ASTM B 247).

Die line. Longitudinal line or scratch resulting from the use of a roughened tool or the drag of a foreign particle between the tool and the product.

Diffusion streaks. Brown colored streaks in copper-bearing clad products, such as alclad 2024, resulting from diffusion of core-alloying constituents to the surface of the coating during thermal treatment.

Direct chemical attack. Corrosion caused by a chemical dissolving of a metal.

Disc. Circular blank fabricated from plate, sheet, or foil from which a central concentric area has been removed.

Draft. Taper on the sides of a die or mold impression to facilitate removal of forgings, castings, or patterns from dies or molds.

Draw dies. Dies made of ductile iron, semisteel, or hardened tool steel for large-scale production. Surfaces contacting the aluminum sheet are highly polished.

Drawing. Process of pulling material through a die to reduce the size, change the cross section or shape, or harden the material.

Drawing stock. Hot-worked intermediate solid product of uniform cross section along its whole length, supplied in coils and in a quality suitable for drawing into wire.

Drawn-in scratch. Scratch that occurs during the fabricating process and is subsequently drawn over; scratch characterized by being drawn over, which makes it relatively smooth to the touch.

Drawn product. Product formed by pulling the material through a die.

Drawn shape. Shape brought to final dimensions by being drawn through a die.

Drawn tube. Tube brought to final dimensions by being drawn through a die (ASTM B 210–B 483).

Drawn wire. Wire brought to final dimensions by being drawn through a die.

Drilled extrusion ingot. Cast solid extrusion ingot that has been drilled to make it hollow.

Dry surface. Foil surface substantially free from oily film and suitable for lacquering, printing, or coating with water-dispersed adhesives.

Ductile. Capable of being drawn out or hammered; able to undergo cold plastic deformation without breaking.

Duct sheet. Coil or flat sheet in specific tempers, widths, and thicknesses, suitable for duct fabrication.

Earing. Characteristic of sheet that causes ears to form when deep drawn or spun.

Ears. Wavy symmetrical projections formed in the course of deep drawing or spinning as a result of directional properties or anisotropy in sheet.

Eccentricity. Deviation from concentricity, difference between the mean wall thickness and minimum (or maximum) wall thickness at any one cross section. Degree of eccentricity can be expressed by a plus and minus wall thickness tolerance.

Edge buckle. Rippled or wavy edge.

Electrical conductivity

1. Capacity of a material to conduct or allow the flow of an electrical current. The international resistivity standard of 10.371 ohms (mil-ft) is equal to 100% conductivity, and conductivity values for aluminum are expressed as percentages of the international resistivity standard for copper.

2. Aluminum has the highest electrical conductivity of all known metals, on the basis of weight, and is used for bus bar conductors and transmission lines.

Electrical finishes. Two types of electrical finishes are possible on aluminum: electrochemical and electroplated. Electrochemical finishing builds up the thickness of the natural aluminum oxide coating, while electroplating deposits a fine layer of another metal on the aluminum.

Electrical metallic tubing. Tubing having standardized length and combinations of outside diameter and wall thickness thinner than those of "rigid conduit," commonly designated by nominal electrical trade sizes and is used with compression-type fittings as a protection for electric wiring.

Electrical resistivity. Electrical resistance of a body of unit length and unit cross-sectional area. The value of 10.371 ohms (mil-ft) at 20°C (68°F) is the resistivity equivalent to the international annealed cooper standard (IACS) for 100% conductivity. This means that a wire of 100% conductivity, 1 ft in length and 1 mil in cross-sectional area, would have a resistance of 10,371 ohms.

Electrochemical finish. Parts to be treated are connected to the anode in an electrolytic bath. Electric current causes oxygen to form on the surface of the casting. Oxygen immediately reacts with the aluminum to form a thick aluminum oxide integral with the metal. The thickness of the coating can be varied widely as can its characteristics. Coatings can be transparent, opaque, or translucent. Formation of the oxide in no way alters the original texture of the surface. A mechanical finish previously applied will remain unchanged, except for the oxide coating. Tiny pores in the oxide coating will absorb dyes and pigments; so an infinite variety of colors can be obtained. Pores also serve to bind corrosion inhibitors and sealers to the aluminum to give improved surface properties to the casting.

Electrolysis. Chemical decomposition by the action of electric current.

Electrolyte. Substance in which the conduction of electricity is accompanied by chemical decomposition.

Electrolytic oxide finishes. Anodic coatings are hard, abrasion- and corrosion-resistant oxide coatings. Those produced in sulfuric acid are thicker and more transparent than films produced in chromic acid. Since these coatings are transparent, the appearance produced by mechanical or chemical finishes is not materially changed by anodizing. Sulfuric-acid-produced coatings vary from 0.0001 to 0.001-in. in thickness, depending on anodizing time. Architectural surfaces that are continuously exposed to weather should have anodic film not less than 0.0008 in. thick. Coatings produced in chromic acid vary from 0.00001 to 0.00009 in. in thickness, depending on anodizing time.

Electroplated finishes. Aluminum can be plated much as other metals are plated. The natural oxide coating on aluminum is first removed, and a zinc coating is produced by immersion in a zincate solution. Other metals can then be plated onto this zinc base without difficulty.

Electroplating

1. Process whereby a casting takes and holds an electroplate applied by the present standard methods.

2. Aluminum alloys are plated with other metals, chiefly for decorative purposes. Zinc, silver, gold, nickel, chromium, and copper are all used to plate aluminum. Silver is used to provide increased surface conductivity, chromium to reduce friction and increase resistance to corrosion by alkalies, and copper to permit assembly by soft soldering. Normally a base coating of zinc is applied to the aluminum by zincating (chemical immersion process) before the other metals are plated.

Electropolishing. Process that produces a highly reflective surface protected by an anodic film.

Elongation. Increase in distance between two gauge marks that results from stressing the specimen in tension to fracture. Original gauge length is usually 2 in. for sheet specimens and round specimens whose diameter is $\frac{1}{2}$ in., or four times the diameter for specimens where that dimensions is under $\frac{1}{2}$ in. Elongation values depend to some extent on the size and form of the specimen.

Embossed foil. Foil on which a pattern has been impressed by means of an engraved roll or plate.

Embossed sheet. Produced by running a commercial mill finish sheet through a pair of rolls having a matched design on their surface; an attractive embossed pattern is rolled into the sheet. Patterns available are diamond, hammered, fluted, leather grain, rib lengthwise, square, stucco, and wood grain. Embossed one-side patterns are pebble grain and fluted.

Embossed tube. Tube whose outside surface has been embossed by rolling, with a design in relief regularly repeated in a longitudinal direction.

Embossing, coining, and stamping. Process easily performed on aluminum using conventional methods and equipment.

Endurance limit. Limiting stress below which a material will withstand an indefinitely large number of cycles of stress. In the case of aluminum alloys, endurance limits are based on 500,000,000 cycles of completely reversed stress using a rotating-beam machine and specimen.

Equivalent round. Diameter of a circle having a circumference equal to the outside perimeter of a given shape or tube.

Etched foil. Foil roughened chemically or electro-chemically to provide an increased surface area.

Expansion and contraction. Aluminum has a comparatively high coefficient of expansion. Architectural aluminum alloys have a coefficient of thermal expansion of 0.000013 per °F per unit of length. For example, a piece of aluminum 10 ft. long will expand or contract 0.166 in. or approximately 5/32 in. in a 100°F temperature change. In building design, provision should be made for expansion and contraction caused by temperature changes. This is especially important where aluminum meets other materials with different coefficients of expansion.

Extrude. Form lengths of shaped sections by forcing plastic material through a hole cut in a die.

Extruded bar. Bar brought to final dimensions by extruding.

Extruded rod. Rod brought to final dimensions by extruding.

Extruded section. Rod, bar, tube, or any other product shape produced by the extrusion process (ASTM B 221).

Extruded shape. Shape brought to final dimensions by extruding.

Extruded structural pipe. Pipe having certain standardized sizes of outside diameter and wall thickness commonly designated by "Nominal Pipe Sizes" and American National Standards Institute (ANSI) schedule numbers (ASTM B 429).

Extruded structural shape. Structural shape brought to final dimensions by extruding.

Extruded structural tube. Hollow product having a round cross section and uniform wall thickness, brought to final dimensions by extruding through a bridge-type die, or by similar methods at the option of the manufacturer (ASTM B 429).

Extruded tube. Tube brought to final dimensions by extruding.

Extruding. Process in which aluminum, heated to a plastic state, is squeezed by hydraulic presses through a die opening. Ejected (extruded) aluminum assumes the same size and cross section as the die opening. Extruded products include shapes of varying cross section, rods, bars, and seamless tubes.

Extrusion. Product formed by pushing the material through an orifice in a die.

Extrusion billet. Solid, wrought, semifinished product intended for further extrusion into rods, bars, or shapes.

Extrusion butt end defect. Longitudinal discontinuity in the extreme rear portion of an extruded product which is normally discarded.

Extrusion defect. Cone-shaped cavity in an extruded product (ring in a hollow shape or tube) formed in the extreme rear portion if extruded too far.

Extrusion ingot. Solid or hollow cylindrical casting used for extrusion into bars, rods, shapes, or tubes.

Extrusion seam. Seam in tube, pipe, or hollow shape resulting from the pressure bonding of two or more edges in the course of extruding through a spider or porthole die.

Fabricating ingot. Cast form suitable for subsequent working by such methods as rolling, forging, or extruding.

Fabrication. Forming and machining operations performed on aluminum products and coatings after they leave the mill or foundry.

Fillet. A concave junction between two surfaces.

Fin. Thin projection on a forging or casting resulting from trimming or from the metal under pressure being forced into hairline cracks in the die or mold or around die or mold inserts.

Finish. Surface appearance of a product.

Finishing. Important characteristic of aluminum is that no protective or decorative finish is required for most building applications. Aluminum does not rust, and under most conditions it does not corrode. In outdoor applications the original bright luster is gradually altered to a soft silvery appearance. If a bright, highly reflective surface is desired permanently, a coating of clear lacquer or wax may be used.

Fin stock. Coiled sheet or foil in specific alloys, tempers, and thickness ranges suitable for manufacture of fins for heat-exchanger applications.

Flag. Marker inserted adjacent to the edge at a splice or lap in a roll of foil.

Flash. Thin protusion at the parting line of a forging or casting which forms when metal, in excess of that required to fill the impressions, is forced between the die or mold interfaces.

Flashless forging. Closed die forging made in dies constructed and operated to eliminate, in predetermined areas, the formation of flash.

Flat roofing sheets. Flat sheets in standardized sizes and of specific temper, width, and thickness intended for the manufacture of corrugated or V-crimp roofing.

Flat sheet. Sheet furnished in rectangular form with sheared, slit, or sawed edges, and which may be flattened by standard industry methods.

Flat-sheet circles. Flat sheet cut into circular form.

Flattened and slit wire. Flattened wire that has been slit to obtain flat edges.

Flattened wire. Solid section having two parallel flat surfaces and rounded edges, produced by flattening round wire by passing it between cylindrical rolls.

Flow lines

1. Lines on the surface of painted sheet, brought about by incomplete leveling of the paint.
2. Line pattern revealed by etching, showing the direction of plastic flow on the surface or within a wrought structure.

Flow through. Forging defect caused when metal flows past the base of a rib, resulting in rupture of the grain structure.

Fluidity. Ability of liquid alloy to flow readily in the mold and fill thin sections at normal pouring temperatures.

Fluted hollow shape. Hollow extruded or drawn shape whose cross-sectional inside periphery is plain, and whose cross-sectional outside periphery has regular, longitudinal, concave corrugations with sharp cusps between corrugations.

Fluted tube. Tube of nominally uniform wall thickness, having regular, longitudinal, concave corrugations with sharp cusps between corrugations.

Foil. Solid sheet section rolled to a thickness of less than 0.006 in.

Foil stock. Semifinished coiled material intended for further rolling into foil (thinner than 0.006 in.).

Forging

1. Metal part worked to a predetermined shape by one or more such processes as hammering, upsetting, pressing, rolling, or other processes selected by the manufacturer (ASTM B 247).

2. Working (shaping) of metal parts by forcing between shaped dies.

Forging plane. A reference plane or planes, normal to the direction of applied force, from which all draft angles are measured.

Forging stock

1. Rod, bar, or other section suitable for subsequent change in cross section by forging.

2. Rod, bar, or other section suitable for further shaping by impact or pressure in a die is called "forging stock." Forging rod is nearly always rolled.

Forging temperatures. All the aluminum forging alloys generally fall in the 750°F to 900°F range. The exact temperature depends on the alloy and type of operation. It should not vary more than ±20°F once the optimum value has been determined.

Formability

1. Characteristic of metals that has no specific scale of reference. Four properties that have a strong influence on formability can be accurately measured and used to evaluate specific formability: hardness, elongation, yield strength, and the spread between yield strength and ultimate strength. Greatest formability is found when the hardness and yield strength are low, elongation is high, and the spread between yield and ultimate strength is large.

2. Aluminum can be fabricated into a variety of shapes by bending, stretching, rolling, drawing, and forging. Formability varies greatly with the aluminum alloy and temper. After a certain stress in the aluminum is reached (varying with the alloy and temper), the metal deforms at an increasing rate. If a load is applied that stresses the metal past this point, the metal takes a permanent set.

3. High-purity aluminum has the best formability. High-strength alloys have less formability. Alloys that contain appreciable amounts of magnesium work harden more rapidly than others and thus require annealing more often in severe forming.

4. 1100 aluminum strain hardens very slowly, and great reductions are possible without intermediate annealing.

Foundry characteristics of an alloy. Foundry characteristics are especially important if a large or intricate casting is involved. The size and shape of the part to be cast determines the required foundry characteristics. For instance, if the casting is to have thin sections, an alloy that has good fluidity must be selected. Moreover, for castings having an intricate shape, together with thin sections, an alloy with both good fluidity and high resistance to hot cracking must be used.

Frac-shot. Spherical product offered in two sizes: 8 to 12 mesh and 12 to 16 mesh in 99+ purity aluminum. Frac-Shot is used as a propping agent in the fracturing process that stimulates oil and gas wells.

Friction scratches. Short, longitudinal scratches introduced during coiling or reeling by relative movement between adjacent wraps of the coil.

Galvanic corrosion. Corrosion produced by electrolytic action between two dissimilar metals in the presence of an electrolyte.

Gauge (gage). Diameter thickness of a solid product or wall thickness of a tubular product expressed in terms of a system of numbers. Gauge is not a measuring device when used in this sense. Dimensions expressed in decimals are usually preferred.

Gouge. Deep scratch or indentation.

Grained aluminum. Very coarse aluminum powder produced by stirring molten aluminum as it freezes, used primarily for deoxidizing steel and exothermic compounds.

Grain size. All metals are crystalline in structure. Crystals are generally referred to as grains. Grain size is a measure of the individual crystal size and is usually reported in terms of grains per unit area (square millimeters) or unit volume (cubic millimeters).

Granules. Aluminum particles in three mesh ranges consisting of 91 to 95% aluminum, with the balance carbon. Used for exothermic reactions, such as "hot tops" in iron and steel foundries.

Hammer forging. Shaping by application of repeated blows as, for example, in a forging hammer.

Hand forging. Forging worked between flat or simply shaped dies by repeated strokes or blows and manipulation of the piece (ASTM B 247).

Hardener. An alloy containing at least some aluminum and one or more added elements for use in making alloying additions to molten aluminum.

Hard foil. Foil fully work-hardened by rolling.

Hardness. Resistance to plastic deformation, usually by indentation.

Heat-exchanger tube. Tube generally designed to be used in apparatus in which fluid inside the tube is heated or cooled by fluid outside the tube. Term is usually not applied to coiled tube or to tubes used in refrigerators or radiators (ASTM B 234).

Heat-treatable alloys

1. Alloys capable of gaining strength by being heat treated. Alloying elements show increasing solid solubility in aluminum with increasing temperature by pronounced strengthening.

2. Alloys have the best machinability in the heat-treated and fully aged temper; non-heat-treatable alloys machine best in the fully hard tempers.

3. Contain elements or constituents that have considerable solid solubility at elevated temperatures and limited solubility at lower temperatures. The heat-treating process used to increase the strength of these alloys consists of two parts: high-temperature-solution heat treatment followed by a drastic quench in a cooling medium, and precipitation or aging treatment at room or slightly elevated temperatures.

Heat treating. Heating and cooling a solid metal or alloy in such a way as to obtain desired conditions or properties. Commonly used as a shop term to denote a thermal treatment to increase strength.

Heat-treat stain. Discoloration of the metal surface caused by oxidation during thermal treatment.

Helical extruded shape. Extruded shape twisted along its length.

Helical-welded tube. Tube formed from sheet and fastened at the seam by welding, with the weld line curved around the tube like an ordinary screw thread.

Herringbone streaks. Elongated, alternate bright and dull markings at an angle to the rolling direction of the sheet. Streaks have the general appearance of a herringbone pattern.

Hollow drawn shape. Hollow shape brought to final dimensions by being drawn through a die.

Hollow extruded shape. Hollow shape brought to final dimensions by extruding.

Hollow shape. Shape, any part of whose cross section completely encloses a void.

Hot-rolling. Shaping of plate metal by rolling heated slabs of metal. Hot-rolling is usually used for rolling metal down to approximately 0.125 in. thickness.

Hot-working. Forming metal at elevated temperatures, at which the metals can be easily worked (for aluminum alloys, usually in the 300°F to 400°F range).

Impact. A part formed in a confining die from a metal slug, usually cold, by rapid single stroke application of force through a punch, causing the metal to flow around the punch and/or through an opening in the punch or die.

Inclusion. Foreign material in the metal.

Ingot

1. Mass of metal cast into convenient shape for storage or transportation, to be later remelted for casting or finishing by rolling, forging, etc.

2. Casting, suitable for working or remelting, that has been poured from either a melting or blending furnace.

Intergrandular corrosion. Corrosion occurring preferentially at grain boundaries, also termed "intercrystalline corrosion."

Interleaving. The insertion of paper between layers of metal to protect from damage or to facilitate shearing and prevent sticking of sheets of foil.

Intermediate temper foil. Foil intermediate in temper between annealed foil and hard foil.

Iron. Iron additions to aluminum are sometimes employed to reduce shrinkage. This element also acts as a grain refiner. However, iron in silicon alloys results in coarse crystals and a brittle structure. Often 0.80% iron is desirable with alloys of 8% or more silicon because it tends to eliminate welding of the dies in pressure castings. Allowable iron content in most other aluminum casting alloys must be controlled and varies from 0.15 to 1.2%, as the mechanical properties of some alloys are materially affected by the iron content.

Kink. Departure from flatness that is great in relation to its extent.

Knock-out mark. Small solid protrusion or circular fin on a forging or a casting, resulting from the depression of a knock-out pin under pressure or inflow of metal between the knock-out pin and the die or mold.

Laminant. The bonding agent used in joining two or more sheets or films.

Lamination. Internal crack aligned parallel to the worked surface of the sheet.

Lap-welded tube. Tube formed from sheet by lapping the edges and joining by welding.

Lateral bow or camber. Deviation of longitudinal edge from straight. The term "lateral bow" is preferred.

Lead. Lead is used as an alloying element because it improves the machinability of the alloy, especially when used along with tin or bismuth. Less than 0.5% is usually ample, the actual amount being carefully controlled

Lip hollow shape. Hollow extruded or drawn shape of generally circular cross section with nominally uniform wall thickness with one hollow or solid protuberance or lip

parallel with the longitudinal axis, used principally for heat-exchange purposes.

Lithographic sheet. Sheet having superior surface on one side with respect to freedom from surface imperfections and supplied with a maximum degree of flatness, for use as a plate in offset printing.

Lock. Condition in which the parting line of a forging is not all in one plane.

Lock-seam tube. Tube formed from sheet with a longitudinal, mechanically locked seam.

Longitudinal bow. Longitudinal curvature in the plane of the sheet.

Lot, heat treat. Material of the same mill form, alloy, temper, section, and size traceable to one heat treat furnace or, if heat treated in a continuous furnace, charged consecutively during an 8-hour period.

Lot, inspection

1. For non-heat-treated tempers, an identifiable quantity of material of the same mill form, alloy, temper, section, and size submitted for inspection at one time.

2. For heat-treated tempers, an identifiable quantity of material of the same mill form, alloy, temper, section, and size traceable to a heat-treated lot or lots and submitted for inspection at one time.

Lubricant residue. The carbonaceous residue resulting from lubricant burned on the surface of a forged part.

Lüder's lines. Surface markings resulting from localized flow, which appear on some alloys after light forming. They lie approximately parallel to the direction of maximum shear stress and appear as depressions when forming is in tension and as elevations when in compression.

Machinability. Aluminum and its alloys possess excellent machinability. Common alloys offer little resistance to cutting, and tool pressure is low. Machinability is improved by cold-working and will be easier to machine to a good finish in the fully hard temper.

Machining aluminum alloys. Aluminum alloys are machined at relatively high feeds and at the highest speeds available on presently used equipment. The rate of machining aluminum alloys is limited only by the capacity of the available machine tools.

Magnesium. Magnesium is one of the most important alloying elements. Like copper, magnesium has the solid solubility characteristics required to make an alloy heat treatable. Aluminum alloys containing over 8% magnesium will respond to heat treatment. Alloys with less magnesium are not heat treatable unless some other alloying elements (such as copper or silicon) are also present.

Manganese. Manganese additions act as grain refiners to reduce shrinkage. When added to both copper and silicon alloys, manganese improves the strength of castings in high-temperature applications. However, manganese must be controlled in combination with iron, or the reverse effects will result; that is, the formation of large particles of primary constituent may result in lower strength properties and no benefits would be derived from the manganese addition.

Mean diameter. Average of two measurements on the diameter taken at right angles to each other.

Mean wall thickness. Average of two measurements on the wall thickness of a tubular product taken opposite each other.

Mechanical finishes

1. Mechanical finishes are used to alter the texture of the aluminum surface. The texture may be changed to provide a more decorative surface or as a treatment

prior to other finishing, such as painting. Grinding, polishing, and buffing result in smoother reflective surfaces. Abrasive blasting gives a rough, matte finish, which can be used as a base for organic coatings or to eliminate minor imperfections in the metal surface. Scratch finishing, mechanical satin finishing, Butler finishing, and spin finishes are scratched-line finishes. These remove minor surface defects while producing distinctive decorative effects. Hammering and burnishing provide other decorative effects. Mechanical finishing removes the original heavy oxide film and leaves only a light film to protect the aluminum from corrosion. For this reason mechanically finished parts are often given a protective coating by anodizing or lacquering.

2. Mechanical methods for cleaning aluminum castings leave a comparatively rough surface. Fine polishing, called "greasing or oiling," is used after rough polishing with a wheel. In fine polishing, the wheel is lubricated with a tallow, oil, beeswax, or similar material to eliminate burning, and a 100- to 200-grit emery is used. The surface may then be refined by buffing. Buffing is done by carrying the abrasive in a heavy vehicle, such as grease, and working it into the surface being finished. The choice of buff and abrasive depends on the finish desired.

3. Aluminum's natural surface can be altered by a variety of mechanical processes—grinding, polishing, buffing, or sandshot-blasting, scratch- or satin-brushing, lubricated compound abrasion, or even hand rubbing with steel wool.

Mechanical properties. Includes specified limits covering tensile ultimate strength, yield strength, and elongation for standard alloys and products.

Mechanical properties of an alloy. The mechanical properties of an alloy will usually determine if the alloy is to be heat-treatable. High-strength requirements necessitate the use of a heat-treatable alloy in order to meet the required mechanical properties. In certain applications special requirements, such as extreme hardness, high impact resistance, or dimensional stability, decide which alloy is to be employed.

Mill finish

1. Uncontrolled finish that may vary from sheet to sheet and within a sheet and that may not be entirely free from stains or oil.

2. Surface condition of a sheet that is finished on rolls that have not been highly polished. Mill finish is relatively uncontrolled within certain limits and varies from a bright to dull surface.

3. Natural finish resulting from the production of the particular form of aluminum involved. Exposed to the weather, this finish becomes dull, gray, and in some atmospheres slightly rough. It will accumulate dirt in relation to the environment. Exposure in an industrial atmosphere, for example, will result in graying down more rapidly than exposure in a rural location. Change in surface appearance does not, for practical purposes, detract from the service life of the aluminum component.

Mill finish foil. Foil having a nonuniform finish which may vary from coil to coil and within a coil.

Mill finish sheet. Sheet having a nonuniform finish that may vary from sheet to sheet and within a sheet, and may not be entirely free from strains or oil (ASTM B 209).

Mill standard sizes. Mill standard sizes are usually produced in large quantities and comprise normal warehouse stock.

Miscellaneous alloying elements. Elements used in aluminum casting alloys include iron, manganese, chromium, nickel, zinc, tin, beryllium, and lead.

Mismatch. Error in register between two halves of a forging or casting caused by opposing die or mold halves not being in perfect alignment.

Modulus of elasticity. Ratio of stress to corresponding strain throughout the range where they are proportional. There are three kinds of stress and three kinds of moduli of elasticity for any material: modulus in tension, in compression, and in shear.

Modulus of rigidity. Modulus of elasticity in shear.

Nick. Surface or edge discontinuity in the form of a slight cut, indentation, or notch.

Nickel. Nickel improves dimensional stability and strength at elevated temperatures and is used in combination with other alloying elements. The addition of 5% nickel (or more) produces high shrinkage. Common aluminum casting alloys employ maximum nickel contents ranging from 0.5 to 3.0%

Nofill. Failure of metal to fill a forging die impression.

Non-heat-treatable alloys

1. Aluminum casting ingots can be identified by groups in specified lists and data according to whether or not the mechanical properties of the material can be improved by heat treatment.

2. Commercially pure 1100 aluminum is more resistant to chemical attack and weathering than any of the other alloys. It is therefore, excellent for chemical processing equipment and other applications where product purity is a factor. Alloy 1100 is very easily worked and is ductile enough to permit deep draws.

3. Alloys not capable of gaining strength by heat treatment but which depend on the initial strength of the alloy or cold-working for additional strength. Also called "common alloys."

4. "Common" or non-heat-treatable alloys contain alloying constituents that remain substantially in solid solution or are insoluble at all temperatures. Group includes high-purity aluminum and the wrought alloys in the 1000, 3000, and 5000 series.

5. Non-heat-treatable alloys contain elements that remain substantially in solid solution or form constituents that are insoluble. Their strength depends on the amount of cold-work introduced after annealing. Strength attained through cold-work is removed by subsequent heating in the annealing range.

Nonmagnetic. Aluminum is nonmagnetic, a property of advantage in the electrical and electronics industries.

Odd-shaped plate blanks. Plate cute into shapes other than circles or rectangles.

Odd-shaped sheet blanks. Sheet cut into shapes other than circles or rectangles.

Off gauge

1. Term applied to thickness outside the specified tolerance.

2. Term applied to deviation of thickness or diameter of a solid product, or wall thickness of a tubular product, beyond the standard or specified dimensional tolerances.

Oil can. Buckle that can be snapped from one position to another. Also referred to as a "snap buckle."

Oil stain. Stain produced by the incomplete burning of the lubricants on the surface of the sheet. Rolling subsequent to staining will change the color (from darker brown to lighter brown down to white).

One-side-bright mill finish. With this finish the sheet has a moderate degree of brightness on one side. The reverse side is uncontrolled and may have a dull nonuniform mill finish appearance (ASTM B 209).

Opaque coatings. For opaque coatings there are many excellent points available. The prime coat should be one of the materials specifically recommended for aluminum.

Open-seam tube. Shape approaching tubular form of nominally uniform wall thickness, but having a longitudinal, unjointed seam or gap of width not greater than 25% of the outside diameter or greatest overall dimension; normally produced from sheet.

Orange peel. Surface roughening encountered in forming products from material with a coarse grain size.

Organic coatings. Organic coatings may be applied by brush, dipping, regular spray, electrostatic spray, hot-spray, flow coating, or roller coating; they may be either air-drying or baking types.

Organic, vitreous, and laminated coatings. Applied finishes on aluminum paint, enamel, lacquer, porcelain enamel, and plastic laminate appear much the same as applied finishes on other materials. Although forming of porcelain-enameled aluminum should not be performed, other applied finishes on aluminum sheet permit forming after application of the finish. Cut edges or accidental surface scratches will not rust, making it possible to expose cut edges to the elements without fear of unsightly staining.

Ovalness. Deviation from a truly circular periphery, usually expressed as the total difference found at any one cross section between the individual maximum and minimum diameter, which usually occurs at or about 90° to each other. Ovalness is not expressed as plus and minus.

Oxide discoloration. Discoloration of the metal surface caused by oxidation during thermal treatment.

Pack marks. Small, densely distributed abrasions on the surface of the sheet resulting from rolling sheets in packs of two or more. Marks occur on the sheet surfaces in contact with each other.

Paint, enamel, and lacquer. In uses where the surface color cannot be applied during processing, it may be desirable to paint or coat the casting with a suitable paint, lacquer, or enamel. Adequate surface treatment is essential if the paint is to adhere to the casting, especially if the coating is to have a protective function.

Panel flat sheet. Sheet which has a higher degree of flatness than a flat sheet.

Parent plate. Plate that has been processed to final temper as a single unit and subsequently cut into two or more smaller plates to provide the required width or length or both (ASTM B 209).

Partial annealing. Thermal treatment given cold-worked metal to reduce the strength to a controlled level.

Patterned or embossed sheet. Sheet product on which a raised or indented pattern has been impressed on either one or both surfaces by the use of rolls.

Permanent casting. Metal object produced by introducing molten metal by gravity or low pressure into a mold

constructed of durable material, usually iron or steel, and allowing it to solidify (ASTM B 108).

Permanent mold castings. Castings made by employing metal molds with a metal core when a large number of castings from the same mold are required. If a sand core is used, the casting is called a semipermanent mold casting. Molten metal is poured into the mold by gravity.

Physical properties of an alloy. The physical properties of an alloy must be considered in special applications where high electrical or thermal conductivity, good corrosion resistance, low thermal expansion, or certain other properties are needed.

Pickup. Small particles of oxidized metal adhering to the surface of a product.

Pinch marks. Elongated markings, generally running in the direction of the grain, resulting from a folding over of the metal during rolling. Such folds occur at the entry side of the rolling mill and are rolled over and smoothed out in the subsequent rolling.

Pinholes. Minute holes in foil.

Pinion hollow shape. Hollow extruded or drawn shape that has regular, accurately spaced, longitudinal serrations on the outside and is round on the inside, used primarily for small gears.

Pipe. Shape having a certain standardized combination of outside diameter and wall thickness commonly designated by "nominal pipe size" and ANSI schedule number.

Pit

1. Sharp depression in the surface of the sheet or plate.

2. Sharp depression in the surface of the metal, resulting in a reduction of the thickness.

Pitting corrosion. Localized corrosion resulting in small pits or craters in a metal surface.

Plate. Solid section rolled to a thickness of 0.250 in. or heavier, in rectangular form, and with either sheared or sawed edges (ASTM B 709).

Plate circles. Plate cut into circular form.

Polishing. Mechanical finishing operation for the purpose of applying a gloss or luster to the surface of a product.

Polishing rating. Composite rating based on the ease and speed of polishing and the quality of the finish provided by typical polishing procedures.

Porcelain enamel finishes. Porcelain enamel finishes can be applied to a wide variety of alloys. Extrusions and sheets of 6061 alloy, as well as 3003 and 1100 alloy sheets, can be porcelainized. Enamel in water suspension is sprayed on the aluminum parts and fired at 90°F to 1050°F. A wide variety of colors are available in porcelain enamel. Finishes have excellent abrasion and corrosion resistance.

Pots. Carbon-lined vessels used in the reduction process, at the bottom of which molten aluminum is collected and siphoned off.

Precision forging. Forging produced to tolerances closer than standard.

Preheating. High-temperature soaking treatment used to change the metallurgical structure in preparation for a subsequent operation, usually applied to the ingot.

Press forging. Shaping by gradually applying pressure in a press.

Primary aluminum rotor ingot. Ingot product for remelting and casting into rotors for induction motors (99.3 to 99.5 minimum purity). Rotor metal is a purity ingot with a special iron-silicon relationship (an iron-silicon ratio of 2 to 1, minimum) to improve casting characteristics, reduce welding to the laminations, and minimize shrinkage.

Published tolerance. Term is sometimes used synonymously with "standard tolerance." In such cases the term "standard tolerance" is preferred.

Pure aluminum. Pure aluminum melts at 1220.4°F while various alloy constituents melt at temperatures as low as 900°F. This contrasts strikingly with the melting point of iron, 2700°F, and indicates that care must be exercised to prevent melting away an aluminum part being welded.

Quenching. Controlled rapid cooling of a metal from an elevated temperature by contact with a liquid, gas, or a solid.

Reamed extrusion ingot. Cast hollow extrusion ingot that has been machined to remove the original inside surface.

Recording sheet circles. Sheet circles for recording and reproducing sound.

Red mud

1. Solid matter impurities collected by either filtering or gravity settling in the Bayer aluminum refining process.
2. Residual containing a high percentage of iron oxide, obtained in purifying bauxite in the production of alumina by the Bayer process.

Redraw rod. Coiled rod of a quality suitable for drawing into wire.

Reduction. Electrolytic process used to separate aluminum from aluminum oxide.

Refined aluminum. Aluminum of very high purity (999.950% or higher) obtained by special metallurgical treatments.

Reflectivity. The natural surface of aluminum has high reflectivity to both light and radiant energy (heat). This property, together with very low emissivity, is of outstanding value in applications where thermal insulation and interior heat gain are important.

Reflector sheet. Sheet of controlled composition suitable for use in the manufacture of reflectors.

Reheating. Thermal operation designed solely to heat stock for hot-working. In general, no metallurgical changes are intended.

Remelt ingot. Cast form suitable for remelting.

Reroll stock. Semifinished rolled product of rectangular cross section in coiled form suitable for further rolling.

Resistance to chemical attack. Aluminum's resistance to attack by many chemicals and foods make it especially valuable for many types of chemical equipment, food containers, kitchenware, etc. Resistance to chemical attack of the aluminum alloys is roughly in proportion to their purity.

Resistance to corrosion. Corrosion resistance in aluminum products is based on the alloy's resistance to corrosive atmospheres.

Resistance to hot-cracking. This property indicates the relative ability of the alloy to withstand contraction stresses while cooling through the short-hot temperature range and is especially important for selecting a permanent mold alloy for certain types of castings.

Rib. An elongated projection to provide stiffening on a shape, forging, or casting.

Rigid conduit. Tube having a certain standardized length and combination of outside diameter and wall thickness greater than electrical metallic tubing, commonly desig-

nated by the nominal size corresponding to ANSI Schedule 40 Pipe, for use with thread-type fittings as a protection for electric wiring.

Ring condition. Groove revealed by caustic etching of the cross section, generally following the outline of the extruded product, and formed from the liquidated surface of an extrusion ingot. This condition, depending on severity, may or may not be considered a defect.

Riveting

1. Riveting is a highly satisfactory method of joining aluminum parts and assemblies. It is the oldest and most reliable method of joining heat-treated aluminum alloy structures because no heat is involved, and it is well understood and highly developed, Modern riveting methods are largely independent of the operator's skill. Principles of joint design and the actual riveting procedures are the same as for iron or steel.
2. Riveting is extensively used in the fabrication of aluminum sheet and plate structures. It is the most reliable method of joining stress-carrying parts of heat-treated aluminum alloy structures because no heat is involved and riveting procedures are highly developed.

Rivets. Whenever practicable, rivets of the same alloy as that being joined should be used to avoid the possibility of galvanic corrosion. Rivets and rivet wire are made in various aluminum alloys.

Rod. Solid round section, $\frac{3}{8}$ in. or greater in diameter, whose length is greater in relation to its diameter.

Rolled bar. Bar brought to final dimensions by hot-rolling.

Rolled-in metal. Extraneous chip or sliver of metal rolled into the surface of the sheet.

Rolled-in scratch. Scratch that occurs during the fabricating process and is subsequently rolled over.

Rolled ring. Cylindrical product of relatively short height circumferentially rolled from a hollow section.

Rolled rod. Rod brought to final dimensions by hot-rolling.

Rolled rod and bar hot-rolling. Hot-rolling is used to produce rod and bar shapes. Cold-finished rods and bars are produced by being hot-rolled to a size larger than specified and then cold-drawn through a die for a better finish and closer dimensional tolerances. Bars are commercially available as squares (round or square edged) and square-edged rectangles and hexagons.

Rolled shape. Shape brought to final dimensions by hot-rolling.

Rolled special shape. Rolled shape other than a structural shape.

Rolled structural shape. Structural shape brought to final dimensions by hot-rolling.

Rollings. Shaping plate and sheet metal by passing metal slabs through steel rollers.

Rolling ingot. Cast form suitable for rolling.

Rolling slab. Rectangular semifinished product, produced by hot-rolling fabricating ingot and suitable for further rolling.

Roll mark. Raised area on the sheet caused by the imprint of a pit or depression in the rolls during the rolling operation.

Roofing sheet. Coiled or flat sheets in specific tempers, widths, and thicknesses suitable for the manufacture of corrugated or V-crimp roofing.

Rub marks. Minor form of scratching consisting of areas made up of a large number of very fine scratches or abrasions.

Sand blasting. Sand blasting produces a finished surface and texture that is determined by the air pressure, grade of sand or abrasive used, rate of feeding the sand, distance and angle of attack of the nozzle, and size of nozzle. The best interrelationship of these variables can only be worked out by trial and error for any given casting.

Sand casting. Metal object produced by pouring molten metal into a sand mold and allowing it to solidify (ASTM B 26).

Sand castings. Castings produced from molds made of sand that permit maximum flexibility in design changes. Sand castings require heavier section thicknesses than other methods, and tolerances are not so close. They are used for spandrel panels, decorative panels, lighting standard bases, and general fittings.

Scalped extrusion ingot. Cast solid or hollow extrusion ingot that has been machined on the outside surface.

Scalping. Mechanical removal of the surface layer from a fabricating ingot or semifinished wrought product so that surface imperfections will not be worked into the finished product.

Scratch. Visible linear indentation caused by a sharp object passing over the surface.

Scratch-brushed foil. Foil abraded, usually with wire brushes, to produce a roughened surface.

Screw machine stock. Bar, rod, and wire in certain standard alloys, tempers, sizes, and shapes suitable for automatic screw-machine applications.

Seam defect

1. Defective design seam.
2. An unbonded fold or lap on the surface of the metal, which appears as a crack, usually the result of defects in casting or working that have not bonded shut.

Seamless pipe. Pipe produced form hollow extrusion ingot.

Seamless tube

1. Tube that does not contain any line junctures resulting from the method of manufacture (ASTM B 210).
2. Tube having an initial continuous periphery as when produced by extrusion or by being drawn through a die.

Selection of alloys. Alloys are selected for a particular application on the basis of a careful evaluation of design requirements and the alloy's mechanical properties and finishing characteristics.

Selection of finishes. Aluminum does not require special finishing. Nevertheless there are numerous applications where decorative values, additional protection from corrosion, or special engineering functions do require finishing operations. Selection of a particular finish takes into consideration the character of the project—whether it is monumental or industrial, its location and the atmospheric conditions to be encountered, the aesthetic effect desired or the need for color, and the probability of periodic maintenance.

Semihollow drawn shape. Semihollow shape brought to final cross-sectional area and contour by being drawn through a die.

Semihollow extruded shape. Semihollow shape brought to final cross-sectional area by extrusion.

Semihollow shape. Shape, part of whose cross section partially encloses a void and in which the area of the void is substantially greater than the square of the width of the gap.

Semi-permanent mold casting. Permanent mold casting that is made using an expendable core such as sand (ASTM B 108).

Shape. Section that is long in relation to its cross-sectional dimensions, has a cross section other than that of a wire, rod, bar, or tube, and is produced by extrusion, rolling, drawing, or cold-finishing.

Shearing strength. Maximum shearing stress that a material is capable of developing. In practice it is considered to be the maximum average stress computed by dividing the ultimate load in the plane of shear by the original area subject to shear.

Sheet. Solid section rolled to a thickness range of 0.006 to 0.249 in., inclusive, and supplied with sheared, slit, or sawed edges.

Shrinkage. Contraction that occurs when metal cools from the hot-working temperature.

Side set. Difference in gauge between the two edges of a sheet.

Silicon. Silicon, in quantities up to 12%, is the most widely used alloying element. It improves the fluidity of the molten aluminum, allowing it to flow farther through thin walls in the mold cavity and reproduce finer details. It also reduces external shrinkage, decreases leaks in the finished casting, reduces the coefficient of expansion, and improves weldability.

Sized tube. Tube which, after extrusion, has been cold-drawn a slight amount to minimize ovalness.

Sliver. Slender fragment or splinter that is a part of the material but not completely attached thereto.

Slug. Metal blank for forging or impacting.

Soft temper. State of maximum workability of aluminum obtained by annealing.

Soldering. Several satisfactory solders and fluxes have been developed for use in joining aluminum. However, soldering is not recommended for applications where mechanical strength is a factor. Soldering is employed only to provide sealing or to ensure electrical contact. The oxide coating is removed and kept from reforming until the surface of the aluminum has been "tinned"; then the parts can be joined by conventional soldering methods.

Solder metal. Various grades of solder metal are produced and consist primarily of tin-lead, tin-lead-antimony, tin-antimony, and silver-lead alloys in any form, formerly known as "soft solder." These solder metal alloys are so formulated as to be usable in molten state at or below 800°F (430°C) (ASTM B 32).

Solid drawn shape. Solid shape brought to final dimensions by being drawn through a die.

Solid extruded shape. Solid shape brought to final dimensions by extrusion.

Solidification shrinkage tendency. Tendency of the alloy to decrease in volume when freezing. This rating gives an indication of the amount of compensating feed metal required in the form of risers.

Solid shape. Shape other than hollow or semihollow.

Solution heat treatment

1. First temperature-raising step in the thermal treatment of a heat-treatable alloy. Also called "heat treatment."

2. Alloy is heated to a specified temperature (close to the solidus), held at this temperature for a specified time, and then quenched rapidly in cold water.

Special tolerance. Tolerance that is closer or wider than "standard."

Specialty sheet. Sheet product supplied to perform specific special functions and usually designated by a name rather than by alloy and temper.

Spinning of aluminum alloy sheet. Operation performed on a spinning lathe. Any shape that can be cut on a lathe from a solid block can be spun from sheet. Normally circular blanks are used for spinning, although spinning may be employed for adding details to parts that were partially formed by drawing or stamping.

Splice. The end joint uniting two webs.

Squareness. Characteristic of having sides straight and parallel with 90° corners.

Stabilizing. Thermal treatment to reduce internal stresses in order to promote dimensional and mechanical property stability.

Standard bright finish. Finish providing the sheet with a relatively bright, uniform appearance on both sides but somewhat less lustrous than the standard one-side-bright finish.

Standard one-side-bright finish

1. Finish providing a sheet with a uniformly bright and lustrous surface on one side. The reverse side is uncontrolled and may have a dull, nonuniform mill finish appearance (ASTM B 209).

2. Surface conditions of a sheet given a bright finish approaching a controlled, uniform mirror finish on one side and having a mill finish or better on the reverse side.

Standard tolerance. Established tolerance for a certain class of product. Term is preferred to "commercial" or "published" tolerance. Maximum limits mill products may be allowed to deviate from the nominal or specified dimensions.

Standard two-sides-bright finish. Sheet having a uniform bright finish on each side (ASTM B 209).

Stepped drawn tube. Drawn tube whose cross section changes in area at intervals along its length.

Stepped extruded shape. Extruded shape whose cross section changes abruptly in area at intervals along its length.

Strain. Measure of the change in size or shape of a body due to force, referred to its original size or shape. Tensile or compressive strain is the change due to force per unit of length in an original linear dimension in the direction of the force. It is usually measured as the change (in inches) per inch of length.

Strain hardening. Method of strengthening nonheat-treatable alloys by either cold-rolling or other physical or mechanical working.

Streak. Line, elongated mark, or stripe causing nonuniformity of surface appearance; in the case of painted sheet, it is visible as a variation in gloss or color.

Streamline hollow shape. Hollow extruded or drawn shape with a cross section of teardrop shape.

Stress. Intensity of force within a body that resists a change in shape, measured in pounds per square inch (psi). Stress is normally calculated on the basis of the original cross-sectional dimensions. The three kinds of stress are tensile, compressive, and shearing. Flexure in-volves a combination of tensile and compressive stress. Torsion involves shearing stress.

Stress corrosion cracking. Failure by cracking resulting from selective directional attack caused by the simultaneous interaction of sustained tensile stress at an exposed surface with the chemical or electrochemical effects of the service environment.

Stress relieving. The reduction of the effects of internal residual stresses by thermal or mechanical means.

Stretch-forming

1. Stretching of large sheets for the purpose of flattening them or for forming them into the shape of a form block.

2. Useful operation for producing long, shallow, three-dimensional curves in large parts. Excellent method for making bends in extruded shapes. Form and size of parts are restricted only by the equipment available.

Strip conductor. Electric conductor, in the form of coiled sheet or foil, with specially controlled or prepared edges.

Structural pipe. Extruded pipe, which may contain an extrusion seam, suitable for applications not involving internal pressure.

Structural shapes

1. Shapes, rolled or extruded, commonly used for structural purposes, but limited to shapes commonly produced by rolling, such as angles, channels, Tees, Zees, I-beams, and H-sections.

2. Though most structural shapes are extruded, some shapes may be hot-rolled by a series of specially shaped rolls. The more complicated structural shapes require a greater number of roller passes to prevent undue straining of the metal in any single pass.

Structural streak. Streak revealed by etching or anodizing and resulting from structural heterogeneities within the product.

Structural tube. Extruded tube, which may contain an extrusion seam, suitable for applications not involving internal pressure.

Suck-in. Defect caused when one face of a forging is sucked in to fill a projection on the opposite side.

Surface finishes. Three general types of finishes can be produced directly on the surface of aluminum castings. These are mechanical, chemical, and electrochemical finishes. Properly applied, these are all permanent in nature. Temporary finishes, such as paints, enamels, and lacquers can also be applied to aluminum castings. These temporary finishes may also be applied on top of any of the surface finishes. The exact treatment given the surface of any aluminum casting will depend on the use of the product and its environment.

Surface finishing. The wide variety of finishes for aluminum vastly extends its usefulness, Color and texture can be varied in a number of ways. Hard, tough coatings that protect, as well as beautify, the surface can be applied.

Sulfuric acid anodizing. Although many acid solutions may be used in anodizing, sulfuric acid is the most often used because the transparent coating it produces enhances the natural color and appearance of architectural aluminum and improves its weatherability.

Tapered extruded shape. Extruded shape whose cross section changes in area continuously along its length or a specified portion thereof.

Telescoping. Transverse slipping of successive layers of a coil of sheet or foil so that the ends of the coil are conical rather than flat.

Temper designation. Designation (following an alloy designation number) that denotes the temper of an alloy.

Thermal conductivity. On a weight basis, aluminum is the most efficient heat conductor of the common metals, of marked advantage when it is desirable to conduct or dissipate heat rapidly and uniformly.

Tin. Tin used to improve the machinability of the copper alloys. It also acts to provide a fine bearing alloy when used in conjunction with copper and nickel additions. Common aluminum casting alloys employ tin additions in varying amounts depending on the alloy.

Titanium. Titanium is usually added as a grain refiner (0.05 to 0.20%) to all casing alloys intended for sand or permanent-mold castings. Titanium is desirable to improve the mechanical properties of the castings.

Tolerance. Allowable deviation from a nominal or specified dimension.

Tool. Term usually used to describe dies, mandrels, and other devices necessary to produce extruded or drawn shapes or tubes.

Tooling pad. Cast or rolled product with rectangular cross section of thickness 0.250 in. or greater, with edges either as-cast, sheared, or sawed, and with internal stress levels controlled to achieve maximum stability for machining purposes in tool and jig applications.

Torn surface. Surface that shows a deep longitudinal rub mark resulting from abrasion by extruding or drawing tools.

Traffic marks. Abrasions that result from metal-to-metal contact and vibration during transit. These abrasions are usually dark in appearance because of the presence of a dark powder consisting of aluminum and aluminum oxide fines produced by the abrasive action of the sheets' rubbing together.

Transvere bow. Concavity or convexity of the sheet across its width.

Tread plate. Sheet or plate product having a raised, figured pattern on one surface, to provide improved traction.

Trim inclusion. Edge trimming accidently wound into a roll of foil.

Tube. Hollow product whose cross section is completely symmetrical. Tube can be round, square, rectangular, hexagonal, octagonal, or elliptical in shape; with sharp or rounded corners and walls of uniform thickness except as affected by corner radii (ASTM B 210–B 483).

Tube drawing. Hollow tube sections are drawn in a process similar to the drawing of bars and rods, except that the tube is threaded over a mandrel (shaper) positioned in the center of the die. When the tube is drawn, close tolerances both inside and out are maintained.

Tube stock. Semifinished tube intended for subsequent reduction in cross section.

Tubular conductor. Bus conductor made from any tube section.

Tubular product. General term encompassing tube, hollow shape, and semihollow shape.

Twist. Winding departure from flatness.

Utility sheet. Mill finished coiled or flat sheet of unspecified composition and properties produced in specific standard sizes and suitable for general building trade use.

Ultimate or tensile strength. Maximum tensile stress that a material is capable of developing under a gradual and uniformly applied load. Tensile strength is calculated from the maximum load carried during a tension test and the original cross-sectional area of the specimen.

Upset forging. Forging having part or all of its cross section greater than that of the stock.

Vent mark. Small protrusion on a forging or casting resulting from the entrance of metal into die or mold vent holes.

Water stain. Superficial etching of the surface from prolonged contact with moisture in a restricted airspace, such as that between layers of the product. Such stain is generally white in appearance.

Wavy edge. Rippling departure of an edge from flat.

Weathering corrosion (galvanic and or chemical). Corrosion produced by atmospheric conditions.

Web

1. Single thickness of foil as it leaves the rolling mill.
2. Connecting element between ribs, flanges, or bosses on shapes and forgings.

Welded joints. Practically all aluminum alloys can be welded. For fusion welding, the inert-gas shielded-arc methods are used exclusively. They offer the advantages of high welding speed and freedom from flux problems. The three methods of resistance welding—spot, seam, and flash butt—are also readily adaptable to joining aluminum.

Welded tube. Tube formed from sheet and fastened at the seam by welding.

Welding aluminum. Aluminum is one of the most readily weldable of all metals. Four outstanding factors affecting the welding of aluminum are its low melting point, the presence of an oxide film, its low strength at high temperatures, and the fact that it shows no color even at temperatures up to the melting point.

Welding rod. Rod (rolled, extruded, or cast) for use in joining metals by welding.

Welding wire. Wire for use in joining metals by welding.

Wettability. Degree to which a metal surface may be wet to determine the absence of or the amount of residual rolling or added lubricants or deposits on the surface.

Whip marks. Elongated markings, generally running in a cross-grain direction, resulting from a whipping of the sheet as it enters the rolling mill.

Wide tolerance. Any special tolerance that is wider than "standard."

Wire. Solid section that is long in relation to its cross-sectional dimensions, having a completely symmetrical cross section that is square or rectangular (excluding flattened wire) with sharp or rounded corners or edges, or is round or a regular hexagon or octagon, and whose diameter, width, or greatest distance between parallel faces is less than $\frac{3}{8}$ in.

Wire brushing. Using a high-speed rotating wire brush on a sandblasted surface results in a pleasing lined effect. Lines can be coarse or smooth, deep or shallow, depending on the size and stiffness of the brush. With a very fine brush, a satin finish results. Satin finishing is also used after buffing or coloring. The finest satin finishes are produced by rubbing the surface with a fiber brush or by hand rubbing with a mixture of oil and pumice.

Wire products. Wire, which must be kept within close tolerances, is cold-drawn by pulling redrawn rod ($\frac{3}{8}$ in. diameter) through a series of progressively smaller dies.

Workability. Aluminum sheet responds readily to bending, stamping, roll-forming, drawing, embossing, cutting, and spinning operations. It may also be extruded, cast, and forged and can be joined by all commonly used methods, such as by welding, adhesives, and mechanical fasteners.

Wrought alloy ingot for remelting. Primary ingot produced in the wrought alloys for remelting for manufacturers who desire to cast their own extrusion and sheet ingot. Wrought alloy ingot is offered in 50- to 1000-lb ingot sizes.

Wrought products. Products formed by rolling, drawing, extruding, and forging.

Yield strength. Stress at which a material exhibits a specified permanent set. The value of the set used for aluminum and its alloys is 0.002 in./in., 2%. For aluminum alloys the yield strengths in tension and compression are approximately equal.

Zinc. Zinc added to aluminum makes the metal very hot and produces high shrinkage when used in large amounts. Exceptionally fast melting is necessary to obtain the best characteristics. Similarly, large risers are important. Zinc additions, particularly in combination with magnesium, in general, produce good impact resistance, high tensile strength, and excellent ductility. Small amounts of zinc in the copper alloys help improve machinability.

COPPER, BRASS, AND BRONZE

Acid copper. Copper electrodeposited from an acid solution of a copper salt, usually copper sulfate.

Admiralty brass (copper, zinc, and tin). Metal typically used for steam powerplant equipment, chemical and process equipment, and marine uses. General properties are excellent corrosion resistance combined with strength and ductility.

Aluminum bronze. Like aluminum, bronzes form an aluminum oxide skin on the surface, which materially improves resistance to corrosion, particularly under acid conditions. Since the color of the 5% aluminum bronze is similar to that of 18-carat gold, it is used for costume jewelry and other decorative purposes. Aluminum-silicon bronzes are used in applications requiring high tensile properties in combination with good corrosion resistance.

Anode copper. Specially shaped copper slabs, resulting from the refinement of blister copper in a reverberatory furnace, used as anodes in electrolytic refinement.

Apparent density (bulk density). Weight of the unit volume of copper powder, of particular importance in the pressing operation. The lower the value, the greater the volume needed for a part of given size.

Architectural bronze. Most important and widely used of the cooper alloys in the architectural metal industry, combining sufficient strength and hardness with free cutting and machining characteristics. It is available in a great many extruded shapes in the form of rounds, squares, flats, angles, channels, moldings, and many other shapes.

Babbitt metal. Alloy of tin, copper, and antimony used for lining bearings and bushings to reduce friction and wear.

Base metal. Metal present in the largest proportion in an alloy; brass, for example, is a copper-base alloy.

Beryllium. Beryllium aluminum silicates are the chief source of the metal, which is prepared by electrolysis. It is hard enough to scratch glass and resembles magnesium in appearance and chemical properties. Its alloys are strong,

light, and resistant to corrosion. Beryllium is used chiefly to alloy with copper, beryllium copper alloys containing about 3% beryllium having greatly increased strength and hardness.

Beryllium bronze. Metal containing over 2% beryllium or beryllium plus metals other than copper. Most of these alloys are heat treatable.

Billet. Refinery shape used primarily for tube manufacture. A billet is circular in cross section, usually 3 to 10 in. in diameter, and produced in lengths up to 52 in., weighing from 10 to 1500 lb.

Blister copper. Impure intermediate product in the refining of copper, produced by blowing copper matte in a converter. The name is derived from the large blisters on the cast surface that result from the liberation of sulfur dioxide and other gases.

Brass

1. Alloy consisting mainly of copper (over 50%) and zinc, to which smaller amounts of other elements may be added.

2. Copper-base alloy containing zinc as the principal alloying component, with or without smaller quantities of other elements. There are alloys that are definitely brasses by definition; yet they have names that include the word "bronze," for example, commercial bronze (90% copper, 10% zinc).

Bronze

1. Copper-rich copper-tin alloy, with or without small proportions of other elements such as zinc and phosphorus. By extension, certain copper-base alloys containing considerably less tin than other alloying elements, such as manganese bronze (copper and zinc plus manganese, tin, and iron) and leaded tin bronze (copper and lead plus tin and sometimes zinc). Also, certain other essentially binary copper-base alloys containing no tin, such as aluminum bronze (copper-aluminum), silicon bronze (copper-silicon), and beryllium bronze (copper-beryllium).

2. Copper-base alloy having tin as the principal alloying constituent. However, the term "bronze" is seldom used alone and the copper-tin alloys have come to be known as phosphor bronze because of the small residual phosphorus content. With a suitable modifier, use of the term "bronze" also extends to a variety of copper-base alloy systems having a principal alloying element other than tin or zinc.

Bronzes. Bronzes are copper alloys in which the major alloying element is one other than zinc or nickel.

Bus bar. Heavy metal conductor, usually copper, for high-amperage electricity (ASTM B 187).

Cake

1. Copper or copper alloy casting rectangular in cross section used for rolling into sheet or strip.

2. Refinery shape for rolling into plate, sheet, strip, or shape, rectangular in cross section and of various sizes. Cast either horizontally or vertically, with range of weights from 140 to 4000 lb or more.

Cartridge brass (copper and zinc). Cartridge brass provides the best combination of ductility and strength of any brass, as well as excellent cold-working properties. Typical uses are products made from deep drawing, stamping, spinning, etching, and rolling, primarily for use in the electrical parts industry.

Casting alloy. Metal used for matching architectural bronzes, Muntz metal, and rich low brass. It may also be used for statuary castings and for engineering purposes where high strength is required.

Casting copper. Fire-refined tough pitch copper usually cast from melted secondary metal into ingot and ingot bars only and used for making foundry castings but not wrought products.

Coalesced copper. Massive copper made from ground brittle cathode copper by briquetting and sintering in a reducing atmosphere at high temperatures with pressure.

Commercial bronze. Alloy available in many forms. The appearance of the metal is quite similar to that of copper, and it is often selected as a substitute for pure copper. Its color, both when new and after oxidation and weathering, is similar to that of some alloys of statuary bronze. It is a rich pleasing color that does not match closely with architectural bronze, Muntz metal, or rich low brass. Designated by the trade as a bronze, it has many uses as an architectural metal.

Compressibility. Ratio of the volume of loose to the volume of compact copper powder. Compressibility is important for both fabrication and end properties and is affected by the physical characteristics of the powder particles and the particle size distribution. Particle size distribution affects press feed, dimensional changes during sintering, porosity of the compact, and final attainable density and strength.

Constantan

1. Alloy of about 45% nickel and 55% copper that is distinguished by its high electrical resistance and unusually constant temperature coefficient of electrical resistivity. Extensive use is made of this property in electrical instruments.

2. Group of copper-nickel alloys containing 45 to 60% copper with minor amounts of iron and manganese and characterized by relatively constant electrical resistivity irrespective of temperature; used in resistors and thermocouples.

Copper. Metallic element, reddish in color with a bright metallic luster, that is malleable, ductile, and a good conductor of heat and electricity (second to silver in electrical conductivity). Copper is obtained from its ores by smelting, leaching, or electrolysis. Its chief use is in copper wire for electrical purposes. Copper does not corrode readily, although it oxidizes on continuous exposure to the atmosphere. Prolonged exposure results in formation of the attractive green patina or verdi-green coatings.

Copper and alloys. Use of copper and its alloys is universal. They are practically synonymous with all things electrical and have a myriad of uses.

Copper brazing. Brazing with copper as the filler metal.

Copper-coated sheet. Full sheet of copper, coated on both sides with a flexible, rubbery bituminous compound.

Copper fabric. Fabric made of full sheet of copper between two layers of asphalt-impregnated cotton fabric. The copper is bonded to both layers of the fabric.

Copper-lead sheet. Single bimetal sheet of copper and lead bonded to a reinforced glass fiber creped kraft backing.

Copper-lead fabric sheet. Single bimetal sheet of copper and lead between two layers of asphalt-impregnated cotton fabric. The full copper-lead sheet is bonded to both layers of the fabric.

Copper-lead fiber glass. Single bimetal sheet of copper and lead bonded on one side with fiber glass reinforced kraft.

Copper-nickels. Alloys with nickel as the principal alloying element, with or without other designated alloying elements present.

Copper-nickel-zinc alloys. Commonly known as "nickel silver," these are alloys that contain zinc and nickel as the principal and secondary alloying elements, with or without other designated elements present.

Copper paper

1. Paper made of copper bonded on one side by asphalt to a waterproofed creped kraft paper.

2. Paper made of copper bonded to and between two layers of waterproofed creped kraft paper.

Copper powder

1. Finely divided copper particles produced by high-velocity atomization of molten copper with a stream of compressed gas, steam, or water; gaseous reduction of finely divided oxides; the precipitation from solutions.

2. Finely divided copper particles produced by electrode-position.

Coppers. Metal which has a designated minimum copper content of 99.3%.

Copper steel. Alloy made by adding a fraction of 1% of copper, 0.20% minimum, to low-carbon steel during the melting process.

Cupronickel (copper and nickel). Cupronickel has high strength and ductility and is resistant to corrosion and erosion. It ranges in color from white to silver. Typical uses are condenser tubes an plates, tanks, vats, vessels, process equipment, automotive parts, meters, and refrigerator pump valves.

Cyanide copper. Copper electrodeposited from an alkali-cyanide solution containing a complex iron made up of univalent copper and the cyanide radical; also, the solution itself.

Deoxidized copper. Copper cast in the form of refinery shapes, freed from cuprous oxide through the use of metallic or metalloidal deoxidizers. By extension, the term is also applicable to fabricators' products made therefrom.

Dezincification. Corrosion of some copper-zinc alloys involving loss of zinc and the formation of a spongy porous copper.

Electrolytic copper. Copper that has been refined by electrolytic deposition, including cathodes that are the direct product of the refining operation; refinery shapes cast from melted cathodes and, by extension, fabricators' products made therefrom. Usually when this term is used alone, it refers to electrolytic tough pitch copper without the presence in significant amounts of elements other than oxygen.

Ferrous alloys. Pure copper and pure iron are miscible in the liquid phase in all proportions. Introducing carbon affects the miscibility, however, so that a molten eutectic iron containing 4.3% carbon can dissolve only about 3% copper. Ductile alloys can be produced throughout the entire composition range, but copper in commercially important iron-base alloys is limited to approximately 2.5% and is usually lower.

Fire-refined copper. Copper that has been refined by the use of a furnace process only, including refinery shapes and, by extension, fabricators' products made therefrom. Usually when this term is used alone, it refers to fire-refined tough pitch copper without elements other than oxygen being present in significant amounts.

Fire scale. Intergranular copper oxide remaining below the surface of silver-copper alloys that have been annealed and pickled.

Flat product. Rectangular or square solid section of relatively great length in proportion to thickness. Depending on width and thickness, flat products include plates, sheets, strips, and bars. Also, included are the products known as "flat wires."

Flow rate. The ease with which copper powder can be fed into the die is determined by its flow rate. Low flow rates retard automatic pressing and may require the use of vibrating equipment.

Forging brass (copper, zinc, and lead). Alloy that is extremely plastic when hot, combining good corrosion resistance with excellent mechanical properties. Typical uses are hot forgings, hardware, and plumbing goods.

Gold-copper-silver alloys. Copper hardens gold principally by forming solid solutions. Other elements are added to impart certain properties or characteristics. Gold-copper-silver alloys are more commonly used than gold-copper alloys.

Green strength. Property of the copper powder to be handled without breakage after compacting and before sintering.

Hard drawn. Temper of copper or copper alloy tubing drawn in excess of 25% reduction in area.

High-conductivity copper. Copper that in the annealed condition has a minimum electrical conductivity of 100% I.A.C.S. as determined in accordance with ASTM B 193 or E 1004.

High-leaded brass (copper, zinc, and lead). Brass whose general properties are free machining and good blanking. Its typical uses are engraving plates; machined parts, instruments (both professional and scientific), and various types of locking devices.

High-leaded tin bronze (copper, tin, zinc, and lead). Commonly used foundry alloy that may be further modified by addition of some nickel or phosphorus, or both.

High-residual phosphorus copper. Deoxidized copper with residual phosphorus present in amounts generally sufficient to decrease appreciably the conductivity of the copper.

High-silicon bronze (copper, silicon, and manganese). Bronze with the corrosion resistance of copper and the mechanical properties of mild steel. Typical uses are tanks, pressure vessels, vats, and forgings.

Lead bronze (lead, zinc, and tin). Bronze used for special bearing applications.

Leaded brass. Lead is added to brass to improve its machinability, particularly in such applications as automatic screw machines where a freely chipping metal is required. Leaded brasses cannot easily be cold-worked by such operations as flaring, upsetting, or cold-heading.

Leaded coppers. Series of cast alloys of copper with 20% or more lead, and usually a small amount of silver present, but without tin or zinc.

Leaded high strength yellow brass. Leaded brass containing over 17% zinc and more than a 2% total of aluminum,

manganese, tin, nickel, and iron. Leaded brass containing more than 5% lead and less than 6% tin is a commonly used foundry alloy.

Leaded nickel brass. Leaded nickel brass contains more than 10% zinc, 0.5% lead, and nickel in sufficient amounts to give it a white color. It is a commonly used foundry alloy and is sometimes called "German silver."

Leaded nickel bronze. Leaded nickel bronze contains over 10% nickel and less zinc than nickel. Less than 10% tin and more than 0.05% lead. A commonly used foundry alloy, it is sometimes called "German silver" or "nickel silver."

Leaded red brass. Leaded red brass contains more than 2% to 8% zinc, and less than 6% tin, (usually less tin than zinc) and over 0.5% lead. It is a commonly used foundry alloy and may be further modified by the addition of nickel.

Leaded semired brass. Leaded semired brass contains more than 8% to 17% zinc, less than 6% tin, and over 0.5% lead. It is a commonly used foundry alloy and may be further modified by the addition of nickel.

Leaded tin bronze. Leaded tin bronze contains up to 20% tin and zinc in amounts equal to or less than the tin, and over 0.5% lead but never exceeding 6%. It is a commonly used foundry alloy and may be further modified by the addition of some nickel or phosphorus.

Leaded yellow brass. Leaded yellow brass contains over 17% zinc, less than 6% tin, less than a 2% total of aluminum, maganese, nickel, and iron, and over 0.5% lead. It is a commonly used foundry alloy.

Light drawn. Temper of copper or copper alloy tubing drawn to between a 10 and 25% reduction in area, corresponding roughly to quarter hard.

Low-residual phosphorous copper. Deoxidized copper with residual phosphorus present in amounts generally too small to decrease the conductivity of the copper appreciably.

Manganese bronze (copper, zinc, iron, tin and manganese). Alloy combining high strength combined with excellent wear resistance. Typical uses are forgings, condenser plates, valve stems, and coal screens.

Monel metal. Registered trademark applied to a technically controlled nickel-copper alloy of high nickel content. Monel metal is mined, smelted, refined, rolled, and marketed solely by one manufacturer. Its composition is approximately two-thirds nickel and one-third copper. This alloy can be formed, machined, cast, forged, spun, drawn, brazed, soldered, welded, and cloisonne and champleve enameled. It can be worked in any way employed with other metals except that it cannot be extruded.

Muntz metal (copper and zinc). Alloy combining high strength with low ductility. Typical uses are sheet form, perforated metal, architectural work, condenser tubes, valve stems, and brazing rods.

Nickel brass. Nickel brass is a commonly used foundry alloy; it is sometimes called "German silver."

Nickel bronze (nickel, zinc, tin, and lead). Nickel bronze is a commonly used foundry alloy; it is sometimes called "German silver" or "nickel silver."

Nickel-copper alloys. Copper is soluble in nickel in both the liquid and solid states in all proportions, and all the alloys have homogenous structures regardless of the heat treatment. Although all of the alloys are ductile, only Monel and constantan have much commercial utilization.

Oxygen-free copper

1. Electrolytic copper free from cuprous oxide, produced without the use of residual metallic or metaloidal deoxidizers.

2. The types of oxygen-free copper are as follows: oxygen free without residual deoxidants (OF), oxygen free phosphorus bearing (OFP), oxygen free phosphorus and tellurium bearing (OFPTE), oxygen free silver bearing (OFS), and oxygen free tellurium bearing (OFTE). The OF and OFS types are high-conductivity coppers.

Phosphorized copper. General term applied to copper deoxidized with phosphorus. Phosphorized copper is the most commonly used deoxidized copper.

Phthalocyanine pigments. Organic pigments of extremely stable chemical configuration resulting in very good fastness properties. Its properties are enhanced by the formation of the copper complex, which is the phthalocyanine blue most used. The introduction of chlorine atoms into the blue molecule gives the well-known phthalocyanine green, also usually in the form of a copper complex.

Pipe. Seamless tube conforming to the particular dimensions commercially known as "standard pipe sizes" (ASTM B 42).

Platinum-copper and palladium-copper alloys. Both platinum and palladium form continuous series of solid solutions with copper. Although the alloys are potentially useful, little actual demand exists for the binary alloys. The 60% palladium, 40% copper alloy has some use as electrical contacts. Ternary palladium-silver-copper alloys with more than 5% copper and not more than 65% palladium are age hardenable and are used in dentistry.

Polling. Step in the fire refining of copper to reduce the oxygen content to tolerable limits by covering the bath with coal or coke and thrusting green wood poles below the surface. There is a vigorous evolution of reducing gases that combine with the oxygen contained in the metal.

Purity. Clean copper powder particle surfaces ensure good contact between particles (needed for optimum mechanical properties). Presence of foreign particles must be held to a minimum, usually specified.

Red or rich low brass. Red or rich low brass is an alloy available in sheet, strip, rod, bar, tube, and pipe. Its reddish yellow color is a close match to that of architectural bronze and Muntz metal, and it is used with those metals, particularly in tube or pipe form, where they are a part of the composition.

Rochelle copper. Copper electrodeposit obtained from copper cyanide plating solution to which Rochelle salt (sodium potassium tartrate) has been added for grain refinement, better anode corrosion, and cathode efficiency; solution from which a Rochelle copper electrodeposit is obtained.

Rod. Round, hexagonal, or octagonal solid section. The round rod used for further processing into wire (known as "hot-rolled rod," "wire-rod," or "redraw wire") is furnished coiled. Rod for other uses is furnished in straight lengths.

Set (pitch). Shape of the solidifying surface of a metal, especially copper, with respect to concavity or convexity.

Set copper. Intermediate copper product containing about 3.5% cuprous oxide, obtained at the end of the oxidizing portion of the fire-refining cycle.

Shape. Solid section, other than rectangular, square, or standard rod or wire section, furnished in straight lengths. Shapes are usually made by extrusion, but may also be fabricated by drawing.

Silicon bronze. High-copper alloys containing percentages of silicon ranging from about 1 to slightly more than 3%. In addition, they generally contain one or more of the four elements, tin, manganese, zinc, and iron. High silicon bronze is typical.

Silver-copper alloys. The silver-copper system is relatively simple, with a eutectic at 28.1% copper that solidifies at 779.4°C (1435°F). The maximum solid solubility of copper in silver at this temperature is 8.8%, decreasing to about 0.1% in equilibrium at room temperature. High-silver silver-copper alloys can be age hardened. All of the alloys of silver and copper are ductile. Of the high-silver alloys, the most important is the 92.5 silver-7.5 copper alloy known as sterling silver.

Sintering properties. Sintering is usually done either in an inert, reducing, or neutral atmosphere or in a vacuum. The sintering temperature of copper powder is critical; when copper powder compacts are sintered, either in the solid state or in the presence of a minor portion of a liquid phase, suitable physical and mechanical properties must be obtained within predictable and reproducible dimensions.

Special alloys. There are many special and proprietary alloys that are essentially defined by their names; included are aluminum-tin bronze, aluminum-silicon bronze, cadmium bronze, chromium copper, beryllium copper, leaded copper, selenium copper, and others.

Statuary bronze. Alloy or series of alloys that have been known for many centuries. The alloys may be of different compositions as produced by various manufacturers, containing varying percentages of copper, zinc, tin, and sometimes lead and nickel. They have individual colors and other characteristics to meet the requirements of each specific project. Bronze alloys are primarily intended for statuary work but may also be used for ornamental castings when these are to be color-matched or contrasted with other bronze.

Tin brass. Tin is added to a variety of basic brasses to obtain hardness, strength, and other properties that would otherwise not be available.

Tin bronze. Bronzes are all alloys of copper and tin. The term "bronze" is generally applied to engineering metals having high mechanical properties, and the term "brass" to other metals. Commercial wrought bronzes do not usually contain more than 10% tin because the metal becomes extremely hard and brittle. When phosphorus is added as a deoxidizer to obtain sound dense castings, the alloys are known as "phosphor bronzes." Two most commonly used tin bronzes contain 5 and 7% tin. Both have excellent cold-working properties.

Tough pitch copper

1. Copper containing from 0.02 to 0.05% oxygen obtained by refining the copper in a reverberatory furnace.

2. Copper that is either electrolytically refined or fire refined, cast in the form of refinery shapes, and contains a controlled amount of oxygen for obtaining a

level set in the casting. The term is also applicable to the products of fabricators' products made therefrom.

Tube. Hollow product of round or any other cross section having a continuous periphery.

Virgin metal. Pure metal obtained directly from ore.

White Metal

1. Group of white-colored metals of relatively low melting points (lead, antimony, bismuth, tin, cadmium, and zinc) and of the alloys based on these metals. A copper matte of about 77% copper obtained from the smelting of sulfide copper ores.
2. Commercial designation of certain cast alloys employed for many purposes other than architectural, such as castings produced in quantity for machinery and other products. Specifications of architectural metal in the light colors employ the definite terms nickel silver, aluminum, Monel metal, or stainless steel.

Wire. Solid section, including rectangular hot wire but excluding other flat products, furnished in coils or on spools, reels, or bucks. Flat wire may also be furnished in straight lengths.

Wire bar. Refinery shape for rolling into rod (and subsequent drawing into wire), strip, or shape. Approximately $3\frac{1}{2}$ to 5 in. in cross section, usually from 38 to 54 in. in length, and from 135 to 420 lb in weight. Wire bar is tapered at both ends when used for rolling into rod for subsequent wire drawing and may be unpointed when used for rolling into strip. It is cast either horizontally or vertically.

Wrought alloys. Alloys made by alloying the various components and casting the alloy into slabs, cakes, or billets for hot- or cold-working by rolling, drawing, extrusion, or forging. Depending on the alloy, the wrought products comprise such diverse shapes as plate, sheet, strip, bar, rod, wire, and seamless tube. The chief elements alloyed with copper are zinc, tin, lead, nickel, silicon, and aluminum and, to a lesser extent, manganese, cadmium, iron, phosphorus, arsenic, chromium, beryllium, selenium, and tellurium.

Yellow or high brass. Alloy available in almost every form except extrusions, and produced in a wide range of tempers from soft to spring. It is usually made of approximately $\frac{2}{3}$ copper and $\frac{1}{3}$ zinc; however, other elements may be introduced for specific purposes as for example, small percentages of lead to improve machinability. It is especially suitable for spinning, bright yellow in color, and characterized by strength, hardness, and excellent cold-working properties.

Zinc-copper alloys. Binary high-zinc zinc-copper alloys are of no great importance commercially. More important are the ternary zinc-copper-aluminum diecasting alloys. The tendency toward expansion is an important property of these alloys, causing them to fill molds completely.

INDUSTRIAL AND COMMERCIAL METALS

Alpha iron. Body-centered cubic form of pure iron, stable below 1670°F.

Alumel. Nickel-based alloy used as a component of pyrometric thermocouples.

Amalgam. Alloy of mercury with one or more other metals.

Antimony. Metallic element occurring native in rare instances but derived chiefly from stibnite or gray antimony ore, kermesite or red antimony, valentinite or white antimony, and certain ores of gold, silver, and lead. Antimony is extracted from the sulfide by roasting to the oxide, which is reduced by salt and scrap iron; it is also prepared by reduction from the oxides with carbon. It is an extremely brittle metal of a flaky, crystalline texture, blue white color, and metallic luster; has a hardness of 3 to 3.5; and is not acted on by air at room temperature, but when heated burns brilliantly with the formation of the white fumes of oxide. It is a poor conductor of heat or electricity. The most important metal alloys include type metal, stereotype metal, and Babbitt metal.

Base bullion. Crude lead containing recoverable silver, with or without gold.

Bismuth. Brittle, white, crystalline metal with a pinkish tinge, which occurs in many places free, as well as in combination as the sulfide, oxide, and carbonate, and is extracted from the ore by melting out the free metal, the oxides and sulfides being composed by the addition of carbon and iron. It is also recovered as a by-product in lead smelting. Bismuth is a poor conductor of electricity, and very diamagnetic; it solidifies with expansion; and when heated in air it burns with a blue flame, forming yellow fumes of the oxide. Its soluble salts form insoluble basic salts on the addition of water—a property sometimes used in its detection. Bismuth forms many alloys with metals, which are often used because of their low melting points and their property of expanding on cooling, thus making them particularly suitable for sharp castings of objects subject to damage by high temperatures.

Bullion. Semirefined alloy containing sufficient precious metal for reuse. Refined gold or silver, uncoined.

Cadmium. Cadmium occurs in small quantities in zinc ores. It distills before zinc in the preparation of that metal, condensing as the brown oxide, which is then reduced with carbon. It tarnishes in air and burns when heated, forming the oxide. Cadmium is a soft, bluish white metal, malleable and ductile. It forms a number of salts of which the sulfate is the most common. Cadmium is a component of one of the lowest melting alloys and is alloyed with silver in electroplating. It is used as the chief constituent in many alloys for machine bearings.

Cast iron. Saturated solution of carbon in iron, with the carbon content varying from $1\frac{1}{2}$ to 4% according to the other impurities contained. Cast iron is hard, brittle, non-malleable, and very fluid when melted and is well adapted for casting into complex forms.

Cast steel. Foundry product made by pouring molten steel into molds that determine its form or shape. Being a steel, its carbon content is relatively low, seldom over 0.40%, and it may be produced by any of the regular steel-making processes.

Cathode. Unmelted flat plate produced by electrolytic refining. Its customary size is about 3 sq ft and about $\frac{1}{2}$ to $\frac{7}{8}$ in. thick, weighting up to 280 lb.

Chromium. Metallic element resembling iron, occurring chiefly in chrome iron ore, and prepared by the reduction of the oxide with aluminum. Chromium is a very infusible hard gray metal used to harden steel. Chromium alloys with iron have become highly important owing to the increased use of the stainless steels, which are remarkably resistant to corrosion. It is one of the most valuable of plating materials because of its hardness, resistance to high temperature, and imperviousness to most acids and salt spray. It can be plated over copper and nickel plating on wrought iron and steel and directly on copper, brass, and other metals with satisfactory results.

Coin silver. Alloy containing 90% silver with copper being the usual alloying element.

Cold-rolled steel. Cold-rolled or cold-finished mild steel or other low-carbon steel is produced by passing the previously cleaned or pickled metal between one or more sets of heavy rolls or through one or more dies to work the metal while cold. This acts to harden and stiffen the steel and to increase its tensile strength, as well as to improve the surface.

Columbium. Very rare metallic element, occurring chiefly in niobite or columbite and prepared by reducing the oxide with carbon in the electric furnace. Columbium is a gray metal that forms an acid oxide from which the salts are derived. A small amount of columbium in steel increases its corrosion resistance.

Common iron. Iron made form rerolled scrap iron or a mixture of iron and steel scrap, no attempt being made to separate the iron and steel scrap.

Cupellation. Oxidation of molten lead containing gold and silver to produce lead oxide, thereby separating the precious metals from the base metal.

Dore silver. Crude silver containing a small amount of gold, obtained after removing lead in a cupelling furnace. Same as dore bullion and dore metal.

Double-refined iron. Iron classed as double refined is considered to be all new wrought iron, which is first rolled into muck bars or slabs. Double-refined iron is free of steel and foreign scrap.

Dow process. Process for the production of magnesium by electrolysis of molten magnesium chloride.

Fineness. Measure of the purity of gold or silver expressed in parts per thousand.

Fine silver. Silver with a fineness of 999, equivalent to a minimum content of 99.9% silver, with the remaining content not restricted.

Gold. Gold occurs native or uncombined, principally in either rock or alluvial deposits, and is obtained form its ores by cyanidation, amalgamation, and smelting. It is a metallic element, having a yellow color when in mass, but when finely divided it may be black, ruby, or purple. Gold is the most malleable and ductile, and also one of the softest, of the metals. It is a good conductor of heat and electricity and is not affected by air and most reagents.

Gold filled. Covered on one or more surfaces with a layer of gold alloy to form a clad metal.

High-strength steels. Specific types of steels that hold attractive possibilities for use in the metal industry have been developed. These steels, variously known as low-alloy high-tensile steels, high-strength low-alloy steels, or low-alloy structural steels, are distinctly different from the older and more widely known types of alloy steels. They combine high strength and at the same time usually possess substantially greater atmospheric corrosion resistance than low-carbon steel.

Hot-rolled steel. Hot-rolled low-carbon steel manufactured for general use. Hot-rolled mild steel has a thin, tight scale over a smooth rolled surface, and all corners not sheared or cut are usually slightly rounded.

Ingot and ingot bar. Refinery shapes employed for alloy production (not fabrication). Both are used for remelting.

Ingots usually weigh from 20 to 35 lb, and ingot bars from 50 to 70 lb. Both are usually notched to facilitate breakage into smaller pieces.

Iridium. Metallic element belonging to the platinum family. Iridium is a very hard, brittle, white metal, occurring in alluvial deposits along with platinum. It is used in apparatus for high temperatures. When alloyed with platinum, it is used for standard weights and measures; when alloyed with osmium, it is used in tipping pens and compass bearings.

Iron. Iron is the most abundant of metals, although aluminum occurs in a larger percentage of the earth's crust. The pure metal, which is practically unknown in the arts (although some grades of soft steel are almost chemically pure), is silver white, very ductile, and magnetic and may be prepared by electrolytic deposition from ferrous sulfate or by reduction of the pure oxide with hydrogen or aluminum. Pig iron is hard, brittle, and fairly fusible, containing about 3% carbon and varying amounts of sulfur, silicon, manganese, and phosphorus.

Kroll process. Process for the production of metallic titanium by the reduction of titanium tetrachloride with a more active metal, such as magnesium, yielding titanium as granules or powder.

Lead. Metallic element of bluish white color and bright luster that is very soft, highly malleable, slightly tenacious, ductile, and a poor conductor of electricity. Lead is used in making pipe and containers for corrosive liquids, and is a constituent of many useful alloys. Lead contracts on cooling and consequently cannot be used for making castings or other products requiring careful measurements. It flows under pressure, and this characteristic makes it useful as a material for anchoring objects to concrete or masonry, but this same characteristic may be unsatisfactory in situations where heavy pressure will cause too great a flowage. Solders contain lead in varying percentages, as do tinfoil and certain copper alloys. Lead is an important alloying element in the lead-bearing brasses, including architectural bronze.

Light metal. One of the low-density metals, such as aluminum, magnesium, titanium, or beryllium, or their alloys.

Magnesium. Magnesium occurs very widely distributed in combination with other elements as magnesite, dolomite, epsom salts, and kainite and is obtained by electrolysis of the fused chloride. It is a light, white, and fairly tough metal that tarnishes slightly in air and that in ribbon, wire, or powder form ignites on heating, burning with a dazzling white heat. Magnesium is the lightest metal used in construction and is used where both lightness and strength are desired.

Malleable iron. Special cast iron that has been subsequently annealed. The production of malleable cast iron consists of two distinct steps: the production of white iron castings that are brittle and hard, and the annealing or graphitizing of the white iron so as to secure the final ductile product. Annealing or graphitizing changes the combined carbon to free carbon.

Manganese. Manganese is obtained by reduction of the oxide with sodium, magnesium, or aluminum or by electrolysis. A gray white metal resembling iron, but harder and very brittle, it is used in the production of alloys with iron, copper, brass, and nickel.

Molybdenum. Molybdenum does not occur native but is obtained from molybdenite and wulfenite. The metal is prepared by reduction of the oxide with carbon in the electric furnace. It is a very hard, silver white metal used chiefly in the manufacture of certain grades of tool steel, boiler plate, rifle barrels, and large cranks, as well as in making filaments, screens, and grids for radios.

Nickel. Nickel is a hard, malleable, ductile, and tenacious metal, white in color, somewhat magnetic, and a fair conductor of heat and electricity. It belongs to the iron-cobalt group of elements and is chiefly valuable for the alloys it forms with other metals—nickel steel and Monel metal. Electrodeposition of nickel plate is used as a protective coating for metals.

Nickel silvers. Alloys of copper, nickel, and zinc. Depending on the composition, they range in color from a definite to slight pink cast through yellow, green, whitish green, whitish blue, to blue. A wide range of nickel silvers is made. Those that fall into the combined alpha-beta phase of metals are readily hot-worked and therefore are fabricated without difficulty into such intricate shapes as plumbing fixtures, stair rails, architectural shapes, and escalator parts. Lead may be added to improve machining.

Osmium. Bluish white, hard, crystalline metal belonging to the platinum family of elements. Osmium is the heaviest known form of matter and is very infusible, oxidizing when heated in the air with an oxide. Osmium is used in making lamp filaments; with iridium it forms the alloy osmiridium, which is used because of its hardness in tipping gold pens and for fine machine bearings.

Pattinson process. Process for separating silver from lead, in which the molten lead is slowly cooled so that crystals poorer in silver solidify out and are removed, leaving the melt richer in silver.

Platinum. Tin-white metal of metallic luster, tenacious, malleable, and ductile. It is welded at a red heat, has a coefficient of linear expansion approximately equal to that of glass, and does not oxidize in air at any temperature but is corroded by halogens, cyanide, sulfur, and caustic alkalies, It forms alloys with lead.

Platinum black. Finely divided from of platinum of a dull black color, usually, but not necessarily, produced by the reduction of salts in aqueous solution.

Precious metal. One of the relatively scarce and valuable metals: gold, silver, and the platinum-group metals.

Refined bar iron or refined wrought-iron bars. Iron bars rolled from a muck bar pile, slab pile, or box pile, of muck bars and wrought-iron scrap bars free from steel, all bars running the full length of the pile.

Rhodium. Silver white metallic element belonging to the platinum family, occurring native with other members of this group in river sands in the Urals and in North and South America. Its salts form red solutions. Rhodium is used chiefly to form alloys with other metals for use in thermoelements and other measuring devices.

Rolled gold. Same as gold filled except that the proportion of gold alloy to the weight of the entire article may be less than one-twentieth. The fineness of the gold alloy may not be less than 10 Karat. Karat is the 24th part by weight of gold in an article.

Silver. Copper ores yield considerable silver, which is obtained from the ores by smelting with lead or copper or by amalgamation with mercury. Silver is a pure white metal having a brilliant luster. It is a little harder than gold and excelled only by that metal in malleability and ductility, while excelling all other metals as a conductor of heat

and electricity. Silver undergoes no change in water or pure air but when melted, absorbs 22 times its volume of oxygen, which it expels again on cooling. It tarnishes in the vapors of sulfur compounds forming a sulfide.

Single-refined iron. All new wrought iron that is first rolled into muck bars or slabs that are then once piled and rolled.

Stainless steel. Stainless steel that is used for architectural metal purposes is commonly understood to be a ferrous alloy containing approximately 18% chromium and 8% nickel. Stainless alloy generally identified as Type 302 Stainless Steel, or more commonly, simply as 18-8.

Standard goal. Legally adopted alloy for coinage of gold. In the United States the alloy contains 10% copper.

Sterling silver. Silver alloy containing at least 92.5% silver, the remainder being unspecified but usually copper.

Tantalum. Gray metal resembling platinum in appearance. Tantalum is resistant to most acids and for this reason is used in industry in acid-handling equipment. Because of its high melting point, it is used for incandescent filaments in electric lights and in radios. Tantalum carbide is used in metal-cutting tools because of its great hardness. It is soluble in fused alkalies, insoluble in acids.

Terne alloy. Alloy consisting of lead and tin.

Tin. Silver white, malleable, and rather ductile metal with low tenacity and a highly crystalline structure. Tin takes a high polish and is used because of its corrosion-resisting properties as a coating for other metals.

Tungsten. Tungsten is obtained by reduction of the oxide with hydrogen, carbon, or aluminum. It is a hard, brittle, gray black metal, resistant too most acids, but soluble in nitric acid and strong alkalies. It forms the oxide when heated in air. Tungsten is used somewhat to form alloys with iron, to which it imparts inc.eased harness; the carbide is used in metal-cutting tools. Because of its very high melting point, tungsten is used extensively in filaments for incandescent lamps.

Wrought iron. Ferrous material aggregated from a solidifying mass of pasty particles of highly refined metallic iron, with which, without subsequent fusion, is incorporated a minutely and uniformly distributed quantity of slag. Wrought iron is produced from pig iron that is heated in a puddling furnace to a pasty consistency, then squeezed and rolled into the desired shapes.

Zinc. Zinc is used for making brass but was not recognized as a separate metal until it was obtained by heating calamine with charcoal. Zinc is a bluish white metal, which is brittle at ordinary temperatures but becomes malleable at 100°C. It is a fair conductor of electricity and burns in air at a high red heat with evolution of white clouds of the oxide. Used to alloy with other metals such as copper, it forms brass. Galvanizing consists in coating other metals with zinc to prevent corrosion. Zinc is used as the negative electrode in various types of electric batteries.

METALLIC POWDERS

Acicular powder. Needle-shaped particles.

Air classification. Separation of powder into particle size fractions by means of an airstream of controlled velocity; application of the principle of elutriation.

Alloy powder. Powder, all particles of which are composed of the same alloy of two or more metals (ASTM B 243).

Apparent density. Weight of a unit volume of powder, determined by a specified method of loading and usually expressed in grams per cubic centimeter.

Atomization. Dispersion of a molten metal into particles by a rapidly moving stream of gas or liquid.

Atomized metal powder. Metal powder produced by the dispersion of molten metals or alloys into particles by the impingement of a rapidly moving gas or liquid stream or by mechincal dispersion.

Binder. Cementing medium, either a material added to the powder to increase the green strength of the compact and expelled during sintering, or a material (usually of relatively low melting point) added to a powder mixture for the specific purpose of cementing together powder particles which alone would not sinter into a strong body.

Blank. Pressed presintered or fully sintered compact, usually in the unfinished condition, requiring cutting, machining, or some other operation to give it its final shape.

Blending. Thorough intermingling of powders of the same nominal composition (not to be confused with mixing) (ASTM B 243).

Bridging. Formation of arched cavities in a powder mass.

Bumping down. Consolidation of a mass of metal powder by vibration before the pressing operation.

Cake. Coalesced mass of unpressed metal powder (ASTM B 243).

Capping. Partial or complete separation of a compact into two or more portions by cracks that originate near the edges of the punch faces and that proceed diagonally into the compact.

Carbonyl powder. Metal powder prepared by the thermal decomposition of a metal carbonyl.

Chemical deposition. Precipitation of one metal from a solution of its salts by the addition of another metal or reagent to the solution.

Chemically precipitated metal powder

1. Powder produced as a precipitate by chemical displacement.
2. Powder produced by the reduction of a metal from a solution of its salts either by the addition of another metal higher in the electromotive series or by another reducing agent.

Classification. Separation of a powder into fractions according to particle size (ASTM B 243).

Coining. Final pressing of a sintered compact to obtain a definite surface configuration (not to be confused with repressing or sizing).

Cold-pressing. Forming a compact at a temperature low enough to avoid sintering, usually room temperature.

Cold-welding. Cohesion between two metal surfaces, generally under the influence of externally applied pressure at room temperature.

Compact. Object produced by the compression of metal powder, generally while confined in a die, with or without the inclusion of nonmetallic constituents. Synonymous with brique (ASTM B 243).

Composite compact. Metal powder compact consisting of two or more adhering layers of different metals or alloys with each layer retaining its original identity.

Compound compact. Metal powder compact consisting of mixed metals, the particles of which are joined by pressing or sintering, or both, with each metal particle retaining substantially its original composition.

Compressibility

1. Reciprocal of the compression ratio when the compact is made following a procedure in which the die, the pressure, and the pressing speed are specified.
2. Density ratio determined under definite testing conditions.

Compression ratio. Ratio of the volume of loose powder to the volume of the compact made from it.

Continuous sintering. Presintering or sintering in such manner that the objects are advanced through the furnace at a fixed rate by manual or mechanical means (ASTM B 243).

Cored bar

1. Bar shaped compact whose interior has been melted by the passage of electricity.
2. Compact of bar shape heated by its own electrical resistance to a temperature high enough to melt its interior.

Core rod. Part of a die used to produce a hole in a compact.

Dendrites. Crystals, usually formed during solidification or sublimation that are characterized by a tree-like pattern composed of many branches—pine tree or fir tree crystals (ASTM E 7).

Dendritic powder. Particles, usually of electrolytic origin, having the typical pine tree structure (ASTM B 243).

Density ratio. Ratio of the determined density of a compact to the absolute density of a metal of the same composition, usually expressed as a percentage.

Die. Part or parts making up the confining form in which a powder is pressed. Parts of the die may be some or all of the following: die body, punches, and core rods.

Die body. Stationary or fixed part of a die.

Die insert. Removable liner or part of a die body or punch.

Die lubricant. Lubricant applied to the walls of the die and punches to facilitate the pressing and ejection of the compact.

Die set. Parts of a press that hold and locate the die in proper relation to the punches.

Disintegration. Reduction of massive material to powder.

Dry density. Weight per unit volume of an unimpregnated sintering.

Electrolytic powder. Powder produced by electrolytic deposition, by the pulverization of an electrodeposit, or from metal made by electrodeposition.

Exudation. Action by which all or a portion of the low-melting constituent of a compact is forced to the surface during sintering.

Fines. The product that passes through the finest screen in the sorting of crushed or ground material.

Flake powder

1. Relatively thin flat or scalelike particles.
2. Flat or scalelike particles whose thickness is small compared with the other dimensions.

Flow rate. Time required for a powdered sample of standard weight to flow through an orifice in a standard instrument according to a specified procedure.

Fraction. Portion of a powder sample that lies between two stated particle sizes.

Fritting. Sintering in the presence of a liquid phase (ASTM B 243).

Gas classification. Separation of powder into particle size fractions by means of a gas stream of controlled velocity.

Grain size. Grain sizes are reported in terms of number of grains per unit area or volume, average diameter, or as a grain size number derived from area measurements.

Granular powder. Particles having approximately equidimensional nonspherical shapes.

Granulation. Production of coarse metal particles by pouring molten metal through a screen into water or by violent agitation of the molten metal while solidifying.

Green. Unsintered, for example, green compact, green density, green strength.

Growth. Increase in the dimensions of a compact that may occur during sintering.

Hot-pressing

1. Forming a compact at a temperature high enough to have sintering.
2. Simultaneous heating and molding of a compact.

Hydrogen loss. Loss in weight of metal powder or a compact caused by heating a representative sample for a specified time and temperature in a purified hydrogen atmosphere. Broadly, a measure of the oxygen content of the sample, when applied to materials containing only such oxides as are reducible with hydrogen and no hydride-forming element (ASTM B 243).

Hydrogen-reduced powder. Powder produced by the hydrogen reduction of a compound.

Impregnation. Process of filling the pores of a sintered compact with a nonmetallic material such as oil, wax, or resin.

Infiltration. Process of filling the pores of a sintered or unsintered compact with a metal or alloy of lower melting point.

Intercommunicating porosity. In a sintered compact, the type of porosity that connects the pores in such a way that a fluid may pass from one to another through the entire compact.

Irregular powder. Particles lacking symmetry.

K factor. Strength constant in the formula for the radial crushing strength of a plain sleeve specimen of sintered metal.

Liquid-phase sintering. Sintering of a compact or loose powder aggregate under conditions where a liquid phase is present during part of the sintering cycle (ASTM B 243).

Loading. Filling the die cavity with powder.

Lower punch. Lower member of a die that forms the bottom of the die cavity. It may or may not move in relation to the die.

Lubricating. Mixing some agent with or incorporating it into a powder, to facilitate pressing and ejecting the compact from the die body; applying a lubricant to the die walls and punch surfaces (ASTM B 243).

Matrix metal. Continuous phase of a polyphase alloy or mechanical mixture; physically continuous metallic constituent in which separate particles of another constituent are embedded.

Mesh. Screen number of the finest screen of a specified standard screen scale through which almost all the parti-

cles of a powder sample will pass. It is also called "mesh size."

Milling. Mechanical treatment of metal powder or metal powder mixtures, as in a ball mill, to alter the size or shape of the individual particles or to coat one component of the mixture with another (ASTM B 243).

Minus sieve. Portion of a powder sample that passes through a standard sieve of specified number.

Mixing. Thorough intermingling of the powders of two or more materials.

Molding. Pressing of powder to form a compact.

Molding press. Press used to form compacts.

Needles. Elongated rodlike particles.

Nodular powder. Irregular particles having knotted, rounded, or similar shapes.

Oversize powder. Particles coarser than the maximum permitted by a given specification for particle size.

Packing material. Material in which compacts are embedded during the presintering or sintering operations.

Particle size. Controlling the lineal dimension of an individual particle as determined by analysis with sieves or other suitable means (ASTM B 243).

Particle size distribution. Percentage by weight or by number of each fraction into which a powder sample has been classified with respect to sieve number of particle size (micrometers). The preferred usage is "particle size distribution by weight," or "particle size distribution by frequency" (ASTM B 243).

Permeability. Rate of passage of a liquid or gas through a compact, measured under specified conditions.

Pine tree crystal. A type of dendrite.

Plates. Flat particles of metal powder having considerable thickness.

Plus sieve. Portion of a powder sample retained on a standard sieve of specified number.

Pore-forming material. Substance included in a powder mixture that volatilizes during sintering and thereby produces a desired kind and degree of porosity in the finished compact.

Pores. Minute cavities in a compact.

Powder. Aggregate of discrete particles that are usually within the size range of 1 to 1000 μm.

Powdered. Metal powder is a powder pressed into shape and then sintered in a furnace. For additional strength it may be sintered again. The process is employed to form a range of products from various powdered metals.

Powder flow meter. Instrument for measuring the rate of flow of a powder according to a specified procedure (ASTM B 243).

Powder lubricant. Agent mixed with or incorporated in a powder to facilitate the pressing and ejecting of the compact.

Powdered metal. The portion of a powder composed of particles that are smaller than 44 μm.

Powder metallurgy. Art of producing metal powders and utilizing them for the production of massive materials and shaped objects.

Preforming. Initial pressing of a metal powder to form a compact which is subjected to a subsequent pressing operation other than coining or sizing; also, the preliminary shaping of a refractory metal compact after presintering and before the final sintering.

Presintering. Heating of a compact to a temperature lower than the normal temperature for final sintering, usually to increase the ease of handling or forming the compact or to remove a lubricant or binder before sintering.

Pressed bar. Compact in the form of a bar; green compact.

Pressed density

1. Density of an unsintered compact, sometimes called "green density."

2. Weight per unit volume of an unsintered compact. The term is synonymous with great density.

Pressing crack. Rupture in the pressed compact that develops during the ejection of the compact from the die.

Puffed compact. Compact expanded by internal gas pressure.

Pulverization. Reduction of metal to powder by mechanical means.

Punch. Part of a die assembly; used to transmit pressure to the powder in the die cavity.

Radial crushing strength. Relative capacity of a plain sleeve specimen of sintered metal to resist fracture induced by a load applied between flat parallel plates in a direction perpendicular to the axis of the specimen.

Rate of oil flow. Rate at which a specified oil will pass through a sintered porous compact under specified test conditions.

Repressing. Application of pressure to a previously pressed and sintered compact usually for the purpose of improving some physical property.

Roll compacting. Progressive compacting of metal powders by the use of a rolling mill (ASTM B 243).

Rolled compact. Compact made by passing metal powder continuously through a rolling mill to form relatively long sheets of pressed material.

Rotary press. Machine fitted with a rotating table carrying multiple dies in which material is pressed.

Segment die. Die made of parts that can be separated for the ready removal of the compact. The term is synonymous with "split die."

Shrinkage. Decrease in dimensions of a compact that may occur during sintering.

Sieve analysis. Particle size distribution, usually expressed as the weight percentage retained on each of a series of standard sieves of decreasing size and the percentage passed by the sieve of finest size.

Sieve classification. Separation of powder into particle size ranges by the use of a series of graded sieves.

Sieve fraction. Portion of a powder sample that passes through a standard sieve of specified number and is retained by some finer sieve of specified number.

Sintering

1. Bonding of adjacent surfaces of particles in a mass of metal powders or a compact by heating; shaped body composed of metal powders and produced by sintering with or without prior compacting.

2. Product made by sintering metal powders.

Sizing. Final pressing of a sintered compact to secure desired size.

Slip crack. Rupture in the pressed compact caused by the mass slippage of a part of the compact.

Specific surface. Surface area of one gram of powder, usually expressed in square centimeters.

Spherical powder. Globular shaped particles.

Sponge iron. Coherent, porous mass of substantially pure iron produced by solid-state reduction of iron oxide (ASTM B 243).

Sponge iron powder. Ground and sized sponge iron, which may have been purified or annealed, or both.

Spongy. Porous condition in metal powder particles usually observed in reduced oxides.

Stocking. Presintering or sintering in such a way that the compacts are advanced through the furnace at a fixed rate by manual or mechanical means. The process is also called "continuous sintering."

Stripper punch. Punch that in addition to forming the top or bottom of the die cavity later moves further into the die to eject the compact.

Subsieve analysis. Size distribution of particles, all of which will pass through a 44-μm standard sieve, as determined by specified methods.

Subsieve fraction. All particles that will pass through a 44-μm standard sieve.

Superfines. Portion of a powder composed of particles that are smaller than a specified size, currently less than 10 μm.

Tap density. Apparent density of a powder, obtained when the volume receptacle is tapped or vibrated during loading under specified condition (ASTM B 243).

Transverse rupture strength. Stress, calculated from the flexure formula, required to break a specimen like a simple beam supported near the ends when the load is applied midway between the fixed line center of the supports.

Upper punch. Member of a die assembly that moves downward into the die cavity to transmit pressure to the powder contained in the die cavity.

Warpage. Distortion that may occur in a compact during sintering (ASTM B 243).

METALLURGY

Acid Bessemer (converter). Suspended, tilting vessel that uses high-speed air and oxygen to oxidize impurities in hot pig iron and other iron-bearing materials to produce high phosphorus steel.

Acid bottom and lining. Inner bottom and lining of a melting furnace consisting of materials like sand, siliceous rock, or silica brick that give an acid reaction at the operating temperature.

Acid embrittlement. Form of hydrogen embrittlement that may be induced in some metals by acid treatment.

Acid flux. Mineral, such as sand, gravel, and quartz rock, used in acid furnaces to make acid (high-phosphorus) steel.

Acid steel. Steel melted in a furnace with an acid bottom and lining and under a slag containing an excess of an acid substance such as silica.

Activation. Changing of the passive surface of a metal to a chemically active state.

Age hardening. Hardening by aging, usually after rapid cooling or cold-working.

Agglomeration. Process for increasing the particle size of iron ores to make them suitable for iron and steel making.

Aging. In a metal or alloy, a change in properties that generally occurs slowly at room temperature and more rapidly at higher temperatures. Aging includes age hardening, artificial aging, interrupted aging, natural aging, overaging, precipitation hardening, precipitation heat treatment, progressive aging, quench aging, and strain aging.

Air furnace. Nonregenerative reverberatory furnace.

Air hardening steel. Steel containing sufficient carbon and other alloying elements to harden fully during cooling in air or other gaseous mediums from a temperature above its transformation range. The term is restricted to steels that are capable of being hardened by cooling in air in fairly large sections, about 2 in. or more in diameter.

Alligatoring. Longitudinal splitting of flat slabs in a plane parallel to the rolled surface.

Allotropy. Ability of a material to exist in several crystalline forms.

Alloy

1. Substance having metallic properties and composed of two or more chemical elements, at least one of which is an elemental metal.

2. Mixture of metals melted together to form a uniform compound.

Alloying element

1. Substance added to metal to purposely change its mechanical or physical properties.

2. Element added in steel making to achieve desired properties.

3. Element added to a metal to effect changes in properties, and remaining within the metal.

Alloy steel

1. Steel containing significant quantities of alloying elements (other than carbon and the commonly accepted amounts of manganese, silicon, sulfur, and phosphorus) added to effect changes in the mechanical or physical properties.

2. Steel in which residual elements exceed the limits prescribed for carbon steel, or to which alloying elements are added within specified ranges.

Amorphous. Noncrystalline, random orientation of the atomic structure.

Anchor pattern. Contour of the surface of a metal, the height of which is generally measured in thousandths of an inch.

Angle of bite. In rolling metals where all the force is transmitted through the rolls, the angle of bite is the maximum attainable angle between the roll radius at the first contact and the line of roll centers. If the operating angle is less, it is called the contact angle or rolling angle.

Anistropy. Variation in properties when a material is tested along axes in different directions.

Annealing

1. Heating metal to high temperatures (1350°F to 1600°F for steel), followed by controlled cooling to make the metal softer or change its ductility and toughness.

2. Type of controlled slow heating and cooling of a material to alter its mechanical and physical properties.

3. Heating to and holding at a suitable temperature and then cooling at a suitable rate, for such purposes as reducing hardness, improving machinability, facilitating cold-working, producing a desired microstructure, or obtaining desired mechanical, physical, or other properties.

Anodizing. To subject a metal to electrolytic action as the anode of a cell in order to coat it with a protective or decorative film; used for nonferrous metals.

Arc furnace. Furnace in which material is heated either directly by an electric arc between an electrode and the work or indirectly by an arc between two electrodes adjacent to the material.

Arc melting. Melting metal in an electric arc furnace.

Artificial aging. Aging above room temperature.

Atomization. Dispersion of a molten metal into particles by a rapidly moving gas or liquid stream.

Austempering. Quenching a ferrous alloy from a temperature above the transformation range, in a medium having a rate of heat abstraction high enough to prevent the formation of high-temperature transformation products, and then holding the alloy, until transformation is complete, at a temperature below that of pearlite formation and above that of martensite formation.

Austenite. Solid solution of one or more elements in face-centered cubic iron. Unless otherwise designated (for example, nickel austenite), the solute is generally assumed to be carbon.

Austenitic steels

1. Tough, strong, and nonmagnetic steels. Austenitic stainless steels have a chromium content of up to 25% and a nickel content of up to 22%, and can be hardened by cold-working.

2. Alloy steel whose structure is normally austenitic at room temperature.

Austenitizing. Forming austenite by heating a ferrous alloy into the transformation range (partial austenitizing) or above the transformation range (complete austenitizing).

Autoradiography. Inspection technique in which radiation spontaneously emitted by the material is recorded photographically.

Baking. Heating to a low temperature in order to remove gases.

Ball mill. Mill in which crushed ores and a variety of other materials are finely ground in a rotating cylinder containing pebbles or metal balls.

Bar

1. Hot-rolled or cold-drawn round, square, hexagonal, and multifaceted long shapes, generally larger than wire in cross section; also, hot- or cold-rolled rectangular flat shapes (flats) generally narrower than sheets and strips.

2. Piece of material thicker than sheet, long in proportion to its width or thickness, and with a width-thickness ratio much smaller than sheet or plate—as low as unity for squares and rounds.

Bark. Decarburized layer just beneath the scale that results from heating steel in an oxidizing atmosphere.

Base metal. One of the parts to be welded together; also called "parent metal."

Basic flux. Mineral (limestone, dolomite) used in basic furnaces to make basic (low-phosphorus) steel.

Basic oxygen vessel. Suspended, tilting vessel that uses high-purity oxygen to oxidize impurities in hot pig iron and other iron-bearing materials to produce low-phosphorus (basic) steel.

Basic steel. Steel melted in a furnace with a basic bottom and lining and under a slag containing an excess of a basic substance such as magnesia or lime.

Bauschinger effect. Phenomenon by which plastic deformation of a polycrystalline metal, caused by stress applied in one direction, reduces the yield strength where the stress is applied in the opposite direction.

Bed. Stationary portion of a press structure that usually rests on the floor or foundation, forming the support for the remaining parts of the press and the pressing load. The bolster and sometimes the lower die are mounted on the top surface of the bed.

Bend test

1. Test for ductility performed by bending or folding, on a specimen but in some instances by blows, a specimen having a cross section substantially uniform over a length several times as great as the largest dimension of the cross section.

2. Test for determining the relative ductility of metal that is usually formed into sheet, strip, plate, or wire and for determining the soundness and toughness of metal. The specimen is usually bent over a specified diameter through a specified angle for a specified number of cycles.

Beneficiation. Concentrating process used to increase the iron content of ores prior to use.

Bessemer (acid) pig iron

1. Low-phosphorus iron used in acid steel-making furnaces.

2. Pig iron with sufficiently low phosphorus (0.100% max) to be suitable for use in the Bessemer process.

Bessemer process. Process for making steel by blowing air through molten pig iron contained in a refractory lined vessel so as to remove by oxidation most of the carbon, silicon, and manganese.

Billet. Solid, semifinished, round or square product that has been hot-worked by forging, rolling, or extrusion. An iron or steel billet has a minimum width or thickness of $1\frac{1}{2}$ in. and the cross-sectional area varies from $2\frac{1}{4}$ to 36 sq in. For nonferrous metals, it may also be a casting suitable for finished or semifinished rolling or for extrusion.

Billet mill. Primary rolling mill used to make billets.

Biscuit. Upset blank for drop-forging; small cake of primary metal, such as uranium made from uranium tetrafluoride and magnesium in a bomb reduction.

Black annealing. Box annealing or pot annealing ferrous alloy sheet, strip, or wire.

Blackplate. Cold-rolled flat carbon steel products thinner than sheets and wider than strips, generally used for coating with zinc, tin, or terne metal.

Blank

1. In forming, a piece of sheet material, produced in cutting dies, that is usually subjected to further press operations.

2. Pressed, presintered, or fully sintered compact, usually in the unfinished condition, requiring cutting, machining, or some other operation to give it its final shape.

Blank carburizing. Simulating the carburizing operation without introducing carbon, usually accomplished by using an inert material in place of the carburizing agent, or by applying a suitable protective coating to the ferrous alloy.

Blank nitriding. Simulating the nitriding operation without introducing nitrogen, usually accomplished by using an inert material in place of the nitriding agent, or by applying a suitable protective coating to the ferrous alloy.

Blast furnace

1. Tall, cylindrical masonry structure lined with refractory materials, used to smelt iron ores in combination with fluxes, coke, and air into pig iron.

2. Shaft furnace in which solid fuel is burned with an air blast to smelt ore in a continuous operation. Where the temperature must be high, as in the production of pig iron, the air is preheated. Where the temperature can be lower, as in smelting copper, lead, and tin ores, a smaller furnace is economical, and preheating of the blast is not required.

Blister. Defect in metal, on or near the surface, resulting from the expansion of gas in a subsurface zone. Very small blisters are called "pinheads" or "pepper blisters."

Bloom. Semifinished hot-rolled product, rectangular in cross section, produced on a blooming mill. For iron and steel, the width is not more than twice the thickness, and the cross-sectional area is usually not less than 36 sq in. Iron and steel blooms are sometimes made by forging.

Bloomer. Mill or equipment used in reducing steel ingots to blooms.

Blooming mill. Primary rolling mill used to make blooms.

Blow hole. Hole in a casting or a weld caused by the entrapment of gas in the metal during its solidification.

Blue annealing. Heating hot-rolled ferrous sheet in an open furnace to a temperature within the transformation range and then cooling in air in order to soften the metal. The formation of a bluish oxide on the surface is incidental.

Blue brittleness. Brittleness exhibited by some steels after they are heated to a temperature within the range of 300°F to 650°F more especially if the steel is worked at the elevated temperature. Killed steels are virtually free of this kind of brittleness.

Bluing. Subjecting the scale-free surface of a ferrous alloy to the action of air, steam, or other agents at a suitable temperature, thus forming a thin blue film of oxide and improving the appearance and resistance to corrosion.

Board hammer. Type of forging hammer in which the upper die and ram are attached to "boards" that are raised to the striking position by power-driven rollers and let fall by gravity.

Bonding. Chemical bonds of metals or other compounds that are related to the valence of the atoms. Bonding in welding is the fusion of two metals.

Box annealing. Annealing a metal or alloy in a sealed container under conditions that minimize oxidation. In box annealing a ferrous alloy, the charge is usually heated slowly to a temperature below the transformation range, but sometimes above or within it, and is then cooled slowly. This process is also called "close annealing" or "pot annealing."

Box pile. In manufacturing wrought-iron bars, a pile, the outside of which is formed of flat bars and the interior of a number of small bars, all bars running the full length of the pile.

Bright annealing. Annealing in a protective medium to prevent discoloration of the bright surface.

Brinell hardness test. Test for determining the hardness of a material by forcing a hard steel or carbide ball of specified diameter into it under a specified load (ASTM E 10).

Brittleness. Characteristic of a material that leads to cracking or breaking without appreciable plastic deformation.

Burning. Permanently damaging a metal or alloy by heating to cause either incipient melting or intergranular oxidation; in grinding, getting the work hot enough to cause discoloration or to change the microstructure by tempering or hardening.

Busheling. Process of heating to a welding heat, in a reverberatory furnace, miscellaneous iron, steel, or a mixture of iron and steel scrap cut into small pieces.

Cake. Refinery shape for rolling into plate, sheet, strip, or formed shape.

Camber. Crown in rolls where the center diameter has been increased to compensate for deflection caused by the rolling pressure.

Capped steel. Semikilled steel cast in a bottle-top mold and covered with a cap fitting into the neck of the mold. The cap causes the top metal to solidify. Pressure is built up in the sealed-in molten metal and results in a surface condition much like that of rimmed steel.

Carbon steel

1. Steel for which no minimum content is specified for any element added to obtain desirable alloying effects. Carbon is the most important element, and its content determines the properties of carbon steel.

2. Steel in which the residual elements are controlled, but alloying elements are not usually added.

3. Steel containing up to about 2% carbon and only residual quantities of other elements except those added for deoxidation, with silicon usually limited to 0.60% and manganese to about 1.65%. It is also termed "plain carbon steel," "ordinary steel," and "straight carbon steel."

Carburizing. Introducing carbon into a solid ferrous alloy in contact with a suitable carbonaceous material. Carburized alloy is usually quench hardened.

Case. In a ferrous alloy, the outer portion that has been made harder than the inner portion or core by case hardening.

Case hardening

1. Hardening of the outer skin of an iron-base alloy by promoting surface absorption of carbon, nitrogen, or cyanide, which is generally accomplished by heating the alloy in contact with materials containing these elements and rapidly cooling it.

2. Hardening a ferrous alloy so that the outer portion or case is made substantially harder than the inner portion or core. Typical processes used for case hardening are carburizing, cyaniding, carbonitriding, nitriding, induction hardening, and flame hardening.

Casting

1. Pouring a liquid or molten material into a mold to produce an object of desired shape.

2. Object at or near finished shape obtained by solidification of a substance in a mold.

Cast iron

1. Series of alloys primarily of iron, carbon, and silicon in which the carbon is in excess of the amount that can be retained in solid solution in austenite at the eutectic temperature.

2. High carbon iron made by melting pig iron with other iron-bearing materials and casting in sand or loam molds. Cast iron is characterized by hardness, brittleness, and high compressive and low tensile strengths.

3. Iron containing carbon in excess of the solubility in the austenite that exists in the alloy at the eutectic temperature. For the various forms of cast iron, such as gray cast iron, white cast iron, malleable cast iron, and nodular cast iron the word "cast" is often left out, resulting in "gray iron," "white iron," "malleable iron," and "nodular iron," respectively.

Cast steel. Steel in the form of castings.

Catalyst. An agent that induces catalysis, observed where a chemical reaction between two or more substances is influenced by the presence of a third substance (the catalyst).

Cathode

1. Unmelted flat plate produced by electrolytic refining.

2. Electrode where electrons enter (current leaves) an operating system such as a battery, an electrolytic cell, an x-ray tube, or a vacuum tube. In the first of these, it is positive; in the other three, negative. In a battery or electrolytic cell, it is the electrode where reduction occurs. Contrast with anode.

Cathodic polarization. In an electrolyte in which hydrogen gas can accumulate around the cathode, very little corrosion can take place on the anode.

Caustic. A chemical base material; caustic soda (a solution of sodium hydroxide used to clean metals).

Cementation. Introduction of one or more elements into the outer portion of a metal object by means of diffusion at high temperature.

Cementite. Compound of iron and carbon, also known as "iron carbide."

Centrifugal casting. Casting made by pouring metal into a mold that is rotated or revolved.

Cermet. Composite material made by bonding a ceramic substance with a metal or alloy at high temperatures and under controlled conditions.

Chemical deposition. Precipitation of one metal from a solution of its salts by the addition of another metal or reagent to the solution.

Chemical machining. Removal cutting process applied to metals, involving the chemical attack of corrosive fluids; form of etching.

Chilled cast iron. Cast iron that would normally solidify as a gray cast iron is purposely caused to solidify as white cast iron locally or entirely by accelerated cooling caused by contact with a metal surface, that is, a chill.

Cladding. Bonding thin sheets of a coating metal with desirable properties (such as corrosion resistance or chemical inertness) over a less expensive metallic core not possessing these properties. Copper cladding over steel may be applied by hot-dipping; stainless steel and aluminum cladding by hot-rolling.

Cluster mill. Rolling mill where each of the two working rolls of small diameter is supported by two or more backup rolls.

Cohesion. Fastening process in which materials are permanently joined by the fusing of like materials by means of molten metal or heat and pressure into a homogeneous, continuous mass.

Coke. Processed form of bituminous coal used as a fuel; reducing agent and a source of carbon in making pig iron.

Cold-drawing. Shaping by pulling through a die to reduce cross-sectional area and impart desired shape. Cold-drawing is generally accompanied by increase in strength, hardness, closer dimensional tolerances, and smoother finish.

Cold-finishing. Cold-working that results in finished mill products.

Cold-forming

1. Plastically deforming metal, usually at room temperature, into desired shapes with compressive force.
2. Forming thin sheets and strip to desired shapes at room temperature, generally with little change in mechanical properties of the metal. The process includes roll, stretch shear, and brake forming.

Cold-reduction. Cold-rolling that drastically reduces sheet and strip thickness with each pass through the rolls. Cold-reduction is generally accompanied by increase in hardness, stiffness, and strength and results in a smoother finish and improved flatness.

Cold-rolling. Gradual shaping between rolls to reduce cross-sectional area or impart desired shape. Cold-rolling is generally accompanied by increase in strength and hardness.

Cold treatment. Cooling to a low temperature, often near $-100°F$ for the purpose of obtaining desired conditions or properties, such as dimensional or structural stability.

Cold work. Permanent strain produced by an external force in a metal below its recrystallization temperature.

Cold-working

1. Shaping by cold-rolling, cold-drawing, or cold-reduction at room temperature. Cold-working is generally accompanied by increase in strength and hardness.
2. Deforming metal plastically at a temperature lower than the recrystallization temperature.

Common iron. Iron made from rerolled scrap iron or a mixture of iron and steel scrap, with no attempt made to separate the iron and steel scrap.

Compact. Object produced by the compression of metal powder, generally while confined in a die, with or without the inclusion of nonmetallic constituents.

Compressive strength (ultimate). Maximum stress that can be applied to a brittle material in compression without fracture.

Compressive strength (yield). Maximum stress that can be applied to a metal in compression without permanent deformation.

Congruent melting. Isothermal or isobaric melting in which both the solid and liquid phases have the same composition throughout the transformation.

Continuous casting. Casting technique in which an ingot, billet, tube, or other shape is continuously solidified while it is being poured, so that its length is not determined by mold dimensions.

Continuous mill. Rolling mill consisting of a number of stands of synchronized rolls (in tandem) in which metal undergoes successive reductions as it passes through the various stands.

Controlled cooling. Cooling from an elevated temperature in a predetermined manner to avoid hardening, cracking, or internal damage, or to produce a desired microstructure. Cooling usually follows a hot-forming operation.

Converter. Furnace in which air is blown through a bath of molten metal or matte, oxidizing the impurities and maintaining the temperature through the heat produced by the oxidation reaction.

Cooling stresses. Residual stresses resulting from nonuniform distribution of temperature during cooling.

Core

1. In a ferrous alloy, the inner portion that is softer than the outer portion or case.
2. Specially formed material inserted in a mold to shape the interior or another part of a casting that cannot be shaped as easily by the pattern; in a ferrous alloy, the inner portion that is softer than the outer portion or case.

Cored bar. Compact of bar shape heated by its own electrical resistance to a temperature high enough to melt its interior.

Corrosion

1. Deterioration or failure of metals and alloys by chemical or electrochemical processes.
2. Deterioration of a metal by chemical or electrochemical reaction with its environment.
3. Gradual wearing away or alteration by a chemical or electrochemical (essentially oxidizing) process of the metal.
4. Deleterious effect on a metal surface due to weathering, galvanic action, or direct chemical attack.

Corrosion embrittlement. Severe loss of ductility of a metal resulting from corrosive attack, usually intergranular and often not visually apparent.

Corrosion resistance. Ability of some metals to form an oxide surface skin that resists further corrosion.

Counterblow hammer. Forging hammer in which both the ram and anvil are driven simultaneously toward each other by air or steam pistons.

Coupon. Piece of metal from which a test specimen is to be prepared, often an extra piece as on a casting or forging.

Creep. Slow plastic deformation in steel and most structural metals caused by prolonged stress under the yield point at elevated temperatures.

Critical strain. Strain just sufficient to cause the growth of very large grains during heating where no phase transformations take place.

Crop. Defective end portion of an ingot which is cut off as scrap.

Cross-country mill. Rolling mill in which the mill stands are so arranged that their tables are parallel with a transfer (or crossover) table connecting them. A cross-country mill is used for rolling structural shapes, rails, and any special form of bar stock not rolled in the ordinary bar mill.

Crucible. Vessel or pot made of a refractory substance or of a metal with a high melting point, used for melting metals or other substances.

Crushing test. Radial compressive test applied to tubing, sintered metal hearings, or similar products for determining radial crushing strength (maximum load in compression).

Crystallization. Formation of crystals when a liquid solidifies and the atoms arrange themselves in a definite lattice structure. For metals, the common arrangements are body-centered cubic, face-centered cubic, and close-packed hexagonal.

Crystal unit structure. Simplest polyhedron that embodies all the structural characteristics of a crystal and makes up the lattice of a crystal by indefinite repetition.

Cupola

1. Furnace for melting metal.
2. Cylindrical vertical furnace for melting metal, especially gray iron, by having the charge come in contact with the hot fuel, usually metallurgical coke.

Cutting. Removing or separating pieces of material from a base or parent material.

Cyaniding. Introducing carbon and nitrogen into a solid ferrous alloy by heating while in contact with molten cyanide of suitable composition. The cyanided alloy is usually quench hardened.

Cyanogen. Colorless gas of characteristic odor, forming hydrocyanic and cyanic acid when in contact with water.

Decarburization. Loss of carbon from the surface of a ferrous alloy as a result of heating in a medium that reacts with the carbon at the surface.

Deformation. Alteration of the form or shape as a result of the plastic behavior of a metal under stress.

Dendrite. A crystal that has tree-like branching patterns that are most evident in cast metals slowly cooled through the solidification range.

Density. The density of a body is the ratio of its mass to its volume.

Deoxidizer. Substance that is used to remove oxygen from molten metals.

Descaling. Removing the thick layer of oxides formed on some metals at elevated temperatures.

Dezincification. Corrosion of a metal alloy that contains zinc.

Diamagnetic. Pertaining to materials that exhibit a slightly negative repulsion with regard to a strong magnetic field. The field causes the electron spins to align.

Die casting. Casting made in a die; casting process where molten metal is forced under high pressure into the cavity of a metal mold.

Die forging. Forging whose shape is determined by impressions in specially prepared dies.

Dieing machine. High-speed vertical press, the slide of which is activated by pull rods extending to the drive mechanism below the bed.

Differential heating. Heating that produces a temperature distribution within an object in such a way that after cooling various parts have different properties, as desired.

Diffusion. Process of intermingling atoms or other particles within a solution.

Direct chemical attack. Corrosion caused by a chemical dissolving of the metal.

Direct chill casting. Continuous method of making ingots or billets for sheet or extrusion by pouring the metal into a short mold. An ingot is usually cooled by the impingement of water directly on the mold or on the walls of the solid metal as it is lowered. The length of the ingot is limited by the depth to which the platform can be lowered; therefore, the process is often called "semicontinuous casting."

Direct quenching. Quenching carburized parts directly from the carburizing operation.

Discontinuous yielding. Nonuniform plastic flow of a metal exhibiting a yield point in which plastic deformation is inhomogeneously distributed along the gauge length. Discontinuous yielding may occur, under some circumstances, either at the onset of or during plastic flow in metals not exhibiting a yield point.

Distortion. Deviation from desired shape or contour.

Double-acting hammer. Forging hammer in which the ram is raised by admitting steam or air into a cylinder below the piston, and the blow intensified by admitting steam or air above the piston on the downward stroke.

Double-action mechanical press. Press having two independent parallel movements by means of two slides, one moving within the other. The inner slide or plunger is usually operated by a crankshaft, whereas the outer or

blank-holder slide, which swells during the drawing operation, is usually operated by a toggle mechanism or cams.

Double-crank press. Mechanical single-action press of such width that the slice is operated by a crank shaft having two crankpins to which two connections are attached.

Double-refined iron. Iron classed as double refined is known as "new wrought iron," which is first rolled into muck bars or slabs. Bars or slabs are then twice piled and rerolled. All iron is then free from steel and from foreign scrap. The manufacturer may use his own mill products of at least equal quality, but only in the first piling. In the final piling all bars or slabs are the full length of the pile.

Drawing. Reducing or changing the cross section of a material by pulling it through a die; forming deeply recessed parts by means of the plastic flow of material between dies.

Drop forging. Forging made with a drop hammer.

Drop hammer. Forging hammer that depends on gravity for its force.

Ductile iron. Cast iron that has been treated in the liquid state so as to cause a substantial portion of its graphitic carbon to occur as spheroids or nodules.

Ductility. The ability of a material to deform plastically without fracturing, being measured by elongation or reduction of area in a tensile test, by height of cupping in an Erichsen test or by other prescribed approved means (ASTM E 643).

Durville process. Casting process that involves rigid attachment of the mold in an inverted position above the crucible. The melt is poured by tilting the entire assembly, causing the metal to flow along a connecting launder and down the side of the mold.

Elasticity. Ability of a material to return to its original form after a load has been removed.

Electric arc. Suspended tilting kettle that melts scrap metal, ore, and sometimes ferroalloys with the heat of an electric arc to produce steels of controlled chemical composition.

Electric induction heater. Steel-encased insulated magnesia pot in which metal scrap and ferroalloys are melted with the heat of an electric current induced by windings of electric tubing. Electric induction used chiefly to produce small quantities of high-grade steels such as alloy, stainless, and heat-resisting steels.

Electrochemical machining. Cutting process whereby metal is removed from a workpiece by dissolving the metal in a chemical solution through which an electric current is passing.

Electrolysis. Decomposition of a substance by means of an electric current.

Electrolyte. Nonmetallic conductor in which electric current is carried by the movement of ions.

Embrittlement. Reduction in the normal ductility of a metal due to a physical or chemical change.

Equilibrium. In metals, a condition of balance where no net change results in an alloy system.

Eutectic. Low-melting-point phase of specific chemical composition in an alloy system. A eutectic has a distinctive structure, which is usually alternating layers of two other phases of the alloy.

Eutectoid. Alloy composition that transforms from a high temperature solid into new phases at the lowest constant temperature.

Explosive-forming operation. Press-forming operation in which the force is supplied by detonating an explosive such as blasting powder, used for large metal parts of sheet or plate.

Extensometer (or strain gauge). Instrument for measuring a more or less minute deformation of a test specimen, caused by tension, compression, bending, twisting, etc.

Extruding. Forming process whereby a stock material billet or pellet is converted into a length with a uniform cross section by forcing the material through a die opening of the same cross section.

Extrusion

1. Forcing of a molten metal through a die by pressure. The term is also used to describe nonferrous metal shapes and sections formed by extrusion processes.
2. Conversion of a billet into lengths of uniform cross section by forcing the plastic metal through a die orifice of the desired cross-sectional outline. In "direct extrusion," the die and ram are at opposite ends of the billet, and the product and ram travel in the same direction.

Fagoting. Making a "fagot," or "box," the bottom and sides of which are formed of muck or scrap bars and the interior of miscellaneous iron scrap or a mixture of iron and steel scrap.

Fatigue in metals. Tendency for a metal to fail by breaking or cracking under conditions of repeated cyclic stressing that takes place well below the ultimate tensile strength.

Ferrimagnetic material. Material that macroscopically has properties similar to those of a ferromagnetic material, but which microscopically also resembles an antiferromagnetic material in that some of the elementary magnetic moments are aligned antiparallel. If the moments are of different magnitudes, the material may still have a large resultant magnetization.

Ferrite. Magnetic form of iron; solid solution in which alpha iron is the solvent, characterized by a body-centered cubic crystal structure.

Ferritic steels. Soft, ductile, and strongly magnetic steels. Ferritic stainless steels usually have a chromium content of between 12 and 27% and cannot be hardened by heat treatment.

Ferroalloy. Alloy of iron that contains a sufficient amount of one or more additional chemical elements to be useful as an agent for introducing these elements into molten metal, usually steel.

Ferroalloys. Iron-based alloys used in steel making as a source of desired alloying elements.

Ferromagnetic material. Material that, in general, exhibits the phenomena of hysteresis and saturation, and whose permeability is dependent on the magnetizing force. The unmagnetized condition of a ferromagnetic material results from the overall neutralization of the magnetization of the domains to produce zero external magnetization.

Ferromagnetism. Strong magnetism resulting from the ability of certain metals and metal oxides to be magnetic in their natural state or to retain their magnetic properties after being magnetized.

Ferrous. Pertaining to an iron-based metallic material.

Ferrous alloy annealing. When applied to ferrous alloys, the term "annealing," without qualification, implies full annealing.

Ferrous alloys. Composite metals whose chief ingredient is iron. Ferrous alloys are metallurgically combined with one or more alloying elements.

File hardness

1. Empirical determination of the comparative hardness of a material made by notching it with a file.
2. Hardness as determined by the use of a file of standardized hardness. The assumption is that a material that cannot be cut with the file is as hard as or harder than the file. Files covering a range of hardnesses may be employed.

Finished mill products. Steel shapes that can be used directly in construction.

Finished steel. Steel that is ready for the market and has been processed beyond the stages of billets, blooms, sheet bars, slabs, and wire rods.

Flame annealing. Annealing in which the heat is applied directly by a flame.

Flame hardening. Quench hardening in which the heat is applied directly by a flame.

Flat product. Rectangular or square solid section of relatively great length in proportion to thickness.

Flux

1. Mineral that due to its affinity to the impurities in iron ores is used in iron and steel making to separate impurities in the form of molten slag.
2. In metal refining, a material used to remove undesirable substances, like sand, ash, or dirt, as a molten mixture. It is also used as a protective covering for certain molten metal baths. Lime or limestone is generally used to remove sand in iron smelting; sand is generally used to remove iron oxide in copper refining.

Fog quenching. Quenching in a fine vapor or mist.

Foil

1. Cold-rolled flat product less than 24 in. wide and less than 0.005 in. thick.
2. Metals in sheet form less than 0.006 in. in thickness.

Forging

1. Plastically deforming metal, usually while it is hot, into desired shapes with compressive force. Forging may be accomplished with or without forming dies.
2. Shaping hot metal between dies with compression force or impact.

Forging press. Press, usually vertical, used to operate dies to deform metal plastically. Mechanical presses are used for smaller closed-die forgings, hydraulic or steam hydraulic presses for flat-die and larger closed-die forgings.

Forming. Pressing metal into shape by mechanical operations other than machining.

Four-point press. Press, whose slide is actuated by four connections and four cranks, eccentrics, or cylinders, its chief merit being that it equalizes the pressure at the corners of the slides.

Fracture. Ruptured surface of metal that shows a typical crystalline pattern.

Fracture test. Breaking a specimen and examining the fractured surface with a microscope to determine the composition, grain size, case depth, soundness, or the possible presence of defects.

Fusion. Merging of two materials while in a molten state.

Galvanic action. Corrosion produced by electrolytic action between two dissimilar metals in the presence of an electrolyte.

Galvannealing. Applying a coat of zinc to a soft steel sheet, then passing it through an oven at about 1200°F to produce a dull, unspangled finish suitable for painting.

Galvanizing. Zinc coating by electroplating or hot-dipping, which produces a characteristic bright spangled finish and protects the base metal from atmospheric corrosion.

Gangue. Commercially undesirable portion of an ore that must be removed before the ore is processed into a metal.

Graphite. An allotropic form of carbon.

Gauge

1. Thickness or diameter used in connection with thin materials usually not more than $\frac{1}{4}$ in. Various standards are arbitrary and differ for ferrous and nonferrous materials and for sheets and wires.

2. Thickness or diameter of sheet or wire. Various standards are arbitrary and differentiate ferrous from nonferrous products and sheet from wire.

Gauge length. Original length of the portion of the specimen over which strain, change of length, and other characteristics are measured.

Grain structure. Microscopic internal crystalline structure (size and distribution of particles) of a metal austenitic, ferritic, and martensitic affects its properties.

Graphitizing. Annealing a ferrous alloy in such a way that some or all of the carbon is precipitated as graphite.

Gray cast iron. Cast iron that contains a relatively large percentage of the carbon present in the form of flake graphite. The metal has a gray fracture.

Hammer forging. Forging in which the work is deformed by repeated blows. Compare with press forging.

Hardenability. In a ferrous alloy, the property that determines the depth and distribution of hardness induced by quenching.

Hardener. Alloy rich in one or more alloying elements, added to a melt to permit closer composition control than would be possible by the addition of pure metals or to introduce refractory elements not readily alloyed with the base metal.

Hardening. Increasing hardness by suitable treatment, usually involving heating and cooling.

Hard-facing. Hard material applied to the surface of a soft metal surface.

Hardness

1. Resistance of a material to localized plastic deformations.

2. Resistance of a material to deformation, particularly permanent deformation, indentation, or scratching.

3. Property of metal by which it will resist indentation, bending, cutting, or scratching.

4. Resistance of metal to plastic deformation, usually by indentation. However, the term may also refer to stiffness or temper, or to resistance to scratching, abrasion, or cutting. Indentation hardness may be measured by various hardness tests.

Heading. Process of upsetting or enlarging the end of a piece of metal.

Hearth. Bottom portion of certain furnaces, such as the blast furnace, air furnace, and other reverberatory furnaces, in which the molten metal is collected or held.

Heat-resisting steel. Low-chromium steel consisting of 4 to 12% chromium. Heat-resisting steel retains its essential mechanical properties at temperatures up to 1100°F.

Heat sink. Large mass of metal that has a high thermal conductivity, used to stabilize the temperature of a part held in contact with it.

Heat treatment

1. Heating and cooling a solid metal or alloy in such a way as to obtain desired conditions or properties.

2. Controlled heating and cooling of steels in the solid state for the purpose of obtaining certain desirable mechanical or physical properties.

High-strength low-alloy steel. Steel with less than 1% of any alloying element, manufactured to high standards for strength, ductility, and partial chemical specifications.

Holding furnace. Small furnace into which molten metal can be transferred to be held at the proper temperature until it can be used to make castings.

Homogenizing. Holding at high temperature to eliminate or decrease chemical segregation by diffusion.

Hot-dip process. Coating with zinc, terne metal, or tin by immersion in a bath of the molten coating metal.

Hot extrusion. Similar to cold extrusion except that a preheated slug is used and the pressure application is slower.

Hot-forming

1. Performing working operations such as bending, drawing, forging, piercing, pressing, and beading, above the recrystallization temperature of the metal.

2. Forming hot plastic metal into desired shapes, with little change in the mechanical properties of the metal.

Hot-press forging. Plastically deforming metal at elevated temperatures between dies in forging presses.

Hot-quenching. Quenching in a medium at an elevated temperature.

Hot-rolled. Metal rolled at a high temperature.

Hot-rolling. Gradual shaping by squeezing hot metal between rolls.

Hot-short. Brittleness in hot metal.

Hot-working

1. Deforming metal plastically at such a temperature and rate that strain hardening does not occur. The lowest temperature for hot-working is the recrystallization temperature.

2. Process of forming hot metal at forging heat.

3. Shaping hot plastic metal by hot-rolling, extruding, or forging, usually at temperatures above 1500°F generally accompanied by increases in strength, hardness, and toughness.

Hydraulic press. Press in which fluid pressure is used to actuate and control the ram.

Hydrogen brittleness. Condition of low ductility resulting from absorption of hydrogen in a metal at a high temperature.

Ihrigizing. Trade name for forming of a corrosion-resistant silicon alloy layer by heating a metal in a gaseous silicon chloride.

Impact extrusion (cold extrusion). Process or resultant product of a punch striking an unheated slug in a confining die. The metal flow may be either between the punch and die or through another opening.

Impact test. Test performed to determine the behavior of materials when subjected to high rates of loading, usually in bending, tension, or torsion.

Inclusions. Particles of impurities that are usually formed during solidification and are commonly in the form of silicates, sulfides, and oxides.

Indentation hardness. The resistance of a material to indentation. This is the usual type of hardness test, in which a pointed or rounded indenter is pressed into a surface under a substantially static load.

Indirect extrusion. In "indirect extrusion" the die is at the ram end of the billet and the product travels through and in the opposite direction to the hollow ram.

Induction hardening. Quench hardening in which the heat is generated by electrical induction.

Induction heating. Heating by electrical induction.

Inert gas. Noble gases, such as helium or argon, that are not reactive with any other elements.

Ingot

1. Cast pig iron or steel shape made by pouring hot metal into molds.
2. Casting suitable for working or remelting.

Ingot and ingot bar. Refinery shapes employed for alloy production.

Ingot iron. Commercially pure open-hearth iron.

Interface contact. Area between two materials or mechanical elements.

Intergranular corrosion. Type of galvanic corrosion that progresses along the grain boundaries of an alloy.

Intermediate annealing. Annealing wrought metals at one or more stages during manufacture and before final thermal treatment.

Intermetallic compound. Chemical compound of two or more elements that are constituents of an alloy system.

Internal friction. The conversion of energy into heat by a material subjected to fluctuating stress.

Interrupted aging. Aging at two or more temperatures by steps, and cooling to room temperature after each step.

Interrupted quenching. Quenching in which the metal object being quenched is removed from the quenching medium while the object is at a temperature substantially higher than that of the quenching medium.

Iron. Term refers to the element Fe and not cast iron, steel, or any other alloy of iron.

Isothermal annealing. Austenitizing a ferrous alloy and then cooling to and holding at a temperature at which austenite transforms to a relatively soft ferrite carbide aggregate.

Isothermal transformation. Change in phase at a constant temperature.

Kaolinite. Prime ingredient of ordinary clay; hydrated aluminum silicate.

Killed steel

1. Steel deoxidized to eliminate the reaction between carbon and oxygen during solidification.
2. Steel deoxidized with a strong deoxidizing agent such as silicon or aluminum in order to reduce the oxygen content to such a level that no reaction occurs between carbon and oxygen during solidification.

Kiln. Large furnace used for baking, drying, or burning firebrick or refractories, or for calcining ores or other substances.

Krause rolling mill. Sheet rolling mill that uses a reciprocating action that accomplishes a high reduction in thickness for a single passage of the sheet through the mill.

Ladle. Receptacle used for transferring and pouring molten metal.

Lamellar. Alternating platelike structure in metals.

Laminate. Composite metal, usually in the form of sheet or bar, composed of two or more metal layers so bonded that the composite metal forms a structural member.

Launder. A channel for conducting molten metal.

Low-shaft furnace. Short-shaft-type blast furnace used to produce pig iron and ferroalloys from low-grade ores, using low-grade fuel. The air blast is often enriched with oxygen. It is also used for making a variety of other products such as alumina, cement-making slags, and ammonia synthesis gas.

Macrostructure. Structure of metals as revealed by macroscopic examination.

Magnetically hard alloy. Ferromagnetic alloy capable of being magnetized permanently because of its ability to retain induced magnetization and magnetic poles after the removal of externally applied fields; alloy with high coercive force. The name is based on the fact that the quality of the early permanent magnets was related to their hardness.

Magnetically soft alloy. Ferromagnetic alloy that becomes magnetized readily upon application of a field and that returns to practically a nonmagnetic condition when the field is removed; alloy with the properties of high magnetic permeability, low coercive force, and low magnetic hysteresis loss.

Magnetic-analysis inspection. Nondestructive method of inspection to determine the existence of such variations in magnetic flux concentration in ferromagnetic materials of constant cross section as might be caused by defects, discontinuities, and variations in hardness. Variations are usually established by a change in pattern on an oscilloscopic screen.

Magnetic particle inspection. Nondestructive method of inspection for determining the existence and extent of possible defects in ferromagnetic materials. Finely divided magnetic particles applied to the magnetized part are attracted to and outline the pattern of any magnetic leakage created by discontinuities.

Malleability

1. Ability to be shaped without fracture either by hot- or cold-working.
2. Characteristic of metals that permits plastic deformation in compression without rupture.

Malleable cast iron. Cast iron made by a prolonged anneal of white cast iron in which decarburization or graphitization, or both, take place to eliminate some or all of the cementite.

Malleable iron. Cast iron of such composition that it solidifies as white iron, which upon proper heat treatment is converted to a metallic matrix with nodules of temper carbon.

Malleableizing. Annealing white cast iron in such a way that some or all of the combined carbon is transformed to graphite; in some instances, part of the carbon is removed completely.

Martempering. Quenching an austenitized ferrous alloy in a medium at a temperature in the upper part of the martensite range, or slightly above that range, and holding it in the medium until the temperature throughout the alloy is substantially uniform. The alloy is then allowed to cool in air through the martensite range.

Martensite. In an alloy, a metastable transitional structure, intermediate between two allotropic modifications, whose abilities to dissolve a given solute differ considerably, the high-temperature phase having the greater solubility.

Martensite steels. Steels that can be made very hard and tough by heat treatment and rapid cooling. Martensitic stainless steels have a chromium content between 4 and 12%.

Mechanical fastening. Linking process whereby materials are joined, permanently or semipermanently, with special locking devices such as screws, rivets, nails, bolts, keys, or retainer rings. The resultant joint is discontinuous.

Mechanical properties. Properties of a material that reveal its elastic and inelastic behavior where force is applied, thereby indicating its suitability for mechanical applications. Examples of mechanical properties are modulus of elasticity, tensile strength, yield point, elongation, hardness, and fatigue limit.

Melting. Heating metal to the molten state in a steel-making furnace to control residual elements and/or add beneficial alloying elements.

Metal. Opaque, lustrous, elemental substance that is a good conductor of heat and electricity and when polished, a good reflector of light. Most elemental metals are malleable and ductile and are, in general, heavier than the other elemental substances.

Metallic bond. Bonding of metal atoms in which the atoms adopt a crystal structure and donate all valence electrons into an electron cloud.

Metallizing. Finishing process whereby a metal coating is sprayed onto the surface of a metal (or in some cases, plastic) part. The process is similar to electroplating and is also called thermal spraying or flame spraying.

Metallurgy. Science and technology of metals. Process (chemical) metallurgy is concerned with the extraction of metals from their ores and with the refining of metals, physical metallurgy with the physical and mechanical properties of metals as affected by composition, mechanical working, and heat treatment.

Metal powder. Aggregate of discrete particles that are usually within the size range of 1 to 1000 μm.

Microstructure. Structure of polished and etched metals as revealed by a microscope at a magnification greater than 10 diameters.

Mild steel. Carbon steel with carbon content between 0.15 and 0.25%.

Mill finish

1. Metal surface resulting from rolling, drawing, and extrusion processes.

2. Surface finish on rolled products characteristic of rolling.

Milling

1. Material removal process employing a cylindrical tool having equally spaced sharpened teeth on its surface, which moves through a fixed work piece.

2. Removing metal with a milling cutter; mechanical treatment of material, as in a ball mill, to produce particles or alter their size or shape, or to coat one component of a powder mixture with another.

Mill scale

1. Black scale of magnetic oxide of iron formed on iron and steel when they are heated for rolling, forging, or other processing. Some scale may be tightly adhered and some bonded loosely.

2. Scale on metal due to hot rolling.

Mold. Form into which molten metal is poured to produce a casting.

Mottled cast iron. Mixed structure of gray and white iron of variable proportions. The fracture has a mottled appearance.

Muck bar. Bar rolled from a squeezed bloom.

Natural aging. Spontaneous aging of a supersaturated solid solution at room temperature.

Neutral flux. Mineral (fluorspar) used to make slag more fluid.

Nitriding. Introducing nitrogen into a solid ferrous alloy by holding the alloy at a suitable temperature in contact with a nitrogenous material, usually ammonia or molten cyanide of appropriate composition. Quenching is not required to produce a hard case.

Noble metal. Metal whose potential is highly positive relative to the hydrogen electrode; metal with marked resistance to chemical reaction, particularly to oxidation and to solution by inorganic acids. The terms as often used is synonymous with "precious metal."

Nodular cast iron. Cast iron that has been treated while molten with a master alloy containing an element such as magnesium or cerium to give primary graphite in the spherulitic form.

Nodular graphite. Graphite in the form of nodules, in iron castings.

Nonferrous. Pertaining to metallic materials in which iron is not a principal element influencing the properties.

Nonferrous alloy annealing. Annealing implies a heat treatment designed to soften a cold-worked structure by recrystallization or subsequent grain growth or to soften an age-hardened alloy by causing a nearly complete precipitation of the second phase in relatively coarse form.

Normalizing

1. Heating a ferrous alloy to a suitable temperature above the transformation range and then cooling it in still air to room temperature.

2. Heating a ferrous alloy to a suitable temperature range and then cooling it in air to a temperature substantially below the transformation range (that is, the range within which the structure of the metal changes).

Open hearth. Masonry structure with a hearth exposed to the sweep of flames, in which hot pig iron, scrap metal, and fluxes are melted and oxidized by a mixture of fuel and air to produce basic or acid steel.

Open-hearth furnace. Reverberatory melting furnace with a shallow hearth and a low roof. Flame passes over the charge on the hearth, causing the charge to be heated both by direct flame and by radiation from the roof and sidewalls of the furnace. In ferrous industry, the furnace is regenerative.

Ore

1. Mineral containing metal either as a mixture or in chemical combination.

2. Natural mineral that may be mined and treated for the extraction of any of its components, metallic or otherwise.

Overaging. Aging under conditions of time and temperature greater than those required to obtain maximum change in a certain property, so that the property is altered in the direction of the initial value.

Overheating. Heating a metal or alloy to such a high temperature that its properties are impaired. When the original properties cannot be restored by further heat treating, mechanical working, or a combination of working and heat treating, the overheating is known as burning.

Oxidation. Combination of oxygen with metal, forming patinas, tarnishes, or coatings of varying depths and color.

Oxidation corrosion. Formation of oxide scale on metals at high temperatures.

Oxide. Compound of oxygen and an element.

Oxidizing. Chemical combining of an element with oxygen, often aided by and sometimes resulting in heat.

Paramagnetic. Pertaining to materials that exhibit a weak magnetic field of their own when subjected to a strong magnetic field.

Patenting. Heat treatment applied to medium-carbon of high-carbon steel before the drawing of wire or between drafts. This process consists in heating to a temperature above the transformation range and then cooling in air or in a bath of molten lead or salt.

Pearlite. Lamellar mixture of ferrite and cementite in slowly cooled iron-carbon alloys, as found in steel and cast iron.

Penetrant inspection. Method of nondestructive testing for determining the existence and extent of discontinuities that are open to the surface in the part being inspected. Indications are made visible through the use of a dye or fluorescent chemical in the liquid employed as the inspection medium.

Permanent set. Plastic deformation that remains upon releasing the stress that produces the deformation.

Permeability. The porosity of foundry sands in molds in casting of metals, and the ability of trapped gases to escape through the sand.

Phase. The following phases occur in the iron-carbon alloy: molten alloy, austenite, ferrite, cementite, and graphite.

Phase change. Change in a metal or alloy from one homogeneous, physically distinct substance to another.

Pickling

1. Removing surface oxides from metals by chemical or electrochemical reaction.

2. Removing the oxide scale formed on hot metal as it air cools by dipping the metal in a solution of sulfuric or hydrochloric acid.

Pig

1. Pig iron ingot.

2. Metal casting used in remelting.

Pig iron

1. Metallic product obtained by the reduction of iron ores in the blast furnace or electric furnace.

2. High-carbon crude iron from the blast furnace used as the main raw material for iron and steel making. Basic pig iron is a high-phosphorus iron used in basic steel-making furnaces.

3. High-carbon iron made by reduction of iron ore in the blast furnace; cast iron in the form of pigs.

Pit. Surface defect in metal.

Plasticity. Material that can be deformed without breaking.

Plate

1. Flat-rolled metal product of some minimum arbitrary thickness and width, depending on the type of material.

2. Rolled-flat product over 0.188 in. in thickness and over 12 in. in width.

3. Heavy, flat piece of metal, the thickness of which is measured in fractions of inches.

4. Hot-rolled flat products generally thicker than sheets and wider than strip.

Postheating. Heating weldments immediately after welding for tempering, stress relieving, or providing a controlled rate of cooling to prevent formation of a hard or brittle structure.

Powder metallurgy. Process by which metal shapes are formed from metal powders by pressure and heat.

Precipitation hardening. Hardening caused by the precipitation of a constituent from a supersaturated solid solution.

Precipitation heat treatment. Artificial aging in which a constituent precipitates from a supersaturated solid solution.

Preheating. Heating before some further thermal or mechanical treatment. For tool steel, it is heating to an intermediate temperature immediately before final austenitizing. For some nonferrous alloys, it is heating to a high temperature for a long time in order to homogenize the structure before working.

Primary mill. Mill for rolling ingots; rolled products of ingots, such as blooms, billets, or slabs. The mill is often called a "blooming mill" and sometimes a "cogging mill."

Process annealing

1. Heating of metals to a point slightly below the critical temperature, followed by regulated cooling as desired, used for stress relieving.

2. In the sheet and wire industries, heating a ferrous alloy to a temperature close to, but below, the lower limit of the transformation range and then cooling it in order to soften the alloy for further cold-working.

Proof load. Specified load that the fastener must withstand in a proof test without any indication of failure.

Proof test. Specified test required of a fastener to verify its suitability for the fastening job.

Progressive aging. Aging by increasing the temperature in steps or continuously during the aging cycle.

Pyrometallurgy

1. Metallurgy involved in winning and refining metals where heat is used, as in roasting and smelting.

2. High temperature process metallurgy.

Quench aging. Aging induced by rapid cooling after solution heat treatment.

Quench annealing. Annealing an austenitic ferrous alloy by solution heat treatment.

Quench hardening. Hardening a ferrous alloy by austenitizing and then cooling rapidly enough so that some or all of the austenite transforms to martensite.

Quenching

1. Rapid cooling by immersion in oil, water, or other cooling medium to increase hardness.

2. Rapid cooling. The following are more specific terms: direct quenching, fog quenching, hot quenching, interrupted quenching, selective quenching, spray quenching, and time quenching.

Radiography. Nondestructive method of internal examination in which metal or other objects are exposed to a beam of x-ray or gamma radiation.

Recrystallization. Change from one crystal structure to another, as occurs on heating or cooling through a critical temperature; formation of a new, strain-free grain structure from that existing in cold-worked metal, usually accomplished by heating.

Recrystallization annealing. Annealing cold-worked metal to produce a new grain structure without phase change.

Reduction. Separation of iron from its oxide by smelting ores in the blast furnace.

Reduction of area

1. Measure of the ductility of a metal computed from the decrease in area divided by the original area and multiplied by 100.

2. Difference between the original cross-sectional area of a tension test specimen and the area of its smallest cross section. Reduction of area is usually expressed as a percentage of the original cross-sectional area of the specimen.

3. Commonly, the difference, expressed as a percentage of original area, between the original cross-sectional area of a tensile test specimen and the minimum cross-sectional area measured after complete separation; difference, expressed as a percentage of original area, between the original cross-sectional area and that after straining the specimen.

Refined bar iron or refined wrought-iron bars. Iron bars rolled from a muck bar pile, slab pile, or box pile of muck bars and wrought-iron scrap bars free from steel, all bars running the full length of the pile.

Refining

1. Purifying crude or impure metals.

2. Melting of pig iron and/or other iron-bearing materials in steel furnaces to achieve desired contents of residual and alloying elements.

Refractories. Nonmetallic materials with superior heat and impact resistance, used for lining furnaces, flues, and vessels employed in iron and steel making.

Refractory

1. Pertaining to nonmetallic ceramic materials designed for uses at temperatures of 2700°F or higher in such applications as furnace linings or foundry crucibles. Also, any metal with a melting point above 3000°F is a refractory metal.

2. Material with a very high melting point and properties that make it suitable for such uses as furnace linings and kiln construction; quality of resisting heat.

Refractory alloy. Heat-resistant alloy; alloy having an extremely high melting point; alloy difficult to work at elevated temperatures.

Refractory metal. Metal having an extremely high melting point. In the broad sense, the term refers to metals having melting points above the range of iron, cobalt, and nickel.

Residual elements. Nonferrous elements such as carbon, sulfur, phosphorus, manganese, and silicon which occur naturally in raw materials and are controlled in steel making.

Retort. Vessel used for the distillation of volatile materials, as in the separation of some metals and in the destructive distillation of coal.

Reverberatory furnace. Furnace, with a shallow hearth, usually nonregenerative, and a roof that deflects the flame and radiates heat toward the hearth or the surface of the charge.

Rivet. Headed fastener of malleable materials used to join parts by inserting the shank through aligned holes in each piece and forming a head on the headlines end by upsetting.

Rockwell hardness. Hardness number indicating the resistance of a material to penetration as determined by a Rockwell hardness tester.

Rockwell hardness test. Test for determining the hardness of a material based on the depth of penetration of a specified penetrator into the specimen under certain arbitrarily fixed conditions of test.

Rolled floor plates. Carbon steel floor plates of flat, hot-rolled, finished steel products which have a rolled, raised figure at regular intervals on the surface of the plate.

Rolling

1. Forming by squeezing a workpiece between rotating cylindrical rolls, by hot- or cold-rolling techniques, generally to produce sheets and strips.

2. Reducing the cross-sectional area of metal stock, or otherwise shaping metal products, through the use of rotating rolls.

Rolling mills. Machines used to decrease the cross-sectional area of metal stock and produce certain desired shapes as the metal passes between rotating rolls mounted in a framework comprising a basic unit called a stand. Cylindrical rolls produce flat shapes; grooved rolls produce rounds, squares, and structural shapes.

Scale. Oxide coating caused by hot-rolling or heating; also the product resulting from the corrosion of metals.

Scaling. Forming a thick layer of oxidation products on metals at high temperatures.

Scleroscope test. Hardness test where the loss in kinetic energy of a falling "tup," absorbed by indentation on impact of the tup on the metal being tested, is indicated by the height of the rebound.

Scrap metal. Source of iron for iron and steel making, consisting of rolled product croppings, rejects, and obsolete equipment from steel mills and foundries and waste ferrous material from industrial and consumer products.

Scratch hardness. Hardness of a metal, determined by the depth of a scratch made by a cutting point drawn across the surface under a given pressure.

Screw press. Press whose slide is operated by a screw rather than by a crank or other means.

Secondary hardening. Tempering certain alloy steels at certain temperatures so that the resulting hardness is greater than that obtained by tempering the same steel at some lower temperature for the same time.

Secondary metal. Metal recovered from scrap by remelting and refining.

Selective heating. Heating only certain portions of an object so that they have the desired properties after cooling.

Selective quenching. Quenching only certain portions of an object.

Semikilled steel. Steel that is incompletely deoxidized and contains sufficient dissolved oxygen to react with the carbon to form carbon monoxide to offset solidification shrinkage.

Sequestering agent. Material that combines with metallic ions to form water-soluble complex compounds.

Shapes

1. Solid sections, other than rectangular, square, or standard rod and wire sections, furnished in straight lengths.
2. Term used in referring to metal sections.

Shaping. Cutting by moving a single-edge tool across a fixed workpiece in a straight-line path; material removal process.

Sheet

1. Hot- or cold-rolled flat product generally thinner than plate and wider than strip.
2. Flat-rolled metal product of some maximum thickness and minimum width arbitrarily dependent on the type of metal.

Single-acting hammer. Gravity drop hammer.

Single-action press. Press with a single slide actuated by means of one or more cranks or eccentrics.

Single-refined iron. Iron consisting entirely of new wrought iron, which is first rolled into muck bars or slabs. The bars or slabs are then piled and rolled once.

Single-stand mill. Rolling mill of such design that the product contacts only two rolls at a given moment. Contrast with a tandem mill.

Sintered. To cause to become a consolidated mass by heating without melting.

Slab. Piece of metal, intermediate between ingot and plate, with its width at least twice its thickness.

Slabbing mill. Primary mill that produces slabs.

Slab pile. Pile built up wholly of flat slabs of iron, all slabs running the full length of the pile.

Slack quenching. Process of hardening steel by quenching from the austenitizing temperature at a rate slower than the critical cooling rate for the particular steel, resulting in incomplete hardening and the formation of one or more transformation products in addition to or instead of martensite.

Slag

1. Impurities or basic iron silicates that are separated from molten metal by flotation.
2. Nonmetallic product resulting from the mutual dissolution of flux and nonmetallic impurities in smelting and refining operations.
3. Nonmetallic impurities that form a layer on top of molten metal.
4. Molten mass composed of fluxes, in combination with unwanted elements, which floats to the surface of the hot metal in the furnace and thus can be removed.

Slug. Short piece of metal to be placed in a die for forging, impacting, or extruding; a small piece of material produced in piercing a hole in sheet material.

Smelting

1. Melting of iron-bearing materials in the blast furnace to separate iron from the impurities with which it is chemically combined or mechanically mixed.
2. Thermal processing wherein chemical reactions take place to produce liquid metal from a beneficiated ore.

Soaking

1. Holding a steel ingot at an elevated temperature for attaining uniform temperature throughout the ingot.
2. Prolonged holding at a selected temperature.

Solid solution. Presence of alloy atoms within the structure of a metal. If the alloy atoms are about the same size and form in the same crystallographic arrangement as those of the base metal, they take the place of some base-metal atoms, forming a substitutional solid solution. If the alloy atoms are very small, they fit between the naturally occurring spaces in the base-metal lattice, forming an interstitial solid solution.

Solubility. Degree to which one substance will dissolve in another.

Solution heat treatment. Heating an alloy to a suitable temperature, holding it at that temperature long enough to allow one or more constituents to enter into solid solution, and then cooling it rapidly enough to hold the constituents in solution. The alloy is left in a supersaturated, unstable state and may subsequently exhibit quench aging.

Solvent. Substance capable of dissolving another substance; the major constituent in a solution.

Spheroidal graphite. Graphite in spheroidal form in iron castings.

Spheroidizing. Heating and cooling to produce a spheroidal or globular form of carbide in steel.

Sponge. Form of metal characterized by a porous condition, which is the result of the decomposition or reduction of a compound without fusion. The term is applied to forms of iron, the platinum group metals, titanium, and zirconium.

Sponge iron. Either porous or powdered iron produced directly without fusion, for example, by heating high-grade ore with charcoal or an oxide with a reducing gas.

Spray quenching. Quenching in a spray of liquid.

Stabilizing treatment. Treatment intended to stabilize the structure of an alloy or the dimensions of a part.

Stainless steel. Steel containing 12 to 27% chromium, with excellent corrosion resistance, strength, and chemical inertness at high and low temperatures.

Steckel mill. Cold-reducing mill having two working rolls and two backup rolls, none of which is driven. The strip is drawn through the mill by a power reel in one direction as far as the strip will allow, then reversed by a second power reel, and so on until the desired thickness is attained.

Steel

1. Iron-base alloy, malleable in some temperature ranges as initially cast. It contains manganese and usually carbon, and often has other alloying elements. Steel is to be differentiated from the two classes of iron, the cast irons on the high-carbon side and the relatively pure irons, such as ingot iron, on the low-carbon side.
2. Iron-carbon alloy, containing residual and sometimes alloying elements, characterized by its strength and toughness. Steel is distinguished from iron by its ability to be shaped by hot- and/or cold-working as initially cast.

Steel forging. Product of a substantially compressive plastic working operation that consolidates the material and produces the desired shape. Plastic working may be performed by a hammer, press, forging machine, or forging rolls and must substantially deform the material to produce a completely wrought structure.

Steel furnace. Masonry or steel structure lined with refractory materials, used to melt pig from scrap metal and sometimes agglomerated ores, ferroalloys, and fluxes into steel.

Stepped extrusion. Single product with one or more abrupt cross-sectional changes. It is obtained by interrupting the extrusion by die changes.

Strain aging. Aging induced by cold-working.

Strength. Ability of a metal to resist external forces.

Stress corrosion. Rapid deterioration of properties resulting from repeated cyclic stressing of a metal in a corrosive medium.

Stress relieving. Heating to a suitable temperature, holding long enough to reduce residual stresses, and then cooling slowly enough to minimize the development of new residual stresses.

Strip

1. Hot- or cold-rolled flat products generally narrower than sheets and thinner than plates.
2. Sheet of metal in which the length is many times the breadth.

Structurals. Hot-rolled steel shapes of special design (such as H-beams, I-beams, channels, angles, and tees) used in construction.

Structural shape. Piece of metal of any of several designs accepted as a standard by the structural branch of the iron and steel industries.

Structural size shapes. Rolled flanged sections having at least one dimension of the cross section 3 in. or greater.

Superalloy. Alloy developed for very high temperature service where relatively high stresses (tensile, thermal, vibratory, and shock) are encountered and where oxidation resistance is frequently required.

Superheating. Heating a phase above a temperature at which an equilibrium can exist between it and another phase having more internal energy; heating molten metal above the normal casting temperature to obtain more complete refining or greater fluidity.

Tandem mill. Rolling mill consisting of two or more stands arranged so that the metal being processed travels in a straight line from stand to stand. In continuous rolling, the various stands are synchronized so that the strip may be rolled in all stands simultaneously. Contrast with a single-stand mill.

Temper. In heat treatment, reheating hardened steel or hardened cast iron to some temperature below the eutectoid temperature for the purpose of decreasing the hardness and increasing the toughness.

Temper brittleness. Brittleness that results when certain steels are held within or are cooled slowly through a certain range of temperatures below the transformation range. Brittleness is revealed by notched-bar impact tests at or below room temperature.

Tempering

1. In heat treatment, reheating hardened steel to decrease hardness and increase toughness.
2. Reheating a quench hardened or normalized ferrous alloy to a temperature below the transformation range and then cooling at any rate desired.
3. Reheating to less than 1350°F, after hardening (as by quenching) and slow cooling, to restore ductility.

Terne plate. Black plate that has been coated with terne metal (lead-tin alloy).

Thermal cutting. Removal or separation cutting involving heat, for example, oxyacetylene and laser cutting.

Thermal stresses. Stresses in metal resulting from nonuniform temperature distribution.

Thermoplastic. Capable of softening or fusing when heated and of hardening again when cooled.

Time quenching. Interrupted quenching in which the duration of holding in the quenching medium is controlled.

Tin plate. Black plate that has been coated with tin.

Tolerance. Specified permissible deviation from a specified nominal dimension; permissible variation in the size of a part.

Toughness. Maximum ability of a material to absorb energy, for example, from sudden shock or impact, without breaking.

Transformation range. Temperature interval within which some phase change occurs in a metal or alloy during heating or cooling.

Transformation temperature. Temperature at which a change in phase occurs. The term is also used to denote the limiting temperature of a transformation range.

Transformation temperature ranges. Ranges of temperature within which austenite forms during heating and transforms during cooling. The two ranges are distinct, sometimes overlapping but never coinciding. The limiting temperatures of the ranges depend on the composition of the alloy and on the rate of change of temperature, particularly during cooling.

Transition metal. Metal in which the available electron energy levels are occupied in such a way that the d-band contains less than its maximum number of 10 electrons per atom, as, for example, in iron, cobalt, nickel and tungsten. Distinctive properties of the transition metals result from the incompletely filled d-levels.

Triple-action press. Mechanical or hydraulic press having three slides with three motions properly synchronized for triple-action drawing, redrawing, and forming. Usually, two slides—the blank holder slide and the plunger—are located above, and a lower slide is located within the bed of the press.

Tubular products. Hollow products of round, oval, square, rectangular, and multifaceted cross sections. In construction, round products are generally referred to as pipe, square or rectangular products with thinner wall sections as tube or tubing.

Turning. Material removal process in which cutting is achieved by revolving a workpiece against a fixed, single-edge tool.

Two-point press. Mechanical press in which the slide is actuated by two connections.

Ultrasonic testing. Nondestructive testing employing high-frequency sound vibrations to detect defects in materials.

Underdrive press. Mechanical press in which the driving mechanism is located within or under the bed.

Unfinished fastener. Fastener made to the same basic dimensions as a finished fastener but having relatively wide tolerances and all surfaces in their formed condition.

Universal mill. Rolling mill in which rolls with a vertical axis roll the edges of the metal stock between some of the passes through the horizontal rolls.

Viscosity. Property in fluids, either liquid or gaseous, that may be described as a resistance to flow; also the capability of continuous yielding under stress.

Void. Cavity or hole in a substance.

Weathering. Galvanic and/or chemical corrosion produced by atmospheric conditions.

Weld metal. Molten area of a weld, either introduced by a filler rod or produced by the fusion of the base metal.

Wenstrom mill. Rolling mill similar to a universal mill, but where the edges and sides of a rolled section are acted on simultaneously.

White cast iron. Cast iron that gives a white fracture because the carbon is in combined form.

White iron. Cast iron in which substantially all of the carbon is in solution and in the combined form. The metal has a white fracture.

Wire

1. Solid section, including rectangular flat wire, but excluding other flat products, finished in coils or on spools, reels, or bucks. Flat wire may also be furnished in straight lengths.

2. Cold-finished products of round, square, or multifaceted cross section, generally smaller than bars, round wire is cold-drawn, 0.005 to less than 1 in. in diameter; flat wire is cold-rolled, generally narrower than bars.

Wire bar. Refinery shape for rolling into rod (and subsequent drawings into wire), strip, or shape.

Work hardening. Increase in hardness and strength in a metal caused by plastic deformation.

Wrought iron

1. Ferrous material aggregated from a solidifying mass of pasty particles of highly refined metallic iron, with which, without subsequent fusion, is incorporated a minutely and uniformly distributed quantity of slag.

2. Relatively pure iron mechanically mixed with a small amount of iron-silicate slag, characterized by good corrosion resistance, weldability, toughness, and high ductility.

3. Commercial iron consisting of slag (iron silicate) fibers entrained in a ferrite matrix.

Wrought products. Products formed by rolling, drawing, extruding, and forging.

Yield point. Stress at which a marked increase in deformation occurs without an increase in load stress, as seen in mild or medium carbon steel.

Yield strength. The stress, observed at the proportional limit of metals, at which a material deviates from that proportionality of stress to strain to a specified amount.

METAL PLATING—COATING, DIPPING, AND FINISHING

Abrasive blasting. Process for cleaning or finishing by means of an abrasive directed at high velocity against the work piece.

Activation. Elimination of a passive condition on a surface.

Activity (ion). The ion concentration corrected for deviations from ideal behavior. The concentration is multiplied by an activity coefficient.

Addition agent

1. Material added in small quantities to plating solution to modify its characteristics.

2. Material added to a plating bath for the purpose of modifying the character of the deposit. The material is usually added in small amounts, is often organic or colloidal in nature, and includes brighteners.

Adhesion. Attractive force exists between an electrodeposit and its substrate that can be measured as the force required to separate an electrodeposit and its substrate.

Alkali metals. Group of elements including lithium, sodium, potassium, rubidium, cesium, and fransium. In plating the term is usually applied to sodium or potassium.

Alkaline cleaning. Cleaning by means of alkaline solutions.

Alloy. Substance that has metallic properties and is composed of two or more elements of which at least one is a metal.

Alloy plate. Electrodeposit containing two or more metals intimately mixed or combined so as not to be distinguishable to the unaided eye.

Alloy plating. Codeposition of two or more metals.

Amorphous. Noncrystalline; devoid of regular structure.

Ampere. Unit of current strength. Current flowing at the rate of 1 C (coulomb)/second equals 1 ampere.

Analytical free cyanide. Free cyanide content of a solution, as determined by a specified analytical method. The "true" value of free cyanide is rarely known with certainty and is therefore usually only dealt with in discussions of theory. The "calculated" or "stated" value is usually used in more practical work.

Anion. Negatively charged ion; ion that during electrolysis is attracted to the anode.

Anionic detergent. Detergent that produces aggregates of negatively charged ions with colloidal properties.

Anode. Electrode at which current enters or electrons leave the solution; positive electrode in electrolysis; electrode at which negative ions are discharged, positive ions are formed, or other oxidizing reactions occur.

Anode compartment. In an electrolytic cell, the enclosure formed by a diaphragm around the anodes.

Anode corrosion. Dissolution of a metal acting as an anode.

Anode effect. Effect produced by polarization of the anode in the electrolysis of fused salts, characterized by a sudden increase in voltage and a corresponding decrease in amperage owing to the anode's being virtually separated from the electrolyte by a gas film.

Anode efficiency. Current efficiency of a specified anodic process.

Anode film. Portion of solution in immediate contact with the anode, especially if the concentration gradient is steep; outer layer of the anode itself.

Anode mud. Deposit of insoluble residue formed from the dissolution of the anode in commercial electrolysis. Sometimes called "anode slime."

Anodic cleaning. Electrolytic cleaning where the work is the anode. It is also called "reverse-current cleaning."

Anodic coating

1. Protective, decorative, or functional coating formed by conversion of the surface of a metal in an electrolytic oxidation process.

2. Film on work resulting from an electrolytic treatment at the anode.

Anodic pickling

1. Electrolytic pickling where the work is the anode.

2. Electrolytic pickling in which the treated metal is made the anode.

Anodized coating. Oxide coating, produced by electrolytic means, for aluminum, magnesium, and their alloys.

Generally, anodized coatings are hard and abrasion resistant, and offer excellent resistance to corrosion.

Anodizing

1. Forming a conversion coating on a metal surface by anodic oxidation; most frequently applied to aluminum.

2. Electrolytic oxidation process in which the surface of a metal, when anodic, is converted to a coating having desirable protective, decorative, or functional properties.

3. Anodic treatment of metals, particularly aluminum, to form an oxide film of controlled properties.

Anolyte. Portion of electrolyte in the vicinity of the anode; in a divided cell, the portion of electrolyte on the anode side of the diaphragm.

Antipitting agent. Addition agent for the specific purpose of preventing gas pits in a deposit.

Armature. Part of a generator or motor that includes the main current-carrying winding.

Automatic machine. Machine for mechanically processing parts through treatment cycles, such as cleaning, anodizing, or plating.

Automatic plating. Plating in which the cathodes are automatically conveyed through successive cleaning and plating tanks.

Auxiliary anode. Supplementary anode placed in a position to raise the current density on a certain area of the cathode to get better plate distribution.

Back-electromotive force (EMF). Potential setup in an electrolytic cell that opposes the flow of current, caused by such factors as concentration polarization and electrode films.

Ball-burnishing

1. Burnishing by means of metal balls or other metal shapes in a rotating container.

2. Same as ball-sizing; removing burrs and polishing small stampings and small machined parts by tumbling.

Barrel-burnishing. Smoothing and brightening of surfaces by means of tumbling the work in rotating barrels in the presence of metallic or ceramic shapes and in the absence of abrasive. In ball-burnishing, the shapes consist of hardened steel balls.

Barrel-finishing (or tumbling)

1. Improving the surface finish of metal objects or parts by processing them in rotating equipment along with abrasive particles, which may be suspended in a liquid.

2. Bulk processing of work in rotating barrels in the presence of abrasive particles, which may be suspended in water or some special solution for the purpose of improving the surface finish.

Barrel-plating (or cleaning)

1. Plating articles in a rotating container, usually a perforated cylinder that operates at least partially submerged in a solution.

2. Mechanical plating (or cleaning) in which the work is processed in bulk in a rotating container.

Base metal. Metal that readily oxidizes or dissolves to form ions—opposite of noble metal.

Basis metal. Original metal to which one or more coatings are applied.

Bath voltage. Total voltage between the anode and cathode of an electrolytic cell during electrolysis; voltage equal to the sum of the equilibrium reaction potential, IR drop, and electrode polarizations.

Bipolar electrode. Electrode that is not directly connected to the power supply but is so placed in the solution between the anode and the cathode that the part nearest the anode becomes cathodic and the part nearest the cathode becomes anodic.

Blue dip. Solution containing a mercury compound used to deposit mercury on a metal by immersion, usually prior to silver plating.

Bright dip (nonelectrolytic). Solution that produces through chemical action a bright surface on an immersed metal.

Brightener

1. Addition agent that leads to the formation of a bright plate or that improves the brightness of the deposit over that obtained without its use.

2. Agent or combination of agents added to an electroplating bath to produce a fine-grained, lustrous deposit.

Bright plate. Electrodeposit that is lustrous in the as-plated condition.

Bright plating. Process that produces an electrodeposit having a high degree of specular reflectance in the as-plated condition.

Bright plating range. Range of current densities within which a given plating solution produces a bright plate.

Bright-throwing power. Measure of the ability of a plating solution or a specified set of plating conditions to deposit uniformly bright electroplate on an irregularly shaped cathode.

Brush. Conductor serving to maintain electric contact between stationary and moving parts of a machine.

Brush-plating

1. Method of plating in which the plating solution is applied with a pad or brush within which is an anode and that is moved over the cathode to be plated.

2. Plating with a concentrated solution or gel held in or fed to an absorbing medium, pad, or brush carrying the anode (usually insoluble). The brush is moved back and forth over the area of the cathode to be plated.

Btu (British thermal unit). Quantity of heat required to raise the temperature of 1 lb of water through $1°F$.

Buffer. Substance that when added to a solution causes it to resist any change in hydrogen ion concentration upon the addition of acid or alkali. Each buffer has a characteristic limited range of pH over which it is effective.

Buffing. Smoothing, by means of a rotating flexible wheel, a surface to which fine abrasive particles are applied in liquid suspension, paste, or grease stick form.

Building up. Electroplating for the purpose of increasing the dimensions of an article.

Buildup. Overplating or undesirable deposit built up on corners or edges during plating.

Burnishing

1. Smoothing of surfaces through frictional contact between the work and some hard pieces of material such as hardened metal balls.

2. Smoothing of surfaces by rubbing, accomplished chiefly by the movement rather than the removal of the surface layer.

3. Smoothing of surfaces by means of a hard tool or object, especially by rubbing.

Burring (deburring). Removal of burrs, sharp edges, fins, etc.

Bus (bus bar). Rigid conducting section for carrying current to the anode and cathode bars.

Butler finish. Finish composed of fine, uniformly distributed parallel lines, having a characteristic high luster usually produced by means of rotating wire brushes on cloth wheels and applied abrasives.

Calculated free cyanide. Concentration of cyanide or alkali cyanide present in solution in excess of that calculated as necessary to form a specified complex ion with a metal or metals present in solution.

Calomel electrode. Half-cell containing a mercury electrode in contact with a solution of potassium chloride of specified concentration and saturated with mercurous chloride, of which an excess is present. Modifying terms refer to the concentration of the potassium chloride, (for example, "saturated" calomel).

Calorie. Quantity of heat required to raise the temperature of 1 g of water through 1°C; 4.1840 J (joules) (absolute).

Carburizing. Heat treatment of metals to increase the carbon content of the surface layer; case hardening.

Cathode. Electrode through which current leaves or electrons enter the solution; the negative electrode in electrolysis; electrode at which positive ions are discharged, negative ions are formed, or other reducing reactions occur. In electroplating, the electrode that receives the deposit.

Cathode compartment. In an electrolytic cell, the enclosure formed by a diaphragm around the cathode.

Cathode copper. Copper deposited at the cathode in electrolytic refining.

Cathode efficiency

1. Current efficiency at the cathode.
2. Current efficiency of a specified cathodic process.

Cathode film

1. Layer of solution in contact with the cathode that differs in composition from that of the bulk of the solution.
2. Portion of solution immediately adjacent to the cathode.
3. Portion of solution in immediate contact with the cathode during electrolysis.

Cathodic corrosion. Corrosion resulting from a cathodic condition of a structure, usually caused by the reaction of alkaline products of electrolysis with an amphoteric metal.

Cathodic or direct cleaning. Electrolytic cleaning in which the work is the cathode.

Cathodic pickling

1. Electrolytic pickling where the work is the cathode.
2. Electrolytic pickling in which the metal being treated is made the cathode.

Cathodic protection. Partial or complete protection of a metal from corrosion by making it a cathode, using either a galvanic or impressed current.

Catholyte

1. Electrolyte adjacent to the cathode in an electrolytic cell.
2. Portion of the electrolyte in the vicinity of the cathode; in a divided cell the portion on the cathode side of the diaphragm.

Cation. Positively charged ion.

Cationic detergent. Detergent that produces aggregates of positively charged ions with colloidal properties.

Caustic dip. Strong alkaline solution into which metal is immersed for etching, neutralizing acid, or removing organic materials such as grease or paints.

Cementation. Displacement reaction in which one metal displaces another from solution.

Chelate compound

1. Compound in which the metal is contained as an integral part of a ring structure and is not readily ionized.
2. Type of complex ion or molecule in which the coordinated group occupies at least two positions in the coordination sphere.

Chelating agent. Compound capable of forming a chelate compound with a metal atom or ion.

Chemical deposition. Precipitation or plating out of a metal from solutions of its salts through the introduction of another metal or reagent to the solutions.

Chemical polishing

1. Improving the surface luster of a metal by chemical treatment.
2. Improvement in surface smoothness of a metal by simple immersion in a suitable solution.

Chemical vapor deposition. Process for producing a deposit by chemical reaction induced by heat or gaseous reduction of a vapor condensing on the substrate.

Chromadizing (chromodizing, chromatizing). Forming an acid surface to improve paint adhesion on aluminum or aluminum alloys, mainly aircraft skins, by treatment with a solution of chromic acid.

Chromate treatment. Treatment of metal in a solution of a hexavalent chromium compound to produce a conversion coating consisting of trivalent and hexavalent chromium compounds.

Chrome. Chromium. Sometimes loosely used to mean to plate with chromium.

Chrome pickle. Produce a chromate conversion coating on magnesium for temporary protection or for a paint base; solution that produces the conversion coating.

Chromium plating

1. Electrolytic process that deposits a hard film of chromium metal onto working surfaces of other metal where resistance to corrosion, abrasion, and/or erosion is needed.
2. Process used where surfaces are required to provide the highest possible resistance to abrasion. The electrolyte is usually a solution of chromic acid with a small amount of chromic sulfate and chromium carbonate.

Chromizing. Surface treatment at elevated temperature, generally carried out in pack, vapor, or salt bath, in which an alloy is formed by the inward diffusion of chromium into the base metal.

Cleaning. Removal of grease or other foreign material from a surface.

Colloidal particle. Electrically charged particle, generally smaller in size than 200 mm, dispersed in a second continuous phase.

Colloidal solution or suspension. Mixture of two or more phases, at least one of which consists of colloidal particles dispersed in one of the other phases.

Color buffing. Light buffing of metal surfaces for the purpose of producing a high luster.

Coloring. Production of colors on metal surfaces by appropriate chemical or electrochemical action.

Commutator. Cylindrical ring or disk assembly of conducting members, individually insulated in a supporting structure, with an exposed surface for contact with current-collecting brushes and ready for mounting on an armature shaft.

Complexing agent. Compound that will combine with metallic ions to form complex ions.

Complex ion. Ion composed of two or more ions or radicals, both of which are capable of independent existence, for example, cuprocyanide.

Composite plate. Electrodeposit consisting of two or more layers of metals deposited separately.

Compound-wound motor. Direct-current motor that has two separate field windings, one connected in parallel and the other in series with the armature circuit.

Concentration polarization. That part of the total polarization caused by changes in the activity of the potential-determining components of the electrolyte.

Conductance

1. Capacity of a medium, usually expressed in mhos, for transmitting electric current.
2. Reciprocal of resistance, measured by the ratio of the current flowing through a conductor to the difference of the potential between its ends.

Conducting salt. Salt added to the solution in order to increase its conductivity.

Conductivity. Specific conductance; current transferred across unit area per unit potential gradient.

Contact plating. Deposition of a metal by the use of an internal source of current by immersion of the work in solution in contact with another metal.

Contact potential. Potential difference at the junction of two dissimilar substances.

Conversion coating

1. Coating produced by chemical or electrochemical treatment of a metallic surface, which gives a superficial layer of a compound of the metal.
2. Coating consisting of a compound of the surface metal, produced by chemical or electrochemical treatment of the metal. Examples are chromate coatings on zinc, cadmium, magnesium, and aluminum; oxide or phosphate coatings on steel.

Copperas. Hydrated ferrous sulfate.

Copper plating. Film of copper deposited by electrical immersion or other means on the surface of another material such as iron or steel.

Corrosion. Gradual destruction of a material, usually by solution, oxidation, or other means attributable to a chemical process. (Of anodes in plating). Solution of anode metal by the electro-chemical action in the plating cell.

Corrundum. Native alumina, used as an abrasive.

Coulomb. Quantity of electricity that passes any section of an electric circuit in 1 second when the current in the circuit is 1 A; quantity of electricity that will deposit 0.0011180 g of silver.

Coulometer. Electrolytic cell arranged to measure the quantity of electricity by the chemical action produced, in accordance with Faraday's law.

Covering power

1. Ability of a plating solution under a specified set of plating conditions to deposit metal on the surfaces of recesses or deep holes.

2. Ability of a plating solution to produce a deposit at very low current densities.

Critical current density

1. Current density above which a new and sometimes undesirable reaction occurs.
2. Value of the current density at which a variable exhibits a sudden change in value.

Crocus. Red ferric oxide used in polishing.

Current density. Average true current per unit area; total current divided by total area of electrode in the solution; value of the current density at a given point on the electrode.

Current efficiency. Proportion, usually expressed as a percentage, of the current that is effective in carrying out a specified process in accordance with Faraday's law. In electrodeposition, the proportion of the current used in depositing or dissolving metal, the remainder usually being used in evolving hydrogen or oxygen.

Cutting down. Polishing or buffing for the purpose of removing roughness or irregularities.

Deburring. Removal of burrs, sharp edges, or fins by mechanical, chemical, or electrochemical means.

Decomposition potential. Minimum potential, exclusive of IR drop, at which an electrochemical process can take place at an appreciable rate.

Degreasing. Removing oil or grease from a surface.

Deionization. Removal of ions from a solution by ion exchange.

Density. Weight per unit volume. Usually expressed in grams per cubic centimeter or pounds per cubic foot.

Depolarization. Decrease in the polarization of an electrode. Change from an excessively to a normally polarized condition.

Depolarizer. Substance or a means that produces depolarization.

Detergency. Cleaning ability of a solution, generally aqueous.

Detergent. Surface-active agent that possesses the ability to clean soiled surfaces.

Diaphragm. Porous or permeable membrane separating anode and cathode compartments of an electrolytic cell from each other or from an intermediate compartment.

Dichromate treatment. Production of chromate conversion coating on magnesium alloys in a boiling solution of sodium dichromate.

Dielectric. Medium having the property that the energy required to establish an electric field is recoverable, in whole or in part, as electric energy.

Dielectric constant. Property of a dielectric that determines the electrostatic energy stored per unit volume for unit potential gradient.

Dielectric strength. Maximum potential gradient that a dielectric material can withstand without rupture.

Diffusion coating

1. Alloy coating produced by applying heat to one or more coatings deposited on a basis metal.
2. Coating produced by immersion at high temperature so that the coating metal diffuses into the basis metal and produces an alloy "case" of graded composition. Composite electrodeposited coatings that are subsequently interdiffused by application of heat.
3. Alloy coating produced at high temperatures by the inward diffusion of the coating material into the basis

metal; composite electrodeposited coatings that are subsequently interdiffused by thermal treatment.

Dipcoat. Method of applying a protective or decorative coating to an article by dipping the piece into the coating material.

Diphase cleaning

1. Removing soil by a composition that produces two phases in the cleaning tank, a solvent phase and an aqueous phase. Cleaning is effected by both solvent action and emulsification.

2. Cleaning by means of solutions that contain a solvent layer and an aqueous layer. Cleaning is effected both by solvent and emulsifying action.

Dispersing agent

1. Material that increases the stability of a suspension of particles in a liquid medium.

2. Substance that increases the stability of a suspension by retarding flocculation.

Dip tank. Tank, vat, or container of flammable or combustible liquid in which articles or materials are immersed for the purpose of coating, finishing, treating, or similar processes.

Divided cell. Cell containing a diaphragm or other means of physically separating the anolyte from the catholyte.

Double salt. Compound of two salts that crystallize together in a definite proportion.

Drag-in

1. Water or solution that adheres to the objects introduced into a bath.

2. Water or solution carried into another solution by the work and the associated handling equipment.

Drag-out. Solution that adheres to the objects removed from a bath.

Ductility. Ability of a material to deform plastically without fracturing.

Dummy cathode

1. Cathode that is usually corrugated to give variable current densities; cathode that is plated at low current densities to remove preferentially impurities from a plating solution; substitute cathode that is used during adjustment of operating conditions.

2. Cathode in a plating tank that is used for working the solution but that is not to be used after plating.

Dummying. Plating with dummy cathodes.

Dynamic electrode potential. Potential measured when current is passing between the electrode and the electrolyte.

Electrical conductivity. Ability of a solution to transmit an electric current—an ability closely related to the concentration of ions in the solution.

Electrochemical corrosion. Corrosion that occurs when current flows between cathodic and anodic areas on metallic surfaces.

Electrochemical equivalent. Weight of an element, compound, radical, or ion involved in a specified electrochemical reaction during the passage of a unit quantity of electricity, such as a Faraday, ampere hour, or coulomb.

Electrochemical series. Table that lists in order the standard electrode potentials of specified electrochemical reactions.

Electrochemistry. Branch of science and technology that deals with transformations between chemical and electrical energy.

Electrode. Conductor through which current enters or leaves an electrolytic cell; conductor at which there is a change from conduction by electrons to conduction by charged particles of matter, or vice versa.

Electrodeposition

1. Deposition of a substance on an electrode by the passage of an electric current through an electrolyte. Electroplating (plating), electroforming, electrorefining, and electrowinning result from electrodeposition.

2. Process of depositing a substance on an electrode by electrolysis—includes electroplating, electroforming, electrorefining, and electrowinning.

Electrode potential. Difference in potential between an electrode and the immediately adjacent electrolyte referred to some standard electrode potential as zero.

Electroendosmosis. Movement of fluids through porous diaphragms caused by the application of an electric potential.

Electroforming. Production or reproduction of articles by electrodeposition on a mandrel or mold that is subsequently separated from the deposit.

Electrogalvanizing

1. Electrodeposition of zinc coatings.

2. Electroplating with zinc to provide greater corrosion resistance.

3. Electroplating of zinc on iron or steel.

Electroless plating

1. Immersion plating where a chemical reducing agent changes metal ions to metal.

2. Deposition of a metallic coating by a controlled chemical reduction that is catalyzed by the metal or alloy being deposited.

Electrolysis. Production of chemical changes by the passage of current through an electrolyte.

Electrolyte. Conducting medium in which the flow of current is accompanied by movement of matter. It is most often an aqueous solution of acids, bases, or salts, but includes many other media, such as fused salts, ionized gases, and some solids. Substance that is capable of forming a conducting liquid medium when dissolved or melted.

Electrolytic cell. Unit apparatus in which electrochemical reactions are produced by applying electrical energy or that supplies electrical energy as a result of chemical reactions and that includes two or more electrodes and one or more electrolytes contained in a suitable vessel.

Electrolytic cleaning. Removing soil from work by electrolysis, the work being one of the electrodes. The electrolyte is usually alkaline.

Electrolytic pickling

1. Pickling where electric current is used, the work being one of the electrodes.

2. Pickling during which a current is passed through the metal and the pickle.

Electrolytic protection. Protection from electrochemical corrosion by use of the protected material as the cathode in the corrosion cell. Should an electromotive force counter to the normal flow of current in a corroding system be impressed on the system circuit, the tendency for the anodic metal to go into solution will be decreased.

Electrophoresis. Movement of colloidal particles produced by the application of an electric potential.

Electroplating

1. Electrodeposition of an adherent metallic coating on an electrode for the purpose of securing a surface with

properties or dimensions different from those of the basis metal.

2. Process that employs an electric current to coat a basis metal (cathode) with another metal (anode) in an electrolytic solution.

3. Electrodepositing metal (may be an alloy) in an adherent form on an object serving as a cathode.

Electropolishing. Improvement in surface finish of a metal by making it anodic in an appropriate solution.

Electrorefining. Anodically dissolving a metal from an impure anode and depositing it cathodically in a purer form.

Electrotinning. Electroplating tin on an object.

Electrotyping. Production of printing plates by electroforming.

Electrowinning

1. Recovery of a metal from an ore by means of electrochemical processes.

2. Production of metals by electrolysis with insoluble anodes in solutions derived from ores or other materials.

3. Electrodeposition of metals from solutions derived from ores or other materials using insoluble anodes.

Element. Collection of atoms of one type that cannot be decomposed into any simpler units by chemical means.

EMF (electromotive force). Electrical potential.

Emulsifying agent. Substance that increases the stability of an emulsion.

Emulsion

1. Suspension of fine particles or globules of one or more liquids in another liquid.

2. Suspension of small droplets of one liquid in another in which it is insoluble. For the formation of a stable emulsion, an emulsifying agent must usually be present.

Emulsion cleaning. Cleaning by means of solutions containing organic solvents, water, and emulsifying agents.

Energy efficiency. The efficiency of an electrochemical process is the product of the current efficiency by the voltage efficiency.

Epitaxy. Situation in which a deposit or overlay assumes the lattice habit and orientation of the substrate.

Equilibrium static electrode potential. Potential existing when the electrode and the electrolyte are in equilibrium with respect to a specified electrochemical reaction.

Equivalent conductivity. In an electrolyte, the conductivity of the solution divided by the number of equivalents of conducting solute per unit volume, that is the conductivity divided by the normality of the solution.

Etch

1. Roughened surface produced by a chemical or electrochemical means.

2. Dissolve a part of the surface of a metal unevenly.

Exciter. Auxilliary generator that supplies energy for the field excitation of another electric machine.

Faraday. Number of coulombs (96,490) required for an electrochemical reaction involving one chemical equivalent.

Filter aid. Inert, insoluble material, more or less finely divided, used as a filter medium or to assist in filtration by preventing excessive packing of the filter cake.

Flash plate. Thin electrodeposit, less than 0.1 mil.

Flocculate. Aggregate into larger particles; to increase in size to the point where precipitation occurs.

Fluorescence. Absorption of radiant energy and emission of the energy during, and only during, the period of excitation as light of various wavelengths characteristic of the substance.

Formula weight. Weight, in grains, pounds, or other units, obtained by adding the atomic weights of all elemental constituents in a chemical formula.

Free cyanide. Actual concentration of cyanide radical, or equivalent alkali cyanide, not combined in complex ions with metals in solution.

Full-automatic plating. Electroplating in which the work is automatically conveyed through the complete cycle.

Full-wave rectification. Rectification in which both halves of the alternating-current cycle are transmitted as direct current.

Galvanic cell. Electrolytic cell capable of producing electrical energy by electrochemical action.

Galvanic corrosion. Corrosion associated with the current of a galvanic cell made up of dissimilar electrodes.

Galvanic series. List of metals and alloys arranged according to their relative potentials in a given environment.

Galvanizing. Application of a deposit of zinc, usually on steel or a ferrous basis metal.

Gas plating. Application of a metallic coating in an enclosed and controlled atmosphere by means of heat-decomposable gaseous metal compounds.

Gassing. Evolution of gases from one or more of the electrodes during electrolysis.

Glass electrode. Half-cell in which the potential measurements are made through a glass membrane.

Gold plating. Gold is deposited on a large variety of metals and alloys from heated solutions of double cyanide of gold and potassium.

Gravimetric analysis. Method of analysis consisting of quantitatively separating a desired constituent from a weighed sample and converting that constituent into a compound of known composition, which is weighed to determine the percentage of the constituent in the original sample.

Grinding. Removal of metal by means of rotating rigid wheels containing abrasive.

Grit blasting. Abrasive blasting with small irregular pieces of steel or malleable cast iron.

Half-cell. Electrode immersed in a suitable electrolyte, designed for measurements of electrode potential, electrode designed to offer a known constant potential, such as the calomel electrode, in which case unknown potentials may be measured against it.

Half-way rectification. Rectification permitting only one-half the alternating-current cycle to be transmitted as direct current.

Hall process. Commercial process for winning aluminum from alumina by electrolytic reduction of a fused bath of alumina dissolved in cryolite.

Hard chromium

1. Chromium deposited for increasing the wear resistance of sliding metal surfaces. The chromium is usually applied directly to the basis metal and is customarily thicker than a decorative deposit.

2. Chromium plated for engineering rather than decorative applications. Not necessarily harder than the latter.

Hard water. Water containing calcium or magnesium ions in solution, which will form insoluble curds with soaps.

Haring cell. Rectangular box of nonconducting material with principal and auxiliary electrodes so arranged as to permit estimation of throwing power or electrode polarizations and potentials between them.

Heat capacity. Amount of heat necessary to raise the temperature of a system or substance by one degree of temperature.

Heat of fusion. Quantity of heat absorbed or liberated by the melting or freezing of a unit quantity of a substance at constant temperature.

Highlights. Those portions of a metal article most exposed to buffing or polishing operations.

Horsepower (HP). Rate of work equal to 550 ft-lb/second or 745.7 W.

Hull cell. Trapezoidal box of nonconducting material with electrodes arranged to permit observation of cathodic or anodic effects over a wide range of current densities.

Hydrogen embrittlement. Embrittlement of a metal or alloy caused by absorption of hydrogen during a pickling, cleaning, or plating process.

Hydrogen overvoltage. Overvoltage associated with the liberation of hydrogen gas.

Hydrophilic. Tending to absorb water; tending to concentrate in the aqueous phase.

Hydrophobic. Tending to repel water; lacking affinity for water.

Immersion cleaning. Cleaning where the work is immersed in a liquid solution.

Immersion coating. Coating produced in a solution by chemical or electrochemical action without the use of external current.

Immersion plate

1. Metallic deposit produced by a displacement reaction in which one metal displaces another from solution.
2. Metallic deposit produced by cementation.

Immersion plating (dip plating). Depositing a metallic coating on a metal immersed in a liquid solution without the aid of an external electric current.

Indicator. Substance that changes color when the pH of the medium is changed. In the case of most useful indicators, the pH range within which the color changes is narrow.

Inert anode. Anode that is insoluble in the electrolyte under the conditions prevailing in the electrolysis.

Inhibitor

1. Substance used to reduce the rate of a chemical or electrochemical reaction, commonly corrosion or pickling.
2. Substance that reduces the rate of attack of acids on a metal surface while not affecting the rate of solution of oxides or other surface compounds.

Interfacial tension. Contractile force of an interface between two phases.

Interpole. Auxiliary pole placed between the main poles of a commutating machine.

Ion

1. Electrified portion of matter of atomic or molecular dimensions.
2. Atom or group of atoms that has lost or gained one or more electrons, thereby acquiring a net electrical charge.

Ion exchange

1. Reversible process by which ions are interchanged between a solid and a liquid with no substantial structural changes of the solid.
2. Exchange of ions, usually between a solution and solid. In practice, the process is most commonly effected by ion-exchange resins.

Insoluble anode. Anode that does not dissolve during electrolysis.

Interrupted-current plating. Plating in which the flow of current is discontinued for periodic short intervals to decrease anode polarization and elevate the critical current density. This process is most commonly used in cyanide copper plating.

IR drop. Voltage across a resistance in accordance with Ohm's law.

Joule. Unit of electrical energy, or work; 1 watt-second. The work done (J) in transferring 1 C (coulomb) (Q) between two points having a potential difference (E) of 1 V: $J = EQ$.

Joule's law. Rate at which heat is produced in an electric circuit of constant resistance is proportional to the square of the current.

Karat. Twenty-fourth part of pure gold; thus 18-karat gold is 18/24 pure.

Lapping. Working or rubbing two surfaces together, with or without abrasives, for the purpose of obtaining extreme dimensional accuracy or superior surface finish.

Leveling action. Ability of a plating solution to produce a surface smoother than that of the substrate or the basis metal.

Limiting current density. Cathodic: maximum current density at which satisfactory deposits can be obtained; anodic: maximum current density at which the anode behaves normally, without excessive polarization.

Liquidus. In a temperature composition diagram, the line connecting the temperatures at which fusion is first completed for the various compositions; referring to a single composition, a point on the above line.

Mandrel. Form used as a cathode in electroforming; a mold or matrix.

Matte dip. Solution used to produce a matte surface on a metal.

Matrix. Form used as a cathode in electroforming; a mold or mandrel.

Mechanical plating

1. Application of an adherent metallic coating by mechanical means involving the compacting of finely divided particles of such metal to coherent coatings.
2. Plating operation in which the cathodes are moved mechanically during the deposition.
3. Plating wherein fine metal powders are peened onto the work by tumbling or other means.

Metal distribution ratio. Ratio of the thicknesses of metal on two specified areas of a cathode.

Metallizing. Application of an electrically conductive metallic layer to the surface of a nonconductor; application of metallic coatings by nonelectrolytic procedures, such as spraying of molten metal and deposition from the vapor phase.

Micron. One thousandth of a millimeter (0.001 mm).

Microthrowing power. Ability of a plating solution or a specified set of plating conditions to deposit metal in pores or scratches.

Mill scale. Heavy oxide layer formed during hot-fabrication or heat treatment of metals.

Molar conductivity. In an electrolyte, the conductivity of the solution, divided by the number of gram molecular weights of conducting solute per unit volume, that is, the conductivity divided by the molarity.

Motor-generator (MG set). Machine that consists of one or more motors mechanically coupled to one or more generators. In plating, a machine in which the generator delivers direct current of appropriate amperate and voltage.

Nitriding. Formation of nitrides of metals by heat treatment in special atmospheres.

Noble metal. Metal that does not readily tend to furnish ions and therefore does not dissolve readily nor enter easily into such reactions as oxidations: opposite of base metal.

Noble potential. Electrode potential of a noble metal. Since there is no agreement over the sign of electrode potentials, the words "noble" and "base" are often preferred because they are unambiguous.

Nodule. Rounded projection formed on a cathode during electrodeposition.

Nonionic detergent. Detergent that produces aggregates of electrically neutral molecules with colloidal properties.

Occlusion. Condition of uniform molecular adhesion between a precipitate and a soluble substance or between a gas and a metal.

Ohm. Unit of electrical resistance.

Orange peel. Finish resembling the dimpled appearance of an orange peel.

Overvoltage. Irreversible excess of potential required for an electrochemical reaction to proceed actively at a specified electrode, over and above the reversible potential characteristics of that reaction.

Oxidation. Reaction in which electrons are removed from a reactant; sometimes, more specifically, the combination of a reactant with oxygen.

Oxidizing agent. Compound that causes oxidation, thereby itself becoming reduced.

Passivity. Condition of a metal that retards its normal reaction in a specified environment, associated with the assumption of a potential more noble than its normal potential.

Peeling. Detachment or partial detachment of an electrodeposited coating from a basis metal or undercoat.

Periodic-reverse plating. Method of plating in which the current is reversed periodically. Cycles are usually no longer than a few minutes and may be much less.

pH. Cologarithm (negative logarithm) of the hydrogen ion activity; less precisely, concentration, of a medium as determined by indicators or electrometric means.

Phase. Homogeneous physically distinct and mechanically separable part of a system, which is separated from other parts (phases) by definite bounding surfaces.

Phosphatizing. Forming an adherent phosphate coating on a metal immersed in a suitable aqueous phosphate solution.

Physical vapor deposition. Process for depositing a coating by evaporating and subsequently condensing an element or compound, usually in a high vacuum.

Pickle

1. Acid solution used to remove oxides or other compounds from the surface of a metal by chemical or electro chemical action.

2. Solution or process used to loosen or remove corrosion products, such as scale and tarnish, from a metal.

Pickle patch. Tightly adhering oxide or scale coating not properly removed during the pickling process.

Pickle stain. Discoloration of metal due to chemical cleaning without adequate washing and drying.

Pickling

1. Removal of oxides or other compounds from a metal surface by means of a pickle.

2. Removing surface oxides from metals by chemical or electrochemical reaction.

Pit. Small depression or cavity produced in a metal surface during electrodeposition or by corrosion.

Plating. Forming an adherent layer of metal on an object.

Plating rack

1. Fixture used to hold work and conduct current to it during electrodeposition.

2. Frame for suspending and carrying current to articles during plating and related operations.

Plating range. Current density range over which a satisfactory electroplate can be deposited.

Plating work. Material being plated or otherwise finished.

Polarization

1. In electrolysis, the formation of a film on an electrode such that the potential necessary to get a desired reaction is increased beyond the reversible electrode potential.

2. Change in the potential of an electrode during electrolysis such that the potential of an anode always becomes more noble and that of a cathode less noble than their respective static potentials. Polarization is equal to the difference between the static potential and the dynamic potential.

Polarizer. Substance or means that produces or increases polarization.

Polishing. Smoothing of a metal surface by means of the action of abrasive particles attached by an adhesive to the surface of wheels or endless belts usually driven at a high speed.

Pore. Minute cavity or channel extending all or part of the way through a coating.

Porosity. Condition of an electrodeposit in which pores are present; the number of such pores in a given area.

Primary current distribution. Distribution of the current over the surface of an electrode in the absence of polarization.

Quinhydrone electrode. Half-cell with a platinum or gold electrode in contact with a solution saturated with quinhydrone.

Rectification. Conversion of alternating into unidirectional (direct) current by means of electric valves.

Rectifier. Device that converts alternating into direct current by virtue of a characteristic permitting appreciable flow of current in only one direction.

Reducing agent. Compound that causes reduction, thereby itself becoming oxidized.

Reduction. Reaction in which electrons are added to a reactant, more specifically, the addition of hydrogen or the abstraction of oxygen. Such a reaction takes place, for example, at the cathode in electrolysis.

Reflowing. Melting of an electrodeposit, followed by re-freezing, so that the surface takes on the appearance of being "hot-dipped" rather than electroplated; especially of tin plate.

Relieving. Removal of material from selected portions of a colored metal surface by mechanical means to achieve a multicolored effect.

Resist. Material applied to a part of a cathode or plating rack to render the surface nonconductive; material applied to a part of the surface of an article to prevent reaction of metal from that area during chemical or electrochemical processes.

Ripple (dc)

1. Regular modulations in the output wave of a rectifier unit or a motor-generator set, originating from the harmonics of the direct-current input system in the case of a rectifier or from the harmonics of the induced voltage of a motor-generator set.

2. Alternating-current component of a pulsating current when this component is small relative to the direct-current component.

Rochelle salt. Sodium potassium tartrate.

Sacrificial protection. Form of corrosion prevention whereby the coating corrodes rather than the basis metal even though the latter is exposed to the corroding medium through pores or scratches.

Salting out. Precipitation of a substance from a solution by the action of added salts.

Sandblasting. Abrasive blasting with sand.

Satin finish. Surface finish that behaves like a diffuse reflector and that is lustrous but not mirrorlike.

Scale. Adherent oxide coating that is thicker than the superficial film referred to as "tarnish."

Sealing of anodic coating. Process that by absorption, chemical reaction, or other mechanism increases the resistance of an anodic coating to staining and corrosion, improves the durability of colors produced in the coating, or imparts other desirable properties.

Semiautomatic plating

1. Plating in which the cathodes are conveyed automatically through only one plating tank.

2. Plating in which the prepared cathodes are mechanically conveyed through the plating baths, with intervening manual transfers.

Sequestering agent. Agent that forms soluble complex compounds with, or sequesters, a simple ion, thereby suppressing the activity of that ion. Thus in water treatment the effects of hardness can be suppressed by adding agents to sequester calcium and magnesium.

Shield

1. Nonconducting medium for altering the current distribution on an anode or cathode.

2. Alter the normal current distribution on an anode or cathode by the interposition of a nonconductor.

Shielding. Placing an object in an electrolytic bath so as to alter the current distribution on the cathode. A nonconductor is called a shield; a conductor is called a robber, thief, or guard.

Shunt-wound motor. Direct-current motor in which the field circuit and armature circuit are connected in parallel.

Silver plating. Silver plating uses a cold solution and consists of the reduction of silver to pure silver nitrate dissolved in distilled water and then converted into double cyanide of silver and potassium.

Slurry. Suspension of solids in water.

Soak cleaning. Cleaning by immersion without the use of current, usually in an alkaline solution.

Soda ash. Anhydrous sodium carbonate.

Solution potential. Electrode potential where the half-cell reaction involves only the metal electrode and its ion.

Solvent cleaning. Cleaning by means of organic solvents.

Solvent degreasing. Degreasing by immersion in liquid organic solvent.

Spotting out. Delayed appearance of spots and blemishes on plated or finished surfaces.

Spray cleaning. Cleaning by means of spraying.

Standard electrode potential. Equilibrium electrode potential for an electrode in contact with an electrolyte in which all the components of a specified chemical reaction are in their standard states. Standard state for an ionic constituent is unit ion activity.

Static electrode potential. Electrode potential measured when no net current is flowing between the electrode and the electrolyte.

Stopping off. Application of a resist to any part of an electrode—cathode, anode, or rack.

Stray current

1. Current through paths other than the intended circuit, for example, through heating coils or the tank.

2. Current flowing in electrodeposition by way of an unplanned and undesired bipolar electrode, which may be the tank itself or a poorly connected electrode.

Strike

1. Thin electrodeposited film of metal to be followed by other plated coatings; a plating solution of high covering power and low efficiency designed to electroplate a thin adherent film of metal.

2. Thin film of metal to be followed by other coatings; solution used to deposit a strike.

3. Plate for a short time, usually at a high initial current density.

Striking. Electrodepositing under special conditions a very thin film of metal that will facilitate further plating with another metal or with the same metal under different conditions.

Strip. Process or solution used for the removal of a coating from a basis metal or an undercoat.

Substrate. Layer of metal underlying a coating, regardless of whether the layer is basis metal.

Superimposed ac. Form of current in which an alternating-current component is superimposed on the direct plating current.

Surface-active agent

1. Substance that affects markedly the interfacial or surface tension of solutions even when present in very low concentration.

2. Surfactant; soluble or colloidal substance having the property of affecting markedly the surface energy of solutions even when present in very low concentration.

Surface tension. Property, due to molecular forces, that exists in the surface film of all liquids and tends to prevent the liquid from spreading.

Tank voltage. Total voltage between the anode and cathode of a plating bath or electrolytic cell during electrolysis; voltage equal to the sum of the equilibrium reaction potential, the *IR* drop, and the electrode potentials.

Tarnish. Dulling, staining, or discoloration of metals due to superficial corrosion. A film is formed.

Thief. Auxiliary cathode so placed as to divert to itself some current from portions of the work that would otherwise receive too high a current density.

Throwing power. Improvement of the coating (usually metal) distribution ratio over the primary current distribution ratio on an electrode (usually cathode). The throwing power of a solution is a measure of the degree of uniformity with which metal is deposited on an irregularly shaped cathode. The term may also be used for anodic processes, for which the definition is analogous.

Titration. Analytical process for determining the quantity of an element or compound in solution by measuring the quantity of some reagent liquid or solution that resets quantitatively with it.

Topochemical. Term applied to localized reactions that take place in the inner or outer fields of force of crystalline material. Topochemistry is the study of localized reactions.

Total cyanide. Total content of cyanide or alkali cyanide, whether present as simple or complex ions; sum of both the combined and free cyanide content of a solution.

Transference number. Proportion of the total current carried by ions of a given kind.

Transference, transport, migration. Movement of ions through the electrolyte associated with the passage of the electric current.

Trees. Branched or irregular projections formed on a cathode during electrodeposition, especially at edges and other high current density areas.

Tripoli. Friable and dustlike silica used as an abrasive.

Tumbling. Rotation of the work, usually castings or forgings, in a barrel with metal slugs or abrasives to remove sand, scale or fins. Tumbling may be done dry or with aqueous solution. Operation is sometimes called "rumbling" or "rattling."

Ultrasonic cleaning. Cleaning by chemical means aided by ultrasonic energy.

Vapor area. Area containing dangerous quantities of flammable vapors in the vicinity of dip tanks, their drainboards or associated drying, conveying, or other equipment, during operation or shutdown periods.

Vapor area ventilation. Vapor areas may be limited to the smallest practical space by maintaining a properly designed system of mechanical ventilation arranged to move air from all directions toward the vapor area origin and thence to a safe outside location. Term is applied to required ventilating systems so arranged that the failure of any ventilating fan automatically stops any dipping conveyor system.

Vapor degreasing

1. Degreasing work in vapor over a boiling liquid solvent, the vapor being considerably heavier than air. At least one constituent of the soil must be soluble in the solvent.

2. Degreasing by solvent vapors condensing on the parts being cleaned.

Vapor plating. Deposition of a metal or compound on a heated surface by reduction or decomposition of a volatile compound at a temperature below the melting points of the deposit and the basis material. Reduction is usually accomplished by a gaseous reducing agent such as hydrogen. The decomposition process may involve thermal dissociation or reaction with the basis material. The term is occasionally used to designate deposition on cold surfaces by vacuum evaporation.

Venturi meter. Flow meter for liquids or gases consisting of a specially tapered constriction in a pipe; the pressure difference between a point before the constriction and a point at the narrowest part is a function of the rate of flow.

Viscosity. Resistance to flow shown by liquids; type of internal friction.

Voltage efficiency vapor deposition. Ratio, usually expressed as a percentage, of the equilibrium reaction potential in a given electrochemical process to the bath voltage.

Volumetric analysis. Method of analysis that depends on measurement of the volume of standard solution consumed in a titration.

Water break. Appearance of a discontinuous film of water on a surface signifying nonuniform wetting and usually associated with a surface contamination.

Wet blasting. Process for cleaning or finishing by means of a slurry of abrasive in water directed at high velocity against the work pieces.

Wetting agent. Substance that reduces the surface tension of a liquid, thereby causing it to spread more readily on a solid surface.

Whisker

1. Microscopic crystalline metallic filament growing out from the surface of a metal; the cause of this phenomenon is not completely understood.

2. Metallic filamentary growth, often microscopic, sometimes formed during electrodeposition and sometimes spontaneously during storage or service after finishing.

Zeolite. Any of a number of hydrous silicates; specifically, one (natural or artificial) used in water softening by cation-exchange reactions.

METALS AND FABRICATION

Abrading. Material removal process, usually accomplished by the action of mineral particles against a fixed or movable work piece. Examples are sanding, grinding, and polishing.

Abrasion. Process of rubbing, grinding, or wearing away by friction.

Abrasive

1. Any of a wide variety of natural or manufactured substances used to scour, polish, or clean metal surfaces by high-speed impact.

2. Substance used for grinding, honing, lapping, superfinishing, polishing, pressure blasting, or barrel-finishing. Term includes natural materials such as garnet, emery, corundum, and diamond and electric furnace products such as aluminum oxide, silicon carbide, and boron carbide.

Abrasive belt. Coated abrasive product in the form of a belt used in production grinding and polishing.

Abrasive blasting. Finishing process in which the work is dry blasted with abrasive particles.

Abrasive disk. Grinding wheel mounted on a steel plate with the exposed flat side used for grinding.

Abrasive finishing. Removing small amounts of material from metal surfaces by grinding, honing, buffing, polishing, and other similar processes.

Accurate. Produced within allowable tolerances.

Adhesion. Fastening process in which materials are permanently joined by the bonding of two materials with a third, dissimilar, substance such as solder, glue, or a brazing rod. The resultant joint is discontinuous; that is, the individual parts do not become one continuous piece of the same material.

Adhesives. Materials or compositions that enable two surfaces to joint together.

Aging. Process of holding metals at room temperature or at a predetermined temperature for the purpose of increasing their hardness and strength by precipitation.

Align. Adjusting to specified points.

Allowance. The specified limits that will be permitted to achieve a satisfactory performance.

Alloy. Substance that has metallic properties and is composed of two or more chemical elements of which at least one is a metal.

Anneal. To heat and cool, usually for softening and making the metal less brittle.

Annealing. The process of heating metal to a prescribed temperature and cooling it slowly to remove stresses and induce softness.

Anodizing. Process for applying an oxide coating to aluminum. Can be varied to produce a light colored, porous coating that allows dyeing in a variety of colors.

Austempering. Heat-treating process consisting of quenching a ferrous alloy at a temperature above the transformation range in a medium such as molten lead.

Austenite. Solid solution of iron, carbon, and other elements where a gamma ion characterized by a face-centered cubic crystal structure is in the solvent.

Austenitizing. Process of forming austenite by heating a ferrous alloy above the transformation range.

Automatic press. Press through which the work is fed mechanically in synchronism with the press action. An automation press is a press that is provided with built-in electrical and pneumatic control equipment.

Beading. Formation of ornamental or strengthening ribs, for example, around the circumference of cylindrical light metal parts.

Bending

1. Forming by uniformly straining a material around a straight axis or die. Types include press-brake, roll, and stationary or fixed-die bending.
2. Changing the plane of a surface.

Beryllium. A metal weighing less than steel but offering equal strength characteristics.

Billet. Heavy piece of rolled metal of substantially square cross section from which various shapes are produced.

Blanking

1. In sheet metal work, the cutting out of a piece of metal, usually by means of a press.
2. Cutting desired shapes out of metal to be used for forming or other manufacturing operations.

Blowhole. Hole or defect in a casting resulting from entrained gases.

Bonderizing. Dipping lightly galvanized objects in a hot phosphate solution to form a surface film of zinc phosphate to improve paint adhesion on steel.

Braking. Mechanical bending operation usually performed on sheets and plates.

Brass. A copper-zinc alloy of varying proportions but typically containing large amounts of copper with zinc added.

Brazing. Jointing metals by a heating process using a copper and zinc alloy.

Broaching. Finishing the inside of a hole to a shape usually other than round.

Buckle. Bend, warp, or kink in a piece of metal, usually in sheets or plates.

Buffing. The process of producing luster on the surface of metals.

Brittleness. The opposite of toughness. Characteristic that causes metal to break.

Burnishing. Process of finishing a metal by compressing its surface.

Burr. Sharp edge remaining on metal after cutting, stamping, or machining.

Burring. Removing the rough edge left after shearing, drilling, or machining.

Carbon. A nonmetalic element.

Carbon steel. Variety of steel classified by the amount of carbon it contains.

Carburizing. Process that introduces carbon to the surface of steel by heating metal below its melting temperature in contact with carbonaceous solids, liquids, or gases and holding at that temperature for a predetermined period after which the metal is quenched.

Casehardening. Process of surface hardening iron base alloys, so that the surface layer or case is made substantially harder than the interior or core.

Casting. Process of producing a metal object by pouring molten metal into a mold.

Casting alloy. An alloy that cannot be forged or rolled and can be shaped only as a casting.

Casting copper. Copper used for foundry castings; obtained from copper ores, it is inferior to electrolytic copper.

Cast iron. Carbon-iron alloy cast to shape, composed of small amount of carbon.

Caustic. Chemical base material; caustic soda, solution of sodium hydroxide used to clean metal surfaces to remove oil, grease, and other foreign particles.

Cementite. Compound of iron and carbon.

Centrifugal casting. Casting where the mold is rotated during pouring and solidification of the metal.

Chemical milling. Controlled removal of metal by chemicals.

Clad metal. Composite metal containing two or three layers that have been bonded together. Bonding may have been accomplished by corolling, welding, casting, heavy chemical deposition, or heavy electroplating.

Coating. Covering a material's surface with a decorative or protective layer of another type of material. Types of coating include painting, plating, and enameling.

Coining. Bringing a stamping or forging to exact size by heavy die pressure.

Cold-drawing. Drawing metal through a die without the application of heat.

Cold-heating. Operation where metal is worked cold.

Cold-rolled. Metal rolled at room temperature, below the softening point. Cold-rolled metal is usually harder, smoother, and more accurately dimensioned than hot-rolled metal.

Cold-working. All processes of changing the form or cross section of a piece of metal at a temperature below the softening point.

Continuous casting. Casting where the ingot is continuously solidified while it is being poured.

Copper. A ductile and malleable nonferrous metal found in various ores and used in industry and the arts in both pure and alloyed form.

Copper steel. Low-carbon steel containing a small amount of copper.

Copper wire. Made from copper by drawing from a hot-rolled rod without annealing.

Core. Portion of a mold used to form the interior of a hollow casting during the casting process.

Corrosion. Gradual electrochemical attack on a metal, caused by galvanic action, such as moisture, in an electrolyte.

Corrosion resistance. Ability of some metals to form an oxide surface skin that resists further corrosion.

Corrugating

1. Forming a surface to consist of a series of alternate ridges and valleys, usually semicircular in section.

2. Forming sheet metal into a series of straight, parallel, alternate ridges and grooves by using a rolling mill equipped with matched roller dies or by using a press bake equipped with specially shaped punch and die.

Countersinking

1. Beveling the edge of a hole for the reception of the head of a bolt, rivet, or screw.

2. Forming a flaring depression around the top of a hole for deburring, for receiving the head of a fastener, or for receiving a center.

Crimping. Process similar to corrugating, but with short and small ridges at edges; process of preparing wire for weaving into wire products.

Curling. Causing the edge of a piece of metal to bend down and around in approximately circular section.

Cyaniding. Process of casehardening a ferrous alloy by treating with molten cyanide, causing the metal to absorb carbon.

Decarburization. Loss of carbon from the surface of a ferrous alloy as a result of heating it in the presence of a medium, such as oxygen, that reacts with the carbon.

Decarburizing. Process of removing carbon from metals.

Deoxidizer. Substance that is used to remove oxygen from molten metals.

Dezincification. Corrosion of a metal alloy that contains zinc.

Die. Tool used to impart a desired shape to a piece of metal.

Die casting. Method of casting metal under pressure by injecting it into the metal dies of a die casting machine.

Drawing. In heat treating, reheating a hardened piece to some temperature below the lower critical temperature. The term "tempering" is now preferred for this form of heat treatment.

Drawing and extruding process die. In the drawing and extruding process the die is the tool through which the metal is drawn or pushed.

Drift pin. Tapered round pin used to align holes in several pieces of metal.

Drilling

1. Round-hole-producing cutting operation employing a cylindrical tool with two spiral cutting edges; material removal process generally involving a fixed work piece.

2. Process of cutting round holes in metal by means of a revolving tool.

Drop forging. A forming operation, done under impact, that compresses the metal in dies designed to produce the specified shape.

Ductility. Property of a metal that permits permanent deformation by hammering, rolling, and drawing without breaking or fracturing.

Embossing. Development of a raised design on a metal surface by die pressure or by stamping or hammering on the reverse surfaces.

Etching. Process of eating away an exposed surface by means of acid, alkaline, or abrasive action to form a pattern or design.

Eutectic. Alloy composition that freezes at the lowest constant temperature, causing a discrete mixture to form in definite proportions.

Extrusion. Forcing a solid metal piece through a shaped die by using an extremely high force.

Fastening. Joining materials together permanently or semipermanently.

Fatigue. Tendency for metal to break or fracture under repeated or fluctuating stresses.

Ferrite. Magnetic form of iron.

Ferrous. Family of metals in which iron is the major ingredient.

Finishing. Treating the surface of a material to increase its serviceability and/or improve its appearance.

Fins. Metal that has flowed into the mold joints and that must be removed from the casting.

Flame harden. Method of surface hardening steel by rapidly heating the surface with flame from an oxyacetylene torch, then quenching.

Flanging. Developing a ridge or rim; turning an edge.

Flat. Rectangular bar whose width is greater than its thickness.

Floor plate. Metal plate having ridges, knobs, or other regular protrusions spaced uniformly on one surface to provide nonslip qualities.

Fluting. Formation of a semicircular groove or series of grooves in sheet metal.

Flux. Substance used to promote the fusion or welding of metals by preventing oxidation of the surfaces being joined.

Foil. Very thin metal, usually not over 0.005-in. thickness, most commonly of aluminum, copper, lead or tin, rolled in sheets.

Forging. Heating and hammering or pressing metal into a desired shape.

Forming. Giving shape to a material without adding or removing any of the material.

Forming process die. Formed tool that is forced against the metal to be formed; shape of the die determining the shape of the finished piece.

Foundry. Building and equipment used for the production of castings and other metal working.

Fracture. Ruptured surface of metal that exhibits a typical crystalline pattern.

Full annealing. Heating metals above their critical range, holding them at temperature until the grain structure is stabilized, and following with controlled slow cooling. The process is used to develop maximum softness in most steels.

Fusion. Merging of two materials while in a molten state.

Galvanize. To deposit zinc on the surface of metal by the process of hot dipping, sheradizing, or electroplating.

Gangue. Worthless rock or aggregates found in ore.

Gate. Part of the mold through which the metal is poured and which is removed after the casting has cooled.

Gauge (Gage). Measure of the thickness of sheets or of the diameter of wire or screws, or an instrument for measuring thickness or diameter; the usual established distance for locating holes for bolts and rivets in structural steel.

Graphite. Allotropic form of carbon.

Grinding. Removing metal from a surface by means of abrasive action.

Grit. Granular abrasive such as aluminum oxide or silicon carbide coated on cloth, paper, or buffing wheels and used in making grinding wheels. The fineness of grit varies and is indicated by numbers.

Gusset plate. Plate used for attaching two or more structural members together by bolting, riveting, or welding.

Hardening. Heating and quenching of certain iron-base alloys for the purpose of producing a hardness superior to the untreated material.

Hardness. Property of a metal to resist being permanently deformed.

Heat-affected zone. That portion of the base metal that has not been melted, but whose mechanical properties and microstructure have been changed by the heat of welding.

Heat treatment. Application of a combination of heating and cooling cycles to a metal or alloy in the solid state to bring about desirable conditions such as hardness and toughness.

High-energy forming. Metal forming technique using the release of a source of high energy, such as electrical, pneumatic-mechanical, or explosive.

Hollow metal. Construction consisting of sheet metal formed, or drawn, or joined by seams into hollow shells of the required shape, to form door stiles, rails, and window sashes. Such shapes generally include moldings, recesses, and rebates not ordinarily found in standard metal door or window construction.

Honing. Process which produces an extremely fine surface.

Hot-forming. Working operations such as bending, drawing, forging, piercing, pressing, and heading performed at temperatures above the recrystallization temperature of the metal.

Hot-rolling. Process of heating metals to temperatures above the transformation range and forming between rolls.

Ingot. Large block of metal usually cast in a metal mold; forms the basic material for further rolling and processing.

Interlocking joint. Formed edge of a metal strip or bar designed to grip a similar piece of metal to produce a continuous locking splice.

Iron. Term "iron" always refers to the element Fe.

Kerf. Channel or groove cut by a saw or other tool.

Knurling. Cross ridging or indentation of a surface by means of a suitably shaped milling cutter or revolving tool for the purpose of improving the hand grip.

Lap seam. Joint made by laying the edges of two metal sheets or plates one on the other and riveting, brazing, soldering, or welding them together.

Lead alloy. An alloy composed of bronze, brass, or steel to which lead is added to improve machinability and mechanical properties.

Lead bronze. An alloy composed of a large amount of copper, small amounts of nickel, and tin.

Lock seam. Seam made by double folding the edges of sheets to permit the edges to interlock.

Lost-wax process. Process using patterns of wax that are melted and drained from the mold before the metal is poured, used in the making of castings involving undercuts and other complications.

Machinability. Ability to be milled, sawed, tapped, drilled, and reamed without excessive tool wear and with ease of chip metal removal and surface finishing.

Machining. Mechanical removal of material by means of a sharp-edged tool.

Malleability. Ability of a metal to deform permanently without rupture when loaded in compression.

Metalloid. Nonmetal that exhibits some, but not all, of the properties of a metal.

Milling. Removal of metal to develop a desired contour by means of a revolving cutting tool.

Nickel. Silver-gray, ductile, malleable, tough metal, used in alloys, plating, ceramics, and electronic circuits.

Nickel-aluminum bronze. An alloy composed of a small amount of aluminum bronze with nickel added to increase strength, corrosion resistance, and heat resistance.

Nickel bronze. An alloy composed of a large amount of copper and small amounts of tin, nickel, and zinc.

Nickel cast iron. An improved-strength alloy cast iron composed of a small percentage of nickel; increasing the nickel produces greater corrosion resistance.

Nickel-chromium steel. Steel containing small amounts of nickel and chromium as alloying elements.

Nickel-molybdenum iron. An alloy composed of a large amount of nickel and a smaller amount of molybdenum with carbon.

Nickel-molybdenum steel. Steel containing small amounts of molybdenum and nickel.

Nickel silver. Silver-white alloy composed of a large amount of copper with smaller amounts of nickel and zinc; also may contain tin and lead.

Nickel steel. Carbon steel containing small amounts of nickel as a major alloying element.

Nickel-vanadium steel. Nickel steel containing small amounts of nickel, manganese, carbon, and vanadium, used for high-strength cast parts.

Nitriding. Casehardening system where a ferrous alloy is heated in an atmosphere of ammonia or in contact with a nitrogenous material to produce surface hardness by the absorption of nitrogen.

Nonferrous. Metals containing no iron.

Normalizing. Process of heating ferrous alloys to approximately $100°F$ above the critical temperature range and cooling slowly in still air at room temperature to relieve stresses that may develop during machining, welding, or forming.

Notching. Cutting a nick or deep indentation in the surface of a piece to be forged so as to facilitate bending.

Nut

1. Block or sleeve having an internal thread designed to assemble with the external thread on a bolt, screw, stud or other threaded part.
2. Piece of metal drilled and tapped to receive the threaded end of a bolt, produced in various shapes.

Oil hardening. Using mineral oil as a quenching medium in the heat treatment of certain alloys.

Pattern. Original from which molds for castings are made.

Pearlite. Lamellar mixture of ferrite and cementite in slowly cooled iron-carbon alloys, found in steel and cast iron.

Peening. Mechanical working of metals by means of hammer blows.

Perforating

1. Piercing holes of desired shapes and arranged patterns in sheets, blanks, or formed parts.
2. Punching or drilling multiple holes in sheet metal or light gauge plates.

Pickling. Cleaning a metal surface by immersion in an acid bath.

Piercing. Forcing a point through a piece of metal to form a hole without the removal of any metal.

Pig iron. Product of a blast furnace. It is a raw iron.

Plaster molds. Casting process using plaster molds instead of sand molds, producing a better surface finish.

Pipe fittings. Elbows, crosses, T's (tees), flanges, and similar connections used to connect pipe or tubing.

Polishing. Process of removing surface roughness of metal to obtain a smooth, bright surface.

Porcelain enamel. Inorganic glass coating for metals made by fusing powdered glass to a metal surface; similar to ceramic glazes.

Pressing. Forcing metal to conform to the shape of a die by means of pressure.

Punching. Forcing a punch through metal into a die, forming a hole the shape of the punch.

Punching-process die. That part of the tool receiving the punch that pierces the metal.

Quenching. The process of rapid cooling of metal alloys for the purpose of hardening. Quenching media include air, oil, water, molten metals, and fused salts.

Reaming. Enlarging a round hole in a piece of metal by means of a revolving-edge tool.

Refractory. Materials that resist change of shape, weight, or physical properties at high temperatures.

Riveting. Forming a permanent connection between two or more pieces of material by passing a rivet through aligned holes and upsetting to form a head.

Rod. Round, square, rectangular, hexagonal, or other cross section of rolled, drawn, or extruded metal.

Rolling. Shaping metal, either hot or cold, by passing it between revolving rolls set a predetermined distance apart. Rolls may be flat over their entire surface or shaped as desired so long as reentrant angles are not used.

Round pipe. Hollow round section of metal, the size of which is determined by the nominal inside diameter in inches and fractions thereof.

Rust

1. Reddish iron oxide formed by the oxidation of iron and steel.
2. Corrosion product consisting of hydrated oxides of iron. Term is applied only to ferrous alloys.

Sandblasting. Blowing particles of sand or other hard granules through a nozzle onto a metal surface to produce a uniform, slightly roughened appearance.

Sanding. Rubbing a surface with an abrasive material.

Sand mold casting. Pouring molten metal into a cavity formed in a sand mold.

Scale. Surface oxidation caused by heating metals in air.

Seaming. Uniting the edges of a sheet or sheets by bending them over or doubling them and pinching them together.

Shaving. Removing a thin slice of metal by a tool.

Shearing

1. Cutting metal by the action of two opposing passing edges.
2. Material separation process in which a single cutting edge or two crossed edges are forced through a fixed work piece.

Sheet

1. Flat-rolled metal product of some maximum thickness and minimum width arbitrarily dependent on the type of metal. Sheet is thinner than plate.
2. Thin, flat piece of metal, the thickness of which is measured by gauge or decimals of inches.

Shielded metal arc welding. Arc welding in which a flux-coated metal electrode is consumed to form a pool of filler metal and a gas shield around the weld area.

Shielded welding. Process using gases or fusible granular materials to shield the weld area from damaging effects of oxygen and nitrogen in the air.

Shipbuilding- and carbuilding-type bulb angles. Angles produced with the elongated leg having a small bulb extending at the end of the leg.

Shrink hole. Defect in a casting due to shrinkage.

Sintering. Method of bonding metal powders that have been compacted by heating them to a predetermined temperature.

Slitting. Cutting sheets or strip metal into long narrow strips, usually by means of revolving shears or slitting rolls.

Smelting. Process of heating ores to a high temperature in the presence of a reducing agent, such as carbon, and of a fluxing agent to remove the gangue.

Soldering

1. Joining metal parts by another metal having a melting point substantially lower than that of the base metal.
2. Joining of two or more pieces of metal with a low-strength, low-melting-point filler metal that bonds chemically with the base metal. No melting of base metal occurs.

Spinning. Shaping sheet metal by bending or buckling it under pressure applied by a smooth hand tool or roller while the metal is being revolved rapidly.

Stock sizes. Sections of metal usually available in warehouse stocks.

Straightening. Eliminating deformations by pressing, rolling, or stretching.

Stretcher leveling. Stretching a metal sheet to produce a straight, flat surface.

Strip. Narrow metal sheets produced either in coil form or cut to length and finished flat. The thickness of the strip is measured by gauge; width is measured in inches and fractions of inches.

Spraying. Coating metal with paint or similar materials by air pressure; coating metal with molten metal by air pressure.

Square pipe. Hollow square section of metal, the size of which is determined by the nominal outside diameter in inches and fractions.

Squares. Referring to square rods. Corners may be slightly rounded or sharp.

Stainless steel. Alloy of iron containing small amounts of chromium and nickel. Resists almost all forms of rusting and corrosion.

Stamping. Bending, shaping, cutting out, indenting, embossing, or forming metal, either hot or cold, by means of shaped dies in a press or power hammer.

Standard pipe size. Round pipe of any metal, in which the nominal inside diameter and wall thickness are the same as that of standard steel pipe.

Steel. An alloy of iron and a small amount of carbon plus impurities and small amounts of alloying elements.

Steel foil. Very thin sheet of steel, the thickness measured in thousandths of an inch.

Steelmaking. Any of various processes for making steel from pig iron.

Swaging. Surface working of a forging by means of repeated blows, usually between dies. Swaging may be done either hot or cold.

Tampico wheel. Polishing wheel with coarse fiber bristles.

Tantalum. Metal that is capable of withstanding temperatures in the 2500 to 4000°F range.

Tapping. Cutting internal threads in a punched or drilled hole.

Tempering. In ferrous metals, the stress relief of steels hardened by quenching for the purpose of toughening them by reducing their brittleness.

Template (templet). Flat pattern used as a guide in making a part of a piece for a particular location.

Threaded fasteners. Group classification of mechanical fasteners that includes screws, bolts, nuts, and washers.

Threading. Developing the helical rib on a bolt or screw and the corresponding groove in a nut or hole whereby they can be screwed together.

Threading-process die. Tool or set of tools that cuts the thread.

Tin. A lustrous silver-white ductile, malleable metal used in alloys for solder, terneplate, and tinplate.

Tin-bronze. A tin-copper alloy.

Titanium. Metal used for applications that require the properties of light weight, high strength, and good temperature and corrosion resistance.

Tolerance. Amount of allowable variation from a specified dimension.

Toughness. Measured in terms of notch toughness, the ability of a metal to resist rupture from impact loading when a notch is present.

Tube. Hollow product of round or any other cross section having a continuous periphery.

Tubing. Hollow section of metal that may be round, square, rectangular, hexagonal, octagonal, or otherwise shaped, measured by the external size in inches and fractions. Wall thickness is measured by gauge or decimals of inches.

Tumbling. Cleaning metal articles by placing them in a revolving container with or without cleaning material.

Turning. Removal of metal by means of an edged cutting tool while the piece is being revolved about its axis.

Upsetting. Building up or thickening the section of a piece of metal by shortening the piece by axial compression. Operation may be performed hot or cold.

Welding

1. Creating a metallurgical bond between metals with heat and sometimes with the use of pressure and filler metal.

2. Joining two metals or alloys by fusion.

Wire. Small-diameter rod measured by gauge, produced by a drawing process.

Work hardening. Hardening of a metal as a result of cold-working.

Wrought alloy. An alloy that has been mechanically worked after casting.

Wrought iron. Iron with a small percentage of slag, distributed as threads and fibers, imparting a tough fibrous structure. Also contains a small percentage of carbon. Tough, malleable, and relatively soft.

Zinc. A shiny, bluish-white, lustrous metal that is ductile when pure, used in alloys, metal coatings, electrical fuses, anodes, and dry cells.

STRUCTURAL METAL FRAMING

AISC specifications. The specification for the design, fabrication, and erection of structural steel for buildings as adopted by the American Institute of Steel Construction.

ANSI. American National Standards Institute.

Assembly. Fitting together and/or attaching by any means, including brazing, welding, bonding, riveting, or bolting.

ASTM. The material standards of the American Society for Testing and Materials.

AWS Code. The Structural Welding Code of the American Welding Society.

Bay (part of a structure). Space between two adjacent piers or columns.

Beam. Horizontal load-bearing structural member, transmitting superimposed vertical loads to other horizontal structural members, walls, or columns.

Beam-column. Structural member whose primary function is to carry loads both transverse and parallel to its longitudinal axis.

Bearing. That portion of a supporting structural member that is bearing on a wall, column, or other structural member.

Bent. Framework of a beam or truss members that are fabricated to support loads and the columns that support these members.

Bolster. Short horizontal structural member bearing on top of a column for the support of other structural members.

Box girder. Girder having a hollow cross-section similar to that of a rectangular box.

Box system. Structural system without a complete vertical load-carrying space frame; in this system the required lateral forces are resisted by shear walls or the braced frames.

Brace. Usually an inclined member forming a triangle in a framework to stiffen the frame.

Braced frame. Frame in which the resistance to lateral load or frame instability is primarily provided by a diagonal, a K-brace, or other auxiliary system of bracing.

Bridging. Diagonal or cross bracing installed between structural members to resist twisting.

Built-up member. Member made of structural metal elements that are welded, bolted, or riveted together.

Cellular steel deck. Structural floor system consisting of two layers of sheet metal shaped to form cells and welded together. Cells can serve as electrical raceways.

Cellular steel floor. Construction consisting of sheet or strip steel formed into an integrated system of parallel steel beams which combine the function of load-bearing members and a continuous floor spanning between main supporting beams, girders, or walls.

Clear span. Distance between the inside faces of supports.

Code. The Code of Standard Practice adopted by the American Institute of Steel Construction.

Cold-formed members. Structural members formed from steel without the application of heat.

Cold-pressing quality plate. Plate made of soft steel suitable for bending and forming at ordinary temperatures.

Column. Structural member whose primary function is to support loads parallel to its longitudinal axis.

Composite beam. Steel beam structurally connected to a concrete slab so that the beam and slab respond to loads as a unit.

Composite column. Steel column fabricated from rolled or built-up steel shapes and encased in structural concrete or fabricated from steel pipe or tubing and filled with structural concrete.

Concrete-encased beam. Beam totally encased in concrete cast integrally with the slab.

Connection. Combination of joints used to transmit forces between two or more structural members.

Contract Documents. Documents that define the responsibilities of the parties involved in bidding, purchasing, supplying, and erecting structural steel.

Crane rails. Rails used for crane runways, using: carbon Steel Crane Rails. (ASTM A 759).

Decking. Fabricated metal forms installed over other structural members to which other materials can be attached.

Diagonal bracing. Inclined structural members carrying primarily axial load, employed to enable a structural frame to act as a truss to resist horizontal loads.

Diaphragm. Floor slab, metal wall, or roof panel possessing a large in-plane shear stiffness and strength adequate to transmit horizontal forces to resisting systems.

Drawing quality plate. Plate produced from low-carbon steel suitable for drawing into identified forms.

Drawings. Shop fabrication and field erection drawings prepared by the fabricator and the erector, required for the performance of the work.

Embedment. Steel component cast in a concrete structure, used to transmit externally applied loads to the concrete structure by means of bearing, shear, bond, friction, or any combination thereof. The embedment may be fabricated of structural-steel plates, shapes, bars, bolts, pipe, studs, concrete reinforcing bars, shear connectors, or any combination thereof.

Encased steel structure. Steel-framed structure where all of the individual frame members are completely encased in cast-in-place concrete.

Erector. The contractor responsible for the erection of the structural members, providing all of the required tools, equipment, and scaffolding.

Fabricator. The contractor responsible for furnishing the fabricated structural steel. Certificates to indicate location of purchase, name of mill, and to certify the structural steel values are to be provided before fabrication begins.

Fastener. Term used for welds, bolts, or other connecting devices.

Flame-cut plate. Plate where the longitudinal edges have been prepared from a large plate by oxygen cutting.

Floor assembly. Combination of materials providing the horizontal separation between stories, including the ceiling, floor, and horizontal structural members supporting the floor, but excluding those primary structural members that serve as part of the structural frame.

Floor plates. Plates having raised patterns using a commercial grade carbon steel except in special uses, for which various types of steel can be provided.

Forging quality plate. Plate intended for forging, heat treating, or similar purposes, requiring uniformity of composition and freedom from injurious defects.

Formed steel construction. Construction composed of sheet or strip steel formed into structural panels, decks, studs, joists, and other fabricated structural members.

Full Circle Plates. Plates available up to 1 in. in thickness, sheared or gas cut, depending on thickness.

Fully composite beam. Composite beam with sufficient shear connectors to develop the full flexural strength of the composite section.

Girder. Usually a long-span structural member composed of one or more members to support other structural members.

Girt. Horizontal structural framing member to aid in providing rigidity to vertical structural members.

Grade beam. Structural member installed at or near ground level, usually performing the functions of a foundation.

Hybrid beam. Fabricated steel beam composed of flanges with a greater yield strength than that of the web.

Hot-pressing quality plate. Plate used for ordinary hot pressing, flanging, or bending work.

Joint. Area where two or more ends, surfaces, or edges are attached by a fastener, such as a bolt, rivet, or weld.

K-bracing. System of struts, used in a braced frame, with a pattern resembling the letter "K", either upright or on its side.

Lally column. Patented type of concrete-filled steel pipe, used as a vertical support.

Lateral bracing member. Structural member utilized individually or as a component of a lateral bracing system to prevent buckling of other members or elements and/or to resist lateral loads.

Light gauge steel construction. Construction composed of sheet or strip steel, less than $\frac{3}{16}$ in. thick, formed into structural panels, decks, studs, joists, and other such fabricated structural members.

Lintel. Horizontal structural member used to support other loads over an opening, door or window.

Long-slotted holes. Permitted in only one of the connected parts of either a slip-critical or bearing-type connection at an individual faying surface.

Maximum size holes. Holes for bolts as described in the *Manual of Steel Construction*, except that larger holes, required for tolerance on location of anchor bolts in concrete foundations, are as indicated in column base details.

MBMA. Metal Building Manufacturers Association.

Mill material. Steel mill products ordered by the fabricator for a specific project.

Oversized holes. Permitted in any or all plies of slip-critical connections, but are not to be used in bearing-type connections. Washers over oversized holes are to be installed in an outer ply.

Partially composite beam. Composite beam for which the shear strength of shear connectors governs the flexural strength.

Pedestal. Upright compression member, the height of which does not exceed three times its least lateral dimension.

Pipe. Fabricated in three types: standard weight, extra strong, and double extra strong. Primarily used for vertical supports.

Plane frame. Structural system assumed to be two-dimensional for the purpose of analysis and design.

Plate girder. Fabricated structural beam.

Primary member. Member of the structural frame of a building, used as a column or a grillage beam, or to support masonry walls or partitions.

Rigid frame. Structure in which connections maintain the angular relationship between beam and column members under load.

Riveted construction. Parts of riveted structural members, well-pinned or bolted and rigidly held together during riveting.

Rivet steel. Rivets should conform to standards described in ASTM A 502 (latest issue).

Secondary member. Any member of the structural framework other than a primary member, including filling-in beams of floor systems.

Sheared plates. Plate material rolled between horizontal rolls and trimmed (sheared or gas cut) on all edges.

Short-slotted holes. Permitted in any or all plies of slip-critical or bearing-type connections.

Sketch plates. Plates with fully detailed dimensions and cuts.

Slip-critical joint. Bolt joint in which the design of the slip resistance connection is required, to assure that slip will not occur at working load.

Space frame. Three-dimensional structural system composed of interconnected members, other than bearing walls, laterally supported so as to function as a complete self-contained unit with or without the aid of horizontal diaphragms or floor-bracing systems.

Space frame—vertical load-carrying. Space frame designed to carry all vertical loads.

Span. Clear horizontal distance between two supports.

Splice. Connection between two structural elements joined at their ends to form a single, longer element.

Standard holes. Provided in member-to-member connections, unless oversized, short-slotted, or long-slotted holes in bolted connections are approved by the engineer.

Steel joist. Any prefabricated secondary member for a building or structure, made from hot or cold-rolled steel sections, designed to resemble a truss.

Stepped column. Column with changes from one cross section to another, occurring at abrupt points within the length of the column.

Stiffener. Structural member, usually an angle or plate, attached to a plate or web of a beam or girder to distribute load, to transfer shear, or to prevent buckling of the member to which it is attached.

Structural alterations. Any change or revision to the supporting structural members of a building or structure, such as footings, bearing walls or partitions, columns, beams or girders, or any substantial change or revision to the roof or exterior walls.

Structural frame. All vertical load-supporting structural members, other than bearing walls, all primary horizontal load-supporting structural members rigidly connected to vertical members, and all other primary members essential to the stability of the structural frame.

Structural members. Structural steel fabricated to include columns, studs, posts, trusses, girders, beams, joists, and other supporting devices that are essential in carrying vertical, horizontal, or torque forces to bearing materials upon which a structure rests.

Structural quality plate. Plate used for bridges, buildings, and miscellaneous other uses.

Structural steel. Structural members and assemblies and the accessory parts and materials included in the fabrication and construction of the assemblies or frames.

Structural system. Assemblage of load-carrying components that are joined together to provide regular interaction or interdependence.

Stub column. Short compression-test specimen, long enough for use in measuring the stress-strain relationship for the complete cross section, but short enough to avoid buckling as a column in the elastic and plastic ranges.

Supported frame. Frame that depends upon adjacent braced or unbraced frames for resistance to lateral load or frame instability.

Truss. Complete or framed structural unit composed of structural members connected at their intersections, where, if loads are applied at their intersections, the stress in each member is in the direction of the length of the member.

Turn-of-nut method. Procedure whereby the specified pre-tension in high-strength bolts is controlled by rotation of the wrench a predetermined amount after the nut has been tightened to a snug fit.

Types of structural steel framing

1. Type 1 designated as "rigid-frame" (continuous frame); assumes that beam-to-column connections have suffi-

cient rigidity to hold virtually unchanged the original angles between intersecting members.

2. Type 2 designated as "simple framing" (unrestrained, free-ended); assumes that, insofar as gravity loading is concerned, ends of beams and girders are connected for shear only and are free to rotate under gravity load.

3. Type 3 designated as "semi-rigid framing" (partially restrained); assumes that the connections of beams and girders possess a dependable and known moment capacity intermediate in degree between the rigidity of Type 1 and the flexibility of Type 2.

Unbraced frame. Frame providing resistance to lateral load by the bending resistance of frame members and their connections.

Universal-mill plate. Plate with longitudinal edges formed by a rolling process during manufacture. Abbreviated as "UM" plate.

Vertical bracing system. System of shear walls, braced frames, or both, extending throughout one or more floors of a building.

Weathering steel. Type of high-strength, low-alloy steel that can be used in normal environments (not marine) and outdoor exposures without protective paint covering. This steel develops a tight adherent to rust at a decreasing rate over time.

WELDING—BRAZING AND CUTTING

Accurate. Performance within the prescribed tolerances.

Acetylene. Colorless gas with a characteristic odor. Very soluble in acetone. Used with oxygen as a welding gas to produce a very high temperature flame.

Admixture. The combining of a base metal with filler metal to alter the characteristics of either.

Air-acetylene welding. Gas welding process wherein coalescence is produced by heating with a gas flame or flames obtained from the combustion of acetylene with air, without the application of pressure and with or without the use of filler metal.

Alloy. Mixture of two or more metals fused or melted together to form a new metal.

All-position electrode. In arc welding, a filler-metal electrode for depositing weld metal in the flat, horizontal, overhead, and vertical positions.

All-weld-metal test specimen. Test specimen of which the portion being tested is composed wholly of weld metal.

Arc blow. Magnetic disturbance of arc that causes it to waver from its intended path.

Arc brazing. Brazing with an electric arc, usually with two nonconsumable electrodes.

Arc cutting

1. Group of cutting processes whereby the severing of metals is effected by melting with the heat of an arc between an electrode and the base metal.

2. Metal cutting with an arc between an electrode and the metal itself. The terms "carbon arc cutting" and "metal arc cutting" refer, respectively, to the use of a carbon or metal electrode.

Arc length. Distance from the end of electrode to the surface of the molten material.

Arc time. Length of time the arc is maintained in making an arc weld.

Arc voltage. Voltage across the welding arc.

Arc welding

1. Welding methods employing an electric arc as the source of heat.

2. Group of welding processes producing coalescence by heating with an electric arc or arcs, with or without the application of pressure and with or without the use of filler metal.

As welded. Condition of weld metal, welded joints, and weldments after welding but prior to any subsequent thermal or mechanical treatment.

Atomic-hydrogen welding

1. Arc welding with heat from an arc between two tungsten or other suitable electrodes in a hydrogen atmosphere. The use of pressure and filler metal is optional.

2. Arc-welding process wherein coalescence is produced by heating with an electric arc maintained between two metal electrodes in an atmosphere of hydrogen. Shielding is obtained from the hydrogen. Pressure may or may not be used, and filler metal may or may not be applied.

Automatic oxygen cutting. Oxygen cutting with equipment that performs the cutting operation without constant observation and adjustment of the controls by an operator. Equipment may or may not perform loading and unloading of the work.

Automatic welding. Welding with equipment that performs the entire welding operation without constant observation and adjustment of the controls by an operator. Equipment may or may not perform the loading and unloading of the work.

Axis of a weld. Line through the length of a weld, perpendicular to the cross section at its center of gravity.

Backfire. Momentary recession of the flame into the torch tip followed by immediate reappearance or complete extinguishment of the flame.

Backhand welding

1. Gas welding technique whereby the flame is directed opposite to the progress of welding.

2. Welding in which the back of the principal hand (torch or electrode hand) of the welder faces the direction of travel. This technique has special significance in gas welding in that it provides postheating.

Backing

1. Material backing up the joint during welding to facilitate obtaining a sound weld at the root.

2. Material placed under or behind a joint to enhance the quality of the weld at the root. It may be a metal backing ring or strip, a pass of weld metal, or a nonmetal such as carbon, granular flux, or a protective gas.

Backing pass. Pass made to deposit a backing weld.

Backing ring. Backing in the form of a ring, generally used in the welding of piping.

Backing strip. Backing in the form of a strip.

Backing weld. Backing in the form of a weld.

Back pass. Pass made to deposit a back weld.

Backstep sequence

1. Longitudinal sequence by which the weld bead increments are deposited in the direction opposite to the progress of welding the joint.

2. Longitudinal welding sequence in which the direction of general progress is opposite to that of welding the individual increments.

Backup. In flash and upset welding, a locator used to transmit all or a portion of the upsetting force to the work pieces.

Back weld. Weld deposited at the back of a single-groove weld.

Bare electrode. Filler metal arc welding electrode in the form of a wire or rod having no coating other than that incidental to the drawing of the wire.

Bare-metal arc welding. Arc welding process wherein coalescence is produced by heating with an electric arc between a bare or lightly coated metal electrode and the work, no shielding being used. Pressure is not used, and filler metal is obtained from the electrode.

Base metal. Metal to be brazed, cut, or welded; after welding, the part of the metal that was not melted.

Base metal test specimen. Test specimen composed wholly of base metal.

Bead weld. Type of weld composed of one or more string or weave beads deposited on an unbroken surface.

Bell-welded. Furnace-welded pipe produced in individual lengths, from cut-length skelp, having its longitudinal butt joint forge welded by the mechanical pressure developed in drawing the furnace-heated skelp through a cone-shaped die (welding bell), which serves as a combined forming and welding die.

Bevel. Type of edge preparation.

Bevel angle. Angle formed between the prepared edge of a member and a plane perpendicular to the surface of the member.

Beveling. Type of chamfering.

Block-brazing

1. Brazing process wherein coalescence is produced by the heat obtained from heated blocks applied to the parts to be joined and by using a nonferrous filler metal having a melting point above 800°F but below that of the base metals. The filler metal is distributed in the joint by capillary attraction.

2. Brazing with heat from hot blocks.

Block sequence

1. Combined longitudinal and built-up sequence for a continuous multiple-pass weld whereby separated lengths are completely or partially built up in cross section before intervening lengths are deposited.

2. Longitudinal welding sequence by which blocks of weld metal are built to a desired thickness with the intervening longitudinal space between them being filled subsequently.

Blowhole. Hole in a casting or weld caused by gas entrapped during solidification.

Blowpipe. Welding or cutting torch.

Bond. Junction of the weld metal and the base metal; junction of the base metal parts when weld metal is not present.

Boxing. Continuing a fillet weld around a corner as an extension of the principal weld; also called an "end return."

Braze. Cover with brass; solder with brass.

Brazed joint. Union of two or more members produced by the application of a brazing process.

Braze welding. Method of welding whereby a groove, fillet, plug, or slot weld is made using a nonferrous filler metal having a melting point below that of the base metals but above 800°F. The filler metal is not distributed in the joint by capillary attraction.

Brazing

1. Joining of two or more pieces of metal with a molten filler metal. No melting of parts to be joined occurs, and the filler metal is distributed along the joint surfaces by its own wetting action.

2. Group of welding processes wherein coalescence is produced by heating to suitable temperatures above 800°F and using a nonferrous filler metal having a melting point below that of the base metals. The filler metal is distributed between the closely fitted surfaces of the joint by capillary attraction.

3. Joining metals by flowing a thin layer, capillary thickness, of nonferrous filler metal into the space between them. Bonding results from the intimate contact produced by the dissolution of a small amount of base metal in the molten filler metal without fusion of the base metal. Sometimes the filler metal is put in place as a thin solid sheet or as a clad layer, and the composite is heated as in furnace brazing. The term brazing is used where the temperature exceeds some arbitrary value, such as 800°F. The term "soldering" is used for temperatures lower than the arbitrary value.

Brazing filler metal. Nonferrous filler metal used in brazing and braze welding.

Brazing sheet. Brazing filler metal in sheet form; a flat-rolled metal clad sheet with brazing filler metal on one or both sides.

Buildup sequence. Order in which the weld beads of a multiple-pass weld are deposited with respect to the cross section of the joint.

Buttering. Depositing weld metal on the face of a joint to increase weldability.

Butt joint. Joint between two abutting members lying approximately in the same plane. A welded butt joint may contain a variety of grooves.

Butt weld. Weld in a butt joint.

Butt welding. Welding a butt joint.

Butt weld joint. Welded pipe joint made with ends of the two pipes butting each other, the weld being around the periphery.

Butt weld pipe. Pipe welded along a seam butted edge to edge and not scarfed or capped.

Capillary attraction. Phenomenon by which adhesion between the molten filler metal and the base metals, together with surface tension of the molten filler metal, distributes the filler metal between the properly fitted surfaces of the joint to be brazed.

Carbon arc cutting. Arc cutting process wherein the severing of metals is effected by melting with the heat of an arc between a carbon electrode and the base metal.

Carbon arc welding

1. Arc welding process wherein coalescence is produced by heating with an electric arc between a carbon electrode and the work, with no shielding used. Pressure may or may not be used, and filler metal may or may not be applied.

2. Welding in which an arc is maintained between a nonconsumable carbon electrode and the work.

Carbon electrode. Non-filler-metal electrode, used in arc welding, consisting of a carbon or graphite rod.

Carbon-electrode arc welding. Group of arc welding processes wherein carbon electrodes are used.

Cascade sequence. Combined longitudinal and buildup sequence in which weld beads are deposited in overlapping layers, usually laid in a backstep sequence.

Chain intermittent fillet welding. Two lines of intermittent fillet welding on a joint wherein the fillet weld increments in one line are approximately opposite to those in the other line.

Chain weld. Two lines of intermittent welding on a joint wherein the weld increments in one line are approximately opposite to those in the other.

Chamfering. Preparing a contour, on the edge of a member for welding, except for a square groove weld.

Coated electrode. Filler-metal electrode, used in arc welding, consisting of a metal wire with a light coating, usually of metal oxides and silicates, applied subsequent to the drawing operation primarily for stabilizing the arc.

Coil weld. Butt weld joining the ends of two metal sheets to make a continuous strip for coiling.

Cold-welding

1. Welding that produces cohesion between two surfaces of metal, generally under the influence of externally applied pressure at room temperature.

2. Solid-phase welding in which pressure without added heat is used to cause interface movements that bring the atoms of the faying surfaces close enough together so that a weld ensues.

Collar. Reinforcing metal of a nonpressure thermit weld.

Color coding. Method of identifying steel. The color coding is painted on the end of the bar or rod.

Commutator-controlled welding. Making a number of spot or projection welds by means of several electrodes, in simultaneous contact with the work, progressively functioning under the control of an electrical commutating device.

Complete fusion. Fusion that occurs over the entire base metal surfaces exposed for welding.

Complete joint penetration. Joint penetration that extends completely through the joint.

Composite electrode

1. Filler-metal electrode used in arc welding, consisting of two or more metal components combined mechanically. The electrode may or may not include materials that protect the molten metal from the atmosphere, improve the properties of the weld metal; or stabilize the arc.

2. Arc welding electrode having two or more metals mechanically combined. It may or may not be flux coated. Tubular electrode having a flux-filled core.

Composite joint. Joint wherein welding is used in conjunction with a mechanical joining process.

Concave fillet weld. Fillet weld having a concave face.

Concavity. Maximum distance from the face of a concave fillet weld perpendicular to a line joining the toes.

Concurrent heating

1. Application of supplemental heat to a structure during a welding or cutting operation.

2. Using a second source of heat to supplement the primary heat in cutting or welding.

Cone. Conical part of a gas flame next to the orifice of the tip.

Continuous sequence. Longitudinal sequence wherein each pass is made continuously from one end of the joint to the other.

Continuous weld

1. Weld that extends without interruption for its entire length.

2. Weld extending continuously from one end of a joint to the other; weld where the joint is essentially circular; weld completely around the joint.

Convex fillet weld. Fillet weld having a convex face.

Convexity. Maximum distance from the face of a convex fillet weld perpendicular to a line joining the toes.

Cool time. In multiple-impulse and seam welding, the time interval between successive heat times.

Corner joint. Joint between two members located approximately at right angles to each other in the form of an _L_.

Corona. Area sometimes surrounding the nugget at the faying surfaces, contributing slightly to overall bond strength.

Covered electrode. Filler-metal electrode used in arc welding, consisting of a metal core wire with a relatively thick covering that provides protection for the molten metal from the atmosphere, improves the properties of the weld metal, and stabilizes the arc. The covering is usually a mixture of mineral or metal powders with cellulose or another binder.

Cover glass. Clear glass used in goggles, hand shields, and helmets to protect the filter glass from spattering material.

Crater

1. Depression at the termination of a weld bead.

2. In arc welding, a depression at the termination of a bead or in the weld pool beneath the electrode.

Crater crack. Crack in the crater of a weld bead.

Cross-wire weld. Projection weld made between crossed wires or bars.

Crushing test. Axial compressive test for determining the quality of the tubing, for example, the soundness of the weld in welding tubing.

Cup weld. Pipe weld where one pipe is expanded on the end to allow the entrance of the end of the other pipe. The weld is then circumferential at the end of the expanded pipe.

Current decay. In spot, seam, or projection welding, the controlled reduction of the welding impulse from its peak current amplitude to a lower value in controlled time to prevent rapid cooling of the weld nugget.

Current regulator. Automatic electrical control device for maintaining a constant current in the primary of the welding transformer.

Cutting attachment. Device that is attached to a gas welding torch to convert it into an oxygen cutting torch.

Cutting tip. That part of an oxygen cutting torch from which the gases issue.

Cutting torch. Device used in oxygen-cutting for controlling and directing the gases used for preheating and the oxygen used for cutting the metal.

Cylinder. Portable cylindrical container used for transportation and storage of a compressed gas.

Deposited metal. Filler metal that has been added during a welding operation

Deposition efficiency. Ratio of the weight of deposited metal to the net weight of electrodes consumed, exclusive of stubs.

Deposition rate. Weight of metal deposited in a unit of time.

Deposition sequence. Order in which the increments of weld metal are deposited.

Depth of fusion. Distance that fusion extends into the base metal from the surface melted during welding.

Deseaming. Removing defects by gas-cutting.

Die opening. In flash or upset welding, distance between the electrodes, usually measured with the parts in contact, but before welding has commenced or immediately upon completion of the cycle.

Die welding

1. Forge welding process wherein coalescence is produced by heating in a furnace and applying pressure by means of dies.
2. Forge welding between dies.

Dip brazing

1. Brazing process wherein coalescence is produced by heating in a molten chemical or metal bath and using a nonferrous filler metal having a melting point about 800°F but below that of the base metals. Filler metal is distributed in the joint by capillary attraction. When a metal bath is used, the bath provides the filler metal.
2. Brazing by immersion in a molten salt or metal bath. Where a metal bath is employed, it may provide the filler metal.

DOT. Department of Transportation.

Double-bevel groove weld. Groove weld in which the joint edge of one member is beveled from both sides.

Double-J groove weld. Groove weld in which the joint edge of one member is in the form of two J's, one from either side.

Double-U groove weld. Groove weld in which each joint edge is in the form of two J's or two half-U's, one from either side of the member.

Double-V groove weld. Groove weld in which each joint edge is beveled from both sides.

Double-welded joint. Butt, edge, T (tee), corner, or lap joint in which welding has been done from both sides.

Down slope time. In resistance welding, time associated with current decrease using slope control.

Drag. Distance between the point of exit of the cutting oxygen stream and the projection, on the exit surface, of the point of entrance.

Drag technique. Method used in manual arc welding where the electrode is in contact with the assembly being welded without being in short circuit. Electrode is usually used without oscillation.

Draw bead. Bead or offset used for controlling metal flow.

Duty cycle. For electric welding equipment, percentage of time that current flows during a specified period. In arc welding the specified period is 10 minutes.

Dynamic electrode force. In spot, seam, and projection welding, the force (pounds) between the electrodes during the actual welding cycle.

Edge joint. Joint between the edges of two or more parallel or nearly parallel members.

Edge preparation. Contour prepared on the edge of a member for welding.

Effective length of weld. Length of weld throughout which the correctly proportioned cross section exists.

Electrical discharge machining (EDM). Cutting process whereby metal is removed from a work piece by the energy of an electric spark that arcs between a tool and the surface of the work piece. Electrical discharge grinding (EDG) is the same as EDM, except that a rotating wheel is used as the tool.

Electrode. Current-carrying rod that supports the arc between two rods as in twin-carbon arc welding. The electrode may or may not furnish filler metal. In resistance welding, the electrode is a part of a resistance welding machine through which current and, in most cases, pressure are applied directly to the work. The electrode may be in the form of a rotating wheel, rotating roll, bar, cylinder, plate, clam, or chuck or modification thereof.

Electrode deposition. Weight of weld metal deposit obtained from a unit length of electrode.

Electrode force. Force between electrodes in spot, seam, and projection welding.

Electrode holder. Device used for mechanically holding the electrode and conducting current to it.

Electrode lead. Electrical conductor between the source of arc welding current and the electrode holder.

Electrode skid. During spot, seam, or projection welding, the sliding of an electrode along the surface of the work.

Electrogas welding. Vertical position method of gas-metal arc welding or flux-cored arc welding where a gas is supplied externally and molding shoes confine the molten weld metal.

Electronic heat control. Device for adjusting the heating value of the current in making a resistance weld by controlling the ignition or firing of the tubes in an electronic contactor. Flow of current is initiated each half-cycle at an adjustable time with respect to the zero point on the voltage wave.

Electroslag welding. Vertical-position welding process where coalescence is produced by molten slag that melts in the filler metal and the surfaces of the work to be welded. The weld pool is shielded by this slag, and molding shoes confine the molten slag and weld metal.

Exothermic welding. Welding process where coalescence is produced by a superheated liquid metal and slag.

Face of weld. Exposed surface of a weld, made by an arc or gas welding process, on the side from which the welding was done.

Faying surface

1. Contact surface between adjacent parts in a joint.
2. Surface of a piece of metal (or a member) in contact with another to which it is or is to be joined.

Ferrous metals. Metals that contain iron in their composition.

Filler metal. Metal added in making a brazed, soldered, or welded joint.

Fillet. Radius (curvature) imparted to inside meeting surfaces; concave corner piece used on foundry patterns.

Fillet weld. Weld of approximately triangular cross section joining two surfaces approximately at right angles to each other in a lap joint, T (tee) joint, or corner joint.

Filter glass. Glass, usually colored, used in goggles, helmets, and hand shields to exclude harmful light rays.

Fixed-position welding. Welding in which the work is held in a stationary position.

Flash. Molten metal that is expelled or squeezed out by the application of pressure and solidifies around the weld.

Flashback. Recession of the flame into or back of the mixing chamber of the torch.

Flashing. Heating portion of the cycle, consisting of a series of rapidly recurring, localized short circuits followed by molten metal expulsions, during which time the surfaces to be welded are moved one toward the other at the proper speed.

Flashing time. In flash welding, the time during which the flashing action is taking place.

Flash welding

1. Resistance welding process wherein coalescence is produced simultaneously over the entire area of abutting surfaces by the heat obtained from resistance to the flow of electric current between the two surfaces and by the application of pressure after heating is substantially completed. Flashing and upsetting are accompanied by expulsion of metal from the joint.

2. Resistance butt welding process in which the weld is produced over the entire abutting surface by pressure and heat, the heat being produced by electric arcs between the members being welded.

Flat fillet weld. Fillet weld having a face that is relatively flat.

Flat position. Position of welding when welding is performed from the upper side of the joint and the face of the weld is approximately horizontal. Also called "downhand welding."

Flow brazing. Brazing process wherein coalescence is produced by heating with molten nonferrous filler metal, poured over the joint until brazing temperature is attained. The filler metal has a melting point above 800°F but below that of the base metals, and is distributed in the joint by capillary attraction.

Flow welding. Welding process wherein coalescence is produced by heating with molten filler metal, poured over the surfaces to be welded until the welding temperature is attained and the required filler metal has been added. The filler metal is not distributed in the joint by capillary attraction.

Fluorescent penetrant inspection. Inspection is based on capillary action. The penetrant solution is applied to the surface by dipping, spraying, or brushing. Capillary action pulls the solution into the defect.

Flux

1. Fusible material used in welding or oxygen cutting to dissolve and facilitate removal of oxides and other undesirable substances.

2. In brazing, cutting, soldering, or welding, material used to prevent the formation of or to dissolve and facilitate removal of oxides and other undesirable substances.

Flux-oxygen cutting. Oxygen cutting process wherein the severing of metal is facilitated by use of a flux.

Forehand welding

1. Gas welding technique wherein the flame is directed toward the progress of welding.

2. Welding in which the palm of the principal hand (torch or electrode hand) of the welder faces the direction of travel. It has special significance in gas welding in that it provides preheating.

Forge delay time. In spot, seam, and projection welding, the time elapsed between the beginning of weld time or

weld interval and the time when the electrode force first reaches the specified pressure for forging.

Forge welding. Group of welding processes wherein coalescence is produced by heating the metal or work in a forge or other type furnace and applying pressure or blows to the weld metal.

Full fillet weld. Fillet weld whose size is equal to the thickness of the thinner member joined.

Furnace brazing. Brazing process wherein coalescence is produced by the heat obtained from a furnace and by using a nonferrous filler metal having a melting point above 800°F but below that of the base metals. The filler metal is distributed in the joint by capillary attraction.

Fusion. Melting together of filler metal and base metal or melting of base metal only that results in coalescence.

Fusion welding

1. Welding, without pressure, in which a portion of the base metal is melted.

2. Process of welding metals in a molten or molten and vaporous state without application of mechanical pressure or blows. Such welding may be accomplished by the oxyacetylene or oxyhydrogen flame or by the electric arc. Thermit welding is also classed as fusion welding.

Fusion zone. Area of the melted base metal as determined on the cross section of a weld.

Gas-metal arc welding. Arc welding process producing coalescence by heating an electric arc drawn between a metal stud or similar part and the other work part until the surfaces to be joined are properly heated, when they are brought together under pressure. Shielding is obtained from an inert gas such as helium or argon.

Gas pocket. Weld cavity caused by entrapped gas.

Gas-shielded arc welding. Arc welding in which the arc and molten metal are shielded from the atmosphere by a stream of gas, such as argon, helium, argon-hydrogen mixtures, or carbon dioxide.

Gas welding. Group of welding processes wherein coalescence is produced by heating with a gas flame or flames, with or without the application of pressure and with or without the use of filler metal.

Groove. Opening provided for a groove weld.

Groove angle. Total included angle of the groove between parts to be joined. Thus, the sum of two bevel angles, either or both of which may be zero degrees.

Groove face. Portion, surface, or surfaces of a member included in a groove.

Groove radius. Radius of a J or U groove.

Groove weld. Weld made in the groove between two members. Standard types are square, single bevel, single, flare bevel, single-flare V, single J, single U, single V, double bevel, double-flare bevel, double flare V, double J, double U, and double V.

Gross porosity. Condition of weld metal or a casting in which pores, gas holes, or globular voids are larger and in greater number than obtained in good practice.

Ground connection. Connection of the work lead to the work.

Hammer welding. Forge welding process wherein coalescence is produced by heating the metal to be welded in a forge or other furnace and applying pressure by means of hammer blows to the weld metal.

Hand shield. Protective device used in arc welding for shielding the face and neck. The hand shield is equipped with a suitable filter glass and is designed to be held by hand.

Hard facing. Depositing filler metal on a surface by welding, spraying, or braze welding, for the purpose of making the surface resistant to abrasion, erosion, wear, galling, and impact.

Heat-affected zone. Portion of the base metal that has not been melted, but whose mechanical properties or microstructures have been altered by the heat of welding, cutting, or brazing.

Heat time. In multiple-impulse welding or seam welding, the time duration of the current flow during any one impulse.

Heating gate. Opening in a thermit mold through which the parts to be welded are preheated.

Helmet. Protective device used in arc welding for shielding the face and neck. The helmet is equipped with a suitable filter glass and is designed to be worn on the head.

High pressure oxygen manifold. A manifold connecting oxygen containers having a DOT service pressure exceeding 200 psig.

High temperature metals. Metals that have unique properties of high strength for extended periods at elevated temperatures.

Hold time
1. In seam, flash, and upset welding, the time during which force is applied to the work after current ceases to flow.
2. In spot and projection welding, the time during which force is applied at the point of welding after the last impulse of current ceases to flow.

Horizontal fixed position. In pipe welding, the position of a pipe joint wherein the axis of the pipe is approximately horizontal, with the pipe not rotated during welding.

Horizontal position for a fillet weld. Position of welding wherein welding is performed on the upper side of an approximately horizontal surface, against an approximately vertical surface.

Horizontal position for a groove weld. Welding position wherein the axis of the weld lies in an approximately horizontal plane, and the face of the weld lies in an approximately vertical plane.

Horizontal-position welding. Making a fillet weld on the upper side of the intersection of a vertical and a horizontal surface; making a horizontal groove weld on a vertical surface.

Horizontal-rolled-position welding
1. In pipe welding, the position of a pipe joint wherein welding is performed in the flat position by rotating the pipe.
2. Topside welding of a butt joint connecting two horizontal pieces of rotating pipe.

Horn. In a resistance welding machine, a cylindrical arm or beam that transmits the electrode pressure and usually conducts the welding current.

Horn spacing
1. In a resistance welding machine, the unobstructed work clearance between horns or platens at right angles to the throat depth. The distance is measured with the horns parallel and horizontal at the end of the downstroke.
2. Distance between adjacent surfaces of the horns of a resistance welding machine.

Hydrogen brazing. Method of furnace brazing in a hydrogen atmosphere.

Impregnated tape-metal arc-welding. Arc welding process wherein coalescence is produced by heating with an electric arc between a metal electrode and the work. Shielding is obtained from decomposition of an impregnated tape wrapped around the electrode as it is fed to the arc. Pressure is not used, and filler metal is obtained from the electrode.

Inadequate joint penetration. Joint penetration that is less than specified.

Incomplete fusion. Fusion that has occurred over less than the entire base metal surfaces exposed for welding.

Indentation. In a spot, seam, or projection weld, the depression on the exterior surface or surfaces of the base metal.

Induction brazing
1. Brazing process wherein coalescence is produced by the heat obtained from resistance of the work to the flow of induced electric current and by using a nonferrous filler metal having a melting point above 800°F but below that of the base metals. The filler metal is distributed in the joint by capillary attraction.
2. Brazing with induction heat.

Induction welding
1. Welding process wherein coalescence is produced by the heat obtained from resistance of the work to the flow of induced electric current, with or without the application of pressure.
2. Welding with induction heat.

Inert gas-carbon arc welding. Arc welding process wherein coalescence is produced by heating with an electric arc between a carbon electrode and the work. Shielding is obtained from an inert gas such as helium or argon. Pressure may or may not be used, and filler metal may or may not be applied.

Inert gas-metal arc welding. Arc welding process wherein coalescence is produced by heating with an electric arc between a metal electrode and the work. Shielding is obtained from an inert gas such as helium or argon. Pressure may or may not be used, and filler metal may or may not be applied.

Inert-gas-shielded arc cutting. Metal cutting by the heat of an arc in an inert gas such as argon or helium.

Inert-gas-shielded arc welding. Arc welding in an inert gas such as argon or helium.

Inspector. Makes sure that welding quality is maintained. Must be familiar with welding standards, and standard welding symbols, have knowledge of operating various types of welding and cutting equipment.

Intermittent weld. Weld in which the continuity is broken by recurring unwelded spaces.

Interpass temperature. In a multiple-pass weld; the lowest temperature of the deposited weld metal before the next pass is started.

Iron powder electrode. Welding electrode with a covering containing up to about 50% iron powder, some of which becomes a part of the deposit.

Joint. Location where two or more members are to be or have been fastened together mechanically or by brazing or welding.

Joint (unwelded). Location where two or more members are to be joined by welding.

Joint brazing procedure. Detailed methods, practices, and materials employed in the brazing of a particular joint.

Joint design. Joint geometry, together with the required dimensions of the welded joint.

Joint efficiency. Strength of a welded joint expressed as a percentage of the strength of the unwelded base metal.

Joint geometry. Shape and dimensions of a joint in cross section prior to welding.

Joint penetration

1. Minimum depth a groove weld extends from its face into a joint, exclusive of reinforcement.
2. Distance the weld metal and fusion extend into a joint.

Joints. Ways of arranging the metal pieces in relation to one another so they can be arc welded. There are five basic joints.

Joint welding procedure. Detailed methods, practices, and materials employed in the welding of a particular joint.

Kerf

1. Space from which metal has been removed by a cutting process.
2. Space that was occupied by the material removed during cutting.

Lap. Surface defect, appearing as a seam, caused by folding over hot metal, fins, or sharp corners and then rolling or forging them into the surface, but not welding them.

Lap joint. Joint between two overlapping members.

Layer. Stratum of one or more weld beads lying in a plane parallel to the surface from which welding was done.

Leg of a fillet weld (actual). Distance from the root of the joint to the toe of a fillet weld, the length of a side of the largest right triangle that can be inscribed in the cross section of the weld.

Lightly coated electrode. Filler-metal electrode, used in arc welding, consisting of a metal wire with a light coating applied subsequent to the drawing operation, primarily for stabilizing the arc.

Local preheating. Preheating a specific portion of a structure.

Local stress-relief heat treatment. Stress-relief heat treatment of a specific portion of a structure.

Longitudinal seam welding. Seam welding in a direction essentially parallel to the throat depth of a resistance welding machine.

Longitudinal sequence. Order in which the increments of a continuous weld are deposited with respect to its length.

Low-hydrogen electrode. Covered arc welding electrode that provides an atmosphere around the arc and molten weld metal that is low in hydrogen.

Low pressure acetylene. Acetylene at a pressure not exceeding 1 psig.

Low pressure oxygen manifold. A manifold connecting oxygen containers having DOT service pressure not exceeding 200 psig.

Machine. A device in which one or more torches using fuel gas and oxygen are incorporated.

Machine oxygen cutting. Oxygen cutting with equipment that performs the cutting operation under the constant observation and control of an operator. The equipment may or may not perform the loading and unloading of the work.

Machine welding. Welding with equipment that performs the welding operation under the observation and control of an operator. The equipment may or may not perform the loading and unloading of the work.

Magnetic particle inspection. Nondestructive testing to detect flaws on or near the surface of ferromagnetic materials.

Manifold. Multiple header for connecting several cylinders to one or more torch supply lines.

Manual oxygen cutting. Oxygen cutting wherein the entire cutting operation is performed and controlled by hand.

Manual welding. Welding wherein the entire welding operation is performed and controlled by hand.

Mash seam weld. Seam weld made in a lap joint, in which the thickness at the lap is reduced plastically to approximately the thickness of one of the lapped parts.

Medium pressure acetylene. Acetylene at pressures exceeding 1 psig but not exceeding 15 psig.

Melting rate. In electric arc-welding, the weight or length of electrode melted in a unit of time. Sometimes called "melt-off rate" or "burn-off rate."

Metal arc cutting. Arc cutting process wherein the severing of metals is effected by melting them with the heat of an arc between a metal electrode and the base metal.

Metal arc welding

1. Arc welding with metal electrodes. Term commonly refers to shielded metal arc welding using covered electrodes.
2. Metal arc welding includes such processes as shielded-metal arc welding, impregnated tape-metal arc welding, submerged arc welding, atomic-hydrogen welding, bare-metal arc welding, inert gas-metal arc welding, stud welding, and shielded stud welding.

Metal electrode. Filler or non-filler-metal electrode used in arc welding, consisting of a metal wire with or without a covering or coating.

Metal-electrode arc welding. Group of arc welding processes wherein metal electrodes are used. Metal-electrode arc welding includes such processes as shielded metal arc welding, impregnated tape-metal arc welding, atomic-hydrogen welding, inert gas-metal arc welding, submerged arc welding, shielded stud welding, stud welding, and bare-metal arc welding.

Mixing chamber. That part of a gas welding or oxygen cutting torch wherein the gases are mixed.

Multiple-impulse welding

1. Making spot, projection, and upset welds using more than one impulse of current. When alternating current is used, each impulse may consist of a fraction of a cycle or a number of cycles.
2. Spot, projection, or upset welding with more than one impulse of current during a single machine cycle; sometimes called "pulsation welding."

Multiple-impulse weld timer. In resistance welding, a device for multiple-impulse welding that controls only the heat and cool times and either the weld interval or the number of heat times.

Multiple-pass weld. Weld made by depositing filler metal with two or more successive passes.

Multiple-spot welding. Spot welding in which several spots are made during one complete cycle of the welding machine.

Neutral flame

1. Gas flame wherein the portion used is neither oxidizing nor reducing.
2. Gas flame in which there is no excess of either fuel or oxygen.

Nondestructive testing. Welds are critical in joining metals together; they must be tested without causing any damage by several methods available.

Nonferrous metals. Metals containing no iron, including copper, lead, tin, etc.

Nonpressure thermit welding. Thermit welding process wherein coalescence is produced by heating with superheated liquid metal resulting from the chemical reaction between a metal oxide and aluminum, without the application of pressure. Filler metal is obtained from the liquid metal.

Nonsynchronous initiation. In resistance welding, the initiation or termination of the welding transformer primary current at any random time with respect to the voltage wave.

Nugget. Weld metal joining the parts in spot, seam, or projection welds.

Off time. In resistance welding, the time that the electrodes are off the work. The term is generally used when the welding cycle is repetitive.

Open-circuit voltage. In arc welding, the voltage between the terminals of a power source when no current is flowing in the circuit.

Overhead-position welding. Welding position wherein welding is performed from the underside of the joint.

Overlap. Protrusion of weld metal beyond the bond at the toe of the weld; in spot, seam, or projection welding, the amount one sheet overlays the other.

Oxidizing flame

1. Gas flame wherein the portion used has an oxidizing effect.
2. Gas flame produced with excess oxygen.

Oxyacetylene cutting

1. Oxygen cutting process wherein the severing of metals is effected by means of the chemical reaction of oxygen with the base metal at elevated temperatures, the necessary temperature being maintained by means of gas flames obtained from the combustion of acetylene with oxygen.
2. Oxygen cutting in which the initiation temperature is attained with an oxyacetylene flame.

Oxyacetylene welding

1. Gas welding process wherein coalescence is produced by heating with a gas flame or flames obtained from the combustion of acetylene with oxygen, with or without the application of pressure and with or without the use of filler metal.
2. Welding with an oxyacetylene flame.

Oxy-arc cutting. Oxygen cutting process wherein the severing of metals is effected by means of the chemical reaction of oxygen with the base metal at elevated temperatures, the necessary temperature being maintained by means of an arc between an electrode and the base metal.

Oxy-city gas cutting. Oxygen cutting process wherein the severing of metals is effected by means of the chemical reaction of oxygen with the base metal at elevated temperatures, the necessary temperature being maintained by means of gas flames obtained from the combustion of city or utility gas with oxygen.

Oxygen cutter. One who is capable of performing a manual oxygen cutting operation.

Oxygen cutting

1. Metal cutting by directing a stream of oxygen on a hot metal. The chemical reaction of oxygen with the base metal furnishes heat for the localized melting and, hence, cutting.
2. Group of cutting processes wherein the severing of metals is effected by means of the chemical reaction of oxygen with the base metal at elevated temperatures. In the case of oxidation-resistant metals the reaction is facilitated by use of a flux.

Oxygen cutting operator. One who operates machine or automatic oxygen cutting equipment.

Oxygen gouging. Oxygen cutting wherein a chamfer or groove is formed.

Oxygen lance. Length of pipe used to convey oxygen to the point of cutting in oxygen-lance cutting.

Oxygen-lance cutting. Oxygen cutting process wherein only oxygen is supplied by the lance, and the preheat is obtained by other means.

Oxyhydrogen cutting

1. Oxygen cutting process wherein the severing of metals is effected by means of the chemical reaction of oxygen with the base metal at elevated temperatures, the necessary temperature being maintained by means of gas flames obtained from the combustion of hydrogen with oxygen.
2. Oxygen cutting in which the initiation temperature is attained with an oxyhydrogen flame.

Oxyhydrogen welding

1. Gas welding process wherein coalescence is produced by heating with a gas flame or flames obtained from the combustion of hydrogen with oxygen, without the application of pressure and with or without the use of filler metal.
2. Welding with an oxyhydrogen flame.

Oxy-natural gas cutting. Oxygen cutting process wherein the severing of metals is effected by means of the chemical reaction of oxygen with the base metal at elevated temperatures, the necessary temperature being maintained by means of gas flames obtained from the combustion of natural gas with oxygen.

Oxypropane cutting. Oxygen cutting process wherein the severing of metals is effected by means of the chemical reaction of oxygen with the base metal at elevated temperatures, the necessary temperature being maintained by means of gas flames obtained from the combustion of propane with oxygen.

Pack rolling. Hot-rolling a pack of two or more sheets of metal. Scale prevents their being welded together.

Partial joint penetration. Joint penetration that is less than complete.

Pass

1. Single longitudinal progression of a welding operation along a joint or weld deposit; the result of a pass in a weld bead.
2. Weld metal deposited in one trip along the axis of a weld.

Peening. Mechanical working of metal by hammer blows or shot impingement.

Penetration. Distance the fusion zone extends below surface of part or parts being welded.

Percussion welding

1. Resistance welding process wherein coalescence is produced simultaneously over the entire area of abutting surfaces by the heat obtained from an arc caused by a rapid discharge of stored electrical energy, with pressure percussively applied during or immediately following the electrical discharge.

2. Resistance welding simultaneously over the entire area of abutting surfaces with arc heat, the pressure being applied by a hammerlike blow during or immediately following the electrical discharge.

Platen. In a resistance welding machine, member with a substantially flat surface to which dies, fixtures, backups, or electrode holders are attached and that transmits the electrode force or upsetting force.

Platen force. In flash and upset welding, the force available at the movable platen to cause upsetting. The force may be dynamic, theoretical, or static.

Plug weld. Circular weld made by either arc or gas welding through one member of a lap or T joint joining that member to the other. The weld may or may not be made through a hole in the first member. If a hole is used, the walls may or may not be parallel, and the hole may be partially or completely filled with weld metal.

Porosity. Condition of metal containing gas pockets or voids.

Portable outlet header. An assembly of piping and fittings used for service-outlet purposes that is connected to the permanent service piping by means of hose or other nonrigid conductors.

Positioned weld. Weld made in a joint that has been so placed as to facilitate making the weld.

Position of welding. Welding positions are designated as follows: flat, horizontal, vertical, overhead, horizontal rolled, horizontal fixed, and vertical pipe.

Postheating

1. Application of heat to a weld or weldment immediately after a welding or cutting in accordance with prescribed standards.

2. Application of heat to a weld or weldment subsequent to a welding or cutting operation.

Postweld interval. In resistance welding, the heat time elapsed between the end of weld time or weld interval and the start of hold time. During this period the weld is subjected to mechanical and heat treatment.

Preheating. Application of heat to the base metal prior to a welding or cutting operation.

Pressure-controlled welding. Process of making a number of spot or projection welds wherein several electrodes progressively function under the control of a pressure sequencing device.

Pressure gas welding. Gas welding process wherein coalescence is produced simultaneously over the entire area of abutting surfaces by heating with gas flames obtained from the combustion of a fuel gas with oxygen and by the application of pressure, but without the use of filler metal.

Pressure thermit welding. Process wherein coalescence is produced by heating with superheated liquid metal and slag resulting from the chemical reaction between iron oxide and aluminum, and by applying pressure. Liquid metal from the reaction is not used as a filler metal.

Pressure welding. Welding process or method wherein pressure is used to complete the weld.

Preweld interval. In spot, projection, and upset welding, the time between the end of squeeze time and the start of weld time or weld interval in flash welding, during which the material is preheated in flash welding; the time during which the material is preheated.

Procedure qualification. Demonstration that welds made by a specific procedure can meet prescribed standards.

Progressive block sequence. Block sequence wherein successive blocks are completed progressively along the joint, either from one end to the other or from the center of the joint toward either end.

Projection welding. Resistance welding process wherein coalescence is produced by the heat obtained from resistance to the flow of electric current through the work parts held together under pressure by electrodes. The resulting welds are localized at predetermined points by the design of the parts to be welded. Localization is usually accomplished by projections, embossments, or intersections.

Puddle. Portion of weld that is molten at place where heat is applied.

Pulsation welding. Sometimes used as a synonym for multiple-impulse welding.

Push welding (poke welding). Spot or projection welding in which the force is applied manually to one electrode and the work or a backing bar takes the place of the other electrode.

Quench time

1. In resistance welding, that part of the postweld interval from the cessation of flow of welding current to the application of a current impulse for postheating.

2. Time from the finish of the weld to the beginning of temper. Also called "chill time."

Radiographic (x-ray) inspection. Involves the use of x and gamma radiation projected through an object under inspection onto a film. Developed film shows an image of the internal structure at the particular location being questioned.

Rare metals. Metals not available in large quantities, such as gold, cerium, europium, holmium, etc.

Rate of flame propagation. Speed at which a flame travels through a mixture of gases.

Reaction stress. Residual stress that could not otherwise exist if the members or parts being welded were isolated as free bodies without connection to other parts of the structure.

Reactor. Device used in arc welding circuits for the purpose of minimizing irregularities in the flow of welding current.

Reducing flame

1. Gas flame wherein the portion used has a reducing effect.

2. Gas flame produced with excess fuel.

Regulator. Device for controlling the delivery of gas at some substantially constant pressure regardless of variation in the higher pressure at the source.

Reinforcement of weld

1. Weld metal on the face of a groove weld in excess of the metal necessary for the specified weld size.

2. In a butt joint, weld metal on the face of the weld that extends out beyond a surface plane common to the members being welded; in a fillet weld, weld metal

that contributes to convexity; in a flash, upset, or gas pressure weld, the portion of the upset left in excess of the original diameter or thickness.

Residual stress.   Stress remaining in a structure or member as a result of thermal or mechanical treatment, or both.

Resistance brazing

1. Brazing process wherein coalescence is produced by the heat obtained from resistance to the flow of electric current in a circuit of which the work is a part, and by using a nonferrous filler metal having a melting point above 800°F but below that of the base metals. The filler metal is distributed in the joint by capillary attraction.

2. Brazing by resistance heating, the joint being part of the electrical circuit.

Resistance welding

1. Group of welding processes wherein coalescence is produced by the heat obtained from resistance of the work to the flow of electric current in a circuit of which the work is a part, and by the application of pressure.

2. Welding with resistance heating and pressure, the work being part of the electrical circuit. Examples: resistance spot welding, resistance seam welding, projection welding, and flash butt welding.

Resistance welding die.   Part of a resistance welding machine that is usually shaped to the work contour, with which the parts being welded are held, and that conducts the welding current.

Resistance welding electrode.   Part or parts of a resistance welding machine through which the welding current and, in most cases, pressure are applied directly to the work. Electrode may be in the form of a rotating wheel, rotating roll, bar, cylinder, plate, clamp, or chuck, or modification thereof.

Reverse polarity

1. Polarity achieved by the arrangement of the direct-current arc welding leads so that the work is the negative pole and the electrode is the positive pole of the welding arc.

2. Polarity achieved by an arc welding circuit arrangement in which the electrode is connected to the positive terminal.

Roll spot welding

1. Making separated spot welds with circular electrodes.

2. Spot welding with rotating circular electrodes.

Roll welding.   Forge welding process wherein coalescence is produced by heating in a furnace and applying pressure by means of rolls.

Root.   For example, root of joint and root of weld.

Root crack

1. Crack in the weld or base metal occurring at the root of a weld.

2. Crack in either the weld or heat-affected zone at the root of a weld.

Root face

1. That portion of the groove face adjacent to the root of the joint.

2. Unbeveled portion of the groove face of a joint.

Root of joint

1. That portion of a joint to be welded where the members approach closest to each other. In cross section

the root of the joint may be either a point, a line, or an area.

2. Location of the closest approach between parts of a joint to be welded.

Root of weld.   Point, as shown in cross section, at which the bottom of the weld intersects the base metal surfaces —may be coincident with the root of joint.

Root opening

1. Separation between the members to be joined at the root of the joint.

2. Distance between the parts at the root of the joint.

Root pass.   First bead of a multiple-pass weld.

Root penetration.   Depth a groove weld extends into the root of a joint, measured on the centerline of the root cross section.

Scarfing.   Cutting surface areas of metal objects, ordinarily by using a gas torch. The operation permits surface defects to be cut from ingots, billets, or the edges of plate that are to be beveled for butt welding.

Scarf joint.   Butt joint in which the plane of the joint is inclined with respect to the main axis of the members.

Seal weld.   Weld used primarily to obtain tightness and prevent leakage.

Seam weld.   Weld consisting of a series of overlapping spot welds, made by seam welding or spot welding.

Seam welding

1. Resistance welding process wherein coalescence is produced by the heat obtained from resistance to the flow of electric current through the work parts held together under pressure by circular electrodes. The resulting weld is a series of overlapping spot welds made progressively along a joint by rotating the electrodes.

2. Arc or resistance welding in which a series of overlapping spot welds is produced with rotating electrodes or rotating work, or both; making a longitudinal weld in sheet metal or tubing.

Seam weld timer.   In seam welding, a device that controls the heat times and cool times.

Selective block sequence.   Block sequence wherein successive blocks are completed in an order selected to create a predetermined stress pattern.

Semiautomatic arc welding.   Arc welding with equipment that controls only the filler-metal feed. The advance of the welding is manually controlled.

Sequence timer.   In resistance welding, a device for controlling the sequence and duration of any or all of the elements of a complete welding cycle except weld time or heat time.

Sequence weld timer

1. In resistance welding, a device for controlling the sequence and duration of any or all of the elements of a complete welding cycle.

2. Same as sequence timer, except that either weld time or heat time, or both, are also controlled.

Series welding.   Making two or more resistance spot, seam, or projection welds simultaneously by a single welding transformer with three or more electrodes forming a series circuit.

Sheet separation.   In spot, seam, or projection welding, the gap that exists between faying surfaces surrounding the weld after the joint has been welded.

Shielded arc welding. Arc welding in which the arc and the weld metal are protected by a gaseous atmosphere, the products of decomposition of the electrode covering, or a blanket of fusible flux.

Shielded carbon arc welding. Arc welding process wherein coalescence is produced by heating with an electric arc between a carbon electrode and the work. Shielding is obtained from the combustion of a solid material fed into the arc or from a blanket of flux on the work, or from both. Pressure may or may not be used, and filler metal may or may not be applied.

Shielded metal arc welding

1. Arc welding process producing coalescence by heating with an electric arc between a covered metal electrode and the work. Shielding is obtained from decomposition of the electrode covering. Pressure is not used, and filler metal is obtained from the electrode.

2. Arc welding in which the arc and the weld metal are protected by the decomposition products of the covering on a consumable metal electrode.

Shielded stud welding. Arc welding process wherein coalescence is produced by heating with an electric arc drawn between a metal stud, or similar part, and the other work part until the surfaces to be joined are properly heated, whereupon they are brought together under pressure. Shielding is obtained from an inert gas such as helium or argon.

Silver brazing. Brazing with silver-base alloys as the filler metal.

Silver brazing alloy. Filler metal used in silver brazing.

Single-bevel groove weld. Groove weld in which the joint edge of one member is beveled from one side.

Single-impulse welding. Making spot, projection, and upset welds by a single impulse of current. When alternating current is used, an impulse may consist of a fraction of a cycle or a number of cycles.

Single-J groove weld. Groove weld in which the joint edge of one member is prepared in the form of a J from one side.

Single-pass weld. Weld made by depositing filler metal with one pass only.

Single-U groove weld. Groove weld in which each joint edge is prepared in the form of a J or half U from one side.

Single-V groove weld. Groove weld in which each member is beveled from the same side.

Single welded joint. Joint welded from one side only.

Size of fillet weld (equal-leg fillet welds). Leg length of the largest isosceles right triangle that can be inscribed within the fillet weld cross section.

Size of fillet weld (unequal-leg fillet welds). Leg lengths of the largest right triangle that can be inscribed within the fillet weld cross section.

Size of groove weld. Joint penetration (depth of chamfering plus the root penetration when specified).

Size of weld. Joint penetration in a groove weld; the lengths of the nominal legs of a fillet weld.

Slag inclusion. Nonmetallic solid material entrapped in weld metal or between weld metal and the base metal.

Sleeve weld. Joint made by butting two pipes together and welding a sleeve over the outside.

Slot weld

1. Weld made in an elongated hole in one member of a lap or T-joint joining that member to the portion of the surface of the other member that is exposed through the hole. The hole may be open at one end and may be partially or completely filled with weld metal.

2. Weld similar to a plug weld, the difference being that the hole is elongated and may extend to the edge of a member without closing.

Slugging. Unsound practice of adding a separate piece of material in a joint before or during welding, resulting in a welded joint that does not comply with design, drawing, or specification requirements.

Socket weld. Joint made by use of a socket weld fitting which has a prepared female end or socket for insertion of the pipe to which it is welded.

Solder embrittlement. Reduction in mechanical properties of a metal as a result of local penetration of solder along grain boundaries.

Soldering. Process similar to brazing, with the filler metal having a melting temperature range below an arbitrary value, generally 800°F. Soft solders are usually lead-tin alloys.

Solid-phase welding. Method of welding in which pressure or heat and pressure are used to consummate the weld without fusion.

Spacer strip. Metal strip or bar inserted in the root of a joint prepared for a groove weld to serve as a backing and maintain the root opening throughout the course of the welding operation.

Spark test. Testing steel by placing the metal at a grinding wheel and observing the resulting sparks, used to determine the grade of steel.

Spatter. In arc and gas welding, the metal particles expelled during welding that do not form a part of the weld.

Spatter loss. Metal lost as a result of spatter.

Specialist welder. Person whose knowledge of metals enables him to decide on the preferred type of weld to be performed. Experimentation may be required to resolve the problem.

Speed of travel. In welding, speed with which a weld is made along its longitudinal axis, usually measured in inches per minute or spots per minute.

Spelter solder. Brazing filler metal of approximately equal parts of copper and zinc.

Spot welding

1. Resistance welding process wherein coalescence is produced by the heat obtained from resistance to the flow of electric current through the work parts held together under pressure by electrodes. The size and shape of the individually formed welds are limited primarily by the size and contour of the electrodes.

2. Welding of lapped parts in which fusion is confined to a relatively small circular area. It is generally resistance welding but may also be gas-shielded tungsten, gas-shielded metal, or submerged arc welding.

Square groove weld. Groove weld in which the abutting surfaces are square.

Squeeze time

1. In spot, seam projection, and upset welding, the time interval between the initial application of the electrode force on the work and the first application of current.

2. In resistance welding, the time between the initial applications of pressure and current.

Stack cutting. Oxygen cutting of stacked metal plates arranged so that all the plates are severed by a single cut.

Stack welding. Resistance spot welding of stacked plates, all being joined simultaneously.

Staggered intermittent fillet welding

1. Two lines of intermittent fillet welding on a joint wherein the fillet weld increments in one line are staggered with respect to those in the other line.

2. Making a line of intermittent fillet welds on each side of a joint so that the increments on one side are not opposite those on the other.

Standard. Accepted base for a uniform system of accepted performances; quality expected.

Standard types of groove welds. Standard types of groove welds are as follows: square groove; single V, single bevel, single U, single J, double V, double bevel, double U, and double J.

Standard welding symbols. Welding symbols used on drawings should be those established by the American Welding Society (AWS).

Static electrode force. In spot, seam, and projection welding, the force between the electrodes under welding conditions but with no current flowing and no movement in the welding machine.

Station outlet. The point at which gas is withdrawn from the service piping system.

Stored-energy welding. Welding with electrical energy accumulated electrostatically, electromagnetically, or electrochemically at a relatively low rate and made available at the higher rate required in welding.

Straight polarity. Polarity achieved by the arrangement of direct-current arc welding leads so that the work is the positive pole and the electrode is the negative pole of the welding arc.

Stress-relief heat treatment. Uniform heating of a structure or portion thereof to a sufficient temperature, below the critical range, to relieve the major portion of the residual stresses, followed by uniform cooling.

Stringer bead. Continuous weld bead made without appreciable transverse oscillation; contrast with weave bead.

Stringer beading. Deposition of stringer beads.

Stud welding

1. Arc welding process wherein coalescence is produced by heating with an arc drawn between a metal stud, or similar part, and the other work part until the surfaces to be joined are properly heated, whereupon they are brought together under pressure. No shielding is used.

2. Welding a metal stud or similar part to another piece of metal, the heat being furnished by an arc between the two pieces just before pressure is applied.

Submerged arc welding. Arc welding process producing coalescence by heating with an electric arc or arcs between a bare metal electrode or electrodes and the work. The welding is shielded by a blanket of granular fusible material on the work. Pressure is not used, and filler metal is obtained from the electrode and sometimes from a supplementary welding rod.

Surfacing. Deposition of filler metal on a metal surface by welding, spraying, or braze welding to obtain certain desired properties of dimensions.

Synchronous initiation. In spot, seam, and projection welding, the initiation and termination of each half-cycle of welding transformer primary current so that all half-cycles of such current are identical.

Synchronous timing. In spot, seam, or projection welding, the method of regulating the welding transformer primary current so that all the following conditions will prevail: The first half-cycle is initiated at the proper time in relation to the voltage to ensure a balanced current wave; each succeeding half-cycle is essentially identical to the first; and the last half-cycle is of opposite polarity to the first.

Tacking. Making tack welds.

Tack welds. Small scattered welds to hold parts of a weldment in proper alignment while the final welds are being made.

Tandem welding. Arc welding in which two or more electrodes are in a plane parallel to the line of travel.

Tee joint. Joint between two members located approximately at right angles to each other in the form of a T.

Tempering. Heat treatment whereby metal is brought to a desired degree of hardness or toughness.

Temper time. In resistance welding, that part of the postweld interval during which a current suitable for tempering or heat treatment flows. Current can be single or multiple impulse, with varying heat and cool intervals.

Theoretical electrode force. In spot, seam, and projection welding, the force, neglecting friction and inertia, available at the electrodes of a resistance welding machine by virtue of the initial force application and the theoretical mechanical advantage of the system.

Thermit crucible. Vessel in which the Thermit reaction takes place.

Thermit mixture. Mixture of metal oxide and finely divided aluminum, with the addition of alloying metals as required.

Thermit mold. Mold formed around the parts to be welded to receive the molten metal.

Thermit reaction. Chemical reaction between metal oxide and aluminum that produces superheated molten metal and aluminum oxide slag.

Thermit welding

1. Group of welding processes wherein coalescence is produced by heating with superheated liquid metal and slag resulting from a chemical reaction between a metal oxide and aluminum, with or without the application of pressure. Filler metal, when used, is obtained from the liquid metal.

2. Welding with heat produced by the reaction of aluminum with a metal oxide. Filler metal, if used, is obtained from the reduction of the appropriate oxide.

Throat depth. In a resistance welding machine, the distance from the centerline of the electrode or platens to the nearest point of interference for flatwork or sheets. In the case of a seam welding machine with a universal head, the throat depth is measured with the machine arranged for transverse welding.

Throat of a fillet weld (actual). Shortest distance from the root of a fillet weld to its face.

Throat of a fillet weld (theoretical). Distance from the beginning of the root of the joint perpendicular to the hypotenuse of the largest right triangle that can be inscribed within the fillet weld cross section.

Through weld. Weld of appreciable length made by either arc or gas welding through the unbroken surface of one member of a lap or T (tee) joint joining that member to the other.

Toe crack. Crack in the base metal occurring at the toe of a weld.

Toe of weld. Junction between the face of a weld and the base metal.

Torch. Gas burner used to braze, cut, or weld. For brazing or welding, it has two gas feed lines, one for fuel, such as acetylene or hydrogen, the other for oxygen. For cutting, there may be an additional feed line for oxygen.

Torch brazing. Brazing process wherein coalescence is produced by heating with a gas flame and by using a nonferrous filler metal having a melting point above 800°F but below the base metal. The filler metal is distributed in the joint by capillary attraction.

Transverse seam welding. Making a seam weld in a direction essentially at right angles to the throat depth of a seam welding machine.

Tungsten arc welding. Inert-gas-shielded arc welding using a tungsten electrode.

Tungsten electrode. Non-filler-metal electrode used in arc welding, consisting of a tungsten wire.

Twin-carbon arc brazing. Brazing process wherein coalescence is produced by heating with an electric arc maintained between two carbon electrodes, and by using a nonferrous filler metal having a melting point above 800°F but below that of the base metals. The filler metal is distributed in the joint by capillary attraction.

Twin-carbon arc welding. Arc welding process wherein coalescence is produced by heating with an electric arc maintained between two carbon electrodes, with no shielding. Pressure is not used, and filler metal may or may not be applied.

Ultrasonic inspection. The use of sound waves to obtain information about the interior of material by observing the length of time required to receive an echo reflected from a flaw inside the material.

Underbead crack

1. Crack in the heat-affected zone not extending to the surface of the base metal.
2. Subsurface crack in the base metal near the weld.

Undercut. Groove melted into the base metal adjacent to the toe of a weld and left unfilled by the weld metal.

Upset

1. Localized increase in volume in the region of a weld resulting from the application of pressure.
2. Portion of the welding cycle during which the volume increases.

Upsetting force. In flash and upset welding, the force exerted at the welding surfaces during upsetting.

Upsetting time. In flash and upset welding, the time during upsetting.

Upset welding. Resistance welding in which the weld is produced simultaneously over the entire area of abutting surfaces or progressively along a joint by the heat obtained from resistance to the flow of current through the area of contact of those surfaces. Pressure is applied before heating is started and is maintained throughout the heating period.

Vertical-position welding

1. Welding position wherein the axis of the weld is approximately vertical.
2. In pipe welding, the position of a pipe joint wherein welding is performed in the horizontal position and the pipe may or may not be rotated.

Voltage regulator. Automatic electrical control device for maintaining a constant voltage supply to the primary of a welding transformer.

Wandering block sequence. Block sequence wherein successive blocks are completed at random after several starting blocks have been completed.

Wandering sequence. Longitudinal sequence wherein the weld bead increments are deposited at random.

Wax pattern. Wax molded around the parts to be welded by a Thermit welding process to the form desired for the completed weld.

Weave bead. A weld bead made with oscillations transverse to the axis of the weld; contrast with string (stringer) bead.

Weave beading. Deposition of weave beads.

Weld

1. Localized coalescence of metal which is produced by heating to suitable temperatures, with or without the application of pressure, and with or without the use of filler metal. The filler metal either has a melting point approximately the same as that of the base metal or has a melting below that of the base metal, but above 800°F.
2. Union made by welding.
3. Union of metallic pieces by melting the place at which such joining occurs. To unite pieces of metal by heating the parts to be joined so that they fuse.

Weldability. Capacity of a metal to be welded under the fabrication conditions imposed into a specific, suitably designed structure and to perform its intended service satisfactorily.

Weld bead

1. Weld deposit resulting from a pass, for example, stringer bead and weave bead.
2. Deposit of filler metal from a single welding pass.

Weld crack. Crack in weld metal.

Weld delay time

1. In spot and projection welding, the length of time the weld time is delayed to ensure proper sequence of mechanical functions in relation to subsequent electrical functions.
2. In spot, seam, or projection welding, the length of time the current is delayed with respect to starting the forge delay timer in order to synchronize the forging pressure and the welding heat.

Welded joint. Union of two or more members produced by the application of a welding process.

Welder. One who performs manual or semiautomatic welding operations or operates machine or automatic welding equipment.

Welder-fitter. Person who is competent to plan and set up the work to be done.

Weld gauge. Device designed for checking the shape and size of welds.

Welding

1. Joining two or more pieces of material by applying heat or pressure, or both, with or without filler material, to produce a localized union through fusion or recrystallization across the interface. The result is one continuous piece of material. The thickness of the filler material is much greater than the capillary dimensions encountered in brazing.
2. Metal-joining process used in making welds, as for example, forge welding, Thermit welding, flow welding, gas welding, arc welding, resistance welding, induction welding, and brazing.

Welding current. Current flowing through the welding circuit during the making of a weld. In resistance welding, the current used during preweld or postweld intervals is excluded.

Welding cycle. Complete series of events involved in making a resistance weld. Term is also applied to semiautomatic mechanized fusion welds.

Welding engineer. Person who has an understanding of metallurgy, structural engineering, mechanical engineering, and welding and its effect on various types of metals.

Welding generator. Generator used for supplying current for welding.

Welding goggles. Goggles with tinted lenses, used during welding or oxygen cutting, that protect the eyes from harmful radiation and flying particles.

Welding lead (welding cable). Work lead or an electrode lead.

Welding leads. Work lead and electrode lead of an arc welding circuit.

Welding machine. Equipment used to perform the welding operation, for example, spot welding machine, arc welding machine, or seam welding machine.

Welding pressure. Pressure exerted during the welding operation on the parts being welded.

Welding procedure. Detailed methods and practices, including joint welding procedures, involved in the production of a weldment.

Welding process. Metal joining process wherein coalescence is produced by heating to suitable temperatures, with or without the application of pressure, and with or without the use of filler metal. Examples are forge welding, Thermit welding, flow welding, gas welding, arc welding, resistance welding, induction welding, and brazing.

Welding rod

1. Filler metal, in wire or rod form, used in gas welding and brazing processes and in those arc welding processes wherein the electrode does not furnish the filler metal.

2. Filler metal in rod or wire form used in welding.

Welding schedule. Record of all welding machine settings plus identification of the machine for a given material, size and finish.

Welding sequence

1. Order of making the welds in a weldment.

2. Order of welding the various component parts of a weldment or structure.

Welding stress. Residual stress caused by localized heating and cooling during welding.

Welding technique. Details of a manual, machine, or semiautomatic welding operation that, within the limitations of the prescribed joint welding procedure, are controlled by the welder.

Welding tip. Replaceable nozzle for a gas torch that is especially adapted for welding, spot welding or projection welding electrode.

Welding torch. Device used in gas welding or torch brazing for mixing and controlling the flow of gases.

Welding transformer. Transformer used to supply current for welding.

Weld interval. In resistance welding, the total of all heat and cool times when making a single multiple-impulse weld.

Weld interval timer. Device used in resistance welding to control heat and cool times and weld interval when making multiple-impulse welds singly or simultaneously.

Weld line. Junction of the weld metal and the base metal; the junction of the base metal parts when filler metal is not used.

Weldment. Assembly whose component parts are joined by welding.

Weld metal. Portion of a weld that has been melted during welding.

Weld metal area. Area of the weld metal as measured on the cross section of a weld.

Weld nugget. Weld metal in spot, seam, or projection welding.

Weld penetration. Examples of weld penetration are joint penetration; root penetration.

Weld time. In single-impulse and flash welding, the length of time that welding current is applied to the work in making a weld.

Weld timer. Device used in resistance welding to control the weld time only.

Work angle. In arc welding, the angle between the electrode and one member of the joint, taken in a plane normal to the weld axis.

Work lead. Electrical conductor connecting the source of arc welding current to the work; also called "welding ground" or "ground lead."

6

Wood and Plastics

ADHESIVES

Abhesive. Material that resists adhesion and is applied to surfaces to prevent sticking.

Acceptance test. Investigation performed on an individual lot of a previously qualified adhesive or by under the observation of the purchaser to establish conformity with a purchase agreement.

Adherend. A body that is held to another body by an adhesive.

Adherend failure. Rupture of an adhesive joint, such that the separation appears to be within the adherend.

Adhesion. The clinging or sticking together of two surfaces. The state in which two surfaces are held together by forces at the interface.

Adhesive. A substance capable of holding materials together by surface attachment.

Adhesive bond. Forces such as dipole bonds that attract adhesives and base materials to each other.

Adhesive bonding. Fastening together of two or more solids by the use of glue, cement, or other adhesives.

Adhesive dispersion. Two-phase system in which one phase is suspended in a liquid.

Adhesive failure. Rupture of an adhesive bond, such that the plane of separation appears to be at the adhesive–adherend interface.

Adhesive strength. The strength of an adhesive bond, usually measured as a force required to separate two objects of standard bonded area by either shear or tensile stress.

Adhesive tape. Tape coated with a substance that binds or sticks to a surface.

Affinity. Attraction or polar similarity between adhesive and adherend.

Aging. The progressive changes in the chemical and physical properties of an adhesive.

Amylaceous. Term pertaining or relating to the nature of starch; starchy.

Anaerobic. Adhesive that cures in the absence of oxygen.

Animal glues (characteristics). Water acts as a solvent for animal glues that are melted and applied hot; liquids are applied cold. These glues are excellent for bonds to wood, leather, glass, and paper, but not to metals. They develop very high strength on wood—up to 10,000 to 12,000 psi in shear. Their temperature resistance is medium for heat and good for cold. They have good creep resistance but very poor water resistance. Initially they are made into a gel and then are air dried at room temperature to cure.

Aromatics. Term refers to chemical compounds that are derivatives of benzene.

Asphalt. Naturally occurring solid or semisolid mineral pitch or bitumen, more or less soluble in carbon disulfide, naptha, and turpentine, and fusible at varying temperatures. Also bituminous residues left from petroleum and coal-tar.

Asphaltic adhesive mixtures (characteristics). Water, aromatics, carbon tetrachloride, and disulfide act as solvents. These mixtures are thermoplastic. Natural asphalts are usually hard and brittle when cold. Asphaltic adhesive mixtures are good for bonding metals to rubber or glass, and installing floor coverings and roofing felts. Their strength is low to fair depending on grade and temperature. Their temperature resistance is poor for heat and good for cold. The melting point may be as high as 200°F or as low as 50°F. They have very poor creep resistance but their water resistance is good to excellent. Elevated temperatures or cooling to room temperature to cure.

Assembly. Group of materials or parts, including adhesive, that has been placed together for bonding or that has been bonded together.

Assembly adhesive. Adhesive that can be used for bonding manufactured parts together. The term is commonly used in the wood industry to distinguish such adhesive from those used in making plywood.

Assembly time. Time interval between the spreading of the adhesive on the adherend and the application of pressure or heat, or both, to the assembly.

A stage. Early stage in the reaction of certain thermosetting resins during which the material is fusible and still soluble in certain liquids.

Bag bonding. Method of bonding involving the application of fluid pressure to a flexible cover, which, usually in connection with a rigid die, completely encloses and exerts pressure on the assembly being bonded.

Batch. Manufactured unit or a blend of two or more units of the same formulation and processing of an adhesive. Also a measured mix of adhesive materials.

Binder. Component of an adhesive composition that is primarily responsible for the adhesive forces that hold two bodies together.

Bite. Ability of an adhesive to penetrate or dissolve the uppermost portions of the adherend.

Blister. Elevation of the surface of a adherend, somewhat resembling in shape a blister on the human skin. Its boundaries may be indefinitely outlined, and it may have burst and become fastened.

Blocking. Undesired adhesion between touching layers of a material, such as occurs under moderate pressure during storage or use.

Blood alumin glues (characteristics). Water acts as solvent for these glues, which are usually in dry powder form. They provide fair bonding for wood, good bonding for leather and paper, and poor bonding for metals or glass. Their average-strength temperature resistance is fair for both heat and cold. They have good creep resistance but poor water resistance. They are dried at room temperature or low heat 150°F to 200°F for cure.

Body. Consistency of an adhesive; thickness; viscosity.

Bond. Unite material by means of an adhesive.

Bondability. Term indicating ease or difficulty in bonding a material with adhesive.

Bonded fabric. Web of fibers held together with an adhesive and not from a continuous sheet of adhesive material.

Bond face. The part or surface of a building component that serves as a substrate for an adhesive.

Bond failure. Rupture of adhesive bond.

Bondline. The layer of adhesive that attaches two adherends.

Bondline slip. Movement within and parallel to the bonding during shear.

Bond strength. Unit load applied in tension, compression, flexure, peel, impact, cleavage, or shear, required to break an adhesive assembly, with failure occurring in or near the plane of the bond.

Borate. To add borax to a starch adhesive to improve tack and viscosity.

B stage. Intermediate stage in the curing of a thermosetting adhesive during which the adhesive softens when heated and swells in contact with certain liquids, but does not entirely fuse or dissolve.

Built-up laminated wood. Assembly made by joining layers of lumber with mechanical fastenings so that the grain of all laminations is essentially parallel.

Butt joint. A joint with the structural units joined to place the adhesive or sealant into tension or compression.

Calk. To fill voids with an adhesive.

Casein glue. Adhesive substance composed of casein (the curd of milk), lime, and sodium salt. It comes as dry powder to which water is added.

Catalysis. Action of any substance that does not undergo any apparent chemical change but accelerates or inhibits a chemical reaction.

Catalyst. Substance that markedly speeds up the cure of an adhesive when added in minor quantity as compared to the amounts of the primary reactants.

Caul. Sheet of material employed singly or in pairs in hot- or cold-pressing assemblies being bonded. It is usually employed to protect either the faces or the press platen, or both, against marring and staining, prevent sticking, and facilitate press loading.

Cellulose cements (characteristics). Water emulsion ethyl acetate or acetone act as solvents. They are thermoplastic and fused by heating. They provide good bonding for glass, wood, and leather but not for rubber. Their strength is good from 1000 to 1400 psi on wood in shear. Their temperature resistance is fair to good for both heat and cold. They have good creep resistance and water resistance poor if mixed with water as a solvent, and fair to medium if mixed with chemical solvents. They are air dried and set for cure.

Chlorinated rubber (characteristics). Ketones or aromatics act as solvents and the mixture is used in liquid form. It provides a medium bond for wood, metals, or glass and a good bond for paper. Its temperature resistance is medium for both heat and cold. It has poor creep resistance and medium to good water resistance. It is dried at room temperature for cure.

Closed assembly time. Time interval between the completion of the assembly of parts for bonding and the application of pressure or heat, or both, to the assembly.

Cobwebbing. During the spray application of an adhesive, the formulation of weblike threads along with the usual droplets as the adhesive leaves the nozzle of a spray gun.

Cohesion. State in which the particles of a single substance are held together by primary or secondary valence forces; as used in the adhesive field, the state in which the particles of the adhesive (or the adherend) are held together. The internal strength of film.

Cohesive failure. Rupture of an adhesive bond such that the separation appears to be within the adhesive.

Cold-pressing. Bonding operation in which an assembly is subjected to pressure without the application of heat.

Cold-setting adhesive. Adhesive that sets at temperatures below 68°F.

Cold-setting resin glue. Resin base glue that comes in powder form and is mixed with water.

Compatibility. Ability of two or more substances to mix or blend without separation or reaction.

Condensation. Chemical reaction in which two or more molecules combine, with the separation of water or some other simple substance. If a polymer is formed, the process is called "polycondensation."

Consistency. Property of a liquid adhesive by virtue of which it tends to resist deformation.

Construction adhesive. Any adhesive used to assemble primary building material into components during building construction—most commonly applied to elastomer-based mastic-type adhesives.

Contact adhesive. Adhesive that is apparently dry to the touch and that will adhere to itself instantaneously upon contact. It is also called "contact bond adhesive" or "dry bond adhesive."

Contact cement. Adhesive used to bond plastic laminates or other thin material, so called because the bond is made on contact, eliminating the need of clamps.

Contact failure. Failure of an adhesive joint due to incomplete contact, during bonding, between the adhesive and adherend or between adhesive surfaces.

Continuous film. Coating of adhesive film that is free of breaks, pinholes, or holidays.

Cottoning. During machine application of an adhesive, the formation of weblike filaments of adhesive between machine parts or between machine parts and the receiving surface during transfer of the liquid adhesive onto the receiving surface.

Crazing. Fine cracks that may extend in a network on or under the surface, of or through a layer of adhesive.

Creep. Dimensional change with time of a material under load, following the initial instantaneous elastic or rapid deformation. Creep at room temperature is sometimes called "cold flow."

Cross laminated. Term describing laminate in which some of the layers of material are oriented at right angles to the remaining layers with respect to the grain or strongest direction in tension.

Cure. Change in the physical properties of an adhesive by chemical reaction, which may be condensation, polymerization, or vulcanization, usually accomplished by the action of heat and catalyst, alone or in combination, with or without pressure.

Curing temperature. Temperature to which an adhesive or an assembly is subjected to cure the adhesive.

Curing time. Period of time during which an assembly is subjected to heat or pressure, or both, to cure the adhesive.

Daluent. Ingredient usually added to an adhesive to reduce the concentration of bonding materials.

D Glass. High boron-content glass that when finely chopped or pulverized is used as a filler for liquid-type adhesive.

Delamination. Separation of layers in a laminate because of failure of the adhesive, either in the adhesive itself or at the interface between the adhesive and the adherend, or because of cohesive failure of the adherend.

Desiccant. Substance that can be used for drying purposes because of its affinity for water.

Doctor bar. Scraper mechanism that regulates the amount of adhesive on the spreader rolls or on the surface being coated.

Doctor roll. Roller mechanism that revolves at a different surface speed or in an opposite direction from the spreader roll, resulting in a wiping action for regulating the adhesive supplied to the spreader roll.

Double spread. Application of adhesive to both adherends of a joint.

Dry. Change in the physical state of an adhesive on an adherend by the loss of solvent constituents by evaporation or absorption, or both.

Drying temperature. Temperature to which an adhesive on an adherend or in an assembly or the assembly itself is subjected to dry the adhesive.

Dry joints. Joints lacking adhesion, more specifically, poor contact of adhering surface.

Dry strength. Strength of an adhesive joint determined immediately after drying under specified conditions or after a period of conditioning in standard laboratory atmosphere.

Dry tack. Ability of certain adhesives, particularly non-vulcanizing rubber adhesives, to adhere on contact to themselves at a stage in the evaporation of volatile constituents, even though they seem dry to the touch, sometimes called "aggressive tack."

Edge joint. A joint made by bonding two pieces of wood together edge to edge, commonly by gluing. The joints may be made by gluing two squared edges, as in a plain edge joint, or by using machined joints of various kinds, such as tongue-and-groove joints.

Elasticity. The extensible property of adhesive films or adhesive interfaces to contract or expand in such a manner as to overcome the differential contraction and expansion rates that the bonded adherends may exhibit.

Elastomer. Material which at room temperature can be stretched repeatedly to at least twice its original length and upon immediate release of the stress will return with force to its approximate original length.

Elongation. Increase in length expressed numerically as a fraction or percentage of initial length.

Epoxy adhesives. Two-part adhesives (one part hardener and one part resin) that can be used to hold almost any kind of material. They create a strong bond and are moisture resistant. Also, they do not set up instantly, which is an advantage if the parts that are to be bonded must remain movable until the final fit is made.

Epoxy resin (characteristics). No solvent is needed for use. Epoxy resin is thermosetting and provides an excellent bond for wood, metal, glass, and masonry. It has high strength from 1000 to 7000 psi on wood in shear. Its temperature resistance is excellent for both heat and cold. Its creep resistance is good to poor, depending on compounding. Its water resistance is fair to excellent, depending on compounding. It is cured by a catalyst and hot-pressing (up to 390°F) or a strong catalyst at room temperature.

Ester gum. One of the classes of compounds obtained by esterfying rosin (solophony) or natural gums with a polyhydric alcohol; specifically, the ester resulting from esterification of rosin with glycerin.

Extender. Substance, generally having some adhesive action, added to an adhesive to reduce the amount of the primary binder required per unit area.

Facing. One of the two outer layers of a material that has been bonded to the core of a sandwich.

Faying surface. Surface of an object that comes in contact with another surface that is to be bonded; the bonding surface.

Filler. Relatively nonadhesive substance added to an adhesive to improve its working properties, permanence, strength, or other qualities.

Filler sheet. Sheet of deformable or resilient material, which, when placed between the assembly to be bonded and the pressure applicator or when distributed within a stack of assemblies, aids in providing uniform application of pressure over the area to be bonded.

Fillet. Portion of an adhesive that fills the corner or angle formed where two adherends are joined.

Film. Thin layer of adhesive material having a thickness not greater than 0.010 in.

Finger joint. An end joint made up of several meshing wedges of fingers of wood bonded together with an adhesive. Fingers are sloped and may be cut parallel to either the wide or narrow face of the piece.

Flow. Movement of an adhesive during the bonding process before the adhesive is set. Fluidity of an adhesive material.

Foamed adhesive. Adhesive whose apparent density has been decreased substantially by the presence of numerous gaseous cells dispersed throughout its mass.

Force dry. Drying of an adhesive by exposing it to circulating air at a temperature between 220°F and 235°F for a specified time.

Freeze resistance. Resistance to the effect of low temperature of an adhesive or assembly specimen subjected to a flexure, torsion, or impact according to prescribed test procedures.

Gap filling adhesive. Adhesive suitable for use where the surfaces to be joined may not be in close or continuous contact, owing either to the impossibility of applying adequate pressure or to slight inaccuracies in matching mating surfaces.

Gel. Semisolid system consisting of a network of solid aggregates in which liquid is held.

Gelation. Formation of a gel.

Glue. Originally, a hard gelatin obtained from hides, tendons, cartilage, bones, etc., of animals; also, an adhesive prepared from this substance by heating with water. The term is now synonymous with the term "adhesive."

Glue joint. Area of a bonded assembly where the adhesive and adherend are in contact.

Glue-laminated wood. Assembly made by bonding layers of veneer or lumber with an adhesive so that the grain of all laminations is essentially parallel.

Glue line. Adhesive layer between two adherends.

Glue line heating. Dielectric heating in which the electrodes are designed to give preferential heating to a thin film of glue or other relatively high-loss material located between layers of relatively low-loss material such as wood.

Gum. Any of a class of colloidal substances exuded by or prepared from plants, sticky when moist, and composed of complex carbohydrates and organic acids that are soluble or swell in water.

Hardboard. Fine pieces of wood bound together with an adhesive and pressed into 4×8-foot sheets. Thicknesses are approximately $\frac{1}{8}$, $\frac{3}{16}$, and $\frac{1}{4}$ in. Thermosetting resins are usually used as the adhesive binder.

Hardener. Substance or mixture of substances added to an adhesive to promote or control the curing reaction by taking part in it. The term is also used to designate a substance added to control the degree of hardness of the cured film.

Haze. Cloudy appearance within or on the surface of an adhesive.

Heat-activated adhesive. Dry adhesive film that is rendered tacky or fluid by application of heat or heat and pressure to the assembly.

Heat endurance. Time of heat aging that an adhesive can withstand before failing at a specified load (physical load).

Hot-melt adhesive. An adhesive that is applied in a molten state and forms a bond on cooling to a solid state.

Heat resistance. Ability of an adhesive to resist the deteriorating effect of elevated temperatures.

Hot-setting adhesive. Adhesive that requires a temperature of 212°F or above to set it.

Impact resistance. Relative susceptibility of an adhesive to fracture by shock.

Impregnate. The dipping or immersion of a fiberous substrate into an adhesive liquid.

Inhibitor. Substance that slows down chemical reaction. Inhibitors are sometimes used in certain types of adhesives to prolong storage or working life.

Insulator. Adhesive material of low electrical or low thermal conductivity.

Intermediate-temperature setting adhesive. Adhesive that sets in the temperature range of 87°F to 211°F.

Izod impact test. Destructive test designed to determine the resistance of an adhesive to the impact of a suddenly applied force.

Joint. Location at which two adherends are held together with a layer of adhesive.

Ketones. Compounds containing the carbonyl group to which are attached two alkyl or aryl groups. Ketones such as methyl ethyl ketone are commonly used as solvent for resins and plastics.

Laminate
1. Product made by bonding together two or more layers of material or materials.
2. Unite layers of material with adhesive.

Lamination. Process of preparing a laminate; also, layer in a laminate.

Lap joint. Joint made by placing one adherend partly over another and bonding together the overlapped portions.

Lay-flat. Industry term for an adhesive material that provides noncurling and distension characteristics.

Liquid resin. Organic polymeric liquid that when converted to its final state for use becomes a resin.

Manufactured unit. Quantity or batch of finished adhesive or finished adhesive component processed at one time.

Marine glue. A form of glue resisting the action of water and containing rubber, shellac, and oil.

Mastic. Material with adhesive properties, usually used in relatively thick sections, which can be readily formed by application with trowel or spatula.

Matrix. Part of an adhesive that surrounds or engulfs embedded filler or reinforcing particles and filaments.

Maturing temperature. Temperature as a function of time and bonding condition, which produces desired characteristics in bonded components.

Mechanical adhesion. Adhesion between surfaces in which the adhesive holds the parts together by interlocking action.

Melamine resins (characteristics). Water or alcohol act as solvents. The resins are thermosetting. They are used as a powder with a separate catalyst and are applied cold. They are colorless and nonstaining and provide an excellent bond for paper or wood but a poor bonding for metals or glass. Their temperature resistance is excellent for both heat and cold. They have a very good creep resistance and excellent water resistance especially when alcohol is used as a solvent. Malamine resins are hot-pressed at 300°F for cure.

Metal bond. Process of joining metals by the use of adhesives, heat, and pressure, generally used to produce a final bond.

Modifier. Chemically inert ingredient added to an adhesive formulation that changes its properties but does not react chemically with the binder.

Moisture absorption. Pickup of water vapor from the air by an adhesive. Does not include water absorption that results from immersion.

Moisture-resistant adhesive. Adhesive with bonds sufficiently durable to withstand occasional exposure to wet and damp conditions. These bonds will offer considerable resistance to severe conditions of exposure for a limited time, but, when permanently exposed to severe conditions, they deteriorate slowly and ultimately fail.

Monomer. Relatively simple compound that can react to form a polymer.

Mucilage. Adhesive prepared from a gum and water; also, in a more general sense, a liquid adhesive that has a low order of bonding strength.

Multiple-layer adhesive. Film adhesive, usually supported, with a different adhesive composition on each side, designed to bond dissimilar materials; as in the core-to-face bond of a sandwich composite. For assemblies involving multiple layers of parts, the assembly time begins with the spreading of the adhesive on the first adherend.

Natural rubber adhesives (characteristics). Water emulsions, aromatics, and various hydrocarbons act as solvents. These adhesives provide good bonding for rubber, glass or leather and fair bonding for wood or ceramics. They have low strength, providing only 340 psi in tension on wood. Their temperature resistance is fair for both heat and cold. They have poor creep resistance but good water resistance. They are dried at room temperature for cure.

Neoprene rubber adhesives (characteristics). Water emulsions or volatile liquids act as solvents. They are thermoplastic with some thermosetting characteristics. These adhesives provide excellent bonding for wood, asbestos board, metal, and glass. They have a strength of up to 1200 psi in shear. Their temperature resistance is good for heat and cold and creep resistance is fair to good with excellent water resistance. Some heat is desirable for cure.

Nitrile rubber adhesives (characteristics). Water emulsions or volatile liquids act as solvents. Thermoplastic and thermosetting types are available. They provide bonding for wood, paper, porcelain, enamel, and polyester skins. They are thermosetting to 4000 psi in shear and thermoplastic to 600 psi. Their temperature resistance is good for both heat and cold. They have good to fair creep resistance and excellent water resistance. Heat curing is preferable.

Nonwarp. Term referring to an adhesive system that does not distend, curl, shrink, wrinkle, or cause any condition that would render a laminated assembly unacceptable.

Novalak. Phenolic-aldehydic resin that, unless a source of methylene groups is added, remains permanently thermoplastic.

Opalescence. Limited clarity of vision through a sheet containing a coating of cured adhesive.

Open assembly time. Time interval between the spreading of the adhesive on the adherend and the completion of assembly of the parts for bonding.

Organic adhesive. Prepared organic material that is ready to use without additional solvents, and cures or sets by evaporation.

Parallel laminated. Term describing a laminate in which all the layers of material are oriented approximately parallel with respect to the grain or strongest direction in tension.

Paste. Adhesive composition having a characteristic plastic-type consistency, that is, a high order of yield value, such as that of a paste prepared by heating a mixture of starch and water and subsequently cooling the hydrolyzed product.

Peel test. Test made on bonded strips of metals by peeling the metal strips back and recording the adhesive strength values.

Penetration. Entering of an adhesive into an adherend.

Permanence. Resistance of an adhesive bond to deteriorating influences.

Permanent set. The amount of deformation that remains in an adhesive after removal of a load.

Phenolic resins (characteristics). Water, alcohol, and ketones act as solvents with dry or liquid resins. Phenolic resins provide good to excellent bonding for wood and paper and medium to poor bonding for glass and metals. They have good strength. Their temperature resistance is excellent for both heat and cold. They have excellent creep and water resistance. Some resins set at room temperature and some require hot-pressing for cure.

Pick-up roll. Spreading device in which the roll for picking up the adhesive runs in a reservoir of adhesive.

Plasticity. Property of adhesives that allows the material to be deformed continuously and permanently without rupture upon the application of a force that exceeds the yield value of the material.

Plasticizer. Material incorporated in an adhesive to increase its flexibility, workability, or distensibility. The addition of the plasticizer may cause a reduction in melt viscosity, lower the temperature of the second-order transition, or lower the elastic modulus of the solidified adhesive.

Plywood. Cross-bonded assembly made of layers of veneer or veneer in combination with a lumber core or plies joined with an adhesive. Two types of plywood are recognized, namely, veneer plywood and lumber core polywood.

Plywood adhesive. Adhesive suitable for bonding wood veneers together in the production of plywood. Two types are used—interior and waterproof exterior.

Polymer. Compound formed by the reaction of simple molecules having functional groups that permit their combination to proceed to high molecular weights under suitable conditions. Polymers may be formed by polymerization (addition polymer) or polycondensation (condensation polymer). When two or more monomers are involved, the product is called a "copolymer."

Polymerization. Chemical reaction in which the molecules of a monomer are linked together to form large molecules whose molecular weight is a multiple of that of the original substance. When two or more monomers are involved, the process is called "copolymerization" or "meteropolymerization."

Polyvinyl resin emulsion adhesives. White glues that come in squeeze bottles and are used for just about everything from gluing paper together to holding wood joints. They are an excellent choice for interior construction.

Polyvinyl resins (characteristics). Water and ketones act as solvents. When the resin is in liquid form, the solvents will usually create an emulsion. They provide good bonding for wood or paper and have a strength of up to 950 psi in shear on wood. Their temperature resistance is fair for heat and good for cold. They fuse at 220°F to 350°F. They have fair to poor creep resistance and fair to medium water resistance. They are air-dried and set at room temperature for cure.

Post cure. A treatment (normally involving heat) applied to an adhesive assembly following the initial cure, to complete cure or to modify specific properties.

Precure. Condition of too much cure, set, or solvent loss of the adhesive before pressure is applied, resulting in inadequate flow, transfer, and bonding.

Premix. Adhesive that has all the ingredients mixed and sealed in containers and frozen or otherwise preserved for future use.

Primer. Coating applied to a surface prior to the application of an adhesive to improve the performance of the bond.

Pressure-sensitive adhesive. Adhesive made so as to adhere to a surface at room temperature by briefly applied pressure alone.

Qualification test. Investigation, independent of a procurement action, performed on an adhesive product to determine whether or not the product conforms to all requirements of the applicable application.

Resin. Solid, semisolid, or pseudosolid organic material that has an indefinite and often high molecular weight, exhibits a tendency to flow when subjected to stress, usually has a softening or melting range, and usually fractures conchoidally.

Resinoid. Any of the class of thermosetting synthetic resins, either in their initial temporarily fusible state or in their final infusible state.

Resite (C stage). Final stage in the reaction of certain thermosetting resins, in which the material is relatively insoluble and infusible. Certain thermosetting resins in a pull cured adhesive layer are in this stage.

Resitol (B stage). Intermediate stage in the reaction of certain thermosetting resins, in which the material softens when heated and swells when in contact with certain liquids but may not entirely fuse or dissolve. Resin in an uncured thermosetting adhesive is usually in this stage.

Resol (A stage). Early stage in the reaction of certain thermosetting resins, in which the material is fusible and still soluble in certain liquids.

Resorcinol glue. A glue that is high in both wet and dry strength and resistant to high temperatures. It is used for gluing lumber or assembly joints that must withstand severe service conditions.

Resorcinol resins (characteristics). Water, alcohol, and ketones act as solvents. These are thermosetting resins, usually applied in liquid form with a separate catalyst. They provide bonding for wood or paper but are poor for glass or metals. They have a strength up to 1950 psi in shear on wood. Their temperature resistance is excellent for cold and they are more heat resisting than wood. They have very good creep resistance. Room temperature or moderate (200°F) heat is adequate for curing.

Restricted adhesive. Adhesive that for any reason cannot be validated and therefore cannot be assigned a usable life. Also such an adhesive cannot be used for structural bonding.

Retarder. A substance added to slow down the cure rate of adhesive.

Retrogradation. Change of starch pastes from low to high consistency on aging.

Room temperature setting adhesive. Adhesive that sets in the temperature range of 68°F to 86°F.

Rosin. Resin obtained as a residue in the distillation of crude turpentine from the sap of the pine tree (gum rosin) or from an extract of the stumps and other parts of the tree (wood rosin).

Scarf joint. Joint made by cutting away similar angular segments of two adherends and bonding the adherends with the cut areas fitted together.

Self-vulcanizing. Term describing an adhesive that undergoes vulcanization without the application of heat.

Separate-application adhesive. Adhesive consisting of two parts, with one part applied to one adherend, the other part applied to the second adherend, and the two brought together to form a joint.

Set. Adhesive converted into a fixed or hardened state by chemical or physical action, such as condensation, polymerization, oxidation, vulcanization, gelation, hydration, or evaporation of volatile constituents.

Setting temperature. Temperature to which an adhesive or an assembly is subjected to set the adhesive.

Shelf life. Period during which the manufacturer guarantees that an adhesive stored at a specified temperature will produce specified mechanical properties when used for designated purposes.

Shortness. Term describing an adhesive that does not string, cotton, or otherwise form filaments or threads during application.

Single spread. Application of adhesive to only one adherend of a joint.

Sizing. Applying a material on a surface in order to fill pores and thus reduce the absorption of the subsequently applied adhesive or coating or to otherwise modify the surface properties of the substrate to improve the adhesion. The material used for this purpose is sometimes called "size."

Slip. Physical property of an adhesive, referring to the ability to move or position the adherends after the adhesive has been applied to their surfaces.

Slippage. Movement of adherends with respect to each other during the bonding process.

Sodium silicate (characteristics). Water acts as a solvent for sodium silicate, which is used in liquid form This substance provides good bonding for wood and metals and excellent bonding for paper or glass. Its temperature resistance is excellent for heat or cold. It has good creep resistance but poor water resistance. It is dried at room temperature or under moderate heat to 150°F to 200°F for cure.

Softener. Material added to an adhesive film to prevent embrittlement of the adhesive layer, and also to increase long-term flexibility properties.

Solids content. Percentage by weight of the nonvolatile matter in an adhesive.

Solvent. Medium within which a substance is dissolved.

Solvent-activated adhesive. Dry adhesive film that is rendered tacky just prior to use by the application of a solvent.

Solvent adhesive. An adhesive having volatile organic liquid as a vehicle. This term excludes water-based adhesives.

Solvent cement. Liquid adhesive utilizing an organic solvent, such as benzene, MEK (methyl ethyl ketone), or toluol, (a coal-tar product), as the vehicle for the deposit of the adhesive polymer.

Soybeam glue (characteristics). Water acts as a solvent for the glue, which is used in a dry powder form. Provides fair bonding for wood or glass and poor bonding for metals or rubber. Its temperature resistance is fair for heat and poor for cold. It has good creep resistance and poor water resistance. It is dried at room temperature for cure.

Specific adhesion. Adhesion between surfaces that are held together by valence forces of the same type as those that give rise to cohesion.

Spread. Quantity of adhesive per unit joint are applied to an adherend, usually expressed in points of adhesive per thousand and square feet of joint area.

Sqeezeout. Bead of adhesive squeezed out of a joint when pressure is applied.

Starch and dextrin glues (characteristics). Water acts as a solvent for these glues, which are available in dry or liquid form. They provide bonding for wood, leather, and paper. Their strength is fair to medium on paper and poor for metals and glass. Their temperature resistance is fair for both heat and cold. They have fair creep resistance and poor water resistance. They are dried at room temperature for cure.

Starved joint. Joint has an insufficient amount of adhesive to produce a satisfactory bond.

Stencil application. Adhesive application method that involves the deposit of adhesive in predetermined patterns, applying adhesive through stencil cutouts.

Storage life. Period of time during which a packaged adhesive can be stored under specified temperature conditions and remain suitable for use.

Strain. Ratio of the extension to the original length of the measured elongating section of the test specimen.

Strength. Maximum stress required to overcome the cohesion of a material.

Stringiness. Property of an adhesive that results in the formation of filaments or threads when the adhesive transfer surfaces are separated.

Structural adhesive. Adhesive having sufficiently high mechanical properties, when cured, that it may safely be used for bonding parts in assemblies where safety is of prime importance.

Substrate. Material on the surface of which an adhesive-containing substance is spread for any purpose, such as bonding or coating. It is a broader term than adherend.

Supported film adhesive. Adhesive supplied in a sheet or film form with an incorporated carrier that remains in the bond when the adhesive is applied and used.

Synthetic rubber (characteristics). Elastomer manufactured by a chemical process as distinguished from natural rubber obtained from plants and trees. Typical examples are buytl, neoprene, nitrile, and thiokol rubbers. Used as an adhesive the product uses aliphatic hydrocarbon as a solvent. It has excellent moisture resistance.

Synthetic rubber-resin adhesive (characteristics). The resin is a fire retardant adhesive using aromatic hydrocarbon as a solvent. It has excellent water resistance. Used at temperature ranges from 30°F to 250°F.

Tack. Property of an adhesive that enables it to form a bond of measurable strength immediately after the adhesive and adherend are brought into contact under low pressure.

Tack range. Period of time in which an adhesive will remain in the tacky-dry condition after application to an adherend, under specified temperature and humidity.

Tacky-dry. Term describing the condition of an adhesive when the volatile constituents have evaporated or been absorbed sufficient to leave it in a desired tacky state.

Telegraphing. Condition in a laminate or other type of composite construction in which irregularities, imperfections, or patterns of an inner layer are visibly transmitted to the surface. Telegraphing is occasionally referred to as "photographing."

Temperatures. The temperature attained by the adhesive during the process of drying (adhesive drying temperature) may differ from the temperature of the atmosphere surrounding the assembly (assembly drying temperature). The temperature attained by the adhesive during the process of curing (adhesive curing temperature) may differ from the temperature of the atmosphere surrounding the assembly (assembly curing temperature). The temperature attained by the adhesive during the process of setting (adhesive setting temperature) may differ from the temperature of the atmosphere surrounding the assembly (assembly setting temperature).

Tensile strength. Capacity of a material to resist a force that tends to stretch it.

Thermoplastic. Material that repeatedly softens when heated and hardens when cooled.

Thermoplastic glues and resins. Glues and resins that are capable of being repeatedly softened by heat and hardened by cooling.

Thermoset

1. Term pertaining to the state of a resin in which it is relatively infusible.
2. Material that will undergo or has undergone a chemical reaction by the action of heat, catalysts, ultraviolet light, etc., leading to a relatively infusible state.

Thermosetting. Having the property of undergoing a chemical reaction by the action of heat, catalysts, ultraviolet light, etc., leading to a relatively infusible state.

Thermosetting glues and resins. Glues and resins that are cured with heat, but do not soften when subsequently subjected to high temperatures.

Thinner. Volatile liquid added to an adhesive to modify the consistency or other properties.

Thixotropy. Ability of adhesive systems to thin upon isothermal agitation and to thicken upon subsequent rest.

Throwing. Characteristic behavior of some adhesives, occurring when they are transferred from rollers or rotary stencil mechanisms, where, owing to peripheral speed, small droplets of adhesive are thrown from the roller or stencil.

Toughness. Energy required to break an adhesive material (cohesively).

Unsupported film adhesive. Adhesive supplied in sheet or film form without an incorporated carrier.

Urea formaldehyde resorcinol glues. These glues create a strong bond when gluing wood and requires clamping. (The excess glue can be wiped away with a damp cloth.) They may come as powders that require mixing with water or a provided catalyst. Drying time ranges from three to ten hours; glues should be used in the area where the temperature is not less than 70°F.

Urea resins (characteristics). Water, alcohol, or alcohol hydrocarbon blends act as solvents. These resins are thermosetting and excellent bonding for wood, leather, and paper but poor bonding for metals or glass. They have good creep resistance and fair water resistance. Some heat is required for curing but some types of resins will cure at room temperature.

Veneer glue. Adhesive used to make plywood laminates, or other products in the wood industry in contrast with "joint glue."

Vinyl acetate (characteristics). Vinyl acetate uses alcohol as a solvent and has good water resistance. Acetone or toluene used for cleanup.

Viscosity. Ratio of the shear stress existing between laminate of moving fluid and the rate of shear between these laminae.

Viscosity coefficient. Tangentially applied shearing stress that will induce a velocity gradient.

Volatile. Property of liquids to pass away by evaporation.

Vulcanization. Chemical reaction in which the physical properties of a rubber are changed in the direction of decreased plastic flow, less surface tackiness, and increased tensile strength by reacting it with sulfur or other suitable agents.

Warp. Significant variation from the original, true, or plane surface. Distortion in an adhesive bonded assembly after bonding.

Water absorption. Ratio of the weight of water absorbed by an adhesive to the weight of the same material in a dry condition.

Weathering. Exposure of adhesives to an outdoor environment.

Webbing. Filaments or threads that may form when adhesive transfer surfaces are separated.

Wet-state strength. In latex adhesives, the joint strength when the adherends are brought together with the adhesive still in the wet state.

Wet strength. Strength of an adhesive joint determined immediately after removal from a liquid in which it has been immersed under specified conditions of time, temperature, and pressure.

Wetting. Relative ability of a liquid adhesive to display interfacial affinity for an adherend and to flow uniformly over the adherend surface.

White latex adhesive. Adhesive used to carpet and resilient flooring installation.

Wood block floor-installation. General-purpose adhesives consisting of nonflammable polyvinyl acetate applied directly to concrete or wood subflooring.

Wood failure. Rupturing of wood fibers in strength tests on bonded specimens, usually expressed as the percentage of the total area involved that shows such failure.

Wood veneer. Thin sheet of wood, generally within the thickness range of 0.01 to 0.25 in., to be used in a laminate.

Working life. Period of time during which an adhesive, after mixing with catalyst, solvent, or other compounding ingredients, remains suitable for use.

Yield value. Stress at which a marked increase in deformation occurs without an increase in load.

ARCHITECTURAL WOODS AND USES

Aboudikrou. African wood resembling mahogany.

Acacia. Figured wood that varies from light brown to shades of red and green.

Advanced decay. The older stage of decay in which the destruction is readily recognized because the wood has become punky, soft, and spongy, stringy, ring-shaked pitted, or crumbly.

African mahogany. One of the true mahoganies, with a very pleasing open grain and heartwood ranging in color from light to medium-dark reddish brown. In lumber form it is available as plain sawed and selectively so as quarter-sawed. In veneer form the quarter or "ribbon-striped" cut predominates, but the plain sliced, as well as many of the exotic "figure" cuts, can be produced.

African zebrawood. Equatorial tree of medium size and hardness, and weighing between 37 and 45 lb/cu ft when dry. The sapwood is pale in color and distinct from the heartwood, which is creamy yellow, color veined, or striped with very dark brown or black. The striped effect is seen when the wood is quartersawed.

Afrormosia. African wood, strong, heavy, and hard, reddish brown in color with bands of golden brown.

Agba. African wood with medium brown color and straight grain.

Alder. Light-colored softwood of the betulaceous genus *Alnus*, a deciduous tree whose wood is used in the manufacture of furniture. Alder is not considered a construction wood.

Almon. Phillipine wood used for veneer. The wood is similar to mahogany but has a coarser texture. Depending on source, color varies from light red to dark red.

American black walnut. Domestic wood species. Its grain pattern variations are extensive, and in veneered form it produces, in addition to its normal plain sliced cut, quartered or "pencil striped" cuts, as well as specialty cuts such as crotches, swirls, burls, and others. The heartwood color varies from gray-brown to dark purplish brown. The cream colored sapwood is very prevalent in solid lumber and is usually completely eliminated.

Andiroba. Tropical American wood. The heartwood color varies from reddish brown to dark reddish brown. The texture is like that of mahogany but heavier than mahogany. Wood is suited for such uses as flooring, cabinetwork, millwork, and decorative veneers.

Angelique. Native to French Guiana and Surinam. Heartwood normally is russet colored when freshly cut but darkens to dull brown with a purplish cast. Texture is coarser than that of black walnut. It is highly resistant to decay and is used for ship decking, planking, boat frames, and underwater members.

Angiosperms. Seed-bearing plants that far outnumber the gymnosperms. Important hardwood families belong to this group.

Anisotropic. Exhibiting different properties when measured along different axes; in general, fibrous materials such as wood are anisotropic.

Apamate. Native to southern Mexico, through parts of Central America and South America, particularly in Venezuela and Ecuador. The heartwood varies through the browns, from a golden to a dark brown. The wood resembles oak. Suitable for use in the production of furniture, interior trim, doors, and flooring.

Applewood. American fruitwood usually used for furniture.

Araucariaceae. Family of the order Coniferae. Contains species of trees known as hoop pine, bunya pine, Chilean pine, and parana pine, all in the Southern Hemisphere.

Aromatic. Term applied to woods having characteristic aromas, usually to cedar, particularly incense and Tennessee cedar.

Ash (southern hardwood)

1. There are two commerical species of ash—white and black. Its principal market is in plywood decorative facings, but it is also used for blocking and stair treads. Ash is a tough wood with a pale brown heart, white sap, and pronounced grain.

2. Heavy, hard, coarse-grained wood with reddish brown heartwood and almost white sapwood. It is used for interior trim, plywood, tool handles, some furniture, and cabinet work.

3. Figured blond wood; extremely tough hardwood used in furniture.

Ash, beech, birch, soft elm, maple, and sycamore. Woods that are good as beams and fair as sills but that require good preservative treatment if they are to be exposed to moist conditions.

Aspen. American wood of the poplar family. The heartwood is grayish white to light grayish brown. The sapwood is lighter in color, lightweight, soft, and low in strength.

Australian clears. Pitchy selects, D and better; pine grade, allowing heavy pitch streaks and open pocket pitch, otherwise clear grade.

Avodire. African blond wood, with strong, dark brown vertical streaking and medium-hard texture.

Ayacahuite. Satiny pinewood with very little pine graining. It grows in Mexico and Central America.

Ayous. African wood, creamy white to pale yellow in color.

Bagtikan. Native to Southeast Asia. The heartwood is gray to straw colored and in some species in pale brown with a pinkish cast. Suited primarily for veneer and plywood purposes.

Balsa. Native to southern Mexico, Brazil, and Bolivia, as well as Ecuador, which has been the principal area for this wood. The wood is lightweight and exceedingly porous. The principal uses are floats, rafts, core stock, insulation, cushioning, sound modifiers, models, and novelties.

Balustra. Dense African hardwood.

Bamboo. Strong, smooth, giant grass.

Basswood. Lightweight wood used for core stock. The heartwoodis pale yellowish brown, creamy-white to pale brown sapwood.

Baywood. Lighter in color and softer than Spanish Mahogany. Also called "Honduras Mahogany." Fine markings make it a desirable veneering wood.

Beech (hard wood). Tough, strong, straight-grained wood that looks like birch or maple and works fairly easily. It is white to brown in color and is used for furniture, tool handles, and plywood.

Beechwood. American hardwood, mostly stave sawed for barrel stock for beer aging.

Bella rose. Moderately hard and heavy Phillipine and Malaya wood, pink to yellowish beige in color.

Benge. Hard-textured African wood, rich brown with tan or darker coloring. The wood is fine textured and similar to birch. Suitable uses are veneer and lumber for furniture.

Betulaceae. Family of trees and shrubs that includes the birch and alder.

Birch (hardwood). Wood that is heavy, very strong, fine grained, and hard. It takes an excellent finish and a beautiful polish. The heartwood is brown, sometimes tinged with red; the sapwood is a creamy yellow. It is used for furnish, flooring, and plywood. Birch has a relatively large volume change due to changes in moisture content. Used in making kitchen cabinets and for other cabinet work.

Black American cherry. Close-grained cabinet and veneer wood. The heartwood color ranges from light to medium reddish brown; the sapwood is a light creamy color and is usually selectively eliminated from the veneer and lumber. American cherry resembles red birch but has a more uniform grain and is further characterized by the presence of small dark gum flecks, which, when sound, are not considered as defects, but add to its interest.

Black bean. Native of eastern Australia. Heartwood dark brown, streaked or mottled with black, and the figure depends on the color variation. Straight grained, moder-

ately heavy, and of excellent strength. Suitable uses are for paneling, decorative work, veneers, and furniture.

Black knot. Knot resulting from a dead branch that the wood growth of the tree has surrounded.

Blackwood

1. Wood similar to acacia, native to Australia. Suitable uses are cabinet work and furniture.

2. Heavy, hard, strong wood. The heartwood is exceptionally durable. It is not a framing timber and is used mostly for posts and poles, as well as for insulator pins.

Bleeding. An exudation of resin, gum, creosote, or other substance in lumber.

Blemish. Any defect, scar, or mark that tends to detract from the appearance of wood.

Blond woods. Blond woods include primavera, avodire, aspen, holly, and birch.

Bloom. Crystals formed on the surface of treated wood by exudation and evaporation of the solvent in preservative solutions.

Blue Stain. Bluish or grayish discoloration of the sapwood caused by the growth of moldlike fungi on the surface and in the interior of the piece, made possible by the same conditions that favor the growth of other fungi.

Borer holes. Voids made by wood-boring insects, such as grubs or worms.

Bosse. African wood with a uniform pink-brown color and a cedarlike aroma.

Bossona. Hard Brazilian wood, red-brown in color with black and brown streaks.

Boxed pith. Pith between the four faces on an end of a piece.

Box heart check. Box heart check occurs in timber that has pith in the piece.

Boxwood. Very dense, light-colored wood. Native to the West Indies. Used for border edge work on satinwood furniture.

Branch knots. Two or more divergent knots sawed length-wise and tapering toward the pith at a common point.

Brashness. Condition of wood characterized by low resistance to shock and by abrupt failure across the grain without normal splintering.

Brazilian rosewood. Cabinet and paneling wood. Its name is derived from the fragrance given off when it is cut or burned. The wood is very hard with dense open grain.

Brazilwood. Reddish wood, similar to mahogany.

Bright. Term denoting the absence of weathering effects. When applied to sapwood, it denotes the absence of stain.

Bright sapwood. Lumber which has bright sapwood is not defined as having any defect. Bright sapwood is permitted in each piece in any amount. Shows no stain and is not limited in any grade except as specifically provided.

Brown ash. Not a framing timber, but an attractive trim wood. The heartwood is brown, the sapwood much lighter. Attractive veneers are sliced from stumps and forks.

Brown stain. Dark, often chocolate brown, discoloration found in the sapwood of some softwoods stored under unfavorable seasoning conditions; it is caused by a fungus.

Brush box. Native to New South Wales and Queensland, Australia. Heartwood color grayish pink, figure is established by curly interlocking fibers. Suitable uses are piling, wharf and dock construction.

Bubinga. African wood, purple in color, with a closely striped grain on a pale to red brown background.

Buckeye. Wood is uniform in texture, straight-grained lightweight. Suitable for pulping for paper and used principally for furniture, boxes, and crates.

Burl—dome shaped. Growth on the trunk of the tree just below the ground from which panels, briar pipes, furniture stocks, etc., are manufactured. It has a wild grain configuration.

Butternut. American wood, resembling black walnut in its fine graining. Heartwood is a light brown and the sapwood nearly white.

Cabinet doors. Due to constant use especially on kitchen cabinets, certain manufacturing requirements have been established for the woods to be used. Basic requirements are: freedom from warp and moderate hardness woods, such as maple, oak, birch, and cherry. These woods are also suitable for natural finishes. Woods suitable for paint finish are Douglas fir, southern yellow pine, gum, ponderosa pine, magnolia, and poplar.

Camphorwood. Native to Malay and Borneo. Heartwood color brownish red, generally straight grained with plain figure. The wood is moderately tough with medium strength value. Suitable for construction lumber, flooring, and railroad car framing.

Capirona. Native to most of Latin America and some areas of Cuba. The heartwood ranges from a light brown to gray. The texture is fine and uniform. The wood is difficult to machine because of its density and hardness. It is used in the textile industry for shuttles and picker sticks.

Cativo. Native to Central America and Columbia. The species exhibit an unusual condition; the sapwood is usually very thick as related to the heartwood. The sapwood has a commercial use and the color range is from very pale pinkish to distinctly reddish. Suitable uses are: furniture, cabinet parts, picture frames, and edge banding for doors.

Cedar. Fragrant, fine-grained wood, used for chests and the linings of closets.

Cedar (Tennessee softwood). Soft, very light, close-grained wood lacking strength. The heartwood is light reddish brown; the sapwood, almost white. It has a strong odor that acts as an insect repellent and is used for closet and trunk linings.

Cedar (western red softwood). Largest cedar native to North America. The heartwood color varies from pinkish red to deep, warm brown; the thin sapwood layer is light yellow. The wood is soft, light, and straight grained but lacks strength. It has high resistance to changes in moisture content and retains its size and shape exceptionally well after seasoning. It is used for shingles, shakes, planking, siding, prefabricated building logs, and yard lumber.

Chemical brown stain. Dark brown discoloration normally confined to the heartwood, which may develop during air-seasoning or kiln-drying.

Chen chen. Light North African softwood, whitish to pale gray in color.

Cherry (hardwood). Light and close-grained wood that works easily and takes a good finish. The heartwood is reddish brown, the sapwood is yellowish white. It is used for cabinet work and interior trim.

Cherry wood. American wood, with a light to dark reddish color, fine grained, and resembling mahogany.

Chestnut. Moderately hard, with coarse grain, resembling oak. Most frequently used now as "wormy chestnut" for paneling and trim.

Chestnut (hardwood). Light, open-grained wood that works easily. The heartwood is dark tan; the sapwood, light brown. It is used for cabinet work.

Circassian walnut. European wood, from the Black Sea area, highly figured, produced from warped walnut trees.

Classification of trees. Trees generally are divided into two groups, hardwoods and softwoods. Botanists refer to the former as "deciduous," that is, with broad leaves that are normally shed in the fall. Softwoods are called "conifers;" that is, they have needles rather than leaves and bear their seeds in cones.

Clear. Lumber and timbers selected for lack of imperfections and knotholes, desirable for finished interior beams and paneling for construction.

Clear all heart (redwood grade). Sound, live heartwood with no sap clear face. Two sound, tight pin knots are allowed on the back.

Clear straight-grained unchecked wood. Wood of a given species whose density and moisture content has been found to be fairly uniform in strength. Tests are made on small, clear specimens to secure strength data, and then separate investigations are made of the effect of the more important growth characteristics.

Clear wood. Wood member free from splits, checks, shakes, knots, or other characteristics.

Cleavage. Splitting or dividing along the grain.

Close grain. Lumber with an average of approximately 6 and not more than 30 annual rings per inch on either one end or the other of the piece.

Close-grained wood. Wood with narrow, inconspicuous annual rings. The term is sometimes used to designate wood having small and closely spaced pores, but in this sense the term "fine textured" is more often used.

Coarse grain. Wood with wide and conspicuous annual rings in which there is considerable difference between springwood and summerwood. Designation is also applied to wood with large pores.

Coco wood. Indian wood, hard and brittle, with purplish brown stripes.

Colored knots. A red knot is one that results from a live branch growth in the tree that is intergrown with the surrounding wood; a black knot is one that results from a dead branch that the wood growth of the tree has surrounded.

Comb grain (rift slices or rift sawn). Result of producing veneer by slicing or sawing at an angle of approximately 45° with the annual rings to bring out certain figures produced by the medullary rays that are especially conspicuous in oak.

Compreg. Wood in which the cell walls have been impregnated with synthetic resin and compressed to give it reduced swelling and shrinking characteristics and increased density and strength properties.

Coniferae. Order of cone-bearing resinous trees or shrubs supplying most of the structural lumber to the world markets.

Courbaril. Native to the West Indies, Central and South America. The heartwood varies through the countless shades of brown with an occasional purplish cast. The sapwood is gray and usually deep in area. Suitable for handle stock, veneer, and furniture.

Crook. Edgewise deviation from a straight line drawn from end to end of a piece and measured at the point of greatest distance from the straight line.

Cross band. Place layers of wood with their grain at right angles to minimize warping.

Cross break. Separation of the wood cells across the grain. Such breaks may be due to internal strains resulting from unequal shrinkage or to external forces.

Cross grain. Wood in which the fibers deviate from a line parallel to the sides of the piece. Cross grain may be either diagonal or spiral grain or a combination of the two.

Crotch swirl. Heartwood at the base of the tree where the grain is extremely distorted and irregular.

Crown edge. Timber having a crook or edge warp and placed so that its ends are bearing on a surface and its midpart is raised off the bearing surface forming an arch or crown.

Cup. Deviation in the face of a piece from a straight line drawn from edge to edge of the piece, measured at the point of greatest distance from the straight line.

Cypress. Southern wood, very resistant to weather, belonging to the family Cupressacea, grown mostly in Louisiana and Florida. It is probably the most durable wood for contact with the soil. The wood is moderately light and close grained; the heartwood, red to nearly yellow; the sapwood, nearly white. It does not hold paint well, but otherwise it is desirable for siding and outside trim.

Cypress (softwood). Light, very durable, close-grained, easily worked wood. The sapwood is light brown; the heartwood, bright yellow. It is used for interior and exterior trim and for cabinet work.

Decayed knot. Knot that is softer than the surrounding wood and that contains advanced decay.

Deciduous. Trees that lose their leaves annually.

Defect. Irregularity occurring in or on the wood that may lower its strength or appearance.

Delignification. Removal of part or all of the lignin from wood by a chemical treatment.

Dense-grain. Term used when cut pieces indicate an average of six or more annual rings with an additional one-third or more summerwood on one end or the other of the piece. The color of the summerwood must be distinct and differs from that of the springwood.

Density rule. Rule for estimating the density (mass per unit volume) of wood based on the proportion of summerwood and rate of growth.

Diamonding. Distortion in drying that causes a piece of wood, rectangular in cross section, to become diamond shaped on account of the differences between the radial and the tangential shrinkage.

Dimensional stabilization. Reduction in swelling and shrinkage of wood through special treatment that minimizes the moisture content changes in wood resulting from changes in relative humidity.

Dip-grained wood. Wood that has single waves or undulations of the fibers like those occurring around knots and pitch pockets.

Discoloration. Stick stain mold or mildew, blue stain, or any other abnormal coloring of the wood, except coloration resulting from weathering, pitch deposits, compression wood, and soft rot. Discoloration through exposure to the elements is admitted in all grades of framing and sheating lumber.

Dote. "Dote," "doze," and "rot" are synonymous with decay.

Douglas fir

1. Curly-grained wood that resembles white pine, used for plywood or laminated sheets.

2. Species native to the northwest that accounts for much of the lumber produced in North America. Its production is developed for structural and construction products, but some of its upper grades are used for stock mill work and specialized woodwork. The heartwood is reddish tan; the sapwood, creamy yellow. Growth rings are conspicuous, and a rather bold grain pattern develops when the wood is either plain sawed for lumber or rotary cut, as is common in plywood. Some lumber and veneer is cut edge or vertical grain, producing a superior form of the product since the tendency to grain-raise is greatly reduced.

Durability. General term for permanence or lastingness, frequently used to refer to the degree of resistance of a species or of an individual piece of wood to decay. In this connection "resistance to decay" is a more specific term.

Eastern hemlock. Brittle, moderately weak wood, not at all durable. The heartwood is pale brown to reddish; the sapwood is not distinguishable from the heart. The wood may be badly wind shaken. It is used for cheap, rough framing veneers.

Eastern red cedar (juniper). Wood with a pungent aromatic odor reported to repel moths. It has a red or brown heartwood, and an extremely rot resistant white sapwood. it is used for lining clothes closets and chests and for fence posts.

Eastern spruce

1. Stiff, strong, hard, tough wood, moderately lightweight, with a light color and little difference between the heart- and sapwood. Commercial eastern spruce includes wood from three related species. It is used for pulpwood, framing lumber, mill work, etc.

2. White spruce and black spruce possess about the same properties. The wood stays in place well, is moderately light and easily worked, has moderate shrinkage and strength and is unilight in color.

Ebony. Hard, dense, heavy, fine-grained brown-black Asian wood.

Elm (hardwood). Very heavy, strong hardwood with a light brown color, tinged with red and gray. It is used for tool handles and heavy construction purposes.

Engleman spruce. Straight-grained, lightweight, low-strength wood, white in color with a red tint. It is used for dimension lumber and boards and for pulpwood and has extremely low rot resistance. The softwood has a fine texture. Works easily and takes paint and glue well.

English ash. Native English wood, usually called "olive burl." It is used for furniture.

English brown oak. (Pollard oak). The tree varies in height from 60 to 130 ft, depending on soil conditions. The wood varies in color from a light tan to a deep brown, with occasional black spots. Burls and swirls that are very brittle and fragile are obtained from trees that have had their tops cut out before reaching maturity.

Exterior trim. The basic manufacturing requirements for exterior trim include medium decay resistance, acceptable painting qualities, maximum freedom from warp, excellent weathering characteristics, and very little effort in tooling and working the wood. The best woods for this purpose are cedar, cypress, and redwood.

Feather grain. Results from the slight separation of springwood and summerwood fibers on veneer surfaces.

Fence posts. The basic manufacturing requirements for exterior fence posts include: high decay resistance, good bending strength, straightness in the entire length of the post, and good nailing or staple holding qualities. Permanent installations requires a preservative treatment for the particular wood used. The best woods are black locus and osage orange, but popular woods are cedar and cypress.

Fiber saturation point. Point in the drying or wetting of wood at which the cell walls are saturated but the cell cavities are free from water.

Fiddleback-grained wood. Figure produced by a type of fine wavy grain found in species of maple.

Fir. Belonging to order of the Coniferae and the family Abies. The true firs are the white fir and Douglas fir.

Fir (Douglas—Softwood). Hard, strong, durable wood. The heartwood color ranges from light red to yellow; the sapwood is whitish. It is used for all types of construction and in the manufacture of plywood.

Firm knot. Knot that is solid across its face but may contain incipient decay.

Firm red heart. Form of incipient decay characterized by a reddish color produced in the heartwood, which does not render the wood unfit for the majority of uses. Firm red heart contains none of the pockets that characterize the more advanced stage of decay (not to be confused with natural red heart).

Fixed knot. Knot that will retain its place in dry lumber under ordinary conditions, but that can be moved under pressure though not easily pushed out.

Foundation plates and floor sleepers. General construction requirements of woods used include the following qualities: good natural decay resistance, or treated with an acceptable preservative for the particular wood used, low moisture content to prevent future dry-out or warp, and good nail-holding qualities. The best wood is white oak, but normally acceptable woods include: Douglas fir, western larch, southern yellow pine, redwood, and other medium-density species for normal conditions. Woods are to be treated when used in damp locations.

Foundation sills. Cedar and redwood are usually used for sills only, as these species do not have the high bending strength desirable for beams. The heartwood has high decay resistance.

Foundation sills and beams. Douglas fir, western larch, southern yellow pine, and rock elm, which are high in strength and nail-holding qualities; are used for sills and beams in basement or dry areas. Under moist conditions they require preservative treatment.

Framing joists, rafters, and headers. Wood used for these purposes based on good construction practices include the following qualities: freedom from pronounced warp, high stiffness, good bending strength, good nail-holding, and limited amount of moisture as established for each type of wood used. The best woods are: Douglas fir, western larch, southern yellow pine, and other woods providing the necessary design stresses.

Framing studs and plates. Wood used for these purposes based on good construction practices include the following qualities: freedom from pronounced warp, moderate stiffness (studs used for bearing partitions should have high stiffness qualities), good nail-holding, and limited amount of moisture content as established for the particular type fo wood being used. The best woods are: Douglas fir, western larch, southern pine, and other woods providing the same qualities.

Free of heart centers. Without pith (side cut). When a piece has been sawed so as to eliminate the pith (heart center), an occasional piece showing pith on the surface for not more than one-fourth the length may be accepted.

French burl. Persian walnut wood with a curly grain.

Fruitwood trees. Fruitwood trees include the cherry, apple, pear, and the like.

Fustic. Light yellow wood from the West Indies, primarily used for marquetry and inlay work.

Gaboon ebony. Rust, brown, or black African wood.

Green timber. Green is the condition of timber as taken from a living tree. Immediately upon being sawed from the tree, lumber begins to lose moisture and otherwise change its condition.

Gum (hardwood). Soft-textured, durable, close-grained wood with color similar to walnut, reddish brown heart wood, and yellowish white sapwood. The wood has a tendency to warp but works easily. It is used for furniture, sometimes as a substitute for walnut.

Gum vein. Local accumulation of gum found in certain hardwoods and occurring in the form of a vein.

Gumwood. Heavy, strong textured wood pink to reddish brown in color.

Hackberry. American wood that resembles elm.

Hard maple. Wood that is very similar in general characteristics to yellow birch. It is heavy, hard, strong, and resistant to shock and abrasion. The heartwood is reddish brown, and the sapwood near white with a slight reddish brown tinge. A natural characteristic is the prevalence of dark mineral streaks (predominantly in the heartwood), which can be minimized in the sapwood by selective cutting.

Hardness. Hardness represents the resistance of wood to wear and marring. Values are presented for end-grain and side-grain surfaces (average of radial and tangential values). It is measured by the load required to imbed a 0.444-in. ball to one-half its diameter in the wood.

Hardwoods. Woods from leaf-bearing trees, such as cherry, oak, maple, mahogany, and walnut, as opposed to that from cone-bearing trees or evergreens.

Harewood. English sycamore, a creamy white wood.

Heart. Portion of the tree contained within the sapwood. Term is sometimes used to mean the pith.

Heart center. Pith or center core of the log.

Heart center decay. Localized decay developing along the pith in some species, readily identifiable and easily detected by visual inspection. Heart center decay develops in the living tree and does not progress further after the tree is cut.

Heart check. Check starting near the pith and extending toward but not to the surface of a piece. Several occurring together are called a "star check."

Heart shake. Heart shake occurs when the pith or center of the log has grain separation, causing a weakening of its structural property. The condition is also called "star shake."

Heartwood and sapwood. Heartwood and sapwood of equivalent character equal strength. No requirement of heartwood need be made when strength alone is the governing factor. Heartwood is more durable than sapwood, and for wood that is to be exposed to decay-producing conditions without preservative treatment, the minimum percentage of heartwood to be present in all pieces in a shipment of any species may be specified.

Heartwood of a durable species. Includes the heartwood of tidewater red cypress, cedar, or redwood or other approved decay-resistant wood.

Heavy stain. Darkest color that develops in lumber.

Heavy-stained sapwood. Sapwood with a pronounced difference in color so that the grain may be obscured; however, the lumber containing it is acceptable for paint finishes.

Hemlock. Wood that resembles white pine and is strong, lightweight, and easy to work. Straight-grained softwood with a fine texture that machines to a smooth satin surface. It is rated very strong in relation to its weight and is good for all construction purposes. The color ranges from light to medium-dark brown.

Hemlock (western softwood). Fine-textured, straight-grained, easily worked wood that is free from pitch and possesses exceptionally good quality for general building purposes when interchanged with fir.

Hickory

1. Hard, tough, heavy wood of the walnut family.
2. Hickory possesses a combination of hardness, weight, toughness, and strength found in no other native wood. It is a specialty wood, almost impossible to nail when dry. However it is not rot resistant. Several related species are marketed together.

Holes. Term is applied to holes from any cause. When knotholes are referred to, they are identified as such. Holes that extend only partially through the piece are classed as surface pits.

Hollow knot. Apparently sound knot in all respects except that it contains a hole over $\frac{1}{4}$ in. in diameter. A through opening a hollow knot may be of a size equal to other permitted holes.

Holly. Native American, found in range that extends along the Atlantic coast, Gulf Coast, and the Mississippi valley. Both heartwood and sapwood are white. The wood has a uniform and compact texture, moderately low in strength when used as a beam or a column. Suitable for use as woods for scientific and musical instruments.

Honduras cedar. Native to British Honduras. Heartwood light red, usually straight grained figure, but frequent wavy, curly, or mottled-grained. Suitable uses are: furniture, veneers, cabinetwork, and finish work for yachts and racing boats.

Honduras mahogany. Mahogany species with open-grained wood and a heartwood color ranging from light tan to a rich golden brown, depending to some extent on the country of its origin. Its outstanding stability and decay resistance expands its potential to include exterior applications for monumental projects. It is generally available as plain sawed lumber and plain spliced veneer with different veneer cuts.

Honeylocust. Native to the eastern United States, except for New England, the south Atlantic, and Gulf coastal plains. The heartwood color is light red to reddish brown and the sapwood is yellowish in color. It is very heavy, very hard, strong in bending, and stiff. Suitable for use as lumber for general construction work, also pallets and crating.

Idaho white pine. Straight-grained, soft, and even wood, ranging in color from pale cream to reddish brown. The wood is light, but strong, and an excellent, highly permanent base for paints and gluing.

Ilomba. Native to the rain forests and ranges of Guinea and Sierra Leone, west topical Africa to Uganda and Angola. The wood is grayish white to pinkish brown. There is no distinction between heartwood and sapwood. The wood has only been utilized for plywood.

Imbuya. Brazilian walnut, with color that varies from olive to a rich red brown.

Impreg. Synthetic resin-treated wood made so as to reduce materially its swelling and shrinking.

Incense cedar. Nonresinous softwood with a fine texture and color ranging from light cream to reddish brown. It is easily machined, suitable for residential and light commercial construction, highly resistant to decay and is permanent.

Interior trim for natural finishes. The basic manufacturing requirements for wood to be machined and worked into uses for trimming include: freedom from warp, pleasing texture and grain, hardness, and free of all knots. The best woods for natural finishing include: oak, birch, maple, cherry, beech, sycamore, and walnut. High-class hardwood interior trim is usually first grade finish (A grade). Special grades of knotty pine, pecky cypress, and others are available to meet special architectural design requirements.

Interior trim for paint finishes. The basic manufacturing requirements for wood to be machined and worked into uses for trimming include: freedom from warp and shrinkage, fine uniform texture and smoothness, moderate hardness, absence of knots and dicoloring pitch pockets, finish nail-holding ability, and a surface that will take a paint finish. The best woods for this purpose include: northern and Idaho white pine, ponderosa pine, sugar pine, and poplar.

Internal sapwood. Zone of wood within the heartwood that retains the light color of the sapwood.

Ironwood. Unusual hardwood olive wood obtained from South America.

Jack pine. Pine that occurs in the Great Lakes states, particularly on sandy soil where it reseeds readily. The wood is of moderate strength, inclined to be coarse, and ranging from white to orange in color. It is used for pulp, ties, and boxes and in boards and dimensions for light frame construction.

Jarrah. Native to the coastal belt of southwestern Australia. The heartwood is uniform pinkish to dark red having in some species the rich dark red mahogany hue. Suitable for use in construction work such as decking, underframing of piers, jetties, and bridges.

Jelutong. Native to the Malaya territory. The wood is white or straw colored and there is no difference between heartwood and sapwood. Texture is moderately fine and even and the grain is straight. It is suitable as an excellent core stock.

Juniper. Small treelike shrub of the Cupressaceae family. Of little or no commercial value other than ornamental.

Kapur. Native to parts of Malaya, Sumatra, and Borneo. The heartwood is a light reddish brown, clearly demarcated from the pale colored sapwood. Wood is coarse textured but uniform. Principal use is lumber for construction purposes.

Karri. Native of western Australia. Heartwood is rated as moderately durable, heavy hardwood possessing mechanical properties. Practical uses is lumber in heavy construction work.

Khaya. Native of west central Africa. Heartwood varies from a pale pink to a dark reddish brown. The grain and the texture is equal to that of mahogany. Principal use is for furniture and veneers.

Kiln drying. Process of drying wood products in closed chamber where the temperature and relative humidity of the circulated air can be controlled.

Knot cluster. Two or more knots grouped together as a unit with the fibers of the wood deflected around the entire unit.

Knots. Formed when the part of a branch or limb embedded in a tree is cut through. They reduce bending strength of beams, having greater influence when near the top or bottom face in the center portion of the span or in any portion where bending moment is large. Knots near the center of the face parallel to the direction of load or near the ends have little or no effect.

Knotty pine. Pinewood showing knots or dark oval shapes.

Kokrodua. Native of West Africa. The heartwood is brownish yellow and darker streaks. The wood is fine textured with straight to interlocked grain. Used for the same purposes that teak is used for.

Lapacho. Native to most Latin American countries. The heartwood is light to dark olive brown, and the sapwood yellowish gray to gray brown. Texture is fine and the grain is closely and narrowly interlocked. The wood is very heavy and strong. Suited for heavy-duty flooring in trucks and box cars.

Larch

1. Softwood species with fine, uniform, and straight grain and distinctive color ranging from pale yellow to light brown. Larch is one of the heaviest of softwoods and one of the strongest. It is very permanent and glues well.

2. Fairly hard and strong wood, resembling Douglas fir more closely than any other softwood. The heartwood is deep reddish brown; the sapwood is much lighter. Suitable for lumber used for construction purposes.

Large knot. Knot more than $1\frac{1}{2}$ in. in diameter.

Larch pitch picket. Picket over $\frac{3}{8}$ in. in width and over 4 in. in length or over $\frac{1}{3}$ in. in width and over 8 in. in length.

Lauan. Reddish brown Philippine hardwood that looks like mahogany.

Laurel wood. East Indian walnut, gray or brown in color, coarse grained, hard, and brittle.

Light-stained sapwood. Sapwood with a discoloration slight that it does not materially affect natural finishes.

Lignin. Principal constituent of wood, second only to cellulose. It encrusts the cell walls and cements the cells together.

Limba. Native of west central Africa. Wood varies from grayish white to creamy brown. Wood is generally straight grained, uniform with a coarse texture. Principal uses include: trim, paneling, and plywood.

Long-leaf southern yellow pine. Term designates not a species, but a grade. All southern yellow pine that has six or more annual rings per inch is marketed as long leaf, and it may contain lumber from any of the several species of southern pine. Wood so designated is heavy, hard, and strong, but not especially durable in contact with the soil. The sapwood takes creosote well. It is useful timber for light framing.

Long split. Split that is longer than a medium split.

Lumber. Wood that is cut from logs and milled into boards, planks, and timber.

Macassar ebony. Very hard, dense wood with a black-brown stripe on reddish background.

Magnolia. Straight-grained, uniform-textured wood, similar to yellow poplar.

Mahogany

1. Highly prized southern hardwood used in paneling, furniture manufacture, etc. It finishes to a blonde or beautiful red-brown luster.

2. Strong, open-pored, durable wood. It works easily and takes a very good finish. Its color is reddish brown. Interior-quality mahogany contains small, gray specks; it is used for furniture, plywood, and interior trim.

Maple. Hard, strong, light-colored wood, similar to birch.

Maple (silver hardwood). Wood that is soft, light, and easily worked, but not very strong or durable. It is used most frequently for wood turning and for interior trim.

Maple (sugar hardwood). One of the hardest and strongest woods in America. It is very close grained and takes an excellent finish but is difficult to work. This maple is light yellow to brown in color. It is used for flooring and furniture.

Medium checks. Checks that are not over $\frac{1}{32}$ in. wide and not over 10 in. long.

Medium grain. Average of four or more annual rings per inch on either one end or the other of the piece, using a radial-line basis of measurement.

Medium knot. Knot over $\frac{3}{4}$ in. but not more than $1\frac{1}{2}$ in. in diameter.

Medium pitch pocket. Pocket not over $\frac{3}{8}$ in. in width and 4 in. in length or not over $\frac{1}{8}$ in. in width and not over 8 in. in length.

Medium roller check. Perceptible opening over 2 in. long but exceeding 4 in. in length.

Medium split. Split equal in length to twice the width of the piece and that in no case exceeds one-sixteenth the length.

Medium stain. Pronounced discoloration.

Medium-stained sapwood. Sapwood with a pronounced difference in coloring that sometimes affects its usefulness for natural but not for painted finishes.

Merulius Lacrymans. Fungus that commonly occasions dry rot.

Mineral streaks. Slight discolorations inherent in certain species.

Moist condition use. Poplar, eastern and West Coast hemlock, and red oak are used for moist conditions. Wood requires good preservative treatment if exposed to moise conditions for long periods of high humidity.

Moisture content. Weight of water contained in wood expressed as a percentage of the weight of the oven dry wood. The moisture content given for "green" wood is the average for specimens taken from the pith to the circumference of the log. It includes the moisture found in both the heartwood and sapwood and is the approximate moisture content of the living tree. The moisture content of seasoned wood varies somewhat among the different species. To facilitate comparison of the strength properties, the test values have been adjusted to conform to a uniform moisture content of 12%.

Moisture equilibrium. The point at which the moisture content of wood is in balance with the humidity of the surrounding air.

Moisture gradient. Graduated moisture content between the inner and outer portions of the wood caused by the loss or absorption of moisture.

Mold or mildew. Superficial fungus growth, usually appearing in the form of a woolly or fuzzy coating of varying color.

Myrtle burl. Highly figured, very blond to golden brown wood. Native to western United States. Principal uses include wood for cabinetwork, inlay work, and veneering.

Narrawood. Philippine hardwood resembles mahogany.

Nogal. Native of northern Central America identified as tropical walnut. The wood is darker than American black walnut. Practical use has been wood for veneers.

Northern white cedar. Wood with light brown heart and a thin, nearly white sapwood. It is light, weak, soft, and decay resistant and holds paint well. It is marketed with Atlantic white cedar.

Northern white fir. Only true fir native to the Great Lakes states. Its wood is soft and coarse, and it is a preferred pulp species. This wood is also used for boxes and for light framing material.

Northern white pine. Variety known also as "genuine white pin," "white pine" or "cork white pine." The wood combines a variety of excellent characteristics that render it superior for many industrial and construction purposes. The wood is light in weight and of fine, uniform soft texture, shrinks and swells little, and is not inclined to distort or split. It is used for window sash, blinds, doors, etc.

Norway pine. Tree sometimes called "red pine" owing to the distinctive red color of the bark. The wood is heavy, hard, and strong and used almost entirely for structural purposes.

Oak. Hard, durable wood, red or whitish in color.

Oak (red hardwood). Variety a little softer than white oak, with coarser grain. The color varies from tan to reddish brown. It is used for furniture, plywood, bearing posts, implement parts and interior trim.

Oak (white hardwood). Very heavy and hard wood, close grained with open pores. It takes an excellent finish, but is difficult to work. It has a tan heartwood with light sapwood and is used for flooring, furniture, plywood, and interior trim.

Obeche. Native to west central Africa. The wood is creamy white to pale yellow with little difference between the heartwood and the sapwood. Fairly soft wood of uniform texture. Suitable for wood in connection with core stock and veneers.

Okoume. Native to west central Africa. The wood has a salmon-pink color with a uniform texture. Texture is slightly coarser than that of birch. There is no difference between the heartwood and the sapwood.

Olive wood. Small Italian tree, light yellow, with greenish yellow figures.

Open defects in wood. Include but are not limited to checks, knots, wormholes, or other defects interrupting the smooth continuity of the surface.

Open-grained wood. Common classification used by painters for wood with large pores, such as oak, ash, chestnut, and walnut. Such wood is also known as "coarse textured."

Open knot in wood. Opening where a portion of the wood substance of the knot has dropped out, or where cross-checks have occurred to present an opening.

Oval knot. Knot cut at slightly more than right angles to the length of the knot (limb).

Pacific Coast Cypress. Variety also known as "Port Orford" or "Alaska Yellow cedar," grown on the Oregon coast and areas in British Columbia.

Paneling lumber. General requirements for natural-finish or lightly stained paneling lumber are pleasing grain, figure or surface treatment, freedom from warp and shrinkage, and some resistance to abrasion. These qualities are obtained in oak, redwood, cypress (pecky), walnut, cedar (knotty), ash, birch, pine (knotty), and cherry or gum, western larch, Douglas fir, beech, southern yellow pine, hemlock, and ponderosa pine. The best grade hardwood for high-quality houses is first grade. Softwood first or second grades are commonly used.

Paneling plywood. Oak, birch, maple, pecan, hickory, and walnut or cedar, pine, Douglas fir, and southern yellow pine, as well as some imported species are used for the manufacture of plywood paneling. Some woods are specially treated to create a variation in the grain for unique surface effects.

Particle boards. Boards produced by gluing small particles of wood together into a panel. Hot-setting resins produce the bond necessary to give the panels form, stiffness, and strength. Particle board is generally classified as low density when it has a density of less than 37 lb/cu ft, medium density when the density is between 37 and 50 lb/cu ft, and high density when the board weight is more than 50 lb/cu ft.

Partridge wood. Brazilian wood with graining resembling partridge feathers.

Pearwood. Pinkish brown, finely grained wood, usually stained to simulate ebony in inlay work.

Pecan. American wood of the hickory family resembling walnut.

Pine (lodge pole softwood). Soft, straight-grained wood with a fine, uniform texture. The wood is light in color with a sapwood that is almost white. It seasons easily and uniformly and is of moderate durability. It is used for lumber and for mine props.

Pine (ocote). Native to Mexico, Guatemala, Nicaragua, and the Honduras. The heartwood is a light reddish brown and the sapwood is a pale yellowish brown. Strength properties comparable to longleaf pine. Practical use is for construction lumber.

Pine (ponderosa). Species that, while not a true pine botanically speaking, is the softwood most commonly used for exterior and interior woodwork components. Its heartwood is tannish pink, and its sapwood a lighter creamy pink. Also, like true pines, the proportion of sapwood is high, and the heartwood has only a moderate natural decay resistance. Its receptivity to preservative treatment is high, and like all pines, it should be treated when used on the exterior.

Pine (sugar softwood). Variety that is very soft, light, and uniform in texture. It works very easily and takes an excellent finish. In the quarter-sawed boards, reddish brown specks are easily recognized. The heartwood is reddish brown, the sapwood light yellow. It is used for interior trim and inexpensive furniture.

Pine (white softwood). Soft, light, durable wood with a creamy white heartwood. It works very easily and is used for furniture.

Pine (yellow or southern softwood). Hard, tough, strong variety with an orange red heartwood and a lighter sapwood. It is very difficult to work and is used for heavy structural purposes, such as floor planks, beams, and timbers.

Pitch. Accumulation of resin in wood. This accumulation may be in the form of pitch pockets, pitch streaks and pitch seams.

Pitch streak. Well-defined accumulation of pitch at one point in the piece.

Pith knot. Sound knot except that it contains a pith hole not more than $\frac{1}{4}$ in. in diameter.

Pith shake (heart shake or heart check). Shake that extends through the growth rings from or through with pith toward the surface of a piece. A pith shake can be distinguished from a season check by the fact that its greatest width is nearest the pith, whereas the greatest width of a season check in a pith-centered piece is farthest from the pith.

Plumwood. Dark-red wood, resembles mahogany.

Plywood. Glued panel made up of layers of veneer (thin sheets of wood) with the grain of adjacent layers at right angles to each other. Two common types are interior and exterior and these names designate their recommended uses.

Pole house construction. Requirements for load-bearing poles and posts used in construction where the butt end of the pole is usually embedded in the soil are high stiffness and strength, freedom from crook, minimum taper, good nail holding, and good decay resistance. Poles used in permanent construction are pressure treated. The best woods for this purpose are western larch, Douglas fir, West Coast hemlock, and southern yellow pine, all preservative treated. Class of pole required in a building is usually determined by the circumference at the top. The lengths vary and include distance above ground, as well as embedment depth, depending on soil conditions.

Pollard oak. English brown oak, nut brown to deep brown in color with a medium hard texture.

Poplar (hardwood). Wood with a light, soft, uniform texture that works easily. It varies in color from yellow to white. It is used for crates, plywood, and occasional pieces of inexpensive furniture.

Pores. Openings on the surface of a piece of wood; openings result when vessels in the wood are severed during sawing.

Porosity. The ratio of the volume of a material's pores to that of its solid content.

Raised grain. Unevenness between springwood and summerwood on the surface of dressed lumber. Slightly raised grain is an unevenness somewhat less than $\frac{1}{64}$ in.

Ramin. Native of southeast Asia, and in particular, from the Malaya peninsula to Sumatra and Borneo. The wood is uniform pale straw or yellowish to whitish in color. The texture is similar to mahogany. Suitable for use in plywood for doors and in solid form for interior trim.

Random grain. Combination of vertical and flat grain.

Rate of growth. Rate at which a tree has laid on wood, measured radially in the trunk or in the lumber cut from the trunk. The unit of measure in use is the number of annual growth rings per inch.

Rays. Strips of cells extending radially within a tree and varying in height from a few cells in some species to 4 in. or more in oak. Rays serve primarily to store food and transport it horizontally in the tree.

Red alder. American hardwood that looks like maple and can be stained to look like mahogany or walnut. Principal use is for furniture, but also used for sash, doors, panel stock, and millwork.

Red cedar. Very aromatic, uniform, soft, medium-brown wood with a straight grain.

Red gum. Wood with a moderately heavy, interlocking grain that warps badly in seasoning. The heart is reddish brown, the sapwood nearly white. The sapwood may be graded out and sold as white gum, the heartwood as red gum, or they may be sold together as unselected gum. It cuts into attractive veneers.

Red knot. Knot that results from a live branch growth in the tree and that is firmly grown into the wood structure.

Red oak. Domestic hardwood possessing strength, wearability, and appealing grain characteristics. It is open grained and when plain sawed or sliced expresses a very strong "cathedral" type grain pattern. The heartwood is reddish tan to brown and very uniform in color; the sapwood is lighter in color and minimal in volume, making its elimination by selective cutting very easy.

Red or Norway pine. Pine resembling the lighter weight specimens of southern yellow pine. It is moderately strong and stiff with a moderately soft heartwood, pale red to reddish brown. It is used for millwork, siding, framing, and ladder rails.

Redwood. Variety that grows in northern California coastal areas. The logs ensure a large volume of soft, close-grained clear lumber, which can be produced in any reasonable length and width. The wood is easily worked, is comparatively free from warping and shrinkage, and has a beauty of grain and texture that makes it suitable for a wide range of decorative purposes, as well as for general-purpose lumber. It makes the most durable and rot-resistant timbers. It has a light, soft, moderately high-strength reddish brown heartwood and a white sapwood. It does not paint exceptionally well, as the paint often "bleeds" through.

Redwood (softwood). Light, easily worked, coarse-grained dull-red wood, lacking strength. It is used for interior and exterior trim, planking, and shingles.

Redwood, flat grain (heartwood). The heartwood is a fairly uniform brownish red color, while the very limited sapwood is lemon colored. In the plain sawed form a medium "cathedral"-type figure develops, while in the illustrated vertical grain a longitudinal striped figure results.

Resin ducts. Intercellular passages that contain and transmit resinous materials.

Resin passages. Intercellular passages or ducts that contain and transmit resinous materials. On a cut surface they are usually inconspicuous. They may extend vertically parallel to the axis of the tree or at right angles to the axis and parallel to the rays.

Ring—annual growth. Growth layer put on in a single growth year.

Rings. Rings are those circular markings around the center of a tree section produced by the contrast in density, hardness, color, etc., between springwood and summerwood. One ring, known as an annual ring, consists of a layer of springwood and a layer of summerwood.

Roof sheathing plywood. Roof sheathing plywood requires adequate stiffness for span and roof loading. Sheathing-grade plywoods are classified into groups by density, hence strength and stiffness. Each grouping sets the distance between supports for proper application and performance. Groups 1 and 2 include such softwoods as Douglas fir, southern yellow pine, and western larch. The standard interior grade (C-D) is used under ordinary conditions; for unusually damp conditions the standard interior grade with exterior glue is used.

Rosewood (Indian). Native of India. The heartwood is a dark purplish brown with denser blackish streaks. The texture is uniform and moderately coarse, and similar in appearance to that of Brazilian and Honduras rosewood. Essentially used for high-grade furniture and cabinet work.

Rosin. A hard resin used in making certain varnishes.

Round knot. Knot cut at approximately a right angle to its long axis so that the exposed section is round or oval.

Round timbers. Timbers used in the original round form.

Salicaceae. Family of trees that includes willows, poplars, cottonwoods, and aspens.

Sande. Native to Ecuador and Columbia. The heartwood and the sapwood show no distinction, being a uniform yellowish white to yellowish brown to light brown. Used as a general utility wood in lumber and plywood.

Satinwood. Highly figured, close-grained, hard, durable wood, light yellow to golden brown in color.

Shakes. Defects originating in the living tree as a result of frost, wind, or other causes or occurring through injury in felling, which later show in the manufactured lumber, most commonly as partial or complete separation between growth rings.

Shelving. Shelving requirements are stiffness and freedom from warp. Lumber or plywood suitable for natural finishes comes from the ash, birch, maple, oak, and walnut. Douglas fir, popular, southern yellow pine, redwood, ponderosa pine, sugar pine, and Idaho white pine are suitable for paint finishes.

Side cut. Cut so that the pith is not enclosed within the four sides of a piece.

Silver grain. Condition where there are conspicuous medullary rays in quartersawed wood.

Single knot. Knot occurring by itself, with the fibers of the wood deflected around it.

Sitka. Light, soft, medium-strong wood with light reddish brown heartwood and nearly white sapwood shading into the heartwood. It is usually cut into boards, planing-mill stock, and boat lumber.

Slight satin. Light-colored, barely perceptible stain.

Slope of grain. Deviation of the line of fibers from a straight line parallel to the sides of the piece.

Small checks. Checks not over $\frac{1}{32}$ in. wide or 4 in. long.

Small knot. Knot over $\frac{1}{2}$ but not more than $\frac{3}{4}$ in. in diameter.

Small pitch pocket. Pocket not over $\frac{1}{8}$ in. in width or 4 in. in length, or not over $\frac{1}{4}$ in. in width or 2 in. in length.

Soft maple. Softer and lighter wood than hard maple, with a lighter color. Box elder is sometimes marketed with soft maple. It is used for much the same purposes as hard maple but is not nearly so desirable.

Softwoods. Botanical group of trees with needle like or scalelike leaves, often referred to as conifers. The term "softwood" has no reference to the softness of the wood.

Sound wood. Wood free from any form of decay, incipient or advanced.

Southern yellow pine. Pine, commonly called "short leaf pine," with a yellowish, noticeably grained, moderately hard, strong, and stiff wood. One cubic foot of air-dried southern yellow pine weighs 36 to 39 lb. It is used extensively in house building for framing, ceilings, weatherboarding, panels, window and door frames, casings, and carved work. The grain shows well in natural finish or when stained. Frames of overstuffed furniture, chairs,

desks, agricultural machinery, wood pulp, mine props, barrels, and crates are also made of this pine.

Spruce (Engelmann softwood). Straight-grained wood, light in color, that works easily and takes a good finish. The trees are usually larger than white spruce and produce a larger percentage of lumber clear of defects. It is used for oars, paddles, sounding boards for musical instruments, and construction work.

Spruce (Sitka softwood). The trees grow very tall with straight grain. The wood is creamy white, sometimes with a pinkish tinge. It is tough, strong, and resists splintering and shattering well. It is used in airframe construction, masts and spars, scaffolding, and general construction.

Spruce (white softwood). Lightweight, softwood that works very easily. The color varies from white to pale yellow with little contrast between heartwood and sapwood. It is a medium-strength softwood, but not very durable in situations favorable to decay. It is used as light construction material and is of particular value as paper pulp.

Straight-grained wood. Wood in which the fibers run parallel or nearly parallel to the edges of the piece.

Stained heartwood and firm red heart. Stained heartwood and firm red heart are characterized by marked variation from the natural color; the color may range from pink to brown. Firm red heart should not be confused with natural red heart. Natural color is usually uniformly distributed throughout certain annual rings, whereas stains usually occur in irregular patches. In grades where stained heartwood or firm red heart is permitted, it has no more effect on the intended use of the piece than other characteristics permitted in the grade.

Stained sapwood. The stain has no effect on the intended use of the pieces in which it is permitted but affects appearance in varying degrees.

Star-checked knot. Knot having radial seasoning checks, usually running from the pitch toward the edge of knot.

Structural insulating board wall sheathing (house). Lumber used as insulating board wall sheathing must have good resistance to water, to nailhead pull-through, and to racking if properly attached. It is applied vertically in sheets 4×8 ft or longer with perimeter nailing. Regular-density structural insulating board (about 18 lb/cu ft in density) is furnished in 2- or 4-ft widths and when applied with long edges horizontal do not provide necessary resistance to racking forces of wind or earthquake, so other bracing must be provided.

Structural insulating roof deck. Structural insulating roof decks are shop fabricated products. The types vary by type of materials, thicknesses, which are usually $1\frac{1}{2}$, 2, and 3 in., depending on span and insulation requirements, surface treatment, and vapor retarder needs.

Subfloor plywood. General requirements for subfloor plywood are moderate stiffness when finish is strip flooring and high stiffness for wood block or resilient finish flooring, and good nail-holding qualities. Suitable woods include group 1 and 2 softwoods such as Douglas fir, southern yellow pine, and western larch.

Sugar pine. Straight grained softwood with a distinctive light creamy color. It is relatively light and not considered a structural wood. It nails without splitting, glues well, is highly workable to a beautiful and smooth surface, and weathers well.

Summerwood. Denser layer of wood in the annual rings of a tree, which is put on in summer or the latter part of the growing season.

Surface check. Check occurring on one surface of the piece but not extending through the piece.

Surface shake. Lengthwise separation of the wood fibers, nowhere extending through from one face of the piece to the other.

Swirls. Irregular grain in wood usually surrounding knots or crotches.

Tamarack or larch. Small- to medium-sized trees. The wood is not usually used for framing lumber but is cut into boards, posts, and poles. The heartwood is yellowish brown, the sapwood white.

Teak. Teak is a wood with extensive figure variations and is available in both lumber and veneered products. Adding to its appeal is its distinctive tawny yellow to dark-brown color, often with light and dark accent streaks. It is most appealing in plain sawed or sliced cuts. Teak has unique stability and weathering properties, making it ideal for exterior applications. It is also used for decorative interior woodwork, most often in veneer form. The wood is available in commercial quantities in India, Burma, Thailand, Laos, Cambodia, Vietnam, and generally in the East Indies.

Timber. Broad term including standing trees and certain products cut from them, including lumber 5 in. or larger in the least nominal dimension.

Texture. Distribution and relative size of wood cells as in coarse texture, fine texture, even texture, and close texture.

Through check. Check extending through the piece from one surface to an opposite or adjoining surface.

Through shake. Shake extending from one surface of a piece to an opposite or adjoining surface.

Twist. Flatwise or a combined flatwise and edgewise deviation in the form of a curl or spiral; the amount is the distance an edge of a piece at one end is raised above a flat surface against which both edges at the opposite end are resting snugly.

Unsound knot. Knot that usually contains decay.

Wall sheathing plywood. General requirements for wall sheathing plywood are good nail-holding qualities, workability, and resistance to racking. Suitable species are Douglas fir, southern pine, and western larch. Most species used for plywood can be employed with satisfactory results. For exterior finishes, such as shingles or shakes, the thickness of softer plywoods should be increased to obtain greater nail penetration. The standard interior grade (C-D) is used under ordinary conditions; the standard interior grade with exterior glue is used in unusually damp locations.

Walnut (hardwood). Very hard, strong, durable, easily worked wood that takes an excellent finish. It is used for fine cabinet work, furniture, flooring, plywood, and interior trim. It has a reddish brown to dark-brown heartwood with lighter sapwood.

Warp. Deviation from a true or plane surface, including bow, crook, cup, and twist or any combination thereof. Warp restrictions are based on the average form of warp as it ccours normally, and any variation from this average form, such as short kinks, is appraised according to its equivalent effect.

Well-scattered knots. Knots that do not occur in clusters. Each knot is separated from any other by a distance at least equal to the diameter of the smaller of the two.

Well-spaced knots. The sum of the sizes of all knots in any 6 in. length of a piece must not exceed twice the size of the largest knot permitted. More than one knot of maximum permissible size must not be in same 6-in. length, and the combination of knots must not be serious.

Western hemlock. Moderately strong wood that is not durable and is used mostly for pulpwood. Its heartwood and sapwood are almost white with a purplish tinge. Some are marketed mixed with Douglas fir.

Western red cedar. Species possessing all the desirable qualities of the other cedars and additional advantages besides. It is exceptionally light in weight and soft textured and has a close, even grain. It is easily worked, finishes to a very smooth silky surface, is free from pitch, and has superior gluing qualities. The coloring is particularly attractive, varying from the almost pure white of the sapwood to the light straw shade or dark reddish brown of the heartwood. It is also called "canoe cedar" or "shinglewood." It has small shrinkage and holds paint well. The heart is light brown and extremely rot resistant; the nearly white sapwood is quite narrow.

Western white pine. Pine possessing creamy or light-brown sapwood; sapwood is thick and white. The moderately light, moderately strong, easy to work wood holds paint well. It is used mostly for mill work and siding. It is also called "Idaho white pine."

White ash. Cream to light brown heartwood with thick light colored sapwood. Grain is similar to oak. The lumber is very strong for its weight and finishes well with hand or machine tools. Used extensively for paneling, furniture, cabinet work and handles for many tools.

White cedar. Light brown heartwood and nearly white sapwood. Heartwood is highly resistant to decay. The sapwood is low in strength, but is easily worked and holds paint very well. Lumber is used extensively for poles, ties, posts, and decorative fencing.

White fir. Softwood that is nonresinous and fine textured with uniformly sized cells and a very narrow summerwood band. The springwood is flat white. The summerwood has a slight reddish brown tinge. The wood is lightweight with an excellent weight-strength ratio. It takes and holds paint well.

White oak

1. Several species are marketed together, but the woods are practically identical. White oak is hard, heavy, tough, strong, and somewhat rot resistant. It has a brownish heart with a lighter sapwood. It is desirable for trim and flooring and is one of the hardwood framing timbers.

2. White oak is used for sills and beams in crawl spaces. The heartwood has high decay resistance; also used for outside doorsills and interior thresholds.

3. Hard, strong wood with a heartwood that has good weathering characteristics, making its use for selected exterior applications appropriate. It is open grained and in its plain sawed form is highly figured. The heartwood varies considerably in color from light grayish tan to brown, making the maintenance of color consistency difficult. The sapwood is much lighter in color and fairly prevalent and its elimination is accomplished by selective ripping.

White oak (plain sawn). Exhibits a grayish tan color with an open grain and a high figure. Exterior uses are for doors, sills, and door frames. Interior uses are for trim, frames, paneling and cabinetwork.

White oak (quarter sawn). Exhibits a grayish tan color, accented with flakes, low figure, and open grain. Used mostly for interior work such as trim, frames, paneling, and cabinetwork.

White oak (rift sawn). Exhibits a grayish tan color, with an open grain and a low figure. Used mostly for interior uses such as trim, frames, paneling, and cabinetwork.

White or butternut walnut. Sapwood light to brown, heartwood light chestnut brown with an attractive sheen. Cut is small, mostly going into cabinet work and interior trim. Moderately light, rather weak, not rot-resistant.

White or gray ash. Hard, heavy, springy, open-grained wood with light reddish brown heartwood and nearly white sapwood. It is too hard to nail when dry and is used for industrial products where hardness, shock resistance, stability, and strength are important. Open grained with a strong and pronounced grain pattern.

White specks. Small white pits or spots in wood caused by the fungus *Fomes pini*. Develops in the living tree and does not develop further in wood in service. Where permitted in the rules, it is so limited that it has no more effect on the intended use of the pieces than other characteristics permitted in the same grade. Pieces containing white speck are no more subject to decay than pieces that do not contain it.

White spruce. Very little contrast between heartwood and sapwood, color varies from white to pale yellow. The lumber is lightweight, of medium strength and not very durable in situations favorable to decay. Used as light construction material and is of particular value as paper pulp.

Wood plank roof decking. Lumber used for this purpose should include moderate stiffness and strength, freedom from warp or wind, low moisture content, and some insulating value. The decking can be of solid stock or laminated wood (edge matched).

Wood sliding. Lumber used for purpose of fabricating exterior siding. Should include medium decay resistance, freedom from warp and wind, easy working qualities in the fabrication procedure, and most important, provides good surfaces for applying paint. The recommended lumber for lap siding, drop siding, matched vertical boards, and battens includes western red cedar, cypress, and redwood. Lumber from heartwood is preferable and includes edge-grained material that provides the best paint holding qualities.

Wood species. Woods are divided into two classes— hardwoods, which have broad leaves, and softwoods or conifers, which have scalelike leaves or needles. The terms "hardwood" and "softwood" do not denote hardness or softness of the wood. Some "hardwoods" like cottonwood and aspen are less dense (or hard) than some "softwoods" like southern pine and Douglas fir.

Xylem. The portion of the tree trunk, branches, and roots that lies between the pith and the cambium.

Yard heart common. All heartwood of redwood common grade, including any number of knots. It is a usable grade of good quality. Bench stock for greenhouse usage, etc., is drawn from this mill grade.

Yellow birch. Tree with a heartwood that varies in color from medium to dark or reddish brown, while the sapwood, which comprises a better than average portion of the tree, is near white. Yellow birch is available as lumber and as veneered products; the wood is adaptable to either paint or transparent finish.

Yellow cypress. Available cypress lumber does not contain the heartwood of inherently high decay resistance once associated with the species; in lumber form it contains a high percentage of sapwood. Thus, as for most softwoods, preservative treatment is imperative if used on the exterior. The wood is more generally utilized for paneling where its strong, bold grain is best displayed.

Yellow or tulip poplar. Easily worked native wood. The old growth has a yellow to brown heart; the sapwood and young trees are tough and white. It is not a framing lumber but is used to a great extent for siding, where it may be marketed with cucumber magnolia, a botanical relative.

Yellow poplar. Sometimes incorrectly called "whitewood," an even-textured, close-grained, stable wood of medium hardness with an inconspicuous grain pattern. The heartwood is pale greenish yellow while the sapwood ls white. Occasional dark purple streaks also occur. The tight, close grain results in outstanding paintability, while its modest figure and even texture permits staining to simulate more expensive hardwood. Owing to its indistinct grain figure, poplar is seldom used for decorative veneered products.

CONSTRUCTION LUMBER AND TIMBER

Actual dimension. The true size of a piece of lumber, after milling and drying.

Actual size. The size of lumber after it has been surfaced.

Air-dried lumber. Lumber that has been dried by being stored in yards or sheds for any length of time. The minimum moisture content of thoroughly air-dried lumber is 12 to 15% and the average is higher.

Air-dried wood. Wood seasoned by exposure to the atmosphere without artificial heat.

Air seasoning. The process of drying green lumber or other wood products by exposure to prevailing atmospheric conditions outdoors or in an unheated shed.

Bastard sawn. Lumber (primarily hardwoods) in which the annual rings make angles of 30° to 60° with the surface of the piece.

Board. Lumber less than 2 in. thick and 1 in. or more wide.

Board foot. A unit of lumber equal to the volume of a board 1 foot square and 1 in. thick.

Board measure. System of designating quantities of lumber in terms of board feet.

Built-up laminated wood. An assembly made by joining layers of laminated wood with mechanical fastenings so that the grain of the laminated wood is essentially parallel.

Built-up timbers. An assembly made by joining layers of lumber together with mechanical fastenings so that the grain of all laminations is essentially parallel.

Check. Lengthwise split in the end of lumber usually resulting from more rapid drying of the end than the balance of the piece.

Clear. Lumber and timbers selected for lack of imperfections and knotholes, desirable for finished interior beams and paneling for construction.

Common lumber. In softwoods, a general term for non-stress graded lumber that has appearance quality less than select grade but is suitable for general construction and utility purposes.

Crosscut. Cut made across the grain of lumber.

Dimension. Lumber from nominal 2-in. through 4-in. thick and 2 or more inches wide.

Dressed lumber. Lumber that has been surfaced to attain smoothness of a surface and uniformity of size on one side (S1S) two sides (S2S), one edge (S1E), two edges (S2E), or any combination of these.

Dressed Size. The dimensions of lumber after surfacing with a planing machine. Usually $\frac{1}{4}$ or $\frac{3}{4}$ in. less than nominal size. The American Softwood Lumber Standard lists standard dressed sizes.

Edge-grained lumber. Lumber that has been sawed so that the wide surfaces extend approximately at right angles to the annual growth rings. Lumber is considered edge grained when the rings form an angle of 45° to 90° with the wide surface of the piece.

End-grained wood. Wood where the grain is on a cut made at a right angle to the direction of the fibers.

Factory or shop lumber. Lumber intended to be cut up for use in further manufacture and graded on the basis of the percentage of the area that will produce cuttings of a given quality and size.

Finish. Lumber suitable for millwork or for the completion of the interior of a building. Chosen particularly because of appearance or ability to accept a high quality finish.

Flat grain. Lumber that has been sawed so that wide surfaces extend approximately parallel to the annual growth rings. Lumber is considered flat grain when the annual growth rings make an angle of less than 45° with the surface of the piece.

Form lumber. Lumber against which concrete is poured to hold a form or shape. It is used to construct forms or retaining walls.

Frame construction. Construction in which the structural parts are of wood or depend on a wood frame for support.

Framing. The unfinished structure of a building, including interior and exterior walls, floors, roof, and ceilings.

Framing lumber. Lumber used for the structural members of a building, such as studs, rafters, and joists.

Glue-laminated wood. Assembly made by bonding layers of veneer or lumber with an adhesive so that the grain of all laminations is essentially parallel.

Girder. Heavy timber or beam designed to support the ends of floor joists.

Grade. Quality or classification of lumber in relation to its adaptability for different uses.

Grading rules. Lumber is manufactured in a sawmill to grading rules of regional associations. As it leaves the saw, the wood is unseasoned. Much lumber is sold and used in this condition, though it is usually partly air-seasoned when it reaches the job.

Green lumber. Unseasoned or wet lumber in which free water still remains within the cells; lumber that has a moisture content above the fiber saturation point (approximately 25 to 30%).

Green timber. Green is the condition of timber as taken from a living tree. Immediately upon being sawed from the tree, lumber begins to lose moisture and otherwise change

its condition. The rapidity of these changes is determined by the species, humidity, and circulation of air, heat, etc.

Horizontally laminated timbers. Laminated timbers designed to resist bending loads applied perpendicular to the wide faces of the laminations.

Kiln. Heated chamber for drying lumber; veneer, and other wood products.

Kiln-dried. Lumber that has been seasoned in a dry kiln, usually, though not necessarily, to a lower moisture content than that of air-seasoned lumber.

Lumber. Product of the saw and planing mill that is not further manufactured than by the process of sawing, resawing, passing lengthwise through a standard planing mill, crosscutting to length, and working. Lumber of thickness not in excess of $\frac{1}{4}$ in. to be used for veneering is classified as "veneer."

Lumber defects. Defects can occur naturally as a result of insects, injury, or disease to the tree. Other defects may result from improper seasoning or the milling process.

Lumber worked to pattern. Lumber that, in addition to being dressed, has been matched, shiplapped, and patterned or molded, bevel siding included.

Manufacturing defects. Includes all defects or blemishes that are produced in manufacturing, such as chipped grain, raised grain, torn grain, skips in dressing, variation in sawing, miscut lumber, machine burn, machine gouge, mismatching, and insufficient tongue or groove.

Matched lumber. Lumber edge dressed and shaped for a tongue-and-groove joint when pieces are laid edge to edge or end to end.

Milling. Machining or planing lumber to determine sizes.

Mill run. All the lumber produced in a mill, without reference to grade.

Nominal size. Commercial size designation of width and depth in standard lumber grades, somewhat larger than the standard net size of dressed lumber.

Off grade. Lumber found to be below the grade shipped and invoiced; lumber pile at the mill that has fallen below the grade sought.

Old growth. Timber growing in or harvested from a mature, naturally established forest.

Patterned lumber. Lumber that is shaped to a pattern, in addition to being dressed, matched, or shiplapped, or any combination of these workings.

Plain end lumber. Worked lumber without end matching or with plain trimming and square ends.

Plain-sawed lumber. Another term for flat-grained lumber, used generally in hardwoods.

Plain sawing. Lumber sawed regardless of the grain; the log is simply squared and sawed to the desired thickness. Sometimes called "slashing" or "bastard sawing."

Planked lumber. Lumber planed to a size in excess of the corresponding standard-dressed size to permit remanufacture or special use.

Precision end-trimmed lumber. Lumber trimmed square and smooth on both ends to uniform length with a manufacturing tolerance of $\frac{1}{16}$ in. over or under in length in 20% of the pieces.

Premium grade. Term meaning "the best of its grade." A better selection from a given grade; the top of a grade.

Quartersawed lumber. Another term for edge-grained lumber used generally in hardwoods.

Quartersawed, quarter cut. Lumber cut in a radial direction, that is, at right angles to the direction of the annual rings. In softwood it is usually called "edge grain"; rift sawed is another term for quartersawed.

Remanufactured lumber. Lumber that has been further processed to change its size or shape after grading.

Resawn lumber. Product of sawing any thickness of lumber to develop narrower sized lumber. Term as used in commercial transactions is mostly to denote the product of resawing dressed and graded lumber.

Ripped lumber. Product of sawing any width of lumber to develop narrower lumber. Term as used in commercial transactions is mostly to denote the product of ripping dressed and graded lumber.

Roof sheating lumber. General requirements for roof sheathing lumber include moderate stiffness, good nail holding, little tendency to warp, and ease of working. These qualities are found in such varieties as Douglas fir, western larch, and southern yellow pine.

Rough lumber. Lumber as it comes from the saw prior to any dressing operation.

Round timber. Lumber used in the original round form, for example, poles, piling, and mine timbers.

Saw-sized lumber. Lumber uniformly sawed to the net size for surfaced lumber, for uses requiring a rough texture. A slight variation in sawing of not more than $\frac{1}{32}$ in. under in 20% of the pieces and $\frac{1}{8}$ in. over is permitted.

Scab. Short piece of lumber used to splice or prevent movement of two other pieces.

Seasoned lumber. Lumber which has been air-dried for at least sixty (60) days, or which has at the time of installation in the structure reached a moisture content approximately equal to that which it will eventually contain to service. Where green or recently cut lumber is used, tabulated bolt values shall be reduced one-third.

Select lumber. In softwoods, a general term for lumber of good appearance and finishing qualities.

Shiplapped lumber. Lumber that is edge dressed to make a lapped joint.

Square end-trimmed lumber. Trimmed square permitting slight manufacturing tolerance of $\frac{1}{64}$ in. for each nominal 2 in. of thickness or width.

Standard matched. Matched lumber that is tongued and grooved to correspond with a specification provided by the grading rules other than for the center of the thickness of the material.

Standing timber. Timber still on the stump.

Stress-graded lumber. Lumber of any thickness and width that is graded for its mechanical properties.

Strips. Yard lumber less than 2 in. thick or 8 in. wide.

Structural glued laminated lumber. Member comprising an assembly of laminations of lumber in which the grain of all laminations is approximately parallel longitudinally and in which the laminations are bonded with adhesives and that conforms to the standards applicable thereto.

Structural lumber. Lumber that is 2-in. thick or more, and 4-in. wide or more, intended for use where the design requires strength.

Structural timbers. Pieces of wood of relatively large size, with a cross section greater than 4 in. × 6 in., the strength of which is the controlling element in their use.

Subfloor lumber. General requirements for subfloor lumber are not exacting, but moderate stiffness, medium shrinkage and warp, and ease of working are desired. Suitable woods include Douglas fir, western larch, and southern yellow pine.

Surfaced lumber. Lumber that has been dressed or planed by running it through a planing machine.

Timbers. Lumber 5 in. or more in least dimension.

Wall sheating lumber. General requirements for wall sheating lumber are easy workability, easy nailing, and moderate shrinkage.

Worked lumber. Lumber that has been run through a matching machine, sticker, or molder. The lumber may be matched, with a tongue-and-groove joint at the edges; shiplapped, with a closely rabbetted or lapped joint at the edges; or patterned shaped to a patterned or molded form.

Yard lumber. Lumber of all sizes and patterns intended for general building purposes. The grading of yard lumber is based on the intended use of the particular grade and is applied to each piece with reference to its size and length if further manufacture is not considered.

FASTENERS AND ACCESSORIES

Aluminum nail. Nails made from an exclusive heat treating process that produces high tensile strength. The nails come in sizes from 6 to 20 penny.

Anchor bolts. Bolts used for anchoring materials to each other, for example, wood plates to masonry or concrete. Anchor bolts with one end threaded for hexagonal nuts. Bolts, available with full body, national coarse-cut threads on one end only, and hook on other end. Bolts manufactured from steel rods in plain high-strength steel, or finished hot-dipped galvanized.

Barbed plywood nail. Nail with flat head and sloped sides under the head, which bite into the plywood, including indentations on the body of the shank for greater holding power.

Barbed roofing nail. Nail designed particularly with a large head and shank provided with indentations and ridges to give it greater holding power for installation of roofing materials.

Blued nail. Nail with oxidized bluish surface as a result of heating.

Bolts. Bolts are used for wood-to-wood fastenings and for fastening steel to wood members.

Box nail. Similar to the common nail with the exception of these changes: the head or nailing surface is thinner and the shank is slimmer than most ordinary common nails.

Buck anchors. Galvanized corrugated steel anchors for fastening to wood or metal bucks and for anchoring into masonry or concrete.

Bugle-head dry-wall screws. Screws used for fastening plasterboard to steel studs. They are made in several styles and many lengths, with or without Phillips recessed heads.

Carriage bolt. Bolt used to fasten two pieces of structural wood together at a joint or elsewhere. Recessing the square shoulder gives the bolt stability in its location. The bolt has a round head, square shoulder, and American

Standard cut thread. It is made of plain or hot-dipped galvanized steel.

Casing nails. Nails for the installation of wood door casings and stair treads where the head of the nail needs to be flush with the material.

Cement-coated nails. Cement-coated nails have up to 100% greater holding power than uncoated nails in withdrawal. However, the increase is usually temporary, and after a short period of time has elapsed, the holding power is the same as that of the uncoated nail.

Channel anchors. Several styles of anchors are available in steel or galvanized steel, with a single or split drop leg, or in flat-plate style with anchor pin.

Clasp nail. Square-section cut nail whose head has two pointed projections that sink into the wood.

Common cut nail. Nail used primarily for installation of framing, in the erection of scaffolding, and other general rough temporary work. Nail is available from 2 to $3\frac{1}{2}$ in. in length and has a tapered shank which provides easy withdrawal.

Common nails. Nails are the simplest type of fastening devices. They are made in hundreds of styles for a multitude of purposes. A standard list of sizes is published with recommended uses for each. Materials for nails vary with use and include steel, aluminum, stainless steel, copper, and bronze. Nails usually are sized from the common 2-penny nail, which is 1 in. long, to the 60-penny nail, which is 6 in. long.

Concrete nail. Nail used primarily in nailing materials to concrete. Nail is oil quench hardened, with $\frac{5}{16}$ in. head and 9 gauge wire shank.

Countersunk finishing washers. Washers used to accommodate the heads of flat- or oval-head wood screws, machine screws, or stove bolts.

Cut washers. Washers made of American National Standard wrought or plain steel. They may be finished by being zinc plated or hot-dipped galvanized. These washers are made to fit many bolt sizes.

Double-style hanger. Hanger designed for hanging joists that are aligned on each side of a girder or beam.

Drift bolts. Drift bolts or drift pins are used to fasten heavy timbers together. These fastenings develop capability of high lateral loads, but are relatively weak in withdrawal loads. Their action is very similar to that of nails in that they prevent lateral movement and separation of the members.

Drivescrew nail. Nail made only in one size, 7 penny, with a $11\frac{1}{2}$ gauge shank that has a swirl cut into the shank for greater holding power. Nail is vinyl coated eliminating temperature and moisture problems.

Duplex nails. Nails used primarily for temporary work, such as formwork and scaffolding. Nail has two heads and the shank tip has a diamond point.

Electroplated nails. Nails that have been electroplated with a thin electro-chemical deposit of brass, cadmium, copper, nickel, tin, or zinc.

Enameled nail. Nail coated with selected color and often baked.

Explosive-driven fastenings. Powder-driven studs used to fasten wood to steel and wood to concrete, and for many other uses. These fastenings are available with threaded head with washer, plain head with disk, and plain head without disk.

Eye bolts. Lag-screw-type bolt, threaded for a hexagonal nut, and finished by zinc plating.

Fender washers. Washers with a large area surrounding the circular opening. They are made to fit many bolt sizes and come in zinc-plated finish.

Flat head wire nail. Nail used for miscellaneous purposes, available from $\frac{5}{8}$ in. to 1 in. in length, with 18 gauge shanks.

Framing clip. Preformed galvanized steel device for connecting wood mullions, posts, or lintels to resist wind and earthquake forces. They are fastened by nailing.

Hardened steel nails. Nails of smaller diameter than common nails of the same length. They are used to advantage under some conditions without decreasing the allowable lateral loads.

Heat-treated. Nails heated to critical temperature and subsequently cooled by quenching in oil or water.

Helically grooved, annularly grooved, and threaded nails. Nails that have been developed for certain uses and conditions. These nails possess the same holding power as common nails, but their strength is greater at larger joint deformations. They are also less sensitive to moisture changes in wood.

Hex (neoprene washer) tapping screw. Screw with integral metal washer that holds neoprene washer captive, preventing spreading when compressed between screw head and metal. These screws are for use with corrugated or flat-sheet metal, flashings, metal panels, etc.

Hex tapping screw. One-piece hex-head, steel-washer-type screw for use in light-gauge metals.

Hi-load nails. Nails designed expressly for fastening plywood diaphragm sheathing to studs and rafters. They are also used for many other purposes, where resistance to shear and withdrawal are essential.

Hot-dipped galvanizing. Method of galvanizing in which metal materials are put through an acid cleaning tank where all scale and foreign materials are removed. The materials are then placed in neutralizing tanks and dipped several times. Hot water dipping, whirl cleaning, and a final rinse are performed before the material is placed in galvanizing solution.

Joist hangers. Metal preshaped hangers with a minimum 2-in. seat fastened to the support by nailing.

Joist-purlin hanger. One-piece, metal, prefabricated, shaped seat hanger with tapered throat for installation. It is fastened to the support by nailing.

Lag-screw expansion shields. Shields used in mortar joints by drilling a hole, placing the shield in the mortar between bricks, attaching the fixture, and tightening up on the lag screw.

Lag screws

1. Lag screws are used in many places for many types of installations and fastenings in lieu of long spike nails for positive fastening. They are made of plain, hot-dipped galvanized, or zinc plated steel.

2. Lag screws are usually used where the fastening of a bolt is difficult due to the inaccessibility of one side of a joint or in places where the presence of a nut on the surface would be objectionable.

Lead wall anchors. Anchors for use in masonry and concrete walls or surfaces. They are made of lead and will expand without turning in soft material.

Lead washers. Washers used with 16-penny common nails, made of pure lead, for use in fastening metal roofing or siding. These washers prevent leaks around the nail head.

Machine bolts. Bolts used in finish metal work, with National coarse threads and hexagonal nuts. Manufactured plain steel or zinc plated, or hot-dipped galvanized. Thread length is optional (ANSI B-18.2 and ASTM D 1761).

Machine screws. Screws manufactured in flat or round slotted heads with a zinc-plated finish. They are made in various types of threading.

Malleable iron washers. These washers have large bearing surfaces for greater strength and come in plain steel or with a hot-dipped galvanized finish. They are made to fit many bolt sizes.

Masonry nails. Nails made of hardened steel, $\frac{9}{16}$ in. checkered head, knurled thread on shank, .250 in. diameter.

Medium split-load washers. Washers used to lock set bolts when tightened. They are made with a zinc-plated finish.

Metal bridging. Bridging made in two types, nail on or no nail. The strong compression type is ribbed for additional strength and oil treated for rust resistance.

Nail heads. Depending on the use of the nail and the manufacturer, many types of nail heads are available: flat, round, double-double, curved, button, oval, checkered, sinker, projection, cupped oval, numeral head, countersink, headless, lettered head, casing, cup-head, oval-countersink, and slotted.

Nail points. Depending on the size of the nail and the manufacturer, several types of nail points are available: regular diamond, long diamond, blunt, round, needle, and sheared bevel.

Nails. Nails that are exposed to the weather and that cannot be set should be monel (such as those holding insulating panels of cement board or metal). All other nails, sash pins, or fasteners on the exterior should be hot-dipped galvanized or of nonstaining metal (ASTM D 1761).

Nonferrous nails. Nails that have been developed for uses in which rusting or corrosion of steel wire nails is objectionable. Nails can be obtained in aluminum, copper, brass, bronze, silicon, monel, and stainless steel.

Nut eyebolts. Eyebolts used for wire and cable fastenings, with the bolt portion used through wood or steel. The bolt is drop-forged and hot-dip galvanized finished.

Penny. Measure of nail length abbreviated by the letter d. A 2-d nail is 1 in. in length.

Plastic screw anchors. Anchors that are ideal for light loads such as household wall-mounted accessories, drapery fixtures, or picture frame hanging. A hole is drilled and the plastic anchor is inserted. The tightening action of the screw expands the anchor in the wall.

Ratchet nail. Nail available in sizes from $1\frac{1}{2}$ in. to 2 in., flat thin head, with reverse annular thread on shank, and terminating in a diamond point. The nail is very difficult to remove due to its reverse thread.

Ringshank nail. Nails available in two sizes, 7 and 8 penny, $11\frac{1}{2}$ gauge, with entire shank with annular thread. Nail is vinyl coated to eliminate temperature and moisture problems.

Rope binding hook. Hook used for halyards, trucks, tarpaulins, awnings, clotheslines, and boats and made with a hot-dip-galvanized finish.

Round bend screw hook. Open-ended sharp hook, used vertically or horizontal.

Scaffold nails. Double-headed diamond point nails used in temporary construction where the nail can be easily pulled.

Screw eyes. Screw eyes are easy to start in wood because their ends are pointed or tapered. The thread is sharp and deep and draws the eye into the hardest of woods. They are usually zinc chromate finished.

Shear plates. Shear plates are intended primarily for wood-to-steel connections or for wood connections in demountable structures when used in pairs. They are used for attaching columns to footings with steel straps, connecting timber members to steel gusset plates, attaching steel heel straps in bowstring trusses, and for other steel-to-wood connections in timber structures.

Snap back hook. Spring-actuated hook that lies flat and out of the way when not in use. The spring holds the hook down under constant tension. The hook pivots 180°, permitting it to be mounted for maximum efficiency and ease of use. It has a hot-dip-galvanized finish.

Spike grids. Grids designed for use with piles and poles in trestle construction, piers, wharves, and transmission lines. They have, however, many other uses. They are installed by pressure with a hand-pressure device.

Split rings. Split rings are the most efficient mechanical devices for fastening joints in timber construction. A power-driven tool is recommended for cutting the groove.

Spring-wing toggle bolt. Zinc-plated steel bolt in spring-actuated wings. It is made with a round, flat, or mushroom head, each slotted, and is for use in hollow walls or suspended ceilings.

Square-bend screw hooks. Hooks made to support heavy loads. The threads are deep and sharp.

Square-plate washers. Washers made in plain steel or with a hot-dipped galvanized finish. They fit many bolt sizes.

Stainless steel nail. Standard common design, from 6 to 16 penny in size.

Steel nailer clips. Slotted, galvanized steel, metal clips for attaching wood nailer or concrete planks to the tops of steel beams.

Stove bolts. Bolts manufactured in flat or round slotted heads, usually with a zinc-plated finish.

Superfast screws. Screws used for panelized roof systems. These special wood screws are waxed for speedy installations and come with Phillips recessed heads.

Tamping anchors. Anchors used for the installation of clip angles to concrete floors by drilling a hole in the concrete and using a setting tool to expand the lead sleeve in the concrete. The clip angle or metal frame is installed and tightened by machine bolt.

Tapping screws. Self-tapping screws used for fastening material to light-gauge sheet metal, asbestos composition sheets, resin-impregnated plywood, etc. The heads are slotted or Phillips head styled; the screws usually have a zinc-plated finish.

Teks. Self-drilling screws made to fit magnetic sockets, used in sheet metal and structural metal installations.

Thread-cutting screws. Screws for use in castings, forgings, structural steel, and very thick plywood sheets. A cutting screw has fine tapping flutes and cuts a standard machine thread as the screw is turned. This type of screw provides security against vibration and resistance to torque and shear and tension loads.

Toggle bolt. Spring-wing toggle bolts used for fastening materials, equipment, and accessories of all kinds to hollow walls and suspended ceilings. Bolt portion available in flat, round, and mushroom heads up to 6 in. in length.

Toothed rings. Toothed rings are used to transfer stress between two timber members, particularly in light, simple structures. No grooves are required since the rings are embedded with a pressure device that is hand-operated.

Turn buckles. Buckles used for take-up purposes and to tighten cables, etc. They are hot-dipped galvanized and made in two styles: eye and eye or hook and eye. The shaft is usually drop-forged or made in aluminum.

U bolts. Bolts principally used to support conduits and piping, but with many other uses. They are made in several styles: standard, extralong, and special extralong for supporting piping. They usually have a zinc chromate finish.

Wafer teks. Self-drilling screws used for $\frac{1}{4}$-in. material.

Wall braces. Prefabricated steel braces used in small-structure construction by attaching bracing across wood studs to prevent racking of framework.

Wall plugs. Galvanized metal wall plugs for building into masonry walls to attach wood furring.

Wire brads. Miscellaneous use nails from $\frac{1}{2}$ in. to $\frac{3}{4}$ in. in length, 18-gauge shanks, with shank terminating in a diamond point.

Wood screws. Screws used in a manner similar to nails, but with greater holding power than nails of similar sizes. They are used widely in attaching fixtures, accessories, and hangers to wood in light structural connections.

FINISH CARPENTRY

Abrasive paper. Paper or cloth covered on one side with a grinding material and used for smoothing and polishing. Materials used for this purpose include crushed flint, garnet, emery, and corundum.

Acoustical board. Material used to control or deaden sound.

Adhesive. Substance capable of holding materials together by surface attachment. It is a general term and includes cements, mucilage, and paste, as well as glue.

Alignment. Course or location of elements of design or construction in relation to a determined line.

Aperture. Opening provided in a wall.

Apron. Flat member of the inside trim of a window placed against the wall immediately beneath the stool. Its function is to cover the rough edge of the interior wall finish.

Architrave. Group of moldings above and on both sides of an opening or panel.

Arris. Meeting of two surfaces producing an angle.

Astragal. Small semicircular molding, plain or ornamented.

Back. Reverse side of the face of a panel; back of a finished mill work piece.

Backband. Molding installed around the outer edge of a plain rectangular casing as a decorative feature.

Backing. Strips or blocks installed in walls or ceiling for the purpose of fastening or supporting trim or fixtures.

Back miter. Angle cut starting from the end and coming back on the face of the stock.

Backset. The distance an object is set back from an edge, side, or end of stock, such as the distance a hinge is set back from the edge of a door.

Band. Low, flat molding.

Banding. Portion of wood extending around one or more sides of a core, usually with its grain parallel to the edge of the panel. The banding of solid wood facilitates shaping the edges of the piece, which may also be finished flat to cover the side and end grains exposed in the core.

Base. Bottom of a column; finish of a room at the junction of vertical surfaces meeting the floor.

Baseboard. Board placed against the wall around a room next to the floor to finish properly between floor and finished wall surface.

Base cap. Molding applied to the top edge of the baseboard.

Base molding. Molding used to trim the upper edge of interior baseboard.

Base shoe. Molding used next to the floor on interior baseboard; sometimes called a "carpet strip."

Batten
1. Narrow strips of wood used to cover joints or as decorative vertical members over plywood or wide boards.
2. Strip of wood placed over vertical joints in external or internal cladding or finish.

Bead. Small molding, semicircular in section.

Bed. Molding used to cover the joint between the plancier and frieze; also used as a base molding on heavy work and sometimes as a member of a cornice.

Bevel. Angular surface across an edge of a piece of wood.

Blank. Piece of wood cut to a size from which the manufactured article is finished.

Blemish. Any defect, scar, or mark that tends to detract from the appearance of wood.

Blind nailing. Nailing in such a way that the nailheads are not visible on the face of the work—usually at the tongue of matched boards.

Blocking. Pieces installed between studs in a wall, usually used to provide fastening for fixtures on the finished wall.

Bond. Adhesion of two surfaces by the action of glue in curing.

Bracket. Projecting support for a shelf.

Brad. Thin, short finishing nail.

Builders' finishing hardware. Hardware that includes manufactured devices for supporting doors, cabinet work, devices and associated hardware, and other miscellaneous finishing hardware generally required to complete the interior of a building.

Bull nose. Rounded corner or edge used in positions where sharp arrises might be damaged.

Butt. Type of door hinge. One leaf is fitted into space routed into the door frame jamb and the other into the edge of the door.

Butt joint. Junction where the ends of two members meet in a square-cut joint.

Cabinet work. Prefinished, shop-assembled cabinet work, such as cases, counters, decks, and cabinets, ready for installation.

Cant strip. Strip of wood placed under a piece to tilt the piece at a slant.

Cap. Upper member of a column, pilaster, door cornice, molding, and the like.

Carpentry abbreviations. See Appendix 3.

Cased opening. Interior opening without a door that is finished with jambs or trim.

Casework (finished carpentry). The term "casework" usually includes the cutting of holes for job-applied vents, weeps, or grilles; cutting for job-applied hardware; linoleum, cork, leather, vinyl, metal, or resilient covering or lining; hinges, pulls, catches, locks, coat and hat hooks, and track assemblies for frameless glass doors; metal angles or brackets for attaching cases or shelves to walls or floors; and glass or glazing.

Casing. Trimming around an opening, either outside or inside; the finished lumber around a post, beam, etc.

Ceiling

1. Narrow, matched boards; sheathing of the surfaces that enclose the upper side of a room.
2. Matched lumber, normally with V-joint pattern, used for ceiling and wainscoting.

Center-matched. Joint between two pieces of lumber. It consists of a groove in the edge of one piece and a tongue in the other. More commonly known as a tongue-and-groove joint.

Center punch. Tool used to make an indentation at the centerlines of holes.

Chair rail. Interior molding applied along the wall of a room to prevent chairs from damaging the finished wall surface.

Chamfer. Corner of a board beveled usually at a 45° angle. Two boards butt-jointed and with chamfered edges form a V-joint.

Cleat. Small strip applied to support a shelf or for similar applications.

Cock bead. Bead raised above the surface.

Compound miter. Bevel cut across the width and also through the thickness of a piece.

Coped cut. Profile cut made in the face of a piece of molding that allows for butting it against another piece at an inside corner.

Coped joint. Type of joint between moldings in which the end of one piece is cut to fit the molded surface of the other.

Corner bead. Strip of formed sheet metal, sometimes combined with a strip of metal lath, placed on corners before plastering to reinforce them; also, strip of wood finish three-quarters round or angular placed over a plastered corner for protection.

Cornice. Horizontal molding that may be a combination of several shaped pieces.

Counterbore. To drive a screw below the surface of the surrounding wood. The void created is filled later with putty or plugged with wood filler.

Countersink. Make a cavity for the reception of a metal plate or the head of a screw or bolt so that it does not project beyond the face of the work.

Cove molding. Molding with a concave face used as trim or to finish interior corners.

Crown molding. Molding used on cornice or wherever an interior angle is to be covered.

Cuts. Grading term used where the piece is to be cut up for further manufacture, as distinguished from other lumber which is intended for use as the original piece.

Dado

1. Rectangular groove across the width of a board or plank; in interior decoration, a special type of wall treatment.
2. Lower portion of an interior wall, often treated or decorated differently from the upper portion of the wall.

Dado joint. Joint in which one piece is grooved to receive the piece which forms the other part of the joint.

Dado molding. Wall molding, similar to a chair molding, installed about 30 in. above the floor to give the effect of a wall wainscot.

Direct nailing. Nail perpendicular to the initial surface or to the junction of the pieces joined. Same as face nailing.

Double acting. Hinges used on doors so they can swing in both directions.

Dovetail. Tenon shaped like dove's spread tail, fitting into corresponding mortise to form a joint.

Dovetail joint. Joint in which one piece has dovetail-shaped pins or tenons that fit into corresponding holes in the other piece.

Dowel. Cylindrical wooden pin used for holding two pieces of wood together.

Dowel joint. Joint made by boring the pieces involved to receive dowels, which are set in glue and assembled under pressure.

Drawboard. Mortise-and-tenon joint with holes so bored that when a pin is driven through, the joint becomes tighter.

Dressed and matched (tongued and grooved). Boards or planks machined in such a manner that there is a groove on one edge and a corresponding tongue on the other.

Driwood moldings. Moldings processed out of solid wood that has been scientifically kiln dried and carefully milled. They are manufactured by the Driwood Moulding Company, Inc., Florence, S.C. 29501.

Dry wall. Interior covering material, such as gypsum board or plywood, which is applied in large sheets or panels.

Dry-wall construction. Construction in which the interior wall finish is applied in a dry condition, generally in the form of sheet materials or wood paneling, as contrasted to plaster.

Dutch door. Door divided horizontally. The bottom can be closed and secured while the top can be left open.

Dutchman. Odd-shaped piece usually used to fill or cover an opening.

Dwarf wall. Framed wall of less than normal full height.

Eased edges. Slightly rounded surfacing on pieces of lumber to remove sharp corners.

Edge. Narrow face of rectangularly shaped pieces; corner of a piece at the intersection of two adjoining faces; in stress grades, that part of the wide face nearest the corner of the piece.

Edge joint. Joint between two pieces of wood glued edge to edge and in the direction of the grain.

End match. Tongue and groove the ends of lumber.

Facade. Main or front elevation of a building.

Face. The best appearing side of a piece of wood or the side which is exposed when installed, such as finish flooring.

Face frame. Framework of narrow pieces on the face of a cabinet containing door and drawer openings.

Face nail. A nail driven perpendicular to the surface of a piece.

Face nailing. Fastening a member by driving nails through it at right angles to its exposed surface.

Face side. Side of a piece that shows the best quality.

Facings. Dressed boards, with or without moldings, used in exposed places.

Factory and shop lumber. Lumber intended to be cut up for use in further manufacture. It is graded on the basis of the percentage of the area that will produce a limited number of cuttings of a specified, or a given minimum, size and quality.

Fascia. Flat member of a cornice or other finish, generally the board of the cornice to which the gutter is fastened.

Feather edge. One edge thinner than the other edge in the width of a board.

Fiberboard. Building board made from fibrous material such as wood pulp.

Filler. Heavily pigmented preparation used for filling and leveling off the pores in open-grained woods.

Filler wood. Heavily pigmented preparation used for filling and leveling off the pores in open-grained woods.

Fillet. Strip of material used with paneling and specified as to pattern; usually a machined narrow strip covering a joint or fitting between two panels, enhancing appearance.

Finger joint. Joint consisting of a series of fingers, precision machined on the ends of two pieces to be joined, which mesh and are firmly held together by an adhesive.

Finish

1. As applied to lumber grades, the term refers to the upper grades suitable for natural or stained finishes.
2. Wood products to be used in joiner work, such as doors and stairs, and in other fine work required to complete a building, especially the interior.

Finish builders' hardware. Fabricated metal devices, with or without plated finishes, used for the support and movement of doors, windows, cabinet work, and other wood manufactured items.

Finish carpentry. That part of the carpentry trade involved with the application of exterior and interior finish.

Finish nail. Thin nail with a small head designed for setting below the surface of finish material.

First and seconds. The best grade of hardwood lumber.

Fished. End butt splice strengthened by pieces nailed on the sides.

Fish joint. Joint formed by pieces that are joined butt end to end and connected on each side by wood or metal plates that are secured to the pieces joined.

Flooring. Material used in the construction of floors. The surface material is known as "finish flooring" while the base material is called "subflooring."

Flush. Even; in the same plane.

Flush bead. Level with the surface.

Flush doors. Door of any size having both surfaces smooth and flat.

Flutes. Hollows or grooves cut longitudinally for ornamental purposes.

Foxtail wedging. Mortising, in which the end of the tenon is notched beyond the mortise and split, and a wedge inserted, which, being forcibly driven in, enlarges the tenon and renders the joint firm and immovable.

Furring. Strips of wood or metal applied to a wall or other surface to plumb or level it and normally to serve as a fastening base for finish material.

Glue. Adhesive used in joining wood parts.

Glue block. Wood block, triangular or rectangular in shape, that is glued into place to reinforce a right angle butt joint. Sometimes used at the intersection of the tread and riser in a stairs.

Glue joint. Joint held together with glue.

Groove

1. Long hollow channel cut by a tool, into which a piece fits or in which it works. Two special types of grooves are the "dado," a rectangular groove cut across the full width of a piece, and the "housing," a groove cut at any angle with the grain and part way across a piece. Dados are used in sliding doors and window frames; housings are used for framing stair risers and threads in a stringer.
2. Sunk channel whose section is rectangular. It is usually employed in the edge of a molding, stile, or rail, into which a tongue corresponding to its section is inserted.

Graphite. Mineral used as pencil lead and also as a lubricant for the working parts of locks and certain tools.

Grounds. Guides used around openings and at the floorline to strike off plaster. They can consist of narrow strips of wood or of wide subjambs at interior doorways. They provide a level plaster line for installation of casing and other trim.

Half-round. Shaped molding used in conjunction with other moldings; it can be figured with decoration or plain.

Halved. Joint made by cutting half of the thickness of the wood away from each piece so as to bring the sides flush.

Hardboard. Man-made board of processed wood produced by converting wood chips into wood fiber, which is then formed into panels under heat and pressure.

Headroom. Clear space between floor line and ceiling, as in a stairway.

Hollow backed. Board in which one or more grooves have been cut from the central part of the back.

Hollow core door. Flush door with a core assembly of strips or other units which support the outer faces.

Horn. The extension of a stile, jamb, or sill that is removed before installation of the finished item.

Housed. Joint in which a piece is grooved to receive the piece that is to form the other part of the joint.

Housing. Groove or trench in a piece of wood made for the insertion of a second piece.

Interior. The inside of a structure; generally associated with the finishing process of the structure.

Interior finish. Material used to cover the interior framed areas, or materials of walls and ceilings.

Interior plywood. General term for plywood manufactured for indoor use or in construction subjected to only temporary moisture. The adhesive used for the assembly may be interior, intermediate, or exterior.

Interior trim. General term for all the molding, casing, baseboard, and other trim items applied within the building by finish carpenters.

Jamb. The top and side frames of a door or window.

Jig. Device that simplifies a hand or machine operation, usually by guiding a tool or serving as a template.

Joinery. Term used by woodworkers when referring to the various types of joints used in a structure.

Joint

1. Space between the adjacent surfaces of two members or components joined and held together by nails, glue, cement, mortar, or other means.

2. Contact surface between two adjacent pieces of wood. An "edge" or "face" joint is parallel to the grain of the wood. An "end" or "butt" joint is at right angles to the grain of the wood. A "scarf" joint is a sloping or bevel joint, where pieces of wood are lapped together.

3. Place where two members meet for fastening.

Joint butt. Squared ends or ends and edges adjoining each other.

Joint cement. Powdered material usually mixed with water and used for joint treatment in gypsum wallboard finish.

Joint end. Joint formed by joining two pieces of lumber end to end with dowels and adhesives.

Joint scarf. End joint formed by joining with adhesives the ends of two pieces tapered to form sloping plane surfaces; when fitted together, they form the original size of the piece scarfed.

Joint stepped. Usually a joint scarfed with a step or notch and machined into the scarfed area so as to position the pieces together for correct alignment, creating a constant size of continuous form with both pieces joined. Also called a "hooked" joint.

Kerfing. Longitudinal saw cuts or grooves of varying depths (dependent on the thickness of the wood member) made on the unexposed faces of millwork members to relieve stress and prevent warping; members are also kerfed to facilitate bending.

Key. Small piece of wood inserted in one or both parts of a joint to align it and hold it firmly together.

Knotting. Quick-drying composition used in joinery to form an impervious covering for knots, usually consisting of shellac in industrial methylated spirit.

Laminated. Term applied to a type of construction in which layers of wood, the grain in all cases running lengthwise of the assembly, are joined by gluing or other means to form a single member. The term is also applied to certain types of flooring made up of pieces of timber laid on edge instead of on their side.

Lap joint. Joint composed of two pieces, one overlapping the other.

Lindermann joint. Glued dovetail joint, shaped by a Lindermann jointer, joining two pieces of wood edge to edge longitudinally.

Lip. Molding with a lip that overlaps the piece against which the back of the molding rests.

Lock block. Block of wood which is joined to the inside edge of the stile of a hollow core door and to which the lock is fitted. Flush doors have a lock block on each side.

Lock set. Assembly consisting of a latching mechanism.

Louver. Slatted, exterior opening for ventilation in which the slats are so placed as to exclude rain, light, or vision.

Machined and knocked down. All pieces of a unit, fully machined, ready for assembly or partially assembled.

Mantel. The ornamental finished wood trim around a fireplace, including the shelf above the opening.

Manufacturing imperfections. All imperfections or blemishes that are the result of the manufacturing process.

Marine glue. Form of glue, containing rubber, shellac, and oil, that resists the action of water.

Marquetry. Art of inlaying wood with wood of other colors or with various other materials.

Mastic. Term applied to caulking preparations used for bedding and setting glazing, caulking window and door frames, bedding wood block flooring, etc.

Matched. Term applied to lumber machined or dressed to any form of joint; tongue-and-groove joint; fancy or ornamental wood matched in figure.

Matched lumber. Lumber that is dressed and shaped on one edge in a grooved pattern and on the other in a tongued pattern.

Matching or tonguing and grooving. Method used in cutting the edges of a board to make a tongue on one edge and a groove on the other.

Mechanical sanders. Equipment, usually electrically operated, with the proper sand paper will sufficiently smooth the wood cutting marks that can then be concealed by painter.

Medallion. A raised decorative piece, sometimes used on flush doors.

Mill run. All the lumber produced in a mill, without reference to grade.

Millwork. Building materials made of finished wood and manufactured in millwork plants; such materials include doors, window and door frames, sash, porch work, mantels, panel work, stairways, special woodwork, and cabinetry. Excluded are finished dressed four sides, siding, or partition, which are items of yard lumber.

Miscellaneous millwork. Interior millwork usually varies a great deal, both in the type and amount used. Uses include doors, kitchen cabinets, shelving, and stairs. In addition, such millwork items as fireplace mantels, wall paneling, ceiling beams, china closets, bookcases, and wardrobes.

Mismatch. Uneven fit in worked lumber when adjoining pieces do not meet tightly at all points of contact or when the surfaces of adjoining pieces are not in the same plane.

Miter. Molding returned upon itself at right angles is said to miter. In joinery, the ends of any two pieces of wood of corresponding form that are cut off at 45° and necessarily abut upon one another so as to form a right angle are mitered.

Miter joint. Joint of two pieces at an angle that bisects the joining angle. A miter joint at the side and head casing at an opening is made at a 45° angle.

Molded edge. Edge of a piece machined to any profile other than square.

Molded plywood. Plywood made to some desired shape other than perfectly flat. Often this shaping is done at the time the layers are glued together. Two ways of molding plywood are with curved forms and by applying fluid pressure.

Molder. Woodworking machine designed to run moldings and other wood members with regular or irregular profiles; also called a "sticker."

Molding

1. Wood strip having a curved or projecting surface, used for decorative purposes.
2. Lumber that has been worked on its side or edge to a uniform cross section, other than rectangular, to give an ornamental effect.

Molding base. Molding on the top of a baseboard.

Molding stock. Softwoods that are suitable for manufacturing into standard molding patterns.

Mortise

1. Cut made in a board to take a tenon.
2. Slot cut into a board, plank, or timber, usually edgewise, to receive the tenon of another board, plane, or timber to form a joint.
3. Hole that is to receive a tenon; any hole cut into or through a piece by a chisel, generally of rectangular shape.

Mortise-and-tenon joint. Projection that is machined on one piece, snugly fits into a rectangularly shaped recessed opening machined in a second piece, and is secured under pressure with an adhesive.

Nail-gluing. Method of fastening where nails hold the wood members until the glue sets.

Nail popping. Protrusion of nail heads due to the shrinking and swelling of wood.

Natural finish. Transparent finish that does not alter the original color or grain of the natural wood. Natural finishes are usually sealers, oils, varnishes, water-repellent preservatives, and similar materials.

Nonbearing partition. Partition extending from floor to ceiling that supports no load other than its own weight.

Nosing. Term applied to the rounded edge of a board.

Notch. Crosswise rabbet at the end of a board.

O.G. (ogee). Molding with a profile in the form of a letter S; molding with the outline of a reverse curve.

Ovola. Convex molding the profile of which is a quadrant of a circle (quarter round).

Panel

1. Piece of wood framed within four other pieces, as in styles and rails of a door, filling up the aperture; also, the ranges of sunken compartments in wainscoting, cornices, ceilings, etc.
2. Thin flat piece of wood, plywood, or similar material framed by stiles and rails, as in a door, or fitted into grooves of thicker material with molded edges for decorative wall treatment.

Panel doors. Any door having decorative panels on either surface, or an entire smooth face on each side of the door.

Parquet. Flooring laid in geometrical designs with small pieces of wood.

Particle board. Manufactured board composed of processed wood chips held together with special adhesives.

Partition. Wall whose primary function is to divide space within a building or structure.

Penny. Term used to indicate nail length; abbreviated by the letter d. Applies to common, box, casing, and finishing nails.

Perforated. Material which has closely spaced holes in a regular or irregular pattern.

Picture molding. Molding shaped to form a support for picture hooks, often placed on the wall at some distance from the ceiling to form the lower edge of the frieze.

Pilaster. Portion of a square column, usually set within or against a wall.

Pilot hole. Small hole drilled into a wooden member to avoid splitting the wood when driving a screw or nail.

Phillips head. Type of screw head with a cross-slot.

Plain sawing. Plain sawed lumber is obtained by making the first saw cut on a tangent to the circumference of the log and the remaining cuts parallel to the first. This method provides the widest boards and least waste. About half of the lumber produced by plain sawing is of tangential grain and the other half of radial grain.

Planing mill products. Products, such as flooring, ceiling, and siding, worked to pattern, usually in strip form.

Plank. Broad board, usually more than 1 in. thick; especially one laid with its wide dimension horizontal and used as a bearing surface.

Plaster grounds. Strips of wood used as guides or strike-off edges around window and door openings and at base, tops, and edges of walls.

Plastic laminate. A very tough, thin material in sheet form used to cover countertops; available in a wide choice of colors and designs.

Plinth block. Small decorative block, thicker and wider than a door casing, used as part of the door trim at the base and at the head.

Plough. Cut a lengthwise groove in a board or plank.

Plumb. Exactly perpendicular or vertical; at right angles to the horizon or finished floor surface.

Plywood. Cross-banded assembly made of layers of veneer or veneer in combination with a lumber core or plies joined with an adhesive. Two types of plywood are recognized: veneer plywood and lumber core plywood.

Pocket door. Type of door that slides into a recess in the wall.

Quarter round. Small molding that has the cross section of a quarter circle.

Quarter sawed lumber. Lumber produced by first quartering the log and then sawing it perpendicular to the growth rings, indicating the radial grain. This method produces relatively narrow boards and creates waste. Red oak, white oak, and the mahoganies are quartersawed lumber.

Rabbet

1. Rectangular, longitudinal groove cut in the corner edge of a board or plank.
2. Longitudinal channel, groove, or recess cut out of the edge or face of any wood member, especially one intended to receive another member.

Rabbetted joint. Lapping joint machined on the meeting portions of the pieces to be connected.

Rake. Cornice on the gable edge of a pitch roof, the members of which are made to fit those of the molding of the horizontal eaves.

Resilient flooring. Vinyl, vinyl-asbestos, linoleum, or other synthetic floor covering that provides a smooth surface.

Return. Continuation of molding or finish of any kind in a different direction.

Reveal. Vertical face revealed in the thickness of an opening or the depth of a recess.

Rubbing compound. Abrasive material used to produce a smooth-finished wood surface.

Sanding. Applying sandpaper or similar abrasive materials to surfaces before coating, brushing, or spraying finish material.

Sawing. Cutting with a tool or blade featuring pointed, equally spaced teeth along the blade edge. Typical methods are hacksawing, band sawing, and circular sawing. It is a material separation process involving both fixed and movable work pieces.

Scarfing. Joint between two pieces of wood, which allows them to be spliced lengthwise.

Scotia. Hollow molding used as part of a cornice.

Screed. Strip of wood, usually the thickness of the plaster coat, used as a guide for plastering.

Scribing. Fitting woodwork to an irregular surface. In moldings, cutting the end of one piece to fit the molded face of the other at an interior angle to replace a miter joint.

Shoe mold. For interior finish, a molding strip placed against the baseboard at the floor; also called "base shoe," or "carpet strip."

Smoothly machined. Free of defective manufacturing, with a minimum of 16 knife marks to the inch.

Smoothly sanded. Sanded sufficiently smooth so that sander marks will be concealed by painter's applied finish work.

Soffit. Underside of a subordinate part or member of a building, such as staircase, archway, or cornice.

Splayed edge. Chamfer extended to the full thickness or depth of the piece.

Splice. Joining of two similar members in a straight line.

Spline. Rectangular strip of wood that is substituted for the tongue.

Standing finish. Term applied to the finish of the openings and the base, and to all other finish work necessary for the inside.

Stile. Upright framing member in a panel door.

Stool. Flat, narrow shelf, forming the top member of the interior trim at the bottom of a window.

Stop bead. Vertical member of the interior finish of a window against which the sash butts or slides.

Stop molding. Small molding that stops the door from swinging through the opening as it is closed; also used to hold window sash in place.

Tenon. End of a piece of lumber formed to fit into a mortise.

Toenailing. Driving a nail so that it enters the first surface diagonally and usually penetrates the second at a slant.

Tongue. Projection on the edge of a board machined to fit into a groove in the adjacent piece.

Torus. Convex molding that generally approaches a semicircle in profile.

Trim. Finish components such as moldings applied around openings or at the floors and ceilings of rooms.

Trimming. Inside and outside finish and hardware on a building.

Wainscoting. Matched boarding or panel work covering the lower portion of a wall.

MANUFACTURED BOARDS AND PANELS

Acoustical board. Low-density, sound-absorbing, structural insulating board having a factory-applied finish and a fissured, felted mineral fiber, slotted or perforated surface pattern provided to reduce sound reflection. It is usually supplied for use in the form of tiles.

Acoustical panels (noncombustible). Acoustical panels are composed of asbestos fibers and cement binder, autoclaved for dimensional stability.

Adhesive. Substance that causes materials to stick to each other through surface attachment.

Air-felting. Forming a fibrous felted board from an air suspension of damp or dry fibers on a batch or continuous forming machine. It is referred to as the dry or semidry process (ASTM D 1554).

Asbestos cement board. A paste of portland cement heavily reinforced with asbestos fibers is used in the manufacture of asbestos cement board. To provide high strength in a short period of time, the boards are steam-cured. The result is a hard, smooth, highly fire-resistant material that has many uses in the construction industry.

Attrition mill. Reduction unit utilizing revolving disks with abrading plates to fiberize lignocellulosic raw materials.

Bark extract. Aqueous extract of various bark species possessing a significant quantity of complex, tanninlike constituents, which are reactive to formaldehyde and have potential value in the manufacture of resins.

Binder. Extraneous bonding agent, either organic or inorganic, used to bind particles together to produce a particle board.

Building board. Natural finished multipurpose structural insulating board.

Catalyst. Substance intended to accelerate hardening in adhesives such as hardener, accelerator, or activator.

Chip board

1. Board made by binding wood chips with phenolic resin or urea formaldehyde glue in a 4-ft-wide form. Chips are produced by feeding pieces of log into a disk waferizer, which cuts the pieces into thin wafers about $1\frac{1}{2}$ in. square and from 0.010 to 0.050 in. thick. They are dried, separated according to thickness into core and face wafers (thick wafers form the core), and fed to the production line. Here they receive a coating of glue and are blown into a steel forming table, first a layer of face wafers, then two layers.

2. Particle-board panel made of chiplike particles.

Chipping. Reduction of solid wood to particles; separation of pieces from the board itself.

Chips

1. Chunky lignocellulosic particles of various sizes.

2. Small pieces of wood chopped off a block by axlike cuts or produced by mechanical hogs, hammermills, etc.

Composite panels. Panels made by combining a reconstituted wood core with veneer faces. Waferboard or flake board usually is used as the core; however, any type of particleboard may be used. Panels are warp-free and have dimensional stability.

Compression ratio. Ratio of the densities of the wood species to the particleboard made from it.

Compression strength. Particleboard strength exhibited by forces tending to shorten the test specimen.

Conditioning. Bringing into equilibrium the moisture content and temperature of particleboard with those of a desired environment.

Core. Inner layer of particleboard.

Core stock

1. Particleboard used for overlaying. The overlay may consist of wood veneer, plastics, etc.
2. Products of flakes or particles, bonded with urea formaldehyde or phenolic resins, with various densities and related properties. It is used for furniture, casework, architectural paneling, doors, and laminated components.

Cork board. Ground cork is mixed with synthetic resin, formed into sheets of from 1 to 6 in. thick, and baked under pressure into rigid boards. The standard board length is 36 in., and widths vary according to the thickness.

Cured panel. Term refers to a solidified particle panel in which the binder has largely hardened.

Curls. Long flat flakes manufactured by the cutting action of a knife in such a way that they tend to be in the form of a helix (ASTM D 1554).

Cut-to-order. Sawing of the standard panel sizes into various shapes and dimensions.

Decorative hardboard. Hardboard scored or engraved after manufacture or by pressing during manufacture on a patterned caul to produce a decorative surface.

Delamination. Partial or total separation along the plane of the board.

Density. Stock panels range in density from 24 to 62 lb/cu ft. An increase in the density of a given product will increase the strength properties proportionately. An increase in density will also make a smoother, tighter edge.

Density profile. Density differentiation over the thickness of the board, which is usually highest at the surfaces and lowest in the innermost layer.

Dilutability. The ability to become diluted with water.

Dimensional stability. Term refers to the lack of dimensional changes in particleboard as the environmental humidity fluctuates or as the product is wetted with water.

Dryer. Equipment utilizing heat to bring the particle moisture content to acceptable low levels.

Embossed. Term applied to surfaces that are heavily textured in various decorative patterns by branding with heated roller. Embossed surfaces are used for doors, architectural paneling, wainscots, display units, and cabinet panels.

Excelsior board. Cement-bonded, relatively low density panel made up of excelsior particles and used mostly for roof decking although there are a number of structural applications.

Extender. Materials used to increase coverage of synthetic resins and improve certain properties of the manufactured board.

Exterior. Boards made with phenolic resins for resistance to weathering. For use as an exterior covering material.

Extruded particleboard. Particleboard manufactured by forcing a mass of particles coated with an extraneous binding agent through a heated die with the applied pressure parallel to the faces and in the direction of extruding.

Extrusion process. Process of particleboard manufacture in which the loose, blended material is forced into two hot parallel plates known as the extrusion die.

Felted mineral core. Rigid, monolithic, mineral fiber construction with a high degree of sound absorption.

Fiberboard. Material made from vegetable fibers and binding agents (such as glue) and pressed into sheets of varying densities. It is also called "hardboard." Its properties are similar to those of plywood.

Fibers. Slender threadlike elements, groups of wood fibers, or similar cellulosic material resulting from chemical or mechanical defiberization, or both, and referred to as "fiber bundles."

Fibrous felted boards. Felted wood-base panel material manufactured of refined or partly refined lignocellulosic fibers, characterized by an integral bond produced by an interfelting of fibers and in the case of certain densities and controlled manufacturing conditions by ligneous bond, and to which other materials may have been added during manufacture to improve certain properties.

Filled. Term applied to a board whose surface is filled and sanded and ready for painting. Such board is used for painted end products requiring firm, flat, true surfaces.

Filler. Inert product applied to the particleboard surface in order to fill the pores before the desired finish is applied.

Fire core. Special mineral fiber core that is highest in fire resistance.

Fire retardant. Particles treated with fire retardants.

Flake. Small wood particle of predetermined dimensions specifically produced as a primary function of specialized equipment of various types, with the cutting action across the direction of the grain, either radially, tangentially, or at an angle between. The cutting action is such as to produce a particle of uniform thickness, essentially flat, and having the fiber direction essentially in the plane of the flakes, cutting in over all character, it resembles a small piece of veneer.

Flake board. Resin-bonded particle board made up of flakes.

Flat-platen-pressed particleboard. Particleboard manufactured by pressing a mass of particles coated with an extraneous binding agent between parallel platens in a hot press with the applied pressure perpendicular to the faces.

Flat-press board. Particleboard manufactured by the flat-press process.

Flax board. Resin-bonded particleboard made of flax sheaves.

Floor underlayment. Panels specifically engineered for floor underlayment; underlay for carpets or resilient floor coverings.

Fluted board. Thick, extruded-type particleboard exhibiting holes along the length and within the thickness designed to lower unit weight. Holes are generated by pipes centrally located between the two hot plates in the extrusion die.

Fungicide. Chemical compound incorporated into the particleboard to forestall fungal attack.

Hardboard

1. Board made from processed wood chips. Chips of controlled size are subjected to high-pressure steam in pressure vessels. When the pressure is released, the chips "explode," and the cellulose and lignin are separated from the unwanted elements. Cellulose fibers and lignin are then mixed into a homogeneous mass and formed into a continuous board which is cut up into convenient lengths. These are pressed into uniform, hard, grainless sheets in heated presses.

2. Fibrous felted, homogeneous, or laminated panel having a density range of approximately 50 to 80 lb/cu ft, manufactured under carefully controlled optimum combinations of consolidating pressure, heat, and moisture so that a softening of lignin occurs and the board produced has a characteristic natural ligneous bond, and to which other materials may have been added during manufacture to improve certain properties.

Hard-pressed particleboards. Particle board having a density range of approximately 50 to 80-lb/cu ft.

Heat treating. Subjecting a wood-base panel material (usually hardboard) to a special heat treatment after hot-pressing to increase some strength properties and water resistance.

Hot-pressing. Increasing the density of a wet-felted or air-felted mat of fibers or particles by pressing the dried, damp, or wet mat between platens of hot-press to compact and set the structure by simultaneous application of heat and pressure.

Humidity control (noncombustible). An unperforated asbestos cement grid panel with a premium grade finish and special moisture-resistant back coating used for noncombustible humidity control.

Hydrolysis. Process of chemical decomposition either in the wood material or the binder.

Hygroscopic. Term characterizes ready absorption and retention of moisture as the environmental humidity increases. The reverse also holds true.

Impregnation. The introduction of chemicals such as fungicides into the manufactured board or particles. Also introduction of resin into the particles prior to consolidation in flapreg production.

Impurities. The inclusion of foreign matter such as inorganic dust, metal objects, and other impurities that are mixed in with wood particles.

Inorganic binder. Binder of inorganic origin, such as portland cement and gypsum.

Insecticide. Chemical compound incorporated into particleboard to forestall insect attack.

Insulating-type particleboard. Particleboard having a density range of approximately 15 to 26 lb/cu ft.

Insulating form board. Specially fabricated structural insulating board designed for use as a permanent form for certain poured-in-place roof constructions.

Insulating roof deck. Structural insulating board product designed for use in open-beam-ceiling roof construction. The product is composed of multiple layers of structural insulating board laminated together with water-resistant adhesive.

Interior-finish board. Structural insulating board with a factory-applied paint finish, fabricated in the form of plank, board, panel, or tile for interior use.

Intermediate density. Density ranging from 37 to 50 lb/cu ft, same as medium density.

Internal bond

1. Tensile strength of particleboard in the direction perpendicular to the surfaces of the board; same as tensile strength perpendicular to the surface.
2. Measure of the force two faces of a panel will withstand before pulling apart, measure of the strength of the bond between individual particles within the panel.

Labeled panels. Panels manufactured with a fire-retardant wood flake core.

Laminate. Single component of a laminar structure of more or less distinct characteristics.

Linear expansion. Change in dimension in the plane of a board in response to a specified change in relative humidity.

Low density. Term applied to particleboard whose density is less than 37 lb/cu ft.

Manufactured boards. Boards made by combining wood flakes or particles with resin binders and hot-pressing them into panels up to 2 in. thick, up to 8 ft wide, and up to 24 ft long. Virtually any dimension may be obtained by cutting or gluing segments together.

Medium-density building fiberboard. Felted, homogeneous, or laminated panel having a density range of approximately 26 to 50 lb/cu ft, manufactured of refined or partly refined lignocellulosic fibers with a primary integral bond, and to which other materials may have been added during manufacture to improve certain properties (ASTM D 1554).

Medium-density particleboard. Particleboard having a density of approximately 26 to 50 lb/cu ft.

Mineral fiberboards. Thick mats of mineral fibers, usually glass or rock wool, are covered with a backing of stiff paper on one or both sides to form rigid boards, ranging in thickness from $\frac{1}{2}$ to 2 in. The usual board size is 24×48 in. These units are used for roof deck insulation and are cemented to the deck with asphalt adhesive.

Modulus of elasticity. Measure of resistance to deflection under an applied load; an indication of the stiffness of a panel.

Modulus of rupture. Measure of the load necessary to break a panel.

Moisture content. It is important to control the moisture content of particleboard panels to be overlaid, as the core material and the overlay material must have near equal moisture content. Panels meeting industry standards have a moisture content of 7 to 9% unless otherwise specified.

Nail-base fiberboard sheathing. Specially manufactured high-density structural insulating board product designed for use in frame construction to permit the direct application of certain exterior siding materials, such as wood or cement asbestos shingles.

Oil contamination. Oil droplets that strike the particle surfaces during the drying operation.

Oriented strand board. Type of particle panel product composed of strand-type flakes that are purposefully aligned in directions that make a panel stronger, stiffer, and with improved dimensional properties in the alignment directions than a panel with random flake orientation.

Outer layer. Term refers to the outermost layer of the particleboard, same as surface layer.

Overlaid. Term applied to board faced with impregnated fiber sheets, hardboard, or decorative plastic sheets and used for applications such as furniture, doors, wall panelings, sink tops, cabinetry, and store fixtures.

Oxboard. An oriented strand board (OSB) that has five layers and is bonded with liquid phenolic resin. It is comparable in strength to plywood and is being used where sheathing plywood traditionally has been used.

Particle. Aggregate component of a particleboard manufactured by mechanical means from wood or other lignocellulosic material (comparable to the aggregate in concrete), including all small subdivisions of wood such as chips, curls, flakes, sawdust, shavings, slivers, strands, wood flour, and wood wool. Particle size may be measured by the screen mesh that permits passage of the particles and another screen on which they are retained or by the measured dimensions, as for flakes and strands.

Particleboard. Resin-bonded product made basically of wood or other lignocellulosic particles and produced in panel form. Particle geometry may vary from fibers to large chips. Particleboard may consist of a single or a number of layers.

Particleboard core stock. Common name given to particleboard manufactured for use as a core for overlaying.

Particleboard panel stock. Common name given to particleboard manufactured primarily for use as panel material and whose surfaces may be treated to obtain decorative effects.

Particle boards. Panel material composed of small discrete pieces of wood or other lignocellulosic materials that are bonded together in the presence of heat and pressure by a synthetic resin adhesive. Particle boards are further defined by the method of pressing. When the pressure is applied in the direction perpendicular to the faces as in a conventional multiplaten hot-press, they are defined as flat-platen pressed; when the applied pressure is parallel to the faces, they are defined as extruded.

Particleboard shelving. Though only one-fourth to one-eighth as stiff as wood or plywood, particleboard is used increasingly where loading is light, extra support is provided, or spans are short. They are frequently veneered or overlaid with higher stiffness materials to provide additional stiffness.

Particle geometry. The size and shape of the individual flakes and particles and the ratio of resin to particles greatly influence the properties of the board. Particle shape and resin content can be controlled to create a given set of physical properties. The size, type, and position of the particles also influence a panel's surface smoothness.

Perforated hardboard. Hardboard with closely spaced factory punched or drilled holes.

Perlite core. Rigid cellular combination of expanded perlite, mineral binders, and fibers. It is highly resistant to airborne moisture and has an exceptionally high sound attenuation value.

Planed-to-caliper hardboard. Hardboard machined to a close thickness tolerance.

Plastic foam boards. Polystyrene and polyurethane plastics are foamed by a patented process to about 40 times their original volume. The foamed material is molded into boards from $\frac{1}{2}$ to 3 in. thick, 12 or 24 in. wide, and from 4 to 12 ft long. This material is used for perimeter insulation for concrete floor slabs, for wall and roof deck insulation, and for roof decks when properly supported.

Prefinished particleboard. Particleboard with a factory-applied finish, such as lacquer, baked-on enamel, or similar finish.

Prefinished wall panel. Hardboard with a factory-applied finish, such as baked-on enamel, lacquer, or similar finish.

Primed or undercoated. Term applied to either filled or regular board with a factory-painted base coat, exterior or interior. Primed or undercoated board is used for many painted products.

Residual moisture. Moisture remaining in the particleboard upon its exit from the hot press.

Resin

1. General term used to refer to synthetic adhesives.

2. Two types of resin, urea-formaldehyde and phenol-formaldehyde are used in the manufacture of boards. Urea-formaldehyde is the most common and is suitable for interior use. Phenol-formaldehyde is used where the panel is subjected to extreme heat or humidity or for exterior applications.

Roof insulation board. Structural insulating board fabricated for use as above-deck-roof insulation.

Sawdust. Wood particles resulting from the cutting and breaking action of sawteeth.

Screen-back hardboard. Hardboard with a reverse impression of a screen on the back produced when a damp or wet mat is hot-pressed into a board and dried in the press.

Shaving. Small wood particle of indefinite dimensions developed incidental to certain woodworking operations involving rotary cutter heads usually turning in the direction of the grain. Because of the cutting action, a thin chip of varying thickness, usually feathered along at least one edge and thick and usually curled at another, is produced.

Sheathing. Structural insulating board for use in building construction, integrally treated, impregnated, or coated to give it additional water resistance.

Shingle backer. Specially fabricated sheathing-grade structural insulating board used as a backer strip in coursed shingle construction.

Size. Asphalt, rosin, wax, or other additive introduced to the stock for a fibrous felted board prior to forming or added to the blend of particles and resin for a particle board to increase water resistance.

Sliver. Particle of nearly square or rectangular cross section with a length parallel to the grain of the wood of at least four times the thickness.

Smooth-two-side hardboard. Hardboard produced from a dry mat pressed between two smooth hot-platens.

Special properties. Panels are given special finishes or treatments at the mill for specific end uses. They may be filled or primed for easy painting or embossed for decorative textured surfaces. Panels may be made edge-banded or given special sanding or overlays. Laminating and edge-gluing are used to achieve unusual sizes.

Standard hardboard. Hardboard substantially as manufactured at the end of hot-pressing, except for humidification to adjust moisture content, trimming to size, and other subsequent machining, and having the properties associated with hardboard meeting specifications for that quality product (ASTM D 1554).

Strand. Relatively long (with respect to thickness and width) shaving consisting of flat, long bundles of fibers having parallel surfaces.

Strawboard. Board made of compressed wheat straw, processed at 350°F to 400°F and covered with a tough kraft paper. Two grades are produced, structural board and insulating board. The insulating grade weighs 3 lb/cu ft in 2-in. thickness; the same thickness of structural board weighs 3.5 lb/cu ft. For exterior use, the paper used is impregnated with asphalt.

Structural insulating board. Fibrous felted homogeneous or laminated panel having a density range of approximately 10 to 26 lb/cu ft, manufactured of refined or partly refined lignocellulosic fibers with a primary integral bond, and to which other materials may have been added during manufacture to improve certain properties (ASTM D 1554).

Tack. Slight stickiness with which blended particles can adhere to each other.

Tempered hardboard. Hardboard subjected to tempering or specially manufactured with other variation in the usual process so that the resulting product has the special properties of stiffness, strength, and water resistance associated with boards meeting specifications for quality products.

Tempering. Manufacturing process of adding to a fiber or particle panel material a siccative, such as a drying oil blend of oxidizing resin, which is stabilized by baking or other heating after introduction.

Three-layer board. Particleboard possessing three distinct layers—two faces and one core.

Toxic-treated board. Board treated with chemicals to resist insects, mold, and decay-producing fungi. Toxic-treated board is used for tropical or other applications where wood products require protection against insect attack or decay.

Urea resin. Synthetic resin obtained by reacting urea with formaldehyde in the presence of acid or alkaline catalysts. It is the most common resin utilized in particleboard manufacture.

Viscosity. Liquid or gas parameter that basically refers to its resistance to flow.

Waferboard. Waferboard is made up of large wood flakes, generally $1\frac{1}{4}$ in. or longer and either square or oblong. The thickness of the flakes is usually $\frac{1}{20}$ to $\frac{1}{40}$ of length. The particles are laid out randomly, which produces a board with strength that is equal in all directions and is not dependent on the direction of the grain. Waferboards are coated with a waterproof resin and wax and then bonded with heat and pressure.

Warp. Any type of distortion in particle-board panels.

Wet felting. Forming a fibrous felted board mat from a water suspension of fibers and fiber bundles by means of a deckle box, fourdrinier, or cylinder board machine.

Wood-base fiber and particle panel materials. Generic term applied to a group of board materials manufactured from wood or other lignocellulosic fibers or particles, to which binding agents and other materials may be added during manufacture to obtain or improve certain properties. Composed of two broad types, fibrous felted and particle boards.

Wood-cement board. Panel material where wood, usually in the form of excelsior, is bonded with inorganic cement.

Wood flake core panels. Wood flake core panels are available in fire-retardant grade, with the label of Underwriters Laboratories, Inc., for flame spread rating of Class I or Class II.

Wood flour. Very fine wood particles generated from wood, reduced by a ball or similar mill until it resembles wheat flour in appearance, and of such size that the particles usually will pass through a 40-mesh screen.

Wood wool (excelsior). Long, curly, slender strands of wood used as an aggregate component for some particle boards.

Wood veneered. Core stock overlaid at the mill with various wood veneers. It is used for furniture, panels, wainscots, dividers, cabinets, etc.

PLASTICS AND USES

Abrasion resistance. Ability to withstand the effects of repeated scuffing or scratching.

ABS plastics

1. Group of amorphous thermoplastics produced by combining the three monomers: acrylonitrile, butadiene, and styrene. Resultant products provide a wide range of properties based on specific compositions, molecular weight, and degree of grafting.
2. Copolymers of acrylonitrile, butadiene, and styrene.

Acetal

1. Strong, stiff, and most resilient thermoplastic with excellent low-temperature and electrical properties. It resists most organic solvents and has superior abrasion resistance.
2. Compound that bridges the gap between metals and plastic. It has superior tensile and flexural properties, resists bending under load, has a low coefficient of friction, and demonstrates dimensional stability.

Acetal plastics. Plastics based on polymers having a predominance of acetal linkages in the main chain.

Acetate. Generic name for cellulose acetate plastics, particularly for the fibers thereof.

Acetone. Clear inflammable liquid capable of dissolving other materials.

Acetyl tributyl citrate. Plasticizer commonly used in vinyl plastics.

Acidolysis. Process of reacting an acid with an ester.

Acrylate. Colorless, unsaturated acid which polymerizes readily.

Acrylic ester. Ester of acrylic acid or of a structural derivative of acrylic acid, such as methyl methacrylate.

Acrylic plastics

1. Thermoplastic molding powders, extruded sections and sheets, and cast sections and sheets produced by the polymerization of the monomeric derivatives of acrylic acid.
2. Plastics formed by the polymerization of an acrylate, such as a derivative of acrylic acid.
3. Plastics that transmit and control light. Acrylic plastics are extremely stable against discoloration and possess superior dimensional stability, desirable structural and thermal properties, and exceptional resistance to weather, breakage, and chemicals.

Acrylic resin. Synthetic resin prepared from acrylic acid or from a derivative of acrylic acid.

Acrylic rubber. Synthetic rubber at least partially made from acrylonitrile, or from ethyl acrylate copolymerized with many of the monomers or block polymers of the synthetic rubber family.

Acrylonitrile. Acrylonitrile is most useful in copolymers. Its copolymer with butadiene is nitrile rubber, and there exist several copolymers with styrene that are tougher than polystyrene. It is also used as a synthetic fiber and chemical intermediate.

Acrylonitrile-butadiene-styrene (ABS). ABS combines high rigidity with high impact strength. Other properties include heat resistance, surface hardness, dimensional stability, chemical resistance, and electrical properties. It is easily thermoformed.

Addition polymerization

1. Production of a polymer molecule by inducing smaller molecules to break an internal double carbon bond and then join together.

2. Polymerization that takes place by the linking of small molecules, such as those of a vinyl compound, without the loss of water or other by-products of chemical reaction.

Additive.   Material added in minor amounts to basic resins or compounds to alter properties.

Adhesion promoter.   Coating that is applied to the substrate before it is extrusion coated with the plastics and that improves the adhesion of the plastics to the substrate.

Adiabatic.   Process in which no heat is deliberately added or removed.

Adipic acid.   Esters of adipic acid are used as plasticizers and lubricants.

Alcohol.   Used directly as solvent and diluent. Many esters of alcohols with organic acids are plasticizers.

Alkyd molding compounds.   Compounds formulated with polyester resins, catalysts, cross-linking monomers, fillers, reinforcements, lubricants, and pigments.

Alkyd plastics.   Plastics based on resins composed principally of polymeric esters, in which the recurring ester groups are an integral part of the main polymer chain, and in which ester groups occur in most cross-links that may be present between chains.

Alkyd resin
1. Polyester convertible into a cross-linked form; polyester requiring a reactant of functionality higher than 2 or having double bonds.
2. Synthetic resin made by the condensation of a polyhydric alcohol, such as glycerol, and a polybasic acid or anhydride, such as phthalic anhydride.
3. Polyester resins made with some fatty acid as a modifier.

Alloy.   Plastic industry term to denote blends of polymers or copolymers with other polymers or elastomers.

Allyl plastics
1. Plastics based on allyl resins.
2. Plastics made by polymerization of substances containing the allyl group.

Allyl resin
1. Resin made by polymerization of chemical compounds containing the allyl group.
2. Synthetic resin formed by the polymerization of chemical compounds. The principal commercial allyl resin is a casting material that yields allyl carbonate polymer.

Aluminum distearate.   White powder used as a lubricant for plastics.

Amino plastics
1. Plastics based on amino resins.
2. Urea and melamine formaldehyde.

Amino resin.   Resin made by polycondensation of a compound containing amino groups, such as urea or melamine, with an aldehyde, such as formaldehyde, or an aldehyde-yielding material.

Amyl oleate.   Solvent and plasticizer for cellulose and vinyl resins.

Anatase.   Naturally crystalized form of titanium dioxide.

Aniline-formaldehyde plastics.   Plastics made by condensation of aniline and formaldehyde.

Antiozonant.   Substance added to elastomers to retard or prevent deterioration caused by exposure to air containing ozone.

A Stage.   Early stage in the production of phenol-formaldehyde-type plastics, consisting of a resin that will melt on heating and is soluble in solvents such as alcohol or acetone. A-stage resins are used for making varnishes and laminates.

Automatic hydrocarbons.   Hydrocarbons derived from or characterized by the presence of unsaturated resonant ring structures.

Bacteriostat.   Agent that, when incorporated in a plastic compound, will prevent the growth of bacteria on surfaces of formed particles.

Bagasse.   Reinforcement fiber from cane sugar, used in laminates and molding powders.

Bakelite.   Hard amberlike substance manufactured from the coal tar derivatives, phenol and formaldehyde.

Barium ricinoleate.   Heat stabilizer imparting good clarity, used most often in vinyl plastisols and organosols.

Benzene.   Solvent and intermediate in the production of phenolics, epoxies, styrene, and nylon.

Benzoyl peroxide.   Catalyst used in the polymerization of styrene, vinyl, and acrylic resins and in curing polyester and silicone resins.

Benzyl alcohol.   Water-white liquid used as a solvent for cellulosics and other resins.

Benzyl cellulose.   Member of the cellulosic family of plastics, a benzyl ether of cellulose.

Binder.   In a reinforced plastic, the continuous phase that holds together the reinforcement.

Bipolymer.   Polymer derived from two species of monomer.

Bisphenol.   Intermediate in the production of epoxy, polycarbonate, and phenolic resins.

Bleeding.   Undesired movement of certain materials in a plastic to the surface of the finished article or into an adjacent material.

Blister
1. Imperfection; a rounded elevation of the surface of a plastic with boundaries that may be more or less sharply defined, somewhat resembling in shape a blister on the human skin.
2. Raised area on the surface of a molding caused by the pressure of gases inside it.

Blocking
1. Unintentional adhesion between plastic films or between a film and another surface.
2. Undesired adhesion between touching layers of a material like that which occurs under moderate pressure during storage or use.

Bloom.   Visible exudation or efflorescence on the surface of a material.

Blow molding.   Method of forming plastics by the application of air pressure to force the plastic against a rigid surface shaped to contours desired in the finished product.

Boric acid esters.   Flame retardants for plastics and plasticizers.

Boss.   Protuberance on a plastic part designed to add strength, facilitate alignment during assembly, provide for fastenings, etc.

Brominated aromatic antimony oxide.   BAAO and halogenated organophosphorus products are used to flame-retard polystyrene. Flame-retarded polystyrene is used in building and construction, transportation, appliances, packaging, and furniture.

B Stage

1. Second stage in the curing or hardening of thermosetting materials such as phenol-formaldehyde resins. Molecular growth has reached the stage at which the material is still thermoplastic but will not melt. The material will soften and swell in solvents such as alcohol or acetone but will not dissolve.

2. Intermediate stage in the reaction of a thermosetting resin in which the material softens when heated and swells in contact with certain liquids, but does not entirely fuse or dissolve. Resins in thermosetting molding compounds are usually in this stage.

Buckling. Crimping of the fibers in a composite material, often occurring in glass reinforced thermosets due to resin shrinkage during cure.

Burning rate. Tendency of plastic articles to burn at given temperatures. Certain plastics, such as those based on shellac, burn readily at comparatively low temperatures. Others will melt or disintegrate without actually burning, or will burn only if exposed to direct flame. The latter are often referred to as "self-extinguishing."

Butanetriol. Colorless liquid, used as an intermediate for alkyd resins and a plasticizer for cellulosics.

Butt fusion. Method of joining pipe, sheet, or other similar forms of a thermoplastic resin whereby the ends of the two pieces to be joined are heated to the molten state and then rapidly pressed together to form a homogeneous bond.

Butyl benzoate. Plasticizer and solvent for cellulosics.

Butylene plastics. Plastics based on resins made by the polymerization or copolymerization of butene with one or more unsaturated compounds, the butene being greater by weight.

Butyl formate. Solvent for several resins, including nitocellulose and cellulose acetate.

Butyl ricinoleate. Plasticizer for vinyl resins and cellulose acetate butyrate, derived from castor oil and butyl alcohol.

Calendering. Method of making thin sheets or film by squeezing the plastic between heavy revolving rollers. Plastic may also be applied to paper or cloth in this manner.

Capric acid. Plasticizer and an intermediate for resins.

Casein. Protein material precipitated from skimmed milk by the action of either rennet or dilute acid. Rennet casein finds its main application in the manufacture of plastics. Acid casein is a raw material used in a number of industries, including the manufacture of adhesives.

Casting. Method of forming a plastic into a desired shape by pouring it into a mold or onto a wheel or belt and allowing it to harden without use of external pressure.

Cast phenolic resins. Cast resins made by combining phenol, formaldehyde, and one or more of a variety of catalysts plus additives, including color, which in liquid form are poured directly into straight draw molds, split molds, or cored molds.

Catalyst. Substance which causes or accelerates a chemical reaction when added to the reactants in minor amounts, without being permanently affected by the reaction.

Cavity. Depression in a mold that, shaped to the contours desired in the finished product, receives the plastic during processing and gives the material its own form.

Cellular plastic. Plastic whose density is decreased substantially by the presence of numerous cells disposed in its mass.

Celluloid

1. Inflammable thermoplastic that tends to yellow and become brittle with age.

2. Plastic made from nitrocellulose and camphor.

3. Thermoplastic material made by the intimate blending of cellulose nitrate with camphor. Alcohol is normally employed as a volatile solvent to assist plasticization and is subsequently removed.

Cellulose

1. Substance used by land plants as a structural material in the walls of cells. Cellulose in the form of cotton linters is the raw material from which plastics such as nitrocellulose and cellulose acetate are made.

2. Natural high polymeric carbohydrate found in most plants; the main constituent of dried woods, jute, flax, hemp, ramie, etc. Cotton is almost pure cellulose.

Cellulose acetate butyrate molding compositions. Combinations of cellulose acetate butyrate, plasticizers, and pigments and dyes supplied as granules and pellets for fabrication by injection, extrusion, blowing, or compression molding.

Cellulose acetate sheets. Thermoplastic sheets made from cellulose acetate, plasticizers, pigments and dyes. Cellulose acetate is produced by esterifying alpha cellulose with acetic acid and anhydride in the presence of a catalyst.

Cellulose derivatives. Cellulose is a naturally occurring high polymer found in all woody plant tissue and in such materials as cotton. It can be modified by chemical processes into a variety of thermoplastic materials, which, in turn, may be still further modified with plasticizers, fillers, and other additives to provide a wide variety of properties. The oldest of all plastics is cellulose nitrate.

Cellulose nitrate

1. A tough plastic, cellulose nitrate is widely used for tool handles and similar applications requiring high impact strength. Its high flammability requires great caution, particularly when it is in the form of film. Cellulose nitrate is the basis of most of the widely used commercial lacquers for furniture and similar items.

2. Derivative of cellulose produced by nitration of cellulose.

3. Nitric acid ester of cellulose manufactured by the action of a mixture of sulfuric and nitric acids on cellulose, for example, on purified cotton linters.

Cellulose propionate molding compositions. Molding compositions consisting of cellulose propionate, plasticizers, and pigments and dyes. Cellulose propionate is made by esterifying purified cellulose with propionic acid and anhydride, with or without acetic acid present, in the presence of a catalyst.

Cellulosic. Classed among the higher impact thermoplastics. Acetate butyrate functions well as steering wheels, tool handles, dispensers, blister packages. Cellulose acetate applications include containers, handles, buttons, displays, etc.

Cellulosic plastics. Plastics based on cellulose compounds, such as esters (cellulose acetate) and ethers (ethyl cellulose).

Cementing. Joining of plastics by means of solvents or adhesives.

Chalking. Specific type of bloom, characterized by a dry, chalk-like appearance of the surface of a plastic article.

Chlorinated polyether

1. Polymer obtained from pentaerythritol by preparing a chlorinated oxetane and polymerizing it to a polyether by opening the ring structure.

2. Polyether that is resistant in a corrosive environment. Its electrical, mechanical, and chemical properties make it possible to precision machine parts that remain extremely stable under severe conditions.

Chlorinated poly(vinyl chloride) plastics. Plastics based on chlorinated poly(vinyl chloride) in which the chlorinated poly(vinyl chloride) is in the greatest amount by weight.

Chlorofluorocarbon plastics. Plastics based on polymers made with monomers composed of chlorine, fluorine, and carbon only.

Chopped reinforcing fibers. Fibers available in various lengths from $\frac{1}{8}$ to 2 in., blended with resins and other additives to prepare molding compounds for compression, transfer, and injection molding and for encapsulation.

Clamping plate. Plate fitted to a mold, used to fasten the mold to a molding machine.

Clarifier. Additive that increases the transparency of a material.

Clicker die. Cutting die for stamping out blanks from plastic sheeting.

Closed cell. Cell totally enclosed by its walls and hence not interconnecting with other cells.

Coagulation. Process of bringing together the tiny droplets or particles of a substance that is in the form of an emulsion or dispersion. The addition of certain chemicals will often bring about coagulation.

Coextrusion. Combining of two or more polymer melt layers during the film or sheet extrusion process.

Cold-drawing. Stretching process employed to improve the tensile properties of thermoplastic filaments and films.

Cold flow. Distortion of a plastic that takes place under continuous stress at low room temperatures.

Cold-molded plastics. Plastics shaped by molding or stamping under high pressure at room temperature.

Cold-pressing. Bonding operation in which an assembly is subjected to pressure without the application of heat.

Cold-setting. Hardening or setting of a plastic that takes place at normal temperatures.

Collapse

1. Inadvertent densification of cellular material during manufacture resulting from a breakdown of the cell structure.

2. Contraction of the walls of a container that, on its cooling leads to a permanent indentation.

Collodion. Solution of cellulose nitrate in alcohol and ether.

Colorfastness. Ability of a plastic to retain its original color.

Combination reinforcing mats. Mats consisting of one ply of woven roving chemically bonded to chopped strand mat are available. These products form a strong and drapable reinforcement that combines the bidirectional fiber orientation of woven roving with the multidirectional fiber orientation of chopped strand mat.

Compatibility. Ability of two or more substances to mix together without objectionable separation.

Composite. Combination of two or more materials that do not dissolve into each other or otherwise merge completely although they act in concert.

Composite mold. Mold in which several different shapes are produced in one cycle.

Compound. Combination of ingredients before being processed or made into a finished product.

Compression molding. Method of forming plastics wherein the material, after being placed in a mold, is made to conform to the outlines of the mold by the use of heat and pressure.

Compression zone. Portion of an extruder barrel in which melting is completed.

Condensation

1. Linking together of two or more molecules by chemical action, which results in the release of simple molecules, such as water, as by-products.

2. Chemical reaction in which two or more molecules combine, with the separation of water or some other simple substance.

Condensation resins. Resins formed by poly condensation, such as the alkyd, phenol-aldehyde, and urea-formaldehyde resins.

Contact pressure resins. Liquid resins that thicken or resinify on heating and when used for bonding laminates require little or no pressure.

Copolymer

1. Complex molecules formed by a reaction between two or more unlike monomers.

2. Polymer formed by the polymerization of a mixture of two or more monomers. Vinyl acetate and vinyl chloride are polymerized in this way to provide plastics in which the molecule is formed by the linking together of both molecules. Copolymers are also described as "interpolymers."

Core. Central member of a sandwich construction (can be honeycomb material, foamed plastic, or solid sheet) to which the faces of the sandwich are attached.

Cotton linters. Short cotton fibers that are unsuitable for textile use, the source of cellulose for making cellulose plastics.

Crater. Small, shallow surface imperfection.

Crazing

1. Fine cracks at or under the surface of a plastic.

2. Fine cracks that may extend in a network on or under the surface or through a layer of a plastic material.

Creep

1. Time-dependent part of strain resulting from stress.

2. Dimensional change with time of a material under load, following the initial instantaneous elastic deformation. Creep at room temperature is sometimes called "cold flow."

Cresol resin. Phenolic-type resin obtained by condensing a cresol with an aldehyde.

Cross-linking. Joining together of threadlike polymer molecules by chemical bonds. Cross-links result in the creation of a network structure, preventing free movement of the long molecules relative to one another. Cross-linking brings about the curing or setting of a polymer. Vulcanization of rubber is caused by cross-linking of the long rubber molecules.

Crystalline. Term applied to a body whose molecules are arranged in regular geometric patterns.

C stage

1. Final stage in the curing (hardening) of a thermosetting plastic. The molecules have formed a network

structure, and the material will not melt on heating nor dissolve in solvents.

2. Final stage in the reactions of a thermosetting resin in which the material is relatively insoluble and infusible. Thermosetting resins in fully cured plastics are in this stage.

Cure

1. Change the properties of a polymeric system into a more stable, usable condition by the use of heat, radiation, or reaction with chemical additives.

2. Change the physical properties of a material by chemical reaction, which may be condensation, polymerization, or vulcanization. The change is usually accomplished by the action of heat and catalysts, alone or in combination, with or without pressure.

Curing. Setting process, during which a plastic material loses its plasticity and becomes infusible and insoluble. Casein, for example, is cured by immersion in formaldehyde; rubber is cured (vulcanized) by heating with sulfur. The curing process takes place as a result of the formation of cross-links between the threadlike molecules, or by the establishment of a network structure by molecular branching.

Curing temperature. Temperature at which a cast, molded, or extruded product, a resin-impregnated reinforcing material, an adhesive, etc., is subjected to curing.

Cycle. Time required to complete a molding operation.

Cyclohexanol acetate. Nonflammable solvent for cellulosics and many other resins.

Debossed. Term describing the results of indent or cut in design or lettering of a surface.

Decorative laminates. Products made by combining different materials to take advantage of the specific property of each component. Laminates are characterized by a hard surface that is highly resistant to damage from scuffing or scratching, unaffected by boiling water and most common household chemicals, and with good resistance to staining and cigarette burns.

Deep drawing. Process of forming a thermoplastic sheet in a mold involving a high draw ratio.

Deflection temperature. Temperature at which a specimen will deflect a given distance at a given load under prescribed conditions of test.

Deformation. Any distortion of the outline or form of a plastic article.

Dehydroacetic acid. Colorless crystalline material with uses as a plasticizer, fungicide, and bactericide.

Delamination

1. Separation of the layers of material in a laminate.

2. Separation of the layers in a laminate caused by the failure of the adhesive.

Deliquescent. Capable of attracting moisture from the air.

Density. Weight per unit volume of a substance, expressed in grams per cubic centimeter, pounds per cubic foot, etc.

Depolymerization. Reversion of a polymer to its monomer or to a polymer of lower molecular weight.

Desiccant. Substance that can be used for drying purposes because of its affinity for water.

Deterioration. Permanent change in the physical properties of a plastic, evidenced by impairment of these properties.

Diallyl ester. Ester made by reaction of a dibasic acid with allyl alcohol, such as diallyl phthalate.

Dibasic acid. Acid containing two acidic groups in the molecule.

Dibenzyl ether. Plasticizer for cellulose nitrate.

Dibutyl adipate. Plasticizer for vinyl and cellulose resins.

Dicapryl adipate. Plasticizer for cellulose and vinyl resins, yielding good low temperature flexibility.

Die. Block, usually metal, containing one or more cavities into which a plastic material is placed for processing and from which it takes its shape upon the application of heat and/or pressure.

Die blades. Deformable members attached to a die body that determine the slot opening and that are adjusted to produce uniform thickness across the film or sheet produced.

Dielectric strength. Relative ability of a plastic to resist passage of an electric current.

Diffusion. Movement of a material, such as a gas or liquid, in the body of a plastic.

Diffusion couple. Assembly of two materials in such intimate contact that each diffuses into the other.

Dimensional stability. Ability of a plastic part to retain the precise shape in which it was molded, fabricated, or cast.

Discoloration. Change from the original color, often caused by overheating, light exposure, irridation, or chemical attack.

Dispersion. Finely divided particles of a material in suspension in another substance.

Draw. To form a sheet of plastic material by stretching it in a hot die or mold.

Dry-coloring. Method used by fabricators for coloring plastics by tumble-blending uncolored particles of the plastic material with selected dyes and pigments.

Dry spot. Imperfection in reinforced plastics; an area of incomplete surface film where the reinforcement has not been wetted with resin.

Durometer hardness. Indentation hardness test used for plastics and rubbers.

Dyes. Synthetic or natural organic chemicals that are soluble in most common solvents.

E glass. Borosilicate glass used in reinforced plastics.

Elastomer

1. Macromolecular material that at room temperature returns rapidly to approximately its initial dimensions and shape after substantial deformation by a weak stress and release of the stress.

2. Material that at room temperature stretches under low stress to at least twice its length and snaps back to the original length upon release of stress.

Electrical properties. Primarily the resistance of a plastic to the passage of electricity.

Electronic treating. Oxidizing a film of polyethylene to render it printable by passing the film between the electrodes and subjecting it to a high-voltage corona discharge.

Embossing. Techniques used to create depressions of a specific pattern in plastics film and sheeting.

Emulsion polymerization. Polymerization that takes place while the monomer is suspended as tiny droplets in another liquid, usually water.

Epoxy. Thermoset with excellent electrical and mechanical properties, chemical resistance, and resilience; it can be made flexible.

Epoxy plastics. Thermoplastic or thermosetting plastics containing ether or hydroxyalkyl repeating units, or both, resulting from the ring-opening reactions of lower molecular weight polyfunctional oxirane resins or compounds with catalysts or with various polyfunctional acidic or basic coreactants.

Epoxy resins

1. Wide class of condensation polymers made by reacting epichlorohydrin with polyhydric compounds, such as phenols, glycols, and novolacs.

2. Based on ethylene oxide, its derivatives or homologs, epoxy resins form straight-chain thermoplastics and thermosetting resins by, for example, the condensation of bisphenol and epichlorohydrin.

Ester. Reaction product of an alcohol and an acid.

Ethers. Organic compounds in which an oxygen atom is interposed between two carbon atoms or organic radicals in the molecular structure.

Ethyl cellulose plastics. Thermoplastic molding powder or sheet consisting of ethyl cellulose, plasticizers, pigments, and dyes. Ethyl cellulose is the ethyl ether of cellulose made by etherifying alkali cellulose with ethyl chloride.

Ethylene plastics. Plastics based on polymers of ethylene or copolymers of ethylene with other monomers, the ethylene being in greatest amount by mass.

Ethylene-vinyl acetate. Copolymers from these two monomers form a class of plastics materials that retain many of the properties of polyethylene but have considerably increased flexibility, for their density-elongation and impact resistance are also increased.

Ethyl formate. Solvent for cellulose acetate and cellulose nitrate.

Extender. Substance, generally having some adhesive action, added to a plastics composition to reduce the amount of the primary resin required per unit area.

Extensibility. Ability of a material to extend or elongate upon application of sufficient force.

Extrusion. Process whereby heated or unheated plastic is forced through a shaping orifice and becomes one continuously formed piece.

Fabricate. Method of forming plastic into a finished article by machining, drawing, and similar operations.

Fadeometer. Apparatus for determining the resistance of resins and other materials to fading. The apparatus accelerates the fading by subjecting the article to high-intensity ultraviolet rays of approximately the same wavelength as those found in sunlight.

Fiber

1. Threads of a small cross section made by chopping filaments.

2. Form of polymer that results from the close packing of long molecules alongside one another, like sticks in a bundle of faggots. Fiber is normally many thousands of times as long as it is wide and displays great strength in the direction of its long axis.

Fiber show. Strands or bundles of fibers that are not covered by resin and that are at or above the surface of a reinforced plastic.

Filaments

1. Individual glass fibers of indefinite length; many together form a strand.

2. Variety of fiber characterized by extreme length, which permits its use in yarn with little or no twist and usually without the spinning operation required for fibers.

Filler

1. Relatively inert material added to a plastic to modify its strength, permanence, working properties, or other qualities.

2. Substance that is mixed with a plastic to modify the plastic's properties. Wood flour is commonly used as a filler with phenolformaldehyde plastics.

3. Inert substance added to a plastic. Fillers may also improve physical properties, particularly hardness, stiffness, and impact strength.

Film

1. Sheeting with nominal thickness no more than 0.010 in. Term may also refer to liquid coating.

2. Very thin sheet of material, such as that used for packing.

Fines. Very small particles, usually under 200 mesh, accompanying larger grains, usually of molding powder.

Fish eye

1. Small globular mass that has not blended completely into the surrounding material.

2. Fault in transparent or translucent plastics materials, such as film or sheet, appearing as a small globular mass and caused by incomplete blending of the mass with surrounding material.

Fissure. Crack, separation, or split in a formed cellular article.

Flake. Term used to denote the dry, unplasticized base of cellulosic plastics.

Flame-resistant plastic. Plastic material that will not support flame when tested in accordance with the ASTM standards for flammability (ASTM E 84).

Flame-retardant resin. Resin that is compounded with certain chemicals to reduce or eliminate its tendency to burn. For polyethylene and similar resins, chemicals such as antimony tiroxide and chlorinated paraffins are useful.

Flame-retarded ABS polymers. Polymers manufactured using additive-type chemicals. Most important are brominated aromatic products in combination with antimony oxide. End uses include appliances and transportation.

Flame-spraying

1. Application of a coating of plastic, such as polythene, by spraying the plastic through a flame. Plastic melts as it passes through the flame, hardening to a coherent coating.

2. Applying a plastics coating in which finely powdered fragments of the plastics, together with suitable fluxes, are projected through a cone of flame onto a surface.

Flame-treating. Rendering inert thermoplastic objects receptive to inks, lacquers, paints, adhesives, etc., by bathing the object in an open flame to promote oxidation of the surface of the article.

Flammability. Measure of the extent to which a material will support combustion.

Flash. Excess plastic material that forms on a molded part where the sections of the mold come together.

Flash point. Lowest temperature at which a combustible liquid will give off a flammable vapor that will burn momentarily.

Flexibilizer. Additive that makes a resin or rubber more flexible, that is, less stiff; also, a plasticizer.

Flexural strength. Strength of a material in bending, expressed as the tensile stress of the outermost fibers of a bent test sample at the instant of failure. With plastics, this value is usually higher than the straight tensile strength.

Fluorocarbon plastics. Plastics based on polymers made with monomers, composed of fluorine and carbon only.

Fluorohydrocarbon plastics. Plastics based on polymers made with monomers, composed of fluorine, hydrogen, and carbon only.

Fluoroplastics

1. Plastics based on polymers with monomers, containing one or more atoms of fluorine or copolymers of such monomers with other monomers, the fluorine containing the monomer(s) being in greatest amount by mass.

2. Most slippery of all solids, with almost complete chemical inertness and excellent electrical and mechanical properties. Fluoroplastics are capable of continuous service up to 500°F and down to 450°F. They are unaffected by outdoor weathering.

Foamed plastics

1. High-molecular-weight plastics expanded by the use of a foaming agent to produce a sheet or block having either unit cells that are hollow spheres or interconnected cells.

2. Resins, either rigid or flexible, in sponge form, useful in sandwich construction and insulation.

3. Resins in sponge form. The sponge may be flexible or rigid, the cells closed or interconnected, the density anything from that of the solid parent resin down to, in some cases, 2 lb/cu ft. The compressive strength of rigid foams is fair, making them useful as core materials for sandwich structures. Both types are good heat retarders.

Foaming agents. Chemical agents that when added to plastics and rubbers generate inert gases on heating, causing the resins to assume a cellular structure.

Foam in place. Term referring to the deposition of foams that requires the foaming machine to be brought to the work that is "in place" as opposed to bringing the work to the foaming machine.

Foil decorating. Printing molding paper or textile or plastic foils with compatible inks directly into a plastic part so that the foil is visible below the surface of the part as an integral decoration.

Former. Shaped object that is dipped into liquid and then withdrawn to leave a film of the liquid on its surface. A former dipped into an emulsion of plastic, for example, will provide a film of plastic in the shape of the former as it dries or hardens.

Formica. Registered trademark of the Formica Company for high-pressure laminates of melamine-formaldehyde, phenolic, and other thermosetting resins with paper, linen, canvas, glass, etc.

Forming. Process in which the shape of plastic pieces such as sheets, rods, or tubes is changed to a desired configuration.

Formulation. Combination of ingredients before being processed or made into a finished product.

Friction calendering. Process whereby an elastomeric compound is forced into the interstices of woven or cord fabrics while passing through the rolls of a calender.

Frosting. Light-scattering surface resembling fine crystals.

Frothing. Technique for applying urethane foam in which blowing agents or tiny air bubbles are introduced under pressure into the liquid mixture of foam ingredients.

FRP. Fiber-glass-reinforced plastic.

Fungi resistance. Ability of plastic pipe to withstand fungi growth or their metabolic products, or both, under normal conditions of service or laboratory tests simulating such conditions.

Furane resins. Dark-colored thermosetting resins available primarily as liquids ranging from low-viscosity polymers to thick, heavy syrups.

Furan plastics. Plastics based on furan resins.

Furan resins. Resins in which the furan ring is an integral part of the polymer chain, the furan being in greatest amount by mass.

Furfural resin. Dark-colored synthetic resin of the thermosetting variety obtained by the condensation of furfural with phenol or its homologs. It is used in the manufacture of molding materials, adhesives, and impregnating varnishes. Its properties include high resistance to acids and alkalis.

Fuse. To join two plastic parts by softening the material by heat or solvents.

Gate. Part of the passage through which plastic material enters the cavity of a mold.

Gauge (gage). Standard of measurement of the thickness of plastic sheet or film, expressed in thousandths of an inch.

Gel

1. Semisolid system consisting of a network of solid aggregates in which liquid is held; the initial, jellylike, solid phase that develops during the formation of a resin from a liquid.

2. In polyethylene, the small amorphous resin particle that differs from its surroundings by being of higher molecular weight and/or cross-linked, so that its processing characteristics differ from the surrounding resin to such a degree that it is not easily dispersed in the surrounding resin. Gel is readily discernible in thin films.

Gel coat. Thin outer layer of resin, sometimes containing pigment, applied to a reinforced plastics molding as a cosmetic.

Glass-bonded mica. Moldable thermoplastic material that uses glass as a binder and mica as a filler.

Glass fiber. Glass filament cut into definite lengths suitable for spinning into yarn.

Glass filament. Form of glass less than 0.005 in. in diameter, long in length.

Glass finish. Material applied to the surface of glass fibers used to reinforce plastics and intended to improve the physical properties of such reinforced plastics over that obtained using glass reinforcement without finish.

Glitter. Group of special decorative materials consisting of flakes large enough so that each separate flake produces a plainly visible sparkle or reflection. These materials are incorporated directly into the plastic during compounding.

Gloss. Shine or luster of the surface of a material.

Glyptal. Synthetic resin made by condensation of phthalic anhydride and glycerol. It is used as a shellac substitute and in paints and varnishes.

Graft copolymers

1. Copolymers in which polymeric side chains have been attached to the main chain of a polymer of different structure.

2. Chain of one type of polymer to which side chains of a different type are attached or grafted, e.g., polymerizing butadiene and styrene monomer at the same time.

Granular structure. Nonuniform appearance of finished plastics material due to retention of or incomplete fusion of particles of composition, either within the mass or on the surface.

Gusset. Piece used to give additional size or strength in a particular location of an object, a reinforcement.

Halocarbon plastics. plastics based on resins made by the polymerization of monomers composed only of carbon and a halogen or halogens.

Hand mold. Mold that is removed from the press after each shot for removal of the molded article, generally used for short runs and experimental moldings.

Hardener

1. Substance that brings about the curing of a plastic. Formaldehyde is a hardener for casein; it links the casein molecules together to form a network structure.

2. Substance or mixture of substances added to plastics composition; an additive to promote or control the curing reaction by taking part in it. The term is also used to designate a substance added to control the degree of hardness of the cured film.

Hardness. Resistance of a plastics material to compression and indentation. Among the most important methods of testing this property are the Brinell hardness, Rockwell hardness, and Shore hardness tests.

Haze

1. Cloudy or turbid aspect or appearance of an otherwise transparent specimen caused by light scattered from within the specimen or from its surfaces.

2. Degree of cloudiness in a plastics material.

Heat resistance. Ability to withstand the effects of exposure to high temperatures.

Heat seal. To fuse together or bond two sheets of plastic through use of heat and pressure.

High polymer. Polymer with molecules of high molecular weight, sometimes arbitrarily designated as greater than 10,000.

High-pressure laminates. Laminates molded and cured at pressures not lower than 1000 psi and more commonly in the range of 1200 to 2000 psi.

High-pressure molding. Molding or laminating process in which the pressure used is greater than 200 psi.

High-pressure spot. An area in reinforced plastics containing very little resin, usually due to an excess of reinforced material.

Homopolymer. Polymer consisting of (neglecting the ends, branch junctions, and other minor irregularities) a single type of repeating unit.

Honeycomb. Manufactured product consisting of sheet metal or a resin-impregnated sheet material (paper, fibrous glass, etc.) that has been formed into hexagonally shaped cells. It is used as core material for sandwich constructions.

Hydrocarbon plastics. Plastics based on resins made by the polymerization of monomers composed of carbon and hydrogen only.

Hydrocarbon resin. Family of thermoplastic compounds based on liquid butadiene-styrene copolymers; polybutadiene with various high-molecular-weight hydrocarbon polymers such as polyethylene and elastomers.

Impact resistance. Relative susceptibility of plastics to fracture by shock, such as indicated by the energy expended by a standard pendulum-type impact machine in breaking a standard specimen in one blow.

Impact strength. Ability of a material to withstand shock loading. The work done in fracturing under shock loading a specified test specimen in a specified manner.

Impregnation. Process of thoroughly soaking a material, such as wood, paper, or fabric, with a synthetic resin so that the resin gets within the body of the material. Process is usually carried out in an impregnator.

Industrial laminates. Class of electrical insulating materials produced by impregnating fibrous webs of materials with thermosetting resins and then fusing multiple layers together under high temperature and pressure.

Infrared. Part of the electromagnetic spectrum between the visible light range and the radar range. Radiant heat is in this range, and infrared heaters are much used in sheet thermoforming.

Inhibitor

1. Substance that prevents a chemical reaction.

2. Substance that slows down chemical reaction. Inhibitors are sometimes used in certain types of monomers and resins to prolong storage life.

Injection molding. Method of forming a plastic to desired shape by forcing heat-softened plastic into a relatively cool cavity where it rapidly solidifies.

Ionomer resin. Polymer that has ethylene as its major component but that contains both covalent and ionic bonds. The polymer exhibits very strong interchain ionic forces. Ionomer resins have many of the same features as polyethylene, plus high transparency, tenacity, resilience, and increased resistance to oils, greases, and solvents. Fabrication is carried out as with polyethylene.

Ionomers. Plastics that are transparent, glossy, and tough at low temperatures. They are resistant to oils, grease, and organic solvents and possess high melt strength, abrasion resistance, and good heat-seal characteristics.

Irradiation. Process of subjecting a substance to radiation. Polythene can be cross-linked and rendered resistant to heat by irradiation with gamma rays.

Irradiation (atomic). As applied to plastics, the term refers to bombardment with a variety of subatomic particles, generally alpha, beta, or gamma-rays. Atomic irradiation has been used to initiate polymerization and copolymerization of plastics and in some cases to bring about changes in the physical properties of a plastic material.

Insert

1. Part consisting of metal or other material that may be molded into position or may be pressed into the molding after the completion of the molding operation.

2. Nonplastic part that becomes an integral part of a molded product during the molding operation, e.g., the metal conductors incorporated into an electric plug molded in plastic.

3. Integral part of a plastics molding consisting of metal or other material, which may be molded into position

or may be pressed into the molding after the molding is completed.

In situ foaming. Technique of depositing foamable plastics (prior to foaming) into the place where it is intended that foaming shall take place. An example is the placing of foamable plastics into cavity brickwork to provide insulation. After being positioned, the liquid mix foams to fill the cavity.

Insulation. Coating composed of a dielectric or essentially nonconducting material whose purpose it is to prevent the transmission of electricity.

Isocyanate resin. Most applications for this resin are based on its combination with polyols, such as polyesters or polyethers. During this reaction the reactants are joined through the formation of the urethane linkage; hence this field of technology is generally known as urethane chemistry.

Isomers. Two or more chemical compounds having similar weights and elements, but differing in properties.

Isophorone. Powerful solvent for vinyl and cellulosic resins, with moderate power to dissolve nearly all common thermosetting and thermoplastic resins.

Isotropic laminate. Laminate in which the strength properties are equal in all directions.

Jig. Tool for holding component parts of an assembly during the manufacturing process or for holding other tools.

Joint. Location at which two pieces of materials are connected together.

Jute. Blast fiber obtained from the stems of several species of the plant corchorus found mainly in India and Pakistan. It is used as a filler for plastics molding materials and as a reinforcement for polyester resins in the fabrication of reinforced plastics.

Kieselguhr. Fine powder consisting of siliceous earth formed from the skeletons of diatoms. It is used as a filler for such plastics as phenol-formaldehyde.

Knockout. Any part or mechanism of a mold used to eject the molded article.

Lac. Natural resin from which shellac is obtained by purification.

Lacquer

1. Varnish that dries by the evaporation of solvent leaving a film (such as nitrocellulose) as a protective coating. The term usually refers specifically to those materials in which the film does not harden and become insoluble by combining with atmospheric oxygen. Lacquer can be removed by redissolving it in solvent.

2. Solution of natural or synthetic resins, etc., in readily evaporating solvents, which is used as a protective coating.

Laminate. Product or assembly made by bonding together two or more layers of material or materials.

Laminated plastic. Material formed by impregnating paper or fabric with a resin, followed by the bonding together of many layers into a consolidated structure. Phenolic resins are commonly used for this purpose, the laminated material being heated under pressure until the resin has set.

Laminated plastics (synthetic resin-bonded laminate). Plastics material consisting of superimposed layers of a synthetic resin-impregnated or coated filler that have been bonded together, usually by means of heat and pressure, to form a single piece.

Latent solvent. An organic liquid that has little or no solvent effect on a particular resin until it is activated by either heat or admixture with a true solvent.

Latex

1. Rubber or resin particles dispersed in an aqueous medium.

2. Emulsion or suspension that takes the form of a milky liquid. Rubber latex, for example, exudes from the rubber tree; it is the emulsion in which the rubber is floating as tiny droplets or particles, like the butterfat in milk.

Lead stearate. White powder used as a vinyl resin stabilizer and lubricant in extrusion compounds.

Leather cloth. Name used to describe fabric spread with a layer of plastic such as PVC or nitrocellulose.

Lift. Complete set of moldings produced in one cycle of a molding press.

Light stability. Ability of a plastic to retain its original color and physical properties after exposure to sunlight or artificial light.

Lignin. Natural resinous material that acts as a binder for the cellulose and other materials involved in the structural skeletons of land plants.

Lignin resin. Resin made by heating lignin or by the reaction of lignin with chemicals or resins, the lignin being in greatest amount by mass.

Linters. Short fibers that adhere to the cotton seed after ginning. Linters are used in rayon manufacture, as fillers for plastics, and as a base for the manufacture of cellulosic plastics.

Low-pressure laminates

1. Laminates made by lamination with resins that set at low temperatures and without forming water or similar by-products (such as polyester resins cured with styrene).

2. Laminated plastic made without the application of high pressure. This technique is commonly used with polyester resins cross-linked with styrene or other substances capable of addition polymerization.

3. In general, laminates that are molded and cured in the range of pressures from 400 psi down to and including pressures obtained by the mere contact of the plies.

Lubricant bloom. Irregular, cloudy, greasy film on a plastic surface.

Luminescent pigments. Special pigments available to produce striking effects in the dark. Basically there are two types: one is activated by ultraviolet radiation, producing very strong luminescence and, consequently, very eye-catching effects; the other type, known as phosphorescent pigments, does not require any separate source of radiation.

Magnesium oxide. A white powder, used as a filler and as a thickening agent in polyester resins.

Mar-resistance. Ability to retain a satisfactory surface appearance when subjected to use, wear, rubbing, scuffing, or scratching.

Mat

1. Fibrous material consisting of randomly oriented chopped or swirled filaments loosely held together with a binder.

2. Randomly distributed felt of glass fibers used in reinforced plastics lay-up molding.

Matrix. A matrix is used in reinforced plastics to bind the materials.

Mechanically foamed plastic. Cellular plastic in which the cells are formed by the physical incorporation of gases.

Melamine-formaldehyde molding materials. Heat-hardening melamine-formaldehyde resin compounds incorporating one or another of the following fillers; alpha cellulose, mineral, chopped fabric, cellulose, glass fiber, and cellulose-mineral.

Melamine-formaldehyde plastics. Plastics made by condensation of melamine with formaldehyde.

Melt. Normally solid thermoplastic material that has been heated to a molten condition.

Metalizing. Applying a thin coating of metal to a non-metallic surface either by chemical deposition or by exposing the surface to vaporized metal in a vacuum chamber.

Metallic pigments. Class of pigments consisting of thin, opaque aluminum flakes (made by ball-milling either a disintegrated aluminum foil or a rough metal powder and then polishing to obtain a flat, brilliant surface on each particle) or copper alloy flakes (known as bronze pigments). When incorporated into plastics, they produce unusual silvery and other metallike effects.

Methyl glucoside. Plasticizer for alkyd, amino, and phenolic resins; also used as a polyol for urethane foam production.

Methyl methacrylate. Colorless, volatile liquid derived from acetone cyanohydrin, methanol, and dilute sulfuric acid and used in the production of acrylic resins.

Methylpentene polymer. Transparent, high heat-resistant polyolefin that retains its form and clarity after repeated sterilization. It is used for labware, special syringes, and respiration equipment.

Migration. The transfer of a constituent of a plastic compound to another contacting substance.

Migration of plasticizer. Loss of plasticizer from an elastomeric plastic compound with subsequent absorption by an adjacent medium of lower plasticizer concentration.

Mixture. Combination of two or more substances intermingled with no constant percentage composition, in which each component retains its essential original properties.

Modified. Containing ingredients such as fillers, pigments, or other additives that help to vary the physical properties of a plastics material.

Modified resins. Synthetic resins, modified by the incorporation of natural resins, elastomers, or oils, which alter the processing characteristics or physical properties of the basic resins.

Moisture resistance. Ability to resist absorption of moisture or water.

Mold. Hollow form or matrix into which a plastic material is placed and that imparts to the material its final shape as a finished product.

Molding powder

1. Plastics material in varying stages of granulation, comprising resin, filler, pigments, plasticizers, and other ingredients, ready for use in the molding operation.

2. Powder consisting of plastic mixed with other necessary ingredients, used as charge for molding.

Mold seam. Line on a molded or laminated piece, differs in color or appearance from the general surface, caused by the parting line of the mold.

Molecular weight. The sum of the atomic weights of all the atoms in a molecule.

Monofil. Monofilament made by the extrusion of plastic to form a single filament thicker than those usually used as textile fibers. Bristles are commonly cut from monofilaments.

Monomer

1. Substance in which the individual molecules are capable of linking together to form the long molecules of a polymer. Ethylene, for example, is the monomer from which the polymer polyethylene (polythene) is made.

2. Relatively simple compound that can react to form a polymer.

Montan wax. Hard white wax derived from lignite, used as a mold lubricant.

Multifilament. Single yarn made up of a number of single plastic fibers.

Nesting. In reinforced plastics, the placing of plies of fabric so that the yarns of one ply lie in the valleys between the yarns of the adjacent ply.

Nitrile resins. Polymers containing high concentrations of nitrile functionality with good barrier properties, as well as the other features needed for their wide-scale use in packaging applications. Properly formulated high-nitrile polymers are clear, rigid, and tough and have low gas and water vapor transmission rates and low creep.

Nitrocellulose. Cellulose nitrate.

Nonrigid plastic

1. Plastic that has a modulus of elasticity either in flexure or in tension of not over 70 MPa (700 kgf/cm^2; 10,000 psi) at 23°C and 50% relative humidity (ASTM D 790: flexural properties of plastics and electrical insulating materials).

2. Nonrigid plastic has a stiffness or apparent modulus of elasticity of not over 50,000 psi at 25°C when determined according to ASTM test procedure (ASTM D 747: stiffness of plastics by means of a cantilever beam). ASTM D 638: Test for tensile properties of plastics. ASTM D 882: Test for tensile properties of thin plastic sheeting.

Novolac

1. Resin made by condensation of phenol and formaldehyde, using an acid catalyst. Novolacs are fusible and soluble; they will set and become infusible and insoluble when neutralized and treated with more formaldehyde.

2. Phenolic-aldehyde resin that, unless a source of methylene groups is added, remains permanently thermoplastic.

Nylon

1. Generic name for all synthetic fiber-forming polyamides, which can be formed into monofilaments and yarns characterized by great toughness, strength and elasticity, high melting point, and good resistance to water and chemicals. Material is widely used for bristles in industrial and domestic brushes, and for many textile applications; it is also used in injection molding gears, bearings, combs, etc.

2. Material having a low coefficient of friction; excellent resistance to impact and fatigue, high abrasion resistance, inertness to aromatic chemicals, and good electrical properties.

Nylon monofilaments. Coarse strands of nylon used for fishing lines, brush bristles, racket strings, surgical sutures, and many other uses.

Nylon plastics. Plastics based on resins composed principally of long-chain synthetic polymeric amide, which has recurring amide groups as an integral part of the main polymer chain.

Oil-soluble resin. Resin that at moderate temperatures will dissolve in, disperse in, or react with drying oils to give a homogeneous film of modified characteristics.

Olefins. Group of unsaturated hydrocarbons named after the corresponding paraffins by the addition of "ene" or "ylene" to the stem. Examples are ethylene and propylene.

Oleoresins. Semisolid mixtures of the resin and essential oil of the plant from which they exude, sometimes referred to as "balsams." Oleoresinous materials also consist of products of drying oils and natural or synthetic resins.

Opalescence. Limited clarity of vision through a sheet of transparent plastic at any angle because of diffusion within or on the surface of the plastic.

Opaque. Term descriptive of a material or substance that will not transmit light. The opposite of transparent. Materials that are neither opaque nor transparent are sometimes described as "semiopaque," but are more properly classified as "translucent."

Open cell foamed. Cellular plastic that has a predominance of interconnected cells.

Orange peel

1. Term applied to an uneven surface somewhat resembling an orange peel.
2. Term applied to moldings that have unintentionally rough surfaces.

Organic matter. Either natural or synthetic matter that is derived from living matter.

Organosol

1. Suspension of a finely divided plastic in a plasticizer, together with a volatile organic liquid. The volatile liquid evaporates at elevated temperatures, and the resulting residue is a homogeneous plastic mass, provided the temperature is high enough to accomplish mutual solution of the plastic and plasticizer.
2. Vinyl or nylon dispersion, the liquid phase of which contains one or more organic solvents.

Osmosis. Passage of solvent from a mass of pure solvent into a solution, or from a less to a more concentrated solution through a membrane which is permeable to the solvent but not to the solute.

Overcure. Thermal decomposition in a thermosetting resin or vulcanizing elastomer due to overheating or excessive molding time.

Overlay sheet (surfacing mat). Nonwoven fibrous mat (either in glass, synthetic fiber, etc.) used as the top layer in a cloth or mat lay-up to provide a smoother finish or minimize the appearance of the fibrous pattern.

Parallel laminate. Laminate in which all the layers of material of assembly are oriented approximately parallel with respect to the grain or the strongest direction in tension.

Parison. Shaped plastic mass, generally in the form of a tube, used in blow molding.

Parting line. Slight line visible on the finished product, identifying the place where the parts of the mold came together.

Parylene. Poly-para-xylene is used in ultrathin films for capacitor dielectrics and as a pore-free coating. Films are formed by heating a monomer and condensing it on a cool surface.

Pearlescent pigments. Class of pigments consisting of particles that are essentially transparent crystals of a high refractive index. The optical effect is one of partial reflection from the two sides of each flake. When reflections from parallel plates reinforce each other, the result is a silvery luster. The effects possible range from brilliant highlighting to moderate enhancement of the normal surface gloss.

Peel ply. Outside layer of a laminate that is removed or sacrificed to achieve improved bonding of additional layers.

Permanent set. Increase in length, expressed in a percentage of the original length, by which an elastic material fails to return to original length after being stressed for a standard period of time.

Permeability

1. If a gas or liquid is absorbed on one side of a piece of plastic and given off on the other side, the phenomenon is called "permeability." Diffusion and permeability are not due to holes or pores in the plastic.
2. Passage or diffusion of a gas, vapor, liquid, or solid through a retarder without physically or chemically affecting it; rate of such passage.

Phenol-aralkyl. Group of resins based on prepolymers formed by the reaction of aralkyl ethers and phenols.

Phenol-formaldehyde plastic. Thermosetting plastic made by condensation of phenol and formaldehyde.

Phenol-furfuraldehyde plastic. Plastic made by condensation of phenol with furfuraldehyde.

Phenolic molding materials. Heat-hardened phenol-formaldehyde and furfuraldehyde-phenol molding materials, usually combined with one or more of the following fillers: wood flour, cotton flock; chopped paper, fabric or cord, asbestos, mica, and silica.

Phenolic plastics. Plastics based on resins made by the condensation of phenols, such as phenol and cresol, with aldehydes.

Phenolic resin. Synthetic resin produced by the condensation of an aromatic alcohol with an aldehyde, particularly of phenol with formaldehyde. Phenolic resins form the basis of thermosetting molding materials, laminated sheet, and staving varnishes. They are also used as impregnating agents and as components of paints, varnishes, lacquers, and adhesives.

Phenolic thermosetting resins. Phenolic thermosetting resins and heat resistant, nonburning, and dust free and have good resistance to high impact, pressure, moisture, and most chemicals.

Phenoxy resins. High-molecular-weight thermoplastic polyester resin based on bisphenol-A and epichorohydrin. Material is available in grades suitable for molding, extrusion, coatings, and adhesives.

Phenylene oxide. Material with good mechanical and electrical properties over a wide temperature range and possessing unusual resistance to aqueous environments. The material can be repeatedly autoclaved. It replaces glass and stainless steel in hospital utensils.

Plasticizer. Liquid or solid substance incorporated in a plastic to develop such properties as moldability, resiliency, elasticity, and flexibility.

Plastify. To soften a thermoplastic resin or compound by means of heat alone.

Phthalate esters. Main group of plasticizers, produced by the direct action of alcohol on phthalic anhydride. Phthalates are the most widely used of all plasticizers and are generally characterized by good stability and good all-round properties.

Pinhole. Very small hole in the extruded resin coating.

Pit. Small crater in the surface of the plastic, with its width of approximately the same order of magnitude as its depth.

Plastic

1. Material containing an organic substance of large molecular weight and which at some stage in processing into finished articles can be shaped by flow.

2. Pliable material capable of being shaped by pressure. Plastic is incorrectly used as the generic word for the industry and its products.

3. Pliable and impressionable material of synthetic nature, capable of being formed to a desired shape under heat or pressure, or both, and processed into parts or articles through molding, extrusion, casting, laminating, and machining operations.

Plastic conduit. Plastic pipe or tubing used as an enclosure for electrical wiring.

Plastic films. Film form of compositions based on thermoplastic resins or polymers having a thickness no greater than 0.010 in.

Plastic flow. Deformation of a polymer that takes place under the effect of a sustained force. Plastic flow is accompanied by movement of the long molecules relative to one another; they take up new positions corresponding to a change in the shape of the mass of the polymer as a whole.

Plastic foam. Cellular plastic.

Plasticize

1. Soften a material to make it kneadable by an addition of plasticizer or heat.

2. Soften a material and make it plastic or moldable, either by means of a plasticizer or the application of heat.

Plastic material. Material having an organic substance of large molecular weight as an essential component. It has a solid finished state and can be shaped to flow for use in manufacturing other articles.

Plastic pipe. Hollow cylinder of plastic material in which the wall thicknesses are usually small when compared to the diameter and in which the inside and outside walls are essentially concentric.

Plastics

1. Material that contains as an essential ingredient one or more organic polymeric substances of large molecular weight, is solid in its finished state, and, at some stage in its manufacture or processing into finished articles, can be shaped by flow.

2. Substances that are plastic at some stage during their fabrication, enabling them to be molded or otherwise shaped, but that can be set, in one way or another, when they have been shaped.

3. Generic term for the industry and its products, which is properly used only as a plural word. Plastics products include polymeric substances, natural or synthetic, and exclude the rubbers.

Plastic tubing. Particular size of plastic pipe in which the outside diameter is essentially the same as the corresponding size of copper tubing, or a small diameter flexible pipe.

Plastisols. Mixtures of resins and plasticizers that can be molded, cast, or converted to continuous films by the application of heat.

Polacrylate. Thermoplastic resin made by the polymerization of an acrylic compound such as methyl methacrylate.

Polyallomers. Crystalline polymers produced from two or more olefin monomers.

Polyamide

1. Polymer made by condensation of diamine and dibasic acid, or related compounds, in such a way as to create long molecules by the linking of small molecules via amide groups.

2. Polymer in which the structural units are linked by amide or thioamide groupings. Many polyamides are fiber forming.

Polybutylene plastics. Plastics based on polymers and made with butene as essentially the sole monomer.

Polycarbonate

1. Thermoplastic resin with high impact strength as its most outstanding property. It has good electrical properties and excellent weatherability. Chemically it resists oil, noncorrosive chemicals, and stains. It is self-extinguishing, has excellent barrier resistance to water, moisture, and humidity, and has 87% light transmission.

2. Polyester polymer in which the repeating structural unit in the chain is of the carbonate type.

Polycarbonate plastics. Polyester plastics based on polymers in which the repeating structural units in the chains are essentially all of the carbonate type.

Polycarbonate resins. Polymers derived from the direct reaction between aromatic and aliphatic dihydroxy compounds with phosgene or by the ester exchange reaction with appropriate phosgene-derived precursors.

Polycarbonates. Material possessing an unmatched combination of toughness, heat and flame resistance, and dimensional stability. It replaces safety glass as glazing.

Polyester

1. Polymer in which the repeated structural unit in the chain is of the ester type.

2. Polymer made by condensation of a glycol and dibasic acid, or related compounds, in such a way as to create long molecules by the linking of small molecules via ester groups.

3. Resin formed by the reaction between a dibasic acid and a dihydroxy alcohol, both organic. Modification with multifunctional acids and/or bases and some unsaturated reactants permits cross-linking to thermosetting resins.

Polyester-reinforced urethane. Poromeric material, which may have a urethane impregnation or a silicone coating, used for shoe uppers and industrial leathers.

Polyesters. Thermosetting resin with or without glass reinforcement; strong, tough material that can be blended as rigid, resilient or flexible, weathers well, has good electrical properties, and resists breakage, water, and most chemicals.

Polyether. Polymer containing recurring ether groups as an integral part of the polymer chain.

Polyethylene

1. Flexible, resilient, and tough with excellent electrical and chemical properties. It is nontoxic, tasteless, and odorless and resists oils and fats. The film has low

water-vapor permeability, flexibility, and toughness even at low temperatures. It is not affected by solvents.

2. Polymer prepared by the polymerization of ethylene as the sole monomer.

3. High-density polyethylene has good chemical resistance and electrical properties, but has a high thermal expansion and tends to fail under mechanical and thermal stress.

4. Thermoplastic material composed by polymers of ethylene. It is normally a translucent, tough, waxy solid that is unaffected by water and by a large range of chemicals.

Polyethylene plastics

1. Thermoplastic resin made by the polymerization of ethylene with crystallinity ranges including low, medium, and high density.

2. Plastics based on polymers made with ethylene as essentially the sole monomer.

Polyimides. Outstanding physical, electrical properties. Film remains tough, flexible at cryogenic temperatures. Most flame resistant of organic materials. Very high radiation resistance. Outstanding electric motor insulator.

Polyisobutylene. Polymerization product of isobutylene. Varies in consistency from a viscous liquid to a rubber-like solid with corresponding variation in molecular weight from 1000 to 400,000.

Polymer

1. Substance consisting of molecules characterized by the repetition (neglecting ends, branch junctions, and other minor irregularities) of one or more types of monomeric units.

2. Substance formed by the linking together of many smaller molecules in such a way as to form long molecules.

Polymers. Organic compounds, natural or synthetic, with the same chemical components and proportions by weight, but differing in molecular weight.

Polyolefin. Polymer prepared by the polymerization of an olefin(s) as the sole monomer(s).

Polyolefin plastics. Plastics based on polymers made with an olefin(s) as essentially the sole monomer(s).

Polyolefins. Chlorinated hydrocarbons, brominated hydrocarbons, and antimony oxide are the major products used to flame-retard these polymers. Polyolefins are particularly hard to flame-retard because of their general molecular structure.

Polyoxymethylene. Polymer in which the repeated structural unit in the chain is oxymethylene.

Polyoxymethylene plastics. Plastics based on polymers in which oxymethylene is essentially the sole repeated structural unit in the chains.

Polypropylene

1. Polymer prepared by the polymerization of propylene as the sole monomer.

2. Material more rigid that polyethylene, with exceptional flex life, good surface hardness, scratch and abrasion resistance, high chemical resistance, and excellent electrical properties. It will not stress crack and is boilable.

3. Tough, lightweight, rigid plastics made by the polymerization of high-purity propylene gas in the presence of an organometallic catalyst at relatively low pressures and temperatures.

Polypropylene plastics. Plastics based on polymers made with propylene as essentially the sole monomer.

Polystyrene

1. Polymer prepared by the polymerization of styrene as the sole monomer.

2. Hard, rigid thermoplastic material that has low impact strength and is soluble in citric acid, cleaning fluids, gasoline, and ketone products but is not affected by cold and can withstand up to 175°F. It is not good for outdoor applications.

3. Material with good dimensional stability, excellent electrical properties, and high optical clarity. A high-impact grade is available. Films have high gas permeability and intermediate water vapor transmission rates.

4. Water-white thermoplastic produced by the polymerization of styrene (vinyl benzene). The electrical insulating properties of polystyrene are outstandingly good, and the material is relatively unaffected by moisture.

Polysulfone

1. Chemical linkage of isopropylidene, ether, and sulfone provides this thermoplastic material with exceptional high-temperature and low-creep properties. It has arc resistance and good electrical properties, is self-extinguishing, and may be molded and extruded.

2. Polysulfone maintains its properties from 150°F to 349°F and is NSF sanctioned for potable water applications. It meets ASTM standards for self-extinguishing compounds. High tensile strength and flexural modulus; excellent chemical, good electrical properties (ASTM D 635).

Polyterephthalate. Thermoplastic polyester in which the terephthalate group is a repeated structural unit in the polymer chain.

Polyterephthalate plastics. Thermoplastic polyester in which the terphthalate group is a repeated structural unit in the chain, the terephthalate being in greater amount than other dicarboxylates which may be present.

Polyterpene resins. Thermoplastic resins obtained by the polymerization of turpentine in the presence of catalysts. These resins are used in the manufacture of adhesives, coatings, and varnishes and in food packaging. They are compatible with waxes, natural and synthetic rubbers, and polyethylene.

Polyurethane

1. Thermosetting resin used for forming foam, both rigid and flexible. It is tough and lightweight; gives good support; has good chemical resistance and sound and shock absorption; can be prefoamed or foamed in place; is resistant to water, moisture, rot, and vermin; and can be made from flameproof.

2. Material available in both thermoplastic and thermosetting types. The most significant properties of the thermoplastic grades are their toughness; outstanding oil, grease, and abrasion resistance; and usefulness over a wide temperature range.

Polyurethane resins. Family of resins produced by reacting diisocyanate with organic compounds containing two or more active hydrogens to form polymers having free isocyanate groups. These groups, under the influence of heat or certain catalysts, will react with each other or with water, glycols, etc., to form a thermosetting material.

Polyvinyl acetal

1. Thermoplastic derived from polyvinyl ester, usually a colorless solid of high tensile strength with good resistance to water, heat, and sunlight.

2. Member of the group of vinyl plastics, polyvinyl acetal is the general name for resins produced from a condensation of polyvinyl alcohol with an aldehyde. There are three main groups: polyvinyl acetal itself, polyvinyl butyral, and polyvinyl formal. Polyvinyl acetal resins are thermoplastics which can be processed by casting, extruding, molding and coating, but their main uses are in adhesives, lacquers, coatings, and films.

Polyvinyl acetate

1. Polymer prepared by the polymerization of vinyl acetate as the sole monomer.
2. Thermoplastic that is slow burning and insoluble in water, fats, waxes, and hydrocarbons. It is used for heat-sealing films and is best known as a milky white household cement that is transparent when dry.
3. Thermoplastic material composed of polymers of vinyl acetate in the form of a colorless solid. It is obtainable in the form of granules, solutions, lattices, and pastes and is used extensively in adhesives, for paper and fabric coatings, and in bases for inks and lacquers.

Polyvinyl alcohol

1. Polymers prepared by the essentially complete hydrolysis of polyvinyl esters.
2. Thermoplastic that is water soluble; impervious to all nonaqueous solvents, gases, and animal and vegetable oils; tough; strong; and abrasion resistant.
3. Thermoplastic material composed of polymers of the hypothetical vinyl alcohol. It is usually a colorless solid, insoluble in most organic solvents and oils, but soluble in water when the content of hydroxy groups in the polymer is sufficiently high.

Polyvinyl butyral. Thermoplastic material derived from a polyvinyl ester in which some or all of the acid groups have been replaced by hydroxyl groups and some or all of these hydroxyl groups replaced by butyral groups by reaction with butyraldehyde. A colorless, flexible, tough solid, it is used primarily in interlayers for laminated safety glass.

Polyvinyl carbazole. Thermoplastic resin, brown in color, obtained by reacting acetylene with carbazole. This resin has excellent electrical properties and good heat and chemical resistance. It is used as an impregnant for paper capacitors.

Polyvinyl chloride

1. Polymer prepared by the polymerization of vinyl chloride as the sole monomer.
2. Thermoplastic material composed of polymers of vinyl chloride. When compounded with plasticizers, it yields a flexible material superior to rubber in aging properties. It is widely used for cable and wire coverings, in chemical plants, and in the manufacture of protective garments.

Polyvinyl chloride acetate. Thermoplastic material composed of copolymers of vinyl chloride and vinyl acetate; a colorless solid with good resistance to water and concentrated acids and alkalies. When compounded with plasticizers, it yields a flexible material superior to rubber in aging properties. It is widely used for cable and wire coverings, in chemical plants, and in protective garments.

Polyvinyl fluoride (PVF). Highly crystalline plastic commercially available as a tough but flexible film. It has good resistance to abrasion and resists staining. PVF film can be laminated to plywood, vinyl, hardboard, reinforced polyesters, galvanized steel, and metal foils. PVF film is sold by Du Pont Co. under the trademark "Tedlar."

Polyvinyl formal. One of the groups of polyvinyl acetal resins made by the condensation of formaldehyde in the presence of polyvinyl alcohol. It is used mainly in combination with cresylic phenolics for wire coatings and for impregnations, but it can also be molded, extruded, or cast and is resistant to greases and oils.

Polyvinyl formal and butyral plastics. Materials that are the product of hydrolysis of polyvinyl acetate and reaction with formaldehyde or butyraldehyde.

Polyvinylidene chloride. Thermoplastic material composed of polymers of vinylidene chloride, a white powder with softening temperature at 185°C to 200°C. This material is also supplied as a copolymer with acrylonitrile or vinyl chloride, giving products that range from the soft flexible type to the rigid type.

Polyvinylidene fluoride resins. Group of plastics that are homopolymers of vinylidene fluoride. These resins are supplied as powders and pellets for molding and extrusion and in solution form for casting. The resins have good tensile and compressive strength and high impact strength. Applications are chemical equipment, such as gaskets, impellers, and other pump parts, and packaging uses, such as drum linings and protective coatings.

Premix. In reinforced thermosetting plastics, an admixture of resin, reinforcements, fillers, etc., not in web or filamentous form, ready for molding.

Prepolymer. Polymer of a degree of polymerization between that of the monomer or monomers and the final polymer.

Preprinting. In sheet thermoforming, the distorted printing of sheets before they are formed. During forming the print assumes its proper proportions.

Primary plasticizer. Plasticizer that has sufficient affinity to the polymer or resin so that it is considered compatible and may therefore be used as the sole plasticizer.

Printing of plastics. Printing of plastics, materials, particularly thermoplastic film and sheet. Basically, the printing processes used are the same as in other industries, but the adaptation of machinery and the development of special inks have been a constant necessity, particularly as new plastics materials have arrived, each within its own problems of surface decoration. Among the printing processes commonly used are gravure, flexographic, inlay (or valley), and silk screen.

Processing. Shaping a plastic to the desired form by molding, casting, fabricating, blowing, and laminating.

Propylene plastics. Plastics based on polymers of propylene or copolymers of propylene with other monomers, the propylene being in the greatest amount by mass.

Propylene-vinyl chloride copolymer resins. Thermoplastics that provide the end-use property advantages of PVC homopolymers plus improvements in stability and flow characteristics.

Proxylin plastic. Plastic product, substance, material, or compound, other than nitrocellulose photographic film, having soluble cotton or similar nitrocellulose as a base, including pyralin, celluloid, fiberloid, viscoloid, zylonite, and similar products, materials, or compounds by whatever name known, whether in the form of raw material or finished product.

Pyroxylin products. Products composed wholly or partly of nitrocellulose or similar flammable material or plastic, such as films, combs, pens, pencils, toilet articles, and other novelties or products.

Pyyrone. Combination of polyimides and polybenzimidazoles that forms a 900°F material.

Quench. Process of shock cooling thermoplastic materials from the molten state.

Quench bath. Cooling medium used to quench molten thermoplastic materials to the solid state.

Rayon. Generic term for fibers, staple, and continuous-filament yarns composed of regenerated cellulose, but also frequently used to describe fibers obtained from cellulose acetate or cellulose triacetate. Rayon fibers are similar in chemical structure to natural cellulose fibers, such as cotton, except that the synthetic fiber contains shorter polymer units. Most rayon is made by the viscose process.

Ream. Layers of unhomogeneous material parallel to the surface in a transparent or translucent plastic.

Regenerated cellulose (cellophane). Transparent cellulose plastics material made by mixing cellulose xanthate with a dilute sodium hydroxide solution to form a viscose. Regeneration is carried out by extruding the viscose, in sheet form, into an acid bath to create regenerated cellulose.

Reinforced plastic. Plastic with high-strength fillers imbedded in the composition, resulting in some mechanical properties superior to those of the base resin.

Reinforced polypropylene. Reinforced polypropylene represents about one-fourth of all the reinforced thermoplastics currently being used. Glass fiber reinforcement improves dimensional stability, resistance to warpage, stiffness, and strength.

Reinforced polystyrene. The addition of glass fiber to polystyrene increases strength, stiffness, dimensional stability, and impact resistance.

Reinforced polyurethane elastomers. Glass fiber reinforcement of polyurethane drops its elongation considerably, lending it to applications where a more rigid material would be desirable.

Reinforced thermoset resins. Reinforced thermoset resins have good high-temperature properties, high modulus, high strength, and a good corrosion resistance. Desirable properties are retained on foaming such systems, and improved thermal insulation is obtained.

Reinforced thermosetting plastic. Thermosetting plastic reinforced with a glass fiber mat having not less than $1\frac{1}{2}$ ozs of glass fiber per square foot.

Reinforcing. Press-forming process in which plastic resin and glass or other fibers are squeezed under heat and pressure between halves of a mold and allowed to cure. Glass-reinforced plastics (fiber glass) are the resultant products. Hand lay-up techniques, requiring no heat or pressure, are also employed.

Reinforcing mats. Mats made either of chopped strands in a random pattern or of continuous strands of glass fiber laid down in a swirl pattern. Strands are generally held together by resinous binders. These are used for medium-strength parts with uniform cross section. Both chopped- and continuous-strand reinforcing mats are available in weights varying from $\frac{3}{4}$ to $4\frac{1}{2}$ oz/sq ft and in various widths.

Reinforcement. Strong inert material bound into plastics to improve its strength, stiffness, and impact resistance. Reinforcements are usually long fibers of glass, sisal, cotton, etc., in woven or nonwoven form. To be effective, the reinforcing material must form a strong adhesive bond with the resin.

Resin

1. Solid or pseudosolid organic material, often of high molecular weight, that exhibits a tendency to flow when subjected to stress, usually has a softening or melting range, and usually fractures conchoidally.

2. Any of a class of solid or semisolid organic products of natural or synthetic origin, generally of high molecular weight with no definite melting point. Most resins are polymers.

Resinoid

1. Name generally applied to phenolic synthetic resins.

2. Any of the class of thermosetting synthetic resins, either in their initial temporarily fusible state or in their final infusible state.

Resite (C-stage)

1. Final stage in the reaction of certain thermosetting resins in which the material is practically insoluble and infusible.

2. C-stage phenolic resin.

Resitol (B-stage)

1. Intermediate stage in the reaction of certain thermosetting resins in which the material swells when in contact with certain liquids and softens when heated, but may not entirely dissolve or fuse.

2. B-stage phenolic resin.

Resole (A-stage)

1. Early stage in the preparation of certain thermosetting resins in which the material is still soluble in certain liquids and fusible.

2. Phenolic resin produced by condensing phenol with formaldehyde using an alkaline catalyst. Further heating causes the resole to set.

Resolite. Thermosetting resin in an intermediate, partially cured form.

Reworked material. Scrap plastic parts such as runners, flaws, and reject parts that have been reclaimed for reprocessing.

Rigid plastic. Plastic with stiffness greater than 100,000 psi at 23°C, determined in accordance with ASTM D 747.

Rigid PVC. Polyvinyl chloride or a polyvinyl chloride acetate copolymer characterized by a relatively high degree of hardness. It may be formulated with or without a small percentage of plasticizer.

Rigid resin. Resin having a modulus of 10,000 psi or greater.

Rockwell hardness. Common method of testing a plastics material for resistance to indentation, in which a diamond or steel ball, under pressure, is used to pierce the test specimen.

Rosin

1. Resin that remains after the distillation of turpentine obtained from pine trees.

2. Resin obtained as a residue in the distillation of crude turpentine from the sap of the pine tree (gum rosin) or from an extract of the stumps and other parts of the tree (wood rosin).

Roving

1. Filament winding; collections of bundles of continuous filaments that are either untwisted strands or twisted yarns.

2. Form of fibrous glass in which spun strands are woven into a tubular rope. The number of strands is variable, but 60 is usual. Chopped roving is commonly used in preforming.

Safety glass. Shatterproof glass made by sandwiching a layer of transparent plastic between two layers of glass.

San. Styrene acrylonitrile thermoplastic copolymer with good stiffness and scratch, chemical, and stress crack resistance.

Sandwich constructions. Panels composed of a lightweight core material, e.g., honeycomb or foamed plastic, to which two relatively thin, dense, high-strength faces or skins are adhered.

Sandwich heating. Method of heating both sides of a thermoplastic sheet simultaneously prior to forming.

Scrap. Plastic material which has been processed or formed, then reground for reuse.

Sealing wax. Thermoplastic material consisting of shellac or other resin, turpentine, and coloring matter.

Secondary plasticizer (extender plasticizer). Plasticizer that has insufficient affinity for the resin to be compatible as the sole plasticizer and must be blended with a primary plasticizer. The secondary acts as a diluent with respect to the primary, and the primary-secondary blend has less afinity for the resin than does the primary alone.

Self-extinguishing. Term describing the ability of a material to cease burning once the source of flame has been removed.

Semirigid plastic. Plastic with stiffness between 10,000 and 100,000 psi at 23°C, determined in accordance with ASTM D 747.

S-glass. Magnesia-alumina-silicate glass used to produce high-tensile-strength glass filaments.

Sheet (thermoplastic). Flat section of a thermoplastic resin considerably greater in length than in width and 10 mils or greater in thickness.

Sheeting. Form of plastic in which the thickness is very small in proportion to length and width and in which the plastic is present as a continuous phase throughout, with or without filler.

Shellac. Natural resin produced by insects.

Shot. All the plastic parts that are produced at one time in a single mold.

Silicone

1. General name given to polymers of silicon and oxygen.
2. One of the group of polymeric materials in which the recurring chemical group contains silicon and oxygen atoms as links in the main chain. At present these compounds are derived from silica (sand) and methyl chloride. Various forms obtainable are characterized by their resistance to heat.

Silicone molding materials. Heat-hardened silicone resins combined with chopped glass fiber and/or other fillers such as titanium dioxide or diatomaceous earth.

Silicone plastics. Plastics based on polymers in which the main polymer chain consists of alternating silicone and oxygen atoms.

Single-stage phenolic resin compound. Phenolic material in which the resin, because of its reactive groups, is capable of further polymerization by application of heat.

Sintering. Welding together of powdered plastic particles at temperatures just below the melting or fusion point.

Skin. Relatively dense layer at the surface of a cellular polymeric material.

Slow-burning plastic (check test). Plastic material that burns no faster than $2\frac{1}{2}$ in. per minute when tested in accordance with the ASTM standards for flammability (ASTM E 84).

Solution coating. Any coating process employing a solvent solution of a resin.

Solvation. Process of swelling, gelling, or solution of a resin by a solvent or plasticizer as a result of mutual attraction.

Solution polymerization. Polymerization that takes place in a solvent.

Solvent. Substance, usually a liquid, that dissolves other substances.

Solvent cement. Adhesive made by dissolving a plastic resin or compound in a suitable solvent or mixture of solvents. Solvent cement dissolves the surfaces of the pipe and fittings to form a bond between the mating surfaces, provided the proper cement is used for the particular materials and proper techniques are followed.

Sorption. Binding of one substance to another by any mechanism, such as adsorption, absorption, or persorption.

Splash. Small pit-like surface defects caused by excessive water in injection-molding resins.

Spray coating

1. Formation of a protective coating by spraying a solution of plastic on to a supporting structure, such as webbing.
2. Spray coating is usually accomplished on continuous webs by a set of reciprocating spray nozzles traveling laterally across the web as it moves.

Stabilizer. Ingredient used in the formulation of some plastics, especially elastomers, to assist in maintaining the physical and chemical properties of the compounded materials at their initial values throughout the processing and service life of the material.

Strength. The mechanical properties of a plastic, such as the ability to withstand sharp blows, and to carry weight.

Stress wrinkles. Distortions in the face of a laminate caused by uneven web tensions, slowness of adhesive setting, selective absorption of the adherends, or by reaction of the adherends with materials in the adhesive.

Styrene butadiene. A class of thermoplastic elastomers with properties indistinguishable from vulcanized rubbers but with the processing advantages of plastics.

Styrene plastics

1. Thermoplastics made from styrene monomer either polymerized alone or copolymerized with other monomers.
2. Plastics based on polymers of styrene or copolymers of styrene with other monomers, the styrene being in greatest amount by mass.

Styrene-rubber pipe and fitting plastics. Plastics containing at least 50% styrene plastics combined with rubbers and other compounding materials, but not more than 15% acrylonitrile.

Styrene-rubber plastics. Plastics based on styrene polymers and rubbers, the styrene polymers being in the greatest amount by mass.

Sulfone polymers. Transparent, high-temperature, engineering thermoplastics.

Surface mat. Thin mat of fine fibers used primarily to produce a smooth surface on a reinforced plastic.

Sweating. Exudation of small drops of liquid, usually a plasticizer or softener, on the surface of a plastic part.

Syntactic foam

1. Material consisting of hollow sphere fillers in a resin matrix.
2. Cellular structure produced by adding preformed hollow spheres into a liquid resin matrix followed by

chemically setting the mix to a fairly rigid state. Syntactic foam differs from gas-blown foam since no blowing agent is needed.

Telomer. Polymer composed of molecules having terminal groups incapable of reacting with additional monomers under the conditions of the synthesis to form larger polymer molecules of the same chemical type.

Tetrafluoroethylene resin. Highly crystalline linear polymer, unique among organic compounds in its chemical inertness, resistance to change at high temperatures, and extremely low dielectric loss factor.

Thermally foamed plastic. Cellular plastic produced by applying heat to effect gaseous decomposition or volatilization of a constituent.

Thermoplastic

1. Plastic that repeatedly can be softened by heating and hardened by cooling through a temperature range characteristic of the plastic, and that in the softened state can be shaped by flow into articles by molding or extrusion.

2. Solid plastic material that is capable of being repeatedly softened by increases in temperature and hardened by decreases in temperature without causing a change in its chemical composition.

Thermoplastic elastomers. Materials possessing the properties of rubber but that can be processed like thermoplastics. These materials can be processed directly into finished parts by conventional thermoplastic techniques, such as injection molding or extrusion. It is these rubber-like properties that distinguish thermoplastic elastomers from traditional flexible thermoplastic materials.

Thermoplastic material. Solid plastic material capable of being repeatedly softened by increase of temperature and hardened by decrease of temperature.

Thermoset

1. Plastic that after having been cured by heat or other means is substantially infusible and insoluble.

2. Material that will undergo or has undergone a chemical reaction by the action of heat, catalysts, ultraviolet light, etc., leading to a relatively infusible state.

Thermosetting. Capable of being changed into a substantially infusible or insoluble product when cured by heat or other means.

Thermosetting alkyd compounds. Compounds with lesser monomer content that do not have the high-impact strength of thermosetting polyester but are useful in an easily handled form for a wide range of electrical, electronic, and appliance uses.

Thermosetting material

1. Solid plastic material that will not soften or harden with temperature changes unless a chemical change occurs.

2. Solid plastic material that is capable of being changed into a substantially infusible and insoluble product when cured under the application of heat or by mechanical means.

Thermosetting plastics (thermosets)

1. Cured plastics that are chemically cross-linked and that degrade or decompose, rather than soften, when heated.

2. Plastics that become permanently rigid on application of heat above a critical temperature and cannot be softened by reheating.

Toughness. Resistance of a plastic to shocks caused by hard blows or dropping.

Transfer molding. Variation on the injection-molding method for use with thermosetting materials.

Translucent. Descriptive of a material or substance capable of transmitting some light, but not clear enough to be seen through.

Transparent. Descriptive of a material or substance capable of a high degree of light transmission, such as glass. Some polypropylene films and acrylic moldings are outstanding in this respect.

Trimer. Molecule formed by the union of three molecules of a monomer.

Two-stage phenolic resin compound. Phenolic material in which the resin is essentially not reactive at normal storage temperature but contains a reactive additive that causes further polymerization upon the application of heat.

Undercure. Condition of inadequate physical properties in a thermosetting resin or elastomer, resulting from too little time and/or temperature during the curing cycle.

Unicellular. Term applied to foamed plastics where each cell is an isolated unit.

Unidirectional laminate. Reinforced plastic laminate in which substantially all of the fibers are oriented in the same direction.

Unsaturated polyester. The essential ingredients of an unsaturated polyester are a linear polyester resin, a cross-linking monomer, and inhibitors to retard cross-linking until the resin is to be used by the fabricator.

Urea-formaldehyde group of molding materials. Heat hardening molding compounds consisting essentially of urea-formaldehyde resin combined with cellulose fillers.

Urea-formaldehyde plastic. Plastic made by condensation of urea with formaldehyde.

Urea-formaldehyde resin (urea resin). Synthetic resin derived from the reaction of urea (carbamide) with formaldehyde or its polymers.

Urea plastics. Plastics based on resins made by the condensation of urea and aldehydes.

Urethane elastomers. Materials usually classified as thermoplastic, castable, and millable gums. Some castables are thermoplastic, although most are thermosetting.

Urethane plastics. Plastics based on polymers in which the repeated structural units in the chains are of the urethane type, or on copolymers in which urethane and other types of repeated structural units are present in the chains.

Vacuum forming. Forming process in which a heated plastic sheet is drawn against the mold surface by evacuating the air between it and the mold.

Vacuum metalizing. Process in which surfaces are thinly coated with metal by exposing them to the vapor of metal that has been evaporated under vacuum (one millionth of normal atmospheric pressure).

Valley printing ink. Ink that is applied to the high points of an embossing roll and subsequently deposited in what becomes the valleys of the embossed plastic material.

Varnish. Liquid, usually a solution of polymer, that forms a protective coating when applied to a solid surface. The coating may be a film of plastic, such as nitrocellulose, or it may be formed by oxidation of a drying oil by oxygen in the air.

Vehicle. Liquid medium in which pigments, etc., are dispersed in coatings such as paint and that enables the coating to be applied.

Veil. Thin mat of very fine, relatively long fibers used at outermost layer of a composite in order to improve surface characteristics.

Vinyl. Class of plastics that contain acrylics, vinyl, and styrene polymers.

Vinyl acetate plastics. Plastics based on polymers of vinyl acetate or copolymers of vinyl acetate with other monomers, the vinyl acetate being in greatest amount by mass.

Vinyl chloride. Vinyls can be compounded to yield any degree of flexibility. Unplasticized PVC is hard and hornlike. Its weathering characteristics are excellent. Vinyls have color, clarity, flex life, and chemical, moisture, and abrasion resistance.

Vinyl chloride plastics. Plastics based on polymers of vinyl chloride or copolymers of vinyl chloride with other monomers, the vinyl chloride being in greatest amount by mass.

Vinyl chloride polymer and copolymer plastics. Thermoplastic materials that include polyvinyl chloride and copolymers that are predominately vinyl chloride. The principal copolymers contain vinyl acetate and vinylidene chloride. Flexible compositions are mixtures of vinyl resins with stabilizers, lubricants, colorants, and plasticizers, such as high-boiling liquids, soft resins, or elastomers.

Vinyl ethyl ether. Colorless monomer which can be polymerized either in the liquid or the gaseous state. In plastics, it is used as a comonomer and intermediate.

Vinylidene chloride. One of the very inert thermoplastics with excellent drug and meat packaging properties. It offers low water vapor and low odor and food-flavoring transmission.

Vinylidene chloride plastics

1. Thermoplastic materials made from unsymmetrical dichlorethylene, which is converted to monomeric vinylidene chloride and subsequently polymerized or copolymerized.
2. Plastics based on polymer resins made by the polymerization of vinylidene chloride or copolymerization of vinylidene chloride with other unsaturated compounds, the vinylidene chloride being in the greatest amount by weight.

Vinylidene fluoride. Thermoplastic chemical and ultra violet (uv) resistant material with a thermal range from 80°F to 300°F. It is nondrip and self-extinguishing and may be molded by all processes.

Vinyl resin. Synthetic resin formed by the polymerization of chemical compounds. In particular, polyvinyl chloride, acetate, alcohol, and butyral are referred to.

Virgin material. Plastic material in the form of pellets, granules, powder, floc, or liquid that has not been subjected to use or processing other than that required for its initial manufacture.

Void. An empty space in any material or medium.

Warp. Dimensional change in a plastic object after molding, fabrication, or aging.

Wash. An area where the reinforcement has moved during closing of the mold, resulting in a resin-rich area.

Weather resistance. Ability of a plastic to retain its original physical properties and appearance upon prolonged exposure to outdoor weather.

Web. Continuous length of sheet material handled in roll form as contrasted with the same material cut into sheets.

Weld-line. Discontinuity in a molded plastic part formed by the merging of the two or more streams of plastic flowing together.

Weld mark. Flaw on a molded plastic article caused by the meeting of two flow fronts during the molding or extrusion operation.

Wet-out. Condition of an impregnated reinforcement wherein substantially all voids between the sized strands and filaments are filled with resin.

Working life. Period during which a compound, after mixing with a catalyst, solvent, or other compounding ingredients, remains suitable for its intended use.

Woven fabrics for plastics. Fabrics made from glass fiber yarns with weights that vary from $2\frac{1}{2}$ to 40 oz/sq yd and thicknesses that vary from 0.035 to 0.048 in.

Woven roving fabric reinforcing. Heavy, drapable fabric made from continuous-strand roving.

Wrinkle. Imperfection in reinforced plastic that has the appearance of a wave molded into one or more plies of fabric or other reinforcing material.

Xenon arc light aging. Test for evaluating the light stability of plastics, employing a xenon gas discharge lamp of special design that emits radiation duplicating the spectra of natural sunlight better than most artificial sources.

Yellowness index. Measure of the tendency of plastics to turn yellow upon long-term exposure to light.

Zinc palmitate. Amorphous white powder used as a lubricant in plastics.

Zinc ricinoleate. White powder used as a stabilizer in vinyl plastics.

PLYWOOD

A-A exterior plywood. Sanded panel grade described by grades of face (A) and back (A) veneers, with minimum C quality inner plies, bonded with exterior glue, and meeting all the performance requirements of exterior-type plywood.

A-A interior plywood. Sanded panel grade described by grades of face (A) and back (A) veneers, with minimum D quality inner plies, bonded with interior or exterior glue, and used where both sides can be painted or stained.

A-B exterior plywood. Sanded panel grade described by grades of face (A) and back (B) veneers, with minimum C quality inner plies, bonded with exterior glue, and used as an alternate for A-A exterior where the appearance of one side is less important.

A-B interior plywood. Sanded panel grade described by grades of face (A) and back (B) veneers, with minimum D quality inner plies, bonded with interior or exterior glue, and used as an alternate for A-A interior where the appearance of one side is less important.

A-C exterior plywood. Sanded exterior panel grade described by grades of face (A) and back (C) veneers with minimum C quality inner plies, bonded with exterior glue, and meeting all the requirements of exterior-type plywood. The face is finish grade and can be painted or stained.

Acoustical characteristics

1. Natural fibrous material provides plywood with good acoustical properties. Many practical floor and wall

assemblies using plywood have been acoustically tested and produced good sound attenuation.

2. Thicker plywood panels provide better sound insulators than thin panels.

A-D interior plywood. Sanded interior panel grade described by grades of face (A) and back (D) veneers, with minimum D quality inner plies, bonded with interior or exterior glue. Finish-grade face can be painted or stained.

Adhesive. Substance capable of holding materials together by surface attachment. It is a general term that includes cements, mucilage, and paste, as well as glue.

Aggregate-coated panel. Plywood panel coated with stone chips.

All-veneer construction. Plywood in which all plies are of veneer. Ordinarily no single ply of veneer will exceed $\frac{5}{16}$ in. in thickness.

Along the grain. In the same direction as the grain; the same direction as the grain of the face ply, normally the long (8-ft) dimension. Plywood is stronger and stiffer along the grain (as in all wood) than it is across the grain.

APA (testing agency). A testing agency is an organization whose function is continuous testing of random samples of plywood from the plywood mills to ensure that the product meets applicable standards. The Division for Product Approval of the American Plywood Association is such an organization.

Back

1. Side opposite the face; the poor side of a plywood panel.

2. Side reverse to the face of a panel; the poor side of a panel in any grade of plywood calling for a face and back.

3. Concealed or unfinished side of plywood and veneered panels.

4. Side of a panel that is of lower veneer quality on any panel whose outer plies are of different veneer grades.

5. Veneer sheet on the underside of a plywood panel corresponding in thickness to the face veneer on the upper or exposed surface; grain running parallel to the grain of the face.

Backing grade No. 4. Grade that permits larger defects. The grain and color are not matched, and the veneer is used primarily as the concealed face. The defects must not affect the strength or serviceability of the panel made from it. At the manufacturer's option, this face can be of some species other than the exposed face.

Backpriming. Application of a coat of primer to the back of a plywood panel. When plywood is used in especially damp locations, the panels should be backprimed to prevent warping.

Balanced construction plywood. Plywood constructed so that the forces induced by uniformly distributed changes in moisture content will not cause warpage. Symmetrical constructions in which the grain directions of the plies are either parallel or perpendicular to each other are balanced constructions.

Banded cores. Cores that have been made with banding on one or more sides.

Banding. Portion of wood extending around one or more sides of a piece of core, usually with the grain extending the long way. The banding of solid wood facilitates shaping the edges of the piece, which may also be finished flat to cover the end or side grain of the core.

Batten. Thin narrow strip of plywood or lumber used to seal or reinforce a joint between adjoining pieces of lumber or plywood.

B-B exterior plywood. Sanded panel grade described by face (B) and back (B) veneers, with minimum C quality inner plies, bonded with exterior glue, and meeting all the performance requirements of exterior-type plywood. This grade has solid paintable faces and is used for outdoor utility panels.

B-B interior plywood. Sanded panel grade described by face (B) and back (B) veneers, with minimum D quality inner plies and bonded with interior or exterior glue. This grade permits circular plugs and is paintable.

Bending. Plywood can be bent without loss of strength, to form curved surfaces for concrete forms.

Bending radii. Simple curves are easy to form with plywood for a continuous uniform curvature.

Bevel. Plywood edges or ends cut at an angle to make smooth mating joints between panels.

Bleed through. Glue or components of glue that have seeped through the outer layer or ply of a glued wood product and that cause a blemish or discoloration on the surface.

Blemish. Anything marring the appearance of the veneer that is not classifiable as a defect.

Blister. Spot or area where veneer does not adhere and bulges like a blister.

Blocking. Light lumber strips nailed between major framing members to support edges of plywood panels where they meet.

Bolt (veneer). Short log cut to length suitable for peeling in a lathe.

Bond

1. Attachment at an interface between an adhesive and an adherend.

2. Attach materials together by means of an adhesive.

3. Glue together. Veneers are "bonded" to form a sheet of plywood. Pressure can be applied to keep matching parts in proper alignment. Some glues used in plywood manufacture require both heat and pressure to cure properly.

4. Grip of adhesive on wood at the line of application, particularly with heat-reactive resins.

Book matching. Turning alternate adjacent sheets of veneer of a flitch over for matching.

Bow. Distortion in a plywood panel so that it is not flat lengthwise.

Broken grain. Separation on veneer surface between annual rings.

Brushed plywood

1. Plywood siding whose surface is treated in manufacturing so that the softer wood is abraded away and the harder grain pattern stands in relief.

2. Plywood produced in exterior and interior types in standard dimensions and $\frac{5}{16}$- and $\frac{3}{8}$-in. thicknesses. The softer grain of the plywood is removed, usually by running the panels through a heavy wire brush.

Bundle. Unit or stack of plywood sheets held together for shipment with metal bands.

Burl. Localized severe distortion of the grain, generally rounded in outline, usually resulting from overgrowth of dead branch stubs, and varying from $\frac{1}{2}$ in. to several inches in diameter; it frequently includes one or clusters of several small continuous conical protuberances, each usually having a core or pith but no appreciable amount of end grain (in tangential view) surrounding it.

Butt joint. Joint formed when two parts are fastened together without overlapping. For end-to-end joints a nailing strip is used. For corner joints, the nails are driven directly into the plywood if it is at least $\frac{3}{4}$ in. thick. If plywood is thinner than $\frac{3}{4}$ in. a wood reinforcing block.

Cauls (gluing). Boards, panels, or metal sheets that are used in gluing operations to provide uniform distribution of the gluing pressure or to prevent precure of glue by slowing up transfer of heat to plywood having thin face veneers.

C-C exterior plywood. Exterior unsanded panel grade described by grade of face (C) and back (C) veneers, with minimum C quality inner plies, bonded with exterior glue, and meeting all the requirements of exterior plywood. It is the lowest grade of exterior plywood, used for rough construction that is exposed to weather.

C-C plugged exterior plywood. Exterior panel grade described by grade of face (C-plugged) and back (C) veneers with minimum C inner plies, bonded with exterior glue, and meeting the requirements of exterior-type plywood. The face is plugged with open defects not larger than $\frac{1}{4} \times \frac{1}{2}$ in. It is used as underlayment primarily for floors, and linings.

C-D interior plywood. Interior unsanded panel grade described by grade of face (C) and back (D) veneers, with minimum D quality inner plies, and bonded with interior or exterior glue. It is used for sheathing and for structural uses.

Centers. Inner plies whose grain direction runs parallel to that of the outer plies.

Chamfer. Flat surface created by slicing off the square edge or corner of a piece of wood or plywood.

Channel groove. Channel grooving design in plywood siding, shiplapped for continuous patterns. Grooving is typically $\frac{1}{16}$ in. deep, $\frac{3}{8}$ in. wide, and 4 in. on center in faces of $\frac{3}{8}$-in. thick plywood.

Check. Lengthwise separation of wood fibers usually extending across the rings of annual growth. Caused chiefly by strains produced in seasoning.

Checking. Plywood exposed, unprotected, to severe conditions of moisture or dryness may eventually develop open cracks or "checks." Checking can be reduced by sealing the edges of panels prior to installing them to minimize moisture absorption, and by using a priming coat or resin sealer on the surfaces.

Chemical resistance. Exterior-type plywood has excellent resistance to a wide range of chemicals. Strength is not significantly affected by organic chemicals, neutral and acid salts, or most acids and alkalies in the pH range of 3 to 10. High-density overlaid plywood has even better resistance.

Chord. Either of the two outside members of a truss connected and braced by the web members. The term also sometimes refers to beam flanges or perimeter members of a plywood diaphragm.

Clear face. Plywood panel with face ply free of knots, knotholes, patches, or resin fills. In some cases a few small, well-matched repairs are permitted. The definition of clear face varies from mill to mill.

Clipper. Shearing machine used to dimension dry or green veneers.

Coefficient of expansion. Rate at which a material expands or shrinks when the temperature changes. Plywood

expands slightly when heated and contracts when cooled. Because of plywood's cross-laminated construction, however, the change is so small that it is seldom important.

Cold-pressed plywood. Interior-type plywood manufactured in a press without external applications of heat.

Comb grain (rift sliced or rift sawed). Method of producing veneer by slicing or sawing at an angle of approximately 45° with the annual rings to bring out certain figures produced by the medullary rays, which are especially conspicuous in oak.

Commercial standard plywood. Plywood produced to meet commercial standards.

Component. Glued or nailed structural assembly of plywood and lumber, such as a box beam or stressed-skin panel. Term is also used to describe prefabricated building sections in panelized construction.

Compregnated wood (compreg). Synthetic, resin-treated, compressed wood with reduced swelling and shrinking characteristics and increased density and strength properties.

Concrete forms. Plywood is used for concrete formwork because it is tough, durable, easy to handle, split resistant, and lightweight.

Condensation. Moisture condensation on the surface of a plywood panel may warp the panel or permit mold to develop. Plywood is always stored in a cool, dry place out of the sunlight and weather. Outdoors, plywood stacks are covered in such a way as to provide good air circulation and ventilation between the panels.

Construction. Arrangement of veneers or lumber in fabrication of plywood.

Contact cements. These cements are useful for applying laminates and edge stripping to plywood. They are not recommended for structural joints.

Conventional or double wall construction. Plywood siding over sheathing applied to wood framing.

Core

1. Innermost portion of plywood, consisting of either hardwood or softwood sawed lumber, veneer, or composition board.

2. Plywood inner plies whose grain runs perpendicular to that of outer plies.

3. Center ply in a plywood panel. It may be composed of either one piece of lumber or of several pieces edge-glued together, or of one or more thicknesses of veneer.

4. In cutting rotary veneer, the portion of the bolt remaining after available veneer has been cut. It is also referred to as a core block.

5. Innermost portion of plywood. It may be of sawed lumber, either one piece or several pieces joined and glued, veneer, or particle board.

Core gap. Open veneer joint extending through or partially through a plywood panel, which results when core (or center) veneer sections are not tightly butted together during manufacture.

Corrosion-resistant hardware. Metal fasteners or hardware with the ability to withstand chemical erosion or wearing away. Untreated nails or staples used to apply exterior plywood siding may corrode or rust, staining the surface. Fasteners must be made of or coated with corrosion-resistant material. Hot-dip galvanizing is one common coating method.

Cross band

1. Layer of veneer in a plywood panel whose grain direction is at right angle to that of the face plies.

2. Layers of veneer whose grain direction is at right angles to that of the face plies, applied particularly to five-ply plywood and lumber-core panels.

3. Place the grain of the layers of veneer at right angles in order to minimize swelling and shrinking.

Cross bands. Inner layers whose grain direction runs perpendicular to that of the outer plies. May be parallel laminated plies. Sometimes referred to as the core.

Crossband gap and center gap. An open joint extending through or partially through a panel, which results when crossband or center veneers are not tightly butted.

Cross banding. Veneer used in the construction of plywood with five or more plies. In five-ply construction, it is placed at right angles between the core and faces.

Cross bar. Type of figure or irregularity of grain resembling a dip in the grain, running at right angles, or nearly so, to the length of the veneer.

Cross break. Separation of the wood cells across the grain. Such breaks may be due to internal strains resulting from unequal longitudinal shrinkage, or to external forces.

Cross lamination. Consecutive veneer layers are placed at right angles to each other (cross laminated) in plywood manufacture. Cross-laminated construction minimizes shrinkage and produces a strong panel.

Cross panel stiffness. Stiffness in the direction perpendicular to (across) the grain of the face ply, normally stiffness in the 4-ft direction. Panel is usually 4 ft wide by 8 ft high.

Cup. Distortion of a plywood panel from flatness across the panel.

Curved panel. Stressed-skin or plywood sandwich panels that are curved to varying degrees.

Decking. Extra-heavy plywood, $1\frac{1}{8}$ in. in thickness. It is manufactured with a minimum of seven plies and interior-type adhesives, and produced in C-D grade with the C side repaired and sanded. Its primary use is for floor systems having supports 4 ft on center, and it may serve as a base for either hardwood or resilient flooring. Plywood may be purchased with tongue-and-groove edges to eliminate the need for blocking and edge support.

Defect. Any irregularity occurring in or on veneer that may lower its strength.

Deflection. Downward bending of a plywood panel between supports when a load is applied to it.

Delamination. Separation of the plies through failure of the adhesive. This condition may be referred to as "durability of the glue line."

Diaphragm. Thin plywood skin of a building working in combination with the framing to withstand wind and earthquake loads imposed on the structure.

Dimensional stability

1. Plywood is dried during manufacture and has no "initial shrinkage;" subsequent movement is reduced by the cross-laminated construction. From oven dry to complete saturation, plywood panels swell an average of only $\frac{2}{10}$ of 1%.

2. Plywood does not swell as much as other materials when it is wet. In normal construction $\frac{1}{8}$ in. spacing between panel edges and $\frac{1}{16}$ in. between panel ends is enough to take care of any expansion. In very wet or humid areas spacing should be doubled.

Discolorations. Common veneer stains are sap stains, blue stains, stains produced by the chemical action caused by the iron in the cutting knife coming in contact with the tannic acid in the wood, and those resulting from the chemical action of the glue.

Drier

1. Tunnellike forced-air oven used to dry veneer for plywood to a predetermined level (usually about 3%) before grading and lay-up.

2. Kiln, chamber, or machine through which the green veneers are passed to remove excess moisture.

Earthquake design. Design of a structure enabling it to withstand the forces imposed during an earthquake. Also referred to as "seismic design." Because of its bracing strength and rigidity, plywood sheathing is commonly used in earthquake-resistant shear wall and diaphragm construction.

Eased edges. Edge surfaces of a piece of lumber or plywood that have been slightly rounded to remove sharp corners.

Edge grain. Grain in wood or veneer sawed so that the annual rings form an angle of 45 to 90°, with the surface of the piece.

Edge splits. Wedge-shaped openings in the inner plies caused by splitting of the veneer before pressing.

Effects of heat. Because of its glue-line durability, plywood performance is not generally degraded by heat. However plywood should generally not be used where temperatures are to exceed 230°F. Panels subjected to dry heat for long periods naturally will dry out, inviting checking, unless properly finished and protected. Low temperatures do not affect plywood adversely.

Effects of humidity. Moisture absorption by wood causes it to swell slightly. Plywood's cross-laminated construction restricts expansion and contraction within the individual plies. From oven dry to soaking wet, plywood panels swell across the grain less than $\frac{1}{8}$ in. When applying plywood always leave $\frac{1}{16}$ in. space between panel ends and $\frac{1}{8}$ in. between edges of adjacent panels. If wet or humid conditions are expected, these spacings should be doubled.

Electrical conductivity. At low moisture content, plywood is normally classified as an electrical insulator (dielectric); however, its resistance to passage of current decreases as moisture content increases.

Embossed plywood. Specially treated product manufactured from regular plywood by indenting or pressing the surface of each sheet with patterned rollers. Embossed plywood has a decorative appearance for variation in paneling and similar uses.

Extender. Additive often combined with adhesive resins in order to reduce the cost of resin in the mix.

Exterior finishing. All edges of plywood to be painted should be sealed with heavy application of exterior oil-base house paint primer or similar sealer. If plywood is not to be painted, edge seal with a good water-repellent preservative should be provided.

Exterior glue. Panels with exterior glue may be used where durability is required in long construction delays. When panels are to be permanently exposed to weather or moisture, only exterior-type plywood is suitable.

Exterior plywood

1. Plywood bonded with highly resistant adhesives so that it is capable of withstanding prolonged exposure to severe service conditions without failure in the glue bonds (ASTM D 1038).

2. Plywood bonded with a type of adhesive that by systematic tests and service records has proved highly resistant to weather; microorganisms; cold, hot, and boiling water; steam; and dry heat.

Exterior-type plywood. Exterior-type plywood is used when the wood will come into contact with excessive moisture and direct contact with the weather or water. This type of plywood is manufactured with phenolic or resorcinol-type adhesives, which are insoluble in water.

Face

1. Better side of a panel in any grade of plywood calling for a face and back; also, either side of a panel where the grading rules draw no distinction between faces.

2. Exposed or finished surfaces of plywood and veneered panels.

3. Veneer used on the exposed side of plywood, usually carefully selected and matched where attractive appearance is required. When required by location or use, face veneer is used on both sides.

4. The face of the plywood panel is the side of a panel that is of higher veneer quality or any panel whose outer plies are of different veneer grades; either side of a panel where the grading rules draw no distinction between faces.

Face checking. Partial separation of wood fibers in the surface of plywood or wood caused chiefly by the strains of weathering and seasoning.

Face grain parallel (or perpendicular). Direction of the grain of the outer ply (face) of a plywood panel in relation to the supports. Plywood's greatest strength is parallel to the face grain. Therefore, in construction the face grain should run across the supports.

Fascia. Wood or plywood trim used along the eave or the gable end of a structure.

Fasteners. Device, such as a nail, screw, clip or staple, used to hold plywood securely in place of application.

Fastening. Ultimate test loads for mechanical fasteners (the maximum load carried by the plywood under test at actual failure) must be reduced by a factor appropriate to the intended use and should take into account duration of stress, moisture content, and variability among panels and between species groups. A load factor of 2 is appropriate for many product design applications; load factors of 4 or 5 are often used for permanent structural applications.

Fiber-glass-reinforced plastic. Product made of glass fibers combined with resins to create tough, nonscuff coating over plywood. The resulting coated panel (composite) is used in truck and trailer bodies.

Field-applied plywood panels. Panels installed at the job site. These panels are used as wall or roof sheathing and are nailed to conventional framing at the construction location, as opposed to "factory applied" where panelized or modular units are finished and ready for installation in completed form.

Field gluing. Procedure whereby tough, especially developed glues are applied to the top edge of the floor joists, with the plywood then laid on and also nailed in place. The glue rather than the nails carries the load, developing a stiffer floor construction that is virtually squeak free.

Figure. Pattern produced in a wood and plywood panel surface by annual growth rings; rays; knots; deviations from regular grain, such as interlocked and wavy grain; and irregular coloration.

Figured. Term applied to veneer containing irregular grain formations that add to its value for furniture and other decorative uses. The various figures are referred to as rift cut, needlepoint comb grained, stripe, rope, roe, mottle, fiddleback, raindrop, finger roll, cross fire, chain, curly, blister, birds eye, feather, crotch, stump, burl, etc.

Fillers. Additives often combined with adhesive resins to change bonding characteristics of the resin mix.

Finger joint. Joint formed by a series of fingers machined on the ends of the two pieces to be joined; the fingers mesh together and are held firmly in position by an adhesive.

Finishing. Finishing improves the appearance and protects the surface of the plywood. Finishes may be paints, stains, or coatings.

Finnish birch plywood. Strong, lightweight, and durable plywood. Extremely high standards of materials and construction, with many more veneer layers, contribute to high performance. Finnish plywood satisfies rigorous requirements of exterior construction and the formability and aesthetic requirements for interior uses. A product of Finnish Plywood Development Association, Falls Church Va.

Fire-rated systems using plywood. Wall, floor, and roof construction of specific materials and designs that has been tested for conformance to fire safety criteria, (such as flame-spread rate and fire resistance), and rated accordingly. Testing and approval is done by agencies such as Underwriters Laboratories, Inc. Plywood is an approved material in a number of fire-rated designs.

Fire-retardant treated. Term applied to wood and plywood that have been chemically treated to retard combustion. Fire-retardant chemicals mixed in water are deposited on the wood under pressure.

Flame spread. Term that relates to the spread of flame along the surface of a material. Flame-spread ratings are expressed in numbers and are used in describing interior finish requirements in building codes.

Flange. Longitudinal ribs that give a member most of its strength. Plywood box beams are fabricated with lumber flanges (top and bottom) and plywood webs (sides).

Flat cut. Term applied to veneer sliced parallel to the pith of the log and approximately tangent to the growth rings. Also termed "plain sliced."

Flat grain

1. Veneer sawed so that the annual rings form an angle of less than 45° with the surface of the piece.

2. Veneer cut so that the growth rings meet the face over at least half the width at an angle of less than 45°. Also called "plain cut," "flat sawed," "slash grain."

Flitch

1. Portion of a log sawed on two or more sides and intended for remanufacture into lumber or sliced or sawed veneer. Term is also applied to the resulting sheets of veneer laid together in sequence of cutting.

2. Portion of a log sawed on two or more faces—commonly on opposite faces—leaving two wavey edges. When intended for resawing into lumber, it is resawed parallel to its original wide faces. Or it may be sliced or sawed into veneer, in which case the resulting sheets of veneer laid together in the sequence of cutting are called a "flitch."

Floors. Plywood for floors may be used as conventional subflooring, underlayment, and combination subfloor and underlayment in a single layer (single floor), depending on support spacing.

Foam core. Center of a plywood "sandwich" panel. Liquid plastic is foamed into all spaces between the plywood panels and serves to insulate, as well as support, the component skins.

Fortified glue. Hot-press protein adhesives to which phenolic resin has been added. Fortified glue is used in interior-type plywood and is not suitable for exterior exposure.

Frame construction. Type of building in which the structural parts are wood or dependent on a wood framework for support. Commonly, lumber framing is sheathed with plywood for roofs, walls, and floor.

Furring. Process of leveling parts of a ceiling, wall, or floor by means of wood strips, called "furring strips," before adding plywood cover.

Gang forming (concrete). Grouping of lumber-framed panels to create large formwork for massive concrete pouring operations.

Gang grooved. Term applied to plywood panels produced by passing under a machine with grooving knives set at certain intervals.

Gap. Open slits in the inner ply or plies; improperly joined veneers when joined veneers are used for the inner plies.

Glue

1. Substance capable of holding materials together by surface attachment. (In woodworking it often implies a substance capable of forming a strong bond with wood).

2. Attach materials together by means of glue.

3. In plywood construction many adhesives are used, preferably in conjunction with nails or other fasteners, to produce strong joints. The type used depends on the purpose and exposure of the finished product.

Glued floor system. Floor construction system in which structural plywood underlayment is glued with elastomeric adhesives and nailed to wood joists. The bond is so strong that floor and joists act like an integral unit, which greatly increases stiffness and virtually eliminates squeaks. Joist sizes can sometimes be reduced or, in some cases, spaced further apart.

Glue line. Line of glue visible on the edge of a plywood panel. Also, the layer of glue itself.

Glue nailed. Term applied to plywood joints and connections that are both glued and nailed for the stiffest possible construction. For most effective fastening the pieces to be joined should contact at all points. Glue is applied to one or both surfaces according to the manufacturer's directions; then surfaces are pressed together and nailed in place. For work such as cabinets or drawers, or whenever possible, the joint should be clamped, as well as nailed, to maintain pressure until the glue sets.

Good grade—No. 1. Veneer on the face is fabricated to avoid sharp contrasts in color and grain.

Good one side. Term for plywood that has a higher veneer on the face than on the back and is used in situations where only one side of the panel will be visible.

Grade

1. Designation of the quality of a piece of plywood. In plywood there are several veneer grades described by letter, which calls out the face and back veneers (B-C, etc.), or by intended end use of the grade.

2. Within each type of plywood there is a variety of appearance and sheathing grades generally determined by the grade of the veneer (A, B, C, or D grade) used on the face and back of the panel.

Grade trademark. In the plywood industry, a mark stamped on the back or branded into the edge of a panel of plywood. It includes the type and grade of the panel, in addition to an identifying symbol.

Grading rules. Standards of grading for the softwood lumber industry. Grading rules were standardized originally about 1925 with the concurrence of the major softwood lumber associations and were issued by the U.S. Department of Commerce. Product Standard PS 20-70, the American Lumber Standard.

Grain

1. Direction, size, arrangement, appearance, or quality of the fibers in wood or veneer. To have a specific meaning the term must be qualified.

2. Natural pattern of growth in wood. Grain runs along the length of the tree; therefore, the strength is greatest in that direction.

Grain character. Varying pattern produced by cutting through growth rings, exposing various layers. The most pronounced is that in veneer cut tangentially or rotary cut.

Grain direction. In the face and back veneers of a plywood panel, the grain usually runs in the long dimension making the panel strongest in that direction.

Grain raise. Condition resulting on the surface of a plywood panel when the harder or more dense fibers of wood swell and rise above the surrounding softer wood.

Grain rupture. Slight breaks in a veneer resulting from improper cutting or irregular grain.

Groove. One of the surface treatments frequently given to textured plywood, in which a series of narrow parallel channels are cut into the surface of the panel. Grooving is available in a variety of widths and spacings on several surface textures.

Grooved plywood. Exterior-type Douglas fir plywood designed primarily as a combination sheathing and siding material and produced in one thickness, $\frac{5}{8}$ in. The $\frac{3}{8}$ in. grooves are $\frac{1}{4}$ in. deep and spaced either 2 or 4 in. apart.

Group

1. Term used to classify species covered by the Product Standard PS-1 in an order that provides a basis for simplified marketing and efficient utilization.

2. Plywood is manufactured from some 50 species that have been classified into five groups under PS-1. The group number appearing in many APA grade trademarks is based on the species used for outer plies. Where face and back veneers are not from the same species group, the number is based on the weaker group. The strongest species are in Group 1.

Gusset plate. Piece of plywood, generally triangular in shape, used to connect lumber members of a truss or other frame structure. A gusset may be applied to one or both sides of the joint. Plywood is used because of its great strength and nail-holding capacity.

Half-round

1. Flitch is mounted on a stay log and is cut on a lathe. It differs from rotary-cut veneer in that the flitch is cut with a wider sweep than when mounted at the lathe center, and the center of the tree is not near the center of rotation.

2. Term refers to a method of cutting veneer to bring out certain beauty of figure, accomplished in the same manner as rotary cutting, except that the piece being cut is secured to a "stay log," a device that permits the cutting of the log on a wider sweep than when mounted with its center secured in the lathe.

Hairline. Thin perceptible line usually showing at the joint.

Hand sawing plywood. For hard sawing plywood it is recommended that a 10- to 15-point (teeth per inch) crosscut saw be used.

Hardboard-faced plywood. Plywood to which hardboard has been glued to one or both faces. It is produced in both exterior and interior types in standard dimensions and in $\frac{1}{2}$-, $\frac{5}{8}$-, and $\frac{3}{4}$-in. thicknesses.

Hardwood

1. General term used to designate lumber produced from broad-leaved or deciduous trees in contrast to softwood produced from evergreen or coniferous trees.

2. Wood of the broad-leaved trees—oak, maple, ash, and walnut—as contrasted to the softwood of the needle-leaved trees, such as pine, fir, spruce, and hemlock. The term has no reference to the actual hardness of the wood.

Hardwood plywood. The species used in the face plies identifies hardwood plywood; that is, black walnut plywood would have one or both face plies of black walnut. Cherry, oak, birch, black walnut, maple, and gum among the native woods and mahogany, lauan, and teak in the imported category are some of the more common species used in hardwood plywood. The major difference in the manufacture of softwood and hardwood plywood is the use of a solid "core" or extra-thick middle ply in some hardwood panels.

Hardwood plywood—Type I. Hardwood plywood—Type I is manufactured with waterproof adhesives and is used in areas where it would come in contact with water.

Hardwood plywood—Type II. Hardwood plywood—Type II is manufactured with water-resistant adhesives and is used in areas where it would not ordinarily be subjected to contact with water. It can be used in areas of continued dampness and excessive humidities.

Hardwood plywood—Type III. Hardwood plywood—Type III is manufactured with moisture-resistant adhesives and is intended for use in areas where it will not come in contact with any water. It can be subjected to some dampness and excessive humidity.

Heartwood

1. Wood extending from the pith to the sapwood, the cells of which no longer participate in the life processes of the tree. Heartwood may be infiltrated with gums, resins, and other materials that usually make it darker and more decay resistant than sapwood.

2. Nonactive core of a tree, distinguishable from the growing sapwood by its usually darker color.

Heavy white pocket. Veneers may contain a great number of pockets, in dense concentrations, running together and at times appearing continuous; holes may extend through the veneer, but the wood between pockets appears firm. Brown cubicle and similar forms of decay that have caused the wood to crumble are prohibited.

High-density overlay. Exterior-type plywood finished with a resin-impregnated fiber overlay to provide extremely smooth hard surfaces that need no additional finishing and have high resistance to chemicals and abrasion. The overlay material is bonded to both sides of the plywood to become an integral part of the panel faces.

High-density plywood. Overlaid plywood having 40% resin by weight in the overlay sheet.

Hot press. In the plywood industry, a press that utilizes both heat and pressure to bond plywood veneers into panels. All exterior-type plywood is manufactured in hot presses. Hot press may also refer to interior-type plywood that has been manufactured in a hot press.

Identification index

1. Set of numbers used in the marking of sheathing grades of plywood. The numbers are related to the species of panel face and back veneers and to the panel thickness in such a manner as to describe the bending properties of a panel. The index is particularly applicable where panels are used for roof sheathing and subflooring to describe recommended maximum spans in inches under normal use conditions. Actual maximum spans are established by local building codes.

2. Identification index numbers appear in Douglas Fir Plywood Association (DFPA) grade trademarks on Standard C-D INT-DFPA, C-C EXT-DFPA, Structural I C-D, and Structural I C-C grades. Identification index numbers refer to maximum recommended spacing of supports in inches when panels are used for roof decking and subflooring with face grain across supports. The left-hand number shows spacing for roof supports; the right-hand number shows spacing for floor supports.

Impact resistance

1. Plywood is splitproof and relatively punctureproof. Even when it is supported on only two edges, cross-laminated construction and large panel size effectively distribute impact loads.

2. Ability to withstand sharp blows or violent contact. Plywood's cross-laminated construction provides great resistance to these blows and distributes the force across both dimensions of the panel.

Inner plies. All layers of a plywood panel except face and back.

Insulation. Softwood plywood has low moisture content and density and is an excellent insulator. Large panel size reduces joints and resultant drafts and air infiltration.

Interior. Type of plywood bonded with adhesives that maintain adequate bonds under conditions usually existing in the interior of buildings in the United States (ASTM D 1038).

Interior-exterior plywood. Plywood is classified as interior or exterior, depending on the type of adhesive employed. Interior-grade plywood must have a reasonable degree of moisture resistance but is not considered to be waterproof. Exterior-grade plywood must be completely waterproof and capable of withstanding immersion in water or prolonged exposure to outdoor conditions.

Interior-type plywood

1. Plywood intended for inside uses or for construction where plywood will be subjected to only occasional wetting. Interior-type plywood is bonded with adhesives that are highly moisture resistant, but not necessarily waterproof.

2. Interior-type plywood can be used in protected areas for wall and roof sheathing, interior wall paneling, and cabinets. Interior plywood is assembled with a protein-type adhesive, such as casein, soybean, or blood-based products. Synthetic adhesives are also used in the production of interior-type plywood; these adhesives are not waterproof and tend to deteriorate or dissolve from contact with excessive moisture. Interior-type plywood should not be used where it will be exposed to excessive moisture or high humidity. If used as roof sheathing, it must be protected from the weather.

Interior-type plywood with exterior-type glue. Interior-type plywood manufactured with waterproof glue for greater ability to withstand exposure to moisture. This type of plywood is not suitable for permanent exterior exposure.

Intermediate glue. Glue used in the manufacture of interior type plywood that is more moisture resistant than interior glue but not intended for permanent exposure. Panels with intermediate glue are suitable where moderate delays in providing protection are expected, or where high humidity or water leakage may occur temporarily.

Jointed core. Core veneer that has had edges machined square. Gaps between pieces of core should not exceed $\frac{3}{8}$ in., and the average of all gaps in the panel should not exceed $\frac{3}{16}$ in.

Joint edge. Joint running parallel to the grain of the wood.

Jointed inner plies. Crossband and center veneer that has had edges machine-squared to permit tightest possible layup.

Joint treatment. Method of covering or sealing space between panels. For subfloor, roof, and wall sheathing, a space of $\frac{1}{8}$ in. should be left between sheathing panel edges and $\frac{1}{16}$ in. between ends to allow for expansion. Exterior siding joints may be sealed with building paper or caulking. Edges may be shiplapped, or horizontal butt joints may be sealed with Z flashing. Battens may be used with vertical textured sidings to double as decorative finish. In all cases exterior joints should be backed with solid lumber framing.

Joist. Horizontal framing member of a floor or ceiling. Plywood is commonly used for subflooring and underlayment over floor joists, and for the manufacture of a joist where the plywood is used as a web member.

Kiln-dried. Wood that has been dried in ovens (kilns) by controlled heat and humidity to specified limits of moisture content. Plywood veneers are kiln dried before lay-up.

Knot. Cross section of branch or limb with grain usually running at right angles to that of the piece in which it occurs.

Knotholes. Voids in veneers produced by the dropping of knots from the wood in which they were originally embedded.

Laminate. Bond together a series of layers to make such products as plywood or laminated structural members. Plywood is laminated with the grain of adjacent layers at right angles for two-way strength. Structural members, such as beams, are laminated with the grain running in the same direction in all layers.

Lap

1. Condition in which adjacent veneers are so misplaced that one piece overlaps the other instead of making a smooth edge joint.

2. Position panels so that one surface extends over the other. Term may be used to designate a plywood exterior lap siding technique, in which each plywood panel overlaps the edge of the next lower panel. A plywood shiplap joint unites two panels when half the thickness of each panel is cut away so that the two pieces fit together with outer faces flush.

Latex paints. Water-base paints that are available for both interior and exterior use over plywood and that have excellent adhesion and blister-resistant characteristics.

Lathe. Machine on which rotary and half-round veneer is cut.

Layer. Single ply or veneer, or two or more plies laminated with grain directions parallel.

Lay-up. Step in the manufacture of plywood in which veneers are "stacked" in complete panel "packages" after gluing and before pressing.

Light white pocket. Term describing stage that is advanced beyond the incipient stain stage to the point where pockets are present and plainly visible. Most of these pockets are small, filled with white cellulose, and generally distributed with no heavy concentrations. The pockets for the most part are separate and distinct with few to no holes through the veneer.

Loose side. In knife-cut veneer, the side of the sheet that was in contact with the knife as the sheet was being cut, containing cutting checks because of the bending of the wood at the knife edge.

Lot. Any number of panels considered as a single group for evaluating conformance to the standard.

Lumber core construction

1. Plywood construction in which the core is composed of lumber strips. Face and back (outer) plies are veneer.

2. Lumber-core plywood contains a thick core made by edge-gluing several narrow strips of solid wood. This core forms the middle section to which veneer cross bands and face plies are glued.

3. Plywood in which the center ply or core is of lumber rather than of veneer. Ordinarily cores that are $\frac{3}{8}$ in. or greater in thickness will be of lumber.

Marine-grade panels. Panels manufactured with the same glue-line durability requirements as other exterior-type panels, but with more restrictive requirements for the wood characteristics. This grade is particularly suitable for marine applications where bending is required.

Marine plywood. Plywood that is very similar to the exterior types. It is produced especially for boat hulls and is available in long lengths. The inner plies are B grade or better and handled to reduce the possibility of defects appearing along cut edges.

Matching. Orienting of sheets of veneer to obtain a particular pattern.

Medium-density overlay. Exterior-type plywood finished with an opaque resin-treated fiber overlay to provide a smooth surface that is an ideal base for paint.

Medium-density plywood. Overlaid plywood having 20% resin by weight in the overlay sheet.

Melamine glue. Glue used in scarf-jointing plywood panels. Heat and pressure are required for proper curing. It is also used in patching panels and in some coatings.

Metal overlaid. Term applied to plywood that has a metal face permanently bonded to it. Panels may be metal overlaid on one or both sides. Surface finishes include pebble texture, as well as assorted colors (baked-on enamel).

Methods of obtaining veneers. Depending on whether plywood is to be used for general utility or for decorative purposes, the veneers employed may be cut by peeling from the log, by slicing, or by sawing. Sawing and slicing give the greatest freedom and versatility in the selection of grain. Peeling provides the greatest volume and the most rapid production because the logs are merely rotated against a flat knife and the veneer is peeled off in a long continuous sheet.

Mildew. Mold or discoloration on wood that occurs under warm, humid conditions or where there is poor air circulation. Mildew begins on plywood as dark spots on the surface and spreads in gray, fan-shaped areas. A dark spot usually can be identified as mildew if it disappears when household bleach is applied.

Mildewed surfaces. Mildewed surfaces should be scrubbed lightly with mild detergent, rinsed with household bleach, rinsed with clear water, and dried. New paint or stain finish should be applied to the prepared surface according to recommendations by the manufacturer.

Mildew protection. Primers are available with mildewcide for use in areas of continuous warm, damp conditions where mildew may discolor paint finishes. Primers for wood should be free of zinc oxide, even if a zinc-containing paint is used for the finish coat. Primers should always be brush-applied to work the paint thoroughly into the wood.

Miter joints. Joint formed by fitting together two pieces of lumber or plywood that have been cut off on an angle.

Moisture content. The amount of water held by the wood. It is usually expressed as a percentage of the weight of the oven-dried wood (a sample of oven-dried wood could absorb 25% of its weight in water and weigh $1\frac{1}{4}$ lbs at 25% moisture content). The softwood plywood Product Standard PS-1 limits the moisture content of plywood to a maximum of 18% of the oven-dry weight at the time of shipment from the mill.

Molded plywood. Plywood that is glued to the desired shape either between curved forms or more commonly by fluid pressure applied with flexible bags or blankets (bag molding), or other means.

Nail coating. Surface treatment, such as galvanizing, to prevent nails from rusting on exposure to weather and thereby prevent staining of the siding material. Coatings are also used to color match nails to the panels with which they are used.

Nail popping. Serious problem that occurs frequently with thin floor coverings. The nails used in applying the plywood underlayment appear to "pop" up so that nailhead impressions are visible on the surface of the finished floor covering. The problem is caused by the shrinkage of floor joists away from the nails after installation. To prevent the problem, all nailheads should be countersunk $\frac{1}{16}$ in. below the surface of the plywood underlayment just prior to laying on the floor covering. The nail holes should not be filled in.

Nail schedules. Nail schedules are recommended by the manufacturers and refer to the size and spacing of nails that should be used with various thicknesses of plywood.

Nail staining. Staining of a plywood surface due to corrosion of the fastener or nail on exposure to weather.

Nail withdrawal. Resistance of a nail being pulled out of wood. Ring shank or spiral thread nails have much higher nailholding power than regular smooth-shank nails or coated nails.

Nominal thickness. Full "designated" thickness. For example, $\frac{1}{10}$ in. nominal veneer is 0.10 in. thick. Nominal $\frac{1}{2}$ in. thick panel is 0.50 in. thick. Commercial size designation is subject to acceptable tolerances.

Noncertified (mill certified). Term that refers to plywood for which there is no industry standard of manufacture and that may bear the mark of the manufacturer rather than one of a recognized agency such as the American Plywood Association or the California Redwood Association.

Open defects. Irregularities such as splits, open joints, knotholes, or loose knots that interrupt the smooth continuity of the veneer.

Open joint

1. Failure of bond or separation of two adjacent pieces of veneer so as to leave an opening. Term usually applied to edge joints between veneers.

2. Joint in which two adjacent pieces of veneer do not fit tightly together.

Overlaid plywood

1. Plywood in which the face veneer is bonded on one or both sides with paper, resin-impregnated paper, or metal.

2. Plywood panels with factory-applied, resin-treated fiber faces on one or both sides. Term may also be applied to metal-overlaid plywood and other composites.

3. Exterior-type plywood that has a resin-impregnated fiber sheet glued to one or both faces. It is produced in standard sizes and is used principally for signs and combination sheathing and siding applications. A medium-density overlay is made to be painted; a high-density overlay, designed for use in concrete forms, requires no further finish.

P&TS. Plugged and touch-sanded face of a plywood panel.

Panel. Sheet of plywood of any construction type.

Panel faces. Outer veneers of a plywood panel.

Panel grades. Appearance or structural grade of a plywood panel, determined by the veneer grades used for the face and back of the panel.

Paneling. Wood panels joined in a continuous surface, especially decorative panels for interior wall finish. Textured plywood in many varieties is often used as interior paneling, either in full wall sections or for accent walls.

Panelized construction. Building components fabricated in the form of wall, floor, ceiling, etc., sections, which will be assembled into the completed structure at the building site.

Panel patch. Repair installed in the face or back veneer of a panel after lay-up and pressing of the panel.

Panel spacing. Gap left between plywood panels when they are installed in a structure. Space should be left between panels in floor, wall, or roof deck construction to allow for expansion due to changes in moisture conditions. When wet or humid conditions can be expected, the normal spacing for sheathing and subfloor panels should be doubled.

Particle-board core. Core that is an aggregate of wood particles bonded together with a resin binder. Face veneers are usually glued directly to the core, although cross banding is sometimes used. In most instances, particleboard-core plywood has greater dimensional stability than the other types.

Patches

1. Insertions of sound wood, having tapered sides, placed and glued into veneers from which defective portions have been removed.

2. Insertions of sound wood in veneers of panels for replacing defective areas. Patches may be in a variety of shapes. Most commonly used are "boat" patches, oval shaped with sides tapering to points or small rounded ends; "router" patches with parallel sides and rounded ends; "router" patches with parallel sides and rounded ends; and "sled" patches, rectangular in shape with feathered ends.

Peel

1. Convert a log into veneer by rotary cutting.
2. Cut wood into thin sheets of veneer in a lathe by rotating the log against a long knife.

Peeler log. Log selected and suitable for cutting into rotary veneer.

Penta (pentachlorophenol). Preservative that gives long-lasting protection against termites, rot, mold, and mildew. Penta-treated plywood is noncorrosive and fairly clean to handle, does not produce skin burn, and may be used for exterior exposure.

Permeability. Readiness with which water vapor moves through the plywood panel thickness. To prolong the life of finishes, a vapor retarder should be installed on the warm side of the wall and ceiling construction. Vapor transmission is also a factor in construction of the all-weather wood foundation system and must be controlled through use of a vapor retarder.

Phenol resorcinol. Synthetic resin-compound adhesive used in the manufacture of plywood and plywood components, in scarf joints, and for exterior-type plywood panel patching. Also called "modified resorcinol."

Pitch pocket. Well-defined opening between rings of annual growth, usually containing, or which has contained pitch, either solid or liquid.

Planing. Smoothing or shaping a wood surface. Planing plywood edges with a plane or jointer is not often necessary if the saw cut has been made with a sharp saw blade.

Platen. Plate of metal, especially one that exerts or receives pressure, as in a press used for gluing plywood.

Plugs

1. Straight-sided insertions of sound wood placed into veneers from which defective portions have been removed.
2. Sound wood of various shapes, including circular, dog-bone, and leaf shapes, for replacing defective portions of veneer; also, synthetic plugs of fiber and resin aggregate used to fill openings and provide a smooth, level, durable surface. Plugs usually are held in veneer by friction until veneers are bonded into plywood.

Ply

1. Single sheet of veneer or several sheets laid with adjoining edges, which may or may not be glued, that form one layer in a piece of plywood.
2. Single veneer in a glued plywood panel.

Plyclips. American Plywood Association trade name for special aluminum H clips used as an economical substitute for lumber blocking in roof construction.

Plyform. American Plywood Association trade name for a concrete form panel.

Plywood

1. Plywood is made of an odd number of thin sheets of wood glued together, with the grain of the adjacent layers perpendicular. The grain of the two outside plies must be parallel to provide stability. This gives the panel nearly equalized strength and minimizes dimensional changes. Thin layers of wood, called plies, usually are "peeled" from a log as veneer. In some instances the veneer is sliced from the log. The veneer is cut into various lengths; dried, selected, or graded; then glued together to make a sheet or panel of plywood.
2. Series of thin layers of wood (veneer) glued together to form a solid sheet. The grain direction runs at right angles in adjacent layers.

3. Composite panel or board made up of cross-banded layers of veneer only or of veneer in combination with a core of lumber or particle board bonded with an adhesive. Generally the grain of one or more plies is roughly at right angles to the other plies, and almost always an odd number of plies are used.
4. Plywood consists of thin sheets or veneers of wood glued together. The grain is oriented at right angles in adjacent plies. To obtain plywood with balance, that is, that will not warp, shrink, or twist unduly, the plies must be carefully selected and arranged to be mirror images of each other with respect to the central plane. All plies are at the same moisture content at the time of manufacture. Outside plies or faces are parallel to each other and are of species that have the same shrinkage characteristics. The same holds true of the cross bands or inner plies and of the core or central ply itself. As a consequence, plywood, as ordinarily made, has an odd number of plies, the minimum being three. Matching the plies involves thickness, the species of wood with particular reference to shrinkage, equal moisture content, and making certain that the grain in corresponding plies is parallel.
5. Cross-banded assembly made of layers of veneer, or veneer in combination with a lumber core, particle-board core, or other type of composition material, or plies joined with an adhesive. Three types of plywood are recognized, namely, veneer plywood, lumber-core plywood, and particle-board-core plywood. (Except for special construction, the grain of one or more plies is approximately at right angles to that of the other plies, and an odd number of plies is used.)

Plywood box beam. Beam built up from lumber and plywood in the form of a long hollow box, which will support more load across an opening than will its individual members alone.

Plywood construction systems. Plywood roof systems range from conventional installation of roof sheathing over lumber joists or preframed panels to stressed-skin panels and folded plates. Wall systems provide resistance to wind and earthquake forces, and are durable and economical. They include single-wall, double-wall, and preframed panel systems. Floor systems include the conventional subfloor and underlayment installation.

Plywood decking. Plywood sheathing used for the deck (flat) portion of a roof.

Plywood I-beam. Beam whose cross section resembles the letter I and are used extensively as structural supporting members.

Plywood panel foundations. All-weather wood foundations; residential building foundation systems in which pressure-preservative-treated exterior plywood panels are used in place of masonry walls under certain soil and weather conditions.

Plywood panel sizes. The most common panel size is 4×8 ft, but many manufacturers produce 4×10 panels. Other sizes are available on special order.

Plywood roof decking. Plywood has an excellent strength-weight ratio, takes and securely holds mechanical fasteners, and can be used under any type of single or built-up roofing.

Plywood shear walls. Because of its strength and nail-holding abilities, plywood sheathing is ideal in structures that require engineered shear walls to transfer lateral loads.

Plywood soffit. Edge of a roof that extends beyond or overhangs the wall. The underside of an eave may form an "open soffit"; plywood, applied face down to eave rafters as roof sheathing, gives a finished surface to the open soffit for painting.

Plywood system. General term to describe a method of construction that incorporates plywood as a major structural material. Plywood construction systems include APA glued-floor system and APA single-wall system.

Plywood types. Plywood is manufactured in two types—exterior, with fully waterproof glue line, and interior, with a moisture-resistant glue line. Veneers in backs and inner plies of exterior-type plywood are of a higher grade than those of interior-type plywood.

Plywood wall sheathing. Plywood wall sheathing delivers strength and rigidity to the structure. Neither let-in bracing nor building paper is required with plywood. Panels may be installed either vertically or horizontally, although horizontal application is recommended when sidings such as shingles are to be nailed directly into the sheathing. Horizontal application delivers greater stiffness under loads perpendicular to the surface; vertical application delivers greater racking resistance.

Postformed plywood. Product formed when flat plywood is reshaped into a curve configuration by steaming or plasticizing agents.

Power sawing. Electrical power sawing equipment with standard type plywood blade gives best cutting and sawing results, but the proper selected combination blade may also be used.

Prefabricated. Term indicating that all parts, including plywood panels, have been constructed or fabricated at the factory, so that final construction consists only of assembling and uniting standard parts at the job site.

Prefinished

1. Term applied to plywood panel that has factory-applied finish (paint, overlays, coatings) so that the panel is ready for use.
2. Several manufacturers are producing factory-finished plywood with a clear surface treatment.

Preframed panels

1. Panels fabricated from precut lumber and plywood.
2. Construction term for panelized building, in which the wall, floor, or roof sections are complete; that is, they are framed and sheathed at the factory.

Premium grade—No. 1. The veneer on the face is fabricated for matched joints, and contrast in color and grain is avoided.

Preprimed. Plywood panel that has a factory-applied primer or undercoat so that only the final finish need be applied after the panel is in place.

Preservatives. Any of a number of treatments to prevent plywood or wood from deteriorating due to exposure to weather, adverse moisture conditions, or insect attack. Treatments range from chemical impregnation under pressure, as for wood foundation or reservoir cover use, to any of the many paints, stains, or sealers.

Press. Apparatus for applying and maintaining pressure on an assembly of veneers and adhesive in the fabrication of plywood. It may be operated mechanically or hydraulically, and the platens may be cold or heated depending on the type of adhesive used.

Pressure treated. Term applied to wood treated with preservatives or fire retardants. The treating solution is forced into the wood cells under pressure.

Prime coat. Most important coat of the paint system, applied as soon as possible after panels are cut to size or installed. The primer provides the necessary bond between the wood and finish paint. Normal drying, high-grade exterior primers, thoroughly brushed, give the best results.

Primer. Undercoat applied to bare woods to serve as a sealer and base for paint.

Product standards. Industry manufacturing quality specifications approved by the U.S. Department of Commerce. American Plywood Association grade trademarks are positive identification by the manufacturer that this plywood has been produced in conformance with U.S. Product Standard PS 1 for "Softwood Plywood" and the quality requirements of the APA.

Puncture resistance. Ability to withstand damage when sharp or extremely heavy objects are dropped on the surface. Plywood is extremely puncture resistant because of its cross-laminated structure, which distributes the impact forces.

Qualified coatings. Coatings that have been tested and approved for use on plywood. A certificate is issued by the American Plywood Association for a qualified coating after it has passed a series of performance tests established by the association. The certificate indicates that when a coating is applied to APA plywood in accordance with stated conditions, it will perform satisfactorily as a finish, as specified. The certificate does not imply that the coating itself or its application are subject to association quality control.

Quality testing program. Testing program administered by APA and designed to maintain quality levels equal to or exceeding that prescribed by the "Softwood Plywood," Product Standard, PS-1. The program is based on a scientific random sampling process. Plywood quality in every American Plywood Association member mill is checked in frequent unannounced calls by trained inspectors. The program includes provisions for withdrawal of grade trademark privilege if quality levels are not maintained.

Quartered. Term applied to veneer manufactured by slicing or sawing to bring out certain figures produced by the medullary or pith rays, which are especially conspicuous in oak. The log is flitched in several different ways to allow the cutting of the veneer in a radial direction.

Rabbet joint. Joint formed by cutting a groove in the surface or along the edge of a board, plank, or panel to receive another piece.

Reaction to chemical exposure. Plywood behaves much like solid wood, with good resistance to chemical exposure. Medium-density- and high-density-overlaid plywoods have very high resistance to chemicals.

Redwood plywood standards. Redwood plywood manufactured by California Redwood Association member mills conforms to U.S. Product Standard PS 1 and the American Plywood Association's 303, Siding Specifications. Face veneers of redwood plywood are exceptionally free of noticeable repairs, knots, and knotholes permitted under the standards.

Refinishing. Application of new surface treatment (paint, stain, or sealer) to renew the appearance and prolong the life of the plywood. Refinishing is necessary only if original finish shows signs of deterioration.

Repainting previously painted plywood. Remove all loose paint, surface dirt, and chalk. On textured plywood use a stiff brush or water blaster. On sanded plywood use moderate sanding, a stiff-bristle brush, or power tools. Feather edges of remaining paint by sanding. Wash and rinse off surface dirt and dust and allow to dry. If plywood has checked, fill checks with a pliable patching compound before repainting.

Repair. Patch, plug, or shim in a plywood panel.

Resawed. Rough sawed; decorative treatment normally provided by scoring the surface of a panel with a saw after manufacture, which imparts a rough, rustic appearance.

Resilient floor covering. Resilient floor covering installed over plywood subflooring and underlayment provides smooth, stiff, rigid floor for comfortable walking.

Resistance to impact loads. Plywood resists impact loads owing to its cross-laminated construction. The grain running crosswise in alternate layers distributes impact forces that would shatter other materials down the length and width of the panel.

Resistance to racking. Ability to resist forces that would "pull" a plywood panel from its rectangular shape. (Tests prove that $\frac{1}{4}$-in. plywood wall sheathing on a standard stud wall has twice the strength and rigidity of 1×8 diagonal board sheathing.)

Resorcinol (waterproof glue). Resorcinol comes as powder plus liquid and must be mixed each time used. It is dark colored, very strong, and completely waterproof. This glue is used with exterior-type plywood for work exposed to extreme dampness.

Restaining previously stained plywood. Remove surface dust and finish chalk using a mild detergent solution followed by thorough rinsing. If surface fibers are loose, remove with light bristle brushing before cleaning. If an opaque stain was the original finish, remove any loosened film using a water blaster.

Reversed matching. Turning alternate adjacent sheets of veneer of a flitch end for end; also called "swing matching."

Rift-sliced (rift sawed). Method of producing veneer by slicing or sawing at an angle of approximately 45° with the annual rings to bring out certain figures produced by the medullary rays, which are especially conspicuous in oak; also termed "comb-grain."

Rigid frame. Structural member in which the studs and rafters are joined together with plywood gussets, so that they act together like an arch. Rigid frame construction eliminates the need for ceiling or tie members.

Roofs. Plywood provides a tight roof deck and high resistance to racking.

Roof systems. These include plywood roof sheathing glued and nailed to joists or rafters.

Rotary-cut veneer. Veneer cut by the rotary cut method, by which the entire log is centered in a lathe and turned against a broad cutting knife, which is set into the log at a slight angle.

Rough grain. Grain characteristics that prevent sanding to a smooth surface.

Rough sawed. Decorative treatment normally provided by scoring the surface of a panel with a saw after manufacture, which imparts a rough, rustic appearance. Same as resawed.

Sanding. Smoothing the surface of wood with sandpaper or some other abrasive. With plywood, sanding should be confined to edges. Most appearance-grade plywood is sanded smooth in manufacture, and further sanding of the surfaces merely removes soft grain. After sealing, sand in direction of grain only. Sanding block will prevent gouging.

Sandwich panels. Sections (as of walls) of layered construction made up of high-strength plywood faces or "skins" attached to or over low-density core materials, such as plastic foam or honeycomb paper fillers.

Sawed veneer. Veneer produced by sawing.

Scarf joint. Angled or beveled joint in plywood where pieces are spliced together. The length of the scarf is 5 to 12 times the thickness.

Scarf jointing. Joint in which the ends of plywood panels are beveled and glued together.

Seasoning. Removal of moisture from wood to improve its serviceability, usually by air-drying, i.e., drying by exposure to air without artificial heat, and kiln-drying, i.e., drying in a kiln using artificial heat. Plywood veneers are seasoned before lay-up and gluing into panels.

Secondary glue line. Glue joint formed in gluing together wood and plywood parts in an assembly such as a component. Glue line between veneers in a plywood panel is called the "primary glue line."

Shear. Plywood's high shear strength and split resistance obtain effective structural connections with mechanical fasteners. Plywood is easy to fasten with glue or a combination of gluing and nailing or screwing. U.S. Forest Service laboratory tests have shown that "walls sheathed even with $\frac{1}{4}$-inch plywood are twice as rigid and over twice as strong as comparable diagonal sheathed walls."

Sheathing. Structural covering, usually of plywood or boards, on the outside surface of the framing. Sheathing provides support for construction, snow, and wind loads and backing for attaching the exterior facing material, such as wall siding or roof shingles. Standard with exterior glue are recommended as sheathing products.

Shim. Long, narrow repair made of wood or suitable synthetic, not more than $\frac{13}{16}$ in. wide.

Shop cutting panel. Plywood panels that have been rejected as not conforming to grade requirements of standard grades. Identification of these panels are by a separate mark that contains the notation "shop cutting panel —all other marks void." Blistered panels are not considered as coming within the category covered by this stamp.

Single and double shear. Term applies to fasteners such as nails. When a gusset plate is nailed to one side of a framing member, the nails are said to act in "single shear" (one place the nail might break). When gusset plates are placed on both sides of a framing member so that the nails penetrate both plates and are clinched (two places a nail might break), the nails act in "double shear" and can carry twice the load of nails in single shear.

Single-wall construction. Plywood panel sidings attached directly to studs.

Size tolerance. Tolerance of plus 0.0 in., minus $\frac{1}{6}$ in. is allowed according to the standards.

Sliced veneer

1. Veneer that is sliced off a log, bolt, or flitch with a knife.

2. Veneer cut by a method in which logs or sawed flitches are held securely in a slicing machine and thrust downward into a large knife, which shears off the veneer in sheets.

Slicer. Machine for producing veneer by slicing.

Slip matching. Laying adjacent sheets of veneer tight side up without turning; also called "slide matching."

Soffit. Underside of the roof overhang; the underside of canopies. Plywood is often used as a finishing material for soffits.

Softwood plywood. Softwood plywood is manufactured from several species of wood, of which Douglas fir is the most common. Some of the other species used in significant quantity include western larch, western hemlock, Sitka spruce, commercial white firs, Alaska and Port Orford cedar, and California redwood (Softwoods are trees with needlelike or scalelike leaves.)

Solid core. Inner ply construction of solid C-plugged veneer pieces. The gaps between pieces of core should not exceed $\frac{1}{2}$ in.—

Sound grade—No. 2. Veneer on the face of this grade plywood is not matched for color or grain. Some defects are permissible, but the face is free of open defects and is sanded and smooth. It is usually used for surfaces that are to be painted.

Sound transmission (control). Passage of sound through a panel or construction assembly. Large panel size reduces the number of joints and cracks that can leak airborne noise. Plywood is also used as a base for resilient coverings that cut impact noise.

Spacing. Small spaces should be left between plywood panels to allow for expansion. Normal spacing for sheathing and subfloors is used, but for areas of high humidity, it is recommended that the spacing be doubled.

Special overlay. Overlay similar to high- or medium-density overlay, which does not exactly fit the product standard description, but which meets the glue-line durability requirements of PS 1 and is overlaid for special uses.

Specialty grade (SP). Plywood of this grade usually entails special matching of the face veneers.

Specialty plywood. Specialty plywood is not typically or necessarily manufactured to meet commercial standards.

Species group. Plywood is manufactured from over 50 different species of varying strength. These species have been grouped on the basis of strength and stiffness, and for purposes of the manufacturing standard PS 1, they are divided into five classifications: Groups 1–5. The strongest woods are found in Group 1. The group number of a particular panel is determined by the weakest (highest numbered) species used in the face and back.

Split. Lengthwise separation of wood fibers completely through the veneer caused chiefly by manufacturing process or handling.

Split resistance. Grain running crosswise in alternate layers allows no line of cleavage for splitting in plywood.

Standard. Name for unsanded interior-type plywood commonly used for construction and industrial applications. Standard is produced with C grade or better.

Standard with exterior glue. Unsanded sheathing grade of plywood. It is known as CDX in the trade and is manufactured with waterproof adhesive; however it is *not* an exterior-type panel.

Staple. Fastener for applying plywood sheathing. Staple sizes and spacing are as recommended by American Plywood Association charts.

Starved joint. Glue joint that is poorly bonded because of an insufficient quantity of glue.

Stay log. Device used on a veneer lathe, to which is fastened a segment of the bolt or flitch to secure desired grain effects in the veneer.

Strength. Strength is an outstanding characteristic of plywood, created by cross-laminating veneers and gluing them together to distribute the wood's high along-the-grain strength in both directions.

Strength and rigidity. The two-way strength of plywood makes it ideal for diaphragm construction and any application like sheathing or siding where it provides major stiffness for the building. Plywood retains this strength under extremes of temperature. It can be exposed to dry temperatures of up to nearly 150°F for a year or more without significant permanent loss of strength, and when

temperature falls below normal, plywood actually becomes stronger.

Strength in bending. Large panel size with high bending strength makes plywood an excellent material for bulkheads and retaining walls.

Stressed-skin panel. Engineered structural flat panel for roof deck or floor applications built up of plywood sheets glued to framing members. The assembly covers quickly with greater combined load-carrying capacity than its individual members would have if installed separately.

Stiffness. Plywood is strongest and stiffest when panels are installed with the face grain across supports.

Striated

1. Term describing plywood with a face veneer that has been grooved or scored parallel to the grain.

2. Term describing texture in plywood siding characterized by closely spaced shallow grooves or striations that form a vertical pattern.

3. Striated texture is produced in exterior and interior types by striating the face of the plywood with planer knives that have been specially ground for the pattern.

Stripping. Removing plywood forms after pouring.

Structural I and II. Unsanded construction grades of plywood with special limitations with respect to species, veneer, grade, and workmanship. These grades are recommended for heavy-load applications where the plywood's strength properties are of maximum importance, as for truss gussets. Structural I and Structural II C-D Interior, and Structural I Exterior are all bonded with exterior glue.

Structural properties. Durability, bending strength, shear strength, and stiffness are plywood qualities that are built in during the manufacturing process with the cross lamination of the veneers.

Stud spacing. Distance between the upright members of a wall's framework, traditionally 16 in. on center, but sometimes 24 in. on center when plywood is used for structural stability, as in the MOD 24 system.

Sub-face. The ply adjacent to the exposed face (or back) of a parallel laminated outer layer.

Subflooring. Plywood applied directly over floor joists that will receive an additional covering, such as underlayment, if tile is to be applied. Plywood is used for strength and to reduce the number of floor joints.

Sunken joint. Depression in the surface of the face ply directly above an edge joint in a lumber core or cross band, usually the result of localized shrinkage in the edge-jointed layer. Sunken joints may also be present in edge-laminated lumber.

Swelling. Expansion of wood caused by the absorption of water. In normal construction $\frac{1}{8}$-in. spacing between panel edges and $\frac{1}{16}$ in. between panel ends are enough to take care of expansion. In very wet or humid areas the spacing should be doubled.

Symmetrical construction. Plywood panels in which the plies on one side of a center ply or core are essentially equal in thickness, grain direction, properties, and arrangement to those on the other side of the core.

Synthetic repairs. Plastic fillers that may be used to repair defects in plywood.

Tanks. Plywood is used for liquid tank construction because of its workability, strength, insulating value, and acid resistance.

Tape

1. Ribbons, usually of paper or cloth, coated with adhesive and used to fasten veneers together for convenience in handling during the gluing operation.

2. Strips of gummed paper or cloth used to hold the edges of the veneer together at the joints prior to gluing.

Tension and compression. Plywood splice plates are used to transfer tension (pulling loads) and compression (pushing loads) between skins of stressed-skin panels.

Termites. Wood-destroying insects. Plywood panels may be treated with pentachlorophenol to increase their resistance to these pests.

Textured plywood. Plywood panels with many different machined surface textures, available in exterior type with fully waterproof glue line for siding and other outdoor uses and for interior wall paneling.

Thermal conductivity. Under normal room temperature and humidity conditions, plywood is a natural insulator.

Thickness. Plywood panels range from $\frac{1}{4}$ through $1\frac{1}{4}$ in. in thickness and are made up of three to seven or more veneers.

303 specialty sidings. Grade designation covering proprietary plywood products for siding, fencing soffits, windscreens, and other exterior applications. Panels have special surface treatments; surfaces may be rough sawed, striated, and brushed, as well as grooved in different styles.

Tight side

1. In knife-cut veneer, the side of the sheet that is farthest from the knife as the sheet is being cut and containing no cutting checks.

2. Term used with its opposite "loose side," to refer to veneer cut with a knife. The product as it is cut by the wedge-shaped or beveled knife may be curved, thus producing small ruptures on the convex side, known as the "loose side;" the opposite surface, strained slightly in compression, but free from any ruptures, is known as the "tight side."

Type. Designation of the moisture resistance of plywood.

UL. Underwriters Laboratories, Inc. Plywood marked with the UL fire-hazard classification label has been treated according to required fire-retardation specifications.

Underlayment. Material applied directly to nonstructural finish flooring such as tile or carpeting. Plywood is available in underlayment and in combination subfloor-underlayment grades.

Unsanded plywood. Sheathing grades of plywood are left unsanded for greater stiffness, strength, and economy.

Utility grade—No. 3. Tight knots, discoloration, stain, worm holes, mineral streaks, and some slight splits are permitted in this grade. Decay is not permitted.

Vapor transmission. Even $\frac{1}{4}$-in. exterior plywood rates as a vapor retarder. Permeance is about 0.72 perm. High-density-overlaid, Medium-density-overlaid, and painted exterior plywood have ratings of about 0.2 perm. Any material with a value below 1 perm is considered a vapor retarder (ASTM E 96).

Veneer

1. Thin layer or sheet of wood cut on a veneer machine.

2. Thin slice of wood cut with a knife or saw from a log and used in the manufacture of plywood.

3. Thin sheets of wood, cross laminated in the manufacture of plywood, glued to form a panel that is split and puncture resistant.

4. Thin sheet of wood rotary cut, sliced, or sawed from a log, bolt, or flitch. Veneer may be referred to as a "ply" when assembled into a panel.

Veneer core. Plywood manufactured with layers of wood veneer joined in the standard manner.

Veneer grades. Standard grades of veneer used in plywood manufacture.

V-grooved wall panel. Interior hardwood plywood designed to simulate lumber plank paneling. Grooves are cut at the joints between mismatched strips of veneer on the face. Material is available in $\frac{1}{4}$-in. panels, generally 4×8 ft long.

Wall construction. Plywood systems for building walls.

Wall sheathing. Plywood nailed to the studs to act as a base for the siding application.

Warp and warping. Bend, twist, or turn from a straight line. A piece of lumber, when improperly seasoned and exposed to heat or moisture, may warp. To reduce the possibility of warping, avoid exposing plywood panels to dampness or moisture. Painting and water-repellent dips will minimize moisture absorption. Also, sealing all edges and back priming will reduce the chances of warping.

Waterproof. Term is synonymous with "exterior," that is, it designates plywood that is bonded with highly resistant adhesives and is capable of withstanding prolonged exposure to severe weather and service conditions without failure in the glue bonds.

Waterproof adhesive. Glue capable of bonding plywood in such a manner as to satisfy the exterior performance requirements recommended by American Plywood Association charts.

Water resistant. Term describing plywood that is bonded with moderately resistant adhesive and is capable of withstanding limited exposure to water or to severe conditions without failure in the glue bonds.

White pocket. Form of decay (Fomes Pini) that attacks most conifers but has never been known to develop in wood in service. In plywood manufacture routine drying of veneer effectively removes any possibility of decay surviving. (Admissible amounts of white pocket were established through a two-year research project at the U.S. Forest Products Laboratory.)

Wood failure (percent). Wood failure is expressed in terms of the area of wood fiber remaining at the glue line following completion of the specified shear test. Determination is by means of visual examination and expressed as a percent of the 1-sq in. test area.

Wood filler. Aggregate of resin and strands or shreds of wood that are used to fill openings and provide a smooth durable surface.

Z Flashing. Z-shaped piece of galvanized steel, aluminum, or plastic used at horizontal joints of plywood siding to prevent water from entering wall cavity.

PREFABRICATED STRUCTURAL WOOD

Adhesive

1. Substance capable of holding materials together by bonding the contact surfaces.

2. Laminated members bonded with a waterproof adhesive.

Adhesives

1. Adhesives are of two general types: dry use (water resistant) and wet use (waterproof). Dry-use adhesives are those that perform satisfactorily when the moisture content of wood does not exceed 16% and are used only when this condition exists. Wet-use adhesives perform satisfactorily for all moisture conditions, including exposure to weather, marine use, and where approved pressure treatments are used either before or after gluing. They may be used for all moisture conditions of service, but their use is required when the moisture content exceeds 16%.

2. Gluing practices take into consideration the characteristics and limitations of the specific glue used and conform to good practices as regards preparation of wood surfaces for gluing, control of temperature and moisture content of materials, mixing and spreading of glue, and compatibility of the glue with any other wood treatments employed. Glues of quality will develop the full strength of the wood used. Glues must provide adequate strength under conditions of service and be durable and permanent for the entire service life of the structure. The glue bond must extend over the entire surface to be glued.

A-frame design. Popular type of laminated timber construction that can be used to create geometrically simple structures.

Air-dried or air-seasoned. Dried by exposure to the air, usually in a yard, without artificial heat.

AITC. American Institute of Timber Construction.

Ambient

1. Surrounding or encompassing.

2. Blanket or box to hold in heat for laminated beams while they are curing in a jig.

Appearance grades

1. Appearance grades apply to the surfaces of glued laminated members and are based on such items as growth characteristics, inserts, wood fillers, and surfacing operations, but not lamination procedures, stain, paint, varnish, or other protective coatings. Appearance grades do not modify the design stresses, fabrication controls, grades of lumber used, and other provisions of the standards for structural glue laminated lumber.

2. Three appearance grades are available, such as premium, architectural, and industrial, and all should conform to American Institute of Timber Construction specifications, AITC 110, and AITC Spec-data sheet on structural glued laminated timber.

Approved plastic fillers. Fillers where and when used are subject to strength and finish requirements. Their application is to fill and conceal natural voids in the material and thus primarily improve its appearance.

Arch. Structural element whose general form is that of a curve and that is so supported that horizontal as well as vertical motion is resisted. When the ends of the arch are fixed in position with respect to the abutments and a member is continuous between the abutments, it is known as a "fixed arch;" when the arch is supported at the abutments by connections incapable of transmitting moment and the member is continuous between hinges, it is known as a "two-hinged arch;" when the arch is supported as a two-hinged arch but is made of two parts joined at an intermediate point, usually the center of the length, by a connection incapable of transmitting moment, it is known as a "three-hinged arch."

Architectural appearance grade

1. Grade for laminated decking ordinarily suitable for construction where appearance is an important requirement. Any small voids are usually filled by others than the fabricator if the final decorative finish so requires.

2. Grade used for laminated decking where attractive appearance is a requirement. It contains normal growth characteristics such as tight knots and medium seasoning checks, but no knotholes except for an occasional edge-sloughed knot. Hem-Fir or cedar face ply equivalent in appearance to tight knotted construction grade boards or ponderosa pine face ply equivalent to tight knotted No. 3 boards is acceptable face ply surface.

3. Grade used for three ply laminated wood decking, where attractive natural appearance is a requirement. Occasional face plies may have some but not all of the following normal growth characteristics: pin holes, short splits or checks, medium torn grain (including shelling), light skips in surfacing, and sound tight knots (no knotholes) up to 2-inch in diameter in Hem-Fir and up to 3 to $3\frac{1}{2}$ for ponderosa pine.

Arch peak shear plates. Back-to-back arch peak shear plates centered on a dowel are used with a tie plate and through bolts. When appearance is important, a bent plate may be dapped into the top of the arch and secured with lags.

Assembly

1. Process of fastening together, by means of hand or portable tools, fabricated components of structural timber framing with nails, bolts, connectors, and/or other fastening devices to form larger components or subassemblies of a structural frame. Term includes assembly in a fabricating plant (shop assembly) or at the job site (field assembly).

2. Pieces of wood and glue that have been placed together for bonding or that have been bonded together.

Assembly clamping pressure. Pressure of not less than 100 psi as determined by suitable mechanical equipment; clamping is rechecked for the original pressure approximately 30 minutes after initial application. Clamp spacing is such that the pressure is as uniform as practicable over the whole area. The nailing of laminations in lieu of clamping for pressure is *not* permitted. Clamping may start at any point, but progress should proceed to an end or ends.

Assembly time

1. Time interval between the spreading of the adhesive on the laminations and the application of final pressure or heat, or both, to the entire package. Assembly time is composed of two parts, open and closed.

2. Time interval between spreading adhesives on laminations and applying final pressure or heat, or both, to the entire assembly.

Basic stress. Working stress for defect-free material. It includes all the factors appropriate to the nature of structural timber and the conditions under which it is used, except those that are accounted for in the strength ratio.

Batch. Term generally used by adhesive manufacturers to identify a "lot," "blending," or "cook" of adhesive.

Beam. Structural element supported at one or more places along its length. Its principal function is to support loads acting more or less transversely to the long dimension.

Beam anchorage in masonry. Beam anchorage is used for anchorages that resist both uplift and horizontal forces. It may have one or more anchor bolts in masonry, and one or more through bolts with or without shear plates through the beam. A 1-in. minimum clearance or impervious moisture retarder is required on all wall contact surfaces, ends, sides, and tops (if masonry exists above beam end).

Beam or purlin hanger for seasoned members. When supported members are seasoned material, the top of the supported member may be set flush with the top of the hanger strap.

Beams-to-glulam-column T plate. A steel T plate is bolted to abutting glulam beams and to a glulam columns. A loose bearing plate may be used where the column cross-sectional area is insufficient to provide bearing for beams in compression perpendicular to the grain.

Beams-to-glulam-column U plate. A steel U plate passes under abutting glulam beams and is welded to steel plates bolted to a glulam column.

Block shear (test). Method of testing wood or glue lines for shear strength.

Board. Lumber less than 2 in. thick and at least 1 in. wide. Boards less than 6 in. wide may be classified as strips.

Bolted joint. Joint in which bolts transmit shear and hold members in position. Included are wood-to-wood and wood-to-steel assemblies.

Bonding. Joining materials together with adhesive.

Bowstring. Generally the most economical of all clear-span construction, bowstring trusses in standard patterns are manufactured up to 120 ft.

Bracing. Bracing consists of structural elements that are installed to provide restraint or support, or both, to other members so that the complete assembly form a stable structure. Bracing may be in lateral, longitudinal, and transverse planes and may consist of sway, cross, vertical, diagonal, horizontal, and like elements to resist wind, earthquake, initial erection, and other forces. Bracing may consist of knee braces, cables, rods, struts, ties, shores, diaphragms, rigid frames, and similar items and combinations.

Break. Fracture of the wood fibers across the grain caused by external forces.

Bridging. Bracing between trusses used to stiffen the structure, spread concentrated loads to adjacent areas, and brace against deflection.

Building codes. Structural glued laminated timber is generally accepted by major building codes.

Butt joint

1. End joint formed by abutting the squared ends of two pieces. Because of the inadequacy and variability of glued butt joints, but joints are not generally glued.

2. A butt joint consists of two abuting square ends. This type of joint is the most economical of material, but it cannot transfer either tension or compression stress. Though ends are tight in lay-up, they may not be at the end of the pressure operation. Steel butt plates could add compressive capacity, but they are difficult to install. Butt joints produced serious sources of stress concentration and are always undesirable in curved members. They should never be used in top or bottom laminations and are not suitable for tension areas. For compression areas, the design should disregard all laminations at a single cross section that have butt joints. In tension areas the design should disre-

gard 1.2 times the area of all such laminations, when acceptable to local building codes.

Calibration. Determination of the relationship between the actual values being measured and the values indicated by the equipment used for the measurement.

Camber

1. Design allowance for deflection that results in a horizontal lower chord when the trusses underload.

2. Glued laminated members may be cambered to offset dead loads or dead loads with live load deflections.

3. Convexity sometimes placed by deliberate design in beams and trusses to prevent them from becoming concave under their loads.

Categories of arches. There are two main categories of arches: the first is the haunched, gothic, or boomerang arch, usually of three-hinged construction in which bending moment is the more critical consideration; the second is the constant-radius arch of either two- or three-hinged construction in which compression is the critical consideration.

Certificate. Document that should be furnished in conjunction with identification marks to provide certification by a laminator or fabricator that his work conforms to a standard and to job specifications; certification by a central organization that the quality control of a plant is under its general surveillance.

Certificate of inspection. A certificate of inspection should accompany a prefabricated assembly. The certificate certifies that the assembly has been inspected and meets all the requirements of the drawings and specifications, and meets all building code requirement.

Check

1. Separation of the wood grain due to internal stresses caused by severe moisture cycling or seasoning.

2. Separation along the grain generally extending across the rings of annual growth.

Chord. One of the principal members of a truss, usually horizontal, braced by the diagonal or vertical web members.

Clamp. Device for application of uniform pressure on a laminating package.

Clamping. Application of pressure to join laminated stock and glue.

Clip angle column anchorage to concrete base. Through-bolts through the clip angles on both sides of a column. Recommended for industrial buildings and warehouses to resists both horizontal forces and uplift. A bearing plate or moisture retarder is recommended.

Closed assembly time. Time elapsed from the assembly of the first laminations of the package into intimate contact until final application of pressure or heat, or both, to the entire package.

Close-grain rules. Rules for the classification of lumber on the basis of rate of growth (rings per inch). The rules at present apply only to Douglas fir and redwood and differ slightly. Structural material of these species meeting the requirements of these rules is assigned somewhat higher working stresses than material not meeting these requirements.

Column. Vertical compression member on which the principal loads are parallel to the axis (a line joining the centers of gravity of all cross sections) of the piece. Short columns are those that fail primarily by shearing or crushing; intermediate columns are those that fail by a combination of shearing or crushing and flexure; long columns are those that fail in flexure; simple columns are those whose cross section at all points is a closed area; spaced columns are those composed of two or more simple columns spaced some distance apart and connected at two or more points, usually with metal connectors at the juncture of the simple column and the spacing element.

Column design

1. Glued laminated columns offer the same advantages as other laminated timber elements, such as higher allowable stresses, improved selected appearance, and the ability to fabricate variable or curved sections.

2. Three types of columns are normally designed and fabricated with laminations, simple rectangular, tapered, and spaced columns.

Composite. Structural element consisting of wood and combinations of other materials in which all pieces are attached together to act as a single unit.

Compressometer. Device for measuring force or pressure. In inspection work it may be used for calibrating torque wrenches.

Concealed electrical wiring. To conceal wiring, only the top of the decking parallel to the span is routed. Routing perpendicular to the decking span may be done only over support members.

Concealed-type purlin hanger. Hanger used for light and moderate loads. The hardware is completely concealed. This type of hanger is not recommended for use with unseasoned purlins. Purlin design includes bearing area and notched beam action. The 50% increase for shear in joint details does not apply.

Connector. Toothed metal plate used to join truss members.

Construction grade. Lumber of this grade is distinguished by its excellent strength and appearance and is used where high-quality construction is of prime importance. It is suitable for use without waste. A serious combination of defects is not permitted in any one piece.

Cross grain. Lack of parallelism between the longitudinal elements of the wood and the axis of a piece. The term applies to either diagonal or spiral grain, or a combination of the two.

Curing time. Period of time for adhesive to attain its full strength.

Curved laminated members

1. No defects in the lumber will be permitted that will interfere with bending to the required curvature without localized irregularities in the curvature that interfere with bringing laminations into intimate contact.

2. When laminations are glued over tapered pieces, the taper should not exceed a slope of 1 to 12. Minimum practicable number of end joints of lamination should occur at any given cross-section of a member. Not more than one end joint in any three successive laminations at a given cross section should be permitted.

Decayed knot. Knot which, due to advanced decay, is softer than the surrounding wood.

Decking. Hem-Fir, cedar and alder decking are manufactured to a nominal 6 in.-width in several thicknesses. Generally, all three plies of cedar decking are cedar; however, cores and backs may be Hem-Fir or Douglas fir at plant option. Alder decking generally will have all three plies alder; cores and backs may be Hem-Fir or fir at plant option. Alder decking face ply has a uniform color, cut

generally will contain at least one finger joint in the face lamination.

Defect. Any irregularity in or on wood that may lower its strength.

Delamination

1. Separation of layers in an assembly because of failure of the adhesive, either in the adhesive itself or at the interface between the adhesive and the lamination. Delamination in glue joints is failure either in the adhesive itself or failure at the wood-adhesive interface.

2. Separation of two laminations due to failure of the adhesive.

Density rule. Rules for classification of lumber based on percentage of summerwood and rate of growth (rings per inch). Rules at present apply only to southern yellow pine and Douglas fir, and differ slightly. Structural material meeting the requirements of these rules is assigned somewhat higher working stresses than is material not meeting these requirements.

Depth. Dimension of the cross section of a member in bending measured parallel to the direction of the principal load.

Diagonal grain. Form of cross grain resulting from sawing at an angle with the bark of the tree.

Diaphragm. Relatively thin, usually rectangular element of a structure that is capable of withstanding shear in its plane. By its rigidity it limits the deflection or deformation of other parts of the structure. Diaphragms may have a plane or curved surface.

Dielectric heating. Generating heat at high frequencies in nonconducting materials by placing them in a strong alternating electric field. Dielectric heating of the glue line is done at very high frequencies (MHz).

Double- and triple-bevel scarves (patented modifications of the plain scarf). In design, double- and triple-bevel scarves are treated as a single plain scarf. Double and triple features act as positioners and seem to bind the joint together in a strong and durable manner. These scarves resist twisting actions and have been successfully used in moderate-span arches to field-join wall arms, haunches, and roof arms.

Double chords. Two leaf members separated by the web system.

Double spread. Application of adhesive to both surfaces of a joint.

Dry use. Normal use conditions where the moisture content in service is less than 16%, as in most covered structures.

Dry-use adhesives. Adhesives that perform satisfactorily when the moisture content of the wood does not exceed 16% for repeated or prolonged periods of service.

Durability. As applied to glue lines, durability means the life expectancy of the structural qualities of the adhesive under the anticipated service conditions of the structure.

Edge grain lumber. Lumber that has been so sawed that the annual growth rings form an angle of 45° or more with the wide surface of the piece.

Edge joint

1. Juncture of two pieces joined edge to edge, commonly by gluing. This is done by gluing two square edges, as in a plain edge joint, or by using machined joints of various kinds, such as tongue and groove.

2. Side joint formed by the use of two or more widths of lumber to make up the full width of a lamination.

Encased knot. Knot whose rings of annual growth are not intergrown with those of the surrounding wood.

End joint

1. Joint formed by joining pieces of lumber end to end with adhesives.
2. Juncture of two pieces joined end to end, commonly by gluing.

Equilibrium moisture content. Any piece of wood will give off or take on moisture from the surrounding atmosphere until the amount of moisture in the wood balances that in the atmosphere. The moisture content of the wood at the point of balance is called the "equilibrium moisture content" and is expressed as a percentage of the oven-dry weight of the wood.

Erection

1. The materials are prefabricated at the plant and arrive at the site ready for immediate erection. Owing to its relatively light weight, glulam can usually be erected with light construction equipment in the field.
2. Erection normally includes the hoisting and/or installing in place in the final structure of fabricated components and usually involves the use of cranes, hoists, and other powered equipment..

Erection bracing. Bracing installed to hold the framework in a safe condition until sufficient permanent construction is in place to prove full stability.

Fabrication. Fabrication consists of boring, cutting, sawing, trimming, dapping, routing, planing and/or otherwise shaping, and/or framing, and/or furnishing wood units, sawed or laminated, including plywood, to fit them for particular places in a final structure.

Finger joint. End joint made up of several meshing tongues or fingers of wood. Fingers are sloped and may be cut parallel to the wide faces or edge faces.

Fire and construction safety. The heavy timber used in laminated wood construction is difficult to ignite. Heavy timber burns slowly and resists heat penetration. The uncharred inner portion of a timber usually maintains its original strength. Laminated timbers do not expand or deform sufficiently to push out supporting walls. Laminated timber usually has the capacity for absorbing impact and temporary overloads; it also provides great safety under high-wind and earthquake conditions.

Fire rating. Glued laminated members are pressure treated with fire-retardant chemicals conforming to the Underwriters Laboratories, Inc., fire hazard classification for a given flame spread (15 for incombustible, or higher as required by project requirements).

Fire resistance. Laminated wood structural members usually are slow burning and provide safety from collapse long after some noncombustible materials have failed. Tests are conducted under controlled conditions, meeting ASTM standards, to determine fire-resistance ratings.

Flat grain lumber. Lumber so sawed that the annual growth rings form an angle of less than 45° with the wide surface of the piece.

Form factor. Factor to be applied to the usual formula for resisting moment in a beam to take account of the difference between the stresses developed in beams having certain cross-sectional shapes and those developed in a beam having a solid 2 × 2 in. cross section when computed by means of the usual engineering formulas.

Frame construction

1. Type of construction in which the structural parts are of wood or are dependent on a wood frame for support. In some codes, if brick or other incombustible material is applied to the exterior walls, the classification of this type of construction is usually unchanged.
2. Type of construction where the building frame is of wood and the structural parts and enclosing walls are also of wood. In many codes if the enclosing walls are veneered, encased or faced with masonry materials or metal the construction may also be termed frame construction.

Girder-to-glulam column connection. Throughbolts in metal straps on both sides of a girder and column provide for uplift resistance. A metal bearing plate is used where the column cross-sectional area is insufficient to provide bearing for the girder in compression perpendicular to the grain.

Glued buildup. Structural elements consisting of wood, plywood, or combinations of the two, in which the grain is not necessarily parallel and in which all pieces are bonded together with adhesive.

Glued laminated structural member. Member in which the grain of all laminations is approximately parallel. The glue line between laminations is as strong as the wood itself.

Glued laminated timber

1. Assemblies of wood laminations bonded together with adhesives so that the longitudinal grain of all laminations are approximately parallel.
2. Product obtained by gluing together a number of laminations having their grain essentially parallel.

Gluing. Joining together edges, faces, or ends of pieces or laminations with adhesives.

Glulam

1. Term used to describe large beams fabricated by bonding layers of specially selected lumber together with strong, durable adhesives. End and edge jointing permits production of longer and wider structural wood members than are normally available. Glulam timbers are used with plywood for many types of heavy-timber construction.
2. Seasoned timber that has been glued together under pressure and fabricated to curved shapes if required.
3. Structural glued laminated lumber that differs from sawed lumber in that it may be fabricated to curved shapes and unusually large sizes and long lengths.

Grade. Designation of the quality of a manufactured piece of wood.

Hardwoods. Botanical group of trees that are broad-leaved. The term has no reference to the actual hardness of the wood. Angiosperms is the botanical name for hardwoods.

Heavy-timber construction. Type of construction in which fire resistance is attained by placing limitations on the minimum sizes, thickness, or composition of all load-carrying wood members; by the avoidance of concealed spaces under floors and roofs; by the use of approved fastenings, construction details, and adhesives; and by providing the required degree of fire resistance in exterior and interior walls.

Height factor. Factor to be applied to the usual formula for resisting moment in a beam to take account of the difference between the stresses developed in beams of various heights and those developed in a beam having a solid 2 × 2 in. cross section when computed by means of the usual engineering formulas.

Hem-Fir. Hem-Fir is generally a light tan wood with a fine textured straight grain. It is noted for strength, stiffness, freedom from pitch, and splintering. Insulation value is high and it takes stain very well.

Horizontally laminated wood. Laminated wood in which the laminations are so arranged that the wider dimension of the lamination is approximately perpendicular to the direction of the loads.

Horizontal shear stress. Horizontal shear stress particularly relates to the stress grades of lumber. This is stress that tends to slide fibers over each other at each support, and the resistance to this stress from the internal force is known as the horizontal shear value of the wood. Horizontal shear stress is the greatest at the center of the depth of any particular piece.

Hot service. Normal use conditions where the wood members may remain at a temperature of 150°F or above continuously or for prolonged periods of time.

Howestring trusses. Most often used in large warehouses, these trusses can be combined with parallel chord trusses or any structural roof member to meet any specific design.

Hydroplat. Hydraulic instrument used for measuring force or pressure.

Identification mark. Stamp, brand, label, or other mark that is owned and administered by a qualified central inspection organization and that is applied to a product by licensed manufacturers to evidence conformance to standards and job specifications. Such marks include identification of a standard, the qualified central inspection organization, and the producing plant. In addition, such mark may include species, allowable unit stresses, wet or dry use, and appearance grade.

Industrial-appearance grade

1. Grade ordinarily suitable for construction in industrial plants, warehouses, garages, and for other uses where appearance is not of primary concern.

2. For industrial-appearance grade inserts or wood fillers are not required. Soffit and face boards can be free of loose knots or open knotholes. Members surface two sides only, permitting an occasional miss along individual laminations.

Inserts. Nonstructural repairs to correct appearance defects.

Inspection. Process of measuring materials and methods for conformance with quality controls by the manufacturer.

Intergrown knot. Knot whose rings of annual growth are completely intergrown with those of the surrounding wood.

Jig. Device for holding truss members in position during fabrication.

Joint area. Area of the glue line, usually expressed in square feet (75 lb of glue spread to 1000 sq ft of joint area indicates that the quantity 75 lb of glue is contained within the 1000 sq ft of glue line area bonding adjacent board surfaces that total 2000 sq ft).

Joint edge. Side joint in laminations formed by use of two or more widths of lumber to make up a full-width lamination.

Joints. Contact surfaces between two adjacent pieces of wood glued together. An edge or face joint is parallel to the grain of the wood. An end joint is at right angles to the grain of the wood.

Joist and plank. Pieces (nominal dimensions 2 to 4 in. in thickness ×4 in. or more in width) of rectangular cross section graded with respect to their strength in bending when loaded either on the narrow face as a joist or on the wide face as a plank.

Kiln-dried. Dried in a kiln with the use of artificial heat.

Kiln-drying. Each 2-in. nominal thickness lamination is kiln-dried prior to surfacing to 12% average moisture content in electronically controlled kilns, the maximum moisture content not to exceed 15%. Controlled drying resists practically all tendency of the fabricated vertically laminated members toward checking or twisting.

Knot. Portion of a branch or limb that has been surrounded by subsequent growth of the wood of the trunk or other part of the tree; also, a cross section of such a branch or limb on the surface of a piece of wood.

Laminated arches. Curved members of any desired contour can be used to achieve strength and spaciousness. Specific types of arches include the Tudor, radial, Gothic, three centered, inverted or reverse curve, A-frame, and parabolic. Spans usually range between 30 and 100 ft, although much longer spans are possible when desired.

Laminated bowstring trusses. Glued laminated top and bottom chords and sawed timber web members.

Laminated decking. Glued laminated timber decking is manufactured from three or more individual kiln-dried laminations into single decking members with tongue-and-groove joints.

Laminated wood. Assembly made by bonding layers of veneer or lumber so that the grain of all laminations is essentially parallel.

Lamination

1. Wood layer contained in a member comprising four or more layers. It extends the full width and full length of the member. It may be composed of one or several wood pieces in width or length, but of only one in depth. Wood pieces may or may not be end or edge glued.

2. One layer in, or to be used in, laminated wood. It may consist of a single piece or of a number of pieces in width or length, but of only one in depth. Wood pieces may or may not be end or edge glued.

Laminating. Bonding laminations together with adhesive. The term applies to the preparation of laminations, the preparation and spreading of adhesives, the assembly of laminations in packages, and pressure and curing.

Laminating species. Softwood species commonly used for laminating are Douglas fir, larch, southern pine, Hem-Fir and California Redwood. Other species, including hardwoods, may be used.

Lumber

1. Laminating lumber should be kiln dried and stress graded to meet the requirements of Standard Specifications for Structural Glued Laminated Timber, AITC 117.

2. Manufactured product derived from a log in a sawmill (or in a sawmill and planing mill), which when rough has been sawed, edged, and trimmed at least to the extent of showing saw marks in the wood on the four longitudinal surfaces of each piece for its overall length and which has not been further manufactured other than by cross cutting, ripping, resawing, joining crosswise and/or endwise in a flat plane, and surfacing with or without end matching and working.

Lumber gauge. Measuring device for determining the thickness of lumber.

Lumber grade

1. The grade of gluded laminated lumber should conform to the grade and species specified. Individual laminations should be not more than 2 in. in thickness, and all such laminations in the stressed portion should be approximately parallel to the neutral plane of the member. Outer laminations at the top and bottom consist of a better grade than the interior members.

2. Lumber for laminating should meet the structural requirements and laminating specifications of Product Standard PS 56-73, structural glued laminated timber. Stress grades for lumber used should be selected by the designer for bending, tension, compression parallel to grain, and compression perpendicular to grain.

Lumber moisture content. At time of factory gluing and/or fabrication, each member should be less than 16% moisture content except that when equilibrium moisture content of member in use is 16% or more. Moisture content, at time of gluing, should not exceed 20%. The range of moisture content in various laminations assembled into a single member should not exceed 5 percentage points if any piece in the assembly exceeds 12%. Moisture content is determined periodically by a moisture meter, by oven tests, or by other standard means.

Lumber used in fabrication. Lumber that is properly selected and prepared for gluing, bending of laminations to the desired shape, general appearance, strength, and moisture content. Material should be free from defects, such as large knots, pitch, cross grain, warp, twist or other characteristics that will prevent intimate contact of adjacent glued faces or interfere with uniform bending to the required curvature under clamping pressure.

Mechanically laminated. Term applied to wood structural element comprised or laminations that are not glued but held together with mechanical fastenings such as nails or bolts and in which all laminations have the grain approximately parallel longitudinally.

Mill-type or heavy-timber construction. Glued laminated members of nominal sizes are required to meet the standards of "heavy-timber construction" and have all of the durability necessary under exposure to fire. Pressure impregnation with fire retardants provides added fire resistance when specified.

Moderate-temperature service. Normal-use conditions for ranges in which the temperature of the wood members remains below 150°F.

Moisture content

1. Amount of water contained in the wood, usually expressed as a percentage of the weight of the oven-dry wood.

2. At the time of gluing, the moisture content should be not more than 15 and not less than 7%. The range of the moisture content of laminations assembled into a single member should not exceed 5% at the time of gluing.

Moment splice. Compression stress is taken in bearing on the wood through a steel compression plate. Tension is taken across the splice by means of steel straps and shear plates. Side plates and straps are used to hold sides and tops of members in position. Shear is taken by shear plates in end grain. Bolts and shear plates are used as design and construction considerations require.

Monochords. Single leaf members with web members normally abutting the chords.

Multiple chords. Three or more leaf members separated by two or more web systems.

One-piece sawed timber. Timber consisting of a single piece of wood and formed to its final dimensions by no other manufacturing operations than sawing, or sawing and subsequent planing, of one or more surfaces.

Open assembly time. Time elapsed between spreading of the adhesive and assembling the spread surfaces into close contact with one another.

Ordinary construction. Timber framing that conforms to all requirements of ordinary timber construction as described in local building codes.

Panelized roof systems. Panelized roof systems consist of glulam, lumber, and plywood elements.

Parallel chord trusses. Parallel chord trusses are manufactured in spans up to 150 ft but are generally limited by economy to 80 ft.

Permanent bracing. Bracing designed and installed to form an integral part of the final structure. Some or all of the permanent bracing may also act as erection bracing.

Physical properties of wood. Wood will not shatter when it is struck; its resilience permits it to absorb shocks. Wood has fine natural insulating qualities. Through laminating, wood can be produced in large sizes and in almost any shape.

Piece mark. Mark placed on an individual piece of timber framing to designate its location in the assembly as indicated in the shop drawings.

Pitch. Ratio of rise to span of a nonparallel chord member of a truss.

Pith knot. Sound knot having a pith hole not over $\frac{1}{4}$ in. in diameter.

Plain sawed. Another term for flat grain.

Plain scarf. Scarf having a straight slope from end to end. Usually, slopes vary from 1–8 to 1–12; the flatter the slope, the stronger the joint.

Plywood. Cross-banded assembly made of layers of veneer in combination with a lumber core or plies joined with an adhesive. Two types of plywood are recognized, namely, veneer plywood and lumber-core plywood.

Pot life. Period of time during which an adhesive, after mixing with catalyst, solvent, or other compounding ingredients, remains suitable for use.

Prefabricated wood truss. Structural truss whose members are made of lumber or laminated lumber and that has been cut and assembled or partially assembled in a shop or place other than its position of final usage. Truss fabrication is completed prior to erection of the truss into the structural assembly.

Premium-appearance grade

1. Grade used when the finest appearance is required. Applications and specifications for appearance grades may be found in the American Institute of Timber Construction standards.

2. In all exposed surfaces voids that cannot be properly filled are replaced with clear wood inserts to match the grain and color. Soffit and face board material is clear and selected to match color and grain at scarf and end joints. Exposed faces are surfaced smooth.

Premium grade. Grade having a face of substantially clear lumber with natural characteristics allowed in prime mixed-grain finish of the species used. Not available in ponderosa pine.

Pressure preservative treatment. Treatment whereby preservatives are introduced into the wood pores under a vacuum (full-cell or empty-cell methods) as opposed to treatment by soaking bath.

Production procedures. All of the manufacturing operations performed in the manufacture of structural glued laminated timber.

Psychrometer. Instrument for measuring relative humidity.

Purlins. Lumber used in laminated purlins is the same quality kiln-dried material as that used in arches and beams. This ensures a uniform appearance of all exposed structural members to the interior space.

Pyrometer. Instrument for measuring temperatures by means of the change in electrical resistance, the production of thermoelectric current, expansion of gases or solids, specific heat of solids, or the intensity of heat or light radiated. Pyrometers used in laminating plants are usually of the thermoelectric or expansion-of-metals type and are commonly used to measure the temperature of interior glue lines.

Quality control

1. Control recommended in accordance with PS 56–73, voluntary product standard for structural glued laminated timber, and the The American Institute of Timber Construction Inspection Manual, AITC 200.

2. System whereby the manufacturer ensures that materials, methods, workmanship, and final product meet the requirements of a standard.

Quartersawed. Another term for edge grain.

Radial. Coincident with a radius from the axis of the tree or log to the circumference.

Radial arch design. Glued laminated radial arches achieve large, unobstructed, clear span enclosures for a variety of uses. They may be either buttressed or tied arches depending on soil conditions and other building requirements. In the buttressed arch-type, horizontal and vertical reactions are taken through concrete abutments. In the tied arch type, the horizontal reaction is taken by steel tie rods located at the ceiling height. This type of arch is usually set on masonry walls or columns. Glued laminated timber radial arches of the two- or three-hinged types are used for structural building members, for interior and/or exterior work.

Rate of growth. Rate at which a tree has grown wood, measured radially in the trunk or in lumber cut from the trunk. The unit of measure in use is the number of annual growth rings per inch.

Relative humidity. Ratio of actual pressure of existing water vapor to maximum possible pressure of water vapor in the atmosphere at saturation at the same temperature, expressed as a percentage.

Representative sample. Portion or section of a glued laminated timber that represents all of the conditions surrounding the manufacturing process of a specific or group of laminated timbers. When the sample is examined and tested and the results determined, it can be factually stated that the level of quality of the individual or group of members is of comparable quality.

Rigid frame. Frame made up of pieces in the same general plane fastened rigidly together at the joints.

Round knot. Knot whose sawed section is oval or circular.

Saddle-type cantilever beam connection with tension tie. Prefabricated metal strap hinge connector for two members edge attached to each other and providing vertical reaction of a supported member carried by side plates and transferred to both members by bearing plates in perpendicular to grain bearing. Saddle rotation due to eccentric loading is resisted by tabs at top and bottom. Separate tension tie resists separation force developed between beams due to rotation.

Sawed. Term applied to wood structural natural element grown and sawed and not laminated or built up.

Scarf joint

1. Sloping or bevel joint, where pieces of wood are lapped together. No joint can be directly over another in adjacent pieces or with scarfed joints sloping in the same direction closer together in adjacent pieces than 6 in. Scarf joints are preglued.

2. End joint formed by joining with adhesive the ends of two pieces of lumber that have been tapered to form sloping plane surfaces, usually to a thin edge and with the same slope of the plane with respect to length in both pieces. In some cases a step or hook may be machined into the scarf to facilitate alignment of the two ends, in which case the plane is discontinuous and the joint is known as a "stepped" or "hooked" scarf joint. Wood pegs or other scarf positioning devices may be used.

Seasoning. Removal of moisture from green wood in order to improve its serviceability.

Selection of grade combinations. For structural use, combinations of grades that provide working stresses intermediate between the highest and lowest should be used.

Shake. Separation along the grain extending, generally, between the rings of annual growth.

Shear plate connector joint. Joint in which shear plates transmit shear between bolts and wood or bolts and steel members. Bolts also hold members in position. Wood-to-wood and wood-to-steel assemblies can be used.

Simple beam masonry anchorage. Anchorage consists of throughbolts for beams with depths of 24 in. and less. Anchorage resists uplift and small horizontal forces. Bolt holes should be field bored $\frac{3}{16}$ in. oversize. A bearing plate and moisture retarder is recommended.

Simple rectangular columns. Formulas for laminated wood columns assume pin-end conditions, but they are also applied to square-end conditions. For simple, rectangular sawed or glued laminated timber columns, the "slenderness ratio," or the ratio of the unsupported length between points of lateral support '1' to the least dimension, 'd' may not exceed 50. Verify with local codes.

Single spread. Application of adhesive to only one surface of a joint.

Smoothness. Relationship of the actual surface to that of a true plane or other surface free from roughness.

Softwoods. Botanical group of tress that have needle- or scalelike leaves and are evergreen for the most part, bald cypress, western larch, and tamarack being exceptions. The term has no reference to the actual hardness of the wood. Softwoods are often referred to as conifers, and botanically they are called "gymnosperms."

Sound knot. Knot that is solid across its face and at least as hard as the surrounding wood, and shows no indication of decay.

Spaced columns. Columns that consist of two or more individual members with their longitudinal axes parallel, separated at the ends and at the midpoint by blocking and joined at the ends by fastenings capable of developing the required shear resistance.

Spike knot. Knot cut approximately parallel to its long axis so that the exposed section is definitely elongated.

Spiral grain. Form of cross grain resulting from the growth of the longitudinal elements of the wood spirally about instead of vertically along the bole of the tree.

Split

1. Separation of the wood fiber caused by external forces.

2. Lengthwise separation of the wood extending from one surface generally across the rings of annual growth through the piece to the opposite surface or to an adjoining surface; a through check.

Split ring connector joint. Joint in which split rings transmit shear and bolts hold members in position; includes wood-to-wood assemblies.

Spread. Quantity of adhesive per unit joint area applied to a lamination, preferably expressed in pounds of liquid or solid adhesive per thousand square feet of joint area.

Springwood. Portion of the annual growth ring formed during the early part of the season's growth. It is usually less dense and weaker mechanically than summerwood.

Squeeze-out. Glue extruded from glue lines when pressure is applied. Though only an indication, it is still an excellent signpost as to the quantity of glue spread and amount of pressure. If the glue spread is adequate and even and attains full coverage out to all edges, then the pressure will force glue out from between the lamination in a pattern of beads and threads. If the proper assembly period is used and spread and pressure are uniform and adequate, squeeze-out will emerge in a uniform pattern along all edges. If either spread or pressure is inadequate in any location, then that area will not show squeeze-out.

Starved joints. Glue lines that do not contain enough adhesive. They may result from low viscosity of the adhesive, excessive pressure, or application of an inadequate amount of adhesive.

Stepped scarfs. Combinations of plain scarfs and steps. A step adds no appreciable strength but is helpful in positioning the scarf ends. Some types of gluing machines permit the leading piece to pull the trailing piece into the machine.

Storage life. Period of time during which a packaged adhesive can be stored under specified temperature and humidity conditions and remain suitable for use.

Strength. The term in its broader sense embraces collectively all the properties of wood that enable it to resist different forces or loads. In its more restricted sense strength may apply to any one of the mechanical properties, in which event the name of the property under consideration should be stated, such as strength in compression parallel to the grain, strength in bending, or hardness.

Strength ratio. Ratio representing the strength of a piece of wood that remains after allowance is made for the maximum effect of the permitted knots, cross grain, shakes, and other defects.

Stress. Force per unit of area.

Structural glued laminated lumber. Member comprising an assembly of laminations of lumber in which the grain of all laminations is approximately parallel longitudinally, in which the laminations are bonded with adhesives, and that conforms to the standards applicable thereto.

Structural glued laminated timber

1. Laminated timber used as load-carrying structural timber framing for roofs and other structural portions of buildings, for other exterior construction uses, and marine installations.

2. Engineered, stress-rated product of a timber laminating plant comprising assemblies of specially selected and prepared wood laminations securely bonded together with adhesives. The grain of all laminations is approximately parallel longitudinally. Laminations should not exceed 2-in. net thickness. The assembly may be comprised of pieces end joined to form any length, of pieces placed or glued edge to edge to make wider ones, or of pieces bent to curved form during gluing.

Structural strength. Laminated engineered timber has a fatigue limit above customary design stress levels and resists fatigue and repeated loading. Glulam is laminated lightweight construction.

Submerged service. Normal use conditions under which timbers in service will be continually submerged in water.

Summerwood. Portion of the annual growth ring that is formed after the springwood formation has ceased. It is usually more dense and stronger mechanically than springwood.

Tapered beams. Beams tapering in one direction only should have the depth required for bending or deflection at the center of the span. Beams tapering both ways from the center should have the required depth at a distance 0.15 of the span each side of the center. The depth at the support should not be less than that required for shear. A tapered curved beam having a roof pitch of more than 4 in 12 requires special consideration because of elongation from deflection.

Tapered columns. The least dimension for laminated tapered columns is taken as the sum of the smaller dimension and one-third the difference between the smaller and greater dimensions.

Thermocouple. In laminating thermocouples are used in conjunction with the determination of the temperature of the innermost glue line.

Thickness of laminations. For practical reasons the two maximum thicknesses of laminations are limited to 2 in. net. Beams are laminated of 2-in. lumber, except that 1-in. boards are sometimes used for architectural reasons. When members are curved, the thickness of laminations is controlled by the radius of curvature. Also, members that are curved to their extremities require greater radius for the same lamination thickness than do members that have either straight ends or longer radius toward the ends.

Tooth value. Unit shear value of connector teeth.

Torque. Product of a force and a lever arm that tends to twist or rotate a body, for example, the action of a wrench turning a nut on a bolt. Torque is commonly expressed as foot pounds, that is, the product of the applied force measured in pounds and the lever arm measured in feet. The same action is also called a "moment."

Torque wrench. Wrench with an attached dial or other indicator that registers the torque applied by the wrench.

Triangular trusses. Triangular trusses are popular for certain industrial and commercial buildings for spans up to 80 ft. Triangular trusses are also recommended for church and school construction for reasons of economy and for special designs.

Truss

1. Framework of members in the same general plane joined only at their ends. Primary stresses in the members are axial, tensile, or compressive stresses. Secondary stresses are generally bending stresses due to either a degree of joint fixity eccentricity and/or beam action of the members under load. Trusses are generally composed of assemblages of triangles for stability.

2. Framework consisting of vertical or diagonal members, with axes all lying in the same plane, so connected as to form a triangle or series of triangles. In some cases, as in a bowstring truss, the chord members may be slightly curved, but the curvature is generally so slight that the definition is essentially complied with.

Truss chord. One of the principal members of a truss, usually horizontal, braced by the web members.

Uniform moisture content. All lumber used in laminated wood products are scientifically kiln-dried. Laminated decking has a maximum moisture content of 19%; laminated beams and arches have a maximum moisture content of 16% before bonding and surfacing. The result is a dimensionally stable product with a permanent appearance that minimizes warping, twisting, and checking.

Unit nail value. Unit shear value of connector nails.

U-strap column anchorage to concrete base. Anchorage assembly consisting of throughbolts, a through U strap, and a column, recommended for industrial buildings and warehouses to resist both horizontal forces and uplift. A bearing plate and moisture retarder is recommended. This type of anchorage may be used with shear plates.

Utility grade. Grade used where a greater number of natural growth characteristics, including knotholes and open defects in the face ply, are acceptable. Utility is appropriate for use wherever appearance is not an important consideration and the effects of such additional growth characteristics are acceptable.

Vertical grain. Another term for edge grain.

Vertically laminated wood. Laminated wood in which the laminations are so arranged that the wider dimension of the lamination is approximately parallel to the direction of the loads.

Wane. Bark or lack of wood or bark from any cause on an edge or corner of a piece.

Waterproof adhesive. All laminated wood products are only bonded with waterproof adhesive conforming to the test requirements of PS 56-73 for waterproof adhesives. This assures the glue line for both interior and exterior locations without the need for separate specifications.

Water-resistant glues. Glues that are used in interior protected surfaces where the moisture content of the materials must not exceed 15% while in use.

Welded and bent-strap-type purlin hanger. Hanger used for moderate and heavy loads. It provides uniform fit where good appearance is desired. Purlins must be raised above the top of the beam to allow the roof sheathing to clear the straps.

Wet use. Normal use conditions, such as those that may occur in exterior construction, under which the moisture content in service is 16% or more.

Wet-use adhesives

1. Adhesives that will perform satisfactorily for all conditions of service including exposure to weather, marine use, and pressure treatment.

2. Laminated timber members bonded with wet-use (waterproof) adhesives are suitable for all applications. Wet-use adhesives are used if the members are subject to occasional or continuous wetting or for applications, either exterior or interior, where the moisture content of the wood will exceed 16%.

Width. Dimension of the cross section of a member in bending measured normal to the direction of the principal load.

Wood failure. Rupturing of wood fibers. The percentage of wood failure is expressed as the percentage of the total area involved that shows such failure.

Wood preservation. Waterborne salts and oil-borne chemicals in mineral spirits or volatile-solvent carriers are the only preservatives recommended for treating individual laminations prior to gluing.

Wood treatment and gluing. Laminated members may be pressure impregnated with chemicals either before or after gluing. When treatment is to be done after gluing, an exterior-type glue must be used, and oil-borne treatments are most often used, especially with large members subjected to severe exposure conditions.

Working stress. Stress for use in the design of a wood member that is appropriate to the species and grade. It is obtained by multiplying the basic stress for the species and the strength property by the strength ratio of the grade.

ROUGH CARPENTRY

Acclimatization. Storage of lumber in a building for a short period so that it acquires the desired interior temperature before installation.

Air-dried. Lumber dried by exposure to air, usually in a yard, without artificial heat.

Approximate shrinkage. Tolerance allowed for normal variance from green to dry.

Awning window. Type of window where the sash are hinged at the top and swing outward.

Backband. Narrow rabbeted molding applied to the outside corner and edge of interior window and door casing to create a "heavy trim" appearance.

Backing. The bevel on the top edge of a hip rafter that allows the roof sheathing to fit the top of the rafter without leaving a triangular space between it and the lower side of the roof covering.

Batten. Narrow strip of wood used to cover joints between boards or panels. "Board and batten" is the term used to describe the type of cladding in wood frame buildings where battens cover the joints between boards.

Batter. The slope, or inclination from the vertical, of a wall or other structure or portion of a structure.

Batter board. Boards set at right angles to each other at each corner of an excavation, used to indicate the level and alignment of the foundation wall.

Bay. One of the intervals or spaces into which a building plan is divided by columns, piers, or division walls.

Bay window. Rectangular, curved, or polygonal window or group of windows, usually supported on a foundation extending beyond the main wall of a building.

Bed molding. Molding in an angle, as between the overhanging cornice, or eaves, of a building and the sidewalls.

Belt course. Horizontal board, around a building. It is usually made of a flat member and a molding.

Bench mark. Mark on a permanent vertical object fixed to the ground from which vertical measurements and elevations are taken.

Bevel. To cut to an angle other than a right angle, such as the edge of a board or door.

Bevel cut. Angle cut through the thickness of a piece of wood.

Bevel siding. Boards tapered to a thin edge and used as exterior wall covering.

Blind nailing. Nailing in such a way that the nailheads are not finally visible on the face of the work.

Blinds (shutters). Lightweight, louvered or flush, wood covers located at each side of a window. Some are made to close over windows to shut out light or give protection from the weather. Others are fastened to the wall and used as a decoration.

Blind stop. Member applied to the exterior edge of the side and head jamb of a window to serve as a stop for the top sash and to form a rabbet for storm sash, screens, blinds, and shutters.

Board and batten. Exterior finish where boards or sheathing are used with strips of wood to cover the joints.

Board foot. The equivalent of a board one foot square and one inch thick.

Boarding in. Process of nailing boards or sheathing on the outside studs of a house.

Bond. Joining or adhering, as of two surfaces. Also, a substance which causes such a joining to take place.

Break joints. To arrange joints so that they do not come directly under or over the joints of adjoining pieces, as in shingling, siding, etc.

Brick molding. Molding for window and exterior door frames. Serves as the boundary molding for brick or other siding material and forms a rabbet for the screens and/or storm sash or combination door.

Buck. Rough framing around any type of opening on which a finished frame is to be installed.

Building paper. Lightweight roofing felt installed on sheathing before the siding of roofing is put on. It is sometimes placed between double flooring to prevent squeaking.

Butt joint. Place where two pieces of wood are joined together end to end.

Cant strip. Triangular-shaped piece of lumber, usually cut from a 2 × 4, generally installed around the perimeter of a flat roof deck, or at the junction of the roofing surface and the adjoining walls. Cant strips are also available in other materials.

Casement window. Type of window where the sash are hinged at the edge and usually swing outward.

Center matched. Matched lumber that is tongued and grooved in the exact center of the material on opposite edges.

Check rails. Meeting rails of a double-hung window, which are made thicker to fill the opening between the top and bottom sash. They are usually beveled.

Clapboard. Type of siding consisting of narrow boards that are usually thicker at one edge than the other.

Cleat. Strip of material, such as wood, fastened to another piece to strengthen it or to furnish a hand grip.

Clinch. To hammer the exposed tip of a nail at an angle, bending its point into the surrounding wood for added joint strength.

Closed valley. Roof valley where the roof covering meets in the center of the valley, completely covering the valley.

Combination doors or windows. Doors or windows with selfstoring or removable glass and screen inserts. The need for handling a different unit each season is thus eliminated.

Corner boards. Built-up wood members installed vertically on the external corners of a house or other frame structure, against which the ends of the siding are butted.

Cornice

1. Overhang of a pitched roof at the eave line, usually consisting of a fascia board, a soffit for a closed cornice, and appropriate moldings.

2. A decorative member, usually molded, placed at or near the top of a wall.

Cornice return. That portion of the cornice that returns on the gable end of a house.

Cricket. Small roof structure at the junction of a chimney and a roof to divert rain water around the chimney.

Cupola. Small vented four-sided structure installed on a roof. Adds decoration to the building and provides ventilation for the attic.

Deck. The flat portion of a roof or floor upon which some type of covering will be placed.

Door frame. Assembly of wood parts that form an enclosure and support for a door. Door frames are classified as exterior and interior.

Door jamb. The surrounding case into which and out of which a door closes and opens. It consists of two upright pieces, called "side jambs," and a horizontal head jamb.

Door stop. Molding nailed to the faces of the door frame jambs to prevent the door from swinging through.

Dormer. Roofed projection in a sloping roof, with framing forming, vertical walls suitable for the installation of windows or other openings.

Double-end trimmed. Term applied to lumber trimmed reasonably square by a saw on both ends.

Dress. Plane one or more sides of a piece of sawed lumber.

Double-hung window. Window in which two sash slide vertically by each other.

Dressed and matched (tongued and grooved). Boards matched in such a way that there is a groove on one edge and a corresponding tongue on the other.

Drip. The projection of a window sill with an extended edge to allow water to drain clear of the wall face.

Drip cap. Molding placed on the top of exterior door and window casings for the purpose of shedding water beyond the outside face of the wall.

Drip groove. Semicircular groove on the underside of a drip cap or the lip of a window sill that prevents water from running back under the member.

Drop siding. Siding, usually $\frac{3}{4}$ in.-thick and machined into various patterns. Drop siding has tongue and groove or shiplap joints.

Eased edges. Slightly rounded surfacing on pieces of lumber to remove sharp corners. Lumber 4 in. or less in thickness is frequently shipped with eased edges, unless otherwise specified. Recommended standard for lumber of 1- and 2-in. thickness is to round to a radius of $\frac{1}{16}$ and $\frac{1}{8}$ in., respectively.

Eaves. The lower part of a roof that projects over an exterior wall. Also called the "overhang."

Eave soffit. The undersurface of the eave.

Edge. There are three meanings for the word "edge:" narrow face of rectangular shaped pieces; the corner of a piece at the intersection of two adjoining faces; usually in stress grades that part of the wide face nearest the corner of the piece.

End check. Check occurring at an end of a piece.

End matched. Lumber that has machined tongues and grooves in the square-cut ends.

Exposure. The amount that courses of siding and roofing are exposed to the weather.

Exterior-grade plywood. Plywood manufactured with 100% waterproof glue. The veneers in back and the inner plies are of a higher grade than those used for interior-grade veneer panels.

Exterior trim. Includes cornices, mouldings, and other ornamental shop-fabricated shapes attached to the outside face of exterior walls, doors, windows, porches, and bay windows.

Face nailing. To nail perpendicular to the surface or to the junction of the pieces joined.

Face width. Width of the face of a piece of dressed and matched or shiplapped lumber, excluding the width of the tongue or lap. The amount of flooring, ceiling, siding, or other matched items required to cover a given area should be computed on the basis of the face width of the pieces.

Fascia. Wood member used for the outer face of a box cornice where it is nailed to the ends of the rafters and lookouts.

Fascia board. Finish member around the face of eaves and roof projections.

Fenestration. The placement or arrangement and sizes of the windows and exterior doors of a building.

Filler. Piece used to fill the space between two surfaces.

Flat roof. Roof that is level or pitched only enough to provide for drainage, as required by the code.

Fireproofing. Making wood resistant to fire so that it is difficult to ignite and will not support its own combustion.

Flush. Continued surface of two continuous areas on the same plane.

Floor sleepers. Wood embedded in or laid directly on a concrete slab that is in direct contact with the ground should be treated with an approved pressure preservative treatment. Sleepers are shimmed level to accept a finish wood floor.

Frieze. Part of the exterior finish applied at the intersection of a overhanging cornice and the wall.

Full sawed. When specified to be full sawed, lumber may be manufactured to the basic oversize tolerance as provided by the standards but may not be undersize at time of manufacture.

Gable. Vertical triangular end of a building from the eaves to the apex of the roof.

Gable end. The entire end wall of a house having a gable roof.

Gable roof. Type of roof that pitches in two directions.

Gambrel. Symmetrical roof with two different pitches or slopes on each side.

Grading rules. Lumber is manufactured in a sawmill to grading rules of regional associations. As it leaves the saw, the wood is unseasoned, though it is usually partly air-seasoned when it reaches the project.

Gutter (eave trough). A shallow channel or trough of metal or wood set below and along the eaves of a house to catch and carry off rainwater from the roof.

Halved joint. Joint made by cutting half the wood away from each piece so as to bring the sides flush.

Header. Pieces placed at right angles to joists, studs, and rafters to form openings in a wood frame.

Hip. The external angle formed by the meeting of two sloping sides of a roof.

Hip roof. Roof that slopes up toward the center from all sides, requiring a hip at each corner.

Hopper window. Type of window where the sash are hinged at the bottom and swing inward.

Horn. Extension of the stiles of doors or the side jambs of window and door frames, which are removed before installation.

Housing. A grove or trench in a piece of wood made for the insertion of a second piece.

Insulation. Material used in walls, floors, and ceilings for the purpose of reducing transmission of hot or cold air.

Jalousie. Series of small horizontal overlapping glass slats, held together by an end metal frame attached to the faces of window frame side jambs or door stiles and rails. The slats or louvers move simultaneously like a venetian blind.

Jamb. The top and two sides of a door or window frame that contact the door or sash; top jamb and side jambs.

Kerf. Cut made by the saw in solid lumber on the back face to prevent buckling and curving.

Knocked down. Unassembled; refers to structural or nonstructural units requiring assembly after being delivered to a job.

Kraft paper. Brown building paper that resists puncturing. Used to face some blanket insulation materials.

Lap joint. A joint of two pieces lapping over each other.

Lap siding. Boards used to cover the sides of buildings, the lower edge of one board being lapped over the upper edge of the board below.

Laminated veneer lumber (LVL). Structural lumber manufactured from veneers laminated into a panel with the grain of all veneers running parallel to each other. The resulting panel is normally manufactured in $\frac{3}{4}$- to $1\frac{1}{2}$-in. thicknesses and ripped to common lumber widths of $1\frac{1}{2}$ to $11\frac{1}{2}$ in. or wider.

Lattice. Thin strips of wood, spaced apart and applied in two layers at angles to each other, resulting in a kind of grillework.

Ledger. Horizontal strip (quite often wood) used to provide support for the ends or edges of other structural framing members.

Light construction. Construction generally restricted to conventional wood stud walls, floor and ceiling joists, and rafters.

Louver. A finned or slatted device installed, usually in the peaks of gables and the tops of towers, to exclude rain and snow while allowing airflow.

Mansard roof. Roof with two slopes on all four sides; the lower slope is steep, while the upper is almost flat.

Marine plywood. Plywood panels manufactured with the same glue line durability requirements as other exterior panels, but with more restrictive veneer quality requirements.

Matching or tonguing and grooving. The method used in cutting the edges of a board to make a tongue on one edge and a groove on the other.

Medium-density overlay. Plywood with moisture resistant glues, mostly for exterior applications, signage, etc. The resin-treated facing on the finished product presents a smooth, uniform, or uniformly textured surface intended for high-quality paint finishes. Some evidence of underlying grain may appear.

Meeting rails. Rails sufficiently thicker than a window to fill the opening between the top and bottom sash made by the parting stop in the frame of double-hung windows. They are usually beveled.

Mill run. All the lumber produced in a mill without reference to grade.

Mixed grain. Lumber may have either or both vertical and flat grain.

Moistureproofing. Making wood resistant to change in moisture content and especially to the entrance of moisture.

Modular construction. Method of construction in which parts are preassembled in conveniently sized units.

Mullion. The construction that divides a window opening to accommodate two or more windows.

Muntin. Vertical member between two panels of the same piece of panel work. The vertical and horizontal sashbars separating the different panes of glass in a window.

Nominal dimension. The stated size of a piece of lumber, such as a 2×4 or a 1×12. The actual dimension is somewhat smaller.

Oriel window. Window that projects from the main line of an enclosing wall of a building and is carried on brackets, corbels, or a cantilever.

Overhang. That part of a roof which extends beyond the outer wall of a building.

Panel siding. Large sheets of plywood or hardboard, which may serve as both sheathing and siding.

Parapet. Low wall or railing along the edge of a roof, balcony, or bridge. The part of a wall that extends above the roof line.

Parting strip. Small strip of wood separating the upper and lower sash of a double-hung window.

Pitch. Inclination or slope, as of roofs or stairs. Rise divided by the span.

Pitch of roof. Amount of slope of a roof. Rise and steepness of a roof. The usual way to express roof pitch is by means of numbers, such as 8 and 12, 8 being the rise and 12 the run.

Plate. Horizontal structural member placed on a wall or supported on posts, studs, or corbels to carry the trusses of a roof or to carry the rafters directly above. Also, a sole or base member of a partition or other frame.

Plywood exterior. General term for plywood manufactured of veneers of Grade C and better and bonded with adhesive that is highly resistant to weather, microorganisms, cold or boiling water, steam, and dry heat.

Porch. Floor extending beyond the exterior walls of a building. It may be covered and enclosed or open.

Precision end-trimmed lumber. Trimmed square and smooth on both ends to uniform length.

Rabbetted joint. A lapping joint machined on the meeting portions of the pieces to be connected.

Rake. The trim members that run parallel to the roof slope and form the finish between the roof and wall.

Reveal. The amount of setback of the casing from the face side of window and door jambs or similar pieces.

Ridge. Top edge or corner formed by the intersection of two roof surfaces.

Roof sheathing. Boards or sheet material, fastened to the roof rafters, on which the shingles or other roof covering is laid.

Roof ridge. The horizontal line at the junction of the top edges of two roof surfaces where an external angle greater than $180°$ is formed.

Rough carpentry. That part of the trade involved with construction of the building frame or other structural work.

Roughing-in. The framing stage of a carpentry project. This framework is concealed later in the finishing stages.

Rough opening. Opening made in the wood framing to install windows, doors, or similar units.

Run. Length of the horizontal projection of a piece such as that of a rafter when in position.

Saddle board. Finish of the ridge of a pitch-roof house. Sometimes called "comb board."

Sash. Frame unit, normally glazed (e.g., a window) that is hung or fixed in a frame set in an opening.

Sash balance. Device, usually operated by a spring or tensioned weather-stripping, designed to counterbalance double-hung window sash.

Saturated felt. Felt that is impregnated with tar or asphalt.

Scaffold or staging. Job-constructed temporary structure or platform enabling workers to reach high places.

Scantling. Lumber with a cross section ranging from 2×4 to 4×4 in.

Scarfing. A joint between two pieces of wood that allows them to be spliced lengthwise.

Seat of a rafter. The horizontal cut on the bottom end of a rafter that rests on the top of the wall plate.

Shakes. Handsplit shingles.

Sheathing paper. The paper used under siding or shingles to insulate the house.

Shed roof. Type of roof that slopes in one direction only.

Shim. Thin strip of wood, sometimes wedgeshaped, for plumbing or leveling wood members. Especially helpful when setting door and window frames.

Shingles. Covering applied in overlapping layers, as for the roof or sides of a building. Shingles can be made of wood, asphalt, asbestos, tile, or slate, among other materials. They are cut fairly small.

Shiplap. Lumber with edges that have been rabbeted to form a lap joint between adjacent pieces.

Shutter. Wood assembly of stiles and rails to form a frame that encloses panels used in conjunction with door and window frames. Also may consist of vertical boards cleated together.

Sidelight. Framework containing small lights of glass placed on one or both sides of the entrance door.

Side of trim. Trim required to finish one side of a door or window opening.

Siding. The finish covering of the outside wall of a frame building. Many different types are available.

Sill. The horizontal member forming the bottom of an opening such as a door or window.

Sill plate. Structural member anchored to the top of a foundation wall, upon which the floor joists rest.

Skylight. Glazing framed into a roof or a glazed metal frame.

Sleeper. Usually, a wood member fastened to a concrete, floor, that serves to support the wood subfloor and flooring.

Sliding window. A window consisting of two sash which slide from side to side past one another.

Sole plate. The lower horizontal member of a wood frame wall nailed to the bottom of the wall studs and to the floor framing members; also called a "bottom plate."

Solid bridging. Solid members of the same thickness and width as floor joists, placed between them to distribute the floor load over a wider area.

Spline. A rectangular strip of wood that is substituted for the tongue.

Stile. The upright or vertical outside pieces of a sash, door, blind, or screen.

Stoop. Small porch, veranda, platform, or a stairway, outside an entrance to a building.

Storm sash. Additional sash placed on the outside of a window to create a dead airspace to prevent the loss of heat from the interior in cold weather.

Structural plywood. Plywood for structural use, such as flooring, siding, and roof sheathing.

Structural sandwich construction. Layered construction comprising a combination of relatively high-strength facing materials intimately bonded to and acting integrally with a low-density core material.

Structural window wall panel. A window unit framed into a wall panel at the factory; also called a "factory-assembled structural wall panel."

Stub tenon. A short tenon intended for insertion in a plow or groove.

Subfloor. Boards or panels laid directly on floor joists over which a finished floor will be laid.

Threshold. Strip of wood, metal, or other material beveled on each edge and used at the junction of two different floor finishes under doors or on top of the door sill at exterior doors.

Toenailing. Nailing at an angle to the first member so as to ensure penetration into a second member.

Tongue-and-groove lumber. Any lumber, such as boards or planks, machined in such a manner that there is a groove on one edge and corresponding tongue on the other.

Transom. Small opening above a door separated by a horizontal member (transom bar). Usually contains a sash or a louver panel hinged to the transom bar.

Trimmer stud. Stud which supports the header for a wall opening. The stud extends from the sole plate to the bottom of the header. It is parallel to and in contact with a full-length stud that extends from sole plate to top plate.

Underlayment. Any material installed on the subfloor that will give a smooth level surface to receive the finish floor covering.

Valley. The internal angle formed by two slopes of a roof.

Verge boards. Boards that serve as the eaves, finish on the gable end of a building.

Wash. Slope on a sill, wall or chimney capping, etc., to allow water to run off easily.

Weatherstrip. Narrow strips of metal, vinyl, plastic, or other material, so designed that, when installed at doors or windows, they will retard the passage of air, water, moisture, or dust.

Window unit. Consists of a combination of the frame, window, weatherstripping, and sash activation device. May also include screens and/or storm sash. All parts are factory assembled as a complete operating unit.

Wood fiber insulation. Composed of wood fibers, with or without binders.

Wood-cement board. Panel material where wood, usually in the form of excelsior, is bonded with inorganic cement.

Yard lumber. Lumber of all sizes and patterns, intended for general building purposes.

STRUCTURAL WOOD FRAMING

Anchor bolts. Bolts that secure wood plates to the construction.

Anchors. Special metal devices or bolts used to fasten timbers to masonry or concrete.

Arch. Construction of any wood material or laminated wood in a form so that it supports its own weight by mutual pressure. An arch is also capable of sustaining superimposed weight.

Ballon frame construction. Type of framing for two-story buildings where both studs and first-floor joists rest on the anchored sill. The second-floor joists bear on a 1×4 in. ribbon strip that has been set vertically into the inside edges of the studs.

Band joist. Joist running perpendicular to the primary direction of the joists in a floor and closing off the floor platform at the outside face of the building.

Basic and working stresses for timber. Basic stresses are the safe stress values of clear, straight-grained wood of structural member size that have been determined by applying a safety factor to ultimate strength.

Beam. Girders, rafters, and purlins acting as a structural member, supported at two ends.

Beam pocket. Notch formed at the top of a foundation wall to receive and support the end of a beam.

Beams and stringers. Lumber having a minimum rectangular cross section, 5×8 in., graded with respect to its strength when loaded on the narrow face.

Bearing. Point of support on a wall, beam, or structural member.

Bearing partition. Interior partition that supports any vertical load in addition to its own load.

Bearing wall. Exterior wall that supports any vertical load in addition to its own load.

Bending. The amount of deflection of a piece of lumber when a load is applied.

Blocking. Pieces of wood installed between joists, studs, or rafters in a building frame to stabilize the structure, or between studs to receive fixtures.

Board. Piece of rough, dressed, or worked lumber less than 2 in. in nominal thickness and 4 in. or more in width.

Bottom plate. The lower horizontal member of a wood frame wall, nailed to the wall plate, which has been anchored to the foundation with anchor bolts.

Braces. Pieces fitted and firmly fastened to two others at any angle in order to strengthen the angle thus treated.

Bracing. In buildings of one and two stories the corner post should be the equivalent of not less than three (3) pieces of two by four (2 × 4) inch studs. Bracing should be accomplished by diagonal wood sheathing or plywood panels, or other sheathing specified in the code, applied vertically in panels of not less than four by eight (4 × 8) feet in area in compliance with approved nailing complying with the code. Ledger members used to support joists should not be less than two by four (2 × 4) inch nominal side, adequately nailed with 16d nails to doubled header joists.

Bridging. Small wood or metal members that are inserted in a diagonal position between the joists acting both as tension and compression members for the purpose of bracing the joists, as well as creating a truss action and spreading the floor loads.

Built-up members. Structural members, the sections of which are composed of combinations or sawn lumber or plywood, in which all parts are bonded or joined together with glue, bolts, nails, metal clips or other similar fastening.

Built-up timber. Timber made of several pieces nailed, bolted, or lag-screwed together, forming one of the larger dimensioned pieces.

Camber. Convexity of a truss or beam to offset weight or pressure that might result in its becoming concave.

Cantilever. Beam or truss or joist extending beyond its last point of support.

Cavity wall. Combination of several wall materials with a continuous airspace between these materials.

Ceiling joists. Series of parallel beams evenly spaced and used to support ceiling loads.

Centering. Temporary framework for the construction of an arch or other shaped form.

Chords. Top and bottom members of a truss. Mono-chords are single-member trusses; upper chords are used on a bowstring truss forming a circular arch or comprising the top member of the truss.

Collar beam. Board, usually 1 × 4, 1 × 6, or 1 × 8 in., fastened to a pair of rafters in a horizontal position at some desired location between the plate and ridge of the roof. A collar beam tends to keep the roof framing from spreading.

Collar brace. Horizontal piece of lumber usually located in the middle third of the rafters, used to provide intermediate support for opposite roof rafters.

Collar tie. A wood member nailed across two opposing rafters near the ridge to resist wind uplift.

Column. Perpendicular supporting member, circular or rectangular in section; vertical structural compression member that supports loads acting in the direction of its longitudinal axis.

Combination frame. Frame combining the principal features of the full and balloon frames.

Combination truss joists. Manufactured with top and bottom wood chords of machine stress-rated lumber, with webs made of $1\frac{1}{8}$-in. diameter tubular steel members, varying in gauges and diameters, depending on design requirements.

Common rafters. Rafters that run square with the plate and extend to the ridge.

Corner brace. Brace used to stiffen the wall frame at the corner of the building.

Corner post. Double studs acting as columns at the corner of the building.

Cripple jack. Rafter that intersects neither the plate nor the ridge and is terminated at each end by hip and valley rafters.

Cripple rafter. Rafters that are cut between valley and hip rafters.

Cripple stud. Wood wall framing member that is shorter than full-length studs because it is interrupted by a header.

Cross bridging. Metal or wood diagonal bridging, set in pairs and crossing each other, between floor joists to distribute loads over a greater area and to stiffen the floor construction.

Cross-grain wood. Wood incorporated into a structure in such a way that the direction of the grain is perpendicular to the direction of the principal loads on the structure.

Cross-wall construction. Type of construction in which floor and roof loads are carried entirely on walls or bearing partitions running across a building.

Curtain wall. Nonbearing wall between columns or piers that is not supported by girders or beams.

Dead load. Refers to type and weight of floor or roof construction in relation to pounds per square foot.

Decking. Prepared wood surfacing material installed over rafters or joists to which other finish materials can be applied.

Diaphragm action. The installation of continuous bracing to stiffen the structural supporting frame.

Dimension. Term applied to all yard lumber except boards, strips, and timbers, that is, to yard lumber from 2 in. to, but not including, 5 in. thick, and of any width.

Dimensional stabilization. Reduction, through special treatment, in the swelling and shrinking of wood caused by changes in its moisture content with changes in relative humidity.

Dimension stock. Square or flat stock, usually in pieces smaller than the minimum sizes admitted by standard lumber grades, that is rough or dressed, green or dry, and cut to the approximate dimension required for the various products of woodworking factories.

Face nail. Nail driven through the side of one wood member into the side of another.

Firecut. Sloping end cut on a wood beam or joist where it enters a masonry wall for bearing. The purpose of the firecut is to allow the wood member to rotate out of the wall without causing any damage to the wall in case of a fire.

Fire resistive. Term applied to construction materials not combustible in the temperatures of ordinary fires and that will withstand such fires without serious impairment of their usefulness for at least 1 hour.

Firestop. Plywood baffle used to close openings between studs or joists in a balloon or platform frame in order to prevent spread of fire through the openings.

Frame. Surrounding or enclosing woodwork of windows, doors, etc.; the timber skeleton of a building.

Frame construction. Wood framing that conforms to all requirements of frame construction as described in building codes.

Framing. Rough timber structure of a building, including interior and exterior walls, floor, roof, and ceilings.

Gable studs. Vertical framing members placed between the end rafters on a gable roof and the two plates of the end wall.

Girder. Large-sized beam used as a main structural member, normally for the support of other beams.

Girt (ribband). Horizontal member of the walls of a full or combination frame house that supports the floor joists or is flush with the top of the joists.

Glued built-up members. Structural elements, the sections of which are composed of built-up lumber, plywood, or plywood in combination with lumber; all parts bonded together with adhesives.

Glue laminated structural lumber. Lumber consisting of laminations in which the grain of all laminations is approximately parallel and where all laminations are bonded together with glue.

Glulam. Term used to describe glue laminated structural wood members.

Grade (lumber). The classification of lumber in regard to strength and utility. A mechanical means of grading may be accepted when approved by the building department.

Grade-stress. A lumber grade defined in such terms that a definite working stress may be assigned to it.

Hardware. All types of fasteners and devices necessary to complete rough carpentry installations, such as nails, screws, lag screws, bolts, and the like.

Header. Short joist supporting tail beams and framed between trimmer joists; piece over an opening; lintel.

Heavy timber construction. Construction composed of planks or laminated floors supported by beams or girders. Exterior walls may be frame, masonry, or metal; however for "mill construction," they must be of masonry (verify with local code requirements).

Heel of a rafter. End or foot that rests on the wall plate.

Hip rafter. Rafter extending from the outside angle of the plates toward the apex of the roof.

Impact bending. In the impact bending test, a hammer of given weight is dropped upon a beam from successively increased heights until complete rupture occurs.

Interior fire retardant treated wood. Some code requirements include that wood treated with fire retardant chemicals bear third-party inspection as provided by Underwriters Laboratories, Inc., who will assure fire performance if approved.

Jack rafter. Rafter that is square with the plate and intersecting the hip rafter.

Joist. Supporting member; solid load-bearing board set on edge and evenly on centers.

Joist lumber. Lumber of rectangular cross sections 2 in. up to, but not including, 5 in. in thickness and 4 in. wide or more, graded with respect to its strength in bending when loaded either on the narrow face (joist) or on the wide face (plank). If 5 in. or more thick, the lumber is known as "beam" or "post" lumber.

Joist plate. Plate at the top of masonry walls supporting rafter or roof joists and ceiling framing lumber.

Joists and planks. Lumber of a rectangular cross section 2 in. to less than 5 in. thick and 4 or more inches wide, graded with respect to its strength in bending, when loaded either on the narrow face as a joist, or on the wide face as a plank.

Joist spacing. Distances (specified by building codes) between joists. Usually specified as o.c. (on center), which is the distance from the center of one joist to the center of the next joist.

Knee brace. Corner brace, fastened at an angle from wall stud to rafter, stiffening a wood or steel frame to prevent angular movement.

Lag screw. Screw used in heavy wood construction. It is longer and heavier than the common wood screw, with coarser threads and a larger head. It is used to connect wood structural members of a building.

Lamella. Short piece of lumber used in the construction of the network arches that form a lamella roof.

Lamella roof structure. Arched roof-framing structure identified by the diamond shaped arrangements of the pieces of plank from which it is formed.

Laminated decking. Random lengths wood lay-up system consisting of continuous planks applied over three or more spans. Each individual plank must bear on at least one supporting member. Glue used should be a dry melamine resin that conforms to ASTM D 2559.

Laminated structural beams. Composed of laminated veneer lumber. Adhesives used for laminating should meet requirements described in ASTM D 2559. Process manufacturer must guarantee that the beams will not twist, shrink, split, or produce checking of any kind.

Ledger board. Board used as a support for the second floor joists of a balloon frame house or for similar uses; ribband.

Lintel (header). Piece of construction material such as stone, wood, or metal, that is placed over an opening; header.

Live load. Weight of all moving and variable loads that may be placed on or in a building, such as snow, wind, occupancy, furnishings, merchandise, and equipment.

Load bearing. Building material or element subjected to or designed to carry loads in addition to its own dead load.

Lookout. End of a rafter or the construction that projects beyond the sides of a house to support the eaves; also, the projecting timbers at the gables that support the verge boards.

Maximum crushing strength. Maximum stress sustained by a compression specimen having a ratio of length to least dimension of less than 11 under a load slowly applied parallel to the grain. This property permits evaluation of the strength of posts or short blocks.

Mechanical properties. Properties of wood that enable it to resist deformations, loads, shocks, or forces. Thus the ability to resist a shearing force is a mechanical property.

Mill construction. Building type with exterior masonry bearing walls and an interior framework of heavy timbers and solid timber decking.

Modulus of elasticity. Measure of stiffness or rigidity. For a beam, the modulus of elasticity is a measure of the resistance to deflection. As determined from bending tests, it includes deflection due to shear distortion.

Modulus of rupture. Measure of the ability of a wood beam to support a slowly applied load for a short time. Modulus of rupture is an accepted criterion of strength, although it is not a true stress, since the formula by which it is computed is only valid to the proportional limit.

Nominal dimension. Dimension of lumber corresponding approximately to the size before dressing to actual size and used for convenience in defining size and computing quantities.

Nominal measure. In worked lumber, the dimensions of the rough board before dressing.

Nominal size. As applied to timber or lumber, the rough-sawed commercial size by which it is known and sold in the market.

Nonbearing wall. Wall supporting no other load than its own weight.

On center (o.c). Measurement of spacing joist, timber, beam, or purlin in a building from its center to the center of the one next to it.

Ordinary construction. Building type with exterior masonry bearing walls and an interior structure of wood framing.

Partition. An interior nonload-bearing wall, unless constructed to be considered as a loadbearing partition.

Physical properties. Properties of wood that have to do with its structure, such as density, cell arrangements, or fiber length. In its broad sense the term physical properties includes all those properties of wood listed as mechanical properties, as well as those pertaining to its structure.

Planks. Material 2 or 3 in. thick and more than 4 in. wide, such as joists or flooring.

Plate. Top horizontal piece on the walls of a frame building upon which the roof rests.

Plate cut. Cut in a rafter that rests on the plate; sometimes called the "seat cut."

Plate line. That part of the wall that supports the rafters.

Platform frame. Structural wood frame building composed of closely spaced members nominally not less than 2 in. thick, in which the wall members do not run past the floor framing members.

Plywood. Wood panel composed of several layers of wood veneer and bonded together under pressure, using several types of adhesives.

Plumb cut. Cut made in a vertical plane; the vertical cut at the top end of a rafter.

Posts. Lumber of square or approximately square cross section, 5×5 in. and larger, graded primarily for use as posts or columns carrying longitudinal loads but adapted for miscellaneous uses in which strength in bending is not especially important.

Prefabricated. Fabricated on or off the site before incorporation into a building or structure.

Pressure processed fire retardant wood. Treated wood pressure impregnated with an approved formulation Type A fire retardant chemical that produces a structurally stable fire retardant wood.

Pressure-treated lumber. Lumber that has been impregnated with chemicals under pressure, to retard decay or fire.

Purlin. Timber supporting several rafters at one or more points or the roof decking directly above it.

Rafter. Lumber used in the framing of a roof on a house to support the sheathing and roofing materials. Flat-roof members are called "joists."

Rafter or joist plate. Plate at top of masonry wall supporting rafter or roof joist and ceiling framing.

Ribbon. Narrow board let into the studding to add support to joists.

Ridge beam. Top horizontal member of a sloping roof, against which the ends of the rafters are fixed or supported.

Ridge board. Board placed on edge at the ridge of the roof to support the upper ends of the rafters.

Ridge framing. Term denoting the members of a wood structure that comprise the frame or form of a ridge that makes up the roof of the building. It usually consists of two chords mitered forming the roof peak and supporting web members with a bottom chord.

Run of rafter. The horizontal distance covered by one rafter.

Rough dimension. Term applied to lumber indicating an actual measurement equal to or larger than the usually specified.

Rough lumber. Lumber that has not been dressed but that has been sawed, edged, and trimmed at least to the extent of showing saw marks in the wood on the four longitudinal surfaces of each piece for its overall length.

Rough stock. Lumber that has been sawed, edged, and trimmed, but not dressed. Rough stock will vary in thickness and width owing to unavoidable variations in sawing, difference in shrinkage, etc.

Seat cut or plate cut. Cut at the bottom end of a rafter to allow it to fit on the plate.

Seat of a rafter. Horizontal cut on the bottom end of a rafter that rests on the top of the plate.

Shear. Stress that tends to keep two adjoining planes or surfaces of a body from sliding, one on the other, under the influence of two equal and parallel forces acting in opposite directions. The force that produces shear (or shearing stress) in a material is called a "shearing force."

Shearing strength parallel to grain. Shearing strength is a measure of the ability of timber to resist one part's slipping on another along the grain.

Sheathing. Boards applied horizontally or diagonally, consisting of 6-, 8-, or 10-in. width, or plywood panels applied to the outside face of the studs.

Sill. Lowest member of the frame of a structure, resting on the foundation and supporting the uprights of the frame.

Sill plate. The plate on top of a foundation wall that supports floor framing.

Sleeper. Timber laid on a concrete slab to support wood flooring.

Span. The distance between the bearings of a timber or arch.

Span rating. Number stamped on a sheet of plywood or other types of wood panels to indicate dimensions between supports.

Static bending. Fiber stress at the proportional limit in static bending or flexure is the computed stress in the wood specimen at which the strain (or deflection) becomes no longer proportional to the stress (or load). Therefore it is the stress in the specimen at which the load deflection curve departs from a straight line.

Strain. Deformation or distortion produced by a stress or force.

Strength properties. In a board sense strength implies all those properties that enable a material to resist forces. In a more restricted sense strength is resistance to stress of a single kind or to the stresses developed in a particular member. Definiteness requires that the name of the specific property be stated, for instance, strength in bending, strength in horizontal shear, or strength in compression parallel to grain. Seldom, if ever, do any two species contain all of the various propertries in the same degree. This accounts for special uses of the different species.

Strength ratio of wood. Percentage of remaining strength left in commercial grades after allowances are made for the effect on an unseasoned piece of the permitted growth characteristics, such as knots, cross grain, and shakes. Strength ratio for the particular grade is applied to basic stress to obtain working stress. Strength ratio is not, however, applied to modulus of elasticity.

Stress. Stress is distributed force. Fiber stress is the distribution force tending to compress, tear apart, or change the relative position of the wood fibers. Stress is measured by the force per unit area. Thus a short column 2×2 in. (4 sq in.) and supporting a load of 2000 lb will be under a stress or fiber stress of 500 lb/sq in.

Stress grades. Lumber grades having assigned working stress and modulus of elasticity values in accordance with accepted basic principles of strength grading.

Stringer. Long horizontal timber in a structure supporting a floor.

Structural insulating board. A structural insulating material, made principally from wood, cane, or other vegetable fibers and preformed into a rigid, fibrous, insulating board, lath, or plank, and used principally in building construction.

Structural lumber. Lumber that is 2 in. or more thick and 4 in. or more wide, intended for use where working stresses are required. Grading of structural lumber is based on both the strength of the piece and the use of the entire piece.

Structural timber. Wood products whose strength is the controlling element in its selection and use, for example, trestle timbers, car timbers, framing for buildings, ship timbers, and cross arms for poles.

Stud. Upright beam in the framework of a building.

Studding. Vertical framework of an interior partition or the exterior wall of a building, usually referred to as "2 by 4's."

Tail beam. A relatively short beam or joist supported in a wall on one end and by a header on the other.

Tensile strength perpendicular to grain. Measure of the resistance of wood to forces acting across the grain that tend to split a member.

Tie beam (collar beam). Beam so situated that it ties the principal rafters of a roof together and prevents them from thrusting the plate out of line.

Timber. Lumber with a cross section of more than 4×6 in., such as used for posts, sills, and girders.

Timber connectors. Rings, grids, plates, or dowels of metal or wood set in adjoining members, usually in precut grooves or holes, to fasten the members together in conjunction with bolts.

Tongue. Projection on the edge of a board machined to fit into a groove in the adjacent piece.

Top plate. Horizontal member at the top of a stud wall.

Trimmer. Beam or floor joist into which a header is framed.

Truss. Factory-built assembly used in lieu of rafters or joists to support a roof load over a wide span.

Trussed rafter. Truss where the chord members also serve as rafters and ceiling joists and are subject to bending stress in addition to direct stress.

Utility

1. Dimension grade in western woods, usually 2- to 4-in. thick, 6 in. and wider. Utility-grade wood is used for studs and framing in light construction and joists and rafters in limited spans.

2. Grade falling just below "standard."

Valley rafters. Rafters extending from an inside angle of the plates toward the ridge or centerline of the building.

Wale. Horizontal beam.

Wood I-Beam. Wood beam with routed top and bottom flanges, bonded to plywood web with waterproof glue.

Wood joist hangers. Metal hangers fully die formed, of galvanized steel, and adapted for nailing according to code. Several types are available: double shear joist hanger, top flange nailing hanger, side-nailing purlin hanger, joist hangers for masonry walls, joist hangers for laminated beams, and many other types.

Wall plate. Plate at top or bottom of wall or partition framing. Further defined as "top plate" at top, and "sole plate," at bottom.

Wood structural members. Each wood structural member should be of sufficient size to carry the design loads without exceeding the allowable unit stress specified in the code. Adequate bracing and bridging to resist wind and other lateral forces should be provided.

Work in bending to maximum load. Work to maximum load in static bending represents the ability of the timber to absorb shock with some permanent deformation and more or less injury to the timber; measure of the combined strength and toughness of wood under bending stresses.

WOOD STAIRS

Angle post. Railing support at landings or other breaks in the stairs. If an angle post projects beyond the bottom of the strings, the ornamental detail formed at the bottom of the post is called the "drop."

Baluster

1. Small pillar or column used to support a rail.

2. Vertical member supporting the railing.

3. Plain or turned piece that is placed in a vertical position to form a guard at the open side of an exposed stairway and around a well hole. The complete guard, consisting of a number of balusters, is called a balustrade.

4. A baluster is usually a small vertical member in a railing used between a top rail and the stair treads or a bottom rail.

Balustrade

1. Coping or handrail with its supporting balusters.

2. Series of balusters connected by a rail, generally used for stairs, porches, balconies, and the like.

3. Railing made up of balusters, top rail, and sometimes bottom rail, used on the edge of stairs.

Box stairs. Stairs built between walls, usually with no support except the wall stringers.

Built-up stair horse. Fabricated stair stringer made at the job. Triangular blocks are sawed from a piece of framing lumber and are nailed on 2×4 or 2×6 in. stock. If a sawed-out horse stair stringer is made first, then the triangular blocks that are cut from the material can be nailed to appropriate stock to make a built-up horse stair stringer.

Bull-nose step. Step that (in plan) is half-round or quarter-round at the front edge.

Buttress cap. Cap made from finish stock with a groove cut in the underside. The bottom ends of the balusters are fitted into the groove at regular intervals. (In some types of construction, the balusters are fitted into the treads and nailed in place.) When a face stringer is used in conjunction with a housed stringer to provide a thicker base for the balustrade, the buttress cap covers the space between two stringers.

Carriages

1. Rough timber supporting the steps of wood stairs.

2. Supports for the steps and risers of a flight of stairs.

Closed and open stairway. Stairway that may have one closed side and one side partially open, or both sides partially open.

Closed stairway. Stairway constructed between two walls and generally used for rear or attic stairs. If the arrangement of the building permits, this type of stairway can be built as a complete unit on the floor and then raised to its final location. In general, however, because rear or attic stairs are located in an area where the surrounding rooms are small, they usually are built into place between the walls.

Continuous handrail. Occasionally, a staircase is designed without newel posts. In this case the handrail is continuous, and specially milled, turned, and shaped pieces—casements, goosenecks, volutes, and quarter-, or half-turns—are required at the various turns of the stairs. A volute or spiral piece can be used at the lower end of the railing and is usually placed above the first tread.

Fillet. Piece of finish material that is fitted into the groove between each baluster so that the top of the buttress cap is smooth.

Flight of stairs. Run of stairs between landings or a set of winders to another run.

Glue blocks. Right-angled triangles of wood, about 4 to 6 in. in length. The blocks are glued and nailed on the underside of the stairway in the right angles formed by the junction of each finished riser and tread.

Handrail. Plain or molded piece of finish stock that serves as a hand rest, guide, and support. The handrail runs parallel to the stringer. In open stairway construction the handrail forms the top member of the balustrade along the open side and around the well hole. The term "handrail" can also mean a rail fastened to the walls of an enclosed stairway. In this case the vertical height range of the handrail above the tread remains at 30 to 34 in. For residential purposes the handrail can be made of wood at least 2 in. in diameter, while handrails for commercial use are made of pipe with an inside diameter of at least 1 in. or from an extruded metal shape. (Verify with OSHA requirements.)

Handrail brackets. Brackets used on closed stairways and generally made of metal. The brackets fasten the handrail to the wall and should be spaced about 8 ft apart or less on centers.

Head room

1. Minimum vertical clearance between a tread nosing in a stairway and the open end of the well hole. Headroom is specified in building codes.

2. Minimum clear height from a tread to overhead construction, such as the ceiling of the next floor, ductwork, or piping. Ample headroom should be provided not only to prevent tall people from injuring their heads but to give a feeling of spaciousness. A person of average height should be able to extend his hand forward and upward without touching the ceiling above the stair. The minimum vertical distance from the nosing of a tread to overhead construction should never be less than $6\frac{1}{2}$ ft. (Verify height with local codes.)

3. Clear space between floor line and ceiling, as in a stairway.

Housed stringer

1. A stairway with at least one open side requires an additional piece of finish lumber to provide additional support for the balustrade on the open side. There are four types of stair stringers: plain, square-cut, mitered, and housed.

2. Stringer made from a piece of finish stock. Dadoes are cut in the stringer to receive the ends of the treads and risers. Tread dadoes are rounded at the closed end so that the nosing of the tread will fit. Both the tread and riser dadoes are wider at the open end that at the closed end to allow for the insertion of wedges to hole the treads and risers in position.

Landing

1. Level platform between two flights of stairs used to break the length of a single flight or to change the direction of a stairway.

2. Platform between flights of stairs or at the termination of a flight of stairs.

3. Flat platform between a series of steps in a staircase.

Mitered stringer

1. Mitered stringers are used on open-finish stairways where it is important that the end grain of the risers is not exposed. The joint between each riser and the stringer is formed by mitering both the riser and stringer at 45° angle and then fitting them together. The miter cuts are made so that the finished corner is exactly a right angle (90°). Treads are nailed on top of the tread cut and extend beyond the face of the stringer and the face of the riser an equal amount to form a projection wide enough to receive a cove molding. The front and ends of each tread are nosed (shaped) to form a rounded edge.

2. Mitered stringers are precision cut from a piece of finish stock in approximately the same shape as a stair horse, except that the vertical cuts are mitered to fit the mitered ends of the risers.

Newel. Principal post at the foot of a staircase; central support of a winding flight of stairs.

Newel post

1. Vertical, plain, or turned piece located as needed in a stairway. The number of posts depends on the shape

and the design of the stairway. Newel posts are commonly located at the bottom tread, and top tread, and at the corners of the balustrade when it is constructed around the open sides of the well hole. Platform stairs require one or more newel posts at each landing. The handrail is fitted to the newel posts and the balusters.

2. Post at which the railing terminates at each floor level.

3. Post to which the end of a stair railing or balustrade is fastened. Also, any post to which a railing or balustrade is fastened.

Nosing

1. Term refers both to the molded front edge of a stair tread and to the milled stock that is fitted against the flooring on the top step and around the well hole.

2. Projection of a tread beyond the riser below.

3. Portion of a tread projecting beyond the face of the riser immediately below.

4. Part of a stair tread that projects over the riser or any similar projection; term applied to the rounded edge of a board.

5. Edge of a board worked into the form of a semicircle.

Open newel stair. Stair having successive flights rising in opposite directions, and arranged about a rectangular well hole.

Open riser. Airspace between the treads of stairways without upright members (risers).

Open stairway. Stairway designed to give a pleasing architectural effect and used where this effect can be viewed, such as in the front or entrance areas of buildings. An open stairway has one or both sides exposed. The exposed side is guarded with an ornamental railing or balustrade. The starting tread for an open stairway in more detailed than the tread required for a closed stairway.

OSHA. The Occupational Safety and Health Administration has issued precise regulations regarding the construction requirements for public and industrial use stairs.

Plain stringer. Stringer used for closed stairways. Only the top and bottom ends of this type of stringer are cut. A plain stringer is nailed to the wall at the angle specified in the plans, and all risers and treads are fitted against it.

Platform. Extended step or landing breaking a continuous run of stairs.

Platform stairway. Stairway constructed when it is necessary to change the direction of the stairs at a given point between a lower level and the next floor level. The platform (landing) must be large enough to accommodate the flight of stairs from the first floor and the flight from the platform to the next floor. Depending on the plans of the building, the second set of stairs may be constructed from any one of the remaining sides of the platform.

Plumb bob. A blumb bob is a machined metal weight that is suspended from a line and is used to indicate a true vertical position. The point of the plumb bob always indicates the lower position of the point from which it is suspended. In other words, it indicates a point that is directly below the point of suspension.

Pitch board. Board sawed to the exact shape formed by the stair tread, riser, and slope of the stairs and used to lay out the carriage and stringers.

Ralling. Protective bar placed at a convenient distance above the stairs for a handhold.

Rise

1. Distance from floor to floor.

2. Vertical distance from the top of a tread to the top of the next higher tread.

3. Perpendicular height of a step or flight of stairs.

4. Vertical distance through which anything rises, as the rise of a stair.

Riser

1. Each of the vertical boards closing the spaces between the treads of stairways.

2. Finished piece of material placed vertically between two treads to close the space and form a suitable finish. Risers are often omitted in the construction of basement or rough stairs.

3. Vertical face of a step. The height is generally taken as the vertical distance between treads.

4. Upright member of a step situated at the back of a lower tread and near the leading edge of the next higher tread.

5. Vertical board between two treads of a flight of stairs.

Rough stairways. Stairways used for access to basements or cellars and usually constructed at a steeper angle than are major stairs between levels. To give the effect of added tread width, the riser material is often omitted in construction; that is, the vertical surface at the back of each tread is open. Treads usually are surfaced on one side only, and the front edge is smoothed.

Run

1. Total length of stairs in a horizontal plane, including landings.

2. Net width of a step or the horizontal distance covered by a flight of stairs.

Sawed-out stair horse. Stair horse stringers are made by sawing out the steps from a single piece of framing lumber. Standard lumber sizes are used for stair horse stringers.

Square-cut stringer. Square-cut stringers are not recommended for finished construction except on an open stairway in an inconspicuous location. The stringer is cut to exactly the same shape as a rough horse stringer, but with greater accuracy. To simplify the finish work, the risers are allowed to project about $\frac{1}{4}$ in. beyond the face of the stringer, and the treads about 1 in. Plain and square-cut stringers are used in conjunction with stair horse stringers.

Stair. Series of steps constructed to provide access between different levels within a building.

Stair carriage. Supporting member for stair treads, usually a 2-in. plane, notched to receive the treads; it is sometimes called a "rough horse stringer."

Staircase. Entire stair structure, including stair stringers, balusters, railing or handrail assembly, moldings, brackets, and panels.

Stair gauges. Gauges that provide the carpenter with a means of reproducing exactly, any number of times, the rise and run measurements for the stair layout. One gauge set to the riser measurement is clamped to the tongue of the framing square, and another gauge set to the run measurement is clamped to the blade. To ensure accurate and repeatable measurements, the gauges must be correctly and securely positioned on the framing square.

Stair head. landing at the top of a flight of stairs.

Stair horse

1. Basic supporting member required in the construction of wood stairs. In general, a stair horse is a stair stringer and is made of 2-in.-thick framing lumber, which is cut in a series of steps on which the finished risers and treads are nailed. The lumber must be of sufficient thickness and width to give adequate support to the entire stairway structure after the steps are cut. A stairway of average width requires two stair horses, while wider stairs require three or even four horse stringers. There are two commonly used types of stair horses—sawed out and built up.

2. Piece of rough lumber with portions cut away to form steps on which the treads and risers are nailed. Also called a "stair stringer."

Stair horse or stringer length. The mathematical length of a horse or stringer is the diagonal of the total rise and total run measurements of the stairs.

Stair railing or balustrade. A stair railing is a complete assembly consisting of balusters, newel posts, handrail, buttress cap, and fillets.

Stair rise. Vertical distance from the top of one stair tread to the top of the next one above.

Stairs

1. Series of steps for passing from one level to another.
2. Series of steps leading from one level or floor to another, or leading to platforms.

Stair stringer

1. A stair stinger is similar to a stair horse except that it is made from finish stock. There are a number of ways in which the stringer can be cut.
2. Support on which the stair treads rest.

Stairway. Way up or down a staircase.

Stair work (finished carpentry). Stair work usually includes the following: structural wood framing, finish treads, risers and nosings, metal handrail brackets, and wood handrails.

Straight stairway. Single set or flight of stairs from a lower level to an upper level.

Stringers. Inclined members along the sides of a stairway. A stringer along a wall is called a "wall stringer." Open stringers are those cut to follow the lines of risers and treads. Closed stringers have parallel tops and bottoms, and treads and risers are supported along their sides or mortised into them. In wood stairs stringers are placed outside the carriage to provide a finish:

String board. Board placed next to the well hole in wood stairs, terminating the ends of the steps. The string board is the piece put under the treads and risers as a support, and forming the support of the stair.

Total rise. The total rise is the vertical distance from finish floor to finish floor and is the basic measurement for all stair layout. (The term "finish floor" means the hardwood flooring material that is nailed over the subfloor.)

Total run. The overall horizontal distance occupied by a stairway is called the "total run." The measurement is obtained by multiplying the tread run dimensions by the number of treads. The total run measurement is used to locate the exact position of the bottom or first riser. A plumb bob is dropped from the stair end of the well hole to the floor below, and the total run is measured from the plumb bob mark. The measurement locates the first riser.

Tread

1. Part of a stairway on which a person steps or "treads." Each tread is usually made of hardwood or material to match the finish flooring, unless it is totally carpeted.
2. Horizontal member of a step.
3. Horizontal surface of a step. The width is usually taken as the horizontal distance between risers.

Tread and riser ratio. Most building codes regulate the height of risers and width of tread. They provide the formula that is based on the sum of 2 risers and a tread, exclusive of projection of nosing, that is not to be less than 24 in. and not more than 25 in. The height of the riser is limited to $7\frac{3}{4}$ in. and the tread to a minimum of 9 in. exclusive of nosing. Verify the formula in the local code.

Tread projection. Portion of the tread that extends beyond the face of the riser. A cove molding is usually placed in the angle formed by the tread projection and the riser to trim the tread and provide a tight joint. The tread projection is usually no more than $1\frac{1}{4}$ in. wide.

Tread rise. Vertical distance from the top of one tread to the top of the next tread. A stairway is mathematically correct only when any one rise dimension is exactly equal to any other rise dimension.

Tread run

1. Horizontal distance from the face of one riser to the face of the next riser, that is, the distance as measured from the vertical riser cut on a stair horse stringer to the next vertical cut.
2. Horizontal distance from the leading edge of a tread to the leading edge of an adjacent tread.

Tread with. Horizontal distance from front to back of tread, including nosing when used.

Wedges. Wedges are preferably made of hardwood and are cut to the exact taper of the dadoes in a housed stringer. The tread wedge is larger than the riser wedge so that it will fit the larger dado provided for the tread.

Well hole. Framed opening in the floor construction of a building through which a stairway passes. The opening is laid out to dimensions somewhat larger than the dimensions of the finish staircase to allow for the fitting and nailing of interior finish lumber around the edges of the well hole. The well hole is guarded with balusters and a handrail if it is an open stairway.

Winders. Steps with tapered treads in sharply curved stairs.

Winder treads. Treads tapered so that a 90° turn in the direction of the stairway can be made without using a landing.

Wood stairways. Building codes regulate the use of wood stairways for residences and most buildings that are not over two stories in height. Residences usually have two stairways, front hall, rear, or stair to the basement. Regardless of use or location, each stairway should be constructed of lumber having qualities for wear resistance to treads, risers, and railings. Lumber should be hard and tough, free from warp or wind, and of pleasing-grained wood for natural finish. Main stair or front hall stair lumber should include oak, birch, maple, walnut, beech, ash, and cherry. Rear or stair to basement could include such lumber as Douglas fir, southern yellow pine, gum, and sycamore.

WOOD TREATMENT

Acid copper chromate (celcure). ACC according to FS TT-W-546 and AWPA standard P5, contains 31.8% copper oxide and 68.2% chromic acid. Equivalent amounts of copper sulfate and potassium dichromate may be used in place of copper oxide (FS) Federal specification, (AWPA) American Wood-Preservers' Association.

Acid copper chromate (celcure). Waterborne preservative consisting of copper sulfate, sodium dichromate, and chromic acid. It is toxic to fungi and insects, paintable, clean, odorless, and corrosive to metal. Wood is usually air- or kiln-dried before acid copper chromate is used on it (ASTM D 1624 and ASTM D 1627).

Advanced decay. Causes decay, and is older stage of decay in which the destruction is readily recognized because the wood has become punky, soft and spongy, stringy, ring shaked, pitted, or crumbly. Decided discoloration or bleaching of the rotted wood is often apparent.

Ammoniacal copper arsenite. ACA according to FS TT-W-549 and AWPA standard P5, contains approximately 49.8% copper oxide or an equivalent amount of copper hydroxide, 50.2% of arsenic pentoxide or an equivalent amount of arsenic trioxide, and 1.7% of acetic acid. Preservative marketed under the name of Chemonite.

Ammoniacal copper arsenite. Waterborne preservative consisting of copper hydroxide, arsenic trioxide, and acetic acid. It is toxic to wood destroying organisms and insects, paintable, clean, odorless, and resistant to leaching and will not bleed through concrete, plaster, or paint. Wood is usually kiln- or air-dried before ammoniacal copper arsenite is used on it (ASTM D 1325).

Antifouling paints. Antifouling paints provide protection but are effective only as long as the paint film remains unbroken. The thorough treatment of the bare wood by brush with a solution of tributyltin oxide and thiodan before an antifouling paint is applied provides the wood with good protection even in case of paint damage.

Bacteria. Most wood that has been wet for any considerable length of time probably will contain bacteria. Mixtures of different types of bacteria in combination with fungi have been found to accelerate decay in some woods.

Beetles and boring insects

1. Bark beetles, ambrosia beetles, roundhead borers, and flathead borers all attack freshly cut logs and do considerable damage to healthy wood. To avoid damage, logs usually are cut at the dormant season (October, November); otherwise logs are dipped in a preservative solution of gamma isomer or benzene hexachloride and piled off ground to dry.

2. Bettles and boring insects infest logs and freshly cut lumber, particularly hardwoods. Softwood logs are attacked by sawflys. Dip treatments, poisons, and fumigation are all used to eradicate these harmful pests.

Bis-tri-*N*-butyltin oxide solution. Clear, odorless, colorless, stable, organotin chemical compound insoluble in water, but readily soluble in most organic solvents. This solution is highly effective in controlling fungi, bacteria, and marine organisms. It has a natural affinity for cellulosic materials, and when wood or fabrics are treated with it, its effectiveness is long lasting even under severe leaching conditions.

Boliden K-33. Boliden K-33 is toxic to fungi and insects, paintable, clean, odorless, and very resistant to leaching. It will not bleed through concrete, plaster, or paint and has good resistance to electrical conductivity. Wood is usually kiln- or air-dried before Boliden K-33 is used on it.

Boliden salts. Boliden salts are relatively resistant to leaching, toxic to fungi and insects, paintable, clean, and odorless. Wood is kiln- or air-dried before boliden salts are used on it.

Borax. Borax is usually effective on hardwood only.

Brown rot. Decay in which the attack concentrates on the cellulose and associated carbohydrates rather than on the lignin, producing a light- to dark brown friable residue, loosely termed "dry rot." The advanced stage in which the wood splits along rectangular planes and shrinks is termed "cubical rot."

Brushing and spraying. Brushing and spraying are usually employed for oil-borne preservative applications but can be used for waterborne preservatives. Penetrations at best will be shallow and retentions are slight. While these treatments do offer some protection to the wood, subsequent abrasions and exposure to the elements will quickly reduce their effectiveness.

Carpenter ants. Carpenter ants are destructive even though they are not attracted to wood for food purposes. They use wood only for shelter and usually prefer wood that is naturally soft or has been made soft by decay. They enter a building directly by crawling in or can be carried into the building in fuel wood.

Causes of wood deterioration. Decay, insects, marine borers, and strong alkaline liquids are all causes of wood deterioration.

Celcure (acid copper chromate). Celcure is toxic to decay and insects; paintable, clean, odorless, and corrosive to metal. It is necessary to air- or kiln-dry wood before celcure is used on it.

Chelura (marine borer). This marine borer is found only along the Atlantic coasts of North America and Europe.

Chemonite (ammonical copper arsenite). Chemonite is toxic to decay and insects; paintable, clean, odorless, and very resistant to leaching. It will not bleed through concrete, plaster, or paint. It is necessary to air- or kiln-dry wood before chemonite is used on it.

Chromated copper arsenate (Type I or Type A—greensalts or erdalith). Chromated copper arsenate, Type I or A consists of potassium dichromate, copper sulfate, and arsenic pentoxide (ASTM D 1625 and ASTM D 1628).

Chromated copper arsenate (Type II or Type B—Osmose K-33 or Boliden K-33). Chromated copper arsenate, Type II or B consists of arsenic acid, chromic acid, copper oxide and other inert ingredients. Osmose products are made by Osmose Wood Preserving Co. of America Inc. Buffalo, N.Y.

Chromated zinc arsenate. Preservative available in solid form or solution. The solid form contains specified amounts of arsenic acid, sodium arsenate, sodium dichromate, and zinc sulfate.

Chromated zinc chloride. Waterborne preservative consisting of zinc chloride and sodium dichromate. It is resistant to leaching, toxic to decay and insects, paintable, clean, and odorless. Wood is usually air- or kiln-dried before use of the material. It has good fire retardancy at high retentions, is corrosive to metal fastenings, and has high electrical conductivity and poor leaching resistance (ASTM D 1032 and ASTM D 1033). CZC according to FS TT-W-551 and AWPA standard P5 contains 80% of zinc oxide and 20% of chromium trioxide. Zinc chloride may be substituted for the zinc oxide and sodium dichromate for the chromium trioxide. Chromated zinc chloride (FR) is included as a fire retarding chemical, in AWPA standard P10.

Coal tar creosote (creosote oil)

1. Coal tar fraction, boiling between 240°C and 270°C. Crude creosote oil is used as a raw material for producing tar acids, etc., or is used direct as a germicide, insecticide, or disinfectant in various connections (ASTM D 390).

2. Distallate of coal tar produced entirely by high-temperature carbonization of bituminous coal. It is heavier than water and has a continuous boiling range of at least 125°C.

Control of mold, stains, and decay. Lumber is kiln-dried to 20% or less moisture content. Dip or spray treatment applied with antiseptic solutions (not preservative) will prevent fungus infection during air-drying.

Copper-8-quinolinolate (solubilized). Odorless, noncrystalline, oil-borne preservative that provides excellent decay resistance to wood. It is normally formulated in conjunction with water repellents to provide dimensional stability. Quinolinolate is an easy-to-handle liquid completely soluble in aliphatic petroleum oils (such as mineral spirits). This preservative finds its most common use in the treatment of wood that may come into intimate contact with foodstuffs, such as the flooring for reefers and refrigerated vehicles.

Copper napthenate

1. Copper napthenate lends itself well to applications in boats to prevent dry rot. While it is a very good preservative, its green color is difficult to cover with paint, the green bleeding through the paint.

2. Preservative highly toxic to fungi and insects. It imparts a green color to the wood, but the color may be somewhat darkened depending on the color of the petroleum used in the treating solution.

Creosote. Most widely used of all preservatives. Wood may be protected in severe exposure with high retentions and in less severe exposures with lower retentions. Creosote is black to very dark brown in color and is distilled from a variety of different tars. It is most commonly used in marine pilings, poles, and cross ties.

Creosote and solutions containing creosote. Creosote and its solutions are used where protection against decay and attack by termites and other wood-destroying organisms is of first importance, and where painting is not required and a slight odor is not objectionable. These mixtures are accepted preservatives for marine or saltwater installations and are ideally suited for pilings for shore installations.

Creosote-coal tar solutions. Mixtures of up to 50% coal tar with coal tar creosote, used primarily to reduce costs. This type of preservative decreases the tendency of the treated wood to check, but is apt to cause "bleeding" (ASTM D 391).

Creosote-petroleum solutions. The primary purpose of adding petroleum to creosote is to reduce the cost of the preservative. The surfaces of the treated wood are more oily and thus checking is reduced. Toxicity is greatly reduced and there is a greater tendency to "bleed." Penetration with this preservative is more difficult (ASTM D 1858).

Decay (rot)

1. Decomposition of wood substances by certain fungi.

2. The fungus organisms that change the physical and chemical properties of the cell walls, thus seriously affecting the strength of the wood, are called "wood-destroying" fungi. These fungi produce the condition in wood referred to as "decay" or "rot." Most of these fungi attack wood after the tree has died or been felled, but there are also a few fungi that attack a living tree after an injury or a weakening of its physical condition by insects.

Decay-causing fungi. Two groups of decay-causing fungi that thrive on wood cells are white rot and brown rot. White rot fungi remove more lignin (the substance that cements the cells of wood together) than cellulose (the cell wall itself) from the wood. Brown rot fungi remove more cellulose than lignin. These two fungi are characterized by the color of the remaining rotten residue.

Diffusion processes. In addition to the steeping processes, diffusion processes are used with green or wet wood. These processes employ waterborne preservatives that will diffuse out of the water of the treating solution or paste into the water of the wood. There are several diffusion methods of preserving wood, such as the Osmose process, the double-diffusion method, and the paste and bandage method.

Dip or soak (nonpressure treatment). This process consists simply of submerging the wood in the preservative and may be effectively used for treating sash and mill work with water-repellent preservatives. The term "dip" usually refers to immersion for periods up to about 15 minutes, "soak" to longer immersion.

Dipping

1. Method of application most generally used by the mill work industry. Mill work lends itself well to this type of treatment because conditions for decay are not generally severe where mill work is used. The wood is most often dipped in a combination of preservative and water-repellent solution to protect it from decay, as well as from dimensional changes.

2. Dipping is not recommended where the wood will be subjected to heavy wear or abrasion in service. As with brush and spray applications, the preservative is not readily retained in the wood, nor does it penetrate deeply. Dipping is used primarily where some protection is desired, a lot of wood is to be treated, and pressure treating is not economically available. Thoroughly dry wood is usually immersed in a bath of oil-borne preservative for a period of from 3 to 15 minutes. Can be expected to add several years of service life to wood not used under severe decay-causing conditions.

Disadvantages of wood and protection. Wood rots or is destroyed by insects and fire. Wood can be processed to meet any situation where it may be required; it can be made to resist decay, insects, fire, and dimensional changes indefinitely if treated properly.

Double-diffusion method. Method that depends on diffusion for penetration. Green or water-saturated wood is soaked in a solution of one chemical for a period of time until the chemical penetrates into the wood. Immediately thereafter, this pretreated wood is soaked in another chemical. After penetration of the second chemical, the two chemicals react with one another inside the wood to form a toxic, insoluble precipitate with a high resistance to leaching.

Dowicide G. Waterborne treatment for lumber and logs. This treatment is not recommended for hand-dipping unless special precautions are taken to protect workers' hands.

Dowicide H.   Waterborne treatment for lumber and logs. In the South it is recommended only for hardwoods.

Drying.   High moisture content may be permitted in wood for treatment with waterborne preservatives by certain diffusion methods; however, for treatment under other methods, drying is essential. Drying the material permits adequate penetration and distribution of the preservative and reduces the risk of checking.

Dry rot.   Term applied to any dry, crumbly rot, especially to that which in an advanced stage permits the wood to be crushed easily to a dry powder. The term, however, is a misnomer, since all wood-rotting fungi require considerable moisture for growth.

Empty-cell pressure process

1. Method used when deeper penetrations with relatively low net retentions of preservative is desired. It is generally used for impregnation of oil-borne preservatives when it is desired to increase the penetration while conserving the preservative.

2. This process leaves the cell cavities empty of solution, but to the extent of the penetration, the cell wall material is effectively treated. Retention is much lower than with the full-cell process. This process is used with solutions of pentachlorophenol or copper napthenate in petroleum oil.

3. In the two empty-cell processes, Lowry and Rueping, the timber is subjected to atmospheric or high initial air pressures prior to introducing and injecting the preservative under somewhat greater pressures. When pressure is released, the expanding entrapped air expels the excess preservative from the wood cells, thus reducing the net retention of preservative. The method gives deeper and more uniform penetrations than are possible with the same absorptions injected by the full-cell process. The empty-cell process should be used for all oil treatments where the stipulated retentions can be obtained by this method.

Environmental factors necessary for decay.   Oxygen (air), moisture (water), favorable temperatures, and a food supply (wood) are all absolutely necessary for decay to take place. Fungi cannot survive if any one of these factors is eliminated. Even the so-called dry rot fungi needs a source of moisture. In this case the fungus attacks dry wood by maintaining contact with a remote moisture supply by means of tiny hairlike growths called "hyphae."

Erdalith or greensalts.   Chemicals that are toxic to fungi and insects; paintable, clean, odorless, and very resistant to leaching. It is necessary for the wood to be air- or kiln-dried before erdalith or greensalts is used on it.

Factors contributing to decay.   Factors contributing to decay include the amount of sapwood present, (the more sapwood the greater the risk of decay), temperature (moderate temperatures favor most fungus organisms), a moisture content above the fiber saturation point, design practices (poor air circulation causes condensation), the presence of fungus spores capable of attacking the species used (many fungi cannot attack all species—one species might show good durability in one area of the country and poor durability in another), the absence of light, and contact with soil, water, or concrete.

Factors influencing decay (untreated wood).   The amount and kind of chemical extractives, the amount of heartwood present, and a moisture content below 20% are factors that influence decay.

Fire-retardant chemicals.   Waterborne fire-retardant chemicals usually are used where clean, odorless, paintable lumber is required and where leaching is not a problem. Chemical combinations in water solution provide recognized fire resistance for wood, and some of them provide protection against decay and insect attack. Many building codes recognize the advantages of these proprietary formulas and various retentions have passed ignition, combustion, flame spread, and glow tests, resulting in rate reductions by insurance underwriters.

Fluor-chrome-arsenate-phenol (Type I or Type A—Wolman Salts or Tanalith).   Type I or Type A of this chemical consists of sodium fluoride, sodium arsenate, sodium chromate, and dinitrophenol (ASTM D 1034 and ASTM D 1035).

Fluor-chrome-arsenate-phenol (Type II or Type B—Osmosalts).   Type II or Type B of this chemical consists of sodium fluoride, sodium arsenate, sodium dichromate, and dinitrophenol. Osmosalts is a product of Osmose Wood Preserving Co. of America; Buffalo, N.Y.

Fluor chrome arsenate phenol.   FCAP according to FS TT-W-535 and AWPA standard P5, includes FCAP type I Wolman Salts and FCAP type II Osmosalts.

Full-cell pressure process

1. Method used when the retention of a maximum quantity of preservative is desired. It is the standard method of pressure-treating marine timbers with creosote and for pressure-treating with all waterborne preservatives and fire-retardant chemicals. In applying waterborne preservatives by this process, the preservative retention can be accurately controlled by regulating the concentration of the treating solution and calculating in advance the exact amount of preservative solution that will be pumped into the wood in order to attain a specified retention.

2. When high retentions are required for creosote or creosote mixtures, such as in marine installation, the full-cell process is used. The preservative is not only absorbed by the cell walls, but the cell spaces also are left full of preservative. The process leaves the maximum amount of preservative in the wood, but when used with creosote or its solutions, it will not leave as clean a surface as the other processes. The full-cell process is also used for treating with waterborne preservatives and fire-retarding chemicals.

Full-cell-process creosote treatment.   This process produces a product that is not paintable, exudes oil when heated, and is dirty. It is, however, the most durable treatment and offers protection against weathering, moisture penetration, and decay.

Fungi.   Decay is caused by microorganisms. These are fungi of the wood-destroying type, a form of parasitic plant life to which unprotected wood is the host and the food supply. Wood-destroying fungi feed on cellulose or lignin, or both. Brown rot fungi feed on cellulose, leaving the brown lignin as residue, while the white rot fungi feed on the lignin.

Fungicide.   Chemical that is poisonous to fungi.

Hot and cold bath.   Both oil-borne and waterborne preservatives can be applied by this method, but oil-borne preservatives are used for the bulk of the wood treated in this manner. Only seasoned wood can be treated. The wood is immersed for hours, first in a hot and then in a relatively cool preservative solution in an open tank. This method of treatment finds its most important use in the butt treatment of western red cedar poles.

Hot-and-cold soak processes.   These processes are used for poles and pilings of certain hard-to-penetrate species where usually pressure treatments are effective, but slower. Only sapwood can be penetrated. Southern pine with a thick sapwood band, treats very well by pressure treatment as compared to hot and cold soaking, boiling in oil, or vacuum treatment. Douglas fir and cedar, with narrow sapwood, are more difficult to penetrate.

Incising. Wood that is resistant to penetration by preservatives is often incised before treatment to permit deeper and more uniform penetration. To accomplish this, sawed or hewed timbers are passed through rollers equipped with teeth that sink into the wood to a predetermined depth, usually $\frac{1}{2}$ to $\frac{3}{4}$ in. Incising is practiced chiefly on Douglas fir, western hemlock, western larch ties, timbers for pressure treatment, and on poles of cedar and Douglas fir.

Insects. Wood-destroying insects are divided into four distinct classes: beetles, termites, carpenter ants, and marine borers.

Lignasan

1. Lignasan is a combination of water and Dowicide G, or Santobrite and water, plus borax for lumber. It is particularly effective under severe conditions.

2. Waterborne treatment for lumber and logs. In bulked pine and in situations where drying is particularly slow, this treatment may allow some molding, but it is effective under severe conditions.

Limnoria, Sphaeroma, and Chelura. These borers do not imprison themselves in the wood they attack. Their burrows are much smaller than those of the molluscan borers, and their galleries are seldom deep into the wood. Limnoria is present almost worldwide. The damage done by the crustacean borers is less spectacular than that of the molluscan borers; crustacean borers usually take at least a year to accomplish the damage done by molluscan borers in a few months.

Liquid creosote, anthracene oil, and carbolineums. These substances are distillates of coal tar and have higher specific gravities and a higher boiling range than ordinary coal tar creosote. The chemical compounds that crystallize at ordinary temperatures are removed to leave a completely liquid oil with a reduction in evaporation when heated. Creosote oils are used primarily in open tanks for dipping, brush-on, or spray treatments.

Lowery pressure process. This process is virtually the same as the Rueping process except that the initial air pressure involved is not above atmospheric pressure. The Rueping tank is not used.

Marine borers (mollusks and crustaceans)

1. Group of wood-boring marine animals that are found in brackish and salt water. They attack any wood between the water line and the mud line. They are divided into two groups: molluscan borers, related to oysters and clams and represented by the shipworms (Teredo, Bankia, and Martesia); and crustacean borers, related to lobsters and crabs and represented by Limnoria, Sphaeroma, and Chelura. No wood is known to be naturally immune to destruction by these animals.

2. Marine borers attack wood in salt water. The only real protection lies in heavy treatment with creosote. Boats are best protected by a well-maintained antifouling paint and copper-containing preservatives. Marine piling can be ably treated against marine borer attack.

Martesia (marine borer). This species is found only along the shores of the Gulf of Mexico. It differs from the shipworm in that its entire body is encased within a bivalve shell, but it too enters the wood and becomes imprisoned in it like the shipworm.

Merulius lacrymans. Fungus that commonly causes dry rot.

Methods of nonpressure process applications. Brushing and spraying the preservative on the wood, dipping, steeping, the hot and cold bath method, diffusion processes, pole spraying, and the vacuum process are all nonpressure methods of preserving wood.

Methods of preservative application. There are two general methods of applying wood preservatives: pressure processes and nonpressure processes.

Moisture content. Amount by weight of water in wood computed as a percentage of the oven-dry weight of the wood.

Molds. Molds are evidenced by a cottony growth on the surfaces of wood . They range in color from white to black and appear when an abundance of moisture is present. When the wood is dry, they can be brushed or dressed off and never seriously affect the strength.

Molds and stains. Molds and stains do not ordinarily significantly affect the strength of wood as they feed on the contents of the cell rather than the cell itself. Fungus organisms generally are unable to attack wood below fiber saturation (27 to 32% moisture content).

Naphthenate solutions. Two naphthenates are commonly used as wood preservatives: copper naphthenate and zinc naphthenate. These preservatives are single, metal-organic, noncrystalline compounds soluble in oil. They are not readily available for application in a pressure plant, but are available in most areas for brush, spray, or dip application. To be fully effective, most specifications call for oil solutions with at least a 2% metallic (copper or zinc) content.

Natural durability. The heartwood of some species, such as cedar, redwood, cypress, and locust is naturally durable. Sapwood is not durable but receives treatment well, and species with wide sapwood rings can be treated extremely well. Southern pine is such a species, as are many hardwoods. Timbers are often incised to make them more penetrable for preservatives.

Natural vacuum (nonpressure treatment). This process is a modified version of the hot and cold bath, used for treatment with waterborne salts.

Nonpressure treating methods. Preservatives most commonly used for nonpressure treatment are pentachlorophenol and copper napthenate. Volatile petroleum solvents (preferably a mineral spirits type when paintable surfaces are desired) containing 5% pentachlorophenol or an amount of copper napthenate equivalent to at least 1% metallic copper (preferably 2% copper for brush or short-period dips) are commonly used.

Nonpressure treatment. Solutions of toxic chemicals in oil or water are often applied by dipping, soaking, brush, or spray treatments. This type of treatment is useful primarily against wood-staining fungi and to protect lumber during distribution.

Nonsubterranean termites. This species is less prevalent than the subterranean kind and is harder to locate and control. Nonsubterranean termites live in the wood which they infect but do not need contact with outside moisture. These termites are found, so far, only in the extreme southern and southwestern parts of the United States.

Noxtane. Waterborne treatment used for lumber only. Control may be somewhat inconsistent where drying is particularly slow.

Oil-borne preservatives. Oil-borne preservatives are used in any type of installation, except those in contact with salt water; the wood is clean, odorless, and paintable after the petroleum oil solvents in the treating solutions have been largely removed, either through air-seasoning or by some other acceptable methods.

Oil-borne treatments. Oil-borne treatments containing 5% penta or 0.75% copper napthenate in petroleum oil are relatively clean treatments, preferred to the creosotes where contact with the public or odor problems are likely to prevail. These treatments are applied with the empty-cell process and are paintable.

Osmosel diffusion process. With this method of preservative application only green or water-saturated wood is instantly immersed in a vat containing a water suspension of the preservative.

Paste and bandage diffusion method. This preservative treatment is perhaps better known as the "ground-line treatment." It has become a standard maintenance treatment for ground-line areas of in-service utility poles and railroad trestle bents.

Peeling. Peeling the bark from round or slabbed wood products is a necessary requirement to enable the wood to dry quickly enough to avoid decay and insect damage and to permit the preservative to penetrate satisfactorily.

Penetration. Complete penetration of sapwood is the goal. The heartwood is more difficult to treat. Some lumber contains no sapwood and penetration may not be very deep; so incising the timber with knives is often a practical way to improve penetration.

Pentachlorophenol

1. Penta is highly toxic to both fungi and insects; it is insoluble in water and thus permanent; and it is the most widely used oil-borne preservative. Depending on the solvent, the treated lumber varies in color from dark brown to colorless.

2. This preservative is a single organic crystalline compound soluble in oil. The term is generally applied to its solutions in petroleum oil. It can be dissolved in light and heavy oils, but in any case it is usually used in a 5% solution by weight (ASTM D 1272 and ASTM D 1274).

3. This preservative is toxic to fungi and insects and resistant to leaching, and has good penetration characteristics. It lends itself well to blending with water repellents for dimensional stability. It has a persistent odor, is cleaner than creosote, but is more difficult than waterborne salts to handle. It is a fire hazard, difficult to paint over, and toxic to humans and plants. The preservative "blooms" on the surface of wood.

Permatox 10s. Waterborne treatment used on lumber and logs. It is not recommended for hand-dipping unless special precautions are taken to protect workers' hands.

Pholads. Group of wood-boring mollusks that resemble clams and are not included in the shipworm classification. The Martesia are the best known species, but a second group is the Xylophaga.

Pole spraying. Special method of applying pentachlorophenol to in-service, above-ground portions of standing utility poles and wood railroad trestles. Like the ground-line treatment, it is a maintenance treatment. After a period of time, the original pressure treatment begins to deteriorate and the wood again becomes susceptible to decay and insect attack. When it is thoroughly dry, the Penta is sprayed onto the surface by means of special applicators and allowed to soak deep into the weathered, dry wood to saturation.

Powder post beetle. Beetle found in all parts of the United States that attacks both hardwoods and softwoods. Of the wood-destroying beetles, it does the most damage. Eggs are laid in the pores of wood from where the larvae burrow, leaving tunnels throughout the interior. Lumber can be protected from beetle attack by pressure treatment. Wood already infected and in use can be protected from further damage by a liberal brush treatment with a suitable preservative, such as a 5% solution of pentachlorophenol.

Preservative. Substance that for a reasonable length of time will prevent the action of wood-destroying fungi, borers of various kinds, and similar destructive agents when the wood has been properly coated or impregnated with it.

Preservative paints and finishes. Protective coatings that temporarily protect the wood from moisture and other weathering agents that might promote conditions for decay and mechanical damage. Such coatings eventually lose their protective ability and flake off. Should decay fungi exist within the wood before such finishes are applied, paint applications are ineffective in killing the decay or preventing its spread.

Preservatives—pressure treatment. Standard wood preservatives are divided into three classifications: creosote, oil-borne preservatives, and waterborne preservatives.

Pressure process equipment. A pressure plant consists of pumps, valves, storage tanks, controls, regulating equipment, a boiler, and a large horizontal cylindrical tank called a "retort." The retort is where the wood is conditioned and treated, and its size determines the capacity of the plant with regard to both the size of the material that can be treated and the total volume that can be treated per day or year.

Pressure treating methods. All pressure preserving processes may be classified as full-cell or empty-cell processes. Standard pressure processing provides the most dependable means of ensuring uniform penetration and distribution of preservative.

Pressure treatments. Full-cell and empty-cell processes. The full-cell process leaves the cell cavities in the portion of the wood that has been penetrated, essentially filled with treating solution. It is the process that provides the greatest retention of chemicals and is used for treating marine piling with creosote. Waterborne preservatives are also applied by this process. Water then is evaporated off in dry kilns or by air-drying in service to leave a heavy concentration of toxic chemical. Waterborne chemicals are used for mine timbers, structural lumber, plywood for dry use, and roof decking where condensation may be a problem.

Quality treated wood. Specifications and standards covering the treatment of woods are available to the public such as FS TT-W-571, "Wood Preservation-treating Practices," FS TT-W-572, "Wood Preservation–Water Repellent." Official quality control standards of the American Wood Preservers' Bureau, and the inspection of material for conformity to the minimum requirements in accordance with the American Wood Preservers' Standard M2, "Standard for Inspection of Treated Timber Products."

Rueping pressure process. In this process air is compressed into the retort and the wood cells. The preservative is then either forced into the retort while maintaining a constant air pressure of up to 100 psi in the retort or by means of a tank mounted over the retort, called a "Rueping tank," where the preservative is held under the same air pressure that is maintained in the retort the preservative is allowed to flow by gravity down into the retort replacing the air in the retort in the process. In either case, when the retort is filled with the preservative, additional pressure is applied which forces the preservative into the wood cells against the entrapped air. The air in the wood cells is compressed so that when the pressure is finally released, some of the preservative is forced out of the wood and is recovered. An excess of preservative is orginally forced into the wood so that after recovery is made, the final desired retention is left in the wood.

Santobrite. Waterborne treatment for lumber and logs. It is not recommended for hand-dipping unless special precautions are taken to protect workers' hands.

Shipworms. The most destructive of the marine borers. The group includes several species of Teredo and several species of Bankia, which are especially damaging. The animal grows within the wood and remains a prisoner in its burrow. It lives on the wood borings and the organic matter extracted from the sea water.

Soil poisoning. Effective chemicals are available for this use, but special knowledge and care is usually required for effective application. The normal life of this type of treatment is about five years.

Solvent recovery. Pressure treatment variation whereby the solvent is removed after conventional pressure treatment with oil-borne preservatives. The major field of application is in above-ground exposures where exceptional cleanliness and paintability are desired.

Stains. There are two kinds of stains: those caused by fungi and those caused by chemical changes in the materials infiltrated into the wood cells. The latter have no effect on the strength properties, but certain stains caused by fungi do slightly affect the strength. Blue stain is one of these, but the strength is not affected enough to make the wood unfit for ordinary commercial uses. Unlike the molds, stains go deep into the wood and cannot be dressed off.

Steeping preservative. The wood, green or dry, is allowed to soak for several days or even weeks in a waterborne preservative solution. During this period the preservative "salt" is absorbed by the wood, and if this period were extended long enough, the absorption and penetration could conceivably be equal to a pressure treatment. Absorption takes place by diffusion over a period of a week to 10 days. Penetrations will be much greater in sapwood than in heartwood.

Subterranean termite. Termites live in the ground and, if they find it necessary, build mud shelter tubes up over foundation walls to reach wood needed for food. Termites live in colonies and infect wood in basementless buildings that are poorly drained and ventilated. Wood that has less than 20% moisture content is relatively free from fungus attack and decay but is highly subject to termite attack, since the termites obtain their necessary moisture from the ground.

Superficial applications. The simplest application of a preservative such as creosote or other oils to woods is to brush it on or with spray equipment. Surface applications of this type require that the oil be flooded on the wood so that the greatest amount of penetration occurs.

Teredo and Bankia (shipworms). The larvae of these animals are free swimming. They attach themselves to wood below the water line and bore into it, making minute holes in the exterior of the wood. As the animal grows, it enlarges its burrow to accommodate its wormlike body and becomes entrapped in the wood for the rest of its life. In less than a year, a timber can be completely honeycombed and its structural strength greatly reduced. Shipworms are found in nearly all the coastal waters of the United States.

Thermal treatment. The hot and cold bath is referred to commercially as "thermal treatment." Coal-tar creosote or pentachlorophenol in heavy petroleum oil is also an effective nonpressure process.

Vacuum process (nonpressure treatment)

1. This process uses a preliminary vacuum, but the preservative is injected under atmospheric pressure. The vacuum process is effective for the application of water-repellent preservative solutions to lumber subjected to relatively low decay hazard and insect attack. This process has been used to treat millwork with water-repellent preservatives and construction lumber with waterborne and water-repellent preservatives. Commercial Standard CS-262.

2. Process that employs the use of a closed vessel. The wood is put into the vessel and placed under vacuum, after which the preservative is introduced and fills the vessel. The vessel holding the wood, having had some of the air removed from them, sucks in the preservative. The excess preservative is then pumped back out to a storage tank, and a second vacuum is drawn, which removes the surplus preservative and dries off the wood.

3. Mild pressure process consisting of evacuating the chamber containing the treating charge and then admitting the toxic chemical liquid at atmospheric pressure when a treating pressure of about 15 psi is achieved. This process is better than dipping and brushing; however it does not produce the results of pressure treatment with cold or hot solutions.

Waterborne preservative

1. Waterborne preservatives are used principally in the treatment of wood for building construction where the treated wood must have no odor, clean handling characteristics, and be paintable. To be effective, waterborne preservatives must be applied by pressure methods. The leaching resistance to some of these preservatives has been developed to the extent that excellent performance can be expected in ground contact or in other wet installations. After treatment, since water is added to the wood, it must be dried to the moisture content required for use.

2. Waterborne preservatives are used for installations off the ground and to a limited extent in ground contact where clean, odorless, paintable wood is required. They are especially adapted to interiors or other places where the wood is not subject to leaching by water, but where it is subjected to decay or exposed to termites. When lumber treated with waterborne preservatives is used in places where shrinkage after placement in the structure would be undesirable, it should be kiln- or air-dried to the proper moisture content after treatment.

Waterborne salts. Waterborne salts are excellent for dry use and also paintable. The use of waterborne salts on dry laminated timber or on dry-use adhesives is not recommended because of the possible damage to the product by swelling. Water-borne salts in general are not likely to produce satisfactory results unless applied before laminating.

Water-gas-tar creosote. Creosote is a distillate of water-gas-tar with petroleum oil. The composition will vary with the character of the petroleum used. This type of cresote differs from coal tar creosote in that there are practically no tar acids or tar bases. Like coal tar creosote, the toxicity varies with the proportion of distillate obtained below 275°C. Water-gas-tar cresote is generally less toxic than coal tar creosote.

Water-repellent preservatives. Preservative systems containing water-repellent components are sold under various trade names, principally for the dip, or equivalent treatment of window sash and other types of millwork. Specifications for the chemicals to be used are covered in FS TT-W-572.

Wolman salts. Wolman salts are toxic to fungi and insects, paintable, clean, odorless, somewhat fire retardant, and noncorrosive to metals. They are subject to moderate leaching under extreme conditions. Wood should be air- or kiln-dried before Woman salts (a registered trademark) are used on it.

Wood. The wood of all species is composed of two distinct parts with relation to where in the tree stem it originates. The drier and more dense center portion of the stem is called "heartwood," while the surrounding moister wood encircling the heartwood is called "sapwood." It is in the sapwood that the liquids and the dissolved foods and extractives are present in sufficient quantities to enable the species to resist decay and insect attack. The extractives act as a natural wood preservative.

Wood-destroying fungi

1. Reproduction of wood-destroying fungi, as well as mold and stain fungi, is by means of microscopic cells called "spores." Spores are produced in fruiting bodies (bracketlike, shelflike toadstools, growths on the exterior for rotting wood, etc.) and are released into the air everywhere. The wind carries them to new wood, and when conditions are right, they germinate and infect the new wood.

2. The fungus organisms that change the physical and chemical properties of the cell walls thus seriously affecting the strength of the wood are called "wood-destroying" fungi. These fungi produce the condition in wood referred to as decay or rot. Most of these fungi attack wood after the tree has died or been felled, but there are also a few fungi that attack a living tree after an injury or a weakening of its physical condition by insects.

Wood-destroying organisms. Three general classes of organisms attack wood and lower its quality: molds and stains, decay fungi, and insects and marine borers.

Wood fungi. Microscopic plants that live in damp wood and cause mold, stain, and decay.

Wood preservatives. Wood preservatives must have the ability to do three things: penetrate the wood, poison the food supply within the wood on which decay fungi and wood-destroying insects live, and be present in sufficient quantities so that its protection outlasts the useful life of the wood product. Preservatives fall into two general classes: oil-borne and waterborne preservatives.

Wood-staining fungi. Fungi that discolor the sapwood and cause moldy growths, but do not harm the wood structurally. The fungi feed on the nutrients in the liquid sap rather than on the cellulse and lignin. Conditions that permit one type of fungi to flourish are not favorable to others.

Wood tar creosote. Creosote made by distilling the wood tar that results as a by-product in the destructive distillation of hardwoods or softwoods. As with the other types of creosote, any refinement of the oil will affect the quality of the creosote. This type of creosote seldom is used because it is not available in large enough quantities. Hardwood tar creosote is highly acidic and will corrode iron and steel.

Working stresses. The same allowable stresses are used for treated wood as for untreated wood. Standard specifications on preservative treatment set limitations on pressures and temperatures in order to prevent possible injury to the wood.

Zinc naphthenate. Zinc naphthenate is easy to cover with paint, but it is not nearly as effective a wood preservative as copper naphthenate. In order to get paintability, toxicity must be sacrificed.

DIVISION

7

Thermal and Moisture Protection

ASBESTOS CEMENT CORRUGATED SHEETS

Accessories. The following accessories are used in conjunction with the installation of corrugated sheets: ridge rolls, battens, corner angles or corner rolls, louver blades, filler or closure strips, fasteners, plastic lap cement, and caulking or sealant.

Color. Natural cement gray or color as specified.

Corner angles or corner rolls. Outside or inside types, with or without overlapping bell ends, of asbestos cement.

Corrugated sheets. Sheets and accessory shapes, except for filler strips, are composed of a combination of asbestos fiber and portland cement or portland blast-furnace slag cement, and not more than 1% by weight of organic fiber, with or without the addition of inert mineral pigments, mineral fillers, coatings, or a curing agent, and are formed under pressure and cured to meet the physical requirements of the standards (ASTM C 221, C 746, and FS SSB-750).

Design. Both types of asbestos cement corrugated sheets are produced in 4.2-in. pitches and 42-in. widths (10 complete corrugations).

Efflorescence. Although efflorescence may appear on asbestos cements sheets, it is not a defect and does not result in a permanent change in color.

Fasteners. Fasteners are usually nonstaining, noncorrosive nails, screws, or lead-head bolts of suitable sizes and types depending on construction of the structure. Lead or neoprenecupped washers are normally used with all types of fasteners to produce a tight weatherproof installation.

Filler or closure strips. Filler or closure strips are of various types for closing flat surfaces or straight edges to corrugations.

Finish. Exposed surface of the sheet is relatively smooth, and requires no paint or preservative.

Fire-resistive materials. Sheets and accessory parts manufactured of portland cement, silica, and asbestos fibers. The material does not burn, contribute to combustion, or produce toxic smoke or fumes. It possesses exceptional resistance to alkalies, smoke, and chemical vapors.

Flexural strength. Average breaking load, in pounds, of dried test specimens, loaded as simple beams, with the load applied at the center and tested by a method in the standard (ASTM C 221).

Lightweight sheet. Sheet approximately $\frac{3}{16}$ in. thick, weighing approximately $1\frac{1}{4}$ to 2 lb/sq. ft, and designed as a roofing and siding sheet for use over sheathing or for direct application on girts and purlins.

Ridge roll. Ridge rolls are manufactured in two types: half-round sections of asbestos cement with battens furnished to fit the ridge roll and two-piece adjustable-pitch asbestos cement of several designs.

Standard sheet. Sheet approximately $\frac{3}{8}$ in. thick, weighing approximately $3\frac{3}{4}$ to 4 lb/sq. ft, designed as a structural roofing and siding sheet, and suitable for use over sheathing or for direct application on girts and purlins.

Thickness. Thickness is defined as the face-to-face or overall thickness between two planes contacting the crests of corrugations on opposite sides of the sheet, and the thickness of the section of the sheet across corrugations, as determined by ASTM C 221.

Types. Sheets are manufactured in two types, standard and utility. The standard sheet weighs 4 lb/sq. ft, and the utility sheet weighs 2 lb/sq. ft. Sheets come in 42-in. widths, and in lengths from 6 in. to 12 ft in multiples of 6 in.

Water absorption. The average of the dried test specimens from a selected lot being tested should not exceed 25 weight percent for all types when calculated with the water absorption test in ASTM C 221.

Workmanship. Exposed surface of the sheet should be free of defects that impair appearance or serviceability, and be smooth or factory textured.

ASBESTOS CEMENT ROOFING SHINGLES

American method. With the American method, shingles are of uniform thickness, generally rectangular in shape, with straight or irregular edges. They are intended to be laid with a minimum 2-in. head lap and with no side lap and to provide double coverage. Shingles are made either as individual or multiple units to simulate the appearance of individual shingles.

Asbestos cement roofing shingles. Shingles are composed of a combination of asbestos fiber and portland cement or portland blast-furnace slag cement, and not more than 1% by weight of organic fiber, with or without the addition of curing agents, water-repellent substances, mineral fillers, coatings, pigments, or mineral granules, and are formed under pressure and cured (ASTM C 222).

Butt.   Overexposed edge of a shingle.

Class A (UL).   Asbestos cement shingles. Roof covering composed of asbestos cement shingles laid to provide two or more thicknesses over a layer of 25 lb asphalt-saturated asbestos felt installed in accordance with manufacturers instructions. Limited to roof decks capable of receiving and retaining nails, and to inclines exceeding 3 in. to the horizontal foot.

Class B (UL).   Asbestos cement shingles. Roof covering composed of asbestos cement shingles laid to provide one or more thicknesses over a layer of 25 lb asphalt-saturated asbestos felt in accordance with manufacturers instructions. Limited to roof decks capable of receiving and retaining nails, and to inclines exceeding 3 in. to the horizontal foot.

Closed valley.   Roof valley where the shingles of the intersecting slopes are continuous without open space.

Color.   The exposed surface of the shingles may be natural cement gray or colored by the addition of mineral pigments, chemical impregnation, pigmented coatings, veneers, or embedded mineral granules.

Coverage.   Number of thicknesses of roofing material applied to the surface of a roof. The kind of shingle and the method of application may furnish single, double, or triple coverage.

Deflection.   The average deflection at midspan of the dried test specimens from the lot being tested should not be less than 0.15 in.

Dutch.   Square in shape.

Dutch or Scotch method.   With the Dutch or Scotch method shingles are of uniform thickness with straight or irregular edges and designed to be laid with a lap at the top and on one side of each shingle; also called "ranch design."

Efflorescence.   The efflorescence that may appear on asbestos cement shingles is not a defect and should not result in a permanent change in color.

Exposure.   Shortest distance in inches between exposed edges of overlapping shingles.

Fiber-cement shingles.   Shingles are reinforced and textured to give the appearance of natural slate or cedar shakes. Shingles conform to ASTM C 222.

Finish.   The surface of the shingles to be exposed are smooth, grained, granuled, coated, or otherwise textured.

Fireproof roof shakes.   Manufactured under a patented process using an organic blend or natural ingredients. The base is perlite, a lightweight mineral aggregate, with portland cement, reinforcement fibers, and iron oxides. No asbestos is used.

Fire rating.   Asbestos cement roofing shingles are classified by the Underwriters Laboratories, Inc. as Class A and Class B.

Flexural strength.   The average breaking load of the dried test specimens from the lot being tested should not be less than 27 lb psf.

French or hexagonal method.   With the French or hexagonal method shingles are of uniform thickness, square in shape, and with at least three corners clipped to give the desired pattern when the shingles are laid with their diagonals perpendicular to the eave of the roof and with their apex sides lapped.

Head lap.   That part of a course of shingles that is covered by the overlapping course. Head lap is defined as the shortest distance in inches from the lower edge of an overlapping shingle to the upper edge of the unit in the second course below. At butted side joints, between individual units, complete and continuous coverage should be provided by the undercourse shingles. Side lap has a similar meaning in respect to the shortest horizontal distance from the exposed side edge to the uncovered roof deck area.

Hip and ridge finishing pieces.   Finishing pieces for application on hips and ridges are available for each type of shingle and should have the same general characteristics as the shingles.

Holes for nails and fasteners.   Holes are provided in the units during manufacture and are so placed as to provide at least the minimum lap, as specified, and to allow for proper application of necessary clips or storm fasteners.

Metal flashing shingles.   Rectangularly shaped pieces of sheet metal used to form base flashing at vertical surfaces or protrusions through the roof.

Nails and fasteners.   Nails and fasteners are made of corrosion-resistant, nonstaining metal and have flat heads substantially larger than the diameter of the holes in the shingles with which they are to be used. Nails should be of such length as to hold securely to the deck.

Nonasbestos simulated slate shingles.   Shingles made without asbestos from a combination of cement, organic fibers, silica, water, and other additives. Slates are colored and of nontextured surfaces; they are also available in textured surfaces.

Open valley.   Roof valley in which the shingles of the intersecting slopes leave an open space covered by metal flashing.

Pitch or slope.   Slope or angle of a roof indicated in inches per foot in relation to a horizontal plane.

Plastic asphalt cement.   Asphalt- or coal-tar-pitch-base waterproof cement, mixed with asbestos fibers and a solvent for cold application.

Ranch shape.   Rectangular oblong, with the greater dimension horizontal.

Ridge roll.   Curved or angular ridge roll for certain types of shingles, having the same general characteristics as the shingles.

Saddle or cricket.   Watershed change in pitched roof slope, as behind a chimney.

Slope.   Incline of a roof expressed as a ratio of vertical rise to horizontal run. For example, a roof that rises at the rate of 4 in. for each 12 in. of run is designated as having a 4 in 12 slope.

Square.   Unit of roofing that equals 100 sq ft.

Starter.   First course of a half-shingle overhanging eave $\frac{3}{4}$ in. and in line with butt with regular size shingles.

Storm anchors or clips.   Anchors or clips made of corrosion-resistant, nonstaining metal, readily bendable, and having flat bases substantially larger than the diameter of the holes in the shingles with which they are to be used;

storm anchors of adequate length to secure the shingles in place.

Strip shingles. Shingles of uniform thickness, in various shapes and designs, to provide top lap only and shingle coverage under the butt joints, with the design such that it provides only single coverage.

Thickness. The average thickness of all units should not be less than 0.150 in., and the average thickness of any one unit should not vary from the average of all units. The minimum average thickness of any shingle should be 0.135 in. For textured shingles the thickness should be construed as the gross overall measurement from the top of the textured surface to the back of the unit.

Underlayment

1. The underlayment is usually composed of asphalt-saturated felt, preferably asbestos felt where maximum fire resistance of the roof covering is desired, suitable for underlayment use. Coal-tar saturated felt is not suitable. The felt used should weigh not less than 25 lb/100 sq ft. Underlayment sheet material for use with asbestos-cement shingles should be water repellent and of the "breather" type, permeable to water vapor.

2. Single thickness of felt parallel to eaves with double thicknesses at hips and ridges.

Water absorption. The average water absorption should not exceed 25 weight percent (ASTM C 222).

ASBESTOS CEMENT SIDING

Backer strips. Backer strips, usually about 3 in. wide, are used for flashing the vertical joint between siding units; they should be asphalt-saturated and coated roofing felt.

Clapboards. Clapboards should be of uniform thickness, generally longer and narrower in shape than siding shingles, and usually with the exposed edge straight.

Color and surface. The exposed surface of the siding should be natural cement gray or colored by the addition of mineral pigments, chemical impregnation, pigmented coatings, veneers, or embedded mineral granules.

Deflection. The average minimum deflection at midspan of the dried test specimens from a selected lot being tested should be not less than 0.15 in. (ASTM C 223).

Efflorescence. Although efflorescence may appear on asbestos cement siding, it is not a defect and should not result in a permanent change in color.

Face nails. Face nails have small, flat heads with shanks sized to enter the face nail holes provided in the units and should be long enough to hold the siding units securely to a wood lumber nailing base. They are made of nonstaining, corrosion-resistant material.

Finish. The exposed surface of the siding should be smooth, grained, granule coated, or otherwise textured.

Flexural strength. The average breaking load of the dried test specimens from the lot being tested should not be less than 25 lb-f.

Holes for nails and fasteners. Holes for nails and fasteners are provided and so placed as to satisfy the top lap requirements.

Sheets. Sheets are of uniform thickness, rectangular in shape, and larger in unit sizes than shingles or clapboards.

Siding. Siding is composed of a combination of asbestos fiber and portland cement or portland-blast-furnace slag cement, and not more than 1% by weight of organic fiber, with or without the addition of curing agents, water-repellent substances, mineral fillers, coatings, pigments, or mineral granules, and is formed under pressure and cured (ASTM C 223).

Siding shingles. Siding shingles are of uniform thickness and generally rectangular in shape, and have wavy, random, thatched, straight, or irregular butts.

Special fasteners. Special fasteners are for attaching siding units to other than wood nailing base; they are made of nonstaining corrosion-resistant material.

Thickness. The average thickness of all units should be not less than 0.150 in., and the average thickness of any one unit should not vary from the average of all units. The minimum average thickness of any one unit of shingle or clapboard should be 0.135 in. For textured siding the thickness should be construed as the gross overall measurement from the top of the textured surface to the back of the unit.

Top lap. Shortest distance between the lower edge of a course of siding and the most proximate area of side wall not covered by the preceding course.

Underlayment. Underlayment is composed of asphalt-saturated felt suitable for underlayment use and weighs not less than 25 lb/100 sq. ft. Underlayment sheet material for use with asbestos-cement siding should be water resistant and of the "breather" type, permeable to water vapor.

Water absorption. Average water absorption should not exceed 25 weight percent (ASTM C 223).

ASPHALT SHINGLES

Abut. To position snugly against the top or side of a shingle. Shingle courses can be abutted when a new layer is applied over worn, but not over warped or buckled, shingles.

Alignment notch. Factory-cut end of a shingle where cutouts meet to form proper shingle alignment.

Asphalt shingles (felt base). Shingles produced by using a single thickness of dry roofing felt impregnated with a hot asphaltic saturant, then coated on both sides with hot asphaltic coating, which may be compounded with a fine substantially water-soluble mineral stabilizer, and finally completely surfaced on the weather side with a colored mineral granule embedded in the hot asphaltic coating (ASTM D 225).

Asphalt shingles (fiberglass base). Shingles produced by using an inorganic fiberglass mat saturated with coating asphalt and top surfaced with colored ceramic granules. The shingles are rated to receive the UL Class A label.

Aviation snips. Tin snips can be used to trim asphalt-fiberglass shingles as well as metal flashing.

Backing down. High nailing courses of roofing material to tie in lower, successive courses.

Blast-furnace slag. Air-cooled blast-furnace slag is crushed to sizes required for granules. It is colored as required. The slag is a nonmetallic product developed simultaneously with iron in a blast furnace and solidified under atmospheric conditions.

Blind nailing. Installing nails so that the nail heads are concealed by roofing material.

Border shingles. Roofing material applied to the outer edges of a roof section to provide protection and to present an even edge.

Boston ridge. Applying asphalt shingles at the ridge or hips of a roof as a finish.

Building paper. Light, asphalt-saturated material used to temporarily waterproof a roof deck; felt.

Butt. Overexposed edge of a shingle to weather.

Capping. One-tab shingles centered and applied horizontally to the ridge or hip of a roof; the last course of roll roofing centered and applied horizontally to the ridge.

Ceramic granules. Fire-hardened material added to asphalt and fiberglass shingles to provide color and for weather resistance.

Closed-valley. Roof valley where the shingles of the intersecting slopes are continuous without open space.

Colors. Colors are established by the various manufacturers and are created by colored mineral, ceramic granules, and crushed slag.

Course. One series of shingles in a horizontal, vertical, or 45° angle from the eaves to the ridge. Shingles, shakes, and roll roofing are laid in successive courses.

Coverage. Number of thicknesses of roofing material applied to the surface of a roof. The kind of shingle and method of application may furnish single, double, or triple coverage.

Cutout. Where alignment notches of three-tab shingles meet to create a pattern; the watermark.

Deck. Materials such as plywood sheathing or planking installed over framing members; the roof surface before shingles or other roofing materials are installed.

Drip edge. Lightweight metal strips designed to fit against rakes and eaves.

Eaves. Edge of a roof that projects over an exterior wall.

Exposure. Shortest distance in inches between exposed edges of overlapping shingles.

Felt. The felt used in asphalt-based shingles is produced by "felting" vegetable or animal fibers, or a mixture of the two.

Fiberglass base-glass mat. The base or mat is composed of a random distribution of fine glass fibers bonded into a sheet with a resin, and suitably prepared for impregnation in the manufacture of bituminous waterproof and roofing membranes (ASTM D 3462).

Fiberglass asphalt shingles. Shingles include incombustible fiberglass base, covered with ceramic coated granules deeply embedded in refined water-resistant asphalt, with self-sealing thermoplastic adhesive.

Fiberglass strip shingles. Shingles are mineral surfaced, self-sealing, laminated multi-ply, overlay construction fiberglass strip shingles. Shingles usually conform to UL Class A fire rating and UL wind resistance rating.

Filling in. Roofing a section by squaring off an angled portion to obtain long vertical runs of shingles.

Fire rated. Asphalt shingles are in the "Prepared Roof Covering Materials" UL classification. This category covers materials in 3 classes of labels, Class A, Class B, and Class C.

Gable. Upper part of a terminal wall under the ridge of a pitch roof; the end or wing of a building so gabled.

Granule. Finely crushed or ground minerals, sand, or rock adhered to the portion of shingles and roll roofing exposed to the weather to provide color and weather resistance.

Head lap. Shortest distance in inches from the lower edges of an overlapping shingle to the upper edge of the unit lapped in the second course below.

High nailing. Driving nails about 1 in. below the top of the shingle (which is well above the standard nailing location) to properly position one course of shingles while allowing one or more courses to be added underneath the high-nailed course. Each shingle eventually must be properly nailed.

Hip roof. Angle formed by the meeting of sloping roof sections. Slopes are angled toward the center from four sides and there are no rakes.

Individual shingle. One-piece shingle without cutout tabs. It can be square, rectangular, or hex shaped.

Interlocking shingle. Shingle so shaped as to have cutouts or tabs in its edges interlocked with adjacent shingles. Interlocking shingles are primarily used to provide effective resistance to strong winds.

Laminated fiberglass shingles. Mineral-surfaced, self-sealing, laminated multi-ply, overlay construction fiberglass strip shingles. Product of Georgia-Pacific Corporation.

Laminated shingles. Composed of either an organic felt base or a fiberglass mat base depending on region for their use. Ceramic coated mineral granules are tightly embedded in carefully refined, water-resistant asphalt. The laminated tabs are firmly adhered in a special tough asphaltic cement. Product of Certainteed Corporation.

Metal flashing shingles. Rectangularly shaped pieces of sheet metal used to form base flashing at vertical surfaces or protrusions through the roof.

Mineral granules. Ceramic baked, fire-resistant mineral granules, acceptable as "grit surfacing" for asphalt shingles for UL approval.

Open valley. Roof valley in which the shingles of the intersecting slopes leave an open space covered by metal flashing.

Organic felt. Previously identified as rag-felt composition, the term is now used to identify felt composed of felted vegetable or animal fibers, or mixtures of both (ASTM D 224).

Pitch or slope. Slope or angle of a roof indicated in inches per foot in relation to a horizontal plane.

Plastic asphalt cement. Asphalt- or coal-tar-pitch-base waterproof cement, mixed with asbestos fibers and a solvent for cold application.

Random spacing. Spacing pattern in which cutouts in the first five courses are never aligned. The pattern is repeated after each five courses.

Ridge. The horizontal junction of the two top edges of two sloping roof sections.

Run. The distance covered by the application of shingles in one pass of the pattern; an inclined course.

Saddle or cricket. Watershed change in pitched roof slope, as behind a chimney.

Saturant. Bitument for the saturant and coatings used on felts should be composed of asphaltic materials.

Seal-down strip. Factory-applied, sunlight-activated adhesive that bonds asphalt-based fiberglass shingles to the course above.

Shingle. Covering made from asphalt, fiberglass, wood, aluminum, tile, slate, or other water-shedding material.

Slope. Incline of a roof expressed as a ratio of vertical rise to horizontal run. For example, a roof that rises at the rate of 4 in. for each 12 in. of run is designated as having a 4 in 12 slope.

Square. Unit of roofing that equals 100 sq ft.

Starter. Shingle of regular width but less than standard length, used at eaves and gutters under the first course to produce a uniform cant.

Starter course. First row of shingles or roll roofing applied at the eaves.

Step flashing. Usually aluminum or galvanized sheet metal cut in L-shaped pieces that are weaved at the joints between the roof surface and the roofing material; installed along walls and masonry.

Straight pattern. Vertical application of three-tab shingles with cutouts aligned every other shingle.

Strip shingle. Shingle that is usually 12×36 in. with or without cutout openings.

Surface finish. The weather surface should be uniform in finish or texture and may be embossed to simulate a grain texture. Mineral granules should cover the entire surface and should be firmly embedded in the asphalt coating.

Top lap. Width of a shingle minus the exposure; portion of the shingle not exposed to weather.

Underlayment. Base or lining between a roof deck and the shingles, usually composed of roofing felt.

Valley. Line of intersection of two roof slopes where their drainage combines.

Weights. Asphalt shingles weights are based on coverage of 100 square ft. Weights vary from approximately 235 to 350 lbs/square.

Wind resistant. The term refers to asphalt-organic felt shingles that are provided with factory applied adhesive or integral locking tabs. Some shingles with factory applied adhesive utilize bands or spots of a heat sensitive adhesive located either on the surface of the shingles or on the back side of each tab. Adhesives of this type are activated by solar heat. This type of shingle is classified as UL Class C label.

Woven valley. Area where shingles continue through the valley, being laid on both roofs at the same time, weaving each course in turn over the valley.

BUILT-UP AND SINGLE PLY ROOFING

Abrasion resistance. Ability of the membrane to resist mechanical abrasion such as foot traffic and windblown particles, which tend to progressively remove materials from its surface.

Adhesion. Ability of the membrane to remain adhered during its service life to the substrate or to itself.

Aggregate

1. Crushed stone, crushed slag, or water worn gravel used for surfacing a built-up roof.

2. Any granular mineral material.

Air-blown asphalt. Asphalt produced by blowing air through molten asphalt or asphaltic flux at an elevated temperature to give it characteristics desired for certain uses.

Alligatoring

1. Cracking in the membrane caused by thermal cycling, material degradation, and exposure to weather.

2. Shrinkage cracking of the bituminous surface of built-up roofing or the exposed surface of smooth-surfaced roofing, in which the loss of volatile oils under solar radiation produces a pattern of deep cracks with the scaly look of an alligator's hide. Alligatoring occurs only in unsurfaced bitumen exposed to the weather.

Aluminized coating. Coating formulated specifically for application on asphalt built-up roofs and other felt roofs. An aluminized roof reflects heat and ultraviolet rays, which are entirely absorbed by ordinary black roofs; applied directly over bitumen.

Application of felt. Plies of felt are laid single fashion. This method produces a strong membrane and also makes it possible to construct a roof of any desired number of plies in a single progressive operation.

Application rate. The quantity (mass, volume, or thickness) of material applied per unit area.

Area divider. A raised, double wood member attached to a properly flashed wood base plate that is anchored to the roof deck. It is used to relieve thermal stresses in a roof system where no expansion joints have been provided.

Asbestos. A group of natural, fibrous, impure silicate materials.

Asbestos-base felt. Asbestos felt thoroughly saturated with asphalt, used as a base sheet on certain asbestos bonded built-up roofs, with or without perforations (ASTM D 250).

Asbestos finishing felt (perforated). Asphalt-impregnated perforated asbestos felt, used for asbestos and combination rag and asbestos built-up roofs. Perforations allow entrapped air to escape from underneath the felt at the time of application and causes better embedment in the asphalt.

Asbestos finishing felt (unperforated). Asphalt-impregnated unperforated asbestos felt, used as a vapor retarder under approved roof insulation.

Asphalt

1. Dark-colored, more or less viscous-to-solid residue obtained from the distillation of certain petroleum crudes.

2. Dark-brown to black highly viscous hydrocarbon produced from the residuum left after the distillation of petroleum and used as the waterproof agent of a built-up roof. It comes in a wide range of viscosities and softening points from about 135°F (dead-level asphalt) to 210°F or more (special steep asphalt).

Asphalt emulsion. Asphalt that has been rendered liquid by emulsification with water, usually with the aid of a small quantity of an emulsifying agent. After application, the emulsion "breaks," allowing the water to evaporate, leaving the desired grade of asphalt behind.

Asphaltene. High molecular weight hydrocarbon fraction precipitated from asphalt by a designated paraffinic naptha solvent at a specified temperature and solvent asphalt ratio used. The asphaltene fraction should be identified by the temperature and solvent-asphalt ratio used.

Asphalt felt. An asphalt-saturated felt or an asphalt-coated felt.

Asphalt mastic. Mixture of asphaltic material, graded mineral aggregate, and fine mineral matter, which can be poured when heated but requires mechanical manipulation to form.

Asphalt organic-felt coverings with hot-mopping asphalt. Classes A and B. These coverings are composed of asphalt-saturated organic felt and/or combinations of cap or base sheets (saturated felts that are asphalt coated on one or both sides) applied with hot-mopping asphalt and surfaced with approximately 400 lb of roofing gravel or crushed stone or 300 lb or crushed slag per 100 sq ft of finished roof embedded in a flood coat of hot-mopping asphalt. The roofing gravel, crushed stone, or slag is to be dry, relatively free from dirt and dust, and graded in size from $\frac{1}{4}$ to $\frac{5}{8}$ in. These coverings are limited to roof decks having inclines not exceeding 3 in. to the horizontal foot. UL rated classes A and B.

Asphalt organic-felt coverings with hot-mopping asphalt. Class C. These coverings are composed of asphalt-saturated organic felt and/or combinations of cap or base sheets (saturated felts that are asphalt-coated on one or both sides) applied with hot-mopping asphalt. The coverings are surface finished with a cold application coating. UL rated class C.

Asphalt primer. Asphalt that has been thinned to a liquid consistency. Where necessary, it is used to prepare surfaces to receive hot asphalt. It is usually applied without heating.

Asphalt roll roofing. Asphalt roll roofing is composed of roofing felt saturated and coated on both sides with asphalt and surfaced on the weather side with powdered mineral materials such as talc or mica (ASTM D 224).

Asphalt roof cement. Cement consisting of an asphalt base, volatile petroleum solvent, and mineral stabilizers including asbestos fibers (ASTM D 2822).

Asphalt roof coatings. Asphalt coating, used for brushing purposes consisting of an asphalt base, volatile petroleum solvent, asbestos fibers, and mineral stabilizers (ASTM D 2823).

Asphalt-saturated asbestos felts. Single thicknesses of asbestos felt saturated with an asphaltic saturant. Asbestos felt contains at least 85 mass percent of asbestos fiber (ASTM D 250).

Asphalt-saturated rag felt. Heavy asphalt-saturated rag felt is used as a base felt on certain rag and asbestos felt bonded built-up roofs.

Asphalt-saturated roofing felt. Single thickness of dry roofing felt saturated with an asphaltic saturant.

Backnailing. Practice of blind-nailing in addition to hot-mopping all the plies to a substrate to prevent slippage on slopes of $1\frac{1}{2}$ in. or more for steep asphalt, $\frac{1}{2}$ in. or more for coal tar pitch and dead-level asphalt.

Base ply. The lowermost ply of roofing material in a roof membrane assembly.

Base sheet

1. Sheet coated on both sides with heavy asphalt-saturated and/or tar-coated felt for use as a base and in the construction of asphalt or coal-tar-pitch built-up roofs. This type of sheet is recommended as a base sheet particularly over wet fill decks, such as poured

gypsum, vermiculite concrete, and perlite concrete; as a base over mineralized shredded wood fiber decks; as the first ply over roof insulation; and as a vapor retarder under all types of above deck roof insulations.

2. Heavy saturated and coated felt placed as the first ply in a multiply built-up roofing membrane.

Bitumen

1. Portion of asphalt or coal tar that is soluble in carbon disulfide. The term is used more loosely to designate any of the asphalt or coal tar products used in the application, repair, or coating of roofs.

2. Generic term for an amorphous, semisolid mixture of complex hydrocarbons derived from petroleum or coal. There are two basic bitumens: asphalt and coal tar pitch. Before application they are either heated to a liquid state, dissolved in a solvent, or emulsified.

Bituminous. Containing or treated with bitumen. For example, bituminous concrete, bituminous felts and fabrics, bituminous pavement.

Bituminous emulsion

1. A suspension of minute globules of bituminous material in water or in an aqueous solution.

2. A suspension of minute globules of water or an aqueous solution in a liquid bituminous material (invert emulsion).

Bituminous grout. Mixture of bituminous material and sand finer than the No. 20 sieve that, when heated, will flow into place without mechanical manipulation.

Blackberry. A small bubble or blister in the flood coating of a gravel-surfaced roof membrane.

Blind nailing. The practice of nailing the back portion of a roofing ply in a manner that protects the fasteners in the finished product from exposure to the weather.

Blister

1. Irregularly shaped raised portion in the top roofing felt resembling a half-domed effect, arising in various sizes, caused by the expansion with heat of vapor or moisture locked in between the felts. Once the expansion takes place causing the blister, it remains in its most extended shape.

2. Spongy, raised portion of a roofing membrane, usually resulting from the pressure of entrapped air or water vapor.

Blister repair. Blisters are cut open and repaired in such a manner as to produce a flat area in place of the blister. No nails are used, and the area should be reinforced with membranes and water-repellent adhesives, depending on the kind and condition of the roof.

Blocking. Wood built into a roofing system above the deck and below the membrane and flashing to stiffen the deck around an opening, acting as a stop for insulation, or to serve as a nailer for attachment of the membrane or flashing.

Board-type insulations. Insulations come in varying materials, thicknesses, and sizes, with provisions or manner of attachment to be recommended by the manufacturer.

Bond

1. Adhesive strength or bonding ability preventing delamination of two roofing components.

2. Standard roofing bond or roofing system guarantee issued by a manufacturer of the materials to the owner for a premium, or guarantee workmanship and materials for a stipulated period of time, with many exclusions for the guarantor for not meeting the requirements of the guarantee.

Brooming. Embedding a ply of roofing material by using a broom to smooth out the ply and ensure contact with the adhesive under the ply.

Built-up roof. Continuous roof covering of laminations or plies of saturated or coated felts, alternated with layers of bitument and surfaced with mineral aggregate or asphaltic materials.

Built-up roofing. Functionally continuous, flexible membrane of saturated or saturated and coated felts, fabrics, or mats assembled on a roof with a bituminous plying cement.

Built-up roof member (BUR). A continuous, semiflexible roof membrane assembly, consisting of plies of saturated felts, coated felts, fabrics or mats between which alternate layers of bitumen are applied. Generally surfaced with mineral aggregate, bituminous materials, or a granule-surfaced roofing sheet.

Cant strip

1. Continuous strip of triangular cross section fitted into the angle formed by a structural deck and a wall or other vertical surface. The 45° slope of the exposed surface of the cant strip provides a gradual transition for base flashing and roofing membrane from a horizontal roof surface to a vertical surface.

2. Impregnated insulation mineral wool, impregnated wood, or lightweight concrete (mortar) placed in the right-angle juncture between the (flat) roofing surface and the vertical surface of the adjacent parapet, wall, or roof curb in order to provide a gradual transition from horizontal to vertical application of roofing and base flashing felts. The slope of the cant is usually at 45°.

Cap sheet

1. Smooth or mineral surfaced roll roofing for use as the top layer on a built-up roof.

2. Mineral surfaced coated felt (or a coated felt without mineral surfacing) used as the top ply of a built-up roofing membrane.

3. Sheet used as the top ply in a built-up roof assembly where gravel surfacing cannot be used. It consists of a single- or double-ply smooth or mineral surfaced heavy roofing felt which replaces the flood coat and gravel of a built-up roof system. Cap sheets are either black or colored, either rag, asbestos, or glass felts, saturated and coated on both sides with asphalt and surfaced on the exposed side with mineral granules, mica, asbestos fiber, or similar materials.

Cement. Substance, such as asphalt cement or asphalt plastic, used to bind two surfaces together.

Cementing. Solidly mopped application of asphalt, cold liquid asphalt compound, coal tar pitch, or other approved cementing material.

Class rating. Roof coverings are of two types, built-up and prepared, and are intended for the protection of roof decks from external fire exposure only. Classification establishing rating of roofs, by testing is performed by the Underwriters Laboratories, Inc. (UL). Assembly roofing installations after testing are classified as Classes A, B, and C.

Class A roof. This UL classification includes roof coverings that are effective against severe fire exposures. Under such exposures, roof coverings of this class are not readily flammable and do not carry or communicate fire. They afford a fairly high degree of fire protection to the roof deck, do not slip from position, possess no flying brand hazard, and do not require frequent repairs in order to maintain their fire-resisting properties.

Class B roof. This classification includes roof coverings that are effective against moderate fire exposures. Under such exposures, roof coverings of this class are not readily flammable and do not readily carry or communicate fire. They afford a moderate degree of fire protection to the roof deck, do not slip from position, possess no flying brand hazard, and may require infrequent repairs in order to maintain their fire-resisting properties.

Class C roof. This UL classification includes roof coverings that are effective against light fire exposure. Under such exposures, roof coverings of this class are not readily flammable and do not readily carry or communicate fire. They afford at least a slight degree of fire protection to the roof deck, do not slip from position, possess no flying brand hazard, and may require occasional repairs or renewals in order to maintain their fire-resisting properties.

Coal tar. Bituminous substance derived as a by-product in the manufacture of coke from bituminous coal. Roofing pitch is produced by distillation of the coal tar. It is a low-melt type bitumen used on most dead-level or nearly dead-level surfaces. The basic properties of coal tar are its cold flow or self-healing properties.

Coal-tar bitumen. Coal tar used as the waterproofing agent in dead-level or low-slope built-up roof membrane, conforming to ASTM D 450 Type III.

Coal tar pitch

1. Black or dark-brown solid cementitious material obtained as a residue in the partial evaporation or fractional distilation of coke oven tar and that gradually liquefies when heated. It is produced in various consistencies.

2. Dark-brown to black solid hydrocarbon obtained from the residuum of the distillation of coke oven tar, used as the waterproofing agent of dead-level or low-slope built-up roofs. It comes in a narrow range of softening points from 140 to 155°F.

3. Pitch suitable for use as a mopping coat in the construction of built-up roofs (ASTM D 1079).

Coal-tar-saturated roofing felt. Single sheet of dry roofing felt saturated with a coal tar saturant.

Coated base sheet. Felt that has previously been saturated with asphalt and later coated with harder, more viscous asphalt, which greatly increases its impermeability to moisture.

Coated sheet felt

1. An asphalt felt that has been coated on both sides with harder, more viscous asphalt.

2. A glass fiber that has been simultaneously impregnated and coated with asphalt on both sides.

Cold-application asbestos felt. Asphalt-saturated and double-coated felt used for cold-application of built-up roofs, to be applied with cold-application cement.

Cold-process roofing. Bituminous membrane comprising layers of coated felts, bonded with cold-applied asphalt roof cement and surfaced with a cutback or emulsified asphalt roof coating.

Combination sheet. Glass fiber felt integrally attached to kraft paper.

Concealed nailing. Nailing whereby nail heads are protected from weather by overlying sheet(s) of roofing material.

Contractor's guarantee. Guarantee issued by the roofer, normally for a minimum of two years, which requires the roofer to make repairs necessary as a result of ordinary wear and maintain the roofing in a watertight condition. The limits of the roofer's responsibilities are usually specified to include the repair of leaks caused by ordinary wear and conditions caused by the elements or defects due to faulty materials or workmanship in the roofing system. Roofing failures, if they should occur, usually occur within the first two years after application of the roofing system.

Cotton fabric saturated with bitumin. Bituminized cotton fabric composed of woven cotton cloth waterproofed with either asphalt or coal-tar pitch and used primarily for membrane systems (ASTM D 173).

Counterflashing. Metal flashing that covers and protects the top edge of the built-up base flashing.

Coverage. Surface area that should be continuously coated by a specified unit of a roofing material after allowance is made for a specified lap.

Crack. Membrane tear produced by bending of heavy felt sheets to a 90° angle or less; also caused by stepping on a blister.

Creep. Permanent elongation or shrinkage of the membrane resulting from thermal or moisture changes, permanent deflection of structural framing or structural deck resulting from plastic flow under continued stress, or dimensional changes accompanying changing moisture content or temperature.

Cutback. Organic, solvent-thinned, soft or fluid cold-process bituminous roof coating or flashing cement.

Cutoff. Roofing detail designed to prevent lateral water infiltration into the insulation at the edges of the exposed felts where they terminate at the end of the day's work. The term applies to felt strip hot-mopped to the stepped contour of the deck, the insulation edges, and felt and the horizontal insulation surface.

Cutoffs and venting. Cutoffs are installed, provided they are constructed to run in one direction continuously to vented edges and provided no section of the roofing is isolated or blocked from venting. The venting of roof insulation at walls, curbs, eaves, etc., is recommended for all but a few special roof structures.

Dampproofing

1. Membrane treatment by use of roofing felts, fabrics, and rubber materials in combination with waterproofing compounds for a surface or structure to resist the passage of water in the absence of hydrostatic pressure.
2. Treatment of a surface or structure that retards the passage of liquid water.

Dead level. Absolutely flat or perfectly horizontal; or zero slope.

Dead-level asphalt. A roofing asphalt conforming to the requirements of ASTM standard D 312 Type 1.

Dead loads. Nonmoving roof top loads, such as mechanical equipment, air conditioning units, and the roof system itself.

Deck. The structural surface to which the roofing or waterproofing system (including insulation) is applied.

Delamination. Built-up roofing membrane failure characterized by separation of the felt plies, resulting in wrinkling, cracking, and fish mouths at the edge of the top ply.

Deterioration of roofing felts. Tar or asphalt that saturates the roofing felt contains volatile ingredients that gradually distill away under the influence of sunlight. Summer heat on an exposed roof is a deteriorating factor. Consequently, roofing felt does not wear out—it dries out. When the waterproofing oils with which it was originally impregnated have evaporated and oxidized, the felt is left with no waterproofing qualities at all. It is now more vulnerable to breaks from wind action, foot marring, or flying debris.

Double pour. The process of applying two layers of aggregate and bitumen to a built-up roof.

Drain. Device that allows for the flow of water from a roof area.

Dropback. Reduction in the softening point of bitumen that occurs when it is heated in the absence of air.

Dry sheet. Uncemented lightweight sheet that is scatter-nailed to wood or other roof decks under the roofing membrane to prevent seepage of bitumen through the cracks in the deck. Rosin-sized paper and unsaturated and saturated felts are commonly used for this purpose.

Edge sheets. Felt strips that are cut to widths narrower than the standard width of the full felt roll, used to start the felt shingling pattern at a roof edge.

Edge stripping. Application of felt strips cut to narrower widths than the normal felt roll width to cover a joint between flashing and built-up roofing.

Edge venting. Practice of providing regularly spaced openings at a roof perimeter to relieve the pressure of water vapor possibly entrapped in some insulation materials.

Elastomer. A macromolecular material that returns rapidly to its approximate initial dimensions and shape, after substantial deformation by a low level stress and the release of that stress.

Elastomeric. Term applied to roofing material having elastic properties, that is, capable of expanding or contracting with the surfaces to which the material is applied without rupturing.

Embedment

1. The process of pressing a felt, aggregate, fabric, mat, or panel uniformly and completely to hot bitumen or adhesive
2. The process of pressing granules into coating in the manufacture of factory-prepared roofing.

Emulsion. Intimate mixture of bitumen and water, with uniform dispersion of the bitumen globules achieved through a chemical or clay emulsifying agent.

Envelope. Continuous membrane edge seal formed at the perimeter and at penetrations by folding the base sheet or ply over the plies above and securing it to the top of the membrane. The envelope prevents bitumen seepage from the edge of the membrane.

Expansion joints. All materials expand and contract owing to temperature and/or moisture; therefore the use, location, and design of building expansion joints, especially the roof structure system, should be taken into consideration at the time of the original building design.

Exposed nailing. Nailing of roofing whereby nail heads are left exposed to the weather.

Exposure. Portion of a roofing element that is exposed to the weather when laid; the least distance in inches from the lowest exposed edge or point of a roofing shingle to a line connecting the corresponding points of the next course.

Fabric. Woven cloth of organic or inorganic filaments, threads, or yarns.

Felt

1. Material fabricated essentially from fibers of vegetable or animal origin.

2. Fabric manufactured by the interlocking of fibers through a combination of mechanical work, moisture, and heat, without spinning, weaving, or knitting. Roofing felts are manufactured from vegetable fibers, rags, and asbestos or glass fibers, or combinations of any of these.

3. Flexible sheet built up by the interlocking of fibers by any suitable combination of mechanical work, moisture, and heat.

4. Lightweight sheets used in multiple-ply construction of built-up roofs, composed of rag, vegetable, and/or other organic components, asbestos, or glass fibers, saturated with asphalt or coal tar.

Felt layer. Machine used for applying bitumen and built-up roofing felts.

Felt mill ream. The mass in pounds of 480 sq ft of dry, unsaturated felt; also termed "point weight."

Fiberglass roofing felt. Composed of inorganic glass fibers that resist moisture absorption. The felt has sufficient porosity to allow the mopping asphalt to penetrate deeply to assure proper interply adhesion.

Fine mineral surfacing. Water-insoluble, inorganic material, more than 50% of which passes No. 35 sieve, used on the surface of roofing.

Fire-resistance rating. Ratings for "Roof-Ceiling Assemblies" are based on the roofing structure, roof covering, and the ceiling attached to the roof construction. The detailed test method and criteria used for establishing fire-resistance ratings are described in the Standard, Fire Tests of Building Construction and Materials. UL263, ANSI A2.1, ASTM E 119, and NFPA 251.

Fire-retardant roof coverings. Roof coverings should be classified on the basis of protection provided against fire originating outside the building or structure on which they have been installed.

Fish mouth

1. Opening in the top sheet at the edge of the ply due to thermal cycling of the roofing surfaces and the possible use of the wrong bitumen with the wrong felts.

2. Opening formed by an edge wrinkle in an exposed felt ply in a built-up roofing membrane.

Flammability. Ability of the membrane to resist combustion and spreading of the flame.

Flashing

1. Connecting device that seals membrane joints at expansion joints, drains, gravel stops, and other places where the membrane is interrupted. Base flashing forms the upturned edges of the watertight membrane. Cap or counterflashing shields the exposed edges and joints of the base flashing.

2. Attachment of built-up roofing to vertical surfaces and the further protection of the top edge by a counterflashing of metal, preferably two-piece, through-wall type; also the waterproof connections between built-up roofing and flanges of roof connections, drains, outlets, gravel guards, etc.

Flashing cement

1. Trowelable, plastic mixture of bitumen and asbestos (or other inorganic reinforcing fibers) and a solvent.

2. Plastic mixture of bitumen and asbestos, reinforced with organic fibers mixed with solvents to soften the material for hand troweling or application.

Flat asphalt. A roofing asphalt conforming to the requirements of ASTM D 312 Type II.

Flood coat

1. Top layer of bitumen in a mineral-aggregate-surfaced built-up roof assembly into which the aggregate is embedded.

2. Top layer of fast-drying, hot-poured bitumen over the final top sheet of the built-up roof assembly.

3. Top layer of bitumen in an aggregate surfaced, built-up roofing membrane. Bitumen is poured, hot-mopped, to a weight of 60 lb per square for asphalt, 75 lb per square for coal tar pitch.

Fluid-applied plastic roof coatings

1. Coverings are composed of a number of layers or coats of plastic roof coating in which a mass of continuous glass fiber filaments may be embedded. Adequate drying time is allowed between successive layers or coats to form a protective membrane.

2. Coverings are composed of a number of layers or coats of plastic roof coating, in which a glass mat may be embedded to form a protective membrane. Roof decks are primed with a mixture of one part neoprene coating to two parts solvent. A quantity of primer is permitted to flow into all voids and cracks on the surface.

Furring strips. Lightweight wood strips applied as supplemental fastening to temporarily prevent wind damage to felt on a roof deck; fastening base for wood roofing materials.

Glass felt. Glass fibers bonded into a sheet with resin and suitable for impregnation in the manufacture of bituminous water-proofing materials, roof membranes, and shingles.

Glass fiber felt. Glass fiber sheet coated on both sides with bituminous compound.

Glass mat. Mat formed by the random distribution of glass fibers bonded into a sheet with a resin and suitable for impregnation in the manufacture of bituminous water-proofing and roofing membranes.

Glaze coat

1. Top layer of asphalt in a smooth-surfaced built-up roof assembly; thin protective coating of bitumen applied to the lower plies or top ply of a built-up membrane when the top pouring and aggregate surfacing are delayed.

2. Top mopping of asphalt in a smooth-surfaced built-up roof assembly of felts, usually asbestos felts. Asphalt is sometimes used as a special type "filler" with certain mineral fillers.

Granules. Finely ground or crushed rock, ceramic coated, used for exposed surfaces of roofing products.

Gravel. Coarse, granular aggregate, with pieces larger than sand grains, resulting from the natural erosion or crushing of rock.

Head lap. Minimum distance, measured at 90° to the eave along the face of a shingle as applied to a roof, from the upper edge of the shingle to the nearest exposed surface.

Hi-Tuff single-ply roofing systems.. The membrane is a 45-mil (nominal) single ply, consisting of a 10×10, 1000 denier polyester scrim reinforcement, which is fully encapsulated by two layers of membrane based on Hypalon synthetic rubber (chlorosulfonated polyethylene) from Du Pont. Manufactured by the Stevens Roofing Systems Div., JPS Elastomerics Corp.

Holiday. An area where a liquid-applied material is missing.

Ice dam. Mass of ice formed at the transition from a warm to a cold roof surface, frequently formed by refreezing melt-water at the overhang of a steep roof, causing ice and water to back up under roofing materials.

Impact resistance. Ability of the membrane to resist hail and falling objects without puncturing.

Incline. Slope of a roof expressed as the number of inches of vertical rise per horizontal foot.

Infrared roof warranty inspection. ASTM standard, published in 1990, presents the steps necessary for arranging infrared roof warranty inspections.

Inorganic. Being or composed of material other than hydrocarbons and their derivatives; not of plant or animal origin.

Insulated roof membrane assembly. Assembly applied directly to the roof deck with insulation on top of it, the membrane being on the bottom side of the insulation. The membrane is not subjected to roof traffic, ultraviolet degradation, thermal cycling, and exposure that causes alligatoring, splitting, ridging, and blistering.

Interlayment. Layer of felt or nonbituminous saturated asbestos felt not less than 18 in. wide, shingled between each course of roof covering.

Knot. Imperfection or nonhomogeneity in the materials used in fabric construction, the presence of which causes surface irregularities.

Kraft-type vapor retarder. Fire-retardant kraft lamination-type vapor retarders applied with steep asphalt, as specified by the fire underwriters.

Lap. Overlapping of surfaces of layers of materials.

Lap cement. Cement composed of bituminous materials dissolved in a volatile solvent, and of such a nature and viscosity as to firmly bind the laps of the various layers or plies of felt without any injurious effect to the materials.

Lightweight aggregate poured concrete. Wet-mix fills, such as poured vermiculite, perlite, zonolite, and foamed cellular concrete retain moisture from mixing for a considerable time. If not properly released, moisture may later cause the formation of blisters, wrinkles, and buckles in the overlying built-up roofing.

Live loads. Moving or nonpermanent loads such as wind, snow, ice, rain, or portable equipment.

Machine application. The use of equipment to apply roofing felt, or the distribution of roofing aggregates.

Maintenance of roofs. Loose gravel is usually removed by vacuum machines, sweeping with heavy street brushes, and then washing with a fire hose to remove the accumulations of dirt, silt, and soot. After patching weak spots and breaks with the proper materials and bitumen, effective maintenance measures can be applied.

Manufacturer's bond. Roofing service guarantee by a security company issued by a manufacturer to finance roofing repairs occasioned by ordinary wear within a stated period. The premium is paid by the owner.

Membrane

1. Any type of functionally continuous flexible structure of felt, fabric, or mat, or combinations thereof, and bituminous plying cement used for roofing waterproofing or dampproofing.

2. Flexible or semiflexible roof covering materials comprising the total weather-resistant component of the roofing system.

Membrane flashings. Membrane flashings are used in a variety of forms depending on the specific problems encountered. The effective "sliding and overlapping" membrane is most commonly used. The "accordion pleat" is sometimes necessary, and the "regleted" type also is used at times. All are equally effective when properly installed.

Mineral fiber felt. Felt with mineral wool as its principal component.

Mineral granules

1. Granular inorganic mineral material more than 50% of which is retained on the No. 35 sieve.

2. Natural or synthetic aggregate ranging in size from 500 μm (microns) to $\frac{1}{4}$-in. diameter, used to surface cap sheets, slate sheets, and shingles.

Mineral stabilizer. Water-insoluble inorganic mineral material that will completely pass a No. 70 sieve, used in admixture with solid or semisolid bituminous materials.

Mineral-surfaced roofing

1. Felt or fabric saturated with bitumen, coated on one or both sides with a bituminous coating, and surfaced on its weather side with mineral granules.

2. Asphalt-saturated felt, coated on one or both sides and surfaced on the weather-exposed side with mineral granules.

Modified bitumen. Composite sheets consisting of a copolymer-modified bitumen, often reinforced and sometimes surfaced with various types of films, foils, and mats.

Modified bitumen roofing. This system applies modified bitumen to reinforcing sheets of fiberglass or polyester in the factory to form a thick waterproof membrane. The finished roof is constructed by rolling 3 ft-wide strips of this membrane over a base sheet. The top or cap sheet often has a factory-applied granular or metal surface to block ultraviolet rays, reflect heat, and improve fire resistance.

Mop and flop. An application procedure in which roofing elements (insulation boards, felt piles, cap sheets, etc.) are initially placed upside down adjacent to their ultimate locations, coated with adhesive, and then turned over and applied to the substrate.

Mopping coat

1. Heavy application of bituminous material applied hot with mop or mechanical applicator to structural surfaces or saturated felts in waterproofing and membrane roof construction.

2. Application of bitumen applied hot with a mop or mechanical applicator to the substrate or to the felts of a built-up roofing membrane.

Nineteen-inch selvage. Prepared roofing sheet with a 17-in. granule-surfaced exposure and a nongranule-surfaced 19-in. selvage edge. This material is sometimes referred to as "SIS" or as "Wide Selvage Asphalt Roll Material Surfaced with Mineral Granules."

Ninety-pound. Prepared organic felt roll roofing with a granule-surfaced exposure that has a mass of approximately 90 lb per 100 sq ft.

Nonnailable deck. Any deck which is incapable of retaining an approved fastener.

Openings. Openings in the structural deck such as: plumbing vents, conduits, duct work. Bases for equipment around such openings should be completely installed before roofing begins.

Organic. Material composed of hydrocarbons or their derivatives; matter of plant or animal origin.

Phased application. The installation of a roof system or waterproofing system during two or more separate time intervals.

Pitch picket. Flanged metal or plastic container placed around a column or other roof-penetrating element and filled with bitumen or flashing cement.

Ply. Layer of felt or fabric in a built-up roofing membrane.

Plying cement. Cementitious bituminous material used for adhering layers of felts, fabrics, or mats to structural surfaces or to each other in roofing membranes.

Pond. Roof surface that is incompletely drained.

Positive drainage. The drainage condition in which consideration has been made for all loading deflections of the deck, and additional roof slope has been provided to ensure drainage of the roof area within 48 hours of rainfall.

Poured concrete decks. Deck surfaces require priming with concrete primer. Surface should be relatively smooth, dry, and unfrozen before roofing felts are applied.

Poured gypsum decks. Poured gypsum decks have a minimum thickness of 2 in. to accommodate the fastener used to secure the base felt to the smooth, dry, and unfrozen surfaces.

Prepared roofing. Any manufactured or processed roofing material other than untreated wood shingles and shakes as distinguished from built-up coverings.

Primer. Liquid bituminous material applied to structural surfaces in order to seal the surface and promote adhesion of subsequently applied heavier applications of bituminous roofing materials.

Protected membrane. Roofing membrane with insulation and protective surfacing or ballasting on top; also called "inverted" or "upside down" roof.

Rag-base felt. Heavy rag felt, thoroughly saturated and coated with asphalt and used as a base felt on certain rag and asbestos felt built-up roofs.

Re-covering. The process of covering an existing roofing system with a new roofing system.

Re-entrant corner. An inside corner of a surface, producing stress concentrations in the roofing or waterproofing membrane.

Reinforced membrane. Roofing or waterproofing membrane reinforced with felts, mats, fabrics or chopped fibers.

Replacement. The process of removing an existing roofing system and replacing it with a new roofing system.

Ridges. Condition caused by weathering, workmanship, materials, or a combination of all three, including thermal shock on the roof deck caused by quick temperature changes.

Ridging. An upward, tenting displacement of a roof membrane, frequently occurring over insulation joints, deck joints, and base sheet edges.

Roll roofing

1. Roll roofing consists of coated or noncoated felts, either smooth or mineral surfaced.

2. Roofing material laid and overlapped from roll material.

3. Roofing material, composed of fiber and saturated with asphalt, that is supplied in rolls containing 108 sq

ft in 36-in. widths and generally furnished in weights of 55 to 90 lb per roll.

Roof assembly. An assembly of interacting roof components including the roof deck, designed to weather-proof and, normally, to insulate a building's top surface.

Roof construction. Combination of materials providing cover for a building including the ceiling (if any) beneath the roof, the roof decking, roofing, and all horizontal or sloping structural members supporting the roof only, but excluding columns and other vertical load-supporting members.

Roof covering. The covering applied to the roof for weather resistance, fire resistance, or appearance.

Roof decks. Roof decks are designed to drain freely to outlets numerous enough and so placed and installed as to remove water promptly and completely from the roof. Poor drainage will cause roof damage or failure at areas where water stands for an extended period of time or damage to the system due to the standing water or freezing.

Roof deck vapor retarder. Consists of a lamination of two layer of high-strength kraft, black polyethylene laminate and two-directional glass reinforcing fibers. It is also edge reinforced for added strength.

Roofing. Covering applied to a roof surface for the purpose of providing protection from the weather. Roofing does not include insulation.

Roofing fabrics. Roofing fabrics are normally made of paper, rag, or asbestos fabrics, or combinations of these. Single-ply felts usually run 15 lb per square (100 sq ft). Double-ply runs 30 lb. Special felts may run as high as 90 lb. Size and area of one roll vary according to the manufacturer. The more common widths are 32 and 36 in.

Roofing square. 100 square feet of roofing surface.

Roofing system. Assembly of interacting roof components designed to weatherproof and insulate a building's top surface.

Roof insulation. The roof insulation usually forms the substrate for the roof membrane. It provides a solid surface free of wide joints and breaks and is of such surface character that the membrane can be firmly attached by some practical method. Insulation should be installed in accordance with the manufacturer's instructions.

Roof moisture inspection. Testing by a infrared scanner to locate wet insulation on a roof. Testing should be made before the expiration of the warranty period. The scanner locates the moisture areas, if any exist, by sensing temperature differences between wet and dry insulation. In order to verify moist conditions, roofing samples of suspected areas should be taken (ASTM C 1153).

Roofs. Flat roof is one that has 4 in or less fall per foot; a pitch roof is one that has a rise of more than 4 in. per foot.

Rooftop equipment. Any equipment located on the exterior roof of a building and/or structure. Such equipment should be installed on a platform and enclosed or screened with incombustible material as approved by the building official. All rooftop equipment over 200 lb in weight should be engineered for roof loading, wind loads, and stability.

Rosin-sized sheating paper. Sheating paper recommended for use over wood decks when roofing felts are applied without insulation.

Saddle. Small structure that helps channel surface water to drains, frequently located in a valley and often constructed like a small hip roof or like a pyramid with a diamond-shaped base.

Saturant. Asphalt of rather low softening point that is used in saturating felt to produce "saturated felt," which is used in built-up roofing, and as a base for the manufacture of asphalt-coated and mineral-surfaced roofing.

Saturated felt

1. Membrane ply that is filled as completely as practicable with bituminous material.
2. Felt that has been impregnated with bitumen.

Scatter-mopping. Random pattern of heated bitumen beads hurled onto the substrate from a broom or mop.

Scuttle. Hatch that provides access to the roof from the interior of the building.

Seal

1. A narrow closure strip made of bituminous materials.
2. To secure a roof from the entry of moisture.

Sealant. Mixture of polymers, fillers, and pigments used to fill and seal joints where moderate movement is expected; it cures to a resilient solid.

Selvage. Edge or edging that is different from the main part. In roofing, an edge of predetermined width that is normally overlaid by another piece of roofing width applied to the roof.

Selvage joint. Lapped joint designed for mineral-surfaced cap sheets. The mineral surfacing is omitted over a small portion of the longitudinal edge of the sheet below in order to obtain better adhesion of the lapped cap sheet surface with the bituminous adhesive.

Shark fin. An upward-curled felt side lap or end lap.

Shingle

1. A small unit of prepared roofing material designed for installation with similar units in overlapping rows on inclines normally exceeding 25%.
2. To cover with shingles.
3. To apply any sheet material in overlapping rows like shingles.

Shingling

1. The procedure of laying parallel felts so that one longitudinal edge of each felt overlaps and the other longitudinal edge underlaps an adjacent felt. Normally, felts are shingled on a slope so that the water flows over rather than against each lap.
2. The application of shingles to a sloped roof.

Single ply. A nominal description of roofing membranes completely installed in one application effort. The single ply membrane may be homogeneous or composite in nature.

Slag. A hard, air-cooled aggregate that is left as a residue from blast furnaces, used as a surfacing aggregate.

Slippage. Relative lateral movement of adjacent components of a built-up membrane. It occurs mainly in roofing membranes on a slope, sometimes exposing the lower plies or even the base sheet to the weather.

Slip sheet. Sheet material placed between two layers of a roofing system to assure that there is no adhesion between them.

Slope. The tangent of the angle between the roof surface and the horizontal, normally measured in inches per foot.

Smooth-surfaced roof. Built-up roofing membrane surfaced with a layer of hot-mopped asphalt, cold-applied asphalt-clay emulsion or asphalt cutback, mineral surfaced cap sheet, or, sometimes, unmopped, inorganic felt.

Smooth-surfaced roofing. Felt or fabric saturated with bitumen, coated on one or both sides with a bituminous coating, and surfaced with fine mineral surfacing.

Softening point. The temperature at which bitumen becomes soft enough to flow, as determined by an arbitrary, closely defined method.

Softening point drift. Change in the softening point of bitumen during storage or application.

Solid mopping. Continuous mopping surface with no unmopped surface areas.

Special steep asphalt. A roofing asphalt conforming to the requirements of ASTM D 312 Type IV.

Split. Membrane tear resulting from tensile strength.

Spot-cementing. Discontinuous application of asphalt, cold liquid asphalt compound, coal tar pitch, or other approved cementing material.

Spot-mopping. Mopping pattern established by the types of felts used, in which the hot bitumen is applied in spot areas, the spot being based on the size of mop used.

Spreading rate. The quantity of bitumen, roofing aggregate, and other roofing materials spread over a roof deck, given in gals, lbs, or other designation "per square".

Sprinkle mopping. Random mopping pattern in which heated bitumen beads are strewn onto the substrate with a brush or mop.

Spudding. The process of removing the roofing aggregate and most of the bituminous topcoating by scraping and chipping.

Square. One hundred square feet (100 sq ft) of roof area.

Stack vent. Vertical outlet in a built-up roof system designed to relieve the pressure exerted by moisture vapor between the roof membrane and the vapor retarder or deck.

Steamblown asphalt. An asphalt produced by blowing steam through molten asphalt to modify its properties.

Steep asphalt. Roofing asphalt conforming to the requirements of ASTM D 312 Type III.

Strip-mopping. Mopping pattern in which the hot bitumen is applied in parallel bands, depending on the type of material used and the number of plies.

Stripping or strip-flashing

1. The technique of sealing a joint between metal and the built-up roof membrane with one or two plies of felt or fabric and hot-applied or cold-applied bitumen.
2. The technique of taping joints between insulation boards or deck panels.

Structural deck. The foundation or base on which the roofing system is installed.

Structural wood fiber decks. Damage to the roofing membrane by movement of units or moisture migration through joints is minimized by nailing an inorganic or coated base sheet to the deck. A $\frac{3}{4}$-in. minimum layer of insulation material must then be solidly mopped to the base felt to receive the balance of the finished roofing membrane.

Substrate. The surface upon which the roofing or waterproofing membrane is applied (i.e., the structural deck or insulation).

Sump. An intentional depression around a drain.

Surety type roofing bond. Roofing bond issued by the manufacturer of roofing materials, when such materials have been installed by an approved roofing contractor. The bond also covers the flashing. The bond is based on the type of built-up roofing installed. The owner pays for the bond, which is based on a limited time period with an aggregate sum, to be paid by the manufacturer during that time period for all repairs to the roofing subject to the exclusions listed in the bond. Not all roofing bonds and service guarantees are surety type bonds. The National Roofing Contractors Association recommends that the roofing contractor establish a maximum term of roofing guarantee not to exceed two years, which can be included in the cost of the roofing.

Surfacing materials. Aggregates for gravel-surfaced built-up roofs usually consist of water-worn or river-washed gravel, crushed rock, or crushed blast-furnace slag.

Tar. Brown or black bituminous material, liquid or semisolid in consistency, in which the predominating constituents are bitumens obtained as condensates in the processing of coal, petroleum, oil-shale, wood, or other organic materials.

Tar asbestos or organic-felt coverings. Classes A and B. These coverings are composed of tar-saturated organic or asbestos felt in combinations applied with hot-mopping coal tar pitch. Top felt covering surfaced with approximately 400 lbs of roofing gravel or crushed stone or 300 lbs of crushed slag per 100 sq ft of finished roof embedded in a flood coat of hot-mopping coal tar pitch. Roofing gravel, crushed stone, or slag is to be dry, relatively free from dirt and dust, and graded in size from $\frac{1}{4}$ to $\frac{5}{8}$ in. This type of covering is limited to roof decks having inclines not exceeding 3 in. to the horizontal foot. It is recommended that wood decks be covered with building paper before application of the roof covering. UL rated class A and B.

Tarred felt. Roofing felt, saturated with coal tar but not coated with refined coal tar. Surface of the felt is not covered with talc or any other substance to prevent its adhesion to any cement or mastic.

Temporary roofing. Temporary roofing is an application of minimal membrane to permit occupancy of a building during the completion of the construction work. The temporary roofing materials are removed before the installation of the permanent roof.

Test cuts. Cuts into a completed roofing system when there is any doubt as to the type or weight of materials used. Cuts are not recommended by manufacturers or roofing contractors.

Thermal conductance value (C). Measure of the relative thermal or insulating efficiency of a homogeneous material expressed in British thermal units (Btu) passing through 1 sq ft of the material of a given thickness.

Thermal cycling. Changes in recorded temperatures during any given year and from year to year.

Thermal insulation. Material applied to retard the flow of heat through an enclosing surface.

Thermal resistance (R). Index of a material's resistance to heat transmission; the reciprocal of thermal conductivity k or thermal conductance C.

Thermal shock

1. Shock caused by rapid changes in temperatures affecting expansion and contraction of the roofing membrane in combination with the structural roof deck.

2. Stress-producing phenomenon resulting from sudden temperature changes in a roof membrane.

Through-wall flashing. Water-resistant membrane or material assembly extending through a wall and its cavities, positioned to direct water entering the top of the wall to the exterior.

Underlay. One or more layers of felt applied as required for a base sheet, over which finish roofing is applied.

Underlayment. Asphalt-saturated or asphalt-coated felt applied over the roof deck and under the roofing material.

Underwriters Laboratories, Inc. Recognized and accepted testing agency for testing fire resistive properties of materials and construction. It tests and classifies various built-up roof assemblies.

Valley. Line of intersection of two roof slopes, where their drainage combines.

Vapor relief vent. Vent is recommended for use on roof decks where an accumulation of moisture occurs due to the materials used for insulation or fills. The metal vent is approximately 18 in. high, 2 to 6 in. in diameter and topped with a metal cap.

Vapor retarder

1. A vapor retarder is normally used where the heat flow direction and or humidity within the building during any climate or occupancy cycle might cause the dew point or condensation to occur under the built-up roofing within the insulation chamber.

2. Material designed to restrict the passage of water vapor through a wall or roof.

Vapor seal. To avoid condensation from accumulating inside of insulation material, vapor seals are placed below the insulation, which may consist of one or two layers of felt.

Vent. An opening designed to convey water vapor or other gas from inside a building or a building component to the atmosphere, thereby relieving vapor pressure.

Vinyl-type vapor retarder. Fire-resistant adhesive and vinyl-type vapor retarder are applied as specified by Underwriters Laboratories, Inc. and the fire underwriters.

Waterproofing. Treatment of a surface or structure that prevents the passage of liquid water.

Weatherability. The ability of the membrane to resist weathering; i.e., degradation due to sun, rain, wind, etc.

Weather-exposed surfaces. All surfaces of walls, ceilings, floors, roofs, soffits, and similar surfaces exposed to the weather.

Weight. The manufacturer's shipping weight in pounds per 100 square feet of roof covering.

Wide-selvage asphalt roll roofing. Roofing surfaced with mineral granules. A single thickness of dry roofing felt is impregnated with a hot asphaltic saturant, then coated on the weather side for approximately one-half the width of the sheet with a hot asphalt coating, which may be compounded with a fine mineral stabilizer substantially insoluble in water, and surfaced by embedding fine mineral granules in the hot asphaltic coating.

Wood decks, board decks. Kiln-dried, tongue-and-groove lumber securely fastened to supporting structural members for adequate strength and rigidity. Dry sheeting paper is installed over decks to prevent bitumen drippage and sticking of the base felt.

Wood nailing strips. Nonnailable decks require wood nailing strips at eaves, edges, walls, roof openings, etc., for proper securing of metal flanges. Nailers are securely and firmly attached to the deck or through it to its supporting members. Nailers extend not less than 1 in. beyond the apron or flange.

Woven-glass fabric bitumen-treated. Bituminized woven glass fabric treated with asphalt, coal-tar pitch, or a bituminous compound suitable for use in a membrane system (ASTM D 1668).

CAULKING COMPOUNDS, SEALANTS, AND GASKETS

Acrylics. Thermoplastic resins formed by polymerging the esters of acrylic acid. They are also available as water emulsions. Acrylic resins are based on methylethyl or other alkyl acrylates. They are the nearest approach to an organic glass.

Acrylic (latex polymer) caulk. Material used for general-purpose interior and exterior caulking and sealing in residential, architectural, and industrial applications where slight to moderate movement is anticipated.

Acrylic terpolymer sealant. Sealant used for construction joints of every type.

Adhesion. Ability of the sealant to adhere to building substrates.

Adhesive failure. Adhesive failure occurs when the caulking compound or sealant pulls away from the contact surface. Separation of two bonded surfaces that occurs within the bonding material.

Angle bead or joint. Bead of a compound whose cross section is a right triangle with the hypotenuse side exposed.

Application equipment. Caulking compounds and sealants may be applied with caulking guns, hand operated pressure guns, air-operated pressure guns, reciprocal pumps and hoses, and hand pointing tools.

Application life. Period of time during which a sealant, after being mixed with a catalyst or exposed to the atmosphere, remains suitable for application; also referred to as "work life."

Application temperature range. Broad application temperature range extends practical working time.

Aroclor. Plasticizer used in some sealants. It is a chlorinated diphenyl.

ASTM conformance. Standards of quality established for one-part chemically curing sealants (ASTM C 920).

Back bed. Bead of sealant or glazing tape applied between the glass or panel and the stationary stop of the sash or frame; it is usually the first bead of sealant or tape applied when setting glass or panels.

Back painting. Permanent coating applied on metal members. This coating must now be allowed to lap over any surface of the metal that will be in contact with the caulking materials unless approved by the sealant manufacturer.

Backup. Product used to pack a cavity formed by a frame unit having a shape or section that would require too much sealant for a good joint or one that would be too expensive.

Backup material

1. Compressible material used with a liquid-applied sealant and placed in a joint to limit the depth of the sealant to ensure that the sealant takes the proper joint configuration.
2. Material placed into a joint prior to sealant application, primarily used to control the depth of the sealant.

Base. General composition of a compound, such as vegetable oil, polysulfide, and polybutane. In a two-part compound, the major unit of the compound to which a curing agent or accelerator is added before use.

Bead

1. Sealant after application in a joint, irrespective of the method of application.
2. Strip of glazing or sealing compound that has been applied to a surface or a joint; also, a molding or stop used to hold glass or panels in position.
3. Strip of sealant applied in a joint, such as caulking bead of glazing bead.

Bed (bedding)

1. Application of a sealant between a light of glass or a panel and the stationary stop or sight bar of the sash or frame; it is usually the first compound to be applied when setting glass or a panel.
2. Bead of glazing compound that is applied between a light of glass or a panel and the integral stop of a sash or frame; it is usually the first bead of compound to be applied in the glazing operation.
3. Glazing compound applied at the base of the glazing recess just before the removable stop is placed in position or buttered on the inside face of the removable stop.

Bedding compound. General-purpose sealant or bulk filler used to fill voids and cavities, as well as acting as a secondary seal behind the sealant.

Beveled bead. Bead of glazing compound applied so as to have a slanted top surface so that water will drain away from the glass or panel.

Bite

1. Width by which the flanges of a lock-strip gasket overlap the edges of supported or supporting material.
2. Amount by which the stop overlaps the edge of the panel or glass.

Bleeding. Bleeding is the absorption of oil or vehicle from a compound by an adjacent porous surface.

Bond. Attachment at an interface between substrate and sealant.

Bond breaker

1. Film of release material placed in joint to prevent undesired adhesion of the sealant to the substrate.
2. Thin layer of material used to prevent the sealant from bonding to the bottom of the joint.
3. Release type of material (such as polyethylene film sheet with adhesive on one side) used to prevent adhesion of the sealant to the backup material or back of the joint.

Bond durability. Test cycle in ASTM C 920 for measuring the bond strength after repeated weather and extension cycling.

Bond face. Part or surface of a building component that serves as a substrate for a sealant.

Bond strength. The force per unit area necessary to rupture a bond.

Building sealant. One part building sealant including a silicone formulation that cures quickly in the presence of normal atmospheric moisture to a low-modulus silicone rubber. It is particularly effective for sealing building joints designed for substantial movement and is used to seal expansion and contraction joints, precast concrete panel joints, curtain wall joints, mullion joints, and many other kinds of building construction joints. It forms permanent, flexible, watertight bonds with most building materials in any combination: stone, masonry, ceramics, marble, wood, steel, aluminum, and many plastics. In most cases no primer or surface conditioning is required. A product of Dow Corning (790) Midland, Mich.

Bulk compound.   Sealants in containers or cartridges, capable of being extruded in place.

Buttering.   Application of compound or sealant to the flat surface of a member before placing it into position, as, for example, the buttering of a removable stop before securing the stop in a place.

Butt joint.   Joint in which the structural units are joined to place the sealant into tension or compression.

Butyl rubber

1. Synthetic rubber prepared by copolymerization of isobutylene with a small amount of isoprene (both ingredients are gaseous hydrocarbons). Butyl rubber has exceptional sunlight resistance and unusually low permeability of gases.

2. Copolymer of isobutene and isoprene. As a sealant it has low recover and slow cure, but good tensile strength and elongation.

Calking.   Same as caulking.

Carbon black.   Finely divided carbon formed by the incomplete combustion of natural gas. May be used as a reinforcing filler in sealant.

Catalyst.   Substance added in small quantities to promote a reaction, while remaining unchanged itself. Sometimes referred to as the curing agent for sealants.

Caulk

1. Sealant with a relatively low (less than 20%) movement capability.

2. Fill the joints in a building with a sealant.

Caulking

1. Process used in filling a void with a sealant. This can be done manually or with pressure equipment.

2. Process of filling or sealing a joint. Materials used for this purpose usually refer to linseed oil compounds rather than the recently developed elastomeric sealing materials.

Caulking compounds.   Most caulking compounds are composed of a variety of drying oils combined with fillers such as ground limestone, asbestos fiber, chalk, silica, and similar materials.

Cellular material.   Foamy material containing many small cells, either open or closed, dispersed throughout. Density is usually described in terms of pounds per cubic foot.

Chain stopper.   Material added during polymerization process to terminate or control the degree of reaction; this could result in soft sealants or higher elongation.

Chalking.   Formation of a powdery surface due to weathering.

Channel

1. Three-sided U-shaped opening in sash or frame for receiving a light of glass or a panel. In some sash or frame units the light or panel is retained by a removable stop. Certain "fixed" or "pocket-glazing channels" do not have a removable stop.

2. Three-sided U-shaped, member in sash or frame for receiving glass or panel inserts.

Channel depth.   Measurement from the bottom of the channel to the top of the stop; measurement from sight line to the base of channel.

Channel glazing.   Method of glazing in which lights or panels are set in a U-shaped channel formed by removable stops. Channel glazing should be limited to lights of glass or panels that do not exceed 125 united inches.

Channel width.   Measurement between stationary and removable stops in a U-shaped channel at its widest point.

Checking.   Formation of slight breaks or cracks in the surface of a sealant.

Chemical cure

1. Curing by chemical reaction, usually involving the cross linking of a polymer.

2. Change in the properties of a material due to polymerization or vulcanization, which may be effected by heat, catalysts, exposure to the atmosphere, or combinations of these.

Clamping pressure.   Pressure exerted by the lip of the lock-strip gasket on material installed in the channel when the lock-strip is in place.

Clips.   Wire spring devices for holding glass in rabbeted sash, without stops, for face-glazing.

Closed-cell foam.   Foam that will not absorb water because all the cells have complete walls.

Coefficient of expansion.   Coefficient of linear expansion is the ratio of the change in length per degree to the length at 0°C.

Cohesion.   The molecular attraction that holds the body of a sealant together is the internal strength of a sealant.

Cohesive failure.   Failure characterized by pulling the body of the sealant apart, or failure of a compound by splitting open when placed under strain.

Compound.   Formulation of ingredients, usually grouped as vehicle and pigment, to form a sealant such as caulking compound or elastomer joint sealer.

Compression seal

1. Type of joint in which weathertightness is obtained by exertion of compressive pressure on the gasket or sealing material.

2. Performed seal that is installed by being compressed and inserted into the joint.

Compression set.   Amount of permanent set that remains in a specimen after removal of a compressive load.

Concrete curing compound.   Liquid applied to fresh concrete surfaces to form an impervious membrane to seal against loss of moisture and retain water for hydration of the cement, also referred to as concrete sealing or membrane-forming compound.

Concrete mold or form release agent.   Material used to prevent sticking of concrete to forms. It may be based on silicon release agents or nondrying oils.

Cone penetrometer.   Instrument for measuring the relative hardness of soft, deformable materials.

Crazing.   Series of fine cracks that may extend through the body of a layer of sealant.

Creep.   Deformation of a body with time under constant load.

Cross linked.   Term applied to molecules that are joined side by side as well as end to end.

Cure.   To set up or harden by means of a chemical reaction.

Curing agent.   One portion of a two-part sealant that, when added to the other part (normally called the "base"), will cause the sealant to set up a chemical reaction.

Cure method.   Important for compatibility decisions.

Cure time.   Cure time determines how long a sealant takes to develop its full properties, important in glazing and curtainwall sealant applications, as well as in field or plant installations.

Curtain wall. Complete exterior closure, including exterior finish, interior finish, insulation, structural independence, and means of attachment to a building.

Depolymerization. Separation of a complex molecule into simpler molecules; also softening of a sealant by the same action.

Durometer

1. Instrument that measures hardness, used to test a sealant. Resistance to the penetration of a blunt indentor point, read from a scale, from zero to 100 to establish hardness.

2. Instrument for measuring the hardness of rubberlike materials. The term is used to identify the relative hardness of rubberlike materials; for example, "low durometer" indicates relative softness or "high durometer," relative hardness.

Dynamic joint movement. Movement capability determines suitability for known building movement conditions. Sealants should always be matched with building design criteria.

Edge clearance. Distance between the bottom of a channel of a lock-strip gasket and the edge of the material installed in the channel.

Efflorescence. Efflorescence is caused by a powdery deposit (a crystalline substance) that adheres to the surface of the masonry after evaporation of the water solution of soluble salts found in masonry materials.

Elastic glazing compound. Compound formulated from selected processed oils (and/or liquid polyisobutylene) and pigments to remain plastic and resilient over longer periods of time than conventional putty.

Elasticity. Ability of a material to return to its original shape after removal of a load.

Elastomer

1. Elastic, rubberlike substance, such as natural or synthetic rubber.

2. Micromolecular material that returns rapidly to approximately its initial dimensions and shape after substantial deformation by a weak stress and release of the stress.

3. Rubbery material that returns to approximately its original dimensions in a short time after a relatively large amount of deformation.

Elastomeric. Term that describes a condition that may either occur naturally or be produced synthetically in elastic substances.

Elastomeric butyl caulk. General-purpose caulking, sealing, and glazing compound ready for application direct from the container. This material is soft, flexible, adhesive, and cohesive; it provides a durable, lasting seal between similar and dissimilar construction materials, such as concrete, masonry, glass, wood, metal, ceramics, and stone; and it is also used to seal glass, in channel glazing and to caulk joints and flashings.

Elongation

1. Amount of stretch exhibited by a compound before rupture.

2. Increase in length (expressed as a percentage of the original length).

Emulsion. Suspension of microscopic particles in water.

Epoxy. Resin formed by combining epichlorohydrin and bisphenols. Requires a curing agent for conversion to a plastic-like solid. Has outstanding adhesion and excellent chemical resistance.

Excessive shrinking. Shrinking caused by a deficiency in the materials of a compound in which the applied bead loses volume and contracts in size. May be caused by evaporation of the solvents or by absorption into adjacent porous surfaces.

Exothermic. Chemical reaction that gives off heat.

Extender. Organic material used to increase the volume and lower the cost of a sealant.

Extensibility. Ability of a sealant to stretch under tensile load.

Exterior glazed. Glazed from the exterior of the building.

Exterior stop. Removable molding or bead used on the outdoor side of glass or a panel to hold it in place.

Extrusion failure. Failure that occurs when a sealant is forced too far out of the joint. The sealant may be abraded by dirt or folded over by traffic.

Face-glazing. Method of glazing in which a rabbeted glazing recess with no removable stops is used and a triangular bead of compound is applied with a glazing knife. Face-glazing should be limited to single lights of glass or panels that do not exceed 75 united inches.

Failure. Change that renders a material no longer fit for its intended purpose.

Fatigue failure. Failure of a material due to rapid cyclic deformation.

Ferrous metal. Iron-containing metal that is generally subject to rusting (rust must be removed just prior to caulking).

Field-molded sealant. Mastic sealant that takes its shape by being placed into the joint.

Filler

1. Material that is pressed into an opening so that the applied sealant will exert pressure and form good contact against the sides of the opening.

2. Finely ground material added to a sealant to change or improve certain properties.

Firestop sealant. Black, one-component, neutral-cure, silicone-rubber sealant that exhibits good performance in applications where joints and apertures in walls and floors need to be sealed to prevent the spread of fire, smoke, toxic gas, and water.

Flange. Part of a lock-strip gasket that extends to form one side of a channel.

Floor hardener. Chemical solution that when applied to new or existing concrete floors reduces dusting and hardens the surface of the concrete.

Foam sealant. Two-component silicone foam that has excellent crack- and void-filling capabilities. The elastic properties of the foam readily accommodate minor vibration of the pipe, conduit, or the like that is in the opening.

Galvanic action. Term used when corrosion is caused by dissimilar metals touching each other and some moisture is present.

Gasket

1. Preformed shapes, such as strips, grommets, etc., of rubber or rubberlike material, used to fill and seal a joint or opening, either by itself or with a sealant.

2. Preformed deformable device designed to be placed between two adjoining parts to prevent the passage of liquid or gas between the parts.

3. Preformed shape of rubber or rubberlike composition used to fill and seal joint or openings, either alone or in conjunction with a supplemental application of a sealant.

4. Deformable material placed between two surfaces to seal the union between the surfaces.

Gasket glazing. Method of setting glass or panels in prepared openings, using a preformed gasket to obtain a weathertight seal.

Glass-centering shim or spacer shim. Small rubber block placed on the edges of glass or panel inserts to keep the insert centered in the sash or frame.

Glazing

1. Installation and securing of glass or panels in prepared openings.

2. Application of sealant in the process of installing glass in prepared openings in windows, door panels, screens, and partitions.

Grid. Structural framework used to hold windows or panels in place in a curtain wall system.

Gunability. Ability of a sealant to extrude out of a cartridge in a caulking gun.

Hardener. Substance added to control the reaction of a curing agent in a sealant.

Hardness. Resistance of a material to indentation. On the Shore A scale, hardness is measured in relative number from 0 to 100.

Head

1. Horizontal overhead member forming the top of the frame for a wall panel or window section.

2. Top member of a window or door frame.

Heel bead

1. Sealant applied at the base of a channel before or after setting a light or panel and before the removable stop is installed. It acts as a secondary seal to prevent leakage past the stop.

2. Glazing compound applied at the base of a glazing channel after setting glass or panel and before the removable stop is installed, one of its purposes being to prevent leakage past the stop. The heel bead in on the removable stop side of the glass.

3. Sealant applied at the base of a channel after setting the glass or panel and before the removable stop is installed.

Hinge. Minimum thickness of gasket material between the channel recess and the lock-strip cavity; the plane at which bonding occurs when the flange is bent open to receive or release installed material.

Hochman test cycle. The bond durability test cycle used in ASTM C 719.

Hypalon. Chlorosulfonated polyethylene synthetic that has been used as a base for making solvent-based sealants.

Identification of compound. A compound is identified on the outside of its package as to its intended usage, that is, whether it is to be used for face glazing or for channel glazing, and its appropriateness for use on wood or metal, or both.

Interface. Common boundary surface between two substances.

Interior glazed. Glazed from the interior of the building.

Interior stop. Removable molding or bead that holds the light or panel in place when it is on the interior side of the light or panel.

Jamb

1. Upright bar or member forming the side of a frame for a curtain wall panel or window section.

2. Vertical side of a window, door opening, or frame.

Joint

1. Space or opening between two or more adjoining surfaces.

2. Opening between component parts of a structure.

3. Space between adjacent surfaces of members to be sealed.

Lacquer. Protective coating used on aluminum, bronze, or other metal for protection during shipping and erection (must be removed before caulking).

Ladder gasket. Lock-strip gasket in the form of a subdivided frame having one or more integrally formed intermediate cross members.

Laitance. Thin coating that sometimes forms on the surface of concrete.

Lap joint. Joint in which the component parts overlap so that the sealant is placed into shear action.

Latex. Sap of the rubber plant; natural rubber emulsion.

Latex caulks. Rubbery emulsion caulking material. Most common latex caulks are polyvinyl acetate or vinyl acrylic.

Light. Term for a sheet or pane of glass.

Lip. Inner face of the tip of a flange on a lock-strip gasket.

Lip pressure. Pressure exerted by the lip of a lock-strip gasket on material installed in the channel when the lock-strip is in place.

Lip seal pressure. Lip pressure required to effect a seal against the passage of water and air.

Liquid polymer. Liquid material having a chemical structure consisting of a long chain of repeated small molecular units.

Liquid polysulfide polymer. One- and two-part liquid polysulfide-polymer-based joint sealant systems are cold-applied, chemically curing, synthetic-rubber-base compounds used for sealing, caulking, or glazing applications in buildings and on other types of construction.

Lock-strip. Strip that is designed to be inserted in the lock-strip cavity to force the lips against the material placed in the channel.

Lock-strip cavity. Groove in the face of a lock-strip gasket, designed to receive and retain the lock-strip.

Lock-strip gasket. Gasket in which the sealing pressure is produced internally by forcing a keyed lock-strip into a groove (lock-strip cavity) in one face of the gasket.

Low-modulus building sealant. Dow Corning 790 low-modulus building sealant is a silicone rubber chemically derived from quartz. Silicone rubbers are the most durable of all flexible construction sealants.

Mastic

1. The term "mastic" usually refers to the consistency of a given material or substance. Mastic is also commonly accepted to mean a type of material that may or may not skin, but will remain plastic for a varying period of time, depending on the base of material used; it is nonelastomeric.

2. Compound that remains elastic and pliable with age. Sometimes an oleoresinous caulking compound is referred to as a mastic compound.

3. Compound that remains elastic and pliable with age.

4. Field-molded sealant or adhesive, including materials that are gunned, poured, or troweled into place.

5. Thick, pasty sealant or adhesive.

Mastic sealers or caulking compounds. Mastic sealers or caulking compounds are used to seal the points of contact between similar and dissimilar building materials that cannot otherwise be made completely tight. Such points include glazing, the joints between windows and walls, the many joints occurring in panelized construction, the copings of parapets, and similar locations.

Maximum joint dimensions. Important as it relates to the sealant's ability to accommodate the required design movement.

Mercaptan. Test method for Mercaptan Content of the Atmosphere (ASTM D 2913).

Migration. Term used when a condition occurs that causes the spreading of oil or vehicle from a compound onto adjacent nonporous surfaces.

Minimum joint dimensions. Important as it relates to the sealant's ability to accommodate the required design movement.

Modulus. Ratio of stress to strain.

Monomer. Material composed of single molecules; building block in the manufacture of polymers.

Mullions

1. Vertical columns separating curtain wall window or panel sections.

2. External structural members in a curtain wall building. They are usually vertical and may be placed between two opaque panels, between two window frames, or between a panel and a window frame.

MVT. Measurement of the moisture vapor transmission through a film, usually expressed in terms of grams of water per square meter per 24 hours.

Neck down. Term used for change in the cross-sectional area of a sealant as it is extended.

Needle glazing. The application of a small bead of sealant using a nozzle not exceeding $\frac{1}{4}$ in. in diameter.

Nonferrous metal. Metal that does not contain iron and that in the construction field is usually aluminum, brass, or copper (these usually have a lacquer protective coating that must be removed prior to caulking).

Non-sag sealant. Sealant formulation having a consistency that will permit application in vertical joints without appreciable sagging or slumping at temperatures between 40°F and 100°F.

Non-skinning. Substance that does not form a hard or nontacky crust on exposure to air or other oxidizing processes.

Oakum. Loose fiber obtained by untwisting and picking old rope, flax, etc. At the present time it is a seldom-used joint filler before caulking.

Oleoresinous. Combination or mixture of natural or synthetic resins, mixed with drying or semidrying oils, as in caulking compounds, paints, or varnishes.

One-part systems. Systems supplied with base and accelerator in single-package ready-to-use cartridges or bulk containers, available in nonsag form suitable for either vertical or horizontal application.

Oxidation. Formation of an oxide; also, the deterioration of rubbery materials due to the action of oxygen or ozone.

Peeling. Term used for the failure of a compound, whereby the skin curls away from the remainder of the material of the compound.

Peel strength. An indication of elastomeric properties and adhesion (ASTM C 794).

Peel test. Test of a sealant using one rigid and one flexible substrate. The flexible material is folded back (usually 180°), and the substrates are peeled apart. Strength is measured in pounds per inch of width.

Performance range. Temperature range over which sealant is expected to maintain elasticity.

Permanent set. Amount of deformation that remains in a sealant after removal of a load.

Phenolic resin. Thermosetting resin. Usually formed by the reaction of phenol with formaldehyde.

Pigment. Coloring agent added to a sealant.

Plasticizer. Material that softens a sealant or adhesive by solvent action.

Plastisol. Physical mixture of resin (usually vinyl), compatible plasticizers, stabilizers, and pigments. Mixture requires fusion at elevated temperatures in order to convert the plastisol to a homogenous plastic material.

Points. Thin, flat, triangular or diamond-shaped pieces of zinc used to hold glass in a wood sash by driving them into the wood.

Poise. cgs unit of viscosity; for example, a polysulfide highway joint sealant has a viscosity of 500 poise at 77°F. Higher numbers indicate a more viscous material.

Polybutene. Light-colored, liquid, straight-chain aliphatic hydrocarbon polymer. It is nondrying and widely used as a major component in sealing and caulking compounds. It is essentially nonreactive and inert.

Polybutene base. Compound made from polybutene polymers.

Polyester. Resins manufactured by reacting a dicarboxylic acid and a dihydroxy alcohol. Polyesters are used in one-part and two-part systems for coatings, molding compounds, and manufacture of Dacron, a polyester fiber.

Polymer

1. Compound of high molecular weight derived by a combination of many small molecules (polymerization) or by a condensation of many smaller molecules, eliminating water, alcohol, and the like.

2. Compound consisting of long chainlike molecules. The building units in the chain are monomers.

Polymerization

1. Chemical reaction in which the molecules of a monomer are linked together to form larger molecules.

2. Oil is raised to a temperature of 500°F to 600°F to promote an increase in molecular size, which at the same time changes the viscosity of the oil. Oils so treated have good package stability and increased resistance to absorption into the capillaries of masonry materials.

Polysulfide

1. Polysulfide liquid polymers are mercaptan-terminated, long-chain aliphatic polymers containing disulfide linkages. They can be converted to rubbers at room temperature without shrinkage upon addition of a curling agent (two-part system). They are also available as one-part sealants.

2. Synthetic elastomer that is produced by the reaction of thylene dichloride and sodium tetrasulfide and is capable of being compounded into a one- or two-component sealant.

Polysulfide rubber. Synthetic polymer usually obtained from sodium polysulfide. Polymer segments are generally terminated with SH-groups. Polysulfide rubbers make very good sealants.

Polyurethane. Synthetic material that is available as a resin or rubber and is capable of being compounded into a one- or two-part sealant.

Pot life (P.O.T., potential open time). Time during which a two-part sealant remains suitable for use after being mixed with a catalyst.

Preformed gasket. Continuous strip, channel, or other shape molded or extruded from an elastomeric compound for use in conjunction with mechanical pressure-fastening devices to provide weathertight seals.

Preformed sealant. Sealant that is preshaped by the manufacturer before being shipped to the job site.

Prime. Seal a porous surface so that a compound will not stain, lose elasticity, or shrink excessively.

Primer

1. Solution applied to the sides of the seam to be caulked and used in some cases to reduce staining and/or obtain better adhesion for the sealant.

2. Preparatory material that is applied to joint faces in order to improve adhesion for the sealant.

3. Special coating designed to enhance the adhesion of sealant systems to certain surfaces.

Primerless adhesion. Type of adhesion that is desirable when testing or when experience has demonstrated that reliable long-term adhesion may be obtained without primer. In many critical applications, priming is an added safety factor.

Protective coating. Film to protect the surface from destructive agents or environments (abrasion, chemical action, solvents, corrosion, and weathering). Such coatings may be either temporary or permanent. Temporary protective coatings include methacrylate lacquers; permanent types include coating to prevent galvanic corrosion.

Psi stress. Stress at a given extension is an indicator of sealant modulus.

Rabbet

1. Two-sided L-shaped recess in a sash or frame to receive lights or panels. When no stop or molding is added, such rabbets are face-glazed. The addition of a removable stop produces a three-sided U-shaped channel; channel glazing is used.

2. Two sided L-shaped member used for face-glazed window sash.

Rake out. Remove a portion of a compound from a joint to permit application of a second, supplemental bead.

Rebate. Seam formed in concrete or other material to give necessary space for the sealant.

Reglet

1. Seam formed usually in poured concrete to give necessary space for the sealant.

2. Groove or recess formed in material such as concrete or masonry to receive the spline or tongue of a reglet-type lock-strip gasket.

Reinforcement. In rubbers, an increase in modulus, toughness, and tensile strength by the addition of selected fillers.

Resilience. Measure of energy stored and recovered during a loading cycle, expressed in percent.

Resilient tape. Preshaped, rubbery sealing material furnished in varying thicknesses and widths, in roll form. It may be plain or reinforced with twine, rubber, or other materials. Nonoily varieties are suitable for use with curing sealants.

Resins. Solid or liquid organic materials, generally not soluble in water, that have little or no tendency to crystallize, for example, epoxy and polyester resins.

Retarder. Substance added to slow down the cure rate of a sealant.

Rope caulk. Preformed, ropelike bead of tacky caulking compound, which may contain twine reinforcement to facilitate handling.

Routing

1. Method used for cleaning or enlarging a seam to which caulking will be applied.

2. Removing old sealant from a joint by means of a rotating bit.

Rubber latex. Water emulsion of an elastomer.

Sagging. Term used for the flow of a sealant within a joint. Can be caused by improper application, or it results from a sealant or a caulking compound that is unable to support its own weight.

Sash. Frame including muntin bars, when used, and rabbets to receive the lights of glass, either with or without removable stops, designed either for face- or channel glazing.

Screw-on bead or stop. Stop, molding, or bead secured in place by machine screws as opposed to those that snap into position without additional fastening.

Seal. Material applied in a joint or on a surface to prevent the passage of liquids, solids, or gases.

Sealant

1. Product used to fill a void between similar or unlike members.

2. Compound used to fill and seal a joint or opening.

3. Formable, adhesive material used to provide a joint seal.

4. Material used to seal joints or openings against the intrusion or passage of any foreign substance, such as water, gases, air, or dirt.

5. Elastomeric material with adhesive qualities that joins components of a similar or dissimilar nature to provide an effective barrier against the passage of the elements.

Sealant backing. In building construction, a compressible material placed in a joint before applying a sealant.

Sealer

1. Liquid primer used to seal pervious surfaces.

2. Liquid primer used to seal porous surfaces.

3. Surface coating generally applied to fill cracks, pores, or voids in the surface.

Sealing material. Any material intended for use in providing weathertight seals in building application.

Sealing tape. Preformed, uncured, or partially cured material that, when placed in a joint, has the necessary adhesive and cohesive properties to form a seal.

Selection of compound. A compound is selected for application in accordance with the manufacturer's specifications and recommendations for usage. Under no circumstances should a compound be used in a manner contrary to the manufacturer's recommendations.

Self-leveling sealant

1. Sealant that is fluid enough to be poured into horizontal joints, forming a smooth, level surface without tooling.
2. Sealant formulation having a consistency that will permit it to achieve a smooth, level surface when applied in a horizontal joint at temperatures between 40°F and 100°F.

Setting block

1. Short length of suitable material placed in the gasket channel to maintain proper edge clearance.
2. Small block of lead, neoprene, or other material placed under the bottom edge of a light or panel to support it and prevent its setting and causing distortion of the glazing compound.

Setting block or shim. Small supporting block placed under the bottom edges of glass or panel inserts.

Shape factor. Width-to-depth proportions of a field-molded sealant.

Shear test. Method of deforming a sealed or bonded joint by forcing the substrates to slide over each other. Shear strength is reported in units of force per unit area (psi).

Shelf life

1. Length of time that packaged materials can be stored under specified temperature conditions and still remain suitable for use.
2. Length of time of sealant or adhesive can be stored and still retain its properties.

Shore "A" hardness

1. Measurement of firmness with a Shore "A" durometer.
2. Value of hardness obtained with a Shore hardness tester (durometer), measuring the resistance of a rubber surface to penetration of a blunt point pressed onto the surface.

Shrinkage. Percentage weight loss or volume loss under specified accelerated conditions.

Sight line. Peripheral boundary line of the visible area of installed glass, formed by the inner edges of the frame and/or stops and the line to which the sealant contacting the lights or panels is sometimes finished off.

Silicone rubber

1. Synthetic rubber, oil, or resin capable of being compounded into one- and two-part sealants having an exceptional temperature section.
2. Synthetic rubber based on silicon, carbon, oxygen, and hydrogen. Silicone rubbers are widely used as sealants and coatings.

Sill. Horizontal bar forming the lowest member of the window or panel section.

Skin. Dry film that forms on the surface of a compound when oil or vehicle oxidizes near the surface.

Solvent. Liquid in which another substance can be dissolved.

Spacer

1. Small block placed on each side of glass or panel insert to center it in the channel and maintain uniform width of sealant bead.
2. Small block or strip of wood, neoprene, or other material, placed on each face of a glass or panel at its edges to center it in the channel, maintain uniform

width of sealant beads, and prevent excessive sealant distortion.

Spacer shims. Devices having a U-shaped cross section, 1 in. or more in length, placed on the edges of glass or panels to serve the function of spacers; also, continuous lengths of solid or closed-cell sponge neoprene or vinyl, usually round or rectangular in cross section, used for the same purpose.

Spalling. Term applied to a failure that results from external conditions, that force pieces off the compound, usually caused by frozen water or moisture collected behind the bead.

Spandrel. Panel of glass, metal, or other material filling the areas between horizontal and vertical members and between windows of two stories.

Spline. Part of a reglet-type lock-strip gasket designed to be installed in a reglet in supporting material.

Staining. Change in color or appearance of masonry adjacent to the sealant.

Stationary or integral stop. Stop or lip of a rabbet formed as an integral part of a frame or sash against which glass or panels are set.

Stop. Stationary, removable exterior, or interior stop.

Stopless glazing. The use of a sealant as a glass adhesive to keep glass in permanent position without the use of exterior stops.

Stop (stop bead). Removable member at the front of a channel, which serves to hold the glass or panel insert in the sash or frame.

Strain. Deformation per unit length, for example, the change in length divided by the original length of a test specimen. Strain is a dimensionless number. Small strains are expressed in units of inches per inch. Strains in rubbery material may be expressed in percent.

Stress. Force per unit area, usually expressed in pounds per square inch (psi).

Stress relaxation. Reduction in stress in a material that is held at a constant deformation for an extended time.

Striking. Working the sealant to give a slightly concave surface.

Structural glazing. System of bonding glass to a building's structural framing system with a high strength, high performance silicone sealant specifically designed and tested for structural glazing.

Structural sealant. Sealant used as an adhesive to bond two materials together, as in stopless glazing.

Substrate. Adherend; surface to which a sealant is bonded.

Surface conditioner. Product used with certain sealants to obtain better adhesion to the surfaces being caulked.

Synthetic rubber sealant. Synthetic rubber sealants are used for glazing windows, sealing joints in curtain wall construction, caulking masonry expansion joints, and sealing joints that require a highly durable elastomeric type of sealant. It provides a durable, permanent, rubberlike seal that maintains its effective bond between building materials with similar and dissimilar porosities, surface textures, and expansions.

Tack-free time. Time it takes before a sealant can withstand incidental touch, important for handling in many applications.

Tackiness. Stickiness of the surface of a sealant.

Tape. Preformed sealant.

Tear strength. Load required to tear apart a sealant specimen. ASTM D 624 expresses tear strength in pounds.

Tensile strength. Resistance of a material to a tensile force (a stretch); cohesive strength of a material expressed in pounds per square inch (psi).

Terpolymer. Product formed when three or more monomers are polymerized at the same time to produce a complex having different properties than the individual monomers.

Thermoplastic

1. Soft and pliable whenever heated (as some plastics) without any change of the inherent properties.
2. Material that can be repeatedly softened by heating. Thermoplastics generally have little or no chemical crosslinking.

Thermosetting

1. Term applied to a type of material that becomes hard (or firm) and unmoldable when it is heated and is resistant to additional applications of heat once it has set.
2. Material that hardens by chemical reaction and is not remeltable. The reaction usually gives off heat.

Thixotropic. Nonsagging; term applied to material that maintains its shape unless agitated. A thixotropic sealant can be placed in a joint in a vertical wall and will maintain its shape without sagging during the curing process.

Toe bead. Glazing compound applied at the base of a glazing channel before glass or panel is set and the removable stop is installed, one of its purposes being to prevent leakage past the stop. (The toe bead is on the stationary or integral stop side of the glass.)

Toluol. Commercial solvent of the aromatic family used for cleaning in the construction industry.

Tool

1. Narrow, blunt-bladed instrument used for tooling and exposed surfaces of the caulking joint.
2. Shape and finish the surface of a sealant in a joint with a specially designed blade.

Tooling. Method used to force the sealant into a rough or textured surface to improve seal. Tooling is also used to give desired smoothness or shape to the sealant.

Tooling time. Time that it takes for a skin to form on the surface of the sealant bead.

Two-part systems. Systems comprised of exact premeasured quantities of base and accelerator, which must be combined and mixed before application. Such systems are available in nonsag and self-leveling forms.

Ultimate elongation. Elongation at failure.

United inches

1. Sum of the dimensions in inches of the length and width of a light of glass or panel.
2. Sum in inches of one width plus one height, usually of glass.

Urethane. Family of polymers ranging from rubbery to brittle. Urethane is usually formed by the reaction of a diisocyanate with a hydroxyl.

Vehicle

1. Liquid portion of a compound.
2. Liquid component of a material; for example, oil paint is composed of a vehicle (linseed oil) and pigment.

Viscosity. Measure of the flow properties of a liquid or paste. Water (the standard of comparison) has viscosity of $\frac{1}{100}$ poise.

Vulcanization. Improving the elastic properties of a rubber by a chemical change.

Water repellant. Solution of silicone or other resins in an appropriate vehicle, which when applied to dry masonry surfaces renders them water repellant for a period of time.

Weatherometer. Environmental chamber in which specimens are subjected to water spray and ultraviolet light.

Web. Part of an H-type lock-strip gasket that extends between the flanges forming two channels.

Weeping. Weeping is the failure of a compound to support its own weight in a joint. It is similar to, but less pronounced than, sagging.

Window walls. Units usually containing a large window or series of windows.

Work life. Period of time during which a sealant, after being mixed with a catalyst or exposed to the atmosphere, remains suitable for application; also referred to as "application life."

Working life. Period of time after mixing during which a sealant or adhesive can be used.

Wrinkling. Wrinkling is characterized by the formation of wrinkles in the skin of a compound during the formation of its surface skin by oxidation after application.

Xylol. Commercial solvent of the aromatic family used for cleaning by the caulking industry.

CLAY TILE ROOFING

Barge tile. Fitting of a shape suitable for forming the vertical face following, and set back under, the roof edge of the gable.

Bark tile shakes. Interlocking roofing tile, installed in shingle fashion. A product of Ludowici-Celadon Company, Chicago, Ill. It is manufactured in five shapes: field tile, undereave tile, detached gable rake tile, end band tile, and V-type hip and ridge tile.

Casa Rancho roofing tile. Roofs tile manufactured in four shapes: roll gable rake tile, field tile, eave closure tile, and end-band, hip and ridge tile. A product of Ludowici-Celadon Company, Chicago, Ill.

Cement and mastic. Asphaltic, nonrunning, heavy-body plastic cement composed of asbestos fibers, asphalt, and other mineral ingredients (ASTM D 2822 Asphalt roof cement. FS-SS-C-153 Type I).

Clay roofing tiles. Tiles made from hard burned clay or shale, with a maximum water absorption of 5% from 5-hour boiling tests established under ASTM C 67. The roofing tiles should be performed with a kiln-fired colored finish and pattern.

Closed valley. Roof valley where the shingles of the intersecting slopes are continuous without open space.

Compressive strength. The roofing tiles should meet the compressive strength tests as required under ASTM C 67 and provide in excess of 4000 psi.

Covers. Convexly shaped tiles that shed water into adjacent trough formed by pan tiles.

End band. Fitting similar to, but narrower than, field tile (generally half-width), for filling out course(s) at the gable.

English tiles. English tiles measure approximately 9×14 in. and are flat with a $\frac{3}{4}$-in. interlocking side joint.

Felt Underlayment. Asphalt-impregnated, 43 lb per square roofing felt, laid over wood roof sheathing (ASTM D 2626).

Fire test. Roofing tiles should afford external protection against fire, and should also pass the (class A) burning test placed on the surface of the roofing tiles and should completely expire without charring or scorching the wood substrate immediately below.

Flashing valleys. Metal compatible with other metal roof accessories. 16 oz. copper is recommended for this use. Alternate metals are 0.020 in. thick zinc alloy, or 0.015 in. thick stainless steel.

Flexure strength. Tests provided by independent laboratories should verify that a concentrated load applied at midpoint on the tiles would indicate that they are capable of supporting more than 400 lbs (ASTM C 67).

Freeze-thaw test. The severe freezing and thawing tests should comply with the requirements of (ASTM C 67).

French interlocking roofing tile. Roof tile manufactured in eight shapes: field tile, ridge tile, end-band tile, gable rake tile, hip roll tile, hip starter tile, closed-ridge end and mating starter tile, and terminal tile.

French tiles. French tiles measure approximately 9×16 in.; they are heavily corrugated and formed with interlocking side joints and a rounded bull nose butt edge to engage the corrugations of the course immediately below.

Greek and Roman tiles. Greek and Roman tiles consist of a flat pan with raised side edges and a separate cover tile. The Roman cover is a tapered roll and the Greek cover is a tapered tent shape (inverted V with raised side edges). Additional fittings required are eave closure, hip starter and hip roll, gable rake, ridge roll, and terminals.

Head lap. That part of a course of tiles covered by the overlapping course.

Mission tiles. Mission tiles consist of tapered or straight barrel-shaped pans and covers measuring 14 to 18 in. long. Any combination of such pans and covers can be laid.

Monray roof tile. Roofing tiles having a nominal overall size of $16\frac{1}{2}$ in. $\times 13$ in. with an interlocking sidelap of $1\frac{1}{4}$ in. Tiles weigh approximately 10 lbs each. The product is manufactured by the Monier-Raymond Company, Corona, CA.

Mortar. Masonry mortar is used in setting ridge angles and hip and ridge junctures, which are mitered on the job. Mortar type as established by (ASTM C 270) type 0. Mortar should be kept back at least $\frac{3}{4}$ in. from exposed edges of tile, and all surfaces of tiles should be sponged with water to remove all excess mortar.

Nails. Nails are noncorrosive (preferably of copper), 11 gauge, large headed, and of a length to give proper penetration. On plywood the ring-type nail point should just penetrate the underside of the deck. On a board, a slater's nail of at least $1\frac{1}{2}$ in. is used but it should not penetrate the underside of the deck.

Norman roofing tile. Norman roofing tiles are installed shingle fashion; all tiles are flat shaped. They are manufactured in four shapes: field tile, end-band tile, header course ridge tile, and undereave tile. A product of Ludowici-Celadon Company Chicago, Ill.

Open valley. Roof valley where the shingles of the intersecting slopes leave an open space covered by metal flashing.

Pans. Tiles used to form troughs between covers for water to descend from ridge to eave.

Roofing membrane. When the pitch of the roof is lower than 5 to 12, a 2 ply felt underlayment is a requirement under clay roofing tiles. The first ply is nailed to the wood deck and the second ply is hot mopped into the first layer.

Roof pitches. The minimum pitch is 3 in 12 for interlocking patterns, 4 in 12 for roll patterns, and 5 in 12 for Norman tile. On extremely steep and vertical applications where wind currents may cause a "chattering," the butt of each tile is set in a dab of mastic or a bead of sealant placed where it will not be seen and will not stain the surface of tile.

Scandia roofing tile. Tile whose shape is the reverse of the ordinary barrel-type clay tile. It is manufactured in four shapes: field tile, detached gable rake tile, circular cover hip and ridge tile, and end-band tile.

Sealant. Caulk all tile joints at the peak of the roof with a sealant comparable to ANSI A 116.1 Class B, colored to match color of tiles.

Shingle tiles. Shingle tiles vary from 5 to 8 in. in width and from 12 to 15 in. in length; they are of $\frac{3}{8}$- to $\frac{5}{8}$-in. uniform thickness or tapered from $\frac{3}{4}$ in. at the butt to $\frac{3}{8}$ in.

Spanish roofing tile. Spanish roofing tiles are manufactured in six shapes: field tile, top fixture tile, detached gable rake tile, circular cover hip and ridge tile, end-band tile, and eave closure tile. A product of Ludowici-Celadon Company. Chicago, Ill.

Spanish tiles. Spanish tiles measure approximately 9×13 in. and are S-shaped with a rounded surface and a flat portion having a $1\frac{1}{2}$-in. interlocking side joint.

Starter. Tile of regular width but less than standard length, used at eaves and built-in gutters under the first course of tiles to produce a uniform cant.

Straight-barrel mission roofing tile. Roofing tile manufactured in five shapes: field tile, 8-in. booster tile, $\frac{3}{4}$-width cover rake tile, eave enclosure tile, and top fixture tile. A product of Ludowici-Celadon Company, Chicago, Ill.

Terminal. Fitting of a shape proper for covering the intersection of and forming the transition between ridge and hips.

Valley. Line of intersection of two roof slopes where their drainage combines.

Valley cut. Regular field tile modified by having one lower corner cut off at the proper angle to fit at the valley.

Williamsburg interlocking roofing tile. Roofing tile installed in shingle fashion. It is manufactured in five shapes: field tile, under eave tile, V-type hip and ridge tile, end-band tile, and detached cable rake tile. A product of Ludowici-Celadon Company, Chicago, Ill.

Wood strips. 1×2 in. strips of pine or fir under the end band and 1-in. stringers of necessary height under hip roll and ridges (height varies with roof pitch). The wood strips should be pressure preservative treated in accordance with FS-TT-W 57.

FABRICATED METALS

Abrasive. Substance having a rough surface, used for removing the surface layers of an object by polishing or grinding.

Acute angle. Angle of less than 90°.

Adhesion. State in which one body sticks fast to another body.

Air. Mixture of gases composed by volume of 78% nitrogen, 21% oxygen, and 1% other gases.

Alloy

1. Substance composed of two or more metals (or sometimes a metal and a nonmetal) that have been intimately mixed by fusion, electrolytic deposition, or the like.
2. Mixture of two or more metals; for example, brass is an alloy containing copper and zinc.

Alloyed zinc. Alloy of zinc having the corrosion resistance and physical properties of an alloy containing 0.15% titanium, 0.74% copper, and 99.11% zinc and so tempered as to be capable of being formed into the shape required for a watertight joint.

Aluminum clad wood skylights. Skylight constructed with a water repellent preservative vacuum treated wood, clad with aluminum members, hinged sash of extruded aluminum, continuous EPDM rubber flashing with aluminum counter flashing, and glazed with fully tempered, heat strengthened float glass, complying with ASTM C 1048.

Aluminum-coated sheet steel uses. Aluminum-coated sheet steel is used for roofing, siding, and roof deck for commercial and industrial buildings; housings for air conditioners and other equipment located outdoors; weather shields for refining and chemical plant equipment; metal insulation lagging; building panels; rolling doors; and other applications that require resistance to atmospheric corrosion.

Annealing. Rendering a metal soft and less brittle by heating and cooling.

Anodizing. Artificial thickening by an electrolytic process of the natural oxide coating that forms on aluminum.

Base box. Unit of value and measure of tin plate. A base box consists of 112 sheets of tin plate measuring 14×20 in.

Bead. Annular bulge that serves to reinforce a cylindrical sheet metal job.

Bit. Copper head of the soldering iron.

Black plate. Iron or steel sheet produced in widths 12 to 32 in. with thicknesses of No. 29 gauge and lighter for tin or terne plating.

Brake. Machine for bending sheet metal; bend in sheet metal.

Bright finish. Surface condition of sheet metals having a smooth and shiny appearance.

Btu. British thermal unit; the quantity of heat required to raise the temperature of 1 lb of water 1°F.

Burr. Turned-up edge of a sheet metal disk; rough edge left on cut or punched metal.

Caulk. Waterproofing material applied to chimney flashing seams or vent-pipe seams. Not a replacement for roofing cement.

Charcoal

1. Black, porous substance obtained by the imperfect combustion of organic matter, such as wood. Used for heating of solder pots.

2. Carbonaceous material obtained by the imperfect combustion of wood or other organic substances.

Cleaning. Removing dirt and scraps and neutralizing excess soldering flux.

Clearance. Adjustment in allowances for seams and in the dimensions of a sheet metal job to permit easy fitting.

Cleat

1. Sheet metal device attached to the substrate to fasten sheet metal in position and permit expansion and contraction.
2. Fitting used to hold two sheet metal edges in place: Drive cleats prevent the edges from pulling apart, S cleats prevent their telescoping one over the other.

Coalescence. Combination or fusion of two or more metals into one metal; ability of two or more metals to combine.

Coke

1. Solid, carbonaceous fuel obtained by heating coal in ovens or retorts to remove its gases.
2. Solid product resulting from the distillation of coal in an oven or closed chamber or by imperfect combustion. Used as a fuel in metallurgy.

Cold-rolled sheet metal. Sheet metal that is relatively cold during the final rolling process, producing a smooth, oxide-free surface.

Conduction. Transfer of heat through a solid material.

Conductor (leader)

1. Round or square pipe, usually corrugated, that carries rain water from the eaves trough to the drain or to the ground.
2. Pipe for conducting water from a roof to the ground or to a receptacle or drain; downspout.

Conical. Cone shaped.

Continuous strip. Unbroken length of sheet steel formed by welding strips of the metal together during the final rolling operations.

Copper-clad stainless steel. For fully annealed copper-clad stainless steel 10% of the thickness on each side is copper, while the remaining 80% is a core of ferritic stainless steel metallurgically bonded at the atomic level. Some areas of use are flashing (base, counter or cap, eave, valley, through-wall, lintel, and spandrel), roofing (batten seam, flat seam, standing seam, and mansard), reglets, gutters, downspouts, termite shields, and expansion joints.

Copper ounce weight. Weight of copper sheet or strip expressed in ounces per square foot.

Copper ounce-weight thickness. Metal thickness that corresponds to the ounce weight.

Copper sheet. Rolled flat product over 20 in. in width and with an ounce-weight thickness from 8 ozs (0.0108 in.) to 48 ozs (0.0646 in.), furnished in flat lengths of not over 10 ft.

Copper strip. Rolled flat product up to and including 20 in. in width and of source-weight thickness from 8 ozs (0.0108 in.) to 48 oz (0.0646 in.). Strip may be furnished in flat lengths of not over 10 ft or in rolls of one continuous length or not less than 25 ft wound in a cylindrical spiral.

Corrosion

1. When dissimilar metals are in contact with one another in an electrolyte, or when the electrolyte, a conductive fluid such as rainwater, washes from one metal surface to another that is more electropositive, galvanic action occurs which results in deterioration of the more electropositive.
2. Rusting or oxidation of steel, which results in the gradual destruction of the metal.

Corrosion of metals. Corrosion of metal is a result of the oxidation of the metal, caused by the combination of the oxygen of the air with the metal. The presence of heat and moisture speeds this process. Rusting is the type of corrosion that affects iron and steel.

Corrosion-resistant. Any nonferrous metal, or any metal having an unbroken surfacing of nonferrous metal, or steel with not less than 10% chromium or with not less than 0.20% copper.

Counterflashing

1. Flashing that covers and protects the top edge of the builtup base flashing.

2. Strip of metal used to prevent water from entering the top edge of the vertical side of a roof flashing; it also allows expansion and contraction without danger of breaking the flashing.

3. Flashing usually used on chimneys at the roof line to cover shingle flashing and to prevent moisture entry.

Course

1. The term used for each application of material that forms the waterproofing system or the flashing.

2. One layer of a series of materials applied to a surface (i.e., a five-course wall flashing is composed of three applications of mastic with one ply of felt sandwiched between each layer of mastic).

Covered walkways. Aluminum framed, structurally supported and glazed with available glass or acrylic materials.

Crease. Line made by bending and straightening sheet metal to facilitate the forming of the job.

Creep. Movement in a sheet metal construction due to the expansion and contraction caused by changes in temperature.

Cricket. Small, sloped structure made of metal and designed to drain moisture away from a chimney. Usually placed at the back of a chimney.

Crimping. Reducing the diameter of the end of a round pipe by fluting or corrugating the metal.

Cross brake. Slight bend made to opposite corners on the sheet metal, which serves to stiffen the metal.

Cylinder. Hollow roller-shaped sheet metal fabricator.

Dead soft. Fully annealed.

Domed skylights. Available in many sizes, with number of framing sections varying depending on design; glazing should be with plastic materials for use with flat or curved lights.

Drawing. Pressing sheet metal to shape in a die.

Ductile. Capable of being shaped or drawn; not brittle.

Durable. Lasting; firm; stable.

Eaves. Part of a roof that projects beyond the wall.

Electrolytic plating. Electrical process by which a coating of one kind of metal is deposited on another metal; most tin plate is manufactured by this process; also called "electroplating."

Ellipse. Regular oval, symmetrical with respect to both its short and its long axis.

Eutetic alloy. Combination of metals in an alloy that has a lower melting point than any other combination or alloy of the same metals. Eutectic alloy of tin-lead solder occurs when the proportions are 63 to 37.

Evaporate. Change from liquid into vapor.

Expansion. Increase, particularly in the dimensions of sheet metals, due to heat.

Female. Larger section of a die into which the smaller male section can be fitted; an internal thread.

Finish. Surface appearance given to sheet metal in the rolling process.

Fire resistant. Made of material that hinders or impedes fire.

Flange. Metal or plastic trim fitted over pipe or other venting unit to waterproof the intrusion in the roof deck or surface.

Flashing

1. Material installed at the intersection of surfaces to prevent entry of water into a structure.

2. Sheet metal or other material used in roof and wall construction to protect a building from seepage of water.

3. Specially formed strip of sheet metal laid around the junction of two surfaces in order to render the junction weathertight. It is laid, for example, between a roof and a chimney projecting above it or between a roof and a parapet wall.

4. Material used and the process of making watertight the roof intersections and other exposed places on the outside of a structure.

5. Strip, usually of sheet metal, used to waterproof the junctions of building surfaces, such as roof peaks and valleys, and the junction of a roof and chimney.

6. Material primarily used in waterproofing roof valleys, hips, intersection of roof with walls chimneys, vents, and the like.

7. Portion of a waterproofing system that connects the main body of the built-up roofing system with the vertical surfaces that adjoin the roof.

Flat finish. Final rolling process to obtain a flat finish, which removes warps and beads from sheets of metal and renders them absolutely flat.

Flat reducer. Rectangular reducer in which at least one side is at 90° to the end.

Flow point. Temperature at which a metal becomes completely liquid.

Flux

1. Substance used to promote the fusion or welding of metals by preventing oxidation of the surfaces to be joined.

2. Substance used to prevent the formation of oxide during the soldering process.

Flux residue. Remains of the flux left on the metal after the soldering is completed.

Forming operations. Steps involved in shaping and seaming a sheet metal job.

Four-ply seam. Seam construction consisting of sheet metal formed into four layers not including cleats.

Framing systems. Consideration must be given to the thermal movements which occur between the aluminum framing and the glazing. The system should control condensation and water infiltration. The opening in the structure and the enclosure should be engineered to resist live and dead loads and wind loads.

Frustrum of a cone. Section of the cone that remains when the top is cut off parallel to the base, for example, the parts of the funnel.

Fuse. Melt, especially at a high temperature, as with metals; to blend metals by melting.

Gable

1. Vertical, triangular part of a building, contained between the slopes of a double-sloped roof; also, a similar part of a building, even though not triangular. Under a single-sloped roof, that vertical part of the building above the lowest elevation of the roof and below the ridge of the roof.

2. Vertical, triangular end of a building from the eaves to the apex of the roof.

Galvanic action. Electrolytic action causing the slow destruction of one metal by an adjoining metal.

Galvanic corrosion. Process whereby one metal is consumed by another through an electric current induced by a chemical reaction between two metals.

Galvanize. Coat steel and iron with zinc in order to protect the metal against rust.

Galvanized sheet steel (galvanized iron). The most widely used method of protecting sheet steel against rust is coating it with a metal that can withstand the action of air and moisture for a long time. Zinc and tin are the most important of the metals used to coat sheet steel to protect it against rust.

Galvanized steel sheets. Flat, formed, and corrugated zinc coated (galvanized) sheets intended for roofing, siding, and flashing. The types of sheet used for roofing and siding include corrugated, V crimp, roll roofing, and many patterns. Zinc coating is 1.25-oz/sq ft class coating, pot yield.

Gambrel. Symmetrical roof with two different pitches or slopes on each side.

Gauge. Standard of measure represented by a system of numbers to establish the thickness of sheet metals. Different gauges are used when measuring ferrous and nonferrous metal.

Gauge (thickness). Thicknesses of sheet steels are indicated by gauge numbers. The gauge standard for measuring the thickness of sheet steel is known as the United States Standard Gauge for Sheet and Plate Iron and Steel.

Gravel stop. Flanged device, frequently metallic, designed to provide a continuous finished edge for roofing material and to prevent loose aggregate from washing off the roof.

Grease pans (galvanized sheet metal). Pans used to catch and contain droppings of grease or oil from mechanical roof units.

Gutter (eave trough). Shallow box or molded channel of metal or wood set below and along the eaves of a building to catch and carry off rainwater from the roof.

Hard solder. Type of solder that has a higher melting point and forms a stronger joint than soft solder (tin-lead), for example, silver solder and other brazing solders.

Heads. Wiring and turning rolls that are also called "wiring and turning heads."

Heat treatment. Method of changing the characteristics of a metal by means of heat. Heat treatment usually hardens a metal.

Heel radius. Outer radius of an elbow.

Hexagon (regular). Figure having six equal sides and six equal angles.

Hip

1. Line of junction where two adjacent slopes of a roof meet; the angle at which they meet.

2. External angle formed by the meeting of two sloping sides of a roof.

Hip roof

1. Roof that slopes up toward the center from all sides, necessitating a hip rafter at each corner.

2. Roof that rises by inclined planes from all four sides of a building.

Hot-dipped galvanized steel sheet. Steel sheet of commercial weight coated with a two-coat system of organic coating. The coating system consists of a fluorocarbon-base top coat which is applied over a quality matched primer. A whitewash coat is applied to the back side for additional protection.

Hot-dipping. Plating sheet metal by immersing it in a bath of molten metal.

Hot-rolled sheet metal. Sheet metal that is rolled while it is red-hot.

Increaser. Sheet metal fitting designed to make the transition from a smaller pipe to a larger pipe.

Jig. Device for holding the parts of sheet metal to be assembled or soldered. It helps to speed the layout of a number of similar sheet metal patterns.

Lead-coated copper sheets. Sheets for architectural design uses supplied according to weight and coating. A lead coating is applied by molten lead or by electrodeposition to two sides of 100 sq ft of copper sheets in three classes: Class A standard—12 to 15 lb, Class B heavy—20 to 30 lb, and Class C extraheavy—40 to 50 lb.

Lock seam

1. Fold the edges of sheets to permit the edges to interlock. Lose lock seaming permits free expansion and contraction of sheet metal.

2. Double-fold the edges of sheets to permit the edges to interlock.

Long terne sheet. Iron or steel sheet coated with lead-tin alloy known as "terne metal" or "terne alloy." Sheets are available in coating classes of 0.55 oz and less.

Loose lock. Type of lock seaming that permits free expansion and contraction movement of roof panels.

Machine screw. Screw having a fine thread, often used for joining parts of sheet metal.

Male. Smaller half of a die used for forming sheet metal; an external thread.

Malleable. Capable of being rolled, pounded, or drawn into shape without breaking; not brittle.

Matte finish. Dull surface resulting from the electrolytic process of coating tin plate. Further finishing is necessary to produce a bright, smooth surface.

Melting point. Temperature at which a solid begins to liquefy.

Metal flashing. Frequently used as thru-wall flashing, cap flashing, counterflashing or gravel stops.

Metal roofing. Metal shingles or sheets for application on solid roof surfaces and corrugated or otherwise shaped metal sheets or sections for application on solid roof surfaces or roof frameworks.

Miter line. Line or lines where pieces of an elbow meet; angle at which the corner of a sheet metal pattern is cut.

Movable section skylights. Skylights with electrically operated movable sections should be provided with structurally designed supports for the tracks, including the bumper stops.

Nickel stainless steel. Thru-wall flashing and membrane waterproofing material, used for masonry structures to prevent leaks or seepage of moisture, thereby avoiding water damage to building walls, foundations, and decks. Thru-wall flashing is a product of Cheney Flashing Company, Trenton, N.J.

Nonferrous. Containing no iron.

Oblong. Figure longer than it is wide.

Obtuse angle. Angle greater than 90° and less than 180°.

Octagon (regular). Figure having eight equal sides and eight equal angles.

Ogee. S-shaped curve which can be formed in sheet metal.

Oil-canning. Whenever a light-gauge reflective material is used over a broad area, optical distortion or "oil-canning" may be a problem. The following steps can be taken to avoid oil-canning when designing in stainless: Use slightly concave panels to eliminate flat reflective surfaces. Use panels with a shallow depression design. Break up the reflective surface by using textured stainless steel or by designing with angles and facets to avoid large reflective surfaces.

Oxidation

1. Combination of oxygen with metal, forming a rust on ferrous metal and petinas, tarnishes, and coatings of various depths and colors on nonferrous metals.
2. Chemical union of oxygen with any substance.

Oxide. Compound of oxygen and another substance.

Patina. Green surface that forms on copper as a result of atmospheric action and moisture (weathering).

Pattern development. Shape to which sheet metal is cut in order for it to be formed into the shape required for the job; method of working out the desired pattern.

Perpendicular. At right angles to a given plane or surface; line drawn at right angles to another line.

Pitch (roof). Incline or rise of a roof. Pitch is expressed in inches of rise per foot of run, or by the ratio of the rise to the span.

Pitch pans or pitch cups. Pans or cups of sheet metal or asphalt roofing materials used to contain asphalt, bitumen, or plastic roof cement.

Plastic range. State through which solder passes between the melting point and complete liquefaction.

Plate. Coat one metal with a thin layer of another metal.

Polygon. Figure having more than four sides.

Pretinning. Precoating sheet metal with tin to provide improved soldering conditions.

Pyramid skylights. Produced in three- or four-sided, slopes, single or double glazed with glass or plastic.

Radial line development. Method of pattern development by which development lines can be positioned on the radii and arc of a circle. This method is used in the development of conical jobs.

Radiation. Transfer of heat from a body in all directions and in straight lines.

Reducer. Sheet metal fitting designed to make the transition from a larger pipe to a smaller pipe.

Reglet. Continuous horizontal groove in the mortar joint in a wall or a prefabricated metal shape at other vertical surfaces adjoining a roof surface for the embedment of counterflashing.

Ridge. Line of junction where the opposite sides of a sloping roof meet; the angle at which they meet.

Rigid. Stiff; not easily bent.

Rise. Upward slope of a roof; also, a measurement of this upward slope, expressed as a comparison between the sloping part and a horizontal line beneath it.

Rivet. Type of fastener used for joining sheet metals.

Rosin flux cored solder. Cored solder containing specified weight percentages of natural, modified, activated rosin or ductile form of the same.

Scupper. Opening in the wall of a building to permit water to drain off a floor or flat roof; a sheet metal device placed in a wall opening to facilitate drainage.

Seam

1. Method of attaching sheet metals.
2. To join.

Seaming. Uniting the edges of a sheet or sheets by bending over or doubling or pinching together.

Shear. Device for cutting sheet metal; the angle given the upper blade of the squaring shear to promote the cutting action.

Sheet. Flat metal $\frac{3}{16}$ in. or under in thickness measured by gauge, weight, or decimals of an inch.

Sheet metal. Metal formed into sheets less than $\frac{1}{4}$ in. thick. Metal $\frac{1}{4}$ in. or more in thickness is regarded as plate.

Sheet metal work. All components of a building employing sheet metal in connection with the application of roofing materials other than metals, such as flashing, gutters, and downspouts.

Sheet-steel long terne coated. Cold-rolled sheet steel in coils and cut lengths, coated with a lead-tin alloy (terne alloy) by the hot-dip process (ASTM A 308).

Short terne plate. Black plate when it is coated with a lead-tin alloy known as "terne metal" or "terne alloy." FS QQ-T-191b and FS QQ-T-201b.

Soft solder. Alloy, usually of tin and lead, that melts at a temperature lower than 800°F.

Soft stainless steel. Soft annealed chrome-nickel stainless steel sheet or strip, generally used in AISI grade type 304. It is easily bent, cut, and soldered and has no springback. It is used for cap and base flashing, roofing and building applications, and general construction requirements specifying permanence roofing: roof edges, gravel stops, reglets, fascia, spandrels and mullions, copings, gutters, scuttles, spout heads and downspouts, termite shields, through-wall flashings, lock joints, chimney flashing, batten seams, and standing seam roofs.

Solderability. Property that enables a metal to be soldered easily.

Soldering. Joining metal parts by a metal alloy having a melting point lower than that of the base metal.

Solder metal. Tin-lead, tin-lead-antimony, tin-antimony, and silver-lead alloys in any form, known as a soft solder. Solder metal alloys are formulated to be usable in molten state at or below 800°F.

Solidification. Process of changing from a liquid to a solid.

Soldification point. Temperature at which a substance changes from a liquid to a solid.

Square. Four-sided figure having equal sides and four equal angles; tool used in pattern layout.

Stainless—18-8. The type of stainless steel containing approximately 18% chromium and 8% nickel is the all-purpose type known as 18-8. This alloy has high all-around corrosion resistance, especially under atmospheric exposure, making it particularly well suited for exterior applications.

Stainless steel

1. An alloy of iron with chromium or iron with chromium and nickel. An alloy must contain chromium in the amount of $11\frac{1}{2}\%$ or more to be classified as stainless. Small amounts of alloying elements, such as titanium, columbium, or molybdenum, may be added to attain better heat resistance, corrosion resistance, or welding stability.
2. Soft, corrosion-resistant chromium-nickel steel plate, sheet, and strip.

Stainless steel sheet. Material under $\frac{3}{16}$ in. in thickness and 24 in. and over in width, with No. 3 to No. 8 finishes, inclusive.

Standing rib panels. Panels which are secured to the structural system with concealed clips. The ribs are 2 in. high and the panels have either 12 in. or 16 in. centers. Panel materials are manufactured from the following metals: aluminized coated steel, galvanized steel, zinc aluminum alloy coated steel with a polyester topcoat.

Standing seam metal roof panels. Panels primarily manufactured from LFQ coil sheet, structural quality per ASTM A 446, grades A, B, C, and D. Other materials that can be used with this process include aluminum, copper, copper-bearing steel coated with terne alloy, gauges as specified. Available coil stock includes mill finish, polyester, and Kynar 500 coatings. Manufactured on-site by Knudson Manufacturing Inc.

Stiffener. Formed edge on sheet metal work designed to add rigidity to the job.

Substrate. Surface that underlies and serves as a foundation to support finished materials.

Swab. Small brush used to apply flux to a seam to be soldered.

Sweating. Filling seams with solder.

Tacking. Soldering a long seam in spots at short intervals to hold the seam in place while it is being soldered throughout its length.

Temper

1. Degree of hardness imparted to sheet metals.
2. Bring a heat-treatable metal to the proper degree of hardness by heating; the degree of hardness or softness of a metal.

Template. Pattern, usually cut from sheet metal, from which can be traced on any surface, for any number of patterns.

Terne

1. Prime copper-bearing steel coated with terne alloy (80% lead, 20% tin), used for roofs (standing seam, batten seam, flat lock and Bermuda seam), flashings (chimney, base, counter, valley, hip, ridge, ledge, dormer, sill, thru-wall, termite shield), perimeters, mansards, fascia, wall coverings, and drainage gutters, downspouts.
2. Sheets in coils and cut lengths, terne alloy coated by the hot-dip process. The product is commonly known as "long terne sheets."

Terne alloy. Alloy of lead and tin with a minimum tin content of 8%. For ease of manufacture a maximum of 20% of tin is permitted (ASTM A 308).

Terne-coated stainless steel. Terne-coated stainless steel is 304—18% chrome, 8% nickel—dead-soft stainless steel covered on both sides with terne alloy of 80% lead and 20% tin.

Terne plate. Sheet iron or steel coated with an alloy of lead and tin.

Thru-wall flashings. Rib-bond thru-wall flashings are manufactured from austenitic nickel-chromium corrosion-resistant stainless steel that remains unaffected by caustic chemicals, such as alkalis in mortar mixes, and will not create unsightly discolorations or stain adjacent building materials. Flashings are factory formed or shop formed for easy installation during construction without tools, eliminating malleting or bending metal on the job site. Inside or outside receiver miters and inserts are also factory or shop formed. Sheet metal cap flashings interlock in slot receiver when inserted after built-up base or roof flashings are completed. Thru-wall flashing is a product of Cheney Flashing Company, Trenton, N.J.

Tinning. Coating the point of the soldering iron with solder to improve the transfer of heat from the soldering iron to the work being soldered; precoating the seam allowance of a sheet metal job in preparation for the final soldering.

Tin plate. Cold-rolled steel coated with pure tin. A thin coating of tin also protects the steel against rust. Tin plate is identified by its bright, silvery appearance. The quality of tin plate is determined by the weight of the coating of pure tin on the surface of the steel. "First quality" tin plate has a heavier coating of tin than "Lower grade." It is usually shipped in boxes containing 112 sheets, each sheet measuring 20×28 in.

Valley. Internal angle formed by the junction of two inclined surface of a roof.

Vaulted skylights. Designed to permit long unsupported spans; usually glazed with plastic materials to reduce dead loads.

Welding. Uniting metals by heating the parts to be joined, with or without filler material, to the temperature at which fusion takes place and forms a strong, homogeneous joint.

Wetting. Action of solder on metal that enables it to form a strong intermetallic compound with the metal.

Zinc-based copper-titanium alloy strip. Strip used for functional-ornamental construction sheet metal applications, such as standing and batten seam roofing and facia, counter, cap, through-wall, eave, and most other types of flashing, gutters, downspouts, and valleys. It is also used for primarily functional applications, such as dovetail and cavity wall ties, gravel stops, expansion joints, reglets, stone anchors, termite shields, control joints, expanded metal lath, and other applications for which an easily formed, strong sheet metal commonly is specified.

Zinc-copper-titanium alloy. Alloy that comes in preweathered sheets and coils and is used for standing seam or batten seam roofing, fascia, gravel stop flashing (counter or cap, through-wall, eave, lintel, spandrel, valleys, gutters and downspouts, reglets, termite shield, coping covers, control joints, stone anchors, cavity wall ties, and other construction sheet metal applications requiring a lightweight, strong, easily formed material.

INSULATION

Acoustical insulation. Insulation used in a building to resist the passage of sound waves either from the outside or from room areas where the noise transmission would be objectionable. Sound travels much faster through a solid substance than through air. Acoustical insulation, like most thermal insulation, depends on millions of tiny air pockets. These pockets, as well as an irregular surface area, absorb much of the sound.

Adhesives. Manufacturers recommend the type of adhesive to be used for the installation of the particular insulation material. Many adhesives are not compatible with foam insulation such as tar, pitch, and solvent based adhesives. Water dispersed rubber based adhesives and solvent dispersed rubber based adhesives are used for installing foam insulation products.

Airspace. Each measurable airspace also adds to the overall resistance. Foil faced surfaces of low emissitives that form the boundaries of the airspace can further reduce the rate of radiant transfer across the space.

All-purpose insulation. All-purpose insulation for home utility use is 1 in. thick and comes in rolls 24 in. wide and 100 ft long (200 sq ft per roll). Is used for wrapping ducts, hot water tanks, and hot or cold water pipes and many other uses where thermal and/or sound insulation is needed.

Amplitude decrement factor. Difference between the maximum and the mean temperatures of a heat wave passing through a wall, roof, or floor. Is dependent on the thickness, mass, specific heat and, orientation of the wall or roof.

Blankets and batts. Flexible, fibrous type of insulation. Blankets come in rolls while the batts most often come in lengths of 24 to 48 in. Both types are available in assorted widths and thicknesses to fit between studs, joists, and rafters of different standard spacings. They may be obtained with vapor-retarder backing, which provides a flange for attachment to the framing members. Some blankets and batts are completely enclosed on all sides, with a vapor retarder on one side and also a flange. A flexible type of insulation is usually made from mineral wools or vegetable fibers. Blankets and batts serve as both thermal and acoustical insulators.

Board. Rigid or semirigid insulation formed into sections, rectangular in both plan and cross section, usually more than 48 in. long, 24 to 30 in. wide, and generally up to 4 in. thick.

Btu. British thermal unit; approximately the heat required to raise 1 lb of water from 59°F to 60°F.

C (Conductance). Thermal conductance; amount of heat expressed in British thermal units (Btu) transmitted in 1 hour from surface to surface of 1 sq ft of material or combination of materials for each degree temperature difference between the two surfaces. It should be noted that this value is not expressed in terms of per inch of thickness, but from surface to surface.

Cavity wall insulation. Insulation in cavity walls are mainly rigid types of insulation such as polystyrene or polyurethane boards. The insulation is usually most effective when it has a relatively high resistance to vapor transmission.

Coefficient of heat transmission. The resistivity per inch times the total thickness of a material.

C (Thermal conductance). Amount of heat (Btu) transmitted in 1 hour through 1 sq ft of a nonhomogeneous combination of materials, for the thickness or type under consideration for a difference in temperature of 1°F between the two surfaces of the material.

C-value (conductance). Amount of heat transmitted through a given thickness of a homogeneous material in 1 hour to create a 1°F difference in temperature between the two surfaces (Btu/hour/sq ft/F).

Dorvon. Brand of bead board; low-density, molded polystyrene insulation with good thermal and physical properties. Molded polystyrene foams exhibit occasional small voids that affect mechanical and thermal performance and should not be used where high moisture or icing conditions may prevail for extended periods. A product of Dow Chemical Company.

Duro-Board roof insulation. A homogeneous insulating product consisting primarily of perlite, a volcanic ore, with selected binders and fibers. When heated during processing, the perlite expands to form numerous air cells that create high-insulating efficiency. The binders and fibers provide the strength needed for compression resistance. A product of Manville Roofing Systems Division.

Expanded polystyrene. Rigid foam insulation manufactured in accordance with FS HH-I-5242A, Type I, Class A.

Extruded polystyrene. Rigid foam insulation manufactured in accordance with FS HH-I-524A, Type II, Class B.

F (Film conductance). Film or surface conductance; amount of heat, expressed in British thermal units (Btu), transmitted in 1 hour from 1 sq ft of a surface to the air surrounding the surface for each degree Fahrenheit temperature difference. The symbols f_i and f_o are used to designate the inside and outside surface conductances, respectively.

Fiber glass insulations. Insulations made from extrafine and long glass fibers. Insulating blankets have greater insulating value with resiliency.

Flexible or semirigid insulation. Blankets with or without paper covering and batts, which include a woollike material.

Foamed-in place insulation. Materials such as polyurethane, urea formaldehyde, or materials of the same grade or type.

Foamglas. Lightweight, rigid insulation composed of millions of completely sealed glass cells. Each cell is an insulating air space. Foamglas insulation is all glass—completely inorganic with no binders or fillers. It is available in several forms (board, block, tapered block) and in a variety of sizes and thicknesses. A product of Pittsburgh Corning Corporation, Pittsburgh, Pa.

Foil-faced fiberglass. Fiberglass rolls and batts faced with an aluminum foil vapor retarder. The foil projects beyond the fiber glass to form a stapling flange. The reflective foil face is a superior vapor retarder and adds up to 15% more insulation value.

Functions of insulation. Thermal insulation is used in a building to retard the transfer of heat from the warm side to the cold. Heat always moves toward a cooler area. Therefore, thermal insulation should be used where artificial heat is required or where the heat from the sun is objectionable. In the winter it prevents the excessive loss of heat from the building to the outside air. In the summer this same insulation will resist the rapid transfer of heat from the outside to the interior of the building. Most insulating materials depend on the multitude of tiny air pockets that they contain for their effectiveness. Heat travels with difficulty through these air pockets.

Hazard classifications. Insulation tested for flame spread, fuel contributed, and smoke developed should be in accordance with ASTM E 84 and UL 723.

High temperatures. Refractory or other specialized insulation materials used in foundry work, nuclear power facilities, the aerospace industry, and other manufacturing processes requiring insulation.

Insulating boards or slabs. Insulating boards or slabs are available in a great variety of sizes from ceiling or wall tiles 8 in. square to sheets 4 ft wide and over 10 ft long. This form is manufactured in various thickness from $\frac{1}{2}$ in. upward, the choice being dependent on the use and the amount of insulation required. Insulating boards or slabs are used structurally as sheathing materials and roof boards and decoratively as ceiling and wall finishes, as well as for thermal and acoustical insulation. They are made from mineral wools, natural and vegetable fibers, and plastics. Structural uses of insulation board have been considered in the unit when placed on sheathing. Insulation board is also used for decorative purposes in conjunction with thermal and acoustical properties. The uneven surface area of the finish boards tends to absorb sound transmitted through the air.

Insulating materials and adhesives. For pipes, ducts, plenums, and other components of heating, air handling, and cooking exhaust systems should be noncombustible or should have fire hazard classification ratings not exceeding flame spread—25; fuel contributed—35; and smoke developed—50; or should be of other approved composition.

Insulation. Substance that is a nonconductor. In building construction both thermal (or heat) and acoustical (or sound) insulations are considered. In conjunction with the most effective use of thermal insulation, consideration must be given to vapor retarders and proper ventilation.

Insulation fastening. Consists of a threaded fastener made from nonmagnetic stainless steel, with a stainless steel hardened carbon steel self-drilling point.

Jet-sulation. Sprayed-on application of top-quality bonded and laminated insulation, consisting of dry stone fibers with a strong efficient binder and primer. It contains no asbestos, and there is no dusting or flaking. A jet gun forces it into a homogeneous cellular blanket that is vapor resistant, verminproof, and fire resistant it is fabricated and applied on the job. A product of Air-O-Therm, Elk Grove Village, Ill.

k-conductivity. Thermal conductivity; the amount of heat, expressed in British thermal units (Btu), transmitted in 1 hour through 1 sq ft of a homogeneous material 1 in. thick for each degree Fahrenheit of temperature difference between the two surfaces of the material.

Keenekote. Asbestos-free, spray-applied fireproofing material that also provides excellent thermal and acoustical insulation and condensation control. Designed, formulated; and tested under Underwriters Laboratories, Inc. supervision, Kennekote is a factory-prepared blend of virgin mineral fibers and other proprietary inorganic ingredients for application to structural steel members, steel deck assemblies, concrete structures, and fire wall cavities. A product of Keene Corporation, Princeton, N.J.

Kraft-faced building insulation. Resilient, pink Fiberglas with a strong kraft vapor retarder on one side and with flanges at the edges for installation. The asphalted kraft vapor retarder has a vapor transmission (permeance) rating of 1 perm or less. Properly installed, it helps keep moisture vapor inside the building from penetrating the wall or ceiling. A product of Owens-Corning Fiberglas Corporation, Toledo, Ohio.

k (Thermal conductivity). Amount of heat (Btu) transmitted in 1 hour through 1 sq ft of a homogeneous material 1 in. thick for a difference in temperature of 1°F between the two surfaces.

K-value (conductivity). Amount of heat transmitted in 1 hour through 1 sq ft of a homogeneous material 1 in. thick

to create 1°F difference in temperature between the two surfaces of the material (Btu/hour/sq ft/in./F).

Loose-fill-type insulation. This type of insulation comes in bales or bags and may be poured or blown into spaces to be insulated. It is used to fill spaces between studs or ceiling joists after the finish material is in place. It is excellent for packing into irregular spaces and for insulating walls of existing homes. This type is made from mineral wools, vermiculite, loose cork, redwood bark, and wood fibers. It serves as both a thermal and an acoustical insulator.

Material layers. Each layer of material contributes to the resistance of heat flow, usually according to density. A layer of suitable insulation is normally many times more effective in resisting heat transfer than the combination of all other materials in the section.

Medium temperatures. Insulation for tanks, pipes, and equipment in industrial process heat applications.

Metallic insulation. Insulation made of very thin tin plate, copper, or aluminum sheets. It is also made of aluminum foil on the surface of rigid fiberboard or plasterboard. The common form is aluminum foil mounted on asphalt-impregnated kraft paper, the strength of which may be increased by the use of jute netting.

Mineral cellular insulation. Material such as foamed glass, calcium silicate, perlite, vermiculite, foamed concrete, or ceramic materials.

Mineral insulation. Insulation made from rock, slag, and glass. Rock and slag wools are made fundamentally by grinding and melting and then blowing the material into a fine, wooly mass by the use of high-pressure steam. Glass wool is another type of mineral wool or fine, fluffy glass fibers. Vermiculite is a form of mica, which when heated expands or explodes into granules about fifteen times the original size.

Moisture resistance properties. Properties of insulation materials should include information on water absorption percent by volume, based on ASTM C 272 and water vapor transmission (perm-inch) average, based on ASTM E 96.

Moisture resistant. Moisture resistance depends on the basic material of the insulation and the type of physical structure.

Organic fibrous insulation. Materials such as wood, cane, cotton, hair, cellulose, or synthetic fibers.

Outside surface films. The outside surface traps a thin film of air, which resists heat flow. This film varies with wind velocity and surface roughness.

Paroc Safing insulation. Insulation manufactured from basaltic rock, utilizing patented technology, with a melting point in excess of 2000°F. The fibers are bonded and preformed into flexible batts. Manufactured by the Partek Insulations Inc., Phenix City, Ala.

Perlite
1. Type of volcanic rock. When heated above 1600°F, the crude perlite particles expand and turn while (much like popcorn) as trapped moisture in the ore vaporizes to form microscopic cells or voids in the glasslike material. The resulting cellular structure accounts for the light weight and excellent thermal insulation of expanded perlite. Perlite aggregate combined with portland cement and water produces an ultra-lightweight concrete that is used for lightweight roof and floor fills, lightweight structural roof decks, cur-

tain wall systems, and a variety of permanent insulating applications. An air-entraining agent is used to improve the workability and to control water content and insulation value. A product of Grefco Inc., a subsidiary of General Refractories Co., Los Angeles, Calif.

2. Natural volcanic glass expanded by a closely controlled application of heat. Used as a loose fill insulation in masonry cores (ASTM C 549).

Physical properties. Physical properties for insulation materials should include information on density (lb/sq ft) average, based on ASTM C 303 compressive strength (psi) to board face based on ASTM D 1621, and linear coefficient of thermal expansion, based on ASTM D 696.

Plastic insulation. Made from polystyrene and rubber. Polystyrene is an inexpensive chemical plastic that is foamed to lightweight rigid shapes. It will not crumble, rot, or disintegrate.

Polystyrene. Resin made by polymerization of styrene as the sole monomer.

Polyurethane

1. Any of various polymers obtained by reaction of a di- or triisocyanate ester with a polyester or glycol and used in making rigid foams.

2. Resins made by the condensation of organic isocyanates with compounds or resins that contain hydroxol groups.

Poured in place insulation. Uses a lightweight insulating concrete.

Rapco foam. Is a modified urea-formaldehyde resin, formulated in accordance with the patented isoschaum process. It is cold-setting and forms a low-density, noncombustible resilient foam. The material has the ability to flow into odd-shaped spaces, around wires and piping, etc. Setting takes place 10 to 60 seconds after it leaves the applicator gun. The material can be troweled before setting. There is no further expansion of the material after it leaves the applicator gun. Voids can be completely filled without danger of subsequent pressure buildup. A product of Rapco Chemical Inc., New York, N.Y.

Reflective insulation. Sheet material with one or both surfaces of comparatively low heat emissivity, such as aluminum foil. When used in building construction, the surfaces face air spaces, reducing the radiation across the air space.

Reflective or metallic insulation. This type of insulation forms barriers of air spaces and possesses the insulating value of reflecting radiant energy. It depends on its bright reflecting surfaces to turn back the heat. The shiny side must be installed so that it does not touch any other surface.

Resistance ($R = 1/c$). Overall resistance; amount of resistance to heat flow between air on the warm side and air on the cold side of the building section.

Reverse flange batts. Insulation with a black draft vapor retarder one side and flanged breather kraft on the other. This type of insulation is used where application from "outside" the structure is required, as in manufactured home construction and crawl spaces. A product of Johns-Manville, Denver, Colo.

Rigid insulation board. Structural building board made of coarse wood or cane fiber in $\frac{1}{2}$- and $\frac{25}{32}$-in. thicknesses. It can be obtained in various sheet sizes, in various densities, and with several treatments.

Rigid urethane. Form of board stock, manufactured in precured bun form and further processed for specific end uses. It has noncritical through cryogenic applications.

Rigid urethane foam. Cellular plastic formed by the reaction of two liquid chemicals (isocyanates and polyols) in the presence of certain additives and catalytic agents. The mixture begins to foam instantly and quickly expands to about thirty times its original volume. Foam completely fills the area, space, or cavity to be insulated and hardens into an airtight mass. It becomes tack free in minutes and is totally cured in 12 hours. The foam is one of the most inert of chemical compounds; it is stable and nontoxic.

Roof fill insulation. Insulation material that is composed of expanded volcanic glass rock combined with a thermoplastic binder. The material is applied hot and rolled to compact the loose fill to proper density.

R (Thermal resistance of resistivity). Reciprocal of transmission, conductance, or conductivity.

R-value (resistance). Reciprocal of transmission, conductance, or conductivity.

Rubber insulation. Insulation made from synthetic rubber containing cells filled with nitrogen.

Spinsulation. Fiber glass rolls and batts faced with a sturdy black kraft paper vapor retarder. The kraft paper projects beyond the fiber glass to form a stapling flange. A product of Johns-Manville, Denver, Colo.

Sprayed mineral fiber and mastic. Asbestos-free spray fiber, consisting primarily of refined white inorganic mineral wool. The water-base, freeze-thaw-stable, PVA-type mastic contains a nonmercuric fungicide.

Sta-Fit. Unbacked fiber glass rolls and batts. It stays in place because of its natural resiliency and because it is slightly wider than the spaces it fills. It is used with separate vapor retarders such as 2-mil polyethelene film or foil-backed gypsum wallboard. A product of Johns-Manville, Denver, Colo.

Styrofoam brand plastic foam. Styrofoam SM, Styrofoam IB, and Styrofoam TG brand plastic foams are made of poystyrene expanded into a closed-cell foam by a unique extrusion process exclusive to the Dow Chemical Company. All are permanently resistant to water and water vapor owing to this closed-cell formation and consequently remain stable. They will not support mold growth and will not deteriorate or crumble inside the walls they are insulating.

Styropor. Beads are composed of a styrene polymer containing an expanding agent that when subjected to heat expands up to forty times its original volume. Following expansion, the beads are molded into blocks or shapes by further heat processing and after an aging or curing time cut into slabs, sheets, or special shapes. A product of BASF Canada Ltd., Montreal, Quebec, Canada.

Tapered edge strip

1. Tapered insulation strip used on the roof at the perimeter and at curbs that extend through a roof.

2. Provides a gradual transition from one layer of insulation to another.

Thermafiber regular blankets. Insulation made from spun mineral fibers, faced on one side with sturdy asphalted vapor retarder, encased on the other side with porous kraft paper. Blankets supplied open-faced without breather paper. Meets ASTM C 665 Standard Type II. Manufactured by the United States Gypsum Company, Chicago, Ill.

Thermal conductance. Time rate of heat flow (Btu/hour/sq ft/F).

Thermal conductivity. Time rate of heat flow through a homogenous material under steady-state conditions, through unit area, per unit temperature gradient in the direction perpendicular to an isothermal surface, expressed in British thermal units per hour per square foot per degree Fahrenheit per inch of thickness.

Thermal functions of insulation. Two basic functions of building insulations are to control temperatures of inside surfaces and spaces that affect the comfort of occupants and aid or deter condensation, and to conserve energy by reducing heat transmission through building construction that determines the energy requirements for both heating and cooling.

Thermal inertia. Property that modifies the effect of the U value on the heat transmission of a building element by expanding the time scale or time lag.

Thermal insulation

1. Material high in resistance to heat transmission that when placed in the walls, ceiling, or floors of a structure will reduce the rate of heat flow.

2. Material having air- or gas-filled pockets that when properly applied will retard the transfer of heat with reasonable effectiveness under ordinary conditions.

Thermal properties of double-glazed reflective plate glass. U value (winter) = 0.32; heat gain for average orientation = 15.92 Btu/hour/sq ft; heat gain for west orientation = 34.08 Btu/hour/sq ft; no appreciable time lag.

Thermal properties of single clear plate glass. U value (winter) = 0.98; heat gain for average orientation = 93.35 Btu/hour/sq ft; heat gain for west orientation = 234.60 Btu/hour/sq ft; no appreciable time lag.

Thermal resistance. Reciprocal of a heat transfer coefficient as expressed by U, thermal conductance, or film or surface conductance. The unit is Fahrenheit degrees per Btu (hour) (square foot); thus a wall with a U value of 0.25 would have resistance value of $1/U = 1/0.25 = 4.0$.

Thurane or Zer-O-Cel. Urethane foams are rigid polyurethane. Polyurethane is expanded into a uniform closed-cell insulation board. Low thermal conductivity, heat resistance, good compressive strength, and solvent resistance are some of the properties. A product of Dow Chemical Co.

Time lag. Delay caused by heat storage and its subsequent release by the structure. The time lag increases as the mass of the wall increases.

Types of insulation materials. Building insulations such as fiberglass, rock or slag, cellulose, molded polystyrene, extruded polystyrene, polyurethane, polyisocyanurate, urea-formaldehyde, perlite (loose fill) vermiculite (loose fill), and insulating concrete.

Urea-formaldehyde. Thermal setting synthetic resin used in producing rigid foams.

Urethane. Synthetic resin similar to polyurethane, used in making rigid foam.

Urethane and isocyanurate foamed plastics. Systems comprised of two liquids that when mixed can form a rigid mass thirty times their original volume, or as rigid performed materials. These materials act as sound attenuation agents for acoustical insulation. Urethane's 60- to 62-lb/sq ft buoyancy affords a flotation medium ideal for marine use. A product of CPR Division. The UpJohn Company, Torrance, Calif.

Urethane foam roofing board. Insulation board consisting of urethane foam mill-formed between 2 layers of high strength asphalt saturated felt membranes. Manufactured to comply with FS HH-1-530, ASTM D 3490.

U. Overall coefficient of heat transmission or thermal transmittance (air to air); time rate of heat flow usually expressed in Btu per (hour) (square foot) (Fahrenheit degree temperature difference between air on the inside and air on the outside of wall, floor, roof, or ceiling).

U-factor. Overall heat transmission coefficient; the amount of heat, expressed in British thermal units (Btu) transmitted in 1 hour through 1 sq ft of a building section (wall, floor, or ceiling) for each degree Fahrenheit of temperature difference between air on the warm side and air on the cold side of the building section.

U (Overall coefficient of heat transmission). Amount of heat (Btu) transmitted in 1 hour/sq ft of the wall, floor, roof, or ceiling for a difference in temperature of 1°F between the air on the inside and outside of the wall, floor, roof, or ceiling.

U-value (transmission). Overall coefficient of heat transmission through composite materials; reciprocal of the sum of the R values for each element between the inside and outside air (Btu/hour/sq ft/F).

Vapor control. Condensation of moisture from the air occurs when the temperature of warm moist air is reduced and the cooler air is unable to hold as much water vapor. Condensation in exterior walls is a common cause of paint failure, warped siding, and rotted studs and sills. Also, moisture penetration can reduce the effectiveness of wall and ceiling insulation. An efficient vapor retarder installed in walls and ceilings usually prevents this damage and is particularly important in climates with wide summer-winter temperature fluctuations and also in humid mild climates.

Vapor retarder

1. Material used to prevent interior water vapor from passing through the insulation and then, under certain atmospheric conditions, condensing inside exterior walls, ceiling, and floor spaces. Vapor retarders are always placed on the warm side of the walls.

2. Material or materials that when installed on the high vapor pressure side retard the passage of the moisture vapor to the lower vapor pressure side.

Vegetable or natural insulation. Insulation made from processed wood, sugar cane, corn stalks, and certain grasses. Cotton in the form of blankets treated to resist fire is a natural insulation. Cork, redwood bark, sawdust, and shavings may be used in their natural state or processed into various shapes and forms.

Ventilation. Process of changing the air in a room by either natural or artificial means. Proper ventilation under roof areas prevents the accumulation of hot air in the summer, thus aiding the insulating ceiling to maintain a cooler interior temperature. Ventilation is accomplished by installing screened louvers at the highest practical location on the roof or in the gable ends of the house.

Vermiculite. Micaceous mineral that expands by heat in a vermicular motion. Used as a loose fill insulation in masonry cores (ASTM C 516).

RUBBER—BY-PRODUCTS AND SYNTHETICS

Age resistance. Rubber normally oxidizes slowly on exposure to air at ordinary temperatures. Oxidation is accelerated by heat and ozone. Hard rubber ages less rapidly than soft, and most of the synthetics age at somewhat the same or greatly retarded rates compared with natural rubber.

Antioxidants. Substances added to rubber to protect it from oxidation or aging.

Autoclave. Vessel used for production purposes, which withstands high pressures.

Bakelite. Generic name for a thermosetting plastic.

Basic rubber. A generic term that includes elastomers and elastomer compounds, regardless of origin.

Bound rubber. The portion of rubber in a mix that is so closely associated with the filler as to be unextractable by the usual rubber solvents.

Butyl. Butyl is made by the copolymerization of isobutylene with a small proportion of isoprene or butadiene. It has the lowest gas permeability of all the rubbers and consequently is widely used for making products in which gases must be held with a minimum of diffusion.

Butyl synthetic rubber sheeting. Rubber moisture retarder sheeting that is a strong dust-free, textured one-ply rubber sheeting made from butyl synthetic rubber. Used as a continuous waterproof retarder for underground structures.

Carbon black. Black pigment produced by the incomplete burning of natural gas or oil. It is widely used as a filler, particularly in the rubber industry. Because it possesses useful ultraviolet protective properties, it is also much used in polyethylene compounds intended for such applications as cold water piping and black agricultural sheet.

Categories of rubber. Two principal categories of rubber are hard and soft. Hard rubber has a high degree of hardness and rigidity produced by vulcanization with high proportions of sulfur (ranging as high as 30 to 50% of the rubber). In soft rubber, however, sulfur may range as low as 1 to 5% and usually not more than 10%.

Centrifuged rubber latex. Latex in which the rubber concentration has been increased by the removal of a serum by centrifugal force.

Characteristics of rubber. In the raw state, rubbers are generally quite plastic, especially when warm, have relatively low strength, are attacked by various solvents, and often can be dissolved to form cements. These characteristics are necessary for processing, assembling, and forming, but are not consistent with the strength, heat stability, and elasticity required in a finished product. The latter desirable properties are obtained by vulcanization, a chemical change involving the interlinking of the rubber molecules and requiring incorporation of sulfur, zinc oxide, organic accelerators, and other ingredients in the raw rubber prior to heat treatment.

Closed-cell cellular rubber. Cellular material in which practically all the individual cells are nonconnecting.

Closed-cell closure strips. Closure strips manufactured from a basic polymer of neoprene or ethylene propylene terpolymer materials. They make ideal sealants for corrugated metal, asbestos, glass, or plastic roofing and siding installations because of the closed-cell structure.

Compressibility. With the exception of sponge rubber, completely confined rubber, as in a tight container, is virtually impossible to compress.

Conveyor belts. Conveyor belts are used for transporting such material as crushed rock, dirt, sand, gravel, slag, and similar materials. A typical conveyor belt consists of cotton duck plies alternated with thin rubber plies; the assembly is wrapped in a rubber cover, and all elements are united into a single structure by vulcanization. In order for the conveyor belt to withstand extreme conditions, it is made with some textile or metal cords instead of the woven fabric. Some conveyor belts are especially arranged to assume a trough form and made to stretch less than similar all-fabric belts.

Crepe. Type of raw natural rubber.

Crumb. Rubber just after it has been coagulated; ground vulcanized rubber.

Curing. Term used for vulcanizing.

Durometer. Instrument for measuring the hardness of rubber or rubberlike materials (ASTM D 2240).

Ebonite. Rubber that has been heavily vulcanized to a point at which it becomes hard and rigid.

Elasticity. Ability of a substance to undergo deformation when subjected to a force and return to its original shape when the force is removed. Polymers, such as vulcanized rubber, display this characteristic to a remarkable degree owing to the folded shape of their long molecules. This type of elasticity is often described as "rubberlike elasticity."

Elastomers. Elastomers, which are sometimes grouped with the rubbers, include polyethylene, cyclized rubber, plasticized polyvinyl chloride, and polybutene.

Elongation. Soft vulcanized compounds may stretch as much as 1000% of their original length, whereas the elongation of hard rubber ranges usually between 1 and 50%. Even after these great elongations, rubber will return practically to its original length.

Energy absorption. The energy-storing ability of rubber is about 150 times that of spring-tempered steel. The resilient energy-storing capacity of rubber is 14,600 ft-lb/lb as compared with 95.3 for spring steel.

Evaporated rubber latex. Latex in which the rubber concentration has been increased by evaporation of some of the water.

Expanded rubber. Cellular rubber having closed cells, made from a solid rubber compound.

Expansion and contraction. Volumetric changes caused by temperature are generally greater for rubber than for metals. Soft vulcanized rubber compounds, for example, have thermal coefficients of expansion ranging from 0.00011 to 0.00005, whereas hard vulcanized rubber compounds range from 0.00004 to 0.000015, as compared with coefficients for aluminum of 0.000012, glass of 0.000005, and steel of 0.000007.

Expansion joint material. Performed rubber compression sealing extrusion that effectively seals moisture, corrosives, and in compressibles from expansion joints between concrete and other building materials. The elastomeric element is made of neoprene to resist deterioration from exposure to weather, sunlight, oils, chemicals, heat, abrasion, and impact.

Flat rubber belting. Laminate is a combination of several plies of cotton fabric or cord, all bonded together by a soft-rubber compound.

Floor coating. Coating formulated from epoxide-stabilized chlorinated rubber, special elastomers, and aromatic solvents. It forms an impervious tough coating and dries to a slip-resistant, colorless film in approximately $\frac{1}{2}$ hour. When used as a curing compound, it retains up to 98% of the water in concrete and thus ensures development of the maximum strength of the concrete.

Flux. Additive that enables a substance to melt at a lower temperature.

Former. Solid replica of an article, for dipping into latex to acquire a coating of rubber.

Formula. List of chemicals in a specific mix.

Friction. The coefficient of friction between soft rubber and dry steel may exceed unity, but when the surfaces are wet, the friction coefficient drops radically and may become as low as 0.02 in water-lubricated rubber bearings.

Gas permeability. Thiokol and butyl compounds have very low rates of permeation by air, hydrogen, helium, and carbon dioxide, in contrast with the natural and GR-S compounds, whose gas permeability is relatively high.

Gel rubber. The portion of rubber insoluble in a chosen solvent.

Gutta-percha

1. Natural plastic obtained from a tree in Malaya and other eastern countries.

2. Rubberlike material obtained from the leaves and bark of certain tropical trees, sometimes used for the insulation of electrical wiring and for transmission belting and various adhesives.

Hardness. Resistance a rubber shows to deformation (ASTM D 2240).

Hard rubber. Vulcanized rubber containing about 30% of sulfur combined with rubber.

Hydrocarbon. Chemical compound made up of the elements hydrogen and carbon only.

Hypalon. (Registered trademark of E. I. DuPont De Nemours Co.) Synthetic rubber often used in conjunction with neoprene in elastomeric roof coverings.

Inert fillers. Fillers that do not improve the physical properties of a rubber.

Insoluble rubber. Material that is capable of recovering from large deformations quickly and forcibly and can be, or already is, modified to a state in which it is essentially insoluble, (but can swell) in boiling solvent, such as benzene, methlethyl ketone, and ethanol-toluene azeotrope.

International rubber hardness degree. Measure of hardness, the magnitude of which is derived from the depth of penetration of a specified indenter into a specimen under specified conditions. The scale is so chosen that 0° would represent a material showing no measurable resistance to indentation, and 100° would represent a material showing measurable indentation.

Laminated rubber. Rubber is often combined with various textiles, fabrics, filaments, and metal wire to obtain strength, stability, abrasion resistance, and flexibility.

Latex. Emulsion of a rubber hydrocarbon in water.

Light resistance. Natural, GR-S, and nitryl rubber compounds under tension are likely to crack when exposed to sunlight. Other rubbers are highly resistant to similar attack. Sunlight discolors hard rubber somewhat and reduces its surface electrical resistivity.

Manufactured fiber rubber. Fiber-forming substance is comprised of natural or synthetic rubber, including the following categories:

1. A manufactured fiber in which the fiber-forming substance is a hydrocarbon such as natural rubber, polyisoprene, polybutadiene, copolymers of dienes and

hydrocarbons, or amorphous (noncystalline) polyolefins.

2. A manufactured fiber in which the fiber-forming substance is a copolymer of acrylonitrile and a diene, composed of not more than 50% but at least 10% by weight of acrylonitrile units.

Mastication. Reduction of rubber to a soft doughlike condition to enable substances to be mixed into it.

Membrane

1. Butyl rubber membrane formulated from a blend of isobutylene-isoprene and ethylene propylene diene monomer. The formulation has been developed specifically for use as a one-ply roofing application, flashing of roof openings, roof expansion joint covers, panels used in waterproofing foundations, plazas, and decks, as well as liners for reservoirs and canals.

2. Blend of chlorinated rubber and special plasticizers used to form a durable, chemical resistant membrane on concrete surfaces.

Methyl chloride. Volatile liquid used as a diluent in the making of butyl rubber.

Natural rubber. The elastic substance obtained by coagulating the milky juice of any of various plants and prepared as sheets and then dried; chemically, essentially polyisoprene. Also called "caoutchouc," or "India rubber."

Neoprene

1. Neoprene is made by the polymerization of chloroprene. It has very good mechanical properties and is particularly resistant to sunlight, heat, aging, and oil. Neoprene is used for making machine belts, gaskets, oil hose, and insulation on wire cable and for other electrical applications subject to outdoor exposure.

2. Special-purpose rubber containing the element chlorine.

Nitrile. Copolymer of acrylonitrile and butadiene. Its excellent resistance to oils and solvents makes it useful for fuel and solvent hoses, hydraulic equipment parts, and similar applications.

Nitrile rubber. Special-purpose rubber containing the element nitrogen.

Oil-extended rubber. Grade of raw rubber containing a relatively high proportion of processing oil.

Perbunan. Nitrile rubber made from 25% acrylonitrile and 75% butadiene.

Permanent set. Amount of rubber that fails to come back to its original position after being released from extension.

Plantation rubber. Natural rubber obtained from specially cultivated trees.

Ply. Single layer of rubberized fabric.

Polyblends. Colloquial term generally applied in the styrene field to mechanical mixtures of polystyrene and rubber.

Polyisoprene. Chemical name for the natural rubber chain.

Preformed joint seal. Crystallization-resistant polychloroprene (neoprene) that is resilient and resistant to heat, oil, and ozone. The material should, when tested by an approved testing agency, conform to the physical requirements provided by the manufacturer for the product.

Preserved rubber latex. Rubber latex treated to inhibit coagulation and accompanying putrefaction.

Prevulcanized rubber latex. Rubber latex in which the particles have been sufficiently vulcanized to produce films and useful articles by drying only.

Raw rubber. Natural or synthetic rubber, usually in bales or packages, the starting material for the manufacture of rubber articles.

Rubber

1. For construction purposes both natural and synthetic rubbers are used. Natural rubber is often called "crude rubber" in its unvulcanized form and is composed of large complex molecules of isoprene.

2. Elastic substance obtained from the sap of various plants that grow in tropical lands.

Rubber accelerator. Substance that increases the speed of curing of rubber, such as thiocarbanilide.

Rubber adhesive. Adhesive made with a rubber base by using natural or synthetic rubber in an evaporative solvent; a tacky mixture of rubber and filler material, as used on pressure-sensitive tapes; or rubber solvent-catalyst mixtures (usually two-part) that cure in place.

Rubber-base paint. Paint in which chlorinated rubber or synthetic latex is the nonvolatile vehicle.

Rubber belt. Conveyor belt that consists essentially of a rubber-covered fabric made of cotton, nylon, or other synthetic fiber, with steel-wire reinforcement.

Rubber blanket. Rubber sheet used as a functional die in rubber forming.

Rubber cement. Adhesive composed of unvulcanized rubber in an organic solvent.

Rubber compound. As used in the manufacture of rubber articles, and intimate mixture of a elastomer(s) with all the materials necessary for the finished article.

Rubber-covered steel conveyor. Steel conveyor band with a cover of rubber bonded to the steel.

Rubber fiber. Fiber composed of natural or synthetic rubber; used to make elastic yarn for clothing.

Rubber flooring. Floor surfacing material in tile or sheet form, consisting of compounded natural rubber or synthetic rubber, or both, in combination with mineral fillers and pigments.

Rubber gasket. Rubber formed and used as a seal in concrete pipe joints.

Rubber hose. Nearly all rubber hose is laminated and composed of layers of rubber combined with reinforcing materials like cotton duck, textile cords, and metal wire. Laminated rubber hose can be made with a large variety of structures. A typical hose consists of an inner rubber lining, a number of intermediate layers consisting of braided cord or cotton duck impregnated with rubber, and outside that, several more layers of fabric, spirally wound cord, spirally wound metal, or in some cases, spirally wound flat steel ribbon. Outside of all this is another layer of rubber to provide resistance to abrasion. Hose for transporting oil, water, and wet concrete under pressure and for dredging purposes is made of heavy-duty laminated rubber.

Rubber hydrochloride. White, thermoplastic hydrochloric acid derivative of rubber in the form of water-insoluble powder or clear film, soluble in aromatic hydrocarbons, softens at 110–120°C. Used for protective coverings, food packaging, shower curtains, and rainwear.

Rubberize. To impregnate or coat, or both, with rubber compound.

Rubber-lined pipes, tanks, and similar equipment. Lining materials include all the natural and synthetic rubbers

in various degrees of hardness, depending on the application. Frequently, latex rubber is deposited directly from the latex solution onto the metal surface to be covered. The deposited layer is subsequently vulcanized. Rubber linings can be bonded to ordinary steel, stainless steel, brass, aluminum, concrete, and wood. Adhesion to aluminum is inferior to adhesion to steel. Covering for brass must be compounded according to the composition of the metal.

Rubber plating. The laying down of a rubber coating onto metals by electrodeposition or by ionic coagulation.

Rubber products. Items of commerce in which the major portion of the filler bearing material is rubber. Typical examples are rubber bands, rubber balls, rubber tires, etc; the rubber composition in the product as distinct from the fabric or metal to which it may be attached.

Rubber solvent. Fast-evaporating petroleum distillate used as a solvent for tackifying rubber during plying (laminating) operations and in compounding rubber cements.

Rubber sponge. Foamed, flexible rubber produced by beating air into unvulcanized latex or by incorporating a gas-producing ingredient (such as sodium bicarbonate) into a strongly masticated rubber stock. Used for comfort cushioning, packaging, and shock insulation. Also known as "cellular rubber," "foam rubber," "rubber foam," and "sponge rubber."

Rubber tree. Tall tree of the spurge family (Euphorbiaceac) from which latex is collected and coagulated to produce rubber.

S.B.R. Styrene-butadiene rubber, also known in the industry as "synthetic rubber."

Silicone rubber. When made in rubbery consistency forms, silicone rubber is a material exhibiting exceptional inertness and temperature resistance. Is used for making gaskets, electrical insulation, and similar products that maintain their properties at both high and low temperatures.

Skim rubber. Rubber coagulated from the serum separated during the concentration of natural rubber latex.

Sol rubber. The portion of rubber soluble in a chosen solvent.

Sponge rubber. Cellular rubber consisting predominantly of open cells and made from a dry rubber compound.

Stabilized rubber latex. Rubber latex treated to inhibit premature coagulation.

Strength. Tensile strength based on original cross section ranges from 300 to 4500 psi for soft rubber stocks and from about 1000 to 10,000 psi for hard rubber. Under compression, soft rubber merely distorts, whereas true hard rubber can be subjected to 10,000 to 15,000 psi before distorting markedly (ASTM D 412).

Sulfide rubbers (thiokols). Polysulfides of high molecular weight have rubbery properties, and articles made from them, such as hose and tank linings, exhibit good resistance to solvents, oils, ozone, low temperatures, and outdoor exposure.

Synthetic rubber. GR-S is the synthetic rubber most nearly like crude rubber and is the product of styrene and butadiene copolymerization. It is the most widely used of the synthetic rubbers. It is not oil resistant.

Synthetic rubber resin-based adhesive. Adhesive compounded and designed to be used with elastic roof flashing. It produces tough, permanent bonds with resistance to fatigue, vibration, hot oil, gasoline, aeromatic (aviation jet) fuels, organic solvents, and prolonged exposures to temperatures up to 300°F and down to 60°F. The adhesive is inflammable.

Synthetic substance. Substance that has been built up or synthesized, the substance synthesized having exactly the same structure as the natural one. In connection with rubber it usually denotes a substance having rubbery properties.

Tensile properties. Properties connected with the strength of rubber (ASTM D 412).

V belts. Belts that consist of a combination of fabric and rubber, frequently combined with reinforcing grommets of cotton, rayon, steel, or other high-strength material extending around the central portion.

Vibration insulators. A vibration insulator usually consists of a layer of soft rubber bonded between two layers of metal. Another type of insulator consists of a rubber tube or cylinder vulcanized to two concentric metal tubes, the rubber being deflected in shear. A variant of this consists of a cylinder of soft rubber vulcanized to a tubular or solid steel core and a steel outer shell, the entire combination being placed in torsion to act as a spring. Heavy-duty mounts of this type are employed on trucks, buses, and other applications calling for rugged construction.

Virgin rubber. Raw rubber that has not been adequately dried, as shown by a white interior.

Vulcanite. Term used for hard rubber.

Vulcanization. A chemical reaction of sulfur, or other vulcanizing agent, with rubber or plastic to cause cross-linking of the polymer chains; it increases strength and resiliency of the polymer.

Vulcanized fiber. A laminated plastic made by chemically treating layers of 100% rag-content paper to gelatinize the paper and fuse the layers into a solid mass.

Water retention and sealing. A high-solids, chlorinated, rubber-based curing and sealing compound is especially formulated for concrete surfaces. It provides a film with exceptionally low moisture vapor transmission rates and penetrates the pores and capillaries sealing out moisture, deicing salts, acids, grease, oil, and other corrosive solutions harmful to the productive life of a concrete floor.

Wild rubber. Natural rubber obtained from trees growing wild, that is, trees that have not been cultivated.

WATERPROOFING AND DAMPPROOFING

Acrylic coating. Pigmented, solvent type acrylic coating, which can be sprayed or brushed on or roller applied. The coating consists of a blend of resins in a solvent system.

Acid sludge. Waste mixture of sulfonated hydrocarbons resulting from the treatment of bitumens with sulfuric acid.

Aggregate. Inert material, such as sand, gravel, shell, slag, or broken stone, or combinations thereof, with which the cementing material is mixed to form a mortar or concrete.

Albertite. Soft jet black mineral (asphaltic hydrocarbon) derived from petroleum by natural oxidation, obtained in Canada.

Alum. White crystalline substance consisting of a hydrated double sulfate of aluminum and potassium.

Anthracene. Waxy crystalline hydrocarbon found principally in coal tars.

Anti-Hydro. Chemical admixture that when used in accordance with manufacturers' directions, with the concrete mix, may provide impermeable concrete. A product of Anti-Hydro Waterproofing Co., Newark, N.J.

Artificial bitumens. Hydrocarbon residues produced by the partial or fractional distillation of bitumen.

Artificial gilsonite. Product obtained from the distillation of a mixture of fish remains and wood and redistillation of the resulting oil.

Asbestos. Mineral of fibrous crystalline structure composed, chemically, of silicates of lime and magnesia, and alumina.

Asbestos felt. Sheets made of asbestos shreds.

Ash water glass. Same as water glass.

Asphalt

1. Solid or semisolid native bitumens, solid or semisolid bitumens obtained by refining petroleums, or solid or semisolid bitumens that are combinations of the bitumens mentioned with petroleums or derivatives thereof, which melt on the application of heat and which consist of a mixture of hydrocarbons and their derivatives of complex structure, largely cyclic and bridge compounds.

2. Derivative obtained from the distillation of crude petroleum.

Asphalt cement. Fluxed or unfluxed asphaltic material, especially prepared with regard to quality and consistency.

Asphalt emulsion systems. Asphalt emulsion systems eliminate heating kettles and the danger of burns when working in closely confined spaces. Work should be scheduled to avoid rain or freezing weather until the emulsion "sets up." Evaporation of moisture during setup requires good ventilation. The emulsion should always be used with glass fabrics for waterproofing.

Asphaltenes. Components of the bitumen in petroleum, petroleum products, malthas, asphalt cements, and solid native bitumens that are soluble in carbon disulfide, but insoluble in paraffin naphthas.

Asphaltic. Similar to or essentially composed of asphalt.

Asphaltic coal. Solid forms of asphalt (originally derived from petroleum), which through loss of their oil content by oxidation resemble glance coal.

Asphaltic concrete. Broken stone bound together with asphaltic cement.

Asphaltic limestone. Limestone or limestone sands naturally impregnated with asphalt or maltha.

Asphaltic oils. Asphaltic petroleums.

Asphaltic petroleums. Petroleums containing an asphaltic base.

Asphaltic sandstone. Sandstone naturally impregnated with asphalt or maltha.

Asphalt mastic. Refined asphalt, particularly that obtained from bituminous rocks; mixture of fine mineral matter and asphalt.

Asphalt primer

1. Asphalt that has been thinned so as to have liquid properties. It is used in preparing surfaces to receive asphalt coatings.

2. Asphalt that is cut back with solvent. It is used in priming and preparing surfaces to receive asphalt coatings.

Asphalt putty. Mixture of a liquid and a solid asphalt (and fine mineral matter, usually) or asphalt and coal tar pitch having a particular consistency.

Bank-run gravel. Normal product of a gravel bank.

Benzene. Colorless volatile, inflammable liquid. A by-product of coal tar.

Benzine. Colorless and volatile fraction of petroleum.

Bentonite. Granular bentonite passes 90% through a 20-mesh sieve and less than 10% through a 200-mesh sieve. Its mineralogical composition is 90% minimum montmorillonite with 10% maximum native sediments of feldspars, micas, and unaltered volcanic ash. Its proximate chemical analysis is silica 60%, alumina 20%, iron oxides 5%, magnesia 2%, soda 3%, lime 1%, chemically bound water 6%, minor 2%.

Bentonite membrane. Membranes consist of granular bentonite sealed inside a smooth face sheet of corrugated kraft paper coated with a temporary water-resistant resin, designed to help block out prehydration from water prior to placement of backfill.

Benzol. Light, volatile, colorless coal tar distillate.

Binder. Bituminous cementing material employed in the membrane system of waterproofing.

Bitumen

1. Natural hydrocarbon mixture of mineral occurrence, widely diffused in various forms that grade by imperceptible degrees from a light gas to a solid. Commercially the term includes only the heavy liquid and solid asphalts.
2. Asphalt or coal tar product used in the application of waterproofing and dampproofing.

Bituminous. Materials containing bitumen.

Bituminous cement. Bituminous material, suitable for use as a binder, having cementing qualities that are dependent mainly on its bituminous character.

Bituminous emulsion. Mixture of a bituminous oil and water made miscible through the action of a saponifying agent or alkaline soap.

Bituminous grouts. Grouts suitable for waterproofing above or below ground level as protective coatings. These grouts also can be used for membrane waterproofing or for bedding and filling the joints of brickwork. Either asphaltic or coal tar pitch materials of dampproofing and waterproofing grade are used together with siliceous sands.

Bituminous paints. Mixtures of liquid paraffin and asphalt; coal tar mixtures of bitumen with some drying oil.

Bituminous putty. Mixture of bituminous materials and whiting or other mineral of a puttylike consistency.

Bituthene waterproofing membrane. Tough, pliable, waterproof sheet of high-quality polyethylene, coated on one side with a thick, factory controlled layer of adhesive-consistency rubberized asphalt. Bituthene membrane is supplied in rolls interwound with a special release paper that protects the adhesive surface until ready for use and promotes easy handling during installation. A product of Grace Construction Products, Cambridge, Mass.

Blisters. Deformations in membranes due to entrapped air or moisture or to insufficient moppings of bitumen between plies of membrane.

Blown asphalt. Asphalt through which air has been blown during the process of refining.

Blown petroleum. Semisolid or solid product produced primarily by the action of air upon originally fluid native bitumens which are heated during the blowing process.

Bonding adhesive. Liquid adhesive expressly compounded for bonding elastomeric sheet to a surface.

Building paper. Paper, usually a heavy, strong grade, sized with rosin to make it water resisting.

Built-up membranes. Membranes consisting of several plies of treated felt cemented with asphalt or coal tar pitch.

Burlap. Woven fabric made of jute.

Caffali process. Proprietary process for applying paraffin to exterior masonry surfaces.

Calcium compounds. Salts of metal calcium or lime.

Caoutchouc. Hydrocarbon possessing properties similar to India rubber.

Carbenes. Components of the bitumen in petroleums, petroleum products, malthas, asphalt cements, and solid native bitumens that are soluble in carbon disulfide but insoluble in carbon tetrachloride.

Carbon bisulfide. Volatile and extremely inflammable compound of carbon and sulfur.

Carbon disulfide. Same as carbon bisulfide.

Carbon tetrachloride. Volatile, noninflammable compound of carbon and chlorine.

Carborundum. Artifical abrasive material resulting from the burning, in an electric furnace, of a mixture of sand, coke, sawdust and salt.

Cement. Adhesive substance used for uniting particles of materials to each other. Ordinarily, the term is applied only to calcined "cement rock" or to artificially prepared, calcined, and ground mixtures of limestone and silicious materials; however it is sometimes used to designate bituminous binder used in waterproofing.

Chinawood oil. Oil pressed from the seeds of the wood-oil tree of China and Japan.

Clay. Finely divided earth, generally siliceous and aluminous, which will pass a 200-mesh sieve.

Coal tar. Mixture of hydrocarbon distillates, mostly unsaturated ring compounds, produced in the destructive distillation of coal.

Coal tar pitch

1. Residue (of a viscous consistency) resulting from the distillation of coal tar.
2. Derivative obtained from the distillation of bituminous coal-tar.

Coat. Total result of one or more surface applications; apply a coat.

Coke-oven tar. Coal tar produced in by-product coke ovens in the manufacture of coke from bituminous coal.

Colloidal material. Gelatinous substance resembling glue or jelly and consisting of microscopically fine particles of matter.

Compressed asphalt. Rock asphalt pavement.

Concrete primer. Thin liquid compound applied as a first coat to a concrete surface preparatory to coating it with a more viscous compound.

Consistency. Degree of solidity or fluidity of bituminous materials.

Continuous membranes. Continuous membranes are essential to prevent penetration of moisture under hydrostatic head. The membranes are traditionally formed of multiple plies of saturated felt or fabric with interply moppings of hot pitch or asphalt.

Corundum. Crystalline mineral abrasive mined in the United States and ground for use for many purposes.

Cotton drill. Woven cotton fabric used for membranes.

Cracked oil. Petroleum residuum that has been overheated in the process of manufacture.

Creosote primer. Refined coal tar creosote oil having liquid properties; used for priming and preparing surfaces to receive coal-tar-base coatings.

Cutback products. Petroleum or tar residue that have been fluxed, each with its own or similar distillate, to a desired consistency.

Dampproofing

1. Treating masonry internally or externally to prevent dampness or moisture from penetrating the masonry.
2. Treatment of a surface or structure to retard passage of liquid water.
3. Process that retards the penetration of moisture into structures above or below grade and differs from waterproofing in that hydrostatic heads are not involved. In below-grade dampproofing, water-resistant coatings are applied to exterior surfaces. These materials penetrate the masonry sealing the pores and forming a continuous protective film.
4. One or more coatings of a compound that is impervious to water, usually a bituminous type, applied either hot or cold.

Dehydrated tar. Crude tar from which all water has been removed.

Drainage. Provision for the disposition of water in or about a structure, necessary for proper waterproofing installation.

Elastomer. Substance that can be stretched at room temperature to at least twice its original length and, after having been stretched and the stress removed, returns with force to approximately its original length in a short time.

Elastomeric. Term describing macromolecular material that returns rapidly to approximately its initial dimensions and shape after substantial deformation by a weak stress and release of the stress.

Elastomeric sheet. Fully cured (vulcanized) sheet compounded of on elastomer and having its properties.

Elaterite. Soft, elastic variety of asphalt, resembling rubber; also, an appropriated name of a proprietary waterproofing compound.

Emulsion

1. Combination of water and oily material made miscible through the action of a saponifying agent.
2. Asphalt or coal tar pitch that has been rendered liquid by suspension of asphalt or pitch particles in water, usually with the aid of a small quantity of an emulsifying agent. After application, the emulsion "breaks," thereby allowing the water to evaporate and leave the desired grade of bitumen behind.

Expansion joint

1. Separation of the mass of a structure, usually the combination of a metal shape or form filled with elastic material, which provides the means for slight movement in the structure.
2. Joint in the structural surface designed to accommodate expansion and contraction.

Fabric

1. Woven cotton or glass fiber cloth saturated with asphalt or coal tar and used in waterproofing. The material is manufactured in several weights, but the most commonly used weight is 15 lb. It is available in both saturant coated and uncoated types.
2. Cotton cloth or burlap treated with asphalt or coal tar pitch, used in membrane installations.
3. Woven cotton or glass fiber cloth saturated with asphalt or coal tar and supplied in various weights, depending on the use as reinforcement for waterproofing bitumens or spandrel flashing.
4. Woven burlap cloth saturated with asphalt or coal tar, depending on the use as reinforcement for waterproofing bitumens.

Felt

1. Soft form of paper sheet composed chiefly of pulp and rags and saturated with coal tar pitch or asphalt, used in membrane installations.
2. Roofing felt saturated with asphalt or coal tar and used in waterproofing.
3. Rag, asbestos, or wood fiber felt saturated with asphalt or coal tar and used in waterproofing. The material is manufactured in several weights, but the most commonly used weight is 15 lb. It is available in both saturant coated and uncoated types.

Filler. Relatively fine cementitious material used to fill the voids in concrete aggregate; the material used to fill the voids in some expansion joints.

Fixed carbon. Organic matter of the residual coke obtained on burning hydrocarbon products in a covered vessel in the absence of free oxygen.

Flashing. Piece of metal or other waterproof material used to keep water from penetrating joints between walls and other parts of the structure.

Floating. Smoothing, with a trowel, the surface of mortar, concrete, or cementitious materials.

Flux. Bitumens, generally liquid, used in combination with harder bitumens for the purpose of softening the latter.

Foundation coating. Heavy-duty coating in a seminastic consistency, used specifically for the protection of masonry walls, both above and below grade. It is manufactured with an asphalt base and fortified with asbestos fibers, and forms a tough, elastic film over the treated surface, creating a tight moisture and vapor retarder (ASTM C 755).

Free carbon. In tars, organic matter that is insoluble in carbon bisulfide.

Fuller's earth. Fine-grained earthy material of cretaceous formation, resembling clay in appearance.

Furring compound. Compound used to bond plaster to masonry.

Gauging water. Water (in measured quantities) used in mixing mortar or concrete to a required consistency.

Gas drip. Condensate from illuminating gas, present to a greater or less degree in all gas mains and tanks and an effective solvent of most bituminous materials.

Gas-house Coal tar. Coal tar produced in gashouse retorts in the manufacture of illuminating gas from bituminous coal.

Gasoline. Volatile distillate of petroleum.

Gates UWM-28. Liquid-applied polyurethane rubber material that provides a seamless retarder. It is a two-part material that cures to a full rubbery membrane at ambient temperatures and does not depend on variable atmospheric moisture for cure. It becomes virtually tack free overnight and will be substantially cured within a few days at temperatures in the range of 80°F. A product of the Gates Engineering, Wilmington, Del.

Gilsonite. Glance pitch; pure, hard, lustrous asphalt mined principally in Utah.

Glance pitch. Pure, solid asphalt or gum asphalt.

Grahamite. Pure, solid, lusterless asphalt.

Graphite. Soft, dark-colored form of carbon with considerable luster.

Gravel. Small stones or pebbles, usually found in natural deposits more or less intermixed with sand, clay, etc., but in which mixture the particles that will not pass a 10-mesh sieve predominate.

Grit. Stone chips, slag chips, small pebbles or rounded rock particles graded or ranging in size between $\frac{1}{8}$ and $\frac{3}{8}$ in.

Ground dampness. When there is no water pressure against foundation walls or floors and no hydrostatic head is expected to build, water-resistant coatings and vapor stop products of mastics can be used to retard the penetration of ground moisture above or below grade. Coatings are applied to the exterior surfaces. They penetrate the masonry, seal the pores, and form a continuous protective film. Some dampproof coating should always be used, even when the use of drainage or tile systems lowers the water level in the ground sufficiently to eliminate the possibility of a hydrostatic head. This will ensure protection against more severe ground dampness which may occur occasionally.

Groundwater. That part of rain, hail, or snow that has percolated through and accumulated in the ground as water, chiefly in consequence of an underlying impervious strata. Location of groundwater is necessary in the proper installation of a waterproofing system.

Groundwater level. Upper surface of groundwater. Highest accumulated source of water nearest to the ground level.

Grout. Mixture of cement and water or cement, sand, and water of thinner consistency than mortar.

Grouting. Injecting grout or mortar to fill small holes and seams in and around subsurface structures.

Gypsum. Hydrated calcium sulfate.

High-carbon tars. Tars containing a high percentage of free carbon.

Hot stuff. Washing soda (carbonate of lime) when used to quicken the setting time of mortar; also colloquially, hot molten asphalt or coal tar pitch, or mastic made from these.

Hydrated lime. Finely divided white powder made of ordinary lime to which has been added just sufficient water to ensure complete slaking, leaving the product dry.

Hydrex compound. Trade name for a proprietary asphalt.

Hydrocarbons. Chemical compounds composed of the elements hydrogen and carbon.

Hydrolithic. Proprietary trade name applied to the integral system of waterproofing.

Hydrolytic. Materials used in integral waterproofing that tend to prevent the percolation of water through the treated masonry.

Hydrostatic pressure. Foundations, walls, or other portions of structures exposed to hydrostatic pressure are subject to horizontal and vertical thrust loads, greatly increasing the risk of moisture penetration unless the structure is waterproofed.

Imitatite. Black, hard variety of bitumen.

Imposmite. Solid bitumen resembling gilsonite, mined in Oklahoma.

Insulated drainage board. Product composed of polystyrene beads bound together with a waterproof adhesive binder and cut into sheets of various thicknesses. The board is factory laminated with geotextile fabric or a vapor-retardant skin. Boards are installed over a cured waterproofing or dampproofing membrane by a special adhesive.

Integral compound. Material incorporated in mortar or concrete, previous to or during mixing, to waterproof same.

Integral system. Process of incorporating waterproofing materials in mass mortar or concrete.

Iron powder. Cast iron or pig iron in powder form.

Joint filler. Compound used for filling joints between moving parts of steel or masonry (structures) subject to expansion, contraction, and vibration.

Kaolin. Fine clay the purity of which gives it a white color.

Karnak membrane system. System that provides a strong, waterproof seal that will not break down under expansion, contraction, or constant water pressure. Its components are Karnak liquid asphalt and Karnak fabric. The fabric is an open-webbed, asphalt-impregnated cotton fabric, which when bonded to a surface with successive coatings of asphalt forms a strong, pliable interlocking membrane that affords protection from water penetration under extreme conditions. A product of Karnak Chemical Corporation, Clark, N.J.

Keeper Kote—M1. One-component, urethane-polymer-based, fluid-applied adhesive rubber system designed to provide a water-impermeable retarder for foundations. This product adheres to and forms a seamless rubber coating on concrete masonry, plywood, insulating materials, terrazzo, quarry tile, and many other construction materials. A product of Keeper Chemical Corporation, Trenton, N.J.

Lake pitch. Plastic, porous, impure asphalt from the asphalt "lake" in the island of Trinidad.

Land pitch. Surface deposit of solid Trinidad lake asphalt, which is tougher and more tenacious than the "lake" asphalt.

Land plaster. Powdered gypsum.

Lap cement. Liquid bituminous compound used for cementing the laps of felts or fabrics used in membrane installations.

Larutan system. Application of a waterproofing membrane in the form of small squares of asphalt-treated cotton fabric.

Layer. Course or coat made in one application.

Lime. White substance resulting from the burning of limestone.

Linseed oil. Oil extracted from the seed of flax.

Lithocarbon. Commercial name for an asphaltic limestone mined in Uvalde.

Low-carbon tars. Tars containing a low percentage of free carbon.

Maltha. Natural or artificial asphalt containing sufficient lighter compounds to be liquid.

Malthene. Those portions of asphalt and similar materials soluble in both carbon bisulfide and petrolic ether and not readily volatile at a temperature of 163°C.

Manjak. Pure, black, lustrous bitumen from Barbados.

Mastic. Mixture of fine mineral matter and asphalt or coal tar pitch, applicable in a heated condition.

Membrane

1. Thin layer or layers of bituminous material, with or without fabric reinforcement, placed on or about a structure.
2. Functional, continuous, flexible structure, made of felt, fabric, bituminous cementing, bituminous rubber material, or combinations thereof, used for roofing or waterproofing.

Membrane system

1. System of applying a combination of elastic, membranous waterproofing materials.
2. Two or more plies of felt or fabric and asphalt or pitch are used where water pressure against walls exists continuously and builds to a head. The number of plies required required increases with the head of the water and may be determined from the recommendations of the manufacturer of materials used.

Membrane waterproofing. Three or more hot-applied coatings and layers of combination of compatible bituminous materials that are impervious to water.

Mineral naptha. Volatile petroleum distillate heavier than gasoline.

Mineral pitch. Popular name for asphalt.

Mineral rubber. Bitumen of rubbery consistency.

Mineral tar. Liquid bitumen, of a viscid nature.

Mortar. Mixture of sand, cement, or lime (or both) and water mixed to a paste consistency.

Mulseal. Emulsified asphalt for spray application. It seals pores in concrete walls against infiltration of ground moisture. Manufactured from asphalt and special emulsifying chemicals. It is fast setting ASTM D 449 Type A.

Mulseal and membrane. When sprayed on simultaneously with chopped fiber glass reinforcement, Mulseal produces a tough, seamless, watertight envelope around the entire foundation wall. The resultant membrane coating, when applied correctly, is tough and elastic enough to bridge hairline cracks in the foundation walls. A product of Emulsified Asphalts Inc.

Multite. Nonflammable compound of emulsified asphalt and selected chemicals (clay type). Manufactured for easy application by spray or brush and is suitable for above-grade work. ASTM D 449 Type A. A product of Emulsified Asphalts Inc.

Naptha. Volatile petroleum hydrocarbon distillate heavier than gasoline.

Napthalene. White, solid, crystalline hydrocarbon, occurring principally in coal tar.

Native bitumens. Bitumens occurring in nature and used for waterproofing purposes, generally as liquids, viscous liquids, or solids.

Natural cement. Fine cementing powder made by burning and grinding a cement rock at a somewhat lower heat than portland cement.

Nautral oil. Neutral mineral oil.

Oil Asphalts. Artificial oil pitches or asphaltic cements produced as a residuum from asphaltic petroleum.

Oil pitches. More or less hard oil asphalts.

Oil-gas tars. Complex hydrocarbon liquids produced by cracking oil vapors at high temperatures in the manufacture of oil gas or carbureted water gas.

Oil tar pitch. Viscous residuum of any desired consistency from the distillation of oil tars.

Ozocerite. Greasy, waxlike, yellow or brown hydrocarbon, occurring in the form of small veins in tertiary rock in Utah.

Paraffin. Commonly, the same as paraffine; hard, white, waxlike substance, chemically of the higher hydrocarbons.

Paraffine. Number of greasy, crystalline hydrocarbons of the paraffin series.

Paraffin naptha. Naptha from paraffin petroleum.

Paraffin oil. Heavy liquid fraction of the manufacture of paraffin from petroleum.

Paraffin petroleum. Petroleum the base of which is principally of the paraffin series of hydrocarbons.

Petrolene. Those portions of asphalt and similar materials soluble both in carbon bisulfide and petrolic ether and volatile at 163°C and below.

Petroleums. Native mineral oils or fluid native bitumens of variable composition.

Petrolic ether. Volatile naptha lighter than gasoline, obtained from petroleum.

Pine oil. Heavy distillate of rosin.

Pine tar. Gum of the pine tree obtained from an incision or by distillation of the wood; common rosin.

Pitch. Resin from pine tar; semisolid or solid residues from the distillation of bitumen. The term is usually applied to residue obtained from tar. It is also short for coal-tar pitch.

Pitch (straight run). Pitch run in the initial process of distillation to the consistency desired without subsequent fluxing.

Plaster bond. Name of various bituminous compounds used for bonding plaster to masonry walls and that also serve as dampproofing mediums.

Plaster of paris. Hydraulic cement; chalky powder resulting from the calcination of pure gypsum (a hydrated calcium sulfate) at a temperature between 250°F and 400°F.

Plaster type waterproofing. The material used is composed of a batch-blended powder formulation, consisting of a blend of hydraulic cements, lime, pigments, plasticizers, dispersants, and aggregates. The powder is mixed with an integral acrylic polymer emulsion bonding agent and water. When thoroughly mixed it can be applied by brush or spray gun.

Plastic cement. Plastic mixture of paint skins, coal tar, pine tar, and soya oil commonly used to seal flashing joints.

Plastic compound. Compound composed of some fine or fibrous inert substance mixed with tar or other bitumen and applied with a trowel.

Plastic slate. Mixture of coal tar and powdered slate.

Ply. Term used to denote the number of thicknesses of saturated felt used in waterproofing systems, for example, three, four, or five ply.

Polyethlene film membrane. Membrane is a self-adhesive cold-applied sheet consisting of cross-laminated polyethylene film and a rubberized asphalt.

Polymer-modified membrane. Membrane consists of a sprayed on polymer-modified asphalt combined with a rigid fiberglass board. To be used for below grade concrete or masonry foundation walls.

Portland cement. Fine cementing powder made by carefully burning and grinding cement rock or an artificial mixture of limestone and clay.

Primer. First coat applied to masonry preparatory to applying the successive coats of material for waterproofing or dampproofing purposes.

Promenade decks. Horizontal slabs over habitable spaces generally exposed to weather and covered with quarry tile, flagstone, or other permanent topping.

Protection course. Course of material, usually treated composition hardboard, laid directly, on the membrane to protect it during backfilling or application of a waring surface.

Puzzolan cement. Fine cementing powder made by mechanically mixing and powdering slaked lime and volcanic ash or slag.

Pyrobitumens. Mineral organic substances forming bitumens upon being subjected to destructive distillation.

Red rope paper. Red variety of building paper partly composed of rope waste.

Reduced petroleums. Residual oils from crude petroleum after removal of water and some volatile oils, but with the base chemically unaltered.

Refined asphalt. Bitumen after it has been freed wholly or in part from its impurities.

Refined tar. Tar freed from water by evaporation or distillation that is continued until the residue is of desired consistency or a product is produced by fluxing tar residuum with tar distillate.

Residual petroleum. Viscous residue from the distillation of crude petroleum with all the burning oils removed.

Residual tars. Tar pitch or viscous residue from the distillation of crude tar with all the light oils removed.

Resin. Dried and hardened pitch from pine and similar trees.

Rock asphalt. Solid asphalt obtained from a naturally impregnated limestone or sandstone; also, the naturally impregnated stone.

Rubberized asphalt membrane. Membrane that is composed of rubberized asphalt laminated to a polyethylene film. The surfaces are to be primed with a elastomeric-based liquid in a solvent solution before membranes are installed. The membrane has adhesive on one side for installation.

Salammoniac. Ammonium chloride; white crystalline soluble substance.

Sand. Finely divided rock detritus the particles of which will pass through a 10-mesh and be retained on a 200-mesh screen.

Sand cement. Fine cementing powder made by grinding together a mechanical mixture of portland cement and pure clean sand.

Sandwish-type construction. System whereby a structural slab is placed first, a membrane material is applied to it, and a subsequent applied finish, such as concrete topping or brick pavers, is placed over the membrane.

Semiasphaltic petroleum. Petroleum with a semiasphaltic base.

Semi-Mastic. Brush or spray-on dampproofing compound for above or below-grade interior and exterior application. The fibered asphalt coating for either the spray or brush application contains a hydrophilic agent to ensure proper adhesion to damp or green surfaces. It is used to protect concrete and masonry foundation walls from moisture penetration, for backup walls in dry-wall construction, and under plywood used over masonry. It may be applied in cool weather, spans holes and cracks;

and withstands temperature changes. A product of W. R. Meadows, Inc., Elgin, Ill.

Sheet mastic. Bituminous mastic in the form of a sheet used for waterproofing purposes.

Soluble glass. Water glass.

Splicing cement. Liquid cement used for making lap splices of elastomeric sheets.

Splicing tape. Uncured (unvulcanized) elastomeric tape used in conjunction with splicing cement in making lap splices.

Spray-Mastic. Spray-on dampproofing compound for above- or below-grade interior and exterior application. The product is of spray consistency, and is composed of nonfibered asphalt containing a hydrophilic (wetting) agent that ensures adhesion to "green" or damp surfaces. It is used to prevent dampness and moisture infiltration through concrete and masonry foundations, retailing walls, and parapet and fire walls. It may be used as a damp course over footings and as a stone backing. A product of W. R. Meadows Inc., Elgin, Ill.

Spray waterproof system. Product consists of a polymer enhanced bentonite that is combined with adhesives. The material is sprayed to vertical surfaces at a minimum dry thickness of $\frac{1}{4}$ in. After spraying, a polyethylene film is installed immediately.

Stearate. Salt of stearic acid.

Stearic acid. Derivative product of the solid fats of the animal kingdom.

Stearic pitch. Black, elastic, nonbrittle, animal by-product obtained form stearic acid in the manufacture of candles.

Substrate. Structural surface on which the membrane is placed.

Sure-Seal neoprene sheet membrane. This product is highly impermeable when in contact with oils and similar contaminates and is self-extinguishing. It has excellent ozone resistance. A product of Carlisle Tire & Rubber Company, Carlisle Pa.

Surface coating. Compound applied to a masonry surface for dampproofing or waterproofing purposes.

Sylvester process. Process of applying alternate coats of soap and alum solutions for waterproofing and dampproofing purposes.

Tar. Bitumen which yields pitch upon fractional distillation and which is produced as a distillate by the destructive distillation of bitumens pyrobitumens, or organic material.

Tar pitches. Semisolid or solid residual tars.

Tex-Mastic cytoplastic vapor retarder membrane. Homogeneous asphalt composition of special oxidized asphalt and highly controlled mineral aggregates free from foreign matter. This product is laminated, weather coated on both sides with asphalt, and formed under pressure and heat to provide a strong, durable, and flexible membrane that is highly moisture resistant. The board is used as a vapor retarder under concrete slabs and on subgrade vertical walls and as a ground cover in crawl spaces and basements. A product of J. & P. Petroleum Products Inc., Dallas, Tex.

Thermoplastic elastomeric membrane. Sheet membrane manufactured from nonplasticized chlorinated polyethylene (CPE), a synthetic elastomer. Has a broad range of resistance to oils, brines, and other severe environmental conditions. Basic use is for below grade walls and foundation waterproofing.

Torpedo gravel. Coarse hard grit.

Trinidad asphalt. Solid or semisolid asphalt, brown to black in color, porous, and about 50% impure, obtained from the island of Trinidad.

Trowel-Mastic. Trowel-applied dampproofing compound for above- or below-grade interior and exterior application. It consists of a heavy-bodied, asbestos-fibered asphalt compound containing a hydrophilic agent to ensure proper adhesion to damp or "green" surfaces. It provides a dampproof and moisture-resistant coating on terra cotta, concrete, brick, and block masonry walls, footings, foundations, retaining walls, and parapet and fire walls. It is ideal for protecting porous or irregular surfaces. A product of W. R. Meadows Inc., Elgin, Ill.

Varnish gum. Any resinous substance, excluding rosin.

Viscosity. Measure of the resistance to the flow of a bituminous material, usually sated as the time of flow of a given quantity of the material through a given orifice.

Volatile. Term applied to fractions of bituminous materials that will evaporate at climatic temperatures.

Volclay. Mined and processed high-sodium mineral bentonite, the product having excellent waterproofing qualities and used in connection with the development of a process for packing granular bentonite in the corrugated flutes of biodegradable kraft panels A product of the American Colloid Company, Skokie, Ill.

Water absorbent. Property of a floor-hardening or waterproofing material that makes it readily miscible with water.

Water gas tar. Liquid hydrocarbon produced by cracking oil vapors in the manufacture of carbureted water gas.

Water glass. Sodium silicate or alkaline silicates soluble in water.

Waterproofing

1. Treatment of a surface or structure that prevents the passage of water.

2. Treating masonry or concrete to exclude or prevent the percolation or penetration of moisture or water through its material.

3. Treatment of a surface or structure to prevent the passage of liquid water.

4. One or more hot- or cold-applied coatings or layers of a material or combination of materials, usually of a bituminous type, that are impervious to water; coating or coatings applied to prevent permeation by water; multiple layers of reinforcing membranes cemented with suitable waterproofing cements.

Waterproofing membrane system. Membranes are composed of polyethylene adhered to bentonite compounded with butyl rubber and offer a unique self-healing capability due to their ability to swell when contacted with water. The membranes adhere to themselves, creating lap seams; adhesives are used for vertical applications.

Water repellent. Property of a waterproofing material that hinders or prevents it miscibility with water.

Water table. Established average underground water level.

Wearing surface. Final outside surface exposed to back-fill below grade.

WOOD SHAKES—SHINGLES AND SIDING

Adhesive. Phenolic resin adhesives are used in bonding shingles to plywood panels.

Backer board. Undercoursing material, typically fiberboard, used beneath siding materials such as mineral fiber shingles. A backer board adds insulation value, increases resistance to impact, and provides a heavier shadow line at the butt than shingle siding applied directly without the use of a backing material.

Batten. Narrow piece of wood used to cover joints in vertical siding.

Battens and corner trim. Lap corner trim strip over siding joint so that there is no continuous joint through corner trim and siding.

Bevel siding (lap siding)

1. Type of finish siding used on the exterior of a house, usually manufactured by resawing a dry, square, surfaced board diagonally to produce two wedge-shaped pieces.

2. Wedge-shaped boards used as horizontal siding in a lapped pattern. The siding varies in butt thickness from $\frac{1}{2}$ to $\frac{3}{4}$ in. and comes in widths up to 12 in. It is normally used over some type of sheathing.

Board and batten siding. Method of installing vertical siding using alternate wide siding and narrow alternate boards.

Boston ridge. Method of applying wood shingles at the ridge or at the hips of a roof as a finish.

Building paper. Lightweight building paper or 15 lb. felt, nailed to sheathing before siding or shingles are installed. The paper or felt is recommended when the sheathing used is other than wood.

Caulking or sealant. Caulking and sealants compounds used should be nonyellowing and resistant to ultraviolet degradation.

Cedar. Cedar is the most popular wood species for shakes and shingles because of its natural, ageless beauty, rugged texture, and durable qualities.

Characteristics of shingles. Knots, wormholes, decay, shakes, checks, crimps, flat grain, cross grain, and sapwood constitute natural characteristics that are not admissible. Defects in manufacturing, including shims, feather tips, diagonal grain, waves, and torn fiber are likewise not admissible.

Clapboard

1. One form of outside covering for a house; siding.

2. Special form of outside covering for a house; horizontal siding.

Corners. Outside corners for shingle installations should be constructed with an alternate overlap of shingles between successive courses. Inside corners should be mitered over a metal flashing, or may be made by nailing a 2 in. square strip in the corner.

Coverage. Amount of weather protection provided by the overlapping of shingles. Depending on the kind of shingle and method application, shingles may furnish one (single coverage), two (double coverage), or three (triple coverage) thicknesses of material over the surface of the roof.

Cross grain. Condition that should not be confused with the terms "flat" or "edge" grain and that might better be termed "cross fiber," since it is a deviation of the wood fibers from the true parallel of the shingle. It is a defect when it runs from one face of the shingle to the other within a longitudinal distance of 3 in. or less in that portion measured 6 in. from the butt.

Curvatures. In the sawed face of hand-split and resawed shakes, curvatures should not exceed 1 in. from a level plane in the length of the shake. Excessive grain sweeps on the split face are not permitted.

Diagonal grain. Condition where the grain of the wood does not run parallel to the edges of the shingle. Diagonal grain is considered a defect when the grain diverges or slants 2 in. or more in width in 12 in. of length.

Dolly Varden siding. Beveled wood siding that is rabbeted on the bottom edge.

Drop siding

1. Exterior wall covering, usually $\frac{3}{4}$ in. thick and 6 in. wide, machined into patterns. Drop siding has tongue-and-groove of shiplap joints, and is heavier and stronger than bevel siding.

2. Siding usually $\frac{3}{4}$ in. thick and 6 and 8 in. wide with tongued-and-grooved or shiplap edges. Often used as siding without sheathing in secondary buildings.

Double coursing. Method of applying shingles over an undercourse of lower grade shingles or other suitable material. Nails are exposed as a result of face- (butt) nailing, and greater shingle exposures are possible than with single coursing. Double coursing usually results in greater wall coverage.

Edges of shakes. The edges of shakes should be parallel within 1 in. of each other.

Exposure. Shortest distance in inches between exposed edges of overlapping shingles.

Exterior nailing. Small-headed, corrosion-resistant, siding nails. Nails should be hot-dipped, zinc-coated, stainless steel or aluminum.

Feather tips. Feather tip or shim is a condition of manufacture found on the thin ends of some shingles where the saw came out of the piece prematurely, producing a thin, flimsy, featherlike edge. The tip ends of the shingle may be uniformly thin and produce a thoroughly satisfactory roof, but when they are uneven or with corners sawed off, the shingles will not lay up evenly.

Fire resistance. Fire retardant treated shingles and shakes are tested according to UL 790, Test Method for Fire Resistance of Roof Covering Materials. Two classifications have been established, Classes B and C.

Flat grain. Condition in shingles or lumber where the growth rings are flat or horizontal, as opposed to edge-grained or quartered material where the growth rings are on edge or vertical to the surface.

Grades used for shingle roofs. With western red cedar, cypress, and redwood, first-grade shingles (all-heart, edge-grained clear stock) are used for the longest life. Other all-heart, but no edge-grained grades, such as second grade in redwood, western red cedar, and cypress, are frequently used for secondary buildings.

Grades used for sidewalls. The same wood species are used for sidewalls as are used for roofs. For best construction on single-course sidewalls, first-grade (all-heart, edge-grained clear) is used. For double-course sidewalls, third grade is recommended for the undercourse, and first grade for the outer course for best construction.

Grain. Redwood siding will contain both flat and vertical grain unless a vertical grain selection is specified.

Grain in shingles. All commercial standard shingles. The grain can be strictly vertical, or edgegrained. Thin lines constitute the annual, or growth rings, and are vertical

when the shingle is laid flat. "Edge grain" is synonymous with quartered or quarter-sawed lumber or flooring, and the condition is considered fulfilled when no portion of the grain slope exceeds 45° from the perpendicular.

Hand split. Term applied to wood shakes split by mallet and froe.

Hand-split and resawed shakes. Shakes with split faces and sawed backs, produced by running cedar blanks or boards of proper thickness diagonally through a band saw. Two tapered shakes are obtained from each blank.

Hand-split and resawed tapered shakes. Shakes produced by running hand-split shakes diagonally through a band saw.

Head lap. Shortest distance in inches from the lower edges of overlapping shingles to the upper edge of the unit in the second course below.

Hardboard siding. Siding characteristics processed or embossed into panels composed of cellulose fibers and a binding agent under pressure.

Length of shakes. Nominal shake lengths are 18, 24, and 32 in. within a minus tolerance of $\frac{1}{2}$ in. A variation, including shims or feather tips, of 1 in. from these nominal lengths is permitted in 5% of the linear inches of shakes in any bundle.

Open valley. Area where the valley flashing is exposed and shingles are generally 3 in. away from the valley's center line.

Pitch. Pitch indicates the incline of a roof as a ratio of the vertical rise to twice the horizontal run, expressed as a fraction. For example, if the rise of a roof is 4 and the run 12, the roof is designated as having a pitch of $\frac{1}{3}$.

Plywood siding. Plywood siding should be exterior type and meet the requirements of product standard PS 1 for construction and industrial plywood.

Quality standard shakes. Quality standard shakes are 100% clear and graded from the split face in the case of hand-split and resawed shakes and from the best face in the case of taper-split and barn shakes. Shakes are 100% heartwood, free of bark and sapwood. Taper-split shakes and barn shakes are 100% edge grain; hand-split and resawed shakes may include no more than 10% flat grain in the linear inches of any bundle.

Red cedar shingles. Premium grade of shingles for roofs and sidewalls are top grade, 100% heartwood, 100% clear, and 100% edge grain.

Redwood grades. Clear all-heart and clear, should be used for siding and trim. Clear all-heart is a completely clear grade, permitting only the reddish-brown heartwood. Clear grade is similar, but permits the white sapwood that is frequently specified.

Shake. Thick hand-split shingle, resawed to form two shakes; usually edge grained.

Sheathing. Structural covering, usually wood boards or plywood, used over studs or rafters of a structure. Structural building board is normally used only as wall sheathing.

Shingle butt. Lower exposed edge of the shingle.

Shingle panel. Sixteen individual cedar shingles are electronically bonded to the plywood sheathing to make an effective 8-ft panel.

Shingles

1. Relatively small individual siding units that overlap each other to provide weather protection. They are typically applied to a nailing base, such as sheathing or horizontal nailing strips, which supports the shingles between structural framing members.

2. Covering applied in overlapping layers, as for the roof or sides of a building. Shingles are commonly made of wood, asphalt, asbestos, tile, and slate, among other materials, and are cut fairly small.

Shingles, and shakes. The basic requirements for shingles and shakes are high decay resistance, little tendency to curl or check and freedom from splitting in nailing. For roof and sidewalls the best species are red cedar, cypress, and redwood. Principal shingle woods; are heatwood only, and edge grained. Shingles are sawed, and shakes are split. Shingles have a relatively smooth surface, while shakes have at least one highly textured natural grain split surface.

Shingle stain. Form of oil paint, very thin in consistency, intended for coloring rough-surfaced wood, such as shingles, without forming a coating of significant thickness or gloss.

Side or end lap. Shortest distance in inches that adjacent shingles or sheets horizontally overlap each other.

Siding. Finish covering of the exterior wall of a frame building. Siding may be made of weatherboards, vertical boards and battens, shingles, or other material.

Siding shingles. Various kinds of shingles, such as wood shingles or shakes and nonwood shingles, that are used over sheathing for exterior sidewall covering of a structure.

Single coursing. Method of applying shingles without the use of an undercourse. In the case of wood shingles, this method results in concealed nails and a smaller exposure than is found in double-coursed walls.

Slope. Incline of a roof as a ratio of vertical rise to horizontal run. It is expressed sometimes as a fraction, but typically as x in 12. For example, a roof that rises at the rate of 4 in. for each foot (12 in.) of run is designated as having a 4 in 12 slope.

Square. Roofing is estimated and sold by the square. A square of roofing is the amount required to cover 100 sq ft of roof surface.

Square pack. Unit providing sufficient shingles for the coverage of an area of 100 sq ft when the shingles are laid at a specified exposure to the weather.

Starter. Shingle of regular width but less than standard length, used at eaves and gutters under the first course to produce a uniform cant.

Straight-split shakes. Shakes manufactured in the same manner as taper-split shakes, except that the splitting is done from one end of the block only, producing shakes that are the same thickness throughout.

Taper sawed. Term applied to shakes produced in the same manner as shingles except that the edges may be split or rough instead of smooth.

Taper split. Term applied to shakes with one end thinner than the other, produced by reversing the block, end for end, with each split.

Taper-split shakes. Shakes produced by hand, using a sharp-bladed steel froe and a wooden mallet. The natural shinglelike taper from butt to tip is achieved by reversing the block, end-for-end, with each split.

Texture split shakes panel. Twenty-fourth inch shakes that have a split face 10 to 12 in. from the butt and are sawed the rest of the way to form a uniform extraheavy tip. Shakes are electronically bonded to plywood sheathing to form an 8-ft panel of bold texture nailed directly to the rafters.

Thickness. Shingles are measured for thickness at the butt ends and designated according to the number of pieces necessary to constitute a specific unit of thickness. For example, $\frac{4}{2}$ indicates that four shingles measure 2 in., while $5/2\frac{1}{4}$ means that five shingles measure $2\frac{1}{4}$ in. in thickness. Shingles should be uniform in thickness, but a minus tolerance of 3% is allowable to compensate for the difference in shrinkage encountered in kiln-drying. This tolerance is based on the total thickness of the bundle.

Thickness of shakes. Shake thickness is determined by measurement of the area within $\frac{1}{2}$ in. from each edge. If corrugations or valleys exceed $\frac{1}{2}$ in. in depth, a minus tolerance of $\frac{1}{8}$ in. is permitted in the minimum specified thickness.

Tin shingle. Small piece of tinned metal used in flashing and repairing a shingle roof.

Top lap

1. Shortest distance in inches from the lower edge of an overlapping shingle to the upper edge of the lapped unit in the first course below, that is, the width of the shingle minus the exposure.

2. Length of the shingle minus the exposure.

Torn fiber. Fuzzy or whiskered appearance usually caused by a dull saw. This condition may also be referred to as "torn grain."

To the weather. Projecting of shingles or siding beyond the course above.

Underlayment

1. Base or lining between a roof deck and the shingles, usually composed of roofing felt.

2. Material placed under finish coverings, such as flooring or shingles, to provide a smooth, even surface for applying the finish.

Valleys. Internal angle formed by the two slopes of a roof.

Wall shingles. Wall shingles shed water, resist weather penetration, and offer a variety of textural and color finishes. Shingles are a suitable siding material for exterior wall surfaces.

Water repellent preservatives. Preservative pressure applied to woods such as red cedar and redwood in particular, to prevent weathering, which results in a leaching and bleaching action.

Waves. Irregularities on the face of a shingle (also referred to as "washboards") usually caused by a wobbling of the saw on its arbor.

Width of shakes. Random widths, none narrower than 4 in. Hand-split and resawed shakes should have a maximum width of 14 in.

Wood quality. All commercial standard wood shingles are manufactured from 100% heartwood.

DIVISION

8

Doors and Windows

FIRE DOORS AND DEVICES

Access door. Access door is a fire-protective door assembly of smaller size than conventional doors and used to provide access to utility shafts, chases, plumbing equipment, doors to service equipment for elevators, and dumbwaiters, or as an opening to gain entry into an attic or space above a ceiling or below a floor.

Astragals

1. Doors swinging in pairs requiring astragals should have at least one astragal attached in place so as to project approximately $\frac{3}{4}$ in. The local authorities have the jurisdiction to permit pairs of doors without astragals.

2. Doors swinging in pairs are provided with astragals in accordance with National Fire Protection Association Standard NFPA No. 80 "Fire Doors and Windows," (unless otherwise noted for Class B and C locations in the individual classifications). Double egress doors (in pairs) and egress doors (in pairs) in same direction bearing the $1\frac{1}{2}$ HR (B) or $\frac{3}{4}$ HR (C) classification markings may be provided without an astragal by some manufacturers as indicated by the individual classifications.

Automatic-closing. As applied to opening protectives such as fire doors, means normally held in an open position and automatically closing upon the action of a heat-actuated releasing or operating mechanism.

Automatic closing door. Doors that are normally open, but will close at the time of fire.

Automatic closing sliding doors. Fire-rated doors, normally operated independently of the automatic closing device; in case of fire they close upon fusing of the 160°F link.

Automatic device

1. Door in an opening normally held in open position and automatically closed by a releasing device actuated by abnormally high temperature, by a predetermined rate of rise in temperature, or by the presence of smoke.

2. Device that functions without human intervention and is actuated as a result of a predetermined temperature rise, rate of rise of temperature, combustion products, or smoke density, such as an automatic sprinkler system, automatic fire door, automatic fire shutter, or automatic fire vent.

3. Device that is activated by exposure to physical or chemical conditions generated by fire. A device is considered "automatic" when a mechanism forces it into a specified position as soon as any restraining force is removed.

Automatic fire door. Fire door equipped with a heat-actuated closing device that will operate at a predetermined temperature of not more than 165°F, with a rate-of-rise temperature operating device, or with approved photocell and/or electronic operating devices.

Chute doors. Intake- and discharge-type doors of formed steel and flush design, with frames, latching, and closing mechanisms. Doors of the intake type are classified for openings not exceeding 36 in. in width and 36 in. in height, or 24 in. in diameter, and discharge doors are classified for openings not exceeding 36 in. in width and 48 in. in height. Requirements for chute doors intended for rubbish handling are described in the National Fire Protection Association Standard, NFPA No. 82, Incinerators and Rubbish Handling.

Closing device (fire door). Device that will close the door and be adequate to latch and/or hold a hinged or sliding door in a closed position.

Composite doors

1. Steel-covered, wood-covered, and plastic-covered composite doors consist of a manufactured core material with steel edges, untreated wood edges, or chemically impregnated wood edges and face sheets of steel, wood veneer, or laminated plastic. Steel-covered doors are of either the swinging type or the sliding type. Each door of the swinging type bears a "composite fire door" classification marking.

2. Composite doors are of flush design and consist of a manufactured core material with chemically impregnated wood edge banding and untreated wood face veneers or laminated plastic faces, or of a core surrounded by and encased in steel.

Coordinator. Device used on a pair of swinging doors that causes the inactive leaf to close before the active leaf closes.

Curtain doors. Curtain doors consist of interlocking galvanized steel blades installed in a galvanized steel frame. The curtain is brought to the closed position by the weight of the blade package upon release of a listed fusible link. Curtain doors are intended for installation on both sides of the opening in a fire wall provided for a duct in accordance with the National Fire Protection Association Standard for Installation of Air-Conditioning and Ventilating Systems Other Than Residence Type (NFPA No. 90A). Each door bears a "curtain fire door" classification marking with the rating 3 HR (A), with no reference to temperature rise. The classification marking covers the design and construction of the door.

Door closer. Closer is a labeled device applied to the door or frame that causes a door to close by mechanical force. The closing speed can be regulated by this device.

Door holder and release device. Device is a labeled fail-safe device, controlled by a detection device, used on automatic closing doors to release the door at the time of fire causing it to close.

Dumbwaiter doors. Counterbalanced and swinging-type doors of flush design, made of forged steel and including wall guides, frame, latching, and counterbalancing mechanisms. Doors are classified for openings not exceeding 4 ft in width and 5 ft 9 in. in height. Doors may be provided with classified $\frac{1}{4}$ in.-thick wired-glass vision lights not exceeding 3 in. in diameter per opening.

Dutch door. Dutch door panels bearing 3 HR (A), $1\frac{1}{2}$ HR (B), or $\frac{3}{4}$ HR (C) classification markings may be provided by some manufacturers, as indicated by the individual listings.

Fire assembly. Assembly of a fire door, fire window, or fire shutter, including all required hardware, anchorage, door frames, and sills.

Fire dampers. Fire dampers constructed in accordance with UL Standard 555 (latest issue), Standard for Fire Dampers, as published by Underwriters Laboratories, Inc. The use of fire dampers shall be as required in NFPA No. 90A.

Fire door

1. Door and assembly, so constructed and assembled in place as to give the specified protection against the passage of fire.

2. Door, assembly, and method of installation that have been approved to prevent or retard the passage of excessive heat, hot gases, or flames and that has a fire-resistive rating as required by the local building code.

3. Door, including its frame, so constructed and assembled in place to prevent or retard passage of flame or hot gases.

Fire door assembly. Assembly of fire door and its accessories, including all hardware, frames, closing devices and their anchors, so constructed as to give protection against the passage of fire.

Fire door hardware. Hardware that is applied to both swinging and sliding doors as described and listed in NFPA No. 80.

Fire doors

1. Standard types of fire doors are UL classified and approved by local building codes for use in interior partitions, corridors, and vertical shafts. In case of fire, they close automatically upon fusing of the links.

2. Doors designed for the protection of openings in walls and partitions against fire when installed in accordance with the instructions in the National Fire Protection Association Standard for Fire Doors and Windows, NFPA No. 80.

Fire doors and walls. Doors and walls constructed of fire-resistive materials designed to prevent the spread of fires.

Fire doors clad with laminate. Fire doors tested and rated by Underwriters Laboratories, Inc. for fire resistance, heat transmission, and structural stability; $1\frac{3}{4}$-in.-thick doors with a metal UL classification marking attached to the hinge stile.

Fire exit hardware. Hardware that consists of exit devices that have been labeled both for fire and panic, as described and listed in NFPA No. 80.

Fire resistance rating. The time, in minutes or hours, that materials or assemblies have withstood a fire exposure as established in accordance with test procedures of "Standard Methods of Fire Tests of Building Construction and Materials," NFPA No. 251 (latest issue).

Fire shutter. Shutter that is a labeled door assembly used for the protection of an opening in an exterior wall.

Frames. Labeled steel door frames are used with labeled fire doors.

Freight elevator (steel panel, hollow-metal, and metal-clad) doors. Freight elevator fire doors of the single-swing or counterbalanced type. The doors consist of either formed steel sheets or hollow-metal or metal-covered wood units. A counterbalanced door assembly includes the door panels, guides, latching device, and counterbalancing mechanism. Each door or door assembly bears a "freight elevator fire door" classification marking with one of the following ratings: $1\frac{1}{2}$ HR (B); temperature rise, 30°F min, 250°F or 650°F max; or $1\frac{1}{2}$ HR (B) with no reference to temperature rise. The classification marking covers the design and construction of the door or door assembly.

Fusible links. Devices used in connection with automatic closing devices for doors and windows and other automatic devices requiring fusible links.

Glass. Labeled wired glass, not less than $\frac{1}{4}$ in. thick, labeled for fire protection rating, and installed in approved steel frames. The glass is to be well inbedded in glazing putty and all exposed joints between the metal and the glass should be struck and pointed.

Glass panels in rated doors. Doors bearing 3 HR (A) or $1\frac{1}{2}$ HR (D) classification markings are not provided with glass lights. All doors bearing $1\frac{1}{2}$ HR (B) classification markings may be provided with classified $\frac{1}{4}$ in.-thick wired-glass vision panels. The sum of the exposed glass areas per door is not to exceed 100 sq in. with no dimension exceeding 12 in. Also for flush doors, the width and height are not to exceed 10 and 33 in., respectively. Doors bearing $\frac{3}{4}$ HR (C) classification markings may be provided with one or more classified $\frac{1}{4}$ in.-thick wired-glass lights. The exposed area of each glass light shall not exceed 1296 sq in. with no dimension exceeding 54 in. Doors bearing $\frac{3}{4}$ HR (E) classification markings may be provided with one or more classified $\frac{1}{4}$ in.-thick wired-glass lights. The exposed area of each glass light is not to exceed 720 sq in. with no dimension exceeding 54 in.

Heat-actuated device. Devices that include fixed temperature releases, rate-of-temperature-rise releases, and door closers with hold open arms embodying a fusible link.

High door hardware. Doors exceeding 8 ft in height are provided with listed three-point locks in a single-swing door or in the active door of a pair, unless a latch throw is shown on the individual door classification marking with listed top and bottom flush bolts in the stationary door to provide the protection indicated.

Hollow-metal doors

1. Doors of the flush and paneled designs, made of No. 20 gauge or heavier formed steel.

2. Fire doors of formed steel of the flush and paneled types. Each door of the swinging type bears a "hollow-metal fire door" classification marking.

Labeled. Symbol or identifying mark of a nationally recognized testing laboratory, inspection agency, or other organization concerned with product evaluation.

Laminate-faced fire doors. Interior fire doors clad with laminate and tested and rated by Underwriters Laboratories, Inc. for fire resistance, heat transmission, and structural strength. They are B labeled with the UL metal label attached to the hinge stile. Vision panels for Class B and C installations are available in standard sizes and are constructed of inorganic core and crossbands with $\frac{1}{28}$-in. face veneers, $\frac{1}{2}$-in. high-density inorganic inner stile, and $\frac{1}{4}$-in. fire-retardant outer stile of treated soft maple. Door faces and stile edges are $\frac{1}{16}$ in. high of pressure plastic laminate.

Listed. Equipment or materials included in an approved and accepted list published by a nationally recognized testing laboratory, inspection agency, other organization concerned with product evaluation.

Locks or latches. Labeled locks and latches or labeled fire exit hardware (panic devices) meeting both life safety requirements and fire protection requirements are to be used for all fire doors.

Louvers in fire doors. Doors bearing $1\frac{1}{2}$ HR (B) or $\frac{3}{4}$ HR (C) classification markings are provided at the factory with automatic louvers by some manufacturers, as indicated by the individual classifications. The maximum size of the automatic louver shall not exceed 576 sq in. with no dimension exceeding 24 in.

Metal-clad (Kalamein) doors

1. Doors of flush and panel design consisting of metal-covered wood cores or stiles and rails and insulated panels covered with steel of 24 gauge or lighter.

2. Fire doors of flush and paneled design consisting of metal-covered wood members. Each door of the swinging type bears a "metal-clad fire door" classification marking. The classification marking covers the design and construction.

National Fire Protection Association. Organization organized in 1896 to promote the science and improve the methods of fire protection and prevention. It has published several hundred NFPA Standards used worldwide. The standards were developed and processed by the Association under regulations intended to assure procedural fairness and that all concerned interests have an opportunity to participate.

Passenger elevator (hollow metal) doors. Passenger elevator fire doors of the single-swing or sliding hollow-metal type. Sliding door panels may be used in single-slide assemblies of the one-, two-, or three-speed type, or they may be used in center-parting assemblies of the one- or two-speed type. Each door panel bears a "passenger elevator fire door" classification marking with the following rating: $1\frac{1}{2}$ HR (B), with no reference to temperature rise. The classification marking covers the design and construction of the door or door panels only.

Power-operated fire door. Power-operated fire door that normally opens and closes by electrical power.

Rolling steel doors. Fire doors consisting of interlocking galvanized or stainless steel slats, bottom bar, wall guides, barrel assembly, automatic release device, governor, and counterbalancing mechanism. They may be provided with a motor drive assembly which does not interfere with the manual or automatic (fusible link, other fixed temperature release, or a rate-of-rise temperature release) closing of the door. Each door bears a "rolling fire door" classification marking. The classification marking covers the design and construction of the door, including the governor and automatic releasing mechanism.

Self-closing device

1. Normally closed and equipped with an approved device that will ensure closing after having been opened for use.

2. Normally kept in a closed position by some mechanical device and closing automatically after having been opened.

3. A self-closing device is one that will maintain the door in a closed position.

4. Device that will ensure that a door to which the device is connected will close immediately after having been opened.

5. Door or other opening protective device that is normally closed and that is equipped with an approved device to ensure its closing after the opening protective device has been opened for use.

Self-closing door. Doors that are normally kept in a closed position by some mechanical device and that are closed automatically after having been opened, except as otherwise provided in the local building code.

Self-closing fire door. Fire door that when actuated by a fire or smoke detector system, fusible link, or other device will automatically be closed by its closing mechanism.

Service counter door

1. Door that has a labeled fire door assembly used for the protection of openings in walls where the primary purpose of the opening is for nonpedestrian use, such as counter service for food, package and baggage transfer, or in protected observation ports.

2. Single- and two-speed counterbalanced-type doors of flush design or rolling-type doors of formed steel, which include wall guides, frame, sill latching, and counterbalancing mechanisms. Doors of the panel type may be provided with classified $\frac{1}{4}$ in.-thick wired-glass vision panels not exceeding 100 sq in. in area with neither dimension exceeding 10 in. Doors are classified for openings not exceeding 6 ft in width and 5 ft in height.

Sheet metal doors

1. Fire doors are of formed steel and of the corrugated flush, and paneled design. Each door of the swinging type bears a "sheet metal fire door" classification marking with the notation "minimum latch throw-in." Some manufacturers furnish doors bearing the notation "fire doors to be equipped with fire exit hardware" in lieu of the notation "minimum latch throw-in." Each door of the sliding type bears a "sheet-metal fire door" classification marking. The classification marking covers the design and construction of the door.

2. Doors usually formed of No. 22 gauge or lighter steel and of the corrugated flush and paneled design.

Single-swing-door hardware. Single-swing doors bearing classification markings and not exceeding 4 ft in width and

8 ft in height must be provided with listed single-point locks or latches with a minimum $\frac{1}{2}$ in. throw to provide the protection indicated.

Smoke detector. Device that senses visible or invisible particles of combustion.

Smoke stop doors. Smoke stop doors may be of ordinary solid wood type not less than $1\frac{3}{8}$ in. thick with clear wired glass panels. Such doors should be self-closing, single swinging type and may be either single or double. They should close the opening completely with only such clearance as is reasonably necessary for proper operation.

Steel doors. Doors of interlocking steel slat design or plate steel construction.

Steel sectional (overhead) doors. Fire doors consisting of hinged steel panels, wall guides, interlock along top edge, vertical and horizontal tracks, roller wheel, counterbalancing, automatic closing mechanism, and governor. Each door bears a "rolling fire door" classification marking. The classification marking covers the design and construction of the door assembly to be installed in accordance with installation instruction provided.

Swinging-pair hardware. Doors swinging in pairs bearing classification markings and not exceeding 8 ft in width or height must be provided with listed single-point locks or latches with minimum $\frac{3}{4}$ in. throw (except as indicated on the individual manufacturer's classification markings) and listed top and bottom manual or self-latching flush bolts or surface bolts to provide the protection required.

Tin-clad doors

1. Fire doors consist of two or three plywood core construction, covered with No. 30 gauge galvanized steel or terne plate (maximum size 14×20 in.). Each door bears a "tin-clad fire door" classification marking.

2. Doors are of two or three plywood core construction, covered with No. 30 gauge galvanized steel or terne plate (maximum size 14×20 in.) or No. 24 gauge galvanized steel sheets not more than 48 in. wide.

Underwriters Laboratories, Inc.. Organization, founded in 1894, maintains and operates laboratories for the examination and testing of devices, systems, and materials relating to public safety: life, fire prevention, casualty hazards, and crime prevention.

Wired glass. Glass having a wire pattern within the glass, installed in a metal sash and frame, and used to resist heat from exposure fires or to protect horizontal or vertical openings against the spread of fire.

Wood-faced fire door. Door comes in sizes up to 4×8 ft and $1\frac{3}{4}$ in. thick. It is made of treated maple edges (UL approved) with a nontoxic composite-type core (UL approved) and type 1 adhesive. It bears UL-rated $\frac{3}{4}$ hour "C" label. A 100 sq in. wired light opening is the maximum size allowable.

FOLDING DOORS

Accordion door

1. Folding door with independent covers separated by an internal mechanism and a dead airspace operating laterally across the face of the opening to a stack or closed position. Accordion doors are not to be confused with serpentine folding doors, which contain no internal operating mechanism or dead airspace between independent covers.

2. Folding door that includes factory finished wood panels, connected continuously along top and bottom by metal strip hinging. Individual panel hinges are riveted to adjoining panels and contain stops to maintain a uniform extended position. Doors are suspended by nylon ball bearing wheels from noncorrosive overhead track.

Ceiling guard. Protection for ceiling finish from sweepstrips attached to doors, fabricated of sheet metal, aluminum, wood, or hardboard.

Cross-over switch. Ceiling mounted or built into the ceiling, a right angle cross-over switch portion of the track to permit the door sections to enclose a room.

Curved switch. Ceiling mounted or built into the ceiling, a switching portion of the track to allow the door sections to divide a room at several designated points.

Curved track. Ceiling mounted or built into the ceiling, a curved portion of the track to permit door panels to achieve a curved wall effect.

Fire retardant cores. Wood panel doors should have fire retardant cores made of materials that will meet the requirements of ASTM E 84, providing the following: 25 flame test, 15 fuel contributed, and 75 smoke developed.

Hangers. Series of nylon-tire ball bearing rollers, usually attached to alternate wood panels of the folding door, that ride in the ceiling mounted track.

Hardware. Metal parts for the operation of the folding door should be of metal either plated or coated with a corrosion resistant finish.

Header. Structure by which a folding door is supported. It may in turn be supported by the superstructure and may be recessed, exposed, or boxed in a false header.

Hinged recess door. Installation enables a door to be stacked into a recessed area and a panel closes to cover the end of the folding door.

Hinge jamb. Door jamb to which the hinge panel of a folding door is attached.

Installation and operation. A folding door and its hardware must have a proper fit, and be plumb and square within the opening. The door must not drag or bind in the track or on the finished floor and clearance should be as recommended by the manufacturer for the type of door installed. Ease of operation, latching, and tieback is a requisite.

Interlocking jamb molding. Vertical full height wood-shaped molding used to provide positive door alignment.

Latch jamb. Opening jamb that the latching panel of a folding door joins to close or lock.

Latches. Depending on the type of folding door, latches can be positive or magnetic, or they can be latches with privacy lock, key lock, or master-keyed to the building system.

Pantograph. Series of rigid links or straight arms in parallelogram form to duplicate in same, larger, or smaller form the action of a given point and its relationship to a fixed point to ensure symmetry of folding action or evenness of fold. Pantographs form is the skeletal system of the fabric-covered folding door and may be horizontal or vertical in nature.

Sliding jamb. False jamb usually consisting of a vinyl-covered plywood board equipped with carriers and fastened to the anchor post, furnished for a folding door that is stored in a pocket where it is difficult to anchor the door at the rear of the pocket. When the partition is pulled out of the pocket, the sliding jamb board moves forward until it contacts stops at the face of the pocket and thus furnishes a false jamb effecting a pocket face enclosure.

Subchannel. Metal structure of varying gauges and sizes fastened to a header and forming a place in the ceiling for later reception of the track.

Supported vinyl. Vinyl fabric preferably formed by laminating a liquid vinyl to a fabric backing, such as jute, cotton drill, or other type of woven material.

Sweepstrips. Strips of resilient substances, such as rubber, vinyl, felt, or combinations thereof, attached to top and/or bottom of doors. Sweepstrips are used to mask sounds of higher frequencies and to prevent leakage of sound around the perimeter of the door.

Tieback. Device used to secure the stacked accordion door against the anchor jamb.

Track. Aluminum or steel support and guide for accordion doors attached to the header and providing a dual or single rail to accommodate the door carrier system.

Vinyl door covering. Vinyl used for the finish material of accordion folding doors, employing the pantograph system, should meet the requirements of ASTM E 84 and FS CCC-W 408, cover the requirements of three types and two classes of continuous vinyl wall-covering materials.

GLASS, PLASTIC, AND GLAZING

AA quality. Best quality of window glass obtainable.

Acid polishing. Polishing of a glass surface by acid treatment.

Acrylic. Common plastic is more glasslike than polycarbonite. Certain acrylics are warranted to withstand sun and weather exposure. Acrylics can be thermoformed or cold-formed and cemented into different shapes.

Acrylic plastic and polycarbonate sheets. Both materials are tough, break-, shatter- or crack-resistant thermoplastics. Acrylics generally weather better than polycarbonates. Polycarbonates have softer surfaces and are more impact-resistant than acrylics. Certain polycarbonate sheets may be used in some bullet-resisting applications.

Actinic glass. Glass that intercepts a large percentage of the ultraviolet and infrared rays of the sun. It transmits a smaller amount of radiant heat than ordinary glass.

Adhesion. State of being attached; result of a molecular force by which bodies stick together.

American National Standards Institute (ANSI). Document ANSI Z97.1 (latest issue) provides specifications and methods of test for safety of glazing materials used in buildings.

Annealing. Cooling glass slowly under controlled reduction of heat.

Annealed glass. Glass that has been subjected to a slow, controlled cooling process during manufacture to control residual stresses so that it can be cut or subjected to other fabrication.

Antique glass. Antique stained glass windows were originally made from hand-blown, painted glass. Also refers to handcast glass manufactured in the U.S.A, painted as required by design, used in conjunction with lead dividers. Colored plastic glazing, in conjunction with plastic dividers, is now substituted for colored or painted glass.

Antique mirror. Mirror that looks like it has deteriorated over the years, created by a manufacturing process.

A quality. Quality of glass that contains no imperfections that will appreciably interfere with straight vision; A-quality grade usually used for commercial purposes.

Backing point. Pigmented primary coat of specially prepared paint applied over silver and copper deposits on the back of a mirror as protection for silvering.

Back putty. Putty placed in the bed of a frame and into which a light of glass is pressed.

Backup ledge. Horizontal support piece on the wall under the bottom of a mirror. It is rabbeted to hold the mirror to the wall face but provides a clearance from the wall.

Batch. Materials used in the making of glass, consisting of sand, limestone, soda ash, cullet, and other chemicals.

Bead. Sealant or compound after application in a joint, irrespective of the method of application, such as caulking bead, glazing bead, etc.; also, molding or stop used to hold glass or panels in position.

Bed or bedding. Bead of compound applied between a light of glass or panel and the stationary stop or sight bar of the sash or frame, usually the first bead of compound to be applied when setting glass or panels.

Bedding of stop. Application of compound at the base of a channel just before the stop is placed in position, or buttered on the inside face of the stop.

Bent glass. Flat sheet glass laid on a mold and heated until it softens and takes the shape of the mold.

Bevel. Angle that one surface makes with another when not at right angles.

Beveled. Inclined or slanting.

Bevel of compound bead. Bead of compound applied so as to have a slanted top surface so that water will drain away from the glass or panel.

Bite. Amount of overlap between the stop and the panel or light.

Blister. Relatively large bubble or gaseous inclusion in the glass. An imperfection in the finished product.

Block. Small piece of wood, lead, neoprene, or other suitable material used to position the glass in the frame.

Blocking. Polishing process used to remove hairline scratches and other defects.

Boil. Small bubbles $\frac{1}{32}$ to $\frac{3}{32}$ in. in size.

B quality. This quality admits the same kind of defects as the A quality, but the defects are larger, heavier, and more numerous.

Brackets. In the plate glass industry this term means the number of square feet to the piece or a particular group of sizes.

Bubble. Gaseous inclusion in glass, practically always spherical and brilliant in appearance. The term applies to all such inclusions larger than $\frac{1}{8}$ in. in diameter; smaller inclusions are called "seeds," and sizes between $\frac{1}{32}$ and $\frac{3}{32}$ in. are called "small bubbles" or "boils."

Bulb edge. Edge thickened out into a bulbous or rounded form.

Bullet-resistant glazing

1. Laminated assembly of several layers of plate glass that provides exceptional tensile and impact resistance to weapons ranging from medium-powered small arms to high-powered rifles.

2. Glazing composed of three to five sheets of polished plate glass cemented together under heat and pressure with a colorless transparent plastic, having the appearance of solid plate glass, and resistant to the penetration of an ordinary bullet.

Bull's eye. Term for a concave motif or a "dish" put in the glass by scratch polishing.

Buttering. Application of putty or compound sealant to the flat surface of some member before placing the member in position, for example, the buttering of a removable stop before fastening the stop in place.

Came. Grooved, usually H-shaped rod of cast lead used, as in stained glass, to hold the panes together.

Cathedral glass. Glass originally designed for church windows, $\frac{1}{8}$ in. thick, with both surfaces smooth or one side smooth and one side slightly hammered, and with a variety of colors and tints.

Cat's eye. Imperfection on the surface of the glass with a distorted area around it found in sheet and rough glass.

Caulking. Filling of the area between the frame and adjacent material, as against sealing at glass puttying.

Central glass area. In plate glass the central glass area is considered to form an oval or circle centered on the sheet whose axes or diameters do not exceed 80% of the overall dimension. This allows a fairly large area at the corners, which may have imperfections not allowed in the central area.

Channel. Three-sided U-shaped opening in a sash or frame for receiving a light or panel, as with sash or frame units in which the light or panel is retained by a removable stop, in contrast to a rabbet, which is a two-sided L-shaped opening, as with face-glazed window sash.

Channel depth. Measurement from the bottom of the channel to the top of the stop or from the sight line to the base of channel.

Channel glazing

1. Sealing of the joints around lights or panels set in a U-shaped channel employing removable stops.

2. Nonhardening, noncorrosive, gun-applied elastomeric sealant, specifically recommended by the sealant manufacturer for the type of glass specified. In no case should glazing be performed with oleoresinous or oil-based compounds, nor should a glazing sealant be diluted or thinned with a solvent.

Channel width. Measurement between stationary and removable stops in a U-shaped channel at its widest point.

Chip. Imperfection due to breakage of a small fragment out of an otherwise regular surface.

Chipping. Producing a uniformly roughened surface by applying hot glue to a ground glass surface. The glue in cooling and drying shrinks and pulls small chips off the surface.

Clear or diffusing plastic sheets. Clear or diffusing plastic sheets or vacuum formed plastic shapes are available for sidewall lighting. Shaped domes for skylighting are also available.

Clear wire. Term often used to designate "polished wire glass."

Clips. Wire spring devices for holding glass in a rabbeted sash without stops; the glass is face glazed.

Color. Property of visible phenomena distinct from form, light, and shade and depending on the effect of light of different wave lengths on the retina.

Compact. Treatment of glass in a manner, such as heat-treatment, to approach maximum density.

Compound. Formulation of ingredients, usually grouped as vehicle and pigment, to produce some form of sealant, such as a glazing compound, caulking compound, or elastomeric joint sealer.

Compression. Pressure exerted on a compound in a joint, for example, by placing a light or panel in place against bedding or a stop in position against a bead of compound.

Compression material (neoprene, vinyl, rubber). Material extruded or molded into various shapes, channels, angles, etc. It may be installed as a continuous gasket or intermittently as spacer shims. The varying thicknesses and cross sections are dependent on the particular type of glazing or glazing material combinations. To establish and maintain a weathertight joint, the gasket must be compressed not less than 15%, assuming no flow or adhesion, to achieve a seal.

Concave bead. Bead of compound with a concave exposed surface.

Conductivity. Ability to conduct heat.

Configurated glass. Glass having a patterned or irregular surface. The surface configuration is usually applied during manufacture. Such glasses are not transparent and are somewhat light scattering. Glasses falling under this classification are often referred to as pebbled, stippled, rippled, hammered, patterned, chipped, crackled, cathedral, etc., depending on the particular type of surface.

Consistency. Degree of softness or firmness of a compound as supplied in the container, varying according to the method of application (gun, knife, or tool).

Contraction. Shortening, shrinking.

Contraflam glass. Glass made of sodium chloride gel sandwiched between two lights of $\frac{1}{4}$ in. tempered glass. When exposed to fire, the inner light which then explodes exposing the gel, absorbs the heat and crystallizes layer by layer. The fire never reaches the other light or heats it to a great extent. Contraflam has good clarity and works especially well in computer rooms and prisons.

Convex and concave mirrors. Mirrors used in blind spots from visual points to provide both security and safety.

Convex bead. Bead of compound with a convex exposed surface.

Copper bronze powder. The addition of a copper bronze powder to vanish or shellac usually enables a manufacturer to classify the mirror as a "copper-backed" mirror. However, this is not within the meaning of "electrocopper backing" or "electrocopper plating," which are galvanic and electroplating processes.

Cords. Heavy strings incorporated in the sheet, occurring without any regularity of direction and appearing to be of considerable thickness rather than on the surface.

Corrugated glass

1. Glass rolled to produce a corrugated contour.

2. Strong structural glass of $\frac{3}{8}$-in. thickness and an overall pitch about 1 in. thick.

3. Plain or wired glass used for interior partitions, skylights, exterior walls, and many other decorative uses.

Crystal. Twenty-six-ounce window glass classed as double strength. Crystal glass has allowable distortion factors.

Cullet. Broken glass mixed with a batch to help the batch melt more rapidly in the furnace.

Cut sizes. Flat glass sheet cut to specific dimensions.

Daylight size. Size of the glass that is visible after it has been glazed in the opening.

Defects. Recognized blemishes and faults in manufacture and treatment; imperfections incorporated during the process of manufacture. Defects include bubble, cord, fire crack, open bubble, ream, sand hole, scratch, seed, short finish, skim, stone, and string.

Desiccant

1. Drying agent.

2. Ample amount of superior drying agent placed in the hollow spacer to absorb moisture at the time of sealing and provide vapor-free performance.

Diffusing. Scattering; dispersing; tendency to eliminate direct beam of light.

Dimensioning

1. The horizontal figure is given first and then the vertical. Dimensioning is very important with heavy sheet and banded glass.

2. In dimensioning first the horizontal is given, second the vertical, and third the thickness. All glass dimensions must be given in inches and fractions of an inch.

Distortion. Draw or wave distortion incurred in sheet glass manufacture runs in one direction. For best appearance glazing should be with the draw horizontal or parallel to the ground.

Document glass. Ultraviolet absorbing glass used for protecting documents.

Door mirror. Generally a plate glass mirror of a size to cover most of a door; full-length mirror.

Double-glazed window units. Window units usually made from two pieces of float, sheet, glass, or tempered safety glass separated by a metal or rubber spacer around the edges and hermetically sealed in a stainless steel enclosure channel. Entrapped air is at atmospheric pressure and is kept dehydrated by a drying agent located in the spacer. Standard airspaces are $\frac{1}{4}$ and $\frac{1}{2}$ in. Special 2-in. airspace units with improved acoustical characteristics are also available.

Double strength. Approximately $\frac{1}{8}$ in. thick.

Drawing. Continuous mechanical process used in the manufacture of sheet glass.

Drawn glass. Glass made by continuous mechanical drawing operation.

D.S. Double-strength glass.

D.S.A. Double-strength A quality.

D.S.A.A. Double-strength AA quality.

D.S.B. Double-strength B quality.

Durometer. Measure of hardness of a rubber surface in terms of the indentation produced in the surface by a special needle subjected to a practically constant indenting force (10 to 25 durometers equals the hardness of a glazing compound; 50 durometer equals the hardness of an inner tube; 90 durometer equals the hardness of a rubber heel). (ASTM D 2240).

Edge finish. Finish given to the edges of a piece of glass. Edge finishes for flat glass include clean-cut and swiped, clean-cut and seamed, ground and seamed, flat ground, flat polished, bull nose polished, round polished, semiround or pencil polished, mitered edges ground or polished, and beveled edges ground or polished.

Elastic glazing compound. An elastic glazing compound differs from wood and metal sash putties both in composition and performance. It is formulated from selected processed oils and pigments that will remain plastic and resilient over a longer period of time than the common hard putties.

Elastomer. Elastic, rubberlike substance, for example, natural or synthetic rubber.

Electrolytic. Term applied to the process of depositing metals; for example, silvering mirrors is an electrolytic process.

Electroplated copper back. Protective layer of copper electroplated over the backing of silver mirrors. The term "copper back" permits use of copper-bearing paints.

Engraving. Carving decoration on glass by abrasive means.

Etching. Marking or decorating glass by attacking the surface with hydrofluoric acid or other agents.

Evacuated windows. Consists of two sheets of glass, one of which has an infrared-reflective, low emissivity coating. A vacuum gap between them is supported with evenly spaced 3 mm glass spheres. A reactive metal gutter in the sealed space traps gases absorbed from the glass during window's lifetime. Other elements include a transparent low emissivity coating on one or both of the internal glass surfaces.

Exterior glazed. Glass set from the exterior of the building.

Exterior stop. Removable molding or bead that holds the light or panel in place when it is on the exterior side of the light or panel, as contrasted to an interior stop located on the interior side of the light.

Face glazing. On a rabbeted sash without stops, a triangular bead of compound is applied with a glazing knife after bedding, setting, and clipping the light in place.

Face lock. Extruded piece of metal of a variety of cross sections that slides along the interior of store sash face members. A portion of this piece is so positioned that when the setscrew on the opposite side is tightened, the wedging action draws the sash face member tight against the glass and gutter or flashing members.

Federal specifications. The following federal specifications establish the quality standards for glass: (a) DD-G-451 sets the quality characteristics as well as the thickness and dimensional tolerances of flat glass products; (b) DD-G-1403 provides standards for tempered glass, heat strengthened glass and spandrel glass; (c) DD-M-411 prescribes the standards for mirrors and mirror frames.

Felt. Fabric made by compacting wool, fur, or hair, or a mixture of all three. Felt is used in back of and under glass to reduce shock.

Fiber-reinforced plastic. Essentially diffusing and usually corrugated plastic available for both horizontal and vertical openings in a variety of colors.

Figured. Flat glass having a pattern on one or both sides.

Finger pulls. Grooves cut in the glass by small grinding wheels and then polished.

Finish. Type of texture or degree of polish of a surface.

Fire cracks. Small cracks penetrating the surface of the sheet, usually in the shape of short hooked crescents, caused by sudden heating or chilling of the surface.

Fire finish. Flames play on the surface of the glass before it enters the lehr, and the surface is not ground and polished.

Fire polishing. Making glass smooth, rounded, or glossy by heating in a fire.

Fixed glazed panel. Portion of a wall, interior or exterior, containing a frame for the support of glazing material.

Flashing. Applying a thin layer of opaque or colored glass to the surface of clear glass, or vice versa.

Flat glass. Sheet glass, plate glass, and various forms of rolled glass.

Float glass process

1. Most recent method of floating an endless ribbon of molten glass on molten tin, resulting in a flat, fire-finished glass with parallel surfaces. A brilliant surface finish and high degree of flatness qualifies float glass for many uses served previously by polished plate glass.

2. Method of manufacturing glass. Select raw materials are melted in a furnace to form molten glass which is continuously cast in the float process over a pool of molten tin. As the ribbon moves over the metal, the glass surfaces are fire polished and irregularities are removed to give the glass its brilliant appearance. The ribbon moves to an annealing lehr for final cooling. At the end of the lehr, the ribbon is cut into shapes for shipment or further fabrication. The result is a ribbon of glass that has perfect plane and parallel surfaces with no optical distortion.

Florentine glass. A raised pattern is worked on one side, practically roughening the whole surface.

Foreign manufacturers. Foreign manufacturers use the number of ounces to the square foot. Thus 16-oz glass would weigh exactly 1 lb/sq ft. When comparing foreign window glass with domestic products, be sure the same product is being compared. For instance, some foreign glass is 18-oz glass and is called "single strength," but the domestic product is 19-oz glass.

Fourcault process. Method of making sheet glass by drawing vertically upward from a slotted, floating clay block (called a "debiteuse block").

Front putty. Putty forming a triangular fillet between the surface of the glass and the front edge of the rabbet.

Frosting. Surface treatment to scatter light or to simulate frost.

Gasket. Preformed shapes, such as strips or grommets, of rubber or rubberlike composition, used to fill and seal a joint or opening, either alone or in conjunction with a supplemental application of a sealant.

Glare-reducing glass

1. Plate and window glass that by composition has a neutral gray or other tint. This allows for "clear vision" as opposed to translucence in the case of surface-treated glass. Patterned and wire glass may also have one or both surfaces acid etched to diffuse or soften the light.

2. Glass with known characteristics of scattering glare or absorbing light.

Glass. Inorganic product of fusion that has cooled to a rigid condition without crystallizing. It may be transparent, translucent, or opaque and colorless or colored.

Glass block

1. Hollow, partially evacuated unit made of clear pressed glass, sealed during manufacture, providing two surfaces on which designs may be pressed.

2. Fused sections of colorless pressed glass formed into structural units.

Glass door. A door composed of a single glass panel, usually tempered glass, that is not set in a frame.

Glazed door. A glass panel of any size that is installed in a wood or metal frame as part of the door assembly.

Glass-clad polycarbonate. Polycarbonate laminated between two pieces of glass.

Glass cutter. Tool for cutting glass using a hard sharp metal wheel or diamond.

Glass edge or glass seal unit. Constructed by fusing edges of two lights of glass together with $\frac{3}{16}$-in. space filled with a dry gas at atmospheric pressure.

Glass size. Overall size of the glass. This dimension allows for clearance between the edge of the glass and the depth of the rabbet.

Glaze

1. Finish a building by glazing.

2. Process of installing glass or plastic panes in frames using putty or other materials, methods, and techniques.

Glazier. One who installs glass or plastic materials.

Glazier's points

1. Small, triangular, flat metal pieces forced into wood frames as nails or brads to hold glass or plastic in place before putty is applied.

2. Galvanized steel devices for securing glass or plastic in wood frames.

Glazing

1. Finishing a building with glass; or plastic materials used in the construction of a building.

2. Securing glass or plastic in prepared openings in windows, door panels, screens, and partitions and for decorative uses.

Glazing bead. Strip of wood or metal nailed or screwed to the frame to hold the glass or plastic in place in lieu of putty.

Glazing materials. Includes plastics, glass, annealed glass, organic-coated glass, tempered glass, laminated glass, wired glass, or combinations of any of these.

Glazing temperatures. Glazing at temperatures below 40°F is likely to be unsatisfactory for both the application and the glazing material.

Grade. The term applies to all flat-drawn glass and expresses strength: SS (single strength) and DS (double strength).

Gray plate glass. Glare-reducing and heat-absorbing glass.

Greenhouse glass. Low-grade double-strength glass furnished in only a few rather small standard sizes. In addition, a heavy sheet glass is obtainable. It is drawn glass and is sometimes used for windows, but it is slightly wavy and may cause a certain slight distortion of images viewed through it.

Ground glass. Double-strength glass that has one side ground to a frosted finish, thus transforming the clear glass to an obscure glass.

Gun consistency. Compound formulated for a degree of softness suitable for application through the nozzle of a caulking gun.

Hard glass. Glass of exceptionally high viscosity at elevated temperatures.

Heat-absorbing glass

1. Color has nothing to do with heat absorbtion but is the by-product of the iron oxides that do the work. Blue-green glass through its chemical composition is able to absorb a high percentage of the sun's radiant energy, thus reducing glare and light transmission. Heavy sheet, polished plate, pattern, or wire glass absorbs much of the energy of the sun that due to its ferrous admixture. A considerable amount of this heat is dissipated externally, thus reducing the rate at which the

heat enters the building. The heat-absorbing quality is such that the glass retains a considerable amount of heat. This type of glass should be glazed with a permanently elastic glazing compound.

2. Heat-absorbing and glare-reducing glass is recommended for exterior glazing in buildings where reduction of solar heat and glare is required. Plate glass is pale bluish green in color. Gray glass provides a reduction of glare and brightness without sacrificing color uniformity. Bronze glass is also glare reducing and heat absorbing. All three may be tempered to comply with local building codes.

Heat cracks. Heat cracks are caused by the application of the sun's heat or other heat to an area of glass while the other areas of the glass remain cool. This sets up a strain in the glass which causes it to break. The break pattern usually follows an irregular line in contrast to a break caused by a blow or pressure, which is usually in a relatively straight line.

Heat-resisting glass. Glass that is able to withstand high thermal shock generally because of a low expansion coefficient.

Heat strengthened and tempered glass. Produced by reheating and rapidly cooling annealed glass. It has greatly increased mechanical strength and resistance to thermal stress.

Heat-strengthened glass. Glass used in the curtain wall design as spandrel glazing of multistoried buildings. It is usually polished plate glass or patterned glass on one surface of which is fused a colored enamel. Reheating to fuse the ceramic surface and subsequent cooling also strengthens the glass considerably. It may be supplied in a variety of surface textures and an almost unlimited color range.

Heat-treated glass. All regular polished plate glass from $\frac{1}{4}$- to $1\frac{1}{4}$-in. thickness may be heat treated or tempered. All working, such as cutting and drilling, must be done before tempering. ANSI-Z97.1 and FS DD-G-1403B.

Heavy sheet glass. Glass manufactured by the flat or vertical drawn process and supplied in several thicknesses of AA, A, and B quality material.

Heel bead. Compound applied at the base of a channel after a light or panel is set and before the removable stop is installed, its purpose being to prevent leakage past the stop.

High-altitude installation. Double-glazed factory units manufactured especially to be installed at high altitudes, provided with assembly techniques and methods for equalizing the atmospheric pressure.

High-transmission glass. Glass that transmits an exceptionally high percentage of the visible light.

Interior glazed. Term applied to glass set from the interior of the building.

Interior stop. Removable molding or bead that holds the light in place when it is on the interior side of the light, in contrast to an exterior stop, which is located on the exterior side of a light or panel.

Iron content. The slight greenish tint to plate glass is due to the iron content.

Jalousie door. A door having the opening glazed with operable, overlapping glass louvers.

Knife consistency. Compound formulated for a degree of firmness suitable for applications with a glazing knife, for example, face glazing and other sealant applications.

Knurled. Glass finish having a knobby or lumpy surface.

Label. Sticker designating quality and grade of light, placed thereon at the factory. No label is usually placed on types of glass of more complexity than flat glass.

Labeled opening. Opening constructed in accordance with Underwriters Laboratories, Inc. tested standards, and so labeled.

Labeling. Each light of safety glazing material manufactured, distributed, imported, or sold for use in hazardous locations must be permanently labeled by such means as etching, sandblasting, or firing of ceramic material. The label must be legible and visible after installation.

Laminated glass

1. Laminated glass consists of two lights of glass with a tough, transparent vinyl interlayer sandwiched between the glass under heat and pressure to form a single unit.

2. Laminated glass is comprised of two or more sheets of glass with a layer or more of transparent vinyl plastic sandwiched between the glass. An adhesive applied with heat and pressure cements the layers into one unit. The elasticity of the plastic cushions any blow against the glass, and if the glass is cracked, it holds firmly to the glass preventing sharp pieces from flying. There is also laminated glare-reducing glass where the pigment in the vinyl plastic laminate provides the glare control quality.

Laminated glazing. Special care must be taken when glazing laminated glass, as laminates will react unfavorably with certain sealant components. Elastomeric sealants made of 100% solid components and containing no solvents must be used.

Laminated reflective glass. Construction consisting of two lights of float glass laminated with interlayers of architectural-grade polyvinyl butyral. The coating is applied to one light and then is permanently sealed between the lights of glass.

Laminated safety glass

1. Laminated glass made from two lights of $\frac{1}{8}$-in.-thick float glass laminated with a special 0.060-in.-thick vinyl interlayer.

2. Construction composed of two plies of float, plate, or sheet glass bonded by an interlayer of 0.015-in.-thick polyvinyl butyral plastic.

3. Construction consisting of two lights of $\frac{1}{8}$-in.-thick clear float glass separated by a 0.015-in.-thick interlayer of polyvinyl butyral. In areas where added sound control is desired, the polyvinyl butyral thickness is increased to 0.045 in.

4. Two or more pieces of glass held together by a layer or layers of plastic. When fractured, the glass particles tend to adhere to the plastic.

Laminated safety plate glass. Construction consisting of two pieces of float plate glass laminated with a vinyl interlayer.

Laminated safety sheet glass. Construction consisting of two pieces of sheet glass laminated with a vinyl interlayer.

Laminated security glass. Construction consisting of sheet or plate glass laminated with a strong, durable, high-tensile plastic interlayer.

Lead glass. Glass containing a substantial proportion of lead oxide.

Leaded glass. Small lights of cathedral glass, usually stained, set in lead cames, used for decorative windows, particularly in churches. Large windows are usually reinforced at intervals with steel rods or bars, since the lead cames have very little stiffness.

Leeway. Number of inches the factory may use to fill an order of stock sheets over and above the size ordered.

Lehr or leer. Annealing furnace or oven.

Light. Piece of glass used to admit light, as in a window or door; a pane of glass.

Lite. Same as light; a single pane of glass.

Low co-efficient of expansion. Certain types of glass with a particular composition, i.e. borosilicates, expand very little, and can resist differences in temperature, and can classed fire retardant for more than one hour, depending on the type of framing.

Low emissivity glass. Commonly called "low-E glass," has a very thin metallic coating that reduces visible light transmission by 10% when compared to equivalent un-coated glass. The main difference between low-E glass and other types of reflective glass is that the low-E coatings reflect what is known as "sensible heat."

Mastic. Term descriptive of heavy-consistency compounds that remain adhesive and pliable with age.

Matte-surface glass. Glass with surface altered by etching, sandblasting, grinding etc., to increase diffusion. Either one or both surfaces may be so treated.

Maze. Form of figured glass. A confusing and baffling network gives wide diffusion in bright sunlight.

Mesh. Galvanized wire embedded within wire glass, which prevents splintering and scattering of the material if broken. It is available in hexagonal twisted or diamond netting.

Metal sash putty. Metal sash putty differs from wood sash putty in that it is formulated to adhere to a non-porous surface. It is used for the glazing of aluminum or steel sash, either inside or outside.

Mirror. Plate or sheet glass silvered on one side to reflect an image.

Mirror clip. Metal or plastic device for securing a mirror to an adjacent surface.

Mirror mastic. Special mastic compound used in setting wall mirrors. It serves as a leveling and adhesive material.

Miter. Bevel or a V miter on the face of the glass.

Mitering. Forming a joint by pieces fitted on a line bisecting the angle of junction.

Monolithic reflective glass. Glass consisting of one light of float glass with a permanent transparent reflective coating applied to one surface.

Multiplate glass. Laminated glass made of three or more individual thicknesses of plate glass; three to five pieces of polished plate glass laminated, with clear plastic sheet at each layer and cemented together under heat and pressure. Also called "bullet-resisting."

Muntin. Wood or metal members that divide a glazed area into lights; secondary framing.

Neoprene setting blocks. Blocks used to position the glass vertically in the gasket. Their length may be calculated with a design load factor of 30 psi for each square inch of setting block. Length may also be calculated by using $\frac{3}{8}$ in. of setting block for each square foot of glass. Durometer hardness of the setting block should be in the 85 to 95 shore A range.

Nondrying. Term descriptive of a compound that does not form a surface skin after application.

Obscure glass. Glass that has been made translucent instead of transparent.

One-way mirror (transparent mirror). Mirror made by depositing an evaporated chromium alloy on the glass. When it is installed between two rooms one of which is brilliantly lighted and the other dimly lighted, an observer in the dark room can see into the lighted one. However, the glass in the illuminated room has the appearance of an opaque mirror.

Opal glass

1. Nontransparent glass that may be white or colored.

2. Highly diffusing glass having a nearly white, milky, or gray appearance. Diffusing properties are an inherent internal characteristic of the glass.

Open bubble. Bubble that has been broken into by grinding, leaving a hemispherical hole in the glass surface.

Open seed. Small bubble at the surface of the glass; one-half of a bubble.

Orange peel. Short finish on plate glass usually caused by insufficient grinding and polishing. The surface looks like the skin of an orange.

Organic-coated glass. A glazing material consisting of a piece of glass bonded on one or both sides with an applied polymeric coating, sheeting, or film.

Organic sealed edge unit. Constructed with two lights of glass separated by a metal or organic spacer (filled with a moisture absorbing material around the edges) and hermetically sealed.

Pane. Piece of glass set in an opening in a window or door.

Patterned glass. Rolled flat glass with an impressed design on one or both sides, accomplished during the rolling process. Patterned glass is available in a variety of designs and finishes.

Pebble-finished polycarbonate. Plastic that offers light diffusion, obscurity, and minimum glare.

Performance specifications and method of test. The quality of safety glazing material such as tempered glass, laminated glass, wire glass, or rigid plastic used in buildings.

Permanent label. A permanently legible identification, which cannot be removed or destroyed, visible after installation of the glazing material.

Plastics. Transparent plastic materials (of a carbon base) that are much lighter than glass, but more likely to be scratched. They are inflammable and cannot be used where a fire-retardant product is required.

Plate glass

1. Flat glass formed by a rolling process, ground and polished on both sides, with surfaces essentially plane and parallel. It may contain such flaws as seeds, short finish, skim, strings, scratches, bubbles, open bubbles, ream stones, fire cracks, and sand holes.

2. Glass manufactured in a continuous ribbon and cut in large sheets. Both sides are then ground to two flat surfaces producing uniformity of thickness. The glass is further polished, giving a polished-plate-quality glass. F.S. DD-G 451c.

Points. Thin, flat, triangular or diamond-shaped pieces of steel used to hold glass in a wood sash by driving them into the wood.

Polished plate glass

1. Glass furnished in thicknesses of from $\frac{1}{8}$ to $1\frac{1}{4}$ in. Thicknesses of $\frac{5}{16}$ in. and over are termed "heavy polished plate." Regular polished plate glass is available in three qualities: silvering, mirror glazing, and

glazing. The glazing quality is that used generally where ordinary glazing is required.

2. Transparent glass whose surfaces have been ground and polished essentially plane and parallel.

3. Rough plate glass following continuous grinding and polishing of both surfaces. The term is sometimes used to designate plate glass; plate glass is polished.

Polished wire. Wire glass ground and polished on both sides but not of the same quality as polished plate glass.

Polybutene base. Compound made from polybutene polymers.

Polybutene tape. Nondrying mastic which is available in extruded ribbon shapes of varying widths and thicknesses. The tape remains plastic or resilient over extremely long periods of time and possesses great adhesion qualities. It should not be used as a substitute or replacement for spacers. It is used as a continuous bed material in conjunction with a polysulfide sealer compound. The tape must be pressure applied for proper adhesion.

Polycarbonates. A tough plastic used in glazing applications such as skylights, walls, greenhouses, solar energy systems, and overhead sash and windows. The plastic comes clear and solar and in opaque colors.

Polysulfide base. Compound made from polysulfide synthetic rubber.

Polysulfide elastomer sealing compound. Material that is a two-part synthetic rubber based on a polysulfide polymer. Its consistency after mixing is similar to a caulking compound.

P.P. Polished plate glass.

Precast concrete glazing reglets. The surface of the precast concrete where the gasket is to be installed must be plaster smooth, waterproofed, on a plane, and free of aggregate or other irregularities for at least $\frac{1}{4}$ in. beyond the edge of the gasket sealing lips. A slope in the panel away from either side of the groove is desirable but must not exceed 8° (1 in. in 8).

Priming. Sealing of a porous surface so that the compound will not stain, lose elasticity, shrink excessively, etc., because of loss of oil or vehicle into the surrounding materials.

Prism. Transparent body in the form of a solid whose ends are triangles and whose sides are parallelograms.

Prism glass

1. Flat glass into whose surface is fabricated a series of prisms, the function of which is to direct the incident light in desired directions.

2. Glass made with sharp prisms that are glazed horizontally in the windows, refracting the light thrown back horizontally into the room.

Processed glass. Glass that has been chipped, acid etched, or sandblasted, finished on one or two surfaces. Its thickness is the same for plate or window glass.

Punte. Round polished area put in flat glass with an engraver's wheel as a decoration; also, rod used by glass makers when making glass in former years.

Pyrolitic coating. Pyrolitic means "chemical change generated by heat." Pyrolitic coatings are produced by applying metal oxides to hot glass, which can be done in a properly equipped heat strengthening oven or on a float glass line while the glass is still hot from the melting and forming process.

Quality. In clear glass and mirrors, a measure of relative freedom from defects or distortion. In glass the term is used to designate the degree of freedom from defects, for example, A or B quality.

Quartz. Special composition that transmits a large percentage of ultraviolet rays.

Rabbet. Two-sided L-shaped recess in sash or frame for receiving lights or panels. When no stop or molding is added, such rabbets are face glazed. The addition of a removable stop produces a three-sided U-shaped channel.

Rail. Bottom, middle, and top member of a door having one or more lights; horizontal member.

Ream. Area of unhomogeneous glass incorporated in the sheet, producing a wavy appearance; optical defect caused by striae within the glass.

Reek. Scratch made by the blocking heads during manufacturing, usually a large curved line.

Reflection. Throwing off or back from a surface light or heat, for example.

Refraction. Refract or turn back light into the space to be lighted.

Regular mirror. A regular mirror has a protective coat of shellac or the equivalent, covered with coat of mirror-backing, moisture-resistive paint.

Residence plate. Term used to designate double-strength plate.

Ripple glass. Glass that is rippled, fretted, or ruffled on both surfaces.

Rock crystal. Colorless, transparent quartz glass.

Rolled figured. Term applied to flat glass with a decorative design impressed in one surface by rolling.

Rosette. Molded ornamental metal cover over the attachment holes of a wall mirror.

Rough plate. Plate glass before grinding and polishing; it is not transparent but is knurled and obscure.

Rough plate blanks. Glass at rough plate stage of manufacture.

Safety glass

1. Standard glass used in all vehicles for public and private transportation. Two lights of plate or sheet glass are bonded together with a layer of polyvinyl butyral resin. When fractured, the particles of glass adhere to the plastic, affording protection from flying glass particles. Safety glass may also be made from tempered glass and has exceptional strength qualities.

2. Glass so constructed, treated, or combined with other materials as, in comparison with ordinary sheet or plate glass, to reduce the likelihood of injury to persons. Laminated, tempered, and wire glass are the three types of safety glass.

Safety mirrors. Mirrors used for full height, hinged, pivoted, or sliding doors, made by silvering fully tempered glass, or silvering the back of laminated glass, or silvering a light of glass and laminating it to another light of glass with the silvering inside the unit.

Salvage. Glass that has been used.

Sash. Frame including muntin bars, when used, and the rabbets to receive lights of glass, either with or without removable stops and designed either for face glazing or channel glazing.

Scratches. Scratches on the surface of the glass. Light scratches, sometimes called "hairlines," may sometimes be polished out on a felt wheel. Heavy or deep scratches cannot be removed with equipment found in distributors' shops.

Scratch polishing. Removing small surface defects on glass using a felt wheel with water and rouge.

Screw-on bead or stop. Stop, molding, or bead fastened by machine screws, as compared with those that snap into position without additional fastening.

Sealant

1. Compound used to fill and seal a joint or opening, in contrast to a sealer, which is a liquid used to seal a porous surface.

2. Two-component, structural rubber sealant is automatically machine mixed and applied to a spacer's sealant cavities by air-pressure-operated extruding equipment.

Sealed-edge protection. A stainless steel edge band protects the factory-fabricated unit during handling and installation. The band is designed to exert continuous pressure against the glass, assuring a permanent seal regardless of temperature ranges or atmospheric conditions. An extra barrier of sealant is placed around all edges before banding, and the bottom edge is further protected against shock.

Sealed insulating glass. Factory-built unit of two lights of S.S. or D.S. window glass fused together with a $\frac{3}{16}$-in. space hermetically sealed and filled with an inert gas.

Seed. Extremely small gaseous inclusion in glass; a small bubble, less than $\frac{1}{32}$ in. in diameter. Seeds about $\frac{1}{64}$ to $\frac{1}{32}$ in. in diameter are usually considered coarse seeds. Minute bubbles are less than $\frac{1}{32}$ in. in diameter. Fine seeds are visible only on close inspection, usually appearing as small specks, and are an inherent defect in the best quality of plate glass.

Setting. Placement of lights or panels in sashes or frames; also, action of a compound as it becomes more firm after application.

Setting block. Small blocks of neoprene, lead, wood, or other suitable material placed under the lower edge of a sheet of glass when setting it in a frame.

Setting blocks and spacers. Two 70- to 90-durometer neoprene setting blocks are installed on the bottom edge of each light. Setting blocks are placed between quarter-points and 6 in. from corners. Neoprene spacer shims of 40 to 50 durometers are spaced along the sides and top of the unit on both faces, not more than 24 in. apart. Continuous spacer materials, such as neoprene or styrofoam rope, a sash with neoprene or vinyl inserts, or structural neoprene gaskets, eliminate the need for spacers.

Setting blocks and spacer shims. Glass is installed on two setting blocks at the quarter-points of the opening. Blocks should be of 70- to 90-durometer neoprene with height adjusted for proper perimeter clearance. Spacer shims of 40- to 50-durometer neoprene on 24 in. centers (or neoprene rope) should be used to maintain proper clearance from channel faces.

Shatterproof laminated sound control glass. Laminated glass consisting of sheet, plate, or float glass laminated with sound-absorbing polyvinyl butyral.

Sheet glass. Flat glass made by continuous drawing (common window glass). F. S. DD-G451c.

Shims. Small blocks of composition, lead, neoprene, etc., placed under the bottom edge of a light or panel to prevent its settling down onto the bottom rabbet or channel to prevent distorting the sealant.

Shock mirror. Window glass mirror having a distorted reflection.

Shore A hardness. Measure of firmness of a compound by means of a durometer hardness gauge. A range of 20 to 25 is about the firmness of an art gum eraser; a range of 90 is about the firmness of a rubber heel (ASTM D 2240).

Sight line. Imaginary line along the perimeter of lights or panels corresponding to the top edge of stationary and removable stops, and the line to which sealants contacting the lights or panels are sometimes finished off.

Single strength. Sheet glass is approximately $\frac{3}{32}$ in. thick.

Size of bead. Normally, the width of the bead. There are, however, many situations in which both the width and depth should be taken into account in design, specification, and application.

Skim. Streaks of dense seed with accompanying small bubbles.

Skylight glass. Glass that has five ribs to the inch on each side, the groove on one side being opposite the rib on the other, giving a sinuous section.

Sound barrier glazing. System consists of an aluminum horizontal head/sill receptor, a dual-sealed insulating unit with a 2-in. airspace, and a vertical T or cross mullion inserted between the insulating units.

Sound control glass. Usually, a single glazing unit consisting of sheet or plate glass laminated with layers of sound-absorbing polyvinyl butyral.

Spacers. Small blocks of composition, wood, neoprene, etc., placed on each side of lights or panels to center them in the channel and maintain uniform width of sealant beads. They prevent excessive sealant distortion.

Spacer shims. Devices that are U-shaped in cross section and an inch or more in length, placed on the edges of lights or panels to serve both as shims to keep the lights or panels centered in the sash or frames, and as spacers to keep the lights or panels centered in the channels and maintain uniform width of sealant beads.

Spandrel glass. Clear glass with colored ceramic enamel fired on one side, used for exterior spandrels and similar applications, usually not a glazing item.

S.S. Single-strength sheet glass.

Stain. Stain is usually caused by glass getting wet; the paper in between the glass accelerates the staining process. When glass is stained, it means that the surface luster has been destroyed by chemical action. Stain marks appear gray. Sometimes stain appears in a pattern caused by the paper wrinkling. If the stain is light, it may be polished out by the use of a blocking machine. If the stain has eaten into the glass very deeply, it cannot be removed. Stain may be so light that it is hardly discernible to the naked eye and will only show up when viewed in a certain light or when the glass is silvered.

Stained glass. Colored glass made by adding oxides of metals to the molten glass. The various pieces of glass are held in position by lead strips or cames to form the design.

Stationary stop. Permanent stop or the lip of a rabbet on the side away from the side on which lights or panels are set.

Stone. Opaque or partially melted particle of rock, clay, or batch ingredient imbedded in the glass.

Stop. Either the stationary lip at the back of a rabbet or the removable molding at the front of the rabbet serving to hold either or both the light or panel in the sash or frame with the help of spacers.

Striae. Cords or wavy lines inside the glass, usually not visible to the human eye, caused by poor batch or batch mixing or unsatisfactory melting conditions.

Striking off. Smoothing off excess compound at the sight line when applying compound around lights or panels.

Strings. Wavy, transparent lines appearing as though a thread of glass had been incorporated into the sheet.

Structural glass. Homogeneously colored opaque glass, polished or with a honed finish, ordinarily used for interior or exterior wall facing, toilet and shower partitions, counter and table tops, etc., usually not a glazing item.

Structural members. Term describing framing members designed for support and expansion without any abnormal mechanical stressing of the glass.

Tapestry glass. Type of figured glass used where it is desirable to admit the light but also to obstruct the view. When objects are close to the glass, the glass seems transparent; as an object moves away; the glass becomes increasingly obscure.

Tempered glass

1. Glass subjected to a process of reheating and sudden cooling, which greatly increases its strength. Any type of flat glass $\frac{3}{16}$ in. thick or more, except wire glass, may be tempered; all cutting and fabricating must be done before tempering. If the glass is broken, it does not splinter like ordinary glass but disintegrates into small rounded particles.

2. Glass that has been reheated to just below its melting point and suddenly cooled. When fully tempered glass shatters, it disintegrates into small pieces. Since it cannot be cut or drilled after tempering, the exact size desired and details must be specified when ordering. The qualities of tempered glass are determined from the characteristics of the glass used in its processing.

Tempered glass breakage

1. Tong fusing to the glass during quenching.

2. Chill cracks occur when checks or leads in either edges or surfaces are introduced during the tempering process between the time the glass leaves the furnace and the first seconds it enters the quench.

3. Partial shading in the quench results in poor tempering in one area of a part or plate. It usually occurs when the glass is stationary during quenching, but is also caused by quench nozzles blocked by an accumulation of broken glass, or from a less-than-normal volume of air from one area of the quench.

4. Localized strain imbalance occurs when the quench air flowing across the glass surface is interrupted.

Tempered safety glass. Tempered safety glass is usually obtained by subjecting annealed glass not less than $\frac{3}{16}$ in. in thicknesses to a special heat treatment, which imparts greatly increased mechanical strength and resistance to thermal stresses.

Thermal breakage. Glass breakage from thermal shock may occur if heat from any source is localized on the glass surface. All heat runs should be baffled so that the heat is directed away from the glass. All stickers and labels should be removed immediately upon glazing. Drapes should never be erected directly against the glass so as to act as a heat trap.

Thermal shock. Glass breakage may occur if the glass is covered with paper, painted with signs, or decorated. Heat outlets or convectors close to the windows should be baffled to direct the heat away from the glass. For similar reasons, draperies, curtains, or blinds must not be hung too closely to the glass and ventilation should be provided at the bottom, top, and sides for air movement on the face of the glass.

Thermal stresses. Thermal stresses may cause breakage of the glass if heat from any source is localized on the glass.

Thinning. Addition of a slight amount of unleaded gasoline to an oleoresinous glazing compound by the glazier to soften its consistency.

Ticket window. A ticket window is usually made of safety glass that closes an opening and has a 3-in.-diameter speech hole at eye level and an opening or wicket at the bottom.

Tight dimension. Dimension from the depth of rabbet to depth of rabbet; horizontal or vertical dimension (or the distance between the two shoulders of the opening).

Tong marks

1. At factory option any size glass may have tong marks, but all glass over 48 in. in both dimensions and over 120 in. in one dimension will have tong marks on one edge.

2. Tong marks are on the short edge of glass in most instances. Glass without tong marks is available in limited sizes and thicknesses.

Toughened glass. These products undergo a heating and cooling process (thermal toughening) or a chemical process (chemical toughening). The latter, although highly resistant to thermal stress, is not considered a safety glazing material.

Transmittance. Ability of glass to transmit solar energy in the visible light, the ultraviolet, and infrared ranges, centrally measured in percentages of each.

Transparent. Admitting the passage of light and allowing a clear view beyond.

Transparent glass door. Doors containing transparent glass in a ratio of eighty (80) percent or more to the total area of the door.

Transparent mirrors. When these mirrors are installed between two spaces, one brightly lighted and the other dimly lighted, an observer in the dimly lighted room can see into the brighter one, but from the brightly lighted room the glass has the appearance of a standard mirror.

U. Overall coefficient of heat transmission or thermal transmittance (air to air).

Ultraviolet. Shortest wavelength in the spectrum and more capable of being refracted than violet.

Ultraviolet glass. Glass that transmits the ultraviolet rays of short wavelength, which are shut out by ordinary glass.

United inches. Sum of width plus length, in inches, of a rectangular piece of glass; also, size limit the sum of height and width of a piece of glass.

U **Value.** Amount of heat (Btu) passing through 1 sq ft of window area per hour for each degree Fahrenheit; temperature difference between exterior and interior air.

Vacuum process. Method of depositing silver or any metal on glass. The method controls the thickness of the metal coating to be applied.

Vegetable oil base. Formulation with a vehicle of vegetable oils, usually processed with resins by the application of heat.

Vehicle. Liquid portion of a compound.

Venetian mirror. Unframed mirror.

Vent. Ventilator sash; operating sash as opposed to the fixed sash; hinged.

Washboard glass. Corrugated glass, with 21 ribs to the inch on one side and 5 ribs to the inch on the other side, the ribs being parallel.

Waves. Unevenness of the surface of the glass causing distortion of vision, found in sheet glass and also in rolled glass. Waves in rolled glass, however, are not important as they are rendered harmless by the pattern of the glass.

Weather sealant. Sealant that contains no solvent or oil that can react with the plastic laminate. Elastomeric sealants are polysulfide, silicone, butyl, or polybutene base.

Weep holes

1. Weep holes must be provided to drain off to the outside any water that enters the metal channel frame. Structural neoprene (zipper) gaskets must also provide such weep holes.

2. Weep holes may be desired in gaskets used with insulating glass or laminated glass. They may be provided in the non-lock-strip side of the gasket, the web or cross member in the H gasket, and through the tongue into the reglet in a tongue and groove gasket. They should not be drilled through the lock strip side of the gasket. Generally, three holes are provided in the sill approximately $\frac{3}{8}$ in. in diameter. Weep holes must have continuity through the frame system to the atmosphere.

Window glass. Sheet glass, $\frac{3}{32}$ (single strength) or $\frac{1}{8}$ in. (double strength) thick. FS-DD. G451c.

Window glass (clear sheet glass). Transparent, relatively thin flat glass having glossy, fire-finished, apparently plane and smooth surfaces, but having a characteristic waviness of surface which is visible when viewed at an acute angle or in a reflected light.

Window glass and heavy sheet glass. Glass manufactured by the flat or vertical drawn process. It is supplied in two thicknesses termed "single strength" and "double strength."

Window wall. Large area of glass, generally set within a framework of tubular mullions or with large units of sash held in place with metal stops.

Wire glass

1. Glass with a layer of wire mesh completely embedded in the glass, but not necessarily in the center. When the glass is broken, the wire mesh holds the pieces together to a considerable extent.

2. Regular rolled flat glass with either a hexagonal twisted or diamond-shaped welded continuous wire mesh placed as near as possible in the center of the sheet cross section. The surfaces may be either patterned, figured, or polished. F.S. DD-G-451c.

Wired glass. The ductile steel wire used in wired glass is subject to rust if exposed to air in the presence of moisture. It is recommended to eliminate this problem by applying a weather-proofing silicone sealant to the cut edge where the wire is exposed, prior to glazing.

Wood sash putty. Generally a mixture of pigment and linseed oil. The putty may contain other drying oils such as soybean and perilla.

HARDWARE

Active door (in a pair of doors). Leaf that opens first and the one to which the lock is applied.

Adjustable ball hinge. Hinge so designed that when the door is closed, only a ball-type knuckle is exposed. The hinge has an adjustment for vertical control within limited scope.

Adjustable key. Key for sliding door locks, having a stem or shank adjustable as to its length to adapt the key to doors of various thicknesses.

AHC. Letters indicating that the individual so identified is a qualified member of the American Society of Architectural Hardware Consultants.

Alarmed hardware. Non-ferrous alloy cover and base. 9V battery-powered, latchbolt monitored by an electronic switch, dual frequency pulsating horn, 90 decibels at 6 ft. Manufactured by Arrow, Brooklyn, N.Y.

Aluminum. Aluminum is used in many ways in builders' hardware and is usually alloyed with about 4% of other elements. Cast, forged, and wrought products are obtained by much the same processes as they are with other metals. Pressure-cast aluminum is frequently used as a substitute for cast iron in miscellaneous items such as door stops, handrail brackets, and hooks.

Anchor hinge. Heavyweight hinge with each leaf extended at its top edge and bent to form a flange that fastens to the top edge of a door or to a head frame. Two conventional-type hinges are required to complete the set.

Anodizing. Anodizing forms a protective and uniform oxide on aluminum, giving it a hard, tough skin. A variety of color-anodized finishes, such as black and oxidized bronze, are available.

Antifriction axle pulley. Sash pulley, the axle of which is carried in roller bearing to reduce the friction.

Antifriction bearing. Bearing having the capability of effectively reducing friction.

Antifriction bolt. Latch bolt of a lock provided with a device for diminishing the sliding friction of the bolt during the closing of the door.

Antifriction latch bolt

1. Latch bolt designed to reduce friction when the bolt starts to engage the lock strike.

2. Latch bolt of a lock provided with a device for diminishing the sliding friction of the bolt during the closing of a door. A small additional latch connected with the regular latch bolt engages the strike and retracts the regular latch.

Apartment house keying. Setup whereby none of the individual apartment change keys are interchangeable, but all will operate the entrance door locks. Keys for the entrance door locks will not operate any of the individual apartment locks.

Armored front

1. Lock front that consists of two plates; the underplate, fastened to the case and unfinished, and the finish plate, fastened to the underplate, which when in place covers the cylinder setscrews thus protecting them from tampering. This type of lock front is used on mortise locks.

2. Construction in which the regular front of a cylinder lock is covered by an armor plate, secured to the regular front by machine screws to guard the setscrew that checks the cylinder and also to protect the front of the lock while the door is being painted or the lock is being mortised. This latter result is effected by removing the armor plate from the front of the lock during these mechanical processes.

3. Lock front for mortise cylinder locks, consisting of two laminated plates, with the underplate permanently riveted or screwed to the lock case and the outer plate

secured to the underplate by means of screws. The outer plate is applied after the installation of the lock and the painting of the door are completed, and serves two purposes: it protects the finish of the lock front during the painting process and serves as a guard or "armor" that covers and prevents tampering with the cylinder setscrew.

Armor plate. Plate similar to a kick plate but covering the door to a greater height, usually 40 in. or more from the bottom.

Astragal. Molding or strip whose purpose is to cover or close the gap between the edges of a pair of doors. Some types overlap; others meet at the center of the gap.

Astragal front. Lock front having a form coinciding in shape with the edges of a door having an astragal molding.

Astragal strip. Molding applied to the surface of the active door of a pair of doors, overlapping the edge so as to conceal the abutting edges of the two doors.

Asylum lock. Lock used on doors of psychiatric institutions providing special protection against tampering.

Automatic bolts

1. Bolts that supply both self-latching and self-releasing action, nonmanually, while the active leaf is in the open position. When used on pairs of doors with an astragal, a coordinator is required.

2. Automatic bolts consist of a bolt at the top of the door latching into the top jamb and a bolt at bottom of the door latching into the floor or sill. The bolts latch automatically when the door closes and release automatically when the active leaf is opened.

Auxiliary dead latch. Supplementary latch that automatically deadlocks the main latch bolt when the door is closed. It is also called "deadlocking latch bolt."

Auxiliary latch bolt (guard latch). Latch bolt separate from the regular latch bolt, which remains retracted when the door is closed and automatically deadlocks the regular latch against end pressure.

Auxiliary rose. Rose equipped with a spring on the underside for the purpose of holding a level handle in a horizontal position.

Auxiliary spring. Spring-operated hub device, applied under a rose or escutcheon, used to supplement the power of the regular lock hub spring in order to hold a lever handle in a horizontal position.

Auxiliary spring rose. Rose containing a built-in hub and a strong spring to prevent the sagging of a lever handle.

Axle pulley. Synonymous with "frame pulley."

Backcheck. Optional feature in hydraulic door closers that slows the speed of the opening of the door.

Back plate. Plate on the inside of a door and surrounding the orifice leading from a letter drop or plate on front of door.

Back plate (rim cylinder). Small plate applied to the inside of a door through which the tailpiece of the rim cylinder passes and to which the connecting screws that hold the cylinder in place are fastened.

Back pulley. Synonymous with "shutter flap."

Backset

1. Offset or horizontal distance from the front of a lock to the centerline of its knob or keyhole.

2. Horizontal distance from the outside center of the latch front to the centerline of the knob or cylinder.

3. Horizontal distance from the face of the lock to the centerline of knob, hub, keyhole, or cylinder. On locks

with a beveled front, this distance is measured from the center of the lock face. On rabbeted doors, it is measured from the lower step at the center of the back face.

Ball-bearing butt. Butt having a roller or ball bearing to reduce the friction.

Ball-bearing hinge. Hinge equipped with ball-bearing raceway between the hinge knuckles to reduce friction.

Bar handle. Door handle consisting of a bar, usually horizontal, supported by one or more projecting brackets.

Barn door hanger. Sheave mounted in a frame or attachment to the bottom of a sliding barn door, traveling on an overhead rail and carrying the door.

Barn door latch

1. Heavy thumb latch.

2. Cylindrical bolt mounted on a plate containing a guide or a case projecting where the wall and floor meet.

Barn door pull. Large cupped pull for heavy doors.

Barn door roller. Sheave mounted in a frame for attachment to the bottom of a sliding barn door, traveling on a rail laid in the floor, and carrying the door.

Barn door stay. Small roller, usually carried on a spike or screw, for guiding a sliding barn door.

Barrel bolt. Cylindrical bolt mounted on a plate, having a case projecting from its surface to contain and guide the bolt.

Barrel key. Synonymous with "pipe key."

Basic door closers. Closers are equipped with a fixed nonadjustable spring that is classified according to size from number 2 to number 6 (the higher the number, the heavier the spring and the greater the closing force). Door closers consist of a spring and a piston, housed inside a cylinder filled with hydraulic fluid.

Bearings for hinges. Ball, oil-impregnated, or antifriction bearings.

Bevel of bolt. The bevel of the bolt indicates the direction in which the bevel of the latch bolt is inclined; a "regular bevel" commonly indicates a lock for use on a door opening inward and a "reverse bevel" one for a door opening outward.

Bevel of door

1. Angle of the lock edge in relation to the face of the lock stile. The standard bevel is $\frac{1}{8}$ in. in 2 in.

2. Angle other than 90° made by the edge of a door with the sides of the lock stile. Any inclination from a 90° angle is known as the "bevel." The bevel normally used is an inclination of $\frac{1}{8}$ in. to each 2 in. of door thickness and is called a "regular bevel." Any deviation would be a special bevel.

Bevel of lock. Bevel indicating the direction in which the bevel of the latch bolt is inclined. A "regular bevel" commonly indicates a lock for use on a door opening in; a "reverse bevel," one for a door opening out.

Bevel of lock front

1. Angle of a lock front when it is not at a right angle to the lock case, allowing the front to be applied flush with the edge of a beveled door.

2. Angle of the front of a mortise lock when inclined at other than a right angle to the case, to conform to the angle of the edge of the door.

Bit-key lock. Lock operated by a key having a wing bit.

Bitting. Cut or indentation on the part of a key that acts upon and sets the tumblers.

Bit of a key

1. Projecting blade cut so as to actuate the tumblers and permit the lock bolts to be operated.

2. Projecting blade that engages with and actuates either or both the bolt and tumblers of a lock; synonymous with "wing."

Blind rivet. Installation requires a rivet tool. The advantage is that they can be installed and clinched from one side. Also called "clinch rivets."

Bolt. Bar or barrier arranged to secure a door or other moving part and prevent its opening.

Bolts. A lock achieves its function by means of various types of bolts. A bolt is a bar of metal that projects out of the lock into a strike prepared to receive it.

Bookcase bolt. Bolt that automatically fastens or releases one-half of a bookcase door when the other half of the door is closed or opened.

Bored lock or latch. Lock or latch whose parts are intended for installation in holes bored in a door.

Bored-type locks. Locks installed in a door that has two round holes at right angles to one another, one through the face of the door to hold the lock body, and the other in the edge of the door to receive the latch mechanism. When these two are joined together in the door they comprise a complete latching or locking mechanism. Bored-type locks have the keyway (cylinder) and/or locking device, such as push or turn buttons, in the knobs. They are made in three weights: heavy, standard, and light duty. The assembly must be tight on the door, without excessive play. Knobs should be held securely in place without screws, and a locked knob should not be removable. Roses are threaded or secured firmly to the body mechanism. The trim has important effects in this type of lock because working parts fit directly into the trim.

Boston sash fast. Sash fast in which the rotating locking bar is held in the locked position by a trigger or thumb piece, pressure on which permits the bar automatically to unlock.

Bottom bolt. Bolt for use on the bottom of a door and having frictional resistance whereby the bolt is prevented from falling into the locked position unless intentionally moved.

Box or square bolt. Square or flat bolt mounted on a plate having a case projecting from its surface to contain and guide the bolt.

Box strike

1. Strike that also provides complete housing to protect the bolt openings.

2. Strike in which the aperture to receive the bolt is enclosed or boxed to prevent access from the rear.

3. Strike in which the latch bolt recess is enclosed or boxed, thus covering the opening in the jamb.

Bracket bearing. Knob thimble or socket that, projecting like a bracket, supports the knob close to its head instead of at the end of the knob shank.

Brass and bronze. Metal alloys, the greatest portion of which are copper but containing smaller amounts of other metals, notably lead and zinc. Bronze differs in that it contains some tin. Differences in color result from the proportions of the various metals included.

Builders' finishing hardware. Term designating the locks, hinges, and other metallic trimmings used on buildings for protective and decorative purposes, as distinguished from "rough hardware," which include such building items as nails, screws, and bolts.

Builders' hardware. All hardware used in building construction, but particularly that used on or in connection with doors, windows, cabinets, and other movable members.

Butt (hinge)

1. Type of hinge designed for mortising into the edge of the door and into the rabbet of a door frame.

2. Hinge intended for application to the butt or edge of a door, rather than a flat or strap hinge for application to the surface of the door.

Cabinet door hook. Hook and its staple, each with a heavy plate for attaching, used on shipboard to hold a door at either end of its swing.

Cabinet lock. Lock for use on cabinet work and furniture.

Cam

1. Rotating piece, whereby the rotary motion of a key or knob imparts reciprocating motion to the bolt of a lock.

2. Rotating piece attached to the end of the cylinder plug to engage the locking mechanism.

3. Rotating piece, either noncircular or eccentric, used to convert rotary into reciprocating motion; wing of a bit key or a cylinder cam, which converts rotary into reciprocal motion when actuating the bolt of a lock.

Cam-life hinge. A mortise hinge that will raise the door upward as it is opened, to approximately $\frac{3}{8}$ in. at 90°. Hinges are available for doors weighing 500 to 3,000 lb.

Canada bolt. Box or other type of bolt, the sliding bar of which is prolonged considerably beyond the back plate and provided with a separate guide near its other end.

Cane bolt. Heavy cane-shaped bolt with the top bent at right angles, used on the bottom of doors.

Cap of lock. Removable part or lid of a lock; also called "cover."

Capped butt. Butt having on each leaf a cap that covers the fastening screws and is itself attached to the butt by one or more smaller screws.

Cardcode electronic lock. Keycard lock recognizes up to 150 encoded cards. Can be validated/voided on any and all locks. Self-contained and is available in mortise or cylindrical locking devices. Manufactured by Yale Security Inc.

Card plate. Plate for use on doors or drawers and designed to hold a label indicating contents.

Carriage bolts. Bolts having round heads that can't be gripped with a tool and are used to fasten together wood or metal and wood. Common diameters are $\frac{1}{4}$ in. to $\frac{1}{2}$ in., common lengths are 1 in. to 6 in.

Casement. Window (sash) hinged in a frame to open horizontally in or out.

Casement adjuster

1. Device for holding a casement open or shut.

2. Hinged or pivoted rod for moving and fastening the hinged sash of a casement or French window.

Casement fastener. Catch for fastening a casement or French window.

Casement hinge. Hinge to swing a casement window. The term is often used to describe a hinge designed to throw the sash out far enough to permit cleaning the outside of the glass from the inside of the room on an outward-swinging casement.

Casement window. Window with side-hinged sash opening either in or out.

Case of lock. Boxlike metal container for the bolt and other mechanisms of a lock. In giving dimensions the vertical should be stated first and the horizontal second.

Cast iron. Cast iron has long-wearing qualities, natural lubrication, brittleness, low tensile strength, and a tendency to rust. It contains approximately 92% iron and small quantities of such other elements as carbon, manganese, silicon, phosphorus, and sulfur. The metal is formed by pouring into sand molds or other type molds to achieve the desired shape.

Cast metal. Metal produced by pouring molten alloy into premolded forms. This method results in a versatile shape that can be machined, etched, or carved to yield a great variety of designs.

Cast trim. Trim made of cast or forged brass, bronze, or aluminum.

Ceiling hook. Hook for use in ceilings, usually having two prongs.

Center hung pivot. Light-duty style for use on toilet stall doors with emergency door stops.

Center pivot sets. Pivots are installed so that the door pivot is located in the center of the thickness of the door set forward a specific distance from the back edge or heel of the door. The pivot set consists of a top pivot and a bottom pivot, floor mounted in a cement case. Maximum door weights vary from 150 to 1,000 lb, depending on type of installation and the pivot type selected.

Chafing strip (sliding doors). Removable vertical strip used with sliding doors to close the wall pocket opening when a door is either in a completely opened or closed position. It is also used at the bottoms of doors to keep the doors in alignment with the overhead hangers.

Chain bolt

1. Bolt applied at the top of a door, and having a chain hanging therefrom, whereby the bolt may be retracted against the resistance of a spring which tends to hold it in the locked position.

2. Spring bolt actuated by a chain attached to the spring bolt, for application at the top of the door.

Chain door fastener

1. Heavy chain, one end of which is secured to a plate, which may be attached to the edge of the door, the other end of the chain carrying a ball or hook, which may be inserted in a slot formed in another plate attached to the jamb or other half of the door, whereby the door cannot be opened, except slightly, until the chain is removed from the slot.

2. Device that limits the opening of a door by means of a chain.

3. Device that permits sliding a chain into a plate on the door, allowing the door to be opened slightly for conversation without intrusion.

Change key

1. Key of a master-keyed lock that differs from all others of the same series and will operate only its own lock (sometimes called "room key"). The term "change key" is used, rather than "master key."

2. Key that operates only one lock or two or more locks keyed alike.

3. Individual key of a lock cut to operate only a lock with a corresponding key setting.

Changes (key). Different bittings or tumbler arrangements in a series of locks.

Checking floor hinge

1. Device placed in the floor that combines top and bottom pivots for hanging the door with a controlled-speed closing mechanism.

2. Hinge containing a spring for closing a compression chamber in which the liquid escapes slowly, thus retarding the closing action to prevent slamming of the door.

3. Device that combines top and bottom points for hanging the door with a controlled-speed closing mechanism. Top, intermediate, and floor pivots make installation possible and enable transfer of weight from the jamb to the floor.

Clevice. Metal loop fastened to a padlock for attaching a chain.

Closet bar rod or pole. Bar, rod, or pole of wood or metal from which clothes may be hung, usually extending across the width of a closet on brackets.

Closet knob. Single knob on one end of a spindle, on the other end of which is a rose or plate to secure the knob and spindle to the door; for use on closet doors.

Closet set. Lock or latch set having a knob outside and thumb turn or closet spindle inside.

Closet spindle. Spindle having a turn that is used in place of a knob. It is usually used on the inside of closet doors.

Coat and hat hook

1. Hook with two projections, one of which is of sufficient length to receive a hat, the other being usually shorter.

2. Hook with two or more prongs, usually with a short prong for coats and a longer prong for the hat.

Coatings. Coatings are used to prevent tarnishing or oxidation of all natural finishes and of plated brass and bronze finishes. The original color and sheen of natural metals are maintained for a long time with the use of synthetic coating treatments.

Combination lock. Lock having changeable tumblers actuated by a dial on the face of the door, permanently connected to the lock mechanism by a spindle.

Combination tumbler. Circular plate of metal consisting of a central disk, containing the driving pin for communicating motion from one tumbler to the next, and an outer or annular disk, enclosing the central one and containing the "gating," these two parts being variably adjustable in relation to each other, thus forming the permutation wheel or tumblers of a combination lock.

Combined escutcheon plate

1. Plate containing both a keyhole and a knob socket.

2. Escutcheon plate cut for both knob and keyhole or cylinder.

Combined store door lock. Lock containing a heavy dead bolt and a latch bolt adapted to be operated by thumb handles instead of knobs.

Communicating door lock. Lock for use on doors between communicating rooms, usually a knob latch with thumb bolts.

Communicating door lock (mortise type). Lock usually having a latch bolt and two dead bolts. One of the dead bolts is controlled by a turn knob or key from one side only, and each dead bolt is controlled independently of the other.

Compensating hub. Lock hub having an elongated spindle hole or of the movable ball type to compensate for the shrinking and swelling of a door and to prevent derangement of the lock and binding due to misalignment.

Concealed hinge.   Hinge so constructed that no parts are exposed when the door is closed.

Concealed-in-door closer.   Door closer completely concealed in the door.

Concealed-in-floor closer.   Single-acting type of door closer that supports the door and is offset hung. Types for doors independently hung on butts or offset pivots are also available. Single-acting closers may also be center hung.

Concealed-screw rose.   Synonymous with "screwless rose."

Connecting bar.   Flat spindle attached to the cylinder of a rim lock to operate a bolt.

Connecting door lock.   Synonymous with "communicating door lock."

Continuous hinge.   Hinge designed to be the same length as the moving part to which it is applied, for example, the hinge on the lid covering the keyboard of a piano; also called "piano hinge."

Conventional locks.   Term commonly applied to mortise locks.

Coordinator

1.   Device used on a pair of doors to ensure that the inactive leaf closes before the active leaf. A coordinator is necessary when an overlapping astragal is present and exit devices or automatic or self-latching bolts are used with closers on both door leaves.

2.   Device used on pairs of doors, usually in connection with panic devices, consisting of two arms to regulate the proper closing of the inactive leaf before that of the leaf having the astragal.

Corrugated key.   Sheet metal key of uniform thickness, corrugated longitudinally; key having a sinuous cross section and not merely grooved on one or both sides.

Cottage latch.   Small lift latch for use on cupboards and light doors.

Crank handle.   Synonymous with "lever handle."

Cremorne bolt

1.   Fastening for casement or French windows, arranged for application to the surface, consisting of a sliding rod engaging at top and bottom with strikes or plates n the window frame, and provided near its center with a handle or knob the rotation of which causes the upper and lower parts of the bolt to move in opposite directions in locking or unlocking; it is sometimes also provided with an additional horizontal bolt, also operating simultaneously, which serves further to secure the sash at or near its center.

2.   Device, of surface application, that by a turn of knob or lever handle locks the door or the sash into the frame, top and bottom.

Cupboard button.   Small turning bar adapted to secure a door.

Cupboard catch

1.   Small spring catch adapted for fastening a light door and operated by a slide knob or thumb piece.

2.   Small lock for use on cabinets, cupboards, and other small hinged doors.

Cupboard lock.   Lock designed for use on doors of cupboards, boxes, etc.

Cupboard turn.   Small spring catch adapted for fastening a light door and operated by a rotating knob or handle.

Cup escutcheon

1.   Door plate, for use on sliding doors, having a recessed panel to afford fingerhold and to contain the knob, or its equivalent, and a key, all of the contained parts being flush with the surface of the plate in order to offer no obstruction to the movements of the door within its recess.

2.   Door plate, for use on sliding doors, having a recessed panel to afford fingerhold and to contain a flush ring and sometimes a cylinder, all being flush with the surface of the plate.

Curved lip strike.   Strike with a lip curved to conform to a detail to protect door casings and prevent the catching of wearing apparel on the projecting lip.

Cycloid knob action.   Arrangement of intergeared pivotal levers for transmitting motion from a lock hub to the latch bolt.

Cylinder.   Cylindrically-shaped device containing the key-controlled mechanism and cam or spindle for actuating the bolts of a so-called cylinder lock. It is a separate mechanism from the lock proper, but it is a unit assembly with the lock.

Cylinder collar.   Plate or ring used under the head of a cylinder.

Cylinder lock

1.   Lock in which the keyhole and tumbler mechanism are contained in a cylinder or escutcheon separate from the lock case.

2.   Lock in which the locking mechanism is controlled by a cylinder.

3.   Lock equipped with a "cylinder" to actuate its bolts.

Cylinder of a lock

1.   Short cylindrical case containing the keyhole and tumbler mechanism of a lock.

2.   Cylindrically shaped assembly containing the tumbler mechanism and the keyway, which can be actuated only by the correct keys.

Cylinder ring.   Rose or washer placed under the head of a cylinder lock to enable a long cylinder to be used on thin doors.

Cylinder screw

1.   Setscrew in the face of a cylinder lock for preventing the unscrewing of the cylinder; also called "setscrew."

2.   Setscrew that holds a cylinder in place by preventing the cylinder from being turned after installation.

Cylinder types.   Cylinders are of three general types: mortise cylinders for use with mortise locks, rim cylinders for use with rim locks, and knob cylinders for use with key-in-knob locks.

Cylindrical locks and latches.   Locks with a cylindrical case that has a separate latch bolt case that fits into the cylindrical lock case.

Dead bolt

1.   Lock bolt having neither spring action nor bevel and that is operated by a key or a turn piece.

2.   Lock bolt having a square head and moved positively by the key in both directions.

3.   Lock bolt having no spring action and activated by a key or a turn. It must be manually operated and provides security; when hardened steel inserts are used the security is greater. The minimum throw is usually $\frac{1}{2}$ in. A dead bolt may be specified with certain functions of mortise preassembled and integral locks.

4. Bolt that is not actuated by springs but that has to be operated manually. In locked position, the bolt cannot be forced back.

5. Lock bolt, usually rectangular in shape, that is thrown into positively locked projected or retracted positions by means of keys or turn knobs.

Dead latch. Synonymous with "night latch."

Deadlock. Lock with a dead bolt only, controlled by key from either or both sides, or by key from one side with or without a turn knob from the other side.

Deadlocking latch. Latch that has an auxiliary plunger that automatically prevents forcing of the latch bolt when the door is closed.

Deadlocking latch bolt. Latch bolt incorporating a plunger that is held in a retracted position when a door is closed, thus preventing the bolt from being retracted by end pressure.

Desk lock. Lock adapted to secure a drawer or a series of drawers.

Detachable key. Key so constructed that the bits or portion that actuates the tumblers may be detached from the shank or handle of the key for convenience in carrying.

Dial lock. Synonymous with "combination lock."

Direction plate. Plate giving information concerning the purpose of the door or opening in which it is used.

Dogging device. In exiting devices, the mechanism that fastens the cross bar in the fully depressed position and also retains the latch bolt or bolts in a retracted position, thus permitting free operation of the door from either side.

Dome door stop. Floor stop with rubber tip providing minimum projection.

Door bolt

1. Manually operated rod or bar attached to a door and providing a means of locking.

2. Sliding rod or bar suitably mounted for attachment to a door and adapted to secure it.

Door bottom. Automatic, insulating, weather seal soundproofing mechanically operated as door moves into closing position.

Door check. Device that combines a spring for closing a door and a compression chamber in which the slow escape of the liquid or air contained therein retards the closing action and prevents the slamming of the door. It is sometimes made without a spring to be used in conjunction with spring hinges.

Door clearance. Doors must be able to open 180°. For typical $1\frac{3}{4}$ in.-thick doors, with the door flush with casing, the hinge should be a minimum of 4 in. wide. For heavyweight hinges, the width should be a minimum of $4\frac{1}{2}$ in. wide. For projecting trim, wider hinges must be used if the door is to open 180°.

Door closer. A door closer controls the door throughout the opening and closing swings. It combines three basic components: a power source to close the door; a checking source to control the rate at which the door closes; and a connecting component (arm) that transmits the closing force from the door to the frame.

Door closer bracket. Device whereby a door closer may be installed on the frame rather than directly on the door.

Door closer mountings. Three ways to mount surface closers are by means of a hinge side, parallel arm, and top jamb. A wide variety of brackets, including corner and soffit types, are available to meet varying door and frame conditions.

Door closer or check. Device combining a spring for closing and a compression chamber into which the liquid or air escapes slowly, thus providing a means of controlling the speed of the closing action.

Door holder

1. Device for fastening a door in an open position.

2. Device that holds a door open at selected positions.

Door mutes. Mutes or silencers designed to cushion the impact of a door against the frame, thus reducing noise.

Door pivot

1. Hinging device embodying a fixed pin and a single joint. Most types include lateral fastening.

2. Pivot designed to fit into the floor and mortise into the heel of a door with a guide pivot at the head. It has no springs.

Door pull

1. Bent handle usually mounted on a plate and adapted for attachment to the surface of a door.

2. Hand grip device applied directly to the surface of the door or a handle applied on a plate, used as a means of pulling a door open.

Door silencers

1. Rubber buttons with spear-shaped portion for insertion into a hole drilled in a metal frame stop.

2. Rubber buttons inserted into the metal frame of a door to cushion the door, thus quieting the noise of a closing door against the frame.

Door stop

1. Device to stop the swing or movement of a door at a certain point; also, architectural term defining that part of a door frame against which the door closes.

2. Device to limit the swing or movement of a door when open.

Door weights. Typical solid wood core doors and hollow metal doors, with faces no thicker than 16-gage metal, measuring 3 ft wide by 7 ft high and weighing up to 150 pounds to 250 pounds total weight, require heavyweight hinges. For doors weighing over 250 pounds, special high-capacity hinges are required.

Double-acting butt. Butt that permits a door to swing in both directions.

Double-acting spring hinge. Hinge having a double set of springs opposed to each other, each tending to move the door into the closed position; the hinges are so constructed as to permit the door to swing in either direction.

Double-bitted key. Key having bittings on both sides, whereby either or both wings or sides of the key may actuate the tumblers.

Double door bolt. Bolt having two sliding bars, moving in opposite directions, to secure a door simultaneously at the top and bottom.

Double throw bolt

1. Bolt that can be projected beyond its first position into a second or fully extended one, thus providing extra security.

2. Bolt controlled by a mechanism that permits extra projection or "throw" of the bolt.

Double weight ball bearing hinge. Hinge is 5 in. high, for full mortise, half surface, half mortise, or full surface applications for doors weighing up to 800 pounds.

Drawer knob. Small knob suitable for use on drawers and cabinet work.

Drawer lock. Lock adapted for use on drawers; also known as "till lock."

Drawer pull. Handle or grip adapted to receive the fingers.

Drawer roller. Device used to ease the sliding of a drawer open or shut, usually with a metal or fiber wheel rotating in a metal frame.

Drawer slides

1. Mechanism employing guides and rollers that guide and support the drawer, permitting easy operation.
2. Series of telescoping slides that support the drawer, permitting easy operation of the drawer.

Drill pin. Round pin projecting from the back plate of a lock and fitted into a hole in the end of the key.

Drivers. Upper set of pins in a pin tumbler cylinder, which, activated by the springs, project into the plug until raised by insertion of the key.

Drop. Distance from the front edge of the selvage to the center of the cylinder or keyhole in cabinet locks.

Drop drawer pull. Pull or handle pivoted at its ends to its attaching plate.

Drop escutcheon (key plate). Escutcheon or key plate provided with a pivoted drop covering the keyhole.

Drop hook. Synonymous with "shutter bar."

Drop key. Key having a bow or handle pivoted to the shank so that it may drop or fall parallel with the surface of the door.

Drop key plate. Swinging cover or drop to protect the keyhole.

Drop ring. Ring handle attached to a spindle that operates a lock or latch. The ring is pivoted but remains in a dropped position when not in use.

Druggists' drawer pull. Drawer pull combined with a plate to contain a label.

Dummy cylinder

1. Mock cylinder without any operating mechanism for use where effect is desired.
2. Limitation cylinder that can be attached with a cylinder collar or escutcheon for use as dummy trim.

Dummy trim

1. Trim only, without lock, usually used on the inactive door in a pair of doors.
2. Trim only, without lock or working parts, applied to the inactive door of a pair of doors and matching the trim on the active door. This is for the purpose of balancing the hardware ornamentation of a pair of doors.

Duplex lock. Master key lock of the cylinder type provided with two cylinders on the same side, both acting on the same bolt but each controlled by a different key, whereby, when used in a series, one of said cylinders may be operated by the master key, which passes every lock in the series, and the other by a change key, which may be different for each lock throughout the series.

Dustproof strike

1. Strike with a spring plunger that completely fills the bolt hole when the bolt is not projected.
2. Strike with spring-controlled shutters that automatically close the strike opening when the lock bolts are withdrawn.

Dutch door. Door cut horizontally through the lock rail so that the upper part of door may be opened independently of the lower door, both parts being equipped with hardware so that the door can be used as a complete total door.

Dutch door bolt. Device for locking together the upper and the lower leaves of a Dutch door.

Easy spring. Construction of a knob lock in which two springs are employed, one of which (the easy spring) acts only on the latch bolt, while the other acts directly or indirectly on the knob spindle. The motion of the latch bolt is opposed by the easy spring, while both springs give resistance to rotation of the knobs, thus giving a lively action to the knobs while permitting the door to close easily. The same action may also be obtained with a single spring if it is suitably connected with the related parts of the lock.

Edge plate. Angle or channel-shaped guard used to protect the edge of a door.

Edge pull. Pull mortised into the edge of a sliding door.

Elbow catch

1. Spring-loaded device embodying a rocker arm and angle strike, for locking the inactive leaf of a pair of cabinet doors.
2. Pivoted fastening for cupboard doors, one end having a hook to engage with a strike or staple and the other end bent to a right angle to form a handle for releasing the catch.
3. Pivoted fastening, usually used to fasten the inactive door of a pair of cupboard doors, which engages a strike.

Electric strike (electric door opener)

1. Electromechanical device that replaces an ordinary strike and makes possible remote electric locking and unlocking of a door. When a control mechanism actuates the electric strike, it allows the door to be opened without a key and to be relocked when closed.
2. Lock strike equipped with an electrically controlled reacting device operated by push buttons in various locations throughout a building, usually in each apartment of an apartment house. By pressing the electric push button a small electromagnet retracts a section of the strike, thus releasing the lock bolts and permitting the door to open.

Electromagnetic holder. Form of door holder that is electrically powered and designed to hold open a door equipped with a self-closing device. When the current is cut off, the door is under the control of the closing device and swings shut. This is generally a smoke or fire door. The electromagnetic holding function may be released by a manual pull on the door, an electric switch, or an alarm device, which is generally controlled by a smoke and/or heat detector.

Ellipsoid knob. Doorknob of oval design.

Emergency door stop. Stop that may be folded or collapsed to permit reverse opening of a door in an emergency, used on private or semiprivate toilet doors.

End-load arms (overhead closers). Center-hung installation where the door is rotated to a 90° position, the closer spindle is also preloaded to a 90° position, and the door is installed perpendicular to the frame.

Escutcheon. Plate containing a keyhole.

Escutcheon (elongated). Plate long enough to span a lock case and having holes for knob bushing bit key, cylinder, turn knob, and similar operating members as required.

Escutcheon knob. Doorknob containing a key escutcheon, the latter actuating the lock or controlling the rotation of the knob.

Escutcheon plate. Protective metal plate applied to the surface of the lock stile, with or without cylinder hole or keyhole but with knob socket.

Escutcheon trim. Ornamental shield or plate mounted behind the knob.

Espagnolette bolt (or bar). Fastening for casement of French windows, arranged for application to the surface thereof, consisting of a rotating rod extending from top to bottom, with hooks at each end that engage with pins or plates in the window frame when the bar is rotated, and having a hinged handle near the center whereby the bar may be rotated to fasten or release the sash and which also engages with a strike or keeper which holds the bar in the locked position and further secures the sash near its center.

Exit device

1. Door-locking device designed to grant instant exit by pressing on a crossbar that releases the locking bolt or latch.

2. Locking or latching device that may always be released by depressing a crossbar. It is sometimes called a "panic bolt," "exit bolt," or "panic exit hardware," however, "exit device" is the preferred term.

Exit device types. There are four types of exit devices: rim, mortise, surface vertical rod, and concealed vertical rod.

Extension bolt. Bolt having a short plate to receive a knob or thumb piece, which latter is connected at the bolt end at the top or bottom of the door by an extension rod inserted through a hole bored in the thickness of the door.

Extension flush bolt

1. Flush bolt in which the connection between the bolt head and operating mechanism is by means of a rod inserted through a hole bored in the thickness of the door.

2. Extension flush bolts are used in pairs, with one at the top and one at the bottom. they are usually installed n the edge of the door. The bolt head is projected or withdrawn manually by finger operation of a lever.

Extension key. Synonymous with "adjustable key."

Extension link

1. Device used to provide long backsets in bored locks.

2. Metal device that can be linked to the latch of a cylindrical lock to increase the backset.

Extruded shapes. Shapes produced by forcing or drawing semimolten metal through dies. Designs having linear characteristics are possible.

Face (of a lock). Exposed surface that shows in the edge of a door after installation.

Fast joint butt. Butt in which the hinge pin is riveted or otherwise secured, and the two parts of the butt permanently fastened together.

Fast pin butt. Butt in which the pin is fastened permanently in place, preventing separation of the two leaves.

Fast pin hinge. Hinge in which the pin is fastened permanently in place.

Fence. Projecting portion of a lock, usually attached to the bolt, which engages with the tumblers and enters or passes through the "gating" of the tumblers when the bolt is retracted.

Finish builders' hardware. Hardware that has a finished appearance, as well as a functional purpose, and that may be considered part of the decorative treatment of a room or building, also termed "finish hardware," builders' finish hardware," and "architectural hardware."

Fire door latch. Spring latch or deadlocking latch that has a $\frac{3}{4}$-in. throw and an antifriction retractor.

Fire exit bolt. Device designed to grant instant exit by pushing a crossbar that releases a locking bolt or latch.

Flat key. Thin flat key made of sheet or plate metal, usually by stamping and sometimes provided with longitudinal grooves or indentations on one or both sides.

Flat washers. Washer acts as a bearing surface to distribute the load of a bolt head or nut. The flat washer is placed between the bolt head or nut and the surface of the object being joined to protect the object from possible damage caused by tightening and turning the nut or bolt.

Floor closer

1. Closing device installed in the floor under a door.

2. Floor hinge with checking action.

Floor door stop. Stop installed on the floor at a selected point to ensure proper contact with the door.

Floor hinge

1. Combined pivot hinge and closing device set either in the floor or in the bottom of the door; it may be spring type only or may be combined with liquid control.

2. Pivot door hinge, set in the floor, usually combining with the door spring, and frequently also acting as a door check.

3. Pivot hinge set in a floor. It may be spring type only or may be combined with liquid control for quiet closing.

Floor holders. There are two types of floor holders: the spring-loaded "step-on" type and the lever or "flip-down" type. Neither type acts as a stop.

Flush bolt

1. Door bolt so designed that when it is applied, it is flush with the face or edge of the door.

2. Door bolt mounted behind a plate adapted to be attached to and let into the surface of a door.

Flush cupboard catch. Catch that is half mortise, for example, let in flush with the face of the door.

Flush cup pull. Pull mortised flush into a door having a recess to receive fingers to actuate the slide of the door.

Flush plate. Door plate of any kind intended to be let into the wood flush with its surface.

Flush ring

1. Flush door pull mortised into a door, having a ring pull that folds flat into the cup of the pull.

2. Flush drawer handle of circular form.

Flush ring cupboard catch. Catch with a flush ring in place of a knob for actuating the bolt.

Folding key. Key having a handle and a blade or shank hinged together, the blade folding into the handle like a jackknife.

Foot bolt

1. Type of bolt applied at the bottom of a door and arranged for foot operation. Generally, the bolt head is held up by a spring when the door is unbolted.

2. Spring bolt for the bottom of a door which, when retracted, is retained by a trigger, the release of which later permits the spring to shoot the bolt into the locked position.

Forged iron. Forged iron is used in builders' hardware in the manufacture of specialized or decorative trim often imitative of early American hand-forged items. Iron forgings are produced by hammering a red-hot bar of iron into the desired shape. Forged iron is almost pure iron, with only about 1% of other elements.

Forged metal. Forged metal is hammered, pressed, or rolled into shape. A smooth, dense product results from this process, the value of which relates to the thickness of the metal.

Frame pulley. Box containing a sheave and adapted to be mortised into a window frame for carrying the sash cord, or chain.

French door lock. Mortise lock usually equipped with French springs to support the level handle, necessarily used because of the narrow backset.

French escutcheon. Small circular key plate containing a keyhole, secured by driving or screwing into the wood.

French hardware. Rim locks and bolts of ornamental character, as used in French construction.

French shank. Term commonly used to designate a thin shaped, cast, forged, or machined knob shank. The shape varies somewhat in different manufacturers' lines.

French spring. Heavy type of spring usually applied to the hub of a lock, intended to offset the overbalancing effect of the lever handle and thereby relieve the strain, which would otherwise be thrown on the sensitive spring of the latch bolt.

French window. Window mounted on hinges to swing like a door; casement window extending to the floor.

French window lock. Mortise knob lock with small backset, for use on French windows or doors with narrow stiles.

Friction catch. Catch that when it engages a strike is held in the engaged position by friction.

Friction hinge. Hinge designed to hang a door and hold it at any desired degree of opening by means of friction control incorporated in the knuckle of the hinge.

Front door lock

1. Lock for use on entrance doors, having a dead bolt and a latch bolt; the former is controlled from the outside by a key and from the inside by a key or knob; the latter is controlled from the outside by a key and from the inside by a knob. It is usually provided with "stop work" whereby the outside knob may or may not be set to actuate the latch bolt, as desired.

2. Lock designed for use on residential entrance doors. Term is usually applied only to locks whose latch bolts are controlled by knobs from both sides, although locks for this purpose having thumb lift latch control would also fall in this classification. Dead bolts of front door locks are controlled from the outside by key only and from the inside by key or turn knob.

Front of lock

1. Plate through which the latching or locking bolts project.

2. Face plate of a mortise lock through which the ends of the bolts are projected.

3. Latch face plate that is recessed in the edge of the door and through which the latch or bolt protrudes.

Full mortise hinge. Hinge used for wood doors and pressed metal or wood jambs, or for hollow metal doors and pressed metal jambs.

Full surface hinge. Hinge used for channel iron frames and metal-covered doors. Hinges should be applied with throughbolts and grommet nuts and jamb leaf with machine screws.

Fusible link arms (closer). Allows a door to self-close in a fire, because the metallic link melts when exposed to heat closer arms constructed of heavy gauge metal and designed to withstand abuse.

Gating. Opening in the tumbler of a lock into or through which the "fence" passes to release the bolt or permit its movement.

Grand master key. Key that operates locks in several groups, each of which has its own master key.

Grille. Ornamental screen of open metal work, wrought or cast.

Guard bar. Series of two or more crossbars, generally fastened to a common back plate to ensure protection of glass or screen in a door.

Guarded front and strike

1. Parts of a lock so constructed that they may interlock to protect the latch bolt from attack through the crevice between the door and jamb.

2. Interlocking construction so that when a door is closed, the latch bolt cannot be manipulated. This type of construction is designed particularly for psychiatric institution and schoolhouse locks.

Guarded front lock. Lock that has a specially constructed interlocking front and strike. When they are interlocked, the latch bolt is protected against retraction by end pressure exerted through the crevice between the door and jamb. Also referred to as a "recessed front lock."

Guestcode electronic lock. Designed for hotel/motel application, microprocessor controlled with internal battery pack. Available in mortise or cylindrical locking devices. Manufactured by Yale Security Inc.

Gun spring. Heavy flat spring of special construction used in locks and latches to hold lever handles in a horizontal position.

Half-mortise hinges. Hinges used for hollow metal doors and channel iron frames.

Half-rabbeted lock. Mortise lock the front of which is turned into two planes at right angles, thus adapting it to use on a door with rabbet on edge; lock having a front in two planes forming a single right angle.

Half-surface hinges. Hinges used for metal-covered doors and pressed-steel jambs. Hinges are applied with throughbolts and grommet nuts and jamb leaf with machine screws.

Hand

1. Hand indicates whether the article is adaptable to either a right-hand or left-hand door.

2. Hand indicates the direction of swing or movement and/or locking security side of a door.

Hand and bevel of locks. Inclination or bevel of the latch bolt and the lock front always corresponds in direction with the bevel of the door. If no bevel is designated, the bevel is understood to be regular. The hand of such a lock is the same as the hand of the door.

Hand determination. When the person is facing the door.

Handed

1. Term indicating that the product is for use only on doors of the designated hand.

2. Term indicating that the article is designed or assembled for use on either a right-hand or left-hand door, but not both.

Handed (not reversible). Used only on doors of the hand for which designed, for example, on most rabbeted front door locks and latches.

Hanging stile. Vertical rail of a door to which the hinges or butts are applied.

Hanging strip. Vertical strip attached to the door casing to which the hinges are applied.

Harmon hinge. Hinge designed to swing a door into a pocket at a right angle with the frame.

Hasp. Fastening device consisting of a loop and a slotted hinge plate, normally secured with a padlock.

Hasp lock. Prison-type lock permanently attached to the hasp of the door and adapted to secure the same when in a closed position.

Hinge. Pair of jointed plates attached, respectively, to a door and its frame, whereby the door is supported and enabled to swing or move.

Hinge plate

1. Synonymous with "hinge strap."
2. Ornamental plate butted against the hinge or stop to give the effect of a strap hinge.

Hinge requirements. Doors up to 7 ft 6 in. in height should use three hinges, and one additional hinge for each 2 ft 6 in. or fraction thereof.

Hinges. There are four basic applications of hinges: full mortise, half-mortise, full surface, and half-surface. Within each type there are a variety of styles, each designed for a particular situation. In addition to these there are other design types, such as olive knuckle, modern paumelle, and pivots.

Hinge stile (of a door). Stile to which the hinges are applied, as distinguished from the lock stile.

Hinge strap. Plate, usually ornamented, adapted for attachments to the surface of a door, fitting at one end against the knuckle of a butt, and intended to give the effect of a strap hinge.

Hinge template. Clamped over the edge of the door, it ensures correct positioning of hinge mortises and permits use of machine router.

Hinge tips. Flat button tips (FBT) are standard with all hinge manufacturers. Hospital-type tips are available from most manufacturers. The tips available as desired by the architect or owner are: ball, button, and steeple.

Holdback feature. Term used to describe a function for a lock whose latch may be held in the retracted position by a catch or other means, creating the holdback feature.

Horizontal lock. Lock whose major dimension is horizontal.

Horizontal spring hinge. Spring hinge mortised horizontally into the bottom rail of a door and fastened to the floor and head frame with pivots.

Hospital latch. Latch that provides positive latching and push-pull operation. Both push and pull sides use the same latch.

Hospital tip. Tip furnished on hinges where it is desired to avoid the possibility of injury due to conventional tip's projection. Hinges with this type of tip are necessarily fast pin.

Hotel lock

1. Type of master-keyed lock, usually having a latch bolt and either one or two dead bolts. The lock may be controlled from the corridor side by the "guest's" or "room" key, or by a floor master key for maid's use, unless the guest has locked the door from the inside. The master key, known as an "emergency key," unlocks the door under all conditions. Additional keys are sometimes used. The "shout-out" key prevents any lock in the series from being opened, except with the emergency key. The "display" key is an individual shut-out key that protects rooms being used for the display of merchandise by preventing the entrance of maids during the absence of the person responsible for the display.
2. Usually a master-keyed knob lock.

Hub

1. Part of a lock through which the spindle passes to actuate the mechanism.
2. Rotating piece within a lock containing a central aperture to receive the knob spindle and engaging with the bolt or tailpiece in the lock whereby the motion of the knob is communicated to the bolt.

Inactive door (or leaf). Leaf of a pair of doors that does not contain a lock, but is bolted when closed, and to which the strike is fastened to receive the latch or bolt of the active door.

Independently hung (overhead closers). Installations have either a slide-arm or a double-lever type arm connecting the closer spindle to the door. The door is hung separately on pivots or hinges.

Indicator button. Device used in connection with a hotel lock to indicate whether or not the room is occupied.

Inside door lock

1. Synonymous with "room door lock."
2. Locks for interior doors. Term usually is used for bit key locks.

Instant lock. Time lock constructed to lock automatically by spring action upon the closing of the door.

Integral lock and latch. Type of mortise lock having cylinder in the knob.

Integral-type lock. Mortise lock with a cylinder in the knob installed in a prepared recess (mortise) in a door. The complete working unit consists of the lock mechanism and selected trim (knob, rose, escutcheon). Roses or escutcheons are bolted together through the lock case.

Intermediate pivot. Heavy-duty-type pivot for use with pivot hinges and floor hinges on lead-lined doors.

Invisible hinge. Hinge so constructed that no parts are exposed when the door is closed.

Jamb. Inside vertical face of a door or window frame.

Jamb joint. Joint, used on the abutting edges of a French sash, in which the edge of one sash is convex and of the other concave to a radius equal to one-half the thickness of the sash, the purpose being to form a weathertight joint.

Jamb lock. Prison-type lock designed to be built into the masonry of the door jamb, the bolt when locked being projected from the jamb and engaging with the door.

Jamb spring hinge. Hinge applied to a jamb and door containing one or more springs.

Janus-face lock. Rim lock both sides of which are similarly molded or ornamented so that either side may be applied to the door, thus making the lock both right and left hand.

Keeper. Synonymous with "strike."

Key cabinet. Cabinet that incorporates a number of hinged leaves with hooks for storing keys. A door pocket holds records.

Key change. Combination of cuts in a key that enables it to operate the lock for which it is intended.

Key changing lock. Lock actuated by a key, the bits and combination of which are changeable at pleasure.

Key control system. Organization of keys within a unit of cabinets in order to regulate the use of and assign responsibility for each individual in the system. The system itself consists of a choice of small-, medium-, or large-capacity cabinets or drawers, which include key hooks, key mark-

ers, receipt holders, and cross-indexed file cards for identification purposes.

Keyhole. Aperture in a lock case or escutcheon plate
through which the key passes in entering the lock.

Key plate

1. Small plate or escutcheon having only a keyhole.

2. Plate, either plain or ornamental, having one or more
   keyholes (but no knob socket) and adapted for attachment to the surface of a door.

Key removable core cylinder. Cylinder having a core containing both the upper pin chamber and plug in an integral
unit that can be removed by a special control key and that
is interchangeable with all other cylinders in the system by
use of a control key.

Key section. Milling of key to match broaching in barrel.

Key shut-out feature. Locks incorporating the shut-out
feature ensure maximum privacy. Pressing the inside button shuts out all keys except the emergency key.

Key tag. Metal or fiber identification tag attached to
keys.

Keyway

1. Aperture in lock cylinders that receives the key and
   closely engages with it throughout its length.

2. Aperture in locks of the cylinder type that receives the
   key and engages closely with it throughout its length,
   as distinguished from the open keyhole of a common
   lock.

3. Aperture throughout the length of a lock cylinder into
   which the key is inserted. The distinction between a
   keyhole and a keyway is that in a bit key lock the
   keyhole is an aperture in the lock case that serves only
   to permit the key to enter the lock and to provide a
   bearing on which the key is rotated, while in a cylinder
   lock the key not only enters the lock through the
   keyway but remains interlocked with and contiguous
   to the keyway throughout the length of the cylinder.

Kick plate. Protective plate applied on the lower rail of
the door to prevent the door surfaces from being marred.

Knee butt. Synonymous with "pocket butt."

Kob

1. Projecting handle, usually round or spherical, for operating a lock.

2. Projecting handle, usually round, oval, or spherical, for
   operating a latch bolt or lock. A small crescent- or
   otherwise shaped knob designed for operation with
   the fingers is called a "turn knob," sometimes known
   as a "thumb turn" or "thumb knob," and is usually
   employed to throw the dead bolt of a lock from the
   inside.

Knob bolt. Door lock having a bolt that is controlled by a
knob or thumb piece from either or both sides of the door;
it is not actuated by a key.

Knob latch. Door lock having a spring bolt operated
from either or both sides of the door by a knob; it is not
actuated by a key.

Knob lock

1. Door lock having both a spring bolt, operated by a
   knob, and a dead bolt, operated by a key; it combines
   in one structure a knob latch and lock.

2. Door fastening or lock combining a knob latch and
   dead bolt.

Knob rose

1. Round plate or washer forming a knob socket and
   adapted for attachment to the surface of a door.

2. Plate or small escutcheon formed to act as a knob
   socket attached to the surface of the door.

Knob shank

1. Projecting stem of a knob into which the spindle is
   fastened.

2. Projecting stem of a knob containing the hole or
   socket to receive the spindle.

Knob top

1. Part of the knob that the hand grasps.

2. Upper and larger part of a knob that is grasped by the
   hand; it is usually made of porcelain, glass, wood,
   plastic, or metal.

Knuckle

1. Enlarged part of a hinge into which the pin is inserted.

2. Enlarged part of a hinge or butt that receives and
   encloses the hinge pin.

Lag bolts. Tapered bolts, coming to a threaded point and
having a square or hexagonal heads. Like screws, they
can't be threaded into a threaded hole. Used for heavy
supporting applications, such as attaching metal to timber
beams.

Latch

1. Lock, the bolt of which is beveled and self-acting by
   the pressure of a spring or by gravity.

2. Fastening device that has a spring latch bolt but is
   without key function or dead bolt. Rim and mortise
   "night latches" are an exception to this definition.

Latch bolt (of a lock)

1. Bolt having a beveled head and actuated by a spring,
   whereby it is retracted by impinging against the strike
   and automatically thrown forward again by the spring.

2. Beveled spring bolt, usually operated by a knob, handle or turn.

3. Bolt of a lock that is held in extended position by a
   spring, except when retracted by knob, lever handle,
   or thumb piece or by contact with the strike as the
   door is being closed.

4. The function of a latch bolt is to hold the door in a
   closed position. A latch bolt is spring actuated and is
   used in all swinging door locks except those providing
   dead bolt function only. It has a beveled face and may
   be operated by a knob, handle, or turn. Friction occurs
   when a latch bolt hits the lip of the strike; therefore to
   ensure easy closing of the door when door closing
   devices are used, an antifriction feature is recommended. The device may be a split latch bolt, a plastic
   insert in the bolt or in the strike, a pivoted bolt, or a
   self-lubricating bolt.

Leafs

1. Two attaching plates that when fastened together by
   the hinge pin form a complete hinge.

2. Two doors forming a pair of doors.

Left-hand door. Hinges on left, locking device on right,
and door swings inward.

Left-hand reverse door. Hinges on left, locking device on
right, and door swings outward.

Letter box back. Synonymous with "letter box hood."

Letter box back plate. Plate, similar to a letter box plate,
attached to the inside of a door to allow the passage of
mail.

Letter box hood

1. Plate attached to the rear of a door to conceal the
   opening through the door from a letter plate and to
   direct letters downward.

2. Cover attached to the inside of the door to conceal the opening through the door and guide the mail downward.

Letter box plate. Plate attached to the door with an opening to permit insertion of mail.

Letter drop plate. Opening, usually closed by drop or flap, to permit the passing of letters.

Lever. An abbreviation of the term "lever tumbler," and inaccurately used as synonymous with "tumbler."

Lever bell pull. Bell pull actuated by lever action instead of by drawing out a knob.

Lever cupboard catch. Catch consisting of a lever pivoted on a plate, through which it passes, its inner end having a hooked form to engage with a staple and its outer end formed into a knob or handle.

Lever handle

1. Bent handle for actuating the bolt of a lock and used in place of a knob.

2. Horizontal handle for operating the bolt(s) of a lock.

Lever tumbler

1. Flat tumbler having a pivoted motion actuated by the turning of the key and controlling the locking function.

2. Lock tumbler having a pivotal action.

Lever tumbler lock. This type of lock is one whose "obstacle" consists of one to five lever tumblers, which must be raised by the key and properly aligned to pass the "fence" of the dead bolt before the key bit can engage the "talon" and actuate the bolt. On the side of the bolt there projects a lug called a "fence." Openings in the tumblers fit over this fence. The "gatings" connect the openings in the lever tumblers and are all cut at different heights. When the key is inserted and rotated, the bittings of the key lift each of the tumblers to exactly the right height so that they are all in proper alignment to permit the fence to pass. The key engages simultaneously with the "talon" of the bolt, and the bolt is thrown; levers then drop from the action of the tumbler springs, engaging the fence and preventing movement of the bolt. Key changes may vary from 4 for a one-tumbler lock, to more than 144 changes in a five-tumbler lock. Also called "bit key lock."

Lever tumblers. Flat tumblers, one or more of which are used in the bolt-controlling mechanism in bit key locks.

Lift latch. Unencased rim latch consisting of a bar pivoted to a plate and engaging with a hook on the jamb, the bar being operated by a thumb piece on the outside of the door and by a lift handle on the inside; it is usually combined with a door pull on one or both sides of the door.

Lip of a strike. Projecting part on which the latch bolt rides.

Lock. Device of any kind operated by a key; specifically, one having a dead bolt, as distinguished from one having a spring latch bolt.

Locker ring. Pull, for mortising into the edge of a sliding locker door, consisting of a plate containing a ring that may be pushed back flush with the plate or pulled forward for use as a pull to open the door.

Lock nuts. Usually regular nuts that have some part of the thread distorted in the manufacturing process, which makes it difficult to remove or for the nut to come off.

Lock rail

1. Horizontal member of a door intended to receive the lock case.

2. Rail on a door located at the proper height to receive the lock and, for that purpose, usually made broader than the other rails.

Lock set

1. Lock complete with trim, such as knobs, escutcheons, or handles.

2. Lock combined with its trim, such as knobs, escutcheon plates, and screws.

3. Complete assembled set of all component parts that constitute the lock and correlated hardware trimming for a door, such as a complete unit consisting of the lock, strike, keys, knobs, escutcheons, and screws.

Lock stile. Stile of a door to which the lock is applied, as distinguished from the hinge stile.

Lock strike. Metal plate mortised into the door jamb to receive and hold the projected latch bolt and, when specified, the dead bolt also. It secures the door and is sometimes called a "keeper." A wrought box installed in back of the strike in the jamb protects the bolt holes from the intrusion of plaster or other foreign material, which would prevent the bolt from projecting properly into the strike.

Lock washer. Washer that serves the same purpose as flat washers except that it also prevents the nut from coming loose. A helical spring-type lock washer looks like a piece of a coil spring. The points of the washer dig into the nut or bolt and the material beneath the washer. Tooth-type lock washers differ in that the grabbing teeth are either on the outside edge, the inside edge, or both.

Long lip strike. Lock strike with a special long lip to protect the frame work of the door.

Loose joint butt. Butt having a single knuckle on each half, one containing the pin and the other a corresponding hole, whereby the two parts of the butt can easily be separated.

Loose joint hinge. Hinge having but two knuckles, to one of which the pin is fastened permanently and the other containing the pinhole, whereby the two parts of the hinge can be disengaged by lifting. These hinges are "handed."

Loose pin butt. Butt having a hinge pin that can be withdrawn to permit the two parts of the butt to be separated.

Loose pin hinge. Hinge having a removable pin to permit the two parts of the hinge to be separated.

Machine bolts. Bolts having either hexagonal (six-sided) or square heads. In heavy-duty applications they are used with nuts. Common sizes range from $\frac{1}{4}$ in. to $\frac{1}{2}$ in. in diameter and from 1 in. to 6 in. in length.

Magnetic catch. Cupboard catch that uses a magnet to hold the door closed.

Magnetic door holder. Magnet in a wall or floor box that holds the door in an open position against the door closer until the electric current is interrupted.

Malleable iron. Cast iron treated by baking or annealing to make it tough and shock resistant. When properly cast and annealed, malleable iron can be bent and even knotted without breaking. Many manufacturers use malleable iron in items like pulls and closer arms to reduce the possibility of breakage and provide resistance to stress.

Master key

1. Key pertaining to a series of master key locks that will actuate any and all of the locks.

2. Key that operates any quantity of cylinders of different individual key changes.

3. Key that operates two or more locks, each keyed differently and each with its own change key.

4. Key designed to operate two or more locks of a master-keyed set, the regular change keys of which are not interchangeable.

Master-keyed lock

1. Lock intended for use in a series, each lock of which may be actuated by two different keys, one capable of operating every lock of the series and the other capable of operating only one or a few of the locks.

2. Lock of a group or series of locks so keyed that in addition to being operated by its regular change key, it may be actuated by another key (master key) that will also operate some or all of the other locks in the group.

Master-keying

1. Arranging cylinders having individual key changes to permit them all to be operated by a simple key called a "master key."

2. Various methods are employed to accomplish functional group arrangements of master-keyed locks. The possible arrangements are innumerable.

Master-keying of cylinder locks. Cylinder locks may be master-keyed by having one or more of the pin tumblers divided into three sections. With the pins divided into three sections instead of the standard two, it is possible to make key bittings such that one key will raise the pins in one position of alignment which permits rotation of the cylinder "plug" and another key will raise the pins to a different position which will also allow the "plug" to rotate. Very complex master-keying requirements may quite easily be fulfilled by the use of cylinder locks.

Metal gage for hinges. Gage (guage) of metal used for architectural hinges that are exposed in finished areas should meet the requirements of ANSI/BHMA A156.1, standard for butts and hinges, and ANSI/BHMA A156.7 standard for template hinge dimensions.

Metals. Basic metals used in builders' hardware are brass, bronze, iron, steel, stainless steel, aluminum, and zinc. Metals may be cast, extruded, forged, or wrought.

Mop plate. Narrow plate, similar to a kick plate, of sufficient height to protect against the use of the mop; a minimum of 8 in.

Mortise

1. Opening to receive a lock or other hardware; to cut such an opening.

2. Cavity made to receive a lock or other hardware; also, the act of making such a cavity.

Mortise bolt

1. Door bolt designed to be mortised into a door instead of being applied to its surface.

2. Miniature deadlock with bolt projected or retracted by turn of the small knob.

Mortise lock or latch

1. Lock designed to be installed in a mortise rather than applied to the door's surface.

2. Lock or latch to be mortised into the edge of a door rather than applied to its surface.

Mortise-type lock. A lock installed in a prepared recess (mortise) in a door. The working mechanism is contained in a rectangularly shaped case with appropriate holes into which the required components, cylinder, knob, and turn-piece spindles, are inserted to complete the working assembly. In order to provide a complete working unit, mortise locks, except for those with deadlock function only, must be installed with knobs, levers, and/or other items of trim.

Mullion. Fixed or movable post dividing a door opening vertically.

Multiple key system. Each keyway has individual barrel broaching and key section for individual master key with grand master keys milled to pass more than one broached keyway.

Mutes (in doors). Spring-cushioned rubber block mounted in a bronze case for mortising.

Necked bolt. Bolt, the projecting end of which has a bend or offset to engage with a strike or keeper not in line with the body of the bolt.

Night key. That one of the two keys of a front door lock that controls the night work and operates the latch bolt.

Night latch

1. Auxiliary lock having a spring latch and functioning independently of, and providing additional security to, the regular lock of the door.

2. Door lock having a spring bolt that cannot be operated from the outside except by a key.

3. Auxiliary lock separate from, but supplementing, the regular lock on a door, having a spring latch bolt operating from the outside by key only and from the inside by turn knob. Some night latches are equipped with stop works that, when engaged, deadlock the latch bolt against end pressure. Others are so constructed that the latch bolt may be held in a retracted position when it is desired to have the night latch inoperative. Other night latches have incorporated both the deadlocking and holdback features.

Night work. Part of the mechanism of a front door or vestibule lock that controls the latch bolt and is actuated by the night key.

Night works (stop works). Interior mechanism of a lock that deadlocks the latch bolt against the outside knob or thumb piece. The mechanism is controlled by buttons in the front of the lock or by the cylinder.

Nose plate. Small plate surrounding the nose or escutcheon of a cylinder lock.

Occupancy indicator. Indicator that provides a visual indication to service personnel that the room is occupied. It consists of a small pin in the outside cylinder which is extended when the shut-out feature is manually set.

Office lock. Lock usually having a latch bolt controlled from both sides by knobs, with the latch arranged to be set by stops so that the outside knob may be rendered inoperative and entrance gained by key only. The inside knob is always operative. If the stop is set, the door locks when closed. A dead bolt or knob bolt may be incorporated.

Offset pivot set with intermediate pivot. The bottom pivot is jamb mounted; however, floor-mounted bottom pivots are also available. The top pivot fastens to the head jamb and the top of the door; so the screws provide better holding power against lateral movement. The intermediate pivot, as in the case of any center-mounted hinge, supports some weight but primarily holds the door in alignment. The bottom pivot supports most of the door weight.

Oilite bearing hinges. Hinge equipped with oil-impregnated bearings between knuckles to reduce friction.

Olive knuckle hinge

1. Paumelle hinge with knuckles forming an oval shape.

2. Hinge, shaped like an olive, designed to show only the knuckle when the door is closed.

Outside. Side from which the hand and bevel of locks are determined, usually the outside of an entrance door, the hall side of a room door, and the room side of a closet door. It is less confusing to determine the hand and bevel of a lock from the side having the more important key function or, if the key function is the same on both sides, from the side on which the butts are invisible.

Overhead concealed closer. Closer concealed in the head frame, with an arm connecting with the door at the top rail.

Panicproof locks. Locks that provide immediate exit from the inside at all times.

Paracentric

1. Term used in connection with cylinder plugs having projections on the sides of the keyway that extend beyond the vertical centerline of the keyway.

2. Term designating a peculiar form of key and keyway, the cross section of which shows ribs projecting from opposite sides of the keyway past its centerline and extending longitudinally throughout its length, thereby preventing the use of picking tools; the opposite sides of the key are grooved to correspond with the contour of the keyway and the key and keyway are thus interlocked throughout their length.

Parliament butt. Butt having T-headed leaves, usually broad.

Parliament hinge. Hinge having T-shaped, usually broad leaves.

Passage set. Term commonly used to describe a latch set with knobs on both sides and no locking feature.

Patio lock. Lock designed with a push button or a turn-locking outside knob. When a door is so locked, there is no entrance by key from the outside.

Paumelle. Hinge style embodying a single pivot-type joint, generally of streamlined design.

Paumelle hinge. Type of loose joint hinge and therefore handed. Only the knuckle shows when the door is closed.

Permutation lock. Lock having changeable tumblers actuated by either a key or a dial.

Piano hinge. Synonymous with "continuous hinge."

Pin tumbler

1. Small pins contained in a cylinder, that dog the plug until they are perfectly aligned by a key, thus permitting the rotating of the plug, which actuates the lock bolt.

2. Small sliding pins in a lock cylinder, working against coil springs and preventing the cylinder plug from rotating until the pins are raised to the proper alignment by bitting of the key.

3. Locking mechanism that is fitted with a number of cylindrical pins and is operated only by the keys with which it has been combined.

4. Small sliding pin actuated by the key and dogging the plug or key hub, thereby transmitting motion to the bolt.

Pipe key. Round key having a hole drilled into its end to fit over a drill pin in the lock, used chiefly for cabinet locks. Synonymous with "barred key."

Pivot-reinforced hinge. Construction that consists of a heavyweight hinge with an added pivot using the same pin. The leaves of the pivot interlock with the hinge leaves at the top. Two conventional hinges are required to complete the set.

Plug

1. Cylindrical piece containing the keyhole and rotated by the key to transmit motion to the bolt.

2. Round part containing the keyway and rotated by the key to transmit motion to the bolt(s).

3. Cylindrical part, housed by the "shell" of a lock cylinder, that contains the cylinder keyway and rotates in the shell to actuate the lock bolts.

Plug retainer. Part that retains the plug in a cylinder lock.

Pocket butt. Hinge or butt for three-ply inside shutters, each leaf of the butt being bent at a right angle near the center, for use on the third leaf of the shutter to permit the latter to enter and leave its pocket without jamming.

Post. Round part of a bit key to which the wing or bit is attached.

Preassembled-type lock. A preassembled-type lock is installed in a rectangular notch cut into the door edge. The lock has all the parts assembled as a unit at the factory and when installed, little or no disassembly is required. Preassembled-type locks have the keyway (cylinder) in the knobs. Locking devices may be in the knob or on the rose or escutcheon.

Preparatory key system. Keying system in which the lock must be set up by a preparatory key before the regular key will operate.

Prison lock

1. Heavy duty lock designed especially for use on jail cells.

2. Lock designed especially for use on cells or other doors in a prison.

Privacy set. Term used for mortise tubular and cylindrical locks, used on bathroom and bedroom doors, having an inside button or turn to lock the outside knob and usually an emergency key function that will unlock the set from the outside.

Protected strike. Strike with a flange fitting against the doorstop, making it impossible to insert a sharp instrument between the casing and the stop for the purpose of sliding the latch bolt to a retracted position.

Pull. Door pull.

Pull bar. Similar to a push bar, sometimes labeled "pull," with a vertical pull attached to the bars.

Pull down handle. Light handle for attachment to the underside of the bottom rail of the upper sash for use in moving the latter.

Pull down hook. Synonymous with "sash hook."

Pulley stile. Vertical sides of a double-hung sash casing to which the pulleys are applied.

Push bar. Bar or series of bars, usually across the door, used as a means of pushing a door open.

Push button. Small movable knob or button within a socket, the movement of which actuates a bell, electrically or otherwise.

Push button (lock). Button in the knob of a cylindrical or bored-type lock used to lock the opposite knob in a fixed position.

Push (or thrust) key. Key that performs its whole function of setting the tumblers by longitudinal motion without rotation.

Push plate

1. Plate applied to the lock stile to protect the door against soiling and wear.

2. Plate for protecting the surface of a door against soiling and wear from handling, frequently made with the word "push" incorporated into the design.

Push-pull cabinet lock. Lock used principally to lock bypassing sliding cabinet doors together by means of a pin extended into locked position by pushing; it is unlocked by inserting a key and pulling.

Quadrant. Device to fasten together the upper and the lower leaves of a Dutch door.

Rabbet

1. Abutting edges of a pair of doors or windows so shaped as to provide a tight fit. One-half of the edge projects beyond the other half, usually $\frac{1}{2}$ in. Also, that portion of a door frame into which the door fits.
2. Offset on the abutting edges of a pair of double doors; also, the corresponding offsets on the fronts and strikes of rabbeted locks.

Rabbeted door. Door whose front edge is not beveled but has a step, usually $\frac{1}{2}$ in. deep.

Rabbeted lock

1. Mortise lock, the front of which is formed with an offset or rabbet conforming to the corresponding rabbet on the edge of the door; lock having a front in three planes, forming two right angles.
2. Lock in which the face conforms to the rabbet found on a rabbeted door.

Rail

1. Horizontal member of a door that joins the stiles. It may be exposed, as in a paneled door, or concealed, as in a flush door.
2. Any of the horizontal members of a door that enclose the panels and that, with the stiles, constitute the framework.

Reach. Distance from the center of the operating rod to the nearest edge of the transom sash.

Recess. Distance inward from the face of the door casing to the face of the transom sash.

Recessed front. Recessed offset in the face of a lock front that fits a correlated offset in the strike. This prevents the reaching and retracting of the latch bolt by means of the insertion of a thin instrument between the door and jamb and thus provides protection against that form of lock picking.

Reinforcing unit

1. Box-shaped metal reinforcement for use in a metal door in which a bored lock is to be installed. It provides both vertical and horizontal latch support.
2. Metal insert that prevents metal door walls from collapsing when locks are firmly installed.

Reinforcement unit. Metal box designed to reinforce a metal door in which a cylindrical lock is to be used.

Removable key. Key that is able to remove active core of cylinders.

Removable mullion. Mullion used on pairs of doors with rim-type exit devices. Surface-mounted strikes are typical.

Reverse bevel. Bevel of a latch bolt that is reversed or inclined in the opposite direction to that which is regular.

Reverse bevel lock. Lock for a door having a reverse bevel; lock for a door opening outward from a space and having key control to prevent entrance to the space.

Reversed. Term applied to articles made of wrought or sheet metal with edges turned back to give the appearance of increased thickness.

Reversed door. Door with opening in the opposite direction to that which is usual or regular. Room doors if opening inward are "regular"; if opening outward they are "reversed." Cupboard doors are regular if opening outward.

Reversible function. Hand can be changed by revolving from left to right or turning upside down, or by reversing some part of the mechanism on many types of locks and latches.

Reversible lock

1. Lock in which the latch bolt can be reversed to adapt the lock to a door of either hand.
2. Lock that by reversing the latch bolt may be used by any hand. On certain types of locks other parts must also be changed.
3. Lock in which the latch bolt and/or hubs can be reversed to adapt the lock to a door of either hand.

Right-hand door. Hinges on right, locking device on left, door swings inward.

Right-hand reverse door. Hinges on right, locking device on right, door swings outward.

Rim. Articles of hardware applied to the surface of doors, windows, etc., rather than those mortised into the wood.

Rim lock or latch. Lock or latch that is applied to the surface of the door, not mortised into it.

Rising pins. Rising pins are usually a feature of nonquality hinges.

Rollback

1. Rotating piece within a lock, permanently attached to the knob spindle, for transmitting motion to the bolt. Term is inaccurately used as synonymous with "hub."
2. Rotating piece within a lock that transmits motion by a rotary motion.

Roller latch

1. Friction door latch employing a roller latch head under spring tension, which engages a strike having a recess formed to receive the roller.
2. A roller engages the depression in the strike, thus holding the door in a closed position until the door is pushed with sufficient force to disengage it.

Roller strike. Strike having a rolling member at the point of latch bolt contact to minimize friction.

Room door lock. Knob lock for doors leading from halls or corridors into rooms; also called "inside door lock."

Rose

1. Trim plate attached to the door under the knob; it sometimes acts as a knob bearing.
2. Circular, square, or oblong plate for attachment to a door and containing a socket for supporting and guiding the shank of a knob.

Rose trim. Term generally used for a round shaped plate or escutcheon.

Rounded front. Lock or bolt front conforming to the rounded edge of a double-acting door. The standard radius is 4 in.

Round key. Key having a round shank or stem.

Sash balance. Spring device used to counterbalance the window sash, eliminating the necessity for pulleys, weights, and cords.

Sash center

1. Pivoted support for the transom of a sash, comprised of two parts, one of which contains a pivot, the other a socket for the pivot.
2. Pin or bearing for a transom light or other sash turning on a horizontal axis, consisting usually of a pair of plates, one carrying a pin and the other a

socket, one plate intended for attachment to the sash and the other to the jamb or frame in which the sash is hung.

Sash chain

1. Metal chain adapted for use with a sliding sash, attached to the sash and to the counterbalancing sash weight.

2. Metal chain used with a sliding sash in place of a cord or rope.

Sash cord

1. Cord or rope used in place of a chain.

2. Small cord or rope used to connect a sliding sash with its counterweight.

Sash cord iron. Small metal holder inserted in the edge of the sash, to which a sash cord or sash chain is attached.

Sash fast

1. Fastener attached to the meeting rail of double-hung windows.

2. Device usually attached to the meeting rail of sashes to prevent their being opened until released.

Sash hook. Metal hook attached to the end of a metal or wood rod and adapted to engage with a hole or socket in the upper sash so that the sash may be raised or lowered.

Sash lift. Plate, bar, or hook adapted for attachment to a window sash so that the sash may be conveniently raised and lowered.

Sash lift and lock. Sash lift provided with a locking lever, which locks the sash by engaging with a strike in the window frame and is released by the raising of the sash.

Sash lock. Fastening controlled by a key and adapted to secure a sash.

Sash pin. Form of a window spring bolt.

Sash pivot. Synonymous with "sash center."

Sash plate. Synonymous with "sash center."

Sash pole. Wood or metal pole to which a sash pole hook is attached.

Sash pole hook. Metal hook attached to a wood or metal pole used to lower or raise a transom or sash beyond hand reach.

Sash pulley

1. Synonymous with "frame pulley."

2. Pulley mortised into the frame of a double-hung sash frame over which the sash cord or sash chain passes.

Sash ribbon. Thin metal band adapted for use with a sliding sash in place of a cord or rope.

Sash socket. Metal plate containing a hole or cup adapted to receive a sash hook.

Sash weight. Weight used to balance a sliding sash, usually of cast iron or, if conditions require, of lead.

Screen door latch. Light knob latch, similar to a cupboard turn, but furnished with a hub, a spindle, and a pair of knobs or lever handles.

Screwless knob

1. Knob attached to a spindle by means of a special wrench, as distinguished from the more commonly used side knob screw.

2. Knob provided with a clamp or vice for attachment to the spindle, thus dispensing with the side screw; knob that eliminates the side screw and substitutes a fastening that obviates all tendency to become loose, even though employing a setscrew.

Screwless rose. Rose with concealed method of attachment.

Secret gate latch. Surface-applied latch operated by a concealed button or other device, usually used on bank gates.

Security hinge. Types of hinges used for security purposes include a nonremovable pin (NRP) that offers protection against the removal of the hinge pin from the exterior side of out-swing doors when the door is closed and locked and a maximum security pin (MSP) that provides the same protection as the NRP hinge pin.

Self-latching bolt. A self-latching bolt provides self-activating engagement with the strike when the leaf in which it is installed (the inactive leaf) is closed, but it must be manually released.

Self-latching bolt (concealed type). The concealed type of self-latching bolt latches automatically when the door closes but requires operation of a lever on the face of the door to release.

Self-latching bolt (flush type). For the flush type of self-latching bolt two bolts are required, one at the top and one at the bottom. It latches automatically but requires finger operation of the slides in the door's edge to release.

Self-latching bolt (surface type). The surface type of self-latching bolt latches automatically when the door closes but requires operation of a lever by hand to release.

Selvage. Distance from the front edge to the center of the keyway of a cabinet lock.

Semiconcealed closer. Door closer mortised into the top rail of a door with the barrel partially extending beyond the surface of the door.

Serrated locking screws. Screws available in hex flange head designs and other standard configurations. They have serrated teeth under the head to grip joint material, very useful in spanning slotted holes.

Serrated lock nuts. Nuts available in hex and hex flange designs, have serrations on bearing surfaces to grip joint material and prevent loosening.

Setscrew

1. Screw that by checking another screw or movable part prevents it from loosening.

2. Type of screw used to attach knobs to spindles and cylinders to locks.

Shank (of a key). Part that connects the bit or wing with the bow or handle.

Shank (of a knob)

1. Part that contains the hole or socket to receive the spindle and that forms a base for the top or enlarged portion of the knob.

2. Projecting stem of a knob into which the spindle is fastened.

Sheet metal screws. Screws having various types of heads, such as flat, oval, pan, bugle, hex, and combination hex washer, produced in thread forming, thread cutting, and self-drilling and taping.

Shelf pin. Metal pin for supporting a book shelf; also called "shelf support" or "shelf rest."

Ship lock. Lock made of bronze for use on ships, usually of heavy construction.

Shutter adjuster. Swinging arm for adjusting and securing shutters in any desired position.

Shutter bar. Fastening for folding blinds consisting of a bar pivoted to a plate and engaging with a hook or stud attached by another plate to the other half of the blind.

Shutter bar or bolt. Fastening to bolt shutters in closed position.

Shutter butt. Small hinge, usually narrow, adapted for use on shutters and light doors.

Shutter dog. Ornamental device to hold shutters open; sometimes called "shutter turnbuckles."

Shutter flap. Small hinge, usually broad, screwed to the surface of a shutter or small door.

Shutter hinge. Usually, ornamental hinges designed to swing shutters.

Shutter knob. Small knob for inside shutters.

Shutter lift. Lift for shutters, similar to a sash lift, but heavier.

Shutter operator. Device incorporating a hinge and a method of opening or closing a shutter by means of a crank or turn from inside without opening the window.

Shutter screw. Heavy thumbscrew for securing one end of a vertical shutter.

Side-load arms (overhead closers). Center-hung installation with a notch in the face of the door and the door is installed in the closed position.

Slide screw. Small screw used for securing a common knob to its spindle.

Signal sash fastener. Sash fastening device to lock double-hung windows that are beyond reach from the floor. The fastener has a ring for a sash pole hook and when it is locked, the ring lever is down; when the ring lever is up, it signals by its upright position that the window is unlocked.

Silencers. Rubber bumpers mounted on a stop or door, three to single doors, two for each leaf of a pair of doors.

Size of hinges. Doors up to 3 ft wide use $4\frac{1}{2}$-in.-high hinges for $1\frac{3}{4}$-in.-thick doors. Doors 3 ft to 4 ft wide, use 5-in.-high hinges, depending on thickness of door; doors 4 ft to 5 ft wide use 6-in. or 8-in.-high hinges, again depending on thickness and weight of door.

Slider. Small sliding tumbler actuated by the key and dogging the plug by which motion is transmitted to the bolt.

Sliding door key. Key adapted for use with a mortise lock and a cup escutcheon on sliding doors; it is usually adjustable as to length.

Sliding door lock. Lock for use on a door that slides; it has hook-shaped bolts to engage with its strike.

Sliding door lock or latch. Device for use in latching or locking a door that slides; it has a hook-shaped bolt to engage the strike.

Sliding door pull

1. Plate or box mortised into the edge of a sliding door and containing a handle or pull for use in moving the door from its recess.

2. Pull used flush on the edge, a flush pull mortised into the face of a lock rail, or a rim pull, all three of which are designed particularly for use on sliding doors.

Sliding door rail. Metallic rail for carrying and guiding the sheaves of sliding doors.

Sliding door stops. Small plate for attachment to floor or ceiling, provided with a projection to limit the motion of a sliding door.

Sliding tumbler. Lock tumbler having a sliding motion.

Smoke and heat detectors. Smoke and heat detectors are placed in the ceiling at proper positions to activate release of doors held open by magnetic holders.

Solid or hollow rivets. Rivet that looks like a bolt without threads, having a rounded head. A hole the size of the rivet must be drilled into the material to be fastened; the rivet is placed in the hole, and the head is held firmly against the material, while the other end of the rivet is hammered and expanded to a point where the rivet cannot be withdrawn.

Solid-rolled. Term designating escutcheon plates and other articles made from rolled or wrought metal of sufficient thickness to show a suitable bevel without turning back the edges.

Spacing. Distance between the center of a knob hub and the center of a keyhole of a lock or its escutcheon plate.

Special heavyweight ball bearing hinge. Hinge is 8-in. high, for full mortise, half-surface, half-mortise, or surface applications. Hinges are for use on doors weighing up to 600 lb.

Spindle (of a knob). Bar or tube connected with the knob or lever handle that passes through the hub of the lock or otherwise engages the mechanism to transmit the knob action to the bolt(s).

Spindle (of a lock). Axis or shaft, usually of square section, that carries the knobs of a lock and communicates their motion to the latch mechanism.

Split astragal. Astragal that is split through the middle, allowing each door leaf to operate independently.

Split dead bolt. Two dead bolts of the same or twin construction, both projecting from a common hole in the lock front and engaging in a common opening in the lock strike. This construction is used in communicating locks.

Spring hinge

1. Hinge containing one or more springs to move the door into the desired position. It may be either single or double acting.

2. A spring hinge contains one or more springs to move the door into a closed position. The energy stored in the spring closes the door. There is no checking control. Spring hinges may be single or double acting and either with or without a hanging strip.

Spring latch. Plain latch generally used on interior doors. It has no locking mechanism in itself.

Square bolt. Rim bolt of rectangular section.

Stainless steel. Iron product of which there are some 40 standard types. Each consists substantial amounts of chromium and small quantities of a number of other elements. The majority of types also contain appreciable percentages of nickel. Because it is highly rust resistant, finishes with a high luster, and is easily maintained, stainless steel is one of the best builders' hardware materials.

Standard mounted closer. Provides a good physical connection between the door and the opening. The closer is fastened to the door while the arm shoe is attached to the header on the pull side.

Steel. Metal, stronger than iron, widely used in builders' hardware. Ordinary carbon steel, used in builders' hardware, contains not only iron but portions of other elements such as carbon, manganese, phosphorus, and sulfur. Exposed to the weather, carbon steel is likely to rust. Most of the builders' hardware items made of wrought steel are formed from flat sheets by dies in heavy presses.

Stem (of a key). Round portion of the bit or wing which forms the trunnion or axis of the key and on which it rotates when in the lock.

Stile

1. Vertical member of the door structure; each door has two, a lock stile and hinge stile.

2. Vertical outside member of a door or window, generally, in builders' hardware, the member to which the butts and locking device are attached.

Stop (of a lock)

1. Button or other small device that locks the latch bolt against the outside knob or thumbpiece or unlocks it if locked. Another type holds the bolt retracted.
2. Mechanism that fastens the bolt of the knob in the locked or unlocked position, usually the latter.
3. Mechanism, button, or lever that fastens a bolt or knob, or both, in the locked or unlocked position. It may or may not take control away from the key.

Stop bead screw

1. Synonymous with "stop screw."
2. Screw used to fasten the stop of a window to the frame.

Stop key. Key for insertion in a keyhole from one side to prevent the entrance of a key from the opposite side.

Stop screw. Screw for fastening the stop bead of a window of the frame.

Store door handle

1. Heavy grip or pull mounted on sectional or elongated plates and provided with a thumbpiece to operate the latch trip of a store door lock.
2. Bent handle, usually mounted on a plate, provided with a lever or thumb handle for actuating a latch bolt and adapted for application to the surface of a door.

Store door latch. Latch containing a spring latch bolt only and adapted to operation by thumb handles.

Store door lock. Heavy lock containing a dead bolt only and usually operated by a key from both sides.

Stove bolts. Bolts have either flat or round heads and unified coarse (UNC) threads and are used with nuts to fasten metal parts. Common diameters range from $\frac{1}{8}$ in. to $\frac{5}{16}$ in.; common lengths range from 1 in. to 6 in.

Strap hinge. Hinge, of which one (or both) of the leaves has considerable length, adapted for attachment to the surface of a heavy or large door.

Strike

1. Metal plate or box that is pierced or recessed to receive the bolt or latch when projected.
2. Metal fastening on the door frame into which the bolt of a lock is projected to secure the door; also, the flat plate used with mortise locks and the projecting box used with rim locks. Synonymous with "striker," "striking plate," and "keeper."
3. Plate recessed in the door jamb into which the latch or dead bolt fits when the door is closed.

Stump. Small piece or projection in a lock for the engagement of one part with another, or to receive a screw or rivet; also, inaccurately used as synonymous with "fence."

Submaster key. Key capable of controlling a subordinate group of master key locks, each having a different key of its own, but all in turn controlled by the main or grand master key. There may be a number of submaster keys under one grand master key.

Subsequent lock. Time lock constructed to lock by the action of a clock at a predetermined hour subsequent to, and irrespective of, the time of closing the door.

Subtreasury lock. Lock for use on metal doors of the small chests or boxes within a fireproof safe, commonly called "subtreasuries."

Surface bolt. Bolt held in place by spring friction. It may be placed at the top or bottom of the door. The strike varies to suit jamb and sill details.

Surface hinge

1. Synonymous with "strap hinge."
2. Hinge having both leaves surface applied.
3. Hinge both leaves of which are applied directly on the surfaces of the jamb and the door in the same parallel planes.

Surface sash center. Sash center adapted for application to the surface of a transom sash.

Swaging. Slight offset of the hinge at the barrel, which permits the leaves to come closer together and improves the operation and appearance of the door.

Swing clear hinge. When doors are required to provide a clear opening at approximately 90°, this type of hinge should be used. It is available in full mortise, half-mortise, full surface, and half-surface.

Swinging latch bolt

1. Bolt that is hinged to a lock front and is retracted with a swinging rather than a sliding action. Sometimes called "hinged latch bolt."
2. Latch called "hinge latch." Type of latch that is hinged at the inner face of the lock front and swings into the lock case on contact with the strike, thus affording easy action with little friction.

Swivel spindle

1. Spindle having a joint midway along its length to permit the knob at one end to be made rigid by the stop works while the other end is free to operate.
2. Spindle having a joint or swivel midway along its length, whereby the knob attached to one end may be made stationary and inoperative, while the knob attached to the other end is left free to rotate and thus to actuate the latch mechanism.

Tail piece

1. Sliding or vibrating piece intermediate between the hub and latch bolt of a lock for transmitting motion from the former to the latter.
2. Sliding part of connecting link through which the lock is operated by hub or key.

Talon. Notch or opening in the bolt of a lock with which the key engages to throw the bolt.

Tempered glass doors. Hardware consists of push-pull handles, center-pivoted floor or overhead closers, standard and custom locking devices, including electric and magnetic locks with exit devices.

Template hardware. Item of hardware that is made to template, that is, exactly matching the master template drawing as to spacing of all holes and dimensions.

T handle. Cross handle for actuating the bolt of a lock and used in place of a knob.

Thimble

1. Socket or bearing on an escutcheon plate to receive the knob shank. Also called "socket."
2. Socket or bearing attached to an escutcheon plate, in which the end of the knob shank rotates.

T hinge

1. Surface hinge of which the chief dimension of one leaf is vertical and of the other leaf horizontal.
2. Surface hinge with the short member attached to the jamb and the long member attached to the door.
3. Surface hinge with one long leaf, similar to a strap hinge, and one smaller rectangular leaf with its long dimensions set at a right angle to the long leaf. It resembles the letter T from which the name is derived.

Thread-cutting screws. Screws that cut their own threads when driven into untapped holes. They require a pilot hole.

Thread escutcheon. Small key plate conforming to the outline of a keyhole and intended to be inserted therein.

Thread-forming screws. Screws that form threads by squeezing and displacing the metal in the material into which they are driven. Usually requires a pilot hole.

Three-ply butt. Synonymous with "pocket butt."

Three-point lock. Device sometimes required on three-hour fire doors to lock the active leaf of a pair of doors at three points.

Threshold. Metal, wood, or marble strip fastened to the floor beneath a door, usually required to cover the joint where two types of floor material meet or to provide an edge for carpeting.

Throw (of a dead bolt or latch bolt). Measurement of the maximum projection when bolt is fully extended.

Thrust pivot unit and hinge set. Hardware set consisting of a pivot unit for the top of the door, having both jamb and top plates for the door and frame and requiring three conventional hinges to complete the set.

Thumb bolt. Door bolt operated by a rotating thumb-piece of a small knob.

Thumb latch. Door fastening consisting of a pivoted bar that crosses the joint of the door to engage with the strike on the jamb, the free end of the bar being raised to disengage it from the strike on the jamb by a transverse pivoted bar passing through the door, and the latter bar operated on one side by the thumb and on the other by the finger.

Thumb piece

1. Small knob, usually flat but sometimes circular in form.

2. Small pivoted part above the grip of a handle to be pressed by the thumb to operate a latch bolt.

Time lock. Lock actuated automatically by a clock and having no keyhole spindle or other connection through the door. Also called "chronometer lock."

Tolerances. Close tolerances, especially in the hinge pin, prevent excessive wear and are in important characteristic of high-quality, heavy-duty hinges.

Touchcode electronic lock. Keypad operated, totally self-contained, battery powered. Available in mortise or cylindrical locking devices. Manufactured by Yale Security Inc.

Tower bolt. Modified form of barrel bolt, in which the locking bar is shortened.

Transom. Small ventilating window or panel over a door or window.

Transom bar

1. Part of a door frame that separates the top of a door or a window from the bottom of the transom.

2. Horizontal member of a transom that divides the opening.

Transom catch

1. Fastener applied to a transom, having a ring by which the latch bolt is retracted.

2. Fastening adapted for use on transom lights.

Transom chain. Short chain used to limit the opening of a transom, usually provided at each end with a plate for attachment.

Transom lift. Vertically operated device attached to a door frame and transom by which the transom may be opened or closed.

Transom lifter. Apparatus for actuating and holding a transom light or panel.

Transom plate. Synonymous with "sash center."

Trim. Exposed portions of the lock, such as the knob, grip handle, lever, or rose.

Triple weight ball bearing hinge. Hinge is 5 in.-high, for full mortise, half-surface, half-mortise, or full surface applications. Hinge is for doors weighing up to 2,000 lb.

Triplex spindle. Lock spindle composed of three triangular rods, which when combined form a rectangle and which give an automatic adjustment by frictional engagement with the knob when expanded by a setscrew.

Tubular lock. Rim lock having a fixed tube containing the tumblers attached to the lock case and usually projecting through the door.

Tubular lock (or latch)

1. Type of bored lock.

2. Lock having a case that requires a bored (round hole) rather than a rectangular mortise. The face plate is attached to the lock case, fastened as a unit.

Tumbler

1. Obstruction or guard in a lock that dogs or prevents the motion of the bolt and that is set by the key during the act of locking and unlocking.

2. Guard or obstruction that prevents operation of a bolt except by insertion of the proper key.

Tumbler lock master-keying. In lever tumbler locks master-keying is accomplished by the introduction of auxiliary tumblers that are not actuated by the regular change keys, but when the proper master key is used, all the tumblers are raised, permitting the bolt to be thrown.

Turnbuckle. Synonymous with "turn button."

Turn button. Rotary bolt or fastening made in various forms. The common form is a simple bar secured by a screw in the center on which it rotates. In another form this bar is mounted on a circular plate. The term is also applied to a catch having a sliding bolt operated by a rotating knob or T handle.

Turn knob. Small knob with a spindle attached for operating the dead bolt of a mortise bolt or lock.

Turn piece. Small knob, lever, or tee turn with a spindle attached for operating the dead bolt of a lock or a mortise bolt.

Two-point latch. Device sometimes required on three-hour fire doors to lock the inactive leaf of a pair of doors at top and bottom.

Unit lock. Lock set so constructed that all of its parts (lock, knobs, and escutcheon plates) are permanently combined in a single construction or unit. Synonymous with "union lock."

Universal. Term that describes a lock, door closer, or other device that can be used on doors of any hand without change.

Upright lock. Lock whose major dimension is vertical.

Veneered front. Lock front or face consisting of two plates, the lower riveted to the lock case and the upper (usually of a more expensive material) permanently fastened to the lower. Used in contradistinction to "armored front."

Vertical spring pivot hinge. Spring hinge mortised into the heel of a door and fastened to the floor and head with pivots.

Vestibule door lock. Mortise latch operated by knob, lever handle, or thumb piece from inside at all times, operated from outside by knob, lever handle, or thumb piece when stop works are not set, and by key from outside when stops are set.

Vestibule latch. Lock resembling a front door lock except that the dead bolt mechanism is omitted. Latch in which the latch bolt is actuated from the outside by a key and from the inside by a knob, the outer knob being controlled by a stop.

Wafer tumbler. Locking mechanism that is fitted with a number of flat disks or wafers and is operated only by the keys with which it has been combined.

Wall-type door stop. Door stop for mounting on base or wall. It may also be mounted directly on door.

Wall-type door stop and holder. Mechanism that provides automatic engagement when door is fully open. It may be disengaged by pulling door toward closed position.

Ward

1. Obstruction projecting from the lock case or side of a keyhole intended to prevent entrance or rotation of an improperly cut key.

2. Projection from the case of a keyhole of a lock tending to obstruct the entrance of the key and necessitating a coincident depression of grooving in the key.

3. Projection in the case or the keyhole of a lock obstructing the entrance or rotation of an unmatched key.

Warded key. Key having grooves or notches, usually in the wing or bit, that coincide with corresponding wards or projections in the lock case or keyhole.

Warded lock. Lock having for obstacles, projections cast in the keyhole opening or on the outside of the case. It requires corresponding grooves in the key to permit it to enter any turn. The number of variation in wards is few. Twelve is the usual limit. Usually only four are employed, which means that every fourth lock can be opened by the same key and four keys would enter and operate every lock in the series. A warded lock is simple to pick. Owing to the limited number of key changes and the lack of security, warded locks are not used when master keying is required.

Wardrobe hook. Hook with a single prong for use on the side walls of closets and wardrobes and in bathrooms.

Window spring bolt. Spring bolt for holding a sliding sash in any desired position, open or shut; used with unbalanced sash.

Wing key. Key having a projection for operating the bolt or tumblers of a lock.

Wood screws. Commonly produced in 20 different stock thicknesses and lengths, ranging in length from $\frac{1}{4}$ in. to 5 in. The screw gauge ranges from 0 to 24 (the higher the number, the larger the diameter).

Working trim. Trim that is threaded for other than cylindrical lock spindles.

Wrought metal. Metal rolled into flat sheets or strips. Products are formed by punching or die cutting into the desired forms. Wrought metal may be thick, as in a hinge, or thin, as in a push plate.

Wrought trim. Trim that is fabricated from heavy-gauge sheet brass, bronze, aluminum, or stainless steel.

Zinc. Zinc is used in builders' hardware as a coating over iron and steel; it resists rust. Many products are made using die-cast zinc as a base metal. It is easily cast, machined, and plated.

METAL AND MISCELLANEOUS DOOR TYPES

Aluminum-clad wood doors. Constructed of western pine, water-repellent preservative treated in accordance with NWWDA standards, randomly laminated blocks with water-resistant adhesive and cross-banded both sides, primed on exterior surfaces and edges, interior surfaces veneered; exterior surfaces clad with aluminum.

Aluminum-clad wood sliding glass doors. Constructed of western pine, water-repellent preservative vacuum treated in accordance with NWWDA standards, primed on exterior surfaces and edges and clad with aluminum lap-jointed and sealed. Sliding panel on two adjustable permanently sealed ball bearing rollers, set on stainless steel track. Glazed with fully-tempered float glass.

Anchor jamb. Metal device inserted in the back of a metal door frame to anchor the frame to the wall. A masonry anchor is used in a masonry wall, a stud anchor is used in a wood or metal stud wall.

Backbend. Return leg section at back of a steel frame.

Base anchor. Fabricated metal piece, either fixed type or adjustable, attached to the base of a steel frame to secure frame to the floor.

Curtain-type doors. Constructed of interlocking steel blades or a continuously formed spring steel curtain in a steel frame.

Cutout. Preparation for hardware and/or accessories in a metal frame.

Darkroom revolving doors. Designed as a light barrier, constructed with metal walls with red vision panels in the metal leaves.

Double-acting swinging doors. Heavy solid or polymer cell core covered with lightweight aluminum alloy facing; kick plates on both sides of stainless steel, galvanized steel, or aluminum; extruded aluminum bumper strips; hot-dipped galvanized jamb guards; clear acrylic vision lights; pivotal hinge mechanism that permits easy operation.

Double-acting traffic doors. Doors fabricated of high density foamed-in-place urethane core insulation with a high-impact cross-linked polyethylene outer skin forming monolithic one-piece hollow shell. Equipped with black spring bumpers, self-closing gravity hinges, and $\frac{1}{8}$ in. polycarbonate viewing windows.

Dustcover box. Metal cover attached to a metal frame behind reinforcement for any mortised or recessed hardware to prevent mortar or plaster from entering the mounting holes.

Electromechanical automatic sliding doors. Metal framing and doors, anodized aluminum, stainless steel, or Muntz metal clad. Glass installation must meet local codes. Low horsepower motor provides the drive mechanism.

Floor anchor. Metal device attached to the back of a metal door frame jamb to secure the frame to the floor; may be either fixed or adjustable in height.

Grouted frame. Metal frame completely filled with mortar during the process of installing the adjoining masonry.

Hollow metal doors. Constructed of formed, 20 gauge or heavier steel of flush and paneled design.

Insulated upward-acting sectional doors. Core made from a continuous foamed-in-place polyurethane lamination process resulting in a homogeneous sandwich of even-textured polyurethane insulation with a cladding of galvanized corrosion-resistant embossed sheet steel that is then factory prepainted with two-coat baked-on polyester paint.

Knocked-down frame. Door frame furnished in several pieces by the manufacturer for assembly in the field.

Labeled door and frame. Door and frame that conforms to the requirements of a nationally recognized testing authority, who has established a rating for the type of door and frame and has issued to the manufacturer a rated approval that permits the manufacturer to attach the label to the door and to the frame.

Masonry anchor. Fabricated metal piece placed inside the throat of a door frame that secures the frame to a masonry wall.

Metal-clad (Kalamein) doors. Constructed of metal-covered wood cores or stiles and rails and insulated panels covered with steel of 24 gauge or lighter in flush or paneled design.

Motorized oval doors. Similar to large-diameter revolving doors, except that the leaves actually travel on an oval-shaped track rather than spinning on a shaft.

NWWDA. National Wood Window and Door Association.

Revolving doors

1. Circular glass enclosures with wood or metal frames. Glazed with annealed $\frac{1}{4}$ in. glass or as required by local codes. Review Consumer Products Safety Commission Standards for public glazing requirements.

2. Doors fabricated from stainless steel, aluminum, or bronze sections. Glazed as required for enclosure glazing.

3. Manually driven or electrically operated at very slow speeds.

Revolving door speed controls. Doors should be equipped with a governing device to prevent them from spinning too fast.

Rolling steel doors. Constructed with interlocking steel slat design or plate steel.

Security revolving doors. Similar to regular motorized revolving doors, except that they are noncollapsible, with provisions for automatic release in case of emergency.

Sheet metal doors. Constructed of 22 gauge or lighter steel and of the corrugated flush or paneled design.

Sliding glass doors. Units consists of metal frames and glazed metal-framed doors. Glass as required by local codes. Aluminum sections used to provide lower weight for easier sliding functions.

Three-leaf revolving doors. Recommended for use in hotels because the larger space between leaves allows patrons to carry luggage through the doors.

Tin-clad doors. Constructed of two or three-ply wood cores, covered with 30-gauge galvanized steel or terneplate having maximum size panels 14 in. by 20 in., or 24-gauge galvanized steel sheets not more th an 48 in. in width.

METAL WINDOWS

Accessories. Clips and anchors necessary for the installation of windows and hardware for the operation of the ventilators, including fasteners required for attaching such items to the window.

Adjustment. All members are adjusted for proper tolerance and weathering in shut and latched position after installation.

Air infiltration. Penetration of air through a window, normally expressed in cubic feet per minute per unit length of crack.

Aluminum finishes. Aluminum finishes consist of the following: etch and lacquer and clear and color anodic finishes.

Anchor

1. Metal clip or device that secures the window to the surrounding construction.

2. Component of the window frame, usually allowing for adjustment in three dimensions, used for holding the window frame to the building structure.

Applied weathering. Sheet metal weather protection member that is applied to a solid section window.

Architectural projected window. Window constructed from industrial-type sections with an unequal leg channel frame to provide an equal margin around the perimeter for receiving the interior finish.

Assembly. Windows, doors, and all component parts that make up an entire opening.

Awning window

1. Window consisting of a number of top-hinged ventilators operated by a single control device that swings the lower edges outward giving an "awning" effect when open. Operation is through a rotooperator, which also serves as the lock. The ventilating area is 100% of the window area.

2. Window that generally includes two or more top-hinged or projected-out ventilators that operate simultaneously.

3. Type of window providing a series of ventilators projecting outward and operating in unison. It may be made of intermediate, heavy intermediate, or heavy custom materials and construction and may be manually or mechanically operated.

Back putty. Putty placed in a window frame into which the glass is pressed, in contrast to putty applied after the glass is installed.

Baffle. Integrally rolled or applied member designed to provide a contact surface for weathering.

Balance. Device for holding a vertical sliding vent in any desired position by the use of a spring or weight to counterbalance the weight of the vent.

Balance arm. Metal strap device used in both jamb sections of a projected window to support the ventilator in the desired position.

Bar lift. Handle for raising double-hung windows.

Basement window. Light residential window, top hinged and opening out or bottom hinged and opening in, usually including a removable vent.

Bay window. Building projection in various shapes and forms formed by two or more windows, supported on a foundation extending beyond the main wall of the building.

Bead glazing. Stop or strip of material, usually metal, but sometimes rubber, used to structurally secure glass into a window.

Bedded. Perimeter of the window set in materials such as mastic or grout.

Blocking. Shims used to level and plumb windows in required position, prior to grouting or fixing in final position. The general term for this work is "blocking."

Bonderize. Treating steel by removing rust, scale, and all foreign matter and converting the metallic surface to a non-metallic phosphate coating to retard corrosion and provide a better base for paint.

Bonderized finish. Crystalline phosphate deposit integral with the metal itself. A rust-inhibiting coating provides a keyed surface for the subsequent primer, bonding it to the metal and increasing the durability of the finish.

Bottom hinged. Term describing a ventilator that hinges from the sill, usually to open in.

Box screen. Screen with subframe, often used in connection with chain-operated, projecting-out ventilators to allow sufficient clearance for hardware.

Brass guide shoe track. Guide used in the jambs of projected ventilators for the sliding shoe in lieu of having the shoe slide in the channel formed by the frame member.

Cam lock. Ventilator locking device that employs an eccentric friction action.

Casement combination window. Window unit made up of side-hinged casement ventilators and projected ventilators assembled in one complete window.

Casement window

1. Window containing a side-hinged ventilator that operates individually and swings outward. It frequently has nonoperating fixed lights that the ventilators close to. Casements may be open 100% for ventilation.
2. Window with side-hinged ventilators that usually swing out but that may swing in instead.

Casing. Finish member that surrounds the window unit or assembly either inside or outside, or both.

Caulking. Applying mastic with a gun or knife between the window and adjacent materials to provide a weather-tight installation.

Caulking pocket. Reglet or recess provided in the wall opening for installation of a caulking compound. The reglet may also be an integral part of the window section.

Caustic etch. A decorative matte finish is produced on aluminum alloys by immersion in a solution of sodium hydroxide and water. The solution actively attacks aluminum, producing a frosted surface, normally silvery in color, which may be highly matte, depending on the duration of the etch.

Cladding. Thin coating of aluminum metallurgically bonded to a core alloy. Through the cladding process, high-strength core alloys may be surfaced with highly corrosion-resistant alloys, often having excellent finishing characteristics, thus providing optimum strength, weatherability, and appearance.

Classroom windows. Windows designed to fill the special requirements of school buildings. Classroom windows feature a large fixed glass area with provision for see-through vision and controlled ventilation by inward or outward projected ventilators.

Cleaner-type hinge. Hinge with the pivot point approximately 3 in. from the outside face of the vent to allow space between the vent jamb and the window jamb so that window may be cleaned from the inside when the vent is open.

Commercial window. Inside, putty-glazed, pivoted or projected window; window of medium weight suitable for commercial structures.

Condensation. Thermal performance of any insulated window product, whether wood or aluminum, affected by the environmental conditions to which the unit is exposed. Factors such as outside and inside temperature, draperies, type of construction, inside relative humidity, design of heating equipment, and the air infiltration characteristics of the window are important and affect window condensation performance.

Condensation gutter. Through section on the inside of a window at the sill to receive and carry off moisture.

Continuous top-hung windows or continuous windows. Assembly of individual windows joined together in such a manner that they form an unbroken expanse or run of fixed and/or openable sash. Windows are top hung, and if openable, must be mechanically operated.

Coped. Removing a portion of window or frame section to allow irregularly shaped horizontal and vertical members to fit tightly together at joints.

Corrosion. The corrosion normally encountered in metal may be caused by atmospheric conditions or it may be chemical corrosion, galvanic corrosion due to the proximity of dissimilar materials, or "poultice"-type corrosion where wet, porous materials are held in close contact with metal over a period of time.

Cover plate. Sheet metal strip, flat or molded, to cover the splice between windows, usually at mullions.

Cremorne bolt. Hardware consisting of vertical slide rods activated by a cam action handle, for use on large casement windows or casement doors to provide for locking at the top, bottom, and intermediate points.

Curtain wall. Design of window and/or panel units that will permit joining in a horizontal run, as with window walls, but with the additional feature of permitting stacking the units two or more stories high to form an integral facade outside the building frame line.

Cut-up. Pattern of a window as determined by the number of vertical and/or horizontal muntins or vent framing bars.

Deflection. Displacement of a structural member due to loading, for example, the inward bending of a window frame member due to horizontal wind loads.

Depth of section. Distance from the front to back of any rolled, formed, or extruded section of fixed glass.

Detention screens. Screens used in mental institutions, constructed of high-tensile stainless steel mesh mounted in a special frame with a spring mechanism to absorb shock.

Detention windows

1. Windows having tool-resistant steel bars at prescribed spacing for maximum detention; windows having mild steel bars at prescribed spacing for moderate detention and fitted with prescribed screens; windows arranged with prescribed bar spacing for minimum detention and fitted with prescribed protection screens.
2. Windows made up of heavy sections and small glazed areas, designed to resist sustained force.

Double glazing. Two panes of glass separated by an airspace; normally with the perimeter of the panes hermetically sealed at the factory.

Double-hung window

1. Window that has two operables sashes; window with a sash that moves vertically within the main frame with the assistance of mechanical balancing mechanisms that minimize the effort required to raise or lower the sash. When both lower and upper sash are opposite each other, 50% of the window is then available for ventilation.
2. Window comprised of sliding upper and lower sashes that bypass each other vertically when operated.

Double-locking device. Device having two cam handles with a connecting rod, providing simultaneous operation for casement windows.

Drip. Projecting fin or groove designed to cause water to drip off instead of running back into undesired places.

Drop window. Window with a single vertical sliding vent that drops into a pocket.

Dust-enclosed power. Gear box totally enclosed for the purpose of keeping dust and dirt out of the working parts of mechanical operators. Used especially in industrial plants.

E label. Rating given to a steel window that conforms to certain prescribed regulations established by ASTM E 163.

Electrically controlled. Term describing operators using electrically driven motors.

Electric motor operation. Electrically operated equipment used to open and close ventilator sections in windows, usually occuring in industrial buildings where the windows are installed in group fashion, one next to each other, divided vertically with flat mullion bars, and cannot be reached easily from the operating floor level. The controls contain limits switches to the operating gear drive line attached to the ventilator, so that when the ventilator reaches the fully opened or closed position it will cut off the power automatically. Reversing starters also can stop or start the movement of the ventilators at any position required by the operator.

Equal-leg frame. Window frame section having inside and outside legs of identical lengths around the perimeter.

Explosion hardware. Hardware that allows windows to open automatically in the event that pressure from an explosion occurs within a building.

Exposed glass area. Glass portion that is not covered by any part of the window frame sections.

Extruded section. Section formed by forcing heated metal through a die under pressure.

Extrusion. Forming or shaping metals by forcing material through a die.

Fin. Continuous metal leg extension secured to the perimeter of the window frame for anchorage and as a weather bar stop.

Finish. Factory-applied surface treatment of the operating hardware and windows.

Fire-rated or labeled windows. Windows used in window openings where it is necessary to restrict the spread of fire and smoke within buildings, whether from interior or external fire. The actual determination as to when a fire-rated window is required for window openings is governed by the building code of the locality in which the building is situated and applicable federal and state statutes.

Fire window

1. Window, frame, sash, and glazing that will successfully resist fire to the degree required by, and that has been constructed and installed in accordance with the requirements of, ASTM E 163.
2. Steel window approved by fire protection codes for use in a hazardous location.

Fixed screen. Stationary screen, used when it will not interfere with window hardware or operation.

Flashing. Metal applied to the perimeter of a window frame to prevent the entrance of water.

Flex-arm operator. Rotooperator equipped with a flexible arm to permit the operation of a projecting-out ventilator.

Folding window. Window with a pair of vents, hinged together on a vertical axis, that fold together and outward like an accordian pleat when opened.

Frame. Perimeter section of a metal sash or window.

Friction hinge. Hinge with lock washer and nut on a pivot shaft to control resistance of movement, thereby holding a window open in any given position under normal conditions.

Friction stay. Device attached to a ventilator to keep it in any given position by means of friction.

Glass or glazing material. Transparent or translucent glass or plastic sheets mounted in a window. Sometimes opaque sheets are used where privacy is desired.

Glazing. Installing glazing material in a window.

Glazing angle. Continuous angle-shaped member used to hold glass and/or panels in place with screw fastenings.

Glazing bead. Continuous formed, extruded, or solid metal section used to hold glass or panels in place. It is attached with screws or snap-on or wedge-type fastenings.

Glazing clip. Formed spring wire or short length of glazing angle used to hold glass in place until putty is placed and has set.

Glazing compound

1. Material used in glazing windows that is applied in a plastic state and does not harden throughout.
2. Soft, cohesive material used to hold glass in the window frame. It must adhere to adjacent surfaces in such a manner as to prevent leakage of moisture and air.

Glazing rebate. Integral ledge or lip forming a seat for glazing compound.

Grout. Sand, cement, and water mixture to fill a "grout pocket" after installation of a frame; install grout.

Grout pocket. Offset or recess in a masonry opening permitting the installation of window, door, or mullion.

Guard bar. Pattern of steel bars welded to the main frame which resists sustained force.

Guard screen. heavy mesh used to protect glass from breakage.

Guard windows. Windows designed for use in jails and correctional institutions, available from manufacturers in standard sizes and three types of construction that provide minimum, moderate, or maximum detention. The basic design employs fixed main frames or grilles with restricted glass sizes and superimposed ventilators attached to the inside or outside of the grilles.

Hand chain. Chain used with window hardware or mechanical operator to manually operate ventilators beyond normal reach.

Hardware. Usually all items handled during the opening, closing, and locking of ventilators; also, every item of mechanism related to movement and locking devices of ventilated sections.

Hardware (sliding windows). Interior movable sash equipped with spring-loaded, self-locking aluminum latch with insert that engages an integral part of the fixed meeting rail.

Head. Uppermost horizontal portion of a window or frame.

Heavy custom section. Solid-section steel window weighing 4.2 lb/ft, combined weight of frame and vent, and with $1\frac{1}{2}$-in. minimum depth of section, as listed by the Steel Window Institute.

Heavy intermediate section. Solid-section steel window weighting 3.5 lb/ft, combined weight of frame and vent, members and with $1\frac{5}{16}$-in. minimum depth, front to back, as listed by the Steel window Institute.

Hopper ventilator. In-swinging or projection ventilator located at the sill of a window.

Horizontally pivoted window. Window with a ventilator operating on pivots, located at the sides, to allow the vent to open out at the bottom and in at the top. The most common type has pivots slightly above the center of the vent and is called "center pivoted." "Top-pivoted" and "bottom-pivoted" types are also available; they have their pivot points near the tops or bottoms.

Horizontally sliding windows

1. Window with a vent that slides or rolls horizontally on tracks.

2. Windows that have one or more operable sashes arranged to move horizontally within a main frame. The ventilating area is 50% of the window area. This type of window is sometimes combined with fixed lights and constructed in large sizes reaching the proportion of a "window wall."

Horizontal shaft. Metal axle or rod used to transmit power to two or more ventilators.

Hot-dip galvanizing

1. Following all fabricating operations, but prior to final assembly, all steel window frames and ventilators are thoroughly cleaned, pickled, fluxed, and completely immersed in a bath of molten zinc. The resulting coating becomes adherent. The normal coating is obtained by immersing the windows in a bath of molten zinc and allowing them to remain in the bath until their temperature becomes the same as the bath. The weight of the coating conforms to ASTMA A 123. Work is then bonderized and painted, and one prime coat of epoxy alkyd primer is baked on.

2. All materials (except screens) should be properly cleaned and hot-dip galvanized; hot-dip galavanized and passivated; or hot-dip galvanized, properly treated, and prime painted. The slab zinc (spelter) used should conform to ASTM B 6. The weight of the coating should conform to ASTM A 123. All fittings, nuts, bolts, etc., should be hot-dip galvanized and centrifuged or should be of nonferrous materials.

Impost. Integral structural member used as a mullion.

Impregnated anodizing. Sulfuric acid anodizing produces an oxide film that is normally used for "clear" or "natural finishes." Prior to sealing, the film is absorptive and can be impregnated with a wide variety of colors by immersion in organic dyes or by coloring with mineral pigment solutions.

Industrial window. Commercial projected, commercial pivoted, or security window fabricated from industrial sections.

Insect screen

1. Fine mesh screen provided over ventilators for the purpose of excluding insects.

2. Screen frame formed of electrogalvanized steel sections having a nominal thickness of not less than 0.032 in. and painted one shop coat of primer, or of formed or extruded aluminum having a thickness of not less than 0.040 in. The screen consists of 18×14 mesh cloth of bronze, 18×16 mesh cloth of aluminum, or fiber glass. Bronze cloth has a nominal diameter of not less than 0.0113 in. and aluminum cloth has a diameter not less than 0.011 in. Wire cloth should be held taut with removable spline. Screens with a sliding or hinged wicket are furnished for projected-out ventilators. Screens are interchangeable for the same size ventilator or similar windows and are of the demountable type, held in place by clips.

Installation. Windows are usually set as the wall construction progresses or are installed without forcing. Windows are set plumb, level, square, and in alignment and properly secured to walls and mullions. Field metal-to-metal joinery within the window materials are sealed at the exterior with an approved sealant furnished and applied by the window erector. Excess sealant is trimmed off. After windows are erected in place and before glazing, hardware is attached and ventilators are adjusted by the erection contractor to operate smoothly and be weathertight when closed.

In-swing side hinged. Term applied to a casement ventilator hinged to swing in.

Integral color anodizing (hard coat). Integral color anodizing utilizes the characteristics of the aluminum alloy employed for color derivation. Anodized coatings are extremely light, fast, and durable.

Intermediate section. Solid-section steel window measuring $1\frac{1}{4}$ in. minimum in depth through the web and weighing (combined weight) 3.0 lb/ft, as listed by the Steel Window Institute. Some companies on the west coast manufacture a frame and sash section, known as a semiintermediate section, that is $1\frac{1}{8}$ in. in depth and weighs less than 2.79 lb/ft (linear).

Jal-awning window. Window that resembles an awning window in appearance but uses different operating and locking mechanisms. The mechanisms are usually separate and require individual operation. The ventilating area is 100% of the window area.

Jalousie window (louver window)

1. Window consisting of a series of overlapping horizontal glass louvers that pivot together in a common frame. The ventilating area is 100% of the window area. Jalousies are not as resistant to air and dust infiltration as other types but are used in moderate climates and as porch enclosures. They should only be used where large heat losses or heat gains can be tolerated. Insulating effectiveness may be improved by inside storm panels, designed to be interchangeable with screens.

2. Operating louvered window with slats of glass, wood, metal, or other material.

Jamb. Vertical side of a frame.

Jamb plate. Metal anchoring extension applied to the jamb.

Jib window. Window with a single vertical sliding vent that slides up into a pocket in the construction.

Keeper. Strike plate.

Key-lock hardware. Locking device that can be operated by authorized personnel.

Kick-out latch. Snap lock type of hardware equipped with a spring-loaded arm which starts the ventilator to a partially open position.

Labeled windows. Special types of windows suitable for fire exposure, built to meet the standards of the NFPA-80.

Lacquer. Organic coating often used for the protection of windows during construction. The lacquers most commonly used are clear, water white methacrylates.

Lead glass. Small panes of glass, colored or plain, joined together and framed with lead, usually in a special design or pattern.

Left-hand-swing vent. Vent hinged at left-hand side when viewed from outside.

Lever arm. Device consisting of a rod and arm with pivoted connection and one end attached to a horizontal shaft, the other to a ventilator.

Lever-under-screen operator. Gearless adjuster for open-out casement windows to operate vents and hold them in an open position.

Lift handle. Handle designed for use in raising the lower sash of a double-hung window.

Limit stop. Stop that prevents the opening of a ventilator beyond a predetermined position.

Lock keeper. Strike plate to engage the locking device on projected windows.

Long-leg frame section. Solid section with an exterior leg extending approximately $\frac{7}{8}$ in. beyond a standard frame section.

Loose-crank operator. Operator with a removable handle to prevent operation of windows except by authorized personnel.

Louver screen. Screen mesh woven or stamped into slats or vanes, constructed to reduce sky glare and solar heat and to exclude insects.

Lug. Metal spacer between window and adjoining construction.

Manual operation. Manual operation is performed by means of a hand chain or vertical shaft. Chains operate over a cast-iron wheel having a chain guard and mounted on the worm shaft. Chains terminate approximately 2 ft above the floor. Where it is impractical to hang a chain vertically from the wheel, idlers are provided. Vertical shafts of pipe or solid rod are coupled directly to the worm shaft and supported by steel or iron brackets spaced not more than 15 ft apart. The operating miter gear is located 4 ft above the floor.

Manual operator. Window operating device for hand operation.

Mastic. Compound for sealing and waterproofing the joints between windows and mullions or adjoining construction.

Material for steel windows. Sections used in the manufacture of steel windows are standard intermediate, heavy intermediate, or heavy custom. Principal members are made of hot-rolled new billet steel and so constructed that the glass in each window is in the same plane within a tolerance of $\frac{1}{4}$ in.

Mechanical joint. Joint held together without welding.

Mechanical operator. Combination of power, shaft, arms, etc., designed to operate a group of ventilators simultane-

ously. It may be operated by chain, vertical rod with miter gear, hand wheel or crank, or detachable pole or be electrically controlled.

Meeting rail. Interior framing bar adjacent to a ventilator.

Mill finish. Exposed surfaces of aluminum members have a mill finish and are uniform in color and clean and free from serious surface blemishes.

Miter gear. Mechanism for changing the direction of operating power.

Miter joint. Joint formed when two pieces of identical cross section are joined at the ends, and where the joined ends are beveled at equal angles, in contrast to a coped joint.

Modular. Window size based on modular coordination using 4 in. as a standard unit of measure.

Monumental window. Window of heavyweight construction, suitable for edifices, institutions, and monumental structures, where the utmost in performance and permanence are desired.

Mortise-and-tenon joint. Interlocking joint with one section coped to permit a portion of the section to pierce the other in order to permit riveting.

Mullion anchor. Clip or plate attached to a mullion for anchorage to adjoining construction.

Mullion cover. Flat or formed sheet metal cover plate over the inside and/or outside face of a mullion.

Mullions

1. Horizontal or vertical dividing units between the frames of windows.
2. Steel mullions are usually of the standard T bar or plate type and not less than $\frac{1}{8}$ in. in thickness. Mullions and window units are bolted or clipped securely together to form a rigid assembly with mullion covers.

Mullion stiffener. Attached member used to stiffen or strengthen a mullion. It may be attached on the outside of the mullion or inside the mullion construction.

Multiple operated window. Window with two or more ventilators linked together to operate in unison.

Muntin

1. Vertical or horizontal division bar between two lights.
2. Small horizontal or vertical dividing bars within the basic framework of a window, subdividing and supporting the panes of glass.

Nonfriction hinge. Hinge without friction washers or other devices to restrict its movement.

Normal reach. Point measured from the finished floor to the operating hardware. It is approximately 5.5 ft above floor for an out-swing casement or projecting-out ventilator and 6 ft above the floor for others.

Oil-enclosed power. Gear box for mechanical operators having a machine-cut worm and bronze gear enclosed in an oil-tight housing with the worm immersed in oil or packed in grease.

Opening dimension. Rough opening in a frame or masonry wall to receive a window or door.

Organic coatings. Various paints, usually composed of pigment, a volatile or nonvolatile vehicle, and a drying agent. Paints most frequently used for windows are alkyd, vinyl, epoxy, or acrylic based.

Oriel window. Window that projects from the main line of an enclosing wall of a building and is carried on brackets or corbels.

Out-swing ventilator. Ventilator hinged to swing to the outside.

Parting strip. Vertical strip on the jambs of a double-hung window frame to keep the top and bottom sash apart.

Pass-through window. Sliding or hinged vent used at counter height.

Pole. Long cylindrical piece of wood or metal tubing with a hook on one end used to reach the operating hardware of windows above normal reach.

Pole ring. Metal ring attached to the head of a ventilator for pole operation.

Prefinished color coatings. Windows, muntins, trim, and miscellaneous material are properly cleaned, coated with a suitable primer, and given an in-plant oven-cured color finish in accordance with the manufacturer's standard specification.

Prime painting. Steel work is cleaned free from rust and loose scale and zinc phosphate treated with Bonderite 37, followed by one coat of DuPont's special epoxy alkyd primer, oven-baked at 325°F for 35 minutes. Field coats of paint, if required, are applied by the painting contractor.

Primer coat. Protective base coat of paint.

Projected window. Window that has a projected vent or vents.

Projected windows. Windows that consist of horizontally mounted ventilators that may project out or in. They differ from awning and hopper windows in that the hinged side of the ventilator moves in a track up or down when the ventilator is operated. Projected windows are used commonly where large glass areas are desired and where it is desired to deflect incoming air in an upward direction.

Project-in vent. Vent whose bottom slides up and top swings in.

Project-out vent. Vent whose top slides down and bottom swings out.

Protective coating

1. Aluminum sash and frame members receive a temporary uniform protective coating of water-clear lacquer after fabrication in accordance with Architectural Aluminum Manufacturer's Association specifications, and AA Aluminum Association specifications.

2. Any one of several types of coatings utilized to provide a barrier between dissimilar materials or to protect a finished surface from damage during construction.

Psychiatric window. Window designed for use in mental institutions. Heavy intermediate sections and construction are used in conjunction with layouts having restricted glass sizes and ventilator heights plus screens and hardware especially designed for this product. Ventilators may be either projected to swing in and manually operated or projected to swing out and mechanically operated.

Pull-down socket. Recess to received a pole hook.

Putty. Glazing material that is applied in a plastic state and hardens to a relatively brittle condition.

Putty glazing. Retaining glazing by means of putty. Putty glazing may be either inside or outside depending on whether sections are designed for glazing from inside or outside.

Push bar. Horizontal operating bar applied to the sill of a vent.

Rack and pinion. Bar with teeth that mesh with a gear that is secured to a torsion shaft. It is held against the shaft by yoke and rollers and used for operating several window ventilators at above normal height.

Rail. Surrounding framework of a ventilator or window.

Ranch-type window. Residential section window with top-hinged, open-out, or project-out ventilators.

Reglet. Groove or pocket in masonry construction to receive the leg of a window section.

Removable handle. Detachable handle to lock or operate a ventilator.

Residential window. Narrow, solid-section metal window of 1-in. minimum depth for casements; also, any type of light-weight window for residential use.

Reversible window. Window with a pivoted ventilator that can be rotated 180° for cleaning. It is usually vertically pivoted.

Right-hand swing vent. Vent hinged on the right-hand side when viewed from the outside.

Rigid vinyl components. Rigid vinyl components are integrated into the window construction to provide an effective thermal break between aluminum members in both frame and sash.

Rolled section. Window section formed by hot-rolling steel bar stock or by cold-rolling sheet steel or other sheet metal.

Roller. Sheave for use in horizontal sliding windows.

Roto-type operator. Worm and gear device used to open the ventilator.

Sash. Operable section of window, usually applicable to double- or single-hung windows.

Sash chain. Galvanized chain used to operate a vent or to hold sash in balance with counterweights.

Screen cloth. Closely interwoven metal or plastic mesh.

Screen flap. Slotted insert in a screen to provide for passage of the operator arm.

Screens

1. Screens cover the ventilating portion of the unit. They are installed from the interior with push-in-type retainers and are easily removable for cleaning or replacement. Screens consist of extruded aluminum tubular frame, 18×16 mesh aluminum screen cloth, and removable splines.

2. Aluminum exterior half or full flat-frame screens are available; a half-screen engages retainer pins in the sill and latches to recessed strikes; a full screen engages hanger pins and latches to a strike pin factory installed in the sill. Both screens are furnished with gray-painted aluminum screen cloth and supplied complete with all necessary hardware.

Screen spline. Metal or plastic strip used to secure screen cloth in a frame groove.

Sealants. Soft cohesion materials used to adhere to adjacent materials and to seal joints so as to prevent the passage of air and water.

Section. Rolled, extruded, or formed metal member.

Security screen. Screen consisting of a standard screen frame with heavy wire mesh, designed to afford limited restraint.

Security windows

1. Windows designed to provide ventilation with protection for commercial and industrial buildings. They have small lights and ventilator guard frames.

2. Windows with grilles over the vent openings. Grilles and muntin spacing are not to exceed 7×16 in.

Setting block. Piece of material used to support a window in position until permanently secured; also, pad to support glass.

Shaft bracket. Metal support that carries the operating drive member of a mechanical operator.

Shop paint finish. Steel surfaces are thoroughly cleaned and freed from rust and mill scale. A protective treatment of hot phosphate or cold phosphate-chromate or a method such as bonderizing or parkerizing is necessary to protect the metal and provide a suitable base for paint. Windows are then given one shop coat or rust-inhibiting primer, baked on.

Shop primer. Steel windows, fins, mullions, cover plates, and other associated parts are cleaned, properly treated, and prime painted after fabrication.

Sill. Lowermost horizontal portion of a window or frame.

Simplex operation. Casement hardware consisting of friction hinges and cam handles

Single-hung window. Vertical sliding window with one operating section.

Slide bolt. Hardware locking device that is formed by a metal bar housed within a sleeved member and moved into a keeper to form a lock.

Sliding friction pivot. Device attached to the pivot point of a projected ventilator to slide in a track or guide in the frame section.

Spring catch or latch. Spring-loaded device for locking a ventilator.

Stay bar hardware. Elbow-type operating and locking device that controls the extent of the opening of a window.

Steel finishes. Finish produced by bonderizing, followed by baked-on shop primer suitable for field painting or galvanizing.

Steel window. Windows made from hot-rolled, structural, grade, new billet steel. However the principal members of double-hung windows are cold-formed from new billet strip steel. The principal manufacturers conform to the specifications of the Steel Window Institute, which has standardized types, sizes, thicknesses of material, depths of sections, construction, and accessories.

Stool. Interior horizontal trim member forming the bottom finish portion of a window assembly.

Stop. Retaining bead holding a pane of glass in its proper position within a frame.

Storm ash, storm windows. Removable, light-framed glazed panels attached to windows or set in frames to provide double glazing for the purpose of reducing heat losses.

Strike plate. Metal piece that receives the locking device.

Subframe. Pressed-metal frame consisting of head, jambs, and sill to receive the frame and metal sash.

Surrounds. Separate or integral trim used to receive exterior wall covering at the perimeter of the window.

Sweep lock. Cam-type locking devices applied to the meeting rail of a double-hung window.

Tape

1. Thin, continuous, flexible material, backed with other material or unbacked, normally used as a sealant in window construction.

2. Preformed stiff mastic strip used in place of mastic where windows contact metal. It is also used in place of back putty.

Tension toggle operator. Shaft that moves horizontally to activate a figure 4 arm, which in turn operates a window.

Tolerance. Specified allowance for dimensional variation resulting from manufacturing limitations of a material or assembly of materials.

Top-hinged in-swinging window. Window that consists of a ventilator, hinged to the main frame at the top, that swings into the room, permitting the glass to be cleaned from the interior. The window is not used for normal ventilation and is operated solely for cleaning purposes and emergency ventilation. Its use is generally in completely air-conditioned high-rise buildings.

Top-hung window. Window with ventilators hinged at the top to swing out.

Transom bar. Horizontal member at the head of a door or window, which forms the supporting sill for a window above.

Transom window. Fixed or operating unit that is located directly over the head of a door or window.

Ultimate strength. Maximum strength that may be developed in a material prior to rupture.

Under screen push bar. Bar type window-operating device that operates a ventilator through the sill of the window.

Unequal lag. Frame section whose exterior leg is longer than its interior leg.

Ventilator. Operable portion of a window.

Vertically pivoted window

1. Window that consists of a ventilator sash mounted on pivots located in the center of the top and bottom main frame members, allowing the ventilator to be reversed by rotating and permitting the glass to be cleaned from the interior. The window is not used for normal ventilation and is operated solely for cleaning purposes or emergency ventilation. It is used generally in completely air-conditioned high-rise buildings.

2. Window operating on or through bearing points at the center of the sill and head, which allow the vent to move to a half-in and half-out position.

Vertical shaft. Metal rod member that transmits power in a mechanical window operator assembly.

Vertical sliding windows. Vertical sliding windows resemble single- or double-hung windows in appearance but use no sash-balancing devices. They are operated manually by lifting and are held in various open positions by mechanical catches that engage in the jamb or hold by friction. The ventilating area is 50% of the window area.

Weatherstrip. Strip of metal, rubber, or other material used to seal the joints between the ventilator and frame against air infiltration.

Weatherstripping. Operable windows should be provided with weatherstripping. Before leaving the factory the window should be tested in the closed and locked position in accordance with ASTM E 283.

Weep holes

1. Small openings designed and located to allow egress of any condensation or water that might otherwise accumulate in a window sill.

2. Small holes provided in the sill section of a sash to allow water to condensation to escape.

Wicket screen. Small sliding or hinged portion of a larger screen that provides access to operating window hardware.

Window. Unit assembly consisting of frame, sash, and hardware, installed in the walls of a building to provide light or vision and optional ventilation.

Window cleaner anchor. Hooks or eye bolts that are secured at the sides of windows to provide for the fastening of window cleaners' safety belt hooks.

Window guard. Hinged or fixed metal frame covered with heavy wire mesh or similar material, which may or may not be a part of a window assembly and provides protection over the glazed portions of the window.

Window wall. Continuously joined windows that cover or almost cover to the extremities of the space or building, extending or almost extending from floor to ceiling.

Yield point. Unit stress at which material continues to deform without further increase in the load.

Yield strength. Maximum unit stress that can be developed in the material without causing a specified permissible deformation. For aluminum the permissible deformation is 0.20% of the original length.

Zipper glazing. Method of holding glass with a continuous plastic gasket that is deformed into final shape by the insertion of a continuous plastic spline.

REFLECTIVE GLASS

Annealed glass. Reflective glass may be annealed glass, which prevents distorted reflections and meets requirements of safety glazing codes if heat treated.

Banded effects. Due to high daylight transmittance characteristics, the use of painted opacifiers, applied films, or light colored backup materials behind reflective glass may alter the exterior appearance and produce a banded effect between vision and spandrel areas.

Cleaning and maintenance. Glass has a hard, durable coating, but care must be taken not to damage this during cleaning processes. Compounds containing fluorine, strong acids, corrosive alkaline detergents, or abrasives should not be used.

Coating test. Glass should be vision-tested against a bright, uniform background.

Coating uniformity. Uniformity within each light.

Condensation. Batt or semirigid insulation contacting the back of reflective glass, or the use of nonuniform color backup materials, may cause objectional "read through" under certain lighting conditions. Moisture condensation, if trapped between these materials and the glass, may also cause gradual etching of the glass and will produce a mottled appearance when viewed from the exterior.

Construction site care. Reflective glass must be protected with clear plastic sheets. Protective material must not touch the glass surface. Airspace between the temporary protection and the glass must be vented to prevent moisture staining or thermal breakage.

Curtain wall. Any building wall, of any material, that carries no superimposed vertical loads; any nonbearing wall.

Dry glazing system. Dry glazing is recommended whenever reflective coatings are glazed to first surface.

Energy efficiency. Reflective glass provides energy cost reductions compared to other types of glass.

Glazing tapes and sealants. Tapes and sealants with high oil content on the coated surface (outside) may cause oily deposits on the reflective coating that may be difficult to clean.

Heat treatable. Glass is heat treatable for heat strengthening, tempering, or bending.

Heat treated. Thermal stress may occur in spandrel glass panels because building structure components prevent adequate dissipation of heat buildup on sunlit elevations. To prevent this stress, reflective glass must be heat treated.

Low heat absorption. Glass has the characteristic of offering a combination of reflectance, daylight transmittance, and resultant low absorption.

Low thermal stress. Reflective glass controls unwanted solar heat gain by reflecting and absorbing a majority of the sun's radiation.

Metal curtain wall. An exterior curtain wall that may consist entirely or principally of metal, or may be a combination of metal, glass, and other surfacing materials supported by or within a metal framework.

Pinholes. Pinholes greater than $\frac{1}{8}$ in. should not be acceptable. Large clusters or close spacing of smaller pinholes, visible from a distance of 6 ft, are also not acceptable.

Pyrolytic coating. Glass can be handled, cut, bent, insulated, laminated, heat strengthened and tempered, using standard annealed float glass techniques.

Reflective glass. Glass pyrolytically coated by chemical vapor deposition in which a gas reacts with the semimolten surface of a ribbon of float glass to form a reflective coating. Several colors are available, including clear.

Sealant. Sealants for reflective glass are compatible with most sealants commonly used in the manufacture of insulating glass.

Scratches. Viewed at a distance of 10 ft, visible scratches longer than 3 in. are not acceptable and are not allowed in an area of the glass through which a person at normal height would be expected to look.

Spandrel applications. Spandrel panels of reflective glass used for cladding over opaque walls may be used to create various designs.

Spandrel construction. Using heat-strengthened insulating glass with ceramic enamel on the room-side surface will eliminate real-through, minimize banding effects, and ensure stable and durable spandrel design.

Ultraviolet transmittance. Reflective glass limits color fading and breakdown of plastic materials, with 90% of the sun's damaging ultraviolet radiation effectively blocked.

Uniform coating. Viewed from a distance of 10 ft, some mottling or streaking of the coating may appear.

Uniformity. Uniformity between vision and spandrel glass is impossible to achieve when viewed from the exterior.

Window wall. Type of metal curtain wall installed between floors or between floor and roof and typically composed of vertical and horizontal framing members containing operable sash or ventilators, fixed lights or opaque panels, or any combination thereof.

SEALED INSULATING GLASS

Active solar heat gain. Solar heat that passes through a material and is captured by mechanical means.

Adhesion. Property of a sealant/compound that produces ability to bond to the surface to which it is applied.

Adhesion failure. The pulling away of a sealant/compound from the surface it is applied to, resulting in water penetration.

Air infiltration. The amount of air that passes between a window sash and frame, or a door panel and frame; for windows it is measured in terms of cubic feet of air per square foot of area, and for doors it is measured in terms of cubic feet of air per minute, per foot of crack.

Bead. Sealant/compound after application in a joint. Also a molding or stop used to hold the glass product in position.

Bite. The dimension by which the edge of a glass product is engaged into the glazing channel.

Block. Rectangular, cured sections of neoprene or other approved materials, used to position the glass product in the glazing channel.

Breather (tube) units. Insulating glass unit where a tube or a hole is factory-placed into the unit's spacer to accommodate elevation of pressure differences encountered in shipping. These tubes or holes are to be sealed on the jobsite prior to unit installation.

BTU. Abbreviation of British thermal unit, defined as the amount of heat needed to raise the temperature of one pound of water one degree Fahrenheit.

Butt glazing. The installation of glass products where the vertical glass edges are without structural supporting mullions.

Capillary tube units. Insulating glass unit where a very small metal tube of specific length and inside diameter is factory-placed into the unit's spacer to accommodate both the pressure differences before installation and also the pressure differences encountered daily after installation. Capillary tubes are not sealed after installation.

Certified IG unit. Insulating glass unit constructed like a unit test model, which has successfully passed the ASTM E 773 and E 774 tests of insulating glass seal durability performance at specific levels.

Channel glazing. The installation and sealing of glass products into U-shaped glazing channels employing removable stops.

Cohesive failure. The splitting and opening of a sealant/compound within its body, resulting in water penetration.

Compound. Formulation of vehicle, fillers, and polymer(s) producing an elastomeric sealant.

Condensation. Moisture that forms on surfaces when they are colder than the dew point.

Conduction. The transfer of heat through matter, whether solid, liquid, or gas.

Convection. Transfer of heat through a liquid or gas when that medium hits against a solid surface.

Curing agent. One part of a two-part sealant that, when added to the base material, causes it to vulcanize by chemical reaction.

Dead loads. Load force due to glass weight.

Desiccants. Porous crystalline substances used to absorb moisture and solvent vapors from the airspace of insulating glass units. More properly called "absorbments."

Dew point. The temperature above 32°F at which visible water vapor or other liquid vapor begins to deposit on the airspace glass surface of a sealed insulating glass unit in contact with the measuring surface of the dew-point apparatus.

Double-glazed units. Units of two lites of glass and one airspace.

Dual-sealed units. Sealed insulating glass units fabricated with an inner primary seal and an outer secondary seal. Generally, each of the two seals has been selected for its special performance characteristic, i.e., adhesion and moisture vapor transmission properties.

Durometer. Gauge to measure the hardness of an elastomeric material.

Edge clearance. Nominal spacing between the edge surface of the glass product and the glazing channel base.

Elastomeric. Having the property of returning to original shape and position.

Emissivity. The relative ability of a surface to radiate heat, with emissivity factors ranging from 0.0 (or 0%) to 1.0 (or 100%).

Emittance. Heat energy radiated by the surface of a body, usually measured per second per unit area.

Equivalent/combined glass load. Combination of the instant applied load of wind and the factored long-term loading of glass weight and snow accumulation.

Exterior glazed. Glass set from the exterior of a building.

Failed IG unit. Installed unit failure exhibits permanent material obstruction of vision through the unit due to a accumulation of dust, moisture, or film on the internal surface of the glass. Surface numbers 2 or 3 in dual-pane units, surface numbers 2, 3, 4 or 5 in triple-pane units.

Float glass. Transparent glass with flat, parallel surfaces formed on the surface of a pool of molten tin.

Fogged unit. Permanent deposit of contaminants on the interior glass surfaces of an insulating glass unit.

Frost point. The temperature below 32°F at which visible frost begins to deposit on the airspace surface of a sealed insulating glass unit in contact with the measuring surface of the frost-point apparatus.

Fully tempered glass. Transparent or patterned glass with a surface compression of not less than 10,000 psi or an edge compression of not less than 9,700 psi.

Gas-filled units. Insulating glass units with a gas other than air in the airspace to decrease the unit's thermal conductivity U-value and to increase the unit's sound insulating value.

Gasket. Preformed shape of rubber or rubber-like compositions, used as a weather seal. Also, a spacer for supplemental application of a sealant.

Glass. Transparent, brittle substance formed by fusing sand with soda or potash or both; it often contains lime alumina or lead oxide.

Glazing. the installation and weather-sealing of a glass product in a prepared sash opening.

Glazing bead. Strip surrounding the edge of the glass in a window or door; applied to the sash on the outside, it holds the glass in place.

Glazing channel. Three-sided, U-shaped sash detail into which a glass product is installed and retained by removable stops.

Glazing channel depth. The measurement from the bottom of the glazing channel to the top of its stops.

Glazing channel width. The measurement between the stationary stop and the removable stop.

Heat-absorbing glass. Glass (usually tinted) formulated to absorb an appreciable portion of solar energy.

Heat-strengthened glass. Transparent or patterned glass with a surface compression of not less than 3,500 psi or greater than 10,000 psi, or an edge compression of not less than 5,500 psi.

Insulating glass

1. Insulating glass units comprise two or more sheets of glass separated by either $\frac{3}{16}$, $\frac{1}{4}$, or $\frac{1}{2}$-in. airspace. Units are factory sealed, and the captive air is dehydrated at atmospheric pressure. Typical units are made up of either window glass or polished plate glass. Special units of varying combinations of heat-absorbing, laminated, patterned, or tempered glass are available.

2. Insulating glass is made from two or three thicknesses of standard glass panes with one or two airspaces between. The contained air layers are dehumidified or dried to prevent fogging in cold weather, and the edges are sealed. Units may be fabricated with polished plate or window glass.

Interior glazing. Glass set from the interior of a building.

Internal muntins. Decorative grids installed between the glass lites that do not actually divide the glass.

Laminated glass. Two or more lites of glass bonded together with a plastic interlayer.

Light-reducing glass. Glass formulated to reduce the transmission of visible light.

Lite (Light). Unit of glass in a window or door; it is enclosed by the sash or by muntins and bars. Also called "pane."

Live load. Load force due to the weight of nonpermanent attachments such as people, glazing rigs, washing rigs.

Low-emissivity glass. Glass with a transparent metallic or metallic oxide coating applied onto or into a glass surface that reflects long-wave infrared energy and thus improves the U-value.

Metal spacers. Roll-formed metal shapes used at the edges of an insulating glass unit to provide the desired spacing of the lites; they allow areas for sealant applications and contain desiccants.

Modulus. Stress at a given strain; also tensile strength at a given elongation.

Moisture vapor transmission (MVT). The steady water vapor flow in unit time through a unit area of a body, normal to specific parallel surfaces, under specific conditions of temperature and humidity at each surface.

Mullion. Horizontal or vertical member that holds together two adjacent lites of glass or sashes or curtain wall sections.

Multiple-glazed units. Units of three glass lites (triple glazed) or four glass lites (quadruple glazed) with two and three airspaces, respectively.

Muntins. Horizontal or vertical bars that divide the sash frame into smaller lites of glass. Similar to mullions but smaller in dimension and weight.

Needle glazing. Application of a small bead of sealant/compound at the slight line by a nozzle gun.

Passive solar heat gain. Solar heat that passes through a material and is captured naturally, not by mechanical means.

Patterned glass. Rolled glass having a distinct pattern on one or both surfaces.

Permeability. The time rate of water vapor or gas transmission through unit area of a material of unit thickness induced by unit vapor pressure differences between two specific surfaces under specified temperature and humidity conditions.

Permeance. The time rate of water vapor or gas transmission through a unit area of a body, normal to specific parallel surfaces, under specific temperature and humidity conditions.

Primary sealant. Sealant applied to the inner shoulders of a spacer for the principal purpose of minimizing moisture, gas, and solvent migration into the unit's airspace.

Priming. Sealing of surfaces to promote the adhesion of sealants.

Purlins. Structural members, generally horizontal, in sloped glazing frames.

Rabbet. Two-sided, L-shaped recess in a sash frame to receive glass products. Addition of a removable stop will convert it to a glazing channel.

Radiation. Energy released in the form of waves or particles, due to a change in temperature within a gas or vacuum.

Rafters. Structural members, vertical in sloped glazing frames.

Reflective coated glass. Glass with metallic or metallic oxide coatings applied onto or into the glass surface to provide reduction of solar radiant energy, conductive heat energy, and visible light transmission.

Relative heat gain. Energy comparison factor for glass products combining the radiant and conductive heat gain under specific conditions.

R-value. The resistance of conductive heat energy transfer in one hour through a 1 sq ft area of a specific insulating glass unit assembly for each 1°F temperature difference between the indoor and outdoor air.

Sash. Frame into which glass products are glazed, i.e. the operating sash in a window.

Sealants. (for IG units) Formulated elastomeric compounds of specific application and vapor transmission properties as well as controlled adhesion, cohesion, and resiliency properties.

Sealed IG units. Units constructed of two or more lites of glass separated and hermetically sealed to spacer frames at the glass edges with the enclosed air chamber(s) dehydrated at the plant's atmospheric pressure.

Sealant spacer. Permanent adhesive sealant extrusion that may contain a structural metal insert and a precompounded desiccant.

Secondary sealant. Sealant applied into the exterior glass-spacer cavity to provide elastic, structural bonding of the assembly. In single-sealed units, this sealant also has low gas and moisture vapor transmission properties to achieve effective unit performance.

Setting blocks. Rectangular, cured extrusions of neoprene rubber or other approved material on which the glass product bottom edge is placed to effectively support the weight of the glass.

Shore hardness. Measurement of the hardness of a cured elastomeric material by means of a durometer hardness gauge.

Sight line. Imaginary line around the perimeter of a glazed glass product defined by the top edge of stationary and removable stops, or the line where the glazing sealant or gasket contacts the glass.

Single-sealed units. Sealed insulating glass units where the structural bonding and moisture sealing is accomplished by a single seal at the edge.

Skylight. Glass or plastic and frame assembly that is installed into a roof of a building.

Sloped glazing. Any installation of glass that is at a slope of 15° or more from the vertical.

Snow load. Load force due to snow accumulation.

Solar energy. Thermal radiation from the sun, as measured by short radiation wavelengths, less than 3μm long.

Solar energy absorptance. The percentage of the solar spectrum energy (ultraviolet, visible, and near infrared), from 300 to 3,000 nm that is absorbed by the glass product.

Spacer corners. Specific methods used in joining the spacer lengths into spacer frames, including interlocking keys, bending, soldering, or welding.

Spacer depth. Dimension of the spacer that is measured parallel to the glass surface.

Shade coefficient. The ratio of the rate of solar heat gain through a specific glass unit assembly to the solar heat gain through a single lite of $\frac{1}{8}$-in. clear glass in the same situation.

Shatterproof insulating glass

1. Insulating glass made of an interior light of sheet, plate, or float glass $\frac{1}{8}$, $\frac{3}{16}$, or $\frac{1}{4}$ in. thick and an exterior light of sheet, plate or float glass $\frac{1}{8}$, $\frac{3}{16}$, or $\frac{1}{4}$ in. thick, separated by a $\frac{1}{4}$-, $\frac{1}{2}$-, or $\frac{5}{8}$-in. dehydrated airspace.

2. Insulating glass consisting of two lights of sheet, plate, or float glass separated by a metal spacer and hermetically sealed to contain dehydrated air and keep out moisture vapor.

Shatterproof insulating sound control glass. Insulating glass consisting of one light of sound control laminated glass and one light of sheet, plate, or float glass, hermetically sealed to contain a dehydrated airspace.

Spacers (shims). Small blocks of neoprene, or other approved material, placed on each side of the glass product to provide glass centering, maintain uniform width of sealant bead, and prevent excessive sealant distortion.

Spacer width. Dimension of the spacer that is measured perpendicular to the glass surface and establishes the unit's air space.

Spandrel. Portion of the exterior wall of a multistory commercial building that covers that area below the sill of the vision glass installation and above the head of the glass installation below.

Stops. The stationary lip of the back of the glazing channel and removable molding (retainer) at the front of the glazing channel.

Structural glazing gaskets. Cured elastomeric channel-shaped extrusions used in place of a conventional sash to install glass products onto structurally supporting subframes, with the pressure of sealing exerted by the insert of separate lock-strip wedging splines.

Structural silicone glazing. A system in which the glass product is bonded to the framing members of a curtain wall, using a structural silicone adhesive/sealant without the presence of outdoor retainers or stops.

Sunlight. The portion of solar energy that is detectable by the human eye; it accounts for about 44% of the total radiation wavelength spectrum.

Tinted glass. Body-colored glass of specific batch ingredient formulation to produce light-reducing and/or heat-absorbing glass products.

Total heat gain/summer/daytime (BTU per hour, per square foot). The sum of the radiant energy and the conductive energy transmitted into a building. (Shade coefficient times ASHRAE solar heat gain factors + summer U-value × the indoor/outdoor temperature differences.)

Total heat gain/summer/nighttime (BTU per hour, per square foot). The conductive energy transmitted into a building. (Summer U-value × the indoor/outdoor temperature difference.)

Total heat loss/winter/daytime (BTU per hour, per square foot). The result of the radiant energy transmitted into a building and the conductive energy transmitted out of a building. (Shade coefficient × ASHRAE solar heat gain factors + the winter U-value × the outdoor/indoor temperature difference.)

Total heat loss/winter/nighttime (BTU per hour, per square foot). The conductive energy transmitted to the outdoors. (Winter U-value × the outdoor/indoor temperature difference.)

Transmittance. The fraction of radiant energy that passes through a given material.

Units. Term used to refer to sealed insulating glass units.

United inches. Total of one width and one length in inches.

U-value. The amount of conductive heat energy (BTUs) transferred through a 1 sq ft area of a specific insulating glass unit for each 1°F difference between the indoor and outdoor air.

Ultraviolet. Type of radiation with wavelengths shorter than those of visible light and longer than those of x-rays.

Visible light transmittance. The percentage of light in the visible spectrum range of 390 to 780 nm that is directly transmitted through the glass products.

Weep holes. Slots or holes in the sill (bottom) member of a sash frame to provide outdoor release of infiltrated water.

Wind load. Load force on glass due to the speed and direction of the wind.

Wired glass. Glass having a layer of meshed wire completely embedded in the glass lite. It may have polished or patterned surfaces.

WINDOW WALLS, CURTAIN WALLS

Abrasion resistance. Property of material enabling it to withstand scratching or rubbing, commonly expressed as depth of penetration, in inches, after testing with silicon carbide wheel for approximately 10,000 wear cycles using an abrasion-testing machine such as the Taber abraser.

Accelerated aging. Artificial aging, within a short period of time, of the finished surfaces of a material through chemical or other means of treatment.

Acid etching. Immersion in various types of chemicals, the effect of which depends on the type of metal used.

Acrylic modified fiberglass panels. Acrylic modified fiberglass-reinforced panels are permanently bonded under an electronically controlled pressure and heat process to a unitized aluminum grid core ±0.003-in. thickness. The bond is waterproof. The exterior translucent face of the unitized grid is laminated with DuPont Tedlar to provide long lasting protection against erosion and other destroying elements caused by general weathering.

Adjustment and alignment. Arranged in line and fitted, as one member to another.

Age hardening (precipitation hardening). Process of hardening metals that occurs at ordinary temperatures, usually resulting in greater strength and less ductility.

Air infiltration. Movement of air through, in, and around the various component parts of an assembly.

Alclad. Aluminum product clad with an aluminum alloy coating that is anodic to the alloy it covers, protecting it both physically and electrolytically against corrosion.

Alodine. Proprietary trade name for a chemical conversion coating used as a protective and/or decorative finish on aluminum. It is applied by either a spray or dip process. The finish may be colorless or one of various shades of green or gold.

Alumilite. Proprietary trade name used by Aluminum Company of America for their anodized finishes on aluminum.

Alumilite finish. Hard oxide coating formed by an electrolytic process. Alumilite finishes have excellent resistance to abrasion, weather, and dirt accumulation.

Aluminize. Apply a surface coating of aluminum to another metal or another base material, usually by spraying or dipping in molten aluminum. On steel, such coatings greatly increase corrosion resistance.

Aluminum curtain wall system. Curtain wall system designed using tubular vertical and horizontal mullions; a "stick" system with all verticals and horizontals the same and all vertical mullions typical. The system is factory fabricated or easily field fabricated from stock. The wall is dry set, outside glazed, with an interior gasket of closed-cell neoprene forming the primary seal. This system is also used as a window wall system between floors, with horizontal expansion handled within each module, or as a curtain wall system bypassing floors with vertical expansion joints. It is reinforced with steel tubes or aluminum I-members as required by wind loads.

Anchor. Device used to secure the metal curtain wall or its parts to the building frame. Anchors should generally be adjustable in three dimensions.

Anneal. Heat metal, glass, or other material above the critical or recrystallization temperature and then cool them to eliminate the effects of cold-working, relieve internal stresses, or improve electrical, magnetic, or other properties.

Annealing. Heat treatment intended to soften metal, thereby removing residual stresses, ordinarily consisting of raising the temperature to a point below that of recrystallization and then cooling slowly.

Anodic coating. Surface finish resulting from anodizing.

Anodize. Provide a hard, noncorrosive oxide film on the surface of a metal, particularly aluminum, by electrolytic action. An electrochemical process produces an anodic coating by conversion of aluminum into essentially aluminum oxide. Appearance depends on both the alloy involved and the surface preparation. Anodic coatings may be transparent, of varying shades of silver, gray, or brown; or colors may be incorporated by the use of dyes or pigments.

Anodizing. Providing an electrochemical finish that produces a hard, integral film or protective oxide.

Applied weathering. Sheet metal weather protection member that is applied to a solid section window.

Architectural effect. Designed appearance of a curtain wall.

Arc welding. Process for joining metal parts by fusion, in which the necessary heat is produced by means of an electric arc struck between an electrode and the metal or between two electrodes.

Asbestos-reinforced panels. Asbestos-reinforced, noncombustible panel designed for a high degree of weather resistance. The finish is a pigmented, multimil acrylic coating, which prevents peeling, blistering, or crazing.

Assembly. Combination of windows, panels, mullions, etc., that comprise a curtain wall design.

Back putty. Bedding of glazing compound, which is placed between the indoor face of glass and the frame or sash containing it.

Backup wall

1. Assembly of materials used on the inside of curtain walls from floor to stool and/or ceiling to floor slab to provide fire-rated construction.

2. Assembly of materials behind a curtain wall to form fire-rated construction.

Baffle. Deflecting surface within a metal wall member, so located as to control or prevent the penetration of air or water into or through the wall. It is commonly used in conjunction with weep holes or slip joints.

Baked-on enamel finish. Factory-applied fluoropolymer enamels containing special formulated resins, highly resistant to chemical attack. The inert base is immune to industrial acids, alkalis, and salts; the finish will not peel, flake, or chip.

Baked-on synthetic resin enamel. Enamel that withstands weathering without appreciable fading, cracking, or peeling.

Batten. Similar to mullion window stop joint.

Bead

1. Strip of metal or wood used around the periphery of a pane of glass to secure it in a frame or sash (also referred to as a "stop") or strip of sealant, such as caulking or glazing compound.

2. Sealant or compound applied to a joint; also, molding or stop used to hold glass or panels in position.

Bedded. Term applied to parts set with a layer of caulking or sealant material.

Bite. Amount of overlap between the stop and the panel or light.

Block. Shim, level, and plumb windows in required position prior to grouting or fixing in final position.

Bonderite. Surface treatment for aluminum.

Bonderize. Remove rust, scale, and all foreign matter from steel surface and convert the metallic surface to a nonmetallic phosphate coating to retard corrosion and provide a better base for paint.

Bonderizing. Type of treatment for iron and steel in which the surface is converted into an insoluble phosphate. Bonderite has little corrosion resistance in itself, but provides an excellent base for paint.

Bottom-hinged ventilator. Ventilator that hinges from the sill member of the window.

Bottom-pivoted ventilator. Ventilator opening on pivots approximately 2 to 4 in. above the bottom of the vent.

Brass track. Guide used in the jambs of projected ventilators for the sliding shoe in lieu of having the shoe slide in the channel formed by the frame member.

Brinell hardness. Measure of resistance to indentation, determined by measuring the area of indentation produced by a hard steel ball under standard conditions of loading.

Butyl rubber (synthetic butyl tape). Synthetic rubber produced by copolymerization of isobutene with a small proportion of isoprene or butadiene.

Burr. Rough or sharp edge left on metal by a cutting tool.

Cam lock. Locking device used on hinged ventilator employing eccentric-type friction action.

Capillary break. Expanded separation between two members designed to stop capillary action.

Carburize. Produce a hard surface layer on steel by heating in a carbonaceous medium to increase the carbon content and then quenching. The process is also referred to as "case-hardening."

Casement window. Window with side-hinged ventilators.

Casing. Finish member that surrounds the window unit or assembly.

Casting. Pouring a liquid form of a substance into a mold to give it the desired shape. Common casting methods for metals include sand casting, permanent mold casting, plaster mold casting, and die casting, all classified by the type of mold.

Caulk. Fill joints, cracks, or crevices in order to make them watertight.

Caulking. Applying mastic with a gun or knife to provide a weathertight installation.

Caulking cartridge. Expendable container filled with a plastic nozzle, made of plastic, fiberboard, or metal, filled with caulking compound, for use in a caulking gun.

Caulking compound. Soft puttylike material intended for sealing joints in buildings and other structures where leakage or structural movement may occur. It is usually available in two consistencies; "gun grade" for use with a caulking gun and "knife grade" for application with a putty knife.

Caulking gun. Device for applying caulking compound by extrusion. In a hand gun the necessary pressure is supplied mechanically by hand; in a pressure gun the pressure is usually greater and is supplied pneumatically.

Caulking pocket. Reglet or recess provided in a wall opening for installation of a caulking compound.

Caustic etch

1. A decorative matte texture is produced on aluminum alloys by an etching treatment in an alkaline solution, generally caustic soda (sodium hydroxide).

2. Chemical treatment consisting of immersion in sodium hydroxide; also known as "frosted finish."

Certificate of test. Written verification issued by a testing laboratory for a manufacturer or supplier for a test conducted on a sample or at the job site.

Cladding. Method of producing a composite material composed of one metal covered with another in order to gain important characteristics not available in either metal alone. Cladding is ordinarily produced by placing the surfaces of two metals in contact with each other under heat and pressure to cause a metallurgical bond between them. Cladding is commonly found in Alclad alloys of aluminum, which usually possess the advantages of high core strength and great corrosion resistance of the exterior metal.

Cladding of grids. Grids are clad to emphasize the lines of the grid itself for aesthetic reasons. A vertical mullion is installed to stand out. Cladding can be installed in the field; it is fitted over the base grid with concealed fasteners and extends the depth of the mullion by as much as 6 in. Cladding is self-draining to prevent the entrapment of condensation or moisture. Cladding also permits the use of more expensive metals such as bronze, stainless steel, or Cor-Ten weathering steel for decorative purposes.

Cleaner-type hinge. Hinge that has the pivot point approximately 3 in. from the outside face of the vent to allow space between the vent jamb and the window jamb so that the window may be cleaned from the outside when the vent is open.

Clearance. Space or distance allowed for anchorage or erection processes or to accommodate dimensional variations in the building structure.

Clip. Small device, usually of metal, for holding larger parts in place, either by friction or by mechanical action; in glazing, a spring device of metal used to hold glass in a metal sash.

Closure. Material used to close a gap between a curtain wall and the surrounding construction.

Cold-welding. Method of joining metals, such as aluminum, by subjecting the thoroughly cleaned joining surfaces to pressure in specialty-shaped dies. When the combined thicknesses of the surfaces are reduced by a specific percentage, a weld occurs at normal temperatures.

Cold-working. Process performed on metals at temperatures below that at which recrystallization occurs. Cold-working causes metals to gain in strength and hardness.

Commercial window. Inside-glazed window fabricated from industrial sections.

Condensation gutter. Trough-like section on the inside of a window at the sill to receive and carry off moisture.

Coped. Term describing a section with a portion removed to allow for the interlocking of members.

Coping. Top closure or cap trim on a wall.

Core adhesives. Cores are laminated to skins with adhesives, which provide for durability, water resistance, and stability at elevated temperatures.

Core fire hazards. The use of polyurethane, polystyrene, and isocyanurate cores in applications presents a major fire hazard under certain circumstances.

Core material. Inner layer or layers of sandwich-type panel.

Corrosion. Deterioration of metal by chemical or electrochemical reaction resulting from exposure to weathering, moisture, chemicals, or other agents or mediums.

Cover plate. Sheet metal strip, flat or molded, to cover splice between windows, usually at mullions.

Curtain wall. Nonload-bearing, windowed and paneled exterior wall surface treatment. A curtain wall is attached to the building structure.

Cut-out. Notch cut from a window perimeter member to clear or fit to an adjacent member.

Cut-up. Pattern of a window as determined by the number of vertical and/or horizontal muntins or vent framing bars.

Depth of section. Distance from the front to back of any rolled, formed, or extruded section.

Detachable pole. Pole having a hook or ring on one end and a crank-operating handle on the other, for working window operating devices that are beyond reach.

Double glazing. Two panes of glass separated by an airspace or sheet of plastic.

Drawing. Any of a number of operations on metal that ordinarily involve forcing the metal into a shape with a punch working into a die.

Drip. Formed sheet metal member that is attached to the head of a fixed or ventilating sash and at the head of an open-out ventilator to shed water.

Duranodic finish. Finish produced by coating an alloy sheet with oxide and treating the coated sheet electrochemically. During processing, distinctive colors are derived from the alloy without dye impregnation; color is determined by controlling the thickness of the oxide coating and by characteristics of the alloy. (Duranodic 300. A product of Construction Specialties Inc., Chicago, Ill.)

Durometer. Instrument that measures the relative hardness of rubber-like materials (ASTM D 2240).

Epoxy resins. Synthetic resins obtained by the condensation of phenol, acetone, and epichlorohydrin.

Equal-leg frame. Section having inside and outside legs of identical lengths.

Exposed glass area. Glass portion that is not covered by any part of the window frame sections.

Extruded section. Section that is formed by forcing heated metal through a die under pressure.

Extrusion. Forming or shaping of metals by forcing through a die. This method is ordinarily used for many aluminum, magnesium, and copper-base alloys, but it can be used, within certain limits, for some ferrous metals.

Fin. Applied continuous metal leg extension secured to the perimeter of the window frame, used for anchoring and support. It is commonly used as a weather bar in masonry construction.

Finish hardware. Operating hardware of a window and door, such as handles, hinges, and locks.

Flashing. Sheet metal strip or bar applied to the perimeter of a window frame to prevent the entrance of water.

Flatness. The flatness of a panel depends on the material bonded to the core. Flat aluminum facings are considered smooth if clearance is less than 0.015 in. along a 6-in. rule. Hardboard, plywood, and other materials should conform to the specifications of their respective manufacturers.

Forging. Shaping of metals by a compressive force, ordinarily at elevated temperatures, by a hammer or other means. The usual types are drop forging, press forging, and upset forging (all classified by the type of equipment used).

Forming. Shaping of metals or other materials without reducing their thickness to any great degree. This method is ordinarily used for sheet materials of various types and is accomplished in single or multiple operations of many combinations. Two major types used for curtain wall components are press forming, consisting of shaping the material in machines designed for bending the material to the desired shape with the aid of dies, and roll forming, consisting of shaping the material on machines with a series of driven rolls of proper conformation.

Frame. Perimeter bar of a metal sash or window.

Framing system. Main grid pattern of vertical and horizontal members.

Gasket. Formed piece of resilient material that seals the joint between two adjoining materials or members.

Glazing. Windows or door sections requiring glazing from either the inside or outside.

Glazing angle. Continuous angle-shaped member used to hold glass and/or panels in place.

Glazing bead. Continuous formed or solid metal section used to hold glass and/or panels in place.

Glazing clip. Formed spring wire or short length of glazing angle used to hold glass in place until putty is placed.

Glazing compound. Plastic material used to hold and/or weatherproof a light of glass in a window.

Glazing rebate. Integral ledge or lip forming a seat for glazing compound.

Grid dimension. Dimension based on modular coordination.

Grid-type curtain wall

1. Wall in which the horizontal and vertical lines are accentuated.
2. Wall with a design in which both horizontal and vertical frame lines are prominent.

Grout. Mortar used to fill a "grout pocket" after installation of the sash.

Heat transmission. Movement of heat by convection, conduction and radiation.

Heat treatment. Any of a number of processes using heat applied to metals to produce changes in their properties. The most common heat treatments given to metals used in curtain walls are annealing and tempering.

Heavy intermediate section. Solid-section steel window with 3.5-lb/ft combined weight and $\frac{15}{16}$-in. minimum depth, front to back.

Hopper ventilator. In-swing or project-in ventilator located at the sill of a window.

Horizontally pivoted window. Window with the ventilators operating on pivots located at jambs approximately 2 in. above the centerline, which allows the vent to open out at the bottom and in at the top.

Hot-dip galvanize. Apply a protective coating to steel by dipping it in a bath of molten zinc.

Hot-platen press. Device employing heat and pressure in the process of laminating materials.

Impact. Force exerted by a body in motion against a fixed structure.

Impact resistance. Ability of materials to resist the damage of a striking force.

In-fill panel system. System suitable for either stucco or masonry exteriors. The stud wall system is supported on the floor slab; windows, floor-to-ceiling window panels, or doors are incorporated. Where the slab extends beyond the wall, the projection may be used as a balcony or walkway area or for sunshading. When panels project beyond the face of the structure, special attention to cap flashing and sealant is required to prevent leaking.

Insect screen. Screen provided over-ventilators for the purpose of excluding insects.

Interlocking joints. Joints commonly used for panels on large wall areas such as those of industrial buildings.

Intermediate section. Solid-section steel window measuring $1\frac{1}{4}$ in. minimum in depth through the web and weighing (combined weight) 3.0 lb/ft.

Insulation materials and values. Insulation materials may be polystyrene, polyurethane, or isocyanurate with $\frac{3}{4}$- to 8-in.-thick insulation cores. Selection depends largely on the insulation value needed to minimize heat gain and loss.

Jalousie. Louvered glass window, stationary or operable.

Jamb plate. Metal extension applied at the jamb.

Joints. Panels are joined by Alcoa's patented Snug Seam or by caulking sealants, splines, battens, or frames.

Laminated panel. Sandwich-type unit made up of various materials bonded together.

Leaded glass. Small panes of glass are joined together and framed with lead.

Left-hand swing vent. Vent hinged at the left-hand side of a window when viewed from outside.

Light. Single pane of glass.

Light reflectance. Amount of light returned by the surface of any material.

Linear expansion. Increase in the overall size of materials due to temperature changes.

Load. Pressure due to applied weight or force.

Long-leg frame section. Solid section with exterior leg extending approximately $\frac{7}{8}$ in. beyond the standard frame section.

Louver. Slatted panel, the slats of which slope outward and downward, providing ventilation and keeping out rain and snow.

Louver screen. Screen made of woven mesh or stamped into slats or vanes constructed to reduce sky glare and solar heat and exclude insects.

Main frame. Portion of the window exclusive of ventilators.

Mastic. Compound for sealing and waterproofing the joints between windows and mullions or adjoining construction.

Materials. When the exterior skin of wall panels are aluminum, the interior skin can be aluminum, hardboard, plywood, cement-asbestos board, or any other solid panel-type material.

Mechanical joint. Joint held together without welding.

Meeting rail. Interior framing bar adjacent to a ventilator.

Mill finish. Term applied to aluminum sections as they are received from the mill, without other finishing processes.

Miter joint. Joint formed when two pieces of identical cross section are joined at the ends and where the joined ends are beveled at equal angles.

Mock-up. Full-size or built-to-scale working model of the design, which includes all materials and parts to be used in final curtain wall construction.

Modular-stack-type wall system. System designed so that no clips are required for the installation of the wall system. Fastenings are unexposed at the joints. Flush glazing is horizontal and vertical with full glass penetration. This system provides positive resistance to water infiltration through its internal drainage system. The system is designed for inside glazing.

Moisture retarder. Caulking, sealant, or closure that prevents the infiltration of air and water.

Mortise-and-tenon joint. Interlocking joint with the vertical section continuous and the horizontal section coped to fit the vertical and then fastened.

Mullion

1. Vertical or horizontal connecting unit between two windows.
2. Vertical framing member in a wall, separating and usually supporting adjacent windows, glass areas, panels, or doors.

Mullion anchor. Clip or plate attached to a mullion for anchorage to adjoining construction.

Mullion cover. Flat or formed cover plate over the inside and/or outside face of a mullion.

Mullion-type curtain wall. Design in which the vertical frame lines only are emphasized by depth or width.

Mullion window stop joint. One of the most common joints, used for materials of various thickness. Assembly is possible from stock store front section. It has many parts that require precision in assembly and installation.

Multistory curtain wall system. This system is designed to be ideal for two- and three-story buildings, particularly where vertical windows or window wall panels are desired. Wall-height studs are installed outside the structural elements, secured to supports at each floor, and, if necessary, spliced at the support. Stud size, spacing, and anchorage are determined by floor-to-floor heights or the distance between supports. Parapet height should be less than one-fourth of the limiting stud height.

Muntin. Bar member supporting and separating panes of glass within a sash or door.

Narrow frame screen. Screen with a formed open-channel frame.

Neoprene. Synthetic rubber made by polymerizing chloroprene.

Noncombustible molded asbestos cement panels. Large-size, lightweight, profiled building enclosures, used as band courses and as complete curtain wall systems.

Oil-canning. Action of metal panels, which, due to temperature and/or pressure changes, move in and out producing noises and an appearance of waviness or unevenness.

Organic coating. Coating, such as paint, lacquer, enamel, or plastic film, in which the principal ingredients are derived from animal or vegetable matter or from some compound of carbon (which includes all plastics).

Out-swing ventilator. Ventilator hinged to swing to the outside.

Oxygen starvation. Localized corrosion of metals, in the presence of an electrolyte, due to a smothering or "poultice" action or resulting from a crevice between the metal and another material.

Panel. Solid filler or facing material, either of one piece or an assembly, for use within a surrounding frame, (e.g., spandrel panel on a wall); preassembled section of wall, including framing (if any), window area, and solid area; length of formed metal sheet, or an assembly of such sheets, usually with insulation between, such as that used for a wall enclosure on industrial buildings.

Parkerizing. Type of treatment for iron and steel in which the clean surface is treated with manganese dihydrogen phosphate. Its primary value is to improve the bonding of paints and lacquers, but it also provides a durable finish, which minimizes corrosion due to porosity or imperfections in the paint film.

Perm. Vapor-transmission rate of one grain of water vapor through 1 sq ft of material per hour when the vapor-pressure difference is equal to 1 in. of mercury (7000 grains equal one lb).

Phosphatized. Term applied to steel with a coating that retards corrosion and produces an etched base to receive paint.

Pickling. Treatment of stainless steel surfaces with a strong oxidizing agent, such as nitric acid, to make them chemically clean and provide a strong inert oxide film that increases corrosion resistance.

Pipe organ effect. Sound produced by the movement of a column of air through hollow mullions.

Pivoted window. Horizontally and/or vertically operated pivoted window.

Polybutenes. General term for polymers of isobutene.

Polysulfide liquid polymer (Thiokol). Thiokol is a polysulfide base sealant, including any one of the polymers produced by the chemical reaction between dichlorodiethylformal and an alkali polysulfide. A product of Thiokol Chemical Division, a division of Thiokol Corporation, Trenton, N.J.

Porcelain enamel. Substantially vitreous or glassy inorganic coating bonded to metal by fusion at a temperature above 800°F.

Porcelain enamel frit. Mixture of minerals and oxides that have been fused at high temperatures and then quenched from the molten condition to produce small friable particles.

Pot life. Period of time during which a sealant, adhesive, or coating, after being mixed with a catalyst, solvent, or other compounding ingredient, remains suitable for use; also referred to as "work life."

Preassembled modules. A system designed to be installed in preassembled modules, provides for separate drainage at each light with concealed and baffled weepage. Water is not allowed to drain through the verticals. The system is designed to provide for expansion and contraction (both horizontal and vertical) outside the glazing area. The curtain wall is dry-glazed from the exterior with neoprene wedges.

Pressed section. Formed section pressed from sheet metal stock.

Projected window. Window that projects in or out as a unit or that has fixed and projecting sections.

Project-in vent. Vent that opens up and in.

Project-out vent. Vent that opens down and out.

Reflectivity. Reflectivity expresses the diffuse reflection of the light falling on a surface. It is commonly expressed as a percentage per unit area for a particular surface.

Reglet. Groove cut or formed in masonry or concrete to receive and hold the edge of flashing material.

Rigidized. Term generally used in reference to lightgauge sheet metal that is embossed or textured by a rolling process; also, name used by one specific manufacturer of this material.

Rolling. Shaping of metal by passing it through successive rolls of decreasing size and varying shapes. Some products, such as ferrous and aluminum alloy structural shapes, are produced by hot-rolling; others are cold-rolled.

Rope caulk. Preformed bead or "rope" of tacky caulking compound, often supplied with twine reinforcement to facilitate handling.

Rotary press (nip roller). Machine for making sandwich panels in which the panel materials are bonded together by running them between two rotating rollers.

Rough hardware. Bolts, anchors, screws, clips, etc., used to secure a window into its frame or opening.

Sag. Measure for the deflection of a material at elevated temperatures, of particular importance in base metals for porcelain enameling. It is commonly expressed in inches at a specified temperature.

Sample testing. Samples of finished panels are tested for stress rupture of the bond interface. The temperature at the surface tested is raised to 180°F and maintained at this temperature throughout the test. When the temperature has been reached, a static load sufficient to develop a stress of 10 lb/sq in. of bond area should be applied and maintained for 10 minutes. The sample should show no delamination or core rupture.

Sandwich panel. Panel made by laminating a core material, usually of low density, between sheets or "skins" of a material or materials of higher density and strength.

Sealant. Mastic or viscous material used to seal joints or openings against the passage of water or air.

Seismic expansion. Series of slip joints incorporated into an assembly to compensate for seismic movement.

Semiintermediate section. Some companies manufacture a frame and sash section known as a semiintermediate section. It is $1\frac{1}{8}$ in. in depth and weighs less than 2.79 lb/ft (linear).

Sheath-type curtain wall. Wall in which no structural elements are indicated.

Sheet metal closure. Formed metal member designed to cover the space between a window and adjacent construction.

Shelf life. Length of time that a packaged material, such as adhesives and sealants, can be stored under specified temperature conditions and still remain suitable for use.

Shop drawings. Detailed graphic representations made expressly to clarify items of construction as an aid to fabrication.

Slip joint. Fastening connection capable of movement, which allows for expansion and contraction.

Snap-on joint. Friction-type joint held by the clamping effect of members so designed as to cause one member to grip another.

Spandrel. Area of an exterior wall between two superimposed windows or openings.

Spandrel panel. Panel covering the spandrel area.

Spandrel-type curtain wall

1. Type of curtain wall that is designed to emphasize the horizontal lines of the structure.
2. Design in which spandrel panels are used to emphasize horizontal lines between window elements at each floor level.

Spandrel wall system. System used when columns are exposed to provide vertical accent and where horizontally placed windows are combined with spandrel panels. A metal stud curtain wall is installed between structural steel girts that support the windows and the wall panels. Tubular girt sections, instead of angles or channels, simplify installation of windows and wall panels.

Splice cover. Similar to mullion cover, except it is used to cover splices in materials.

Spline joints. Joints that allow flexibility for replacement, removal, and expansion. The disadvantage is the concentration of metal at one point, requiring insulation against thermal short circuit.

Split mullion. Vertical separation member composed of two or more interlocking shapes usually covered by separate closures.

Squareness. The squareness of a panel is governed by the type of facing specified. For commercial squareness, the length of either diagonal across the panel may vary $\pm \frac{1}{4}$ in. on a 5×18 ft panel.

Stack effect. Vertical design employing elements one above the other.

Stamping. Shaping of materials by impressing the desired conformation into the sheet material.

Starved joint. Adhesively bonded joint in which the amount of adhesive is insufficient to produce a satisfactory bond.

Steel grid system. Grid system designed to be made up of large prefabricated curtain wall modules instead of many field-assembled parts. Grids are all steel and factory welded. Steel has the strength to deliver greater wind load capacities and the ability to support larger glass areas than do lighter framing metals.

Stick system. System designed for high-rise applications for which a vertical accent is desired. The horizontal member includes a front cover and a stiffener. The cover makes possible the advantage of concealed and baffled drainage that prevents water from backing into the system. The vertical mullion design distributes the metal more efficiently and delivers more strength per pound of aluminum. The horizontals run continuously through the verticals, reducing cuts and shelf clips. Installation is further simplified through a design of bolts that fasten the inside and outside mullion components to each other.

Stretcher level. Flatten and level sheet metal stock by a mechanical stretching process.

Stretcher leveling

1. Flattening metal sheets by stretching them mechanically.
2. Removing deformations, such as warping and bow, from sheet materials by applying uniform tension at the ends of the sheet. A definite amount of permanent set is induced into the sheet, causing it to flatten considerably.

Structural clearance. Space required between the curtain wall construction and the structure.

Structural grid elements of extruded shapes. Connectors fix a panel to either the metal or masonry of a building's frame; transitional shapes connect a panel to ceilings, floors, or partitions and to corners, copings, sills, or interior trim; other shapes meet special architectural needs.

Stucco exterior-ledger-supported system. Metal stud walls are supported by ledger angles placed outside and secured to the structural framework of the building. The concept provides a broad expanse of smooth stucco surface. Vertical fins to accentuate appearance are constructed from studs and channels to suit job requirements. Windows are used as individual units or as part of a window wall assembly with spandrel panels included.

Styrene butadiene copolymer. Highly refractive colorless liquid.

Sunshade. Series of louvers or other devices that afford protection from sky glare.

Synthetic butyl tape. Polybutene extruded tape on crepe paper removable backings.

Tack weld. Weld for temporarily holding metal parts in position.

Theodolite. Instrument for measuring horizontal and vertical angles, used in curtain wall construction to establish elevations in reference to fixed bench marks.

Thermal conductivity. Measure of the amount of heat a material will conduct through itself, commonly expressed in British thermal units per hour per square foot of surface per degree Fahrenheit difference in the temperature of the surfaces per inch of thickness.

Thermal expansion. Increase in length of a material caused by heat, commonly expressed in inches of expansion per inch of original material size per range of temperature in degrees Fahrenheit.

Thermal limits. Useful temperature limits of a material, ordinarily assumed for metals as temperatures at which metal has lost 50% of its strength, commonly expressed in degrees Fahrenheit. Thermal limits affect creep and fatigue strength greatly.

Thermal separator or thermal break. Nonheat-conductive element used to separate conductive materials.

Thermal setting adhesive. Bonding agent that requires the application of heat to cause it to adhere to other materials.

Thermal stress. Internal force within a material due to temperature variations.

Tolerance. Permissible deviation from a nominal or specified dimension or value.

Tongue-and-groove joint. Very simple joint for erection of panel sections that speeds up installation. A principal disadvantage is the alignment of adjacent panels if smooth surfaces are required. Smooth surfaces accentuate the difference in surface planes.

Ultimate set. Final degree of firmness obtained by a plastic compound after cure, evaporation of volatiles, or surface polymerization.

U-value. Overall coefficient of heat transmission, air to air, through building materials, either singly or in combination; the time rate of heat flow expressed in British thermal units per hour per square foot per degree Fahrenheit temperature difference between air on the inside and outside surfaces.

Ventilator. Portion of a window that can be opened.

Vinyl plastics (copolymer resins). Polymers and resins derived by polymerization or copolymerization of vinyl monomers.

Wall system. The term "wall system," can be used in a technical or commercial sense. In a technical sense any wall system is either custom or standard. In a standard wall system of the architectural type the panels include formed framing members and are usually used with fenestration.

Wall system installation. Wall systems are classified by the system or method used to attach them to the building structure. Classical attachment systems include stick, unit, unit and mullion, panel, and column cover and spandrel.

Weatherstrip. Piece of metal, rubber, or other material used to seal joints between ventilator and frame.

Window wall

1. Nonstructural continuous facing consisting of combinations of windows, panels, and mullions filling the space between structural elements.
2. Nonstructural system of metal curtain wall in which the windows are the most prominent element.

WOOD DOORS

Accordion folding door. Door assembled from narrow wood strips or single wood slats, 5 in. wide with fabric, plastic, or metal hinges, resembling long, drapelike doors.

Acoustical door. Consists of an acoustical damping material, hardboard crossband, and hardboard facing material. To further prevent sound transmission, gasketed stops and neoprene bottom seals can be provided.

Active door. The operating door of a pair of doors; usually the one that has a locking device.

Adhesive. Adhesives used for doors are classified into two types; adhesives used for interior doors, and adhesives used for exterior doors. Exterior doors assembled use Type 1, fully waterproof bond. Interior doors assembled use Type II, water-resistant bond. Both comply with Commercial Standard CS-35.

Astragal. Molding attached to one of a pair of swinging doors, against which the other door strikes.

Batten door. Door made of sheathing, secured by strips of board, placed crossways, and nailed with clinched nails.

Bifold doors. Pair of doors, hinged at the longitudinal sides between the pair and at one jamb, with a track at the head carrying the free end of the pair with a nylon roller or support. Bifold doors are used primarily for closets.

Biparting door. Door that slides vertically and consists of two or more sections or pairs of sections that open away from each other and are so interconnected that two or more sections operate simultaneously.

Blueprint matching. Term used to describe matching of architectural-grade veneers. Matched for continuity of grain and color for various sizes of panels, doors, and transoms.

Bottom rail. Bottom crosspiece or horizontal piece of a door.

Bow. Flatwise deviation from a straight line drawn top to bottom; curvature along the length of the door.

Buck. Job terminology for a rough wood door frame.

Bumper sill. Special sill used in out-swing doors that utilizes a stop with a weatherstrip at the sill, thus preventing moisture penetration.

Butted frame. Door frame which fits against wall structures rather than around it. Frame jamb depth is normally equal or less than the wall thickness.

Casing. Mill work used as interior trim around door frames.

Classified types of doors. Type 1, fully waterproof bond for exterior and interior doors, Type II, water-resistant bond for interior doors only.

Closer reinforcement. Metal plate applied internally to a door frame or door to provide additional strength for the attachment of a door closer.

Combination door. Door with interchangeable storm sash and screen insert.

Composite doors. Primarily of a flush design, consisting of a manufactured core material with chemically impregnated wood edge banding and untreated wood face veneers, or laminated plastic faces, or surrounded by and encased in metal.

Composition core. Solid core of wood or other lignocellulose particles bonded together with a suitable binder, cured under heat, and pressed in a flat platen press.

Core material for custom grade. Grace AC exterior, rotary-cut Douglas fir plywood with no voids in the second ply or exterior-grade particle board meeting the physical property requirements set for economy grade.

Core material for economy grade. Grade AD group I, interior, rotary-cut Douglas fir plywood or particleboard manufactured to the following minimum standards: modulus of rupture, 3000 psi; screw holding power, 250 lb; and face density, 200 lb.

Core material for premium grade. Rotary-cut Philippine mahogany or other close-grain plywood with no voids in second ply or equivalent nontelegraphing material.

Core stock. Stock that is used to form the inner portion of a solid door. Core stock is covered on both sides by the type of wood paneling desired.

Corner seal. Unit used at connection of jamb and threshold to block passage of light and/or moisture.

Cross banding. Veneer used in the construction of flush doors placed between the core and face veneers with the direction of the grain at right angles to that of the face veneer.

Cup. Deviation from a straight line drawn from side to side; curvature along the width of the door.

Custom grade. Grade that includes all the requisites of a high-quality product and is suitable for all normal uses in high-grade construction in such buildings as better-class residences, schools, and commercial buildings.

Door frame

1. Finished mill work frame on which a door is supported with devices that allow it to swing or to slide.
2. Surrounding case into and out of which the door shuts and opens; supporting portion of door opening.

Door gradings

1. Grade 10, flush wood, particle core, $1\frac{3}{4}$ in. thick, wood panel door with minimum thickness of panel at $\frac{1}{2}$ in., including rebate, stiles having a minimum dimension of $1\frac{3}{4}$ in. by 6 in.
2. Grade 20, flush wood, particle core, $1\frac{3}{4}$ in. thick, lock block of dense wood at least 6 in. wide by 24 in. high. Hinge blocks, 6 in. wide by 12 in. high.

Door hanging. The swing of the door, determined by standing on the hinge pin side and noting on which side the hinges are located, either right-hand hinged or left-hand hinged.

Door type cores. Cores vary with the individual manufacturer and consist of cores such as grid hollow core of interlocking horizontal and vertical wood strips, honeycomb hollow core of expanded corrugated kraft fiber, particleboard core, inert mineral composition, edge-glued wood block, implanted blanks core, mesh core, and ladder core.

Double-acting door. Door equipped with hardware that permits it to swing to either side of the plane of its frame.

Double rabbet frame. Metal frame having recesses capable of receiving doors on both sides of a stop. Normally only one recess is provided for a door.

Dutch door. Door divided into two parts, each part separately hinged and locked. When used in residences at the kitchen, the lower part has a shelf for serving purposes.

Economy grade. Grade that establishes a standard for meeting the requirements of lower cost residential and commercial construction wherein economy is the principal factor.

Edge band. Strip of wood along the outside edges of two sides and the top and bottom of the door. It may be either a separate strip applied to the edges of the stiles or rails or it may be applied directly to the edges of the core.

Edge banding. Doors may be finished with or without edge banding. Edge-banded finished doors cannot be trimmed to fit on the job.

Edge margin. Perimeter margin around a prehung door that is actually the clearance between the door and the frame.

Entrance door. Closure of an opening for passage of persons that may be removed by sliding laterally, swinging in or out, or being raised above the opening and that may be again positioned as a closure.

Exterior door. Door in an entrance that provides access from the outside to the inside of a building.

Face panels. Plywood, hardboard, or plastic panels used for the face of the door.

Factory finishing

1. Sealing coats applied and the specified finish is to be completed at the job.
2. Complete finishing at the factory, requires prefit conditions and final premachining.

Filler. Material used in slab door construction that fills the area between the core framing.

Fire door. Four wood-type fire doors are presently manufactured and labeled, or UL classified as wood fire doors: the 20 min. door, $\frac{3}{4}$ hour "C" label, 1 hour "B" label, and the $1\frac{1}{2}$ hour "B" label.

Five-ply door. Door with two plies of veneer on each side of the core.

Flitch. Portion of a log sawed on two or more sides and intended for remanufacture into lumber or into sliced or sawed veneer. The term is also applied to the resulting sheets of veneer laid together in sequence of cutting.

Flitch matching. Premium-grade veneers from several flitches may be used for the plywood surface. If more than one flitch is used, grain and color of the several flitches may not be similar. Doors may not have similar grain or color.

Flush door

1. Door assembly consisting of a core and face panels of either cross banding and flat face veneers, plastic, or hardboard.
2. Door constructed from hollow or solid wood or composition cores supporting face panels of wood veneers, hardboard, or plastic laminates. It presents a flush, smooth appearance unless interrupted by glass lights or louver openings.

Flush door (slab door). Door of composite construction having flush surfaces. It may be prepared to take glass.

Framed-construction hollow door. Door utilizing a mesh or cellular-type core and $1\frac{3}{8}$-in. stiles, $2\frac{7}{8}$-in. rails, and $2\frac{7}{8} \times 21$ in. solid wood lock blocks, with stiles running full vertical height.

Framed-construction solid door. Door using a mineral core consisting of expanded Perlite particles, mineral binders, and fibers and $1\frac{3}{8}$-in. stiles, $2\frac{7}{8}$ in. rails, and $2\frac{7}{8} \times 21$ in. solid wood lock blocks, with stiles running full vertical height.

French door. Door consisting entirely or in large part of divided glass openings.

Glass and louver beads. Beads that hold glass lights in doors should be of wood to match the door finish. Beads that hold the louver in the door can be of metal to match a metal louver, or of wood to match the wood louver.

Glass door. Solid or veneered door having one or more openings for glass.

Glazed door (sash). Door that is similar in construction and appearance to a panel door, except that one or more panels are replaced with glass. Completely glazed doors without panels are casement (French) doors.

Grid core. Type of core construction made up of strips of a suitable material arranged so as to form a mesh, lattice, or grid throughout the core area.

Grounded doors. Grounding is required in certain electrical, radio, and television installations. Recommended grounding can consist of wire mesh located at center of the core, grounded with copper wire through the hinges to the frame, and the frame grounded.

Guarantees. Variety of guarantees offered by most manufacturers vary from one year to lifetime depending on the construction of the door and its uses. The National Woodwork Manufacturers Association has issued standard NWMA Guarantees.

Head jamb. Cross or horizontal jamb member forming the top of the frame.

High density overlays. Recommended by manufacturers for exterior doors. It is an overlay of a phenolic resin cellulose fiber applied to the faces of the door. The overlay is applied under heat and pressure to the hardwood veneer face. Some manufacturers will issue lifetime guarantees for doors with this overlay.

High pressure plastic laminated door. Plastic laminated flush doors of solid or hollow core-type doors. The plastic faces should conform to Type 1 or Type 4 of the National Electrical Manufacturers Association NEMA standard LP 2.

Hollow core

1. Core assembly of strips or other units of wood, wood derivatives, or insulation board, with intervening hollow cells or spaces. Core assembly supports the outer faces.
2. Type of slab door construction in which the core filler consists of a spaced ribbing or grid fabricated to support the facing surfaces.

Implanted core. Hollow core composed of a series of blanks, forms, tubular columns, or shapes of wood, wood derivatives, or insulation board, which may or may not be joined together, implanted between and supporting the outer faces of the doors. Air cells or spaces are between the blanks, but having a density of not less than 15 lb/cu ft.

Inactive door. Part of a pair of doors that does not have the locking hardware, but does have top and bottom bolts to hold the door in a closed position.

Interior door. Any door other than a door that is designed as an exterior door.

Interior door jamb. Surrounding case into and out of which a door closes and opens. It consists of two upright pieces, called "side jambs," and a horizontal head jamb.

Jamb

1. Side of an aperture, door, or window frame.
2. Sidepiece or post of an opening; sometimes, door frame.

Kiln-dried lumber. Lumber dried in a closed chamber in which the removal of moisture is controlled by artificial heat and usually by relative humidity.

Ladder core. Hollow core composed of strips of wood, wood derivatives, or insulation board, with the strips running either horizontally or vertically throughout the core area and with air cells and/or spaces between the strips.

Lead core door. Framed construction utilizing a flakeboard or particle-board core of 40- to 50-lb density and with $2\frac{7}{8}$-in. rails and $1\frac{3}{8}$-in. stiles, the stiles running the full vertical height. A lead lining $\frac{1}{16}$ or $\frac{1}{8}$ in. thick, depending on the degree of protection required, is available.

Lead-lined doors. Verify latest developments on lead-lined doors with Underwriters Laboratories, Inc. standards. Optional location within door construction for installation of $\frac{1}{32}$ in. to a maximum of $\frac{1}{2}$ in. continuous lead sheet from edge to edge to cover the entire area of the door, which may be reinforced to support the load with lead bolts or approved adhesive.

Leaf. An individual door used either singly or in multiples.

Light and louver openings. Installed no closer in any case than 5 in. between opening and edge of door. In hollow core doors, it is recommended that the cutout area maximum be not over one-half of the door height. For exterior doors, weatherproofing is required to prevent moisture from penetrating the core material.

Lock block

1. Block of wood (solid or made by gluing up blocks) the thickness of the door stile, which is attached to the inside edge of the stile and into which the lock is fitted.

2. Piece of solid material in hollow-case door construction providing for installation of lock and latch sets.

Louver door. Door composed of a stile and rail frame with integral louver construction, mortised into stiles or vertical dividing bars.

Louvered cafe doors. Doors $1\frac{1}{8}$ in. thick, usually with scalloped tops. Doors are built solid or with mesh or cloth fillers, and are usually about 39 in. in height, hung on double-acting hinges.

Louvered door. Door with one or more sections of angled slats providing openings for ventilation.

Mesh or cellular core. Hollow core composed of strips of wood, wood derivatives, or insulation board, interlocked and running horizontally, vertically, or diagonally throughout the core area, forming air cells and/or spaces between the strips and supporting the outer faces.

Mineral composition core. The core consists of a mineral composition material of calcium silicate with asbestos fibers. The core has very poor screw holding ability and should include solid blocking to receive hardware.

Moisture content. All wood used to fabricate doors should be kiln dried, having a moisture content between 6 and 12%.

Opening size. Size of a door frame opening, measured horizontally between jamb rabbets and vertically between the head rabbet and the finished floor. The opening size is usually the nominal door size and is equal to the actual door size plus the recommended clearances and threshold height.

Panel door

1. Door constructed of stiles, rails, and one or more panels of wood or glass.

2. Door assembled from stiles and rails (vertical and horizontal components) that frame and support one or more panels. It has a characteristic paneled appearance resulting from raised and recessed areas, patterned molds, etc.

Particleboard solid core door. Particleboard cores are engineered-laminated wood panels consisting of predetermined size wood chips bonded together under heat and pressure with synthetic resins. Core should meet the requirements of ASTM D 1037.

Phenolic facing finish. High or medium low-density overlay of phenolic resins and cellulose fibers fused to inner faces of hardwood in lieu of final veneers as base for final opaque finish only.

Plank door. Door constructed of a number of solid individual planks joined vertically and assembled with horizontal blind battens or splines. It may be prepared to take glass.

Plastic finish. $\frac{1}{16}$ in. minimum plastic sheet bonded to $\frac{1}{16}$ in. wood backing of two or more plies, or $\frac{1}{8}$ in. hardboard, smooth on one or two sides.

Prefit doors. Doors usually $\frac{3}{16}$ in. less in width, and $\frac{1}{8}$ in. less in height than nominal size, $\pm\frac{1}{32}$ in. tolerance, with vertical edges eased.

Premachining. Doors factory mortised for locks and cut out for hinges (when so specified).

Premium grade. Premium grade entails superior-quality workmanship and materials with corresponding increase in cost.

Rabbet. The recess or offset formed in the frame to receive the door.

Rail. The horizontal structural member forming the top and bottom edges of a door.

Reveal. The distance from the face of a door to the face of the frame on the pivot side.

Screens. Screen doors, blinds, and shutters (finished carpentry) usually include metal screening and miscellaneous hardware.

Sequence matching. Architectural-grade veneers are matched for color, and all panels of the same size should have continuity of grain. Other size panels must be cut during installation, which may interrupt grain continuity. Doors are made from veneers of a similar color but do not have continuity of grain.

Seven-ply door. Door having three plies of veneer on each side of the core.

Shim. Wood shingle piece placed between the back of the wood frame and the adjacent wall, allowing the frame to be nailed in place after it has been plumbed and leveled.

Shutter. Louvered, flush wood, or nonwood frames in the form of doors located at each side of a window. Some are made to close over the window for protection; others are fastened to the wall as a decorative device.

Side jambs. The upright or vertical jamb members forming the sides of the frame.

Sidelight. Fixed glazed frame adjacent to the door frame on one or both sides of a door.

Sill. Lower boundary of a door.

Solid core

1. Innermost layer in veneered door construction.

2. Term pertaining to slab door construction wherein solid materials extends the full width and height of the door.

Solid wood flake door. Door with a 30-lb density core that complies with U.S. Department of Commerce Standard CS-236 and National Woodwork Manufacturers Association interim standards.

Solid wood staved core door. Door with core of kiln-dried, low-density, finger-jointed blocks or edge- and end-glued blocks bonded together with end joints staggered in adjacent rows and joined with full-length edge banding of clear lumber bonded to all four edges.

Sound insulating door. Doors produced to meet the requirements of ASTM E 90 established at Riverbank Acoustical Laboratory. Doors are furnished complete with special door stops at the door frame, gaskets, and automatic threshold closing devices.

Square edge door. Door having vertical edges that are 90° perpendicular to the horizontal edges.

Stiles. Upright or vertical pieces of the framework of a door or the jamb in which it is hung.

Storm and screen door. Lighter, thinner stile and rail construction supporting screening (screen doors), glass panels (storm doors), or interchangeable screen and storm panel inserts (combination doors).

Strike jamb. Vertical member of the door frame prepared for the installation of the lock strike.

Sweep strip. Weatherstrip or sound insulation strip at the bottom of a door.

Threshold. Beveled piece over which the door swings. It acts as a weather stop at exterior doors.

Tolerance standards. Refer to NWMA Standards for requirements applying to height, width, thickness, squareness, and warp. Standards vary for solid and built-up construction.

Top rail. Top crosspiece or horizontal piece of a door.

Transom. Fixed or operable solid or glazed frame above a door.

Twist. Deviation in which one or more corners of the door is out of the plane of the other corners of the door.

Veneer cutting methods. Face veneers for doors are cut in several ways: plain or flat slicing, quarter slicing, rotary cutting, half round cutting, and $\frac{1}{8}$ in. and $\frac{1}{4}$ in. sawed veneers.

Vision panels. Allowed if the vision panel meets the requirements of the local codes. Location may decide the type of glass used. Wired glass, tempered glass, or laminated glass may be used, installed as per code.

Warp. Distortion in the plane of the door itself and not in relationship to the frame or jamb in which it is to be hung.

Weatherstripped doors. Doors for locations requiring maximum protection against weather and for the retention of heated or cooled air. These doors also minimize the infiltration of wind, water, sound, and dust.

Wood blocks or strips-solid core door. Cores are manufactured in two types: the glued-up core or the frame block core. The glued-up core is recommended. It consists of blocks only of one species of wood that are of varying lengths and well staggered (like flooring is laid), not more than $2\frac{1}{2}$ in. wide, and there should be no voids left between the blocks. Recommended lumber for the core includes white pine, cedar, or basswood.

Wood facings. Standard thickness face veneers should be in the range of $\frac{1}{16}$ in. to $\frac{1}{32}$ in., bonded to hardwood,

crossband in the range of $\frac{1}{10}$ in. to $\frac{1}{16}$ in. $\frac{1}{8}$ in. sawn veneers, bonded to crossband, are easily repaired and refinished.

Wood panel doors. Constructed of West Coast hemlock, Sitka spruce, or equal domestic woods. All joints use mortise and tenon construction, are glued with water-resistant casein glue and steel pinned under pressure.

Wood flush door. Door assembly consisting of a core and face panels of either cross bounding and flat face veneers, plastic, or hardboard.

WOOD WINDOWS

Apron. Flat part of the inside trim of a window, placed against the wall directly beneath the stool of the window.

Awning windows. Windows that have one or more top-hinged out-swinging sash. Single awning sash is often combined with fixed and other types of sash into larger window units. Several sashes may be stacked vertically and may close on themselves or on meeting rails that separate the individual sash. When awning windows have sliding friction hinges, which move the top rail down as the sash swings out, they have "projected action." The ventilating area is considered to be 100% of the operating sash area.

Bar

1. Either a vertical or horizontal member that extends the full width or length of the glass opening.

2. Horizontal or vertical member that divides glass openings and that extends the full width or height between stiles or rails. A filler or muntin bar is a short bar intersecting horizontal or vertical bars.

Basement windows. Basement windows are generally single sash units of simplified design. Sash may be of the awning, hopper, or top-hinged in-swinging type; 100% of the window is available for ventilation.

Bay window. A rectangular, curved, or polygonal window, supported on a foundation extending beyond the main wall of the building.

Blind stop

1. Thin strip of wood machined so as to fit the exterior vertical edge of the pulley stile or jamb and keep the sash in place.

2. Rectangular molding, usually $\frac{3}{4} \times 1\frac{3}{8}$ in. or more in width, used in the assembly of a window frame. It serves as a stop for storm and screen or combination windows and to resist air infiltration.

Casement. Window sash that opens on hinges fastened to a vertical side of the frame. Windows with such a sash are called "casement windows."

Casement frames and sash. Frames of wood or metal enclosing part or all of the sash, which may be opened by means of hinges affixed to the vertical edges.

Casement windows. Casement windows have a side-hinged sash, mounted to swing outward. They may contain one or two operating sashes and sometimes a fixed light between the pair of sashes. When fixed lights are used, a pair of casements may close on a mullion or against themselves, providing an unobstructed view when open. An operating sash may be opened 100% for ventilation.

Casing. Molding of various widths and thicknesses used to trim window openings.

Center-hung sash. Sash hung on its center so that it swings on a horizontal axis.

Check rails. Meeting rails sufficiently thicker than the window to fill the opening between the top and bottom sash made by the check strip or parting strip in the frame. They are usually beveled and rabbeted.

Clerestory. Part of a building rising clear of the roofs or other parts; its walls contain windows for lighting the interior of the building.

Combination windows. Combination windows are used over regular openings and provide winter insulation and summer protection. They often have self-storing or removable glass and screen inserts.

Dormer window. Substantially vertical window and its enclosing structure erected as an appendage to a sloping roof.

Double-hung windows

1. Windows that have two operating sashes (single-hung windows have only the lower sash operative). The sashes move vertically within the window frame and are maintained in the desired position by friction fit against the frame or with balancing devices. Balancing devices also assist in raising the sash. Fifty percent of the window area is available for ventilation.
2. Window having top and bottom sash, each balanced by springs or weights to be capable of vertical movement in its own grooves with relatively little effort.

Double window. Window arranged with a double sash, enclosing an airspace intended to act as a sound and heat insulator.

Drip cap. Molding placed above the exterior of a door or window causing water to drip beyond the frame.

Extension blind stop. Molded piece, usually of the same thickness as the blind stop and tongued on one edge to engage a plow in the back edge of the blind stop, thus increasing its width and improving the weathertightness of the frame.

Face measure. Distance across the face of a wood part, exclusive of any solid mold or rabbet.

Fenestration. Any opening or arrangement of openings for the admission of daylight and ventilation.

Finished size. Overall dimension of a wood part, including the solid mold or rabbet.

Fixed sash. Sash permanently fixed in a solid frame.

Fixed windows. Windows usually consisting of a frame and a stationary sash, glazed, and often flanked with double-hung and casement windows or stacked with awning and hopper units to make up windows of custom design. To keep the sight lines (width or height of view area) consistent, fixed sash members are made to the same appropriate cross-sectional dimensions as adjacent operating sash.

Frames and sash. Material requirements for frames and sash are good to high decay resistance, good paint holding, moderate shrinkage, freedom from warping, good nail holding, and ease of working. The best woods are cypress, cedar, redwood, northern and Idaho white pine, ponderosa pine, and sugar pine. The majority of door and window frames and sashes are treated with water-repellent preservative at the time of manufacture. Decay-resistant species of wood are used for basement frames and sashes where resistance to moisture and decay is important. Under severe moisture conditions, pressure-treated material is desirable.

French window. Glazed casement serving as both window and door.

Full bound. Term applied to window sash having a similar width of wood in the stiles and top and bottom rails, usually described as "same rail all around."

Glazing. Glazing is usually done with a proven glazing compound, of knife consistency, and chocolate brown or gray in color. The glass is bedded with the same material. At the bottom of the fixed glass in the frame or sash, a back bed and pin is placed every 8 in. with $\frac{7}{8}$ in. pure zinc points. No bottom outside bead or compound is necessary on a fixed sash.

Glazing bead. Small molding fastened onto a sash or door rebate to hold glass in its opening.

Hardware for pulley-type frames. Pulleys are installed at the factory and closed faced with bronze face plates. Wheels are sized to center weights in the pockets and have steel axles, oilite wood bushing, and wheels for the chains. The chains are copper plated over galvanized steel, usual size No. 30. Pendulum strips of sheet galvanized iron are provided. Weight pockets are conveniently located, and pulley arrangements are designed to accommodate practical sizes of iron or lead weights and result in smooth operation.

Head jamb. Horizontal member forming the top of the frame.

Hopper windows. Hopper windows have one or more bottom-hinged, in-swinging sash. A hopper sash is similar in design and operation to the awning type and may actually be an inverted awning sash with minor hardware and weather-stripping modifications. For this reason, windows with a hopper sash are sometimes referred to as "awning windows." An operating sash provides 100% ventilating area.

Horizontal sliding windows. Horizontal sliding windows have two or more sashes of which at least one moves horizontally within the window frame. In a three-sash design, the middle sash is usually fixed; in two-sash units, one or both sashes may be ventilating. This type of window sometimes is increased in size to door proportions. The ventilating area is 50% of the window area in most designs.

Jalousie. Adjustable glass louver. Also refers to doors or windows containing jalousies.

Jamb. Part of a frame that surrounds and contacts the window or sash the frame is intended to support.

Jamb liner. Small strip of wood, either surfaced on four sides or tongued on one edge, which, when applied to the inside edge of a jamb, increases its width for use in thicker walls.

Light (Lite). Space in a window sash for a single pane of glass; also, pane of glass.

Margin lights. Narrow lights of glass near the edges of a sash.

Measurements between glass. Distance across the face of any wood part that separates two sheets of glass.

Meeting rail. Bottom rail of the upper sash of a double-hung window. Sometimes called the "check rail."

Meeting rails. Rails of a pair of sash that meet when the sash are closed.

Mullion

1. Construction between the windows on a frame that holds two or more windows.

2. Vertical bar or divider in the frame between windows, doors, or other openings.

Muntin

1. Short light bar, either vertical or horizontal.
2. Small member that divides the glass or openings of sash or doors.
3. Vertical or horizontal member between two panels of the same piece of panel work; vertical or horizontal sash bars separating the different panels of glass.

Nails. Nails exposed to the weather that cannot be set (such as those holding insulating panels of cement board or metal) should be monel. All other nails or fasteners on the exterior should be hot-dipped galvanized or of non-staining metal.

Oriel bay window. Extension of a room suspended or projected beyond the general wall line of an enclosed structure in one or more vertically aligned spaces but not extending to the foundations, whereby floor space is gained in the room.

Painted frames. The following woods are listed in the approximate order of their ability to stay in place and adaptability: redwood, African mahogany, northern white pine, Idaho pine, sugar pine, vertical grain fir, ponderosa pine, Sitka spruce, and southern yellow pine.

Parting stop or strip

1. Thin strip of wood let into the jamb of a window frame to separate the sash.
2. Small wood piece used in the side and head jambs of double-hung windows to separate the upper and lower sash.

Prefabricated frame

1. A prefabricated frame includes the head, jambs, and sill. The wood parts are treated with a toxic water-repellent preservative, and exterior surfaces are primed. The jamb liner sash slide is of high-impact polyvinyl chloride backed by continuous hard-temper aluminum springs.
2. Group of wood parts so machined and assembled as to form an enclosure and support for a window or sash.

Prefabricated window hardware. Galvanized steel spiral spring-type balances are connected to the sash with 300-lb test nylon cord. Electrogalvanized steel balance hardware is concealed within the head frame. A self-aligning recessed sash lock and strike are factory installed. The sash lock, strike, and lifts are furnished in baked bronze enamel.

Pulley stile

1. Side jamb into which a pulley is fixed and along which the sash slides.
2. Member of a window frame that contains the pulleys and between which the edges of the sash slide.

Putty

1. Soft, pliable type of cement.
2. Type of cement usually made of whiting and boiled linseed oil, beaten or kneaded to the consistency of dough, and used in sealing glass in sash and filling small holes and crevices in wood and for similar purposes.

Rabbeted jamb. Jamb with a rectangular groove along one or both edges to receive a window or sash.

Rails. Crosspieces or horizontal pieces of the framework of a sash or screen.

Removable muntins. Aluminum muntin bars of hexagonal cross section. Joints are secured with $\frac{1}{4} \times 4$ in. steel pins. Muntin assemblies snap into the sash by means of concealed fittings. They are available in horizontal, vertical, or diamond arrangements.

Sash

1. Single assembly of stiles and rails made into a frame for holding glass, with or without dividing bars, and supplied either open or glazed.
2. Framework that holds the glass in a window.

Sash balance. Device, usually operated by a spring or tensioned weather-stripping, designed to counterbalance double-hung window sash.

Sash-balance-type hardware. Where narrow mullions are desired, Unique sash balances such as Caldwell or Pullman tape types are recommended.

Sash fabrication. Sash are $1\frac{3}{4}$ in. thick and treated with toxic water-repellent preservative. All exterior surfaces are primed. Corners are mortised and tenoned, glued, and nailed; the edges of stiles are factory painted.

Sash lift. Hardware installed on the bottom rail of the operating window to facilitate vertical lifting.

Sash lock. Locking device installed on the meeting rails.

Sash woods. The following woods are adaptable to sash manufacture: African mahogany, sugar pine, Idaho pine, northern white pine, ponderosa pine, redwood, vertical grain fir and Sitka spruce.

Side jamb. Upright member forming the vertical side of the frame.

Sill. Horizontal member forming the bottom of the frame.

Spring-pin sash lock. Locking device installed in the vertical rail of a sash. The pin engages the window jamb through the sash.

Stained frames. The clear heart of redwood is desirable where natural or stained finished wood is used. When a door opening adjacent to a window is subjected to heavy traffic or abrasive action, the following woods are recommended: white oak, African mahogany, and vertical grain fir; these woods may be stained to match redwood.

Stiles. Upright or vertical outside pieces of a sash or screen.

Stool

1. Flat, narrow shelf that forms the top member of the interior trim at the bottom of a window.
2. Flat molding fitted over the window sill between jambs and contacting the bottom rail of the lower sash.

Storm sash or storm window. Extra window usually placed on the outside of an existing one as additional protection against cold weather.

Water bar. Bar set in the joint between the wood and masonry sill of a window to prevent penetration of water.

Water-repellent pentachlorophenol preservative treatment. Exterior windows, frames, and the outside finish of all species of wood are treated with nonswelling water-repellent paintable preservative according to the 3-minute immersion process defined by the National Woodwork Manufacturers Association.

Weatherstrip. Narrow strips of a combination of materials or metal, installed around doors and windows to retard passage of air, water, moisture, or dust.

Weatherstripping. Spring-type weather stripping makes tight contact at the check rail and sill. Polyvinyl chloride bulb weather stripping at the head engages the aluminum extruded frame lip. Silicone-treated woven pile strips set into sash stiles seal at vinyl jamb liners.

Window. One or more single sash made to fill a given opening, supplied either open or glazed.

Window unit. Combination of window frame, window, weather strip, balancing device, and at the option of the manufacturer, screen and storm sash, assembled as a complete and operating unit.

Wood moisture content. Exposed portions of frames, wood dried to moisture content averaging from 7% to a maximum of 15% depending on location of area.

Wood allowance. Difference between the outside opening and the total glass measurement of a given window or sash.

Wood trim treatment. Wood parts are treated with toxic water-repellent preservative; the exterior surfaces are primed. Interior trim; mullion covers, apron, stool, and casings, prepared for finish.

Wood windows. The woods commonly used to resist shrinking and warping for the exposed parts of windows are white pine, sugar pine, ponderosa pine, fir, redwood, cedar, and cypress. Hard pine or some relatively hard wood is used for stiles against which double-hung windows slide. The parts of a window exposed to the inside are usually treated with trim and made of the same material.

9

Finishes

ACOUSTICS AND SOUND CONTROL

Absorbers. Materials that have the capacity to absorb sound, such as acoustical tile and panels, carpeting, draperies, and upholstered furniture. People absorb sound.

Absorption

1. Ability of a material to absorb rather than reflect the sound waves striking it by converting the sound energy to heat energy within the material.

2. Taking up and holding or dissipating of matter or energy in the same way as a sponge takes up water.

Absorption coefficient

1. Ratio of the sound-absorbing effectiveness (at a specific frequency) of 1 sq ft of a material to 1 sq ft of perfectly absorptive material, usually expressed as a decimal value (e.g., 0.70) or in percent.

2. Measure of the efficiency of an absorbing material or method. It is the ratio of the sound energy absorbed to the incident sound energy.

Absorption conversion. Conversion of acoustic energy to heat or another form of energy within the structure of sound-absorbing materials.

Absorptive. Porous materials such as acoustical plaster, acoustical tile, woods, carpets, draperies, furniture, and people are primarily absorptive.

Acoustic

1. Adjective used in conjunction with a basic property of sound, for example, acoustic energy.

2. Containing, producing, arising from, actuated by, related to, or associated with sound. "Acoustic" is used when the term being qualified designates something that has the properties, dimensions, or physical characteristics associated with sound; "acoustical" is used when the term being qualified does not designate explicitly something that has such properties, directions, or physical characteristics.

Acoustical. Adjective used in conjunction with apparatus, design, or other nouns or verbs connected with sound control, e.g., acoustical tile, acoustical analysis.

Acoustical analysis. Detailed study of the use of a structure, the location and orientation of its spaces, and a determination of noise sources and the desirable acoustical environment in each usable area.

Acoustical array. Sound-transmitting or receiving system with elements arranged to give desired directional characteristics.

Acoustical correction. Special planning, shaping, and equipping of a space to establish the best possible hearing conditions for faithful reproduction of wanted sound within the space.

Acoustical "dead" environment. Condition occurring in rooms containing a large amount of absorption. There will be little reflected sound, resulting in minimum reverberation; thus the sound will die out very quickly, and the room is considered dead.

Acoustical Doppler effect. Change in pitch of a sound observed when there is relative motion between source and observer.

Acoustical double glass. Two monolithic glass panels set into a frame, with an airspace between the two panels that is usually larger than 1 in. and generally not hermetically sealed.

Acoustical environment. All of the factors, interior or exterior, that affect the acoustic conditions of the location, space, or structure under consideration.

Acoustical fiberboard. Insulating fiberboard, not less than 1 in. thick, matte surfaced, perforated, slotted, or of other surface finish providing the desired acoustical absorption, painted or unpainted. Formboard is treated to provide effective resistance to attack from mildew, rot, and termites.

Acoustical holography. Technique for using sound to form visible images, where acoustic beams form an interference pattern of an object, a beam of light interacts with this pattern and is focused to form an optical image.

Acoustical impedance. Resistance to the flow of acoustical energy, measured in rayls at specific frequencies. Acoustical impedance is affected by density and fiber diameter. It is tested in accordance with ASTM C 384.

Acoustical "live" environment. In rooms lacking adequate absorptive surfaces, the sound will keep bouncing and die out very slowly, thus increasing the reverberation time.

Acoustical material. Material that is manufactured or natural considered in terms of its acoustical properties. A material that absorbs sound.

Acoustical stiffness. Acoustic reactance associated with the potential energy of a medium multiplied by 2π times the sound frequency.

Acoustical tile. Acoustical absorbents produced in the form of sheets or units resembling tiles. They are usually 12×12 in. or multiples thereof.

Acoustical treatment. Use of acoustical absorbents, acoustical isolation, or any changes or additions to the structure to correct acoustical faults or improve the acoustical environment.

Acoustic center. Center of the spherical sound waves radiating outward from an acoustic transducer.

Acoustic clarifier. System of cones loosely attached to the baffle of a loudspeaker and designed to vibrate and absorb energy in order to suppress sudden loud sounds.

Acoustic comfort index. Scale designed to indicate the noise inside passenger cabin of an aircraft; on this scale, +100 represents ideal conditions or zero noise, 0 represents barely tolerable conditions, and −100 represents intolerable conditions.

Acoustic compliance. The reciprocal of acoustic stiffness. Also known as "acoustic capacitance."

Acoustic dispersion. Complex sound wave's separation into its frequency components as it passes through a medium; usually measured by the rate of change of velocity with frequency.

Acoustic domain. Concentration of crystal lattice vibrations traveling at the speed of sound; used to generate light from an array of pn junctions.

Acoustic doublet. Double sound source, such as the front and rear surfaces of a diaphram or loudspeaker cone.

Acoustic energy. Total energy of a given part of the transmitting medium minus the energy that would exist in the same part of the medium with no sound waves present; energy added as a result of sound vibrations.

Acoustic feedback. Transfer of physical vibrations from a loudspeaker to other apparatus in the reproducing chain in such a way that spurious electrical signals are fed back to the loudspeaker.

Acoustic generator. Transducer that converts electrical, mechanical, or other forms of energy into sound.

Acoustic grating. Series of rods or other suitable objects of equal size placed in a row a fixed distance apart; causes sounds with different wavelengths to be diffracted in different directions.

Acoustic homing. Following of a path of acoustic energy to or toward its source or point of reflection.

Acoustic image. Geometric space figure that is made up of the acoustic foci of an acoustic lens, mirror, or other acoustic optical system and that is the acoustic counterpart of an extended source of sound.

Acoustic impedance. Complex ratio of the sound pressure on a given surface to the sound flux through that surface, expressed in acoustic ohms.

Acoustic interference. Combined action of two waves moving simultaneously through the same region. Two waves of the same frequency in phase with each other and moving in the same direction produce reinforcement. Two waves of the same frequency in phase opposition and moving in the same direction produce destructive interference. If they also have equal amplitudes, the result is a complete annulment.

Acoustic mass reactance. That part of the acoustic reactance associated with kinetic energy of a medium.

Acoustic measurement. Process of quantitatively determining one or more properties of sound.

Acoustic noise. Undesired sound. The frequencies involved include at least the band from 15 to 20,000 Hz.

Acoustic ohm. Unit of acoustic impedance, also known as "acoustic reactance unit" or "acoustic resistance unit."

Acoustic plaster. Plaster having good acoustic absorbing properties.

Acoustic pressure. Instantaneous pressure at a point as a result of the sound vibration minus the static pressure at that point; change in pressure resulting from sound vibration.

Acoustic radiometer. Instrument for measuring sound intensity by determining the undirectional steady state pressure caused by reflection or absorption of a sound wave at a boundary.

Acoustic ratio. Ratio of the intensity of sound radiated directly from a source to the intensity of sound reverberating from walls of an enclosure at a given point in the enclosure.

Acoustic insulation. Material used to diminish sound energy that passes through it or strikes its surface.

Acoustic interferometer. Device for measuring by an interference method the velocity and attenuation of sound waves in a gas or liquid.

Acoustic labyrinth. Special baffle arrangement used with a loudspeaker to prevent cavity resonance and to reinforce bass response.

Acoustic lens. Selected materials to refract sound waves in accordance with the principles of geometrical optics, as is done for light.

Acoustic levitation. Use of a very intense sound wave to keep a body suspended above a device producing the sound wave.

Acoustic reactance. Imaginary component of the acoustic impedance.

Acoustic receiver. Complete equipment required for receiving modulated radio waves and converting them into sound.

Acoustic refraction. Variation of the direction of sound transmission due to spatial variation of the wave velocity in the medium.

Acoustic resistance. The real component of the acoustic impedance.

Acoustic resonator. Enclosure that produces sound-wave resonance at a particular frequency.

Acoustics. Science of sound, including its production, transmission, and effects.

Acoustic scattering. Irregular reflection, refraction, and diffraction of sound in many directions.

Acoustic shielding. Sound barrier that prevents the transmission of acoustic energy.

Acoustic spectrometer. Instrument that measures the intensities of the various frequency components of a complex sound wave.

Acoustic spectrum. Range of acoustic frequencies, extending from subsonic to ultrasonic frequencies, that is approximately from zero to at least one megahertz.

Acoustic stiffness. Product of the angular frequency and the acoustic stiffness reactance.

Acoustic tile. Plain or decorative tile, available in various sizes, with sound-absorbing properties, used primarily to cover wall and ceiling areas.

Acoustic transformer. Device, such as a horn or megaphone, for increasing the efficiency of sound radiation.

Acoustic treatment. Application of absorbent or reflecting material to the walls, floor, or ceiling of a room to alter its acoustic properties.

Acoustic velocity. Speed of sound or similar pressure waves through the atmosphere.

Acoustic wave velocity. Vector quantity that specifies the speed and direction with which a sound wave travels through a medium. Sonic speed is sometimes used to describe the speed of a body when it is equal to the acoustic wave velocity in the medium in which the body is moving.

Airborne noise. Noise produced by a source that radiates sound directly into the air, such as people talking, radio or television playing, or a dog barking.

Airborne sound

1. Sound that reaches the point of interest by propagation through air (ASTM E 90).
2. Sound produced by vibrating sources that radiate sound directly into the air. It is transmitted through air as a medium rather than through solids or the structure of the building.

Airborne sound transmission. Sound transmitted when a surface is set into vibration by the alternating air pressures of incident sound waves.

Air column. Airspace enclosed on at least five sides that produces specific sound wave characteristics. A typical example is the pipes of a pipe organ, each tuned by virtue of size and length to a specific pitch.

Air conditioning. Process of delivering sound-controlled air simultaneously with its temperature, humidity, cleanliness, and distribution to meet the comfort requirements of the occupants of the conditioned space.

Ambient. Existing surrounding conditions.

Ambient noise. Noise associated with a given environment, usually a composite of sounds from many sources near and far.

Ambient or background noise. Total of all other noise in a system or situation, independent of the presence of the desired signal.

Ambient sound. Continuous existing sound level (background noise) in a room or space, which is a composite of sounds from both exterior and interior sources, none of which generally are indentifiable individually by the listener.

Amplitude

1. Peak or maximum value of a vibration or wave motion.
2. Maximum displacement of the conductor molecules during the passing of a sound wave.
3. Amplitude of the wave represents the intensity of the sound pressure produced by the energy source. Amplitude is the maximum displacement to either side of the normal or "rest" position of the molecules, atoms, or particles of the medium transmitting the vibration.

Anechoic. Without echo. An anechoic chamber is a chamber or room where the walls are lined with a material that completely absorbs sound.

Anechoic chamber or room. Free-field acoustic testing environment in which all the sound emanating from a source is essentially absorbed in the walls or materials. Free from all reflections, rated as an "echoless" or acoustically "dead" room.

Antinode. Point, line, or surface in a standing wave system at which the amplitude is a maximum, such as an antinode of pressure or an antinode of particle velocity.

Architectural acoustics. Acoustics of buildings and structures.

A-scale sound level. Quantity in decibels, read from a standard sound level meter switched to the weighting scale labeled A. The A-scale discriminates against the lower frequencies, as does the human ear at moderate sound levels. It measures the relative noise and annoyance levels of many common sounds.

Attenuation. Reduction of sound-pressure level, usually expressed in decibels.

Audible. Capable of producing the sensation of hearing.

Audible sound. Sound containing frequency components lying between about 15 and 20,000 Hz with sufficient sound pressure to be heard.

Audio. Pertaining to frequencies of audible sound waves between about 15 and 20,000 Hz.

Audiofrequency. Rate of oscillation corresponding to that of sound audible to the human ear, approximately 15 to 20,000 Hz.

Auditory fatigue. Moderately loud noises create fatigue in human beings and reduce their output and efficiency.

Average deviation in pressure. The average deviation in pressure above or below the static value is called "sound pressure," which is proportional to the intensity and related to the loudness of the sound.

Average transmission loss (TL). The numerical average of the transmission loss values of a construction measured at nine frequencies. It is a single number rating for comparing the airborne sound transmission through walls and floors.

A-weighted sound level. Sound level in decibels using a frequency filter similar to human hearing.

Background level. Normal sound level present in a space, above which speech, music, or similar specific wanted sound must be presented.

Background noise. Normal sound always present in a space, created either by outdoor sounds, such as street traffic, or indoor sounds, such as ventilating noise or appliances.

Baffle. General expression for the wall, board, or enclosure carrying the loudspeaker. The purpose of the baffle is primarily to separate the front and back radiations from the cone or diaphragm. Otherwise, they would cancel each other.

Bass. Name given to the lower frequencies in the audio range.

Bassy. Sound reproduction that overemphasizes low-frequency tones.

Beats

1. Two tones of slightly different frequencies sounded together interfere to give a sound of regularly varying intensity. The number of beats per second is the difference in frequency of the two tones.
2. Audible difference tones produced when two signals of nearly the same frequency are sounded together.

Bimorph. Two thin slabs of crystal cemented together.

Broadband noise. Noise with a wide frequency range, such as airflow noise.

B-scale. Filtering system that has characteristics that roughly match the response characteristics of the human ear at sound levels between 55 and 85 dB.

Capacitor. Device for storing an electric charge, analogous to the compliance of a spring that may be used to store mechanical energy.

Cardioid microphone. Class of microphone having a heart-shaped directivity pattern in the horizontal plane.

Characteristic impedance of air. The specific acoustic impedance at a point in a plane wave in a free field. It is a pure specific resistance since the sound pressure and the particle velocity are in phase, and is equal in magnitude to the product of air and the speed of sound.

Coincident. Term applied to microphone arrangements in stereophony. Two microphones are said to be coincident if they are placed immediately adjacent to each other so that any difference in the times of arrival of the sound are negligible.

Communication. Signals or stimuli (or their transmission) that produce reactions, and furnish information. The sounds for communication are controlled with reflective surfaces.

Compatibility. Stereophonic system that provides a signal capable of giving satisfactory results on monophonic equipment.

Complex waves. Sound waves combining two or more frequencies, pure tones, overtones, and harmonics.

Compliance. Ease with which a panel of a material can be flexed by application of a force or pressure.

Compression

1. Condition that occurs in part of a sound wave where the pressure is above the standing atmospheric pressure.

2. In sound, the concentrating of the conductor molecules to produce a high-pressure layer.

Concave. Curved toward the observer. Concave surfaces, such as those behind a speaker or a musical or instrumental group, are useful in fortifying sound.

Conductor. Material that carries or transmits energy from one location to another. A conductor of sound must be an elastic material.

Consonance. Combination of two tones that is generally accepted as producing a satisfying effect.

Constant. Value that remains the same under changing conditions.

Convex. Curved away from the observer; rounded. Convex wall surfaces are highly effective in diffusing sound.

Coupling. Means of joining separated masses of any media so that sound energy is transmitted between them.

Criteria. Standards by which performance can be judged.

Criterion. Statement of the desired acoustical environment given in an appropriate numerical system of values.

Critical frequency. Lowest frequency at which the wavelength of a bending wave, traveling in a structure, is the same as the wavelength in air at that frequency.

Crystal. In most pickups, a bimorph of Rochelle salt. Two crystal slices are cemented together with a conducting cement.

Crystal microphone. Type of microphone in which a fluctuating voltage is generated by applying pressure to a slab of material such as Rochelle salt or barium titanate.

C-scale sound level. Quantity in decibels, read from a standard sound level meter switched to the weighting scale labeled C. The C-scale weights the frequencies between 70 and 4000 Hz uniformly, but below and above these limits frequencies are slightly discriminated against. C-scale measurements are essentially the same as overall sound pressure levels, which require no discrimination at any frequency.

Cycle

1. One complete series of changes in a periodically varying quantity.

2. One complete phase of an action, such as one revolution of a wheel or one full swing of a pendulum. In relation to sound, one to-and-fro movement of the

vibrating object or one high- and low-pressure sequence of a sound wave.

3. Entire sequence of movement of a particle (during periodic motion) from rest to one extreme of displacement, back through the rest position to the opposite extreme of displacement, and back to the rest position.

Damping

1. Addition of sound attenuation materials within the construction to effectively dissipate sound energy by converting it to heat.

2. Dissipation of structure-borne noise (vibrational energy), usually by conversion to heat.

Dead acoustic. Dull acoustic effect of an enclosed space with little reverberation.

Dead room. Room that is characterized by an unusually large amount of sound absorption.

Decay rate. The rate of decrease of sound pressure level after the source of sound has stopped.

Decibel

1. Division of a uniform scale based on 10 times the log of the relative intensity of sound levels being compared.

2. Logarithmic unit of measure for sound pressure (or power) calculated according to an applicable formula. Zero on the decibel scale corresponds to a standardized reference pressure of 0.0002 μbar.

3. Logarithmic unit expressing the ratio between a given sound being measured and a reference point.

4. Unit ratio of power, voltage, or current.

5. Unit of measure for sound pressure (or power) in representing vastly different sound intensities. It is 10 times the logarithm to the base 10 of the ratio of the sound intensity to a reference intensity of 10-16 W/sq cm. This reference intensity is considered the lowest value that the ear can detect.

6. A division of the logarithmic scale customarily used to express the ratio of two like quantities proportional to power or energy (ASTM C 634).

Decibel scale. A logarithmic scale used to express the ratio of two like quantities proportional to power or energy. The ratio is expressed on the decibel scale by multiplying its common logarithm by 10.

Deflect. To bend from a straight course or an original shape.

Density. Ratio of the mass (or weight) of a body to its volume. A common unit of measure is pounds per cubic foot.

Diaphragm. Thin body that separates two areas; in sound, the skin of a partition or ceiling that separates the room from the structural space in the center of the partition or ceiling assembly.

Diaphragmatic absorbents. Materials that flex under sound pressures and vibrate as a diaphragm, dissipating acoustic energy within their structure as heat and mechanical energy of vibration.

Diffraction

1. Change in direction that occurs when a wave contacts a space object, surface, or edge smaller than the wavelength.

2. Ability of a sound wave to "flow" around an object or through openings with little energy loss.

3. Bending of waves around solid objects. Low-frequency sound waves bend around the edge of the loudspeaker enclosure and produce some reradiation from the

edges which may provide irregularities in the frequency response.

Diffuse. Spread out evenly and thus become less dense or concentrated.

Diffuse sound field

1. Region in which the average rate of flow or sound energy is equal in all directions.
2. Region in which sound intensity is independent of direction.

Diffusion. Dispersion of sound within a space so that a uniform density of energy exists throughout.

Directional microphone. Microphone with response dependent on the direction of the sound wave.

Direct sound. Sound travels in all directions from the sources but each listener will hear just the one segment of the overall sound wave that is traveling in a direct line to his ear (in a space free from reflecting surfaces). As the direct sound loses intensity, the importance of reflected sound increases. In a room the ear will always hear direct sound before it hears reflected sound.

Direct sound field. Region in which all but a negligible part of the sound arrives directly from the source without reflection.

Discontinuous construction. Any of several construction methods, such as the use of staggered studs, double walls, or resilient mounting of surfaces, used to break the continuous paths through which sound may be transmitted.

Discrimination. Ability of the hearing apparatus to discern discrete, particular signals in a complex sound field.

Dispersion. Scattering or distribution of sound in a space.

Displacement

1. Distance of a particle from its normal or mean position.
2. Forced movement away from an original location.

Dissonance. Combination of two tones that does not form a consonance.

Distortion. Change in the transmitted sound that alters the character of the energy-frequency distribution within the signal so that the sound being received is not a faithful replica of the source sound.

Distribution. Pattern of sound intensity levels within a space; also, pattern of sound dispersion as the sound travels within the space.

Dominant pitch. Subjective response of the ear to the fundamental frequency of a sound. The dominant pitch is usually louder than the harmonic overtones and of lower frequency.

Dynamic range. Ratio (in phons) between the softest sound and the loudest sound in a live performance; ratio (in decibels) of the largest signal a given system can handle without distortion to the smallest signal that it can reproduce successfully (that is, without the inherent background noise masking it).

Dyne

1. Unit of force that if acting on a mass of 1 g would accelerate it 1 cm/second/second.
2. Unit of force in the metric system; force that will give to a moving mass of 1 g an acceleration of 1 cm/second/second.

Doppler effect. The change in frequency (or wavelength) of a wave disturbance, sensed by an observer, due to relative motion of the source and the observer.

Doppler's principle. The apparent frequency of a sound as affected by the motion of the hearer, the source, and the medium is given by formula.

Echo. Reflected sound that is loud enough and received late enough to be heard as distinct from the source.

Elastic. Term applied to a material with the capacity to return to its original shape after deflection.

Elastic medium. Substance in which strain or deformation is directly proportional to stress or loading.

Electronic amplification. Increasing the intensity level of a sound signal by means of electrical amplification apparatus.

Electrostatic microphone. Class of microphone in which a fluctuating electric current is produced by the movement of a diaphragm relative to a rigid plate.

Energy. Ability to perform work; in sound, the capacity to compress the conductor molecules.

Fidelity. Faithful reproduction of the source sound.

Field insertion loss (of a barrier or enclosure). The reduction in sound pressure level produced at an observation point by the introduction of the barrier or enclosure between the observation point and the source of sound.

Field sound transmission class. The sound transmission class of a partition installed in a building derived from values of field sound transmission loss.

Field transmission coefficient (of a partition installed in a building, in a specified frequency band). The fraction of the airborne sound power incident on the partition that is transmitted by the partition and radiated on the other side.

Field transmission loss (of a partition installed in a building, in a specified frequency band). The ratio expressed on the decibel scale, of the airborne sound power incident on the partition and radiated on the other side (ASTM C 634).

Flanking paths

1. Wall or floor/ceiling construction that permits sound to be transmitted along its surface; opening that permits the direct transmission of sound through the air.
2. Lightweight constructions along which sound impulses are transmitted.
3. Sound is transmitted along the path of least resistance, which is usually through the structure and around sound barrier walls and floor/ceilings.

Flanking transmission. Transmission of sound from the source to the receiving room by paths other than directly through the partition under test (ASTM C 634).

Flow resistance

1. Ratio of the pressure differential across a sample of porous material to the air velocity through it. Flow resistance is one of the most important quantities determining the sound-absorbing characteristics of a material.
2. The quotient of the air pressure difference across a specimen, divided by the volume velocity of airflows through the specimen (ASTM C 634).

Flow resistivity. The quotient of its specific flow resistance divided by the thickness of a homogeneous material.

Flutter. Rapid reflection or echo pattern between parallel walls, with sufficient time between each reflection to cause a listener to be aware of separate, discrete signals.

Focusing. Concentration of acoustic energy within a limited location in a room as the result of reflections from concave surfaces.

Frequency

1. Number of complete cycles of a vibration performed in 1 second, measured in cycles per second (cps) and expressed in hertz (Hz).

2. Number of vibrations of a body or particle completed in a second.

3. Number of complete cycles per second of a vibration or other periodic motion.

4. Number of times a sine wave repeats itself in each sound. In acoustics, the units of frequency is the hertz (Hz) which is numerically equal to the cycle per second (cps).

Frequency analyzer. Electrical apparatus capable of measuring the acoustic energy present in various frequency bands of a complex sound.

Frequency of a wave. Number of times per second that the wave motion repeats itself. The unit of frequency is the hertz.

Fundamental. Lowest frequency component of a complex wave.

Fundamental frequency. Most intense and usually lowest frequency of a sound. The fundamental frequency establishes the frequencies of the harmonics.

Fundamental frequency of a standing wave. Lowest frequency at which wave motion can be set up in a particular situation. Possible higher frequencies are called "harmonics" or "overtones."

Harmonic. Component of a complex wave that is an integral multiple of the fundamental. It can be twice the fundamental, three times the fundamental, etc.

Harmonic response. Periodic response of a vibrating system exhibiting the characteristics of resonance at a frequency that is a multiple of the excitation frequency.

Harmonics. Secondary frequencies that are whole number multiples of the fundamental frequency of a sound. Harmonics combine with the fundamental to produce the complex sound wave, giving timbre or quality to the total sound as perceived.

Hearing. Subjective response to sound, involving the entire mechanism of the internal, middle, and external ear and including the nervous and cerebral operations that translate the physical operations into meaningful signals.

Hearing loss. Increase in the threshold of audibility, at specific frequencies, as the result of normal aging, disease, or injury to the hearing organs.

Helmholtz resonator. Acoustic resonator comprising a cavity having an aperture open to free air. This aperture may be fitted with a duct or tunnel either internally or externally.

Hertz (Hz)

1. Unit of frequency of a periodic process equal to 1 cycle per second.

2. Unit of measure of sound frequency, representing 1 cycle per second; named for Heinrich R. Hertz, noted German physicist.

Homogeneous. Of uniform composition and structure.

Human ear. The ear is not equally sensitive to all sound frequencies, and some it does not detect at all. Even though two different sounds produce the same sound pressure intensity, one sound may be judged to be louder than the other if its energy is concentrated in a frequency to which the ear is more sensitive.

Impact. Sharp, rapid contact between two solid bodies.

Impact insulation class

1. Single number rating developed by the Federal Housing Administration to estimate the impact sound isolation performance of floor/ceiling systems.

2. Whole, positive number rating, based on standardized test performance, for evaluating the effectiveness of assemblies in isolating impact sound transmission.

Impact noise. Noise that is structure borne, caused by a direct mechanical impact. The impact or vibration is transmitted through the structure and radiated from surfaces as airborne sound.

Impact noise rating

1. Single number rating used to compare and evaluate performance of floor/ceiling constructions in isolating impact noise transmission.

2. Single number rating, based on standardized test performance, for evaluating the effectiveness of assemblies in isolating impact sound transmission. This rating method is being replaced by the impact insulation class.

Impact sound pressure level. Sound level (in decibels), measured in the "receiving" room, resulting from the transmission of sound through the floor/ceiling construction, produced by a standard tapping machine.

Impedance. Complex ratio related to the sound absorption characteristics of acoustical absorbents. It is similar to electrical or mechanical impedance.

Impedance ratio. The ratio of the specific normal acoustic impedance of a surface to the characteristic impedance of air.

Incident sound. Sound striking a surface, contrasted to reflected sound. The angle of incidence equals the angle of reflection.

Incident wave. Wave traveling toward a reflecting or refracting surface.

Inertia. Tendency of a mass to resist any change in its state of motion or rest.

Infrasonic frequencies. Frequencies below those by which the human hearing apparatus is stimulated, usually about 16 Hz.

Intensity

1. Rate of flow of sound energy per unit of area normal to the direction of propagation.

2. Rate of sound energy transmitted in a specified direction through a unit area.

Intensity level. Ten times the log of the ratio of intensity of the sound to a reference intensity.

Intensity—loudness. There is an important distinction between sound intensity and loudness. Intensity results from the maximum high-low pressure variations of the wave. It is a physical property that can be measured precisely, but it is sensed as loudness by the ear.

Intensity of sound. The intensity of sound depends on the energy of the wave motion. Intensity is measured by the energy in ergs transmitted per second through 1 sq cm of surface.

Intermolecular forces. Attraction that exists between molecules, similar in effect to a magnetic attraction.

Interval. In music, the pitch interval between two notes of a scale.

Isolation. The separation of opposite surfaces of a construction will improve sound isolation. Surfaces may be separately supported with no structural connections between them, or they may be resiliently mounted to common supports.

Isolation of vibration. Any of several means of preventing transmission of sound vibrations from a vibrating body to the structure in which or on which it is mounted.

Kilohertz. 1000 Hz or 1000 cps.

Lateral. To the side.

Lateral flow resistivity. The flow resistivity when the direction of airflow is parallel to the face of an anisotropic homogenous material in sheet or board form, from which the test specimen is taken.

Leaks. Cracks, penetrations, or any openings, however small, readily conduct airborne sound.

Level

1. Strength of a continuous signal used for test purposes. It is measured by comparison with a standard reference level, which is usually a power of 1 mW.

2. The ratio, expressed on the decibel scale, of two like quantities, the second of which is a standard reference quantity.

Limpness. Inelastic motion; motion in which the material under stress does not recover from the strain upon removal of stress.

Line source. Ideally, an acoustic radiator having the dimension of length only; array of loudspeakers arranged in a line.

Live acoustic. Bright acoustic effect of a room with considerable reverberation.

Live room. Room that is characterized by an unusually small amount of sound absorption.

Lobes. Variations from a minimum value, through a maximum, and back to minimum in the polar response of a radiator.

Logarithm. Mathematical device consisting of the power to which a base number has been raised.

Longitudinal. Term applied to wave motion in which the particles of the medium vibrate along the line of propagation of the wave.

Longitudinal wave

1. Wave in which the motion or disturbance of the medium is in the same direction as the direction of propagation of the wave. Sound waves are longitudinal waves.

2. Waves characterized by molecular movement parallel to a straight line from the source. Sound waves are a typical example: high-low pressure combinations are propagated in a direction parallel to the motion of the wave, and thus produce longitudinal waves.

Loudness

1. Subjective response to sound indicating the magnitude of the hearing sensation, which is dependent on the listener's ear.

2. Magnitude of the auditory sensation produced by a sound. The unit used in modern work is the sone. The loudness scale is linear in that, for example, 2 sones sound twice as loud as 1 sone. More generally, x sones sound x times as loud as 1 sone.

3. Effect on the hearing apparatus of varying sound pressures and intensities.

4. Intensive attribute of an auditory sensation in terms of which sounds may be ordered on a scale extending from soft to loud. Loudness depends primarily on the sound pressure of the stimulus, but it also depends on the frequency and wave form of the stimulus.

Loudness level

1. The sound pressure level in decibels (relative to 0.0002 microbar) of a simple tone of 1000 Hz frequency.

2. The loudness level, expressed in phons, is equal to the average sound pressure level, in decibels, of a free progressive sound wave of 1000 Hz, that, in a number of trials, is judged by listeners while they are facing the source.

Loudspeaker. System for converting electrical energy into sound energy.

Masking

1. Process by which the threshold of audibility for one sound is raised by the presence of another (masking) sound.

2. Effect produced by background noise, which appears to diminish the loudness of transmitted noise.

3. Effect on the hearing of one sound when another sound is present; shift of the intensity threshold of audibility of one sound due to the presence of a second one.

4. Increase in the threshold of audibility of a sound necessary to permit its being heard in the presence of another sound.

Mass

1. Property of a body that resists acceleration. The weight of a body is the result of the pull of gravity on the body's mass.

2. Quality of matter that permits it to resist acceleration; quality of matter that produces the effect of inertia.

3. The heavier the construction, the greater its resistance to sound transmission.

Mass or density. The frequency of the source is constant because it is a function of the mass or density and the elasticity of the vibrating source object. The frequency of the originating sound depends on the mass of density and the elasticity of the conductors. The greater the mass, the longer the period of each vibration; consequently, the waves will be less frequent. The decrease in frequency is detected by the ear as a lower tone.

Microphone. Kind of electric ear that is finely sensitive to the sound impinging on its diaphragm. It is used to convert acoustic energy into electrical energy. A vibration generator is an example.

Mode. Manner or fashion. Resonant modes are the manners in which a complex system resonates.

Modulation. Moving from one key to another.

Molecule. Smallest particle of a material that retains the material's identity. A molecule is a combination of atoms held together by a mutual attraction similar to magnetism. Molecules themselves are held together by a similar attraction, the degree of which determines the physical state of the material. For instance, the molecules of air are only loosely attracted while those of steel at moderate temperature are rigidly locked together. Molecules do not move, but only vibrate within their immediate areas.

Natural frequency. Rate at which a freely vibrating body oscillates.

Node. Point, line, or surface in a standing wave system at which the amplitude is zero, such as a node of pressure, or a node of particle velocity.

Noise. Undesired sound. By extension, noise is any unwanted disturbance within a useful frequency band, such as undesired electric waves in a transmission channel or device. Erratic, intermittent, or statistically random oscillation.

Noise-criterion curves. Series of criterion curves that portray sound pressure levels for background noises which generally should not be exceeded, or should be maintained, in various human environments.

Noise exposure. Measure of noise duration as well as noise level.

Noise isolation class. A single number rating derived in a prescribed manner from the measured values of noise reduction. It provides an evaluation of the sound isolation between two enclosed spaces that are acoustically connected by one or more paths (ASTM C 634).

Noise level. Level of noise, the type of which must be indicated by further modifier or context.

Noise reduction

1. Reducing the level of unwanted sound by any of several means of acoustical treatment.
2. Difference in the sound pressure levels, expressed in decibels, on either side of a structural configuration. Noise reduction is essentially synonymous with attenuation.

Noise reduction coefficient. Single number index of the noise reducing efficiency of acoustical materials. It is found by averaging the sound absorption coefficients at 250, 500, 1000, and 2000 Hz.

Normal mode (of a room). One of the possible ways in which the air in the room, considered as an elastic body, will vibrate naturally when subjected to an acoustical disturbance. When each normal mode is associated a resonance frequency and in general, a group wave propagation direction comprising a closed path (ASTM C 634).

Octave

1. Interval between two sounds that have a frequency ratio of 2 to 1.
2. Interval between a sound of one frequency and a sound with a frequency that is exactly double the first.

Octave band. Frequency spectrum that is one octave wide. Bands of one-third octave are used for recording sound test results and are designated by the center frequency of the band.

Optimum reverberation time

1. Empirically determined reverberation time, varying directly with room volume, which produces hearing conditions considered "ideal" by an average listening audience.
2. The duration of reflections is controlled by the amount of absorption in a room. The optimum duration time is long enough for proper blending and reinforcement of sounds, but short enough to prevent excessive overlapping and the resulting confusion. The amount of acoustical treatment required to obtain this goal usually results in only a moderate decrease in loudness of the wanted sound.

Orientation. The location of a building or a site can affect the sound control conditions encountered. The shape of the building is also important.

Oscillograms. Graphic representations of sound waves, recorded with the aid of electronic equipment.

Overall sound pressure level. Sound pressure level measured in a broad frequency band covering the frequency range of interest. This generally extends across the whole audible frequency range of 20 to 16,000 Hz.

Overtone

1. Component of a complex wave that may or may not be an integral multiple of the fundamental.
2. Subjective response of the ear to harmonics.

Partial. Component of a complex sound. Overtones and harmonics are partials.

Partition. An assembly of materials into a building component used to divide spaces, and installed from floor to the ceiling of the space.

Peak noise. Maximum pressure of an impulsive noise.

Perceived noise level. Single number rating of aircraft noise in decibels, used to describe the acceptability or noisiness of aircraft sound. Noise level is calculated from measured interior and exterior noise levels and correlates well with subjective responses to various kinds of aircraft noise.

Perimeter. Outer edge of a figure or area.

Period or periodic time. Time taken by an alternating quantity to perform one cycle.

Phase. Two waves are in phase if they are always exactly in step.

Phase angle. Measurement of the phase difference between two waves.

Phon. Measure of loudness level (on a logarithmic scale) that compares the effect of a sound to the effect of a 1000-Hz tone of a given sound pressure level.

Piezoelectric. Term applied to phenomena by which certain crystals (Rochelle salt, quartz, tourmaline, etc.) expand along one axis and contract along another, when subjected to an electrostatic field, or when subjected to twisting and bending, certain crystals develop a potential difference between opposite faces.

Pitch

1. Subjective quality of a sound that determines its position in the musical scale.
2. Highness or lowness of a sound as perceived by the ear. While the frequency of the sound determines the highness, pitch is the subjective response to it.
3. Attribute of auditory sensation in terms of which sounds may be ordered on a scale extending from low to high. Pitch depends primarily on the frequency of the sound stimulus, but it also depends on the sound pressure and wave form of the stimulus.

Pitch of sound. The pitch of sound is determined by the frequency or number of vibrations per second.

Polar response. Plot of the variation in radiated energy with the angle relative to the axis of the radiator.

Pressure gradient operation. The sound has access to both sides of the diaphragm.

Pressure operation. The sound wave has access to one side of the diaphragm only.

Pressure waves. Layers of high and low pressure that radiate out in all directions from a sound source.

Pure pitch sounds. Very few pure pitch sounds are heard, since most sounds consist of combinations of numerous frequencies.

Pure tone. Sound wave, the instantaneous sound pressure in which is a simple sinusoidal function of the time;

sound sensation characterized by the singleness or pitch; sound wave of only one sinusoidal frequency.

Radiate. Travel in straight lines away from a center, such as sound waves moving out from a source.

Radius. Straight line distance from the center of a circle or sphere to the edge.

Rarefaction

1. Part of a sound wave where the pressure is below the pressure due to standing atmospheric pressure.
2. In sound, the drawing apart of the particles (molecules) of the conductor, producing a low-pressure area.

Rayl. Unit of specific acoustical impedance, equal to a sound pressure of 1 dyne/sq cm divided by a sound particle velocity of 1 cm/second.

Receiving room. The room in which the sound transmitted from the source room is measured. The measurement is usually of sound transmission loss or noise reduction.

Reflect. Rebound off an object and travel in a new direction.

Reflected sounds. Reflected sounds can be multiple and frequent in a room with few absorptive materials. The simplest forms of sound reflection occur when sound waves strike a flat nonabsorbtive surface.

Reflecting surfaces. Room surfaces from which significant sound reflections occur; special surfaces used particularly to direct sound throughout the space.

Reflection. Return from surfaces of sound not absorbed on contact with the surfaces.

Reflection of sound wave. Return of energy due to the wave striking some discontinuity in its supporting medium.

Reflective ceilings. Splays or tilted portions near the speaker can reflect sound to the sides or rear of the room.

Refract. Deflect at the point of traveling from one conductor into another. An example of refraction is the bending that occurs when light rays pass into air from water.

Refraction. Bending or changing of direction of a wave when it passes from one medium into another in which the speed of wave propagation is different. All types of waves can exhibit refraction.

Residual noise. In acoustics, the A-weighted sound level exceeding 90% of a noise-monitoring period.

Resilient mounting. Mounting, attachment system, or apparatus that permits room surfaces or machinery to vibrate normally without transmitting all of the energy of vibration to the structure.

Resonance

1. Sympathetic vibration, resounding, or ringing of enclosures, room surfaces, panels, etc., when excited at their natural frequencies.
2. Effect when the vibrations of a body reach maximum amplitude on being caused to vibrate by a force having a particular frequency.
3. Natural, sympathetic vibration of a volume of air or a panel of material at a particular frequency as the result of excitation by a sound of that particular frequency.
4. Abnormally large response of a system to a relatively small stimulus that has the same, or very nearly the same, frequency (or vibration period) as the natural frequency (or free vibration period) of the system; phenomenon of amplification of a free wave or oscillation of a system by a forced wave or oscillation of exactly equal period. The forced wave may arise from an impressed force on the system or from a boundary

condition. The growth of the resonant amplitude is characteristically linear in time.

Resonant frequency dip. Dip that occurs in the performance graph of a sound attenuation assembly at the resonant frequency of the assembly.

Response. Motion or other output resulting from an excitation or stimulus under specified conditions.

Reverberant sound field. An enclosed or partially enclosed space in which all but a negligible part of the sound has been reflected repeatedly or continuously from the boundaries (ASTM C 634).

Reverberation

1. Persistence of sound in an enclosed space as a result of multiple reflections after the sound source has stopped; sound that persists in an enclosed space as a result of repeated reflection or scattering after the source of the sound has stopped.
2. Continuing travel of sound waves between reflective surfaces after the original source is stopped. The reverberation time (T) is the time in seconds required for a sound to diminish 60 dB after the source is stopped.
3. Prolongation of sound in a room due to reflections from the walls, ceiling, and floor.
4. Continuation of sound reflections within a space after the sound source has ceased.

Reverberation room. Room designed so that the reverberant sound field closely approximates a diffuse sound field both in the steady state, when the sound source is on, and during decay, when the sound source has been stopped.

Reverberation time

1. Time taken for the sound intensity in a room to fall from its steady state by 60 dB (for example, to a millionth).
2. Time in seconds required for a sound to decay to inaudibility after the source ceases (strictly, the time in seconds for the sound level at a specific frequency to decay 60 dB after the source is stopped.)
3. Time that would be required for the mean-square sound pressure level, originally in a steady state, to decrease 60 dB after the source is stopped.

Room acoustics

1. Branch of architectural acoustics dealing with both acoustical correction and noise reduction.
2. Shaping and equipping of an enclosed space to obtain the best possible conditions for faithful hearing of wanted sounds, and the reduction and absorption of unwanted sounds.

Room criteria. System of maximum noise levels indicated by octave bands, used to specify the appropriate background noise level.

Room location. Orientation of a room with respect to other rooms on a given floor, other surrounding spaces, or other outside factors, including spaces outside the building under consideration.

Room shape. Configuration of enclosed space resulting from the configuration, orientation, and arrangement of surfaces defining the space.

Room volume. Cubic feet of volume space enclosed by the room surfaces.

Sabin

1. Measure of sound absorption of a surface, equivalent to 1 sq ft of a perfectly absorptive surface.

2. Unit of sound absorption. One sabin corresponds to the absorption that would be produced by 1 sq ft of an infinitely absorbing surface such as an open window. The average person has an absorption of 4.2 sabins.

3. Unit of measure of sound absorption; the amount of sound absorbed by a theoretically perfect absorptive surface of 1-sq ft area; named for Wallace C. W. Sabine, noted American physicist.

Shielding. Attenuating of sound by placing a barrier between a source and a receiver.

Short circuit. Bypassing connection or transmission path that tends to nullify or reduce the sound-isolating performance of a building construction.

Simple harmonic motion. Periodic oscillatory motion in a straight line in which the restoring force is proportional to the displacement. If a point moves uniformly in a circle, the motion of its projection on the diameter (or any straight line in the same plane) is simple harmonic motion.

Simple sound source. Source that radiates sound uniformly in all directions under free-field conditions.

Simple tone. A sound wave, the instantaneous sound pressure of which is a simple sinusoidal function of the time.

Sine wave. Wave shape of a pure note.

Sone. Measure of loudness (on a linear scale) that compares the effect of a sound to the effect of a 1000-Hz tone of 40-dB sound pressure level.

Sound

1. Vibration in an elastic medium within the frequency range capable of producing the sensation of hearing; chain reaction of vibrations.

2. Change in the ambient air pressure that has an amplitude and frequency content within the response range of the human ear.

Sound absorbents. Materials that absorb sound readily, usually building materials designed specifically for the purpose of absorbing acoustic energy.

Sound absorption

1. Change of sound energy into some other form, usually heat, in passing through a medium or on striking a surface.

2. The process of dissipating or removing sound energy (ASTM C 634).

3. The property possessed by materials, both natural and manufactured, objects, and structures such as rooms, of absorbing sound energy.

Sound absorption coefficient. Fraction of incident sound energy absorbed or otherwise not reflected by a surface.

Sound attenuation

1. Reduction of the energy or intensity of sound.

2. Reduction of sound energy as it passes through a conductor, resulting from the conductor's resistance to the transmission.

3. Reduction in the intensity of a sound signal, measured in decibels.

4. Diminution of the pressure or velocity, from one point to another due to spherical propagation, reflection, or dissipation.

Sound behavior. Sound always behaves predictably; it takes the physical forms of radiating pressure waves, and as such obeys specific laws of physics.

Sound condition. Generated sound waves travel through the air in the form of small pressure changes, alternately

above and below a static atmospheric pressure, the ever-present air pressure that is measured by a barometer.

Sound conditioning. Designing and equipping a space for a faithful retention of desirable sounds and maximum relief from undesirable acoustical effects.

Sound control. Application of the science of acoustics to the design of structures and equipment to permit them to function properly and to create the proper environment for the activities intended.

Sound detection. Discrimination of a sound from background noise, either by the ear or by an electronic instrument such as a volume indicator.

Sound effects. Mechanical devices or recordings used to provide lifelike imitations of various sounds.

Sound energy. Difference between the total energy and the energy that would exist if no sound waves were present.

Sound image. Photographic image of a sound, as on a film sound track.

Sound intensity. The quotient obtained when the average rate of sound energy flow in a specified direction and sense is divided by the area, perpendicular to that direction, through or toward what it flows. The intensity at a point is the limit of the quotient as the area which includes the point approaches zero (ASTM C 634).

Sound isolation

1. Condition created by materials or methods of construction designed to resist the transmission of airborne and structure-borne sound through walls, floors, and ceilings.

2. Descriptive term that denotes the performance of wall or floor-ceiling construction assemblies to reduce sound transmission.

Sound lag. Time necessary for a sound wave to travel from its source to the point of reception.

Sound leak. Leaks caused by openings that permit airborne sound transmission.

Sound level. Measure of sound pressure level as determined by electrical equipment meeting American National Standards Institute requirements.

Sound level meter. Electrical instrument for determining sound pressure level.

Sound masking. Ability of one sound to make the ear incapable of perceiving another sound.

Sound oscillation. Oscillation in pressure, stress, particle displacement, particle velocity, etc., in a medium with internal forces, such as the superposition of such propagated oscillations; auditory sensation evoked by the oscillation described above.

Sound power. The rate at which acoustic energy is radiated. In general, the rate of flow of sound energy, whether from a source, through an area, or into an absorber.

Sound power level

1. The ratio expressed on the decibel scale, of the sound power under consideration to the standard reference power of 1 picowatt (1 pW) (ASTM C 634).

2. In decibels, 10 times the logarithm to the base 10 of the ratio of a given power to a reference power. The reference power must be indicated.

Sound pressure

1. Instantaneous pressure at a point as a result of the sound vibration minus the static pressure at that point; change in pressure resulting from sound vibration. Sound pressure is measured in dynes per square centimeter. The sound pressure of 1 dyne/sq cm is about that of conversational speech at close range and is approximately equal to one millionth atmospheric pressure.

2. Instantaneous change in pressure resulting from vibration of the conductor in the audible frequency range. Conversational speech at close range produces a sound pressure of about 1 dyne/sq cm.

Sound pressure level

1. In decibels, 20 times the logarithm to the base 10 of the ratio of the pressure of a sound to the reference pressure. The reference pressure must be indicated.

2. Value equal to 20 times the logarithm of the ratio of the pressure of the sound to the reference pressure.

3. Sound pressure when measured on the decibel scale; ratio in decibels between measured pressure and a reference pressure.

4. Decibel quantity that equals 20 times the logarithm to the base 10 of the ratio of the pressure of a sound to the reference pressure. The common reference pressure for acoustics in air is 0.002 μbar.

Sound production. Conversion of energy from mechanical or electrical into acoustical form, as in a siren or loudspeaker.

Sound propagation. Origination and transmission of sound energy.

Sound reception. Conversion of acoustical energy into another form, usually electrical, as in a microphone.

Sound recording. Process of recording sound signals so they may be reproduced at any subsequent time, as on a phonograph disk, motion picture sound track, or magnetic tape.

Sound source. When a sound source vibrates in a conductor (such as air), it creates a high-pressure layer in the conductor.

Sound spectrograph. Instrument that records and analyzes the special composition of audible sound.

Sound stage. The term is used in reference to stereophony. The sound stage is the region in which the sound images appear. For each source of sound in the studio there will be a corresponding point from which the reproduced sound appears to come in the region between the two spaced loudspeakers. These points are termed "images."

Sound transmission

1. Passage of sound through a material construction or other medium.

2. Transfer of sound energy from one place to another, through air, a structure, or other conductor.

Sound transmission class (STC)

1. Single number rating for evaluating the efficiency of constructions in isolating airborne sound transmission. The higher the STC rating, the more efficient the construction.

2. Single number rating system that compares the sound transmission loss of a test sample with a standard contour.

3. Method of rating the acoustic efficiency of various wall and floor systems. Sound transmission losses at various frequencies for different types of construction are compared with class contours established by ASTM E 90. The sound transmission class is determined for each partition or floor by its location on the contour chart according to criteria set forth in the standard.

Sound transmission coefficient. The fraction of the airborne sound power incident on the partition that is transmitted by the partition and radiated on the other side of a partition in a specified frequency band.

Sound transmission loss. The sound transmission loss is the reduction in the sound pressure level between two stated points. The transmission loss is usually expressed in decibels.

Sound travel speeds. Sound waves travel in air at a uniform rate of about 1128 ft/second, at 70° F. In a solid conductor, such as steel, the rate if 18,000 ft/second; in wood, the rate is 11,700 ft/second.

Sound vibrations. Sound vibrations are conducted by any elastic material, such as air, metal, water, or wood.

Sound wave

1. Mechanical disturbance advancing with finite velocity through an elastic medium and consisting of the longitudinal displacement of the ultimate particles of the medium, such as compressional and rarefactional displacements parallel to the direction of advance of the disturbance; a longitudinal wave.

2. One complete cycle or vibration from pressure, through high pressure, back to normal, through low pressure, and back to normal pressure again to complete the cycle.

3. Disturbance that is propagated in a medium in such a manner that at any point in the medium the displacement is a function of the time.

4. A sound wave reaches the human ear causing the eardrum to vibrate, producing the sensation of hearing.

5. Longitudinal waves in air or some other medium that are in the audible range of frequencies.

Source room. The room that contains the noise source or sources from which the measurement of sound transmission loss or noise reduction is made.

Spaced microphones. Microphone arrangements in stereophony. Microphones are said to be spaced when the stereophonic effects are produced by differences in time of arrival at the microphones.

Specific flow resistance. The product of the flow resistance of a specimen and its area. This is equivalent to the quotient of the air pressure difference across the specimen divided by the linear velocity of airflow through the specimen, measured outside the specimen (ASTM C 634).

Specific frequency. Every sound, musical or otherwise, has a specific frequency—the number of complete vibrations that occur in 1 second.

Spectrum. Description of the resolution of a sound wave into components, each of different frequency and (usually) different amplitude and phase.

Speed of sound in air. The speed of sound in air is 344 m/second or 1128 ft/second at 25° C or 77° F. The speed of sound is an important consideration in large-room acoustics where the relative timing of sound fronts (direct and reflected) has a strong bearing on sound quality.

Splay. Slight offset in angle from a flat plane such as a wall or ceiling.

Standing or stationary wave system. Interference pattern characterized by nodes and antinodes.

Standing wave. Periodic sound wave having a fixed distribution in space. Standing waves are prevalent in small, rigidly geometric rooms and detract from the usually desired acoustic uniformity.

Static atmospheric pressure. Air pressure around us (approximately 14.7 psi) that results from the weight of the air above.

Stationary or standing waves. Waves produced in a medium by the simultaneous transmission in opposite directions of two similar wave motions. Fixed points of minimum amplitude are called "nodes." A segment extends from one node to the next. An antinode or loop is the point of maximum amplitude between two nodes.

Structure-borne sound

1. Sound that reaches the point of interest, over at least part of its path, by vibrations of a solid structure (ASTM C 634).
2. Sound energy transmitted through the solid mediums of the building structure.

Structure-borne sound transmission. Sound transmitted as a result of direct mechanical contact or impact caused by vibrating equipment, footsteps, object dropped, etc.

Subjective. Related to conditions of the brain and sense organs rather than to direct physical actions.

Symmetrical. State of being identical or balanced on each side of a real or imaginary dividing line.

Threshold of audibility

1. Sound pressure at which the average normal hearing apparatus begins to respond (usually 0.0002 μbar).
2. Minimum sound pressure level of a specified signal that is capable of evoking an auditory sensation.

Threshold of pain. Sound intensity level sufficiently high to produce the sensation of pain in the human ear (usually above 120 dB).

Tile mounting. Method of attaching acoustical tile to the building structure or building surfaces.

Timbre

1. Subjective response of the ear to the quality of richness of a sound, produced by the number and relative energy of the harmonics and other frequencies present in the sound.
2. Tone color that enables a listener to recognize the difference between two sounds having the same loudness and pitch.

Tone

1. Term that describes subjective response of the ear to the pitch of a sound.
2. Sound wave capable of exciting an auditory sensation having pitch; sound sensation having pitch that provides the tone to the human ear.

Transducer. Device that converts power of one type into power of another. A loudspeaker is such a device.

Transient. Effect on a vibrating system when there is a sudden change of conditions that persists for only a short time after the change has occurred.

Transmission. Propagation of a vibration through various mediums.

Transmission loss

1. Decrease or attenuation in sound energy (expressed in decibels) of airborne sound as it passes through a building construction.
2. Absolute measure of sound isolation of a structural configuration, expressed in decibels. Transmission loss is derived from the configuration's noise reduction, with corrections for the transmitting surface area and for room absorption on the "receiving" side. A decrease in power during transmission from one point to another.

Transverse. In the transverse of a wave motion the particles vibrate at right angles to the direction of propagation.

Transverse wave

1. Wave in which the motion or disturbance of the medium is perpendicular to the direction of propagation of the wave. Electromagnetic waves are always transverse waves.
2. Waves characterized by molecular movement perpendicular to a straight line from the source; water ripples produced by a falling stone are a typical example.

Traveling wave. Periodic disturbance that moves forward in a medium.

Ultrasonic frequencies. Frequencies above those that stimulate the human hearing apparatus, usually 18,000 to 20,000 Hz.

Ultrasonics. Physical science of the acoustic waves that oscillate in the range of about 18 to 80 kHz.

Unwanted sound. Noise or interfering sound, whatever its source or nature.

Vacuum. Absence of all matter; total void.

Variable. Value that changes with changes in conditions.

Velocity. Time rate of change of position of a reference point moving in a straight line. Rate at which the sound waves progress through a conductor.

Velocity of sound in air. The velocity of sound in air is 3.44×10^4 cm/second.

Vibration. Uniform, rapid movement of an elastic material in a back-and-forth direction.

Volume unit meter. Volume indicator with a decibel scale and specified dynamic and other characteristics, used to obtain correlated readings of speech power necessitated by the rapid fluctuations in level of voice currents.

Wanted sound. Audible signals that communicate necessary and desirable information or stimuli to the listener.

Watt. Unit of power equal to 1×10^7 dyne/cm/second, which is the basic expression of the flow of sound energy.

Wave. Propagating disturbance in a medium. All waves carry energy and momentum, but not matter.

Wave form

1. Configuration of a sound wave.
2. Shape of the graphic representation of a sound wave.

Wave front. Spherical surface of the wave as it travels out in all directions from the source.

Wavelength

1. Minimum distance between two points of a wave that are in phase.
2. Perpendicular distance between analogous points on any two successive periodic waves. The wavelength of sound is inversely proportional to the frequency of the sound.
3. Distance traveled by the sound in the time of one complete vibration.
4. The distance between successive crests or troughs of the wave.

Wave motion. Progressive disturbance propagated in a medium by the periodic vibration of the particles of the medium. Transverse wave motion is that in which the vibration of the particles is perpendicular to the direction of propagation. Longitudinal wave motion is that in which the vibration of the particles is parallel to the direction of propagation.

Wave packet. A bundle of waves that retains its oscillatory character but is localized in space.

Wave sound. Disturbance that is propagated in a medium in such a manner that at any point in the medium the displacement is a function of the time.

White noise. Noise with a spectrum density (or spectrum level) substantially independent of frequency over a specified range.

CARPETING

Acrylics. Generic term including acrylic and modified acrylic (modacrylic) fibers. Acrylic is a polymer composed of at least 85% by weight acrylonitrile; modacrylic is a polymer composed of less than 85% but at least 35% by weight acrylonitrile.

Adhesives. Waterproof adhesives such as latex adhesives and carpet cements.

All-hair pad. Cushion pad consisting of 100% cleaned animal hair, treated for moth protection.

Axminster

1. The Axminster loom is highly specialized and nearly as versatile as hand weaving. Color combinations and designs are limited only by the number of tufts in the carpet. Almost all the yarn appears on the surface, and a characteristic of this weave is a heavy ribbed back allowing the carpet to be rolled lengthwise only. Axminsters produce single-level cut-pile textures.

2. Carpet made on an Axminster loom, capable of intricate color designs, usually with level cut-pile surface.

Backing

1. Carpet foundation of jute, kraftcord, cotton, rayon, or polypropylene yarn that secures the pile yarns and provides stiffness, strength, and dimensional stability.

2. Foundation construction that supports the pile yarn.

Backing compound. Coating of a heavy layer of latex that holds the tufts of a tufted carpet in place.

Backing material. In the Wilton carpet the weave itself forms part of the backing. Backing can be wool, kraftcord, jute, cotton, or other fabric, depending on the type and quality of the carpet.

Back seams. While all carpet seams are located on the back or underside of the carpet, those made when the carpet is turned over or facedown are called "back seams," while those made with the carpet faceup are called "face seams."

Bearding. Long fiber fuzz occurring on some loop pile fabrics, caused by fibers snagging and loosening due to inadequate anchorage.

Binding. Strip sewed over a carpet edge for protection against unraveling.

Bonded rubber cushioning. Rubber or latex cushioning adhered to the carpet at the mill.

Broadloom

1. Carpet woven in widths wider than 27 or 36 in., usually in 6-, 9-, 12-, 15-, and 18-ft widths, and up to 30 ft in chenille. "Broadloom" is not a type of weave of carpet, nor a pattern nor color, but simply a designation of width.

2. Carpet woven on a broad loom in widths of 6 ft or more.

Buckling. Wrinkling or ridging of the carpet after installation, caused by insufficient stretching, dimensional instability, or manufacturing defects.

Bulked continuous filament. Continuous strands of synthetic fiber made into yarn without spinning, often extruded in modified cross section, such as multilobal mushroom or bean shape, and/or texturized to increase.

Burling

1. Removing surface defects such as knots, loose threads, and high spots to produce acceptable quality after weaving; filling in omissions in weaving.

2. Hand-tailoring operation after weaving to remove any knots and loose ends, to insert missing tufts of surface yarn, and check otherwise the condition of the fabric; repair operation on worn or damaged carpet.

Carpet. Fabric constructions that serve as soft floor coverings, especially those that cover the entire floor and are fastened to it, as opposed to rugs.

Carpet testing. Performance certification by a manufacturer of fibers and yarn.

Carpet tiles. Preassembled tiles or modules fabricated of nylon yarn fusion bonded to a primary vinyl backing, which in turn is fused to a fiberglass reinforced secondary vinyl backing. The combination of materials prevents the tile from curling, buckling, warping, and other distortions. A product of Carpets International-Georgia, Inc., LaGrange Ga.

Carpet yarn. Combination of fiber and yarn blending.

Chain binders. Yarns running warpwise (lengthwise) in the back of the carpet. As the name implies, they bind all construction yarns together. Chain binders run alternately over and under the weft binding and filling yarns, thereby pulling the pile yarn down and the stuffer yarns up for a tightly woven construction.

Chain warp. Zigzag warp yarns that work over and under the shot yarns of the carpet, binding the backing yarns together.

Chamber test. The chamber test (UL 992) evaluates the fire behavior of flooring systems. This test is a variation of the tunnel test in that the sample is exposed to a high-intensity burner in a tunnel. The specimen is placed on the floor and the forced draft, supply air to the flame, sweeps the carpet in the direction opposite to that observed on the floor in full-scale corridor fires.

Chemicals. Basic raw materials for nylon 6 fibers are caprolactam and polycaprolactam.

Chenille. Pile fabric woven by the insertion of a prepared weft row of surface yarn tufts in a "fur" or "caterpillar" form through very fine but strong cotton "catcher" warp yarns and over a heavy woolen backing yarn.

Commercial matching. Matching of colors within acceptable tolerances or with a color variation that is barely detectable to the naked eye.

Construction

1. Method by which a carpet is made, i.e., by which the pile fibers are combined with the backing materials. The term applies to woven, tufted, and knitted carpet.

2. Method by which the carpet is made (loom or machine type) and other identifying characteristics, including pile rows per inch, pitch, wire height, number of shots, yarn count and plies, pile yarn weight, and density.

Cotton count. Number of 840-yd lengths per pound. Thus 1.00 cu cm has 840 yd/lb, 2.00 cu cm has 840×2- 1680 yd/lb.

Count. Number identifying yarn size or weight per unit of length (or length per unit of weight) depending on the spinning system used (such as denier, woolen, worsted, cotton, or jute).

Crab. Hand device usually used for stretching carpet in a small area where a power stretcher or knee kicker cannot be used.

Crimping. Method of texturizing staple and continuous-filament yarn to produce irregular alignment of fibers and increase bulk and covering power. Crimping also facilitates interlocking of fibers, which is necessary for spinning staple fibers into yarn.

Cross seams. Seams made by joining the ends of carpet together.

Cushion (pad). Cushioning material for installation under rugs and carpet, made of felted cattle hair, jute, wool, and combinations of those materials or rubber and plastic foam.

Cushioning. Soft, resilient layer provided under carpeting to increase underfoot comfort, absorb pile-crushing forces, and reduce impact sound transmission.

Cut-loop pile. Pile surface in which tufts have been cut to reveal the fiber ends.

Cut pile. Carpet that has a surface composed of cut ends of pile yarn.

Delustered nylon. Nylon on which the normally high sheen has been reduced by surface treatment.

Denier

1. System or yarn count 9000 m used for synthetic fibers; number of grams per 9000 m of yarn length. One denier equals 4,464,528 yd/lb or 279,033 yd/oz.
2. Unit of weight for the size of a single filament yarn. The higher the denier number, the heavier the yarn.

Density. Calculation used to measure the compactness of face yarns in a carpet. Increased density generally results in better performance.

Dimensional stability. Ability of a fabric to retain its dimensions in service and wet cleaning.

Direct glue-down. Gluing of the carpet directly to the floor with adhesives.

Double back. Webbed backing cemented to the backing of tufted, knitted, and some woven carpets as additional reinforcement, to provide greater dimensional stability. Also known as "scrim back."

Double backing. Second backing added to many carpets to provide dimensional stability. The material used is usually jute, and prevents the carpet from stretching and buckling after it is installed.

Dry-compound cleaning method. This is more a maintenance than a cleaning method, but it is especially effective between cleanings where oily staining or tracked-in grease is a problem. This method utilizes a dry powder impregnated with various combinations of grease solvents, detergents, and brighteners.

Dry-foam cleaning method. This method employs a machine that applies a hot, fairly dry detergent shampoo through a reel brush, with a vacuum pickup immediately behind the brushing section. Holding tanks for clean shampoo and soiled suds and water are mounted on the machine. The higher temperature (150° F) and alkalinity (pH 9.5 to 11) is recommended for the dry-foam method and contribute to its effectiveness.

Dutchman. Colloquial name for a narrow strip of carpet side-seamed to standard-width broadloom to compensate for unusual offsets, sloping walls, etc., but never used as a substitute for good planning and proper stretching techniques.

Embossed. Type of pattern formed when heavy twisted tufts are used in a ground of straight yarns to create an engraved appearance. Both the straight and twisted yarns are often of the same color.

Face seams. Seams, either sewed or cemented, that are made without turning the entire carpet over or facedown. These seams are made during installation when it is not possible to make back seams.

Face weight. Total weight of pile yarns in the carpet measured in ounces per square yard, excluding backing yarns or fabric.

Felting. Process of pressing or matting together various types of hair or fibers to form a continuous fabric, known as "felt."

Fibers. Typical fibers are nylon, acrylic, and modacrylic.

Filling yarn. Yarns, usually of cotton, jute, or kraftcord, running across the fabric and used with the chain yarns to bind the pile tufts to the backing yarns.

Flameproofing. Materials in carpets and rugs treated by special chemicals, as required by many fire codes.

Flame retardant carpet cushion. Carpet cushion or pad for carpet installations. Materials consist of compounded synthetic, high density latex foam rubber with a reinforced cellulose backing. Flame retardant additives provide the safety requirements of the U.S. Bureau of Standards ZZ-C-0811B.

Flocked carpet. Single-level velvety pile carpet composed of short fibers embedded on an adhesive-coated backing.

Flooring radiant panel test. Full-scale test designed to simulate a set of conditions likely to lead to fire spread in a carpet system. The test method determines a critical radiant flux measured in watts per square centimeter or the critical radiant energy necessary for a fire to continue to burn and spread.

Fluffing

1. Lint and fuzz appearing on newly installed carpet, which is merely the factor-sheared pile ends working their way to the surface—not the tufts or pile yarns themselves. Fluffing disappears as the carpet is used.
2. Bits of carpet come off, forming fluffy balls of carpet fiber and lint on the surface. Fluffy balls are formed by bits of the pile that were cut off and not entirely removed from the rug after manufacture.

Frames. Racks at the back of a jacquard loom, each holding a different color of pile yarn. In Wilton carpets two to six frames may be used, and the number is a measure of quality, as well as an indication of the number of colors in the pattern, unless some of the yarns are buried in the backing.

Frieze carpet. Rough, nubby-textured carpet using tightly twisted yarns.

Frieze yarn. Tightly twisted yarn that gives a rough, nubby appearance to the pile. In addition to its use in plain colors, it is employed to form designs against plain grounds and thus gives an engraved effect.

Fusion bonding. Process that produces a complete carpet by imbedding pile yarns and adhering backing to a viscous vinyl paste which hardens after curing.

Fuzzing. Temporary condition on a new carpet consisting of irregular fuzzing appearance caused by slack yarn twist, fibers snagging, or yarn breaking. Fuzzing is remedied by spot shearing.

Gauge (gage)

1. Distance, in fractions of an inch, between two needle points.

2. Distance between tufts across the width of knitted and tufted carpets, expressed in fractions of an inch.

Graying. Carpet changes color slightly; additions to carpet do not match. This is caused by the settling of fine dust, but it often looks as if the carpet has faded. Carpet added later will not match but will gradually gray to the same tone. Cleaning will not correct this.

Grin

1. Condition where the backing shows through sparsely spaced pile tufts. Carpets may be grinned (bent back) deliberately to reveal the carpet construction.

2. Backing of the carpet shows between the rows of pile tufts.

Ground color. Background color against which the top colors create the pattern or figure in the design.

Heat-set nylon. Nylon fiber that has been heat treated to retain a desired shape.

Installation systems. Installation systems are rated in the following order: direct glue-down with attached cushion; tackless strip stretched over cushion; and direct glue-down with no cushion.

Jacquard

1. Pattern control on a Wilton loom. A chain of perforated cardboard "cards" is punched according to the design elements, which when brought into position activates the mechanism by causing it to select the desired color of yarn to form the design on the pile surface. Unselected colors are woven "dormant" through the body of the fabric.

2. Mechanism for a Wilton loom that uses punched cards to produce the desired color design.

Jaspe Carpet. Carpet surface characterized by irregular stripes produced by varying textures or shades of the same color.

Jute

1. Strong, durable yarn spun from fibers of the jute plant, used in the backings of many carpets.

2. Fibrous skin between the bark and stalk of a plant native to India and the Far East. Shredded and spun, it forms a strong, durable yarn used in carpet backing to add strength, weight, and stiffness.

Knee kicker. Short tool with gripping "teeth" at one end and a padded cushion on the other, used in making small stretches during carpet laying.

Knitted carpet

1. Type of nonwoven carpet construction in which the backing and stitching yarns are looped together with three sets of "knitting needles," producing an uncut loop pile.

2. Carpet made on a knitting machine by looping together backing, stitching, and pile yarns with three sets of needles, as in hand knitting.

Knitting. Knitting process resembles weaving in that the face and back are made simultaneously. Backing and pile yarns are looped together with a stitching yarn on machines with three sets of needles. Knitted carpets are

usually solid or tweed in color with a level loop pile texture.

Kraftcord. Tightly twisted yarn made from wood pulp fiber, used as an alternate for cotton or jute in carpet backing.

Latex foam. Cushion (pad) reinforced with fiber mesh plus cellulose with a connecting latex cell structure.

Lay-fit method. Method used exclusively for carpet tiles.

Leno weave. Weave in which warp yarns, arranged in pairs, are twisted around one another between "picks" of weft yarn.

Lip. Chain and/or stuffer left on the edge of carpet after it has been cut.

Loom

1. Machine in which yarn or thread is woven into a fabric by the crossing of the warp or chain by other threads, called the weft or filling, at right angles to the warp threads.

2. Machine on which carpet is woven, as distinguished from other machines on which carpets may be tufted, flocked, or punched.

Loomed. Term applied to carpet made on a modified upholstery loom with characteristic dense low-level loop pile, generally bonded to cellular rubber cushioning.

Loop pile. Face yarns of a carpet in which the loops are woven into the body and not cut.

Luster fabric. Cut-pile fabric woven with surface yarns spun from special types of staple and chemically washed, like hand woven oriental fabrics, to give a bright sheen or luster.

Matting. Pile flattens as a result of foot traffic or the weight of furniture. Pile should be brushed and furniture rearranged occasionally.

Miter. Junction of two pieces of carpet, wood, or other material at an angle, usually 45° to form a right angle, but it may be any combination of angles.

Moresque

1. Special coloring or textural effect created by winding together in the spinning process two or more yarns of different colors or tones of the same color.

2. Multicolored yarn made by twisting together two or more strands of different shades or colors.

Multilevel carpet. Carpet with a texture or design created by different heights of tufts of either cut or uncut loop.

Nap. Pile on the surface of a carpet or rug.

Narrow carpet. Fabric woven in widths of 27 and 36 in., as distinguished from broadloom.

Natural gray yarn. Unbleached and undyed yarn spun from a blend of black, brown, or gray wools.

Noil. By-product in worsted yarn manufacture consisting of short wool fibers, less than a determined length, which are combed out.

Nylon carpet fiber. Nylon was first tufted in 1947 and has now been developed to be antistatic and antibacterial, to resist fading, bleaching agents, and cleaning solutions.

Olefins. Long-chain synthetic polymers composed of at least 85% by weight ethylene, propylene, or other olefin units. Currently, only polypropylene has been produced in fiber form for carpet manufacture.

Omalon carpet foundation (cushion pad). Proprietary name of a high-density, high-bulk, homogeneous, unfilled polymeric foam with a tightly laminated fabric slip surface.

Package dyeing. Placing spun and wound yarn on large perforated forms and forcing the dye through the perforations.

Padding

1. Cellular rubber, felted animal hair, or jute fibers in sheet form, used as cushioning under carpet.

2. Cushioning material for installation under rugs and carpet, made of felted cattle hair, jute, wool, and combinations of these or of rubber and plastic foam.

Patent-back carpet. Carpet so constructed that the fabric can be cut in any direction without the edges raveling. Edges are joined by tape and adhesives instead of being sewed.

Pattern. Decorative feature of a carpet that distinguishes it from being plain; sketch to scale of a design.

Piece dyeing. Immersing an entire carpet in a dyebath to produce single or multicolor pattern effects.

Pile

1. Raised yarn tufts of woven, tufted, and knitted carpets that provide the wearing surface and desired color, design, or texture. In flocked carpets, the upstanding, nonwoven fibers.

2. Upstanding fibers or wearing surface of a carpet.

Pile crushing. Bending of pile due to foot traffic or the pressure of furniture.

Pile height

1. Height of the loop or tuft from the surface of the backing to top of the pile, measured in fractions or decimals of an inch.

2. Height of pile measured from the top surface of the back to the top surface of the pile, not including the thickness of the back.

Pile setting. Brushing after shampooing to restore the damp pile to its original height.

Pile warp. Lengthwise pile yarns in Wilton carpets that form part of the backing.

Pile yarn. Yarn used to form the loops or tufts of a pile fabric.

Pile yarn density. Weight of pile yarn per unit of volume in carpet, usually stated in ounces per cubic yard.

Pilling

1. Appearance defect associated with some staple fibers where balls of tangled fibers are formed on the carpet surface and are not removed readily by vacuuming or foot traffic; pills can be removed by periodic clipping.

2. Condition in certain fibers in which strands of the fiber separate and become knotted with other strands causing a rough, spotty appearance. Pilled tufts should never be pulled from carpet but may be cut off with a sharp scissors at the pile surface.

Pill test. Federal regulations require that all carpet sold in the United States pass the pill test. This test screens out carpet easily ignited by a small incendiary source; it measures the response of carpet exposed to a timed burning tablet in the absence of an imposed external radiation field.

Pitch

1. Number of tufts or pile warp yarns in 27-in. widths of woven carpet.

2. Number of pile ends per inch of width. Actually, in practical floor covering specifications, it is taken as the number of pile ends per unit of standard 27-in. width. The terms of pitch used commonly in the industry are 180, 189, 192, 216, and 256.

3. Numbers of loops or yarns across a 27 in. width of carpet usually varying between 162 and 256.

Pitch or gauge. Pitch and gauge are equivalent and represent the number of ends of pile yarns across the width or carpet. In woven carpet pitch is the number of ends of yarn in 27 in. of width: 216 pitch ÷ 27 in. = 8 ends per inch. In tufted carpet gauge is the spacing of needles across the width of the tufting machine expressed in fractions of an inch: $\frac{1}{8}$ gauge = 8 ends or needles per inch.

Planting. Method of placing spools of different colors of surface yarn in the frames in back of jacquard Wilton looms so that more colors will appear in the design than are supplied in the full solid colors used. These extra "planted" colors are usually arranged in groups of each shade to give added interest to the pattern.

Plush carpet. Smooth face-cut pile surface that does not show any yarn texture.

Ply

1. Number of single ends of yarn twisted together to form a heavier, larger yarn, for example, 1300/3 (3 ends of 1300 denier yarn plied together). Ply is not a measure of quality.

2. Ply designates the number of single strands used in the finished yarn.

3. Layer or thickness of yarns used in carpet. If the pile yarn is described as "four-ply," it means that each tuft is made of four yarns spun together. A ply is one strand of yarn thickness.

Point. One tuft of pile.

Polypropylene. Olefin used in carpet manufacturing.

Polyester pneumacel (cushion pad). Generic material made of pneumatic cellular fibers. Resilient pneumatic strand, sheet, or cushion composed of a cellular polymeric structure having predominantly closed cells that are inflated to higher than surrounding pressure with two or more gases, one of which is impermeable to the cell walls.

Power stretcher. Extension-type version of the knee kicker with larger "teeth" arranged in a patent head, which can be adjusted for depth of "bite," used to stretch larger areas of carpet than can be handled by the knee kicker.

Primary and complementary colors. Primary colors are the principal colors that make up "white" light. Complementary colors are colors that when combined produce a neutral color or whitish gray, as when orange and blue are combined or mixed.

Primary backing. In tufted carpets, a woven or nonwoven fabric into which pile yarns are attached, usually jute or polypropylene. In woven carpets backing yarns are usually kraftcord, cotton, polyester, jute, or rayon.

Print dyeing. Screen printing a pattern on carpet by successive applications of premetalized dyes, which are driven into the pile construction by an electromagnetic charge.

Puckering. Condition in a carpet seam due to poor layout or unequal stretching wherein the carpet on one side of the seam is longer or shorter than that on the other side, causing the long side to wrinkle or develop a "pleated" effect.

Punched carpet. Carpet made by punching loose, unspun fibers through a woven sheet which results in a pileless carpet similar to a heavy felt. Such a carpet usually consists entirely of synthetic fibers.

Quarter. Quarter of a yard or 9 in., formerly used as a unit of carpet-width measure. A 27-in. carpet is designated $\frac{3}{4}$ carpet, and a 36-in. carpet is known as 4/4 carpet. Actual feet and inches are usually given in describing carpet width.

Quarter width. Unit of yard measure one quarter used in referring to carpet or loom widths. Carpets are woven in widths of 27 in. or $\frac{3}{4}$ yd.

Repeat. Distance lengthwise from one point in a figure or pattern to the same point at which it again occurs in the carpet.

Resilience. Ability of a carpet fabric or padding to spring back to its original shape or thickness after being crushed or walked upon.

Resist printing. Placing a dye-resist agent on carpet prior to piece dyeing so that the pile will absorb color according to a predetermined design.

Restretch. Remedial step necessary for the correction of improperly laid carpet resulting from the application of wrong stretching techniques, carpet defects, or undetermined causes.

Reverse coloring. Changing yarn frames in Jacquard weaves to cause the interchanging of ground and top colors.

Riser. Vertical or front surface of a step, rising from the back of a tread. (Data for computing carpeting for stairs.)

Rotary shampoo cleaning method. Rotary machines feed detergent dissolved in tepid water from a tank mounted on the machine through a revolving brush or mop. A rotating brush works suds into the pile. A separate wet-vacuum is then used to extract the suds-dissolved dirt from the pile.

Round-wire or loop-pile carpet. Wilton or velvet carpet woven with the yarn uncut.

Rows or wires. Number of lengthwise yarn tufts in 1 in. of carpet. In Axminster of chenille, they are called "rows," but in Wilton and velvet they are known as "wires."

Rows per inch or wires per inch. In all woven carpets except knitted, the yarn is looped over a wire, which can vary in thickness to create the pile. The thickness of the wire determines the height of the pile.

Rubberized cushion. Cushion consisting of conventional cleaned animal hair and India fiber with rubberized backing, or India fibers and rubberized process on both sides.

Rug. Soft loose floor coverings laid on the floor but not fastened to it. A rug does not cover the entire floor.

Scallops. Up-and-down uneven effect along the edge of carpet caused by indentations where tacks are driven.

Scribing. Transferring the exact irregularities of a wall or other surface onto a piece of paper, which is then cut to fit those irregularities and then used as a pattern to cut the carpet.

Scrim. Rough, loosely woven fabric often used as a secondary backing on tufted carpets.

Scrim back. Double back made of light, coarse fabric, cemented to a jute or kraftcord back in tufted construction.

Sculptured. Type of pattern formed when certain tufts are eliminated or pile yarns drawn tightly to the back to form a specific design in the face of the carpet. The resulting pattern simulates the effect of hand carving.

Sculptured carpet. Carpet with surface designs created by combinations of cut and loop pile and/or variations in pile height.

Secondary backing. Extra layer of material laminated to the underside of the carpet for additional dimensional stability and body, usually latex, jute, foam, sponge rubber or vinyl.

Selvage

1. Finished lengthwise edge of woven carpet that will not unravel and will not require binding or serging.

2. Edge of a carpet so finished that it will not ravel or require binding or hemming.

Serging

1. Method of finishing a lengthwise cut edge of carpet to prevent unraveling, distinguished from finishing a cut end, which may require binding.

2. Serging, also known as "oversewing," is a method of finishing the edge of carpet where it has been cut. It is customary to serge the side and bind the end.

Set or drop match. In a set-match carpet pattern, the figure matches straight across on each side of the narrow carpet width; in a drop match, the figure matches midway of the design; in a quarter-drop match, the figure matches one-quarter of the length of the repeat on the opposite side.

Sewing pole. Piece of wood or other material, more or less rounded, over which carpet may be laid prior to opening up the fabric in order to facilitate sewing and other related operations. Most carpet layers prefer a wood pole, about 4 in. in diameter, that has been slightly flattened on one side.

Shading

1. Carpet appears to be lighter or darker in certain spots as a result of the bending of cut pile fibers so that light is reflected from the side.

2. Crushing or bending out of cut pile fibers so that reflected light from the side (rather than from the top) gives the illusion of a light spot on the rug; viewed from the opposite side of the rug or carpet, the spot will appear dark. Shading is not a defect but rather a characteristic of cut pile fabrics.

Shag carpet. Carpet with a surface that consists of long twisted loops.

Shearing

1. Carpet finishing operation that removes stray fibers and fuzz from loop pile and produces a smooth, level surface on cut pile.

2. Process in manufacture in which the fabric is drawn under revolving cutting blades, as in a lawn mower, in order to produce a smooth face on the fabric.

Shedding. Temporary condition of dislodged loose short fibers in new carpeting after initial exposure to traffic and sweeping.

Shot. Number of weft yarns in relation to each row of pile tufts crosswise on the loom. A two-shot fabric has two weft yarns for each row of pile tufts; three-shot fabric has three weft yarns for each row of tufts.

Side seams. Seams running the length of the carpet; also called "length seams."

Skein-dyed yarn. Surface yarn spun from white fiber staple and dyed in kettles or vats by immersion in skein form.

Skein dyeing. Immersing batches of yarn (skeins) in vats of hot dye.

Solution-dyed yarns. Man-made fibers dyed in liquid form before becoming solid threads. In this method the dye becomes part of the fiber.

Solution dyeing. Adding dye or colored pigment to synthetic material while it is in liquid solution before its extrusion into fiber.

Space dyeing. Applying alternating bands of color to yarn by rollers at predetermined intervals prior to tufting.

Spot removal. The single most difficult problem in carpet maintenance is that of stains, usually the result of spills. Immediate action following a spill is essential in removing stains that might be extremely difficult if not impossible to remove completely after drying.

Sprouting
1. Condition in new carpet caused by long ends of pile protruding above the surface of the rug. These ends were unclipped during the factory shearing process; they should never be pulled out but may be clipped off with a scissors level with the pile surface.
2. Temporary condition on new carpets where strands of yarn work loose and project above the pile. The condition can be remedied by careful clipping or spot shearing.

Staple fibers
1. Fibers in their natural state before being spun into yarns.
2. Relatively short natural (wool) or synthetic fibers, ranging from approximately $1\frac{1}{2}$ to 7 in. in length, which are spun into yarn.

Static control. Control of static and the production of antishock carpet with interwoven stainless steel fibers or static control yarn.

Static electricity. Persons who receive an electrical shock when touching metal after being in contact with a rug. This occurs with any carpet when it is new or when the humidity is low (in the winter, for instance, in a centrally heated house). Rugs now can be antistatic finished or treated fibers are used to eliminate or reduce this problem.

Stay tacking. Temporary tacking of the carpet in stages during a long stretch with the power stretcher or knee kicker to "hold the stretch" until the end of the carpet can be kicked over the pins.

Steam extraction cleaning method. Most steam extraction machines have separate twin tanks for the cleaning solution and for the dirty water vacuumed from the carpet. The water temperature is kept between 130° F and 160° F depending on fiber type, and this hot solution is sprayed deep into the pile. There is never any actual live steam, although steamlike vapor will rise from the carpet as it is being cleaned. The soil-laden water, flushed to the surface, is immediately picked up by the vacuum section of the machine.

Stitches, or rows, or wires. These are equivalents and represent the number of tufts, counting lengthwise, per 1 in. of carpet.

Stitches per cushioning inch. Number of stitches counted along the length of a tufted carpet.

Stock-dyed yarn. Surface yarn spun from fibers that have already been dyed in staple form in large quantities.

Stock dyeing. Dyeing raw fibers before they are carded, combed or spun.

Stria carpet. Striped surface effect obtained by loosely twisting two strands of one shade of yarn with one strand of a lighter or darker shade.

Stuffers. Extra yarns running lengthwise through the fabric to increase weight and strength.

Stuffer warp. Yarn that runs lengthwise in the carpet but does not intertwine with any filling (weft shot) yarns; it serves to give weight, thickness, and stability to the fabric.

Stuffer yarns. Series of extra yarn running lengthwise in a fabric to add weight and bulk to form the backing.

Supportive cushioning. Attached or detached, thick or thin, soft or semirigid cushioning to extend the wear life of most carpets.

Tackless strip. Pin frames are fastened around the perimeter of an area, and the carpet is secured by stretch-tensioning and attached to the anchoring pins.

Tapestry. Looped pile fabric woven on the velvet loom.

Template. Pattern, generally of paper or cardboard, for shaping carpet to be cut.

Texture. Surface effect obtained by using different heights of pile or two or more forms of yarn, by alternating the round and cut pile wires, by "brocade" engraving, by simulated or actual carving or shaving with an electric razor, or by specially treating the design in another way to give added interest beyond that provided by the woven design or tones.

Top colors. Colors of the yarn used to form the design, as distinguished from ground color.

Tone on tone. Carpet pattern made by using two or more shades of the same color.

Total weight. Includes the face weight and backing.

Traffic. The passing to and fro of persons, with special reference to carpet wear resulting therefrom.

Tuft density. The total tufts per square inch. Ends × tufts = tuft density.

Tufted carpet
1. Carpet made by inserting face yarn or tufts through premanufactured backing by the use of needles, similar in principal to a sewing machine. Yarns are held in place by coating the back with latex, and a secondary back is applied to add body and stability. A variety of textures is possible.
2. Carpet or rug fabric that is not woven in the usual manner but formed by the insertion of thousands of needles that punch tufts through a fabric backing on the principle of the sewing machine.

Tufts. Cut loops of a pile fabric. Term applies to both woven and tufted carpets.

Tufts per square inch. Tufts per square inch are calculated by multiplying the number of ends across the width (gauge or pitch) by the number of tufts lengthwise (stitches or rows) per inch.

Tunnel test. ASTM E 84 was developed to evaluate the fire hazards of wall and ceiling materials. There is agreement now in the fire protection community that the tunnel test is not adequate for testing the fire hazards of carpet as a flooring material.

Uncut loop carpet. Pile yarns are continuous from tuft to tuft, forming visible loops.

Vacuuming. A systematic vacuuming or maintenance program is essential not only to maintain appearance, but also to control the abrasive effect of soil and grit below the surface of the pile. The frequency of vacuuming largely depends on the volume of traffic.

Vat dyes. Dyes formed in fabrics by oxidation and precipitation of the original dye liquor, such as indigo. Vat dyeing refers to a kind of dye rather than to a method of dyeing. Raw stock dyeing, skein dyeing, or solution dyeing can be performed with vat dyes.

Velvet

1. Pile fabric woven on a velvet loom. It is the simplest of all carpet weaves and is used mostly for solid colors. The carpet is woven with wires, a looped pile being created when the wires are withdrawn. A cut pile results when knife blades on the ends of the wires cut the loops. Tightly twisted yarns in cut pile provide a frieze surface. Varied heights of looped pile can be achieved by using shaped wires.

2. Simplest of all carpet weaves. Pile is formed as the loom loops warp yarns over wires inserted across the loom. Pile height is determined by the height of the inserted wire. Velvets are traditionally known for smooth cut-pile plush or loop pile textures, but they can also create hi-lo loop or cut-uncut textures. Usually the carpet is solid, moresque, or striped in color.

3. Carpet made on a simple loom, usually of solid color or moresque, with cut or loop pile of either soft or hard twisted yarns.

Vinyl foam cushioning. Carpet cushioning made from a combination of foamed synthetic materials.

Waffled sponge cushion. Carpet cushion or pad used under carpet installations compounded from specially developed elastomers for maximum tensile strength and durability.

Warp

1. Series of yarns in the backing that extend lengthwise in the loom.

2. In a fabric, threads or yarn running lengthwise between which the cross threads (or weft, filling) are woven.

3. Backing yarns running lengthwise in the carpet.

Warp chain. Warp yarn woven over the weft.

Weaving

1. Process of forming carpet on a loom by interlacing the warp and weft yarns.

2. Formation of carpet on a loom by interlacing the warp and filling threads.

Weft. The yarn used across the loom sometimes called a "woof." A weft shot is a yarn that crosses the loom on a shuttle usually at high speed.

Weft or woof

1. Backing yarns that run across the width of the carpet. In woven carpets the weft shot (filling) yarns and the warp chain (binder) yarns interlock and bind the pile tufts to the backing. In tufted carpets, pile yarns that run across the carpet are also considered weft yarns.

2. Threads running across a fabric from selvage edge to selvage edge, binding in the pile and weaving in the warp threads.

Wilton

1. Fabric woven on a loom controlled by a jacquard pattern device, which raises one of from two to six surface yarns over a bladed pile wire that is then withdrawn to cut the tufts and give a plushlike face. Other yarns run "dormant" through the center and back of the fabric in the warp direction.

2. The operation of a Wilton loom is basically the same as that of a velvet loom with the addition of a jacquard mechanism with up to six colors or frames. Owing to one color being utilized in a surface at a time, other yarns remain buried in the body of the carpet until utilized. Wilton looms can produce cut-pile, level loop, multilevel or carved textures.

3. Carpet made on a loom employing a jacquard mechanism, which selects two or more colored yarns to create the pile pattern.

Wire height. Pile height expressed in thousandths of an inch.

Wires (pile wire, gauge wire, standing wire)

1. Metal strips over which the pile tufts are formed in woven carpets.

2. Metal strips inserted in the weaving shed under the surface yarns to form loops when the yarns are bound by the weft shuttle in the velvet and Wilson weaves. A round wire will withdraw, leaving uncut pile loops for round-wire velvet and Wiltons, while bladed knife ends on flat wires will cut these loops, forming a plushlike surface of tufts. The term has come to indicate the number of rows of tufts per inch of warp; thus 11 wires means 11 rows of tufts to each inch of length.

Woolen yarn

1. Soft, bulky yarn spun from both long and short wool fibers that are not combed straight but lie in all directions so they will interlock to produce a feltlike texture.

2. Yarn spun from short fibers that are interlocked as much as possible in the spinning operation and twisted. In a cut-pile fabric, woolen yarn resists the penetration of dirt particles but will "shed" in use.

Wool fibers. Fibers are elastic and can be stretched without rupture and still recover their original dimensions. The outer layer of wool fiber sheds water while water vapor passes through its microscopic pores, which makes wool suitable for climatic extremes.

Worsted yarn

1. Yarn spun from longer types of staple, carded to lay the fibers as nearly parallel as possible, and then combed to extract the shorter fibers as "noil." Fibers will stand independently erect in a cut-pile fabric, with no short staple to "shed" in use.

2. Strong, dense yarn made from long staple fibers that are combed to align the fibers and remove extremely short fibers.

Woven carpet

1. Woven carpet will be either velvet, Axminster, or Wilton. The face and back are formed by the interweaving of the warp and weft yarns. Warp yarns run lengthwise and usually consist of chain, stuffer, and pile yarns. The weft yarns bind in the pile and weave in the stuffer and chain yarns which form the carpet back.

2. Carpet made by simultaneously interweaving backing and pile yarns on one of several types of looms from which the carpets derive their names.

Wrinkling. Carpet, apparently smoothly installed, suddenly shows wrinkles. Wrinkling is caused by changes in humidity and should correct itself when the air becomes drier.

Yarn. Spun, threadlike material of natural or synthetic origin prepared for use in weaving.

Yarn size. Carpet yarn is measured by the number of yards in length to the ounce of weight in a single ply. Two to four plies of this single strand will be twisted to form the best balance for any particular weave and quality of fabric desired. It is necessary to know yarn sizes and plies in order to make fair comparisons.

Yarn weight or face weight. Measured in ounces per square yard and should be distinguished from total weight. Face weight may be only 26 oz/sq yd, while the total weight is 151 oz/sq yd. The face weight is the actual surface yarn or the yarn exposed to wear.

CERAMIC TILE

Abrasive tile. Nonslip tile having abrasive grain within the body, as in ceramic mosaics, or pressed into the surface, as in quarry or pavers.

Abrasive wear index. When tested according to pre-scribed requirements, tiles should have an abrasive wear index of not less than 25.

Absorption. Relationship of the weight of the water absorbed to the weight of the dry specimen, expressed in percent.

Accessories. Ceramic or nonceramic articles affixed to or inserted in tile work, as exemplified by towel bars, paper, soap and tumbler holders, grab bars, and the like.

Acid-alkali resistant grout. Grout that is resistant to the effects of prolonged contact with acids and alkalis.

Acid cleaning. Cleaning using a liquid consisting of 1 part muriatic acid (36% strength) to 9 parts of water, only if recommended by manufacturer.

Adhesives. Water-resistant organic adhesives used for bonding tile consist of resin- or rubber-type substances in combination with a solvent.

Aggregate. Noncementitious materials, such as the sand in the mortar and sand, gravel, and stone in the concrete fill and in other such similar operations used in connection with the tile work.

Air-slack. Condition where soft-body clay, after absorbing moisture and being exposed to the atmosphere, will spall a piece of clay and/or glaze.

Angle divider. The angle divider is used by the tilesetter to determine the degree of an angle to cut tile. Used for fitting trim, moldings, and floors into corners. Corner angle is measured by adjusting the divider to fit the corner.

ANSI. American National Standards Institute. Organization established 1918 to promote knowledge and voluntary use of approved standards in industry, including engineering and safety design.

Asphalt emulsion. Plastic water suspension of asphalt and clay or other fillers.

Back-butter. Application of adhesive on the back of a tile to supplement the adhesive applied on the setting bed. All button-backed tiles, sheet-mounted mosaic tiles, small cut tiles, and any tile with an uneven back require back-buttering to ensure a strong bond between them and the setting bed.

Backing (backup). Material used as a base, over which a finished material such as tile is to be installed.

Back-mounted tile. Tile packaged in sheet format with the mounting material applied to the back of the tile.

Backs of tiles. Backs of tiles are manufactured either plain, raised, or with depressed designs.

Balanced cuts. Cuts of tile at the perimeter of an area that will not take full tiles. The cuts on opposite sides of such an area, after careful measurements, should be the same size.

Ball clay. Secondary clay, commonly characterized by the presence of organic matter, high plasticity, high dry strength, long vitrification range, and a light color when fired.

Base. One or two rows of tiles installed above the floor.

Basic types. Glazed and unglazed tile.

Beating. Process by which a wooden block and hammer are used to firmly embed tile in the mortar and to bring their faces into the same plane.

Biscuit chips. Glazed-over chips on the edge or corner of the body.

Biscuit cracks. Fractures in the body of the tile visible on both face and back.

Bisque (biscuit). Twice-fired tile after the first firing. Bisque is subsequently coated with glaze and again fired.

Blister. Bloated effect in which an area of the surface of a tile has been raised by gas formation in the interior of the tile.

Block angle. Square of tile specially made for changing direction of the trim.

Bloom. Visible exudation or efflorescence on the surface.

Blots. Green marks or stains on the face.

Body. Structural portion of a ceramic article or the material or mixture from which it is made.

Bond. Adherence of one material to another, as mortar to scratch coat, tile to mortar, or adhesive to backing.

Bond coat. Thin coat of pure portland cement, used as the bond between tile and mortar.

Bond strength. When tested according to prescribed requirements, tile should develop a bond strength of not less than 50 lb/sq in.

Brick-veneer tile. Tile produced by several methods that simulate the appearance of real brick.

Bridge. Straightedge used as a starting line for the laying of tile.

Bright glaze. Colored glaze having a high gloss and a highly reflective surface.

Brown coat. Straightening-up or leveling coat of cement mortar applied to the scratch coat or other backing to prepare the surface to receive tile.

Bull nose. Trim tile having one edge with a convex radius for the general purpose of finishing the top of a wainscot or turning an outside corner.

Bull nose corner. Type of bull nose trim with a convex radius on two adjacent edges.

Bush hammer. Hammer having a rectangular head and corrugated or toothed face, used for roughing concrete to provide a masonry bond.

Butterball. Apply a heavy amount of mortar known as "butterball" to the back of each tile to compensate for unevenness of the scratch coat of plaster.

Buttering. Applying a thin layer of neat cement with a small buttering trowel to the back of each tile before pressing it into the mortar setting bed.

Button-backed tile. Tile manufactured with raised dots or squares on the back that serve to separate the individual tiles when they are stacked in the kiln and ensure that heat circulates uniformly among them.

Calcine. Ceramic material or mixture fired to less than fusion for use as a constituent in a ceramic composition.

Cap. Tile normally used for the top course on a tile wall; bull nose trim tile.

Caulking. Filling in of joints or crevices with a type of mastic, oakum, or other wadding.

Caulking compound. Sealant used for filling cracks such as those between tile and adjacent materials.

Cement grouts. Grouts consist of portland cement blended with ingredients making them smoother, waterproof, white, uniform in color, and more resistant to shrinking than plain cement. They are suitable for grouting all walls subject to ordinary use. Wetting of the tile and sustained moisture of the area are necessary for proper cure.

Cement tile. Tile whose body consists of vibrated portland cements, aggregate, and mineral pigments. The face is buffed and highly polished and all edges are square.

Ceramic adhesive. Used for bonding tile to a surface. Rubber solvents and rubber- and resin-based emulsions can be used as adhesive.

Ceramic mosaic. Tiles that measure less than 6 sq in. in face area and about $\frac{1}{4}$ in. in thickness. The surface is usually unglazed, but glazed ceramic mosaics are often made. They are impervious or have fully vitrified or fairly dense bodies. There are two distinct types of ceramic mosaics, the porcelain and the natural clay types.

Ceramics. General term applied to the art or technique of producing articles by a ceramic process.

Ceramic tile. Tile whose body is made of clay or a mixture of clay, sometimes known as "bisque." Materials are fired by intense heat, about 2000° F to produce a material that can be either nonvitreous, semivitreous, vitreous, or impervious.

Chips. Fragments that scale or break off at the edges of a tile, from rough handling, for example.

Classes of ceramic tile. There are three basic classes of ceramic tile: glazed wall tile, ceramic mosaic and quarry tile, and pavers.

Cleaning. Removal of all excess grout, checking for air holes or gaps. Use of acid for cleaning is not recommended.

Clear glaze. Colorless glaze capable of reflecting color or property of the tile body.

Cleavage membrane. Membrane that provides a separation and slip sheet between the mortar setting bed and the backing or base surface.

Cold joint. Any point in a tile installation where tile and setting bed have terminated and the balance of the surface has lost is plasticity before work is continued.

Colors. Manufacturers offer a wide range of colors. Production processes may cause color changes and may differ from original selection.

Commercial cement grout. Mixture of portland cement and other ingredients to produce a water-resistant, uniformly colored, usually white material.

Compressive strength. The ability of tile adhesive to withstand a heavy load without fracturing.

Composition tile. Hard tile surfacing unit made from a mixture of chemicals. The finished surface can be the mixture of chemicals or can be marble chips to create a terrazzo finish. The unit is made hard by the set of the chemicals; the product is not fired as in the manufacture of ceramic tile.

Conductive dry-set mortar. Water-retentive, presanded, moderately electrically conductive portland cement mortar for installation of conductive ceramic tile by the thin-bed method.

Conductive mortar. Mortar to which specific electrical conductivity is imparted through the use of additives and that has the other normal physical properties of conventional tile mortar.

Conductive setting beds. Conductive setting beds are provided in inorganic electrically conductive cement mortar mixtures of various compounds of electrically conductive organic adhesives.

Conductive tile. Tile made from special body compositions or by methods that result in specific properties of electrical conductivity while other normal physical properties of tile are retained.

Conductor mortars and adhesives. Mortars and adhesives designed for use with conductive ceramic tile in floors where it is necessary to dissipate static electricity and provide a path of moderate electrical conductivity between all persons and conductive equipment making electrical contact with the floor.

Conventional portland cement mortar. Mixture of portland cement and sand, roughly in proportion of 1:6 on floors and of portland cement, sand, and hydrated lime in proportions of $1:5:\frac{1}{2}$ to $1:7:1$ for walls. Cement mortar is the conventional material for setting ceramic tile and is suitable for most surfaces and ordinary types of installation. A thick bed, $\frac{1}{8}$ to 1 in. on walls and $\frac{1}{4}$ to $1\frac{1}{4}$ in. on floors, facilitates accurate slopes or planes in the finished tile work. When membrane and wire mesh are required, additional thickness should be allowed.

Cove. Trim tile forming a concave junction, generally between the bottom wall course and the floor.

Cove base. Trim tile having a concave radius on one edge and a convex radius with a flat landing on the opposite edge, often used as the only course of tile above the floor tile.

Cove—internal. Trim tile having one edge with a concave radius for the purpose of forming a junction between the bottom wall course and the floor and at the same time forming an inside corner.

Cracks. Hairline fissures.

Crazing

1. Minute cracking sometimes visible on the surface of glazed tile, commonly due to a difference of coefficients of expansion and contraction between the materials in the body and those of the glaze.

2. Evidence of crazing consists of fine minute cracking, caused by the manufacturing process.

Crooked edges. Curvature of the sides, either convex or concave, measured along the sides. Degree of crook is the departure from the straight line between two corners, expressed in percentage of the tile length.

Crystalline glaze. Sparkling textured glaze with extreme resistance to abrasion. Crystalline glazed tile may be used for light-duty floors.

CTI (Ceramic Tile Institute of America). Organization established in 1954, works to upgrade installation standards and to test new materials.

Curtain wall. Ceramic tile is used in curtain wall construction in many types of frostproof and styles. Backing surfaces are waterproof and structurally sound. Most tile panels are precast or processed prior to erection.

Cushion edge. Tile edge that is slightly rounded rather than square.

Cushion edge tile. Tile whose facial edges have a distinct curvature, which when the tile is installed results in a slightly recessed joint.

Cut mosaic. Trade term for vitreous unglazed or semivitreous or nonvitreous glazed tiles that are pressed or molded with grooves or notches to facilitate breaking or cutting into irregular tesserae for ungeometrical design or picture work.

Darby. Straight-edge tool used to smooth the surface of mortar, especially on walls and ceilings.

Decorative tile. Tile having special decorative appearance, such as colored designs or pictures in glaze, or tiles having crude, rough facial texture simulating handmade or "faience" tile. All types of tile are included.

Die. Metal or plaster-of-paris plates used to shape dust-pressed tiles.

Domestic manufacture. The following materials are manufactured domestically: various kinds of clay, silica, flux, metallic oxides for coloring glazes, cements, and marble. Methods of manufacturer are dust-pressed, plastic, and cast. There are two basic kinds of ceramic tile, glazed and unglazed.

Dot-mounted tile. Tile packaged in sheet format and held together by plastic or rubber dots between the joints.

Down angle. Trim tile with two rounded or curved edges that serve to finish off an outside corner.

Drying. Removal by evaporation of uncombined water or other volatile substance from a ceramic raw material or product, usually expedited by low temperature heating.

Dry-set grout. Mixture of portland cement and additives providing water retentivity. Dry-set grout has the same characteristics as dry-set mortar. It is suitable for grouting all walls and floors subject to ordinary use.

Dry-set mortar

1. Portland cement containing a water-retentive additive that permits setting dry tile on either wet or dry backing, thus eliminating soaking either the tile or backup during installation.

2. Portland cement, mixed with sand and additives imparting water retentivity, that is used as a bond coat for setting tile. Suitable for use over a variety of surfaces. It is used in one layer, as thin as $\frac{3}{32}$ in., has excellent water and impact resistance, is water cleanable, nonflammable, good for exterior work, and requires no presoaking of tile.

Dry spots. Small areas on the face of tile that have been insufficiently glazed.

Dry-wall grout. Grout with the same additives and characteristics as dry-set mortar, suitable for grouting all walls, subject to ordinary use. It obviates the soaking of wall tile, although dampening is sometimes required under very dry conditions.

Double bull nose. Type of trim having the same convex radius on two opposite edges.

Dust-pressing. Production technique invented by Richard Prosser and patented by Herbert Minton in 1840. Tiles are made from pressed dust, which eliminates shrinkage. The process facilitated tile production and enabled tiles to be made thinner, smoother, and more uniform in appearance.

Dutchman. Cut tile used as a filler in the run of a wall or floor installation.

Edge-mounted tile. Type of mounted tile where tile is assembled into units or sheets and bonded to each other at the edges or corners of the back of the tiles by an elastomeric or resinous material that becomes an integral part of the tile installation.

Efflorescence. Whitish powder or crust that occasionally forms on masonry and sometimes on tile floors and walls. The condition is caused by water contained in the pores of the tile and mortar that dissolves soluble salts which may be present. As the water evaporates, it deposits these salts on the surface. Efflorescence commonly contains sodium and potassium carbonates. Calcium and magnesium salts are often present together with chlorides, nitrates, hydroxides, sulfates, and silicates.

Embossed-back tile. Tile with back embossed with marks to identify manufacturing specifications for a particular production run.

Enameled tiles. Tiles finished by dipping the bisque in glaze of the desired color and refiring. Certain "decorator" colors can only be obtained in this manner.

Encaustic tile. Decorative tile made of clay stamped in relief, glazed, and fired to hardness; used principally for ornamentation.

Epoxy adhesive. Two-part tile adhesive system consisting of epoxy resin and epoxy hardener portions.

Epoxy grout. Epoxy resin and hardener system producing stainproof, chemically resistant joints. It produces smooth, permanently hard, impermeable grout lines and is highly desirable for counters subjected to food and other stains.

Epoxy grout for quarry tiles and pavers. Two-part grout system employing an epoxy resin and hardener, often containing coarse silica filler, especially formulated for industrial and commercial installations where chemical resistance is of paramount importance. Grout has excellent bonding characteristics.

Epoxy grout or mortar. Grouting material or setting mortar made with epoxy resin.

Epoxy mortar

1. Compound developed by the Tile Council of America for use with ceramic and quarry tile. It is especially formulated for use where a high degree of chemical resistance or waterproofness is needed.

2. Two-part mortar system employing epoxy resin and hardener. Mortar is suitable for use where chemical resistance of floors or high bond strength is an important consideration. Acceptable subfloors when properly prepared include concrete, wood, and plywood, steel plate, and ceramic tile. Application is made in one thin layer. Pot life, adhesion, water cleanability before cure, and chemical resistance vary with manufacturer.

Epoxy resin. Synthetic resin of the thermosetting type, used with a catalyst or hardener. When properly used for grouting and/or setting tile, it imparts a very high chemical resistance, together with extreme hardness and permanence of bond.

Epoxy thinset adhesive. One of three types of sand and cement based, thinset tile adhesives, which must be mixed with liquid epoxy resin before use.

Expansion joint. Control joint to allow for expansion and contraction wherever backup is not dimensionally stable.

Expansion strips. Strips of pliable or resilient material inserted between expanses of tile, concrete, or other masonry to provide room for expansion and contraction.

Extra-duty glaze. Durable glaze for tile that is suitable for light-duty floors and all other surfaces on interiors where there is no excessive abrasion or impact.

Extra-heavy loading. Term used to describe tile installation supporting capacity for extra-heavy commercial and industrial uses based on 300-lb loads when using carriers having steel wheels. Equivalent to passing test cycles 1–14 of ASTM test method C 627.

Extruded tile. Tile shaped or formed by extrusion from a clay preparation that contains enough moisture to make it plastic.

Facial defect. Portion of the facial surface of the tile that is readily observed to be nonconforming and that will detract from the aesthetic appearance or serviceability of the installed tile.

Face-mounted tile. Tile packaged in sheet format, with the mounting paper applied to the face of the tile.

Faces of tile. Faces of tile are manufactured as plain with either square or cushion edges.

Factory-mixed mortar. Mechanically preblended carbon and portland cement shipped to the job where sand and water are added in specified ratio.

Faience mosaics. Faience tiles that are less than 6 sq in. in facial area, usually $\frac{5}{16}$ to $\frac{3}{8}$ in. thick, and mounted on paper or webbing to facilitate installation.

Faience tile. Glazed or unglazed tile, generally made by the plastic process, showing characteristic variations in the face, edges, and glaze that give a handicrafted, nonmechanical, decorative effect. FS SS-T-308b.

Fan or fanning. Spacing tile joints to widen certain joints so that they will conform to a section that is not parallel.

Feather edging. Chipping away the body from beneath a facial edge of a tile in order to form a miter.

Feature strip. Narrow strip of tile having contrasting color, texture, or design to the field in which it is used as accent.

Field tile. Tile in the area covering walls or floor, bordered by tile trim.

Filler. Dry mixture of cement and sand, brushed into open tile joints of ceramic mosaics; also called "spacing mixture."

Flexible grouts. Rubber latex, organic, or other proprietary materials that are flexible or that make hydraulic cement more flexible and enable it to cure under dry conditions.

Float coat. Same as "setting bed" when the floating method is used; final mortar coat onto which the bond coat is applied.

Floated bed. Bed of mortar applied to a floor or wall to serve as a setting surface for tile.

Flux. Substance that promotes fusion in a given ceramic mixture.

Foreign manufacture. Imported ceramic and glass tile units do not necessarily conform to domestic standards. They are available in many sizes and shapes not found in domestic selections.

Freeze-thaw stability. Ability of a tile installed outdoors to withstand, without cracking, the cycle of freezing and thawing in colder climates. Porous nonvitreous tiles, which readily absorb water, are less freeze-thaw stable than nonporous vitreous tiles.

Frit. Glass that contains fluxing material and is used as a constituent in a glaze, body, or other ceramic composition.

Fritted glaze. Glaze in which a part or all of the fluxing constituents are prefused.

Frostproof (glazed tile). Tile with a nonabsorbent body, $\frac{5}{16}$-in. thick, and an impervious glaze. It is normally used for counter tops, refrigerators and chemical tank linings, and exterior areas subject to freezing.

Fully integrated thin-bed mortar. Setting bed of carbon, portland cement, and aggregate; a preblended and pretested compound used in $\frac{1}{2}$-in. thickness. It is machine mixed on the job with the addition of water only.

Furan mortar. Two-part system consisting of a furan resin and hardener. Furan mortar is suitable for use where the chemical resistance of floors is an important consideration.

Furan mortar and grout. Compound of furan resins especially formulated for a high degree of resistance to nonoxidizing inorganic and organic acids, alkalies, oils, grease, solvents, water, and steam, at temperatures up to 380° F. It is used extensively with acidproof tile installations.

Furan resin adhesive. Dark-colored thermosetting adhesive liquids used in setting and/or grouting tiles where high acid-alkali resistance is desired.

Furan resin grout for quarry tiles and pavers. Two-part grout system consisting of a furan resin and hardener.

Furring. Stripping used to build out a surface such as a stud wall where strips of suitable size are added to the studs to accommodate vent pipes or other fixtures.

Gel. Chemical change of state in portland cement that occurs 10 to 30 minutes after the addition of water. Cement or mortar should not be used before this reaction has taken place.

Glass mosaic. Small pieces of molded fused silica glass, usually less than 1 sq in. in surface area.

Glass mosaic tiles. Tiles made of glass, usually in sizes not over 2-in. square and $\frac{1}{4}$-in. thick, mounted on sheets of paper that are usually 12-in. square.

Glaze

1. Bright matte or crystalline finishes.

2. Impervious ceramic material produced by heat and used to cover the surface of a tile to provide a variety of color and other properties, such as the prevention of absorption of liquids and gases and resistance to abrasion and impact.

Glazed ceramic mosaics. Ceramic mosaics with glazed faces.

Glazed interior tile. Glazed tile with a body that is suitable for interior use and that is nonvitreous and is not required or expected to withstand excessive impact or to be subjected to freezing and thawing conditions. FS SS-T-308b.

Glazed tile. Glazed tile has a fused impervious facial finish composed of ceramic materials fused onto the body of the tile, which may be a nonvitreous, semivitreous, vitreous, or impervious body. The glazed surface may be clear or colored. FS SS-T-308b.

Glaze fit. The stress relationship between the glaze and body of a fired ceramic product.

Grade mark seals. Tiles are shipped in sealed packages with the grade of the contents indicated by grade seals. Grade seals consist of strips of paper, in distinctive colors, on which the grade name is printed in plain, clearly legible type. Standard paper colors are blue for standard grade and yellow for seconds. Grade seals are placed in such a position that the container cannot be opened without breaking the seals. Grade seals of the applicable paper color are placed on all types of containers.

Grade marking practices. Grade marking is considered fundamental to an understanding of tile grades and to the proper and satisfactory use of tiles. Various grade names are used by architects, manufacturers, and contractors without definition, and the new designations applied from time to time to establish grades or to mixed grades of tiles are indefinite or confusing.

Grades. There are only two recognized grades of tile, standard and seconds, although there are other grades of tile on the market of varied quality. Warpage and manufacturing blemishes on the face of tile are considered in grading. As all grades of tile are made from similar raw materials and by similar processes, uniformity of practice on the part of producers in the sorting and grading of tiles is important. Grade names have significance only when they are defined in terms that are carefully and accurately explained.

Green bisque. The unfired bisque or body of a tile.

Grout. Rich or strong cementitious or chemically setting mix used for filling tile joints.

Grouting. Finishing the joints of all tile work or the spaces between tiles by filling with cement, mortar, or thermosetting material. Commercial cementitious grouts differ from ordinary mortars in that most contain hardeners, bleaches, and shrinkage control fillers.

Grout, plain and sanded. Cement-based material used to fill in joints between tiles. Plain grout is used for joints less than $\frac{1}{16}$-in. wide, while grout to which sand has been added for strength is used for wider joints.

Hand-decorated tile. Hand-decorated tile includes feature strips, picture tile, geometric inserts, allover patterns on a glazed interior tile body. Also available in raised line and hand-decorated tiles under this term.

Hang. Tile adhesive's ability to hold a tile in place on a vertical surface before the adhesive has cured.

Hard tile. Term used to designate types of tile, such as ceramic, glass mosaic, marble tile, and like materials.

Hawk. Square metal or wooden plate with central handle attached to the undersurface, used to hold a supply of mortar.

Heavy-duty tile. Natural clay tiles ideal for service in high traffic areas. This tile is unglazed, slip resistant, highly vitreous, and frostproof.

Heavy loadings. Term used to describe tile installation supporting capacity for heavy commercial and industrial uses, based on 200-lb loads when using carriers having rubber wheels. Equivalent to passing test cycles 1–12 of ASTM test method C 627.

Impervious ceramic tile. Tile with water absorption of 0.5%.

Impervious tile. Extremely dense-bodied tile that, by comparison with nonvitreous tile, is fired at a very high temperature for a long period of time. The bisque of this tile contains less than 0.5% air pockets, is therefore almost waterproof, and is freeze-thaw stable. Some impervious tiles have high compressive strength, while others are very fragile, as indicated by manufacturer's markings.

Initial setting time. Time required for a freshly mixed cement paste, mortar, or concrete to achieve initial set.

Inlaid tile. Same as encaustic tile, except inlaid tile has the figure, ornament, or decoration formed into the tile before glazing or firing.

Inside corner. The joint at the internal angle of two intersecting surfaces.

Interior glazed tile. Glazed tile with a body that is suitable for interior use and that is usually nonvitreous. It is not required or expected to withstand excessive impact or to be subject to freezing and thawing conditions.

Isolation membrane. Covering used on top of a setting bed to prevent seasonal movement in the bed from cracking the tiles.

Jagged edges. Irregularities left on the edges of the tile mainly due to the use of handcutting tools.

Job-mixed mortar. Setting bed consisting of portland cement and sand mortar with carbon black added to mortar as specified.

Leather-hardened tile. Bisque, ready for firing in a kiln, that has dried sufficiently to lose its pliability.

Latex grout. Portland cement with latex additive, suitable for use in all installations subject to ordinary use. It is somewhat more flexible and less permeable than regular cement grout.

Latex portland cement mortar. Mixture of portland cement, sand, and special latex additive. The uses of latex portland cement mortar are similar to those of dry-set mortar. This mixture is somewhat more flexible than portland cement.

Layout lines. Lines chalked on a setting bed to guide the accurate setting of tiles.

Leadless glaze. Ceramic coating matured to a glassy state on a formed article, or the material or the mixture from which the coating is made, to which no lead has been deliberately added.

Leveling coat. Mortar coat applied when needed to produce a plumb and level surface so that subsequent coats can be applied in uniform thicknesses.

Light traffic and loading. Term used to describe tile installation supporting capacity for light commercial and residential uses based on 200-lb loads when using carriers having hard rubber wheels. Equivalent to passing test cycles 1–6 of ASTM Test Method C 627.

Manufactured ceramic tile surfacing units. Units that are usually relatively thin in relation to the facial area. They are made from clay or a mixture of clay and other ceramic materials and are known as the body of the tile. They may have either a glazed or unglazed face, which is fired above red heat in the course of manufacture to a temperature sufficiently high to produce specific physical properties and characteristics.

Manufacturing methods. Tile may be dry-pressed, dust-pressed, plastically extruded, hand or machine molded, or cast in a mold or form. FS SS-T-308b.

Marble mosaic tile. Tile made of small marble tesserae varying slightly in size (usually about $\frac{1}{2}$-in. square) and mounted on sheets of paper to facilitate installation.

Marble tiles. Marble cut into tile sizes 12-in. square or less, usually $\frac{1}{2}$-in. to $\frac{3}{4}$-in. thick. Several types of finishes are available.

Mastic grout. One-part grouting composition that is used directly from the container.

Materials in tile. Tile is composed of various kinds of clay, silica, flux, metallic oxides for coloring glazes, cements, and marble.

Matte glaze. Dull glaze providing a uniformly smooth glazed surface that is almost lacking in gloss and not reflective of an image.

Membranes. Membranes are used under tile floors in wet locations to prevent possible leaking to floors below. A membrane consists of several layers of bituminous coated felts or fabrics or metal pans.

Metal lath. Lightweight expanded, painted, or galvanized lath, depending on location of use.

Mexican paver tiles. Terra cotta-like tile, handmade, used mainly for floors, varying in color, texture, and appearance from tile to tile. Available up to 12-in. square, hexagon, octagon, and many other shapes. Tiles are coated with sealers to prevent absorption and to provide a wearing surface.

Mixed shades. Decided variations in color values.

Moderate loadings. Term used to describe tile installation supporting capacity for normal commercial light institutional and industrial uses based on 300-lb loads when using carriers having rubber wheels, and occasional use of 100-lb loads, when using carriers having steel wheels. Equivalent to passing test cycles 1–10 of ASTM test method C 627.

Mortar. Cementitious mixture of hydraulic cement and water, which usually contains sand and may contain lime or other additives such as dry-curing additive, and which subsequently hardens to perform a bonding, leveling, and filling function.

Mortar hawk. Square metal or wooden plate with central handle attached to the undersurface, used to hold a supply of mortar.

Mosaics. Small tiles or bits of tile, stone, glass, etc., which are used to form a surface design or intricate pattern. They are often laid in mortar or adhesive and usually have the joints grouted.

Mosaic tile. Glass or vitreous porcelain or clay tiles that are 2-in. square or smaller, usually unglazed, and generally packaged in sheet-mounted format.

Mounted tile. Tiles assembled into units or sheets by suitable material to facilitate handling and installation. Tiles may be face-mounted, back-mounted, or edge-mounted. Face-mounted tile assemblies may have paper or other suitable material applied to the face of the tiles, usually water-soluble adhesives that can be easily removed after installation, but prior to grouting of the joints. Back-mounted tile assemblies may have perforated paper, fiber mesh, or other material bonded to the back and/or edges of each tile, becoming an integral part of the tile installation.

Natural clay tile

1. Tile made by either the dust-pressed method or the plastic method, from clays that produce a dense body having a distinctive, slightly textured appearance. FS SS-T-308b.
2. Tile made from natural clay or shales by the dust-press method. Restricted in color owing to the limited colors of the raw clays, but is strong and dense. It can be rendered nonslip by the addition of alundum or carborundum to the mix. The tile is generally obtainable with either cushion or square edges.

Nippers. End-cutting pliers used for making irregular cutouts in tile.

Nominal size. The nominal size is approximate but lies within a narrow range of sizes, due to the variation in processes and systems employed by various manufacturers.

Nonslip tile. Tile having greater nonslip characteristics due to an abrasive admixture, abrasive particles imbedded in the surface, or grooves or patterns in the surface or because of a natural nonskid surface characteristics.

Nonstaining grout. Epoxy or resin-type nonstaining grouts are used for surfaces and areas subject to excessive staining action.

Nonvitreous ceramic tile. Tile with water absorption of over 7%.

Notched trowel. Trowel having a serrated or notched edge for spreading tile mortar or adhesive in ridges of a definite thickness.

Opaque glaze. Nontransparent colored or colorless glaze.

Open time. Period of time during which the bond coat retains its ability to adhere to the tile and bond the tile to the substrate.

Orange peel. Pitted texture of a fired glaze resembling the surface of an orange peel.

Organic adhesive. Organic adhesive consists of a rubber or resinous substance in combination with a solvent. The solvent gives a degree of fluidity that allows spreading, while the rubber or resinous material provides a film or layer adherent to the backing and tile. Organic adhesive, after its application over a backing material, is transferred from a liquid to a solid condition by evaporating the solvent to the surrounding atmosphere and to the backing, when this is relatively porous.

Organic adhesive (mastic). Prepared organic material, ready to use with no further addition of liquid or powder, that cures or sets by evaporation.

Packing house tile. Similar to quarry tile, but usually of greater thickness.

Paver. Unglazed porcelain or natural clay tile formed by the dust-pressed method and similar to ceramic mosaics in composition and physical properties but relatively thicker and with 6 or more square in. of facial area. FS SS-T-308b.

Pavers. Pavers are larger and thicker than ceramic mosaics but have the same characteristics. Tile is available with nonslip grain. Pavers are normally used for heavy-duty service.

Penetrating oil. Type of sealer used to protect the surface of unglazed tile, that penetrates the bisque of the tile.

Perimeter joint. Expansion joint placed at the margin of an interior floor between 16-ft squares and 24-ft squares, or at a wall 12 ft to 24 ft long. The joint allows for seasonal expansion in the floor or wall while preventing damage to the tiles.

Pimples. Small surface bubbles or blowouts resulting from the explosion of gas during firing.

Pinholes. Imperfections resembling pinpricks in the surface of a ceramic body or glaze.

Pitted. Indentations in the finished surface of individual tiles other than at the corners and edges. These are caused by sharp corners or trowels and other tools of the workmen and are different from manufacturing defects.

Pointing mix. Mortar of stiff paste consistency that is forcibly compressed or tooled into the tile joints where it hardens.

Pointing trowel. Small trowel used for forcing pointing mortar into tile joints.

Pool deck. Natural clay ceramic mosaics recommended for use on pool decks because of its natural coarse-grain, slip-resistant surface.

Pool liner. Porcelain ceramic mosaics are ideal for pool liners because of their $\frac{1}{2}\%$ absorption factor and resistance to stain and chemical abuse.

Porcelain tile. Ceramic mosaic or paver, generally made by the dust-pressed method, consisting principally of china clay and feldspar, resulting in a tile that is dense, fine grained, and smooth, with a sharply formed face, usually impervious. Colors of the porcelain type are usually clear and luminous, granular, or a texture blend thereof.

Porcelain tile mosaic. Tile made by the dust-press process from carefully blended ceramic materials, with a vitreous and impervious body that is resistant to freezing, thawing, and the abrasive wear of foot traffic. Porcelain tiles have a wide range of colors, which are produced by adding mineral oxides to the mix. Colors extend throughout the body, and the tiles have a plain, mottled, textured, or flashed surface.

Precast floor tiles. Floor tiles that best withstand heavy traffic or frost conditions.

Pregrouted ceramic tile sheets. Individual ceramic tile factory assembled into pregrouted sheets of various sizes for various interior floor and wall installations. Such sheets, which also may be components of an installation system, are generally grouted with an elastomeric material such as silicone, urethane, or polyvinyl chloride (PVC) rubber, each of which is engineered for its intended use.

Premounted tile sheets. Tiles premounted in a sheet to make installation easier and faster. Instead of the installation of one tile at a time, each sheet allows up to 2 sq ft to be set all in one easy movement.

Pressure-sensitive adhesive. Adhesive made so as to adhere by briefly applied pressure alone to a surface at room temperature.

Pulls. Small depressions or scratches in the body, noticeable through the glaze at a distance of more than 3 ft.

Pure coat. Thin coat of pure cement and water mixed to a pasty consistency.

Quarry tile. Tile usually 6 sq in. or more in surface area and $\frac{1}{2}$ to $\frac{3}{4}$ in. in thickness, made by the extrusion process from natural clay or shales. It is normally supplied only in unglazed form, but some producers do apply glazes for special effects.

Rack. Metal grid that is used to properly space and align floor tiles.

Radius bullnose. Curved tile used to border and complete a field of tile installed on a raised setting bed.

Ragging off. Procedure of spreading and pulling damp cheese cloth over the tile surface during the tile grouting process in order to remove excess grout and clean the tile.

Raw glaze. Glaze compounded primarily from raw constituents. It contains no prefused materials.

Receptor. Metallic or nonmetallic subpan for a tile shower stall or an entire tile bathroom.

Red body swimming pool tile. Brightly glazed, frostproof tile designed primarily for depth markers for outdoor installations.

Reference lines. Pair of lines, chalked on a setting bed, that intersect at a 90° angle and establish the starting point for plotting a grid of layout lines to accurately guide the setting of tiles.

Reinforcing mesh. Wire mesh used over a mortar substrate being floated or at midpoint in a mortar bed to strengthen the bed surface.

Residential loadings. Term used to describe tile installation supporting capacity for a normal residential foot traffic and occasional 300-lb furniture or equipment loads when using carriers having soft or inflated rubber tire wheels. Equivalent to passing test cycles 1–3 of ASTM test method C 627.

Ridge-backed tile. Tile manufactured with a series of ridges on the back that increase the surface area contacting the tile adhesive to improve the adhesive's grip.

Rodding. Leveling the brown coat or the float coat on walls to which tile is to be installed.

Salt glaze. Glaze produced by the reaction, at elevated temperature, between the ceramic body surface and the salt fumes in the kiln atmosphere.

Sand

1. Noncoherent rock particles smaller than $\frac{1}{8}$ in. in maximum dimension.

2. Grains of sand imbedded in the glaze.

Sand holes. Tiny pits visible in the surface of the tile.

Sand-portland cement grout. On-the-job mixture of one part portland cement to one part fine graded sand, used for ceramic mosaic tile; on-the-job mixture of one part portland cement to two parts fine graded sand, to which one-fifth part of lime may be added, used for quarry tile and pavers.

Scarifier. Piece of thin sheet metal with teeth or serrations cut in the edge (like a saw blade); piece of metal lath used to roughen mortar surfaces to receive fresh mortar.

Scarred faces. Surface blemishes caused by scrapping or other marring of the tile.

Scratch coat. First coat of mortar applied directly to the metal lath or other backup to provide a sufficiently strong, level, and rigid surface to receive the mortar setting bed or float coat.

Scratched. Tiles that have surface scratches caused from sand, tools, or rough handling.

Screed. Straight-edge tool drawn along and upon screed strips to work mortar into a level, plumb, and true plane.

Screed strips. Wood, metal, or plastic strips placed in the same plane by plumb, rule, and straight edge, forming gauges or guides upon which the screed is worked to obtain a true mortar surface.

Sculptured tile. Tile with a decorative design of high and low areas moulded into the finished face of the tile.

Scum. Lack of gloss, crystalline or frosted in effect, appearing on the face of the tile.

Seconds (glazed wall tiles). Tiles of this grade may have minor blemishes and defects that are not permissible in standard grade, but they are free from blots and biscuit cracks.

Seconds (grade tolerance). A tolerance of 5% of seconds in the standard grade is permissible.

Seconds (unglazed ceramic mosaic tiles). Tiles of this grade may have minor blemishes and defects that are not found in the standard grade, such as mixed shades, warpage, chipping, sand holes, pimples, scarred face, chips, cracks, and/or other imperfections.

Self-spacing tile. Tile with lugs, spacers, or protuberances on the sides which automatically space tie tile for mortar joints.

Semimatte glaze. Surface with some gloss but that does not reflect a clear image; medium-gloss ceramic glaze with or without color.

Semivitreous tile. Ceramic tile with water absorption of between 3 and 7%.

Setting bed. Layer of mortar into which the tile is installed; also, final coat of mortar when bond coat is not used.

Setting methods. Portland cement mortars, dry-set portland cement mortar, water-resistant organic adhesives, and epoxy mortars are used for setting tile.

Shade. Gradation of color.

Ship and galley tile. Special quarry tile having an indented pattern such as a diamond grid or corduroy scored into the face to produce a nonslip effect.

Shivered edge. Minute fracture of the glaze running along the edge, which appears as a fine, silvery thread when struck by light from an angle.

Shivering. Splintering that occurs in fired glazes or other ceramic coatings due to critical compressive stresses.

Silicone rubber grout. Engineered elastomeric grout system for interior use employing a single-component of nonslumping silicone rubber which upon curing is resistant to staining, moisture, mildew, cracking, crazing, and shrinking.

Sink cap. Tile trim used on the front edge of a countertop or drainboard or around the edge of a sink.

Sink trim. Special tile trim used on the edges around a sink set within a drainboard.

Slaking. Slaking denotes a period of time after the addition of water during which the portland cement granules acquire a gelatinous envelope thereby causing a gel, a chemical reaction within cementitious mortar.

Slip-resistant tile. Tile having greater slip-resistant characteristics due to an abrasive admixture, abrasive particles in the surface or grooves, or patterns in the surface.

Soaping. Method of applying a soapy film to newly tiled walls to protect them from paint, sizing, and plaster droppings during construction.

Spacers. T-shaped and Y-shaped, used in installation to separate tile on walls and floors. Manufactured in various thicknesses from $\frac{1}{16}$ in. to $\frac{1}{2}$ in.

Spacing lugs. Built-in projections on the edges of a tile that, like tile spacers, enable the setter to install tiles with a consistently sized grout joint.

Special methods of setting. Epoxy, furan, and conductive mortars and adhesives are used for setting tile. Epoxy or furan resins are principally used as highly acid-resistant mortars and grouts.

Special-purpose glaze. Some glazes are used for such special purposes as resisting abrasion or imparting special textures.

Special-purpose tile. Tile, either glazed or unglazed, made to meet or have specific physical design or appearance characteristics such as size, thickness, shape, color, or decoration. Some have keys or lugs on backs or sides; special resistance to staining, frost, alkalies, acids, thermal shock, or physical impact; high coefficient of friction; or electrical properties. FS SS-T-308b.

Special tile-setting mortars. In connection with other tile-setting materials, instructions for use and installation specifications should be obtained from the manufacturer and followed carefully.

Speckled glaze. Glaze containing granules of oxides or ceramic stains that are of contrasting color.

Specks. Dark dots on the face less than $\frac{1}{64}$ in. in diameter and noticeable at a distance of more than 3 ft.

Spots. Dark dots on the face more than $\frac{1}{64}$ in. in diameter.

Square-edge tile. Tile whose facial edges are square and produce joints that are flush with the face of the tile.

Stacking. Method of installation whereby glazed wall tiles are placed on the wall so that they are in direct contact with the tile below, as well as with the adjacent tile, without the use of string or other means to form a joint width. Stacking is normally done only with lugged tile, which may be set with either straight or broken joints.

Standard-grade glazed wall tiles. Tiles of this grade are as perfect as is commercially practicable to manufacture. They are harmonious in color, although they may vary in shade, and they are free from warpage, exceeding 0.4% convexity and 0.3% concavity when measured along the edges. They are free from wedging and crooked edges, biscuit cracks, ragged and shivered edges but may have glazed-over biscuit chips, spots, specks, blots, pulls, dry spots, scum, sand, or stickers visible at a distance of more than 3 ft.

Standard-grade unglazed ceramic mosaic tiles. Tiles of this grade are as perfect as is commercially practicable to manufacture. Colors and shades are reasonably uniform. The face of the tile presents a smooth, even surface (excepting natural clay tiles, which have a more rugged face) and is uniform in texture, without chips noticeable at more than 3 ft and free from sand holes or blisters. Slight warpage, not to exceed 1.5% on the edges and 1.0% on the diagonal, is permissible. Blemishes on the backs of the tiles are permitted if they do not affect the appearance or permanence of the installed surface. Joints between all tiles mounted on the permanent backing material are uniform. All unglazed ceramic mosaics fall into one of three classes: impervious, vitreous, and semivitreous.

Standard-grade unglazed quarry tiles. Tiles of this grade are as perfect as is commercially practicable to manufacture. Colors and shades are reasonably harmonious. The face of the tile is a smooth, even surface, uniform in texture, without chips noticeable at more than 3 ft, and free from sand holes or blisters. Reasonable variation in size, thickness, and warpage is permissible.

Stickers. Small rough or raised spots in the glaze.

Stretcher. Running length of a trim tile in contrast to the angles of the same pattern.

Striking (raking). Removing excess grout from between the edges of tile by wiping with a sponge or cloth or scraping with a tool of desired contour.

Structural defects. Cracks or laminations in the body of the tile which detract from the aesthetic appearance of the tile installation.

Substrate. The underlying support for the ceramic tile installation.

Surface trim. Field tile with one rounded edge used to border and complete an installation.

Tapping tile. Inspection technique whereby a coin, key, or other small metallic object is tapped against an installed tile to determine by sound whether the tile is properly bonded to its backing.

TCA (Tile Council of America). Organization established in 1945, composed of companies that produce tile and related products, with the purpose and object of promoting the industry and establishing installation specifications.

Terrazzo tile. Portland cement and sand body made by a mixture of marble chips and portland cement and usually ground smooth.

Textured glaze. Clear, colorless glaze applied to white-bodied, nonvitreous tiles. It is not an opaque white glaze.

Thermal shock. When tested by prescribed requirements, glaze should display no shivering, and the body of the tile should show no dusting or other evidence of disintegration.

Thick-bed installation. Installation using a floated bed of mortar as its setting bed.

Thin-bed or thinset installation. Installation using a non-mortar setting bed, which may or may not use a thinset adhesive.

Thinset. The bonding of tile with suitable materials applied approximately $\frac{1}{8}$-in. thick.

Thinset adhesive. One of three types of powdered, cement-based tilesetting adhesives that must be mixed with a liquid before use. Whether water-mixed, latex or acrylic, or epoxy, this type of adhesive, by comparison with an organic mastic, has greater bond and compressive strength, sets up more quickly, is more flexible when dry, and is more water and heat resistant.

Thinset mortar. The use of mortar-based tilesetting adhesives.

Threshold. Acts as a transitional piece between two different finished floor levels.

Tile. Ceramic surfacing unit, usually relatively thin in relation to facial area, made from clay or a mixture of clay and other ceramic materials, called the "body" of the tile, having either a glazed or unglazed face, and fired above red heat in the course of manufacture to a temperature sufficiently high to produce specific physical properties and characteristics. According to the Federal Trade Commission, the word "tile," when standing alone, means a baked clay or ceramic product.

Tile accessories. Ceramic or non-ceramic articles affixed to or inserted in conjunction with tile work, such as paper, soap, tumbler holders, and towel bars.

Tile adhesives (mastic). Organic adhesives of several types, such as synthetic latex, rubber solvents, and resin and rubber emulsions, used for bonding tile to a surface.

Tile spacer. Plastic device, made in various shapes and thicknesses, used to ensure a consistent grout joint when setting tile.

Tile trim units. Variously shaped units such as angles, bases, beads, bull noses, butterflies, corners, coves, curbs, moldings, nosings, plinths, shoes, sills, stair treads, stops, etc. All shapes necessary or desirable to make a complete installation and to achieve sanitary purposes, as well as architectural design, for all types of tile work are available. Trim tiles are identified by a universal numbering system developed by the Tile Council of America.

Top-coat sealer. Product that seals and protects the surface of unglazed tiles.

Trimmers. Units of various shapes consisting of such items as bases, caps, corners, moldings, angles, and all other shapes necessary to complete an installation.

Underlayment. Applied to subfloors to provide a smooth surface on which tile will be set.

Unglazed tile. Hard, dense tile of homogeneous composition, deriving color and texture from the materials of which it is made. Colors and characteristics of the tile are determined by the materials used in the body, the method of manufacture, and the thermal treatment. The term applies to porcelain and natural clay ceramic mosaic, pavers, and quarry tiles.

Vellum glaze. Semi-matte glaze having a satin-like appearance.

Vertical broken joint. Style of installing tiles with each vertical row of tile offset for half its length.

Vitreous tile. Dense-bodied, nonporous tile that is fired at high temperature for a longer period of time than nonvitreous tile.

Wainscot. Lining or inner surface of the lower portion of walls, usually the tiled lower part of inner walls, as in bathrooms, kitchens, and corridors.

Wall tile. Glazed tile with a body that is suitable for interior use and that is usually nonvitreous, and is not required nor expected to withstand impact.

Warpage. Surface curvature, either convex or concave, measured on the face of the tiles along the edges or on the diagonal; concave or convex curvature of a tile so that the surface is not perfectly flat. The degree of warpage is the variation from the plane expressed in percentage of the tile length.

Water cleaning. Clean water is used for cleaning all tile surfaces while the cement is still in a plastic state.

Water-mixed thinset adhesive. One of three types of sand- and cement-based thinset tile adhesives that must be mixed with water before use.

Wedging. Difference in the lengths of opposite sides, expressed in percentage of the tile length.

Welts. Usually heavy accumulations of glaze in the form of a ridge along the edge.

Wet location. Area or surface prepared for tile work subject to wetting conditions, such as showers or tub and shower recesses.

CLEANING OF SURFACES

Abrasive blast cleaning. Process used for the removal from a concrete surface of all laitance, oil, grease, curing compounds, coatings, and paint by action of a high-velocity stream of abrasive in air or water followed by a blast of air to provide a concrete surface free of abrasive dust or other particles. The abrasive used may be clean silica sand, fine mineral grit, steel shot, steel grit, or mineral slag.

Brushes and sanders. Wire brushes and disc sanders are not very effective in removing tightly adherent coatings.

Chemical stripping. Method in which a chemical paste containing alkaline solvents or acid materials is applied to a surface. The paste dissolves the paint and is removed and disposed of as a monolithic sheet.

Dry ice pellets (solid carbon dioxide). Used to remove paint and rust. The dry ice volatilizes, leaving only the paint or oxide residue.

Environmental Protection Agency. The Resource Conservation and Recovery Act (RCRA) of 1976 provides specific rules for removing, handling, and disposing of potentially hazardous waste.

Power tool cleaning. Power tools reduce the quantity of waste debris generated.

Power tools. Developed to include abrasive embedded rotary flap wheels, nonwoven abrasive discs, and needle guns.

Pressurized water jetting. High-pressure water jets dislodge paint, rust, and contaminants from surfaces to be painted. Small amounts of abrasive can be added to increase efficiency.

Recyclable abrasives. Method using metallic shot and metallic grit or aluminum oxide and silicone carbide.

Sand blasting. Method or system for cutting, abrading, or scarifying concrete, masonry, or steel by a stream of wet or dry sand ejected at high speed from a nozzle by compressed air.

Sodium bicarbonate pellets. Used to remove soft coatings adhered to paint or rust, such as grease and wax. The abrasive is usually wetted to prevent excessive dusting.

Vacuum blasting. An abrasive recovery head connected to and surrounding the blast nozzle. The vacuum suction captures the abrasive and paint residue directly from the surface.

Wet abrasive blasting. Water added to the abrasive stream or spraying a water ring around the nozzle.

COATINGS—PAINTING AND FINISHING

Abrasion. Wearing away of paint film by some external force.

Abrasion resistance. When a surface resists wearing by rubbing, it is abrasion resistant. Abrasion resistance relates to toughness rather than merely hardness and is a necessary quality for floor finishes, enamels, and varnishes.

Absorb. To suck in an applied material such as through the pores in wood grains, or through soft surfaces of plaster and certain stones.

Accelerate. To quicken or hasten the natural or usual course.

Acetone. Water-white volatile solvent with ether-like odor.

Acid. Corrosive chemical substance that attacks decorative finishes and coatings.

Acid number. Value obtained in the analysis of oils, fats etc; number of milligrams of potassium hydroxide required to neutralize the free fatty acid in a gram of the substance.

Acoustic paint

1. Paint formulated to coat acoustical material without materially reducing the acoustical or sound-absorbing qualities.
2. Paint that will not affect the acoustical properties of a surface.

Acrylate resins. Group of thermoplastic resins formed by polymerizing the esters or amides of acrylic acid, used chiefly when transparency is desired. Lucite and plexiglass are in this group.

Acrylic. Monomer or polymer that is characterized by good durability, gloss retention, crystal clarity, and color retention.

Acrylic resin (emulsion type). Resin from which plexiglass is made. A preoxidized resin precluding further oxidation. It especially gives to house paints long life, flexibility, and resistance to water and alkali.

Acrylic resin (solvent type). The solvent type is used in industrial finishes and automobile lacquers for good color retention and resistance to weather.

Actinic rays. Invisible rays in sunlight that cause chemical changes to take place in oil, paint, and varnish coatings, resulting in yellowing, chalking, and disintegration.

Adhesion

1. The bonding of force that makes two materials stick together. Paint with paint is called "intercoat adhesion." Epoxies have great adhesion to most surfaces.
2. Property that causes one material to adhere to another by mechanical and/or chemical force.

3. Satisfactory adhesion of the first coat of paint to the underlying surface is extremely important. The success of a paint job depends to a very large extent on the bonding of the first coat to the substrate. Insufficient or unskilled preparation of the surface may result in defective adhesion.

Adulteration. Dilution in value by the addition of base ingredients; addition of inferior components to those of greater value.

Advancing colors. Warm colors; colors that contain red or yellow to advance or stand out. Advancing colors make an object seem nearer and larger.

Agglomeration. Gathering together of small particles of pigments to form a larger mass.

Aging. Permitting a material to stand for some time and grow old.

Agitator. Mechanical paint mixer.

Air-dry. Dry a coating to its ultimate hardness under normal atmospheric conditions.

Air-drying. Exposing to air at moderate atmospheric temperatures to form a solid film.

Airless spray. Method of forcing a coating through a sprayer by hydraulic pressure rather than air.

Alcohol. Flammable solvent miscible with water. The alcohols commonly used in painting are ethyl alcohol as a shellac solvent, and methyl alcohol or wood alcohol in paint removers.

Aliphatic. Term used to describe a major class of organic compounds, many of which are used as solvents.

Aliphatic solvents. Straight-chain hydrocarbons, such as gasoline, mineral spirits, naptha, and kerosene, usually petroleum products derived from a paraffin-base crude oil.

Alkali

1. Soluble salt that is the opposite of acid. It has the ability to neutralize acid. Also called a "base."
2. Chemical substance characterized by its ability to combine with acids to form neutral salts, damaging to many paints and coatings.
3. Caustic or basic substance which releases hydroxyl ions in aqueous medium. Lye is the most common alkali.

Alkali-refined oil. Drying oil refined with alkali to reduce its acid number.

Alkyd

1. Synthetic resin, usually made with phthalic anhydride, glycerol, and fatty acids from vegetable oils.
2. Chemical combination of alcohol, acid, and oil, used as a vehicle for paint.
3. Most versatile of synthetic resins used in paint chemistry; available in a full range of sheens for interior and exterior use, and made by combining synthetic materials with various vegetable oils (linseed, soya, tung, etc.) to produce clear, hard resins. It possesses good self-sealing properties on many surfaces, eliminates the need for special primer-sealers or undercoaters, and has good weather resistance and gloss retention for exterior exposure. Because of their oil content, alkyds tend to darken with age.

Alkyd-chlorinated paraffin-linseed oil. This substance combines the positive features of alkyd and linseed oil formulations (when properly pigmented) plus excellent color retention in exterior use. It possesses outstanding film resistance to blistering, fumes, and dirt collection and excellent gloss retention and will not stain from rust or copper washdown. Its improved flexibility gives better resistance to cracking. It may be used as its own primer on bare spots when repainting wood surfaces.

Alkyd resin

1. Synthetic resins made by reacting polyhydric alcohols, such as glycerin and the glycols, with dibasic organic acids, such as phthalic, maleic, succinic, and sebacic. A modifying agent is generally present to impart certain properties. Some of these agents are drying, semidrying, and nondrying oils, fatty acids of the oils, natural resins such as rosin; synthetic resins, and other substances.

2. A considerable number of alkyds can be reacted with a wide variety of drying oils in combinations ranging from mostly alkyd to mostly oil to provide a large choice of alkyd-oil combinations useful for numerous air-drying and heat-hardening finishes.

Alkyds. Oil-modified resins that dry faster and are much harder than ordinary oils, used in greater volume than any other paint vehicle. Drying results from both evaporation of the solvent and oxidation of the oil. The more oil there is in a formula, the longer it takes to dry; the lower the gloss is, the better the wetting properties and the better the elasticity. Alkyds are used for making fast-drying enamels for both interior and exterior architectural and industrial paints. They are superior to phenolics in color retention and recoatability. They should not be used directly on masonry or other alkaline surfaces except over an alkali-resisting primer or sealer.

Alkyd varnish. Medium-oil-alkyd clear-gloss varnish for interior wood floors, trim, paneling, and other areas where a clear-gloss-varnish finish is desired. It is resistant to abrasion and direct wear, hot or cold water, soap and common household detergents, fruit juices and alcohol.

Alligatoring

1. Development of interlacing lines over relatively large areas of a paint film, giving the appearance of an alligator's skin. One cause of alligatoring is the application of a hard-drying coat of paint or varnish over a comparatively soft undercoat (for example, a flat white paint applied over a bituminous coating). The outer coat tends to oxidize and harden during drying, after which it contracts and shrinks. By applying the hard topcoat over the soft undercoat, the oxidation and consequent hardening of the undercoat is stopped. Good practice is to allow the priming coat and all subsequent coats of paint or varnish to dry hard before the next coat is applied.

2. Paint film failure that resembles an alligator's skin or a dried-out river bed.

3. Alligatoring is caused by low-quality paint, insufficient drying time between coats, or a hard coating applied over a soft oil-base paint.

4. Paint film condition in which the surface is cracked and develops an appearance somewhat similar to the skin on the back of an alligator.

5. Coarse checking pattern characterized by a slipping of the new paint coating over the old coating to the extent that the old coating can be seen through the fissures.

Aluminum. Prepare aluminum surfaces for painting by removal of corrosion by wire brushing and oil and foreign matter by cleaning with solvent, vapor, alkali, emulsion, or steam. Unweathered metal is treated with surface conditioner before painting.

Aluminum paint

1. Mixture of finely-divided aluminum particles in flake form combined in a vehicle.

2. Fine metallic aluminum flakes suspended in drying oil plus resin or in nitrocellulose.

Aluminum pigment

1. Largely pure metallic aluminum containing appreciable amounts of polishing lubricant, a mixture of stearic and other fatty acids.

2. Nonleafing aluminum paints, usually colored finishes with flakes of aluminum.

Aluminum stearate. Metallic soap used as a flatting agent in varnishes and lacquers and simulating a hand-rubbed appearance in some varnishes. It waterproofs flat paints, reduces the settling of pigments, and regulates the false body of some products. Because it is a "soap," it is used in small quantities.

Amber. Pale brownish yellow color of varnishes and resins.

American gallon. 231 cu in.

American turpentine. Light colored, volatile essential oil obtained from resinous exudates or resinous woods associated with coniferous trees.

Amine resins. Reaction products of formaldehyde and alcohols, water-white thermosetting resins, which are commonly called "melamine" or "urea resins," used to modify properties of alkyd resins, particularly nondrying alkyd resins, and to produce baking finishes for automobiles, refrigerators, etc.

Angular sheen. Sheen given from flat wall paint when viewed from nearly parallel to the surface.

Aniline colors. Organic dyes made from a coal tar base and used as colorants for penetrating wood stains. Pigmented stains get their colors from pigments, not dyes.

Anodizing. Electrolytic surface treatment for aluminum that builds up an aluminum oxide coating.

Anticorrosive paint. Metal paint designed to inhibit corrosion and rusting; applied directly to the metal. It is usually a primer for finish coats and often contains pigments such as zinc chromate or red lead.

Antifoam. Antifoam is a material added to paints, especially water-base paints, to prevent them from bubbling excessively when agitated.

Antifouling paint. Bottom paints containing poisonous substances, such as mercuric or cuprous oxides, to prevent barnacles and marine growth from attaching to the hull of a ship; final coat on a ship's bottom.

Antimony oxide. Pure white pigment that provides the same hiding power as lithopone.

Antique finish. Finish used to simulate age. Chains and buckshot are used to make scratches and indentations on furniture to look used and old, and an antique-looking finish is then applied.

Antiskinning agent. To reduce a paint's tendency to skin over, guiacol (nonvolatile), dipentene (volatile), or other antioxidants are often added.

Aromatic

1. Derived from or belonging to a major class of organic compounds, many of which are used as solvents.

2. Derived from or characterized by the presence of the benzene ring, as applied to a large class of cyclic organic compounds, many of which are odorous.

Aromatic solvents

1. Group of organic compounds derived from coal or petroleum, such as benzene and toluene.
2. Hydrocarbon solvents with a benzene ring nucleus, such as xylol or toluol.

Asbestine. Natural fibrous magnesium silica that is pure white in color, and is used as an extender pigment in paints.

Asbestos siding (preparation for painting). Allow asbestos siding to weather at least two years before painting and then treat with a marketed masonry conditioner prepared for this purpose.

Asphalt

1. Black resin or gum that occurs naturally or that can be derived from petroleum.
2. Natural asphalt is a waterproofing material. The vehicle includes both petroleum and natural asphalts. Asphalt coatings are widely used for the protection of concrete, steel, and wood. Asphalt has good water resistance, is thermoplastic, and tends to cold-flow even at moderate temperatures.

   On exterior exposures it is quickly degraded by sunlight, which results in cracking or alligatoring. The addition of aluminum flake pigment to asphalt reduces both cold flow and alligatoring because the aluminum reflects the heat rays of the sun. Asphalt aluminum coatings are recommended for roofs because they can lower the temperature inside the building. The addition of epoxy resin to asphalt minimizes the cold flow and alligatoring and maximizes the chemical resistance of asphalt.

Asphalt-asbestos roof coating. Black paint, usually composed of asphalt base (with or without fatty oils), a volatile solvent, and a mineral filler such as asbestos fiber.

Asphaltum. Paints formulated with asphaltum cannot normally be recoated without bleeding.

Asphalt varnish. Asphalt varnish is used for interior and exterior surfaces and is particularly suitable for painting indoor water and gas pipes. It dries with a smooth, black, lustrous finish similar to black enamel. Its composition is hard native asphalt or asphaltite fluxed and blended with drying oils, then thinned with solvents and driers.

Atomization. Breaking up of paint into finely divided tiny droplets, usually accomplished in spray painting.

Back priming. Application of paint to the back of woodwork and exterior siding to prevent moisture from getting into the wood and causing the grain to swell.

Baking finish

1. Finish whose properties are measurably changed through the application of heat, as compared to those properties obtained through normal or forced drying.
2. Finish obtained by baking paint or varnish at temperatures above 150°F to dry and develop desired properties.

Ball mill. Large cylindrical revolving mill that utilizes the principle of falling bodies. As the mill turns, steel balls or spherical stones cascade down over each other, dispersing the pigment with the vehicles in the manufacture of paint.

Bar finish. Alcohol-proof, clear varnish or lacquer used to protect bar tops.

Barium sulfate. Extender pigment made from mineral and barite, and unaffected by acids or alkalies.

Base. Colored or white paste or heavy liquid added to paints to tint them and to which is added a vehicle to make a finished paint.

Basic lead silico chromate. Quality pigment used in anti-corrosion finishes, principally for steel.

Batch. Unit of production. It may be measured by the size of a special order or limited to the capacity of the equipment.

Benzene. Aromatic solvent used for many materials. Its use is restricted due to its toxicity and the fact that it is a fire hazard.

Benzine. Petroleum product used by the painting industry as a thinning solvent and dilutent. It is highly flammable.

Binder

1. Part of a paint that holds the pigment particles together, forms a film, and imparts certain properties. It is usually a polymer and is also called "vehicle solids."
2. Nonvolatile portion of a paint vehicle that serves to bind or cement the pigment particles together. Oils, varnishes, and proteins are examples of binders.
3. Nonvolatile component of a paint or varnish that is responsible for binding the pigments within the film for cohesion and for adhesion to the undersurface.

Biting. Solvent in topcoat dissolves or bites into the coat below.

Bituminous plastic cement. Black cement composed of a bituminous binder, asphaltic or coal tar base, and inorganic filler such as asbestos filler, and a solvent.

Black graphite paint. Paint for structural, steel such as that used for bridges and tanks. It is furnished in 2 types. Type 1 is a natural flake graphite pigment, which is dark steel gray with a metallic luster. Blackness is obtained by the addition of lampblack or carbon black. Graphite has, to a certain extent, the property of "leafing," a characteristic that probably accounts for its durability as paint. Type II is generally darker in color and may include all four forms of graphite: natural crystalline vein or lump, natural crystalline flake, artificial graphite, and natural "amorphous."

Black pigments. Four black pigments are used for tinting and solid colors: bone black, carbon black, lampblack, and synthetic black iron oxide. High-color carbon black and lampblack are stronger than bone black in tinting strength, and lampblack is the pigment most frequently used for tinting purposes. However, bone black and carbon black are blacker than lampblack. All these color pastes produce paints suitable for trim colors. Black iron oxide is used in black metal protective paints and for tinting and blending.

Blast cleaning. Cleaning of materials accomplished by the use of a air and a grit such as sand, contained in a vehicle under pressure and applied by a gun. Cleaning with propelled abrasives.

Bleaching

1. Restoring discolored or stained wood to its normal color or making it lighter with acids or other bleaching agents.
2. The use of bleaching agents to restore discolored or stained wood to its original normal color or to make it lighter.

Bleaching agent. Color lightener usually applied to wood to obtain a permanently lighter color.

Bleeding

1. Property of some paints to transfer their color upward into the topcoats that are applied over them; diffusion of coloring matter through a coating from a substrate.

2. When coloring material from either the wood or an undercoat works up into succeeding coats and imparts to them a certain amount of color, paint is said to "bleed."

3. Migration of a dye or stain from stained wood or an undercoat into subsequent coats. Bleeding usually occurs as the result of solubility in the vehicle portion of the topcoat.

4. Vehicle in a topcoat dissolving the dye, stain, or pigments in an undercoat and allowing the color to float to the surface of the topcoat. Bleeding is most common with certain reds—hence the name. It is frequently stopped or prevented by an intermediate coat of either shellac or aluminum paint.

5. Painting over knots or other resinous wood causes a dissolving action of some of the resin. As a result, white paint turns yellowish. Painting over creosote, bitumen, or colors soluble in oil (mahogany stain, for example) likewise may cause some of the material to dissolve into the paint coat. The remedy is to treat the surface with a sealer prior to applying paint.

Blending. Mixing one color with another so that the colors mix and blend gradually.

Blistering

1. Formation of bumps or pimples on a paint film, normally caused by moisture or heat.

2. Formation of bubbles on the surface of a paint film, usually caused by moisture behind the film.

3. Common type of paint failure in which the paint film forms blisters, which are often caused by a vapor expanding between the coated surface and the film. The vapor may be the paint solvent when the paint is applied in strong sunlight or water vapor when the paint is applied to a moist surface.

Blistering and peeling. Blistering of paint is caused by fluid (gas or liquid) pressure beneath an airtight coating. Water usually plays a secondary part, chiefly that of making it possible for fluid pressure to be developed within the wood. Wood can be very wet without the paint blistering; otherwise, marine painting would be largely impracticable. Peeling paint may be caused by the application of paint to a glossy surface. A glossy surface should be sanded before applying paint. It is not good practice to apply a glossy paint over a glossy undercoat.

Block cinder and concrete-exterior or interior. Block cinder and concrete are ready for painting when free from dirt, loose or excess mortar is removed, and the mortar is thoroughly dry. Painting with a latex paint may proceed in seven days under normal drying and painting conditions.

Bloom

1. Haze or clouded effect that appears on the surface of dried enamel or varnish, affecting the gloss of the film.

2. Bluish cast formed on the surface of some films. Bloom may be caused by such foreign material as smoke, oil, or dust during the drying process.

3. Clouded appearance on a varnished surface.

4. Bluish white haze that sometimes appears on a varnish film, also, a hazy appearance on any film.

Blue lead. Slate-gray sublimed pigment that consists of approximately 75% basic lead sulfate and generally more than 20% other lead compounds. Oil paints brush easily and are durable for priming structural steel. Blue lead is also used in special vehicles designed for marine use.

Blue-lead paint. Dark slate-gray paint intended for priming and body coats on iron and steel. It is characterized by very good working and keeping properties and is durable.

Blue pigments. The two blue pigments are iron blue (also called "Prussian blue") and ultramarine blue. Iron blue has a relatively low specific gravity and high bulk and is stronger than ultramarine blue. For tints, iron blue is very susceptible to the action of alkalies, however weak; whereas ultramarine blue is not affected by alkalies, but it is affected by even weak acids. Although both colors, in either tints or solid colors, have fair permanency when exposed to the weather, a blue pigment known as "copper-phthalocyanine blue" is more durable in light or "shutter blue" blue.

Blush. Lacquer finish is said to "blush" when it takes on white or grayish cast during drying. Blushing is usually caused by precipitation or separating of a portion of the solid content of the material, causing an opaque appearance.

Blushing

1. Hazing or whitening of a film caused by absorption and retention of moisture in a drying paint film.

2. Condition in which a cloudy film appears on a newly lacquered surface. It is caused directly by the precipitation of a portion of the solid content or the material usually due to oil or water mixed in the lacquer, a relatively high humidity condition, or too rapid drying.

Bodied linseed oil

1. Bodied linseed is oil that has been thickened by heat treating or blowing, thus increasing the "body" or consistency. It may be obtained in various degrees of body or viscosity, indicated on the Gardner Holdt scale by letters of the alphabet.

2. Linseed oil heated to acquire a specific viscosity. Bodied linseed oil will decrease the moisture permeability.

Body. Thickness or thinness of a liquid paint. Other terms are "viscosity" and "consistency."

Body coat. Intermediate coat of paint between the priming or first coat and the finishing or last coat.

Boiled linseed oil

1. Raw linseed oil that has been heated in the presence of metallic drying compounds.

2. Linseed oil that has been heated with chemicals. The term "boiled" is a misnomer, as the oil is not actually boiled.

Boiled oil. Oil that is heat treated until it is partly oxidized. The treatment gives the oil quicker drying properties.

Bonderizing. Proprietary name for a chemical process for phosphate-coating iron, steel, or zinc surfaces as a rust-preventing base for painting.

Bonding primer. Clear, medium oil length primer available for both nonporous and porous surfaces. It is used as a primer over nonporous surfaces, such as glass, steel, galvanneal, galvanized iron, stainless steel, and aluminum, and as a sealer on porous surfaces, such as pressed wood and masonite. It offers excellent adhesion to the support surface and also to the following topcoat. Its flexibility adds resistance to popping and peeling. This primer offers excellent adhesion to both substrate and topcoat and superior sealing and holdout qualities; it adds flexibility for improved weather resistance.

Boxing

1. Pouring paint from one container into another repeatedly, not only for mixing, but to ensure uniformity.

2. Mixing paint by pouring it from one bucket to another several times to ensure the most uniform consistency and smoothness.

Breaking up. Adding thinner to heavy paste paint and stirring until the proper consistency is obtained.

Breather film. Type of paint that lets water vapor from within pass through (eliminating blisters) and yet holds out surface moisture, such as rain and condensation.

Brick-exterior or interior. Brick is ready for painting when it is free from dirt, and loose excess mortar or foreign material is removed by brush, air, or steam-cleaning. Glazed brick is roughened by sandblasting and treated with a masonry conditioner. Before painting with a latex material, new brick is allowed to weather one month; then the surface is treated with masonry conditioner before finish coats are applied.

Bridging. Ability of paint to span small gaps or cover cracks through its elastic qualities. Bridging is a desirable quality for some coatings, which are so formulated. It is not desirable in coatings for screens and acoustical tile.

Brightness. Total amount of light reflected by a surface, usually used as a measure of the whiteness or lightness of a surface.

Brittle. Easily cracked or flaked when bent.

Brittleness. Lack of flexibility, usually combined with a lack of toughness.

Bronzing. Formation of a metallic-appearing haze on a paint film.

Bronzing liquid. Liquid binder for gold, bronze, or aluminum powders, usually varnish or lacquer especially formulated for this purpose.

Brown pigments. The four brown pigments are metallic brown, burnt sienna, burnt umber, and raw umber. Metallic brown is adaptable as a trim color but is used more as a solid paint and is durable for painting barns, tin roofs, and freight cars. Other brown pigments are suitable for tinting purposes and for stains. They consist of oxides of iron and are durable when exposed to the weather.

Brush. Bristle applicator for applying paint.

Brushability. Ability or ease with which a paint can be brushed under practical conditions; ability of a paint to be applied by a brush.

Brushing

1. Applying paint by a brush.
2. Best method for applying stain. Brushing tends to work the stain into the surface better and gives a more uniform distribution on the surface.

Brush-off blast-cleaning. Cleaning of all except tightly adhering residues of mill scale, rust, and coatings, exposing numerous evenly distributed flecks of underlying metal.

Bubbling. Presence of air bubbles in a drying film caused by excessive brushing or vigorous stirring just before application. When the bubbles break, they may form pinholes in the film.

Butadiene. Petroleum derivative capable of forming polymers such as artificial rubber.

Butyl acetate

1. Solvent for paint and commonly used in lacquers.
2. Lacquer solvent made from butyl alcohol by reaction with acetic acid.

Calcimine. Water-thinned paint composed essentially of calcium carbonate or clay and glue.

Calcimine (kalsomine). Inexpensive powdered mixture of pigments, such as whiting, china clay, and glue, made ready for application by the addition of water. It is intended for temporary interior use and is easily removed by sponging the surface with warm water.

Calcination. Oxidation by heating; expulsion of volatile ingredients with the aid of heat.

Carbon. Toxic solvent generally used for cleaning.

Carbon black. Black pigment manufactured from carbon.

Carnauba wax. Yellowish or greenish wax obtained from the young leaves of the Brazilian wax palm tree, used in various waxes and polishes.

Casein

1. Component of emulsion paints.
2. Protein from milk used for making water paints. Synthetic casein resins are made by mixing formaldehyde with casein.
3. Protein of milk and the principal constituent of cheese. Casein is used extensively in the manufacture of water paints.

Casein-powder paints. Cold-water paints, also packaged in powder form. All types require the addition of water, and one kind is furnished with a "mixing liquid" composed of raw or boiled linseed oil or spar varnish to be used in addition to the water.

Castor oil

1. Natural vegetable oil used in the manufacture of alkyd resins and as a plasticizer.
2. In its natural state castor oil does not dry; its satisfactory drying properties are achieved by dehydrating it under heat.

Catalyst

1. Component that speeds up a chemical reaction without entering into the reaction.
2. Substance that by its presence accelerates velocity of reaction between substances.

Caulking compound. Semi- or slow-drying plastic material used to seal joints or fill crevices around windows, doors, etc., usually made in two grades, gun type and knife type.

Cellulose. Natural polymer, generally of wood or cotton.

Cement-base paint. Paint composed of portland cement, lime pigment, and other modifying ingredients and sold as dry powder. It is mixed with water for application.

Cement powder paint. Paint made from white portland cement, pigments, and (usually) small amounts of water repellent, mixed with water just before application. Painted surfaces should be kept damp by sprinkling with water until the paint film is well cured. This paint does not provide a good base for other types of finishes. It is applied with a fiber brush.

Cement water paints. Cement water paints are packaged in powder form and require only the addition of water to prepare them for use. In mixing them, the recommended procedure is to make a stiff paste by adding water in small portions to the dry material, stirring constantly. After this, additional water may be gradually stirred into the paste to the desired consistency. The proper amount of water varies with the fineness of the dry materials.

Chalk. Form of natural calcium carbonate.

Chalking

1. Result of the weathering of a paint film, characterized by loose pigment particles on the surface of the paint.
2. Development of a chalky powder on the surface of a paint film, usually by weathering, which occurs in the majority of exterior paints to a greater or lesser degree. White exterior paints with controlled chalking have been developed.

3. Chalking or powdering of the paint film is generally caused by the destruction of the outside binding material in the film. In normal weathering of outside paints, chalking follows loss of gloss. Certain pigments (for example, titanium dioxide) cause more chalking of the paint film than others (for example, zinc oxide).

Check. Paint failure characterized by small cracks on the surface of the paint film.

Checking

1. Tiny breaks in the surface of the paint film, usually V shaped. The underlying surface is not visible.

2. Slight breaks in the surface of a paint film. Breaks are called "cracks" if the underlying surface is visible.

3. Checking starts at the exposed surface, works progressively deeper into the coating, tends to take on a V-shaped cross section with the open part at the exposed surface, but gives no sign of widening at the bottom by contraction of the coating, though the checks may be widened slightly by erosion. Slight checking is not a serious defect, as it indicates a relieving of the shrinkage stresses in a paint film. If the film does not check because of its great tensile strength, it may crack with the expansion and contraction of the surface to which the paint is applied; scaling may then result.

Chemical pigments. Synthetic pigments as distinguished from natural pigments.

Chinawood oil (tung oil). Oil originally imported from China and used in very hard chemical-resistant finishes, such as quality spar varnishes.

Chipping. Separation of paint from previous coats in chips or flakes.

Chlorinated rubber

1. This vehicle produces coatings that are especially resistant to alkalies and acids and have outstanding chemical and water resistance. Rubber coatings dry exclusively by solvent evaporation like lacquers. A film is resoluble in aromatic (coal tar) solvents no matter how old it is; it is easy to remove and easy to recoat. Rubber coatings have poor resistance to aromatic solvents, esters, and ketones and to animal and vegetable oils and fats; they have limited heat resistance and will deteriorate if exposed above 150°F for prolonged periods. They are durable on exteriors but chalk slightly more than alkyds.

2. Resin made with natural rubber and chlorine. Paints made of chlorinated rubber have good chemical resistance and dry rapidly.

Chlorinated rubber enamel. Stabilized chlorinated rubber finish pigmented with chemically resistant pigments. It is used as a topcoat on interior and exterior iron, steel, plaster, brick, and concrete and is also part of a system recommended for the protection of surfaces exposed to excessive moisture and chemicals. It is excellent for use on underwater surfaces and will not support the growth of mildew or fungus. Its fast-drying properties make it well suited for areas where there is chemical fallout. Its inertness to alkalies and high resistance to acids, both dilute and concentrated, make it ideally suited for use in paper mills, chemical plants, and related applications. Its imperviousness to water makes it well suited for use in bleacheries, laundries, and shower rooms. It is not resistant to animal or vegetable fats and oils; thus it is not to be used in dairies and packing houses where it would be exposed to these conditions.

Chrome orange pigments. Two shades of chrome orange pigments may be used for tinting and for solid or trim colors. Chrome orange pigments are relatively permanent when exposed to the weather. The pigments are basic lead chromate obtainable in shades designated as light and dark, the latter shade known as "international orange."

Clear. Term applied to paint containing no pigment or only transparent pigments.

Clear film. Transparent film. It may be water clear to amber in color, but does not materially change the appearance of the surface coated.

Clear varnish finishes for wood. Clear varnish provides a durable and attractive finish and forms a tough, transparent coat that will withstand frequent scrubbings and hard use. It tends to darken the wood surface and give the impression of visual depth. It readily shows scratch marks, which are difficult to conceal without redoing the entire surface. Some varnishes turn yellow with age. An extra coat is recommended on new work. Clear varnish can provide a flat, satin, semiglossy, or glossy finish.

Cloudy film. Hazy appearing film on varnish or enamel.

Coagulate. To change from a liquid into a dense mass; solidify; curdle.

Coagulation. Resin particles curdle into a rubberlike mass as the result of freezing, excessive heat, or chemical change.

Coal-tar-base enamels—applied hot. Coal-tar-base enamels are used on steel submerged in water. These enamels are composed of processed coal tar usually combined with an inert mineral filler. Coal-tar-base enamels are applied hot in a uniform coating approximately $\frac{1}{16}$ in. thick.

Coal-tar epoxy paint. Paint in which the binder or vehicle is a combination of coal tar with epoxy resin.

Coal-tar solvent. Solvent derived from distillation of coal tar, and consists mainly of products for the paint industry. These are benzene, solvent naphta, toluene, and xylene.

Coal-tar urethane paint. Paint in which the binder or vehicle is a combination of coal-tar with polyurethane resin.

Coarse particles. Particles of pigment larger than necessary for securing desirable properties in paint film. Particles are considered coarse if they will not wash through a 325-mesh screen.

Coat. Single layer of paint spread at one time and allowed to harden.

Coating

1. Applying paint or the actual film left on a substrate by a paint.

2. Mastic or liquid-applied surface finish, regardless of whether a protective film is formed or merely decorative treatment results.

Coating materials. Coating materials for application to building surfaces are divided into two categories: fire-retardant coatings for application to interior combustible surfaces (and occasionally interior noncombustible surfaces) for the purpose of reducing the fire hazard, and general-purpose coatings for various purposes. The purpose of the classification is to express the degree of fire hazard of the coating.

Coconut oil. Natural vegetable oil used in the manufacture of alkyd resins.

Cohesion. Molecular attraction by which the particles of a body, whether like or unlike, are united together throughout a mass, such as a paint film.

Cold application. Coal-tar-base enamels for cold application are composed of a processed coal tar vehicle and an inert inorganic filler thinned with a solvent.

Cold-checking. Checking caused primarily by low temperature.

Cold water paint

1. Latex paint; paint having a binder and vehicle that are soluble in water.

2. Paint in which the binder or vehicle portion is composed of casein, glue, or other protein material dissolved in water, usually employed on concrete, masonry, or plaster surfaces.

3. Paint furnished in powder and paste form, intended chiefly for decorating plaster and other dry interior masonry surfaces but sometimes used for temporary exterior work. It is not recommended for use on basement walls if dampness is present because of its tendency to mildew. The binder in the paint is milk casein, soya bean protein, or some other form of protein.

Color. Property of visible phenomena in which certain impressions or effects are formed on the retina of the eye by the light of different wave lengths. Color is divided into three principal parts—hue, tint, and shade.

Colored pigments. Inorganic materials, especially used outdoors, where the brilliant but fugitive organic pigments soon fade.

Color in oil. Pigment ground in linseed oil to a paste form, usually used for tinting.

Color retention

1. Permanence of a color under a set of conditions.

2. Ability of a paint to retain its original color for a long period of time.

Commercial blast cleaning. Blast cleaning used for rather severe conditions of exposure and continued until at least two-thirds of each element of a surface is free of all visible residues.

Compatibility. Ability of materials to get along together without separation or reaction. Latex paints can usually be mixed with other latex paints, and most oil paints can be mixed with oil paints; but oil paints and latex paints are not compatible.

Concrete and masonry paint. Oil-base exterior paint ready-mixed as a finishing coat for suitably primed concrete, brick, and stucco surfaces, except floors. The paint dries to an eggshell finish. Old coatings of organic or cement-base water paint in good condition need not be removed. Peeling, scaling, or flaking paint and whitewash should be completely removed. As moisture under the paint film seriously impairs the life of oil paint coatings, new masonry should not be painted until the walls are dry.

Concrete and masonry surfaces. Porch and deck paints are made with natural rubber base. They are good flowing and leveling and have good resistance to rain, moisture, and detergents.

Concrete floors (preparation). Prior to painting, concrete floors are cleaned and etched, with a prepared etch solution or muriatic acid. They must be rinsed thoroughly and allowed to dry before paint is applied.

Concrete, poured, exterior or interior. Concrete is prepared for painting by removal of dirt, loose or excess mortar, film left from an incompatible form of oil, or concrete curing compounds. The surface must be allowed to weather 1 month under normal drying conditions before painting with latex paint.

Conductive primer. Primer that when electrodeposited and cured will not act as an insulator of the coated metallic object but will conduct current for the electrodeposition of a second coating or topcoat film.

Consistency

1. Fluidity of a system.

2. Thickness, body, or the resistance a paint offers to stirring or flow.

Contrast ratio. The hiding power of paint is usually determined by coating both white and black areas of a drawdown sheet. The brightness of the film over the black area is divided by the brightness over the white area to determine the ratio, normally expressed in percent. The higher the number, the better the hiding ability.

Cool colors. Receding colors, predominantly blue or green, as opposed to warm colors containing red and yellow (the colors of fire and the sun).

Copolymer. Polymer made from two monomers.

Copolymerization. Union of unlike molecules. Two or more substances polymerize to make a substance different from both, for example, vinyl acrylate emulsion.

Copper (prepared for painting). Copper is prepared for painting by being cleaned free of dirt and foreign matter. It is then painted with an oil-base paint.

Copper and bronze fly screens. Copper and bronze fly screens cause unsightly discoloration on white and light-tinted paints. Discoloration is a result of corrosion products from the copper or bronze and may be avoided by varnishing or painting the screens before corrosion takes place.

Copper stains. Copper staining is usually caused by the corrosion of copper screens, gutters, flashing and downspouts. Corrosion products are washed down by rain and usually cause yellow or brown stains on painted surfaces. Painting or varnishing the copper can prevent the stains.

Corrosion

1. Term describing interaction between a material and the environment, such as decay or oxidation, in which the material is slowly eaten away.

2. Chemical action; oxidation process (rusting).

Coverage

1. Amount of area a volume of paint will cover at a certain thickness.

2. Capacity of a pigment material to hide the surface beneath and produce a uniform, opaque surface.

Covering power. Ability of a coating to completely cover a surface, usually expressed in square feet per gallon.

Crack. Break extending through a paint film so that the undersurface is visible.

Cracking

1. Larger than hairline breaks in the surface of a film. There are usually curled edges, and the underlying surface is exposed.

2. Phenomenon manifest in coatings by a breaking extending through to the surface painted.

Cracking and scaling. Cracking and scaling are common types of paint failure. If the tension of the paint film is not broken in some way when the surface to which it is applied contracts and expands, cracking and subsequent scaling is apt to result. Cracking differs from checking or alligatoring in that the cracks extend all the way through the coating to the underlying surface. Subsequently, the coating may separate from the edges of the crack and curl outward without widening because of the contraction of the coating. On wood the cracking may occur at right angles to or parallel with the grain. Moisture enters through the cracks, works under the paint film, and causes scaling. Cracking and scaling usually take place when a paint has very little elasticity, particularly when a thick coat builds up with continued repaintings. Certain paints are inclined to fail by cracking, curling, and scaling, whereas others fail by moderate chalking with or without checking. Failure of a paint by moderate chalking is the least serious type of paint failure.

Crawling

1. Wet film defect that results in the paint film's pulling away from certain areas or not wetting certain areas, leaving those areas uncoated.
2. Tendency of some liquids to draw themselves into beads or drops, caused by high surface tension or applying paint to a high-gloss surface.
3. Tendency of a paint or other liquid to draw itself into drops after application to a surface, leaving bare surface areas; also called "cissing."
4. Tendency of a paint or varnish to form a discontinuous film by drawing up into drops or globules shortly after application, resulting from surface tension caused by either the surface or the paint. Oil paint applied at room temperature over a cold and greasy surface may also cause crawling. Other causes are the application of a glossy coat of paint over a fresh glossy coat, particularly when an outside house paint contains an excessive amount of linseed oil; the application of paint or varnish over previous coats that are not hard and dry; the presence of moist finger marks on a surface prior to varnishing; the mixing together of various brands of varnish; the use of varnish that has become thick and viscous; the presence of a thin film of wax on the surface left by the use of liquid paint removers; and the application of paint in cold or foggy weather.

Crazing

1. Film failure that results in surface distortion or fine cracking.
2. Fine lines or minute surface cracks occurring on painted or enameled surfaces as a result of unequal contraction during the drying process.

Creosote. Liquid with a penetrating odor, distilled from coal tar and used as a wood preservative (used in shingle stain).

Creosote bleaching oil. Refined and clarified creosote oil containing a small amount of gray pigment and a chemical ingredient that actually bleaches the wood.

Creosote oil. Distillate from coal-tar that is heavier than water, used as a wood preservative.

Creosote stain. Creosote made from wood and coal tar is mixed with linseed oil, dried and thinned with benzine or kerosene.

Creosote stains for siding and shingles. Stains dye the wood and penetrate well into the surface, giving a permanent color. A high proportion (over 60%) of highly refined creosote oil protects the wood and prolongs its life. Stains will not crack, peel, or blister.

Crocking

1. Removing surface color by rubbing. Most latex finishes exhibit crocking when rubbed with a wet rag.
2. Removal of color pigment from a film by harsh abrasion or rubbing as, for example, in a chalked film.

Curdling. Coagulation. Curdling also refers to lumpy solids in paints or varnishes.

Cure. Harden a coating by applying heat or a catalyst.

Curing. Chemical reaction that takes place in the drying of paints that dry by a chemical change.

Curing agent. Catalyst to produce drying or hardening effect, or both.

Curtaining. Sagging of paint in a curtain effect, usually caused by applying too heavy a coat of paint.

Cut. Number of pounds of resin per gallon of solvent.

Cut back. Add liquid to thin paint.

Cutting in. Operation calling for the most careful workmanship to keep a clean edge, for example, cutting in on a window sash with a brush called a sash tool, which permits the painter to get a clean painted edge. The basic idea is to keep the paint off of the glazed portions.

Damar. Pine resin from trees in South America, New Zealand, and Australia, used as a paint vehicle in liquid form.

Damar varnish. Spirit varnish applied to indoor surfaces, both as a transparent finish and as a vehicle for white or light-colored enamels exposed to high temperatures. It is not intended for use under conditions that require an abrasion-resistant, moisture-resistant, or indentation-resistant varnish.

Deck paint. Enamel with a high degree of resistance to mechanical wear, for use on such surfaces as porch floors.

Degradation. Gradual or rapid disintegration of a paint film.

Degreaser. Combination of solvents manufactured for the purpose of removing grease and oil from a surface in preparation for painting.

Degreasing. Cleaning a substrate (usually metallic) by removing grease, oil, and other surface contaminants.

Dehydrated castor oil. Dehydrated castor oil is made by heating castor oil under vacuum with catalysts to remove water. Natural castor oil is not necessarily a drying oil, but dehydrated castor oil is excellent with good flexibility, adhesion, and rapid-drying characteristics.

Density. Weight of any material per unit of volume.

Depth of finish. Glassy appearance of multiple coats of varnish.

Diluent. Liquid for reducing varnish or lacquer for the purpose of increasing the bulk.

Dipping

1. Applying paint to an article by immersing the article in a container of paint and then withdrawing the article and allowing excess paint to drain from the part.
2. Dipping is recommended for rough surfaces only. The dipping trough is lined with 10-mil polyethylene film. Lumber is submerged in the stain and allowed to drain thoroughly. Each piece of lumber is dried without any object touching the surface to be exposed.

Dipping compound. Bituminous compound used for coating pipes and iron tunnel segments to preserve them against rust.

Dirt and foreign matter. Dirt and foreign matter should be removed thoroughly by bristle brushes or by being blown clean with air pressure or steam cleaned. Surface deposits of oil or grease must be prevented from spreading over additional areas in the cleaning process.

Discoloration. Paints containing lead pigments will discolor if there is hydrogen sulfide in the atmosphere. Discoloration appears as a brownish or grayish film because of the formation of lead sulfide. Discoloration may be "bleached out" with hydrogen peroxide.

Dispersed. Scattered or completely integrated, as pigment separated in a binder.

Dispersion. Distribution of solid particles, uniformly throughout a liquid, commonly, the dispersion of pigments in a vehicle.

Dispersion agent. Substance that aids the holding of pigments in dispersion.

Drag. Resistance a paint offers to brushing.

Drier

1. Catalyst added to a paint to speed up the cure or dry.
2. Material that hastens the drying or hardening of a drying or semidrying oil.
3. Accelerating the action of drying yet undergoing no change in themselves, driers are metallic compounds known as "catalysts." Lead cobalt, manganese, iron, zinc, calcium, and zirconium all have their own specific drying purpose.
4. Catalyst that hastens the hardening of drying oils. Most driers are salts of heavy metals, especially cobalt, manganese, and lead, to which salts of zinc and calcium may be added. Iron salts, usable only in dark coatings, accelerate hardening at high temperatures. Driers are normally added to paints to hasten hardening, but they must not be used too liberally or they cause rapid deterioration of the oil by overoxidation.

Dripless enamel (special alkyd base). This enamel does not drip from a brush or roller. It is made with special alkyd resins to form a soft gel that liquefies with agitation but gels again on standing. It is soft, buttery, easily brushed, and self-sealing. It has excellent color retention and low odor and is solvent and water resistant.

Dry

1. Change from a liquid to a solid that takes place after paint is deposited on a surface. Included in drying is the evaporation of the solvents and any chemical changes that occur.
2. Film formation by evaporation of solvent or oxidation of unsaturated compound or catalytic action.

Dry color. Pulverized coloring substance (pigment).

Dry film thickness. Measurement, usually in mills, of the paint film after drying (one mil is a thousandth of an inch).

Drying. Act of changing from a liquid to a solid state by evaporation of volatile thinners and by oxidation of oils.

Drying oils

1. Natural oils, like linseed oil or tung oil, or synthetic oils that air-dry within 48 hours under normal atmospheric conditions.
2. Paint vehicle ingredients, such as linseed oil, that when exposed to the air in a thin layer oxidizes and hardens to a relatively tough elastic film.

Drying-oil vehicles or binders. Most vehicles contain drying oils. These are vegetable and animal oils that harden or dry by absorbing oxygen.

Drying power. Certain kinds of paints and varnishes lose drying power if stored too long in the original containers. This happens more frequently to the "fast-drying" types of enamels and varnishes than to the older types of paints. If paint is old stock, its drying time should be tested before use. If the drying power has been lost, paint drier should be added to restore it.

Drying time

1. A paint is dry to the touch when the paint film does not adhere to the fingertips on being touched very lightly. A paint has dried to a dust-free condition when dust blown onto a paint film may be completely removed after the film has dried hard. A surface is tack free when it no longer sticks to the thumb after being squeezed between the thumb and forefinger. The paint film is hard when the film is not permanently marred by being pressed as hard as possible with the heel of the hand.
2. Expected time for a paint film to harden. Varying by product, temperature, and atmospheric conditions. Drying time is usually expressed as dust free and completely dry, or dry for recoat.

Dry pigments. Dry pigments are almost always mixed to pastes at the factory, but when it is necessary to mix paint from dry pigments on the job, the workroom should be free from drafts and properly ventilated.

Durability

1. Length of life. The term usually applies to a paint that is used for exterior purposes.
2. Lasting qualities or wearability of paint under the conditions for which it was designed.

Earth pigments. Color pigments that are mined, such as iron oxides. Earth pigments are stable in everything but color consistency. Greater color control can be maintained by synthetic pigments.

Efflorescence

1. Become covered with a white, hard crust formed by deposits of salts upon the evaporation of water—most common in damp basements.
2. Water-soluble salts that show as a white powder on masonry surfaces. This condition is caused by moisture bringing the salts to the surface.
3. Deposit of water-soluble salt on the surface of masonry or plaster caused by the dissolving of salts present in masonry. The solution migrates to the surface and deposits the salts when the water evaporates.

Eggshell finish

1. Low-luster finish with a gloss very little higher than a flat paint.
2. Degree of gloss between flat finish and semigloss; sometimes, a color.

Egg shell luster. Luster closely resembling that of an eggshell.

Electric holiday detector. Pinholes or cracks in bituminous coatings can be readily detected by means of an "electric holiday detector." The unit usually consists of a high-voltage transformer, designed to operate on a 105 to 115-V, 60-cycle power line, and an adjustable Tesla coil. A metal chain, band, or brush, whichever best conforms to the contour of the surface to be tested, is attached to the discharge terminal of the Tesla coil.

Elasticity. Ability to change size and return to normal without breaking. Most thermoplastic finishes are more elastic than oxidized coatings.

Electrodip coating. Process in which a paint film is deposited on all surfaces of a metallic product by the passage of an electric current between the object being coated and another electrode, while the metallic object is immersed in a bath of water-thinned paint. This process is also called "electrodeposition," "electrocoating," or "electropainting."

Electroendosmosis. Process by which occluded water is forced from a deposited film during the paint film deposition process.

Electrolysis. Passage of current via charged particles (ions) that are discharged on the relevant electrodes (applies to electrodeposition of paint films.)

Electrostatic spray. Application of paint through electricity. The spray is charged as it leaves the gun, while the surface to be coated has the opposite charge. The spray particles are thus attracted to the object with a minimum of overspray and more uniformity.

Emulsion

1. Suspension of fine polymer particles in liquid. The dispersed particles may be binder, pigments, or other ingredients.

2. Preparation in which minute particles of one liquid, such as oils, are suspended in another, such as water.

Enamel

1. Actually "pigmented clear finish," free-flowing finish that dries hard to the degree of sheen formulated, gloss to flat.

2. Opaque coating, usually with a varnish vehicle, designed to flow and level well and give a smooth finish. It is usually gloss but may be semigloss or, occasionally, flat.

Enamel holdout. Property that prevents the penetration of subsequent enamel coats to underlying surfaces; it prevents unequal absorption and uneven gloss.

Enamel undercoater (alkyd base). Hard, tight films. This undercoat provides a good base for enamel and makes for easy brushing and smooth leveling. It dries in about 12 hours.

End seal. Paint applied to the ends of boards in order to seal the pores.

Epoxides. Epoxides bond well to a variety of surfaces, including metals, and are used for rough, durable, and transparent or pigmented coatings.

Epoxy. Class of resins characterized by good chemical resistance.

Epoxy adduct. Epoxy resin having all of the required amine incorporated, but requiring additional epoxy resin for curing.

Epoxy amine. Amine-cured epoxy resin.

Epoxy—catalyzed. Catalyzed epoxies are two-component coatings produced by adding an activator or catalyst to a pigmented enamel or primer. The catalyst reacts chemically to produce a film as hard as most baked enamels. The mixture has a limited time of workability (referred to as "pot life"), which may vary from a few minutes to a day or more, depending on formulation. Catalyzed epoxy coatings, when properly cured, have excellent solvent and chemical resistance and are widely used in chemical plants. They make excellent coatings for both concrete floors and walls, producing a surface that is highly resistant to traffic, abrasion, chemicals, and cleaning. All epoxy coatings—both catalyzed epoxies and epoxy esters—chalk rather freely on exterior exposure, but otherwise they have good durability on exterior exposures.

Epoxy enamel. Hard film with a wide gloss range and low odor, ideal where vigorous and frequent cleaning is done. It has excellent adhesion and resistance to abrasion, water, solvents, greases, and dirt. It is packaged in two containers with enamel in one and the curing agent in the other. The contents of both containers are mixed together prior to use.

Epoxy ester

1. Epoxy modified oil; single package epoxy.

2. These epoxies are modified with the fatty acids of drying oils and require no catalyst. They require solvents of less strength than catalyzed epoxy and have

good adhesion and good color retention. They are less resistant to chalking on exteriors than alkyds but are not as chemical resistant as catalyzed epoxy.

3. Epoxy esters are epoxies modified with oil to produce a coating that will dry by oxidation. Neither as hard nor as chemically resistant as catalyzed epoxies, epoxy esters have a good intermediate degree of chemical resistance and can be used on areas subjected to occasional spillage of chemicals or chemical fumes. They are easy to apply and produce hard, tough films. They appear as conventional paints in a single package. Epoxy esters do not require the use of a catalyst, therefore there is no pot life restriction.

Epoxy resin

1. Synthetic resin resulting from the chemical combination of epichlorohydrin and bisphenol, resulting in a very chemically wet film.

2. Condensation of epichlorohydrin and bisphenol. The film made from epoxy resins is extremely durable and solvent resistant. Pure epoxies require strong solvents and a catalyst.

Erosion. Wearing away of the finish coat to expose the substrate or undercoat.

Ester. Organic compound formed from an alcohol and an organic acid by eliminating the water.

Ester gum

1. Hard brittle resin used in lacquer.

2. Resin of rosin and polyhydric alcohol, such as glycerin. It is a hard, brittle resin used in lacquer.

Evaporation

1. Change from liquid to gas. When solvents leave a wet paint film, they do so by evaporation.

2. Drying by the removal of moisture. Lacquers, vinyls, and most latex finishes dry by the evaporation of solvents.

Evaporation rate. Speed with which a liquid evaporates.

Excessive chalking. Excessive chalking may be the result of paint applied in rain, fog, or mist, paint applied too thin, or low-quality paint.

Exposure tests. By exposing applied film to the conditions the film will have to withstand, its wearability can be determined. Conditions are simulated in the laboratory where the action is accelerated in a weatherometer.

Extender

1. Inert pigment added to paint to increase the bulk. It is also used to vary the consistency or dilute colored pigment of great tinting strength. To control gloss and sheen of finish, products such as calcium carbonate, magnesium silicate, aluminum silicate, and silica are used.

2. Pigment used not for hiding but to improve the film structure.

3. Compatible substance that can be added to a more valuable substance to increase the volume of material without substantially diminishing its desirable properties; in paints, extender pigments improve storage and application properties.

Extender pigments

1. Inert, usually colorless, and semitransparent pigment used in paints to fortify the pigment systems.

2. Pigment added to extend the opaque pigments, increase durability, and provide better spreading characteristics. Principal extender pigments are silica, china

clay, talc, mica, barium sulfate, calcium sulfate, calcium carbonate, magnesium oxide, magnesium carbonate, and barium carbonate plus others used for specific purposes.

Exterior black gloss paint (slow drying). Paint for use on wood and structural steel. It is one of the most durable paints when exposed to the weather. When applied to structural steel primed with two coats of a rust-inhibitive paint (for example, red lead paint), it gives very good service to a topcoat or cover coat and retains its color well. This paint has good working properties and excellent hiding power.

Exterior stain. Pigmented water-repellent preservative. The preservative component consists of an organotin compound that is effective protection against wood discoloring mildews and woodrotting fungi. The stain must meet the regulations of the Environmental Protection Agency.

Exterior traffic paint. Ready-mixed traffic paint for centerline, zone-marking, and road-marking paint, obtainable in white and yellow colors. The yellow is called "highway marking yellow" (formerly "federal yellow"). Both paints are intended for application at a wide range of temperatures to bituminous and concrete highways bearing heavy traffic. The paints are applied by a brush or machine.

Factory primed. Term applied to materials that have received a first coat at the factory.

Fading. Loss of color through sunlight, heat, or other conditions.

Fading of color. Most paints change color on prolonged exposure to the weather. This may be a true fading of color, but generally it is the result of either a chemical change in the pigment on exposure to the weather or excessive chalking of the paint. Some paints chalk unevenly, which results in a blotchy effect on the color, particularly in dark grays. Other paints chalk evenly and self-clean themselves evenly. This is the preferred type of color "fading." Blotching or so-called color fading occurs in spots where the porous surface has not received sufficient coats of paint or a suitable primer.

False body. Puffing up in the container. It is possible through formulation to give abnormal (heavy) body to paint that can be lost through thinning to brushing consistency. Present-day thixotropic or homogenized paints are false-bodied.

Fastness. Ability to withstand exposure to heat, light, and weather without losing color.

Fatty acid. Acid derived from a natural oil.

Fatty acids. Glycerides in animal and vegetable fats and oils used in resin manufacture.

Feathering. Disappearing or blending-out of the edge of a paint film.

Filler

1. In a paint formula, inert pigments such as barytes, silica, or powdered mica. Also, a product to fill open-grained wood, made with silica and linseed oil.
2. Heavily pigmented paint used to fill imperfections or pores in a substrate.
3. Inert material used to fill or level porous surfaces such as open-pore woods like oak and walnut.
4. Pigmented composition used for filling the pores or irregularities of a surface prior to the application of other finishes.

Film. Very thin continuous sheet of material. Paint forms a film on the surface to which it is applied.

Film thickness. Thickness of a coating measured in mils.

Fineness of grind. The fineness of grind is measured from 0 (very coarse) to 8 (perfect dispersion). On a Hegman grind gauge enamels normally require a 6 to 8 grind and flat finishes, a 1 to 3 grind.

Finish coat. Last coat applied in a paint or wood-finishing job.

Fireproof paint. Paint that resists burning through foaming or bubbling action, referred to as an intumescent coating.

Fireproof paints. Paint products are not flammable; water-thinned type.

Fire-retardant paint. Paint that will significantly reduce the rate of flame spread, resist ignition at high temperatures, and insulate the underlying material so as to prolong the time required for the material to reach ignition, melting, or structural weakening temperature.

Fire retardation. Retardation of the flame includes the possible sacrifice of the paint film as such in retarding the spread of the flame over the paint and the wood on which the paint is applied.

Fish oil

1. Fish oil has good wetting properties; its chief disadvantages are its yellowing, poor flexibility, poor durability, and poor weather resistance; paints containing fish oil are usually inferior.
2. Fish oil is used in paints where softness and aftertack are not undesirable. The oil must be refined, by refrigeration, to remove nonhardening constituents.
3. Fish oils are obtained mostly from menhaden, sardines, and herring, and are claimed to be rust preventive. Actually, newer formulas have better rust-wetting properties without the slow drying of fish oils.

Flagging. Split end of a bristle in a paint brush.

Flaking

1. Paint film separates from the substrate and flakes off.
2. Protective coating failure associated with paints, varnishes, lacquers, and allied formulations, characterized by actual detachment of pieces of the coating either from its substratum or from paint previously applied. Flaking is sometimes referred to as "scaling" and is generally preceded by checking or cracking. Flaking is attributed to loss of adhesion of the coating.

Flame-cleaning. Cleaning by means of a special torch through which an oxyacetylene flame is applied to the steel surface. The process removes loosely adherent scale and rust and, in addition, drives moisture from the surface (dehydrates it). If the paint is applied while the steel is still warm, a good bond between the priming coat and the steel should result.

Flame-cleaning of new steel. Dehydrating the surface and removing rust, loose mill scale, and some tight mill scale by use of flame, followed by wire brushing.

Flashing

1. Uneven degree of gloss over a surface due to poor prime coat, poor applications, or the early exposure of a film to condensation (moisture).
2. Nonuniform appearance of walls or other surfaces on which a coating dries with spotty differences in color or gloss, usually due to improper sealing of the porous surface.
3. Uneven gloss apparent in a dried paint film. Flashing is usually due to a porous undersurface.

Flash point

1. Temperatures at which the vapor of a thinner or solvent will ignite in the presence of sparks or open flame.

2. Degree of temperature at which a liquid emits gases that will ignite in an open flame.

Flat.   Without gloss, sheen, luster, matte finish.

Flat alkyd enamel.   Enamel made with alkyd resins and having a flat finish practically free of sheen. It is used in the same way as latex wall paint but has slightly better washability and abrasion resistance, dries in about four hours, and has practically no odor.

Flat paint.   Interior paint that contains a high proportion of pigment and dries to a flat or lusterless finish.

Flats.   Term applied to flat paints and coatings.

Flatting agent

1. Ingredient added to paints and coatings to reduce the gloss of the dried film.

2. Ingredient used in lacquers and varnishes to give a flat or hand-rubbed effect (calcium, aluminum, or zinc stearate). Silicate flatting pigments give a better product than these metallic soaps.

Flatting oil.   Complete vehicle for thinning white lead paste and similar paste paints, composed of processed drying oils and drier and thinned with turpentine, mineral spirits, or a mixture thereof. When added to white lead paste and similar paste paints, it produces flat, washable, interior finishes for use on plaster, wood, wallboard, and fabric surfaces.

Flexibility

1. Quality of film that is easily bent without cracking or losing adhesion.

2. Ability of a paint film to withstand dimensional changes.

Floating

1. Tendency of some pigments to separate and float to the surface. Floating results in a streaked or spotty application. Also called "flooding."

2. Nonuniform rising to the surface of pigments of colors.

Flocculation

1. Formation of masses of particles either by settling or by forming a gel.

2. Dispersed pigment particles coming together and being held in a cluster. Flocculation causes a loss of color but can often be broken down by brushing. This sometimes produces an uneven color.

Flow

1. Degree of leveling without brush marks. Excess flow may cause sagging.

2. Leveling characteristics of a wet paint film.

3. Degree of leveling or ability to level, eliminating brush marks produced during the film application.

Foots.   Mucilaginous matter that settles to the bottom of a container. Settlings in vegetable oils.

Forced drying.   Drying at increased temperatures, but usually not over 150°F.

Ford cup.   Viscosimeter that works like a liquid hourglass. The viscosity of a liquid is determined by the time it takes to flow through the opening in the cup.

Foreign matter.   Visible material unrelated to the true origin of the liquid specified.

Fresco.   Ornamental plaster that is decorated while it is still drying.

Frosting.   Frostlike appearance of a semiopaque or translucent coating.

Fungicides

1. Agents for destroying molds or mildews.

2. Additives for paint used to prevent the growth of mold or fungus in the container or on a dry paint film.

Fugitive.   Not permanent. The term generally refers to pigments, etc., that fade in sunlight or under heat.

Fungus.   Group of living plants like mildew, mold, rust, and smut. They can actually feed on paint ingredients and attack most organic material under the right conditions of moisture and temperature.

Galvanized iron.   Galvanized iron is prepared for painting by a thorough cleaning to remove grease, residue, and corrosion products on the surface, with solvent or with chemical washes, used as directed by the manufacturer. New galvanized iron with galvanized iron that has been exposed to the weather for long periods of time and shows rust is primed with a corrosion-inhibiting metal primer.

Galvanizing primers.   Primers with a high percentage of zinc dust that provides good antirust protection and adhesion. Galvanizing/zinc dust primers give excellent coverage, one coat usually being sufficient on new surfaces. Two coats are ample for surfaces exposed to high humidity.

Ghosting.   Surface coated with a flat paint appears to be flatter in some areas than others.

Glaze.   Very thin, semitransparent paint film, usually tinted with a Vandyke brown, burnt sienna, or similar pigment and generally applied over a previously painted surface to produce a decorative effect.

Glazing

1. Adding a thin layer of transparent color to modify the tone of painting, like antiquing.

2. Applying transparent or translucent coatings over a painted surface to produce blended effects.

Glazing Compound

1. Puttylike compound used for securing panes of glass or sealing cracks and crevices prior to painting.

2. Doughlike material, consisting of vehicle and pigment, which retains its plasticity over a wide range of temperature and an extended period of time.

Glazing liquid.   Special varnish to which pigment is added for antiquing, blending, and special effects.

Gloss.   Shine, luster, or sheen of a dried film.

Gloss (luster, sheen).   Property of a surface by which it reflects light specularly. The terms "high," "enamel," or "mirror" indicate the highest gloss or luster, "semigloss," "eggshell," and "flat" indicate decreasing degrees of gloss in the order given.

Gloss (paint or enamel).   Paint or enamel that contains a relatively low proportion of pigment and dries to a sheen or luster.

Gloss enamel.   Finishing material made of varnish and sufficient pigments to provide opacity and color, but little or no pigment of low opacity. Such an enamel forms a hard coating that has a maximum smoothness of surface and a high degree of gloss.

Gloss meter.   Instrument that measures the degree of gloss of a film by its reflectance, also called a "glossimeter." The most commonly used is a 60-degree meter which measures gloss at an angle of 60°.

Gloss oil. Inexpensive varnish made by dissolving rosin in petroleum thinner.

Gloss retention. Ability of a finish to retain its gloss without flatting or dulling.

Glyceryl phthalate resin (alkyd). Synthetic resin of the alkyd group, used principally in paints, varnishes, and lacquers—sometimes called "phthalic alkyd resin." It is made by reacting glycerin and phthalic anhydride. Alkyd resins as a group are made by reacting polyhydric alcohols, such as glycerin and the glycols, with dibasic organic acids, such as phthalic, maleic, succinic, and sebasic.

Grain checking. Tiny breaks in the paint film parallel to the grain of the wood.

Graininess. Gritty appearance of a film due to the lumping of pigment.

Graining. Simulating the grain of wood by means of specially prepared colors or stains and graining tools.

Grain raising. Feathery fibers of wood standing up from the absorption of water. Light sanding is the corrective measure.

Green pigments. The two green pigments are chrome green and chromium oxide green. Chrome green consists of a mixture of two pigments, chrome yellow and iron blue; it covers a variety of light, medium, and dark chrome greens, used both for tinting purposes and as solid trim colors. Chrome greens are brighter and stronger than chromium oxide greens, but they are not as permanent and are sensitive to alkalies. Chromium oxide green consists of a single pigment, chromium oxide, and is used with white for tinting purposes to give soft gray-green tints, for trim purposes as a durable, solid color, and for tinted bridge paints. It is somewhat weaker in tinting strength and lower in hiding power than chrome green but is more permanent; it will withstand high temperatures and is unaffected by alkalies and acids.

Ground coat. Usually a buff color used under graining colors for contrasts when graining.

Gum

1. Natural resins that are the gummy substances from various trees or fossils.

2. Viscous vegetable secretion that hardens but, unlike a resin, is water soluble. The name is often applied in the varnish industry to natural resins as, for example, kauri "gum." A more appropriate term is "gum resin."

3. Solid resinous material that can be dissolved and will form a film when the solution is spread on a surface and the solvent is allowed to evaporate; ingredient of varnishes.

Gum spirits and wood turpentine. Turpentine is one of the oldest and most widely used volatile thinners for oil paints and oleoresinous varnishes. However, in prepared paints, volatile thinners derived from petroleum and having the general characteristics of turpentine are used. Destructively distilled wood turpentine, used in the manufacture of paint and varnish, is a good solvent and, in general, can be used in place of gum spirits or wood turpentine with good results.

Gum turpentine. Oleoresinous material obtained from living pine trees. Gum turpentine when distilled provides gum rosin and gum spirits of turpentine.

Hairlines

1. Fine lines that appear in some coatings as a result of weathering or temperature changes. Hairline cracks also appear in plaster.

2. Very narrow cracks in paint or varnish film.

Hand tool cleaning. Hand tool cleaning includes removal of loose rust, mill scale, and paint by hand chipping, scraping, sanding, and wire brushing.

Hardness

1. Quality of a dry paint film that gives the film resistance to surface damage deformation.

2. Cohesion of particles on the surface as determined by ability to resist scratching or identation.

Haze. Development of a cloud in a film or in a clear liquid.

Heat Resistance. Ability of a finish to withstand high temperatures without ill effects.

Heavy-bodied oil. Oil of high viscosity.

Heat-resisting black enamel. There are two types of black enamel, bituminous base (unpigmented) and resin base (pigmented). Both types are intended for use on surfaces exposed to temperatures of up to 400°F including steam pipes and boiler fronts. Neither type should be applied when the surfaces are hotter than 140°F, and 48 hours of drying time at room temperature should elapse before the coating is subjected to the maximum temperature of 400°F. Because of the nature of the volatile thinners, these enamels should be applied in well-ventilated areas free from open lights or flames.

Hiding power

1. Ability of a paint product to hide previous coats or the surface beneath; opacity.

2. Power of a paint or paint material to obscure the surface to which it is applied.

3. Obliterating or obscuring power of a paint, sometimes referred to as "covering power."

Holdout. Ability to seal a surface so that the finish coat will be of even gloss and color. Primers have holdout.

Holidays

1. Area or surfaces that were accidentally missed when painted.

2. Skipped or missed areas unintentionally left uncoated with paint.

Homogeneous. Uniform or similar throughout.

Hot spots. Alkali bleed through. Hot spots are free lime spots that have not totally cured with the rest of the wall and attack the applied paint.

Hot spray. Reduce the varnish or paint to spray consistency with heat, rather than solvent.

House paint. Pigmented paint designed to be applied to the exterior surfaces of residences.

Hue

1. Characteristic by which one color differs from another.

2. Name of a color (red, blue, yellow, etc).

3. Specific quality distinguishing one color from another.

Incompatible

1. Term applied to paints that should not be mixed together or applied one over the other. Water paints and oil paints should not be mixed. Epoxies might "lift" certain finishes.

2. Term applied to paints when they will not mix homogeneously or without impairing their original properties.

Inert. Without active properties. Inert extender pigments like barytes, asbestine, silica, and mica are added to paint. They usually add desirable properties but sometimes are only extending the quality of the product.

Infrared rays. Invisible rays that are used for their heating effect.

Inhibitor

1. Additive to a paint that slows up some process, such as yellowing or skinning.
2. Material used to retard rusting, corrosion, and chemical reaction.
3. Material that has a tendency to prevent chemical action; for example, zinc chromate in a primer inhibits rust formation.

Inorganic zinc-rich primers. Silicate vehicles, usually sodium silicates or ethyl silicates.

Iodine number. In the analysis of oils, fats, etc., the number of centigrams of iodine absorbed by 1 g of the substance.

Intercoat adhesion. Adhesion between two coats of paint.

Interior oil stain. Quick-drying pigmented alkyd resin wood stains for interior use.

Intumescence. Property of fire-retardant paint that makes it swell and foam when heated.

Iron. Metallic substance that requires painting to prevent corrosion and rust.

Iron oxide paint. Durable paint used widely on exterior wood and metal, particularly on tin roofs, red metallic paint, metallic brown paint, mineral red and brown paint, or freight car red. It may be used both as a priming and a finish or cover coat on structural steel.

Japan

1. Black varnish, term that once indicated giving a surface a hard black gloss.
2. Solutions of metallic salts in drying oils; varnishes containing asphalt and opaque pigments.

Japan drier. Varnish gum with a large proportion of metallic salts added to hasten drying. It is used in paints, varnishes, and enamels.

Japan finishes. Coatings usually made of bituminous materials, with oil and resins, applied to metal and heated.

Jelled. Term applied to products that are allowed to thicken by adding certain other products; formula failure.

Kalsomine (calcimine). Residue left from water paint coatings made with whiting, glue, etc.

Kauri reduction test. Method of determining the relative flexibility of certain oleoresinous varnishes by using a solution of properly treated kauri resin in turpentine.

Kerosene. Distillate obtained in petroleum refining, which evaporates slowly.

Ketone. Colorless volatile thinner or solvent, such as acetone.

Kettle-boiled oil. Union or blending of linseed oil and driers by boiling in an open kettle.

Knots in lumber. No further preparation, such as spot painting of knots with shellac varnish or aluminum paint, is necessary, unless pitch has not been completely removed.

Krebs-Stormer viscometer. Instrument for measuring the consistency of a paint in Krebs units, often abbreviated to KUs.

Krebs units

1. Unit measurement of viscosity as determined by a Stormer viscosimeter. Most paints are between 65 and 100 KU.

2. Arbitrary values representing viscosity or consistency of paints based on tests with the Krebs-Stormer viscometer.

Label directions. Every reliable manufacturer usually states concisely and clearly how his product is to be used.

Lac. Natural resin secreted by certain insects that live on the sap of trees in India and other bordering countries. The product is marketed in various forms such as seed lac, button lac, and shellac.

Lacquer

1. Synthetic resin film former, usually nitrocellulose plus pasticizers, volatile solvents, and other resins.
2. Surface coating that dries only by the evaporation of solvents, unlike a paint or varnish, which usually dries by evaporation and chemical reactions.
3. Finish or protective coating consisting of a resin or a cellulose ester dissolved in a volatile solvent (sometimes pigment is added). Drying occurs when the solvents evaporate.
4. Finishing material that dries by the evaporation of the thinner or solvent. There are many different types of lacquer, the most important being that based on cellulose nitrate. Besides the cullulosic compound, lacquers contain resins, plasticizers, solvents, and diluents.
5. Coating material that dries by evaporation. There are many types including cellulosic lacquers, Chinese lacquers, sanitary or tin-plate lacquers, and spirit lacquers, for example, solutions of shellac in alcohol. Cellulosic lacquers are the most important, and the term "lacquer" is applied almost exclusively to this type. Cellulosic lacquers, either transparent or pigmented, contain cellulose esters or ethers and plasticizers, with or without natural or synthetic resins as the basic film-forming ingredients. Lacquer dries rapidly by solvent evaporation.

Lacquer thinner. Volatile thinner intended for use with spraying lacquer consisting of butyl acetate, butyl alcohol, ethyl acetate, petroleum naptha, and toluene.

Laminate. Bond layers of material together; product made by so doing.

Latex

1. Microscopic dispersion of droplets of synthetic resin in an aqueous medium. Synthetic resin may be acrylic, polyvinyl acetate, butadiene styrene, etc.
2. Synthetic resin, developed similarly to synthetic rubber, now used as a base for water emulsion paints.
3. Natural latex is a milky substance from a rubber tree; synthetic resins are those from which emulsion or water-base paints are made (polyvinyl acetate, styrene butadiene, or acrylic).

Latex emulsion. Latex binders are synthetic materials that can be varied in hardness, flexibility, gloss development, and retention. Common types of latex are styrene butadiene, polyvinyl acetate, acrylics, and vinyl acetate-acrylics. Drying results from coalescence of latex particles as the water evaporates from the film.

Latex house paints. Exterior latex paints have durability comparable to oil-base paints. They are resistant to weathering and yellowing and so quick drying that they can be recoated in 1 hour. They can be applied in damp weather over a damp surface. They are easy to apply, and the brush or roller can be cleaned quickly with water. They are free from fire hazard.

Latex paint

1. Low-sheen exterior acrylic emulsion paint, for use on exterior surfaces on properly primed wood, masonry, and metal surfaces. It requires no special undercoat or additions for repainting over secure surfaces.

2. Latex paint is easy to clean up, quick drying, and easy to apply; it offers long-term durability, alkali resistance, excellent adhesion, and exceptional color retention.

Latex primer-sealer (water thinned). Latex primer-sealer dries quickly and can be recoated in about two hours. It is not flammable and almost odorless. One coat is usually sufficient. Thinning is unnecessary unless recommended by manufacturer.

Latex wall paint (water thinned). Durable paint with excellent coverage and good washability; it is quick drying, easy to touch up, safe to use and store, and nontoxic. It has practically no odor.

Latex water-thinned acrylic. Paint with excellent adhesion to interior surfaces of all types, including plaster, wood, and preprimed metal, excellent gloss and color retention with proper preparation, good flow, easy application, and excellent hiding and self-sealing properties.

Latex water-thinned polyvinyl acetate (PVA). Latex paint with excellent adhesion to nonchalky masonry surfaces of all types, good alkali resistance, excellent color retention permitting easy touchup of missed or patched areas at a later date, and excellent self-sealing properties (thus no primer is needed except on bare metal surfaces).

Latex water-thinned polyvinyl chloride (PVC). Latex paint used as exterior finish on wood or masonry. It is approximately four times as flexible as polyvinyl acetate. It is excellent for use in repaint work, even over chalky surfaces.

Leaching. When stains are applied over a cresote-based stain, there is a potential for the leaching of the creosote through the stain. Leaching is less likely the longer the creosote stain has been exposed to the weather. Creosote leaching, while not detrimental to the life of a stain finish, tends to be unsightly and is more apparent with the lighter stain colors than with the darker ones.

Lead. Metal commonly used in the manufacture of driers and pigments.

Lead drier. Additive to paint to accelerate drying, made of lead and organic acid.

Leaded zinc. Oxides of zinc are produced from zinc ore by transforming lead sulfide into basic lead sulfate and united with zinc oxide. These pigments are opaque and add body to paint because they contain-lead sulfate.

Leading. Action peculiar to aluminum paints, in which the flat flakes in the pigment overlap each other.

Leafing. Ability of aluminum or gold paint particles to align themselves more or less parallel with the coated surface. Leafing gives the paint its brilliant or silvery appearance. Good leafing is produced with coated pigments and the proper bronzing liquid.

Let down. Add white paint to a colored paint to produce a lighter color.

Leveling. Spreading out into a smooth, level film. Paint with good leveling properties will dry without brush marks or the appearance of an orange peel.

Lifting

1. Solvents of the topcoat penetrate the coat underneath and cause wrinkling or a break in the adhesion.

2. Attack by the solvents in a topcoat on the undercoat which results in distortion or wrinkling of the undercoat.

3. Softening and penetration of a dried paint film by the vehicle of another coating, causing raising and wrinkling of the first film.

4. Buckling of a finish coat when it is applied over a previous coat that is not yet dry or where solvents in the second coat are too strong.

Lightfastness. Ability of a color or film to withstand exposure to light rays without changing color.

Lightness. Whiteness of a paint.

Linseed oil

1. Drying oil, obtained by pressing flax seed, used in paints, varnishes, lacquers, etc.

2. Vegetable oil widely used in the manufacture of alkyd resins and also as a binder by itself.

3. Oil derived from flaxseed by crushing, cooking, and pressing. Like other drying oils, this raw oil hardens slowly and is therefore usually treated by "boiling" or blowing. Boiling consists of heating the oil at elevated temperatures to promote polymerization, which is accompanied by increased body and viscosity and increased readiness to react with oxygen. Blowing consists of heating the oil to moderate temperature and blowing air through to cause bodying and increased viscosity by oxidation.

4. Linseed oil is relatively slow in drying and an excellent wetter.

Linseed replacement oil. Linseed replacement oil is a mixing or thinning oil blended before packaging and used widely in thinning paste white lead on the job and ready-prepared paints. This material is a replacement for either raw linseed oil, thinner and drier, or for boiled linseed oil and thinner.

Liquid driers. Driers consist of metallic soaps or salts of organic acids dispersed in suitable mediums thinned to a liquid with petroleum spirits, turpentine, or a mixture thereof, and used in paints or similar finishing materials. Resinates, linoleates, and napthenates are permitted. The addition of insufficient drier to a paint results in the paint's drying too slowly; the addition of too much drier may also cause the paint to dry too slowly or may produce other undesirable results, such as wrinkling of the film.

Liquid Wax. Liquid wax is intended for use on bituminous floors, rubber flooring, and linoleum. It is the type of wax that does not have to be polished after application. It is a water-wax emulsion.

Livering

1. Causing a coagulated mass partly by reacting a reactive pigment with an acid vehicle actually producing a soap formation.

2. Formation of a rubbery mass due to a chemical reaction between a pigment and a vehicle.

Long oil. Varnish or paint vehicle with more than 55% of the resin consisting of oil or fatty acid.

Long-oil gloss paint. Long-oil gloss paint is used as a priming and body coat on steel and iron.

Long-oil varnish

1. Type of varnish that possesses greater elasticity or flexibility at a sacrifice of faster drying power and greater hardness quality.

2. Varnish with a large percentage of oil to gum resin. Usually more than one gal of oil to each four lbs of resin. Long-oil varnish is more elastic, and more durable than short-oil varnish. Spar varnish is an example of long-oil varnish.

Loss of gloss. On outdoor paints the first sign of weathering is loss of gloss, followed by chalking. Gloss is produced by an excess of oil or other binding material (linseed oil or varnish) and forms a smooth, glasslike film on the surface. Air, moisture, and sunlight cause these organic binding materials to deteriorate rapidly, resulting in loss of gloss. Among other conditions causing loss of gloss are inadequate preparation of the surface, insufficient drying time between coats, and painting in cold weather, incorrect use of paint removers and alkaline cleaners on paint, or exposure of the freshly painted surface to frost, fog, or moisture.

Luster (gloss). Appearance of depth obtained by multiple coats of varnish.

Maleic resins. Maleic resins are used to increase the value and body of alkyds; they also contribute to nonyellowing properties.

Masking. Protecting areas by means of masking tape, etc., from paint application where it is not wanted.

Malamine and urea. Melamine and urea are used in a large variety of industrial finishes like those for automobiles and refrigerators.

Metallic zinc coatings. Coatings for protecting iron and steel in uncontaminated atmospheres are galvanized, sherardized, and electroplated zincs.

Metal primer. First coat of paint or preserving compound applied to iron or steel.

Mica. Reflective extender that is a platy form of hydrated aluminum silicate.

Micrometer. Device for measuring minute distances. A micrometer caliper is an instrument for measuring thickness with precision. The micrometer is used for measuring the thickness of paint films.

Mil. Unit of measuring film thickness (1 mil = 0.001 in.).

Mildew

1. Fungus growth which appears on substrates in warm, humid areas.

2. Plant disease that causes discoloration, produced by various parasitic fungi (mold). It appears as dirty spots on paint film.

3. Fungus or mold growth. Mildews that feed on paint are usually black, brown, or purple.

4. It is extremely important that mildew-infected surfaces be thoroughly scrubbed with water containing some trisodium phosphate (phosphate cleaner). The most efficient method of combating badly mildewed areas is to wash the areas with a solution of 1 part of bichloride of mercury dissolved in 300 parts of water. When mildew is particularly severe, the priming and finishing paint should contain a fungicide.

Mildew prevention. Mildew generally forms in black splotches on a paint film, causing an unsightly change of color. The condition is fairly common in the south. Commercial fungicides for the prevention or retarding of mildew are available in paste form to be added by the user.

Mileage. Coverage obtained from a coat of paint; synonymous with "coverage."

Mills. Machines for grinding pigment and vehicles, such as Kady, Ball, Cowles Hi-Speed, or dispersion mills.

Mill white. Enamels of high reflectance used for the interior of industrial, office, and school buildings. It is usually offered in gloss, semigloss, and flat.

Mineral spirits

1. Paint thinner originally formulated to match the evaporation of turpentine.

2. Petroleum solvent that has a high solvency and a reasonably fast evaporation rate.

Miscible. Two liquids are miscible if they can be mixed without reacting chemically, turning cloudy, appearing turbid, or separating into layers; for example, alcohol and water are miscible, while oil and water are immiscible.

Mist coat. Semitransparent spray coat of paint.

Mixer. Container that includes an agitator.

Mixing of pastes and powders. Powder and paste pigments should be properly thinned if the desired effect is to be obtained. Some pigments furnished in paste form are to be mixed with oil by the purchaser; others, such as resin emulsion paints, are to be blended with water.

Mixing oil. Complete vehicle for thinning white lead paste and similar paste paints, composed of processed drying oils with or without resins, driers, turpentine, and mineral spirits. It is used for finish coat paints for exterior surfaces where low-gloss or eggshell finishes are desired and for interior decoration where finishes having a gloss higher than those obtainable with ordinary flatting oil are desired.

Moisture. Water vapor or liquid.

Moisture-cured urethane. Urethane that has not been modified. It has excellent flexibility and chemical and water resistance. Its abrasion resistance is outstanding. Curing is by evaporation of solvents and reaction with moisture in the air; it must have a relative humidity between 30% and 90%. Moisture-cured urethane is packaged either pigmented or clear, and has a limited pot life after the package is opened.

Moisture curing. Curing by reaction with moisture from the air in contrast to the conventional method for curing oil-based paints and varnishes, which is by reaction with oxygen from the air.

Moisture resistance. Ability to resist moisture.

Mold or mildew removal. Mold or mildew must be removed completely from the surface. A solution of detergent and Clorox mixed with water is used. The surface is rinsed thoroughly with clean water and allowed to dry before painting.

Monomer. Chemical compound, usually simple, capable of reacting with itself or other monomers to form polymers.

Mottling. Film defect appearing as blotches.

Mud cracking. Broken network of cracks in the film.

Mulling. Thorough dispersion of a pigment or pigments in a vehicle by hand or a mill.

Multicolor finish. Coating that can disguise wall imperfections. Small particles of color appear when coating has dried.

Natural bristle brushes. Brushes made with hogs' hair. They were originally the only kind of brush recommended for applying oil-base paints, varnishes, lacquers, and similar finishes because natural fibers are more flexible, leave fewer brush marks, and hold more paint than synthetic bristles. They also resist most strong solvents but are not as easy to clean as synthetic bristles.

Natural finish. Transparent finish, usually a drying oil, sealer, or varnish, applied on wood for the purpose of protection against soiling or weathering. Such a finish may not seriously alter the original color of the wood or obscure its grain pattern.

Natural resin

1. Natural resin may be a fossil of ancient origin, such as amber, or the extract of certain pine trees, such as rosin, copal, and damar.

2. Nonvolatile solid or semisolid organic substance obtained from certain plants and trees.

Naval stores. Group of products derived from the pine trees such as turpentine, rosin, and pine oil.

Near-white blast cleaning. Blast cleaning to nearly white metal, until at least 95% of each element of surface area is free of all visible residues.

Neutral. Term applied to dull or grayed color; term applied to the material that is neither acid nor alkaline.

Neutralization level. Level of amine neutralizing agent added, based on the acid value and resultant molecular weight of the water-thinnable polymer being used.

Nitrocellulose. Major ingredient of most lacquers, made by a reaction of nitric and sulfuric acids with cotton.

Nondrying oil. Oil that does not readily oxidize and harden when exposed to air.

Nontoxic. Term applied to a coating that is nonpoisonous, to a degree.

Nonvolatile. Does not evaporate.

Nonvolatile matter. Remaining portion of a chemical material after its exposure to certain arbitrary conditions of atmosphere, temperature, and time.

Nonvolatile vehicle. Liquid portion of paint, excluding volatile thinner and water.

Ochres. Natural earths; mixtures of hydrated oxide of iron with various earthy materials, producing pigments ranging from yellow, to brown, to red.

Oil and grease. Oil and grease are removed with mineral spirits xylol, used if the area is well ventilated.

Oil-base primers. Oil-base primers have good adhesion and sealing, are resistant to cracking and flaking when applied to unprimed wood, and have good brushability and leveling, controlled penetration, and low sheen. They are unsuitable as a topcoat and should be covered with finish paint within a week or two after application.

Oil color. Single-pigment dispersion in linseed oil, used for tinting paints.

Oil length. Oil length is determined by the number of gallons of oil cooked with 100 lb of resin (short, medium, or long.)

Oil length of varnish. Number of gallons of drying oil with which 100 lb of resin or gum is heated. "Long-oil" varnish contains 25 gal of oil or over; "short-oil," 10 gal or less; "medium-oil," from 10 to 20 gal. An example of long oil is spar varnish.

Oil-modified urethanes. Urethanes that have been modified with drying oils and alkyds; resulting in film properties that are more nearly those of the modifying resin. They dry by evaporation of the solvent and oxidation of the oil and are available as one-package, clear, air-drying types. Their performance is similar to epoxy esters, but they have a much harder film with excellent abrasion resistance.

Oil of (pine) tar. Volatile oil recovered by distilling pine-tar oil to convert it into pine tar.

Oil or oil-alkyd-base house paints. Paints made with drying oil combined with alkyd resin. They have excellent brushing and penetrating properties, provide good adhesion, elasticity, durability, and resistance to blistering on wood and other porous surfaces, and are often modified with alkyd resins to speed drying time. They are applied with a brush to obtain a strong bond, especially on old painted surfaces.

Oil paint

1. Drying oil vehicles or binders plus opaque and extender pigments.

2. Paint with drying oil or oil varnish as the basic vehicle ingredient.

Oils

1. Commonly vegetable oils obtained from various natural sources. Oils are relatively viscous liquids that have a slippery feel. They are used as modifiers for alkyd resins, paint vehicles, varnish constituents, and plasticizers.

2. Oils are used in the manufacture of alkyds.

Oil-type and alkyd-modified linseed oil. Standard oil for coating exterior wood. It has excellent wetting, penetration, and adhesion and can be applied to many surfaces with a minimum of surface preparation. It remains flexible for a long period.

Oil varnish

1. Varnish that contains resin and drying oil as the basic film-forming ingredients and is converted to a solid film primarily by chemical reaction.

2. Blend of drying oil and resins, natural or synthetic, with driers and volatile thinners. Spirit varnishes do not contain drying oils.

Oiticica oil. Very similar to tung oil and used as a substitute for it.

Oleoresin. Pine gum, the nonaqueous secretion of resin acids dissolved in a terpene hydrocarbon oil, which is mainly produced or extruded from the intercellular resin ducts of trees.

Oleoresinous

1. Vehicles composed primarily of oil but modified with various resins. They generally take on the characteristics of the oil. They are harder than oil coatings, dry faster, possess higher gloss, and are more durable.

2. Paints and varnishes made of oil and resin.

Oleoresinous varnish. Varnish composed of resin or gum dissolved in a drying oil, which hardens as it combines with oxygen from the air.

Oleoresins and oil. Varnishes containing resin and drying oil as the chief film-forming ingredients.

One coat. Under normal conditions one coat of material gives maximum coverage and durability usually at 400 sq ft/gal, depending on the type and viscosity of material.

One-package (one-part) formulation. Paint or coating formulated to contain all the necessary ingredients in one package and generally not requiring any field additions except pigment and thinner.

Opacity. Degree of obstruction to the passage of visible light.

Opaque

1. Opposite of transparent. A coating that has unusual hiding power is said to be opaque.

2. Material is opaque if light rays cannot pass through it.

Orange peel

1. Film that has the physical appearance of an orange peel, caused by improper spray application or the application of some finishes by roller or spray.

2. Pebbled film surface similar in appearance to the skin of an orange.

Orange shellac. Manufactured product of stick lac, the secretion of the *Laccifer lacca kerr*.

Organic zinc-rich primers. Organic zinc-rich primers are made with chlorinated rubber, polystyrene, catalyzed epoxy, or other vehicles that have sufficient film strength and are not reactive with the zinc.

Overspray. Spray paint on areas surrounding target objects.

Oxidation

1. Reaction of oxygen with another substance. Oils are dried by oxidation.

2. Chemical reaction involving combination with oxygen.

Oxidize. To unite with oxygen.

Paint

1. Pigmented material generally producing an opaque film, as distinguished from varnish and stain, which are transparent or translucent.

2. Substance that can be put on a surface to make a layer or film of material.

3. Mixture of pigment with vehicle intended to be spread in thin coats for decoration or protection, or both.

4. Mixture or dispersion of pigments or powders in a liquid or vehicle.

Paint and varnish removers. Paint and varnish removers are intended principally for application to interior surfaces but may be utilized for exterior work. They are for use in places that are relatively free from fire hazards.

Paint drier. Paint driers are usually oil-soluble soaps of such metals as lead, manganese, or cobalt, which in small proportions hasten the oxidation and hardening (drying) of the drying oils in paints.

Paint failures. Paint failures occur because of the lack of suitable surface preparation prior to painting, but also because of improper application or unsuitable composition of paint. Factors to be considered in avoiding failures are careful selection and correct proportioning of materials for each coat, allowing adequate drying time between coats, proper spreading rates, and suitable weather and temperature conditions for painting.

Paint-holding properties of woods. Woods that hold paint longest and suffer least when repainting is neglected are cedar, redwood, and cypress; next in order are northern white pine, western white pine, and sugar pine, and then, in order are ponderosa pine, spruce, and hemlock. The woods that have the poorest paint-holding properties are Douglas fir, western larch, and southern yellow pine.

Painting in cold weather. When painting in cold weather (45°F to 50°F) weather forecasts should be consulted daily. When a marked drop in temperature (20°F) is predicted for the evening of the day that paint is to be applied, it is recommended that painting be discontinued early in the afternoon.

Paint liquids. Varnishes, resin solutions, or other liquids used in the manufacture of paints, enamels, and lacquers.

Paint remover. Mixture of active solvents used to remove paint and varnish coatings.

Paint sprayers. Paint sprayers are particularly useful for large areas. Spraying is much faster than brushing or rolling, and although some paint will likely be wasted through overspraying, the savings in time and effort may more than compensate for any additional paint cost.

Paint system. A paint system includes one or more coatings selected for compatibility with each other and the surface to which they are applied, as well as their suitability for the expected exposure and decorative requirements.

Paste. Paint with sufficiently concentrated pigment to permit substantial thinning before use.

Pastes in oil. Pigments can be broken up by stirring with a strong wooden paddle in the original container, taking care that no heavy sediment is left in the bottom. It may be necessary, even after careful stirring, to pour off some of the partly mixed paste into another clean can in order to thoroughly break up the bottom layer by more vigorous stirring. The part that has been poured off can then be returned to the can and the entire quantity again stirred until uniform. Mechanical mixers are available for this purpose.

Paste wood filler. Compound supplied in the form of a stiff paste, which is then applied to the surface of woods to fill the open grains such as oak, walnut, mahogany, and others.

Pearl lacquer. Lacquer into which has been suspended fish-scale crystals found attached to the skin of cold water fish.

Peeling

1. Loss of adhesion of a paint film, which results in large pieces of film splitting away from the surface.

2. Detachment of a paint film in relatively large pieces. Paint applied to a damp or greasy surface often "peels." Peeling is sometimes caused by moisture under the painted surface.

Permeability

1. Ability of a film, membrane, or other form of material to allow another material, usually a vapor or liquid, to pass through.

2. Ability to pass through a substance or mass. Most latex finishes are permeable to moisture vapor.

Penetrating stain. Stain made by dissolving oil-soluble dyes in oil or alcohol.

Phenolic. Class of resins characterized by good chemical resistance.

Phenolic resins. Oil-soluble resins of phenol and formaldehyde. Oil-modified resins make durable chemical-resistant varnishes.

Phenolics

1. The vehicles for phenolics are phenolic resins modified with oil in varying amounts. The coatings made with these vehicles dry from both the evaporation of the solvent and from the oxidation of the oil. They have good resistance to water and chemicals but are softened by strong solvents. Phenolics discolor with age, and the film becomes very hard. Aged phenolic film is difficult to recoat because of poor adhesion. Intercoat adhesion can be improved if the old surface is properly sanded.

2. In varnishes, phenolics are used for outdoor and other severe applications on wood and metals. They are especially durable when baked.

Phenylmercury compounds. Class of materials used as fungicides.

pH value

1. Numerical expression describing the alkalinity or acidity of a solution.

2. Chemical symbol that together with a number describes the alkalinity or acidity of a solution, 7 being neutral. A value below 7 indicates an acid condition.

Pickling method. Complete removal of rust and mill scale by acid pickling, duplex pickling, or electrolytic pickling. Pickling may pacify the surface.

Pigment

1. Insoluble solid having a small particle size, incorporated into a paint system by a dispersion process. Pigments are used to color paints.

2. Fine solid particles used in the preparation of paint and substantially insoluble in the vehicle.

3. Finely ground, dry substance that when mixed with a vehicle in which it is insoluble becomes paint.

4. Pigments form the solid portion of paint and are practically insoluble in the vehicle or liquid portion.

Pigment oil stain. Used mostly as a wiping stain and consisting of finely ground insoluble color pigments, such as those used in paints, in solutions with linseed oil, varnish, and mineral spirits or in accordance with a manufacturer's formula.

Pigment volume. Percentage by volume of pigment in the nonvolatile portion of a paint as calculated from bulking value and composition data.

Piling. Too much coating in one spot; defect of not leveling out.

Pinholing. Occurrence of tiny round breaks in a paint film giving the appearance of a pinhole.

Piping colors. To designate the contents of pipes it is recommended that color bands be painted on the pipes, preferably adjacent to valves or fittings. If desired, the entire length of the piping system may be painted with the correct identifying color. Piping colors are recommended for better identification and protection and preservation of the pipes.

Plaster preparation for painting. Plaster must be allowed to dry thoroughly for at least 30 days before painting. The room should be ventilated while drying and heated in cold, damp weather. Damaged places are repaired with patching paste. Bare plaster, either new or old, must be dry, cured, and hard. Existing texture or swirl type plaster and soft, porous or powdery plaster should be treated with a solution of a mild vinegar solution in water. Repeat treatment until the surface is hard, then rinse off with plain water and allow to dry.

Plaster primer and sealer. Pigmented primer and sealer in white or tints for use as a priming or sizing coat on interior walls and ceilings of plaster, brick, and cement wallboard. It is also used for priming interior woodwork. Pigment is 56% titanium dioxide by weight, 24% calcium sulfate, and 20% magnesium silicate.

Plasticity. Capability of being molded or brought to a definite form.

Plasticizer

1. Material that is added to a paint system to make the film more flexible.

2. Ingredient added to a plastic to soften, increase toughness, or otherwise modify the properties.

3. Any of a group of substances used in plastics or paints to impart softness and viscous quality to the finished product. Resins can be plasticized with vegetable oil fatty acids, and emulsion paint can be plasticized with glycols.

Plastisol. Film former containing resin and plasticizer without solvents.

Plastic wood. Mixture designed for the repair of woodwork. Consists of wood flour, resins, volatile solvents, and plastic binding material such as cellulose nitrate.

Poise. The cgs (centimeter-gram-second) unit of absolute viscosity.

Poisons. Volatile liquids (turpentine, mineral spirits, and gasoline) used by painters are skin irritants. The frequent use of such thinners to remove paint from the hands should be discouraged. Prolonged inhalation of toxic fumes such as those emitted from benzol, used in paint removers, and carbon tetrachloride or carbon disulfide, used in removing old bituminous coatings, constitutes a health hazard. Some of the pigments, particularly those containing lead and chromates, are poisonous.

Polishing. Wall paints with shiny spots or surfaces resulting from washing or wiping.

Polyamide-epoxy (two component)

1. Chemical-resistant finish with excellent hardness, abrasion resistance, and adhesion. It is also resistant to alkalies and acids. It is used as a concrete floor finish where heavy traffic wears through an alkyd finish in a short time, but it loses gloss and chalks on prolonged exterior exposure. Film integrity is not adversely affected.

2. Tilelike finish applied to any firm interior surface. Available in material to produce gloss and semigloss sheens. The two components are mixed prior to application. Pot life is a full working day. The finish combines the physical toughness, adhesion, and chemical resistance of peoxy with the color retention and permanent clarity of polyester. The film has outstanding stain resistance and film is also impervious to moisture.

Polyester resin. Synthetic resin produced with polyhydric alcohols and polybasic acid. It requires a catalyst for curing. Oxygen-inhibited types usually require peroxide catalysts. The "oil-free" polyesters used extensively in industrial finishes are cross-linked with amino resins.

Polymer. Chain or network of repeating units combined chemically, formed from monomers by polymerization. For example, polyvinyl acetate is a polymer formed by hooking together many vinyl acetate units. Vinyl acetate is a monomer.

Polymerization

1. Formation of a polymer from monomers. The two types of polymerization are addition and condensation.

2. Union of two or more molecules of a compound to form a more complex compound with higher molecular weight.

3. Combination of two or more molecules of the same substance, such as that occurring in the heat-bodying of oil.

4. Reaction in which two or more molecules of the same substance combine to form a product of higher molecular weight without changing the chemical composition of the original material. In the protective coating field, the term is applied to various materials, including drying oils, such as linseed oil and tung oil, and resins, such as rosin. In most varnishes the oil is

polymerized by careful heating. The presence of phenolic resin along with the oil during the varnish cooking greatly accelerates polymerization. Such varnishes dry largely by polymerization rather than by oxidation. Varnish films produced mainly by polymerization are characterized by improved water resistance and resistance to sunlight and weathering. Tung oil, a pure phenolic resin spar varnish, is a good example.

Polyurethane. Hard, highly abrasion- and chemical-resistant coating, often used as a floor varnish and bar top finish.

Polyurethane (one component). Polyurethane provides abrasion resistance on wood floors, furniture, paneling, cabinets, etc., and resistance to normal household materials, such as alcohol, water, and grease.

Polyurethane gloss clear plastic coating. Clear polyurethane gloss for interior and exterior wood surfaces. It is ideal for use when abrasion, heat, or household chemicals are a consideration and possesses excellent wearing qualities.

Polyurethane resins. Finishes made with isocyanate resin have tremendous abrasive resistance and are noted for their elastic qualities. Industrial floor finishes are made of these resins. Some polyurethanes approximate the qualities of epoxy coatings. Most retailed finishes are oil modified.

Polyurethane satin clear plastic coating. Polyurethane clear satin finish for interior application. Toughness and fast-drying qualities make it ideal for areas where excellent wear resistance is required. It resists household chemicals, soaps, detergents, heat, water, alcohol, oil, and grease.

Polyvinyl acetate. Vehicle for latex paint; synthetic resin made by polymerizing vinyl acetate, either by itself, or with other resins.

Porosity. State of a film having noncontinuous voids; quality of exterior latex paint that allows it to breathe. Porous materials allow the passage of liquid and vapor. Porous flat wall finish will not be stain resistant.

Power tool cleaning methods. Loose rust, loose mill scale, and loose paint are removed, to the degree specified, by power tool chipping, descaling, sanding, wire brushing, and grinding.

Preparation of surfaces and application of painting materials. The success of a painting job depends principally on four factors: condition of the surface to be painted, the condition of the preceding paint coat in a repainting job, prevailing atmospheric conditions, and quality of the paint and its suitability for the service expected.

Preventing butt end watermarks. Where the butt ends of lumber, plywood, or shingles are exposed to the weather or other source of moisture and water, it is essential to stain those edges to prevent the "wicking" of water into the wood, which would invariably result in unsightly watermarks.

Preventing lap marks. Lap marks result when an attempt is made to cover too large an unbroken area at one time. When using stain, maintain a so-called wet edge so that the freshly applied stain can be uniformly merged into the last applied stain.

Primary colors. Yellow, red, and blue. These are the basic colors from which all paint colors are made.

Primer

1. First coat of paint applied to a substrate. This is often the most important coat for controlling such things as corrosion, adhesion loss, and blisters.

2. Paint or analogous substance applied next to the surface of the material being painted; priming paint.

Primers. Undercoat that binds the topcoat to the substrate.

Primer-sealer. Product formulated to possess the properties of both a primer and a sealer.

Priming paints for steel. Inhibitive washes and paints are usually applied to structural steel by brushing, but they may also be applied by spraying. Paints used for spraying are mostly of the oil type, which require considerable time for drying. A widely used priming paint for structural steel exposed to the weather is red lead meeting ASTM D 83 standards.

Pumice. Type of stone commonly used by painters in pulverized form mixed with water to reduce gloss on an enamel or vanish and give a "hand rubbed" effect.

Putty

1. Cement of doughlike consistency, made of whiting and linseed oil, used for securing window glass and filling holes, that sets hard upon aging.

2. Doughlike mixture of pigment and oil (usually whiting, and linseed oil, sometimes mixed with white lead). It is used to set glass in window frames and fill nail holes and cracks.

PVA

1. Polyvinyl acetate.

2. Polyvinylacetate; synthetic latex used in some water emulsion paints.

P.V.C. Pigment Volume Concentrate, term used to express high or low pigment content of a paint in proportion to the amount of free available binder or oil vehicle.

Quick drying. Architectural paints and enamels that will dry in about 4 hours.

Rain spots. Condition caused by rain falling on a newly applied finish before it has set.

Raw linseed oil

1. Oil just as it is extracted from the flaxseed.

2. Material from which boiled and refined linseed oils are made and the most important of the liquids used for mixing paint on the job. The quality of raw linseed oil is probably more influenced by the presence of "foots." Paint made with oil containing considerable foots, but otherwise of high quality, dries very slowly and may easily be washed off long after it appears to be dry.

Raw tung and China wood oil. Tung oil is one of the chief oils used in the manufacture of fast-drying, waterproof oil varnishes. The most important characteristics for identification of tung oil are its high specific gravity and high refractive index. Raw tung oil dries rapidly but not to a smooth film and is not used in exterior house paints where linseed oil is almost universally used. Some tung oil, prepared by heat-treating the raw oil so as to overcome its defects, is used in house paints. However, the great use of this oil is in the manufacture of oil varnishes used as such, in varnishes and liquids for the manufacture of interior paints, floor and deck enamels, and water-resisting enamels, and in other specialized finishes.

Ready-mix paints. New paints are usually ready for use when purchased and require no thinning except when they are to be applied with a sprayer. Check the label before mixing or stirring. Some manufacturers do not recommend mixing as it may introduce air bubbles.

Receding colors. Light colors are better adapted to receding then are dark colors.

Red label goods. Products requiring a red label, for shipment, according to I.C.C. regulations. This requirement is for products having a flash point below 80°F.

Red lead. Primer for structural steel or iron. A heavy lead oxide substance, orange to red in color, it is used as a paint pigment.

Red pigments. The five red pigments are bright-red iron oxide, Indian red, mineral red, toluidine red, and Venetian red. With the exception of toluidine, these red pigments may be used for tinting purposes. Bright-red iron oxide and Indian red are used in all types of paints, enamels, and stains, and are quite permanent; toluidine red, a bright, organic red, is permanent when exposed to the weather but is not permanent in a light tint, such as pink, when used outdoors.

Reduce

1. Thin in viscosity by adding a thinner or solvent.

2. Decrease the consistency of a product by the addition of a thinner, such as mineral spirits or linseed oil.

Reducer. Volatile ingredients used to thin or reduce the viscosity of a finishing material.

Refined soybean oil. Refined soybean oil is used largely in edible products rather than in paint. It is slower drying than linseed oil and generally is used in combination with faster drying oils, such as tung, perilla, or linseed oils. It is mainly used in varnish vehicles for interior paints and enamels because paints made with soybean oil do not yellow as much as those made with many other oils. Likewise, it is used in some of the best interior architectural white enamels based on synthetic resins of the alkyd type.

Reflection. Return of light from a surface.

Relative dry hiding power of a paint. Ability of a paint to reduce the contrast of a black and white surface to which it is applied and allowed to dry.

Relative humidity. Scientific way of measuring moisture in the air; percentage ratio of water vapor in the air to the amount required to saturate it at the same temperature.

Resin

1. Solid or semisolid material, usually polymeric, that deposits a film and is the actual film-forming ingredient in paint.

2. Semisolid or solid complex amorphous mixture of organic compounds with no definite melting point and insoluble in water. Resins either are usually partly soluble in alcohols, ethers, and other organic solvents or can be made so by heating. On heating, resins soften, melt, and burn with a smoky flame.

3. Nonvolatile solid or semisolid exudation from pine trees and plants. Resins are also made synthetically by polymerizing molecules. Examples of natural resins are rosin and damar.

4. Mixture of organic compounds with no sharply defined melting point and no tendency to crystallize, and soluble in certain organic solvents but not in water.

Resin-emulsion paints

1. Paints usually packaged in paste form to be thinned with water approximately in the proportion of two parts by volume of paste to one part of water.

2. Water-thinnable paints usually supplied to the customer in paste form. The binder can be of either the alkyd or oleoresinous type, finely dispersed in water to form an emulsion. On application and drying the emulsion breaks to form a cohesive, adherent film.

3. Paint whose vehicle (liquid part) consists of resin or varnish dispersed in fine droplets in water, analogous to cream (which is butterfat dispersed in water).

4. Water-thinned paste paint of the resin-emulsion type that can be applied as a decorative coating on interior walls and ceiling of plaster. In addition to plaster, it may be used on concrete, brick, masonry, wallboard, and fabric. This paint is not intended for wood and metal but may be used on such surfaces if they are suitably primed.

Resins for paints. Natural and synthetic resins are used in air-drying and baked finishes. Natural resins include both fossil resins, which are harder and usually superior in quality, and recent resins tapped from a variety of resin-exuding trees. The most important fossil resins are amber kauri, Congo, Boea Manila, and Pontianak. Resins include damar, East India, Batu, Manila, and rosin. Shellac, the product of the lac insect, may be considered to be in this class of resins.

Respirator. Mask, usually of gauze, that prevents the inhalation of noxious substances during spraying.

Retarder. Solvent added to paint to reduce the evaporation rate.

Rich. Adjective used to describe the quality of deep, dark, and warm colors.

Roller care. Rollers used with alkyd or oil base paints should be cleaned with turpentine or mineral spirits. Rollers used with latex paints should be cleaned immediately with luke-warm water and soap.

Roller covers. Like brushes, roller covers are available with either natural or synthetic fibers. Natural fiber covers, made with wool or mohair, are used for oil-base paints, varnishes, stains and similar finishes. Synthetic fiber covers are used for applying latex paints. Newer blends of synthetic fibers combine the best features of both synthetic and natural, leaving a smooth "unstippled" finish more like that attained with a brush.

Roller mill. Grinding mill that disperses pigments in a vehicle. Rollers revolve in opposite directions to each other and at varying speeds.

Rolling. Roller applications are used only on rough and resawed surfaces, where brushing does not penetrate the deeper crevices or surfaces. Rollers are now made in various coverings to add texture to smooth surface wall and ceiling.

Roof coatings. Bituminous roof coatings are made of asphalt (chosen for good weather resistance) dissolved in a suitable solvent. Asbestos and other fillers are added to prevent sagging on sloping roofs and to permit application of relatively thick coatings. They are basically made in gray and black; however, the addition of aluminum powders provides for other colors. Asphalt emulsion roof coatings can be applied over damp surfaces.

Ropiness. Stickiness inherent in some paints that results in an irregular pile up of their film, which has the appearance of the texture of a rope. In the Western states only, this same term applies to a film having poor flow or showing brush marks.

Ropy. Description of applied paint showing brush marks.

Rosin

1. Solid resin obtained as the residue from the preparation of turpentine from the crude resin of the pine tree.

2. Gum rosin is the sap that exudes from living pine trees. Wood rosin is steam-distilled from stumps and other dead wood.

Rosin oil. Relatively viscous, oily portion of the condensate obtained when rosin is subjected to dry, destructive distillation.

Rosin spirits. Relatively light, volatile portion of the condensate obtained in the first stages when rosin is subjected to dry, destructive distillation.

Rottenstone. Similar to pumice, but softer. It is used chiefly by piano finishers.

Rubbed effect. Desirable sheen, similar to low luster, hand-rubbed on fine furniture.

Rubber-emulsion paint. Paint whose vehicle consists of rubber or synthetic rubber dispersed in fine droplets in water.

Rubbing oil. Neutral, medium-heavy mineral oil used as a lubricant for pumice stone in the rubbing of varnish or lacquer.

Rubbing varnish. Hard drying varnish that can be rubbed with an abrasive and water or oil.

Running and sagging. Running and sagging may occur when the paint contains too much oil or is applied too freely. If an old paint surface is too glossy, the fresh paint may show sagging.

Runs. Runs are usually caused by improper paint consistency or by applying paint too heavily.

Rust. Corrosion product that forms on iron or steel exposed to moisture.

Rust-inhibitive pigments. Pigments that react chemically on metal to prevent or retard further corrosion.

Rust-inhibitive washes. Rust-inhibitive washes are solutions that when correctly formulated etch the metal and form a dull gray coating of uniformly fine texture, thus producing a rust-inhibitive surface receptive to the priming paint.

Safflower oil

1. Drying oil that is similar to soybean oil but dries faster.

2. One of the best nonyellowing oils available; it is slower drying than linseed oil but slightly faster than soybean.

Sagging. Excessive flow on a vertical surface resulting in drips and other imperfections on the painted surface.

Sags. Sagging of paint (a curtain effect). Usually sags are caused by applying too heavy a coat of paint or too much thinning.

Saponification. Conversion of a fatty acid into soap with alkali; the breakdown of a vehicle as a result of reaction with alkali.

Saponify. Convert a fat or oil into soap by the action of an alkali. When esters are boiled with strong bases, soaps are formed. Linseed oil contains the glyceryl ester of linoleic acid. When a linseed oil paint comes in contact with a surface that contains strong alkali and water, such as damp concrete basement floors, oil is saponified and thus loses its bonding properties.

Sandblasting. Method of cleaning a building prior to painting. A blast of air or steam, laden with sand, is used to clean surfaces.

Sanding. Smoothing of a surface with an abrasive paper or cloth.

Satin finish. Paint finish resembling the luster of satin.

Sealer

1. Liquid coating composition, usually transparent, such as varnish, that also contains pigment for sealing porous surfaces, especially plaster, preparatory to the application of finish coats.

2. Primer that does not allow succeeding coats to penetrate and that seals in material that might otherwise bleed through the surface.

3. Coat of paint, or the like, intended to close or seal the pores in a surface.

Sealer and primer

1. Sealer and primer, a clear penetrating alkyd resin sealer for both interior and exterior wood priming, penetrates wood surfaces, coats wood fibers, and equalizes the surface porosity of new wood.

2. Sealer and primer may also be used over concrete to retard dusting, is ideal for all woods, and is of particular value for porous soft woods.

Secondary colors. Orange, violet, and green (colors made from mixing two primary colors).

Sediment. Solid that can settle or be centrifuged from the main portion of the liquid.

Seeding

1. Lumping of pigment; vehicles, becoming gelatinous, forming relatively large particles in the coating. Seeding most often occurs with zinc oxide in alkyd finishes.

2. Result of a chemical reaction that produces aggregates of pigment or vehicle particles, which show up as a "sandy" effect when the material is brushed out.

Self-cleaning. Controlled chalking of a paint film so that dirt and small amounts of paint film do not adhere to the surface.

Self-sealing. Term applied to a product that can be used as a primer and finish coat on porous surfaces.

Semigloss

1. Intermediate gloss level between high and low gloss.

2. Degree of surface reflectance midway between gloss and eggshell; also, paints and coatings displaying these properties.

Semigloss and full-gloss enamel (alkyd base). Enamel made with alkyd resins having good gloss retention, grease and oil resistance, and better washability and resistance to abrasion than flat alkyd enamel.

Semigloss and full-gloss latex enamel (water thinned). Enamel that has most of the properties of alkyd enamels plus the usual advantages of latex paints—easy application and cleanup, rapid drying, low odor, and nonflammable. It has a good leveling, but lapping does not compare favorably with alkyd enamels.

Semigloss paint or enamel. Paint or enamel made with a slight insufficiency of nonvolatile vehicle so that its coating, when dry has some luster but is not very glossy.

Settling. The pigment separates from the vehicle and settles to the bottom of container.

Sheen

1. Gloss or flatness of a film when viewed at a low angle.

2. Luster, gloss, semigloss, eggshell, etc.

Shelf life. Length of time a paint product may remain on the shelf, or be stored and still be usable.

Shellac

1. Shellac is available in clear and "orange" finishes. It is fast drying. The thinned first coat provides an excellent seal for new wood and can be overcoated in about 30 minutes. It should be lightly sanded between coats. Paste wax, as the final coat, provides luster and some protection against scratches.

2. Transparent coating made by dissolving lac, a resinous secretion of the lac bug (a scale insect that thrives in tropical countries, especially India), in alcohol.

3. Fast-drying varnish consisting of lac resins, produced by the lac insect, dissolved in alcohol.

4. Natural gum that is useful in the manufacture of certain types of paint, chiefly lacquers; also a solution of this natural gum, which is widely used to paint hardwood floors.

5. Resinous material commonly known as "flake shellac," secreted by the insect *Laccifer lacca kerr*. Shellac is obtainable in two forms, "orange" and "bleached."

Shop primed. Article primed at the factory.

Short oil varnish. Harder or faster drying type of varnish that may be brittle owing to its property of being short in oil and long in resin content.

Sienna. Iron-bearing earth pigment that is brownish yellow naturally but when roasted becomes orange red or reddish brown.

Silicone alkyds. Silicone alkyds can be modified with oils to dry in the same way as any conventional alkyd, and it may be thinned with either mineral spirits or xydol, depending on the formulations. Many formulations of these coatings are on the market, but it is generally recognized that silicone alkyds must contain at least 25% silicone resin to provide the outstanding color and gloss retention for which silicones are noted. Coatings are recommended for heat-resistance applications with the advantage that the alkyd portion protects the surface during erection or standby and that when equipment is fired or heated, the coating is heat-cured and the silicone portion becomes effective.

Silicone resin. Synthetic resins using silicone as a component. Silicones are water repellent, and they are used in clear liquid form as a coating for the exterior of buildings. Silicones are also resistant to heat and chemicals.

Silicones. Silicones are used when temperatures higher than can be borne by the other finishes are encountered.

Silking. Lines in a paint film resulting from the draining off of excess paint in a dip or flow coating process.

Size

1. Water-based formulation with glue or starch binders, intended as a sealer over existing wall paint or plaster; now seldom used.

2. Liquid coating used for sealing porous surfaces.

3. Sealer used to seal the pores of a surface so that subsequent coats of paint or varnish will not penetrate.

Skin

1. Tough, skinlike covering that forms on paints, varnishes, etc., when left exposed to air for long periods; it is formed by oxidation or polymerization.

2. Partial solid layers of material that may form from the material itself or otherwise.

Skinning. Development of a solid layer on the top of the liquid in a container of paint.

Skips

1. Areas unintentionally left bare by a painter.

2. Skips, also called "holidays," are small areas that are left unpainted unintentionally.

Slow drying. Slow drying may be caused by insufficient drier in the paint or varnish, the use of poor-quality linseed oil, too liberal an application of paint, the application of paint or varnish over an undercoat that is not dry or during damp, wet, or foggy weather, and the application of paint during cold weather.

Softness. Film property displaying low resistance to scratching or indentation; opposite of hardness.

Solid covering. Covering having high opacity.

Solids

1. Percentage, on a weight basis, of solid material in a paint after the solvents have evaporated.

2. Pigment and nonvolatile vehicle components of paint that remain on the surface (they can be measured by weight and volume).

Soluble. Term applied to a product that can dissolve into another.

Solvency. Measure of the ability of a liquid to dissolve a solid.

Solvent

1. Volatile portion of the vehicle, such as turpentine and mineral spirits, that evaporates during the drying process.

2. Liquid that will dissolve something, commonly resins or gums or other binder constituents.

3. Component of a solution that dissolves other components; in paint, the liquid is usually volatile. There are two major categories of paints: solvent based, which refers to oil or resin, and water based, in which water is the solvent.

4. A liquid capable of dissolving a mateiral is a solvent for the material.

Solvent cleaning. Removal of oil, grease, dirt, soil, salts, and contaminants by cleaning with solvent, vapor, alkali, emulsion, or steam.

Solvent thinned. Term applied to a formulation in which the binder is dissolved in the thinner, as in oil paint, rather than emulsified, as in latex paint.

Soybean oil

1. Soybean oil is obtained from the seeds of the soya plant. It has excellent flexibility and is nonyellowing but dries slowly and is usually combined with tung, perilla, or linseed oil.

2. Oil pressed from soybeans, used widely in alkyd resin vehicles. Its nonyellowing qualities make it a preferred ingredient for white enamels. By itself, it has poor drying qualities.

3. Films have good flexibility and excellent nonyellowing characteristics. Soybean oil is considered semidrying oil and is slower drying than both linseed oil and tung oil.

Spackling compound. Type of plaster used to fill surface irregularities and cracks in plaster. The compound when mixed with paste paint makes what is known as "Swedish putty."

Spar varnish

1. Very durable, water-resistant varnish for severe service on exterior exposure. It consists of one or more drying oils, such as linseed, tung, or dehydrated castor; one or more resins; such as rosin, ester gum, 100% phenolic resin, or modified phenolic resins; one or more volatile thinners, such as turpentine or petroleum spirits; and driers; such as linoleates, resinates, or naphthanates of lead, manganese, and cobalt.

2. Top-quality interior-exterior varnish, resistant to rain, sun, and heat. It usually contains phenolic resin.

3. Clear varnish that is useful on surfaces that are exposed to the exterior.

4. Clear phenolic resin wood finish, transparent gloss. It produces a rich transparent gloss that provides a protective film for all wood surfaces without hiding natural grain beauty. It is one of only a few varnishes designed primarily for wood siding and is also used for outdoor furniture, fences, etc. and interior woodwork.

Specifications. Written instructions on details of paint applications, types of products to be used, areas to be painted, and painting procedures.

Specific gravity. Weight of a material, ordinarily relative to water, in which case specific gravity is the weight of a specific volume of the material divided by the weight of an equal volume of water under standard conditions.

Spirit stain. Stain produced by dissolving a dye in an alcohol.

Spirit varnish

1. Varnish that is converted to a solid film by solvent evaporation; damar varnish.

2. Shellac (gum or resin dissolved in solvent). This type of varnish dries by evaporation, rather than oxidation.

3. Spirit varnish is intended (originally) to replace shellac varnish, which it resembles in appearance. It can be applied to wood, metal, paper, and textiles and may also be used in place of oil-resin varnishes where rapid drying is more important than length of service.

Spray. Application of a paint by spraying (spray gun application).

Spray booth. Small room with spray equipment designed with proper ventilation and dust control.

Spraying. Applying a coating by means of a spray gun. Skill is necessary to provide an adequate coating. Spraying usually requires a follow-up with a dry brush to ensure an even coating.

Spraying lacquer. Spraying lacquer should contain celluosic derivatives. This lacquer is intended for indoor or outdoor use and is applied by a spray gun; it is not intended to be a brushing lacquer. It consists of cellulose nitrate, alkyd resin, dibutyl phthalate, fish oil, and lacquer thinner.

Spreading rate. Area that may be covered by a paint expressed as the number of square feet per gallon per mil: 1 gal equals 231 cu in. and can be spread to 1604 sq. ft as a wet film in a thickness of 1 mil, depending on the material.

Stabilizer. Something that is added to paint to prevent degradation.

Stain

1. Transparent composition of colored penetrating liquid that leaves little or no surface buildup.

2. Penetrating formulation intended primarily for wood surfaces that changes the color of the wood without obscuring the grain; depending on the amount of pigment, stains may be more or less transparent and may leave little or no surface film.

Staining of shingles and wood. Shingle stains are thin, fluid paints in which the shingles are dipped before being laid. Suitable stains property applied to shingles or rough siding give a durable finish on exterior woodwork. With the exception of some dark-brown stains, which are refined coal tar or creosote and volatile thinner, shingle stains are usually made from very finely ground pigments, drying oils, and volatile thinners. Many commercial shingle stains contain some creosote oil.

Stains. Stains are available in natural finish and in a variety of colors that provide attractive, natural appearance. Several coats are required for bare wood, with light sanding between coats. A final coat of paste wax provides luster and some protection against scratches, particularly furniture. "Thick" stains can be thinned with turpentine or mineral spirits.

Steel structural plate. Steel structural plate is cleaned by surface preparation methods established by the Steel Structures Painting Council.

Storage of painting materials. Paints, varnishes, lacquers, thinners, and other painting material should be stored in well-ventilated places where they will not be exposed to excessive heat, smoke, sparks, flame, or direct rays of the sun. Packages should be kept tightly closed when not in use. Powder paints should be stored in moistureproof containers. Ready-prepared paints and pastes stored in suitable cans or drums should be inverted very month to retard settling of the solid matter. However, unpigmented liquid materials (linseed oil, for example) should be left undisturbed.

Strong. Term denoting the strength of pigment, intensity of a color, or degree of odor.

Strontium chromate. Rust-inhibitive pigment of less tinting strength than zinc chromate, but similar in color.

Stucco. Stucco is prepared for painting by being cleaned and freed of any loose mortar. Manufacturer's recommended procedures for applying paint to stucco should be followed, and normal drying conditions prevail. The surface may be painted with a latex paint in 5 to 7 days after installation.

Styrene butadiene (emulsion type). Latex paints made with styrene butadiene usually have slightly more sheen than acrylic or PVA flats and dry practically odorless in less than an hour. Butadiene cures by oxidation, making the film water resistant in 30 days. There is also a solvent type that is a copolymer of styrene and butadiene. It is nonoxidizing, resistant to strong alkalies, moisture, and mildew, and tends to yellow more than PVA or acrylic latex. The solvent type is often used in masonry coatings.

Substrate. Layer lying under another; material on the surface on which a coating is applied.

Surface preparation

1. New wood should be clean and dry for best results from either solid color or semitransparent stains. Surface dirt, pencil marks, and grade stamp marks must be removed.

2. Conditioning of a surface to receive a coating. Surface must be free from dirt, grease, and dust and must be properly sanded, etc.

Surfacer. Pigmented formulation for filling minor irregularities before a finish coat is applied; it is usually applied over a primer and sanded for smoothness.

Surface tension. Property of liquid or solid matter due to unbalanced molecular forces near the surface; also, a measurement of this property.

Swedish putty. Slow-drying mixture of whiting, flour paste, and boiled linseed oil, used for filling hairline cracks and uneven spots on plastered walls.

Synthetic. Chemically man-made.

Synthetic bristle brushes. Brushes made from a synthetic fiber, usually nylon. Nylon brushes are recommended for both latex (water soluble) and oil-base paints because this tough synthetic fiber absorbs less water than natural bristles do, while resisting most strong paint and lacquer solvents. Nylon bristles are easier to clean than natural bristles.

Synthetic resins. Chemical reproductions of natural resins as far as color and appearance are concerned but of superior performance in the manufacture of paints due to better uniformity and other properties.

Tack. Stickiness of a paint film.

Tacky. Term applied to a sticky condition that exists during the drying process between the wet and dry stages.

Tempera. Pigments ground in egg whites or glue, used by watercolor artists on posters, etc.

Tensile strength. Load necessary to break a film when pulled in the direction of length; stress necessary to break a film.

Texture. Rough or uneven physical appearance produced by a tool, of the surface structure.

Thermoplastic. Soft and pliable when heated, returning to solid when cooled.

Thermosetting. Term describing a type of plastic that becomes hard and unmoldable when heated and thereafter is heat resistant.

Thinner

1. One or a mixture of several solvents or diluents that are used to reduce the viscosity of paint or to lower the solids.

2. Volatile liquid with which the viscosity of a paint product can be modified. Thinner evaporates when the coating is drying.

Thinners. Volatile constituents added to coatings to promote their spreading qualities by reducing viscosity. They should not react with the other constituents and should evaporate completely. Commonly used thinners are turpentine and mineral spirits, such as derivatives of petroleum and coal tar.

Thixotropic paint. Paint consisting of jellylike materials that change to flowing liquids when agitated by stirring, shaking, etc. However, when the agitation ceases, these materials revert to their former jellylike state in due time.

Thixotropy. Property of becoming liquid exhibited by certain gels when shaken or stirred. Term generally refers to a heavy-bodied (buttery) paint.

Throwing power. Ability of an electrodeposit to penetrate into hard-to-reach areas, such as a hollow metal object.

Tinner's red. Inexpensive red oxide paint; oil for painting roofs and gutters.

Tint

1. Add color; let down with white; pastel shade of a color.

2. Very light color, also, to add color to another color or to white.

3. Light color obtained by the addition of small quantities of coloring materials to a white base.

Tinting strength. Coloring power of paint or pigment.

Titanium dioxide

1. Extremely white inert pigment. It has greater hiding power than lead or zinc and retains its whiteness.

2. White pigment that has the greatest hiding power of all white pigments.

Toluol. Solvent used in industrial finishes and as a diluent in lacquer.

Tooth

1. Roughen a surface with sandpaper or clean with a liquid degreaser in preparing a surface for painting, which helps adhesion of succeeding coats of paint.

2. Term describing a flat or roughened surface that readily allows a topcoat of paint to adhere.

3. Roughened or absorbent quality of a surface that affects adhesion and application of a coating.

Topcoat. Final layer of paint applied to a substrate.

Total solids. Nonvolatile ingredients of paint after the volatile portion has evaporated.

Toughness. Ability of a hard film to resist scratches, abrasion, and breaking.

Traffic paint. Reflective marking paint for designating traffic lanes, safety zones, and intersections, but usually available in nonreflecting paints for parking designations and traffic lanes in industrial buildings.

Transparent. Clear enough to see through (such as window glass).

Treatment of aluminum and magnesium. Aluminum and aluminum-magnesium alloys should be anodized before being given one or more coats of a suitable high-grade paint. Zinc chromate primers are widely used and known to give good results. Trim paint, aluminum paint, iron oxide paint, graphite paint, or the usual outside house paints can be used over the primer with good results.

Trim enamel paint. Trim enamel paint is a subdivision of surface coatings known as house paints and differs from ordinary house paint body colors by faster drying, having more gloss, and showing fewer brush marks. It is principally designed for use on trim, screens, and shutters.

Trim paint. Paint usually made with oil-modified alkyds. It is slow drying (overnight) and is made in high-sheen bright colors, have good gloss and color retention. They are substantially more durable than conventional oil-alkyd enamels.

Tung oil. Tung oil dries faster, has greater reactivity with varnish resins, and possesses greater water and weathering resistance than most other oils.

Tung or china wood oil

1. This oil is derived from the tung tree nut and is especially waterproof, fast drying, and durable. It is usually used with less reactive oils, such as linseed or soya. In the raw state it tends to wrinkle on hardening and is used for wrinkle finishes; otherwise it is treated or combined with other oils.

2. Drying oil extracted from the seeds of the tung tree, used in making varnish and giving extra water resistance. It is also used to plasticize phenolic resin for marine paints that are resistant to chemicals.

Turbid. Term describing a liquid containing a relatively great amount of nonsettling suspended matter or other insoluble or separated matter even though the liquid is translucent and transmits at least a little light.

Turpentine

1. Distilled product from pine trees, it is both colorless and volatile.

2. Solvent obtained from the distillate of the exudation of pine trees.

Two-coat paint system. System developed for painting new exterior woodwork that consists of special primer and finish paint for application in relatively thick coats.

Two-package (two-part) formulation. Paint or coating formulated in two separate packages and requiring that the two ingredients be mixed before the characteristic properties can be obtained and the material can be applied.

Ultraviolet light

1. Portion of the spectrum that is largely responsible for the degradation of paints.

2. Invisible rays of the spectrum lying outside the violet end of the spectrum that are responsible for film failure in exterior exposure.

3. Invisible part of sunlight that causes skin to become "suntanned" and similarly causes paints and varnishes to fade, yellow, darken, crack, etc.

Umber. Earth pigment consisting chiefly of hydrated oxide of iron and some oxide of manganese, brown in its raw state, but reddish brown when heated to prescribed temperatures.

Undercoat

1. Coating applied prior to the finishing coats or topcoats of a paint job. It may be the first of two or the second of three coats. Synonymous with "priming coat."

2. Primer over which a topcoat will be applied.

3. First coat, primer, sealer, or surfacer.

Undercoater. Coat applied in preparation for a finish coat to be applied over it.

Undertone. Color of a pigment that shows up when that pigment is mixed with much white pigment.

Underwater paints. The vehicle for underwater paints is generally a pure phenolic resin-tung oil varnish, with or without linseed oil. Usually the same vehicle is used in both the primer and topcoat or finish coats.

Urea-formaldehyde resins. Resins requiring strong solvent whose main value is to improve their flow and leveling of epoxies without diminishing their chemical resistance.

Urea-melamine resins. Resins which produce a tough finish which approaches porcelain. Product of melemine and formaldehyde.

Urethane resins. Particular group of film formers; isocyanate.

Urethanes. Polymers introduced with characteristics similar to the epoxies.

Value. Light or dark. As the color is let down toward white, the value gets lighter. As it approaches black, the value gets darker.

Varnish

1. Clear resin solution made by applying heat to a mixture of hard gum and a vegetable oil, and then dissolving the product formed in a suitable organic solvent.

2. Protective coating made from resins dissolved in oil for oil varnish or in alcohol or other volatile liquid for spirit varnish. Varnish is a transparent coating made in varying degrees of gloss.

3. Homogeneous liquid generally composed of resin, drying oil, volatile thinner, and drier. When applied in a thin layer and exposed to the air, the varnish is converted to a transparent or translucent solid film.

4. Transparent combination of drying oil and natural or synthetic resins.

5. Thickened preparation of drying oil, or drying oil and resin, suitable for spreading on surfaces to form continuous, transparent coating or for mixing with pigments to make enamels.

Varnish-type paints. The vehicle for underwater paints is generally a pure phenolic resin-tung oil varnish, with or without linseed oil. Usually the same vehicle is used in both the primer and topcoat or finish coats.

Vehicle

1. All of a paint except the pigment.

2. Liquid (such as oil) that is mixed with pigment to make paint.

3. Liquid portion of any paint or enamel, including anything, such as a resin, that is dissolved in the liquid.

4. Liquid portion of a paint, except water, that is volatile in a current of steam at atmospheric pressure.

Vegetable oils. Oils pressed from seeds and nuts of vegetable growth. Vegetable oils used in paint manufacture include linseed, soybean, perilla, hempseed, tung, and castor. They produce paint binders with good drying qualities.

Venetian red. Red pigment made by calcining a mixture or lime and iron sulfate.

Vinyls

1. Class of monomers that can be combined to form vinyl polymers, characterized by their toughness, flexibility, and durability.

2. Solution vinyl coatings are plasticized copolymers of vinyl chloride and vinyl acetate dissolved in strong solvents such as ketones and esters. They dry very rapidly by solvent evaporation; thus spray application is preferred.

Vinyl resins. Resin or polymer produced by reaction between acetylene and acetic acid in the presence of a mercuric catalyst.

Viscosity

1. Thickness or thinness of a liquid; measure of flow resistance.

2. Internal friction of a fluid; resistance to flow; opposite of fluidity. Linseed oil is more viscous than turpentine; bodied linseed oil is more viscous than raw linseed oil. It is an important physical property of oil, varnish, and lacquer.

3. Property of fluids, either liquid or gaseous, that may be briefly described as "resistance to flow."

Volatile. Term applied to material that passes off in the form of vapor.

Volatile mineral spirits. Petroleum distillate known as mineral spirits or petroleum spirits. It evaporates at about the same rate as turpentine and is frequently used in its stead for thinning oil paints. Chemically, it is a mixture of aliphatic, naphthenic, and aromatic compounds. It evaporates completely from the film during the drying period.

Volatile thinner. Liquid that evaporates during the drying of the film.

Volatile thinners and driers. Liquids that evaporate plus drying catalysts that assist in film formation.

Volume solids. Volume or nonvolatile material expressed as percent of a gallon (25% volume solids would equal 1 qt of solids).

Warm color. Advancing colors that contain red or yellow.

Washing

1. Paint failure that allows the film to be rubbed off as a soapy emulsion when wet.

2. Leaching out of the soluble part of a pigment in a paint during rainstorms. Washing sometimes occurs when a paint contains a pigment that is either water soluble or is converted from a water-insoluble to a water-soluble material through chemical action as the result of weathering.

Water emulsion paints. Mixtures of pigment and synthetic resin in water. When water evaporates, the resin particles form a film. They are more durable than oil paints, do not become brittle, are resistant to ultraviolet rays, and have less tendency to blister or peel. Ease of cleanup with water is another desirable factor. They are now called "latex paint."

Water paint

1. Pigments plus vehicles based on water, casein, protein, oil emulsions, and rubber or resin latexes, separately or in combination.

2. Paint with a vehicle that is a water emulsion or water dispersion or with ingredients that react chemically with water.

Water-repellent preservative (silicone type). Silicone water repellents are transparent liquids that help repel water without changing the surface appearance. They must be applied strictly in accordance with instructions to ensure adequacy of film and water repellency. They should not be top-coated with paint until surface has weathered for at least two years.

Water-resisting red enamel. Bright-red enamel of high grade intended for general exterior use. It is used widely on fire alarm boxes and fire hydrants. It is water resistant, has a high gloss, and is durable. Toluidine pigment used in the enamel makes the color of different lots less variable than the color of other bright-red paints that do not contain toluidine.

Water-resisting spar varnish. Durable general-utility varnish, suitable for both interior and exterior use where high gloss or initial hardness of film is not required. Tung oil and phenolic resins are used widely in this type of spar varnish.

Water spotting. Presence of spots or stress of a nonuniform appearance in color, gloss, or sheen, of such shape as to be assumed to have been caused by drops or rivulets of water.

Weathering. Changes caused in a paint film by natural forces, such as sun, rain, airborne dust.

Weatherometer

1. Laboratory instrument designed to produce weathering effects, such as heat, sunlight, cold, and rain.

2. Electrical testing instrument that accelerates various weather conditions on panels exposed to ultraviolet light and water spray.

Wet-film thickness. Thickness of the coating when applied, but before the evaporation of the solvent.

Wetting. Process by which a liquid forms intimate contact with the substrate to which it is applied.

Wetting agent. Substance to lower the surface tension of water as an aid to dispersion.

White lead

1. Oldest white pigment. Basic lead carbonate imparts adhesion, toughness, elasticity, durability, and chalking properties, and is widely used, especially in exterior paints. It is the only pigment that by itself will produce a durable exterior paint. Basic lead sulfate is widely used as a replacement for lead carbonate in mixed-pigment paints.

2. Basic lead carbonate used in exterior house paints and primers.

3. Component of almost all white and light colored paints, and one of the most important white paint pigments. Both kinds, basic carbonate and basic sulfate, are frequently used in ready-mixed paints. For paint mixed on the job, paste in oil is used, whereas the dry pigment is used by the paint manufacturer. The term "white lead" paste is used to designate the usual paste. White lead is composed of about 91% basic-carbonate white lead and 9% linseed oil, or 89% basic-carbonate white lead, 2% turpentine, and 9% linseed oil.

White pigments. White lead (basic carbonate and basic sulfate), lead titanate, zinc oxide, leaded zinc oxide, lithopone, titanated lithopone, zinc sulfide, titanated zinc sulfide, zinc sulfide-barium, zinc sulfide-calcium, zinc sulfide-magnesium, antimony oxide, titanium dioxide (anatase and rutile), titanium-barium, titanium-calcium, and titanium-magnesium.

Whitewash. One of the oldest water paints used for both exterior and interior surfaces. The principal ingredients in whitewash is lime paste. A satisfactory paste can be made with hydrated lime, but better results are obtained by using quicklime that has been slaked with enough water to make a moderately stiff paste and kept in a loosely covered container for several days or, preferably, months.

Whiting. Pure white chalk (calcium carbonate) used in making putty and whitewash and as an extender in paints.

Wiping. Application of stain with a cloth or wiping with a cloth after application by brush will produce shades lighter than those shown on the color card.

Wood affected by water. Wood cells are organic materials that are always taking on or giving up water. Wood swells and shrinks. Under these conditions paint creates a film over the wood that blisters, cracks, and lifts.

Wood affected by mildew. Mildew and molds cause wood discoloration. They thrive on moisture and live only on damp wood.

Wood alcohol. Methyl alcohol (very toxic).

Wood clear finishes. These finishes are not as durable as pigmented paint. Alkyd varnishes have good color and color retention but may crack and peel. Some synthetics, such as polyurethane varnishes, have good durability but may darken on exposure. Spar varnish (marine varnish) is quite durable but will also darken and yellow. It is recommended that thin penetrating coats on the bare wood are applied followed by the unreduced varnish.

Wood exterior. Wood exteriors are prepared for painting by being cleaned and dried. Primer and paint applied as soon as possible. No painting should be done immediately after a rain, during foggy weather, or when the temperature is below 50°F. Knots and pitch streaks should be scraped or burned, spot sanded, and primed before full priming coat is applied. All nail holes or small openings must be filled after the priming coat is applied.

Wood interior. Wood interiors are prepared for painting or finishing. Finishing lumber and flooring, when delivered, are protected from the weather and stored on the premises in dry, warm areas to prevent the absorption of moisture, shrinkage, and roughening of the wood before finishing. Surfaces are sanded smooth with the grain and never across it. Surface blemishes are corrected. Areas are cleaned and free of dust before proceeding with finishing coats. Enamels and varnish finishes are lightly sanded between coats to a smooth finish.

Wood paste filler. Wood filler is a mixture of silex or ground quartz and quick-drying varnish and is used to fill the pores of open-grained wood, such as oak, before varnish is applied and sometimes before painting. Paste filler is thinned to brushing consistency with turpentine or mineral spirits. Wood paste filler is intended for use on open-grained floors and flooring. When thinned, it can be substituted for a pigmented liquid wood filler.

Wood-rotting-fungi preventative. Effective water-repellent material that also includes a fungicide to protect against rotting caused by fungi, used on boats and other water contact objects and items that are almost constantly in the water.

Wood stains. Semitransparent stains are available for exterior wood but are not as durable as house paints. Stain improves the appearance of wood by highlighting the grain and texture of the surface. Stains are available in many colors, the most popular being cedar, light redwood, and dark redwood.

Wrinkle. Pattern formed on the surface of a paint film by improperly formulated coatings; appearance of tiny ridges or folds in the film.

Wrinkle finish. Varnish or enamel film that purposely looks like fine wrinkles.

Wrinkling. When a paint is applied too liberally, particularly during cold, damp weather, the top of the film surface dries first, leaving the paint beneath this skin soft. The result is that the finished surface has a wrinkled appearance.

Yellowing

1. Condition usually caused by the dried vehicle turning amber with age. In white paints this presents a yellowish cast. In colored paints it affects the color and may change blue to a greenish cast.

2. Discoloration to the yellow, which is commonly caused by smoke, grease, certain gases, and sunlight.

Yellow pigments. There are a variety of yellow pigments: lemon chrome yellow, medium chrome yellow, primrose chrome yellow, yellow ocher, raw sienna, and yellow iron oxide. For tinting purposes, the chrome yellows are suitable for outside use. Another yellow pigment, zinc yellow, is used to some extent as a color pigment, but its main use is as a rust-inhibitive pigment in metal primers.

Zinc chromate

1. Effective rust-inhibitive pigment, greenish yellow in color, and usually combined with some high-hiding pigment.

2. Yellow rust-preventing pigment useful on steel.

Zinc dust. Medium-gray pigment with extreme hiding power. Zinc-rich paints adhere unusually well to galvanized metal. Other desirable qualities are one-coat hiding, weather and heat resistance, and rust resistance.

Zinc dust primers. Also known as "galvanized iron primers." They are oleoresinous and generally contain less than 50% zinc dust by weight in the dried film and should not be confused with zinc-rich primers, that usually contain 85 to 95% zinc in the dried film. In addition, these primers generally incorporate some zinc oxide with the zinc dust to improve the characteristics of the paint.

Zinc oxide

1. Zinc oxide is widely used by itself or in combination with other pigments. Its color is unaffected by many industrial and chemical atmospheres. It imparts gloss and reduces chalking but tends to crack and alligator instead.

2. White pigment that is useful in preventing mold or mildew on paint film.

3. Compound of zinc and oxygen widely used in interior and exterior paints. Along with the hiding power of white lead, it reduces the tendency toward yellowing, increases drying, and promotes hardness and gloss. It resists sulfur fumes and mildew and is used in self-cleaning exterior paints in combination with linseed oil.

4. Zinc oxide with varying properties is made from ore (American process) or from spelter (French process). Made by either process, it may differ in color and oil absorptive qualities.

Zinc-rich primers. Coatings are formulated with zinc dust dispersed in various vehicles. Zinc is higher than iron in the electromotive series and these coatings protect steel cathodically just as with hot-dip galvanizing. Zinc becomes the anode and iron the cathode in electrochemical reactions of corrosion. In order for this cathodic protection to take place, there must be contact between the zinc pigment particles and the steel substrate. Therefore the zinc content must be 85 to 95%, and the steel must be absolutely clean. Poorly cleaned steel only results in poor galvanic protection and early failure of the coating. Zinc-rich primers are generally available in two types—organic and inorganic, depending on the nature of the vehicle.

Zinc silicate. Inorganic binder that contains no resins, plasticizers, or drying oils. This is an entirely different way of formulating chemically resistant coatings. When pigmented with zinc dust, it takes on a quality like that produced by fusing ceramics or hot-dip galvanizing and is resistant to weather, intense heat, alcohol, and solvents.

Zinc sulfide. Compound of zinc used as a white pigment in paints.

Xylol (xylene). Solvent used in industrial finishes having a higher boiling point than toluol.

GYPSUM CONSTRUCTION

Acid-etched nails. Chemically deformed nails (ASTM C 514).

Adhesive

1. Casein rubber, or latex-base adhesive specifically intended for the application of gypsum wallboard may be classed according to use as laminating adhesive for bonding layers of wallboard, stud adhesive for attaching wallboard to wood supports (ASTM C 557) and contact adhesive for bonding layers of wallboard or bonding wallboard to metal studs.

2. Organic adhesive intended for bonding the back surface of gypsum wallboard to wood framing members. Adhesive essentially free of foreign matter and of a uniform consistency suitable for application in accordance with accepted commercial standards. Adhesive should remain as applied until application of the gypsum wallboard arrives at an ambient temperature range of 40°F to 110°F.

Adhesive application. Surfaces of gypsum board and framing members to be adhered to by an adhesive should be free from dust, dirt, grease or any other foreign matter that could impair bonding.

Adhesive nail-on method

1. Continuous bead of adhesive applied to wood framing plus supplemental nailing improves bond strength.

2. System of application providing for adhesive application of wallboard to wood framing with supplemental fastening.

3. Application of adhesive to the face of studs or joists in continuous or intermittent beads. Bead sizes and permanent and temporary fasteners (nails or screws) and their spacing are as required by manufacturer.

Aggregate

1. Inert material used as a filler in stucco, plaster, mortar, concrete, etc., without regard to its function as a binding material.

2. Aggregates most commonly used in gypsum plaster include perlite, sand (natural and manufactured), and vermiculite. Other aggregates may be employed provided tests have demonstrated them to yield plaster of satisfactory quality.

Alkali discoloration. Painted surfaces on gypsum wallboard may inherit discoloration or burning in spots on the finished surface due to the alkali sensitive pigments used in the paints. The high alkaline joint compounds affect these paints and it is recommended that water-base paints be used since they have alkali resistant pigments.

Angles and channels. Angles and channels are formed of not less than 22-gauge coated steel with legs not less than 1 in. wide.

Annular ringed nails

1. Nails for use in the application of $\frac{3}{8}$-in, $\frac{1}{2}$-in and $\frac{5}{8}$-in gypsum wallboard. The steel wire used in the manufacture of the nails is of hard-drawn low-carbon or medium-low carbon steel. Before fabrication wire is sufficiently ductile to withstand coldbending, without fracture, through 180°F over radius not greater than the diameter of the wire (ASTM C 514).

2. Nail developed to reduce or eliminate "nail popping." It is the shortest nail recommended with maximum holding power. The nail has greater holding ability than a cement-coated nail of the same length. The annular ring nail's high carbon-steel content meets the requirements of ASTM C 514 Nails for Application of Gypsum Wallboard.

Automatic taper. Device that applies tape and proper amount of joint compound simultaneously to flat joints or corners. It is designed for high-volume machine tool application.

Back blocking. Application of an additional thickness of board by adhesive at the back of a joint, usually on ceilings for reinforcing butt-end and/or edge joints to minimize surface imperfections such as ridging.

Backer board

1. Base layer in laminated construction.

2. Backer board is available in many thicknesses. It is recommended for the base layer for 2-layer wallboard application; $\frac{1}{2}$-in backer board and $\frac{5}{8}$-in. fire-resistant backer board are used for acoustical tile application.

Backing boards. Boards that serve as a base to which gypsum wallboards or acoustical ceiling tiles are attached or laminated. Backing boards differ from regular gypsum wallboards mainly in that the paper covering on both faces and edges is a gray liner paper and is not suitable for decorating. Backing boards may be secured to framing with staples, nails, screws, or adhesive nail-on methods (ASTM C 442).

Base layer. First layer next to the vertical supporting member (in multilayer construction).

Bedding. Installation of the tape in the first application of joint cement.

Bedding cement. First application for joint cement for installation of bedding tape.

Bedding compound. Compound used for first application for bedding tape.

Beveled edge. A beveled edge on a gypsum board gives a paneled effect. After fasteners are covered with a compound, the board is ready for paint with grooves exposed.

Blisters in tape. Blisters in the tape are caused by insufficient compound used under the tape, or tape is not initially pressed into good contact with the compound.

Butt-end joint. Joint in which mill or job cut (exposed core) wallboard ends or edges are butted together during installation.

Calcined gypsum

1. Sulfate of lime, $CaSO_4 + 2H_2O$ heated (calcined) to form $CaSO_4 + \frac{1}{2} 2H_2O$, which mixed with water at normal temperature returns to $CaSO_4 + 2H_2O$.

2. Gypsum partially dehydrated by means of heat, having the approximate chemical $CaSO_4 + \frac{1}{2} H_2O$. Calcined gypsum plaster has a purity of not less than 66.0% by weight.

Cant. Angular gypsum concrete cant against parapet walls and equipment bases, poured integral with the gypsum concrete roof slab.

Carrying channel. Main supporting member of a suspended ceiling system to which furring members or channels attach.

Caulking

1. Resilient, nonhardening, nonskinning compound used to improve acoustical performance of drywall construction. It is applied around cutouts, under runners, and around the perimeter of partitions and ceilings to prevent leaks and transmission of airborne sound.

Caulking should have flexibility to absorb slight movement and maintain an air-tight seal without adhesive failure. It should be longlasting, nonstaining, nonbleeding, and easily applied with mechanical or pneumatic caulking apparatus.

2. For sound rated partitions, acoustical sealant is recommended to prevent sound leaks in areas not adequately sealed by normal construction methods. Sealant material must be nonhardening, nonstaining, and easily applied with a caulking gun.

Cavity shaft walls. Cavity shaft walls provide excellent sound and fire resistance plus a vertical chaseway for installation of electrical services and sound attenuation blankets.

Chase partition. Partition constructed of two rows of vertical supporting members, with sufficient space between each row to allow the installation of large size conduits or plumbing stacks, including one layer of sound absorption material in the pipe chase, and then closed up with one or two layers of plasterboard on each side of the partition. Partition of this type should provide a STC rating of 55.

Circle cutter. Device consisting of an adjustable arm with a cutting blade supported by a post with a puncture point which establishes the radius length of the arm. Device cuts holds in gypsum wallboards up to 12 in. in diameter.

Cold weather. It is not recommended that gypsum concrete be mixed or poured during weather when the temperature and job exposure will allow the concrete to freeze before it takes its complete chemical set. Since the setting action of gypsum concrete generates considerable heat within the slab, it is usually safe to pour slabs in the coldest weather during which mechanics normally work.

Column fireproofing. System of metal channel furring at the flanges of a steel column combined with the screw attachment of the plasterboard and metal vertical corner angles, which may provide up to 4 hours rating depending on the thickness of the plasterboard and the amount of layers.

Condensation. Moisture conditions trapped behind gypsum wallboard installations can introduce water vapor transmission into the wallboard causing large wet spots that will tend to disengage wallpapers attached to the wallboard by affecting the adhesive. Discoloration on the wallpaper will also occur due to this problem.

Consistency. Property of a material determined by the complete flow-force relation.

Control joints. Control joints are designed to relieve stress of expansion and contraction transverse to the joint in large ceiling and wall areas. They are made from roll-formed zinc with a small open slot that is protected with plastic tape which is removed after finishing. They are highly resistant to corrosion in interior and exterior uses.

Control joints and isolation details. Drywall systems are non-load bearing but need structural consideration of their ability to retain integrity over a period of time and be relatively rigid. When other elements, such as doors, fixtures, or cabinets are affixed or joined to a drywall system, adequate provision must be made to ensure that these loads are adequately supported. Structural framing, completely separate from the drywall system, is necessary to support eccentric or heavy loads. Control joints are necessary to prevent cracking in the gypsum wallboard facing of drywall systems. Isolation should be considered where structural elements, such as slabs, columns, or exterior

walls can bear directly on non-load-bearing partitions. Gypsum wallboard is subject to some form of movement due to the structure.

Core board. Solid gypsum board product fabricated for use with solid, double solid, and triple solid partitions. Additional layers of gypsum board are generally laminated to the core board to provide the completed wall assembly.

Corner bead. Corner reinforcements provide true and straight lines for smooth finishing at outside wallboard corners. The exposed bead also helps prevent damage from impact.

Corner finisher. Corner finisher distributes excess compound evenly over tape and feathers edges.

Corner roller. A corner roller is used to embed tape in corners and force excess compound from under the tape prior to finishing.

Cover coat drywall compound. Vinyl-base product designed for filling and smoothing monolithic concrete ceilings and columns located above grade; no extra bonding agents are needed. The compound is supplied in ready-mixed form (sand can be added) and is easily applied with drywall tools in two or more coats. It dries to a fine white surface.

Cracking problems. On the basis of research, two factors can be estimated—the "critical height" above which type A cracking may occur in flat-plate buildings of various heights and the amount of slab deflection, resulting from column movement. Type A cracking usually occurs in the form of cracks and gaps at the opposite corners of partitions that connect an exterior and interior column. Diagonal cracks may also show up on the face of the partition if the edges are not well reinforced.

Cracks in inside corners. Cracks in inside corners are the result of too much compound applied over tape at the apex of the angle or excessive structural movement at corners where dissimilar surfaces join.

Cross tees. Sheet metal tees (T's), usually painted or galvanized, used to support form board at joints.

Crushed gypsum

1. Run-of-mine gypsum further reduced so that all of it will pass in a 3-in. ring and not more than 25% shall pass a No. 100 sieve.

2. Gypsum subjected to a primary crushing operation.

Cutting wallboard. Gypsum wallboards cut by scoring and breaking or by sawing, working from the face side where the board meets projecting surfaces, and neatly scribed.

Damaged edges to wallboard. Damaged or abused paper-bound edges may result in ply separation along the edges or in the loosening of paper from the gypsum core. Damaged edge may also fracture or powder the core itself; edges are more susceptible to ridging when the joint system is applied.

Decoration. Strong, highly calendered face paper on wallboard is suitable for any type of decorative treatment, such as paint, surface applied texture, or wallpaper, and permits repeated redecorations.

Decorative applications. Decorative applications and treatment to gypsum wallboard should not be started until the adhesive between gypsum wallboard and backing board has dried or cured. This is especially important in the application of water-based texture finishes.

Dimple. Impression in the board at the nail head, made by a hammer or driving mechanism.

Discoloration to finish. Differences in suction of board paper and joint compound may lighten paint color or change gloss or sheen in higher suction areas; discoloration is most common when conventional oil paints are used and is also caused by overthinning of paint. Suction differences may also cause greater amounts of texturing material to be deposited over high-suction areas causing color differences when viewed from an angle.

Double-nailing system

1. Double-nailing system is used for minimizing defects due to loosely nailed wallboard by placing the second nails within 2 in. of the first nails.

2. Method of attachment devices to guard against nail pops. This system requires doubling up on the field nails. The total quantity of nails used does not double, however, since maximum nail spacing is increased to 12 in. on center and conventional nailing is used on the perimeter.

Double pour. Apply gypsum concrete in more than one layer over substantial areas.

Double pouring. Application of gypsum concrete over substantial areas in more than one layer to complete the full slab thickness, as distinguished from the common practice of "scratching in" gypsum concrete ahead of the general pour to "seal" joints, etc.

Double solid laminated (non-load bearing). Lamination consists of $\frac{1}{2}$-in. gypsum board to a 1-in. tongue-and-groove core board, homogenous or laminated on each side of a 3-in. airspace. The system has a 2-hour fire rating, 47 dB STC.

Double-layer wallboard—metal stud partition. The double-layer system offers a fire-rated partition excellent for corridors and gives the high STC value necessary for partitions.

Double wallboard layer—wood stud partition. Wood stud framing for drywall partitions, a base layer of wallboard, backing board, or wood fiber sound-deadening board nailed or screwed to the wood studs. The wallboard face layer is job-laminated to the base layer.

Drainage slope fills. Fills of quick-setting gypsum concrete are poured in place over structural roof decks of poured or precast concrete or wood. They provide smooth, monolithic sloped surfaces that are easily roofed. Lightweight fills offer rapid installation, even in cold and freezing weather.

Drywall. Generic term referring to the installation of various types of gypsum wallboards to framing supporting members to form a complete wall product, including metal and other accessories.

Drywall adhesive applicator. Caulking-type gun applicator that has a trigger mechanism and offers minimum resistance to a large bulk load of adhesive.

Drywall adhesive gun. Cartridge-type caulking gun. The trigger mechanism's bearing is mounted on the gun body and offers minimum resistance to a $\frac{1}{4}$-gal cartridge for adhesive.

Drywall and rigid foam insulation furring. Exterior wall furring assembly consists of gypsum wallboard adhesively bonded to polystyrene or urethane rigid foam insulation.

Drywall screws. Drywall screws are used to apply all types of gypsum wallboard, whether the framing is metal or wood or the application is gypsum to gypsum. They are made of high-strength carbon steel. A phosphate overcoating is used as a rust inhibitor. It is recommended that all screws be mechanically driven, for best results.

Drywall system. System consisting of the installation of wood or metal vertical supports and faced on both sides of the supports with one or more layers of plasterboard sheets.

Edge cracking. After completion of joint treatment, straight narrow fissures or cracks may appear along the edges of tape. They result from too rapid drying because of high temperature accompanied by low humidity, excessive drafts, improper application, such as overdilution of the joint compound, excessive joint compound under the tape, or failure to follow embedding with a skim coat over the tape. Cold, wet application conditions may also cause poor bond.

Edges. Board extremities that are paperbound and run the long dimension of the wallboard as manufactured.

Electric radiant ceiling system. System comprised of electric heating cables embedded within the thickness of a plasterboard panel. The panels are shop-fabricated and delivered to the job ready for installation. Construction of panels should meet the requirements of the National Electrical Code NFPA No. 70, and listed by the Underwriters Laboratories, Inc.

Elevator bolt. Flat, plain, circular countersunk head bolt with a square neck to prevent rotation. The head diameter is slightly over three times that of bolt body. The large low head shape provides a flush, wide bearing connection. It is suitable in all laminated gypsum partitions for the installation of medium or heavy fixtures.

Ends. Board extremities that are mill- or job-cut, exposing the gypsum core, and run the short dimension of the wallboard as manufactured.

Expansion and contraction. Gypsum concrete roof decks are subject to expansion and contraction owing to temperature changes. Bulb tees welded to steel framing limit slab movement that would exert itself at right angles to the direction of the bulb tees.

Face layer. Outermost or finish layer in laminated construction, single base layer next to supporting members.

Fastener. Nails, screws, or staples used for the mechanical application of the gypsum board.

Fastener treating. Method of concealing fasteners by successive applications of compound until a smooth wallboard surface is achieved.

Feather edging (feathering). Tapering the joint compound to a very thin edge to ensure inconspicuous blending with the adjacent wallboard surface.

Field. Surface of the board exclusive of the perimeter.

Finish cement. Finish application of joint over cement tape and for spotting nails.

Finish compound. Finish application over tape and for spotting nails.

Finishing. Finishing includes joints, concealment with joint treatment compound of such joints, heads of fasteners, and edges of corner protective devices, and the sanding of such areas to prepare them to receive the field application of priming, painting, coating, decorative coating, and coverings such as wallpaper and vinyl materials.

Finishing or topping compound adhesive. Adhesive with or without fillers.

Fire block gypsum board. Proprietary name for a gypsum board used over roof rafters beneath the sheathing, providing resistance to fire from without. It is of particular value beneath strip sheathing and combustible roof covering.

Fire fighter gypsum board. Proprietary name for a product designed for use on roofs, over rafters, and under sheathing and roof covering. Tests indicate that it blocks roof fires from entering the attic and living areas. It has a special core and paper.

Fireproof sheathing. Fireproof gypsum board with an asphalted gypsum core encased in specially formulated brown water-repellent paper on both sides and long edges. It provides weather resistance, water repellence, and fire resistance.

Fire-rated board (type X). Board rated as to fire resistance by the Underwriters Laboratories, Inc. as a result of controlled tests. Panels are labeled. Board that has a core of Perlite with glass fibers (ASTM C 37).

Fire rated form board. Rigid, highly insulative board suitable for 2-hour fire-rated construction. Its mineral fibers will not contribute to mildew growth. It reduces reflected noise and provides performance up to 50 NRC (Noise Reduction Coefficient).

Fire resistance. Ability of a construction, such as a partition, floor/ceiling, or protected column, to contain a fire. The degree to which such assemblies prevent the spread of fire or damaging heat is indicated in intervals of time. If a construction assembly contains the fire and heat for 2 hours during the test, it is given a 2-hour fire resistance rating. Tests are in accordance with procedures outlined in ASTM E 119, Fire Tests of Building Construction and Materials.

Fire-resistance rating. Rating that denotes the length of time a given assembly can resist the passage of intense heat and flames while supporting the imposed design loads. Fire ratings are correlated with all components of a given assembly—not with the ceiling or partition membrane alone.

Fire resistant. Gypsum will not support combustion.

Fire-resistant board. Any gypsum wallboard, since gypsum is a naturally fire-resistant material.

Fire-resistant (type X) wallboard. This product is similar in exterior covering, appearance, edge treatment, and installation to regular wallboard, but it has a core specially formulated with additives and glass fibers for greater fire resistance. When prescribed installation methods are followed, fire-resistant assemblies of 45 and 60 minutes are possible with $\frac{1}{2}$- and $\frac{5}{8}$-in. boards, respectively, over wood frame construction. Fire-resistant wallboard is also available with an aluminum foil vapor retarder for use in exterior walls and roof/ceiling assemblies.

Flat metal trim. Flat metal trim is used to conceal unfinished joints.

Floating angle method. Gypsum wallboard application that eliminates the use of the perimeter nails at interior corners and where ceilings and walls meet. This method reduces the stress and strain on the board if framing settles or moves.

Floating end joints. Placing of boards so that end joints occur off framing.

Foil-back gypsum board. Regular or Type X gypsum wallboard or gypsum backing board with foil laminated to the back surface. The foil is a vapor retarder and provides thermal insulation when installed in conjunction with a $\frac{3}{4}$-in. minimum enclosed airspace next to the foil.

Form board. Sheet material spanning between subpurlins or across primary framing members and used as a form to receive the gypsum concrete. Form boards are

permanent, remaining in place as part of the finished structure.

Form board installation. Form boards are replaced on subpurlin flanges with all end or cross joints supported, and the forms to fit neatly on all four edges. Forms are cut to fit at walls, curves, and openings as required. Cross tees installed to support end joints of square-edge form boards not supported by roof framing. Form board is installed only if it can be covered by the completed slab on the same day.

Framing member. That portion of the framing, furring, blocking, or base material to which gypsum board is attached.

Furring. A member or means of supporting a finished surfacing material away from the structural wall or framing. Used to level uneven or damaged surfaces, or to provide space between substrates.

Furring channel. Channel formed of not less than 0.021-in. nominally thick hot-dipped galvanized steel. It is hat shaped with a face not less than $1\frac{1}{4}$ in. wide to receive the gypsum wallboard. The furring channel is not less than $\frac{7}{8}$ in. deep and $2\frac{3}{4}$ in. wide.

Gypsum. Principal core material in wallboard, consisting chemically of hydrous calcium sulfate.

Gypsum backing board

1. Board used as a base layer or backing material in multilayer construction, consisting of an incombustible core, essentially gypsum, with or without fiber, but not exceeding 15% fiber by weight, and surfaced with paper firmly bonded to the core. The back surface of insulating gypsum backing board is covered with aluminum foil. Type X designates gypsum backing board complying with specifications that provide for at least 1-hour fire-retardant ratings for boards $\frac{5}{8}$ in. thick or $\frac{3}{4}$-hour fire-retardant ratings for boards $\frac{1}{2}$-in. thick, applied in single-layer nailed application on each face of load-bearing wood framing members, when tested in accordance with the requirements of ASTM E 119, Methods of Fire Tests of Building Constructions and Materials (ASTM C 442).

2. Board designed for use in two-ply construction (double wall) or as a base for acoustical tile. Gray face paper permits easy lamination of gypsum face panels to backing board.

Gypsum base for veneer plaster. Essentially a gypsum noncombustible core, not exceeding 15% fiber by weight, surfaced with paper firmly bonded to the core. The back surface of the insulating gypsum base is covered with pure brightly finished aluminum foil.

Gypsum board

1. Gypsum wallboard, gypsum backing board, or water-resistant gypsum board.

2. Sheet, slab or panel having a noncombustible core essentially of gypsum and surfaced with fibrous material suitable for its purposes.

Gypsum board components. Gypsum ore is mined or quarried, crushed, dried, ground to flour fineness, and heated or calcined to drive off the greater part of the chemically combined water as steam. This produces calcined gypsum or plaster of paris, which is then mixed with water and other ingredients and sandwiched between two sheets of specially treated paper to form a ribbon of gypsum board. After the gypsum core has set, the board is cut to length, dried, and prefinished as required.

Gypsum board enclosure systems. Gypsum board partitions for enclosing elevator shafts, stairs, smoke towers, and other vertical openings providing non-load-bearing walls, designed for resistance to uniform wind loads without excess deflection.

Gypsum bond plaster. Calcined gypsum mixed at the mill with other ingredients to control working quality and setting time and to adapt it for application as a bonding scratch coat over monolithic concrete. The addition of water only is required on the job. Gypsum bond plaster contains not less than 93.0% calcined gypsum by weight and not less than 2.0 nor more than 5.0% hydrated lime.

Gypsum cement plaster. Plaster, sometimes called "hard wall" or "neat gypsum plaster," sold in powder form and mixed with an aggregate and water at the construction site. Mixed with no more than three parts sand by weight, it makes a strong base coat. Scratch coats generally consist of one part plaster powder to two parts sand by weight, fibered or unfibered; the base coat in two-coat work usually is a $1:2\frac{1}{2}$ mix; brown coats are $1:3$ mixes.

Gypsum concrete

1. Combination of aggregate or aggregates with calcined gypsum as a binding medium, which after mixing, with water sets into a conglomerate mass (ASTM C 317).

2. Calcined gypsum to which may be added aggregates, wood chips, or wood shavings in proportion to meet applicable requirements. Calcined gypsum conforms to Specifications for Gypsum Plasters (ASTM C 28). Aggregates conform to Specifications for Inorganic Aggregates for Use in Gypsum Plaster (ASTM C 35). Wood chips or wood shavings, when used, should be of dry wood, uniform and clean in appearance, and pass a 1-in. sieve. They should not be more than $1\frac{1}{16}$ in. in thickness (ASTM C 317).

Gypsum concrete mill mixture. Mixture to which water only is added at the site. The amount of water added is recommended by the manufacturer and the consistency of the mixture must be accurately measured. Either manual or mechanical mixing is employed. In either case the concrete is thoroughly mixed but not beyond setting time limits.

Gypsum core board. Core material in solid, double, and triple solid partitions, as well as shaft wall assemblies. Additional layers of gypsum wallboard are laminated to core board to complete the assembly.

Gypsum drywall. Generic term referring to the installation of various types of gypsum wallboard.

Gypsum fiber concrete. Gypsum concrete in which the aggregate consists of shavings, fiber, or chips of wood.

Gypsum form board

1. Sheet or slab having an incombustible core, essentially gypsum, surfaced on the exposed side with a fungus-resistant paper and on the reverse side with paper suitable to receive poured-in-place gypsum concrete.

2. Form board consisting of an incombustible core, essentially gypsum, with or without fiber, but not exceeding 15% fiber by weight, and surfaced with sheets of fibrous material. The exposed surface is treated specifically to resist fungus growth. Gypsum form board is used for poured-in-place reinforced gypsum concrete roof decks.

3. Rigid gypsum board treated to resist mildew effectively where adequate ventilation is provided. Boards

with 2-hour fire ratings and 46 STC (sound transmission class) are available with 2-in. gypsum slabs and exposed tees. Form board is $\frac{1}{2} \times 32$ or $\frac{1}{2} \times 48$ in. wide, treated, in lengths equal to main purlin spacings (12 ft max) (ASTM C 442).

4. Boards used as permanent forms and as finished ceilings for gypsum concrete roof decking. Form boards are surfaced on the exposed face and longitudinal edges with calendered manila paper specially treated to resist fungus growth. The back face is surfaced with gray liner paper as in regular wallboard. Vinyl-faced board, which provides a white, highly reflective, and durable surface, is also available.

Gypsum gauging plaster

1. Plaster prepared for mixing with lime putty for the finish coat. It may contain materials to control setting time and working quality.

2. Plaster for finish coat containing not less than 66.0% $CaSO_4 \frac{1}{2} H_2O$ by weight.

Gypsum lath

1. Lath that consists of an incombustible core, essentially gypsum, with or without fiber, but not exceeding 15% fiber by weight, and surfaced with paper firmly bonded to the core.

2. Class of board products intended as a base for plastering. These products consist of a gypsum core enclosed by a multilayered, fibrous paper covering designed to ensure good bond with gypsum plaster. A suction bond is created by tiny crystals of plaster forming in the porous paper covering (ASTM C 37).

3. Sheet or slab having an incombustible core, essentially gypsum, surfaced with paper suitable to receive gypsum plaster.

4. Sheet, generally $\frac{3}{8}$- or $\frac{1}{2}$-in. thick, composed principally of calcined gypsum that has been mixed with water, hardened and dried, and sandwiched between two paper sheets.

Gypsum molding plaster. Material consisting essentially of calcined gypsum for use in marking interior embellishments, cornices, as gauging plaster, etc. Gypsum molding plaster contains not less then 80% calcined gypsum (ASTM C 59).

Gypsum neat plaster

1. Calcined gypsum plaster mixed at the mill with other ingredients to control working quality and setting time. Neat plaster may be fibered or unfibered. The addition of aggregate is required on the job. Gypsum neat plaster contains not less than 66.0% calcined gypsum by weight.

2. Plastering material in which not less than 85% of the cementitious material is calcined gypsum mixed at the mill with other materials.

Gypsum neat plaster—Type R. Calcined gypsum plaster mixed at the mill with other ingredients to control working quality and setting time. This plaster is for use with sand aggregate only. Gypsum neat plaster, Type R, contains not less than 66.0% calcined gypsum by weight.

Gypsum partition tile or block

1. Gypsum building unit in the form of tile or block for use in nonbearing construction in the interior of buildings and for the protection of columns, elevator shafts, etc., against fire.

2. Gypsum tile for nonbearing fire walls, partitions, and enclosures, and for furring the inside or outside of

other walls, consists of hollow or solid tile or blocks which are manufactured at the mills and delivered for erection on the building site. Generally gypsum tile consists of about 95% finely ground calcined gypsum uniformly mixed with from 2 to 5% fibrous material by weight, this compound being mixed to a plastic state with water, and molded in the form of the desired units.

Gypsum plaster. Gypsum when heated to a temperature between 250°F and 400°F loses about three-fourths of its combined water, and the calcined product is known commercially as "plaster of paris." This product, when finely powdered and mixed with water, takes up in combination as much water as it lost through calcination and becomes rigid or set through recrystallization.

Gypsum plaster board

1. Plaster board used as a sheet lath or base for gypsum plaster on walls, ceilings, and partitions on the interior of buildings.

2. Sheet or slab having an incombustible core of gypsum and designed to be used as a lath or backing for gypsum plaster on the walls, ceilings, and partitions in the interior of buildings. It consists either of sheets or slabs of gypsum with not more than 15% fiber, by weight, intimately mixed; or of an incombustible core of gypsum, with or without fiber, surfaced with paper or other fibrous material firmly bonded to the core, or with intermediate layers of such material within the core.

Gypsum products. Gypsum plasterboard, wallboard, partition tile or block, roof slabs, and other formed building products are made from calcined gypsum mixed with various aggregates, such as fiber or wood pulp. These products are molded as units by various processes and supplied ready to place on the job.

Gypsum ready-mixed plaster. Calcined gypsum plaster, mixed at the mill with a mineral aggregate, designed to function as a base to receive various finish coats. It may contain other materials to control setting time and other desirable working properties.

Gypsum ready-sanded plaster. Plastering material in which the predominating cementitious material is calcined gypsum, and which is mixed at the mill with all the constituent parts, including sand, in their proper proportion. It requires only the addition of water to make it ready for use.

Gypsum ribwall wallboard partition. Proprietary name for a gypsum partition incorporating interior airspace between opposite facings for improved fire and sound resistance. It also accommodates electrical services and plumbing, and serves efficiently as dividing partitions for vent shafts. This non-load-bearing construction inherently provides back-blocking and minimizes ridging.

Gypsum sheathing board

1. Sheet or slab having an incombustible core, essentially of gypsum, surfaced with water-repellent paper.

2. Gypsum sheathing board consists of an incombustible core, essentially gypsum, with or without fiber, but not exceeding 15% fiber by weight, and surfaced with water-repellent paper firmly bonded to the core. Core-treated water-repellent gypsum sheathing, in addition, has a water-repellent material incorporated in the core (ASTM C 79).

3. Sheathing board that provides fire resistance, wind bracing, and a base for exterior finishes in wood and metal frame construction. It is usually made of water-resistant gypsum core completely enclosed by a firmly bonded water-repellent paper, which eliminates the need for sheathing paper.

Gypsum studs and ribs. Gypsum studs and ribs serve as non-load bearing internal vertical member partition systems.

Gypsum tile

1. Precast kiln-dried gypsum tile laid up with gypsum mortar to make non-load-bearing column fireproofing and interior partitions.

2. Tile that are solid or cored and rectangular in shape with straight and square edges and true surfaces. They may also be of special shape, free from cracks and other imperfections that would render it unfit for use, and is dried before shipment.

Gypsum veneer plaster. Construction that provides an overall monolithic trowel finish or textured surface free of surface defects. All angles are cut sharply and are well detailed. The plaster is applied in accordance with manufacturers' instructions and exhibits joint strength between paperbound edges such that, when subjected to tensile test, specimens break within the field of the board and not at the joint, leaving the plaster securely bonded thereto. Gypsum veneer plaster exhibits sufficient bond strength between plaster and the gypsum base and, where applicable, between plaster base coat and finish coat to resist shock delamination when tested (ASTM C 587).

Gypsum wallboard

1. Sandwich-type material; gypsum plaster with a heavy paper coating on both sides. When fastened directly to studs, it forms a wall surface.

2. Sheet, slab or panel, having an incombustible core, essentially gypsum, surfaced with paper suitable to receive decoration (ASTM C 36).

3. Wallboard consisting of an incombustible core, essentially gypsum, with or without fiber, but not exceeding 15% fiber by weight, surfaced with paper firmly bonded to the core. It is designed to be used without the addition of plaster for walls, ceilings, or partitions and affords a surface suitable to receive decoration.

4. Sheet, slab or panel having an incombustible core of gypsum and that without the addition of plaster furnish to interior walls, ceilings, or partitions provides a surface that may receive decoration. It is surfaced with paper or other fibrous material firmly bonded to the core.

Gypsum wallboard—stud wall partitions. Gypsum stud wall partitions are incombustible interior dividers. Lightweight, non-load-bearing assemblies, with built-in chase for electrical services. System inherently provides back-blocking.

Gypsum wood-fibered plaster

1. Gypsum plaster in which wood fiber is used as an aggregate. Gypsum wood-fibered plaster contains not less than 66.0% calcined gypsum by weight and wood fiber made from nonstaining wood.

2. Gypsum plaster in which wood fiber is used as an aggregate. It contains not less than 80% calcined gypsum by weight and not less than 1% wood fiber by weight made from a nonstaining wood. The remainder may consist of materials to control the working quality and setting time.

3. Plaster containing wood fiber to give the mix more bulk and coverage and requiring only the addition of water at the site. It is used where a high fire rating is required and is about three times stronger in compression and tension than sanded plaster and about 50% harder.

Hammer. Hammer for nailing wallboard has round edges and slightly crowned head, which forms a "dimple" in the face of the board as the nail is "driven home."

Hurricane area installations. Standard gypsum roof decks resist uplift action by nearly four times the normal requirements of 35 lb/sq ft when constructed with bulb or truss tee subpurlins welded to the primary framing.

Impact noise rating (INR). Impact noise rating designates the ability of a floor-ceiling construction to resist impact sound transmission. INR is measured on a plus or minus scale in relation to a standard performance curve, INR−O. The higher the positive number, the better the assembly resists impact sound transmission.

Improperly fitted wallboard. Board wedged into place or improperly nailed which cannot be brought into natural contact with framing. Also, it may buckle causing one end or edge to override the other.

Inorganic aggregates. Inorganic aggregates used in gypsum plaster includes perlite, sand (natural or manufactured), and vermiculite (ASTM C 35).

Installation. Usually up to 30,000 sq ft of gypsum deck can be poured in one day under normal clear dry weather conditions. The quick-setting action of gypsum concrete permits roofing almost immediately after a short drying out period.

Insulating (foil-back backing board). Aluminum foil is applied to the back of regular gypsum backing board and provides an effective vapor retarder and excellent insulating qualities.

Insulating (foil-back gypsum wallboard). Insulating gypsum wallboard provides an insulation (C) factor of 0.25 to 0.30 when used with an airspace for inside walls and a perm rating of less than 0.3.

Insulating gypsum lath. Lath meeting the requirements for plain gypsum lath except that, in addition, the back surface is covered with a continuous sheet of pure bright-finish aluminum foil. By cementing a sheet of bright aluminum foil to the back of a plain lath, insulating gypsum lath is made. It is used for vapor control and insulation against heat loss.

Insulating gypsum wallboard. Insulating wallboard is regular wallboard with a bright-finish aluminum foil bonded to the back. The foil backing serves as a vapor retarder and reflective thermal insulator. It eliminates the need for a separate vapor retarder in an exterior wall assembly and provides increased insulating performance. Installation and edge treatment of insulating wallboard is the same as for regular wallboard.

Insulating Wallboards. Wallboards that consist of various gypsum-base wallboards with aluminum foil laminated to their back surface. They provide effective insulation. The vapor permeability is 0.061 to 0.277 perms and their C factor range for $\frac{1}{2}$-in. board on sidewalls is 0.25 to 0.32 (including airspace).

Insulating wool blankets. Insulating wool blankets consist of mineral fibers mechanically formed into a uniform mat of definite dimension and controlled density. One side

is enclosed with a strong asphalted paper that also forms nailing flanges. The breather side is enclosed in a durable, fire-resistant, porous kraft paper; the open-faced batts are also available.

Interior metal studs. Nonbearing channel-type studs roll-formed from 20 to 25 gage galvanized steel, or truss type studs, made from cold drawn seven gage steel wire rods.

Joint cement. Bedding cement applied to gypsum board joint, for the first application for bedding tape followed by the final cement application over tape. Also used for spotting nails.

Joint compound adhesive. Adhesive with or without fillers.

Joint compound taping. Polyindurate-type material for embedding tape and metal accessories. It is also ideal for heavy fills because it chemically hardens in 3 to 6 hours. It is virtually unaffected by high humidity and changes in humidity. It is not to be used as finishing coat.

Joint compound—topping. Smooth-sanding material used for second and third coats over taping compound.

Joint ridging or beading. Post raising of the joint. This condition can occur after joints are properly finished.

Joint tape. Strip of reinforcing material designed to be embedded in the joint compound to reinforce the joints in gypsum wallboard construction.

Joint tees

1. Steel tee sections or other steel shapes used to support form board edges at right angles to the subpurlins or to the primary framing where subpurlins are not used.

2. Hot-rolled or sheet metal tees shop-painted with one coat of rust-inhibitive paint or galvanized.

Joint treatment. Method of reinforcing and concealing wallboard joints with tape and successive layers of joint compound (ASTM C 475).

Keene's cement

1. Anhydrous calcined gypsum, the set of which is accelerated by the addition of other materials. It sets in not less than 20 minutes nor more than 6 hours, having a compressive strength of not less than 2500 psi (ASTM C 61).

2. Cement obtained by calcining pure gypsum at red heat, immersing it in an alum bath, and then drying and calcining it again.

Keyhole saw. Saw used for cutting small openings and right-angle cuts in gypsum wallboard.

Laminated to unpainted masonry walls. Gypsum wallboard or prefinished gypsum wallboard may be laminated to unpainted masonry walls of monolithic concrete or block or interior partitions and the interior or exterior walls above or below grade with waterproof adhesives recommended by the manufacturer.

Laminating adhesive. Either a joint compound of the type used for embedding tape or an adhesive or laminating compound recommended by the manufacturer of the gypsum board.

Lamination. Two or more plies either mechanically or adhesively bonded.

Limitations. Gypsum wallboards are not recommended in locations exposed directly to water.

Long-span precast gypsum slabs. Steel-edged tongue-and-groove slabs designed for support on primary structural framing. The slabs are rectangular in shape and uniform in thickness and contain a wire reinforcing mat.

Manufactured sand. Fine material resulting from the crushing of rock, gravel, or blast furnace slag and classified by screening or otherwise.

Mastic spreader. A mastic spreader is used to apply adhesive in vertical strips to base layer gypsum boards or core-board in strip-laminated partition construction.

Metal-backed tape. Regular tape with metal backing. It is used to halt beading and also in lieu of back blocking.

Metal casings. Edge finish at doors, windows, ceiling-wall junctures, and built-up column enclosures.

Metal-edged gypsum plank. Plank having a water-resistant core, precast at the factory for installation in any weather on flat or steeply pitched roofs. The planks are laid over steel, or concrete, to form a noncombustible, structurally reinforced roof deck ready for roofing.

Metal furring channel. Drywall metal-furring of exterior masonry walls using furring channels to which the wallboard is screw-attached. An effective vapor retarder and significant insulating value are added to walls by using insulating (foil-back) wallboard.

Metal lock fastener. Fastener specially designed to rigidly attach metal studs to drywall runners and metal trim to drywall studs.

Metal runners. Roll-formed galvanized steel channel-type sections designed to receive metal studs and to secure various types of partitions to floor and ceiling.

Metal stud curtain wall. Metal stud curtain wall systems used on exterior walls to enclose a structure. Fabricated quickly on the job with conventional components, these lightweight, non-load-bearing systems can be made highly insulative and suitable for concrete and steel frame structures.

Metal studs. Non-load-bearing channel-type studs roll-formed from galvanized steel. Standard 25-gauge studs are recommended for interior partitions; 20-gauge studs are designed for exterior, non-load-bearing curtain wall systems but are also used for interior partitions to provide more rigidity or when greater heights occur.

Metal trims. Flat trim used to conceal unfinished joints. Metal casing trim used as edge finish at doors, windows, and other types of openings. Metal corner trim used for reinforcing and finishing exterior corners.

Method of testing gypsum. The chemical and physical properties of gypsum are determined in accordance with the Methods for Chemical Analysis of Gypsum and Gypsum Products (ASTM C 471) and Methods for Physical Testing of Gypsum Plasters and Gypsum Concrete (ASTM C 473).

Methods of testing gypsum plaster. Chemical and physical properties of gypsum plaster are determined in accordance with the Methods for Chemical Analysis of Gypsum and Gypsum Products (ASM C 473) except that the methods for chemical analysis contained therein are not applicable to gypsum ready-mixed plaster or gypsum wood-fibered plaster.

Mill-mixed gypsum concrete. Concrete consisting essentially of calcined gypsum and suitable aggregate, requiring the addition of water only at the job. Gypsum concrete is intended for use in the construction of poured-in-place

roof decks or slabs. Two classes, based on compressive strength and density, are covered.

Mill mixture. Factory-proportioned ingredients that when job-mixed with water forms a slurry of gypsum concrete that can be pumped into the framework.

Mineral fiber sound-deadening board. Incombustible rigid core board used behind gypsum wallboard to reduce sound transmission through a stud partition.

Minimum thickness. The minimum thickness of gypsum concrete used structurally should be 2 in. from the top of the form boards, except in the suspension system, where it should be not less than 3 in. The measurement for minimum thickness is taken over the primary framing members and at the centerline between subpurlins where such members are used.

Mixes. Gypsum concrete is mixed with clean water only, using $8\frac{1}{2}$ gal/80 lb of pyrofill concrete and 8 gal/67 lb of thermofill concrete.

Modified beveled edge. A modified beveled edge is available in some panels. This type of edge more readily accommodates the vinyl surface decoration, which is delivered completely edge wrapped.

Moisture-resistant gypsum wallboard

1. Wallboard specially processed for use as a base for ceramic and other nonabsorbent wall tiles in bath and shower areas. The board is tapered on the edge so that joints above the area to be tiled can be treated in the usual manner.

2. Wallboard manufactured with a specially treated water- and moisture-resistant gypsum core and face paper. This wallboard provides a permanent backing for ceramic, metal, and plastic tile and prefinished panels. Wallboard $\frac{1}{2}$ or $\frac{5}{8}$ in. is available for assemblies requiring added fire protection.

Moldings (mouldings). Moldings used to cover joints and edges, protect corners made of extruded aggregate, vinyl plastic, and steel with vinyl laminated surfaces.

Molly bolt. Molly bolt, $\frac{1}{4}$ in., installed in wallboard only. One advantage of this type of fastener is that threaded section remains in the wall when the screw is removed. Also, wide-spread spider supported formed by the expanded anchor spreads the load against the wall material, increasing the load capacity.

Monolithic unit. Concrete gypsum roof deck systems form a monolithic unit that structurally integrates the roof deck with the roof framing. The roof deck is a rigid diaphragm that provides firm resistance to harmonic wave action (as uplift is applied and relaxed).

Mortar. Material used in a plastic state, which can be troweled (and becomes hard in place).

Mud. Common term for joint cement.

Mud pan. Pan used for containing small portions of joint cement ready for the applicator, made of nonbreakable plastic, unaffected by temperature changes or joint cement chemicals. Pan is also used for mixing powder joint compounds.

Multi purpose compound. Compound used for bedding and finishing.

Nails for drywall application. Nails suitable for wallboard application are typically bright, coated or chemically treated low-carbon steel nails with flat, thin, slightly filleted and countersunk heads approximately $\frac{1}{4}$ in. in diameter and medium diamond point.

Nails for the application of gypsum wallboard. Steel wire nails suitable for use in the application of $\frac{3}{8}$-, $\frac{1}{2}$-, and $\frac{5}{8}$-in. thick gypsum wallboard and gypsum backing board directly to wood nailing members. The steel wire used in the manufacture of nails is made of hard-drawn low or medium-low carbon steel. Before fabrication the wire is sufficiently ductile to withstand cold-bending, without fracture, through 180° over a radius not greater than the diameter of the wire (ASTM C 514).

Nail popping. Nail popping is caused by improper installation of nails, loose nails, lumber shrinkage.

Nail pops. A protrusion directly over the had of a nail results from outward movement of the concealed nail head in relation to the smooth finished surface of the wallboard. The protrusion is usually only slightly larger than the nail head but may be about the size of a hammer dimple. Nail pop may occur with any type of material secured to wood with nails.

Notched spreader. Notches should be $\frac{3}{8}$ in. high and $\frac{5}{16}$ in. wide at the base, slightly tapered, and spaces $3\frac{1}{2}$ in. on center (o.c.).

Partition end cap. End cap designed as a one-piece terminal for any free-standing and section or trimmed header. This section ensures a strong, crack-resistant finished product. The flush solid-metal face eliminates problems due to joint cement shrinkage.

Partition heights. Non-load-bearing partitions are limited in height due to deflection resulting from size, spacing, and gage of studs, thickness of plasterboard, and the method of application.

Perforated gypsum lath

1. Perforated lath has $\frac{3}{4}$-in. holes drilled 4 in. on center in each direction to improve the lath-to-plaster bond. In addition to the suction bond, perforated lath develops a mechanical bond by keying the plaster to the lath as the plaster is squeezed through the holes.

2. Perforated lath is produced during the manufacturing process by punching holes through the lath at regular intervals.

3. Lath meeting the requirements for plain gypsum lath except that it has perforations not less than $\frac{3}{4}$ in. in diameter with one perforation for not more than 16 sq in. of lath, and except that the breaking loads when tested with the bearing edges parallel to the fiber of surfacing are not less than 20 lb for $\frac{3}{8}$-in. thickness and not less than 30 lb for $\frac{1}{2}$-in. thickness (ASTM C 37).

Perimeter isolation. The perimeters of non-load-bearing dry wall surfaces should be isolated from all structural members except the floor. Isolation is important to reduce possibilities for cracking in partitions, ceilings, wall furring, and column and beam fireproofing.

Perlite. Volcanic glass that can be expanded by heating. It is used as an insulating material, and as a lightweight material mixer.

Perlite aggregate. Siliceous volcanic glass properly expanded by heat. The weight of perlite aggregate is not less than $7\frac{1}{2}$ nor more than 15 lb/cu ft.

Pistol-type stapler. Stapler used to attach base layer wallboard to wood studs and insulating and sound attenuation blankets.

Plasterboard. Board consisting of a core of gypsum faced with two sheets of heavy paper.

Plasterboard form board. Rigid gypsum board treated to resist mildew.

Pole drywall blade. Blade used for smoothing and finishing cover coat applications. The tool has a handle and a replaceable spring steel blade.

Pole sander. Tool used for sanding ceilings and walls from the floor. It has a handle, aluminum sanding block and a universal joint.

Portable caulking applicator. Applicator used to apply caulking to a seal under runners and around openings. The compound is pumped from a 5-gal pail under air pressure through a nozzle at a constant rate of flow. It has a trigger cutoff.

Powder-type compound for joint taping. Compound used to embed tape and conceal corner beads and fasteners in the application of gypsum wallboard.

Power-driven screwdrivers. Electric power-driven screwdrivers are used to drive screws for attachment of gypsum wallboard to wood or metal framing or to gypsum core materials in laminated construction.

Precast reinforced gypsum slabs. Metal reinforced gypsum slabs containing wood chips, wood shavings, or reinforcing fibers. There are two types—short-span slabs and long-span slabs.

Predecorated gypsum board. Gypsum board with a decorative wall covering or coating applied in-plant-by the gypsum board manufacturer.

Predecorated wallboard. Wallboard with a factory-applied decorative finish on the face side of the board. Common finish materials are paper and vinyl in a variety of colors, patterns, and simulated wood finishes. Some manufacturers provide a loose flap of the decorative covering material, which can be cemented down after installation to conceal the joint.

Predecorated wallboard panels. Panels manufactured in standard and custom factory-laminated vinyl-faced wallboards, with a range of colors, patterns, and textures.

Prefinished wallboard panels. Panels that require no painting or joint finishing. Wall are complete as soon as the panels are installed. Fire-resistant panels usually have a durable finish that is easily maintained.

Primary framing. Structural members installed to receive the reinforced gypsum roof deck assemblies.

Protrustions. Bridging, headers, firestops, or mechanical lines improperly installed so as to project beyond the face of the framing, preventing the board from contacting nailing surface. The result will be loose board, and nails driven in the area of protrusion will puncture the face paper.

Pyrofill gypsum concrete. Mill-formulated concrete composed of calcined gypsum and wood chips or shavings, mixed with clean water only at the job site, and poured in place over permanent formboards or other decks as a drainage fill. A product of United States Gypsum, Chicago, Ill.

Pure gypsum. Calcium sulfate combined in crystalline form with two molecules of water.

Rasp. Tool made on the job and used to trim or smooth cut wallboard ends or edges. It consists of an 8-in.-long 2×4 in. block with a metal lath stapled to the face.

Ready-mixed joint compound

1. Vinyl-base formulation widely used for embedding and finishing. It comes premixed with a creamy, lump-free plasticity that produces excellent slip and bond. It is available in either machine or hand-tool consistency.

2. Compound used to embed tape, fill and finish joints, coat corner beads and fasteners, and texture gypsum wallboard.

Ready-mixed plaster with perlite aggregate. Mix containing not more than 2 cu. ft of aggregate per 100 lbs of calcined gypsum plaster and having a compressive strength of not less than 600 psi. Mix used over lath base.

Ready-mixed plaster with sand aggregate. Mix containing not more than $2\frac{1}{2}$ cu ft of aggregate per 100 lb of calcined gypsum plaster and having a compressive strength of not less than 700 psi. Mix is used over lath base.

Ready-mixed plaster with sand or perlite aggregate. Mix containing not more than 3 cu ft of aggregate per 100 lbs or calcined gypsum plaster and having a compressive strength of not less than 400 psi. Mix is used over masonry base.

Ready-mixed plaster with vermiculite aggregate

1. Mix containing not more than 2 cu ft of aggregate per 100 lb of calcined gypsum plaster, and having a compressive strength of not less than 450 psi. Mix is used over lath base.

2. Mix containing not more than 3 cu ft of aggregate per 100 lb of calcined gypsum plaster and having a compressive strength of not less than 325 psi. Mix is used over porous masonry base.

Regular gypsum wallboard. Wallboard with a tapered edge that permits smooth joint treatment; the surface takes any decoration. Basic recommendations are for $\frac{1}{2}$- and $\frac{5}{8}$-in. board for a single layer, $\frac{3}{8}$-in. board for two layers, and $\frac{1}{4}$-in. for remodeling and sound deadening.

Reinforcement placement. Reinforcing mesh is placed with the heaviest wires at right angles to subpurlins. If keydeck is used, place 16-gauge wires at right angles to subpurlins. Lap mesh ends at least 6 in, but do not lap sides of mesh. For fire-rated assemblies, lap mesh sides 4 to 6 in. Cut the mesh to fit at the wall, curbs, and openings and carry the mesh into all areas where gypsum concrete is poured. Keydeck is a product of Keystone Steel & Wire, Peoria, Ill.

Reinforcing fabric. Galvanized welded wire fabric used for gypsum concrete reinforcement having an effective cross-sectional area of not less than 0.026 sq in./ft of width, equivalent to No. 12 gauge longitudinal wires spaced not over 4 in. on centers and transverse wires equivalent to No. 14 gauge and spaced not over 8 in. on centers (ASTM A 185).

Reinforcing fabric installation

1. Fabric is placed with the longitudinal wires at right angles to the subpurlins. The ends of the fabric are lapped not less than 6 in. and wired together. The sides of the fabric are butted or spaced not more than 4 in. apart. Reinforcing fabric is cut to fit at all walls, curbs, openings, etc.

2. The fabric is installed over the form boards with the longitudinal wires running at right angles to the primary framing members. The ends of the fabric are lapped not less than 6 in. The sides of the fabric are butted or spaced not more than 4 in. apart. Reinforcing fabric must fit at all walls, curbs, openings, etc. Reinforcing fabric is carried into all areas where gypsum concrete is placed, including poured curbs.

Reinforcing mesh

1. Keydeck galvanized wire mesh, woven with 16-gauge straight wires and 19 gauge diagonal wires.

2. Galvanized welded wire mesh with 12-gauge longitudinal wires at 4 in. and 14-gauge transverse wires at 4 in. on center.

Resilient furring channel. Galvanized steel channel that provides for the resilient attachment of gypsum wallboard to wood framing. It is widely used to improve sound transmission loss in partitions and ceilings.

Resilient furring channels. Furring channels used for the attachment of wallboard to wood studs with the furring channels providing limited sound isolation.

Ridging. Surface defect resulting in conspicuous wrinkling of the joint tape at treated joints.

Roofing application. Built-up roof covering installed as soon as practical, but not later than two days after pouring to protect the gypsum from excessive wetting from rain or snow and to develop optimum nail-holding power.

Rubber mallet. Mallet for impacting wallboard in the adhesive nail-on method of application.

Runners. Runners are formed of not less than 0.021-in. nominal thickness hot-dipped galvanized steel; with legs not less than 1 in. high and slightly bent in to hold the studs by friction.

Run-of-mine gypsum. Form in which the gypsum comes from the mine or quarry.

Sand. The sand used for plastering in which gypsum is employed consists of fine granular material, naturally or artificially produced by the disintegration of rock containing by weight not less than 80% silica, feldspar, dolomite, magnesite or calcite and that is free from saline, alkaline, organic, or other deleterious substances.

Sanded gypsum plaster. Sanded gypsum plaster requires only the addition of water at the site, since it is sold in bags containing the proper proportions of aggregate and putty.

Sanding cloth. 320 grit (fine) and 220 grit (medium) usually used for sanding successive layers of joint compound.

Scratch in. Seal the joints between form boards (with gypsum concrete) ahead of the main pour.

Screeds. Metal bars or wood strips placed as guides to screed or guide the installer as to the level of the gypsum concrete to the specified thicknesses. Immediately on pouring, the gypsum concrete is spread and screeded. The surface of the gypsum concrete is left smooth enough to receive the roof covering.

Screw applications. Screws have been developed for wallboard applications, usually with a self-tapping self-threading point, special-contour flat head, and deep Phillips recess for use with a power screwdriver.

Screws. See ASTM C 1002. Specification for Steel Drill Screws for the Application of Gypsum Sheet Metal to Light-Gage Steel Studs. Specially designed metal screws are used for gypsum board application to wood framing or to gypsum studs, as recommended by the manufacturer of the gypsum board.

Seismic resistance

1. Poured gypsum roof decks with truss tees and reinforcing mesh offers construction to resist seismic shock. Gypsum fill flows through the truss tee to form a rigid diaphragm with excellent resistance to shear and uplift.

2. Poured gypsum roof decks with bulb tees or truss tees structurally tie the framing system together to reinforce the building and provide resistance to wind and seismic loads.

Semi-solid laminated partition. Non-load-bearing partition composed of a center vertical strip of 2 layers of laminated gypsum board spaced 24 in. on center on which $\frac{5}{8}$ in. gypsum board is laminated to each side. The space between the 1×6 in. vertical strips provides space for the installation of electrical wiring. The partition has been rated as one hour, and an STC rating of 30.

Shaft wall systems. Non-load-bearing partitions, designed to provide resistance to fire, sound, and horizontal loads. Shaft wall systems are used around elevator shafts, stairways, smoke towers, and other such service shafts. They are constructed of steel studs, steel angles, gypsum core board, and gypsum wallboard.

Sheet lamination. In multilayer construction, adhesive is applied in parallel strips spaced 16 to 24 in. apart.

Sheetrock. Mill-fabricated gypsum wallboard composed of a fireproof gypsum core encased in a heavy manila-finished paper on the face side and a strong liner paper on the back side. The face paper is folded around the long edges to reinforce and protect the core, and the ends are square-cut and finished smooth. A product of United States Gypsum, Chicago Ill.

Shoring. Once wallboard is in place, pieces of wood or strips of wallboard are temporarily nailed to the frame with 6-penny double-headed nails to hold wallboard firmly while the compound dries.

Short-span slabs of precast gypsum. Slabs designed for end support on subpurlins. The slabs are rectangular in shape and uniform in thickness, except for grouting grooves along each upper longitudinal edge. A wire reinforcing mat is contained in the slab.

Single layer partition. Partition consisting of vertical supporting members with one layer of plasterboard on each face of the member.

Single nailing. Conventional attachment for wood framing.

Single wallboard layer—metal stud partition. Metal stud vertical framing system for wallboards, screw fastened to assure positive attachment of wallboard and freedom from nail pops.

Single wallboard layer—wood stud partitions. Wood stud framing, drywall partitions wallboard is either horizontally or vertically applied directly to conventional 2×4 in. wood studs. Attachment is by nails, screws, or the adhesive nail-on method.

Sized gypsum

1. Gypsum that can pass a $1\frac{1}{2}$-in. ring and of which not more than 10% can pass a $\frac{1}{4}$-in. sieve.

2. Crushed gypsum of prescribed size of individual particles.

Skimming. Light finish coat of cement.

Sling psychrometer. Used to determined temperature and humidity for figuring joint compound drying time.

Soffit board. Used on exterior soffits where it is not exposed directly to the weather. Finishes and decorates like interior gypsum-board with no nails or joints showing.

Solid and semisolid gypsum partitions. Partition systems composed entirely of gypsumboard having added values of increased fire resistance.

Solid laminated partitions. Factory assembled partition consisting of an inner core of two layers of laminated gypsum coreboard and faced on both sides with a single layer of gypsum wallboard. The partition has been rated as two hours when the facing is not less than $\frac{1}{2}$-in. thick and the partition has been given a sound transmission class (STC) rating of 30.

Solid wallboard partition. Solid partitions of the 2-hour fire-rated gypsum drywall assemblies. This non-load-bearing construction is used for vent shaft or elevator shaft enclosures.

Sound-absorbent surfaces. Roof deck systems with gypsum concrete poured over exposed glass-fiber form board offer a large sound-absorbent surface. Glass-fiber form boards effectively absorb sound and reduce reverberation. Interior noise levels can be reduced with these form boards.

Sound attenuation blankets. Paperless, semirigid, spun mineral fiber blankets of uniform dimension and controlled density especially developed for efficient sound retarder use. When inserted in the partition cavity from floor to ceiling, these blankets help provide excellent fire ratings and substantially improve sound transmission class (STC) ratings of partitions.

Sound classification. Sound ratings are based on carefully controlled laboratory tests conducted in accordance with ASTM E 413.

Sound control. Partitions, floors, and ceiling of a building that provide resistance to the transmission of airborne and structure-borne sound. Their efficiency as sound retarders is dependent on drywall construction.

Sound ratings. Drywall construction systems are laboratory tested to establish their sound isolation characteristics. Airborne sound isolation is reported as the sound transmission class (STC), whereas impact noise, tested on floor-ceiling systems only, is reported as the impact insulation class (IIC). Tests are conducted in accordance with ASTM E 90 and the STC is determined per ASTM E 413.

Sound transmission class (STC). Rating that has become the most widely accepted means of indicating a partition's sound resistance. It is the result of comparing the sound transmission loss of a tested assembly with a "standard contour" of known sound loss performance.

Square edge. Gypsum board square edge was the original wallboard edge, designed initially to be a base with a final covering such as wallpaper, paneling, or tile. Square edge can be finished with joint compounds to form a clean monolithic wall suitable for paint. It is also used where an exposed joint is desired.

Staggered stud wallboard partitions. Staggered stud systems offers sound isolation and fire protection for drywall wood frame partitions used as party walls.

Standard application (single nailing). Gypsum wallboard is usually applied directly to wood framing members. Ceilings are applied first and then sidewalls. Boards are accurately cut and joints abutted but not forced together. Horizontal application, with the long edges at right angles to nailing members, is preferred for it minimizes joints and strengthens the wall or ceiling. Nails are spaced not to exceed 7 in. on ceilings or 8 in. on sidewalls, a minimum of $\frac{3}{8}$ in. and a maximum of $\frac{1}{2}$ in. from edges and ends of wallboard. Annual ring nails are recommended for nailing application.

Standard metal studs. Non-load-bearing channel-type studs formed from galvanized steel.

Staples. Number 16 USS (The United States Standard gauge) flattened galvanized wire staples with a $\frac{7}{8}$-in. wide crown outside measure and divergent point are used for the first ply only by two-ply gypsum board application.

Steel drill screws. Steel drill screws are used for the application of gypsum sheet material to light-gauge steel studs, and provides minimum requirements for steel drill screws for use in fastening gypsum sheet metal to light-gauge steel members (ASTM C 1002).

Steel studs. Steel studs are formed of not less than 0.021-in. nominal thickness hot-dipped galvanized steel, with flanges not less than $1\frac{1}{4}$ in. wide to receive the gypsum wallboard. The edge of the flange is bent back 90° and doubled over to form a $\frac{3}{16}$-in. minimum return.

Steep roofs. On steep roofs, where slate, clay tile, or rigid-type shingle roof coverings are required, metal edge gypsum flank can be used.

Steel subpurlins

1. Bulb tests, standard structural or bar size tee, A.A.R. or A.S.C.E. rails, special flanged sections, or other sections having the required structural characteristics to carry the dead and live loads. Steel subpurlins are shop-painted with one coat of rust-inhibitive paint.

2. Steel subpurlins vary in size, weight, and shape and are selected according to required span and loading. They provide lateral bracing and anchorage against uplift, and restrict deck movement due to temperature change. Subpurlin spacing accommodates 24-, 32-, or 48-in. form board widths with a slight tolerance for ease of form board placement. Subpurlins are spaced approximately $24\frac{5}{8}$, $32\frac{5}{8}$, or $48\frac{5}{8}$ in. on center and are welded to the structural framing members. When 48-in.-wide form board is used with light subpurlin sections, supporting steel spacing should not exceed 36 in. on center.

Straight edge. Calibrated metal straight edge and T square used as a guide in making cuts across the full 4-ft. width of the wallboard. It eliminates the need for drawing lines, and thus speeds accurate cutting. A metal edge is preferred because it prevents the knife from cutting into the edge as the wallboard is scored.

Structural quicklime. All classes of quicklime, such as crushed lime, granular lime, ground lime, lump lime, pebble lime, and pulverized lime. Its chemical composition consists of calcium oxide, magnesium oxide, silica, alumina, iron oxide and carbon dioxide.

Structural strength. Gypsum concrete decks have high structural strength and a hard surface. In tests, standard assemblies supported uniform roof loads over 450 lb/sq ft wet and 700 lb/sq ft when dry. At dry densities of 48 to 50 lb/cu ft for pyrofill and 38 to 40 lb/cu ft for thermofill, the compressive strength of the slab is 500 psi min. Gypsum concrete exceeds the strength of other insulating fills; it provides a better base for roofing and adequate support for normal roofing equipment.

Stucco. Material used in a plastic state, which can be trowled to form, when set, a hard covering for the exterior walls or other exterior surfaces of any building or structure.

Subpurlin installation. Each subpurlin is placed and welded to the main purlins at each contact point, using fillet welds and $\frac{1}{2}$-in. minimum length placed on alternate sides of the subpurlins where accessible. All end joints bear on the roof supports.

Subpurlins. Steel members applied transversely to the primary faming and used to support the form boards and the gypsum concrete roof slabs.

Supporting structure. Framing that supports subpurlins, framing that (when directly spaced) supports the form boards directly.

T&G edge. Tongue-and-groove edge used on 2-ft-wide sheathing and backer boards.

Tape. Fiber-reinforced paper material used to cover joints. It can be punch or spark perforated or plain.

Tape corner tool. Tool that permits application of tape and joint compound to both sides of a corner at once.

Tape joint compounds. Phosphoprotein (casein-type) power products. The product should be easily mixed, smooth working, provide ample working time and controlled minimum shrinkage, without alkali burning of paint. It should not be used in combination with noncasein or hardening-type joint compounds.

Tapering edge. Originally, a "recessed edge." It is now called "tapered" because of the incline from the edge to about 2 in. into the board. The taper allows for tape and joint treatment to be applied and leaves the completed job flat, smooth, and monolithic.

Tapered edge wallboard. Tapered edge wallboard has long edges tapered on the face side in order to form a shallow channel for the joint reinforcement, which provides smooth, continuous wall and ceiling surfaces.

Tapered wallboard with rough edge. Designed to reduce the beading and ridging problems commonly associated with standard-type gypsum board. Edge formation provides a stronger, more rigid joint and results in a smoother wall surface.

Tapering end joints. Floated end joints with a depression to receive the tape in back-blocked work.

Taping. Applying joint tape over embedding compound in the process of joint treatment.

Temperature. Temperatures between 55°F and 70°F must be maintained both day and night. This temperature should be maintained 24 hours before, during, and after the entire wallboard and joint treatment application.

Thermofill gypsum concrete. Mill-formulated concrete composed of calcined gypsum and graded perlite aggregate, mixed only with clean water at the job site and poured in place over permanent form boards. A product of United States Gypsum, Chicago, Ill.

Toggle bolt. One-quarter-inch toggle bolt installed in wallboard for attaching and hanging of fixtures. The disadvantage of toggle bolt is that when the bolt is removed, the wing fastener on back will fall down into a hollow wall. Another disadvantage is that a large hole is required to allow wings to pass through wall facings to install the bolt.

Tongue-and-groove edge. Actually a V joint used for tight taped or untaped joints.

Topping compound. Compound specifically designed for the second and third coats over the bedding compound. Since it is used only for topping, the amount of adhesive can be reduced and other materials incorporated to reduce the shrinkage and make it easier to sand and work.

Transmission of sound. Exterior noise is efficiently attenuated by the high sound transmission loss of the gypsum deck. Pyrofill gypsum concrete, poured $2\frac{1}{2}$ in. thick over sheetrock form board, developed a sound transmission class (STC) of 46.

Treated joint. Joint between gypsum boards that is reinforced and concealed with tape and joint treatment compound or covered by strip moldings.

Triple solid laminated partition—(non-load bearing). The center piece is a 1-in. gypsum core panel running the height and length of the wall, one side of which is dead space and the other side filled with fiber-glass insulation. The outer section of the wall (on both sides) is $\frac{1}{2}$-in. gypsum board laminated to 1-in. core board; this gives a $1\frac{1}{2}$-in. wall on both sides of the center section. The system has a 3-hour fire rating, and a 57-dB sound transmission class (STC) rating.

Troweled spot. Insulation mastic is applied using spots of adhesive $1\frac{1}{2}$ in. in diameter, 1-in. high, and spaced 12 in. on center. Spots are also applied 2 in. from the edges to be taped and spaced 6 in. on center.

Two ply. Two thicknesses of gypsum wallboard, nailed, screwed or adhesively bonded.

Type "X" lath. Special fire-retardant lath that, when it complies with the requirements for plain gypsum lath, will endure the type "X" lath test for 30 min. as outlined in ASTM C 473 (ASTM C 37).

Untreated joint. Joint that is left exposed.

Uplift. All roof decks are subject to uplift forces and must be anchored to support to resist uplift. Although a dead load can be considered as part of the total resistance, the chief resistance is obtained by securely welding the subpurlin to the main purlin at bearing points to transmit slab loads. Reinforcing mesh also absorbs tensile stress, distributes it across the slab, and transmits it to the framing.

Ventilation. Provide proper ventilation and dry joint compound in about 24 hours. During hot dry weather avoid drafts that cause too rapid drying of compound.

Vermiculite aggregate. Micaceous mineral properly expanded by heat. The weight of vermiculite aggregate is not less than 6 and not more than 10 lb/cu ft.

Vinyl-covered gypsum wallboard. Wallboard with vinyl fabric surface. The standard product with 6-mil vinyl has a flame spread of 25. It is treated with plasticizers, pigments, and stabilizers and is chip or peel resistant.

Vinyl-surfaced gypsum. Fire-resistant panels, factory laminated with a washable vinyl sheet providing a permanent scuff-, crack-, and chip-resistant surface.

Wallboard. Factory-fabricated product composed of a fibered gypsum core between two layers of tough paper. The two long edges are also covered with paper.

Wallboard corner beads. Galvanized steel with perforated flanges provides a protective reinforcement of straight corners.

Wallboard hammer. Hammer that has a symmetric convex face designed to compress wallboard and leave a perfect dimple.

Wallboard installation. Wallboard may be applied horizontally (long edges of the board at right angles to the framing) or vertically (long edges parallel to framing).

Fire-rated partitions generally require vertical application.

Wallboard—masonry. Gypsum wallboard, adhesively applied directly to exterior or interior masonry walls or partitions above grade.

Wallboard nails. Nails have been developed to concentrate maximum holding power over the shortest possible length—notably the annular ring-type nail, which has 20% greater holding power than a cement-coated nail of the same length.

Water

1. Water must be potable and must not contain impurities that affect the setting of gypsum.

2. Water must be clear and free from injurious amounts of oil, acid, alkali, organic matter, or other deleterious substances.

Water-damaged gypsum boards. Gypsum boards that acquire water or moisture damage may be develop paper bond failure and possibly mildew or fungus growth on the surface after installation.

Water ratio. Water ratio determines to a major degree the ultimate compressive strength of the dry gypsum concrete. Precautions must be taken to be sure that excessive water is not added. Proper devices for accurate measurement or water or consistency of mixture are used.

Water-resistant gypsum backing board (ASTM C 630). Board for use in bath and shower areas as a base for the application of ceramic or plastic tile. It consists of a water-resistant gypsum core, with or without fiber or aggregates and with a surface of water-resistant paper firmly bonded to the core. Water-resistant Type X gypsum backing board designates a product complying with requirements of ASTM C 36, Gypsum Wallboard, Section 2.

Water resistant gypsum board. For use as a base for the adhesive application of ceramic, metal, or plastic tile, gypsum core consists of incombustible gypsum rock combined with an emulsion giving high water resistance. The product is ideal for bathrooms, kitchens, powder rooms, and all high moisture areas.

Wet density. Weight of a specific volume of gypsum concrete before drying.

Wood fiber. Material produced by grinding or shredding wood.

Wood fiber sound-deadening board. Low-density board used behind gypsum wallboard to reduce sound transmission through a partition. This structural board is easily nailed to wood studs and provides a resilient base for the application of wallboard. It is limited to interior uses only.

Wood-grained board. Board with various simulated wood grains superimposed on paper which is laminated to the board.

Wood shrinkage. Wood shrinks as the moisture content of the wood is reduced below 30%, and shrinks in the direction of the growth rings (edge grain) and very little, as a rule, along the grain (longitudinally). Wood shrinkage has the most effect on the nailing of the gypsumboard. As the wood shrinks the nail emerges and creates a space between the wood and the back of the gypsumboard, causing the board to loosen and quite often tearing the taped joint. Small nail pops are also associated with wood shrinkage (U.S. Department of Agriculture publication, *Wood Handbook No*. 72).

Yield. Volume of gypsum concrete obtained from a given amount of gypsum mill mixture.

LATH AND PLASTER

Accelerator. Material added to gypsum plaster or portland cement that speeds the natural set.

Accessories. Metal shapes that are embedded in plaster for special purposes ranging from protection of exterior corners to providing for expansion and contraction of plaster and providing the means of attaching light fixtures, pictures, etc.

Acoustical plaster. Finishing plaster designed to correct sound reverberations or reduce noise intensity.

Additive. Admixture that is added to a product at the mill during manufacture.

Admixture. Material other than water, aggregate, or basic cementitious material that is added to the batch immediately before or during its mixing for the purpose of improving flow and workability or imparting particular qualities to the mortar.

Aggregate

1. Inert graded material, such as sand, vermiculite, or perlite, used as a filler for mixing with a cementitious material.

2. Collection of granulated particles of different substances into a compound or conglomerate mass.

Air entrainment. Intentional introduction into portland cement plaster in its plastic state a controlled number of minute disconnected air bubbles well distributed throughout the mass to improve flow and workability or to impart other desired characteristics to the mortar.

Alpha gypsum. Class of specially processed calcined gypsum having properties of low consistency and high strength.

Angle irons. Metal sections sometimes used as main runners in lieu of channels in suspended ceiling construction.

Arch corner bead. Corner bead so designed that it can be job-shaped for use on arches.

Arris. Sharp edge forming an external corner at the junction of two surfaces.

Asbestos fibers

1. Short fiber mineral added to gypsum plaster to provide a dense hard gypsum plaster and a fire rated material.

2. Asbestos fibers are sometimes mill- or job-added to prevent segregation in portland cement-lime plasters.

Atomizer. Device by which air is introduced into material at the nozzle to regulate the texture of machine-applied plaster.

Autoclaved lime. Thoroughly hydrated lime, manufactured by hydrating (slaking) in a pressure chamber rather than at normal atmospheric pressure.

Backing. Lath used as a base for plaster.

Band. Flat molding (moulding).

Base coat. Plaster coat or combination of scratch and brown coats applied before the finish coat.

Base coat floating. Spreading, compacting, and smoothing plaster to a reasonably true plane.

Base screed. Performed metal screed with perforated or expanded flanges that provides a ground for plaster and separates areas of dissimilar materials.

Bead. Strip of metal usually formed with a projecting nosing and perforated or expanded metal wings and used as a plaster ground and edge protection at external corners.

Beaded molding (moulding). Cast plaster string of beads set in a molding or cornice.

Beam clip. Formed wire section used to attach lath to flanges of steel beams.

Bed mold. Flat area in a cornice in which ornamentation is placed.

Bevel. Slanted surface.

Blister. Protuberance on the finish coat of plaster caused by application over too damp a base coat, troweling over too damp a base coat, or troweling too soon.

Bond. Adherence between the plaster and base produced by adhesive and cohesive properties of plaster or special supplementary materials.

Bonding agent. Nonoxidizing, noncrystallizing, resinous water emulsion providing bond for plaster to concrete, masonry, steel, or existing plaster.

Bonding plaster. Base coat material especially formulated for application directly to roughened interior monolithic concrete surface.

Boss. Gothic ornament set at the intersection of moldings.

Brackets. Formed shapes of channel or pencil rod, used in erecting furred assemblies.

Break. Interruption in the continuity of a plastered surface.

Bridging. Section sized to fit inside the flanges of studs and channels to stiffen construction.

Brown coat. Coat of plaster directly beneath the finish coat. In two-coat work, brown coat refers to the base coat plaster applied over the lath. In three-coat work, the brown coat refers to the second coat applied over a scratch (first) coat.

Brown out. Complete application of base coat plastering.

Bull nose. External angle that is rounded to eliminate a sharp corner, used largely at window returns and door frames.

Butterflies. Color imperfections on a lime putty finish wall caused by lime bumps not put through a screen or insufficient mixing of the gaging.

Butterfly reinforcement. Strips of metal reinforcement placed diagonally over the plaster base at the corners of openings before plastering.

Caisson. Panel sunk below the normal surface in flat or vaulted ceilings.

Calcine. Make powdery or oxidize by removing chemically combined water by the action of controlled heat.

Calcining. Producing lime by the heating of limestone to a high temperature.

Capital. Ornamental head of a column or pilaster.

Carrying channel. Heaviest integral supporting member in a suspended ceiling. Carrying channels or main runners are supported by hangers attached to the building structure and in turn support various grid systems and/or furring channels or rods to which the lath is fastened.

Case mold (mould). Plaster shell used to hold various parts of a plaster mold in correct position; also used with gelatin and wax molds to prevent distortions during pouring operations.

Casing

1. Beam used at the perimeter of a plaster membrane or around openings to provide a stop and separation from adjacent materials.

2. Sometimes called a "plaster stop," this bead is used where plaster is discontinued, around openings (thus providing a ground), where the plaster adjoins another material, and to form the perimeter of a plaster membrane or panel.

Casing clip. Formed metal section that puts pressure on a casing bead to ensure rigid positioning.

Casting plaster. Preparation of plaster of paris or whiting, glue, and fiber, used mainly for casting ornaments in gelatin or rubber molds.

Cat face

1. Pitlike imperfection in the white coat finish.

2. Flaw in the finish coat comparable to a pockmark. Sometimes, base coat knobs showing through the finish cost are referred to by this term.

Ceiling runner track. Formed metal section, anchored to the ceiling, into which metal studs for hollow or solid partitions are set; formed metal section to which lath is attached for studless partitions; metal channel or angle used for anchoring the partition to the ceiling.

Ceiling track. Channel used as a ceiling runner to set prefabricated steel studs; long lath for partitions.

Cement. Material or mixture of materials that when in a plastic state possesses adhesive and cohesive properties and that will set in place. The word "cement" is used without regard to the composition of the materials.

Cementitious material

1. Material binding aggregate particles together in a heterogeneous mass.

2. Mixture of materials that when mixed with water provides binding action to hold aggregate particles together in a solid mass when set.

Chamfer. Beveled corner or edge.

Channel—carrying. Heaviest integral supporting member in a suspended ceiling. Carrying channels or main runners are supported by hangers attached to the building structure and in turn support various grid systems and/or furring channels to which the lath is fastened.

Channel—furring. Smaller horizontal member of a suspended ceiling, applied at right angles to the underside of carrying channels and to which the lath is attached; the smaller horizontal member in a furred ceiling; in general, the separate members used to space the lath on any surface or member over which it is applied.

Channels. Hot-rolled or cold-rolled steel channels used for furring and not carrying channels or runners but as studs.

Chase. Space in a masonry wall to provide for pipes, ducts, or conduits.

Check cracks. Random cracks caused in plaster by shrinkage when plaster is bonded to its base, sometimes referred to as "craze," "alligator cracks," or "wind checks."

Chemical bond. Adherence of one plaster to another or to the base, which implies formation of interlocking crystals or fusion between the coats or to the base.

Chip cracks. Similar to check cracks, except that the bond is partly destroyed; also referred to as "fire cracks," "map cracks," "crazing," "fire checks," and "hair cracks" (eggshelling).

Clip for control of movement. Flexible, resilient metal section separating the plaster membrane from supports to reduce sound transmission and plaster cracking due to structural movement.

Coat. Thickness or covering of plaster work done at one time. The first coat of plastering is called the "scratch coat," the second coat (when there are three coats) is called the "brown coat," and the last coat is known as the "finish coat."

Coffered ceilings. Ornamental ceilings made up of sunken or recessed panels.

Colored finishes. Plaster finish coats containing integrally mixed color pigments or colored aggregates.

Consistency. Degree of density, fluidity, or vicosity of a mortar or cementitious paste.

Contact ceiling. Ceiling composed of lath and plaster and secured in direct contact with the construction above without runner channels or furring.

Contact fireproofing. Application of fire-resistive material direct to structural members to protect them from fire damage.

Control joint (expansion). Formed metal section limiting the areas of unbroken plaster surfaces to minimize possible cracking due to expansion, contraction, and initial shrinkage in portland cement plaster.

Copper alloy steel. Steel containing a minimum of 0.2% pure copper.

Corner bead. Small projecting molding built into plastered corners to prevent accidental breaking of the plaster; strip of formed metal, sometimes combined with the strip of metal lath, placed on corners before plastering to reinforce the corners and used to protect arrises.

Corner bead clip. Metal section used, where necessary, to provide an extension for attachment of various types of corner beads.

Corner reinforcement

1. Reinforcement metal mesh lath cut into strips and bent to form a right angle for use in interior corners of walls and ceilings to prevent cracks in plaster.

2. Corner reinforcement for interior plastering where the plaster base is not continuous around an internal corner or angle.

Corner reinforcement—exterior. Metal section, usually shaped of wire, for the reinforcement of exterior plaster arrises.

Corner reinforcement—interior. Flat or shaped reinforcing units of metal or plastic mesh.

Cornice. Molding (moulding) with or without reinforcement.

Cove. Curved concave or vaulted surface.

Cross furring. Furring members that are attached at right angles to the underside of main runners or other structural supports.

Cross scratching

1. Scratching of the semidry base coat in two directions with a relatively sharp instrument to provide mechanical bond for the successive coat.

2. Scratching the plaster scratch coat in two directions to provide mechanical bond for the brown coat.

Curing. Methods employed to keep cement plasters moist by preventing too rapid loss of the moisture necessary for its hydration; chemical action by which the plaster obtains its set and strength.

Cut nail. Square, tapered fastening, used where great holding power is desired.

Dado. Lower part of a wall usually separated from the upper by a molding or other accessory.

Darby. Flat wood tool with handles about 4 in. wide and 42 in. long, used to smooth or float the brown coat. It is also used on the finish coat to give a preliminary true and even surface.

Dash bond coat. Thick slurry of portland cement, sand, and water dashed on concrete or masonry surfaces by hand or machine to provide a mechanical bond for succeeding plaster coats.

Dash coat. Fine- or coarse-textured coat applied either by hand or by machine.

Dead burn. Removal of all water content during calcining of gypsum.

Decibel. Unit measure of sound intensity that can be used in expressing sound volume or loudness.

Dentils. Small rectangular blocks set in a row in the bed mold of a cornice.

Diamond mesh flat expanded metal lath. Metal lath slit and expanded from metal sheets or coils into such a form that there will be no rib in the lath; lath weighs 2.5 lb. painted, 3.4 lb painted or galvanized per sq. yd.

Direct-attachment clips. Clips used to attach the lath directly to furring or steel studs.

Dope. Additives mixed with plaster to accelerate or retard set.

Dot. Small projection of base coat plaster places on a surface and faced out between grounds to assist the plaster in obtaining the proper plaster thickness and surface plane. Occasionally, pieces of metal or wood are applied to the plaster base at intervals as spot grounds to gauge plaster thickness.

Double hydrated lime. Lime that achieves 92% hydration (normal hydrated lime 62%), which minimizes spalling of the putty coat.

Double up. Plaster in successive operations without setting and drying interval between coats.

Dryout. Soft, chalky condition in gypsum plaster caused by loss of water due to excessive evaporation or suction before the setting action has taken place.

Efflorescence

1. White fleecy deposit on the face of plastered walls caused by salts in the sand or backing. It is also referred to as "whiskering" or "saltpetering."
2. White deposit sometimes found on plaster surfaces, usually caused by water-soluble salts in the plaster ingredients or in the concrete or masonry base.

Eggshelling. Chip-cracking of plaster in the base or finish coat. Its form is concave to the surface, and the bond is partially destroyed.

End clip. Metal section used to secure ends and edges of gypsum lath.

Enrichment. Cast ornament that cannot be executed by a running mold.

Expanded metal. Sheet metal that has been slit and expanded to produce diamond or rib lath.

Expanded metal lath. Metal lath is one of three types: diamond mesh (also called flat expanded metal lath), rib, or sheet. Metal lath is slit and expanded, or slit, punched, or otherwise formed, with or without partial expansion, from copper alloy or galvanized steel coils or sheets. Metal

lath is coated with rust-inhibitive paint after fabrication or is made from galvanized sheets.

Expanded metal—stucco mesh. Reinforcement similar to metal lath but cut and expanded from heavier metal sheets and especially suitable for exterior stucco work.

Exterior corner reinforcement. Metal section, usually shaped of wire, for the reinforcement of exterior plaster arrises.

Fat. Material accumulated on the trowel during the finish coat operation, which is often used to fill small imperfections, also, a term used to describe working characteristics of highly plastic mortars or mortars containing a high percentage of cementitious material.

Feather edge. Bevel-edged tool used to straighten the finish coat in reentrant angles.

Fiber. The addition of sisal, glass, wood, or asbestos fibers to gypsum plaster provides fire-rated and dense, hard gypsum plaster. The fibers are usually added at the mill to the gypsum plaster.

Fibered. Term applied to plaster containing a reinforcing fiber, such as sisal, glass, wood, or asbestos fiber.

Fine aggregate. Sand or other inorganic aggregate for use in plastering.

Fines. Small aggregate passing a No. 200 sieve.

Finish coat. Last and final coat of plastering that receives the particular finish required for the installation.

Finish coat floating. Bringing the aggregate to the surface to produce a uniform texture.

Finishing coat. White coat, putty coat, sand finish, or acoustical plaster, as applied to plastering, means the third or last coat applied.

Fire-retardant lath. Type X lath, the same as a plain gypsum lath, except that the core has increased fire-retardant properties to improve its fire-resistive rating.

Fish eyes. Spots in finish coat about $\frac{1}{4}$ in. in diameter caused by lumpy lime due to age or insufficient blending of material.

Flat-rib metal lath. Combination of expanded metal lath and ribs in which the rib has a total depth of less than $\frac{1}{8}$ in. measured from the top inside of the lath to the top side of the rib.

Float. Rectangular tool with a handle, faced with cork, felt, wood, carpet, plastic, or rubber.

Floating. The act of spreading, compacting, and smoothing plaster to a reasonably true plane, and bringing the aggregate to the surface to produce a uniform texture.

Floor runner track. Formed metal section, anchored to the floor, into which metal studs for hollow or solid partitions are set; formed metal section into which lath is inserted for studless partitions; wood member into which lath is inserted for studless partitions; metal channel used for anchoring the partition to the floor.

Foam plaster base (rigid type). Rigid foamed backing that acts as a plaster base.

Framing. Structural members such as columns, beams, girders, studs, joists, headers, or trusses.

Fresco. Art of decoration achieved by applying a water-soluble paint to freshly spread plaster before it dries.

Furred ceiling. Ceiling composed of lath and plaster attached by means of steel channels, rods, or wood furring strips indirectly to the construction above.

Furring

1. Wall or ceiling construction beyond or below the normal surface plane; also, the members used in such construction.
2. Framework of wood, hollow tile, or metal, not a part of the structure of the building, employed to provide airspaces for insulation, to even or level surfaces, to cover unsightly construction or equipment, or by aligning surfaces and balancing elements, to achieve the requirement of the architectural design.

Furring channels. Smallest horizontal members of a suspended ceiling, applied at right angles to the underside of carrying channels, to which the lath is attached; also, the smallest horizontal members in a furred ceiling; also, in general, the separate members used to space the lath on any surface over which it is applied.

Furring clip. Metal section for attaching cross furring to main runners.

Furring insert. Formed metal section that is inserted in concrete or masonry walls for the attachment or support of wall furring channels.

Gauging. Addition of another cementitious material (usually gypsum or Keene's or portland cement) to lime putty to accelerate its setting.

Gauging plaster

1. Plaster mixed with finish lime to make lime putty for the finish coat of plaster. The initial strength and proper setting time of the lime putty is determined by the gauging plaster.
2. Specially ground gypsum plaster that mixes easily with lime putty and Type S hydrated lime, available in fast or slow-setting formulations.

Gesso. Composition of gypsum plaster, whiting, and glue, used as a base for decorative painting.

Glazing. Condition created by the fines of a machine dash texture plaster traveling to the surface and producing a flatened texture and shine or discoloration. Glazing may be caused by the base coat's being too wet or the acoustical mortar's being too moist. Glazing occurs in hand application when the mortar being worked is excessively wet.

Glitter finish. Finish produced by blowing mica or metallic flakes onto the wet exterior or interior finish coats.

Gradation. Particle size distribution of aggregate as determined by separation with standard screens.

Green plaster. Wet or damp plaster.

Grillage. Framework composed of main runners and/or carrying and furring channels to support the lath.

Ground. Strip of wood or metal that acts as a straight edge and thickness gauge to which the plasterer works to ensure a straight plaster surface of proper thickness.

Grout

1. Mortar made so thin by the addition of water that it will run into joints and cavities of masonry; fluid cement used to fill crevices.
2. Gypsum or portland cement plaster mortar used to fill crevices or hollow metal frames.

Gypsum

1. Hydrous calcium sulfate, in crystalline form, having the approximate chemical formula $CaSO_4 + 2H_2O$ (ASTM C 22).
2. Natural mineral consisting of hydrous sulfate of calcium.

Gypsum bond plaster. Mill plaster specially formulated with a small percentage of lime for direct application to roughened monolithic concrete surfaces (ASTM C 28).

Gypsum cement plaster. Gypsum cement that is prepared to be used, with the addition of aggregate, as a base coat plaster.

Gypsum gauging plaster. Plaster for mixing with lime putty to control the setting time and initial strength of the finish coat. It is classified as either quick setting or slow setting (ASTM C 28).

Gypsum lath

1. Plaster base made in sheet forms, and composed of a core of fibered gypsum, faced on both sides with a treated paper.
2. Plaster base manufactured in the form of sheets of various thicknesses and sizes, having a gypsum core and surfaced with a special paper suitable for receiving gypsum plaster (ASTM C 37).

Gypsum molding plaster. Specially formulated plaster used in casting and ornamental plaster work; it may be used neat or with lime (ASTM C 59).

Gypsum neat plaster. Plaster requiring the addition of aggregate on the job. It may be unfibered or fibered (with animal, vegetable, or glass fibers) (ASTM C 28).

Gypsum plaster. Ground calcined gypsum combined with various additives to control the set; also, applied gypsum plaster mixture.

Gypsum ready-mixed plaster. Plaster that is mixed at the mill with a mineral aggregate. It may contain other ingredients to control the time of set and working properties. Similar terms are "mill-mixed" and "premixed." Only the addition and mixture of water are required on the job (ASTM C 28).

Gypsum wood-fibered plaster. Mill-mixed plaster containing a small percentage of wood fiber as an aggregate, used for fireproofing and high strength (ASTM C 28).

Hanger insert. Formed metal section inserted in concrete members for the attachment of hangers.

Hangers. Vertical members that carry the steel framework of a suspended ceiling; vertical members that support furring under concrete construction; wires used in attaching the lath directly to the concrete construction.

Hardening. Gain of strength of a plastered surface after setting.

Hard wall. Gypsum base coat plaster. Regionally the term differs; in some areas it refers to plaster to which aggregate has been added; in others to neat plaster.

Hard white coat. Surface coat of sufficient hardness to require 70 kg of pressure to force a 10-mm ball 0.01 in. into the plaster face.

Hawk. Flat wood or metal tool, 10 to 14 in. square, with a handle used by the plasterer to carry plaster mortar.

High-calcium lime. Quicklime in which the principal chemical constitutent, calcium oxide, is normally in excess of 95%.

High-magnesium lime. Quicklime in which the magnesium oxide content is normally approximately 35% with a corresponding reduction in the calcium oxide content. The combined oxides normally are 98% plus in high-quality lime.

High-strength gypsum gauged plaster. High-strength gypsum gauging plaster for use with lime putty to produce surfaces of extreme hardness and durability.

Hog ring. Heavy galvanized wire staple applied with a pneumatic gun which clinches it in the form of a closed ring around a stud, rod, pencil rod, or channel.

Hydrated lime

1. Relatively stable lime produced by slaking quicklime with water.

2. Lime after having been slaked with the optimum quantity of water to yield a dry, fine powder.

Hydraulic cement. Cement, such as portland cement, that will set and harden under water; also, quick-setting expansion-type cement compound used to fill cracks and to waterproof.

Hydraulic hydrated lime. Type of lime used for scratch or brown coat of plaster, stucco, mortar, or in portland cement concrete either as a blend, amendment, or admixture.

Individual stud clip. Formed metal section for use where a floor runner is impractical.

Inorganic aggregates. Inorganic aggregates most commonly used in gypsum plaster include perlite, sand (natural or manufactured), and vermiculite (ASTM C 35).

Insulating gypsum lath. Plain gypsum lath with the addition of a sheet of aluminum foil attached to its back side to act as a vapor retarder and also as reflective thermal insulation.

Interior corner reinforcement. Flat or shaped reinforcing unit of metal or plastic mesh.

Interior stucco. Regional term designating a finish plaster for walls and ceilings that finishes smooth or textured; mechanically blended compound of Keene's cement lime (Type S), and inert fine aggregate. Color pigment may be added to produce integrally colored interior stucco.

Joining. Juncture of two separate plaster applications of the same coat, usually within a single surface plane.

Keene's cement

1. Anhydrous calcined gypsum (dead-burned), used principally as gauging for lime putty to obtain smooth finish coat (ASTM C 61).

2. Quick-setting, white, hard-finish plaster, which produces a wall of extreme durability. It is made by soaking plaster of paris in a solution of alum or borax and cream of tartar.

Key. Grip or mechanical bond of one coat of plaster to another coat or to a plaster base. Keying may be accomplished physically by the penetration of wet mortar or crystals into paper fibers, perforations, scoring irregularities or by the embedment of the lath.

Lath. Perforated or expanded metal, gypsum board, insulation board, or wire used as a plaster base.

Lath clip (generic). Metal section to secure lath to supports.

Lead back lath. Plain gypsum lath to which sheets of lead have been laminated.

Lightweight aggregate. Vermiculite or perlite, as distinct from sand aggregate.

Lime

1. Calcium oxide, chemically known as calcia, occurring chiefly in combination with carbon dioxide as a calcium carbonate in limestone, marble, chalk, coral, and shells. It is obtained by heating limestone in a furnace or kiln to about 1000°F to burn out the carbonic acid gas.

2. General term usually referring to quicklime, formed by burning (calcining) limestone at temperatures sufficient to drive off the natural carbon dioxide content.

Lime plaster. Interior base coat plaster containing lime, aggregate, and sometimes fiber. Lime base coat plaster is slow-setting, and should not be applied to gypsum lath.

Lime putty

1. Mixture of gauging plaster and finish lime used for the finish coat of plaster.

2. Product produced from quicklime or hydrated lime to which an excess of water has been added to form a plastic putty of "hodable" consistency.

Liquid bonding agent. Proprietary material used to improve adherence of the plaster coat to the surface receiving plaster.

Load-bearing metal studs. Studs formed from 18-gauge, minimum, structural-grade strip steel, with punched webs. They are also available in double form widths to permit attachment of the lath by nailing or other means.

Long lath. Gypsum lath used in solid plaster partitions. Generally a long lath is 24 in. (600 mm) wide and the length of the wall height.

Low-consistency plaster. Neat (unfibered) gypsum base coat plaster especially processed so that less mixing water is required than in standard gypsum base coat plaster to produce workability. This type of plaster is particularly adapted to machine application.

Lump lime. Quicklime in large chunks as produced in vertical kilns. Lime in this form is rarely used in modern plastering techniques.

Main runner or carrying channel

1. Heaviest horizontal member supported by hangers in a suspended ceiling, to which the furring channels or rods are attached; it may also be directly attached to the construction above.

2. Heaviest integral supporting member in a suspended ceiling. Main runners or carrying channels are supported by hangers attached to the building structure and in turn support furring channels or rods to which the lath is fastened.

Manufactured sand. The fine material resulting from the crushing and classification by screening, or otherwise, of rock, gravel, or blast furnace slag (ASTM C 35).

Marble chips. Graded aggregate of maximum hardness made from crushed marble to be thrown or blown onto a soft plaster bedding coat to produce marblecrete.

Marblecrete. Surface-bedded materials, such as marble chips, glass, crushed ceramic tile, etc., thrown forcibly onto a bedding coat by hand or machine and then tamped lightly to give uniform embedment.

Marezzo. Imitation marble formed with Keene's cement to which colors have been added.

Masking. Affixing paper, plastic, or any other flexible protective material or coating to protect adjacent work. Masking is used particularly in plastering machine applications.

Masonry base. Masonry surface over which plaster is applied.

Mechanical application. Application of plaster mortar by mechanical means, generally pumping and spraying, as distinguished from hand placement.

Mechanical bonding. Physical keying of one plaster coat to another or to the plaster base.

Mechanical trowel. Motor-driven tool with revolving blades used to produce a denser finish coat than by hand troweling.

Medium-hard white coat. Surface coat of sufficient hardness to require 50 kg of pressure to force a 10 mm ball 0.01 in. into the plaster face.

Membrane fireproofing. Lath and plaster system that is separated from the structural steel members, in most cases by furring or suspension, to provide fireproofing.

Metal arch casing. Sheet-steel-formed arch for use as a base lath or corner reinforcement at arched opening in partitions.

Metal base clip. Formed metal section to which is attached the metal base for partitions or walls.

Metal corner bead. Fabricated metal with flanges and nosings at the juncture of the flanges, used to protect or form arrises.

Metal laths. Sheets of copper alloy or galvanized steel that are slit and expanded or slit, punched, or otherwise formed, with or without partial expansion, on which the plaster is spread. A metal lath is coated with rust-inhibitive paint after fabrication or is made from galvanized sheets.

Metal lath clip. Formed wire section for fastening a metal lath to the flanges of steel joists.

Metal lath—diamond mesh. Metal sheet, slit and expanded in two directions to form diamond-shaped openings in the sheet.

Metal lath—expanded—ribbed and sheet. Copper-bearing steel sheet, coated with rust-inhibitive paint after cutting; lath cut from zinc-coated steel sheet.

Metal lath—flat rib. Combination of expanded metal lath and ribs in which the rib has a total depth of less than $\frac{1}{8}$ in. measured from the top inside of the lath to the top side of the rib.

Metal lath—paper-backed. Factory-assembled combination of metal lath or expanded metal reinforcing with paper, fiber, or other backing, the assembly being used as a plaster or stucco base.

Metal lath—self-furring. Metal lath so formed that portions of it extend from the face of the lath so that it is separated at least $\frac{1}{4}$ in. from the background to which it is attached.

Metal partition base. Fabricated integral metal section, which may also serve as a ground for the plaster (attached to framing member or masonry).

Metal studs. Upright members of the framework of a wall fabricated as channels from strip steel, with punched openings along the entire length of web between flanges; or fabricated from doubled round rods on each side, with a single round rod forming an open-web trusslike pattern and all points of contact welded; or fabricated from steel angles with a single round rod forming an open-web trusslike pattern between flanges and all points of contact welded.

Miter. Diagonal joining of two or more moldings at their intersection.

Moist cure. Process of keeping the plastered surface damp in order to ensure the hydration of the material to prevent excessive cracks.

Mortar. Material used in a plastic state, which becomes hard or sets in place.

Moulds. Consists of two types. Running, such as cornices and other continuous mouldings, and casting, such as standard designs or custom designed ornamental plaques that are applied at the job.

Nailing channel. Channel fabricated from not lighter than 25-gauge steel so as to form slots to permit attachment of the lath by means of ratchet-type annular nails or other satisfactory attachments.

Natural sand. Fine granular material resulting from the natural disintegration of rock or from crushing of friable sandstone.

Neat

1. Material that does not contain aggregate but may be fibered or unfibered.

2. Term applied to plaster material requiring the addition of aggregate.

Neat plaster. Base coat plaster to which sand is added at the job.

Niche. Curved or square recess in a wall.

Noise reduction coefficient (NRC). The noise reduction coefficient is a standard of measurement of sound control in the design of acoustical materials.

Normal finishing hydrated-lime. Type N—normal finishing hydrated-lime that is suitable for use in scratch, brown, and finish coats of plaster, stucco, mortar, and as an addition to portland cement concrete (ASTM C 206).

Nozzle. Attachment at the end of a plastering machine material hose, that regulates the fan or spray pattern.

Ogee. Curved section of a molding (moulding) partly convex and partly concave.

Orifices. Attachments to the nozzle on the hose of a plastering machine of various shapes and sizes, which may be changed to help establish the pattern of the plaster as it is projected onto the surface being plastered.

Paper-backed metal lath. Factory-assembled combination of many types of metal lath with absorbent slot perforated paper or a waterproofed building paper, the assembly being used as a plaster base.

Partition (lath and plaster). Bearing or nonbearing interior wall that may be hollow or solid, fire rated, and/or sound rated.

Partition cap. Formed metal section for use at the end of a free-standing solid partition to provide protection for plaster. It is also used as a stair rail cap, mullion cover, light cove cap, etc.

Pencil rods. Mild steel rods used as hangers for suspended ceilings.

Perforated gypsum lath

1. Same as a plain lath, except that it has perforations not less than $\frac{3}{4}$ in. in diameter, with one perforation for not more than each 16 sq in.

2. Gypsum lath made in sheet form, composed of a gypsum core, faced on the sides and long edges with treated paper, and having round holes punched through the lath that provide a mechanical "key" for plaster.

Perlites

1. Nonmetallic siliceous volcanic mineral containing particles of chemically bound water. When heated rapidly, the mineral expands and flakes off, forming a lightweight aggregate consisting of a spongelike mass of rock-enclosed air bubbles.

2. Siliceous volcanic glass properly expanded by heat and weighing not less than $7\frac{1}{2}$ nor more than 15 lb/cu ft, used as a lightweight aggregate in plaster.

Pinhole. Small hole appearing in a cast because of excess water.

Plain lath. Sheet or slab having an incombustible core, essentially gypsum, surfaced with paper suitable to receive gypsum plaster. One face may be variously treated by, for example, mechanical pricking or indenting, or impregnation with a catalyst.

Plaster. Cementitious material or a combination of cementitious materials and aggregates that when mixed with a suitable amount of water forms a plastic mass, which when applied to a surface adheres to it and subsequently sets or hardens, preserving in a rigid state the form or texture imposed during the period of plasticity. The term "plaster" is used with regard to the specific composition of the material and does not explicitly denote either interior or exterior use.

Plaster bead. Member of run or precast plaster moldings.

Plastering machine. Mechanical device by which plaster mortar is conveyed through a flexible hose and deposited in place; also known as a "plaster pump" or "plastering gun." It is distinct from "gunite" machines in which the plaster or concrete is conveyed, dry, through the flexible hose and hydrated at the nozzle.

Plastering on monolithic concrete. Specially prepared bond plaster is used on concrete. Aggregate is not added.

Plaster stop. Piece of flat metal formed in a U shape with one edge used as a plaster ground and the other edge used to attach it to the base.

Plaster screeds. Narrow strips of mortar are applied on a wall or ceiling surface and faced out straight and true to serve as thickness and plane guides for plastering the space between them.

Plastic cement. Portland cement to which small amounts of plasticizing agents, not more than 12% by weight, have been added at the mill.

Plasticity. Workability and water-retentive characteristics are imparted to plaster mortars by such agents as natural cement, lime, asbestos, flour, clays, air-entraining agents, or other approved lubricators or fatteners. Duration of mixing time may be a factor in the plasticity of some mortars.

Pops or pits. Ruptures in finished plaster or cement surfaces, which may be caused by expansion of improperly slaked particles of lime or by foreign substances.

Portland cement. Product obtained by finely pulverizing the clinkers produced by calcining to incipient fusion an intimate and properly proportioned mixture of argillaceous and calcareous materials, with no additions subsequent to calcination except water and calcined or uncalcined gypsum (ASTM C 150).

Portland cement-lime plaster. Portland cement and lime (either Type S hydrated lime or properly aged lime putty) combined in proportion as outlined in building codes.

Portland cement plaster. Plaster mix in which portland cement is used as the cementitious material, usually designed for exterior surfaces or interior areas where wet or moist conditions exist or that may be hosed down at regular and frequent intervals.

Processed quicklime. Quicklime that has been pulverized to a very fine, powdery form.

Puddling. Mechanical dash textures resulting in glazing; texture deviation or discoloration caused by holding the plastering machine nozzle too long in one area.

Pumice. Lightweight volcanic rock, which, when crushed and graded, may be used as a plaster aggregate.

Putty. Product resulting from slaking, soaking, or mixing lime and water together.

Putty coat. Troweled finish coat composed of lime putty or Type S hydrated lime gauged with gypsum gauging plaster or Keene's cement. Fine aggregate may be added.

Quicklime

1. Solid product remaining after limestone has been heated to a high temperature.

2. The quicklime classification includes all types of quicklimes such as crushed, granular, ground, lump, pebble, and pulverized (ASTM C 5).

Ready-mixed. Term applied to aggregate that has been incorporated in the dry mixture as shipped by the manufacturer. Ready-mixed plaster requires the addition of water only, in the field, for use.

Relief. Ornamented figures above a plane surface.

Resilient clips. Clips used in sound-conditioned walls to hold the lath base away from furring members or steel studs.

Resilient clips and pencil rods. Clips and pencil rods combined are used in sound-conditioned walls to hold the lath base away from furring members or steel studs.

Resilient system. Method of attaching plaster bases to a surface to provide for flexibility between the surface and the plaster.

Retarder. Material added to gypsum plaster that slows up its natural set.

Retempering. Addition of water to portland cement plaster after mixing but before the setting process has started. Gypsum plaster must not be retempered.

Return. Terminal of a cornice or molding that takes the form of an external miter and stops at the wall line.

Reveal. Vertical face of a door or window opening between the face of the interior wall and the window or door frame.

Rib. Unexpanded portion of a metal lath, which leaves the plane of the lath at a certain angle and returns at the same angle; separately attached stiffening member.

Rock gun. Device for throwing aggregate onto a soft bedding coat in applying marblecrete, or small gravel chips.

Saddle tie. Specific method of wrapping hanger wire around main runners; also, method of wrapping tie wire around the juncture of a main runner and cross furring.

Sand. Mineral aggregate that passes a $\frac{1}{4}$-in. mesh sieve and is retained on a 200-mesh sieve.

Sand-float finish. Coarse surface finish achieved by adding sand to the finish coat material and floating to a rough but true surface.

Scoring. Grooving, usually horizontal, of portland cement plaster scratch coat to provide a mechanical bond for the brown coat; also, decorative grooving of the finish coat.

Scratch coat. First coat of plaster. The term originates from the practice of cross raking or scratching the surface of this coat with a comblike tool to provide a mechanical key for the brown or finish coats.

Screed. Formed galvanized steel strip used to divide different types of plaster finishes and as a separation between plaster and a cement base; permanent incombustible strip to control plaster thickness and alignment beneath wood or other rigid base trim.

Self-furring metal lath. Metal so formed that portions of it extend from the face of the lath so that it is separated at least $\frac{1}{4}$-in. from the background to which it is attached; it is painted or galvanized.

Set. Change in mortar from a plastic, workable state to a solid, rigid state.

Sgraffito. Decoration generally consisting of two or more layers of differently colored plaster. While still soft, part of the top layer is removed by scratching, exposing part of the base or underlying layer.

Sheet lath. Metal lath, slit, or punched, or otherwise formed from metal sheets.

Shielding. Method of protecting adjacent work by positioning temporary protective sheets of rigid material, particularly used for machine applications.

Shoe. Formed metal section used in attaching metal studs to floor and ceiling tracks, also the end section of a channel turned to an angle (usually 90°) to permit attachment, generally to other channels.

Sisal fiber. Glass mill additives to gypsum plaster.

Slaking. Adding water to quicklime to produce hydrated lime or lime putty.

Soffit. Underside of a subordinate part or member of a building, such as staircase, archway, cornice, or eave.

Sound transmission clip. Flexible, resilient metal clip used to decrease sound transmission through partition and floor assemblies. It also serves to lessen plaster cracking resulting from structural movement.

Special finishing hydrated lime. Type S-special finishing hydrated lime that is suitable for use in the scratch, brown, and finish coats of plaster, stucco, mortar, and as an addition to portland cement concrete (ASTM C 206).

Splay angle. Where two surfaces come together forming an angle of more than 90°.

Spray texture. Surface finish achieved by application of finish coat material with a plastering machine or gun.

Staffs. Plaster casts of ornamental details made in molds and reinforced with fiber. They are usually wired, nailed, or stuck into place.

Starter clip. Metal section used at the floor; initial course of a gypsum lath.

Sticker. Piece of metal channel inserted in concrete or masonry walls for the attachment or support of wall furring channels.

Stiffener. Horizontal metal shape tied to vertical members (studs or channels) of partitions or walls to brace them.

Stipple perforated. Surface texture achieved by rolling an absorptive roller over a tacky plaster surface and then puncturing the surface with a pointed round tool in an irregular pattern.

Straight edge. True flat tool or rod, used to strengthen the brown coat or plaster screeds.

String wire. Soft annealed steel wire placed horizontally around a building of open-stud construction to support waterproofing paper or felt.

Strip lath. Narrow strip of diamond mesh metal lath applied as a reinforcement over joints between sheets of nonmetallic lathing bases or at junctures between such bases and where dissimilar bases join.

Strip lath or stripite. Strips of metal, wire, or wire fabric lath used over continuous joints of gypsum lath at junctions of dissimilar materials, used to reinforce door and window openings at intersections of the jamb with the head and sill, and used at arches and other locations to reinforce plaster subject to localized stresses. In veneer plastering plastic mesh or paper is used for these purposes.

Stucco. Mortar used for exterior portland cement plaster. Depending on locality, stucco may be the combined base and finish coat or the colored finish coat only.

Studs (metal, load bearing). Metal load-bearing studs are formed from minimum 18-gauge, structural-grade strip steel with punched webs. They are available also in double form widths to permit attachment of the lath by nailing or other means.

Suction. Absorptive quality of surfaces, such as concrete, masonry, or gypsum lath, to be plastered. A plaster base coat must also have suction in order to absorb water from the succeeding coat and so induce a bond. Excessive suction should be controlled so that sufficient water will remain in each plaster coat to assure hydration.

Suspended ceiling. Ceiling composed of lath and plaster and steel channels suspended from and not in direct contact with the floor or roof construction above.

Suspended framing. Furring members are suspended below the structural members of the building.

Sweat out. Soft, damp gypsum plaster caused by poor drying conditions which delay or impede setting of plaster.

Tape. Plastic reinforcing mesh or paper used to reinforce angles and bridge lath joints in veneer plastering.

Temper. Mix plaster to a workable consistency.

Template. Gauge, pattern, or mold used as a guide to produce arches, curves, molds, and other shapes of a repetitive nature.

Textured finishes. Finishes obtained by special methods such as stippling, dashing, troweling, floating, or a combination of these.

Thermal shock. Stress created by an extreme change in temperature that may result in the cracking of plaster that has not yet attained its ultimate strength.

Three-coat work. Plastering consisting of two base coats and one finish coat.

Three-eighths-($\frac{3}{8}$) in. rib metal lath. Combination of expanded metal lath and ribs of a total depth of approximately $\frac{3}{8}$ in. measured from the top inside of the lath to the top side of the rib, or another metal lath of equal rigidity.

Three-quarters-($\frac{3}{4}$) in. rib metal lath. Combination of expanded metal lath and ribs of a total depth of approximately $\frac{3}{4}$ in. measured from the top inside of the lath to the top side of the rib.

Ties. Two types are used for the attachment of the lath: the butterfly tie, which is formed by twisting the wire and cutting so that the two ends extend outward oppositely, and the stub tie, which is twisted and cut at the twist.

Tie wire. Soft annealed steel wire used to join lath supports, attach the lath to supports, attach accessories, etc.

Trowel. Flat, steel tool used to spread and smooth plaster.

Troweled finish. Finish produced by water troweling with a steel trowel to produce a smooth and dense surface.

Turtle back. Term used synonymously with blistering. It is used regionally to denote a small localized area of wind crazing.

Two-coat work. Plastering consisting of one base coat and a finish coat.

Type S hydrated lime. Special finishing hydrated lime, distinguished from Type N (normal finishing hydrated lime) by restrictions on the amount of unhydrated oxides. The plasticity requirements of Type S hydrated lime may be determined after a shorter soaking period than with Type N.

Veneer plaster. Specially formulated high-strength plaster for thin-coat application to large-size veneer plaster lath.

Veneer plaster lath. Large-size base for veneer plasters having an incombustible core, essentially gypsum, surfaced with a special face paper suitable to receive veneer plaster.

Vermiculite. Micaceous mineral expanded by heat and used as a lightweight aggregate.

V-stiffened wire lath. Wire lath not lighter than No. 20 W. & M. gauge wire, $2\frac{1}{2}$ meshes per in. with No. 24 U.S. gauge V-rib stiffeners spaced not to exceed 8 in. apart, coated with zinc or rust inhibitive paint.

Wadding. Hanging staff by fastening wads made of plaster of paris and excelsior or fiber to the casts and winding them around the framing.

Wainscot. Lower portion of an interior wall when it is finished differently from the remainder of the wall.

Wall furring base clip. Formed metal section used to attach a metal base to furred walls.

Wash-out. Lack of proper coverage and texture buildup in machine dash textured plaster caused by the mortar's being too soupy.

Waste mold (mould). Precast plaster mold made for the forming of decorative monolithic or cast-in-place concrete. The mold cannot be removed without being destroyed.

Waterproof cement. Portland cement to which waterproofing agents, such as surface repellents, have been added at time of blending materials at the mill.

Welded wire fabric. Plaster reinforcement of copper-bearing soft annealed wire not lighter than 16 gauge, zinc coated, electrically welded at all intersections forming openings not to exceed 2×2 in.. It may have an absorptive paper separator and an additional paper or foil backing for purposes of waterproofing or insulation. It is flat or self-furring.

White coat

1. Hard-gauged lime putty troweled finish; white topcoat on a plastered surface.

2. Surface coat of sufficient hardness to require 70 kg of pressure to force a 10-mm ball 0.01 in. into the plaster face.

White coat—medium hard. Surface coat of sufficient hardness to require 50 kg of pressure to force a 10-mm ball 0.01 in. into the plaster face.

Wire cloth lath. Plaster reinforcement of wire not lighter than No. 19 gauge, $2\frac{1}{2}$ mesh per in., coated with zinc or rust-inhibitive paint (not to be used as reinforcement of exterior portland cement plaster).

Wire lath

1. Woven or welded wire lath.

2. Wire lath not lighter than No. 19 W & M gauge wire, $2\frac{1}{2}$ mesh per in., coated with zinc or rust-inhibitive paint.

Wood fiber aggregates. Mill-added grained or shredded nonstaining wood fiber aggregated in gypsum plaster. The addition of wood fiber provides fire-rated and dense hard gypsum plaster.

Wood fiber plaster. Plaster that is formulated for use either neat or with sand. It contains shredded wood fiber.

Woven wire fabric. Plaster reinforcement of zinc-coated wire, not lighter than No. 18 gauge when woven into 1-in. openings, nor lighter than No. 17 gauge when woven into $1\frac{1}{2}$-in. openings. Lath may be paper backed, flat, or self-furring.

RESILIENT FLOORING

Above grade. Suspended or supported floor construction with a minimum of 18 in. of cross-ventilated air space beneath the floor construction.

Abrasion. Form of wear in which a gradual removal of flooring surface is caused by frictional action of foot traffic introducing relatively fine, dense, and hard particles.

Adhesive. Material used as a bonding agent.

Air-conditioning influence. Usually an air-conditioning system will favorably affect the drying of subfloors because large quantities of water are removed from the air by such systems. The effect on the drying of subfloors or adhesives may be quickly ascertained by learning whether the system is removing water from or adding it to the air. The latter circumstance is likely during the winter months for most systems. These should not be unfavorable drying conditions, for normally the relative humidity would be controlled at 50% or less.

Alkali. Resilient flooring installed over concrete slabs may be endangered by the existence of alkali in the soil or in the slab, which may penetrate the slab surface and attack the adhesive and the flooring itself.

Application of adhesive. Most manufacturers produce and recommend the adhesive to be used for each type of flooring. The application of the adhesive usually is required by the size of area covered to the maximum working area recommended by the manufacturer. Adhesive that films over or shows signs of having dry spots must be removed by scraping or solvent.

Asphalt cutback adhesives. Solutions of asphalt in hydrocarbon solvents containing fibrous fillers to permit application by trowel. Because of the solvents they present some flammability and toxicity hazards. Like asphalt emulsions, cutbacks require at least 30 minutes open time before the installation of tile; they become too dry for the installation of tile after 4 to 7 hours. Cutbacks are still superior to emulsions for installing service-gauge vinyl-asbestos tile.

Asphalt emulsion adhesive. Asphalt emulsions are water based, and present no flammability or toxicity hazards. Adhesives contain from 50 to 60% asphalt and are applied to the subfloor by a notched trowel. After application the adhesives must be permitted to become substantially dry, which requires a waiting period of 30 minutes to 1 hour before tile can be installed. Drying time may be longer on humid days. Once the film dries, the tile may be installed up to 24 hours later, provided it remains dry and free from dust and dirt. Pressure is not generally required. Asphalt emulsion adhesives are used on concrete both above and below grade and on suspended wood subfloors with or without lining felt.

Asphaltic adhesives. There are two types of asphaltic adhesives, water emulsion and cutback. Both are suitable for installing asphalt tile or vinyl-asbestos tile. Although they eventually become hard, asphaltic adhesives will remain soft and tacky for a long period of time. Asphalt and vinyl-asbestos tile are relatively rigid with a high degree of cold flow, and require adhesives with long tack retention to ensure intimate contact.

Asphalt rubber. A blend of asphalt cement, reclaimed tire rubber and certain additives in which the rubber component is at least 15% by weight of the total blend and has reacted in hot asphalt cement sufficiently to cause swelling of the rubber particles.

Asphalt rubber adhesive. Adhesive having a water base and consisting of asphalt and rubber, which are brush and troweling grade materials, for use in the installation of asphalt and vinyl-asbestos tile. It can be used over all subfloors and over lining felt. Allow the adhesive to set 30 minutes before installing tile, or as recommended. Use brushing grade with $\frac{1}{16}$-in. vinyl-asbestos tile. Keep the adhesive from freezing and clean it with fine steel wool and a cleaner recommended by manufacturer after completion.

Asphalt tile

1. Tile consists of asbestos fibers, finely ground limestone fillers, mineral pigments, and asphaltic or resinous binders. Asphaltic binder is usually a blackish, high-melt asphalt mined principally in Utah and Colorado; the resins are derived from the distillation of petroleum or coal tars. Asphalt tile is made in several color and cost groups; group B includes dark background and marbleized colors group C; includes intermediate colors; group D includes light colors, and group K covers special pattern effects.

2. Thoroughly blended composition of thermoplastic binder, asbestos fibers, pigments, and fillers formed under heat and pressure, and then cut to size.

3. Gilsonite and petroleum asphalts are used as binders in dark-colored asphalt tiles. In the lighter shades, cumarone-indene resins, with very little or no asphalt, are used as binders. Various synthetic thermoplastic resins are used in grease-resistant asphalt tile. Fine asbestos fibers, ground limestone, and silica are used as fillers, along with mineral pigments.

4. Floor surfacing unit composed of thermoplastic binder, asbestos fibers, mineral fillers, and pigments. The binder is essentially asphalt or hydrocarbon resins or both, or coal tar and petroleum origin, compounded with suitable plasticizers and stabilizers (ASTM F 141).

Backed vinyl. Vinyl resin stock bonded to scrap vinyl, rubber, or other nonvinyl backing.

Base. Flat or shaped, extruded or molded material used to provide a functional and/or decorative border between walls and floor.

Battleship linoleum. Heavy gauge linoleum at least $\frac{1}{8}$-in. thick and manufactured in plain dark colors.

Beveled edging. Transition piece of wood or metal used at exposed edges of resilient flooring, usually at doorways.

Border. Flooring at the perimeter of the room adjacent to the walls, which is installed separately from the field. Flooring is installed so that the border is approximately the same width at all walls.

Brushable adhesives. Adhesives applied by brush are most often used by nonprofessionals. Although designed for brush application, many of these adhesives have characteristics similar to those of trowel-applied asphaltic adhesives. Brushable adhesives usually are combinations of asphalt and rubber, latex emulsions, or rubber in a solvent vehicle.

Brush-on latex emulsion-type adhesive. This type of adhesive dries to a transparent film and is nonstaining, nonflammable, and nontoxic. It has no offensive odor. It is used with reinforced vinyl, vinyl polymer and asphalt floor tiles over wood or concrete and is easy spreading. Tile can be laid almost immediately.

Capping. Finished protective edging material used as a stop for resilient flooring.

Conductive resilient flooring. Floor surfacing which serves as a convenient means of electrically connecting persons and objects together to prevent the accumulation of electrostatic charges (ASTM F 150).

Cork base. Usually the same cork material used for the cork floor covering.

Cork tile

1. Cork tile consists of particles of cork thoroughly and uniformly bonded into small, thin sheets by a baking process. Tile may contain thermosetting binder.

2. Tile composed chiefly of the granulated bark of the cork oak tree, native to Spain, Portugal, and North Africa. Synthetic resins are added to the granulated cork, which is pressed into sheets or blocks and baked. Surfaces are finished with a protective coat of wax, lacquer, or resin applied under heat and pressure. Sheets are then cut to tile sizes. Vinyl cork tile has a film of clear PVC vinyl fused to the top surface to improve durability, water resistance, and ease of maintenance.

3. Floor surfacing unit made from natural cork, thoroughly and uniformly bonded together (ASTM F 141).

4. Tiles are usually made from clean cork shavings with no added binder or inert fillers or from cork granules with some added binder. Cork particles are bonded together by subjecting them to a high compressive load and a baking process.

Cove base. Flat extruded or molded material used to provide a functional permanent border between walls and floor.

Cushioned vinyl flooring

1. Flooring composed of a vinyl wearing surface, a center section of a composite of fiberglass-vinyl, and the bottom section of vinyl-foam. The composite flooring does not meet standards for smoke requirements.

2. Usually vinyl sheet floor covering in which a foam layer is incorporated as part of the product thickness (ASTM F 141 and F 387).

Cutback asphalt adhesive. Adhesive consisting of asphalt, having a solvent base, and made from troweling-grade materials; for use in the installation of asphalt tile and vinyl-asbestos tile. It can be used over all concrete subfloors and panel board underlayment primed with asphaltic primer. Allow the adhesive to set 30 minutes; install tile within 4 to 18 hours as recommended. It is not recommended for use under or over lining felt or asphaltic underlayments. Material is combustible.

Edging. Finished, protective edge material used as a stop for resilient covering.

Embossed linoleum. Embossed inlaid linoleum of 0.090-in. gauge thickness. Embossing tends to conceal subfloor irregularities and traffic indentations.

Emulsion. Adhesives and mastic underlayments consisting of cementitious binders and fillers suspended in a liquid carrier such as water.

End stop. Resilient base used for finishing the base run at flush openings.

End wall. Wall along the short dimension of the room.

Epoxy. A multicomponent resin grout that usually provides very high, tensile, compressive, and bond strengths.

Epoxy adhesive. Adhesive consisting of resin and catalysts with troweling-grade materials, for use in the installation of solid vinyl and rubber tile on or below grade. It can be used over all types of subfloors. Mix only as much adhesive as can be spread within 30 minutes and covered within $2\frac{1}{2}$ hours, or as recommended by manufacturer. It is advisable to guard against tile slipping while the adhesive sets.

Epoxy adhesives. Epoxies are two-component adhesives that may have a pot life as long as several hours. Although they must be applied to relatively dry surfaces, they exhibit superior bonding on subfloors that subsequently become wet. They are noted for their durability, good adhesion, high strength, and good aging characteristics. Among their drawbacks are the necessity of mixing, difficulty in spreading and extensive movement of tile after application.

Exposure to strong sunlight. Strong exposure to sunlight may affect the performance and appearance of some types of resilient floors by causing fading, shrinking, blisters, and brittleness.

External corner. Resilient base preformed for external corner application.

Feature strip. Decorative or functional inlay in resilient flooring.

Field. Middle portion of the flooring installation, exclusive of the border.

Fillet strip. Structural backing for a flash cove.

Fire resistance. Vinyl asbestos tile and asphalt tile must meet the requirements of the Underwriters Laboratories, Inc. "Test method for measuring the flame propagating characteristics of flooring and floor covering materials"

designated as UL Subject 992. Some state authorities and federal agencies demand the UL 992 test method, requiring an FPI (flame propagation index) of 4.0 or less.

Flash cove and base. Combined base and border which is coved at the junction of horizontal and vertical surfaces.

Flashing. Bending up sheet material against a wall or a projection, either temporarily for the purpose of fitting, or permanently so as to form a one-piece resilient base.

Flexibility. That property of a resilient flooring that allows it to be deformed by bending or rolling without cracking, breaking, or showing other permanent defects (ASTM F 137).

Gauge. Form of wear, consisting of a wide groove deformation accompanied by material removal and penetrating a considerable distance below the immediate flooring surface (ASTM F 141).

Hardboard. Dense panel board manufactured of wood fibers with the natural lignin in the wood reactivated to serve as a binder for the wood fibers.

Heating efficiency. The temperature difference between the surface of concrete subfloors and the resilient flooring surface is not enough to result in significantly increased fuel consumption.

Homogeneous vinyl tile

1. Thoroughly blended composition of thermoplastic binders, fillers, and pigments. The binder consists of a polyvinyl chloride resin or copolymer resin.

2. Tiles are unbacked and usually have uniform composition throughout. The flooring trade uses the term "homogeneous" rather loosely to mean vinyl formulations, usually neither backed by nor laminated to nonvinyl compositions.

Homogeneous vinyl tile flooring. Floor surfacing units composed of vinyl-plastic binder and pigments with or without mineral fillers. The vinyl-plastic binder is an essentially poly (vinyl chloride) resin, or a poly (vinyl chloride) resin, or a poly (vinyl chloride) copolymer resin compounded with suitable plasticizers and stabilizers (ASTM F 141).

Inlaid patterned linoleum. Inlaid patterns in many colors in gauges .070 in. and up.

Inlaid sheet flooring. Floor surfacing material in which the pattern is formed by colored areas that extend from the surface through to a backing, and are bonded together (ASTM F 141).

Internal corner. Resilient base preformed for internal corner application.

Laminated resilient vinyl. Properly compounded vinyl resin stock bonded to a scrap vinyl, rubber, or other nonvinyl backing.

Latex. Milky colloid in which natural or synthetic rubber or plastic is suspended in water; an elastomer product made from latex.

Latex adhesive

1. Adhesive consisting of rubber and water-base brush and troweling-grade materials, for use in the installation of vinyl tile or with asbestos-vinyl or rubber backing, rubber tile, solid vinyl tile, and lining felt. It can be used over all subfloors. Install flooring immediately; keep adhesive from freezing.

2. Latex adhesives permit the installation of rubber and vinyl tiles on and below grade. The first latex adhesives introduced were two-component mixtures (NR) which are quite difficult to apply after 30 to 40 minutes from mixing time. Once the adhesive is applied, the tile must be installed within less than 10 minutes.

Latex flooring and tile adhesive. Adhesive for the installation of rubber flooring (tiles or rolls) over suspended wood and concrete floors and over suitable on-grade concrete floors under conditions where excessive moisture exists.

Latex leveling compound. Waterproof self-curing latex and dry powder are mixed to form a compound used for rapid patching and filling of worn or dished-out stairs. This compound will feather-edge without crumbling and bonds well to wood and most other building materials. It cures to a tough resilient surface and will not shrink, check, crack, or break up.

Latex underlayment

1. Material for smoothing rough subfloors on, above, or below grade. It is easily troweled, dries rapidly to a concretelike hardness and applies from the skin coat to $\frac{1}{8}$ in. and thicker. The latex liquid is mixed with powder just prior to use.

2. Cementitious product with liquid rubber additive, used to smooth or fill subfloor irregularities.

Lignin paste. Lignin or so-called linoleum paste is used to cement felt underlays to wood subfloors. It is most generally used to bond flexible vinyl tiles, rubber tiles, cork tiles, linoleums, and felt-backed floor coverings to suspended concrete and wood subfloors and to underlayments on such subfloors. Lignin paste is partly water soluble and special resin cements or alumina cement-latex paste are recommended by some manufacturers where the surface of the floor tiles is apt to be exposed to appreciable water or dampness.

Lining felt

1. Felt that may be saturated or semisaturated with asphaltic material.

2. Semisaturated asphalt felt specifically manufactured for use under resilient flooring. Generally, lining felt is recommended under burlap backed linoleum and may be an acceptable substitute for underlayment for the installation of most tile products over suitable existing strip-wood floors.

Linoleum

1. Thoroughly blended composition of oxidized oleoresinous binders, pigments, and fillers on a supporting backing of burlap or felt.

2. Composition of high-quality natural raw materials, including oxidized linseed oil, cork, wood, flour, and colorstable mineral pigments.

3. Floor surfacing material composed of oxidized linseed oil, fossil, or other resins or rosin, or an equivalent oxidized oleoresinous binder, mixed with ground coal or wood flour, mineral filler and pigments, that is bonded to a burlap fiber or other suitable organic backing. Linoleum is made in either sheet or tie units (ASTM F 141).

Linoleum floor coverings. The binder in linoleum is linoleum cement, consisting of a mixture of oxidized linseed oil, kauri gum, rosin, and other resins. The cement is intimately mixed with ground cork, wood floor, and pigments to form the linoleum composition. The composition is calendered onto a burlap backing or an asphalt-saturated felt backing. It is then cured at a constant temperature of about 140°F for a period of from 2 to 4 weeks, depending on the thickness of the linoleum.

Linoleum paste

1. Paste consisting of clay with water-base, troweling-grade materials, for use in the installation of lining felt, all sheet flooring, vinyl tile with backing, rubber tile, cork, and vinyl cork tile, which can be used over suspended wood or concrete subfloors, and panel board and latex underlayments. The flooring is installed immediately after troweling the paste. The paste is not recommended for use on or below grade, over suspended subfloors subject to moisture, or for solid vinyl, asphalt, and vinyl-asbestos tile.

2. Linoleum pastes are sulfite-based liquids which are used to install linoleum, rubber tile, some vinyls, cork tile, lining felt, and some other materials. They are water based emulsions that remain forever water soluble, making them unsuitable for use on concrete slabs or on ground or suspended slabs that are wet. Once linoleum paste has been applied, the flooring must be installed and rolled with a 100-lb roller within about 15 minutes to ensure intimate contact between flooring and subfloor. The adhesive film will normally dry to a firm film within 24 hours and will become hard and rigid within a few days. Rigidity restrains sheet flooring and prevents tiles from curling.

Linoleum tile and sheets

1. Linoleum products consisting of oxidized linseed oil and resin binders, mixed with wood flour, pigments, mineral fillers, and/or ground cork. The composition is bonded to an organic backing of burlap or asphalt-saturated rag felt.

2. Light and standard gauges are suitable for residential and light commercial work, and are adaptable for counter and work tops in commercial and institutional projects. Linoleum tile and sheets are manufactured with felt backs and are available in a wide range of marbleized and textured effects. The heavy gauge is manufactured with a burlap back and is recommended for commercial and institutional work.

Marbleized linoleum. Directional or nondirectional grained-effect linoleum, in many colors and patterns, .070 in. gauge and up.

Mastic. A glasslike, brittle, yellow to greenish yellow resinous exudation of the mastic tree (Pistacia lentiscus).

Mastic-type underlayments. Mastic underlayments are of several kinds. The best contain a binder of latex, asphalt, or polyvinyl acetate resins in the mix. Those that consist simply of a powdered mixture (cement, gypsum, and sand), to which only water is added, all too often break down under traffic when applied in thin coats or feather edges. For some installations, all types will be satisfactory, but the mastic type is best where a thin film is required; moreover portland cement can be made mastic by commercially available additives of polyvinyl acetate. For best results the maximum thickness of latex underlayments is $\frac{1}{8}$ in. Any thickness greater than $\frac{1}{8}$-in. should be applied in two or more applications. For satisfactory union with any type of mastic underlayment, it is important that the subfloor be free from paint, oil, and varnish.

Moisture. Resilient flooring installed over concrete slabs, which have been poured on the ground, may be endangered by the penetration of moisture through the slab and finally may be trapped under the flooring, which may result in buckled or warped tiles in due time. Installation of polyethylene film, .004 in. (4 mil) is recommended over the ground surface.

Neoprene. A synthetic rubber with outstanding resistance to ozone, weathering, various chemicals, oil and flame, made by polymerization of chloroprene.

Neoprene adhesive. Water resistant adhesive for installing vinyl and rubber stair treads, rubber nosings, and rubber and vinyl corner guards.

Nonslip surface. Surface specially prepared to prevent slipping.

Nosing. Finished protective formed edge material used as a stop for resilient stair tread covering.

No-wax flooring. Resilient flooring that does not require wax or polishing due to manufacturing processes used for the protection of the finished surfaces. Washing and cleaning will suffice but the floor must be protected from unusual abrasion, deep scratches, and gritty sharp particles being ground into the flooring surface.

On grade. Floor construction at ground level, on the ground, or less than 18 in. above the ground.

Particle board. Panel board manufactured of wood chips and/or flakes or other lignocellulosic material, together with a synthetic resin adhesive.

Particle-board underlayment. Underlayment for resilient flooring, frequently referred to as "chipboard" or "flake board."

Plywood underlayment

1. Several types and grades of plywood are suitable for this use. Guarantees of quality and performance are the responsibility of the plywood manufacturer and should always be qualified in underlayment specifications in accordance with standards developed by the American Plywood Association in conjunction with the U.S. Department of Commerce.

2. Underlayment, $\frac{1}{4}$ in. or thicker, depending on construction of the subfloor. It must be underlayment grade or better. Panels are slightly butted and installed with staggered joints (none of which coincide with the joints of the material under them), nailed securely every 2 in. on edges and 6 in. apart throughout the panel with ring-shanked nails driven flush with the surface of the plywood. If "power cleats" are used, slightly closer spacing is used than for nails. Thicker panels require less nailing. Panel joints are sanded and filled with leveling and patching compound. Cracks and depressions are also filled with leveling and patching compound. The floor should be clean, smooth, dry, free of grease, grit, paint and solvent spills, and loose particles.

Polymeric-poured (seamless) floors. Floor covering composed of polymeric material applied to the substrate in a liquid form alone, or in combination with mineral or plastic chips, pigments, desiccants, or fillers, which finally converts to a thick built-up covering (ASTM F 141).

Polyvinyl acetate. A thermoplastic, polymer insoluble in water; gasoline, oils, and fats. Soluble in ketones, alcohols, benzene; esters, and chlorinated hydrocarbons; used in adhesives, lacquers, and latex paints.

Polyvinyl acetate underlayments. Polyvinyl acetate resins are the main ingredients of concrete bonding compounds for securing concrete topping to old concrete, terrazzo, marble, concrete block, brick, ceramic tile, painted or nonpourous surfaces such as metal, or other materials providing no suction or absorption.

Poured latex topping. Several types of topping used for leveling and preparing a subbase for receiving resilient flooring.

Primer. Brushable, solvent-base, asphaltic preparation recommended as a first coat over porous or dusty concrete floors and panel underlayments for the purpose of sealing of pores and improving the bond with asphaltic adhesives used for the installation of asphalt and vinyl-asbestos tile.

Printed sheet vinyl flooring. Floor surfacing material in which the pattern is printed on a backing and protected with a wearing layer of transparent or translucent vinyl-plastic (ASTM F 141).

Resilience. Resilience is a property that involves the elastic energy in a material which causes it to regain its original shape when an external load is withdrawn. For practical purposes resilience in its broadest sense, consists of properties beyond recovery from indentation.

Resilient flooring

1. Flooring manufactured in tile and sheet form, in different thicknesses, and from a variety of ingredients. Tile sizes range from 9×9, 12×12, up to 36×36 in.; sheets are generally made in 6-ft widths, although some products are made in 4-ft., 6-in. widths. Matching resilient accessories, such as wall base, thresholds, stair treads, and feature strips, are available to complete flooring installation.

2. Organic floor surfacing material made in sheet or tile form, or formed in place as a seamless material of which the wearing surface is nontextile. The resilient floor covering classification by the common usage includes, but is not limited to asphalt, cork, linoleum, rubber, vinyl, vinyl asbestos, and polymeric-poured seamless floors. Resilient in this sense is used as a commonly accepted term, but does not necessarily define a physical property (ASTM F 141).

Resilient flooring products. Resilient flooring products may be classed according to basic ingredients: asphalt, vinyl, vinyl-asbestos, rubber, cork, and linoleum.

Resinous waterproof adhesive. Adhesives containing alcohol as a solvent. They have approximately the same characteristics as linoleum paste, although the dried films are insoluble in ordinary water and likely to be attacked by alkaline solutions, which makes them unsuitable for use on or below grade.

Resistance to impact indentation. The momentary indentations caused by ordinary walking traffic are of the utmost importance. The entire weight of a person acting dynamically on a specific restricted contact area could produce pressures of from 12,000 to 60,000 lbs sq in. Resilient floors normally tend to recover from such impact indentations.

Resistance to static loads. Indentations that are caused by concentrated loads that remain on the flooring in a stationary position for long periods of time. It is recommended that flooring should not be exposed to more than 75 lbs sq ft. The standard method of test for indentation of resilient flooring (McBurney Test) (ASTM F 142).

Rubber. A natural, synthetic, or modified high polymer with elastic properties and, after vuncanization, has elastic recovery; the generic term is "elastomer."

Rubber base. Extruded or molded material for use against walls.

Rubber floor coverings. Floor coverings consisting of vulcanized compounds of rubber, mainly synthetic rubber of the butadiene-styrene type. To many of the floorings, reclaimed rubber is added. Mineral fillers, such as zinc oxide, magnesium oxide, and various clays, are added in considerable amounts, along with mineral pigments. Rubber floorings are obtained in tile or sheet form in thicknesses ranging from $\frac{3}{22}$ to $\frac{1}{4}$ in. In sheet form the rubber compound is usually keyed to a cotton cloth backing.

Rubber flooring. Floor surfacing material in tile units or sheet form, consisting of compounded natural or synthetic rubber, or both, in combination with mineral fillers and pigments. (ASTM F 141).

Rubber sheet or tiles. Sheets or tiles composed of natural rubber, synthetic rubber, or reclaimed rubber, alone or in combination.

Rubber tile. Natural or synthetic rubber is the basic ingredient of rubber flooring. Clay and fibrous talc or asbestos fillers provide the desired degree of reinforcement; oils and resins are added as plasticizers and stiffening agents. Color is achieved by nonfading organic pigments, and chemicals are added to accelerate the curing process. Ingredients are mixed thoroughly and rolled into colored sheets. Sheets are calendered to uniform thickness and vulcanized in hydraulic presses under heat and pressure into compact, flexible sheets with a smooth, glossy surface. Backs are then sanded to gauge, ensuring uniform thickness, and sheets are cut into tiles.

Saddle. Insert in the flooring at an interior opening.

Scratching. Form of wear in which a minute groove-like break in a flooring surface is made by a rubbing contact with a tool or dense hard particle, the total deformation being confined to the most immediate level (ASTM F 141).

Scribing. Transferring the profile of an obstruction, projection, or flooring edge to a piece of flooring so that it can be accurately cut and fitted.

Scuff. Form of wear in which a mark, gall, roughness, or other damage is caused by the rubbing of foot-traffic bodies against a flooring surface and may involve deposition of a foreign material onto the flooring surface (ASTM F 510).

Sealer. Solution of equal parts of wax-free shellac and denatured alcohol, recommended as a first coat over existing stripwood floors from which finish has been removed and intended to seal the wood pores, prevent excessive moisture absorption, and provide a dimensionally stable base for direct application of lining felt or resilient flooring.

Sealer and primer. Base coats used to prepare surfaces for the application of underlayments or adhesive.

Seam masters. Machine that cuts seams by automatically lifting the flooring material off the subfloor, allowing the electric saw blade to slice through it. The lifting of the material is achieved with a shoe or guide fitted to the bottom of the machine and adjusted to the thickness of the flooring material.

Sheet vinyl. The vinyl content allows them to be made especially colorful and adds to their ability to resist wear,

grease, and alkalies. Sheet vinyl floors are available in varying thicknesses.

Sidewall. Wall along the long dimension of the room.

Solid vinyl. "Solid vinyl" is a misnomer because a flooring of pure vinyl would have poor wearing qualities. Vinyl is combined with asbestos fiber and inert fillers to produce the vinyl asbestos tile, which provides superior wearing qualities and resistance to fire and cigarette burns.

Solvent-base adhesives and primers. Adhesives and primers consisting of cementitious binders and fillers dissolved in a volatile hydrocarbon carrier such as alcohol or cutback.

Standard cove wall base cement or adhesive. Cement or adhesive having a solvent base, with troweling-grade materials, combined with for use in the installation of vinyl and rubber cove base. They are used over dry walls above grade. The base is installed within 15 minutes after adhesive has been applied. The material is combustible.

Static conductive homogenous vinyl. Vinyl tile containing electrically conductive ingredients formulated to be a conductor of static electricity.

Static conductive linoleum. Burlap-backed sheet linoleum containing electrically conductive ingredients formulated to be a conductor of electricity.

Statically conductive flooring. Electrically conductive resilient flooring.

Stick tile. Tile with adhesive already on the back. Protective paper is removed and the tile is pressed in place over concrete, wood, or old floor coverings in suitable condition.

Subfloor. Structural material or surface that supports floor loads and finish floor, and serves as a working platform during construction. If the subfloor is sufficiently dense, smooth, stiff, and dimensionally stable and possesses adequate bonding properties, resilient flooring may be applied directly without the use of underlayment.

Subfloor locations

1. Subfloors when built at least 18 in. above grade over a ventilated crawl space are considered suspended.

2. Selection of resilient flooring products and adhesives is determined largely by subfloor location: below grade, on grade, or suspended. These designations are based mainly on the relative likely exposure to moisture and alkali, which adversely affect many flooring materials and adhesives.

Subfloor tolerances. In order to accomplish an acceptable smooth, level, and tight-jointed resilient flooring, installation depends on the condition of the subfloor. The subfloor surfaces must not vary more than $\frac{1}{4}$ in. above or $\frac{1}{4}$ in. below the established required plane. In no case should there be more than $\frac{1}{16}$ in. variation within any 12 in.

Surface finishes. Protective coating in paste or liquid form for resilient floors; usually natural waxes, but may be synthetic solutions, particularly for homogenous vinyls.

Suspended wood floors. A suspended wood floor is one with at least 18 in. of well-ventilated space below it. The ground under a crawl space must be covered with 6-mil polyethylene sheeting. The wood floor should be double construction or the equivalent, with a minimum thickness of 1 in.

Synthetic rubber cement for wall base. Cement consisting of synthetic rubber resin and solvent-base, brush-grade materials, for use in the installation of vinyl cove base and metal nosing and edges. It can be used with all resilient tile, except asphalt and vinyl-asbestos to metal surfaces. It can be used for applications to both walls and floors. The material must be installed within 20 minutes after the adhesive is applied.

Temperatures. Most adhesives for resilient flooring are designed for use at about 70°F. Temperatures up to 90°F may not be harmful, but the higher temperatures will accelerate set. Temperatures below 70°F may cause slow setting and permit indentation of the flooring. Asphaltic adhesives tend to lose the ability to bond at low temperatures.

Underlayment

1. Composition board or plywood firmly attached to the underfloor as a base for resilient flooring; trowel-applied material for smoothing or filling subfloor irregularities.

2. Underlayment grade hardboard $\frac{1}{4}$-in. thick installed over the subfloor to provide a suitable base for resilient flooring when the subfloor does not possess the necessary properties for direct application of the flooring.

Underlayment boards. Resilient flooring is not recommended to be laid on particle board, chipboard, and flake board. These boards tend to swell, buckle, and warp under certain moisture conditions.

Vinyl acetate. A colorless, water-soluble, flammable liquid that boils at 73°C, used as a chemical intermediate and in the production of polymers and copolymers.

Vinyl-asbestos tile

1. Thoroughly blended, semiflexible composition of asbestos fibers, vinyl resins, plasticizers, color pigments, and fillers, formed under heat and pressure and then cut to size. FS SS-T-312, Type IV.

2. Flooring in tile form, composed of vinyl resins and asbestos fillers. Because of outstanding physical properties, its exceptional durability and ease of cleaning, its performance is excellent over suspended subfloors, on-grade slabs, and below-grade concrete with conventional adhesives. It has exceptionally good resistance to alkalies and grease.

3. Tile composed of asbestos fibers, ground limestone, plasticizers, pigments, and polyvinyl chloride (PVC) resin binders. These resins permit lighter colors and greater variety of designs than the binders used in asphalt tile. The ingredients are mixed under heat and pressure and rolled into blankets to which decorative chips may be added. Additional rolling and calendering produce smooth sheets of desired thickness. After waxing, sheets are cut into tiles.

4. Floor surfacing unit composed of vinyl plastic binder, asbestos fibers, mineral fillers, and pigments. The vinyl plastic binder is an essentially poly(vinyl chloride) resin or a poly(vinyl chloride) copolymer resin compounded with suitable plasticizers and stabilizers (ASTM F 141).

Vinyl sheet and tile. The chief ingredient of vinyl products is polyvinyl chloride (PVC) resin. Other ingredients include mineral fillers, pigments, plasticizers, and stabilizers. Plasticizers provide flexibility; stabilizers fix the mixture to ensure color stability and uniformity. Vinyl products sometimes are referred to as "flexible" vinyl to distinguish them from vinyl-asbestos products, which also are made with PVC resins and are termed semiflexible vinyl.

Vinyl tile. Vinyl tile may be of homogenous solid composition or it may be backed with other materials, such as organic felts, asbestos fibers, or scrap vinyl. Ingredients for solid vinyl tiles are mixed at high temperature, hydraulically pressed, and/or calendered into homogeneous sheets of required thickness; the sheets are cut into tile sizes. Backed products are essentially vinyl sheet flooring cut into tile sizes.

Waterproof resin adhesive. Adhesive consisting of resin and solvent-base, troweling-grade materials; for use in the installation of linoleum; vinyl with rag felt, vinyl, or rubber backing; rubber tile; cork tile; and solid vinyl. It can be used over all suspended subfloors. Flooring should be installed within 15 minutes after adhesive has been troweled on. This adhesive is not recommended for use on or below grade, for asphalt or vinyl-asbestos tile or to install lining felt. The adhesive is combustible.

Waxing. After wax stripping or as often as needed after light-duty cleaning, the floor should be waxed. It is preferable to apply a thin coat so that the polish film can dry properly in a reasonable length of time.

Wear. The accumulative and integrative action of all the deleterious mechanical influences encountered in use which tends to impair a material's serviceability. Such influences include, but are not limited to abrasion, scratching, gouging, and scuffing (ASTM F 510).

TERRAZZO

Abrasion resistance. The wearing ability of poured-in-place or precast terrazzo is based on the hardness value of marble chips in the terrazzo. It is recommended that the marble chips have a minimum Ha (abrasive hardness) 10 value, based on ASTM C 241. Test for abrasion resistance of stones subjected to foot traffic. Exterior installations require that all types of chips used have a Ha 50 minimum value.

Abrasive aggregate. Aggregate combined with terrazzo mixtures to provide nonslip surfaces. The aggregate material usually is aluminum oxide or silicon carbide. The aggregate can also be broadcast applied to the top of the fresh terrazzo topping surface.

Absorption. Marble chips with a 24-hour absorption rate in excess of 0.75% should be used with caution. High absorption can cause an unsightly appearance and resulting maintenance problems. Chips used for exterior installations must have a maximum absorption rate of 25%.

Acetylene carbon black. Additive material used in poured-in-place terrazzo mixtures to reduce static electricity explosion hazards. Used primarily in hospital operating rooms.

Aggregate. The principal aggregate in most poured-in-place and precast terrazzo consists of marble chips, although granite, quartz, slate and other dense hard stones are used in combination with marble. Abrasive aggregates are also listed as aggregates.

Art marble. Terrazzo product produced in a shop. The basic ingredients are usually the same as poured-in-place terrazzo.

Binder. Ingredient that holds the chips in position. Binders can be cementious, modified cementious, and resinous. The binder is the matrix.

Bonded terrazzo. System for installing a terrazzo floor where the underbed is structurally bonded to the concrete substrate. The total thickness can vary from $1\frac{1}{2}$ to $1\frac{3}{4}$ in. The minimum thickness of the topping is usually not less than $\frac{1}{2}$ in.

Broken marble. Fractured slabs of marble not crushed by machines into chips.

Byzantine. Art of mosaics expressed in vitreous materials.

Channels for abrasive strips. Zinc channels designed to form grooves in terrazzo stair treads and floor areas, which are considered hazardous for foot traffic, and takes a silicon carbide abrasive strip, installed in the channel.

Chemical matrices. Terrazzo system including chemical matrices usually applied in a thin cross-section in which small chips are used. The matrix is composed of resinous or chemical materials sometimes in addition to portland cement and are often highly resistant to acids, alkalis, and other harmful materials.

Chips

1. Chips used in terrazzo can be defined to include all calcareous, serpentine, and other rocks capable of being ground and taking a satisfactory polish, such as marble, onyx, and other mineral of this same character. Quartz, ceramic coated granules, granite, quartzite, and silica pebbles are suitable for these finishes that do not require polishing, such as rustic terrazzo and textured mosaics.

2. Chips are crushed uniformly so that all dimensions are reasonably close to the limits of the recommended sizes. The percentages of flats or flaky chips should be held to a minimum. Dust content should be limited to less than 1% by weight.

3. Marble granules screened to various sizes. Range in size from No. 0 to 8, as established by the National Terrazzo Mosaic Association Inc.

Cleaner. Neutral liquid cleaners should be used on a regular cycle.

Cleaning materials. The liquid cleaner selected must be neutral with a pH as near 7 as possible and free from any harmful alkali, acid, etc. Soaps and scrubbing powders containing water solubles, inorganic salts, or crystalizing salts should never be used in the maintenance of terrazzo.

Color pigments. Inorganic matter used in the terrazzo mix to vary the color; powdered substance that when blended with a liquid vehicle gives the cement its coloring.

Concrete slab surface. Slab that is level and of uniform thickness. A finished surface should not vary more than $\frac{1}{8}$ in. All laitance is removed from the slab and it is left with a roughened or broomed finish. Slab must be clean of all foreign materials.

Conductive terrazzo

1. Acetylene carbon black added to the topping and underbed and mixed to meet or exceed the requirements of Bulletin 56A of the National Fire Protection Association.

2. Terrazzo flooring system that will conduct static electricity within prescribed resistance levels. It eliminates the build-up of static electricity and is therefore a safe floor for use in area subject to explosive hazards. Chip sizes should be no larger than No. 1 chip and the matrix color is black, since the carbon black is used as the conductive vehicle in the floor.

Control joints. Allowance made by use of a metallic strip, or other device to allow for movement without damaging the terrazzo.

Cove base dividing strip. Dividing strip of zinc or brass usually matching the floor strip and used to divide the wall base into sections not to exceed 6 ft.

Crushing. Uniform crushing of marble chips is desirable within the tolerances specified, and the bagged chips should have a dust content of less than 1% by weight.

Curing. Proper moisture and temperature conditions maintained for normal hydration and portland cement.

Divider strips

1. Divider strips are made of white alloy of zinc, brass, or plastic for use in portland cement terrazzo systems. They are used in thin-set systems as logical stop strips. Brass and plastic may react with some resinous materials and should be used only if deemed safe by the supplier of the resin.

2. All-metal or plastic-top metal strips provided in the terrazzo finish to control cracking due to drying, shrinkage, temperature variations, and minor structural movements. They are also used for decorative purposes and convenience in placing the topping.

3. Terrazzo floor dividing strips are made of half-hard brass, white alloy zinc (99% zinc), and plastic in various colors. The thickness of the strips should be specified using Brown and Sharpe gauges or fractions of an inch. Strips less than $\frac{1}{8}$-in. thick are made of uniform thickness for their entire depth. Strips $\frac{1}{8}$ in. and thicker are of the "heavy-top" type with the top member having a minimum wearing depth of $\frac{1}{4}$ in. and a thin bottom member.

Dividing strips for thin-set terrazzo. Dividing strips provide a leveling device for thin-set terrazzo. Thin-set strips are made in various configurations and in brass, zinc, aluminum and in plastic. Depths range in sizes from $\frac{1}{8}$ to $\frac{1}{4}$ in.

Edging strip. Strip used where terrazzo floors adjoin composition or resilient floors such as vinyl, vinyl asbestos, asphalt tile and linoleum. Strip is made in zinc or brass.

Epoxy. Two-component thermosetting resinous material that is an excellent binder for use in thin-set terrazzo. The minimum physical properties are stipulated in National Terrazzo Mosaic Association, Inc. Epoxy terrazzo specifications.

Epoxy conductive terrazzo (floor matrix). Terrazzo consisting of colorful aggregates and an epoxy matrix, designed to eliminate static buildup, and meeting NFPA requirements, as set forth in Bulletin 56A. It polishes smooth to a gleaming finish.

Epoxy-modified cement. Composite resinous material that is an excellent binder for use in thin-set terrazzo.

Epoxy terrazzo. Terrazzo consisting of colorful aggregates and an epoxy matrix. It polishes to a smooth gleaming finish.

Expansion strips. Double divider strip separated by resilient material and provided generally for the same purpose as divider strips, but used where a greater degree of structural movement is expected.

Expansion thin-set dividing strip. Expansion strip at expansion or control joints consisting of two angles with an exposed neoprene strip. The metal portions of the strip are made in zinc or brass and are also available in plastic.

Flats or flakes. Flats or flakes are produced when crushing marble for chips.

Gray portland cement. Some gray portland cements may not be able to meet the National Terrazzo and Mosaic Association's modifications of ASTM C 150 requirements for white cement. During manufacture gray portland cement is not color controlled, and it may, therefore, cause terrazzo to be unevenly colored.

Grout. Paste used as the binder, sometimes mixed with color pigments, applied to the floor to fill the voids and pits after rough grinding.

Heavy dividing top strip. Dividing strip for terrazzo floor topping with a wide top member in sizes for $\frac{1}{8}$ to $\frac{1}{2}$ in. The strip is made of solid zinc or brass to withstand grinding and polishing.

Integral abrasive edging. Edging used at the housing portion of a stair tread by inserting the nonslip pattern material in the form before casting.

Isolation membrane. Membrane such as asphalt-saturated roofing felt, building paper, or polyethylene film, installed between subfloor and underbed to prevent bond and permit independent movement of each.

Liquid cutting compounds for the curing of concrete. A liquid cutting compound should not be used on slabs to receive terrazzo, as it will act detrimentally to the bonding of the terrazzo.

Marble. Metamorphic (recrystallized) limestone, composed predominantly of crystalline grains of calcite or dolomite, or both, having interlocking or mosaic texture.

Marble sizes. Marble chips used in terrazzo are graded by number according to sizes adopted by the National Terrazzo and Mosaic Association.

Matrix
1. Ingredient in a terrazzo floor that acts as a binder to hold the chips in position. There are three basic types of matrix: cementitious, modified cementitious, and resinous.
2. Topping mortar consisting of binders, and sometimes pigments and inert fillers, which fills the spaces between chips and binds them into a homogeneous mass.
3. Portland cement and a water-mix or noncementitous binder used to hold the marble chips in place for the terrazzo topping.

Mineral pigments. Pigments used as colorants in the matrix in combination with portland cement should be alkali-resistant and non-fading.

Monolithic terrazo. Terrazzo topping that is bonded directly to the structural slab without an underbed only if the slab has not been thoroughly cured.

Mosaic
1. Artistic finish for walls, ceilings, floors, and other elements composed of small hand-cut pieces of smaltite glass, or marble called "tesserae." Tesserae are mounted on paper by hand to form patterns, designs,

or murals. Sheets of mosaic are then set in mortar on the job site.
2. Art, requiring high skill, of placing small, thin, and often hand-cut pieces of marble, colorful stone, or vitreous enamel material, in a manner so as to form various designs or pictures.

Neoprene control strips. Sandwich-type strip in which a neoprene expansion strip is fastened between white alloy, brass or zinc strips.

Noncementious matrices. Matrices such as epoxy, polyacmilate, or polyester.

Nonslip terrazzo. Standard or Venetian small chip terrazzo, used in areas where a smooth, highly slip-resistant surface is desired.

Oxychloride terrazzo. Flooring resembling portland cement terrazzo, but which incorporates oxychloride cement.

Oxalic acid. Acid sometimes used to give ornamental and precast terrazzo a highly honed finish.

Palladiana terrazzo. Form of terrazzo in which $\frac{1}{2}$-in. thick random fractured slabs of marble are set in an underbed and the joints between these slabs are filled with terrazzo or a low shrinkage white or tinted jointing material.

Panels. Spaces in terrazzo topping formed by the divider strips.

Plate numbers. Numbers appearing adjacent to the various terrazzo illustrations in the National Terrazzo and Mosaic Association's color catalog.

Polyacrylate modified cement. Composition resinous material that is excellent binder for use in thin-set terrazzo.

Polyester. Two-component thermosetting resinous material that is an excellent binder for use in thin-set terrazzo. Minimum physical properties are stipulated in the National Terrazzo and Mosaic Association's polyester terrazzo specifications.

Portland cement (Type I—ASTM C 150)
1. Portland current is the product obtained by pulverizing clinker consisting essentially of hydraulic calcium silicates, to which no additions have been made subsequent to calcination other than water and/or untreated calcium sulfate, except that additions not to exceed 1.0% of other materials may be interground with the clinker at the option of the manufacturer.
2. Either white or natural (gray) Type I portland cement may be used for terrazzo finishes. Natural portland cement is less expensive and generally acceptable if a vivid or light-colored matrix is not essential to the color scheme. White portland cement offers a lighter background for displaying the decorative chips and produces clearer, truer colors with mineral pigments. Type IA air-entraining portland cement is recommended for exterior applications subject to freezing and thawing.

Precast terrazzo. Terrazzo that is custom fabricated in watertight molds, in a shop of factory, by a compression and vibratory method.

Production of marble chips. Marble of various types and colors is quarried and selected to avoid off-color or contaminated material. It is crushed by a process that largely eliminates flat or silvery chips and is accurately sized to yield marble chips for terrazzo. Marble is defined as a metamorphic rock formed by the recrystallization of limestone, but has been redefined to include all calcareous rocks capable of taking a polish, including onyx, travertime, and attractive serpentine rocks.

Resinous matrices (epoxy). Two-component, thermo-setting, resinous material used as a binder in thin-set terrazzo.

Rustic terrazzo

1. Variation of stone or marble chips used in the topping where the surface is washed in lieu of grinding and polishing.

2. Terrazzo in which decorative quartz, quartzite, onyx, and granite chips have been substituted for marble chips and whose surface, in lieu of being ground and polished, is washed with water or otherwise treated to expose the stone chips.

3. Form of terrazzo in which the topping is washed or treated in such a way as to expose the chips, creating a uniform textured surface. In exterior work where the rustic terrazzo will be exposed to freezing conditions, a dense hard chip is used in an air entraining matrix.

Samples. Slight color variations in chips can be anticipated, and where the interior design of a room may depend on the terrazzo flooring, a sample should be requested.

Sand. Small noncoherent rock particles.

Sand cushion base. Base used where structural movement that may cause injury to the terrazzo topping is anticipated, either from settlement, expansion, contraction, or vibration.

Screeding. Leveling the top of the mortar bed with a wood or metal strip.

Sealer. Protective coating or treatment applied to terrazzo to prevent any foreign liquid or matter from being absorbed by closing off the pores of the cement.

Setting bed

1. Backing of cement plaster on vertical surfaces to receive terrazzo wainscot or base.

2. A setting bed consists of mortar and used as a bond, base, and level for the terrazzo topping. It is applicable to vertical surfaces.

Sprinkle. Practice of broadcasting or densifying the compaction of marble on the surface of the terrazzo. Sprinkling is accomplished by using the same marble chips as used in the matrix, wetting them, and placing them after installing the terrazzo topping, but prior to the rolling operation.

Standard terrazzo. Standard terrazzo toppings incorporates No. 1 and No. 2 size marble chips, in equal parts, of the same marble. It is the usual practice to combine different colored marbles in the topping.

Standard topping. A standard topping has a minimum thickness (finished) of $\frac{5}{8}$ in., and is composed of Nos. 1 and 2 size marble chips, and sometimes No. 3.

Subfloor. Structural material or surface intended to serve as a working platform during construction and to support the finish floor and design loads.

Surfacing. Surfacing includes the grinding, grouting, filling and finishing operations on terrazzo topping.

Thinset terrazzo

1. Thinset terrazzo (modified cementitious and resinous terrazzo) may, if the cross section is less than $\frac{3}{8}$ in., require the use of No. 1 and smaller chips only.

2. Terrazzo systems that can be applied in $\frac{1}{2}$ in. or less.

3. Relatively thin toppings ($\frac{1}{4}$ to $\frac{1}{2}$ in.) installed directly over any suitable subfloor, generally with resinous binders only.

Terrazzo

1. Type of Venetian marble mosaic, using a portland cement matrix. Its mixture is composed of two parts marble to one part portland cement, to which color pigment and water may be added.

2. Composition material, poured in place or precast, that is used for floor and wall treatments. It consists of marble chips, seeded or unseeded, with a binder or matrix that is cementitious, noncementitious, or a combination of both. Terrazzo is poured, cured, and then ground and polished or otherwise finished.

3. Terrazzo is derived from the Italian "terrace" or "terrazza" and by definition over the centuries is "a form of mosaic flooring made by embedding small pieces of marble in mortar and polishing."

Terrazzo finishes. To acquire the patina and luster associated with fine marble finishes, terrazzo must be given an application of a penetrating sealer to fill the pores of the cement matrix to prevent, in large measure, absorption of traffic dirt and stains, as well as to facilitate their removal in routine care with neutral cleaners.

Tesserae. Thin slices of marble, colorful stone, or glasslike highly colored vitreous enamel material cut into squares or other shapes of any size and used in mosaic work.

Topping

1. Decorative wear layer consisting of mineral chips embedded in a suitable matrix and requiring grinding, filling, polishing, or washing to form the finished terrazzo surface.

2. Wearing surface of the terrazzo floor.

Types of terrazzo installations. Installations commonly include the following: sand cushion, bonded, monolithic, palladiana, epoxy, polyester, polyacrylate modified, epoxy modified, terrazzo over permanent forms, vertical terrazzo, stair treads, and conductive epoxy terrazzo.

Underbed

1. Layer of mortar or grout used as a bearing surface for the installation of the terrazzo topping.

2. Subsurface to accept terrazzo strips.

3. Layer of nonstructural portland cement mortar sometimes used over the subfloor to provide a suitable base for portland cement terrazzo and to minimize cracking.

Venetian mosaic. Same as Byzantine mosaic. Venetian mosaic is so called to distinguish it from mosaics that are formed by tesserae of marble.

Venetian terrazzo

1. Toppings incorporating marble chips larger than the intermediate sizes are usually referred to as "Venetian toppings." The minimum topping thickness for Venetian terrazzo is $\frac{5}{8}$ in.

2. Same as standard terrazzo except that large chips sizes Nos. 3 through 8 are utilized. Venetian terrazzo therefore requires a thicker topping to accommodate the larger sized chips.

Venetian topping. Venetian topping has a minimum thickness that varies from $\frac{5}{8}$ to 1 in. and is composed of marble chip sizes Nos. 3 through 8. It requires a minimum of $1\frac{1}{2}$ in. deep strips, depending on the thickness of the terrazzo topping.

White portland cement. White portland cement provides a good background for marble chips. It can be tinted to produce clear colors. White cement is carefully color controlled during manufacture. For use in terrazzo, white portland cement should exceed the minimum standards of ASTM C 150.

WALL COVERINGS

Abrasion resistance. Federal Standard 191b, Method 5304: Type I—200 + double rubs, Type II—300 + double rubs, Type III—1000 + double rubs.

Adhesion of coating to fabric. Minimum requirements ASTM D 751; Type I—4 lbs per 2 in. width. Type II—6 lbs per 2 in. width, Type III—6 lb per 2-in. width.

Adhesive. GC-460A at full strength, containing mildew inhibitors, or GC-120, depending on material used.

Breaking strength (W × F). Minimum requirements ASTM D 751, grab method: Type I—40×30 lb, Type II—50×55 lb, Type III—100×95 lb.

Coating weight. Minimum requirements CFFA-W-101A, paragraph 7.1.1: Type I—5 oz per sq yd, Type II—7 oz per sq yd, Type III—12 oz per sq yd.

Color fastness to light. Federal Standard 191b Method 5660: There was no evidence of any changes after 200 hours, for Types I, II, and Type III.

Colors. Selection is the option of the decorator or owner.

Concrete block (new construction). Masonry contractor should be notified that wall coverings are to be applied directly to the block, and should adhere to the following: joints to be flush or bagged, not tooled, all burrs of excess mortar and other protusions must be removed, and the wall must present a clean smooth surface. Walls will then be sealed with Right Arm Primer sealer or a coat of straight acrylic point. GC-460A adhesive is then applied.

Drywall. New drywall surfaces including newly taped seams and all color identification marks on the drywall should be primed with a white pigmented alkyd or latex primer.

Enameled painted walls. Walls should be sanded and washed with trisodium phosphate and then rinsed with clean water.

Exterior use. Wall coverings should not be used on exterior surfaces.

Fabric backed vinyl (maintenance). Recommended cleaning materials for some materials can be a mild soap or detergent in warm water. Stubborn surface stains, depending on the material and color, can be removed by using isopropyl alcohol or a very mild solution of household bleach. Testing of a small area is advised before proceeding with the entire surface area. Rinse all cleaned areas with clean lukewarm water. Manufacturers usually suggest cleaning methods for their product.

Fire hazard classifications. As tested and classified by Underwriters Laboratories, Inc. and ASTM E 84 (Surface Burning Characteristics of Building Materials).

Flame spread rating. When applied to reinforced cement board, lightweight material—10, medium-weight material—15, heavyweight material—20, unless other ratings have been established by state or local authorities.

Gypsum substrate. A fabric which is impregnated with uncrystallized gypsum. The gypsum is formulated so that when it is applied to the substrate surface with an adhesive, it will crystallize and form a secure bond with the substrate.

Heat aging. Federal Standard 191b Method 5850 specifies the requirements for all types that the materials shall not become stiff, brittle, or discolored or show loss of grain.

Heavy duty vinyl wall covering. Material 22 oz/sq yd. For use on wall surfaces where a wainscot occurs, and wall surfaces in areas that are exposed to rough treatment and heavy traffic (Type III).

Light duty vinyl wall covering. Material, 7 oz/sq yd. To be used in areas and on surfaces not subjected to abrasion or wear (Type I).

Material and patterns. Various materials and patterns are available and include the following; textures, florals, stripes, silks, damasks, woods, leathers, suedes, linens, foils, burlaps, grasscloth, and metallic types.

Medium duty vinyl wall covering. Material 13 oz/sq. yd. To be used in areas and surfaces subjected to average traffic and minor scuffing (Type II).

Method of application. Brush or roll adhesive, using approximately 1 gal for every 10 to 12 lin yd, depending on the material used; unless otherwise specified, seams should be overlapped on the wall and double cut with a straight edge to ensure pattern uniformity. Other installation instructions are usually supplied by the manufacturer.

Other wall surfaces. Before application of vinyl wall coverings to surfaces other than the standard wall surfaces, consult the manufacturer for their recommendations to apply the product to the wall surface to be used.

Painted surfaces. Verify walls for pigment bleeding, sand surfaces to dull the painted coat, apply GC Right Arm Primer.

Plastered walls. Applications on new plastered walls require the testing of the moisture content, which should not exceed 5%. Remove all crystals due to efflorescence. prime walls with GC Wall Prep Primer.

Stain resistance. ASTM D 1308-b Standard requires that all types of materials shall show no appreciable effect of staining.

Size. Most patterns available in 53/54 in. widths. Type I products—10.5 to 15 oz in 60 yd bolts, Type II products in 30 yd bolds in 54 in. widths with weight ranging from 20 to 38 oz per lin yd.

Smoked developed ratings. When applied to reinforced cement board, lightweight material—5, medium-weight material—10, heavyweight material—10, unless other ratings have been established by state or local authorities.

Standards. Chemical Fabrics and Film Association (CFFA) Quality Standard for vinyl-coated fabric wallcoverings, CFFA-W-101-A, Federal specification CCC-W-408A, and various standards and specifications established by state and local authorities.

Substrate fabrics. The material that lies between the wallcovering and the wall itself becomes a substrate for the wallcovering, allowing a smoother application over surfaces that may be imperfect.

Tear strength (W × F). Minimum requirements ASTM impulse method: Type I—14 × 12 scale reading, Type II—25 × 25 scale reading, Type III—50 × 50 scale reading.

Total Weight. Minimum requirements Federal Standard 191b, Method 5041: Type I—5 oz. per sq yd, Type II—13 oz. per sq yd, Type III—22 oz. per sq yd.

Underwriters Laboratories, Inc. Testing laboratory is available to manufacturers for testing of their products.

Vinyl wall covering. Materials may consist of three layers, the first, the supporting material of cotton cloth, nonwoven fiberglass, asbestos, or other suitable base material. The base material should be mildew-resistant. The second layer is a containing compound of specialized vinyl chloride resin is laminated to the supporting base material in a continuous film. The third layer could be a clear coating or as required by the specifications.

Wallpaper. Printed patterns and designs produced on many weights of paper in all colors. Some are treated to make them waterproof and stain proof. Many are washable with mild detergents and many types of cleaning materials. Manufacturers issue instructions on installations according to the type of wall surface and the adhesive to be used.

Wall surface preparation. Surfaces are to be absolutely clean and smooth. All openings of any size, including minute pinholes, should be filled with a spackling compound and sanded to a smooth surface.

Warranty. Most manufacturers warrant their products for a period of five years from date of installation. Warranty of the products are usually for separation of the backings, staining caused by bleeding of impurities or supporting the growth of mildew; warranty depends on the product's being installed on a proper base that was treated as suggested by the manufacturer and used under normal interior conditions.

WOOD FLOORING

Acrylic/wood flooring. Flooring made of acrylic-impregnated wood, a composite material combining the advantages of acrylic plastic with the aesthetic appeal of fine hardwood.

Adhesive

1. Adhesive for the installation of deep wood block floors consists of coal-tar pitch coating approximately $\frac{1}{8}$-in. thick.

2. Parquet hardwood blocks are usually installed over concrete or wood subfloor with a PVA mastic that provides an excellent bond.

Adhesives. Latex rubber and solvent base for installation of prefinished wood flooring over wood, concrete, or terrazzo subfloors. Nonflammable, nonsolvent based adhesives are available to satisfy fire ratings. Two part epoxy adhesive for areas with moisture problems.

Air-dried lumber. Lumber that has been stored to reduce moisture content to the 12–15% range. American hardwoods air-dried lumber may be no lower than 19%.

American Parquet Association standards. Publications indicate that these standards meet or exceed all building standards. APA patterns comply with FHA minimum property standards. It also meets Voluntary Product Standard PS 27, issued by the U.S. Bureau of Standards.

Anchors. Metal $\frac{3}{8}$-in.-diameter flat headed drive pins and are a minimum $1\frac{1}{2}$ in. long.

Asphalt coating compound. Compound used in the installation of deep wood blocks, requiring a coating compound that is bituminous solvent type, and acid-resistant black asphalt coating compound. FS TT-C 494.

Asphalt-saturated roofing felt. Felt used in the construction of combination flooring systems over concrete slabs and as a waterproofing membrane (ASTM D 226).

Bevel-corner end-grain wood blocks. Bevel-corner-type blocks with all four corners beveled. Blocks are manufactured from carefully selected, air-dried southern yellow pine lumber meeting the standard grading rules of the Southern Pine Assocaition specifications for medium-grain lumber. Blocks are a minimum of 2 in. in depth, approximately $2\frac{1}{2}$ to 4 in. in width, and vary from $3\frac{1}{2}$ to $6\frac{1}{2}$ in. in length. Season checks are not considered defects. Blocks are impregnated under pressure with creosote oil or pentachlorophenol for protection against decay, moisture, vermin, and termites.

Black karuni. South American wood flooring panels in shades from deep brown to almost black with pronounced golden grain. The hardness of this species make it almost impervious to wear and damage. Easy-to-install 19 × 19 in. paper-faced panels are available, unfinished.

Blind nailing. Nailing in such a way that the nailheads are not visible on the face of the work, usually at the tongue of matched boards.

Block flooring. Wood flooring made of blocks, in square or rectangular pieces, rather than the normal strips.

Channel anchors. Modified steel drivepins $\frac{5}{16} \times 1\frac{1}{2}$ in. developing a tensile (pull-out) strength in excess of 900 psi and a horizontal shear strength of over 1300 lb. The drivepins penetrate an average of 1 in. in 4000-psi concrete.

Channels. 16-gauge hot-dipped galvanized steel.

Cold-type floor mastic. Mastic that needs no heating and trowels directly from the container at 70°F or above. It is recommended for installing wood block floors and dampproofing slabs on grade using polyethylene film moisture retarder.

Direction of the flooring. Flooring strips are usually laid to run along the longest dimension of the room, when possible. Strips should run continuously through doorways into adjoining rooms, without thresholds, giving the advantages of a flush floor throughout. The shortest pieces of flooring are used inside closets and scattered over the floor area. The shortest pieces should not be used at entrances and doors. The length of strips elsewhere in a floor is not important, as the process of sanding and finishing a floor blends all pieces into one surface whether the desired effects is a "natural" finish or a dark stained one.

First-grade beech, birch, and hard maple (unfinished). The face is practically free of all defects, but the varying natural color of the wood is not considered a defect.

First-grade pecan (unfinished). The face is practically free of defects, but the varying natural color of the wood is not considered a defect.

First-grade red pecan (unfinished). Same as first grade except that the face is all heartwood.

First-grade white pecan (unfinished). Same as first grade except that the face is all bright sapwood.

Floating floor system. The system consists of $\frac{1}{8}$-in.-thick closed-cell resilient foam applied directly to the dry sub-floor. Two $\frac{1}{2}$-in.-thick layers of fir plywood or pine particle board are stapled together to form an integral floating unit and are finished off with wood floor tiles glued to the plywood with a waterproof adhesive.

Floor ventilation. Adequate provisions must be made for the free movement of air currents under the wood floor assembly if no basement exists.

Foam-cushioned hardwood flooring. Four 6-in. blocks of wood tile are glued together by an automated process, making a 12×12 in. panel with milled tongues and grooves, and the $\frac{1}{8}$-in. layer of closed-cell foam is adhered directly to the back of each panel.

Glued floor system. System based on gluing techniques and elastomeric adhesives that adhere the structural plywood to wood joists firmly and permanently. Floor and joists are fused into an integral T-beam unit and floor stiffness is increased when compared with conventional floor construction. Glue rather than nails carries the stress. Field gluing also eliminates the squeaks that can result from shrinking lumber and nail popping.

Grooved end-grain wood blocks. Grooved blocks with at least two grooves at the side and one groove at the end. Lugs at the side project $\frac{1}{16}$ in. Blocks are manufactured from carefully selected air-dried southern yellow pine lumber meeting the standard grading rules of the Southern Pine Association specifications for medium-grain lumber. Blocks are a minimum of 2-in. in depth, approximately $2\frac{1}{2}$ to 4 in. in width, and vary from $3\frac{1}{2}$ to $6\frac{1}{2}$ in. in length. Season checks are not considered defects. The blocks are impregnated under pressure with creosote oil or pentachlorophenol for protection against decay, moisture, vermin, and termites.

Hardboard underlayment. Produced in 4-ft. squares, 0.220-in. thick and planed to uniform thickness, hardboard underlayment should be installed to manufacturers' specifications for proper performance. It is mainly used in remodeling or in new construction where minimum thickness buildup is desired.

Hardwoods. The hardwoods most commonly used for flooring are oak, maple, beech, birch, and pecan. Oak, the most plentiful, is by far the most extensively used for the residential flooring in the United States.

Isolation strip. Strip 1 in. wide and 0.10-in. thick of material having a maximum compression set of 25%.

Joints. All pieces of the flooring are tongued and grooved on the sides and on the ends; this is called "side and end matching." Bundles are made up of strips of varying lengths and pieces fit perfectly when laid side to side and end to end. End jointing may come anywhere in the floor, without regard to whether there is a joist or sleeper bearing at the point, as the end and side tongue support the flooring, but it should be staggered so as to avoid having two or three joints clustered together.

Kandatawood (registered trademark of Bangkok, Industries, Inc.). Wood known as "South American walnut." Panels make up into a parquet floor that is a light walnut tone in its natural state. The color will not wear off because it goes through the wood. The wood is used for residential and commercial areas. The $\frac{5}{16} \times 19 \times 19$ in. panels are unfinished.

Karpawood (registered trademark of Bangkok Industries, Inc.). Wood also known as "Asian ironwood." It is used for commercial and institutional areas subjected to unusually heavy traffic. It has an extremely low coefficient of expansion and is ideal for on-grade and radiant-heated floors. It also has an exceptionally low flame spread index of 22.5. It possesses vivid grain and comes in a variation of tones from tawny light to deep warm brown. It is available in many patterns, prefinished and unfinished. Panel sizes are $\frac{5}{16} \times 18 \times 18$ in.

Kerriwood (registered trademark of Bangkok Industries, Inc.). The hardest of floorings, possessing more than double the hardness of oak or maple. Its dense grain structure makes it almost naturally fire resistant without further treatment. Its rich warm red brown natural color mellows to a deeper tone as it ages. Panel size $\frac{5}{16} \times 18 \times 18$ in.

Light Karuni (registered trademark of Bangkok Industries, Inc.). South American wood floor panels that are extremely hard. The finished floor reveals a neutral tawny tan that is available in $\frac{3}{16} \times 19 \times 19$ in. panels, unfinished.

Lock-tite floor system. Combination flooring system using a method of anchoring a premium hard maple floor to a concrete slab by means of $\frac{3}{8}$-in. 16 gauge channels held to concrete slab with steel anchors, locking each strip of flooring with a heavy-duty 16 gauge clip. Product of E. L. Bruce Co., Inc., Memphis Tenn. Product is listed as Robbins Hard Maple Flooring Systems.

Lug end-grain wood blocks. Blocks with integral lugs projecting from one side and one end. There are at least two lugs at one side and one lug at the end. Blocks are manufactured from carefully selected air-dried southern yellow pine lumber meeting the standard grading rules of the Southern Pine Association specifications for medium-grain lumber. Blocks are a minimum of $1\frac{1}{2}$ in. in depth and approximately $2\frac{1}{2}$ to 4 in. in width, and vary from $3\frac{1}{2}$ to $6\frac{1}{2}$ in. in length. Season checks are not considered defects. Blocks are impregnated with creosote oil or pentachlorophenol, under pressure, for protection against decay, moisture, vermin, and termites.

Metal channels and clips. Zinc-coated 16-gauge channels and clips developing an average holding power in excess of 600 lb. Tension is perpendicular to the surface. Clips are solid, 1-in. wide, and hold-down wings and $\frac{3}{16}$-in. locking spikes.

MFMA standard specifications. Specifications concerning the official grading rules for northern hard maple, beech, and birch flooring. Publication issued by the Maple Flooring Manufacturers Association, Oshkosh, Wis. Publication contains information about grading and tolerances of the wood concerned, and information about installation of adhesive-applied floors. The specification can be used as standards for species of woods.

National Oak Flooring Manufacturers' Association (NOFMA). The association is the accredited source of the specifications, technical data, and manufacturing and grading rules for oak flooring. In addition it provides grading for beech, birch, hard maple, and pecan flooring.

Northern maple. A very hard, dense, fine-fibered wood with tight grain structure. Wood is nonsplintering, long-wearing, and resilient.

Oak. In clear-grade oak the amount of sapwood is limited; otherwise, variations in color are disregarded in grading. Red and white oak ordinarily are separated, but that does not affect their grading. In most cases the average length of strip flooring is greater in the higher grades. Oak is classified into two grades of quarter-sawed stock and four of plain sawed. In descending order the quarter-sawed grades are clear, select, No. 1 common, and No. 2 common.

Oak, finished, No. 2 common. This grade of oak may contain sound natural variations of the forest product and manufacturing imperfections. The purpose of this grade is to furnish general, utility use flooring or where character marks and contrasting appearance are desired.

Panels. Wood flooring panels are assembled into square or rectangular shapes at the factory. Fabricated panels are laid as units, either in mastic or by nailing.

Parquet solid hardwood floors. Parquet solid hardwood floors come in a wide variety of patterns and derive from such wood species as white oak, red oak, pecan, black walnut, hard maple, cherry, cedar, panga-panga, angelique, teak, and Rhodesian leak.

Particle-board underlayment. Produced in the same thicknesses as plywood, particle-board underlayment is often preferred because its uniform surface and somewhat higher density make it more resistant to indentation than plywood when thin resilient flooring is applied over it. Because it tends to change more in length and width with changes in moisture content than plywood, manufacturers' directions for installation conditions and specifications for adhesives must be followed for good performance.

Pattern (not parquet). Floors manufactured in short lengths of individual pieces. Each piece is cut to exact dimension of another piece, or multiples thereof. Customarily tongued and grooved and end matched, the pieces are laid separately either by nailing or setting in mastic.

Pecan. Pecan is processed in six standard grades, two of which specify all heartwood, and one bright sapwood. Otherwise color variation is not considered. The grades are first grade, first grade red, first grade white, second grade, second grade red, and third grade.

Plain rectangular end-grain wood blocks. Plain rectangular blocks manufactured from carefully selected air-dried southern yellow pine lumber meeting the standard grading rules of the Southern Pine Association specifications from medium-grain lumber. Blocks shall be a minimum of 2 in. in depth and approximately $2\frac{1}{2}$ to 4 in. width, and vary from $3\frac{1}{2}$ to $6\frac{1}{2}$ in. in length. Season checks are not considered defects. Blocks are impregnated under pressure with creosote oil or pentachlorophenol for protection against decay, moisture, vermin, and termites.

Plain-sawed clear oak (unfinished). The face is practically clear, admitting an average of $\frac{3}{8}$ in. of bright sap. Color is not considered.

Plain-sawed select oak (unfinished). The face may contain sap, small streaks, pinworm holes, burls, slight imperfections in working, and small tight knots that do not average more than one to every 3 ft.

Planks. Planks are usually random-width pieces, tongued and grooved, with square edges and ends. Frequently the edges of planks are beveled to reproduce the effect of large cracks, which characterize early hand-hewn plank floors.

Plywood underlayment. Plywood underlayment is a special grade produced for this purpose from group 1 woods (for indentation resistance). It is produced in $\frac{1}{4}$-, $\frac{3}{8}$-, $\frac{1}{2}$-, $\frac{5}{8}$-, and $\frac{3}{4}$-in. thicknesses, and the face ply is C plugged grade (no voids) with a special C or better veneer underlying the face ply to prevent penetration from such concentrated loads as high heels.

Preservatives. Pressure treated waterborne preservatives for softwood lumber, timber, and plywood above ground use, as described in the requirements of the American Wood Preservers Institute publication A WPB LP-2.

Prime-grade oak (finished). The face is selected for appearance after finishing, but sapwood and the natural variations of color are permitted.

Quarter-sawed clear oak (unfinished). The face is practically clear, admitting an average of $\frac{3}{8}$ in. of bright sap. Color is not considered.

Quarter-sawed select oak (unfinished). The face may contain sap, small streaks, pinworm holes, burls, slight imperfections in working, and small tight knots that do not average more than one to every 3 ft.

Second-grade beech, birch, and hard maple (unfinished). These woods will admit of tight, sound knots, and slight imperfections in dressing, but must lay without waste.

Second-grade pecan (unfinished). This wood has tight, sound knots or their equivalent, pinworm holes, streaks, light stains and slight imperfections in working.

Second-grade red pecan (unfinished). Same as second grade except that the face must be all heartwood.

Shorts—oak (unfinished). Pieces 9 to 18 in. long are to be bundled together and designated as $1\frac{1}{4}$ ft. shorts. Pieces grading No. 1 Common, Select, and Clear, are to be bundled together and designated No. 1 common and better with pieces grading No. 2 common bundled separately and designated as such.

Sleeper

1. Timber laid on a slab to support flooring materials.
2. Usually, a 2×2 in. wood member fastened to the concrete slab, that serves to support and fasten the subfloor or flooring.

Special-grade beech, birch, and hard maple (unfinished). Second and better grade are combinations of first and second grade developing in the strip without cross-cutting for each grade. The lowest grade pieces admissible may not be less than standard second grade.

Sta-Loc floor system. Combination flooring using a method of anchoring a hard maple flooring to a concrete slab by means of steel channels, locking each piece of floor in place with a heavy-duty type of flooring steel clip. Product of Horner Flooring Co., Dollar Bay, Mich.

Standard-grade oak (prefinished). This grade may contain sound wood characteristics, may be even and smooth after filling and finishing, and will lay a sound floor without cutting.

Strip and wood block flooring. The requirements for strip and wood block flooring are high resistance to wear, attractive figure or color, and minimum warp and shrinkage. Material should be used at a moisture content near the level it will average in service, such as maple, red and white oak, beech, and birch.

Strip flooring

1. This type of flooring is usually laid over boards nominally 1-in. thick because the boards must be think enough to hold the nail. For best results, boards for subfloors are laid diagonally and in nominal widths no greater than 6 or 8 in. Plywood $\frac{5}{8}$ to $\frac{3}{4}$-in. thick is also satisfactory. $\frac{1}{2}$-in. plywood is satisfactory for the subfloor when strip flooring is nailed to floor joists.

2. Wood flooring consisting of narrow, matched strips.

Strip oak flooring over concrete slab. Flooring system using no subfloor, consisting of a double layer of 1×2 in. wood sleepers nailed together, with a moisture barrier of 4 mil polyethylene film between them. The bottom layer of the wood sleepers is secured to the slab by mastic and by concrete nails. The strip hardwood flooring is then nailed to the sleepers with one nail at each point of bearing. Before the slab is poured it is recommended to install a 4 mil polethylene film over the ground.

Subfloor. Particle-board or plywood laid on joists or sleepers over which a finished wood floor is to be laid.

Tavern and better beech and pecan (prefinished). A combination of prime, standard, and tavern grade containing the full product of the board except that no pieces are to be lower than tavern grade.

Tavern-grade oak (prefinished). This grade must be able to make and lay a serviceable floor without cutting but purposely contains typical wood characteristics that are to be properly filled, such as flags, heavy streaks and checks, worm holes, knots, and minor imperfections in working.

Teakwood. Teak from Thailand can be finished in a natural golden tone or stained-sealed to any degree of darkness to meet specifications. It is available in many patterns, prefinished and unfinished. Panel sizes are $\frac{5}{16} \times 18 \times 18$ in.

Teak T&G plank (manufactured by Bangkok Industries, Inc.). Flooring $\frac{25}{32}$ in. thick in lengths 3 ft. and up. It is tongue and grooved and end matched. Face widths are $1\frac{3}{4}$, $3\frac{1}{2}$, $5\frac{1}{2}$, and $7\frac{1}{2}$ in. Flooring must be genuine *Tectona Gradis* (botonical name) of first quality with no sapwood or knots or beeholes in its face. The floor should be finished in accordance with the manufacturer's specifications.

Third-grade beech, birch, and hard maple (unfinished). This grade must be of such character as will lay and give a good serviceable floor.

Third-grade pecan (unfinished). This grade must be of such character as will give and lay a good serviceable floor.

Tongue and groove. Boards or planks machined in such a manner that there is a groove on one edge and a corresponding tongue on the other.

Vapor retarder. 4 or 6-mil carbonized polyethelene film. Installed over a firm base to receive the concrete slab.

Wood block flooring. Wood block flooring requires an even and uniform base; therefore for best results, a plywood subfloor is frequently used. A $\frac{5}{8}$- or $\frac{3}{4}$-in. thickness should be used if block flooring is installed by nailing. Laminated block flooring $\frac{1}{2}$-in. thick or less may be used over a $\frac{1}{4}$- or $\frac{3}{8}$-in. plywood or particle-board underlayment that has been nailed to a wood subfloor.

Wood floors. The hardwoods most commonly used for flooring are maple, beech, oak, and pecan. Softwoods are yellow pine, Douglas fir, and western hemlock. Hardwoods are available as strips and unit wood blocks; softwoods are available as strips and as end-grain blocks for industrial establishments. Strip floorings over wood subfloors placed diagonally on wood joists are widely used to serve as both structural and surface floors. For satisfactory performance, wood flooring require a protective coating, such as penetrating sealer, varnish, or shellac, in conjunction with wax.

CHALKBOARDS AND TACKBOARDS

Aluminum trim. Aluminum trim includes aluminum chalktrays, trim, map and display rails, end caps, and fasteners. Exposed aluminum trim No. 6063-T5 has an anodized satin aluminum finish and comes in extruded shapes.

Baked-on finish to hardboard. Chalkboard surface baked on temper-treated hardboard. This finish withstands rigorous use, misuse, and normal neglect and is dentproof, colorfast, and easy to maintain in perfect condition by washing with any good soap, detergent, or household cleaner. Hardboard $\frac{1}{4}$ in., single ply, or $\frac{1}{2}$ in. two ply.

Cement asbestos board-composition chalkboard. Steam-cured, $\frac{1}{4}$-in. cement asbestos board. A writing surface finish is applied to the board.

Coil steel chalkboards. Light gauge, pretreated, cold-rolled steel coil stock prepared for the porcelain finish. Coil chalkboards are available up to 300 ft in length.

Combination clothes rack and space divider. This clothes rack, a hat and boot rack unit, is ideal for schools, churches, clubs, etc. One side of the unit includes a chalkboard; the other side includes a tackboard plus a coat rack fitted with sliding coathooks. The panel is permanently attached to a heavy-gauge extruded aluminum frame.

Core stock for porcelain enamel steel chalkboards. Core stock includes $\frac{1}{4}$-in. gypsum board, $\frac{3}{8}$-in. fiber honeycomb, and $\frac{3}{8}$-in. and $\frac{3}{8}$-in. plywood, $\frac{1}{4}$ and $\frac{7}{16}$-in. hardwood, $\frac{3}{8}$- and $\frac{1}{4}$-in. composition wood, $\frac{3}{8}$-in. particle board.

Electroplated steel face chalkboard. Chalkboard writing surface applied to 26-gauge electroplated steel. The surface is baked on at high temperatures for a harder, smoother, and tougher nonglaring surface. Face sheets are mounted on various types of core materials and three types of backing.

Extruded aluminum display board. Display installed in wood or metal trim frame directly on the wall, made of heavy-gauge extruded aluminum with $\frac{1}{4}$-in. thick cork insert. The rail is approximately 1 in. wide. Accessories include display hooks, roller brackets, flag holders, and map rail.

Extruded aluminum map rail. Extruded aluminum map rail with cork insert, spring clip map hook, roller brackets, flag holders, and end stops.

Fabric-natural cork tackboard. Natural cork product composed of large particles of pure cork compressed into a cellular cork layer and securely fused to heavy fabric producing a tough sheet that is easily mounted on any flat surface.

Felt-tip dry markers. Markers used and recommended for semipermanent drawings, sketches, graphs, charts, etc., in areas where chalk dust might be objectionable. Felt-tip markings are easy to wipe off a specially treated high-gloss, stain-resistant porcelain on steel surface with a clean felt eraser or clean cloth.

Fire resistance. Fire retardant characteristics of materials used for bulletin boards, such as cork, vinyl, fabrics, wood, and plastic frames are usually tested under the requirements of ASTM E 84.

Gloss measurement. Porcelain enamel gloss measurements are made by a 45 degree gloss meter in accordance with the Porcelain Enamel Institute Bulletin T-18 "Gloss test for porcelain enamels" (ASTM C 346).

Heavy-duty factory-built units. These units include a map rail with cork insert, chalk tray under a chalkboard with cast aluminum end closures, steel hanging clips, $\frac{1}{4}$-in. plywood or hardboard backing.

Horizontal sliding chalkboards. Assemblies available in any length and various heights. Standard installations consist of two or four sliding chalkboards and a fixed chalkboard, a cork bulletin board, or a projection screen, or a combination of all three. The trim and chalkboard edging are extruded aluminum in a dull satin finish. For ease of operation chalkboards are top-suspended and glide easily over the extruded aluminum track on molded nylon rollers.

Horizontal sliding chalkboard or tackboard. Chalkboard or tackboard consisting of heavy tubular aluminum casings, reinforced aluminum, frames, reinforced aluminum tracks, ball bearing nylon rollers, and overhead hardware of anodized aluminum.

Lighting system. Light fixture designed to be mounted above the chalkboard, which has a parabolic-reflecting finished dome mounted inside the fixture frame. The intent of the fixture is to provide shadowless light beam on the chalkboard without reflection problems. Flourescent tubes are used for the light source.

Masonite-tempered panel-composition chalkboard. Panel manufactured with tempering liquid that is polymerized by baking. The process reduces the rate of moisture absorption, and increased resistance to abrasion. A finished writing surface is applied to panel.

Metal trim factory-built units. Boards are furnished complete and ready to install with 2-in. clip angle hangers 2 ft on center, top and bottom, or 1-ft Z bars 2 ft on center for the top and 2-in. clip angle hangers 2 ft on center for the bottom.

Motor-operated horizontal sliding chalkboards. These units come complete with motors, sprockets, roller chain, tracks, rotary types, limit switches, operating switches, reversing controls, and necessary mounting brackets. Chalkboard is 24 gauge, in standard colors, with a honeycomb core, and backed with 24-gauge galvanized steel or 0.015 aluminum, as required, to provide a rigid panel of sufficient strength to eliminate all vibration and flexing. The tackboard comes in standard colors. Panels are trimmed with 6063-T5 aluminum channels and have an anodized satin finish and corners reinforced with steel angles. The unit is complete with support brackets for the motors, sprockets, and a horizontal top rack for attachment to the structural framework. Overhead rolling hardware consists of two hangers with four nylon wheels and a heavy aluminum track for panels weighing up to 200 lb, panels over 200 lb having a formed steel track. In all cases rolling hardware is designed to ensure smooth, easy, and safe operation of the panels.

Oak trim panels. Panels with oak wood trim perimeter frame with mitered corners. Included are an oak chalk tray under the chalkboard, steel hanging clips, and $\frac{1}{4}$-in. hardboard backing.

Porcelain enamel steel chalkboard. The enamel is applied automatically to a uniform thickness and fired under controlled temperatures to fuse the porcelain permanently to the steel. The steel comes in various gauges and in stretcher-leveled steel sheets manufactured in accordance with the performance specifications for porcelain enamel steel chalkboards established by the Porcelain Enamel Institute. The steel sheet is laminated to the core stock with hot-type neoprene contact adhesive applied to both surfaces automatically. Each substrate must have a minimum of 80% covering with 1.5 to 2.0 dry mils of adhesive. Panel components have uniform pressure applied mechanically over the entire area. Panel backing is available in aluminum foil, sheet aluminum, and steel.

Porcelain layer gauge. Magnetic porcelain layer thickness gauge, O'Hommel Model M-894 or its equivalent.

Porcelain-on-steel chalkboards. A porcelain writing surface consists of three uniform coats: a nickel deposition coat of 2 g/sq ft, a cobalt primer coat of 0.003-in. minimum thickness, and a chalkboard surface coat of 0.003-in. minimum thickness. The side opposite the writing surface consists of two uniform coats: a nickel deposition coat of 2 g/sq ft and a cobalt primer coat of 0.003-in. minimum thickness. If a steel chalkboard sheet is to be laminated to other material, the surface texture of reverse side coat shall be No. 50 grit coarseness. The base metal should be a special-purpose "enameling iron or steel" of low metaloid

and copper content, especially manufactured and processed for the high temperature (over 1500°F) used in coating porcelain on steel units for architectural purposes, and of 18 gauge minimum, uncoated, or 22 gauge, uncoated. The reflectance factor must be not more than 20% or less than 15% and guaranteed not to vary as a result of wear and use.

Porcelain-on-steel laminated chalkboard. A base metal of 28- or 24-gauge steel, suitable for performing as described by the Porcelain Enamel Institute Standard PEI: S-104, is used. Ceramic coatings on chalkboard are manufactured at the lowest possible temperatures. Lower temperatures put less stress on the steel and thereby achieve the highest degree of surface flatness in the finished product. The base metal receives a nickel dip coating of sufficient thickness to assure complete bond between the steel and the subsequent vitreous enamel coatings. Porcelain enamel completely covers all exposed surfaces of the finished chalkboard. The steel sheet is laminated to hardboard, gypsum board, or other core material. Moistureproof, fungusproof contact adhesives are used. Direct heat drives off excess solvents. Lamination is under controlled pressure and time to produce the ultimate in bond strength.

Porcelain-steel-masonite chalkboard. Chalkboard manufactured of 24-gauge enameling steel with the frit fired and fused at high temperature to make the writing surface. The steel chalkboard is laminated under extreme pressure to $\frac{1}{4}$-in. masonite and backed with aluminum foil.

Porcelain-steel-plywood chalkboard. Porcelain-enameled steel is manufactured by firing and fusing the frit at high temperature on 18-gauge special enameling steel. The writing surface is manufactured in strict accordance with the Porcelain Enamel Institute's "Standards for Architectural Porcelain." An 18-gauge porcelain steel chalkboard is laminated under extreme pressure to $\frac{1}{4}$-in. exterior-grade plywood with a backing of 0.015-gauge aluminum sheet.

Portable chalkboards or tackboards. Chalkboard that comes in a choice of standard colors on both sides with a tackboard on one side and chalkboard on the other. The entire frame, including struts, is of anodized satin-finish aluminum tubing or available in wood. Swivel ball-bearing rubber-wheeled locking casters permit ease of movement.

Revolving chalkboard and/or tackboard. Panel framed with aluminum channel and perfectly balanced and pivoted on a steel pin in a brass bushing. The panel locking device is beneath chalk tray; simply pull catch to rotate. Standards are of sturdy $2 \times 2 \times \frac{1}{8}$ in. aluminum tubing.

Snap-on aluminum trim. Trim designed for permanent installation on chalkboards and bulletin boards, with a satin anodized finish. The grounds, trim, chalk troughs, and map rails are completely coordinated and matched. All parts join together in a neat precision fit.

Swinging panel chalkboards. Units hung on plated steel brackets to prevent sagging, consisting of reinforced anodized aluminum frames. Units of two to six panels can be hung on frame. Each panel may be positioned 180°; panels have corkboard or chalkboard on one or both sides.

Swing leaf boards. Panels may be turned like pages of a book; when not in use, the entire unit swings 180° to either side and lies flat against the wall. Standard units consist of four or six panels. Panels are faced on both sides with chalkboard, tackboard, corkboard, or pegboard.

Swing leaf units. Each unit is furnished with panels trimmed with heavy, satin-finished, anodized aluminum channels. The corners of panels are reinforced with concealed steel angles and secured with machine screws. Brackets and hardware are finished to match frames. All panels swing 180° in either direction or fold flat against the wall to either side.

Vertical sliding chalkboard. Chalkboard consisting of heavy-gauge steel or aluminum casings, reinforced aluminum frames and tracks, ball-bearing sheaves, roller-bearing chain, lead weights (counterbalanced), and vertical tracks with nylon rollers.

Vertical sliding units. Vertical sliding units may be single chalkboard or bulletin board panels or double or triple panels, which operate one in front of the other. They can be wall-mounted or extended to the floor for greater panel movement. Panels are counterbalanced by lead weights moving over roller-bearing sheaves at each end of the panel. Units glide freely and smoothly in the channel guides. The maximum travel of the panel is approximately equal to the panel's overall height. The operating mechanism's housing and all trim are of heavy-gauge extruded aluminum in etched and anodized satin finish. The housing of the single-panel unit is $2\frac{5}{8}$ in. deep, and the face is $2\frac{5}{8}$ in. wide. Any number of single-panel units may be installed in front of each other to obtain any desired number of sliding panels.

Vinyl-cork tackboards. Fabric-covered small-particle cork underlay, available in many colors and some fabrics and patterns. Assembly conforms to Federal Specification CCC-W-408, Type 2, Class 1 and 2, UL rating flame spread 25. Fuel contributed 10, smoke developed 5.

Vitreous porcelain enamel chalkboards. The facing sheet is 0.0003 in. thick, and consists of a vitreous enamel writing and erasing coat on a 0.0025-in. cobalt ground coat, with a nickel deposition over 2g/sq ft on the front and back of sheet. Eleven types of cores are available to be backed up with aluminum backer sheet when required by the manufacturer.

Wood fiberboard-composition chalkboard. Board treated with tempering oil polymerized by baking to increase hardness and reduce absorption and with the finished writing surface applied to board.

Wood trim factory-built units. Units substantially built and designed to fit into any area. The chalkboard may be any one of various colors with applied writing surfaces, and various cores with vitreous enamel steel chalkboard. The perimeter trim and chalk trough are oak and walnut finish. Other finishes are available.

COMPUTER ROOM AND ACCESS FLOOR

Access flooring. Flooring should conform to NFPA Standard 75, "Computer/Data Processing Equipment", Chapter 2.

Access floor supporting structure. The supporting system consists of individual pedestals, which are attached or carry the ends of four panels. The pedestal consists of a steel base plate to which a threaded steel stud is welded, a locking nut that supports a hollow metal tube or a special hollow tubing that is welded to the plate, and the threaded steel rod is supported by the locking device; either is acceptable. The threaded rod and its locking device allow height adjustment of the panel floor.

Air-conditioning design. The following design criteria for air supply are recommended to be considered for heat gain from equipment: assumed possible additional equipment, heat generated from the under-floor plenum, maximum personnel, lighting system, and infiltration from adjacent areas (including the assumed times the entrance to the room is opened, unless a double entry door enclosure is installed). All suggestions from equipment manufacturers should be followed in order for the equipment to function properly.

Air supply. Computer room heat gains are highly concentrated, due to the location and type of equipment. The air distribution system installed should be flexible to accommodate possible relocation of equipment or equipment to be added in the future. Air-conditioning units using the under-floor plenum should provide adequate conditioned air for the room above the plenum.

Alarm systems. The computer room should be monitored by alarm systems, such as temperature and humidity, smoke and fire, electric power shutdown, communication failure, and equipment failure.

ASTM. American Society for Testing and Materials.

Carpet. Patterns, weights, and colors are optional and should have static control requirements that have less than 1.8 KV when tested according to AATCC-134 at 20% relative humidity; should be installed with a permanent adhesive, having mildew resistance.

CISCA. Ceiling and Interior Systems Construction Association.

Clamped stringer system. System is used for computer rooms and provides high lateral stability, complete access to the space below, and electrical continuity for grounding and static control. The system's contact between panel edge and stringer provides a plenum seal.

Conductive plastic laminate. Conforming to NEMA LD 3, high-wear type, thickness of 0.062 in. (minimum), color selection optional, and finished on the panel with a noncombustible vinyl edging. Surface to ground resistivity should meet requirements of NFPA 99, Chapter 3, with suggested modifications. Surface resistivity to meet requirements of ASTM D 257.

Conductive vinyl tile. Conforming to FS SS-T-312, Type III. Surface resistivity in accordance with NFPA 99, Chapter 3. Provide one single piece for each panel, colors optinal, edge panels provided with a noncombustible vinyl edging.

Dampers. Dampers used with air supply grilles from the under-floor plenum should have automatically controlled fire dampers.

Debris and dust. All openings in the access floor should be installed with protection to prevent debris and dust from entering the under-floor plenum.

Electrical outlets. All materials used to be UL listed and labeled. Provisions for under-floor metal boxes to be capable of providing space for duplex receptacles, data wiring, and telephone wiring.

Electrical resistance. Floor panels and floor coverings when tested should be in accordance with NFPA 99, Chapter 3, and specified modifications.

Electrical service. The electric service to the computer room should be separate from the building service, with its own distribution and circuit breakers.

Exhaust system. Exhaust system should be independent of building exhaust systems.

Finished floor surfaces. Floor surfaces must be conductive, grounded to avoid accumulation of static electricity, dust free, and nonmagnetic.

Fire detection system. Under-floor plenum should be protected by a fire detection system with an alarm system, in case of fire from overheated wiring systems.

Fire-resistant construction. Computer rooms should be separated from all other occupancies within a building by fire-resistant rated walls, floors, and ceilings, with a minimum resistance of not less than 1 hour, or to meet local codes.

Heat transmission. The temperature of the top surface should not increase more than 150°F above the ambient temperature when the bottom surface is exposed to 1,600°F for 15 minutes.

Impact load. 120 ft/lb impact applied to 1 sq in. area at any location on the panel. Deflection should be limited to 0.020 in.

NEMA. National Electrical Manufacturers' Association.

NFPA. National Fire Protection Association.

Panel fill materials. Fill materials for the panels can be lightweight concrete or high-density particleboard core.

Panel floor finishes. Panels may be covered with carpet, plastic laminate, conductive plastic laminate, vinyl composition tile, or conductive vinyl tile. Colors, textures and patterns are optional.

Plastic laminate. Conforming to NEMA LD 3, high-wear type, having a minimum thickness of 0.062 in.; colors, patterns, and material weights are optional. Panels to be edged with a noncombustible vinyl edging.

Panel lock type. System used in general construction and designed without a stringer connection at the edge. Bolted at the corner and at midpanel, it provides added rigidity and flexibility over stringless systems.

Pedestals. Either standard or corner-lock type, constructed of galvanized steel. Includes pedestal base, 16 sq in. to be set in adhesive, threaded steel rod, having a minimum of $\frac{5}{8}$ in. diameter, adjusting mechanism with a locking device that will provide a 2-in. vertical adjustment for 6-in.-high floors, and over 2 in. for 7 in. and higher. Supporting head either will lock panels in place or panels will be bolted to the pedestal head with flush fasteners.

Ramps. Due to the access floor being raised above the normal floor level, a ramp must be installed. The ramp should be adequately constructed to support the same loads as the floor and be surfaced with nonslip material.

Rigid grid system. Used in computer rooms and areas of heavy loading. Provides maximum rigidity for seismic or dynamic loading, electrical continuity for grounding or static control, and plenum seal. Maximum loading capacity is 400 lbs per sq ft and a supporting concentrated load of 1250 lbs.

Snap-on-grid system. Used in general construction and computer rooms where access is required to the space below. Provides improved lateral stability.

Standard floor panels. Nominal size 24 in. × 24 in. fabricated from 24-gauge galvanized steel, as a pan encloses the bottom, sides, and laps the top $\frac{1}{4}$ in., filled with a lightweight concrete fill, having a minimum compressive strength of 3,000 psi and is finished smooth to receive a

floor covering. For a corner-lock system, panels should accept a flush fit captured metal fastener that securely fastens the panel corner to the pedestal head.

Stringer construction. Fabricated from 16-gauge (minimum) galvanized steel, bolted to the pedestal head, supports all four edges of the floor panels.

Temperature and humidity control. In order for the equipment to function at its best, it is very important to provide precision temperature and humidity controls.

Uninterruptible power supply. Large computer rooms with large numbers of computers and accessory equipment should have power systems that closely control power supply voltage and frequency modulations. UPS installations can be served by the local utility company or by standby generator sources.

Under-floor plenum. Plenums should not exceed 10,000 sq ft in area and must be divided by noncombustible bulkheads to prevent spread of smoke or fire.

Vibrationproof. Computer room should be located in a building that can provide vibrationproof conditions. If possible, the room should be constructed as a separate entity, not connected to the building. The floor slab should be separated from the structural floor slab by a sand bed.

Vinyl composition tile. Conforming to FS SS-T-312, Type IV, composition 1, minimum thickness $\frac{1}{8}$ in., colors optional, and panel tile edged with a noncombustible vinyl edging.

Water drainage. Drainage of under-floor area should be provided in order to reduce damage by water from any source.

IDENTIFYING DEVICES

Acrylic injection molded letters. These letters come with plain or adhesive back. Adhesive-back letters are easily applied; simply press into position. They are excellent for "do-it-yourself" jobs.

Acrylic plastic cubes. Cubes offer a larger visual mass to gain attention, are excellent for counter top identification, and are easily moved or changed. Several messages may be incorporated into one cube, which can be turned to reveal the desired information. Countertop identification cubes have fully mitered corners.

Acrylics. Acrylics are clear or colored for contrast, and lettering may be enameled after engraving. Clear lettering may be used in edge-lighted signs. Acrylics are not as scratch resistant as the ES plastic.

Add-on (registered trademark). Add-on directories feature an acrylic insert holder that is untarnishable and fadeproof. The material is $\frac{1}{4}$-in. thick, having machine-grooved slots for accepting ES plastic inserts. Inserts are easily removed for engraving or rearranging. Acrylic holders are available in black and pearl white.

Adhesives. Adhesives are available on glass, glazed or ceramic tile, porcelain and vinyl surfaces; special adhesives are available for installations on all other surfaces, such as painted wood or metal, brick, concrete block. General-purpose cement is a rubber-based paste-type adhesive. Silicone adhesive is available for permanent installations only. Silicone installations cannot be removed without damaging both the parts and the surfaces. Epoxy adhesive is available but not recommended.

AIGA. American Institute of Graphic Arts.

Aluminum

1. Architectural aluminum sheets of 0.064-, 0.090-, and 0.125-gauge thickness, heliarc-welded, are required to give a structurally strong sign that will survive the most adverse weather conditions.

2. Aluminum is available in regular metallic color or in anodized gold (or black) finish, with contrasting letter colors.

Aluminum alloy

1. Number 356 aluminum alloy has been recommended to be a sound base material for the natural satin lacquered finishes, as well as acrylic baked enamel finishes.

2. Number 214 aluminum alloy is recommended for natural anodized finishes.

Aluminum cast letters. Letters cast of certified No. 214 alloy aluminum using ingots only (gates and risers prohibited), of sufficient thickness to ensure rigid structures, and poured at temperatures not exceeding those recommended for this alloy. The casting surface is free of all imperfections and only sound and advanced foundry practices employed. Letter faces are ground smooth and buffed to a mirrorlike finish or polished to a fine grain satin finish; edges are filed and ground smooth and all traces of sand texture removed. After thorough degreasing, alumilite finish No. 704 is applied.

Aluminum heliarc-welded letters. Welded aluminum letters fabricated from architectural sheet aluminum of 0.090 gauge thickness. Letters are welded by the latest "Welded Aluminum" methods, mechanically sanded and etched and degreased to receive their specific finish. Letters are anodized or enamel finished as required.

Aluminum letters

1. Letters cast from aluminum ingot No. 214 if alumilited, duranodic, or anodized and from aluminum ingot B.443.0 if baked enamel.

2. Letters with satin polish faces and matte finish sides, cast of F-214 alloy. Alumilite finish is available.

Aluminum welded. Aluminum letters fabricated and welded from alloy 5005-H-14 in 0.090 gauge thickness.

Anodic Type H. Letters requiring duranodic or kalcolors are cast in No. 214 alloy and mechanically sanded and etched to a fine texture finish. Although anodic finishes build up an extremely hard oxide surface and are highly resistant to weathering and abrasion, variation in color will exist in all aluminum material in sheets, extrusions, or cast owing to the uncontrollable characteristics.

Baked enamel finish. Exterior satin smooth nonglare finish available in many colors, the usual available standard colors being red, ivory, light gray, light or dark blue, green, fire red, yellow, orange, black, brown, and white. Other colors are available.

Baked enamel finish on aluminum

1. Letters are cast from prime F-214 alloy aluminum ingots. Sides are filed smooth and sandblasted, and faces are belt polished to a smooth surface. Letters are degreased, acid etched, and then given a two-coat primer application, after which two heavy coats of high-grade porcelainlike acrylic baking enamel are applied and baked at 350°F.

2. Finish consisting of three coats of baked enamel on both faces and sides. Standard colors are available from all manufacturers.

Baked enamel letters. The letters consist of aluminum alloy, surfaced smooth before priming. The letters are sprayed with two coats of baking enamel with each coat baked separately.

Baked enamel on metal. Metal is immersed in hot alkaline cleaner to remove contamination and oxidation metal is prime coated to improve paint adhesion and inhibit corrosion, then enameled with conventional enamels by spray application, and baked in a box-type oven for required time.

Baked-on enamel finish. Letters are welded mechanically, sanded, and etched to receive baked-on enamel finish. Paint is bonderized to aluminum at 350°F.

Baked-on enamel letters

1. Letters are cast, mechanically sanded, and etched to receive baked-on enamel finish. Paint is bonderized to aluminum at 350°F. Many colors are available, including matching colors to anodized, duranodic, and Kalcolors.

2. Letters are welded, mechanically sanded, and etched to receive baked-on enamel finish. Paint is bonderized to aluminum at 350°F.

3. Letters are sprayed with heavy coats of aluminum primer before receiving the finish coats of enamel, then conveyed to an oven at approximately 325°F to bind the primer and enamel permanently to the metal. Colors are available from the manufacturer.

Base stanchions. Signs supported by stanchions consisting of 30-lbs, 18-in.-diameter cast-iron base with 4-ft-long, $1\frac{1}{4}$-in. pipe, painted black, and drilled with standard holes for 18- and 24-in. signs.

Basic metal treatment. Letters are mechanically sanded, etched, and degreased prior to receiving finish.

BM-6-mounting. Free-standing letter is mounted on a bar or channel base, which is bolted to the structure.

BM-7-mounting. Free-standing letter is mounted on aluminum bars or channel.

BMB-6-mounting. Letters are mounted on two parallel-back bars or channels for attachment to masonry or other surface.

BMT-1-mounting. Letters are mounted on two parallel back bars or channels for attachment to masonry or other surface.

BPM-1-mounting. Projected mounting using U brackets in aluminum or stainless steel.

Bracket plaques. Plaques that incorporate a slide-in aluminum bracket for wall and ceiling units, which slide into an aluminum retainer channel that attaches to the mounting surface. There are no exposed fastening devices.

Bright satin aluminum finish. Letters are cast from prime F-214 alloy aluminum ingots. The sides are filed smooth and sand blasted, and faces are belt polished to a fine satin finish. Letters are then degreased and etched before receiving two coats of high-grade clear lacquer protective coating.

Bright satin anodize. The faces of the letters are given a satin finish and anodized by specification. Two coats of clear retardant applications are applied to anodized letters.

Bright satin anodized letters. Letters cast in virgin No. 214 aluminum, mechanically sanded, filed, etched, and given a fine satin texture to receive a patented anodic oxide coating that retards corrosion. The letter finish is preserved by two coats of colorless chemical for additional protection. Finishes are available in color.

Bronze letter finish. Letters with satin polish faces and matte finish sides. (Oxidized sides are furnished when specified.)

Bronze or brass letters. Polished or satin finish. With or without enamel coating for engraved portion of the letter.

Bronze soldered. Bronze letters are fabricated from 20-gauge Muntz metal and finished in a satin and lacquer finish or oxidized statuary finish.

Building dedication tablets. Tablets of cast bronze with deep statuary finish, flat-face Roman letters, and flat-band border or of modern aluminum with Gothic letter styles, and straight edge or deep bevel border, and background finished in charcoal gray. Building plaques have satin finished letters and borders and are furnished with concealed mounting devices.

Caducei symbols. Symbols of the medical-health profession are cast in aluminum or bronze, coated with clear lacquer or baked enamel, and anodized or gold leaf finished. Special sizes and styles are fabricated to specifications.

Cast aluminum and bronze plaques. The background surface is of pebble texture. Plaques are etched and degreased to receive baked-on enamel finish for background.

Cast aluminum letters

1. Letters are mechanically finished to satin texture, etched, and degreased to receive Type A baked-on enamel finish. The mounting method is PMC-1 projected $\frac{1}{2}$ in. from the wall.

2. Letters made of solid aluminum, cast of certified Aluminum Association's alloy designation B443.0 (formerly alloy 43) conforming to Federal Specifications QQ-A-596d. Where anodic coatings are specified, the alloy shall conform to Aluminum Association's alloy designation 514.0 or A514.0.

3. Letters cast of No. 214 aluminum ingots recommended for anodizing and sound castings. This material is best suited for anodized finishes, as well as baked enamel finishes.

Cast aluminum tablet. Tablet is sand cast of corrosion-resistant aluminum; no junk remelt or scrap aluminum is used. Castings are free of pits and blemishes, and all letters are sharp and of clear character. Letters are raised the proper height above the background to give good three-dimensional appearance. The borders and faces of raised letters are of satin finish. The background is of leatherette design and gray aluminum finish. A colored background is available of weatherproof vinsynite with vinyl color and finish. At least four coats of weatherproof vinyl permanent clear coatings are pressure sprayed on the complete plaque to form a weathertight seal against atmospheric contamination.

Cast bronze and aluminum letters and plaques. Letters and plaques cast from virgin bronze ingots or aluminum alloy F-214. The lettering is raised or recessed, or both. They are bronze, oxidized, spray painted with lacquer, or aluminum, anodized or spray painted in many colors and shades. The fastening of letters and plaques is by means of screws and rosettes or concealed fasteners.

Cast bronze and aluminum tablet. Cast of bronze virgin ingots 85-5-5-5/aluminum alloy F-214. The casting must be free of pits and gas holes. All letters are hand-chased and satin finished. The background is flat satin texture. The plaque is colored and coated with two coats of clear lacquer to prevent tarnishing.

Cast bronze letters

1. Letters cast of certified 85-5-5-5 alloy bronze using ingots only (gates and risers prohibited), of sufficient thickness to ensure rigid structure, and poured at temperatures not exceeding those recommended for this alloy. The casting surface must be free of all imperfections and only sound and advanced foundry practices employed. The letter faces are ground smooth and polished to a satin finish. Edges are filed and ground smooth, and all traces of sand texture are removed. After thorough degreasing, two coats of high-grade clear lacquer are applied.

2. Solid bronze letters, cast of certified copper-tin-zinc 85-5-5-5 alloy using only virgin ingots (no scrap permitted). The metal must conform to industry standards.

Cast bronze letters and plaques. Letters and plaques cast from grade architectural bronze ingots (85% copper, 5% lead, 5% tin, 5% zinc) and finished as satin or oxidized bronze (85-5-5-5).

Cast letters of aluminum/bronze. Cast letters available in many styles and sizes. Aluminum is cast from virgin 214 alloy and bronze from architectural bronze alloy (85-5-5-5).

Cast metal letters. Letters must be free of all porosity and with sharp corners, flat faces, and accurate profiles. Remove burrs and rough spots. Belt-polish faces to a uniform high-luster finish. Sides are filed smooth with all tool marks removed by fine abrasive grain air-blasting. Each letter is cleaned ultrasonically in a special degreasing bath using high-frequency sound waves.

Cast tablets. Cast tablets are used for dedication plaques, commemorative tablets, building identification, religious and fraternal insignia, or service and merit awards. Cast metal tablets of bronze and aluminum will stand most weather conditions.

Ceiling flush mounting. Mounting useful for information and directional interior signage, mounted above traffic.

Ceiling suspended mounting. Where high ceilings exist, suspended mountings are advisable to keep the information and directional-type interior signage in eyesight limits.

Changeable letter boards. Sculptured $1\frac{1}{2}$ inches extruded aluminum frame, natural satin or optional, simulated bronze, or chrome or stainless steel anodized finishes. Indoor use only.

Changeable letter directory board. Board that features removable letter panels, precision grooved every $\frac{1}{4}$ in. to accept all changeable letters to 3 in. in height, and a choice of rich black or brightly colored felt or nonfading Koroseal (vinyl) backgrounds. Doors are mounted on full-length piano hinges and equipped with heavy-duty locks and keys.

Changeable wall signs. Signs that feature aluminum or acrylic holders that accept ES $\frac{1}{16}$-in. plastic inserts, machine engraved. Signs are ideal where frequent change of information is required.

Channel-shaped aluminum mounting. Extruded aluminum standard clear anodized, gold, and bronze mounting with two flat-head screws that accepts $\frac{1}{8}$-in. thick sign.

Church bulletin boards. Outdoor boards illuminated or nonilluminated. Cases are thoroughly weatherproofed with watertight corners and reinforced corner plates and come in satin silver, bronze, duronodic, or black finish. Illuminated cases are $5\frac{1}{2}$ in. deep with lighting fixtures mounted inside the top portion of the case. Message panel backgrounds come in a wide variety of background colors. Plastic changeable letters are available in various styles, sizes, and colors. The church nameplate is mounted on plexiglas with raised red, black, or blue plexiglas letters. Mounting posts and hardware are available.

Clear anodized aluminum finish. Aluminum is belt polished and then given a clear anodic finish.

CM3-mounting. Concealed method of fastening whereby threaded rods on the back of the tablet are inserted into cement-filled holes.

CMB-5 mounting. Concealed mounting on brick, stone or concrete. A threaded rod in the tapped boss on the reverse side of the plaque is inserted into a drilled hole filled with quick-setting cement.

Colored aluminum finish letters. Satin polished faces with etched sides or background. The entire letter is color anodized in a selected color. Two coats of high-grade clear lacquer are applied for protection.

Colored anodic finishes

1. Letter finish is the same as Satin Alumilite, except that the finish is Kalcolor, Duranodic, or Permanodic in light, medium, or dark bronze. Although anodic finishes are very durable, a variation in color may exist, especially in casting and weldments, owing to uncontrollable characteristics. Kalcolor, Duranodic and Permanodic are proprietary names.

2. Colored anodic finishes, such as gold, Duranodic, Kalcolor, and Permanodic are available as special finishes on cast and fabricated aluminum letters.

3. Special finishes on cast letters are subject to considerable color variation. Color variation between letters (as well as within the same letter) is possible. Baked enamel colors simulating earth tones are strongly suggested in lieu of anodic finishes.

Combination metal letters. Illuminated letters with channel backs and plexiglass faces. Letters are fabricated of stainless steel (or aluminum) in a gauge recommended by the manufacturer and conform accurately to the plastic letter faces. The plastic faces are engaged to the metal letter channel backs with alumilite aluminum retainer rings. The retainer rings are fitted with neoprene gasketing to prevent entry of the elements. Neon tubing, transformers, and erection should be as recommended by the manufacturer.

Concealed fastening. Studs on back set in cement.

Console directory. The console directory comes with table top mounting; it may be nonilluminated, back-lighted or illuminated and of either the removable or fixed type. It comes with or without a heading and has a $\frac{1}{4}$-in. face and an aluminum or bronze frame.

Cross symbol (religious). Fabricated crosses for houses of worship, religious schools, memorials, parish houses, convents, or other religious areas where display of the cross is required. Crosses are fabricated in stainless steel, aluminum, bronze, or wood (in natural, baked enamel, painted, varnished, or gold leaf finish). There are special sizes and designs to meet specifications.

Custom directory. Directory with a background of walnut veneered plywood with walnut (wood tape) edges, stained and to have flat wood lacquer. Directory heading plate and name strips are selected with spacing and in size in accordance with the furnished layout. Mounting is by mitered wood hanger clip with a screw lock at the bottom.

Cut-out aluminum plate. Cut-out aluminum letters are available in many thicknesses.

Dark oxidized bronze finish. Bright polished faces, contrasting dark statuary etched sides or background.

Desk and counter nameplates. Nameplates are machine engraved, enabling freedom of design and styles. The aluminum bar nameplate offers a distinctive appearance with natural aluminum lettering on a black or bronze anodized aluminum bar. Holders are of solid aluminum in standard frosted anodized gold, silver, and bronze.

Description of finishes. Letters that are not anodized, plated, or enameled are spray-coated with a weather-resistant clear material that will not discolor. The edges of all letters are carefully belt polished. The faces of the letters are furnished in finishes as desired.

Die-embossed signs. Signs are sharply drawn to $\frac{100}{1000}$ to $\frac{125}{1000}$ of an inch, making the entire sign free from wind, twist, or buckle, increasing readability, and simplifying refinishing when necessary. Signs are fabricated from heavy-gauge steel, zinc chromated, and protected with three coats of baked enamel, $\frac{5}{16}$-in. diameter holes are punched at top and bottom ready for mounting on U-channel posts. Embossed signs are reflectorized with Ultra-Brite glass beads covering the entire face of the sign. Ultra-Brite is a proprietary name.

Directional arrow sign. Sign used with name sign for a complete directional system. The module may be oriented for left, right, up, and down.

Directories. Directories are designed to help people locate the area of their destination within a space zone. Directories are highly functional by themselves or as part of a complete sign system.

Directory. A directory utilizes a brief building layout and conveys to its users various rooms from a particular entrance. A "You Are Here" marker on the layout (usually in red) identifies that certain location. This type of directory is usually in multilevel entrance situations.

Duranodic finish. Aluminum finish, a trademark of the Aluminum Company of America, includes light bronze No. 311E, medium bronze No. 312E, dark bronze No. 313E, and black No. 335E.

Durolite (phenolic). Durolite signs are washable and nontarnishable; no paints or lacquers are used. They are ideal for double-faced signs. Their thickness is standard $\frac{1}{8}$ in., but they also come in $\frac{1}{16}$- and $\frac{3}{16}$-in. thicknesses. The standard polished finish is also satin. They have rigid high-impact density. Engraving is done through the surface layer showing the exposed white or black core. Available woodgrains and colors include black, white, blue, green, red, yellow, brown, gray, charcoal, tan, walnut, mahogany, blond, maple, and aqua. Durolite is a proprietary name.

Engraved letters. The sign is made in a plastic laminated base. Letters are machine engraved and evenly cut into a contrasting core.

Engraved signs. These signs are recommended where the signs are up to 18 × 24 in. (indoors). Each is individually designed and cut into durolite ES plastic, the acrylic, brass, aluminum, or stainless steel. Durolite and plexiglas are available in a wide choice of colors. Signs are ideally suited for color coding to help direct the flow of traffic in a more efficient manner. Durolite and Plexiglas are proprietary names.

Epoxy enamel letters and tablets in color. Hard-fired epoxy enameled tablets and letters are baked with brilliant colors and require no maintenance or polishing, being as hard as granite and smooth as glass. Lettering on the plaque can be either raised or incised and is bronze or aluminum finished to provide the desired contrast.

ES. Engraver's stock.

E. S. plastic. A thermoset plastic, rated to be self-extinguishing, meets U.S. specification No. L-P-387A type NDP.

Exit indicators. Opaque images on acrylic plaques, 0.125-in. thick. Where applicable, the unit incorporates a slide-in aluminum bracket for wall and ceiling units, which slides into an aluminum retainer channel that attaches to the mounting surface. There are no exposed fastening devices.

Extruded aluminum directories. Aluminum directories are manufactured from extruded stock designs. All structural members are heavy gauge and have a satin anodized finish. Special metal tones, such as duranotic finish, can be made on request. Duranotic is a proprietary name.

Extruded aluminum mounting. T-shaped extruded aluminum available in several lengths for wall or ceiling application. This mounting takes any $\frac{1}{8}$-in.-thick sign. It is installed with two flat-head screws, with the extrusion tapped into place. The sign is press fit with clear vinyl extrusion furnished and is available in clear anodized gold and bronze finishes. The design is patented.

Fabricated aluminum letters

1. Letters fabricated of 0.090-gauge select alloy aluminum. Faces and edges are joined by the heliarc process. The welds are carefully ground and a fine satinlike finish is applied to the faces. The faces and edges are free of distortion due to welding and free of discoloration after anodizing.

2. Letters fabricated from 0.090-gauge aluminum sheet, 6061 alloy. All joints are heliarc welded in conformance with the American Welding Society's and the Aluminum Association's specifications. Where necessary, letters are braced internally so that they may be free from waves, buckles, or warps.

Fabricated backlighted letters. Letters fabricated from 0.090 gauge sheet aluminum, heliarc welded, mechanically finished to a satin texture, are etched and degreased to receive Type A baked enamel. Letters are backlighted by two tubes, 13-mm white neon, and powered by normal power factor transformers installed in the letters.

Fabricated illuminated letters. Letters fabricated from 0.090-gauge sheet aluminum, heliarc welded, mechanically finished to satin texture, and etched and degreased to receive Type A baked-on enamel finish. Letters have flat plastic faces and are illuminated by two 13-mm neon tubes and powered by normal power factor transformers, installed in the letters. The letters are UL approved.

Face fastening. Fasteners attached through the face of the tablet with wood screws, toggle bolts, or expanding screw anchors, depending on wall material. Rosettes are placed over the fastener.

Fiber glass. Three-ply laminated $\frac{1}{4}$-in.-thick fiber glass. Fiber glass is both break and weather resistant (flexural strength is 38,000 psi, its tensile strength 28,000 psi).

Fiber-reinforced polyester sheets. Sheets made from acrylic-modified polyester resin reinforced with high solubility strand fiberglass mat.

FMM-1 mounting. Flush mounting to masonry. The threaded stud of aluminum or stainless steel is set with nonstaining cement.

FM mounting. Flush mounting with threaded studs.

FMC-2 mounting. Flush mounted to masonry wall, casts integral spurs drilled and tapped for threaded rods.

FMT-1 mounting. Flush mounting to masonry. The threaded stud of aluminum or stainless steel is set with nonstaining cement.

FMT-2 mounting. Flush mounting to wood. The pointed end of the threaded stud of aluminum or stainless steel is forced into undersized holes with a small amount of nonstaining cement.

FMT-3 mounting. Flush mounting to wood or metal panel. The threaded stud is of aluminum or stainless steel with a nut and washer.

FMW-2 mounting. Flush mounting to wood. The pointed end of the threaded stud of aluminum or stainless steel, with a small amount of nonstaining cement on the thread, is forced into undersized holes.

FSC-3 mounting. Free standing, grinding, and tapping for free-standing letters on inverted channel base.

FST-1 mounting. Free-standing letters mounted on an aluminum bar or channel.

General letter finishes. Letters are mechanically sanded, etched, and degreased before receiving specified finish.

Glass-fiber-reinforced plastic sign. The sign fabricated from material with a high strength-to-weight ratio, designed for interior and exterior use, and possessing exceptional resistance to wear, cleaning agents, scratching, and shattering. A sign medium in which the graphics are integral with the fabricated materials and become a subsurface strata. There are 12 type styles, furnished with a matte finish surface as standard. Special colors, type styles, graphics, and additional sizes and configurations are available.

Glazed indoor-outdoor changeable letterboard. Letterboard with distinctive "reveal effect." It comes with a $1\frac{1}{4}$-in. extruded aluminum frame in natural satin or optional bronze anodized finishes and is for indoor or outdoor use.

Glazed outdoor changeable letterboard. Letterboard with contemporary flat $1\frac{1}{4}$-in. extruded aluminum frame in natural satin or optional bronze anodized finishes. This letterboard is for outdoor use.

Graphic blast. (Registered trademark of Best Manufacturing Company, Kansas City, Mo.) Exclusive engraving process developed by Best, which utilizes a unique photomechanical technique to reproduce any graphic image on virtually any durable material with absolute fidelity. This process produces a permanent engraved image of the highest quality, while offering total design freedom, outstanding durability, and ease of maintenance.

Ground-with-skirt mounting. Sign supports are concealed by an aluminum skirt.

Ground-with-supports mounting. Ground mounting of sign with exposed supports. The size, number, and material of the posts depend on sign weight, length, etc.

Heat-stamped signs. Modern in appearance and produced in a wide variety of decorator colors, heat-stamped signs have acquired wide acceptance for office and commercial use. The letter colors are fused to satin-matte-finished acrylic plastic and will not fade or damage under ordinary maintenance procedures. These signs can be used outdoors without fading or being damaged by the elements but should not be used where they will be subject to surface abrasion.

Honor rolls, donor plaques. Plaques showing a full complement of names. They can be cast in one piece to any size. Honor rolls or past president plaques with solid cast backgrounds are available in any size or shape and allow for plates to be added at a later date by drilling and tapping the desired number of holes in the provided blank area.

High-temperature baked enamel finish. Letters are cast from prime S-214 alloy aluminum ingots. The sides are filed smooth and sandblasted, and faces are belt polished to a smooth surface. Letters are degreased, acid etched, and then given a two-coat primer application, after which two heavy coats of high-grade porcelainlike acrylic baking enamel are applied and baked at 350°F in a chest-type oven.

Horizontal and vertical dividers (directories). White horizontal plastic strip and vertical nylon cord charts. The board is manufactured exactly as desired for directory columns and organizational charts, which can be easily inserted and rearranged.

Illuminated indicator. Compact electrical fixture consisting of a permanently mounted exterior housing and a removable and interchangeable interior cartridge unit. Graphics in these indicators are accomplished utilizing photographically precise film negatives. Standard exit units have 6-in.-high graphics, complying with OSHA requirements. Ceiling and projecting wall mount units are available with either single- or double-sided graphics. Units are also available for surface wall mount use.

Illuminated outdoor letterboard. Letterboard with satin anodized extruded aluminum framing, S3000 brushed stainless steel frame and changeable letter panel covered with nonfading washable vinyl. The panel is removable from its case. Sturdy black or white letters align automatically. A glass door is mounted on a rugged full-length hinge and has heavy-duty tumbler-type lock and key. The case has a metal back, completely weatherproofed. The white translucent header plate is at the top with raised black Old English letters. Fluorescent lighting is concealed.

Illuminated sectional directory—backlighted. This directory is available for surface, freestanding, or in-the-wall mounting. It comes in single, double, or triple tier, with or without a heading, and with aluminum, bronze, or stainless steel frames. Has a fixed or removable illumination feature.

Imprinted letters. Signs made from letters made with heat and pressure on colored foil directly to a solid plastic base.

Imprinted signs with plexiglas (registered trademark). Acrylic sheet utilizes acrylic plastic with either a satin matte or polished finish. The color selection in the acrylic plastic is large and varied. Plates can be imprinted with black, white, red, gold, or silver. The background material is $\frac{1}{8}$-in.-thick solid plastic with the imprinted image fused to the surface to a depth of about $\frac{3}{1000}$ by pressure and heat. Signs may be mounted with 3M $\frac{1}{32}$-in. Scotch-foam tape. $\frac{1}{8}$ in. thick pads are sometimes used to give the sign some relief from the wall.

Individual metal letter installation. Letters up to 3 in. high are furnished with integral cast spurs for easy driving into wood or other soft backgrounds. Small letters are furnished with sanded backs for gluing into position with clear epoxy. All letters over 3 in. are high drilled, tapped and furnished with 2-in. threaded rods ready for mounting.

Individual relief letters. Letters used on doors, walls, paneling, etc., adhered by epoxy or by pieces of scotch

mount so they can be removed at a later date. Letters average from $\frac{1}{16}$- to $\frac{1}{8}$-in. thickness. Letters can be sprayed with any standard krylon color.

K2-alloy. Special formulated K2-alloy for bronze-toned anodized finishes.

Laminated plastic ES (engravers stock). Versatile permanent sign material widely adopted for directional signs, number and nameplates, control panels, plaques, diagrams, directories, and a myriad of commercial, institutional, and industrial uses (flexible material is also available).

Lawn sign. Aluminum message panel with cast aluminum letters and vertical steel tube supports, anchored in concrete foundation.

Letter accessories (directories). White plastic figures and symbols are available to punctuate the message and hold the readers' attention.

Letter storage box (directories). Durable fiberboard storage box that accommodates 1000 letters and numbers to 1 in. and smaller quantities of larger sizes or 1500 letters and numbers to 1 in. and smaller quantities of larger sizes.

Light-gold bronze finish. Bright polished faces with contrasting light etched sides or background.

Medium oxidized bronze finish. Bright polished faces with contrasting medium statuary etched sides or background.

Memorial plaque. Engraved plate in aluminum, brass, or bronze attached to a wood panel and framed with $\frac{1}{4}$-in. extruded aluminum angle. It is flexible as to size, and any number of plates can be affixed to the background.

Metal plaques. Product presents a means of accomplishing photographically precise graphics in metal, which are an integral part of the plaque. Graphics are etched and are available in relief or sunken-form lettering styles. Plaques can also be furnished with pictographs, corporate symbols, and special alphabets. Plaques are standard in a polished or oxidized bronze finish with other metals and finishes available.

Metals. Anodized aluminum claims the same freedom from maintenance enjoyed by thermoset plastics. Brass, bronze, and copper-base alloys depend on clear lacquer for tarnish resistance. Stainless steel is both durable and tough. All of the metal groups depend on separate coloring of letters, with enamels or by oxidizing methods, to produce high-visibility signs. A striking effect may also be obtained with the natural Graphic Blast (registered trademark) finish in metals.

Metal surface-mounted directory. Directory that consists of a bronze laminated glass cover with aluminum returns that hold the edge of the glass. The unit has a concealed hinge and is designed for surface mounting. Standard aluminum finishes are natural anodized, medium-bronze duranodic, and black duranodic.

MFH mounting. Mounting frames provided with mechanical fastening holes.

MM-1 mounting (flush). A noncorrosive pin is set in expanding cement.

MM-2 mounting (keyhole). Fabricated letter held firm by noncorrosive double nut on noncorrosive pin set in expanding cement or a lead anchor.

MM-3 mounting (extended sleeve). Aluminum or stainless steel spacer and noncorrosive pin set in expanding cement.

MM-5 mounting (expansion). The threaded end of the sleeve is screwed into the rear of the letter. A noncorrosive pin is screwed into a lead anchor that is set into the wall. A stainless steel set screw bites into the pin for positive and theftproof fastening. This also allows for removal to clean and service.

MM-6 mounting (base). A noncorrosive machine screw is screwed through the mounting bar or channel into drilled and tapped holes in the bottom of each letter.

MM-7 mounting (back bar). A noncorrosive machine screw is screwed through the mounting bar to drilled and tapped holes in the back of each letter. The mounting bar is then secured to the wall with noncorrosive screws into lead anchors.

Monolithic fabricated letters. Letters consisting of metal faces with fabricated metal returns. The letters are unique, as the method of fabrication presents the visual appearance of a solid monolithic form. Standard metals and finishes include polished chrome, satin brass, polished oxidized brass, polished brass, polished bronze, satin bronze, polished copper, and cast steel with a linear polyurethane coating in many standard colors. Monolithic fabricated letters are suitable for exterior and interior use.

MT mounting. Mounting by the use of magnetic tape.

MTF-4 mounting. Molly or toggle bolt fastening for mounting into hollow masonry, plaster, or wood walls, flat-head screw finished off with a rosette.

Natural aluminum and baked enamel finish. Bright, polished, natural aluminum faced with a three-coat baked enamel finish on the sides and two coats of high-grade clear enamel for protection.

Natural aluminum finish letters. Bright, polished, natural aluminum faces with etched sides or background and two coats of high-grade clear lacquer for protection.

Natural aluminum satin finish letters. Aluminum receives two coats of clear retardant application.

Natural satin finish letters. Cast aluminum letters are mechanically sanded, filed, and chemically etched to a satin texture. Letters are finished with two coats of colorless chemicals for protection against oxidation.

Nonilluminated projected sign. Sign consisting of an aluminum panel with cast or fabricated letters mounted either flush or projected from the face.

Nonilluminated screen printed signs. Sign consisting of an aluminum panel for silk-screened copy or pressure-sensitive vinyl letters. The panel is particularly suited for parking, directional, and information signage, both interior and exterior.

Nonilluminated sectional directory. Directory available for surface, freestanding, or in-the-wall mounting in single, double or triple tiers with or without heading, and with an aluminum, bronze, or stainless steel frame. It can be used with name strips or is available with a one-piece grooved background for use with individual changeable letters or plastic name bases.

Number sign. Sign that may be used individually for room numbering and combined with a name sign.

Opaque signs. Sign consisting of opaque graphics with opaque background. It is single or double sided, with integral molded side returns and encapsulated core material up to $2\frac{1}{2}$-in. thick and comes in sizes to 48×96 in. Standard core materials are plywood and high-density polyurethane foam.

Oxidized bronze. Letters finished same as "Satin Bronze." Letters oxidized to dark statuary color and hand finished before given the two coats of colorless chemical to preserve finish and retard oxidation.

Oxidized bronze finish

1. Light-, medium-, or deep-brown statuary finish, conforming to trade industry specifications.
2. Letters with the same finish as satin bronze, except they are chemically oxidized to a dark statuary color, hand-rubbed, and protected with two coats of high-grade clear lacquer.

Oxidized bronze letters. Letters that have been immersed in vats to receive the medium or dark oxidized finish, as specified, and two coats of the best grade clear retardant application.

Patient directories and directory systems. Accurate at a glance patient directory available in a variety of sizes with matte-finish plastic plates easily marked. Free-standing and wall-mounted units are available.

Permabrass signs (a proprietary name). Distinguished-looking signs preferred for banking, insurance, and professional buildings. They are also available as individual tablets, memorials, etc., and come with satin brush-finished surface baked with synthetic lacquer. The lettering is precision engraved into solid brass and filled with black lacquer. Signs never need polishing or refinishing.

Permalon (a proprietary name). Chemical milling process by which it is possible to reproduce "line copy" in metal. Any part of the metal may be chemically removed by etching, thus leaving the line copy in bold relief. The copy area or artwork is finished in rich lustrous gold, and the background color is baked-on permanent black crackle finish.

Permalum signs (a proprietary name). Aluminum signs suited for all interiors. These signs have a clear anodized satin brush finish. All lettering is cut into solid aluminum and filled with black lacquer unless otherwise specified. Signs need no polishing or refinishing.

Pictorial directories. Directories indicating to people the zone in which their destination may be found. Such directories have become a vital part of a well-designed sign system, with color-coded floor plans and simple graphic reproductions and maps showing buildings, grounds, streets, and landmarks.

Plain (no glass face) directory. Directory with a background of grooved corkboard or chalkboard, or a combination of the two, available in single- and double-face styles. It comes with a fixed or removable background and may be mounted on standards, wall hung, or recessed.

Plastic framed plaque. Plaque material is 0.125-in. solid acrylic plastic with a satin matte finish and integral permanent color. The type image is fused to the plaque material under heat and pressure to a 0.003-in. depth. The frame is fabricated of solid acrylic plastic with permanent integral matching plaque colors. Corners have a $\frac{1}{16}$-in. outer radius, and frame is separated from the insert plaque by a $\frac{1}{16}$-in. reveal. The insert plaque is applied to the $\frac{1}{4}$-in.-thick black acrylic plastic reveal area with foam mounting tape.

Plexiglas-lucite (acrylic). Letters are cut into solid stock and filled with any color or shade lacquer. Standard colors are black, white, and clear, and other colors are available. This material is used for electric signs and edgelite effects. Specific color combinations can be achieved with clear Lucite by spraying background any desired shade and then cutting letters through the coloring. Plexiglas is a trademark of the Robin and Haas Company.

PMB-4 mounting. Projected mounting to masonry using stainless steel U bracket, cap screw, and expansion bolt, standard length.

PM mounting. Projected mounting with sleeve covering stud and spacing letter from face of surface.

PMC-1 mounting. Projected mounting to masonry wall, letters supplied with aluminum separators and threaded rods.

PMM-1 mounting. Bronze dowels are set in cement for concealed fastening to brick or stone without rosettes.

PMM-2 mounting. Angle clips used with wood screws and toggle or expansion bolts for concealed fastenings.

PMM-3 mounting. Toggle bolts are used for hollow tile or cement block walls with rosette covers.

PMM-4 mounting. Expansion bolts are used for brick, stone, or cement with rosette covers.

PMS-3 mounting. Projected mounting to masonry with aluminum or stainless steel studs set with nonstaining mastic. Aluminum sleeve spacers in standard lengths.

Polished aluminum finish

1. Finish that is buffed to a high gloss before lacquering.
2. Finish that is buffed to high gloss, conforming to trade industry specifications.

Polished bronze finish. Finish that is buffed to high gloss, conforming to trade industry specifications.

Porcelain enamel. Porcelain coatings are high-grade organic glass frits, fused to the aluminum at temperatures over 800°F. They are available in manufacturers' standard colors.

Porcelain enamel on steel. Surfaces are finished in hard-fired vitreous enamel. All standard "Porcelain Enamel Institute" colors are available by most manufacturers.

Portable changeable letterboards. These letterboards come with bright chrome or satin anodized aluminum frames and matching pedestals. A glass front provides complete protection. The letter panel removes from the rear and is covered in black felt or an optional-color felt or vinyl.

Precision cut-out sign. Vandal-resistant aluminum panel with precision cut-out copy backed up with plastic and illuminated by concealed lighting.

Precision profiled letters. Each letter is individually profiled out of 0.187 gauge sheet aluminum, bronze, or plastic and made in accordance with designed patterns.

Pressure-sensitive letters. Letters adaptable to such diverse uses as simple door identification and giant signing requirements. They are made of Plytex, which is a 0.004-in.-thick film that can easily stand up to several years of outdoor exposure. Its lifetime can be lengthened significantly by the application of a protective coat of polyurethane varnish. The film will resist grease, oil, mild acids and alkalies, salt, and most common solvents. (Plytex is a proprietary name.) All letters and numerals are die-cut and can be applied to any nonporous surface.

Pressure-sensitive vinyl letters. Applied letters that consist of a vinyl film 3.5 mils in thickness. Letters are suitable for interior and exterior use and are standard in black or white, with special colors available.

PRM-1 mounting. Projected mounting to masonry or wood. The threaded stud of aluminum or stainless steel is set with nonstaining cement. An aluminum sleeve spacer is furnished in standard lengths of $\frac{1}{4}$ through 3 in.

PRM-2 mounting. Projected mounting to wood or metal panel. The threaded stud of aluminum or stainless steel comes with nut and washer. Aluminum sleeve spacer is furnished in standard lengths.

PRM-3 mounting. Projected mounting to masonry using stainless steel U bracket, cap screw, and expansion bolt.

Projected cut-out sign. Vandal-resistant aluminum panel with precision cut-out copy. Precision-cut plastic letters are inserted into an aluminum cut-out face and illuminated by concealed lighting.

Projection wall signs. Painted or walnut-oiled-finish wood ceiling brackets that will accept an engraved plastic plate or three-dimensional letters.

Protective coating. Finishes of metal protected with exclusive protective coating developed by the Battelle Institute.

Raised letters. Three-dimensional metal-faced letters, also supplied on either matte-finished acrylic plastic or material. A wide variety of custom letter styles are available.

Red bronze alloy. In the bronze family the 85-5-5-5 red bronze alloy is the basic metal for the red to brown colors.

Reflective outdoor signs. Reflective sheeting applied to treated aluminum or GPX paneling emits tremendous reflective power. It makes any sign visible at great distance as soon as a car light is directed at it. Signs can be made from 12×18 to 48×96 in. in one piece. All signs are custom-made in accordance with specifications.

Reflective signs. Reflective sheeting is applied to treated aluminum or GPX wood paneling. The sheeting is composed of finely powdered glass and silver, which gives it tremendous reflective powers and requires absolutely no repainting or maintenance of any kind. Rain, snow, or ice do not impair its reflective quality. It needs no electricity or "recharging" from sunlight.

Relief letters. Cast styrene characters permanently affixed to a colorful plastic background.

Relief letter signs (with Plexiglas—a registered trademark). Relief letter signs have letters that cannot be removed since they are chemically fused to $\frac{1}{8}$-in. acrylic plastic background material. These signs are childproof, theftproof, and weatherproof. All background colors are nonfading. White-lettered signs are guaranteed weatherproof.

Reverse-engraved. Engraved on the reverse side of acrylic plastic. This process achieves a three-dimensional effect on a smooth surface. Acrylic plastic permits a crystal clear view of the contrasting metallic letters. Reverse-engraved letters are impervious to damage by scratching or abrasion. Reverse-engraved acrylic nameplates are weather resistant; however, darker colors will tend to fade if exposed to excessive direct sunlight.

SAM mounting. Mounting that uses silastic adhesive.

Sandwichlike construction. Plastic signs have a sandwich-like construction with layers of different colors. Engraving through the top layer exposes the contrasting layer. No paint is used, therefore there is nothing to chip or peel. All stock colors and wood grains are available by manufacturers of nameplates and holders.

Satin alumilite finish. Letters are cast from prime F-214 alloy aluminum ingots. Sides are filed and ground smooth and faces polished to a fine satin finish. Letters are degreased, etched, and given a 204R1 alumilite-patented clear anodized finish. A gold anodized finish is available from some manufacturers.

Satin aluminum finish

1. Satin bright faces with etched sides of background and protected with a natural anodized finish over the entire surface.

2. Natural metal color, with subtle horizontal or vertical lines and lacquer coated.

3. Finish with vertical or horizontal grain lines and conforming to trade industry specifications.

Satin bright aluminum finish. Letters are cast from prime S-214 alloy aluminum ingots. Sides are filed smooth and sandblasted, and faces are belt polished to a fine satin finish. Letters are degreased and etched before receiving two coats of high-grade clear-lacquer protective coating.

Satin bronze finish

1. Cast bronze letters are cast, mechanically sanded, filed, and chemically etched to a beautiful satin finish. Letters are given two coats of colorless chemicals to retard oxidation of the bronze finish.

2. Finish with vertical or horizontal grain lines, conforming to industry trade specifications.

3. Letters cast from bronze allot (85-5-5-5). Sides are filed, ground smooth, and sandblasted to a fine satin finish. Faces are belt polished to a fine satin finish. Letters are then given two coats of high-grade clear lacquer to prevent oxidation and preserve the fine finish.

Sawed aluminum plate. Letters or other designs may be cut from selected quality aluminum plate. Faces are satin finished and alumilited; sides are filed and ground smooth. Aluminum plate provides a material ideal for baked enamel, color anodizing, or porcelainlike acrylics.

Screen-printed signs. Plastic panel for silk-screened copy. These signs are ideal for plot plans and directional and other graphic media. The panel is illuminated by concealed lighting. These signs are excellent for exterior as well as interior use.

Sculptural indigenous flora tablet. Tablet that incorporates realistic imprints of the actual flowers, leaves, etc., in the bas-relief sculpture, in addition to the linear pictorial qualities.

Sculptural linear pictorial tablet. Tablet that incorporates all the design elements of bas-relief sculpture and finished by selective tonal oxidation to enhance the three-dimensional linear pictorial qualities.

Sculptured letters. Molded with epoxy resins impregnated with integral permanent color. The integrally colored letters are impervious to weather.

Sculptured plaques. Hand-molded portraits from photographs and cast in bronze from proper sculptured likenesses.

SCW-4 mountings. Screws and rosettes are used for installations on wood.

Seals. Sculptural bas-relief of public, educational and private designs.

Self-adhesive PVC letters. Single letters fabricated from a permanent grade of polyvinyl chloride to which a self-adhesive material has been applied.

Sheet aluminum letters. Corrosion-resistant letters, fabricated from architectural sheet aluminum 0.091 to 0.125 gauge in thickness, depending on height, to minimize any distortion or waves. The seams are heliarc-welded and cleaned, with the sides filed and ground smooth. The letters are degreased to receive specified finish. Letters can be anodized in any standard color.

Signage system. System that provides the directional information that guides and controls the flow of traffic. Communication of this type of information is vital to the operations of any complex.

Silk-screened wall panel. Panel made of fabricated plastic with dark-brown acrylic lacquer finish with screened letters.

Single-border-style memorial plaques. Rectangular signs with number and nameplates featuring flat-band border or straight- and bevel-edge designs, cast from stock patterns with one or many lines of copy.

Stainless steel letters

1. Letters fabricated from 18- to 20-gauge stainless steel with No. 4 polish or satin finish. Letters are of soldered construction, braced where necessary, and free of warpage.

2. Lifetime durability. Letters are made of 18- to 22-gauge 18-8 (Type 302) stainless steel with No. 4 finish. Letters are well braced with faces flat and buckle free.

3. Letters fabricated from 18-gauge, 18-8 alloy, type 302 stainless steel with No. 4 finish. Letters have straight edges and flat, buckle-free faces with all surfaces carefully polished.

4. Made from natural metallic color or with enamel of contrasting color in the engraved portion of the lettering.

Stainless steel soldered letters. Stainless steel letters fabricated from 20-gauge (or heavier, depending on size of letter) 18-8 alloy, type 302, with satin finish.

Standing signs. Signs that are ideal where movable signs are required. Framer and stands are chrome. Sign faces are type "ES plastic" in selected color. All may be engraved on one or both sides. Standing signs are applicable to most sign systems.

Steel embossed signs. Heavy-duty steel signs for parking and traffic control needs. Embossing (raised letters) is a die-stamping process which adds rigidity, visibility, and durability to all outside signs. Steel blanks are embossed, zinc chromated, and triple coated with white baked enamel. Messages are red or black. Holes are punched to fit U-channel rail posts.

Strip directory. Directory with laminated Micarta (a proprietary name) background and adhesive-mounted engraved aluminum strips.

Strip-type modular directories. Illuminated or nonilluminated 9-in. wide extruded aluminum glass-enclosed modules with removable name-strip retainers. Directories can be surface recessed-mounted. Directories may have satin silver, bronze duronodic, or black finishes. Spacing between modules is $\frac{1}{4}$ in. allowing flexibility for later additions.

Subsurface printed sign. Signs for interior and exterior use having a matte finish surface, with the graphics subsurface printed onto either vinyl or acrylic plastic prior to the application of a background color and laminated to a plastic base material.

Supersine. Supersines are signs fabricated from 0.102-in. H-14 aluminum sheet, bright dipped and alumilited prior to fabrication. Borders and characters are formed by a unique punch and die that produces a combination of semishear and extrusion to form evenly raised characters and sharp, well-defined edges. This process is used exclusively by the Supersine Company.

Surface-mounted directory. Directory consisting of a bronze acrylic plastic face with mitered bronze acrylic plastic returns to present a monolithic form. The unit has a concealed hinge and is designed for surface mounting.

Symbol signs. Passenger and pedestrian-oriented symbol signs created for the U.S. Department of Transportation, Office of Facilitation, by contract with the American Institute of Graphic Arts. The document on this information is available at National Technical Information Service, Springfield, Va.

TBI mounting. Toggle bolts for hollow tile and similar conditions.

Thermoset plastic. Melamines, such as "ES plastic," are tough and scratch resistant. Letter and design durability is ensured because there is no separate need for paint filling images when they are machine and/or Graphic Blast (a registered trademark) engraved through the surface to the contrasting core color. This high-pressure laminate is suitable for both interior and exterior use.

Triangular wood-base desk sign. Triangular shaped sign of oiled walnut or black painted sign with plastic or metal nameplate mounted on one side.

Type B mounting. Free-standing channel bar.

Type F mounting. Flush mounting with pins or bolts.

Type P-B mounting. Projected mounting with brackets.

Type P-C mounting. Angle mounting for fabricated letters only.

Type P-S mounting. Projected mounting with sleeves.

Unitized directory. Illuminated or nonilluminated directory with surface or in-wall mounting. No foundation frame is needed. It is used as a single unit or in series, with or without a heading.

Veneered and precision-engraved sign. Engraving is done on multicolored or one-color front. Plastics and metals can be integrated on the face side of the sign, to achieve contrasting or complementary display of the message. Sides and back are protected with veneer over outdoor GPX plywood for moisture protection.

VTM mounting. Mounting of frames by using vinyl tape.

Wall mounting—flush. This mounting method is excellent for lobbies, hallway junctions, and areas beside or above doorways and can be used for informational and directional exterior and interior signage.

Wall mounting—projected. Mounting of this type is normally reserved for short-copy signage, such as "exit," "stairs," or "enter."

Welded aluminum letters. Letters fabricated from architectural sheet aluminum of 0.090 gauge thickness. Letters are continuously welded by "welded aluminum" methods, mechanically sanded, etched, and degreased to receive their specified finish. Letters are anodized or enameled finished.

White bronze or nickel silver finish. This finish closely resembles the stainless steel color. There is great stability and retention of finish in all atmospheric conditions.

WMP-1 mounting. The wood mounting of the plaque is drilled and countersunk for flat-head screw and finished off with rosette coverings.

WS-2 mounting. Wood screws are set into lead anchors or Rawl plugs. Ornamental rosettes cover the heads of the screws.

WSM-2 mounting. Wood or sheet metal mounting with self-tapping or wood screws. Round, oval, or pan-head screws are furnished for small plaques only.

Yellow brass and gold bronze alloys. Available for metal letters.

MANUAL EXTINGUISHING EQUIPMENT

All-purpose extinguisher. Chemical fire extinguisher tested and listed as capable of handling fires in ordinary combustibles, flammable liquids, and electrical equipment.

Antifreeze extinguisher. Extinguisher using calcium chloride solution or a proprietary compound having a very low freezing point.

Carbon dioxide extinguisher. Compressed gas, used primarily on small fires, with an effective range of 3 to 8 ft maximum on Class B and C fires. The extinguisher is rated Class 2B, C. Extinguisher will operate at $-40°F$ and will not conduct electricity.

Cartridge extinguisher. Water base, gas cartridge, used primarily for small fires, must be used close up and directly on the blaze on Class A fires only. The extinguisher is rated Class 2A. Extinguisher will freeze if kept outdoors or in cold areas.

Class A fires. Incipient fires, containing wood, paper, miscellaneous rubbish, and textiles, where quenching with water is of primary importance.

Class B fires. Incipient fires where extinguishing by blanketing or smothering to reduce the oxygen content is of primary importance. Fires of gasoline, oil, grease, and tar are in this class.

Class C fires. Incipient fires that occur in electrical equipment and electrical devices and require a nonconducting extinguishing agent.

Combination cabinets. Metal cabinet with glazed metal framed doors, of size to contain a hose rack for 75 ft of $1\frac{1}{2}$ in. lined hose and a $2\frac{1}{2}$ gal extinguisher.

Dry chemical extinguishers. Portable hand, wheel-mounted, or apparatus-mounted appliances of stored-pressure or cartridge-operated types. In the cartridge-operated type, the dry chemical is expelled by carbon dioxide or nitrogen stored in a separate gas expellant cartridge. In the stored-pressure type, the dry chemical is expelled by air, nitrogen, or other gases stored in a single chamber with the dry chemical. Dry chemical extinguishers may be charged with sodium bicarbonate base dry chemical, multipurpose dry chemical with ammonium phosphate base, or with other base ingredients, such as potassium bicarbonate. The dry chemical may also be treated to make it compatible with mechanical form.

Extinguisher cabinets. Metal cabinet with glazed metal framed door, of size to contain two $2\frac{1}{2}$ gal extinguishers or one $2\frac{1}{2}$ gal extinguisher.

Extinguishers. Portable first-aid fire-fighting devices approved for use on certain types and classes of fires.

Extra hazard occupancy. One unit of extinguishing capacity for use on Class A fires for every 1,000 sq ft of floor area. Verify with existing codes.

Factory test pressure. The pressure at which the shell was tested at time of manufacture. The pressure is indicated on the nameplate.

Fire extinguisher. Portable device, the contents of which are for extinguishing a fire.

Fire protection appliance. Apparatus or equipment provided or installed close at hand for immediate use in the event of fire.

First-aid hose. Hose that is permanently attached to a standpipe outlet and provided primarily for use in fire fighting by the occupants of the building.

First-aid hose station. Hose connection with valve in a system of piping adequately supplied with water, hose, and nozzle for use by building personnel in extinguishing a fire.

Gas cartridge. Extinguisher used for Class A fires only, water base, $2\frac{1}{2}$ gal, Class 2A. Must not be mounted in cold areas or it will freeze.

Gas cartridge or cylinder. Expellent gas is confined in a separate pressure vessel until the operator releases it to pressurize the extinguisher shell.

Halogenated agent extinguisher. Pressurized type for Class B and C fires, with an effective range up to 30 ft, must be recharged after every use, can operate at $-40°F$, and will not conduct electricity. The extinguisher is rated 5B, C.

Hand fire extinguishing equipment. Portable equipment intended for the control of small or incipient fires and designed for manual operation.

Hand propelled. The material is applied with scoop, pail, or bucket.

Horizontal fire line. A fire line installed around the interior walls and columns of a building, pier, or wharf, with hose outlets located so that every part of the floor area is within reach of at least one fire stream.

Hoses. Lined synthetic fiber plastic hose is usually recommended for use on standpipe installations. Cotton rubber-lined hose is standard for fire department and heavy equipment use.

Humpback swing rack. Hose rack for $1\frac{1}{2}$ and $2\frac{1}{2}$ in. lined hose. By using a reducing coupling of $1\frac{1}{2}$ in., hose can be attached to the $2\frac{1}{2}$ in. valve. Valves may be located 5 ft 6 in. above floor, depending on the requirements of local and state codes.

Inspection. A check to see that the extinguisher is fully charged, and operable, that it has not been tampered with or damaged.

Light hazard occupancy. One unit of extinguishing capacity for use on Class A fires for every 3,000 sq ft of floor area, subject to local or state codes.

Loaded stream extinguisher. Water based alkametal salt solution, used primarily for small fires. Used on Class A and B fires, with an effective range of a maximum of 40 ft under pressure. The extinguisher is rated 2A, $\frac{1}{2}$B. Due to the salt content, extinguisher will operate at $-40°F$.

Maintenance. A check to see that the extinguisher will operate safely and effectively.

Manual fire-extinguishing equipment. All hand operated auxiliary fire-extinguishing equipment of an approved type suitable to the occupational use of the building and installed in the corridors or other locations, visible and readily accessible to the occupants of the building in accordance with the requirements of the local and state building and fire departments.

Mechanically pumped. The operator provides expelling energy by means of a pump and the vessel containing the agent is not pressurized.

Mild steel shell. Except for stainless steel and steel used for compressed gas cylinders, all other steel shells are "mild steel" shells.

Multipurpose extinguisher. Chemical fire extinguisher rated and listed as capable of handling fires in ordinary combustibles, flammable liquids, and electrical equipment.

Ordinary hazard occupancy. One unit of extinguishing capacity for use on Class A fires for every 1,500 sq ft of floor area. Verify with local and state codes.

Pressurized extinguisher. Water base, compressed air type, used primarily for quenching small fires, with an effective range of a maximum of 55 ft. Used on Class A fires only. The extinguisher is rated Class 2A. Extinguisher will freeze if kept outdoors or in cold areas.

Pump tank extinguisher. Water base, hand pump type, used primarily for small fires, with an effective range of a maximum of 40 feet. Used on Class A fires only. The extinguisher is rated 2A for $2\frac{1}{2}$ gal capacity and 4A for 5 gal capacity. The extinguisher will freeze if kept outdoors or in cold areas.

Recharging. The replacement or replenishment of the extinguishing agent.

Self-expelling. The agents have sufficient vapor pressure at normal operating temperatures to expel themselves.

Self-generating. Actuation causes gases to be generated that provide expellent energy.

Service pressure. The normal operating pressure as indicated on the gage and nameplate.

Standard fire extinguisher. Portable fire extinguisher that bears the label of approval of a national testing laboratory acceptable to the fire prevention bureau.

Stored pressure. The extinguishing material and expellent gas are kept in a single container.

Swing rack semiautomatic. Fire hose rack and reels, recommended hose size of $1\frac{1}{2}$ in., for use with $1\frac{1}{2}$ in. in diameter building standpipes, and 100 ft of hose. Connection for $2\frac{1}{2}$ in. hose should be available to each station for use of firemen. Verify local and state codes regarding standpipe requirements.

Swing reel rack. Rack and reels are made for $1\frac{1}{2}$ and $2\frac{1}{2}$ in. lined hoses only. Verify use of this type of hose rack with local and state authorities.

Typical portable extinguisher identifications

1. Extinguishers suitable for Class A fires should be identified by a triangle containing the letter "A." If colored, the triangle should be colored GREEN.

2. Extinguishers suitable for Class B fires should be identified by a square containing the letter "B." If colored, the square should be colored RED.

3. Extinguishers suitable for Class C fires should be identified by a circle containing the letter "C." If colored, the circle should be colored BLUE.

4. Extinguishers suitable for fires involving metals should be identified by a five-pointed star containing the letter "D." If colored, the star should be colored YELLOW.

NOTE: Colors recommended for identifications as described are in accordance with Federal Color Standards:

GREEN -No. 14260
RED   -No. 11105
BLUE   -No. 15102
YELLOW-No. 13655

Extinguishers suitable for more than one class of fire may be identified by multiple symbols.

PROTECTIVE COVERINGS

Abrasion resistance. All fabrics to have normal good abrasion resistance.

Acrylic-painted cotton duck. Acrylic coated fabric, 12 oz per sq yd, solids and stripes available in various color combinations, opaque material.

Acrylic woven fabric. Woven fabric made of 100% acrylic solution-dyed fibers with a fluorocarbon finish. Typical weight is 9.25 oz per sq yd. Resistant to ultra-violet light, color degradation, water, and mildew. Solids and stripes in various color combinations.

Aluminum frames. Frames should be welded construction using all aluminum extruded components 6063-5 alloy. Extrusions are mill-finish aluminum; painted frames are available.

Awning shapes available. Frame is fabricated to fit the condition or situation, but typical shapes are available, such as concave, convex, elongated bullnose, halfdome, and typical awning type.

Backlit awnings. Basic fabric is polyester, available in solid colors and white underside. Fabric decorated by silkscreen printing, spraying, pressure sensitive lettering, or heat transfer systems. The fabric has dramatic day and night identification.

Dimensional stability. Most fabrics have been treated to resist stretch.

Durability / average life span. Most awning fabrics have a life span from 5 to 8 years, depending on environment and proper maintenance.

Flame retardant. Fabrics are available as flame retardant, if treated accordingly. Manufacturer advises on fabric quality and flame resistance.

Flexible polyester. The basic fabric is polyester, with a colored, eradicable, flexible, pre-coated membrane, providing maximum brilliance and consistent translucency. Weight is 20 oz per sq yd. PVC is combined with ultraviolet stabilizers, fungicides, and whiteners. Translucent, formulated for backlighting. All colors are flame retardant. California Fire Marshall UL 48, UL 94.

Modacrylic woven fabric. Woven fabric made of 100% modacrylic solution-dyed fibers with a fluorocarbon finish. Typical weight is approximately 9.25 oz per sq yd. Resistant to ultraviolet rays and color degradation. Somewhat translucent, depending on color. All colors are flame retardant.

Nontracking polyester. The basic fabric is polyester with wick-resistant scrim, extrusion-coated with optically consistent polyvinyl chloride resins. Weight is 16.5 oz per sq yd.

Maximum ultraviolet and mildew resistance. Solid colors only available, underside white for light transmission. California Fire Marshall Certification F-102.08. Translucent, formulated for back-lighting.

Perlgard woven fabric. Woven fabric made of 100% acrylic solution-dyed fibers with Perlgard finish. Typical weight is 9.50 oz per sq yd. Resistant to ultraviolet rays and color degradation. Solid colors and patterns available in various color combinations.

Pigmented polyester. Fabric of pigmented vinyl resin-coated polyester plain weave. Weight is 12.5 oz per sq yd. Maximum resistance against wear and fading from wind, water, mildew, and sunlight. Glossy textured appearance. Opaque and all colors are flame retardant.

Polyester base fabric. Tri-layer fabric with vinyl top and bottom layers, middle layer of polyester scrim. Typical weight is 15 oz per sq yd. Resistant to ultraviolet light, mildew, and water (ASTM E 84). Solids and stripes available in many color combinations.

Stitch type fastening. Patented fastening where fabric is stapled into a channel in a special frame, in lieu of normal stitch sewing.

Surface finishes. Finishes such as matte, smooth, nonglare, woven texture, and glossy vary with coating materials and the fabric.

Underside colors. Colors vary from pearl gray, tan, same as top color, or optional patterns. Backlit fabrics have white undersides for light transmission.

Vinyl-coated cotton duck. Vinyl coated fabric 15 oz per sq yd, solids and stripes available in various color combinations, opaque material (ASTM E 84).

Warrant. Most manufacturers offer five-year limited warranty. Warranty should delineate color fastness, mildew, frame stability, and normal allowable wear and tear.

Wick-resistant polyester. The basic fabric is with a wick-resistant polyester weft insertion scrim, extrusion coated with PVC resin. Weight is 15 oz per sq yd. Fabric is mildew inhibiting, heat and dielectrically sealable, soft and pliable. California Fire Marshall Certification F-102.08. Fabric is translucent, formulated for backlighting.

SIGNS AND OUTDOOR DISPLAY STRUCTURES

Accessory sign. Computed and based on the area of the principal front of the building only and may be subdivided into not exceeding four units and not exceeding the permissible signs area. Accessory signs as permitted, should be used only to advertise the principal use of the building.

Advertising device. A device other than a recognized or standard type of sign that is placed or affixed to advertise or to attract attention or to promote publicity for an individual, firm, organization, product or event, and includes devices of a decorative nature.

Advertising or billboard signs. Advertising sign or billboard is a sign that directs attention to a business, commodity service, or entertainment conducted, sold, or offered elsewhere than upon the same zoning lot.

Advertising sign. A sign that directs attention to a profession, business, commodity, service, or entertainment other than one conducted, sold, or offered upon the premises where such sign is located, or on the building to which the sign if affixed.

Advertising structure. Any structure erected for advertising purposes, with or without any advertisement display thereon, situated upon or attached to real property, upon which any poster, bill, printing, painting, device, or other advertisement of any kind whatsoever may be placed, posted, painted, tacked, nailed, or otherwise fastened, affixed, or displayed.

Animated sign. Any sign having a conspicuous and intermittent variation in the illumination or physical position of any part of the sign, provided, however, that a slow rotation of the sign not be considered animation.

Approved combustible plastic. Plastic material more than one-twentieth ($\frac{1}{20}$) in. in thickness which burns at a rate of not more than two and one-half ($2\frac{1}{2}$) in. per minute when subjected to standard tests for flammability of plastics in sheets of six-hundredths (0.06) in. thickness.

Area of a sign. Consists of one or more letters, symbols, or other parts, affixed to or mounted upon a building or affixed to any other approved mounting, which sign does not have a border or frame, is all of the area of the surface to which the sign is attached lying within the extremities of the sign.

Artisans' signs. Signs of mechanics, painters, and other artisans may be erected and maintained during the period such persons are performing work on the premises on which such signs are erected.

Awning, canopy, roller curtain, or umbrella sign. Any sign painted, stamped, perforated, or stitched on the surface area of an awning, canopy, roller curtain or umbrella.

Awning sign. An awning is any structure made of cloth or metal with a metal frame attached to a building and projecting over a public area, having sufficient surfaces for sign work, which can be raised to a position flat against the wall of the building when not in use, but having the sign work visible to the public.

Banjo sign. Sign having a total area of not more than fifty (50) sq ft, the advertising content of which is not closer than ten (10) ft to the surface of the ground unless otherwise required by code.

Banner. Any sign having the characters, letters, illustrations or ornamentations applied to cloth, paper, balloons, or fabric of any kind with only such material for a foundation.

Banners, balloons, posters, etc.. Includes signs that contain or consist of banners, balloons, posters, pennants, ribbons, streamers, spinners, or other similarly moving devices.

Billboard. Includes all structures, regardless of the material used in the construction of the same, that are erected, maintained or used for public display of posters, painted signs, wall signs, whether the structure be placed on the wall or painted on the wall itself, pictures or other pictorial reading matter that advertise a business or attraction that is not carried on or manufactured in or upon the premises upon which said signs or billboards are located.

Billboard (poster panel). Board, panel, or tablet used for the display of printed or painted advertising matter.

Billboards. Any framework for signs advertising merchandise, services, or entertainment, sold, produced, manufactured or furnished at a place other than the location of such structure.

Building master identification signs. Building master identification signs are signs which identify the name of a multiple-tenant commercial building.

Building or wall sign. Sign, other than a roof sign, that is supported by a building or wall.

Business identification sign. A sign that is affixed to or placed or mounted upon a commercial establishment or property solely to name and identify the business conducted on that property. The sign must be affixed to the establishment or be placed upon the same property as the establishment. A business identification sign may display the registered or commonly used name or the registered trademark, or both, of a commercial establishment. No other message, symbol or device shall appear upon such sign.

Business sign. A sign that directs attention to a profession or business conducted, or to a commodity, service, or entertainment sold or offered upon the premises where such sign is located, or on the building to which such sign is affixed.

Change-panel signs. A sign designed to permit an immediate change of copy, which may be other than the name of the business.

Closed sign. Display sign in which the entire area is solid or rightly enclosed or covered.

Combination sign. Any sign incorporating any combination of the features of pole, projecting, and roof signs.

Combustible ground sign. Any sign constructed in whole or in part of combustible material erected or maintained upon the ground and not attached to any building.

Commercial sign. Any sign belonging to or controlled by the owner or occupant of a building or premises which is used to identify the building or premises or the products or services sold therein or thereon.

Construction sign. Sign identifying the contractor, developer, architect, and financing agency that is erected temporarily on the lot in which the premises under construction is located.

Detached sign. Any sign not attached to or painted on a building, but which is affixed to the ground.

Direction sign. Any sign permanently or temporarily erected to denote the route to any city, town, village, historic place, shrine or hospital; signs directing and regulating traffic; notices of any railroad, bridge, ferry or other transportation or transmission company necessary for the direction or safety of the public; signs, notices or symbols for the information of aviators as to locations, directions and landings; and conditions effecting safety in aviation; signs, notices or symbols as to the time and place of civic meetings, and signs or notices erected or maintained upon private property giving the name of the owner, lessee, or occupant of the premises or the street number.

Directional sign. Sign for the purpose of traffic control, located on private property.

Directory sign. Sign containing the name of a building, complex, or center.

Directory signs. Directory signs are used to guide pedestrians to individual businesses within a multiple-tenant commercial building.

Display sign. Structure that is arranged, intended, designed, or used as an advertisement, announcement, or direction, and includes a sign, sign screen, billboard, and advertising devices of every kind.

Display surface. The surface made available by the structure, either for the direct mounting of letters and decoration or for the mounting of facing material intended to carry the entire advertising message.

Electric sign. Fixed or portable, self-contained electrically-illuminated appliance with words or symbols designed to convey information or attract attention.

Entry-way sign. A freestanding sign used to identify the entrance to a project or facility.

Erect. To build, construct, attach, hang, place, suspend, or affix and also includes the painting of wall signs.

Facia sign. Single-faced building or wall sign that is parallel to its supporting wall.

Facing. The surface of the sign upon, against, or through which the message of the sign is exhibited.

Facing surface. The face or surface of the sign upon, against, or through which the message is displayed or illustrated on the sign.

Fence signs. Signs painted on the surface of enclosure or division fences or on picket or other ornamental fences.

Field advertising sign. A sign that advertises a business not conducted, or a product or service not available, upon the property on which the sign is located.

Fin sign. A sign that is supported wholly by a one-story building of an open-air business or by poles placed in the ground or partly by such pole or poles and partly by a building or structure.

Fixed projecting sign. Any sign projecting at an angle from the outside wall or walls of any building and rigidly affixed thereto.

Flashing sign. An illuminated sign on which the artificial light is not maintained constant or stationary in intensity or color at all times when such sign is in use. A revolving sign or any advertising device that attracts attention by moving parts operated by mechanical equipment or movement caused by natural sources, whether or not illuminated with artificial lighting, is considered a flashing sign.

Flat or wall sign. Any sign erected parallel to the face or outside wall of any building and supported throughout its length by the wall of the building.

Flat sign. Any sign attached to and erected parallel to the face of, or erected or painted on the outside wall of a building and supported throughout its length by such wall or building, or any sign in any way applied flat against a wall.

For sale or lease sign. Sign on a lot denoting that same lot or the premises thereon are for sale, lease, or rent.

Freestanding or attached signs. The area includes all lettering, wording, and accompanying design and symbols, together with the background, whether open or enclosed on which they are displayed.

Freestanding sign. A permanent or portable sign that is neither attached to a wall of a building nor within a building, including a sign that is installed upon the roof of a building.

Freestanding wall sign. A sign consisting of individual letters on a wall that is integrated architecturally with the building.

Government building signs. Signs erected on a municipal, state or federal building that announce the name, nature of the occupancy, and information as to use of or admission to the premises.

Grand opening sign. A sign advertising the introduction, promotion, announcement of a new business, store, shopping center, office or the announcement, introduction, promotion of an established business changing ownership or management.

Gross surface area of sign. Entire area within a single continuous perimeter enclosing the extreme limits of such and in no case passing through or between any adjacent elements of same. However, such perimeter does not include any structural or framing elements lying outside the limits of the sign and not forming an integral part of the display.

Ground sign. Freestanding sign supported by one or more uprights, braces, or pylons located in or upon the ground or to something requiring location on the ground including "billboards" or "poster panels" so called.

Height of a sign. The height of a sign with a border or frame is the vertical distance from the ground on which it stands to the highest extremity of the sign.

Holiday sign. Sign in the nature of a decoration, clearly incidental and customary and commonly associated with any national, local, or bona fide religious holidays, provided that there be on the sign no names of firms or products.

Home occupation signs. Signs advertising home occupations, bearing the name and occupation of the practitioner.

Horizontal projecting sign. Any projecting sign that is greater in width than in height.

Identification sign. A sign, design, or symbol that clearly defines or illustrates the name and/or primary nature of the business establishment.

Illuminated sign. Sign designed to give forth artificial light or through transparent or translucent material from a source of light within such sign, including but not limited to neon and exposed lamp signs.

Incombustible sign material. Incombustible material is any material that will not ignite at or below a temperature of 1,200° F during an exposure of five minutes, and that will not continue to burn or glow at that temperature.

Indirect lighting. A source of external illumination located a distance away from the sign, which lights the sign, but which is itself not visible to persons viewing the sign from any normal position of view.

Indirectly-illuminated sign. Illuminated, non-flashing sign whose illumination is derived from an external artificial source so arranged that no direct rays of light are projected from such source into any residential district or public street.

Indirectly lighted sign. Sign illuminated by artificial light reflecting from the sign face, the light source not visible from any street right-of-way.

Individual letter sign. Letters or figures individually fashioned from metal or other approved materials and attached to the wall of a building, but not including a sign painted on a wall or other surface.

Information sign. A sign that conveys a message other than advertising, which sign is placed solely for the guidance of individuals or groups.

Insignias and flags. Includes such insignias, flags and emblems of the United States, the state, and municipal and other bodies of established government, or flags that display the recognized symbol of a nonprofit and/or non-commercial organization.

Institutional bulletin board.   On-premises sign containing a surface area upon which is displayed the name of a religious institution, school, library, community center, or similar institution and the announcement of its services or activities.

Internal–indirect lighting.   A source of illumination entirely within the sign (generally a freestanding letter) which makes the sign visible at night by means of lighting the background upon which the freestanding character is mounted. The character itself should be opaque and thus will be silhouetted against the background. The source of illumination should not be visible.

Internal lighting.   A source of illumination entirely within the sign that makes the contents of the sign visible at night by means of the light being transmitted through a translucent material but wherein the source of the illumination is not visible.

Internally illuminated sign.   Sign illuminated by an artificial light source that is not visible, but that reaches the eye through a diffusing medium.

Length of a sign.   The length of a sign, without border or frame, affixed to or mounted upon a building or other approved mounting, is the horizontal distance between the first and last extremities of the lettering, symbols, or other parts of the sign.

Letters and decorations.   Include the letters, illustrations, symbols, figures, insignia and other devices employed to express and illustrate the message of the sign.

Location.   A lot, premises, building, wall, or any place whatsoever upon which a sign is erected, constructed and maintained.

Maintenance.   The replacing or repairing of a part or portion of a sign made unusable by ordinary wear, tear, or damage beyond the control of the owner or the reprinting of existing copy without changing the wording, composition, or color of said copy.

Mansard and parapet signs.   A sign permanently affixed to a wall or surface designed to protect the edge of a roof, constructed no more than 20° from vertical unless otherwise required by code.

Marqueee sign.   Any sign projecting from, attached to, or hung from a marquee; "marquee" means a canopy or covered structure projecting from and supported by a building, when such canopy or covered structure extends beyond the building, building line, or property line.

Menu board.   A permanently mounted sign displaying the bill of fare of a drive-in or drive-thru restaurant.

Monumental sign.   Freestanding sign affixed to a sign monument.

Monument sign.   Structure, built on grade, that forms an integral part of the sign or its background is in conformance with the zoning requirements of the district in which it is located.

Nameplate.   A sign indicating the name and/or address of a building or the name of an occupant thereof and the nature of a permitted occupation therein.

Name plate sign.   Sign which states the name or address or both of the profession or business on the lot where the sign is located.

Neon strip lighting.   Neon strip lighting is prohibited above the roof level of any building except on approved pylons and parapets.

Nonaccessory sign.   Sign that is not accessory to the principal use of the premises.

Noncombustible ground sign.   Any sign constructed of noncombustible materials, including the foundation, supporting framework, poles, or posts erected or maintained upon the ground and not attached to, and independent of, any part of the building.

Nonconforming sign.   Sign, outdoor advertising structure, or display of any character, that was lawfully erected or displayed, but that does not conform with standards for location, size, or illumination for the district in which it is located by reason of adoption or amendments of the ordinance, or by reason of annexation of territory to the city.

Nonflashing sign.   An illuminated sign on which the artificial light is maintained stationary and constant in intensity and color at all times when such sign is illuminated. For the purpose of the ordinance, any moving, illuminated sign is not considered a nonflashing sign.

Nonstructural trim.   Nonstructural trim is the molding battens, caps, nailing strips, latticing, and walkways that are attached to the sign structure.

Number of signs.   "Sign" should be considered to be a single display surface or display device containing elements organized, related, and composed to form a unit. Where matter is displayed in a random manner without organized relationship of elements, or where there is reasonable doubt about the relationship of elements, each element should be considered to be a single sign.

Obsolete sign.   A sign that advertises an activity, business, product or service that no longer is conducted on the premises on which the sign is displayed.

Off-premise sign.   A structure bearing a sign that is not appurtenant to the use of the property where the sign is located, or a product sold or a service offered upon the property where the sign is located, and that does not identify the place of business where the sign is located as a purveyor of the merchandise or services advertised upon the sign. Permanent off-premise signs are absolutely prohibited.

On-premises sign.   Sign the primary purpose of which is to identify and/or direct attention to a profession, business, service, activity, product, campaign, or attraction manufactured, sold, or offered upon the premises where such sign is located.

Open sign.   Display sign in which at least 50% of the area is uncovered or open to the transmission of wind.

Outdoor advertising display sign.   Any fabricated sign, including its structure, consisting of any letter, figure, character, mark, point, plane, marquee sign, design, poster, pictorial picture, stroke, stripe line, trademark, reading matter, or illuminating device, constructed, attached, erected, fastened, or manufactured in any manner whatsoever so that the same shall be used for the attraction of the public to any place, subject, person, firm, corporation, public performance, article, machine, or merchandise whatsoever, and displayed in any manner whatsoever outdoors for recognized advertising purposes.

Outdoor advertising signs.   An attached or freestanding structure constructed and maintained for the purpose of conveying information, knowledge, or ideas to the public.

Outdoor advertising structure. Anything constructed or erected, either freestanding or attached to the outside of a building, for the purpose of conveying information, knowledge, or ideas to the public about a subject either related or unrelated to the premises upon which located.

Outdoor commercial advertising device. Visible, immobile contrivance or structure in any shape or form, the purpose of which is to advertise any product or service, campaign, event, etc.

Outdoor display structure. Any structure erected or attached outdoors used for advertising or display, or for the affixment, attachment, or support of a sign or signs, or for any similar purposes; includes billboards and displays for special occasions such as Christmas displays.

Outdoor sign. Any arrangement of letters, figures, symbols, or other devices used for advertising, announcement, direction, or declaration, intended to attract or inform the public, affixed or attached to the exterior walls of a building or other structure, or upon constructed surfaces erected, attached, or supported outdoors. The word "sign" includes the structure of such signs.

Outdoor signs. Includes all fabricated signs and their supporting structures erected on the ground or attached to or supported by a building or structure.

Painted wall sign. Any sign painted on the outside wall of any building.

Permanent sign. A sign that is firmly attached to a structure supported by a foundation.

Point of purchase sign. Any structure, device, display board, screen, surface or wall with characters, letters, or illustrations placed thereto, thereon, or thereunder by any method or means whatsoever where the matter displayed is used for advertising on the premises a product actually or actively offered for sale or rent thereon or therein or services rendered.

Point of sale sign. A sign that advertises a product or service available, or a business conducted, upon the property on which the sign is located.

Pole or ground sign. Any sign erected upon a pole or poles and that is wholly or totally independent of any building for support.

Pole sign. Freestanding sign other than a portable sign or a monumental sign.

Political sign. Any advertising structure or banner used in connection with a local, state, or national election campaign.

Portable display sign. A display surface temporarily fixed to a standardized advertising structure that is regularly moved from structure to structure at periodic intervals.

Portable sign. Freestanding sign not permanently anchored or secured.

Posted sign. Tablet, card, or plate that defines the use, occupancy, fire grading, and floor loads of each story, floor, or parts thereof for which the building or part thereof has been approved.

Price sign. A permanently mounted sign displaying the retailing cost of a gallon of gasoline on the premises of a service station.

Private identification sign. A sign that is affixed to or placed or mounted upon a private or residential property solely to name or identify the property, the occupant, or the owner. A sign placed by a private club or association to name or identify the club, club premises, etc., is also classified as a private identification sign.

Projecting sign. Any projecting sign or marquee sign extending beyond the building wall or affixed to any hood, canopy, or marquee.

Projection. The distance by which a sign extends over public property or beyond the building line.

Public sign. A sign of a noncommercial nature and in the public interest, erected by or upon the order of a public official in the performance of his public duty, such as safety signs, danger signs, trespassing signs, traffic signs, memorial plaques, signs of historical interest, and all other similar signs, including signs designating hospitals, libraries, schools, airports, and other institutions or places of public interest or concern.

Pylon sign. An advertising structure projecting from the wall or extending over the roof of any building, comprising a framework and display surface, the structural members of which are an integral part of the building upon which such sign is erected.

Raised or embossed sign. Includes any lettering, characters, or numbers affixed to the building or structure.

Real estate sign. Any sign erected by the realtor or owner, advertising the real property for rent or for sale upon which the sign is located, but does not include rooming housing signs.

Roof sign. Outdoor advertising display sign erected, constructed, or maintained above the roof of any building.

Sale and rental signs. Ground signs or wall signs used exclusively for the sale or lease of property on which they are erected, having an area not exceeding twenty four (24) sq ft, may be constructed entirely of combustible materials, provided that such ground signs are located not less than ten (10) ft from any building or public way, or as required by the local code.

Sandwich sign. Any sign that is not permanently affixed to any structure on the site or permanently ground-mounted; any portable sign.

Service sign. Sign identifying rest rooms and other service facilities.

Sign. Any structure or part thereof, or device attached thereto, or painted thereon, located outside a building that is arranged intended, designed for, or used for displays or includes any letter, word motto, banner, flag, pennant, insignia, device, or representation that is in the nature of an advertisement, announcement, direction, or attraction.

Sign area. Area of a sign should be computed as the entire area within a single continuous rectilinear perimeter of not more than eight straight lines enclosing the extreme limits, writing, representation, emblem, or design, together with a material or color forming an integral part of the display or used to differentiate the sign from the background against which it is placed. Sign supports should not be included in determining sign area unless they are an integral part of the display.

Signboard. Any structure or part thereof on which lettered or pictorial matter is displayed for advertising or notice purposes.

Signfacing. The opaque or transparent surface or surfaces of the sign, upon, against, or through which the message of the sign is exhibited.

Sign frontage. The length in feet of the ground floor level of a building front or side facing a street (or facing a right-of-way accessible from a street) that is occupied by an individual business.

Sign-on-site. Sign relating in its subject matter to the premises on which it is located or to products, accommodations, services, or activities on the premises. On-site signs do not include signs erected by the outdoor advertising industry in the conduct of the outdoor advertising business.

Sign structure. Any structure which supports or is capable of supporting any sign as defined in the code. A sign structure may be a single pole and may or may not be an integral part of the building.

Sign surface area. Entire area within a parallelogram, triangle, circle, semicircle or other geometric figure, including all of the elements of the matter displayed, but not including blank masking, frames, or structural elements outside the advertising elements of the sign and bearing no advertising matter.

Size of sign. The smallest rectangle in which the sign, including decorative borders or symbols, will fit.

Sky or roof sign. Deemed to be any letter, word, model, sign, or device in the nature of an advertisement, announcement or direction supported wholly or in part over or above any wall, building, or structure.

Small professional or announcement signs. Should be limited to not over one (1) sq ft in area if fixed flat to the main wall of a residence building. Name and announcement signs firmly fixed to the main wall of public or semipublic buildings, and which are not over six (6) sq ft in area; signs having an area of not exceeding six (6) sq ft on horticultural or agricultural buildings, and real estate signs are displayed behind the prevailing front building line of that block and further provided that there should be but one (1) such sign on each lot, which should be used only to advertise the premises upon which it is erected, or as required by local code.

Snipe sign. Any sign under twenty (20) sq ft in area made of any material, including paper, cardboard, wood and metal, when such sign is tacked, nailed, posted, pasted, glued, or otherwise attached to trees, poles, fences, or other objects, and the advertising matter appearing thereon is not applicable to the premises upon which said sign is located, or as required by local code.

Special displays. Special decorative displays on which there is no commercial advertising, used for holidays, public demonstrations, or promotion of civic welfare or charitable purposes, when authorized by the municipal authorities.

Spectacular sign. Any sign of noncombustible materials, erected or maintained independently of any building, upon a structure of structural steel or concrete and exceeding twenty-five (25) ft in height, or as required by local codes.

Stop or danger signs. Any sign that uses the word "stop" or "danger" or presents or implies the need or requirement of stopping, or the existence of danger, or that is a copy of imitation of official signs. Red, green or amber (or any color combination thereof) revolving or flashing light giving the impression of a police or caution light is considered a prohibited sign, whether on a sign or on an independent structure.

Street banners. Signs advertising a public event providing that specific approval is granted under regulations established by the local authorities.

Street clock. Any timepiece erected upon a standard upon the sidewalk or on the exterior of a building or structure placed and maintained by some person for the convenience of the public and for the purpose of advertising their place of business.

Structural trim. Includes the molding battens, cappings, nailing strips, latticing, and platforms that are attached to the sign structure.

Structure. That which is built or constructed, an edifice or building of any kind or any piece of work artificially built up or composed of parts joined together in some definite manner.

Subdivision sign. Any sign located either on or off a subdivision tract that indicates the direction to or advertises the location, existence, or sale of a subdivision or any part thereof.

Surface area of a sign. Computed as including the entire area within a regular geometric form or combination of regular geometric forms comprising all of the display area of the sign and including all of the elements of the matter displayed. Framed and structural members not bearing advertising matter should not be included in computation of surface area.

Swinging projecting sign. Any sign projecting at an angle from the outside wall or walls of any building that is supported by only one rigid support, irrespective of the number of guy wires used in connection therewith.

Swinging signs. Signs that swing or otherwise noticeably move as a result of wind pressure because of the manner of their suspension or attachment.

Temporary sign. A sign or cloth or other combustible material, with or without a frame, that is usually attached to the outside of a building on a wall or storefront, intended for a limited period of display.

Temporary window signs. Allowed by local code only if they advertise special sales or events lasting no more than 15 days. They may cover no more than 30% of the area of the window in which they appear.

Traffic sign. A temporary or permanent sign designed solely for the purpose of regulating vehicular or pedestrian traffic, the warning of danger, or the providing of directions to public facilities or streets, provided that, where possible, such signs should be of the same type as in use by the state and local authorities.

Vehicle signs. Signs accessory to the use of any kind of vehicle, providing the sign is painted or attached directly to the body of the vehicle.

Vertical projecting sign. Any projecting sign that is greater in height than in width.

Wall area sign. The width multiplied by the height of that section of the building occupied by the advertiser, facing any one public way.

Wall sign. Any sign attached to or erected against the wall of a building or structure, with the exposed face of the sign in a plane parallel to the plane of the wall.

Wind loads. For the purpose of design, and except for roof signs, wind pressure should be taken upon the gross area of the vertical projection of all signs at not less than fifteen (15) lb per sq ft for those portions less than sixty (60) ft above the ground, and at not less than twenty (20) lb per sq ft for those portions more than sixty (60) ft above the ground. (Verify wind loads with local authorities).

Wind sign. Any sign in the nature of a series of two or more banners, flags, pennants, or other objects of material that call attention to a product or service, fastened in such a manner as to move upon being subjected to pressure by wind or breeze.

Window area sign. The width multiplied by the height of the total window area facing any one public way.

TOILET AND BATH ACCESSORIES

Aluminum shelf. Shelf for articles such as purses, combs, or gloves, which would provide the person the use of his or her hands while at the lavatory. Shelf is fabricated of 6063-T3 aluminum, polished and anodized.

Aluminum shelf, single-roll toilet paper holder and ash tray. Combination of items installed on toilet stall partitions. The shelf is fabricated of 6063-T3 polished and anodized heat-treated aluminum. The unit holds a standard core tissue roll and a one-piece aluminum ash tray with integral cigarette holder. The installation is meant to be vandal resistant.

Ash receptacle. Receptacle used for cigarette butts to prevent their being discarded in other receptacles or on the floor. The receptacle is finished in satin chrome and has a corrosion-resistant steel body and a removable heat-proof glass inner liner. Its tight-closing tip-action blades prevent smoldering of cigarettes or cigars.

Bottle opener. Bottle openers are installed in motel and hotel rooms for the occupants' convenience. They are usually finished in solid brass or are nickel-chrome plated.

Bumper hook. Combination hook used not only as a door stop, but also as a holding device for clothes hangers or towels on the bathroom door. It comes finished in chrome-plated cast brass in various projections.

Combination mirror/shelf. Convenience shelf of seamless stainless steel, 5 in. deep, mounted to the wall mirror of $\frac{1}{4}$-in. plate glass, edge-bound with polished stainless steel, with concealed hanger on a full-steel backplate.

Combination toilet tissue cabinet. Cabinet fabricated of 20-gauge-type 18-8 stainless steel with a satin finish. The unit should hold approximately 1250 single- or 975 double-fold tissues.

Concealed scale. Standard-size bath scale mounted in a recessed unit to permit in-wall storage when not in use. The recessed box is made of rustproofed steel with a brass frame and a rubber-cushioned floor rest that also serves as an access handle.

Concealed toilet paper holders. These holders also provide recessed compartment for spare roll, face panel pulls out and down to provide access to the spare roll. The panel is solid brass or bright nickel-chrome plated.

Double-roll toilet fixture. Fixture fabricated of zinc die-casting, triple chrome plated with a concealed hinge at end for servicing and a tension spring to control the flow of paper.

Dual napkin/tampon dispenser. Dispenser fabricated of 22-gauge, Type 304L 18-8 stainless steel with a prepolished satin finish. The door is 18-gauge stainless steel with a piano hinge of heavy duty stainless steel.

Facial tissue holders—surface unit. Unit fabricated of solid brass with a bright nickel-chrome finish. It holds a standard tissue package and can be mounted on either a vertical or horizontal surface.

Foot-operated soap dispenser. Dispenser used primarily in hospitals, medical clinics, and doctors' offices and in food plants under the jurisdiction of health departments or other government agencies. The body and spout are fabricated of Type 316 stainless steel. The pumping mechanism is heavy duty with 72-in. long heavy nylon tubing. The soap level indicator is unbreakable.

Foot-operated surgical-type soap dispenser. Dispenser used primarily in hospitals, medical clinics, and doctors' offices. The rectangularly shaped reservoir is formed of 22-gauge, satin-finished, Type 316 stainless steel. The hinged filling cover is equipped with a lock. The soap level indicator is located on front of the tank. The soap outlet spout can be rotated 190°. There is a foot pump and 6-ft vinyl tubing. The unit is mechanically locked to a surface-mounted concealed wall plate.

Framed mirrors and mirror/shelf combinations. The mirror frame is of one-piece roll-formed construction with a continuous integral stiffener on all sides, and Type 304 satin-finished stainless steel angles. The edges of the angles in contact with the mirror have a bevel design. The mirror is $\frac{1}{4}$-in. plate glass, electrolytically copper plated, and protected by filler strips and shock absorbing padding. A wall hanger is furnished for locking the mirror to the wall. The galvanized steel back is equipped with concealed locking setscrews for securing the mirror the the wall hanger. The framed mirror is equipped with Type 304 satin-finished stainless steel shelf, 5 in. deep with return edges on all four sides and the front edge hemmed.

Grab bars

1. Safety railings securely anchored to walls, located in showers, at toilets for the handicapped, and in the hallways of hospitals, nursing homes, and similar locations.

2. Grab bars are installed in hospital and hotel bathrooms to aid getting in and out of tubs and supply support in showers. They are especially useful for the handicapped. Bars are available to withstand 600-lb downward pull. Bars consist of solid brass tubing with $\frac{7}{8}$-in. outside diameter and a wall thickness of 0.045 in. They have cast posts and a bright chrome finish on the exposed parts.

Hand hair dryer. Automatic electrically operated dryer, with a $\frac{1}{8}$ hp motor protected by safety circuit breaker against overheating and burnout. A fan/blower directs air flow through heating element with a range between 1,800 and 2,300 W. Timer has electronic sensor, automatic shut-off, and a time range from 30 to 90 seconds. Tamper proof cover fabricated from 18-gauge stainless steel or baked enamel on die-cast metal. 360° rotating die-cast zinc alloy nozzle.

Heavy-duty hook. Hook for support of clothes hangers or towels in the bathroom. A hook of chrome-plated brass with a bright or satin finish will support 300 lb. A bright or satin-finished stainless steel hook will support 500 lb.

Heavy-duty stainless steel soap and grab bar. Unit usually installed at the tub or shower. The grab bar is made of stainless steel tubing held by two formed stainless steel brackets.

Heavy-duty stainless steel soap and one-piece grab bar. Unit is considered safe for the handicapped. The grab bar is of $\frac{1}{4}$-in thick and 1-in. high stainless steel welded to the supporting body.

Heavy-duty stainless steel soap dish. Soap dish fabricated of seamless, deep-drawn, 20-gauge Type 302 stainless steel with a satin finish. The retaining lip extends $\frac{3}{4}$ in.

and is welded to the body. The patterned dimples keep soap dry and prevent it from slipping.

Heavy duty towel hook. Polished chromium-plated brass hook with a base plate and anchor plate for installation.

Hook strip. A hook strip is used in large shower rooms. The number of hooks is based on the number of shower heads in the room. The hook is constructed of Type 302 (18-8) satin-finished stainless steel. Hooks are 14-gauge stainless steel, the strip is 18-gauge (0.50) stainless steel, channel formed.

Hospital solution dispenser. Dispenser used in medical clinics and doctors' offices. It dispenses aseptic solutions, surgical-type synthetics, and other liquid soaps. Each stroke of the valve delivers a measured amount. The dispenser body is polished chromium-plated brass. The $\frac{1}{2}$-qt polyethylene globe unscrews for filling.

Hotel towel holders. Stem- or tree-type towel holder with a polished chrome-plated steel spring with spaces to provide convenient storage for two to four sets of bath towels, hand towels, and wash cloths. It is easily installed on guest bathroom walls. The holder is of heavy-gauge metal with a polished chrome finish.

Illuminated mirror side. Fully recessed double-compartment steel cabinet with sliding plate glass mirror doors. It comes with built-in light fixtures with opaque glass panels, a convenience outlet, and switch.

Large-capacity waste receptacles. Recessed waste receptacle constructed entirely of Type 304 stainless steel, all-welded construction, with satin finish on exposed surfaces. It is flange drawn, of one-piece seamless construction. The top and bottom edges of the removable waste container are hemmed for safe handling and are secured to the cabinet by a tumbler lock, keyed like the other accessories. The waste receptacle has a 12-gal minimum capacity.

Lather-forming liquid soap dispenser. Dispenser fabricated with the following operating features: each push of the valve delivers a measured amount of lather with the stainless steel valve above the soap level preventing continuous flow. All moving metal components are of Type 302 stainless steel. A separate wall plate prevents theft.

Lather-type soap dispenser. Dispenser fabricated with the following operating features: push of the valve delivers a measured amount of creamy lather. Continuous flow is impossible because the valve is always above soap level. It will not leak, as the soap must be pumped. The unit is equipped with a self-cleaning piston: air is drawn in on the return stroke. The soap container is removable for cleaning and is enclosed in a protective chromium-plated brass case. The soap supply can be instantly checked from top to bottom through viewing slots on both sides of the dispenser, which fills from the top by unlocking with the service key. The filler cap is permanently chained to the dispenser. All moving metal parts are of 18-8 stainless steel. A separate wall plate, with screw heads concealed, prevents tampering or theft. The flat-against-the-wall design eliminates side sway.

Lavatory soap dispenser. Dispenser valve and body are of chrome-plated brass with all internal working parts of stainless steel. The container has a capacity of 16 fl oz and is top filled.

Liquid-type dispenser. All metal parts in contact with the soap are of 18-8 stainless steel. The unit dispenses liquid soaps and lotions without leaking or dripping. Each push of the valve delivers a measured amount of soap. A

wide-mouthed jar unscrews for easy filling. The body of dispenser is of polished chromium-plated brass.

Liquid-type soap dispenser

1. Straight-liquid, push-in valve soap dispenser. The metal parts in contact with the soap are formed of 18-8 stainless steel. The valve is located above the soap level. The body of the dispenser is constructed of polished chromium-plated brass. The dispenser is filled by unscrewing the globe. The dispenser is mechanically locked to a surface-mounted concealed wall plate.

2. Dispenser installed in public bathrooms. The moving metal parts are fabricated of 18-8 stainless steel. The valve is located above the soap level. The unit is equipped with a self-cleaning piston with an adjustable corprene seal. A glass globe is contained in a chromium-plated brass tapered cylindrical housing with twin slotted sides to indicate soap level. The chained top-filling cap is unlocked with a special key. The dispenser has a 12-oz usable soap capacity. The dispenser is mechanically locked to a surface-mounted concealed wall plate.

Mirror-slide. Fully recessed steel cabinet with smooth-sliding mirror doors and a larger mirror area than hinged-door types. The baked enamel interior is finished in hammertone textured metallic color. Provided with two bulb edge glass shelves. Aluminum frame polished to match chrome.

Multifold towel cabinet. Towel cabinet of 22-gauge, Type 18-8 stainless steel with a satin finish. It dispenses 315 C-fold or 700 multifold towels. It has a sloping top, a refill indicator slot, and a keyed tumbler lock.

Multiroll toilet tissue dispensers. The recessed multiroll toilet tissue dispenser is constructed of Type 304 stainless steel, all-welded construction, with satin finish on exposed surfaces. It is of flange(s)-drawn, one-piece seamless construction. The unit is equipped with an adjustable flange for installation in the toilet partition. The unit automatically drops an extra toilet tissue roll in place after the bottom roll has been used up. The door(s) is equipped with a tumbler lock.

Partition-mounted toilet seat cover dispenser. This dispenser serves two toilet compartments. Seat covers are dispensed from both sides, but the unit is serviced from one side. It mounts through $\frac{3}{4}$- to $1\frac{5}{8}$-in. thick partitions. Its capacity is 1000 seat covers.

Polished stainless steel mirrors. Mirrors are used primarily in industrial plants where glass breakage is possible. Steel mirrors are hand selected, hydrogen bright, annealed stretcher level, continuous controlled tolerance, polish, and good reflectivity.

Powdered soap dispenser. The dispenser is of all-stainless-steel construction. The entire container rocks back and forth to shake out powder. There is no waste because it will not deliver a steady flow. The container is filled through 4-in. opening. The dispenser has a locked cover and a separate wall plate.

Pull-down utility shelf. This shelf provides a convenient surface for gloves, parcels, briefcases, purses, etc. After use, the shelf must return to upright position before the door can be opened, making it impossible to forget belongings. The unit is sturdily cast of heavy-duty metal. The stippled satinlike tray has a lustrous chrome edging.

Push-in-type soap dispensers. Dispenser fabricated of a heavy-cast chrome-plated body with a separate wall bracket. No screws are visible when it is mounted. It cannot drip or leak because the valve is above the fluid level. The mechanism allows the user to regulate the output as desired. It has a stainless steel plunger and a precision-engineered valve mechanism.

Push-up type liquid soap dispensers. The translucent, white, 16-oz plastic globe clearly and instantly shows the soap level. It is inert to reactions of most alkalies and acids and is noncorroding. The unit is mounted on the wall and is filled through a large opening.

Razor blade slot. Metal slot for the disposal of razor blades in a hollow wall interior. It is furnished with a slide for used blades and a perforated mounting plate. It is fabricated of 16-gauge, 18-8 stainless steel with a satin finish.

Recessed cup dispenser and disposal. Unit fabricated of 20-gauge, Type 304L 18-8 alloy stainless steel with satin finish. The face frame is of one-piece construction with no miters or welding, 1 in. wide and returning $\frac{1}{4}$ in. to the wall. The door is of solid stainless steel with a concealed pivoting rod; it is locked by a theftproof spring-tensioned cam lock. A cup level indicator is provided in the lower section of the dispenser door.

Recessed electric hand and hair dryers. Recessed electric dryer and one-piece cast-iron cover with acid-resistant porcelain enamel finish. It has a universal-type motor of $\frac{1}{10}$ hp with sealed-lubricant ball bearings and resilient mounting, and is protected by a 20-A fuse. The heating element is protected by an automatic circuit breaker. After actuation of the push button, the timing mechanism is designed to operate 30 seconds. The dryer UL listed.

Recessed facial tissue dispenser. Face plate fabricated from 22-gauge stainless steel that snaps into wall housing, which is heavy gauge preplated steel. Its capacity is 250 two-ply tissue.

Recessed feminine napkin disposals. Unit constructed entirely of Type 304 stainless steel, all-welded construction, with satin finish on exposed surfaces. It is of flange(s)-drawn, one-piece seamless construction. The single stainless steel receptacle is completely removable for servicing.

Recessed feminine napkin vendors. Unit constructed of stainless steel, all-welded construction, with satin finish on exposed surfaces. The 18-gauge door is equipped with two tumbler locks keyed like the other accessories. The mechanism dispenses all popular brands of feminine napkins including tampon types. It has two operating handles so that either napkins or tampons can be dispensed at the user's option.

Recessed hinge retractable shower seat. The seat and housing are fabricated of cast aluminum with an anodized flange. The hardware is made from Type 304L stainless steel, the mounting frame from cadmium plated steel.

Recessed medicine cabinet. Cabinet and body fabricated from 22-gauge stainless steel; the mirror frame is prepolished stainless steel with satin finish. The cabinet is provided with magnetic catches and three adjustable shelves.

Recessed paper cup dispenser. Unit fabricated of Type 304 stainless steel, welded construction with exposed surfaces of 22-gauge stainless steel, satin finish. The flange is drawn and seamless. The door has a stainless steel piano hinge and double-pan back. The tumbler lock is keyed. It dispenses all makes of cone-shaped flat bottomed or pleated paper cups.

Recessed paper towel dispenser. Stainless steel cabinet with laminated plastic door. It dispenses all makes of paper towels and has a capacity of 600 C-fold, 800 multifold, or 1100 single-fold towels. It has a concealed latch.

Recessed paper towel dispenser and waste receptacle. Stainless steel cabinet with laminated plastic door. This is a dispenser for all makes of paper towels and has a capacity of 600 C-fold, 800 multifold, or 1100 single-fold towels. The unit incorporates a removable $10\frac{1}{2}$-gal rigid, leakproof, plastic waste container. It has a concealed latch.

Recessed powdered soap dispenser. This unit, including the mounting box, is constructed of stainless steel. The door is of 22-gauge, Type 304 stainless steel with satin finish, attached to the mounting box with a concealed, stainless steel piano hinge, and is equipped with a keyed tumbler lock. The spillage tray locks in position, and is easily removed for cleaning. An adjustable output mechanism dispenses any free-flowing powdered-soap from the finest to the coarsest. An agitator spring prevents soap buildup. A vessel attached to the back of the door swings out for easy refilling without removing from the wall.

Recessed soap dispenser. The cabinet with all its exposed surfaces are of Type 304 stainless steel, satin finished, with tumbler lock(s), keyed like the other accessories. The face plate is attached to a stainless steel mounting box with stainless steel piano hinge. The unit has a minimum capacity of 48 fl. oz and an unbreakable refill indicator. The soap dispenser is a fully adjustable output mechanism equipped with an agitator to minimize soap clogging. The soap dispenser has a mechanism that dispenses individual soap leaves and has a capacity of 1000 leaves. It has a double locking device.

Recesses toilet seat cover dispenser. Stainless steel cabinet with laminated plastic door, designed for installation in the back wall of the toilet compartment. It has a concealed latch and a capacity of 500 paper toilet seat covers.

Recessed towel dispenser and waste receptacle. Unit whose exposed surfaces are fabricated of 22-gauge, Type 18-8 stainless steel, satin finished. The flange is of one-piece construction with a $\frac{1}{4}$-in. return and no mitered corners. The doors are of warp-free double-pan construction, hung on continuous stainless steel piano hinges, and fitted with tumbler locks (keyed alike). The towel dispenser has a capacity of 340 C-fold, 725 multifold, or 875 single-fold towels. The waste receptacle is equipped with a removable 10-gal stainless steel and plastic inner container. This dispenser is designed for use in public restrooms.

Recessed vanity shelves. Recessed vanity shelf with brass frame. The recessed steel box is finished with hammertone metallic baked enamel protected by a bottom glass plate. It is equipped with a convenience outlet, tooth-brush holder, and sliding glass doors with metal knobs.

Recessed vertical soap dispenser. Dispenser with a face fabricated of 22-gauge, Type 304L, 18-8 alloy stainless steel with satin finish. The face trim is of one-piece construction returning $\frac{1}{2}$ in. to the wall. The door is locked by a cam lock and mounted on full stainless steel piano hinge. The unit has a liquid level indicator.

Recessed wall urn ash trays. Ash trays constructed of Type 304 stainless steel, all welded construction, with satin finish on the exposed surfaces. The flange is drawn and of one-piece seamless construction. The satin-finished stainless steel flip-top ash tray is secured to the cabinet with a full-length stainless steel piano hinge. The unit is furnished with a removable aluminum inner ash receptacle.

Recessed waste receptacle. Stainless steel cabinet with laminated plastic door; a self-closing panel, lettered "Push," which covers the receptacle opening. The unit has a concealed latch and removable $10\frac{1}{2}$-gal rigid, leakproof, molded plastic waste container.

Retractable clothes line. Clothes line for bath, kitchen, or utility room and provides up to 10 ft of line between any two facing walls and then retracts out of the way by being rewound in a protective container.

Retractable shower seats. Units constructed of 1×0.049 in. (18 gauge) stainless steel tubing. When they are in position for use, the legs angle into the base of the wall for rigid support. When they are retracted, the legs fold against wall providing a low silhouette, and conserving space. No extra fittings are required to hold the seats in position. They are ideal where space is limited. Foam rubber cushions are covered in off-white Naugahyde.

Robe hook. Single or double hook types, fabricated from solid forged brass, with nickel and copper plating and chrome coating. Die-cast plates for mounting.

Sanitary napkin disposal. Unit fabricated of 22-gauge, Type 304L stainless steel with satin finish. The piano hinge is made of heavy-duty stainless steel. The bottom door has a hidden spring-tensioned self-catching lock.

Sanitary napkin disposal with folding purse shelf. Unit fabricated of 22-gauge, Type 304L stainless steel with satin finish.

Shelf with paper towel dispenser and soap dispenser. Unit fabricated of 20-gauge, Type 304L, 18-8 alloy stainless steel with satin finish. The unit has a spring-loaded towel dispenser.

Shower curtain rod. Type 302 stainless steel bar with stainless steel flange, bright or satin finished. It comes in 2 straight sections only with chrome-plated brass curtain hooks.

Shower stall seat. Permanent installation, surface mounted, fabricated of 16-gauge stainless steel, all edges beveled, welded, ground smooth, and polished.

Single-fold towel cabinet. Cabinet with door and flange of 20-gauge, Type 18-8 stainless steel, satin finished. The sides are of 24-gauge stainless steel. The unit holds approximately 200 single-fold towels.

Single or double robe hook. Brass hook with highly polished chrome-plated finish over nickel. It has concealed screw fastenings and projects $2\frac{1}{2}$ in. from the back of the flange.

Soap dispenser—surface mounted. Unit fabricated of 20-gauge steel with all surfaces copper nickel and chrome plate with a satin finish. The positive shutoff valve is of chrome-plated brass and stainless steel. The unit is fitted with a tamper-proof refill cap and liquid level indicator.

Stainless shower seat. Shower seat fabricated of 16-gauge, 18-8 stainless steel, satin finished. It is of welded construction and polished smooth. It has a full-length stainless piano hinge and mounts left or right with 11-gauge stainless full-length mounting brackets.

Stainless steel double-roll toilet paper holder and ash tray. Self fabricated of $\frac{1}{8}$-in. thick stainless steel polished to a satin finish with stainless steel mounting brackets. Toilet paper holder, double roll.

Stainless steel framed mirrors. Frame is fabricated of 18-gauge, 18-8 stainless steel, satin finished with corners welded and ground smooth. Mirror is of $\frac{1}{4}$-in. polished plate glass with two coats of silver, electrocopper clad. The

backing is made of 20-gauge galvanized steel and supported by concealed heat-proof brackets.

Stainless steel lather-type soap dispenser. Lather-forming liquid soap dispenser. Each stroke of the valve delivers a measured amount of soap. It cannot leak or drip because the valve is located above the liquid level and soap is pumped from the bottom of the tank. It has a separate wall plate. The rectangularly shaped dispenser is formed with two front corners beveled vertically at 45°. The metal components in contact with liquid are fabricated of 18-8 stainless steel, with exposed surfaces satin finished. The tank is constructed of 22-gauge and the cover of 18-gauge stainless steel. The pump-up-, push-in-type soap valve is located approximately 1 in. from the top of the dispenser. The locked cover is removable for filling. The dispenser is mechanically locked to a surface-mounted concealed wall plate.

Stainless steel shelf. Shelf fabricated of 18-gauge, Type 304L stainless steel with satin finish. Brackets are 16-gauge stainless steel welded to turned-down front and back edges. No spot welds appear on surface.

Stainless steel theft-proof framed mirror. Frame of welded one-piece construction with ground and polished corners. The spring-lock device guarantees a theftproof installation and easy removal by maintenance personnel. It has a stainless steel satin finish.

Surface-mounted bedpan rack. Unit fabricated from 18-gauge stainless steel, with satin finish on all exposed surfaces. Holding straps are of 16-gauge stainless steel. Backplate is provided with internal holding brackets.

Surface-mounted facial tissue dispenser. Unit fabricated of chrome-plated steel with back mounting plate fabricated of heavy gauge galvanized steel. Its capacity a minimum of 300 single-ply or 150 two-ply tissues.

Surface-mounted multiroll toilet tissue dispenser. This unit holds and dispenses two rolls of standard toilet tissue, as well as 5-in.-diameter, 1500-sheet rolls. An extra toilet tissue roll automatically drops in place after the bottom roll is used up. Dispenser is fabricated of Type 304 stainless steel, satin finished and has a keyed tumbler lock.

Surface-mounted paper towel dispenser. Paper towel dispenser with tumbler lock, for installation where use of single-fold paper towels is desired. The cabinet is constructed of 22-gauge stainless steel with satin finish. The front of the cabinet is equipped with a full-length piano hinge at the bottom and swings down for convenient refilling. The tumbler lock on top of the cabinet is keyed. A slot on each side of the cabinet indicates refill time. The unit should have a minimum capacity of 400 single-fold paper towels and accommodates all brands.

Surface-mounted powdered soap dispenser. Unit fabricated of amytl butyl styrene in pure white. The dispenser is mounted on a theftproof device.

Surface-mounted sanitary napkin dispensers. Unit that dispensers all popular brands of external napkins. The coin box is separately locked and keyed. The unit has a capacity of 22 napkins. It is fabricated of Type 18-8 stainless steel, satin finished. The door is constructed of 22-gauge steel, reinforced to the equivalent of 16-gauge. The cabinet is reinforced 22-gauge stainless steel.

Surface-mounted stainless steel combination sanitary napkin and tampon vendor. Dual surface-mounted vendor equipped with two coin mechanisms for dispensing both sanitary napkins and tampons simultaneously. It holds 12 napkins and 19 tampons. The unit is fabricated of Type

18-8 stainless steel, satin finished. The door is constructed of 22-gauge stainless steel, reinforced to the equivalent of 16-gauge steel. The cabinet is fabricated of reinforced 22-gauge stainless steel.

Surface-mounted toilet seat cover dispenser. Unit that dispensers popular makes of single- or half-fold paper toilet seat covers. The entire unit is heavy-gauge, Type 304 stainless steel, welded construction. The exposed surfaces are satin finished. The dispenser fills from the bottom through a concealed opening and mounts to the wall or toilet partition with two mounting screws. The capacity of the unit should have a minimum of 250 paper toilet seat covers.

Tank-type soap dispenser. The tank-type soap dispenser is formed with two recessed panels on an angled front surface. The tank is constructed of 20-gauge cold-rolled steel with a triple-plated polished chrome exterior. The inside of the tank is copper-nickel plated. A view-level gauge is located on the front of the dispenser. A chain top-filling cap unlocks with a special key. The dispenser is mechanically locked to an adjustable self-leveling wall plate fabricated of 16-gauge cadmium-plated cold-rolled steel.

Three-piece mirror frame. Frame fabricated from 6063 T5 aluminum alloy having hairline miters and screw-locking cadmium-plated corner keys and continuous center section wall hangers at the top and bottom. It has a clear anodized finish.

Three-rolled surface-mounted tissue dispenser. The automatic self-locking mechanism will release spare rolls only when the exposed roll is complete used. The dispenser is available in white baked enamel or burnished aluminum finish.

Toilet paper holder. Holder made of brass with a highly polished chrome-plated finish over nickel. It has a chrome spring roller, concealed screw fastenings, and contoured flange(s).

Towel ladder. Solid brass unit with all exposed parts chrome finished.

Towel pull-out rods. Convenience rod for bath, powder room, kitchen, or dressing room. Pull-out rods extend 14 in. and have die-cast flange and solid brass tubing with bright nickel-chrome-plated finish.

Towel racks. Surface mounted guest towel rack in polished chrome-plated steel. Comes in several sizes to accommodate from two to four guests.

Towel rings. Clear Lucite rings with supporting flange of die-cast zinc, copper-nickel-chrome plated. (Lucite is a proprietary name).

Towel supply shelves. Shelves having zinc die-cast posts, copper-nickel-chrome plated, and polished stainless steel square tubing.

Towel tree. Unit designed primarily for use in hotel bathrooms. It has a polished brass finish. The towel bars slide laterally and permit various arrangements. Bars are spaced on vertical 24 in. trunk.

Triple paper cup dispenser. Dispenser is fabricated of 20-gauge Type 304L, 18-8 stainless steel with satin finish. The door is locked by a theftproof spring-tensioned cam lock. A cup level indicator is provided on the dispenser door.

Twin-roll toilet tissue dispenser. Unit that dispensers one roll of tissue while holding another in reserve. The reserved roll automatically moves down into position as the first roll is exhausted. It is fabricated of 18-gauge Type 18-8 stainless steel, satin finished. The door has a full-length stainless steel piano hinge and tumbler lock.

Valet towel holder. Unit that holds wash cloths, face towels, bath towels, and a bath mat. Hooks on the bottom hold shoeshine cloth or other items. It is made in two-, three-, or four-guest sizes.

Wall corner guards. Finished wall-mounted corner guards fabricated from 18-gauge stainless steel with satin finish and vinyl protective coating. Guards can be mounted with an adhesive or screws.

Wall mirror. Mirror consisting of $\frac{1}{4}$-in. plate glass, edge-bound with polished stainless steel, with a concealed hanger on the full steel backplate.

Wall-mounted shelf. Shelf fabricated from 18-gauge stainless steel, with brackets fabricated from 16-gauge stainless steel welded to shelf. Available in several depths and lengths.

Wall-mounted waste receptacle. Stainless steel wall-mounted waste receptacle. The unit is fabricated of 22-gauge Type 18-8 stainless steel, satin finished. The liner is quickly removed for waste disposal. The receptacle has rounded edges for safety.

Wall urn. Flip-top ash try and wall urn of 22-gauge Type 304 stainless steel, satin finished. The flip-top ash tray is secured to the wall urn with a hinge. The mounting bracket is of anodized aluminum, satin finished. The aluminum receptacle removes for servicing.

Waste receptacle for use under counter tops. Specially designed waste receptacle for installation between two lavatories or to the right or left of a single lavatory. It mounts to the underside of the counter top, off the floor, permitting easy floor maintenance. It is constructed of Type 304 stainless steel with a satin finish. The unit is equipped with a self-closing panel lettered "Push." It is furnished with a heavy-gauge vinyl liner, removable for servicing. It has a 5-gal capacity. Two stainless steel drawer slides are provided to allow for easy servicing.

Equipment

DARKROOM EQUIPMENT

Air-tempering systems. A fully automatic heating/refrigeration system with remote electronic control, water-cooled compressor, and heater, located in storage cabinets. Total unit size: 16-in. wide, 20-in. long, and 14-in. high.

Burst grids

1. Built-in grid for 1 gal tanks.
2. Built-in grid for $4\frac{1}{2}$ gal tanks.

Cabinet construction. Doors, drawer fronts, and finished sides fabricated from $\frac{11}{16}$ in. 45 lb particleboard, composed of a sandwich of decorative high-pressure laminate on the face side and liner high-pressure laminate on the back side, permanently bonded under pressure to both surfaces with a PVC self edge.

Cabinet hardware. Consists of concealed hinges, black plastic pulls attached from back of doors or drawers, pin tumbler locks (cam type where specified), ball bearing drawer slides, full extension file drawers, swivel casters 3 in. in diameter with 90-lb load capacity.

Combination graphic arts sink. Used for litho-film, a tray processing sink, washer molded in one continuous unit, mounted on a stand, preplumbed and fully equipped with hot and cold water faucet, wash water jet control, and vacuum breaker. Shatterproof "safety glass" in the light fixture. Unit requires hot and cold water supply, drain connection, and power outlet. Sink is one-piece glass fiber reinforced resin, impervious to rapid fixers, color bleaches, iron chloride, and other such chemicals. Heavy duty trays are molded glass fiber sized from $12 \times 14 \times 3$ in. deep to the largest size $35 \times 45 \times 4$ in. deep.

Combination sink. Unit contains a tank processing sink, a tray processing sink, and a round print washer, molded in one continuous unit and mounted on a particleboard core plastic laminated on both sides of the exposed surfaces. The unit is preplumbed and fully equipped with the accessories necessary to use the unit for processing.

Deep tank processing sink. One-piece molded glass fiber, coated with a corrosion-resistant resin, 10-in. deep with 6-in. utility ledge on 7-in.-high backsplash, with drain-channeled bottom.

Dot etch and wet negative viewer. Unit that has a one-piece molded top surrounded by a drain trough, with a heavy-duty safety glass diffuser covering viewing area. Provided with a drain in the drain trough and a smooth working surface. Double valve mixing faucet with spray or other types of water supply available with unit.

Electrical. Factory-wired units requiring electrical outlets are completely wired for 115 V service, using either junction boxes for direct wiring on site, or three-prong male grounded connections.

Electron microscope. Tank deck is a one-piece molded glass fiber coated with a blend of resins; tanks are recessed in deck and suspended in air-tempered cabinet base. Chemical tanks have built-in burst grids and lids, plus floating lids on developer tanks. Wash tanks have built-in inlets and vacuum breaker. Adjustable temperature blender, built-in burst agitation system, adjustable timer, individual on-off control for each chemical tank, separate distribution manifolds for nitrogen to developer tanks and air or nitrogen to other chemical tanks.

Faucets

1. Single valve faucet with integral vacuum breaker and laboratory nozzle spout end, 4-in. back-splash space.
2. Single valve faucet with elevated vacuum breaker, 4-in. backsplash space.
3. Double valve faucet, pedestal mount, 10-in. backsplash space.
4. Double valve faucet, hot and cold water, 12-in. back-splash space.
5. Double valve faucet with spray for hot and cold water, 16-in. backsplash space.

Faucets for 45° mount. Double valve faucet with spray for graphic arts sink, having 45° backsplash. 16-in. back-splash space.

Film-print dryer. Cabinet is fabricated from wood particleboard and plastic laminate and includes film hangers. The dryers have filtered air circulation with mechanical refrigeration. The cabinet can be fabricated for pass-thru doors for thru-wall installation between dark and light rooms.

Gas burst systems and equipment. Primary burst control system with timer, for nitrogen and compressed air, can be used as a control for a single-tank system with built-in grid, for a multi-tank operation with carry-along grid, or for the first tank in a multi-tank system with built-in grids. Timer serves all additional units used in the operation.

Gas valve unit. On-off solenoid gas valve unit without timer. One unit is required for each tank in a multi-tank burst system, in addition to a primary control unit for first tank. Unit includes solenoid valve with waterproof coil, remote on-off switch for timed burst, bulkhead fittings, and 5-ft plastic tubing.

Graphic arts sink. Molded glass fiber sinks, preplumbed and equipped with jet wash system with vacuum breaker, jet control valve and 5-in. standpipe, drain and 7-in. standpipe, and trays. Sinks have dump trough in front of 7-in.-high backsplash. Trays are suspended in a water jacket. Backsplash is angled 45° to make controls easy to reach.

Laboratory table. Fabricated of particleboard laminated with plastic, with or without drawers, 48 in. wide and 5 or 6 ft long.

Negative file cabinet. Cabinet with eight standard drawers including dividers and followers; stores sheets up to 4×5 in. and film strips up to 11-in. long. Unit is 15-in. wide, 52-in. high and 24-in. deep.

Pass-thru unit. Stainless steel wall pass-thru unit for darkroom wall, designed to pass a tray containing prints up to 16×20 in.

Photochemical mixing unit. Mobile chemical mixers that afford means of mixing, transferring, and draining photochemicals. Mixer is designed to minimize chemical oxidation through turbo-blending action within a closed circulatory system. It contains a corrosion-resistant mixing bowl of high density polyethylene material, an electric pump for chemical recirculation, and casters to allow for movement of mixer.

Platemaking table (in-line table). Consists of a one-piece molded sink with $1\frac{1}{2}$-in. lips and $5\frac{1}{2}$-in. backsplash. Requires water supply for faucets or spray headers, either hand control or pedal foot valve.

Plumbing. Manufacturer provides all sink units, preplumbed, including all faucets, valves, spouts, vacuum breakers, circulation systems, temperature blenders, chilled water systems, compressed air systems, nitrogen systems, and necessary accessories to complete the installation of the mechanical work on the units.

Remote control valves

1. Single pedal foot valve with volume control

2. Single remote hand valve with vacuum breaker.

Rotary darkroom doors. Inner and outer cylinders are fabricated in one piece of black-matte-finished petroleum hydrocarbon. Has a flash point of 700°F, a flame class rating UL 94 HB and does not emit toxic fumes. Made with a two- or three-way door, with reinforced floor and rubber gasket. Sizes from 28 to 48 in. 54-in. size with a two- or four-way door will accommodate wheelchairs and wheelchair ramps can be supplied. Doors are equipped with breakaway hardware in case of emergency.

Rotary tube film processors. Fabricated to accommodate a wide range of color and black and white processing applications. Engineered with an automatic temperature-controlled water jacket to maintain chemicals at a constant temperature. Provides daylight processing capabilities.

Sink unit construction. Sink unit composed of a molded glass fiber coated with a blend of resins that resists rapid fixers, color bleaches, iron chloride, and other photographic chemicals. The wide back ledge completely seals plumbing pipes and serves as a shelf for graduates and glasswear. The fluted bottom provides fast drainage and eliminates duckboards, while trays remain level. Water service and drainage in accordance with manufacturer's instructions.

Special purpose washer. One-piece molded sink includes water jet system with control and vacuum breaker, overflow drain connected to drain with strainer on round washer. This type of unit is recommended for large quantities of prints.

Spray header system

1. Spray header and single remote hand control,

2. Spray header and single pedal foot valve with volume control.

Squeegee board. One-piece wall-mounted molded board with slopes to drain trough at bottom and hose connection.

Storage cabinet. Fabricated of particleboard laminated with plastic with four individual spaces, each space having one fixed shelf, and two pairs of stacked doors.

Tanks and accessories. Molded of hard rubber or plastic, having from 1 to $4\frac{1}{2}$-gal capacities, with floating lids, and light-tight covers; used as a developing tank or wash tank.

Tank processing sink. A 10-in. deep tank with fluted bottoms, designed with a 7-in.-high backsplash and 5-in. wide top to accommodate accessories.

Temperature blender. Adjustable temperature blender mounted in sink backsplash with jet control, circulation jets, vacuum breaker, large dial thermometer, and check stops.

Tray processing sink. Sink is 5-in. deep with a fluted drain channel for a flat bottom, a 7-in. backsplash, and a 5-in. wide top to mount accessories, conceal plumbing pipes, and serve as a utility shelf.

Utility sink. Drop-in molded sink with stainless steel mounting rim to fit a 29-in.-deep cabinet with a 30-in.-wide top.

Walk-around sinks. Designed for island installation. One-piece molded sink with drain and combination strainer, standpipe, and tailpipe. Sink is 5-in. deep, installed in standard or special type cabinets. Endsplash is 6-in. high with $1\frac{1}{2}$ in. lips on sides. Water supply to distribution outlets required.

Water jet circulation system. Sinks to be used as washers or water-jacketed processors, equipped with jet control with vacuum breaker.

Water temperature blender. The blender used with combination sinks (5-in. deep section only) has jet inlets, jet control, shut-off valve, vacuum breaker, and thermometer.

Workstations. Free standing baffled station with two drawers and a $3\frac{1}{2} \times 4\frac{1}{2}$ in. safe viewlight. Top drawer is provided with lock and bottom drawer has a light-tight lid.

Wheelchair accessible equipment. To accommodate persons confined to wheelchairs, products and equipment have been designed to meet or exceed the requirements of the Architectural and Transportation Barriers Compliance Board, American National Standards Institute, and state and local codes. Units mounted with countertops are no higher than 34 in. from the finished floor and provide leg and knee clearance and rotary doors to have ample clearance to pass a wheelchair.

FOOD PREPARATION EQUIPMENT AND FOOD SERVICES

Adulterated. Condition of food. Term describing food that bears or contains a poisonous or deleterious substance rendering it injurious to health.

Approved. Term indicating acceptance by health authorities; term indicating conformance to appropriate standards and good public health practice.

Automatic fire protection system. System installed at the exhaust hood, in kitchen area over stoves and other cooking, baking, and frying equipment in accordance with local fire department regulations.

Backsplashes. Backsplashes occur at the backs and sides of counter tops. The turnup meets the wall surface in an enclosing condition to provide a tight fit between the counter and the wall as required by most public health authorities.

Baker's table. Maple topped table with maple backsplash on back and sides. The drawers are galvanized metal with stainless steel front and legs. The table usually contains portable ingredient bins.

Banquet carts. A banquet cart has a capacity of up to 96 plates and includes a removable humiture heat unit; the cart is wired and provides its own electrical cord and grounded plug. The cart is portable and has perimeter rubber bumper wheels with brakes.

Beam scale. Scale used in receiving areas. Beam scales are accurate, easy to use, and moveable. Many accessories available, depending on manufacturer.

Bench scale. Countertop scale with beam and/or dial head. Accessories include cabinet base, wheels, extra pans, and a heater in head to protect from moisture and dust.

Beverage dispensing. Use of the common drinking cup is prohibited in public areas. Use of any fountain, cooler, or dispenser for filling glasses or other drinking receptacle where the top rim of receptacle comes in contact with any part of the appliance is usually prohibited under most health codes.

Booster water heaters. Manufactured piece of equipment used to keep the water in dish machines at the temperature required by public health departments. Types and sizes vary.

Cabinet bases. Cabinets installed below countertops and kitchen equipment for the support of the countertop. Cabinets usually house utensils or the miscellaneous storage of small kitchen equipment. The cabinet is usually constructed of galvanized steel parts and stainless steel shelves and front.

Cabinet hinged doors. Doors, usually constructed of galvanized or stainless steel, of the sound-deadened double-pan type. The doors are mounted on stainless steel piano hinges extending the full length of the hinged side. Each hinged door is equipped with a magnetic catch and recessed grip.

Cabinet sliding doors. Doors, usually constructed of galvanized or stainless steel, of the sound-deadened pan type. The doors are removable and operated on tracks, having nylon tires with stainless steel ball-bearing wheels with guide pins, door stops, and recessed-type handles.

Carhop service. Delivery service between a restaurant or other place where cooked food or beverages, or both, are served and delivered by the owner or operator, its agents or employees, to the occupants of automobiles parked on, in or about the restaurant and the like, premises, or adjacent premises or thoroughfare, for consumption there.

Carry-out restaurant. A restaurant where the majority of the food prepared and served is "carried out" for consumption off the premises.

Chef table. Length of table depends on size of operation. Generally contains a bain-marie, self-leveling dish dispensers, overshelf, cutting board, undercounter shelving, and heat lamps to keep food warm. Other options include a heated or refrigerated base, shelving, and other countertop accessories.

Closed. Fitted snugly together, leaving no openings large enough to permit the entrance of vermin.

Codes and regulations. Food service equipment must be furnished and installed to comply with all governing regulations and health codes.

Coffee urns. Urns are usually automatic with twin coffee liners and unlimited water capacity for brewing. A swivel spray arm and heated coils for incoming water are suggested options. Many options are available.

Coffee warmer. Available in various station quantities, electrically operated. Water lines are required for some models. Warmers are fully automatic or manually operated.

Commercial cooking appliances. Ranges, ovens, broilers, and other miscellaneous cooking appliances of the types designed for use in restaurants, hotel kitchens, and similar commercial establishments.

Commercial cooking hood. Hood for the collection of cooking odors, smoke, steam, or vapors from commercial food heat-processing equipment.

Commercial food heat-processing equipment. Equipment used in a food establishment for heat-processing food or utensils and which produces steam, vapors, smoke, or odors that are required to be removed through a local exhaust ventilation system.

Commercial food-processing establishment. Establishment in which food is processed or otherwise prepared and packaged for human consumption and that is subject to sanitary regulations and periodic sanitary inspection by a federal, state, or local governmental inspection agency.

Commissary. Any establishment or concession preparing, serving, or selling food products to occupants of buildings or employees or commercial or industrial enterprises for consumption upon the premises.

Compartment steamer. Steamer that has two compartments with automatic controls for each compartment. The steamer has pull-out shelves for holding perforated pans. It has automatic boiler blowdown and a cold-water condenser. It comes with an enamel exterior finish.

Container. The package, wrapper, or other receptacle in which food may be placed and includes, but is not limited to, any cup, mug, glass, jar, can, bottle, box, or bag.

Convection oven. Single deck oven, finished in gray baked enamel with a stainless steel front and french doors with full-size double-pane thermal-type windows. The oven interiors are of aluminized steel usually provided with oven shelves and placed on adjustable legs (adjusts for height of cooking utensils).

Convenience food. Food item whose condition greatly simplifies preparation and portion control. It may be fresh (an egg, for example), frozen, dried, or canned.

Conveyor dishwasher. Partially or fully automatic rack type dishwasher available with single tank or two tanks or with prewash tank. Accessories include inspection door, automatic temperature control with visible thermometers, splash shields, and other associated parts.

Corrosion-resistant material. Material that maintains its original surface characteristics under prolonged exposure to food, cleaning compounds, and sanitizing solutions, which may contact it.

Countertops. Countertops are fabricated in one piece with welded joints. If one-piece construction is impractical, joints should be of the hairline butt type, having continuous field weld, ground smooth, and polished to the original finish. Stainless steel is the recommended metal for its wearing ability and for sanitary and health requirements. The underside of the tops is usually undercoated with a hard drying mastic material sprayed on after all work has been completed.

Cutouts. Openings made available for drop-in-type equipment. Edges allow the flange of the drop-in unit to set flush with the countertop. Where lids or insulated covers are used a spillage edge is provided around the opening.

Cutting and chopping block. Reversible block, 6-in. thick, made from selected northern hard rock maple. The cutting surfaces are finished in hot paraffin. The block is bolted to wood or steel legs.

Deck oven. Available in one, two, or three decks and many sizes, depending on requirements. Used to bake or roast. Equipped with legs for mounting on floor or on cabinet bases.

Deep-fat fryer. Available in several capacities, countertop models, or floor types. Fry basket lift may be manual or automatic. For custom-fabricated units, drop-in units are available. Fat filter may be self contained. Gas or electrically operated.

Dish dispensers. Drop-in-type tubes have capacities of up to 72 saucers. The dispenser can be adjusted to raise the last saucer to shelf height automatically.

Dishwasher. Regulated electric heat with thermostatic temperature control and low water cutoff. High-velocity jets and automatic hot water final rinse. Stainless steel wash chamber. Equipment must meet public health department approval.

Dishwater vent stacks. Vent stacks provide ventilation for dishwasher. Usually stainless steel vents are the only type that resist acids, and grease detergent solutions used in the machine.

Disposables. Term applied to plates, cups, saucers, and utensils that are made of material such as paper or plastic and designed for one-time use.

Disposer. Manufactured piece of equipment used in connection with all sinks where food scraps are deposited. Adapted collar or cone, vacuum breaker, solenoid valve, flow control valves, and drum-type manual reversing switch are important to installation. Types and sizes vary.

Drainboards. Drainboards are usually integrally fabricated with sinks. They are pitched a minimum of $\frac{1}{8}$ in./ft to permit drainage and have a continuous rim and backsplash matching those provided on the adjoining sink. The front and free ends have an approximately 3-in. high rolled rim. Where drainboards exceed 3 ft in length or are to be provided with a disposer, they should be supported with a set of legs having a cross rail and adjustable bullet feet.

Drawers. Drawers consist of a removable die-drawn coved liner constructed of galvanized metal set in a galvanized frame mounted on ball-bearing wheels. Drawer fronts are made of stainless steel and are equipped with a convenient handle or a flush stainless recessed pull. The drawer housing with enclosures is provided to meet Na-

tional Sanitation Foundation and health department approval.

Drink. Includes any and all liquid used or intended to be used as a beverage for human consumption, irrespective of any nourishing quality thereof.

Drive-in restaurant

1. Establishment where food or beverages are dispensed and where such food or beverages are consumed on the premises, but not within a building.

2. Prepared food and beverage sales enterprise that is operated in such a manner that all or a portion of its patrons consume their food and beverages while situated in automobiles parked on the premises.

3. Restaurant that is laid out and equipped to serve food and beverages to patrons in automobiles and/or to allow consumption of food and beverages by patrons in automobiles on the premises.

4. Restaurant or public eating business so conducted that food, meals, or refreshments are brought to the motor vehicles for consumption by the customer or patron.

5. Restaurant designed to permit or facilitate the serving of meals, sandwiches, ice cream, beverages, or other food directly to, or permit their consumption by, patrons in automobiles or other vehicles parked on the premises or patrons elsewhere on the site, outside the main building.

Drive-in restaurant or refreshment stand. Place or premises used for the sale, dispensing, or serving of food, refreshments, or beverages in automobiles, including those establishments where customers may serve themselves and may eat or drink the food, refreshments, or beverages on the premises.

Duct system. Ducts serving a hood should be constructed of No. 16 standard gauge steel, or stainless steel not lighter than No. 18 standard gauge steel.

Easily cleanable. Readily accessible and of such material and finish and so fabricated that residue may be completely removed by normal cleaning methods.

Employee. Any person, including the proprietor, working in a food service establishment who transports food or food containers, engages in food preparation or service, or comes in contact with food utensils or equipment.

Employees' hand-washing facilities. Employees' hand-washing facilities should be separate from utensil-washing facilities and should be located in or immediately adjacent to the food preparation area whenever possible in existing restaurants.

Equipment. Stoves, ranges, barbecue facilities, hoods, meat blocks, tables, counters, refrigerators, sinks, dishwashing machines, steam tables, and similar items, other than utensils used in the operation of a food service establishment.

Equipment drawers. Drawers consisting of a removable die-drawn coved liner constructed of galvanized metal set in a frame mounted on stainless steel ball-bearing wheels. Drawer fronts are made of stainless steel and are equipped with a convenient handle or a flush stainless recessed pull. Drawer housing with enclosures must meet National Sanitation Foundation and local health department requirements.

Equipment legs. Equipment legs are constructed of stainless steel or galvanized metal and have bullet-type adjustable feet. Legs should position equipment a minimum of 10 in. off the floor, and should be spaced no further than 6 ft apart, depending on the size and weight of the equipment.

Exhaust ventilators. Exhaust ventilators are required over equipment producing grease-ladened vapors. Installed from ceiling or wall mounted, single depth for one line of equipment or double-sided for use with island equipment. Many types of filters are available.

Fabrication. Equipment fabrication must meet the standards established by the National Sanitation Foundation (NSF) in order to qualify for approval in most states.

Fan scale. Countertop scale, generally up to 6 lb capacity. Usually automatic in operation, should have well-defined figures and graduations for accurate reading. May be supplied with stainless steel platters, scoops, or square and round pans.

Fast food. Generic term applied to the limited-menu, quick-service type of restaurant.

Flight-type dishwasher. Generally automatic with visible water temperature thermometers, inspection doors, and detergent dispensers. Optional equipment includes blow-dryers, control sensors, and energy-saving connections. Many other devices and parts available. Electrically powered with gas, electrical, or steam tank heaters.

Food. Any raw, cooked, or processed edible substance, beverage, or ingredient used or intended for use or for sale, in whole or in part, for human consumption.

Food contact surfaces. Those surfaces of equipment and utensils with which food normally comes in contact, and those surfaces with which food may come in contact or that drain back onto surfaces normally in contact with food.

Food cutter. Electrically operated food cutter for multiple uses such as salads and chopped steaks and converted for baking mixes. Knives are made of special-cutlery stainless steel. Cutter comes with a stainless steel bowl and a removable knife guard.

Food service establishment. Usually any fixed or mobile restaurant; coffee shop; cafeteria; short-order cafe; luncheonette; grille; tearoom; soda fountain; sandwich shop; hotel kitchen; smorgasbord; tavern; bar; cocktail lounge; nightclub; roadside stand; industrial feeding establishment; school lunch project; hospital kitchen; nursing home kitchen; private, public, or non-profit organization or institution routinely serving the public; catering kitchen; commissary or similar place in which the food or drink is prepared for sale or for service on the premises or elsewhere; and any other eating and drinking establishment where food is served or provided for the public with or without charge.

Food service operation. A food operation in which food is served, or prepared and served, for consumption in or about the food establishment, or in which food is prepared for service and consumption elsewhere.

Food service system design. Planning, layout, and design of volume feeding kitchens and service areas, which generally includes menu making, staff charting, information systems, internal controls, and facilities design.

Food slicer. Countertop equipment, electrically operated, for slicing meats and cheese. Adjustable speeds and for thickness of product. Various sizes and capacities are available.

Food warmer strips. Food warmer strips are mounted above an open-dispenser food shelf with infinite control, pilot light, and electric cord and plug.

Food waste disposer. Vertical unit with its exterior housing cast from heat-treated, corrosion-resistant alloy, a shredder of specially heat treated steel, a heat-treated rotor, and a totally enclosed electric motor. There is a waterproof fitting at the electrical connection.

Free-standing sinks. Stainless steel sinks with the interior vertical and horizontal corners of the sink bowls coved to a minimum of $\frac{1}{4}$-in. radius. The front rim has a rolled edge, the backsplash is usually $9\frac{1}{2}$-in. high with the ends enclosed. Holes are punched in the backsplash for each faucet. The bottom of the compartment is die-stamped with tapered grooves to a crumb cup or lever handle drain.

Frozen desserts. Includes ice cream, frozen custard, French ice cream, French custard ice cream, sherbet, fruit sherbet, ice milk, ice, water ice, quiescently frozen confection, quiescently frozen dairy confection, whipped cream confection, bisque tortoni, artifically sweetened ice cream, or artificially sweetened ice milk.

Fryer. Floor-model-type fryer with syphon filter and strainer bags, stainless steel front and top adjustable legs, twin baskets, and stainless steel tank.

Galvanized iron. Copper-bearing-type steel with tight-coat exposed surfaces painted, metal cleaned, properly primed with a rust-inhibiting primer, degreased, and finished with two coats of epoxy-base gray hammertone paint.

Garbage and refuse disposal. Prior to disposal, garbage containing food wastes is kept in leakproof, nonabsorbent containers covered with tight-fitting lids. When filled, the containers are usually stored in refrigerated areas until disposed of.

General fabrication methods. Integral construction methods of fabrication where possible, within the sheet size. Where extra sizes are required, and sheets are welded together welds, where exposed, should be ground smooth and reinforced with a like metal. "Integral construction" means design from front to rear of fabricated unit, to sheet width or length, with integral breaks, processed coved or as required with ends of contour specified, welded in, and finished.

Grease collector. Device other than a filter used to remove grease and other contaminants from the air before it enters the duct system.

Grease filter. Device used to capture by entrapment, impingement, adhesion, or similar means, grease and similar contaminants before they enter a duct system.

Grease hood. Commercial cooking hood that is at or over equipment that produces or may produce grease vapors.

Hardware. Heavy-duty, high-quality, standard hardware with parts readily available for maintenance and, where at all possible, from the same manufacturing source.

Hinged doors. Stainless steel or galvanized metal sound, deadened, double-pan-type doors with corners closed and welded. Doors are mounted on stainless steel piano hinges extending the full length of the hinged side. Each hinged door should be equipped with a magnetic catch and recessed grip.

Hood. Any air-intake device connected to a mechanical exhaust system for collecting grease vapors, fumes, smoke, steam, heat, or odors from the area over the food preparation equipment.

Hot chocolate dispenser. Manufactured piece of equipment, with enameled or stainless steel finish. Capacity is based on cups per hour. Types vary.

Hot food servers. Unit that keeps food warm or hot, ready for serving. Units are free standing and electrically operated. The food table cabinet holds several drawers.

Hot food table. Usually four-section electric-type food table with blanked-out openings, a front portion with a carving board, and enamel-finished base set on adjustable stainless steel legs. The table includes aluminum spillage pans.

Ice cream cabinet. Manufactured piece of equipment with enamel or stainless steel exterior finish. Capacities and sizes vary.

Ice machine. Manufactured piece of equipment, for making small cubes of ice, rated at pounds per 24 hours. Storage bin capacities and sizes varies. Unit comes with adjustable legs.

Ice tea dispenser. Manufactured piece of automatic equipment for tea products that have 100% cold-water solubility. Sizes and types vary.

Ingredient spice bins. Stainless steel, die-drawn, fully coved bins on ball-bearing wheels, usually mounted on the underside of the overshelf.

Inset sinks. Die-drawn, fully coved sink constructed of the same gauge stainless steel as the work surface top. The minimum depth of $8\frac{1}{2}$ in. is soundproofed, and is integrally welded in top. The sink is usually equipped with a basket-type strainer drain and swing-type spout faucets.

Island serving shelf. Shelf that is furnished with sets of stainless steel tubular supports, welded or properly secured to underside of shelf, and provided with concealed flanges for mounting to work surfaces. Supports are to be positioned at approximately 5- to 10-in. spacing or as mounting work surfaces allow.

Itinerant eating and drinking establishment. Any push-cart, wagon, truck, or other food-and-drink vending vehicle from which food or drink is prepared, dispensed, or sold.

Juice dispenser. Manufactured piece of equipment with delivery systems, using concentrated juices. Systems and dispensers vary.

Kitchen equipment. Equipment for use in connection with the preparation of food. It includes kitchen sink, or other device or equipment used as a kitchen sink; range, stove, hot plate, broiler, heater, or other device or equipment used for the cooking or warming of food; refrigerator and cabinets used for food storage or utensil or dish storage in connection with the preparation or service of food, when installed within the same room, recess, or space for other kitchen equipment.

Kitchenware. All multiuse utensils other than tableware used in the storage, preparation, conveying, or serving of food.

Lunch wagon. Any prefabricated structure brought in complete form or assembled on the site designed to be used for the purposes of a restaurant, whether standing on its own wheels or on a fixed foundation, whether or not connected with sewer or water mains.

Maintenance. All utensils and food-contact surfaces of equipment, used in the preparation or serving of food or drink, including food-storage utensils, must be thoroughly cleaned before each use.

Major kitchen. A major kitchen has been established in most codes to be a kitchen with an area of 500 sq ft, or more, in any building except a single-family or two-family dwelling.

Malt mixer. Manufactured piece of equipment with one to three spindles and three speeds. The unit comes with stainless steel and enamel combination finishes.

Meat chopper. Countertop equipment, electrically operated, for chopping fresh and frozen meats, located on counter for easy cleaning. Various sizes and capacities and attachments available.

Meat saw. Electrically operated two-speed saw for cutting fresh and frozen fish and meats.

Microwave oven. Fully automatic microwave oven with a push-button timer and preset button controls.

Milk dispenser. Manufactured piece of equipment whose capacity may vary. It is all stainless steel, inside and out, with milk and creamer valve arrangements and adjustable legs.

Mixer. Manufactured piece of equipment whose capacity may vary. It comes with a slicer attachment, shredder plate, and meat chopper assembly and is available in floor or table models. Types and sizes vary.

Mobile food service operation. One that may be moved without significant alteration of the structure or equipment after the structure and equipment has been moved from one location to another.

Mobile restaurant. Restaurant operating from a movable vehicle, trailer, or boat that periodically or continuously changes location and wherein meals or lunches are prepared or served or sold to transients or the general public. Vehicles used in the delivery of preordered meals or lunches prepared in a licensed restaurant are not included. The term "mobile restaurant" does not include a common carrier regulated by the state or federal government.

Multiple compartment sinks. Bodies designed like single-compartment bowls. Partitions between the bowls consist of two pieces of the same gauge metal. Where partitions occur, no straps or beads occur on the outside of the sink bowls. Where ends are without drainboards and are free of walls, they terminate in a rolled edge. Bull-nosed rim corners are formed and welded into rolled edges.

Multi-service items. Includes all utensils that are reused for food preparation, eating, and drinking, and must be of materials that are nontoxic in nature.

Off-own premises food caterer. One who provides prepared bulk food, including beverages, if any, for off-own premises consumption, with or without other service incidental thereto.

Overhead pot rack. Rack constructed of galvanized iron flat bars having radius corners at the ends on the triple-bar type and square ends on the double-bar type. Hooks are the malleable or corrosion resistant double-duty, sliding type on 6-in. centers. Mounting supports have an epoxy-base hammertone finish. Each end of the island-type rack has a piece of tubing welded to flat bars. Type of mountings in a kitchen for overhead pot racks can be an island, wall, or ceiling; triple or double bar rack.

Overshelves. Shelves constructed of stainless steel or galvanized metal with $1\frac{1}{2}$-in. flange down front edge. Back and ends are constructed of 1-in. minimum flat stainless steel risers with free corners and edges rounded. Shelves are provided with applicable stainless steel supports.

Pass-through opening. Stainless steel shelf the full length of the opening. Where required by fire regulations, there is an automatic door, frame, and closing device.

Perishable food. Food of such type or in such condition that it may spoil unless the proper refrigeration is provided.

Permanent ware. China, glassware, silver, and linen made of reusable and long-lasting materials.

Pipe and tubing. Seamless or welded stainless steel or galvanized pipe and tubing of true roundness, properly annealed, pickled, and ground smooth. Stainless tubing, where exposed to view, should have a finished grind with not less than 150 grit emery.

Portable silver soak sink. Rack with chute, lever handle waste, casters, and setting brake.

Potentially hazardous food. Any perishable food that consists wholly or in part of milk products, eggs, meat, poultry, fish, shellfish, or other ingredients capable of supporting rapid and progressive growth of infectious or toxigenic microorganisms.

Pot sink with overflow basket. Three-compartment pot sink with integral drainboards, overflow compartment, and basket. Drainboards and sink compartments should have sufficient slope for complete drainage. Lever or rotary handle wastes are recommended. Strainer basket and plug also available. Recirculating water agitator may be installed on soiled dish compartment for ease in waste removal. Overflow basket may be installed next to soiled dish compartment with the suggested perforated basket. Final rinse sink supplied with a sanitizing solution or heater to raise water temperature to required level. All exposed surfaces and parts fabricated from stainless steel.

Pot shelving. Shelving, usually of open wire construction, with zinc plating on a baked protoxy coating on shelves and uprights.

Prerinse spray. Deck-mounted spray with a temperature control valve, usually mounted on top of the backsplash ledge on the soiled dish table over the prerinse sink.

Range. Ranges are available for gas or electric and are manufactured with various numbers of burners and cook tops. Fry top is equipped with grease trough, burner top with drip tray, griddle top with spillage channels. Ranges can be installed in countertops or mounted on a cabinet or oven base with flue riser and shelves. Range should be installed under vented hood with a fire protection system.

Ready foods. Derivative of the phrase "table-ready foods"; also called "convenience foods."

Receiving scale. Manufactured piece of equipment, usually with a single aluminum weighbeam and a 1000-lb (minimum) capacity. Graduations are from $\frac{1}{4}$ to 50 lb, or as selected. Types and sizes vary.

Refrigerator. Manufactured in many types and capacities, such as roll-in, pass through, mobile, and undercounter. Available with sliding or hinged full or half doors. Units with dual temperature control for combination refrigerator/freezer compartments. Refrigeration compressor may be self-contained or free standing.

Restaurant

1. Building in which food is prepared and sold for consumption within the building, as opposed to a drive-in restaurant establishment where food may be taken outside of the building for consumption either on or off the premises.

2. Place where meals are served to the public.

3. Retail establishment offering food or beverages, or both, for consumption on the premises. Restaurants do not usually include barrooms, nightclubs, or lounges.

4. Establishment principally offering food for consumption on the premises.

5. Establishment, including tearoom, at which food is sold for consumption on the premises. However, a snack bar or refreshment stand at a public or non-profit community swimming pool, playground, or park, operated solely for the convenience of patrons of the facility, is not usually deemed to be a restaurant in most codes.

6. Establishment selling prepared food and drink for consumption on the premises, but not providing dancing or entertainment.

7. Establishment, usually including cafes, coffee shops, ice cream parlors, soda fountains, hamburger and hot dog stands, and any other store of a similar nature, where prepared foods or convenience foods are sold.

8. Building, room, or place where meals or lunches, including soft drinks, ice cream, milk, milk drinks, ices, and confections, are prepared, served, or sold to transients or the general public.

Roll-in freezer. Manufactured self-contained freezer with stainless steel interior and exterior. Door hinging is by selection. Types and sizes vary.

Roll-in refrigerator. Manufactured self-contained refrigerator with enamel or stainless steel finish. Door hinging is by selection. Types and sizes vary.

Roll warmer. Three-drawer stainless steel cabinet unit, with automatic electric control.

Salamander broiler. Installed in the shelf area above the range top where possible. Used for light broiling and finished with drip shield to protect range top. Optional gas or electrically operated.

Sandwich grille. Electrically operated countertop unit, available in many sizes.

Sandwich unit. Unit used for the preparation of salads and sandwiches. Refrigerated pans are accessible from top preparation area. Base or cabinet is refrigerated for storing food supplies. Refrigeration compressor is self-contained. Many sizes and types are available.

Sanitize. Make sanitary by effective bactericidal treatment of clean surfaces of equipment and utensils by hot water or chlorine or by any other method that has been approved by the local or state health officer as being effective in destroying microorganisms, including pathogens.

Sealed. Free from cracks or other openings that permit the entry or passage of moisture.

Single compartment sinks. Sinks whose ends are without drainboards, free of walls, and terminated in rolled edges. Bull-nosed rim corners are formed and welded into rolled edges.

Single-service articles. Cups, containers, lids, or closures; plates, knives, forks, spoons, stirrers, toothpicks, paddles, straws, place mats, napkins, doilies, wrapping materials, and all similar articles that are constructed wholly or in part from paper, paperboard, molded pulp, foil, wood, plastic, synthetic, or other readily destructible materials and that are intended by the manufacturers and generally recognized by the public as for one use only, then to be discarded.

Single tank dishwasher. Generally used in small food service operations. Manual operation for straight-through or corner installations. Available with low temperature control for use with sanitizing solution.

Sliding doors. Stainless steel or galvanized metal pan-type, sound-deadened doors with perimeter channel reinforcing on the interiors. Doors are removable and operate on a track with nylon tires on stainless steel ball-bearing wheels having bottom guide pins, and door stops. Doors to have recessed type handles.

Soft ice cream dispenser. Manufactured piece of equipment with enamel or stainless steel finish. Capacities and types vary.

Soft serve. Available in floor or counter models, electrically operated, for mixing yogurt or ice cream products. Various sizes and capacities available.

Soiled and clean dish tables. Units adjacent to the dishwashing machine having a quick-drain trough the full width of the table. Recessed scrapping troughs are sloped to the disposer sink. The table contains a prerinse sink whose minimum depth is 8 in. with a removable perforated stainless steel scrap basket and basket drain.

Soup wells. Automatically controlled electric warming wells with a minimum of two spaces. Sizes vary.

Stainless steel. Stainless steel used in fabrication must be new material, low carbon, non-magnetic, austenitic 18% chrome-8% nickel, Type 302, No. 3 mill finish, as listed by the American Iron and Steel Institute (ASTM A 167).

Stationary steam kettle. Usually 40-gal-type kettle with a modular base, baked-enamel finish, and stainless steel adjustable legs. It comes with steam connections and hot and cold fill faucets.

Steam jacketed kettle. Units available in various sizes and capacities. Tilting or stationary types, with tri-leg pedestal, floor mounted. Units may be gas, electric, or steam.

Steamers. Available with one, two, or three compartments. Used to rewarm and heat bread, rolls, pies, and cakes and for many other purposes. Available in various sizes and types.

Step-in utility cooler. Prefabricated, job-installed cooler, complete with air-cooled compressor. Types, sizes, and equipment vary.

Structural steel. Material used for framing or bracing consisting of angle iron, bands, bars, channels, etc. It is ductile and free of hard spots, runs, checks, cracks, and other defects. Excess welds are ground and rounded to eliminate sharp protrusions. Structural steel surfaces are primed with rust-inhibiting primer and a heavy coat of high-grade aluminum paint.

Table-service restaurant. Food service operation in which patrons are served only at tables.

Tableware. All multiuse eating and drinking utensils, including flatware (knives, forks, spoons).

Take-out service. A food establishment that serves or dispenses food through a window or door directly to the public for consumption on or off the premises is offering a take-out service; any food establishment that serves or dispenses food inside its premises, but for consumption of such food off the premises only is offering a take-out service.

Temporary food service establishment. Usually any food service establishment that operates at a fixed location for a temporary period of time, not to exceed several weeks, in connection with a fair, carnival, circus, public exhibition, promotional activity, or similar transitory gathering.

Tilting skillet. Skillet with a minimum 9-in. depth, made of polished stainless steel pan. The base has a baked enamel finish. The unit comes with stainless steel adjustable legs.

Toaster. Manufactured piece of electrical equipment available as a popup or continuous ladder type. Sizes, capacities, and types vary.

Undercoating. Sound-deadening, hard-drying mastic material on the underside of all tops.

Undershelves. Stainless steel or galvanized metal undershelves. Corners are die formed and fitted with a slotless contour head bolt for a tight sanitary contact with legs. Edges are rolled down 90°, the flange having rounded corners. The top surfaces of undershelves are usually positioned 10 in. above the floor.

Utensils. Tableware and kitchenware used in the storage, preparation, conveying, or serving of food, including such items as straw dispensers, silverware containers, and napkin holders.

Vegetable preparation sink. Stainless steel or galvanized metal sink with two compartments, each at least 14-in. deep. There are drainboards on each side, with one drainboard accepting the disposer cone. The sink is to have lever handle waste, with basket strainer and tail piece, including faucet and spout. The supporting structure for the sink should have stainless steel legs with adjustable bullet feet.

Vending machine. Any self-service device offered for public use that, upon insertion of a coin, coins, or token, or by other means, dispenses unit servings of food or beverage, either in bulk or in package, without the necessity of replenishing the device between each vending operation, but not including devices dispensing peanuts, wrapped candy, gum, or ice exclusively.

Waffle baker. Manufactured piece of equipment with teflon-finished grids. Sizes and types vary.

Waitress stations. Custom or standard designed manufactured items. Finishes, types, and sizes vary.

Walk-in cooler-freezer

1. Prefabricated, job assembled, modular sandwich panel, cold storage room. Floor insulation, finished hardware, light switch assemblies, light fixtures vapor proof, freezer door heater wire, finished trim, heated pressure relief vents, coil supports, utility penetration openings and closures, interior ramps, dial thermometers. Types, sizes and equipment vary.

2. Prefabricated, installed-at-job freezer combination, with metal interior and exterior and high-density inplace or bonded urethane foam (4-in. minimum thickness) insulation. Either floor or floorless types are available. Wire shelving is zinc plated with baked-on protoxy coating. Types and sizes vary.

Waste compactor. Generally located at the point of waste generation area or near central accumulation area. Waste compacted in bags or boxes.

Welds. All welding performed for the fabrication of the equipment should provide welds that are full and homogeneous, using noncorrodible rods for stainless steel work, square-formed without pits, piping, chilled or disintegration and with all excess weld to be ground to original gauge. Fabrication marks and discoloration should be removed and finished same as original material.

Wholesome. In sound condition, clean, free from adulteration, and otherwise suitable for use as human food.

Wire shelving. Stationary or portable shelving, assembled usually in five-shelf units. Finishes are chrome with protoxy, stainless steel, zinc, or zinc with protoxy. It is designed to carry weights of from 115 to 500 lb.

Yogurt dispensers. Manufactured piece of equipment with enameled or stainless steel finish. Capacities and types vary.

12

Furnishings

DECORATIVE TREATMENTS AND MATERIALS

Accordion shades

1. Shades made of accordion pleats sharply creased at regular intervals horizontally across their width.
2. Fabric shades that open and close like an accordion, made from fabric bonded to a stiff backing.

Acele. Trademark of E.I. DuPont de Nemours & Co. (commonly known as DuPont) for acetate fiber.

Acetate

1. Synthetic fiber made from cellulose acetate. Solution- and spun-dyed acetates are colorfast to sunlight, and air pollution. Acetate is often used for very luxurious fabrics as it resembles silk; it is occasionally mixed with other fibers to give additional sheen.
2. Acetate burns and melts while in a flame and after removal from the flame it leaves a hard, brittle, black bead. It may smell like vinegar when burning.
3. Synthetic fiber with good sun resistance.

Acrilan. Trademark of Monsanto Textiles Co. for acrylic fiber.

Acrylic

1. Synthetic fiber made from acrylonitrile, which comes from coal, air, water, petroleum, and limestone. It is lighter in weight for the warmth it gives than other fibers and is extremely popular for blankets as a substitute for wool.
2. Acrylic burns and melts while in a flame and after removal from the flame. Burning leaves a hard, brittle, black bead.

Alencon lace. Fabric with a solid design outlined in cord on a shear net background.

Alpaca

1. Soft, silky, woollike hair from the South American animal, alpaca, woven into a fabric.
2. Type of llama that has very long hair which is considered a wool. Alpaca fabric is a soft, silky, and fairly lightweight fabric resembling mohair.

Angora

1. Yarn made from the fleece of angora sheep.
2. Long and soft wool of the angora goat which is called "mohair" when made into fabric or yarn. It is usually blended with other fibers. The angora rabbit has soft, silky hair which is also made into fabrics.

Anidex

1. Synthetic elastic fiber with exceptional resistance to sunlight and heat.
2. Synthetic fiber made from a monohydric alcohol and acrylic acid which gives permanent stretch and recovery to fabrics and resists gas, oxygen, sunlight, chlorine bleaches, and oils.

Anim. Trademark of Rohm and Haas Co. for anidex fiber.

Antimacassar. Piece of cloth that was originally pinned to the back of a chair to protect the upholstery from hair oil.

Antique satin

1. Satin weave fabric primarily used for draperies. It can be used on either side. The face is a classic lustrous satin; the reverse has a slubbed look similar to shantung.
2. Heavy lustrous fabric woven with uneven yarns.

Antistatic. Buildup of static electricity is a problem with many of the synthetic fibers. Antistatic finishes have been developed for materials made from these fibers.

Antron. Trademark of E.I. DuPont de Nemours & Co. for a type of nylon.

Applique´

1. Design stitched or glued onto a background fabric.
2. Decorations consisting of a piece of fabric cut out and added to another fabric by sewing, embroidering, gluing.

Arnel

1. Trademark of Celanese Corporation for triacelate fiber.
2. Synthetic fiber distinguished by its tolerance of heat.

Arras. Handwoven tapestry.

Armure. Raised satin pattern on a fabric with a repp background.

Asbestos. Mineral fiber that is nonmetallic and whose greatest virtue is that it is nonflammable. It is used in combination with other fibers for theater curtains and where flameproofing is essential.

Astrakhan. Wool from karakul lambs. The term is also used to describe fabric woven or knitted to look like this wool, i.e., curly and fairly heavy.

Astrakhan cloth. Heavy pile fabric with curled loops.

Aubusson. Tapestries made in Aubusson, France, and designed for use as wall hangings. The term is also applied to patterned rugs with little or slight rib and no pile.

Austrian drape. Shirred fabric treatment for windows, which gives the effect of vertical rows of swags from top to bottom.

Austrian shade. Shade treatment achieved by gathering vertical sections of the shade to produce a swagged effect.

Austrian shade cloth. Crinkled, woven, stripped cotton, silk, or synthetic fabric.

Austrian shades. Shades made of fabric that is shirred across the width of the shade. When drawn up, Austrian shades hang in graceful loops of fabric.

Avlin. Trademark of FMC Corporation for polyester.

Avril. Trademark of FMC Corporation for high-wet-modulus rayon.

Azion. Synthetic fiber made from regenerated, naturally occurring proteins. It gives a soft feeling when blended with other fibers.

Backed fabric. Fabric with an extra warp or filling, or both, to make it heavier and thicker.

Bagheera

1. Uncut-pile velvet fabric which is crease resistant because the surface is not smooth.

2. Fine uncut-pile velvet with a rough, crush-resistant finish.

Baize

1. Fabric made from wool or cotton, originated in Baza, Spain.

2. Loosely woven fabric originally made from cotton or wool but now made of other fibers.

Baku. Dull-finish straw fabric.

Baline. Plain-woven coarse fabric used for stiffening in upholstery work.

Ball fringe. Trimming consisting of round fluffy balls (pompoms) attached by threads to a band of fabric by which the trimming is sewn to fabric. It is often used on curtains and upholstery.

Ban-lon. Trademark of Bancroft Licensing for a texturizing process that uses heat-setting to add bulk and a small amount of stretch to the filament yarns of thermoplastic fibers.

Barathea. Fine cloth with a broken rib pattern, originally made of silk or wool but now made in many fibers.

Barege. Sheer, gauzelike fabric of wool combined with cotton or other fibers.

Bargello. Needlepoint stitch.

Barkcloth. Fabric found throughout the South Pacific, made from the inner bark of certain trees. Bark is beaten into a paperlike fabric and is then dyed or otherwise colored. Tapa cloth is one of the best known types of true barkcloth. Fabric that resembles barkcloth is used extensively for draperies, slipcovers, and other home furnishings.

Bark crepe. Fabric designed to resemble bark, but the effect is more exaggerated than in the case of barkcloth. Usually, one fiber is used for the warp and another for the filling to help create the textured look of bark.

Barre. Fabrics either knit or woven, in which stripes run in the crosswise direction; flaw in fabric, which appears as unwanted crosswise stripes of texture or color.

Basket weave

1. Textile woven with large, similarly sized warps and wefts.

2. Decorative loose weave; monk's cloth; oxford cloth.

3. In basket weave, pairs of warp yarns pass over pairs of filling yarns in a plain weave. Gives a more open, slightly coarser looking fabric than does conventional plain weave with its single yarn.

Baste. Stitch loosely to hold fabric sections together before final stitching.

Bast fibers. Fibers obtained from stalk plants, for example, jute, flax, and ramie.

Batik. Brightly colored fabric, usually of linen or rayon, used for curtains, bedspreads, and wall hangings.

Batiste

1. Fine sheer fabric, usually made of cotton, with a lengthwise striation.

2. Sheer or semisheer curtain fabric made of cotton, silk, or polyester.

3. Fine, sheer, lightweight fabric, that may be made of almost any fiber. The degree of sheerness depends on the fiber. Cotton and synthetic batiste are probably the lightest and sheerest, wool batiste the heaviest.

Batting

1. Filling material that can be used to stuff pillows and quilts. It used to be made of cotton but now is usually polyester fiberfill.

2. Carded cotton used for stuffing and padding upholstered furniture.

Bayadere. Fabric of strong contrasting multicolored horizontal stripes.

Beach cloth. Imitation linen crash made of lightweight cotton warp and mohair filling.

Bead curtains. Individual strings of beads, either glass, wood, plastic, or ceramic, joined together to form a curtain.

Bedford cloth. Strong woven fabric with lengthwise ribs, used extensively for upholstery. It may be made from any fiber.

Bedford cord. Strong, durable rib-weave fabric, made of wool, silk, cotton, rayon, or a combination of these.

Bedspread. Fabric covering for tops and sides of a bed.

Beetling. Finishing process in which linen is pounded to produce a hard, flat surface with a sheen.

Belting. Heavyweight, fairly stiff fabric that comes in various widths.

Bengaline. Strong fabric with clearly defined crosswise ribs.

Bestpleat. Woven pleater tape with spaced pockets for hooks.

Bias

1. True diagonal of a piece of fabric.

2. Diagonal of a woven fabric between the warp and the filling threads; part of a woven fabric that has the greatest amount of stretch.

Biconstituent fiber. Fiber made by mixing two different man-made fiber materials together in their syrupy stage before forcing it through a spinneret.

Binding. Narrow fabric used to enclose (bind) edges, usually raw edges. It can also be used for purely decorative purposes. Bias tape is often used as binding.

Bird's-eye. Fabric woven with a pattern that has a center dot and somewhat resembles the eye of a bird; popular pique weave.

Bleaching. Process used to remove impurities in fabrics to obtain a pure color.

Blend

1. Combination of fibers that produces a fabric that has the good qualities of both fibers. The development of blends of polyester and cotton, producing fabrics that require a minimum of ironing, has been one of the most significant developments in fabrics. The term "blend" refers only to fabrics made from yarns that have been spun to combine the two fibers in one yarn. The term "mixture" should be used to describe fabrics in which, for instance, the warp thread is polyester and the filling thread is cotton.

2. Two or more fibers mixed together to make one yarn.

Blind. Shade or screen device used over a window area to control light.

Blind stitching. Method of sewing in upholstery to achieve no final visual effects.

Blister. Bump on a fabric. Blisters are often used to give additional depth to a design. Flowers may be blistered to make them stand out from the rest of a fabric. Blister crepe is technically a fabric produced chemically by shrinking some of the yarns and leaving others unshrunk in a crepe pattern after the fabric is manufactured.

Block print. Fabric printed by hand.

Block printing. Hand-printing process in which a design is carved on a block of wood or linoleum. Dye is placed on the surface, and the block is then placed on the fabric, transferring the dye. Every color requires a different block, making this type of printing both tedious and expensive.

Bolster. Long, narrow pillow of round or rectangular shape used for decoration and support.

Bolt. Term referring to a quantity of fabric. Fabric and the board on which the fabric is wrapped, together are called a bolt. A bolt usually has between 15 and 20 yd of fabric.

Bonded fabric

1. Bonding is the term used for the joining together of layers of fiber, usually through the use of glue or the melting of some of the fibers. Nonwoven fabrics are produced this way. The term "bonded" is occasionally used as a synonym for "laminated."

2. Lightweight, usually knitted cloth that is strengthened and thickened by having a lining material fixed to it permanently by means of heat and pressure. Often a very thin layer of foam interlining is placed between two layers of the fabric.

Bonding. Joining two layers of fabric with glue or a web of fibers that melts when heat is applied. Nonwoven fabrics are made in this way. The term is occasionally used as a synonym for "laminating", but this is technically incorrect.

Boucle

1. Rough, fairly thick, quite slubby yarn. Fabric made from boucle yarn, also called "boucle," has a textured nubby surface which is usually dull unless shiny yarns are used. Fabrics may be woven or knit by hand or machine.

2. Fabric or trim distinguished by small spaced loops on the surface.

3. Nubby fabric made of various fibers with little loops on the surface, used in draperies, slipcovers, and bedspreads.

Box-edged pillow. Pillow that is three dimensional rather than two sided, as are most bed pillows. The pillow fabric covering is shaped like a round or rectangular box and has a fabric band (boxing) that covers the edges of the pillow and joins the top and bottom sides.

Boxing. Straight strip of fabric that covers the sides of a three-dimensional round or square pillow. Boxing is joined to the rest of the cover with seams, which occasionally include a decorative trimming such as welting.

Boxing strip. Narrow strip joining top and bottom sections of a pillow cover.

Box pleating. Technique in which fabric is folded back on itself and folded back again in an opposite direction.

Box pleats

1. Heading used to give controlled fullness to draperies, slipcover skirts, and dust ruffles. At spaced intervals sections of fabric are folded and pressed or sewn flat.

2. Box pleats are made by folding fabric so that the edges of two pleats face in opposite directions on the right side of the fabric.

Bracket. Device used to support a curtain rod.

Braid

1. Narrow strip made by intertwining several strands of silk, cotton, or other fabric.

2. Tape or ribbon that is woven the same on both edges.

Braiding. Method of making fabric by interlacing three or more yarns or strips of fabric; also known as "plaiting."

Broadcloth

1. Lustrous cotton cloth with a right, plain weave and a crosswise rib.

2. Originally broadcloth was any fabric made on a loom of a certain width, but now it is a fine, tightly woven fabric with a faint rib. Originally, it was made of mercerized cotton, but now it is made of any fiber. Wool broadcloth usually has a soft, slightly napped surface and is of medium weight.

Brocade

1. Fabric resembling embroidery, woven on a jacquard loom.

2. Fabric made with a jacquard weave. It has a prominent and raised design.

3. Rich satin, silk or velvet fabric with a heavy texture and elaborate decoration, used in draperies and upholstery.

Brocatelle

1. Heavy fabric resembling damask, with the pattern appearing to be embossed.

2. Fabric resembling stiff brocade, with an embossed effect, and used for drapery fabric and upholstery.

Broche. Silk or satin ground fabric similar to brocade.

Brussels curtains. Curtains made of net with an embroidered design, done either by hand or machine, over the net. The net may be of one or two layers.

Buckram. Stiffly finished, heavily sized, plain-weave fabric.

Buckskin. Fabric made in a form of satin weave with a napped finish. Originally made of wool, buckskin is now made of varios synthethic fabrics and has a smooth surface with or without a napped finish.

Bullion

1. Twisted, cordlike fringe used on upholstery instead of a fabric skirt.
2. Twisted, shiny, cordlike fringe used primarily in upholstery.

Burlap

1. Plain-weave cotton, jute, or hemp that is coarse in finish and loosely woven.
2. Rought-textured fabric made from jute or hemp, used for bedspreads, draperies, and wall coverings with paper backing.
3. Coarse, heavy fabric made of jute and used for upholstery, wall coverings, commercial items, and occasionally, fashion items. It dyes well but may have a disagreeable odor unless treated.

Butcher's linen. Strong, heavy, plain-weave fabric.

Butcher rayon. Medium-weight rayon fabric woven in a plain weave.

Cafe curtains

1. Short curtains, usually made with a scalloped top.
2. Curtains that are hung in tiers so that one row covers the top half of a window, a second row the bottom. They are hung on wood or metal poles placed across the top and center of the window. The curtains are often finished with scalloped edges through which the poles slide; there are also cafe curtain rings available for hanging them.
3. Short curtains deriving from the style in French coffee houses.

Calendering. Finishing process for fabrics, producing a flat, smooth, glazelike finish.

Calico

1. Plain-weave printed cotton fabric, similar to percale.
2. Lightweight plain-weave cotton cloth that originated in Calcutta, India.
3. Smooth-surfaced plain-weave cloth. Term is applied to fabric with bright, sharply contrasting, usually small-print designs. Traditionally, calico is a popular fabric for patchwork.

Cambric

1. Plain-weave fabric finished with a slightly glossy surface. The fabric is traditionally made from cotton or linen but can be made from any fiber. Its major use is in upholstering, where it is used on the underside of chairs and sofas.
2. Soft, white, loosely woven cotton or linen fabric.

Cambric finish. Glossy finish applied to fabrics.

Camel's hair. True camel's hair, a luxury fiber, is considered a wool and comes from the camel. It is almost always blended with another fiber—sometimes sheep's wool or man-made fibers.

Camlet. Fabric made from camel's hair.

Candlewick

1. Fabric with a chenille effect.
2. Thick, soft yarn used to form tufts by pulling it through a base fabric and then cutting it. Term also describes the fabric made by this method.

Canvas

1. Heavy, strong, usually plain-weave fabric, which historically was made of flax, hemp, or cotton. It is usually made of cotton, but some fabrics made of man-made fibers or blends are also called "canvas." Canvas is heavier than duck or sailcloth, although the three names are often used interchangeably.
2. Heavy cotton or linen fabric with an even weave.

Caprofan. Trademark of Allied Chemical for nylon.

Carding. Process in the conversion of cotton, wool, some silks, and man-made staple fibers into yarn. Carding separates the fibers and causes them to lie parallel to each other.

Carpet brocade. Fabric in which the pattern is formed by heavy, twisted yarn tufts on a ground of straight fiber yarns.

Cartridge pleats

1. Rolled heading used to give controlled fullness to draperies. At spaced intervals, sections of fabric are rounded and stuffed with crinoline to hold their shape.
2. Unpressed, very narrow pleats, usually used more as decoration than to control fullness.

Casement cloth

1. Lightweight sheer drapery cloth of cotton, silk, rayon, mohair, or a mixture of any of these.
2. Term applied to lightweight, closely woven, opaque fabric used for cutrains.
3. Sheer fabric made of various fibers, used for curtains and draperies.
4. Term for fabrics, usually sheer, that can be used for cutrains, draperies, or shades, although in practice the term is usually limited to open-weave curtain fabrics.

Cashmere

1. Wool of the cashmere (or Kashmir) goat, noted for its softness. Cashmere is one of the luxury fibers and is usually blended with sheep's wool or man-made fibers.
2. Soft wool textile or yarn made from Indian goat hair.

Cellulose. Naturally occurring polymer (giant molecule) that forms the solid framework of plants. Cellulose from wood pulp is the base for rayon and acetate, both of which are man-made fibers. Cotton is more than 90% cellulose before it is cleaned (scoured).

Cellulosic fibers (cotton, linen, rayon). Cellulosic fiber burns rapidly with a yellow flame and continues to glow after it is removed from the flame. It smells like burning paper and leaves a soft, gray ash.

Chainette fringe. Fine fringe made of yarn resembling a chain, used on Austrian shades.

Chair tie. Cord with a tassel at each end, used to tie seat cushions to a chair.

Challis

1. Soft, lightweight fabric made of wool, cotton, or man-made fibers. It is traditionally printed with vivid floral patterns on dark grounds or with paisley designs.
2. Soft fabric woven of wool silk, rayon, or cotton.
3. Fabric resembling soft, lightweight wool, now also made of cotton and rayon.

Chambray. Lightweight cloth with a colored warp and a white filling thread, originally made of cotton, but now of any fiber.

Channel. Passage stitched into the top of a curtain to allow the rod to slide through.

Chantilly lace. Bobbin lace with a delicate ground.

Charmeuse. Soft, silken luster produced on fine cotton-warp sateens.

Check. Small, regular pattern of squares that is woven or knitted into or printed on a fabric.

Cheesecloth. Loosely woven, plain-weave fabric fashionable for curtains.

Chenille

1. Pile yarn originally made by weaving a pile fabric and subsequently cutting it into strips. Chenille is popular in rugs, bedspreads, and bathroom accessories.
2. Yarn covered with short cut fibers of pile, used chiefly in bedspreads and rugs.
3. Yarn made of several fibers that has a short cut pile.

Chiffon. Sheer, lightweight, drapable woven fabric, originally made of silk, but now man-made fibers. Chiffon is available in a wide range of colors from soft pastels to bright bold colors.

China silk

1. China silk has been replaced with lining fabrics of man-made fibers.
2. Sheer plain-weave silk fabric.

Chine. Speckled; term applied to fabrics in which the warp threads are printed before weaving while the filling threads are left plain, giving a shadowy effect to the finished fabric.

Chintz

1. Fine cotton cloth with printed design.
2. Gaily printed or solid colored, highly glazed fabrics.
3. Cotton fabric with or without a glazed finish, used for slipcovers, curtains, and draperies.
4. Closely woven, plain-weave fabric printed in bright designs that are most often floral. Most chintz have a glazed finish. It is used extensively for draperies and upholstery and is one of the few fabrics still made almost exclusively of 100% cotton; however, it can be made in man-made fibers.

Cire. Extremely shiny, glossy surface given to fabrics as part of the finishing process. Cire fabrics have a much higher shine than glazed fabrics and are usually somewhat slippery.

Cloth. Fabric or material. Implicit in the word "cloth" and not in "fabric" or "material" is the use of fibers to produce the resulting product.

Colorfast. Term that implies that the color in a fabric will not wash out or fade upon exposure to sunlight or other atmospheric elements. There are no standards established for the use of this term.

Combing. Process in the manufacture of cotton and man-made yarns in which the fibers are combed to remove short lengths of fiber leaving only longer ones. Combed fibers are finer than those not combed.

Continuous filament. Term that emphasizes the long, uncut nature of a filament of fiber, always man-made except in the case of silk.

Cord. Heavy, round string consisting of several stands of thread or yarn twisted or braided together.

Corded fabric. Fabrics with a lengthwise rib, often woven in stripes. Any fabric with a lengthwise rib is a corded fabric.

Cordelan. Kohjin Company's biconstituent, flame-retardant fiber of 50% vinyl and 50% vinyon.

Cord gimp. Gimp with cord superimposed for added decorative effect.

Cording

1. Cotton cord covered with bias strips to form welting.
2. Round decorative edging. Term also used to describe white cord which can be covered with bias strips of fabric to form welting or piping.

Cord tension pulley. Device to prevent tangling of traverse-rod cords.

Corduroy

1. Sturdy cotton fabric with closely spaced soft pile cords, used in upholstery and bedspreads.
2. Corded fabric in which the rib has been sheared or woven to produce a smooth, velvetlike nap. Usually made of cotton, it now can be made of many different fibers.

Cornice

1. Decorative heading for window draperies, often covered with fabric to match. It has corners and usually juts out into the room. The framework is usually made of wood.
2. Decorative strip of wood or buckram used to conceal curtain and drapery hardware.

Cotton

1. Most important of all textile fibers. It is strong, easily dyed, and holds color.
2. Fiber from the cotton plant and also the fabric made from this fiber. Different types of cotton plants produce cotton of higher or lower quality usually associated with staple length and fineness of the fiber.

Cotton ball tassel. Ball fringe with a tassel effect on each ball.

Cotton loop ball fringe. Flat heading hung with alternating loops and pompons.

Coverlet. Short bedspread, usually used with a skirt.

Crash

1. Cotton, jute, or linen fabric having coarse, uneven yarns and a rough texture.
2. Durable, soft, rough-textured fabric, used in linens, draperies, and some upholstery.
3. Coarse-woven fabric with a rough surface, used as a curtain fabric.

Crease resistant fabric. Fabric has been treated so that it will wrinkle less than it would normally. Fabrics are usually made crease resistant as part of the finishing process.

Crepe

1. Dull-surfaced fabric with an allover crinkled surface. A crinkled surface may be obtained through the use of crepe yarns (yarns that have such a high twist that the yarn kinks), chemical treatment with caustic soda, embossing, or weaving (usually with thicker warp yarns and thinner filling yarns). Although crepe is traditionally woven, crepe yarns are also used to produce knit crepes.
2. Fabrics having a crinkled surface obtained by hard twisting of yarns, chemical treatment, type of weave, or embossing.

Crepe-backed satin. Satin that can be used on either side. The face is satin; the back is crepe.

Crepe de chine

1. Traditionally, a very sheer silk fabric; lightweight crepe, usually of man-made fibers.
2. Soft, lustrous silk fabric with a crinkled surface, used for curtains.

Cretonne

1. Printed fabric, resembling chintz, but lacking the glazed finish.
2. Firmly woven cotton, linen, or rayon fabric, used for slipcovers and draperies.

3. Printed drapery and home furnishings fabric, similar to unglazed chintz, traditionally made of cotton.

Crewel

1. Embroidery in wool on almost any fabric.

2. Embroidery that utilizes almost every embroidery stitch and is worked with a fairly thick wool yarn called "crewel yarn." Designs are often quite large and often extremely stylized.

Crinkle crepe. Fabric with an uneven surface created by the use of caustic soda which causes it to shrink unevenly. Plisse´ is an example of a crinkle crepe fabric. Crinkle crepe and plisse´ usually have a larger pattern to the surface irregularities than crepe.

Cross-dyed fabrics. Fabrics subjected to several dyeing processes for multicolor effect.

Cross-dyeing. Method for coloring fabrics made from more than one kind of fiber, for example a wool and cotton blend. Each fiber in a fabric designed for cross-dyeing will take a specific dye in a different color or in variations of one color. A fabric that is cross-dyed is more than one color. Cross-dyeing is often used to create heather effects (soft, misty colorings), but strongly patterned fabrics can also be achieved.

Curtain

1. Movable covering of lightweight materials for windows, doors, or alcoves.

2. Unlined, often sheer, window hanging.

Curtains and draperies. Curtains are window coverings, usually unlined, that hang within the framework of the window, ending at the windowsill. Draperies are almost always lined and are usually made of fairly heavy fabrics, such as satin or velvet. They normally hang to the floor, but occasionally in very formal rooms they may even lie on the floor. Curtains and draperies can be made of almost any fabric. Fabrics made of glass fiber for window coverings as curtains or draperies are now available.

Dacron

1. Trademark for a polyester fiber manufactured by Dupont.

2. Synthetic fiber distinguished by extreme wrinkle resistance.

Damask

1. Firm, glossy, patterned fabric, with a jacquard weave. Damask can be woven in cotton, rayon, linen, silk, wool, or a combination of any of these.

2. Fabric woven on a jacquard loom with a flat design.

3. Fabric made from various fibers into which a design is woven, used in upholstery, draperies, bedspreads, and table linens.

4. Heavy jacquard-weave fabric used for tablecloths and home furnishings. Linen damask is the traditional fabric for fine tablecloths.

Decorative fabrics. Fabrics used in home decorating for upholstery, slipcovers, curtains, and draperies. They are also called "decorator fabrics" and "home furnishings fabrics."

Delustering. Process that dulls the characteristic shine of man-made fibers. Particles of a chemical are added to the fiber mixture before it is "spun," resulting in fibers with softer, muted color tones.

Denier. Size or number of filaments of fibers in a yarn or thread.

Denim

1. Heavy, tough cotton twill fabric produced in blends of cotton and synthetics, used in slipcovers, bedspreads, and draperies.

2. Firm, heavy, twill-weave cotton fabric.

3. Twill-weave fabric with a colored warp and a white filling thread. When the fabric became popular, the name was given to many other types of fabric, including cross-dyed fabrics and brushed fabrics, both knit and woven, that resemble true denim.

Dimity

1. Lightweight, moderately sheer fabric that often has fine woven stripes or other patterns, such as small flowers. The fabric was traditionally made of cotton, but now is often made of man-made fibers. It is used for curtains.

2. Double- or multiple-thread sheer cotton fabric woven in a corded, striped, or checked pattern.

Dobby weave

1. Dobby fabric is one with small geometric figures incorporated in the weave, made with a dobby attachment on the loom. Less elaborate than a jacquard attachment, which also produces geometric designs, the dobby is used to produce geometric designs such as those found in pique fabrics.

2. Small geometrically patterned weave made with a special loom attachment, somewhat similar to a jacquard but less complicated and resulting in a smaller design. The pattern is usually raised. A typical dobby weave is pique.

Doeskin. Usually the skin of a white sheep, although originally it was the skin of a deer, hare, or rabbit; also, any fabric made of wool or man-made fibers with a soft, often napped, finish.

Doily. Piece of fabric, round, square, or rectangular in shape, that is used under plants and decorative objects partly to protect furniture surfaces and partly as decoration.

Domestics. Such household items as sheets and towels. Domestics are also items made domestically, that is, in the United States.

Dotted Swiss

1. Fine sheer fabric of almost any fiber; fabric with very small dots on it, often woven in. The dots may be flocked or even printed. Some knitted fabrics made with a thread on the surface that forms a dot are also called dotted Swiss, although they are not the traditional Swiss fabric.

2. Crisp, sheer cotton, woven, embroidered, or printed; plain-weave fabric.

Double cloth. Fabric made of two fabrics woven one above the other and joined at the center with threads. True double cloth can be split into two distinct layers of fabric by cutting the threads between the layers. Velvet is often made as a double cloth and then cut to form the pile.

Double damask. Rich traditional tablecloth fabric, made in a heavier weight than ordinary damask.

Double-faced. Term applied to double cloth that can be used on either side. It is also used to describe any fabric with two right sides.

Double-faced satin. Satin fabric that has the satin appearance on both sides unlike ordinary satin which has a definite right and wrong side.

Double knit fabric. Fabric knit of two interlocking layers that cannot be separated. Except when the double knit is made with a pattern, the face and back are identical.

Double-woven. Term applied to two different cloths woven together, such as drapery and lining materials.

Drape. Term that describes the way a shaped fabric hangs or falls.

Drapeable fabrics. Fabrics that are soft and flowing, and can be arranged in soft gathers. Drapeable fabrics are made from a variety of yarns and in a variety of ways, including both knitting and weaving. They must be fairly lightweight to drape properly.

Drapery. Decorative window hanging, usually lined.

Draw curtain. Curtain that may be drawn along a rod or rail by means of a traverse arrangement of cords and pulleys.

Dry cleaning. Method of cleaning certain fabrics accomplished by means of organic solvents instead of water. Manufacturer's suggestions are recommended.

Duck. Originally, a fabric lighter in weight than canvas. The terms are now synonymous.

Dust ruffle

1. Part of a bedspread; separate ruffles placed over the box spring. A dust ruffle can be removed from the bedspread for separate cleaning.

2. Gathered or pleated skirt covering the legs and springs of a bed.

Dyeing. Giving color to a fabric, yarn, or the solution from which the fiber is made. Manufacturer's suggestions are recommended.

Dynel (a proprietary name). Synthetic fiber, mainly used as a blend with other fibers.

Embossing. Method of producing an indented design on a fabric. Embossing is usually done with a heated roller having a raised section that forms the design and is part of the finishing process. Most embossing is permanent.

Embroidery

1. Term for a group of decorative, usually nonfunctional stitches done with thread or yarn on fabric.

2. Art of decorating a fabric with a raised design or pattern.

End pin. One-prong pin used at the end of a section of pleater tape.

Extender plates. Plates added to a window frame to hold a rod and gain height or width as needed.

Fabric. Any braided, felted, woven, knitted, or nonwoven material, including not only cloth, but lace, also referred to as "cloth," "goods," "material," and "stuff."

Fabric wall covering. Fabric applied to interior walls consisting of such fabrics as burlap, felt, grasscloth, shiki silk, velvet, moire, linen, and many synthetic fabrics.

Faille. Silk or man-made fiber fabric that has a very narrow crosswise rib. Ottoman is similar to faille but has a wider rib.

Felt (felted). Fabrics made from fibers that are joined by heat, moisture, and mechanical action. Wool is the traditional felted fabric, and wool is usually required in mixtures of other fibers to form the bond that results in a felt.

Festoon. Decorative cord, usually tasseled, used to create a valance effect over draperies. It is also used on table covers, bedspreads, etc.

Fiber. The basic material from which fabrics are made is called "fiber." Fibers are much longer than they are wide. The term used to be limited to materials that could be spun into yarn but is now used to include filaments that do not requrie spinning, such as silk and man-made fibers.

Fiberfill. Man-made fluffy material used, among other things, to stuff pillows and make quilts. Most fiberfill is polyester.

Fill. Crosswise yarns in a piece of fabric.

Filler. Composition used to fill wood pores before applying paint or finish.

Filling. The crosswise thread that interlaces with the warp threads on a woven fabric is called "filling." It is also called "weft," "woof," "shoot," and "shute." Filling, however, is the most common term used in the textile industry, partly because it describes the function of the yarn so well.

Fire foe. Spring Mills' name for their flame-retardant fabrics.

Fireproof fabric. Fabric that literally will not burn. To be labeled "fireproof," the Federal Trade Commission requires that a fabric must be 100% fireproof. If the fiber or fabric has been treated to prevent flames from spreading, it must be labeled as "fire resistant."

Fire resistant. Fire resistant refers to a fabric or fiber that has been treated to discourage the spreading of flames.

Fire stop. Name given by Cotton Incorporated to 100% cotton or cotton blend fabrics treated to meet government or industry flammability standards.

Flame-retardant fabric. Fabric that resists or retards the spreading of flames. Flame-retrdant fabric can be made by using fibers that are in themselves flame retardant or by using special finishes on fabrics.

Flame-retardant finishes. Flame-retardant finishes may be applied to any fabric.

Flange

1. Flat border. In fabrics the term usually applies to a flat border on a pillow.

2. Flat border. On pillows it is an unstuffed decorative edging surrounding a stuffed pillow. It differs from a ruffle in that it is flat rather than shirred.

Flannel

1. Soft woven fabric made from wool, cotton, or of wool and cotton.

2. Soft fabric, usually with a brushed surface. Flannel may be made of just about any fiber although the traditional fibers used for flannel are wool and cotton.

Flat. Description of both a loom or knitting machine and the finished product. A flat machine weaves or knits a fabric that is all in one plane as opposed to circular looms and machines which produce tubular fabrics.

Flat screen printing. Mechanical method by which the cloth is printed using a number of stationary flat screens of a rectangular shape, working in a line. Screens have color pressed through the mesh, then they are lifted, the cloth moves on the space of one repeat and the process begins again.

Flax

1. Fiber from the inner bark of the flax plant used in the manufacturer of linen.

2. Fiber from which linen yarn and fabric is produced. It is a product of the flax plant. The word "linen" is derived from "linum," part of the scientific name for the flax plant.

Fleck. Spot, usually of color, included in a fabric to add visual and textural interest to it. Flecks are often made by the addition of small pieces of colored fiber to the base fiber during the process of spinning it into yarn.

Flocking

1. Method of adding design with texture to a fabric. Flocking involves the use of an adhesive (either on its own or as part of a printing dye) which is printed onto a finished fabric in a pattern. Small pieces of fluffy material are then sprinkled over the fabric and stick to the adhesive in the selected pattern. Flocked fabrics are often intended to imitate more expensive fabrics such as cut velvets. Dotted Swiss can be made with the dots flocked rather than woven.

2. The technique for applying flock consists of scattering finely powdered wool over the entire surface to be decorated; particles adhere only to the tacky surface.

3. Finely powdered wool or synthetic applied to fabric or wallpaper in a design.

Flock paper. Wallpaper that has a velvetlike or suede surface.

Floral. Wallpaper or fabric designed and patterned with flowers.

Flounce. Wide strip of fabric used to decorate a bedspread or slipcover, similar to a ruffle.

Foam. Material with bubbles as part of its basic structure. Foam rubber and foam polyurethane are two of the most common foams. The foam structure gives a springy, bouncy effect to the basic material, making foam items suitable for pillows, floor padding, backings, and upholstery.

Foam back. Layer of foam (usually polyurethane) that is laminated to another fabric.

Foam rubber. Rubber made in foam form and used for pillows, floor padding, backings, and upholstery.

Fortrel. Trademark of Fiber Industries, Inc., for polyester fiber marketed by Celanese Fibers Marketing Co.

Frieze. Fabric used primarily for upholstery and slipcovers. The fabric is looped, and the loops are often sheared to varying heights to form the pattern. Originally made of cotton, the fabric is now usually made in blends of cotton and man-made fibers.

Fringe

1. Border, but when it refers to fabrics for home furnishings it means a shaggy edging.

2. Ornamental edging used to finish or trim drapery, upholstery, etc.

Frog. Ornamental triple-loop fastening used as a decorative device in home furnishings.

Fusible fabric. Fabric that can be joined to another fabric in a fairly permanent bond through the application of heat, moisture, and pressure, accomplished with a pressing iron. A fusible fabric has dots of polyamide resins (polyamids are the bases of many synthetic fibers) on the wrong side, which is placed against the wrong side of the outer fabric; the fusing agent melts and fuses to the other fabric when the pressing iron is applied.

Galloon

1. Flat, closely woven, wide braid used for trimming draperies and upholstered furniture, same as gimp, but wider.

2. Narrow edging or a narrow lace made with scallops on both edges.

3. Closely woven, flat braid used for accenting draperies and furniture and also called "braid." The term "galloon" is also used for any narrow fabric with decorative edges, such as scallops, which are finished in the same way on each side.

Gauze

1. Thin, sheer, woven fabric that is quite open. It can be made in many fibers and used for many purposes including curtains.

2. Transparent fabric, netlike or plain weave, or combination of both.

3. Sheer, loosely woven fabric of natural or synthetic yarns.

Georgette. Sheer fabric, very similar to chiffon, made with a crepe yarn that gives the fabric a crepe appearance.

Gimp

1. Edging that often has small scallops of fine cord along its edges. Gimp was originally designed to hide such things as upholstery tacks on chairs and sofas but is now used for other decorative purposes.

2. Narrow ornamental trimming designed to cover upholstery nails, popular for general trimming purposes.

3. Ornamental braid used to trim drpaeries and furniture.

Gingham

1. Lightweight, yarn-dyed cotton material

2. Yarn-dyed, plain-weave fabric with check pattern.

3. Fabric made of crisp cotton or a blend of cotton and synthetic fibers, printed in checks, strips, or plaid designs.

4. Plain-weave fabric with a pattern made from dyed yarns. Traditionally made of cotton (although other natural fibers have been used in ginghams and given that name), gingham now is usually made of a blend or of man-made fibers.

Glass curtains

1. Sheer window coverings which hang in front of a window affording a degree of privacy without cutting off an excessive amount of light. Glass curtains are often used behind draperies.

2. Curtains made of sheer, semitransparent fabrics.

Glass fiber. Fiber made from glass, used extensively for cutrains and draperies. Glass fiber fabrics are very strong and wash well, but care should be taken to avoid getting small splinters of the glass yarns in the hands. Glass fiber is stiff and has poor resistance to wear and abrasion; it is, however, fireproof.

Glass fiber curtains. Curtains made of glass fiber yarns. Sheer glass fiber curtains are often used behind draperies, but glass fiber curtains are also available in heavier, opaque fabric constructions.

Glazed chintz. Chintz or plain colored fabric that is glazed by means of calendering.

Glazing. Fabric treated with starch, glue, shellac, or parafin.

Godet

1. Inset sewn into an article to give it fullness, often used in bedspreads.

2. Piece of fabric, tapering from wide to very narrow, inserted into another fabric section, often at a seam,

for additional fullness either for function or appearance. It is often used in home decorating at corners of beds, chairs, sofas, and slipcovers.

Gooseneck bracket. Small curved bracket for round curtain rods.

Grass cloth

1. Wall-covering material glued onto a paper or fabric backing.

2. Fabric made of vegetable or synthetic fibers, or a combination of both, laminated to a paper backing.

3. Plain-weave, loosely woven fabric made from such fibers as hemp, ramie, and even nettle. True grass cloth is relatively rare, but the appearance of grass cloth is copied in wallpaper and fabrics of man-made fibers.

Grosgrain. Fairly heavy ribbed fabric, often made in narrow widths for use as trimming. Grosgrain is used for ribbons in which the ribs are usually quite narrow, but it can be made with larger ribs.

Grosspoint. Needlepoint made with large stitches.

Ground. Background of a fabric design or print, as when red flowers are printed on a black ground.

Gypsum-coated wall fabric. Fabric impregnated with gypsum and hung like wall paper, for application on concrete block walls.

Haircloth

1. Stiff fabric made from a combination of natural or man-made fibers and animal hair, usually either goat or horsehair, used in upholstery and as interfacing for its strength.

2. Stiff, wiry fabric made of cotton with horsehair, mohair, or human hair filling and plain woven.

3. Fabric made of cotton or linen and horsehair and used for furniture upholstery. It is usually stiff, glossy, and black.

Hand-blocked. Term applied to a hand-pressed motif on wallpaper.

Hand-blocking. Method of printing a design on a piece of paper, fabric, or other surface.

Hand-printed wallpaper. Wallpaper produced by the silk-screen process, one roll at a time.

Hand prints. Wallpapers, murals, fabrics, accessories, etc., produced by a hand-screening process.

Hang-rites (proprietary name). Perforated nylon strips inserted inside pleats to hold drapery pins.

Headboard. Board or frame at the pillow end of a bed. The headboard stands perpendicular to the floor. It may be made of wood or slipcovered or upholstered to match a bedspread or other decorative area of a room.

Heading. Top portion of a curtain or drapery. It is usually decorative and is often made from trimmings such as braid.

Hem

1. Finish on the edge of a fabric or garment designed to prevent it from running or raveling. The hem is made by turning the fabric up to the inside and stitching it in place. It may be finished by adding a decorative trimming of some kind to protect the edge.

2. Turned-under section forming a finished edge at the side or bottom of drapery.

Hemp. Tough vegetable fiber woven into small squares and sewn together for porch rugs.

Herringbone weave

1. Ornamental variation of basic twill weave.

2. Twill weave in which the lines of the twill change direction, forming a pattern of V's. The weave is often described as a "broken twill" weave.

Holland. Plain-weave fabric that is used in the home primarily for window shades.

Homespun

1. Fabric made from yarns that were spun by hand. The term is now used for fabric that imitates this look. Homespun has a fairly rough surface and is made from nubby, uneven yarns.

2. Loosely woven fabric, irregular in texture, rough in appearance, and moderately heavy, usually made of wool or cotton.

Hopsacking. Coarse, durable fabric made of cotton or rayon, used for draperies and slipcovers.

Horsehair

1. Hair from the manes and tails of horses. It is occasionally used for upholstery but is used in interfacings for stiffening and strength. When it is used it is always combined with other man-made fibers.

2. Furniture covering woven from horses hair.

Hygenic finishes. Finishes that retard the growth of bacteria and fungi and stop mildew.

Indian blanket. Woolen blanket hand-woven by American Indians in the western part of the United States. Indian blankets are usually made in bright colors or in earth colors. The term is used for any blanket that resembles an authentic Indian blanket.

Indian head. Trade name for a smooth cotton fabric of plain weave.

Insect-repellent finishes. Insect-repellent finishes are used on fabrics to protect them against moths and carpet beetles.

Interfacing. Stiffening fabric which may be made of horsehair, often goat hair, wool, man-made fibers, or combinations of these fibers.

Inverted pleats. Pleats formed in the same way as box pleats, but with the edges meeting on the right side of the garment.

Jabot

1. Fabric hung lengthwise and folded, slanted at the bottom. It is used with a swag.

2. Decorative ruffled drapery heading.

Jacquard weave

1. Term used to describe fabrics with a woven or knitted pattern, whether or not they are made with a jacquard attachment of the loom. A jacquard attachment for weaving and knitting machines makes possible the manufacture of complicated, repeated geometrical designs in knits and wovens.

2. Name of an attachment for a knitting or weaving machine that enables the machine to make complicated and quite large patterns in the course of weaving or knitting the fabric itself. Damask is an example of a woven jacquard; many patterned double knits are made on machines with jacquard attachments.

3. Woven design made on special loom; damask and brocade.

Jaspe cloth. Hard, firm, durable cotton or rayon fabric used in draperies and upholstery work.

Jute. One of the natural fibers still used extensively for fabrics. It comes from jute plants grown in India, Pakistan, and Bangladesh. Jute is used in many ways including the manufacture of burlap, twine, and rope and as trimmings and backing for rugs.

Jute border. Braid made of rough-textured jute yarn.

Kapok

1. Fluffy fiber that comes from the seed pods of the kapok tree found in the tropics. Kapok was used at one time for stuffing pillows and life preservers, as it is naturally buoyant. However, man-made fibers have replaced kapok in many cases.
2. Silky fiber obtained from the seed pods of the kapok tree.
3. Natural seed-pod fiber used in cushion fillings.

Kettlecloth. Cotton and polyester fabric producing a crisp, grainy-textured gingham.

Kick pleat. Small pleat at the bottom of a decorative skirting.

Knife pleats. Narrow, straight pleats similar to accordion pleats, with each pleat facing in the same direction.

Lace

1. Decorated fabric made either on a background fabric of net or without a background fabric. The pattern in lace is usually open and most often floral in design. Machine-made lace is now produced in many patterns that were formerly only made by hand.
2. Netlike fabric, usually with an ornamental design, made from silk, cotton, acetate, or nylon.

Lambrequin

1. Decorative wooden construction projecting several inches from the wall, framing a window along the sides and top, and usually painted or fabric covered.
2. Short drapery hung across the top of a door or window.
3. Structure at the top and sides of a window that frames the window and is usually part of the window decoration. It is usually made of wood and is often covered with fabric and trimmed, or it may be simply painted.

Lamé

1. Fabric woven or knitted with all metallic yarns or with a combination of metallic and other fiber yarns. Most lamé is made from one of the nontarnishable metallic fibers. "Glitter" is sometimes used to describe this type of fabric instead of lamé.
2. Silk or rayon fabric with metallic threads in the design, used for draperies.

Laminated fabric. Double-thickness fabric made by bonding one fabric to another or to another material.

Laminating. Joining two different fabrics together by use of a fusing material that melts and forms a bond between the two fabrics. Synthetic foam is often laminated to fabric in home furnishings fabrics. The permanency of a laminate depends on the quality of the manufacturing process.

Lasta firm. Lightweight crinoline used for headings on sheer curtains.

Latex. Liquid form of natural or man-made rubber. It can be formed into thread for use as an elastic yarn.

Lauhala. Leaves of an Hawaiian tree folded into strips. Their color is yellow brown, and the material is glossy. The strips are interwoven to make mats and screens.

Lawn

1. Sheer fabric made from cotton or linen material. It is soft, light, and crispy.
2. Fairly sheer, lightweight, plain-weave fabric made originally from linen but now usually made from combed cotton or blends of cotton and man-made fibers. Lawn is slightly stiffer than batiste but can be used for similar purposes.

Leno. Open, somewhat lacelike fabric made when warp yarn pairs are crossed over each other and around the filling yarn. The airy quality of leno weaves makes them especially popular in curtains. A special attachment for the loom is required for this weave.

Linen

1. Made from flax, which comes from the flax plant, linen is one of the oldest fabrics known. Strong man-made fibers are often blended with it to improve its wrinkle resistance and give the fabric other desirable qualities. It is woven in various weights for different purposes.
2. Linen is a natural fiber like cotton, silk, and wool and has no trademarks in the way that man-made fibers do. Terms such as "Irish" or "Belgian" linen, used as indications of quality, actually refer to where the linen is produced.
3. Strong, durable fabric made from the fibers of the flax plant.

Lining. Fabric made in the same shape as the outer fabric, a lining supports and protects the outer fabric and hides seams as well. Linings are found in draperies and occasionally in curtains and bedspreads. Items that are lined tend to last longer, and the appearance of a lined item is usually better. The lining should be of the same construction as the outer fabric.

Lining paper

1. Wallpaper used as a base for the finish wallpaper. It is used primarily to hide rough uneven surfaces and provide a smooth surface for the final product.
2. Inexpensive wallpaper that is first applied to a wall before the final wallpaper is applied.

Linum. Part of the scientific name for the flax plant.

Lyons velvet. Velvet originally made of silk in Lyons, France. Lyons is a thick, rather stiff velvet with a very short pile. This type of velvet is now made of man-made fibers. It is used for home furnishings.

Machine-printed wallpaper. Printed on high-speed presses, this wallpaper is easy to hang and usually durable.

Macrame. Method of forming open fabrics by knotting string, yarn, or other forms of thread. Macrame can be used to make anything from delicate trimmings to sturdy items like hammocks.

Man-made fibers. Overall term referring to all fibers not found naturally. It includes rayon and acetate, which are made from cellulose, a natural product. The term "synthetic fibers" applies only to man-made fibers produced entirely in the laboratory from such substances as petroleum (polyester, for instance).

Marquisette

1. Term used for a group of lightweight open fabrics extremely popular for curtains and mosquito netting. Marquisette is made of many different fibers, including cotton and nylon.
2. Fine, sheer fabric with a lacy effect, made of polyester. It can be soft or crisp depending on the finish of the material, and is usually used for curtains.

Matelasse

1. Double fabric with quilted or crinkled effect, used for draperies and upholstery.
2. One of the fabrics that, like cloque, has a blistered or quilted look to the design. The term "matelasse" implies the use of two different yarns that when finished react differently to the finishing, resulting in a puckered effect in the fabric. Matelasse is popular for upholstery.

Matte

1. Flat paint finish without gloss or luster.
2. Dull surface on a fabric. Since one of the characteristics of fabrics made from man-made fibers is a shiny surface, matte-finished fabrics have become popular, and matte looks for man-made fabrics are achieved in yarn processing or finishing.

Mattress cover. Fairly thick quilted pad placed on top of a bed mattress and beneath a bottom sheet to protect the mattress and make the bed more comfortable. It often has elastic at all corners to hold it on the bed, and it should completely cover the top of the mattress.

Metallic fiber

1. Manufactured fiber composed of metal, plastic-coated metal, or metal-covered or plated plastic.
2. Generic name for a manufactured fiber that may be metal, metal coated with a synthetic, or a man-made fiber core covered with metal. When the metal is coated with a man-made film, the metal does not tarnish.

Mildew resistant. Among the many properties that can be given to fabrics in the finishing is resistance to traditional enemies. Such fabrics as canvas, which are exposed to the damp conditions that encourage the growth of mildew fungus, can be treated with finishes to resist this fungus, making them mildew resistant.

Modacrylic. Fabric that shrinks away from the flame and then burns very slowly and melts; it is self-extinguishing when removed from the flame. Burning leaves a hard, brittle, black bead.

Mohair. Hair from the angora goat makes a fabric that is durable and resilient and used for upholstery work.

Moire

1. Fabric woven from silk, cotton, or rayon fibers, used for bedspreads and draperies.
2. Wavy, rippling pattern somewhat like a watermark produced in the finishing on certain fabrics by calendaring. On acetate a moire made this way is permanent; on most other fabrics it is not. Moire effects can also be achieved by printing and in the weaving of the fabric but the finishing method is the most common.

Monk's cloth

1. Heavy, coarse, loosely woven basket weave fabric. The fabric is brownish beige and is made of cotton, sometimes with the addition of flax or jute. Monk's cloth may also be made of man-made fibers. It is used for home furnishings such as draperies and slipcovers.
2. Heavy, coarse-woven cotton fabric, also called "friar's cloth."

Moss crepe. Fabric made in a plain or dobby weave with rayon yarns that produce the mosslike effect. The term refers to any crepe, including polyester, that can be considered to have a mosslike surface.

Mousseline. Name for a broad category of fabrics, usually fairly sheer and lightweight and made in a variety of fibers, including man-made silk, cotton, and wool. It usu-ally has a crisp feel and is often used today for a fabric resembling mousseline de soie.

Mull. Soft, sheer, lusterous cotton, silk, or rayon fabric used for draperies.

Multipocket bestpleat. Woven pleater tape with extra pockets for extra pleating versatility.

Muslin

1. Name for a very large group of plain-weave fabrics originally made of cotton but now made of blends of cotton and man-made fibers. These fabrics range from sheer to very heavy weights. Muslin sheets made of cotton and man-made fiber blends approach the softness of percale sheets after a few washings. Muslin is used as a furniture covering that is subsequently covered with upholstery fabric or slipcovers.
2. Plain-weave fabric that may be bleached or unbleached. It is used for undercovering on upholstered pieces.
3. Sturdy cotton fabric, usually rather coarse but sometimes woven very sheer, used as lining for wallpapers and for upholstery undercovering.

Nap. Raised fibers on the surface of a fabric that create a downy or fuzzy appearance.

Napping. Finishing process, used to produce a true napped surface, in which the fabric is passed over fine wires, brushes, or burrs which raise some of the fibers to the top, producing the characteristic soft or fuzzy napped surface. Shearing may be a part of the napping process; it cuts the raised nap to a uniform height, very much in the way a lawn mower cuts grass. The process is used for fabrics in which a striped nap effect is considered desirable.

Natural fibers. Fibers found in nature and that can be made into yarn. The best known fibers include cotton, silk, wool, linen, hemp, jute, ramie, and nettle. The term "natural fibers" is used since the development of man-made fibers to distinguish between them.

Needlepoint. Hand embroidery made on a mesh canvas, usually with wool or cotton yarn.

Net. Net is produced on machines that imitate the handwork that originally produced this fabric. Originally net consisted of geometrically shaped holes outlined by yarn. Net fabric ranges from the sheerest and lightest of weights to the heavy fishnet, used for decorative as well as industrial purposes.

Ninon. Chiffonlike fabric made from various natural and synthetic fibers and used for draperies and curtains.

Nip-Tite pleater hooks (a proprietary name). Hooks that lock into pleater-type pockets to create perfect pinch pleats.

Nonwoven. Fabric that is neither knit nor woven. The category of nonwoven fabrics usually includes fabrics that are felted, as well as those in which fibers are joined by glue or heat. Occasionally a distinction is made between felted fabrics, paper, and other nonwovens.

Nylon

1. Nylon, the first of the synthetic fibers, is very strong and resists abrasion and wrinkles. It has a natural luster, holds body heat, and resists moths. Nylon dyes well but fades in the sunlight, may pill, and melts under high heat. It is naturally mildew resistant and is used for such things as mosquito netting in the topics. Nylon is a thermoplastic man-made fiber synthesized from petroleum.

2. Nylon shrinks away from the flame and then burns slowly and melts. It is usually self-extinguishing when removed from the flame. It smells like celery and leaves a hard gray bead.

3. Strongest of man-made fibers, nylon has good dimensional stability, resiliency, and abrasion resistance.

Olefin

1. Man-made fiber that is durable, fadeproof, moisture-resistant. It is used in awning fringe, upholstery, and carpets.

2. Generic name for fibers derived from polyethylene or polypropylene. Olefin is primarily used in home furnishings for inexpensive rugs and upholstery as it has good bulk and coverage and resists chemicals, mildew, and weather.

3. Olefin shrinks away from the flame and then burns and melts. It is self-extinguishing when removed from the flame. Burning leaves a hard tan bead. It may smell like a candle burning.

Organdy. Thin, transparent, crisp cotton fabric.

Orlon

1. Synthetic fiber with excellent shape retention.

2. Trademark of E.I. DuPont de Nemours & Co. for acrylic and modacrylic fibers.

Osnaburg

1. Coarse, plain-weave cotton fabric. It is quite strong and used extensively for bags and other industrial purposes. It is occasionally used for draperies and upholstery.

2. Rough-textured, open-weave cotton fabric that is strong and durable.

Ottoman. Fabric with wide horizontal ribs similar to faille. It is usually made in wool, wilk, or man-made fibers and appears in upholstery and draperies.

Paisley

1. Colorful comma-shaped printed or woven design

2. Light wool fabric with woven design, on a silk fabric, design is printed to imitate wool.

Panel. In drapery, a width of fabric with hemmed edges and finished top and bottom.

Patchwork. Fabric resulting from the joining of small pieces of fabric together in a pattern or in random fashion to make one large piece. Patchwork was originally sewn by hand, and often a complete piece of patchwork became the top of a quilt. Printed fabric is now made to imitate the real thing.

Pattern. Design that is repeated. On fabric it is the design that is usually repeated several times in every yard of fabric. The term "pattern" is used as a synonym for "design" and "motif."

Pebble. Fabric with a somewhat bumpy, grainy appearance. The term is used to describe crepes.

Pellon

1. Nonwoven stiffening material for adding firmness to a fabric or drapery heading.

2. Nonwoven fabric used as a lining material for curtains and drapes.

Percale

1. Printed in plain-weave cotton with smooth finish.

2. Plain, closely woven, cotton fabric, that is strong, lightweight, and durable and used for curtains and bed linens.

Perma-Crin (a proprietary name). All-purpose crinoline used for curtain headings.

Permanent press. Factory-applied finish that makes fabric wrinkle and crease resistant and helps it to hold its shape.

Permette. Stiffened buckram used in sewing cornices.

Piece-dyed fabric. Fabric dyed after it is woven.

Pile

1. Fabric surface formed by yarns that are brought to the right side of the fabric in the course of making the fabric. Some fabrics have pile surfaces on both sides. Looped fabrics, such as terry cloth; tufted fabrics, including candlewick and many rugs; and fabrics such as velvet, made by the double-cloth construction method in which the heavy joining yarns are subsequently split to make two fabrics are all pile constructions. In certain specialized areas pile fabrics are considered to have a nap.

2. Fabric in which yarns are left on the surface of the fabric either in loops (as in the case of terry cloth) or clipped to form a hairlike surface (as in the case of velvets and many rugs).

Pilling. Formation of small balls of fiber on the surface of certain fabrics. Pills develop on woolen fabrics but tend to disappear when the fabrics are cleaned; man-made fiber pills remain on the fabric. Man-made fibers also tend to develop picks (small loose threads that can snag). The tendency of man-made fibers to pill and pick can be reduced by steps taken in the processing of the yarn or the finishing of the fabric.

Pillow. Cloth bag (often made of ticking) that has been stuffed with feathers, down, kapok, rubber, synthetic foam, fiberfill, or a similar substance and is used to support some part of the body and to make furniture comfortable.

Piping. Decorative tubular edging used to trim upholstery, draperies, etc.

Pique. Fabric woven with small raised geometric patterns on a loom with a dobby attachment. It is usually a crisp medium- or heavy-weight fabric made of cotton or a blend of cotton and synthetic fiber, often printed with colorful designs. The look of woven pique can be duplicated with embossing and heat-setting.

Plaid. Fabric or wall-covering with a pattern of colored stripes at right angles to one another.

Plain weave. Simplest form of over- and underweaving; chambray and percale are examples of plain weave.

Pleating

1. Way of folding cloth. Pleats differ from tucks in that tucks are usually stitched on the fold line, while pleats are not.

2. Folding or doubling over of fabric to create a fullness below the pleats.

Pleats

1. Decorative, regularly spaced fabric folds.

2. Folds of cloth arranged in a certain way. Pleats can be made as the fabric is produced (usually through heat-setting). Pleats are used to control fullness or for decorative purposes. The thermoplastic nature of man-made fibers (they change their shape under heat) means that permanent pleating which does not have to be renewed when a pleated item is cleaned is now available.

Plisse. Puckered fabric made by printing plain fabric, usually cotton, with a chemical (caustic soda). The printed area shrinks causing the unprinted area of the fabric to pucker. The pucker is permanent. The same effect can be achieved in thermoplastic man-made fiber fabrics by using heat to set the puckers.

Plush. Long-pile velvet, usually used as an upholstery fabric.

Polyester

1. Widely used man-made fiber that is extremely strong with excellent wrinkle and abrasion resistance. It also resists mildew and moths. Polyester may pill and attract lint.

2. Polyester shrinks away from the flame and then burns slowly and melts, giving off black smoke. It is usually self-extinguishing when removed from the flame. It smells slightly sweet and leaves a hard black bead.

Polypropylene (olefin). Man-made fiber that is durable, fadeproof, and moisture resistant. It is used in carpeting and awning fringe.

Polyurethane. Man-made material used for foam that is laminated to other fabrics to provide warmth, as well as for mattresses and stuffing. Polyurethane foam tends to yellow from exposure to air, but it is claimed that this does not affect its performance. It will not harden and is not affected by mildew, moisture, or strong sunlight. Spandex fibers are based on polyurethane.

Pom-Decor (a proprietary name). Fluffy pompons strung on cord for use as trim or hung from rods to replace curtains or shades. Pom-Decor is also used for room dividers.

Pompon. Fluffy ball, usually made from yarn, used as a decorative accent.

Pongee. Plain-woven silk fabric, now made from cotton and synthetic fibers.

Poplin

1. Plain-weave fabric with fine cross ribs.

2. Tightly woven cotton or rayon fabric with crosswise ribs, used in draperies.

3. Fabric with a fine horizontal rib, usually made of cotton, a combination of cotton and man-made fibers, silk, or man-made fibers designed to imitate silk.

Prepasted wallpaper

1. Paper that requires only moistening, not pasting.

2. Machine-printed wallpaper coated with an adhesive at the factory. It is best used over a layer of old wallpaper.

Preshrunk. Term applied to shrinking of fabrics at the factory.

Pretrimmed wallpaper. Selvages are removed at the factory so that the paper is ready to hang.

Protein fibers (silk, wool). Protein fibers burn slowly, sizzling and curling away from the flame. They are sometimes self-extinguishing when removed from the flame. They smell like burning hair or feathers and leave a crushable black ash.

Quilted fabric. Thick, soft fabric consisting of two layers of cloth with a soft material filling between the layers, stitched into a definite design or pattern.

Quilting. Stitching through two or more layers of fabric to form a design or pattern. The most common design today is a diamond pattern, but quilting stitches (usually a short running stitch) may also be done in other geometric or floral patterns. Stitches are often used to outline patchwork or applique designs on a quilt.

Ramie

1. Strong, lustrous, natural fiber from the ramie plant grown in Asia.

2. Fiber yielded by a shrubby Chinese and East Indian perennial of the nettle family, used for cordage and certain textiles.

Rattail fringe. Edging made with a satinlike cord used in a garland effect.

Rattan. Strong, slender, light-brown woody strips woven into panels and used in furniture. Synthetic rattan is also available.

Ravel. Tendency of fabric to come unwoven or unknitted at unfinished edges. The term "unravel" means the same as "ravel." Loosely woven fabrics tend to ravel more than those made of tight weaves. Occasionally the tendency to ravel is desirable in order to create a fringed edge.

Rayon

1. Extremely versatile man-made fiber. Rayon resembles silk and is used for rugs and carpets.

2. First successful man-made fiber. Rayon was originally called "artificial silk." It is made from cellulose and is weak when wet. It is soft and comfortable and dyes well but is weakened by exposure to sunlight. Because of its low wet strength, rayon may shrink or stretch unless treated. Several different types of rayon are made.

Repeat

1. Design that appears over and over again on a fabric; also, the amount of space the design takes before it starts over again. Since it is desirable to center the design in a fabric on such things as sofas, the size of the repeat must be known in order to determine the yardage needed. The larger the pattern, the larger the repeat and the more fabric needed.

2. Size of a single pattern motif on drapery material.

Repp. Silk, cotton, rayon, or wool fabric having a crosswise rib.

Rib weave

1. Weave that uses a heavier warp than fill, or heavier fill than warp, as, for example, in corduroy.

2. Plain weave in which the yarns in either the warp or the filling direction are thicker or further apart than those in the other direction, producing a regularly furrowed bumpy surface. Grosgrain and faille are common examples of rib-weave fabrics.

Roller blinds. Shades that are wound around a roller or dowel when the window is exposed. Originally made only in neutral colors, are also made in colors or matched and coordinated with the draperies in a room.

Roller shades. Shades made from slender wood, bamboo, or synthetic materials which are spaced to admit a small amount of light between the strips but are otherwise opaque.

Roman shades. Shades similar to Austrian shades. When the window is exposed, the fabric of Roman shades hangs in graceful folds at the top of the window. Austrian shades are shirred throughout when they cover the window, but Roman shades hang straight and only form folds when drawn up to bare the window.

Ruffle. Gathered strip of fabric used for trimming purposes.

Rush. Long grass that is twisted and woven to make seats for chairs.

Sailcloth

1. Heavy, strong, plain-weave fabric made of cotton, linen, or jute.

2. Smooth, stiff cotton or cotton and rayon fabric.

3. Firmly woven cotton canvas, now made of man-made fibers as well as cotton. The terms "duck," "sailcloth," and "canvas" are often used interchangeably.

Saran. Generic name for a man-made fiber derived from vinylidene chloride. It is strong and resists common chemicals, sunlight, and weather. Saran is used in the fabric field primarily for upholstery on public transportation vehicles and for garden furniture.

Sateen

1. Fabric with a luster finish like satin, made from various fibers and used for bedspreads.

2. Strong, lustrous, satin-weave fabric made of cotton. The term is also used to distinguish between the cotton satin-weave fabric and satin-weave fabrics made of silk or man-made fibers.

Sateen-weave fabric. Fabric made by floating fill yarns over several warp yarns. It is used for drapery linings.

Satin

1. Fabric originally made of silk imported from China but now made from various natural and synthetic fibers.

2. One of the basic weaves. Satin weave has proved so popular that various tyeps of stain-weave fabrics have developed.

Satin-weave fabric. Fabric with a smooth, lustrous surface produced by laying warp yarns over several fill yarns.

Scalloped bestpleat. Pleater tape for cafe curtains; it alternates scallops and pinch pleats.

Scrim. Plain-weave, open-mesh fabric used for curtains, bunting, and as a supporting fabric for some laminated fabrics. It has been traditionally made of cotton but is now usually made of nylon or other man-made fibers.

Seam binding. Usually a narrow strip of twill-weave fabric finished at each edge. It is stitched to a raw edge of fabric to cover it and prevent it from raveling. Strips of bias fabric can be used in the same way, and when they are, they are also called "seam binding."

Seam roller. Tool used to smooth seams when hanging wallpaper.

Seersucker

1. Lightweight cotton or cotton-polyester fabric with lengthwise crinkled strips.

2. Puckered fabric. The puckered look appears in the alternating stripes and is achieved in the actual weaving process. Groups of tight warp yarns alternate with groups of slack or loose warp yarns so that when the filling thread is woven in, the loose yarns pucker. Seersucker is available in plain colors and stripes and is also popular in plaids and prints. Seersucker effects can be imitated in knits.

Selvage

1. Edge of a woven fabric that does not ravel because the filling yarns wrap around the warp yarns. It may also be called "self-edge" or "selvedge."

2. Woven edge of a piece of fabric.

Shade cloth. Name for any fabric used for window shades.

Shades. Window coverings that play a double role. They provide both light control and privacy and can also lend a decorative accent. Shades range from the traditional roller blinds, available in versions that exclude light completely and in versions that permit some light to come in, to some with a more decorative purpose like Austrian shades.

Shantung

1. Rough, plain-weave, silk fabric in which slubs in the yarn provided a textural effect. Shantung is usually made of man-made fibers or combinations of man-made and natural fibers.

2. Heavy-gauge pongee made originally in Shantung, China.

Sheer

1. Term applied to lightweight, gauzy, transparent fabrics like China silk, marquisette, and net. Sheers are woven of natural or synthetic fibers.

2. Opposite of opaque. Sheer fabrics are usually made in an open weave which creates fabrics with varying degrees of transparency. Batiste, organdy, and voile are examples of sheer fabrics.

Shiki silk. Horizontal-weave, delicately textured fabric resembling grasscloth, mounted on lining paper and used as a wall covering.

Shir-Rite (a proprietary name). One inch flat cord with two woven-in cords for shirring.

Shirred. Gathered. A shirred heading is made by pulling a sewn-on thread or cord to gather it.

Shower curtain. A shower curtain is a length of fabric hung around a bathroom shower or shower-tub combination to keep water from splashing onto the floor. It should be waterproof. When decorative nonwaterproof shower curtains are used, a waterproof liner, usually made of plastic, is placed inside between the curtains and the shower.

Showerproof. One of the many terms used to describe varying degrees of imperviousness to water. Showerproof fabric will repel water to a limited extent but is not waterproof.

Shrinkage-control finishes. Finishes used on fabrics that require frequent laundering.

Shrinkage-controlled fabric. Fabric that has been treated in some way to prevent it from shrinking more than a specified amount. Shrinkage control is usually achieved by shrinking the fabric in the finishing steps or by the addition of finsihing agents to the fabric.

Silk

1. Product of the silkworm and the only natural filament fiber (it is produced in a long thread), silk has been a leading luxury fiber for thousands of years. There are many types of silk and many ways of making it into cloth. Man-made fibers have to a very large extent replaced silk.

2. Silk is a natural fiber like cotton, linen, and wool and has no trademarks the way man-made fibers do.

3. Natural fiber produced by the larvae of silkworms. Other insects spin a similar fiber.

Silk-screening. Method of producing a wallpaper print through a silk screen.

Size. Sealer used to seal pores in a surface before papering.

Sizing. Starch or gelatine that is added to fabrics in the finishing stages to give them additional body and a smoother appearance. Cotton fabrics are those most commonly treated in this manner. Sizing is rarely used and fabrics usually retain their initial appearance through cleaning. A few fabrics such as needlepoint canvas are still sized so that they can be handled more easily; this in no way affects their final performance.

Slipcover

1. Unattached covering for a sofa or chair. Slipcovers are made with openings so that they can be removed for cleaning. Slipcovers are also called "loose covers."

2. Removable covers, cut and made to fit over upholstered furniture to protect the original materials. Fabrics usually are sailcloth, chintz, linens, or tightly woven cottons.

Slubbed fabric. Fabric having thick, uneven sections in the yarn.

Slubs. Uneven area in a yarn which gives the fabric made from it a degree of texture. Slubs can be produced naturally (as in hand-spinning, which often has slubs) or artificially, by deliberately making them in spinning. In manmade fibers slubs are usually produced by making parts of the fiber thicker than other parts. Short staple fibers mixed with other fibers in the yarn will produce slubs. "Nub" is another word for "slub."

Soil and oil repellents. Soil and oil repellents are applied to upholstery fabrics to repel water and resist staining.

Soil-release finishes. Factory-applied finishes that help to retard soiling.

Spread. Any kind of covering. A bedspread is usually a decorative covering that covers the blanket and pillows on a bed during the day. Spreads are available in many styles from simple throws arranged casually over the bed to tailored box spreads. A box spread is a shaped and fitted bedspread with a tailored appearance; corners are square, giving the spread its name.

Spring socket. Curtain fixture designed for recessed windows; it uses no screws.

Spun-fiber yarn. Yarn made from staple lengths of manmade fibers rather than the long filaments in which manmade fibers are formed. To accomplish this, long, filament fibers are chopped into staple lengths and then spun to imitate natural fiber yarns.

Spun yarn. Spun yarn is less smooth with lower luster than filament yarns.

Stock-dyed fabrics. Fabrics made from fibers that are dyed before they are spun into yarn.

Strie. Striped fabric with color lines relating closely to the background color.

Stripe. Band of color, usually on a plain ground and used in multiples. Stripes can be very narrow or wide, in all one color, in patterns of alternating colors, or in multicolored patterns.

Strippable. Descriptive term applied to wallcoverings designed to be easily removed.

Sunburst pleats. Pleats that begin at a central point and move out to the edge of a fabric. They are often narrow at the top of the fabric and wider at the edge.

Swag

1. Draped fabric section placed above a window.

2. Piece of fabric hung horizontally over a window and allowed to droop. It is usually held at the sides by ties.

Swag ball fringe. Large version of ball-and-loop fringe, woven with alternating balls and swags.

Swatch. Small piece of material or paper used as a sample.

Taffeta

1. Plain-weave fabric with a crisp, somewhat shiny surface. Originally made of silk, taffeta is now also made of rayon or acetate.

2. Fabric made of silk, rayon, nylon, or wool, finished in a fine, crisp, lustrous fabric for use in curtains, draperies, and bedspreads.

Tapa cloth. Papery cloth made by pounding and flattening the inner bark of certain trees found in the Pacific Islands. Often used for decorative wall hangings. Also called "barkcloth."

Tapestry

1. Handwoven fabric with ribbed surface.

2. Usually, a decorative wall hanging, traditionally woven to depict a scene. The filling threads are changed in color to fit the design. Some rugs are made in tapestry weaves. Machine-made fabrics, also called tapestry, have regular designs on the surface and a slightly looped pile.

Tassel. Several strands of yarn loops joined together shortly below the top and cut at the end. Tassels are used in rows as home furnishings trimmings and singly for such uses as zipper pulls or on the corners of pillows.

Tassel and slide. Large decorative tassel for use on tiebacks. The cord from which the tieback is made slips through a hollow bead or slide on a loop from which the tassel hangs.

Texture. Look and feeling of depth on the surface of fabric.

Ticking

1. Closely woven cotton fabric with a twill or satin weave.

2. Strong, durable cotton or linen fabric used for upholstery and other coverings.

3. Broad term for extremely strong woven fabrics that are used as coverings for pillows, mattresses, and box springs. It usually has a pattern of woven stripes, jacquard or dobby designs, or printed patterns.

Tieback. Device made of fabric or cord that is looped around a drapery and tacked to the window, wall, or tieback block, which is covered with the drapery material, the block being fastened to the wall.

Tieback curtain. Full-length (either to the windowsill or to the floor) curtain or drapery that is looped back at the side of the window with a band of trimming or self-fabric. The curtain or drapery is closed at the top of the window and almost entirely open at the point of the tieback.

Thermoplastic. Term used to describe fibers that are heat-sensitive. Most man-made fibers are thermoplastic. A thermoplastic fiber is one whose shape can be changed when heat is applied. This can be both an advantage and a disadvantage. It is advantageous because in fabrics made of thermoplastic fibers, certain features like pleats can be made permanent through heat-setting. However, care must be taken in drying and ironing fabrics made of thermoplastic fibers because of their sensitivity to heat.

Thermosetting. Process for giving thermoplastic fibers or fabrics certain characteristics such as crimp or permanent pleats through the application of heat. The process is also used to develop certain finishes in a fabric to produce desirable characteristics like durable press.

Thread. Thin, continuous length of twisted fibers. Thread is used primarily in sewing. The term thread is occasionally used instead of yarn, as, for example, "warp thread" and "filling thread."

Toile. French word for cloth. Toile is also a woven fabric that has been printed, usually in one color only, with a scenic design. It is occasionally called "toile de Jouy." It is used as a home furnishing fabric.

Toile de Jouy

1. Printed cotton or linen fabric used for matching wallpapers.

2. Sheer linen with allover scenic design printed in one color.

Torque yarns. Yarns that tend to rotate or twist when they hang free.

Trapunto. Form of quilting in which the fabric is quilted only in certain areas. The design to be quilted is first worked through two layers of fabric, and then the backing fabric is slit so that the quilted areas can be padded with yarn or cord or with a filling like fiberfill.

Traverse rod. Curtain rod with a mechanism allowing curtains to be drawn across the window.

Tuck

1. Fold made by doubling a fabric back on itself and stitching it parallel to the edge of the fabric.

2. Small, narrow section of fabric that is folded and then stitched, either down the fold of the tuck or across each end. Pin tucks are the narrowest of tucks. They usually serve a decorative function and are narrow, while pleats made in much the same way are usually used to control fullness as well.

Tufted. Term applied to fabric with small clumps of fibers on the front side.

Tufting. Carpet construction method of inserting tufts into a backing.

Tweed. Heavy, coarse, rough-textured wool fabric.

Twill weave. Fabric woven with diagonal ridges.

Unpressed pleats. Pleats whose edges (the folds) have not been set by pressing. The term is usually used for wide unpressed pleats, while the term "cartridge pleats" is used to describe narrower, decorative pleats which are also unpressed.

Valence

1. Decorative top for curtains or draperies, usually hung from a rod and made of fabric or fabric over a stiffening material such as buckram. It differs from a ruffle in that it is absolutely flat.

2. Shallow decorative strip, usually of fabric, that conceals curtain and drapery hardware.

3. Short drapery hung from a rod over the top of the window draperies or curtains.

Vat dyeing. Vat dyeing is considered the most satisfactory of dyeing techniques and usually makes fabric colorfast.

Vellus (a proprietary name). Fabric made from urethane foam covered with nylon flock on one side and with another fabric on the other side. It is used in making bedspreads.

Velour

1. Soft, closely woven smooth fabric with short thick pile.

2. Soft, heavy, velvetlike fabric.

3. Knit or woven fabric with a thick, short pile, Terry velour cloth has cut loops to produce the velour effect and is often used for bath towels. Terry velour has a rich look like velvet or velveteen but is not as effective in drying as conventional terry cloth.

Velvet

1. Fabric with a short, closely woven pile. There are two methods of making velvet. One method uses double-cloth construction in which two layers of fabric are woven with long threads joining them. After the double fabric is woven, the center threads that join them are cut, producing two pieces of velvet. The other method of making velvet utilizes wires. In the weaving process the yarn is lifted over the wires to form the pile. When the wires are removed, the yarn is cut to form the velvet surface. Velvet was originally made of silk but is now made of many other fibers, nylon being one of the most popular.

2. Fabric with a thick, short pile, less than $\frac{1}{8}$ in. on the top surface, and a plain back.

3. Soft, lustrous, low-pile fabric.

Venetian blinds

1. Horizontal wood, metal, or synthetic louvers, which can control light by cord action.

2. Popular window coverings for controlling light and privacy made of strips of fabric, metal, or plastic. The strips can be tipped to shut out light completely or opened to varying degrees to filter light to the desired intensity. They can also be raised to the top of the window to bare it completely. Conventional venetians hang with the slats or strips horizontal to the windowsill but vertical venetians are also available and are often used as room dividers as well as window coverings. Venetians are available in various colors and widths.

Vinyl. Any fabric made with a base of vinyl, including those listed as vinyl and vinyon. The term usually refers to thick fabrics coated with a vinyl-based coating and used for upholstery.

Vinyl wall covering. Vinyl wall covering printed on presses or by hand and hung like wallpaper. Made with paper or cloth backing, it comes in a variety of weights for the use intended.

Voile

1. Sheer, soft fabric made from cotton or various other types of natural and synthetic fibers and used for curtains.

2. Lightweight, crisp, sheer fabric of a plain weave. Voile is made of cotton or blends and is often printed to match heavier fabrics. Voile is used for curtains.

Wallpaper. Paper printed by hand or machine methods in a variety of patterns, textures, and colors.

Warp. Lengthwise yarns in a piece of fabric.

Warp knit. A warp knit is made on a machine in which parallel yarns run length wise and are locked into a series of loops. Fabrics have a good deal of stretch in the crosswise direction.

Washable wallpaper. Machine- and hand-printed washable papers are available. They are treated with a clear plastic coat at the factory.

Weft knit. A weft knit is made on a machine that forms loops in a circular direction and has one continuous thread running across the fabric.

Welting

1. Decorative edging that lends a certain degree of strength to the area in which it is sewn. Welting is made by covering cord with bias strips of matching or contrasting fabric and is a popular finish for seams on upholstery and slipcovers. Welting and piping are synonyms.

2. Fabric-covered cord inserted in seams of slipcovers or upholstery. It often edges bedspreads.

Wool. Natural crimped fiber from the fleece of lambs, sheep, and goats.

Yarn-dyed fabrics. Fabrics made from yarns dyed before weaving.

FURNITURE AND ACCESSORIES

Accessories. Items both useful and necessary for a room's comfort, often the accents to the general color scheme, including such items as ashtrays, vases, plants, lamps, books, pictures, throw pillows, and small sculptured works.

Airfoam. Goodyear's trade name for a rubber foam latex material used for cushions and padding.

Akari lamps. Folding rice-paper Japanese lanterns.

Alabaster. Fine-textured, compact variety of sulfate of lime or gypsum. The finished material is used for applied ornaments or sculpture.

Andirons. Pair of upright metal supports with a transverse rod that holds the logs for buring on an open hearth.

Antique finish. Furniture finish used on wood to give it an aged look.

Antiquing

1. Applying finish to furniture to simulate a soft, worn, aged appearance.
2. Method of treating wood to lend an old appearance or "patina." Antiquing may be done with chemicals, paint, or stain. It adds a worn-away look to the color and reduces the brilliance of the surface. Specking, instead of continuous color, adds a soft look.

Apothecary chest. Chest with many small (or simulated drawer fronts) drawers adapted from early American druggists' chests.

Apron

1. Structural support placed at right angles to the underside of a table or chair and can be decorative trim.
2. Structural part of a table directly beneath and at right angles to the top, connecting with the legs.

Armchair

1. Chair with armrests or arm supports.
2. Chair with arms, designed to be used at dining tables.

Armoire. Tall, deep piece of furniture for hanging clothes. Also made with bottom drawers.

Artifact. Article of great antiquity made by man.

Bachelor chest. Chest of drawers, sometimes used in pairs.

Backplate. Metal mount or escutcheon on which the drawer handle or pull is mounted.

Bail handle. Handle made of metal in the form of a half-loop.

Banister-back chair. Chair with split turned spindels or flat bars for the uprights of the chair back.

Banjo clock. Wall clock whose contour resembles an inverted banjo.

Banquette. Upholstered bench usually used for dining.

Barrel chair. Semicircular chair, usually upholstered.

Base wood. Basic construction wood for a piece of furniture.

Basset table. Gaming table for playing a five-handed game popular in the eighteenth century.

Bean bag chair. Shapeless chairs made from plastic or leather and filled with polystyrene beads.

Bed frame. Steel or wood frame with supports for the spring, mattress, and headboard.

Bedroom chair. Light-frame side chair.

Bedside table. Small table placed at one side of the bed; also called "night table" or "night stand."

Bench. Rectangular seating device most often backless.

Bentwood. Furniture made by steaming and bending beech wood.

Bergere. Upholstered chair with closed arms extending only part way to the front.

Beta fabric. Fiberglass fabric that is soft, pliable, strong, and durable. It is used in curtains and draperies.

Biscuit tufting. Method of tying back upholstery and padding to create plump square tufts on a chair back and seat.

Bolster

1. Long, cylindrical stuffed pillow or cushion.
2. Deep oval or triangular deep cushion used for armrest or backrest.

Bonnet bed. Bed that has a short extended canopy on tall headboard posts.

Bookcase. Piece of furniture that resembles a china cabinet and that can be made with or without glazed doors and with fixed or adjustable shelving.

Box bed. Bed enclosed on three sides.

Box spring. Sleeping unit made of spiral steel springs encased in a boxlike frame.

Brad. Small, thin, nail with a small head, less than 1 in. in length.

Breakfront

1. Piece of furniture whose front is formed on two or more planes.
2. Large, high storage piece for china, silver, and glassware.
3. Tall cabinet or secretary with a front divided into several vertical sections, one section projecting forward from the others.

Broken pediment. Triangular top, from classical architecture, sometimes used on headboard, mirrors, tall chests, or cabinets. The sloping lines stop short of the peak, leaving a gap for an ornamental finial such as a turned knob or urn shape.

Bucket armchair. Armchair similar to the spoon-bucket chair.

Buffet. Cupboard or sideboard.

Bunk bed

1. Two or three single beds, built in a tier, one over the other.
2. Single bed with matching bed fitted on top (with ladder). Some types can be rearranged into a trundle or twins.

Butler's table. Low table with a top rim in which holes are cut for handles.

Butterfly table

1. American-designed drop-leaf table.
2. A butterfly table is comparable to a gateleg table, except that the drop leaves are supported on wing-shaped brackets that swing out from the stretchers.
3. Table with rounded drop leaves supported by wing brackets.

Cabriole leg.   Furniture leg with a graceful S-curve shape.

Candlestand table.   Small tripod, pedestal or four-legged table used to hold candles and their drippings.

Cane.   Reedlike material plaited or woven into a mesh, used in the back of seating portion of chairs as a panel insert in a wood frame.

Cannonball bed.   Large balls cap each of its four low posts.

Canopy.   Covering over a bed, usually supported on high posts.

Canopy bed

1. Bed with a fabric roof supported usually by the corner posts of the bed.

2. Tall four-poster with fabric-covered wood frame over it. Originally designed with vertical hangings to keep out drafts in unheated bedrooms.

Captain's bed.   Single bed with drawers or cabinets built under the bed frame.

Captain's chair.   Type of chair with a saddle seat and often with low bentwood back and arms. This type of chair was once used on ships and later in taverns.

Carcass.   Base assembled wood framework of a piece of furniture ready for finishing and upholstery.

Card table

1. Table top with folding legs for storage. Tops are finished with many types of materials.

2. Folding table used for gaming.

Cedar chest.   Low chest with hinged top.

Center drawer guide.   The channel under the center of the drawer rides over a guide on the case frame to prevent jamming and ensure smooth operation.

Chair bed.   Chair or settee that converts into a bed.

Chaise lounge.   Reclining chair with a very elongated seat.

Chest-on-chest

1. Furniture piece consisting of two complete chests of drawers.

2. Two-sectioned chest. The section mounted on top is usually slightly smaller.

Cheval mirror.   Mirror that swings between posts and is adjustable. Small versions, often with drawers in the base, are used on chests or tables.

China cabinet.   Display cabinet, usually with glazed doors and glass or wood shelving.

Club chair

1. Large, roomy, upholstered easy chair.

2. Small-scale upholstered arm chair.

3. Low-backed upholstered chair with fairly low seats.

Cobbler's bench.   Originally a shoemaker's work seat with drawers for tools, a cobbler's bench is now designed for use as a cocktail or coffee table.

Coffee table.   Usually long low table used in front of a sofa or couch, also called a "cocktail table."

Comb-back chair.   Type of Windsor chair with spindles resembling an old-fashioned high comb.

Commode

1. Originally an enclosed bedroom "chamber box," later combined with a wash stand. The term also means "night stand" or "console chest."

2. Chest of drawers or a cabinet, usually low and squat.

3. Small drawer cabinet.

Console table

1. Narrow, rectangular table placed against a wall.

2. Shelflike table, attached to a wall.

Contour chair.   Chair that is molded, shaped, and contoured to conform to the human body.

Corner chair.   Chair whose back forms a semicircle around its two sides. The seat is diagonal to the middle of the back. The occupant straddles the front leg.

Cornice.   Projecting molding used to give an architectural finish to the top edge of a chest, dresser, or cabinet.

Credenza.   Buffetlike storage cabinet, usually with doors.

Crown bed.   The canopy is suspended over the bed.

Cupboard.   Storage cabinet with doors.

Cushion.   Shaped, flexible bag of fabric or leather filled with feathers or other filling materials.

Davenport bed.   Sofa that converts into a bed.

Daybed

1. Studio couch, restbed, or narrow bed.

2. A day bed converts from a couch to a bed.

Dentil.   Small-shaped rectangular block used in a row and projecting like teeth under a cornice; classic Greek form of ornamentation.

Desk.   Type of table having writing surface area with or without drawers.

Dinette set.   Table and chairs for a small breakfast room. Furniture should be in scale to the room size.

Dinette table.   Usually a plastic topped, metal legged table for kitchen eating areas. However, many dinette tables are available in fine woods.

Divan.   Armless and backless upholstered settee.

Divider.   Furniture or screen arrangement that separates one area of a room from the rest of the room.

Double bed.   The standard-size double bed is a minimum of 53 in. wide × 75 in. long.

Dovetail

1. Furniture joint; interlocking two pieces of wood without nails.

2. Joint made by hard tongues of wood that interlock with shaped pieces of wood to hold front and back drawer corners securely.

Doughboy table.   Deep slope-sided box used for "raising" bread, adapted for use as an occasional table with roomy storage compartment.

Dowel.   Round peg of wood that fits into a corresponding hole to form a strong joint, used instead of nails.

Down.   Soft, fluffy feathers from fowl, used for stuffing pillows or cushions.

Dresser.   Sideboard with drawers for storage space in combination with mirrors.

Drop handle.   Pear or tier-shaped pull, usually brass.

Drop-leaf table.   Table with end leaves that drop to the sides.

Drum table.   Round library or lamp table with a deep open or closed top (which suggests a drum) that sometimes revolves on a pedestal base.

Dry sink.   Cupboard with open well on the top, often lined with copper tray. Originally water was added for washing dishes.

Easy chair.   Roomy, comfortable, upholstered chair.

Elderdown.   Soft, fluffy feathers obtained from large sea ducks.

End table. Small table placed at the ends of a sofa or couch to hold a lamp or a few books or magazines.

Extension table. Tabletop that separates in the center and extends outward in both directions.

Fiddleback chair. Chair back whose design resembles a violin.

Finial. Tip of an ornamental pinnacle, as atop a lamp, bedpost, curtain rod.

Flap table. Fixed center slab and two side flaps, which can be lowered by folding back the legs that support them.

Floating construction. Construction used for furniture made of solid woods; it permits the top and side panels to expand or contract with changes of temperature and humidity, thus avoiding warping or cracking.

Fluting. Vertical channels carved into columns, as seen in ancient Greek architecture.

Foam rubber. Rubber material made of latex; the sap of the rubber tree whipped with air to create a light, porous rubber composition.

Folding bed. Cot, on casters, that folds up, end to end, and locks into an upright position.

Folding chair. Straight chairs that have hardware to enable them to fold up for storage.

Folding furniture. Collapsible furniture that can fold into a compact unit.

Folding screen. Screen comprised of two or more vertical panels, hinged together, for forming a divider in rooms.

Four-poster bedstead. Bed with two posts in front and two in the back.

Gallery. Decorative railing around edge of table, shelf or tray.

Gateleg table

1. Table with two drop leaves and legs that swing out like a gate to support the leaves when extended.
2. Table with out-swinging gate-shaped legs, which support the end leaves of the table top when they are raised and placed over the gate legs.

Grandfather clock. Floor-standing clock in a wood case.

Grill. Metal or wood combined with glass for some cabinet doors.

Hall tree. Floor-standing hat and coat rack made of metal or wood.

H and L hinge. Design resembling the letters H and connecting L, made originally of iron or brass.

Harvest table. Long drop-leaf table, usually with narrow leaves running the entire length, originally designed for use for feeding harvest hands.

Hassock. Heavy cushion or thick mat used as a footstool.

Headboard. Board or panel that rises above the mattress at the head of the bed.

Highboy

1. Tall chest of four or five drawers, on legs.
2. Tall chest of drawers supported by legs and usually crowned with cornice moldings or a pediment.
3. Two-piece chest of drawers with the upper chest taller, narrower, and shallower than the bottom one.

Hollywood bed

1. Bed without a footboard. The spring and mattress are set on a metal bed/frame unit.
2. Box spring attached to four legs, to which a headboard and footboards may be attached.

Hurricane lamp. Tall glass cylinder shade set over a candlestick to protect the flame.

Hutch

1. Cabinet or cupboard placed over a buffet unit.
2. Small cupboard, chest, or bin.
3. Open-shelved storage piece on a base with cupboards and/or drawers.

Isinglass. Translucent sheets of mica, originally used for fenestration and the glazing in old parlor stoves.

Kidney desk. Ornamental kidney-shaped table with the concave side toward the sitter.

King-size bed. Bed that is usually 72 to 78 in. wide by 76 to 84 in. long and has one large headboard.

Kneehole desk. Desk with a central space below the writing surface for leg room.

Knocked down. Term applied to furniture that is shipped unassembled or in several parts which can be put together at the point of use.

Ladder-back chair. Chair with a ladder-shaped back and horizontal slats.

Lampshade. Covering or shield over a light bulb that is set on top of a lamp base.

Lampshade trim. Narrow edging, often interwoven with metallic thread.

Lawson chair. Upholstered low-backed chair with arms that are often set back at the front to accommodate a T cushion (seating cushion that extends forward and across the front of the arms).

Lazy Susan. Round, revolving wood tray for the center of a dining table.

Leather. Upholstery covering material, usually the cured hide of animals.

Leatherette (artificial leather). Nitrocellulose material with heavy cotton backing.

Limed oak. Special finish applied to oak to give it a frosted or silver gray appearance.

Lounge. Sofa or couch

Love seat. Upholstered settee for two persons.

Lowboy. Serving table or low chest of drawers.

Mate's chair. Chair that is smaller than a Captain's chair, with shorter arms.

Mattress. Fully filled pad placed over the springs or slats of a bed frame.

Modular furniture

1. Modern concept in furniture design.
2. Size-matched pieces of furniture that can be stacked or placed next to each other.

Morris chair. Overstuffed easy chair.

Mortise and tenon. A tongue or projecting part of wood fits into a corresponding rectangular hole or mortise.

Moss fringe. Short, thick fringe, with narrow, tailored head, that can be used in slipcover seams or as drapery edging.

Nail heads. Nails or brads with plain or decorated oversized heads, made of brass, copper, or other metals.

Naugahyde. Trade name for a vinyl upholstery and wall covering of the United States Rubber Company.

Nested tables. Series of small tables, graduating in size, so that one can be set inside the other.

Occasional chair. Armchair with fairly thick upholstered back and seat.

Occasional table. General term for a small table used for many purposes.

Ogee bracket foot. Foot shaped like a bracket, with a double curve.

Ottoman. Long, backless, cushioned seat, couch, or divan.

Overstuffed furniture. Heavy upholstered pieces of furniture.

Pedestal table

1. A single turned center pedestal supports the table, or the pedestal may have a tripod base. Extension tables often have two pedestals for firm support when extended.

2. Table supported by a column rather than legs.

Pembroke table. Small rectangular table with two drop leaves and a shallow drawer.

Portable server. Movable serving cart on casters.

Poster bed. Bed that has four tall posts, decoratively turned.

Pouf. Hassock resembling a big, thick, puffy cushion.

Queen-size bed. Bed with a single mattress approximately 60 in. wide by 76 to 84 in. in length.

Quilting. Two layers of fabric with padding between the layers, which is held in place by stitches that follow a design or pattern.

Reclining chair

1. Mechanically operated chair with a back that lets down and a footrest.

2. Type of upholstered chair with a back that can be pushed and locked in several positions. The chair is also mechanically designed to have the back act in conjunction with the support for the occupant's legs.

Rocking chair. Chair designed to rock back and forth.

Saddle seat. Seat that is scooped away to the side and back from a central ridge for more comfort. It resembles the pommel of a saddle.

Sconce. Decorative light fixture; wall bracket with one or more arms holding candle sockets for fancy-shaped bulbs.

Secretary

1. Desk surface with space for writing appliances, containing a drawer base below and a bookcase cabinet above.

2. Desk with drawers below and open or closed bookshelves above.

Sectional furniture. Upholstered or case furniture made in modules or small units which can be pulled together in many ways and seating arrangements.

Settee. Seating device; long seat with a carved or upholstered back, arms, and a soft seat.

Settle chair. All wood chair, often of Windsor design; bench with arms, sometimes called a "deacon's bench."

Sideboard. Auxiliary case piece in a dining room, having drawers and cupboard space.

Side chair. Essentially straight-backed chairs without arms.

Single bed. Bed usually 39 in. wide by 78 to 84 in. long.

Slat-back chair. Chair back that has several horizontal rails or crossbars.

Sling chair. Chair with a metal frame on which hangs a hammocklike seat of canvas, vinyl, or leather.

Slipcover. Removable cover for a chair, sofa, or other furniture piece.

Sofa. Upholstered daybed or couch.

Sofa bed. Sofa whose back drops down and becomes parallel with a seat.

Sofa sleeper. Convertible sofa with a concealed sleeper beneath the seat.

Spindle bed. Bed that has decorative turnings, like some chair backs.

Spool bed. Bed whose turnings are shaped like thread spools.

Standard-size bed. Double bed, usually 53 to 54 in. wide by 75 in. long.

Stenciling. Applying a pattern to furniture by painting over thin metal or heavy paper cutouts.

Step table. The step table was originally made for reaching high bookshelves. It is now used as a two- or three-level end table with the steps retained as part of the design.

Stick table. Combination pedestal table and lamp.

Straight chair. Similar to a side chair, but specifically designed for dining.

Stretcher

1. Wood or metal strip connecting furniture legs.

2. Supports connecting the legs of any piece of furniture. They are frequently tuned for decorative interest.

Studio couch. Seating device that converts into a sleeping unit.

Swag light. Light fixture that is supported at the ceiling, with an electric cord entwined in a metal chain. The electric outlet may or may not be at the ceiling.

Swedish modern. Contemporary style of furniture and furnishings.

Swing bed. Twin beds attached to a single large headboard. The beds can be separated and positioned apart to serve as single beds; together they function as a king-size bed.

Swivel chair. Desk chairs mechanically set to spin 360°.

Tester. Four-poster bed fitted with a canopy frame.

Tilt-top table. Pedestal table with a hinged top that can be dropped vertically when not in use.

Trestle table. Long rectangular top on two vertical supports.

Trundle bed

1. Single bed with a second single bed stored underneath. The latter can be pulled out at night and raised to the height of the top bed.

2. Low bed on rollers that fit under a single bed.

Tub chair. Small, circular upholstered chair, rather like half a barrel.

Tufting. Upholstery finishing technique.

Tuxedo chair. Upholstered chair with arms the same height as the back.

Upholstering. Act of stuffing, padding, and covering chairs, sofas, couches, settees, etc.

Upholstery. Fabric covering, cushioning, springs (if any), and all materials attached to the rough frame.

Wall bed. Bed that folds upright into a wall recess or closet enclosed behind doors.

Water bed. Bed with a mattress made of plastic and filled with water.

Webbing

1. Strips of tightly woven burlap used in upholstery construction.
2. Interwoven strips that give support to upholstered furniture.

Windsor chair

1. Chair so called because the shapes forming spindles were turned near Windsor, England.
2. Chair characterized by slender turned spindles, a wooden saddle seat, and turned, splayed, or raked legs usually joined by a stretcher at the bottom. Also referred to as "Duxbury."

Wing chair

1. Upholstered chair with high side pieces designed to ward off drafts.
2. Upholstered chair with high wings flanking the back.

RUGS AND CARPETS

Abrash. Irregularities in the same color, which results when all the wool is not dyed at the same time. The effect is often deliberately produced in machine-made carpets and rugs.

Abrasion resistance. Degree to which material resists wearing due to foot traffic.

Acrylic. Acrylic resembles wool; it is resilient, resistant to soiling and crushing, durable, and easy to clean.

Animal fur. Animal fur is used for throw rugs or wall hangings. Simulated and synthetic man-made furs are also used.

Area rug

1. Rug, sometimes shaped irregularly but that is mostly rectangular.
2. Small, usually decorative rug often placed on a carpet as an accent in a room or to define an area such as a dining section in a living room.

Aubusson

1. Rug without a pile; rug that is woven like a tapestry.
2. Woven rug with little or no rib and a low pile.

Axminster

1. Rug woven on a loom that makes fairly complicated designs possible.
2. Type of carpet. The term refers to rugs woven originally at Axminster, England. The carpet is woven tightly and the pile is usually cut.
3. Type of loom used in the manufacture of carpet.

Backing. Material into which carpet yarns are tufted, made of jute, olefin, or polypropylene fiber, often latex coated.

Bashrah. Term used for Anatolian rugs with a white ground.

Braided fabrics. Fabrics formed by interlacing three or more strips of fabric or yarn to form a flat or tubular, usually narrow, length of fabric. Braided rugs are made from strips of braid that are subsequently joined together.

Braid rug

1. The rug is made by joining strips of braid together with stitches. Rugs may be either rectangular, oval, or round.
2. Rug made of strips of fabric that are braided together and then stitched together to form a rug.

Broadloom

1. Seamless carpet woven in many widths and a variety of textures, weaves, and colors.
2. Carpet manufactured in wide strips 9, 12, or 15 ft or over.
3. Carpet made on a wide loom.

Broadloom carpet. Carpet woven on a loom at least 9 ft wide or wider. The term is also used to describe wide, tufted carpeting in which tufts of yarn are pulled through a backing to form the rug surface.

Brocade. Rug in which a pattern is formed by using yarns of the same color with different twists. Light strikes the yarns differently, giving a shaded design effect.

Brussels carpet. Uncut-wool-loop pile fabric woven on the Wilton loom.

Carpet. Term interchangeable with rug. Carpet usually refers to a heavy fabric floor covering that covers the entire floor and is most often fastened to it in a somewhat permanent fashion.

Carpet tile. Small pieces of carpet with adhesive backing.

Carved rug. Rug on which the pattern is created by having the pile cut at various levels.

Chain. Total of the warp threads stretched on the two beams of the loom.

Chenille carpet

1. Thick, soft, cut-pile fabric woven on two looms.
2. Plush carpet with a deep pile.

Cotton. Durable, but soils and crushes easily. It is easy to clean.

Count. Length per pound of yarns, varying with the intrinsic weight of the yarn and its thickness.

Cut-pile

1. Fabric woven with an extra set of warp of filler yarns.
2. Many fabrics are formed with loops on the surface. When these are cut, they form a cut pile. Some velvets and many pile rugs are made in this manner.

Drugget. Coarse, felted floor covering made from mixtures of such fibers as cotton, jute, and wool. A drugget is usually napped on one side and is traditionally a floor covering used by institutions.

Felt-base rug. The covering material has an enameled design print and is similar to linoleum.

Fiber rug. Reversible rug woven of kraft or sisal fibers, sometimes combined with wool or other fibers.

Floor cloth. Area rug made of canvas.

Flossa. Scandinavian word denoting a particular closely tufted type of rug.

Frieze carpet. Carpet with tightly twisted yarns in a rough, nubby cut pile.

Garden carpet. Carpet reproducing more or less distinctly the layout of a Persian garden with its camels, pools, and flower beds.

Grass rug. Rug produced in many sizes from several types of long grass, in natural colors or dyed. The rug comes with a thin rubber padding or backing.

Ground. Principal color or background.

Hand-knotted rug. Hand-knotted rugs, including oriental and Persian rugs, are among the most expensive made. Intricate designs are possible. The higher the number of knots to the inch, the finer is the rug.

Hooked rug

1. Pile-surfaced rug made of threads or strips of cloth pushed through a canvas backing.

2. Rug made by hand or machine using a hook to pull loops of yarn or fabric through a coarse backing or canvas to form a pile.

Indoor-outdoor carpeting. Carpeting that can be used outdoors where it will be exposed to weather conditions, as well as indoors. Most of this type of carpeting is made of olefin.

Interlock. Method of creating a pattern used in Röllakan and Khelim rugs.

Khelim. Rugs of oriental origin in tapestry weave with inlay.

Kilim. Woven carpet; web ends of a knotted carpet.

Knotted rugs. Rugs in which tufts are knotted on a foundation weave to form a raised pile.

Loop

1. Uncut yarns of carpet or fringe.

2. Yarn forms loops.

Loop rug. Usually, a rug with an uncut loop pile. Types of loops include high-low loop (a rug in which some loops are higher than others giving a sculptured effect), one-level loop, and random-sheared loop (a rug with some loops cut and others uncut to create a sculptured effect).

Medallion. Form of design widely used in all types of rugs, especially in Persian carpets of the Safavid period. The medallion generally appears in the middle of the field, often with one or two pendants.

Modacrylic. Similar to acrylic, but not as resilient. Modacrylic fiber is blended with acrylic to increase fire resistance.

Narrow carpet

1. Narrow carpet is used as the opposte of broadloom carpet. Narrow carpet is woven in narrow widths like 36 in. and is popular for stair coverings.

2. Carpet made on a loom about 3 ft wide. The term is used to ditinguish this carpeting from broadloom carpeting.

Navajo rug. Rug woven in a tapestry weave in a geometric pattern by Navajo Indians in the western part of the United States. Navajo rugs are usually brightly colored and have become true collectors' items.

Nylon. Strongest and longest wearing of the man-made fibers. Nylon cleans well, is mothproof, and remains mildew free.

Oriental rugs. Hand-made rugs produced in both the Middle and Far East. They are either hand-woven, by the tapestry method, or hand-knotted.

Padding. Any item that provides a degree of support to a fabric. It is usually the layer of fabric placed underneath a carpet or rug to provide it with longer life and to give it a more luxurious appearance and feeling. Carpet padding is made of the cattle hair, rubberized hair, rubber, and combinations of jute and cattle hair, as well as some of the man-made fibers. Some rugs and carpets have a bonded foam, sponge, rubber, or man-made backing, in which case no separate padding is needed. Padding is also called "cushion" and "underlay."

Persian rug. Oriental rug made in Iran (formerly called Persia).

Pile. Raised surface, silk or wool, of a knotted carpet. The length of the pile varies greatly, depending, apart from the condition of the carpet, on the place of origin.

Pile weave. Fabric with clipped yarns on the right side; examples are velvet and corduroy.

Plush carpet. Carpet made from straight yarns that are sheared and used in a tight pile.

Polonaise rug. Certain type of seventeenth-century silk rug, often brocaded in gold and silver.

Polyester. Polyester rugs and carpets are soft, durable, mothproof, and mildew free. They are resistant to crushing and also clean well.

Portuguese carpets. The corners of these carpets show sailing ships. Since the Portuguese were the first Europeans in India, the carpets were at first thought to be Indian. The tendency now is to attribute them to Persia.

Prayer rug. Rug on which devout Moslems say their prayers, either kneeling or standing.

Rag rug. Rug woven with strips of cotton, wool, or synthetic fabrics used as the filling on a cotton or synthetic yarn warp. Rag rugs are made both by hand and machine and, with the exception of some hand-made antique rag rugs, they are usually the most inexpensive rugs.

Random-sheared carpet. The pile is a combination of sheared and looped yarns.

Random-sheared rug. Pile rug in which some sections of the pile are cut and other sections are not.

Rayon. Rayon gives a very dense pile. It will improve soil resistance, abrasion resistance, and resiliency.

Rug. Loose floor covering of any size.

Rugs and carpets. Rugs and carpets are usually the most expensive fabrics most people ever buy. The words "rug" and "carpet" are often used interchangeably. A rug usually refers to a floor covering that does not cover the entire floor and is not fastened to it, while carpet is used for fabric that does cover the entire floor and is fastened to it.

Rya rug (rye-ah)

1. Rya rugs are popular as area rugs, largely because of their dramatic color combinations. The highest quality rya rugs are hand-knotted.

2. A rya rug is a high-pile shaggy rug. It is the Scandinavian version of the tufted rug, generally having rather long tufts.

Saran (a proprietary name). Saran has good soil and stain resistance and only fair abrasion resistance. It will not burn, but will melt at certain temperatures.

Scatter rug. Small area rug used as an accent on the floor.

Sculptured carpet

1. Carpet with a plush surface cut at various heights to achieve a design.

2. Carpet (or rug) in which the pile is cut in different lengths to form a pattern or design.

Shag

1. Long-haired carpet surface.
2. Long yarn used in a loose pile.

Shag rug or carpet. Rug or carpet with an extremely long pile.

Short-pile method. Method of tufting rugs in which a continuous weft yarn is used instead of cut lengths.

Splush carpet. Combination of plush and shag.

Straw. Naturally grown material used for weaving into rugs. The backing material may be paper or rubber.

Tufted rug

1. Rug with a raised-pile surface.
2. Type of rug which the tufts are inserted in the backing.
3. Type of rug construction. Tufted rugs are formed by needles rapidly punching yarn into the rug backing to form a pile that can be left looped as it is, cut, or sheared.

Tufting. Method of making rugs. Groups of yarns are forced through a backing fabric. The yarns are held in place permanently when the underside of the rug is coated, often with liquid latex.

Twist. Technical term referring to the way in which yarn is turned during the course of its manufacture. In capeting, twist is a corkscrewlike cut pile that has a pebbly appearance.

Twist rug. Rug or carpet made of twist, a strong, long-wearing yarn that has been tightly twisted in its manufacture. Twist rugs and carpets are recommended for high-traffic areas.

Vase carpet. Carpet of a rich floral design which usually includes a vase.

Vinyon. Synthetic fiber used for embossed carpets, pressed felts, and other nonwoven fabrics.

Weft. Threads or yarn crossing a weave from selvedge to selvedge.

Wilton. Type of loom used in the manufacture of carpet.

Wilton rug. Woven cut-pile rug with a velvety texture. The designs in Wilton rugs often show an oriental influence.

Wool. Basic carpet and rug fiber. It is very resilient and resists crushing and soil. Its durability is very high and cleaning is simple. Wool must be mothproofed.

FACTORY FABRICATED STRUCTURES

Air-inflated structure. A building where the shape of the structure is maintained by air pressurization of cells or tubes to form a barrel vault over the usable area. Occupants of such a structure do not occupy the pressurized area used to support the structure.

Air-supported structure. A building wherein the shape of the structure is attained by air pressure and occupants of the structure are within the elevated pressure area. Air-supported structures are of two basic types.

1. Single skin—Where there is only the single outer skin and the air pressure is directly against that skin.

2. Double skin—Similar to a single skin, but with an attached liner which is separated from the outer skin and provides an air space which serves for insulation, acoustic, aesthetic, or similar purposes.

Building component. Any subsystem, subassembly, or other system designed for use in, or as a part of, a structure, including but not limited to: structural, electrical, mechanical, fire protection, and plumbing systems, and other systems affecting health and safety.

Building system. Plans, specifications, and documentation for a system of manufactured building or for a type or a system of building components, including but not limited to, structural, electrical, mechanical, fire protection, and plumbing systems, and including such variations thereof as are specifically permitted by regulation, and which variations are submitted as part of the building system or amendment thereof.

Cable-restrained air-supported structure. One in which the uplift is resisted by cables or webbing that is anchored to either foundations or dead men. Reinforcing cable or webbing may be attached by various methods to the membrane or may be an integral part of the membrane. This is not a cable-supported structure.

Cable structure. A nonpressurized structure in which a mast and cable system provides support and tension to the membrane weather retarder and the membrane imparts structural stability to the structure.

Closed construction. Any building, component, assembly, or system manufactured in such a manner that all portions cannot be readily inspected at the installation site without disassembly, damage to, or destruction thereof.

Factory-built building. Building or component part thereof that is either wholly or in substantial part manufactured at a location other than the building lot.

Factory-built chimney. Factory-made chimney, listed by an accredited authoritative agency, for venting gas appliances, gas incinerators, and solid or liquid-fuel-burning appliances.

Flame retardant or flame resistant. Fabric or material resistant to flame or fire to the extent that it will successfully withstand standard flame resistance tests adopted and promulgated by the state fire marshal. Membrane is a thin, flexible, impervious material capable of being supported by an air pressure of 1.5 inches of water column.

Frame-covered structure. A nonpressurized building wherein the structure is composed of a rigid framework to support tensioned membrane, which provides the weather barrier.

Installation. The process of affixing or assembling manufactured buildings or building components on the building site, or to an existing building.

Label. Approved device affixed to a manufactured building or building component by an approved agency, evidencing code compliance.

Manufactured building. Any building that is of closed construction and that is made or assembled in manufacturing facilities, on or off the building site, for assembly and installation, on the building site. Manufactured building may also mean, at the option of the manufacturer, any building of open construction, made or assembled in manufacturing facilities away from the building site, for installation, or assembly and installation, on the building site.

Manufactured home. Structure built in accordance with the National Manufactured Home Construction and Safety Standards Act.

Mass and industrialized production. Prefabrication applies to all prefabricated forms of building elements and assembled construction units intended for both structural and service equipment purposes in all buildings of all use groups. Prefabrication covers the precutting and assembling of individual elements either in the shop or at the site before erection in the building structure. Prefabricated shop assemblies are shipped in structurally complete

585

units ready for installation in the building structure or in knock-down and packaged form for assembly at the site.

Membrane. A thin, flexible, impervious material capable of being supported by an air pressure of 1.5 in. of water column.

Mobile home. Structure built on a permanent chassis, capable of being transported in one or more sections, and designed to be used, with or without a permanent foundation, as a dwelling when connected to on-site utilities.

Modular home. Dwelling, factory-produced as a complete unit, ready to install at the home site on a prepared permanent foundation. Modular homes, constructed of plywood components, are fabricated under factory quality-inspection procedures.

Noncombustible membrane structure. A membrane structure in which the membrane and all component parts of the structure are noncombustible as defined by the code.

Open construction. Any building, component, assembly or system manufactured in such a manner that all portions can be readily inspected at the installation site without disassembly, damage to, or destruction thereof.

Plans and specifications. Complete legible dimensioned drawings and specifications covering every type of prefabricated construction submitted to the building department for approval. Permit application usually describes all essential elements of the structure or assembly.

Portable cabana. Any prefabricated cabana that is designed to be readily assembled and disassembled and adapted to ready transportation from place to place.

Portable grandstand. Assembly of prefabricated units, readily erected, dismantled, and transported, and used or intended for use as movable permanent or temporary support of audiences.

Prefabricated

1. Fabricated prior to erection or installation on a building or structure foundation.
2. Term applied to construction materials or assembled units fabricated prior to erection or installation in a building or structure.

Prefabricated assembly. Structural unit, the integral parts of which have been built up or assembled prior to incorporation in the building.

Prefabricated building. Completely assembled and erected building or structure, including the service equipment, of which the structural parts consist of prefabricated individual units or subassemblies using ordinary or controlled materials, and in which the service equipment may be either prefabricated or at-site construction.

Prefabricated building assembly. Building unit, the parts of which have been built up or assembled prior to incorporation in the building.

Prefabricated subassembly. Built-up combination of several structural elements designed and fabricated as an assembled section of wall, ceiling, floor, or roof to be incorporated into the structure by field erection of two or more such subassemblies.

Prefabricated unit. Built-up section forming an individual structural element of the building, such as a beam, girder, plank, strut, column, or truss, the integrated parts of which are prefabricated prior to incorporation into the structure, including the necessary means for erection and connection at the site to complete the structural frame.

Prefabricated unit service equipment. Prefabricated assembly of mechanical units, fixtures, and accessories comprising a complete service unit of mechanical equipment, including bathroom and kitchen plumbing assemblies, unit heating and air-conditioning systems, and loop-wiring assemblies of electric circuits.

Relocatable structure. Factory-built unit for use without a permanent foundation.

INCINERATORS

Breechings. Flue connections and breechings usually constructed of not less than No. 16 U.S. gauge sheet metal when less than 12 in. in diameter and of not less than No. 12 U.S. gauge sheet metal when more than 12 in. in diameter.

By-product waste. Tar, paints, solvents, sludge, and from other industrial operations creating such wastes.

Charging chute (incinerator). An enclosed vertical passage through which refuse is fed to an incinerator.

Charging gate (incinerator). A gate in an incinerator used to control the flow of combustion gases into the charging chute and the entry of refuse into the combustion chamber.

Commercial and industrial incinerator

1. Incinerator designed to burn waste matter incidental to any class of occupancy.
2. Incinerator other than a domestic-type or flue-fed incinerator.

Domestic gas-fired-type incinerator. Direct-fed, gas-fired type, generally located within a building or structure and designed primarily for use in specific occupancies and for the burning of ordinary waste material with a capacity of no more than four bushels.

Domestic-type incinerator

1. Incinerator with a combustion chamber of not over 5 cu ft or a capacity not exceeding a 25-lb/hour burning rate.
2. Direct-fed incinerator with a fire-box or charging compartment 5 cu ft designed to incinerate materials that burn with no higher intensity than wood or paper.

Domestic outside-type incinerator. Free-standing incinerator does not form an integral part of a building or structure.

Draft regulators. Incinerators provided with an automatic draft control that will limit the draft in the primary combustion chamber to a maximum of 0.15 in. of water.

Firebrick. Refractory fireclay brick that meets the standards of the local building code.

Flue-fed apartment-type incinerator. Incinerator having a chimney that also serves as a charging chute from one or more floors above the incinerator.

Flue-fed incinerator

1. Incinerator arranged and constructed so that refuse can be fed into it through hopper doors in the flue that serves it.
2. Incinerator having a combined refuse chute and smoke flue and designed to feed waste materials directly into the combustion chamber.

Fuel-fired incinerator. Incinerator in which combustion of the material to be incinerated is aided by gas, liquid, or solid fuel. Gas or liquid fuel may be introduced by a blower.

Garbage. Consists mainly of animal and vegetable wastes from restaurants, cafeterias, hotel dining rooms, hospital food services, markets, and similar installations where food is prepared and served.

High-heat-duty firebrick. Refractory-fired clay brick that meets the standards of the local building code.

Human and animal remains. Carcasses, organs of all kinds, organic wastes from hospitals, animal pounds, laboratories, autopsy departments, and similar sources.

Incinerator

1. Device intended or used for the reduction of garbage, refuse, or other waste material by burning or incineration.
2. Appliance or combustion chamber used for burning rubbish, garbage, and other combustible waste material.
3. Device, using heat, for the reduction of garbage, refuse, or other waste materials.
4. Special type of furnace for burning combustible refuse in order to reduce it to stable gases and inert solids. The term, as generally used, includes the furnace, building, chimney or stack, storage space, auxiliary equipment, and accessories used in controlling combustion.
5. Combustible apparatus designed for high-temperature operation in which solid, semisolid, liquid, or gaseous combustible wastes are ignited and burned efficiently and from which solid residues contain little or no combustible material.
6. Approved device intended or used primarily for the reduction of easily combustible dry waste materials by burning or incineration, or other approved special types of incinerators also designed to reduce garbage, refuse, or other waste material by the same process.

Incinerator (domestic gas-fired). A domestic appliance used to reduce combustible refuse material to ashes and which is manufactured, sold, and installed as a complete unit.

Industrial incinerator. Incinerator for rubbish and waste material resulting from industrial or office occupancy and suitable for any occupancy.

Large domestic or small commercial incinerator. Incinerator with a combustion area not over 12 cu ft.

Leg-supported incinerator. Incinerators set on legs that provide open space under the base of the appliance. They may be mounted on floors, provided the appliance is so constructed that the flame or hot gases do not come in contact with its base.

Low-temperature incinerator. Low-temperature non-fuel-fired incinerator for garbage and other waste material incidental to residential occupancy and suitable only for use in dwellings, apartment, tenements, hotels, clubs, hospitals, dormitories, churches, schools, and restaurants.

Medium-temperature incinerator. Incinerator with a fuel-fired combustion chamber, used for the destruction of garbage from occupancies established by code and for industrial waste material.

Non-fuel-fired incinerator. Incinerator that uses no fuel for combustion other than the refuse to be incinerated, except that a gas flame (without blower) or other approved means may be used to accomplish primary ignition.

Packaged incinerators. Incinerators consisting of an extra-heavy steel enclosure with insulated walls, high-heat-alloy cast-iron grates, and refractory linings that will withstand approximately 3000° F. Incinerator can be gas or oil fired and have water spray wash to scrub flue gases and remove particulates, such as fly ash.

Paper. Newspapers, periodicals, cardboard, and all wastepaper.

Private incinerators. Incinerators for the burning of refuse and garbage produced on the same premises, provided that the construction is such as to assure immediate and complete combustion and freedom from offensive smoke, ash, unburned particles, and odors.

Refuse. An even mixture of rubbish and garbage accumulated from persons living in residential and apartment occupancies.

Residential incinerator. Incinerator used by not more than three families. It is gas-fired with a charging compartment of not over 5 cu ft.

Rubbish. A mixture of combustible waste such as paper cardboard cartons, wood scraps, foliage, and combustible floor sweepings from domestic, commercial, and industrial activities. Also contains small amounts of cafeteria and restaurant waste.

Solid by-product waste. Rubber, plastics, wood, and other such solid products incinerated from industrial operations.

Trash. A mixture of highly combustible waste such as paper, cardboard cartons, wood boxes, and combustible floor sweepings from commercial and industrial activities. Also contains a small percentage of plastic bags, coated paper, laminated paper, treated corrugated cardboard, oily rags, and plastic or rubber scraps.

RADIATION SHIELDING

Absorbed dose. Quantity of energy acquired by a mass of material exposed to radiation. The unit of measurement is the rad.

Areaways. Passageways, window wells, or acrossways below ground level and adjacent to an exterior wall.

Areaway walls. For an areaway, the exterior wall mass thickness of the building part adjacent to the areaway is recorded, not the mass thickness of the retaining wall at the outside of the areaway.

Atmospheric radiation. Electromagnetic radiation emitted by the atmosphere.

Attic. Space beneath the roof whose floor is at or above the level of the eaves and that has head-room of at least 7 ft in some area. An attic is considered a story.

Azimuthal sector. Portion of a circle represented by a plane angle at a detector location subtending a shield or plane of contamination. The angle is determined by drawing a line from the detector position to the left and right extremities of the shield or contaminated plane. This plane angle is designated azimuthal sector (degrees).

Background count. Evidence or effect on a detector of radiation other than that which is desired to detect, caused by any agency. In connection with health protection, the background count usually includes the radiation produced by naturally occurring radioactivity and cosmic rays.

Backward scatter. Scattering of radiant energy into the hemisphere of space bounded by a plane normal to the direction of the incident radiation and lying on the same side as the incident ray; the opposite of forward scatter.

Barrier shielding. Shielding that places mass between the shelter occupant and the radioactive source.

Basement

1. Lowest story of a building or building part with no single wall completely above ground; therefore if one or more walls are completely above the ground, that story is not the basement. The basement is always numbered 00.

2. The basement is measured from the basement floor to the top of the floor of the first story, or to the average height of the roof, if there is no first story. If there is no basement, a 0 designation is used.

Basement extension. Portion of a basement that extends beyond the exterior wall of the first story.

Basement walls. The mass reported for the exterior walls of basements is that of the exposed portion of the basement walls. If a building contains a partial basement, the exterior walls reported are those located directly below the first-story exterior walls. If a crawl-space exists under the portion of the building occupied by the partial basement, an interior partition may be recorded at the location of the partial basement walls that are adjacent to the crawl space.

Biological shielding. Shielding provided to attenuate or absorb nuclear radiation, such as neutron, proton, alpha, beta, and gamma particles; the shielding is provided mainly by the density of the concrete, except that in the case of neutrons the attenuation is achieved by compounds of some of the lighter elements, such as hydrogen and boron.

Building part. Rectangular portion of a building obtained by passing vertical planes through a building with a nonrectangular floor plan. Each part is reported as a complete and separate building on a separate form.

Building population. The figure given for this item is an estimate. Estimates can be made as follows: An apartment with 50 units could house an average of 3.60 persons per unit; thus the population would be 180. An office building could be estimated for one person per 100 sq ft of area.

Building sides. Four sides of a building or building part are identified by the letters A, B, C, and D. Side A is designated the address side. Sides B, C, and D are assigned clockwise from side A.

Crawl space. If there is a crawl space, the wall weight is to be recorded so that the contribution through that wall to the first story can be calculated.

Delayed fallout. Fallout resulting from high-yield atomic or thermonuclear detonations that thrust their clouds into the stratosphere. Storage times vary from less than 1 year to as high as 10 years, depending on the latitude and altitude of detonation. Particles are of submicron size.

Distance out from side. Dimension of the basement extension perpendicular to the wall it adjoins.

Dosage (radiation). Amount of nuclear radiation received by a person under a given set of circumstances.

Dose (radiation). Absorbed dose from any ionizing radiation; also, exposure dose, properly expressed in roentgens, which is a measure of the total amount of ionization that x-rays or gamma rays produce in the air.

Dosimetry. Measurement of radiation doses. It applies to both the devices used (dosimeters) and to the techniques.

Early fallout. Fallout near the site of an explosion.

Emergency shelter. Structure or portion of a structure above or below the ground, so constructed as to provide emergency shelter from hazards of storms or destructive forces of an enemy attack.

Exterior walls—mass thickness. The mass thicknesses reported for the exterior walls are used to determine the amount by which radiation is attenuated when it passes through the exterior walls.

Fallback. That part of the material carried into the air by a surface or subsurface atomic explosion that ultimately drops back to the earth or water at the site of the explosion.

Fallout. Process of precipitation to earth of radioactive particulate matter from a nuclear cloud; the term is applied to the particulate matter itself.

Fallout contours. Lines joining points that have the same radiation intensity and that define a fallout pattern, represented in terms of roentgens per hour.

Fallout pattern. Distribution of fallout as portrayed by fallout contours.

Fallout shelter

1. Building, structure, or other real property, or an area or portion thereof, constructed, altered, or improved to afford protection against harmful radiation resulting from radioactive fallout, including such plumbing, heating, electrical, ventilating, conditioning, filtrating, and refrigeration equipment and other mechanical additions or installations, if any, as may be an integral part thereof.

2. Structure of portion of a structure intended to provide protection to human life during periods of danger to human life from nuclear fallout, air raids, storms, or other emergencies.

First floor. Floor at the bottom of the first story.

First story

1. Story of a building or building part immediately above the basement. If the building part has no basement, the lowest story is the first story.

2. The first story is measured from the floor of the first story to the top of the floor of the story (or average height of the roof) above. If there is no first story, a 0 designation is used.

Floor. Top surface of any horizontal structural plane (except the roof) of a building or other structure normally designed as a base for storage, equipment, operations, personnel activities, or personnel and vehicular traffic.

Floor to head height dimension. Dimension obtained by measuring, to the nearest foot, the vertical distance from the floor to the top of the aperture. This measurement is made for each story on each side. As with sill height, the reported dimension represents an average of the aperture heights in a wall.

Geometric shielding. Geometric shielding places people out of the direct path of radiation or at some distance from it.

Ground contribution. All similar radiation from fallout originating from the ground source plane. The ground contribution is further subdivided into ground direct, wall scattered, skyshine, and ceiling shine.

Ground contribution—ceiling shine. Some radiation interacts with particles in the ceiling and is deflected or scattered to the interior.

Ground contribution—direct. Radiation that reaches the interior directly from the ground source plane without being deflected or scattered.

Ground contribution—skyshine. Some radiation is scattered to the interior by interaction with molecules of the air.

Ground contribution—wall scatter. Some radiation interacts with particles in the wall and is deflected or scattered to the interior.

High-density concrete. Concrete of exceptionally high density, usually obtained by use of heavyweight aggregates, used especially for radiation shielding.

Initial radiation. Nuclear radiation accompanying a nuclear explosion and emitted from the resultant fireball; immediate radiation.

Nuclear airburst. Explosion of a nuclear weapon in the air at a height greater than the maximum radius of the fireball.

Nuclear cloud. All-inclusive term for the volume of hot gases, smoke, dust, and other particulate matter from the nuclear bomb itself and from its environment, which is carried aloft in conjunction with the rise of the fireball produced by the detonation of the nuclear weapons.

Nuclear surface burst. Explosion of a nuclear weapon at the surface of land or water, or above the surface, at a height less than the maximum radius of the fireball.

Nuclear underground burst. Explosion of a nuclear weapon in which the center of the detonation lies at a point beneath the surface of the ground.

Offset. Overhang or setback.

Overhang. Substantial increase in building width or length occurring along one or more sides of a building at given story.

Overhead mass thickness. Since the basement extension is quite often covered with a large slab (such as a sidewalk), a special entry is recorded for this mass thickness in pounds per square feet.

Partial basement. Basement that is smaller in area than the first story.

Partition. Interior wall extending at least 75% of the length or width of the building (part).

Protection factor (PF). The protection factor expresses the relation between the amount of gamma radiation that would be received by an unprotected person compared to the amount that would be received by one in a shelter.

Radiation. Stream of particles such as electrons, neutrons, protons, α-particles, or high energy photons, or a mixture of these.

Radiation area. Area in which the level of radiation is such that a major portion of an individual's body could receive in any one hour a dose in excess of 5 millirem or in any five consecutive days a dose in excess of 150 millirem.

Radiation burn. Burn caused by overexposure to radiant energy.

Radiation corrosion. Accelerated corrosion of a metal caused by radiation.

Radiation counter. Instrument used for detecting or measuring nuclear radiation by counting the resultant ionizing events.

Radiation damage. Harmful changes in the properties of liquids, gases, and solids caused by any type of radiation.

Radiation detection instrument. Device that detects and records the characteristics of ionizing radiation.

Radiation dose. Total amount of ionizing radiation absorbed by material or tissues, in the sense of absorbed dose (expressed in rads), exposure dose (expressed in roentgens), or dose equivalent (expressed in rems).

Radiation dose rate. Radiation dosage absorbed per unit of time; a radiation dose rate can be set at some particular unit of time: H-hour plus 1 hour would be called "H-hour plus 1 radiation dose rate."

Radiation effects. The harmful effects of ionizing radiation on humans and other animals.

Radiation gauge. Instrument for measuring radiation quantity and intensity.

Radiation hazard. Health hazard arising from exposure to ionizing radiation.

Radiation monitoring. Continuous or periodic determination of the amount of radiation present in a given area.

Radiation protection. Measures to reduce exposure to radiation.

Radiation protection guide. Standards established by the Federal Radiation Council.

Radiation safety. Protection of personnel against harmful effects of ionizing radiation by making provisions to ensure that people will not receive excessive doses of radiation.

Radiation scattering. Diversion of radiation (thermal, electromagnetic, or nuclear) from its original path as a result of interactions or collisions with atoms, molecules, or larger particles in the atmosphere or other media between the source or radiation, such as a nuclear explosion and at a point some distance away. As a result of scattering, radiation (especially gamma rays and neutrons) will be received at such a point from many directions instead of only from the direction of the source.

Radiation shield. Shield or wall of material interposed between a source of radiation and the absorbing sensitive body.

Radiation shielding. Shelters with high protection factors are achieved by the planning and control of geometric and barrier relationships between the radioactive source and sheltered enclosure.

Radiation standards. Exposure standards, permissible concentrations, rules for safe handling, regulations for transporting, and regulations for industrial control of radiation.

Radiation survey meter. Portable device used to measure the intensity of nuclear radiations in a given region.

Radiation warning symbol. Standard symbol used on posters displayed in locations where radiation hazards exist.

Roof contribution

1. Radiation that reaches the interior from fallout that may accumulate on an overhead source plane.

2. Radiation originating from radioactive particles (dust and debris) that may accumulate on an overhead source plane.

Setback. Substantial decrease in building length or width occurring along one or more sides of a building at given story.

Shielding concrete. Concrete employed as a biological shield to attenuate or absorb nuclear radiation, usually characterized by high specific gravity or high hydrogen (water) or boron content and having specific radiation attenuation effects.

Sill height dimension. Dimension obtained by measuring the vertical distance from the floor to the sill of the predominant apertures. "Floors" refers to the building floor and not the areaway level when an areaway is involved. This measurement is made for each story on each side.

Story. That part of a building between one floor and the next higher floor or roof. An attic is considered a story.

Story number. All stories of a building or building part are numbered in sequence, beginning with 00 for the basement and 01 for the first story, etc.

Subbasement

1. Story below the basement. The first subbasement is numbered 1, the second 2, etc.

2. The subbasement is measured from the subbasement floor to the top of the basement floor. If there is more than one subbasement, the average height of the subbasements is used in the calculations.

Sum of aperture width. This value is the sum of the widths of the windows and doors in a side. If an adjoining building shields part of the building side, the width of apertures is calculated using that portion of the wall not shielded by the adjoining building.

Thermal radiation. Heat and light produced by a nuclear explosion; electromagnetic radiation emitted by any substance as a result of thermal excitation of its molecules.

Upper stories. Average height of all upper stories (including the attic, if any). If there are no upper stories, a 0 designation is used.

SOLAR ENERGY SYSTEMS

Absorber. Surface in a collector that absorbs solar radiation and converts it to heat energy.

Absorptance. Ratio of the radiation absorbed by a surface to the radiation incident on that surface.

Absorption. Process in which radiation is converted within a material into excitation energy.

Absorption coefficient. Measure of the absorbing strength of a material for radiant energy per unit length.

Absorptivity. Capacity of a material to absorb radiant energy. Absorptance is the ratio of the radiant energy absorbed by a body to that incident on it.

Active residential solar heating system. Solar heating system for heating homes that utilizes forced circulation of the collection and distribution transfer medium. It is a system that combines the means for collecting, controlling, transporting, and storing solar energy with the primary heating system in the house.

Active solar energy systems. Uses outside energy to operate the system and to transfer the collected solar energy from the collector to storage and distribute it throughout the system.

Active solar system (flat plate or concentrating collector based). A system characterized by the use of powered mechanical equipment to move the heat transfer fluid (liquid or gas) through a collector and from a collector to load or storage.

Albedo. The percentage of radiant energy reflected by a surface.

Altitude. Angular distance of a heavenly body measured on the great circle that passes perpendicular to the plane of the horizon through the body and the zenith. It is measured positively from the horizon to the zenith, from 0 to 90°.

Ambient air temperature. Temperature of the outside air or air surrounding a space or building.

Antireflective coating. Absorbing surface that is coated to increase the amount of light penetration. Usually applied to the surface of solar cells or to the glass or plastic surfaces of collectors.

Apparent extraterrestial irradiation at air mass zero. Variable, expressed in British thermal units per square foot per hour, used to calculate the direct normal solar radiation incident at the earth's surface on a clear day. It accounts for the seasonal variation of the distance between the earth and sun.

Apparent solar day. Time required for the sun to cross a given meridian twice.

Atmospheric extinction coefficient. Dimensionless variable used to calculate the direct normal solar radiation incident at the earth's surface on a clear day. It accounts for the seasonal variation of the water vapor content of the atmosphere.

Auxiliary energy subsystem. Equipment utilizing conventional energy sources both to supplement the output provided by the solar energy system and to provide full energy backup during periods when the solar H or DHW systems are inoperable.

Back-up energy system. Conventional fuels and systems used when the demand is beyond the capability of the solar energy system to provide the energy required.

Bioconversion. Conversion of solar energy to fuel by the natural process of photosynthesis.

Btu (British thermal unit). Basic heat measurement, equivalent to the amount of heat needed to raise one pound of water 1°F.

Charge. Directing heat into storage through radiant absorption or convective heat transfer.

Clearness number. Ratio between the actual clear-day direct solar radiation intensity at a specific location and the intensity calculated for the standard atmosphere for the same location and date.

Clear sky. Sky with less than 30% cloud cover.

Cloudy sky. Sky with more than 70% cloud cover.

Collection. Act of trapping solar radiation and converting it to heat.

Collection circuit. Path followed by the collection transfer medium as it removes heat from the collector and transfers it to storage.

Collector angle. Angle between the surface of a solar collector and the horizon.

Collector efficiency. Ratio of the amount of heat usefully transferred from the collector into storage to the total solar radiation transmitted through the collector covers. Some authorities define collector efficiency as the ratio of the amount of heat usefully collected to the total solar radiation incident on the collector.

Collector subsystem. The assembly for absorbing solar radiation, converting it into thermal energy, and transferring the thermal energy to a heat transfer fluid.

Collector tilt. Angle at which a solar collector is inclined with respect to a horizontal plane.

Combined system (combined collectors and storage devices). A combined component system characterized by a system with integral construction and operation of the components such that the solar radiation collection and storage phenomena cannot be measured separately in terms of flow rate and temperature changes.

Concentrating collector. Means of raising the energy density of an area by funneling the received energy from a larger area.

Concentrating focusing collector. Collector that concentrates the solar radiation incident on the total area of the reflector onto all absorbing surfaces of smaller area, thereby increasing the energy flux.

Concentrating thermal collectors. Devices that reflect or retract incoming direct solar radiation to a special focal point.

Concentration ratio. The ratio of the heat flux within the image to the actual flux received on earth at normal incidence.

Concentrator. Device used to intensify the solar radiation striking a surface.

Conductivity. Property of a material indicating the quality of heat that will flow through 1 ft of a material for each degree of temperature difference.

Controls. Installation of devices to regulate the processes of collecting, transporting, storing, and utilizing solar energy.

Control subsystem. An assembly of devices and its electrical, pneumatic, or hydraulic auxiliaries used to regulate the processes of collecting, transporting, storing, and utilizing energy.

Conversion efficiency. The electrical power output of a solar cell expressed as a percentage of the incident radiant power.

Cooling internal spaces. Solar cooling of internal closed spaces can be achieved with the use of absorption cooling methods.

Design life. The period of time during which a solar energy system or component is expected to perform without major maintenance or replacement.

Diffuse sky. Solar radiation received from the sun after its direction has been changed by reflection and scattering by the atmosphere.

Direct gain. Passive solar heating system whereby solar radiation is admitted directly into the conditioned space.

Direct radiation. Solar radiation received from the sun without undergoing a change of direction.

Direct solar radiation or direct solar insulation. Portion of the sun's heat energy that penetrates the atmosphere without being absorbed.

Discharge. Removing stored heat by radiation or convective heat transfer.

Distribution circuit. Path followed by the distribution transfer medium as it leaves storage, below which useful heat cannot be delivered.

Downpoint temperature. Temperature of the distribution transfer method as it leaves storage, below which useful heat cannot be delivered.

Drawdown. Removal of all useful heat from storage.

Electric conversion collectors. Collectors utilizing special solar cells that convert solar energy into electricity.

Emittance. The ratio of the radiant energy emitted by a body to the radiant energy emitted by a black body at the same temperature.

Eutectic salt. Combination of two (or more) mutually soluble materials that requires a large amount of heat to melt or an equally large amount of heat to solidify and that melts or freezes at constant temperature and with constant composition. The phase change temperature is the lowest achievable with these materials.

Evacuated tubular collector. Collector manufactured from specially coated concentric glass tubes with an evacuated space between the outer two tubes.

Extraterrestrial radiation. Radiation (electromagnetic and charged particles) originating in space beyond the earth's atmosphere, that is, solar radiation and primary cosmic rays; solar electromagnetic radiation received at the outer limits of the atmosphere.

Facility. A building or structure including appliances, heating or cooling equipment, industrial or manufacturing processes to be served by the solar energy system.

Fixed-plate collector. Collector that is stationary and does not concentrate the solar radiation; that is, the absorbing area is the same size as the area intercepting the incoming radiation.

Flat face collector. Any nonfocusing flat-surfaced solar collector.

Flow condition. The condition existing in the solar energy system when the heat transfer fluid is flowing through the collector under normal operating conditions.

Heating load. Rate of heat flow required to maintain indoor comfort in Btu/hr.

Heat storage. Device or medium that absorbs collected solar heat and stores it for use during periods of inclement or cold weather.

Heliodon. Device used to simulate the effect the sun's position has on models of buildings and other objects. It is primarily used to conduct shadowing studies.

Heliostat. Instrument consisting of a mirror mounted on an axis, mechanically rotated to steadily reflect the sun in one direction.

Heliotropism. Property of being able to follow the sun's apparent motion across the sky.

Hybrid system. System incorporating a major passive aspect, where at least one of the significant thermal energy flows is by natural means and at least one is by forced means.

Incident radition. Quantity of radiant energy incident on a surface per unit time and unit area.

Indirect solar radiation. Portion of reradiation and reflected radiation that eventually reaches the earth's surfaces.

Insolation. Incident solar radiation; amount of solar radiation striking a surface during a specified period of time.

Internal mass. Massive materials with heat storage potential contained within a building, such as walls, floors, or other absorbing materials and assemblies.

Ionosphere. Part of the earth's upper atmosphere that is sufficiently ionized by solar ultraviolet radiation.

Ionospheric disturbance. Temporal variation in electron concentration in the ionosphere that is caused by solar activity.

Isolated gain. Type of passive solar heating system where heat is collected in one area and transferred elsewhere for use.

Langley

1. Measure of irradiation in terms of langleys per minute, where 1 langley equals 1 calorie/cu cm. The langley is named for American astronomer, Samuel P. Langley.
2. Unit of measurement of solar radiation equal to 1 g calorie/cu cm.

Mean solar day. Duration of one rotation of the earth on its axis, with respect to the mean sun. The length of the mean solar day is 24 hours of mean solar time or 24 hours, 0.03 minutes, 56.555 seconds of mean sidereal time. The mean solar day beginning at midnight is called a "civil day," and one beginning at noon, 12 hours later, is called an "astronomical day."

One-step thermal collectors. Uncontrolled inexpensive heating devices.

Operating energy. The conventional energy required to operate the H, HC and HW systems, excluding any auxiliary energy which supplements the solar energy collected by the systems (e.g., the electrical energy required to operate the energy transport and control subsystems).

Overcast sky. Sky having 100% cloud cover, no sun is visible.

Partly cloudy sky. Sky having 30 to 70% cloud cover.

Passive solar system. Integral energy system or assembly of natural and architectural components such as collectors, thermal storage devices, and transfer fluids, in which no appreciable off-site energy is used to accomplish the transfer of thermal energy.

Peak load. Maximum energy demand placed on a system.

Photosynthesis. Production of chemical compounds in plants using solar radiant energy, specifically the production of carbohydrates by plants containing chlorophyll using sunlight, carbon dioxide, and water.

Potential energy. Stored energy that is available for release.

Pyranometer. Instrument used to measure the total hemispherical solar radiation incident on a surface. This includes direct radiation from the sun, diffuse radiation from the sky, and reflected shortwave radiation (albedo) from the surroundings.

Pyrheliometer. Instrument used to measure direct solar radiation incident on a surface located normal to the sun's rays.

Radiation. Direct transport of energy through a space by means of electromagnetic waves.

Reflected radiation. Solar radiation reflected by light-colored or polished surfaces, used to increase solar gain.

Retrofit. Installation of a solar heating system on an existing building.

Selective surface

1. Solar collecting surface consisting of a thin coating having high absorptance for solar radiation and a substrate with low emittance for long-wave radiation.
2. Surface produced by the application of a coating to an absorber plate (solar collector) with spectral selective properties, which maximizes the absorption of incoming solar radiation (0.3- to 3.0-μm range) and emits

much less radiation (3.0- to 30.0-μm range) than an ordinary black surface at this same temperature would emit.

3. Coating with high solar radiation absorptance and low thermal emittance, used on the surface of an absorber to increase system efficiency.

Sky light. Visible radiation from the sun redirected by the sky.

Sky radiation. Portion of reradiation and reflected radiation that eventually reaches the earth's surface.

Skyshaft. Multichambered plexiglass device that penetrates a roof and is used to provide natural interior illumination with a minimum of heat loss or gain.

Sky temperature. Effective temperature of the sky, used in determining the heat lost from the earth's surface by radiation to the sky.

Solar absorption index. Relation of the sun's angle at various latitudes and local times with the ionospheric absorption.

Solar activity. Disturbances on the surface of the sun.

Solar air mass. Optical air mass penetrated by light from the sun for any given position of the sun.

Solar altitude. Angle of the sun above the horizon.

Solar array. Number of individual solar collection devices arranged in a suitable pattern to effectively collect solar energy.

Solar azimuth. Horizontal angle between the sun and due south.

Solar battery. Array of solar cells, usually connected in parallel and series.

Solar cell

1. Device made from semiconductor materials that absorbs solar radiation and converts it into electrical energy.
2. Photovoltaic cell. Device employing crystals (silicon, for example). When exposed to solar radiation, it generates an electric current.

Solar climate. Hypothetical climate that would prevail on a uniform solid earth with no atmosphere. Climate of temperature alone, determined only by the amount of solar radiation received.

Solar collector. Device used to collect solar radiation and convert it to heat.

Solar constant

1. Power per unit area received from the sun outside the earth's atmosphere on a surface normal to the direction of the sun at the earth's mean distance from the sun; solar irradiance prior to attenuation by the earth's atmosphere.
2. Amount of solar radiation incident on a unit area of surface located normal to the sun's rays outside the earth's atmosphere at the earth's mean distance from the sun.

Solar cooker. Device for cooking that uses the sun as an energy source.

Solar day. A day whose length is measured between successive appearances of the sun on a given meridian.

Solar declination. Angle of the sun north or south of the equatorial plane. It is positive if north of the plane and negative if south.

Solar degradation. Process by which exposure to sunlight deteriorates the properties of materials and components.

Solar energy. Energy in the form of electromagnetic radiation received from the sun.

Solar fixed plate. Stationary plate that does not concentrate the solar radiation. The absorbing area is the same size as the area intercepting the incoming radiation.

Solar flare. A great eruption of glowing gas into the photosphere.

Solar furnace. Unitized, self-contained, solar heating system. A device for achieving high temperatures by concentrating the sun's energy using suitable optical systems.

Solar gain. Absorption of heat from the sun. Amount of solar radiation (Btus) received on an identified surface.

Solar generator. Electric generator powered by radiation from the sun.

Solar heating. Conversion of solar radiation into heat for technological or comfort heating and for the preparation of foods.

Solar heat storage. Storage of solar energy for later use, usually accomplished by heating of water or fusing a salt, although sand and gravel have been used as storage media.

Solar intensity. Direct irradiance normal to the solar beam.

Solar noon. Time of day when the sun is due south, that is, when the solar azimuth is zero and the solar altitude at maximum.

Solar pond. Shallow body of water that can collect solar radiation. Heat exchanges located at the bottom of the pond are used to extract the heat generated.

Solar power. Conversion of the energy of the sun's radiation to useful work.

Solar power farm. Installation for generating electricity on a large scale using solar energy, consisting of an array of solar collectors, steam or gas turbines, and electrical generators.

Solar radiation

1. Radiant energy emitted form the sun in the wavelength range between 0.3 and 3.0 μm. Of the total solar radiation reaching the earth, approximately 3% is in the ultraviolet region, 44% in the visible region, and 53% in the infrared region.

2. Total electromagnetic radiation emitted by the sun. To a first approximation, the sun radiates as a blackbody at a temperature of about 5700 K; hence about 99.9% of its energy output falls within the wavelength interval from 0.20 to 11.0 μm, with peak intensity near 0.47 μm.

Solar sail. Surface of a highly polished material upon which solar light radiation exerts a pressure.

Solar spectrum. Spectrum of the sun's electromagnetic radiation extending over the whole electromagnetic spectrum, from wavelengths of 10^{-9} cm to 30 km.

Solar thermal electric conversion. The conversion of solar energy to thermal energy that powers turboelectric generators.

Solar time. Time based on the rotation of the earth relative to the sun.

Solar wind. Streams of protons and electrons that are blown out from the sun in all directions.

Solstice. The solstice occurs twice a year when the sun is farthest north or south of the equator. In the Northern Hemisphere, the summer solstice occurs about June 21 and the winter solstice about December 21.

Stagnation. Condition in which heat is not added to or removed from storage mechanically, but only as a result of natural heat transfer from the storage container.

Storage. Device or medium that absorbs collected solar heat and stores it for future use.

Sun. Source of fuel for all solar energy applications, in the form of radiation.

Sun factor. Average amount of measured solar radiation divided by the total possible solar radiation for a given month and location. The number indicates the amount of cloudiness that occurs at a given location.

Sunlight. Direct visible radiation from the sun; also, solar illumination.

Sunshine. Direct radiation from the sun, as opposed to the shading of a location by clouds or by other obstructions.

Sun time. Time of day at a specific location as determined by the position of the sun.

Sun tracking. Ability to follow the apparent motion of the sun across the sky.

Terrestrial radiation. Total infrared radiation emitted from the earth's surface, to be carefully distinguished from effective terrestrial radiation, atmospheric radiation, and insolation.

Thermal mass. Thermally absorptive building component used to store heat energy. In a passive solar system, the thermal mass absorbs the sun's heat during the day and radiates it at night as the temperature drops. Thermal mass is the amount of potential heat storage capacity available in a given system or assembly.

Thermal storage wall. Passive system in which the heat storage mass is a wall located between a glazed window wall and the interior space to be heated. The mass can be a variety of materials, including water or masonry.

Time-lag heating. Process of heating a building's interior by using the heat loss properties of massive materials to delay the requirements of solar heat.

Torrid zone. Wide belt encompassing 47 degrees between the tropics over which the vertical rays of the sun pass.

Transfer medium. Substance that carries heat from the collector to storage and from storage to the house. The medium is typically a fluid such as air, water, or a water-ethylene glycol solution.

Transmissivity. Capacity of a material to transmit radiant energy. Transmittance is the ratio of the radiant energy transmitted through a body to that incident on it.

Trombe wall. Passive heating concept consisting of a south-facing masonry wall with glazing in front. Solar radiation is absorbed by the wall, then converted into heat and conducted and radiated into the building. Vents may be used to circulate the warm air from the space between the glass and the wall to the building.

Troposphere. Convective region of the atmosphere that extends from the earth's surface to the stratosphere. Its height ranges from 5 miles at the poles to 11 miles at the equator.

Ultraviolet radiation

1. Radiant energy of wavelengths from 0.1 to 0.4 μm.

2. Electromagnetic radiation of shorter wavelength than visible radiation, but longer than x-rays; roughly, radiation in the wavelength interval from 10 to 4000 Å.

Useful heat. Heat delivered by the solar system on demand by the thermostat that contributes to a reduction in the conventional fuel heating bill. Useful solar heat replaces an equivalent amount of heat that otherwise would have to be provided by the primary heating system.

***U*-value.** Coefficient that indicates the time rate at which energy (British thermal units per hour) passes through a component for every degree (Fahrenheit) of temperature difference between one side and the other under steady-state conditions.

Water wall. Passive solar heating system, implementing water-filled containers, for collecting and storing solar energy.

Wavelength conversion. Direct solar radiation arrives from the sun in the form of shortwave radiation (0.3 to 3 μm in length). After the shortwave radiation strikes matter and is absorbed, it is radiated in longwave form (3 to 30 μm in length).

SWIMMING POOLS

Acid. Chemical used to lower pH of pool water.

Acid demand. Amount of acid a pool water requires to adjust the pH to the correct range.

Air pump. The filter media of a diatomaceous earth filter are cleaned by releasing compressed air into the water on the low-pressure side of the media.

Algae
1. Tiny plants that grow in the pool and discolor the water, create slimy conditions, and give off unpleasant odors.
2. Microscopic plant growth found in all bodies of water. Algae is introduced into swimming pool water by bather's feet, wind, and rain. It causes the water to discolor and starts serious water pollution.

Algicide. Natural or synthetic chemical used for killing algae.

Algistat. Substance that inhibits the growth of algae but is incapable of killing it.

Alkalinity. Amount of all the soluble alkalies in the water, such as hydrotides, carbonates, or bicarbonates.

Alum
1. An acid salt for floccing or settling to the pool floor, matter suspended throughout the pool water.
2. Flocculating agent. Potassium and ammonium alum are the most common types used for the treatment of pool water.

Aluminum sulfate. Filter alum; term generally applied to the aluminum compounds of sodium, potassium, and ammonia. Alum is commercially available in pieces, crystals, powder, liquid, or in granular form. Alum aids in the formation of floc to settle out small impurities on a filter bed. It is easily dissolved by the swimming pool water. Alum is not used with diatomaceous earth filters.

Artificial swimming pool. Structure intended for wading, bathing, or swimming, made of concrete, masonry, metal, or other impervious material, located either indoors or outdoors, and provided with a controlled water supply. No filter system is required if approved by health department regulations.

Available chlorine. Term used in rating the total oxidizing power of chlorine products. The higher the percentage of available chlorine in a product, the more powerful it is.

Backwash. Process of reversing water flow to clean a filter.

Backwash cycle. Operating time (following the filter cycle) required to completely clean the filter.

Backwash piping
1. Piping that extends from the backwash outlet of the filter to its terminus at the point of disposal.

2. Pipe going from the backwash outlet of a filter to a disposal point.

Backwash rate. Rate (speed and volume) of water flow during the postfilter cleaning cycle, expressed in gallons per minute per square foot of effective filtration area.

Bacteria. Minute unicellular organisms of various forms, some of which can cause disease.

Bacteriological quality. Local health departments may demand to take water samples of the swimming pool at their discretion. The bacteriological quality of the water from the recirculating type pool must meet the standards for drinking water as established by the U.S. Public Health Service.

Bad odor. Bad odor may be due to hydrogen sulfide in the water supply or excess of algae formations.

Baking soda. Sodium bicarbonate that reacts with air and liquid to form carbon dioxide, used to raise the alkalinity of the water, but does not raise the pH noticeably.

Balanced water. Achieved by proper adjustment of pH level, total alkalinity, calcium hardness, and total dissolved solids. When water has more minerals than required, the excess minerals are deposited on the pool surfaces and the pipes and equipment as scale.

Bather. A person dressed for bathing or swimming and permitted to be in the pool area, on the pool deck, or in the pool.

Bathing beach. Natural bathing place, together with its buildings and appurtenances, if any, at a pond, lake, or tidal or nontidal stream or other body of water that is open to the public for bathing or swimming purposes.

Bathing load. The number of persons using the pool at any given time.

Black algae. Algae that forms tough, dark-colored spots on pool surfaces, particularly visible on pool plaster.

Bleach. Sodium hypochlorite; an active bleach available in powder or liquid form. Both forms carry an amount of chlorine usually offering 5 to 15% available chlorine in strength. When added to pool water it breaks down to produce hypochlorous acid, the active agent of a disinfectant. It purifies because of its strong oxidizing power.

Body feed
1. Continuous addition of small amounts of filter aid during the operation of a diatomaceous earth filter.
2. Filter aid fed into a diatomite-type filter throughout the filtering cycle.

Breakpoint chlorination. Addition of sufficient chlorine to water in the pool to destroy the combined chlorine present.

Bromide. Compound of bromine. Two such compounds, sodium bromide and potassium bromide, are salts sometimes used to produce disinfectants and algicides.

Bromine
1. Element sometimes used for pool water purification. Bromine is a dark, heavy, reddish brown liquid (its normal state) closely related to chlorine.
2. Bromine is a halogen not unlike chlorine or iodine. It exists as a reddish-brown liquid or gas.

Calcium carbonate. Combination of carbonate ions and calcium ions.

Calcium hardness. Term to indicate the mineral content of pool water. Hardness helps to protect pool surfaces from corrosive effects of water.

Calcium hypochlorite

1. White granular powder that is soluble in water and generally offers up to 70% in available chlorine strength.
2. Compound of chlorine and calcium used in powder or granulated form, usually containing 70 to 80% available chlorine by weight, which is released in water to act as a germicide or algicide.

Cartridge. Replaceable porous element that traps dirt from entering the media.

Cartridge filter. Filter using filter cartridges.

Caustic. Sodium hydroxide.

Chemical feeder. Device used to feed chemicals into a pool, but usually one that introduces an alum, acid, filter aid, algicide, or soda ash. Included in this category are proportioning pumps, injection feeders, pot feeders, and dry feeders.

Chemical residual. Concentration at which a chemical must be maintained in solution in the pool so that the chemical can do its work effectively.

Chlormaine. Compound formed when chloride combines with nitrogen or ammonia, has a strong objectionable odor, is a poor disinfectant, and is a skin and eye irritant.

Chlorination. Addition of chemicals containing chlorine to sanitize pool water.

Chlorinator. Device that dispenses, regulates the flow of and meters the chlorine introduced into water being treated.

Chlorine

1. A gemicide used primarily in residential swimming pools.
2. Elemental gas that can be stored in cylinders when liquefied under pressure, and used as a disinfectant and algicide in the treatment of pool water.

Chlorine demand. Quantity of chlorine required to destroy pollutants in the pool water, such as bacteria, algae, and chloramines.

Chlorine gas. Halogen compressed as a liquid. Commercially available in cylinders from 100 to 150 lbs, also in one ton containers for large installations.

Chlorine residual. Amount of chlorine available for sanitizing after the initial chlorine demand of the water has been met. Chlorine residual should be maintained at 1.0 to 1.5 ppm in a stabilized pool.

Clean water. Water added to a swimming pool after treatment in the pool recirculation system.

Cloudy water. Condition caused by the introduction of foreign matter in the pool water, usually by improper filtration, or too high an alkalinity (8.0 pH or higher).

Colored water. Condition caused by the introduction of raw water, which may include as its mineral constituents, iron, manganese, or copper.

Combined chlorine. Portion of the total chlorine existing in pool water in chemical combination with ammonia, nitrogen, or organic compounds. Combined chlorines are poor disinfectants, have bad odors, and are skin and eye irritants.

Controlled water supply. Quantity and sanitary quality of the water supplied to the pool can be regulated and maintained by the swimming pool operating personnel.

Copper sulfate. Copper sulfate available in crystals or powder. Manufactured by combining copper salts and sulphuric acid. Used for combating algae formations.

Corrision. Etching or oxidation of a material.

Corrision-resistant material. Material with exceptional resistance to the effects of corrosion.

Cross-connection. Poorly insulated connection between a domestic water system and a pool that can cause water backflow to the domestic system; appropriate protective measures include providing vacuum breakers or air gaps.

Cyanuric acid. Chemical used for pH stabilization.

Cyanuric acid compounds. Compound available in tablet, powder, or packaged packet form. The compound is used for reducing the decomposition of residual chlorine by ultraviolet rays of the sun.

Deep area. Area on the deep side of the transition point in the swimming pool.

Depth-type cartridge. Filter cartridge, with media not less than $\frac{3}{4}$-in thick, that relies on penetration of dirt into the media to achieve its removal and to provide adequate holding capacity for the cartridge.

Design load. The maximum number of persons permitted in the pool at any given time and to be determined on dividing the total square footage of the swimming pool water surface area by twenty-seven (27).

Design rate. Average flow rate over the filtration cycle.

Diatomaceous earth

1. Filter material of petrified diatoms and other unicellular algae used for diaomite filtration.
2. Minute, variously shaped silica skeltons of diatoms.
3. Odorless, tasteless, talclike powder made up of the petrified skeletal remains of microscopic single-celled marine plants.

Diatomite (diatomaceous earth). Type of filter aid.

Diatomite filter. Filter designed to filter water through a thin layer of diatomaceous earth; diatomite filters may be of the pressure, gravity, suction, or vacuum type.

Directional inlet fitting. Inlet fitting that allows the adjustment of water direction and flow rate within a pool.

Discharge head. Total head (pressure of a fluid), including the static and friction heads, on the discharge side of a pump.

Distributor. Device in a filter that diverts incoming water to prevent erosion of the filter medium.

Diving area. Area of a pool reserved for diving.

Diving board. Flexible board, having a nonslip surface finish, that is provided for aquatic diving. "Board" is synonymous to "platform."

Diving platform

1. Rigid platform, having a nonslip surface finish, that is provided for aquatic diving. "Platform" is synonymous to "board."
2. Platform normally used for (standard) 5- and 10-m official diving competitions.

Diving pool. Swimming pool intended for use exclusively by divers.

Diving stand. Stand or support for a springboard or diving board.

Diving tower. Support usually associated with the 3-m springboard.

Drain. Outlet at the deep end of a pool through which waste water can exit.

Dry chlorine. Calcium hypochlorite, lithium hypochlorite, or a cyanuric acid compound.

Dry niche. Weatherproof fixture placed in an opening in the pool wall and protected from water by a watertight window.

Effluent. Water that flows out of a filter.

Electrolysis. Migration of ions between submerged dissimilar metals, caused by an electrical potential between them.

Eye irritations. Condition caused by pool water that may be on the acid side of the pH range.

Face piping

1. Piping with all valves and fittings used to connect the filter system together as a unit.

2. Piping, valves, and fittings used within the filter system, including all the hardware necessary to the filter's functioning.

3. Piping that connects the vacuum fitting to the pump suction.

Feet of head. Basis for indicating the resistance in a hydraulic system, equivalent to the height of a column of water that would cause the same resistance (100 ft of head equals 43 psi). The total dynamic head is the sum of all resistances in a complete operating system. The principal factors affecting a head are vertical distances and the resistance caused by friction between the fluid and pipe walls. (Friction head is the head loss due to friction only).

Fill-and-draw swimming pool

1. Swimming pool that has no provisions for recirculating and filtering the water.

2. Dependence for maintaining water of sanitary quality is placed on the complete removal and replacement of the water at periodic intervals.

3. Swimming pool so operated that the water is completely drained to waste intermittently and replaced by make-up water.

Filter. Material or apparatus by which water is clarified.

Filter aid

1. Diatomite type of filter medium.

2. Fine medium used to coat a septum filter, usually diatomaceous earth or vulcanic ash. (Alum used on a bed of sand filters is also a filter aid.)

Filter cartridge

1. Disposable-filter-medium package. There are two types: the surface or area type that removes suspended matter at the surface and the depth type.

2. Disposable or renewable filter element that employs no filter aid.

Filter cycle. Operating time between backwash cycles.

Filter element

1. Part of a filter device that retains the filter medium. Removes the suspended particles in the water.

2. Part of a filter on which the filter aid is deposited (usually found in diatomite filters).

Filter media. The fine material which entraps the suspended particles.

Filter pool. Swimming pool having a recirculating filter system.

Filter rate. Rate (speed and volume) of water flow through a filter, expressed in gallons per minute per square foot of effective filtration area.

Filter rock

1. Graded rock and gravel used to support filter sand.

2. Graded, rounded rock or gravel used to support a filter medium.

Filter sand. Type of filter medium.

Filter septum

1. Part of a filter element consisting of a cloth or wire mesh (or other porous material) on which the filter cake is deposited.

2. Part of the filter element in a diatomite filter, on which a cake of diatomite is deposited.

Filtration rate. Rate of filtration of water through a filter during the filter cycle expressed in US gallons per minute per square foot of effective filter area.

First-aid equipment. Equipment required for the emergency treatment of any pool participant, usually recommended by the local Red Cross. Telephone numbers of a physician in the immediate area, the fire department, and the police department should be listed in the equipment container.

Floccing. Floccing swimming pool water consists of adding a chemical to the water for the purpose of settling suspended dirt and other solids to the pool floor.

Flocculating agent. Compound, such as one of the alums, causes small suspended particles to become a flocculent (fluffy aggregate).

Floor slope. Slope of a pool floor, usually expressed in vertical rise or horizontal distance.

Flow rate. Volume of flow per unit of time expressed in gallons per minute (GPM).

Flow-through swimming pool

1. The sanitary quality of the water is maintained by the water flowing through the pool from some natural or developed source, but the overflowing water from the pool is wasted.

2. Swimming pool in which, when the pool is in use, the water is undergoing continuous displacement to waste by makeup water.

3. Pool having a continuous flow of fresh water into and out of the pool.

Free available chlorine. Portion of total chlorine in chlorinated water that is not combined with ammonia or nitrogen compounds.

Galvanic action. Creation of an electrical current by chemical action.

Gravity-sand filter. Filter with a layer of filter medium (usually silica sand) supported on graded gravel, through which water is made to flow safely by its weight.

Green water. Condition caused by the overabundance of algae formations.

Gunite concrete. Sprayed concrete usually on wire mesh reinforcing.

Gutter fitting. Drainage fitting used in an overflow gutter.

Hardness. Amount of calcium and magnesium dissolved in water. High levels of calcium and magnesium in balanced water contributes to the protection of plaster and metals.

Hard water. Hard water is caused by the presence of calcium and magnesium. The condition is related to the total alkalinity of the water. High alkalinity combined with hard water produces scaling.

Heaters. Heaters maintain comfortable water temperatures, between 80°F and 84°F i.e., temperatures.

High-rate permanent-medium filter. Filter using a fast water flow, made possible by the uniform distribution and collection of incoming and outgoing water.

Hose connector. Fitting connecting a hose to a vacuum wall fitting (usually a combination hose sleeve/nut).

Hydrotherapy inlet fitting. Special air-entraining inlet fitting that produces a massaging effect.

Hydrotherapy spa. Not used for swimming purposes, but as a recreational and therapeutic installation that is not drained, cleaned, or refilled for each user. It may include hydrojet circulation, hot water, cold water mineral solutions, air-induction bubbles, or any combination thereof.

Hypobromous acid. A very powerful disinfectant form of bromide in water.

Hypochlorinator. Device used to feed, control, and meter a solution of sodium or calcium hypochlorite into the water being treated. There are three general types: the positive displacement type, which is usually motor driven; the aspirator type, actuated by a pressure differential created within the hydraulic system; and the metering type, connected to a suction pump and regulated by an orifice that is opened and closed by a timing mechanism.

Hypochlorite. Any salt form of hypochlorous acid (chlorine).

Hypochlorite acid. An aqueous solution of approximately 38% hypochloric acid. Also known as "muriatic acid." The commercial product is yellow in color and has a pungent odor.

Hypochlorous acid. Most active sanitizing agent in chlorine. Formed when chlorine is added to water, hypochlorous acid production is controlled by the pH level of the water.

Indoor pool. Swimming pool where the pool and pool deck are totally enclosed within a building or structure covered by a roof.

Influent. Water entering a filter or other device.

In-ground swimming pool. Pool whose sides rest in partial or full contact with the earth.

Inlet

1. Fitting or opening through which water enters the pool.

2. Fitting through which filtered water passes to the pool (filtered water inlet), or the fitting through which "raw" water passes to the pool (raw water inlet).

Iodine

1. Element related to chlorine and bromine, used as a disinfectant both in its natural form and in its combined forms. When iodides are used, chlorine is normally employed to free the elemental iodine.

2. A halogen not unlike chlorine or bromine.

Isocyanurates. Groups of self-stabilizing pool sanitizer products that contain cyanuric acid.

Ladders. Single-unit stainless-steel-constructed stepladder that is grounded in accordance to most local codes.

Lifeguard. A qualified, certified person appointed by the owner or operator to maintain surveillance during published and established time periods, over bathers or swimmers while they are in the pool area or enclosure, to supervise and control behavior and safety.

Lifeguard service. The attendance, at all times that persons are permitted to engage in water contact sports, of one or more lifeguards who hold Red Cross or YMCA senior lifeguard certificates or other equivalent qualifications and who have no duties to perform other than to superintend the safety of participants in water contact sports.

Lifeline anchors. Rings at the transition point.

Lithium hypochloride. White granular powder yielding approximately 35% available chlorine.

Main outlet

1. Outlet(s) at the deep portion of the pool through which the main flow of water leaves the pool when being drained or circulated.

2. Outlet at the bottom of a swimming pool through which water passes to a recirculating pump.

Main suction. Line connecting the main outlet to the pump suction.

Make-up water

1. Fresh untreated water that is periodically added to the swimming pool.

2. Water added from an external source to a swimming pool.

Markings, lines, and decorative designs. Lane lines or other markings on the bottom of a pool in accordance with local code requirements.

Milky water. Condition caused by defective filter screens where diatomaceous earth is used as the filter media. Excessive use of calcium hypochlorite may also cause this condition.

Modified swimming pool. Swimming pool that is not an indoor pool and that has the form of a basin-shaped depression in the earth, the floor of which slopes downward and inward toward the center from the rim, and in which water recirculation, water disinfection, or water filtration, or in which water displacement on the flow-through or fill-and-draw principle or any combination of these is part of this pool operation.

Multiple filter-control valve. Multiport valve having at least four control positions for various filter operation, thus combining in one unit the function of two or more single valves.

Muriatic acid. Commercial form of hydrochloric acid used to lower the pH (alkalinity) of pool water.

Nonconforming pool. Public swimming pool that has been designated by the local health department as a nonconforming pool because of a condition found therein that constitutes a hazard to health or safety.

Nonpermanently installed swimming pool. Pool constructed so that it may be readily disassembled for storage and reassembled to its original integrity.

Nonslip decks. Areas surrounding spas and pools to prevent slipping on leaving the spa or pool.

Nonswimming area. Portion of a pool wherein water depths, offset ledges, or similar hindrances would prevent normal swimming activities.

Onground swimming pool. Pool whose sides rest fully above the surrounding earth.

OTO (orthotolidine). Solution (reagent) used in a pool water test kit to measure the amount of residual or free chlorine present.

Outlet line. Pipe or pipes from the main drain and/or skimmers used to pump water from the pool to the pump and filter.

Overflow gutter

1. Gutter around the top of a pool, used to carry away waste on the surface of the water or collect it for return to the filters.

2. Continuous trough in the top of the pool wall which may be used as an overflow and also skims the pool water surface.

Overflow system. Includes perimeter-type overflows, surface skimmers, and surface-water collection systems.

Oxidation. Action or union of oxygen with another substance. In pool water, hypochlorous acid oxidates algae, bacteria, and suspended dirt particles.

Ozone. Gaseous molecule, composed of three atoms of oxygen, that is generated on-site and used for oxidation of water contaminants.

Permanently installed swimming pool. Pool constructed in the ground, on the ground, or in a building in such a manner that the pool cannot be readily disassembled for storage.

pH

1. Chemical symbol denoting the degree of alkalinity or acidity in the pool water.

2. Value expressing the relative acidity or alkalinity of a substance; indication of the hydrogen-ion concentration.

Phenol red. Chemical reagent used for determining the pH of pool water.

pH plus blocks. Fused cakes of sodium carbonate (soda ash) used to counteract acidity in the pool water and to restore pH to the alkaline side.

pH scale. Scale that denotes the range of alkalinity or acidity of a solution. The pH scale ranges from 0 to 14. The neutral point on the scale is 7. Any reading below 7 indicates the water is on the acid side. Any reading above 7 indicates the water is on alkaline side.

Pool boiler. Indirect type of pool heater that uses steam (instead of hot water) in a closed system.

Pool deck. The finished nonslip materials used around the swimming pool area.

Pool depth

1. Vertical distance between the floor and the waterline.

2. Distance between the floor of the pool and the maximum operating level when the pool is in use.

Pool floor. Portion of the pool that is horizontal or inclined less than 45° to the vertical from the horizontal.

Pool heater. Device through which pool water is circulated to increase its temperature. In the direct type the heat is transferred directly to the pool water via circulation tubes. The indirect type uses a separate enclosed system directly exposed to a heat generator that heats the pool water by circulating steam or hot water around the tubes of a heat exchanger (through which the water circulates).

Pool shell. Made of reinforced concrete, or other material equivalent in strength and durability; it is designed and built to withstand anticipated stresses and is of watertight construction with smooth and impervious surfaces. It has a white or light-colored waterproof interior finish, which will withstand repeated brushing, scrubbing, and cleaning procedures.

Pool slope. Slopes at the bottom of swimming pools are required based on the use of the pool. Diving and diving board uses control depths and slopes and are carefully established by state or local building codes.

Pool walks. Finished nonslip areas around the pool.

Pool wall. Portion of the pool that is vertical or inclined more than 45° to the vertical from the horizontal.

Potable water

1. Water fit for human consumption.

2. Approved domestic water supply, bacteriologically safe and otherwise suitable for drinking.

Precoat

1. Initial coating of filter aid on the septum of a diatomaceous earth filter.

2. In a diatomite filter, the initial coating of filter aid placed on the filter septum at the start of the filter cycle.

Precoat feeder. Device used to feed a measured amount of filter aid into a diatomaceous earth filter upon the start of its cycle, following the cleaning operation.

Pressure differential. Pressure difference between two parts of a hydraulic system (influent pressure versus effluent pressure of a filter, for example).

Pressure sand filter. Sand filter enclosed within a pressure-operated tank.

Private residential swimming pool. Pool located on a private residential property under the control of the owner or occupant, the use of which is limited to swimming or bathing by members of the family and their visitors.

Private swimming pool. Pool established or maintained on any premises by an individual for his own or his family's use or for guests of his household.

Public pool. Every swimming or wading pool admission to which may be gained by the general public with or without the payment of a fee.

Public swimming pool. Pool, other than a private residential swimming pool, intended to be used collectively by numbers of persons for swimming or bathing.

Pump strainer. Device placed on the suction side of a pump that contains a removable strainer basket designed to trap debris in water flowing through it with a minimum of flow restriction.

Quaternary ammonia compounds

1. Compounds of ammonia used as algicides and germicides, which reduce the surface tension of water.

2. Essentially wetting agents that exhibit properties that make them suitable as algaecides. Organic chemicals of the amine group.

Rate of flow. Volume of flow per unit of time expressed in gallons per minute.

Rate-of-flow indicator. Device that indicates the rate of flow in a pipeline.

Recirculating piping. Piping from the pool to the filter and returned to the pool, through which the water circulates.

Recirculating skimmer. Device connected with the pump suction used to skim the pool over a self-adjusting weir and return the water to the pool through the filter.

Recirculating system. Complete water system, including suction piping, pump, strainer, filter, face piping, and return piping.

Recirculating-type swimming pool. The circulation of the water is maintained through the pool by pumps, the water being drawn from the pool, clarified by filtration, and disinfected before being returned to the pool.

Red water. Condition caused by raw water with precipitated iron content.

Residential-apartment pool. Constructed pool, permanent or portable, intended for noncommercial use as a swimming pool usually by no more than three owner families and their guests, and that is over 24 in. in depth and has a surface area exceeding 250 sq ft or a capacity of over 3250 gal.

Residual. Term refers to residual chlorine: the amount of chlorine that can be measured after water has been treated with chlorine. Free residual chlorine, in contrast to the combined residual chlorine, is not combined with ammonia or other compounds and is a more effective disinfectant.

Return piping. Part of the pool piping between the filter and the pool through which filtered water passes.

Safety equipment. The kind of equipment required at the swimming pool is based on the size of the pool, and usually determined by state or local codes.

Sand filter. Filter medium composed of hard, sharp silica, quartz, or similar particles which have been graded for size and uniformity.

Scale. Precipitate that forms on surfaces on contact with water when calcium hardness, pH, or total alkalinity is too high.

Semipublic pool. Swimming or wading pool on the premises of or used in connection with a hotel, motel, trailer court, apartment house, country club, youth club, school, camp, or similar establishment where the primary purpose of the establishment is not the operation of the swimming facilities, and where admission to the use of the pool is included in the fee or consideration paid or given for the primary use of the premises. A semipublic pool also means a pool constructed and maintained by groups for the purposes of providing bathing facilities for members and guests only.

Septum. Part of a filter element consisting of cloth, wire screen, or other porous material on which the diatomaceous earth filter medium is deposited.

Service factor. Indicator of how much over an electric motor's specified output it can be operated without danger of failure due to overloading.

Shallow area. Area on the shallow side of the transition point in the swimming pool.

Shock treatment. Addition of pool chemicals in larger than normal amounts in order to eliminate unusual pool water conditions such as infestations of algae, the presence of chloramines, or colored water.

Silver protein. Solution containing protein and silver ions, used as an antiseptic.

Skimmer. Device to remove leaves and other floating debris from the water surface.

Skimmer filter. Recirculating skimmer with a filter forming an integral part of the device.

Skimmer weir. Hinged floating dam within a skimmer.

Slurry. Watery mixture (not solution) of diatomaceous earth used to body feed diatomaceous earth filters.

Slurry feeder. Device that feeds a variable amount of filter aid into a filter during the filter cycle.

Soda ash

1. Dry chemical used to increase the pH (alkalinity) of pool water.

2. Chemical name for sodium carbonate. White powder or cake used to raise the pH of the pool water to an optimum range of 7.2 to 8.2 pH where disinfecting agents work most effectively.

Sodium bisulfate. Colorless crystals soluble in water. Used for lowering the pH of highly alkaline waters. Also known as "sodium acid sulfate."

Sodium carbinate. Soda ash.

Sodium hydroxide. Chemical used for off-setting a low pH factor. Used to prevent the corrosion due to excessive acidity.

Sodium hypochlorite

1. Compound containing 5 to 15% (sometimes more) available chlorine by weight in a caustic soda solution, which releases chlorine when added to pool water.

2. Commercial bleach offering 5 to 15% available chlorine in solution form.

Special-use pools. Pools designed and used primarily for a single purpose, such as wading, instruction, or diving.

Springboard. Diving board installed with an anchor (base) and fulcrum that provides a spring action to the driver.

Stabilization. Addition of a stabilizer or conditioner, such as cyanuric acid, to pool water to extend the effective life of chlorine by protecting it from the dissipating effects of sunlight.

Static head. Vertical distance between the waterline of water supply and the point of discharge, or to the surface of the discharged water.

Suction head. Total head on the suction side of a pump.

Sulfamic acid. Acid that is commercial, sold as granular pellets, powder, or tablets. Used for lowering the alkalinity of pool water. It titrates in the range of 4.5 to 9.0 pH. It will stabilize chlorine, but is less effective than cyanuric acid compounds.

Superchlorination. Addition of larger-than-normal amounts of chlorine during periods of excessive heat or rainfall or heavy pool use to convert chloramines into free available chlorine by destroying ammonia.

Surface skimmer. Device designed to continuously remove surface film from water and return it to the filter. Considered part of the recirculation system, it usually incorporates a self-adjusting weir (given) a collection tank, and a means of preventing the pump from locking (sometimes referred to as a "mechanical" or "automatic" skimmer).

Surface-type cartridge. Filter cartridge with media less than $\frac{3}{4}$-in. thick that relies on retention of dirt on the surface of the cartridge to achieve its removal.

Surge chamber. Storage chamber within a pool's recirculation system used to absorb water displaced by bathers.

Swimming area. Area of a pool more than 3 feet deep that is used for swimming.

Swimming pool

1. Artificial swimming pool or wading pool, together with the buildings and appurtenances used with it.

2. Every artificial pool of water having a depth of 2 ft or more at any point and used for swimming or bathing, located indoors or outdoors, together with the bathhouses, equipment, and appurtenances used in connection with the pool. The term does not include any residential pool nor does it include any pool used primarily for baptismal purposes or the healing arts.

3. Structural basin, chamber, or tank containing an artificial body of water for swimming, diving, or recreational bathing and having a water depth of 2 ft, 6 in. or more at any point.

4. Constructed pool, used for swimming or bathing, over 24 in. in depth or with a surface area exceeding 250 sq ft.

5. Artificial basin, chamber, or tank constructed of impervious material and used, or intended to be used, for swimming, diving, or recreational bathing. The term does not include baths where the main purpose is the cleaning of the body, or individual therapeutic tubs.

TDS (total dissolved solids). The sum total of all dissolved materials in the pool water.

Total alkalinity. Measurement of all alkaline chemicals in pool water. A total alkalinity that is too high causes the pH to resist adjustment to the desired range; if the total alkalinity is too low, it is difficult to maintain the pH within the desired range.

Training pool. Swimming pool not normally in excess of 3 ft. deep at its point of maximum depth and usually reserved for use by persons learning to swim.

Transition point

1. Point between shallow and deep area.

2. Place in the floor of the pool between water depths of 4 ft, 6 in., and 5 ft, 6 in., where an abrupt change in slope occurs.

Turbidity. Cloudy condition of pool water due to the presence of extremely fine particulate materials in suspension.

Turnover

1. Rate of water interchange in recirculating or flowing-through pool, expressed as the ratio of the volume of clean water entering the pool in 24 hours to the total pool volume. All swimming pools should have a minimum of three turnovers per day (24 hours), which means that the total volume of water recirculated or flowing through the pool in 24 hours should be at least three times the total pool volume.

2. Time required to recirculate the volume of water the pool contains through the filtration system and back to the pool.

Turnover rate. Period (usually stated in hours) required to circulate a volume of water equal to the pool's capacity.

Underdrain

1. Distribution system at the bottom of a filter that collects water during the filter cycle and distributes backwash water during the cleaning operation.

2. Appurtenance at the bottom of the filter to assure equal distribution of water through the filter medium.

Underwater lights. Light designed to illuminate a pool from beneath the water surface.

Vacuum (suction) filter. Filter that operates by means of a vacuum (from the suction side of a pump).

Vacuum fitting. Fitting in the wall of the pool that is used as a convenient outlet for connecting the underwater suction cleaning equipment.

Vacuum piping. Piping that connects the vacuum fitting to the pump suction.

Vacuum wall fitting. Fitting in the pool wall just below the waterline to which is attached the underwater suction cleaner's hose.

Venturi tube. Tube with a constricted throat that causes differences in pressure and can be used to operate feeding devices and instruments.

Vertical. Slope of no more than 11° (1 ft horizontally for each ft vertically) from plumb.

Wading area. Area less than 3 ft deep reserved for nonswimmers.

Wading pool. Artificial basin, chamber, or tank constructed of impervious material used, or intended to be used, for wading by small children and having a maximum depth not to exceed 18 in. at the deepest point or more than 12 in. at the side walls.

Water clarity. The clarity of the water in any pool is established by the visual observation of a 6-in. black disc placed at the deepest point in the pool and to be clearly visible from the pool deck at a minimum distance up to 30 ft.

Wall slope. Inclination from the vertical in a pool wall, expressed in degrees, or in feet or inches of horizontal distance at a specific depth.

Water depth markers. Markers placed in the side walls of the pool to indicate the depths by ft, required by most local and state codes.

Waterline. The waterline is deemed to be at the midpoint of the operating range of the skimmers. On pools with overflow systems, the waterline is deemed that established by the height of the overflow rim.

Water testing kit. Commercial devices used to test the pool water for determining chlorine residual and pH range.

Weir. The weir contains a buoyant section so that it floats high in the water, allowing only about $\frac{3}{16}$ in. of surface water to flow into the skimmer box.

Wet niche. Watertight, water-cooled submersible housing placed in a niche in the pool wall.

Width and length. Measurements which are determined by actual water dimensions at the surface of the pool.

X-RAY RADIATION AND SHIELDING

Absorbed dose. Energy imparted by x-rays or other ionizing radiation to a part of the body exposed to radiation. The special unit of absorbed dose is the "rad", which represents 100 "ergs" of energy absorbed by 1 gr of material.

Absorption. Process whereby radiation is stopped or reduced in intensity as it passes through matter. Lead, which is denser than most materials, is a good absorber of x-rays.

Added filtration. Filtration placed in the x-ray beam to absorb the less penetrating radiations, which do not contribute to the quality of the x-ray image. The use of appropriate filtration prevents unnecessary x-ray exposure. (Filtration materials may be aluminum, copper, or lead).

Alpha particle. Form of ionizing radiation consisting of two neutrons and two protons. Alpha particles are often emitted from radioactive materials.

Anode. The positive electrode within an x-ray tube, toward which electrons are accelerated from the cathode. The kinetic energy possessed by the high-speed electrons is converted to heat and x-rays when the electrons strike the anode.

Atom. Fundamental unit of which matter is composed; it consists of a heavy nucleus and surrounding electrons.

Beta particle. Form of ionizing radiation consisting of a fast electron. Beta particles are often emitted from radioactive materials.

Cardboard film holder. Used in certain types of x-ray examinations. This type of film holder has no mechanism for intensification of the x-rays.

Cathode. The negative electrode in a tube where electrons are produced. It consists of one or two filaments and focusing cups.

Chronic exposure. Irradiation that is spread out over a period of years. Those who are exposed to occupational radiation can suffer from chronic exposure.

Collimation. Restriction of the size of the x-ray beam.

Cone. Round shield placed in front of an x-ray tube to limit the size of the beam.

Cosmic rays. Energetic radiations from outer space, consisting of x-rays and charged particles.

Creep. The horizontal or vertical movement of fluoroscopic equipment during an x-ray examination.

Dead-man switch. A switch so constructed that a circuit-closing contact can only be maintained by continuous pressure by the operator.

Depth dose. A radiation dose delivered to a point at a given distance below the skin.

Diagnostic-type tube housing. An x-ray tube housing so constructed that the leakage radiation at a distance of 1 m from the target cannot exceed 100 milliroentgens in 1 hour when the tube is operated at any of its specified ratings.

Dose. Expresses the amount of energy absorbed in a unit of mass.

Dose rate. Dose delivered per unit of time.

Dosimeter. Instrument which measures dose.

Electron. Negatively charged part of an atom or a molecule. When electrons strike materials at high energy, x-rays are produced.

Exposure. Measure of the number of ions produced in air by radiation. The unit of exposure is the "roentgen," which can be easily measured with an ionization chamber. "Exposure" should not be confused with "dose," which is not as easy to measure and represents energy deposited in a certain amount of material.

Film badge. Lightproof film packet used for estimating radiation exposure of personnel who work with x-rays or other forms of ionizing radiation.

Filter. Material placed in the useful beam to absorb preferentially the less penetrating radiations.

Fluoroscope. Fluorescent screen coated with a special substance that emits light when exposed to x-rays.

Gamma rays. Photons or bundles of electromagnetic radiation of high energy, which are usually more penetrating than x-rays.

Grid. Device similar to a grating; its purpose in radiology is to absorb scatter radiation that would impair the clarity of the image on the x-ray film.

HVL (half-value layer). The thickness of a specified metal material required to decrease the dosage rate of a beam of x-rays to one-half its initial value.

Hard x-rays. x-rays of high penetrating power.

Hardness. Term used to describe the penetrating quality of radiation. The higher the energy of the radiation, the more penetrating the radiation.

Heel effect. The unequal intensity of the x-ray beam; the intensity is greatest on the cathode side of the beam and least intense on the anode side of the beam.

High-voltage x-rays. Voltage range from 140 to 250 kv peak.

Inherent filtration. Filtration effect of the materials, such as glass and oil, making up the wall of the x-ray tube.

Ion. An atom or molecule that has one or more of its surrounding electrons separated from it and therefore carries an electric charge.

Ion pair. Positively charged atom or molecule (ion) and an electron formed by the action of radiation upon a neutral atom or molecule.

Ionization. Process whereby one or more electrons are removed from a neutral atom by the action of radiation.

Ionization radiation. Radiation such as x-rays, gamma rays, beta particles, and alpha particles that are capable of producing ions in matter.

Interlock. A device for precluding access to an area of radiation hazard either by preventing entry or by automatically removing the hazard.

Kilovolt (kV). Unit of 1,000 volts, used to describe the energy of x-rays. Most x-ray machines (medical) in use generate 20 to 150 kV x-rays.

Lead-plastic sheet. Lead-impregnated transparent plastic sheet that contains 30% lead by weight. Combines light transmission and effective radiation shielding. Sheet is shatter-resistant and is the latest innovation in effective radiation shielding.

Leakage radiation. All radiation coming from within the tube housing except the useful beam.

mA (milliamperage). Electron current flowing across the x-ray tube in milliamps (1,000 milliamps = 1 ampere).

Millirad. One-thousandth of a rad.

Millirem. One-thousandth of a rem.

Milliroentgen. One-thousandth of a roentgen.

Molecule. Group of atoms bonded together by electrostatic (chemical) forces.

Penetrating power. The x-ray's power to penetrate, a solid substance depends on the density of the substance and the speed at which the electrons are traveling.

Photons. Bundle of electromagnetic energy. Each photon carries a fixed amount of energy. An x-ray beam consists of photons having energies in the x-ray region of the electromagnetic spectrum.

Primary protective barrier. A barrier sufficient to attenuate the useful beam to the required degree.

Protective barrier. A barrier of attenuating materials used to reduce radiation exposure.

Quality. Term that describes the penetrating power of x-rays or gamma rays, which is related to the energies of the photons in the beam.

Quantity. Term used to describe the number of photons in a beam of x-rays or gamma rays.

Rad. Special unit of absorbed dose equal to 100 ergs of energy deposited by ionizing radiation-like x-rays in one frame of matter.

Radiation. Energetic subatomic particles of electromagnetic waves that move at high speeds.

Rem. Unit of dose equivalent that is the same as a rad for x-rays.

Roentgen. Amount of radiation required to produce ions that carry a charge of .000258 coulomb in a kilogram of air.

Scattered radiation. Radiation that, during passage through matter, has been deviated in direction.

Secondary protective barrier. A barrier sufficient to attenuate stray radiation to the required degree.

Shielding. Material interposed between a radiation source and an irradiated site for the purpose of minimizing the radiation hazard. Shielding has usually been made of lead, which is dense and absorbs radiation easily.

Shutter. A device, generally of lead, fixed to an x-ray tube housing to intercept the useful beam.

Soft x-rays. X-rays of low energy and penetrating power.

Stray radiation. Radiation not serving any useful purpose. It includes leakage and secondary radiation.

Target. Part of the metal anode or plate that faces the cathode and is struck by the beam of electrons. X-rays are produced when the electrons strike the anode.

Total filtration. Filtration of the x-ray beam provided by both the inherent filtration and the added filtration.

Tube. The glass tube within the head of the x-ray unit, wherein x-rays are produced as the result of high-speed electrons striking a metallic target (anode).

X-ray diffraction. Since all crystals act as three-dimensional gratings for x-rays, the pattern of diffracted rays is characteristic for each crystalline material. This method is of particular value in determining the presence or absence of crystalline silica in an industrial dust.

X-ray field. The area of the intersection of the useful beam and any one of the set of planes parallel to and including the plane of the image receptor, whose perimeter is the locus of points at which the exposure rate is $1/4$ of the maximum in the intersection.

X-ray machine. The x-ray tube, power supply, and associated equipment required for producing x-ray photographs.

X-ray room. Room where walls, floor, and ceiling construction are protected from radiation transmission by the installation of sheet lead. In ceilings the lead sheet is placed above the structural slab, in wood framing, above the plasterboard or lath and plaster.

X-rays. Penetrating photons of electromagnetic radiation having wavelengths shorter than visible light. Usually produced by bombarding a metallic target with fast electrons in a vacuum. These rays are sometimes called "roentgen rays" after their discoverer, Wilhelm K. Roentgen.

Conveying Systems

ELEVATOR ACCESSORY EQUIPMENT, CONTROLS, AND OPERATIONS

ANSI. American National Standards Institute. Publication, ANSI Code for Elevators, recent edition.

Automatic operation

1. Operation by means of buttons or switches at the landings, with or without buttons or switches in the car, the momentary pressing of which will cause the car to start and automatically stop at the landing corresponding to the button pressed.
2. Operation wherein the starting of the elevator car is effected in response to the momentary actuation of operating devices at the landing, and/or of operating devices in the car identified with the landings, and/or in response to an automatic starting mechanism, and wherein the car is stopped automatically at the landings.

Auxiliary rope-fastening device. Device attached to the car or counterweight or to the overhead dead-end rope-hitch support that will function automatically to support the car or counterweight in case the regular wire-rope fastening fails at the point of connection to the car or counterweight or at the overhead dead-end hitch.

Blind hoistway. Portion of a hoistway that passes floors or other landings at which no normal landing entrances are provided.

Blower. Ventilating fan located above the car ceiling or outside the car enclosure.

Bottom car clearance. Clear vertical distance from the pit floor to the lowest structural or mechanical part, equipment, or device installed beneath the car platform, except guide shoes or rollers, safety jaw assemblies, and platform aprons or guards, when the car rests on its fully compressed buffers.

Bottom clearance. Vertical distance between any obstruction in the pit, exclusive of the compensating device, buffer, and buffer supports, and the lowest point of the understructure of the car, exclusive of the safeties, car frame channels, guide shoes, and other necessary equipment attached to the underside of the platform, when the car floor is level with the bottom terminal landing.

Bottom elevator car run-by

1. Distance the car floor can travel below the level of the lower terminal landing until the car strikes its buffer. The bottom run-by of an elevator counterweight is the distance the counterweight can travel below its position when the car floor is level with the upper terminal landing until the counterweight strikes its buffer.
2. Distance between the car buffer striker plate and the striking surface of the car buffer when the car floor is level with the bottom terminal landing.

Bottom overtravel of the counterweight. Distance the counterweight can travel below its position when the car platform is level with the upper terminal landing until the full weight of the counterweight rests on the buffers; the resulting buffer compression is included.

Bottom overtravel of the elevator car. Distance the car floor can travel below the level of the lower terminal landing until the weight of the fully loaded car rests on the buffers; the resulting buffer compression is included.

Bottom terminal elevator landing. The lowest landing served by the elevator that is equipped with a hoistway door and hoistway door locking device that permits egress from the hoistway side.

Buffer

1. Device designed to absorb the impact of the car or counterweight at the extreme lower limits of travel.
2. Device designed to stop a descending car or counterweight beyond its normal limit of travel by storing or by absorbing and dissipating the kinetic energy of the car or counterweight.

Buffer spring. The buffer stores in a spring the kinetic energy of the descending car or counterweight.

Bumper. Device, other than an oil or spring buffer, designed to stop a descending car or counterweight beyond its normal limit of travel by absorbing the impact.

Cab. Finished car enclosure.

Cable lock. Device installed and maintained so that the operating cable can be locked at any landing.

Car. Load-carrying unit that moves through the hoistway, including the platform, car frame enclosure, and car door or gate.

Car annunciator. Electrical device in the car that indicates visually the landings at which an elevator landing signal registering device has been actuated.

Car enclosure

1. Top and walls of the car resting on and attached to the car platform.

2. The enclosure or cab of an elevator consists of the walls and the top or cover built up on the platform.

Car frame (sling). Supporting frame to which the car platform, upper and lower sets of guide shoes, car safety, and the hoisting ropes or hoisting-rope sheaves or the plunger of a direct plunger elevator are attached.

Car-leveling device. Mechanism or control that will move a car within a limited zone toward, and stop the car at, the landing. For an elevator, the device may also be used for emergency operation of the car throughout its entire travel and for safe lifting purposes.

Car or counterweight safety. Mechanical device attached to the car frame, an auxiliary frame, or the counterweight frame to stop and hold the car or counterweight in case of predetermined overspeed or free fall, or if the hoisting ropes slacken.

Car platform. Structure that forms the floor of the car and directly supports the load.

Car-switch automatic floor-stop operation. Operation in which the stop is initiated by the operator from within the car with a definite reference to the landing at which it is desired to stop, after which the slowing down and stopping of the elevator is automatically effected.

Car-switch operation. Operation wherein the movement and direction of travel of the car are directly and solely under the control of the operator by means of a manually operated car switch or continuous-pressure buttons in the car.

Centering rope. Used in connection with hand cable control, which when pulled with throw the operating device to the stop position.

Class A freight elevator. General freight loading, by hand truck. Single items may not exceed 25% of the car-rated load. Rated load is based on 50 pounds per square foot of net inside platform area.

Class B freight elevator. Elevator designed to carry automobiles or automobile trucks. Rating is based on a load of 30 psf of platform area.

Class C freight elevator. Industrial truck loading, based on a maximum loading, 150% of rated capacity, and on a figure of 50 psf of net inside platform area.

Compensating chain or rope. Chain or rope attached to the bottom of the car at one end and the bottom of the counterweight at the other which compensates for the hoist rope's moving from car to counterweight, and vice versa.

Compensating rope sheave switch. Device that automatically causes the electric power to be removed from the elevator driving-machine motor and brake when the compensating sheave approaches its upper or lower limit of travel.

Composite fixture. Signal fixture indicating more than one function (such as position and direction) in a single unit.

Continuous-pressure operation. Operation by means of buttons or switches in the car and at the landings, any one of which may be used to control the movement of the car as long as the button or switch is manually maintained in the actuating position.

Contract load or rated load (capacity). Approved safe live load specified.

Control. System governing the starting, stopping, direction of motion, acceleration, speed, and retardation of the moving member.

Control equipment. Devices used by those operating or servicing elevators to control the starting, stopping, and direction of motion of the car.

Controller. Device or group of devices designed to control in a predetermined manner the apparatus to which it is connected. The system governs the starting, stopping, direction of motion, acceleration, speed, and retardation.

Counterweight. Calculated weight attached to the hoisting cable on the opposite end from the elevator to balance the car's weight and facilitate motion.

Cylinder well. Steel casing extending below the bottom of the pit in a hydraulic elevator hoistway, designed to receive the hydraulic cylinder of the hoisting unit.

Digital indicator. Car position indicators with numerals on a rotating drum or readout slide viewed through a small lens. The numeral indicates the floor that the elevator car is passing.

Direction of passage. Passage from a landing to a treadway, or vice versa, in the direction of treadway travel at the point of passenger entrance or exit.

Double-deck elevators. Elevators designed to provide additional service for high-rise buildings. The elevator increases shaft capacity and decreases the number of stops.

Dual operation. System of operation whereby the elevator controller is arranged for either automatic operation by means of landing and car buttons or switches, or for manual operation by an operator in the car, who may either use a car switch or the buttons provided in the car. When operated by the operator, upon the throwing of a suitable switch or switches, the car can no longer be started by the landing buttons; the buttons may, however, be used to signal the operator that the car is desired at certain landings.

Electric elevator machine room. Located at the top of the hoistway, includes the hoisting machine and control, adequate ventilation or air conditioning, soundproofing, and structural support across the hoistway to support the hoisting equipment.

Elevator arrangement. The hoistway location is a matter of planning to provide accessible service to the passengers and tenants in a building.

Elevator automatic dispatching device. Device whose principal function is either to operate a signal in the car to indicate when the car should leave a designated landing or actuate its starting mechanism when the car is at a designated landing.

Elevator automatic signal transfer device. Device by means of which a signal registered in a car is automatically transferred to the next car following, in case the first car passes a floor for which a signal has been registered without making a stop.

Elevator cab. Enclosure consisting of walls and top built upon a car platform.

Elevator car. Guided by guide rails on each side of the hoistway. Design of the car should include a finished ceiling, walls, floor covering, ventilation or air conditioning, electric lighting, car controls, floor indicators, and equipment for the blind and the deaf.

Elevator car flash signal device. Signal light in the car, which is illuminated when the car approaches the landing at which a landing signal registering device has been actuated.

Elevator car interior. Design should include nonslip floor or carpeting, telephone for emergency use, call buttons and control panel, car position indicator, frame that displays inspector permit, ventilation or air conditioning, hand rails on each side of the car, provisions for blind and deaf controls, and sufficient lighting.

Elevator car leveling device. Mechanism that either automatically, or under control of the operator, moves the car within the leveling zone toward the landing only and automatically stops it at the landing.

Elevator landing. Portion of a floor, balcony, or platform used to receive and discharge passengers or freight.

Elevator landing signal registering device. Button or other device, located at the elevator landing, that when actuated by a waiting passenger causes a stop signal to be registered in the car.

Elevator landing stopping device. Button or other device, located at the elevator landing, that when actuated causes the elevator to stop at that floor.

Elevator landing zone. Zone extending from a point 18 in. below a landing to a point 18 in. above the landing.

Elevator lobbies. The waiting side of the lobby, facing the elevator doors, should have car direction indicators, indicator that locates which floor the elevator is serving, doors and frames, entrance safety device light mounted on car door, call buttons, and a traffic director's panel with fire control provisions on the first floor.

Elevator nonstop switch. Switch that when operated will prevent the elevator from making registered land stops.

Elevator or dumbwaiter hoistway. Shaftway, hatchway, well hole, or other vertical opening or space in which an elevator or dumbwaiter is designed to operate.

Elevator or power dumbwaiter hoistway. Shaftway for the travel of one or more elevators or power dumbwaiters. It includes the pit and terminates at the underside of the overhead machinery space floor or grating, or at the underside of the roof where the hoistway does not penetrate the roof.

Elevator pit. Portion of a hoistway extending from the threshold level of the lowest landing door to the floor at the bottom of the hoistway.

Elevator potential switch. Switch that disconnects the power from the elevator apparatus when the supply voltage fails or decreases below a definite value and that is usually opened by various electrical safety devices. Switches are of the magnetic type.

Elevator selection. Selection depends on the need to service the habitants of the building, type of service according to height of building, and the size of the elevator cab. Local and state codes will govern most installations.

Elevator separate signal system. System consisting of buttons or other devices, located at the landings, that when actuated by a waiting passenger illuminate a flash signal or operate an annunciator in the car indicating floors at which stops are to be made.

Electric service equipment. Electrical equipment, located near the point of entrance of electric supply conductors to a building, that constitutes the main control of supply and means of cutoff of electricity for the building and includes switches, fuses, and electrical accessories.

Elevator signal transfer switch. Manually operated switch, located in the car, by means of which the operator can transfer a signal to the next car approaching in the same direction, when he desires to pass a floor at which a signal has been registered in the car.

Elevator slack-cable switch. Device for automatically cutting off the power in case the hoisting cables become slack.

Elevator starter's control panel. Assembly of devices by means of which the starter may control the manner in which an elevator or group of elevators functions.

Elevator truce zone. Limited distance above an elevator landing within which the truck-zoning device permits movement of the elevator car.

Elevator truck-zone device. Device that permits the operator in the car to move a freight elevator within the truck zone with the car door or gate and hoistway door open.

Emergency stop switch (safety switch). Device in the car used manually to cut off the power from the elevator machine independently of the operating devices.

Emergency stop switch. Device, located in the car, that when manually operated causes the electric power to be removed from the driving machine motor and brake of an electric elevator or from the electrically operated valves and/or pump motor of a hydraulic elevator.

Emergency terminal stopping device. Device that automatically causes the power to be removed from an electric elevator driving-machine motor and brake, or from a hydraulic elevator machine, at a predetermined distance from the terminal landing, and independently of the functioning of the operating device and normal terminal stopping device, if the normal terminal stopping device does not slow down the car as intended.

Entrance. Door system for access to the hoistway and elevator car.

Factor of safety. Ratio of the ultimate strength of the material or part divided by the actual load imposed on it.

Fascia. Falsework or steel sheeting running down the sill on one entrance to the hanger pocket of the door below. Sometimes done in wall construction materials where elevator entrance units are spaced more than one floor apart.

Fascia plate. Metal plate not less than $\frac{1}{16}$ in. in thickness, securely fastened, and extending flush from the top of the hoistway landing door frame to the landing sill above and running the full width of the door opening.

Final terminal stopping device

1. Automatic device for stopping the car and counterweight from rated speed within the top clearance and bottom overtravel, independently of the operation of the normal terminal stopping device and the operating device.

2. Device that automatically causes the power to be removed from an electric elevator or dumbwaiter driving-machine motor and brake, or from a hydraulic elevator or dumbwaiter machine, independently of the functioning of the normal terminal stopping device, operating device, or any emergency terminal stopping device, after the car has passed a terminal landing.

Flooring. Interior finish of the platform.

Freight elevator cabs, gates, and doors. Cabs for freight service are designed for hard service. Built of heavy gauge steel, multilayer wood floors, guarded ceiling light fixtures. Cab gates slide vertically, hoistway doors usually vertical lift, center-opening doors, operated manually or by power.

General purpose freight elevator. Elevators designed to carry loads up to 20,000 lb, and can be traction or hydraulic types.

Governor. Safety device set to trip at excess speed.

Group automatic operation. Automatic operation of two or more no-attendant elevators equipped with power-operated car and hoistway doors. Operation of the cars is coordinated by a supervisory control system, including automatic dispatching means whereby selected cars at designated dispatching points automatically close their doors and proceed on their trips in a regulated manner. There is one button in each car for each landing served and up and down buttons at each landing (single buttons at terminal landings). Stops set up by the momentary actuation of the car buttons are made automatically in succession as a car reaches the corresponding landings irrespective of its direction of travel or the sequence in which the buttons are actuated, automatically by the first available car that approaches the landing in the corresponding direction.

Guide rails (car and counterweight). A metal rail with safety devices guides the vertical movement of the car.

Hall and corridor. Hall where the elevator stops.

Hall button. Signaling device at a landing outside the hoistway enclosure for calling a car.

Hall lantern. Signal light at a landing that indicates the arrival of an elevator car and its direction of travel.

Handhold (handgrip). Device attached to the belt that can be grasped by a passenger for maintaining balance while on a manlift.

Hanger pocket. Space required for the hangers of door panels of passenger-type entrances.

Hatchway. Space in which an elevator or dumbwaiter is designed to operate.

Hatchway enclosure. Structure that separates the hatchway, either wholly or in part, from the floors or landings through which the hatchway extends.

Hatchway unit contact system. Contact system that prevents the operation of the car unless all hatchway doors are closed.

Hatchway unit interlock system. Interlock system that prevents the operation of the car unless all hatchway doors are locked in the closed position.

Hoistway

1. Enclosed shaft in which one or more elevators is designed to travel.
2. Vertical opening, space, or shaftway in which an elevator or dumbwaiter is installed.
3. Shaftway, hatchway, well hole, or other vertical opening or space in which an elevator, escalator, or dumbwaiter operates.
4. Shaftway for the travel of one or more elevators or dumbwaiters. It includes the pit and terminates at the underside of the overhead machinery space floor or grating, or at the underside of the roof where the hoistway does not penetrate the roof.

Hoistway access switch

1. Switches are located at the lower and upper terminal landings to permit access to the pit and top of the car. Car travel is limited to a zone sufficient for the full door opening.
2. Switch, located at a landing, whose function it is to permit operation of the car with the hoistway door at this landing and the car door or gate open, in order to permit access to the top of the car or to the pit.

Hoistway unit system

1. Interlock system that, in addition to fulfilling the requirements given under the definition of interlock, will also prevent the operation of the car unless all hoistway doors are locked in the closed position.
2. Series of hoistway door interlocks, hoistway door electric contacts, hoistway door combination mechanical

locks and electric contacts, or a combination thereof, whose function it is to prevent operation of the driving machine by the normal operating device unless all landing doors are locked in the closed position.

Hydraulic elevator. Low-rise structures are usually equipped with hydraulic elevators. These elevators provide accurate control, smooth operation, and very accurate automatic leveling. Hydraulic units normally exceed 60 ft in height, and operate at 125 fpm.

Hydraulic elevator machine room. Located next to the hoistway at or near the bottom terminal landing. The room is designed to accommodate the pump and motor drive unit, hydraulic fluid storage tank, control panel, adequate lighting and electrical service, and outlets. Ventilation or air conditioning should be provided.

Hydraulic plunger. Supporting member of a hydraulic hoisting unit that is attached to the bottom of the elevator car.

Inspection of elevators. Periodic inspection of all types of elevators and escalators, together with proper maintenance and safe operation of equipment and appurtenances as required by local building codes.

Landing. Portion of a floor or platform used to receive and discharge passengers or freight.

Landing zone. Space from a point not more than 18 in. below the landing to a point not more than 18 in. above the landing.

Lantern. Lighted signal fixture indicating which car will be dispatching next; it may indicate in which direction the car is traveling.

Leveling zone. Limited distance above or below an elevator landing within which the leveling device may cause movement of the car toward the landing.

Limit switch for a manlift. Device whose purpose it is to cut off the power to the motor and apply the brake to stop the carrier in the event that a loaded step passes the terminal landing.

Load-weighing device. Each power elevator car (not hydraulic) is equipped with an automatic load-weighing device that permits the car to operate up to 125% of capacity passenger load and activates a signal at near capacity load to bypass all landing calls until the car has reduced its load by discharging passengers or weight. After a car is filled at the lower dispatching terminal the device, when activated, will cause this car to be dispatched without waiting for the lapse of the dispatching interval.

Machine beam. Structural member directly supporting the elevator machine.

Machine final terminal stopping device. Stopping device operated directly by the driving machine.

Manlift step platform. Passenger-carrying unit.

Manlift travel. Distance between the centers of the top and bottom pulleys.

Multiple hoistway. Hoistway for more than one elevator or dumbwaiter.

Multivoltage control. System of control that is accomplished by impressing successively on the armature of the driving-machine motor a number of substantially fixed voltages such as may be obtained from multicommutator generators common to a group of elevators.

Nonselective collective automatic operation. Operation by means of one button in the car for each landing level served and one button at each landing, wherein all stops registered by the momentary pressure of landing or car buttons are made irrespective of the number of buttons pressed or of the sequence in which the buttons are pressed. With this type of operation the car stops at all landings for which buttons have been pressed, making the stops in the order in which the landings are reached after the buttons have been pressed, but irrespective of its direction of travel.

Normal terminal stopping device

1. Device or devices to slow down and stop an elevator or dumbwaiter car automatically at or near a terminal landing independently of the functioning of the operating device.

2. Automatic device for stopping the car within the overtravel independently of the operating device.

Observation car elevators. By the arrangement of the traction lifting mechanism and the attachment of the car, the glass enclosure of the building and the glass panel in the car can provide a visual sight and openness to the passenger.

Oil buffer. Buffer using oil as a medium that absorbs and dissipates the kinetic energy of the descending car or counterweight.

Oil buffer stroke. Oil-displacing movement of the buffer plunger or piston, excluding the travel of the buffer plunger accelerating device.

One-way automatic leveling device. Device that corrects the car level only in the case of underrun of the car but that will not maintain the level during loading and unloading.

Operating device

1. Car switch, push button, rope, wheel, lever, treadle, etc., employed to enable the operator to actuate the controller.

2. Car switch, push button, lever, or other manual device used to actuate the control.

Operation. Method of actuating the control.

Overhead structure. Structural members, platforms, etc., supporting the elevator machinery sheaves and equipment at the top of the hoistway.

Overslung car frame. Frame to which the hoisting rope fastenings or hoisting rope sheaves are attached to the crosshead or top member of the car frame.

Pit. Portion of a hatchway extending below the level of the bottom landing to provide for bottom overtravel and clearance and for parts that require space below the bottom limit of car travel.

Platform. Structural unit bolted to the sling and stabilized in a horizontal plane.

Position indicator. Device indicating the position of the elevator car in the hoistway. It is called a "hall position indicator" when placed at a landing and a "car position indicator" when placed in the car.

Preregister operation. Signals to stop are registered in advance by buttons in the car and at the landings. At the proper point in the car travel the operator in the car is notified by a signal, visual, audible, or otherwise, to initiate the stop, after which the landing stop is automatic.

Raceways. Channels for holding wires or cables, designed and used solely for this purpose. Raceways are of metal, and the term includes rigid metal conduit, flexible metal conduit, or electrical metallic tubing.

Rated load. Load that the elevator, dumbwaiter, escalator, or private residence inclined lift is designed and installed to lift at the rated speed.

Rated speed

1. Speed at which the elevator, power dumbwaiter, escalator, or moving walk or moving ramp is designed to operate.

2. Speed for which the device is designed and installed.

3. Speed that the car is designed to attain when carrying its rated load in the "up" direction.

4. Speed at which the elevator, dumbwaiter, escalator, manlift, or inclined lift is designed to operate.

Rated speed of elevator or dumbwaiter. Speed in the "up" direction with the rated load in the car.

Rated speed of escalator or inclined lift. Rate of travel of the steps or carriage, measured along the angle of inclination, with the rated load on the steps or carriage. In the case of a reversible escalator, the rated speed is the rate of travel of the steps in the "up" direction, measured along the angle of inclination, with the rated load on the steps.

Rheostatic control. System of control accomplished by varying the resistance or reactance in the armature or field circuit of the driving machine induction motor, which is arranged to run at two different synchronous speeds by connecting the motor windings so as to obtain different numbers of poles.

Roller guide. Guide composed of three rollers mounted at the top of the car frame which grips the guide rails.

Run-by of the bottom elevator counterweight. Distance between the counterweight buffer striker plate and the striking surface of the counterweight buffer when the car floor is level with the top terminal landing.

Safety. Device attached to the car or counterweight that applies sufficient gripping or clamping action to the guide rails to retard and stop the elevator car or counterweight when it has passed its designed rate of speed in the downward direction.

Safety devices. Mechanical device attached to the car frame, auxiliary frame, and counterweight designed to stop the car or the counterweight during free fall, overspeed, or by slacking the hoisting ropes.

Second-level machine room. Room a half-level below the overhead machine room in the shaft area to accommodate hoisting machines separated from motor-generator sets in machine rooms of larger elevator installations.

Selective collective automatic operation. Automatic operation by means of one button in the car for each landing level served and by up and down buttons at the landings, wherein all stops registered by the momentary actuation of the car buttons are made as defined under nonselective collective automatic operation, but wherein the stops registered by the momentary actuation of the landing buttons are made in the order in which the landings are reached in each direction of travel after the buttons have been actuated. With this type of operation all "up" landing calls are answered when the car is traveling in the up direction, and all "down" landing calls are answered when the car is traveling in the down direction, except in the case of the uppermost and lowermost calls which are answered as soon as they are reached, irrespective of the direction of travel of the car.

Sensitive edge. Vertical rubber edge on a passenger door or a horizontal rubber edge on a freight elevator gate.

Shaft. Vertical opening through a building for an elevator, dumbwaiter, light, ventilation, or similar purposes.

Sheaves (drive and deflection). Wheel with grooves that route the hoist ropes (deflecting), assist in providing traction, or provide driving power for the vertical motion of the car in traction-type elevators.

Signal control. Device required to slow, stop, and level elevators automatically.

Signal equipment. Message-relaying devices that tell the operator or passenger that an event or sequence of events involving the elevator equipment can be anticipated. The primary signal method consists of lighted buttons, numerals, or arrows; hall lanterns; and gongs, buzzers, bells, dial indicators, and readout indicators.

Signal operation. Operation by means of single buttons or switches (or both) in the car, and up or down direction buttons (or both) at the landings by which predetermined landing stops may be set up or registered for an elevator or for a group of elevators. Stops set up the momentary pressure of the car buttons are made by automatically in succession as the car reaches those landings, irrespective of its direction of travel or the sequence in which the buttons are pressed. Stops set up by the momentary pressure of the up and down buttons at the landing are made automatically by the first available car in the group approaching the landing in the corresponding direction, irrespective of the sequence in which the buttons are pressed. With this type of operation the car can be started only by means of a starting switch or button in the car.

Sill. Finished threshold of an elevator hoistway entrance unit.

Single automatic operation. The operation by means of one button in the car for each landing level served and one button at each landing is so arranged that if any car or landing button has been actuated, actuation of any other car or landing operation button will have no effect on the operation of the car until the response to the first button has been completed.

Single hoistway. Hoistway for a single elevator or dumbwaiter.

Slant elevators. Type of elevator used primarily at sloped hill sides and for conditions in the design of structures that provide space adjacent to the shaft which are required for other purposes.

Sling. Structural unit guided in the hoistway.

Spring buffer. Buffer that stores in a spring the kinetic energy of the descending car or counterweight.

Spring buffer load rating. Load required to compress the spring an amount equal to its stroke.

Spring buffer stroke. Distance the contact end of the spring can move under a compressive load until all coils are essentially in contact.

Structural load rating. Load design (minimum) rating of 100 lb/sq ft of exposed treadway for a moving walk (usual safety code requirements).

Subpost car frame. Frame all of whose members are located below the car platform.

Suspension rope equalizer. Device installed on an elevator car or counterweight to equalize automatically the tensions in the hoisting wire ropes.

Terminal landings. Highest and lowest standings served by the elevator.

Toe guard

1. Final fascia at the bottom terminal landing (passenger elevator).

2. Beveled metal furring plate below the door sill at each hoistway opening between the car and face of the hoistway well.

Top clearance of a car

1. The overhead or top clearance of the elevator car is the shortest vertical distance between the car crosshead and appurtenances and the nearest part of the overhead structure or any other obstruction when the car floor is level with the top landing.

2. The shortest vertical distance between the top of the car crosshead or between the top of the car where no crosshead is provided and the nearest part of the overhead structure or any other obstruction when the car floor is level with the top terminal landing.

3. Distance the car floor can travel above the level of the top terminal landing without any part of the car or devices attached thereto coming in contact with the overhead structure.

Top clearance of a counterweight. Shortest vertical distance between any part of the counterweight structure and the nearest part of the overhead structure or any other obstruction when the car is level with the bottom terminal landing.

Top counterweight clearance. Shortest vertical distance between any part of the counterweight structure and the nearest part of the overhead structure or any other obstruction when the car floor is level with the bottom terminal landing.

Top overtravel of an oil hydraulic elevator car. Distance provided for the car floor to travel above the level of the upper terminal landing until the car is stopped by the normal terminal stopping device.

Top overtravel of a traction elevator. Distance the car platform can travel above the level of the upper terminal landing until the counterweight buffer is fully compressed.

Top overtravel of the counterweight. Distance the counterweight can travel above its position when the car platform is level with the bottom terminal landing until the car buffer is fully compressed.

Top terminal elevator landing. Highest landing served by the elevator that is equipped with a hoistway door and hoistway door locking device which permits egress from the hoistway side.

Transfer floor. Floor at which a passenger leaves one car and transfers to another car, usually between low-rise and higher-rise elevators.

Travel. Vertical distance between the bottom terminal landing and the top terminal landing of an elevator, dumbwaiter, escalator, and private residence inclined lift.

Traveling cable. Cable made up of electrical conductors which provides the electrical connection between an elevator and a fixed outlet in the hoistway. The cable provides power for lights, fans, telephone, panel control, and similar equipment.

Travel or rise. Vertical distance between the bottom terminal landing and the top terminal landing.

Truckable sill. Part of a freight elevator hoistway door that forms the bridge between the hoistway sill and car sill.

Two-way automatic maintaining leveling device. Device that corrects the car level on both underrun and overrun and maintains the level during loading and unloading.

Two-way automatic nonmaintaining leveling device. Device that corrects the car level on both underrun and overrun but will not maintain the level loading and unloading.

Underslung car frame. Frame to which the hoisting rope fastenings or hoisting rope sheaves are attached at or below the car frame.

Waiting passenger indicator. Indicator that shows at which landings and for which direction the elevator hall stop or signal calls have been registered and are unanswered.

Weatherproof. Construct or protect in such a way that exposure to the weather will not interfere with the successful operation of the elevator.

Working pressure. Pressure measured at the cylinder of a hydraulic elevator when lifting the car and its rated load at rated speed.

ELEVATOR AND HOISTWAY DOORS AND GATES

Astragal. Small rubber strip used on the leading edges of center-opening passenger-type doors; buffer used on the leading edge of the upper section of a biparting freight-type door. The strip is made of special material. It also acts as a fire seal and is part of the labeled door construction.

Automatic opener. Hoistway power-operated door or gate. The door or gate is opened by power, the opening of the door being initiated by the arrival of the car at or near the landing. The closing of such a door or gate may be under the control of the elevator operator or may be automatic.

Biparting door. Vertically or horizontally sliding door consisting of two or more sections so arranged that the sections or groups of sections open away from each other and so interconnected that all sections operate simultaneously.

Biparting panels. Two-panel arrangement generally for freight elevators. The panels move vertically, one up and the other down. Variations of the biparting gates may range from a single panel moving up to a single panel moving down with intermediate one-third/two-third panel combinations.

Car door or gate

1. Door or gate in or on the elevator car that closes the opening ordinarily used for entrance and exit.
2. Door or gate in or on an elevator or dumbwaiter car.

Car door or gate electric contact

1. Electrical device whose function it is to prevent operation of the driving machine by the normal operating device unless the car door or gate is in the closed position.
2. Device whose purpose it is to open the control circuit or an auxiliary circuit, unless the car door or gate is in the closed position, and thus prevent operation of the elevator by the operating device in a direction to move the car away from the landing.

Car door or gate power closer. A device or assembly of devices that closes a manually-opened car door or gate by power other than by hand, gravity, spring, or the movement of the car.

Car door or gate switch. As applied to an elevator, means an electrical device, the function of which is to prevent operation of the driving machine by the normal operating device unless the car door or gate is in the closed position.

Car or hoistway door or gate. Sliding portion of the car or the hinged or sliding portion in the hoistway enclosure that closes the opening giving access to the car or to the landing.

Center-opening door panels. System of two panels that meet in the center and open by one panel's moving to the left and the other to the right.

Channel jamb and head. Steel channel frame at the sides and top of the freight-type elevator hoistway entrance.

Collapsing car gate. Gate that is distorted in opening and closing.

Door. Vertically or horizontally operated hollow metal panels of a passenger or freight elevator car entrance.

Door closer. Device, operated by gravity or other means, that will automatically close a door when released by the operator or by suitable automatic means.

Door or gate closer. Device that closes a manually opened hoistway door or car door or gate by means of a spring or gravity.

Door or gate (manually operated). A door or gate that is opened and closed by hand.

Door or gate power-operator. A device or assembly of devices that opens a hoistway door and/or a car door or gate by power other than by hand, gravity, springs or the movement of the car, and that closes them by power other than by hand gravity or the movement of the car.

Door or gate—self-closing door or gate. A manually-opened hoistway door or a car door or gate that closes when released.

Door unit contact system. System that requires that the hatchway door or gate, at which the elevator is standing, must be closed before the elevator can leave the landing, but that does not prevent the operation of the car if other doors in the hatchway are not closed.

Emergency release. Device to make the door or gate electric contacts or door interlocks inoperative in case of emergency.

Fully automatic door or gate. Vertically moving door or gate opened directly by the motion of the elevator car approaching the terminal landings and closed by gravity as the car leaves the landing.

Gate. Expanded metal closure of a freight car entrance whose operation is usually vertical. The car gate of a freight elevator may or may not have a sensitive reversing edge if power operated.

Hatchway door interlock. Device that prevents the car's moving away from a landing unless the hatchway door at that landing is locked in the closed position, and that prevents the opening of the hatchway door from the landing side, except by special key, unless the car is at rest within the landing zone or is coasting through the landing zone with its operating device in the stop position.

Hatchway door or gate. Hinged or sliding portion of the hatchway enclosure for access to the car at any landing.

Hatchway door or gate electric contact. Device to open the control circuit or an auxiliary circuit, unless the hatchway door or gate at which the car is standing is in the closed position, and thus prevent the car's moving away from the landing.

Hatchway door unit interlock system. The interlock system requires the hatchway door at which the elevator is standing to be locked in the closed position before the elevator can leave the landing, but it does not prevent the operation of the car if other doors in the hatchway are not locked.

Hoistway biparting door. Vertical or horizontal sliding door consisting of two or more sections so arranged that the sections or pairs of sections open away from each other, and so interconnected that both sections operate simultaneously.

Hoistway door. Landing door that completely fills the door opening giving access to the elevator or dumbwaiter car at any landing. It is of solid construction, with or without vision panels, regardless of design or method of operation.

Hoistway door combination mechanical and electrical contact or contact locks. Combination mechanical and electric device whose two related but entirely independent functions are to prevent the operation of the driving machine by the normal operating device unless the hoistway gate is in the closed position, and to lock the hoistway gate in the closed position and prevent its being opened from the landing side unless the car is in the landing zone.

Hoistway door interlock. Device whose purpose it is to prevent the opening of the hoistway door from the landing side, unless the car is at rest within the landing zone or is coasting through the landing zone with its operating device in the stop position, and to prevent the operation of the elevator machine by the operating device in a direction to move the car away from a landing, unless the hoistway door at which the car is stopping or is at rest locked in the closed position.

Hoistway-door interlock retiring cam device. A device that consists of a retractable cam with its actuating mechanism and that is entirely independent of the car-door or hoistway-door and power-operator.

Hoistway door or gate electric contact

1. Electrical device whose function it is to prevent the operation of the driving machine by the normal operating device unless the hoistway door or gate is in the closed position.
2. Device whose purpose it is to open the control circuit or an auxiliary circuit, unless the hoistway door or gate at which the car is standing is in the closed position, and thus prevent operation of the elevator by the operating device in a direction to move the car away from the landing.

Hoistway door or gate locking device. Device that secures a hoistway door or gate in the closed position and prevents it from being opened from the landing side except under certain specified conditions.

Hoistway enclosure. Fixed structure, consisting of vertical walls or partitions, that isolates the hoistway from all other parts of the building or from an adjacent hoistway and in which the hoistway doors and door assemblies are installed.

Hoistway entrance. Opening through the hoistway enclosure from a landing closed off by gates or doors.

Hoistway fully automatic door or gate. Vertically moving door or gate that is opened directly by the motion of the elevator car approaching the landing and closed by gravity as the car leaves the landing.

Hoistway gate. Gate that gives access to the elevator car at any landing and that consists of slats, bars, spindles, wire screen or expanded metal regardless of the method of operation.

Hoistway-gate separate mechanical lock. A mechanical device, the function of which is to lock a hoistway gate in the closed position after the car leaves a landing and prevent the gate from being opened from the landing side unless the car is within the landing zone.

Hoistway manually operated door or gate. Door or gate opened and closed by hand.

Hoistway power-operated door or gate. Door or gate opened and closed by power other than by hand, gravity, springs, or the movement of the car.

Hoistway semiautomatic door or gate. Door or gate that is opened manually and closes automatically as the car leaves the landing.

Horizontal slide type entrance. An entrance in which the panel(s) or door(s) slides horizontally.

Hoistway telescoping gate. Gate in which the sections slip together without distortion of the section.

Hoistway-unit system. A series of hoistway-door interlocks, hoistway-door electric contacts or hoistway-door combination mechanical locks and electric contacts, or a combination thereof, the function of which is to prevent operation of the driving machine by the normal operating device unless all hoistway-doors are in the closed position and, where so required by the code, are locked in the closed position.

Manually controlled door or gate. Hoistway power-operated door or gate, which is opened and closed by power, but whose door movement in each direction is controlled by the elevator operator.

Manually operated door or gate. Door or gate that is opened and closed by hand.

Pass-type biparting gates. Gates that run in a two-track system allowing all the gate panels to pass by the gate panel of the floor above and below. These gates are sometimes required owing to the short distance between floors.

Power-closed car door or gate. Door or gate closed by a car door or gate power closer or by a door or gate power operator.

Power-closed door or gate. Door or gate manually opened and closed by power other than by hand, gravity, springs, or the movement of the car.

Power closer for a car door or gate. Device or assembly of devices that closes a manually opened car door or gate by power other than by hand, gravity, springs, or the movement of the car.

Power-opened self-closing door or gate. Door or gate that is opened by power other than by hand, gravity, springs, or the movement of the car, and that when released by the operator is closed by energy stored during the opening operation.

Power-operated door or gate

1. Hoistway door and/or a car door or gate opened and closed by a door or gate power operator.
2. Device or assembly of devices that opens a hoistway door and/or car door or gate by power other than by hand, gravity, springs, or the movement of the car, and that closes them by power other than by hand, gravity, or the movement of the car.

Power-operated door or gate—automatically opened. Door or gate opened by means other than by hand, gravity, springs, or the movement of the car, the opening of the door being initiated by the arrival of the car at or near the landing. The closing of such a door or gate may be under the control of the operator, or it may be automatic.

Power-operated door or gate—manually controlled. Door or gate that is opened and closed by power other than by hand, gravity, springs, or the movement of the car, the door movement in each direction being controlled by the operator.

Regular type biparting gates. Biparting gates run in a continuous vertical, in line track.

Self-closing door or gate. Manually opened hoistway door and/or car door or gate that closes when released.

Semiautomatic gate. Gate that is opened manually and that closes automatically as the car leaves the landing.

Side jamb. Side of the entrance where the door stops in its opening motion.

Single-speed door panel. Single panel door moving horizontally to the left or right.

Strike jamb. Side of the entrance where the door stops its closing motion.

Three-speed door panels. Three telescoping door panels moving horizontally to either the left or right.

Two-speed center opening. Combination of the two-speed and center opening systems involving four horizontally moving panels with two telescoping to the right and two to the left.

Two-speed door panels. Two telescoping door panels moving horizontally to either the left or right.

ELEVATOR TYPES AND MACHINE TYPES

Auxiliary power elevator. Elevator having a source of mechanical power in common with other machinery.

Belt-driven elevator. Elevator in which the driving mechanism is connected to a prime mover by a single or multiple belts, and where multiple belts are used, the direction of motion of the elevator car is changed without reversal of the prime mover.

Car elevator. Load-carrying unit, including its platform, car frame, enclosure, and car door or gate.

Carriage elevator. Elevator supported by cables attached to the platform at four or more points in such a manner that the supporting cables are relied on to maintain the platform substantially level.

Chain drive machine. Indirect-drive machine having a chain as the connecting means.

Chain-driven elevator. Elevator having its machine connected to a reversible motor, engine, or turbine by a chain.

Cylinder well hydraulic elevator. This type of elevator requires a well the depth of the hydraulic piston, which is the height of the travel distance.

Direct-drive machine

1. Electric driving machine whose motor is directly connected mechanically to the driving sheave, drum, or shaft without the use of belts or chain either with or without intermediate gears.
2. Machine whose power is transmitted directly to the driving sheave or sheaves without intermediate mechanism or gears.

Direct-plunger driving machine. Machine in which the energy is applied by a plunger or piston directly attached to the car frame or platform and that operates in a cylinder under hydraulic pressure.

Direct-plunger elevator. Hydraulic elevator having a plunger or piston directly attached to the car frame or platform.

Double-belted elevator. Auxiliary power elevator whose direction of travel is changed without reversal of the prime mover.

Driving machine. Power unit which applies the energy necessary to raise and lower an elevator or dumbwaiter car or to drive an escalator, moving stairway, or an inclined lift.

Dumbwaiter. Hoisting and lowering mechanism equipped with a car, which moves in a substantially vertical direction, whose floor area does not usually exceed 9 sq ft, whose compartment height does not usually exceed 4 ft, whose capacity does not usually exceed 500 lb, and which is used exclusively for carrying materials.

Electric driving machine. Machine whose energy is applied by an electric motor. It includes the motor, brake, and driving sheave or drum, together with its connecting gearing, belt, or chain, if any.

Electric elevator. Elevator operated by an electric motor directly applied to the elevator machinery.

Electric geared traction elevator. System designed to operate in the range of 100 to 350 fpm.

Electric gearless traction elevator. Systems are available in preengineered units with speeds of 500 to 1200 ft per minute. Systems with greater speeds are available. Gearless elevators provide a smoother ride.

Electrohydraulic elevator. Direct-plunger elevator where liquid is pumped under pressure directly into the cylinder by a pump driven by an electric motor.

Elevator

1. Hoisting and lowering mechanism equipped with a car or platform that moves in guides in a substantially vertical direction and that serves two or more floors of a building or structure.
2. Hoisting and lowering mechanism equipped with a car or platform that moves in guides in a substantially vertical direction and that is designed to carry passengers and/or freight; hosting mechanism, such as a portable hoist or tiering machine, when used to elevate material between two or more permanent levels and when the device is fixed in a permanent location.
3. Device within or in connection with a building used for carrying persons or things upward or downward; it includes escalators, moving stairs, and similar devices.
4. Hoisting and lowering mechanism equipped with a car or platform that moves in guides in a substantially vertical direction, but not including tiering or piling machines which operate within one story, or endless belts, conveyors, chains, buckets, or similar devices used for the purpose of elevating materials.
5. Machinery, construction, apparatus, and equipment used in raising and lowering a car, cage, or platform either vertically or substantially vertically on or between permanent rails or guides, including all elevators, power dumbwaiters, escalators, gravity elevators, and other lifting or lowering apparatus permanently installed on or between rails or guides, but not including hand-operated dumbwaiters, manlifts of the platform type with a platform area not exceeding 900 sq in., construction hoists, or other similar temporary lifting or lowering apparatus.

Elevator machinery. Machinery and its equipment used in raising and lowering the elevator car or platform.

Freight elevator

1. Elevator designed and used for the carrying of freight and such persons only as are necessary for its safe operation or the handling of the freight carried by it.
2. Elevator normally used for carrying freight and on which, in addition to the operator, only employees in pursuit of their duties and with permission from their employer are allowed to ride.

Geared-drive machine. Direct-drive machine in which the energy is transmitted from the motor to the driving sheave, drum, or shaft through gearing.

Geared traction machine. Geared-drive traction machine.

Gearless traction machine. Traction machine without intermediate gearing that has the traction sheave and the brake drum mounted directly on the motor shaft.

Generator field control. System of control accomplished by the use of an individual generator for each elevator or dumbwaiter wherein the voltage applied to the driving-machine motor is adjusted by varying the strength and direction of the generator field.

Grade-level elevator. Freight elevator whose hoistway is located partially outside the building, in an area not used by people or vehicles as a place of travel, and having no opening into the building at the upper terminal landing.

Hand elevator. Elevator utilizing manual energy to move the car.

Hoisting machine. Power unit that applies the energy necessary to raise or lower an elevator.

Holeless hydraulic elevator. This type of elevator has a telescoping hydraulic piston as the driving mechanism, eliminating the need for a cylinder well.

Hydraulic driving machine. Machine in which the energy is applied by means of a liquid under pressure in a cylinder equipped with a plunger or piston.

Hydraulic elevator. Power elevator where the energy is applied by means of a liquid under pressure in a cylinder equipped with a plunger or piston.

Hydroelectric elevator. Direct-plunger elevator where liquid is pumped under pressure directly into the cylinder by a pump driven by an electric motor.

Inclined lift. Hoisting and lowering mechanism equipped with a car or platform that moves in guides installed at a degree not substantially vertical, that is designed to carry passengers and or freight, and that serves two or more landings of a building or structure.

Indirect-drive machine. Electric driving machine, the motor of which is connected indirectly to the drive sheave, drum, or shaft by means of a belt or chain through intermediate gears.

Light-duty freight elevators. Freight capacity of 1,000 to 2,500 lb, may utilize hydraulic or traction drives.

Machine. Machinery and its equipment used in raising or lowering the car or platform.

Material-handling elevators. Hoisting and lowering mechanism equipped with a car platform used in conjunction with manual or automatic loading and/or unloading devices, that moves in guides in a substantially vertical direction, and whose travel exceeds 56 in.

Miscellaneous hoisting and elevating equipment. All power-operated hoisting and elevating equipment for raising, lowering, and moving persons or merchandise from one level to another, including inclined elevators, slings and hooks, tiering and piling machines not permanently located in a fixed position, mine elevators, skip hoists for blast furnaces, stage and orchestra lifts, lift bridges, temporary builders' hoists, and similar equipment.

Motor generator set. Combination unit that provides variable voltage from ac service for operation of a dc motor driving unit on an elevator machine.

Passenger elevator. Elevator used primarily to carry persons other than the operator and persons necessary for loading and unloading.

Private residence elevator

1. Power passenger electric elevator, installed in a private residence, that has a rated load usually not in excess of 700 lb, a rated speed usually not in excess of 50 ft/minute, a net inside platform area usually not in excess of 12 sq ft, and a rise usually not in excess of 50 ft.
2. Power passenger elevator serving a single family, installed in a dwelling, and having a rated capacity of usually not more than 700 lb and a rated speed of usually not more than 50 ft/minute.

Plunger elevator. Hydraulic elevator having a ram or plunger directly attached to the underside of the car platform.

Power elevator

1. Elevator utilizing means other than gravity or manual energy to move the car.
2. Elevator in which the motion of the car is obtained through the application of electrical energy.

Roped hydraulic driving machine. Energy is applied by a piston connected to the car with wire ropes, which operates in a cylinder under hydraulic pressure. The machine includes the cylinder, piston, and multiplying sheaves, if any, and their guides.

Roped hydraulic elevator. Hydraulic elevator having its piston connected to the car with wire ropes.

Roped-geared hydraulic elevator. Movement of the car is obtained by multiplying the travel of a piston or ram by a system of sheaves over which the hoisting ropes operate.

Screw machine. Electric driving machine, the motor of which raises and lowers a vertical screw through a nut, with or without suitable gearing, and in which the upper end of the screw is connected directly to the car frame or platform. The machine may be of direct- or indirect-drive type.

Service elevators. Usually standard passenger elevators that are modified for service use.

Sidewalk elevator

1. Freight elevator with a hatch opening at sidewalk level, located partially or wholly outside of the structure served and with no opening into the structure at its upper terminal landing.
2. Freight elevator that operates between a sidewalk or other area exterior to the building and floor levels inside the building below such area that has no landing opening into the building at its upper limit of travel, and that is not used to carry automobiles.
3. Freight elevator having a speed of not more than 50 ft/minute and having the top landing not more

than 4 ft above grade level at the point where the elevator is located. The platform of the elevator is suspended or supported at or below the platform level and in such a manner as will not permit tipping of the platform.

Slack rope switch. Device that automatically causes the electric power to be removed from the elevator driving machine motor and brake when the hoisting ropes of a winding drum machine, become slack.

Special hoisting and conveying equipment. Manually or power-operated hoisting, lowering, or conveying mechanisms, other than elevators, moving stairways, or dumbwaiters, for the transport of persons or freight in a vertical, inclined, or horizontal direction on one floor or in successive floors.

Spur-geared machine. Power is transmitted to the driving sheaves or drum through spur gearing.

Traction machine

1. Direct-drive machine in which the motion of a car is obtained through friction between the suspension ropes and a traction sheave.

2. Machine in which the movement of the car and counterweight is obtained by means of traction between the driving drum, sheave or sheaves, and the hoisting cables.

Two-speed alternating-current control. Two-speed driving machine induction motor arranged to run at two different synchronous speeds by connecting the motor windings so as to obtain a different number of poles.

Undercounter dumbwaiter. Dumbwaiter that has its top terminal landing located underneath a counter and that serves only this landing and the bottom terminal landing.

Winding-drum machine. Geared-drive machine in which the hoisting ropes are fastened to and wind on a drum.

Worm-geared machine. Direct-drive machine in which the energy from the motor is transmitted to the driving sheave or drum through worm gearing.

GENERAL USE CONVEYANCE EQUIPMENT AND SYSTEMS

Amusement device

1. Device or structure open to the public by which persons are conveyed or moved in a manner intended to amuse, divert, or excite.

2. Contrivance or structure for use by the public by which persons are conveyed in an unusual manner for diversion.

3. Mechanically operated device used to convey persons in any direction as a form of amusement.

4. Merry-go-rounds, revolving wheels, shooting the chutes, giant swings, etc., carrying passengers.

5. Manually or power-operated devices used to convey persons in any direction as a form of amusement.

Automated people movers. System used in airports, shopping centers, college campuses, and other stretched-out facilities that need automatic transportation. Vehicles are controlled automatically and are programmed to operate on a schedule or on a demand mode. Vehicles operate continuously, stopping at each station for a predetermined dwell time, before moving on to the next station. Vehicles operate on a electric rail at the bottom of shuttles. Passenger capacity varies with size of shuttles and varies from 24 to 124 persons.

Automobile parking lift. Mechanical device for parking automobiles by movement in any direction.

Automotive lift. Fixed, mechanized device for raising an entire motor vehicle for maintenance or repair purposes above the ground or grade floor level, but not through successive floors of the building or structure.

Belt conveyor system. Power-driven conveyor consisting of heavy-duty combination rubber and fiber belts. Idlers support the belt and can be adjustable. Used primarily at airport baggage pick ups, manufacturing assembly plants, and for food inspection prior to wrapping. Has an extensive speed range, which can be controlled.

Bolster. Supporting member of the platform on which the lifting piston acts.

Circular conveyor lift. Used primarily to transport cartons between operating levels and between work stations within a level. Installation depends on height and travel of feed and exit conveyors.

Closed-type manlift. Cup-shaped device, open at the top in the direction of travel of the step for which it is to be used and closed at the bottom, into which the passenger may place his fingers.

Console lift. A section of the floor area of a theater or auditorium that can be raised and lowered.

Controls. Devices for actuating the up and down movement of a platform.

Counterbalance truck forklift. Electrically operated, by industrial, batteries, used to move large volumes of material where maneuvering area is not limited. 17 ft 9 in. maximum lifting height, 2,000 to 15,000 lbs-load capacities, travels up to 11 mph.

Conveyor

1. System of machinery and manual or mechanized devices, other than elevator and dumbwaiter equipment, consisting of belts, chains, rollers, buckets, aprons, slides and chutes, and other miscellaneous equipment for hoisting, lowering, and transporting materials and merchandise in packages or in bulk in any direction in a building or structure.

2. Mechanical device to transport material from one point to another, often continuously.

Cylinder assembly. Hydraulic cylinder with a piston inside.

Cylinder platform lift. Used to facilitate lifting bulky loads in crates or otherwise from the floor or ground level to vehicles. Designed for loads ranging from 2,000 to 30,000 lb, operated by hydraulic power cylinder.

Electric hand pallet truck forklift. Used to transport unitized loads for loading docks and production areas. 6 to 7 in. maximum lifting height, 1,500 to 6,000 lb load capacity, travel speed up to 5 mph, truck weight 1,000 to 2,000 lb.

Electrohydraulic type lift. Type in which the operating liquid is pumped directly into the lifting cylinder by an electrically driven pump. Lowering may be by either a mechanically or electrically operated valve.

Elevator parking device. Electrical or mechanical device whose function it is to permit the opening from the landing side of the hoistway door at any landing when the car is within the landing zone of that landing. The device may also be used to close the door.

Escalator

1. Power-driven inclined assembly of parts and equipment consisting of a horizontal tread-type operating in the direction of travel, angle of inclination within the limitation from horizontal and vertical travel contained in a maximum distance of 35 ft.

2. Stairway or incline arranged like an endless belt so that the steps or treads ascend or descend continuously while the device is in service.

3. Moving stairway for transporting passengers from one level to another.

Freight platform hoist. Hoist structurally installed in a fixed place, equipped with a direct plunger and a platform, and used for the purpose of elevating or lowering material or freight from one level to another.

Full-hydraulic lift. Automotive life of the plunger type that employs a liquid under pressure as the direct lifting and load-sustaining agent. Such a lift is so designed and constructed that the full weight of the load and lifting assembly rest on a continuous column of liquid, which extends from the cylinder to the liquid control valve.

Hinged loading ramps. Hinged loading ramps may be made in full-hydraulic, hydropneumatic, or pneumatic types. In hydropneumatic or pneumatic types the air is used as the positioning agent only, and means other than the direct air pressure are used to support the platform when loads are rolled across it.

Hinged loading ramps (nonportable type). The nonportable type of hinged loading ramp is either mechanical or hydraulic, hand or power operated, and is used for spanning gaps and/or adjusting heights between loading surface and carrier or between loading surface and loading surface. It is not used for marine applications.

Hydraulic leveler. Leveler of which a liquid under pressure is the direct lifting and load-sustaining agent.

Hydropneumatic leveler (semihydraulic). Type in which compressed air is employed as the primary lifting agent, acting continuously against a column of liquid to provide the lifting effort. Lowering is accomplished by release of air pressure.

Inclined lift. A hoisting and lowering mechanism equipped with a car or platform that moves in guides, installed at a degree not substantially vertical, that is designed to carry passengers and freight.

Industrial lift (material lift). Nonportable, power-operated raising or lowering device for transporting freight vertically, operating entirely within one story of the building or structure.

Lip. Hinged extension at the outer (truck) end of a platform.

Loading dock. Structure for receiving items unloaded from parked motor vehicles or railroad cars preparatory to moving said items to interior locations or reloading them on other motor vehicles or railroad cars.

Loading ramp. Hinged nonportable device, either mechanical or hydraulic, hand or power operated, used for spanning gaps and/or adjusting heights between loading surface and carrier or between loading surface and loading surface. It is not used for marine applications.

Load pad. Steel plate used to transmit a load from cylinder base to foundation.

Manlift. Device consisting of a power-driven endless belt provided with steps or platforms and handholds attached to it for the transportation of persons in a vertical position through successive floors or levels of a building or structure.

Material lift (industrial lift). Power-operated raising or lowering device for transporting freight vertically, operating entirely within one story of a building of structure.

Mechanical amusement device. Device designed or used to move a person or to permit the movement of a person by mechanical means in any direction for amusement and operated within a space or over a route devoted exclusively to such use.

Mechanical life. Automotive lift so designed that the motive power is transmitted to the lifting frame by mechanical means. There are three principal types: cable-and-drum, rack-and-pinion, and screw type.

Metal roller conveyor system. Skate wheel type, gravity driven, designed for light or heavy loading, and in sizes to accommodate products or materials to be delivered.

Moving stairway (escalator). Moving inclined continuous stairway or runway used for raising or lowering passengers.

Open-type manlift. Manlift that has a handgrip surface fully exposed and capable of being encircled by the passenger's fingers.

Order picker truck forklift. Electrically operated by industrial batteries, designed to allow access to multiple level pick slots. 30 ft 6 in. maximum lift height, 1,500 to 3,000 lb load capacity, travels up to 5 mph.

Platform assembly. Load deck including hinge and lip.

Platform lift. Raising or lowering mechanism in a fixed position, designed for loading or unloading, with a travel up to 5 ft but not to exceed 7 ft, equipped with an open platform or a platform hinged at one end.

Pneumatic lift. Type in which compressed air is employed directly as the lifting agent to provide the lifting effort.

Pneumatic tube down discharge terminals. System where the down discharge terminals can be recessed in walls with only dispatching and receiving doors exposed. These terminals can be used for all automatic selective systems.

Pneumatic tube pedestal terminal. Pedestal base unit located in convenient location to receive tube system. Pedestals vary in size depending on type of system and operation.

Pneumatic tube vacuum—combination line type. System dispatches carriers from the central station to all substations via separate lines, but return lines are common. Computer controlled systems consist of control center stations.

Pneumatic tube vacuum—pressure type. System utilizes both vacuum and pressure and provides fast service. Its use is restricted to mercantile houses, drug, grocery, and meat packing plants, and to similar buildings where there are great distances between departments and buildings. This is a combination line system.

Pneumatic tube vacuum system (twin line). System may dispatch carriers from all stations simultaneously with continuous, nearly unlimited transactions. Lines exposed to weather or refrigerated areas must be protected and insulated to prevent condensation.

Private residence inclined lift. Power passenger lift installed on a stairway in a private residence for raising and lowering persons from one floor to another.

Pump unit. Package consisting of electric motor, hydraulic pump, and oil reservoir.

Reach truck forklift. Forklift, cylinder operated. 20 ft maximum lift height, 2,000 to 6,000 lb load capacity, travels up to 7 mph.

Residential wheelchair lifts. Usually installed adjacent to a stairway and operated off standard household current. Platform is screw driven, lifted along a threaded rod that is rotated by the power unit.

Residential chair lift. Usually installed adjacent to a stairway and operated off standard household current. Chair is mounted on a support that is guided by a guide rail and a track for the chair. Power unit that controls the chair movement is under the armrest, the power unit that propels the chair is mounted on the back of the chair or elsewhere. Call boxes are at the top and bottom of chair lift travel sequence.

Segmental moving surface. Power driven metal pan slats, usually used in airports for assembly and delivery of baggage from the unloading station to the unloading area, in the airport baggage public area. Moves in circles or around curves.

Scissor lift. Lift used to raise and lower unit loads to delivery vehicles from the ground floor or levels above or below the vehicles. Operated by hydraulic cylinders.

Semihydraulic (hydropneumatic) lift. Automotive lift of the plunger type that employs compressed air as the primary lifting and loading-sustaining agent; such compressed air acts continuously against a column of liquid to provide the lifting and load-sustaining effort.

Sideloader truck forklift. Electrically operated by industrial batteries, used for narrow aisle operations. 30 ft maximum lifting height, 2,000 to 10,000 lb load capacity, travel speed up to approximately 6 mph. Truck weight 9,000 to 12,000 lb.

Skirt. Vertical shield at three sides of a ramp to create a safe condition by closing the space between the platform and ramp when in the raised position.

Special hoisting, elevating, and conveying equipment. Permanently or semipermanently located device, manually or power operated, used for raising, lowering, and moving material, equipment, or persons, other than elevators, dumbwaiters, and moving stairways, and including but not limited to amusement devices; automotive lifts; inclined elevators for one or two persons or for materials; manlifts; builders' material hoists; temporary elevators to transport workers during building construction; belt, bucket, scoop, roller, or similarly inclined freight conveyors; tiering or piling machines; skip hoists; stage and orchestra lifts; lift bridges; elevators of capacity exceeding 30,000 lb and platform areas exceeding 300 sq ft when suspended by cables near each corner of the hoistway and at additional positions; and similar equipment.

Split-rail switch for a manlift. Electric limit switch operated mechanically by the rollers on the manlift steps. The switch consists of an additional hinged or "split" rail mounted on the regular guide rail, over which the step rollers pass. It is spring-loaded in the "split" position. If the step supports no load, the rollers will "bump" over the switch; if a loaded step should pass over the section, the split rail will be forced straight, tripping the switch and opening the electrical circuit.

Straddle truck forklift. Cylinder operated forklift. Approximately 20 ft maximum lift height, 2,000 to 6,000 lb load capacity, travels up to 6 mph.

Travel. Vertical movement of a ramp, measured at the lip end; also vertical movement of a lift.

Vertical pallet lift. Lift used to transport loads from one level to another within a conveyor system or for manual loading or unloading at each level. Lift speed 20 fpm. Installation floor to floor with platform either flush with floor or above floor level, depending on loading or unloading operations.

MOVING WALKS AND MOVING RAMPS

Belt pallet-type moving walk. Moving walk with a series of connected and power-driven pallets to which a continuous belt treadway is fastened.

Belt type. Moving walk or moving ramp with a power-driven continuous belt treadway.

Grooving. Treadway surface of a moving walk grooved in a direction parallel to its travel for the purpose of meshing with combplates at the landing.

Landing. Stationary area at the entrance to or exit from a moving walk or moving walk system.

Moving ramp. Moving ramp having a slope or angle exceeding 3° with the horizontal.

Moving sidewalk. Type of passenger-carrying device, other than steps, on which passengers stand or walk and in which the passenger's carrying surface remains parallel to its direction of motion and its movement is uninterrupted. The angle of inclination does not exceed 15° from the horizontal.

Moving walk. Moving walk having a slope or angle not exceeding 3° with the horizontal.

Moving walk and moving ramp landing. Stationary area at the entrance to or exit from a moving walk or moving ramp.

Moving walk or moving ramp. Type of passenger-carrying treadway on which passengers stand or walk and in which the passenger-carrying surface remains parallel to its direction of travel and its movement uninterrupted.

Moving walk system. Series of moving walks in an end-to-end or side-by-side relationship with no landing between treadways.

Pallet. One of a series of rigid platforms that together form an articulated treadway or the support for a continuous treadway (moving walk or moving ramp).

Pallet-type moving walk or moving ramp. Series of connected and power-driven pallets that together constitute the treadway.

Roller type moving walk or moving ramp. Moving walk or ramp that is belt supported by a succession of rollers with their axes at right angles to the direction of the treadway motion.

Slider-bed-type moving walk or moving ramp. Treadway sliding on the supporting surface.

Slope of a moving walk. Angle that the treadway makes with the horizontal.

Threshold comb. Toothed portion of a threshold plate designed to mesh with a grooved treadway surface (moving walk or moving ramp).

Threshold plate. Portion at the entrance or exit to the treadway consisting of one or more stationary or slightly movable plates (moving walk or moving ramp).

Treadway. Exposed passenger-carrying member of a moving walk or moving ramp.

Width. The width of a moving walk is the exposed width of the treadway.

POWER WORK PLATFORMS

Angulated roping. System of platform suspension in which the upper wire rope sheaves or suspension points are closer to the plane of the building face than the corresponding attachment points on the platform, thus causing the platform to press against the face of the building during its vertical travel.

Babbitted fastenings. Wire rope attachments in which the ends of the wire strands are bent back and are held in a tapered socket by means of poured molten babbitt metal.

Brake (disk type). Brake in which the holding effect is obtained by frictional resistance between one or more faces of disks keyed to the rotating member to be held and fixed disks keyed to the stationary or housing member (the pressure between the disks being applied axially).

Brake (self-energizing ban type). Essentially unidirectional brake in which the holding effect is obtained by the snubbing action of a flexible band wrapped about a cylindrical wheel or drum affixed to the rotating member to be held, the connections and linkages being so arranged that the motion of the brake wheel or drum will act to increase the tension or holding force of the band.

Brake (shoe type). Brake in which the holding effect is obtained by applying the direct pressure of two or more segmental friction elements held to a stationary member against a cylindrical wheel or drum affixed to the rotating member to be held.

Building face rollers. Specialized form of guide roller designed to contact a portion of the outer face or wall structure of the building and to assist in stabilizing the operators' platform during vertical travel.

Continuous pressure. Operation by means of buttons or switches, any one of which may be used to control the movement of the working platform or roof car, only as long as the button or switch is manually maintained in the actuating position.

Control. System governing starting, stopping, direction, acceleration, speed, and retardation of moving members.

Controller. Device or group of devices, usually contained in a single enclosure, that serves to control in some predetermined manner the apparatus to which it is connected.

Direction relay. Electrically energized contactor responsive to an initiating control circuit, which in turn causes a moving member to travel in a particular direction.

Electrical ground. Conducting connection between an electrical circuit or equipment and the earth or some conducting body that serves in place of the earth.

Guide roller. Rotating, bearing-mounted, generally cylindrical member, operating separately or as part of a guide shoe assembly, attached to the platform, and providing rolling contact with building guideways or other building contact members.

Guide shoe. Assembly of rollers—slide members or the equivalent—attached as a unit to the operators' platform and designed to engage with the building members provided for the vertical guidance of the operators' platform.

Interlock. Device actuated by the operation of some other device with which it is directly associated to govern succeeding operations of the same or allied devices.

Operating device. Pushbutton, lever, or other manual device used to actuate a control.

Potential for vertical travel relay. Electrically energized contactor responsive to the initiating control circuit, which in turn controls the operation of a moving member in both directions. The relay usually operates in conjunction with direction relays.

Powered platform. Equipment to provide access to the exterior of a building for maintenance, consisting of a suspended power-operated working platform, a roof car, or other suspension means and the requisite operating and control devices.

Rated load. Combined weight of employees, tools, equipment, and other material that the working platform is designed and installed to lift.

Roof car. Structure for the suspension of a working platform, providing for its horizontal movement to working positions.

Roof-powered platform. Powered platform having the raising and lowering mechanism located on a roof car.

Self-powered platform. Powered platform having the raising and lowering mechanism located on the working platform.

Traveling cable. Cable made up of electrical or communication conductors, or both, and providing electrical connection between the working platform and the roof car or other fixed point.

Weatherproof. Term applied to equipment so constructed or protected that exposure to the weather will not interfere with its proper operation.

Working platform. Suspended structure arranged for vertical travel which provides access to the exterior of the building or structure.

Zinc fastenings. Wire rope attachments in which the splayed or fanned wire ends are held in a tapered socket by means of poured molten zinc.

RUNWAY CRANES AND HOISTS

Automatic crane. Crane that when activated operates through a preset cycle or cycles.

Auxiliary hoist. Supplemental hoisting unit of lighter capacity and of higher speed than usually provided for the main hoist.

Brake. Device used for retarding or stopping motion by means of friction or power.

Bridge. Part of a crane, consisting of girders, trucks, end ties, footwalks, and drive mechanism, that carries the trolley or trolleys.

Bridge conductors. Electrical conductors located along the bridge structure of a crane to provide power to the trolley.

Bridge travel. Crane movement in a direction parallel to the crane runway.

Boom guards. Cage-type boom guards, insulating links, or proximity warning devices used on cranes.

Bumper (buffer). Energy-absorbing device for reducing impact when a moving crane or trolley reaches the end of its permitted travel or when two moving cranes or trolleys come in contact.

Cab. Operator's compartment on a crane.

Cab-operated crane. Crane controlled by an operator in a cab located on the bridge or trolley.

Cantiliever gantry crane. Gantry or semigantry crane in which the bridge girders or trusses extend transversely beyond the crane runway on one or both sides.

Choker. Sling hitch that is self-tightening.

Clearance. Distance from any part of the crane to a point of the nearest obstruction.

Clearance between parallel cranes. When the runways of two cranes are parallel, and there are no intervening walls or structure, adequate clearance is provided and maintained between the two bridges.

Clearance from obstruction. Minimum clearances of 3 in. overhead and 2 in. laterally are provided and maintained between the crane and obstructions.

Control braking. Method of controlling the crane motor speed when in an overhauling condition.

Countertorque. Method of control by which the power to the motor is reversed to develop torque in the opposite direction.

Crane. Machine for lifting and lowering a load and moving it horizontally, with the hoisting mechanism an integral part of the machine. Cranes, whether fixed or mobile, are driven manually or by power.

Current collectors. Contacting devices for collecting current from runway or bridge conductors.

Double beam crane. Used to handle heavy loads in storage warehouses and manufacturing plants. The crane has two-direction travel, vertical lift over the entire length of the beams and rails. Load capacities are based on spans between rails and working heights.

Drag brake. Brake that provides retarding force without external control.

Drift point. Point on a travel motion controller that releases the brake while the motor is not energized. This allows for coasting before the brake is set.

Drum. Cylindrical member around which the ropes are wound for raising or lowering the load.

Dynamic. Term describing the method of controlling crane motor speeds when in the overhauling condition to provide a retarding force.

Emergency stop switch. Manually or automatically operated electric switch to cut off electric power independently of the regular operating controls.

Equalizer. Device that compensates for unequal length or stretch of a rope.

Exposed. Capable of being contacted inadvertently. Term applied to hazardous objects not adequately guarded or isolated.

Fail-safe. Provision designed to automatically stop or safely control any motion in which a malfunction occurs.

Fire extinguishers. Carbon dioxide, dry chemical, or an equivalent fire extinguisher is kept in the cab or in the vicinity of the crane.

Floor-operated crane. Crane that is pendent or nonconductive rope controlled by an operator on the floor or an independent platform.

Footwalk. Walkway with a handrail, attached to the bridge or trolley for access purposes.

Gantry crane. Crane similar to an overhead crane except that the bridge for carrying the trolley or trolleys is rigidly supported on two or more legs running on fixed rails or other runway.

Hoist. Apparatus that may be a part of a crane, exerting a force for lifting or lowering.

Hoist chain. Load-bearing chain in a hoist.

Hoist motion. Motion of a crane that raises and lowers a load.

Holding brake. Brake that automatically prevents motion when the power is off.

Hot-metal-handling crane. Overhead crane used for transporting or pouring molten material.

Limit switch. Switch that is operated by some part or motion of a power-driven machine or equipment to alter the electric circuit associated with the machine or equipment.

Load. Total superimposed weight on the load block or hook.

Load block. Assembly of hook or shackle, swivel, bearing, sheaves, pins, and frame suspended by the hoisting rope.

Magnet. Electromagnetic device carried on a crane hook to pick up loads magnetically.

Main hoist. Hoist mechanism provided for lifting the maximum rated load.

Main switch. Switch controlling the entire power supply to the crane.

Main trolley. Trolley having an operator's cab attached thereto.

Master switch. Switch that dominates the operation of contactors, relays, or other remotely operated devices.

Mechanical control. Method of control by friction.

Overhead crane. Crane with a movable bridge carrying a movable or fixed hoisting mechanism and traveling on an overhead fixed runway structure.

Overhead wires. Any overhead wire is considered to be an energized line until the person owning such line or the electric utility authorities indicate that it is not an energized line.

Personnel safety. Passageways or walkways are provided obstructions and placed so that the safety of personnel will not be jeopardized by movements of the crane.

Power-operated crane. Crane whose mechanism is driven by electric, air, hydraulic, or internal combustion means.

Pulpit-operated crane. Crane operated from a fixed operator station not attached to the crane.

Rated load. Maximum load for which a crane or individual hoist is designed and built by the manufacturer and shown on the equipment nameplate(s).

Rated load marking. The rated load of the crane is plainly marked on each side of the crane, and if the crane has more than one hoisting unit, each hoist has rated load marked on it or its load block, and this marking is clearly legible from the ground or floor.

Regenerative. Term describing the form of dynamic braking in which the electrical energy generated is fed back into the power system.

Remote-operated crane. Crane controlled by an operator not in a pulpit or in the cab attached to the crane, by remote control or any method other than pendency or rope control.

Rope. Usually a wire rope.

Running sheave. Sheave that rotates as the load block is raised or lowered.

Runway. Assembly of rails, beams, girders, brackets, and framework on which the crane or trolley travels.

Runway conductors. Electrical conductors located along a crane runway to provide power to the crane.

Semigantry crane. Gantry crane with one end of the bridge rigidly supported on one or more legs that run on a fixed rail or runway, the other end of the bridge being supported by a truck running on an elevated rail or runway.

Side pull. That portion of the hoist pull acting horizontally when the hoist lines are not operated vertically.

Span. Horizontal distance center to center of runway rails.

Spring return controller. Controller that when released will return automatically to a neutral position.

Stacker crane. Stacker cranes allow storage retrieval above conventional forklift truck heights. Cranes can be remote computer controlled. Load capacity depends on overall lifting height and travel speed.

Standby crane. Crane that is not in regular service but that is used occasionally or intermittently as required.

Stop. Device to limit travel of a trolley or crane bridge. The device is normally attached to a fixed structure and normally does not have energy-absorbing ability.

Storage bridge crane. Gantry-type crane of long span, usually used for bulk storage of material. The bridge girders or trusses are rigidly or nonrigidly supported on one or more legs. It may have one or more fixed or hinged cantilever ends.

Switch. Device for making, breaking, or changing the connections in an electric circuit.

Trolley. Unit that travels on the bridge rails and carries the hoisting mechanism.

Trolley travel. Trolley movement at right angles to the crane runway.

Truck. Unit, consisting of a frame, wheels, bearings, and axles, that supports the bridge girders or trolleys.

Wall crane. Crane having a jib, with or without a trolley, and supported from a sidewall or line of columns of a building. It is a traveling type and operates on a runway attached to the sidewall or columns.

Wind indicators and rail clamps (exterior cranes). Wind-indicating device that gives a visible or audible alarm to the bridge operator at a predetermined wind velocity, with automatic rail clamps.

AIR-CONDITIONING AND REFRIGERATION SYSTEMS AND EQUIPMENT

Absolute humidity

1. Grains of moisture in the air per cubic foot.
2. Weight of water vapor per unit volume (pounds per cubic foot or grams per cubic centimeter).

Absolute pressure

1. Pressure measured above absolute vacuum.
2. Gauge pressure plus 14.7-psi atmospheric pressure.
3. Total pressure measured from absolute zero, that is, from an absolute vacuum. It equals the sum of the gauge pressure and barometric (atmospheric) pressure. Gauge pressure is expressed in pounds per square inch.

Absolute temperature

1. Temperature measured from absolute zero.
2. Degrees Fahrenheit plus 459.69 or degrees Celsius plus 273.16. Absolute temperatures are referred to on the Fahrenheit scale as "degrees Rankine" and on the Celsius scale as "degrees Kelvin."

Absolute zero

1. Temperature where all molecular motion ceases ($-459.7°$F).
2. Zero from which absolute temperature is reckoned, approximately $-273.2°$C or $-459.7°$F.
3. The temperature at which no heat exists in a substance.

Absorber. Equipment in which a gas is absorbed by contact with a liquid.

Absorber (adsorber). Part of the low side of an absorption system used for absorbing (adsorbing) vapor refrigerant.

Absorption. Occurrence during which the absorbent undergoes a physical or chemical change (or both).

Absorption refrigeration

1. Refrigeration process whereby a secondary fluid absorbs the refrigerant, and in so doing, gives up heat, then releases the refrigerant, during which it absorbs heat.
2. Refrigeration in which cooling is effected by the expansion of the liquid refrigerant into gas and absorption of the gas by water. The refrigerant is reused after the water evaporates.

Absorption system

1. Refrigerating system in which the gas evolved in the evaporator is taken up by an absorber or adsorber.
2. Equipment intended or installed for the purpose of heating or cooling air by an absorption unit, either by direct or indirect means, and discharging such air into any room or space.

Absorption unit. Factory-tested assembly of component parts producing refrigeration for comfort cooling by the application of heat. This definition applies to absorption units that also produce comfort heating.

Across-the-line motor starter. Motor starter or switch that when engaged, impresses full line voltage on the motor windings.

Accumulator

1. Prevents the refrigerant from entering the suction line; sometimes called suction accumulator.
2. Pressure vessel whose volume is used in a refrigerant circuit to reduce pulsation.

ACR tubing. Tubing for refrigeration that has been cleaned, sealed, and charged with dry nitrogen. ACR means "air-conditioning refrigeration."

Adiabatic. Process or action during which no heat is added or subtracted.

Adiabatic compression. Adiabatic compression of a gas is effected when no heat is transferred to or from the gas during the compression process. In compressor practice, this definition refers only to the reversible adiabatic or isoentropic compression process. The characteristic equation relating pressure p and volume v during adiabatic compression is $pv^k = c$, in which c is constant and k is the ratio of the specific heat at constant pressure to the specific heat at constant volume.

Aftercoolers. Devices for removing the heat of compression of the air or gas after compression is completed. An aftercooler is one of the most effective means of removing moisture from compressed air.

Air cleaner. Device used to remove dirt, lint, and airborne impurities from the air.

Air coil. Coil used in refrigeration as a condenser or evaporator to remove heat or pick it up.

Air compressors. Includes all stationary power-driven compressors.

Air conditioner. Device used to reduce temperature, humidity, and impurities in the air.

Air-conditioner unit. Unit that provides a means for moving, heating, cooling, humidifying, dehumidifying, and filtering the air.

Air-conditioning

1. Process by which the temperature, humidity, movement, and quality of air in buildings and structures used for human occupancy are controlled and maintained to secure health and comfort.

2. Simultaneous control of all or at least the first three of those factors affecting both the physical and chemical conditions of the atmosphere within any structure. These factors include temperature, humidity, motion, distribution, dust, bacteria, odors, and toxic gases, most of which affect in greater or lesser degree human health or comfort.

Air-conditioning or comfort-cooling equipment. All of that equipment intended or installed for the purpose of processing the treatment of air so as to control simultaneously its temperature, humidity, cleanliness, and distribution to meet the requirements of the conditioned space.

Air-conditioning system. System of a building that provides conditioned air for comfort cooling by the lowering of temperature, requiring a total of more than 15 motor horse power or a total of more than 15 tons of mechanical refrigeration, in single or multiple units, and air distribution ducts.

Air-conditioning system or apparatus. A system or apparatus that ventilates, heats, and humidifies in winter and/or cools and humidifies in summer the space under consideration and provides the desired degree of air motion and cleanliness.

Air-conditioning unit. Factory assembled equipment in compact formation used for the treatment of air, which is passed through the equipment so as to control simultaneously its temperature, humidity, cleanliness, and purity for distribution for a given space.

Air-cooled condenser

1. Condenser designed to remove heat from gas by transferring the heat to the air surrounding or passing through the condenser.

2. Device used to remove heat from the vapor or refrigerant by passing air through the unit.

3. Heat exchanger that transfers heat from the vapor as the surrounding air is moved through.

Air cooler. Factory assembly of devices whereby the temperature of air passing through the elements is reduced.

Air diffuser. Air distribution outlet, generally located in the ceiling, consisting of a frame containing blades or louvers to guide the discharging supply of air in various directions and planes.

Air distribution system. System that includes supply and return air ducts and plenum chambers.

Air duct

1. Tube or conduit used for conveying air. Air passages of self-contained systems are not to be considered as air ducts.

2. Tube or conduit or an enclosed space or corridor within a wall or structure used for conveying air.

Air filter. Device used to remove dust and other undesirable solids from the air.

Air filter units. Both washable and throw-away types used for removal of dust and other airborne particles from air circulated mechanically in equipment and systems.

Air-handling unit. Blower or fan used for the purpose of distributing a conditioned air supply to a room, space, or area.

Air-indicated horsepower. Horsepower calculated from compressor indicator diagrams. The term applies only to power delivered at the piston(s) of the compressors.

Air washer. Equipment designed to allow the passage of air that is washed from sprays for the purpose of humidifying or cleaning.

Ambient. Encompassing on all sides. Ambient air is the air surrounding.

Ambient temperature. Temperature of the air surrounding the object under consideration.

Ammonia refrigeration. Refrigeration accomplished by the use of a mechanical refrigeration machine that uses ammonia as a refrigerant.

Anemometer. Instrument for measuring the velocity of air speed. Usually used at the face of air supply grilles or diffusers.

Apparatus—heating and cooling. A device that utilizes fuel or other forms of energy to produce heat, refrigeration, or air conditioning.

Appliance. Device that utilizes fuel or other forms of energy to produce light, heat, power, refrigeration, or air conditioning. This definition includes a vented decorative appliance.

Approach. Term used to describe the difference in degrees between the temperature of cold water leaving the cooling tower and the wet-bulb temperature of the surrounding air.

Atmospheric pressure

1. Pressure due to the weight of the atmosphere; pressure indicated by a barometer. Standard atmospheric pressure or 1 standard atmosphere is equivalent to 14.696 psi or 29.921 in. of mercury at 32°F.

2. The pressure at any point in an atmosphere due solely to the weight of the atmospheric gases above the point concerned.

3. Pressure exerted by the earth's atmosphere. Under standard conditions, at sea level, atmospheric pressure is 14.7 psia or 0 psig.

Automatic expansion valve

1. Expansion valve designed to maintain a constant pressure in the evaporator regardless of superheat. It is seldom used in air-conditioning where loads generally fluctuate and cannot be used on systems with multiple-valve installations.

2. Device used to control the flow of a refrigeration system by pressure.

Autotransformer. Automatic compensator used with alternating-current motors, in which the motor is fed from different points in an impedance coil placed across the supply circuits.

Auxiliary contact. The pilot or auxiliary contact is operated whenever the main contact or contacts are closed; an example is the holding circuit contact in a magnetic starter. Also, an interlock contact.

Auxiliary receiver. Extra vessel used to supplement the capacity of the receiver when additional storage volume is necessary.

Auxiliary switch. Accessory switch available for most damper motors and control operators. The switch can be arranged to open or close a circuit whenever the control motor reaches a certain position.

Axial flow fan. Fan with a disc or air-foil shaped blade mounted on a shaft, that moves the air in the general direction of the shaft axis.

Back pressure. Pressure in the low side or suction pressure.

Back-seat. Seat against which a valve disc may seat when the valve is in full open position. By "back-seating," the pressure on the valve is prevented from reaching the valve packing.

Back-seat port or tapping. Port entering the valve body behind the back-seat. Since no pressure can reach this port when the valve is back-seated, the port can be used for attaching test gauges or making other connections to the system while under pressure.

Ball-bearing motor. Type of motor having ball bearings, usually arranged for grease lubrication.

Ball-check. Device consisting of a ball and orifice. Pressure on one side of the device will seat the ball across the orifice and stop the flow. Pressure in the opposite direction will lift the ball from seat permitting flow. The device is generally used in liquid level gauges to seal off the gauge glass in case of breakage.

Bending spring. Spring that is placed over or inside tubing to keep the tubing from collapsing while bending.

Bimetal strip. Two different metals fused together so that when heated, the strip will bend and open and close points (as on a thermostat).

Blower or fan. Assemblage comprised of blades or runners and housings or castings which cause a motion of air when in operation.

Blower shutoff. Approved manually reset, heat-actuated, thermostatic device installed in the duct system and arranged to shut off the supply of power to the blower whenever the temperature of the air at the point of installation reaches a point not in excess of 200°F. The device is nonadjustable above 200°F.

Boiling. Change of state from a liquid to a gas or vapor.

Boiling point or boiling temperature. Temperature at which a fluid will change from a liquid to a gas. The boiling point depends on the pressure exerted on the surface of the liquid.

Boiling temperature. Temperature at which a fluid changes from a liquid to a vapor.

Booster compressors. Machines for compressing air or gas from an initial pressure that is considerably above atmospheric pressure to a still higher pressure.

Brake (or shaft) horsepower
1. Unit of power equal to 1 hp delivered at the shaft of an engine or motor; also, the actual power required to drive a machine. The term "brake horsepower" is used to avoid confusion with other horsepower definitions, such as electrical horsepower, which would be based on the power input to the driving motor and would not consider the efficiency of the driver.
2. Measured horsepower input of the compressor. Horsepower, either indicated or brake, for any displacement-type compressor varies with the compression ratio, as well as absolute intake and discharge pressure. Performance guarantees are expressed in terms of horsepower-per-cubic-foot capacity. In comparing test results with performance guarantees, corrections should be made for any deviation from specified values of absolute intake pressures and ratio of compression.

Brazed joint. Gas-tight joint obtained by the joining of metal parts with alloys that melt at temperatures higher than 1000°F but less than the melting temperatures of the joined parts.

Brazing. Method of joining two metals together.

Brine. Liquid, used for the transmission of heat without a change in its state, having no flash point or a flash point above 150°F as determined in an approved manner.

Brine cooler. Evaporator for cooling brine in an indirect system.

Brine thickness insulation. Commercial classification applied to molded cork covering and similar insulation for refrigerant and brine lines. It is somewhat thicker than "ice water thickness."

British thermal unit. Classically the Btu is defined as the quantity of heat required to raise the temperature of 1 lb of water 1°F. The exact value depends on the initial temperature of the water. Several values of the Btu are in more or less common use, each differing from the others by a slight amount. One of the more common of these is the mean Btu, which is defined as $\frac{1}{180}$ of the heat required to raise the temperature of 1 lb of water from 32°F to 212°F at a constant atmospheric pressure of 14.696 psia.

Btu rating. Listed maximum capacity of any appliance, absorption unit, or burner expressed in British thermal unit input per hour.

Burner. Device in which combustion of fuel takes place.

Bypass
1. Means of circumventing an object; connection around a coil for the purpose of reducing the capacity of the coil.
2. Passage to one side or around regular passage.

Calorie. Unit of heat energy equal to 4.184 joules. Unit of energy equal to the heat required to raise the temperature of 1 g of water from 14.5° to 15.5°C at a constant pressure of 1 standard atmosphere, equal to 4.1855 ± 0.0005 J.

Capacitor. Condenser used for improving the power factor of inductive loads. Capacitors are designed for indoor and outdoor use and are made in single-, two-, and three-phase types.

Capacitor start motor
1. Induction motor having a separate starting winding similar to the split-phase motor except that the capacitor start motor has an electrical condenser connected into the starting winding for added starting torque.
2. Motor that has a capacitor in the starting circuit.

Capacitor start and run motor. Motor similar to the capacitor start motor except that the capacitor and starting winding are designed to remain in the circuit at all times, thus eliminating the switch used to disconnect the starting winding.

Capacity. The capacity (actual delivery) of an air or gas compressor is the actual quantity of air or gas compressed and delivered, expressed in cubic feet per minute at conditions of total temperature, total pressure, and composition prevailing at the compressor inlet. Capacity is always expressed in terms of air or gas at intake conditions rather than in terms of standard air or gas. The capacity of a positive displacement-type compressor, working under a given compression ratio, is not affected by barometric pressure or temperature at the compressor intake.

Capacity control actuator. Device that responds to variations in the suction pressure and proportions a fluid pressure to a mechanism on the cylinder liner that loads or unloads the compressor cylinders.

Capacity modulation. Method of varying the capacity of a compressor.

Capillary tube

1. Tube having a very small internal and external diameter, sometimes referred to as a "restrictor tube." Capillary tubes are used between the bulb and power element of thermostatic expansion valves and remote bulb temperature controllers.

2. Very small tube to meter refrigerant into the evaporator. Usually it is several feet in length.

Cap seal valve. Manual valve having the stem protected by a tightly fitting cap.

Cap wrench valve. Cap seal valve having a seal cap which serves also as a wrench to operate the valve stem for opening or closing the valve.

Celsius. Metric system temperature scale. In this scale, water freezes at 0°C and boils at 100°C. (Also centigrade.)

Centigrade. Thermometer scale in which 0 degrees represents the freezing point and 100 degrees represents the boiling point of water at a pressure of 1 atmosphere. Centigrade is generally used with metric units of measure.

Centimeter. Metric unit of length (1 cm = 0.394 in.); $\frac{1}{100}$ m.

Central air conditioner. Means an air conditioner which is not a room air conditioner.

Central cooling plant. Comfort cooling equipment installed in a manner to supply cooling by means of ducts or pipes to areas other than the room or space in which the equipment is located.

Central fan system. Mechanical indirect air delivery system in which the air is treated or handled by equipment located outside of the rooms or areas being served, usually in a central location, and conveyed to and from the rooms or areas by means of a fan and a system of distributing ducts.

Central plant. A refrigeration system utilizing refrigerant containing components that are interconnected by piping in the field.

Centrifugal compressor. Compressor that imparts motion, and therefore pressure, to a gas by means of high-speed impellers. It is not a positive displacement machine.

Centrifugal fan. Fan rotor or wheel within a scroll type housing and including driving mechanism supports for either belt drive or direct connection motor.

Change of state. Change in the physical characteristic of a substance, such as a change from a liquid to a solid upon freezing, or the change from a liquid to a gas upon evaporation, or vice versa.

Charge. Amount of fluid forced or drawn into a closed system, such as the refrigerant in a refrigerating system, the fluid in the bulb and power element of a thermostatic expansion valve, or oil in the crankcase of a compressor.

Charging valve. Valve, located on the liquid line, through which refrigerant may be charged into the system.

Check valve. Valve designed to permit flow in one direction only. The valve is designed to close against backflow.

Chilled water system. Closed circuit system that recirculates water between a mechanical refrigeration water chilling unit and remote cooling equipment, usually operating with water temperatures in the range between 40°F and 55°F.

Circulating air supply. Air conveyed from a conditioned area or from outside the building through openings, ducts,

plenums, or concealed spaces to a heat exchanger of a comfort heating, cooling, absorption, or evaporative cooling system.

Closed cycle. System where the fluid is used over and over again without the introduction of additional fluid.

Coefficient of heat transmission. The quantity of heat (Btu) transmitted from fluid to fluid per unit of time (1 hour) per unit of surface (1 sf.) through a material or assembly of materials under a unit temperature differential (1°F) between fluids.

Coil. Cooling or heating element made of pipe or tubing.

Comfort air-conditioning

1. Process by which the temperature, moisture content, movement, and quality of the air in enclosed spaces intended for human occupancy may be simultaneously maintained within required limits.

2. Mechanical conditioning of air for the comfort or well-being of human beings, as distinguished from conditioned air used in manufacturing or industrial processes.

Comfort chart. Chart used in air-conditioning work to show dry bulb temperature.

Comfort cooling

1. System used to cool a design area to a comfortable temperature.

2. Air cooling to 50°F or above.

Comfort cooling system. All the equipment intended or installed for the purpose of cooling air by mechanical means and discharging such air into any room or space. This definition does not include any evaporative cooler.

Comfort cooling unit. Self-contained refrigerating system, factory assembled and tested, installed with or without conditioned air ducts and without connecting any refrigerant-containing parts. This definition does not include a portable comfort cooling unit or an absorption unit.

Comfort line. Line on the comfort chart showing the relation between the effective temperature and the percentage of adults feeling comfortable.

Comfort zone. That area on the psychrometric chart covered by combinations of temperatures and humidities at which, in subjective tests, more than 50% of the subjects were comfortable.

Comfort zone (average). Range of effective temperatures over which the majority (50% or more) of adults feel comfortable.

Commercial refrigerating system. Refrigerating system assembled or installed in a building used for business or commercial purposes.

Compound gauge. Instrument used to measure pressures above and below atmospheric pressure.

Compound wound motor. Direct-current motor with relatively constant speed characteristics.

Compression. Increase of pressure (especially on a gas) by using a mechanical device.

Compression efficiency (adiabatic). Ratio of the theoretical horsepower in isentropic compression to the horsepower imparted to the air or gas actually delivered by the compressor. The power imparted to the air or gas is brake horsepower minus mechanical losses.

Compression gauge. Gauge used to measure pressure above atmospheric.

Compression ratio. Ratio determined by dividing the discharge pressure (psia) by the suction pressure (psia).

Compression system. Refrigerating system in which the pressure-imposing element is mechanically operated.

Compressor

1. Specific machine, with or without accessories, for compressing a given refrigerant vapor.

2. Machine designed to pump a gas from a low-pressure space to a high-pressure space.

3. Mechanical device used in a refrigerating system for the purpose of increasing the pressure on the refrigerant.

4. Mechanical device used to move a refrigerant through a system by removing the refrigerant from the low side and raising the pressure on the high side.

5. Device having one or more pressure-imposing elements, used in a refrigerating system to increase the pressure of the refrigerant in its gas or vapor state for the purpose of liquefying the refrigerant.

6. Machine designed for compressing air or gas from an initial intake pressure to a higher discharge pressure.

Compressor capacity modulation. Mechanism to control the capacity of a compressor by rendering one or more cylinders ineffective.

Compressor displacement. Volume swept by the pistons or impellers or a reciprocating or rotary machine, usually stated in cubic feet per shaft revolution or in cubic feet per minute at some definite rpm.

Compressor efficiency (adiabatic). Ratio of the theoretical horsepower in isentropic compression to the shaft horsepower. It is equal to the product of compression efficiency and mechanical efficiency.

Compressor lubricating oil. Highly refined lubricant made especially for refrigeration compressors.

Compressor relief device. Valve or rupture member located between the compressor and the stop valve on the discharge side, arranged to relieve the pressure at a predetermined point.

Compressor seal. Device to prevent the leakage of refrigerant gas at the point where the crankshaft must pass through the crankcase. It usually takes the form of two finely finished surfaces separated only by a thin film of oil and maintained under pressure by means of a spring.

Compressor unit

1. Condensing unit less the condenser and liquid receiver.

2. Unit consisting of a compressor, motor, drive, and frequently the essential compressor controls, all mounted on a common base.

Condensate

1. Water that forms on the cool evaporator coil.

2. Liquid formed by the condensation of a vapor in steam heating; water condensed from steam.

Condensate pump. Pump used to remove the moisture from underneath the evaporator coil.

Condensation

1. Process of changing a vapor into liquid by the extraction of heat.

2. Formation of liquid or droplets when the vapor is cooled below a certain point.

Condense. Change a gas or vapor back to a liquid.

Condenser

1. Vessel or arrangement of pipe or tubing in which vaporized refrigerant is liquefied by the removal of heat.

2. Device for removing heat from gas for the purpose of causing the gas to condense to a liquid.

3. Part of a refrigeration system that removes the heat from the high-pressure vapor, causing the vapor to change back to a liquid.

4. Vessel or system of tubing in which the compressed refrigerant gas is liquefied by the removal of heat.

Condenser drain line. That part of the refrigerant piping between the condenser and the liquid receiver.

Condenser fan. Motor with a blade that moves the air through the condenser.

Condenser shutoff valve. Valve located in the hot gas or discharge line at the inlet to the condenser.

Condensing temperature. Temperature of the fluid in the condenser at the time of condensation.

Condensing unit

1. Specific refrigerating machine combination for a given refrigerant, consisting of one or more power-driven compressors, condensers, liquid receivers (when required), and the regularly furnished accessories.

2. Unit consisting of a compressor, motor, drive, condenser, and frequently the essential controls, all mounted on a common base.

3. Specific refrigerating machine combination consisting of a motor-driven compressor, a condenser, a liquid receiver, and the regularly furnished accessories.

4. Unit containing the compressor, condenser, fan motor, service valves, lines, and devices used to start the compressor.

Condensing water. Water supplied to cool a water-cooled condenser or condensing unit.

Conditioned air supply. Air conveyed to a conditioned area through ducts or plenums from a heat exchanger of a comfort heating, cooling, absorption, or evaporative cooling system.

Conditioned area. Area, room, or space normally occupied and heated or cooled for human comfort by any equipment.

Conditioned space. Space within a building that is provided with a positive heat supply or a positive method of cooling, either of which has a connected output capacity in excess of ten Btu/hr per sq ft.

Condition line. The infinite number of combinations of wet and dry bulb temperatures that will satisfy the requirements of an air supply for a given room condition from what is known as the condition line on the psychrometric chart.

Conductance (C). Thermal conductance is the amount of heat, expressed in British Thermal Units (Btu), transmitted in 1 hour from surface to surface of 1 sq ft of material or a combination of materials for each degree of temperature difference between the two surfaces. It should be noted that this value is not expressed in terms of per inch of thickness, but of transmission from surface to surface.

Conductance of air space (a). The thermal conductance of an air space is the amount of heat, expressed in British thermal units (Btu), transmitted in 1 hour across an airspace of 1 sq ft area for each degree Fahrenheit temperature difference.

Conductivity (K)

1. Thermal conductivity is the amount of heat expressed in British Thermal Units (Btu), transmitted in 1 hour

through 1 sq ft of a homogeneous material 1 in. thick for each degree Fahrenheit of temperature difference between the two surfaces of the material.

2. Ability of a material to transmit or conduct heat or electricity.

Container. Cylinder for the storage and transportation of a refrigerant.

Control. Regulation of the system equipment to maintain the conditions desired.

Control cycle. Sequence of operations under automatic control intended to maintain the desired conditions at all times.

Control relay. Electromagnetic device that closes or opens contacts when its coil is energized.

Cooling tower. Device for cooling water by evaporation in air. Water is usually sprayed into an airstream where part of the water evaporates thus reducing the temperature of the remaining water.

Cooling unit. Unit that provides a means for moving and cooling air.

Crankcase. Casing or covering for the crankshaft of the reciprocating compressor.

Crankcase double-pipe equalizer. System in which both the gas pressure and the oil level in all compressors connected in multiple are equalized.

Crankcase pressure. Pressure that exists in the crankcase of a reciprocating compressor.

Crankcase single-pipe equalizer. Pipe or tube connection between the crankcases of the compressors of a multiple compressor system, the function of the pipe or tube being to equalize the pressure within the crankcases.

CTI. The Cooling Tower Institute is a non-profit, self-governing technical association of manufacturers, suppliers, owners, operators, and specifiers.

Cut-in point. Temperature or pressure at which a controller will function to start the equipment controlled.

Cut-out point. Temperature or pressure at which a controller will function to stop the equipment controlled.

Cycle

1. Sequence of operations or functions.

2. Series of events that have a tendency to repeat.

Cylinder. Chamber in a reciprocating compressor in which a piston is impelled by the crankshaft to compress the refrigerant gas.

Cylinder bore. Finished internal diameter of a cylinder.

Cylinder capacity modulation. Method used for modulating the capacity of a compressor cylinder by interrupting the normal action of the suction valve. It is usually activated by a device controlled by electricity or oil pressure.

Cylinder discharge valve. Valve in the compressor through which the gas leaves the cylinder.

Cylinder head. Upper part or cap of the cylinder.

Cylinder spring. Spring located between the cylinder head and discharge valve assembly of a reciprocating compressor to ensure protection against damage from slugs of liquid refrigerant or oil. The entire discharge valve assembly will "lift" against the spring and pass the slug and then return to its normal position. It is often referred to as a "safety head."

Cylinder suction valve. Valve in the compressor through which the gas enters the cylinder.

Damper

1. Valve of vane, leaf, or butterfly type for controlling the flow of air.

2. Adjustable gate, usually located in the flue, to restrict the flow of combustion gas.

Dampers, face and bypass. A set of coordinated dampers, arranged to direct the air through an evaporator, around an evaporator, or partly through and partly around an evaporator in any desired proportion, in response to control demand.

Dehumidifier. Apparatus for removing moisture content from a substance either by precipitation or reduction of temperature or by the use of a hygroscopic substance.

Dehumidification. Reduction of moisture in a given volume of air.

Dehydration. Removal of water vapor from the air by the use of absorbing or adsorbing materials.

Dehydrator. Device containing a desiccant for the purpose of removing moisture from the refrigerant.

Demand factor. Ratio of the coincident maximum demand of a group of apparatus to the sum of the individual maximum demands of the apparatus making up the group.

Density. Weight per unit volume of a substance.

Depressor fork. Device that when energized moves to hold open the cylinder suction valves in certain compressor capacity modulation systems.

Desiccant. Chemical agent used for moisture removal.

Design temperature (indoor). The temperature to be maintained within the conditioned space.

Design temperature (outdoor). The outdoor temperature arbitrarily established as the maximum against which the system must be able to maintain the desired indoor conditions. For economic planning, it is somewhat lower or higher than the actual maximum.

Design working pressure

1. Maximum allowable working pressure for which a vessel is designed.

2. Maximum allowable working pressure for which a specific part of a system is designed.

Desuperheat. Cool a gas to saturation temperature.

Dew point. Temperature at which water vapor will condense.

Dew point temperature

1. Temperature at which water vapor begins to condense when a constant mixture of air and water vapor is cooled.

2. Temperature at which the condensation of water vapor in a space begins for a given state of humidity and pressure as the temperature of the vapor is reduced; temperature corresponding to saturation (100% relative humidity) for a given absolute humidity at constant pressure.

Dichlorodifluoromethane. Refrigerant commonly known as "refrigerant-12" or "Freon-12" (R-12).

Differential pressure control. Method of maintaining a given pressure difference in two pipelines or spaces.

Direct absorption unit. Unit in which the refrigerant evaporator is in direct contact with the air to be conditioned.

Direct expansion evaporator. Evaporator designed to cool a medium in direct contact with it.

Direct heater. Furnace or heating device installed in a duct system for conditioning purposes, whose heating effect on the air therein is supplied directly from the heater to the air in the duct by radiation and convection.

Direct refrigeration expansion. Refrigeration system or apparatus in which the evaporator is in direct contact with the refrigerated material or space or is located in air circulating passages communicating with such spaces.

Direct refrigerating system. System in which the refrigerant evaporator is in direct contact with the material or space to be refrigerated or is located in air-circulating passages communicating with such spaces.

Direct system. System in which the evaporator is in direct contact with the material or space refrigerated or is located in air-circulating passages communicating with such spaces.

Direct system of refrigeration. System in which the evaporator is located in the material or space refrigerated or in air-circulating passages communicating with such space.

Discharge gas. Refrigerant gas leaving the compressor.

Discharge line. Part of the refrigerant piping between the compressor and the condenser, usually called the "hot-gas line."

Discharge manifold. Piping fitting used for collecting the compressed refrigerant from the various cylinders of a compressor.

Discharge pressure

1. Pressure against which the compressor must deliver the gas.
2. Absolute total pressure at the discharge flange of the compressor; commonly stated in terms of gauge pressure. Unless the associated barometric pressure is included, this is an incomplete statement of discharge pressure.

Discharge pressure gauge. Device used to measure fluid pressures above atmospheric pressure.

Discharge shutoff valve. Valve in the discharge passage of a compressor. It is frequently mounted directly on the compressor body or discharge manifold.

Discharge temperature

1. Total temperature at the discharge flange of the compressor.
2. Temperature of the gas leaving the compressor.

Displacement. In a compressor, the volume displaced per unit of time, usually expressed in cubic feet per minute; in a reciprocating compressor, the net area of the compressor piston multiplied by the length of the stroke and by the number of compression strokes per minute. The displacement rating of a multistage compressor is the displacement of the low-pressure cylinder only.

Diversified load. The diversified load of a group of apparatus is the sum of the individual demands of the group times the demand factor.

Double-acting compressors. Compressors in which compression takes place on both strokes per revolution in each compressing element.

Double indirect vented open-spray system. System in which a liquid, such as brine or water, cooled by an evaporator located in a vented enclosure, is circulated through a closed circuit to a second enclosure where it cools another supply of a liquid, such as brine or water, and this liquid in turn is circulated to a cooling chamber and is sprayed therein.

Double (or secondary) refrigerant system. System in which an evaporative refrigerant is used in a secondary circuit. Each system enclosing a separate body of an evaporative refrigerant is considered as a separate direct system.

Draft hood. Device placed in and made part of the vent connector from an appliance, or in the appliance itself, which is designed to ensure the ready escape of the products of combustion in the event of no draft, backdraft, or stoppage beyond the draft hood; prevent a backdraft from entering the appliances; and neutralize the effect of stack action of the chimney flue on the operation of the appliance.

Drier. Device used to remove moisture from a refrigeration system.

Drift. Fine spray of water carried away from a cooling tower by the wind.

Dry air cooler. Equipment designed to remove sensible heat from the dehydrated air whenever it leaves the dehydrator at an elevated temperature.

Dry bulb. Instrument used to measure air temperature.

Dry-bulb temperature. Temperature of the air measured with an ordinary thermometer and indicative only of sensible heat changes.

Dry expansion chiller. Device designed to utilize the evaporation of refrigerant within a tube bundle, the refrigerant being metered to the inside of the tubes by means of one or more thermostatic expansion values. The liquid to be chilled passes through the chiller between the shell and tubes.

Dry expansion evaporator. Evaporator designed to evaporate liquid refrigerant as rapidly as it is fed by the expansion valve. The feed must be so regulated that no liquid enters the evaporator unless there is sufficient heat available (load) for evaporation.

Dry nitrogen—oil pumped. Nitrogen pumped through oil to remove moisture.

Dry ton. Sensible heat load expressed in tons.

Dual pressure controller. Controller consisting of two pressure bellows that operate a switch within a single enclosed case. When applied to compressor control, one bellows functions on a change in suction pressure, the other on a change in discharge pressure.

Dual riser system. System in which two vertical risers are used when a single vertical riser sized for 1000 ft/minute at a minimum load will have an excessive pressure drop at maximum load. The smaller riser is sized for a minimum of 1000-ft/minute velocity at minimum load, while the larger is sized so the velocity through both risers will not be less than 1000 ft/minute at maximum load.

Duct. Tube, pipe, conduit, or continuous enclosed passageway used for conveying air, gases, or vapors.

Duct fan. Axial flow fan mounted in a section of duct.

Duct lining. Sound-absorbing or insulating material placed in the interior of duct walls or plenum chambers.

Duct systems. Ducts, duct fittings, plenums, and fans assembled to form a continuous passageway for the distribution of air.

Effective temperature. Arbitrary index that combines in a single value the effect of temperature, humidity, and air movement on the sensation of warmth or cold felt by the human body. The numerical value is the temperature of still, saturated air that would induce an identical sensation.

Electrical control circuit. Wiring diagram showing how the various pieces of the control circuit are to be connected for proper operation.

Electrostatic filter

1. Electrical filtering device for the removal from air of smoke and other particles too small for the usual mechanical filter. Very small particles are forced to adhere to collector plates because of the electric charges imparted by the filter.

2. Device in which particles of dust are given an electrical charge; oppositely charged plates collect the particle.

Elementary chilled water system. The system consists of a refrigeration water chilling unit, a chilled water recirculating pump, terminal cooling equipment, and an expansion tank.

Eliminator. Device containing many stationary vanes or louvers designed to remove entrained water particles from an airstream.

Emergency relief valve. Manually operated valve for the discharge of refrigerant in case of fire or other emergency.

End play. Slight longitudinal movement of a shaft; movement of a motor shaft along the centerline.

Entering temperature. Temperature of a substance as it enters a piece of apparatus.

Entrained oil. Oil droplets carried by high-velocity refrigerant gas.

Entrained water. Small particles of water carried along by a rapidly moving airstream.

Equalizer. Pipe connection between two or more pieces of equipment made in such a way that the pressure in each piece is maintained equally.

Equalizer line. Line that equalizes the gas or oil pressure in two or more pieces of equipment.

Equipment. Equipment includes such items as materials, fittings, devices, appliances, and apparatus used as part of or in connection with installations.

Estimated design load. In a heating or cooling system, the sum of the useful heat transfer occurring in any auxiliary apparatus connected to the system in British thermal units per hour—or in heating, equivalent direct radiation (EDR).

Evaporating temperature. Temperature at which the fluid boils under the existing pressure.

Evaporation. Process whereby a liquid substance is converted into a vapor or a change of state.

Evaporative condenser. Condenser designed to remove heat from gas by utilizing the cooling effect of evaporating water.

Evaporative cooler. Device used for reducing the sensible heat of air for comfort cooling by the evaporation of water into an airstream.

Evaporative cooling. The adiabatic exchange of heat between air and a water spray or wetted surface. The water approaches the wet-bulb temperature of the air, which remains constant during its traverse of the exchange.

Evaporative cooling system. Equipment intended or installed for the purpose of comfort cooling by an evaporative cooler, from which the conditioned air is distributed through ducts or plenums to the area to be conditioned.

Evaporator

1. Device in which a refrigerant is evaporated for the purpose of extracting heat from the surrounding medium.

2. Part of the refrigeration system that absorbs heat as the refrigerant boils into a vapor.

3. Part of a system in which the refrigerant is expanded or vaporized to produce refrigeration.

Evaporator condenser. Compact form of condenser in which air is positively directed over the surface by mechanical means.

Evaporator temperature. Temperature of the evaporator due to the boiling refrigerant.

Expansion coil. Evaporator constructed of pipe or tubing.

Expansion valve

1. Valve designed to meter the flow of liquid refrigerant to an evaporator.

2. Device to meter the refrigerant from the high side to the low side.

3. Valve for controlling the flow of liquid refrigerant to the evaporator.

Exploring tube. Small flexible tube attached to a halide torch in such a manner that air is continually drawn through the tube to the torch flame. When the free end of the searching tube is placed near a refrigerant leak, some of the refrigerant is carried to the flame where its presence is indicated by coloring the flame.

Explosionproof motor. Motor so constructed that a fire originating within the motor cannot be transmitted to the outside.

External equalizer. Tube connecting the chamber under the diaphragm of the power element of a thermostatic expansion valve to the low-pressure side of an evaporator to eliminate the effect of pressure drop through the evaporator on superheat response.

External pilot control. The internal connection of the pilot is plugged and an external connection provided to make it possible to use an evaporator pressure regulator as a suction stop valve as well.

Face and bypass control. Device, usually a valve or damper, to divert the flow of air over the face of an extended surface evaporator or through a passage around the evaporator.

Face and bypass dampers. Set of coordinated dampers arranged to direct the air through an evaporator, around an evaporator, or partly through and partly around an evaporator in any desired proportion, in response to control demand

Fahrenheit. Temperature scale in which $+32$ and $+212$ degrees are the freezing point and boiling point, respectively, of water under standard atmospheric conditions. This system is generally used where other units of the English system are employed.

Fan. Motor driven device comprising a wheel or blades within a housing or orifice plate.

Film conductance (F). Film or surface conductance; the amount of heat, expressed in British thermal units (Btu) transmitted in 1 hour from 1 sq ft of a surface to the air surrounding the surface for each degree Fahrenheit temperature difference. The symbols f_i and f_o are used to designate the inside and outside surface conductances, respectively.

Filter. Device used to remove dust and lint from the air.

Final temperature. Temperature of a substance as it leaves a piece of apparatus.

Fire damper. Approved automatic or self-closing non-combustible barrier designed to prevent the passage of air, gases, smoke, or fire through an opening, duct, or plenum chamber.

Flammable refrigerant. Refrigerant that will burn when mixed with air.

Flare fitting. Device used to connect two lines.

Flash gas. Gas generated whenever pressure is reduced on a liquid held at boiling temperature.

Float expansion valve. Valve designed to maintain a constant liquid level in a flooded evaporator.

Float valve. Valve designed to maintain automatically a constant liquid level. Valve operation is governed by a float mechanism.

Flood back. Carry-over of liquid refrigerant from the evaporator to the suction line, frequently caused by faulty expansion valve operation or "slop over" from a flooded evaporator.

Flooded chiller. Device designed to utilize the evaporation of refrigerant on the outside surface of a tube bundle, the refrigerant level being maintained by a float valve. The liquid to be chilled passes through the tubes of the chiller.

Flooded evaporator. Evaporator designed to contain a definite quantity of liquid refrigerant at all times. Any refrigerant evaporated owing to load is replaced by means of a float valve.

Flooded system. System in which only part of the refrigerant passing over the heat transfer surface is evaporated and the portion not evaporated is separated from the vapor and recirculated.

Fluid. Substance that is a liquid or gas in its normal state.

Flywheel. Heavy wheel whose weight resists sudden changes of speed, thus securing uniform motion. It is usually the driven wheel in a drive combination.

Forced draft cooling tower. Tower in which the flow of air is created by one or more fans discharging air into the tower.

Forced vibration. Vibration created by an artificial means, for example, by a motor in a refrigeration system.

Force-feed lubrication. Lubrication system in which the lubricant is forced to the various bearing surfaces by an oil pickup or pump, as opposed to splash lubrication systems.

Fouling factor. Factor that determines the loss of heat transfer due to deposits of foreign material in the water side of the tubing in refrigeration condensers or chillers.

Free area. The total area of open space in a grille or diffuser through which air can pass without restrictions or friction.

Freezing. Change of state from a fluid to a solid by cold; to harden into ice.

Freezing point. Temperature at which freezing occurs.

Freon. Refrigerant manufactured by E. I. DuPont De Nemours & Company.

Friction loss. Loss of pressure in a system due to frictional resistance to flow.

Front seat. Part of a refrigeration valve that forms the seal with the valve button when the valve is in the closed position.

Fused (or fusible) disconnect switch. Electric switch designed to isolate part of the system when required. It is provided with fuses for the protection of the equipment.

Fusible plug

1. Device having a predetermined-temperature fusible member for the relief of pressure.
2. Safety device having an insert of low-melting-point alloy. At excessive temperature the alloy will melt and release the refrigerant.
3. Device arranged to relieve the pressure in a container by operation of a fusible member at a predetermined temperature.
4. Device for the relief of pressure, having a fusible metal that will melt at a maximum temperature of 200°F.

Gas. Vapor phase or state of a fluid.

Gas at saturation temperature. Pressure equilibrium. No superheat is present; if any heat is removed, some of the gas will condense to liquid.

Gas specific weight. Weight of air or gas per unit of volume at conditions of total pressure, total temperature, and composition prevailing at the inlet of the compressor.

Gas suction. The gas entering the suction side of the compressor.

Gas tracer. A gas having a powerful odor. Sometimes used in small quantities with odorless refrigerants to give warning of a leak.

Gas velocity. Speed of the gas in the piping or equipment, usually stated in feet per minute or feet per second.

Gauge port plug. Plug screwed into the front- or back-seat port of a packed-type shut-off valve.

Gauge pressure

1. Pressure existing above atmospheric pressure. Gauge pressure is, therefore, 14.7 psi less than the corresponding absolute pressure.
2. Pressure measured from atmospheric pressure as a base. Gauge pressure may be indicated by a manometer that has one leg connected to the pressure source and the other exposed to atmospheric pressure.

Generator. Device equipped with a heating element, used in the refrigerating system to increase the pressure of the refrigerant in its gas or vapor state for the purpose of liquefying the refrigerant.

Genetron. Refrigerant manufactured by Allied Chemical Corporation, General Chemical Division.

Halide. Chemical compound of the type MX, where X is fluorine, chlorine, iodine, bromine, or astatine, and M is another element or organic radical.

Halide torch

1. Device used to detect refrigerant leaks in a refrigeration system. The burner is equipped with a source of fuel, a mixing chamber, reactor plate, and an "exploring tube." The reactor plate surrounds the flame. When the open-end exploring tube is held near a refrigerant leak, some of the refrigerant is drawn to the mixing chamber where its presence changes the color of the flame.
2. Propane torch used to detect refrigerant leaks.

Handhole. Small-sized opening capable of accommodating the insertion or placement of the hand to maintain and repair machinery, equipment, or controls in small vessels.

Head pressure. Pressure that exists on the high side of a refrigeration system.

Head pressure control. Control that is used to disconnect the circuit if the head pressure becomes excessive.

Head pressure gauge. Device used to measure the discharge pressure of a pump or refrigeration compressor.

Heat content. Amount of heat, usually stated in British thermal units (Btu) per pound, absorbed by a refrigerant in raising its temperature from a predetermined level to a final condition and temperature. Where a change of state is encountered, the latent heat necessary for the change is included.

Heat exchanger. Device that removes heat from one fluid and adds it to another.

Heat gain. Amount of heat that must be removed from a space in order to achieve the desired temperature and humidity.

Heat—latent. Heat removed or added that cannot be measured by a change in temperature but that accomplishes a change in state.

Heat load. Amount of heat removed from an area in 24 hours.

Heat loss. Amount of heat lost through conduction, leakage, etc., that must be compensated for in winter air conditioning.

Heat of compression. Heat developed within a compressor when a gas is compressed, as in a refrigeration system.

Heat of the liquid. Heat content of the liquid; the heat necessary to raise the temperature of the liquid from a predetermined level to a final temperature.

Heat of the vapor. Heat content of the gas; the heat necessary to raise the temperature of the liquid from a predetermined level to the boiling temperature plus the latent heat of vaporization necessary to convert the liquid to a gas.

Heat pump

1. Factory packaged units that provide reverse-cycle air-conditioning for cooling and heating.

2. Equipment, used in a system to heat or cool an area, that has a compression cycle and a reverse refrigeration cycle.

Heat pump system. The system usually consists of a series of water source heat pump units connected by a two-pipe water circuit. The units get heat from the circuit on the heating cycle and give it to the water circuit on the cooling cycle. Heat transfer occurs when some areas need cooling. The heat in the water resulting from the cooling process in one area is absorbed from the water and used in another area. The common water circuit becomes the medium for the heat exchange.

Hermetic compressor. Unit in which compressor and motor are sealed in a dome.

Hermetic reciprocating compressor. Sealed reciprocating compressor and motor combination with no external coupling.

High-pressure cut-out. Controller, such as a switch, for breaking a circuit when pressure exceeds a predetermined point.

High side

1. Parts of a refrigerating system that contain the refrigerant at the condensing pressure of the system.

2. Part of the refrigeration system that is under pressure. Heat is given up, and the refrigerant changes back to a liquid.

High-side charging. Introducing liquid refrigerant into the high side of the refrigerating system. It is the acceptable way of placing the refrigerant into the system.

High-vacuum pump. Mechanism that will pull a deep vacuum and is used to remove the moisture and air from a system.

Holding charge. Partial charge of refrigerant placed in a piece of refrigeration equipment after dehydration and evacuation either for shipping or testing purposes.

Holding coil. Part of a magnetic starter or relay that causes the device to operate when energized.

Horsepower. Unit of power; the effort necessary to raise 33,000 lb a distance of 1 ft in 1 minute.

Horsepower (boiler horsepower). Heat required for the evaporation of 34.5 lb of water per hour at 212°F or the delivery of 33,475 Btu/hour to the water, steam, or other liquid in a boiler. Where data is not obtainable, the horsepower rating is obtained either by dividing the calorific value (in Btu) of fuel burned per hour by 50,000, or by dividing the area in square feet of boiler heating surface exposed to the products of combustion by 12.

Hot-gas bypass. Connection from the discharge to either the evaporator inlet or the suction side of the compressor. It is used to sustain compressor operation at loads that are less than the minimum stage of compressor capacity.

Humanly occupied space. Space normally frequented or occupied by people but excluding machinery rooms and walk-in coolers used primarily for refrigerated storage.

Humidifier. Device used to add moisture to a confined space.

Humidistat. Device used to control a humidifier; it senses the humidity in a confined space.

Humidity controller. Controller sensitive to changes in humidity.

Humidity ratio (HR). The weight of the actual water vapor in a mixture per pound of dry air.

Hunting. Fluctuation caused by the controls attempting to establish an equilibrium against difficult conditions.

Hydraulic cylinder. Part of the valving mechanism that is used in the capacity control actuator of the capacity modulation system.

Ice water thickness insulation. Commercial classification applied to molded cork covering and similar insulation for refrigerant and chilled water lines. It is somewhat thinner than "brine thickness."

Indirect absorption unit. Unit in which the refrigerant evaporator is not in direct contact with the air to be conditioned.

Indirect closed-surface system. System in which a liquid, such as brine or water, cooled by an evaporator located in an enclosure external to a cooling chamber, is circulated to and through such a cooling chamber in pipes or other closed circuits.

Indirect open-spray system. System in which a liquid such as brine or water, cooled by an evaporator located in an enclosure external to a cooling chamber, is circulated to such a cooling chamber and is sprayed therein.

Indirect refrigerating system. System in which brine, cooled by a refrigerating system, is circulated to the material or space to be refrigerated or is used to cool air so circulated.

Indirect system. System in which a liquid, such as brine or water, cooled by the refrigerant, is circulated to the material or space refrigerated or is used to cool air so circulated.

Indirect system of refrigeration. System in which a liquid, such as brine or water, cooled by the refrigerant, is circulated to the material or space refrigerated or is used to cool air applied to such space.

Indirect vented closed-surface system. System in which a liquid, such as brine or water, cooled by an evaporator located in a vented enclosure external to a cooling chamber is circulated to and through such cooling chamber in pipes or other closed circuits.

Indoor design temperature

1. Temperature that a heating or cooling system is designed to maintain in an indoor space.

2. Temperature to be maintained within the conditioned space.

Induced draft cooling tower. Tower in which the flow of air is created by one or more fans drawing the saturated air out of the tower.

Industrial refrigerating system. Refrigerating system used in the manufacture, processing, or storage of materials located in a building used exclusively for industrial purposes.

Inlet pressure. Absolute total pressure at the inlet flange of the compressor.

Inlet temperature. Total temperature at the inlet flange of the compressor.

Inspection plate. Handhole covers which, when properly removed, make possible the inspection of the internal parts of a compressor.

Insulation. Material intended to reduce the flow of heat.

Intercoolers. Devices for removing the heat of compression of the air or gas between consecutive stages of multistage compressors.

Intercooling. Removal of heat from the air or gas between stages or stage groups of compression. The degree of intercooling is the difference in air or gas temperature between the inlet of the compressor and the outlet of the intercooler. Perfect intercooling prevails when the temperature of the air leaving the intercooler is equal to the temperature of the air at the compressor intake.

Interlocking system. Arrangement of interlocks to ensure that all interrelated parts of an air-conditioning or refrigeration system are operating properly together.

Interlocks. Interlocks are used to prevent certain parts of an air-conditioning or refrigeration system from operating when other parts of that system are not operating.

Internal equalizer. Port connecting the chamber under the diaphragm of the power element of a thermostatic expansion valve to the suction side of the valve passage.

Internal heat gain. Amount of heat generated within an enclosed space. Typical sources are human bodies, cooking, electric lights, television sets, washing machines, and other heat- and moisture-producing appliances.

Inverted trap. Trap in refrigeration piping to prevent the slugging of the refrigerant or oil back to the compressor.

Irritant refrigerant. Refrigerant that has an irritating effect on the eyes, nose, throat, or lungs.

Isothermal compression. Isothermal compression is effected when interchange of heat between air or gas and surrounding bodies occurs at a rate precisely sufficient to maintain the air or gas at a constant temperature during compression. It may be considered a special case of polytropic compression.

Isotron. Refrigerant manufactured by the Pennsalt Chemicals Corporation.

K. Symbol for thermal conductivity for a unit thickness of a test specimen, usually 1 in., expressed as Btu·in./h·sq ft·deg F. This value is identified as the specific thermal conductivity of a material.

K (bulk density). Symbol for thermal conductivity for a unit thickness of a test specimen, usually 1 ft, expressed as Btu·ft/h·sq ft·deg F. It is the value commonly used by engineers in connection with the bulk density of materials.

K exponent. Exponent k occurs in the equation of adiabatic compression.

Latent heat

1. Heat energy involved in making the change of state, for example, in changing water vapor to liquid water. In air conditioning, specifically, it is that part of the heat gain produced by the water vapor that comes from outside air, human bodies, cooking, bathing, laundering, etc. The cooling system removes latent heat by removing some of the humidity from inside air.

2. Term used to express the energy involved in a change of state of a body without changing its temperature.

Latent heat of fusion. Latent heat absorbed when a solid melts to a liquid; latent heat liberated when a liquid freezes.

Latent heat of vaporization. Latent heat absorbed when a liquid evaporates to a gas; latent heat liberated when a gas condenses to a liquid.

Leak detector. Device that is typically a form of propane torch, used to find a leak in a refrigeration system.

Leak test. Test made in various ways to determine the existence and location of refrigerant leaks in a refrigeration system.

Lift. Elevate a fluid from one level to a higher level.

Limit control

1. Thermostatic device installed in the duct system to shut off the supply of heat at a predetermined temperature of the circulated air.

2. Control used to open or close a set of points on temperature rise or fall.

Limited charged system. A system in which, with the compressor idle, the internal volume and total refrigerant charge are such that the design working pressure will not be exceeded by complete evaporation of the refrigerant charge.

Liquid charge. Usually, a liquid charge is applicable to the power element of temperature controls and thermostatic expansion valves. The power element and remote bulb are sometimes charged with liquid rather than gas.

Liquid header. Manifold of the evaporator into which the liquid refrigerant is introduced for distribution through the evaporator.

Liquid indicator. Device located in the liquid line where the liquid flow can be observed.

Liquid level gauge. Gauge mounted on or in a vessel to indicate the liquid level within the vessel.

Liquid line

1. Line carrying liquid refrigerant from the receiver or condenser to the evaporator.

2. Tube that carries the liquid refrigerant from the condensing unit to the refrigerant control.

Liquid receiver

1. Vessel permanently connected to a system by inlet and outlet pipes for storage of a liquid refrigerant.

2. Cylinder for liquid storage, located after the condenser.

3. Vessel permanently connected to the high-pressure side of a system for the storage of refrigerant.

Liquid shut-off valve. Valve in the liquid line, usually located immediately at the condenser or receiver liquid outlet.

Liquid sight glass. Glass "bulls-eye" installed in the liquid line permitting visual inspection of the liquid refrigerant, primarily for the purpose of detecting bubbles in the liquid, indicating the shortage of refrigerant in the system.

Listed and listing. Terms referring to equipment that is shown in a list published by an approved testing agency that is qualified and equipped for experimental testing and maintaining an adequate periodic inspection of current productions and whose listing shows that the equipment complies with the standards and the code where the equipment is to be installed.

Load factor. Ratio of the average compressor load during a given period of time to the maximum rated load of the compressor.

Louver. Overlapping and sloping formed metal strips fabricated into a frame to allow the passage of air, but shaped to prevent the inclusion or passage of rain or snow.

Low-limit control. Protective device to prevent a system, or any part of it, from operating at a point below the setting of that device. It is used to prevent freeze-up in a water chiller or excessively low temperatures in duct work.

Low-pressure control. Pressure-operated switch in the suction side of a refrigeration or air-conditioning system that opens its contacts to stop the compressor at a given cutout setting.

Low side

1. Parts of a refrigerating system in which the refrigerant pressure corresponds to the evaporator pressure.

2. Parts of a refrigeration system that are under low pressure.

Low-side charging. Introducing a refrigerant into the low side of the system. Low-side charging is usually reserved for the addition of a small amount of refrigerant after repairs.

Machinery

1. Refrigerating equipment forming a part of the refrigerating system, including any or all of the following: compressor, condenser, generator, absorber (adsorber), liquid receiver, connection pipe, or evaporator.

2. Refrigerating equipment including any or all of the following: compressor, condenser, generator, absorber, receiver, connecting pipe, evaporator, or complete unit system.

Machinery room

1. Room in which a refrigerating system is permanently installed and operated, excluding evaporators located in a cold-storage room, refrigerator box, air-cooled space, or other enclosed space. Closets contained within and opening only into a room are not considered machinery rooms, but a part of the machinery room in which they are contained or open into.

2. Room reserved for mechanical equipment, including, but not necessarily restricted to, refrigerating and air-conditioning apparatus, heating boilers, and associated equipment and closed off from the remainder of the building by fire-resistive walls of not less than $\frac{3}{4}$-hour

fire-rating having no inside openings other than fire-resisting, self-closing doors of approved UL class C type.

Magnetic motor starter. Motor switch operated by a magnetic power unit or holding coil and equipped with overload relays for protection of the motor.

Manifold

1. Fabricated fitting that includes several evenly spaced outlets to allow connections to branch piping. Acts in the same capacity as a header.

2. Portion of refrigerant main in which several branch lines are joined together; also, single piece in which there are several fluid paths.

Manifold gauge. Device constructed to hold the low- and high-pressure gauges and two valves and connect them to a refrigeration unit.

Manifold service. Device equipped with gauges, service hoses, and valves, used for servicing refrigeration equipment.

Manual motor starter. Motor switch operated by hand.

Manual shutoff valve. Hand-operated device to stop the flow of liquids in a piping system.

Master switch. Main switch that controls starting and stopping of the entire system.

Mechanical efficiency. Ratio of the horsepower imparted to the air or gas to the brake horsepower; in a displacement-type compressor, the ratio of the air- or gas-indicated horsepower to the indicated horsepower of the power cylinders of a steam-engine-driven or internal-combustion-engine-driven compressor, or to the brake horsepower delivered to the shaft in a power-driven compressor.

Methyl chloride. Chlorinated hydrocarbon refrigerant.

Microfarad. Practical unit of measure for a capacitor; one-millionth of a farad.

Mixer. Vessel or device for mixing the refrigerant with another substance.

Modulating thermostat. Temperature controller employing a potentionmeter winding instead of switch contacts.

Moisture indicator. Instrument used to measure the moisture content in a refrigeration system.

Moisture separators. Devices for collecting and removing moisture precipitated from the air or gas during the process of cooling.

Monochlorodifluoromethane. Common refrigerant known as "Refrigerant-22" or "Freon-22" (R-22 or F-22).

Motor-driven sequence controller. Program device to control progressively the solenoid valves of certain types of compressor capacity modulation systems.

Motor-operated valve. Valve and motor combination in which the motor is operated to open and close the valve.

Muffler. Device installed in hot-gas line to silence discharge surges.

Multiple compressors. Two or more compressors installed in parallel.

Multiple-dwelling system. Refrigerating system employing the direct system in which the refrigerant is delivered by a pressure-imposing element to two or more evaporators in separate refrigerators or refrigerated spaces located in rooms of separate tenants in multiple dwellings.

Multiple end use chilled water system. The system consists of multiple chillers and pumps and a differential pressure-controlled bypass valve arrangement.

Multiple system. System employing the direct system of refrigeration in which the refrigerant is delivered to two or more evaporators in separately refrigerated spaces.

Multispeed. Term usually applied to a machine designed to operate at more than one speed.

Multistage compressors or compound compressors. Compressors in which compression from initial to final pressure is completed in two or more distinct steps or stages.

Multistep compressor capacity modulation. Compressor capacity modulation system arranged to reduce capacity in two or more consecutive steps.

Natural draft cooling tower. Tower in which the flow of air depends on natural air currents or a breeze, generally applied where the spray water is relatively hot and will cause some convection currents.

Natural frequency. Frequency that exists naturally in a spring or rubber-in-shear.

Natural ventilating system. Ventilating system whose effectiveness depends on natural atmospheric conditions and on the operation of windows, transoms, and other openings, the operation of which is in control of the person or persons in the room or space that is ventilated.

N **exponent.** *N* occurs in the equation of polytropic compression.

Noncombustible air filter. Filter that in itself or by treatment is of such construction and composition that fire spreading over its surface when it is loaded with dust and under operating conditions will not be fed by the burning of the filter itself nor cause the generation of any smoke or toxic gases.

Noncondensable gas. Gas that is combined with the refrigerant and that cannot be condensed at temperatures near the condensing temperature of the refrigerant—usually air or impurities.

Nonpositive displacement compressor. Compressor in which increase in vapor pressure is attained without changing the internal volume of the compression chamber.

Nonrecycling control relay. Electrical device installed in the control circuit to prevent the refrigeration system from operating when temperature controls are not calling for cooling.

Nonrecycling pump-down circuit. Refrigerant circuit, including a nonrecycling control relay, in which a liquid line solenoid valve is operated by a control thermostat. When the liquid line solenoid valve is closed, the compressor pumps down the low side of the refrigeration system to a predetermined setting on a low-pressure cut-out switch which stops system operation.

Off-and-on thermostat. Thermostat designed to open or close an electric circuit in response to temperature change.

Off cycle. Period when equipment, specifically a refrigeration system, is not in operation.

Oil check valve. Valve of the check valve type, installed between the suction manifold and the crankcase of a compressor, intended to permit oil to return to the crankcase but prevent exit of the oil from the crankcase on starting.

Oil filter. Device in the compressor to remove foreign matter from the crankcase oil before it reaches the bearing surfaces.

Oil level. Level in a compressor crankcase at which oil must be carried for proper lubrication.

Oil loop. Loop placed at the bottom of a riser for the purpose of forcing oil to travel up the riser.

Oil pressure failure control. Device that acts to shut off a compressor whenever the oil pressure falls below a predetermined set point.

Oil pressure gauge. Device to show the oil pressure developed by the pump within a refrigeration compressor.

Oil pump. Device that provides the source of power for force-feed lubrication systems in reciprocating compressors.

Oil return line. Line carrying the oil collected by an oil separator back to the compressor crankcase.

Oil separator. Device for separating out oil entrained in the discharge gas from a compressor and returning it to the compressor crankcase.

Oil sight glass. Glass "bull's-eye" in the compressor crankcase permitting visual inspection of the compressor oil level.

Oil trap. Low spot, sag in the lines, or space where oil will collect; also, mechanical device for removing entrained oil.

On cycle. Period when equipment, specifically a refrigeration system, is in operation.

Open motor. Motor in which air is circulated directly over the motor windings.

Open reciprocating compressor. Reciprocating compressor coupled externally to an open motor.

Operating charge. Total amount of refrigerant required by a system for correct operation.

Outdoor design temperature. Outdoor temperature arbitrarily established as the maximum against which the system must be able to maintain the desired indoor conditions. For economic planning it is somewhat lower or higher than the actual maximum.

Outlet or supply opening. An opening, the sole purpose of which is to deliver air into any space to provide heat, ventilation, or air-conditioning.

Outside air. Air that is taken from outside the building and is free from contamination of any kind in proportions detrimental to the health or comfort of the persons exposed to it.

Outside intake. Includes the ducts and outdoor openings through which outside air is admitted to a ventilating, air-conditioning, or heating system.

Overall efficiency. Ratio obtained by dividing the theoretical power required by the actual power required.

Overload. Load greater than that for which the system or machine is intended.

Overload protection. Device designed to stop the motor should a dangerous overload occur.

Overload protector. Device used to protect the system from overload by excess pressure, temperature, or current.

Package cooling towers. Several types are available. The most frequently used:

1. Small crossflow induced draft package cooling tower, designed to provide from 5 tons to 500 tons of refrigeration,
2. Counterflow forced draft package cooling tower, designed to provide from 20 tons to 1600 tons of refrigeration. Manufacturers may vary in capacities for their specific cooling towers.

Packaged terminal air conditioner. A room air conditioner consisting of a factory-selected combination of heating and cooling components, assemblies, or sections, intended to serve an individual room or zone and constructed in a manner that complies with the definition contained in the Standard for Packaged Terminal Air Conditioners approved by the Air-Conditioning and Refrigeration Institute in 1976, known as ARI-76, revised and updated.

Packed angle receiver valve. Angle valve of the packed stem type, usually used to close off the outlet of a refrigerant receiver. Because of its design and construction, it lends itself to many shutoff valve applications in refrigeration work.

Packing. Resilient impervious material placed around the stems of certain valves to prevent leakage; also, slats or surface in cooling towers designed to increase the water-to-air contact.

Packless diaphragm valve. Manual valve in which the stem packing is eliminated by the use of a diaphragm through which the stem motion is transmitted.

Pilot control. Valve arrangement in an evaporator pressure regulator to sense the pressure in the suction line and regulate the main valve.

Pipe duct. Tube or conduit used for encasing pipe.

Piping. Pipe or tube mains for interconnecting the various parts of the refrigerating system.

Piston. Disk fitted to slide in a cylinder and connected with a rod for exerting pressure on a fluid in the cylinder.

Piston clearance. Space between the top face of piston and the valve assembly when the piston is at the top of its stroke.

Pitch. Slope of a pipeline for the purpose of enhancing drainage.

Plenum. Air compartment or chamber to which one or more ducts are connected and that forms part of either the conditioned air supply, circulating air supply, or exhaust air system, other than the occupied space being conditioned.

Plenum chamber. Air compartment or enclosed space to which one or more distributing air ducts are connected.

Plug-in air-conditioning appliance. A complete factory-tested air-conditioning unit or a window air-conditioning unit, in a suitable frame or enclosure, that is fabricated and shipped in one complete assembly in a ready-to-operate condition, requiring no refrigerant containing parts to be connected in the field and requiring no field expertise to place it in operation.

Polytropic compression. Polytropic compression is effected when heat is transferred to or from the gas during the compression process at a precise rate. The relation between pressure p and volume v can be expressed by the equation $pv^n = c$, in which n is constant. When the actual compression path for a particular compressor is known, and when the heat transfer to or from the gas is at the proper rate, the value of n may be determined from the equation.

Portable comfort cooling unit. Self-contained refrigerating system, with not over 3-hp rating, that has been factory assembled and tested and installed without conditioned air ducts and without connecting any refrigerant-containing parts.

Portable compressors. Compressors consisting of compressor and driver so mounted that they may be readily moved as a unit.

Portable evaporative cooler. Evaporative cooler that discharges the conditioned air directly into the conditioned area without the use of ducts and can be readily trans-

ported from place to place without the dismantling of any portion thereof.

Positive displacement compressor. Compressor in which increase in vapor pressure is attained by changing the internal volume of the compression chamber.

Potential relay. Voltage-controlled switch used to disconnect the starting windings of a compressor motor.

Power element. Actuating mechanism of a temperature control or thermostatic expansion valve.

Pressure control

1. Control used to stop a motor at a certain pressure, sometimes used as a safety device.
2. Pressure-triggered device for controlling the operations of a machine, engine, or motor.

Pressure controller. Controller sensitive to changes in pressure.

Pressure drops. Loss of pressure due to friction or lift.

Pressure drop due to lift. Difference in pressure between the top and bottom of a column of fluid due to the weight of the fluid.

Pressure-imposing element. Device or portion of the equipment used for the purpose of increasing the refrigerant vapor pressure.

Pressure-limiting device. Pressure- or temperature-responsive mechanism for automatically stopping the operation of the pressure-imposing element at a predetermined pressure.

Pressure or compression ratio. Ratio of the absolute discharge pressure to the absolute inlet pressure.

Pressure relief device. Pressure-actuated valve or rupture member designed to relieve excessive pressure automatically.

Pressure relief valve. Pressure-actuated valve held closed by a spring or other means and designed to relieve pressure in excess of its setting automatically.

Pressure rise. Difference between the discharge pressure and the inlet pressure.

Pressure-sensing device. Part of the capacity control actuator of the capacity modulation system that is sensitive to variations in suction pressure.

Pressure tube. Small line carrying pressure to the sensitive element of a pressure controller.

Pressure vessel

1. Any refrigerant-containing receptacle of a refrigerating system other than expansion coils, headers, and pipe connections.
2. Any refrigerant-containing receptacle of a refrigerating system other than evaporators (each separate section of which does not exceed $\frac{1}{2}$ cu ft of refrigerant-containing volume), expansion coils, compressors, controls, headers, pipe; and pipe fittings.

PSI. Pressure in pounds per square inch.

Psychometer. Device for measuring the humidity in the air and employing a wet bulb and a dry bulb thermometer.

Psychrometric chart. Graph indicating the properties of air-stream mixtures and used primarily in the design of air-conditioning systems.

Pulling down. Removing refrigerant from a part or removing all of the refrigerant from a system; literally, creating a vacuum in a closed system.

Pump down. Reduction of pressure with a system.

Pumping down. Removing the refrigerant from one part of the system to another part; removing the refrigerant from a system.

Purge. Discharge impurities and noncondensable gases to the atmosphere.

Purge valve. Valve through which noncondensable gases may be purged from the condenser or receiver.

Purging. Removing the refrigerant from a system and letting it bleed into the atmosphere. This is followed by recharging the system.

Push button station. Switching or controlling device equipped with marked buttons. Depression of the buttons opens or closes contacts in electrical circuits.

Quick-connect or quick-disconnect coupling. Device used for fast, easy connection of refrigeration lines.

Readily accessible. Capable of being reached safely and quickly for operation, repair, or inspection without requiring those to whom ready access is requisite to climb over or remove obstacles.

Receiver. Vessel for storing the refrigerant liquefied by the condenser.

Receiver shutoff valve. Valve in the line that connects the condenser to the receiver, usually located at the inlet to the receiver.

Reciprocating compressor

1. Compressor that imparts motion and pressure to a gas by means of reciprocating pistons; positive-displacement machine.
2. Compressor that uses a piston and cylinder to provide the pumping action.
3. Compressor in which each compressing element consists of a piston moving back and forth in a cylinder.

Reduced-voltage motor starter. Motor starter, either magnetic or manual, having a means for reducing the voltage temporarily at starting. It is usually equipped with a timing device to increase the voltage to full line as the motor approaches full speed.

Refrigerant

1. Fluid used to produce a cooling effect by means of evaporation under controlled conditions.
2. Substance used in a refrigeration system to absorb heat in the evaporator and release it in the condenser as it changes in physical state; sometimes called "gas."
3. Medium for conveying heat in a refrigerating system. It is evaporated by absorbing heat at a lower temperature and liquefied by surrounding heat at a higher temperature.
4. Substance used to produce refrigeration by expansion or vaporization of such substance in a closed thermodynamic cycle.
5. Medium used to produce cooling or refrigeration by the process of expansion or vaporization.

Refrigerant charge. Amount of refrigerant a system holds, usually expressed in pounds per square inch gauge.

Refrigerant coil. Evaporator or condenser made up of tubing either with or without extended surface (fins).

Refrigerant control

1. Device used to control the amount of refrigerant that flows through the system.
2. Device that meters the correct amount of refrigerant into the evaporator and maintains a certain pressure.

Refrigerant drum-valve. Shutoff valve to control the flow of refrigerant from the drum.

Refrigerant filter. Very fine strainer for removing foreign matter and dirt from the refrigerant.

Refrigerant gas. Refrigerant in the gaseous state.

Refrigerant pressure vessel. Refrigerant-containing receptacle that is a portion of a refrigeration system but does not include evaporators, headers, or piping.

Refrigerant receiver. Vessel permanently connected to a system by inlet and outlet pipes for storage of a liquid refrigerant.

Refrigerant tables. Tables that show the properties of saturated refrigerants at various temperatures.

Refrigerant-12. Dichlorodifluoromethane—a refrigerant commonly used in refrigeration and air-conditioning systems.

Refrigerant-22. Monochlorodifluoromethane—a refrigerant commonly used in refrigeration and air-conditioning systems.

Refrigerant velocity. Movement of the gaseous refrigerant required to entrain oil mist and carry it back to the compressor.

Refrigerating capacity. Rate at which a system can remove heat, usually stated in tons or British thermal units (Btu) per hour.

Refrigerating circuit. Course followed by the refrigerant in passing through the evaporator, compressor, condenser, and back to the evaporator.

Refrigerating effect. Amount of heat a given quantity of refrigerant will absorb in changing from a liquid to a gas at a given evaporating pressure.

Refrigerating fluid. Fluid used to transfer heat between cold refrigerant and the substance or bodies to be cooled, by circulation of the fluid without change of state, or by evaporation-condensation process at essentially equal pressures.

Refrigerating systems. Combination of interconnected refrigerant-containing parts constituting one closed refrigerant circuit in which a refrigerant is circulated for the purpose of extracting heat.

Refrigeration

1. Complete system used for process space, or product cooling and humidity control, other than for human comfort.
2. Mechanical process of extracting heat from the air or other medium in an enclosed space in a building or structure.

Refrigeration condenser. A vapor condenser in a refrigeration system, there the refrigerant is liquefied and discharges its heat to the environment.

Refrigeration cycle. Complete operation involved in providing refrigeration.

Refrigeration duty motor. Squirrel cage motor wound for high starting torque and low starting current.

Refrigeration oil. Special oil used in refrigeration systems.

Refrigerator. Room or space in which an evaporator or brine coil is located for the purpose of reducing or controlling the temperature below 50°F.

Refrigerator system ratings. Refrigerator system ratings are expressed in terms of 1 hp, 1 ton, or 12,000 Btu per hour, all equal to the same quantity.

Relative humidity

1. Amount of moisture in the air stated in terms of percentage of total saturation at the existing dry-bulb temperature.
2. Ratio of the weight of water vapor actually present in a unit volume of air to the weight that would be

present if the air were saturated with vapor at its actual temperature.

Relay. Device that is operative by a variation in the conditions of one electric circuit to effect the operation of other devices in the same or another electric circuit.

Relief bypass. Direct connection from the discharge to the suction side of the compressor, port that opens at a set point and relieves abnormally high discharge pressure to the suction side of the compressor.

Relief valve

1. Valve designed to relieve the pressure from a vessel or system whenever the pressure exceeds the setting of the valve.

2. Valve designed to be normally closed or opened, to open or close at a predetermined pressure to relieve pressure in a system, and to return to its original condition when a predetermined safe pressure is attained.

Remote bulb. Part of the expansion valve. The remote bulb assumes the temperature of the suction gas at the point where the bulb is secured to the suction line. Any change in the suction gas superheat at the point of bulb application tends to operate the valve in a compensating direction to restore the superheat to a predetermined valve setting.

Remote system

1. Refrigeration system that has the condensing unit outside and the evaporator inside the area to be cooled.

2. Refrigeration system in which the compressor or generator is located in a space other than the cabinet or fixture containing the evaporator.

Repulsion induction motor. Motor with commutated bars in the rotor and a centrifugal mechanism that when the motor approaches full running speed short-circuits the commutator.

Residual gas. Gas remaining the clearance space of a compressor cylinder after the piston has reached the top of its stroke.

Resistance ($R = 1/c$). Overall resistance; amount of resistance to heat flow between air on the warm side and air on the cold side of the building section.

Resistor. Device offering electrical resistance, used in an electrical circuit for protection or control.

Return air. Air returning to a room conditioner or a duct system from the conditioned space.

Return duct. Duct for conveying air from a space being heated, ventilated, or air-conditioned back to the heating, ventilating, or air-conditioning appliance.

Reversing valve. Device used to reverse the flow of refrigerant in a heat pump system.

Riser. Vertical tube or pipe that carries refrigerant in any form from a lower to a higher level.

Room air conditioner. A factory encased air conditioner designed as a unit for mounting in a window or through a wall or as a console. It is designed for delivery of conditioned air to an enclosed space without ducts. "Room air conditioner" includes packaged terminal air conditioners.

Rotary compressor

1. Compressor that imparts motion and pressure to a gas by means of a rotating impeller usually sealed with a sliding blade; positive-displacement machine.

2. Compressor that uses vanes instead of pistons for the pumping action.

Rupture member

1. Pressure relief device that operates by the rupture of a diaphragm within the device.

2. Mechanical device that will rupture at a predetermined pressure to control automatically the compressor or maximum pressure of operation of the refrigerant.

R-12 (dichlorodifluoromethane). Refrigerant-12 or Freon-12, often used in small air-conditioning units.

R-22 (monochlorodifluoromethane). Refrigerant-22 or Freon-22, a common refrigerant.

Saddle valve (line tap valve). Valve that can be installed around the refrigeration lines so that a pressure reading can be taken. It can also be used to tap a water line.

Safety factor. Ratio of extra strength or capacity to the calculated requirement. A safety factor ensures freedom from breakdown and ample capacity.

Saturation pressure. Pressure at which gas at any specific temperature is saturated.

Saturation temperature. Boiling point of a refrigerant at a given pressure. In refrigeration it is considered the evaporator temperature.

Schraeder valve. Service valve that can be installed in a refrigeration system to read the pressure (it has a spring-loaded core).

Sealed absorption system. Unit system for refrigerants only in which all refrigerant-containing parts are made permanently tight by welding or brazing against refrigerant loss.

Sealed unit. Pressure-imposing element that operates without a stuffing box, or that does not depend on contact between moving and stationary surfaces for refrigerant retention.

Self-contained. Having all essential working parts except energy and control connections so contained in a case or framework that they do not depend on appliances or fastenings outside the machine.

Self-contained air-conditioning unit. Air conditioner containing a condensing unit evaporator, fan assembly, and complete set of operating controls within its casing.

Self-contained refrigeration system. Complete factory-made and factory-tested refrigerating system in a suitable frame or enclosure, fabricated and shipped in one or more sections, and in which no refrigerant-containing parts are connected in the field other than by companion or block valves.

Semicombustible air filter. Filter that in itself or by treatment is sufficiently fire-resistant so that fire spreading over its surface when loaded with dust and under operating conditions will not be materially fed by the burning of the filter itself, nor cause the generation of quantities of smoke or toxic gases.

Sensible cooling effect. Term used to describe the difference between the total cooling effect and the dehumidifying effect.

Sensible heat

1. Term used in heating and cooling to indicate any portion of heat that changes only the temperature of the substances involved.

2. Heat that when added or subtracted results in a change of temperature, as distinguished from latent heat.

Serpentining. Arrangement of tubes in a coil to provide circuits of the desired length. It is intended to keep pressure drop and velocity of the substance passing through the tubes within the desired length.

Service valve. Device used on a system so that a pressure reading can be obtained.

Shaded pole motor. Small induction motor having a shading pole for the purpose of starting and very low starting torque.

Shell and coil. Designation for heat exchangers, condensers, and chillers consisting of a tube coil within a shell or casing.

Shell and tube. Designation for heat exchangers, condensers, and chillers consisting of a tube bundle within a shell or casing.

Shell-type apparatus. Refrigerant-containing pressure vessel having tubes for the passage of a heating, cooling, or refrigerating fluid.

Short cycle. Too frequent starting and stopping; short on and off cycles.

Shutter control. Damper assembly and pressure-operated actuator used on systems with non-unloading-type hermetic compressors in air-cooled condensing unit to maintain minimum required condensing pressure during low ambient operation.

Sight glass. Glass window or tube installed in a refrigeration system to see the refrigerant flow or to check the oil.

Single-acting compressors. Compressors in which compression takes place on but one stroke per revolution in each compressing element.

Single package. Complete factory-made and factory-tested refrigeration system in a suitable frame or enclosure that is fabricated and shipped in one or more sections and in which no refrigerant containing parts are connected in the field.

Single-stage compressors. Compressors in which compression from initial to final pressure is completed in a single step or stage.

Single-step compressor capacity modulation. Compressor capacity modulation system arranged to reduce capacity in one step. It usually reduces capacity to one-half of full capacity.

Sleeve-bearing motor. Motor with a sleeve-type babbit or bronze bearing which requires oil lubrication.

Slip ring motor. Polyphase induction motor whose rotor has a winding welded to the end ring and provided with slip rings through which electrical connections are made to the rotor windings.

Smoke detector. Device installed in the plenum chamber or in the main supply air duct of an air-conditioning system to detect the presence of smoke. Upon so doing it automatically shuts off the blower and closes a fire damper.

Soap bubble test. Method of detecting leaks by coating with a soap solution. Any major leak will cause bubbles to form.

Soldered joint. Gas-tight joint obtained by the joining of metal parts with metallic mixtures or alloys that melt at temperatures below 1000°F and above 400°F.

Solenoid coil. Electrical winding having an open core, used for operating solenoids.

Solenoid valve. Magnetically operated valve generally used to control the flow of liquid to an evaporator. It may also be used wherever off-on control is permissible. Solenoid electrical winding controls the action of the valve.

Specific gravity. The specific gravity of a given air or gas is the ratio of the weight of this air or gas to the weight of dry air at same pressure and temperature.

Specific heat. The number of British thermal units required to raise 1 lb of a substance 1°F. For air 0.241 Btu may be used and for water vapor 0.444 Btu.

Speed. In air compression, speed refers to the revolutions per minute (rpm) of the compressor shaft.

Splashproof motor. Motor protected against splashing water or against rain when exposed to the weather.

Split-phase motor. Induction motor having a separate winding for starting.

Split system. Air-conditioning system that places the condensing unit outside and the evaporator inside the house.

Spray-type air-cooler. Air-cooler having a forced circulation wherein the coil surface capacity is augmented by a liquid spray during the period of operation.

Spray water. Water sprayed over the coils of an evaporative condenser or the packing of a cooler tower.

Squirrel cage motor. Straight induction motor.

Standard air

1. Air with a density of 0.075 lb/cu ft. This is substantially equivalent to dry air at 70°F and 29.92-in. (HG) barometer.

2. Air at a temperature of 68°F, a pressure of 14.70 psia and a relative humidity of 36% (0.0750-lb/cu ft density).

Standby. Additional machine that is used as a substitute in times of emergency.

Starter holding coil circuit. Circuit in a magnetic starter that is energized when the start button is pressed to close the main contacts and remains energized as long as the main current flows through the starter to the motor.

Starting torque. Turning effort of an electric motor on starting.

Static head. Pressure due to the weight of a fluid in a vertical column; more generally, resistance due to lift.

Static pressure. Pressure measured in air or gas in such a manner that no effect on the measurement is produced by the velocity of the air or gas.

Static temperature. Actual temperature of a moving gas stream. It is related to total temperature in the same way that static pressure is related to total pressure and measured in such manner that no effect on the measurement is produced by the velocity of the air or gas.

Stop valve. Shutoff for controlling the flow of refrigerant.

Storage capacity. Volume of fluid that may be safely stored in a vessel such as a receiver.

Stroke. Length of travel of the piston from top to bottom of the cylinder; maximum travel without reversal of direction.

Subcooling. Cooling of liquid below its condensing temperature.

Subcooling coil. Supplement coil in an evaporative condenser, usually a coil or loop immersed in the spray water tank, for the purpose of reducing the temperature of the liquid leaving the condenser.

Suction gas. Gas entering the suction side of the compressor.

Suction hold-back valve. Regulating valve in the suction line designed to prevent the suction pressure at the compressor from rising above the predetermined setting of the valve, used to prevent overloading the compressor.

Suction line

1. Pipe to conduct refrigerant vapor from the evaporator to the compressor.

2. Line used to carry the vapor that has boiled off in the evaporator back to the compressor; low-side line.

Suction manifold. Device to distribute suction gas equally from a common suction line to multiple compressors connected in parallel.

Suction pressure. Pressure forcing the gas to enter the suction inlet of the compressor.

Suction pressure gauge. Gauge connected in the suction side of a refrigerating system for the purpose of measuring the pressure within the system at that point.

Suction pressure regulator. Automatic valve or control device designed to maintain the pressure, and therefore the temperature, in an evaporator above a predetermined minimum.

Suction riser. Vertical tube or pipe that carries suction gas from an evaporator on a lower level to a compressor on a higher level.

Suction shutoff valve. Valve in the suction line to the compressor; usually mounted on the compressor body or compressor suction manifold.

Suction stop valve. Service valve to open and close the suction line.

Suction temperature. Temperature of the gas as it enters the compressor.

Sump. Reservoir in which a fluid is collected before recirculation.

Superheat. Temperature increase above the saturation temperature or above the boiling point.

Surface cooling. Air cooled by passing it over or through cold surfaces.

Switch disconnect. Switch usually provided for a motor that will completely disconnect the motor from the source of electric power. It is particularly advantageous to disconnect the refrigeration system from the source of power.

Synchronous motor. Motor having a stationary armature winding that is connected on the polyphase power line, a moving field winding that is connected through slip rings to a direct-current supply, and a damper winding on the rotor that is short-circuited, similar to the rotor of a squirrel cage motor.

Take-up ring. In the capacity modulation mechanism, the ring that raises and lowers the lift pins to transmit the response of the pressure-sensing device to the compressor suction valves.

Temperature control. Mechanism used to control the temperature that starts and stops the compressor.

Temperature controller. Controller sensitive to changes in temperature; thermostat.

Temperature rise efficiency. Ratio of the theoretical horsepower to the horsepower computed from the measured inlet temperature and the measured discharge temperature of the air or gas. Under certain conditions it may be used to determine compression efficiency (adiabatic). These conditions include constancy of specific heats of the air or gas, determination of loss of heat from the compressor, and extreme accuracy of air temperature measurement to avoid errors caused by velocity and radiation. Temperature rise efficiency is useful principally to designers and is not satisfactory as the basis of a guarantee or commercial statement of performance.

Test charge. Amount of gas forced into a refrigerating system for leak-testing purposes.

Theoretical horsepower. Horsepower required to compress adiabatically the air or gas delivered by the compressor through the specified pressure range.

Thermal conductance. The time rate of heat flow of a body between two definite surfaces, under steady state conditions, divided by the difference of their average temperatures and by the area of one of the surfaces.

Thermal conductivity. The basic unit used in measuring heat flow is thermal conductivity, defined as the number of British thermal units (Btu) that will flow through a material 1 sq ft and 1 in. thick owing to a temperature difference of $1°F$ in 1 hour.

Thermal overload elements. Alloy piece holding an overload relay closed that melts when the current drawn is too great.

Thermal overload relay. Thermal device that opens its contacts when the current through a heater coil exceeds the specified value for a given time.

Thermal resistance. Reciprocal or thermal transmittance; U value.

Thermal transmittance. (U overall coefficient of heat transfer). The ratio of the steady state of heat flux from the surroundings on one side of a body, through the body, to the surroundings on its opposite side. Time rate of heat flow per unit area of a surface that must be identified to the temperature difference between the two surroundings.

Thermometer. Instrument for measuring temperature.

Thermometer well. Small pocket or recess in a pipe or tube designed to provide good thermal contact with a test thermometer.

Thermostat

1. Device for controlling equipment in response to temperature change; temperature-sensitive controller.

2. Device used to control the temperature in a confined area.

Thermostatic expansion valve

1. Expansion valve designed to meter the flow of liquid to a dry expansion evaporator at a rate sufficient to maintain a constant superheat in the gas leaving the evaporator.

2. Control valve that meters refrigerant into the evaporator and is controlled by pressure and temperature.

Throttling valve. Small valve used primarily in gauge lines to shut off the line when readings are not to be made and in the throttling line to prevent fluctuations when readings are being made.

Time delay relay. Relay that is activated after a predetermined time has elapsed from the point of impulse.

Ton of refrigeration

1. Unit of refrigeration capacity corresponding to the removal of 200 Btu/minute, 12,000 Btu/hour, or 288,000 Btu/day; so named because it is equivalent in cooling effect to melting 1 ton of ice in 24 hours.

2. Heat removal at the rate of 12,000 Btu/hour. Compressor capacity is based on $5°F$ evaporator temperature and is $86°F$ condenser temperature, except that the capacity of a compressor when used for comfort cooling or air-conditioning purposes is based on $40°F$ evaporator temperature.

Torque. That which tends to produce rotation or torsion.

Total cooling effect. Term used to describe the difference between the total heat content of the airstream mixture entering a conditioner per hour and the total heat of the mixture leaving per hour.

Total heat load. Sum of latent heat load and sensible heat load.

Total load. "Sum of the diversified group loads times the usage factor.

Totally enclosed fan-cooled motor. Motor built with internal fans to circulate air between an interior shell and an exterior shell.

Totally enclosed motor. Motor in which motor heat is liberated through the motor shell and that may be used where dust conditions are severe.

Total pressure. Pressure that would be produced by stopping a moving air or gas stream; pressure measured by an impact tube. In a stationary body of air or gas, the static and total pressures are numerically equal.

Total temperature. Temperature that would be measured at the stagnation point if a gas stream were stopped, with adiabatic compression from the flow condition to the stagnation pressure.

Tracer gas. Gas having a powerful odor, sometimes used in small quantities with odorless refrigerants to give warning of a leak.

Transmittance. Term used to describe the rate of heat flow per unit area per unit temperature difference.

Tubeaxial fan. Propeller fan usually with adjustable blades within a cylinder. The driving mechanism can be direct connected or by a belt. Motor usually supported on top of the cylinder.

Tube within a tube. Term used to identify a heat-exchange surface or condensers constructed of two concentric tubes.

Two-pipe dual temperature system. The system serves two functions: Hot water is circulated through the terminals during the cold weather season, and chilled water is circulated during the hot weather season. The distribution system may be divided into zones, each of which is capable of changeover from heating to cooling of the other zones.

Two-state compressors. Compressors in which compression from initial to final pressure is completed in two or more distinct steps or stages.

U **factor**

1. Overall heat transmission coefficient; the amount of heat, expressed in British thermal units (Btu) transmitted in 1 hour through 1 sq ft of a building section (wall, floor, or ceiling) for each degree Fahrenheit of temperature difference between air on the warm side and air on the cold side of the building section.

2. Time rate of heat flow (Btu/hour) for 1 sq ft of surface for a temperature difference of 1°F between the fluids (air) on the two sides of this surface.

Unit compressor. A unit consisting of a compressor, motor, drive, and frequently the essential compressor controls, all mounted on a common base.

Unit cooler. Factory assembled, direct-cooling encased unit including a cooling element, fan, and motor and air directional outlet.

Unit refrigeration system. Refrigerating unit, not to exceed a 3-hp rating, that has been factory assembled and tested prior to its installation. The unit is not connected to any duct work. It is a complete one-unit package without remote parts.

Unit system

1. Self-contained system that has been assembled and tested prior to its installation and that is installed without connecting any refrigerant-containing parts. The unit system may include factory-assembled companion or block valves.

2. System that can be removed from the user's premises without disconnecting any refrigerant-containing parts, water connections, or fixed electrical connections.

Usage factor. Ratio of the coincident diversified demand of several groups of apparatus to the sum of the individual diversified demands of the groups.

Useful oil pressure. Difference in pressure between the discharge and suction sides of a compressor or pump.

Vacuum

1. Reduction in pressure below atmospheric pressure, usually stated in inches of mercury.

2. Pressure below atmospheric pressure; negative pressure.

Vacuum pump

1. Pump for exhausting a system; pump designed to produce a vacuum in a closed system or vessel.

2. Pump used to remove the air and moisture from the inside of a refrigeration system.

Vacuum pumps. Vacuum pumps are machines for compressing air or gas from an initial pressure that is below atmospheric to a final pressure that is near atmospheric.

Valve lift pin. Part in a capacity modulating cylinder.

Valve port. Passage in a valve that opens and closes to control the flow of a fluid in accordance with the relative position of the valve button to the valve seat.

Valve (purge). A valve through which noncondensable gases may be purged from the condenser or receiver.

Valve (receiver shutoff). A valve in the line that connects the condenser to the receiver. Usually located at the inlet to the receiver.

Valve (refrigerant drum). A shut-off valve to control the flow of refrigerant from the drum.

Valve (relief). A valve designed to relieve the pressure from a vessel or system, whenever the pressure exceeds the setting of the valve.

Valve (solenoid). A magnetically operated valve generally used to control the flow of liquid to an evaporator, but may be used wherever off-on control is permissible. A solenoid electrical winding controls the action of the valve.

Vaneaxial fan. Propeller fan with as many as seven blades within a cylinder having air guide vanes either before or after the blades. The motor is mounted on the cylinder or behind the cylinder.

Vapor. Fluid in the gaseous state following evaporation.

Vaporization. Evaporation or boiling.

Velocity chart. Charts constructed for each type of refrigerant to show gas velocity in feet per minute through various sizes of tubing at standard refrigeration conditions.

Velocity pressure. Total pressure minus the static pressure in an air or gas stream, generally measured by a Pitot tube. In displacement-type compressor practice, it is usually so small that it can be neglected.

Vent

1. Port or opening through which pressure is relieved.

2. Listed factory-made vent pipe and vent fittings for conveying flue gases to the outside atmosphere.

Ventilating openings. In any room or space ventilating openings are defined as apertures opening onto a public way, yard, court, public park, public waterway, or onto a roof of a building or structure in which the room or space is situated. Included are windows, skylights, transoms, or other openings that are provided for ventilating purposes and that are equipped with adjustable louvers, dampers, or other devices to deflect or diffuse the air currents.

Ventilation

1. Process of supplying or removing air by natural or mechanical means to or from any space. Such air may or may not have been conditioned.

2. Providing and maintaining in rooms or spaces, by natural or mechanical means, air-conditioning that will protect the health and comfort of the occupants thereof.

Volatile fluid. Fluid that vaporizes easily.

Volumetric efficiency

1. Ratio obtained by dividing the actual volume of gas delivered by the displacement.

2. Ratio of the capacity of the compressor to the displacement of the compressor. The term does not apply to centrifugal compressors.

Water (condensing). The water supplied to cool a water-cooled condenser or condensing unit.

Water-cooled condenser. Condenser designed to remove heat from gas by transferring the heat to water flowing through the condenser.

Water-cooling induced draft tower. Enclosed device utilizing one or more fans to move the air through the tower, the fans being an integral part of the tower.

Water-cooling tower. Enclosed device for evaporatively cooling water by contact with air.

Water make up. Water added to a cooling tower or evaporative condenser to replace that lost through evaporation or other causes.

Water-regulating valve. Pressure-operated valve used to control the flow of condenser water in proportion to condenser requirements as reflected by the condensing pressure.

Water treatment. Treatment of water with chemicals to reduce its scale-forming properties or change other undesirable characteristics.

Welded joint or seam. Joint or seam obtained by the joining of metal parts in the plastic or molten state.

Wet-bulb temperature. The thermodynamic wet-bulb temperature is the temperature at which liquid or solid water, by evaporating into air, can bring the air to saturation adiabatically at the same temperature. The wet-bulb temperature (without qualification) is the temperature indicated by a wet-bulb psychrometer constructed and used according to specifications.

Wet compression. Refrigeration system in which some liquid refrigerant is mixed with vapor entering the compressor so as to cause the discharge vapors from the compressor to be saturated rather than superheated.

Window air conditioner. Factory assembled unit designed to be installed in window openings, providing temperature control, cooling or heating cycles, ventilation, and filtered air.

Year-round air-conditioning system. System which provides ventilation and heating, humidification in winter and dehumidification in summer, and also provides the desired degree of air motion and cleanliness.

Zimmerli gauge. Device used to measure a vacuum in inches of mercury.

COMPRESSED GAS SYSTEM

Approach channel. Passages through which gas must pass from the cylinder to reach the operating parts of the safety relief device.

Cargo tank. Container designed to be permanently attached to any motor vehicle or other highway vehicle and which is to transport any compressed gas.

Combination frangible disc and fusible plug. A frangible disc in combination with a low-melting-point fusible metal is intended to prevent bursting at a predetermined bursting pressure unless the temperature is also high enough to cause yielding or melting of the fusible metal.

Combination safety relief valve and fusible plug. Safety relief device utilizing a safety relief valve in combination with a fusible plug. The combination device may be an integral unit or separate units and is intended to open and close at predetermined pressures or to open at a predetermined temperature.

Compressed gas in solution (acetylene). Nonliquefied gas that is dissolved in a solvent.

Containers. All vessels, such as tanks, cylinders, or drums, used for transportation or storing liquefied petroleum gases.

Compressed gas. Gas, other than acetylene, oxygen, or carbon dioxide, that when in the container is subject to an absolute pressure of more than 40 psi at 70°F or an absolute pressure of more than 104 psi; flammable liquid that has an absolute pressure of more than 40 psi at 100°F.

Discharge channel. Passages beyond the operating parts through which gas must pass to reach the atmosphere, exclusive of any piping attached to the outlet of the device.

DOT regulations. U.S. Department of Transportation regulations for transportation of explosives and other dangerous articles by land and water in rail freight and express and baggage services and by motor vehicle (highway) and water, including specifications for shipping containers. Code of Federal Regulations, Title 49, Parts 171 to 178.

Excess flow valve. Valve designed to prevent a flow rate greater than the maximum preset flow rate.

Filling density formula. The ratio of the weight of gas in a container to the weight of water the container will hold at 60°F, expressed as a percentage.

Flow capacity. Safety relief device having the capacity in cubic feet per minute of free air discharged at the required flow rating pressure.

Flow rating pressure. Pressure at which a safety relief device is rated for capacity.

Frangible disc. Operating part in the form of a disc, usually of metal, so held as to close the safety relief device channel under normal conditions. The disc is intended to burst at a predetermined pressure to permit the escape of gas.

Free air or free gas. Air or gas measured at a pressure of 14.7 psia and a temperature of 60°F.

Fusible plug. Operating part in the form of a plug of suitable low-melting material, usually a metal alloy, that closes the safety relief device channel under normal conditions and is intended to yield or melt at a predetermined temperature to permit the escape of gas.

Liquefied compressed gas. Gas under charging pressure is partially liquid at a temperature of 70°F. Flammable compressed gas is normally nonliquefied at 70°F but is partially liquid under the charging pressure and temperature following the requirements for liquefied compressed gases.

Nonliquefied compressed gas. Gas, other than a gas in solution, that under the charging pressure is entirely gaseous at a temperature of 70°F.

Operating part. Part of a safety relief device that normally closes the safety discharge channel, but that when moved from this position as a result of the action of heat or pressure, or a combination of the two, permits escape of gas from the cylinder.

Portable tank. Container designed primarily to be temporarily attached to a motor vehicle, railroad car other than a tank car, or marine vessel, and equipped with all of the necessary equipment and accessories to facilitate safe handling procedures for the container in which compressed gas is transported.

Pressure opening. Orifice against which the frangible disk functions.

Pressurized liquid compressed gas. Compressed gas, other than a compressed gas in solution, that cannot be liquefied at a temperature of 70°F and that is maintained in the liquid state at a pressure not less than 40 psia at a temperature less than 70°F.

Rated bursting pressure. Maximum pressure for which the frangible disk is designed to burst when in contact with the pressure opening for which it was designed when tested.

Reinforced fusible plug. Fusible plug consisting of a core of suitable material having a comparatively high yield temperature surrounded by a low-melting-point fusible metal of the required yield temperature.

Safety relief device. Device intended to prevent rupture of a cylinder under certain conditions of exposures.

Safety relief device channel. Channel through which gas released by operation of the device must pass from the cylinder to the atmosphere, exclusive of any piping attached to the inlet or outlet of the device.

Safety relief valve. Safety relief device containing an operating part that is held normally in a position closing the safety relief device channel by spring force and is intended to open and close at predetermined pressures.

Set pressure. Safety relief valve set pressure is the pressure marked on the valve and at which it is set to start the discharge.

Start to discharge pressure. For a safety relief valve, the pressure at which the first bubble appears through a water seal of not over 4 in. on the outlet of the safety relief valve.

Test pressure of the cylinder. Minimum pressure at which a cylinder must be tested as prescribed in Department of Transportation specifications for compressed gas cylinders.

Yield temperature. Fusible plug yield temperature is the temperature at which the fusible metal or alloy will yield when tested.

FIXTURES AND ACCESSORIES

Accessibility of fixtures. All plumbing fixtures should be installed and spaced so as to be reasonably accessible for their intended use and maintenance.

Accessible. "Accessible" when applied to a fixture connection, appliance, or equipment means having access thereto, but which first may require the removal of an access panel, door, or similar obstruction; "readily accessible" means direct access without the necessity of removing or moving any panel, door, or similar obstruction.

Allen wrench. Hexagonal-end wrench used to remove and install faucet valve seats.

Antisiphon ball cock. Device consisting essentially of a float valve equipped with a flow splitter to provide for tank and trap refill. It has an integral vacuum breaker and is used in conjunction with flush tanks.

Ball cock

1. Assembly inside a toilet tank that connects with the water supply and controls flow of water into tank.
2. Ball or device that floats on the surface of an enclosed container and opens or closes a water supply device in accordance with a predetermined release procedure.

Bar sink. Receptacle for the disposal of liquid wastes only.

Bathroom. Room containing at the least a water closet, lavatory, bathtub, and/or shower.

Battery of fixtures. Any group of two or more similar, adjacent fixtures that discharge into a common horizontal waste or soil branch.

Bedpan washer. A fixture designed to wash bedpans and to flush the contents into the soil drainage system. It may also provide for steaming the utensils with steam or hot water.

Bedpan washer hose. A device supplied with hot and cold water and located adjacent to a water closet or clinic sink to be used for cleansing bedpans.

Bidet. Bowl equipped with cold and hot running water, used for bathing the external genitals and posterior parts of the body.

Bonnet. Casing for wall-mounted bath and shower faucets. It screws into the faucet body behind the wall.

Bracket hanger. Preshaped metal hanger that supports a wall-hung sink or lavatory.

Chemical toilet. Toilet arranged to receive non-water-carried human waste directly into a deodorizing and liquefying chemical solution contained in a watertight tank.

Clinic sink or bedpan hopper. A sink designed primarily to receive wastes from bedpans, provided with a flush rim, integral trap with a visible trap seal, and having the same flushing and cleansing characteristics as a water closet.

Closet bolt. Bolt used to attach a water closet securely to the closet flange.

Closet spud. Connector between the base of the ball cock assembly in a water closet tank and the water supply pipe.

Combination fixture

1. Fixture combining one sink and tray or a two- or three-compartment sink or tray in one unit.
2. Fixture that is an integral combination of one sink and one or two laundry trays in one fixture or of a two- or three-compartment sink or laundry tray in one fixture.

Construction of fixtures. Plumbing fixtures should be made of smooth nonabsorbent material, free from concealed fouling surfaces, and the fixture enclosures should be ventilated.

Drainage fixture unit (d.f.u.). Measure of the probable discharge into a drainage system by various types of plumbing fixtures, 1 DFU being equal to 7.5 gal per minute discharge. The drainage fixture-unit value for a particular fixture depends on its volume rate of drainage discharge.

Drinking fountain. Free-standing or wall-mounted fixture that delivers a stream of water by actuating the release valve through a nozzle.

Escutcheon. Decorative piece that fits over the faucet body or pipe coming out of the wall.

Faucet. Valve that permits controlled amounts of water from a water pipe.

Filler tube. On a ball cock assembly, the tube through which water enters the toilet tank.

Fixture. A receptacle or device that is either permanently or temporarily connected to the water distribution system of the premises and demands a supply of water therefrom, or it discharges used water, liquid-borne waste material, or sewage either directly or indirectly to the drainage system of the premises, or that requires both a water supply connection and a discharge to the drainage system of the premises.

Fixture branch. A pipe connecting several fixtures.

Fixture drain. The drain from the trap of a fixture to the junction of that drain with any other drain pipe.

Fixture outlet pipe. Pipe that connects the waste opening of a fixture to the trap serving the fixture.

Fixtures. Includes water closets, bathtubs, sitz tubs, catch basins, slop sinks, kitchen sinks, urinals, wash trays, wash basins, lavatories, pantry sinks, showers, drinking fountains, floor drains, laundry tubs, and all other appliances requiring running water or connection to a sewer.

Fixture supply. The water supply pipe connecting a fixture to a branch water supply pipe or directly to a main water supply pipe.

Fixture unit. The fixture unit is the rate of discharge through a plumbing fixture of 7-$\frac{1}{2}$ gal per minute. This is termed "one fixture unit."

Fixture unit, drainage or d.f.u. A measure of the probable discharge into the drainage system by various types of plumbing fixtures. The drainage fixture unit value for a particular fixture depends on its volume rate of drainage discharge, on the time duration of a single drainage operation, and on the average time between successive operations.

Fixture unit, supply or s.f.u. A measure of the probable hydraulic demand on the water supply by various types of plumbing fixtures. The supply fixture unit value for a particular fixture depends on its volume rate of supply, on the time duration of a single supply operation, and on the average time between successive operations.

Float-arm. Wire arm that connects the float ball at one end to the ball cock assembly at the other in a toilet tank.

Float ball. Large copper or plastic ball that floats on the surface of the water in a toilet tank and descends and ascends with the water level.

Float valve

1. Valve in a ball cock assembly that controls the flow of fresh water into the toilet tank.
2. Valve used to control the water level in a tank or other container. It is operated by a float and is considered a positive operating valve.

Flush ball. The part of a water closet assembly that controls the flow of water into the water closet bowl.

Flush tank. Receptacle designed to discharge, either manually or automatically, a predetermined quantity of water to fixtures for flushing purposes.

Flush valve seat. Opening between the tank and the bowl in a water closet against which the flush ball is fitted.

Freezeless water faucet. Faucet designed to be installed through an exterior wall that prevents freezing.

Frostproof closet

1. Closet without an integral trap that has its trap and the control valve for its water supply installed below the frost line.
2. Hopper that has no water in the bowl and the trap and control valve for its water supply installed below the frost line.

Full bath. A bathroom that includes a water closet, lavatory, and a bathtub with or without shower provisions.

Garbage disposal. Electric device that, with the use of water, grinds food wastes into a fine pulp, which is washed down into the drainage system.

Handicapped and disabled provisions. Access to fixtures should be provided with heavy-duty chrome-plated brass 1-$\frac{1}{2}$ in. rails as required by the American with Disabilities Act (ADA), PL 101-336. Provisions should also be applicable to state and local codes.

Latrine. Sewer-connected receptacle or privy that provides water closet facilities for two or more persons.

Laundry tray. Deep fixed tub installed in laundry rooms, equipped with hot and cold running water, connected to a drain, and used for washing household items or clothes.

Lavatory. A basin or other similar vessel that is fixed in place and is plumbed with water, used for washing the hands, arms, face, and head.

Mixing faucet. Separate faucets having a common spout permitting the control of water temperature.

Mop basin. Basin set close to the floor, used for the washing of mops and the dumping of pail water.

Overflow tube. Vertical tube in a water closet tank that prevents overfilling of the tank.

Packing nut. Nut screwed down onto faucet stem, holding packing tight.

Plumbing appliance. Any one of a special class of plumbing fixtures that is intended to perform a special plumbing function. Its operation or control or both may be dependent upon one or more energized components, such as motors, controls, heating elements, or pressure or temperature sensing elements. Such fixtures may operate automatically through one or more of the following actions: time cycle, temperature range, pressure range, measured volume or weight; or the fixture may be manually adjusted or controlled by the user or operator.

Plumbing fixture

1. Receptacle or device that is either permanently or temporarily connected to the water distribution system of the premises and demands a supply of water therefrom, or that discharges used water, liquidborne waste materials, or sewage either directly or indirectly to the drainage system of the premises, or that requires both a water supply connection and a discharge to the drainage system of the premises. Plumbing appliances are defined as a special class of fixture.

2. Approved type of installed receptacle, device, or appliance that is supplied with water or that receives liquid or liquid-borne wastes and discharges such wastes into the drainage system to which it may be directly connected. Industrial or commercial tanks, vats, and similar processing equipment are not plumbing fixtures but may be connected to or discharged into approved traps or plumbing fixtures when and as otherwise provided for.

3. Water-supplied receptacle intended to receive and discharge water, liquid, or water-carried wastes into a plumbing or drainage system with which is connected.

Plumbing fixture requirements. The number and type of fixtures for men, women, and children are determined by the use of the building and the maximum number of persons in the facility at one time and are based on local and state building codes.

Plumbing fixtures. Plumbing fixtures are installed receptacles, devices, or appliances that are supplied with water or that receive or discharge liquids or liquid-borne wastes, with or without discharge into the drainage system with which they may be directly or indirectly connected.

Porcelain. White ceramic material used in the finishing of bathroom fixtures; also called "vitreous enamel".

Port control faucet. Single-handle, noncompression faucet that contains within the faucet body a port for both cold and hot water and some method of opening and closing these ports.

Public or public-use plumbing fixtures. The term "public" is applicable to fixtures in general toilet rooms or schools, gymnasiums, hotels, railroad stations, airports, public buildings, bars, public comfort stations, or places to which the public is invited or that are frequented by the public without special permission or special invitation and also to other installations, whether pay or free, where a number of fixtures are installed so that their use is similarly unrestricted.

Public washroom. Room that contains one or more sanitary units and to which employees of a business or institution; patrons of or visitors to a place of business; students, patients, inmates, or visitors of an institution; the traveling or transient public; or all tenants of an apartment building or condominium would expect to have the right of access without any special permission from management.

Receptor. An approved plumbing fixture or device of such material, shape, and capacity as to adequately receive the discharge from indirect waste pipes, constructed and located to be readily cleaned.

Refill tube. Rubber or copper tube extending from the ball cock to the overflow tube in the water closet operating assembly.

Rim. Unobstructed open edge of the receptacle section of a plumbing fixture.

Rotating ball faucet. Single-handed faucet that controls water flow and temperature with a channeled rotating plastic ball. Holes in the ball are aligned with orifices for hot and cold water.

Sanitary fixtures. Each dwelling unit abutting on a public sewer or with a private sewage disposal system should have at least one (1) water closet, one (1) lavatory, one (1) tube or shower bath and one (1) kitchen-type sink. All other structures for human occupancy or use abutting on a sewer or with a private sewage disposal system should have at least one (1) watercloset and one (1) fixture for cleansing purposes. In hotel and dormitory residential buildings, there should be not less than one (1) toilet room for each sex containing not less than (1) watercloset, one (1) lavatory, and one (1) tube or shower bath for every six (6) occupants depending on code requirements.

Sanitary unit. Water closet, urinal, bidet, or bedpan washer.

Shower pan lining. Fabricated from nonplasticized chlorinated polyethylene (CPE), a synthetic elastomer. The material is guaranteed against rotting, cracking, and microorganism deterioration when installed according to manufacturer's instructions.

Sill cock. Faucet used on the outside of buildings, to which a hose can be connected.

Single lever faucet. Several types of washerless faucets using a single control are available.

Sink. Receptacle for general washing or receiving liquid wastes.

Spout. End of faucet that serves as a passageway for water.

Sterilizer instrument. A nonpressure type fixture used for boiling for disinfecting instruments, utensils, or other equipment, which may be portable or connected to the plumbing system.

Stopper. Plug that controls wastewater drainage from a lavatory or bathtub.

Tailpiece. A connection used from outlet of fixture strainer to trap connection.

Toilet compartment. No water closet should be located in a room or compartment that is not properly lighted and ventilated.

Toilet facility. A fixture, maintained within a toilet room, that may be used for defecation or urination or both.

Toilet room. Enclosed space containing one or more water closets, which may also contain one or more lavatories, urinals, and other plumbing fixtures.

Trim. Water supply and drainage fittings that are installed on the fixture to control the flow of water into the fixture and the flow of wastewater from the fixture to the drainage system.

Urinal. A toilet facility that is used only for urination.

Vanity. Cabinet that supports a lavatory.

Wall-hung water closet. Water closet installed in such a way that no part of the water closet touches the floor.

Wash basin. Receptacle for washing any part of the human body.

Washrooms. Enclosed space containing one (1) or more bathtubs, showers, or both and that should also include toilets, lavatories, or fixtures serving similar purposes.

Water closet. A toilet facility (which may be used for both defecation and urination) in which the waste matter is removed by automatic flushing with water.

Water closet bowls. Water closet bowls and traps should be glazed vitreous earthenware made in one piece and of such form as to hold sufficient quantity of water, when filled to the trap overflow, to prevent fouling of surface, and should be provided with integral flushing rims constructed so as to flush the entire interior of the bowl. The use of water closet bowls with side inlets or of the valve-in-bowl type is prohibited.

Water closet combination. Water closet bowls may be siphon-jet, washdown, reverse-trap, or blowout type with wall outlet. Water closet bowls and traps should be made in one piece and should be provided with integral flushing rims constructed to flush the entire interior of the bowl.

Water closet compartment. An enclosed space containing one (1) or more toilets or one (1) or more urinals and other plumbing appliances.

GAS-BURNING EQUIPMENT AND APPLIANCES

Absolute pressure. Pressure above that of a perfect vacuum; sum of gauge pressure and atmospheric pressure (psia).

Accessory gas appliance. Unit designed to be attached to a gas appliance, such as a pilot light, regulator, safety device, control valve, or relief valve. A regulator may or may not be used as an appliance accessory.

a / conductance of airspace. Thermal conductance of an airspace; amount of heat, expressed in British thermal units transmitted in 1 hour across an airspace of 1 sq ft area for each degree Fahrenheit temperature difference.

Adjustable spring-type regulator. Regulator in which the regulating force acting upon the diaphragm is derived principally from a spring, the loading of which is adjustable.

Air mixer. Portion of an injection-type burner (Bunsen) into which the primary air is introduced.

Air shutter. Adjustable device for varying the size of the opening of the primary air inlet.

Air supply. With respect to the installation of an appliance, the air for combustion, ventilation, and flue gas dilution.

Appliance. Device using gas for a fuel to produce heat, light, power, or refrigeration.

Appliance accessory. A unit designed to be attached to a gas appliance, such as pilot lights, regulators, safety devices, control valves, and relief valves.

Appliance flue. Flue passages within an appliance.

Appliance fuel connector. Assembly of listed semirigid or flexible tubing and fittings to supply fuel between the fuel piping outlet and a fuel-burning appliance.

Appliances automatically controlled. Appliances equipped with automatic devices that accomplish complete turn-on or shutoff of the gas to the main burner(s) but do not effect complete shutoff of the gas.

Appurtenance. An accessory or adjunct to a gas appliance, intended to be used in connection with it.

Atmospheric injection-type burner. Burner into which the air at atmospheric pressure is injected by a jet of gas.

Automatically lighted boiler. Gas to the main burner is normally turned on automatically.

Automatic control. Device(s) installed on an appliance to accomplish, without manual attention, a complete turn-on

or shutoff of gas to the main burner or burners, or to regulate the supply of gas to the main burner or burners.

Automatic draft regulator. Device attached to or made a part of the vent outlet from an appliance and designed to govern the effect of stack action of the flue on the operation of the appliance.

Automatic gas control valve. Automatic device consisting essentially of a gas valve for controlling the main gas supply to a boiler.

Automatic gas shutoff device. Device constructed so that the attainment of a water temperature in a hot-water supply system in excess of some predetermined limit acts in such a way as to cause the gas to the system to be shut off.

Automatic gas shutoff valve. Valve used in conjunction with an automatic gas shutoff device to shut off the gas supply to a gas-fired water heating system. It may be constructed integrally with the shutoff device, or it may be a separate assembly.

Automatic ignition. Ignition of the gas at a burner when the gas burner valve controlling the gas to that burner is turned on. Reignition will be effected if the flames on the burner have been extinguished by any means other than closing the gas burner valve.

Automatic input control valve. Gas valve for controlling the gas supply to the main burner without manual attention.

Automatic instantaneous water heater. Water heater that heats the water as it is drawn.

Automatic operation. Operation, sequence, or cycle of operations that is performed by a device or combination of devices without manual attention.

Automatic pilot. Automatic pilot device and pilot burner securely assembled in a fixed functional relationship.

Automatic pilot device. Device employed with gas-burning equipment that will either automatically shut off the gas supply to the burner(s) being served or automatically actuate electrically or otherwise a gas shutoff device when the pilot flame is extinguished. The pilot burner may or may not be constructed integrally with the device.

Automatic pilot device—complete shutoff type. Device for shutting off automatically the gas supply to the main burner and the pilot in event of pilot or gas failure.

Automatic shutoff valve

1. Valve designed to shut off the gas flow to the burner without requiring manual attention.

2. Automatic shutoff valve that will automatically shut off the gas flow and remain closed until manually reopened.

Automatic storage water heater

1. Self-contained gas, oil, or electric heating element in a water storage tank.

2. Water heater that combines a water-heating element and water storage tank in which the supply of gas to the main burner is controlled by a thermostat.

Automatic valve for gas appliances. Automatic or semi-automatic device consisting essentially of a valve and operator that controls the gas supply to the burner(s) during normal operation of an appliance. The operator may be actuated by application of gas pressure on a flexible diaphragm, by electrical means, by mechanical means, or by some other means.

Backflow valve. Valve used when liquefied petroleum or other standby gas is interconnected with the regular gas piping system. The valve prevents backflow into either system.

Baffle. Object placed in an appliance to change the direction or retard the flow of air, air–gas mixtures, or flue gases.

Base. Lowest supporting frame or structure of the appliance, exclusive of legs.

Blast furnace gas. Gas produced as a by-product of the iron blast furnace, and containing a high carbon dioxide content.

Blast heater. Set of heat transfer coils or sections used to heat air that is drawn or forced through it by a fan.

Bleed venting. Expirating or inspirating of air or gas from or to one side of a diaphragm of a valve or regulator.

Blue gas. Gas produced by introducing steam through a bed of incandescent carbon, resulting in a mixture of 50% carbon monoxide and 50% hydrogen.

Boiler

1. Self-contained gas-burning appliance for supplying hot water or low-pressure steam, primarily intended for domestic and commercial space-heating application.

2. Closed pressure vessel in which a liquid, usually water, is raised in temperature by the application of heat. It may be used for power or process work, heating by either steam or hot water, or the heating of water for commercial or domestic supply.

Boiler blow-off. Outlet on a boiler to permit emptying or discharge of sediment.

Boiler blow-off tank. Vessel designed to receive the discharge from a boiler blow-off outlet and to cool the discharge to a temperature that permits its safe discharge to the drainage system.

Boiler heating surface. Boiler surfaces in contact with hot gases, including water walls.

Boiler horsepower. Boiler horsepower is equivalent to the evaporation of 34.5 lb water per hour from and at 212°F 33475 Btu/hour, 140 ft of steam radiation, 224 sq ft of water radiation at 150°F or 10-kWh electrical input to boiler.

Boiler horsepower rating. The largest rating is determined by dividing the square feet of boiler heating surface by 10. Where the boiler rating is based on other than 10 sq ft of heating surface per horsepower, the manufacturer's output rating in horsepower is used; where the manufacturer's output rating is expressed in terms other than horsepower, such rating is converted into horsepower by the use of one of the factors given above.

Boiler input control valve. Automatic gas control valve for regulating boiler input.

Boiler room. Room containing a steam or hot-water boiler.

Branch line. Part of a piping system that conveys gas from a common supply line or common header to an appliance(s).

Breeching. Conduit connecting the boiler with the flue.

Breeching (vent connector). Connection from a boiler or furnace to the stack. The term "breeching" is normally used for larger equipment; "flue" or "vent connector" is used for gas-fired equipment. The term "breeching" applies to any type of equipment connection.

British thermal unit (Btu). Amount of heat required to raise the temperature of 1 lb water 1°F at 60°F.

Broiler. General term including broilers, salamanders, barbecues, and other devices for cooking primarily by radiated heat, excepting toasters.

Btu. Abbreviation for British thermal unit or the quantity of heat required to raise the temperature of 1 lb of water 1°F.

Btuh. Number of British thermal units per hour.

Building gas piping. System of piping within a structure or a building, either exposed or concealed, that conveys gas from the outlet of the service meter or line to appliances at various places throughout the building. Any piping underground that contains measured gas is also building piping.

Built-in domestic cooking unit range. Gas appliance for domestic food preparation, providing at least one function of top or surface cooking, oven cooking, or broiling and designed to be recessed into, placed on, or attached to counters, cabinets, walls, or partitions.

Buried pipe storage. System of storage in especially designed high-pressure pipe sections or bottles capable of storing natural gas at pressures near or equal to the pressure of maximum supercompressibility. It is not storage in ordinary steel pipe.

Burner. Device for the final conveyance of gas or a mixture of gas and air to the combustion zone.

Burner head. Portion of a burner beyond the outlet end of the mixer tube that contains the ports.

Burner valve—gas. Manually or mechanically operated valve that permits control of the flow of gas.

Butane-air gas. Manufactured gas composed mainly of the two hydrocarbons, butane and isobutane.

Bypass (thermostat). Passage in the valve body of a thermostat that permits a flow of gas from the inlet to the outlet connection entirely independent of the action of the thermostatic valve.

By-products (residuals). Secondary products that have commercial value and are obtained from the processing of a raw material. They may be the residues of the gas production process, such as coke, tar, and ammonia, or the result of further processing of such residues, such as ammonium sulfate.

Carbon black. Almost pure amorphous carbon consisting of extremely fine particles, usually produced from gaseous or liquid hydrocarbons by controlled combustion with a restricted air supply or by thermal decomposition.

Carburetted blue gas. Gas consisting of a mixture of blue gas and oil gas formed by the cracking of oil in a chamber through which the blue gas is passing.

Central heating gas appliance. Vented gas-fired appliance comprising the following classes: boiler, central furnace, floor furnace, and vented wall furnace.

Chimney. That part of a building constructed so as to provide the spaces to include a flue or flues for transmitting products of combustion to the outer air.

Chimney connector. Pipe that connects a fuel-burning appliance to a chimney; a breeching.

Circulating space heater

1. Space heater designed to convert the energy in fuel gas to convected heat or radiant and convected heat by the circulation of the products of combustion and room air or room air only.

2. Space heater having an outer jacket surrounding the casing around the combustion chamber, arranged with openings at top and bottom so that air circulates between the inner casing and the outer jacket. Space heaters with openings in the outer jacket to permit

some direct radiation from the inner casing are classed as radiating types.

Circulating-type room heaters. Room heaters having an outer jacket surrounding the combustion chamber, arranged with openings at top and bottom so that air circulates between the inner and outer jacket, and without openings in the outer jacket to permit direct radiation. Heaters usually have clearances at the sides and rear of not less than 12 in.

Circulator. Room heater designed to convert the energy in fuel gas to convected heat or radiant and convected heat. It may be one of the vented or unvented type and may be equipped with radiants. There may be an external jacket surrounding the burner and heating elements and the jacket may be open in front of the radiants.

Clock controls. A clock mechanism operates to shut off or start the gas supply to an oven at any predetermined time. Clock controls are used on range ovens having oven ignition systems.

Clothes dryer. Device used to dry wet laundry by means of heat derived from the combustion of fuel gas.

Coal gas. Manufactured gas made by distillation or carbonization of coal in a closed coal gas retort, coke oven, or other vessel.

Cock. Valve of the plug-and-barrel type that controls the supply of gas.

Combustible material

1. The definition pertains to materials adjacent to or in contact with heat-producing appliances, vent connectors, gas vents, chimneys, steam and hot-water pipes, and warm-air ducts. These are materials made of or surfaced with wood, compressed paper, plant fibers, or other materials that will ignite and burn. Such materials are considered combustible even though flameproofed, fire retardant treated, or plastered.

2. Material that will ignite at or below a temperature of 1200°F and will continue to burn or glow at that temperature.

Combustion. Rapid oxidation of fuel gases accompanied by the production of heat and/or light.

Combustion chamber

1. Portion of an appliance within which combustion occurs.

2. Metal or refractory chamber located within the fire box and used to contain the combustion flame.

Combustion products. Constituents resulting from the combustion of a fuel gas with the oxygen of the air, including the inert substances but excluding excess air.

Combustion zone. Zone where combustion is intended to occur.

Component. Essential part of an appliance; it may be certified separately from the appliance.

Compressed gas. Any material or mixture having in the container an absolute pressure exceeding 40 psia at 70°F or regardless of the pressure at 70°F having an absolute pressure exceeding 104 psia at 130°F.

Compression tank. Closed tank used in hot-water heating systems, the air in which serves as a cushion to allow for the expansion of the water in the boiler when heated; expansion or cushion tank.

Concealed. Made inaccessible to view, inspection, adjustment, and/or repair by a permanent cover that cannot be removed except by digging and/or demolition.

Concealed gas piping. Gas piping access to which when it is in place in the finished building would require removal of permanent construction.

Condensate. Liquid resulting when a vapor is subjected to cooling and/or pressure reduction; also, liquid hydrocarbons condensed from gas and oil wells.

Condensation. Separation of liquid from a gas, including flue gases, due to a reduction in temperature.

Conductance C. Thermal conductance is the amount of heat, expressed in British thermal units (Btu), transmitted in 1 hour from surface to surface of 1 sq ft of material or a combination of materials for each degree Fahrenheit of temperature difference between the two surfaces. The value is not expressed in terms of per inch of thickness but from surface to surface.

Consumer gas piping. All pipe and fittings installed on any premise or in any building or other structure on the outlet side of the gas meter, or extending from gas piping anywhere beyond that location and ending at capped or plugged outlets ready to connect with fixtures or gas appliances.

Consumer gas service line. All piping, fittings, and accessories, including regulators used to convey unmetered gas from the point of connection to the utility's service pipe to the inlet side of the meter.

Continuous or standby pilot. Gas pilot that operates at all times regardless of whether the main burner is on or off.

Control. Device designed to regulate the gas, air, water, and/or electrical supply to an appliance. It may be manual or automatic.

Control action. Temperature at which the control acts to turn on the gas to the main burner(s).

Control cock. Cock used in piping to control the gas supply to any section of a system of piping or to an appliance.

Controls. Devices designed to regulate the gas, air, water, or electrical supplies to a gas appliance. It may be manual, semiautomatic, or automatic.

Control valve or cock. Valve or cock used in piping to control the supply to any section of a system of piping.

Conversion burner

1. Burner designed to supply gaseous fuel to an appliance originally designed to utilize another fuel.

2. Burner designed to burn gas in an appliance in which another fuel can be burned.

Converter. Vessel containing a heating coil used to heat water for either domestic purposes or hot-water heating, but not both at the same time. It is not a storage tank.

Cubic foot of gas. Amount of gas that would occupy 1 cu ft when at a temperature of 60°F if saturated with water vapor under a pressure equivalent to that of 30 in. of mercury.

Cutoff point of control. Temperature at which the control acts to shut off the gas to the main burner(s).

Corrosion protection. Protection required for ferrous piping installed in soils that have been determined to cause corrosion. Field wrapping is required in most situations when approved by local codes.

Crude oil gas. Gas manufactured by cracking topped crude oil that is atomized with steam in a chamber containing hot checker brick.

Cubic foot. Most common unit of measurement of gas volume; amount of gas required to fill a volume of 1 cu ft under stated conditions of temperature, pressure, and water vapor.

Damper

1. Valve or plate for regulating the draft.
2. Fixed or movable plate for regulating draft and controlling the flow of air, or for controlling the flow of gases.

Dayton process gas. Manufactured gas produced from crude oil that is atomized with air and introduced into the hot chamber.

Deadweight-type regulator. Regulator in which the regulating force acting upon the atmospheric side of the diaphragm is derived from a weight or combination of weights.

Demand. Maximum amount of gas per unit of time, usually expressed in cubic feet per hour or British thermal units (Btu) per hour, required for the operation of the appliance(s) supplied.

Design temperature. Indoor temperature that the heating system of a building is designed to maintain.

Design temperature difference. Difference between the design temperature and the normal wintertime low outdoor temperature for the locality.

Dial (thermostat). Part of the thermostat by which the position of the thermostatic valve in reference to the valve seat may be manually adjusted. The dial is provided with temperature markings.

Diaphragm automatic valve. Device consisting essentially of an automatic valve actuated by the application of gas pressure on a flexible diaphragm.

Diaphragm valve. Device consisting essentially of a gas valve actuated by the application of gas pressure on a flexible diaphragm.

Diffusion. Intermixture of gases or liquids, regarded in warm-air heating as the intermingling of warmer with colder air to produce a comfortable temperature.

Dilution air. Air that enters a draft hood or draft regulator and mixes with the flue gases.

Direct-fired oven. Oven in which the flue gases flow through the oven compartment.

Direct gas-fired air heaters. A gas heating device in which gas is burned and in which the products of combustion are mixed with the air that is to be heated in passing through the heater. The term means the unit and equipment from its outside air inlet to the exit where the heated air leaves the unit.

Diversity factor. Ratio of the maximum probable demand to the maximum possible demand.

Domestic gas appliance. Pressure regulator; device, either adjustable or nonadjustable, for controlling and maintaining a uniform outlet gas pressure.

Domestic gas-fired incinerator. Domestic appliance used to reduce combustible refuse material to ashes and that is manufactured, sold, and installed as a complete unit.

Draft diverter. Device built into an appliance or made a part of the flue or vent connector from an appliance, designed to ensure the ready escape of the products of combustion in the event of no draft, backdraft, or stoppage beyond the draft hood; to prevent a backdraft from entering the appliance; and to neutralize the effect of stack action of the flue or vent on the operation of the appliance.

Draft hood. Device built into the appliance or made a part of the vent connector from an appliance, designed to ensure the ready escape of the products of combustion in the event of no draft, or stoppage beyond the draft hood; to prevent a backdraft from entering the appliance; and to neutralize the effect of stack action of the chimney or gas vent on the operation of the appliance.

Draft regulator. Device that maintains a desired draft in the appliance by automatically reducing the draft to the desired volume.

Drip

1. Container placed at a low point in a system of piping to collect condensate and from which container may be removed.
2. Short piece of pipe, capped at one end and connected at the other end to low points in fuel lines for the purpose of catching condensation formed in the fuel lines.

Drip leg. Chamber of ample volume, with suitable cleanout and drain connections, into which gas is discharged so that liquids and solids are trapped.

Drip pocket. Designed pocket placed at a low point in any part of a piping system for the collection of condensate and designed so that the collected condensate may be removed.

Drop. Vertical pipe or nipple that conducts the gas downward.

Dry gas. Gas having a moisture and hydrocarbon dew point below any normal temperature to which the gas piping is normally exposed.

Dual fuel burning. A gas burner firing into the same combustion zone into which another fuel is utilized.

Duct. Tube, pipe, conduit, or continuous enclosed passageway used for conveying air, gases, or vapors.

Dust pocket. Designed pocket placed at a low point in any part of a piping system for the collection of dust or dirt and designed so that the collected dust or dirt may be removed.

Electrical-type automatic valve. Device actuated by electrical energy for controlling the gas supply.

Exposed piping. Gas piping that will be in view in the finished structure.

Excess air. Air that passes through the combustion chamber and the appliance flues in excess of that theoretically required for complete combustion.

Explosion door. Door or cover that will open to relieve sudden excessive pressure from the furnace gases.

Explosive mixture. Flammable mixture in a confined space.

External inspection. Inspection of the exterior surfaces and appurtenances of a boiler or unfired pressure vessel.

Fail-safe valve. Approved, automatic, fast-closing, safety shutoff valve with positive closure against a pressure equivalent to at least 150% of the valve rating in the event of failure of the operating medium(s).

Film conductance, f. Film or surface conductance; the amount of heat, expressed in British thermal units (Btu), transmitted in 1 hour from 1 sq ft of a surface to the air surrounding the surface for each degree Fahrenheit temperature difference.

Fire box. Metal enclosure in which gas is burned and that forms a portion of the heat exchanger.

Fire damper. Approved automatic or self-closing non-combustible barrier designed to prevent the passage of air, gases, smoke, or fire through an opening, duct, or plenum chamber.

Fireplace insert. Open-flame radiant-type gas heater mounted on a decorative metal panel designed to cover the fireplace or mantel opening.

Firing door. Conversion burner designed specifically for the boiler or furnace ashpit door installation.

Firing test valve. Manually operated, lubricated, plug-type quarter-turn valve that has stops in the open and closed positions, that has an attached handle, loose-fitting key, or extended handle wrench, and that is located downstream of all automatic safety shutoff valves on the valve train and as close to the burner as is practicable.

Fixed appliance. Appliance that is fastened or otherwise secured at a specific location.

Fixed liquid level gauge. A type of liquid level gauge using a relatively small positive shutoff valve and designed to indicate when the liquid level in a container being filled reaches the point at which this gauge or its connecting tube communicates with the interior of the container.

Fixed maximum liquid level gauge. A fixed liquid level gauge that indicates the liquid level at which the container is filled to its maximum permitted filling density.

Flame. Rapid oxidation of gas.

Flame (Bunsen). Flame produced by premixing some of the air required for combustion with the gas before it reaches the burner port or point or point of ignition.

Flame safeguard

1. Device that will automatically shut off the gas supply to a main burner or group of burners when the means of ignition of such burners become inoperative.

2. Safety device that is sensitive to the properties of gas flame, that detects the presence or nonpresence of a gas flame, and that causes the gas supply to be shut off in the event of flame or ignition failure.

Flexible tubing

1. Gas conduit other than that formed by a continuous, one-piece metal tube.

2. Gas conduit that depends for tightness on joint packing or that has a wall structure other than that formed by a continuous one-piece metal tubing member.

Floor furnace. Completely self-contained furnace unit attached to the floor of the space being heated, taking air for combustion from outside this space, and with means for observing the flames and lighting the appliance from such space.

Flue. Primarily vertical passageway used to remove products of combustion and suitable for devices or appliances using any type of fuel.

Flue collar

1. Projection or recess provided to accommodate the flue pipe.

2. Portion of an appliance that is designed for the attachment of the draft hood, chimney breeching, or vent connector.

Flue exhauster. Device that is installed and made part of a gas vent or vent connector and that will provide a positive induced draft.

Flue gas baffle. Object in the path of the gases that is exposed to flue gases or radiant heat and is intended to restrict or modify flue gas flow. It may be a projection from the heat exchanger or suspended in the flue gas passages by some other means.

Flue gases. Products of combustion and excess air.

Flue losses. Above-room-temperature and latent heat of the flue gases leaving the appliance.

Flue outlet (vent). Opening provided in an appliance for the escape of the flue gases.

Flue products. Products of combustion and excess air.

Fuel gas. Includes acetylene, hydrogen, natural gas, LP gas, methylacetyle-propadiene, stabilized and other lique-fied and nonliquefied flammable gases, which are stable because of their composition or because of the conditions of storage and utilization stipulated in the code.

Fuel line. Gas-piping system extending from the meter to outlets supplying appliances or other gas-burning equipment.

Gas

1. Natural, manufactured, and mixed gas and liquefied petroleum products.

2. Natural gas, manufactured gas, or a mixture of the two, which may be used to produce light, heat, power, or refrigeration.

Gas alarm. Alarm device used in connection with a gas sensing detector.

Gas appliance. Fixture or apparatus manufactured and designed to use natural, manufactured, or mixed gas or any gas as a medium for developing light, heat, or power, including but not limited to gas ranges, gas room heaters; gas, steam and hot-water boilers; and gas burners of all kinds, together with any attachments or apparatus designed to be attached to any gas appliance, such as solid tops, pilot lights, governors, regulators, so-called fuel savers, and safety devices.

Gas appurtenance. Adjunct to a gas appliance and intended to be used with it.

Gas building line. Piping system extending from the outlet of the utility meter to the building.

Gas building service line. Pipe that brings gas from the main to the meter.

Gas burner. Factory-fabricated metal chamber or pipe in which one or more holes occur to allow the passing of gas or gas–air mixture for cooking or heating.

Gas burner valve. Manually or mechanically operated valve that controls the flow of gas to the main burner.

Gas cleaning equipment. A device or process designed for removing particulate matter from the gas or air in which it is entrained.

Gas counter appliances. Appliances such as gas-operated coffee, brewers or coffee urns and appurtenant water-heating equipment, food and dish warmers, hot-water immersion sterilizers, hot plates, griddles, and waffle bakers.

Gas distribution piping. All piping from the building side of the gas meter; building distributes gas supplied by a public utility to all fixtures and apparatus used for illumination or fuel in any building.

Gas fitting. Installation of all gas and oil house piping and all fuel piping, fittings, controls, burners, and venting for all gas- and oil-fired appliances and equipment.

Gas hose. Gas conduit that depends for tightness on joint packing or on a wall structure other than that formed by a continuous one-piece metal tubing member.

Gas input. Amount of gas in cubic feet per hour or in British thermal units per hour that an appliance consumes.

Gas log. Unvented open-flame room heater consisting of a metal frame or base supporting simulated gas logs or designed for installation in a fireplace.

Gas main. A street main (or main) is a portion of the system used for distributing gas, generally located entirely outside of the premises, and designed to supply gas to the service pipes of one or more units. A main is generally parallel to the line of the roadway in which it lies.

Gas meter. Instrument installed to measure the volume of gas delivered through it.

Gas outlet

1. Threaded or bolted flange connection in a house gas-piping system to which a gas-burning appliance is or may be attached.

2. Connection in the gas-piping system of a building or structure to which a gas-burning appliance is attached; a capped or plugged connection ready for the attachment of a gas-burning appliance.

Gas pipe. Rigid conduit of iron, steel, copper, or brass.

Gas piping. Installation of pipe, valves, or fittings that are used to convey fuel gas.

Gas-piping system. Piping from the meter or service regulator when a meter is not provided for an appliance or appliances.

Gas pressure. Pressure in piping or appliances that is imparted by the source of the gas supply, usually by the gas company, from outside the building. It is usually stated in inches of water and measured with a U-tube gauge.

Gas pressure regulator

1. Device that maintains a substantially uniform gas pressure at its outlet.

2. Device for controlling and maintaining a uniform gas supply pressure.

Gas ranges and plates. Gas-fired units used for domestic and commercial cooking or heating of food.

Gas service. A pipe conveying as to a building from the distribution system of a public service corporation.

Gas service line. Pipe and fittings used to convey unmeasured gas from the main to the premises to be supplied. The line extends underground to the inside face of the foundation wall.

Gas service line extension. All pipe and fittings, including service pressure regulators, meter headers, etc., that are installed inside the premises to connect the end of the service with the fitting to which the inlet piping for the meter installation is to be attached, and that contains unmeasured gas.

Gas service piping. The supply pipe from the street main through the building wall and including the stopcock or shutoff valve inside the building.

Gas shutoff relief valve. Device so constructed that the attainment of a temperature in the medium being heated in excess of some predetermined limit acts upon a chemical or metallic element in such a way as to cause the gas to the appliance to be shut off and remain shut off.

Gas tubing. Semirigid conduit of copper, brass, steel, or aluminum.

Gas utility. The duly enfranchised public utility supplying the gas from its street mains.

Gas vent

1. Conduit or passageway, vertical or nearly so, for conducting flue gases to the outer air, including any offset section in the gas vent that inclines not more than 60° from the vertical.

2. Vent pipe, primarily vertical, used for conveying to the outer air the products of combustion from gas-fired appliances, but which is not suitable for use with other fuels.

Gas vent connector. Portion of a gas-venting system that connects a listed gas appliance to a gas vent.

Gas vents (Type B)

1. Listed factory-made gas vents for venting listed or approved appliances equipped to burn only gas, except those specifically listed for use with chimneys only.

2. Vent piping of combustible, corrosion-resistant material of sufficient thickness, cross-sectional area, and heat-insulating quality to avoid excess temperature on adjacent combustible material and that has been approved as Type B gas vent by the Underwriters Laboratories, Inc., or other approved testing laboratory.

Gas vents (Type BW). Listed factory-made gas vents for venting listed or approved gas-fired, vented, recessed heaters.

Gas vents (Type C). Vent piping of sheet copper of not less than No. 24 U.S. standard gauge, galvanized iron of not less than No. 20 U.S. standard gauge, or other approved corrosion-resistant material.

Graduating thermostat. Thermostat in which the motion of the thermostatic valve is in direct proportion to the effective motion of the thermal element induced by temperature change.

Hazardous location. Area or space where combustible dust, ignitable fibers, or flammable, volatile liquids, gases, vapors, or mixtures are or may be present in the air in quantities sufficient to produce explosive or ignitable mixtures.

Heat exchanger

1. Fire box and any auxiliary heat transfer surfaces within the casing of an appliance.

2. Device used in a warm-air furnace to transfer the heat of the fire to the air being circulated.

3. Chamber in which heat resulting directly from the combustion of fuel, or heat from a medium such as air, water, steam, or a refrigerant is transferred through the walls of the chamber to or from the air entering the exchanger.

Heating appliance

1. Appliance intended primarily to convert gas to heat energy.

2. Device designed or constructed for the generation of heat from solid, liquid, or gaseous fuel or electricity.

Heating boiler. Low-pressure steam boiler; hot-water heating boiler.

Heating elements. All parts that transmit heat from the flames or flue gases to the medium being heated and in contact with both.

Heating surfaces. All surfaces that transmit heat from the flames or flue gases to the medium to be heated.

Heating value (total). Number of British thermal units (Btu) produced by the combustion at constant pressure of 1 cu ft of gas when the products of combustion are cooled to the initial temperature of the gas and air, when the water vapor formed during combustion is condensed, and when all the necessary corrections have been applied.

Heat transmission. Transfer of heat from one point to another due to the tendency of heat to flow from an area of higher temperature to an area of lower temperature.

Heat transmission factor. Number used in calculating the rate of heat loss from a building; the number is a guide to the heat-transmitting properties of any given type of wall, window, door, etc.

High-gas-pressure switch. Pressure-actuated device arranged to effect a safety shutdown of the burner or prevent it from starting when the gas supply pressure exceeds the normal supply pressure by 20%.

High heat appliances. Billet and bloom furnaces, blast furnaces, brass furnaces, brick kilns, coal gas retorts, cupolas, earthenware kilns, glass furnaces, open-hearth furnaces, porcelain baking and glazing kilns, and water gas retorts.

High-pressure system. System that operates at a pressure higher than the standard service pressure delivered to the customer; thus, a pressure regulator is required on each service to control pressure delivered to the customer. The system is sometimes referred to as "medium pressure."

High-static-pressure-type gas-burning appliance. Self-contained, automatically controlled, vented gas-burning appliance, limited to the heating of nonresidential space. Appliances having integral means for circulation of air against 0.2-in. or greater static pressure and are designed for installation in the space to be heated unless they are equipped with provisions so they can be attached to both the inlet and outlet air ducts.

High-static-pressure-type unit heater. Self-contained, automatically controlled, vented gas-burning appliance, limited to the heating of nonresidential space. Appliances have integral means for circulation of air against 0.2 in. or greater static pressure and are designed for installation in the space to be heated unless they are equipped with provisions so they can be attached to both inlet and outlet air ducts.

Hood. Canopy or similar device connected to a duct for the removal of heat, fumes, or gases.

Hotel and restaurant range. Self-contained gas range providing for cooking, roasting, baking, or broiling, or any combination of these functions, and not designed specifically for domestic use.

Hot plate. Gas-burning appliances consisting of one or more open-top burners mounted on either short legs or a base.

Hot-water heating boiler. Boiler used primarily for space heating where the medium is water, to be operated at a pressure not in excess of 160 psi and a temperature not in excess of 250°F.

Hot-water storage tank. Closed tank used for the storage of hot water, with or without a heating element in the tank.

Hot-water supply boiler. Boiler used primarily for heating water for such general uses as baths and washing, and similar purposes, and in which the heating medium is water, to be operated at a pressure not in excess of 160 psi and a temperature not in excess of 250°F.

House piping. Gas piping downstream from the house meter outlet.

Humidistat. Device that registers the changes in the amount of humidity.

Humidity. Moisture contained in the air.

Hydrocarbon. Compound that contains only hydrogen and carbon. The simplest and lightest forms of hydrocarbon are gaseous. With greater molecular weights they are liquid, while the heaviest are solids.

Hydrogen. Largest component by volume of coal gas, blue gas, carburetted blue gas, and crude oil gas. It is a combustible property.

Ignition pilot. Gas pilot that operates during the lighting cycle and discontinues during main burner operation.

Incinerator. Device using heat, for the reduction of garbage, refuse, or other waste materials.

Indirect oven. Oven in which the flue gases do not flow through the oven compartment.

Individual main burner valve. Valve that controls the gas supply to an individual main burner.

Induced-draft burner. Burner that depends on the draft induced by a fan beyond the appliance for its proper operation.

Industrial gas boiler. Gas appliance designed primarily to furnish steam for industrial or commercial processes, as distinguished from central heating.

Industrial heating equipment. Appliance, device, or equipment used or intended to be used in an industrial, manufacturing, or commercial occupancy for applying heat to any material being processed. Water heaters, boilers, or portable equipment used by artisans in pursuit of a trade are not included in this definition.

Inerting. Scavenging of the furnace and boiler gas passes of a fuel rich or unknown gas mixture by dilution with an inert atmosphere.

Injection-type (Bunsen) burner. Burner employing the energy of a jet of gas to inject air for combustion into the burner and mix it with the gas.

Input rating

1. Gas-burning capacity of an appliance in British thermal units (Btu) per hour as specified by the manufacturer. Appliance input ratings are based on sea-level operation, and need not be changed for operation up to 2000-ft elevation. For operation at elevations above 2000 ft, input ratings are reduced at the rate of 4% for each 1000 ft above sea level.

2. Amount of gas fuel in British thermal units (Btu) per hour that can be safely burned in an appliance at the altitude of the location.

In-shot type burner. Conversion burner normally designed for boiler or furnace ashpit installation and fired in a horizontal position.

Insulation. Materials that are poor conductors of heat, used to reduce heat loss from a building.

Intermittent pilot. Gas pilot that operates during the ignition cycle of the main burner and continues during main burner operation but is shut off at other times.

Internal inspection. Inspection of the interior surfaces of a boiler or unfired pressure vessel.

Interrupted pilot. Gas pilot that operates during the ignition cycle of the main burner but is shut off at other times.

Interruptible service. Low-priority service offered to customers under schedules or contracts that anticipate and permit interruption on short notice, generally in peak-load seasons, by reason of the claim of firm service customers and higher priority users. Unlike off-peak service, gas is available at any time of the year if the supply is sufficient.

Journeyman gas fitter. Person engaging in the trade of installing, constructing, or repairing gas piping, gas appliances, or other apparatus using gas as a fuel.

Laundry stove. Gas-burning appliance consisting of one or more open-top burners mounted on high legs or having a cabinet base.

Limit control

1. Device responsive to changes in pressure, temperature, or liquid level for turning on, shutting off, or throttling the gas supply to an appliance.
2. Device that is sensitive to changes in flow, pressure, temperature, or liquid level and that shuts off or allows to be turned on to the gas supply to an appliance as a result of such changes.

Liquefied petroleum gas (LP gas). Material that is composed predominantly of the following hydrocarbons or mixtures thereof: propane, propylene, butane, (normal butane or isobutane), and butylenes.

Liquefied petroleum gas equipment. All containers, apparatus, piping (not including utility distribution piping systems), and equipment pertinent to the storage and handling of liquefied petroleum gas. Gas-consuming appliances are not to be considered liquefied petroleum gas equipment.

Listed

1. Entered in a directory as being approved by a nationally recognized authority in accordance with their standards.
2. Appliances, accessories, and appurtenances that are shown in a list published by an approved nationally recognized testing agency (such as the American Gas Association Laboratories or the Underwriters Laboratories, Inc.).

Listed and listing. Terms used in the definitions when specifically referring to equipment. This equipment must be indicated on a list published by an approved testing agency that is qualified and equipped for experimental testing and maintaining an adequate inspection of current equipment production and whose published listing indicates that the equipment complies with the standards set forth in the applicable code.

Low-gas-pressure switch. Pressure-actuated device arranged to effect a safety shutdown of the burner or prevent it from starting when the gas supply pressure falls below 50% of the normal gas supply pressure.

Low-heat appliances. Bakery ovens, candy furnaces, coffee-roasting ovens, core ovens, cruller furnaces, lead-melting furnaces, rendering furnaces, steam boilers operating at not over 50-psig pressure, steam boilers of not over 10 boiler hp regardless of operating pressure, stereotype furnaces, and wood-drying furnaces. Appliances otherwise classed as medium-heat appliances may be considered as low-heat appliances if furnace volume is not larger than 100 cu ft in size.

Low-static-pressure-type gas-burning appliance. Self-contained, automatically controlled, vented gas-burning appliance, limited to the heating of nonresidential space in which it is installed. Such appliances have integral means for circulation of air, normally by a propeller fan or fans, and may be equipped with louvers or face extensions made in accordance with the manufacturer's approved specification.

Low-static-pressure-type unit heater. Self-contained, automatically controlled, vented gas-burning appliance, limited to the heating of nonresidential space in which it is installed. Such appliances have integral means for circulation of air, normally by a propeller fan or fans, and may be equipped with louvers or face extensions made in accordance with the manufacturer's approved specifications.

Low water cut off

1. Device so constructed as to automatically cut off the gas supply when the surface of the water in a boiler falls to the lowest safe water level. This point should not be lower than the bottom of the water glass.
2. Device or control for shutting down the fuel-feeding apparatus when the water level in the boiler reaches a predetermined point.

Lubricated plug-type valve. Valve of the plug and barrel type provided with means for maintaining a lubricant between the bearing surfaces.

Luminous or yellow-flame burner. Burner in which secondary air only is depended on for complete combustion of the gas.

Main burner

1. Device or group of devices essentially forming an integral unit for the final conveyance of gas or a mixture of gas and air to the combustion zone, and on which combustion takes place to accomplish the function for which the appliance is designed.
2. Device or group of devices forming an integral unit for the final release of gas or mixtures of gas and air to the combustion zone for ignition.

Main burner and pilot load application regulator. Regulator capable of controlling the flow of gas to the main and pilot burners. In such applications the pilot is taken off downstream from the regulator valve.

Main burner control valve. Valve that controls the gas supply to the main burner manifold.

Main burner load application regulator. Regulator capable of controlling the flow of gas to main burners only. In such applications the pilot is taken off upstream from the regulator.

Manifold

1. Conduit of an appliance that supplies gas to the individual burner.
2. Conduit of a gas appliance downstream of the last valve in the valve train.

Manifold gas pressure. Gas flow pressure downstream of the valve train, taken at right angles to the direction of flow.

Manually lighted boiler. Boiler in which gas to the main burners is turned on only by hand.

Manual main gas shutoff valve. Manually operated valve in the gas line for the purpose of completely turning on or shutting off the gas supply to the boiler.

Manual main shutoff valve. Manually operated valve in the gas line for the purpose of completely turning on or shutting off the gas supply to the appliance, except to the pilot(s), which are provided with independent shutoff valves. In floor furnaces the valve operated above the floor to control gas to the burner is regarded as the manual main gas control valve.

Manufactured gas. Gas obtained by destructive distillation of coal, the thermodecomposition of oil, or the reaction of steam passing through a bed of heated coal or coke. Examples are coal gas, coke oven gas, producer gas, blast furnace gas, blue (water) gas, and carbureted water gas. The Btu content varies widely.

Master gas fitter. Person who has a valid certificate of registration issued by an authorized agency to operate as a master gas fitter.

Measured gas. Gas that has passed through and the volume of which has been recorded by a meter.

Mechanical joint. Gas-tight joint, obtained by the joining of metal parts through a positive-holding mechanical construction.

Mechanically fired fuel-burning equipment. Any device by means of which fresh fuel is mechanically fired from outside the furnace into the zone of combustion, the same being actuated by automatic control.

Medium-heat appliances. Annealing furnaces (glass or metal), charcoal furnaces, galvanizing furnaces, gas producers, steam boilers of over 10 boiler hp operating at over 50-psig pressure. Appliances otherwise classed as high-heat appliances may be considered as medium-heat appliances if furnace volume is not larger than 100 cu ft in size.

Meter. Device installed by the utility company for measuring the quantity of gas used by the consumer.

Meter set assembly. Piping and fittings installed by the serving gas supplier to connect the inlet side of the meter to the gas service and the outlet side of the meter to the customer's house or yard piping.

Methane. Product of natural decomposition of organic matter.

Miniature boiler. Boiler used mainly for the operation of a pressing machine in a service establishment.

Mixed gas. Gas in which manufactured gas is comingled with natural or liquefied petroleum gas (except where the natural or liquefied petroleum gas is used only for "enriching" or "reforming") in such a manner that the resulting product has a Btu value higher than that previously produced by the utility prior to the time of the introduction of natural or liquefied petroleum gas.

Mixer. Combination of mixer head, mixer throat, and mixer tube.

Mixer face. Air inlet end of the mixer head.

Mixer head. Portion of an injection-type (Bunsen) burner, usually enlarged, into which primary air flows to mix with the gas stream.

Mixer throat. Portion of the mixer that has the smallest cross-sectional area between the mixer head and the mixer tube.

Mixer tube. Portion of the mixer that lies between the throat and the burner head.

Modulating

1. Modulating or throttling is the action of a control from its maximum to minimum position in either predetermined steps or increments of movements as caused by its actuating medium.

2. Infinite variance of the volume of the flow of gas or air, or both, between predetermined minimum and maximum limits.

Multijet-type burner. Conversion burner consisting of a number of individual burner tubes or jets mounted on one or more manifolds. Each jet in itself is a complete burner with an orifice, primary air openings, and flame ports.

Natural gas

1. Gas obtained from wells in a manner similar to petroleum. Composed mainly of methane together with smaller amounts of higher hydrocarbons of the marsh gas series.

2. Naturally occurring mixture of hydrocarbon and non-hydrocarbon gases found in porous geologic formations beneath the earth's surface, often in association with petroleum. The principal constituent is methane.

Natural gas shrinkage. Reduction in volume of wet natural gas due to the extraction of some of its constituents, such as hydrocarbon products, hydrogen sulfide, carbon dioxide, nitrogen, helium, and water vapor.

Needle (adjustable). Tapered projection, coaxial and movable with respect to an orifice, the position of which is fixed to regulate the flow of gas.

Negative pressure. Less than atmospheric pressure.

Neutral pressure point. Point or plane where pressure in the furnace or boiler changes from negative to positive. With the firing door cracked open it can be demonstrated by the flowing outward of a match flame above this point and its drawing inward below this point.

Neutral pressure point adjuster. Restricting device installed between the draft and the flue outlet of the heating plant, used to establish the position of the neutral pressure point of the flame.

Nonadjustable spring-type regulator. Regulator in which the regulating force acting upon the diaphragm is derived principally from a spring, the mounting of which is not adjustable.

Normal test pressures. Pressures specified for testing purposes at which adjustment of burner ratings and primary air adjustments are made.

Oil gas. Gas resulting from the thermal decomposition of petroleum oils, composed mainly of volatile hydrocarbons and hydrogen. The true heating value of oil gas may vary between 800 and 1600 Btu/cu ft depending on the operating conditions and feedstack properties.

On-and-off-type boiler control. Boiler input control valve that intermittently opens or closes the main gas supply.

One-hundred percent safety shutoff valve. Valve with a pilot connection designed in such a way that only the pilot can receive gas when it is being lighted. A knob must be held down for about 30 seconds during and after lighting the pilot; only then can the main gas valve be turned on. Should the pilot be extinguished for any reason, the main gas and pilot are both shut off.

Orifice

1. Opening in a cap, spud, or other device whereby the flow of gas is limited and through which the gas is discharged to the burner.

2. Machined opening in a cap, plug, spud plate, or other device through which gas is discharged and by which its flow is limited.

Orifice cap (hood). Movable fitting having an orifice that permits adjustment of the flow of gas by the changing of its position with respect to a fixed needle or other device.

Orifice spud. Removable plug or cap containing an orifice that permits adjustment of the flow of gas either by substitution of a spud with a different-size orifice or by motion of a needle with respect to it.

Outlet. End of a particular branch of a fuel line at which an appliance is connected or is to be connected.

Output rating. Amount of heat in British thermal units (Btu) per hour that an appliance will deliver for useful service when operating at rated input.

Oven heat control. Device actuated by temperature changes designed to control the gas supply to the oven burner in order to maintain a fixed oven temperature. The controls operate to cut the flame down to a small size when the predetermined oven temperature is reached.

Oven ignition systems. When a constant-burning pilot is used in the oven, a device is supplied to prevent the flow of gas to the main burner unless the pilot light is burning.

Overhead heater. Room heater designed for suspension from the ceiling in the room being heated.

Pilot. Flame that is utilized to ignite the gas at the main burner or burners.

Pilot light. Flame from a small burner having control independent from the main burner or burners and that is utilized to ignite the main burner or burners of an appliance; flame from the main burner or burners of an appliance resulting from the diversion of a limited amount of gas through a bypass when the main gas supply is shut off; flame utilized to maintain ignition of the main burner or burners.

Pilot loaded regulator. Type in which the loading is accomplished by applying gas pressure to the diaphragm of the main regulator through a pilot regulator.

Pilot valve. Manual valve to control the supply of gas to a pilot burner.

Pipe dope. Standard compound used as a pipe and fitting thread lubricant for gas service.

Pipe fitting. Pipe elbow, return bend, tee, union, bushing, coupling, cross, reducing coupling, nipple, and similar items.

Piping. Either pipe or tubing, or both.

Piping drop. Vertical pipe that conducts gas down to an appliance.

Piping extensions or additions. Additional system piping installed to supply added appliances.

Piping in concrete. Piping installed in concrete is placed in properly constructed ducts or thoroughly encased in portland cement or its equivalent. In no case should the piping system be installed so as to form a part of the floor reinforcement.

Piping riser. Vertical pipe that conducts gas upward to one or more floors.

Piping system. Piping that conducts gas from the meter to each outlet, including a plugged or capped manual gas valve.

Plenum chamber. Chamber for distributing warm air from a furnace to the supply ducts (supply plenum) or for receiving air to be heated by the furnace (return plenum).

Port. Opening in a burner head through which gas or an air-gas mixture is discharged for ignition.

Portable appliance. Appliance that is actually moved or can be easily moved from one place to another in normal use.

Power boiler. Steam boiler designed for a pressure of more than 15 psi.

Power burner. Burner in which either gas or air, or both, are supplied at pressures exceeding the line pressure for gas, and the atmospheric pressure for air, this added pressure being applied at the burner. A burner for which air for combustion is supplied by a fan ahead of the appliance is commonly designated as a forced-draft burner.

Premixing burner. Power burner in which nearly all of the air for combustion is mixed with the gas as primary air.

Pressure burner. Burner that is supplied with an air-gas mixture under pressure (usually from 0.05 to 14 in. of water and occasionally higher).

Pressure relief valve. Automatic device that opens or closes a relief vent, depending on whether the pressure is above or below a predetermined value.

Primary air. Air that when introduced into the burner mixes with the gas before it reaches the port or ports.

Primary air inlet. Opening(s) through which primary air is admitted into a burner.

Propane

1. Member of the marsh gas series of hydrocarbons and usually associated with methane in natural gas.

2. Heavy, colorless, gaseous petroleum hydrocarbon gas of the paraffin series.

Psig. Pounds per square inch gauge.

Purge

1. Clear of air, water, or other foreign substance.

2. Free gas piping or an appliance of air, gas, or a mixture of gas and air.

Purging of gas piping. After the piping and meter have been checked, all piping receiving gas through the meter should be fully purged at the end of the pipeline to the outside of the building.

Radiant heater

1. Room heater designed primarily to convert the energy in fuel gas to radiant heat by means of refractory radiants or similar radiating materials. A radiant heater has no external jacket, but it is equipped with an exposed back wall.

2. Heater designed to transfer heat primarily by direct radiation.

3. Space heater designed primarily to convert the energy in fuel gas to radiant heat.

4. Room heater utilizing ceramic, asbestos, clay-back, or the equivalent as the radiating media.

5. Suspended heaters should be safely and adequately supported with due consideration given to their weight and vibration characteristics.

Radiant room heaters. Radiant room heaters should have clearances at the sides and rear of not less than 18 in. except that heaters making use of metal, asbestos, or appliance should have a clearance of 36 in. in front, and if constructed with a double back of metal or ceramic they may be installed with a clearance of 18 in. at the sides and 12 in. at rear, or as required by local codes.

Radiation shield. Separate panel or panels interposed between heating surfaces and jackets to reduce heat losses through radiation.

Radiator. Heat source that transmits heat by radiation and conduction; heat exchanger.

Range. Gas cooking appliance consisting of an oven(s) along with additional top or side burners for other cooking purposes.

Range boiler. Tank with a capacity of 120 gal. or less for the storage of water at a pressure not in excess of the maximum working pressure stamped on the tank at a temperature not in excess of 200°F, and having no self-contained means of heating.

Rated heat input. Measured amount of fuel, expressed in British thermal units (Btu) per hour, that a gas appliance is designed to burn completely and that is marked on every appliance meeting the requirements and specifications of a recognized agency, such as the American Gas Association.

Readily accessible. Capable of being reached quickly for operation, renewal, servicing, or inspection without requiring the climbing over or removal of obstacles or the use of portable ladders.

Recessed heater. Completely self-contained gas-fired heating unit, usually recessed in a wall and located entirely above the floor of the space it is intended to heat.

Recessed wall heater. Space heater designed for installation within a wall or partition and approved for such use.

Reformed gas. Gas produced by the reforming of refinery oil gas and natural gas, and composed almost entirely of hydrocarbons.

Register. Warm-air outlet in a room equipped with a damper and a grill to aid in the diffusion of the warm air.

Regulator. Device for controlling and maintaining a uniform gas supply pressure.

Relative humidity

1. Amount of moisture contained in the air, expressed as a percentage of the amount of moisture that air at the same temperature could hold if saturated.

2. Percentage of moisture in the air compared with the maximum amount of moisture that air will hold at a given temperature. Increase in temperature increases the amount of moisture that air will hold. When relative humidity is below 30% out of doors, the fire danger is considered to be serious. During the winter heating season relative humidity in buildings may be very low, contributing to rapid ignition and spread of fire. High relative humidity acts as a fire barrier in the air and when humidity is very high active combustion is unlikely unless air is dried by a brisk deliberate fire.

Relief devices. Safety devices designed to forestall the development of a dangerous condition in the medium being heated by relieving either pressure, temperature, or vacuum built up in the appliance or by permanently shutting off the main gas supply.

Relief opening. Opening provided in a draft hood to permit the ready escape to the atmosphere of the flue products from the draft hood in the event of no draft, backdraft, or a stoppage beyond the draft hood and to permit infiltration of air into the draft hood in the event of a strong chimney updraft.

Relief valve. Valve of the spring-loaded type, without disk guides on the pressure side of the valve, used on a hot-water heating boiler, a hot-water supply boiler, or a water storage tank to relieve excess pressure.

Residential-type appliance. Appliance commonly used in but not restricted to one- or two-family dwellings.

Residue gas. Natural gas remaining after the extraction of various liquid hydrocarbons.

Resistance, $R = I/C$. Overall resistance; amount of resistance to heat flow between air on the warm side and air on the cold side of the building section.

Return systems. Assembly of connected ducts, plenums, fittings, registers, and grilles through which air from the space or spaces to be heated or cooled is conducted back to the heat exchanger.

Reversible flue. Passage for flue gases in which the normal upward direction of the products is reversed at some point in their travel, the outlet usually being at a level only slightly above the burner.

Riser

1. Vertical pipe that conducts the gas upward.

2. Fuel gas supply pipe that extends vertically one full story or more.

Riser heat pipe. Duct that extends at an angle of more than 45° from the horizontal.

Room heater. Self-contained, free-standing, nonrecessed, gas-burning air-heating appliance intended for installation in the space being heated and not intended for duct connection. The heater may be of either the gravity or mechanical air-circulation type, vented or unvented, except that unvented heaters do not have a normal input rating in excess of 50,000 Btu hour.

Room or space heater. Space heaters are above-the-floor devices for direct heating of the space in and adjacent to that which the device is located without heating pipes or ducts.

Safety circuit. Circuit or portion thereof, involving one or more safety controls, in which failure due to grounding, opening, or shorting of any part of the circuit can cause unsafe operation of the controlled appliance.

Safety shutoff valve. Valve that automatically shuts off the supply of gas through the functioning of a flame safeguard control or safety limiting device.

Safety shutoff valve at boiler. Automatic gas control valve of the "on" and "off" type that is actuated by the safety control system or by an emergency device.

Safety valve. Valve of the spring-pop type such as that used on a steam boiler, an air receiver, or a hydropneumatic tank to relieve excess pressure.

Sag. Low place in a horizontal pipe where liquid may collect.

Sealed combustion chamber appliances. Appliances so constructed and installed that all air for combustion is derived from outside the space being heated and all flue gases are discharged to the outside atmosphere.

Sealed combustion unit. Appliance where all the air required for combustion is taken directly from outdoors and the combustion products are exhausted directly outdoors.

Secondary air. Air externally supplied to the flame after ignition or at the point of combustion.

Semiautomatic valve. Valve that is opened manually and closed automatically or vice versa.

Semirigid tubing. Gas conduit having a semiflexible metallic wall structure.

Service pipe. Pipe that brings the gas from the main to the inside of the building.

Service piping. Piping, valves, fittings, and equipment between the street gas main and the gas piping system inlet, which is installed by, and under the control of the gas supplier.

Shut-off valve. Valve used in the piping to fully turn on or fully shut off the gas supply to any section of a system or piping or to an appliance.

Snap-acting thermostat. Thermostat in which the thermostatic valve travels instantly from the closed to the open position or vice versa.

Sour gas. Gas that in its natural state contains such amounts of sulfur compounds as to make it impractical to use without purifying it because of its corrosive effect on piping and equipment.

Space heater (room heater)

1. Above-the-floor device for direct heating of the space in and adjacent to that in which the device is located without external heating pipes or ducts.

2. Individual gas heating appliance, vented or unvented, used for heating a room that is self-contained and connected to a gas outlet. Examples are radiant heaters, reflector heaters, gas steam radiators, gas warm-air radiators, gas unit heaters, gas floor heaters, and floor furnaces.

3. Above-the-floor free-standing heating unit burning solid gas or liquid fuel, for direct heating of the space in and adjacent to that in which the unit is located without external heating pipes or ducts.

Special equipment. Special gas-fired commercial and industrial equipment consisting of boilers for other than central heating plants, various types of ovens, processing equipment, sterilizing equipment, gas-illuminated fixtures, clothes dryers other than the domestic type for private residences, and other such commercial and industrial gas-fired equipment.

Special gas-fired commercial and industrial equipment. Consists of boilers for other than central heating plants, various types of ovens, processing equipment, sterilizing equipment, gas-illuminated fixtures, clothes dryers other than domestic type for private residences, and other commercial and industrial gas-fired equipment.

Specific gravity. Ratio of the weight of a given volume of gas to that of the same volume of air, both measured under the same conditions.

Spring-type regulator. Regulator in which the regulating force acting on the atmospheric side of the diaphragm is derived from a compressed spring.

Stationary appliance. Appliance that is not easily moved from one place to another in normal use because of its size or weight.

Steam radiator. Space heater in which all the energy in the fuel gas (with the exception of that lost from the flue by radiation or convection from the combustion chamber) is transmitted to the surrounding atmosphere through the medium of steam or hot water generated within the appliance. The combustion chamber is that part of the appliance in which combustion of the gas takes place and does not include the flue passages.

Stove. Appliance intended for cooking and space heating.

Street gas main. Portion of the system of a public gas utility used for distributing gas, generally located entirely outside the premises, and designed to supply gas to the utility gas service pipe, consumer's gas service line, and the consumer's gas piping.

Sweat joint. Gastight joint obtained by the joining of metal parts with metallic mixtures or alloys that melt at temperatures below the melting point of the metal parts being joined.

Sweet gas. Gas that in its natural state contains such small amounts of sulfur compounds that it can be used without being purified, with no deleterious effects on piping and equipment.

Temperature relief valve (fusible plug type). Valve that opens and keeps open a relief vent by the melting or softening of a fusible plug or cartridge at a predetermined temperature.

Temperature relief valve (resetting or self-closing). Automatic device that opens and closes a relief valve, depending on whether the temperature is above or below a predetermined value.

Thermal element. Part of a thermostat that is directly acted on by temperature changes and that through the physical change thus produced originates the motion directly or indirectly controlling the action of the thermostatic valve.

Thermostat. Automatic device actuated by temperature changes, designed to control the gas supply to a burner(s), in order to maintain temperatures between predetermined limits.

Thermostat valve. Part of a thermostat that controls gas flow by the position of a movable member with respect to the valve seat.

Throttling-type boiler control. Boiler input control valve that regulates the main gas supply throughout the entire range from the open to the closed position.

Trap. Section of piping in a horizontal line that is below the general level and from which condensate cannot naturally be removed by grading.

Type B gas vents. Factory made vent piping of noncombustible, corrosion-resistant material, which has been approved for use, as a result of tests and listing by a nationally recognized testing laboratory, for venting of approved gas appliances equipped to burn only gas.

Type BW gas vents. Factory-made vent piping of noncombustible, corrosion resistant material, which has been approved for use as a result of tests and listing by a nationally recognized testing laboratory, for venting of approved gas-fired wall furnaces.

Type C gas vents. Vents constructed of sheet copper usually not less than No. 24 U.S. standard gauge, galvanized iron usually not less than No. 20 U.S. standard gauge, or other approved noncombustible corrosion-resistant material.

Type L venting systems. Systems using factory-made vent piping and fittings of noncombustible material, which has been approved for use as a result of tests and listing by a nationally recognized testing laboratory, for use with fuel-burning appliances when used with such systems.

Unfired pressure vessel

1. Closed metal vessel that contains air, steam, gas, or liquid pressure in excess of 50 psig supplied from an external source. Hot-water storage tanks are defined as unfired pressure vessels if the water temperature exceeds 125°F and the tank capacity exceeds 120 gal.

2. Tank or pressure vessel in which a gas or a liquid is stored under pressure with the pressure obtained from a source external to the vessel itself or from the indirect application of heat.

Unit heater

1. Suspended space heater with an integral air-circulating fan, intended for the heating of nonresidential space in which it is installed.

2. Appliance that consists of an integral combination of heating element and fan within a common enclosure and that is located within or adjacent to the space to be heated.

Unit heater—high-static type. Self-contained, automatically controlled, vented appliance limited to the heating of nonresidential space. The appliances incorporate integral means for circulation of air against 0.2-in. water gauge or greater external static pressure.

Unit heater—low-static type. Self-contained, automatically controlled, vented appliance limited to the heating of nonresidential space in which it is installed. The appliances incorporate integral means of circulation of air, normally by propeller or fans, and may be equipped with louvers or face extensions as supplied by the manufacturer.

Unmeasured gas. Gas that has not passed through and has not been recorded by a meter.

Unvented appliance. Appliance designed or installed in such a manner that products of combustion are not conveyed by direct connection from the appliance into a flue or chimney and conducted to the outside atmosphere.

Unvented circulator. Circulator in which air is heated by direct mixing with the combustion products and excess air inside the jacket.

Unvented gas appliance. Gas appliance designed and installed in such a manner that the products of combustion are not conveyed directly to an approved chimney or flue.

Unvented, open-flame radiant-type wall heater. Room heater of the open-front type, designed for insertion in or attachment to a wall or partition, having exposed flames the heat from which is reflected by fireclay or similar radiating materials. It incorporates no venting arrangements in its construction and discharges all products of combustion through the open front into the room being heated rather than to the outside air.

Upshot-type burner. Conversion burner normally designed for boiler or furnace ashpit installation and fired in a vertical position at approximately grate level.

Utility gases. Natural gas, manufactured gas, liquefied petroleum (LP) gas-air mixtures, or mixtures of any part of these gases.

Utility gas service line. Pipe and fittings installed by a public gas utility, used to convey unmeasured gas from the main to a service stop installed at the end of the utility gas service pipe or, where no service stop is installed by the utility, at the point where the service pipe of the utility connects with the service pipe of the consumer.

Vacuum relief valve. Automatic device that opens or closes a relief vent depending on whether the vacuum is above or below a predetermined value.

Valve train. Combination of valves, controls, and piping of an appliance through which gas is supplied to the appliance and by which the gas flow is controlled.

Valve (shutoff). A valve used to shut off either individual equipment (valve to be located in the piping system and readily accessible and operable by the consumer) or the entire piping system (valve to be located between the meter or source of supply and the piping system.)

Valve (tamperproof). Valves designed and constructed to minimize the possibility of the removal of the core of the valve accidentally or willfully with ordinary household tools.

Vent connector. (vent connector pipe). Portion of the vent system that connects the gas appliance to the gas vent or chimney.

Vented appliance. Gas appliance designed and installed in such a manner that all the products of combustion are conveyed directly to an approved chimney or flue.

Vented circulator. Circulator in which room air is heated by contact with surfaces that enclose the flue gases, there being no mixing of flue gases with the heated air in the room.

Vented decorative appliance. Vented appliance whose only function lies in the aesthetic effect of the flames.

Vented gas appliance. Gas appliance designed and installed in such a manner that all the products of combustion are conveyed directly to an approved chimney or flue.

Vented wall furnace. Self-contained vented appliance complete with grilles or the equivalent, designed for incorporation in or permanent attachment to a wall, floor, ceiling, or partition, and furnishing heated air circulated either by gravity or by a fan directly into the space to be heated through openings in the casing.

Vent gases. Products of combustion from gas appliances plus excess air plus dilution of air in the vent connector, gas vent, or chimney above the draft hood or draft regulator.

Venting. Removal of flue gases to the outer air by means of roof openings, chimneys, gas vents, or mechanical exhaust systems.

Vent limiting means. Means that limit the flow of air or gas from the atmospheric diaphragm chamber of a gas pressure regulator to the atmosphere.

Vent system. Gas vent or chimney and vent connector, if used, assembled to form a continuous unobstructed passageway from the gas appliance to the outside atmosphere for the purpose of removing vent gases.

Wall heater. Unit heater supported from or recessed in the wall of the room or space to be heated.

Water gas (blue gas). Gas manufactured by passing steam through an incandescent bed of coke or coal. It consists chiefly of equal percentages of carbon monoxide and hydrogen and generally has a heating value of about 320 Btu/cu ft.

Water gas—carbureted water gas. Gas manufactured in a water gas set by the concurrent production of blue gas in the generator of the set and a high Btu gas by the chemical cracking of gas oil or other hydrocarbon in the carburetor and superheater of the set. The resultant mixture of gases is made to have a heat content of between 525 and 550 Btu/cu ft, although other values in excess of 300 Btu/cu ft are possible, depending on the amount of enriching agents used.

Water heater

1. Appliance for supplying hot water for domestic or commercial purposes other than space heating.

2. Tank or a system of coils, or a combination of both, containing water that is heated by the combustion of gas.

Weight-type regulator. Regulator in which the regulating force acting on the atmospheric side of the diaphram is derived from a weight or a combination of weights.

Wet natural gas. Unprocessed or partially processed natural gas produced from strata containing condensable hydrocarbons. The term is subject to varying legal definitions as specified by certain state statutes.

Zero governer regulator. Regulating device that is normally adjusted to deliver gas at atmospheric pressure withins its flow rating.

HEATING SYSTEMS AND EQUIPMENT

Alteration of a heating system. A change, addition or modification of an existing heating system, or of any part thereof; replacement of an existing furnace with a new furnace; substitution of another furnace for an existing furnace; moving of an existing furnace, and conversion of a furnace to the use of fuel other than that previously used.

Atmospheric tank. A storage tank designed to operate at pressures from atmospheric through 0.5 psig.

Automatic boiler. When applied to any class of boiler, such boiler must be equipped with specified controls and limit devices, including all safety accessories, as required by local or state codes.

Automatic boiler burner. Burner is a device to convey fuel into the combustion chamber in proximity to it, with a combustion air supply to permit a stable controlled heat release compatible with the burner design, and equipped with an ignition system to reliably ignite the entire heat release surface of the burner assembly.

Auxiliary heating equipment. Equipment used in connection with combustion equipment, such as stokers, burners, draft regulations, etc.

Baseboard convector. A low type of convector normally installed at the floor of the space to be heated.

Baseboard heating. Heating where the heating element, usually an electric resistance or forced hot water, is located at the base of the wall.

Boiler and furnace room. Every boiler or furnace room, including the breeching and fuel room, in places of indoor assembly should be enclosed with a 2-hour fire-resistive enclosure or better and all interior openings in walls forming such enclosures should be protected by self-closing, fire-resistive doors. Gas-fired appliances for heating water should be installed in a boiler or furnace room. Chimneys should be constructed in conformity with the requirements of the code.

Boiler blow-off. Controlled outlet on a boiler that permits discharge of sediment or the emptying of the unit.

Boiler blow-off tank. Vessel designed to receive the discharge from a boiler blow-off outlet and to cool the discharge to a temperature of 130°F or less, which permits its safe discharge to the drainage system.

Boiler burning fuel in suspension. Fuel-burning device in which fuel is conditioned or pulverized previous to admitting the fuel into the furnace for combustion. The combustion process is completed with the fuel in suspension.

Boiler (high pressure). A closed vessel in which steam or other vapor (to be used externally to itself) is generated at a pressure of more than fifteen (15) psig by the direct application of heat.

Boiler (low pressure). A boiler operated at pressures not exceeding fifteen (15) psig steam or at water pressure not exceeding 160 psig and temperatures not exceeding 250°F.

Boiler room. Any room containing a fuel-fired steam or hot-water boiler or furnace.

Boilers. The size, number, and location of power or heating boilers to be installed should be marked on the plans, and, except in single dwellings, should be approved by the department for the inspection of steam boilers, unfired pressure vessels, and cooling plants, and by the department of smoke inspection and abatement, before a permit is issued by the department of buildings for the erection of such building.

Ceiling-type direct-fired unit heaters. Direct-fired unit heaters that are suspended from the ceiling, mounted between uprights, or on wall or column brackets. Unit heaters are appliances consisting of a combination of heating element and fan, having a common enclosure, and placed within or adjacent to the space to be heated.

Central heating boilers and furnaces. Heating furnaces and boilers include warm air furnaces, floor mounted direct-fired unit heaters, hot water boilers, and steam boilers operating at not in excess of 15 psig used for heating of buildings or structures.

Central heating plant

1. Equipment centrally located, designed to heat buildings in part or completely by means of pipes or ducts, water, steam, or air used as a medium for conveying the heat generated to radiators or registers in the portions of the building to be heated.

2. Comfort heating equipment installed in a manner to supply heat by means of ducts or pipes to areas other than the room or space in which the equipment is located.

Central heating system. A system whereby heat is furnished from a central source of supply by a heat-producing mechanism or device that is completely separated from those parts of the building to which heat is supplied.

Closed water-piping system. System of water piping where a check valve or other device prevents the free return of water or steam to the water main.

Continuous pilot. Pilot that burns without turndown throughout the entire period that the boiler is in service, whether or not the main burner is firing.

Controls. Required electrical, mechanical, safety, and operating controls must have received approval from a nationally recognized testing agency and be installed in accordance with rules and regulations of the building department.

Convector. Radiator for either hot water or steam heat with many radiation surfaces, such as fins, to increase contact with air moved either by natural or forced convection.

Direct-fired low static unit heater. Direct-fired suspended, self-contained automatically controlled, vented heating appliance, having integral means for circulation of air by a propeller fan or fans.

Direct-fired unit heater. Unit heater that uses liquid or gas fuel for heating of the heat-emitting element.

Direct heating system. System in a building that produces heat to raise the temperature of the space within the building for the purpose of human comfort in which electric heating elements, or products of combustion exchange heat either directly with the building supply air or indirectly through a heat exchanger and using an air distribution system of ducts.

Electric boiler. Any heating apparatus that provides for heat transfer from an electric heating element to a liquid conduction medium such as water.

Expansion tank, closed system. An airtight tank that provides a means of pressurizing the system over a wide range of conditions.

Fuel-burning equipment. Any furnace, boiler, water heater, device, mechanism, stoker, burner, stack, structure, oven, stove, kiln, still, or other apparatus, or a group or collection of such units used in the process of burning fuel, refuse or other combustible material.

Fumes. Air-borne colloidal systems that are formed by chemical reactions or physical processes, such as, but not limited to, combustion, distillation, sublimation, calcination, or condensation.

Furnace

1. The furnace proper comprises an enclosed metal structure partially lined with firebrick and equipped with grates, drying hearth, ash pit, and chambers for mixing, ignition, and combustion.

2. Enclosure for the combustion of fuel.

3. Completely self-contained direct heating unit with burners, combustion chamber, heat exchanger, and casing.

Gravity or circulating type space heater. Vented, self-contained freestanding or wall recessed heating appliance using liquid or gas fuels.

Hand-fired fuel-burning equipment. Any fuel-burning equipment in which fresh fuel is manually thrown directly on the hot fuel bed.

Handling of solid fuel. Includes but is not limited to its transport by water on boats, barges, car ferries, and motor vehicle ferries; its transport by land, by railroad, truck, or trailer; its transfer from water transport to land transport and vice versa; its transfer to and from storage bins, silos, hoppers, or piles; and its transfer to or from the equipment in which it is processed or burned.

Heater room. Space containing central heat-producing or heat-transfer equipment.

Heating. Heating system of a building, which requires the use of high or low pressure steam, vapor or hot water, including all piping, ducts, and mechanical equipment appurtenant thereto, within, adjacent to, or connected with a building, for comfort heating.

Heating appliance. Any device designed or constructed for the generation of heat from solid, liquid, or gaseous fuel or electricity to be used for the purposes other than heating a building or structure.

Heating boiler. Any boiler carrying not in excess of 15 psi steam or 30 psi water pressure.

Heating degree day. Measure of the coldness of the weather experienced, based on the extent to which the daily mean temperature falls below a reference temperature, usually 65°F. The daily mean temperature usually represents the sum of the high and low readings divided by 2.

Heating surface. All surfaces in contact with hot gases for the purpose of transferring the heat by conduction, radiation, or convection.

Heating system. Any combination of building construction, machinery, devices, or equipment, proportioned, arranged, installed, operated, and maintained to produce and deliver in place the required amount and character of heating service.

Heating value. Amount of heat produced by the complete combustion of a unit quantity of fuel; gross higher heating value obtained when all of the products of combustion are cooled to the temperature existing before combustion. The water vapor formed during combustion is condensed, and all the necessary corrections are made. The net or lower heating value is obtained by subtracting the latent heat of vaporization of the water vapor formed by the combustion of the hydrogen in the fuel from the gross or higher heating value.

Heat-producing appliance. An appliance or device used for the production of heat by the combustion of fuel.

Heat pump. A system in which refrigeration equipment is used in such a manner that heat is taken from a heat source and given up to the conditioned space when heating service is wanted, and is removed from the space and discharged to a heat sink when cooling and dehumidification is desired.

High-capacity heating equipment. Containing equipment having an individual or combined rated gross capacity of 1,000,000 Btu per hour or more, or capable of operating at more than 15 psi for steam or more than 30 psi or 250°F for hot water.

High-pressure boiler

1. Boiler furnishing steam at pressures in excess of 15 psi or hot water at temperatures in excess of 250°F or at pressures in excess of 160 psi.

2. Closed vessel in which steam or other vapor to be used externally to itself is generated at a pressure of more than 15 psig by the direct application of heat. Direct-fire hot-water supply heaters and hot-water heating boilers are each defined as high-pressure heating boilers if the water temperature is in excess of 250°F or if the water pressure is in excess of 160 psi.

High-pressure systems. Systems operating in excess of 6-in. water column pressure.

High-static pressure-type heating appliance. Direct-fired suspended or floor standing, self-contained, automatically controlled and vented heating appliance having an integral means for circulation of air against 0.2 in. or greater static pressure.

High-temperature water boiler. Boiler used for heating water or liquid to a pressure exceeding 160 psi or to a temperature exceeding 250°F.

Horizontal-type central furnace. Furnace designed for low headroom installation, with airflow through the appliance essentially in a horizontal path.

Horse power. Boiler horse power is figured as equivalent to the evaporation of $34\frac{1}{2}$ lb of water per hour from and at 212°F.

Humidifier. Device designed to accumulate and discharge water vapor into a confined space for the purpose of increasing or maintaining the relative humidity in an enclosure.

Hydronics. Science of heating and cooling with liquids.

Industrial furnaces and power boilers—stationary type. For the purpose of the standards, stationary-type industrial furnaces and power boilers must be classified as low; medium- or high-heat appliances in accordance with their character and size and the temperatures developed in the portions thereof where substances or materials are heated for baking, drying, roasting, melting, vaporizing, or other purposes.

Interlock. Device that senses a limit or off-limit condition or improper sequence of events and shuts off the offending or related piece of equipment or prevents proceeding in an improper sequence in order to prevent a hazardous condition from developing.

Intermittent pilot. Pilot that burns during light-off and while the main burner is firing and that is shut off with the main burner.

Interrupted pilot. Pilot that burns during light-off and that is shut off during normal operation of the main burner.

Jacketed stove. Vented, self-contained freestanding non-recessed heating appliance, using solid, liquid, or gas fuels. The effective heating is dependent on a gravity flow of air circulation over the heat exchanger.

k conductivity. Thermal conductivity; amount of heat, expressed in British thermal units (Btu), transmitted in 1 hour through 1 sq ft of a homogeneous material 1 in. thick for each degree Fahrenheit of temperature difference between the two surfaces of the material.

Large room or space in comparison with the size of the appliance. For boilers, a room or space having a cubic volume at least 16 times the total volume of the boiler (usually dictated by local codes).

Limit control. A thermostatic device installed in the duct system to shut off the supply of heat at a predetermined temperature of the circulated air.

Liquid piping system. A system in which water or other liquid is used as the medium by which heat is carried through pipes from the supply source to or from the heating or cooling units.

Low-capacity heating equipment. Containing equipment having a rated gross capacity of less than 250,000 Btu per hour and operating at less than 15 psi for steam or less than 30 psi or 250°F for hot water.

Low-pressure boiler

1. Boiler constructed of ferrous or nonferrous metal in which the maximum allowable gauge working pressure is limited to 15 psi for steam and 160 psi at 250°F for hot-water heating boilers.

2. Steel or cast-iron heating boiler in which the maximum allowable gauge working pressure is limited to 15 psi for steam and 30 psi for hot-water heating boilers.

Low-pressure hot-water or low-pressure steam boiler. Boiler furnishing hot water at pressures not exceeding 160 psi and at temperatures not more than 250°F or steam at pressures not more than 15 psi.

Low-pressure steam boiler. Steam boiler designed to operate at a pressure not in excess of 15 psi.

Low-pressure systems. Systems operating not in excess of 2-in. water column pressure.

Low-pressure tank. Storage tank that has been designed to operate at pressures above 0.5 psig but not more than 15 psig.

Low-static-type unit heater. Direct-fired suspended, self-contained automatically controlled, vented heating appliance, having integral means for circulation of air by means of a propeller fan or fans.

Low-temperature gas-oil-burning equipment. Fuel-burning appliances listed as exhausting low-temperature combustion products containing a minimum of excess air and listed for use with Type L low-temperature vent systems.

Maximum allowable working pressure. Pressure at which a boiler is permitted to be operated or used under code.

MBH. 1000 Btu/hour.

Medium-pressure systems. Systems operating with pressures over 2- and up to and including 6-in. water column pressure.

Metal housing for heating appliances. The burner of the appliance should be enclosed with a metal housing so constructed that there will be no open flame, and the burner housing should be effectively guarded against personal contact. The arrangement should be such that the shield will prevent any combustible material in the vicinity of the appliance from coming in contact with the flame or with the housing that encloses the burner.

Miniature boiler. Boiler having an internal shell diameter of 16 in. or less and a gross volume of 5 cu ft or less for steam or vapor under pressures exceeding 15 psi and not exceeding 100 psi.

Moderate capacity heating equipment. Containing equipment having an individual or combined rate gross capacity from 250,000 to 1,000,000 Btu per hour, and operating at less than 15 psi for steam or less than 30 psi or 250°F for hot water.

Net rating of heating boiler. Rating specified by the Institute of Boiler and Radiator Manufacturers for cast-iron boilers and by the Steel Boiler Institute for steel boilers.

One-pipe system. A one-pipe system employs a single pipe main with special fittings installed at riser connections to the heating elements.

Package boiler. Boiler shipped equipped complete with fuel-burning equipment, automatic controls and accessories, with mechanical draft equipment when specified.

Panel heating. A method of space heating in which radiant heat is supplied by large heated areas of the room surface operating at low surface temperatures, 80°F to 140°F.

Permit requirements. All boilers and pressure vessels and their installations must conform to the minimum requirements for safety from structural and mechanical failure and excessive pressures, established by a nationally recognized agency and local and state building codes.

Pilot. Device ignited by a spark or other independent and stable ignition source and that provides ignition energy required to immediately light the main burner.

Prefabricated heating panels. A factory-built heating unit. Heat may be supplied by hot water, steam, or electricity.

Pressure piping. Includes piping for power, refrigeration, hydraulic, liquefied petroleum gas, and heating piping as listed in the code.

Pressure vessel. Storage tank or vessel designed to operate at pressures above 15 psig.

Purge. Acceptable method of scavenging the combustion chamber, boiler passes, and breeching to remove all combustible gases.

Radiant heaters. Reflector heaters, gas steam radiators, gas warm-air radiators, gas unit heaters, gas floor heaters, or floor furnaces.

Radiant heating system. A heating system in which only the heat radiated from panels is effective in providing the heating requirements.

Radiator. A heating unit exposed to view within the room or space to be heated, that emits heat to objects within visible range by radiation and to the surrounding air by convection.

Reversible-flue furnace. Furnace in which the course of the flue gas is reversed before it reaches the vent outlet.

Room or space heater. Space heaters are above-the-floor appliances for direct heating of the space in and adjacent to that which the appliance is located, without heating pipes or ducts.

Room sizes for furnaces. For furnaces, usually a room or space having a volume at least 12 times the total volume of the furnace.

Small power boiler. Boiler with pressure exceeding 15 psi but not exceeding 100 psi and having less than 350,000 Btu/h heat output.

Space heater. Space-heating appliance for heating the room or space within which it is located, without the use of ducts.

Space-heating appliance. Appliance intended for the supplying of heat to a room or space directly, such as a space heater, fireplace, or unit heater, or to rooms or spaces of a building through a heating system, such as a central furnace or boiler.

Stack dampers. Boilers fired with oil or solid fuel should not close off more than 80% of the stack area when closed, except on automatic boilers with prepurge, automatic draft control, and interlock. Operative dampers should not be placed within any stack, flue, or vent of a gas-fired boiler, except an automatic boiler with prepurge automatic draft control and interlock.

Standard air. Air that is equivalent to dry air at 70°F and 29.92 inches of mercury.

Stationary industrial furnaces and power boilers. Furnaces and boilers classified as low-, medium- or high-heat appliances in accordance with their character and size and the temperatures developed in the portions thereof where substances or materials are heated for baking, drying, roasting, melting, vaporizing, or other purposes.

Tempered air. Treated air heated or cooled, compatible with the space to which it is being introduced.

Tempered outside air. Outside air heated before distribution.

Two-pipe direct return system. A system in which the heating medium, after it has passed through a heat exchanger unit, is returned to the boiler by the shortest direct path, resulting in considerable differences in the lengths of the several circuits composing the system.

Two-pipe reversed return system. A system in which the heating medium from each heat transfer unit is returned along paths arranged so that all circuits composing the system are of equal length.

***U* factor.** Overall heat transmission coefficient; the amount of heat, expressed in British thermal units (Btu) transmitted in 1 hour through 1 sq ft of a building section (wall, floor, or ceiling) for each degree Fahrenheit of temperature difference between air on the warm side and air on the cold side of the building section.

Unassisted gravity return. Steam return line to a boiler in which all of the condensate returns to the boiler without the assistance of a pump, injector, trap, or similar apparatus.

Unfired pressure vessels. Includes jacketed kettles, steam cookers, stills, digesters, compressed air tanks, and other pressure vessels used for storage of gases under pressure exceeding 15 psi.

Unit heater. Heating appliance other than a floor-mounted space heater that consists of a heat-emitting element and fan contained within a common enclosure, designed and installed for delivery of warm air for space heating directly into the space in which or adjacent to which the appliance is located.

Unit heater (infrared). A heater consisting of an element heated to a high temperature electrically or by burning fuel and the radiant heat so produced is directed by a reflector.

Up-feed system. A heating system in which the supply mains are below the level of the heating units that they serve.

Volume damper. Any device that when installed will restrict, retard, or direct the flow of air, in any duct, or the products of combustion in any heat-producing equipment, its vent connector, vent, or chimney therefrom.

Wall furnace. A vertical furnace designed to be installed in or against a wall. Cool air enters at bottom and is discharged at the top. Air may be circulated by gravity or by a fan.

Welding. Welding on pressure vessels should be by approved welders who have received a welder's certificate of competency, and the work to be done should be in conformance with nationally recognized testing agencies and be subject to inspection and approval by local building officials.

HOT WATER AND STEAM HEATING SYSTEMS AND EQUIPMENT

Blower unit heater. Assembly of heating coils, finned tubes, or electric elements in a housing, but with quiet centrifugal-type fans to move air through the heating element.

Boiler. Heating appliance intended to supply hot water or steam for space heating, processing, or power.

Boiler blowoff. Outlet on a boiler to permit emptying or discharge of sediment.

Boiler heating surface. Portion of the surface of the heat transfer apparatus in contact with the fluid being heated on one side and the gas or refractory being cooled on the other in which the fluid being heated forms part of the circulating system; the surface measured on the side receiving heat, including the boiler, water walls, water screens, and water floor.

Boiler horsepower. One boiler horsepower is equivalent to the evaporation of 34.5 lb of water per hour at 212°F.

Breeching. Casing at the (end of a boiler) opening of a device carrying products of combustion where a flue is connected as part of a system to convey the products of combustion (from the boiler) to a chimney or vent.

Burner unit. One or more burners that can be ignited safely from one source of ignition.

Central fan system. Mechanical indirect system of heating, ventilating, or air-conditioning where the air is treated or handled by equipment located outside the rooms served, usually at a central location, and conveyed to and from the rooms by means of a fan and a system of distributing ducts.

Central heating plant. Steam or hot-water boiler located outside the spaces served with piping arranged to convey heat to one or more rooms or spaces.

Chimney effect. Tendency of air or gas in a duct or other vertical passage to rise when heated owing to its lower density compared with that of the surrounding air or gas; in buildings, the tendency toward displacement (caused by the difference in temperature) of internal heated air by unheated outside air due to the difference in density of outside and inside air.

Circulating air. Heating medium being moved through the furnace from the air inlet opening to the air outlet opening.

Closed hot-water system. A forced hot-water system in which the circulating water is completely enclosed, under pressure above atmospheric, and closed to the atmosphere.

Coefficient heat transmission. Any one of a number of coefficients used in the calculation of heat transmission by conduction, convection, and radiation through various materials and structures.

Condensate. When steam loses sufficient heat, it returns to water. This liquefied steam is condensate.

Convection. In a fluid, the motion resulting from the difference in density and the action of gravity. In heat transmission this definition has been extended to include both forced and natural motion or circulation.

Convector

1. Assembly of ferrous or nonferrous tubes with metal fins attached, mounted in a metal enclosure, designed to transfer heat to the air, principally by convection.

2. Agency of convection. In heat transfer, the surface designed to transfer its heat to a surrounding fluid largely or wholly by convection. The heated fluid may be removed mechanically or by gravity (gravity convector). The surface may or may not be enclosed or concealed. When it is concealed and enclosed, the resulting device is sometimes referred to as a "concealed radiator."

Corner joints. Corner joints are similar to tee joints except that one member does not extend beyond the other.

Degree day. Unit based on temperature difference and time, used in estimating fuel consumption and specifying nominal heating load of a building in winter. For any one day, when the mean temperature is less than 65°F there exist as many degree days as there are degrees Fahrenheit difference in temperature between the mean temperature for the day and 65°F.

Direct-return system (hot water). Hot-water system where the water, after it has passed through a heating unit, is returned to the boiler along a direct path so that the total distance traveled by the water is the shortest feasible. There are considerable differences in the lengths of the several circuits composing the system.

Down-feed system (steam). Steam heating system where the supply mains are above the level of the heating units which they serve.

Draft head (side outlet enclosure). Height of a gravity convector between the bottom of the heating unit and the bottom of the air outlet opening (top outlet enclosure); height of a gravity convector between the bottom of the heating unit and the top of the enclosure.

Draw off. Valved connection from the return header of a boiler, usually piped to the drain. Scale and sediment are blown from the boiler through the draw off.

Drip. Pipe (or a steam trap and a pipe considered as a unit) that conducts condensation from the steam side to the return side of a steam heating system.

Dry-bulb temperature. Temperature indicated by an ordinary thermometer, usually expressed in degrees Fahrenheit.

Drying hearth. Inclined deck located slightly above and forward of the grates, on which wet garbage is deposited.

Dry return. Return pipe in a steam heating system that carries both water of condensation and air. The dry return is above the level of the water line in the boiler in a gravity system.

EDR (equivalent direct radiation). The term EDR is the result of originally rating heat output of direct radiation (free-standing cast-iron radiators and pipe coils) on the basis of the actual square feet of heating surface. Radiators were found to vary in heat output per square foot of surface depending on height, depth, and width of sections. A standard was established, namely, 240-Btu/hour output per square foot of heating surface, and has been adopted for rating present-day radiation used with steam. The term in use is equivalent direct radiation or EDR: 1-sq-ft EDR equals 240 Btu/hour.

Effective temperature. Arbitrary index that combines into a single value the effect of temperature, humidity, and air movement on the thermal sensations of the human body. Effective temperatures vary from summer to winter.

Emissivity. Ratio of the energy emitted from a radiator that is not a blackbody to that emitted from a blackbody at the same temperature.

Equivalent direct radiation (EDR). Unit of heat delivery of 240 Btu/hour. It does not imply 144 sq in. of surface.

Equivalent evaporation. Amount of water a boiler would evaporate, in pounds per hour, if it received feed water at 212°F and vaporized it at the same temperature and corresponding atmospheric pressure.

External heat gain. Amount of heat added to an inside space from outside sources. External heat is transmitted primarily through walls, ceilings, window, and doors when the outside temperature is higher than the inside temperature and through sun effect on windows, walls, and roofs and infiltration through cracks.

Factory-built chimney. Chimney consisting entirely of factory-constructed metal parts, each part designed to be assembled with the others without requiring field fabrication, and certified as to assembly methods performed by a designated testing organization.

Fillet weld. Weld of approximately triangular cross section joining two surfaces approximately at right angles to each other in a lap joint, tee joint, or corner joint.

Fin tube radiation. Term describing an assembly of ferrous or nonferrous tubes with metal fins attached, square or rectangular, usually installed continuous along a wall or ceiling, and always provided with a metal cover to provide convection where required and safety to the user.

Flywheel effect. Time lag between a peak external heat gain of short duration and a peak internal temperature. Time lag is due to the absorption of the heat gain by building materials and furnishings.

Forced hot-water heating system. A system in which water is heated in the boiler and is forced through the pipes by the action of a circulating pump.

Forced hot-water system. System in which circulation is created by means of a pump, usually driven by an electric motor.

Furnace volume (total). The total furnace volume for horizontal-return tubular boilers and water tube boilers is the cubical contents of the furnace between the grate and the first plane of entry into or between tubes. It includes the volume behind the bridge wall in ordinary horizontal-return tubular boiler settings, unless manifestly ineffective (for example, no gas flow taking place through it), as in the case of waste-heat boilers with auxiliary coal furnaces, where one part of the furnace is out of action when the other is being used. For Scotch or other internally fired boilers it is the cubical contents of the furnace, flues, and combustion chamber, up to the plane of first entry into the tubes.

Fusion joint. Joint formed by melting together filler metal and base metal or base metal only, which results in coalescence.

Gauges. Steam boilers must be equipped with a pressure gauge and a water gauge glass. Water boilers must be equipped with a pressure gauge and a temperature indicator.

Grate area. Area of the grate surface, measured in square feet, to be used in estimating the rate of burning fuel. The area is construed to mean the area measured in the plane of the top surface of the grate, except that with special furnaces, such as those having magazine feed or special shapes, the grate area is the area of the active part of the fuel bed taken perpendicular to the path of the gases through it. For furnaces having a secondary grate, such as those in double-grate downdraft boilers, the effective area is the area of the upper grate plus one-eighth of the area of the lower grate, both areas being estimated as previously defined.

Gravity hot-water heating system. A system in which water is heated in the boiler and, as the water temperature rises, it flows out through supply pipes to the space distribution units; the cooled water flows downward to the return pipes to the boiler.

Gravity hot-water systems. Heating systems where circulation of water is due to the head created by the difference in density of the water between the supply and return risers.

Gravity low-pressure steam heating system. One in which the condensate is returned to the boiler by gravity due to the static head of water in the return mains. The elevation of the boiler waterline must consequently be sufficiently below the lowest heating units and steam main and dry return mains to permit the return of condensate by gravity. The waterline difference must be sufficient to overcome the maximum pressure drop in the system and the operating pressure of the boiler when radiation and drip traps are used as in two-pipe vapor systems. This applies only to closed circuit systems, where the condensation is returned to the boiler. If the condensation is wasted, no waterline difference is required.

Hartford Connection. Manner of connecting pipe and pipe fittings in the bleeder or equalizer between the steam and return headers of a boiler. The connection into the bleeder or equalizer is made at a height that would prevent the boiler water level from lowering dangerously below normal level (because of backward flow into the return mains or pump discharge lines).

Heating and hot-water supply installation. A fuel burning installation used only for space heating or hot water supply.

Heating boiler

1. Boiler operating at less than 15 psi.
2. Heating boilers include hot-water boilers operating at temperatures not in excess of 240°F and steam boilers operating at pressures not in excess of 15 psi, both types used for heating buildings or structures.

Heating surface. Exterior surface of a heating unit. An extended heating surface (or extended surface) consists of fins, pins, or ribs that receive heat by conduction from the prime surface. A prime surface is a heating surface having the heating medium on one side and air (or medium to which heat is transferred) on the other.

High-pressure hot water. Any boiler, generator, pressure vessel, system, piping, or equipment used for the purpose of heating or distributing hot water for heating or processing, operating at pressures in excess of 45 psig and temperatures in excess of 250°F classified as high pressure.

High-pressure steam

1. Any boiler, generator, pressure vessel, system, piping, or equipment used for the purpose of heating or distributing steam for heating, power, and processing, operating at pressures in excess of 15 psig, is classified as high pressure.
2. Boiler furnishing steam at pressures in excess of 15 psi or hot water at temperatures in excess of 250°F or at pressures in excess of 160 psi.

Hot water. Any boiler, generator, pressure vessel, system, piping, or equipment used for the purpose of heating or distributing hot water for heating or processing, operating at pressures in excess of 160 psig and/or temperatures in excess of 250°F is classified as high pressure.

Hot-water forced circulation system. System that has a booster or circulating pump installed to circulate the water in the system mechanically.

Hot-water gravity system. Gravity circulated system that depends on the difference in the weight or in the density between the hot water and the cold water to create circulation within the piping system.

Hot-water heating system. A heating system in which water is used as the medium by which heat is carried from the boiler to the heating units.

Hot-water supply. Devices for heating and storing water in boilers or hot-water tanks should be designed and installed to prevent danger form explosion through overheating.

Hot-water supply boiler. Boiler having volume exceeding 120 gal or a heat input exceeding 200,000 Btu/h or an operating temperature exceeding 200°F that provides hot water to be used externally to itself.

Lift fitting or lift connection. Casting or assembly of pipe fittings that provides a seal between a horizontal return main and a vertical connection to another return main at a higher level, used in piping between the outlet taping of an accumulator tank and the suction connection of a vacuum pump if this distance exceeds 5 ft in height.

Low-pressure hot water

1. Any boiler, generator, pressure vessel, system, piping, or equipment used for the purpose of heating or distributing hot water for heating or processing, operating at pressures of 45 psig or less and temperatures of 250°F or less is classified as low pressure.
2. Boiler furnishing hot water at pressures not exceeding 160 psi and at temperatures not more than 250°F.

Low-pressure steam. Any boiler, generator, pressure vessel, system, piping, or equipment used for the purpose of heating or distributing steam for heating, power, and processing, operating at pressures of 15 psig or less is classified as low pressure.

Mechanical return low-pressure steam heating system. One in which the condensate flows to a receiver and is then forced into the boiler against the boiler pressure. The lowest parts of the supply side of the system must be kept sufficiently above the waterline of the receiver to insure adequate drainage of water from the system, but the relative elevation of the boiler waterline is unimportant in such cases except that the discharge head on the mechanical return device becomes greater as the height of the boiler waterline above the pump increases.

Mixing chamber. Space or passage connecting the ignition and combustion chambers.

One-pipe steam heating system. A system where the pipe that carries the steam to the distribution units also returns the condensed steam to the boiler.

One-pipe systems. Those in which the flow of the steam supply to the radiation and the return of condensation flow are in opposition to each other.

Outdoor design temperature. For heating load calculations, the outdoor temperature that is equaled or exceeded during $97\frac{1}{2}\%$ of the hours in December through March; for cooling load calculations, the outside temperature that is equalled or exceeded by $2\frac{1}{2}\%$ of the total hours of June through September.

Overhead system. Any steam or hot-water system in which the supply main is above the heating unit. In a steam system the return must be below the heating units;

in a hot-water system the return may be above or below the heating units.

Panel heating

1. Heating system in which heat is transmitted by both radiation and convection from panel surfaces to both air and surrounding surfaces.

2. Radiant heating system of any type. The heating medium may be hot water in pipes buried in floors, walls, or ceilings, or electric resistance heating elements buried in walls or ceilings in panels mounted on the walls or ceilings.

Panel radiator. Heating unit placed on or flush with a flat wall surface and intended to function essentially as a radiator.

Portable heating appliance. Approved, unvented air-heating appliance designed for human comfort and not secured or attached to a building by any means other than fuel piping or electrical wiring.

Power boiler plant. One or more steam boilers or power hot-water boilers and connecting piping and vessels on the same premises.

Power steam boiler. Boiler in which steam or other vapor is generated at pressures exceeding 15 psi.

Pressure vessel. Tank, jacketed vessel, or other unfired pressure vessel used for transmitting steam for power or for using or storing steam under pressure for heating or steaming purposes.

Process equipment. Equipment used for processing, such as canning equipment, tannery equipment, clothing manufacturing equipment, cleaning and pressing equipment, laundry equipment, or hospital equipment, which usually requires high-pressure steam.

Radiant heat. Heat transmitted through an object that may not itself be hot; the heat transmitted through glass on which the sun is shining.

Radiant heating

1. Heating system with warm or hot surfaces used to radiate heat into a cooler area.

2. Heating system in which only the heat radiated from panels is effective in providing the heating requirements. The term "radiant heating" is frequently used to include both panel and built-in radiant heating.

Radiation

1. Heat transfer by means of rays traveling in direct lines from the source to another body; also, heat transfer units such as convectors or radiators.

2. Flow of heat through space from a warm surface to a cooler one. The flow is like that of light and is of practical importance in connection with reflective insulation.

Radiator. Assembly of cast-iron sections connected to form a space for steam or hot water and having a large area for transmission of heat.

Return header of a boiler. Horizontal piping connected to the return tapping(s) of the boiler. The bleeder or equalizer of the steam header is connected to the return header. Condensate from the steam header and also from the heating system piping returns to the boiler through the return header.

Return main. Horizontal piping through which the heating medium is conveyed from the various return pipes from the radiation to the boiler, return trap, or pump.

Reversed-return system. System in which the heating or cooling medium from several heat transfer units is returned along paths arranged so that all circuits composing the system or a major sub-division of it are of practically equal length.

Riser. Vertical pipe carrying either steam or condensate from floor to floor in a building.

Runout. Horizontal connection from a convector or radiator to the riser.

Split system. System in which the heating is accomplished by means of radiators or convectors and mechanical circulation of air from a central point supplies the ventilation. Mechanically circulated air does not supply the heat to balance the heat loss.

Springpiece. Horizontal connection from a main to a riser or radiator.

Square foot of heating surface (equivalent). Synonymous with equivalent direct radiation (EDR).

Stack height (convector). Height of a gravity convector between the bottom of the heating unit and the top of the outlet opening.

Steam. Any boiler, generator, pressure vessel, system, piping, or equipment used for the purpose of heating or distributing steam for heating, power, and processing, operating at pressures of 15 psig or less, is classified as low pressure.

Steam header of a boiler. Horizontal piping connected to the boiler steam outlet or outlets. The header serves as a steam reservoir, and the steam main(s) connected to it. The bleeder or equalizer connection between the steam and return header also equalizes the boiler pressure to keep the boiler waterline steady.

Steam heating system. The steam is generated in the boiler and rises to the space radiators, convectors, and coils, where it condenses and forms water and returns to the boiler.

Steam piping system. A system in which steam is transferred from a source to a steam-utilizing device at, above or below atmospheric pressure for a purpose other than for heating a building.

Steam trap. Device for allowing the passage of condensate, or air and condensate, and preventing the passage of steam.

Stratification. Arrangement in strata or layers. In heating, stratification of air may occur in a room with a high ceiling, resulting in a marked temperature difference between floor and ceiling.

Stub. Vertical piping connection to a radiator valve or trap from the steam and return riser runouts or springpieces.

Superheated steam. Steam at a higher temperature than that at which water would boil under the same pressure.

Supply main. Horizontal pipe through which the heating medium flows from the boiler or source of supply to the springpieces, risers, and runouts leading to the heat transfer units.

Total effective grate surface. Cast-iron grate area plus the hearth area that is effective as a grate surface.

Two-pipe steam heating system. A system where the steam rises through a supply main to the distribution units; air in the system and the condensed steam are forced through thermostatic traps at the bottom outlets of the units to the return main. An air eliminator in the return main expels the air through a vent and allows the water to return to the boiler.

Two-pipe system. A steam or hot-water heating system in which one pipe is used for the supply of the heating medium to the heating unit and another pipe for the return of the heating medium to the source of the heat supply. The essential feature of a two-pipe system is that each heating unit receives a direct supply of the heating medium, which medium cannot have served a preceding heating unit.

Two-pipe systems. Those in which one pipe is used for the supply of steam to the radiator and another for the return of condensation.

Unit heater

1. Assembly of hot water or steam heating coils or electric elements with a semienclosed housing and a fan to recirculate room air through the elements, usually designed for ceiling support and for a horizontal or downward blast of heated air.

2. A steam or hot-water unit heater is a heater in which the heating element is supplied heat from a steam or hot-water heating system.

Vacuum heating system. A two-pipe steam heating system equipped with the necessary accessory apparatus that will permit operating the system below atmospheric pressure when desired.

Vapor heating system. A steam heating system that operates under pressures at or near atmospheric and that returns the condensation to the boiler or receiver by gravity. Vapor systems have thermostatic traps or other means of resistance on the return ends of the heating units for preventing steam from entering the return mains; they also have a pressure-equalizing and air-eliminating device at the end of the dry return.

Water hammer. Noise resulting from steam coming in contact with condensate in pocketed or back-graded piping.

Wet return. Part of a return main of a steam heating system that is filled with water of condensation. A wet return is usually below the level of the waterline in the boiler, although not necessarily so.

Zoned unit. Rooms or unit of the building with auxiliary spaces, with heat supply arranged so that the zoned unit can be controlled independently from the rest of the building.

LIQUEFIED PETROLEUM GAS SYSTEMS

Adequate ventilation. During normal operation ventilation is considered adequate when the concentration of the gas in a gas–air mixture does not exceed 25% of the lower flammable limit.

API. American Petroleum Institute.

Cargo tank. Tank designed to be permanently attached to motor vehicle and in which liquefied petroleum gas is to be transported. A fuel tank used solely for the purpose of supplying fuel for the propulsion of a vehicle is not a cargo tank.

Container assembly. Assembly consisting essentially of the container and fittings for all container openings, including shutoff valves, liquid level gauging devices, safety relief devices, and protective housings.

Containers. All vessels, such as tanks, cylinders, or drums, used for transporting or storing liquefied petroleum gases.

Cylinder. Vessel having a capacity not exceeding that of 1000 lb of water under normal conditions.

Filled by volume. Amount of liquefied petroleum gas in the vessel as determined by volumetric measurements.

Filled by weight. Net amount of liquefied petroleum gas in the vessel as determined by weight.

Filling density. Percent ratio of weight of liquefied petroleum gas to the weight of water the vessel will hold at 60°F.

Float gauge. A gauge constructed with a float inside the container resting on the liquid surface that transmits its position through suitable leverage to a pointer and dial outside the container indicating the liquid level. Normally the motion is transmitted magnetically through a nonmagnetic plate so that no LP gas is released to the atmosphere.

Gallon. U.S. standard gal = 0.83 imperial gal, 231 cu. in. and 3.785 liters.

Gas. Liquefied petroleum gases in either the liquid or gaseous state.

Liquefied petroleum (LP) gas

1. Gas–air mixture. Liquefied petroleum gases are distributed at relatively low pressures and normal atmospheric temperatures. They are diluted with air to produce desired heating value and utilization characteristics.

2. Material in the liquid or gaseous phase that is composed predominantly of any of the following hydrocarbons or mixtures thereof: propane, propylene, butanes (normal butane or isobutane), and butylenes.

3. Fuel gases, including commercial propane (predominantly propane and/or propylene) or commercial butane (predominantly butane, isobutane and/or butylene). Liquefied petroleum gases are supplied in cylinders and used without liquid vaporizers. They must be either commercial propane or commercial butane as defined above to avoid major variations in heating value of the gas as it is released from the cylinder.

Liquefied petroleum gas equipment. All containers, apparatus, piping (not including utility distribution piping systems), and equipment pertinent to the storage and handling of liquefied petroleum gas. Gas-consuming appliances are not to be considered as being liquefied petroleum gas equipment.

LP gas system. An assembly consisting of one or more containers with a means for conveying LP gas from the container(s) to dispensing or consuming devices (either continuously or intermittently) and that incorporates components intended to achieve control of quantity, flow, pressure, or state (either liquid or vapor).

Maximum allowable working pressure. Internal working pressure for which the vessel was designed or the maximum internal working pressure allowable according to the conditions of the vessel when last inspected, whichever is less.

Mobile vehicle. Easily movable, readily portable, or self-propelled vehicle driven by an internal combustion engine.

Motor fuel tank. Vessel used to supply liquefied petroleum gas to the motor of a mobile vehicle.

Movable fuel storage tenders or farm carts. Containers with a capacity not in excess of 1200-gal water capacity, equipped with wheels to be towed from one location of usage to another. They are basically nonhighway vehicles but may occasionally be moved over public roads or highways. They are used as a fuel supply for farm tractors, construction machinery, and similar equipment.

Outage. Portion of the vessel not permitted to be filled with liquid, providing for the expansion of the liquid from an increase in temperature.

Point of delivery. End of the delivery hose or pipe where it is connected to the receiving vessel or the fill line connected thereto.

Saddle. Part of the supporting foundation upon which a horizontally installed vessel rests.

Skid tank. Portable tank used to transport or store liquefied petroleum gas and equipped with fixed skids or with feet or lugs to which skids are attached when used to transport liquefied petroleum gas.

System. Assembly of equipment consisting essentially of the vessel or container and major devices such as vaporizer, relief valves, excess flow valves, regulators, and piping connecting such parts.

Tank. Vessel of more than 1000-lb water capacity (nominal) used for the storage, transportation, or utilization of liquefied petroleum gas.

Tank motor vehicle. Motor vehicle designed or used for the transportation of liquefied petroleum gases in a cargo tank.

Trailer tank. Transportation tank permanently installed on a trailer or semitrailer.

Vaporizer. Device used to convert liquefied petroleum gas from the liquid to the gaseous state.

Vaporizer-burner. Integral vaporizer-burner unit dependent on the heat generated by the burner as the source of heat to vaporize the liquid used for dehydrators or dryers.

Vessel. Container used for the storage, transportation, or utilization of liquefied petroleum gas.

OIL-BURNING EQUIPMENT AND APPLIANCES

Accessibility. The installation of the equipment or appliances should be such as to provide reasonable accessibility for cleaning heating surfaces, removing burners, replacing motors, adjusting controls, replacing air filters, adjusting draft regulators, and the cleaning and lubrication of all moving or stationary parts that will require maintenance.

Accessory. Part capable of performing an independent function(s), certified separately from and contributing to the operation of the appliance(s) that it serves.

Air change. Quantity of air, provided through the burner, equal to the volume of furnace and boiler gas passes (air volume to be calculated at 14.7 psia and 70°F.).

Air rich. Ratio of air to fuel supplied to a furnace that provides more air than that required for an optimum air/fuel ratio.

Air supply. Air for combustion, ventilation, and flue gas dilution.

Air temperature rise. Difference in temperature, in degrees Fahrenheit, between the average temperature of the air discharged from the air outlet opening, such as outlet air temperature, and the average temperature of the air entering the air inlet opening, such as the inlet air temperature.

Air—theoretical. Chemically correct amount of air required for complete combustion of a given quantity of a specific fuel.

Alarm. Audible or visible signal indicating an off-standard or abnormal condition.

Antiflooding device. Primary safety control that causes the fuel to be shut off upon a rise of fuel above the normal level or upon the receipt of excess fuel.

Appliance. Device to convert fuel into energy, including all components, controls, wiring, and piping required by the applicable standard to be part of the device.

Appliance control system. Combination, as applicable, of primary safety control, limit controls, and operating controls that are used to control the burner of an appliance.

Appliance supply piping. Fuel supply piping leading to the burner of an appliance from a supply tank or, in the case of a central oil distribution system, from the subatmospheric or demand valve immediately downstream of the meter.

Atomizer. Device in an oil burner that emits liquid fuel in a finely divided state with the assistance of an atomizing medium, such as stream or air.

Atomizing medium. Supplementary fluid, such as steam or air, that assists in breaking down oil into a finely divided state.

Atomizing-type oil burner. Oil burner that breaks up the fuel into liquid fuel droplets prior to vaporization.

Attic furnace. Horizontal furnace intended for installation in attics or other unoccupied spaces with low headroom.

Automatically lighted burner. Burner where fuel oil to the main burner is normally turned on and ignited automatically.

Automatically operated damper. Damper operated by an automatic control.

Automatic appliance. Appliance equipped with an automatic burner.

Automatic burner. Burner equipped with an automatic appliance control system.

Automatic burner control. When the burner is started, it will continue an unlimited number of operating cycles without manual attention unless shut down by the combustion safety control.

Automatic burner control (nonrecycling). System by which a furnace is purged and a burner is started, ignited, modulated, and stopped automatically but does not recycle automatically.

Automatic burner control (recycling). System by which a furnace is purged and a burner is started, ignited, modulated, and stopped automatically and that recycles on a preset pressure range.

Automatic input control valve. Valve for modulating the fuel supply to the main burner without requiring manual action.

Automatic operation. Operation or sequence or cycle of operations performed by a device or combination of devices without manual attention.

Automatic valve. Valve designed to turn on or shut off the fuel flow to the burner without requiring manual action.

Automatic valve of the manual reset type. Automatic valve that remains closed until manually reopened.

Auxiliary tank. Supply tank installed in the fuel supply between a burner and its main fuel supply tank

Baffle (fixed damper). Stationary device used to divert the flow of fluid (air, water, or steam) or flue gases (flue baffle) or to shield parts of an appliance from the effects of flame (flame baffle) or heat (heat baffle).

Barometric tank. Supply tank that automatically maintains by barometric pressure a definite level of oil in a sump.

Blast tube. Round, pipe-shaped casting in which the oil nozzle and electrodes are located, extends to the rear and supports the motor, blower, and oil pump. Also holds the air turbulator to spin the air for the flame.

Boiler. Appliance intended to supply hot water or steam for space heating, processing, or power purposes.

Boiler-furnace explosions. Furnace explosions occur when sufficient quantities of fuel and air are accumulated in explosive mixtures and then ignited. Ignition energy may be supplied by the regular ignition devices, from another flame envelope, or from other sources within the furnace and boiler passes. Such accumulations of fuel and air can result from improper operating sequences, inability to ignite fuel as it enters the furnace, and fuel-rich or unstable air/fuel rations. Explosions can result from improper design, application, operating procedures, maintenance, or mechanical failure of equipment components.

Boiling point. The boiling point of a liquid is the temperature of the liquid at which its vapor pressure equals the atmospheric pressure.

Bonnet. Part of a furnace casing that forms the supply plenum or to which the supply plenum is attached.

Breeching. Flue pipe or chamber for receiving flue gases from one or more flue connections and for discharging these gases through a single flue connection.

Btu. British thermal unit or the quantity of heat required to raise 1 lb of water 1°F.

Btuh. Btu per hour.

Burner. Device for the introduction of fuel and air into a furnace at the required velocities, turbulence, and concentration to establish and maintain proper ignition and stable combustion of fuel within the furnace.

Burner unit. One or more burners that can be ignited safely from one source of ignition.

Central oil distribution system. System by which oil is supplied through piping from a central supply tank(s) to buildings, mobile homes, travel trailers, or other structures.

Certified (listed) with respect to appliances, components, and accessories. Investigated and suitably marked by a recognized testing agency as conforming to recognized standards or requirements, or accepted test reports.

Cleaner-filter. Device to remove foreign matter from fuel.

Clothes dryer. Appliance intended to supply heat for the drying of wet laundry.

Combustion air. Air required for satisfactory combustion of fuel, including excess air.

Combustion chamber. Metal or refractory chamber located within the firebox of an appliance and used to contain the combustion flame.

Combustion products. Constituents resulting from the combustion of fuel with the oxygen of the air, including the inerts but excluding excess air.

Combustion safety control (flame safeguard). Primary safety control sensing the presence of flame and causing fuel to be shut off in the event of flame failure or ignition failure.

Combustion zone. Zone where combustion is intended to occur.

Commercial-type appliance. Appliance other than a residential-type appliance.

Commercial-type equipment. Equipment other than residential-type equipment.

Component. Essential part of an appliance, and which may be certified separately from the appliance.

Condensate (condensation). Liquid that separates from gas or from combustion products because of a reduction in temperature.

Connector. Tube or hose with a fitting at each end, for connecting combinations of appliances, fuel containers, and piping.

Constant-level valve. Device for maintaining, within a reservoir, a constant level of fuel for delivery to the burner.

Construction heater. Portable oil-fired appliance intended for temporary space heating during construction.

Continuous pilot. Pilot that burns without turndown throughout the entire time the burner is in service, whether the main burner is firing or not.

Cooking appliance. Appliance intended for supplying heat for cooking purposes.

Damper. Movable plate baffle or valve for regulating the flow of air or flue gas.

Direct-fired appliance. Appliance in which the combustion products or flue gases are intermixed with the medium being heated.

Direct service water heater. Service water heater that derives its heat directly from either an electrical resistance element or the combustion of fuel.

Draft. Flow of air or combustion gases, or both, through an appliance and its venting system.

Draft regulator (barometric damper). Draft control device intended to stabilize the natural draft in an appliance by admitting room air to the venting system.

Draft stabilizer. Hinged damper installed in a flue pipe tee directly after the flue connection of an oil-burning appliance that provides a constant draft over the fire. The damper swings freely on hinges and has an adjustable weight, providing a "stabilized" draft over the fire.

Drop pipe. Vertical pipe that conducts fuel down to an appliance.

Duct furnace. Furnace intended for installation in the air distribution ducts, with air circulation provided by a blower that is not an integral part of the furnace.

Electrical relay. Device that closes electrical contacts when an electromagnetic coil is energized, such as a 120 V pump motor started upon a signal from a 24 V thermostat.

Electrodes. Rounded, metal-end wires, supported by porcelain holders, used in pairs to provide a spark between the two ends separated $\frac{5}{32}$ in. to make a spark gap. The spark is used to light oil in an oil burner in a furnace or boiler.

Excess air. Air supplied to the combustion zone in excess of that theoretically required for complete combustion.

Expanding pilot. Pilot that normally burns at a low turndown throughout the entire time the burner is in service whether or not the main burner is firing, except that upon a call for heat the pilot is automatically expanded so as to reliably ignite the main burner.

Fan-assisted burner. Burner in which the combustion air is supplied by a fan or blower at sufficient pressure to overcome the resistance of the burner only.

Fast-closing valve. Valve that has a maximum closing time of 5 seconds upon being deenergized.

Firebox. Metal enclosure in which fuel or gaseous derivatives of fuel are burned and forming a portion of the heat exchanger.

Fire-resistance rating. Rating (hours) assigned by a nationally recognized authority to a material or assembly of materials in accordance with standard fire test methods.

Fire valve. Automatic fuel line shutoff valve in an oil line to the burner. Should the temperature rise in the vicinity of this valve, it snaps shut, stopping any oil spread.

Firing rate (flow rate). Fuel input rate to a burner.

Flame. Quantity of burning gas or vapor, not necessarily visible.

Flame-establishing period (trial-for-ignition period). Length of time fuel is permitted to be delivered before the flame-sensing device is required to detect the flame.

Flame failure reaction time (response time). Interval between flame extinguishment and the deenergizing of the devices that are used for stopping or reducing the fuel supply to the burner.

Flame-out. Sudden disappearance of the flame in an oil burner combustion chamber from lack of oil or stoppage of the burner, due to malfunction.

Flame-sensing device. Component of a combustion safety control that senses flame.

Flammable (explosive) range. The range of flammable vapor or gas-air mixtures between the upper and lower flammable limits is known as the "flammable range," also often referred to as the "explosive range."

Flammable or explosive limits. In the case of gases or vapors which form flammable mixtures with air, oxygen, or other chemicals, there is a minimum concentration of vapor in air or oxygen below which propagation of flame does not occur on contact with a source of ignition.

Flash point. Lowest temperature of a liquid fuel at which application of the test flame, under specified test conditions, causes the vapors above the surface of the liquid to ignite but not continue to burn.

Flexible connector. Connector made of flexible tubing.

Flexible coupling. Coupling used on an oil burner, between the motor and fan and pump. Usually made of rubber with separate or connected metal couplings. Three-piece coupling ends can be interchanged for others of different design or size.

Flexible tubing. Tubing that can be easily bent without the use of special tools, does not necessarily retain its bend, and is not subject to damage if rebent several times.

Flue. Enclosed passageway for conveying flue gases.

Flue collar. The portion of fuel-fired appliance designed for the attachment of the flue pipe or breeching.

Flue damper. Damper located in a flue.

Flue gases. Combustion products and excess air.

Flue gas loss. Heat loss escaping in the flue gases.

Flue outlet. Opening of a fuel-fired appliance through which the flue gases pass to the flue pipe or breeching.

Flue outlet pressure (flue outlet draft). Pressure difference between the pressure at the flue outlet of the appliance and the pressure of the surrounding air, the latter being used as the datum.

Flue pipe (chimney connector). Conduit connecting the flue collar of an appliance to a chimney

Forced draft. Mechanical draft created by a device upstream from the combustion zone of an appliance.

Fuel oil

1. Kerosene or any hydrocarbon oil as classified in the standards meeting the general requirements for oil-burning equipment.

2. Kerosene or any hydrocarbon oil conforming to nationally recognized standards and having a flash point not less than 100°F.

Fuel oil grade. Classification number for a particular fuel oil as specified in the standards meeting the general requirements for oil-burning equipment.

Fuel rich. Ratio of air to fuel supplied to a furnace that provides less air than that required for an optimum air/fuel ratio.

Gallon. Unit of liquid measure. A U.S. gallon contains 231 cu in. or 8.3359 lb avdp of distilled water at its maximum density and with the barometer at 30 in.

Gravity tank. Supply tank from which the oil is delivered to the burner by gravity.

Gauge (thickness of sheet metal). For uncoated and galvanized sheet steel the nominal thickness corresponding to the gauge number, together with mill tolerance, is implied unless otherwise given.

Heat exchanger. Firebox and auxiliary heat transfer surfaces within the casing of an appliance.

Heat reclaimer (flue pipe type). Device intended to be installed in the flue pipe between an appliance and the chimney to transfer heat from the flue gases through metal to air or water.

Heat transfer surface. Surface of a heat exchanger designed to transfer heat between two physically separated fluids.

Heating surface. Fireside of the walls of the heat exchanger exposed to the flue gases.

Heavy oils. Heavy, thick, and viscous oils, usually refinery residuals commonly specified as grades 5, 6, and Bunker C.

High-limit control. Safety switch that operates to stop oil burner in the event of overheating or other possible malfunctions of the equipment.

High-steam-pressure switch. Pressure-actuated device arranged to effect a normal burner shutdown when the steam pressure exceeds a preset pressure.

Hose. Flexible tubing that does not consist of a single continuous metal wall.

Hose connector. Connector made of flexible tubing hose.

Hot plate. Appliance consisting of one or more open-top burners mounted on short legs or a base.

Ignition temperature. The minimum temperature required to initiate or cause self-sustained combustion independently of the heating or heating element, whether the substance is solid, liquid, or gaseous.

Ignition transformer. Transformer that steps up the voltage from 120 V to 10,000 V. Provides an electric spark to ignite the oil spray in an oil burner.

Incinerator. Appliance complete with firebox and chimney in which combustible wastes are ignited and burned.

Indirect fire appliance. Appliance in which the combustion products or flue gases are not mixed with the air being heated within the appliance.

Indirect service water heater. Service water heater that derives its heat from a heating medium such as warm air, steam, or hot water.

Induced draft. Mechanical draft created by a device downstream from the combustion zone of an appliance.

Instantaneous-type (tankless) service water heater. Service water heater designed to supply hot water directly to the outlets without storage facilities.

Installation. Consists of a complete setting in place, ready for operation, of oil-burning equipment together with its accessories, oil supply, and controls.

Integral tank. Supply tank that is a component part of the appliance on which it is mounted.

Interlock. Device that senses a limit or off-limit condition or improper sequence of events and shuts down the offending or related piece of equipment or prevents proceeding in an improper sequence in order to prevent a hazardous condition.

Intermittent ignition. Electrically ignited pilot that is automatically lighted each time there is a call for heat and burns during the entire period that the flame is present.

Intermittent pilot
1. Pilot that burns during light-off and while the main burner is firing, and that is shut off with the main burner.
2. Electrically ignited pilot that is automatically lighted each time there is a call for heat and burns during the entire period that the main burner is firing.

Interrupted ignition. Electric ignition that ceases to function after the flame-establishing period.

Interrupted pilot
1. Pilot that burns during light-off and that is shut off (interrupted) during normal operation of the main burner.
2. Electrically ignited pilot that is automatically lighted each time there is a call for heat and in which the pilot fuel is cut off automatically at the end of the flame-establishing period of the main burner.

Kerosene (kerosine). Oil or liquid product of petroleum that does not emit a flammable vapor below a temperature of 115°F, when tested in a Tag closed-cup tester.

Kerosene stove. Appliance that is a nonflue connected, self-contained, self-supporting kerosene-burning range, room heater, or water heater equipped with an integral tank not exceeding 2 gal kerosene capacity. Review local codes for regulations governing installation.

Labeled. Equipment that may be required by local codes to bear the inspection label of a nationally recognized testing agency.

Light off. Establish combustion of fuel entering the furnace.

Light-off time limit timer. Device used on supervised manual systems that limits the allowable time between completion of purge and light-off. The time is usually not more than 5 minutes.

Light oil. Generally, all oils lighter than residual fuel oils Nos. 5 and 6; oils that have a low specific gravity, usually products of controlled distillation of crude oil, but also including by-products benzol and toluol.

Limit control. Safety control intended to prevent unsafe condition of temperature, pressure, or liquid level.

Low-oil-pressure switch. Pressure-actuated device arranged to effect a safety shutdown of the oil burner or prevent it from starting when the oil supply pressure falls below that recommended by the burner manufacturer.

Low-oil-temperature switch. Temperature-actuated device arranged to effect the safety shutdown of the oil burner or prevent it from starting when the oil temperature falls below the limits required to maintain the viscosity range recommended by the burner manufacturer.

Low water cutout. Device arranged to effect a safety shutdown of the burner when the water level in the steam drum falls to a predetermined low level.

Main burner. Burner unit exclusive of the pilot burner.

Manual appliance. Appliance equipped with a manual burner.

Manual burner. Burner equipped with a manual appliance control system.

Manually-lighted burner. Burner where fuel oil to the main burner is turned on only by hand and ignited under supervision.

Manually operated damper. Adjustable damper manually set and locked in the desired position.

Manual oil shutoff valve. Manually operated valve in an oil line for the purpose of turning on or completely shutting off the oil supply to the burner.

Manual system
1. System in which the burner, when started, will complete only one cycle of operation without manual attention.
2. System by which a furnace is purged and a burner is started, ignited, modulated, and stopped manually.

Mechanical draft. Draft created by a mechanical device, such as a fan, blower, or aspirator, which may supplement natural draft.

Mechanical sprayer. Device in an oil burner that emits liquid fuel in a finely divided state without using an atomizing medium.

Metering valve. An oil control valve for regulating burner input.

Modulate. Vary gradually the fuel and airflows to the burner in accordance with load demand.

Modulating. Term used to describe the desired adjustment to control infinite variance of the volume of the flow of fuel or air, or both, between predetermined minimum and maximum limits.

Monitor. Sense and alarm a condition requiring attention without initiating corrective action.

Natural draft. Draft other than a mechanical draft.

Natural draft burner. Burner not equipped with a mechanical device for supplying combustion air.

Negative pressure. Pressure less than atmospheric pressure.

Normal fuel supply pressure. Pressure at the fuel service connection for which the fuel-burning system has been designed.

Normal shutdown. Stopping burner operation by shutting off all fuel and ignition energy to the furnace.

No. 2-grade fuel oil. Lightweight, nearly colorless oil from petroleum. Used for domestic pressure-type oil burners.

Oil burner. Piece of assembled equipment used as a heating device in conjunction with other equipment or appliances such as boilers, furnaces, water heaters, ranges, and the like.

Oil burner—one pipe. Piping arrangement using only one oil line between the oil tank and the burner. The oil tank is usually slightly above the level of the oil burner.

Oil-burning assembly. Assembly comprising an oil burner and such devices as may be required to control the supply of fuel and air to the burner.

Oil-burning equipment

1. One or more oil-burning appliances, together with their fuel tanks, fuel piping, wiring, controls, and accessories.

2. Oil burner of any type together with its tank, piping, wiring, controls, and related devices, including all conversion oil burners, oil-fired units, and heating and cooking appliances.

Oil control valve. Automatically or manually operated device consisting essentially of an oil valve for controlling the fuel supply to a burner.

Oil distribution main. Pipe intended to convey fuel oil from a central supply tank(s) to oil service pipes.

Oil-fired unit. Heating appliance equipped with one or more oil burners and all the necessary safety controls and electrical and related equipment manufactured for assembly as a complete unit. Kerosine or oil stoves are not included.

Oil pump. Pump of an oil-burning furnace that draws oil from a tank and forces it through the oil nozzle at 100 psi to produce an oil flame.

Oil service pipe. Pipe intended to convey fuel oil from an oil distribution main to the appliance supply piping.

Oil transfer pump. Pump, automatically or manually operated, that transfers oil through continuous piping from a supply tank to an oil-burning appliance or to an auxiliary tank and that is not designed to stop pumping automatically in case of total breakage of the oil supply or return lines.

Operating control. Control used to regulate or control the normal operation of the equipment.

Operating range. Region between the maximum and minimum fuel inputs in which the burner flame can be maintained continuously and stably. The range is determined by test.

Operational testing. Final installation of all of the equipment and the control system for oil-burning equipment and appliances should require a test run before it is judged as meeting the established standards and the requirements of the local codes.

Optimum ratio of air to fuel. Minimum ratio of air to fuel supplied to a furnace that will provide complete combustion of the fuel with sufficient range of excess air to maintain a stable flame envelope.

Overfire pressure (over fire draft). Pressure difference between the pressure in the combustion chamber directly over the fire and the pressure of the surrounding air, the latter being used as the datum.

Outlet draft. Flue gas pressure at the outlet of the last convection pass of the boiler.

Package unit. Appliance supplied by one manufacturer as a complete unit including burner, controls, and integral wiring.

Petroleum

1. Complex mixture of various hydrocarbons existing as a liquid in the upper strata of the earth.

2. Naturally occurring complex liquid hydrocarbon that after distillation yields combustible fuels.

Pilot. Flame that is utilized to ignite the fuel at the main burner(s).

Pilot burner. Burner at which the pilot is established.

Pilot-establishing period. That interval of time during light-off that a safety control circuit permits the pilot fuel safety shutoff valve(s) to be opened before the flame safeguard is required to prove the presence of the pilot flame.

Pilot valve. Valve to control the supply of fuel to a pilot burner.

Pipe fitting. Item in a piping system that is used as a connector, such as an elbow, return bend, tee, union, bushing, coupling, cross, or nipple, but not including such functioning items as a valve or regulator.

Piping. Fuel conduits of circular cross section that are of sufficient wall thickness and suitable outside diameter for threading to required standards and that are specified by nominal inside diameter (ID).

Piping outlet. Termination of fuel piping near or at the location of an appliance or proposed appliance.

Piping riser. Vertical pipe that conducts fuel upward.

Portable equipment (or appliance). Equipment (or appliance) that is readily moved from place to place. Such equipment may be flue connected.

Power burner (forced draft burner). Burner in which the combustion air is supplied by a fan or blower at sufficient pressure to overcome the resistance of the burner and the appliance.

Pressure relief valve. Valve that opens automatically to relieve a pressure in excess of a predetermined setting and closes after such relief.

Primary air. Portion of the combustion air that is supplied for the initial stages of the combustion process. It is supplied upstream from the point of ignition.

Primary heating surface. Surfaces of those portions of the heat exchanger exposed to direct radiation from the fire or combustion chamber, including the portion that encloses the combustion chamber.

Primary safety control. Automatic safety control intended to prevent abnormal discharge of fuel at the burner in the event of ignition or flame failure.

Process application. Application of heat for other than space heating or service water heating.

Proof. Establish by measurement or test the existence of a specified condition, such as flame, level, flow, pressure, or position.

Proved pilot. Pilot flame supervised by a primary safety control that senses the presence of the pilot flame prior to permitting the main burner fuel to be delivered for combustion.

Purge

1. Replace the existing fluid, gaseous or liquid, in the fuel piping or the appliance with the desired fuel.

2. Flow of air through the furnace, boiler, gas passages, and associated flues and ducts that will effectively remove any gaseous combustibles and replace them with air.

Radiator. Metal enclosure located within the furnace casing, consisting of secondary heating surfaces and forming a portion of the heat exchanger.

Readily accessible. Capable of being reached quickly for operation, renewal, servicing, or inspection without requiring the climbing over or removal of obstacles or the use of portable ladders.

Recycle. Start-up initiated by steam pressure following a normal shutdown.

Repeatability. Ability of a device to maintain a constant set point characteristic.

Residential-type equipment. Equipment commonly used in, but not restricted to, one- or two-family dwellings.

Restart. Manually initiated start-up.

Rigid tubing. Tubing that normally cannot be bent without the use of special tools.

Safety circuit. Circuit used in a safety control system.

Safety combustion control. Safety control responsive directly to flame properties, sensing the presence of flame and causing fuel to be shut off in event of flame failure.

Safety control. Automatic control of a safety control system.

Safety control system. System of automatic controls intended to prevent unsafe operation of the controlled equipment automatically. It may include relays, switches, and other auxiliary equipment and interconnecting circuitry.

Safety shutdown. Stopping burner operation by shutting off all fuel and ignition energy to the furnace by means of a safety interlock(s) and requiring a manual restart.

Safety shut-off valve. Valve that automatically shuts off the supply of fuel in response to the action of a combustion safety control or limit control.

Safety valve. Automatic oil control valve of the "on" and "off" type (without any by-pass to the burner) that is actuated by a safety control or by an emergency device.

Sealed combustion unit. Appliance where all the air required for combustion is taken directly from outdoors and the flue gases are exhausted directly outdoors.

Secondary air. Portion of the combustion air supplied for the intermediate and final stages of the combustion process. It is supplied externally downstream from the point of ignition.

Secondary heating surfaces. Surfaces of those portions of the heat exchanger conveying the combustion products or flue gases from the enclosure forming the primary heating surfaces to the flue outlet.

Self-energized control system. Burner control system where part of the heat energy of the pilot is converted through a thermopile to electrical energy sufficient to operate the control system and automatic valves.

Self-energized pilot. Pilot used in a self-energized control system.

Self-energized valve. Valve used in a self-energized control system

Self-generating millivolt circuit. Circuit in which an electromotive force is generated by the effect of the heat of a flame on a thermopile element.

Semiautomatic appliance. Appliance equipped with a semiautomatic burner.

Semiautomatic burner

1. Burner that when started will continue an unlimited number of operating cycles without manual attention unless shut down by the combustion safety control or limit control.

2. Burner equipped with a semiautomatic appliance control system.

Semirigid connector. Connector made of semirigid tubing.

Semirigid tubing. Tubing that normally can be bent without the use of special tools, retains its bend, and is subject to appreciable damage if rebent several times.

Service compartment. Normally enclosed compartment of an assembly that is accessible for occasional operations, such as "lighting-up," or the adjustment, cleaning, or servicing of such parts of the equipment as air filters, blowers, motors, and controls.

Service connection. Point at which fuel, atomizing medium, or power is connected to the boiler, firing equipment, or controlled devices.

Service water heater. Appliance intended for the heating of water for plumbing services (as distinct from water for space heating).

Set point. Predetermined value to which an instrument is adjusted and at which it performs its intended function.

Shutoff valve. Manual valve used in the piping to fully turn on or shut off the fuel supply to any section of a piping system or to an appliance.

Space heater (room heater). Space-heating appliance for heating the room or space within which it is located without the use of ducts.

Space-heating appliance. Appliance intended for the supplying of heat to a room or space directly (e.g., space heater, fireplace, unit heater) or to rooms or spaces of a building through a heat-distributing system, such as a central furnace or boiler.

Spark gap. The distance between the spark electrodes of an oil burner, $\frac{5}{32}$ in., across which a spark is formed. This ignites the oil sprayed into the combustion chamber.

Stack relay. Electrical-mechanical device to control the operation of an oil burner in response to the stack temperature. The relay must be manually reset; this alerts the operator to investigate the cause of the lockout.

Storage tank. Tank for the storage of fuel and from which the fuel-burning equipment is not intended to be fed automatically.

Storage-type service water heater. Service water heater with an integral hot-water storage tank.

Sump. Receptacle used in conjunction with a vacuum tank.

Supervise. Sense and alarm a condition requiring attention and initiate corrective action.

Supply tank. Tank for the storage of fuel and from which the fuel-burning equipment is intended to be fed automatically.

Temperature relief valve—fusible plug type. Device that opens and keeps open a relief opening by the melting or softening of a fusible plug or cartridge at a predetermined temperature.

Temperature relief valve—reseating or self-closing type. Valve that opens automatically when a predetermined temperature is exceeded and closes automatically when the temperature falls below a predetermined lower value.

Temperature (total). Actual measured temperature, including the room ambient temperature.

Trial-for-ignition period (main-burner-establishing period). Interval of time during light-off that a safety control circuit permits the main burner fuel safety shutoff valves to be opened before the flame safeguard is required to supervise the main burner flame only.

Tubing. Fuel conduits of circular cross section that are not of sufficient wall thickness or suitable outside diameter to permit threading to iron pipe size (IPS) standards and that are specified by outside diameter (OD).

Two-pipe oil supply. Piping arrangement whereby two oil lines run between the oil pump and the oil tank. This is necessary if the oil tank is underground or at a great distance from the oil burner operation.

Unvented appliance. Appliance not intended to be connected to a venting system.

Vacuum or barometric tank. Tank not exceeding 5 gal capacity that maintains a definite level of oil in a sump or similar receptacle by barometric feed. Fuel is delivered from the sump to the burner by gravity.

Valve. Device by which the flow of a fluid may be started, stopped, or regulated by means of a movable part that opens or obstructs passage.

Vapor density. The relative density of a vapor or gas (with no air present) as compared with air.

Vaporizing-type oil burner. Oil burner in which oil is vaporized from a film on a surface of the burner.

Vented appliance. Appliance intended to be connected to a venting system.

Venting system. System for the removal of flue gases or vent gases to the outside air by means of chimneys, gas vents, or exhaust systems, natural or mechanical.

Wall-flame-type oil burner. Oil burner in which oil in a liquid stream is fed against a ring of wall inside the combustion chamber.

Water solubility. Information on the degree to which a flammable liquid is soluble in water is useful in determining effective extinguishing agents and methods.

PIPES, FITTINGS, AND DEVICES

ABS. Rigid plastic pipe; acrylonitrile-butadiene-styrene.

Adapter. Fitting for pipes of two different materials.

Aerator. Sievelike filter device on the spigot end that mixes air with the water flow.

Alloy pipe. Steel pipe with one or more elements other than carbon, which give it greater resistance to corrosion and more strength than carbon steel pipe.

Aluminum pipe. Used in installations where no chemical actions could occur and primarily used for water service.

Angle of bend. Angle of the center of the bend between radial lines from the beginning and end of the bend to the center.

Angle valve. Valve, usually of the globe type, in which the inlet and outlet are at right angles.

Annealing and temper. Brass pipe, annealed sufficiently to indicate complete recrystallation and to enable the pipe

to meet the test requirements prescribed by acceptable standards, and local codes.

Antisiphon valve. Valve installed on a supply line to prevent siphoning of contaminated water back into the watersupply system.

Average diameter. The average of the maximum and minimum outside diameters or the maximum and minimum inside diameters, whichever is applicable as determined by any one cross section of a pipe or tube, or as listed by the manufacturer.

Backflow connection. Arrangement whereby backflow can occur. Synonymous terms include "cross connection," "interconnection," "backsiphonage," and "gravity flow."

Backflow preventer (vacuum breaker). Device installed in a water supply pipe to prevent backflow of water into the potable water supply system.

Backing ring. Metal strip used to prevent melted metal from the welding process from entering a pipe when making a butt-welded joint.

Backsiphonage. Flowing back of used, contaminated, or polluted water from a plumbing fixture or vessel into a water supply pipe due to a negative pressure in such pipe.

Backsiphonage preventer

1. Device or means to prevent backsiphonage. It is not to be used under continuous pressure.

2. Device or means for preventing backsiphonage, type designed to be used under continuous pressure.

Backwater traps. When there is a possibility that a plumbing drainage system will be subject to backflow of sewage, suitable provisions must be made to prevent its overflow into the building.

Backwater valve

1. Valve installed in a building drain or building sewer to prevent sewage from flowing back into the building.

2. Device installed in piping to prevent the back or reverse flow of storm or sewage into a drainage system or its branches.

Ball valves. Valve in which the flow is controlled by a rotating drilled ball that fits tightly against a flexible seat in the valve body.

Bell-and-spigot cast-iron pipe. One end of the pipe has a bell-shaped hub, the other end a lip called a "spigot." The spigot end of one pipe fits into the bell of another, and the joint is caulked with oakum and sealed with molten lead.

Bell and spigot joint. Commonly used joint in cast-iron pipe. Each piece is made with an enlarged diameter or bell at one end into which the plain or spigot end of another piece is inserted. The joint is then made tight by cement, oakum lead, or rubber caulked into the bell around the spigot.

Bell (or hub). Pipe section that is enlarged for a short distance to receive a portion of another pipe of identical diameter to form a joint.

Black and hot-dipped zinc-coated (galvanized) welded and seamless steel pipe for ordinary uses. Pipe that meets the standard of ASTM A 53.

Black pipe. Steel pipe that has not been galvanized.

Black steel. Steel that is not coated with any metallic substance.

Black wrought iron. Wrought iron that is not coated with any metallic substance.

Blank flange. Flange in which the bolt holes have not been drilled.

Bonnet. Part of a valve used to guide and support the valve stem.

Branch tee. Tee having many side branches.

Brass pipe. Brass pipe should contain 84 to 86% copper and not more than 0.06% lead, 0.05% iron. The remainder is zinc.

Building or house trap. Running trap installed in the building drain to prevent circulation of air between the drainage system of the building and the building sewer.

Building trap. Device, fitting, or assembly of fittings installed in the building drain to prevent circulation of air between the drainage system of the building and the building sewer.

Bull head tee. Tee the branch of which is larger than the run.

Bushing. Pipe fitting for connecting a pipe with a female fitting of larger size; hollow plug with internal and external threads.

Butt-weld steel pipe. Used for steam, water, gas, or air services. It is not intended for medium or high pressure installations.

By-pass. Supplementary line leaving the main run and rejoining it at some point beyond a valve or other apparatus so that service is not interrupted when the valve or apparatus is not usable.

Cap. Female pipe fitting that is closed at one end. It is used to close off the end of a piece of pipe or tubing.

Carbon steel. Steel that contains high percentages of carbon as distinguished from the other elements.

Carbon steel pipe. Steel pipe that owes its properties chiefly to the carbon it contains.

Cast iron. Composite of alloys primarily of iron, carbon, and silicon in which the carbon is in excess of the amount that can be retained in solid solution in austenite at eutectic temperature.

Cast iron, malleable and ductile iron pipe fitting material. Materials used in the manufacture of pipe fittings in accordance with ASTM standards.

Cast-iron screw pipe. Cast-iron metal or alloy pipe with threaded joints. Pipe having same inside and outside diameters as occurs in extra strong wrought-iron or steel pipe, with uniform wall thickness, and full pipe area and capable of withstanding threading in a satisfactory manner.

Cast iron soil pipe and fittings. Pipe and fittings that meet the standards of ASTM A 74.

Cast-iron soil pipe support. Cast-iron soil pipes are supported and secured at not less than every story height and at its starting point.

Caulked joints. All caulked joints should be firmly packed with oakum or hemp and secured only with pure lead, not less than 1 in. deep, and caulked tight.

Caulking. Operation or method of rendering a joint tight against water or gas by means of plastic substances, such as lead and oakum.

Centrifugally cast pipe. Pipe formed from the solidification of molten metal in a rotating metal or sand mold.

Check valve

1. Valve designed to allow a fluid to pass through in one direction only. The common type has a plate so suspended that the reverse flow aids gravity in forcing the plate against a seat, shutting off the reverse flow.
2. Automatically operated device that will permit the flow of fluids in one direction and will close if there is a reversal of flow.

Close nipple. Nipple with a length twice the length of a standard pipe thread.

Common system. Part of a plumbing system designed and installed to serve more than one appliance, fixture, building, or service.

Common waste. Drain from a fixture containing multiple compartments connected to a single trap.

Companion flange. Pipe flange to connect with another flange or with a flanged valve or fitting; flange attached to the pipe by threads, welding, or other method and differing from a flange that is an integral part of a pipe or fitting.

Compression joint. Multipiece joint with cup-shaped threaded nuts that when tightened compress tapered sleeves so that they form a tight joint on the periphery of the tubing they connect.

Continuous-welded. Furnace-welded pipe produced in continuous length from coiled skelp and subsequently cut into individual lengths, having its longitudinal butt joint forge-welded by the mechanical pressure developed in rolling the hot-formed skelp through a set of round pass welding rolls.

Copper drainage tube (DWV). Tubing that meets the standard ASTM B 306.

Copper pipe. Lightweight, rigid pipe joined by soldered or threaded joints. The weight and ease of fitting make it a good choice for water piping.

Copper tubes for waste and vent lines. Copper tube should be seamless, cold-drawn commercially pure hard copper tubing of standard U.S. government or ASTM types.

Copper tubing support. Tubing supported and secured at each story for piping $1\frac{1}{2}$ in. and larger and at not more than 4-ft intervals for piping $1\frac{1}{4}$ in. and smaller.

Corporation stop. Valve installed in the building water service line at or near the water meter, or as required by the local authorities.

Coupling. Threaded sleeve used to connect two pipes; sleeve having internal threads at both ends to fit external threads on a pipe.

Cross. Pipe fitting with four branches in pairs, each pair on one axis and the axes at right angles.

Crossover. Small fitting with a double offset or shaped like the letter U with the ends turned out. It is only made in small sizes and is used to pass the flow of one pipe past another when the pipes are in the same plane.

Cross valve. Valve fitted on a transverse pipe so as to open communication between two parallel pipes.

Curb cock or curb stop. Valve placed on the water service, usually near the curb line.

Developed length. Length of a pipeline measured along the centerline of the pipe or fittings.

DHP. Phosphorized copper, high-residual phosphorous.

Diameter. Unless otherwise specifically stated in a particular plumbing code, the term "diameter" is the nominal commercial designation, normally the inside diameter of the pipe.

Diameter of pipe. Nominal diameter as designated commercially, unless otherwise specified.

Direct flush valve. Device designed to discharge a predetermined quantity of water to fixtures for flushing purposes.

DLP. Phosphorized copper, low-residual phosphorous.

Double extrastrong pipe. Schedule of steel or wrought iron pipe weights in common use.

Double hub. Cast-iron sewer pipe having a bell at both ends.

Double offset. Two changes of direction installed in succession or series in continuous pipe.

Double-sweep tee. Tee made with easy (long-radius) curves between body and branch.

Drain. Pipe that carries waste water or waterborne wastes in a building drainage system.

Drain cock. Simple valve connection at the lowest part of a water supply system that can be opened to drain the system.

Drop elbow. Small ell used in gas fittings. Fittings have wings cast on each side, the wings having countersunk holes so that they may be fastened by wood screws to a ceiling, wall, or framing timbers.

Drop tee. Tee having the same type of wings as the drop elbow.

Drum trap. Trap occasionally used for tubs and showers instead of curved pipe sections; cylindrical drum with inlet and outlet at different levels.

Ductile iron. Cast iron that has been treated in the liquid state so as to cause substantially all of its graphitic carbon to occur as spheroids or nodules in the as-cast condition.

Eccentric fitting. Pipe fitting in which the centerline of the opening is offset.

Eight-bend. Pipe fitting that causes the run of pipe to make a 45° turn.

Elastic limit. Greatest stress that a material can withstand without permanent deformation after stress release.

Elbow. Fitting used for making turns in pipe runs (for example, a 90° elbow makes a right-angle turn). A street elbow has one male end and one female end.

Elbow (ell). Fitting that makes an angle between adjacent pipes. The angle is 90° unless otherwise indicated or specified by local codes.

Electric-fusion-welded pipe. Pipe having a longitudinal butt joint wherein coalescence is produced in the preformed tube by manual or automatic electric arc-welding.

Electric-resistance-welded pipe. Pipe produced in individual lengths or in continuous lengths from coiled skelp and subsequently cut into individual lengths, having a longitudinal butt joint wherein coalescence is produced by the head obtained from resistance of the pipe to the flow of electric current in a circuit of which the pipe is a part and by the application of pressure.

Expansion joint. Joint whose primary purpose is not to join pipe but to absorb longitudinal expansion in the pipeline due to heat.

Expansion loop. Large radius bend in a pipeline to absorb longitudinal expansion in the line due to heat.

Extra-heavy. Term used to designate the heaviest and strongest grades of cast-iron and steel pipe.

Female. Pipes, valves, or fittings with internal threads.

Ferrule. Metallic sleeve or fitting used to connect dissimilar plumbing materials.

Field tile. Short lengths of pipe that are perforated as subsurface drains to emit drainage.

Fitting. Device used to join sections of pipe.

Fixture carrier. Special fitting for supporting an off-the-floor water closet or other plumbing fixture.

Fixture trap. Trap integral with or serving a fixture and includes an interceptor serving as a trap for a fixture.

Flange. Ring-shaped plate on the end of a pipe at right angles to the end of the pipe and provided with holes for bolts to allow fastening the pipe to a similarly equipped adjoining pipe. The resulting joint is a flanged joint.

Flange faces. Pipe flanges that have the entire face of the flange faced straight across and use either a full face or ring gasket are commonly employed for pressures less than 125 lb on stream and waterlines.

Flare fitting. Female fitting used on copper tubing.

Float valve. Positive operating valve operated by a float and used to control the water level in a vessel, tank, or other container.

Floor flange. Fitting attached at the floor level to the end of a water closet bend so the water closet can be bolted to the drainage piping.

Flush bushing. Pipe fitting used to reduce the diameter of a female-threaded pipe fitting.

Flushometer valve

1. Device actuated by direct water pressure in such manner as to discharge to a fixture for flushing purposes a predetermined quantity of water.

2. Automatically operating metered valve.

Flush valve

1. Valve for water closets, urinals, bidets, and similar fixtures.

2. Device located at the bottom of the tank for the purpose of flushing water closets and similar fixtures.

3. Valve for flushing a sanitary unit.

Furnace lap-welded pipe. Pipe having a longitudinal lap joint made by the forge-welding process wherein coalescence is produced by heating the preformed tube to welding temperature and passing it over a mandrel located between two welding rolls that compress and weld the overlapping edges.

Galvanized pipe. Steel pipe coated with zinc to resist corrosion.

Gas piping. Term applied to the installation, repair, replacement, and relocation of pipes, fixtures, and other apparatus for distributing a gas supply for illuminating or fuel purposes in any premises.

Gate valve. Valve employing a gate, often wedge shaped, allowing fluid to flow when the gate is lifted from the seat. Gate valves have less resistance to flow than globe valves.

Globe valve. Valve with a somewhat globe-shaped body with a manually raised or lowered disk that when closed rests on a seat so as to prevent passage of a fluid.

Gray iron castings for valves, flanges, and pipe fittings. Castings that meet the standard of ASTM A 126.

Ground joint. Parts to be joined are precisely finished and then ground in so that the seal is tight.

Hose bib (or hose cock). Valve with male-threaded outlet for accepting the hose fitting.

Hubless cast-iron pipe. Cast-iron pipe that is joined by means of rubber gaskets and clamps, making it much easier to work with than a bell and spigot, which uses molten lead.

Hydrant. Water supply outlet with a valve located below or above ground level. Five hydrant as required by the local fire department with type and outlets as meet the fire code.

Increaser. Fitting having larger opening at one end for accepting the larger diameter pipe.

Increasers and reducers. Where different sizes of pipe or pipes and fittings in the drainage system are to be connected, proper size increasers or reducers are to be used.

Inside copper water tube. Copper tubing for inside water supply distribution system should be seamless, cold-drawn, commercially pure hard copper tubing of standard U.S. government type K or L except that copper tubing used for water supply under or in concrete slabs should be soft or hard drawn Type K copper. On that part of inside water piping extending from the floor or wall to the plumbing fixture, a fixture supply of the type known as "Flexible" supplies may be used, such supplies to be of not less than $\frac{1}{4}$ in. I.D. copper tubing.

Internal valve. A primary shutoff valve for containers that has adequate means of actuation and that is constructed in such a manner that its seat is inside the container and that damage to parts exterior to the container or mating flange will not prevent effective seating of the valve.

Interceptor

1. Device designed and installed so as to separate and retain deleterious, hazardous, or undesirable matter from normal wastes while permitting normal sewage or liquid wastes to discharge into the drainage system by gravity.
2. Receptacle designed and constructed to intercept or separate and prevent the passage of oil, grease, sand, or similar materials into the drainage system to which it is directly or indirectly connected.

Lapped joint. Pipe joint made by using loose flanges on lengths of pipe whose ends are turned over or lapped over to produce a bearing surface for a gasket or metal-to-metal joint.

Lap weld pipe. Welding along a scarfed longitudinal seam in which one part is overlapped by the other.

Lead joint. Joint made by pouring molten lead into the space between a bell and spigot and making the lead tight by caulking.

Lead pipe joints. Joints in a lead pipe or between a lead pipe and brass or copper pipes, ferrules, soldering nipples, bushings, or traps, in all cases on the sewer side of the trap and in concealed joints on the inlet side of the trap, are full-wiped joints.

Lead pipe supports. Lead pipes are supported and secured at intervals usually not exceeding 4 ft.

Lead to cast-iron steel or wrought iron. Joints made by means of a caulking ferrule, soldering nipple, or bushing.

Lead water service. Lead water service pipe should be of best quality of not less weight per linear foot than defined in the code and in accordance with federal specifications. Lead water supply pipe shall be AAA double extra strong.

Lip union. Form of union characterized by the lip that prevents the gasket from being squeezed into the pipe so as to obstruct the flow.

Locknut. Nut fixed onto one piece (for instance, a flexible connector for a water heater) and screwed onto another piece to join the two.

Long sweep fitting. Drainage fitting that has a long radius curve at the bends.

Long quarter-bend. 90° fitting with one section longer than the other.

Male. Pipes, fittings, and valves with external threads.

Malleable iron. Cast-iron heat treated to reduce its brittleness. The process enables the material to stretch to some extent and to withstand greater shock.

Manifold. Fitting with a number of branches in line connecting to smaller pipes. The term is used largely as in interchangeable term with "header."

Mechanical joint. Joint for the purpose of mechanical strength or leak resistance, or both, where the mechanical strength is developed by threaded, grooved, rolled, flared, or flanged pipe ends or by bolts, pins, compounds, gaskets, rolled ends, caulking, or machined and mated surfaces. Mechanical joints have particular application where ease of disassembly is desired.

Medium pressure. When the term is applied to valves and fittings, it implies that they are suitable for a working pressure of from 125 to 175 psi.

Meter stop. Valve used on a water main between the street and a water meter.

Mild steel pipe. All steel pipe should conform to the ASTM Standards for welded and seamless steel pipe A 53 and should be galvanized.

Mill length. Also known as "random length." Run-of-mill pipe is 16 to 20 ft in length. Some pipe is made in double lengths of 30 to 35 ft.

Miter. Two or more straight sections of pipe matched and joined on a line bisecting the angle of junction so as to produce a change in direction.

Needle valve. Valve provided with a long tapering point in place of the ordinary valve disk. The tapering point permits fine graduation of the opening.

Nipple. Tubular pipe fitting usually threaded on both ends and under 12 in. in length. Pipe over 12 in. long is regarded as cut pipe.

No-hub pipe. Soil pipe that has smooth ends, but doesn't have a spigot or hub.

Nominal size. The approximate dimensions of standard materials.

Nonrising stem valve. Gate valve in which the stem does not rise when the valve is opened.

Nozzle. Fitting, attached to the outlet of a pipe or hose, that varies the volume of water.

OD pipe. Smaller size pipe is usually designated by its inside diameter. For pipe over 14 in. however, the nominal size is the outside diameter, and such pipe is termed "OD."

OF. Oxygen-free copper without residual metallic deoxidants.

Outside copper water tube. Copper tubing used for underground water supply or water service should be K type only and should be fully annealed, and meet requirements of local codes.

Petcock. Small ground key type valve used with soft copper tubing to control water flow.

Pipe

1. Tube, usually cylindrical, used for conveying a fluid or transmitting fluid pressure, and normally designated as "pipe" in the applicable specification; also, similar

components designated as "tubing," used for the same purpose.

2. Closed conduit used to convey fluids, air, or gas.

Piping fittings of wrought carbon steel and alloy steel for moderate and elevated temperatures. Fittings that meet the standards of ASTM A 234.

Plug. Closed-end, male-threaded fitting for closing off a pipe end that has female threads.

Plug valve. Short section of a cone or tapered plug through which a hole is cut so that fluid can flow through when the hole lines up with the inlet and outlet; when the plug is rotated 90°, flow is blocked.

Plumbing appurtenance. Manufactured device, prefabricated assembly, or on-the-job assembly of component parts that is an adjunct to the basic piping system and plumbing fixtures. An appurtenance demands no additional water supply, nor does it add any discharge load to a fixture or the drainage system. It is presumed that it performs some useful function in the operation, maintenance, servicing, economy, or safety of the plumbing system.

Polyethylene tubing. Flexible plastic pipe, often used for underground sprinkling systems.

Pop-off valve. Safety valve that opens automatically when pressure and temperature exceed the predetermined limits.

Pressure-reducing valve. Safety valve for a water heater. It lets water and steam escape.

Pressure regulator. Valve that reduces water pressure in the supply piping.

Pressure-type vacuum breaker. Vacuum breaker designed to operate under conditions of static line pressure.

P trap. Fixture trap in the shape of the letter P.

PVC (polyvinyl chloride). Type of plastic used in fabricating pipes and fittings for water distribution, irrigation, and natural gas distribution.

Quarter-bend. Drainage pipe fitting that makes a 90° angle.

Receptor. Device receiving the discharge of a waste pipe or pipes and discharging it by gravity into the storm or sanitary drainage system.

Reduced-pressure-zone-type backflow preventer. Assembly of differential valves, including an automatically opened spillage port to the atmosphere, designed to prevent either backsiphonage or backflow due to any superior pressure on the downstream side of the assembly.

Reducer. Fitting with a larger size at one end than at the other, with the larger size designated first. Reducers are threaded inside, unless specified flanged or welded, or for some special joint.

Relief valve. Valve designed to open automatically to relieve excess pressure.

Resistance weld pipe. Pipe made by bending a plate into circular form and passing electric current through the material to obtain a welding heat.

Return offset. Double offset installed so as to return the pipe to its original alignment.

Rigid copper tubing. Hard copper used for the installation of water lines.

Rising stem. Type of valve stem that moves up and down as the valve is opened and closed.

Rolling offset. Same as an offset, but used where the two lines are not in the same vertical or horizontal plane.

Roof drain. Drain installed to receive water collecting on the surface of a roof and to discharge it into a leader or a conductor.

Roof joints. Joints at the roof made watertight by the use of a sheet lead or copper plate with a sleeve turned over the top of the ventilating pipes at least 1 in. and dressed tightly against the inside of the pipe, or approved galvanized roof flashing of usually not less than No. 20 gauge.

Rotary pressure joint. Joint for connecting a pipe under pressure to a rotating machine.

Run. Length of pipe made of more than one piece of pipe; portion of a fitting having its ends in the line or nearly so, in contradistinction to the branch or side opening of, for example, a tee.

Saddle fitting. Fitting used to install a branch from an existing run of pipe.

Saddle flange. Flange curved to fit a boiler or tank and attached to a threaded pipe. The flange is riveted or welded to the boiler or tank.

Saddle tee. Fitting for copper or galvanized pipe that is bolted onto the pipe, eliminating cutting and threading or soldering.

Safety valve. Combination temperature and pressure relief valve generally installed in a hot-water tank to prevent an explosion caused by overheating or excessive pressure in the tank.

Sanitary fittings. Fittings that have no inside shoulders to block flow of waste.

Screwed flange. Flange screwed on the pipe that it is connecting to an adjoining pipe.

Screwed joint. Pipe joint consisting of threaded male and female parts screwed together.

Screwed pipe supports. Screwed pipes are supported and secured at not less than every other story height.

Screw joints. Joints having American Standard screw joints.

Seamless carbon steel pipe. Manufactured from carbon steel for high-temperature, high-pressure service. An open-hearth steel that is fabricated in Grades A and B. Grade A can be used for forming or welding; B has a higher carbon and manganese content and greater tensile strength, but is less ductile (then required by standard ASTM A 106).

Seamless copper water tube. Tubing that meets the standards of ASTM B 88.

Seamless pipe. Pipe or tube formed by piercing a billet of steel and then rolling.

Seamless steel pipe. A general service pipe, suitable for bending, coiling, fusion welding, lapping, or flanging. Meets the standards of ASTM A 53. Grade B does not lend itself to close coiling, forge-welding, or cold bends.

Seat. Valve part into which washer or other piece fits, stopping the flow of water.

Service fitting. Street ell or street tee with male threads at one end and female threads at the other.

Set. Same as offset, but also used in place of offset where the connected pipes are not in the same vertical or horizontal plane; rolling offset.

Setback. In a pipe bend, the distance measured back from the intersection of the centerlines to the beginning of the bend.

Short nipple. Nipple whose length is a little greater than that of two threaded lengths or somewhat longer than a close nipple so that it has some unthreaded portion between the two threads.

Shoulder nipple. Nipple of any length that has a portion of pipe between two pipe threads. As the term is generally used, it is a nipple halfway between the length of a close nipple and a short nipple.

Shutoff valve. Valve installed in a waterline whenever water must be stopped from flowing to a fixture or leak in the piping.

Siphon breaker. Valve, device, or appurtenance constructed and installed to prevent backflow in the plumbing system or any portion thereof.

Size and length. A given caliber or pipe size is for a nominal internal diameter with the exception of iron pipe and brass pipe that is measured by outside diameter. The developed length of a pipe is its longitudinal length along the center line of the pipe and fittings.

Size of pipe or tubing. Usually the nominal size by which pipe or tubing is commercially designated. The actual dimensions of the different kinds of pipe and tubing are indicated or specified.

Sleeve weld. Joint made by butting two pipes together and welding a sleeve over the outside.

Slip coupling. Coupling that has no stop to prevent it from slipping over a pipe.

Slip joint. Connection where one pipe slides into another making a tight joint with a threaded retainer or an approved gasket.

Slip joints. Slip joints are permitted only in trap seals or on the inlet side of the trap. Ground joint brass connections, which allow adjustment of tubing but provide a rigid joint when made up, are not considered as slip joints.

Slip nuts. Nuts that are not fixed but can move up and down the pipes, allowing for adjusting length for proper connection.

Slip-on flange. Flange slipped over the end of the pipe and then welded to the pipe.

Socket weld. Joint made by use of a socket weld fitting that has a prepared female end or socket for insertion of the pipe to which it is welded.

Solder joint. Joint made by joining two tubes by use of lead/tin solder.

Spigot. Plain end of a cast-iron pipe. The spigot is inserted into the bell end of the next piece of pipe to make a watertight joint.

Spiral pipe. Pipe made by coiling a plate into a helix and riveting or welding the overlapped edges.

Stack cleanout. Plugged fitting located at the base of all soil or waste stacks.

Stainless steel pipe. Alloy steel pipe with corrosion-resisting properties, usually imparted by nickel and chromium.

Standard pressure. The term was formerly used to designate cast-iron flanges, fittings, valves, etc., suitable for a maximum working steam pressure of 125 lb.

Stop and waste valve. Gate or compression-type valve that has a die opening, or port, and may be opened to allow water to drain from the piping supplied by the valve.

Stopcock. Underground valve near or on the property line used for shutting off water in emergencies. Location determined by the city engineer.

Stop valve. Device used to stop or regulate the flow of fluids in a pipe. It is normally manually operated.

S-trap. S-shaped, water-sealed trap used primarily in the installation of water closets.

Street elbow. Elbow with a male thread on one end and female thread on the other.

Supports. Supports, hangers, and anchors are devices for supporting and securing pipe, equipment, and fixtures to walls, ceilings, floors, or structural members.

Swing joint. Arrangement of screwed fittings and pipe to provide for expansion in pipe lines.

Swivel joint. Joint employing a special fitting designed to be pressuretight under continuous or intermittent movement of the machine or part to which it is connected.

Tee. Fitting with three openings, shaped like a T.

Tempering valve. Valve that mixes a small amount of hot water with the cold water entering the toilet tank in order to prevent a sweating tank.

Trap

1. Fitting or device so designed and constructed as to provide, when properly vented, a liquid seal that will prevent the backpassage of air without materially affecting the flow of sewage or waste water through it.

2. Device (most often a curved section of pipe) that holds a water seal to prevent sewer gases from escaping into a home through a fixture drain.

3. Fitting or device so constructed as to prevent the passage of air or gas through a pipe or fixture by means of a water seal.

4. Designed fitting normally shaped with a U-type part that continuously retains a liquid seal preventing the backpassage of air without affecting the flow of liquids in the system.

Trap arm

1. Section of a trap that connects a J-bend with a drainpipe behind the wall.

2. That portion of a fixture drain between a trap and its vent.

Trap (building). A building (house) trap is a device, fitting, or assembly of fittings installed in the building drain to prevent circulation of air between the drainage system of the building and the building sewer.

Trap dip. Lowest part of the upper interior surface of a trap.

Trap primers. A device or system of piping to maintain a water seal in a trap.

Trap (resealing). A resealing trap is a trap constructed and installed to retain a satisfactory seal when subjected to siphonic or aspiratory effects of wastes discharging through or past the branch into which the trap is connected.

Trap standard. Trap for a fixture that is integral with the support for the fixture.

Type ACR copper tubing. Tubing used for the installation of air-conditioning and refrigeration services.

Type K copper tubing. Tubing used for underground and interior water service.

Type L copper tubing. Tubing used for interior water service only.

Type M copper tubing. Nonpressure water tubing used for above-ground installations.

Union. Device used to connect pipes and usually consisting of three pieces; a thread end fitted with exterior and interior threads; a bottom end fitted with interior threads and a small exterior shoulder; and a ring that has an inside flange at one end and an inside thread like that on the exterior of the thread end on the other end. Unions are extensively used because they permit connections with little disturbance of the pipe positions.

Union ell. Ell with a male or female union at one end.

Union joint. Pipe coupling, usually threaded, that permits disconnections without disturbing other sections.

Union tee. Tee with male or female union at one end of the run.

Vacuum breaker. Device designed to prevent back-siphonage by providing an opening through which air may be drawn to relieve negative pressure (vacuum) in the water supply pipe.

Vacuum relief valve. Device to prevent excessive vacuum in a water storage tank or heater.

Valve. Device that controls the flow of water or any other type of liquid or gas.

Valve tags. Tags are available in brass, color-coded anodized aluminum, fiberglass-reinforced plastic, polyester laminated paper, and engraved phenolic.

Volume of a pipe. Measurement of the space within pipe walls.

Water hammer arrester. Device other than an air chamber or calculated air chamber designed to provide protection against excessive surge pressure without maintenance.

Welded and seamless steel pipe. Pipe that meets the standard ASTM A 53.

Welding end valves. Valves without end flanges and with ends tapered and beveled for butt welding.

Welding fittings. Wrought or forged steel prefabricated elbows, tees, reducers, saddles, and the like, beveled for welding to pipe.

Welding neck flange. Flange with a relatively long neck beveled for butt welding to the pipe.

Wrought iron. Iron refined to a plastic state in a puddling furnace. It is characterized by the presence of about 3% slag irregularly mixed with pure iron and about 0.5% carbon.

Wrought pipe. The term refers to both wrought-steel and wrought-iron pipe. Wrought in this sense means worked, as in the process of forming furnace-welded pipe from skelp or seamless pipe from plates or billets. The expression "wrought pipe" is used in contradistinction to "cast pipe." When wrought-iron pipe is referred to, it should be designated by its complete name.

Wye (Y). Fitting, either cast or wrought, that has one side outlet at any angle other than 90°.

Y. Fitting with three outlets in the shape of the letter Y.

PLUMBING INSTALLATIONS AND ACCESSORIES

Accessibility of services. Public sanitary sewer system, storm water system, or water supply system accessible to dwelling units on any lot or parcel of ground when any point of the lot or parcel abuts on or is within a distance of a street, alley, or parcel of land in which such a sanitary sewer, storm sewer, or water supply service has been constructed and is in operation.

Accessible. Approachable by person or tools as required, without undue hindrance or impediment, and where all obstacles may be removed and replaced without the cutting or breaking and subsequent patching or replacing of the materials.

Air break

1. Physical separation, which may be a low inlet into the indirect waste receptor from the fixture, appliance, or device indirectly connected.

2. Drain piping arrangement from a mixture or device discharging indirectly into another fixture or device above the trap seal but below the flood level rim.

Air break (drainage system). Piping arrangement in which a drain from a fixture, appliance, or device discharges indirectly into a fixture, receptacle, or interceptor at a point below the flood level rim of the receptacle, so installed as to prevent backflow or siphonage.

Air chamber (or air cushion or water hammer arrester). Device attached to supply pipes near outlets to prevent water hammer.

Air gap

1. Unobstructed vertical distance through the free atmosphere between the lowest opening from any pipe or faucet conveying water or waste to a tank, plumbing fixture receptor, or other device and the flood level rim of the receptacle.

2. Vertical distance between the supply fitting outlet (spout) and the highest possible water level in the receptor when flooded. If the plane of the end of the spout is at an angle to the surface of the water, the mean gap is the basis for measurement.

Air gap (drainage system). Unobstructed vertical distance through the free atmosphere between the outlet of waste pipe and the flood level rim of the receptacle into which it is discharging.

Anchor. Special metal fastener used to attach pipes, fixtures, and other such parts to the building structure.

Appliance. Receptacle or equipment that receives or collects water, liquids, or sewage and discharges water, liquids, or sewage directly into an indirect waste pipe or fixture.

Approved

1. Accepted or acceptable under an applicable requirement stated or cited in a code or accepted as suitable for the proposed use under procedures and powers of an administrative authority.

2. Accepted as satisfactory to the authority having jurisdiction over plumbing.

3. Work, materials, methods, and procedures that are acceptable in plumbing by the approving authority for a given locality and a type of work or job.

Area drain

1. Receptacle designed to collect surface or rainwater from an open area.

2. Receptacle provided with a strainer that permits surface or rainwater to flow through at the specified rate.

Asbestos joint runner. Runner made of an asbestos rope and a clamp that holds molten lead in the bell of a cast pipe until it has cooled.

Auger. Springlike tool forced into waste lines to break up blockages.

Automatic flushing cistern. Water cistern mainly used for flushing urinals automatically at predetermined periods.

Backflow

1. Flow of water or other liquids, mixtures, or substances into the distributing pipes of a potable supply of water

from any source(s) other than its intended source. Backsiphonage is one type of backflow.

2. Reverse flow of liquids in a pipe.

Backflow connection. Arrangement whereby backflow can occur.

Backflow preventor (reduced pressure zone type). Assembly of differential valves and check valves, including an automatically opened spillage port to the atmosphere. Gauges are installed where pressure reduction is mandatory by code.

Backflow siphonage. Flowing back of used, contaminated, or polluted water from a plumbing fixture or vessel into a water supply pipe due to a negative pressure in such a pipe.

Back pressure

1. Force exerted causing or tending to cause water or air to flow in a pipe opposite to the normal direction flow.

2. Air pressure in plumbing pipes that is greater than the surrounding atmospheric pressure.

Backsiphonage

1. Flowing back of used, contaminated, or polluted water from a plumbing fixture, vessel, or other sources into a water supply pipe due to a negative pressure in such pipe.

2. Flowing back by negative pressure of contaminated or polluted water from a plumbing fixture into a potable water system.

Back-to-back. Any two fixtures connected at the same level to a stack and complying with the vent requirements.

Back vent

1. Branch vent installed primarily for the purpose of protecting fixture traps from self-siphonage.

2. Pipe installed to vent a trap or waste pipe and connected to the vent system at a point above the fixture served by the trap or waste pipe.

Backventing (reventing or secondary venting). Connecting a fixture with a nearby main vent instead of venting directly out the roof.

Bend. Change in direction in piping.

Branch

1. Any part of the piping system other than a main, riser, or stack.

2. Any part of a plumbing system other than a main.

3. The branch of any system of piping is the part of the system that extends horizontally at a slight grade, with or without lateral or vertical extension or vertical arms, from the main to receive fixture outlets not directly connected to the main.

Branch arm. Horizontal waste extending from the vent and/or waste stack to a fixture trap or traps. It should not exceed 8 ft in length.

Branch interval. Length of soil or waste stack corresponding in general to a story height, but in no case less than 8 ft, within which the horizontal branches from the floor or story of a building are connected to the stack.

Branch vent

1. Vent connecting one or more individual vents with a vent stack or stack vent.

2. Vent pipe connecting from a branch of the drainage system to the vent stack.

Branch water-distributing pipe. Pipe connected to a distributing or riser pipe and conveying the water therefrom to the plumbing fixture.

Branch water supply pipe. Pipe connected to a principal supply pipe and conveying the water therefrom to the riser or distributing pipe.

Braze. The joining of metal with an alloy having a melting point higher than common solder but lower than the metal being brazed.

Building connection. Sewer extending from the main sewer to the curb line.

Building drain

1. Part of the lowest piping of a drainage system that receives the discharge from soil, waste, and other drainage pipes inside the walls of the building and conveys it to the building sewer outside the building wall (with the distance determined by contract).

2. That part of the lowest horizontal piping of a building drainage system from the stack or horizontal branch, exclusive of storm sewer, extending outside the building wall (with the distance determined by contract).

3. Horizontal piping of drainage piping in or adjacent to a building or other structure that receives the discharge from the drainage piping and conveys it to the building sewer, including offsets.

Building drain branch. Soil or waste pipe that extends horizontally from the building drain and receives only the discharge from fixtures on the same floor as the branch.

Building drainage system. All piping for carrying waste water, sewage, or other drainage from the building to the street sewer or place of disposal.

Building gravity drainage system. Drainage system that drains by gravity into the building sewer.

Building main. Water supply pipe, including fittings and accessories, from the water (street) main or other source of supply to the first branch of the water-distributing system.

Building sewer. Extension from the building drain to the public sewer or other places of disposal.

Building storm drain. Building drain used for conveying rainwater, groundwater, subsurface water, condensate, cooling water, or other similar discharge to a building storm sewer or a combined building drain and building storm sewer.

Building storm sewer. Extension from the building storm drain to the public storm sewer, combined sewer, or other place of disposal.

Building subdrain. Portion of a drainage system that usually cannot drain by gravity into the building sewer.

Building sanitary drain. Part of the drainage system that extends from the end of the building drain and conveys its discharge to the public sewer system.

Building service supply. It is unlawful to connect water piping supplied directly from city water mains or other approved sources with or to piping from underground storage tanks or other unapproved sources, and no cross connection is permitted between the potable water distributing system and any portion of the waste or soil system or fixtures or devices that may contaminate, pollute, or otherwise render the water unsafe.

Building sewer

1. Part of the horizontal piping of a drainage system that extends from the end of the building drain receives the discharge of the building drain and conveys it to a public sewer, private sewer, individual sewage disposal system, or other point of disposal.

2. Pipe extending from the building drain to the building connection or other point of disposal.

3. Combined building sewer that receives storm and sewage.

4. That part of the horizontal piping of a building drainage system extending from the building drain to the outside of the inner face of the building wall to the street sewer or other place of disposal (a cesspool, septic tank, or other type of sewage treatment device or devices) and conveying the drainage of but one building site.

5. Part of the drainage piping outside a building or other structure that connects a building drain to the main sewer or, where the place of disposal of the sewage is on the property, to the place of disposal on the property and that commences at a point from the outer face of the wall of the building or other structure and terminates at the property line or place of disposal on the property.

6. Pipe that begins outside the inner face of the building wall and extends to a public sewer, septic tank, or other place of sewage disposal.

Building storm drain

1. Building drain used for conveying rainwater, surface water, groundwater, subsurface water, condensate, cooling water, or other similar discharge to a building storm sewer or a combined building sewer, extending to a point outside the building wall.

2. Horizontal piping of storm drainage piping in or adjacent to a building that receives the discharge from storm drainage piping and conveys it to the building sewer and includes offsets.

3. Part of the piping of the building drainage system that takes surface water, groundwater, subsurface water, cooling water, or similar discharge to a public or approved discharge point.

Building storm sewer

1. Extension from the building storm drain to the public storm sewer, combined sewer, or other disposal system.

2. Part of storm drainage piping outside a building or other structure that connects the building storm drain to the main storm sewer or, where the place of disposal is on the property, to the place of disposal on the property, and that commences from the outer face of the wall of the building or other structure and terminates at the property line or place of disposal on the property.

Building subdrain

1. Portion of a drainage system that cannot drain by gravity into the building sewer.

2. Portion of a building drainage system that cannot drain by gravity into the building drainage system.

Building supply. Pipe carrying potable water from the water meter or other source of water supply to a building or other point of use or distribution on the lot. Building supply also means water service.

Building trap

1. Device fitting or assembly of fittings installed in the building drain to prevent circulation of air between the drainage system of the building and the building sewer.

2. Running trap installed in the building drain to prevent circulation of air between the drainage system of the building and the building sewer.

3. Running hand-hole trap installed in a building drain to prevent circulation of air between the building drain and the building sewer.

Burr. Protruding metal or roughness on the walls of a pipe resulting from pie cuts or threading.

Catch basin. Receptacle that separates and retains greases, oil, dirt, gravel, and all other substances lighter or heavier than the liquid waste that bears them in order to prevent their entrance into the house sewer. A catch basin may perform the functions of a gravel or grease basin, or both, except that the liquid waste that it receives should not contain fecal matter.

Caulking lead. Lead used for caulking should consist of not less than 99.73% lead.

Cesspool. Covered pit with open-joined lining in its bottom portions into which raw sewage is discharged, the liquid portion of the sewage being disposed of by seeping or leaching into the surrounding porous soil, and the solids or sludge being retained in the pit to undergo partial decomposition before occasional or intermittent removal.

Chain wrench. Adjustable tool for holding and turning large pipe up to 4 in. in diameter. A flexible chain replaces the steel jaws of standard pipe wrenches.

Circuit or loop vent. System of vent pipes arranged as a substitute for individual revents when two or more fixtures are located on a branch between a soil or waste stack and a vent riser.

Circuit vent

1. Branch vent that serves two or more traps and extends from in front of the last fixture connection of a horizontal branch to the vent stack.

2. Group vent extending from in front of the last fixture connection of a horizontal branch to the vent stack.

3. Vent that functions for two or more traps and extends to a vent stack from a point on a horizontal branch in front of the last connected fixture.

4. Branch vent that serves two or more traps or fixtures with integral traps that are battery wasted. The vent extends from the top of the horizontal soil and/or waste branch in front of the last fixture waste to a vent stack adjacent to the upstream end of the horizontal branch.

Cistern. Covered tank in which rainwater from roof drains is stored for household or other purposes.

Cleanout

1. Opening providing access to the drain line or trap under the sink and closed with a threaded plug.

2. Device that has a removable cap or plug securely attached to it and so constructed that it can be installed in a pipe so that the cap or plug can be removed to permit pipe cleaning apparatus to be inserted into the pipe.

Clear water waste. Cooling water and condensate drainage from refrigeration and air-conditioning equipment; cooled condensate from steam heating systems; cooled boiler blowdown water; waste water drainage from equipment rooms and other areas where water is used without an appreciable addition of oil, gasoline, solvent, acid, etc.; and treated effluent in which impurities have been reduced below a minimum concentration considered harmful.

Closed system. Water-piping system where a pressure-regulating device, check valve, or a backflow preventer is installed between the street main or other source of supply and a water heater or a water heater connected to a storage tank and having water shutoff valves between the heater and tank. Water shutoff valves at the water meter, building supply, or cold-water inlet to a water heater do not constitute a closed system.

Closet bend. Drainpipe that joins with the toilet bowl outlet at one end, the drain line or soil stack at the other.

Combination waste and vent system. Specially designed system of waste piping embodying the horizontal wet venting of one or more sinks or floor drains by means of a common waste and vent pipe adequately sized to provide free movement of air above the flow line of the drain.

Combined building drain. Building drain that receives both storm water and sewage.

Combined building sewer. Building sewer that also receives storm water.

Combined sewer. Sewer or drain that receives storm water, other liquid wastes, and sewage.

Common or continuous waste. Waste that has several compartments, such as a double laundry tray connected to a single trap.

Common trap. Trap having a water seal or not less than 2 in. or not more than 4 in.

Common vent (dual vent)

1. Vent connecting at the junction of two fixture drains and serving as a vent for both fixture and drain.

2. Vent that serves two fixtures by a connection at the junction of the two fixture drains.

Compressive stress. Compressive stress is stress that resists a force attempting to crush a body.

Conductor. Pipe inside a building that conveys storm water from the roof to a storm or combined building drain or sewer.

Conductor or leader. The conductor, often termed a "leader", is part of roofing and/or gutter system taking water from a roof or above-surface area to a storm drain or other disposal area or system.

Conductors. Pipes that carry the storm or rainwater from the roofs of buildings to the building sewer to storm water drains. "Downspouts" is a term that is frequently used to denote the vertical portion of conductors or roof leaders.

Confined groundwater. Body of groundwater overlaid by material sufficiently impervious to serve free hydraulic connection with overlying groundwater.

Continuous vent

1. Vent that is a continuation of and in a straight line with the drain to which it connects. A continuous vent is further designated by the angle the drain and vent make with the horizontal at the point of connection, such as a vertical continuous waste and vent, 45° continuous waste and vent, and flat (small-angle) continuous waste and vent.

2. Vertical vent is a continuation of the vertical drain to which it connects.

3. Pipe extending vertically above the soil or water branch.

4. Continuation of a vertical soil or waste-pipe above the point of entrance of the pipe from a fixture trap.

Continuous waste

1. Drain from two or three fixtures connected to a single trap.

2. Waste pipe from two or more fixtures using the same trap.

3. Drain connecting the compartments of a set of fixtures to a trap or connecting other permitted fixtures to a common trap.

Copper pipe straps. Straps used to secure copper pipe.

Continuous waste and vent

1. Vent that is a continuation of and in a straight line with the drain to which it connects. A continuous waste and vent is further defined by the angle the drain and vent make with the horizontal at the point of connection, such as, vertical continuous waste and vent, 45° continuous waste and vent, and flat (small angle) continuous waste and vent.

2. Vent pipe that is a vertical extension of a vertical waste pipe and includes the vertical waste pipe.

Critical level. Marking on a backflow prevention device or vacuum breaker conforming to approved standards and established by the testing laboratory (usually stamped on the device by the manufacturer) that determines the minimum elevation above the flood level rim of the fixture or receptacles served at which the device may be installed. When a backflow prevention device does not bear a critical level marking, the bottom of the vacuum breaker, the combination valve, or the bottom of any such approved device constitutes the critical level.

Cross-connection

1. Physical connection or arrangement of pipes between two otherwise separate water supply systems, one of which contains potable water and the other water of unknown or questionable safety, whereby water may flow from one system to the other, the direction of flow depending on the pressure differential between the two systems.

2. Any connection or arrangement, physical or otherwise, between a potable water supply system and any plumbing fixture or any tank, receptacle, equipment or device through which it may be possible for nonpotable, used, unclean, polluted, or contaminated water or other substances to enter any part of such potable water system under any condition.

3. Physical arrangement whereby one system of piping is connected to another system of piping in such a way that the contents of the two systems may become mixed.

Crown of a trap. Point in a trap where the direction of flow changes from upward to downward.

Crown weir. Point in the curve of a trap directly below the crown.

Cup weld. Pipe weld where one pipe is expanded on the end to allow the entrance of the end of the other pipe. The weld is then circumferential at the end of the expanded pipe.

Curb box. Cylindrical casting placed in the ground over a corporation stop. It extends to ground level and permits a special key to be inserted to turn off the corporation stop.

Dead end

1. Branch leading from a soil, waste, vent, building drain, or building sewer that is terminated at a developed distance by means of a cap, plug, or other fitting not used for admitting water to the pipe.

2. Extended portion of a pipe that is closed at one end. No connections are made on the extended section. A dead end is without free air circulation.

Dead ends. In the installation or removal of any part of a drainage system, dead ends are avoided except where necessary to extend a cleanout so as to be accessible.

Deep seal. Trap having a water seal or more than 4 in.

Developed length. Drainage or vent piping has its length measured along the centerline of the pipe and fitting.

Die. Tool used to cut external threads by hand or machine.

Die stock. Tool used to turn dies when cutting external threads.

Dip of a trap. The lowest portion of the inside top surface of the trap.

Distance. The distance or difference in elevation between two sloping pipes is the distance between the intersection of their centerlines with the centerline of the pipe to which both are connected.

Distributing pipes. Pipe for conveying water from a service pipe to a fixture or to an outlet, including the control valves and fittings connected in it but not including a meter or control valve or other device owned and controlled by the supplier of the water.

Distributing water pipe. Pipe that is connected to a riser pipe or branch supply pipe and conveys the water therefrom to the branch distributing pipe.

Domestic sewage. Liquid and waterborne wastes derived from the ordinary living processes, free from industrial wastes, and of such character as to permit satisfactory disposal, without special treatment, into the public sewer of by means of a private sewage disposal system.

Dope. Pipe joint compound.

Double offset. Two offsets installed in succession or series in the same line.

Downspout. Leader or conductor pipe that carries water from the roof or gutter to the ground or to any part of the drainage system.

Drain. Pipe that carries waste water or waterborne wastes in a building drainage system.

Drainage pipe. Any pipe in drainage piping.

Drainage piping

1. All or any part of the drainpipes of a plumbing system.
2. All the connected piping that conveys sewage to a place of disposal, including the building drain, building sewer, soil pipe, soil stack, waste stack and waste pipe, but not including a main sewer and piping used for sewage in a sewage plant.

Drainage system

1. All the piping within public or private premises that conveys sewage, rainwater, or other liquid wastes to a legal point of disposal, but not including the mains of a public sewer system or a private or public sewage treatment or disposal plant.
2. All the piping within a public or private premise that conveys sewage, storm water, or other liquid wastes, including the building sewer.
3. Piping within public or private premises that conveys sewage, rainwater, or other liquid wastes to a point of disposal, and that includes the building drain and building sewer, but does not include the building connection, public sewer system, or private or public sewage treatment or disposal plant.
4. Systems of piping or conduits through which are conveyed liquids or liquid-borne solids from all plumbing fixtures and appurtenances in buildings and structures and that discharges such liquid or solids into the house sewer. The drainage system of buildings or structures includes the house drain and its branches and the sewer and its branches.

Dry vent

1. Vent that does not carry water or waterborne wastes.
2. Vent pipe that is not a wet vent.

Dual vent

1. Vent connected at the junction of two fixture drains and serving as a vent for both fixtures.
2. Group vent connecting at the junction of two fixture branches and serving as a back vent for both branches. Sometimes called a "unit vent."
3. Vent pipe connecting at a junction of waste pipes serving two fixtures and as a common vent pipe for both fixtures.

Dry well. Pit or receptacle having porous walls that permit the contents to seep into the ground.

Durham system

1. Term to describe soil or waste systems where all piping is of threaded pipe, tubing, or other such rigid construction, using recessed drainage fittings to correspond to the type of piping.
2. Drain-waste-vent system using threaded galvanized pipe.

DWV (drain-waste-vent). System that carries away waste water and solid waste, allows sewer gases to escape, and maintains atmospheric pressure in drainpipes.

Effective opening

1. Minimum cross-sectional area at the point of water supply discharge, measured or expressed in terms of the diameter of a circle or if the opening is not circular, the diameter of a circle of equivalent cross-sectional area.
2. Minimum cross-sectional area between the end of the supply fitting outlet (spout) and the inlet to the controlling valve or faucet. The basis of measurement is the diameter of a circle of equal cross-sectional area. If two or more lines supply one outlet, the effective opening is the sum of the effective openings of the individual lines or the area of the combined outlet, whichever is the smaller.
3. Cross-sectional area of a faucet, fitting, or pipe at the point of discharge.

Effluent. The outflow from sewage treatment equipment.

Ejector. Electrically or mechanically operated device used to elevate sewage and liquid wastes from a lower level to a point of discharge into a sewer or other disposal system.

Fall. Amount of slope given to horizontal runs of pipe.

Fire line. System of pipes and equipment used exclusively to supply water for extinguishing fires.

First. With reference to the connection of a fixture to a horizontal branch, the nearest to the waste stack of soil stack.

Fixture. Receptacle or equipment that receives water, liquids, or sewage and discharges water, liquids, or sewage directly into drainage piping.

Fixture branch

1. Water supply pipe from the water-distributing pipe to the wall or floor line.
2. Water supply pipe between the fixture supply pipe and the water-distributing pipe.
3. Water supply pipe serving more than one fixture.

Fixture drain. Drain from the trap of a fixture to the junction of that drain with any other drainpipe.

Fixture supply pipe

1. Water supply pipe connecting the fixture with the fixture branch.

2. Water supply pipe connecting a fixture to a branch water supply pipe or directly to a main water supply pipe.

3. Water supply source connecting the fixture and supply pipe.

4. Water supply pipe connecting the fixture with the fixture branch at the wall or floor line.

Fixture unit

1. Design factor so chosen that the load-producing values of the different plumbing fixtures can be expressed approximately as multiples of that factor.

2. Quantity in terms of which the load-producing effects on the plumbing system of different kinds of plumbing fixtures are expressed on some arbitrarily chosen scale.

3. Unit in which the hydraulic load produced by fixtures is expressed and determined under requirements of the local plumbing code.

4. The rate of discharge, usually through a plumbing fixture $7\frac{1}{2}$ gal/minute, is termed "one fixture unit."

5. Flow of waste equal to 1 cu ft/minute, usually the measuring stick for sizing pipes in a plumbing system.

Fixture unit flow rate. Total discharge flow in gallons per minute of a single fixture divided by 7.5, which provides the flow rate of that particular plumbing fixture as a unit of flow. Fixtures are rated as multiples of this unit of flow.

Fixture vent. Part of the piping system that connects with the drainage piping near the point where the fixture trap is installed and extends to a point above the roof of the structure.

Flashing (usually lead fabricated). Device that fits over the vent pipe on the roof to prevent water from entering the house through the roof opening for the vent.

Flood control device. Mechanical device consisting of backwater valve or valves; motorized unit of sufficient capacity to overcome backwater pressures and housing, the bottom of which is the invert of the sewer it serves, permitting gravity flow under normal conditions. The motorized unit lifts and ejects the contents of the house drain without overheating or failure by means of a bypass to the outlet side of the backwater valve.

Flooded

1. A fixture is flooded when the liquid therein rises to the flood level rim.

2. A fixture is flooded when the liquid in a fixture or receptacle rises to the flood level rim.

Flood level. Flood level in reference to a plumbing fixture is the level at which water begins to overflow the top or rim of the fixture.

Flood level rim. Top edge of a fixture or receptacle from which water overflows.

Floor drain

1. Drain to receive water from a floor of a building. In its simplest form it consists of a strainer or grate set flush with the upper surface of a floor so that water passing down through the strainer or grate enters a connected drainage pipe. Without limiting the generality of the foregoing, it includes, when located between the strainer and the connected pipe or nipple, any ancillary part, such as a floor drain body, water stop, trap, backwater valve, or primer connection.

2. Receptacle fitted with a strainer or grate and a trap or seal and connected to the plumbing or drainage system.

Flow pressure. Pressure in the water supply pipe at the faucet or water outlet while the faucet or water outlet is wide open and flowing.

Flow rate. Volume of water used by a plumbing fixture in a given amount of time, usually expressed in gallons per minute (gpm).

Fluorocarbon tape. Special tape used as a joint sealer in place of pipe joint compound.

Flux. Paste applied to copper pipe tubing before soldering to prevent oxidation when heat is applied to metal.

Foundation drain. Drain installed below the surface of the ground to collect and convey water from the foundation of a building or other structure.

Free groundwater. Groundwater in the zone of saturation extending down to the first impervious barrier.

Freezing. No water, soil, or waste pipe should be installed or permitted outside a building or in an exterior wall unless adequate provision is made to protect such pipe from freezing where necessary.

Front main cleanout. Plugged fitting located near the front wall of the building where the building drain leaves the building. The front main cleanout may be inside or directly outside the building, depending on local codes.

Frostline. Depth of frost penetration, depending on region of the country and the normal average temperature range.

Fusion weld. Joining metals by fusion, using oxyacetylene.

Gang trapped. The waste piping from a group of two or more fixtures or other drainage openings is so arranged that all the fixtures or other drainage openings drain to a common trap. The term does not apply when the trap is a secondary trap, such as a building trap or the trap of a fixture that receives waste from one or more indirect waste pipes.

Gasket. Device (usually rubber) used to make the joint between two valve parts watertight.

Generally accepted standard. Specification, code, regulation, rule, guide, or procedure in the field of construction or related thereto, recognized and accepted as authoritative.

Grade

1. Slope or fall of a line of pipe in reference to a horizontal plane. In drainage it is usually expressed as the fall in a fraction of an inch per foot length of pipe.

2. Where the grade of a line of pipe in reference to a horizontal plane is intended, the term "slope" applies. Where elevation or ground level is intended, the term "grade" applies.

Graded. With reference to a pipe, it is the slope of the pipe with reference to the true horizontal.

Graphite packing. Wirelike material to be wrapped around the faucet stem to prevent leaking.

Gravel basin. Receptacle through which roof water flows and that is designed to retain sediment.

Grease basin or intercepter. Receptacle designed to cause separation and retention of oil or grease from liquid wastes.

Grease interceptor (grease trap)

1. Receptacle designed to intercept and retain grease or fatty substances contained in kitchen or other wastes.

2. Container designed to intercept and hold grease or fatty substances from wastes from a line to which it is connected.

Groundwater

1. Water obtained from natural storage areas beneath the surface of the ground.
2. Subsurface water occupying the zone of saturation.

Groundwater supply. Groundwater supply includes well, spring, water suction pipe, water pressure pipe, or similar structure or device used to obtain groundwater.

Group vent. Branch vent that performs its functions for two or more traps.

Hanger. Prefabricated metal device used to support pipes or fixtures.

Header. With reference to drainage piping or vent piping, a pipe that receives the flow from two or more pipes.

Heel or side inlet bend. A heel or side inlet quarter-bend is not to be used as a vent when the inlet is placed in a horizontal position.

Horizontal

1. Not departing from the true horizontal plane by more than 45°.
2. Level or with a slope of not more than 1 in./ft.

Horizontal branch

1. Branch drain extending laterally from a soil or waste stack or building drain, with or without vertical sections or branches, that receives the discharge from one or more drains and conducts it to the soil or waste stack or to the building drain.
2. Part of a waste pipe that is horizontal and installed to convey the discharge from more than one fixture.

Horizontal branch drain. Drain branch pipe extending laterally from a soil or waste stack or building drain, with or without vertical sections or branches, that receives the discharge from one or more fixture drains and conducts it to the soil or waste stack or to the building drain.

Horizontal pipe

1. Pipe or fitting that is installed in a horizontal position or with a slope less than 3 in./ft. of length.
2. Installed pipe or fitting making an angle of less than 45° with the horizontal.

Hot water. Water at a temperature of not less than 120°F.

House drain. Part of the horizontal piping of a plumbing system that receives the discharge from soil, waste, downspout, and other drainage pipes inside the walls of a building and conveyed to the house sewer terminating 3 ft outside the building walls or as regulated by the city or utility company.

House sewer. That part of the horizontal piping of a plumbing or drainage system extending from the house drain to its connection with the main sewer or other place of sewage disposal.

Hydraulic gradient. Amount of inclination of a drainage line between the trap outlet and the vent connection, not exceeding one pipe diameter in this total length.

Hydropneumatic tank. Closed tank used for the storage of water under air pressure obtained from an external source. It is used for providing pressure to raise water to the upper stories of buildings when the city water pressure is inadequate.

Identification of piping. All service piping that is accessible for maintenance operations should be identified, by color and printed information including the direction of flow.

Indirect connection. Connection in which there is a break in a line of pipe through which the water, sewage, or other liquid may be discharged from one pipe to another by gravity and that is open to the atmosphere for a sufficient altitude to permit visibility, of such discharge and to prevent a backflow into the pipe above the connection.

Indirect service water heater. Heater that derives its heat from a heating medium such as warm air, steam, or hot water.

Indirect waste. Waste that is not discharged directly into drainage piping.

Indirect waste pipes

1. Pipe that does not connect directly with the drainage system but conveys liquid wastes by discharging into a plumbing fixture or receptacle that is directly connected to the drainage system.
2. Waste pipe that does not connect with the drainage system, but that discharges into it through a properly trapped fixture or receptacle.
3. Waste pipe that does not connect directly with the drainage system, but that discharges into the drainage system through an air break or air gap into a trap, fixture, receptacle or interceptor that is properly wasted and vented.

Individual sewage disposal system. System for the disposal of domestic sewage by means of a septic tank, designed for use apart from a public sewer to serve a single establishment or building where a public sewer is not available.

Individual vent

1. Pipe installed to vent a fixture and that is connected to the vent system above the fixture it serves or terminates in the open air.
2. Pipe installed to vent a fixture trap and connected with the vent system above the fixture it serves.
3. Pipe installed to vent a fixture trap and connected to the general vent system at a point above the fixture.
4. Pipe installed to vent a fixture drain. It connects with the vent system above the fixture served or terminates at a point above the roof level.
5. Vent that is installed in a pipe to vent a fixture trap connected to the vent system above the fixture it serves.

Individual wastes. Liquid, gaseous, solid, or other waste substance, or a combination thereof, resulting from any process of industry, manufacturing, trade, or business or from the development, processing, or recovery of any natural resources.

Individual water supply. Supply other than an approved public water supply that serves one or more families.

Industrial sewers and sewage works. Sewer or sewage works for conveying or treating industrial wastes from any industry or industrial process.

Industrial wastes

1. Liquids and wastes resulting from the processes employed in industrial establishments and free of fecal matter.
2. Wastes that are detrimental to the public sewer system or to the functioning of the sewage treatment plant.
3. Liquid wastes resulting from the process employed in an industrial establishment.
4. All liquid or waterborne wastes from industrial or commercial processes except domestic sewage.

5. Sewage and/or the liquids, solids, or other wastes from an industrial establishment or resulting from any process of industry, manufacture, trade, or business, or from the development of any natural resource.

In front of. With reference to the point of connection of a fixture to a horizontal branch, in the direction of discharge.

Insanitary. Contrary to sanitary principles and injurious to health.

Installed. Completion of work to accomplish alterations, revisions, changes, and installation of new work.

Interconnection. Physical connection or arrangement of pipes between two otherwise separate building water supply systems whereby water may flow from one system to the other, the direction of flow depending on the pressure differential between the two systems. Where such connections occur between the sources of two such systems and the first branch from either, whether inside or outside the building, the term "cross-connection" applies and is generally used.

Invert. Floor or lowest part of an internal cross section of a conduit or pipe.

J-bend. J-shaped piece of drainpipe used in traps.

Joint. Point at which two sections of pipe are fitted together.

Joint runner. Tool composed of asbestos rope and a clamp used in the leading of joints in horizontal runs of bell and spigot cast-iron pipe.

Journeyman plumber. A person, certified as such under the terms of the ordinance, skilled in the practice of installing plumbing and drainage systems, who alters, repairs, dismantles and maintains plumbing and drainage systems or parts thereof as an employee, but who does not furnish materials or supplies or contracts work.

Labeled. Equipment or materials bearing a label of an approval agency.

Last. With reference to the point of connection of a fixture to a horizontal branch, farthest from the waste stack, soil stack, or building drain to which the horizontal branch is connected.

Leach bed. System of underground piping that permits absorption of liquid waste into the earth.

Leader (downspout)

1. Water conductor from the roof to the building storm drain or other piping serving as a storm drain.
2. Exterior drainage pipe for conveying storm water from roof or gutter drains.

Licensed plumber. Person engaged in the vocation of plumbing who has successfully met the qualifications or passed an examination so that the issuance of a license qualifies the holder as a licensed plumber.

Liquid waste

1. Discharge from any fixture, appliance, area, or appurtenance in connection with a plumbing system that does not receive fecal matter.
2. Discharge from any fixture, appliance, area, or appurtenance that does not contain human or animal waste matter.

Listed. Term applied to equipment or materials included in a list published by a listing agency that maintains periodic inspection on current production of listed equipment or materials and whose listing states either that the equipment or material complies with approved standards

or has been tested and found suitable fur use in a specified manner.

Listing agency. Agency, accepted by authority, that is in the business of listing or labeling, that maintains a periodic inspection program on current production of listed models, and that makes available a published report of such listing in which specific information indicating that the product has been tested to approved standards and found safe for use in a specified manner is included.

Load factor. Percentage of the total connected fixture unit flow rate that is likely to occur at any point in the drainage system. Load factor varies with the type of occupancy, the total flow unit above this point being considered, and with the probability factor of simultaneous use.

Local ventilating pipe. Pipe on the fixture side of the trap through which vapor or foul air is removed from a room or fixture.

Local venting pipe. Pipe through which foul air is removed from a room or fixture.

Loop or circuit vent. Pipe that vents a series of fixture traps on the same soil or waste branch and that is continued through the roof or reconnected into a vent stack above all fixture trap branches.

Loop vent

1. Same as a circuit vent except that it loops back and connects with a stack vent instead of a vent stack.
2. Branch vent that functions for two or more traps and loops back or extends to a stack vent from a point in front of the last connection of a fixture to a horizontal branch.

Main

1. The main of any system of continuous piping is the principal artery of the system, to which branches may be connected.
2. The main of any system of horizontal, vertical, or continuous piping is that part of the system that receives the wastes, vents, or revents from fixture outlets or traps, directly or through branch pipes.
3. Part of a system of piping that receives the soil or waste pipes, vent pipes, or revent pipes from fixture outlets or traps, directly or through branch pipes as tributaries.

Main sewer

1. Public sewer under the jurisdiction of the municipality in a street, alley, or other premises.
2. Public sewer, including its branches.

Main vent. Principal artery of the venting system, on which vent branches may be connected.

Main water supply pipe. Pipe that is connected to the service pipe of any building, structure, or premises and conveys the water therefrom to the principal supply pipe.

Male thread. Threads on the outside of a pipe, fitting, or valve.

Manhole. Opening in a plumbing system or sewer large enough to permit a person access to the system for the purposes of inspection and/or cleaning.

Master plumber. Plumber who is licensed to engage in the business of installing plumbing as a contractor.

Minor repairs

1. Repair of an existing fixture; replacement of faucets, valves, or parts thereof with like material or material serving the same purpose; clearance of stoppages; stopping of leaks; relieving of frozen pipes; and other

minor replacements or repairs classified as "minor repairs" by the local plumbing codes.

2. Mending of leaks in drains, soil, vents, water supply piping, faucets, and valves; replacement of a fixture in an old location; forcing out an obstruction and the replacement of a faucet, valve, or not more than 10 ft of soil, waste, or vent piping (length dictated by local plumbing code).

Municipal water system. Plant, pumps, pipes, valves, treatment, storage, and distribution facilities and appurtenances by which water is supplied to the inhabitants of a municipality.

Negative pressure. Pressure within a pipe that is less than atmospheric pressure.

Neoprene. Synthetic rubber material with a superior resistance to oils and grease; often used as a gasket and washer material.

Nonpotable water

1. Water not safe for human or animal use.
2. Water not safe for drinking or personal or culinary use.

Nonpressure drainage. Condition in which a static pressure cannot be imposed safely on the building drain. This condition is sometimes referred to as "gravity flow" and implies that the sloping pipes are not completely filled.

Nonpressure-type (atmospheric) vacuum breaker. Vacuum breaker that is not designed to be subjected to static line pressure.

Nuisance

1. The term "nuisance" includes, but is not limited to, inadequate or unsafe water supply or sewage disposal system.
2. The term "nuisance" embraces public nuisance as known at common law or in equity jurisprudence. Whatever is dangerous to human life or detrimental to health; whatever building, structure, or premises is not sufficiently ventilated, sewered, drained, cleaned, or lighted, in reference to its intended or actual use; and whatever renders the air or human food or drink or water supply unwholesome are considered nuisances.

Oakum. Stranded hemp used in making bell-and-spigot joints watertight.

Offset. An offset in a line of piping is a combination of elbows or bands that brings one section of the pipe out of line, but into a line parallel with the other section.

One-pipe system. System of soil and waste water disposal using two vertical pipes. Waste and soil fittings discharge into the one stack connected directly to the drain; the other pipe is used for ventilation and antisiphonage.

Open air. Atmosphere outside a building.

Open plumbing. Plumbing system in which no plumbing fixtures, except a built-in bathtub, is so enclosed as to form a space in which air does not circulate.

Open trench. Trench dug in earth or rock to make it accessible from the adjacent ground level. It may have either sloping or vertically braced sides.

O-ring. Narrow rubber ring used in some faucets instead of packing to prevent leaking around the stem. It is also used with swivel-spout faucets to prevent leaking at the base of the spout.

Outlet. With reference to the distributing pipe, an opening at which water is discharged from the pipe from a faucet into a boiler or a heating system, into a device or equipment that is operated by water and that is not part of

the distributing system or into the open air, but not into an open tank forming part of the supply system.

Oxidized sewage. Sewage that has been exposed to oxygen to make the organic substances stable.

Packing. Loosely-packed waterproof material installed in the packing box of the valve to prevent leaking around the stem.

Penetrating oil. Penetrating oil is used to help loosen a threaded joint in which corrosion has fused the fittings.

Pipe alignment. Piping installed in a straight line, vertically, horizontally, or at a given angle.

Pipe chase. Space or recess in the walls of a building where pipes are run vertically.

Pipe cutters. Various tools designed specifically for making perfectly square cuts on pipe. Cutters for copper, plastic, and galvanized and cast-iron are available.

Pipe joint compound. Sealing compound used on threaded fittings (applicable to male threads).

Pipe markers. Fabricated from a plastic-coated cloth tape with a rubber-based pressure-sensitive adhesive. Available in a variety of colors and printed in contrasting inks.

Pipe trenches. The water service pipe or underground water pipes should not be run or laid in the same trench as building sewer or drainage piping.

Pipe wrench. Tool with adjustable, slightly curved, toothed jaws, designed to firmly grip pipe as pressure is applied to the handle.

Plumbing

1. Plumbing includes all piping, fixtures, appurtenances, and appliances for a supply of water for all personal and domestic purposes in and about buildings, structures, and public places where persons live, work, or assemble. It also includes all piping, fixtures, appurtenances, and appliances for a sanitary drainage and related venting system within a building, and all piping, fixtures, appurtenances, and appliances outside a building, connecting the building with the source of water supply on the premises or the water main in the public way, or at the curb. It includes all piping, fixtures, appurtenances, appliances, drains, or waste pipes carrying sewage from the foundation walls of a building to the sewer service lateral at the curb or in the public way or other disposal terminal holding private or domestic sewage, excepting any underground house drain or other underground sewer of vitrified tile or masonry construction including any connection with any catch basin or cesspool of masonry.

2. System of connected piping, fittings, valves, and appurtenances that receives water from a source of supply on a property or from a public water main and conveys the water into and within a building or to a place of use on the property, or where the source is on the property, that commences at the source of supply or at the property line and includes all tanks, pumps, heaters, coils, strainers, and treatment devices designed to make physical, chemical, or bacteriological changes in the water being conveyed to fixtures, including drainage piping with all traps, fittings, and appurtenances with storm drainage piping with all traps, fittings, and appurtenances, and a vent system with all fittings and appurtenances.

3. Business, trade, or work having to do with the installation, removal, alteration, or repair of plumbing and drainage systems, or parts thereof.

4. Gas pipes and gas-burning equipment, waste pipes, water pipes, water closets, sinks, lavatories, bathtubs, catch basins, drains, vents, and other provided fixtures, together with the connections to the water, sewer, and gas lines.

5. Work and/or practice, materials and fixtures used in the installation, removal, maintenance, extension, and alteration of a plumbing system; all piping, fixtures, fixed appliances, and appurtenances in connection with sanitary drainage, storm drainage facilities, special wastes, the venting system, and the public or private water supply systems, within or adjacent to any building, structure, or conveyance and their connections to the public disposal mains or other acceptable terminals within the property lines.

6. Art and science of creating and maintaining sanitary conditions in buildings where people live, work, or assemble by providing permanent means for a supply of safe, pure, and wholesome water, ample in volume and of suitable temperature and pressures for drinking, cooking, bathing, washing, and cleaning and by cleansing all waste receptacles and like means for reception and speedy and complete removal from the premises of all fluid or semifluid organic wastes and other impurities incident to human life and occupation.

Plumbing appliance. Any one of a special class of plumbing fixtures intended to perform a special function. Operation and/or control may be dependent on one or more energized components, such as motors, controls, heating elements, or pressure- or temperature-sensing elements. Fixtures may operate automatically through one or more of the following actions: time cycle, temperature range, pressure range, measured volume or weight; or the fixture may be manually adjusted or controlled.

Plumbing branch. Part of a piping system connected to a riser main or stack.

Plumbing inspections supervisor. To meet the requirements of codes, all references to the plumbing inspections supervisor means the building inspections superintendent or his authorized representatives.

Plumbing system

1. Water supply and distribution pipes; plumbing fixtures and traps; soil, waste and vent pipes; building drains and building sewers including their respective connections, devices, and appurtenances within the property lines of the premises; and water-treating or water-using equipment.

2. All potable water supply and distribution pipes, all plumbing fixtures and traps, all drainage and vent pipes, and all building drains, including their respective joints and connections, devices, receptacles, and appurtenances within the property lines of the premises and potable water piping, potable water-treating or using equipment, fuel gas piping, water heaters and vents for same.

3. All water supply, drainage, and venting systems and all fixtures and their traps complete with their connections.

Plumbing systems appliance. Receptacle or equipment that receives or collects water, liquids, or sewage and discharges, water, liquids, or sewage either directly or indirectly to a plumbing system.

Plumbing systems tests. The plumbing systems must be subjected to such tests as will effectively disclose all leaks and defects in the work, and to satisfy local building codes.

Pollution. Addition of sewage, industrial wastes, or other harmful or objectionable material to water. Sources of sewage pollution may be privies, septic tanks, subsurface irrigation fields, seepage pits, sink drains, and other types of wastes.

Positive pressure. Pressure within the sanitary drainage or vent piping system that is greater than atmospheric pressure.

Potable water

1. Water that is satisfactory for drinking, culinary, and domestic purposes and meets the requirements of the health authority having jurisdiction.

2. Water fit for human consumption.

3. Water free from impurities that could cause disease or harmful physiological effects. Bacteriological and chemical quality must conform to the regulations of applicable codes.

Potable water system. Plumbing that conveys potable water.

Pressure drainage. Condition in which a static pressure may be imposed safely on the entrances of sloping building drains through soil and waste stacks connected thereto.

Pressure head. Amount of force or pressure created by a depth of one foot of water.

Primary branch drain. Single sloping drain from the base of a soil or waste stack to its junction with the main building drain or with another branch thereof.

Principal water supply pipes. Such pipes are the water supply arteries in buildings and structures. They are connected to the main water supply pipe and convey the water therefrom to pumps, tanks, filters, heaters, and other equipment, together with all their appurtenances, and to the branch supply pipes.

Private or private use. Term applicable to plumbing fixtures in residences and apartments, to private bathrooms in hotels and hospitals, to rest rooms in commercial establishments containing restricted-use single fixtures, and to similar installations where the fixtures are intended for the use of a family or an individual.

Private sewer

1. Sewer privately owned and not directly controlled by public authority.

2. Sewer serving two or more buildings, privately owned, and not directly controlled by public authority.

3. Sewer built in a street, alley, or granted easement not dedicated for public use and serving or intended for the service of two or more pieces of property.

Private water supply. A private source of water supply may be used provided samples are submitted periodically to the proper authorities for analysis and approval.

Private water supply system. Water supply system that derives its water from a source other than the municipal water system.

Process water. Water from unapproved sources for industrial processing or fire protection, identified by an outlet (provided that it has an outlet) with an approved sign stating that the water is unfit and that its use is prohibited for drinking purposes. Piping carrying potable water is identified and distinguished from water piping from unapproved sources by distinctive painting and appropriate signs.

Public sewer

1. Common sewer directly controlled by public authority.

2. Sewer built by or constructed under the authority of the municipality in a public place, such as a street or alley, or in and through land for which an easement has been granted for the common use of the property abutting on such public place of easement.

Public sewers and public sewage works. Sewerage system or sewage works, whether publicly or privately owned, for conveying or treating sewage from a community, subdivision, municipality, or two or more private premises.

Public water main. Water supply pipe for public use controlled by public authority.

Putty. Prepared soft mixture used to seal sink rims, water closet bases, bathtub edges, and other places needing a caulking-type compound.

Rainwater leader. Conductor inside a building or other structure that conveys storm water from the roof of the building or other structure to a building storm drain or other place of disposal.

Reamer. Tool that fits into pipe ends and is used to grind off internal burrs caused by cutting pipe.

Relief vent

1. Vent whose primary function is to provide circulation of air between drainage and vent systems or to act as an auxiliary vent on a specially designed system.

2. Branch from the vent stack, connected to a horizontal branch between the first fixture branch and the soil or waste stack, whose primary function is to provide for circulation of air between the vent stack and the soil or waste stack.

3. Vent pipe discharging into a vent stack and connected to a horizontal branch between the first fixture connection and the soil or waste stack.

4. Vent so planned as to permit additional circulation of air between drainage and vent systems.

Revent (or back vent pipe). Part of a vent pipeline that connects directly with an individual trap underneath or in back of the fixture and extends either to the main or branch pipe.

Revent pipe

1. Part of a vent pipeline that connects directly with an individual waste or group of wastes underneath or in back of the fixture and extends either to the main or branch vent pipe. (It is sometimes called an "individual vent.")

2. Pipe that connects directly at or near the junction of an individual trap outlet with a waste or soil pipe underneath or in back of a fixture and extends to a connection with the main or branch vent above the top of the fixture.

Rim

1. Unobstructed open edge of a fixture.

2. Stainless steel device that fits around the outside edge of some kinds of sink, holding the sink onto the countertop.

Riser

1. Vertical run of pipe.

2. Water supply pipe that extends vertically one full story or more to convey water to branches or to a group of fixtures.

Riser water pipe. Pipe installed perpendicular to the horizontal through the floors, stories, and other open spaces of buildings and structures that conveys the water from the main or branch supply pipes to the distributing pipes or branch distributing pipes.

Roof gutter. Receptacle either suspended from the edges of a roof or constructed in a roof to convey roof water to the downspout rain leader or conductor pipe.

Roof jacket or flange. Installed on the roof terminals of vent stacks to seal the opening and prevent rainwater from entering the building. Jacket fabricated from lead or copper.

Roughing in. Installation of all parts of the plumbing system that can be completed prior to the installation of fixtures, including drainage, water supply, and vent piping and the necessary fixture supports.

Running trap. Stretch of pipe in which the inlet and outlet are at the same height and the waterway between them is lower than the bottom of either.

Safe pan. Pan or other collector placed beneath a pipe or fixture to prevent leakage from escaping onto the floors, ceilings, or walls.

Sand filter. Treatment device or structure constructed above or below the surface of the ground, for removing from septic tank effluent solid or colloidal material of a type that cannot be removed by sedimentation.

Sand interceptor (sand trap). Watertight receptacle designed and constructed to intercept and prevent the passage of sand or other solids into the drainage system to which it is directly or indirectly connected.

Sanitary building drain. Drain that conveys the discharge of plumbing fixtures.

Sanitary piping. The piping of the plumbing system should be of durable material, free from defective workmanship and designed and installed to give satisfactory service for the reasonable life expectancy of the building.

Sanitary sewer

1. Pipe that carries sewage and excludes storm, surface, and groundwater. A sanitary sewer might also carry processing wastes, acid wastes, and clear-water wastes from special equipment, which because of their chemical components should not be discharged into a storm sewer.

2. Sewer designed or used only for conveying liquid or waterborne waste from plumbing fixtures.

3. House drain or house sewer designed and used to convey only sewage.

Sanitary sewer system. Sewerage system that carries exclusively sewage or industrial wastes, or both.

Seal of a trap. The depth of water in a trap under normal operating conditions.

Secondary branch. The secondary branch of the building drain is any branch of the building drain other than a primary branch.

Second hand. As applied to material or plumbing equipment, that which has been installed, used, removed, and passed to another ownership or possession.

Self-siphonage. Loss of the seal of a trap as a result of removing the water from the trap. Usually caused by the discharge of the fixture to which the trap is connected.

Septic tank. Watertight receptacle that receives sewage.

Service pipe

1. Pipe that conveys water between the main shutoff valve on the public water system and the control shutoff valve in a supply system.

2. Water supply line from the source of supply to the building served.

Sewage

1. Sewage includes any or all of the following, whether untreated or insufficiently treated: human excreta; food wastes disposed of through sewers; wash water; liquid wastes from residences, institutions, business buildings, and industrial establishments; and such diluting water as may have entered the waste disposal system. Sewage consists of the water-carried waste products or discharge from human beings or other wastes from residences, public, or private buildings, together with such ground, surface, or storm water as may be present.

2. Liquid waste containing animal or vegetable matter in suspension or solution and that may include liquids containing chemicals in solution.

Sewage ejectors. Device for moving sewage by entraining it on a high-velocity steam, air, or water jet.

Sewage treatment plant. Structures and appurtenances, including septic tanks and cesspools, that receive the discharge of a sanitary drainage system and are designed to bring about a reduction in the organic and bacterial content of the waste so as to render it less offensive or dangerous.

Sewage works. All devices and appurtenances that treat or were designed to treat sewage or industrial waste by changing the nature of or removing any of its constituents before its final disposal into any waters or on any lands.

Sewer gas. Mixture of vapors, odors, and gases created in sewers by the mixture of wastes.

Sewers or sewerage systems. All structures, conduits, or pipelines by which sewage or industrial waste is collected, transported, and discharged to the point of disposal, except plumbing within and in connection with buildings and service pipes from buildings to street sewers.

Sewer tape. Flexible metal tape forced into sewers to break up blockages.

Shearing stress. Stress that resists a force that would make one layer of a body slide across another layer.

Side vent

1. Vent connecting to the drainpipe through a fitting at an angle not greater than 45° to the vertical.

2. Vent connecting the drainpipe through a 45° wye.

Single-stack system. System of soil and waste water disposal using a single pipe and waste branches with deep-seal traps to prevent antisiphonage. A single-stack system is so named to distinguish it from one- and two-pipe systems.

Siphonage. Suction resulting from the flow of liquids in pipes.

Siphoning. When vacuum occurs in a section of pipe, nearby water is pulled into it. A toilet functions on a siphoning principle. Siphoning is not desirable in other places because it could cause draining of fixture traps. Venting acts to prevent siphoning.

Slope. Grade of a line of pipe in reference to a horizontal plane. In drainage it is usually expressed as the fall in a fraction of an inch per foot length of pipe.

Soil-or-waste pipe. Pipe in a sanitary drainage system.

Soil-or-waste stack. Vertical soil-or-waste pipe that passes through one or more stories, and includes any offset that is part of the stack.

Soil or waste vent. Part of the main, soil or waste pipe that extends above the highest installed branch or fixture connection.

Soil pipe

1. Pipe that conveys the discharge of water closets, urinals, or fixtures having similar functions, with or without the discharge from other fixtures, to the building drain or building sewer.

2. Pipe that conveys the discharges of one or more water closets or bedpan sterilizers, with or without the discharges from other fixtures, to the house drain.

Soil stack

1. Large DWV pipe that connects the toilet to the house drain and also extends up and out the house roof, its upper portion serving as a vent.

2. Stack that conveys the discharge of one or more sanitary units, with or without the discharge from any other fixture.

Soil vent. Part of a stack that extends above the highest installed water closet.

Solder. Soft metal wire (tin and lead) used as a bonding agent to join copper pipe.

Solvent cement. Compound used to join rigid plastic pipe.

Special waste pipe. Pipes that convey special wastes.

Special wastes. Wastes that require some special method of handling, such as the use of indirect waste piping and receptors, corrosion-resistant piping, sand, oil, or grease interceptors, condensers, or other pretreatment facilities.

Special wastes, piping, or treatment. Wastes that require special treatment before entry into the normal plumbing system.

Stack

1. Vertical main of a system of soil, waste, or vent piping extending through one or more stories.

2. Part of drainage piping that is vertical and that runs from the building drain or sewage tank to the open air and includes offsets not exceeding 5-ft perpendicular distance.

3. Vertical line of soil, waste, vent, or inside conductor piping.

Stack group. Location of fixtures in relation to the stack so that by means of proper fittings, vents may be reduced to a minimum.

Stack vent

1. Extension of a soil or waste stack above the highest horizontal drain connected to the stack terminating at a point above the roof level.

2. Vertical extension of a soil, waste, or vent system above the highest horizontal drain connected to the stack.

Stack venting. Method of venting a fixture(s) through the soil or waste stack.

Stop box or curb box. Cast-iron box and removable cover, installed below grade, with the cover at grade line.

Storm building drain. Drain that conveys storm water or other drainage allowed by some codes.

Storm building sewer. Sewer that conveys storm water or other drainage but no sewage.

Storm drain. Drain used for conveying rainwater, surface water, condensate, cooling water, or similar discharges.

Storm drainage pipe. Pipe in storm drainage piping.

Storm drainage piping. All the connected piping that conveys storm water to a place of disposal, including the building storm drain, building storm sewer, rainwater leader, area drain installed to collect surface water from the area of a building, and the piping that drains water from a swimming pool or from water-cooled air-conditioning equipment, but not including a main storm sewer and the subdrain or foundation drain.

Storm drainage system. System that is used for conveying rainwater, surface water, condensate, cooling water, or similar liquid wastes, exclusive of sewage or industrial waste, to the storm sewer or other legal place of disposal.

Storm sewer. Sewer used for conveying rainwater, surface water, condensate, cooling water, or similar liquid wastes, exclusive of sewage and industrial waste.

Storm sewer system. Sewerage system that carries only storm water surface drainage or other clean water and does not carry any sewage or industrial wastes.

Storm water. Rainwater, melted snow, or ice and water leached into the subsoil.

Storm water drain. Pipe or drain that receives the discharge or rainwater from buildings or premises. It may include the discharge of seepage or groundwater which it conveys to public sewer, building sewer, or water course.

Strain. Change of size or shape of a body produced by the action of a stress.

Stress. When external forces act on a body (in this case metals), they are resisted by reactions within the body which are termed "stresses."

Stub-outs. Portions of the supply pipe that stick out of the wall.

Subdrain. Drain that is at a level lower than the building drain and the building sewer.

Subsoil drain

1. Drain that receives only subsurface or seepage water and conveys it to a place of disposal.

2. Part of a drainage system that conveys the subsoil, ground water, or seepage water from the foundations of walls or from below the basement floors under buildings to the building drain, storm water drain, or building sewer.

3. Part of a drainage system that conveys subsoil ground or seepage water to the house drain or house sewer.

Subsurface disposal field. Approved open-jointed system of pipes or drains through which sewage effluent is distributed beneath the surface of the ground for absorption into the subsoil.

Subsurface drain. Drain, other than a foundation drain, installed to collect water from subsoil.

Sump

1. Tank or pit that receives the discharge from drains or other wastes located below the normal grade of the gravity system, and that must be emptied by mechanical means.

2. Approved airtight tank or pit that receives sewage or liquid waste, is located below the normal grade of the gravity system, and that must be emptied by mechanical means.

3. Watertight tank that receives the discharge of drainage water from a subdrain, and from which the discharge flows or is ejected into drainage piping by pumping.

4. Tank or pit below the normal grade of gravity receiving liquid wastes or sewage from which the wastes or sewage must be mechanically pumped to a higher level receiving point.

Sump pump. Mechanical device other than an ejector or bucket for removing sewage or liquid waste from a sump.

Supply system. Service pipe, distributing pipe, and all connecting pipes, fittings, control valves, and devices.

Supports. Devices for supporting and securing pipe and fixtures to walls, ceilings, floors, or structural members.

Supports, hangers, anchors. Devices for supporting and securing pipe, fixtures, and equipment to walls, ceilings, floors or structural members.

Sweating. Accumulation of moisture on pipes and tanks caused by condensation when the cooler surfaces of the pipe or tank meets warm air.

Sweat soldering. Method of soldering in which the parts to be joined are first coated with a thin layer of solder and then joined while exposed to a flame.

Tempered water. Water at a temperature of not less than 90°F and not more than 105°F.

Tensile strength

1. Overall tensile stress that a material will develop. Tensile strength is considered to be the load in pounds per square inch at which test materials rupture.

2. Stress that resists a force that tends to pull a body apart.

Terminal. Upper portion of a soil, waste, or vent pipe that projects above or through the roof of the building.

Testing. Requires that the plumbing system be filled with water, air, smoke or as otherwise required by the local plumbing code.

Threading die. Tool used for cutting threads into pipe.

Tile sewer or drain. Sewer or drain of tile, terra cotta, or cement pipe.

Trade size. Size designation traditionally used by the trade, but restricted to products or classes of products manufactured to a standard or specification, so that the designated trade size may be referred to an industry accepted table or chart which then provides the true dimensions of the item in question.

Trailer park sewer. Part of the horizontal piping of a drainage system that begins 2 ft downstream from the last trailer site connection and that receives the discharge of the trailer site and conveys it to a public sewer, private sewer, individual sewage disposal system, or other point of disposal.

Trap seal

1. Maximum vertical depth of liquid that a trap will retain, measured between the crown weir and the top of the dip of the trap.

2. Vertical distance between the overflow and the dip separating the inlet and outlet arms of the trap.

Trap weir. Highest part of the lower interior surface of a trap.

Turbulence. Deviations from parallel flow in a pipe due to rough inner walls, obstructions, or directional changes.

Underground piping. Piping in contact with earth below grade. Pipe in a tunnel or a watertight trench is not included.

Union. Fitting that joins two lengths of pipe and permits assembly and disassembly without taking the entire section apart.

Unit vent. One vent pipe that serves two or more traps.

Vacuum. Pressure less than that exerted by the atmosphere.

Vacuum breaker. Device designed to protect a water supply system against backsiphonage by providing an opening through which air may be drawn to relieve negative vacuum pressure.

Vent. Pipe installed to provide a flow of air to or from a drainage system or to provide a circulation of air within such system to protect trap seals from siphonage and backpressure.

Vent pipe

1. Pipe provided to ventilate a building drainage system and prevent trap siphonage and backpressure.
2. Part of the vent system.

Vent pipe or vent. Pipe provided to ventilate a plumbing system, prevent trap siphonage and backpressure, and equalize the air pressure within and without the piping system.

Vent stack

1. A vent stack, sometimes called a "main vent," is a vertical vent pipe installed primarily for the purpose of providing circulation of air to or from any part of the building drainage system.
2. Continuous run of vent pipe connected to a soil stack, waste stack, or building drain and terminating in the open air.

Vent system

1. System of piping installed to provide a flow of air to or from drainage piping or storm drainage piping.
2. Pipe or pipes installed to provide a flow of air to or from the drainage system or to provide a circulation of air within such system to protect trap seals from siphonage and backpressure.

Vertical. Not departing from the true vertical plane by more than 45°.

Vertical pipe. Pipe or fitting that makes an angle of 45° or less with the vertical.

Waste pipe

1. Pipe that conveys only liquid waste free of fecal matter.
2. Pipe that receives the discharge of any fixture, except water closets or fixtures having similar functions, and conveys it to the building drain or to a soil or waste stack.
3. Drain pipe that receives the discharge of any fixture other than water closets or other fixtures receiving human excreta.
4. Part of the drainage piping that runs from a fixture to a waste stack, soil stack, building drain, or sewage tank.

Waste pipe and special waste pipe. Pipe that receives the discharge from any fixture or device, except water closets and bedpan sterilizers, and conveys it to the building drain, sewer, or waste pipe. When such pipe does not connect directly with the house drain or soil pipe, it is termed "special waste pipe."

Waste stack. Stack that is not a soil stack.

Waste vent. Part of a stack that extends above the highest installed fixture trap other than a water closet.

Water conditioning or treating device. Device that conditions or treats a water supply so as to change its chemical content or remove suspended solids by filtration.

Water-distributing pipe

1. Pipe that conveys water from the water service pipe to the plumbing fixtures branch or other plumbing outlets.
2. Pipe within the building or on the premises that conveys water from the water service pipe to the point of usage.

Water hammer. Sound of pipes vibrating and banging, resulting from the lack of air cushion.

Water main

1. Water supply pipe for public or community use.
2. Pipe used to convey public water supply.

Water mains. Pipes through which city water is distributed from the city waterworks system's pumping stations to all water service connections.

Water outlet. In connection with the water-distributing system, a water outlet is the discharge opening for the water to a fixture, to atmospheric pressure (except into an open tank that is part of the water supply system), to a boiler or heating system, or to any water-operated device or equipment that requires water to operate but that is not a part of the plumbing system.

Water service pipe

1. Part of a building main installed by or under the jurisdiction of a water department or company.
2. Pipe from the water main or other source of water supply to the water-distributing system of the building served.

Water storage facility. Reservoir, cistern, storage tank, water supply tank, pressure tank, or similar facility utilized to store water in a private water supply system.

Water supply system. System of a building or premises that consists of the water service pipe, the water-distributing pipes, and the necessary connecting pipes, fittings, control valves, and appurtenances in or adjacent to the building or premises.

Weather protection. Drainage and water piping should be protected against freezing temperatures and water supply systems should be installed to permit complete drainage when necessary.

Wet vents

1. Vent that receives the discharge from wastes other than water closets.
2. Drainpipe that also serves as a vent.
3. Waste pipe also functioning as a vent pipe.
4. Waste pipe that also serves as a vent, on the same floor level.
5. Drainpipe that in addition to receiving discharge waste also serves as a vent.

Wiped joint. Refers to the joint made by the adhesion of metal with solder at a finish thickness of at least $\frac{1}{4}$ in. at the point where the pipes are joined. The joint is smoothly finished with a wiping cloth.

Workmanship. Work of an acceptable nature and character that will fully secure the expected results of the local codes for the safety, welfare, and health protection of all persons.

Yoke vent

1. Pipe connecting upward from a soil or waste stack to a vent stack for the purpose of preventing pressure changes in the stacks.
2. Vertical or 45° relief vent of the continuous-waste-and-vent type formed by the extension of an upright wye-branch or 45° wye-branch inlet of the horizontal branch to the stack. It becomes a dual yoke vent when two horizontal branches are thus vented by the same relief vent.
3. Vent pipe connecting a soil stack or a waste stack to a vent stack.

REGISTERS AND GRILLES

Air circulation. Distribution of air depends on location of supply and return outlets, high and low ceiling changes, leakage through doors, windows, and elevator shafts, and the volume of air supplied.

Air quantity. The quantity of air used to ventilate a given space during period of occupancy should always be sufficient to maintain the standards of air temperature, air quality, air motion, and air distribution.

Air supply. The supply and distribution of the air required for heating, ventilating, and air-conditioning.

Air throw. Distance from the air supply outlet measured in the direction of the air flow, from the opening or grille to the point where the air velocity is 50 ft/minute.

Blow throw. In air distribution, the distance an airstream travels from an outlet to a position at which air motion along the axis reduces to a velocity of 50 ft/minute.

CFM. Measure of volume of air in cubic feet per minute.

Core area. The total plane of the portion of a grille, face, or register bounded by a line tangent to the outer opening through which air can pass. The core area is less than the register size.

Damper. Device used to control the volume of air passing through a duct by varying the cross-sectional area.

Diffuser. Outlet discharging supply of air in a spreading pattern.

Drop. The vertical distance (in feet) between the base of the outlet and the bottom of the air stream at the end of the horizontal throw.

Effective area. The calculated area of an outlet based on the average measured velocity between the fins.

FPM. Measure of air velocity in feet per minute.

Free area. The actual measured perpendicular area between the fins of a grille or register.

Grille. Framed metal or wood treatment at the conclusion of a duct run, including fins, blades, or louver slats used for air diffusion and direction.

Induction. The process of drawing room air into the projected air stream because of the velocity of the projected air stream.

Jet velocity. The average measured velocity of air passing between the fins.

Lattice-type return grille. Fabricated from aluminum, an all square grid-type grille for ceiling and wall installations.

Linear supply diffuser. Fabricated from extruded aluminum with anodized, duranodic, or special enamel finishes. One-way or two-way opposite direction air pattern, can be used for return air if so desired. Available up to 5 ft in length.

Outlet velocity. The average velocity of air emerging from the outlet measured in the plane of the outlet.

Pressure loss. The total pressure required to move air through a register.

Radius of diffusion. Horizontal distance in feet from the diffuser to a point where the terminal velocity of 50 fpm occurs.

Rectangular louvered face diffuser. Fabricated in aluminum or baked enamel steel, to fit the tile modules of lay-in ceilings, and used for supply and return outlets.

Rectangular perforated face diffuser. Fabricated in aluminum or baked enamel steel, to fit tile modules of lay-in ceilings, and can be used for supply and return air outlets.

Register. Combination grille and damper assembly covering the outlet of an air distribution opening.

Return. Any opening through which air is removed from a conditioned space.

Round louvered face diffuser. Fabricated normally of steel with baked enamel finish, surface-mounted for all types of ceilings.

Round perforated face diffuser. Fabricated from aluminum or steel baked enamel finish, in several patterns, and can be used for supply or return air outlets.

SPRINKLER SYSTEMS AND STANDPIPES

Air compressor

1. Device for supplying air under pressure to the cushion tank of an inside standpipe system.
2. Device for supplying air under pressure to the pressure or cushion tank of a sprinkler system, to an inside standpipe system, or to a dry-pipe sprinkler system.

Air-filling connection. Connection pipe from the air compressor that enters the system above the priming water level of the dry-pipe valve should not be less than 3/4 in. In this air line, there should be installed a check valve and on the supply side of this check valve, a shutoff valve of the renewable disc type is installed.

Air test. In dry-pipe systems an air pressure of 40 psi is pumped up and allowed to stand for 24 hours. All leaks that allow a loss of pressure of over $1\frac{1}{2}$ lb for the 24 hours are stopped.

Aquamatic sprinkler head. The only approved fully automatic on-off sprinkler. The aquamatic discharges water only when required; after the fire is controlled, it shuts off and resets itself automatically. The sprinkler is ready to operate again, and provides fail-safe protection.

Automatic air compressor. Dry-pipe system supplied by an automatic air compressor plant air system, having a device or apparatus used for automatic maintenance of air pressure. Should be of a type specifically approved for such service and capable of maintaining the required air pressure on the dry-pipe system. More than one dry-pipe system should not be connected on a single automatic air maintenance device where the air supply piping to the systems is subdivided only by check valves. Otherwise when one dry-pipe valve operates, leakage past the check valves could water column other dry-pipe valves.

Automatically controlled fire pump. Pump that starts automatically when the pressure in the system drops to a predetermined point and stops automatically when the pressure in the system rises to a predetermined point.

Automatic dry-pipe sprinkler system. System in which the piping up to the sprinkler heads is filled with air, either compressed or at atmospheric pressure, with the water supply controlled by a Type A or Type B dry-pipe valve.

Automatic dry standpipe system. A standpipe system in which all piping is filled with air, either compressed or at atmospheric pressure. Water enters the system through a control valve actuated either automatically by the reduction of air pressure within the system or by the manual activation of a remote control located at each hose station.

Automatic fire pump. A pump that maintains a required water pressure in a fire extinguishing system and which is actuated by a starting device adjusted to cause the pump to operate when the pressure in the system drops below a predetermined pressure, and to stop the pump when the pressure is restored.

Automatic sprinkler. Equipment for fire control and extinguishment whereby water is piped to specially designed orifices or sprinkler "heads" distributed throughout a property and operated automatically in the event of fire.

Automatic sprinkler head. Device connected to a water supply system that opens automatically at a predetermined fixed temperature and disperses a stream or spray of water.

Automatic sprinkler system. Standard and approved system consisting of one or more sources of a water supply; piping, valves, controls, and devices for automatically distributing water on a fire in sufficient quantities either to extinguish it or to hold it in check until it can be manually extinguished.

Automatic water supply source. Water supplied through a gravity or pressure tank, or automatically operated fire pumps, or from a direct connection to an approved city water main.

Automatic wet-pipe sprinkler system. A sprinkler system in which all piping and sprinkler heads are at all times filled with water under pressure, which is immediately discharged when a sprinkler head operates, with the water continuing to flow until the system is shut off.

Branch line. Horizontal pipe that conveys the water from a branch main to the sprinkler heads.

Branch lines. Lines of pipe in which the sprinklers are placed.

Branch main. Horizontal pipe that conveys the water from a system main or a system riser to the branch lines.

Brazing filler metal (classification BCuP-3 or BCuP-4). Brazing filler metal should meet the standards of AWS A 5.8. (American Welding Society).

Carbon dioxide

1. Colorless, odorless, electrically nonconductive inert gas that is a suitable medium for extinguishing fires. It extinguishes fire by reducing the concentrations of oxygen or the gaseous phase of the fuel in air to the point where combustion stops.

2. A carbon dioxide fire extinguishing system consists of a fixed supply of carbon dioxide normally connected to fixed piping with nozzles arranged to discharge carbon dioxide directly on the burning material. The system is arranged to operate automatically or manually by those normally in the vicinity of the hazard.

Chemical sprinkler system. System of automatic sprinklers controlled by thermostatic operating devices for the diffusion of approved fire-extinguishing chemicals or gases.

Chemical total flooding systems. Total flooding systems should meet NFPA 12A, requirements. The standard covers the designing, installing, testing, inspecting, approving, listing, operating, and maintaining halogenated agent extinguishing systems.

Class I standpipe system. System for use by fire department and those trained in handling heavy fire streams ($2\frac{1}{2}$ in. hose).

Class II standpipe system. System for use primarily by the building occupants until the arrival of the fire department ($1\frac{1}{2}$ in. hose).

Class III standpipe system. System for use by either fire department and those trained in handling heavy hose streams or by the building occupants.

Combination standpipe. Fire line system with a constant water supply, installed for the use of the fire department and the occupants of the building.

Combined dry pipe and preaction sprinkler system. System employing automatic sprinklers attached to a piping system containing air under pressure with a supplemental heat-responsive system of generally more sensitive characteristics than the automatic sprinklers themselves, installed in the same areas as the sprinklers. Operation of the heat-responsive system, as from a fire, actuates tripping devices that open dry-pipe valves simultaneously and without loss of air pressure in the system. Operation of the heat-responsive system also opens approved air exhaust valves at the end of the feed main, which facilitates the filling of the system with water, which usually precedes the opening of sprinklers. The heat-responsive system also serves as an automatic fire alarm system.

Corrosion-resistant sprinklers. Sprinkler heads specially coated or treated so they can be used in locations where chemicals, moisture, or other corrosive vapors exist.

Corrosive chemicals. Brine or other corrosive chemicals should not be used for testing systems.

Cross-connection (fire-extinguishing system). Piping between risers and siamese connections in a standpipe or sprinkler system.

Cross mains. Pipes directly supplying the lines in which the sprinklers are placed.

Cushion tank. Metal container holding water under air pressure, connected to the discharge from a fire pump supplying an inside standpipe system or a sprinkler system.

Deluge system. System employing open sprinklers attached to a piping system connected to a water supply through a valve that is opened by the operation of a heat-responsive system installed in the same areas as the sprinklers. When this valve opens, water flows into the piping system and discharges from all sprinklers attached thereto.

Deluge valve. Valve controlling a deluge system whereby an entire area subject to high fire or explosion hazard can be deluged by open sprinklers.

Differential dry-pipe valve. Valve that utilizes two seat rings, one to control entry of water and a second to seal air pressure in the sprinkler piping. Differential ratios of water pressure to air pressure when this type of valve operates may be nominally 6 to 1.

Domestic connections

1. Sprinkler piping should not be used in any way for domestic water service. Circulation of water in sprinkler pipes is objectionable, owing to increased corrosion, deposit of sediment, and condensation drip from pipes.

2. Connections for domestic use should not be taken from the fire protection piping. If permitted, such connections should be made on the supply side of the city check valve in the city connection near the point of entrance to the property.

Dry chemical. Especially useful on electrical and flammable liquid fires. Powdered extinguishing agent, under pressure of dry air or nitrogen, commonly discharged over cooking surfaces.

Dry hydrant. Permanently installed suction pipe or connection with the proper thread to permit a fire department pumper to take suction from a static water source, for example, at a bridge, pier, or farm pond.

Dry-pipe system

1. Sprinkler system in which all pipes and sprinkler heads are filled with air or other gas under pressure and in which the supply of water or other approved extinguishing agent is controlled by an automatic valve which, when opened, permits free flow of the extinguishing agent through the sprinkler heads.

2. System employing automatic sprinklers attached to a piping system containing air under pressure, the release of which, as from the opening of sprinklers, permits the water pressure to open a valve known as a dry-pipe valve. Water then flows into the piping system and out the opened sprinklers.

Dry-pipe valves. The alarm apparatus for a dry-pipe system consists of approved alarm attachments to the dry-pipe valve. When a dry-pipe valve is located on the system side of an alarm valve, the actuating device of the alarms for the dry-pipe valve may be connected to the alarms on the wet-pipe system.

Dry standpipe. Pipe extending full height of a building to which a fire hose can be attached and which, when not in operation, is free of water to prevent freezing.

Dry standpipe system

1. Fire line system without a constant water supply, equipped with fire department inlet and outlet connections and installed exclusively for the use of the fire department.

2. Standpipe system without permanent connection to any water supply.

Dry system. Dry-pipe automatic sprinkler system having air under pressure in the sprinkler piping installed in areas that might be subject to freezing. Operation of one or more sprinklers releases the air pressure actuating the control valve allowing water to flow through the piping and out opened sprinklers.

Dry valve. Valve of a dry-pipe automatic sprinkler system in which air under pressure keeps the valve sealed and keeps water out of the sprinkler piping until heat operates a sprinkler to release the air.

Extended sprinkler head. Head is designed for light hazard areas where an extra-long water spray pattern is required. The head is a solder-type, designed for horizontal wall installation.

Extinguish. Put out flames; essentially, to completely control the fire so that no abnormal heat or smoke remain.

Extra hazard maximum coverage. Sprinkler head coverage is as follows: 90 sq ft for all types of construction, 100 sq ft if hydraulically calculated, or as required by code.

Feed mains. Bulk mains supplying risers or cross mains.

Filling pump. Device used to supply water to the gravity or pressure tanks, or both, of a sprinkler system.

Fire

1. Usually unintentional or undesired burning; rapid oxidation of combustible materials resulting in light and heat.

2. Most fires in solid combustibles develop in four stages: incipient, smoke, flame, and intense heat.

Fire department connection. The piping between the check valve in the fire department inlet pipe and the outside connection should be tested in the same way as the balance of the system.

Fire protection equipment. Apparatus, assemblies, or systems, either portable or fixed, for preventing, detecting, controlling, or extinguishing fire.

Fire protection systems. Fire sprinkler extinguishing systems, fire alarm systems, fire detection systems, standpipe systems, carbon dioxide extinguishing systems, foam extinguishing systems, and water supplies or other extinguishing mediums suitable for the specific purpose for which they are designed and installed.

Fire sprinkler systems. Arrangement of open or closed sprinkler heads attached to piping containing an approved extinguishing agent.

First-aid hose. Hose permanently attached to a standpipe outlet, provided primarily for use in fire fighting by the occupants of the building.

First-aid standpipe. Auxiliary vertical or horizontal fire line designed primarily for emergency use by the occupants of the building or by the private fire brigade before the arrival of the municipal fire department.

Flame. Light from burning gases and incandescent particles during a fire.

Foam. Used to suppress flammable liquid fires. Foam can be distributed by a piping network to nozzles or other discharge outlets.

Foam-water sprinkler head. Heads available in upright or pendant-type nozzles, which discharge mechanical foam in a spray pattern. The nozzles also have water distribution patterns that meet standard sprinkler requirements.

Freezing of systems. Systems, or portions of a system, subject to freezing should be maintained dry, filled with antifreeze solution, heated by electric strip heater, or otherwise protected against freezing by any other method.

Fusible plug or link. Safety device consisting of a low-melting-point alloy designed to release pressure at a predetermined temperature.

Gravity tanks. Size of tanks hold the number of U.S. gallons to supply the sprinkler system in the building. Elevated tanks are used as gravity links.

Grid. Water main or piping system with lateral feeders or arteries to strengthen the supply and improve the distribution of water with minimum pressure loss due to friction.

Halon (haloginated hydrocarbon). Used where water damage to building contents or equipment would be unacceptable. Piping network connects fixed supply of Halon to nozzles that discharge uniform, low-concentration spray throughout room. An alternate system, to avoid the use of the network, is the use of discharge cylinders.

High-temperature sprinklers. Automatic sprinklers normally set to operate at 212°F (boiling point of water), 286°F or 360°F for solder-type sprinklers and at somewhat lower temperatures for nonsolder types. Sprinklers are designated as "intermediate," "high," and "extra high" and are colored white, blue, and red, respectively. Some very high-temperature, non-solder sprinkler heads are available for 400°F and 500°F operation and are colored green and orange.

Horizontal fire line. Fire line installed around the interior walls and columns of a building, pier, or wharf, with hose outlets located so that every part of the floor area is within reach of at least one fire stream.

Hose. Standpipes inside buildings and structures should have not more than 100 feet of $1\frac{1}{2}$-in. diameter hose equipped with the proper accepted and approved nozzle and couplings at each outlet and hung in an approved rack or cabinet to meet the requirements of the code.

Hose cabinet. Identifiable cabinet to house folded hose and valve, wall mounted or partially recessed.

Hose streams (solid or straight). Used to cool tanks and other equipment exposed to flammable liquid fires, or for washing burning spills away from danger points.

Hydrant. Valved outlet to a water supply system with one or more threaded outlets to supply fire department hose and pumpers with water, usually a post hydrant, but also a wall hydrant and chuck hydrant.

Hydraulically designed sprinkler system. Sprinkler system in which pipe sizes are selected on a pressure loss basis to provide a prescribed density (gallons per minute per unit area) distributed with a reasonable degree of uniformity over a specified area.

Hydrostatic test

1. New dry-pipe systems are tested hydrostatically, except that at seasons of the year that will not permit testing with water, they are tested for hours with at least 50-psi air pressure. The clapper of a differential-type dry-pipe valve is held off its seat during any test at a pressure in excess of 50 psi to prevent injuring the valve.

2. Test of strength and leak-resistance of a vessel, pipe, or other hollow equipment by introducing internal pressure with a test liquid, air, or air-gas mixture.

Latched-clapper dry-pipe valve. Valve that usually utilizes a diaphragm sensor to release the single water-controlling clapper at a preselected air pressure.

Life safety suppression system. Fire detection and suppression system primarily intended to protect human life from fire. A secondary benefit is fire protection for the structure and interior contents. The system is a combined hydraulic and electrical system. The hydraulic components are intended to deliver automatically the fire-extinguishing medium to the base of the fire. The electrical portion is intended to supervise the hydraulic system in order to detect and/or prevent system faults and to supplement the hydraulic system as necessary to accomplish the overall objectives of the system.

Light hazard maximum coverage. Sprinkler head coverage is as follows: 200 sq ft for smooth ceiling, and beam and girder construction (225 sq ft if hydraulically calculated), 130 sq ft for open wood joists.

Limited water supply system. System employing automatic sprinklers conforming to standards but supplied by a pressure tank of limited capacity. Sprinkler systems employing limited water supplies, reduced pipe sizes, and other departures from the requirements for standard systems are not classified as standard sprinkler systems.

Linen hose. Unlined fire hose, formerly used for first-aid standpipes, consisting of a linen, or flax fabric without a rubber lining. Under present practice, cotton or synthetic fiber is more commonly used.

Loop. Water supply main system having a feeder main extending in a loop making it possible to supply a given point from two directions and, when adequately valved, minimizing the danger of interruption to supply.

Low-differential dry-pipe valve. Valve that utilizes a single clapper which is held shut by air pressure in excess of the water pressure. It is equipped with a pilot valve or split seat ring to provide the fire alarm feature upon operation. The water pressure to air pressure ratio at the time of tripping is usually between 1.0 and 1.2 to 1.

Low flash points. Water may be ineffective in fighting fires with low flash points. The lower the flash point the less effective water will be.

Manually controlled fire pump. Pump that must be started and stopped by hand.

Mechanical dry-pipe valves. The earliest designed valves that achieved their differential through external lever and escapement mechanisms.

Microfast sprinkler head. A product of the Viking Corporation, is a very sensitive sprinkler head, smaller than other sprinkler-actuating mechanisms, with an element as resistant to damage as any other, but with a faster response time.

Nonautomatic sprinkler system. System in which all pipes are maintained dry and that is equipped with a siamese fire department connection.

Nonautomatic standpipe system. System in which all piping is maintained dry and which is supplied with water through a fire department siamese connection.

One-source sprinkler system. Automatic sprinkler system that is supplied from one of the approved automatic sources of water supply.

Open-head system. Fire protection sprinkler system having open sprinklers without fusible links and controlled by a valve that may be operated manually or by a thermostatic device.

Operating test of dry-pipe valve. Working test of the dry-pipe valve and quick opening device.

Ordinary hazard maximum coverage. Sprinkler head coverage is as follows: 130 sq ft for all types of construction except 100 sq ft for high-piled storage, or as required by code.

Outside sprinklers. Sprinkler system with open heads, manually operated, and used to protect a structure and window openings against a severe exposure hazard.

Outside standpipe. Standpipe riser on the exterior of a building, usually adjacent to an exterior stair provided for use by the fire department and equipped with a fire department siamese at the base.

Partial sprinkler system. Automatic sprinkler system consisting of a limited number of automatic sprinkler heads serviced from the building water supplies with one or more fire department siamese connections as required, for use in exitway facilities and isolated hazardous locations when approved by the building official.

Pendant sprinkler. Automatic sprinkler pointing downward from concealed piping rather than placed in an upright position above the supply pipe.

Permissible leakage. Inside sprinkler piping is installed in such a manner that there is no visible leakage when the system is subjected to the hydrostatic pressure test.

Pipe, steel, black and hot-dipped, zinc-coated welded and seamless. Steel pipe should meet the standard specifications of ASTM A 53.

Pipe, steel, black and hot-dipped zinc-coated (galvanized) welded and seamless for ordinary uses. Steel pipe should meet the standards of ASTM A 53.

Preaction and deluge valves. The alarm apparatus for preaction and deluge systems consists of approved electric alarm attachments actuated by a thermostatic system independently of the flow of water in the system. A mechanical alarm (water motor gong) may also be required.

Preaction system. System employing automatic sprinklers attached to a piping system containing air that may or may not be under pressure, with a supplemental heat-responsive system of generally more sensitive characteristics than the automatic sprinklers themselves, installed in the same areas as the sprinklers. Actuation of the heat-responsive system, as from a fire, opens a valve that permits water to flow into the sprinkler piping system and to be discharged from any sprinklers that may be open.

Pressure gauges. Approved pressure gauges are installed in sprinkler risers above and below each alarm check valve.

Private hydrant. Hydrant provided on a private water system for the protection of private property.

Protective systems—equipment or apparatus. Automatic sprinklers, standpipes, carbon dioxide systems, automatic covers, and other devices used for extinguishing fires and for controlling temperatures or other conditions dangerous to life and property.

Pump suction pipe. Pipe that conveys the water from the city main or other source of supply to the fire pump.

Relief valve. An approved relief valve is provided between the compressor and controlling valve and is set to relieve at a pressure 5 lb in excess of the maximum air pressure that should be carried in the system.

Return bends. Where piping on wet systems is concealed with sprinklers installed in pendent position below a ceiling, return bends are required when the water supply to the sprinkler system is from a raw water source, millpond, or open-top reservoirs. Return bends are connected to the tops of branch lines in order to avoid accumulation of sediment in the drop nipples. In new systems the return bend pipe and fittings should be 1 in. in size. In revamping existing systems, where it is not necessary to retain sprinklers in the concealed space, $\frac{1}{2}$-in.-close nipples inserted in the existing sprinkler fittings may be used with 1-in. pipe and fittings for the other portion of the return bend. Where the water supply is potable, return bends are not required.

Risers. Vertical pipes supplying the sprinkler system.

Seamless copper tube. Tubing should meet the standards of ASTM B 75.

Seamless copper water tube. Tubing should meet the standards of ASTM B 88.

Siamese. Device on the exterior wall of a building or in another approved location, connected to a standpipe and equipped with multiple inlets for attachment of fire department hose.

Siamese connection. Two or more multiple inlet fittings installed on the outside of a building and connected to the standpipe main of an inside standpipe system for the use of the fire department only, to supply water to the system.

Sidewall sprinkler. Sprinkler designed to be installed on piping along the sides of a room instead of with the normal spacing.

Single-source supply. Automatic sprinkler system that is supplied from the public water supply or from a pressure tank.

Size of risers. Each system riser should be of sufficient size to supply all the sprinklers on the riser on any one floor of one fire section as determined by the standard schedules of pipe sizes. There should be one or more risers in each building and in each section of the building divided by fire walls. Where conditions warrant the sprinklers in an adjoining building or section cut off by fire walls may be fed from a system riser in another fire section or building.

Smoke. Combination of gases, carbon, particles, and other products of incomplete combustion hindering respiration and obscuring visibility and access to the seat of a fire.

Solder metal. Solder metal should meet the standards of ASTM B 32.

Special sprinkler systems. Systems employing limited water supplies, reduced pipe sizes, and other departures from the requirements for standard systems should not be classified as standard sprinkler systems. Systems of this type may include those pressurized with air or nitrogen. Authorities having jurisdiction may recognize the degree of protection afforded by special types of sprinkler systems.

Spray sprinkler. Automatic sprinkler designed to control fire by the spray principle of heat absorption. Sprinklers discharge water in a horizontal pattern to absorb heat below the ceiling.

Sprinkler. Automatic sprinkler or an automatic sprinkler system for extinguishing fires.

Sprinklered

1. Equipped with an approved and properly maintained automatic sprinkler system.

2. Term that refers to a building or portion thereof that is protected by an automatic sprinkler system.

Sprinkler head. Device connected to a piping system. When it is activated by physical or chemical conditions generated by a fire, it will open to permit the discharge of an extinguishing agent on the fire.

Sprinkler spacing. Distribution of automatic sprinklers to provide the number of square feet of coverage specified for light hazard, ordinary hazard, and extrahazardous locations.

Sprinkler stopper. One of several devices for stopping the flow from individual sprinklers.

Sprinkler system

1. Complete automatic sprinkler system installed in compliance with generally accepted standards.

2. For fire protection purposes a sprinkler system is an integrated system of underground and overhead piping designed in accordance with fire protection engineering standards. The system includes a suitable water supply, such as a gravity tank, fire pump, reservoir, or pressure tank, and/or connection by underground piping to a city main. The portion of the sprinkler system above ground is a network of specially sized or hydraulically designed piping installed in a building, structure, or area, generally overhead, to which sprinklers are connected in a systematic pattern. The system includes a controlling valve and a device for actuating an alarm when the system is in operation. The system is usually activated by heat from a fire and discharges water over the fire area.

3. Arrangement of pipes and sprinkler heads designed and constructed to release a fire-extinguishing agent automatically when a fire occurs.

4. System of piping connected to one or more approved sources of water supply and provided with distributing devices so located and arranged as to discharge or

diffuse an effective stream or spray of water over the interior of a building, part of a building, or the exterior of a building.

Sprinkler system—thermostatic. An open- or closed-head sprinkler system operated through an auxiliary thermostatic device that functions at a predetermined rate of temperature rise.

Sprinkler system water flow alarm and supervisory signal service. Sprinkler system with signaling attachments for indicating the flow of water in the system and off-normal condition of the sprinkler system components that may adversely affect the performance of the system. Alarm and supervisory signals are transmitted to a signal reception system.

Standpipe

1. Pipe exclusively for fire-fighting purposes extending vertically throughout all stories of a building, including the basement, having one or more inlets close to the grade level and on or adjacent to those exterior walls of a building that face a street or alley, and having outlets on every story and basement.

2. Vertical water pipe riser used to supply fire hose outlets in buildings. First-aid standpipes provided for use by building occupants have hose and small nozzles and are under domestic or private water pressure. Fire department standpipes have $2\frac{1}{2}$-in. hose outlets and are supplied by fire department siamese connections near the ground level.

3. Wet or dry fire line installed exclusively for fighting fires, extending from the lowest to the topmost story of a building or structure with hose outlets at every floor, equipped with reducing valves, and designed to operate at required working pressures.

Standpipe during building construction. Every building six stories or more in height should be provided with one standpipe for temporary fire department use during construction, when topmost construction reaches several stories in height. Such a standpipe should be provided with fire department inlet connections and approval for the accessible location and located adjacent to usable stairs.

Standpipe—first-aid. An auxiliary vertical or horizontal fire line designed primarily for emergency use by the occupants of the building or by the private fire brigade before the arrival of the local fire department.

Standpipe main. Pipe that conveys the water from the fire pump to the standpipe risers.

Standpipe risers or standpipe. Vertical pipe that extends upward through a building and conveys the water from the standpipe main to the hose outlets.

Standpipe system

1. Approved installation of piping and appurtenances whereby all parts of a building can be quickly reached with an effective stream of water.

2. System of wet or dry piping, including the necessary appurtenances, within a building or structure.

3. Pipeline installed exclusively for fighting fire, extending from the lowest to the topmost story of a structure, with hose outlets at every story, and equipped with manually operated valves to permit the flow of water at pressures required by code.

Supervised sprinkler system. A system in which all water supply, valves, and accessory equipment are provided with electrical contact devices to transmit signals to an outside central supervisory station.

System main. Pipe that conveys the water from the tank riser or water supply to the system risers.

System riser. Vertical pipe that extends upward through a building and conveys the water supply or system main to the branch lines.

Tank heater. Device for heating the water in a gravity tank, pressure tank, or tank riser to prevent the water in those portions of a sprinkler system from freezing.

Tank riser. Pipe that conveys the water from a gravity or pressure tank to the system main or sprinkler system.

Test gasket. In testing extensions to old systems a special type of self-indicating blank is used whenever a blank gasket has been used for testing purposes. A testing blank has lugs painted red protruding beyond the flange in such a way as to clearly indicate its presence.

Test pressure. New systems, including yard piping, are tested hydrostatically at not less than 200-psi pressure for 2 hours, or at 50 psi in excess of the maximum static pressure when the maximum static pressure is in excess of 150 lb.

Thermostatic sprinkler system. Open- or closed-head sprinkler system operated through an auxiliary thermostatic device that functions at a predetermined rate of temperature rise.

Two-source supply. An automatic sprinkler system can be supplied from a combination of public water supply and pressure tank, from two pressure tanks, or by direct connections to the public water supply on two streets in which the water mains are separately controlled.

Wall hydrant. Two or more multiple-outlet fittings installed on the outside of a building and connected to the standpipe main of an inside standpipe system for the use of the fire department only, to obtain an additional supply of water from the fire pump in the building.

Wall sprinkler. Automatic sprinkler designed for placement near a wall and arranged to project a spray into the area protected. It is technically known as a "sidewall sprinkler."

Water curtain. System of sprinkler heads installed above openings in floors, walls, or partitions. When the system is activated, a supply of water is dispensed adjacent to the opening.

Waterspray sprinkler head. Nozzle has an external deflector type head that produces a filled cone of small water droplets.

Water hammer. Where connections are made from water mains, subject to severe water hammer (especially where pressure is in excess of 100 lb), it may be desirable to provide either a relief valve, properly connected to a drain, or an air chamber in the connection. If an air chamber is used, it should be located close to where the pipe comes through the wall and on the supply side of all other valves and so located as to take the full force of the water hammer. Air chambers having a capacity of not less than 4 cu ft should be controlled by an O. S. & Y. gate valve and provided with a drain at the bottom and also an air vent with control valve and plug to permit inspection by local authorities.

Water supply piping in an inside standpipe system. Water supply piping from the source of supply to hose outlets.

Water supply piping in a sprinkler system. Water supply piping from the sources of supply to sprinkler heads.

Wet-pipe sprinkler system. Sprinkler system in which all piping is maintained filled with water and connected to an approved source or sources of water supply.

Wet-pipe system

1. System employing automatic sprinklers attached to a piping system containing water and connected to a water supply so that water discharges immediately from sprinklers opened by a fire.

2. System of automatic sprinklers in which all pipes are filled with water or another extinguishing agent at all times.

Wet standpipe

1. Standpipe connected to an open or automatically provided source of water with water available at all outlets at all times.

2. Auxiliary fire line system with a constant water supply, installed primarily for emergency fire use by the occupants of the building.

3. Standpipe fire line having a primary water supply constantly available at every hose outlet, or made available by opening the hose outlet or by automatic functioning of a control station.

Wet standpipe system. Standpipe system that is connected to the public water supply system or to a source of water conforming to fire code requirements.

Wrought seamless copper and copper-alloy tube. Tubing should meet the standards of ASTM B 251.

Wrought steel pipe. Pipe should meet the standards of ANSI B36.10.

Zone. A vertical division of a building fire standpipe system used to establish the water working pressures within the system and also to limit the pressure at the lowest hose outlet in the zone.

VENTILATION AND AIR DISTRIBUTION

Air change. Quantity of ventilation air in cubic feet per hour of minute divided by the volume of the space, gives the number of air changes during the selected interval of time. The recommended number of air changes is usually required by local codes.

Air cleaner. Device which is designed to remove airborne impurities.

Air diffuser. Circular, square, or rectangular shaped air distribution outlet, mounted in a frame to fit in a recessed space or be surface mounted in a wall, ceiling or floor. The diffuser includes fins for deflecting and directing the flow of air.

Air filter. A screen that prohibits the passage of fine airborne particles from infiltrating the air stream delivered to a space.

Air outlet. Any opening through which air is delivered to a space.

Attic fan. Primarily an exhaust fan installed in the topmost story of a building to discharge the accumulation of hot air.

Attic ventilators. Openings provided in gable ends of roofs with louvers or the use of exhaust fans placed behind the louvers.

Auxiliary ventilating openings. The free area when louvers, dampers, or other devices are in position to deflect or diffuse the air currents in such a manner that there will be no objectionable drafts.

Axial flow fan. Fan unit with a disc or air-foil shaped blade mounted on a shaft and which moves the air in the general direction of the shaft axis.

Backdraft damper. Multiblade damper to prevent outside air from passing through the unit in reverse direction when the fan is not operating.

Bird screen. Screen ($\frac{1}{2}$-in. mesh) installed behind the intake louver to prevent entrance of birds into the housing.

Blower. A fan used to force air under pressure into an affected area.

Blowers. Centrifugal or squirrel cage blowers installed in ceilings or soffits, using a round duct for air passage through the roof or horizontally through a wall.

Borrowed air. Room air drawn by fan or gravity ventilation from another room or space.

Ceiling fans. Fan units that are mounted in ceiling space between joists and discharge vertically, through a round duct through the roof construction.

Central fan system. Fan unit located at a central location and conveying air supply through a duct system.

Centrifugal fan. Fan rotor or wheel within a housing and including motorized driving mechanism, either directly connected to the shaft or by belt drive.

CFM. Cubic feet per minute, used to designate the volume rate of air flow.

Court vent. Inner court solely for providing light and ventilation for rooms facing the court.

Damper. Bladed device used to vary the volume of air passing through an inlet or outlet and also to direct the air currents.

Duct. A fabricated conduit used to deliver air to a given space.

Duct fan. Fan unit mounted inside of a duct.

Duct systems. Duct systems employing mechanical means for the movement of air and used for heating and ventilating, including warm-air heating systems, plain ventilating systems, combination heating and ventilating systems, air-cooling systems, air-conditioning systems, and exhaust systems.

Duct (ventilation). A pipe, tube, conduit, or an enclosed space within a wall or structure, used for conveying the air.

Effective area. The net area of an outlet or inlet device through which air can pass; it is equal to the free area of the device times the coefficient of discharge.

Exhaust air. Air removed from a ventilated space and discharged to the outside without recirculating any of it.

Exhaust duct. Duct through which air is conveyed from a room or space to the outdoors.

Exhauster. A fan used to withdraw air from an affected area under suction.

Exhaust fan-heater. Combination exhaust fan and heater used primarily in bathrooms.

Exhaust fan-heater-light. Combination exhaust fan unit that includes a heater and provides a lighting fixture for use in bathrooms.

Exhaust opening. Opening through which air is removed from a space.

Exhaust system. Any combination of building construction, machinery, devices, or equipment, so proportioned, arranged, maintained, and operated that gases, dusts, fumes, vitiated air, or other materials injurious to health are effectively withdrawn from the breathing zone of employees and frequenters, and disposed of in a proper manner.

Exhaust Ventilation. Ventilation of a space by drawing exhaust air from it by mechanical or gravity pull, thus causing outside air or other air supply to be drawn into the space for ventilation.

Exhaust ventilation system. Any combination of construction, machinery, devices or equipment, designed and operated to remove harmful gases, dust, fumes, or vitiated air, from the breathing zone of the persons using the space.

Exterior fan. Fans mounted on the exterior of buildings, that pull rather than push the air.

Fan

1. Assembly comprising blades or runners and housings or casings. It is either a blower or exhauster.
2. Equipment providing air movement by the action of a wheel or shaft with blades enclosed in a housing or mounted on an orifice plate.

Fire damper. Damper arranged to seal off the airflow automatically through part of an air duct system, so as to restrict the passage of heat. A fire damper may also be used as a smoke damper if the location lends itself to the dual purpose.

Free area. Total minimum area of openings in an air inlet or outlet through which air can pass.

Free delivery-type fan unit. Unit that takes in air and discharges it directly to the space without external elements that usually impose air resistance.

Fresh air. Outdoor air.

General ventilation. Ventilation in which air is supplied to or removed from any area.

Gravity exhaust ventilation. A process of removing air by natural means, the effectiveness depending on atmospheric condition, such as difference in relative density, difference in temperature or wind motion.

Gravity ventilation. Exhaust ventilation of a space by utilizing the exhaust pressure of a column of air having a different temperature than the outside air or the induction effect of wind velocity across a roof ventilator, or both.

Impeller. Device that rotates and causes the air to move through it.

Infiltration. Leakage of air into a building through doors and windows and through the cracks around them.

Local exhaust ventilation. Ventilation in which dusts, fumes, vapors, gases, and mists are removed from the atmosphere near the sources of their generation.

Louver. Assembly of sloping vanes or blades designed to allow passing or air through its openings and prevent the intake of snow or water.

Mechanical ventilating exhaust system. System for removing air from a room or space by artificial means combined with a supply of air through windows, skylights, transoms, grilles, shafts, ducts, or other relief openings direct to the outside.

Mechanical ventilating supply system. System for forcing air into a room or space by artificial means combined with the removal of air through windows, skylights, transoms,

undercut doors, grilles, shafts, ducts, or other relief openings direct to the outside.

Mechanical ventilation. Mechanical process for introducing fresh air or for providing changes of air in a building or structure.

Natural ventilation. Air supplied to a space, usually through open windows, by natural air currents or low-thermal-difference movement. The air is generally exhausted by other windows or exfiltration, without special provision. There may be casual circulation of room air.

Odor removal. Dilution of odors by direct exhaust ventilation with the use of air washer and carbon filters. Continuous odors, which may be unavoidable, may be masked by use of ozone treatment and aerosol automatic spray systems.

Openings—screened. Any openings in exterior walls of dwellings such as those used for ventilation should be completely closed off with noncorrodible 16 mesh wire cloth in metal frames.

Open shaft. Shaft extending through the roof of a structure and open to the outer air at the top.

Outlet velocity. Average discharge velocity that is discharged at the outlet.

Outside air. Uncontaminated new or fresh air introduced from outside points into a building for ventilating spaces.

Outside air opening. Opening used as an entry for air supply from the outdoors.

Plenum. Air compartment or chamber to which one or more ducts are connected and that forms part of an air distribution system.

Plenum chamber. An air compartment or enclosed space to which one or more distributing air ducts are connected.

Plenum floor system. The use of space between the floor structural system with earth covered by non-porous vapor barrier for the distribution of conditioned air.

Portable ventilating equipment. Ventilating equipment that can be readily transported from place to place without any portion thereof being dismantled. It has no duct connection.

Power-type venting system. System that depends on a mechanical device to provide a positive draft within the venting system.

Propeller fan. Propeller or disc type wheel within a mounting ring or plate and including motorized driving mechanism, either directly connected or by belt drive.

Recirculated air. The transfer of air from a space through the air-handling equipment and back to the space.

Required window. Window that provides all or part of the code required natural light and ventilation in the room or space in which it is located.

Return air. Air returned to a ventilating unit for the purpose of reuse by recirculating it to the same or other occupied spaces.

Return (or exhaust opening). Any opening for the sole purpose to remove air from any space being heated, ventilated or air-conditioned.

RPM. Revolutions per minute, used to designate motor or impeller speed.

Saturated air. Air containing saturated water vapor with both air and vapor at the same dry-bulb temperature.

Shaft. An enclosed vertical or inclined space for the transmission of light, air, or persons through one or more stories of a building, and connecting two or more openings in successive floors or in a floor or floors and a roof; but not including ducts forming an integral part of a heating or ventilating system, or of a blower or exhaust system.

Skylights. Glazed metal frames manufactured in various shapes and sizes, or custom-made to fit into the roof of a building to provide light, and air. Glass or plastic may be used for glazing, subject to the approval of the building department.

Smoke damper. Damper arranged to seal off airflow automatically through a part of an air duct system so as to restrict the passage of smoke. The smoke damper may be a standard louvered damper serving other control functions if the location lends itself to the dual purpose. A smoke damper does not need to meet all the requirements of a fire damper unless required by code.

Squirrel cage fan. Centrifugal blower with forward-curved blades.

Standard air. Air with a density of 0.075 lb/cu ft substantially equivalent to dry air at 70°F and a barometric pressure of 29.92 in. of mercury.

Supply air. Air introduced into a space for ventilation, heating, or cooling. It may be moved by fan or gravity and may be all outside air or a mixture of return (inside) air and outside air.

Tempered air. Air transferred from a heated or cooled area of a building.

Tempered outside air. Outside air heated or cooled before distribution.

Throw. The horizontal or vertical axial distance an air stream travels after leaving the air outlet.

Transoms. Solid or glazed operating sash above windows or door openings to provide ventilation between spaces.

Tubeaxial fan. Propeller or disc type wheel within a cylinder including motorized driving mechanism for direct drive or by belt drive with motor supports mounted on the cylinder.

Vaneaxial fan. Disc type wheel within a cylinder, set of guide vanes located either before or after the wheel, and including motorized driving equipment mounted on the cylinder.

Vaned outlet. Register or grille equipped with vertical and/or horizontal adjustable vanes.

Ventilated. Provided with a means to permit circulation of air sufficient to remove an excess of heat, fumes, or vapors.

Ventilating plenum ceiling. Suspended ceiling containing many small apertures through which air, at low pressure, is forced downward from an overhead plenum dimensioned by the concealed space between the suspended ceiling and the floor or roof above.

Ventilating openings. In any room or space, apertures opening upon a public way, yard, court, public park, public waterway, or onto a roof of a building or structure in which the room or space is situated. They may be windows, skylights, transoms, or other openings that are provided for ventilating purposes and that are equipped with adjustable louvers, dampers, or other devices to deflect or diffuse the air currents. Windows and doors are considered ventilating openings in living quarters.

Ventilating system. Equipment used for the purpose of supplying air to or removing air from room or space.

Ventilation. Process of supplying or removing air by natural or mechanical means to or from any space. Such air may or may not have been conditioned.

Ventilation exhaust system. A complete system, including all hoods, ducts, fans, separators, and receptacles where required, and any other part necessary to the receptacles, and any other part necessary for the proper installation and operation thereof.

Ventilation exhaust system refuse receptacle. Part of the exhaust system into which dust or other material separated from the air is deposited.

Ventilation exhaust system separator. Part of an exhaust system in which the contaminant or entrained material is separated from the air that conveys it.

Ventilation fan. The machine that creates the movement of air in a mechanical ventilation system.

Ventilation gravity system. Ventilation depending wholly upon relative air density.

Ventilation (mechanical). The process of supplying or removing air by power-driven fans or blowers.

Ventilation of basements. Basements, and all rooms located therein except storage rooms should be lighted and ventilated by windows in exterior walls having both a glass and ventilation area of not less than the code requirements for the room or space floor area.

Ventilation or exhaust duct. Any tube or duct forming a part of an exhaust or ventilation system, used to convey air, dusts, fumes, mists, vapors or gases.

Ventilation system. All equipment intended or installed for the purpose of supplying air to or removing air from a room or space by mechanical means, other than equipment that is a portion of a comfort heating, cooling, absorption, or evaporative cooling system.

Vent shaft. Court used only to provide ventilation or light for indoor rooms.

Wall fan. Fan installed through the exterior wall of a building with weatherproof hood at the exterior and a face outlet grille on the interior side.

WARM-AIR HEATING SYSTEMS AND EQUIPMENT

Air bunching. Accumulation of air around a ventilator due (usually) to the inadequacy of the extractor.

Air-circulating blower. Complete blower assembly, including the blower wheel (or fan), blower housing, and motor and drive, used to provide the means for the circulation of air in a forced-air furnace.

Air duct. Tube, conduit, or enclosed space used for conveying air.

Air filter. Device designed to remove solids from the heating, cooling, or ventilating air.

Attic furnace. Forced-warm-air furnace designed and approved specifically for installation in a normally unoccupied attic.

Auxiliary equipment. Equipment used in connection with combustion equipment such as stokers, burners, or draft regulations.

Blower control (fan control). Temperature-actuated switch controlling only the on-off operation of an air-circulating blower or fan.

Breeching. Metal connector to chimney or stack to provide exhaust facilities for medium- and high-heat appliances.

Bridge wall. Baffle wall at the forward end of the ignition chamber.

Btu. Abbreviation for British thermal unit, the amount of heat required to raise the temperature of 1 lb of water 1° F.

Burner. Device for the final conveyance of a gas-air mixture to the combustion zone.

Casing. Jacket of a heating or cooling unit.

Central fan system. Indirect system of heating air. Air is heated by steam or hot water at a central location and distributed to the rooms to be heated by a fan and a system of ducts.

Central furnace

1. Self-contained direct-fired appliance intended primarily to supply heated air through ducts.

2. Furnace utilizing ducts, intended for heating rooms or spaces separate from the room in which the furnace is located.

Central heating boilers and furnaces. Heating furnaces and boilers include warm-air furnaces, floor-mounted direct-fired unit heaters, hot-water boilers, and steam boilers operating at not in excess of 15-psig pressure, used for heating buildings or structures.

Central warm-air heating system. Heating system consisting of a heat exchanger with an outer casing or jacket; electrical heating unit together with its supply and return duct system.

Chimney connector. Pipe or breaching that connects the heating appliance to the chimney.

Combustion air. Sufficient air for the complete combustion of fuel.

Combustion chamber.

1. Space forward of the mixing chamber where combustion is completed.

2. Space in a heating appliance provided for the combustion of fuel.

Comfort heating equipment. All warm-air furnaces, warm-air heaters, combustion product vents, comfort heating air distribution ducts and fans, steam and hot-water piping, together with all control devices and accessories installed as part of or in connection with any comfort heating system or appliance regulated by local code.

Comfort heating system. Warm-air heating plant consisting of a heat exchanger enclosed in a casing, from which the heated air is distributed through ducts to various rooms and area. A comfort heating system includes the circulating air supply, conditioned air supply, and all accessory apparatus and equipment installed in connection therewith.

Convection. Transfer of heat from one place to another by virtue of warm air rising.

Conversion burner. Burner provided to convert a heating appliance from its designed fuel to a different fuel.

Counterflow furnace. Furnace through which the circulating air flows in the opposite direction to the flue gases.

Direct space-hating furnace. Furnace intended for heating the space in which the furnace is located.

Downflow-counterflow furnace. Forced-warm-air furnace designed with the airflow through the furnace in an essentially vertical path downward, with the air discharged at or near the bottom of the furnace.

Downflow furnace. Furnace in which the circulating air flows downward, discharging at or near the bottom of the casing.

Downflow-type central furnace. Furnace designed with the airflow in an essentially vertical path, discharging air at or near the bottom of the furnace.

Draft gauge. Instrument used to measure air movement.

Draft hood or draft diverter. Device attached to or made part of the vent outlet from an appliance and designed to ensure the ready escape of the products of combustion in the event of no draft, backdraft, or stoppage in the vent or flue beyond the draft hood. It prevents a backdraft from entering the appliance and neutralizes the effect of stack action of the flue upon operation of the appliance.

Draft indicator. Instrument used to indicate or measure chimney draft and combustion gases.

Draft regulator. Device that functions to maintain a desired draft in the appliance by automatically reducing the draft to the desired value.

Dry heating system. Heating systems that utilize air as a medium for conveying heat.

Duct

1. Tube or other provision for the passage of air, gases, or services.

2. Pipe for the transmission of air.

Duct and vent material. Ducts and vents should be constructed of aluminum, copper, monel metal, galvanized steel, cement-asbestos, or other approved noncombustible, corrosion-resistive materials of adequate strength, durability, and for the temperatures involved; and the seams must be made secure and substantially air and gas tight.

Duct furnace

1. Warm-air furnace normally installed in an air distribution duct to supply warm air for heating. The definition applies only to a warm-air heating appliance that depends for air circulation on a blower not furnished as part of the furnace.

2. Furnace designed for insertion or installation in a duct of an air distribution system to supply warm air for heating. The duct furnace depends for air circulation on a blower that is not a part of the furnace.

Duct system. Continuous passageway for the transmission of air, which includes the ducts, duct fittings, dampers, plenums, fans, and accessory air-handling equipment.

Enclosed furnace. Specific heating or heating and ventilating furnace incorporating an integral total enclosure and using only outside air for combustion.

Equivalent friction loss. Loss of pressure in a duct system due to changes in size or shape, bends, takeoff, or obstructions, expressed in terms of the length of straight pipe in which an equivalent friction loss would take place.

Fan-type floor furnace. Floor furnace equipped with a fan that provides the primary means for circulating air.

Fan-type vented wall furnace. Wall furnace equipped with a fan.

Fire damper. Damper assembly arranged to seal off the airflow automatically through part of an air duct system so as to restrict the passage of smoke, heat or fire.

Floor furnace. Self-contained flue-connected or vented furnace taking air for combustion outside this heated space, and with a means for observing flame and igniting the appliance from the space being heated.

Flue. Passage leading from the combustion chamber to the stack.

Flue pipe. Conduit or pipe, vertical or nearly so in direction, designed to convey all the products of combustion to the outside atmosphere.

Forced-air central furnace. Central furnace equipped with a fan or blower that provides the primary means for circulation of air.

Forced-air furnace. Furnace equipped with a blower that provides the primary means for circulation of air.

Forced-air heating system. Central warm-air heating system equipped with a fan or blower that provides the primary means for circulation of air.

Forced-air-type central furnace. Central furnace equipped with a fan or blower that provides the primary means of air circulation.

Forced-warm-air furnace. Furnace equipped with blower to provide the primary means for circulating air.

Forced warm-air heating plant. A plant that consists of one or more warm air furnaces, enclosed within casings, together with necessary appurtenances thereto, consisting of warm air pipes and fittings, cold air or recirculating pipes, ducts, boxes and fittings, smoke pipes and fittings, registers, and grilles, intended for means of providing warm air, intended for the heating of buildings in which they may be installed.

Free air. Air at atmospheric conditions at any specific location. Because altitude, barometric pressure, and temperature vary at different localities and at different times, it follows that this term does not mean air under identical or standard conditions. Free air as a measure of volume may be applied either to displacement or capacity and in no way distinguishes between these two.

Furnace. Completely self-contained direct-fired, automatically controlled, vented appliance for heating air by transfer of heat of combustion through metal to the air and designed to supply heated air through ductwork to spaces remote from the appliance location.

Furnace (warm-air furnace). Space-heating appliance using warm air as the heating medium and usually having provision for the attachment of ducts.

Furnace room. Room primarily used for the installation of heating equipment.

Gauge (gage). U.S. standard for measuring the thickness of sheet plate iron and steel.

Gravity furnace. Furnace that depends on the difference in density of warm and cool air for circulation. The furnace may also use an integral fan or blower to overcome the internal furnace resistance to airflow.

Gravity heating system

1. Central warm-air heating system through which air is circulated by gravity.
2. Comfort heating system consisting of a gravity-type warm-air furnace, together with all air ducts or pipes and accessory apparatus installed in connection with it.

Gravity-type central furnace. Central furnace depending primarily on circulation of the air by gravity.

Gravity-type central furnace with booster fan. Central furnace equipped with a booster fan that does not materially restrict free circulation of the air by gravity flow when the fan is not in operation.

Gravity-type central furnace with integral fan. Central furnace equipped with a fan or blower as an integral part of its construction and operable on gravity systems only. The fan or blower is used only to overcome the internal furnace resistance to airflow.

Gravity-type floor furnace. Floor furnace depending primarily on circulation of the air by gravity. This classification also includes floor furnaces equipped with booster-type fans which do not materially restrict free circulation of the air by gravity flow when such fans are not in operation.

Gravity-type vented wall furnace. Wall furnace depending on circulation of air by gravity.

Gravity-type venting system. System that depends entirely on the heat from the fuel being used to provide the energy required to vent an appliance.

Gravity-type warm-air furnace. Warm-air furnace depending primarily on circulation of air through the furnace by gravity. This definition also includes any furnace approved with a booster-type fan, which does not materially restrict free circulation of air through the furnace when the fan is not in operation.

Gravity warm-air heating plant. One or more warm-air furnaces, enclosed within casings, together with the necessary appurtenances, consisting of warm-air pipes and fittings, cold-air or recirculating pipes, ducts, boxes and fittings, smoke pipes and fittings, registers, borders, faces and grilles, intended for the heating of buildings, in which they may be installed.

Gravity warm-air heating system. Warm-air heating system that depends on the difference in density between warm and cold air for the circulation of air throughout the system.

High-velocity or high-pressure systems. Air systems in which the duct velocities and static pressures are such that special control and acoustic equipment are required for proper introduction of the air into the space to be served.

High limit control. Device responsive to change in temperature and used for the purpose of interrupting the fuel supply when plenum or bonnet temperatures exceed predetermined settings.

Hood

1. Air-intake device connected to a mechanical exhaust system for collecting vapors, fumes, smoke, dust, steam, heat, or odors from, at, or near the equipment, place, area where generated, produced, or released.
2. Canopy device placed over a stove, range, or other heating installation connected to a ventilating duct.

Horizontal furnace. Furnace in which the circulating air flows horizontally with the air inlet and discharge openings at opposite ends of the casing.

Horizontal-type central furnace. Furnace designed for low headroom installation with the airflow through the appliance in a horizontal path.

Humidifier. Device used to add moisture to the air.

Ignition chamber. Portion of the furnace directly above the grates and hearth.

Incombustible (noncombustible) material. Material that will not ignite at or below a temperature of 1200° F during an exposure of 5 minutes and that will not continue to born or glow at that temperature, according to tests as required.

In-line furnace. Furnace in which the air inlet and discharge openings are in the top of the casing.

Limit control. Thermostatic device installed in the duct system to shut off the supply of heat at a predetermined temperature of the circulated air.

Mechanical warm-air furnace. A warm-air furnace that is equipped with a fan to circulate the air.

Mechanical warm-air heating plant. One or more warm-air furnaces enclosed within casings, together with necessary appurtenances thereto, consisting of warm-air supply pipes and fittings, cold air or recirculating pipes, ducts, boxes and fittings, smoke pipes, dampers and registers, grilles, fans, or blowers, intended for heating the building in which they may be installed. The circulation of air within such a system is dependent upon the motive power furnished by a fan or blower, and the duct work in connection therewith is designed especially for such system. However, the incorporation of a booster fan, blower, or any power-driven device for the purpose of accelerating the air circulation in a gravity warm-air heating plant is construed as changing the classification of such gravity system to a mechanical system.

Mechanical warm-air heating system. A warm-air heating system in which circulation of air is effected by a fan. Such a system may include air cleaning devices, such as removeable or permanent type filters.

Pipeless furnace. Gravity furnace in which the entire heat output is delivered through one opening directly above the combustion chamber. Return air enters the unit through grille work around the perimeter of the outlet grille.

Plenum. Air compartment or chamber to which one or more ducts are connects and that forms part of either the supply or return systems.

Plenum chamber

1. Air compartment maintained under pressure and connected to one or more distributing ducts.

2. Compartment or chamber to which one or more ducts are connected and that forms a part of either the supply or return air system.

3. Chamber used to connect the furnace and ducts together.

Register. Combination grille and damper used on the end of a duct system to direct the airflow into a room.

Revertible-flue furnace. Furnace in which the upward direction of the flow of the flue gas is reversed before reaching the flue collar.

Safety pilot control. Device provided to shut off automatically the fuel supply to the main burner when the means of ignition of such fuel supply becomes inoperative.

Sealed combustion system appliances. Appliances constructed and installed so that all air for combustion is derived from the outside atmosphere and all flue gases are discharged to the outside atmosphere.

Shaft. Enclosed vertical or inclined space for the transmission of light and air, through one or more stories of a building, connecting two or more openings in successive floors or in a floor(s) and a roof, but not including ducts forming an integral part of a heating or ventilating system or of a blower or exhaust system.

Smoke damper. Damper arranged to seal off the airflow automatically through a part of an air duct system so as to restrict the passage of smoke.

Smoke pipe. Pipe or breeching connecting a heating appliance and flue.

Supply system. Assembly of connected ducts, plenums, fittings, registers, and grilles through which air, heated or cooled, is conducted from the exchanger to the space or spaces to be heated or cooled.

Suspended furnace. Horizontal furnace having provision for suspension.

Tuyere. Adjustable opening in a furnace door or wall that admits and controls combustion air.

Underfloor horizontal furnaces. Horizontal-flow forced-warm-air devices located under the floor of the heated structure and connected to a duct system.

Upflow furnace. Furnace in which the circulating air flows upward, discharging at or near the top of the casing.

Upflow-type central furnace. Furnace designed with the airflow in an essentially vertical path, discharging air at or near the top of the furnace.

Vent. Pipe designed to convey the products of combustion from an appliance to a flue or chimney.

Vented appliance. Appliance designed or installed in such a manner that the products of combustion are conveyed directly from the appliance, connected directly into a flue or chimney, and conducted to the outside atmosphere.

Venting collar. Outlet opening of an appliance provided for connection of the vent system.

Venting system. Vent or chimney and its connectors assembled to form a continuous open passageway from an appliance to the outside atmosphere for the purpose of removing products of combustion. This definition also includes the venting assembly, which is an integral part of an appliance.

Volume damper. Device that when installed will restrict, retard, or direct the flow of air in any duct or the products of combustion in any heat-producing equipment, vent conductor, vent, or chimney therefrom.

Wall furnace. Furnace installed in or on a wall or partition, supplying warm air through grilles, boots, or the equivalent without the use of ducts.

Warm-air all-year air-conditioning system. Includes a mechanical warm-air heating plant, together with such other devices and such automatic controls as will secure the simultaneous control of the temperature, motion, humidity, and a reduction in the dust and odor content, of the air employed in the ventilation of rooms. This includes both warming and humidifying in winter and cooling and dehumidifying in summer.

Warm-air furnace. Solid, liquid, or gas-fired appliance for heating air to be distributed with or without duct systems to the space to be heated.

Warm-air heating system. Heating plant consisting of an air-heating appliance from which the heated air can be distributed by means of ducts or pipes. It includes accessory apparatus and equipment installed in connection therewith.

Warm-air heating, ventilation, and air-conditioning systems in buildings. Air duct systems employing mechanical means, regardless of the type of fuel used, for the movement of air and used for warm-air heating, ventilating, and air-conditioning systems and combination heating and air-conditioning systems.

Warm-air winter air-conditioning system. Includes a mechanical warm-air heating plant, together with such other devices and such automatic controls as will secure the simultaneous control of the temperature, motion, humidity, and a reduction in the dust and odor content, of the air employed in the ventilation of rooms, but not provided with such devices and automatic controls as will provide for cooling and dehumidifying in summer.

ALARM AND WARNING SYSTEMS

Accidental alarm. Alarm device set off and transmitted through accidental operation of an automatic or manual fire alarm device, frequently caused by low air pressure on automatic sprinkler dry valves, excessive heat from industrial processes, or cold weather trouble.

Actuate. Set a fire alarm signal device or fire protection device into operation either automatically or manually.

Alarm. Signal device indicating an emergency requiring immediate action, as an alarm for fire from a manual box, a water-flow alarm, or an alarm from an automatic detection system; also, other emergency alarms.

Alarm services. Services rendered as a result of manual pulling of a fire alarm box or the transmission of an alarm indicating the operation of protective equipment or systems, such as an alarm for water flow from a sprinkler system, the discharge of carbon dioxide, the detection of smoke, or the detection of excess heat by a thermostat; also, alarms of other protective systems.

Alarm signal. Signal device indicating an emergency requiring immediate action as an alarm for fire from a manual box, a water-flow alarm, and an alarm from an automatic fire alarm system; other emergency signal.

Alarm signaling devices. Consist of three basic types: gongs, which are made in three standard sizes; horns, which are suitable for use in high-noise areas; and chimes, used in areas where a lower sound is necessary. Signaling devices may be surface mounted or recessed in suitable enclosures.

Annunciator. A unit containing two or more identified targets or indicator lamps in which each target or lamp indicates the circuit, condition, or location to be annunciated.

Audible alarm indicating devices. Should be of such character and so distributed as to be effectively heard above the ambient noise level obtained under normal conditions of occupancy, and should produce signals that are distinctive from audible signals used for other purposes in the same area, section or building.

Automated control console. When fire is detected by smoke, heat, infrared detectors, or water-flow indicators in sprinkler system piping, automated control systems immediately summon the fire department.

Automatic. Term applied to a fire protection device that functions without human intervention, such as an automatic sprinkler or an automatic fire alarm system.

Automatic alarm. Alarm device usually actuated by thermostats, sprinkler alarm valves, or other automatic devices, received over private alarm circuits that may go through a central station and be received on separate tapper or identified by a special signal. Automatic alarm signal; assignment dispatched to an automatic alarm.

Automatic alarm initiating device. Alarm initiating devices such as fire detectors, smoke detectors, and water-flow switches that automatically transmit an alarm signal when a condition indicative of a fire occurs to which they respond.

Automatic closing. Term describing a device that is normally held in an open position and automatically closes upon the action of a heat-actuated releasing or operating mechanism.

Automatic fire alarm system. System that automatically detects a fire condition and actuates a fire alarm signal device.

Automatic smoke alarm service. Smoke alarm service that consists of equipment designed and installed, whereby abnormal products of combustion other than heat results in the transmission of a distinctive smoke alarm signal to a signal reception system.

Auxiliary box. Fire alarm signal situation on public or private property used to actuate a master box that transmits an alarm to the fire department.

Auxiliary fire alarm system. System that is maintained and supervised by a responsible person or group having alarm initiating devices that, when operated, cause an alarm signal to be transmitted over a municipal fire alarm system to the fire station or to the fire alarm headquarters for retransmission to the fire station.

Auxiliary system. Fire alarm signal system wherein fire alarms are transmitted to a community fire alarm headquarters on the same equipment and by the same alerting methods as alarms transmitted from public fire alarm boxes located on streets.

Auxiliarized local system. A local system (NFPA 72A) that is connected to the municipal alarm facilities.

Auxiliarized proprietary system. A proprietary system (NFPA 72D) that is connected to the municipal alarm facilities.

Box circuit. Electrical signal circuit connecting manually or automatically operated fire alarm boxes and fire alarm detection devices with a fire alarm central station. In very small systems a box circuit may also serve as an alarm circuit to operate devices such as tape recorders, alarm bells, sirens, and air horns.

Breakglass. False alarm deterrent available in fire stations; a glass rod is placed across the pull-lever and breaks easily when the lever is pulled.

Central station system

1. Automatic sprinkler or fire alarm system in which all equipment is supervised by a central or proprietary station to which all alarm signals are transmitted and relayed to the municipal fire department.

2. System of electrically supervised circuits employing a direct circuit connection between signaling devices at the protected premises and signal-receiving equipment in a remote station, such as the city fire alarm headquarters or other location acceptable to the fire department.

3. System or group of systems, whose operation is signaled to, recorded in, maintained, and supervised from an approved central station, in which there are competent and experienced observers and operators in attendance at all times whose duty upon receipt of a signal is to take such action required under the rules established for their guidance. Systems controlled and operated by a person, firm, or corporation whose principal business is the furnishing and maintaining of supervised protective signaling service and that has no interest in the protected properties.

Class A fire alarm system. A closed circuit electrically supervised fire alarm system. System may be coded or selective code ringing type, a noncoded type with or without annunciators (automatic detection type), or combination fire and sprinkler alarm.

Class B fire alarm system. A fire alarm system that may consist of one or more bells, sirens, or other sounding devices electrically or mechanically operated so that all sounding-devices will sound when any fire alarm station is operated.

Closed circuit system. System in which all circuits between alarm sending stations, sounding devices, and annunciators are maintained as normally closed electrical circuits.

Coded alarm signal. Alarm signal that represents a one-, two-, three-, or four-digit number indicative of the location of the fire alarm station operated.

Coded closed circuit. Circuit used for an interior fire alarm signal system consisting of sending stations and signaling devices operated on supervised closed electrical circuits where rounds of automatically sounded coded signals are transmitted to indicate the floor or portion of the same from which the alarm was sent.

Coded or selective code ringing. Fire alarm system where each sending station, when operated will cause sounding devices to signal a predetermined number of strokes, which is indicative of the location of the sending station.

Coded system. System in which not less than three rounds of coded alarm signals are transmitted, after which the fire alarm system may be manually or automatically silenced.

Combination system. A local protective signaling system for fire alarm, supervisory, or watchman service whose components may be used in whole or in part in common with a nonfire-emergency signaling system, such as a paging system, a musical program system, or a process monitoring service system, without degradation of or hazard to the protective signaling system.

Continuous ringing. Arranged in coded systems to the signal that sounds after the completion of the normal number of rounds (usually four) of identifying coded alarm signal.

Control panel with zone annunciator. Annunciator provides a visual indication of station or zone from which a fire alarm signal was initiated.

Direct circuit auxiliary alarm system. An auxiliary alarm system connected by a municipally controlled individual circuit to the protected property, to interconnect the actuating devices and the municipal fire alarm switchboard.

Double supervised system. System in which the source of power for the trouble signal is supervised, in addition to the circuitry.

Dual-coded system. System in which a unique coded alarm is sounded for each separate fire box or fire zone to notify owner's personnel of the fire location, while noncoded or common coded alarm signals are sounded on separate signals to notify other occupants to evacuate the building.

Early detection and alarm. System of fire detection that will give an alarm in the earliest stages of combustion.

Electrically supervised. A system so designed, installed, and maintained that, upon break or ground fault of its circuits or a failure of its main operating current supply source, a trouble signal will sound or the signal sounding devices of the system will be activated.

Fire alarm. A system, automatic or manual, arranged to give a signal indicating a fire emergency.

Fire alarm box. Manually operated alarm initiating device; may be equipped to generate a continuous signal (noncoded station) or a series of coded pulses (coded station).

Fire alarm—coded or selective code-ringing. A fire alarm system where each sending station, when operated, will cause sounding devices to signal a predetermined number of strokes, which indicates the location of the sending station.

Fire alarm—noncode system. A fire alarm system in which all sending stations when operated will cause sounding devices to sound the same signal.

Fire alarm panel control unit. Consists of the controls, relays, switches, and associated circuits necessary to furnish power to a fire alarm system, receive signals from alarm initiating devices and transmit them to indicating devices and accessory equipment, and electrically supervise the system circuitry.

Fire alarm system

1. Integrally supervised, electrically operated system consisting of manual stations that will actuate audible and/or visual alarm signals, or both, throughout the building or structure.

2. Approved installation of equipment for sounding a fire alarm.

Fire-indicating unit. Power control unit for zone system that indicates alarm and controls alarm devices. It supervises detector and audible device wiring, as well as critical internal circuits. It controls supplementary equipment such as blowers, dampers, and door releases and transmits trouble and fire signals to supervising station.

Indicator panel. Alarm panel indicating the source of a signal from automatic alarm or water-flow alarm in a specified area of a protected property.

Interior fire alarm system. System installed within a building and provided and intended to warn the occupants of the building in the event of a fire by means of an audible signal transmitted from manual trip, stations, signal stations, or boxes, or automatic fire detecting devices.

Leased line annunciator and power supply. Device that transmits fire and trouble signals from protected property to the central fire station via leased telephone lines. Telephone lines are supervised for shorts or openings.

Local alarm system. A local system sounding an alarm as the result of the manual operation of a fire alarm box or of the operation of protection equipment or systems, such as water flowing in a sprinkler system, the discharge of carbon dioxide, the detection of smoke or the detection of heat.

Local energy auxiliary alarm system. An auxiliary alarm system which employs a locally complete arrangement of parts, initiating devices, relays, power supply, and associated components, to automatically trip a municipal transmitter or master box over electric circuits which are electrically isolated from the municipal system circuits.

Local fire alarm system. Electrically operated system producing signals at one or more places at the premises served, primarily for the notification of the occupants.

Local noninterfering coded station. Fire alarm station that, once actuated, will transmit not less than four rounds of coded alarm signals and cannot be interfered with by any subsequent actuation of that station until it has transmitted its complete signal.

Local supervisory system. A local system arranged to supervise the performance of watch patrols, or the operative condition of automatic sprinkler systems, or of other systems for the protection of life and property against fire hazard.

Local system. Fire alarm signal system in which the operation and alarm signals are on the protected premises and that is designed to warn the occupants of the premises.

Manual alarm initiating device. Fire alarm station that will transmit an alarm signal when manually operated.

Manual fire alarm service. Manually actuated fire alarm service.

Manual fire alarm system. Interior alarm system composed of sending stations and signaling devices in a building, operated on an electric circuit, and so arranged that the operation of any one station will ring all signals throughout the building or at one or more approved locations. Signals may be either noncoded or coded to indicate the floor area in which the signal originated and may be transmitted to an outside central station.

Manual fire station. Station operated on the modified pull level principle. Instead of pulling the handle out and down, the entire face panel pushes down to set off the alarm.

Manual system. System in which the alarm initiating device is operated manually to transmit or sound an alarm signal.

Master box. A municipal fire alarm box that may also be operated by remote means.

Master coded system. System in which a common coded alarm signal is transmitted for not less than three rounds, after which the fire alarm system may be manually or automatically silenced. The same code is sounded regardless of the location of the alarm initiating device. This system can be programmed to ring continuously.

Mechanical fire alarm service. A manually actuated and operated mechanical device for sounding an alarm, and consists primarily of a vertical rod, tripping mechanism, spring, clapper, and bell, and wherein the alarm does not have continuity after actuation.

Municipal fire alarm box. A specially manufactured enclosure housing a transmitting device that can only be operated manually.

NFPA. National Fire Protection Association.

Noncoded closed circuit. Interior fire alarm signal system, consisting of sending stations and signaling devices operated on supervised closed electrical circuits so arranged that operation of any station will automatically sound the sounding devices throughout all portions of the building.

Noncoded fire alarm system. All sending stations when operated will cause sounding devices to sound the same signal.

Noncoded system. System in which a continuous audible alarm is transmitted for a predetermined length of time, after which it may be manually or automatically silenced.

Open circuit interior fire alarm system. System in which all circuits are maintained as electric circuits normally open at the alarm sending stations and that is operated by closing the circuit at any alarm sending station.

Positive noninterfering and succession coded station. Fire alarm station that, once actuated, will transmit not less than four rounds of coded alarm signals without interference from any other station on the circuit. One or more of these stations, if subsequently operated, will transmit not less than four rounds of their coded signals without interference with each other or with the first station actuated.

Power alarm annunciator. Device that provides alarm in case of power failure by means of battery-operated alarm bell and indicator lamp.

Presignal system. System in which the operation of an automatic detector or the first operation of a manual fire alarm station actuates only a selected group of alarm indicating devices for the purpose of notifying key personnel. A general alarm may be sounded on these same indicating devices and on an additional group of devices from any manual station, to warn all occupants.

Process monitoring alarm system. An alarm system used to supervise the functioning of a commercial process, such as manufacturing operations, heating or refrigerating systems temperature control, etc., when failure of the supervised process could result in fire or explosion endangering life or property.

Proprietary fire alarm system. System with supervision by competent and experienced observers and operators in a supervising station at the property to be protected.

Proprietary system. Fire alarm signal system supervised by competent and experienced personnel in a central supervising station at the property protected, in which the personnel take such action as required under the rules established for their guidance.

Protective signaling system. Electrically operated circuits, instruments, and devices, together with the necessary electrical energy, designed to transmit alarms and supervisory and trouble signals necessary for the protection of life and property.

Punched tape recorder. Device for tape recording the actuation of a fire alarm initiating device. The station or zone code is punched on the tape.

Remote station fire alarm system. System of electrically supervised devices employing a direct-circuit connection between alarm initiating devices or a control unit in protected premises and signal-indicating equipment in a remote station, such as a fire or police headquarters.

Remote station system. Fire alarm system employing a direct-circuit connection between the alarm signal initiating devices in protected premises and signal-receiving equipment in a remote station, such as fire or police headquarters or a fire station.

Selective coded system. System in which each manual fire alarm station and each group of automatic detectors has its own individual code, which sounds on all alarm indicating devices in the system when the manual station or automatic detector is actuated.

Shunt auxiliary alarm system. An auxiliary alarm system electrically connected to an integral part of the municipal alarm system extending the municipal circuit into the protected property to interconnect the actuating devices, which, when operated, open the municipal circuit shunted around the trip coil of the municipal transmitter or master box, which is thereupon energized to start transmission, without any assistance whatsoever from a local source of energy.

Signaling equipment. Equipment intended to give an audible signal such as bells, horns, chimes, buzzers, and the like.

Single-stroke bell. Device whose gong is struck only once each time operating energy is applied to the bell.

Siren. A motor-driven siren produces a high-pitched wailing sound, readily distinguishable from horns, whistles, or other monotone audible devices and easily recognized through a confusion of other sounds.

Standard fire alarm system. Manually operated fire alarm system equipped with automatic detectors; system installed in a building for the purpose of notifying the occupants of the building of conditions due to fire or other causes that necessitate the building's being vacated immediately by the occupants. The system and all equipment and devices used in the installation of such a system must be tested by and bear the label of approval of a nationally recognized testing laboratory. Equipment and quality of installation must conform to the electrical requirements of the applicable code.

Stations. Devices for initiating the signal. Stations are either manually operable or automatically operable by temperature changes or rate of rise of temperature.

Supervised fire alarm system. That accidental interruption of current flow will be signaled as prescribed by the regulations.

Supervised sprinkler system. System instrumented in such a way that a supervisory signal will be transmitted when conditions essential to the proper operation of the sprinkler system are in jeopardy. Supervision includes but is not limited to control valves, water level and pressure in pressure tanks, water temperatures, fire pump and booster pump power; it does not include conditions related to public water mains, tanks, reservoirs, and other containers of water controlled by a municipality or a public utility.

Supervised system. System in which a break or ground in the wiring that prevents the transmission of an alarm signal will actuate a trouble signal.

Supervisory service. Service required to ensure performance of watch patrols and the operative condition of automatic sprinkler systems and of other systems for the protection of life and property.

Supervisory signal. Signal indicating the need for action in connection with the supervision of watchmen and sprinkler and other extinguishing systems or equipment, or with the maintenance features of other protective systems.

Trouble signal. Signal indicating trouble of any nature, such as a circuit break or ground, occurring in the devices or wiring associated with a protective signaling system.

Voice communication system. An electrically supervised communicating system provides a two-way emergency communication system and a one-way address communication system.

Watchman's tour supervisory service. Consists of one or more watchmen with specific routes having reporting and signaling stations and facilities for keeping a permanent record at a supervising location of each time a signal transmitting station is operated.

Zone coded system. System in which the building has been divided into zones. Alarm initiating devices in each zone activate a zone code that indicates only the location of the affected zone.

AUTOMATIC DETECTION SYSTEMS

Air duct detectors. Detectors designed for detecting fire and smoke in air-handling systems through an air-sampling arrangement. They are used to prevent the spread of fire and smoke throughout the system by shutting down fans and/or activating smoke exhaust dampers.

Automatic detection. Automatic detection may be by any listed or approved method or device capable of detecting and indicating heat, flame, smoke, combustible vapors, or airborne materials.

Automatic fire-detecting system. System that utilizes thermostatic or other approved detecting elements for the detection of fire and the automatic transmission of an alarm.

Automatic fire detection service. Fire detection service that consists of heat detection and other equipment installed throughout sufficient parts of the protected premises to detect automatically any fire condition on said premises and transmit a signal or alarm to a specific signal reception system.

Combination detector. Device that either (a) responds to more than one of the fire phenomena classified in the code or (b) employs more than one operating principle to sense one of these phenomena. Typical examples are (a) a combination of a heat detector with a smoke detector, or (b) a combination rate-of-rise and fixed temperature heat detector.

Combination door-holder and closer. Device provides door-holding automatic release by remote signal from a detector, in conjunction with a fire alarm system.

Cloud chamber smoke detector (sampling). The air pump draws a sample of air into a high humidity chamber within the detector. After the air is in the humidity chamber, the pressure is lowered slightly. If smoke particles are present, the moisture in the air condenses on them, forming a cloud in the chamber. The density of the cloud is measured by the photoelectric principle. When density is greater than a predetermined level, the detector responds to the smoke.

Detection systems. Each type of detector and each subtype has its own unique advantages and disadvantages. The correct choice depends on the use of the space and the location of the detectors. In most installations, both heat and smoke detectors generally are necessary, and in certain conditions, flame detectors should be installed. Manufacturers should be consulted before selection is made.

Duct detectors. The function of air duct smoke detectors is to detect smoke for the primary purpose of controlling blowers and dampers of air-conditioning and ventilating systems in an attempt to prevent possible panic and damage from distribution of smoke and gaseous products.

Electromagnetic holder / release. Device that holds doors in the open position until the circuit is broken by a signal from a detector, and the doors are released to close.

Filter dust detector. Device that detects the presence of products of combustion in industrial lint collectors and dust filters.

Fire and smoke-detecting system. An approved installation of equipment that automatically actuates a fire alarm when the detecting element is exposed to fire, smoke, or abnormal rise in temperature.

Fire characteristics. Flame detectors are sensitive to glowing embers, coals, or actual flames, which radiate to the detectors energy of sufficient intensity and spectral quality to initiate action.

Fire-detecting system. Approved installation of equipment that automatically actuates a fire alarm when the detecting element is exposed to fire or abnormal rise in temperature.

Fire detection system

1. System for detecting the presence of abnormal fire or heat. Various means are employed including rate of temperature rise, fixed temperature settings, a combination of these, or electrical detection of smoke or flame.

2. Integrally supervised, electrically operated system of either smoke- or heat-responsive devices, so connected by wiring or tubing that either will operate an audible and/or visual alarm system throughout the building or structure.

Fixed temperature detector. Device that will respond when its operating element becomes heated to a predetermined level.

Flame detector. A device that detects the infrared, or ultraviolet, or visible radiation produced by a fire. A device that responds to the appearance of radiant energy visible to the human eye (approximately 4000 to 7700 Å) or to radiant energy outside the range of human vision.

Flame fire detector. Detector that instantly senses infrared radiation emanating from flames, intended for areas of fast-developing fires where ignition is almost instantaneous, high-ceiling areas, or high-air-movement areas. To prevent false alarms, the unit is sensitive only to flame flickering sustained 3, 10, or 30 seconds, depending on the detector selected.

Flame flicker detector. A photoelectric flame detector including means to prevent response to visible light unless the observed light is modulated at a frequency characteristic of the flicker of a flame.

Heat. Added energy that causes substances to rise in temperature. Also, the energy liberated by a burning substance.

Heat detector. Device that detects abnormally high temperature or rate-of-temperature rise.

Infrared detector. A device whose sensing element is responsive to radiant energy outside the range of human vision (above approximately 7700 Å).

Ionization detector. Detector has a small amount of radioactive material that ionizes the air in the sensing chamber, thus rendering it conductive and permitting a current flow through the air between two charged electrodes. This gives the sensing chamber an effective electrical conductance. When smoke particles enter the ionization area, they decrease the conductance of the air by attaching themselves to the ions, causing a reduction in mobility. When the conductance is less than a predetermined level, the detector circuit responds.

Ionization fire detectors. Fire detectors that operate on the ionization principle and react to combustion gases in the first stage of fire. They do not require heat, flame, or visible smoke to operate.

Line-type detector. Device in which detection is continuous along a path. Typical examples are rate-of-rise pneumatic tubing detectors, projected beam smoke detectors, and heat sensitive cable.

Nonrestorable detector. Device whose sensing element is designed to be destroyed by the process of detecting a fire.

Paddle-type detectors. Water-flow indicators (paddle type) should not be installed in dry-pipe, preaction, or deluge systems as the surge of water when the valve trips would seriously damage the device.

Photoelectric beam-type detector. The photoelectric beam-type detector consists of a light source that is projected across the area to be protected into a photosensing cell. Smoke between the light source and the receiving photosensing cell reduces the light reaching the cell, causing actuation.

Photoelectric fire detectors. Detectors that response directly to visible smoke, intended for areas where it is not practical to use ionization detectors owing to normal high ambient level of combustion gases or where the material protected will produce heavy smoke.

Photoelectric flame detector. Device whose sensing element is a photocell, which either changes its electrical conductivity or produces an electrical potential when exposed to radiant energy.

Photoelectric spot-type detector. Contains a chamber with either overlapping or porous covers that prevent the entrance of outside sources of light but that allow the entry of smoke. The unit contains a light source and a special photosensitive cell in the darkened chamber. The cell is either placed in the darkened area of the chamber at an angle different from the light path or has the light blocked from it by a light stop or shield placed between the light source and the cell. With the admission of smoke particles, light strikes the particles and is scattered and reflected into the photosensitive cell. This causes the photosensing circuit to respond to the presence of smoke particles in the smoke chamber.

Pneumatic rate-of-rise tubing. Line-type detector comprising small diameter tubing, usually copper, which is installed on the ceiling or high on the walls throughout the protected area. The tubing is terminated in a detector unit, containing diaphragms and associated contacts set to actuate at a predetermined pressure. The system is sealed except for calibrated vents that compensate for normal changes in temperature.

Projected beam-type smoke detector. Device that consists of a separate light source, projected across the area to be protected, and a photosensing cell. Smoke between the light source and receiving photosensing cell reduces the light reaching the cell, causing actuation.

Rate compensation detector. Device that will respond when the temperature of the air surrounding the device reaches a predetermined level, regardless of the rate of temperature rise.

Rate of rise. Method of detecting fire by equipment sensitive to an abnormal rate of increase in heat, frequently operating on the pneumatic expansion principle; i.e., a normal amount of heated air is allowed to escape, but an abnormal amount exerts pressure on an electrically operated alarm diaphragm.

Resistance bridge smoke detector. Device responds to an increase of smoke particles and moisture, present in products of combustion, which fall on an electrical bridge grid. As these conductive substances fall on the grid, they reduce its resistance and cause the detector to respond.

Restorable detector. Device whose sensing element is not ordinarily destroyed by the process of detecting a fire. Restoration may be manual or automatic.

Sampling smoke detector. Consists of tubing distributed from the detector unit to the area or areas to be protected. An air pump draws air from the protected area back to the detector through air sampling ports and piping. At the detector, the air is analyzed for smoke particles.

Self-contained air duct detector unit. Completely self-contained ionization detector unit for mounting directly to the air duct of any air-conditioning, heating, or ventilating system to prevent the spread of fire or smoke throughout the system. It contains its own power supply and supervision and control circuitry, along with visual indicators for alarm and trouble conditions.

Self-contained fire detector unit. A single ionization detector unit with its own power supply and control circuitry for isolated and limited areas, such as valve or remote stations. The unit is capable of automatically activating external devices such as bells, extinguishers, and central station transmitters. An audible signal is provided for trouble or alarm conditions.

Self-restoring detector. Restorable detector whose sensing element is designed to be returned to normal automatically.

Smoke. Totality of the airborne visible or invisible particles of combustion.

Smoke detector. Device that senses visible or invisible particles of combustion.

Spot-type detector. Device whose detecting element is concentrated at a particular location. Typical examples are bimetallic detectors, fusible alloy detectors, certain pneumatic rate-of-rise detectors, certain smoke detectors, and thermoelectric detectors.

Spot-type pneumatic rate-of-rise detector. Device consisting of an air chamber, diaphragm, contacts, and compensating vent in a single enclosure.

Temperature classification. Heat detectors of the fixed temperature type or rate-compensated spot-pattern type are classified as to the temperature of operation and marked with the appropriate color code.

Thermal detection. A thermal device operates only after the temperature of the thermal element itself reaches the rated operating temperature.

Thermal (heat) detectors. There are several types of thermal detectors available, such as:

1. Electrically operated, based on the use of a thermostat and temperature rising above a fixed temperature, the rate of temperature rise, or a combination of the two.

2. Mechanically operated (spring wound), based on fixed temperature only.

3. Operated by compressed gas, based on fixed temperature.

4. Thermally sensitive wires, based on fixed temperature.

5. Device used with pneumatic tubing, based on rate of temperature rise, usually temperatures in excess of 165°F or a rate of temperature rise of 15°F or more per minute, which will activate the fire alarm system.

Thermal lag. Fixed temperature device operates when the temperature of the surrounding air will always be higher than the operating temperature of the device itself. This difference between the operating temperature of the device and the actual air temperature is "thermal lag" and is proportional to the rate at which the temperature is rising.

Thermoelectric effect detector. Device whose sensing element comprises a thermocouple or thermopile unit that produces an increase in electric potential in response to an increase in temperature. This potential is monitored by associated control equipment, and an alarm is initiated when the potential increases at an abnormal rate.

Ultraviolet detector. Device whose sensing element is responsive to radiant energy outside the range of human vision (below approximately 4000 Å).

Water-flow detector. The installation of a flow switch in sprinkler system piping will cause a water-flow alarm to be initiated when a single sprinkler head is activated.

COMMUNICATIONS

Accentuation. Emphasis of certain band of frequencies, to the exclusion of all others, in an amplifier or electronic device.

Active transducer. Transducer whose output signal is dependent on power that is controlled by one or more actuating signals.

Actuating system. Manually or automatically operated mechanical or electrical device that operates electrical contacts to effect signal transmission.

Adapter. Fitting designed to change the terminal arrangement of a jack, plug, socket, or other receptacle so that connections other than the original electrical connections are possible; intermediate device that permits attachment of special accessories or provides special means for mounting.

Ambient temperature. Temperature of air or liquid surrounding an electrical part or device.

Ampere. Unit of electrical current or rate of flow of electrons: 1 V across 1 ohm of resistance causes a current flow of 1 A. The flow of 1 C/second = 1A.

Amplification. Increase in signal magnitude from one point to another or the process causing this increase.

Amplifier. Device that enables an input signal to control power from a source independent of the signal, thus rendering it capable of delivering an output that bears some relationship to and is generally greater than the input signal.

Amplitude

1. Measure of the magnitude of the maximum deviation from the rest position of a parameter. Amplitude may be expressed in either a positive or negative direction, polarity, or sense, depending on the parameter.
2. Magnitude of variation in a changing quantity from its zero value. The term must be modified with an adjective such as "peak," "rms," or "maximum," which designates the specific amplitude in question.

Amplitude-frequency distortion. Distortion due to an undesired amplitude-frequency characteristic.

Amplitude-frequency response. Variation of gain, loss, amplification, or attenuation as a function of frequency.

Amplitude range. Ratio, usually expressed in decibels, of the upper and lower limits of program amplitudes which contain all significant energy contributions.

Antenna. Portion, usually wires or rods, of a radio transmitter or receiver station used for radiating waves into or receiving them from space; also called "aerial."

Attack time. Interval required, after a sudden increase in input signal amplitude to a system or transducer, to attain a stated percentage (usually 63%) of the ultimate change in amplification or attenuation due to this increase.

Attenuation. Decrease in signal magnitude from one point to another or the process causing this decrease.

Attenuator

1. Adjustable passive network that reduces the power level of a signal without introducing appreciable distortion.
2. Resistive network that provides reduction of the amplitude of an electrical signal without introducing appreciable phase or frequency distortion.

Audio frequency. Frequency corresponding to a normally audible sound wave. Audio frequencies range roughly from 15 to 20,000 Hz.

Audio-frequency noise. Unwanted disturbance in the audio-frequency range.

Audio-frequency oscillator. Nonrotating device for producing an audio-frequency sinusoidal electric wave whose frequency is determined by the characteristics of the device.

Audio-frequency spectrum. Continuous range of frequencies extending from the lowest to the highest audio frequency.

Automatic dialing unit. Device capable of generating dialing digits automatically.

Automatic gain control. Process or means by which gain is automatically adjusted as a function of input or other parameters.

Automatic volume control. Process or means by which a substantially constant output volume is automatically maintained in a system or transducer.

Available power. Maximum power obtainable from a given source by suitable adjustment of the load.

Babble. Aggregate cross talk from a large number of interfering channels.

Baffle. In a speaker a baffle is used to increase the acoustic loading of the diaphragm.

Balanced. In communication practice in the term usually signifies electrically alike and symmetrical with respect to a common reference point, usually ground, or arranged to provide conjugacy between certain sets of terminals.

Balanced amplifier. Amplifier in which there are two identical signal branches connected so as to operate in phase opposition and with input and output connections each balanced to ground.

Balanced transmission line. Transmission line having equal conductor resistances per unit length and equal impedances from each conductor to earth and to other electrical circuits.

Balun. Acronym from balanced to unbalanced; for device used for matching an unbalanced coaxial transmission line to a balanced two-wire system. Also called "balanced converter" or "bazooka."

Band. Range of frequencies that lies between two defined limits.

Band elimination filter. Filter that has a single attenuation band, neither of the cutoff frequencies being zero or infinite.

Bandpass filter. Filter that has a single transmission band, neither of the cutoff frequencies being zero or infinite.

Bandwidth. Range within the limits of a band.

Bass boost. Accentuation of the lower audio frequencies in the amplitude-frequency response of a system or transducer.

Beat frequency. Either of the two additional frequencies obtained when signals of two frequencies are combined, equal to the sum or difference, respectively, of the original frequencies.

Boom. Mechanical support for a microphone, used in a television studio to suspend the microphone within range of the persons and voices, but out of camera range.

Branch. In an electronic network, the section between two adjacent branch points; portion of a network consisting of one or more two-terminal elements in a series.

Bridging. Shunting of one signal circuit by one or more circuits, usually for the purpose of deriving one or more circuit branches.

Bridging amplifier. Amplifier with an input impedance sufficiently high so that its input may be bridged across a circuit without substantially affecting the signal level of the circuit across which it is bridged.

Bridging gain. Ratio of the signal power a transducer delivers to its load to the signal power dissipated in the main circuit load across which the input of the transducer is bridged.

Bridging loss. Ratio of the signal power dissipated in the main circuit load across which the input of a transducer is bridged to the signal power the transducer delivers to its load; ratio of the signal power delivered to that part of the system following the bridging point before the connection of the bridging element to the signal power is delivered to the same part after the connection of the bridging element.

Broad band. As applied to data transmission, the term denotes transmission facilities capable of handling frequencies greater than those required for high-grade voice communications.

Broad-band amplifier. Amplifier that has an essentially flat response over a wide frequency range.

Cable. Assembly of one or more conductors, usually within a protective sheath, so arranged that the conductors can be used separately or in groups.

Cable splice. Connection between two or more separate lengths of cable. Conductors in one length are individually connected to conductors in the other lengths, and the protecting sheaths are so connected that protection is extended over the joint.

Cgs Electromagnetic System of Units. Coherent system of units for expressing the magnitude of electrical and magnetic quantities.

Cgs Electrostatic System of Units. Coherent system of units for expressing the magnitude of electrical and static quantities.

Channel. Portion of the spectrum assigned for the operation of a specific carrier and the minimum number of side bands necessary to convey intelligence; single path for transmitting electric signals.

Characteristic. Inherent and measurable property of a device; also, a set of related values.

Characteristic impedance. Driving-point impedance of a line if it were of infinite length; ratio of voltage to current at every point along a transmission line on which there are no standing waves. Also called "surge impedance."

Clipper. Device whose output is zero or a fixed value for instantaneous input amplitudes up to a certain value, but a function of the input for amplitudes exceeding the critical value.

Clipper amplifier. Amplifier designed to limit the instantaneous value of its output to a predetermined maximum.

Coaxial cable. Transmission line in which one conductor completely surrounds the other, the two being coaxial and separated by a continuous solid dielectric or by dielectric spacers. Such a line has no external field and is not susceptible to external fields from other sources. Also called "coaxial line," "coaxial transmission line," and "concentric line."

Columnar speakers. In-line array of speakers in a finished cabinet enclosure.

Communication. Transmission of information from one point, person, or piece of equipment to another.

Communication system. Electrical system whereby intelligence or signals may be transmitted to or through a central station, including telephone, telegraph, district messenger, fire and burglar alarm, watchman or sprinkler supervisory system, and other central station systems of a similar nature, which commonly receive the power supply necessary for their operation from central office or local power sources, but not including radio communication equipment.

Compandor. Combination of a compressor at one point in a communication path for reducing the amplitude range of signals, followed by an expander at another point for a complementary increase in amplitude range.

Composite picture signal. Television signal produced by combining a blanked picture signal with the sync signal.

Compressor. Transducer that for a given input amplitude range produces a smaller output range.

Conjugate branches. Of a network, any two branches such that an electromotive force inserted in one branch produces no current in the other branch.

Conjugate impedances. Impedances having resistance components that are equal and reactance components that are equal in magnitude but opposite in sign.

Connection. Attachment of two or more component parts so that conduction can take place between them; point of such attachment.

Connector. Coupling device that provides an electrical and/or mechanical junction between two cables or between a cable and a chassis or enclosure; device that provides rapid connection and disconnection of electrical cable and wire terminations.

Control. In any mechanism, one or more components responsible for interpreting and carrying out manually initiated directions.

Conversion. Process of changing from one form of representation to another.

Converter. Facsimile device that changes the type of modulation delivered by the scanner; device capable of converting impulses from one mode to another, such as analog to digital, or parallel to serial, or one code to another, device that changes audio-frequency shift modulation.

Corner reflector antenna. Antenna consisting of a primary radiating element and a dihedral corner reflector formed by the elements of the reflector.

Coulomb (C). Quantity of electricity that passes any point in an electric circuit in 1 second when the current is maintained constant at 1 A.

Crossover network. Electrical filter that separates the output signal from an amplifier into two or more separate frequency bands for a multispeaker system.

Cross talk. Undesired energy appearing in one signal path as a result of coupling from other signal paths.

Cue circuit. One-way communication circuit used to convey program control information.

Current amplification. Increase in signal current magnitude in transmission from one point to another, or the process thereof; of a transducer, the scalar ratio of the signal output current to the signal input current.

Current attenuation. Decrease in signal current magnitude in transmission from one point to another, or the process thereof; of a transducer, the scalar ratio of the signal input current to the signal output current.

Cutoff frequency. Frequency that is identified with the transition between a pass band and an adjacent attenuation band of a system of transducer.

Cycle. Change of alternating wave from zero to a negative peak to zero to a positive peak and back to zero. The number of cycles per second (hertz) is called "frequency."

dBm. Unit for expressing power level in decibels with reference to a power of 1 mW (0.01 W).

Decade. Interval between any two quantities having the ratio of 10:1.

Decibel (dB). Standard unit for expressing transmission gain or loss and relative power levels.

Deemphasis. Process that has an amplitude-frequency characteristic complementary to that used for preemphasis.

Device. Single discrete conventional electronic part, such as a resistor or transistor, or a microelectronic circuit—also called an "item"; any subdivision of a system, mechanical, electrical, and/or electronic contrivance intended to serve a specific purpose.

Diode. Electron tube having two electrodes, a cathode and an anode.

Diplexer. Coupling unit that allows more than one transmitter to operate together on the same antenna.

Dipole antenna. Straight radiator usually fed in the center. Maximum radiation is produced in the plane normal to its axis. Also called "dipole."

Dissipation. Loss of electrical energy as heat.

Distortion. Of a signal, an undesired change in waveform.

Distribution amplifier. Power amplifier designed to energize a speech or music distribution system and having sufficiently low output impedance so that changes in load do not appreciably affect the output voltage.

Disturbance. Irregular phenomenon that interferes with the interchange of intelligence during transmission of a signal.

Dividing network. Frequency-selective network that divides the spectrum into two or more frequency bands for distribution to different loads.

Dynamic range. Difference in decibels between the overload level and the minimum acceptable signal level in a system or transducer.

Dynamic speaker. Speaker in which the moving diaphragm is attached to a coil that is conductively connected to the source of electric energy and placed in a constant magnetic field. The current through the coil interacts with the magnetic field, causing the coil and diaphragm to move back and forth in step with the current variations through the coil. Also called a "moving-coil speaker."

Earphone. Electroacoustic transducer intended to be placed in or over the ear. Also called "receiver."

Echo. Wave that has been reflected or otherwise returned with sufficient magnitude and delay to be perceived in some manner as a wave distinct from that directly transmitted.

Effective bandwidth. Bandwidth of an ideal transmission system, which has uniform transmission in its pass band equal to the maximum transmission of the specified system and transmits the same power as the specified system when the two systems are receiving equal input signals having a uniform distribution of energy at all frequencies.

Electric wave. Another term for the electromagnetic wave produced by the back-and-forth movement of electric charges in a conductor.

Electroacoustic. Pertaining to a device that involves both electric current and sound frequency pressures.

Electromagnetic. Having both magnetic and electric properties.

Electromotive force. Force that causes electricity to flow when there is a difference of potential between two points. The unit of measurement is the volt.

Electronic. Pertaining to the branch of science that deals with the motion, emission, and behavior of currents of free electrons, especially in vacuum, gas, or phototubes and special conductors or semiconductors.

Element. Portion of a part that cannot be renewed without destruction of the part; constituent of a microcircuit or integrated circuit that contributes directly to the operation.

Energy. Capacity for performing work. A particle or piece of matter may have energy because it is moving or because of its position in relation to other particles or pieces of matter.

Envelope delay distortion. Of a system or transducer, the difference between the envelope delay at one frequency and the envelope delay at a reference frequency.

Equalizer. Device designed to compensate for an undesired amplitude-frequency or phase-frequency characteristic, or both, of a system or transducer.

Expander. Transducer that for a given input amplitude range produces a larger output range.

Fax machine. Machine connected to a telephone system that produces copies of documents, which are quickly and efficiently relayed to other fax machines, to 1,000 destinations if necessary.

Filter (wave filter). Transducer for separating waves on the basis of their frequency.

Fitting. Accessory, such as a locknut or bushing, to a wiring system. Its function is primarily mechanical rather than electrical.

Flat frequency response. The response of a system to a constant-amplitude function that varies in frequency is flat if the response remains within specified limits of amplitude, usually specified in decibels from a reference quantity.

Frequency. Number of recurrences of a periodic phenomenon in a unit of time. Electrical frequency is specified as so many hertz. Radio frequencies are normally expressed in kilohertz at and below 30,000 kHz and in megahertz above this frequency.

Frequency band. In communications and electronics, a continuous range of frequencies extending between two limiting frequencies. The term may also be applied to those frequencies that are encountered in shock and vibration excitation.

Frequency modulation. Type of modulation in which the frequency of a continuous carrier wave is varied in accordance with the properties of a second (modulating) wave.

Frequency response. Measure of how effectively a circuit or device transmits the different frequencies applied to it; portion of the frequency spectrum that can be sensed by a device within specified limits of amplitude error.

Frequency separator. Circuit that separates the horizontal-scanning from the vertical-scanning synchronizing pulses in a television receiver.

Front-to-back ratio. Ratio of power gain between the front and rear of a directional antenna. Also called "front-to-rear ratio."

Function. Quantity whose value depends on the value of one or more other quantities; specific purpose of an entity or its characteristic action.

Gain. Increase in power when a signal is transmitted from one point to another, usually expressed in decibels. The term is widely used for denoting transducer gain.

Gain control. Device for adjusting the gain of a system or transducer.

Gamma (γx). Unit of magnetic intensity equal to 10 microoersteds or 0.00001 oersted, number indicating the degree of contrast in a photograph, facsimile reproduction, or received television picture.

Ghost image. Undesired duplicate image, offset somewhat from the desired image as viewed on a television screen, due to a reflected signal traveling over a longer path and therefore arriving later than the desired signal.

Ground. Metallic connection with the earth to establish a ground potential voltage reference point in a circuit.

Harmonic distortion. Nonlinear distortion of a system or transducer characterized by the appearance in the output of harmonics other than the fundamental component when the input wave is sinusoidal.

Hertz (Hz). Unit of frequency equal to 1 cycle/second.

High frequency (hf). Frequency band from 3 to 30 MHz.

High-pass filter. Filter having a single transmission band extending from some cutoff frequency, not zero, up to infinite frequency.

High _Q_. Having a high ratio of reactance to effective resistance; factor determining the efficiency of a reactive component.

Hiss. Audio-frequency noise having subjective characteristics analogous to prolonged sibilant sounds.

Horn. Tubular or rectangular enclosure for radiating or receiving acoustic waves.

Hotel/motel control systems. Signaling system controlling room status, occupancy, maid occupancy, messages received while the guest is out, and wake-up alarm system. Many other controls are available for security.

Hybrid coil. Single transformer having effectively three windings. It is designed to be connected to four branches of a circuit so as to render these branches conjugate in pairs.

Hybrid set. Two or more transformers interconnected to form a network having four pairs of accessible terminals to which may be connected four impedances so that the branches containing them may be made conjugate in pairs when the impedances have the proper values, but not otherwise.

Ideal transducer. Hypothetical linear passive transducer that transfers the available power of the source to the load.

Ideal transformer. Hypothetical transformer that neither stores nor dissipates energy and has unity coefficient of coupling.

Image impedances. Impedances that will simultaneously terminate all inputs and outputs of a transducer in such a way that at each of its inputs and outputs the impedance in both directions will be equal.

Impedance. Total opposition a circuit offers to the flow of alternating current at a given frequency. It is measured in ohms, and its reciprocal is called "admittance."

Incidence angle. Angle between an approaching light ray or emission and the perpendicular to the surface in the path of the ray.

Infinite. Boundless; having no limits whatsoever.

Infrasonic frequency. Frequency lying below the audio-frequency range.

Input. Current, voltage, power, or other driving force applied to a circuit or device.

Input impedance. Impedance presented by the transducer to the source.

Insertion gain. Ratio of the power delivered to the part of a transmission system following a transducer to the power delivered to the same part of the system before the insertion of the transducer.

Insertion loss. Ratio of the power delivered to the part of a transmission system following a transducer to the power delivered to the same part of the system after the insertion of the transducer.

Interference. Electrical or electromagnetic disturbance, phenomenon, signal, or emission, man-made or natural, that causes or can cause undesired response, malfunctioning, or degradation of the electrical performance of electrical and electronic equipment.

Intermodulation distortion. Nonlinear distortion of a system or transducer characterized by the appearance in the output of frequencies equal to the sums and differences of integral multiples of the two or more component frequencies present in the input wave.

Isolation amplifier. Amplifier employed to minimize the effects of a following circuit on the preceding circuit.

Isolation transformer. Transformer inserted in a system to separate one section of the system from undesired influences of other sections.

Iterative impedance. Impedance that when connected to one pair of terminals of a transducer produces an identical impedance at the other pair of terminals.

Joule (J). Work done by a force of 1 N acting through a distance of 1 m.

Lavalier microphone. Small microphone suspended from the neck by a cord or wire or fastened on a person's clothes.

Level (in audio). Magnitude of a quantity considered in relation to a reference value.

Line amplifier. Amplifier that supplies a transmission line or system with a signal at a stipulated level.

Line transformer. Transformer connecting a transmission line to terminal equipment used for such purposes as isolation, line balance, impedance matching, or additional circuit connections.

Load. Device that receives power; power delivered to such a device.

Load impedance. Impedance presented by the load.

Local speed of sound. Velocity of propagation of acoustic waves over a small region as determined by the conditions there. It is principally a function of temperature.

Loss. Decrease in power suffered by a signal as it is transmitted from one point to another, usually expressed in decibels. Energy is dissipated without accomplishing useful work.

Low-pass filter. Filter having a single transmission band extending from zero to some cutoff frequency, not infinite.

L pad

1. Volume control that has practically the same impedance at all settings. It consists essentially of an L network in which both elements are adjusted simultaneously.

2. An L pad is designed for use in adjusting speaker volume at a remote location while keeping a constant load on the audio line.

Magnitude (size). Quantity assigned to one unit so that it may be compared with other units of the same class; ratio of one quantity to another.

Mast. Pole on which an antenna is mounted.

Matching transformer. Transformer used for matching impedances.

Maximum. Highest value occurring during a stated period.

Megahertz (MHz). One million hertz.

Microphone. Electroacoustic transducer that responds to sound waves and delivers essentially equivalent electric waves.

Microphonics. Noise caused by mechanical shock or vibration of elements in a system.

Microvolt (μV). One millionth of a volt.

Microwave. Very short radio wave, usually shorter than 1 m; wave shorter than 1 m. A wave shorter than 10 m is called an "ultrashort wave."

Mixer. Device having two or more inputs and a common output. The latter combines the separate input signals linearly in the desired proportion to produce an output signal. Circuit that generates output frequencies equal to the sum and difference of two input frequencies.

Mixer (in audio techniques). Device having two or more inputs, usually adjustable, and a common output. It operates to combine the separate input signals linearly in a desired proportion to produce an output signal.

Modulation. Process of modifying some characteristic of a wave (called a "carrier") so that it varies in step with the instantaneous value of another wave (called a "modulating wave" or "signal").

Monitor. Listen to a communication service, without disturbing it, to determine its freedom from trouble or interference; device used for checking signals.

Monitoring amplifier. Power amplifier used primarily for evaluation and supervision of a program.

Motorboating. Undesired oscillation in an amplifying system or transducer, usually of a pulse type, occurring at a subaudio or low audio frequency.

Narrow band. Waves contained within a relatively small portion of the spectrum. If a narrow-band pass filter is used, the intensity at the two cutoff wavelengths is a certain percentage of the maximum intensity within the interval.

Network. Combination of elements.

Newton (N). Unit of force that will impart an acceleration of 1 m/second to a mass of 1 kg.

Noise. Unwarranted disturbance within a dynamic electrical or mechanical system.

Noise level. Noise power density spectrum in the frequency range of interest; average noise power in the frequency range of interest; indication on a specified instrument.

Nonlinear distortion. Distortion caused by a deviation from a linear relationship between specified measures of the input and output of a system or transducer.

Octave. In communication, the interval between the two frequencies having a ratio of 2:1.

Oersted (Oe). In the cgs electromagnetic system, the unit of magnetizing force equal to $1000/4\,\vartheta$ ampere-turns per meter.

Ohm. Unit of resistance: 1 ohm is the value of resistance through which a potential difference of 1 V will maintain a current of 1 A.

Operate time characteristic. Relation between the operate time of an electromagnetic relay and the operate power.

Operating voltages. Direct voltages applied to the electrodes of a vacuum tube under operating conditions.

Oscillator. Electronic device the generates alternating-current power at a frequency determined by the values of certain constants in its circuits.

Output. Current, voltage, power, or driving force delivered by a circuit or device.

Output impedance. Of a device, the impedance presented by the device to the load.

Output power. Power delivered by a system or transducer to its load.

Overload level. Of a system or component, the level above which operation ceases to be satisfactory as a result of signal distortion, overheating, or damage.

Pad. Nonadjustable passive network that reduces the power level of a signal without introducing appreciable distortion.

Parabolic antenna. Antenna with a radiating element and a parabolic reflector that concentrates the radiated power into a beam.

Parameter. Constant or element whose value characterizes the behavior of one or more variables associated with a given system; measured value that expresses performance.

Passive. Term describing inert component that may control but does not create or amplify energy.

Passive transducer. Transducer that has no source of power other than the input signal(s), and whose output signal power cannot exceed that of the input.

Peak limiter. Device that automatically limits the magnitude of its output signal to approximate a preset maximum value by reducing its amplification when the instantaneous signal magnitude exceeds a preset value.

Peak-to-peak amplitude. Of an oscillating quantity, the algebraic difference between the extremes of the quantity. Two times the amplitude of a simple harmonic oscillatory quantity is its peak-to-peak amplitude or double amplitude.

Percent harmonic distortion. Measure of the harmonic distortion in a system or transducer numerically equal to 100 times the ratio of the square root of the sum of the squares of the root-mean-square voltages (or currents) of each of the individual harmonic frequencies to the root-mean-square voltage (or current) of the fundamental.

Phase. Angular relationship between current and voltage in alternating-current circuits; number of separate voltage waves in a commercial alternating-current supply.

Phase delay. Ratio of the total phase shift (radians) experienced by a sinusoidal signal in transmission through a system or transducer to the frequency (radians per second) of the signal.

Phase delay distortion. Of a system or transducer, the difference between the phase delay at one frequency and the phase delay at a reference frequency.

Plane. Screen of magnetic cores. Planes are combined to form stacks.

Playback. Reproduction of a recording.

Power amplifier. Amplifier that drives a utilization device such as a loudspeaker.

Power gain. Ratio of the signal power that a transducer delivers to its load to the signal power absorbed by its input circuit.

Power level. Magnitude of power averaged over a specified interval of time.

Power loss. Ratio of the signal power absorbed by the input circuit of a transducer to the signal power delivered to its load.

Power supply. Unit that supplies electrical power to another unit. It changes alternating current to direct current and maintains a constant voltage output within limits.

Power supply hum. Interference from a power system characterized by the presence of undesired energy at power supply frequency or harmonics thereof.

Preamplifier. Amplifier connected to a low-level signal source to present suitable input and output impedances and provide gain so that the signal may be further processed without appreciable degradation in the signal-to-noise ratio.

Preemphasis. In a system, the process that increases the magnitude of some frequency components with respect to the magnitude of others in order to reduce the effects of noise introduced in subsequent parts of the system.

Process control. Automatic control of continuous operations, contrasted with numerical control, which provides automatic control of discrete operations.

Program. Sequence of signals transmitted for entertainment or information.

Program level. Magnitude of a program in an audio system expressed in volume units (vu).

Public-address system. One or more microphones, an audio-frequency system, and one or more speakers used for picking up and amplifying sounds to a large audience, either indoors or out.

Radio. The transmission of signals through space by means of electromagnetic waves; usually applied to the transmission of sound and code signals, although television and radar also depend on electromagnetic waves.

Range. Maximum useful distance of a radar or radio transmitter; difference between the maximum and the minimum value of a variable; set of values that may be assumed by a quantity or function.

Ratio. Value obtained by dividing one number by another. This value indicates their relationship to each other.

Receiver. Portion of a communications system that converts electric waves into a visible or audible form; electromechanical device for converting electrical energy into sound waves.

Reception. Listening to, copying, recording, or viewing any form of emission.

Record changer. Device that will automatically play a number of phonograph records in succession.

Recovery time. Time interval required, after a sudden decrease in input signal amplitude to a system or transducer, to attain a stated percentage (usually 63%) of the ultimate change in amplification or attenuation due to this decrease.

Reference volume. Volume which gives a reading of zero on a standard volume indicator.

Reflection. Phenomenon in which a wave that strikes a medium of different characteristics is returned to the original medium with the angles of incidence and reflection equal and lying in the same plane.

Remote control. System of control performed from a distance. The control signal may be conveyed by intervening wires, sound, light, or radio.

Remote line. Program transmission line between a remote pickup point and the studio or transmitter site.

Resistance. Property of conductors that, depending on their dimensions, material, and temperature, determines the current produced by a given difference of potential; property of a substance that impedes current and results in the dissipation of power in the form of heat. The unit of resistance is the ohm.

Resistor. Device connected into an electrical circuit to introduce a specified resistance.

Ripple. Portion of the output voltage of a power supply harmonically related in frequency to the input power and to any internally generated switching frequency. Ripple is expressed as an rms percentage, but it can also be expressed as peak-to-peak.

Roll-off. Gradually increasing loss or attenuation with increase or decrease of frequency beyond the substantially flat portion of the amplitude-frequency response characteristic of a system or transducer.

Root mean square (rms). Square root of the average of the squares of the values of a periodic quantity taken throughout one complete period; effective value of a periodic quantity.

Separator. Insulating sheet or other device employed in a storage battery to prevent metallic contact between plates of opposite polarity within a cell; insulator used in the construction of convolutely wound capacitors.

Series. Components connected end-to-end in a circuit to provide a single path for the current.

Shunt. Precision low-value resistor placed across the terminals of an ammeter to increase its range. A shunt may be either internal or external to the instrument.

Side bands. Frequency bands on both sides of the carrier frequency. Frequencies of the wave produced by modulation fall within these bands.

Signal. Visual, aural, or other indications used to convey information; information to be conveyed over a communication system; wave in a communication system that conveys information.

Signal level. Magnitude of a signal, especially when considered in relation to an arbitrary reference magnitude.

Singing. Undesired self-sustained oscillation in a system or transducer.

Singing margin (gain margin). Ratio of the singing point to the operating gain of a system or transducer.

Singing point. Minimum value of the gain of a system or transducer that will cause singing to start.

Single channel. Carrier only for single-tone modulated radio control transmitter and matching receiver installation.

Single-edged push-pull amplifier circuit. Amplifier circuit having two transmission paths designed to operate in a complementary manner and connected so as to provide a single unbalanced output without the use of an output transformer.

Single-ended amplifier. Amplifier in which each stage normally employs only one active element (tube, transistor, etc.), or if more than one active element is used, in which they are connected in parallel so that operation is asymmetric with respect to ground.

Sinusoidal. Varying in proportion to the sine of an angle or time function.

Solid state. Pertaining to circuits and components using semiconductors; pertaining to the physics of materials in their solid form.

Sound effects filter. Filter used to adjust the frequency response of a system for the purpose of achieving special aural effects.

Source. That which supplies signal power to a transducer.

Source-impedance. Impedance presented by a source of energy to the input terminals of a device.

Speaker. Electroacoustic transducer that radiates acoustic power into the air with essentially the same wave form as that of the electrical input.

Spectrum. Continuous range of electromagnetic radiations from the longest known radio waves to the shortest known cosmic rays; band of frequencies necessary for transmission of a given type of intelligence; range of frequencies considered in a system.

Speed of sound. Speed at which sound travels in a given medium under specified conditions. The speed of sound at sea level in the international standard atmosphere is 1108 ft/second, 658 knots, or 1215 km/hour. In the United States standard atmosphere, 1962, and the atmosphere adopted by the International Civil Aeronautics Organization, the speed of sound in air at zero altitude is 340.294 m/second or 1116.45 ft/second.

Splice. Device used for joining two or more conductors.

Splitter. Passive device similar to an antenna coupler, but designed to match a 75-ohm impedance.

Standard volume indicator. Standardized instrument having specified electrical and dynamic characteristics and read in a prescribed manner, for indicating the volume of a complex electric wave such as that corresponding to speech or music.

Subharmonic. Sinusoidal quantity having a frequency that is an integral submultiple of the fundamental frequency of a periodic quantity from which it is derived.

Sync signal. Signal employed for synchronizing the scanning. In television it is composed of pulses at rates related to the line and field frequencies. Also called a "synchronizing signal."

System. Assembly of component parts and devices linked together by some form of regulated interaction into an organized whole.

Tape recorder. Mechanical electronic device for recording voice, music, and other audio-frequency material.

Telegraph bandwidth. Difference between the limiting frequencies of a channel used to transmit telegraph signals.

Telegraph cable. Uniform conductive circuit consisting of twisted pairs of insulated wires or coaxially shielded wires or combinations of each, used to carry telegraph signals.

Telegraph carrier. Single-frequency wave that is modulated by transmitting apparatus in carrier telegraphy.

Telegraph circuit. Complete wire or radio circuit over which signal currents flow between transmitting and receiving apparatus in a telegraph system.

Telegraph code. System of symbols for transmitting telegraph messages.

Telegraph emission. Signal transmitted by a telegraph system, classified by type of transmission, type of modulation, bandwidth, and other characteristics.

Telegraph grade. Class of communication circuits that can transmit only telegraphic signals, comprising the lowest types of circuits in regard to speed, accuracy, and cost.

Telegraph interference. Any undesired electrical energy that tends to interfere with the reception of telegraph signals.

Telegraph transmitter. Device that controls an electric power source in order to form telegraph signals.

Telegraphy. Communication at a distance by means of code signals consisting of current pulses sent over wires or by radio.

Telephone. System of converting sound waves into variations in electric current that can be sent over wires and reconverted into sound waves at a distant point, used primarily for voice communication.

Telephone answering system. Special type of private branch exchange system used by a telephone answering service bureau to provide secretarial service.

Telephone channel. One-way or two-way path suitable for the transmission of audio signals between two stations.

Telephone circuit. Complete circuit in a telephone system over which audio and signaling currents travel between the two telephone subscribers in communication with each other.

Telephone data set. Equipment interfacing a data terminal with a telephone circuit.

Telephone line. Conductors extending between telephone subscriber stations and a central office.

Telephone set. Assembly including a telephone, transmitter, telephone receiver, and associated switching and signaling devices.

Telephone signal. Electrical signal transmitted by a telephone system, classified by type of transmission, type of modulation, bandwidth, and supplementary characteristics.

Teletypewriter. Special electric typewriter that produces coded electric signals corresponding to manually typed characters and automatically types messages when fed with similarly coded signals produced by another machine.

Teletypewriter exchange service. Service furnished by telephone companies to subscribers in the U.S., whereby any of the subscribers can communicate directly with any other subscriber via teletypewriter.

Televise. To pick up a scene with a television camera and convert it into corresponding electric signals for transmission by a television station.

Television. System for converting a succession of visual images into corresponding electric signals and transmitting these signals by radio or over wires to distant receivers at which the signals can be used to reproduce the original images.

Television antenna. Antenna suitable for transmitting or receiving television broadcasts; a receiving antenna is a horizontally mounted half-wave dipole.

Television bandwidth. Difference between the limiting frequencies of a television channel.

Television broadcast band. Several groups of channels, each containing a number of 6-MHz channels, that are available for assignment to television broadcast stations.

Television broadcasting. Transmission of television programs by means of radio waves, for reception by the public.

Television pickup station. Land mobile station used for the transmission of television program material and related communications to a television broadcast station from the scene of an event occurring at a remote point.

Television receiver. Radio receiver for converting incoming electric signals into television pictures and the associated sound.

Television satellite. Orbiting satellite that relays television signals between ground stations.

Television screen. Fluorescent screen of the picture tube in a television receiver.

Television station. Installation, assemblage of equipment, and location where radio transmissions are sent or received.

Telewriter. System in which writing movement at the transmitting end causes corresponding movement of a writing instrument at the receiving end.

Telex. Worldwide teleprinter exchange service providing direct send and receive teleprinter connections between subscribers.

Terminal. Point of connection for two or more conductors in an electrical circuit.

Terminating. Closing of a circuit at either end of a line or transducer by connection of some device.

Thump. Low-frequency transient disturbance in a system or transducer characterized audibly by the onomatopoeic connotation of the word.

Transducer. Device capable of being actuated by signals from one or more systems or media and of supplying related signals to one or more other systems or media.

Transducer gain. Ratio of the power transducer delivers to its load to the available power of the source.

Transducer loss. Ratio of the available power of the source to the power the transducer delivers to its load.

Transformer. Electrical device that by electromagnetic induction transforms electric energy from one or more circuits to one or more other circuits at the same frequency, but usually at a different voltage and current value.

Transformer loss (in communication). Ratio of the signal power an ideal transformer would deliver to a load to the power delivered to the same load by the actual transformer, both transformers having the same impedance ratio.

Transistor. Active semiconductor device, usually made of silicon or germanium, having three or more electrodes. The three main electrodes used are the emitter, base, and collector.

Transition loss. At a junction between a source and a load, the ratio of the available power to the power delivered to the load.

Transmission. Conveyance of electrical energy from point to point along a path; transfer of a signal, message, or other form of intelligence from one place to another by electrical means.

Transmission gain. Increase in signal power in transmission from one point to another.

Transmission loss. Decrease in signal power in transmission from one point to another.

Trap. Selective circuit that attenuates undesired signals but does not affect the desired one.

Treble boost. Accentuation of the higher audio frequencies in the amplitude-frequency response of a system of transducer.

Tuner. Packaged unit capable of producing only the first portion of the functions of a receiver and delivering radio frequency, intermediate frequency, or demodulated information to some other equipment.

UHF. Ultra high frequency.

Unbalanced. Not balanced.

Underwriters Laboratories, Inc. (UL). Independent laboratory that tests equipment to determine whether it meets certain safety standards when properly installed and used.

Very high frequency. (vhf). Frequency band of 30 to 300 MHz and a wavelength of 10 to 1 m.

Vibration. Continuous reversing change in the magnitude of a given force; mechanical oscillation or motion about a reference point of equilibrium.

Voltage amplification. Increase in signal voltage magnitude in transmission from one point to another or the process thereof; of a transducer, the scalar ratio of the signal output voltage to the signal input voltage.

Voltage attenuation. Decrease in signal voltage magnitude in transmission from one point to another or the process thereof; of a transducer, the scalar ratio of the signal input voltage to the signal output voltage.

Volume. In an electric circuit, the magnitude of a complex audio-frequency wave as measured on a standard volume indicator.

VU. Unit of volume in which the standard volume indicator is calibrated.

Volume unit meter. Volume indicator with a decibel scale and specified dynamic and other characteristics, used to obtain correlated readings of speech power necessitated by the rapid fluctuations in level of voice currents.

Watt (W). Unit of the electric power required to do the work at the rate of 1 J/second.

Waveform. Shape of an electromagnetic wave.

Wave trap. Device used to exclude unwanted signals or interference from a receiver. Wave traps are usually tunable to enable the interfering signal to be rejected or the true frequency of a received signal to be determined.

Wide band. Waves contained within a relatively large portion of the spectrum. If a wide bandpass filter is used, the intensity at the two cutoff wavelengths is a certain percentage of the maximum intensity within the interval.

ELECTRICAL EQUIPMENT AND DEVICES

AC-DC general-use snap switch. Form of general-use snap switch suitable for use on either direct or alternating-current circuits for controlling resistive loads not exceeding the ampere rating at the voltage involved; inductive loads not exceeding one-half the ampere rating at the voltage involved, except that switches having a marked horsepower rating are suitable for controlling motors not exceeding the horsepower rating of the switch at the voltage involved; and tungsten filament lamp loads not exceeding the ampere rating at 125 V when marked with the letter T.

AC general-use snap switch. Form of general-use snap switch suitable only for use on alternating-current circuits for controlling resistive and inductive loads (including electric discharge lamps) not exceeding the ampere rating at the voltage involved, tungsten filament lamp loads not exceeding the ampere rating at 120 V and motor loads not exceeding 80% of the ampere rating of the switches at the rate voltage.

Adjustable. As applied to circuit breakers, a qualifying term indicating that the circuit breaker can be set to trip at various values of current and/or time within a predetermined range.

Adjustable-speed motor. Motor whose speed varies gradually over a considerable range but that once adjusted remains unaffected by the load, for example, a shunt motor with field-resistance control.

Adjustable variable-speed motor. Motor whose speed can be adjusted gradually, but that once adjusted for a given load will vary with change in load, for example, a compound-wounded motor with field control or slip-ring induction motor with rheostat speed control.

Air circuit breaker. Circuit breaker in which the interruption occurs in air.

Air switch. Switch in which air is the insulating medium between contacts in the open position.

Alternate power source. One or more generator sets intended to provide power during the interruption of the normal electrical service, or the public utility electrical service intended to provide power during interruption of service normally provided by the generating facilities on the premises.

Ambient temperature. Temperature of the surrounding medium that comes in contact with the device or equipment.

Amp-trap or hi-cap. High-interrupting-capacity, current-limiting, fusible device that anticipates dangerous short-circuit currents and breaks the circuit.

Armature. Revolving part of an electric motor or relay.

Attachment plug

1. Device that by insertion in a receptable establishes connection between the conductors of the attached flexible cord and the conductors connected permanently to the receptacle.

2. Plug that consists of projecting blades enclosed in an insulated enclosure. It is connected to the electrical wiring cord leading to an electrical appliance. Receptables are classified as flush mounting, surface mounting, and for mounting in special enclosures. Plugs are specified according to current rating and the number and arrangement of blades, such as two-pole parallel, two-pole tandem, two-pole polarized, three-pole, and four-pole blades.

Automatic circuit recloser. Self-controlled device for automatically interrupting and reclosing an alternating-current circuit, with a predetermined sequence of opening and reclosing followed by resetting, hold-closed, or lockout operation.

Automatic throw-over switch. An air switch and automatic means responsive to changes in circuit conditions to change a conductor connection from one circuit to another.

Auto-transformer. A transformer in which part of the winding is common to both the primary and secondary circuits.

Auxiliary power plants. Used for lighting or power purposes, may be located in boiler or heating equipment rooms if the fuel to be used is natural gas or diesel fuel. If the fuel to be used is gasoline, such equipment should be located on the exterior of the building and housed in a ventilated room with at least eight (8) in. thick masonry walls. The use of liquified petroleum gas for fuel is prohibited.

Auxiliary section. Any section other than the main, distribution, or combination section.

Auxiliary switches. Switches that are mechanically operated by the main switching device for switching, interlocking, or other purposes. Auxiliary switch contacts should be designated as "a" or "b," described as follows:

1. "a" contacts. Contacts that are open when the switching device contacts are open or tripped and closed when the switch contacts are closed.

2. "b" contacts. Contacts that are closed when the switching device contacts are open or tripped and open when the switch contacts are closed.

Back-connected switch. Switch in which the current-carrying conductors are connected to the back studs of the mounting base.

Branch bus. Bus that usually originates at a section bus and terminates in one or more overcurrent devices.

Branch circuit device. Device that is the final overcurrent device protecting a circuit.

Break distance of a switch. The minimum open gap distance between the stationary and movable contents, or live parts connected thereto, when the blade is in the open position.

Bus structure. Assembly of bus conductors, with associated connecting joints and insulating supports.

Butt-contact switch. Two contacts are butted together when the switch is closed.

Cabinet. Enclosure of adequate mechanical strength, composed entirely of fire- and absorption-resistant material, designed either for surface or flush mounting, and provided with a frame, mat, or trim in which swinging doors are hung.

Capacitor

1. Electrical condenser used on some fractional horse-power single-phase motors to provide added starting torque; also, an electrical condenser used for power factor correction.

2. Device for electricity storage, used in starting and running the circuits of many electrical motors.

Cartridge fuses. Fuses are manufactured in ferrule type for capacities not exceeding 60 A and knife blade type for capacities of 61 to 600 A for voltages not to exceed 600 V. They are manufactured for one-time operation or are renewable. When renewable, the blown fuse links in the casing are replaceable. When for one-time operation, the entire fuse has to be replaced.

Circuit bonding jumper. The connection between portions of a conductor in a circuit to maintain required ampacity of the circuit.

Circuit breaker

1. Device designed to interrupt the flow of electric current when the circuit becomes overloaded.

2. Device designed to open and close a circuit by nonautomatic means, and to open the circuit automatically on a predetermined overload of current, without injury to itself when properly applied within its rating.

3. Switching device capable of making, carrying, and breaking currents under normal circuit conditions, and also making, carrying for a specified time, and breaking currents under specified abnormal circuit conditions, such as those of short circuit.

4. Electromechanical device designed to open a current-carrying circuit, under both overload- and short-circuit conditions, without injury to the device. The term applies only to the automatic type designed to trip on a predetermined overcurrent.

5. Protective device used against overloading of circuits and equipment. It is also used to protect smaller conductors where the size of the conductors is changed.

Circuit breakers

1. Circuit breakers, as well as fuses, possess a time element of operation. They operate instantaneously on short circuits and with time lag on overload.

2. Circuit breakers have an advantage over fuses in that they can be reset in less time than required for replacing blown fuses. Circuit breakers are desirable where service is an important factor and where fuse replacement is inconvenient.

Closed-circuit batteries. Batteries that can be discharged continuously.

Combination section. Switchboard section that performs the functions of both the distribution and the main sections.

Commutator. Segmented conductor used as a collector on the armature of a dc motor or generator.

Compartment. Area within a section that is so constructed as to isolate devices in that compartment from the surrounding area, except for openings used for interconnections, control, or ventilation.

Compound-wound motor. Direct-current motor that has two separate field windings, one connected in parallel with the armature circuit and the other connected in series with the armature circuit.

Constant-potential transformer. Transformer that consists essentially of three parts—a primary coil, core (usually iron), and secondary coil. The primary winding of the transformer is connected to the power source. The secondary winding delivers the power to the load.

Constant-speed motor. Motor with a constant or practically constant speed at normal operation, such as an ordinary direct-current shunt-wound motor, a synchronous motor, or an induction motor with small slip.

Continuous current. Amount of current a conductor, device, or a piece of equipment can carry continuously for an indefinite period of time without exceeding its allowable temperature rise.

Continuous load. Load in which the current is expected to continue for three hours or more.

Controlled vented power fuse. Fuse with provision for controlling discharge circuit interruption such that no solid material may be exhausted into the surrounding atmosphere. Discharge gases should not ignite or damage insulation in the path of the discharge, nor should these gases propagate a flashover to or between grounded members or conduction members in the path of the discharge when the distance between the vent and such insulation or conduction members conforms to manufacturer's recommendations.

Controller. Device or group of devices that serves to govern in some predetermined manner the electrical power delivered to the apparatus to which it is connected.

Control wiring. Wiring for the circuit or circuits of a piece of equipment that carries the electrical signals directing the performance of the devices in that equipment, but which does not carry power current.

Convenience receptacle. Device used as a safe means of connecting a portable electrical appliance to an electric circuit; receptacle provided with openings for insertion of the blades of an attachment plug and mounted in an insulated enclosure of porcelain or composition arranged for flush or surface mounting.

Coupling capacitor. The coupling between circuits or components by means of a capacitor.

Current balance relay. Relay that operates on a difference in current input or output of two circuits.

Current-carrying part. Conducting part intended to be connected in an electric circuit to a source of voltage. Non-current-carrying parts are those not intended to be so connected.

Current-limiting overcurrent protective device. Device that when interrupting a specified circuit will consistently limit the short-circuit current to a specified magnitude substantially less than that obtainable in the same circuit if the device were replaced with a solid conductor having comparable impedance.

Current rating of a switchboard. Designated maximum direct current of alternating current in rms amperes at rates frequency that the switchboard bus can carry continuously without exceeding its temperature rise limits when subjected to specified heating tests.

Current relay. Device used to open or close a circuit on the basis of change of current flow strength.

Current transformer. Transformer used to charge the current of a system.

Cutout. Assembly of a fuse support with either fuse holder, fuse carrier, or disconnecting blade. The fuse holder or fuse carrier may include a conducting element (fuse link) or may act as the disconnecting blade by the inclusion of a nonfusible member.

Cutout box. Enclosure of adequate mechanical strength, composed entirely of fire- and absorption-resistant material, designed for surface mounting, and having swinging doors or covers secured directly to and telescoping with the walls of the box proper.

Dead front

1. Term applied to switches, circuit breakers, switchboards, control panels, and panelboards that are designed, constructed, and installed so that no current-carrying parts are normally exposed on the front.

2. Equipment constructed so that all live parts are enclosed in such a manner as to be inaccessible to unauthorized persons.

Dead-front switchboard. Switchboard with no live parts mounted on the front of the board, used in systems limited to a maximum of 600 V for direct current and 2500 V for alternating current. Bench, desk, pedestal and post, and combustion dead-front boards are used for the support of control and metering equipment for remote control electrically operated circuit breakers in central stations.

Dead rear. Switchboard or panelboard where there is no exposed portion of any current-carrying conductor, part, device or other equipment on the rear of such board.

Device. Single discrete conventional electronic part, such as a resistor or transistor, or a microelectronic circuit—also called an "item"; any subdivision of a system; mechanical, electrical, and/or electronic contrivance intended to serve a specified purpose.

Dielectric. Material separating the plates of a capacitor.

Dielectric withstand tests. Tests that determine the ability of the insulating materials and spacings to withstand overvoltages.

Differential current relay. Fault-detecting relay that functions on a differential current of a given percentage or amount.

Diffuser. Device used to redirect the luminous flux from a source, primarily by the process of diffuse transmission.

Disconnecting means. Device, group of devices, or other means whereby the conductors of a circuit can be disconnected from their source of supply.

Disconnecting switch. Switch used for disconnecting a light or power circuit from its source of supply.

Disconnecting (or isolating) switch (disconnector, isolator). Mechanical switching device used for isolating a circuit or equipment from a source of power.

Disconnector. Switch that is intended to open a circuit after the load has been thrown off by some other means. Manual switches designed for opening loaded circuits are usually installed in circuit with disconnectors to provide a safe means for opening the circuit under load.

Distribution section. Switchboard section having branch or feeder circuit switching and overcurrent protective devices.

Dripproof motor. Upper parts of the motor are shielded to prevent vertical overhead dripping.

Drawout mounted device. Device that may be removed from the stationary portion of a switchboard without un-bolting connections or mounting supports.

Draw-out-type switchboard. Metal-clad switchgear consisting of a stationary housing mounted on an angle iron framework and a horizontal draw-out circuit-breaker structure. The equipment for each circuit is assembled on a frame, forming a self-contained and self-supporting mobile unit.

Dynamo
1. Electric generator with a permanent magnet.
2. Large-capacity generator.

Electrical center. Points approximately midway between the ends of an inductor or resistor that divides the inductor or resistor into two equal electrical values.

Electrical equipment
1. Any apparatus, appliance, device, instrument, fitting, fixture, machinery, material, or thing used in or for, or capable of being used in or for, the generation, transformation, transmission, distribution, supply, or utilization of electric power or energy, and that without restricting the generality of the foregoing, includes any assemblage or combination of materials or things that is used or is capable of being used or adapted to serve or perform any particular purpose or function when connected to an electrical installation, notwithstanding that any of such materials or things may be mechanical, metallic, or nonelectric in origin.
2. Installations of electrical conductors, fittings, devices, and fixtures within or on public and private buildings.

Electrical service equipment. Equipment located at a point of entrance of supply conductors to a building that constitutes the main control of supply and means of disconnecting electricity, including circuit breaker, switches, fuses, and electrical accessories.

Electrical supply equipment. Equipment that produces, modifies, regulates, controls, or safeguards a supply of electrical energy.

Electrical connector. Device that joins electric wires or conductors mechanically and electrically to other wires or conductors and to the terminals of devices, equipment, or apparatus.

Electric control. Control of equipment, machinery, lighting fixtures, or device by switches, relays, or rheostats.

Electric controller. Device or a group of devices for controlling in some predetermined manner the electric power delivered to the apparatus to which it is connected.

Electric demand limiter. Device that selectively switches off electrical equipment whenever total electrical demand rises beyond a predetermined level.

Electric distribution center. A terminal at which electric energy is received from the transmission system and is delivered to the distribution system only.

Electric equipment standards. Materials, fittings, appliances, devices and other equipment listed in publications of inspected electrical equipment of the Underwriters Laboratories, Inc., and other nationally recognized testing organizations, and installed in accordance with the recommendations of accredited agencies.

Electrical protective device. Particular type of equipment used in electrical power systems to detect abnormal conditions and to initiate appropriate corrective action.

Electric substation. A terminal at which electric energy is received from the transmission system and is delivered to other elements of the transmission system and, generally, to the local distribution system.

Electrolier switch. Multiple-circuit switch. One lamp or group of lamps may be turned on alone or in combination with other lamps.

Emergency system. System of feeders and branch circuits meeting the requirements of the National Fire Protection Association or local codes, connected to the alternate power source by transfer switch and supplying energy to an extremely limited number of prescribed functions vital to the protection of life and patient safety, with automatic restoration of electrical power within ten seconds after power interruption. Only circuits serving those areas or functions specifically required by code or the NFPA.

Enclosed panelboard. Assembly of buses and connections, overcurrent devices, and control apparatus, with or without switches or other equipment, installed in a cabinet.

Enclosed-type switch. Switch so designed and constructed that no current-carrying parts are normally exposed and the operator cannot come into contact with the current-carrying parts during ordinary operation. The term includes all approved types of externally operable switches and circuit breakers that meet the above requirements.

End bell. End structure of an electrical motor which usually houses the bearings.

Equipment. Fittings, devices, appliances, fixtures, apparatus, and the like, used as part of or in connection with an electrical power transmission and distribution system or communication systems.

Equipment bonding jumper. Connection between two or more portions of the equipment grounding conductor.

Equipment grounding conductor. Conductor used to connect non-current-carrying metal parts of equipment, raceways, and other enclosures to the system's grounded conductor at the service and/or the grounding electrode conductor.

Equipment system. System of feeders and branch circuits arranged for delayed, automatic or manual connection to the alternate power source and that serves primarily three-phase power equipment.

Expulsion fuse unit. Vented fuse unit in which the expulsion effect of gases produced by the arc and lining of the fuse holder, either alone or aided by a spring, extinguishes the arc.

Explosionproof apparatus. Apparatus enclosed in a case that is capable of withstanding an explosion of a specified gas or vapor which may occur within it, and of preventing the ignition of a specified gas or vapor surrounding the enclosure by sparks, flashes, or explosion of the gas or vapor within, and that operates at such an external temperature that a surrounding flammable atmosphere will not be ignited thereby.

Externally operable. Capable of being operated without exposing the operator to contact with live parts. The term is applied to equipment, such as a switch, that is enclosed in a case or cabinet.

Filter. Device that changes, by transmission, the magnitude and/or spectral composition of the flux incident on it. Filters are called "selective" (or "colored") or "neutral," according to whether or not they alter the spectral distribution of the incident flux.

Frame size. Term applied to a group of circuit breakers that are physically interchangeable with each other. Frame size is expressed in amperes and corresponds to the largest ampere rating available in the group.

Front accessible. Bus and device connections are accessible from the front. If required, a limited number of devices can be removed to permit accessibility.

Fuse. Protective device with a current-sensitive fusible part that, during a continuous overcurrent condition, will melt open and interrupt the flow of current. A fuse may or may not be the complete device necessary to connect into an electrical circuit.

Fusible switch. Switch in which one or more poles have a fuse in series in a composite unit.

General-purpose motor. Motor having a continuous rating and designed for use without restriction to a particular application.

Generator. Electromagnetic machine that converts mechanical energy to electrical energy.

General-use switch. A switch intended for use as a switch in general distribution and branch circuits. It is rated in amperes and is capable of interrupting its rated current at its rated voltage.

Ground bus. Bus to which the equipment grounding conductors from the individual pieces of equipment are connected and that, in turn, is connected to the grounding electrode conductor at one point. It grounds each switchboard section through which it passes.

Ground fault protector. Device or system that provides protection for equipment by opening the circuit in case of a predetermined ground fault current. The ground fault protector includes a ground fault current sensing device and relaying equipment or combination of ground fault current sensing device and relaying equipment that will operate to cause a disconnecting means to function at a predetermined value of ground fault current.

Hermetic motor. Motor compressor and housing sealed in a dome.

Indicating switch. Switch of such design or so marked that the fact as to whether it is "on" or "off" may be readily determined by inspection.

Individually mounted device. Device that is not panel-mounted and that may or may not be enclosed in its own compartment.

Instantaneous trip. As applied to circuit breakers, a qualifying term indicating that no delay is purposely introduced in the tripping action of the circuit breaker.

Insulation level. Insulation strength of a material expressed in terms of a rms to withstand voltage.

Interlock. Electrical or mechanical component actuated by the operation of a device or other means with which it is directly associated, to govern succeeding operations of the same or allied devices.

Interrupter switch. Switch capable of making, carrying, and interrupting specified currents.

Interrupting capacity. Devices intended to break current must have an interrupting capacity sufficient for the voltage employed and for the current that must be interrupted.

Interrupting rating. An overcurrent protective device whose rating is based upon the highest available rms symmetrical current, or direct current at a specified voltage that the device is capable of interrupting under prescribed test conditions.

Inverse time. As applied to circuit breakers, a qualifying term indicating there is purposely introduced a delay in the tripping action of the circuit breaker, with the delay decreasing as the magnitude of the current increases.

Isolated device. Device that is segregated from other devices by metal or insulating barriers or enclosures that is not readily accessible to personnel unless special means for access are used and provided for.

Isolating switch

1. Switch intended for isolating an electric circuit from the source of power. It has no interrupting rating and is intended to be operated only after the circuit has been opened by some other means.

2. Switch intended for isolating either a circuit or some equipment from its source of supply. It is not intended either for establishing or interrupting the flow of current in any circuit.

Jack. Plug-in spring terminal used to facilitate connections between internal and external circuits.

Knifeblade switch. Switch that makes contact when movable copper blades are inserted into forked contact jaws.

Labeled. Equipment or materials having a label, symbol, or other identifying mark of a nationally recognized testing laboratory, inspection agency, or other organization concerned with product evaluation that maintains periodic inspection of production of labeled equipment or materials and by whose labeling is indicated compliance with nationally recognized standards or tests to determine suitable usage in a specified manner.

Life safety branch. System of feeders and branch circuits meeting the requirements of the NFPA or local codes connected to alternate power source by a transfer switch, and functioning as a component of the emergency system.

Lighting and appliance branch circuit panel board. Panelboard having more than 10% of its overcurrent devices rated 30 A or less, for which neutral connections are provided.

Listed. Equipment or materials included in a list published by a nationally recognized testing laboratory, inspection agency, or other organization concerned with product evaluation that maintains periodic inspection of production of listed equipment or material, and whose listing states that the equipment or material meets nationally recognized standards or has been tested and found suitable for use in a specified manner. Means for identifying listed equipment may vary for each testing laboratory, inspection agency, or other organization concerned with product evaluation, some of which do not recognize equipment as listed unless it is also labeled. The authority having jurisdiction usually utilizes the system employed by the listing organization to identify a listed product.

Live-front switchboards. Current-carrying parts of the switch equipment mounted on the exposed front of the vertical panels and usually limited to systems not exceeding 600 V. Generally they are installed in restricted areas where permitted by code.

Low-voltage protection. Effect of a device that on the reduction or failure of voltage causes and maintains the interruption of power supply to the equipment protected and the main circuit.

Low-voltage release. Effect of a device that on the reduction or failure or voltage causes the interruption of power supply to the equipment but does not prevent the reestablishment of the power supply on the return of the voltage.

Main bonding jumper. Connection between the grounded circuit conductor and the equipment grounding conductor at the service.

Main device. Single device that disconnects all ungrounded switchboard conductors, other than control power conductors when used, from the supply bus.

Main section. Portion of a switchboard where the main or service disconnect device or devices are located. This section is also permitted to contain utility meters or other measuring instruments. Incoming line conductors are usually terminated in this section.

Manually operable. Designed and intended for operation by the hand directly to a handle, lever, push button, or other suitable contrivance that is an integral part of the equipment.

Mercury contact switch. Switch consisting of a mercury tube with stationary contacts located in its opposite ends. When the tube is tilted by the switch handle, the mercury moves from one contact to the other, thereby making and breaking the circuit.

Metal-clad switchgear. Metal structure completely enclosing a circuit breaker and associated equipment, such as current and potential transformers, interlocks, controlling devices, buses, and connections.

Mimic bus. Single-line diagram on the face of the switchboard indicating the principal connections of the system.

Momentary-contact switch. Switch that opens or closes a circuit for only a short period. The switch returns to its normal position when a handle or button is released.

Motor. Electromechanical machine that converts electrical energy to mechanical energy.

Motor circuits switch

1. Switch, rated in horsepower, capable of interrupting the maximum operating overload current of a motor of the same horsepower rating at the rated voltage.

2. Manually operated knife or snap switch, rated in horsepower, fused or unfused.

Motor control. Control used to start and stop a motor at the proper time.

Motor efficiency. Rate at which a motor converts electrical energy. The greater the efficiency, the lower the energy cost. A well-designed induction motor may have an efficiency of between 75 and 95% full load.

Motors. Motors are generally divided into three groups: direct-current, single-phase alternating current, and polyphase alternating current.

Multiple fuse. Assembly of two or more single-pole fuses.

Multiplier. Resistor in series with the coil of a voltmeter to increase the range.

Multispeed motor. Motor that can be operated at different speeds, independently of the load, as, for example, a dc motor with two armature windings or an induction motor with windings capable of various pole groupings.

Multiwinding motor. Motor having multiple and/or tapped windings, intended to be connected or reconnected in two or more configurations, for operation at any one of two or more speeds and/or voltages.

Mutual inductance. Flux linkage in one coil resulting from a current in another coil.

Neutral assembly. Assembly consisting of an appropriate number of terminals for connecting neutral conductors.

Neutral bus. Bus having the appropriate number of terminals to provide for the connection of the neutral line and load conductors.

Neutral conductor. Grounded conductor of a three-wire, single-phase alternating current or direct current supply system or a four-wire, three-phase alternating current supply system.

Nonadjustable. As applied to circuit breakers, a qualifying term indicating that the circuit breaker does not have any adjustment to alter the value of current at which it will trip or the time required for its operation.

Nonautomatic. Implied action requires personal intervention for its control. As applied to an electric controller, nonautomatic control does not necessarily imply a manual controller, but only that personal intervention is necessary.

Nontamperable overcurrent device. Circuit breakers so designed that the seal must be broken or the entire circuit breaker replaced to alter the current-carrying capacity of the device.

Nonvented power ruse. Fuse without intentional provision for the escape of arc gases, liquids, or solid particles to the atmosphere during circuit interruption.

Notching relay. Relay that functions to allow only a specified number of operations of a given device or equipment, or a specified number of successive operations within a given time of each other. It is also used to allow periodic energizing of a circuit.

Oil cutout (oil-filled cutout). Cutout in which all or part of the fuse support and its fuse link or disconnecting blade are mounted in oil with complete immersion of the contacts and the fusible portion of the conducting element (fuse link), so that arc interruption by severing of the fuse link or by opening of the contacts will occur under oil.

Oil switch. Switch having contacts that operate under oil (or askarel or other suitable liquid).

Open. When applied to electrical equipment, the term indicates that moving parts, windings, or live parts are exposed to accidental contact.

Open-type motor. Motor that allows maximum ventilation; motor that has full opening in the frame and end bells.

Out of reach. Term applied to equipment that is located more than 5 ft horizontally or more than 8 ft vertically from any floor, platform, or other surface from which it would otherwise be readily accessible.

Overcurrent device. Device capable of automatically opening an electric circuit under both predetermined overload and short-circuit conditions either by the fusing of metal or electromechanical means.

Overload device. Device affording protection from excess current, but not necessarily from short circuit, and capable of automatically opening an electric circuit either by the fusing of metal or electromechanical means.

Panelboard

1. Single panel or group of panel units designed for assembly in the form of a single panel, including buses, and with or without switches and/or automatic overcurrent protective devices for the control of light, heat, or power circuits of small individual, as well as

aggregate capacity. It is designed to be placed in a cabinet or cutout box placed in or against a wall or partition and accessible only from the front.

2. Assembly of buses and connections, overcurrent devices, and control apparatus, with or without switches or other equipment, constructed for installation as a complete unit in a cabinet.

Panel-mounted device. Closely grouped assembly of devices that are mounted on a common base or mounting surface utilizing panelboard-type construction. The total combination is then mounted in a switchboard combination or distributing section.

Part-winding start motor. Motor arranged for starting by first energizing part of its primary winding and subsequently energizing the remainder of this winding in one or more steps, both parts then carrying current.

Peak let-through current. Maximum instantaneous current through an overcurrent protective device during its total clearing time.

Plug fuses. Plug fuses are available in sizes up to 30 A for installation in circuits not exceeding 150 V. they are compact and easy to inspect.

Plug-in mounted device. Device that can be plugged in to make electrical connections to a line bus bar. Device need not be self-supporting when withdrawn.

Pole of a switch. Part that makes or breaks one connection. A single-pole switch makes and breaks the connection in one leg of a circuit; a two-pole makes and breaks connections in two legs.

Portable. Term applied to equipment specifically designed not to be used in a fixed position and to receive current through the medium of a flexible cord or cable and usually a detachable plug.

Portable ground fault circuit interrupter. Ground fault circuit interrupter specifically designed to receive current by means of a flexible cord or cable and an attachment plug cap. It incorporates one or more receptacles for the connection of equipment and is provided with a flexible cord or cable and an attachment plug cap.

Position switch. Switch that makes or breaks contact when the main device or apparatus, which has no function number, reaches a given position.

Potential transformer. Transformer used to change the voltage of a system.

Potentiometer. Variable resistor designed for voltage control, characterized by three contacts—two fixed and one variable.

Power fuse unit. Vented, nonvented, or controlled vented fuse unit in which the arc is extinguished by being drawn through solid material, granular material, or liquid, either alone or aided by a spring.

Pressure connector (solderless). A pressure wire connector is a device that establishes the connection between two or more conductors or between one or more conductors and a terminal by means of mechanical pressure and without the use of solder.

Protected. Equipment is constructed so that the electrical parts are protected against damage from foreign objects entering its enclosure.

Rating of a device. Designated limit of operating characteristics based on definite conditions; such operating characteristics as current, voltage, frequency, and so forth are permitted to be given in the rating.

Rear accessible. Incoming and outgoing cable or bus connections that are accessible from the rear of the cabinet.

Rectifier. Device for converting alternating current to direct current.

Regulator bypass switch. Specific device or combination of devices designed to bypass a regulator.

Relay. Electrically operated switch that has one or more sets of points and a coil. A small current through the coil will close the points of the relay. In this way a small amount of voltage or current can control the power in a high-voltage or high-power circuit.

Remote control circuit. Electrical circuit that controls another circuit through a relay or an equivalent device.

Resistor

1. Electrical device that introduces resistance into an electrical circuit.

2. Circuit component designed to limit the current.

Rheostat. Resistor with varying resistance. Rheostats have a variety of applications: Theater dimmers, which consist of a varying number of rheostats mounted in a bank and are controlled by interlocking levers, used for dimming the auditorium and stage lighting to various desired levels of illumination; field rheostats, which are used to regulate the voltage of a generator and the speed or power factor of a motor; dc motor-starting rheostats, which are used to start motors in steps; dc speed-regulating rheostats, which are used for motors requiring varying conditions; and dc battery-charging rheostats, which are used for charging storage batteries.

Running windings. Electrical windings of a motor that have current flowing through them continuously in normal operation.

Safety-type switch. Switch mechanism mounted inside a metal box and operated by a handle outside the box.

Sealable equipment. Equipment enclosed in a case or cabinet that is provided with means for sealing or locking so that live parts cannot be made accessible without opening the enclosure. The equipment may or may not be operable without opening the enclosure.

Sealed (mermetic type) refrigeration compressor. Mechanical compressor consisting of a compressor and a motor, both of which are enclosed in the same sealed housing, with no external shafts or shaft seals. The motor operates in the refrigerant atmosphere.

Section bus. Portion of a bus structure that serves an overcurrent device or devices in the switchboard section and comprises the part of the bus between the through bus and the branch bus.

Section disconnect. Device by which all ungrounded power conductors in a switchboard section can be disconnected from the supply.

Section rating. Each switchboard section is assigned a continuous current section rating, which should not exceed the ampacity of the section bus or section disconnect.

Self-exciting dynamo. System involving a rotating electrical conductor that creates and sustains its own magnetic field.

Semiprotected motor. Openings are lined with screens to protect the motor from falling particles.

Sequencer control (heat relays). Control for the elements in an electrical heating system.

Service disconnect. Device or group of devices by which all ungrounded power conductors of a circuit can be disconnected from the supply.

Service entrance equipment. Assembly of switches and switchlike devices that permits disconnection of all power or its distribution to various branch circuits through overcurrent devices such as fuses or circuit breakers; assembly of fuses or circuit breakers, with or without a disconnecting means; also termed a "distribution panel" or "panelboard."

Service equipment. Necessary equipment, usually consisting of a circuit breaker or switch fuses, and their accessories, located near the point of entrance of supply conductors to a building, another structure, or otherwise defined area, and intended to constitute the main control and means of cutoff of the supply.

Service factor. Safety factor designed into a motor to deliver more than its rated horsepower. A 10-hp motor with a service factor of 1.10 is capable of delivering 11 hp.

Setting (of circuit breaker). Value of current and/or time at which an adjustable circuit breaker is set to trip.

Shaded-pole motor. Type of motor, used for fans, that does not have a start winding and can be used at different speeds.

Short circuit current rating (equipment). Rating that indicates the ability of equipment to withstand the effects of a short circuit current without exceeding specified damage criteria.

Shunt resistor. Precision resistor parallel to the coil of an ammeter.

Shunt-wound motor. DC motor in which the field and armature circuits are connected in parallel.

Single-throw switch. A circuit is made and broken when the switch is thrown in one position only. A double-throw switch makes and breaks a circuit when the switch is thrown in either of two positions.

Special-purpose motor. Motor specifically designed and listed for a particular power application.

Splashproof motor. The bottom parts of the motor are shielded to prevent splashing of particles at an angle not over 100° from vertical.

Splice bus. Bus that electrically connects switchboard sections.

Squirrel-cage induction motor. Motor in which the secondary circuit consists of a squirrel cage winding suitably disposed in slots in the secondary core. Squirrel-cage induction motors are further classified according to the torque and starting current characteristics.

Starter. Electric controller for accelerating a motor from rest to normal speed and for stopping the motor. Usually overload protection is included.

Starting relay. Electrical device used to connect and disconnect the starting winding at the proper time.

Start winding. Auxiliary windings that are used to help the motor get started and are disconnected at about three-fourths the running speed.

Stationary-mounted device. Device that can be removed only by unbolting electrical connections and mounting supports.

Step-down transformer. Transformer connected so that the delivered voltage is lower than the supplied voltage.

Step-up transformer. Transformer connected so that the delivered voltage is higher than the supplied voltage.

Storage battery. Battery consisting of positive plates connected together electrically, negative plates connected together electrically, separators, sulfuric acid, electrolyte for lead-acid battery, connected straps and terminals, and a suitable non-leak container.

Switch

1. Device for making, breaking, or changing the connection in a circuit.

2. Device for opening and closing or changing the connection of a circuit. A switch is usually understood to be manually operable, unless otherwise described.

3. Device used to connect and disconnect an electrical circuit from the source of power.

Switchboard

1. Large single-panel frame or assembly of panels on which are mounted on the face or back, or both, switches, overcurrent and other protective devices, buses, and usually instruments. Switchboards are generally accessible from the rear as well as from the front and are not intended to be installed in cabinets.

2. Panel or assembly of panels on which is mounted any combination of switching, measuring, controlling, and protective devices, buses, and connections, designed with a view to successfully carrying and rupturing the maximum fault current encountered when controlling incoming and outgoing feeders.

Switchboard section. Portion of a switchboard that is prevented by the structural framework from being physically separated into smaller units.

Switching device. Device designed to close and/or open one or more electric circuits.

Supply bus. Bus that conducts electrical power from the source terminations to the main disconnect device or devices.

Symmetrical current. Alternating current having no offset or transient component and therefore having a waveform essentially symmetrical about the zero axis. Symmetrical current is expressed in terms of rms amperes.

Synchronous motor. An alternating-current motor that is essentially an alternator operated as a motor.

Temperature limits for switchboards. Maximum allowable temperature limit is based on an allowable temperature rise over a prescribed ambient temperature as described in Underwriters Laboratories, Inc. Publication No. UL 891.

Temperature rise. Difference in temperature between the temperature of the part under consideration and the ambient temperature.

Thermal cutout. Device affording protection from excessive current, but not necessarily short-circuit protection, and containing a heating element in addition to and affecting a fusible member that opens the circuit.

Thermal protection. As applied to motors, the words "thermal protection" appearing on the nameplate of a motor indicate that the motor is provided with a thermal protector.

Thermal protector. Protective device for assembly as an integral part of a motor or motor-compressor, which when properly applied protects the motor against dangerous overheating due to overload and failure to start. The thermal protector may consist of one or more sensing elements integral with the motor or motor compressor and an external control device.

Thermocouple.   Device that converts heat energy to electrical energy. It is constructed by joining two dissimilar metals and activated by heating the junction.

Three-way switch.   Switch that controls lights from two different locations. A four-way switch is used with three-way switches for controlling lights from three or more different locations.

Through bus.   Bus that extends through a switchboard section, also described as a "horizontal," "cross," or "main bus."

Totally enclosed motor.   Nonventilated motor that may be used in hazardous atmospheres, or that may be explosion-proof.

Train.   Angle between the vertical plane through the axis of the searchlight drum and the plane in which this plane lies when the searchlight is in a position designated as having zero train.

Transformers.   Transformers are used primarily to convert electrical power in an alternating-current system from one voltage and current to some other voltage and current.

Unit safety-type switchboard.   Metal-enclosed switchgear consisting of a completely enclosed, self-supporting, metal structure containing one or more circuit breakers or switches.

Universal motor.   Series-wound motor that may be operated on direct current or single-phase alternating current at about the same speed and output.

Utilization equipment.   Equipment that utilizes electrical energy for mechanical, chemical, heating, lighting, testing, or similar purposes and is not a part of supply equipment, supply lines, or communications lines.

Variable-speed motor.   Motor in which the speed varies with the load. The speed generally decreases as the load increases, as in a series motor or an induction motor with large slip.

Vented power fuse.   Fuse with provision for the escape of arc gases, liquids, or solid particles to the surrounding atmosphere during circuit interruption.

Warning labels.   Switchboard sections are required to have warning labels to inform the user of the potential hazards of the equipment.

Wire connector.   Device that connects two or more conductors together or one or more conductors to a terminal point for the purpose of connecting electrical circuits.

Wiring switches.   Switches used for controlling branch circuits and individual lights or appliances in interior wiring systems. They are manufactured with push-button, rotary, toggle, locking, quiet, or mercury types of operating mechanisms, arranged for flush or surface mounting.

Wireway.   Raceway consisting of a completely enclosing system or metal troughing and fittings so formed and constructed that insulated conductors may be readily drawn in and withdrawn, or laid in and removed, after the system has been completely installed, without injury either to conductors or their covering.

ELECTRICAL INSTALLATIONS

Accessible

1. As applied to wiring methods, capable of being removed or exposed without damaging the building structure or finish; not permanently closed in by the structure or finish of the building.

2. As applied to equipment, admitting close approach because not guarded by locked doors, elevation, or other effective means.

Aluminum-sheathed cable.   Cable consisting of one or more conductors of an approved type assembled into a core and covered with a liquid- and gas-tight sheath of aluminum or aluminum alloy.

Aluminum-sheathed (ALS) cable.   Factory assembled cable consisting of one or more insulated conductors enclosed in an impervious, continuous, closely fitting tube of aluminum. It is used with approved fittings for terminating and connecting to boxes, outlets, and other equipment.

Apparent sag at any point.   Departure of the wire at a particular point in the span from the straight line between the two points of support of the span, at 60°F with no wind loading.

Apparent sag of a span.   Maximum departure of the wire in a given span from the straight line between the two points of support of the span, at 60°F with no wind loading.

Appliance

1. Utilization equipment, generally other than industrial, normally built in standardized sizes or types, and installed or connected as a unit to perform one or more functions such as clothes washing, air conditioning, food mixing, or deep frying.

2. Appliances are current-utilizing equipment, fixed or portable—for example, heating, cooking, and small motor-operated equipment.

Appliance branch circuit.   Branch circuit supplying energy to one or more outlets to which appliances are connected. Such circuits must have no permanently connected lighting fixtures that are not a part of an appliance.

Approved

1. Acceptable to the administrative authority enforcing applicable code.

2. When the term is used with reference to particular electrical equipment, it means that such equipment has been submitted to an acceptable certification agency for examination and testing, that an acceptable certification agency has given formal certification that it conforms to standards as usually established under the provisions of the local electrical code, and that the certification report has been adopted by the inspection authorities.

Approved for the purpose.   Approved for a specific purpose, environment, or application described in a particular code requirement.

Arc resistance.   Time required for a given electrical current to render the surface of a material conductive because of carbonization by the arc flame.

Arc sources.   Carbon arc sources radiate because of both the incandescence of the electrodes and the luminescence of vaporized electrode material and other constituents of the surrounding gaseous atmosphere. Considerable spread in the brightness, total radiation, and spectral energy distribution may be achieved by varying the electrode materials.

Armature.   Moving part of a magnetic circuit.

Armor.   Wrapping of galvanized interlocking steel strip or other approved metal, forming an integral part of the assembly of certain insulated cables, wires, or cords.

Armored cable (BX)

1. Cable provided with a wrapping of metal tape, other than lead, that forms an integral part of the assembly.

2. Rubber-insulated wires are protected by a flexible steel armor, the combination being furnished as a unit. Armored cable may be used in concealed locations when approved by local authorities.

3. Flexible metallic sheathed cable used for interior wiring.

Askarel

1. Nonflammable synthetic insulating liquid that when decomposed by the electric arc evolves only nonflammable gaseous mixtures.

2. Generic term for a group of nonflammable synthetic chlorinated hydrocarbons used as electrical insulating mediums. Askarels of various compositional types are used. Under arcing conditions the gases produced, while consisting predominantly of noncombustible hydrogen chloride, can include varying amounts of combustible gases depending on the askarel type.

Automatic. Self-acting; operating by its own mechanism when actuated by some impersonal influence, such as a change in current strength, pressure, temperature, or mechanical configuration.

Auxiliary gutter. Raceway consisting of a sheet metal enclosure, used to supplement the wiring space of electrical equipment and enclose interconnecting conductors.

Bare conductor. Conductor having no covering or insulation whatsoever.

Barricade. Physical obstruction, such as tapes, screens, or cones, intended to warn and limit access to a hazardous area.

Barrier. Physical obstruction that is intended to prevent contact with energized lines or equipment.

Barrier layer or photovoltaic cell. Barrier layer or photovoltaic cell, when illuminated, generates voltage even though not connected to an external power source. The cell comprises a metal plate coated with a semiconductor (selenium on iron or cuprous oxide on copper, for example). On exposure to light, electrons liberated from the metal surface are trapped at the interface unless an external circuit is provided, through which they may escape. In photographic and illumination meters, this circuit includes a small microammeter calibrated in units of illumination.

Battery

1. Device for transforming chemical energy into electrical energy.

2. Two or more voltage cells connected together for a direct-current power source.

Bond. Electrical connection from one conductive element to another for the purpose of minimizing potential differences or providing suitable conductivity for fault current or for mitigation or leakage current and electrolytic action.

Bonding. Permanent joining of metallic parts to form an electrically conductive path that will ensure electrical continuity and the capacity to conduct safely any current likely to be imposed.

Bonding jumper. Reliable conductor to ensure the required electrical conductivity between metal parts required to be electrically connected.

Box connector. Device for securing a cable, via its sheath or armor, where it enters an enclosure such as an outlet box.

Brake horsepower. Horsepower actually required to drive a fan, including the energy losses in the fan, but not including the drive loss between the motor and fan. The name is derived from the Pony brake, a common method of testing mechanical output of motors.

Bus. Conductor that serves as a common connection for the corresponding conductors of two or more circuits.

Bushing. Insulating structure that includes a through conductor or provides a passageway for such a conductor, with provision for mounting on a barrier, conducting or otherwise, for the purpose of insulating the conductor from the barrier and conducting current from one side of the barrier to the other.

Bus or bus bar. Heavy solid wire or bars that connect points at the same potential.

Busway

1. Raceway consisting of a system of metal troughing (including elbows, tees, crosses, in addition to straight runs), containing conductors, the conductors being supported on insulators.

2. Assembly of rigidly fixed conductors or bus bars suitably braced and arranged within an approved enclosure of sheet metal or other material having substantially equivalent mechanical strength. The term "busway" includes one totally enclosed type, enclosed ventilated type, and trolley type.

Busway and bus duct. Power is transmitted through bare rods or bars insulated from each other, supported and protected by a sheet-metal housing. Bus ducts are of the plug-in, non-plug-in, and movable trolley contact type.

BX cable. Insulated wires enclosed in flexible metal tubing, used in electrical wiring systems where permitted by code.

Cable

1. Combination of conductors that are bound together and insulated from each other; also, single conductors having the same insulation and outside protective covering as commonly used in multiconductor cables.

2. Conductor with insulation, or a stranded conductor with or without insulation and other coverings (single-conductor cable), or a combination of conductors insulated from one another (multiple-conductor cable).

3. Number of wires twisted or braided to form a conductor.

4. Standard conductor; group of solid or standard conductors laid together, but insulated from one another.

Cable sheath. Protective covering applied to cables. A cable sheath may consist of multiple layers of which one or more is conductive.

Cable trough. Raceway consisting of a system of metal troughing and fittings so formed and constructed that insulated conductors and cables may be readily laid in or removed after the system has been completely installed, without injury either to the conductors or their covering.

Cell. As applied to a cellular floor raceway, the hollow space used as a raceway.

Cellular-metal floor raceway. Wiring installed in a floor constructed of metal deck containing hollow spaces that are used as raceways for the wiring.

Center of gravity. Point inside or outside a body, around which all parts of the body balance each other.

Center of mass. On a line between two bodies, the point around which the two bodies would revolve freely as a system.

Certificate of approval. Upon satisfactory completion of electrical wiring and installation of equipment and after final inspection, tests, and when an administrative official issues a certificate of approval to the persons to whom the permit was issued for delivery to the owner.

C.G.S.. Centimeter-gram-second.

Charge. Displacement of electrons. A positive change is a deficiency of electrons, and a negative charge is a surplus of electrons.

Circuit bonding jumper. Connection between portions of a conductor in a circuit to maintain required ampacity of the circuit.

Clearance. The minimum distance between two conductors, between conductors and supports or other objects, or between conductors and ground.

Climbing space. Vertical space reserved along the side of a pole structure to permit ready access by personnel to equipment and conductors located on the pole structure.

Common use. Simultaneous use of facilities by two or more agencies supplying the same type of service.

Concealed. Rendered inaccessible by the structure or finish of the building. Wires in concealed raceways are considered concealed, even though they may become accessible by being withdrawn. Raceways or wiring materials that are unexposed or not normally visible are considered concealed.

Conductor

1. Wire, cable, or other form of metal installed for the purpose of conveying electric current from one piece of electrical equipment to another or to ground.
2. Materials, usually in the form of a wire, cable, or bus bar suitable for carrying an electric current.

Conductor shielding. Envelope that encloses the conductor of a cable and provides an equipotential surface in contact with the cable insulation.

Conductor spacing conflict. Conductor so situated with respect to a conductor of another line at a lower level that the horizontal distanced between them is less than the sum of the following values: 5 ft plus one-half the difference of the level between the conductors concerned plus the values required in the electrical code for horizontal separation between conductors on the same support for the highest voltage carried by either conductor concerned.

Conduit

1. Tube especially constructed for the purpose of enclosing electrical conductors.
2. Raceway of circular cross section into which it is intended that conductors be drawn. The term includes rigid conduit (metallic and nonmetallic) and flexible conduit.
3. Metal pipe that houses electric wiring.

Conduit body. Separate portion of a conduit or tubing system that provides access through a removable cover(s) to the interior of the system at a junction of two or more sections of the system or at a terminal point of the system.

Conspicuity. Capacity of a signal to stand out in relation to its background so as to be readily discovered by the eye.

Consumer's service. All that portion of the consumer's installation from the service box or its equivalent up to and including the point at which the supply authority makes connection.

Cord set. Assembly consisting of a suitable length of flexible cord or power supply cable, provided with an attachment plus at one end and a cord connector at the other end.

Counter mounted cooking unit. Assembly of one or more domestic surface heating elements for cooking purposes,

designed for flush mounting in, or supported by, a counter. The assembly is complete with inherent or separately mountable controls and internal wiring.

Cover. In wire coating, a coating whose primary purpose is to "weatherproof" or prevent casual grounding such as contact with a wet tree branch), or to otherwise protect a conductor.

Covered conductor

1. Conductor having one or more layers of nonconducting materials that are not recognized as insulation under the electrical code.
2. Conductor encased within material of composition or thickness that is not recognized in the electrical code as electrical insulation.

Damp location

1. Location subject to a moderate degree of moisture, such as basements and cold-storage warehouses.
2. Partially protected locations under canopies, marquees, roofed open porches, and the like and interior locations subject to moderate degrees of moisture.
3. Location that is normally or periodically subject to the condensation of moisture in, on, or adjacent to electrical equipment.

Dead. Free from any electrical connection to a source of potential difference and from electric charge; not having a potential different from that of the earth. The term is used only with reference to current-carrying parts that are sometimes alive.

Dead end. The point in a transmission line where all the strain in the conductors is carried by the support.

Distribution system. System to provide overhead and/or underground wiring services (electrical or communications) to individual lots. This definition applies to the system as a whole, or any part thereof (cables, conduits or wires) used as feeders, primaries, secondaries, or similarly designated conductor systems forming a part of such distribution system and not including transmission systems.

Dry location. Location not normally subject to dampness or wetness. A location classified as dry may be temporarily subject to dampness or wetness, as in the case of a building under construction.

Duct. In underground work, a single tubular runway for underground cables.

Dust and ignitionproof. Enclosed in a manner that will exclude ignitable amounts of dust or amounts that might affect performance or rating and that when installation and protection are in conformance with these regulations, will not permit arcs, sparks, or heat otherwise generated or liberated inside of the enclosure to cause ignition of exterior accumulations or atmospheric suspensions of a specified dust on or in the vicinity of the enclosure.

Dustproof. Constructed or protected so that dust will not interfere with successful operation.

Dusttight. Constructed so that dust will not enter the enclosing case.

Electric. Containing, producing, arising from, or actuated by electricity; often used interchangeably with "electrical."

Electrical. Related to or associated with electricity, but not containing it or having its properties or characteristics; often used interchangeably with "electric."

Electrical code. A body of rules and regulations that govern the application and installation of electrical wiring systems, electrically operated equipment, and control devices.

Electrical conductors. Made from materials of comparatively low resistance, through which electricity flows in appreciable quantities.

Electrical conduit. Metal pipes in which electrical wires or conductors are installed.

Electrical construction. Electrical construction includes and governs all work and materials used in installing, maintaining, and extending a system of electrical wiring for light, heat, or power, all appurtenances thereto, and all apparatus or equipment used in connection therewith, inside of or attached to any building, structure, lot, or premises.

Electrical contractor. Any person engaged in the business of installing or altering by contract, electrical equipment for the utilization of electricity supplied for light, heat, or power, not including radio apparatus or equipment for wireless reception of sound and signals, conductors, and other equipment installed for or by public utilities including common carriers that are under state jurisdiction for use in their operation as public utilities. The term does not include employees employed by such contractor to do or supervise such work.

Electrical installation. Installation of any system of wiring in or on any land, building, or premises from the point or points where electric power or energy is delivered by the supply authority or from any other source of supply to the point or points where such power or energy can be used by any electrical equipment including the connection of any such wiring with any of the equipment and any part of a wiring system, and also including the maintenance, alteration, extension, and repair of such wiring.

Electrical installation standards. Conformity of installations of electrical equipment to the applicable standards of the city electrical code, National Electrical Code NFPA 70 latest edition and other accepted engineering standards listed in the code, should be *prima facie* evidence that such installations are reasonably safe for use in the service intended and in compliance with the provisions of the several codes

Electrical metallic tubing (thin-wall conduct)

1. Raceway of metal having a circular cross section into which it is intended that conductors be drawn and having a wall thinner than that of rigid metal conduit and an outside diameter sufficiently different from that of rigid conduit to render it impracticable for anyone to thread it with standard pipe thread.

2. Wiring installed in thin-wall conduit with outlet boxes provided in the conduit system at points as required. Thin-wall conduit is similar to rigid steel conduit but is of thinner material and lighter in weight.

Electrical noise. Noise or unwanted sounds generated by motors, equipment, certain lighting fixtures, and lamps, received by the human ear and senses.

Electrical supply lines. Conductors and their necessary supporting or containing structures that are located entirely outside of buildings and are used for transmitting a supply of electrical energy. Electrical supply lines do not usually include communication lines or open wiring on buildings, in yards, or in similar locations where spans are less than 20 ft and all the precautions required for stations or utilization equipment are observed.

Electrical supply station. Building, room, or separate space within which electrical supply equipment is located and the interior of which is accessible only to properly qualified persons. Included are generating stations and substations and generator, storage battery, and transformer rooms, but manholes and isolated transformer vaults on private premises are excluded.

Electrical wiring. Includes wiring methods, electrical conductors, conduits, insulators, junction boxes, switches, outlets, gas tubes, and other devices used for the functioning and safety of an electrical installation.

Electric cell. Single unit of primary or secondary battery that converts chemical energy into electrical energy.

Electric comfort heating appliance. Device that produces heat energy to create a warm environment by the application of electric power to resistance elements, refrigerant compressors, or dissimilar material junctions.

Electric contact. Physical contact that permits current flow between conducting parts.

Electric heating. Any method of converting electric energy to heat energy by resisting the free flow of electric current.

Electrician. Person who is engaged in the trade or business of electrical construction and who is qualified and has been registered under the terms of the local electrical code.

Electricity. Interaction between particles of positive and negative charge, utilized as a flow of electrons, producing electric current.

Electric lamp. Any type of a lamp in which light is produced by electricity, such as the incandescent lamp, fluorescent lamp, mercury-vapor lamp, arc lamp and glow lamp.

Electric meter. An electricity measuring device that totalizes with time, such as a watt-hour meter or ampere-hour meter.

Electric power meter. Device that measures electric power consumed, either at an instant, as in a wattmeter, or averaged over a time interval, as a demand meter.

Electric sign. Fixed, stationary, or portable self-contained electrically illuminated appliance with words or symbols designed to convey information or attract attention.

Electric supply lines. Conductors used to transmit electric energy and their necessary supporting or containing structures. Signal lines of more than 400 V to ground are always supply lines within the meaning of the rules, and those of less than 400 V to ground may be considered as supply lines if so run and operated throughout.

Enclosed

1. Surrounded by a case that will prevent a person from accidentally contacting live parts.

2. Surrounded by a case, housing, fence, or walls that will prevent persons from accidentally contacting energized parts.

Enclosure. Case or housing of apparatus or the fence or walls surrounding an installation to prevent personnel from accidentally contacting energized parts or to protect the equipment from physical damage.

Explosionproof. Enclosed in a case that is capable of withstanding an explosion of a specific gas or vapor which may occur within it, and/or preventing the ignition of a specific gas or vapor surrounding the enclosure by sparks, flashes, or explosion of the gas or vapor within. An explosionproof device must operate at such an external temperature that a surrounding flammable atmosphere will not be ignited thereby.

Exposed

1. As applied to circuits or lines, in such a position that in case of failure of supports or insulation contact with another circuit or line may result.

2. As applied to live parts, capable of being inadvertently touched or approached nearer than a safe distance by a person. The term is applied to parts not suitably guarded, isolated, or insulated.

3. As applied to wiring methods, not concealed.

4. As applied to wiring methods, on or attached to the surface or behind panels designed to allow access.

Feeder

1. Conductors of a wiring system between the service equipment or the generator switchboard of an isolated plant and the branch circuit overcurrent device.

2. All circuit conductors between the service equipment or the generator switchboard of an isolated plant and the final branch circuit overcurrent device.

3. Conductor or group of conductors that transmits electrical energy from a service supply, transformer, switchboard, distribution center, generator, or other source of supply to branch circuit overcurrent devices.

4. Set of conductors in a distribution system that extends from building switches or switchboards to distribution centers or panelboards.

Feeders. Conductors connecting a distribution panel to service equipment or disconnect switch.

Final unloaded conductor tension. Longitudinal tension in a conductor after the conductor has been stretched by the application for an appreciable period and subsequent release, of the heavy combination loading of ice and wind, and temperature decrease, required by the electrical codes or equivalent loading.

Final unloaded sag. Sag of a conductor after it has been subjected for an appreciable period to the loading prescribed, or equivalent loading, and the loading is removed.

Fished. Term applied to electrical wiring installed or fished through existing inaccessible hollow spaces of building with a minimum damage to the building finish.

Fish tape. Flat spring steel wire used to pull wires through walls or conduit.

Fitting. Accessory, such as a locknut, bushing, or other part of a wiring system, that is intended primarily to perform a mechanical rather than an electrical function.

Fixed appliance. Appliance that is fastened or otherwise secured at a specific location.

Flame arcs. Flame arcs are obtained by enlarging the core in the electrodes of a low-intensity arc and replacing part of the carbon with the chemical compounds known as flame materials, capable of radiating efficiently. Typical flame materials are iron for the ultraviolet, rare earths of the cerium group for white light, calcium compounds for yellow, and strontium for red.

Flame retardant. Material in ordinary locations will not burn for more than a specified period of time, nor will the flame travel or extend beyond a specified distance.

Flexible conduit. Conduit of metallic material which may be easily bent without the use of tools.

Flexible conduit (Greenfield). Wiring installed in flexible steel armor with outlets provided in the conduit system. The installation is similar to rigid conduit except that the conduit is flexible. It is used concealed or for short, flexible extensions of rigid conduit to vibrating equipment. This is not considered BX.

Flexible metallic conduit. Flexible raceway of circular cross section, especially constructed for drawing in or withdrawing wires and cables after the conduit and its fittings are in place. It is made of metal strip, usually of steel, with a metallic corrosion-resistant coating, helically wound, and with interlocking edges.

Flexible tubing. Flexible nonmetallic tubing for the mechanical protection of insulated wires.

Free electron emitted from the cathode. A free electron emitted from the cathode collides with one of the two valence electrons of a mercury atom and excites it by imparting to it part of the kinetic energy of the moving electron, thus raising the valence electron from its normal energy level to a higher one.

Fuse

1. Overcurrent protective device with a circuit-opening fusible part that is heated and severed by the passage of overcurrent through it. The fuse comprises all the parts that form a unit capable of performing the prescribed functions. It may or may not be the complete device necessary to connect into an electrical circuit.

2. A fuse consists of a housing in which there is a wire or strip material that will melt when carrying abnormal current and thus break the circuit.

3. Electric safety device that melts and interrupts current flow when the current becomes dangerously large.

Fuseway. Electrical circuit with its own main fuse.

Grounded. Connected to earth or to some conducting body which serves in place of earth.

Grounded conductor

1. Conductor that is intentionally grounded, either solidly or through a current-limiting device.

2. System or circuit conductor that is intentionally grounded.

Grounded system. System of conductors in which at least one conductor or point (usually the middle wire or neutral point of the transformer or generator windings) is intentionally grounded, either solidly or through a current-limiting device. The ground connection may be at one or more points.

Ground electrode. Buried metallic water piping system, or metal object or device buried in or driven into the ground so as to make intimate contact to which a grounding conductor is electrically and mechanically connected.

Ground fault circuit interrupter. Device whose function it is to interrupt, within a predetermined time, the electrical circuit to the load when a current to ground exceeds some predetermined value that is less than that required to operate the overcurrent protective device of the supply circuit.

Grounding conductor

1. Conductor used to connect equipment or the grounded circuit of a wiring system to a grounding electrode or electrodes.

2. Path of copper or other suitable metal specially arranged as a means whereby electrical equipment is electrically connected to a ground electrode.

Grounding electrode conductor

1. Conductor used to connect the grounding electrode to the equipment grounding conductor and/or to the grounded conductor of the circuit at the service.

2. Conductor used to connect equipment or the grounded circuit of a wiring system to a grounding electrode.

Grounding electrode resistance. Resistance of the grounding electrode to earth.

Grounding receptacle. Convenience outlet designed to receive a three-prong plug, the third prong permitting the appliance or equipment to be grounded, thus minimizing the shock hazard from faulty or loose wiring.

Grounding system. All conductors, clamps, ground clips, ground plates or pipes, and ground electrodes by means of which the electrical installation is grounded.

Ground protective relay. Relay that functions on failure of the insulation of a machine, transformer, or other apparatus to ground, or on flashover of a direct-current machine to ground.

Ground radiation. Visible radiation from the sun and sky reflected by surfaces below the plane of the horizon.

Ground wire. Electrical wire that conducts electricity from the structure to the ground or a grounding acceptable substitute.

Guarded

1. Covered, shielded, fenced, enclosed, or otherwise protected by means of suitable covers, casings, barriers, rails, screens, mats, or platforms to remove the liability of dangerous contact or approach by persons or objects to a point of danger. Wires that are insulated, but not otherwise protected, are not considered guarded.

2. Protected by personnel, or enclosed by means of suitable casings, or other methods or devices in accordance with standard barricading techniques designed to prevent dangerous approach or contact by persons or objects.

Guard zone. Space at minimum clearance from guards to electrical parts where guards may be installed by workers without definite engineering design.

Guy. One or more braces or cables used to stiffen a pole and keep it erect and in position.

Handhole

1. Enclosure installed in the earth, deck, floor of a building, or similar location and used as a pull or junction box for underground electrical or communication conductors. The enclosure is provided with a removable cover and so designed that the conductors may be pulled, spliced, or otherwise handled without requiring a person to enter the enclosure.

2. Opening in an underground system into which workers reach but do not enter.

Hazardous location. Area where ignitible vapors or dust may cause a fire or explosion created by excessive heat energy emitted from lighting or other electrical equipment.

Header. Transverse raceway for electrical conductors, providing access to predetermined cells of a cellular floor made of metal or concrete and permitting the installation of conductors from a distribution center to the cells.

Hot-line tools and ropes. Tools and ropes that are especially designed for work on energized high-voltage lines and equipment; insulated aerial equipment especially designed for work on energized high-voltage lines and equipment.

Hot wires. Power-carrying live wires (usually black or red).

Identified

1. In reference to a conductor or its terminal, the conductor or terminal is recognized as grounded.

2. As applied to a conductor, the white or natural gray covering and raised longitudinal ridge(s) on the surface of the extruded covering on the conductors of certain flexible cords.

3. As applied to other electrical equipment, terminals to which grounded or neutral conductors are to be connected are for identification by being tinned, nickel plated, or otherwise suitably marked.

Inaccessible. Covered by the structure or finish of the building, sufficiently remove from access, or so placed or guarded that unauthorized persons cannot inadvertently touch, interfere with, or enter the equipment room or compartments to which the term is applied.

Insulated. Separated from other conducting surfaces by a dielectric substance or air space permanently offering a high resistance to the passage of current and to disruptive discharge through the substance or space. When an object is insulated, it is understood to be insulated in a suitable manner for the conditions to which it is subjected; otherwise, it is uninsulated.

Insulated conductor. Conductor encased within material of a composition and thickness that is recognized by code as electrical insulation.

Insulating. When the term is applied to the covering of a conductor or to clothing, guards, rods, and other safety devices, it means that a device, interposed between a person and current-carrying parts protects the person making use of it against electric shock from the current-carrying parts with which the device is intended to be used.

Insulation. Material characterized by very few free electrons and high resistance. It is used to confine current to the desired path.

Insulation shielding. Envelope that encloses the insulation of a cable and provides an equipotential surface in contact with cable insulation.

Intrinsically safe. Term indicating that sparking that may occur in normal use or under any condition of fault likely to occur in practice is incapable of causing an ignition of the prescribed flammable gas or vapor.

Isolated. Term applied to an object not readily accessible to persons unless special means for access are used.

Isolated plant. Private electrical installation deriving energy from its own generator driven by a prime mover.

Joint use. Simultaneous use of facilities by two or more agencies not furnishing like services, but having use for similar facilities.

Journeyman electrician. Person who possesses the necessary qualifications, training, and technical knowledge to install apparatus or equipment for light, heat, or power, in accordance with plans and specifications furnished to him and electrical code rules and regulations governing such work.

Junction. Point in a circuit where two or more conductors are connected.

Knockout. Circular die-cut impression, made in electrical boxes, which may be removed to accommodate wiring.

Ladder cable trough. Cable trough with openings exceeding 2 in. in a longitudinal direction.

Lateral conductor. In pole wiring work, the wire or cable extending in a generally horizontal direction approximately at right angles to the general direction of the line conductors.

Lateral work space. Space reserved for working between conductor levels outside the climbing space, and to its right and left.

Lead acid cell. Voltage cell composed of lead plates with a diluted sulfuric acid electrolyte.

Line conductor. One of the wires or cables carrying electric current, supported by poles, towers, or other structures, but not including vertical or lateral connecting wires.

Liquidtight flexible metal conduit. Flexible metal conduit having an outer liquidtight jacket.

Main. Supply circuit to which branch or service circuits are connected through circuit breakers, switches, or fuses at different points along its length. The wire size of the main is the same for its entire length.

Manhole

1. Opening in an underground system through which workers or others may enter for the purpose of installing cables, transformers, junction boxes; and other devices and for making connections and tests.

2. Subsurface enclosure that personnel may enter and that is used for the purpose of installing, operating, and maintaining equipment and/or cable.

Manual. Capable of being used by personnel for instructions to operate the electrical system and equipment.

Master electrician. Person who possesses the necessary qualifications, training, and technical knowledge to plan, lay out, and supervise the installation of electrical wiring, apparatus, or equipment for light, heat, or power in accordance with furnished plans and specifications and standard rules and electrical code regulations governing such work.

Maximum total sag. Total sag at the midpoint of the straight line joining the points of support of the conductor.

Metal-clad cable. Fabricated assembly of insulated conductors and one or more adequate grounding conductors in a flexible metallic enclosure.

Metal wireway and raceway. Wiring installed in sheet steel raceways exposed in walls or ceilings. Outlet boxes are installed in the runs at points as required.

Meter. Device installed by the utility company for measuring the quantity of electric current used by the consumer.

Mineral-insulated cable. Cable having bare solid conductors supported and insulated by a highly compressed refractory material enclosed in a liquid- and gas-tight metallic tube sheathing. The term includes both the regular type (MI) and the lightweight type (LWMI), unless otherwise qualified.

Multioutlet assembly

1. Type of surface or flush raceway designed to hold conductors and attachment plug receptacles, assembled in the field or at the factory.

2. Series of plug-in receptacles, usually spaced at regular intervals and contained in a protective raceway (channel). Such assemblies also are termed "plug-in strips"; they may be substituted where convenience outlets are called for.

3. Surface or flush enclosure carrying conductors for extending one branch circuit to two or more receptacles of the grounding type which are attached to the enclosure.

Multiple-section mobile unit. Single structure composed of separate mobile units, each towable on its own chassis. When towed to the site, they are coupled together mechanically and electrically to form a single structure.

N.E.C. National Electrical Code (latest edition) published by the National Fire Protection Association. NFPA 70.

Nonmetallic sheathed cable. Rubber-insulated wires are bound together and covered with a cotton-bound paper sheath, which is protected with an outer cotton braid treated with a heat-resisting and moistureproof compound. The cable is installed exposed or concealed in hollow spaces in partitions, floors, or ceilings.

Nonmetallic waterproof cable. Rubber-insulated wires covered with a rubber sheath mounted on the surface of walls and ceilings in wet locations.

Nonventilated cable trough. Cable trough in which there are no ventilating openings in the bottom or sides.

Open. When applied to electrical equipment, the term indicates that moving parts, windings, or live parts are exposed to accidental contact.

Open circuit. Interrupted electrical circuit (electricity cannot flow).

Open-circuit batteries. Batteries intended to be used only on intermittent service, such as door bells, call bells, and in some cases, short-distance telephone lines.

Open wire. Conductor or pair of conductors separately supported above the surface of the ground.

Open wiring. Conductors not in a wiring enclosure.

Ordinary location (dry). Location in which at normal atmospheric pressure and under normal conditions of use, electrical equipment is not unduly exposed to injury from mechanical causes, excessive dust, moisture, or extreme temperatures, and in which electrical equipment is entirely free from the possibility of injury through corrosive, flammable, or explosive atmosphere.

Outlet

1. Point on the wiring system at which current is taken to supply utilization equipment.

2. Point on the wiring system at which current is taken to supply fixtures, lamps, heaters, motors, and electrical equipment generally.

3. Device that provides a power source for lights or appliances.

Outside wiring. Electrical wiring located outside of buildings, not including wiring for signs or any extensions of circuits supplying any load within the building.

Overhead system service entrance conductors

1. Service conductors between the terminals of the service equipment and a point usually outside the building, clear of building walls, where they are joined by tap or splice to the service drop.

2. Portion of the service conductors that connects the service drop to the service equipment.

Permit. Official written permission of the permit department on a form provided for the purpose, authorizing work to be commenced on any electrical installation and informing the electrical inspection department.

Plans and specifications. Plans and specifications and schedules submitted with the application for a permit to indicate or explain adequately the scope or manner of a proposed installation, alteration, replacement, or repair of electrical wiring or equipment.

Pole (electrical). Each of the lines or terminals between which a relatively large voltage exists.

Pole face. Side of a pole on which cross arms are attached or that is so designated by the companies owning or operating the pole.

Pole piece. Piece of high-permeability material used to reduce the air gap between magnetic poles.

Portable appliance

1. Appliance capable of being readily moved where established practice or the conditions of use make it necessary or convenient for it to be detached from its source of current by means of a flexible cord and attachment plug.

2. Appliance that is actually moved or can easily be moved from one place to another in normal use.

Portable electric space heaters. Heaters not intended for permanent connection to a structure or electric wiring and usually provided with a cord and plug.

Power supply cord. Assembly consisting of a suitable length of flexible cord or power supply cable provided with an attachment plug cap at one end.

Power outlet. Enclosed assembly that may include receptacles, circuit breakers, fuseholders, fused switches, buses, and watt-hour meter mounting means, intended to supply and control power to mobile homes, recreational vehicles, or boats or to serve as a means for distributing the power required to operate mobile or temporarily installed equipment.

Premises wiring (system). Interior and exterior wiring, including power, lighting, control, and signal circuit wiring, together with all of its associated hardware, fittings, and wiring devices, both permanently and temporarily installed, that extends from the load end of the service drop or of the service lateral conductors to the outlet(s). The wiring does not include wiring internal to appliances, fixtures, motors, controllers, motor control centers, and similar equipment.

Pulling tension. Longitudinal force exerted on a cable during installation.

Raceway. Any channel for holding wires, cables, or bus bars that is designed expressly, and used solely for this purpose. Raceways may be of metal or insulating material, and the term includes rigid metal conduit, rigid nonmetallic conduit, flexible metal conduit, electrical metallic tubing, underfloor raceways, cellular concrete floor raceways, cellular metal floor raceways, surface metal raceways, structural raceways, wireways, and busways.

Rainproof. So constructed, protected, or treated as to prevent rain from interfering with the successful operation of the apparatus.

Raintight. So constructed or protected that exposure to a beating rain will not result in the entrance of water.

Readily accessible. Capable of being reached quickly for operation, renewal, or inspection without requiring those to whom ready access is requisite to climb over or remove obstacles or to resort to portable ladders, chairs, etc.

Receptacle

1. Contact device installed at the outlet for the connection of a single attachment plug. A single receptacle is a single-contact device with no other contact device on the same yoke. A multiple receptacle is a single device containing two or more receptacles.

2. Contact device installed in an outlet for the connection of a portable lamp or appliance by means of a plug and flexible cord.

3. A convenience receptacle is a device used as a safe means of connecting a portable electrical appliance to an electric circuit. A receptable is provided with openings for the insertion of the blades of an attachment plug and is mounted in an insulated enclosure of porcelain or composition arranged for flush or surface mounting.

4. Baseboard or wall outlet into which an electric cord can be plugged.

Receptable outlet. Outlet where one or more receptacles are installed.

Rigid conduit. Rigid conduit of metallic or nonmetallic material.

Rigid metal conduit

1. Rigid conduit of metallic material made to the same dimensions as standard pipe and suitable for threading with standard pipe threads.

2. Tubular raceway with threaded ends, for electric wires and cables. If of ferrous metal, conduit should have a corrosion-resistant coating on all surfaces except threads and if of corrosion-resistant material, conduit should be identified; in either case conduit should be finished with a uniformly smooth interior coating of enamel or like material. Conduit may be made of mild steel tubing of circular cross section having walls which in the various electrical trade sizes comply with the measurements set forth in the electrical codes.

Rigid nonmetallic conduit. Rigid conduit of nonmetallic material which is not permitted to be threaded.

Rigid PVC conduit. Rigid nonmetallic conduit of unplasticized polyvinyl chloride.

Romex. Trade name for nonmetallic sheathed cable used for indoor wiring where permitted by electrical codes.

Separate built-in cooking unit. Stationary cooking appliance, including its integral supply leads or terminals, consisting of one or more surface elements or ovens, or a combination of these, and constructed so that the unit is permanently built into a counter or wall.

Service. Conductors and equipment for delivering energy from the electricity supply system to the wiring system of the premises served.

Service box. Approved assembly consisting of a metal box or cabinet constructed so that it may be effectually locked or sealed, containing either service fuses and a service switch or a circuit breaker. It is designed so that either the switch or circuit breaker may be manually operated when the box is closed.

Service cable. Service conductors made up in the form of a cable.

Service conductor

1. Supply conductors that extend from the street main or transformers to the service equipment of the premises supplied.

2. Portion of the supply conductors that extends from the supply main, duct, or transformers of the serving agency to the service equipment of the premises supplied. For overhead conductors this includes the conductors from the last line pole to the service equipment.

Service drop

1. Overhead service conductors between the last pole or other aerial support and the first point of attachment to the building or other structure.

2. Overhead service conductors from the last pole or other aerial support to and including the splices, if

any, connecting to the service entrance conductors at the building or other structure.

Service entrance. System of service conductors bringing electricity into the building and service equipment that controls and distributes it where needed in the building.

Service entrance conductors—overhead system. Service conductors between the terminals of the service equipment and a point usually outside the building, clear of building walls, where they are joined by tap or splice to the service drop.

Service entrance conductors—underground system. Service conductors between the terminals of the service equipment and the point of connection to the service lateral. Where service equipment is located outside the building walls, there may be no service entrance conductors, or they may be entirely outside the building.

Service feeder. Set of entrance conductors from the utility company's network to the main building switches or switchboard.

Service lateral. Underground service coinductors between the street main, including any risers at a pole or other structure or from transformers, and the first point of connection to the service entrance conductors in a terminal box, meter, or other enclosure with adequate space, inside or outside the building wall. Where there is no terminal box, meter, or other enclosure with adequate space, the point of connection is considered to be the point of entrance of the service conductors into the building.

Service raceway. Rigid metal conduit, electrical metallic tubing, or other raceway that encloses service entrance conductors.

Shockproof. As applied to x-ray and high-frequency equipment, the term indicates that the equipment is guarded with grounded metal so that no person can come into contact with any live part.

Short circuit. Power interruption usually caused by contact between a live wire and a ground wire in the same 115- or 230-V circuit.

Short-circuit ground. Fault in an electrical circuit in which current passes to ground instead of through the desired parts..

Short cycle. System tendency to start and stop more frequently than it should.

Short time duty. Requirement of service that demands operation at a substantially constant load for a short and definitely specified time.

Slow-burning. As applied to conductor insulation, the term indicates that the insulation has flame-retarding properties.

Soldered. Uniting of metallic surfaces by the fusion thereon of a metallic alloy, usually of lead and tin.

Span length. Horizontal distance between two adjacent supporting points of a conductor.

Special-purpose outlet. Point of connection to the wiring system for a particular piece of equipment, normally reserved for the exclusive use of that equipment. Such outlets may be plug-in receptacles to which the equipment is connected with a flexible cord, as with electric dryers or ranges, or they may be protective enclosures (junction boxes) with removable covers containing "permanent" connections. Such junction boxes can be mounted either on the structure or on the equipment.

Splices. All conductors spliced or joined are made mechanically and electrically secure without solder, if an approved splicing device is used. Otherwise conductor should be soldered with a fusible metal or alloy or brazed or welded. All splices and joints and the free ends of conductors should have mechanical protection and be covered with an insulation equal to that on the conductors.

Split-receptacle outlet. Duplex receptacle that has been wired so that each plug-in position is on a separate 120-V circuit; two separate line wires of a single three-wire 240-V circuit.

Splitter. Enclosure containing terminal plates or bus bars having main and branch connectors.

Starting torque. Torque available to start a load from a standstill; torque exerted by the starting current of the motor to overcome the static friction at rest. Also known as "locked-rotor" or "breakaway torque."

Stationary appliance. Appliance that is not easily moved from one place to another in normal use.

Stationary electric space heaters. Heaters permanently mounted in a structure and permanently connected to their electric wiring.

Structure conflict. As applied to a pole line, the line is so situated with respect to a second line that the overturning (at the ground line) of the first line will result in contact between its poles or conductors and the conductors of the second line, assuming that no conductors are broken in either line. Lines are not considered conflicting where one line crosses another or where two lines are on opposite sides of a highway, street, or alley and are separated by a distance not less than 60% of the height of the taller pole and not less than 20 ft.

Subfeeder. Extension of a feeder from one distribution center to another.

Substantial. Constructed and arranged so as to be of adequate strength and durability for the service to be performed under the prevailing conditions.

Supply authority. Person, firm, corporation, company, commission, or other organization-supplying electric energy.

Supply service. Any one set of conductors run by a supply authority from its mains to a consumer's service.

Surface raceway. Raceway in the form of a channel with a backing and capping for loosely holding conductors in wiring on a surface or pendent from the surface.

Surge. Sudden variation in current or voltage; pulse.

Tag. System or method of identifying circuits, systems, or equipment for the purpose of alerting persons that the circuit, system, or equipment is being worked on.

Temperature rise test. Measure of the motor's capacity to dissipate the heat generated by its electrical power losses. If the nameplate shows a temperature rise of 40°C and the ambient temperature is 90°F (32.2°C), the motor could operate at 162°F (72°C). If the motor becomes hot, hang one thermometer (0° to 110°C or 0.230°F in the room near the motor, and place one thermometer in firm contact with the bearing bushes, stator, or other hot stationary part of the motor, and check the rise.

Terminal. Element that connects electrical devices to a circuit.

Temperature radiator. Radiator whose flux density (radiant emittance) is determined by its temperature and the material and character of its surface and is independent of its previous history.

Temperature rise. Difference between the temperature of the surrounding air and that of different parts of a motor.

Temporary service for construction purposes. Installation of a temporary service to a temporary pole for construction purposes, provided that all requirements of the applicable local codes are complied with before a permanent service is installed.

Thin-wall conduit (electrical metallic tubing). Thin-walled steel or corrosion-resistant metal raceway of circular cross section, constructed for the purpose of pulling in or withdrawing wires after it is installed in place. It is coated inside and out, is corrosion resistant, and is connected by means of threadless fittings. The interior diameters should be the same as for the corresponding trade sizes of rigid conduit.

Total sag. Distance measured vertically from any point of a conductor to the straight line joining its two points of support, under conditions of ice loading equivalent to the total resultant loading.

Transformer vault. Isolated fire-resistant enclosure, either above or below ground, in which transformers and related equipment are installed and that is not continuously attended during operation.

Underfloor raceway

1. Raceway suitable for use in the floor.
2. Wiring is installed in sheet metal or fiber duct, which is embedded in the concrete or cement fill of the floor.

Underground system service entrance conductors. Service conductors between the terminals of the service equipment and the point of connection to the service lateral. Where service equipment is located outside the building walls, there may be no service entrance conductors, or they may be entirely outside the building.

Unloaded sag. The unloaded sag of a conductor at any point in a span is the distance measured vertically from the particular point in the conductor to a straight line between its two points of support, without any external load.

Vaportight. Enclosed so that vapor will not enter the enclosure.

Vault. Enclosure above or below ground, which personnel may enter, used for the purpose of installing, operating, and/or maintaining equipment and/or cable.

Vault (transformer vault or electrical equipment vault). Isolated ventilated enclosure, either above or below ground, with fire-resistant walls, ceilings, and floors, for the purpose of housing transformers and other electrical equipment.

Ventilated. Provided with means to permit circulation of air in an enclosed space sufficiently to remove excess heat, fumes, or vapors.

Ventilated cable trough. Cable trough having adequate ventilating openings, with no opening exceeding 2 in. in a longitudinal direction.

Ventilated flexible cableway (VFC). Ventilated metal raceway into which conductors may be drawn. It is designed to be rigid in one plane and flexible in one at 90° to it, and it is constructed with ventilating openings comprising approximately 30% of the surface.

Vertical conductor. In pole wiring work, a wire or cable extending in an approximately vertical direction.

Wall-mounted oven. Oven for cooking purposes designed for mounting in or on a wall or other surface and consisting of one or more heating elements, internal wiring, and built-in or separately mountable controls.

Watertight. Constructed so that moisture will not enter the enclosing case.

Weatherproof

1. Constructed or so protected that exposure to the weather will not interfere with successful operation. Raintight or watertight equipment may fulfill the requirements for "weatherproof." Weather conditions resulting from snow, ice, dust, or temperature extremes.
2. As applied to the protective covering on a conductor, a covering made up of braids of fibrous material that are thoroughly saturated with a dense moistureproof compound after they have been placed on the conductor, or an equivalent protective covering designed to withstand weather conditions.

Wet locations

1. Installations underground or in concrete slabs or masonry in direct contact with the earth; locations subject to saturation with water or other liquids, such as vehicle washing areas; locations exposed to weather and unprotected.
2. Locations in which uncontrolled liquids may drip, splash, or flow on or against electrical equipment.

Wire television distribution system. Distribution system in a building having coaxial or other suitable cable or wire, together with any necessary amplifiers, that is used in the transmission of television signals.

Wireway. Raceway consisting of a completely enclosing system of metal troughing and fittings so formed and constructed that insulated conductors may be readily drawn in and withdrawn, or laid in and removed, after the system has been completely installed, without injury either to conductors or their covering.

Wiring enclosure. Any raceway, cabinet, box, fitting, enclosed switchboard compartment, or other case approved for the protection of electric wiring and permitted by the local code.

Wiring method. Type of wiring and equipment, and details of installation required or used in an electrical installation.

Wiring switches. Switches used for controlling branch circuits and individual lights or appliances in interior wiring systems. They are manufactured with push-button, rotary, toggle, locking, quiet, or mercury types of operating mechanisms, arranged for flush or surface mounting.

Working standard. Any calibrated device for daily use in measurement work.

ELECTRONICS

Alternating current. Electrical current that reverses its direction of flow periodically.

Atom. An atom consists of a central nucleus possessing a positive charge about which rotate negatively charged electrons. In the normal state these electrons remain in particular orbits or energy levels and radiation is not emitted by the atom.

Capacity. Capacity is measured by the charge that must be communicated to a body to raise its potential one unit. Electrostatic unit capacity is that which requires one electrostatic unit of charge to raise its potential one electrostatic unit. A capacity of 1 F requires 1 C of electricity to raise its potential 1 V.

Cathode. Negative electrode in an electrical circuit. Cathode rays are negatively charged particles (electrons) that are repelled from the cathode and stream through the partially evacuated space of a cathode-ray tube.

Conductance. Reciprocal of resistance, measured by the ratio of the current flowing through a conductor to the difference of potential between its ends. The practical unit of conductance, the mho, is the conductance of the body through which 1 A of current flows when the potential difference is 1 V. The conductance of a body in mhos is the reciprocal of the value of its resistance in ohms.

Conductivity. Conductivity is measured by the quantity of electricity transferred across a unit area per unit potential gradient per unit time. Reciprocal of resistivity.

Conductor

1. Material through which electrical current will readily flow. Metals are generally good conductors.
2. Substance capable of transmitting electricity, heat, or sound.
3. Class of bodies that are incapable of supporting electric strain. The charge given to a conductor spreads to all parts of the body.

Current (electric). Rate of transfer of electricity. Transfer at the rate of 1 electrostatic unit of electricity in 1 second is the electrostatic limit of current. An electromagnetic unit of current is a current of such strength that 1 cm of the wire in which it flows is pushed sideways with a force of 1 dyne when the wire is at right angles to a magnetic field of unit intensity. The practical unit of current is the ampere, the transfer of 1 C/second, which is one-tenth the electromagnetic unit. An ampere-turn is the magnetic potential produced between the two faces of a coil of one turn carrying 1 A.

Diamagnetic. Term applied to bodies that tend to set the longest dimension across the magnetic field. The permeability of a diamagnetic substance is less than unity.

Dielectrics, insulators, or nonconductors. Class of bodies supporting an electric strain. The charge on one part of a nonconductor is not communicated to any other part.

Diode. Device that allows electrical current to pass in only one direction. Diodes are often constructed from semiconductor materials.

Electrical current. Flow of electrical charge (usually electrons) around a circuit. The unit of current is the ampere.

Electrical field. Condition in space set up by electrical charges to which other electrical charges react. A free-charge particle will experience a force and be accelerated in an electric field.

Electrical force. One of the four basic forces in nature. The electrical force between two objects is directly proportional to the product of their electrical charges and inversely proportional to the square of the distance between them. The force is repulsive if the charges have the same sign and is attractive if the charges have opposite signs.

Electrical power. The electrical power delivered to a particular part of an electrical circuit is equal to the product of the voltage across the element and the current flowing through it. The unit of electrical power is the watt.

Electric conduction. Ability of a material to conduct or transmit electric current. In a metal conduction occurs by

electron motion of the electron cloud. Nonmetals conduct by a flow of ions within the structure.

Electric field intensity. Electric field intensity is measured by the force exerted on a unit charge. The unit field intensity is the field that exerts the force of 1 dyne on a unit positive charge.

Electric field strength. Electric field strength gives the force per unit charge exerted on an electrical charge in the field.

Electrochemical equivalent. The electrochemical equivalent of an ion is the mass liberated by the passage of a unit quantity (1 C), of electricity.

Electrode. Either terminal of an electric source; either conductor by which the current enters or leaves an electrolyte.

Electrodynamics. The phenomena of electricity in motion; the science of the action of electric currents on themselves and on one another, and of the interaction of currents and magnets.

Electrokinetics. Electrodynamics treating of the laws of distribution of electric current.

Electrolysis. Conduction of an electric current by the charged particles (ions) in an electrolyte so that molecules are formed at the electrodes and liberated as gases or deposited as solids. Chemical decomposition by an electric current.

Electrolyte

1. Liquid conductor containing ions, the passing of an electric current through which causes a liberation of matter at the electrodes, either the giving off of a gas or the depositing of a solid.
2. Substance, such as sodium chloride or an acid, whose water solution produces ions and thus makes the liquid an electrical conductor.

Electrolytic conductance. Tranport of electric charges, under electric potential differences, by charged particles (ions) of atomic or larger size.

Electrolytic conductivity. Conductivity of a medium in which the transport of electric charges, under potential differences, is by particles of atomic or larger size.

Electromagnet. Magnet whose temporary magnetism is due to an electrical current flowing in a wire that is wound around a part of the magnet.

Electromagnetic. Pertaining to the combined electric and magnetic fields associated with radiation or with movements of charged particles.

Electromagnetic current. Motion of charged particles giving rise to electric and magnetic fields.

Electromagnetic field. Magnetic forces developed by the passage of an electric current through a conductor.

Electromagnetic induction. Generation of an electrical current by a changing magnetic field.

Electromagnetic line of sight. Maximum distance at which direct wave transmission is possible between transmitting and receiving antennas of given height, neglecting propagation anomalies.

Electromagnetic radiation

1. Radiation produced by accelerated charges. It takes the form of propagating waves.
2. Energy propagated through space or through material mediums in the form of an advancing disturbance in electric and magnetic fields existing in space or in the mediums. The term "radiation," alone, is used commonly for this type of energy, although it actually has a broader meaning.

Electromagnetic spectrum

1. Ordered array of all known electromagnetic radiations, extending from the shortest gamma rays through x-rays, ultraviolet radiation, visible radiation, and infrared radiation, and including microwave and all other wavelengths of radio energy.

2. The electromagnetic spectrum consists of radiations of all frequencies from low-frequency radio waves to high-frequency rays and includes microwaves, radar waves, x-rays, as well as visible light.

Electromagnetic susceptibility. Tolerance of circuits and components to all sources of interfering electromagnetic energy.

Electromagnetic testing. Nondestructive test method for materials, including magnetic materials, that uses electromagnetic energy having frequencies less than those of visible light to yield information regarding the quality of testing material.

Electromagnetic wave. Propagating disturbance in the electromagnetic field. Such waves require no material medium and can propagate through empty space. Light, radio waves, and x-rays are all electromagnetic waves.

Electromagnetism. Magnetism developed by a current of electricity.

Electromechanical coupling factor. Factor used to characterize the extent to which the electrical characteristics of a transducer are modified by a coupled mechanical system, and vice versa.

Electrometer. Instrument for measuring the difference between the amounts of electrification of two (2) points.

Electromotive. Relates to motion of, or produced by, electricity.

Electromotive force (emf)

1. Force that can cause electric charge to move and thereby give rise to an electrical current. A battery is a source of emf.

2. Force that causes flow of current. The electromotive force of a cell is measured by the maximum difference of potential between its plates. The electromagnetic unit of potential difference is that against which 1 erg of work is done in the transfer of electromagnetic unit quantity. A volt is that potential difference against which 1 J of work is done in the transfer of 1 C; 1 V is equivalent to 10^8 electromagnetic units of potential.

Electromotor. Apparatus for generating a current of electricity.

Electron. The most elementary charge of negative electricity. Electrons are constituents of all atoms and are given off by hot bodies.

Electron accelerator. Device that accelerates electrons to high energies.

Electron attenuation length. Average distance that an electron with a given energy travels between successive inelastic collisions as derived from a particular model in which elastic scattering is assumed to be insignificant.

Electron beam. Narrow stream of electrons moving in the same direction, all having the same velocity.

Electron-beam generator. Velocity-modulated generator, such as a klystron tube, used to generate extremely high frequencies.

Electron charge. Basic unit of electrical charge. The charge carried by any particle or object is an integer number of electron charges.

Electron efficiency. Power that an electron stream delivers to the circuit of an oscillator or amplifier at a given frequency, divided by the direct power supplied to the stream.

Electron emitter. Electrode from which electrons are emitted.

Electron escape depth. Distance (in nanometers) normal to the surface at which probability of an electron escaping without significant energy loss due to inelastic scattering processes drops from its original value.

Electron flooding. Irradiation of a specimen with low-energy electrons in order to change or stabilize the charging potential.

Electron flow. Current produced by the movement of free electrons toward a positive terminal; the direction of electron flow is opposite to that of current.

Electronic. Pertaining to electronic devices or to circuits or systems utilizing electronic devices, including electron tubes, magnetic amplifiers, transistors, and other devices that do the work of electron tubes.

Electronic circuit. Electric circuit in which the equilibrium of electrons in some of the components is upset by means other than an applied voltage.

Electronic component. Component that is able to amplify or control voltages or currents without mechanical or other nonelectrical commands, such as electron tubes, transistors, and other solid-state devices.

Electronic control. Control of a machine or process by circuits using electron tubes, transistors, magnetic amplifiers, or other devices having comparable functions.

Electronic engineering. Engineering that deals with the design and practical applications of electronics.

Electronic fuse. Fuse that is set off by an electronic device that is incorporated in it.

Electronic interference. Any electrical or electromagnetic disturbance that causes undesirable response in electronic equipment.

Electronic jamming. Deliberate radiation, reradiation, or reflection of electromagnetic signals with the object of impairing the use of electronic devices by others.

Electronic motor control. Control circuit used to vary the speed of a direct-current motor operated from an alternating-current power circuit.

Electronics. Broad field pertaining to the conduction of electricity through a vacuum, gases, or solids and the circuits associated therewith.

Electronic specific heat. Contribution to the specific heat of a metal from the motion of conduction electrons.

Electronic structure. Arrangement of electrons in an atom, molecule, or solid, specified by their wave functions, energy levels, or quantum numbers.

Electronic surge arrester. Device used to switch to ground high-energy surges.

Electronic switch. Vacuum tube, crystal diodes, or transistors used as an on-off switching device.

Electronic switching. Use of electronic circuits to perform the functions of a high-speed switch.

Electronic voltage regulator. Device that maintains the direct current power supply voltage for electronic equipment nearly constant in spite of input alternating current line voltage variations and output load variations.

Electron trajectory. Path of an electron.

Electron velocity. Rate of motion of an electron.

Electron volt. Kinetic energy gained by an electron after passing through a potential difference of 1 V.

Electrophorus. Instrument for producing electric charges by induction.

Electroscope. Instrument for detecting an electric charge on a body.

Faraday's law. The mass of a substance decomposed by the passage of equal quantities of electricity through different electrolytic cells is, for the same electrolyte, equal, and for different electrolytes, proportional to the combining weights of the elements or radicals that are deposited.

Field lines. Field lines represent the map of a field, such as an electrical field or a gravitational field. The direction of a field line through a point indicates the direction of the force that a particle will experience at that point owing to the field. The density of the field lines indicates the magnitude of the force.

Galvanometer. Instrument used to measure electrical currents. By combining various resistances with a galvanometer, a voltmeter or an ammeter can be constructed.

Henry (H). The mks unit of self and mutual inductance, equal to the self-inductance of a circuit or the mutual inductance between two circuits if there is an induced electromotive force of 1 volt when the current is changing at the rate of one ampere per second.

Induced electromotive force. In a circuit, induced electromotive force is proportional to the rate of change of magnetic flux through the circuit.

Inductance. Change in magnetic field due to the variation of a current in a conducting circuit itself. The phenomenon is known as "self-induction." If an electromotive force is induced in a neighboring circuit, the term "mutual induction" is used. Inductance may thus be distinguished as self or mutual and is measured by the electromotive force produced in a conductor by unit rate of variation of the current. Units of inductance are the centimeter (absolute electromagnetic) and the henry, which is equal to 10^9 cm of inductance. The henry is that inductance in which an inducted electromotive force of 1 V is produced when the inducing current is changed at the rate of 1 A/second.

Induction. Change in the intensity or direction of a magnetic field causes an electromotive force in a conductor in the field. Induced electromotive force generates an induced current if the conductor forms a closed circuit.

Insulator. Material through which electrical current will not readily flow. Glass and plastics are good insulators.

Line of force. Description of an electric or magnetic field. A line such that its direction at every point is the same as the direction of the force that would act on a small positive charge (or pole) placed at that point. A line of force is defined as starting from a positive charge (or pole) and ending on a negative charge (or pole). A line (of force) is also used as a unit of magnetic flux, equivalent to the maxwell.

Magnetic field due to a current. Intensity of the magnetic field in gauss at the center of a circular conductor of radius in which a current in absolute electromagnetic units is flowing.

Maxwell (Mx). A centimeter-gram-second electromagnetic unit of magnetic flux that produces an electromotive force of one abvolt in a circuit of one turn linking the flux reduced to zero in one second at a uniform rate.

Neutrons. Electrically neutral elementary particles found in the nuclei of all atoms (except the lightest isotope of hydrogen). Neutrons are very similar to protons, the main difference being the lack of any electrical charge on a neutron.

Nuclear reactor. Device in which fission reactions are self-sustaining. The energy released in the continuing fission reactions can be converted into useful electrical energy.

Potential (electric). At any point potential is measured by the work necessary to bring unit positive charge from an infinite distance. Difference of potential between two points is measured by the work necessary to carry unit positive charge from one to the other. If the work involved is 1 erg, we have the electrostatic unit of potential.

Resistance. Property of conductors, depending on their dimensions, material, and temperature, that determines the current produced by a given difference of potential. The practical unit of resistance, the ohm, is that resistance through which a difference of potential of 1 V will produce a current of 1 A.

Resistance of conductors in series and parallel. The total resistance of any number of resistances joined in series is the sum of the separate resistances.

Resistivity (electrical). The electrical resistance offered by a material to the flow of current times the cross sectional area of current flow and per unit length of current path; the reciprocal of the conductivity.

Semiconductors. Materials with electrical conductivity properties intermediate between those of conductors and insulators.

Specific inductive capacity. The ratio of the capacity of a condenser with a given substance as dielectric to the capacity of the same condenser with air or a vacuum as dielectric is called the specific inductive capacity. Ratio of the dielectric constant of a substance to that of a vacuum.

Superconductor. Metallic element or alloy that loses all resistance to the flow of electrical current at some temperature near absolute zero.

Surface density of electricity. Quantity of electricity per unit area.

Surface density of magnetism. Quantity of magnetism per unit area.

Surface resistivity. Resistance of unit length and unit width of a surface.

Susceptibility (magnetic). Susceptibility is measured by the ratio of the intensity of magnetization produced in a substance to the magnetizing force or intensity of field to which it is subjected. The susceptibility of a substance will be unity when unit intensity of magnetization is produced by a field of 1 G.

Temperature resistance coefficient. Ratio of the change of resistance in a wire due to a change of temperature of 1°C to its resistance at 0°C.

Thermoelectric power. Thermoelectric power is measured by the electromotive force produced by a thermocouple for unit difference of temperature between the two junctions. It varies with the average temperature and is usually expressed in microvolts per degree Celsius. It is customary to list the thermoelectric power of the various metals with respect to lead.

Thomson thermoelectric effect. Designation of the potential gradient along a conductor which accompanies a temperature gradient. The magnitude and direction of the potential varies with the substance. The coefficient of the Thomson effect or specific heat if electricity is expressed in joules per coulomb per degree Celsius.

Transformer. Device for increasing or decreasing the voltage in an alternating-current circuit.

Transistor. Three-element device consisting of semiconductor layers that can control and amplify electrical signals.

Volt (V). Unit of measure of potential difference.

Voltage. The electrical pressure or potential difference that causes current to flow. The unit of potential difference is the volt.

LIGHTING AND ILLUMINATION

Absorptance. Ratio of the flux absorbed by a medium to the incident flux.

Absorption. Process by which incident flux is dissipated within a medium.

Absorption spectrum (or dark line spectrum). Series of lines of definite wavelength produced when a beam of white light (all colors) passes through a medium that selectively absorbs some of the light. The lines of an absorption spectrum are characteristic of the medium through which the light passes, not the source of the light.

Accent lighting

1. Directional lighting to emphasize a particular object or draw attention to a part of the field of view.
2. Directional lighting, such as spot lighting, to emphasize a particular object.

Access to boxes. Electric discharge lighting fixtures, surface mounted over concealed outlet, pull, or junction boxes should be installed with suitable openings in the back of the fixture to provide access to the boxes.

Accommodation. Process by which the eye changes focus for objects at various distances, involving changes in shape of the crystalline lens.

Achromatic. Lenses signifying their more or less complete correction for chromatic aberration.

Adaptation

1. Process by which the retina becomes accustomed to more or less light than it was exposed to during an immediately preceding period. Adaptation results in a change in the sensitivity of the eye to light.
2. Process by which the eye adapts itself to light, involving primarily a change in the sensitivity of the photoreceptors.

Adaptive color shift. Change in the perceived color of an object caused solely by change of chromatic adaptation.

Aeronautical beacon. Light specifically provided as an aid to air navigation, visible at all azimuths either continuously or intermittently, and used to designate a particular location on the surface of the earth.

Aeronautical light. Luminous sign or signal accepted, approved, and recognized by competent authority, that is established, maintained, exhibited, or operated as an aid to air navigation.

Altitude. Angular distance of a heavenly body measured on the great circle which passes perpendicular to the plane of the horizon through the body and through the zenith. It is measured positively from the horizon to the zenith, from $0°$ to $90°$.

Angle of collimation. Angle subtended by a light source at a point on an irradiated surface.

Angular aperture. The angular aperture of an objective is the largest angular extent of wave surface that it can transmit.

Apochromat. Photographic and microscope objectives indicating the highest degree of color correction.

Apostilb. $1/\pi$ candela/m^2. A unit of luminance.

Apparent candlepower at a specified distance. Candlepower of a point source that would produce the same illumination at that distance.

Apparent candlepower of an extended source. Candlepower of a point source that would produce the same illumination at that distance.

Approach lights. Configuration of aeronautical ground lights located in an extension of a runway or channel to provide visual approach and landing guidance to pilots.

Arc lamp. Application of an electric arc to produce a brilliant light. Arc lamps are used in large spot lights and motion-picture projectors.

Astigmatism. Error of spherical lenses peculiar to the formation of images by oblique pencils. The image of a point when astigmatism is present will consist of two focal lines at right angles to each other and separated by a measurable distance along the axis of the pencil. The error is not eliminated by reduction of aperture as is spherical aberration.

Aurora borealis (northern lights). Hazy horizontal patches or bands of greenish light on which white, pink, or red streamers sometimes appear 60 to 120 miles above the earth. Apparently, they are caused by electron streams spiraling into the atmosphere, primarily at polar latitudes. Some of their spectrum lines have been identified with transitions from metastable states of oxygen and nitrogen atoms.

Average luminance of a luminaire. Luminous intensity at a given angle divided by the projected area of the luminaire at that angle.

Average luminance of a surface. The average luminance (average photometric brightness) of a surface may be expressed in terms of the total luminous flux (lumens) actually leaving the surface per unit area. It is identical in magnitude with luminous existance, which is the preferred term.

Azimuth. Angular distance between the vertical plane containing a given line or a celestial body and the plane of the meridian.

Back light. Illumination from behind the subject in a direction substantially parallel to a vertical plane through the optical axis of the camera.

Bactericidal (germicidal) effectiveness. Capacity of various portions of the spectrum to destroy bacteria, fungi, and viruses.

Baffle. Single opaque or translucent element to shield a source from direct view at certain angles or to absorb unwanted light.

Ballast

1. Circuit element that serves to limit an electric current or to provide a starting voltage, as in certain types of lamps such as in fluorescent lighting fixtures.
2. Devices used with an electric discharge lamp to provide the necessary circuit conditions for starting and to limit the operating current.
3. For direct-current operation, ballasts are composed of resistance which involves appreciable power loss, and

lamp starting voltage is limited to the maximum line voltage available. For these reasons direct-current operation is practical only for special applications.

4. For alternating-current operation ballasts are composed of transformers, inductive and capacitive reactors, and resistors. Ballasts provide starting and operating voltages, cathode heating, lamp current limitation, shock hazard protection, and radio interference suppression.

Base light. Uniform, diffuse illumination approaching a shadowless condition, sufficient for a television picture of technical acceptability. It may be supplemented by other lighting.

Beacon. Projector designed for communication by light signal or to indicate a geographical location.

Beam spread

1. Angle enclosed by two lines that intersect the candlepower distribution curve at the points where the candlepower is equal to 10% of its maximum.

2. In any plane, the angle between the two directions in which the candlepower is equal to a stated percent (usually 10%) of the maximum candlepower in the beam.

Blackbody. Temperature radiator of uniform temperature of which the radiant emittance in all parts of the spectrum is the maximum obtainable from any temperature radiator operating at the same temperature. A blackbody will absorb all radiant energy falling upon it. It is practically realized in the form of a cavity with opaque walls at uniform temperature and with a small opening for observation purposes.

Blackbody (Planckian) locus. Locus of points on a chromaticity diagram representing the chromaticities of blackbodies having various temperatures.

Blackbody radiation. Light from practical light sources, particularly that from incandescent sources, often described by comparison with that from a blackbody or complete radiator.

Black light. Popular term for ultraviolet energy near the visible spectrum.

Bracket or mast arm. Attachment to a lamp post or pole from which a luminaire or lighting fixture is suspended.

Brewster's law. The tangent of the polarizing angle for a substance is equal to the index of refraction. A polarizing angle is that angle of incidence for which the reflected polarized ray is at right angles to the refracted ray.

Bright-field. Having a brightly lighted background.

Bright-line spectrum. Emission spectrum made up of bright lines on a dark background.

Brightness. Luminous intensity of any surface in a given direction per unit of projected area of the surface as viewed from that direction; property of a surface.

Brightness control. Control that varies the luminance of the fluorescent screen of a cathode-ray tube, for a given input signal, by changing the grid basis of the tube and the beam current.

Brightness of a perceived light source color. Attribute of a perceived light source color in accordance with which the source seems to emit more or less luminous flux per unit area.

Brightness ratio. Ratio between the photometric brightness of any two relatively large areas in the visual field.

Candela

1. Formerly candle.

2. Unit of luminous intensity; 1 candela is defined as the luminous intensity of one-sixtieth of 1 sq cm of projected area of a blackbody radiator operating at the temperature of solidification of platinum.

Candle

1. Primary standard of light is a blackbody radiator operated at the temperature of solidification of platinum. The candle (unit of luminous intensity) is defined as one-sixtieth of the luminous intensity of 1 sq cm of such a radiator. Values for standards having other spectral distributions are derived by the use of accepted spectral luminous efficiency data.

2. Unit of luminous intensity; 1 lumen per unit solid angle (steradian).

Candle (or international candle). Unit of luminous intensity; specified fraction of the average horizontal candlepower of a group of 45 carbon-filament lamps preserved at the Bureau of Standards.

Candle per square centimeter. Brightness of a surface that has, in the direction considered, a luminous intensity of 1 candle/sq cm.

Candle per unit area. Unit of photometric brightness (luminance).

Candlepower

1. Luminous intensity expressed in candles. The mean horizontal candlepower is the average candlepower in the horizontal plane passing through the luminous center of the light source.

2. Luminous intensity expressed in candelas.

Candlepower distribution curve

1. Curve, generally polar, representing the variation of luminous intensity of a lamp or luminaire in a plane through the light center.

2. Curve showing the variation of luminous intensity of a lamp or luminaire with the angle of emission. A vertical candlepower distribution curve is obtained by taking measurements at various angles of elevation in a vertical plane through the light center, unless the angle of azimuth is specified, a vertical curve is assumed to represent an average such as would be obtained by rotating the unit about its vertical axis. A horizontal candlepower distribution curve represents measurements made at various angles of azimuth in a horizontal plane through the light center.

Carbon arc lamp. Electric discharge lamp in which light is produced by an arc discharge between carbon electrodes. One or more of the electrodes may contain chemicals that contribute importantly to the radiation.

Ceiling area lighting. General lighting system in which the entire ceiling is, in effect, one large luminaire.

Central vision. Seeing of objects in the central or foveal part of the visual field, approximately 2° in diameter. Central vision permits seeing much finer detail than does peripheral vision.

Chemiluminescence. Emission of light during a chemical reaction.

Chrome. Relative purity, strength, or saturation of a color, directly related to the dominance of the determining wavelength of the light and inversely related to grayness; one of the three variables of color.

Chromatic aberration. Owing to the difference in the index of refraction for different wavelengths, the light of various wavelengths from the same source cannot be focused at a point by a simple lens. This phenomenon is called chromatic aberration.

Chromaticity coordinates of a light. Ratio of each of the tristimulus values of the light to the sum of the three tristimulus values. Chromaticity coordinates in the CIE system of color specification are designated by x, y, and z.

Chromaticity diagram. Plane diagram formed by plotting one of the three chromaticity coordinates against another.

Chromaticity of a color. Dominant or complementary wavelength and purity aspects of the color taken together; aspects specified by the chromaticity coordinates of the color taken together.

Circuit conductors. Branch circuit conductors, having an insulation suitable for the temperature encountered, permitted to terminate in a fixture.

Clear sky. Less than 30% cloud cover.

Clerestory. Part of a building that rises clear of the adjacent roofs of other parts of the structure and whose walls contain windows for lighting the interior.

Cloudy sky. More than 70% cloud cover.

Coefficient of utilization. Ratio of luminous flux (lumens) received on the work plane to the rated lumens emitted by the lamps.

Coffer. Recessed panel or dome in the ceiling.

Coffered lighting ceiling. Standard module used to allow lighting arrangements with either flat or coffered lighting modules.

Cold-cathode lamp

1. Lamp that uses a rugged iron thimble-type electrode at each end and is not affected by the number of lamp starts. The lamp starts instantly by voltage across its ends, which draws off enough electrons to establish the arc.
2. Lamp with an iron-shell electrode. It starts instantly and operates at a lower temperature.

Color. Characteristics of light by which a human observer may distinguish between two structure-free patches of light of the same size and shape.

Color code. System of colors to indicate resistor values, capacitor characteristics, and circuit connections.

Colorimetric shift. Change of chromaticity and luminance factor of an object color due to change of the light image.

Color of a light source. Characteristics of the source determined by its spectral composition and the spectral properties of the average normal human eye.

Color of an object. Color of the light reflected or transmitted by the object when illuminated by a standard light source.

Color of light emitted by the electroluminescent lamp. The color is dependent on frequency, while the brightness is strongly dependent on both voltage and frequency. The effects of both voltage and frequency change with the specific phosphors. Present efficacy is low compared even to incandescent lamps, though comparable to colored incandescent lamps with filter coats giving blue or green light.

Color rendering. Effect of a light source on the color appearance of objects in conscious or subconscious comparisons with their color appearance under a reference light source.

Color rendition. Property of light that controls the color appearance of objects viewed under the specific light source. There is, at present, no generally accepted method of evaluating this property numerically. The term should not be confused with color appearance of the light source itself.

Color temperature. The radiation characteristics of a blackbody of unknown area may be specified, by fixing only two quantities: the magnitude of the radiation at any given wavelength and the absolute temperature. The same type of specification may be used with reasonable accuracy for tungsten filaments and other incandescent sources. However, in the case of selective radiators, the temperature used is not that of the filament but a value called the "color temperature."

Color temperature of a light source. Temperature at which a blackbody radiator must be operated to have a chromaticity equal to that of the light source.

Color value. Relative lightness or intensity of color, approximately a function of the square root of the total amount of light; one of the three variables of color.

Coma. Aberration of lenses, occurring in the case of oblique incidence, similar to spherical aberration of the axial rays. The image of a point is comet shaped; hence the name.

Commercial electric lampposts. Street lampposts on public property; not owned, operated, or maintained by the city.

Complementary wavelength of a light. Wavelength of radiant energy of a single frequency that matches the color of the reference standard when combined in suitable proportion with the light.

Compound or mixed reflection. Simultaneous occurrence of regular and diffuse reflection in any proportion.

Conjugate foci. Under proper conditions, light divergent from a point on or near the axis of a lens or spherical mirror is focused at another point. The point of convergence and the position of the source are interchangeable and are called "conjugate foci."

Contrast rendition factor. Ratio of visual task contrast with a given lighting environment to the contrast with sphere illumination.

Contrast sensitivity. Ability to detect the presence of luminance differences. Quantitatively, it is equal to the reciprocal of the contrast threshold.

Contrast threshold. Minimal perceptible contrast for a given state of adaptation of the eye. Luminance contrast is detectable during some specific fraction of the times it is presented to an observer, usually 50%.

Cord pendant. Cord hanging freely in the air in a vertical position, which has a fixed connection to a permanent wiring enclosure at the upper end and a suitable cord connection body (receptacle) or lamp holder attached to the lower end.

Cornice lighting. System comprising light sources shielded by a panel parallel to the wall and attached to the ceiling, and distributing light over the wall.

Course light. Aeronautical ground light supplementing an airway beacon, used to indicate the direction of the airway and to identify by a coded signal the location of the airway beacon with which it is associated.

Cove lighting. System comprising light sources shielded by a ledge, a vertical or horizontal recess, and distributing light over the ceiling and upper and sidewalls.

Corpuscular theory. Luminous bodies emit radiant energy in particles. The particles are intermittently ejected in the straight lines and act on the retina of the eye, stimulating the optic nerves to produce the sensation of light.

Cosine law. Lambert's cosine law states that the illumination of any surface varies as the cosine of the angle of incidence.

Cross light. Equal illumination in front of the subject from two directions at substantially equal and opposite angles with the optical axis of the television camera and a horizontal plane.

Cutoff angle of a luminaire. Angle, measured up from nadir, between the vertical axis and the first line of sight at which the bare source is not visible.

Cutoff starter (fluorescent lamps). Starters made to reset either manually or automatically and designed to prevent repeated blinking or attempts to start a deactivated lamp.

Daylight factor. Ratio of the illumination at a given point inside a building to the simultaneous exterior illumination on a horizontal plane from the whole of an unobstructed sky of assumed or known luminance distribution. Direct sunlight is excluded from both interior and exterior values of illumination.

Daylighting design principles (sun). The amount of solar light, heat, and other components of solar energy depends on the position of the sun in the sky. For clear-day daylighting design, the factor of greatest importance is the position of the sun in relation to the room in question.

Decorative lighting. Superfluous light, not used as part of an advertising display, intended to increase attractiveness or other incidental use.

Decorative street lighting equipment. Lamps attached to wires or structures that extend over any public property, does not apply to lamps attached to commercial electric lampposts or to any sign, canopy, or structure.

Diffraction

1. Modification that light or other radiation undergoes when it passes by the edge of an opaque body. The analogous modifications produced upon sound waves when passing by the edge of a building.

2. If the light source were a point, the shadow of any object would have its maximum sharpness; a certain amount of illumination would be found within the geometrical shadow owing to the diffraction of the light at the edge of the object.

Diffused. Illumination by means of light that travels through a material other than the bulb or tubing necessary to enclose the light source so that the light is spread evenly over the surface of the diffusing material.

Diffused lighting. Light that is not predominantly incident from any particular direction.

Diffuser. Device to redirect or scatter the light from a source, primarily by the process of diffuse transmission.

Diffuse reflection. Process by which incident flux is redirected over a range of angles.

Diffuse transmission. Process by which the incident flux that passes through a surface or medium is scattered.

Dimmers. Varying number of rheostats mounted in a bank and controlled by interlocking levers, used for dimming large rooms or spaces and stage lighting to various desired levels of illumination.

Direct glare

1. Glare resulting from high luminances, insufficiently shielded light sources in the field of view, or reflecting areas of high luminance. It is usually associated with bright areas, such as luminaires, ceilings, and windows, that are outside the visual task or region being viewed.

2. Glare resulting from high brightness, insufficiently shielded light sources in the field of view, or reflecting areas of high brightness and large area.

Directional lighting

1. Illumination on the work plane or on an object that is predominantly from a single direction.

2. Lighting designed to illuminate the work plane or an object, predominantly from a preferred direction.

Disability glare

1. Glare resulting in reduced visual performance and visibility, often accompanied by discomfort.

2. Reflected light that impairs a viewer's ability to accurately discern an object, perceive color, or to read printed matter.

Discomfort glare. Glare producing discomfort. It does not necessarily interfere with visual performance or visibility.

Dominant wavelength of a light. Wavelength of radiant energy of a single frequency that matches the color of the light when combined in suitable proportion with the radiant energy of the reference standard.

Effect of illumination. Photoelectric current in vacuum varies directly with the illumination over a very wide range (spectral distribution, polarization, and cathode potential remaining the same). In gas-filled tubes the response is linear over only a limited range.

Effect of material characteristics. In practical lamps the rate of evaporation and the melting point of the filament limit the extent of gains obtainable with operation at elevated temperatures. The melting point of tungsten is 3655 K, the highest of all metallic elements.

Effect of spectral distribution. A large proportion of the energy radiated by incandescent sources is in the infrared and ultraviolet wavelengths, which are invisible. The achievable efficacies of these sources are low compared with the theoretical maximum (680 lumens/W) that would be obtained if all the power input were emitted as green light of 555-mμ wavelength to which the human eye is most sensitive.

Efficacy. The luminous efficacy of a light source is defined as the ratio of the total luminous flux (lumens) to the total power input (watts or equivalent).

Efficiency of a source of light. The efficiency of a source is the ratio of the total luminous flux to the total power consumed. In the case of an electric lamp it is expressed in lumens per watt.

Electric discharge lamp

1. Lamp in which light (or radiant energy near the visible spectrum) is produced by the passage of an electric current through a metallic vapor or a gas.

2. All electric discharge lamps have negative volt-ampere characteristics and must therefore be operated in conjunction with current-limiting devices, commonly called "ballasts."

Electroluminescent lamp. Lamp in which light is produced by the excitation of a phosphor in an electric field.

Electromagnetic theory. Luminous bodies emit light in the form of radiant energy, and this radiant energy is transmitted in the form of electromagnetic waves. Electromagnetic waves act upon the retina of the eve thus stimulating the optic nerves to produce the sensation of light.

Electrometeor. Visible or audible manifestation of atmospheric electricity. Electrometeors either correspond to discontinuous electrical discharges (lightning, thunder) or occur as more or less continuous phenomena (Saint Elmo's fire, polar aurora).

Emergency and exit lights. All lights required by municipal regulations for the purpose of facilitating safe exit in case of fire or other emergency.

Emergency lighting. Lighting system designed to supply illumination essential to the safety of life and property in the event of failure of the normal supply.

Equivalent sphere illumination. Level of sphere illumination that would produce task visibility equivalent to that produced by a specific lighting environment.

Excitation purity of a light. Ratio of the distance on the CIE chromaticity diagram between the reference point and the light point to the distance in the same direction between the reference point and the spectrum locus or purple boundary.

Exterior lamp post. Standard support provided with necessary internal attachments for wiring and external attachments for bracket and luminaire.

Exterior lighting unit. Assembly of pole or post with bracket and luminaire.

Exterior pole mounting height. Vertical distance between the roadway surface and the center of the light source in the luminaire.

Extinction. Attenuation of light; that is, the reduction in illuminance of a parallel beam of light as the light passes through a medium wherein absorption and scattering occur.

Eye light. Illumination in television on a person to produce a specular reflection from the eyes without adding a significant increase in light on the subject.

Fenestration. Opening or arrangement of openings (normally filled with mediums for control) for the admission of daylight.

Festoon lighting. Aerial span of conductors installed outdoors and supplying only weatherproof lamp holders attached thereto.

Fill light. Supplementary illumination to reduce shadow or contrast range during television filming.

Filter. Device that changes, by transmission, the magnitude and/or the spectral composition of the flux incident on it. Filters are called "selective," "colored," or "neutral," according to whether or not they alter the spectral distribution of the incident flux.

Fixed light. Light having a constant luminous intensity when observed from a fixed point.

Fixture disconnection. Fixtures or lamp installations should be controlled either singly or in groups by an externally operable switch or circuit breaker that opens all ungrounded primary conductors.

Flame arcs. Flame arcs are obtained by enlarging the core in the electrodes of a low-intensity arc and replacing part of the carbon with the chemical compounds known as flame materials, capable of radiating efficiently. Typical flame materials are iron for the ultraviolet, rare earths of the cerium group for white light, calcium compounds for yellow, and strontium for red.

Flashing light. Light operated to have luminous periods alternating with dark periods.

Floodlight. Projector designed for lighting a scene or object to a brightness considerably greater than its surroundings. It usually is capable of being pointed in any direction and is of weatherproof construction.

Fluorescence. In the fluorescent lamp and in the improved-color mercury lamp ultraviolet radiation resulting from luminescence of the mercury vapor due to a gas discharge is converted into visible light by a phosphor coating on the inside of the tube or outer jacket. If this emission continues only during the excitation, it is called "fluorescence."

Fluorescent lamp. Low-pressure mercury electric-discharge lamp in which a fluorescing coating (phosphor) transforms some of the ultraviolet energy generated by the discharge into light.

Fluorescent mercury lamp. Electric-discharge lamp having a high-pressure mercury arc in an arc tube and an outer envelope coated with a fluorescing substance (phosphor) which transforms some of the ultraviolet energy generated by the arc into light.

Flush mounted or recessed. Term applied to a luminaire mounted above the ceiling (or behind a wall or other surface) with the opening of the luminaire level with the surface.

Flux. Number of photons that pass through a surface per unit time, expressed in lumens or watts.

Footcandle

1. Unit of illumination when the foot is the unit of length; illumination on a surface 1 sq ft in area on which a flux of 1 lumen is uniformly distributed. A footcandle equals 1 lumen/sq ft.

2. Luminance on a surface of 1 sq ft on which a flux of 1 lumen is uniformly distributed; luminance at a surface 1 ft from a uniform surface of 1 candlepower.

Footlambert

1. $1/\pi$ candle/sq ft. Relation of emittance (or exmittance) to luminance: Theoretically a perfectly diffusing surface emitting, transmitting, or reflecting flux at a rate of 1 lumen/sq ft which gives it an emittance (or exmittance) of 1 lumen/sq ft, has a luminance of $1/\pi$ candle/sq ft in all directions. No actual surface completely fulfills this condition.

2. Unit of luminance (photometric brightness) equal to $1/\pi$ candela/sq ft or to the uniform luminance of a perfectly diffusing surface emitting or reflecting light at the rate of 1 lumen/sq ft.

3. Unit of brightness of a perfectly diffusing surface emitting or reflecting light at the rate of 1 lumen/sq ft.

General lighting. Lighting designed to provide a substantially uniform level of illumination throughout an area, exclusive of any provision for special local electrical code requirements.

Glare

1. The sensation produced by luminance within the visual field that is sufficiently greater than the luminance to which the eyes are adapted to causes annoyance, discomfort, or loss in visual performance and visibility.

2. Effect of brightnesses or brightness differences within the visual field sufficiently high to cause annoyance, discomfort, or loss in visual performance.

Glazing effect on daylighting. Daylight illumination at any point in a given room depends on the amount of light transmitted by the glazing and brightness control mediums, among other factors.

Globe. Enclosing device of clear or diffusing glass, that is used to protect the lamp, diffuse or redirect its light, or modify its color.

Glow lamp. Electric-discharge lamp whose mode of operation is that of a glow discharge, and in which light is generated in the space close to the electrodes.

Ground light. Visible radiation from the sun and sky reflected by surfaces below the plane of the horizon.

High-intensity discharge lamps. General group of lamps consisting of mercury, metal halide, and high-pressure sodium lamps.

High-key lighting. Type of lighting for television that when applied to a scene results in a picture having graduations falling primarily between gray and white. Dark grays and blacks are present, but in very limited areas.

Horizontal plane of a searchlight. Plane that is perpendicular to the elevation axis, and on which lies the train axis.

Hot-cathode instant-start (slimline) lamp. Lamp that is started without a separate starting device by applying an increased voltage across the ends of the lamp to draw off electrons required to start the arc discharge.

Hot-cathode preheat lamp. A small coated filament produces the electrons required to sustain the mercury-arc discharge. The life of the lamp depends on filament coating. Each time the lamp is started, some part of the filament coating is used up. The starter permits current to flow through the lamp filaments before the full lamp voltage is applied across the ends of the lamp, thus heating the electrodes and initiating the arc at a relatively low voltage.

Hot-cathode rapid-start lamp. An auxiliary coil heats the filament by passing a small quantity of current continuously through it. It takes about 1 second for the lamp to start, and the filament is continuously heated thereafter with a reduced current.

Hot-cathode tube. Tube in which the cathode is heated in order to agitate and eject electrons out of the electrode.

Hue. One of the three variables of color. It is caused by light of certain wavelengths and changes with the wavelength.

Hue of a perceived light source color. Attribute of a perceived light source color that determines whether it is red, yellow, green, blue, or the like.

Illuminant A. Tungsten lamp operated at 2854-K color temperature.

Illuminants B and C. These illuminants consist of illuminant A plus a filter. Illuminant B approximates a blackbody source operating at 4800 K and is used by the British as their daylight standard. The filter for illuminant B consists of a layer 1 cm thick of selected solutions contained in a double cell constructed of nonselective optical glass.

Illuminant C. Illuminant C approximates daylight provided by the combination of direct sun and clear sky light having a color temperature of approximately 6500 K. The filter for illuminant C uses a cell identical with that for illuminant B.

Illumination

1. Density of the luminous flux incident on a surface; quotient of the luminous flux by the area of the surface when the latter is uniformly illuminated.

2. Density of the luminous flux incident on a surface; flux divided by the area over which the flux is distributed.

Illumination climate. Worldwide distribution of natural light from the sun and sky as received on a horizontal surface.

Illumination control. Photoelectric control that turns on circuits to lighting fixtures when outdoor illumination decreases below a predetermined level.

Illumination design. Design of sources of lighting and illumination systems that distribute light in order to effect a comfortable and satisfactory environment for the human eye.

Illumination distribution. Manner in which light is dispersed on a surface.

Illumination meter. Instrument for measuring the illumination on a surface. The instrument consists of one or more barrier-layer cells connected to a meter calibrated in footcandles.

Incandescence

1. Familiar physical objects are simple or complex combinations of chemically identifiable molecules, which in turn are made up of atoms. In solid materials the molecules are packed together, and the substances hold their shape almost indefinitely over a wide range of physical conditions. In contrast, the molecules of a gas are highly mobile and occupy only a small part of the space filled by the gas.

2. Emission of visible radiation by a hot body.

Incandescent. Luminous or glowing with intense heat, or shining with intense brilliance.

Incandescent filament lamp. Lamp in which light is produced by a filament heated to incandescence by the flow of an electric current through it.

Incident light. Direct light that falls on a surface.

Index of refraction. For any substance, the ratio of the velocity of light in a vacuum to its velocity in the substance; also, the ratio of the sine of the angle of incidence to the sine of the angle of refraction. In general, the index of refraction for a substance varies with the wavelength of the refracted light.

Indirect illumination. Illumination by means of light cast upon an opaque surface from a concealed source.

Instant-start hot-cathode lamps. Lamp that has filament electrodes and starts without auxiliary starters. The type is referred to as a "slimline lamp."

Intensity. Shortening of the terms "luminous intensity" and "radiant intensity."

Interflectance. Ratio of the lumens received on the work plane to the lumens emitted by the luminaires.

Interior mounting height. Distance from the floor to the light center of the luminaire.

Internal reflection of a light ray. Internal reflection results when a ray is incident on the boundary between two mediums from the side of the more dense medium. If the angle of incidence exceeds the critical angle, the ray will be totally reflected and no light will be transmitted into the less dense medium.

Interreflectance. Portion of the lumens that has been reflected one or more times on reaching the work plane, as determined by the Flux Transfer Theory.

Isocandle line. Line plotted on any appropriate coordinates to show all the directions in space about a source of light in which the candlepower is the same. For a complete exploration the line is a closed curve. A series of such curves, usually for equal increments of candlepower, is called an "isocandle diagram."

Isolux (isofootcandle) line. Line plotted on any appropriate coordinates to show all the points on a surface where the illumination is the same. For a complete exploration the line is a closed curve. A series of such lines for various illumination values is called an "isolux (isofootcandle) diagram."

Kelvin (K). SI (International System of Units) unit of thermodynamic temperature. Correct notation, for example, is 273 K and not 273°K.

Key light. Apparent principal source of directional illumination falling on a subject or area during television filming.

Lambert

1. Unit of luminance equal to $1/\pi$ candle/sq cm; unit emitting or reflecting light as the rate of 1 lumen/sq cm.

2. Unit of photometric brightness equal to $1/\pi$ candle/sq cm.

Lambert's cosine law. Distribution of flux such that the flux per solid angle in any direction from a plane surface varies as the cosine of the angle between that direction and the perpendicular to the surface. The photometric brightness of such a surface is uniform at all angles of view.

Lambert surface. Surface that emits or reflects light in accordance with Lambert's cosine law. Lambert surfaces appear equally bright from all viewing angles.

Lamp

1. Artificial source of light; by extension the term is also used for artificial sources radiating in regions of the spectrum adjacent to the visible. A portable lighting unit consisting of a lamp(s) with housing, shade, reflector, or other accessories is also commonly called a "lamp." In order to distinguish between such a complete luminaire and the light source within it, the latter is sometimes called a "bulb."

2. Generic term for a man-made source of light. By extension, the term is also used to denote sources that radiate in regions of the spectrum adjacent to the visible.

Lamp bank. Number of incandescent lamps connected in parallel or series to serve as a resistance load for full-load tests of electrical equipment.

Lamp efficacy versus filament temperature. Lamp efficacy increases as the filament temperature is raised. This is true for a lamp of any given design and construction.

Lamp holder. Device intended to support an electric lamp mechanically and connect it electrically to the circuit conductors.

Lamp lumen depreciation factor. Multiplier to be used in illumination calculations to relate the initial rated output of light sources to the anticipated minimum rated output based on the relamping program to be used.

Lamp shielding angle. Angle between the plane of the baffles or louver grid and the plane most nearly horizontal that is tangent to both the lamps and the louver blades.

Laser. Acronym for light amplification by stimulated emission of radiation. Laser produces a highly monochromatic and coherent beam of radiation. Steady oscillation of nearly a single electromagnetic mode is maintained in a volume of an active material bounded by highly reflecting surfaces, called a "resonator." The frequency of oscillation varies according to the material used and the methods of initially exciting or pumping the material.

Lateral width of a light distribution. Lateral angle between the reference line and the width line, measured in the cone of maximum candlepower. The angular width includes the line of maximum candlepower.

Least mechanical equivalent of light. At the wavelength of maximum visibility 1 lumen equals 0.00161 W; 1 W at the same wavelength equals 621 lumens.

Lens. Shaped piece of optically transparent material capable of focusing a beam of light or of diverging the light from an apparent source.

Life test. Test in which lamps are operated under specified conditions for a specified length of time, for the purpose of obtaining information on lamp life. Measurements of photometric and electrical characteristics may be taken at specified intervals.

Light

1. Light is psychophysical, i.e., neither purely physical nor purely psychological. "Light" is not synonymous with "radiant energy," however restricted, nor is it merely sensation. In a general nonspecialized sense, light is the aspect of radiant energy of which a human observer is aware through the stimulation of the retina of the eye.

2. For the purposes of illuminating engineering, light is visually evaluated radiant energy. Evaluation is accomplished by multiplying the energy radiated at each wavelength by the spectral luminous efficiency data for that wavelength and adding the results.

3. Radiant energy of the wavelengths to which the human eye is sensitive.

Light absorption. Process in which energy of light radiation is transferred to a medium through which it is passing.

Light adaption. Disappearance of dark adaption; chemical processes by which the eyes, after exposure to a dim environment, become accustomed to bright illumination, which initially is perceived as quite intense and uncomfortable.

Light amplifier. Electronic device that, when activated by a light image, reproduces a similar image of enhanced brightness, and that is capable of operating at very low light levels without introducing spurious brightness variations (noise) into the reproduced image.

Light and the energy spectrum. Wave theory permits a convenient graphical representation of radiant energy in an orderly arrangement according to its wavelength or frequency. The arrangement is called a "spectrum." It is useful in indicating the relationship between various radiant energy wavelength regions.

Lighting fixture. Assembly having one or more lamp holders therein or thereon, or a lamp holder used in lieu of such an assembly.

Lighting fixture raceway. Surface-mounted or pendant metal raceway, primarily intended to form a part of or the support for lighting fixtures, and to hold conductors, some of which may supply the fixtures.

Lighting outlet

1. Outlet intended for the direct connection of a lamp holder, lighting fixture, or pendant cord terminating in a lamp holder.

2. Connection to a branch circuit, made in a protective box, to which a light fixture or lamp holder is directly attached or from which wires are extended to fixtures in coves, valances, or cornices.

Lighting track. A manufactured assembly designed to support and energize lighting fixtures capable of being readily repositioned on the track.

Light loss factor. Factor used in calculating the level of illumination after a given period of time and under given conditions. It takes into account temperature and voltage variations, dirt accumulation on luminaire and room surfaces, lamp depreciation, maintenance procedures, and atmospheric conditions.

Lightness of a perceived object color. Attribute of a perceived object color by which the object seems to transmit or reflect a greater or lesser fraction of the incident light.

Light-source color. Color of the light source. The color of a point source may be defined by its luminous intensity and chromaticity coordinates; the color of an extended source may be defined by its luminance and chromaticity coordinates.

Linear light. Luminous signal having a perceptible physical length.

Live parts. Fixtures, lampholders, lamps, and receptacles will not be permitted to have live parts normally exposed to contact.

Local lighting

1. Illumination provided over a relatively small area or confined space.
2. Lighting designed to provide illumination over a relatively small area or confined space without providing any significant general surrounding lighting.

Localized general lighting. Lighting that utilizes luminaires above the visual task and contributes also to the illumination of the surround.

Louver

1. Series of baffles used to shield a source from view at certain angles or to absorb unwanted light. Baffles usually are arranged in a geometric pattern.
2. Opaque or translucent member used to shield a source from direct view at certain angles or to absorb unwanted light.

Louverall ceiling. General lighting system comprising a wall-to-wall installation of multicell louvers shielding the light sources mounted above it.

Louvered ceiling. Ceiling area lighting system comprising a wall-to-wall installation of multicell louvers shielding the light sources mounted above it.

Louver shielding angle. Angle between the horizontal plane of the baffles or louver grid and the plane at which the louver conceals all objects above.

Low-intensity arcs. Of the three principal types of carbon arcs in commercial use, the low-intensity arc is the simplest. In this arc the light source is the white-hot tip of the positive carbon. The tip is heated to a temperature near its sublimation point ($3700°C$) by the concentration of a large part of the electrical energy of the discharge in a narrow region close to the anode surface.

Low-key lighting. Type of lighting for television that when applied to a scene results in a picture having graduation from middle gray to black, with comparatively limited areas of light grays and whites.

Lumen

1. Unit of luminous flux; basic unit of light measurement used to express the total output of a light source. It is equal to the luminous flux on a unit surface area all points of which are a unit distance from a uniform point source of 1 candle.
2. Unit of luminous flux equal to the flux produced by a uniform point source of 1-candle intensity on a unit area of surface in which all points are a unit distance from the source.

3. Unit of luminous flux equal to the flux emitted through a unit solid angle (1 steradian) from a uniform point source of 1 candle.

Lumen-hour. Unit of quantity of light (luminous energy); quantity of light delivered in 1 hour by a flux of 1 lumen.

Lumen per watt. Unit of luminosity factor and of luminous efficacy.

Lumen-second. Unit of quantity of light (luminous energy), equal to the quantity of light radiated or received for a period of one (1) second by a flux of one (1) lumen.

Luminaire

1. Complete lighting unit consisting of a lamp(s) together with the parts designed to distribute the light, position and protect the lamps, and connect the lamps to the power supply.
2. Complete lighting unit consisting of a light source, globe, reflector, refractor, housing, and such support as is integral with the housing.

Luminaire dirt depreciation factor. Multiplier to be used in illumination calculations to relate the initial illumination provided by clean, new luminaires to the reduced illumination that they will provide owing to dirt collection on them at the time at which it is anticipated that cleaning procedures will be instituted.

Luminaire efficiency. Ratio of luminous flux (lumens) emitted by a luminaire to that emitted by the lamp or lamps used therein.

Luminair modular ceiling. Basic configuration of ceiling modules to allow the installation of recessed troffers.

Luminance (brightness) coefficient. Coefficient, similar to the coefficient of utilization, used to determine wall and ceiling luminances.

Luminance contrast. Relationship between the luminances of an object and its immediate background.

Luminance difference. Difference in luminance between two areas. It is usually applied to contiguous areas, such as the detail of a visual task and its immediate background, in which case it is quantitatively equal to the numerator in the formula for luminance contrast.

Luminance factor. Ratio of the luminance of a surface or medium under specified conditions of incidence, observation, and light source, to the luminance of a perfectly transmitting or perfectly diffusing surface or medium under the same conditions.

Luminance (photometric brightness). Luminous flux per unit of projected area and unit solid angle either leaving a surface at a given point in a given direction or arriving at a surface at a given point from a given direction; luminous intensity of any surface in a given direction per unit of projected area of the surface as viewed from that direction.

Luminance ratio. Ratio between the luminances (photometric brightness) or any two areas in the visual field.

Luminance threshold. Minimum perceptible difference in luminance for a given state of adaptation of the eye.

Luminescence

1. Emission of light other than incandescence.
2. Radiation from luminescent sources results form the excitation of single-valence electrons of an atom, either in a gaseous state, where each atom is free from interference from its neighbors, or in a crystalline solid or organic molecule, where the action of its neighbors exerts a marked effect.

Luminescent material. Material that emits light not ascribable directly to incandescence and that is emitted without elevation in the temperature of the material.

Luminous ceiling. Lighting system comprising a continuous surface of diffusing material with light sources mounted above it.

Luminous density. Quantity of light (luminous energy) per unit volume.

Luminous efficacy of a light source. Ratio of the total luminous flux emitted by the source to the total power input to the source. In the case of an electric lamp, efficacy is expressed in lumens per watt.

Luminous emittance. Density of luminous flux emitted from a surface.

Luminous flux

1. Time rate of the flow of the luminous parts of the radiant energy spectrum measured in lumens.
2. Time rate of flow of light (luminous energy).
3. Time rate of flow of light, usually measured in lumens.

Luminous flux density at a surface. Luminous flux per unit area of the surface.

Luminous intensity. Luminous flux per unit solid angle in a given direction; quotient of the luminous flux on an element of surface normal to that direction by the solid angle (in steradians) subtended by the element as viewed from the source.

Luminous intensity (or candlepower). Property of a source of emitting luminous flux. It may be measured by the luminous flux emitted per unit solid angle. The accepted unit of luminous intensity is the international candle. The Hefner unit, which is equivalent to 0.9 international candles, is the intensity of a lamp of specified design burning amyl acetate, called the "Hefner lamp."

Lux

1. International System unit of illumination; illumination on a surface 1 m^2 in area on which there is a uniformly distributed flux of lumen; illumination produced at a surface all points of which are at a distance of 1 m from a uniform point source of 1 candela.
2. Unit of illumination when the meter is the unit of length, equal to 1 meter-candle or 1 lumen/m^2.
3. Practical unit of illumination in the metric system, equivalent to the meter-candle. It is the illumination on a 1-m^2 area, with a uniformly distributed flux of 1 lumen.

Magnifying power. The magnifying power of an optical instrument is the ratio of the angle subtended by the image of the object seen through the instrument to the angle subtended by the object when seen by the unaided eye. In the case of a microscope or simple magnifier, the object as viewed by the unaided eye is supposed to be at a distance of 25 cm (10 in.).

Maintenance factor

1. Factor formerly used to denote the ratio of the illumination on a given area after a period of time to the initial illumination on the same area.
2. Product of the lamp lumen depreciation factor.
3. The accumulation of dirt on window surfaces results in a decrease in the daylight illumination. The amount of reduction varies with the location, the angle at which the glass is mounted, and the cleaning schedule. If it is known that the windows are to be used in a particularly clean or particularly dirty location, other maintenance factors should be used in the design of required lighting.

Matte surface. Surface from which the reflection is predominantly diffuse, with or without a negligible specular component.

Maximum attainable brightness. The maximum attainable brightness on an illuminated surface is limited by the brightness of the available light sources, the top limit depending on the optical arrangement. If the arrangement does not return significant amounts of radiation to the sources, the maximum brightness attainable will be that of the sources. If radiation is returned to the sources, the top limit will approach the brightness of a blackbody operating at the true temperature of the sources.

Maximum attainable efficacy. the maximum attainable efficacy of any white light source (whether it is a blackbody, tungsten, gaseous discharge, or fluorescent type) with its entire output distributed uniformly with respect to wavelength within the visible region is of the order of 200 lumens/W.

Mean horizontal candlepower. Average intensity measured in a horizontal plane passing through the source.

Mean spherical candlepower. Average candlepower measured in all directions and equal to the total luminous flux in lumens divided by 4π.

Mercury lamp. Electric discharge lamp in which the major portion of the radiation is produced by the excitation of mercury atoms.

Mercury-vapor lamps. Lamps that produce large amounts of ultraviolet radiation. If special glass is used, mercury lamps become efficient producers of ultraviolet for therapeutic purposes, skin tanning, and special fluorescent effects. Mercury lamps cannot be dimmed. Correct voltage and ballasting are obtained by autotransformers; two lamps usually may be operated from one transformer.

Metamers. Lights of the same color, but of different spectral energy distribution.

Microlambert. Unit of brightness; brightness of a perfectly diffusing and completely reflecting surface illuminated by 1 microlumen/sq cm of surface.

Microlumen. Total visible energy received by 1 sq cm of surface when illuminated by a point source of unit candlepower placed at a distance of 10 m.

Minimum deviation. The deviation or change of direction of light passing through a prism is a minimum when the angle of incidence is equal to the angle of emergence.

Moonlight. The moon shines purely by virtue of its ability to reflect sunlight. Since the reflectance of its surface is rather low, its brightness is approximately 1170 footlamberts.

Mottling. Spots or blotches of different color or shades of color interspersed with the dominant color.

Mounting height above the floor. Distance from the floor to the light center of the luminaire or to the plane of the ceiling for recessed equipment.

Mounting height above the work plane. Distance from the work plane to the light center of the luminaire or to the plane of the ceiling for recessed equipment.

Munsell chroma. Index of saturation of the perceived object color defined in terms of the Y value and chromaticity coordinates (x, y) of the color of the light reflected or transmitted by the object.

Munsell color system

1. System of surface color specification based on perceptually uniform color scales for the three variables: Munsell hue, Munsell value, and Munsell chroma. For an observer of normal color vision, adapted to daylight and viewing the specimen illuminated by daylight surrounded with a middle gray to white background, Munsell hue, value, and chroma of the color correlate well with the hue, lightness, and saturation of the perceived color.

2. Color designation system that specifies the relative degrees of the three simple variables of color: hue, value, and chroma.

Munsell hue. Index of the hue of the perceived object color defined in terms of the Y value and chromaticity coordinates (x, y) of the color of the light reflected or transmitted by the object.

Munsell value. Index of the lightness of the perceived object color defined in terms of the Y value. (Munsell value is approximately equal to the square root of the brightness [luminance] factor expressed in percent.)

Neon bulb. Glass envelope filled with neon gas and containing two or more insulated electrodes.

Night. Hours between the end of evening civil twilight and the beginning of morning civil twilight. (Civil twilight ends in the evening when the center of the sun's disk is 6° below the horizon and begins in the morning when the center of the sun's disk is 6° below the horizon.)

Nit. 1 candela/m².

Nodal points. Two points on the axis of a lens such that a ray entering the lens in the direction of one, leaves as if from the other and parallel to the original direction.

Numerical aperture. The sine of half the angular aperture, used as a measure of the optical power of the objective.

Object color. Color of the light reflected or transmitted by the object when illuminated by a standard light source.

Occulting light. Flashing light that has luminous periods of longer duration than the dark periods.

Orbit. The orbit described by a particular electron rotating about the nucleus is determined by the energy of that electron. Particular electron rotating about the nucleus is determined by the energy of that electron. Particular energy associated with each orbit. The system of orbits or energy levels is characteristic of each element and remains stable until disturbed by external forces.

Orientation. Positioning of a building with respect to compass directions and the major source of daylight.

Outline lighting. Arrangement of incandescent lamps or gaseous tubes to outline and call attention to certain features, such as the shape of a building or the decoration of a window.

Overcast sky. Sky with 100% cloud cover, sky with no sun visible.

Overvoltage. Lamp life is shortened by the application of voltage in excess of rated.

Partly cloudy sky. Sky with 30 to 70% cloud cover.

Pendant suspended. Hung from the ceiling or on brackets or supports from a wall or other surfaces.

Perceived light source color. Color perceived to belong to a light source.

Perceived object color. Color perceived to belong to an object and resulting from characteristics of the object, of the incident light, and of the surround, viewing direction, and observer adaptation.

Perfect diffusion. Diffusion in which flux is scattered in accord with Lambert's cosine law.

Peripheral vision. Seeing of objects displaced peripherally from the primary line of sight and outside the central part of the visual field.

Phosphorescent material. Material that after exposure to a light source, either visible or invisible, will continue to emit light in the visible spectral range for an appreciable period of time after excitation of the material has ceased.

Phot. Unit of illumination when the centimeter is the unit of length, equal to 1 lumen/sq cm.

Photoelectric effect. Emission of electrons from a material when light of sufficiently high frequency is incident on the surface.

Photometric brightness (luminance). Luminous flux per unit of projected area per unit solid angle either leaving a surface at a given point in a given direction or arriving at a given point from a given direction; luminous intensity of a surface in a given direction per unit of projected area of the surface as viewed from that direction.

Pilot house control. Mechanical means of controlling the elevation and train of a searchlight from a position on the other side of the bulkhead or deck on which it is mounted.

Point of fixation. Point or object in the visual field at which the eyes look and on which they are focused.

Polarization. Phenomenon in which the transverse vibrations of light waves are oriented in a specific plane.

Polarized light

1. Polarized electromagnetic radiation whose frequency is in the optical region.

2. Light that exhibits different properties in different directions at right angles to the line of propagation is said to be polarized. Specific rotation is the power of liquids to rotate the plane of polarization. The various properties are stated in terms of specific rotation or the rotation in degrees per decimeter per unit density.

Pole. Standard support generally used where overhead lighting distribution circuits are employed.

Portable lighting. Lighting by means of equipment designed for manual portability.

Portable luminaire. Lighting unit that is not permanently fixed in place.

Positive electrode of the low-intensity arc. Electrode that may contain a core consisting of a mixture of soft carbon and a potassium salt. Potassium does not contribute to the light but does increase the steadiness of the arc by lowering the effective ionization potential of the arc gas.

Preheat hot-cathode lamps. Lamps that have filament-type electrodes and require a starting device.

Primary line of sight. Line connecting the point of observation and the fixation point. The point of observation is the midpoint of the baseline connecting the centers of rotation of the two eyes.

Principal focus. The principal focus of a lens or spherical mirror is the point of convergence of light coming from a source at an infinite distance.

Prism. Triangular piece of glass that by refracting light passing through it separates the light into its various component colors.

Quality of lighting. Term that pertains to the distribution of luminance in a visual environment. It is used in a positive sense and implies that all luminances contribute favorably to visual performance, visual comfort, ease of seeing, safety, and aesthetics for the specific visual tasks involved.

Range lights. Groups of color-coded boundary lights provided to indicate the direction and limits of a preferred landing path, normally on an airport without runways but exceptionally on an airport with runways.

Rapid starting. Lamps designed to be operated with continuously heated cathodes. Cathode heating is accomplished through low-voltage heater windings built into the ballast or through separate low secondary voltage transformers designed for the purpose.

Rapid-start lamps. Lamps that are similar to preheat lamps, but require no starters.

Rated voltage. Lamps are operated in the specified burning position at labeled voltage or current held within $\pm 0.25\%$ of rated value.

Recessed. Mounted above the ceiling or behind a wall or other surface.

Recessed incandescent fixtures. Incandescent fixtures must have thermal protection and should be identified as thermally protected.

Reflectance. Ratio of the flux reflected by a surface or medium to the incident flux. The quantity reported may be total reflectance, regular (specular) reflectance, diffuse reflectance, or spectral reflectance, depending on the component measured.

Reflectance of a surface or medium. Ratio of the reflected flux to the incident flux.

Reflected glare. Glare resulting from specular reflections of high luminances in polished or glossy surfaces in the field of view. It usually is associated with reflections from within a visual task or areas in close proximity to the region being viewed.

Reflection. General term for the process by which a part of the incident flux leaves a surface or medium from the incident side.

Reflection coefficient or reflectivity. Ratio of the light reflected from a surface to the total incident light. The coefficient may refer to diffuse or to specular reflection. In general it varies with the angle of incidence and with the wavelength of the light.

Reflector

1. Device used to redirect the luminous flux from a source, primarily by the process of reflection.
2. Device for redirecting the light of a lamp by reflection in a desired direction.

Reflectorized lamp. Lamp in which part of the bulb is coated externally or internally with a specular reflecting material for the purpose of redirecting some of the emitted flux.

Refraction. Bending of a ray of light as it passes obliquely from one medium to another in which its velocity is different.

Refractor. Device used to redirect the luminous flux from a source, primarily by the process of refraction.

Regular or specular reflection. Process by which a portion of the incident flux is reemitted at the specular angle without scattering.

Regular transmission. Process by which incident flux (lumens) passes through a surface or medium without scattering.

Resolving power of a telescope or microscope. Resolving power is indicated by the minimum separation of two objects which appear distinct and separate when viewed through the instrument.

Resultant color shift. Difference between the perceived color of an object illuminated by a test source and that of the same object illuminated by a reference source, taking account of the state of chromatic adaptation in each case; i.e. the resultant of colorimetric shift and adaptive color shift.

Rotary power. Power of rotating the plane of polarized light.

Runway lights. Aeronautical ground lights arranged along a runway indicating its direction or boundaries.

Searchlight. Projector designed to produce an approximately parallel beam of light, mounted on a fixed or moving platform, and having an optical system with an aperture of 8 in. or more.

Service period. Number of hours per day for which the daylighting system provides a specified illumination level, often stated as a monthly average.

Set light. Separate illumination of background or sets for television other than that provided for principal subjects or areas by the types of high lighting.

Shielding angle of a luminaire. Angle between a horizontal line through the light center and the line of sight at which the bare source first becomes visible.

Short-arc lamp

1. High-pressure electric discharge lamp in which the arc length is comparable to the arc diameter.
2. High-pressure argon-mercury, mercury-xenon, and xenon lamps with short arc lengths, used where an extremely high brightness source can be used to good advantage.

Show window. Window requiring special lighting, used or designed to be used for the display of goods or advertising material, fully or partly enclosed or entirely open at the rear, and with or without a platform raised higher than the street floor level.

Side back light. Illumination from behind the subject in a direction not parallel to a vertical plane through the optical axis of the television camera.

Sky. Visible radiation from the sun is redirected by the sky.

Sky factor. Ratio of the illumination on a horizontal plane at a given point inside a building, due to the light received directly from the sky, to the illumination due to an unobstructed hemisphere of sky of uniform luminance, equal to that of the visible sky.

Skylight. A considerable amount of light is scattered in all directions by the earth's atmosphere. On theoretical grounds the scattering should vary inversely as the fourth power of the wavelength when the size of the scattering particles is small compared to the wavelength of light, as in the case of the air molecules themselves.

Solar constant. The irradiance from the sun at the mean distance between the earth and the sun, before modification by the earth's atmosphere.

Solar radiation simulator. Device designed to produce a beam of collimated radiation that has a spectrum, flux density, and geometric characteristics similar to those of the sun outside the earth's atmosphere.

Solar time. Time measured by the daily motion of the sun. Noon is taken as the instant in which the center of the sun passes the observer's meridian. (This is the time measured by sun dials.)

Source A. Source A is a tungsten lamp operated at 2854 K color temperature.

Source B. Source B consists of Source A plus a filter. Source B has a correlated color temperature of approximately 4800 K. Used by British scientists as their daylight standard.

Spacing-to-mounting height ratio. Ratio of the distance between luminaire centers to the mounting height above the work plane. (Usually considered at 30 in. above floor.)

Spectral color. Color corresponding to light of a pure frequency; basic spectral colors are violet, blue-green, yellow, orange and red.

Spectral emissivity. No known radiator has the same emissive power as a blackbody. The ratio of the output of a radiator at any wavelength to that of a blackbody at the same temperature and the same wavelength is known as the spectral emissivity of the radiator.

Spectral extinction. Selective absorption of different wavelengths of light as a function of depth in water.

Spectral luminous efficacy

1. Ratio of the luminous flux at a given wavelength to the radiant flux at that wavelength, expressed in lumens per watt and equal to 680 (lumens per watt) times the spectral luminous efficiency data, in published available data. The spectral luminous efficiency for a particular wavelength is the ratio of the spectral luminous efficacy for that wavelength to the value at the wavelength (555-mμ-nm) of maximum spectral luminous efficacy.

2. Standard measure of an eye to monochromatic light at various wavelengths; the function is normalized to unity at its maximum value.

3. Accepted internationally as representing the spectral luminous efficiency of radiant flux of the wavelengths between 380 and 760 mμ.

Spectral luminous efficacy of radiant flux. Quotient of the luminous flux at a given wavelength by the radiant flux at that wavelength, expressed in lumens per watt.

Spectral series. Spectral lines or groups of lines that occur in an orderly sequence.

Spectrum. Series of colors or radiations with various wavelengths from a source of waves.

Spectrum locus. Locus of points representing the colors of the visible spectrum in a chromaticity diagram.

Specular angle. Angle between the perpendicular to the surface and the reflected ray that is numerically equal to the angle of incidence and that lies in the same plane as the incident ray and the perpendicular, but on the opposite side of the perpendicular.

Specular reflection factor. The ratio of the specularly reflected light to the incident light.

Speed of light. The speed of all radiant energy, including light in vacuum, is approximately 186,000 miles per second. In all material media the speed is less and varies with the material's index of refraction, which itself varies with wavelength.

Speed of vision. Reciprocal of the duration of exposure required for something to be seen.

Sphere illumination. Illumination on a task from a source providing equal luminous intensity (candelas) from all directions, such as an illuminated sphere with the task located at the center.

Spherical aberration. When large surfaces of spherical mirrors or lenses are used, the light divergent from a point source is not exactly focused at a point. This phenomenon is known as spherical aberration. For axial pencils the error is known as axial spherical aberration; for oblique pencils, coma.

Spherical candlepower. Average candlepower of a source in all directions in space, equal to the total luminous flux (lumens) of the source divided by 4π.

Standards. Standards of candlepower, luminous flux, and color are used and established by national testing laboratories.

Starter. Device used in conjunction with a ballast for the purpose of starting an electric-discharge lamp.

State of chromatic adaptation. Condition of the eye in equilibrium with the average color of the visual field.

Stilb

1. 1 candela/cm.

2. Unit of photometric brightness equal to 1 candle/cm^2.

Stray light. Light from a source that is scattered onto parts of the retina lying outside the retinal image of the source.

Street lighting luminaire. Complete lighting device consisting of a light source together with its direct apurtenances, such as globe, reflector, refractor, housing, and such support as is integral with the housing. The pole, post, or bracket is not considered a part of the luminaire.

Subjective brightness. Subjective attribute of any light sensation giving rise to the percept of luminous intensity, including the whole scale of qualities of being bright, light, brilliant, dim, or dark.

Sun bearing. Angle measured in the plane of the horizon through which a vertical plane at a right angle to the window wall must be rotated to contain the sun.

Sunlight

1. Direct visible radiation from the sun (solar illumination).

2. Energy of color temperature, about 6500 K received from the sun at the outside of the earth's atmosphere at an average rate of about 0.135 W/sq cm. About 75% of this energy is transmitted to the earth's surface at sea level (equator) on a clear day. The apparent brightness of the sun is approximately 160,000 candles/sq cm viewed from sea level. The illumination of the earth's surface by the sun may be as high as 10,000 footcandles; on cloudy days the illumination drops to less than 1000 footcandles.

Sunlight control. The brightness ratios resulting from direct sunlight in rooms on a clear day usually will be well above the optimum required for good seeing. It is important that some method of sunlight control be used. Control can be either fixed or variable, incorporated in the glazing medium or located outside or within the room.

Sun's path. The position of the sun is expressed in terms of two angles. One of these is altitude, the vertical angle of the sun above the horizon. The second is solar azimuth, the horizontal angle of the sun from due south. Solar altitudes and solar azimuths for different latitudes are published in many handbooks.

Supplementary lighting. Lighting used to provide an additional quantity and quality of illumination that cannot readily be obtained by a general lighting system and that supplements the general lighting level, usually for specific work requirements.

Surface-mounted luminaire. Luminaire mounted directly on the ceiling, wall, or any other surface.

Suspended luminaire. Luminaire hung from a ceiling by supports.

Talbot. Unit of light, equal to 1 lumen-second.

Total reflection. When light passes from any medium to one in which the velocity is greater, refraction ceases and total reflection begins at a certain critical angle of incidence.

Transmission. Process by which a part of the incident flux leaves a surface or medium on a side other than the incident side.

Transmittance. Ratio of the flux transmitted by a medium to the incident flux. The quantity reported may be total transmittance or regular transmittance, depending on which of the components is measured.

Troffer. Long recessed lighting unit usually installed with the opening flush with the ceiling.

Train. Angle between the vertical plane through the axis of the searchlight drum and the plane in which this plane lies when the searchlight is in a position designated as having zero train.

Ultraviolet radiation. Any radiant energy within the wavelength range 10 to 380 nm.

Undulating light. Light operated to alternately increase and decrease in luminous intensity while remaining continuously luminous.

Utilance (utilization factor). Ratio of the luminous flux (lumens) received on the work plane to that emitted by the luminaire.

Valance lighting

1. Longitudinal shielding member mounted across the top of a window or along a wall and usually parallel to the wall, to conceal light sources giving both upward and downward distributions.

2. System comprising light sources shielded by a panel parallel to the wall at the top of a window.

Veiling brightness. Brightness superimposed on the retinal image which reduces its contrast. This veiling effect is produced by bright sources or areas in the visual field that results in decreased visual performance and visibility.

Veiling reflection. Regular reflections superimposed on diffuse reflections from an object that partially or totally obscures the details to be seen by reducing the contrast; also called "reflected glare."

Velocity of light. The velocity of all radiant energy, including light, is 2.997925×10^6 m/second in vacuum (approximately 186,000 miles/second). In all material mediums the velocity is less and varies with the material's index of refraction and with wavelength.

Vertical plane of a searchlight. Plane that is perpendicular to the train axis and on which the elevation axis lies.

Visibility. Quality or state of being perceivable by the eye. In many outdoor applications visibility is defined in terms of the distance at which an object can be just perceived by the eye. In indoor applications it usually is defined in terms of the contrast or size of a standard test object, observed under standardized viewing conditions, having the same threshold as the given object.

Visible radiation. Radiant energy of wavelengths from 0.4 to 0.76 μm that produces a sensation defined as "seeing" when it strikes the retina of the human eye.

Visual acuity. Measure of the ability to distinguish fine details; quantitatively, the reciprocal of the angular size in minutes of the critical detail that is just large enough to be seen.

Visual angle

1. Angle subtended by an object or detail at the point of observation, usually measured in minutes of arc.

2. Angle that an object subtends at the optical center of the eye.

Visual field

1. Locus of objects which at a given moment can be seen by one or the other of the two eyes. The portion where the fields of the two eyes overlap is called the "binocular visual field."

2. Locus of objects or points in space that can be perceived when the head and eyes are kept fixed. The field may be monocular or binocular.

Visual perception. Interpretation of impressions transmitted from the retina to the brain in terms of information about a physical world displayed before the eye.

Visual performance. Quantitative assessment of the performance of a task taking into consideration speed and accuracy.

Visual surround. The visual surround includes all portions of the visual field except the visual task.

Visual task. Term that conventionally designates those details and objects that must be seen for the performance of a given activity and includes the immediate background of the details or objects.

Wave theory. Light is the resultant of molecular vibration in the luminous material, and vibrations are transmitted through the ether as wavelike movements (comparable to ripples in water). Also, vibrations thus transmitted act on the retina of the eye, stimulating the optic nerves to produce visual sensation.

Wide band. Waves contained within a relatively large portion of the spectrum. If a wide band-pass filter is used, the intensity at the two cutoff wavelengths is a certain percentage of the maximum intensity within the interval.

Width line. Radial line (the one that makes the larger angle with the reference line) that passes through the point of one-half maximum candlepower on the lateral candlepower distribution curve plotted on the surface of the cone of maximum candlepower.

Work plane. The plane at which work usually is done, and at which the illumination is specified and measured. It is assumed to be a horizontal plane 30 in. above the floor.

Zonal constant. Factor by which the mean candlepower of a source of light in a given angular zone is multiplied to obtain the lumens in a zone.

LIGHTNING PROTECTION

Air terminal

1. Combination of elevation rod, brace, or supports placed on the upper portions of structures together with a tip or point if used.

2. Combination of roof saddle or base, placed on upper portions of structures together with a discharge point affixed thereto.

Alloys. Alloys should be substantially as resistant to corrosion as copper under similar conditions.

Aluminum. When aluminum is used, care should be taken not to use it in contact with the ground or where it will rapidly deteriorate, and precautions should be observed at connections with dissimilar metals. Cable conductors are of electrical conductor grade aluminum.

Antenna. Device used for radiating or receiving radio waves. An aerial, radio antenna.

Arrestors. Device used to prevent powerline surge damage.

Bonding. The two types of equipment on a roof that require bonding are:

1. Metal bodies of conductance such as exhaust fan ventilators in sheet metal housings, metal-framed skylights, metal roof hatches, and metal chimneys or stacks.
2. Metal bodies of inductance such as roof drain covers and gratings, metal flashings, metal copings, and expansion joint caps.

Branch conductor. Conductor that branches off at an angle from a continuous run of conductor.

Cable. Conductor formed of a number of wires stranded together.

Cable holders or fasteners. Typical clip and loop type, installed with expansion-type bolts to concrete surfaces, or with lug screws on wood surfaces.

Class I protection. Equipment and system for ordinary buildings under 75 ft in height.

Class II protection. Equipment and system for buildings over 75 ft in height.

Class III modified protection. Equipment and systems for specialty areas covering large heavy-duty stacks and chimneys.

Conductor. Portion of a lightning protection system designed to carry the lightning discharge between the air terminal and ground.

Conductor installation. Conductors should be secured with cable fasteners compatible to the thickness of the conductor. Fasteners are set in special-type adhesive. Spacing of fasteners is based on local code regulations.

Copper. The grade ordinarily required for commercial electrical work is generally designated as being of 98% conductivity when annealed.

Copper-clad steel

1. The copper covering should be permanently and effectively welded to the steel core, and the proportion of copper should be such that the conductance is not less than 30% of the conductance of an equivalent cross section of solid copper.
2. Steel with a coating of copper welded to it, as distinguished from copper-plated or copper-sheathed material.
3. Steel with a heavy coating of copper bonded to it.

Down conductor. Vertical portion of a run of conductor that ends at the ground.

Elevation rod. Vertical portion of a conductor in an air terminal by means of which it is elevated above the object to be protected.

Fastener. Device used to secure the conductor to the structure that supports it.

Flat roof terminals. Terminals installed on the clean surface of the last layer of roofing felt and set in a special adhesive.

Ground. Conducting path, intentional or accidental, between an electric circuit or equipment and the earth or some other conducting body serving in place of the earth.

Ground connection. Buried body of metal with its surrounding soil and a connecting conductor which together serve to bring an object into electrical continuity with the earth.

Grounded system. Conducting apparatus connected to the ground.

Ground resistance. Opposition of the earth to the flow of current through it; its value depends on the composition and moisture content of the soil.

Ground rod. Rod that is driven into the earth to serve as a ground terminal, such as a copper-clad rod, solid copper rod, galvanized iron rod, or galvanized iron pipe.

Ground terminal. Portion of a lightning protection system extending into the earth, such as a ground rod, ground plate, or the conductor itself, serving to bring the lightning protection system into electrical contact with the earth.

Ground-to-cloud discharge. Lightning discharge in which the original streamer processes start upward from an object located on the ground.

Inspection. Equipment and systems accepted and approved by local codes, including criteria to evaluate installation procedures.

Lightning. Meteorological phenomenon arising from the accumulation, in the formation of clouds, of tremendous electrical charges, usually positive, which are suddenly released in a spark type of discharge. The lightning spectrum corresponds closely with that of an ordinary spark in air, consisting principally of nitrogen bands, though hydrogen lines may sometimes appear, owing to dissociation of water vapor.

Lightning arrester. Device that protects equipment from the destructive effects of lightning surges. It allows the high-voltage surge to pass directly to ground without allowing ordinary current to follow.

Lightning channel. Irregular path through the air along which a lightning discharge occurs.

Lightning conductor. Conductor designed to carry the current of a lightning discharge from a lightning rod to ground.

Lightning discharge. Series of electrical processes by which charge is transferred within the atmosphere along a channel of high ion density between electric charge centers of opposite sign.

Lightning flash. In atmospheric electricity, the total observed luminous phenomenon accompanying a lightning discharge.

Lightning mast. Steel mast or extension of the steel structure that is designed to divert all lightning that might otherwise strike a bus, disconnecting switch, or other equipment.

Lightning protection system. Integrated arrangement of air terminals, bonding connections, arrestors, splicers, and other accessories and fittings to safely conduct to ground any lightning discharge.

Lightning rod. Metallic rod installed on an exposed elevation or roof of a structure and connected to a low-resistance ground to intercept lightning discharges and to provide a direct conducting path to ground.

Lightning strike. Series of repeated discharges comprising a single lightning discharge, or lightning flash, specifically in the case of the cloud-to-ground discharge.

Lightning surge. Transient disturbance in an electric circuit due to lightning.

Lightning switch. Manually operated switch used to connect a radio antenna to the ground during electrical storms.

Main conductors. Conductors that tie the air terminals together and interconnect with the grounding system.

Materials. The materials of which protection systems are made must be resistant to corrosion or acceptably protected against corrosion. No combination of materials may be used that forms an electrolytic couple of such nature that in the presence of moisture corrosion is accelerated, but where moisture is permanently excluded from the junction of such metals, contact between them may be permitted.

Parapet air terminals. Where parapets occur on buildings, the air terminal is usually secured to the parapet with anchors and bolts or screws, depending on the design of the parapet.

Point. Pointed piece of metal used at the upper end of the elevation rod to receive a lightning discharge.

Roof conductor. Portion of the conductor above the eaves running along the ridge, parapet, or other portion of the roof.

Roof installation. Air terminals should be located around the perimeter of a flat roof building and along the ridge of sloped roof buildings. Review local codes for spacings and approval of the installation.

Sideflash. Spark occurring between nearby metallic objects or from such objects to the lightning protection system or to ground.

Structural framework. Protection in tall structures can take advantage of the steel frame to fasten the flexible conductor to a plate that is fastened to the steel column and then to the grounded copper rod, or as required by local code.

Utility and service leads. Utility regulations will govern the installation of devices to afford lightning protection for electric and telephone lead-ins, and also protection to radio and television antenna leads.

Water service ground. For most installations, based on local regulations, it is usually mandatory to fasten the conductors from the lightning protection system to the incoming water service pipe.

A

Professional Services

* The INFORM DIVISIONS presented here have been created by the author and are not in any way connected or related directly or indirectly with the Masterformat, published by the Construction Specifications Institute, Inc.

CONSTRUCTION CONTRACTING

Addenda. Statements or drawings prepared by the architect/engineer that modify the basic contract documents after the latter have been issued to the bidders, but prior to the taking of bids.

Agreement. Document that identifies the owner and contractor and states the contract price and certain other information. It must be signed by both parties and dated.

Alteration work. New work done to and in conjunction with existing work, other than complete demolition.

Alternates. Proposals required of bidders reflecting amounts to be subtracted from or added to basic proposals in the event specific changes in the work are ordered.

Application for payment. Statement of amounts claimed by the contractor as payments due on account of work performed and/or materials stored at the job site. Alternates may be either deductive or additive.

Arbitration. Method of settling claims or disputes between parties to a contract, other than litigation, under which an arbitrator or a panel of arbitrators, selected for their specialized knowledge in the field in question, hears the evidence and renders a decision. (Refer to Construction Industry Arbitration Rules of the American Arbitration Association (AAA).

Architect–Engineer–Designer. Party to a contract to provide professional design and other services to an owner (the other party to that contract). In certain cases, services may be part of more comprehensive design and construction services provided to an owner, which may also include land acquisition as well as construction work, as in so-called "package deals" or turnkey projects.

Architect's consultant. Party to a contract to provide professional design or other services to an architect who, in turn, has undertaken to provide design and other services to an owner, of which the services of the architect's consultant are a part.

Area estimates. Approximations of probable project construction cost based on reliable published information on unit costs per square foot.

Award. Notification to a bidder that the submitted bid was accepted.

Award contracts (general contractor). As soon as the general contractor has received the owner's contract, after a review of the subbids, awards the contract to the subcontractor who meets the estimate established for the work, not necessarily the low bidder.

Basic items (of work). Items of work that are described and included in the specifications and cost code, that occur in most of the construction jobs done in any particular field of construction activity, and that constitute part of a basic specification or basic cost code, in that field, as opposed to particular items of work.

Bid

1. Offer to perform construction work for payment, the acceptance of which constitutes a contract between an owner (who has accepted the bid) and a contractor (who has made the bid to the owner). It is sometimes referred to as a general bid (made by a general contractor) to distinguish it from a subbid.

2. Proposal prepared by prospective contractor specifying the charges to be made for completing the work

according to the requirements of the contract documents.

Bid date. Advertised date, or official date that the bid is due.

Bidder. Individual, company, partnership, or corporation who submit a bid or proposal to provide labor, materials, supplies, or equipment for a proposed job, building, or project.

Bidding. The combined action of a bidder or bidders to estimate or secure estimates to prepare a bid or proposal.

Bidding documents. Bidding documents should be the same as the contract documents with the addition of the instructions to bidders, the bid form, and any other informative documents related to the bidding. Instructions to bidders should contain only information and instructions regarding bidding. In some instances the bid form is identified as one of the contract documents because of its contents, such as requiring a list of proposed subcontractors and unit prices to be submitted by the bidder which will be made part of the contract.

Bid form. Form of a proposal prepared by the architect setting forth the basis of the price quotations required and containing spaces to be filled in by all bidders. Unless otherwise called for, the contractor has the option to prepare his bid on his own bid form or on his letterhead.

Bid period (subcontractor). Subcontractor submits finalized bid at the time specified in the invitation to bid.

Bid security

1. Bid bond, certified check, or other forfeitable security usually required by public agencies guaranteeing that a bidder will sign a contract, if offered, in accordance with his proposal.

2. Security usually required in connection with submitting of a bid. The bid bond attached to the bid must be executed by a recognized national corporate surety.

Bonus and penalty clause. Provision in the proposal form for payment of a bonus to the contractor for each day the project is completed prior to the time stated, and for a charge against the contractor for each day the project remains uncompleted after the time stipulated. Bonus and penalty amounts are agreed to at the signing of the contract.

Building permit. Permit issued to the general contractor by authorities in the jurisdiction permitting the construction of a building or structure in accordance with submitted plans and specifications.

Cash allowances. Sums that the contractor is required to include in his bid, which becomes part of the contract for specific use purposes.

Certificate for payment. Statement by an architect informing the owner of the amount due a contractor on account of work accomplished and/or materials stored at the job site. The contractor may have the option of preparing the request for payment on his own form or on a form accepted by various contractor's organizations.

Certificate of insurance. Statement by insurance company as evidence that an insurance policy is written to provide the coverage that was requested in the contract documents, is in effect, date of anticipated start of work in the field, and also should give the expiration date.

Change order. Written work order for the contractor, usually prepared by the architect/engineer signed by the owner or his agent, authorizing a change in the scope of the work.

Codes. The contractor should become familiar with the codes governing any construction job he has been awarded.

Collusion affidavit. Statement by the bidder that he has not entered into collusion with another bidder or other person in the preparation of his bid. It is not recommended unless required by the owner.

Competitive bidding. The owner seeks to obtain the lowest bid, and, after examining all of the proposals, will decide with the aid of his architect and attorney whom the contract should be awarded to, which may not necessarily be the low bidder.

Construction coordination (general contractor). The general contractor will provide a competent superintendent who, in coordination with the appointed project engineer, will be in complete charge and direction of the project. The major subcontractors will provide supervision or competent foremen.

Construction documents. Includes the contract, working drawings, specifications, general conditions, and supplementary general conditions. Any other documents issued by the architect/engineer or the owner may also be included.

Construction management. Group of management activities, over and above normal architectural and engineering services related to a construction program, that are carried out during the conceptual planning, predesign, design, and construction phases and that contribute to the cost, control of time, and quality desired by the owner in the construction of a new facility.

Construction manager. The construction manager contract is awarded to a single entity who will be responsible for providing controls over contracts awarded, seeing that they are within the estimated budget, and administering the construction without having to engage any field employees or tradesmen.

Contingency allowance. Established amount of money included in the contract to cover unpredictable changes in work or unforeseen conditions arising at the site during construction. It is not intended to cover additions that increase the scope of the job.

Contract. Agreement that is enforceable by law.

Contract amount (sum). Amount of money paid by the owner to the contractor for the work in accordance with the terms of the contract. In stipulated sum contracts it is the amount stipulated by the bidder and later stated in the articles of the contract. In some other types of contract (such as cost plus fee) the contract amount is implicit in the articles and conditions of the contract, but it is not explicit until it is determined according to the amount of work done and the terms of the contract.

Contract date. Date shown on the contract as the official date of execution. It may not always be possible for all parties to sign the contract on the same day, and the official date is the "contract date."

Contract documents

1. Includes the contract, drawings, specifications, and other documents prepared by the architect/engineer (first as bidding documents) that illustrate and describe the work, the terms and conditions under which the work is to be done for the owner, and the terms and conditions of payment for the work by the owner to the contractor.

2. Working drawings, specifications, general conditions of the contract, supplementary general conditions, and the owner-contractor agreement.

Contract forms. Standard published forms on which the contract between the owner and contractor is written, usually consisting of the "agreement," "guaranty bond," "power-of-attorney," and, sometimes, "wage rates,"

Contract limit. Limit or perimeter line established on the drawings or stated elsewhere in the contract documents defining the boundaries of the site available to the contractor for construction purposes. It is the owner's responsibility to identify the limits of the site.

Contractor

1. Person who is to perform, is performing, or who has performed the work under consideration in or on a building, and who is or was under a contract with the owner to perform such work.

2. Party to a construction contract who provides construction work for the owner (the other party). The contractor may have subcontracts with subcontractors to do parts of the work that he has undertaken to perform. The term "contractor" is sometimes used in a comprehensive sense to include by implication the subcontractors (and their sub-subcontractors) of the contractor, because that which applies to the contractor usually applies to some or all of the subcontractors and sub-subcontractors.

3. That party to the owner-contractor agreement who is charged with the responsibility of executing the work contracted for.

4. Party signing the contract with the owner to do the work called for in the drawings and specifications. The bidder is not a contractor until the contract is signed.

5. Person or organization responsible for performing the work and identified as such in the owner-contractor agreement.

Contractor's affidavit. Certified and notarized statement from the contractor attesting to having made certain payments to subcontractors, suppliers, installers, and supply and equipment companies with evidence to satisfy the owner.

Contractor's estimate

1. Determination of probable project construction cost prepared by the contractor as a basis for his proposal.

2. Forecast of construction cost, as opposed to a firm proposal, prepared by a contractor for a project or a portion thereof. The term is sometimes used to denote a contractor's application or request for a progress payment.

Contractor's liability insurance. Insurance purchased and maintained by the contractor to protect him from specified claims that may arise out of or result from his operations, whether such operations be by himself, or by anyone directly employed by him.

Contractor's option. Options stated in the specifications under which the contractor may select with approval certain specified materials, methods, or systems at his own option without change in the contract sum.

Contractor shops and yards. Fabrication, repair shops and yards for contractors who require assembly work of their product or materials and also require space and areas, whether inside or outside of enclosed buildings, to store raw material or equipment to be used at a job site.

Contractor's yard. Any space, whether inside or outside a building, used for the storage or keeping of construction equipment, machinery, or vehicles, or parts thereof, which are for use by a construction contractor.

Contract price. Basic total sum shown in the agreement which is the consideration given to the contractor for doing the work required by the contract.

Contract sum. Price stated in the owner-contractor agreement, which is the total amount payable by the owner to the contractor for the performance of the work under the contract documents. The contract sum can be adjusted only by change order.

Contract time

1. Official amount of time allowed the contractor to complete the work of the contract. It begins with the "contract date" or date of "notice to proceed" as required by the terms of the contract. It ends on the "completion date" as called for.

2. Period of time established in the contract documents within which the work must be completed. The contract time can be adjusted only by change order, and time may be added or deducted based on the adjustment of the contract.

Cost accounting. That part of construction management by which actual costs of work are segregated and attributed to specific items of work and to specific construction jobs, after which the cost data are analyzed and formulated for use in job planning and cost control and in estimating the costs of other jobs.

Cost plus contract. Contract in which the contractor keeps a record of the cost of construction and the owner pays this cost plus an agreed profit to the contractor. This is opposed to the "lump-sum contract."

Cost plus fee agreement. Agreement under which the contractor (in an owner-contractor agreement) or the architect/engineer (in an owner-architect/engineer agreement) is reimbursed for his direct and indirect costs and, in addition, is paid a fee for his services. The fee is usually stated as a stipulated sum or as a percentage of cost.

Cost-plus-fee contract. Agreement under which the contractor is reimbursed his costs and, in addition, is paid a professional services fee.

Costs in use. All the costs incurred by an owner through his ownership of a building, apart from and subsequent to the initial costs of constructing the building, including depreciation, maintenance, financing, taxes, insurance, and building operating costs.

Costs of the construction work. All the direct and indirect costs of work, generally classified as labor costs, material costs, plant and equipment costs, job overhead costs, operating overhead costs, and profit; these are the costs to the owner for the work done, and in place.

Cost planning. Estimating the costs of work during the design stage (usually by units or elements or other such recognized means), including, if necessary, selecting materials and construction methods and adjusting the design so as to complete the work within a budgeted cost.

Cube. Item in an estimate with its quantity expressed as a volume, usually in cubic feet or cubic yards (or cubic meters). With a cube item, no dimensions need appear in the item's description, although in some cases it is more informative if some dimensions are included.

Date of agreement. Date usually stated on the first page of the agreement. If no date is stated, it could be the date on which the agreement is actually signed, if recorded, or it may be the date established by the award. It is also sometimes referred to as the "contract date."

Date of commencement of the work. Date established in a notice to proceed or, in the absence of such notice, the date of the agreement or such other date as may be established therein or by the parties thereto, or by the lender, that an account has been established for the job.

Date of substantial completion. Date certified by the architect/engineer when the work or a designated portion thereof is sufficiently complete, in accordance with the contract documents, so that the owner may occupy the work or designated portion thereof for the use for which it is intended, acknowledged by the permit authorities that occupancy is approved.

Design-award-build. Process is as follows: the architect designs the project, producing a set of construction documents that are used to select a contractor and to determine the cost of construction. The architect is the administrator, reviewing all construction documents, observing the work in the field, reviewing shop drawings and samples of materials submitted, processing the contractor's application for payment, and administering final check lists and inspection for the final payment.

Design-build (prime contractor). Agreement with one single entity, who provides the design and construction under one contract.

Detailed estimate of construction cost by the contractor. Forecast of construction cost prepared on the basis of a detailed analysis of materials and labor for all items of work, as contrasted with an estimate based on current area, volume, or similar unit costs.

Developer. Person who develops land for improvement through construction work.

Direct costs of work. The direct costs of work are generally classified as labor costs, material costs, plant and equipment costs, and job overhead costs, all of which are directly attributable to a specific construction job.

Direct expense. Expense directly incurred and assignable to the account of a particular project.

Direct labor costs. Labor costs paid by a contractor directly to an employee, usually based on a wage agreement.

Direct select contractor. Many corporations, having worked with a contractor previously on another construction program and being satisfied with his cooperation and quality of work, will in most instances select that firm for the new project. Individual owners having had the same experience will make their selection on the same basis. In most cases, the construction contract will be negotiated based on preliminary estimates.

Divided contract. System of dividing the work of the job into several prime contracts, such as general, plumbing, heating, and electrical, as opposed to "single contract," which is the system of letting all work on a job to one general contractor.

Division of a specification. One of the 16 divisions in the Construction Specifications Institute format or the Masterformat (latest edition). Divisions are subdivided into sections by the specification writer according to the nature and extent of the work specified to facilitate the production of the specifications and their use in bidding and construction.

Duodecimals. Number system based on twelfths, used in calculating quantities from dimensions given in feet and inches.

Element of a building. An element of a building is defined as the "part of a building that always performs the same functions irrespective of building type." A building may be made up of any number of elements within practical limits.

Estimated cost. Reasonable value of all services, labor, materials, equipment and use of scaffolding and other appliances or devices entering into and necessary to the prosecution and completion of the work ready for occupancy.

Estimates. Forecasts of probable project construction costs to be used as a guide. In preparing contract documents, these include area estimates, cubage, volume estimates, in-place unit cost estimates, quantity and cost estimates, semidetailed and detailed estimates, and statement of probable project construction cost.

Estimating. The estimator and the quantity surveyor, or both, estimate the work to be done, so that its probable costs can be calculated, with the intention that the owner pay for what he receives from the contractor.

Existing work. Work that already exists when a contract is made for other work, and which by definition has some physical relationship or connection with the work of the contract.

Extra over item. Feature of part of an item of work measured and priced separately from the main item to facilitate and simplify the measurement, pricing, and subsequent costing of the work. Establishing extra over items eliminates the need for many minor deductions and adjustments. It also enables the theoretical isolation of basic items by separating them from their minor and variable features, thus effectively increasing the available and useful cost data of basic items of work.

Extras. Construction costs over and above those specified in the agreement.

Fast-track (no fixed price). Process is as follows: contractor starts construction on available architect/engineer documents, subletting work to subcontractors (being familiar with their work on other projects), and reports to owner weekly on costs as of that date. When all of the contract documents have been completed by the owner's architect/engineer, the contractor prepares an estimate to be used as a guide for future contracts to be awarded. The owner at this time can award a maximum-not-exceed contract to the contractor. The original authorization for the contractor to begin construction is based on a previous letter of intent.

Field order. Written order effecting a minor change in the work that may or may not involve an adjustment in the contract sum or amount, or an extension of the contract time, issued to the contractor during the construction phase.

Final acceptance

1. Owner's acceptance of a completed project from the contractor.

2. Owner's acceptance of the project from the contractor upon certification by the lender or architect/engineer that it is complete and in accordance with the contract requirements. Final acceptance is confirmed by the making of the final payment, unless otherwise stipulated at the time such payment is made.

Final completion. The work is complete and all contract requirements have been fulfilled by the contractor, and approved by the local authorities for occupancy.

Final payment. Payment made by the owner to the contractor, upon issuance by the architect/engineer of the final certificate for payment, of the entire unpaid balance of the contract sum as adjusted by change orders.

Force account. Work ordered to be done without prior understanding of either the lump sum or unit price cost thereof; work ordered to be done and billed for at the cost of labor, materials, insurance, taxes, etc., plus an agreed percentage in lieu of overhead expense and profit. (A hazardous agreement.)

General conditions of the contract for construction. Part of the contract documents that sets forth many of the rights, responsibilities, and relationships of the parties. Standard forms published by many organizations.

General contract

1. Agreement between the owner and a general contractor for the construction of a project, including all phases of the work.

2. Under the single-contract system, the contract between the owner and the contractor for construction of the entire work; under the separate contract system, the contract between the owner and a contractor for construction of architectural trades and structural work.

General contractor. Contractor who has subcontractors to do some part (or all) of the work that he has undertaken to do for an owner. Previously, the term referred to a contractor who employed workers of several different trades, and who undertook to do most (or all) parts of the work, as distinct from a specialist (trade) contractor, who normally undertakes the work of only one trade.

General requirements. Temporary services and other requirements for work provided by a contractor, which because of their general nature are related to the work as a whole rather than to specific items of work. Usually found in Division I (of the Construction Specifications Institute Masterformat) for job specifications. Certain general conditions of some contracts are more accurately called general requirements in that they more directly relate to the work rather than to the items of the contract.

Guaranteed maximum cost. Amount established in an agreement between owner and contractor as the maximum cost of performing specified work on the basis of cost of labor and materials plus overhead expense and a stipulated profit.

Indirect costs of work. Costs of work generally classified as operating overhead costs and profit as opposed to direct costs.

Indirect expense. Expenses incurred by but not chargeable to a particular project, sometimes referred to as "overhead expense."

Indirect labor costs. Labor costs paid by an employer on an employee's behalf for such things as insurance, social security, and pension plan.

In-place unit cost estimates. Forecasts of project construction cost based on the cost of units of construction installed.

Instruction to bidders. Document prepared by the lender, architect/engineer, owner, or agency stating the procedures to be followed by bidders.

Invitation to bid. Invitation to a selected list of contractors usually prepared by the architect/engineer, owner, agency, or lending institution, furnishing information on the submission of bids for a project.

Item of work. Part of the work that by its nature can be observed, identified, and distinguished from other parts of work for the purposes of estimating, cost accounting, and construction management. Costs can be segregated and allocated to specific items of work.

Job overhead costs. Direct costs of work that because of their general nature cannot be allocated to specific items of work but can be allocated to a specific job, as distinct from operating overhead costs.

Labor and material payment bond. Bond guaranteeing to the owner that the contractor will satisfy all obligations incurred by him on the project.

Labor costs. Part of the total costs of work expended on labor that is dependent on the labor rates paid for workmen and their productivity. They include both direct and indirect labor costs. The costs of work, other than material costs, plant and equipment costs, overhead costs, and profit.

Labor rate. Total costs per hour for labor paid by the employer, including all direct and indirect labor costs for a specific period of time (and place) divided by the number of hours worked during that period.

Laps. Additional material required by and incorporated into the work because of the dimensions of the material product and of the work, and the resultant need for joining the material by overlapping. Consequently, the actual amount of additional material required for laps usually depends on the dimensions of the material product and the dimensions of the work, the more the additional amount of material required for laps becomes a constant. Laps are sometimes measured as work, but in many cases allowance is made for them on a percentage basis, based on previous experience with the same unit item on other projects, particularly if the amount required for laps is practically a constant.

Licensed contractor. Where required by law, person or organization licensed and certified by governmental authority, to engage in construction contracting.

Liens. Legal claims against an owner for amounts due those engaged in or supplying materials for the construction of his building.

Liquidated damages

1. Agreed-to sum in the contract chargeable against the contractor as reimbursement for damages suffered by the owner because of the contractor's failure to fulfill his contract obligations.

2. Sum established in a construction contract, usually as a fixed penalty per day, as the measure of damages suffered by the owner due to failure to complete the work within a stipulated time.

Lowest acceptable bona fide bid. Proposal reflecting the lowest dollar amount and fulfilling all the stipulated requirements.

Lump sum contract. Agreement in which a stipulated sum forms the basis of payment.

Material costs. Costs of all materials, products, building components, fixtures, and building equipment required to be incorporated and installed in the work, including delivery to the site and taxes; costs of work, other than labor costs, plant and equipment costs, overhead costs, and profit.

Maximum-not-to-exceed contract. Contract price that has been established by bid or negotiated, in which the owner shares in the savings when lower bids are obtained by the prime contractor. Usually there is a percentage split in the savings.

Multiple construction contracts. Contract, which does not include any of the pipe trades, is awarded to a contractor. The pipe trades are individually awarded by the owner. The matter of site cleanup and coordination of the several groups is the responsibility of the owner.

Mutual responsibility of contractors. Obligation of a contractor for damages caused by any separate contractor.

New work. Opposite of existing work. It includes all new materials and equipment.

Number. Item in an estimate with its quantity expressed by enumeration. With a number item, all three dimensions should appear on the item's description.

Open bid. Usually required for public construction, and have strict bidding requirements, which generally reduce the number of bidders who can qualify under the established restrictions.

Operating overhead costs. With profit, the indirect costs of work; costs of operating a construction business that because of their general nature cannot be allocated to specific jobs. They are distinct from job overhead costs, plant and equipment costs, material costs, and labor costs.

Overhead and profit. Usually, operating overhead costs and profit, which are indirect costs.

Overhead costs. Job overhead costs and operating overhead costs. Usually, the more specific terms should be used.

Owner. Party to a construction contract who pays the contractor (the other party) for construction work; the party who owns the right to the land on which the work is done and therefore owns the completed work itself; also, the client of the architect/engineer.

Owner-contractor agreement. Agreement between an owner and a contractor for a construction project. Standard printed forms available from several organizations.

Owning and operating costs. Total plant and equipment costs. Owning costs are those incurred by ownership of the plant and equipment and consist primarily of the costs of investment, maintenance, and depreciation. Operating costs are those incurred by operating and using plant and equipment (over and above the owning costs) and consist mainly of materials, fuel, lubricants, and operator's labor costs. Mobilization and demobilization costs must also be included in an estimate.

Particular items of work. Items of work that are not basic items of work and therefore require an original description in the specifications or cost code for the jobs in which they occur. (In construction specifications, basic items are sometimes prescribed in so-called "master specifications," which may be extant in a printed form or stored in a computer.) Particular items are originated as required for particular projects and are interleaved among the basic items to create the project specifications. Similarly, particular items are originated in an estimate and given a code number specially for that project, so that the item can be identified in cost accounting and other construction management procedures.

Plant and equipment costs. All the owning and operating costs of plant and equipment; costs of work other than material costs, labor costs, overhead costs, and profit, including costs of tools other than those provided by the workers for their own use.

Plumbing contractor. Person, corporation, or firm that undertakes to construct, extend, alter, renew, or repair any part of a plumbing system.

Pre-bid phase (general contractor). General contractor issues bid invitations to subcontractors, allowing sufficient time for the subcontractors to prepare their bids.

Preconstruction meeting (general contractor). General contractor arranges a meeting of the subcontractors, after the construction schedule has been approved by the owner. The rules for the project are discussed and the construction schedule is the mandated document. The architect/engineer may attend.

Prequalification of bidders or contractors. Process of investigating and qualifying bidders as acceptable contractors on the basis of their skills, integrity, and responsibility relative to the project contemplated prior to the award of contracts. Prequalification has been established as a standard procedure by some public lending agencies and owners.

Prime contractor. One who has a direct contract with an owner.

Producers. Manufacturers, processors, or assemblers of building materials and equipment in shops, off site.

Profit. Excess of income over total expenditure from doing work. For the owner the profit is a cost of the work. For the contractor, it is a motive to do the work. The contractor's total expenditures are material costs, labor costs, plant and equipment costs, and overhead costs. These must be accounted for before the amount of profit can be ascertained, which is one reason for proper estimating and cost accounting.

Progress payment. Payments based on contractor's requests made during progress of the work on account of work completed and/or materials suitably stored.

Progress schedule. Diagram prepared by the contractor showing proposed and actual starting and completion dates for the various branches of the work included in a project.

Protection of work and property. Protection by the contractor of his work, employees, equipment, the owner's property, adjacent property, and the public.

Punch list. List prepared by the architect/engineer or the owner, of the contractor's uncompleted or uncorrected work. The list is made before substantial completion date has been established.

Quantity and cost estimates. Contractor's forecasts of probable project construction cost based upon the sum of the estimated costs of materials, labor, equipment, overhead, and profit.

Quantity surveys. Bills of material estimates prepared by specially qualified estimators known as "quantity surveyors." Unit costs are not necessarily included in the estimate.

Restricted bid. Owner's selection of qualified bidders, who may have been recommended by the architect, engineer, lawyer, or by his friends, or the bank that may finance the project.

Retainage. Sum withheld from each payment to the contractor in accordance with the terms of the owner-contractor agreement.

Run. Item in an estimate with its quantity expressed as a linear measurement, usually linear feet or linear yards (meters). With a run item, two dimensions must appear in the item's description.

Schedule of values. Statement furnished by the contractor to the architect/engineer reflecting the portions of the contract sum allotted for the various parts of the work and used as the basis for reviewing the contractor's applications for progress payments.

Section of a specification. Distinct part of a specification division with its own title and reference. The work specified in a section depends on the job's requirements as interpreted by the specification writer.

Separate contract. Contract between the owner and contractor, other than the general contractor, for the construction of portions of a project.

Shop drawings. Drawings to illustrate how specific portions of the work are to be fabricated and/or installed.

Single contract. Contract for building construction where all work is contracted under the responsibility of a single general contractor.

Specifications. One of the contract documents, prepared by the architect/engineer is a written description of administration and procedure, products, and execution.

Standard forms of contracts. Prescribed forms containing articles of agreements and general conditions for different types of construction contracts, usually written (in conjunction with construction associations), published, and sold by architectural and engineering professional organizations.

Stipulated or lump sum (fixed price). Contract where the total cost of construction has been established and a price arrived at in advance, either by bidding or by negotiation.

Subbid. Offer made by a subtrade firm to a contractor to do construction work for payment, the acceptance of which constituted a subcontract between the subcontractor (who made the subbid) and the contractor for part of the work that the contractor has undertaken to perform for the owner.

Subcontract. Agreement between a separate contractor and a prime contractor.

Subcontractor

1. Person who contracts with a contractor for the work on part of a project. (A person who contracts with a subcontractor is also a subcontractor.)

2. Party to a construction subcontract who does construction work for the contractor (the other party to the subcontract) that is part of the work the contractor has undertaken to perform for the owner.

3. Separate contractor for a portion of the work.

4. Individual, firm, partnership, or corporation who, with the written consent of the contractor, assumes obligation for performing specified pay items.

Subcontractor list. List of subcontractors whose proposals were used by a prime contractor when preparing his bid; list of subcontractors proposed to be employed by the prime contractor, to be submitted to the architect for approval after the contract award.

Substantial completion. The date of substantial completion of a project or specified area of a project is the date when the construction is sufficiently completed in accordance with the contract documents, as modified by any change orders agreed to by the parties, so that the owner can occupy the project or specified area of the project for the use for which it was intended, subject to permit approval for occupancy.

Sub-subbid. Offer made by a subtrade firm to a subcontractor to do construction work for payment, the acceptance of which constitutes a sub-subcontract between the sub-subcontractor (who made the sub-subbid) and the subcontractor for part of the work that the subcontractor has undertaken to perform for the contractor.

Sub-subcontractor

1. Party to a construction sub-subcontract who does construction work for a subcontractor (the other party to the sub-subcontract) that is part of the work that the subcontractor has undertaken to perform for the contractor, who, in turn, has undertaken to do all the work for the owner.

2. Person or organization who has a direct or indirect contract with a subcontractor to perform a portion of the work at the site.

Super

1. Item in an estimate with the quantity expressed as a superficial area, usually in square feet or yards (or meters). With a super item, one dimension must appear in the item's description.

2. Abbreviation of the word "superintendent" on the project.

Superintendence. Contractor's responsibility for a representative at the site. No professional service normally rendered by an architect/engineer should be construed as superintendence.

Superintendent. Contractor's representative at the site who is responsible for continuous field supervision, coordination, and completion of the work and, unless another person is designated in writing by the contractor, to the owner and the architect/engineer.

Supervision. Direction of the work by contractor's representative. Supervision is neither a duty nor a responsibility of the architect/engineer as part of his basic professional services.

Supplier

1. One who supplies stock materials, manufactured products, or building components for construction work to a contractor, subcontractor, sub-subcontractor, or owner. The supplier does not usually do any of the work, at the site or elsewhere, as a subcontractor, but the distinction is not always applicable.

2. Person or organization who supplies materials or equipment for the work, including that fabricated to a special design, but who does not perform labor at the site.

Time of completion. Number of days (calendar or working), or the actual date as stated in the contract by which the work is required to be completed.

Trade. Classification or type of work usually done by workers who restrict themselves to a specific type of work, such as plumbers, masons, and carpenters.

Turnkey. Agreement with one single administrative entity, the prime contractor, who provides the site, financing, construction work, furniture, furnishings, fixtures, office equipment, start-up of all systems, maintenance for a stipulated period, and training of the maintenance staff.

Unit price

1. Unit price of an item of work is an average price of a unit of the work, established prior to the bid or contract documents. The unit of work usually includes labor and materials, and all other related costs.

2. Amount stated by the contractor in his proposal, as price per unit of measurement for material to be added to or omitted from the work.

Unit price contract. Contract awarded to a contractor who has presented a proposal with a quantity estimate of materials to be used on the project, including a unit price for each. Units of work, which include installation, are used as a multiplier against total units of each material, which then resolves the contract price. This type of contract allows the owner to add or deduct work from the contract under stipulated conditions.

Unit rate

1. Unit rate is expressed in man-hours per unit to indicate the average rate of labor productivity for an item of work. Since the unit rate is independent of the wage rate, it can be compared with other unit rates for work done at other times and for different wage rates.

2. Unit rate of an item of work is the average productivity rate for a unit of the work.

Upset price. Amount established as a maximum cost to perform a specified work.

Vendor. Person or organization who furnishes materials or equipment not fabricated to a special design for the work.

Void. Deduction made for an opening (door, window, etc.) or for a minor area within a major area.

Volume estimates. Contractor's forecasts of probable project construction cost based on unit cost per cubic foot.

Wage rate. Direct labor costs per hour exclusive of indirect labor costs.

Want. Deduction made for a deliberate overmeasurement of work in an estimate, usually in the case of an irregular area with want(s) adjoining the area's perimeter.

Warranty. Contractor's or manufacturer's guarantee of the quality, workmanship, and performance of his work or equipment.

Waste. Construction material that is additional to the actual quantity required in the work, as indicated in the contract documents, but that is, nevertheless, required by or used in performing the work and therefore contributes to the material costs of the work. Some waste may be inevitable and necessary, as in cutting a standard sheet of material to a smaller size required by the job, thus rendering the offcut unsuitable for any other use. It is essentially variable, and usually it is more easily accounted for (by cost accounting) if it is stated and kept separate from the net measured quantity of work by allowing for it in the synthesis of the unit price.

Withholding payments. Retaining certain sums otherwise due the contractor from payments to the contractor, because of work that is uncompleted or uncorrected, or for other reasons stated by the owner or the agency in writing.

Work

1. Labor, materials, and the use of tools, plant, and equipment, and all other things and services required of the contractor in a contract for which the owner pays.

2. Materials and/or labor required for a project.

Working drawings. One of the contract documents prepared by the architect/engineer illustrating the construction for the project.

CONSTRUCTION INDUSTRY ARBITRATION RULES

AS AMENDED AND IN EFFECT JANUARY 1, 1991

1. Agreement of Parties. The parties shall be deemed to have made these rules a part of their arbitration agreement whenever they have provided for arbitration by the American Arbitration Association (hereinafter AAA) or under its Construction Industry Arbitration Rules. These rules and any amendment of them shall apply in the form obtaining at the time the demand for arbitration or submission agreement is received by the AAA. The parties, by written agreement, may vary the procedures set forth in these rules.

2. Name of Tribunal. Any tribunal constituted by the parties for the settlement of their dispute under these rules shall be called the Construction Industry Arbitration Tribunal.

3. Administrator and Delegation of Duties. When parties agree to arbitrate under these rules, or when they provide for arbitration by the AAA and an arbitration is initiated under these rules, they thereby authorize the AAA to administer the arbitration. The authority and duties of the AAA are prescribed in the agreement of the parties and in these rules, and may be carried out through such of the AAA's representatives as it may direct.

4. National Panel of Arbitrators. In cooperation with the National Construction Dispute Resolution Committee, the AAA shall establish and maintain a National Panel of Construction Industry Arbitrators and shall appoint arbitrators therefrom as hereinafter provided.

5. Regional Offices. The AAA may, in its discretion, assign the administration of an arbitration to any of its regional offices.

6. Initiation under an Arbitration Provision in a Contract. Arbitration under an arbitration provision in a contract shall be initiated in the following manner: (a) The initiating party (hereinafter claimant) shall, within the time period, if any, specified in the contract(s), give written notice to the other party (hereinafter respondent) of its intention to arbitrate (demand), which notice shall contain a statement setting forth the nature of the dispute, the amount involved, if any, the remedy sought, and the hearing locale requested, and (b) Shall file at any regional office of the AAA three copies of the notice and three copies of the arbitration provisions of the contract, together with the appropriate administrative fee as provided in the Administrative Fee Schedule. The AAA shall give notice of the filing to the respondent or respondents. A respondent may file an answering statement in duplicate with the AAA within ten days after notice from the AAA, in which event the respondent shall at the same time send a copy of the answering statement to the claimant. If a counterclaim is asserted, it shall contain a statement setting forth the nature of the counterclaim, the amount involved, if any, and the remedy sought. If a counterclaim is made in the answering statement, the appropriate fee provided in the Administrative Fee Schedule shall be forwarded to the AAA with the answering statement. If no answering statement is filed within the stated time, it will be treated as a denial of the claim. Failure to file an answering statement shall not operate to delay the arbitration.

7. Initiation under a Submission. Parties to any existing dispute may commence an arbitration under these rules by filing at any regional office of the AAA three copies of a written submission to arbitrate under these rules, signed by the parties. It shall contain a statement of the matter in dispute, the amount of money involved, if any, the remedy sought, and the hearing locale requested, together with the appropriate administrative fee as provided in the Administrative Fee Schedule.

8. Changes of Claim. After filing of a claim, if either party desires to make any new or different claim or counterclaim, same shall be made in writing and filed with the AAA, and a copy shall be mailed to the other party, who shall have a period of ten days from the date of such mailing within which to file an answer with the AAA. After the arbitrator is appointed, however, no new or different claim may be submitted except with the arbitrator's consent.

9. Applicable Procedures. Unless the AAA in its discretion determines otherwise, the Expedited Procedures shall be applied in any case where no disclosed claim or counterclaim exceeds $50,000, exclusive of interest and arbitration costs. Parties may also agree to the Expedited Procedures in cases involving claims in excess of $50,000. The Expedited Procedures shall be applied as described in Sections 53 through 57 of these rules, in addition to any other portion of these rules that is not in conflict with the Expedited Procedures. All other cases shall be administered in accordance with Sections 1 through 52 of these rules.

10. Administrative Conference, Preliminary Hearing, and Mediation Conference. At the request of any party or at the discretion of the AAA, an administrative conference with the AAA and the parties and/or their representatives will be scheduled in appropriate cases to expedite the arbitration proceedings. In large or complex cases, at the request of any party or at the discretion of the arbitrator or the AAA, a preliminary hearing with the parties and/or their representatives and the arbitrator may be scheduled by the arbitrator to specify the issues to be resolved, to stipulate to uncontested facts, and to consider any other matters that will expedite the arbitration proceedings. Consistent with the expedited nature of arbitration, the arbitrator may, at the preliminary hearing, establish (i) the extent of and schedule for the production of relevant documents and other information, (ii) the identification of any witnesses to be called, and (iii) a schedule for further hearings to resolve the dispute. With the consent of the parties, the AAA at any stage of the proceeding may arrange a mediation conference under the Construction Industry Mediation Rules, in order to facilitate settlement. The mediator shall not be an arbitrator appointed to the case. Where the parties to a pending arbitration agree to mediate under the AAA's rules, no additional administrative fee is required to initiate the mediation.

11. Fixing of Locale. The parties may mutually agree on the locale where the arbitration is to be held. If any party requests that the hearing be held in a specific locale and the other party files no objection thereto within ten days after notice of the request has been mailed to it by the AAA, the locale shall be the one requested. If a party objects to the locale requested by the other party, the AAA shall have the power to determine the locale and its decision shall be final and binding.

12. Qualifications of an Arbitrator. Any neutral arbitrator appointed pursuant to Section 13, 14, 15, or 54, or

selected by mutual choice of the parties or their appointees, shall be subject to disqualification for the reasons specified in Section 19. If the parties specifically so agree in writing, the arbitrator shall not be subject to disqualification for those reasons. Unless the parties agree otherwise, an arbitrator selected unilaterally by one party is a party-appointed arbitrator and is not subject to disqualification pursuant to Section 19. The term "arbitrator" in these rules refers to the arbitration panel, whether composed of one or more arbitrators and whether the arbitrators are neutral or party appointed.

13. Appointment from Panel. If the parties have not appointed an arbitrator and have not provided any other method of appointment, the arbitrator shall be appointed in the following manner: immediately after the filing of the demand or submission, the AAA shall submit simultaneously to each party to the dispute an identical list of names of persons chosen from the panel. Each party to the dispute shall have ten days from the mailing date in which to cross off any names objected to, number the remaining names in order of preference, and return the list to the AAA. If a party does not return the list within the time specified, all persons named therein shall be deemed acceptable. From among the persons who have been approved on both lists, and in accordance with the designated order of mutual preference, the AAA shall invite the acceptance of an arbitrator to serve. If the parties fail to agree on any of the persons named, or if acceptable arbitrators are unable to act, or if for any other reason the appointment cannot be made from the submitted lists, the AAA shall have the power to make the appointment from among other members of the panel without the submission of additional lists.

14. Direct Appointment by a Party. If the agreement of the parties names an arbitrator or specifies a method of appointing an arbitrator, that designation or method shall be followed. The notice of appointment, with the name and address of the arbitrator, shall be filed with the AAA by that party. Upon the request of any appointing party, the AAA shall submit a list of members of the panel from which the party may, if it so desires, make the appointment. If the agreement specifies a period of time within which an arbitrator shall be appointed and any party fails to make the appointment within that period, the AAA shall make the appointment. If no period of time is specified in the agreement, the AAA shall notify the party to make the appointment. If within ten days thereafter an arbitrator has not been appointed by a party, the AAA shall make the appointment.

15. Appointment of Neutral Arbitrator by Party-Appointed Arbitrators or Parties. If the parties have selected party-appointed arbitrators, or if such arbitrators have been appointed as provided in Section 14, and the parties have authorized them to appoint a neutral arbitrator within a specified time and no appointment is made within that time or any agreed extension thereof, the AAA may appoint the neutral arbitrator, who shall act as chairperson. If no period of time is specified for appointment of the neutral arbitrator and the party-appointed arbitrators or the parties do not make the appointment within ten days from the date of the appointment of the last party-appointed arbitrator, the AAA may appoint the neutral arbitrator, who shall act as chairperson. If the parties have agreed that their party-appointed arbitrators shall appoint the neutral arbitrator from the panel, the AAA shall furnish to the party-appointed arbitrators, in the manner prescribed in Section 13, a list selected from the panel, and the appointment of the neutral arbitrator shall be made as prescribed in that section.

16. Nationality of Arbitrator in International Arbitration. Where the parties are nationals or residents of different countries, any neutral arbitrator shall, upon the request of either party, be appointed from among the nationals of a country other than that of any of the parties. The request must be made prior to the time set for the appointment of the arbitrator as agreed by the parties or set by these rules.

17. Number of Arbitrators. If the arbitration agreement does not specify the number of arbitrators, the dispute shall be heard and determined by one arbitrator, unless the AAA, in its discretion, directs that a greater number of arbitrators be appointed.

18. Notice to Arbitrator of Appointment. Notice of the appointment of the neutral arbitrator, whether appointed mutually by the parties or by the AAA, shall be mailed to the arbitrator by the AAA, together with a copy of these rules, and the signed acceptance of the arbitrator shall be filed with the AAA prior to the opening of the first hearing.

19. Disclosure and Challenge Procedure. Any person appointed as neutral arbitrator shall disclose to the AAA any circumstance likely to affect impartiality, including any bias or any financial or personal interest in the result of the arbitration or any past or present relationship with the parties or their representatives. Upon receipt of such information from the arbitrator or another source, the AAA shall communicate the information to the parties and, if it deems it appropriate to do so, to the arbitrator and others. Upon objection of a party to the continued service of a neutral arbitrator, the AAA shall determine whether the arbitrator should be disqualified and shall inform the parties of its decision, which shall be conclusive.

20. Vacancies. If for any reason an arbitrator is unable to perform the duties of the office, the AAA may, on proof satisfactory to it, declare the office vacant. Vacancies shall be filled in accordance with the applicable provisions of these rules. In the event of a vacancy in a panel of neutral arbitrators after the hearings have commenced, the remaining arbitrator or arbitrators may continue with the hearing and determination of the controversy, unless the parties agree otherwise.

21. Date, Time, and Place of Hearing. The arbitrator shall set the date, time, and place for each hearing. The AAA shall mail to each party notice thereof at least ten days in advance, unless the parties by mutual agreement waive such notice or modify the terms thereof.

22. Representation. Any party may be represented by counsel or other authorized representative. A party intending to be so represented shall notify the other party and the AAA of the name and address of the representative at least three days prior to the date set for the hearing at which that person is first to appear. When such a representative initiates an arbitration or responds for a party, such notice is deemed to have been given.

23. Stenographic Record. Any party desiring a stenographic record shall make arrangements directly with a stenographer and shall notify the other party of these arrangements in advance of the hearing. The requesting party or parties shall pay the cost of the record. If the transcript is agreed by the parties to be, or determined by the arbitrator to be, the official record of the proceeding, it must be made available to the arbitrator and to the other parties for inspection, at a date, time, and place determined by the arbitrator.

24. Interpreters. Any party wishing an interpreter shall make all arrangements directly with the interpreter and shall assume the costs of the service.

25. Attendance at Hearings. The arbitrator shall maintain the privacy of the hearings unless the law provides to the contrary. Any person having a direct interest in the arbitration is entitled to attend hearings. The arbitrator shall otherwise have the power to require the exclusion of any witness, other than a party or other essential person, during the testimony of any other witness. It shall be discretionary with the arbitrator to determine the propriety of the attendance of any other person.

26. Postponements. The arbitrator for good cause shown may postpone any hearing upon the request of a party or upon the arbitrator's own initiative, and shall also grant such postponement when all of the parties agree thereto.

27. Oaths. Before proceeding with the first hearing, each arbitrator may take an oath of office and, if required by law, shall do so. The arbitrator may require witnesses to testify under oath administered by any duly qualified person and, if it is required by law or requested by any party, shall do so.

28. Majority Decision. All decisions of the arbitrators must be by a majority. The award must also be made by a majority unless the concurrence of all is expressly required by the arbitration agreement or by law.

29. Order of Proceedings and Communication with Arbitrator. A hearing shall be opened by the filing of the oath of the arbitrator, where required; by the recording of the date, time, and place of the hearing, and the presence of the arbitrator, the parties, and their representatives, if any; and by the receipt by the arbitrator of the statement of the claim and the answering statement, if any. The arbitrator may, at the beginning of the hearing, ask for statements clarifying the issues involved. In some cases, part or all of the above will have been accomplished at the preliminary hearing conducted by the arbitrator pursuant to Section 10. The complaining party shall then present evidence to support its claim. The defending party shall then present evidence supporting its defense. Witnesses for each party shall submit to questions or other examination. The arbitrator has the discretion to vary this procedure but shall afford a full and equal opportunity to all parties for the presentation of any material and relevant evidence. Exhibits, when offered by either party, may be received in evidence by the arbitrator. The names and addresses of all witnesses and a description of the exhibits in the order received shall be made a part of the record. There shall be no direct communication between the parties and a neutral arbitrator other than at oral hearings, unless the parties and the arbitrator agree otherwise. Any other oral or written communication from the parties to a neutral arbitrator shall be directed to the AAA for transmittal to the arbitrator.

30. Arbitration in the Absence of a Party or Representative. Unless the law provides to the contrary, the arbitration may proceed in the absence of any party or representative who, after due notice, fails to be present or fails to obtain a postponement. An award shall not be made solely on the default of a party. The arbitrator shall require the party who is present to submit such evidence as the arbitrator may require for the making of an award.

31. Evidence. The parties may offer such evidence as is relevant and material to the dispute and shall produce such evidence as the arbitrator may deem necessary to an understanding and determination of the dispute. An arbitrator or other person authorized by law to subpoena witnesses or documents may do so upon the request of any party or independently. The arbitrator shall be the judge of the relevance and materiality of the evidence offered, and conformity to legal rules of evidence shall not be necessary. All evidence shall be taken in the presence of all of the arbitrators and all of the parties, except where any of the parties is absent in default or has waived the right to be present.

32. Evidence by Affidavits and Post-hearing Filing of Documents or Other Evidence. The arbitrator may receive and consider the evidence of witnesses by affidavit, but shall give it only such weight as the arbitrator deems it entitled to after consideration of any objection made to its admission. If the parties agree or the arbitrator directs that documents or other evidence be submitted to the arbitrator after the hearing, the documents or other evidence shall be filed with the AAA for transmission to the arbitrator. All parties shall be afforded an opportunity to examine such documents or other evidence.

33. Inspection or Investigation. An arbitrator finding it necessary to make an inspection or investigation in connection with the arbitration shall direct the AAA to so advise the parties. The arbitrator shall set the date and time and the AAA shall notify the parties. Any party who so desires may be present at such an inspection or investigation. In the event that one or all parties are not present at the inspection or investigation, the arbitrator shall make a verbal or written report to the parties and afford them an opportunity to comment.

34. Interim Measures. The arbitrator may issue such orders for interim relief as may be deemed necessary to safeguard the property that is the subject matter of the arbitration without prejudice to the rights of the parties or to the final determination of the dispute.

35. Closing of Hearing. The arbitrator shall specifically inquire of all parties whether they have any further proofs to offer or witnesses to be heard. Upon receiving negative replies or if satisfied that the record is complete, the arbitrator shall declare the hearing closed and minutes thereof shall be recorded. If briefs are to be filed, the hearing shall be declared closed as of the final date set by the arbitrator for the receipt of briefs. If documents are to be filed as provided in Section 32 and the date set for their receipt is later than that set for the receipt of briefs, the later date shall be the date of closing the hearing. The time limit within which the arbitrator is required to make the award shall commence to run, in the absence of other agreements by the parties, upon the closing of the hearing.

36. Reopening of Hearing. The hearing may be reopened on the arbitrator's initiative, or upon application of a party, at any time before the award is made. If reopening the hearing would prevent the making of the award within the specific time agreed on by the parties in the contract(s) out of which the controversy has arisen, the matter may not be reopened unless the parties agree on an extension of time. When no specific date is fixed in the contract, the arbitrator may reopen the hearing and shall have thirty days from the closing of the reopened hearing within which to make an award.

37. Waiver of Oral Hearing. The parties may provide, by written agreement, for the waiver of oral hearings in any

case. If the parties are unable to agree as to the procedure, the AAA shall specify a fair and equitable procedure.

38. Waiver of Rules. Any party who proceeds with the arbitration after knowledge that any provision or requirement of these rules has not been complied with and who fails to state an objection thereto in writing shall be deemed to have waived the right to object.

39. Extensions of Time. The parties may modify any period of time by mutual agreement. The AAA or the arbitrator may for good cause extend any period of time established by these rules, except the time for making the award. The AAA shall notify the parties of any extension.

40. Serving of Notice. Each party shall be deemed to have consented that any papers, notices, or process necessary or proper for the initiation or continuation of an arbitration under these rules; for any court action in connection therewith; or for the entry of judgment on any award made under these rules may be served on a party by mail addressed to the party or its representative at the last known address or by personal service, in or outside the state where the arbitration is to be held, provided that reasonable opportunity to be heard with regard thereto has been granted to the party. The AAA and the parties may also use facsimile transmission, telex, telegram, or other written forms of electronic communication to give the notices required by these rules.

41. Time of Award. The award shall be made promptly by the arbitrator and, unless otherwise agreed by the parties or specified by law, no later than thirty days from the date of closing the hearing, or, if oral hearings have been waived, from the date of the AAA's transmittal of the final statements and proofs to the arbitrator.

42. Form of Award. The award shall be in writing and shall be signed by a majority of the arbitrators. It shall be executed in the manner required by law.

43. Scope of Award. The arbitrator may grant any remedy or relief that the arbitrator deems just and equitable and within the scope of the agreement of the parties, including, but not limited to, specific performance of a contract. The arbitrator shall, in the award, assess arbitration fees, expenses, and compensation as provided in Sections 48, 49, and 50 in favor of any party and, in the event that any administrative fees or expenses are due the AAA, in favor of the AAA.

44. Award upon Settlement. If the parties settle their dispute during the course of the arbitration, the arbitrator may set forth the terms of the agreed settlement in an award. Such an award is referred to as a consent award.

45. Delivery of Award to Parties. Parties shall accept as legal delivery of the award the placing of the award or a true copy thereof in the mail addressed to a party or its representative at the last known address, personal service of the award, or the filing of the award in any other manner that is permitted by law.

46. Release of Documents for Judicial Proceedings. The AAA shall, upon the written request of a party, furnish to the party, at its expense, certified copies of any papers in the AAA's possession that may be required in judicial proceedings relating to the arbitration.

47. Applications to Court and Exclusion of Liability. (a) No judicial proceeding by a party relating to the subject matter of the arbitration shall be deemed a waiver of the party's right to arbitrate. (b) Neither the AAA nor any arbitrator in a proceeding under these rules is a necessary party in judicial proceedings relating to the arbitration. (c) Parties to these rules shall be deemed to have consented that judgment upon the arbitration award may be entered in any federal or state court having jurisdiction thereof. (d) Neither the AAA nor any arbitrator shall be liable to any party for any act or omission in connection with any arbitration concluded under these rules.

48. Administrative Fee. As a not-for-profit organization, the AAA shall prescribe an Administrative Fee Schedule and a Refund Schedule to compensate it for the cost of providing administrative services. The schedule in effect at the time the demand for arbitration or submission agreement is received shall be applicable. The administrative fee shall be advanced by the initiating party or parties, subject to final apportionment by the arbitrator in the award. When a claim or counterclaim is withdrawn or settled, the refund shall be made in accordance with the Refund Schedule. The AAA may, in the event of extreme hardship on the part of any party, defer or reduce the administrative fee.

49. Expenses. The expenses of witnesses for either side shall be paid by the party producing such witnesses. All other expenses of the arbitration, including required travel and other expenses of the arbitrator, AAA representatives, and any witness and the cost of any proof produced at the direct request of the arbitrator, shall be borne equally by the parties, unless they agree otherwise or unless the arbitrator in the award assesses such expenses or any part thereof against any specified party or parties.

50. Neutral Arbitrator's Fee. Unless the parties agree otherwise, members of the National Panel of Construction Industry Arbitrators appointed as neutrals will serve without compensation for the first day of service. Thereafter, compensation shall be based on the amount of service involved and the number of hearings. An appropriate daily rate and other arrangements will be discussed by the administrator with the parties and the arbitrator. If the parties fail to agree to the terms of compensation, an appropriate rate shall be established by the AAA and communicated in writing to the parties. Any arrangement for the compensation of a neutral arbitrator shall be made through the AAA and not directly between the parties and the arbitrator. The terms of compensation of neutral arbitrators on a panel shall be identical.

51. Deposits. The AAA may require the parties to deposit in advance of any hearings such sums of money as it deems necessary to defray the expense of the arbitration, including the arbitrator's fee, if any, and shall render an accounting to the parties and return any unexpended balance at the conclusion of the case.

52. Interpretation and Application of Rules. The arbitrator shall interpret and apply these rules insofar as they relate to the arbitrator's powers and duties. When there is more than one arbitrator and a difference arises among them concerning the meaning or application of these rules, it shall be decided by a majority vote. If that is unobtainable, either an arbitrator or a party may refer the question to the AAA for final decision. All other rules shall be interpreted and applied by the AAA.

53. Notice by Telephone. The parties shall accept all notices from the AAA by telephone. Such notices by the AAA shall subsequently be confirmed in writing to the parties. Should there be a failure to confirm in writing any notice hereunder, the proceeding shall nonetheless be valid if notice has, in fact, been given by telephone.

54. Appointment and Qualifications of Arbitrator. Where no disclosed claim or counterclaim exceeds $50,000, exclusive of interest and arbitration costs, the AAA shall submit simultaneously to each party an identical list of five proposed arbitrators drawn from the National Panel of Construction Industry Arbitrators, from which one arbitrator shall be appointed. Each party may strike two names from the list on a peremptory basis. The list is returnable to the AAA within seven days from the date of the AAA's mailing to the parties. If for any reason the appointment of an arbitrator cannot be made from the list, the AAA may make the appointment from among other members of the panel without the submission of additional lists. The parties will be given notice by telephone by the AAA of the appointment of the arbitrator, who shall be subject to disqualification for the reasons specified in Section 19. The parties shall notify the AAA, by telephone, within seven days of any objection to the arbitrator appointed. Any objection by a party to the arbitrator shall be confirmed in writing to the AAA with a copy to the other party or parties.

55. Date, Time, and Place of Hearing. The arbitrator shall set the date, time, and place of the hearing. The AAA will notify the parties by telephone, at least seven days in advance of the hearing date. A formal Notice of Hearing will be sent by the AAA to the parties.

56. The Hearing. Generally, the hearing shall be completed within one day, unless the dispute is resolved by submission of documents under Section 37. The arbitrator, for good cause shown, may schedule an additional hearing to be held within seven days.

57. Time of Award. Unless otherwise agreed by the parties, the award shall be rendered not later than fourteen days from the date of the closing of the hearing.

ELECTRICAL ENGINEERING

Abampere. Centimeter-gram-second electromagnetic unit of current. One abampere is equal to 10 A.

Abcoulomb. Centimeter-gram-second electromagnetic unit of electrical quantity. One abcoloumb is equal to 10 C.

Abfarad. Centimeter-gram-second electromagnetic unit of capacitance. One abfarad is equal to 10^9 farads.

Absolute maximum supply voltage. Maximum supply voltage that may be applied without the danger of causing a permanent change in the characteristics of a circuit.

Absolute value. Numerical quantity without regard to polarity, direction, or arithmetic sign.

Absorptance. Ratio of the flux absorbed by a medium to the incident flux. The sum of the total reflectance, total transmittance, and absorptance is 1.

Absorption. General term for the process by which a part of the incident flux at a surface or medium is dissipated within the medium.

Abvolt. Centimeter-gram-second electromagnetic unit of potential difference. One abvolt is equal to 10^8 V.

AC (alternating current). Flow of electricity that rises to a maximum in one direction, then decreases to zero, then reverses itself, rising to a maximum in the opposite direction, and so on. The number of such cycles per second (hz) is the frequency of alternation.

Acoustical ohm. Measure of acoustic resistance, reactance or impedance.

Alive or live (energized). Electrically connected to a source of potential difference; electrically charged so as to have a potential significantly different from that of the earth in the vicinity. The term "live" is sometimes used in place of "current-carrying," where the intent is clear, to avoid repetition of the longer term.

Alternating current

1. Electrical current in which the direction of electron flow is reversed at regular intervals. In the United States, 60 Hz (cycles per second) constitute the standard frequency; in Europe, 50 Hz are standard.

2. Current that changes direction every one-twentieth of a second.

3. Electric current that alternately reverses its direction in a circuit in a periodic manner.

Alternating-current ampere. Current that when flowing through a given ohmic resistance will produce heat at the same rate as a direct-current ampere.

Ammeter. Instrument for measuring the intensity of an electric current in units that are known as amperes.

Ampacity. Current-carrying capacity of electric conductors expressed in amperes.

Amperage—electrical rate of flow. The amperage is measured in amperes and is comparable to gallons per minute in a fluid medium.

Ampere (A)

1. Measure of the quantity of electric current flowing in a circuit: 1 V applied across a resistance of 1 ohm will cause 1 A to flow.

2. Unit for measuring the amount of electrical current flowing through a wire (comparable to gallons of water per minute).

3. Current of 1 C of electricity per second past a given point.

4. Watts divided by volts equal amperes ($W/V = A$).

Ampere turn. Unit of magnetomotive force equivalent to amperes times turns.

Angstrom $\overset{\circ}{A}$

1. Unit of length used in the measurement of the wavelength of light, x-rays, and other electromagnetic radiation and in the measurement of molecular and atomic diameters: 1 $\overset{\circ}{A}$ equals 10^{-8} cm or 10^{-4} μm. The wavelength of visible light ranges from about 4000 to 7000 $\overset{\circ}{A}$, whereas x-ray wavelengths and atomic diameters are of the order of a few angstroms. The unit is named in honor of Swedish spectroscopist A. J. Angstrom.

2. A unit of length, 10^{-10} m, used to express wavelengths of optical spectra.

Angstrom unit. A unit of measurement of wavelength and other radiation.

Appliance branch circuit. Circuit supplying energy to one or more outlets to which appliances are to be connected; such circuits have no permanently connected lighting fixture unless it is part of an appliance.

Asymmetrical current. Alternating current having a waveform that is offset with respect to the zero axis due to a transient condition. The offset occurs at the initiation of a short circuit or other change in current. The offset usually decays quickly until steady-state conditions are reached and the current becomes symmetrical. Asymmetrical current is composed of symmetrical and direct current components; it is expressed in rms total amperes or rms asymmetrical amperes at a specific time (normally one-half cycle) after initiation of a short circuit or other change in current.

Atmospheric transmissivity. Ratio of the flux incident on a surface after passing through a unit thickness of the atmosphere to the flux that would be incident on the same surface if the flux had passed through a vacuum.

Automatic. Self-acting, operating by its own mechanism when actuated by some impersonal influence.

Available short-circuit current. Maximum current in rms symmetrical amperes that a circuit is capable of delivering at the system terminals ahead of the apparatus being supplied.

Avogadro's number. Number (N) of molecules in a gram-molecule or atoms in a gram-atom of a substance $N = 6.02 \times 10^{23}$ molecules.

Back-connected switch. Switch in which the current-carrying conductors are connected to the back studs of the mounting base.

BeV. Billon electron volts. An electron possessing this much energy travels at a speed close to light at 186,000 miles/second.

Branch circuit

1. Portion of a wiring system extending beyond the final overcurrent device protecting the circuit. A device not approved for branch circuit protection, such as a thermal cutout or motor overload protective device, is not considered the overcurrent device protecting the circuit.

2. Circuit conductors between the final overcurrent device protecting the circuit and the outlet(s).

3. Portion of a circuit extending beyond the final overcurrent devices in the circuit.

Branch circuit (general purpose). Branch circuit that supplies a number of outlets for lighting and appliances.

Branch circuit (individual). Branch circuit that supplies only one utilization equipment.

Break distance of a switch. The minimum open gap distance between the stationary and movable contents, or live parts connected thereto, when the blade is in the open position.

Breakdown voltage. The voltage at which the dielectric strength of an insulator breaks down and arcing begins.

Calibration. Method of adjusting an instrument to cause it to conform to a specified standard.

Capacitance. Measure of the ability of a circuit to store electrical energy. When two conducting materials are separated by an insulating material, they have the ability to store electrical energy when the circuit carried alternating current. Such an arrangement of material is called a "condenser" or "capacitor." The unit of capacity is the farad.

Capacitive reactance. Opposition to the flow of alternating current due to capacity. Measured in ohms, it makes the voltage lag behind the current in phase.

Carrier current. High frequency current superimposed on the normal frequency of a power transmission line for communication or telemetering control.

Cell line. An assembly of electrically interconnected electrolytic cells supplied by a source of direct-current power.

Circuit. Complete path of an electric current. Electric current is permitted to flow from a wire connected with a generating source through energy-consuming units and return through another wire to the source. Positive wire is known as such because of its potential or voltage above or below zero. Negative wire may be a neutral maintained at a potential of zero voltage. Neutral wire is ordinarily grounded to the earth at the point of service entrance. A ground is carried through the electric system as a third wire or through the enclosing metallic raceway of the conductors.

Circuit with capacitance only. When a direct-current voltage is impressed across the plates of a perfect capacitor, there is an initial rush of current that charges the capacitor to the impressed voltage. After this there is no further current if the impressed voltage remains constant.

Circular mil. Area of a circle 1 mil or 0.001 in. in diameter. The area of any solid conductor of circular cross section can be found by squaring the diameter expressed in mils.

Coincidence factor. Ratio of the maximum demand of a system or part under consideration to the sum of the individual maximum demands of the subdivisions.

Commercial frequencies. In the United States frequencies are standardized at 60 cycles and 25 Hz, although other frequencies are used. In parts of California and Mexico 50 cycles are used on some large transmission systems.

Concealed. Rendered inaccessible by the structure or finish of the building.

Conductance. The reciprocal of resistance.

Conductivity. The reciprocal of resistivity.

Continuous current. Current that flows continuously in one direction (direct current) and at a steady value.

Continuous duty. Requirement of service that demands operation at a substantially constant load for an indefinitely long time.

Continuous load. Load where the maximum current is expected to continue for 3 hours or more.

Coulomb (C)

1. Quantity of electricity that passes any point in an electric circuit in 1 second when the current is maintained constant at 1 A.

2. Quantity of electrical charge equivalent to 6.24×10^{18} electrons.

Creepage. Conduction of electricity across the surface of a dielectric.

Creep distance. Shortest distance on the surface of an insulator between two electrically conductive surfaces separated by the insulator.

Current

1. The rate of flow of electricity.

2. The movement of electrons through a conductor, measured in amperes.

3. The current taken by the light source is measured by means of an ammeter. The location of the ammeter in the circuit should be such that all of the current to the light source goes through the ammeter. When the current taken by the voltmeter is included in the ammeter indication, a correction must be made to obtain true lamp current.

Cycle. Complete set of positive and negative values through which an alternating current repeatedly passes.

Delay line. Circuit that produces a time interval between input and output.

Demand factor

1. Ratio of the maximum demand of the system or part of the system to the total connected load of the system or part of the system under consideration.

2. Ratio of the maximum demand to the total connected load.

Dielectric constant. Ratio comparing the insulative quality of a material to that of air.

Dielectric strength. Electric voltage gradient at which an insulating material is broken down or "arced through," in volts per mil of thickness.

Different systems. Systems that derive their supply from different sources, from individual transformers or banks of transformers that do not have their secondary windings interconnected, or from individual service switches.

Diffuse reflection. Process by which a portion of the incident flux is reemitted in a non-image-forming (diffused) state.

Diffuse transmission. Process by which a portion of the incident flux is reemitted from a surface or medium on the nonincident side in a non-image-forming (diffused) state.

Direct current

1. Flow of electrons in only one direction through a circuit.

2. Electric current flowing in one direction only and free from pulsation.

3. Current that maintains a constant direction.

Direct-current circuit. The power in a direct-current circuit under steady conditions is given by the product of the volts across the circuit and the current in amperes in the circuit.

Direct-current power. Power determined by observing the current and the voltage and computing the power as their product.

Discharge. Act of neutralizing a charge.

Distributing substation. Assembly of equipment designed to receive electric energy at a high supply voltage and convert it to a form desirable for local distribution. Electric energy may be received as alternating current and distributed as direct current, or as alternating current at one frequency and distributed at another frequency.

Distribution. The supply of power to points of utilization from the source of generation.

Diversity factor. Reciprocal of the coincidence factor.

Diversity factor of an installation. The ratio of the sums of the maximum demands of the various loads in the installation to the maximum demand of the whole installation.

Drude equation. The values of spectral emissivity at wavelengths greater than 2000 nm may be calculated with reasonable accuracy by means of the Drude equation.

Dry cell. Primary voltage cell with an electrolyte of paste.

Dynamic electricity. Electricity in motion. The term is usually applied to practical electricity capable of sustaining a useful current.

Duty. Requirement of service that demands the degree of regularity of the load.

Eddy currents. Stray local currents induced in those metal parts of electrical apparatus that are rapidly changing magnetic fields.

Effective. Capable of performing a specified function with safety.

Effectively grounded

1. Permanently connected to earth through a ground connection of sufficiently low impedance and having sufficient ampacity to prevent the building up of voltages that may result in undue hazard to connected equipment or to persons.

2. Permanently connected to earth through a ground connection of sufficiently low impedance and having sufficient ampacity that ground fault current which may occur cannot build up to dangerous voltages.

Effective value. Square root of the sum of the square of all the instantaneous values in a sine wave, equivalent to $0.707 \times$ peak value.

Effect of polarization. When radiant energy is polarized, the photoelectric current will vary as the orientation of the polarization is changed.

Effect of wavelength. The more electropositive the metal, the longer the wavelength of its maximum photoelectric emission and the lower the frequency threshold below which electrons are not liberated.

Efficiency. Ratio of output to input. The output is the useful energy delivered by a machine. Input is the energy supplied to the machine.

Electrical angle. An angle that specifies a particular instant in an alternating-current cycle or expresses the phase difference between two alternating quantities.

Electrical degree. The three-hundred sixtieth part of the angle subtended at the axis of the machine by two consecutive field poles of like polarity. One mechanical degree is thus equal to as many electrical degrees as there are pairs of poles in the machine.

Electrical energy

1. The work done in a circuit or an apparatus by current flowing through it.

2. Electrical energy is generated by forcing electrons to move in certain paths or circuits. By convention, electric current is assumed to flow from a positive terminal to a negative.

Electrical engineer. Person whose training includes a degree in electrical engineering, or who has comparable knowledge and experience in computations to prepare that person for dealing with the generation, transmission, utilization, and values of standards and codes related to installation of electrical service.

Electrical impedance. The total opposition that a circuit presents to an alternating current, equal to the complex ratio of the voltage to the current in complex notation.

Electrical input. Electrical input is measured at the motor terminals. For synchronous motors with separately driven exciters, excitation input as measured at the sliprings is added to the input of the stator. For synchronous motors with direct-connected exciters, exciter losses are deducted from the measured stator input.

Electrical instability. Persistent condition of unwanted self-oscillation in an amplifier or other electric circuit.

Electrical measurement. Measurement of any one of the many quantities by which electricity is recorded or computed.

Electrical power. The rate of expending electrical energy or the rate of doing work per unit of time.

Electrical properties. Properties of a substance that determine its response to an electric field, such as its dielectric constant or conductivity.

Electrical resistivity. Electrical resistance offered by a material to the flow of current times the cross-sectional area of the current flow and per unit length of current path; the reciprocal of the conductivity.

Electrical symbol. Geometrical indication or symbol used on drawings or other descriptive literature to represent a component of a circuit, device, or equipment that has been included in a schematic circuit diagram.

Electrical system. System of wiring, switches, relays, or other devices associated with receiving and distributing electrical current.

Electrical unit. Standard in terms of which some electrical quantity is evaluated.

Electric arc. A discharge of electricity through a gas, normally characterized by a voltage drop approximately equal to the ionization potential of the gas.

Electric circuit. Path or group of interconnected paths capable of carrying electric currents.

Electric connection. Direct wire path for current between two points in a circuit.

Electric current. Rate at which electricity flows through a conductor or circuit. The practical unit is the ampere, which is a current of 1 C/second. A coulomb is a basic quantity of electricity.

Electric energy measurement. Measurement of the integral, with respect to time, of the power in an electric circuit.

Electric field. Space in the neighborhood of a charged body or of varying magnetic field throughout which an electric charge would experience a mechanical force.

Electric instrument. An electricity-measuring device that indicates flow of electric current, such as a voltmeter or ammeter.

Electric heat. Heat produced by a flow of electrons through a resistive element.

Electricity

1. General name for all phenomena that arise out of the flow or accumulation of electrons in matter. The unit of quantity is the coulomb; that of pressure, the volt; and that of flow, the ampere.

2. Atomic energy capable of producing a flow of electrons.

Electric power. Rate of doing electrical work. The unit is the watt or kilowatt (1000 W); 746 W = 0.746 kW = 1 hp.

Electric power generation. Large scale production of electrical power for industrial, residential, commercial, and rural use, generally in stationary generation plants designed for that purpose.

Electric power plant. Power plant that converts a form of raw energy into electricity, such as a hydro, steam, diesel, or nuclear generating station for stationary or transportation service.

Electric power substation. Installation of equipment for an electric power system for the transmittal, distribution, voltage transformations, and switching as required for service, of electric energy.

Electric power system. Assemblage of equipment and circuits for generating, transmitting, transforming, and distributing electric energy.

Electric power transmission. Process of transferring electric energy from one point to another in an electric power system.

Electrodynamometer. Meter movement characterized by three coils: two fixed and one movable.

Electrolysis. Action that occurs when electric current (direct) is passed through a solution (called "electrolyte"); it is decomposed.

Electrolyte. Chemical ingredient between the plates of a voltage cell.

Electrolytic conduction. In general, solutions of most inorganic salts, bases, acids, certain fused salts, and a few solid substances permit passage of electricity by electrolytic conduction.

Electromagnet. Magnet created by current through the loops of an inductor.

Electromagnetic system. System based on the force exerted between a magnetic field and conductor carrying a current placed in that field.

Electromotive force

1. Force that causes current to flow, such as the electric potential difference between the terminals of any device used as a source of electrical energy.

2. An agency that tends to produce motion of electrons through a circuit is an electromotive force (emf). It is not a force in the mechanical sense.

3. Difference in potential between two points; also, voltage and electrical pressure.

Electron. Negatively charged particle with negligible mass orbiting around the nucleus of an atom.

Electrons of an atom. The electrons of an atom are divided into two classes. The first includes the core electrons, which are not readily removed or excited by radiation other than x-rays. The second includes the valence electrons, which cause chemical bonding into molecules.

Electroscope. Instrument that indicates the presence of small charges.

Electrostatic charge. Difference in potential caused by a displacement of electrons.

Electrostatic field. Area around an electrostatic charge where its influence can be felt.

Electrostatics. Science of static electricity (electricity at rest).

Electrostatic system. System based on the force exerted between two charges of electricity.

Element. Primary substance that cannot be broken down by chemical means.

Energy. Ability to do work. Energy can be converted into different forms but cannot be created or destroyed.

Erythemal exposure. Product of erythemal flux density on a surface by the duration of the exposure. It equals the amount of effective radiant energy received per unit of area exposed.

Erythemal flux. Radiant flux evaluated according to its capacity to produce erythema of the untanned human skin.

Erytheme E-viton. Recommended unit of erythemal flux; amount of radiant flux that will give the same erythemal effect as 10 mW of radiant flux at 296.7 nm.

Extralow potential. Potential up to and including 30 V.

Extralow-voltage power circuit. Circuit, such as valve operator and similar circuits, that is neither a remote control circuit nor a signal circuit, but that operates at not more than 30 V and is supplied from a transformer or other device restricted in its rated output to 1000 VA and is approved for the purpose, but in which the current is not limited in accordance with the requirements for a type circuit.

Farad (F)

1. Capacitance of a capacitor in which a charge of 1 C produces a charge of 1 V in the potential difference between its terminals.

2. Unit of capacitance equivalent to 1 C/V.

Fault. Electrical short circuit or leakage path to ground or from phase to phase inadvertently created.

Ferromagnetic. Term applied to a substance strongly attracted to magnets and characterized by high permeability.

Finsen. Recommended unit of erythemal flux density on a surface, equal to one unit of erythemal flux per square centimeter.

Frequency. Measure of the number of periods of alternating current per unit time, usually expressed in hertz (Hz), which are periods per second.

Frequency of an alternating-current circuit. The number of complete reversals, or cycles, of the current per second.

Galvanic electricity. Electricity is produced when 2 unlike metals are immersed in an acid or salt solution an electromotive force will be generated between them.

Galvanometer. Sensitive direct-current meter that registers current in either direction.

Gauss (G). Centimeter-gram-second electromagnetic unit of magnetic induction: 1 G represents one line of flux (one maxwell) per square centimeter.

General purpose branch circuit. Circuit that supplies a number of outlets for lighting and appliances.

Gilbert. Unit of magnetomotive force, equivalent to flux times reluctance.

Graybody. Temperature radiator whose spectral emissivity is less than unity and the same at all wavelengths.

Graybody radiation. When the spectral emissivity is constant for all wavelengths, the radiator is known as a "graybody." No known radiator has a constant spectral emissivity for all visible, infrared, and ultraviolet wavelengths, but in the visible region a carbon filament exhibits very nearly uniform emissivity; that is, it is nearly a graybody.

Ground

1. Conductive connection, whether intentional or accidental, between an electrical circuit or equipment and earth, or to some conducting body that serves in place of the earth.

2. Connection to earth obtained by a ground electrode.

3. Conductive connection, whether intentional or accidental, by which an electric circuit or equipment is connected to reference ground.

4. Connection of the electrical system to the ground to minimize damage from lightning and protect from electrical shock.

5. Conductive body, usually earth, to which an electric potential is referenced.

Henry (H). Unit equal to the inductance of a circuit in which the variation of a current at the rate of 1 A second induces an electromotive force of 1 V (named after physicist Joseph Henry).

Hertz (Hz). Unit of frequency equal to one cycle per second.

High-intensity arcs. Arcs obtained from the flame arc by increasing the size and the flame material content of the core of the anode, and at the same time greatly increasing the current density, to a point where the anode spot spreads over the entire tip of the carbon resulting in a rapid evaporation of flame material and carbon from the core so that a crater is formed. The principal source of light is the crater source and the gaseous region immediately in front of it.

High potential. Potential above 750 V.

Horsepower

1. Work performed equivalent to 33,000 ft-lb/minute. The term "horsepower" was first used by James Watt to measure the power of his steam engine against the power of a horse.

2. Unit of electric power equal to 746 W.

Hysteresis. That property of magnetic material which causes the induction corresponding to a given magnetizing force to be greater when the latter is decreasing than when it is increasing.

Impedance

1. Total combined opposition to the flow of alternating current. The impedance consists of the resistance, and inductive and capacitive reactance.

2. Combination of reactance and resistance, measured in ohms.

Impedance of a circuit. The resistance offered to the passage of alternating current.

Individual branch circuit. Branch circuit that supplies only one utilization equipment.

Inductance. Inductance occurring in a circuit carrying alternating current (ac) produces opposition to the flow of the current that makes the current lag behind the voltage in time or phase. It is measured in a unit called a "henry."

Inductive reactance. Opposition to the flow of changing current due to inductance. It is measured in ohms.

Initial conductor tension. Longitudinal tension in a conductor prior to the application of an external load.

Initial unloaded sag. Sag of a conductor prior to the application of an external load.

Instantaneous value. Amplitude of a sine wave at any selected instant.

Insulation resistance. Resistance of insulating material measured in megohms.

Intensity. Shortening of the terms "luminous intensity" and "radiant intensity." The term is often misused for level of illumination.

Ion. Charged atom.

Irradiance H. Density of radiant flux incident on a surface.

Intermittent duty. Requirement of service that demands operation for alternate intervals of load and no load; load and rest; or load, no load, and rest.

Kilo. Prefix meaning 1000 (10^3).

Kilowatt (kW). Electrical unit of work or power. Equal to 1000 watts, 1.34 horsepower, and 1.18 KVA.

Kilowatt hour (kWh). Unit of work equal to the consumption of 1000 W in 1 hour.

KVA (kilovolt-ampere). Approximately 89/100 kW.

Live parts. Parts that are electrically connected to points of potential different from that of the earth.

Load factor. Ratio of the average demand to the maximum demand over a designated period of time.

Loss factor. Product of the power factor and the dielectric constant.

Low-energy power circuit

1. Circuit that is not a remote control or signal circuit, but has its power supply limited in accordance with the requirements of Class 2 remote control circuits. Such circuits include electric door openers and circuits used in the operation of coin-operated phonographs.

2. Circuit that is not a remote control or signaling circuit but has its power supply limited in accordance with the requirements of Class 2 and Class 3 circuits.

Low potential. Potential from 31 to 750 V inclusive.

Low-voltage circuit. Circuit having a difference of potential of 600 V (750 V where specified in certain regulations) or less between any two conductors of the circuit.

Magnetomotive force. Magnetic potential difference measured in gilberts and symbolized by mmf.

Maximum demand. Integrated demand for a specified time interval, that is, instantaneous, 5 minutes, 10 minutes, 15 minutes, etc.

Maxwell. One line of magnetic force.

Mega. Prefix meaning one million.

Megawatt. Unit of power equal to one million watts or one thousand kilowatts.

Molecule. Smallest particle of matter that contains all the characteristics of the original substance.

Multiple, parallel, or shunt circuits. Circuits in which the components are so arranged that the current divides among them. In multiple circuits the current through the generator varies with the load, and the generator electromotive force is maintained practically constant. Multiple circuits generally are employed for the distribution of electrical energy for all lighting and power in buildings.

Multiwire branch circuit. Branch circuit consisting of two or more ungrounded conductors having a potential difference between them and an identified grounded conductor having equal potential difference between it and each ungrounded conductor of the circuit, and which is connected to the neutral conductor of the system.

Neutral. Conductor (when one exists) of a polyphase circuit, or of a single-phase, three-wire circuit, that is intended to have a potential such that the potential differences between it and each of the other conductors are approximately equal in magnitude and are equally spaced in phase.

Nominal voltage. Nominal value assigned to a circuit or system for the purpose of conveniently designating its voltage class (120/240, 480Y/277, or 600, etc.). The actual voltage at which a circuit operates can vary from the nominal within a range that permits satisfactory operation of equipment.

Nonincentive circuit. Circuit or part of a circuit in which any sparking that may be produced by normally arcing parts is incapable, under normal operating conditions, of causing an ignition of the prescribed flammable gas or vapor.

Nucleus

1. The nucleus is made up of protons that carry the positive charge and neutrons that are approximately equal in mass to the protons, but uncharged. The number of protons in the nucleus is always the same

for a given element and gives that element its atomic number.

2. Center of an atom consisting of protons, neutrons, and other subatomic particles.

Oersted (Oe). Unit of field strength: 1 Oe is equivalent to 1 dyne of force on a unit pole.

Ohm

1. Unit of resistance: 1 ohm is the value of resistance through which a potential difference of 1 V will maintain a current of 1 A. Equal to 10^9 cgs units.

2. Unit of electrical resistance to current flow. It is equal to a fall in potential of 1 V when a current of 1 A flows.

Ohmmeter. Instrument used to measure resistance.

Ohm's law. The voltage required to force a current through a circuit of constant resistance R is proportional to the current.

Open circuit. Interrupted electrical circuit (electricity cannot flow).

Overload. Use of more current than a circuit or equipment is designed for.

Parallel circuit. Circuit in which devices are connected across one another in a ladderlike arrangement.

Periodic duty. Type of intermittent duty in which the load conditions are regularly recurrent.

Permeability. Ease with which a material passes magnetic flux.

Phase

1. The time relation between the current and potential in an alternating-current circuit, or the time between potentials in two or more circuits.

2. Position of a wave shape at any given instant during its period.

Phase relation. Comparison of the phase of two signals at a given instant.

Photoelectric cell. Device that converts light energy to electrical energy.

Photoelectron. The maximum value of the initial velocity of a photoelectron, and therefore its maximum kinetic energy, decreases as the wavelength of the radiant energy increases.

Piezoelectric. Pressure electricity; property of a crystal that causes it to produce a voltage when placed under mechanical stress.

Planck's equation. The equation for blackbody radiation was developed by the introduction of the concept of radiation of discrete quanta of energy to represent the radiation curves obtained in 1900 by Lummer and Pringsheim, who used the open end of a specially constructed and uniformly heated tube as their source.

Polarity. Direction with respect to an axis.

Power

1. In alternating-current circuits the power is best determined by a wattmeter measurement. The current coil of the wattmeter is connected into the line in a manner similar to an ammeter, and the voltage or potential circuit of the wattmeter is connected directly across the load, as for a voltmeter. Certain wattmeters are designed to compensate automatically for the power taken by the wattmeter potential coil. In such cases the wattmeter is referred to as a "compensated wattmeter," and no correction need be made.

2. Time rate of doing work expressed in watts.

Power factor

1. In a perfect condenser, the current leads the voltage by 90°. When a loss takes place in the insulation, the absorbed current, which produces heat, throws the 90° relationship out according to the proportion of current absorbed by the dielectric. The power factor is the cosine of the angle between the voltages applied and the current resulting. Measurements are usually made at million-cycle frequencies.

2. Ratio of actual to apparent power of an alternating current, found by measuring the current with a wattmeter as against the indicated voltmeter reading; measure of the loss in an insulator, capacitor, or inductor.

Power factor of a circuit. The ratio of the true power passing through it to the product of the volts and amperes.

Precise method. For the measurement of dc current and voltage, a deflection potentiometer, equipped with suitable multipliers and shunts, provides a rapid and accurate method of measuring current and voltage. Adjustment of a single dial gives an approximate balance of the unknown value against a standard cell. Residual unbalance is read from the deflection of a sensitive millivoltmeter, which replaces the galvanometer of the null potentionmeter.

Primary standard. Standard by which a unit of measurement is established and from which the values of other standards are derived. A satisfactory primary standard must be reproducible from specifications.

Pulsating direct current. Current with a constant direction and varying amplitude.

Pulse. Sharp surge of current.

Quantity of electricity. The product of current and time.

Radiance. Radiant flux per unit solid angle (radiant intensity) in a given direction per unit of projected area of the source as viewed from that direction.

Radiant density. Radiant energy per unit volume.

Radiant emittance. Density of radiant flux emitted from a surface.

Radiant energy U. Energy traveling in the form of electromagnetic waves; the time integral of radiant flux.

Radiant flux

1. Time rate of the flow of any part of the radiant energy spectrum measured in ergs per second or in watts.

2. Time rate of flow of radiant energy.

Radiant flux density at a surface. Radiant flux per unit area of the surface.

Radiant intensity. Radiant flux per unit solid angle in a given direction.

Reactance. Opposition to a change in current, measured in ohms.

Reactance of a circuit. That part of the resistance offered to the passage of alternating current that is due to the inductance and capacity of the circuit.

Reflectance. Ratio of the flux reflected by a surface or medium to the incident flux. The quantity reported may be total reflectance, regular (specular) reflectance, diffuse reflectance, or spectral reflectance, depending on the component measured.

Regular transmission. Process by which a portion of the incident flux is reemitted from a surface or medium on the nonincident side without scattering (no diffusion).

Relative erythemal factor. Factor that gives the relative erythemal effectiveness of radiation of a particular wavelength as compared with that of wavelength 296.7 nm, which is rated as unity.

Reluctance. Opposition to magnetic flux, measured in rels.

Residual magnetism. Magnetism that remains in a material after the magnetizing force is removed.

Resistance

1. Opposition offered by a material to the flow of an electric current in it. The electrical resistance is expressed in ohms. It determines for a given current the average rate at which electrical energy is converted into heat.

2. Opposition to current, measured in ohms.

Resistivity. Ability of a material to resist the passage of electrical current either through its bulk or on a surface. The unit of volume resistivity is the ohm-centimeter; the unit of surface resistivity is the ohm.

Resonant frequency. Frequency that produces resonance.

Retentivity. Ability of a material to hold magnetism.

Scalar. Quantity that is completely determined by its magnitude alone.

Secondary network. Distribution system wherein the secondary mains of an alternating-current system are interconnected and supplied through transformers connected in parallel on the secondary side through fuses or automatic switching devices arranged to prevent the feeding of fault current on the primary side of the transformers through the secondary mains. Such a system is also often called an "alternating-current automatic low-voltage secondary network."

Secondary standard. Standard that is calibrated by comparison with a primary standard.

Selective radiators. The emissivity of all known materials varies with wavelength. Therefore, materials are called "selective radiators."

Self-induction of a circuit. The property that opposes any change in the value of the current flowing through it.

Series circuit

1. All components are connected in tandem. The current at every point of a series circuit is the same, and the generator electromotive force varies with the load. The most important commercial application of a series conduit is in street lighting. This type of circuit has the disadvantage that a break any place in the circuit disrupts the current. Thus, if one lamp in a group connected in series goes out, they all go out.

2. Circuit in which the total current passes through each part of the circuit.

Signal circuit

1. Electrical circuit that supplies energy to an appliance that gives a recognizable signal.

2. Electrical circuit, other than a communication circuit, that supplies energy to a device that gives a recognizable signal, such as circuits for doorbells, buzzers, code-calling systems, and signal lights.

Signaling circuit. Electric circuit that energizes signaling equipment.

Short. Low-resistance path through or around a circuit component.

Spectral emissivity. Ratio at a given wavelength of the radiant flux density per unit wavelength interval (at that wavelength) of a temperature radiator to that of a blackbody at the same temperature.

Steradian (unit solid angle). Solid angle subtending an area on the surface of a sphere equal to the square of the sphere radius.

Surface resistivity. Electrical resistance between opposite edges of a unit square of insulating material, commonly expressed in ohms.

System. Electrical system in which all the conductors and apparatus are capable of being readily connected electrically by metallic contact to a common source of potential difference.

Theory of wave mechanics. The theory of wave mechanics is an attempt to reach a harmonious compromise between the quantum and the wave theories. It utilizes wave characteristics and quanta particles as the need arises in the solution of problems.

Three-phase current. Current that consists of three different alternating currents 120° out of phase with each other.

Three-wire circuit. Multiwire circuit. Its advantages over the two-wire circuit are that only three wires are required to supply a load that would require four wires when two-wire circuits are used, and that when other conditions are the same, the percent voltage drop is only half as great as in a two-wire circuit.

Total emissivity. Ratio of the radiant flux density (radiant emittance) at an element of a temperature radiator to that at an element of a blackbody at the same temperature.

Tracking. A high-voltage source current creates a leakage or fault path across the surface of an insulating material by slowly but steadily forming a carbonized path.

Transmission. General term for the process by which a part of the incident flux leaves a surface or medium on a side other then the incident side.

Transmittance. Ratio of the flux transmitted by a medium to the incident flux. The quantity reported may be total transmittance, regular transmittance, diffuse transmittance, or spectral transmittance, depending on the component measured.

Two-phase current. Current that consists of two different alternating currents 90° out of phase with each other.

UHF. Ultrahigh frequency.

Underwriters Laboratories, Inc. (UL). Independent laboratory that tests equipment to determine whether it meets certain safety standards when properly used.

Unit pole. Standard unit of magnetic pole strength. A force of 1 dyne is exerted between two unit poles separated by 1 cm of air.

Vaportight. Approved for installation in damp or wet locations.

Varying duty. Requirement of service that demands operation at loads, for intervals of time, both of which may be subject to wide variation.

Vector. Quantity having both magnitude and direction.

Very high frequency (vhf). Frequency band of 30 to 300 MHz and a wavelength of 10 to 1 m.

Vibration. Continuously reversing change in the magnitude of a given force; mechanical oscillation or motion about a reference point of equilibrium.

Volt (V)

1. Unit of electrical pressure that will push 1 A through a resistance of 1 ohm.

2. Electromotive force that will cause a current of 1 A to flow through a resistance of 1 ohm.

3. Unit of electrical pressure.

4. Unit of electromotive force.

Voltage

1. Direct-current voltage applied to the light source is measured by means of a voltmeter. To avoid correction for voltage drop in the ammeter, the voltmeter is generally connected directly across the load. Where precise measurements are desired, separate voltage leads are connected to the base of the lamp through special lampholders.

2. Voltmeters for use in alternating-current circuits may be self-contained or arranged for use with either a multiplier or a potential transformer. With the voltmeter connected directly across the load, the current taken by the voltmeter is included in the ammeter indication, and the power taken by the voltmeter is included in any wattmeter indication. As in the case of direct-current measurements, the voltmeter current is computed by dividing the voltmeter indication in volts by the voltmeter resistance in ohms.

3. Difference in potential between two points, measured in volts.

Voltage amplification. Increase in signal voltage magnitude in transmission from one point to another or the process thereof; of a transducer, the scalar ratio of the signal output voltage to the signal input voltage.

Voltage attenuation. Decrease in signal voltage magnitude in transmission from one point to another or the process thereof; of a transducer, the scalar ratio of the signal input voltage to the signal output voltage.

Voltage divider. Tapped resistor or a multiresistor circuit used to divide the applied voltage among the various loads.

Voltage drop. Loss of electrical current caused by overloading wires or by using excessive spans of undersized wires, evidenced by dimming of lights, or slowing down of motors.

Voltage of a circuit. Greatest root-mean-square (effective) difference of potential between any two conductors of the circuit concerned. Some systems, such as three-phase four-wire, single-phase three-wire, and three-wire direct-current may have various circuits or various voltages.

Voltage of a circuit not effectively grounded. Voltage between any two conductors. If one circuit is directly connected to and supplied from another circuit of higher voltage (as in the case of an autotransformer), both are considered of the higher voltage, unless the circuit of lower voltage is effectively grounded, in which case its voltage is not determined by the circuit of higher voltage. Direct connection implies electric connection as distinguished from connection merely through electromagnetic or electrostatic induction.

Voltage of an effectively grounded circuit. Voltage between any conductor and ground unless otherwise indicated.

Voltage to ground

1. For grounded circuits, voltage between the given conductor and that point or conductor of the circuit that is grounded; for ungrounded circuits, the greatest voltage between the given conductor and any other conductor of the circuit.

2. Voltage between any given live ungrounded part and any grounded part in the case of grounded circuits, or the greatest voltage existing in the circuit in the case of ungrounded circuits.

Voltage to ground in grounded circuits. Voltage between the given conductor and that point or conductor of the circuit that is grounded.

Voltage to ground in ungrounded circuits. Greatest voltage between the given conductor and any other conductor of the circuit. Where one circuit is directly connected to another circuit of higher voltage (as in the case of autotransformers), both are considered of the higher voltage, unless the circuit of lower voltage is effectively grounded.

Voltaic. Production of electricity by chemical action or galvanism.

Voltaic cell. Device to produce electrical energy by chemical action.

Volt-ampere. Unit of apparent power, without regard to phase.

Voltmeter. High-resistance instrument used to check the voltage in a circuit.

Volts or voltage. When the term "volts" or "voltage" is used without qualification, it means the voltage between conductors if no grounded conductor capable of carrying load is present. If such a grounded conductor is present, "volts" or "voltage" means volts to ground.

Volume. In an electric circuit, the magnitude of a complex audio-frequency wave as measured on a standard volume indicator.

Volume resistivity (specific insulation resistance). Electrical resistance between opposite faces of a 1-cm cube of insulating material, measured under prescribed conditions using a direct-current potential after a specified time of electrification, and commonly expressed in ohm-centimeters. The recommended test is as stated in ASTM D 257.

VU. Unit of volume in which the standard volume indicator is calibrated.

Watt (W)

1. Unit of power; the amount of energy expended for 1 A to flow through 1 ohm.

2. Unit of measure for electric power ($V \times A = W$ or electrical energy consumed): 1 W used for 1 hour is 1 Wh; 1000 Wh = 1 Wh.

Wattage. Electrical power measured in watts; a single unit combining the effect of both voltage (pressure) and rate of flow (amperage) by multiplying these quantities (volts times amperes equals watts).

Watt-hour. Unit of energy consumed, consisting of watts multiplied by time in hours. The result often is expressed in thousands of watt hours, termed "kilowatt-hours."

Wave. Periodic variation in voltage or current.

Wavelength

1. Length of one period of a wave in meters.

2. The red cadmium spectrum line (643,84696 nm in air under standard considerations) has been established as the primary standard for the measurement of all optical wavelengths.

ENERGY ENGINEERING

Absolute humidity. In a system of moist air, the ratio of the mass of water vapor present to the volume occupied by the mixture, that is, the density of the water vapor component.

Absolute pressure. Pressure referred to that of a perfect vacuum; sum of the gauge pressure and atmospheric pressure.

Absorption refrigeration. Cooling system that uses heat as its primary source of energy and evaporation as the cooling means.

Activation energy. The minimum or "threshold" energy necessary for a reaction to occur. The energy of the activated complex minus the energy of the reactant molecules.

Adiabatic. Term applied to a process in which there is no heat flow between a substance or system and its surroundings.

Adiabatic change. Temperature change within a substance caused only by its own expansion or compression.

Adsorption. Adherence of molecules of dissolved substances (gases or liquids) to the surface with which they are in contact, thereby reducing the quantity of fluid in the original substance.

Air change. Replacement of the air contained within an enclosed space within a given period of time.

Air-conditioning

1. Process of treating air to control its temperature, humidity, flow, cleanliness, and odor.
2. Artificial control of humidity, temperature "purity," and motion of the air within buildings and other enclosed spaces; also, the operation of equipment for such controls. The objective is to secure either maximum human comfort or the best environment for a given industrial operation.

Air density. Ratio of the mass of air to the volume occupied by it. In a continuous medium the density is defined by a limiting process and is a point function.

Air mass

1. Large section of the troposphere in which temperature and humidity are fairly uniform at a given level.
2. Path length of solar radiation through the earth's atmosphere, with the vertical path at sea level considered unity.

Air movement. Velocity or speed of air moving adjacent to a body.

Angstrom. Unit of measurement of length equivalent to 1×10 μm.

Annual range of temperature. Difference between the highest and lowest temperatures recorded at a station in any given year.

Askarel. Synthetic nonflammable insulating liquid that when decomposed by an electric arc evolves only nonflammable gaseous mixtures.

Atmosphere. Envelope of air surrounding the earth and bound to it, more or less permanently, by virtue of the earth's gravitational attraction. The earth's atmosphere extends from the solid or liquid surface of the earth to an indefinite height, its density asymptotically approaching that of interplanetary space. A unit of pressure is equal to 101,325 N/m^2 (14.70 psi), representing the atmospheric pressure of mean sea level under standard conditions.

Atmospheric pressure. Pressure exerted by the atmosphere as a consequence of gravitational attraction exerted on the "column" of air lying directly above the point in question. Pressure is usually given in millibars, inches of mercury, pounds per square inch, or pounds per square foot. Its standard value at sea level is about 14.7 psi (101,323 N/m^2).

Atmospheric temperature. Degree of heat or cold in the envelope of air surrounding the earth as measured on some definite temperature scale, usually Celsius or Fahrenheit, by means of any of various types of thermometers.

Automatic controller. Device used to regulate a system on the basis of its response to changes in the magnitude of some property of the system, such as pressure or temperature.

Auxiliary furnace. Supplementary heating unit used to provide heat to a space when its primary heat source cannot do so adequately.

Azimuth. Angular distance between the vertical plane containing a given line or a celestial body and the plane of the meridian.

Baffle. Surface used for deflecting fluids, usually in the form of a plate or wall.

Barometric pressure. Atmospheric pressure as indicated by a barometer. This atmospheric pressure is the pressure exerted by the atmosphere as a consequence of gravitational attraction exerted upon the "column" of air lying directly above the point in question.

Barrel. Measure of a quantity of fluid, usually equal to 42 U.S. gal, 5.6 cu ft, or 159 liters.

Battery. Device to store electrical energy as chemical potential energy. Within limits, most batteries can be repeatedly charged and discharged.

Berm. Man-made mound or small hill of earth.

Binding energy. The energy that holds the protons and neutrons together in the nucleus of an atom.

Biofuels. Renewable fuels and energy sources derived from organic material (wood, methane, etc.).

Biosphere. Zone of air, land, and water, above and below the earth's surface, that is occupied by plants and animals.

Blackbody. Body that absorbs all incident radiation and reflects or transmits none. Additionally, a blackbody is a perfect radiator. It emits or radiates the maximum amount of radiant energy for any surface at any given temperature.

Blackout. Complete shutoff of electrical energy from a power-generating source due to overloads, power shortages, or outages caused by equipment or transmission breakdown.

Boiler. Device used to heat water or produce steam for space heating or other uses, including power generation.

Breeder reactor. Nuclear fission power plant that creates additional nuclear fuel from its original uranium supply by nuclear reaction within the nuclear power plant.

British thermal unit (Btu). Unit of measurement equivalent to the amount of heat energy required to raise the temperature of 1 lb of water 1°F.

Brownout. Reduction in line voltage planned to alleviate overloads on power-generating equipment. Reduced voltage diminishes the brightness of incandescent lamps.

Building envelope. Exterior components of construction that enclose an interior space.

Calorie. Amount of heat necessary to raise the temperature of 1 g of water 1°C.

Cathodic protection. Corrosion protection against electrolytic reactions.

Chain reaction. Reaction in which the material or energy that imitates the reaction is also one of the products.

Change of state (phase change). Change from the solid, liquid, or gaseous state to either of the other two.

Chemical compatibility. The ability of materials and components in contact with each other to resist mutual chemical degradation, such as the chemical degradation caused by electrolytic action or plasticizer migration.

Chemical energy. Chemical energy consists of plants, food, wood, fossil fuel, and man's energy.

Cleaning by water. The movement of water across horizontal surfaces acts as a natural mechanism by dropping heavier objects from its mainstream.

Climate. Long-term manifestations of weather, however they may be expressed. More rigorously, the climate of a specified area is represented by the statistical collective of its weather conditions during a specified interval of time, usually several decades.

Climatic stresses. Stresses that are components of the climatic phase of the environment, such as temperature, moisture, solar radiation, atmospheric pressure, wind, or rain.

Coal

1. An impure form of carbon formed by decay of plant materials over centuries.

2. A sedimentary rock composed of combustible matter derived from the partial decomposition and alteration of celulose and lignin of plant materials.

3. Advanced-stage hydrocarbon. It has the highest density of a naturally occurring solid fuel.

Coal gas. Fuel gas obtained by the destructive distillation of bituminous coal.

Coal gasification. The conversion of coal, charcoal, or coke to a gaseous product by reaction with air, oxygen, steam, carbon dioxide, or a mixture of these.

Coal liquefaction. The process of preparing a liquid mixture of hydrocarbons by destructive distillation of coal.

Coefficient of expansion. Relative increase of the volume of a system (or substance) with increasing temperature in an isobaric process.

Coefficient of linear expansion. Change in length per unit length per degree change in temperature.

Coefficient of performance (COP). Ratio of the energy output of a device such as a heat pump to the energy input.

Coke. Coal from which all volatile and incomplete decomposed plant matter has been driven by destructive distillation. The product is nearly pure carbon.

Cobustion air. Term used to describe the oxygen in the air that promotes the ignition by fire of various substances or their reduction by heat.

Combustion of fossil fuels. The combustion of fossil fuels adds the following to the atmosphere: carbon dioxide, carbon monoxide, water vapor, solid particles, hydrocarbons, sulfur, nitrogen, mineral oxides, and waste heat.

Comfort zone. Range of temperatures and humidities over which the majority of adults feel comfortable under normal living and working conditions.

Compression. Energy can be stored in air by means of compression.

Condensation

1. Physical process by which a vapor becomes a liquid or solid; opposite of evaporation. In meteorological usage this term is applied only to the transformation from vapor to liquid; any process in which a solid forms directly from its vapor is termed "sublimation," as is the reverse process. It is indispensable to avoid confusing condensation with precipitation, for the former is by no means equivalent to the latter, though it must always precede the latter.

2. Process of changing a vapor into a liquid by the extraction of heat.

Condenser. Component of a system in which a working fluid undergoes a change of state from gas to liquid by the rejection of latent heat to a cooling medium, producing a heating effect.

Conduction

1. Transmission of energy between two bodies that are in direct contact.

2. Transfer of energy within and through a conductor by means of internal particle or molecular activity, and without any net external motion. Conduction is to be distinguished from convection (of heat) and radiation (or all electromagnetic energy). Heat is conducted by molecular motion within a few centimeters of the heat source (that is, the earth's surface). Distribution of heat away from that source is accomplished by convection and (in analogy to molecular conduction) by eddy heat conduction.

Continental climate. Climate that is characteristic of the interior of a land mass of continental size, marked by large annual, daily, and day-to-day ranges of temperature, low relative humidity, and (generally) by a moderate or small and irregular rainfall. The annual extremes of temperature occur soon after the solstices.

Continuous air circulation (CAC). Mode of operation of a forced-air heating system in which the blower operates continually.

Controller—automatic. Device used to regulate a system on the basis of its response to changes in the magnitude of some property of the system, such as pressure or temperature.

Convection

1. Mass motions within a fluid resulting in the transport and mixing of the properties of that fluid. Convection, along with conduction and radiation, is a principal means of energy transfer.

2. Heat transfer by convection is accomplished by movement of air or liquid.

Converter reactor. Reactor in which one type of fissionable material is used to produce another type of fissionable material.

Cooling load. Amount of heat that must be removed from a building to maintain a comfortable temperature, measured in British thermal units per hour or tons of air conditioning (1 ton = 12,000 Btu/hour).

Cooling mediums. Air that is cooler than an object in the airstream will carry away heat being given off by the object. The rate at which heat is carried away increases as the velocity of the air increases.

Cooling pond. Body of water that dissipates heat by evaporation, convection, and radiation.

Cooling power. In the study of human bioclimatology, one of several parameters devised to measure the air's cooling effect on a human body. Essentially, cooling power is determined by the amount of applied heat required by a device to maintain it at a constant temperature (usually 34°C); the entire system should be made to correspond as closely as possible to the external heat exchange mechanism of the human body.

Cosmic rays. Radiations coming to the earth from outer space.

Critical humidity. Relative humidity above which the atmospheric corrosion rate of a given metal increases sharply.

Crooke's radiometer. Partially evacuated hollow glass sphere containing vanes that are black on one side and silvered on the other side. These spin in the presence of thermal radiation because of the differing rates of radiation absorption between the silvered and the blackened surfaces.

Crude oil. Natural mixture of liquid hydrocarbons extracted as petroleum from under the earth's surface. Crude oil may also include oil extracted from tar sands and oil shale.

Daily mean. Average value of a meteorological element over a period of 24 hours. The "true daily mean" is usually taken as the mean of 24 hourly values between midnight and midnight, either as continuous values taken from an autographic record or as point readings at hourly intervals. When hourly values are not available, approximations must be made from observations at fixed hours. Long-period mean value of a climatic element on a given day of the year.

Damper. Device used to vary the volume of air passing through an air outlet, inlet, or duct.

Declination of the sun. Number of degrees the sun's vertical ray is north or south of the equator.

Degree day

1. Measure of the departure of the mean daily temperature from a given standard: there is one degree day for each degree (Celsius or Fahrenheit) of departure above (or below) the standard during one day. Recently, degree days have been applied to fuel and power consumption, with the standard being 65°F.

2. Unit of measurement based on temperature difference and time, used in estimating the average heating requirement for a building. For any one day, when the mean outside temperature is less than 65°F there exist as many degree days as there are degrees Fahrenheit difference in temperature between the mean temperature and 65°F. The base of 65°F assumes that no heat input is required to maintain the inside temperature at 70°F when the outside temperature is 65°F.

Degree day—Heating. Unit used to estimate the heating fuel consumption and the nominal heating load of a building. For any one day, when the mean temperature is less than 65°F, there exist as many degree days as there are degrees Fahrenheit difference in temperature between the mean temperature for the day and 65°F. The sum of the degree days constitutes the annual degree day heating requirement.

Degrees of frost. Number of degrees below 32°F or 0°C, the freezing point.

Dehumidifcation. Condensation of water vapor from air by cooling below the dew point; removal of water vapor from air by chemical or physical (adsorption) methods.

Dehumidify. Reduce, by any process, the amount of water vapor in a space.

Density. Ratio of the mass of a substance to its volume.

Design conditions. Selected indoor and outdoor wet-bulb and dry-bulb temperatures for a specific location that determine the maximum heating and cooling loads of a building.

Design outside temperature. Lowest temperature that usually occurs during the heating season at a given location. This temperature is approximately 15°F above the lowest temperature ever recorded by the meteorological station in the area.

Dew point. Temperature to which a given weight of air must be cooled at constant pressure and constant watervapor content in order for saturation to occur. When this temperature is below 0°C, it is sometimes called the "frost point."

Dielectric fitting. An insulating or nonconducting fitting used to isolate electrochemically dissimilar materials.

Diffuse radiation. Portion of reradiation and reflected radiation that eventually reaches the earth's surface.

Direct energy. Energy used in the most immediate available form, that is, natural gas, electricity, and oil.

Dry-bulb temperature. Local air temperature as indicated by a dry temperature measuring sensor.

Duct

1. Conduit or tube through which air or other gases flow.

2. Passageway used for transporting air or other gas at low pressures to the atmosphere.

Earth energy. Energy due to the natural mass and heat of the planet.

Economizer cycle. Cooling mode that uses cool outdoor air instead of an energy-consuming cooling device to offset heat gains in buildings.

Effective temperature. Temperature at which motionless, saturated air induces in a sedentary worker wearing ordinary indoor clothing the same sensation of comfort as that induced by the actual conditions of temperature, humidity, and air movement. With respect to radiation ascribed to an imperfectly radiating body, the temperature at which a perfect radiator (blackbody) would emit radiation at the same rate. Thus, the effective temperature is always less than the actual temperature.

Effective terrestrial radiation. Difference between the outgoing (positive) terrestrial radiation of the earth's surface and the downcoming (negative) counterradiation from the atmosphere. It is to be emphasized that this difference is a positive quantity, of the order of several tenths of a langley per minute, at all times of day, except that under conditions of low overcast clouds, it typically attains its diurnal maximum during the midday hours when high soil temperatures create high rates of outgoing terrestrial radiation. However, in daylight hours the effective terrestrial radiation is generally much smaller than the insulation while at night it typically dominates the energy budget of the earth's surface.

Effluent. Discharged waste from sewage and industrial sources, suspended in liquid or gas.

Electrical energy. Energy generated by forcing electrons to move in certain paths or circuits.

Electric demand limiter. Device that selectively switches off electrical equipment whenever the total electrical demand rises beyond a predetermined level.

Electric resistance heat. Conversion of electrical energy to heat energy by means of an electrical resistor.

Electromagnetic spectrum

1. Arrangement of electromagnetic radiation (infrared, visible, or ultraviolet) on a wavelength or frequency scale.

2. Entire range of wavelengths of electromagnetic radiation, extending from gamma rays to the longest radio waves.

Emissivity

1. Ratio of the rate of emission of radiant energy in a given wavelength interval from a given surface to the rate of emission of a blackbody at the same temperature in the same wavelength interval, with the radiation emitted by the surface due solely to its temperature; that is, excluding transmitted radiation, heat generated by chemical or other reactions, etc.

2. Capacity of a material to emit radiant energy. The emittance is the ratio of the total radiant energy emitted by a body to that emitted by a blackbody at the same temperature. The emissivity of a surface is numerically equal to its absorptivity when the radiating source is a blackbody at the same temperature as the surface.

Emittance. Ratio of the amount of energy radiated by a surface to the energy striking the surface.

Energy

1. The capacity for producing motion and for doing work. Energy holds matter together. It can become mass or can be derived from mass. It takes such forms as kinetic, potential, heat, chemical, electrical, and atomic energy, and it can be changed from one of these forms to another.

2. Capacity for doing work. Energy takes a number of forms, which may be transformed from one into another. The various forms of energy include thermal (heat), mechanical (work), electrical, and chemical. Energy may be measured in customary units, such as kilowatt hours (kWh) or British thermal units (Btu), or in SI units, such as joules (J) where 1 J = 1 W-second.

Energy level

1. Permitted value of energy for an atom or molecule, or for an electron in an atom or molecule.

2. The distance from an atomic nucleus at which electrons can have orbits.

Energy storage. Retention of energy by converting it to a form (gravitational potential, chemical, etc.), from which it can be retrieved for useful purposes.

Equinox. Two times of the year when the sun crosses the equator, thereby making day and night of equal length. The spring equinox occurs about March 21 and the fall equinox about September 21.

Ethyl alcohol. Colorless, volatile, flammable liquid that is produced by the fermentation of grains and fruits.

Eutectic. Mixture of two or more substances that melts or freezes at constant temperature and with constant composition. The term is usually applied to the mixture of a given substance that has the lowest melting point.

Evaporation. Process by which a substance is changed from the liquid into the vapor state. In hydrology, evaporation is vaporization that takes place at a temperature below the boiling point.

Evaporator. Component of a system in which a working fluid undergoes a change of state from liquid to gas, taking on latent heat and thereby providing a cooling effect.

Fall equinox. The beginning of fall that occurs annually approximately September 23.

Fan coil. Heating or cooling device that forces air through heating or cooling coils.

Filter. Device used to remove solid or airborne materials from a fluid or air.

Fin. Extended surface used to increase the heat transfer area.

Fission. Splitting of an atomic nucleus, releasing large amounts of energy. Usually the uranium-235 atom is split, producing heat for the generation of electricity with steam.

Flow rate. Volume or weight per unit time of a fluid flowing through an opening or duct.

Flue. Exhaust channel through which gas and fumes produced by combustion exit a building.

Fluid. Gas, vapor, or liquid.

Forced air heating system. Heating system in which air, circulated mechanically by either a blower or fan, is the transfer medium.

Forced convection. Convection that results from forced circulation of a fluid, as by a fan or pump.

Fossil fuel. Class of organic compounds formed by the decay of matter under the influence of heat and pressure over millions of years and which when burned liberate large quantities of energy. Coal, oil, and natural gas are fossil fuels.

Free energy. The energy of a reaction that can be converted to useful work outside the reaction.

Freezing. Change from a liquid to a solid state, usually by the abstraction of heat.

Frost line. Depth of frost penetration in the earth. The depth varies from one geographic location to another.

Fuel. Substance that can be burned to produce heat.

Fuel cell. Device in which hydrogen and oxygen are combined in an electrochemical reaction to generate electricity and produce water as a by-product.

Fusion (nuclear). Release of energy by the formation of a heavier nucleus from two lighter ones.

Geothermal energy. Heat energy contained in large underground reservoirs of steam and hot water, produced by molten material from the earth's interior.

Greenhouse effect. Term used to describe the ability of air to absorb long heat waves from the earth after allowing the sun's short waves to pass through it.

Ground frost. Freezing condition injurious to vegetation, which is considered to have occurred when a thermometer exposed to the sky at a point just above a grass surface records a minimum temperature of 30.4°F or below.

Heat

1. Form of energy transferred between systems by virtue of a difference in temperature and existing only in the process of energy transformation. By the first law of thermodynamics, the heat absorbed by a system may be used by the system to do work or to raise its internal energy.

2. Form of energy transferred from one mass to another by virtue of a temperature difference.

Heat balance. Equilibrium that exists, on the average, between the radiation received by the earth and its atmosphere from the sun and that emitted by the earth and its atmosphere. That the equilibrium does exist in the mean is demonstrated by the observed long-term constancy of the earth's surface temperature. Equilibrium that is known to exist when all sources of heat gain and loss for a given region or body are accounted for. In general this balance includes advective, evaporative terms as well as a radiation term.

Heat capacity

1. Ratio of the heat absorbed or released by a system to the corresponding temperature rise or fall.

2. Quantity of heat required to raise a system one degree in temperature in a specified way.

Heat conduction. Heat transferred from one part of a body to another part of the same body or from one body to another in physical contact with it without displacement of the matter within the body.

Heat convection. Heat transferred from one point to another by being carried along as internal energy with the flowing medium, which can be either a gas, vapor, or liquid.

Heat equator. Line that circumscribes the earth and connects all points of highest mean annual temperature for their respective longitudes.

Heat exchange flow pattern. Relative flow arrangement of the collection and distribution circuits in storage.

Heat exchanger. Device used to transfer heat from one temperature level to another.

Heat gain. Increase in the amount of heat contained in a space, resulting from solar radiation and the heat given off by people, lights, equipment, machinery, and other sources.

Heating areas. Both passive and active solar energy is being used to heat internal spaces and external areas.

Heating degree day. Unit used to estimate the heating fuel consumption and the nominal heating load of a building. For any one day, when the mean temperature is less than 65°F there exist as many degree days as there are degrees Fahrenheit difference in temperature between the mean temperature for the day and 65°F. The sum of the degree days constitutes the annual degree day heating requirement.

Heat longwave. Radiant energy emitted from bodies at wavelengths longer than 3.0 μm.

Heat loss. Decrease in the amount of heat contained in a space, resulting from heat flow through walls, windows, the roof, and other components of the building envelope and from the infiltration of cold outdoor air.

Heat pipe. Closed pipe containing a liquid and a wick that will transfer heat from one end to the other without any input of work.

Heat pump

1. Reversible refrigeration system that delivers more heat energy to the end use than is input to the compressor. The additional energy input results from the absorption of heat from a low-temperature source.

2. Refrigerating system installed where the heat discharged from the condenser, rather than the heat absorbed by the evaporation, is desired.

Heat radiation. Heat transferred from one body to another by the passage of radiant energy between the two. The radiant energy is then converted back to internal energy when it is absorbed by the receiving body.

Heat resistance. Ability of a material to show little or no deterioration on continuous or intermittent exposure to a predetermined elevated temperature.

Heat sink. Body or substance that is capable of accepting or rejecting heat.

Heat transfer

1. System by which heat may be propagated or conveyed from one place to another.

2. Transfer or exchange of heat by radiation, conduction, or convection in a fluid or between the fluid and its surroundings.

Heat transfer coefficient

1. Nondimensional number arising out of the problem of heat transfer in fluids; the rate of heat transfer per unit area per unit temperature distance, a quantity having the dimensions of reciprocal length.

2. Unit surface thermal conductance for convection or radiation that describes the rate by which heat can be transferred by either mode.

Heat wave. Period of abnormally and uncomfortably hot and usually humid weather. To be a "heat wave," such a period should last at least one day, but conventionally it lasts from several days to several weeks.

Heat wheel. Device used in ventilating systems that tends to bring incoming air into thermal equilibrium with existing air. As a result, hot summer air is cooled and cold winter air is warmed.

Horsepower. Standard unit of power equal to 745.7 W, 2545 Btu/hour, or 550 ft-lb/second.

Hot-water heating. Collection system designed for absorption to supply domestic hot water.

Human contribution. The input of food and water results in an output of solid wastes, human wastes, and heat wastes. The body of an average adult human male gives off a minimum of 250 Btu/hour by convection, radiation, and evaporation while at rest.

Humidification

1. Process for increasing the water content of air or other gases.

2. Using the sun's energy, water evaporates to provide humidification for internal spaces.

Humidistat. Device that continuously monitors the humidity in a space and activates the humidity control equipment to ensure that the humidity is maintained at a preset level.

Humidity

1. Amount of water vapor in the air.

2. The increase or decrease of humidity in the air can provide temperature changes necessary for natural heating and cooling. The sensation of human comfort is greatly influenced by air speed and humidity level.

Hurrican-force wind. In the Beaufort wind scale, a wind whose speed is 64 knots (73 mph) or higher.

HVAC. Abbreviation for heating, ventilating, and air-conditioning.

Hydroelectric plant. Electrical power generating plant in which the kinetic energy of water is converted to electrical energy by means of a turbine generator.

Hydrogen. Water can be reduced to its components of hydrogen and oxygen by electrolytic action. Hydrogen fuel cells are an advanced technology used for energy conversion.

Hydronic heating system. Heating system in which water is the transfer medium.

Infiltration. Uncontrolled flow of air into a building through cracks, openings, doors, or other areas that allow air to penetrate.

Infrared radiation

1. Radiant energy of wavelengths longer than those corresponding to red light, that is, longer than approximately 0.8 μm.
2. Electromagnetic radiation in the wavelength region between visible radiation and microwave radiation, usually considered to begin at 7600 A (0.76 mm) and extend to 1000 mm. (1 mm).

Insulate. Separate or isolate a conducting body from its surroundings by means of a nonconductor to prevent the transfer of electricity, heat, or sound.

Insulation. Material having relatively high resistance to heat flow, electricity, or sound.

Insulation resistance. Measured resistance of the insulation of a device or product. The measurement is taken along the path over which the insulation is intended to be effective.

Internal energy. A characteristic property of the state of a thermodynamic system. It includes intrinsic energies of individual molecules, kinetic energies of internal motions, and contributions from interaction between molecules, but excludes the potential or kinetic energy of a system as a whole.

Kinetic energy. Moving or dynamic energy. Energy of motion.

Latent heat

1. Heat required by a body to change its state or phase without changing temperature.
2. Heat released or absorbed per unit mass by a system in a reversible, isobaric-isothermal change of phase. In meteorology, the latent heats of vaporization (or condensation), fusion, and sublimation of a water substance are of importance.
3. Heat that is necessary to produce a change of state of a material at a constant temperature.

Local winds. Winds that, over a small area, differ from those that would be appropriate to the general pressure distribution, or possess some other peculiarity.

Maximum "flow" temperature. The maximum temperature obtained in a component when the heat transfer fluid is flowing through the system.

Maximum "no-flow" temperature. The maximum temperature obtained in a component when the heat transfer fluid is not flowing through the system.

Maximum service temperature. The maximum temperature to which a component will be exposed in actual service, either with or without the flow of heat transfer fluid.

Mean annual range of temperature. Difference between the absolute maximum and minimum temperatures for a year, averaged over a given number of years.

Mean daily maximum (minimum) temperature. Average of the maximum (minimum) temperatures for each day within a given period, usually a month, over a period of years.

Mean hourly temperature. Average of the daily temperatures at a given hour for an indicated period, generally a month, averaged over a period of years.

Mean monthly cloudiness. Average of the mean cloud cover of each day within a month, averaged over a period of years.

Mean monthly dew point. Average of the mean dew points of each day (generally computed from observations taken at equal time periods of 6 hours or less apart) within a month, averaged over a period of years.

Mean monthly maximum (minimum) temperature. Highest (lowest) temperature for a month, averaged over a period of years.

Mean monthly relative humidity. Average of the mean relative humidities for each day of the month, averaged over a period of years.

Mean monthly temperature. Average of the mean temperatures of each day within a month, averaged over a period of years. Some weather services require that a prescribed length of record be available before "mean" is used.

Mean radiant temperature. Temperature at which an object gives out as much radiation as it receives from its surroundings. In a room it is approximately the mean temperature of the walls, floor, and ceiling.

Methane. Colorless, odorless, flammable gaseous hydrocarbon, which is the product of the decomposition of organic matter. It is a major component of natural gas.

Micrometer. Unit of measurement of length equivalent to one millionth of a meter.

Modulus of elasticity. Ratio of stress to corresponding strain below the proportional limit.

Moistureproof. Able to resist the transmission of water vapor.

Nameplate rating. Statement attached to the equipment or product by the manufacturer of a heating system that the equipment or product will provide the performance of the system under specified operating conditions.

Natural convection. Convection caused by difference in density resulting from temperature changes.

Natural gas. Most efficient direct-combustion hydrocarbon fuel extracted from fossil fuels.

Net energy. Energy remainder or deficit after the energy costs of extracting, concentrating, and distribution are subtracted.

Net radiometer. Radiation balance meter used to measure all radiation (both shortwave and longwave) components.

Net reserves. Estimate of the net energy that can be delivered from a given energy resource.

Nuclear fission. One "fission event" (with a uranium-235 atom) releases 50 million times as much energy as the burning of a hydrocarbon (such as coal) atoms.

Nuclear fusion. Fusion proceeds by the forceful combining or fusing of very light hydrogen atoms.

Oil. Liquid hydrocarbon.

Oil shale. Shale from which oil is extracted.

Outgassing. The emission of gases by component materials, usually during exposure to elevated temperature or reduced pressure.

Overall coefficient of heat transfer. Time rate of heat flow through a body per unit area for a unit temperature difference between the fluids on the two sides of the body under steady-state conditions.

Partial pressure. Pressure exerted by one constituent of a mixed gas. The sum of the partial pressures of each constituent of the mixture equals the total gas pressure.

Percentage of saturation. Ratio of the actual weight of water present in a given weight of dry air to the weight of water that would be present in the same weight of dry air at saturation (100% relative humidity) expressed as a percentage.

Permafrost table. Surface that represents the upper boundary of perennially frozen ground; sometimes called the "frost line."

Photolysis. Chemical decomposition caused by the action of solar radiant energy.

Plenum. Compartment for the passage and distribution of air.

Power. Rate at which work is performed or energy expended.

Pressure

1. Type of stress, characterized by uniformity in all directions.
2. Force exerted by a substance on a unit area of its boundary.

Prevailing wind direction. Wind direction most frequently observed during a given period. The periods most frequently used are the observational day, month, season, and year.

Primary heating system. System used to heat a building when the solar heating system cannot provide useful heat, usually using conventional oil, gas, or electric furnace or heat pump that will provide the additional heat.

Quad. 10^{15} Btu (one quadrillion Btu).

Quanta. The indivisible units of energy in electromagnetic radiation.

Radiant energy. Energy in the form of electromagnetic waves that is continually emitted from the surface of all bodies.

Rain. Precipitation of liquid water particles, either in the form of drops of more than 0.5-mm (0.02-in.) diameter or of smaller widely scattered drops.

Raw energy source. Original, unrefined source of energy.

Recovered energy. Utilized energy that would otherwise have been wasted.

Recycle. Process of recovering resources from waste material for reprocessing and reuse.

Refinery. Chemical processing plant in which crude oil is separated into more useful hydrocarbon compounds.

Reflectance. Ratio of the amount of radiation reflected by a surface to the amount of radiation incident on the surface.

Reflectivity. Capacity of a material to reflect radiant energy. Reflectance is the ratio of the radiant energy reflected from a body to that incident on it.

Reflector. Mirror used to increase shortwave radiation input into the collector.

Relative humidity

1. Percentage of water vapor in the air in relation to the maximum amount of water vapor it can hold at a given temperature.

2. Ratio, expressed in percent, of the amount of water vapor in a given volume of air to the amount the volume of air would contain if saturated; ratio, expressed in percent, of the actual partial pressure of the water vapor in a given mixture of air and water vapor to the saturation pressure of pure water at the same temperature.

Sensible heat

1. Heat that produces a change of temperature in a body.
2. Heat that a body gives off without changing its state or phase and measured as a change in temperature.

Sensible heat storage. Heat storage medium in which the addition or removal of heat results in a temperature change only (as opposed to phase change, chemical reaction, etc.). The medium is typically water or gravel (pebble bed).

Sensible temperature. Temperature at which the "average indoor air" of moderate humidity would induce, in a lightly clothed person, the same sensation of comfort as that induced by the actual environment.

Space heating. Interior heating of a building or room.

Specific heat

1. Amount of heat that has to be added to or taken from a unit of weight of a material to produce a change of 1 degree F in its temperature.
2. Heat capability of a system per unit mass, such as the ratio of the heat absorbed (or released) by unit mass of the system to the corresponding temperature rise or fall.

Specific humidity. In a system of moist air, the ratio of the mass of water vapor to the total mass of the system. Specific humidity may be approximated by the mixing ratio for many purposes.

Stefan-Boltzmann constant. Numerical constant for a perfect radiator (blackbody).

Stratification. Existence of persistent temperature gradients in storage mediums.

Surface temperature. In meteorology, the temperature of the air near the surface of the earth, almost invariably determined by a thermometer in an instrument shelter.

Temperate climate. Very generally, the climate of the "middle" latitudes; variable climate between the extremes of tropical and polar climate.

Temperature. Measure of heat intensity or the ability of a body to transmit heat to a cooler body.

Temperature detector resistance. Temperature-measuring device that employs a sensitive element of extremely pure platinum, copper, or nickel wire which provides a definite value at each temperature within its range.

Therm. Quantity of heat equal to 100,000 Btu.

Thermal. Relatively small-scale, rising current of air produced when the atmosphere is heated enough locally by the earth's surface to produce absolute instability in its lowest layers.

Thermal barrier. Insulated wall, layer, blanket, enclosure, or heat exchanger designed to protect equipment from the effects of high temperatures.

Thermal conductance. Time rate of heat flow through a body per unit area for a unit temperature difference between the body's surfaces under steady-state conditions.

Thermal conductivity

1. Time rate of heat flow through a homogeneous material per unit area and thickness, under steady-state conditions, when a unit temperature gradient is maintained in the direction normal to the cross-sectional area.

2. Intrinsic physical property of a substance, describing its ability to conduct heat as a consequence of molecular motion.

Thermal deterioration. Impairment of physical properties due to effects of high or very low temperatures.

Thermal diffusivity. Property of a material equivalent to its thermal conductivity divided by the product of its density and specific heat.

Thermal efficiency

1. Ratio of the useful heat at the point of use to the thermal energy input for a designated time period, expressed as percent.

2. Expression of the effectiveness of temperature in determining the rate of plant growth, assuming sufficient moisture.

Thermal energy. Energy that is characteristic for thermal neutrons at room temperature, about 0.025 electron volt.

Thermal equilibrium. State of a system at which there are no variations in temperature from one point to another in the system.

Thermal exposure. Total normal component of thermal radiation striking a given surface throughout the course of a detonation, expressed in calories per square centimeter.

Thermal lag. Amount of heat necessary to reachieve downpoint temperature after collection is resumed following a period of stagnation.

Thermal resistance. Reciprocal of thermal conductance.

Thermal transmission. Passage of heat through a material.

Thermistor. Temperature-measuring device that employs a resistor with a high negative temperature coefficient of resistance. As the temperature increases, the resistance goes down and vice versa.

Thermocouple. Temperature-measuring device that utilizes the principle that an electromotive force is generated whenever two junctions of two dissimilar metals in an electrical circuit are at different temperatures.

Thermometer. Instrument for measuring temperature.

Thermostat

1. Device to monitor continuously the temperature in a space and activate temperature control equipment to ensure that the temperature in the space is maintained at a preset level.

2. Instrument that controls temperature by responding to changes in temperature.

Ton (of air conditioning). Thermal refrigeration energy required to create 1 ton of ice (2000 lb) in 1 day. It equals 12,000 Btu/hour.

Torque. Turning or twisting force.

Vacuum. Enclosed space from which air is evacuated to reduce the atmospheric pressure inside the enclosure to a very low value.

Vapor. Substance existing in the gaseous state at a temperature lower than that of its critical point—that is, a gas cool enough to be liquefied if sufficient pressure were applied to it.

Vapor retarder

1. Moisture-impervious layer applied to prevent moisture from traveling to a point where it may condense owing to lower temperatures.

2. Component of construction that is impervious to the flow of moisture, used to prevent moisture traveling to a point where it may condense.

Vapor pressure. Pressure exerted by the vapor of liquid in a confined space such that vapor can form above it; pressure of water vapor in the air; part of the total atmospheric pressure that is due to water vapor.

Vent. Penetration through the building envelope that is specifically designed for the flow of air into or from the building.

Ventilation air. Outside air that is intentionally caused to enter an interior space.

Vicosity. Internal resistance of fluids to shear.

Warm air. Heat energy is converted to molecular kinetic energy, measured as an increase in temperature.

Water distillation. Purification, demineralization, and desalination using solar energy to evaporate and purify water.

Water energy. Movement and cycling of water, influenced by the sun, atmospheric conditions, and weather.

Water vapor. Water in vapor form—one of the most important of all constituents of the atmosphere. Its amount varies widely in space and time due to the great variety of both "sources" of evaporation and "sinks" of condensation that provide active motivation to the hydrologic cycle.

Weather. State of the atmosphere, mainly with respect to its effects on life and human activities; in the making of surface weather observations, a category of individual and combined atmospheric phenomena that must be drawn on to describe the local atmospheric activity at the time of observation.

Weather code. *Pro forma* code used for describing weather conditions.

Weatherstripping. Foam, felt, neoprene, metal, or rubber strips used to form an airtight seal around windows, doors, or openings to reduce infiltration.

Wet-bulb temperature

1. Lowest temperature to which air can be cooled at any given time by evaporating water into it at constant pressure, when the heat required for evaporation is supplied by the cooling of the air. Temperature is indicated by a well-ventilated wet-bulb thermometer.

2. Local air temperature as indicated by a wet temperature measuring sensor.

Wind. Air in motion relative to the surface of the earth. Since vertical components of atmospheric motion are relatively small, especially near the surface of the earth, meteorologists use the term to denote almost exclusively the horizontal component.

Wind energy. Kinetic energy of air motion over the earth's surface caused by the sun's heating of the atmosphere.

Wind machine. Device used to convert the kinetic energy of wind to another form of energy for useful purposes.

Work. Transfer of energy from one physical system to another; in mechanics, the transfer of energy to a body by the application of force.

MECHANICAL ENGINEERING

Absolute humidity

1. Weight of the water vapor in a given quantity of air in pounds per cubic foot.
2. Weight of water vapor per unit volume.

Absolute pressure. Pressure referred to that of a perfect vacuum; sum of gage pressure and atmospheric pressure.

Absolute temperature. Temperature expressed in degrees above absolute zero.

Absolute viscosity. Force per unit area required to produce unit relative velocity between two parallel areas of fluid a unit distance apart.

Absolute zero. Zero point on the absolute temperature scale; 459.69 degrees below the zero of the Fahrenheit scale ($-459.69°F$); 273.16 degrees below the zero on the Celsius scale.

Absorbent

1. Material that, due to an affinity for certain substances, extracts one or more such substances from a liquid or gaseous medium with which it is in contact, and which changes physically or chemically, or both, during the process. Calcium chloride is a solid absorbent, while solutions of lithium chloride, lithium bromide, and the ethylene glycols are liquid absorbents.
2. Material that has the ability to cause molecules of gases, liquids, or solids to adhere to its internal surfaces without changing physically or chemically. Certain solid materials such as silica gel, activated carbon, and activated alumina have this property.

Absorber. Device containing liquid for absorbing refrigerant vapor or other vapors; in an absorption system, that part of the low side used for absorbing refrigerant vapor.

Absorption

1. Accumulation of water in a material or its cells or fibers, accompanied by a physical or chemical change, such as softening of the fibers, the dissolving of a binding agent, or the swelling of wood.
2. Process whereby a material extracts one or more substances present in an atmosphere or mixture of gases or liquids, accompanied by physical or chemical change, or both, of the material.
3. Accumulation of water in or on the surface of a material, or on its fibers or cell walls, without any chemical or physical change.
4. Action, associated with surface adherence, of a material in extracting one or more substances present in an atmosphere or mixture of gases and liquids, unaccompanied by physical or chemical change.

Absorption coefficient. Specific factor characteristic of a substance on which its absorption radiation depends.

Absorption system. Refrigeration system in which the refrigerant gas evolved in the evaporator is taken up in an absorber and released in a generator upon the application of heat.

Absorptivity. Capacity of a material to absorb radiant energy. Absorptance is the ratio of the radiant flux absorbed by a body to that incident to it.

Acceleration. Time rate of change of velocity, that is, the derivative of velocity with respect to time.

Acceleration due to gravity. Rate of increase in velocity of a body falling freely in a vacuum. The value varies with latitude and elevation. The international standard taken at sea level and 45° latitude is 980.665 cm/second/second or 32.174 ft/second/second.

a **(Conductance of airspace).** Thermal conductance of an airspace; the amount of heat, expressed in Btu transmitted in one hour across an airspace of 1 sq ft area for each degree F temperature difference.

Actual displacement. Actual volume of gas or vapor at compressor inlet conditions, moved by a compressor per revolution or per unit of time.

Adiabatic. Term that describes a change in a system in which the system neither receives nor gives out heat.

Adiabatic process. Thermodynamic process during which no heat is added to or taken from a substance or system.

Aeration. Exposure of a substance or area to air circulation.

Aerosol. Assemblage of small particles, solid or liquid, suspended in air. The diameters of the particles may vary from 100 μ down to 0.01 μm or less. Examples are dust, fog, and smoke.

Air changes. Method of expressing the amount of air leakage into or out of a building or room in terms of the number of building volumes or room volumes exchanged.

Air circulation. Natural or imparted motion of air.

Air-conditioning. Process of treating air so as to control simultaneously its temperature, humidity, cleanliness, and distribution to meet the requirements of the conditioned space.

Air temperature. Heat is lost by convection to the surrounding atmosphere at a rate that is dependent on the air temperature.

Air velocity. Air motion affects the rate of heat loss from the body both by convection and evaporation.

Ambient. Near or in contact with.

Ambient air. Air surrounding an object.

Amplitude decrement factor. Difference between the maximum and the mean temperatures of a heat wave passing through a wall, roof or floor, dependent upon the thickness, mass, specific heat and orientation of the wall or roof.

Anemometer. Instrument for measuring the velocity of air.

Approved. With respect to materials, workmanship, and types of construction, the term indicates approval by the plumbing inspector as the result of investigations, inspections, and/or tests conducted or by reason of accepted principles or tests by nationally recognized testing agencies.

Aspiration. Production of movement in a fluid by suction created by fluid velocity.

Atmospheric pressure

1. Pressure of air enveloping the earth, averaged as 14.7 psi at sea level, or 29.92 of mercury as measured by a standard barometer.
2. Pressure due to the weight of the atmosphere; pressure indicated by a barometer. Standard atmospheric pressure or standard atmosphere is the pressure of 76 cm of mercury having a density of 13.5951 g/cu cm under standard gravity of 980.665 cm/second/second. Atmospheric pressure is equivalent to 14.696 psi or 29.921 in. of mercury at 32°F.
3. Pressure of air at sea level, usually 14.7 psia (1 atmosphere) or 0 psig.

Babo's law. The vapor pressure over a liquid solvent is lowered approximately in proportion to the quantity of a nonvolatile solute dissolved in the liquid.

Backpressure. Force exerted causing or tending to cause water or air to flow in a pipe opposite to the normal direction of flow.

Barometer. Instrument for measuring atmospheric pressure.

Boiler horsepower. Boiler horsepower is equivalent to the evaporation of 34.5 lb of water per hour from and at 212°F. This is Btu/hr equal to a heat output of 970.3 × 34.5 = 33,475 Btu/hr.

Boiling point. Temperature at which the vapor pressure of a liquid equals the absolute external pressure at the liquid-vapor interface.

Boyle's law. At constant temperature, the volume of a given mass of gas varies inversely as the pressure.

British thermal unit (Btu)

1. Unit of energy in the form of heat approximately equal to the heat required to raise the temperature of 1 lb of water 1°F. The symbol "Btu" refers to the heat unit alone. The symbol "Btuh" refers to the units of heat transmitted or resisted in 1 hour.

2. Measure of the quantity of heat in a substance: Two Btu is the amount of heat energy required to raise the temperature of 1 lb of liquid water 1°F from 62°F to 63°F.

Buoyancy. Power of supporting a floating body, including the tendency to float an empty pipe (by exterior hydraulic pressure).

British imperial gallon. Fluid gallon equal to 1.2 U.S. gal approximately. It contains 27.42 cu in. There are 6.23 such gallons per cubic foot.

Calorie. Heat required to raise the temperature of 1 g of water 1°C actually from 4°C to 5°C. A mean calorie is 1/100 part of the heat required to raise 1 g of water for 0° to 100°C.

Calorimeter. Instrument for measuring heat quantities, such as machine capacity, heat of combustion, specific heat, vital heat, and heat leakage; also, instrument for measuring quality for moisture content of steam or other vapor.

Capillarity. Action by which the surface of a liquid where it is in contact with a solid (as in a slender tube) is raised or lowered.

Capillary action. Action that occurs when a liquid is drawn into a razor-thin space between two almost-touching solid surfaces.

Ceiling panel system. Heating system in combination with a ceiling construction consisting of gypsum board panels with electric tape or cables embedded in the gypsum. Some panels are listed by Underwriters Laboratories, Inc., and meet the requirements of the NFPA National Electrical Code.

Celsius

1. Thermometric scale in which the freezing point of water is called 0 degree C and its boiling point 100 degrees C at normal atmospheric pressure (14.696 psi).

2. Temperature degree scale on which the freezing point is indicated as 0 degree C and its boiling point as 100°C.

Centrifugal force. Outward force exerted by a body moving in a curved line; force that tends to tip a car over in going around a curve.

Change of state. Change from one phase, such as solid, liquid, or gas, to another.

Charles' law. At constant pressure, the change of volume of a given mass of gas is proportional to the change of temperature.

Chimney effect

1. Tendency of warm air to rise in any vertical passage, such as a duct, stack, air shaft, elevator shaft, or stairwell, or from low to higher ventilators in an attic space, and thus to draw in unheated air.

2. Tendency of air or gas in a duct or other vertical passage to rise when heated owing to its lower density compared with that of the surrounding air or gas. In buildings, the tendency is toward displacement (caused by the difference in temperature) of internal heated air by unheated outside air due to the difference in density of outside and inside air.

Closed system. Heating or refrigerating piping system in which the circulating water or brine is completely enclosed under pressure above atmospheric and shut off from the atmosphere except for a vented expansion tank at the high point of the system.

Clothing. Clothing insulates the body and reduces the rate of heat loss from it.

Code. The word "code" or "the code" when used alone means the regulations, subsequent amendments thereto, or any emergency rule or regulation that the administrative authority having jurisdiction may lawfully adopt.

Coefficient of discharge. For an air diffuser, the ratio of net area or effective area at *vena contracta* of an orificed airstream to the free area of the opening.

Coefficient of expansion. Change in length per unit length or the change in volume per unit volume per degree change in temperature.

Coefficient of heat transmission. Any one of a number of coefficients used in the calculation of heat transmission by conduction, convection, and radiation, through various materials and structures.

Coefficient of performance (heat pump). Ratio of the rate of heat delivered to the rate of energy input, in consistent units, for a complete operating heat pump plant or some specific portion of that plant, under designated operating conditions.

Coefficient of performance (refrigerating plant). Ratio of the rate of heat removal to the rate of energy input, in consistent units, for a complete refrigerating plant, under designated operating conditions.

Comfort air-conditioning. Process of treating air so as to control simultaneously its temperature, humidity, cleanliness, and distribution to meet the comfort requirements of the occupants of the conditioned space.

Comfort chart. Chart showing effective temperatures with dry-bulb temperatures and humidities (and sometimes air motion) by which the effects of various air conditions on human comfort may be compared.

Comfort line. Line on the comfort chart showing the relation between the effective temperature and the percentage of adults feeling comfortable.

Comfort zone. Average: range of effective temperatures over which the majority (50% or more) of adults feels comfortable; extreme: the range of effective temperatures over which one or more adults feels comfortable.

Compression efficiency. Ratio of work required to compress, adiabatically and reversibly, all the vapor delivered by a compressor (per stage) to the actual work delivered to the vapor by the piston or blades of the compressor.

Compressor heating effect (heat pump). Rate of heat delivery by the refrigerant assigned to the compressor in a heat pump system, which is equal to the product of the mass rate of refrigerant flow produced by the compressor and the difference in specific enthalpies of the refrigerant vapor at its thermodynamic state leaving the compressor and the saturated liquid refrigerant at the pressure of the vapor leaving the compressor.

Compressor refrigeration coefficient of performance. Ratio of the compressor refrigerating effect to the rate of energy input to the shaft of the compressor, in consistent units, in a complete refrigerating plant, under designated operating conditions.

Condensation

1. Process of changing water vapor in the air to liquid water by taking away heat; opposite of evaporation (which requires the addition of heat).

2. Process of changing a vapor into liquid by the extraction of heat. Condensation of steam or water vapor is effected in either steam condensers or dehumidifying coils, and the resulting water is called "condensate."

Condensing unit capacity. Refrigerating effect generally measured in British thermal units per hour or tons, produced by the difference in total enthalpy between refrigerant liquid leaving the unit and the total enthalpy of the refrigerant vapor entering the unit.

Conduction. Transfer of heat through a body or substance itself.

Convection

1. Transfer of heat by the wiping effect and movement of air from warm surfaces to coller surfaces. Natural convection is caused by the expansion of air as it is warmed; hence it becomes lighter and rises. Air in contact with cold surfaces gives up its heat, shrinks, becomes denser, and falls. The circuit of air motion and heat transfer exists in practically all spaces large enough for air movement.

2. Transfer of heat by heating a substance and then moving the substance. A warm-air heating system is known as a "convection heating system" because air is heated by the furnace and then transferred or moved to the rooms to be heated.

3. When the air surrounding the body is less than 98.6°F it will pick up heat as it passes over the body, and it carries this heat away. The faster the air moves over the body, the faster the rate of carrying heat away by convection.

Convection heating systems. There are five basic types of convection heating systems: gravity warm air, forced warm air (winter air conditioning), gravity hot water, forced hot water, and steam. Each is called a "convection heating system" because it heats to a large extent by convection. In warm-air jobs the air is heated at the furnace and transferred to the space to be heated. In the "wet heat" systems (hot water and steam) convection radiators in the various rooms are heated by a remotely located boiler and room air then circulates over the hot radiators where it becomes heated and circulates in the space.

Convector. Agency of convection. In heat transfer, a surface designed to transfer its heat to a surrounding fluid largely or wholly by convection. The heated fluid may be removed mechanically or by gravity (gravity convector). Such a surface may or may not be enclosed or concealed.

Cooling (heating) air-conditioning unit. Specific air-treating combination consisting of means for ventilation, air circulation, air cleaning, and heat transfer, with control means for cooling (or heating).

Cooling of air. Reduction in air temperature due to the abstraction of heat as a result of contact with a medium held at a temperature lower than that of the air. Cooling may be accompanied by moisture addition (evaporation), by moisture extraction (dehumidification), or by no change whatever of moisture content.

Critical level. The critical level marking on a backflow prevention device or vacuum breaker is a point established by the testing laboratory and usually stamped on the device by the manufacturer that determines the minimum elevation above the flood level rim of the fixture or receptacle served at which the device may be installed. When a backflow prevention device does not bear a critical level marking, the bottom of the vacuum braker, combination valve, or any approved device shall constitute the critical level.

Critical point. Of a substance, the state point at which liquid and vapor have identical properties: "Critical temperature," "critical pressure," and "critical volume" are the terms given to the temperature, pressure, and volume at the critical point. Above the critical temperature or pressure there is no line of demarcation between liquid and gaseous phases.

Critical temperature. Saturation temperature corresponding to the critical state of the substance at which the properties of the liquid and vapor are identical.

C (Thermal conductance). Applied to specific material as used, either homogeneous or heterogeneous, solid or gaseous, for the thickness of construction stated (not per inch of thickness), and the time rate of heat flow is expressed in Btus per hour per square foot per Fahrenheit degree average temperature difference between two surfaces.

Dalton's law of partial pressure. Each constituent of a mixture of gases behaves thermodynamically as if it alone occupied the space. The sum of the individual pressures of the constituents equals the total pressure of the mixture.

Degree day

1. Measure of the severity of a heating period (usually an entire heating season) based on climatic conditions and established by determining from local weather records the difference between 65°F and the mean temperature for each day. The sum of these differences for all the days in the heating season is the degree days for that locality.

2. Unit based on temperature difference and time, used in estimating fuel consumption and specifying nominal heating load of a building in winter. For any one day, when the mean temperature is less than 65°F there exist as many degree days as there are Fahrenheit degrees difference in temperature between the mean temperature for the day and 65°F.

Degree of saturation. Ratio of the weight of water vapor associated with a pound of dry air to the weight of water vapor associated with a pound of dry air saturated at the same temperature.

Dehumidifying effect. Condition that occurs when heat is removed in reducing the moisture content of air, passing through a dehumidifier, from its entering to its leaving condition.

Density. Ratio of the mass of a specimen of a substance to its volume; mass of a unit volume of a substance. When weight can be used without confusion, as synonymous with mass, density is the weight per unit volume.

Design working drawings. Drawings which contain all schematics of every system with other additional data.

Design working pressure. Maximum allowable working pressure for which a specific part of a system is designed.

Developed length. The developed length of pipe is its length along the centerline of the pipe and fittings.

Dew point rise. Increase in moisture content (specific humidity) of air expressed in terms of rise in dew point temperature.

Dew point temperature

1. Temperature at which the condensation of water vapor in a space begins for a given humidity condition as the temperature of the vapor is reduced. It is the temperature at which dew forms. It is also the temperature at which the air is fully saturated with water vapor.

2. Temperature at which the condensation of water vapor in a space begins for a given state of humidity and pressure as the temperature of the vapor is reduced; temperature corresponding to saturation (100% relative humidity) for a given absolute humidity at constant pressure.

Diameter. Unless specifically stated, diameter is the normal diameter as designated commercially.

Direct method of cooling. System in which the evaporator is in direct contact with the material or space refrigerated or is located in air-circulating passages communicating with such spaces.

Direct-return system. Piping arrangement for a heating, air-conditioning, or refrigerating system, in which the heating or cooling fluid, after it has passed through a heat-exchange unit, is returned to the boiler or evaporator by the shortest direct path, resulting in considerable differences in the lengths of the several circuits composing the system.

Down-feed system. Piping arrangement for a heating, air-conditioning, or refrigerating system in which the heating or cooling fluid is circulated through supply mains that are above the levels of the heating or cooling units that they serve.

Draft. Current of air, when referring to the pressure difference that causes a current of air or gases to flow through a flue, chimney, heater, or space or to a localized effect caused by one or more factors of high air velocity, low ambient temperature, or direction of air flow, whereby more heat is withdrawn from a person's skin than is normally dissipated.

Dry air. Air without water vapor; air only.

Dry-bulb temperature

1. Temperature recorded by a standard shaded thermometer.

2. Temperature of a gas or mixture of gases indicated by an accurate thermometer after correction for radiation.

Duct system. Series of ducts, elbows, and connectors to convey air from one location to another.

Durham system. Soil or waste system where all piping is of threaded pipe, tubing, or other such rigid construction, using recessed drainage fittings to correspond to the types of piping.

Effective opening. Effective opening is the minimum cross-sectional area at the point of water supply discharge, measured or expressed in terms of the diameter of a circle or if the opening is not circular, the diameter of a circle of equivalent cross-sectional area.

Effective temperature. Arbitrary index that combines into a single value the effect of temperature, humidity, and air movement on the sensation of warmth or cold felt by the human body. The numerical value is that of the temperature of still, saturated air that would induce an identical sensation.

Emissivity

1. Measure of the emission of heat from a surface, used primarily with reflective materials. The heat emitted from and the heat reflected by any surface add up to 1.0.

2. Capacity of a material to emit radiant energy. Emittance is the ratio of the total radiant flux emitted by a body to that emitted by an ideal blackbody at the same temperature.

Emittance. Ratio of the radiant energy emitted by a surface to that emitted by a perfect radiator (a blackbody) at the same temperature.

Equivalent direct radiation (EDR). Unit of heat delivery of 240 Btu/hour.

Equivalent temperature differential (ETD). Value determined by research that represents the effect of some nine different factors on the rate at which heat enters a building through its weather-exposed shell. The principal components are solar heat, air heat, mass, and color.

Evaporation. Heat is consumed in the process of evaporation. Latent heat is the heat required to change the form of a substance without changing its dry-bulb temperature. In cold weather some moisture is evaporated from the body, from the body surfaces, and from the lungs as one breathes. In summer when air temperature and surrounding surface temperatures are high, perspiration increases so as to increase the heat loss from the body by evaporation. This is nature's method of compensating for changes in the surroundings so that the rate of total heat loss from the body can remain reasonably near the comfort level.

Expansion valve capacity. Refrigerating effect in British thermal units per hour, or tons, each of 12,000 Btu/hour, produced by the evaporation of refrigerant passed by the valve under specified conditions.

Factor of safety (in pressure vessels). Ratio of ultimate stress to design working stress.

Fahrenheit. Thermometric scale in which 32°F denotes freezing and 212°F the boiling point of water under normal pressure at sea level (14.696 psi).

F (Film or surface conductance). The time rate of heat exchange by radiation conducted and convection of a unit area of surface with its surroundings, usually expressed in Btus per hour per square foot of surface per Fahrenheit degree temperature difference.

Fixture unit. Quantity in terms of which the load-producing effects on the plumbing system of different kinds of plumbing fixtures are expressed on some arbitrarily chosen scale.

Fixture unit (dfu). Common measure of the probable discharge into the drainage system by various types of plumbing fixtures. The drainage fixture-unit value for a particular fixture depends on its volume rate of drainage discharge, on the time duration of a single drainage operation, and on the average time between successive operations.

Fixture unit (sfu). Common measure of the probable hydraulic demand on the water supply by various types of plumbing fixtures. The supply fixture unit value for a particular fixture depends on its volume rate of supply, on the time duration of a single supply operation, and on the average time between successive operations.

Fixture unit flow rate.   Total discharge flow in gallons per minute of a single fixture divided by 7.5, which provides the flow rate of that particular plumbing fixture as a unit of flow. Fixtures are rated as multiples of this unit of flow.

Flash point.   Temperature of combustible material, like oil, at which there is a sufficient vaporization to ignite the vapor, but not sufficient vaporization to support combustion of the material.

Floor radiant systems.   Any system that places the heat in the floor is known as a "floor panel" or "floor radiant" system. Warm air can be conducted through hot water pipes embedded in the floor or air channels created in the floor.

Flow pressure.   Pressure in the water supply pipe near the faucet or water outlet while the faucet or water outlet is wide open and flowing.

Food.   Fuel for the human body. Food "burns" or oxidizes in the body and supplies heat.

Force.   Action on a body that tends to change its relative condition as to rest or motion.

Forced hot water.   An electrically operated water pump is put in the return line of a hot-water heating system and forces water circulation through longer and smaller pipes than can be used with gravity systems. The "circulator" does for a gravity hot-water system exactly what the blower does for a warm-air furnace. Correctly controlled, this type of system can carry heat great distances. Hot-water heat has the advantage over steam of being able to circulate at low temperature in mild weather and at higher temperatures in colder weather.

Freezing point.   Temperature at which a given liquid substance will solidify or freeze upon removal of heat. The freezing point for water is 32°F.

Generally accepted standard.   Specification, code, rule, regulation, guide, or procedure in the field of construction, or related thereto, recognized and accepted as authoritative in the locality of acceptance.

Gravity hot water.   Water is heated in the boiler and then circulated by gravity or natural forces through water pipes to radiators in the various rooms. Hot water flows through such a system for exactly the same reasons that warm air flows through a gravity warm-air system. It is the difference in temperature between the hot-water line and the return line and the height of the radiator above the boiler that supplies the natural force, or "gravity" action, that causes water to flow through such a system.

Gravity warm-air heating.   Air circulates over the furnace or heat exchanger and through the pipes by natural forces. As air is heated, it expands. As it expands, each cubic foot of air becomes lighter in weight than it was at the lower temperature. With air in the warm-air pipes being lighter than the cooler air in the return air pipes, the natural force of gravity causes the air to circulate. Warm air rises, and the heavier return air drops into the base of the furnace and helps push the warm air up and out of the furnace.

Gross refrigerating capacity.   Total rate of heat removal from all sources by the evaporator of a refrigerating system at stated conditions. It is numerically equal to the system refrigerating effect.

Heat.   Form of energy that is the result of molecular action within a substance.

Heat-actuated cooling.   The use of thermal energy to initiate a thermodynamic cycle that results in a local decrease in temperature.

Heat buildup.   Accumulation of thermal energy generated within a material as a result of hysteresis, evidenced by an increase in temperature.

Heat capacity.   Amount of heat necessary to raise the temperature of a given mass one degree F; numerically, the mass multiplied by the specific heat.

Heat curve determination.   Technique in which the temperature of a substance is measured as a function of the program temperature while the substance is subjected to a controlled-temperature program in the heating mode.

Heat durability.   The extent to which a material retains its useful properties at ambient air conditions, following its exposure to a specified temperature and environment for a specified time and its return to the ambient air conditions.

Heat flux transducer.   Device containing a thermopile that produces an output that is a function of the heat flux.

Heat pump.   System employed to transfer heat into a space or substance. The condenser provides the heat while the evaporator is arranged to pick up heat from air, water, etc. By shifting the flow of air or other fluid, a heat pump system may also be used to cool the space.

Heat pump coefficient of performance compressor.   Ratio of the compressor heating effect (heat pump) to the rate of energy input to the shaft of the compressor in consistent units, in a complete heat pump, under designated operating conditions.

Heat transfer.   Heat always transfers in one direction— from the hotter to the cooler substance. It is frequently referred to as being transferred "downhill." Heat transfers by three methods: radiation, conduction, and convection.

Heat transmission.   Any time rate of heat flow; usually conduction, convection, and radiation combined.

Heat transmission coefficient.   Any one of a number of coefficients used in the calculation of heat transmission by conduction, convection, and radiation through various materials and structures.

High-pressure steam heating system.   System employing steam at pressures above 15 psig.

High-temperature water heating system.   System in which water having supply temperatures above 350°F is used as a medium to convey heat from a central boiler, through a piping system, to suitable heat-distributing means.

Horsepower

1.   Measurement of power that includes the factors of force and speed; also, force required to lift 33,000 lb 1 ft in 1 minute.

2.   Unit of power in foot-pound-second system; work done at the rate of 550 ft-lb/second or 33,000 ft-lb/minute.

Hot-water heating system.   System in which water having supply temperatures less than 250°F is used as a medium to convey heat from a central boiler, through a piping system, to suitable heat-distributing means.

Hydraulics.   Branch of science or engineering that treats of water or other fluids in motion.

Hydronics.   Science of heating and cooling with liquids.

Hydrostatic.   Relating to pressure of equilibrium of fluids.

Humidifying effect

1. Latent heat of vaporization of water at the average evaporating temperature times the weight of water evaporated per unit of time.

2. Latent heat of vaporization of water at the average evaporating temperature times the number of pounds of water evaporated per hour, in British thermal units per hour.

Humidity. Water vapor within a given space.

Humidity ratio. The ratio of the mass of water vapor to the mass of dry air in a given air–vapor mixture.

Ice-making capacity. Actual productive ability of a system making ice. This is less than the rated (ice-making) capacity, as some refrigeration is used in cooling the water to the freezing point, cooling the ice below the freezing point, and overcoming heat leakage.

Ice-melting capacity. Refrigeration equal to the latent heat of fusion of a stated weight of ice at 144 Btu/lb.

Ice-melting equivalent capacity. The amount of heat absorbed by 1 lb of ice at 32°F in liquefying to water at 32°F is 144 Btu.

Inch of water. Unit of pressure equal to the pressure exerted by a column of liquid water 1 in. high at a temperature of 4°C or 39.2°F.

Indirect system. Heating, air-conditioning, or refrigerating system in which a fluid, such as air, water, or brine, heated or cooled by electric heating elements, products of combustion, or a refrigerant, is circulated to the material or space to be heated or cooled or is used to heat or cool air so circulated.

Internal energy. Sum of all the kinetic and potential energies contained in substance due to the states of motion and separation of its several molecules, atoms, and electrons, including sensible heat (vibration energy) and that part of the latent heat that is represented by the increase in energy during evaporation.

Isotherm. A line on a graph or map joining points of equal temperature.

Isothermal. Temperature that remains constant; in general, heat is absorbed or rejected.

Joule's law. The intrinsic energy of a gas is independent of the volume of the gas and depends upon temperature only.

Joule-Thomson effect. Ratio of temperature change to pressure change (dT/dp) of an actual gas in a process of throttling or expansion without work done or heat interchanged.

Kelvin. The unit of thermodynamic temperature; the SI unit of temperature for which an interval of one (1) Kelvin (K) equals exactly an interval of one (1) degree Celsius (1°C) and for which a level of 273.15 K equals exactly 0°C.

K (Thermal conductivity). The time rate of heat flow through a homegeneous material under steady-state conditions through unit area per unit temperature gradient in the direction perpendicular to the isothermal surface, expressed in Btus per hour per square foot per Fahrenheit degree per inch of thickness.

Latent heat

1. Change of enthalpy during a change of state, usually expressed in British thermal units per pound. With pure substances, latent heat is absorbed or rejected at constant pressure.

2. Amount of heat required to change the physical form of a substance without changing its temperature as measured by a dry-bulb thermometer. It takes heat to convert ice into water at 32°F. To convert 1 lb of ice into 1 lb of water, apply 144 Btu. To convert 1 lb of water at 212°F to 1 lb of steam or water vapor at 212°F add 972 Btu.

Latent heat of fusion. The amount of heat that must be added to the unit mass of a solid substance to change it to a liquid without any change in temperature.

Latent heat of vaporization. The amount of heat required to change the unit mass of a substance from a liquid to saturated vapor without any change in temperature.

Load. Amount of heat per unit time imposed on a refrigerating system; required rate of heat removal.

Load factor

1. Percentage of the total connected fixture unit flow rate that is likely to occur at any point in the drainage system. It varies with type of occupancy, total flow unit above the point being considered, and probability factor of simultaneous use.

2. Ratio of actual mean load to a maximum load; maximum production capacity in a given period.

Low-pressure steam heating system. System employing steam at pressures between 0 and 15 psig.

Low-pressure system. In air-conditioning, a distributing system delivering air to ordinary ventilating grilles at low velocities with low static losses through the supply grilles.

Manning's formula. Equation for the value of coefficient C in the Chezy formula, the factors of which are the hydraulic radius and a coefficient of roughness.

Mass

1. Weight or quantity of matter in the shell of a building; its inertia with respect to heat.

2. Quantity of matter in a body as measured by the ratio of the force required to produce given acceleration to the acceleration.

MBh. 1000 Btuh.

Mean radiant temperature. Temperature of a uniform black enclosure in which a solid body or occupant would exchange the same amount of radiant heat as in the existing nonuniform environment.

Mechanical efficiency. Ratio of the useful horsepower available at the flywheel or power takeoff to the horsepower developed in the engine cylinders, expressed in percent.

Mechanical equivalent of heat. Energy conversion ratio of 778.177 ft/lb = 1 Btu.

Medium-temperature water heating system. System in which water having supply temperatures between 250°F and 350°F is used as a medium to convey heat from a central boiler, through a piping system, to suitable heat-distributing means.

Micron. Unit of length, the thousandth part of 1 mm or the millionth of a meter. Also known as "micrometer."

Millimeter of mercury. Unit of pressure equal to the pressure exerted by a column of mercury 1 mm high at a temperature of 0°C.

Negative gauge pressure. Amount by which atmospheric pressure within a space is less than the atmospheric pressure immediately outside the space.

Net capacity cooler refrigerating. Rate of heat removal from a fluid flowing through a cooler (air, water, brine, etc.) at stated conditions; difference in specific enthalpies of the cooling fluid entering and leaving the cooler. In case frosting occurs within the cooler, the latent heat of fusion and the subcooling heat of the ice (frost) must be added in determining the net cooler refrigerating capacity.

Net refrigerating capacity. Remaining rate of heat removal from all sources by the evaporator of a refrigerating system, at stated conditions, after deducting internal and external heat transfers to the evaporator that occur before distribution of the refrigerating medium and after its return.

One-pipe system. System in which the fluid withdrawn from the supply main passes through a heating or cooling unit and returns to the same supply main.

Outdoor air. Air taken from outdoors and therefore not previously circulated through the system.

Outside air. External air; atmosphere exterior to refrigerated or conditioned space; ambient (surrounding) air.

Overall coefficient of heat transfer (thermal transmittance). Time rate of heat flow through a body per unit area, under steady conditions, for a unit temperature difference between the fluids on the two sides of the body.

Overall dimensions. Projected dimensions of a device, usually on horizontal and vertical planes, that can be used to determine whether the device will fit in an assigned space or can be moved through a designated passageway.

Overhead system. Heating, air-conditioning, or refrigerating piping system in which the supply main is above the heating or cooling units supplied.

Panel heating system. System in which heat is transmitted by both radiation and convection from panel surfaces to both air and surrounding surfaces.

Pascal's law. The pressure exerted at any point on a confined liquid is transmitted undiminished in all directions.

Peltier effect. Evolution or absorption of heat that occurs when an electric current is passed across the junction of two different metals.

Percentage humidity. Ratio of the weight of water vapor associated with a pound of dry air to the weight of water vapor associated with a pound of dry air saturated at the same temperature.

Performance factor. Ratio of the useful output capacity of a system to the input required to obtain it. The units of capacity and input need not be consistent.

Perimeter warm-air heating system. System of the combination panel and convection type. Warm-air ducts embedded in the concrete slab of a basementless house, around the perimeter, receive heated air from a furnace and deliver it to the heated space through registers placed in or near the floor. Air is returned to the furnace from registers near the ceiling.

Perm. Unit of measurement of water vapor permeance of a material. Value of 1 perm is equal to 1 grain of water vapor per square foot per hour per inch of mercury vapor pressure difference.

Permeability. Water vapor permeability is a property of a substance that permits passage of water vapor, and is equal to the permeance of 1-in. thickness of the substance. When permeability varies with psychrometric conditions, the spot or specific permeability defines the property at a specific condition. Permeability is measured in perm-inches.

Permeance

1. Ability of a sheet of material of any thickness to transmit water vapor in proportion to the difference in vapor pressures between the surfaces.

2. The water vapor permeance of a sheet of any thickness (or assembly between parallel surfaces) is the ratio of water vapor flow to the vapor pressure difference between the surfaces. Permeance is measured in perms.

Permissible relative humidity (PRH). Highest relative humidity that may exist within a space of design temperature before condensation begins to form on the vapor retarder within the construction or on any other nonabsorptive critical surface when the outside is at winter design temeprature. It does not take into account the absorptive capacity of some surfacing materials.

Physical activity. As physical activity increases, food energy in the body is burned at a higher rate; so it is necessary to increase the rate of heat loss from the body for comfort. This is why a person can be confortable in a cooler place or with fewer clothes when active. The average adult loses about 400 Btu/hour at rest but can lose as much as 1000 Btu/hour when active.

Power. Rate of performing work. Common units are horsepower, British thermal units per hour, and watts.

Pressure. Normal force exerted by a homogeneous liquid or gas per unit of area on the wall of its container.

PSI. Abbreviation for pounds per square inch. Water pressure is rated at so many psi's.

PSIG. Pounds per square inch gauge.

Psychrometric chart. Graphical representation of the thermodynamic properties of moist air.

Radiant heating system. System in which only the heat radiated from panels is effective in providing the heating requirements. The term "radiant heating" is frequently used to include both panel and radiant heating pipes.

Radiation

1. Transfer of heat through space without the need of any intervening medium. Heat reaches the earth from the sun entirely by radiation because there is no intervening medium. Radiant heat is very much like light; it travels in straight lines in all directions from the heated body.

2. When the body is surrounded by surfaces that are cooler than body temperature, heat will transfer to these cooler surfaces by radiation. Radiant rays travel in all directions like light.

3. In thermal design, the transmission of heat through the air from warm to cooler surfaces that can "see" each other by means of electromagnetic waves of very long wavelength. In the process the air is not warmed except as it may take heat from warmed surfaces over which it passes. It moves in any direction.

Rated input. Maximum amount of fuel, expressed in Btu per hour, that a heating appliance is designed to burn completely, and which is marked on any appliance meeting the requirements and specifications of the American Gas Association or other similar recognized agency.

Recirculated air. Return air passed through the conditioner before being again supplied to the conditioned space.

Reflectance. The ratio of the radiant energy reflected by a surface to the energy incident upon the surface.

Reflectivity. Ability of a material to reflect heat waves; opposite of emissivity and its complement.

Refrigerant. Fluid used for heat transfer in a refrigerating system. It absorbs heat at a low temperature and low pressure of the fluid and rejects heat at a higher temperature and higher pressure of the fluid, usually involving changes of state of the fluid.

Refrigerating capacity. Rate of heat removal from a medium or space to be cooled, at stated conditions. The term "refrigerating effect" is used to denote heat transfer to or from the refrigerant itself in a refrigerating system, whereas the term "refrigerating capacity" is used to denote the rate of heat removal from a medium or space to be cooled.

Refrigerating circuit. Assembly of refrigerant containing parts and their connections used in a refrigerating cycle.

Refrigerating compressor capacity. Rate of heat removal by the refrigerant assigned to the compressor in a refrigerating system; this is equal to the product of the mass rate of refrigerant flow produced by the compressor and the difference in specific enthalpies of the refrigerant vapor at its thermodynamic state entering the compressor and refrigerant liquid at saturation temperature corresponding to the pressure of the vapor leaving the compressor.

Refrigerating effect. Rate of heat removal by a refrigerant in a refrigerating system, equal to the product of the mass rate of refrigerant flow in the system and the difference in specific enthalpies of the refrigerant at two designated points in the system, or two designated thermodynamic states of the refrigerant. The term "refrigerating effect" is used to denote heat transfer to or from the refrigerant itself in a refrigeration system. The term "refrigerating capacity" is used to denote the rate of heat removal from a medium or space to be cooled.

Refrigerating engineering. Technique of design, manufacture, application, and operation of refrigerating machinery and its primary equipment. Refrigeration (except as exact measure in heat units) refers here to a more general science, concerned with the use of cold temperatures for commercial and other useful purposes.

Refrigerating system capacity. Cooling effect produced by the change in total enthalpy (formerly called "heat content") between the refrigerant entering the evaporator and the refrigerant leaving the evaporator.

Relative humidity

1. Ratio of the mol fraction of water vapor present in the air to the mol fraction of water vapor present in saturated air at the same temperature and barometric pressure. Approximately, it equals the ratio of the partial pressure or density of the water vapor in the air to the saturation pressure or density, respectively, of water vapor at the same temperature.

2. As the percentage of moisture in the air increases, its ability to evaporate moisture from the body decreases. So heat loss from the body by evaporation decreases as the air becomes more moist. This is why a person can be comfortable at a lower temperature in properly humidified air.

3. In effect, the amount of water in vapor form in a given volume of air at a fixed temperature as a percentage of the amount of water vapor the same air could hold

if saturated. Since warm air can hold more water vapor at saturation than cooler air, temperature is a critical factor.

Repeated load. Force applied thousands or perhaps millions of times, for example, the forces in various parts of a running engine.

Resistance, thermal (*R*, ru, *R*-factor, *R*-value). Reciprocal of a heat transfer coefficient ($R = 1/C$, $1/U$, $1/F$). It is the most useful factor in thermal design calculations because resistances are additive, whereas conductances and conductivities are not. The word "ru" is an abbreviation for resistance unit.

***R* (Thermal resistance).** The reciprocal of a heat transfer coefficient is expressed by U, C or F. It is expressed in Fahrenheit degree per Btu's per hour per square foot.

Reversed-return system. System in which the heating or cooling medium from several heat transfer units is returned along paths arranged so that all circuits composing the system or a major subdivision of it are of practically equal length.

Reynolds number (aeronautic). Nondimensional coefficient used as a measure of the dynamic scale of a flow.

Room dry bulb. Dry-bulb (dew point, etc.) temperature of the conditioned room or space.

Roughness coefficient. Factor in the Kutter, Manning, and other flow formulas representing the effect of channel (or conduit) roughness on energy losses in the flowing water.

Saturated air. Moist air in which the partial pressure of the water vapor is equal to the vapor pressure of water at the existing temperature. This condition occurs when dry air and saturated water vapor coexist at the same dry-bulb temperature.

Saturated vapor. Vapor in equilibrium with its liquid; that is, the number, per unit time, of molecules passing in two directions through the surface dividing the two phases are equal.

Saturation. Condition for coexistence in stable equilibrium of a vapor and liquid or a vapor and solid phase of the same substance.

Saturation coefficient. Ratio of absorption by 24-hour submersion in water at room temperature to that after 5-hour submersion in boiling water.

Saturation temperature. For a fluid, the boiling point corresponding to a given pressure; evaporation temperature; condensation temperature.

Sensible heat

1. Heat associated with a change in temperature; specific heat exchange of temperature; in contrast to a heat interchange in which a change of state (latent heat) occurs.

2. Amount of heat in a substance that can be measured with a dry-bulb thermometer.

Solar constant. Solar intensity incident on a normal surface located outside the earth's atmosphere at a distance from the sun equal to the mean distance between the earth and the sun. Its value is 415, 445, or 430 Btu/hour/ cu ft as the July, January, or mean value, respectively. At sea level in July the solar intensity value is about 300 Btu/cu ft/hour since about 28% is absorbed in the earth's atmosphere.

Solid state. One of the three states or phases of matter, characterized by stability of dimensions, relative incompressibility, and a molecular motion held to limited oscillation.

Specific gravity. Ratio of the weight of a body to the weight of an equal volume of water at 4°C or other specified temperature.

Specific heat. Ratio of the quantity of heat required to raise the temperature of a given mass of any substance to the quantity required to raise the temperature of an equal mass of a standard substance (usually water at 59°F) 1 degree.

Specific heat of a substance

1. Ratio of the amount of heat required to raise the temperature of 1 lb of that substance 1°F to the amount of heat required to raise the temperature of 1 lb of water from 62°F to 63°F. For example, the specific heat of steel is 0.118. This means that 1 lb of steel can be heated up 1°F by the expenditure of 0.118 Btu whereas it takes 1.00 Btu to heat a pound of water 1°F. Conversely, in cooling, 1.0 lb of water will give up 1 Btu/1 degree and 1 lb of steel will give up 0.118 Btu as it cools 1°F.

2. The number of calories per degree celsius required to raise the temperature of 1 gram of the substance.

Specific humidity. Weight of water vapor (steam) associated with 1-lb weight of dry air.

Specific volume. Volume of a substance per unit mass; reciprocal of density.

Standard air. Dry air at a pressure of 760-mm (29.92-in.) Hg at 69.8°F temperature and with a specific volume of 0.833 m^3/kg.

Standard conditions. Set of physical, chemical, or other parameters of a substance or system that defines an accepted reference state or basis for comparison.

Standard rating. Standard rating is a rating based on tests performed at standard rating conditions.

Static pressure. Normal force per unit area that would be exerted by a moving fluid on a small body immersed in it if the body were carried along with the fluid. Practically, it is the normal force per unit area at a small hole in a wall of the duct through which the fluid flows (piezometer) or on the surface of a stationary tube at a point where the disturbances, created by inserting the tube, cancel. It is supposed that the thermodynamic properties of a moving fluid depend on static pressure in exactly the same manner as those of the same fluid at rest depend on its uniform hydrostatic pressure.

Steam heating system

1. System in which heat is transferred from the boiler or other source of heat to the heating units by means of steam at, above, or below atmospheric pressure.

2. Like hot water, a steam heating system uses radiators in the various rooms. Water is converted to steam in the boiler, and steam is then carried up to the radiators. Before circulators were developed for hot-water systems, steam could be carried much farther and through smaller pipes than were required for hot-water heating. Therefore, steam was the only way to get heat into distant radiators in large buildings.

Summer air-conditioning. Comfort air-conditioning carried out primarily when outside temperature and humidity are above those to be maintained in the conditioned space.

Sun effect. Solar energy transmitted into space through windows and building materials.

Superheated vapor. Vapor at a temperature that is higher than the saturation temperature at the existing pressure.

Surface conductance. Time rate at which heat is exchanged between the surface of a material and the surrounding air. It is commonly called "film resistance" because the effect is that of a thin insulating material. The unit is British thermal units per hour per square foot per degree Fahrenheit. The symbols are f, which represents the inside surface film, usually in still air, and f^0, the outside surface film, often subject to the wiping effect of winds.

Surface cooling. Method of cooling air or other gas by passing it over cold surfaces.

Surface film conductance. Time rate of heat flow per unit area under steady conditions between a surface and a fluid for unit temperature difference between the surface and the fluid.

Surface heat transfer coefficient. The rate of heat transfer from a unit area of a surface to the adjacent air and environment caused by a temperature difference of one (1) degree F between the surface and the air.

Temperature

1. Thermal state of matter with reference to its tendency to communicate heat to matter in contact with it. If no heat flows upon contact, there is no difference in temperature.

2. Degree or intensity of the heat in a substance at any given time, expressed in degrees. There are two commonly used temperature scales: Fahrenheit, marked F, and Celsius, marked C. The Fahrenheit scale is most commonly used by heating engineers. On the Fahrenheit scale water freezes at +32° and boils at +212° at sea level altitude and normal atmospheric pressure.

Temperature of surrounding surfaces. Heat is lost by radiation to the surfaces surrounding the body. When surrounding surfaces are warm, air temperature can be cool and a person will still be comfortable because one counterbalances the other. Satisfactory temperature comfort is obtained when the numerical sum of the air temperature and the average temperature of the surfaces around the body approximates 140°F. If surface temperatures are 70°F then an air temperature of 70°F will create comfort. But if the average surface temperature is 65°F then it will take approximately 75°F air temperature for comfort.

Theoretical displacement. Total volume displaced by the working strokes of all the pistons of a compressor per revolution or unit of time.

Therm. Quantity of heat equal to 100,000 Btu.

Thermal conductance

1. Time rate at which heat flows through 1 sq ft of a material of known thickness in 1 hour when the temperature difference between the two surfaces of the material is 1°F. it is expressed in British thermal units per hour per square foot per degree Fahrenheit. By industry agreement, roof insulations are rated on their conductance rather than their thickness; hence the term "C-value."

2. Time rate of heat flow through a body (frequently per unit area) from one of its bounding surfaces to the other for a unit temperature difference between the two surfaces, under steady conditions.

Thermal conduction

1. Movement of heat through a material, as through a metal rod held with one end to a fire.

2. Process of heat transfer through a material medium in which kinetic energy is transmitted by particles of the

material from particle to particle without gross displacement of the particles.

Thermal conductivity

1. Time rate at which heat flows through a homogeneous material 1 in. thick and 1 sq ft in area in 1 hour, when the difference in temperature between the two surfaces is 1°F. This is the measure of the effectiveness of most insulation and individual building materials other than reflective insulations. Composite insulation, however, is rated by its conductance per inch or with the actual thickness stated. It is expressed in British thermal units per hour per square foot per degree Fahrenheit per inch.

2. Time rate of heat flow through unit area and unit thickness of a homogeneous material under steady conditions when a unit temperature gradient is maintained in the direction perpendicular to area. Materials are considered homogeneous when the value of the thermal conductivity is not affected by variation in thickness or size of sample within the range normally used in construction.

Thermal inertia. Property which modifies the effect of the U value on the heat transmission of a building element by expanding the time scale or time lag.

Thermal insulation. Material having a relatively high resistance to heat flow and used principally to retard the flow of heat.

Thermal insulation value. The heat conductivity of a material is a measure of the insulating value of the material; the lower the conductivity, the greater the insulating value.

Thermal radiation. Transmission of heat through space by wave motion; passage of heat from one object to another without warming the space between.

Thermal resistance. Reciprocal of thermal conductance.

Thermal resistivity. Reciprocal of thermal conductivity.

Thermal transmittance (u factor)

1. Time rate of heat flowing through 1 sq ft of material in 1 hour from the air on one side to the air on the other when the difference in temperature between the air on the two sides is 1°F. It is sometimes called the "overall coefficient of heat transfer" and is expressed as British thermal units per hour per square foot per degree Fahrenheit. Its symbol is U; hence its value is commonly called the "U-value" of the given construction assembly.

2. Time rate of heat flow per unit area under steady conditions from the fluid on the warm side of a barrier to the fluid on the cold side per unit temperature difference between the two fluids.

Time lag. Delay caused by heat storage and its subsequent release by the structure; increases as the mass of the wall increases.

Ton-day of refrigeration. Heat removed by a ton of refrigeration operating for a day; 288,000 Btu. It is approximately equal to the latent heat of fusion or melting of 1 ton (2000 lb) of ice, from and at 32°F.

Ton of Refrigeration

1. Useful refrigerating effect equal to 12,000 Btu/hour, 200 Btu/minute.

2. Removal of heat at a rate of 12,000 Btu/hour.

Total cooling effect. Difference between the total enthalpy of the dry air and water vapor mixture entering a unit per hour and the total enthalpy of the dry air and

water vapor (and water) mixture leaving the unit per hour, expressed in British thermal units per hour.

Total thermal resistance. Total resistance to heat flow through a complete building section or construction assembly, generally expressed as the temperature difference in degrees F needed to cause heat to flow at the rate of one (1) Btu per hour per square foot of area.

Transfer of heat by conduction. Conduction occurs through continuous materials in which energy is transferred directly between adjacent molecular aggregates without mass motion of the materials.

Transfer of heat by convection. Convection occurs in liquids and gases, the transfer being accomplished by the motion of the fluid from a locality where it receives heat, to a locality where it gives up heat.

Transmissivity. Capacity of a material to transmit radiant energy. Transmittance is the ratio of the radiant flux transmitted through a body to that incident to it.

Two-pipe system. System in which the fluid withdrawn from the supply man passes through a heating or cooling unit to a separate return main.

Up-feed system. Piping arrangement for a heating, air-conditioning, or refrigerating system, in which the cooling fluid is circulated through supply mains that are below the levels of the heating or cooling units that they serve.

U (**Overall coefficient of heat transmission**). As applied to the usual combinations of materials and also to single materials such as window glass and includes the surface conductance on both sides. It is expressed in Btu's per hour per square foot per Fahrenheit degree temperature differance between air on the inside and air on the outside of a wall, floor, roof, or ceiling.

Useful refrigeration capacity. Capacity available for the specific ultimate cooling function for which the system was designed.

Useful sensible capacity—air conditioner. Available refrigerating capacity of an air conditioner for removing sensible heat from the space to be conditioned.

Useful total capacity—air conditioner. Available refrigerating capacity of an air conditioner for removing sensible and latent heat from the space to be conditioned.

U (**Overall coefficient of heat transmission**). Amount of heat (Btu) transmitted in one hour per square foot of the wall, floor, roof, or ceiling for a difference in temperature of one degree F between the air on the inside and outside of the wall, floor, roof, or ceiling.

U (**u-factor**). Overall heat transmission coefficient; the amount of heat, expressed in Btu transmitted in one hour through one square foot of a building section (wall, floor, or ceiling) for each degree F of temperature difference between air on the warm side and air on the cold side of the building section.

***U*-value (transmission).** The overall coefficient of heat transmission through composite materials. The reciprocal of the sum of the R-values for each element between the inside and outside air, expressed in Btu's per hour per square foot per degree Fahrenheit.

Vacuum heating system. Two-pipe steam heating system equipped with the necessary accessory apparatus that will permit it to operate below atmospheric pressure.

Vapor. Gas, particularly one near to equilibrium with the liquid phase of the substance and that does not follow the gas laws. Usually used instead of gas for a refrigerant, and in general for any gas below the critical temperature.

Vapor retarder

1. Material that does not readily permit the passage of water vapor. A material, film, or coating made for this use is expected to have a permeance of 1 perm or less.

2. Moisture-impervious layer applied to the surfaces enclosing a humid space to prevent moisture travel to a point where it may condense owing to lower temperature.

Vapor heating system. Steam heating system that operates under pressures at or near atmospheric and that returns the condensate to the boiler or receiver by gravity.

Vapor permeability. The property of a material that permits migration of water vapor under the influence of a difference in vapor pressure across the material.

Vapor permeance. The ratio of the water vapor flow rate, in grains per hour, through a material of any specified thickness to the vapor pressure difference between the two surfaces of the material, expressed in inches of mercury. The unit is the perm.

Vapor pressure

1. Part of the total atmospheric pressure that is exerted by the water vapor in the air.

2. Pressure exerted by a vapor. If a vapor is kept in confinement over its liquid so that the vapor can accumulate above the liquid, the temperature being held constant, the vapor pressure approaches a fixed limit called the "maximum" or "saturated" vapor pressure, dependent only on the temperature and the liquid.

Vapor resistance. Reciprocal of vapor permeance; rating of the resistance of a material to the passage of water vapor.

Vapor resistivity. Reciprocal of vapor permeability; measure of the resistance of a 1-in. thickness of material to the passage of water vapor.

Vapor system. Refrigerating system employing a condensable vapor as the refrigerant.

Velocity. Vector quantity that denotes at once the time rate and the direction of a linear motion.

Velocity head. For water moving at a given velocity, the equivalent head through which it would have to fall by gravity to acquire the same velocity.

Ventilation. Process of supplying or removing air, by natural or mechanical means, to or from any space. Such air may or may not have been conditioned.

Viscosity. Property of semifluids, fluids, and gases by virtue of which they resist an instantaneous change of shape or arrangement of parts. It is the cause of fluid friction whenever adjacent layers of fluid move with relation to each other.

Volatile fluid. A fluid that vaporizes easily.

Volumetric refrigeration capacity. Of a system, the per unit volume of refrigerant circulated at the compressor suction.

Water vapor

1. Water in the state of an invisible gas diffused in the air.

2. Term used commonly in air-conditioning parlance to refer to steam in the atmosphere.

Wet-bulb temperature

1. The thermodynamic wet-bulb temperature is the temperature at which liquid or solid water, by evaporating into air, can bring the air to saturation aciabatically at the same temperature. The wet-bulb temperature

(without qualification) is the temperature indicated by a wet-bulb psychrometer constructed and used according to specifications.

2. Temperature recorded by a thermometer in vigorously moving air when the thermometer bulb is wetted by a suitable wicking device. It is the temperature at which that air would be saturated when cooled.

Winter air-conditioning. Heating, humidification, air distribution, and air cleaning where outside temperatures are below the inside or room temperature.

Winter air-conditioning system. System that consists of a warm-air furnace with an electrically driven blower to force the air circulation through piping that is smaller in size than can be used with gravity systems. Systems are called "forced air" or "forced warm-air" systems because of the mechanical force needed to push air through the system. Systems usually installed with a humidifier, filter, and automatic controls and that really condition the air during the heating season are not called "winter air-conditioning" systems.

Warm-air heating system. Heating plant consisting of a heating unit (fuel-burning furnace) enclosed in a casing, from which the heated air is distributed to various rooms of the building through ducts.

Water vacuum system. In refrigeration, a system that employs a vacuum to boil water at the temperature desired; system that employs evaporating water vapor as the refrigerant.

PROFESSIONAL PRACTICE

Access to the project. The professional or his representatives have the prerogative to enter the project during the progress of the work for any purpose that will benefit the project.

Acknowledgment. Signed form that acknowledges that the bidder has received the proper bidding documents. This acknowledgment protects the owner against any action the bidder may take in possible withdrawal of his proposal.

Addendum. Revised, changed, or corrected document or addition to the contract documents, usually issued during the time of bidding to all prime bidders; however, it may be issued after the bid opening and before the contracts are signed to cover certain modifications negotiated with the low bidder (plural: addenda).

Additional services. In addition to the basic services professional services are as follows: project analysis, programming, land use studies, feasibility investigations, financing, construction management, scheduled on-site observations, job inspections, supervising associate, and special consulting services.

Additional to building. Addition to an existing building or structure.

Addition to contract amount. Addition to the contractor's basic proposals, either in the form of an accepted additive alternate, or as a change order amending the contract.

Advertisement for bids. Published notice in the trade periodicals; both private and public soliciting proposals for a construction project.

Agency. Relationship between agent and principal.

Agency construction management (ACM). Construction management by an agent who acts for the owner in a fiduciary role during entire tenure of the project without becoming engaged in other functions, such as providing design services, contracting, or doing construction work with its own forces.

Agent. One authorized by another to act in his stead or behalf.

Agreement

1. Document stating the essential terms of the construction contract, incorporating by reference the other contract documents; document that sets forth the terms of a contract between the professional and the owner, or between the professional and a consultant.

2. Instrument of the various documents, which are related to each other by inference, statements, and direct reference, legally binding the signing parties. The agreement usually defines the relationship and obligations between the parties.

3. Contract between parties.

Allowance. Cash allowance, contingency allowance, etc.

Alterations. Revisions within or to a prescribed list of elements in an existing structure.

Alternate. Variation in contract requirements described in the specifications and possibly indicated on the drawings, on which a separate price is to be received as a part of the bid.

Alternate bid. Amount stated in the bid to be added to or deducted from the amount of the base bid if the corresponding change in project scope, alternate materials, or methods of construction is accepted by the owner.

Alternatives. Potential solutions to meet a project's requirements. Professional is entitled to be paid for the preparations of alternatives unless otherwise stated in the contract.

Alternatives analysis. Analysis of the various alternatives to determine validity and impact on project cost, project appearance, project schedule, and socioeconomic and environmental conditions.

Ammendments and supplements to the general conditions. Documents stating the changes and modifications from the General Conditions.

Announcement of bidding. General announcement to the various trade periodicals, bidders, and material suppliers that bids will be received from a selected list of bidders on a specific job, or that the project is open to all bidders.

Application for payment. Statement of amounts on proper acceptables or forms claimed by the contractor as payment due on account of work performed or materials suitably stored.

Appraisal

1. Evaluation or estimate by a qualified professional appraiser of the market, replacement value, demand, utility, or other attribute of the land or the improvement.

2. Approximation of replacement cost, utility, or other attribute of existing facilities.

Appraisers. Persons qualified through experience and judgment to prepare written documentations as to values at the time of the appraisal.

Approved equal. Term used to indicate that the material, equipment, or product finally supplied or installed must be equal to that specified and approved by the architect/engineer and acceptable to the owner, in some cases, and the lender.

Arbitration. An impartial group chosen by the parties to solve a dispute between them. The arbitrators are vested with the power to make a final determination concerning the controversy, bound only by their own discretion and not by rules of law or equity. (Construction Industry Arbitration Rules—American Arbitration Association)

Architect. Designation reserved, usually by law, for a person qualified and duly licensed to perform architectural services, including analysis of project requirements, creation and development of the project design, preparation of working drawings, specifications and bidding requirements, and general administration of the construction contracts. Is entitled under the professional-owner contract to engage other professionals to work jointly on the project.

Architect-Engineers. Individuals or firm offering professional services as architect and/or engineer.

Architect's / Engineer's approval. Written acknowledgment that a material, equipment, or method is acceptable or in accordance with the contract requirements, without increase in cost to the owner.

Architecture. The art and science of designing, planning, writing specifications and the preparation of related documents, for buildings and structures.

Area estimates. Approximations of probable project construction costs based on assumed unit cost per square foot.

Articles of incorporation. Instrument that creates a private corporation, pursuant to the general corporation laws of the state.

As-built drawings. Records kept at the site during the progress of construction, or subsequent thereto, illustrating how various elements of the project were actually installed including changes that were made. Then the information is transferred to a drawing, or the original drawing is revised based on the professional's contract.

Associate or associated architect. Architect working with another architect in a temporary agreement or partnership of joint venture.

Attorney-in-fact. Person having power-of-attorney to represent another person legally as his agent in accordance with defined terms.

Award. Act of presentation or the offering of the proposed contract to the successful bidder for proper signing.

Base bid

1. Bid before any alternates are considered.

2. Amount stated in the bid as the sum for which the bidder offers to perform the work.

Base bid documents. Documents listing or describing materials, equipment, and methods of construction on which the base bid must be predicated.

Base bid price. Bidder quoted price or monetary consideration for the base bid.

Beneficial occupancy. Use of a structure or portion thereof for the purpose intended.

Bid

1. Complete and properly signed proposal to perform the work outlined n the proposal for the sums stipulated therein.

2. Bidder's offered price to do the work covered by the contract documents; executed bid form showing the bid price and other data.

Bid bond. Form of bid security executed by the bidder as principal and by a surety to guarantee that the bidder will enter into a contract within a specified time and will furnish any required performance bond and labor and material payment bond.

Bid date. Advertised time set for the receipt of the bids.

Bidder. Anyone who submits a bid, usually a prime bidder who submits a bid to the owner, rather than a subbidder who submits a bid to the prime bidder. A bidder is not a contractor until he signs a contract.

Bidder's prequalifications. List of general prequalification information desired, including such items as banking connections, experience, labor force, bonding capabilities, and references.

Bidding documents. Documents that include the invitation to bid, instructions to bidders, bid form, bid bond, etc. They are not actually a part of the specifications. The contract documents deal with the activities of construction after the signing of the contract.

Bidding period. Advertised calendar period beginning at the time of issuance of bidding requirements and contract documents and ending at the prescribed bid time.

Bid form

1. Description of an acceptable form prepared on the bidder's letterhead; printed form prepared by a professional or technical institute or association.

2. Form furnished to the bidder on which he should prepare his bid.

Bid guarantee. Deposit of cash, check, money order, or bid bond when the bid is submitted by the bidder to guarantee that he will sign the contract and furnish the required surety if awarded; same as bid security.

Bid opening

1. Opening ceremony on a given date at a given time and location and public reading of the bids that have been submitted.

2. Official opening and tabulation of bids submitted by the prescribed bid time, in conformity with the prescribed procedures.

Bid price

1. Sum stated in the bid for which the bidder offers to perform the work.

2. "Base bid price," not including any alternates or substitutions.

Bid security. Bid guarantees by the bidder; by bond, certified check, or cashier's check that he will accept the contract if awarded to him.

Bona fide bid

1. Bid submitted in good faith, complete and in prescribed form, which meets the conditions of the bidding requirements and is properly signed.

2. Bid honestly submitted on a bid form properly executed, with all necessary guarantees.

Bonus and penalty clause. Provision in the proposal form for payment of a bonus to the contractor for each day the project is completed prior to the time stated, and for a charge against the contractor for each day the project remains uncompleted after the time stipulated.

Book of specifications. Bound volume of specifications, describing the work by divisions and sections, usually containing other documents relating to methods of bidding and other information; the "Project Manual."

Boundary survey. Mathematically perfect closed diagram of the complete peripheral boundary of a site, provided by the owner, reflecting dimensions, compass bearings, and angles. It should bear a licensed land surveyor's signed certification and include a metes and bounds or other legal written descriptions.

Budget for construction project. Total sum established by the owner for the construction project.

Building inspector. Representative of a governmental authority who issued the construction permit and who is employed to inspect construction for compliance with applicable codes, regulations, and ordinances. The building inspector may inspect all phases of the construction, or there may be a separate inspector for the structure, plumbing, electrical installations, and fire protection, health, and safety features.

Building permit. Permit issued by appropriate governmental authority, usually paid for by the contractor, allowing construction of a project in accordance with approved drawings, specifications, and special stipulation, if any to proceed with construction in the field.

CAD. An acronym for "computer-aided design" or for "computer-aided drafting."

CADD. An acronym for "computer-aided design and drafting."

Cash allowance

1. All-inclusive cost to provide a work unit at a predetermined allowance included in the contract. The savings usually revert to the owner.

2. Sum recommended by the architect/engineer for the purchase or contracting of a construction material item, equipment, furnish item, or work yet to be described or generally described in the specifications. The cash allowance or maximum limit established is usually set by an owner or his financial institution.

3. Amount established in the contract documents for inclusion in the contract sum to cover the cost of prescribed items not specified in detail, with provision that variations between such amount and the finally determined cost of the prescribed items will be reflected in properly endorsed change orders appropriately adjusting the contract sum.

Certificate of substantial completion. Certificate prepared by the responsible design professional on the basis of an inspection, stating that the work or a designated portion of the work is substantially complete.

Certificate of occupancy

1. Certificate issued by governmental authority certifying that all or a designated portion of a building complies with the provisions of applicable statutes and regulations and can be occupied.

2. Certificate issued by the governing authority granting permission to occupy a project or parts of a project for a specific use. It is procured by the prime contractor.

Change order

1. Revision to the contract after it has been officially awarded.

2. Work order, usually prepared by the contractor, submitted to the architect-engineers, and signed by the owner or his agent, authorizing a change, revision, or addition.

3. Written order to the contractor signed by the owner, issued after the execution of the contract, authorizing a change in the work or an adjustment in the contract sum and the contract time.

Changes in the work. Changes ordered by the owner, which could consist of additions, deletions, or other revisions within the general scope of the contract, the contract sum, and the contract time.

Class 1 shop drawings. Shop drawings for structural components that will be a part of a completed structure and are fabricated or constructed according to specific requirements as defined by the design professional.

Class 2 shop drawings. Shop drawings for structural components that will be part of a completed structure and are provided as manufactured items according to the performance requirements as defined by the design professional in the contract documents.

Class 3 shop drawings. Shop drawings for structural components that will not be part of a completed structure, but are to be used temporarily during construction for such purposes as erection, shoring, reshoring, bracing, scaffolding, or access.

Clerk of the works. On-site clerk who keeps records of the men on the project, deliveries made, and progress of the job. The clerk makes reports, receives and catalogs samples, and keeps a log of job activities. Employed by the contractor, or can be employed by the owner based on original contract agreements.

Closed specifications. Specifications requiring a particular brand or trade name with no substitutions permitted.

Code of ethics. Professional practice standards estblished by a national organization to which the majority of professionals subscribe.

Codes. Municipal, state, or federal regulations governing private actions for the protection of the public health, safety, and welfare.

Collusion affidavit. Statement by the bidder that he has not entered into collusion with another bidder or other person in the preparation of his bid. It is not recommended on private projects unless required by the owner.

Commercial standards. Standards that provide technical requirements for materials, equipment, construction, dimensions, tolerances, testing, grading, marking and identifying, or other related details. The objective is to define quality levels for products in accordance with the principal demands of the trade or industry.

Competitive bidding. A method, often mandated by law, for public projects of selecting contractors for construction projects by price competition between qualified bidders subjected to various rules and procedures.

Completion bond. Bond given by the contractor to the owner and lending institution, guaranteeing that the work will be completed and that funds will be provided for that purpose.

Completion date. Date set by the contract when all required work shall be completed or substantially completed.

Computer. Sophisticated electronic machine for storing, processing, and manipulating data, regarding job progress and costs.

Computer hardware. Physical components of a computer.

Computer software. Computer programs and instructions that control the way information is processed by the computer hardware.

Conditions of the bid. Instructions to bidders, notice to bidders, advertisement for bids, invitation to bidders, or similar documents prescribe the conditions under which bids are to be prepared, submitted, received, and date of acceptance.

Conditions of the contract. Those portions of the contract documents that define the rights and responsibilities of the contracting parties and others involved in construction.

Constructability analysis. Review of the ability to construct a project, covering economics, availability of materials, site restrictions, and local conditions that may affect the construction process.

Construction contract. Agreement or contract between the owner and constructor for construction of a project, or portions of a project, in accordance with contract documents.

Construction cost. Cost of all the construction portions of a project, generally based on the sum of the construction contract(s) and other direct construction costs.

Construction documents. Agreements, working drawings, specifications, general conditions, and possibly supplementary general conditions.

Construction estimates. Forecasts of probable project construction costs to be used as a guide to the owner and the design professionals in preparing contract documents.

Construction management

1. A special management service performed by the architect/engineer or others during the construction phase of the project under separate or special agreement with the owner. This is not part of the architect/engineer's basic services but an additional service.

2. Generic label for a method of total control of project development, design, and construction under a single entity.

Construction manager (CM). Individual or entity that provides the construction management services with a fiduciary duty to the owner during the design and/or construction phases of a project.

Construction progress photographs. Photographs of the site or the structures taken during the construction phase of the job if required by owner's agreement with the architect/engineer, by the owner or his financial institution, or by the general contractor under contract. Photographs are always related to a plan, sketch of the project, and the directions from which they are taken. Photographs are dated and identified when negatives are printed.

Construction project budget. Sum established by the owner as available for construction of the project, stipulated highest acceptable bid price or, in the case of a project involving multiple construction contracts, the stipulated aggregate total of the highest acceptable bid prices.

Construction quality manager. Individual member of the construction team responsible for quality-assurance and quality-control activities of a construction project.

Constructor

1. Prime contractor or general contractor.

2. Individual or entity responsible for performing and completing the construction of a project as required by the contract documents. Constructor has a direct contract with the owner.

Consultants. Professionals engaged by the owner or the financial institution to complement the services of the architect/engineer.

Contingency allowance. Established amount of money included in the contract to cover unpredictable small changes in work due to site conditions during construction. It is not intended to cover additions that increase the scope of the job.

Contract

1. Agreement that is enforceable by law.

2. Legally enforceable promise or agreement between two or among several persons.

Contract date. Date shown on the contract as the official date of execution. It may not always be possible for all parties to sign the contract on the same day or start construction on that day and the official date is the "contract date."

Contract documents. Working drawings, specifications, general conditions of the contract, supplementary general conditions, and the agreement.

Contract forms. Standard or prepared forms or documents on which the contract is written. They usually consist of the agreement, guaranty bond, and power-of-attorney if required.

Contractor. Party signing the contract with the owner to perform the work called for in the drawings and specifications. The bidder is not a contractor until the contract is signed.

Contract price. Basic amount indicated in the agreement as the consideration given to the contractor for performing the work required by the contract.

Contract time. Official amount of time allowed the contractor to complete the work of the contract. It should begin with the contract date or date of the notice to proceed, as required by the terms of the contract. It should end on the completion date as called for.

Correction of work. Process of correcting work that does not conform to contract documents.

Cost. Amount or equivalent paid or charged for materials and services provided or performed.

Cost breakdown. Schedule of costs for each trade. It could include other costs, such as general conditions, overhead, and profit.

Cost control. Systematic identification of problem areas affecting cost in all phases of design and construction, with appropriate steps taken to mitigate and correct escalating cost trends.

Cost plus contract. Contract in which the contractor keeps a record of the cost of construction and the owner pays this cost plus an agreed profit to the contractor.

Cost plus fee. System used for negotiated contracts where a stipulated sum does not form part of the agreement. The contractor is reimbursed for the entire cost expended and is paid a fee for his services, either fixed or a percentage of the reimbursed amount.

Cubage

1. Enclosed total volume measurements of a structure.
2. Architectural volume of a building, which includes the sum of the products of the areas and the height from the underside of the lowest floor construction system to the average height of the surface of the finished roof above for the various parts of the building.

Deduction. Amount deducted from the contract sum by change order.

Deductive alternate. Alternates listed in the documents that results in a deduction from the bidder's base bid.

Defective work. Work judged by inspection as not complying with the contract documents.

Description of work. Description of the major characteristics of the construction and of the project type and size to provide the bidder the information as to whether the project is within his construction capability and financial and insurance bonding capacity.

Design. Process in which drawings and specifications and the other parts of the contract documents are prepared under the direction of a design professional.

Design-build. Form of contract where the constructor is responsible for the design and construction of a facility.

Design discipline. Specific category of related professional services such as structural engineering, architecture, mechanical engineering, electrical engineering, civil engineering, interior designer, etc.

Design professional. Designation reserved, usually by law, for a person or organization professionally qualified and duly licensed to perform architectural or engineering services, which may include but not necessarily be limited to development of project requirements; creation and development of project design; preparation of drawings, specifications, and bidding requirements; and providing of professional services during the construction phase of the project.

Design team. Group of individuals or entities representing all design disciplines and required for execution of a design contract.

Design team leader. Individual responsible for the coordination of design discipline activities of a project. The design team leader also is responsible for monitoring progress and reporting to the owner.

Detail. Drawing, to a larger-scale, or a part of a drawing, indicating in dimensioned detail the composition, construction, and correlation of the assembly and materials specified.

Detailer. Individual or entity who prepares shop drawings for a constructor or subcontractor.

Developer. Individual or group that arranges for the financing and construction of a project.

Disclaimers of responsibility. Contract language that disclaims responsibility for certain actions.

Discrepancies and ambiguities. Instructions on how the discrepancies and ambiguities in the bidding documents will be resolved during the bidding period.

Disqualification of proposals. List of conditions or causes under which a bidder will be officially disqualified before the proposal is received or opened and after the proposal is opened.

Divided contract. System of dividing the work of the job into several prime contracts such as general, mechanical, and electrical, as opposed to a single contract, which is the system of awarding all work on a job to one general contractor.

Division. Major division of the specifications, similar to chapters in a book. All related work of the same type throughout the job should be grouped together in one division. The divisions are placed in the specification in somewhat chronological order of the progress of the work. If the work within a division falls into more than one trade, each trade should be placed into a separate "section" of the division. It is not usually the intention of the specification writer to divide the work according to jurisdictional trade agreements. The recommended division headings are contained in the CSI Masterformat.

Division of the specifications. One of the sixteen basic organizational divisions used in the specifications format developed by the Construction Specification Institute and published in the Masterformat.

Document deposit. Deposit that is required of a bidder to ensure prompt return of drawings and specifications in good condition.

Drafting practices. Usually standards adopted by each professional office, which are revised or corrected from job experiences or from new information provided by manufacturers. Trade associations also provide details of new installation procedures.

Drawing clarification. Graphic interpretation of the drawings issued by the architect-engineer as part of an addendum or change order.

Drawing deposit. Document deposit to guarantee return of the bidding documents.

Drawings. Portion of the contract documents indicating in graphic or pictorial form the design, location, and dimensions of the elements of a total project.

Elevation. Two-dimensional graphic representation of the design, location, and certain dimensions of the project, or parts thereof, viewed in a vertical plane from a given direction; also, distance above or below a prescribed datum or reference point.

Engineer. A professional engineer who, by reason of special knowledge of the mathematical and physical sciences and the principles and methods of engineering analysis and design, acquired by professional education or practical experience, is qualified to practice engineering as attested by his registration as a professional engineer.

Engineer-in-training. Designation prescribed by statute for a person qualified for professional engineering registration in all respects except the required professional experience.

Engineer of record (EOR). Prime professional engineer or organization who is legally responsible for the engineering design.

Environmental assessment. Document that identifies and asesses the influence of a proposed project on significant economic, social, and physical elements of the environment.

Environmental impact statement

1. Report on the anticipated impact of a proposed project on surrounding conditions. Environmental, engineering, esthetic, and economic aspects are included.

2. A detailed document meeting the goals of the National Environmental Policy Act, discussing alternatives to avoid or minimize adverse impacts or enhance the quality of the human environment.

Estimate. Opinion, based on technical expertise and experience, of the probable construction cost of a project or any part of a project.

Estimate proposals. Determination of probable project construction cost prepared by the contractor as a basis for his proposal.

Estimates—construction. Forecasts of probable project construction cost to be used as a guide for the owner and the design professionals in preparing contract documents.

Examination of documents and site. Description of the bidder's responsibilities to review the bidding documents and his visit to the site before the bid is prepared.

Execution of contracts between the owner and contractor. List of requirements and conditions of contract execution, including the proper preparation and signators, seals, and possible notary attesting.

Expert system. Computer system that separates the decision-making process (the inference mechanism) from the body of knowledge and logic upon which the decisions are made (the knowledge base).

Extras. Construction costs over and above those specified in the agreement. An extra does not exist unless it is acceptable to the owner.

Fabricator. Entity responsible for furnishing fabricated construction components.

Fast-track construction. Process whereby design and construction are performed simultaneously. As design is completed for a portion of the project, construction work commences and proceeds for that particular phase. Thus site work will commence as soon as drawings and specifications are available, while at the same time structural, mechanical, and electrical designs are under way and construction proceeds in the field under a temporary permit.

Field engineer. Resident project administrator.

Field representative. Person in the field designated to represent a responsible party during the construction phase of a project.

Final acceptance. Owner's acceptance of a completed project from the contractor.

Final project cleanup. Final removal of all debris, including the removal of all grease, dust, dirt, stains, labels, fingerprints, and other foreign materials from interior and exterior surfaces.

Fixed limit of construction cost. Maximum allowable cost of the construction work as established by the owner.

General Conditions. Standard form or prepared document that states the general conditions of the contract that describes the rights, responsibilities, and the relations of the parties to the contract.

General contract. Agreement between the owner and a general contractor for the construction of a project.

General requirements. General Conditions, amendments and supplements to General Conditions, special conditions, and alternates are grouped together under this title.

Geological practice. Any professional service or work requiring geological education, training, and experience, and the application of special knowledge of the earth sciences to such professional services as consultation, evaluation of mining properties, petroleum properties, and ground water resources, professional supervision of exploration for mineral natural resources including metallic and nonmetallic ores, petroleum, and ground water, and the geological phases of engineering investigations.

Geologist. A person, not of necessity an engineer, who by reason of his special knowledge of the earth sciences and the principles and methods of search for and appraisal of mineral or other natural resources acquired by professional education and practical experience is qualified to practice geology as attested by his registration as a professional geologist. A person employed on a full-time basis as a geologist by an employer engaged in the business of developing, mining, or treating ores and other minerals is not deemed to be engaged in "geological practice".

Government regulations. General rules relating to a subject made by a governmental body that has been legally established and legally conferred with the power to make such regulations.

Guarantee. Legally enforceable assurance of the duration of satisfactory performance or quality of a product or work.

Guaranteed maximum price (GMP). Sum established in an agreement between the owner and constructor as the maximum cost of performing specified work on the basis of cost and labor and materials plus overhead expenses and profits.

Guaranteed maximum price construction management (GMPCM). Construction management by a constructor who, at some point in design, provides the owner with a guaranteed maximum price.

Impact analysis. Determination, not as formal as an environmental assessment, that identifies and assesses the influence of a proposed project on significant economic, social, and physical elements of the environment.

Indemnification. Collateral contract or assurance by which an individual agrees to secure another against an anticipated loss or to prevent the other individual from being damaged by the legal consequences of an act of forbearance on the part of one of the parties or of some third party.

Information available to bidders. Supplementary information, not usually included in the contract documents, provided for use by bidders, subject to their own interpretation and evaluation.

In-place unit cost estimates. Forecasts of project construction cost based on the cost of units of construction installed.

Inspection. Examination of work completed or in progress to determine its compliance with contract requirements. The architect/engineer ordinarily makes inspections as agreed upon in the contract, of a construction project. Inspections should be distinguished from the more general observations made by the architect-engineer on visits to the site during the progress of the work. The term is also used to mean examination of the work by a public official, union representative, the owner, the owner's representative, utility representatives or others.

Inspection punch list. List of items of work prepared by the architect/engineer with the owner's representative, to be completed or corrected by the contractor by a given date.

Instructions to bidders. Instructions contained in the bidding requirements for preparing and submitting bids for a construction project.

Invitation to bid. Public or private notice to prime bidders that they may bid on a proposed job and to sub-bidders and material suppliers that they may bid on their part of the work. It may be directed toward certain selected bidders, or it may be a notice to all interested and qualified bidders.

Invited bidders. Bidders selected from an approved list, as the only ones from whom bids will be received.

Job site. Place or location of the work.

Joint venture. Collaborative undertaking by two or more persons, groups, or organizations for a specific project or projects.

Labor and material payment bond. Bond guaranteeing to the owner that the contractor will satisfy all obligations incurred by him on the project.

Landscape architect. A person who, by reason of his professional education, practical experience, or both, is qualified to engage in the practice of landscape architecture as attested by his registration as a landscape architect.

Landscape architectural practice. The performance of professional services such as consultations, investigation, reconnaissance, research, planning, design, or responsible supervision in connection with the development of land and incidental water areas where the dominant purpose of such services is the preservation, enhancement or determination of proper land uses, natural land features, ground cover and planting, naturalistic and esthetic values, the settings and approaches to buildings, structures, facilities, or other improvements, natural drainage and the consideration and the determination of inherent problems of the land relating to erosion, wear and tear, light or other hazards. This practice can include the location and arrangement of such tangible objects and features as are incidental and necessary to the purposes outlined in this paragraph, but does not include the design of structures or facilities with separate and self-contained purposes for habitation or industry, such as are ordinarily included in the practice of engineering or architecture; and does not include the making of cadastral surveys or final land plats for official recording or approval, nor manditorially include planning for governmental subdivisions.

Land surveying. Within the meaning and intent, any service comprising the determination of the location of land boundaries and land boundary corners; the preparation of maps showing the shape and area of tracts of land and their subdivisions into smaller tracts; the preparation of maps showing the layout of roads, streets, and rights of way of same to give access to smaller tracts; and the preparation of official plats, or maps, of said land in the state.

Land surveyor. Person who has a license to engage in the practice of surveying tracts of land for the determination of their correct locations, areas, boundaries, and description, for the purpose of conveyancing and recording, or for establishment or reestablishment of boundaries and plotting of lands and subdivisions.

Large-scale details. Construction sections and large-scale details of individual areas on the working drawings that clarify the intent.

Letter agreement. Letter as a temporary measure stating the terms of future formal agreement between two parties, usually prepared to be signed by the party receiving the letter, to indicate acceptance of those terms as legally binding.

Letter of intent. Letter signifying intention to enter into a future formal agreement and setting forth the general terms.

Liens. Legal claims against an owner for amounts due those engaged in or supplying materials for the construction of the building.

Life-cycle cost. Total cost of developing, owning, operating, and maintaining a constructed project for its economic life, including its fuel and energy costs.

Life-cycle costing. Systematic evaluation of life-cycle cost and comparison of alternative systems.

Liquidated damages. Agreed-to sum chargeable against the contractor as reimbursement for damages suffered by the owner, because of the contractor's failure to fulfill contract obligations.

Loss prevention. Use of safety programs and insurance by a constructor to mitigate financial losses resulting from loss of life and personal injuries on a construction project.

Low bid. Bid stating the bid price, including selected alternates, and complying with all bidding requirements, that will be determined by review of all bids to be the low bid.

Low bidder. Bidder who submits the lowest bid price after alternates have been considered but who does not become the lowest responsible bidder until so qualified.

Lowest acceptable bona fide bid. Proposal reflecting the lowest dollar amount and fulfilling all the stipulated requirements.

Lowest-responsible bidder. Bidder who submitted the lowest bid price and has shown himself to be qualified as a "responsible bidder."

Lump sum contract

1. Contract with one fixed basic contract price.
2. Total contract price to deliver a completed project.

Maintenance. Corrective and preventive program, incorporating a record-keeping system, to evaluate a constructed facility with respect to its intended purpose and the arrangement for any warranty or general repairs or replacements to sustain the facility.

Manage. To direct or administer specific activities.

Management. To control, supervise, or guide by direction or regulation, an enterprise, organization, or activity.

Material suppliers. Person or business that bids on or supplies material for the job but does not supply the labor of installing it.

Material testing standards. Standards established such as those published by the American Society for Testing and Materials. (ASTM does not perform physical testing.)

Mechanic's lien. Lien on real property created by statute in all states in favor of persons supplying labor or materials for a building or structure for the value of labor or materials supplied by them. In some jurisdictions a mechanic's lien also exists for the value of professional services. Clear title to the property cannot be obtained until the claim for the labor, materials, or professional services if resolved.

Metes and bounds. Boundaries, property lines, or limits of a parcel of land, defined by distances and compass directions.

Minor change in the work. Change in the work of a minor nature, not involving an adjustment in the contract sum or contract time. It may be authorized by written directive from the owner.

Modification to the contract documents. Written amendment to the contract signed by both parties; change order; written or graphic interpretation issued by the architect/ engineer; written authorization for a minor change in the work from the owner.

Modular measure. System of measurement based upon a stipulated unit.

Module. Repetitive dimensional or functional unit used in planning or constructing buildings or other structures; distinct component forming part of an ordered system.

Mutual responsibility of contractors. Obligation of a contractor for damages caused by any separate contractor.

Negligence. Failure to exercise the degree of care that a reasonable and prudent person would exercise to protect the interest of others.

Noncollusion affidavit. Notarized statement by a bidder that he has prepared his bid without collusion of any kind.

Nonconforming work. Work that does not meet the requirements of the contract documents.

Notice to proceed

1. Written communication issued by the owner to the contractor authorizing him to proceed with the work and establishing the date of commencement of the work. It usually occurs immediately after all financial arrangements have been made with the lender.

2. Document issued to the contractor by the architect-engineer, owner, or lender, which informs the contractor he may proceed with the work of the contract (usually already signed) on a given date. "Contract time" shall begin on that given date with that day counting as the first day. This is only used when the time for the start of work must be delayed for some reason. Usually the contract directs the contractor to begin on the contract date.

Objectives. Detailed and specific statement by the owner of his requirements for a project, including function, operation, schedule of design and construction, construction cost, technical aspects, quality, esthetics, and administrative, fiscal, and management requirements.

Occupancy permit. Certificate of occupancy issued by local authorities, after a final inspection of the project.

Office practice. Standardized program for a design or constructor firm, covering general management of the firm, organization for projects, owner relationships, office procedures, filing and storing material, and operational rules.

On-site observation. Design professional's representative, whose primary function is to observe the progress at the site and report to the office any situation or condition that needs immediate attention. The office will report to the owner after problems have been resolved.

Opening time. Hour set for the bid opening.

Option. Choice given to the contractor to provide specified alternates without prior approval of the owner or architect/engineer.

Or equal. Phrase used in a specification after a brand or trade name to indicate that a substitution will be considered when requested by the contractor.

Outline of construction procedure. Written or graphically explained, the procedure of construction prepared by the contractor, usually by a bar chart of scheduled dates by trades or a critical path chart.

Outline specifications. Specifications containing adequate technical information but written in an abbreviated manner.

Owner. The party signing the contract with the contractor is usually the owner of the structure to be constructed.

Owner/developer. Individual or group that combines the functions of both owner and developer.

Owner's agent. One empowered by the owner to act in his behalf.

Package dealer. Person or company assuming a single contract for total responsibility for the delivery of a complete project, including all services, such as architectural, engineering, construction, furnishings, and financing, if required.

Partnership practice. A partnership is an association of two or more persons who direct an organization as co-owners. Each partner assumes a legal responsibility for his associates, the associates of each partner, and the work of the employees.

Payment request. Application by the architect/engineer for payment by the owner.

Penal sum. Amount to be paid by the contractor of the bonding company, based on an agreement to complete a phase of work.

Percentage agreement. Agreement between the owner and the architect/engineer under which the professional fee is based on a percentage of the construction cost of a project.

Perspective drawing. Graphic representation of the project, or part thereof, as it would appear three dimensionally.

Pert schedule. Acronym for "project evaluation review technique." It charts the activities and events anticipated in a work process.

Plan. Two-dimensional graphic representation of the design, location, and dimensions of the project, or parts thereof, seen in a horizontal plane viewed from above.

Plan rooms. Trade offices and firms distributing job information to contractors set aside plan rooms for contractors to inspect the drawings and specifications so that if interested they can prepare a bid for their work. Advertisements for bids or notice to bidders can document where these rooms are available.

Power-of-attorney. Legal authority for a person to sign for and obligate another person or corporation. The person so impowered is an "attorney-in-fact." Also, legal documents that give this power. The document must be signed by the person or an officer of the corporation conferring this power.

Practice of architecture. Includes any professional service, such as consultation, investigation, evaluation, planning, architectural design, or supervision of construction, if so authorized by the owner, in connection with the construction of any private or public buildings, structures, projects, or the equipment thereof, or addition to or alterations thereof, wherein the public welfare or the safeguarding of life, health, or property is concerned or involved.

Practice of professional engineering. Includes any professional service, requiring the application of engineering principles and data, wherein the public welfare or the safeguarding of life, health, or property is concerned.

Prebid conference. Conference prior to receipt of bids, arranged by the owner and attended by the design professionals and prospective constructors, for the purpose of explaining the project, the design professional's intent, and the owner's objectives and expectations, and for responding to questions by the bidders.

Preconstruction meeting. Meeting arranged by the owner, prior to construction, for the design professional, constructor, and subcontractors to facilitate an uncomplicated start-up process.

Preliminary estimates. Forecasts of probable project construction cost; budgets made before construction documents are complete.

Preoperational test. Test conducted prior to operation of a component of a constructed system to determine if it meets performance requirements.

Prequalification of bidders. Process of investigating and qualifying bidders as acceptable contractors, prior to the award of contracts, on the basis of their skills, integrity, and responsibility relative to the project contemplated.

Prime bidder. One who submits a bid directly to the owner.

Prime contractor. Contractor having a contract with the owner.

Producers. Manufacturers, processors, or assemblers of building materials and equipment.

Professional basic services. Professional basic services consist of the following five phases: schematic design, design development, construction documents, bidding or negotiations, and construction contract administration.

Professional engineer. Designation reserved, usually by law, for a person or organization professionally qualified and duly licensed to perform such engineering services as structural, mechanical, electrical, sanitary, or civil engineering.

Professional engineering practice. Rendering or offering to render any service that requires or would require the application of engineering in any of its branches and fields.

Professional practice. Registered practice of one of the environmental design professions, in which services are rendered within the framework of recognized professional ethics and standards and applicable legal requirements described in state statutes.

Program. Written statement of the owner's conditions, criteria, and requirements of a project.

Progress payments. Payment requests prepared on prescribed forms made during progress of the work on account of work completed and/or materials suitably stored.

Progress schedule. Diagram showing proposed and actual times of starting and completion of the various branches of the work included in a project.

Project

1. At a given location, the contracted work that has been and will be done on the site for the owner by one or several contractors.

2. Detailed documents for the total construction, of which the work performed under the contract may be the whole or a part.

Project cost. Total cost of the project, including professional compensation, land costs, furnishings, equipment, financing, and other charges, as well as the construction cost.

Project evaluation. Critical evaluation by project team members during both design and construction to assess design, schedule, objectives, costs, legal ramifications, and trends that impact cost, quality, and schedule.

Project inspector. Person who provides continuous inspection of the project.

Project management. Planning, organizing, staffing, directing, controlling and coordinating design and construction activities for a construction project under the direction of a single project manager, who has direct responsibility and represents the owner.

Project manager. Person or organization, representing the owner, responsible for overall coordination and management of the project activities. The manager may be a member of the owner's, design professional's, or constructor's staff, or an independent contractor employed by the owner.

Project manual. Documents prepared by the design professional containing the bidding documents, conditions of the contract, and specifications.

Project photographs. Photographs taken at the project at regular time periods to indicate the progress of the work.

Project plan. Work activity diagram and other documents depicting features of a project's requirements.

Project schedule. Diagram, graph, or written listing showing proposed and actual times of starting and completion of various elements of design or construction.

Project team. Entities primarily responsible for completing a constructed project: the owner, design professional, and constructor.

Proposal form. Bid form customarily bound into the project manual form recommended by the instruction to bidders.

Proprietorship. Professional office consisting of one principal.

Protection of work and property. Protection by the contractor of his work, employees, and equipment; the owner's property; adjacent property; and the public.

Provide. Furnish materials or equipment and the labor to install complete.

Punch list. List prepared by inspection of the contractor's uncompleted or uncorrected work.

Qualifications. Attributes of an individual or entity that are judged or reviewed to determine conformity to predetermined standards and requirements.

Quality. Conformance to predetermined requirements.

Quality assurance. All planned and systematic actions necessary to provide adequate confidence that a structure, system, or component will perform satisfactorily and conform with project requirements.

Quality control. Specific procedures involved in the quality assurance process.

Quality management. Control, supervision, or guidance of a quality assurance program.

Quantity and cost estimates. Forecasts of probable project construction cost based on the sum of the estimated costs of materials, labor, equipment, overhead, and profit.

Quantity survey. Detailed analysis and listing of all items of material and equipment necessary to construct a project, with tentative applicable unit costs.

Quantity surveyor. One who prepares a complete list of quantities of materials in conjunction with estimated unit costs.

Quantity surveys. Bills of materials prepared by specially qualified estimators.

Quotation. Price (bid proposal) quoted by a contractor, subcontractor, material supplier, or vendor to furnish materials or labor, or both.

Real estate broker. Includes a person, firm or corporation who, for a fee, commission, or other valuable consideration, or by reason of promise or reasonable expectation thereof, lists for sale, sells, exchanges, buys or rents, or offers or attempts to offer a sale, exchange, purchase, or rental of real estate or an interest therein.

Reasonable interpretation. Process of discovering and expounding the meaning of a written document with reasonable care.

Record drawings. Drawings kept at the site; information secured at the site to be transferred to permanent as-built drawings.

Records. Written account documenting data, activities, transactions, and memorandum of verbal communications, and usually including the contract documents.

Registered architect. Architect licensed in good standing and legally authorized to practice architecture in the state of registration.

Rendering. Perspective or elevation drawing of an interior or exterior view with an artistic delineation of materials, usually in color.

Resident project representative (RPR). Individual representing the owner, sometimes selected from the design professional's firm, who administers the construction contract and monitors progress and relationships among the project site personnel.

Responsibilities. Those duties and activities of project team members defined by a contractual arrangement and the legal obligations to the public required by both statute and common law.

Responsible bidder. Bidder who is acceptable to the owner and has shown himself to be qualified to meet the requirements as set forth in the instructions to bidders.

Restricted list of bidders. Select list of bidders established by lender or owner.

Retainage. Sum withheld from each payment to the contractor in accordance with the terms of the owner-contractor agreement.

Risk transfer. Contractual clauses that transfer the risk of project team members to other parties by means of bonds or insurance.

Sample. Material or assembly submitted by the contractor to the architect/engineer prior to manufacture or delivery to the project.

Samples. Physical examples submitted by the contractor for review and approval that illustrate materials, equipment, assembly, or workmanship, and that establish standards by which the work will be judged at the project.

Schedule of values. Statement furnished to the lender and the owner by the contractor reflecting the amounts to be allotted for the principal divisions of the work. It serves as a guide for reviewing the contractor's periodic application for payment.

Section in a specification. Subdivision of the "Division." It usually covers the work of one trade only but does not outline its jurisdictional direction as to trade selection.

Selected bidder. Bidder selected by the owner for discussions relative to the possible award of a contract.

Selection committee. Committee, established by the owner and guided by preestablished criteria and administrative policy, comprised of qualified individuals, including professionals, to make recommendations on selection of design professionals after conducting necessary investigations, interviews, and inquiries.

Select list of bidders. List of bidders who may submit a bid. The list is usually prepared by the architect-engineer, lender, and owner. It is sometimes called a "restricted list of bidders."

Semidetailed and detailed estimates. Statements of probable project construction cost prepared by specially qualified estimators or select contractors.

Separate contract. Contract between the owner and a contractor other than the general contractor for the construction of portions of a project or specialized work on the project.

Shop drawings

1. Drawings, diagrams, illustrations and picture cuts, schedules, performance charts, brochures, and other data prepared by the contractor or any subcontractor, manufacturer, supplier, or distributor, that illustrate how specific portions of the work will be fabricated and/or installed.

2. Usually larger scale drawings by a contractor or shop or materials supplier to illustrate how specific portions of the work will be fabricated and/or installed.

Single contract. Contract for building construction where all work is contracted under the responsibility of a single or prime general contractor.

Site. Geographical location of the project, usually defined by legal boundary lines.

Special conditions

1. Added section to the conditions of the contract, other than General Conditions and supplementary conditions, which may be prepared for a particular project.

2. Document that states the exceptional conditions that are general enough to cover all of the trade divisions.

Specifications

1. Written technical requirements of the job consisting of the CSI divisions and sections format.

2. Part of the contract documents consisting of written descriptions of administration and procedure, products, and execution.

Speculative builder. One who engages in the construction of building projects for sale or lease.

Square foot estimates. Cost estimates based on usable floor space.

Standard. Prescribed set of rules, conditions, or requirements concerned with the definition of terms; classification of components; delineation of procedures; specification of materials, equipment, performance, design, or operation; descriptions of fit and measurement of size; or measurement of quality and quantity in describing materials, products, systems, services, or practices.

Start-up. Preparing the project or facility for occupancy or use and testing the systems in that facility for operation.

Statements of probable construction cost. Cost forecasts prepared by reliable sources for the guidance of the owner and issued before construction contracts are awarded.

Stipulated sum agreement. Contract in which a specific amount is set forth as the total payment for the performance of the contract.

Stipulated sum contract. Lump sum amount of a contract.

Subbidder

1. One who tenders to a bidder or a prime contractor a proposal to provide materials and/or labor.

2. One who submits a bid to a prime bidder or other subcontractor.

Subconsultant. Person or organization who has a direct contract with the design professional.

Subcontract. Agreement between a separate contractor and a prime contractor.

Subcontractor

1. Individual or organization who has a direct contract with a prime contractor to perform a portion of the work at the site.

2. Subcontractors are only those having a direct contract with the contractor and who furnish material worked to a special design according to the plans and specifications of the work. The term does not include those who merely furnish material.

3. Separate contractor for a portion of the work.

Subcontractor bond. Bond given by the subcontractor to the prime contractor, guaranteeing performance of his contract and payment for all labor and materials.

Subcontractor list. List of subcontractors whose proposals were used by a prime contractor when preparing his bid; list of subcontractors proposed to be employed by the prime contractor, to be submitted to the architect/engineer for approval after the contract award.

Subrogation

1. Transfer of financial responsibility and right of recovery from one party to another.

2. Substitution of one person for another with respect to legal rights such as a right of recovery. Subrogation occurs when a third person, such as an insurance company, has paid a debt of another or claim against another and succeeds to all legal rights that the debtor or person against whom the claim was asserted may have against other persons.

Substantial completion. The date of substantial completion of a project or specified area of a project is the date when the construction is sufficiently completed, in accordance with the contract documents as modified by any change orders agreed to by the parties, so that the owner can occupy subject to permission from the local authorities the project or specified area of the project for the use for which it was intended.

Substitution. Material, equipment, or process offered in lieu of and as being equivalent to a specified material or process.

Substitutions. List of requirements and procedures by which a bidder will be permitted to request permission to substitute materials or equipment and/or methods of construction before the proposal has been submitted.

Subsurface investigation. Soil boring and sampling program, together with the associated laboratory tests supplied by the owner, necessary to establish subsurface profiles and the relative strengths, compressibilty, and other characteristics of the various strata encountered within the depths likely to have an influence on the design of the substructure for the project.

Successful bidder. Bidder to whom the contract is awarded.

Superintendence. Work of a contractor's representative at the site. No professional service normally rendered by an architect should be construed to be superintendence.

Supervising architect-engineer. Architect or engineer who guides, reviews, or approves the work of other architects or supervisory personnel; engaged to be the job representative.

Supervision. guidance of the work by the contractor's personnel during the construction phase.

Supplementary general conditions. One of the contract documents, prepared by the architect/engineer that may modify provisions of the General Conditions of the contract.

Supplier. Person or company who supplies materials, equipment, and prefabricated materials or parts but does not provide the job labor to install.

Survey. Boundary and/or topographic mapping of a property; measuring of an existing building; analyzing a building for existing use of space; determining the owner's requirements for a proposed building project; investigation and reporting of any critical information for a project.

System construction. Combining prefabricated assemblies, components, and parts into single integrated units utilizing industrialized production, assembly, and methods.

Table of contents. List of all parts of the project manual.

Team. Individuals or entities associated together to execute design and construction.

Time of completion. Number of days (calendar or working) or the actual date by which the work is required to be completed.

Title block. Area of a sheet of working drawings in which is recorded the name of the project, the name and address of the architect/engineer, the name and address of the owner, the sheet number, and the total of sheets comprising the set.

Title page. First page in the project manual on which the title of the job is given and the name of the owner, architect/engineer, and consultants.

Title sheet. First sheet in a set of working drawings, containing a variety of information.

Topographic survey. Configuration of a surface, including its relief and the locations of its natural and man-made features, usually recorded on a drawing (site plan) showing surface variations by means of contour lines and indicating height above or below a fixed datum. Original information provided by registered surveyor.

Trade. Classification or type of work usually done by workers who restrict themselves to this type of work, established by jurisdictional agreements.

Transmittal. Form used to transmit items of a standard nature between the parties on the project.

Uncorrected work. Work that must be corrected; work that is not in conformance with the contract documents or the local ordinances and codes.

Unit prices. Amounts stated by the contractor in his proposal as prices per unit of measurement for materials to be added to or omitted from the work.

Upset price. Amount established as a maximum cost to perform a specified work.

Utility supplies. Utility services provided to a project by vendors or suppliers of such services.

Value engineering. Organized, creative approach to identify unnecessary costs efficiently, especially those that do not measurably contribute to a project's quality, utility, durability, or appearance, or to the owner's requirements.

Vendor. Person or company who represents a manufacturer and supplies materials and equipment but does not provide job labor to install.

Verbal agreements. No verbal agreement or conversation with any officer, agent, or employee of the owner or the contractor or his representatives, either before or after execution of the contract, should be made to infer that the intention is to modify any of the terms or obligations contained in any of the documents comprising the proposed contract without being confirmed in writing.

Volume estimates. Forecasts of probable project construction cost based on unit cost per cubic foot.

Warranty. Contractor's or manufacturer's guarantee of the quality, workmanship, and performance of his work or equipment.

Withholding payments. Retaining certain sums, otherwise due the contractor, from payments to the contractor because of work that is uncompleted or uncorrected, or for other legal reasons.

Work

1. All labor necessary to produce the construction required by the contract documents, and all materials and equipment incorporated or to be incorporated in such construction.

2. Materials, products, equipment, and/or labor required for a project.

Working drawings. Descriptive contract document, prepared by the architect/engineer illustrating the design, location, and dimensions of the elements of the project.

Written notice. Written notice is duly served if delivered in person to the individual, a member of the firm, or an officer of the corporation for whom it is intended, or if delivered or sent by registered mail to the last business address known to him who gives the notice.

Zoning approval. Zoning approval is issued by appropriate governmental authority, authorizing land to be used for a specific purpose, building, or use.

REAL ESTATE

Abandonment. The voluntary relinquishment of rights of ownership or another interest, such as an easement, by failure to use the property coupled with an intent to abandon (to give up interest).

Abeyance. Suspension. In property, where there is no existing person in whom a property can vest, it remains in abeyance until a proper owner appears.

Abode. Home; residence; place of habitation.

Abrogate. To annul; to abolish; to destroy.

Absolute fee simple title. Title that is unqualified; best title one can obtain.

Abstract of title

1. Summary of the conveyances (such as deeds or wills) and legal proceedings, giving names of parties, description of the land, and agreements, arranged to show the continuous history of ownership.

2. Condensed title of a parcel of land, documenting conveyances, interests, lists, and charges against the property.

3. Condensed history of the title, consisting of a summary of the various links in the chain of title, together with a statement of all liens, charges, or encumbrances affecting a particular property.

4. History of the recorded instruments, such as deeds, wills, and lawsuits, affecting ownership of a piece of property.

Abut. To touch, border on, or to be contiguous.

Acceleration. Shortening of a time for the vesting of an expected interest, performance on a contract, or payment of a note.

Acceleration clause. Clause in a mortgage, land purchase contract, or lease stating that upon default of a payment due, the balance of the obligation should at once become due and payable.

Acceptance. Voluntarily agreeing to the price and terms of an offer. Offer and acceptance create a contract.

Accession

1. Process of acquiring title to improvements made on one's land without approval before installation.

2. Right to any natural or artificial addition to one's property.

Accessory buildings. Buildings or structures on the same lot with the main or principal building.

Access right. Right of an owner to have ingress and egress to and from his property.

Accretion

1. Addition to the land through natural causes, usually by change in water flow.

2. Gradual and imperceptible accumulation of land, generally by water at a shoreline washing soil onto the shore.

3. Physical addition to land by natural causes, such as deposits left by movement of waters.

Acknowledgment. Formal declaration made before a notary public or other person similarly empowered by the signatory to an instrument as to the genuineness of the signature.

Accrued depreciation

1. Total depreciation accumulated in a property up to the present.

2. Difference between the cost or replacement new as of the date of the appraisal and the present appraisal value.

Acquisition. Process of acquiring or purchasing a piece of property.

Acre. Measure of land 160 square rods, 4840 sq yd, 43,560 sq ft.

Action to quit title. A court action to establish ownership to real property.

Addition. Extension or increase in floor area or height added to the original building or structure.

Administrator. Person appointed by the court to administer the estate of a deceased person who left no will, that is, who died intestate.

Advance commitment. Agreement by which the lender agrees to grant a mortgage loan, at a specified interest rate, to the home buyer. It becomes effective when the buyer takes title to the property.

Advance fee. Fee paid in advance of any service rendered in the sale of a property or in obtaining a loan.

Ad valorm. Tax according to valuation.

Adverse possession

1. Occupancy or enjoyment of property in spite and in defiance of someone else's legal title. If such possession continues for 20 years, legal title can be claimed by the one in possession (depending on state statutes). Adverse possession must be actual possession, continuous, visible, and hostile to the one who has legal title.

2. Right of an occupant of land to acquire title against the real owner, where possession has been actual, continuous, hostile, visible, and distinct for the statutory period.

Affadavit. Statement or declaration reduced to writing and sworn or affirmed before some officer who has authority to administer an oath or affirmation.

Affiant. One who makes an affidavit.

Age-life method. Method based on empirical tables purportedly reflecting past experience with regard to various types of structures.

Agent. One who represents another from whom he has derived authority.

Agrarian. Relating to land, or to a division or distribution of land.

Agreement of sale

1. The word "agreement" is identical in meaning with the word "contract."

2. Written agreement whereby the purchaser agrees to buy certain real estate and the seller agrees to sell on terms and conditions set forth therein.

AIR. American Industrial Real Estate Association.

Air-conditioning. Process of treating air so as to control simultaneously its temperature, humidity, cleanliness, and distribution.

Air rights

1. Rights granted by fee simple, lease agreements, or other conveyance to occupy or use all or any portion of the space above the ground.

2. Ownership of the right to use, control, or occupy the airspace over a designated property.

Alienation. Transfer of real property by one person to another.

Alley. Public way that affords a secondary means of access to abutting property.

Allotment. Parcel of land that has been divided into small parts; subdivision.

Alluvium. Soil deposited by accretion; increase in land on the shore or bank of a river due to change in flow of a stream.

ALTA. American Land Title Association.

Alteration. Any change, rearrangement, or modification of an existing building or lot, other than additions or extension.

Amenities

1. Qualities that attach to property and the benefits derived from ownership that are other than monetary —architectural excellence, scenic beauty, and social environment.

2. Those qualities that increase the pleasure of ownership and are not necessarily related to monetary values.

3. Satisfaction of enjoyable living to be derived from a home; beneficial influence arising from the location of a property.

Amortization

1. Liquidation of a financial obligation on an installment basis.

2. Paying back a debt by installment payments. Most of today's mortgages are monthly amortizing, which means interest is paid monthly along with repayment of the principal.

3. Extinguishing a debt by equal payments or regular intervals over a specific period of time.

Amortize. To reduce a debt by regular payments of both principal and interest.

Ancillary. Portion of a document that is subordinate to or in aid of another primary or principal one.

Annuity. Sum of money or its equivalent that constitutes one of a series of periodic payments.

Apartment. Room or suite of rooms that is occupied by one family for living and sleeping purposes.

Apartment hotel. Building containing both dwelling units for permanent tenants and rooming units for transient guests.

Apartment house. Building that is designed, built, rented, leased, let, or hired-out to be occupied as the home or residence of three or more families living independently of each other in dwelling units.

Appraisal

1. Estimate of quantity, quality, or value; process through which conclusions of property value or property facts

are obtained; also, commonly the report setting forth such estimate and conclusions.

2. Written statement of estimated value made by a trained and experienced person.

Appraisal by capitalization. Estimate of value by capitalization of productivity and income.

Appraisal by comparison. Estimate of value by comparison with the sale prices of other similar properties.

Appraisal by summation. Adding together of parts of a property separately appraised to form the whole. For example, the value of the land considered as vacant is added to the cost of reproduction of the building less depreciation.

Appraisal inventory. Detailed tabulation of the separate items comprising property that is included in an appraisal report and evaluated by the appraiser.

Appraisal report. A written report by an appraiser containing is opinion as to the value of the property and all of the factual data supporting his opinion.

Appraisal surplus

1. Difference between book values and actual values when the latter is in excess of the former as established by an appraisal.

2. Excess of appraised values over book values.

Appraiser. Professional who, for a fee, estimates the value of property.

Appreciation

1. Increase in the value of property due to factors unrelated to the improvements on the land.

2. Increased value of property due to economic or related causes. "Appreciation" is the antonym of "depreciation" where the latter is used to denote shrinkage in value. The term is also applied to gain in the upgrading of physical property.

Appreciation rate. The "index figure" used on computing the cost of reproduction new as of a different date or at a higher price level.

Appurtenance

1. That which belongs to something else; an adjustment; an appendage; something annexed to another thing more worthy as principal, and which passes as incident to it, as a right-of-way or other easement to land or an outhouse, barn, garden, or orchard to a house.

2. That which is attached to the land so as to become a part thereof. Buildings and improvements are typical appurtenances; a property right may also be one.

Aquatic rights. Individual's right to the use of the waterways for the purpose of navigation, fishing and other regulated uses.

Arbitration. Method, in lieu of litigation, for settling claims or disputes between parties to a contract; an arbitrator or a panel of arbitrators selected for specialized knowledge in the field in question, hears the evidence and renders a decision. (See Construction Industry Arbitration Rules of the American Arbitration Association AAA).

Arcade. Range of arches supporting a roofed area along with a column structure, plain or decorated, over a walkway adjacent to or abutting a row of retail stores on one side or both.

Arterial highway. Major highway with heavy traffic count. Usually connecting several largely populated areas.

As-is condition. Premises that include the land and the improvements, accepted by a buyer or tenant in the condition existing at the time of sale or lease.

Assemblage. Cost or estimated cost of assembling two or more parcels of land under a single ownership over the normal cost or current market price of the parcels held individually.

Assessed valuation. Assessment of real estate by a unit of government for taxation purposes.

Assessed value

1. Value set on property by local assessors, normally a fraction of the actual market value. The property tax imposed is a rate of the assessed value.

2. Estimated worth of a piece of property as ascertained by an authorized appraising agency.

3. Value of real or personal property as established by an assessor for the purpose of levying taxes.

Assessment

1. Determination of the value of property for the purposes of fixing a tax. The term is also used to denote a special tax on property based on benefits to the property itself.

2. Valuation of property for taxation; also, the value so assigned.

3. Charge against real estate made by a unit of government to cover the proportionate cost of an improvement, such as a street or sewer.

Assessment ratio. For purposes of property taxation, the relationship between the assessed value of property and the same property's market value.

Assessor. Public official who assesses property for taxation.

Assets. Property possessed by one ownership.

Assignee. Person to whom an agreement or contract is assigned.

Assignment

1. Transfer of rights from one person to another.

2. Written transfer of interest in any instrument.

3. Method or manner by which a right, a specialty, or contract is transferred from one person to another.

Attachment. Taking into custody property belonging to a person involved in a proceeding. Property attachment is sometimes the beginning of a lawsuit against someone who is not within the state, but whose property is within the jurisdiction of the court.

Attached building. Building that has any part of its exterior or bearing walls in common with another building or that is connected to another building by a roof.

Attic. Space between the ceiling construction and the roof construction, and not used for habitation.

Attorney-in-fact

1. Person authorized to act for or in behalf of another person or organization, to the extent prescribed in a written instrument known as a power-of-attorney.

2. One who is appointed in writing to perform a specific act for and in place of another.

Auction. Public sale of property to the highest bidder.

Authentication. Certification of a document by the signature of an officer whose seal is usually affixed to validate the procedure.

Authorization to sell. Another term for "listing."

Avigation rights. Avigation is aerial navigation. Avigation rights are rights to use the air above the land.

Avulsion. Removal of land from one owner to another when a stream or river suddenly changes its direction of flow.

Bachelor apartment. Apartment designed for occupancy by one or more persons, primarily limited in area and containing one large room for sleeping, cooking, eating, and living, with bathroom facilities and closet space.

Backfill. Approved material used to replace ground removed during a construction job. Buyer must check for backfill used other than approved materials.

Backwater. Waterway that, because of a dam or other natural obstruction, is stopped in its course or flows toward its source.

Balloon payment. Amount due when a note or mortgage that is in excess of normal installment payments is due.

Bankruptcy. Legal proceeding under federal statutes whereby an insolvent debtor may be ruled incapable of meeting his obligations and his properties may be sold or distributed to satisfy his creditors. A petition in bankruptcy may be filed by either the debtor or the creditors.

Bargain. Mutual undertaking, contract, or agreement. Contract or agreement between two parties, the one to sell goods or lands, and the other to buy them.

Bargain and sale deed. Deed which conveys the property for valuable consideration.

Basement. The portion of a building that is wholly or partly below grade or partly below and partly above grade and is not used for habitation.

Base title. The result of a review and examination of title for the internal use by a developer, for the anticipation for future sales or subdividing of a large area.

Basic floor area. Total amount of gross floor area a building contains, expressed as a percentage of the total area of the property.

Bathroom. Room in a dwelling unit with the required ceiling height, ventilation, wall and floor finishes, and not less than three plumbing fixtures of which there must be a water closet, lavatory, and a bathtub or shower.

Bench marks

1. Identification symbols on stone, metal, or other durable matter permanently fixed in the ground, from which differences of elevation are measured. They are used to prepare contour and topological surveys and to establish the grade level or first floor of a building.

2. Location indicated on a durable marker by surveyors.

Betterment. Improvement in property that is considered to add to the investment or capital cost, as distinguished from repairs or replacements where the original character or cost is unchanged.

Bilateral contract. Both parties expressly enter into mutual engagements (reciprocal).

Bi-level. Usually a residence containing two levels of living quarters.

Binder

1. Agreement by which the buyer and seller tentatively agree on the terms of a contract. In some states, if for some valid reason the contract cannot be drawn, the agreement is no longer enforceable.

2. Agreement to cover a down payment for the purchase of real estate as evidence of good faith on the part of the purchaser; in insurance: a temporary agreement given to one having an insurable interest and who desires insurance subject to the same conditions that will apply if, as, and when a policy is issued.

Blacktop. Asphaltic paving surface usually at the driveway to the garage.

Blanket mortgage. Single mortgage that covers more than one piece of real estate.

Blight

1. Partial or complete eradication of the desirable features of real estate, such as its productivity, beauty, or social amenities.

2. Reduction in the productivity of real estate due to a variety of causes, which have a harmful effect on the appearance of the property area affected.

Blighted area. Portion of a community where a variety of adverse conditions have partially or wholly destroyed real estate desirability; unsightly, unsavory, unhealthful neighborhood.

Block. Land area shaped in a square or rectangular form that is a portion of a city, town, or village, enclosed by streets and alleys.

Boarding house. Same as a rooming house wherein lodging and meals may or may not be served, not necessarily for transients, but for not less than three or more permanent guests or as restricted by local zoning regulations.

Bona Fide. In good faith; without fraud.

Bond

1. Originally, any obligation under seal. A real estate bond is a written instrument promising to pay a specified sum of money at a specified date. It is usually issued on security of a mortgage or a trust deed.

2. Agreement to pay money (penalty or penal sum) on or before some future date. The agreement is payable only if some other person does or fails to do something described in the bond. Security for faithful performance by another.

Book depreciation. Amount recorded as necessary for replacement of an asset.

Boom. Rapid increase in prices and demand; in community economics, a sudden popularity, increase in population. It is usually accompanied by a period of excessive profits.

Boundary. Legal recorded property line dividing two parcels of land.

Breach of contract. Violation by one of the contracting parties of an obligation or duty under a contract signed by him or in his name. This usually excuses the other party from performing his contract obligations.

Broker

1. One who arranges transactions involving real or personal property for others. The broker does not usually have possession of the property, nor does he buy or sell for his own account. A commission (brokerage) is earned when he has brought together a seller, ready and willing to sell, and a buyer, ready and willing to buy.

2. One who, for compensation, acts as the agent for another in buying or selling property. In some state laws, the broker performs other functions ordinarily associated with the real estate business.

3. One employed by another, for a fee, to carry on any of the activities listed in the license law of the city or state.

Brokerage. Business conducted by a broker; also, sometimes the fee or commission paid a broker.

Builder. Individual, company, or corporation who engage in building construction.

Building. Structure, wholly or partly enclosed within exterior walls, or within exterior walls and party walls and a roof affording shelter to persons, animals, chattels, or property of any kind.

Building code. Code created by an ordinance regulating the construction of buildings within a locality.

Building line. Line fixed at a certain distance from the front and/or sides of a lot, beyond which no building can project.

Building and loan company. Same as savings and loan company; an institution organized to make real estate loans with the funds received from depositors, paying interest to the latter for use of their money.

Building permit. Document issued by the governing authorities in charge of building construction for an individual, company, or corporation to proceed with the physical aspects of work at a site.

Built-up-roof. Composition roofing consisting of several materials laid on a flat or near flat roof. Roofing affords a weatherproof roof deck and can be guaranteed for various time periods depending on the materials used.

Bundle of legal rights. Rights that establish real estate ownership and consist of the right to sell, to mortgage, to lease, to will, to regain possession at the end of a lease (reversion), to build and remove improvements, and to control use within the law.

Built-ins. Term applied to all appliances included in the construction of a property that become real property by the nature of their installation.

Burdensome property. Under bankruptcy laws the trustee is not required to take title to property that is unprofitable or subject to excessive liens and may obtain release from this property upon action by the court.

Business chance broker. One who negotiates the sale of a mercantile business for another, for a fee.

Buy and sell agreement. Contracts between owners of a business executed to protect the surviving partner(s) in the event of the death of one or more parties to the agreement, in which the survivor(s) has the right to acquire, at predetermined amounts, the interests of the deceased party.

Buy back agreement. Term used for a special trade-in housing contract which permits the one who accepted the trade-in to repurchase his property for a specified amount within a limited period of time at his option.

Capital. Wealth that is set aside for the production of additional wealth; specifically, funds belonging to a business, invested with the expressed intention of remaining permanently in the business.

Capitalization. Act or process of computing potential future incomes into current equivalent capital value; the amount so determined.

Capitalization rate. Rate of interest or return adopted in the process of capitalization.

Capitalized value. The value of the property after use of the capitalization approach of the appraisal.

Carport. Supported roof adjacent to a house for the primary purpose of shelter for one or more automobiles.

Carrying charges

1. Cost incidental to home ownership, as taxes, insurance, maintenance.

2. Expenses necessary for holding property, such as taxes on idle property or property under construction.

Caveat emptor. "Let the purchaser beware." The buyer is duty bound to examine the property he is purchasing, and he assumes conditions that are readily ascertainable upon view.

Cavity wall. Exterior wall usually constructed of two separate walls tied or bonded together.

Cellar. Room or area under a building for storage purposes and has the same meaning as "basement" in many localities.

Certification of no defense. Instrument executed by the mortgagor, upon the sale of the mortgage to the assignee, certifying as to the validity of the full mortgage debt.

Certificate of occupancy. Statement signed by a building authority certifying that a structure meets all the requirements set forth in existing building codes for the area.

Certificate of title. Attorney's review and examination of the abstracts or chains of title, for the preparation of a written report and opinion, declaring that the title is vested as stated in the updated abstract.

Cestui que trust. Person who has a beneficial interest in an estate, the legal title to which is vested in another person.

Chain of title

1. Succession of conveyances whereby the present holder derives his title.

2. History of conveyances and incumbrances affecting the title.

Chain store. One of a number of retail stores under the same ownership and a central management, selling uniform merchandise and following a uniform policy.

Chapel. Room in a building, not necessarily a church, used for prayer and retreat.

Chattel mortgage. Mortgage or loan on personal property as contrasted to a mortgage on real property.

Chattels

1. Articles of personal property as distinguished from real property. A chattel mortgage is a right in personal property given as security for a loan.

2. Items of property other than real estate.

3. Personal property, such as household goods or removable fixtures.

Chimney. Self-standing or part of a building of fire-resistant materials containing passageways or lined flues for combustion gases and smoke.

City. City, as a general term, means the entire city, including the central city and suburbs. It is the whole urban area, which constitutes a social, cultural, and economic unit, aside from whatever arbitrary political boundaries may subdivide this entity.

Client. Party whose interests one is engaged to protect or advance.

Closing. In real estate transactions, the time at which the buyer completes payment to the seller and the seller delivers the deed to the buyer.

Closing costs. Expenses incidental to the sale of real estate, such as loan fees, title fees, appraisal fees, and all other fees incidental to the cost of closing.

Closing statement. Accounting of funds in a real estate sale made by a broker to the seller and buyer, respectively.

Cloud on the title. Outstanding claim or encumbrance which, if valid, would affect or impair the owner's title; judgment; dower interest.

Color of title

1. Appearance of validity in an imperfect title to property. A faulty deed or other written instrument which gives apparent title is said to give "color of title."

2. That which appears to be good title, but as a matter of fact, is not good title.

Commission

1. Payment for the performance of specific duties, usually measured by a percentage of another sum, as of the price paid for a property.

2. Administrative and enforcement tribunal of real estate license laws.

3. Sum or percentage allowed an agent for his services. In real estate, it is a payment to a broker for selling property.

Common law. Unwritten set of legal principles that have evolved from judicial proceedings, court precedents, and customs, which were originally introduced into this country by English settlers.

Common law liability. Obligation established under common law, the breach of which may require compensation for damages.

Common property. Land considered as the property of the public in which all persons enjoy equal rights.

Common wall. Wall that separates adjacent dwelling units within an apartment building.

Community garage. Series of private garages located jointly on a common lot.

Community property

1. Property accumulated through joint efforts of husband and wife living together.

2. Property owned in common, or together, by husband and wife.

Compound interest. Interest paid on original principal and also on the accrued and unpaid interest.

Condemnation

1. Judgment by authorities affecting or confiscating of property.

2. Taking private property for public use, with compensation to the owner, under the right of eminent domain.

Condominium

1. Individual ownership units in a multifamily structure, combined with joint ownership of common areas of the building and ground.

2. Form of complete ownership by which the buyer has the entire undivided interest in an apartment or similar type of dwelling, as opposed to ownership of the land on which the dwelling sits.

Confession of judgment. Entry of judgment on the debtor's voluntary authority to any attorney to do so in his or her behalf.

Confirm. Complete or establish that which was imperfect or uncertain; to ratify what has been done without authority or insufficiently.

Conforming use. Lawful use of a building or lot that complies with the provisions of the applicable zoning ordinance.

Consideration. Something of value that induces a contract, usually the price for which property is bought or sold.

Construction. Includes all labor, materials, and equipment for new construction, alterations, enlargements, additions, moving, conversion, razing, or demolition of buildings or structures.

Construction loan. Loan that provides for progressive payments of the loan proceeds during erection of the building.

Constructive eviction. Breach of a covenant of warranty or quiet enjoyment; for example, the inability of a purchaser or lessee to obtain possession by reason of a paramount outstanding title.

Contingent fees. Remuneration based on future occurrences or upon results of services to be performed.

Contour map. A map which uses lines to outline the configuration and elevation of the surface areas.

Contract

1. Voluntary and lawful agreement between two or more parties to do or not to do something.

2. Agreement mutually entered into by two or more parties to do certain things for a consideration.

3. Enforceable agreement between two or more competent persons by which each promises to do or not to do a particular thing.

Contract for deed. Agreement of seller to deliver the deed when certain conditions have been fulfilled; usually the completion of certain payments. Similar to a mortgage.

Contract for sale. Also familiarly known as "land sales contract," "contract to purchase real estate," or a "conditional sales contract."

Contractor. Individual, company, or corporation engaged in the business of construction work of any kind or type, furnishing or installing fixtures or floor covering or incidental items involved in the completion of a building for use.

Convector. Equipment used for the transfer of heat by the use of steam or hot water through radiation surfaces such as fins, or pipes through contact with the air.

Conventional mortgage. Mortgage that is not insured by the Federal Housing Administration or guaranteed by the Veterans Administration.

Conversion value. Value created by changing from one state, character, form, or use to another.

Convey. To pass or transmit the title to property from one to another; to transfer property by deed or instrument under seal.

Conveyance

1. Transfer of title.

2. Written instrument that transfers interest in real estate from one person to another.

3. Transfer of ownership of real property.

4. Means or medium by which title to real estate is transferred.

Cooling tower. Structure, usually on the roof of a building, used for the cooling of liquids by exposure to the open air.

Cooperative. Apartment house where the occupants participate in ownership and cost of occupancy in proportion to the rental value of the space they occupy; also called "co-op."

Corner influence. Change in value caused by greater usefulness of real estate located at or near street or road intersections.

Corner lot. Lot abutting on two or more streets at their intersection.

Corporeal. Pertaining to a right or group of rights of a visible and tangible nature.

Corridor. Enclosed passageway.

Cost of reproduction. Cost as of today of exact duplication of a property with the same or similar materials.

Court. An open unoccupied space bounded on two or more sides by the walls of a building, on the same lot with a building or a group of buildings.

Covenant

1. Any promise or agreement, usually specific promises within a deed.

2. In the law of contracts, an agreement, convention, or promise of two or more parties, by deed in writing, signed, sealed, and delivered, by which either of the parties pledges to the other that something either is done or shall be done, or stipulates the truth of certain facts.

3. Agreement between two or more persons, by deed, whereby one of the parties promises the performance or nonperformance of certain acts, or that a given state of things does or does not exist.

Covered patio. Attached or detached structure not exceeding 12 ft in height, and enclosed on not more than three sides except for posts required for support of the roof.

Crawl space. Shallow space below living quarters of a house. Generally not excavated but accessible for storage.

Creditor's position. The part of real property market value that is represented by a first mortgage or that can be financed by a prime loan or mortgage.

Credit report. Report required by the lender to check the credit rating of the possible owner who pays for the report.

Cubage. Front or width of a building multiplied by the depth and the height of the building, figured from basement floor to the outer surfaces of walls and roof, depending on use of space.

Cubical content. Space contained within a structure as measured from the outer surfaces of walls and roofs and the upper surface of the lowest basement floor, and expressed in cube measurements.

Cubic content. Area within the outside limits of the exterior walls of a building, multiplied by the average height of the portion being considered.

Cul-de-sac

1. Street with one outlet.

2. Street with access from one end only and closed at the other with a curved, bulblike turn-around.

Curb. Substantially vertical member along the edge of a paved street being considered.

Current assets. Property readily converted from one form to another.

Current liabilities. Indebtedness that is not past due.

Curtain wall. Nonbearing, nonsupporting wall of a building between columns for the enclosure of a building.

Damages. Monetary reparation for loss or injury of rights in a parcel of real estate.

Dance hall. Any room or space in a building in which dancing is carried on and to which the public may gain admission, either with or without payment of a fee.

Datum line. A horizontal line from which heights and depths are measured.

Day nursery. Place where children can be left by their parents for care and educational purposes for a period of the day.

Dead end street. Street or portion of a street with only one vehicular outlet.

Debentures. Promises to pay, usually without specific collateral and sometimes without any fixed date of redemption, but with a fixed interest rate.

Decentralization. Movement from a center toward a periphery.

Decree of foreclosure. Decree by a court upon the completion of foreclosure of a mortgage, lien, or contract.

Dedicate. Appropriate and set apart one's private property to some public use; for example, to make a private way public by acts evincing an intention to do so.

Dedication. Appropriation of land by an owner to some public use, together with acceptance for such use by or on behalf of the public.

Deed

1. Written instrument that transfers ownership of property. The two most common types of deeds are quitclaim deed, which relinquishes or releases to another only the seller's present interest in the land, and the warranty deed, in which the seller warrants the title against all claims of all persons whatsoever.

2. Writing by which lands, tenements, and hereditaments are transferred, which writing is signed, sealed, and delivered by the grantor.

3. Written document by means of which real property is transferred.

4. One of several types of written instruments that convey rights in real property.

Deed of trust. An instrument used in many states in place of a mortgage.

Default

1. Failure to meet an obligation when due.

2. Failure to perform an act or obligation legally required, such as to meet payments on a mortgage loan or to comply with provisions of a sales agreement.

3. Nonperformance of a duty, whether arising under a contract or otherwise; failure to meet an obligation when due.

Defeasance

1. Provision that performance of certain specified acts will render an instrument of contract void.

2. Instrument that nullifies the effect of some other deed or of an estate.

Defend. To protect, maintain, or keep secure; to guaranty; to agree to indemnify.

Deferred maintenance. Existing, but unfulfilled requirements for repairs and rehabilitation.

Deferred payments. Payments to be made at a future date or dates; installment payments.

Deficiency judgment. Difference between the indebtedness sued upon and the sale price or market value of the real estate at the foreclosure sale.

Demise. In conveyancing, to convey an estate to another for life, for years, or at will; to surrender rights, as in a lease to surrender the right of occupancy to the tenant; to quit the property.

Depreciation

1. Loss in value, brought about by deterioration through ordinary wear and tear, action of the elements, or functional or economic obsolescence.

2. Loss in value caused by obsolescence, physical deterioration, blight, and other factors.

3. Loss in value of property resulting from age, physical decay, changing neighborhood conditions, or other causes.

Depreciation allowance. Allowance that permits the owner of real property to shelter all or part of his income so that he pays no current tax on it.

Depreciation at accelerated rate. The owner is allowed to take larger depreciation write-offs in the early years and smaller allowances in the later years.

Depreciation rate. The annual percentage of change in property values resulting from depreciation.

Depreciation reserve. Capital amount that summarizes the annual charges to operation by reason of depreciation.

Depth of lot. The mean horizontal distance between the front and rear lot lines, measured in the general direction of its side lot lines.

Depth tables

1. Tables of factors for computing front foot values of lots of varying depths by comparison with a given standard.

2. Tabulation of factors for computing front foot values of lots of varying depths by comparison with a given standard.

3. Tabulation of factors representing the rating of value per front between a selected "standard" depth (usually 100 ft) and other lots of greater or lesser depth.

Detached building. Building surrounded by open space.

Deterioration. Reduction of usefulness and value by physical causes such as the elements and use.

Developer. Individual, company, or corporation engaged in the development and improvement of land for construction purposes.

Devise. Gift of real property by last will and testament.

Devisee. One who inherits property by will.

Discount. Loan placement charge made by the lending institution to the seller, by increasing the yield on the investment (also known in the trade as points).

Discrimination. In real estate, prejudice or refusal to rent or sell to a person because of race, color, religion, or ethnic origin.

Dispossess

1. Deprive one of the use of real estate.

2. Eject; to legally oust from the land.

Distance separation. Open space between buildings or between a building and an interior lot line, which is provided to prevent the spread of fire.

District. Portion of the city within which the use of land, buildings, structures, or other improvements, and the location, height, and bulk of buildings and structures are governed by ordinance. Any section of the applicable city in which the zoning regulations are uniform.

District boundaries. Boundaries of the zoning districts established on the official zoning map of the city, which is available to the public.

Division wall. Any interior wall in a building, dividing a building into separate areas.

Documentary stamp. Stamp issued for the payment of a tax.

Domicile

1. Home; place that is the permanent and real home of a person, where he intends to live permanently. This differs from his residence where he may be for the time being.

2. Place where one has his permanent residence and, usually, is a registered voter.

Dormer. Structure projecting from a slanting roof to accommodate a window.

Double frontage lot. Lot extending between and having frontage on two parallel streets.

Dower

1. Widow's legal rights in the real estate of her deceased husband; amounts vary according to statutory provisions.

2. Right that a wife has in her husband's estate at his death.

Driveway. Private way for use of vehicles.

Dump. A lot or parcel of land or part thereof used primarily for the purpose of the disposal of abandoned materials, burial, burning (if allowed), or any other means of disposal of trash, refuse, junk, discarded machinery, vehicles, or parts thereof or waste material of any kind allowed by the applicable ordinance.

Duplex. Single two-story structure designed for two-family occupancy.

Dwelling. House or building or portion thereof that is occupied in whole or in part as a home, residence, or sleeping place for one or more families.

Earnest money

1. Deposit; token sum, the payment of which constitutes a binder; indication of an intention to go through with a deal.

2. Money paid as evidence of good faith or actual intent to complete a transaction, usually forfeited by willful failure to complete the transaction.

3. Down payment made by a purchaser of real estate as evidence of good faith.

Easement

1. Right, liberty, advantage, or privilege that one individual has in the lands of another (a right-of-way).

2. Legal term denoting the use of land in a certain way by someone other than the land owner.

3. Agreement under which certain privileges or rights of the owner of land are granted to another.

4. Right or privilege of a person, other than the owner or tenant, to use a piece of property.

Economic approach. Approach based on the building residual method employed in the income approach to value.

Economic life. Period over which a property may be profitably utilized; period during which a building is valuable.

Economic obsolescence

1. Loss in value created by factors outside the property, such as changes-in-neighborhood factors. This is one of the principal considerations in depreciation.

2. Economic obsolescence is caused by changes external to the property, such as neighborhood infiltration of inharmonious people or property users, or by legislation or economic changes.

3. Reduction of desirability, usefulness, and value caused by nonphysical forces, such as adverse legislation, style changes, booms, and depressions.

Economic rent. Commonly the justifiable amount payable for use or occupancy of real estate, not including any utilities or services; basic rent. Also, in academic usage, complicated formulas of philosophic reasoning are so termed.

Effective gross revenue. Total income less allowances for vacancies, contingencies, and sometimes collection losses, but before deductions for operating expense.

Efficiency living unit. Any room in a building having cooking facilities used for combined dining and living purposes.

Ejectment

1. Action to repossess real property unlawfully held by another.

2. Form of action to regain possession of real property, with damages for the unlawful retention.

Eluviation. The movement of soil materials, either down-stream or in a horizontal direction, caused by excessive water in the soil.

Eminent domain

1. Right of the people or government to take private property for public use upon payment of compensation.

2. Power of appropriating private property for public use or for public welfare.

3. Power of a municipal, state, or federal government to take property for public use by condemnation proceedings.

4. Right of a governmental body to condemn and/or acquire property for public use.

Enclosed arcade. Arcade for pedestrians with less than 25% of its perimeter abutting a street or plaza.

Enclosed court. Court bounded on all sides by the exterior walls of a building or exterior walls and lot lines on which walls are allowed by the ordinance.

Enclosed porch. Porch having at least 50% of the horizontal section of the exterior walls in glass.

Encroachment

1. In real property, the use of the property of another. Building a house and extending beyond the building line into the public street.

2. Illegal and unauthorized use of the property of another, usually by building or constructing something of a permanent nature in part or whole on such property without permission.

3. Invasion of private rights by persons or economic forces; trespass.

4. Part of a building or an obstruction that intrudes on or invades a highway or sidewalk or trespasses on the property of another.

Encumbrance

1. Right in real property that diminishes its value but does not prevent its transfer from one person to another. Examples are leases, liens, easements, and rights to timber. Such rights are generally held by third persons (not the grantor or the grantee), and the deed can be taken, subject to them.

2. Interest or right in real property that diminishes the value but does not prevent conveyance of the fee by the owner thereof. Mortgages, taxes, and judgments are encumbrances known as "liens." Restrictions, easements, and reservations are encumbrances.

3. Claim, lien, charge, or liability attached to and binding upon real property, such as a judgment, unpaid taxes, or a right-of-way. An encumbrance is defined in law as any right to or interest in land that may subsist in another to the diminution of its value but that is consistent with the passing of the fee.

4. Claim against a piece of property that lessens its value but does not prevent its being transferred, sold, or deeded to someone by the owner. Typical encumbrances are taxes, mortgages, and judgments. These are known as "liens." Easements are also encumbrances but are not liens.

5. Any outside interest in or right to property founded on legal grounds, such as a mortgage or lien for work and materials.

Engineer. Registered and licensed individual practicing lawfully in the state of registration for the designing structures and systems.

Entirety. The phrase "by the entirety" describes the ownership of all of a piece of real property by two or more people on such terms that it cannot be divided between them. Upon the death of one, the whole property goes to the other. This is commonly the way in which property is owned by husband and wife.

Equipment. Fixed assets, such as furniture, that are not attached to real estate physically or legally.

Equity

1. Value of a piece of property, usually expressed in terms of money, over and above all liens.

2. Interest in or value of real estate in excess of mortgaged indebtedness.

3. Value of property less any outstanding mortgages, liens, or other changes.

4. Interest or value that an owner has in real estate over and above the mortgage against it.

Equity of redemption

1. Right of a mortgagor to regain his property after he has lost the legal right to it by failure to make payment in time.

2. Right of the original owner to reclaim property sold through foreclosure proceedings on a mortgage by payment of debt, interest, and costs.

Erosion. Wearing away of land through the processes of nature as by streams and the forces of nature.

Escheat

1. Reversion of property to the state by reason of failure of persons legally entitled to it or by lack of heirs.

2. Reversion of property to the sovereign state owing to lack of any heirs capable of inheriting.

Escrow

1. System by which the buyer submits the purchase price to a disinterested third party who disburses it to the seller after the title has been correctly passed to the purchaser.

2. Deed or other instrument placed in the hands of a disinterested person for delivery upon the performance of certain conditions or the happening of certain contingencies.

3. Deed delivered to a third person for the grantee, to be held by him until the fulfillment or performance of some act or condition.

4. Depository for papers, funds, and instructions with a third party who is obligated to carry out all the instructions, providing they are in accordance with the agreement.

Estate

1. Right in property. Estate in land is the degree, nature, or extent of interest in it.

2. Degree, quantity, nature, and extent of interest that a person has in real property.

Estate in reversion. Residue of an estate left in the grantor, to commence in possession after the termination of some particular estate granted by him. In a lease, the lessor has the estate in reversion after the lease is terminated.

Estoppel. Bar raised by law that precludes a party from using a defense he may have had in consequence of his previous denial, conduct, or admission; similar to but not the same as a waiver.

Estoppel certificate. Certificate that shows the unpaid principal sum of a mortgage and the interest thereon, if the principal or interest notes are not produced or if the seller asserts that the amount due under the mortgage that the purchaser is to assume is less than shown on record.

Estovers. Materials that a tenant is allowed to take from the landlord's premises for necessary fuel, implements, repairs, etc.

Eviction

1. Removal from possession of real property; the dispossess. Eviction is not necessarily accomplished by legal process, nor does it always compel removal. Constructive eviction occurs when a landlord deprives the tenant of the use of the property by some act such as obstructing the entrance or failing to supply water or elevator service. It gives the tenant the right or a cause to move, but if he stays, he must pay the rent.

2. Dispossession by process of law; act of depriving a person of the possession of land that he has held in pursuance of the judgment of a court.

3. Violation of some covenant in a lease by the landlord, usually the covenant for quiet enjoyment; also, the process instituted to oust a person from possession of real estate.

Exchange agreement. Contract that covers the understandings of two or more owners who agree to transfer their properties to each other with or without additional consideration.

Exclusive agency. Appointment of one real estate broker as sole agent for the sale of a property for a designated period of time.

Exclusive listing

1. Contract to sell property as an agent, according to the terms of which the agent is given the sole right to sell the property or is made the sole agent for its sale; also, the property so listed.

2. Right given to a broker by a property owner to sell the property to the exclusion of any and all other brokers.

Executor's deed. Deed given by an executor.

Existing. That which is in existence as of the date of the purchase.

Exit. Means of egress from a building.

Expropriation. Process whereby private property is taken for public use or purposes of public welfare; right of eminent domain.

Extender clause. Clause in an exclusive listing contract that carries the original exclusive period over an additional period to protect the broker if a sale is made to a prospect he obtained during the original listing period.

Extension agreement. Agreement between mortgagee and mortgagor to extend the maturity date of the mortgage after it becomes due.

Exterior balcony. Narrow, open platform on the outside of a building wall, surrounded by a rail or balustrade.

Exterior lot line. Boundary line between a lot and a street alley, public way or railroad right of way.

Family. Group of persons living together who share at least in part their living quarters and accommodations.

Fanny Mae

1. Common name in the mortgage industry given to the Federal National Mortgage Association operated by the government as a secondary market for federally insured loans.

2. Secondary mortgage market. Provides a market for mortgages held by primary lenders, such as banks and savings and loan associations, and provides the primary market with a ready market for mortgages, so as to permit a greater turnover of money for loans.

Fair market value. Price that probably would be negotiated between a willing seller and a willing buyer.

Fee. Remuneration for services; inheritable estate in land.

Fee simple

1. Absolute ownership of land.

2. Estate by which the owner is entitled to the entire property, with unconditional power of disposition during his life. The estate descends to his heirs and legal representatives upon is death intestate.

3. Largest estate or ownership in real property; free from all manner of conditions of encumbrances.

Fee simple absolute. Largest possible interest or estate in property; inheritable estate.

Fee simple limited. Estate giving the owner thereof fee rights as long as certain conditions obtain, termination being governed by the occurrence of some state event.

Fee tail. Estate limited to some particular class of heirs of the person to whom it is granted.

Fee tail estate. Estate of inheritance given to a person and the heirs of his body. If the grantee dies without leaving issue, the estate terminates and reverts to the grantor.

Fence. Independent structure forming a barrier at grade.

FHA (Federal Housing Administration)

1. Agency in the Department of Housing and Urban Development that insures private loans for the financing of new and existing housing and home repairs and improvements. It also administers rent supplements to low-income families in private housing and many recent programs for housing low- and moderate-income families.

2. Agency of the federal government that insures real estate loans.

Finder's fee. Fee or commission paid to a broker for obtaining a mortgage loan for a client or for referring a mortgage loan to a broker; also, a commission paid to a broker for locating a property.

Fire area. Floor area enclosed and bounded by fire walls or exterior walls of a building to restrict the spread of fire.

Fire district. Territory defined and limited in the ordinance, for the restriction of types of buildings construction.

Fire separation. Construction of specific fire resistance separating parts of a building.

Firm commitment. Commitment by the FHA or by a lending institution to insure a mortgage on specified property with a specified mortgagor.

First mortgage. Mortgage that has precedence over all other mortgages.

First story of a building. Floor of the first story. Lowest story in a building wholly above ground.

Fixed assets. Property owned by one ownership whose value is unvarying or varies by regular and recorded degree; furniture, fixtures, equipment, machinery etc.

Fixed charges. Regularly recurring expense for protection of capital investment as opposed to physical maintenance. Insurance premiums are fixed charges; repairs are maintenance expenses.

Fixtures

1. Appurtenances affixed to structures or land, usually in such manner that they cannot be removed without damage to themselves or the property.
2. Articles that were once personal property but have become real estate by reason of their permanent attachment to an improvement.

Floor area of a building. The sum of the gross horizontal area of the floors, mezzanine, and interior balconies, at the established grade of the lowest abutting street, bounded by the outside faces of all exterior walls.

Floor area ratio. Floor area of a building on any lot divided by the area of the lot.

FNMA (Fannie Mae). Federal National Mortgage Association which buys and sells FHA-insured and Veterans Administration guaranteed loans to improve distribution of home mortgage funds. Special assistance purchases support FHA section 220 and 221 programs that are designed to help urban renewal development, rehabilitation, and relocation activities. The Housing and Urban Development Act of 1968 divided FNMA into two separate corporations, one to manage the special assistance functions and the other to administer the secondary market operations. Until 1968 all of FNMA was part of the Department of Housing and Urban Development. The new FNMA is a "Government sponsored private corporation."

Forced sale. Sale of property under compulsion as to time and place; usually a sale made by virtue of a court order, ordinarily at public auction.

Forced sale value. Amount that may be realized at a forced sale; price that could be obtained at immediate disposal.

Forcible entry and detainer. Legal action to recover possession of premises that are unlawfully held.

Foreclosure

1. Proceeding against a property owner who has failed to meet the obligation of a mortgage loan or other liens like tax liens. The method used is to force a sale of the property to pay the debt.
2. Mandatory transfer of property ownership by order of court in satisfaction of an unpaid debt, such as a mortgage or delinquent property taxes.
3. Court process instituted by a mortgagee or lien creditor to defeat any interest or redemption that the debtor-owner may have in the property.

Forfeiture. Reversion to the state of private property and property rights in satisfaction of unpaid taxes.

Foundation. Construction, below or partly below grade, that provides support for the exterior walls or other structural parts of the building.

Free and clear. Real property against which there are no liens or mortgages.

Freehold

1. Tenure of land held in fee simple absolute, fee simple limited, or in fee tail, unencumbered by lease.
2. Estate in fee simple or for life.

Freeway. Divided arterial highway for through traffic to which access from abutting properties is prohibited.

Frontage. All the property abutting on one side of a street between two lot lines measured along the right-of-way line.

Front foot

1. Measure (1 ft in length) of the frontage of real estate on a street.
2. Standard of measurement, 1 ft wide, extending from the street line for a depth generally conceded to be 100 ft.

Front foot cost. Cost of a parcel of real estate expressed in terms of front foot units.

Front lot. That boundary of the lot that abuts on a street.

Front yard. Yard across the full width of the plot facing the street extending from the front line of the property to the front line of the building.

Functional obsolescence

1. Obsolescence that may be due to a poor plan or to functional inadequacy or overadequacy due to size, style, age, or otherwise. It is evidenced by conditions within the property.
2. Impairment of functional capacity or efficiency.

General mortgage bond. Written instrument representing an obligation secured by a mortgage but preceded by senior issues.

General warranty. Covenant in the deed whereby the grantor agrees to protect the grantee against the world.

G.I. loan. Loan guaranteed by the Veterans Administration under the Servicemen's Readjustment Act of 1944, as amended, for honorably discharged veterans and their widows.

Give. Transfer, yield to, or bestow upon another. One of the operative words in deeds of conveyance of real property importing, at common law, warranty or covenant for quiet enjoyment during the lifetime of the grantor.

Graduated lease. Lease that provides for a certain rent for an initial period, followed by an increase or decrease in rent over stated periods.

Grant. Generic term applicable to all transfers of real property; technical term made use of in deeds of conveyance of land to import a transfer.

Grantee. Person to whom real estate is conveyed; buyer.

Grantor. Person who conveys real estate by deed; seller.

Gross acre. An acre, 43,560 sq. ft.

Gross earnings. Revenue from operating sources before deduction of the expenses incurred in gaining such revenues.

Gross floor area. Area of the plan projection of all floors of whatever nature within or attached to a building.

Gross income. Total receipts during a given period; total revenue that, although not necessarily actually received, has accrued from all sources during a specified time.

Gross leasable area. Total floor area designed for tenant occupancy and exclusive use, including basements, mezzanines, upper floors, and storage area, and excluding areas devoted to the housing of mechanical equipment, electrical substations, and mail areas, and also excluding general office floor area having no connection to any of the retail establishments.

Gross lease. Lease of property under the terms of which the lessor is required to meet all property charges regularly incurred through ownership.

Gross multiplier. Number used to determine approximate selling price for income property by multiplying the gross income times this number (income × multiplier = selling price).

Gross profits. Profits computed before the deduction of general expenses.

Gross revenue. Total revenue from all sources before deduction of expenses incurred in gaining such revenue.

Ground rent

1. Net rent paid for the right of use and occupancy of a parcel of unimproved land; portion of the total rental paid that is considered to represent a return on the land only.

2. Rent reserved by a grantor to himself, his heirs, and assigns in conveying land in fee.

Guaranteed mortgage. Mortgage on which payment of principal and interest is guaranteed, usually by insurance or surety.

Guaranteed trade-in. Contract to purchase real property executed by a trade-in company, builder, or dealer with the seller in which a set time is allotted for the performance of the guarantee and during which the seller may sell to someone else for more money than the trade-in company guaranteed.

Guarantee of title. Opinion rendered regarding the condition of title based upon a review of the official records and backed by a fund to compensate those damaged by negligence.

Habendum clause. The "to have and to hold" clause that defines or limits the quantity of the estate granted in the premises of the deed.

Habitable room. Room in a residential unit used for living, sleeping, eating, or cooking purposes, but excluding baths, toilet rooms, storage spaces, and corridors.

Hand money. Same as an earnest money deposit; money in hand.

Hereditaments

1. Every sort of inheritable property, such as real, personal, corporeal, and incorporeal.

2. Largest classification of property, including lands, tenements, and incorporeal property, such as rights-of-way.

Highest and best use. Use of or program of utilization of a site that will produce the maximum net returns over the total period that comprises the future; optimum use for a site.

Holder in due course. One who receives in the course of business and in good faith a note for value without prior knowledge of any defects.

Holdover tenant

1. One who retains possession after the expiration of his term without the consent of the landlord.

2. Tenant who remains in possession of leased property after the expiration of the lease term.

Homestead

1. Home place; place where the home is; the home, the house and the adjoining land, where the head of the family dwells; the home farm. Fixed residence of the head of a family, with the land and buildings surrounding the main house.

2. Real estate occupied by the owner as a home. The owner enjoys special rights and privileges.

Hot spot. Section of a community in which the greatest proportion of retail trade is transacted.

Housing authority. Agency set up by the federal government and by some state and local governments to construct and manage housing presumably for lower income group tenants.

Housing code. Locally adopted ordinance, regulation, or code, enforceable by police powers under the concept of health, safety, and welfare, that specifies the minimum features that make dwellings or dwelling units fit for human habitation or that controls their use or occupancy.

HUD. Department of Housing and Urban Development.

Hundred percent location. Location or site in a city that is best adapted to carrying on a given type of business.

Hypothecate

1. Pledge property as security; to mortgage.

2. Give a stock as security without giving up possession of it.

Implied. Not expressly stated or written but understood to be included.

Implied warranty. A warranty assumed by law to exist in an instrument, although it may not be specifically stated.

Improvement acts. Laws that authorize installation of street and other improvements, which may then be assessed directly to the properties involved.

Improvement bond. A bond issued by a district, city, or state for the installation of improvements such as highways or streets, and which is then sold to investors to finance the projects covered.

Improvements on land. Structures, of whatever nature, erected on a site, for example, buildings, fences, driveways, and retaining walls.

Improvements to land. Facilities, usually public utilities, such as sidewalks or sewers, added to land that increase its usefulness.

Inchoate. Not yet vested or completed. The right to dower is inchoate until the husband dies.

Income. Receipts, usually measured in money, flowing from services rendered by persons or property.

Income property. Real estate that produces monetary income as a direct return for its use; usually rental property.

Incompetent. Mentally incapable (legally) of making reasoned decisions.

Increment. Addition to value, usually the increase due to population growth and other causes of community prosperity.

Indenture

1. Deed to which two or more persons are parties, and in which these enter into reciprocal and corresponding grants or obligations toward each other.

2. Formal written instrument made between two or more persons in different interests. The name comes from the practice of indenting or cutting the deed on the top or side in a waving line.

Industrial property. Property used in industry; in real estate the land and buildings used in manufacturing, processing, and fabricating products.

Inner city. Urban area that does not necessarily have political, geographic, racial, or economic outlines or boundaries but that, in general, was a popularly recognized central shopping and residential part of a city prior to World War II.

Inside property line. Property line dividing lots.

Installment contract. Purchase of real estate on an installment basis. Upon default, payments are forfeited.

Installment payments. Periodic payments, usually equal amounts of money, to discharge an indebtedness.

Insurance rate. Cost of insurance expressed in dollars per thousand dollars of protection or sometimes in percentage.

Intangible property. Nonphysical items of property; for example, good will is intangible in contrast to the real estate owned by the corporation.

Interest. In property, the most general term that can be employed to denote a property in lands or chattels. In its application to lands or things real, the term is frequently used in connection with the terms, "estate," "right," and "title."

Interest rate. Amount, scale, or rate of payment for the use of a principal sum of money, usually expressed as a percentum in relation to the principal amount.

Interim financing. Temporary financing, usually for construction funds.

Interior lot. Side lines of which do not abut on a street.

Intestate. Death of an individual who does not leave a valid will.

Intrastate. Within a particular state.

Intrinsic value. True vale; actual value.

Insure. Serve to the use or benefit of someone.

Inventory. Tabulation of the separate items comprising an assembled property.

Investment. Money spent for purchase, improvement, and development of property in the expectation of regular and permanent return of profits and safety of principal.

Investment property. Property purchased for investment rather than for speculative possibilities. Such a purchase connotes permanency of the capital investment as opposed to quick turnover.

Involuntary lien. Lien on real property imposed without the owner's authorization, such as taxes.

Irrevocable. Without the right to cancel or void the act involved.

Joinder. Joint action with one or more persons; joining.

Joint and several note. Same as a joint note, except that makers may be sued together or individually in the event of a default.

Joint note. Note signed by more than one person, each with equal responsibility for payment, who must be sued together if action is necessary.

Joint tenancy

1. Tenancy shared equally by two or more parties with the right of survivorship.

2. Property held by two or more persons together with the distinct character of survivorship.

Judgment. Final determination of order of a court of law that sets forth the decision of the judge in a lawsuit.

Junior lien. Lien placed on property after a previous lien has been made and recorded and having rights enforceable only after previous lines have been satisfied.

Junior mortgage. Mortgage second in lien to a previous mortgage.

Kennel. Any lot or premise on which animals are boarded.

Key lot. Lot that is located in such a manner that one side adjoins the rear of another lot.

Kitchenette. Space in a dwelling having a floor area of not less than 60 sq ft. Verify with local zoning ordinances.

Land. Solid substance composing part of the earth's surface. In law, land includes natural and man-made appurtenances. In economics, land includes the elements of natural value (as distinguished from elements added by man) as the wealth of the nation.

Land contract

1. Contract to deliver a deed to real property given to a purchaser of real estate who pays a small portion of the purchase price when the contract is signed and agrees to pay additional sums at intervals and in amounts specified in the contract until the total purchase price is paid.

2. Contract for the purchase of real estate on an installment basis. On payment of last installment, deed is delivered to purchaser.

Land coverage. Percentage of a lot covered by the main and accessory buildings as compared to the total lot area.

Land development. Division of land into lots for the purpose of conveying such lots singly or in groups to any person, company, or corporation for the purpose of the erection of buildings by such person, company, or corporation.

Land economics

1. Analysis of the functions, causes of variations in value, utilization, and other elements affecting or affected by the use of land by man.

2. Branch of the science of economics that deals with the classification, ownership, and utilization of land and the buildings erected thereon.

Land improvements. Physical changes that increase utility and value, including improvements to land and improvements on land.

Landlord

1. One who rents property to another.

2. One who owns land that he rents to others.

Landscape architect. Professional landscape architect licensed by the state for the practice of planning and developing of landscaping work.

Land trust certificate. Instrument granting participation in benefits of ownership or real estate, but with title remaining in a trustee.

Land use. Term referring to the use of a lot or parcel of property; for example, a lot occupied by a factory has an industrial land use. The general categories of land use include residential, industrial, commercial, public, semipublic, and institutional. Also, "land use policy" describes a spectrum of policies related to the assembly and use of land.

Latent. That which is concealed, hidden. A defect in property or title may be "latent."

Lease

1. Contract between the owner of real estate and another person stating the conditions under which the tenant may occupy and use the property and the responsibilities of owner and tenant.
2. Contract, written or oral, for the possession of lands and tenements on the one hand and a recompense of rent or other income on the other hand.
3. Agreement by which the lessor (landlord) grants possession for a period of time to the lessee (tenant) in return for rental payments.

Leaseback. Arrangement under which the tenant agrees to pay an amount of yearly rent that is calculated to pay off the mortgage and leave the property free and clear in the owner's hands after a stipulated number of years.

Lease fee. Property held in fee with the right of use and occupancy conveyed under lease to others; property consisting of the right to receive ground rentals over a period of time and the further right of ultimate repossession.

Leasehold

1. Rights held by virtue of a lease.
2. Estate in realty held under a lease.

Leasehold estate. Right in property terminating at a date specified by a contract, usually a lease.

Legal description

1. Designation of boundaries of real estate in accordance with one of the systems prescribed or approved by law.
2. Description recognized by law, which is sufficient to locate and identify the property without oral testimony.

Lessee

1. Tenant; one who has right in land by virtue of a lease.
2. One who possesses the right to use or occupy a property under lease agreement.
3. Person to whom property is rented under a lease.

Lessee's interest. Leasehold estate.

Lessor

1. Landlord; one who grants a right in land by virtue of a lease.
2. One who conveys the right to use and occupy a property under lease agreement.

Let. In conveyancing, to demise or lease.

Leverage. Use of other people's money to help buy a piece of property.

Leveraged. Property is said to be highly leveraged when the owner's cash equity is small in relation to the total value of the property.

License or registration. Privilege or right granted by the state to operate as a real estate broker or salesman; authority to go upon or use another person's land or property without possessing any estate therein.

Lien

1. Claim on the property of another resulting from some charge or debt.
2. Charge against property whereby the property is made security for the payment of a debt.
3. Hold or claim that one person has on the property of another as security for a debt or charge; judgments, mortgages, and taxes.
4. Incumbrance against property, which becomes the security for the obligation.

Life estate

1. Freehold interest in land, the duration of which is confined to the life of one or more persons or is contingent on certain happenings.
2. Estate or interest held during the term of some certain person's life.

Line fence. Fence placed on a boundary line.

Listing

1. Record of property for sale by a broker who has been authorized by the owner to sell; also, the property so listed.
2. Oral or written employment of a broker to sell or lease real estate.

Littoral

1. Shore and the country contiguous to it. Zone between high- and low-water marks.
2. Belonging to shore as of the sea or Great Lakes; corresponding to riparian rights.

Loan constant. The yearly percentage of interest which remains the same over the life of an amortized loan.

Lot. Plotted lot of a recorded subdivision, or a parcel of land, occupied or capable of being occupied by one building or use and including accessory buildings.

Lot area. Total horizontal area within the lot lines of the lot.

Lot depth. Horizontal length of a straight line drawn from midpoint of the front lot line to the midpoint of the rear lot line.

Lot line. Legally defined line dividing one tract of land from another.

Lot width. Distance parallel to the front of a building erected or to be erected, measured between side lot lines at the building line.

MAI. Member of the American Institute of Real Estate Appraisers.

Main building. Any building having the predominant land use that is not an accessory building.

Maintenance. Act of keeping or the expenditures required to keep a property in condition to perform adequately and efficiently the service for which it is used.

Maintenance reserve. Amount reserved to cover costs of maintenance.

Major street. Streets identified by the city street plan as having large volumes of traffic.

Mall. Area composed of sidewalks and landscaping that serves as a pedestrian thoroughfare between buildings, but is not dedicated to public use.

Marginal land. Land that barely repays the cost of working or using it; land whereon the costs of operating approximately equal the gross income.

Marina. A place for docking or storage of pleasure boats.

Marketable title

1. Title to real property that a reasonable purchaser would accept; title that has no defects restricting or reducing its use of value.

2. Title a court would compel a purchaser to accept; title free from any encumbrances or clouds.

Market price. Price paid for a property; amount of money that must be given or can be obtained in exchange under the immediate conditions existing at a certain date.

Market value

1. As applied to existing structures or buildings, the price as of the date of the permit at which a prudent seller should be willing to sell and a prudent buyer willing to buy, determined by the administrative official in accordance with generally accepted methods of estimating prescribed by the approved rules.

2. Price at which the property might be reasonably sold if offered for sale in a fair market.

3. Quantity of other commodities a property would command in exchange; specifically, the highest price estimated in terms of money that a buyer would be warranted in paying and a seller justified in accepting, provided both parties were fully informed and acted intelligently and voluntarily, and, further, that all the rights and benefits inherent in or attributable to the property were included in the transfer.

4. Highest price that a buyer, willing but not compelled to buy, would pay, and the lowest a seller, willing but not compelled to sell, would accept.

Master's deed. Deed issued by a master in chancery, under court order, to satisfy a judgment.

Mechanic's lien

1. Lien or claim by a contractor, subcontractor, or worker for labor done or materials furnished which must be paid by a builder before he can sell to a home buyer.

2. Species of lien created by statute that exists in favor of persons who have performed work or furnished materials in the erection or repair of a building.

3. Charge against property by those who provide labor or materials for improvements whereby the property is security for payment.

Metes and bounds

1. System of describing land by measurements and boundaries of a parcel of real estate, most frequently found in states not having the U.S. rectangular survey system of locating and describing real estate.

2. Description in a deed of the land location, in which the boundaries are defined by directions and distances.

Mill. One-tenth of 1 cent; measure used to state the property tax rate. That is, a tax rate of 1 mill on the dollar is the same as a rate of one-tenth of 1% of the assessed value of the property.

Minimum lot. The smallest lot size allowed by local zoning regulations for development.

Misplaced improvements. Improvements on land or to land that do not conform to the best utilization of the site.

Mixed occupancy. Building used for two or more occupancies, classified within different occupancy groups.

Mobile home park. Lot, parcel, or tract of land used as a site for mobile homes, providing the necessary utilities as required by the applicable code.

Monument

1. Fixed object such as a stone or concrete marker, used to establish real estate boundaries.

2. Artificial or natural landmark.

Mortgage

1. Lien on real property that an owner gives a lender as security for the repayment of the money that the owner borrowed from the lender. The buyer or owner is called the "mortgagor," the lender is the "mortgagee."

2. Conditional conveyance of property contingent upon failure of specific performance, such as the payment of a debt; the instrument making such conveyance.

3. Conditional transfer of real property as security for the payment of a debt or the fulfillment of some obligation.

Mortgage certificate. Fractional interest in a mortgage, evidenced by an instrument that certifies the agreement between the holder of the mortgage and the holder or holders of the certificates as to terms, interests, etc.

Mortgagee

1. Source of the funds for a mortgage loan and in whose favor the property serving as security is mortgaged.

2. Person to whom property is conveyed as security for a loan made by such person (the creditor).

Mortgagee in possession. Mortgage creditor who takes over the income from the mortgaged property upon a default on the mortgage by the debtor.

Mortgage processing charge. The fee may be called a mortgage service fee, initial service fee, brokerage fee, originating fee, or it may be known by some other designation. Primarily it reimburses the lender for the paperwork, time, and expense involved in handling the mortgage for the new owner.

Mortgagor

1. Owner of property who borrows money and mortgages the property as security for the loan.

2. Owner who conveys his property as security for a loan (the debtor).

Motel. Building or group of buildings including either separate unit or a row of units that have individual entrances and contains living and sleeping accommodations.

Multiple dwelling. Building containing three or more dwelling units.

Multiple listing

1. Arrangement among real estate boards or exchange members whereby each broker brings his listings to the attention of the other members so that if a sale results, the commission is divided between the broker bringing the listing and the broker making the sale, with a small percentage going to the board of exchange.

2. Agreement by which one of a group of agents has an exclusive right to sell for a specified period; however, any one of the group can sell the property, but he must share the commission with the exclusive broker.

Municipality. Elected corporate government unit.

Muniments. Muniments of title are those elements that strengthen or fortify the rights in property.

NAREB. National Association of Real Estate Boards.

Net earnings. Revenue from operating sources after deduction of operating expenses, maintenance, uncollectible revenues, and taxes, but before deduction of financial charges and generally before deduction of provision for depreciation and retirements.

Net income. Synonymous with net earnings, but considered a broader and better term; balance remaining after deducting from the gross income all operating expenses, maintenance, taxes, and losses pertaining to operating properties, excepting interest and other financial charges on borrowed or other capital.

Net lease

1. Lease where in addition to the rental stipulated, the lessee assumes payment of all property charges such as taxes, insurance, and maintenance.
2. Lease under which lessor receives a fixed rental and the lessee pays taxes, utilities, and all other operating expenses.

Net listing. Price that must be expressly agreed upon, below which the owner will not sell the property, and at which price the broker will not receive a commission. The broker receives the excess over and above the net listing as his commission.

Net option. Option granting the right to purchase property at a stated or "net" price to the owner.

Net profits. Balance remaining after deducting all expenditures from all receipts.

Nonconforming building. Building or structure or portion thereof that does not conform to the height and area regulations of the zone in which it is located.

Note

1. Instrument given to attest a debt.
2. Written instrument in which the signer promises to pay a specified sum to a specific person on a specific date. Usually, a home owner executes a note as part of a mortgage loan.

Nuisance. Use of one's property in such a way as to cause damage, annoyance, or inconvenience to another. It is a private nuisance when it affects only one or a few persons, a public nuisance when it affects the community.

Nuisance value. Amount that would be paid to remove a property from ownership or use that is felt to be detrimental or annoying to a prospective purchaser.

Obsolescence

1. Impairment of desirability and usefulness brought about by changes in the public taste, art, design, or process, or from external circumstances that make a property less desirable and valuable.
2. Impairment of desirability and usefulness brought about by physical, economic, fashion, or other changes.

Occupancy. Purpose for which a building or portion thereof is used or intended to be used.

Occupancy certificate. Official document certifying that a building or parcel of land is in compliance with the provisions of the applicable codes.

Offset statement. Statement by the owner of a property or owner of a lien against a property, setting forth the present status of liens against the subject property.

Off street parking. Space provided for vehicular parking outside the dedicated street right-of-way.

Open-end mortgage. A mortgage permitting the mortgagor to borrow additional money under the same mortgage.

Open listing

1. Permission granted a broker to offer property for sale on a nonexclusive basis.
2. Oral or general listing.

Open porch. A porch that has no walls or windows other than those of the main building to which it is attached.

Open shed. Any structure that has no enclosing walls.

Option

1. Agreement granting exclusive right to buy, sell, or lease property at a given price and (usually) within a stated period of time.
2. Right to purchase or lease a property at a certain price for a certain designated period, for which right a consideration is paid.

Original cost. Actual cost of a property to its present owner, but not necessarily the first cost at the time it was originally constructed and placed in service.

Original tract. Contiguous body of land under the same ownership.

Overall property tax limitation. Constitutional or statutory limitation on the total amount of taxes that may be levied for all purposes against any parcel of real estate within any one year, such overall limit to be a fixed percentage of the true value of such parcel of real estate. The overall limitation is so called to distinguish it from the limitations on separate portions of the real estate tax now in effect in nearly all states, such as specific limitations on the taxes levied for schools or parks.

Overimprovement. Improvement that is not the best use for the site on which it is placed by reason of excess size or cost.

Owner. Owner of the land as recorded in the registry of deeds for the country, or as registered in the land court.

Package mortgage. Mortgage that includes real and personal property.

Parcel of land. Contiguous quantity of land in the possession of, or owned by, or recorded as the property of, the same person.

Parking area. Lot or part thereof used for the storage or parking of motor vehicles, on impervious, open hardsurface area, with or without the payment of rent or charge.

Parking space. Stall or berth that is arranged and intended for the parking of one motor vehicle in a parking area or a garage.

Parties in interest. Persons, other than mortgagees or holders of vendor's liens, who have an interest of record in or who are in possession of a residence.

Partition

1. Division of property into separate parts by co-owners; also an action to divide the real estate into shares when one (or more) of the joint owners or owners in common desires his separate share.
2. Act of dividing property among the several owners thereof who may hold it either in joint tenancy or as tenants in common.
3. Division made of real property among those who own it in undivided shares.

Partition proceedings. Legal procedure by which an estate held in joint or common tenancy is divided and title passed to each of the previous tenants.

Party wall

1. Wall constructed between two adjoining pieces of property and used in common by both owners.
2. Dividing wall erected on a line separating two adjoining properties and to whose use the owners of the respective parcels have common rights.
3. Wall erected on the line between two adjoining properties, belonging to different persons, for the use of both properties.

Patent. Conveyance of title to government land.

Penthouse. Structure located on the roof of the main building for the purpose of living accommodations or mechanical equipment.

Percentage lease

1. Lease under which the tenant is required to pay as rental a specified percentage of the gross income from total sales made on the premises.

2. Lease of property in which the rental is a stated portion of the receipts from the business conducted by the lessee. It is frequently expressed as a percent of the gross revenues.

Perpetual easement. Easement constantly maintained.

Perpetuity. Continuing forever. Legally pertaining to real property.

Person. Natural person, his or her heirs, executors, administrators, or assigns, also including a firm, partnership, or corporation, its or their successors or assigns, or the agent of any of the aforesaid.

Personal chattel. Item of movable property.

Physical depreciation. Physical deterioration; adverse changes in the physical composition of property, such as loss of topsoil, wear and tear on buildings, or damage caused by storm.

Plan. Drawing or drawings illustrating the work to be done.

Plat

1. Map developed indicating survey planned use of land.

2. Plan or map of land showing its subdivision into smaller parcels or lots.

Plat book

1. Record showing the location, size, and name of the owner of each plot of land in a stated area.

2. Public record of various recorded plans in the municipality or county.

Plot. Parcel of land consisting of one or more lots or portions thereof, which is described by reference in a recorded plat or by metes and bounds.

Plottage. Increment in unity value of a plot of land created by assembling smaller ownerships into one ownership.

Plotting increment. Appreciation in unit value created by joining smaller ownerships into one large single ownership.

Point. One-time charge of 1% or agreed on percentage of the amount loaned, paid by the seller.

Porch. Roofed structure projecting from a building and separated from the building by the walls thereof.

Postponement of lien. Subordination of a presently prior lien to a subsequent judgment or mortgage.

Potential value. Value that would or will exist if and when future probabilities would become actualities.

Power-of-attorney. Instrument authorizing another to act as one's agent.

Preliminary subdivision plans. Complete and exact subdivision plans defining the property rights, proposed streets and other improvements presented for the purpose of securing preliminary approval.

Price level. Relative position in the scale of prices as determined by a comparison of the prices (of labor, materials, capital, etc.) as of one time with prices as of other times.

Principal. Sum employed as a fund or investment as distinguished from its income or profits; original amount of a loan due and payable at a certain date; party to a transaction as distinguished from an agent.

Principal building. Building that houses the main use or activity occurring on a lot or parcel of ground.

Principal note. Promissary note that is secured by the mortgages or trust deed.

Private garage. Accessory building or a portion of a principal building designed or used solely for the storage of motor vehicles owned and used by the occupants of the building to which it is accessory.

Private sewer. Sewer privately owned on private property, connecting to a main sewer.

Private street. Street held in private ownership, installed and maintained by the owner.

Privity. Mutual or successive relationship to the same rights of property.

Property

1. Any object of value subject to ownership by persons. The term is incorrectly used as a synonym for "real estate."

2. Right or interest that an individual has in lands and chattels to the exclusion of all others.

Property management. Real estate specialty that includes the care, leasing, and maintenance of property for a fee.

Proration. Proportionate division among the parties involved on the basis of a fixed rate of computation.

Prospect. Potential customer for real estate purchase or lease.

Prospectus published. Advertisement for a new enterprise, such as rural property or subdivision.

Publication date. In foreclosure proceedings, the date when the notice of sale was first published as prescribed by law.

Public land ownership. Any land use operated by or through a unit or level of government, either through lease or ownership, such as municipal administration and operation, county buildings and activities, state highway offices, and similar land uses—and federal uses such as post offices, bureau of public roads and internal revenue offices, military installations, and the like.

Public parking area. Any land used or intended to be used for the parking of motor vehicles and for which a fee is charged.

Public property. Property the ownership of which is vested in the community.

Public report. As pertains to subdivisions, a report issued by the real estate commissioner (or his equivalent) setting forth the known facts about a proposed subdivision. The report must be given to each buyer before he signs a purchase agreement.

Public use. Use of any land, water, or buildings by a public body for a public service or purpose.

Purchase money mortgage

1. Mortgage loan providing a purchaser with the money necessary to buy property.

2. Mortgage executed by the purchaser as a part of the purchase price.

3. Mortgage given by a grantee to the grantor in part payment of the purchase price of real estate.

Quadrangle. Tract of the land in the U.S. governmental survey system measuring 24 miles on each side of the square.

Quarter section. One quarter of a section contains 160 acres.

Quasi-public use. Use serving a community or public purpose, and operated by a noncommercial entity, or by a public agency.

Quiet enjoyment. Right of an owner to the use of property without interference of possession.

Quiet title. Court action brought to establish title and to remove a cloud on the title.

Quit claim

1. Deed to real property that is no more than a release of whatever rights the grantor may have.

2. In conveyancing, to release or relinquish a claim; to execute a deed of quit claim.

Quit claim deed

1. Deed of conveyance whereby whatever interest the grantor possesses in the property described in the deed is conveyed to the grantee without warranty.

2. Deed given when the grantee already has or claims complete or partial title to the premises and the grantor has a possible interest that otherwise would constitute a cloud upon the title.

Quit notice. Notice to a tenant to vacate rented property.

Range. Strip of land six miles wide, determined by government survey, running in a north-south direction.

Raw land. Acreage that is not in use and for which no specific use has yet been determined. It is undeveloped, unzoned, and usually surrounded by other land in the same category.

Real chattel. Item of property annexed to or concerned with real estate.

Real estate. Land rights and whatever is made part of or attached to it by nature or man.

Real estate broker. Licensed person, firm, partnership, copartnership, association, or corporation who for a compensation or valuable consideration sells or offers for sale, buys or offers to buy, or negotiates the purchase, sale, or exchange of real estate, or who leases or offers to lease, or rents or offers for rent any real estate or the improvements thereon for others as a whole or partial vocation. Definitions differ under licensing and other laws of various states.

Real estate investment trust (REIT). Method of investing in real estate in a group, with certain tax advantages.

Real estate salesman. Licensed person who for a compensation or valuable consideration is employed, either directly or indirectly, by a real estate broker to sell or offer to sell, buy or offer to buy, or to negotiate the purchase or sale or exchange of real estate, or to lease, rent, or offer for rent any real estate, or to negotiate leases thereof or of the improvements thereon as a whole or partial vocation. Definitions differ under licensing and other laws of various states.

Real estate tax. Amount of money levied annually against the ownership of real estate for maintenance of government functions.

Real property

1. The term is synonymous and used interchangeably with real estate and includes not only the land, but buildings, fences, sidewalks, and other improvements on the land. Real property also includes all water within the property boundaries and all minerals, ores, and oils in natural deposit beneath the surface of the land.

2. Land and everything growing or erected on it, including things permanently attached to it.

Realtor

1. Usually a member of the National Association of Real Estate Boards either by virtue of being a member of a constituent local real estate board or as an individual member of the national organization, and a subscriber to a prescribed code of ethics in his relations with other real estate operators, with his clients, and with the general public.

2. Coined word used to designate an active member of a local real estate board affiliated with the National Association of Real Estate Boards.

3. Real estate broker who is licensed or registered by the state.

Rear lot line. Lot line that is generally opposite the lot line along the frontage of the lot.

Rear yard. Yard extending across the full width of the lot and measured between the rear property line and the rear line of the building.

Recapture clause. Clause in an agreement providing for retaking or recovering possession. As used in percentage leases, it permits to taking a portion of earnings or profits above a fixed amount of rent.

Record plan. Exact copy of the approved final plan, reproducible of standard size prepared for necessary signatures and recording with the recorder of deeds.

Redemption

1. Recovery of property that has been lost through foreclosure of a mortgage, tax forfeiture, or other legal process.

2. Right of a mortgagor to redeem the property by paying the debt after the expiration date; right of an owner to reclaim his property after a sale for taxes.

Redevelopment. Development or improvement of cleared or undeveloped land in an urban renewal area.

Reduction certificate. Certificate showing the balance due on a mortgage at the time of closing the sale.

Reformation. Action to correct a mistake in a deed or other instrument.

Rehabilitation. Restoration to good condition of deteriorated structures, neighborhoods, and public facilities, which may include repair, renovation, conversion, expansion, remodeling, or reconstruction.

Release. Relinquishment of some right or benefit to a person who already has some interest in the property.

Release of lien. Discharge of certain property from the lien of a judgment, mortgage, or claim.

Relocation. Title I of the Housing Act of 1949, amended numerous times, requires the preparation of a feasible plan for relocation of families or individuals to decent, safe, and sanitary dwellings.

Remainder. Relinquishment, concession, or giving up of a right, claim, or privilege by the person in whom it exists or to whom it accrues to the person against whom it might have been demanded or enforced.

Remainder estate

1. Estate in property created simultaneously with other estates by a single grant and consisting of the rights and interest contingent upon and remaining after the termination of the other estates.

2. An estate that vests in one, other than the grantor, after the termination of an intermediate estate. An estate limited to take effect and to begin after another estate is determined.

Remise. Remit or give up; formal word in a deed of release and quit claim, the usual phrase being "remise, release, and forever quit claim."

Remodeling. Any change, addition, or modification in construction that involves a change in structure.

Rent. Compensation, either in money, provisions, chattels, or labor, received by the owner of the land from the occupant.

Rental value. Amount of money for which a property does or would rent. It is sometimes used as a base for determining a supposed capital value.

Repair. Reconstruction or renewal of any part of an existing building for the purpose of its maintenance.

Replacement cost. Cost of replacing the subject property new with one having exactly the same utility.

Reproduction cost

1. Cost of replacing the subject improvement with one that is an exact replica.
2. Normal cost of exact duplication of a property as of a certain date.

Required setback line. Line beyond which a building is not permitted to extend under the provisions of applicable zoning ordinance which has established minimum depths and widths of yards.

Required yard. Yard having a depth and width set forth in applicable zoning ordinances that are measured perpendicular to the lot lines.

Reservation. Right reserved by an owner in the sale or lease of a property.

Residence. Place where a person physically lives. This differs from his "domicile." The laws of the various states set up different requirements as to the period of residence for different purposes, for example, maintaining a divorce action.

Residence farm. Usually a parcel of land not less than 1 acre in area on which is located a building used as a dwelling and which land is worked on by a single family who are the occupants of the dwelling.

Residence lot. Lot whose front boundary line abuts on a public street.

Residence street. Portion of a street between intersections with two other streets where the majority of the frontage is occupied by residences and is within a zoned residence district.

Residential purposes. Building used for residential purposes such as one or two family residences, apartment houses, and multiple dwellings.

Restriction covenant. Clause in a deed limiting the use of the property conveyed for a certain period of time.

Restrictions

1. Limitations on the use or occupancy of real estate defined by covenant in deeds, by private agreements, or by public legislative action.
2. Device in a deed for controlling the use of land for the benefit of the land.

Reversed corner lot. Corner lot the side line of which is substantially a continuation of the front property line of the first lot to its rear, or its rear lot line which abuts the side lot line of another lot.

Reversion

1. Return of real estate to the original owner or his heirs after the termination of some temporary grant.
2. Recovery by the lessor of possession of leased property upon the termination of the lease, with all the subsequent rights to use and enjoyment of the property.
3. Residue of an estate left to the grantor, to commence after the determination of some particular estate granted out by him.

Reversionary right. Right of a lessor to recover possession and use of property upon termination of a lease.

Right. Interest or title in property; just and legal claim to hold, use, or enjoy or to convey or donate property.

Right of occupancy. Privilege to use and occupy a property for a certain period under some contractual guarantee, such as a lease or other formal agreement.

Right of survivorship. Right contained in a joint tenancy deed for the surviving tenants to acquire the interest of the deceased tenant.

Right-of-way

1. Privilege to pass or cross; an easement over another's land. The term is also used to describe the strip of land that railroad companies use for a roadbed or land dedicated to public use for a roadway, walk, or other way.
2. Easement over another's land. The term is also used to describe a strip of land used for a public utility for a public purpose.
3. Easement that grants to its receiver the right to pass over or maintain use of a parcel of property belonging to another.

Riparian. Pertaining to the banks of a river, stream, or waterway at a shore line.

Riparian grant. Conveyance of riparian rights.

Riparian lease. Written instrument setting forth the terms, conditions, and date of expiration of the rights to use lands lying between the high-water mark and the low-water mark.

Riparian owner. One who owns lands bounding on a waterway, or at a shore line.

Riparian rights

1. Rights of owners of land adjoining a waterway.
2. All phases of right and title (of the upland owner) in and to the water and land below high-water mark. These rights vary and depend on local legal regulations.

River frontage yard. Yard extending across the rear or along the side of a lot that abuts an established bulkhead line of a waterway.

Roadway. Public thoroughfare devoted to vehicular traffic, or that part included between curbs.

Rooming house. Building, or portions thereof that contain sleeping rooms and which is regularly used or available for permanent occupancy.

Rooming unit. Habitable room or group of habitable rooms forming a single habitable unit not used for cooking or eating purposes.

Row dwelling. Dwelling of which the walls on two sides are party or lot line walls.

Running with the land (easement). Easement that inures to the benefit and advantage of subsequent owners of the land, for which (the easement) was originally created.

Rural districts. All places not urban, usually in unincorporated areas, but in some cases within city limits where sparse population density permits.

Sales contract. Contract embodying the terms of agreement of a sale.

Satisfaction piece. Instrument for recording and acknowledging payment of an indebtedness secured by a mortgage.

Section. Section of land established by government survey and containing 640 acres.

Section of land. One square mile; 640 acres; one thirty-sixth of a township. Part of the survey system of public lands in the United States.

Seisin

1. Possession with an intent on the part of the individual who holds it to claim a freehold interest.
2. Possession of real estate by one entitled thereto.

Separate property. Property owned by a husband or wife that is not community property, acquired by either spouse prior to marriage or by gift or devise after marriage.

Septic tank. Tank in which the solid matter of sewage is deposited and retained until it has been disintegrated by bacteria, used in areas where sewer systems have not been installed.

Septic tank system. Private sewage disposal section for an individual home.

Service property. Property devoted to or available for utilization for a special purpose, such as a clubhouse, church property, public museum, or public school.

Setback. Distance from the curb or other established line within which no building may be erected.

Severalty ownership. Real property owned by one person only; sole ownership.

Severance damage. Impairment in value caused by separation; commonly, the damage resulting from the taking of a fraction of the whole property, reflected in a lowered utility and value of the land remaining brought about by reason of the fractional taking.

Sheriff's deed. Instrument drawn under order of court to convey title to property sold to satisfy a judgment at law.

Side lot line. Dividing line between two lots or a lot and a street.

Side property line. Lot lines connecting the front and rear property lines of a lot.

Sidewalk. The area of a street or other public way that is located between the curb line and the street line.

Side yard. Yard between the building and the side line of the lot and extending from the street line of the lot to the rear yard.

Simple listing. Listing of property with a broker for sale or rent other than through an exclusive agency or exclusive right-to-sell contract; an open listing, usually verbal.

Single and separate ownership. Ownership of property by one or more persons whose ownership is separate and distinct from that of any adjoining property.

Single family dwelling. Detached building designed for or used exclusively by one family.

Single ownership. Ownership by one person, or by two or more persons whether jointly, as tenants by the entirety, or as tenants in common, of a separate parcel of real property.

Sinking fund. Fund set aside from property that, with accrued interest, will eventually pay for replacement of the improvements.

Site

1. Parcel of land or a portion thereof, considered a unit, devoted to or intended for use or occupied by a structure or a group of structures that are united by a common interest or use.
2. Plot of ground set aside for specific purpose and use.

Sky lease. Lease for a long period of time of space above a piece of real estate. The upper stories of a building are erected by the tenant and upon the termination of lease, the improvement belongs to the lessor.

Slum. Primarily a residential area in which rundown housing provides shelter for the poor and the deprived.

Special assessment. Tax levied against real estate ownership to pay (usually) for public improvements of assumed specific benefit to the property.

Special districts. Myriad of school, water, highway, and sewer districts, and other units of government with power to tax and spend for particular purposes. Their boundaries are seldom identical with the political boundaries of cities, townships, or counties.

Special warranty deed.

1. Warranty only against the acts of the grantor himself and all persons claiming by, through, or under him.
2. Deed wherein the grantor limits his liability to the grantee to anyone claiming by, from, through, or under him, the grantor.

Specific performance. Remedy in a court of equity compelling the defendant to carry out the terms of the agreement or contract that was executed.

Specific risk guarantee. In trade-in housing, a guarantee that obligates someone to carry a specific amount of any loss that might result from the trade.

Split lot. Lot that is divided by a zone boundary.

Spot zoning. Zoning on a parcel by parcel basis, rather than an established master plan.

Square foot content. Sum expressed in square feet of the area of all rooms on all floors of a building.

Squatter's rights

1. Rights of occupancy of land created by virtue of long and undisturbed use, but without legal title or arrangement; in the nature of right at common law.
2. Occupancy of land by virtue of long use against the recorded title owner.

Standard depth. As applied to land (such as urban lots), the depth chosen as standard, usually the depth that is most common in the neighborhood.

Statute. An act of the legislature, adopted pursuant to its constitutional authority.

Statute of frauds. Statute that requires certain contracts relating to real estate, such as agreements of sale, to be in writing in order to be enforceable.

Statute of limitations. Statute fixing and limiting certain periods of time beyond which rights of action cannot be enforced.

Statutory warranty deed. Warranty deed form prescribed by state statutes.

Step-up lease. Lease that provides for a certain rent for an initial period, followed by an increase in rent over stated periods.

Story. That vertical portion of a building included between the surface of any floor and the surface of the floor next above.

Straight line method. Cost of the improvement is depreciated equally each year for the economic life of the improvement.

Straight-term mortgage. A mortgage calling for principal to be paid in a lump sum (balloon) at maturity.

Street. Public thoroughfare legally dedicated for public use.

Subdivision. Tract of land divided into lots suitable for home-building purposes.

Sublease. Agreement by a tenant transferring all or part of his rights to another.

Subletting. Leasing by a tenant to another, who holds occupancy under the responsibilities of the tenant.

Subordination clause. Clause in a mortgage or lease, stating that the rights of the holder shall be secondary or subordinate to a subsequent encumbrance.

Surety bond. Instrument in writing and under seal assuring the performance of specific acts and at stated times and indemnifying against monetary loss for any failure so to perform.

Surrender. Cancellation of a lease by mutual consent of lessor and lessee.

Survey

1. Statement, report, map, or plat of the courses, distances, and quantity of land.
2. Process by which a parcel of land is measured and its area ascertained.

Surveying. Ascertaining the quantity, location, and boundaries of a piece of land.

Switch site. Industrial property that has railroad switching facilities or is so located that such facilities can be installed.

Syndicate. Group of investors who pool their funds for the purpose of investment, each sharing the loss or gain for tax purposes.

Tangible property. Property that by its nature is susceptible to the senses. Generally, it includes the land, fixed improvements, furnishings, merchandise, and cash.

Tax

1. Levy (usually in terms of money) against citizens or their property for the support of government functions.
2. Charge assessed against persons or property for public purposes.

Tax abatement. Rebate or reduction of a tax, particularly a tax improperly levied.

Taxable value. Assessed value; value on which the tax levy is computed.

Tax credit. Credit against the actual taxes owed, as opposed to a deduction from taxable income before the taxes are calculated. It is generally viewed unfavorably by tax experts, both because of the loss of revenue and because of the imprecision with which it can be applied.

Tax deed

1. Deed issued by public authority subsequent to a tax sale of property.
2. Deed for property sold at public sale by a political subdivision, such as a city, for nonpayment of taxes by the owner.

Tax foreclosure. Seizure of property by taxing officials for unpaid taxes.

Taxpayer. One who pays a tax; building erected for the primary purpose of producing revenues to meet the taxes on the land.

Tax penalty. Forfeiture of a sum because of nonpayment of taxes.

Tax receivership. Office or function of a receiver appointed by a court or under a statute upon default of taxes.

Tax redemption. Recovery of property by payment of delinquent taxes and penalties.

Tax sale. Sale of property, usually at public auction, for nonpayment of taxes.

Tenancy. Estate less than freehold; holding of real estate under a lease.

Tenancy at suffrance. Tenancy that arises when a tenant holds over after expiration of his lease.

Tenancy at will

1. Tenancy that may be terminated at the will of either the lessor or lessee.
2. License to use or occupy lands and tenements at the will of the owner.

Tenancy in common

1. Tenancy shared by two or more parties.
2. Form of estate held by two or more persons, each of whom is considered as being possessed of the whole of an undivided part.

Tenant

1. Technically one who has possession of land or space in a building by virtue of ownership or any other kind of right. Popularly, one who temporarily, and on certain conditions, possesses the property of another.
2. Person who holds real estate under a lease (lessee).

Tenant at sufferance. One who comes into possession of lands by lawful title and keeps it afterward without any title at all.

Tenement

1. Term commonly applied to houses and other buildings. In its original, proper, and legal sense it signifies everything that may be owned, provided it be of a permanent nature, whether it be of a substantial and sensible kind or of an unsubstantial, ideal kind. It is of greater extent than land, including not only land but rents, commons, and other rights and interests issuing out of or concerning land.
2. Everything of a permanent nature that may be held.

Terre tenant. One who has the actual possession of land.

Thoroughfare. Street, road, way, or other space customarily used for travel.

Through lot. Interior lot having frontage on two streets.

Tidelands. Property covered from time to time by ocean or lake waters.

Tier. Strip of land six miles wide running in an east-west direction, as determined by government survey.

Title

1. Evidence by which the owner proves his ownership, as well as his right to possession.
2. Union of all the elements that constitute proof of ownership. In real property law the title is the means whereby the owner of lands has the just possession of

his property. A title is the means whereby a person's rights to property is established.

3. Evidence of ownership that refers to the quality of the estate.

4. Evidence of ownership and lawful possession of property.

Title by adverse possession. Title acquired by occupation and recognized, as against the claim of the paper title owner.

Title certificate. Written statement prepared by an attorney or title company stating who has ownership of property.

Title guarantee policy. Title insurance furnished by the owner, provided as an alternative for an abstract of title.

Title insurance

1. Insurance written by a title company to compensate the lender or owner for any loss if the property is owned by a person other than the seller, subject to the conditions of the policy.

2. Agreement binding the insurer to indemnify the insured for losses sustained by reason of defects in title to the real estate.

3. Policy of insurance that indemnifies the holder for any loss sustained by reason of defects in the title.

Torrens certificate. Documents issued by a public authority called a "registrar" acting under the provisions of the Torrens Law, indicating the party in whom title resides.

Torrens system. System of title records provided by state law.

Torrens system of land registration. System of state insurance for land titles.

Tort. Wrong committed by one person to another and not arising from contract.

Township. Territorial subdivision, 6 miles long, 6 miles wide, and containing 36 sections, each 1 mile square.

Tract. A parcel of land. In some states synonymous with a "subdivision."

Trespass. Entry without permission on the property of another.

Trover. Action to recover the value of property wrongfully taken by another, more commonly called an "action for conversion."

Trust. Ownership rights transferred by the owner to another party, usually under temporary or conditional terms —for example, ownership transferred to a trustee until an heir becomes of legal age.

Trust deed

1. Conveyance of real estate to a third person to be held for the benefit of a cestui que trust (beneficiary).

2. In some states, a trust deed is a substitute for a mortgage. The buyer deeds the property to a third party, usually a title or escrow company, who holds it as a guarantee to the lender that the buyer will repay what he borrowed.

3. Deed that establishes a trust; generally, an instrument that conveys legal title of property to a trustee and states the authority of and the conditions binding upon the trustee in dealing with the property held in trust. Frequently trust deeds are used to secure lenders against loss; in this respect they are similar to mortgages.

Trustee

1. One who holds title to property for the benefit of another.

2. Person in whom an estate, interest, or power, in or affecting property, is vested or granted for the benefit of another person.

Trustor. One who deeds his property to a trustee.

Turnkey. Term applied to public housing provided by a housing authority's purchase of privately produced construction from the builder, who follows general requirements instead of minutely detailed federal specifications. The term is also used in the provision of rehabilitated private housing for public housing tenancy.

Unbalanced improvement. Improvement that is not the highest/best use for the site on which it is placed.

Underimprovement. Improvement that is not the highest/best use for the site on which it is placed by reason of being smaller in size or cost than one that would bring the site to its highest and best use.

Undeveloped land. Land in parcels sufficiently large enough for the planning of subdivisions, presently used for agriculture or woodland.

Unearned increment

1. Addition or increase in value said to be "unearned" because it results from population increases, community expansion, greater desirability, and more active market in real estate, development of land, and so forth.

2. Increase in value of real estate due to no effort on the part of the owner but often due to increase in population.

Uniform commercial code. A code which regulates the transfer of personal property.

Unilateral contract. One in which one party makes an express undertaking, without receiving in return any promise of performance from the other.

Unimproved. Vacant or lacking in essential improvements required to serve a useful purpose. Term applied to raw land.

United States governmental survey system. Method of describing or locating real property by reference to the governmental survey; also known as the "rectangular survey system."

Unlawful detainer. Statutory proceedings by which a landlord removes a tenant who holds over after his lease has expired, after his tenancy is terminated by notice, or after default in payment of rent or other obligations.

Unit of housing. Entire dwelling unit occupied by a person or family, whether a house or apartment, single room or multiroomed, owned or rented.

Urban district. Densely settled area within the city limits.

Use value. Value (in money) computed on the basis of the amount paid for the use of property.

Usury

1. Illegal interest for a loan. Excess interest over the legal rate is usury. In many places usurious contracts will not be enforced in the courts, and even the person who agreed to pay illegal rates may set up the defense of usury. In some places, courts will enforce the contract by allowing the principal to be paid with legal interest.

2. Charging more than the legal rate of interest for the use of money.

Valuation. Act or process of estimating value; amount of estimated value.

Value. Pertaining to a building or structure, the estimated present cost to replace the building or structure in kind, but not including the value or cost of the foundations, or land.

Variance. Deviation granted to allow a change in the ordinance.

Vendee

1. Purchaser of real estate under an agreement.
2. One who buys property.

Vendor

1. One who disposes of property by sale.
2. Seller of real estate, usually referred to as the "party of the first part" in an agreement of sale.

Waiver

1. Intentional abandonment of a right.
2. Renunciation, abandonment, or surrender of some claim, right, or privilege.

Warrant. In conveyancing, to assure the title to the property sold by an express covenant to that effect in the deed of conveyance; to stipulate by an express covenant that the title of a grantee shall be good, and his possession undisturbed. In contracts, to engage or promise that a certain fact or set of facts, in relation to the subject matter, is, or shall be, as it is represented to be.

Warranty deed

1. Guarantee that the title the property owner transfers to a buyer has not been given previously to someone else, and that the property being transferred is free of all previous claims and debts, except as noted in the bill of sale, mortgage, or deed.
2. Instrument, in writing, by which a freehold is guaranteed by the grantor, his heirs, or successors.
3. Deed that contains a covenant that the grantor will protect the grantee against any claimant.

Waste

1. Abuse of property by a tenant or someone having a temporary interest in the property, resulting in a loss to the owner.
2. Willful destruction of any part of the land or improvements so as to injure or prejudice the estate of a mortgagee, landlord, or remainderman.

Water rights. Property consisting of the rights to a water supply.

Waterway. Any body of water, including any bayou, creek, river, lake, or bay or any other body of water, natural or artificial, on a single lot or parcel of land.

Waterway line. Line marking the normal division between land and water, established by the agency having jurisdiction.

Way. Street, alley, or other thoroughfare or easement permanently established for passage of persons or vehicles.

Width of lot. The mean width of a lot measured at right angles to its depth.

Without recourse. Words used in endorsing a negotiable instrument to denote that the endorser will not be liable to a future holder in the event of nonpayment.

Yard. Open unoccupied space on all sides of a building, based on the required space for such yards.

Yield. Annual percentage rate of return on an investment in real estate, stocks, or bonds.

Zone. Area set off by a governing body for specific use, such as residential, commercial, or industrial use.

Zoning

1. Governmental ordinance regulating the use that may be made of each parcel of land.
2. Public regulation of character and intensity of use of property through the employment of police power.
3. "Spot zoning" occurs when the tract in question is singled out for treatment differing unjustifiably from that of similar surrounding land, thereby creating an island having no relevant differences from its neighbors.
4. Control of land usage by village, city, township, county, or state authorities, with power to limit the property use by established standards.

Zoning ordinance. Exercise of police power by a government in regulating and controlling the character and use of property.

RECORDING AND TESTING EQUIPMENT AND DEVICES

Abrasion resistance index. A measure of the abrasion resistance of a vulcanized rubber relative to that of a standard vulcanized rubber under the same specified conditions.

Abrasion tester. A machine for determining the quantity of material lost by friction wear under specific conditions.

Absolute manometer. Instrument whose calibration can be calculated from the measurable physical constants of the instrument and which is the same for all gases.

Absorption test. A test made to determine the absorption of materials.

Accelerated adsorption tests. Tests in which the end point is hastened by testing at conditions more severe than those anticipated in service.

Accelerated life test. Method designed to approximate, in a short time, the deteriorating effect of normal long-term service conditions.

Acceptance testing. Testing performed to affirm if a material meets acceptance criteria.

Angstrom. A unit of length equal to 10^{-10} meter.

Angstrom compensation pyrheliometer. Instrument developed by K. Angstrom for the measurement of direct solar radiation.

API gravity. Arbitrary scale developed by the American Petroleum Institute.

Arid chamber. Environmental test facility simulating high temperature, low humidity, and solar radiation heat and light as found in arid areas of the earth.

Atmometer. Instrument that measures the evaporation rate of water into the atmosphere. Four main classes of atmometers may be distinguished: large evaporation tanks sunk into the ground or floating in water, small evaporation pans, porous porcelain bodies, and porous paper-wick devices.

Atmosphere for testing. Air at ambient conditions of relative humidity and temperature.

Atterberg test. Test performed to determine the plasticity of soils.

Autographic dilatometer. Instrument that automatically records instantaneous and continuous changes in dimensions and some other controlled variable such as temperature or time.

Back pressurizing testing. Method of testing sealed units in which the units are subjected to a tracer gas pressure for a period of time. The tracer gas is flushed from outside the unit and the tracer gas leakage from the unit is measured.

Ball test. Test to determine the consistency of freshly mixed concrete by measuring the depth of penetration of a cylindrical metal weight with a hemispherical bottom.

Barograph. Recording barometer. Barographs are classified on the basis of their construction into the following types: aneroid barograph (including microbarograph), float barograph, photographic barograph, and weight barograph. The aneroid barograph, which is the least complicated and possibly the least accurate of the barographs, is the most commonly used in weather stations.

Barometer. Instrument for measuring atmospheric pressure. A mercurial barometer employs a column of mercury supported by the atmosphere. The aneroid barometer has a partly exhausted thin-metal cylinder somewhat compressed by atmospheric pressure.

Barothermograph. Instrument that automatically records temperature and pressure.

Barothermohygrograph. Instrument that automatically records pressure, temperature, and humidity of the atmosphere.

Bead test. Test used in mineral identification.

Bearing test. Test made to compute the bearing values of soils.

Beaufort wind scale. System of estimating and reporting wind speeds, invented in the early nineteenth century by Admiral Beaufort of the British Navy. In its present form for international meteorological use, it equates Beaufort force (or Beaufort number), wind speed, a descriptive term, and visible effects upon land objects or sea surface. As originally given, Beaufort numbers ranged from 0, calm, to 12, hurricane. They have now been extended to 17.

Becke test. Microscopic test in which indices of refraction are compared for minerals.

Belgium block course. Test facility for simulating the conditions of transportation. The course is a specially prepared roadbed having varying degrees of roughness, waviness, and other controlled characteristics over which wheeled equipment is moved at varying speeds to study the effects of shock and vibration caused by transportation. Belgium block is only one section of the Munson test course, which also includes a course washboard, a radial washboard, and a single corrugation section.

Bench-scale testing. Testing of methods and materials on a small calibrated scale.

Bend test. A test for ductility performed by bending or folding, usually by steadily applied forces, but in some instances by blows, the specimen having a cross section substantially uniform over a length several times as great as the largest dimension of the cross section.

Beranek scale. Scale that measures the subjective loudness of noise.

Berthon dynamomometer. Instrument for measuring the diameter of small objects.

Bevameter. Mobile or portable instrument developed by the Land Locomotion Laboratory for measuring horizontal and vertical stress-deformation curves of natural soils or soil-simulating materials, and consisting of one or more rotating horizontal shear heads and one or more vertical displacement penetrometers.

B-H meter. Device used to measure the intrinsic hysteresis loop of a sample of magnetic material.

Bierbaum scratch hardness test. Test for the hardness of a solid sample by microscopic measurement of the width of the scratch made by a diamond point under preset pressure.

Black-bulb thermometer. Thermometer whose sensitive element has been made to approximate a blackbody by covering it with lamp black. The thermometer is placed in an evacuated transparent chamber that is maintained at constant temperature. The instrument responds to insulation, modified by the transmission characteristics of its container.

Blaine test. Test performed for determining the fineness of cement and other fine materials on the basis of permeability to air of a sample prepared under specific conditions.

Boiler hydrostatic test. Test that employs water under pressure in a new boiler before use.

Bolometer

1. Radiation detector that converts incident radiation into heat, which in turn causes a temperature change in the material used in the detector. The change is measured to give the amount of incident radiant energy.

2. Instrument that measures the intensity of radiant energy by employing a thermally sensitive electrical resistor.

Bourdon tube. Closed curved tube of elliptical cross section used in some temperature-sensing and pressure-sensing instruments. Expansion of the fluid due to a temperature change causes an increase in the radius of curvature of the tube. Curvature may then be measured by the travel of the tip of the tube. Curvature is a measure of the difference between the pressure inside the tube and that outside.

Brinell hardness test. An indentation hardness test using calibrated machines to force a hard ball, under specified conditions, into the surface of the material under testing and to measure the diameter of the resulting impression after removal of the load.

Calcimeter. Instrument for measuring the amount of lime in soil.

Calorimeter

1. Instrument for measuring heat quantities, such as machine capacity, heat of combustion, specific heat, vital heat, and heat leakage; also, an instrument for measuring quality or moisture content of steam or other vapor.

2. Instrument for measuring heat exchange during a chemical reaction, such as the quantities of heat liberated by the combustion of a fuel or hydration of a cement.

Capacitance meter. Instrument used to measure capacitance values of capacitors or of circuits containing capacitance.

Carbometer. Instrument for measuring the carbon content of steel by measuring magnetic properties of the steel in a known magnetic field.

Cathode-ray voltmeter. Instrument consisting of a cathode-ray tube of known sensitivity, whose deflection can be used to measure voltages.

Caulometer. Electrolytic cell that measures a quantity of electricity by the amount of chemical action produced.

Ceilometer. Automatic, recording, cloud-height indicator.

Celsius temperature scale (same as Centigrade). By recent convention, the Ninth General Conference on Weights and Measure (1948) replaced the designation "degree Centigrade" by "degree Celsius."

Centigrade temperature scale. Temperature scale with the ice point at 0 degrees and the boiling point of water at 100 degrees. Conversion to the Fahrenheit temperature scale is according to formula.

Centrifugal tachometer. Instrument that measures the instantaneous angular speed of a shaft by measuring the centrifugal force on a mass rotating with it.

Centrifuge. Machine used in environmental testing to subject material to steady-state rotational acceleration about a fixed axis.

Chemical dosimeter. Instrument in which the accumulated radiation exposure dose is indicated by color changes accompanying chemical reactions induced by the radiation.

Chronometer. Portable timekeeper with compensated balance, capable of indicating time with precision and accuracy.

Chronometric tachometer. Instrument that repeatedly counts the revolutions during a fixed interval of time and presents the average speed during the last timed interval.

Clean chamber. Enclosed area in which airborne contamination (particulate matter) and, if necessary, temperature, humidity, and air pressure are controlled to a far higher degree than in conventional air-conditioned areas. It is commonly referred to as a "clean room."

Climatic test. Test designed to evaluate the effects of climatic conditions on the equipment undergoing the test. Climatic tests usually include sunshine, rain, hail, snow, sleet, wind, humidity, sand, dust, temperature, fungus, salt, and spray.

Clinometer. Any one of various instruments for measuring angles of slope, inclination, elevation, or the like, such as the angle between the horizontal and the line of sight to the spot of light thrown by a ceiling projector, or between the horizontal and a ship's axis.

Color comparator. Photoelectric instrument that compares an unknown color with that of a standard color sample for matching purposes.

Colorimeter. Optical instrument designed to compare the color of a sample with that of a standard sample or a synthesized stimulus.

Colorimetric photometer. Instrument that can measure light intensities in several spectral regions, using color filters placed in the path of the light.

Compass declinometer. Instrument used for magnetic distribution surveys.

Compression test. Machine-produced test on a test concrete sample that calculates the compressive strength of the concrete in pounds per square inch.

Consistometer. Apparatus for measuring the consistency of cement pastes, mortars, grouts, or concretes.

Consolidation test. Test performed in which the specimen is confined laterally in a ring and is compressed between porous plates which are saturated with water.

Continuity test. Test used to determine the location of a broken electrical connection.

Controlled-strain test. Test in which the load is so applied that a controlled rate of strain results.

Controlled-stress test. Test in which the stress to which a specimen is subjected is applied at a controlled rate.

Core sample analysis. Sample core retrieved by drilling, to analyze the composition, porosity, moisture content, and other values of the sample.

Carrodkote test. An accerated corrosion test for electrodeposits.

Corrosion fatigue limit. Maximum repeated stress endured by a metal without failure in a stated number of stress applications under defined conditions of corrosion and stressing.

Corrosion test. Test designed to determine the adequacy of a part for withstanding corrosion under specified conditions for a known length of time.

Coulombmeter. Instrument that measures quantity of electricity in coulombs.

Crash safety. Type of shock test intended to determine the mechanical integrity of equipment hardware under simulated aircraft crash landing loads.

Cure meter. Testing device that measures the progress of vulcanization.

Current type flowmeter. Device that measures liquid velocity in open and closed channels.

Cylinder test. A cast cylindrical specimen of concrete tested for its values.

Daniell hygrometer. Instrument used for measuring dew point.

DB meter. Meter having a scale calibrated to read directly in decibel values at a specified reference level.

Decibel meter. Instrument for measuring the electric power level, in decibels, above or below an arbitrary reference level; also called a "dB meter."

Declinometer. Device for measuring the direction of a magnetic field relative to astronomical or survey coordinates.

Decremeter. Instrument used for measuring the logarithmic decrement of a train of waves.

Deflectometer. Instrument for measuring minute deformations in a structure under transverse stress.

Densimeter. Instrument that measures the density or specific gravity of a liquid, solid, or gas.

Densitometer. Instrument for measuring the optical density of a material.

Depth micrometer. Instrument used to measure the depths of holes, slots, distances of projections, and shoulders.

Desiccator. Enclosed apparatus in which substances can be kept in a dry atmosphere. The latter is obtained by the inclusion of drying agents such as phosphorus pentoxide or concentrated sulfuric acid.

Destructive test. Test in which material or equipment is subjected to environmental conditions that are inherently damage-producing or destructive; test in which material or equipment is intentionally damaged to determine its damage or fatigue resistance.

Dew cell. Instrument used to determine the dew point. It consists of a pair of spaced bare electrical wires wound spirally around an insulator and covered with a wicking wetted with a water solution containing an excess of lithium chloride. The electrical potential applied to the wires causes a flow of current through the lithium chloride solution, which raises the temperature of the solution until its vapor pressure is in equilibrium with that of the ambient air.

Differential voltmeter. Instrument that measures only the difference between a known voltage and an unknown voltage.

Diffusion hygrometer. Hygrometer based on the diffusion of water vapor through a porous membrane.

Dilatometer. Instrument used in dilatometry for measuring length or volume changes.

Directional gyroscope. Gyroscopic instrument for indicating direction, containing a free gyroscope that holds its position in azimuth and thus indicates angular deviation from a preset heading.

Disk colorimeter. Device for comparing standard and sample colors.

Displacement pickup. Device that converts a detectable change in a medium, such as that produced by a sound wave or an electromagnetic wave, into some form of electrical energy.

Distillation test. Standardized test for finding the initial, intermediate, and final boiling points in the boiling range of petroleum products.

Distortion meter. Instrument that measures the deviation of a complex wave from a pure sine wave.

Distributed impact test. Apparatus or method that produces a spatial distribution of impacts by liquid or solid bodies over an exposed surface of a specimen.

Dose-rate meter. Instrument that measures radiation dose rate.

Dosimeter

1. Instrument that measures the amount of exposure to nuclear or x-ray radiation by utilizing the ability of such radiation to produce ionization of a gas.

2. Instrument for measuring and registering total accumulated exposure to ionizing radiations.

Drag-body flowmeter. Instrument to meter liquid flow.

Drop test. Type of shock test in which a test specimen, or a guided structure to which the specimen is mounted, is released from a specified height and after free fall is decelerated by a specified medium.

Drosometer. Instrument used to measure the amount of dew formed on a given surface.

Dry-bulb thermometer. Thermometer with an uncovered bulb, used with a wet-bulb thermometer to determine atmospheric humidity. The two thermometers constitute the essential parts of a psychrometer.

Durometer. Instrument used to measure hardness of rubber and rubber-like material plastics.

Dynamic impact test. Load displacement test simulating free fall of an object.

Dynamic test. Test conducted under active or simulated loading conditions.

Dynamometer. Instrument in which the force between a fixed and a moving coil provides a measure of current, voltage, or power.

Electrical impedance meter. Instrument that measures the complex ratio of voltage to current in a given circuit at a given frequency.

Electric energy meter. Device that measures the integral, with respect to time, of the power in an electric circuit.

Electric hygrometer. Instrument for indicating by electrical means the humidity of the ambient atmosphere.

Electrometer. Electrostatic instrument that measures a potential difference or an electric charge by the mechanical force exerted between electrically charged surfaces.

Emanometer. Instrument for the measurement of the radon content of the atmosphere. Radon is removed from a sample of air by condensation or absorption on a surface. It is then placed in an ionization chamber and its activity determined.

Endurance test. Dynamic fatigue test, such as a vibration test, usually conducted at accelerated stress levels.

Environmental field test

1. Test or program of tests in which an item of material is subjected to storage and functional testing in one or more specific environments.

2. Test in which a piece of equipment or an entire system is exposed or operated under natural environmental conditions.

Environmental simulation test. Test in which a piece of equipment or an entire system is exposed or operated under simulated service conditions, usually in a laboratory.

Environmental test. Test of equipment, supplies, and techniques under a specific set of environmental conditions in which each is intended to be used. Such a test will normally be an integral part of such other tests as engineering design, engineering, and service tests.

Erichsen test. A cupping test to measure the ductility of a piece of sheet metal and to determine its suitability for deep drawing.

Eudiometer. Instrument for measuring changes in volume during the combustion of gases.

Evaporation pan—type of atmometer. Pan used in the measurement of the evaporation of water into the atmosphere. An evaporation pan (class-A pan) is a cylindrical container fabricated of galvanized iron or monel metal with a depth of 10 in. and a diameter of 48 in.

Evaporimeter. Instrument for measuring the rate of evaporation of water into the atmosphere.

Exhaust-gas analyzer. Instrument that analyzes the gaseous products to determine the effectiveness of the combustion process.

Exposure meter. Instrument used to measure the intensity of light reflected from an object, for the purpose of determining the proper camera exposure.

Extensometer. Device for measuring linear strain.

Faradmeter. Instrument for measuring electric capacitance.

Fire-danger meter. Graphical aid used in fire-weather forecasting to calculate the degree of forest fire danger.

Flame detector. Sensing device that indicates whether or not a fuel is burning in a fuel-fired appliance, or if ignition has been lost, by transmitting a signal to a control alarm system.

Flame photometer. Instrument used to determine elements (especially sodium and potassium in Portland cement) by the color intensity of their unique flame spectra resulting from introducing a solution of a compound of the element into a flame; flame spectrophotometer.

Flattening test. Quality test performed by flattening metal tubing between parallel plates that are a specified distance from each other.

Flexometer. Instrument used for measuring the flexibility of suitable materials.

Flood-frequency curve. Graph showing the number of times per year on the average, plotted as abscissa, that floods of magnitude, indicated by the ordinate, are equaled or exceeded.

Flowmeter. Instrument used to measure pressure, flow rate, and discharge rate of a liquid, vapor, or gas flowing in a contained pipe or vessel.

Fluxmeter. Instrument for measuring magnetic flux.

Fog chamber. Confined space in which supersaturation of air or other gas is produced by reduction of pressure, cooling, or other means, producing an artificial fog.

Fracture general test. Production of a fracture in a metal sample to determine such values as discontinuities, grain size, and composition.

Fracture steel test. Test that utilizes a hardened steel disk section prepared from billet or bar stock that is fractured parallel to the grain flow so that, among other discoveries of the test, discontinuities due to inclusion segregates can be visually detected.

Frequency meter. Instrument for measuring the frequency of alternating current.

Frequency test. Test indicating the number of complete input cycles per unit time of a periodic quantity such as alternating current.

Frigorimeter. Instrument for measuring the physiological cooling power in millicalories per square centimeter and minutes. It consists of a blackened copper sphere, 7.5 cm in diameter, the surface of which is maintained electrically at 36.5°C (97.7°F) against the heat losses due to all meteorological conditions of the ambient air. A temperature of 36.5°C corresponds to the constant deep body temperature of man. (An older model was set at 33°C.)

Galvanometer. Instrument for measuring the magnitude of a small electric current or for detecting the presence or direction of such a current by means of motion of an indicator in a magnetic field.

Gas meter. An instrument for measuring the quantity of a gas passing thru the meter.

Gasometer. Apparatus employing a calibrated volume that is used to calibrate gas measuring devices.

Gas thermometer. Thermometer that utilizes the thermal properties of gas. There are two forms of this instrument: the type in which the gas is kept at a constant volume, and pressure is the thermometric property; and the type in which the gas is kept at constant pressure, and volume is the thermometric property. A gas thermometer is the most accurate of all thermometers and is used as the standard instrument for measurement of temperature.

Gauge. Instrument for measuring the size or state of anything.

Gauge strain. Device for measuring strain, which is the deformation produced in a solid as a result of stress.

Gaussmeter. Magnetometer whose scale is graduated in gauss or kilogauss, and measures the intensity and not the direction of the magnetic field.

Geomagnetic electrokinetograph. Device for measuring the lateral component of the speed of an ocean current by means of two pairs of electrodes towed astern and suitable registering apparatus.

Gillmore needle. Device used in determining the time of setting of hydraulic cement.

Goniometer. Instrument devised for measuring the angle through which a specimen is rotated.

Goniophotometer

1. Instrument used to obtain geometric distribution of reflected or transmitted radiant flux.

2. Instrument for measuring the directional light distribution characteristics of sources, lighting fittings, media, and surfaces.

Halogen leak detector. Detector that responds to halogen tracer gases (ASTM E 425).

Harmonic detector. Voltmeter circuit that measures only a particular harmonic of the fundamental frequency.

Heliograph. Instrument that records the duration of sunshine and gives qualitative measure of the amount of sunshine by the action of the sun's rays on blueprint paper; a type of sunshine recorder.

Helium leak detector. Detector using helium as the tracer gas (ASTM E 425).

High-frequency voltmeter. Instrument designed to measure currents alternating at high frequencies.

High-resistance voltmeter. Instrument having a resistance considerably higher than 1000 ohms per volt.

High-temperature chamber. Enclosed facility for producing thermostatically controlled high temperatures (usually by resistance heaters), used to determine the effect of high temperatures on a test item.

Hood test. An overall test in which an object under vacuum test is enclosed by a hood that is filled with tracer gas so as to subject all parts of the test object to examination at one time (ASTM E 425).

Hot-wire ammeter. Ammeter in which the expansion of a wire moves a pointer to indicate the amount of current being measured.

Hot-wire anemometer. Instrument that measures the velocity of wind or gas by its cooling effect on an electrically heated wire.

Humidity chamber. Laboratory facility constructed with a conditioning device, used to maintain a specified humidity at a specified temperature. The control range may vary with specific applications and construction purposes. Normal construction ranges are as follows: humidity, frost point (F.P.) 15°C to dew point 85°C; dry-bulb temperature, 15°C to 93°C; barometric pressure, 14.90 to 30.27 in. mercury. The chamber is a specialized facility with difficult control requirements.

Hydrometer. Instrument for measuring the specific gravity of a liquid, such as the electrolyte of a storage battery.

Hyetograph. Chart showing rainfall intensity against time; map showing the area distribution of rainfall.

Hygrometer

1. Instrument for measuring the water vapor content of the air. The most common type is a psychrometer, consisting essentially of dry-bulb and wet-bulb thermometers.

2. Instrument that indicates directly or indirectly the relative humidity of the air (ASTM E 41).

3. Any instrument for measuring the humidity of the atmosphere (ASTM E 41).

4. Any properly calibrated instrument that indicates directly or indirectly the relative humidity of the air (ASTM E 337).

Hygroscope. Instrument that indicates variation in atmospheric moisture.

Hygrothermograph. Recording instrument combining on one record variation of atmospheric temperature and humidity content as a function of time. The most common hygrothermograph is a hair hygrograph combined with a thermograph.

Hypsometer. Instrument for measuring atmospheric pressure by determining the boiling point of a liquid at the station; instrument for determining the height of trees or other objects.

Icing-rate meter. Instrument for the measurements of the ice accretion on an unheated body.

Illumination (footcandle) meter. Instrument for measuring the illumination on a surface. Most such instruments consist of one or more barrier layer cells connected to a meter calibrated in footcandles.

Illuminometer. Portable photometer for measuring the illumination on a surface.

Immersion pyrometer. Instrument for determining molten-steel temperature.

Immersion refractometer. Device to measure refractive indices by immersing the prism portion in the sample being checked.

Impedometer. Instrument used to measure impedances in wave guides.

Indicator. Part of electronic equipment in which the data are obtained by observation, usually in the form of a scope or dial; part of an instrument from which the reading is made. It may be at the instrument or at a remote location, or both.

Inductance meter. Device that measures the self-inductance of a circuit or the mutual inductance of the circuits.

Induction motor meter. Meter containing a rotor that moves in reaction to a magnetic field and the currents induced into it.

Infrared bolometer. Bolometer adapted to detecting infrared radiation, as opposed to microwave radiation.

Infrared detector

1. Device for observing and measuring infrared radiation, such as a bolometer, radiomicrometer, thermopile, pneumatic cell, photocell, photographic plate, and photoconductive cell.
2. Device responding to infrared radiation, used primarily in detecting fires.

Infrared spectroscopy. Use of a spectrophotometer for determination of infrared absorption spectra (2.5- to 18-nm wavelengths) of materials. It is especially used for detection, determination, and identification of organic materials.

Integrating meter. Instrument that totalizes electric energy or some other selected quantity consumed over a period of time. The electric watt-hour meter is an example.

Integrating photometer. Instrument that, with a single reading, indicates the average candlepower from a source in all directions or at all angles in a single plane.

Interferometer. Apparatus used to produce and show interference between two or more wave trains coming from the same luminous area and also to compare wavelengths with observable displacements, reflectors, or other parts by means of interference fringes. An interferometer is frequently used to obtain quantitative information on flow around bodies in wind tunnels.

Ionization chamber. Instrument consisting essentially of a closed chamber or tube of air or gas with two electrodes, used for detecting and measuring nuclear radiation. Radiation passing through an ionization chamber ionizes the air or gas in the chamber, permitting detection and measurement of the radiation by electrical means.

Izod test. An impact test in which a falling pendulum strikes a fixed, usually notched specimen with 120 ft-lb of energy at a velocity of 11.5 ft/second; the height of the pendulum swing after striking is a measure of the energy absorbed and thus indicates impact strength.

Jolt test. Application of repeated shocks to equipment.

Jumble test. Application of repeated tumbling to equipment in a box that is rotated around its diagonal axis.

K. Abbreviation for Kelvin temperature scale.

Katharometer. Instrument for detecting the presence of small quantities of gas in air.

Kelly ball test. Apparatus used for indicating the consistency of fresh concrete.

Kelvin temperature scale (same as absolute temperature). In the Kelvin scale, the freezing point of water is 273.15 K (0°C) and the boiling point of water is 373.15 K (100°C).

Kilovoltmeter. Instrument that measures potential differences on the order of several kilovolts.

Knoop hardness test. An indentation hardness test using calibrated machines to force a rhombic-based pyramidal diamond indenter having specified edge angles, under specified conditions, into the surface of the material under test and to measure the long diagonal after removal of the load (ASTM E 140).

Konimeter. Instrument for determining the dust content of a sample of air. One form of the instrument consists of a tapered metal tube through which a sample of air is drawn and allowed to impinge on a glass slide covered with a viscous substance. The particles caught are counted and measured with the aid of a microscope.

Laser anemometer. Instrument in which the wind being measured passes through two perpendicular laser beams, and the resulting change of velocity of one or both beams is measured.

Lateral extensometer. Instrument used in photoelastic studies of the stresses on a plate.

Leak test. Device for detecting, locating, and measuring leakage (ASTM E 425).

Light distribution photometer. Device that measures the luminous intensity of a light source in various directions.

Light meter

1. Portable device for measuring illumination.
2. Electronic device that contains a photosensitive cell and calibrated meter for the measurement of light levels.

Liquid-in-glass thermometer. Thermometer in which the thermally sensitive element is a liquid contained in a graduated glass envelope. The indications of such a thermometer depend on the difference between the coefficients of thermal expansion of the liquid and the glass. Mercury and alcohol are liquids commonly used in meteorological thermometers.

Liquid penetrant test. Method employed in nondestructive testing used to locate defects open to the surface of nonporous materials.

Los Angeles abrasion test. Test to obtain the abrasion resistance of concrete aggregates.

Lovibond tintometer. Colorimeter that compares a solution or object under examination with a series of slides each of three colors.

Low-temperature chamber. Enclosed, thermally insulated facility with equipment and controls to produce an internal temperature below the ambient temperature. Refrigeration can be accomplished by mechanical single-stage systems, mechanical two-stage compound systems, multistage cascade systems, dry ice, or liquid carbon dioxide. Low temperatures can be attained down to $-150°F$ depending on the system and the refrigerant used. The chamber maintains internal temperature conditions by convection and/or radiation.

Macrotech testing. Method of examining section of forgings or billets, blooms, etc., for the detection of such defects as bursts, pipe, excessive segregation, flakes, etc., that employs the action of an acid or other corrosive agent to develop the characteristics of a suitably prepared specimen (ASTM E 340).

Macrometer. Instrument that has 2 meters and a focusing telescope with which the ranges of distant objects can be found.

Magnetic detecting device. Device for detecting cracks in iron or steel.

Magnetic field meter. Instrument designed to measure the flux density of magnetic fields (ASTM E 269).

Magnetic potentiometer. Instrument that measures magnetic potential differences.

Magnetic spectrometer. Device for measuring the momentum of charged particles.

Magnetic test coil. Coil that is connected to a suitable device to measure a change in the magnetic flux linked with it.

Magnetic variometer. Instrument for measuring the differences in a magnetic field with respect to space or time.

Magnet meter. Instrument for measuring the magnetic flux produced by a permanent magnet.

Magneto anemometer. Cup-type anemometer with its shaft mechanically coupled to a magnet.

Magnetomer. Device for measuring and comparing the direction and magnitude of magnetic fields.

Manocryometer. Instrument for measuring the change of a substance's melting point with change in pressure.

Manometer

1. Gauge/gage for measuring the pressure of gas.
2. Instrument for measuring differences of pressure. The weight of a column of liquid enclosed in a tube is balanced by the pressure applied at its opposite ends, and the pressure difference is computed from the hydrostatic equation. The mercury barometer is a type of manometer.

Mass spectrometer

1. Instrument that permits rapid analysis of chemical compounds.
2. Mass spectroscope in which a slit moves across the paths of particles with various masses, and an electrical detector behind it records the intensity distribution of masses.

Maximum thermometer

1. Thermometer that registers the maximum temperature attained during an interval of time.
2. Thermometer in which the mercury, or the indicator used for registering temperature, remains at the highest point reached since its last setting.

Mechanical testing. Testing done to accomplish the determination of mechanical properties (ASTM E 6, E 28).

Megohmmeter. Instrument used for measuring the high resistance of electrical materials of the order of 20,000 megohms at 1000 volts.

Mercury barometer

1. Instrument for measuring atmospheric pressure.
2. Glass instrument, employing mercury in its vertical column, used to measure atmospheric pressure.

Mercury thermometer. Liquid-in-glass or liquid-in-metal thermometer using mercury as a liquid.

Metal detector. Electronic device for detecting metallic objects, concealed or buried.

Metal test specimens. The metals used to evaluate the corrosion properties of engine coolants.

Microdensitometer. Instrument used in spectroscopy to measure lines in a spectrum by light transmission measurement.

Microhardness test. Microindentation hardness test using a calibrated machine to force a diamond indenter of specific geometry, under a test load of 1 to 1000 gf, into the surface of the test material and to measure the diagonal or diagonals optically (ASTM E 384).

Microinterferometer. Combination of a microscope and an interferometer; used primarily to study platings, transparent coatings, and other such thin films.

Micro penetration tester. Testing machine capable of applying low loads, usually in the range from 1 g to 5 kg, to form an indentation or a scratch or both, as a basis for measuring hardness (ASTM E 384).

Microvoltmeter. Voltmeter whose scale is calibrated to indicate voltage values in microvolts.

Microwave detector. Device that demonstrates the presence of a microwave by a specific effect that the wave produces.

Microwave refractometer. Device for measuring the refractive index of the atmosphere or microwave frequencies, usually in the 3-cm region.

Milliammeter. Electric current meter calibrated in milliamperes.

Minimum thermometer. Thermometer that automatically registers the lowest temperature occurring since its last setting.

Mullen test. A test for bursting strength made on a specific type of machine.

Murray loop test. Test method of localizing a fault in a cable.

Nanovoltmeter. Instrument sufficiently sensitive to give readings in thousandths of microvolts.

Nephelometer. Instrument that measures, at more than 1 angle, the scattering function of particles suspended in a medium.

Nichol's radiometer. Instrument used to measure the pressure exerted by a beam of light.

Noise analyzer. Device used for noise analysis that includes data for determining the frequency components that are part of special noises.

Nondestructive test. Test in which material or equipment is subjected to environmental conditions that are inherently nondamaging or nondestructive; test in which material or equipment is subjected to inherently damage-producing conditions, but with stress levels or exposure times intentionally reduced to prevent equipment damage or destruction.

Notched-bar test. Test performed in which a notched metal specimen is bent with the notch in tension.

Olemeter. Device for measuring the specific gravity of oils.

Olsen ductility test. Cupping test in which a piece of sheet metal is deformed at the center by a steel ball until fracture occurs; ductility is measured by the height of the cup and the time of the failure.

Opacimeter. Photoelectric instrument used for measuring the amount of sediment in a liquid.

Orifice meter. Form of gas or liquid flowmeter consisting of a diaphragm in which there is an orifice placed transversely across a pipe. The difference in pressure on the two sides of the diaphragm is a measure of flow velocity.

Oscillator. Electrical device that generates alternating currents or voltages. An oscillator is classified according to the frequency of the generated signal.

Oscillograph. Device for graphically recording or indicating oscillations or changes in an electric current.

Oscilloscope. Instrument for producing a visual representation of oscillations or changes in an electric current. The face of the cathode ray tube used for this representation is called a "scope" or "screen."

Output meter. Alternating-current voltmeter that measures the signal strength at the output of a receiver or amplifier.

Pallograph. Low-frequency vibrograph.

Pantometer. Instrument designed to measure all the angles necessary for determining elevations and distances.

Partial pressure gage. Ionization gage that indicates the partial pressure of any gas in a mixture irrespective of the partial pressure of other gases in the mixture (ASTM E 296, E 21).

Particle-size analysis. Determination of various amounts of the different separates in a soil sample, usually by sedimentation, sieving, micrometry, or a combination of these methods.

Peak detector. Device whose output voltage approximates the true peak value of an applied signal.

Peak-to-peak voltmeter. Instrument that indicates the overall difference between the positive and negative voltage peaks.

Peel test. Test performed to ascertain the adhesive strength of bonded strips of metals by peeling or pulling the metal strips back and recording the adherence values.

Pendant drop test. Test made for the measurement of liquid surface tension by the elongation of the drop of liquid.

Penetration probe. Device for obtaining a measure of the resistance of concrete to penetration, customarily determined by the distance that a steel pin is driven into the concrete from a special gun by a precisely measured explosive charge.

Penetrometer. Portable device used by an individual to test the resistance of soils materials by penetrating the material and reading the resistance value on a calibrated scale on the device.

Phase meter. Instrument for measuring the difference in phase between two alternating quantities of the frequency.

Photoelectric densitometer. Electronic instrument used to measure the density of opacity of a film or other materials.

Photoelectric photometer. Instrument that incorporates a phototube or photoelectric cell for measurement of light.

Photoelectric pyrometer. Instrument used to measure radiant energy given off by a heated object—primarily for measuring high temperatures.

Photoelectric tristmulus colorimeter. Instrument using photoelectric detectors in which the source-filter detector response characteristic are adjusted so that the instrument reads directly the tristimulus values or related quantities.

Photometer. Instrument used for measuring photometric quantities (ASTM E 685).

Physical testing. Term generally includes all types of testing to determine the physical properties of materials based on measurements and visual observation.

Picoammeter. Instrument whose scale is calibrated to indicate current values in picoamperes.

Piezometer

1. Instrument for measuring pressure head.
2. Device installed for measuring the pressure head of pore water at a specific point within the soil mass.

Pitot tube. Instrument for measuring the relative speed of a fluid. It consists of a concentric pipe arrangement in which the inner pipe is open at one end and the outer pipe is perforated and closed at both ends, with each pipe connected to a manometer. The unit is operated with the open end pointing upstream, so that the inner pipe measures the total pressure and the outer pipe measures the static pressure. The difference in these pressures, the dynamic pressure, is proportional to the square of the fluid speed.

Pitot-tube anemometer. Pressure tube anemometer consisting of a Pitot tube mounted on the windward end of a wind vane and a suitable manometer to measure the developed pressure and calibrated in units of wind speed.

Plug gage. Metal gage used to test the dimension of a drilled hole.

Pneumatic test. Pressure testing of a closed vessel by the use of air pressure.

Polyphase meter. Instrument that measures electrical quantity, such as power factor or power, in a polyphase circuit.

Potentiometer. Instrument for measuring differences in electric potential. Essentially, the instrument balances the unknown voltage against a variable known voltage. Potentiometers are frequently used in conjunction with thermocouples for measuring temperature.

Powder flow meter. Instrument for measuring the rate of flow of a powder according to a specified procedure (ASTM B 243, B 9).

Precipitation gauge. Any device that measures the amount of rainfall or snowfall.

Pressure gauge. Pressure above atmospheric (barometric) pressure.

Procter needle plasticity test. Test performed to obtain a measure of the degree of compaction of a soil by measuring its resistance to penetration; also a method of determining soil moisture.

Profilograph. Instrument designed to measure and record the roughness of a surface over which it travels.

Profilometer. Instrument used for measuring the profile of a surface by moving a stylus over the surface and recording the calibrated amplified motion and the stylus.

Proving ring. Device for calibrating load indicators of testing machines, for indicating the magnitude of the deformation under load.

Psychrometer. Instrument for measuring the water vapor content of air; a type of hygrometer with two thermometers, one a wet bulb and the other a dry bulb.

Psychrometric chart. Nomograph for graphically obtaining relative humidity, absolute humidity, and dew point from wet- and dry-bulb thermometer readings.

Puncture test. Test involving the tear and stiffness strength of a material.

Putty Gauge. Type of step gauge that provides an indication of peak acceleration. It consists of a number of spring-loaded masses whose motion results in the indentation of a puttylike material.

Pycnometer. Vessel for determining the specific gravity of liquids or solids.

Pyranometer. Instrument used to measure the total hemispherical solar radiation incident on a surface, including direct radiation from the sun, diffuse radiation from the sky, and reflected shortwave radiation (albedo) from the surroundings.

Pyrheliometer

1. Instrument used to measure direct solar radiation incident on a surface located normal to the sun's rays.

2. Class of instruments that measure the intensity of direct solar radiation. The instrument consists of a radiation-sensing element in a casing that is closed except for a small aperture through which the direct solar rays enter, and a recorder unit.

Quadrant electrometer. Instrument for measuring voltages and charges by means of electrostatic forces.

Qualification test. Series of tests conducted to determine conformance of materials and material systems, to meet the requirements of a specification.

Quantity meter. Type of fluid meter used to measure volume of flow.

Radar. Radio detection and ranging equipment that determines the distance and usually the direction of objects by transmission and return of electromagnetic energy.

Radiac dosimeter. Instrument used to measure the ionizing radiation it absorbs by that instrument.

Radiant energy thermometer. Instrument that determines the blackbody temperature of a substance by measuring its thermal radiation.

Radiation pyrometer. Instrument that uses the radiant power from the object or source whose temperature is being measured.

Radiometer. Any of a variety of instruments used to measure thermal radiant energy or the energy of electromagnetic radiation at wavelengths longer than visible radiation, that is, in the infrared, microwave, and radiowave regions.

Radiosonde. Instrument carried aloft by a free, unmanned balloon and equipped with elements for determining temperature, pressure, and relative humidity and automatically transmitting the measurements by radio.

Rain gauge. Instrument designed to measure the amount of rain that has fallen. Rain gauges are classified according to their operation in the following manner: recording rain gauge, nonrecording rain gauge, and rain-intensity gauge.

Rankine temperature scale. Temperature scale using the degrees of the Fahrenheit scale and the zero point of the Kelvin scale. The ice point is thus 491.67 K and the boiling point of water is 671.67 K.

Raob. Observation of temperature, pressure, and relative humidity obtained by means of a radiosonde. The name "raob" is derived from the words "radiosonde observation."

Ratio meter. Instrument that measures electrically the quotient of two quantities.

Reaumur temperature scale. Scale with the ice point at zero degrees and the boiling point at 80 degrees, with a pressure of 1 atmosphere.

Recording ammeter. Instrument that provides a permanent recording of the value of either an alternating or a direct current.

Recording demand meter. Instrument that records the average value of the load in a circuit during successive short periods.

Recording voltmeter. Instrument that provides a permanent record of the value of either alternating or direct voltage.

Reed gauge (Fahm's reeds). Instrument that measures the frequency at which an object is vibrating.

Reflectometer. Microwave system arranged to measure the incidental and reflected voltages and indicate their ratio.

Reheat test. Prescribed heat treatment of a fired refractory free of externally applied stresses to determine its linear or volume stability by measurements before and after the heating (ASTM C 71, C 8).

Relaxation test. Creep test in which the decrease of stress with time is measured while the total strain is maintained constant.

Resistance magnetometer. Instrument that depends on its operation on the variation in the electrical resistance of a material immersed in the field to be measured.

Resistance thermometer. Instrument for determining temperature by measuring the electrical resistance of a standardized material exposed to that temperature.

Reverberant chamber. Type of acoustical testing facility in which a specimen is subjected to simultaneous impingement of acoustical energy from many directions. It is characterized by highly reflective walls and may have nonparallel opposing walls or multiple energy sources.

Ringelmann chart. Chart used in air pollution evaluation for assigning an arbitrary number, referred to as smoke density, to smoke eminating from any source. This chart is designed specifically for measuring the density of black smoke and is not applicable to other emissions (ASTM D 1356).

Rockwell hardness test. An indentation hardness test using a calibrated machine to force a diamond spheroconical penetrator (diamond penetrator) or a hard steel ball under specified conditions into a surface of the material under test in two operations, and to measure the difference in depth of the impression under the specified conditions of minor and major loads.

Roentgen. Unit of measure of the total quantity of x-/or gamma radiation absorbed in air. Technically, it is defined as the amount of x-/or gamma radiation that, as a result of ionization, will produce in 1 cu cm of dry air, at standard conditions of temperature and pressure, ions carrying one electrostatic unit of electricity of either sign.

Roentgen meter. Instrument for measuring the quantity or intensity of roentgen rays.

Rotameter. Tapered-tube measuring device containing a float whose position indicates the rate of fluid flow.

Rotary voltmeter. Electrostatic voltmeter used for measuring high voltages.

Rotation anemometer. Device in which the rotation of an element serves to measure the wind speed.

Salinometer. Instrument for determining the salinity of a liquid. In its most common form it consists of a hydrometer graduated to indicate the percentage of salt in the solution.

Salt spray chamber. Environmental test chamber for accelerated corrosion testing. The salt solution is atomized by the use of suitable nozzles in conjunction with a compressed air supply. The chamber is equipped for heating the system.

Salt spray test. Accelerated corrosion test in which a piece of metal is subjected to a controlled spray of a solution of sodium chloride for a given time period.

Scratch hardness test. Test in which a cutting point under a given pressure is drawn across the surface of a metal plate and the width and depth of the indentation is measured.

Screen test. Standard test for the fineness of porcelain enamel slip or powder (ASTM C 286, C 22).

Secondary force standard. Instruments or mechanism, the calibration of which has been established by comparison with primary force standards (ASTM E 74, E 28).

Seismic detector. Instrument designed to receive seismic impulses.

Seismochronograph. Instrument for determining the time at which an earthquake shock occurs.

Seismograph

1. Instrument for recording earthquake activity. The electromagnetic type was constructed in Italy by L. Palmieri in 1855 and installed in the observatory on Vesuvius.
2. Instrument used in recording vibrations in the earth.

Seismometer

1. Detector to indicate seismic waves. When combined with a graph or recording mechanism, it is called a "seismograph."
2. Instrument that detects movements in the earth.
3. Instrument that picks up linear or rotational displacement, or acceleration.

Seismoscope. Instrument used for recording only the occurrence or time of an occurrence of an earthquake.

Sensitivity of leak test. The smallest leak rate that an instrument, method, or system is capable of detecting under specified conditions (ASTM E 425).

Sensitometer. Instrument used for measuring the sensitivity of light-sensitive materials.

Shaker-reaction-type vibration machine. Machine consisting of a softly suspended table to which is attached one or more unbalanced rotating masses. It vibrates owing to the unbalanced force reaction of the table, without force reaction against the machine base.

Shaking test. Test used to indicate the presence of significant amounts of rock flour, silt, or very fine sand in a very fine-grained soil (ASTM D 653, D 18).

Shock machine. Device for subjecting a system to controlled and reproducible mechanical shock.

Shock tube. Test device consisting of a controlled-atmosphere tube in which a shock wave is used as a driving force to produce a high mach number of very short duration (order of milliseconds). The shock tube is of interest in environments other than high-speed propulsion since the shock wave causes a tremendous increase in the temperature of the gas.

Shore hardness. Established method of rating the hardness of metal, plastic or rubber material.

Sieve analysis. Performing analysis to determine the proportion of particles lying within certain size ranges in a granular material by separation on sieves of different size openings.

Simulated service test. Controlled test, usually conducted in a laboratory, designed to produce results having a meaningful relationship to those produced in service under natural environmental conditions.

Sling psychrometer

1. Device used to determine temperature and humidity for figuring joint compound drying time.
2. Psychrometer in which the wet- and dry-bulb thermometers are mounted on a frame connected to a handle at one end by means of a bearing or length of chain. The psychrometer is whirled by hand in order to provide the necessary ventilation.

Slump test. Test made in the field to establish the consistency of the concrete. Concrete is poured into a metal mold, and when the mold is removed, the distance that the concrete has settled or slumped below the original height of the mold is termed the "slump."

Solar radiation chamber. Enclosed facility provided with a means for producing a simulated solar radiation environment, including thermostatically controlled temperature. There are two types of solar simulation, terrestrial and space.

Sonic altimeter. Absolute altimeter that determines the height above the terrain by measuring the time interval between transmission of a sound and the return of the echo.

Sonic thermometer. Thermometer based on the principle that the velocity of a sound wave is a function of the temperature of the medium through which it passes. The velocity of a sound wave also depends on the velocity of the medium through which it passes; therefore, this quantity must be known.

Sound-level meter. Instrument that consists of a microphone, amplifier, output-meter, and frequency-weighting networks, for the measurement of noise and sound levels.

Spectrometer. Test instrument that determines the frequency distribution of the energy generated by any source and indicates all components simultaneously.

Spectrophotometer

1. Instrument for measuring the intensity of radiant energy of desired frequencies absorbed by atoms of molecules. Substances are analyzed by converting the absorbed energy to electrical signals proportional to the intensity of radiation.
2. Instrument for measuring the ratio of two special radiometric quantities (ASTM E 349).

Spectropyrheliometer. Instrument that measures the spectral distribution of direct solar irradiance (ASTM E 491).

Spectroradiometer. Instrument for measuring the special concentration of radiant energy or radiant power (ASTM E 349).

Stalag mometer. Apparatus for determining surface tension. The mass of a drop of a liquid is measured by weighing a known number of drops or by counting the number of drops obtained from a given volume of the liquid.

Standard curing. Exposure of concrete test specimens to exact specified conditions of moisture, humidity and temperature.

Step gauge. Type of instrument that indicates that motion of a specified severity has occurred. Generally, the severity of the motion is defined in terms of a value of maximum acceleration that has been reached or exceeded.

Stereocomparagraph. Stereoscopic instrument used for the preparation of topographic maps to determine ground elevations by measuring the displacement of their images on photographs.

Stereophotogrammetry. Photogrammetry with the aid of stereoscopic equipment and methods.

Stereoplanigraph. Very accurate stereoscopic photogrammetric mapping instrument with mechanical drafting attachment, capable of providing a stereoscopic picture from overlapping photographs, regardless of the angle at which they were taken.

Stream gauging. Process and art of measuring the depths, areas, velocities, and rates of flow in natural or artificial channels.

Survey meter. Portable instrument, such as a Geiger counter or ionization chamber, used to detect nuclear radiation and to measure the dose rate.

Synoptic weather chart. Chart of an extended portion of the earth's surface on which are delineated the weather conditions at different points observed at the same moment of actual time.

Tachometer

1. Instrument used to measure the frequency of mechanical systems by the determination of annular velocity.
2. Instrument that measures the revolutions per minute or the angular speed of a rotating shaft.

Tectonometer. Device used on the ground surface level to obtain information of the structure of underlying rocks.

Telemeter

1. Apparatus that transmits, receives, and measures or records the value of quantity at a distance.
2. Measuring, transmitting, receiving, and indicating apparatus for obtaining the value of a quantity at a distance. The radiosonde system is a meteorological example of a telemeter or telemeteorograph.

Televoltmeter. A telemeter that measures voltage.

Telewattmeter. A telemeter that measures power.

Temperature chamber. Enclosed, thermally insulated space with equipment and controls to produce a chamber temperature differing from ambient.

Tensile test. Test in which a specimen is subjected to increasing longitudinal pulling stress until the fracture or break occurs.

Tensiometer. Device for measuring the negative pressure (or tension) of water in soil in situ; porous, permeable ceramic cup connected through a tube to a manometer or vacuum gauge.

Teraohmmeter. Instrument used to measure extremely high resistance.

Terrain analyzer. Mobile instrument developed by the Land Locomotion Laboratory for rapid measurement and automatic reduction of soil strength and terrain profile data, and consisting of a mounting vehicle, Bevameter, gyro-referenced two-point profile follower, and electronic components.

Testing. An element of inspection that generally denotes the determination by technical means of the properties or elements of supplies, material, equipment, or components thereof, and involves the application of established scientific and engineering principals and procedures.

Testing chamber. Enclosed environmental laboratory facility used for duplicating, accelerating, or simulating one or more natural environmental phenomena singly or in various combinations.

Testing machine

1. Mechanical device for applying a load (force) to a specimen (ASTM E 4, E 28).
2. Device for applying tests on specimens and accurately recording the results.

Test method. Definitive, standardized set of instructions for the identification, measurement, or evaluation of one or more qualities, characteristics, or properties of a material.

Test result. The result of most tests determine a value of a quality, strength, characteristic, performance, and other properties which then can be compared with the specifications of the material or equipment.

Test specimen. A specific specimen or a portion thereof on which the test is to be performed.

Test unit. Usually a fraction of a unit product from which one or more test specimens are to be taken for testing.

Thermistor. Device whose electrical resistance varies markedly and monotonically and that possesses a negative temperature coefficient of resistivity. Thermistors used in meteorology are composed of solid semiconducting materials whose resistance increases $4\%/°C$. It is constructed in a variety of sizes and may be obtained with thermal time constants of a millisecond or less. Meteorological applications include thermometers, anemometers, and bolometers.

Thermocouple. Temperature-sensing element that converts thermal energy directly into electrical energy. In its basic form it consists of two dissimilar metallic electrical conductors connected in a closed loop. One pair of junctions forms a thermocouple, several pairs form a thermopile. If electrical energy is passed through a thermocouple it creates "cold" (Peltier effect), which can be used for refrigeration.

Thermogram. Record of a thermograph.

Thermograph. Self-recording thermometer. The thermometric element is most commonly either a bimetal strip or a Bourdon tube filled with a liquid.

Thermometer. Instrument for measuring temperature by utilizing the variation of the physical properties of substances according to their thermal states.

Tide gauge. Device for measuring the height of the tide. It consists of a graduated staff in a sheltered location where visual observations can be made at any desired time; or an elaborate recording instrument (sometimes called "marigraph") making a continuous graphic record of tide height against time. Such an instrument is usually actuated by a float in a pipe communicating with the sea through a small hole that filters out the shorter waves.

Torquemeter. Instrument that is designed to measure torque.

Torque viscometer. Apparatus used for measuring the consistency of slurries, in which the energy required to rotate a device suspended in a rotating cup is proportional to viscosity.

Torsiograph. Type of vibrograph designed for the measurement of torsional vibration.

Torsion hygrometer. Instrument in which the rotation of the hygrometric element is a function of the humidity.

Transducer. Device for converting energy from one form to another. For example, a thermocouple transduces heat energy into electrical energy.

Turbidimeter. Device for measuring the particle size distribution of a finely divided material by taking successive measurements of the turbidity of a suspension in a fluid.

Unconsolidated-undrained test (quick test). A soil test in which the water content of the test specimen remains practically unchanged during the application of the confining pressure and the additional axial (or shearing) force (ASTM D 3017).

Underwriters Laboratories, Inc.. An independent laboratory that tests materials and equipment to determine whether they meet certain safety standards when properly used.

Vacuum gauge. Instrument that indicates the absolute gas pressure in a vacuum system.

Vane anemometer. Instrument used to measure wind speeds and airspeeds in ductwork.

Vane shear test. An in-place shear test in which a rod with thin radial vanes at the end is forced into the soil and the resistance to rotation of the rod is determined (ASTM D 2573).

Varhour meter. An electricity meter that measures and registers the integral of the reactive power of the circuit into which the meter is connected.

Varmeter. Instrument for measuring reactive power in either vars, kilovars, or megovars.

Velocimeter. Instrument for measuring the sound in water.

Venturi meter. Instrument for measuring fluid flow rate in a piping system.

Venturi tube. Tube designed to measure the rate of flow of fluids. It consists of a tube having a construction or throat at its midsection. The difference between the pressure measured at the inlet and at the throat is a function of the fluid velocity.

Vertical anemometer. Instrument that records the vertical component of the wind speed.

Vibration machine. Device for subjecting a mechanical system to controlled and reproducible mechanical vibration.

Vibration meter. Apparatus consisting of a calibrated amplifier, and output meter for vibration pickup, for the measurement of displacement, velocity, and acceleration of a vibrating body. Also called "vibrometer."

Vibratory reed meter. Device for measuring the frequency of an excitation by detecting the resonance of one or more vibratory reeds.

Vibrograph. Mass-spring type of displacement pickup with self-contained means for recording the relative motion between the mass and the case, consisting of a recording medium contained on a supply spool, transport mechanism for the recording medium, and a take-up spool.

Vibrometer. Instrument used to measure the amplitude of a vibration.

Vicat apparatus. Penetration device used in the testing of hydraulic cements and similar materials.

Vicat needle. Weighted needle for determining setting time of hydraulic cements.

Vickers hardness test. An indentation hardness test using calibrated machines to force a square-based pyramidal diamond indenter having specified face angles, under a predetermined load into the surface of the material under test and to measure the diagnosis of the resulting impression after removal of the load (ASTM E 92).

Viscometer

1. Instrument for measuring flow properties.
2. Instrument for determining the viscosity of slurries, mortars, or concretes.
3. Instrument for measuring the viscosity of a fluid.

Viscometer gauge. Vacuum gauge in which the gas pressure is determined from the viscosity of the gas.

Volt breakdown test. Test whereby a specified voltage is applied between given points in a device, to ascertain that no breakdown will occur at a specified voltage.

Voltammeter. Instrument that may be used either as a voltmeter or ammeter.

Volt-ammeter. Instrument calibrated to read both voltage and current.

Volt-ampere-hour meter. Electricity meter that measures the integral, usually in kilovolt-ampere-hours, of the apparent power in the circuit where the meter is connected.

Volt-ampere meter. Instrument for measuring the apparent power in an alternating current circuit. Its scale is graduated in volt-amperes or kilovolt-amperes.

Voltmeter. Instrument for measuring potential difference. Its scale is usually graduated in volts.

Voltmeter-ammeter. A voltmeter and an ammeter combined into a single case, but with separate circuits.

Volt-ohm-milliammeter. Test instrument with several ranges, for measuring voltage, current and resistance.

Volume-unit meter. Instrument having a volume indicator with a decibel scale and specified dynamic and other characteristics. It is used to obtain correlated readings of speech power necessitated by the rapid fluctuations in level of voice currents.

Watt-hour demand meter. Combined watt-hour meter and demand meter.

Watt-hour meter. Electricity meter that measures and registers the integral, usually in kilowatt hours, of the active power of the circuit into which the meter is connected.

Water meter. Instrument for measuring the amount of water passing a specified point in a piping system.

Wattmeter. Instrument that measures electric power in watts.

Weatherometer

1. Instrument that is utilized to subject articles to accelerated weathering conditions, such as rich ultraviolet sources and water sprays.
2. Device used to subject products and finishes to accelerated weathering conditions.

Wet-bulb thermometer. Thermometer having the bulb covered with a cloth, usually muslin or cambric, saturated with water.

Wind component indicator. Device that mechanically determines the range and deflection components of the computed wind, which is equivalent to all true winds encountered by a projectile in flight.

Wind corrector. Mechanical device that computes the correction necessary for the effect of wind, used in sound ranging and artillery fire control.

Wind-resolving mechanism. Device similar to a wind component indicator, which is mounted on a deflection board or is part of a computer. It mechanically determines the range and deflection components of the ballistic wind.

Wind tunnel. Tunnel through which a stream of air is drawn at controlled speeds for aerodynamic tests and experimentation.

X-ray monochromator. A device used to eliminate photons of energies other than those in a narrow band.

SEISMIC TECHNOLOGY

Acceleration. Rate of change in the velocity of a moving body. High accelerations are the most damaging to buildings, which must try to follow the rapid changes in ground movement.

Accelerogram. Record from an accelerograph showing acceleration as a function of time.

Accelerograph

1. Instrument designed to record ground acceleration and ground displacement.

2. Strong-motion earthquake instrument recording ground (or base) acceleration.

Accelerograph record. Record that indicates the high-frequency tremors of an earthquake. They are the waves that have the highest acceleration, are most often felt, and do the most damage. In terms of actual ground motion (displacement), they are the waves of smallest amplitude. All of the longer period waves have much larger amplitudes.

Accelerometer. A seismograph for measuring ground acceleration as a function of time.

Active fault. Any fault that is seen as likely to generate earthquakes in the forseeable future.

Aftershock

1. Earthquake, usually a member of an aftershock series often occurring within the span of several months following the occurrence of a large earthquake (main shock). The magnitude of an aftershock is usually smaller than the main shock.

2. An earthquake creates a "main shock" and many smaller earthquakes called "aftershocks."

Alaska seismic belt. Alaska is one of the important links in the great seismic belt that circumscribes the Pacific. In the Aleutians, many submarine shocks occur in and near the deeps that parallel the islands on the south side. This belt moves inward on the Alaskan Peninsula, widens out in the region of the Kenai Penninsula, and extends northward into central Alaska, the region of Fairbanks marking the northern terminus.

Alpide zone. Zone that begins in the Azores, passes through the Mediterranean and Near East, skirts the northern border of India, and passes through Sumatra and Indonesia to join the Circum-Pacific belt in New Guinea.

Amplification. Increase in earthquake motion as a result of resonance of the natural period of vibration with that of the forcing vibration.

Amplitude. Maximum deviation from the mean or centerline of a wave.

Amplitude of ground motion. Extreme range of a fluctuating quantity.

Andean arc. Line drawn through the Andes Mountains in Chile along the 70° parallel.

Andesite line. Line stretching from New Zealand through the Kermadec and Tonga Islands. There appears to be an oceanic feature that actually follows the so-called "Andesite" line.

Anisotropic. Material whose physical properties (for example, seismic wave velocity) vary quantitatively with the direction in which they are measured.

Aseismic region. Region that is relatively free of earthquakes (all areas show some seismicity over a sufficiently long interval).

Asthenosphere. Worldwide layer below the lithosphere that is marked by low seismic wave velocities and high seismic attenuation. The atmosphere is a soft layer probably partially molten. It may be the site of convection.

Atomic bombs. Seismic waves from atomic bombs have been recorded at great distances.

Attenuation. Reduction of amplitude or change in wave due to energy dissipation over distance with time.

Axial load. Force coincident with the primary axis of a member.

Barrier (fault). An area of fault surface resistant to slip because of geometrical or structural changes.

Basaltic layer. There are two layers in the earth's crust. The upper layer is granitic. The lower layer is the "basaltic" or intermediate layer and occurs in the oceans where granite does not occur.

Basement rocks

1. Igneous or metamorphic rocks, including partly metamorphosed sediments, underlying the sedimentary blanket.

2. Oldest rocks recognized in a given area; complex of metamorphic and igneous rocks that underlies all the sedimentary formations, usually Precambrian or Paleozoic in age.

Base shear

1. Total shear force acting at the base of a structure.

2. Lateral force applied to the bottom of a structure at the foundation. Magnitude is the principal concern of the designer.

Basin. In rectonics, the circular, synclinelike depression of strata.

Bilinear. Representation by two straight lines of the stress versus strain properties of a material, with one straight line to the yield point and the second line beyond.

Block fault. Structure formed when the crust is divided into blocks of different elevation by a set of normal faults.

Block faulting. System of failure that occurs quite clearly in California, Philippines, and New Zealand.

Box system. Structural system without a complete vertical load-carrying space frame. In this system the required lateral forces are resisted by shear walls.

Braced frame. Frame that is dependent on diagonal braces for stability and the capacity to resist lateral forces.

Brittle failure. Failure in material that generally has a very limited plastic range; material subject to sudden failure without warning.

Caldera. Large circular depression in a volcanic terrane, typically originating in collapse, explosion, or erosion.

Capable fault. A fault along which it is mechanically feasible for a sudden slip to occur.

Cataclastic rock. Breccia or powdered rock formed by crushing and shearing during rectonic movements.

Cliff. High steep face of rock. A cliff of considerable length is often called an "escarpment" or "scarp." Cliffs are usually produced by erosion, less commonly by faulting.

Compression. Stress that resists the tendency of two forces acting toward each other.

Compression and dilatation

1. In connection with longitudinal waves, the nature of the motion at a given point, usually at a recording station.

2. The term is used in connection with longitudinal waves, as in acoustics. It refers to the nature of the motion at a given point, usually at a recording station. When the ray emerges to the surface, displacement upward and away from the hypocenter corresponds to compression, the opposite to dilatation.

Continental drift

1. It has been proved that the continents are moving. The assumption is that the continents represent large blocks of granitic material floating on a sea of basalt.

2. Horizontal displacement of rotation of continents relative to one another.

Contraction hypothesis. The contraction hypothesis assumes that the earth is cooling but makes certain simplifying assumptions.

Convergence zone. Band along which moving tectonic plates collide and area is lost either by shortening and crustal thickening or subduction and destruction of crust; site of volcanism, earthquakes, trenches, and mountain building.

Core

1. The complete picture of the earth consists of three distinct cores and one cover. The cover is made of crustal layers, the next zone is the "mantle," and the intermediate zone is the "metallic liquid core." The inner core is believed to be solid material.

2. Central part of the earth below a depth of 2900 km. It is thought to be composed of iron and nickel and to be molten on the outside with a central solid inner core.

Crater. Abrupt circular depression formed by extrusion of volcanic material and its deposition in a surrounding rim.

Creep (along a fault). Very slow periodic or episodic movement along a fault trace unaccompanied by earthquakes.

Critical damping. Minimum damping that will allow a displaced system to return to its initial position without oscillation.

Critical frequency. Frequency at which maximum or minimum amplitudes of excited waves occur.

Crust

1. Part of the earth above the Mohorovicic discontinuity. It is not necessarily identified with the lithosphere.

2. Lithosphere, the outer 80 km of the earth's surface made up of crustal rocks, sediment, and basalt. Its general composition is silicon, aluminum, iron.

3. Outer layer of earth's surface which consists of two layers. The thickness varies from 22 miles, regardless of the number of layers, to approximately 40 miles under some mountain ranges.

Damping

1. Effect of opposing forces causing the free vibration of an elastic body to cease its movements and come to a position of rest. Principal causes of damping are frictional resistances, dissipating the energy of vibration

in the form of heat. During an earthquake, partial damage to structures tends to absorb the destructive energy and thus hasten the damping of vibrations.

2. Rate at which natural vibration decays as a result of absorption of energy.

Dead load. Weight of the structure itself or equipment.

Degrees of freedom. When a body is restrained in such a manner that it may oscillate only in one plane, it is designated as having 1 degree of freedom. It may have as many degrees of freedom as there are planes in which it can oscillate.

Deep-focus earthquakes. Deep-focus earthquakes commonly do not register surface waves with periods near 20 seconds and measurable amplitudes. Even when such motion is recorded in the expected part of a seismogram, much of it appears to be due to body waves, especially S-waves, repeatedly reflected at the surface.

Deep-focus teleseisms. Regularly recorded teleseisms with large S-waves having amplitudes at least as large, relative to those of the P-wave group, as on seismograms of shallow shocks.

Deflection. Displacement of a member due to application of external force.

Depths of foci. Earthquakes are commonly classed by the depth of the focus or hypocenter beneath the earth's surface: shallow (0 to 70 km), intermediate (70 to 300 km), and deep (300 to 700 km).

Descriptive reports on earthquakes. One of the important functions of the Coast and Geodetic Survey is to collect descriptive information on earthquakes. This is done mostly from the San Francisco office of the Survey because most earthquakes occur in the Pacific coast and western mountain regions.

Design criteria. Set of codes or laws established by the local, state, or authorized agency that sets the minimum standards of engineering criteria to design a structure.

Diaphragm

1. A diaphragm is used to resist horizontal forces to horizontal and vertical or resisting elements.

2. Generally a horizontal girder composed of a web (such as a floor or roof slab) with adequate flanges, which distributes lateral forces to the vertical resisting elements.

Dilatations. Almost all earthquakes give the quadrant distribution of compressions and dilatations. An explosion, on the other hand, gives a push outward in all directions.

Dip-slip fault. Fault in which the relative displacement is along the direction of dip of the fault plane; either a normal or a reverse fault.

Dispersion. Increase in velocity of observed surface waves with wavelengths.

Displacement. Particular movements of earthquakes result in displacing solids or fluids from one area to another. Difference between the initial position of a body and a later position.

Divergence zone

1. Belt along which plates move apart and new crust and lithosphere are created; site of midocean ridges, earthquakes, and volcanism.

2. Belt along which tectonic plates move apart and new crust is created.

Double amplitude. Total excursion or overall height of a wave (peak to peak, crest to trough); for a sinusoidal wave, twice the amplitude.

Drag. The effects of elastic rebound should not be confused with those of drag, which are often more conspicuous in the field, and are actually in the opposite sense.

Drift. In buildings, the horizontal displacement of basic building elements due to lateral earthquake forces.

Ductility

1. If a structure is ductile so that it can dissipate energy in alternating cycles of inelastic deformation without fracture, then it can survive an earthquake. (Inelastic: Not elastic, hence, inflexible, unyielding.)

2. Ability to withstand inelastic strain without fracturing.

Duration. The duration of an earthquake seldom exceeds 1 minute. Minor shocks are often observed before and after major earthquakes; these have little destructive power.

Dynamic. Having to do with bodies in motion.

Earth avalanches. Avalanches started by earthquakes and causing flows of relatively dry material consisting of rock and soil from cliffs or bluffs.

Earth flows. Flows that occur following an earthquake. A sudden burst of water in a locality where underground springs exist spouts sand and mud with it downgrade in a flow.

Earth lurches. Earth lurches are caused by earthquakes and produce cracks and fissures parallel to streams or gulches.

Earthquake

1. Physical phenomenon characterized by the shaking of the ground with violence of varying intensity. An earthquake causes horizontal and vertical ground vibrations.

2. Violent oscillatory motion of the ground caused by the passage of seismic waves radiation from a fault along which sudden movement has taken place.

3. Perceptible trembling to violent shaking of the ground produced by the sudden shift of part of the earth's crust.

4. Shocks caused when rocks that have been distorted beyond their strength finally break.

Earthquake belts. Earthquake belts around the world are the Circum-Pacific, Alpine, mid-Atlantic, Artic Oceanic, and Indian Ocean.

Earthquake classification. Earthquake effects may be classified as primary, due to the causative process, such as faulting or volcanic action, or secondary, due to shaking or, more generally, to the passage of elastic waves generated by the primary process.

Earthquake forces

1. In order to determine forces exerted by an earthquake, a scale measuring the violence of the ground motion during an earthquake had to be developed. A number of widely used intensity scales were devised, all of them expressing the force of the ground motion in a particular area in terms of the effect that such ground motion has on people and on material things.

2. Forces that are vibrational or dynamic in character and cannot be treated the same as static or steady forces.

Earthquake fountains. Where excessive groundwater exists, a strong earthquake often produces fountains, spouts, or geysers, which occur during the strong shaking and for some time after the incident.

Earthquake ground motion science. Earthquake forces are vibrational or dynamic in character and cannot be treated the same as static or steady forces.

Earthquake motions. Earthquakes consist of horizontal and vertical ground vibrations. The horizontal motion is usually much greater than the vertical, the latter being one-tenth to one-fifth of the former. The most destructive force is caused by horizontal earth motion.

Earthquake occurrences. One million earthquakes occur each year throughout the world annually. Most earthquakes originate beneath the sea. Seven hundred of the million are classified as strong, capable of causing considerable damage.

Earthquakes measured. Earthquakes are measured in terms of either energy (magnitude) or actual effects (intensity). The first measurement is based on instrument records, the second on personal observations. They are completely separate in intent and results, but they are often confused by the general public.

Earthquake sounds. When an earthquake occurs, vibrations in the ground often disturb the air above it and produce sound waves that are within the range of the human ear.

Earthquake spectra. Graphs exhibiting the effects of an individual earthquake on structures with different free periods of vibration.

Earthquake swarms. In certain localities earthquake swarms occur before and during volcanic eruptions. They consist of a long series of large and small shocks with no one principal event.

Earth slumps. Result of earthquake motion accelerating processes which go on at all times wherever there is unconsolidated material on steep slopes.

Earthwaves. Earthwaves are caused by earthquakes and travel through the body of the earth and on its surface. There are two basic seismology body waves called "P" and "S" waves. P waves are longitudinal or compression and rarefaction waves, and S waves are transverse or shear waves through the earth.

Eastern region earthquake. The earliest recorded earthquake of destructive intensity felt in the United States occurred in the valley of the St. Lawrence River in 1663, centering apparently between the present sites of Montreal and Quebec. Great landslides kept the river muddy for a month.

Eccentric loading. Forces that occur off center of the load on a vertical member.

Elasticity. Ability of a material to return to its original form or condition after a displacing force is removed.

Elastic limit. Maximum stress that can be applied to a body without resulting in permanent strain.

Elastic-rebound theory

1. In most earthquakes stress is gradually accumulated and suddenly relieved by the breaking of a fault.

2. A fault is incapable of movement until strain has built up in the rocks on either side. Strain is accumulated by the gradual shifting of the earth's crust. Rocks become distorted but hold their original positions. When the accumulated stress finally overcomes the resistance of the rocks, the earth snaps back into an unstrained position. The "fling" of the rocks past each other creates the shock waves we know as earthquakes.

3. Theory of fault movement and earthquake generation that holds that faults remain locked while strain en-

ergy accumulates in the country rock, and then they suddenly slip and release this energy.

Elastic Waves. Ground movements are elastic vibrations generally following the laws of simple harmonic motion.

Elastoplastic. Term describing the total range of stress, including expansion beyond the elastic limit into the plastic range.

Elsinore fault in California. This fault has been instrumental in the development of many mountain ranges and valleys near the southern coast. However, earthquake activity during the past 150 years has been insignificant.

Empirical. Depending on experience or observation rather than science or theory.

Energy absorption. Energy is absorbed as a structure distorts inelastically.

Energy dissipation. Reduction in intensity of earthquake shock waves with time and distance or by transmission through discontinuous materials with different absorption capabilities.

Energy-magnitude equation. Relationship between the magnitude of an earthquake and the amount of energy it releases.

Environment of deposition. Geographically limited area where sediments are preserved, characterized by its land forms, relative energy of currents, and chemical equilibria.

Epeirogeny. Large-scale, primarily vertical movement of the crust. It is characteristically so gradual that rocks are little folded and faulted.

Epicenter

1. The epicenter of an earthquake is commonly taken to be near the center of the meizoseismal area.
2. Point on the earth's surface vertically above the focus or hypocenter of an earthquake.
3. Place on the earth's surface located directly over the focus; area on the surface of the earth directly above the subterranean origin.

Escarpment. Long cliff or steep slope facing in one general direction and continuing for a considerable distance. Escarpments may be produced by faulting or by erosion. The abbreviated form, scarp, is usually limited to cliffs formed by faulting.

Failure mode. Manner in which a structure fails (column buckling, overturning of structure, etc.)

Fault

1. Planar or gently curved fracture in the earth's crust across which relative displacement has occurred.
2. Fracture in the earth's crust, accompanied by a displacement on one side with respect to the other in a direction parallel to the fracture.
3. Dislocation in rocks as a result of crustal movements. Faults differ from joints in being not simply cracks but fissures created when differential movement of the two sides has taken place. The three principal types are known as "normal faults," "reversed faults" or "overthrusts," and "transcurrent faults."
4. Dislocation caused by a slipping of rock masses along some plane of fracture; dislocated structure resulting from such slipping.

Fault-block mountain. Mountain or range formed as a horst when it was elevated (or as the surrounding region sank) between parallel normal faults.

Faulting

1. Movement along a fault. Faulting is the most usual cause of earthquakes.

2. Movement that produces relative displacement of adjacent rock masses along a fracture.

Fault-line scarp. Scarp that is not the exact original face created by faulting, but that has been modified by erosion from the original scarp. It may have the same relief as the original scarp but is usually behind it and much more irregular in outline.

Fault movements. Fault movements are always expressed in relative terms, since it is impossible to tell which side actually does the moving. Horizontal movement can be either left or right lateral. Vertical displacement can be created by either an upthrust of one block of a downthrow of the other. A combination of vertical and horizontal components is not uncommon.

Fault plane. Plane that best approximates the fracture surface of a fault.

Fault zone. Instead of being a single clear fracture, a zone is hundreds or thousands of feet wide. A fault zone consists of numerous interlacing small faults.

Felt area. Total extent of the area where an earthquake is felt.

Field act. State of California law (1933) enacted for the protection of the lives of children in public school buildings against the danger of earthquakes.

First motion. On a seismogram, the direction of ground motion at the beginning of the arrival of a P wave. Upward ground motion indicates a compression, downward motion, a dilatation.

Fissure. A crack in the earth's surface, commonly produced by earthquakes.

Flexible structures. Simple 1-degree-of-freedom structure, with its mass concentrated at the top and supported on elastic columns.

Flexible system. System that will sustain relatively large displacements without failure.

Focal depth. Depth of the earthquake focus or hypocenter below the ground surface.

Focus

1. Used as a synonym for "hypocenter" (pl., foci).
2. Source of a given set of elastic waves; true center of an earthquake, within which the strain energy is first converted to elastic wave energy.
3. Center of an earthquake disturbance. The focal depth can vary from nearly zero to more than 400 miles.

Focus (of an earthquake). Point at which the rupture occurs; synonymous with "hypocenter." It marks the origin of the elastic waves of an earthquake.

Folding. The pattern formed where compression squeezes the crust into a wavy pattern without thrusting or trenching.

Forced vibration. Oscillation movement of a body caused by an externally applied vibration force. Forced vibration has no connection with the free vibration of the elastic body itself and depends entirely on the amplitudes, periods, and accelerations of the external force causing it. During an earthquake many of the vibrations are forced vibrations. This is particularly true regarding the movement of the structure caused by the oscillations of a foundation.

Foreshocks. Foreshocks caused by minor movements that sometimes precede and may even provide part of the triggering device for a main shock.

Free oscillation. "Ringing" or periodic deformation of the whole earth at characteristic low frequencies after a major earthquake.

Free vibration. Vibration assumed by an elastic structure or an elastic body when subjected to the shock of a suddenly applied force. It consists of an oscillation movement about a neutral position like a tuning fork. The vibration gradually dies out due to the dampening effect of internal friction, resistance of air, and other causes.

Frequency. When referring to vibrations, the number of wave peaks that pass through a point in a unit of time, usually measured in hertz.

Friction breccia. Breccia formed in a fault zone or volcanic pipe by the relative motion of two rock bodies.

Fundamental period. The longest period (duration in time of one full cycle of oscillatory motion) for which a structure or soil column shows a response peak, commonly the period of maximum response.

Garlock fault. Second largest fault in California. It has made several contributions to the landscape, including the mountain ranges that form the northern edge of the Mojave Desert. There has not been a single earthquake during recorded history that can be blamed on this huge fracture.

Geocentric latitudes. Angles between the radii and the plane of the equator.

Geodimeter. A surveying instrument that measures the distance between two points on the earth's surface.

Geosyncline. Major downwarp in the earth's crust, usually more than 1000 km in length, in which sediments accumulate to thicknesses of many kilometers. The sediments may eventually be deformed and metamorphosed during a mountain-building episode.

Graben. Block, generally long compared to its width, that has been downthrown along faults relative to the rocks on either side.

Graben (rift valley). Long, narrow trough bounded by one or more parallel normal faults. Downdropped faults are caused by tensional crustal forces.

Gravity anomaly. Deviation of the observed acceleration of gravity from an expected value calculated from the general gravitational field of the earth, considering latitude and elevation.

Ground acceleration. Acceleration of the ground due to earthquake forces.

Ground displacement. Distance that the ground moves from its original position during an earthquake.

Ground failure. Situation in which the ground does not hold together, such as landsliding, mud flows, and liquefaction.

Ground motion. Earthquake forces result from erratic vibratory motion of the ground on which the structure is supported. The ground vibrates both vertically and horizontally, but it is customary to neglect vertical components, since most structures have considerable excess strength in the vertical direction through the effect of safety factor requirements.

Ground movement

1. During earthquakes ground movement, as determined by instruments, has been found to be complex and irregular. Seismologists analyze graphs into simple trains of elastic waves. Ground movements are studied as elastic vibrations, generally following the laws of simple harmonic motion.

2. Ground movement includes all aspects of motion (acceleration, particle velocity, displacement).

Ground velocity. Velocity of the ground during an earthquake.

Guided waves. Guided waves exist in a layered medium in the general sense, such that their properties change only vertically and horizontally.

Guttenberg-Richter scale. Scale of magnitude proposed by Richter and later improved by Gutenberg and Richter. This scale gives quantitative values characterizing earthquake shocks in terms of released energy.

Hammering. Result of buildings being built too close to each other. During earthquakes they could hammer each other.

Harmonic motion. The response of a structure depends on how close the period of the structure is to the period of the assumed motion, and the resulting resonance effects could indicate that any structure would fail regardless of construction. If the structure and the motion have the same period of resonance, they are in harmonic motion.

Hayward fault. Branch of the San Andreas zone. This fault has played a significant role in the geologic development of the San Francisco Bay area, and it has also given birth to several large tremors.

Hertz. The unit of frequency equal to one cycle per second.

Hooke's law. The stress within a solid is proportional to the strain. This principle holds only for strains of a few percent or less.

Horizontal members. Beams, girders, trusses, joists, purlins, and slabs.

Hypocenter. Point below the epicenter at which an earthquake actually begins; focus.

Imperial fault. Branch of the San Andreas zone. The exact route of the fault was not exposed until the 1940 earthquake which ruptured the surface on both sides of the United States-Mexico border.

Inelastic behavior. Behavior of an element beyond its elastic limit.

Inertia. Property of matter by which it will remain at rest or in a uniform motion in a straight line or direction, unless acted upon by some external force.

Instrumental epicenter. As derived from seismograms, the epicenter often is near one end or side of the meizoseismal area and in some cases outside of the area.

Intensity

1. Degree of shaking and movement at a specified place.

2. Subjective measure of the force of an earthquake at a particular place as determined by its effects on persons, structures and earth materials. Intensity is a measure of effects as contrasted with magnitude, which is a measure of energy. The principal scale used in the United States today is the modified Mercalli.

3. As measured by the modified Mercalli scale, intensity is a measure of visible destructive effects and other observed phenomena accompanying an earthquake. There are many intensities in each earthquake, but usually the maximum is observed. The violence of earthquake motion in any part of the perceptible area of an earthquake is based on the effects observed on people and objects.

Intensity magnitude. Intensity is used to describe the severity or violence of an earthquake disturbance in any part of a shaken area and is based entirely on the earthquake effects reported on people and inanimate objects, including damage. Magnitude depends on the amount of energy released at the focus of an earthquake and is determined from the amplitudes of the ground vibrations registered at various seismological stations. Magnitudes are not reported in terms of energy but in terms of a mathematical function of the energy. A magnitude scale was developed by seismologists of the Pasadena Seismological Laboratory of the California Institute of Technology.

Interior waves. Two types of waves travel at different speeds through the earth's interior and are known as interior waves. The faster one alternately compresses and dilates the rock as it travels forward; the slower one shakes the rock sidewise as it advances—like the vibration of a violin string. Seismological tables, based on many thousands of seismograph readings, show to the nearest second just how long it takes each of these wave groups to travel to points on the earth's surface at various great circle distances from an earthquake origin.

Isoseismals. Map contours drawn to define limits of estimated intensity of shaking for a given earthquake.

Land mass movement detection. Recognition of mass movement of the earth's surface and its importance in the shaping of the land has lagged far behind the knowledge of the action of glaciers, winds, waves, and running water. Landslides, creeps, and other types of mass movement are occurring constantly without much attention being given them, yet they have great significance to the geologist and engineer.

Landslides. In regions where there are many hills with steep slopes, large earthquakes are often accompanied by landslides.

Lateral. Of or relating to the side; situated or directed toward or coming from the side.

Lateral force coefficients. Factors applied to the weight of a structure or its parts to determine lateral force for aseismic structural design.

Left-lateral fault. Strike-slip fault on which the displacement of the far block is to the left when viewed from either side.

Liquefaction. Transformation of a granular material from a solid state into a liquefied state as a consequence of increased pore-water pressure induced by vibrations.

Lg wave. Wave used for finding the boundary of continental structure; relatively small short-period disturbance easily seen superimposed on the longer surface waves.

Lithosphere. Outer rigid shell of the earth, situated above the asthenosphere and containing the crust, continents, and plates.

Live load. Weight of movable objects such as people, snow, rain, and wind.

Longitudinal wave

1. When an earthquake occurs in an elastic solid, the longitudinal and transverse wave start out together, but the longitudinal wave (sound wave) travels nearly twice as fast. This wave is identified as P, for primus.
2. Pure compressional wave with volume changes.

Long wave

1. Slow surface wave that is usually distinguishable only at a great distance.
2. Transverse vibration of a seismic surface wave.

Low-velocity zone. Region in the earth, especially a planar layer, that has lower seismic wave velocities than the region immediately above it.

Lumped mass. For analysis purposes, an assumed grouping of mass at specific locations.

L waves. Surface waves over the earth, generally of long period.

Maar volcano. Volcanic crater without a cone, believed to have been formed by an explosive eruption of trapped gases.

Macroseismic. Effects of earthquakes that can be observed on the large scale, in the field, without instrumental aid.

Macrozones. Large zones of earthquake activity such as zones designated by state or local building code maps.

Magma. Molten rock material that forms igneous rocks upon cooling.

Magnification factor. Increase in lateral forces at a specific site for a specified factor.

Magnitude

1. Measure of earthquake size that describes the amount of energy released.
2. Rating of a given earthquake independent of the place of observation.
3. Magnitude is related to the energy that is radiated from the earthquake source in the form of elastic waves.
4. Measure of energy released by an earthquake. The ground motion at a fixed distance from the epicenter is stated in terms of a magnitude scale. The magnitude scale is exponential in character so that an increase of one unit in magnitude signifies a 10-fold increase in energy release. Zero of the scale represents the smallest recorded earthquakes. The largest known earthquake magnitudes are about $8\frac{3}{4}$.
5. Measure of earthquake size, determined by taking the common logarithm (base 10) of the largest ground motion during the arrival of a P wave or seismic surface wave and applying a standard correction for distance to the epicenter.

Mantle. Main bulk of the earth between the crust and core, ranging from depths of about 40 to 3480 km. It is composed of dense mafic silicates and divided into concentric layers by phase changes that are caused by the increase in pressure with depth.

Meizoseismal area

1. Area within which the shock is the strongest, bounded by the isoseismal line.
2. Area within the isoseismals of higher intensity.

Mercalli scale. In 1902 the Italian seismologist Mercalli set up a new scale, which was based on a I to XII range and provided for more refined analysis of major damage. The Mercalli scale was modified in 1931 by two American seismologists, Harry O. Wood and Frank Neumann, to take into account modern features such as tall buildings, motor cars and trucks, and underground water pipes. It is this modified Mercalli scale (frequently abbreviated to MM) that is still used today.

Microregionalization. Breaking up of macrozones into much smaller zones of specific earthquake intensity and activity.

Microseism

1. Weak vibration of the ground that can be detected by seismographs and is caused by waves, wind, or human activity but not by an earthquake.

2. A microseism is not a small earthquake; it is a more or less continuous disturbance in the ground that can be recorded by a seismograph.

3. Minute waves called "microseisms" are continuously moving through the rocks over the entire surface of the earth, as can be seen by examining a sensitive seismograph record obtained in any part of the world. The most prominent and important waves have periods from about 4 to 7 seconds.

Microseismic. Effects are small scale and can only be observed with instruments.

Microseismic waves. Waves of meteorological origin just like ocean waves; in fact, ocean waves may play an important part in their generation. Regardless of the mechanism of their origin, it is a fact that storms and low-pressure areas at sea are always accompanied by great increases in the amplitudes of microseismic waves, recorded at seismographic stations in the surrounding coastal areas.

Mississippi Valley region earthquake. Although this region is not generally considered seismic, in 1811 it was the scene of the greatest earthquake this country ever experienced. The earthquake was centered near New Madrid in southeastern Missouri and was felt over two-thirds of the United States.

Modal analysis. Determination of design earthquake forces based on the theoretical response of a structure in its several modes of vibration to excitation.

Mode. Shape of the vibration curve.

Modes of vibration. During an earthquake, a building is deflected in relatively simple shapes called the "natural modes of vibration."

Modified Mercalli intensity scale

1. Scale of 1931, used to evaluate earthquake intensity. It separates the violence of earthquake motions into 12 different grades.

2. Scale originally devised by Mercalli in an attempt to improve on the Rossi-Forel scale and modified in 1931 by Wood and Neumann. It is believed to be more uniformly graded than other scales.

Mohorovicic discontinuity. Boundary between crust and mantle, marked by a rapid increase in seismic wave velocity to more than 8 km/ps. It has a depth of 5 to 45 km. Its abbreviation is "Moho" or "M-discontinuity."

Moment. Tendency or measure of the tendency to produce motion about a point or axis.

Moment frame. Frame that is capable of resisting bending movements in the joints, enabling them to resist lateral forces or unsymmetrical vertical loads through the frame's overall bending action. Stability is achieved through bending action rather than bracing.

Moment resisting space frame. Moment resisting vertical load-carrying space frame in which the members and joints are capable of resisting design lateral seismic forces by bending moments. The system may or may not be enclosed or adjoined by more rigid elements that could tend to prevent the space frame from resisting lateral seismic forces.

Mud flow. Mass movement of material finer than sand, lubricated with large amounts of water.

Mylonite. Very fine lithified fault breccia commonly found in major thrust faults and produced by shearing and rolling during fault movement.

Natural frequency. Constant frequency of a vibrating system in the state of natural oscillation.

Nature of earthquakes and seismic waves. Strong earthquakes are usually due to the rupturing of great masses of rock many miles beneath the surface of the earth. The rupture generally takes the form of slipping of sliding along a rupture plane, called a "fault."

Newport-Inglewood fault in California. This fault was unknown until 1920, when a small earthquake disclosed that there was an active break along the coast. If there were any doubts about its activity, they were dispelled in 1933, when the disastrous Long Beach earthquake occurred along the coast.

Nonstructural components. Building components that are not intended primarily for the structural support and bracing of the building.

Normal fault

1. Dip-slip fault in which the block above the fault has moved downward relative to the block below.

2. Fault under tension where the overlying block moves down the dip or slope of the fault plane.

Normalization. Method of standardizing characteristics of vibration.

Oblique-slip fault

1. Fault that combines some strike-slip motion with some dip-slip motion.

2. Combination of normal and slip or thrust and slip faults whose movement is diagonal along the dip of the fault plane.

Origin time. Instant at which the earthquake event commences at the hypocenter. (It does not include foreshocks.)

Orogeny. Tectonic process in which large areas are folded, thrust-faulted, metamorphosed, and subjected to plutonism. The cycle ends with uplift and the formation of mountains.

Oscillate. Swing backward and forward; to vibrate like a pendulum; to fluctuate between fixed limits. The strength of the complete structure is affected by the nonstructural components, even though these are not taken into consideration in the design. Structural materials behave somewhat differently under dynamic conditions than they do under static loads. The foundation beneath the structure deforms while the structure oscillates.

Out of phase. The state in which a structure in motion is not at the same frequency as the ground motion or the equipment in a building is at a different frequency from the structure.

Overturning. Seismic codes have generally required the calculation of and provision for the overturning moment on the basis that the building is a fixed-end cantilever beam loaded with the static lateral earthquake forces accounting for the shears and acting simultaneously in the same direction. Although this concept is not entirely correct, it is simple in application and for most low- or medium-height buildings it produces results that are fairly reasonable.

Pacific coast and Nevada seismic activity. Two-thirds of the seismic activity of the country is centered in this area, most of it in the coastal ranges of California.

Pacific ocean seismic belt. The rim of the Pacific Ocean outlines the world's greatest seismic belt. This belt includes the Pacific Coast, and western mountain region of the United States, and a large part of Alaska.

Pelean eruption. Volcanic eruption accompanied by great explosions and emanations of hot gas and Nuees ardentes, named for Mont Pelee on Martinique.

Pendulum seismograph. Instrument that measures the relative motion between the ground and a loosely coupled inertial mass.

Period (wave)

1. Time interval between the arrival of successive crests in a homogeneous wave train. The period is the inverse of the frequency of a cyclic event.

2. Time for a wave crest to traverse a distance equal to one wavelength or the time for two successive wave crests to pass a fixed point.

Phreatic eruption. Volcanic eruption of mud and debris caused by the expansion of steam formed when magma comes in contact with confined groundwater.

Plastic deformation. Deformation that proceeds to large strains at constant stress without fracturing.

Plate. One of the dozen or more segments of the lithosphere that are internally rigid and move independently over the interior, meeting in convergence zones and separating at divergence zones.

Plate tectonics. Theory and study of plate formation movement, interaction, and desctruction; theory that explains seismicity, volcanism, mountain building, and paleomagnetic evidence in terms of platemotions.

Plutonic. Term applied to deep focus earthquakes.

Principal shock. When earthquakes occur in series the most important is the principal shock; others are divided into foreshocks and aftershocks.

P waves

1. Longitudinal or compression rarefractional waves through the earth.

2. Primary or fastest waves traveling away from a seismic event through the earth's crust, consisting of a train of compressions and dilatations of the material.

Rayleigh wave

1. Discovered by Lord Rayleigh, a British physicist, the Rayleigh wave has particle motion purely in the plane of propagation. This discovery was the start of surface wave explanation.

2. Forward and vertical vibration of seismic surface waves.

Refraction (wave). Departure of a wave from its original direction of travel at the interface with a material of a different index of refraction (light) or seismic wave velocity.

Resonance

1. Condition of forced vibration such that the frequency of the disturbing force approaches the natural frequency of free vibration of the elastic body subjected to the forced vibrations.

2. Induced oscillations of maximum amplitude produced in a physical spectrum when an applied oscillatory motion and the natural oscillatory frequency of the system are the same.

Response. Effect produced on a structure by earthquake ground motion.

Return period of earthquakes. Time period (years) in which the probability is 63% that an earthquake of a certain magnitude will recur.

Richter magnitude scale

1. Scale that measures the size of an earthquake at its source. Measurements are based on records made on a standard type of seismograph at a distance of 62 miles from the epicenter. Seismograms from several different stations are normally used in computing the magnitude of a quake. Since most stations are bound to be some distance from the source other than the standard 62 miles, many records are compared and complex conversion tables are used to arrive at the final figure.

2. Measure of earthquake size that describes the amount of energy released. The measure is determined by taking the common logarithm (base 10) of the largest ground motion observed during the arrival of a P wave of seismic surface wave and applying a standard correction for distance to the epicenter.

Rift. Break or split in the earth's surface or crust.

Rift valley

1. Fault trough formed in a divergence zone or other area of tension.

2. Elongated narrow trough or valley formed by the sinking of a strip of the earth's crust between two more or less parallel normal faults.

Right-lateral fault. Strike-slip fault on which the displacement of the far block is to the right when viewed from either side.

Rigidity. Relative stiffness of a structure or element. In numerical terms, it is equal to the reciprocal of displacement caused by a unit force.

Rigid structures. Structural members must have ductility as well as strength. Deformation beyond the elastic range will occur in a strong earthquake, even though the structure is designed in full compliance with the provisions of the seismic building code. Ductility may mean the difference between readily repairable damage and catastrophic damage.

Roosi-Forel scale. Scale with 10 grades of intensities. M. S. De Rossi of Italy and F. A. Forel of Switzerland invented the scale, which dates back to 1874.

Rotation. Horizontal diaphragms are used in buildings with masonry walls and walls arranged so that torsional moments are minimized.

Rupture strength. Greatest stress that a material can sustain without fracturing at a pressure of 1 atmosphere.

Sag pond. Pond occupying a depression along a fault. The depression is due to uneven settling of the ground or other causes.

San Andreas fault

1. Most publicized rift in California. It is the longest in the state, and it annually produces dozens of earthquakes.

2. Surface fault, practically vertical. It extends 600 miles from Humbolt County, through the city of San Francisco, and south to the Colorado Desert. The most active areas along this fault are at Hollister and Cape Mendocino.

San Francisco earthquake. California earthquake of April 18, 1906, was the most notable of all earthquakes in this country. There were 700 lives lost, and San Francisco was practically razed by fires that followed shortly after the earthquake. The city burned unchecked for days due to the wrecking of the water supply system. Actual earthquake damage probably did not exceed $24 million (1906 dollar value), but the fire increased this perhaps twenty-fold.

San Jacinto fault. Part of the San Andreas zone, perhaps the most active branch. It has been the source of many important quakes, and the land forms along its route give mute testimony to its long-term significance on the California topography.

Santa Ynez fault. Largest of a group of related breaks that form a large seismic area around the Santa Barbara channel. The most spectacular earthquake to originate in this region was the 1925 Santa Barbara shock.

Scarp

1. Cliff, escarpment, or steep slope of some extent formed by a fault or a cliff; steep slope along the margin of a plateau, mesa, terrace, or beach.
2. A vertical or dip-slip displacement is expressed in the formation of scarps.
3. Deep descent or declivity; as a result of overlapping ground action due to earthquake activity. To cut down vertically or to a steep slope.

Scarplet. Often small enough in throw and linear extent to have originated in a single earthquake.

Scissoring. The vertical component of relative displacement reverses along the fault; such reversal along a fault is called "scissoring," and the points of zero throw where the reversals begin are "scissor points." Scissoring is a common characteristic of strike-slip faulting.

Second-order after shocks. Aftershocks of the second order exist; that is, a large aftershock may be closely followed by a train of small ones falling off more rapidly than the general aftershock activity.

Sedimentary blanket. Blanket that consists of cretaceous and younger strata, generally less consolidated and less competent than the basement rocks.

Sedimentary rock. Rock formed by the accumulation and cementation of mineral grains transported by wind, water, or ice to the site of deposition or chemically precipitated at the depositional site.

Seiche

1. Phenomenon involving the natural oscillation of a lake. Earthquakes cause seiches with amplitudes of as much as 3 ft. Seiches may occur in closed or partly closed bodies of water such as harbors, lakes, ponds, and channels; winds, current changes, or tides will also cause seiches.
2. The word "sieche" originates in Switzerland, where F. A. Forell introduced its general use. A typical seiche is a standing wave set up on the surface of an enclosed body of water.

Seismic. Pertaining to earthquake activities.

Seismic belt. An elongated seismic zone similar to the alpide zone.

Seismic coefficient. Ratio of the lateral acceleration from the earthquake to the vertical acceleration of gravity.

Seismic discontinuity. Surface within the earth across which P wave or S wave velocities change rapidly, usually by more than 0.2 km/second.

Seismicity

1. The phenomena of earth movements, or the general earthquake activity.
2. Worldwide or local distribution of earthquakes in space and time; general term for the number of earthquakes in a unit of time or for relative earthquake activity.

Seismic profile. Data collected from a set of seismographs arranged in a straight line with an artificial source, especially the times of P wave arrivals.

Seismic reflection. Mode of seismic prospecting in which the seismic profile is examined for waves that have reflected from near-horizontal strata below the surface.

Seismic refraction. Mode of seismic prospecting in which the seismic profile is examined from waves that have been refracted upward from seismic discontinuities below the profile. Greater depths may be reached than through seismic reflection.

Seismic sea waves. Submarine earthquakes occasionally generate seismic sea waves that travel thousands of miles over the oceans and cause great damage to shore property when they either pile up and break or simply flood shore line areas.

Seismic sea wave warning service. One of the important services of the Coast and Geodetic Survey is the maintenance of a seismic sea wave warning program. The principal objective is to alert military installations on Pacific islands and public officials in such areas as Hawaii, the West Coast of the United States, and other Pacific countries and islands whenever seismographic records reveal the occurrence of an earthquake that might generate a destructive sea wave.

Seismic support. A mass (heavy) is supported on springs (weak) so that the mass remains almost at rest when the free end of the springs is subjected to sinusoidal motion at the operating frequency.

Seismic surface wave. Seismic wave that follows the earth's surface only, with a speed less than that of S waves. There are Raleigh waves (forward and vertical vibrations) and Love waves (transverse vibrations).

Seismic transition zone. Seismic discontinuity, found in all parts of the earth, at which the velocity increases rapidly with depth, especially at 300 to 600 km.

Seismic vibrations or waves. In a great earthquake seismic vibrations or waves penetrate the entire structure of the earth and travel all over its surface. While great earthquakes are seldom felt farther than a thousand miles from their source, sensitive seismographs have registered these unfelt vibrations in all parts of the world for more than 50 years.

Seismic wave. An elastic wave in the earth usually generated by an earthquake source or explosion.

Seismogram. The record written on a seismograph is called a "seismogram." It is a continuous line that fluctuates according to the pendulum movements. The greater the earth's movement, the greater the variation in the record. Degree of amplitude is the basis for assigning a Richter magnitude to the shock. The arrival time of the P, S, and L waves enables seismologists to fix the distance between the epicenter and the recording seismograph. By comparing the records taken at several different stations, the source of the waves can be accurately pinpointed in both direction and distance.

Seismograph

1. Instrument used to measure and record earthquake vibrations and other earth tremors.
2. Instrument that writes or tapes a permanent continuous record of the earth's motion; seismogram.
3. A seismograph typically consists of three connected units; pendulum, chronograph, and recording device, which writes a line representing the motion of the pendulum on the chronograph drum.

4. Instrument for recording earthquake activity. The electromagnetic type was constructed in Italy by L. Palmierei in 1855 and installed in the observatory on Vesuvius.

5. Instrument for magnifying and recording the motions of the earth's surface that are caused by seismic waves.

Seismographs. Seismographs are basically nothing more than carefully constructed pendulums that appear to swing back and forth as the earth vibrates under them.

Seismology

1. Science of earthquakes from the Greek "seismios" (a shaking); observation and data dealing with the earth's vibrations.

2. Study of earthquakes and seismic waves, and their propagation through the earth.

3. Science of observing and recording the generation and propagation of elastic waves in the earth regardless of whether these are of earthquake or artificial origin.

Seismometer

1. Instrument to pick up linear (vertical, horizontal) or rotational displacement, velocity, or acceleration (mostly on soil surface).

2. Detector to indicate seismic waves; when combined with a graph or recording mechanism, it is called a "seismograph."

3. Seismograph whose physical constants are known sufficiently for calibration, so that actual ground motion can be calculated from the seismogram.

Seismoscope. Device that indicates the occurrence of an earthquake but does not write or tape a record.

Shallow earthquakes. The great majority of so-called shallow earthquakes originate at depths of 10 or 20 miles, but some are deeper, extending to a maximum depth of 450 miles. Earthquakes due to volcanic activity may seem violent locally, but they are never very deep or felt at great distances. Compared with tectonic earthquakes that are frequently registered on instruments all over the world, they are very superficial.

Shear. Stress that resists the tendency of two parallel forces acting in opposite directions to cause the planes of the body to slide onto each other.

Shear distribution. Distribution of lateral forces along the height or width of a building.

Shearing force. During an earthquake a structure is subjected to a forced vibration, imposed on it by the movements of its foundations. The inertia of the structure tends to resist the movements of its foundation. Therefore, at the foundation there is a shearing force, (base shear) imparted to the structure in order to make it move.

Shear strength. Stress at which a material fails in shear.

Shear wall

1. Wall designed to resist lateral forces parallel to the wall. Braced frames subjected primarily to axial stresses are considered shear walls for the purpose of this definition.

2. Wall designed to resist lateral forces parallel to the wall. A shear wall is normal vertical, although not necessarily so.

Shock pulse. Form of shock excitation characterized by a rise and decay of acceleration in a relatively short period of time. The major values of accelerations for a shock pulse are in one direction.

Shock pulse duration. Time required for the acceleration of the pulse to rise from some stated fraction of the maximum amplitude and to decay to this value.

Side hill furrows. Side hill furrows are among the more conspicuous of the permanent smaller features characteristic of strike-slip fault zones.

Sierra Nevada fault. Earthquake-caused movements created the escarpment that forms the eastern edge of the Sierra. The Owens Valley branch of the system was responsible for the 1872 quake—the largest in California's recorded history.

Simple harmonic motion. Oscillatory motion of a wave; single frequency. Essentially, a vibratory displacement like that described by a weight that is attached to one end of a spring and allowed to vibrate freely.

Slickensides. Parallel grooves, ramps, and scratches on one or both of the inside faces of a fault, showing the direction of slip.

Slip (fault). Relative motion of one face of a fault relative to the other.

Soil-bearing capacity. Bearing capacity or resistance of the ground measured in pounds per square foot, determined by test.

Soil-structure interaction. Effects of the properties of both soil and structure on the response of the structure.

Space frame. Three-dimensional structural system composed of interconnected members, other than bearing walls, laterally supported so as to function as a complete self-contained unit with or without the aid of horizontal diaphragms or floor bracing systems.

Spatter cone. Conical deposit of cooled lava congealed around a volcanic vent that disgorges mostly gas with occasional globs of molten rock.

Special seismographs. Seismographs made to record destructive earthquake motions usually have very short-period pendulums. In fact, they are more like weighted springs. They record the acceleration or force of the motion on photographic paper. Instead of operating continuously like regular seismographs, they remain inoperative until a strong earthquake starts them. The starting pendulum closes an electrical circuit that causes the entire recording mechanism to operate for about 1.5 minutes.

Spectra. Plot indicating maximum earthquake response with respect to the natural period or frequency of the structure or element. The response can show acceleration, velocity, displacement, shear, or other properties.

Spectrum. The spectrum can be visualized by assuming a movable base with a series of cantilevered pendulums with varying periods. With the ground motion or force moving to the left, the pendulum moves to the right, with the length of the pendulum determining the period of increase in time. The maximum response is plotted against the period of the pendulum. This develops a curve.

Stability. Resistance to displacement or overturning.

Stiffness

1. Stiffness of a member having the ability to resist bending.

2. Rigidity; reciprocal of flexibility.

Strain. Deformation of the body in the vicinity of a given point.

Strain release. Movement along a fault plane. It can be gradual or abrupt.

Strain seismograph. Instrument that measures changes of strain in surface rocks to detect seismic waves.

Stress. A point in the interior of a body stress is determined by the system of forces acting in the vicinity of that point.

Strike-slip fault. Fault whose relative displacement is purely horizontal.

Strong ground motion. The shaking of the ground near an earthquake source made up of large-amplitude seismic waves of various types.

Subduction. Sinking of a plate under an overriding plate in a convergence zone.

Subduction zone. Dipping planer zone descending away from a trench and defined by high seismicity, interpreted as the shear zone between a sinking oceanic plate and an overriding plate.

Subsidence

1. Land movement that differs from other mass movements, being essentially vertical instead of a superficial displacement down a topographic slope.
2. Sliding and guiding of large blocks of earth.
3. Gentle epeirogenic movement where a broad area of the crust sinks without appreciable deformation.

Surface waves. Elastic waves propulgated along the surface of a bounded elastic solid.

S wave

1. Secondary seismic wave traveling slower than the P wave and consisting of elastic vibrations transverse to the direction of travel. It cannot penetrate a liquid.
2. Shear wave produced essentially by the shearing or tearing motions of earthquakes at right angles to the direction of wave propagation.
3. Transverse or shear wave through the earth.

Tectonic earthquakes. Tectonic earthquakes are often accompanied by the fracturing of earth materials, called "faults." When there is no visible rupture on the earth's surface during an earthquake, it is still believed that an interval rupture takes place in the form of an overthrust or in some other manner. The connection between faulting and earthquakes has has established by observations on the surface and rationalized further by geological studies.

Tectonics. Study of the movements and deformation of the crust on a large scale, including epeirogeny, metamorphism, folding, faulting, and plate tectonics.

Teleseism. Earthquake recorded by a seismograph at a great distance.

Tension. Stress that resists the tendency of two forces acting in opposite directions.

Thermal contraction. Principal reason for the folding and thrusting of the earth's crust.

Thermal expansion. Increase in volume as a result of an increase in internal temperature.

Theory of plate tectonics. The theory of plate tectonics asserts that the crust and upper mantle of the earth are made of internally rigid plates that slowly, continuously, and independently slide over the interior of the earth.

Thrust fault. Dip-slip fault in which the upper block above the fault plane moves up and over the lower block, so that older strata are placed over younger.

Thrust (reverse) fault. Fault under compression where the overlying block moves up the dip of the fault plane.

Tidal waves. Tidal waves are more correctly referred to as "sea waves" or by the Japanese word "tsunami." Any earthquake that causes an abrupt change in the level of the sea bottom, or extensive submarine landslides may generate a tsunami.

Time-dependent response analysis. Study of the behavior of a structure as it responds to a specific ground motion.

Torsion

1. Stress that resists the tendency of a body to twist.
2. Twisting around an axis.

Torsional movement. Torsion is normally assumed to occur when the centroid of rigidity of the various vertical resisting elements in a story fails to coincide with the center of gravity. Distance between the two, called the "eccentricity," times the amount of lateral force is a torsional moment that must be resisted in addition to and simultaneously with the normal design lateral forces. Torsion is simply a twisting about the vertical axis. Building has natural torsional modes in addition to its translational modes.

Transit time (or travel time). Elapsed time between origin time and the arrival of a given seismic wave at a specified point, such as a seismograph station.

Transverse wave. When an earthquake occurs, this wave arrives behind the longitudinal wave. Since it got there second, it is called S or secundus.

Trench. Long and narrow deep trough in the sea floor; interpreted as marking the line along which a plate bends down into a subduction zone.

Triple junction. Point that is common to three plates, and which must also be the meeting place of three boundary features, such as divergence zones, convergence zones, or transform faults.

Tsunami

1. Ocean wave produced by a submarine earthquake, landslide, or volcanic eruption. Tsunamis steepen and increase in height on approaching shallow water, inundating low-lying areas, and where local submarine topography causes extreme steepening, they may break and cause great damage. Tsunamis have no connection with tides.
2. Sea wave caused by an underwater earthquake; great wave; tidal wave.
3. Large destructive wave caused by the sea floor movements in an earthquake.
4. Sea wave produced by large area displacements of the ocean bottom, the result of earthquakes or volcanic activity.
5. Tsunamis are commonly called "tidal waves," although they have nothing to do with tides. Most follow strong earthquakes and are called "seismic sea waves."

Turbidity currents. Earthquake reaction to the currents of the ocean caused by landslide material moving into sea canyons or down the continental slope.

T wave. Sound wave in the sea that is propagated from the shore into the continental interior with the velocity of ordinary seismic waves.

Vertical load transfer. Space frame designed to carry all vertical loads.

Vertical members. Columns, piers, pilasters, posts, struts, and walls.

Vibration. Periodic motion that repeats itself after a definite interval of time.

Vibratory motion. The vibratory motion of the ground sets up inertia forces in the structure. For a rigid structure rigidly coupled to its foundation, force equals the mass of the structure times the acceleration of ground motion at any instant. If the structure yields slightly, that is, if it is flexible, then for short periods of time force may be somewhat less because yielding of the structure absorbs some of the energy, storing it for some later time. However, if a very flexible structure is subjected to a ground motion whose period is near that of the structure, a much greater force may result.

Vibrogram. Vibration measurements.

Volcanic earthquake

1. Earthquake caused by gas explosions at erupting volcano vents.
2. Gas explosions at an erupting volcanic vent cause vibrations in the ground.

Volcanic eruption. Appearance, usually sudden and violent, at the surface of the earth of lava, ash, vapor, steam, and other materials from a volcano.

Volcano

1. Mountain that has been built up by the materials ejected from the interior of the earth through a vent.
2. Opening through the crust that has allowed magma to reach the surface, including the deposits immediately surrounding this vent.

Warping. In tectonics, the gentle, regional bending of the crust, which occurs in epeirogenic movements.

Wavelength

1. Distance between two successive peaks, or between troughs, of a cyclic propagating disturbance.
2. Distance between successive similar points on two wave cycles.

Wave steepness. Maximum height or amplitude of a wave divided by its wavelength.

Western mountain region seismic zone. Excluding Nevada, the principal seismic zone in this area extends in a broad band from Helena, Montana, southward to northern Arizona.

White Wolf fault. Short, relatively insignificant fault that unexpectedly generated the greatest earthquake to hit California since 1906. Its unimpressive size and apparent inactivity was very deceiving.

World earthquakes. It is estimated that more than a million earthquakes occur throughout the world annually. They range from minor tremors that are barely perceptible locally to catastrophic shocks.

Zones. Four classified earthquake zones, zero through four, exist in the United States.

SOIL INVESTIGATION AND SOIL TESTING

Active coefficient of earth pressure. Minimum ratio of minor principal stress to the major principal stress, applicable where the soil has yielded sufficiently to develop a lower limiting value of the minor principal stress.

Active earth pressure. The minimum value of earth pressure. This condition exists when a soil mass is permitted to yield sufficiently to cause its internal shearing resistance along a potential failure surface to be completely mobilized.

Adobe. Natural mixtures of fat clays and sand; exceedingly cohesive and sticky soil found in the southwest.

Airfield borings. Auger borings usually extended 10 ft below the top of the pavement in cuts 10 ft below existing ground in shallow fills, or to a depth at which the California bearing ratio for proposed loading is 1, whichever is greater.

Amorphous structure. Soil of fine texture having a massive or uniform arrangement of particles. Individual grains cannot be recognized.

Angle of repose. Maximum slope or angle, measured from a horizontal plane, at which unconsolidated material (soil or loose rock) remains stable. When exceeded, mass movement by creeping, slipping, or sliding may be expected.

Approximate yield of water. Approximate yield of water that may be anticipated from a previous stratum may be determined by bailing or pumping the bore and observing the corresponding stabilized draw-down level. The yield of a single, small-diameter boring cannot be relied on, however, to furnish data on which a trustworthy estimate of the pumping requirements for a proposed foundation excavation may be based. Better results may be obtained by means of a test pit or caisson.

At-rest coefficient of earth pressure. Ratio of the minor principal stress to the major principal stress; applicable where the soil mass is in its natural state without having been permitted to yield or without having been compressed.

Atterberg limits. Indicators of structural properties of soils.

Atterberg tests. Tests that determine minimum water contents of cohesive soils causing specified conditions of consistency. The water contents of the Atterberg limits are usually expressed as percentage of the weight of the dry soil.

Auger borings

1. Hand- or power-operated augering with periodic removal of material. In some cases a continuous auger may be used, requiring only one withdrawal. Casing generally is not used. They are ordinarily used for shallow explorations above the water table in partly saturated sands and silts and soft to stiff cohesive soils.
2. Samples of cohesive soils and some types of noncohesive soils may be obtained by this method if the soils are not too soft. However, samples brought up on the auger are invariably badly disturbed and consequently of limited value. Fully saturated noncohesive soils will not be retained on an auger.
3. An ordinary wood auger is used where more definite information is required. An auger will often penetrate 100 ft or more and bring up fairly reliable samples. An auger, however, is chiefly of use in fine sand or clay and stops on the first obstruction encountered.
4. Auger borings are primarily utilized for shallow exploration above groundwater. Although materials recovered are disturbed, auger borings furnish continuous samples of soils encountered.

Auger samples. They are samplers designed to completely remove soil from the hole. They are advanced by hand or power rotation and the soil is completely churned upon removal. They are used for continuous identification of materials in the profile and where tube samples are not required. Ordinarily they are used for shallow explorations above the water table but can be taken to depths of about 50 ft.

Base exchange. Adsorption that involves reactions that are essentially chemical or ionic in character.

Bearing capacity. The ultimate bearing power of a soil mass is the minimum load that causes failure of the mass. The safe-bearing capacity is evaluated by the use of a safety factor. Failure of the mass may, however, be determined by a specified settlement.

Bedrock—igneous or sedimentary. The material may contain fissures, folds, and fault zones. It may be filled with compressible soils, or it may be open as the result of scour. Fissures in the upper surfaces of bedrock are common in glaciated regions. Cavernous conditions are common in soluble limestones.

Bedrock surface. Borings are extended 5 ft into sound, unweathered rock. Where the character of the rock is not known or where boulders or irregularly weathered material overlie bedrock, coring 10 ft into sound rock and 20 ft into one or more selected areas is recommended. In cavitated limestone, borings are extended through strata suspected or containing solution channels.

Borings in rock. Borings in rock are made by percussion drilling or core boring with a diamond drill, shot drill, or steel sawtoothed bit. They have extremely limited value as an exploratory method in rock as they yield poor samples. Core boring, on the other hand, if properly done, can yield practically continuous samples from which reliable information may be obtained as to the character and soundness of the rock penetrated.

Boring without sampling. Only the depth to rock or the existence of cavities is first determined. Borings may be made without sampling; so the utility of a boring for sampling is not important.

Boulders. Detached and rounded or worn pieces of rock, the smallest of which are from 6 to 8 in.

Brittle. Soil that will break with a sharp, clean fracture, when dry. If struck a sharp blow, it will shatter into cleanly broken, hard fragments.

Buildings and structure sites. Borings are extended to a depth such that the increase in the vertical soil stress will not exceed 10% of the vertical stress existing prior to the proposed construction, but the depth should not be less than the width of the structure and should extend to not less than 25 ft below the bottom of the deepest part of the proposed foundation.

Calcareous soil. Soil containing sufficient calcium carbonate to effervesce when tested with a weak hydrochloric acid.

Caliche. Layer near the surface, more or less cemented by secondary carbonates of calcium or magnesium precipitated from the soil solution. Caliche may occur as a soft, thin soil horizon, as a hard, thick bed just beneath the solum, or as a surface layer exposed by erosion.

California bearing ratio (CBR) test. Generally the effect of various compaction efforts on CBR values is determined. The test entails compaction of samples at energies ranging from standard to modified Proctor efforts.

California liner—sampler. Sampler obtained by penetration of a thick, solid-wall sampler containing 2-in. liner tubes.

Capillarity of a soil. Ability to raise water in thin films above the level of the water table. Both particle size and the physicochemical properties of the soil, particularly adhesion, influence its value.

Casing. Casing is used in the upper 10 ft of all borings and throughout borings in soils with cavities, or in soils where a full head of drilling fluid cannot otherwise be maintained at all times. Casing is used whenever it is required to obtain groundwater observations at intermediate depths for extended periods and in general whenever successful boring operations make it necessary.

Cavities—fractures and joint zones. In the bedrock being investigated, wash borings or rotary borings without sample recovery, or soundings and probing are spaced as close as 10 ft.

Chalk. Very soft, white to light gray, fine-grained variety of limestone composed largely of the calcareous shells of small marine organisms.

Check borings

1. Geophysical surveys are supplemented by borings to recover representative samples and to check stratification interpreted from the survey. These borings are used to clearly illustrate correlation or lack of it between test borings and geophysical information in presenting subsurface data on contract drawings.

2. During final exploration, at least one boring should extend well below the zone involved in the apparent stability, settlement, or seepage problem to make sure no unusual conditions exist at great depth.

Churn drilling. Method used in drilling through soil, boulders, and rock. In removing the cuttings, only enough water is used to fill the bottom of the hole. Cuttings mix with the water as the churning proceeds, and at intervals the drill stem is withdrawn; cuttings and water are removed by a sand pump or bailer.

Clay

1. Soil separate consisting of particles finer than 0.002 mm in equivalent diameter. Soil material containing more than 40% clay, less than 45% sand and less than 40% silt.

2. Fine-textured soil that forms hard lumps or clods when it is dry.

3. Extremely fine-grained inorganic soil, the individual particles of which are so small as to be impalpable when rubbed between the fingers. Many of the particles are in fact of colloidal size.

4. Combination of silica and alumina with all sorts of impurities mixed with it. When mixed wet and then dried out, it becomes very hard and shrinks in volume. Being of particles so much finer than sand, it is held in suspension and carried much farther out to sea than the coarser-grained sand or gravel, which is deposited first. The finest particles of all are often carried far out into the ocean as mud. Fine material may become shale by pressure or some other means. Shale may be uplifted and exposed to weather where it will disintegrate and again become mud or clay.

Clayey. Containing large amounts of clay; having properties similar to those of clay.

Clay loam. Soil material that contains 27 to 40% clay and 20 to 45% sand.

Clay pan

1. Accumulation or stratum of stiff, compact, and relatively impervious clay. Clay is not cemented and if immersed in water, can be worked into a soft mass.

2. Compact, slowly permeable layer in the subsoil having a much higher clay content than the overlying material, from which it is separated by a sharply defined boundary. Claypans are usually hard when dry, and plastic and sticky when wet.

Clays. These are inorganic clays of high placticity, as well as fat clays. Organic clays are medium to high plasticity.

Coarse sand. Sand the greater portion of which is retained by a 28-mesh sieve.

Coarse sandy loam. Loam that consists of 25% or more very coarse and coarse sand and less than 50% of any other one grade of sand.

Coarse texture. Texture exhibited by sands, loamy sands, and sandy loams (except very fine sandy loam).

Coefficient of consolidation. Coefficient utilized in the theory of consolidation, containing the physical constants of a soil that affect its rate of volume change.

Coefficient of earth pressure. Principal stress ratio at a point in a soil mass.

Coefficient of subgrade reaction. Ratio of load per unit area of horizontal surface of a mass of soil to the corresponding settlement of the surface. It is determined as the slope of the secant, drawn between the point corresponding to zero settlement and the point of 0.05-in. settlement, of a load-settlement curve obtained from a plate load test on a soil using a 30-in. or greater diameter loading plate.

Cohesion. Maximum tensile value of the soil; complicated coordination of many factors, such as gravitational grain attraction, colloidal adhesion of the grain covering, capillary tension of the moisture films, atmospheric pressure where gases in the void space have been dissolved or absorbed, electrostatic attraction of charged surfaces, and many others.

Cohesionless soil. Soil that when unconfined and air-dried has little or no strength and that when submerged has little or no cohesion.

Cohesion of a soil. Property by which resistance to displacement of particles is developed by the forces of attraction that act between them.

Cohesive soil. Soil that when unconfined and air-dried has considerable strength and that when submerged has significant cohesion.

Combination drilling rigs. Rigs equipped for auger boring, percussion drilling, rotary drilling, core boring, operation of dry samplers, dewatering holes, and making field permeability tests. "All-purpose" drilling rigs are most useful in making the extensive subsurface explorations required for major projects.

Compact. Soil packed together in a dense firm mass, but without any cementation. A relative degree of compaction may be expressed by terms such as "slightly compact" or "very compact."

Compacted sample tests. In prospecting for borrow materials, index tests or tests specifically for compacted samples may be required in a number proportional to the volume of borrow involved or the number of samples obtained. Structural properties tests are assigned after borrow materials have been grouped in major categories.

Compaction. Densification of a soil by means by mechanical manipulation.

Compaction curve (proctor curve). Curve showing the relationship between the dry unit weight (density) and the water content of a soil for a given compactive effort.

Compaction of soils. Permanent soil alteration is accomplished by physical means. Most soils can be compacted or densified by the application of pressure sufficient to reduce the water content and eliminate the voids.

Compaction test (moisture-density test). Laboratory compacting procedure whereby a soil at a known water content is placed in a specified manner into a mold of given dimensions and subjected to a compactive effort of controlled magnitude, and the resulting unit weight determined. The procedure is repeated for various water contents to establish a relation between water content and unit weight.

Compressibility

1. Property of a soil pertaining to its susceptibility to decrease in volume when subjected to load.
2. Property by which a decrease in volume results when an external pressure is applied.

Compression index. Slope of the linear portion of the pressure-void ratio curve on a semilog plot.

Compressive strength (unconfined compressive strength). Load per unit area at which an unconfined prismatic or cylindrical specimen of soil will fail in a simple compression test.

Compression wave (irrotational). Wave in which an element of the medium changes volume without rotation.

Concentration factor. Parameter used in modifying the Boussinesq equations to describe various distributions of vertical stress (Joseph Valentin Boussinesq).

Cone index. Index of the shearing resistance of soil obtained with the zone penetrometer; a number representing resistance to penetration into the soil of a 30° cone with a $\frac{1}{2}$-in. base (actually, a load in pounds on a cone base area in square inches).

Cone penetrometer. Instrument used to measure the ability of a soil to support traffic movements.

Consistency

1. Relative ease with which a soil can be deformed.
2. The number of blows required to drive a split spoon sampler 1 in. through the clay gives a rough indication of the consistency or strength of the clay.

Consolidated-drained test (slow test). Soil test in which essentially complete consolidation under the confining pressure is followed by additional axial (or shearing) stress applied in such a manner that even a fully saturated soil of low permeability can adapt itself completely (full consolidate) to the changes in stress due to the additional axial (or shearing) stress.

Consolidated-undrained test. Soil test in which essentially complete consolidation under the vertical load (in a direct shear test) or under the confining pressure (in a triaxial test) is followed by a shear at constant water content.

Consolidation

1. Gradual reduction in volume of a soil mass resulting from an increase in compressive stress.
2. Test in which the specimen is laterally confined in a ring and is compressed between porous plates.
3. The test physically loads an undisturbed sample of clay and can determine if a clay has been preloaded or not by measuring the compression that occurs with successive loading increments.

Consolidation tests. One-dimensional consolidation tests with complete lateral confinement determine the total compression of fine-grained soil under an applied load and the time rate of compression due to gradual volume decrease, which accompanies squeezing of pore water from the soil.

Contour survey (entire site). Contour survey should be made having adequate number of spot elevations in flat areas and on topographical features such as depressions, displacementage swales, saddles, top of high area, or ridges. The survey should meet the requirements of national map accuracy standards. Contour intervals should be established and extended 10 ft beyond all property lines.

Core barrel samples. Samples are taken in rock or hard cohesive soils (materials relatively insensitive to sampling disturbance). The suitability of cores for structural properties tests depends on the quality of individual samples. The percentage of core recovery is an indication of soundness and degree of weathering of rock.

Core barrel sampler (Dension sampler). Sample that is similar to the double-tube core barrel except that the cutting edge of the inner barrel may be extended 3 in. beyond the outer coring bit by exchanging bits. The sampler is adaptable to sampling hard clays, cemented coarse-grained soils, hardpan, weathered or soft rock. The sampler is used in conjunction with rock coring when the overburden consists of materials that are difficult to sample by other methods.

Core lifter. The core lifter automatically grips the core so that it is removed in the barrel. The core provides a plausible record of the material penetrated. Gaps usually exist in the record because core recovery is often incomplete. The core size is commonly $1\frac{1}{8}$ in. with a 2-in. hole.

Core run. The length of the core run is defined as the length of penetration between retrievals of the core barrel from the bore hole, expressed in feet and tenths of feet. Core recovery expresses the length of core recovered from the core barrel per core run, in percent.

Dams and levees borings. Borings extended to penetrate not only the soft and unstable but also the permeable strata, as consideration of seepage conditions, as well as foundation loading, governs design.

Dams and water retention structure borings. Borings to a depth of one-half the base width of earth dams or one to one and one-half times the height of small concrete dams in relatively homogeneous foundations. Borings may terminate after penetration of 10 to 20 ft in a hard and impervious stratum if continuity of this stream is known from reconnaissance.

Dams and water retention structures boring locations. Preliminary borings are spaced approximately 200 ft centerlines over the foundation area. Spacing is decreased to 100 ft centerlines for immediate borings. Included are borings at the cut-off location and critical spots in the abutment.

Deep-cut borings

1. Borings to a depth of between three-quarters and one times the base width of narrow cuts. Where the cut is above groundwater in stable materials, a depth of 4 to 8 ft below the base may suffice. Where the base is below groundwater, it is necessary to determine the extent of pervious strata below the base.

2. Borings extended to a depth below the deepest part of the proposed cut equal to its bottom width, where this width is less than one-half the depth of the cut. Where the bottom of the cut will be below groundwater level, the extent of previous strata and the pore water pressure should be investigated, as the stability of the slopes may be affected by seepage.

Degree of consolidation (percent consolidation). Ratio, expressed as a percentage, of the amount of consolidation at a given time within a soil mass to the total amount of consolidation obtainable under a given stress condition.

Dense structure. Soil mass having a minimum of pore space and an absence of large pores or cracks.

Densification of soils. Loose soils are grouted for densification by the addition of cement (often mixed with fly ash or bentonite to provide better mobility) into the pores, or with slit intrusion or chemical or asphalt emulsion. These procedures replace the void content, which is partly or wholly filled with moisture, with more solid filler.

Density. The density of a soil is its weight per unit volume. Its value is therefore dependent on the state of compaction. The term "mass density" is sometimes used and refers to the density or unit weight of the soil mass or body.

Depth. Depth below reference elevations, usually ground surface unless otherwise shown.

Depths of test borings. Required depths depend to some extent on sizes and types of proposed structures. Depths are controlled to a greater degree by the character and sequence of subsurface strata.

Desert soil. Zonal great soil group consisting of soils with a very thin, light-colored surface horizon, which may be vesicular and is ordinarily underlaid by calcareous material. Soil is formed in arid regions under sparse shrub vegetation.

Detailed analysis. Minimum of one boring to obtain undisturbed samples of critical strata. Sufficient preliminary dry sample borings are provided to determine the most representative location for undisturbed sample borings.

Determination of the free groundwater level. Water levels and any marked excess hydrostatic pressure in pervious strata are essential steps in subsurface explorations. Groundwater conditions may be regular, with a single closely defined water table which is the contact surface between the free groundwater and the capillary zone, or irregular, with one or more bodies of groundwater having water tables perched above impervious strata or held beneath such strata under artesian pressure.

Development of site on soft compressible strata. Borings are spaced 100 to 200 ft on centers at possible building locations. Intermediate borings are added when building sites are determined.

Diamond drilling. Diamond drilling is suitable only for rock drilling. The drill is a hollow steel cylinder with embedded black diamonds outside and inside the bottom edges, forming a bit. The drill is rotated by power-driven equipment cutting into the rock by abrasion. Water forced down the drill rod to the bottom rises in the annular space between the drill rod and the sides of the hole. The water removes the cuttings and cools the bit. The bit is not attached directly to the drill rod, but a cylinder called a "core barrel" is intermediate, providing space for a core length of up to 10 ft.

Diamond drill or core borings. It is necessary to be absolutely sure as to the depth and nature of the bedrock. Borings are obtained by having a cutter that is hard enough to cut out a core of even the hardest rock and bring it to the surface. The cutting tool is made of diamond, shot, or fragmented chilled cast iron. Cores are sometimes about 1 in. in diameter and from a fraction of an inch to 5 or 10 ft long.

Dilatancy or volume change. Dilatancy is the result of applied external forces. It is a complicated property, since it is difficult to separate true dilatancy from shrinkage (sometimes expansion) as a result of variation in fluid content. Dilatancy is a reversible phenomenon; unfortunately, so is the effect of fluid change. If either action is disregarded, prediction of movements of structures on volume-changing soils can lead to inaccurate values.

Direct shear test

1. Shear test in which soil under an applied normal load is stressed to failure by moving one section of the soil container (shear box) relative to the other section.

2. Test to determine the maximum shearing strength and angle of friction of soils for use in stability analysis.

Displacement-type boring. Boring by repeatedly driving or pushing a tube or spoon sampler into the soil and withdrawing the recovered materials. Changes are indicated by examination of the materials and resistance to a driving or static force for penetration. No casing is required. This type of boring is used in loose to medium compact sands above the water table and in soft to stiff cohesive soils. It is limited to holes less than 3 in. in diameter.

Disturbed sample. Sample obtained from auger cuttings or wash water for classification purposes only.

Double-tube core barrel. A tube is enclosed within and attached to a core barrel by means of a swivel. As the outer barrel rotates, the inner tube remains stationary and receives the core. Cuttings are removed by circulating water between the outer and inner barrels. The double-tube core barrel is available with an outside diameter from 2 to 18 in. It is used primarily in nonuniform, fissured, friable and soft rock and obtains samples $1\frac{1}{3}$ to $15\frac{5}{8}$ in. outside diameters up to 20 ft long.

Double-tube auger. Open spiral sampler, containing a thin-wall liner, that is rotated into the soil. The liner receives samples, and the open-top cylinder retains material displaced by the spiral. No water is used. It obtains samples 46 in. long with outside diameters of $1\frac{1}{4}$ and $2\frac{1}{4}$ in. It is used for taking tube samples at relatively shallow depths and may be advanced by hand in soft to medium clays and loose to medium compact sandy soils.

Driving resistance. Number of blows of a pile-driving hammer required to advance the point of a pile a specific distance into the subsoils.

Dry sample. Drill rod with bit with an open-ended pipe or special forms of samplers. The apparatus is placed in the hole, and the pipe is driven into the soil at the bottom. The pipe or sampler is brought to the surface, and the sample removed and examined. Although not dry, the sample is called a "dry sample" to distinguish it from the samples brought up in wash water. The sampling procedure is not adaptable to rock, but the boring method is used for penetrating sand, gravel, clay, boulders, and solid rock.

Dutch cone penetration method. A hardened steel cone is forced vertically into the soil by a static thrust. The thrust required to cause a bearing capacity failure of the point is measured and recorded.

Earth pressure. Pressure or force exerted by soil on any boundary.

Earth pressure at rest. Value of the earth pressure when the soil mass is in its natural state without having been permitted to yield or having been compressed.

Elasticity of a solid. Ability of a solid to return to its original shape and size when the external force causing deformation is removed, provided the deforming force is within the elastic range.

Elastic modulus. Value only where the soil acts as a solid, with complete rebound on the release of the load. Most soils do act in this fashion over limited ranges of load application and favorable external conditions. Reliance on elasticity is limited by the possibility of unfavorable change in conditions during the life of a structure.

Elastic state of equilibrium. State of stress within a soil mass when the internal resistance of the mass is not fully mobilized.

Electroosmosis. The control of moisture is feasible only in fine-grained soils.

Fat clays. Finer grained than lean clays. When wet, fat clays may have a distinctly soapy feel. They are strongly cohesive and when dry, can be broken but not crushed or pulverized by hand. Partially dried fat clays are plastic.

Field expedients. On certain projects, expedient materials (rails or rods with detachable cone points) may be driven to determine stratification and make typical soundings adjacent to borings to correlate penetration resistance with soil type.

Fill. Man-made deposits of natural soils and waste materials.

Final borings. Final borings are arranged so that geological sections may be determined at the most useful orientations. Borings in slide areas should establish geological sections necessary for stability analysis.

Fines. Portion of a soil finer than soil that can pass a No. 200 U.S. standard sieve.

Fine sand

1. Consists of 50% or more fine sand; less than 25% very coarse, coarse, and medium sand and less than 50% very fine sand.

2. Sand that passes a 65-mesh and is retained by a 200-mesh sieve.

Fine sandy loan. Consists of 30% or more fine sand and less than 30% very fine sand; between 15 and 30% very coarse, coarse, and medium sand.

Firm. Soil that is moderately hard. It shows a resistance to forces tending to produce rupture or deformation.

Flood plain deposits. Deposits placed during high-water seasons alongside the lower reaches of rivers. They usually consist of fairly uniform and continuous layers of silt or clay separated by equally uniform layers of sand and even coarser material.

Footing loadings. The load to be used for footing design must be at least the dead load, including the weight of all fixed partitions, furniture, finishes, etc. However, it is doubtful whether the live load has any effect on settlements, except in such structures as grandstands, where the live-load increment may be over 150 lb/sq ft on certain occasions.

Free water (gravitational water). Water that is free to move through a soil mass under the influence of gravity.

Friable

1. Pertaining to the ease of crumbling of soils.

2. Structure that may be easily pulverized to a granular state. Aggregates of a friable soil are readily crushed or ruptured with the application of a moderate force.

Frost action. Freezing and thawing of moisture in materials and the resultant effects on these materials and on the structures of which they are a part or with which they are in contact.

General shear failure. Failure in which the ultimate strength of the soil is mobilized along the entire potential surface of sliding before the structure supported by the soil is impaired by excessive movement.

Gilgai. Microrelief of soils produced by expansion and contraction with changes in moisture. It is found in soils that contain large amounts of clay, which swells and shrinks considerably with wetting and drying. Usually a succession of microbasins and microknolls in nearly level areas of microvalleys and microridges parallel to the direction of the slope.

Glacial. Pertaining to the presence, size, composition, or activities of tensive masses of land ice; pertaining to alterations or distinctive features of terrain resulting from the actions of glaciers.

Glacial deposits. Glacial deposits are notoriously erratic, having been transported, plowed, and mixed by glacial ice and the water running from melting ice to such an extent that irregular pockets and lenses of materials of different types may follow each other heterogeneously.

Gleization. Process of soil formation leading to the development, under the influence of excessive moistening, of a glei (gley) horizon in the lower part of the solum. A soil horizon in which the material ordinarily is bluish gray, more or less sticky, compact, and often structureless, is called a "glei horizon."

Gradation

1. Distribution and size of grains in a soil, determined by gradation analysis of soils or passing the soil through a series of screens of increasing fineness. The result is usually presented in the form of cumulative grain-size curve in which particle sizes are plotted to a logarithmic scale with respect to percentage retained (or passing) by weight of the total sample, plotted to a linear scale.

2. In addition to its use in classification, grain-size analysis may be applied to seepage and drainage problems, filter and grout design, and evaluation of frost heave.

Granular structure. Aggregates varying in size to 2 in. in diameter, of medium consistency, and more or less subangular or rounded in shape.

Gravel

1. Loose, rounded fragments of rock ranging in size from $\frac{1}{4}$ in. to boulder size.

2. Rounded or semirounded particles of rock that will pass a 3-in. and be retained on a No. 4 U.S. standard sieve.

Groundwater. Observations are made at the times indicated on logs. The porosity of soil strata, weather conditions, site topography, etc., may cause changes in the water levels indicated on the logs.

Hand augers. Manually operated augers used for shallow subsurface exploration.

Hand-cut samples. Samples properly taken of materials exposed in test pits should be of the highest quality. Certain materials such as loose sands, highly sensitive cohesive soils, or brittle, weathered rock may not furnish truly undisturbed samples in borings, and hand-cut samples may be necessary.

Hand-operated augers. Augers especially made for subsurface explorations. Hand augers are screwed into the earth a few inches at a time and then pulled out, bringing up a sample of the material on the bit. The process is continued until the exploration is extended to the desired depth or, more probably, until the limiting depth attainable by a hand auger has been reached. The limiting depth is usually about 10 or 15 ft, although under favorable conditions it may be as much as 25 or 30 ft.

Hard. Soil structure resistant to forces that tend to cause rupture or deformation.

Hard or stiff clays. Clay that cannot be molded with the fingers nor excavated without a pick. Such clays when compressed to a low water content may be good foundation materials.

Hardpan

1. Accumulation that has been thoroughly cemented into a rocklike layer that will not soften when wet. True hardpan definitely and permanently limits the downward movement of water in nature. The distinction between hardpan and claypan is an important one in soil classification.

2. Thoroughly compacted mixture of clay, sand, gravel, and boulders or a cemented mixture of sand, or sand and gravel, with or without boulders, that is difficult to remove by hand-tool excavation. The term is applied to different materials in different localities.

3. Hardened soil layer in the lower A or in the B horizon caused by cementation of soil particles with organic matter or with materials such as silica, sesquioxides, or calcium carbonate. Hardness does not change appreciably with changes in moisture content, and pieces of the hard layer do not slake in water.

4. Mixture of sand, clay, and gravel. The proportions and consistency of hardpan vary from mud to a natural concrete that is so hard it has been mistaken for good portland cement concrete; however, it can be removed by pick and shovel. Most hardpan is much harder when dried out than when in its original bed, under water. Some hardpans are watertight, others water bearing.

5. Layer of extremely dense soil.

Harpoon-type sampler. Thick-walled split-barrel tube with stabilizing fins that is dropped through water under its own weight to sample lower sediments. A sampler with an outside diameter of 2 in. obtains 6-ft-long samples with $1\frac{1}{2}$-in. outside diameter. The sampler was developed by the United States Coast Guard for sampling river or harbor bottom muds and silts.

High-embankment borings. Borings to a depth of between one-half and one and one-quarter times the horizontal length of the side slope in a relatively homogeneous foundation. Where deep or irregular soft strata are encountered, borings should reach hard materials.

Highway and airfield borings. Auger holes are usually bored to extend 6 ft below the top of pavement in cuts, 6 ft below the existing ground in shallow fills.

Highway, railroad, and airfield borings. The borings are extended through weathered strata and disclose general stability, drainage conditions, and the maximum effect of frost action. Except where major fills and cuts are necessary, a depth of from 5 to 10 ft is usually sufficient provided unsatisfactory strata are fully explored when encountered.

Honeycomb structure. Natural arrangement of the soil mass in more or less regular five- or six-sided sections separated by narrow or hairline cracks, usually found as a surface structure arrangement.

Horizon (soil horizon). One of the layers of the soil profile, distinguished principally by its texture, color, structure, and chemical content.

Humus. More or less stable fraction of the soil organic matter remaining after the major portion of added plant and animal residues have decomposed. Usually it is dark colored.

Hung piston sampler. A thin-wall sample tube is advanced beyond the piston by driving or pushing the drill rod. The piston is held stationary by an outer barrel which surrounds sample tube and seats on a special casing drive shoe, eliminating the need for a separate set of piston extension rods. The ratchet system holds the piston in place for full penetration or any amount of partial penetration of the sampler. The sampler is used for undisturbed sampling in cased holes only.

Hydraulically activated piston sampler. A thin-wall sampler tube is advanced by pumping water or drilling mud through a drill rod which holds the piston in a stationary position, eliminating the need for a separate set of piston rod extensions. The hydraulic pressure is confined to the top of the sample tube head by the outer tube until it builds up sufficiently to push the sample tube down beyond the piston. The drill rod, holding the piston, is maintained stationary either by the drill chuck or another separate device.

Identification. Identification of soil type is made on the basis of an estimate of particle sizes, and in the case of fine-grained soils also on the basis of plasticity.

Impervious. Term applied to soil that is highly resistant to the penetration of water and usually resistant to penetration by air and plant roots. In field practice the term is applied to strata or horizons that are very slowly penetrated by water and that retard or restrict root penetration.

Inclined borings. Inclined borings are required in special cases when surface obstructions prevent the use of vertical holes or when subsurface irregularities such as buried channels, cavities, or fault zones are to be investigated.

Index properties test. Test used for classification, to group soils in major strata, and to extrapolate results from a restricted number of structural properties tests to determine properties of other similar materials.

Indurated. Cemented into a very hard mass that will not soften or lose its firmness when wet and requires much force to cause breakage; rocklike.

Initial consolidation (initial compression). Comparatively sudden reduction in the volume of a soil mass under an applied load due principally to explusion and compression of gas in the soil voids preceding primary consolidation.

Inorganic silt. Inorganic silts have little cohesion when dried, and they dust off easily when allowed to dry on the hands. Some fine-grained silts are difficult to distinguish from clays when wet. When such silts are shaken in the palm of the hand, however, their surface becomes glossy due to the explusion of water from the pores of the soil.

Internal friction. Coulomb pure friction corresponds to simple shear in the theory of elasticity. It can never exceed the value of internal resistance.

Internal resistance. This is possibly the most important soil property. Results obtained vary with the testing method. In a perfectly dry, graded sand of similarly shaped grains, shear resistance is a definite value; but in other soils, the value varies, being larger at the beginning of motion than after the first step has started. Variations in shear value can result from temperature, chemicals in the soil, moisture, direction of moisture flow, vibration, and time of application of release of loading. Internal resistance is a combination of frictional and cohesive forces. Both can vary, but usually not in parallel relationships.

Internal stability of a soil mass. Property by which particles too large to be affected by molecular attraction attain mechanical stability through the mutual support of the particles.

Isolated rigid foundations. Generally all borings should extend no less than 30 ft below the lowest part of the foundation unless rock is encountered at the shallower depth.

Laboratory tests. Laboratory tests have limited usefulness in evaluating natural soils except for elaborate seepage investigations where undisturbed samples of coarsegrained materials are obtained. Secondary structure in situ, stratification, cracks, or heterogeneous character of the soil have an overwhelming influence on permeability. Their effect may be evaluated only by field permeability tests.

Laminated structures. Arrangement of the soil mass in very thin plates or layers, less than 1 mm in thickness, lying horizontal or parallel to the soil surface.

Large-site preliminary borings. Preliminary borings are spaced so that the area between any four borings includes approximately 10% of the total area. In detailed exploration, borings are added to establish geological sections at the most useful orientations.

Large structure foundation borings. Generally all borings usually extend no less than 30 ft below the lowest part of the foundation unless rock is encountered at a shallower depth.

Large structure footing borings. Borings are spaced approximately 50 ft in both directions and are included at possible exterior foundation walls, at machinery or elevator pits, and to establish geologic sections at the most useful orientations.

Leaf mold. Accumulation on the surface of more or less decomposed organic remains; usually the leaves of trees and remains of herbaceous plants.

Lean clays. Lean clays are coarser grained than fat clays, but the particles are, nevertheless, too fine to be palpable. They are strongly cohesive, but less so than fat clays.

Liquid and plastic limits. Liquid and plastic limits are determined by laboratory tests and indicate the compressibility and plasticity of a clay. The higher the liquid limit, the more compressible and plastic the clay will be. The higher the water content, the softer the clay, and the lower the water content, the stiffer the clay and the greater the probability that the clay is preconsolidated.

Liquid limit. Least water content at which a cohesive soil is practically liquid, determined by finding the water content at which a groove cut in a pat of the soil will close when the sample is subjected to the shock of 25 blows.

Load tests on soils. Loads may be applied by jacks reacting against a loaded truck or by the weight of castings set up on a frame or jacking against a beam held in place by screw anchors.

Loam

1. Soil material that contains 7 to 27% clay, 28 to 50% silt, and less than 52% sand.
2. Soil material that contains more than 50% silt and clay, or sand and clay, but less than 20% clay.
3. Mixture of decomposed organic matter with sand, clay, etc., treacherous even when not full of wormholes. It is not compacted by nature as most sands and clays are by the glacial or other floods, and does not extend to any great depths.

Loamy. Soil is intermediate in texture and properties between fine-textured and coarse-textured soils. This type of soil includes all textural classes with the words "loam" or "loamy" as a part of the class name, such as clay loam or loamy sand.

Loamy coarse sand. Soil material that contains 25% or more very coarse and coarse sand and less than 50% any other grade of sand.

Loamy fine sand. Soil material that contains 50% or more fine sand or less than 25% very coarse, and medium sand and less than 50% very fine sand.

Loamy sand

1. Soil material that contains 25% or more very coarse, coarse, and medium sand and less than 50% fine or very fine sand.
2. Soil material that contains at the upper limit 85 to 90% sand; the percentage of silt plus one and one-half times the percentage of clay is not less than 15. The lower limit contains not less than 70 to 85% sand; the percentage of silt plus twice the percentage of clay does not exceed.30

Local shear failure. Failure in which the ultimate shearing strength of the soil is mobilized only locally along the potential surface of sliding at the time the structure supported by the soil is impaired by excessive movement.

Location and ground elevation. The exact location from property lines and the elevation of each boring hole should be shown on a boring drawing.

Loess

1. Fine-grained, granular, slightly cemented, and slightly compacted soil that was deposited by wind. Loess is characterized by a very uniform grain size, by the presence of vertical root holes, and the ability to stand with nearly vertical slopes.
2. Geological deposit of relatively uniform fine material, mostly silt, presumably transported by wind. Many unlike kinds of soil in the United States have developed from loess blown out of alluvial valleys and from other deposits during periods of aridity.

Long bulkhead or wharf wall test borings

1. Preliminary borings are made on the line of wall at 400-ft space. The spacing of intermediate borings decreases to 100 or 50 ft. Certain intermediate borings are made inboard and outboard of the wall line to determine materials in the scour zone at the toe and in the active wedge behind wall.
2. Borings to a depth below the dredge line of between three-quarters and one and one-half times the unbalanced height of wall. Where stratification indicates a possible deep stability problem, selected borings should reach the top of the hard stratum.

Loose. Soil mass in which the soil particles are independent of each other or cohere very weakly with a maximum of pore space and a minimum resistance to forces tending to cause rupture.

Low-load warehouse building of large area. A minimum of four borings at the corners plus intermediate borings at interior foundations are sufficient to define the subsoil profile.

Machine excavation. Large-diameter rotary bucket augers are used to drill caisson holes that may be inspected and sampled. Pits or trenches are made with a backhoe, bulldozer, or clamshell bucket. Mechanically excavated pits or trenches are made in lieu of equally expensive borings, where shallow conditions are essential in design. The locations of excavations must be clearly identified and recorded for reference during construction. These pits are located so as not to disturb bearing materials at intended positions of shallow foundations.

Major waterfront structure borings. Borings are generally spaced not farther than 100 ft with intermediate borings added at critical locations, such as deep pumpwells, gate seats, tunnels, or culverts.

Marine deposits. Marine deposits frequently comprise fine materials, such as clays deposited in bays, seas, and oceans. Thick clay beds are generally of such formation. Coral rocks are a form of marine deposit.

Marl

1. Marine deposit consisting of clay, sand, and calcium carbonate that may occur in the form of a crumbly and very soft rock or a fairly stiff to very stiff grayish-green claylike soil.
2. Material composed of clay and carbonate of lime in different proportions, the carbonate of lime often making it valuable as a fertilizer. Like clay and sand, it contains many impurities, fossils, etc. Soft marl is called "earthy"; hard marl, "indurated."

Marlacious shales, limestones, and clays. Materials that contain enough marl to give them some of its characteristics. It is exceedingly difficult to define specific distinctions between these materials and marl. In different parts of the country the terms are applied to somewhat different materials.

Marsh. Periodically wet or continually flooded area with the surface not deeply submerged. Marshes are mainly covered with sedges, cattails, rushes, or other hydrophytic plants. Subclasses include freshwater and saltwater marshes.

Mean sea level. Average height of the surface of the sea for all stages of the tide, used as a reference for elevations.

Medium sand. Sand the greater portion of which passes a 28-mesh but is retained by a 65-mesh sieve.

Methods for improving soil-bearing capacity. In normal soil conditions, increasing the width of the footing is the simplest method of increasing bearing capacity, especially if the concrete is poured directly against the soil.

Moderately coarse texture. Texture of soil consisting predominantly of coarse particles (in soil texture classification, all the sandy loams except the very fine sandy loam).

Moderately fine texture. Texture of soil consisting predominantly of intermediate size particles or with relatively small amounts of fine or coarse particles (in soil textural classification, clay loam, sandy loam, sandy clay loam, and silty clay loam).

Modified proctor test. Test applied specifically to a heavily compacted base coarse and subgrade for airfield pavement. Modified Proctor tests should not be used for mass earthwork compacted by ordinary methods because it yields an optimum moisture content generally drier than desirable.

Mohr circle. Graphical representation of the stresses acting on the various planes at a given point.

Mohr envelope. Envelope of a series of Mohr circles representing stress conditions at failure for a given material. According to Mohr's rupture hypothesis, a rupture envelope is the locus of points the coordinates of which represent the combinations of normal and shearing stresses that will cause a given material to fail (Christian Otto Mohr).

Moisture content, unit weight, specific gravity. Tests are used to compare soil volume and weight components. Ordinarily, moisture content is determined for all representative samples, disturbed or undisturbed, for classification and to group materials in principal strata.

Muck. Highly decomposed organic soil material developed from peat. Generally, muck has a higher mineral ash constant than peat and is decomposed to the point that the original plant parts cannot be identified.

Mull. Humus-rich layer of forested soils consisting of mixed organic and mineral matter. Mull blends into the upper mineral layers without an abrupt change in soil characteristics.

Nominal bearing pressures. Allowable bearing pressures for spread foundation on various soil types, derived from experience and general usage, which provide safety against shear failure or excessive settlement.

Normal consolidation. Condition where a soil deposit has never been subjected to an effective pressure greater than the existing overburden pressure and where the deposit is completely consolidated under the existing overburden pressure.

Normally loaded clay. Clay that has not been strengthened by preconsolidation. Any additional load will cause compression of the clay layer. The amount of compression that could occur in a normally loaded clay can be roughly estimated knowing the liquid limit of the material.

Organic materials. In organic materials, secondary compression is particularly important and may dominate the time-compression curve, accounting for more than one-half the total compression.

Organic silt. Fine-grained, more or less plastic soil containing finely divided particles of decomposed organic matter. It ranges from light to dark gray in color and often contains gaseous products of the decomposition of organic matter, which give it a characteristic odor.

Organic soil

1. Soil that contains a high percentage of organic matter throughout the solum.

2. Soil or a soil horizon that consists primarily of organic matter, such as peat soils, muck soils, and peaty soil layers.

Overcompaction. Overcompaction is dangerous where later changes in temperature and moisture may cause serious heaving. An increase in the density of soils improves their strength, but too high a compression will produce an incipient unstable internal structure.

Overconsolidated soil deposit. Soil deposit that has been subjected to an effective pressure greater than the present overburden pressure.

Pans. Horizons or layers, in soils, that are strongly compacted, indurated, or very high in clay content, such as caliche, claypan, duripan, fragipan, and hardpan.

Particle-size analysis. Determination of the various amounts of the different separates in a soil sample, usually by sedimentation, sieving, or micrometry or a combination of these methods.

Passive coefficient of earth pressure. Maximum ratio of the major principal stress to the minor principal stress, applicable where the soil has been compressed sufficiently to develop an upper limit value of the major principal stress.

Passive earth pressure. Maximum earth pressure. Condition exists when a soil mass is compressed sufficiently to cause its internal shearing resistance along a potential failure surface to be completely mobilized.

Peat

1. Highly organic swampy soil.

2. Highly fibrous soil composed predominantly of organic material with easily recognized plant remains.

3. organic soil consisting largely of decomposed vegetable matter. It is plastic and highly compressible. When dried it is light enough to float and will also burn.

4. Vegetable matter not fully carbonized.

5. Acid, dark-colored, soft, usually coarsely fibrous, unconsolidated soil with a 96 to 99% content of partly decomposed, somewhat carbonated plant material accumulated under conditions of excessive moisture.

Peat bog. Area of soft, wet, spongy ground, consisting chiefly of decayed or decaying moss and other vegetable matter, where peat is formed.

Peat soil. Organic soil containing more than 50% organic matter, usually slightly decomposed or undecomposed deposits and much to the highly decomposed materials.

Penetration resistance

1. Variations in penetration resistance indicate strata changes. Dynamic resistance is measured by the number of blows of a particular weight dropped a specific height necessary for penetration of a sounder. The variations in static resistance of a rod pushed or jacked into soil may be determined with greater accuracy than the dynamic resistance.

2. Number of blows of a specified weight hammer, falling a given distance, required to produce a specific penetration of a sampling or penetration device into subsoils.

Percent compaction. Ratio, expressed as a percentage, of dry unit weight of a soil to maximum unit weight obtained in a laboratory compaction test.

Perched water table. Water table usually of limited area maintained above the normal free water free elevation by the presence of an intervening relatively impervious confining stratum.

Percussion drilling (churn drilling)

1. Percussion drilling is used for advancing a boring between the taking of dry or undisturbed samples. Percussion drilling rigs are usually truck mounted. Actual drilling is done by a heavy bit that is churned up and down in such a way as to disintegrate the material at the bottom of the hole. Water is introduced in limited quantities, if it is not present in sufficient amount as groundwater. This method can be used to penetrate many types of soil and rock and is therefore useful in differentiating between large boulders and ledge. It is very slow in sticky clays and shales, and impossible in fine, loose, water-bearing sands.

2. Power chopping with a limited amount of water at the bottom of the hole. Water becomes a slurry which is periodically removed with a bailer or sand pump. Changes are indicated by the rate of progress, action of drilling tools, and composition of the slurry removed. Casing is required except in stable rock. Percussion drilling is sometimes used in combination with auger or wash borings for penetration of coarse gravel, boulders, and rock formations.

Permeability-compression test. For routine control, tests may be performed in devices large enough to accommodate samples in compaction molds.

Permeability of a soil mass. Property that allows a fluid under a hydrostatic head to flow through it.

Piggot coring tube. Tube 10 ft long containing thin-wall liner, shot into the ocean bottom by an explosive charge that is triggered when the sampler makes contact with the bottom. The sampler has an outside diameter $2\frac{1}{2}$ in. and obtains $1\frac{7}{8}$-in. outside diameter samples. It is successful in sampling stiff to hard ocean soils in water depths exceeding 20,000 ft.

Piston sampler. Sampler that contains a piston or plug usually mounted on a long rod that extends the full length of the hollow drill rod. It is lowered to the bottom of the casing with the piston closing the opening like a plug and preventing the entry of extraneous material and scrapings from the walls of the casing or drill hole while it is being lowered. When the sampler come to rest at the bottom of the drill hole, the piston rod is released from the drill rod and is clamped or held at the top in such a way that it cannot descend as the sampler is forced down into the soil. After the sampler has been forced down, the piston rod is again clamped to the drill rod and the whole affair withdrawn.

Pitcher sampler. A thin-wall tube is forced into soil while the core barrel outside tube reams out the hole. Tube leads bit of core barrel an amount depending on consistency of soil. The tube is prevented from rotating by a ball-bearing connection between the tube and the outer rotating barrel. The sampler can be used in medium to soft clays and silts but is best adapted to hard clays and soft rock which are difficult or impossible to sample by other thin-wall samplers.

Plastic. Term describing soil that may be readily deformed without rupture; it is pliable but cohesive. The term applies to those soils in which, at certain stages of moisture, the grains will readily slip over each other without the masses cracking or breaking apart.

Plastic equilibrium. State of stress of a soil mass that been loaded and deformed to such an extent that its ultimate shearing resistance is mobilized at one or more points.

Plastic flow. Deformation of a plastic material beyond the point of recovery, accompanied by continuing deformation with no further increase in stress.

Plasticity. Property of a soil that allows it to be deformed beyond the point of recovery without cracking or appreciable volume change.

Plasticity index

1. Difference between the liquid and the plastic limit. The index is an indication of highly plasticity if numerically high and low plasticity if low.
2. Numerical difference between the liquid limit and the plastic limit.

Plasticity of a solid. Ability to deform continuously without rupture under as force greater than the one that caused the yield. It follows that a force less than the one that caused the yield can be sustained without deformation resulting.

Plastic limit

1. Water content corresponding to an arbitrary limit between the plastic and the semisolid states of consistency of a soil; water content at which a soil will just begin to crumble when rolled into a thread approximately $\frac{1}{8}$ in. in diameter.
2. Minimum water content at which a cohesive soil is plastic. It is determined by finding the minimum water content at which a small sample of the soil can be rolled into threads of about $\frac{1}{8}$ in. in diameter without crumbling. No special apparatus is required, the soil samples being rolled under the hand on a glass plate.

Plastic state. Range of consistency within which a soil exhibits plastic properties.

Poisson's ratio. Poisson's ratio measures the lateral effects produced by a linear strain, in a continuous series of values of reversible sign. With a value of unity, as occurs in a fluid, any force is transferred undiminished in all directions, simultaneously in a pure liquid and with a lag in other fluids.

Pore. Minute orifice or opening in a rock.

Porosity

1. The porosity of a soil mass is the ratio of the volume of voids to the total volume expressed as a percentage.
2. Ratio, usually expressed as a percentage, of the volume of voids of a given soil mass to the total volume of the soil mass.

Potentially compressible strata. Borings in potentially compressible fine-grained strata of great thickness are extended to a depth where stress from a superimposed load is so small that corresponding consolidation will not significantly influence surface settlements.

Power-driven augers. Power-driven augers were developed for boring holes in earth for piles and are used to a considerable extent for making relatively shallow exploratory borings. Continuous helical augers are available for use with equipment of this type. The material is brought to the surface without the interruptions caused by the need to frequently withdraw the auger from the hole. Under favorable conditions, these augers may reach a depth of 100 ft.

Preconsolidated clay. Preconsolidation caused by either preloading or dessication (drying out) of clay. The result of preconsolidation is that the clay becomes stronger. Preconsolidated clay may be capable of supporting a load greater than the overburden of soil that it presently carries and undergo slight or no additional compression.

Preconsolidation pressure. Value that forms the boundary between recompression and virgin compression ranges and is the maximum normal stress to which the material in situ has been subjected by a past loading.

Preliminary borings. For large sites, preliminary borings are so located as to furnish an overall subsoil survey instead of following a rigid geometric pattern.

Preliminary field classification. Identification of material encountered, as reported in the records of the exploration, is based on preliminary field classification. Work subsequently done in the laboratory may, however, lead to the reclassification of materials of doubtful character.

Preliminary subsurface exploration. Preliminary subsurface exploration comprises at least one dry sample boring and possibly a few soundings or probings, made during the initial examination of the site of any major project. Dry sample boring is made and carried to the maximum depth at which a deposit of soft or compressible material, such as clay or peat, might cause troublesome settlement of the structure under consideration. Borings are carried to a depth equal to the least horizontal dimension of the largest proposed structure to be erected on the site.

Pressure distribution under spread footings. Even though the settlement characteristics of a soil are determined, actual expected settlement cannot be approximated unless the distribution of pressure at the base of the footing is known. Without such information, the design of a footing is impossible.

Primary consolidation. Reduction in the volume of a soil mass caused by the application of a sustained load to the mass and due principally to a squeezing out of water from the void spaces of the mass and accompanied by a transfer of the load from the soil water to the soil solids.

Probing. Methods of subsurface exploration of minor importance. Probing is a useful supplement to borings for determining the approximate elevation of the top of a known underlying firm stratum. Probings are useful in large areas to determine the extent and spacing of borings. Probing consists of forcing or driving a rod or pipe into the ground.

Proctor penetration resistance. Number of blows of a hammer of specified weight, falling a given distance, required to produce a given penetration into the soil for a pile, casing, or sampling tube; unit load required to maintain a constant rate of penetration into the soil by a probe or instrument; unit load required to produce a specified penetration into the soil at a specified rate by a probe or instrument. For a Proctor needle, the specified penetration is $2\frac{1}{2}$ in., and the rate is $\frac{1}{4}$ in.

Quay or retaining wall borings. Borings are extended to a depth beyond which soil failure or sliding will not develop. The depth cannot be determined until the results of the initial phases of the subsurface exploration are available.

Quicksand

1. Submerged, saturated sand into which a heavy object easily sinks. Sand is held in a very loose, unstable packing such that a shock moves the grains to a smaller bulk volume. The lack of bearing power may be due to seepage pressure of water percolating through the sand in an upward direction or to inherent instability of the structure of the sand, unaided by seepage pressure.

2. Sand rendered unstable by an upward flow of groundwater. Quicksand is a condition rather than a material. Any sand can be made quick by an upward flow through it of sufficient velocity.

Raymond sampler. Method of performing a test in which the number of blows that are required to drive a 2-in. diameter sampler 1 ft with a weight of 140 lb, at a height of fall of 30 in. The test is made after cleaning the bore with a water jet or auger, lowering the sampler on the drill rod, and driving it about 6 in. into the soil at the bottom of the hole.

Recompression and swell. Depending on the magnitude of preconsolidation, pressures applied by new construction may lie partly or wholly in the recompression range. If the load is decreased by excavation, fine-grained soil will undergo a volumetric expansion in the stress range below preconsolidation.

Records. Exploratory borings and subsurface explorations must be complete and accurate. It is essential that records be kept up to date and events be recorded promptly as the work progresses.

Recovery length. Recovery length is expressed as a ratio of the length recovered to the total length pushed or driven.

Relatively uniform subsurface conditions. In large sites where subsurface conditions are relatively uniform, preliminary borings at spacings of 100 to 500 ft may be adequate. Spacing is decreased in detailed exploration by intermediate borings as required to define variations in the subsoil profile. Final spacing of 25 ft usually suffices for even erratic conditions.

Representative dry samples. Representative dry samples are generally obtained at vertical intervals no less than 5 ft center to center of the sample location and at every change in strata.

Requirements for sampling program. The number, type, and distribution of boring samples for testing depend on strata arrangement and sample usage.

Resistance. Resistance occurs when the sample is pushed in one continuous movement by hydraulic rig action, maximum hydraulic pressure, is required where pertinent. The numbers indicate blows of a 140-lb hammer falling freely 30 in. and recorded per 6 in. of sampler penetration. Standard penetration resistance is the number of blows for the last 12 in. of penetration of the split-spoon sampler.

Resistance of soils to pressure. One factor determining resistance of a soil is the coefficient of friction. Friction is a resistance independent of the area loaded, but directly proportional to the intensity of loading. A second factor is cohesion, which is a characteristic of the material.

Retractable plug sampler. Sampler with an inside plug that is hand-driven into the soil to the desired depth of sampling. The plug is then retracted and the sampler driven into the soil, which enters a series of brass liners. The sampler is removed from the soil by jacking. It is available with outside diameters of 1.4 and 3 in., which obtain samples 42 and 60 in. long with outside diameters of 0.9 and 1.9 in., respectively.

Ring shear. Essentially a classification test whose results have been correlated with sheer failures in the field.

Rock borings or drillings. Boring or drillings made by rotating a hard steel tube with a bit or lower end specially treated to cut the rock along an annular ring. The bits are surfaced with some extrahard, tough alloy, such as tungsten carbide, or with commercial diamond chips known as "borts." High-pressure water jets are always necessary to dissipate the frictional heat. The lower tube section or recovery barrel contains part of the rock freed by the annular cutting.

Rock core drilling. Rock core drilling is effected by power rotation of a core barrel as the circulating water removes ground-up material from the hole. The water also acts as a coolant for the core barrel bit. Generally, the hole is cased to the rock. This type of drilling is used along and in combination with boring types to drill weathered rocks, bedrock, and boulder formations.

Rock flour. Extremely fine-grained inorganic silt, the particles of which are so fine that the material may have a strong superficial resemblance to clay when wet. Rock flour can be distinguished from clay by the shaking test.

Rod test. The rod test is selected where the ground is more or less soft and it is advisable to ascertain the approximate depth of the soft strata. If the rod only penetrates a few feet, more definite means should be taken to ascertain the nature of the material under the surface, whereas if it penetrates many feet, foundations should be set on a hard bottom at that site. The rod may be driven 30 ft or more.

Rotary and percussion borings. Borings used for deep exploration or to penetrate hard soils or strata containing boulders and rock seams.

Rotary core drilling. Rotary core drilling is used in bedrock to recover continuous core or to pass obstructions in the overburden.

Rotary drilling

1. Rotary drilling is effected by power rotation of the drilling bit as circulating fluid removes cuttings from the hole. Changes are indicated by rate or progress, action of drilling tools, and examination of cuttings in the drilling fluid. Casing is usually not required except near the surface. Rotary drilling is applicable to all soils except those containing much large gravel, cobbles, and boulders. It is difficult to determine changes accurately in some soils.

2. Rotary drilling is used in lieu of drilling and washing in making subsurface explorations. boring in such a case is done by a rotary bit lubricated with a drilling fluid or "mud" consisting of a suspension of fat clay, bentonite, or a similar material in water. When samples are to be obtained, the bit and drill rod are removed and the sample is taken with the same type of equipment and by the same methods.

Saline soil. Soil containing excessive amounts of the neutral or nonalkaline salts.

Sampling spoon. Open-ended cylinder especially designed for either sand or clay. The cylinder is driven into the ground and filled. When extracted, it provides a soil test specimen.

Sampling spoons or samplers. Spoons driven into the soil to secure so-called dry samples. Undisturbed samples cannot be obtained with them.

Sand

1. Material consisting of 25%or more very coarse, coarse, and medium sand and less than 50% fine or very fine sand.

2. Individual grains can be seen and felt readily. Squeezed in the hand when dry, it will fall apart when the pressure is released. Squeezed when moist, it will form a cast that will hold its shape when the pressure is released, but will crumble when touched.

3. Cohensionless granular soil consisting of particles smaller than gravel, often occurring as a mixture with gravel.

4. Sand may vary from pure silica in very fine particles, to gravel, or it may be mixed in various proportions with many different materials, such as clay, loam, decayed vegetable matter, minerals, and, most important of all, water.

5. Soil material that contains 85% or more sand. The percentage of silt plus one and one-half times the percentage of clay is not to exceed 15.

6. The material composed of loose granular grains, containing less than 20% silt and clay.

Sandy clay. Soil material that contains 35% or more clay and 45% or more sand.

Sandy clay loam. Soil material that contains 20 to 35% clay, less than 28% silt, and 45% or more sand.

Sandstone. Consolidated rock composed of sand grains cemented together. The size range and composition of the constituents are the same as for sand, and the particles may be rounded or angular.

Sandy loam

1. Soil material that contains 20% clay or less and for which the percentage of silt plus twice the percentage of clay exceeds 30 and 52% or more sand; soil material that contains less than 7% clay, less than 50% silt and between 43 and 52% sand.

2. Soil material that contains 30% or more very coarse, coarse, and medium sand but less than 25% very coarse sand and less than 30% very fine or fine sand.

Satisfactory bearing material. Soil or fill that meets the requirements of code, and does not contain or overlay appreciable quantities of organic material and therefore can be assumed to have a bearing value sufficient to prevent damaging settlements from occurring when subjected to design load.

Saturated samples. When moisture content and dry weight are measured, all volume-weight parameters may be computed by assuming a specific gravity. When moisture content and specific gravity are measured, all volume-weight parameters may be computed directly. The volume-weight of fine-grained soils may be determined with sufficient accuracy for computation of loads by assuming saturation.

Secondary compression. After completion of primary consolidation under a specific load, the semilogarithmic time-compression curve continues approximately as a straight-line which is the range of secondary compression.

Secondary consolidation. Reduction in volume of a soil mass caused by the application of a sustained load to the mass and due principally to the adjustment of the internal structure of the soil mass after most of the load has been transferred from the soil water to the soil solids.

Securing borings. Borings made in foundation areas that eventually will be excavated below groundwater, or where artesian pressures are encountered, must be plugged or grouted unless used for continuing water level observations. In bore holes for groundwater observations, casing is placed in tight contact with the walls of the hole.

Selection of boring method. The choice depends on the efficiency of the boring procedure in prospective materials, the ability to determine, strata changes and material type, and the possible disturbance of materials to be sampled.

Settlement. Settlement of structures during construction must always be watched. Measurement is usually by surveying methods, referring the elevation of fixed points to a bench mark sufficiently remote so as not to be affected by the loaded structure.

Shale

1. Shale is considered a rock and sometimes a soil. It is material in a state of transition from clay to slate. Some shales decompose rapidly when exposed to the weather.

2. Lithifield muds, clays, and silts that are fissile and break along planes parallel to the original bedding. Typical shale is so fine grained as to appear homogeneous to the unaided eye.

Shallow depths. Where stiff or compact soils are encountered at shallow depths, one or more boring must be extended through this material to a depth where the presence of an underlying weaker strata cannot affect stability or settlement.

Shearing strength of a soil mass. Property by which the individual particles resist displacement with respect to one another when an external force is applied.

Shore deposits and delta deposits. Deposits formed where rivers entered seas or lakes and where the resulting sediments may have been eroded by waves and shifted by coastal currents. Such deposits may contain pockets of silt, clay, or even peat.

Shot core barrel. Chilled steel shot is fed to a rotating soft-steel bit through drill rods and a single barrel. The bit and rock are worn away as cuttings are washed above the barrel by circulating water. Cuttings are deposited in a sludge barrel or calyx attachment.

Shot drilling. Shot drilling obtains cores from rock. A hollow cylindrical bit, rotated by an attached drill rod, cuts a circular groove in the rock. The cutting is by means of chilled steel shot, which are fed into the hole and which find their way under the rotary bit. A core barrel is provided between the bit and the sides of the hole, as in the diamond drill. Cuttings are washed from the cut surface by pumped water surging through the drill rod and bit orifices. Water rises in the annular space between the drill rid and the sides of the hole. The presence of cavities delays drilling because the shot disappears into the spaces, losing contact with the lower edge of the bit. After filling the spaces with cement grout, drilling is resumed with the bit cutting through the hardened grout.

Shrinkage limit. The minimum water content for completely saturating a cohesive soil is determined if the specific gravity of the soil is not known; a container of known volume and weight, usually a dish about $1\frac{3}{4}$ in. in diameter and 4 in. deep, is filled with the soil in a plastic condition and weighed. The sample is then dried to a constant weight, the drying being accompanied by considerable shrinkage of the plastic soil. The shrunken sample is accurately weighed and its volume determined by mercury displacement. The shrinkage limit is then computed by formula.

Shrinkage of a soil. Loss of volume due to a decrease in moisture content and to forces resulting from tension in the water films.

Silt

1. Material passing the No. 200, U.S. standard sieve that is nonplastic or very slightly plastic and that exhibits little or no strength when air-dried.
2. Soil material that contains 80% or more silt and less than 12% clay.
3. Soil material that contains less than 20% sand and clay.

Silt loam. Soil material that contains 50% or more silt and 12 to 27% clay; soil material that contains 50 to 80% silt and less than 12% clay.

Silty clay

1. Soil material that contains 40% or more clay and 40% or more silt.
2. Soil having 40% or more of clay and less than 20% sand.

Silty clay loam. Soil material that contains 27 to 40% clay and less than 20% sand.

Silty sand. Mixture of sand with fine-grained materials having some plasticity. It may be of organic or inorganic origin.

Single-grained structure. Incoherent condition of the soil mass with no arrangement of individual particles into aggregates. This type of structure is usually found in sands of coarse texture.

Single-tube core barrel. A tube with a coring bit is rotated down into the rock receiving the core, while circulating water removes cuttings. The barrel has an outside diameter of $1\frac{1}{2}$ to $2\frac{15}{16}$ in. and obtains cores up to 10 ft. long with outside diameters of $\frac{7}{8}$ to $2\frac{1}{8}$ in. It is used primarily in sound rock not vulnerable to erosion, silting, or fracturing.

Slime. Soft, fine, oozy mud or other substance of similar consistency.

Slope stability—boring locations. Provide three to five borings on line in the critical direction to establish a geological section for analysis. The number of geological sections depends on the extent of its stability problem. For an active slide, usually locate at least one boring upslope of the sliding area.

Slope stability test borings. To obtain slope stability use the elevation below estimated active or potential failure surface and auger deep into hard stratum, or to a depth for which failure is unlikely because of geometry of cross-section.

Soft. Yielding to any force causing rupture or deformation.

Soft clays. Soft clays can be molded easily with the fingers and excavated with a spade. The clays are relatively compressible, having low shearing strength, and fail under heavy loads.

Soft ground. The soil about underground openings that does not stand well and requires substantial timbering.

Softly cemented. Cementing material is not strong or evenly diffused through the mass. Aggregates are readily crushed, but do not break with a clean fracture.

Soft rock. Rock which can be removed by air-operated hammers but cannot be removed by a handpick.

Soil. Unconsolidated mineral material on the immediate surface of the earth that serves as a natural medium for the growth of land plants.

Soil auger boring. Holes bored into the soil with soil augers are rotated by hand or power driven. Selection depends on the material to be penetrated and the power system employed. If the penetrated material is damp sand or contains considerable clay, the hole may not cave; but if caving occurs, the auger is operated inside a metal casing which sinks or is driven as the boring progresses. If the soil adheres to the auger, indications of the nature of the soil can be obtained by examining the soil. Below the groundwater level in sand or in silt, it may be necessary to bail out the soil.

Soil bin. Laboratory container in which various soils and soil vehicle relationships can be investigated under controlled conditions.

Soil bulk density. Mass of dry soil per unit bulk volume. The bulk volume is determined before drying to a constant weight at 105°C.

Soil characteristic. Feature of a soil that can be seen and/or measured in the field or in the laboratory on soil samples. Examples include soil slope and stoniness, as well as the texture, structure, color, and chemical composition of soil horizons.

Soil description on boring logs. Description of material should be according to the unified soil classification: a word description that should give soil constituents, consistency or density, and other appropriate classification characteristics. Unified Classification symbols usually shown on "stratification log" column or use geologic names where appropriate. A solid line usually indicates stratigraphic change; a dashed line usually indicates approximate stratigraphic change.

Soil failure model. Mathematical equation describing the stress and strain states developed in natural soils or laboratory soil-simulating materials, when subjected to various external loadings that exceed the failure strength.

Soil identification. Field investigations to identify soils can be made by surface surveying, aerial surveying, or geophysical or subsurface exploratory analysis. Complete knowledge of the geological structure of an area permits definite identification from a surface reconnaissance. Tied together with a mineralogical classification of the surface layers, reconnaissance can at least recognize the structures of some soils.

Soil moisture. When the water content of soil is chiefly film or adsorbed moisture, the mass will not act as a liquid. All solids tend to adsorb or condense on their surfaces any liquids (and gases) with which they come into contact.

Soil profile. Vertical section of the soil through all its horizons and extending into the parent material.

Soil sampler. The simplest type of sampler, suitable only for cohesive soils such as clay, is a thin-walled seamless steel tube not less than 2 in. in diameter and from 2 to 3 ft long; the tube is filled with soil by forcing it into the bottom of the hole. In the testing laboratory, the tube with its contents are cut into lengths of about 6 in. The soil is forced from these short lengths to obtain specimens for testing.

Soil samples. Soil samples are taken at $2\frac{1}{2}$-ft intervals and at each change of strata or at directed intervals.

Soil sampling. Information obtained about the soil penetrated by some exploratory methods is usually inadequate for foundation design. To obtain information about the properties of the soil at various depths, the bits attached to the drill rods are replaced at intervals by various types of samplers forced or driven into the bottom of the hole. The objective of the sampling operation is to obtain an undisturbed soil sample.

Soil stabilization. Chemical or mechanical treatment designed to increase or maintain the stability of a mass of soil or otherwise to improve its engineering properties.

Soil survey. Systematic examination, description, classification, and mapping of soils in an area. Soil surveys are classified according to the kind and intensity of field examination.

Soil tests. Soil tests are required where there is doubt as to the character of the soil or application to impose loads in excess of intended design loading.

Soil tractionability. Capacity of a soil to provide traction for vehicles.

Soil trafficability. Capacity of a soil to withstand traffic, especially the traffic of heavy vehicles.

Soil values. Set of empirically determined mathematical parameters that describe the measured horizontal and vertical stress-deformation curves of natural soils or laboratory soil-simulating materials.

Sounding rod. Steel rod or pipe about $\frac{3}{4}$-in. in diameter, in about 5-ft lengths, joined by standard couplings and provided with a pointed lower end. It is hand driven with a maul or drop weight, lubricated with water, and turned with a pipe wrench to reduce sticking. The number of blows required to drive the rod indicates the nature of the underlying soil. Driving continues until the rod "refuses" further penetration. After driving has been completed, the rod is removed by a lever and chain. The sounding rod yields no soil samples, but experienced operators can estimate the soil character by observing the manner in which the rod penetrates the soil.

Special drilling fluids. Special drilling fluids consisting of suspensions or emulsions of fat clays or bentonites are sometimes used in lieu of water and are useful in minimizing the caving of bore holds which cannot be effectively driven with water alone as a drilling fluid.

Specific gravity of a soil. Ratio of the weight in air of a given net volume of the particles comprising the soil to the weight of an equal volume of distilled water at 20°C.

Specific gravity of a material. Ratio of its weight per unit volume to the weight of a unit volume of water.

Specific surface of a soil. Surface are of the soil particles per unit of absolute volume.

Split-liner piston-type sampler. Thin-wall stationary piston sampler with the inside liner split longitudinally, advanced by driving or pushing. The liner removed after sampling, and the sample examined by separating the liner halves, It is utilized in stratified or varved sands with organic or clayey lenses. Other sampler types are utilized for a wide range of soil types.

Split-spoon sample. The sampler is obtained by driving a 2-in. split spoon to determine penetration resistance and allow classification.

Split spoon sampler. Thick-wall, split-barrel driven or pushed into the soil. A ball check in the sampler head prevents water pressure from forcing the sample out of the tube. The split barrel is opened in the field for examination of the sample. It is used to obtain "dry samples" in practically all soils but it not adapted to coarse gravel and rock.

Standard penetration test. Test that records the number of blows required by a 90-lb weight falling 30 in. on a 2-in. outside diameter soil sampler to penetrate the ground 1 ft.

Standard penetration tests

1. A 2-in. outside-diameter, $1\frac{3}{8}$-in. inside-diameter sampler is driven a distance of 1 ft into undisturbed soil with a 140-lb hammer free falling a distance of 30 in. The number of hammer blows for seating the spoon and making the tests are recorded for each 6 in. of penetration on the drill log (example, 6/8/9). Standard penetration test results can be obtained by adding the last two figures such as $8 + 9 = 17$ blows per foot.

2. Ordinary sounding procedure to determine compactness or hardness in situ.

Standard Proctor test. The standard Proctor test is used for ordinary embankment compaction control. In preparing for control, a family of compaction curves representing principal borrow materials is obtained.

State-of-the-ground code. Standardized surface synoptic observation that describes the condition of the ground surface. Basically, the states of the ground are recognized as dry, moist, wet, frozen, and ice or snow covered.

States of matter affecting soil behavior. Soils can exhibit solid, viscous, plastic, or liquid action. If the true state can be predicted, the structural design can be prepared accordingly.

Static resistance devices. Devices reserved for special exploration. Resistance readings require correlation with conventional test properties of subsoils.

Stationary piston sampler with piston rods. A thin-wall sample tube is advanced beyond piston, generally by pushing drill rod while the piston is held stationary by a clamp to the piston rod extension which runs through the sampler head and inside the drill rod to the surface. The cone lock in the sampler holds the piston to the drill rod and prevents downward motion of the piston. This type of sampler is satisfactory for undisturbed sampling in soft to medium clays and silts from cased and uncased holes.

Sticky. Term applied to soils showing a decided tendency when wet to adhere to other materials and foreign objects.

Stone. Crushed or naturally angular particles of rock that will pass a 3-in. sieve and be retained on a No. 4 U.S. standard sieve.

Structural properties tests

1. Tests are selected by the engineer for particular design problems. Rigid standardization of test procedures is inappropriate. Tests are performed only on undisturbed samples obtained as specified or on compacted specimens prepared by standard procedures. In certain cases completely remolded samples are utilized to estimate the effect of disturbance. Tests are made to determine typical properties of major strata rather than to distributor tests arbitrarily in proportion to the number of undisturbed samples obtained. A limited number of high-quality tests on carefully selected undisturbed samples is preferred to many mediocre tests on specimens selected at random.

2. Permeability, consolidation, and strength characteristics may be determined using methods similar to those for undisturbed samples. Usually the standard compaction sample is trimmed to desired size the sample is compacted in a mold of the test specimen size using standard compaction energy.

Structure. The structure of a soil mass or body refers to the arrangement of the individual soil particles.

Subsoil exploration. Geophysical or seismic exploration of subsoils is a carryover from practice standardized in oil-field surveys to determine discontinuities in soil structure. The principles involved are the well-known characteristics of sound-wave transmission, reflection, and refraction in passing through materials of different densities. The method charts the time and intensity of sound-wave emergence at various points, induced by the explosion of the submerged charge. A similar technique uses the variation in electric conductivity of various densities and discontinuities in layer contacts.

Subsurface logs. Logs contain observations and mechanical data collected by the driller while at the site, supplemented by classification of the materials removed from the borings as determined through visual identification by technicians in the laboratory. Data presented on the subsurface logs together with the recovered samples will provide a basis for evaluating the character of the subsurface conditions relative to the proposed construction. Evaluation must consider all the recorded details and their significance relative to each other. Often analysis of standard boring data indicates the need for additional testing and sampling procedures to evaluate the subsurface conditions more accurately.

Supervision

1. Recovery of undisturbed samples is a specialized operation to be supervised by an experienced soil engineer rather than the boring contractor.

2. Supervision of a subsurface exploration should be in the hands of a competent soils engineer under whose personal observation the work should be done. Leaving the supervision of such work to technically untrained personnel, even though they may be experienced members of drilling crews, leads to the misinterpretation of data and may lead to consequent serious error.

Swedish foil sampler

1. Sampler employed to recover long, continuous samples of soft soil.

2. Method to retrieve samples by using wide thin-metal strips housed inside the sampler head above a thin, sharp cutting edge envelops the sample as it enters the tube, thus minimizing friction between sample and tube. The sampler is advanced by pushing or jetting. The piston is held fixed during penetration by a chain which passes along inside of sample tube to the surface and is fixed to a stationary frame. Sample tubes are added until the sampling run is completed. The sampler is used for taking long, continuous, undisturbed samples in cohesive soils free of gravel, sand layers, or excessive shells, etc., which may break or rupture the foils and samples. Theoretically, it can recover samples 40 ft. or more in length.

3. Samples that can be used where continuous sampling is necessary for sand drain design, elaborate settlement, or stability analysis.

Swell of a soil mass. Increase in volume caused by a change in water content.

Terraces and fills. Borings are extended to depths beyond which surfaces of failure will not develop. For preliminary estimates of the required depths of exploration, it is suggested that one and one-fourth times the horizontal projection of the slope distances of terraces and one-half the horizontal projection of the slope distances of triangular or trapezoidal fills be used.

Terrain. Area considered as to its extent, and man-made and natural features in relation to its use.

Terrain analysis. Process of interpreting a geographical area to determine the effects of its natural and man-made features on intended uses.

Terrain estimate. Portion of an analysis of the area of operations that concerns the description of the terrain for the construction uses of the terrain and the effects of the characteristics of terrain on approaches for construction equipment.

Terrain evaluation. Valuation and interpretation of an area for probable uses and determination of the effect of the terrain of the lines of approach open to construction in the area.

Terrain factor. Specific attribute of the terrain that can be described in quantitative terms.

Test boring logs. Records of each boring from ground elevation to established or directed depth, or to excessive resistance. The log is an accurate description of the soils or rocks encountered, depth of the various soil materials and their descriptions, water table and water levels, driving resistance, and water content.

Test of soil for bearing capacity. Where the local conditions are not well understood, it is advisable to make special tests of the soil by putting a platform on the ground and loading it. The larger the area covered by the testing platform, the more reliable the results, but even the most careful experiments of this nature require a great deal of personal judgments. This type of testing is used where testing equipment is not available.

Test pit

1. Definite method of showing what exists but limited as to depth. Test pits are especially useful where boring records are not consistent with predictions from surface and geological records.

2. A test pit usually constitutes a satisfactory method for securing reliable information concerning subsurface soil conditions. The procedure permits soil examination in its natural undisturbed state. Pits are shored and lined by properly supported horizontal and vertical members to prevent caving.

3. Digging a small test pit will often take the place of borings to supplement the information obtained. But test pits are not usually made under ground water level or to more than a few feet in depth.

4. Test pits for examining or sampling soils in situ range from shallow manual or machine excavations to deep, sheeted, and braced pits. After the test pit is excavated a probe can be made with a pressure measuring device into the subsoils to determine resistance to penetration.

Test schedule. Gradations of a large number of samples are usually not required, for identification. Samples should be grouped in principal strata by visual classification before performing grain-size analysis on specimens of major strata.

Thin-walled tube sampler. Sampler using open or fixed-piston sampling head.

Thin-wall shelby tube sampler

1. Thin-wall sampler that is pushed or driven into the soil. The sampler head contains a ball check to prevent water pressure from forcing the sample into the tube during withdrawal. The sampler accommodates tubes with outside diameters of 2 to $4\frac{1}{2}$ in. This type of sampler is used in soft to hard clays and silts or silty and clayey sands. It is utilized for undisturbed samples of hard cohesive soils when driving is necessary to advance the tube. It recovers samples 24 in. long, with outside diameters of $1\frac{7}{8}$ to $4\frac{3}{8}$ in.

2. Sampler that is pushed into the soil to secure so-called undisturbed samplers. Undisturbed samples $3\frac{1}{2}$ in.or more in diameter are preferable to 2-in samples, which may be of doubtful value if the soil is sensitive to disturbance.

Tough. Soil that is tenacious or shows a decided resistance to rupture. The soil mass adheres firmly.

Triaxial test

1. Test to determine the maximum shearing strength of a soil. It is also called a "confined compression test."

2. Tests used in analysis of slope stability and bearing capacity and to compute earth pressures where sheer strength is mobilized.

True specific gravity. Specific gravity of the soil particles themselves. Mass specific gravity refers to the specific gravity of the soil mass or body.

Tunnel. Long, nearly horizontal or horizontal passageway used for aqueducts, drainage or sewage, vehicular traffic, mines, and other underground uses.

Tunnels. Preliminary explorations for tunnels should be extended to a depth below the proposed invert elevation at least equal to the gross width of the proposed tunnel.

Ultimate bearing capacity. Average load per unit of area required to produce failure by rupture of a supporting soil mass.

Uncased borings. Borings used in some localities where very stable cohesive soils make it unnecessary to extend casings further down the bore hole than is required to recover the wash water and protect the top of the hole from caving and abrasion.

Unconfirmed compression test

1. Test in which a cylindrical clay specimen without lateral support is subjected to axial loading.

2. When cohesive soils are encountered in the field (job site) by the soils engineer the use of the penetrometer is permissible in most cases, to produce a simulated unconfined compression test.

Underconsolidation. Condition where a soil deposit is not fully consolidated under the existing overburden pressure and hydrostatic excess pore pressures exist within the material.

Underwater samplers. Free-fall gravity coring tube; 10- to 15-ft-long tube containing a thin-wall linear, released at a fixed height above the ocean floor by utilizing pilot weight and lever mechanism. The sampler is equipped with lead weights and stabilizing fins. The sampler has an outside diameter of $2\frac{1}{4}$ in. and obtains 1.9-in. outside-diameter samples. It is used for ocean bottom sampling and is successful in medium to stiff clays and in sand and gravel up to $1\frac{1}{2}$ in. in size at depths exceeding 13,000 ft.

Undisturbed samples

1. Soil sample that has been obtained by methods under which every precaution has been taken to minimize disturbance to the sample.

2. The recovery of undisturbed samples is preceded by dry sample borings to determine the thickness and extent of critical strata. The number and spacing of undisturbed samples depend entirely on related design problems and the necessary testing program. The distribution of undisturbed samples may range from only one or two in a boring to practically continuous undisturbed samples in a critical stratum.

3. Samples in which the material has been subjected to little disturbance so that it is suitable for all laboratory tests and thereby for approximate determination of the strength, consolidation, and permeability characteristics and other physical properties of the material in situ.

4. Undisturbed samples are obtained by the penetration of a minimum 3-in.-diameter, thin-wall tube using an open- or, where indicated, fixed-piston sampling head. Sample suitable for laboratory testing of any type necessary.

Undisturbed sampling. Method used for obtaining samples at least $3\frac{1}{2}$ in. in diameter for determination of the strength in shear and the consolidation characteristics of cohesive soils and for determination of certain other characteristics of noncohesive soils.

Unsaturated samples. Samples used for measuring moisture content, dry weight, specific gravity, and total volume of specimen to compute volume-weight relationships.

Unsuitable foundation strata. Borings are extended through unsuitable foundation strata, such as unconsolidated fill, peat, highly organic materials, soft fine-grained soils, and loose coarse-grained soils to reach hard or compact materials of suitable bearing capacity.

Vane test (vane shear test). In-place shear test in which a rod with thin radial vanes at the end is forced into the soil and the resistance to rotation of the rod is determined.

Varved clay. Material that consists of alternating thin strata of clay and silt deposited, respectively, in the winter and summer by water from melting ice at the close of a glacial epoch.

Varved silt or clay. Fine-grained glacial lake deposit with alternating thin layers of silt or fine sand and clay, formed by variations in sedimentation from winter to summer during the year.

Vehicle cone index. Index assigned to a given vehicle indicating the minimum soil strength required for 40 to 50 passes of the vehicle.

Vehicle ground mobility. Measure of ability of an automotive vehicle to traverse the variety of terrain conditions (including inland waterways) found on the surface of the earth, and in a minimum time with minimum support and remaining capable of performing its design function.

Very fine sand. Material that consists of 50% or more very fine sand.

Very fine sandy loam. Material that consists of 30% or more very fine sand; material that consists of more than 40% fine and very fine sand, at least half of which is very fine sand and less than 15% very coarse, coarse, and medium sand.

Virgin compression. Compression comprising the range of pressures exceeding that to which the sample has been subjected in the past.

Void ratio. Volume of the voids divided by the volume of the solids in the soil. In a saturated clay the voids are filled with water.

Volcanic formations and coral formations. Uncovered on tropical islands, these deposits are usually more erratic and unpredictable than glacial deposits.

Wash borings

1. Method for boring test holes into unconsolidated materials. A bit is mounted on the lower end of a pipe called a "drill rod," through which water is forced. The drill rod is worked up and down or churned and rotated slowly. The bit strikes the soil or rock, gradually penetrating it. Cuttings are washed by water rising to the surface in the annular space between the rod and sides of the hole. The process derives its name from this operation. The nature of the soil penetrated is often judged by examining the borings or cuttings surfaced by the wash water.

2. Test hole from which samples are brought up mixed with water.

3. Method of making subsurface explorations. The equipment with which borings are made is similar to that used for making dry-sample borings except that no sampling spoons or samplers are used. Instead, samples of wash water are caught as the boring proceeds, and an attempt is made to identify and classify the material which settles out of the water. Samples thus obtained are often misrepresentative and misleading, and the method is no longer used by progressive practiconers in the fields of soil mechanics and foundation engineering.

Wash boring. Borings used for recovery of either disturbed dry samples or undisturbed samples. Ordinarily this is the type most suitable for locations with difficult access.

Wash samples. Samples are taken from wash water circulated from bore holds and should not be relied on for identification of subsoils.

Wash-type boring for undisturbed or dry samples. Boring effected by the chopping, twisting, and jetting action of a light bit as circulating drilling fluid removes cuttings from hole. Changes are indicated by the rate of progress, action of the rods and examination of cuttings in drilling fluid. Casing is used as required to prevent caving. This method is used in sands, sand and gravel without boulders, and soft to hard cohesive soils. It is the most common method of subsoil exploration and usually can be adapted for inaccessible locations, such as over water, in swamps, or slopes, or within buildings.

Water content

1. Analysis of soil and the determination of water content for each sample.

2. Water content and its relationship to the liquid and plastic limits may be used to estimate the strength of a clay and can be used as an indicator of preconsolidation in the clay.

Water-holding capacity of a soil. Soils having the ability to retain water in thin films against the action of forces tending to drain it. A centrifugal force, usually 1,000 times gravity, is generally used to remove water. Results are evaluated in terms of centrifugal moisture equivalent, which is the moisture content expressed as a percentage of the dry weight of the soil.

Water levels. Water levels shown on the boring logs are the levels measured in the borings at the time of boring. In sand, indicated levels can be considered reliable ground water levels. In clay soil it is not possible to determine the groundwater level within the normal scope of a test boring investigation, except where layers of more pervious water-bearing soil are present, and then a long period of time may be necessary to reach equilibrium. Therefore, the position of the water level symbol for cohesive or mixed-texture soils may not always indicate the true level of the groundwater table.

Water line. In all boring holes the water line must be indicated as accurately as possible.

Water table

1. Surface defined by the upper limit of the zone of saturation; surface of unconfined groundwater.

2. Upper surface of groundwater; level below which the soil is saturated with water; locus of points in soil water at which the hydraulic pressure is equal to the atmospheric pressure.

Weakly cemented. Term applied when the cementing material of a soil mass is not strong, and the aggregates can be readily broken into fragments with a more or less clean fracture.

Wet clay. Wet clay sticks to the hands and when allowed to dry on them does not brush off easily. A rather large range of water content clay can be rolled into threads that have appreciable tensile strength. Clay has pronounced cohesive qualities, and when a lump or pat is allowed to dry, it develops considerable strength.

Weight or density. The amount of solid material on a unit volume is called the "dry weight." Usually, weight is taken as a measure of solidity or compressive strength. The ratio of actual weight as found in nature or in a prepared soil to the corresponding weight of an artificially compacted sample is the measure of density relative to the standard optimum. Of the various standards used, the technique developed by Proctor was first accepted but has been replaced by a modification known as the AASHO test.

Wind-laid deposits. Deposits such as dune sand and loess are usually remarkably uniform in character but are likely to prove troublesome because they are not densely compacted.

STRUCTURAL ENGINEERING

Allowable. Term applied to a live load, a wind load, or other loads; maximum load that can be applied without causing the allowable stresses or deflections in a structural member of assembly to be exceeded or a building or structure to become unsafe or unstable.

Allowable load. Ultimate load divided by a factor of safety.

Allowable soil pressure. The maximum stress permitted in soil of a given type and under given conditions.

Allowable stress

1. Unit stress permitted for a design under code.

2. Maximum permissible stress used in the design of members of a structure, based on a factor of safety against rupture or yielding of any type.

Allowable stress design. Method of proportioning structures based on "allowable" or working loads, such that stresses do not exceed prescribed values. Allowable stresses incorporate a factor of safety against one of the limits of structural usefulness.

Allowable unit stresses. Maximum unit stress considered desirable in a structural member subjected to loads; working unit stress; safe working unit stress.

Amplification factor. Multiplier of the value of moment or deflection in the unbraced length of an axially loaded member to reflect the secondary values generated by the eccentricity of the applied axial load within the member.

Anchorage bond stress. Reinforcing bar forces divided by the product of the bar perimeter(s) and the embedment length.

Apron wall. Part of a skeleton building below a window sill and supported on a spandrel beam.

Aspect ratio. In any rectangular configuration, the ratio of the lengths of the sides.

Assembly. Fitting together and/or attaching by any means, including brazing, welding, bonding, riveting, bolting, and screwing.

Atterberg limits. Arbitrary water contents (shrinkage limit, plastic limit, liquid limit), determined by standard tests, that define the boundaries between the different states of consistency of plastic soils.

Atterberg test. Method for determining the plasticity of soils.

Axial and combined stresses. A force acting parallel to the axis of a member and at the center of gravity of its cross section produces axial stress. Such stress is uniformly distributed over the cross section.

Axial force. Push (compression) or pull (tension) acting along the length of a member. Usually measured in pounds.

Axial load. Force or force system whose resultant passes through the centroid of the section on which the force acts.

Axially loaded. Tension or compression loaded only, with no bending, twist, or shear.

Axial stress. The axial force acting at a point along the length of a member, divided by the cross section area of the member. Usually measured in pounds per square inch (psi).

Balanced load. Load capacity at simultaneous crushing of concrete and yielding of tension steel.

Balanced moment. Moment capacity at simultaneous crushing of concrete and yielding of tension steel.

Balanced reinforcement. Amount and distribution of reinforcement in a flexural member such that in working stress design the allowable tensile stress in the steel and the allowable compressive stress in the concrete are attained simultaneously, or such that in strength design the tensile reinforcement reaches its specified yield strength simultaneously with the concrete's reaching in compression its assumed ultimate strain of 0.003.

Batten plate. Plate element used to join two parallel components of a built-up column, girder, or strut, rigidly connected to the parallel components and designed to transmit shear between them.

Bay (part of structure). The wall space between two columns; the whole space between column centers.

Beam

1. Structural member subjected primarily to flexure.

2. Structural member, ordinarily subject to bending, usually a horizontal member carrying vertical loads. In a framed floor, beams are members on which the floor plank, slab, or arch rest directly.

Beam—action shear. Shear force carried by a plate girder web panel at the theoretical shear buckling stress. The shear stress distribution is that predicted by simple beam theory.

Beam and slab floor. Reinforced concrete floor system in which the floor slab is supported by beams of reinforced concrete.

Beam—column. Structural member that is subjected to forces producing significant amounts of both bending and compression simultaneously.

Beams and stringers. Lumber of rectangular cross section, 5 or more in. thick and 8 or more in. wide, graded with respect to its strength in bending when loaded on the narrow face.

Beam test. Method of measuring the flexural strength (modulus of rupture) of concrete by testing a standard unreinforced concrete beam.

Bearing. Structural support for a beam, girder, truss, etc., usually walls, hangers, or columns.

Bearing capacity. Maximum unit pressure that a soil or other material will withstand without failure or without settlement to an amount detrimental to the integrity or functioning of the structure.

Bearing stratum. Soil or rock stratum on which a footing or mat bears or that carries the load transferred to it by a pile, pier, caisson, or similar deep foundation.

Bearing surface. Contact surface between a foundation unit and the soil or rock upon which it bears.

Bearing walls. Wall that supports a vertical load in addition to its own weight.

Bench mark. Datum point of known elevation that serves as a reference in establishing other levels or locations.

Bending moment

1. At any section in the length of a beam the bending moment is the effect of the various forces acting on the beam that tend to cause it to bend.

2. At any section of a beam the bending moment is equal to the algebraic sum of the moments of the forces on either the right or the left of the section.

3. At any section in the length of a beam, the bending moment equals the moment of the reactions minus the moments of the loads to the left of the section.

4. Algebraic sum of the couples or the moments of the external forces, or both, to the left or right of any section on a member subjected to bending by couples or transverse forces, or both.

5. Bending effect at any section of a structural element. It is equal to the algebraic sum of all moments to the right or left of the section.

Bending moment diagram. Graphical representation of the variation of bending moment along the length of the member for a given stationary system of loads.

Bending strength. Maximum flexible stress developed in such specimens as cast iron, wood, and concrete before they break under bending or flexure.

Bending stress. The force per square inch (psi) of area acting at a point along the length of a member, resulting from the bending moment applied at that point. Usually measured in psi.

Bending stress and modulus of rupture. Bending stresses are stresses induced by loads perpendicular to the member. The modulus of rupture is the maximum bending stress computed on the assumption that elastic conditions exist until failure.

Bend test. Test for measuring relative ductility, soundness, and toughness of certain materials.

Bent. Transverse system of framing composed of columns and beam(s).

Biaxial bending. Simultaneous bending of a member about two perpendicular axes.

Biaxiality. In a biaxial stress state, the ratio of the smaller to the larger principal stress.

Biaxial stress. State of stress in which only one of the principal stresses is zero, the other two usually being in tension.

Bifurcation. Phenomenon whereby a perfectly straight member may either assume a deflected position or else may remain undeflected; buckling.

Blaine test. Method for determining the fineness of cement or other fine material on the basis of the permeability to air of a sample prepared under specified conditions.

Bolster. Short horizontal structural member used to give larger bearing area for supported beams, etc., at top of columns or column caps.

Bond. The combined action of steel and concrete is dependent on the grip of concrete on steel.

Bond area. Area of interface between two elements across which adhesion develops or may develop, as between concrete and reinforcing steel.

Bond strength. Resistance to the separation of mortar and concrete from reinforcing steel and other materials with which they are in contact; collective expression for all forces such as adhesion, friction due to shrinkage, and longitudinal shear in the concrete engaged by the bar deformation that resist separation.

Bond stress. Force of adhesion per unit area of contact between two such bonded surfaces as concrete and reinforcing steel or any other material such as foundation rock; shear stress at the surface of a reinforcing bar, preventing relative movement between the bar and the surrounding concrete.

Bottom chord. Horizontal or inclined (scissors truss) member that establishes the lower edge of a truss, usually carrying combined tension and bending stresses.

Box girder. A girder having a hollow cross-section similar to that of a rectangular box.

Box system. Structural system without a complete vertical load-carrying space frame. In this system the required lateral forces are resisted by shear walls.

Brace. Structural member used to support another. A brace is always designed for compression and sometimes for tension under special load conditions.

Braced frame. Frame in which the resistance to lateral load or frame instability is primarily provided by a diagonal, a K-brace, or other auxiliary system of bracing.

Breaking load. Load required to fracture a specimen under tension, compression, bending, or torsion.

Breaking (or fracture) strength. Breaking load of a material divided by the original cross-sectional area of that material.

Bridging. Diagonal or crossbracing between joists to resist twisting.

Brinell hardness test. Test for determining the hardness of materials by forcing a hard ball penetrator of known size into the surface of the material under specified load.

Brittle fracture. Abrupt cleavage with little or no prior ductile deformation.

Buckling. Failure by lateral or torsional instability of a structural member, occurring with stresses below the yield or ultimate values.

Buckling load. Load at which a perfectly straight member assumes a deflected position.

Built-in (or fixed) beam. Beam rigidly fixed at both ends.

Built-up beam. Single unit composed of two wood members having the same thickness but not necessarily the same depth, which provides greater load-carrying capability as well as greater resistance to deflection.

Built-up member. Member made of structural metal elements that are welded, bolted, or riveted together.

Bulk density. Weight of a material (including solid particles and any contained water) per unit volume, including voids.

California bearing ratio. Ratio of the force per unit area required to penetrate a soil mass with a 3-sq in. circular piston at the rate of 0.05 in./minute to the force required for corresponding penetration of a standard crushed-rock base material. The ratio is usually determined at 0.1-in. penetration.

Camber

1. Slight vertical curve built into beams, trusses, etc., to offset load deflection.
2. Deviation from edge straightness, usually referring to the greatest deviation of side edge from a straight line.

Cantilever beam

1. Beam that projects beyond the support(s) and is loaded on the projecting part.
2. Beam having one end rigidly fixed and the other end free. Extending a simple beam beyond either support gives a combination of a simple beam and a cantilever beam. Beam with both ends free and balanced over a support is also called a "cantilever beam."

Cast iron construction. Construction that utilizes cast iron for structural purposes. The cast iron product of good foundry mixture produces a clean, tough, gray iron free from serious blowholes, cinder spots, or cold shuts, and has a minimum tensile strength of 20,000 psi.

Cast steel. Cast steel used in buildings or structures should be of such quality as to conform to the "Standard Specifications for Carbon Steel Castings" (ASTM A 216).

Cellular steel deck. Structural floor system, consisting of two layers of sheet metal shaped to form cells and welded together. Cells serve as electrical raceways.

Cellular steel floor. Construction consisting of sheet or strip steel formed into an integrated system of parallel steel beams that combine the function of load-bearing members and a continuous deck spanning between main supporting girders, beams, or walls.

Cement. Construction material in powder form that when mixed with water sets into a hard, solid mass. Portland cement is the most common type.

Cement finish. Finish placed on top of the floor arch, slab, or other structural floor element.

Centerline. Line dividing a space into two equal spaces.

Center of gravity. Point in a body about which all the weights of all the various parts balance.

Center to center. Measurement between centers of two adjoining parallel structural elements.

Centrifugal force. Outward force exerted by a body moving in a curved line.

Centripetal force. Force or restriction exerted inward to keep a body moving in a curved line.

Centroid. The center of gravity of a solid is an imaginary point at which all its weight may be considered to be concentrated or the point through which the resultant weight passes.

Charpy impact test. Test used to determine impact strength, that is, the energy absorbed by a specimen when it is supported at both ends as a single beam and fractured by a moving pendulum-type hammer.

Chords of a truss. The upper or top chord consists of the upper line of members. The lower chord consists of the lower line of members. The web members connect the joints of the upper chord with those of the lower chord.

Cladding. Composite plate made of base metal with a plate of corrosion- or heat-resistant metal on one or both sides.

Clear span. Horizontal distance between interior faces of supports.

Cold-formed members. Structural members formed from steel without the application of heat.

Cold-formed steel structural member. Structural member cold-formed to shape from carbon or low-alloy sheet or strip steels and used for load-carrying purposes in buildings and structures.

Column. Member used primarily to support axial compression loads, with a height at least three times its least lateral dimension.

Column, strut, post. Structural member that is compressed endwise. A strut is usually of smaller dimensions than either a column or post.

Combined mechanism. A mechanism determined by a plastic procedure that combines elementary beam, panel, and joint mechanisms.

Control joint. Spaces between the abutting portions of a building or structure to relieve longitudinal stresses due to contraction of materials.

Corrosion. Chemical action that causes gradual destruction of a metal.

Corrosion resistant material. Material that maintains its original surface characteristics under prolonged exposure and use.

Combined stresses. A state of stress that cannot be represented by a single component of stress; that is, one that is more complicated than simple tension, compression, or shear.

Compact shape. Cross-sectional shape that will not experience premature local buckling in the inelastic region.

Components of a force. Any number of forces whose combined effect is the same as that of a single force. The process of finding the components is resolution.

Composite beam. A member consisting of a reinforced concrete slab supported by a structural steel shape, interconnected in such a way that the two act together to resist bending.

Composite column. Concrete compression member reinforced longitudinally with structural steel shapes, pipe, or tubing, with or without longitudinal reinforcing bars.

Composite concrete flexural members. Concrete flexural members consisting of concrete elements constructed in separate placements but so interconnected that the elements respond to loads as a unit.

Composite construction. Type of construction using members produced by combining materials such as concrete and structural steel or cast-in-place and precast concrete, such that the combined components act together as a single member.

Compression. Action of a force on a body in a manner that tends to shorten the body or to push the parts of the body together.

Compression flange. Widened portion of an I, T, or similar cross-sectional beam which is shortened or compressed by bending under normal loads, such as the horizontal portion of the cross section of a simple-span T-beam.

Compression load. Load that tends to squash a material.

Compression member. Member in which the primary stress is longitudinal compression.

Compression or compression stress. Force that tends to shorten a member.

Compression reinforcement. Reinforcement designed to carry compressive stresses.

Compression test. Test used to determine the behavior of materials under compression. A specimen is compressed and deformed or crushed by a load. The various properties determined are elastic limit, proportional limit, yield point, yield strength, and, for brittle material, compressive strength.

Compressive strength

1. Maximum stress a material can withstand before it is crushed by squeeze forces. The compressive strength of brittle materials is calculated by dividing the breaking load by the original cross-sectional area of the specimen.

2. Maximum compressive stress that a material is capable of developing, based on the original area of cross section. In the case of a material that fails in compression by a shattering fracture, the compressive strength has a very definite value. In the case of materials that do not fail in compression by a shattering fracture, the value obtained for compressive strength is an arbitrary value depending on the degree of distortion that is regarded as indicating complete failure of the material.

Concentrated force. Force whose place of application is so small that it may be considered to be a point.

Concentrated load

1. Load applied at a point.

2. Load that extends over a short length of a beam. It is assumed acting at a point.

3. Force applied at a point.

Concrete-encased beam. Structural member totally encased in concrete, cast integrally with the slab.

Concrete filling. Type of short-span floor construction in fireproof and fire-resistive buildings installed between structural steel framing to serve as a combination structural floor slab or arch and fireproof protection of the framing.

Concurrent and noncurrent forces. Forces are concurrent when their lines of action meet in a point. They are nonconcurrent when their lines of action do not meet in this manner.

Connection. Combination of joints used to transmit forces between two or more members. Categorized by the type and amount of force transferred (moment, shear, end reaction).

Contact pressure. Pressure acting at and perpendicular to the contact area between the footing and the soil, produced by the weight of the footing and all the forces acting on it.

Continuous beam

1. Beam that rests on more than two supports.

2. Beam that is continuous and capable of carrying moment over an interior support.

3. Beam having more than two points of support.

Continuous slab or beam. Slab or beam that extends as a unit over three or more supports in a given direction.

Contraction. Decrease of volume occurring as the result of any or all processes affecting the bulk volume.

Control joint. Spaces between the abutting portions of a building or structure to relieve longitudinal stresses due to contraction of materials.

Coplanar and noncoplanar forces. Forces may lie in the same plane or in different planes; that is, they may be either coplanar or noncoplanar forces.

Counter. Slender truss diagonal that resists only tension, usually one of a pair that acts alternately as the shear on a panel changes direction.

Cover. Thickness of concrete between the outer surface of any reinforcement and the nearest surface of the concrete.

Crazing. Development of fine, random cracks caused by shrinkage.

Creep

1. Permanent deformation of a material at a given temperature under sufficiently high sustained loading, continuing with time, but without increasing the load.

2. Time-dependent strain occurring under stress. Creep strain occurring at a diminishing rate is called "primary creep"; that occurring at a minimum and almost constant rate, "secondary creep"; that occurring at an accelerating rate, "tertiary creep."

3. Slow deformation under stress.

4. Permanent deformation occurring over a period of time in a material subjected to constant stress at elevated constant temperatures.

Creep limit. Maximum stress that will cause less than a specified quantity of creep in a given time; maximum nominal stress under which the creep strain rate decreases continuously with time under constant load and at constant temperature. Sometimes used synonymously with "creep strength."

Creep recovery. Time-dependent strain after release of the load in a creep test.

Creep strength

1. Rate of continuous deformation under stress at a specific temperature; commonly expressed as the stress, in pounds per square inch, required to produce a certain percentage of elongation in a specified number of hours.

2. Constant nominal stress that will cause a specified quantity of creep in a given time at constant temperature; constant nominal stress that will cause a specified creep rate at constant temperature.

3. Load that will produce only a specified size change when constantly applied for a given period of time at a specific temperature.

Creep test. Test for determining the creep behavior of materials subjected to prolonged constant tension or compression loading at constant temperature.

Curling. Distortion of an essentially straight or flat member into a curved, warped, or dished shape due to creep or to internal differences in temperature or moisture content.

Curvature. The rotation per unit length due to bending.

Dead load

1. Weight of materials forming a permanent part of the structure.

2. Weight of a structure itself plus any permanent loads. In design, the weight of the structure must be assumed and the design corrected later if the assumed weight is very much in error. Masonry construction has the largest dead load relative to the total load.

3. Sum of all loads that are static and permanent in nature.

4. Weight of all permanent construction, including walls, framing, floors, roof, partitions, stairways, and fixed building service equipment.

5. Weight of walls, partitions, framing, floors, roofs, and all other permanent stationary construction entering into a building.

6. Weight of all of a building's permanent construction and fixed service equipment.

Dead man. Anchorage below the ground surface.

Decking. Fabricated metal forms installed over other structural members on which other materials can be installed.

Deflection. Movement from the original position of a structural element because of bending or shear deformation caused by its weight, applied loads, and temperature or moisture changes.

Deflection of beams. The vertical distance that a point moves on a neutral surface is the deflection of the beam at that point.

Deformation

1. Whenever a body is subjected to a force, there is a change in its shape or size.

2. A material changes shape when subjected to the action of a force. This change in shape is deformation or strain. The deformation per unit of length is the unit deformation.

Density. Mass of a substance in a unit volume. When expressed in the metric system, it is numerically equal to the specific gravity of the same substance.

Design bearing pressure. Maximum allowable net pressure on soil or rock.

Design capacity. Load that a foundation is designed to transfer to the supporting soil or rock.

Design load

1. Total load that a structure is designed to sustain.

2. The design load as applied to a live load, dead load, impact, wind load, or other load is the load used in the design of a building.

Design of steel joist. An open-web steel joist is built up of bars or other sections. A joist can be fabricated by expanding a rolled section designed as a truss.

Design properties. Properties of soil or rock used in proportioning and determining the design capacity of a foundation.

Design span. Span used to calculate the strength and deflection of a member, usually measured from center to center of the bearing surface at each support.

Design strength. Resistance provided by element or connection; the product of the nominal strength and the resistance factor.

Design stress. Specified unit stress used in design.

Diagonal bracing. Inclined structural members carrying primarily axial loads employed to enable a structural frame to act as a truss to resist horizontal loads.

Diaphragm. Floor slab, metal wall, or roof panel possessing a large in-plane shear stiffness and strength adequate to transmit horizontal forces to resisting systems.

Distributed force. Force whose place of application is an area. A distributed force may often be considered as a concentrated force acting at the center of the contact area.

Distributed load

1. A uniformly distributed load exerts an equal downward force for each linear unit of beam length.
2. Force or force system spread over an area, either uniformly or nonuniformly.

Double curvature. Bending condition in which end moments on a member cause the member to assume an S shape.

Drift. Lateral sway of a building due to wind.

Drift bolts. Bolts used to connect members by being driven into a prebored hole of smaller diameter.

Drift index. The ratio of lateral deflection to the height of the building.

Ductility

1. Property of a material that permits it to undergo plastic deformation when subjected to a tensile force.
2. Ability of a material to withstand plastic deformation without fracturing. The percent of elongation and reduction of areas are measures of ductility in metals.
3. Property allowing deformation under tensile stress without rupture.

Ductility factor. The ratio of the total deformation at maximum load to the elastic-limit deformation.

Duration of load. The period of continuous application of a given load, or the aggregate of periods of intermittent applications of the same load.

Dynamic. Term applied to forces tending to produce motion.

Dynamic balance. Condition of rest created by equal strength of forces tending to move in opposite directions.

Dynamic creep. Creep that occurs under conditions of fluctuating load or fluctuating temperature.

Earthquake action. Kinetic action of earthquakes causing lateral load acting in any horizontal direction on the structural frame.

Eccentric force. Force parallel to the axis of a member but not acting along this axis. It is equivalent to an axial force of like amount and a couple whose moment is equal to the product of the force by the normal distance from the force to the axis of the member. Thus an eccentric force as described produces combined stresses.

Eccentricity. Normal distance between the centroidal axis of a member and the parallel resultant load.

Eccentric loads. Walls supporting eccentrically applied loads including eccentric loads produced by the deflection of floor and roof members should be be analyzed for stability and strength. Maximum unit stresses must not exceed those specified in the code.

Effective height. Height of a member assumed for calculating the slenderness ratio.

Effective length. Equivalent length used in the Euler formula for computing the strength of a framed column.

Effective moment of inertia

1. The moment of inertia of the cross section of a member that remains elastic when partial plastification of the cross section takes place, usually under the combination of residual stress and applied stress.
2. The moment of inertia used in the design of partially composite members.
3. The moment of inertia based on effective widths of elements that buckle locally.

Effective stiffness. Stiffness of a member computed using the effective moment of inertia of its cross section.

Effective thickness. Thickness of a member assumed for calculating the slenderness ratio.

Effective width. Reduced width of the plate or slab that, having a uniform stress distribution, is assumed to produce the same effect on the behavior of a structural member as the actual plate width with its nonuniform stress distribution.

Elastic constants. Modulus of elasticity, either in tension, compression, or shear, and Poisson's ratio.

Elastic deformation. Change of dimensions accompanying stress in the elastic range, the original dimensions being restored upon release of stress.

Elasticity

1. Property of a material that enables it to return to its original size and shape when the load to which it has been subjected is removed. Unit stress is called "elastic limit."
2. Ability of a material to return to its original shape upon removal of a deforming load.
3. Property of a material by virtue of which it tends to recover its original size and shape after deformation.

Elastic limit

1. Maximum stress a material can withstand without permanent deformation.
2. Maximum stress to which a material may be subjected without any permanent strain remaining upon complete release of the stress.
3. Maximum unit stress that can be developed in a material without causing permanent deformation, commonly expressed in pounds per square inch.
4. Maximum unit stress that can be applied to metal without causing permanent deformation.

Elastic limit and yield point. The elastic limit is the stress at which the ratio of stress to deformation ceases to be constant. The yield point is the stress at which deformation increases.

Elastic strength. Greatest unit stress a material can resist without a permanent change in shape.

Elements of a force

1. The force acting upon a body is completely known when its general direction, point of application, and magnitude are given.

2. Force has three elements: magnitude, direction, and line of action.

Elongation

1. Amount of permanent stretch in a material after failure in tension, commonly expressed as a percentage of the specimen's original length.

2. In tensile testing, the increase in the gauge length, measured after fracture of the specimen within the gauge length, usually expressed as a percentage of the original gauge length.

3. Increase in gauge length of a tension test specimen, usually expressed as percentage of the original gauge length.

4. Increase in gauge length of a specimen under tensile testing. Elongation is estimated from the increase in gauge length divided by the original gauge length and multiplied by 100.

Embedment. Steel component cast in a concrete structure that is used to transmit externally applied loads to the structure by means of bearing, shear, bond, friction, or any combination thereof. The embedment may be fabricated of structural-steel plates, shapes, bars, bolts, pipe, studs, concrete reinforcing bars, shear connectors, or any combination thereof.

Encased steel structure. Steel-framed structure in which all of the individual frame members are completely encased in cast-in-place concrete.

End restraint. Conditions occurring in structures in which the structure, when designed on the assumption of full or partial end restraint, due to continuous, semicontinuous, or cantilever action, the beams, girders, and trusses, as well as the sections of the members to which they connect, should be designed to carry the shears and moments so introduced, as well as all other forces, without exceeding at any point the unit stresses established for the materials, except that some nonelastic, but self-limiting deformation of a part of the connection may be permitted when this is essential to the avoidance of overstressing of fasteners.

Endurance limit. Maximum stress to which a material can be subjected for an indefinitely large number of cycles without causing rupture, commonly expressed in pounds per square inch. In practice, values are taken after a specified number of cycles. Endurance limit is ordinarily considered to be "fatigue strength."

Equilibrium

1. State of rest or of uniform motion. When a system of forces acting on a body produces no motion in the body, the system of forces is said to be "in equilibrium." Two equal forces opposite and parallel, having the same line of action. The two forces balance each other; the body does not move. The force necessary to produce motion must be a force slightly greater in magnitude than the force required for equilibrium.

2. Dynamic conditions of balance between atomic movements where the resultant is zero and the condition appears to be one of rest rather than change.

Equilibrium of forces. If a number of forces act on a body and the body does not move or if moving does not change its state of motion, then the forces considered are in equilibrium. If any one of the forces balances all the other forces, then it is an equilibrant of those other forces.

Equivalent uniform load. A conventionalized representation of an element of dead or live load, used for the purposes of design in lieu of the actual dead or live load.

Expansion. An increase in volume occurring as the result of any or all processes affecting the bulk volume.

External load. Force acting on the outside of a structure set in equilibrium.

Factored load. Product of the nominal load and a load factor.

Factor of safety

1. Quotient that results from dividing the ultimate strength of a material by the unit stress.

2. Ratio of the breaking strength to the actual load carried by a piece of metal.

3. Quotient obtained by dividing the breaking load or ultimate strength of a material or device by the allowable design load.

4. As used in allowable stress design, the factor by which a designated limit of structural usefulness is divided to determine an allowable stress.

Factor of safety and allowable stress. Stress used in design is the allowable stress. It is obtained by dividing the ultimate stress by the factor of safety.

Failure. Condition of distress in a structural member that upon removal of a load does not react to recovery of the member to its original physical shape and state and causes a permanent and substantial reduction in the load-carrying capacity of the member.

Fastener. Term for welds, bolts, rivets, or other types of connecting devices.

Fatigue

1. Fracture phenomenon associated with a repetitive stress condition.

2. Type of failure in metal resulting from repetitive loading.

3. Condition occurring in a material subjected to fluctuating or cyclic stresses and strains and leading to permanent deformation.

4. Failure of metals by repeated or alternate stresses.

5. Phenomenon leading to fracture under repeated or fluctuating stresses having a maximum value less than the tensile strength of the material. Fatigue fractures are progressive, beginning as minute cracks that grow under the action of the fluctuating stress.

Fatigue life. Number of cycles of stress that can be sustained prior to failure for a stated test condition.

Fatigue limit

1. Maximum fluctuating stress a material can withstand for an indefinite number of cycles without permanent deformation.

2. Maximum stress below which a material can presumably endure an infinite number of stress cycles. If the stress is not completely reversed, the value of the mean stress, minimum stress, or stress ratio should be stated.

Fatigue ratio. Ratio of the fatigue limit for cycles of reversed flexural stress to the tensile strength.

Fatigue strength

1. Stress at which a material fractures by fatigue.

2. Maximum stress that can be sustained for a specified number of cycles without failure, the stress being completely reversed within each cycle unless otherwise stated.

Fatigue test. Test for determining the behavior of materials under fluctuating stresses.

Faulting. Differential vertical displacement of a slab or other member adjacent to a joint or crack.

Faying surface. Contact area of adjacent parts of a connection.

Fiber stress. Local stress through a small area (a point or line) on a section where the stress is not uniform, as in a beam under a bending load.

Fireproof construction. Structural members of approved noncombustible construction having the necessary strength and stability and fire resistance ratings of not less than four hours for exterior nonbearing walls, wall panels, columns, and wall-supporting girders and trusses, and not less than three hours for floors, roofs, and floor and roof supporting beams, girders, and trusses, and in which exterior and interior bearing walls, if any, are of approved masonry or reinforced concrete. The term "fireproof construction" is not acceptable in some building codes.

Flexible connection. Connection permitting a portion, but not all, of the simple beam rotation of a member end.

Flexural strength (Flexural rigidity). Resistance of a material to flexural stress or bending, commonly expressed in pounds per square inch.

Floor assembly. Combination of materials providing the horizontal separation between stories including the ceiling, floor, and horizontal structural members supporting the floor, but excluding those primary structural members that serve as part of the structural frame.

Flow. Time-dependent irrecoverable deformation.

Footing. Structural unit of a substructure used to distribute loads to the underlying strata.

Force. That which tends to change the state of motion of a body; that which causes a body to change its shape if it is held in place by other forces.

Force of gravity. Force by which all bodies are attracted toward the center of the earth. The magnitude of the force of gravity is the weight of the body. The amount of material in a body is its mass. Force action exists in pairs.

Forces. Force produces or tends to produce motion or a change of motion of bodies.

Formed steel construction. Type of construction used in floor and roof systems consisting of integrated units of sheet or strip steel plates, which are shaped into parallel steel ribs or beams with a continuous connecting flange deck and generally attached to and supported on the primary or secondary members of a structural steel or reinforced concrete frame.

Foundation. Wall below the floor nearest grade serving as a support for a wall, pier, column, or other structural part of a building.

Fracture test. Testing by breaking a concrete specimen and examining visually the fractured surface to determine such things as grain size and the presence of defects.

Fracture toughness. Measurement of the ability to absorb energy without fracture, determined by impact loading of specimens containing a notch having a prescribed geometry.

Frame buckling. Condition under which bifurcation may occur in a frame.

Frame construction. Walls and interior construction are wholly or partly of wood.

Fulcrum. Bearing point of a lever as distinguished from the lifting point.

Fully composite beam. Composite beam with sufficient shear connectors to develop the full flexural strength of the composite section.

General engineering set. Working drawings in which engineering design is shown. These drawings are used for estimating and field construction purposes and contain all the important tables and data giving dimensions, load requirements, and material strengths.

Girder

1. Beam that receives its load in concentrations. In a framed floor it supports one or more cross beams, which in turn carry the flooring. The term "girder" is also applied to any large heavy beam, especially a built-up steel beam or plate girder.

2. As distinguished from "beam," the heavier supporting member that carries the load of the beams into the column, as in "beam and girder construction."

Girt. Secondary horizontal member in a sidewall, designed to resist wind pressure.

Guy. Cable for holding a structure in a desired position.

Hardness. Resistance of material to deformation by penetration or indentation, commonly expressed as a hardness number on the scale of a machine used for testing.

Header. Beam or other structural support member that spans across an opening in the normal framing.

Heavy timber construction. Walls of approval masonry or reinforced concrete, in which the interior structural elements, including columns, floors, and roof construction, consist of heavy timbers with smooth, flat surfaces assembled to avoid thin sections, sharp projections, and concealed or inaccessible spaces; and in which all structural members that support masonry walls have a fire-resistance rating of not less than three hours, and other structural members of steel or reinforced concrete, if used in lieu of timber construction, have a fire-resistance rating of not less than one hour.

High-rise building. Building with its height considerably greater than its floor plan dimensions; skyscraper.

Honeycomb. Voids in concrete.

Hooke's law. Stress is proportional to strain. The law holds up only to the proportional limit.

Horizontal force. A horizontal force caused by wind pressure or earthquake effect.

Horizontal shear. The tendency to move laterally is called the "horizontal" or "longitudinal shear." The magnitude of the horizontal shear at any section of a beam is equal to the magnitude of the total vertical shear.

Hybrid beam. Fabricated steel beam composed of flanges with a greater yield strength than that of the web. Whenever the maximum flange stress is less than or equal to the web yield stress, the girder is considered homogeneous.

Impact

1. Stress in a structure caused by the force of a vibratory, dropping, or moving load, generally a percentage of the live load.

2. Load resulting from moving machinery, elevators, craneways, vehicles, and other similar forces and kinetic loads.

3. Force exerted by a body in motion against a fixed structure.

Impact load

1. Force applied within a short period of time, such as a weight dropping to the floor; also described as an "energy load."

2. Dynamic effect of live loads when suddenly applied.

3. Loading resulting from moving machinery, elevators, craneways, vehicles, and other similar forces and kinetic loads.

Impact resistance. Ability of materials to resist the damage of a striking force.

Impact strength

1. Stress required to fracture a notched specimen of a material at one blow; measure of toughness. Impact strength is commonly expressed in foot-pounds of energy required for fracture.
2. Energy required to fracture a specimen under an impact (very rapidly applied) load.

Impact testing. An impact test is a dynamic test in which a selected concrete specimen, machined and surface ground and usually notched, is struck and broken by a single blow of a specially designed testing machine, and the energy absorbed in breaking the specimen is measured.

Imposed load. All loads, exclusive of dead load, that a structure is to sustain.

Inelastic action. Material deformation that does not disappear on removal of the force that produced it.

Inertia. Property of matter by which it will remain at rest or in uniform motion in a straight line, unless acted upon by an external force.

Initial drying shrinkage. The difference between the as-cast length of a specimen and its length when first dried.

Inner forces. The internal or inner forces in a structure are the stresses in the different members that are brought into action by the outer forces and hold the outer forces in equilibrium.

Instability. Condition reached in loading of an element or structure in which continued deformation results in a decrease of load-resisting capacity.

Internal load. Force effect within the entity or any component part of a structure set in equilibrium.

Izod. Type of impact test in which a specimen is struck by a swinging pendulum and the energy absorbed in the fracture is measured.

Joint. Area where two or more ends, surfaces, or edges are attached.

Joists and planks. Lumber of a rectangular cross section 2 in. to less than 5 in. thick and 4 or more inches wide, graded with respect to its strength in bending when loaded either on the narrow face as a joist, or on the wide face as a plank.

Johnson's proportional limit. Stress in a material at which the deformation increases with respect to stress at a rate of 50% greater than at any stress below the proportional limit, commonly expressed in pounds per square inch.

K-bracing. System of struts used in a braced frame in which the pattern of the struts resembles the letter K, either normal upright or on its side.

Kip

1. Load of 1000 lb. The term is derived from "kilopound."
2. Unit of 1000 lb, used in calculating strengths of structural materials.

Laminate

1. Composite material usually in the form of a sheet or bar, composed of two or more layers so bonded that the composite material forms a structural member.

2. Plies of wood joined together with an adhesive and/or mechanical fastening.

Lateral brace. Member installed and connected at right angles to a chord or web member of a beam, girder, or truss to resist lateral movement.

Lateral bracing members. Member utilized individually or as a component of a lateral bracing system to prevent buckling of members or elements and/or to resist lateral loads.

Lateral force resisting system. Part of the structural system to which the lateral forces prescribed in the code are assigned.

Lateral load. Load applied horizontally owing to wind pressure or seismic or blast effects.

Lateral (or Lateral-torsional) buckling. Buckling of a member involving lateral deflection and twist.

Lateral soil load. Lateral pressure in pounds per square foot due to the weight of the adjacent soil, including due allowance for hydrostatic pressure.

Light-gauge, cold-formed stainless steel structural member. Structural member cold-formed to shape from sheet or strip stainless steel, annealed and strain-flattened, and used for load-carrying purposes in buildings and structures.

Light-gauge steel construction. Type of construction in which the structural frame consists of studs, floor joists, arch ribs, rafters, steel decks, and other structural elements that are composed and fabricated of cold-formed sheet or strip steel members less than $\frac{3}{16}$ in. thick.

Light noncombustible construction. Construction in which all structural members, including walls, floors, roofs, and their supports, are of steel, iron, concrete, or other noncombustible materials, and in which the exterior enclosure walls are of masonry, concrete, or other fire-resistive materials or assemblies of materials that have not less than two-hour fire-resistance ratings.

Limit design. Design based on a chosen limit of usefulness.

Limit state. Condition in which a structure or component becomes unfit for service and is judged either to be no longer useful for its intended function or to be unsafe.

Linear expansion. Increase in the overall size of materials due to temperature changes.

Lintel. Beam especially provided over an opening for a door, window, or the like, to carry the wall over the opening.

Live load

1. Load imposed solely by the occupancy.
2. All loads other than dead loads and wind pressure.
3. Load to which a building or structure is subjected that results from the use and occupancy of the building or structure. A live load includes all loads other than dead loads, wind loads, and earthquake loads.
4. Any load other than a dead load or a lateral load.
5. Any moving or variable load that may act upon the structure as, for example, the weight of people, merchandise, or portable equipment.

Load. Pressure due to applied weight or force.

Load-bearing wall. Wall that supports vertical load in addition to its own weight.

Load factor. As used in plastic design, the factor by which the working load is multiplied to determine the ultimate load.

Local buckling. Buckling of a plate component, which may precipitate the failure of the whole member.

Long column. Column whose load capacity is limited by buckling rather than strength.

Main reinforcement. Steel reinforcement designed to resist stresses resulting from design loads and moments, as opposed to reinforcement intended to resist secondary stresses.

Malleability. Property of a material that permits plastic deformation when the material is subjected to a compressive force.

Maximum load design. Design method in which members are selected on the basis of their maximum strength at ultimate strength at ultimate load (determined by multiplying the expected loads by a load factor). In steel structures it is utilized in plastic design.

Mechanical properties

1. Properties manifested in a material's reactions to applied forces and loads. Some mechanical properties are tensile strength, compressive strength, and fatigue strength.

2. Properties of a material that reveal its elastic and inelastic behavior where force is applied, thereby indicating its suitability for mechanical applications—for example, modulus of elasticity, tensile strength, elongation, hardness, and fatigue limit.

Mechanical shear connector. In composite construction, a metal device attached to the top flange of a steel beam that is capable of transmitting shear forces between the concrete slab and the steel beam.

Mechanism. System of members containing a sufficient number of plastic (or real) hinges to be able to deform without a finite increase in load.

Mechanism method. Method of plastic analysis in which equilibrium between external forces and internal plastic hinges is calculated on the basis of an assumed mechanism.

Member. Piece of a structure that is a single unit of the structure, for example, a beam, column, or a web member of a girder.

Method of design. Any system or method of construction that admits rational analysis in accordance with well-established principles of mechanics.

Modulus of elasticity

1. Ratio, within the elastic limit, of stress to strain for a material in tension, compression, and shear, commonly expressed in pounds per square inch. The modulus of elasticity in tension and compression is generally called "Young's modulus" (E); in shear it is called the "modulus of rigidity" (G).

2. Measure of the rigidity of metal; ratio of stress, within the proportional limit, to corresponding strain. Specifically, the modulus obtained in tension or compression is Young's modulus, stretch modulus, or the modulus of extensibility; the modulus obtained in torsion or shear is the modulus of rigidity, shear modulus, or modulus of torsion; the modulus covering the ratio of the mean normal stress to the change in volume per unit volume is the bulk modulus. The tangent and secant moduli are not restricted within the proportional limit; the former is the slope of the stress-strain curve at a specified point; the latter is the slope of a line from the origin to a specified point on the stress-strain curve. Also called "elastic modulus" and "coefficient of elasticity."

3. Stress required to produce unit strain, which may be a change of length (Young's modulus); twist or shear (modulus of rigidity), or a change of volume (bulk modulus), expressed in dynes per square centimeter.

4. The modulus of elasticity of a material is the ratio of the unit stress to the unit deformation and indicates the degree of stiffness.

5. Within the elastic zone, the ratio of stress to corresponding strain. The modulus of elasticity is unique for each material.

6. The ratio between stress and deformation is the modulus of elasticity.

Modulus of rupture

1. Unit fiber stress calculated from the beam formula $f = M/S$ and used as an extreme fiber stress in bending for a material such as plain concrete, which has a different strength in tension and in compression, and for which because of certain phenomena in change of location of the neutral axis, the fiber stress in neither tension nor compression can be used in figuring bending.

2. Nominal stress at fracture in a bend or torsion test. In bending, the modulus of rupture is the bending moment at fracture divided by the section modulus. In torsion, the modulus of rupture is the torque at fracture divided by the polar section modulus.

Moment. Bending or twisting effect; force multiplied by distance. Moment is usually measured in units of pound-feet.

Moment of a force

1. With respect to a point the moment of a force is the measure of the tendency of the force to produce rotation about that point. It is equal to the magnitude of the force multiplied by the perpendicular distance of its line of action from the given point. The point about which the moment is taken is the origin (or center) of moments, and the perpendicular distance from the origin to the line of action is the lever arm (or arm) of the force. When a force tends to cause rotation in the direction of the hands of a clock, the moment is positive, and in the opposite direction, negative.

2. Moment is the tendency of a force to cause a rotation about a given point or axis.

Moment of inertia

1. Function of some property of a body or figure, such as weight, mass, volume, area, length, or position, equal to the summation of the products of the elementary portions by the squares of their distances from a given axis.

2. The moment of inertia of an area is the sum of the products of all the elementary areas multiplied by the square of their distances from an axis.

Moment-resisting space frame. Vertical load-carrying space frame in which the members and joints are capable of resisting design lateral forces by bending moments.

Motion. Force that produces or tends to produce motion or a change of motion of bodies. Motion is a change of position with respect to some object regarded as in a fixed position.

Necking. Localized reduction of cross-sectional area of a tensile specimen under load.

Negative bending moment. Moment which exists between a support of a slab or beam and the point of inflection on either side of the support.

Neutral axis

1. Axis of no stress.

2. Position somewhat above midheight in a reinforced concrete beam where there is neither tension nor compression.

Nominal loads. Magnitudes of the loads as required by the applicable code.

Nominal span. Horizontal distance between outside edges of the outermost supports.

Nonbearing wall. Wall that carries no load other than its own weight.

Noncorrodible metal. Metal that, under the conditions of its use, may reasonably be expected, without unusual or excessive maintenance, to serve its purpose throughout the probable life of the structure in which it is used as determined by the building department.

Nondestructive testing. Testing methods that do not destroy the part to determine its suitability for use.

Non-load-bearing wall. Wall that supports no vertical load other than its own weight.

Notch sensitivity. Measure of the reduction in the load-carrying capacity of a material caused by the presence of stress concentration.

Open-web joists. Open-web load-carrying structural members and assemblies in buildings and structures, including the accessory parts and materials included in the fabrication and construction of the assemblies.

Open-web steel joist. A steel joist made of hot- or cold-formed sections, strip or sheet steel, riveted or welded together, or by expanding, which is used for supporting floors and roofs between girders, trusses, beams, or walls in buildings or structures.

Outer forces. The external or outer forces acting upon a structure consist of the applied loads and the supporting forces, called "reactions."

Panel wall. Nonbearing wall in skeleton construction, built between columns or piers and wholly supported at each story.

Panel zone. Zone in a beam-to-column connection that transmits moments by a shear panel.

Party wall. Wall used or adapted for joint service between two buildings.

Pedestal. Upright compression member, the height of which does not exceed three times its least lateral dimension.

Permanent set. Residual deformation after the removal of all loads (aside from creep effects).

Physical properties. Properties that serve to characterize the describe matter and to distinguish the different kinds of matter. These properties include specific gravity or density, electrical and thermal conductivities, and coefficient of thermal expansion.

Pile. Slender structural element that is driven, jetted, or otherwise embedded on end in the ground for the purpose of supporting a load or for modifying the characteristics of the soil.

Pitch of a truss. Ratio of the rise of the truss to its span.

Plane frame. Structural system assumed for the purpose of analysis and design to be two-dimensional.

Plastic analysis. Determination of load effects on members and connections based on the assumption of rigid-plastic behavior, i.e., that equilibrium is satisfied throughout the structure and yield is not exceeded anywhere.

Plastic deformation. Alteration of form or shape that remains permanently in a material after removal of the force that caused it.

Plastic design. Design method for continuous steel beams and frame which defines the limit of structural usefulness as the "ultimate load" associated with the formation of a mechanism. The term "plastic" comes from the fact that the ultimate load is computed from a knowledge of the strength of steel in the plastic range.

Plastic flow. Gradual time-dependent deformation due to sustained load.

Plastic hinge. Yielded zone found in a structural member when the plastic moment is applied.

Plasticity

1. Opposite quality to elasticity. With this quality a material does not return to its original dimensions when the load causing deformation is removed.

2. Ability of a material to be nonelastically deformed without fracturing.

3. Ability of a metal to deform nonelasticity without rupture.

Plastic modulus. Resisting modulus of a completely yielded cross section; combined static moment about the neutral axis of the cross-sectional area above and below the axis.

Plastic moment. Maximum moment of resistance of a fully yielded cross section.

Plastification. Gradual penetration of yield stress from the outer fiber toward the centroid of a section under increase of moment.

Plate girder. Built-up structural beams and plates.

Plywood. Laminated board or panel consisting of veneer sheets of wood placed alternately crosswise and bonded together with either water-resistant or waterproof adhesives, each stronger than the wood itself.

Poisson's ratio

1. Absolute value of the ratio of the transverse strain to the corresponding axial strain in a body subjected to uniaxial stress, usually applied to elastic conditions.

2. Whenever bodies elongate under stress, they shrink laterally; conversely, when they are compressed, under a load, they expand at right angles to the direction of the load. The ratio of deformation normal to stress to deformation parallel to stress is Poisson's ratio.

Positive bending moment. Moment that exists at all points in beams or slabs, except where the negative moment exists.

Posts and timbers. Lumber of square or approximately square cross section, 5 by 5 in. and larger, graded primarily for use as posts or columns carrying axial loads but adapted for miscellaneous uses in which strength in bending is not especially important.

Post-buckling strength. Additional load or stress that can be carried by a plate or structural member after buckling.

Pounds per lineal foot (plf). Units of loading on a member, measured as average load in pounds per foot of member length.

Pounds per square foot (psf). Units of loading on a surface such as a floor or roof.

Pounds per square inch (psi). Measure of loads distributed over a square inch of surface.

Primary member. A member of the structural frame of a building used as a column, or a grillage beam, or to support masonry walls or partitions; including trusses, isolated lintels spanning an opening of eight (8) feet or more and any other structural member required to brace a column or a truss.

Principal design sections. Vertical sections in a flat slab on which the moments in the rectangular directions are critical.

Principal stresses. Normal stresses on three mutually perpendicular planes on which there are no shear stresses.

Professional engineer or architect. Person authorized by law to practice or to engage in the business of professional engineering or architecture, and who is engaged in such practice or business.

Property of materials. Strength, stiffness, elasticity, ductility, malleability, and brittleness.

Proportional limit

1. Highest stress at which strain is directly proportional to stress.
2. Maximum stress at which strain remains directly proportional to stress.
3. Maximum stress in a material at which the strain is directly proportional to the stress, commonly expressed in pounds per square inch.

Protected construction. Construction in which all structural members are so constructed, chemically treated, covered, or protected that the individual unit or the combined assemblage of all such units has the required fire-resistance rating specified for its particular use or application. Protected construction includes protected frame, protected ordinary, and protected noncombustible construction.

PSI

1. Pressure in pounds per square inch.
2. Abbreviation of pounds per square inch used in measuring load or stress.

Purlin. Horizontal framing members supporting the roof rafters or spanning between trusses to support the roof.

Racking load. Load applied in the plane of an assembly in such manner as to lengthen one diagonal and shorten the other.

Radius of gyration

1. Distance from the reference at which all of the area that still produces the same moment of inertia can be considered concentrated. Numerically it is equal to the square root of the moment of inertia divided by the area.
2. The radius of gyration is used in the design of steel columns. Also used is the term "slenderness ratio."

Ratio of reinforcement. Ratio of the effective area of the reinforcement cut by a section of a beam or slab to the effective area of the concrete at that section.

Reaction. Forces acting on a beam, girder, or truss or other member through its supports that are equal but opposite to the sum of the dead and live load.

Real load. Actual load applied to the structure as opposed to an assumed load.

Redistribution of moment. Process that results in the successive formation of plastic hinges until the ultimate load is reached. Through the transfer of moment which results from the formation of the plastic hinge, the less highly stressed portion of a structure also may reach the Mp value.

Redundant member. Member in a frame or truss that may be omitted in the structure without affecting the ability to analyze the frame or truss by ordinary static methods (such as a counter-diagonal in a truss).

Reinforced concrete. Concrete in which reinforcement other than that provided for shrinkage or temperature changes is combined in such a manner that the two materials act together in resisting forces.

Required design load. Design load that is required to be used in the design of buildings and structures.

Required strength. Load effect acting on an element or connection determined by structural analysis from the factored loads.

Residual deflection. Deflection resulting from an applied load, remaining after removal of such load.

Residual stresses. Stresses that are left in a member after it has been formed into a finished product.

Resistance. The capacity of a structure or component to resist the effects of loads.

Resistant. When used as a suffix (such as absorption-resistant, moisture-resistant, etc.) means material constructed, protected, or treated so that it will not be injured readily when subjected to the specific material or condition.

Resonance. Condition reached when the frequency of the applied dynamic load coincides with natural frequency of the load support.

Restrained beam

1. Beam that is more or less fixed at one or both points of support.
2. Beam of which one or both ends are fixed or continued over a support.
3. In structures, beams that are usually restrained by being riveted to a column or other beam, or poured with a concrete slab or beam.

Restrained support. Flexural member where the supports and/or the adjacent construction provides complete or partial restraint against rotation of the ends of the member and/or partial restrained against horizontal displacement when subject to a gravity load and/or temperature change.

Restraint. Restriction of free movement of hardened concrete; restraint can be internal or external and may act in one or more directions.

Resultant of forces. A single force that would produce the same effect as a number of forces is the resultant of those forces. The process of finding the single force is composition.

Rigid frame

1. In a plane, any structure made up of beams and columns and so constructed that the joints are rigidly fixed to transmit moment and thus to reduce moment in other parts of the frame.
2. Rigid joint structure in which moments and shears in joints maintain the equilibrium of the structure.

Rise of a truss. The rise of a truss is the distance from the highest point of a truss to the line joining the points of support.

Riveted construction. All parts of riveted members should be well pinned or bolted and rigidly held together while riveting. Drifting done during assembling should not distort the metal or enlarge the holes.

Roof loads. Roofs sustain, within stress limitations set by code, all "dead loads" plus unit "live loads." Live loads are to be assumed to act vertically on the area projected on a horizontal plane.

Rotation capacity. Angular rotation that a given cross-sectional shape can accept at the plastic moment value without prior local failure.

Safe. Adequate in strength and stability to withstand the required design loads and applied loads without exceeding the allowable stresses.

Scalar. Quantities that involve magnitude but no sense of direction. Energy, time, and mass are scalar quantities.

Secondary member. Any member of the structural framework other than a primary member, including filling-in beams of floor systems.

Secondary stress. Stress resulting from causes other than the initial effects of design loads including, but not limited to, stresses due to dimensional charges tending to occur in a structure because of the application of design loads or because of temperature changes or possible foundation settlements.

Sectional properties. End area per unit of width, moment of inertia, section modulus, and radius of gyration.

Section modulus

1. Moment of inertia of the area of a section of a member divided by the distance from the center of gravity to the outermost fiber.
2. One of the properties of cross-sections frequently used.

Seismic design. Construction designed to withstand earthquake forces.

Semifireproof construction. Construction with structural members of approved noncombustible construction having the necessary strength and stability with fire-resistance ratings of not less than four hours for exterior walls and wall panels; not less than three hours for columns and wall-supporting girders and trusses; not less than two hours for floors, roofs, and floor- and roof-supporting beams, girders, and trusses; and with exterior and interior bearing walls, if any, of approved masonry or reinforced concrete. Building codes may differ in this definition.

Semirigid framing. Semirigid, partially restrained framing assumes that the connections of beams and girders possess a dependable and known moment capacity intermediate in degree between complete rigidity and complete flexibility.

Service load. Load expected to be supported by a structure under normal usage; often assumed as the nominal load.

Setting shrinkage. Reduction in volume of concrete prior to the final set of cement, caused by settling of the solids and by the chemical combination of water and cement when an external source of curing water is not present.

Shape factor. The ratio of the plastic moment to the yield moment, or the ratio of the plastic modulus to the section modulus for a cross section.

Shear

1. Cut off, as by two equal and opposed forces. Pure shear is a condition of stress consisting of equal tension and compression at right angles and in which no bending exists.
2. Shear occurs when two parallel forces having opposite direction act on a body, tending to cause one part of the body to slide past an adjacent part.
3. Force that causes or tends to cause two contiguous parts of the same body to slide relative to each other in a direction parallel to their plane of contact.

Shear castellation. A set of interrupted keys is cut into wood to provide shear strength at the junction between dissimilar materials in composite construction.

Shear strength. Stress required to produce fracture in the plane of cross section, the conditions of loading being such that the direction of force and of resistance are parallel and opposite although their paths are offset by a specified minimum amount.

Shear stress. Result of forces acting parallel to an area but in opposite directions, causing one portion of the material to "slide" past another.

Shear wall. Wall designed to resist lateral forces parallel to the wall. Braced frames subjected primarily to axial stresses are considered shear walls.

Shoring. Props of timber or other material in compression, placed temporarily to support beams, etc.

Short column. Column whose load capacity is limited by strength rather than buckling; column that is customarily so stocky and sufficiently restrained that at least 95% of the cross-sectional strength can be developed.

Shrinkage. Volume decrease caused by drying or chemical changes; a function of time but not of temperature nor of stress due to external load.

Shrinkage and temperature stresses. Shrinkage is a function of materials that are poured in a semiliquid state and then harden by cooling or chemical action. Such materials are cast iron and concrete.

Sidesway. Lateral movement of a structure under the action of lateral loads, unsymmetrical vertical loads, or unsymmetrical properties of the structure.

Simple beam

1. Beam freely supported at both ends, theoretically with no restraint. The restraint of a wall is considered insufficient to change the beam from "simple" to "restrained."
2. Beam resting on a support at each end, there being no restraint against bending at the supports; the ends are simply supported.

Simple frame construction. Construction of the unrestrained type for which the design is predicted on the assumption that the ends of beams and girders are connected for shear only and are free to rotate under load.

Simple spans. Beams, girders, and trusses designed on the basis of simple spans whose effective length is equal to the distance between centers of gravity of the members to which they deliver their end reactions.

Simple support. Flexural member where the supports and/or the adjacent construction allows free rotation of the ends of the member and horizontal displacement when subject to a gravity load and/or a temperature change.

Skeleton construction. Construction in which the loads from all parts of the structure are carried to the foundation by means of beams and columns and not by walls.

Slanting construction. Type of construction in which blast- and/or bomb-resistant features are incorporated into a new structure without appreciable reduction in efficiency.

Slender column. Column whose load capacity is reduced by the increased eccentricity caused by secondary deflection moments.

Slenderness ratio. Ratio of the effective height of a member to its effective thickness.

Slip-critical joint. Bolt joint in which the slip resistance of the connection is required by code.

Slow-burning construction (mill). Exterior and bearing walls are of approved masonry of reinforced concrete. Structural elements are wholly or partly of wood of smaller dimensions than required for heavy-timber construction, or of steel or iron not protected as required for fireproof or semifireproof construction.

Solid masonry unit. Masonry unit whose net cross-sectional area in every plane parallel to the bearing surface is 75% or more of its gross cross-sectional area measured in the same plane.

Solid masonry wall. Wall built of solid masonry units laid contiguously, with the joint between units filled with mortar or grout.

Space frame. Three-dimensional structural system composed of interconnected members other than bearing walls, laterally supported so as to function as a complete self-contained unit, with or without the aid of horizontal diaphragms or floor-bracing systems.

Space frame–ductile moment resisting. A space frame moment resisting, complying with the requirements for a ductile moment resisting space frame as required in the code.

Space frame—moment resisting. A vertical load-carrying space frame in which members and joints are capable of resisting design lateral forces by bending moments.

Space frame—vertical load-carrying. A space frame designed to carry all vertical loads.

Span. Distance between structural supports, such as walls, columns, piers, beams, girders, and trusses.

Spandrel beam

1. Beam in the building frame that extends between exterior columns at a floor level.

2. Beam from column to column, carrying an exterior wall in a skeleton building.

Span of a roof truss. Horizontal distance in feet between the centers of supports.

Splice. Connection between two structural elements joined at their ends to form a single, longer element.

Static load. Force applied gradually and slowly and not repeated many times.

Statically determinate structure

1. Structure, all the elements of which may be computed by ordinary methods of computation.

2. Structures for which both outer and inner forces may be determined by the aid of statics. If all the outer forces may be found by statics, the structure is statically determinate with respect to the other forces whether or not it is possible to determine the inner forces by the same means.

Statically indeterminate structures

1. Structures, all the elements of which cannot be computed by ordinary static methods of computation.

2. Structures that cannot be statically determined are those which the equations of statics will not suffice to design. All rigidly connected building frames are statically indeterminate.

Statics. Science that treats forces in equilibrium.

Steel joist.

1. Secondary steel member of a building or structure made of hot- or cold-formed solid or open-web sections; riveted or welded bar, strip, or sheet steel members; slotted and expanded or otherwise deformed rolled sections.

2. Open-web structural member fabricated from hot-rolled or cold-formed steel and suitable for the direct support of floors and roof decks in buildings and structures.

Stepped column. Column with changes from one cross section to another occurring at abrupt points within the length of the column.

Stiffener. Member, usually an angle or plate, attached to a plate or web of a beam or girder to distribute load, to transfer load, or to prevent buckling of the member to which it is attached.

Stiffness

1. The stiffness of a material is the property that enables it to resist deformation.

2. Resistance of materials to bending or flexure.

3. Ability of a metal or shape to resist elastic deflection. For identical shapes, stiffness is proportional to the methodulus of elasticity.

4. Rigidity of structural members. In columns or struts stiffness refers to their lateral stability; in beams stiffness refers to the lack of deflection rather than to strength.

Story. Portion of a building included between the surface of the floor and the floor immediately above.

Strain

1. Measure of the linear change in the original size or shape of a solid due to an applied stress.

2. Measure of the change in the size or shape of a body, referred to its original size or shape. Linear strain is the change per unit length of a linear dimension. True (or natural) strain is the natural logarithm of the ratio of the length at the moment of observation to the original gauge length. Conventional strain is the linear strain referred to the original gauge length. Shearing (or shear) strain is the change in angle (expressed in radians) between two lines originally at right angles. When the term strain is used alone, it usually refers to the linear strain in the direction of the applied stress.

Strength. The strength of a material is its ability to resist forces. Three basic stresses are compression, tension, and shear. The strength of the material is considered to be the unit stress that causes failure or rupture.

Strength design. Method of proportioning structural members using load factors and resistance factors such that no applicable limit state is exceeded.

Stress

1. Load per unit of area. Types of stress include tensile, compressive, and shear stresses.

2. Force per unit area, often thought of as force acting through a small area within a plane. Stress can be divided into components normal and parallel to the plane, called "normal stress" and "shear stress," respectively. Truss stress denotes the stress where force and area are measured at the same time. Conventional stress, as applied to tension and compression tests, is force divided by the original area. Nominal stress is the stress computed by simple elasticity formulas, ignoring stress raisers and disregarding plastic flow; in a

notch bend test, for example, it is bending moment divided by minimum section modulus.

3. Cohesive force in a body that resists the tendency of an external force to change the shape of the body.

4. Tension, compression, or shear in metal, usually measured in pounds per square inch.

Stress diagram. Graphical solution of axial forces as they interact within the members of a truss.

Stressed skin. A design in which frame and skin or sheathing are joined so that the skin may aid in resisting strains.

Stress-grade lumber

1. Lumber classifications known as beams and stringers, joists and planks, and posts and timbers, to each grade of which is assigned proper allowable unit stresses, depending on the type of lumber.

2. Lumber 2 or more in. thick and 4 or more in. wide that has been graded for strength by lumber grading, inspection bureau, or other agency.

Stress raiser. Notch that has the effect of concentrating the applied stresses at the point of the notch.

Strong axis. Major principal axis of a cross section.

Structural alterations. Any change in the supporting members of a building, such as footings, bearing walls or partitions, columns, beams, or girders, or any substantial change in the roof or in the exterior walls, excepting such repair as may be required for the safety of the building.

Structural aluminum. Structural member cast or wrought from aluminum alloys and used for load-carrying purposes in buildings and structures.

Structural damage. Loosening, twisting, warping, cracking, distortion, or breaking of any piece, fastening, or joint in a structural assembly, with loss of sustaining capacity of the assembly. The following is not deemed to constitute structural damage: small cracks in reinforced concrete perpendicular to the reinforcing bars or deformation of sheet material when the structural assembly is under applied load, which increases as such load increases, but which disappears when such load is removed.

Structural failure. Rupture, loss of sustaining capacity or stability, marked increase in strain without increase in load, or the more rapid increase in deformation than in imposed load.

Structural feature. Any part of a structure that is designed for or indicative of the intent to accommodate any given use.

Structural frame. All vertical load-supporting members, other than bearing walls; all primary horizontal load-supporting members rigidly connected thereto; and all other primary members essential to the stability of the structural frame.

Structural insulating board. Structural insulating material made principally from wood, cane, or other vegetable fibers and performed into a rigid, fibrous, insulating board, lath, or plain, used principally in building construction.

Structural lumber. Structural lumber consists of lumber classifications known as beams and stringers, joists and planks, and posts and timbers, to each grade of which is assigned the proper allowable unit stresses.

Structural member. Columns, studs, posts, bearing walls or partitions, trusses, girders, arches, beams, joists, rafters, roofing, flooring, or other supporting devices that are essential in carrying vertical, horizontal, or torque forces to bearing materials upon which a structure rests.

Structural metal. Structural member cast, wrought, or otherwise manufactured from a metal or metal alloy and used for load-carrying purposes in buildings and structures.

Structural steel. Structural member hot-rolled to shape from carbon steel or fabricated from hot-rolled carbon steel and used for load-carrying purposes in buildings and structures.

Structural steel construction. Construction composed of hot-rolled structural (carbon) steel shapes and plates, with supplementary members of cast steel, cast iron, and other metals, assembled and erected by bolting, riveting, or welding.

Structural steel member. Primary or secondary member of a building or structure consisting of a rolled steel structural shape other than formed steel, light-gauge steel, or steel joist members.

Structural system. Assemblage of load-carrying components that are joined together to provide regular interaction or interdependence.

Structure

1. Part of assemblage of parts constructed to support certain definite loads. Structures are acted upon by external forces, and these external forces are held in equilibrium by internal forces called "stresses."

2. Assembly of materials forming a construction frame of component structural parts for occupancy or use, including buildings.

Stub column. Short compression-test specimen, sufficiently long for use in measuring the stress-strain relationship for the complete cross section, but short enough to avoid buckling as a column in the elastic and plastic ranges.

Subassemblage. Connected group of beams and columns that form part of a multistory frame.

Supported frame. Frame which depends upon adjacent braced or unbraced frames for resistance to lateral load or frame instability.

Sustained load. Force acting for a long period of time. The dead weight of a structure is an example of such a load. For some materials under certain conditions of temperature and load, this force has an appreciable effect on the structure, such as permanent deflection of a beam or permanent shortening of a column.

Swelling. Volume increase in soil caused by wetting or chemical changes, or both; a function of time but not of temperature nor due to external load.

Tendons. Prestressing strands and wires used in prestressed concrete.

Tensile load. Load that tends to pull a material apart.

Tensile strength

1. Maximum load that can be sustained by metal in tension measured in pounds per square inch.

2. In tensile testing, the ratio of maximum load to original cross-sectional area, also called "ultimate strength."

3. Quality that resists direct pull, usually expressed as allowable pounds per square inch.

4. Maximum tensile stress that a material is capable of sustaining. Tensile strength is calculated from the maximum load during a tension test carried to rupture and the original cross-sectional area of the specimen.

5. Resistance of a material subjected to tensile loading.

Tensile test. Test for determining the behavior of materials under axial tension loading. In a tensile test, the specimen is gripped from its two ends and pulled apart.

Tension. Force that tends to lengthen the body on which it acts.

Tension—field action. Resistance of a plate girder to externally applied shear forces, in which diagonal tensile stresses develop in the web and compressive forces develop in the transverse stiffeners in a manner analogous to a Pratt truss.

Tension or tensile stress. Force that tends to stretch a member.

Thermal stress. Internal force caused within a material due to temperature variations.

Tie. Structural member that tends to lenghten under stress.

Time-dependent deformation. Deformation of concrete occurring with appreciable time (as days, weeks, or months); includes creep and characteristics affected by age and strength changes such as elasticity, drying, shrinkage, and temperature effects.

Tolerance. Allowable variation for practical reasons from the theoretical standards desired.

Top chord. An inclined or horizontal member that establishes the upper edge of a truss, usually supporting combined compression and bending stresses.

Torsion. Twisting action resulting in shear stresses and strains.

Torsion strength. Maximum stress that a material sustains before fracturing under twisting.

Torsion test. Test for determining the behavior of materials subjected to twisting loads.

Toughness. Ability of a metal to absorb energy and deform plastically before fracturing. It is usually measured by the energy absorbed in a notch impact test, but the area under the stress-strain curve in tensile testing is also a measure of toughness.

Transition temperature. Temperature at which the failure mode changes from ductile to brittle.

Transverse. Literally "across," usually signifying a direction or plane perpendicular to the direction of working.

Triaxial stress. State of stress in which none of three principal stresses is zero.

Trimmers. Full-length beams on each side of an opening that support the header, into which are framed the ends of the beams cut to form the opening.

Truss. Framed or jointed structure composed of straight members that are connected only at their intersections, so that if the loads are applied at these intersections, the stress in each member is in the direction of its length. Each member of a truss is either a tie or strut.

Ultimate and rupture stress. The load on a member is increased until the member fails, and the highest unit stress sustained is called the "ultimate stress." Some materials, notably steel, after being stressed to the ultimate, sustain a gradually lessening load until failure. The unit load at failure is called the "rupture stress."

Ultimate load. Largest load a structure will support. In plastic design it is the load attained when a sufficient number of yield zones have formed to permit the structure to deform plastically without further increase in load.

Ultimate strength

1. Maximum stress that can be developed in a material as determined by cross section or original specimen, commonly expressed in pounds per square inch. A brittle material breaks when it reaches stress equal to ultimate strength; a ductile material continues to stretch.

2. Maximum conventional stress—tensile, compressive, or shear—that a material can withstand.

Unbraced frame. Frame in which the resistance to lateral load is provided by the bending resistance of frame members and their connections.

Unbraced length. Distance between braced points of a member.

Unbuttoning. Premature, sequential failure of fasteners, progressing from the ends of the joint inward.

Uniform load. Load distributed evenly over a surface or member.

Unprotected metal construction. Structural supports are unprotected metal. The roofing and walls or other enclosures are of sheet metal or other noncombustible materials, or of masonry deficient in thickness or otherwise not conforming to approved masonry. Building codes may differ in some areas of this definition.

Unsymmetrical bending. Bending in both principal directions (biaxial bending) and simultaneous torsion.

Vectors. A quantity that combines both magnitude and direction is a vector quantity. Forces, velocity, and acceleration are vector quantities. Any vector quantity may be represented graphically by a drawn line.

Veneered wall. Wall having a masonry facing that is not attached and bonded to the backing and so does not form an integral part of the wall for the purposes of load bearing and stability.

Vertical bracing system. System of shear walls, braced structural frames, or both, extending throughout one or more floors of a building.

Vertical shear. Tendency for the beam to fail by dropping vertically between the supports, the fibers of the beam failing by vertical shear. The magnitude of the vertical shear at any section of a beam is equal to the reactions minus the loads to the left of the section.

Virtual eccentricity. Eccentricity of resultant axial loads required to produce axial and bending stresses equivalent to those produced by applied axial and transverse loads.

Viscoelastic properties. Combination of viscous and elastic responses of plastics to applied loads. The mechanical model of this behavior, resembling a spring and shock absorber combination, is called a "Voigt model."

Volume change. An increase or decrease in volume (length, width, and thickness).

Warping torsion. Portion of the total resistance to torsion that is provided by resistance to warping of the cross section.

Web crippling. Local elastic-plastic failure of the web plate in the immediate vicinity of a concentrated load or reaction.

Weldability. Ease of making a sound and serviceable joint by welding.

Width-thickness ratio. Ratio of the width of the outstanding portion of an element to its thickness, an indication of local instability.

Wind load

1. Load resulting from wind blowing in any direction.

2. Lateral pressure on the building or structure in pounds per square foot due to wind blowing in any direction.

Working load or safe load. The product obtained by multiplying the cross-sectional area of a column or tie by the working or allowable unit stress is called the "working load" or "safe load" of a member. For a beam, the safe load is the load that will stress the most processed fibers to the allowable unit stress.

Yard lumber. Lumber that has not been graded for strength, but that is suitable for general building purposes.

Yield moment. In a member subjected to bending, the moment required to initiate yielding.

Yield point

1. Stress at which a material continues to deflect, stretch, or bend without greater load.

2. Point at which strains increase without corresponding increase in stress.

3. Point at which a material continues to deform without any further increase in the load; commonly expressed in pounds per square inch. Only ductile materials have a well-defined yield point. In practice, the yield point is sometimes assumed to be the point at which stress produces an elongation in the specimen of 0.50% of its original length.

4. First stress in a material, usually less than the maximum attainable stress, at which an increase in strain occurs without an increase in stress. Only certain metals exhibit a yield point. If there is a decrease in stress after yielding, a distinction may be made between upper and lower yield points.

Yield strength

1. Stress at which a material exhibits a specified limiting deviation from proportionality of stress and strain. Deviation is expressed in terms of strain.

2. Load at which a specified limited permanent deformation occurs.

3. Stress at which a material exhibits a specified deviation from proportionality of stress and strain. An offset of 0.2% is used for many metals.

4. The term is used to describe the maximum unit stress in a material to produce a permanent deformation of 0.10 or 0.20% of the original length; commonly expressed in pounds per square inch. Term ordinarily used for materials that have no well-defined yield point.

Yield stress

1. Generic term to denote either the yield point or the yield strength.

2. Stress at which a material exhibits a specified limiting deviation from the proportionality of stress to strain.

Yield stress level. Average stress during yielding in the plastic range; stress determined in a tension test corresponding to a strain of 0.005 in./in.

SURVEYING AND SURVEYS

Accuracy. Correctness or freedom from mistakes or carelessness.

Acre. Unit for measuring land, equal to 43,560 sq ft, 4840 sq yd, or 160 rods^2.

Angle. Difference in the direction of two lines that meet or tend to meet, usually measured in degrees.

Angular course. Compass direction in degrees, minutes, and seconds, stated as a deviation eastward or westward from due north or south, and used in metes-and-bounds surveys and descriptions.

Angular measure. Deviation between two lines that meet at a point, expressed in degrees, minutes, and seconds.

Azimuth

1. True bearing, commonly measured from point south moving clockwise in a right-handed direction 360°.

2. Direction expressed as horizontal angle, usually in degrees or mils; and measured clockwise from north. Thus, azimuths will be true azimuths, grid azimuths, or magnetic azimuths, depending on which north is used.

Back tangent. Tangent previous to the curve.

Baseline. Parallel of specified latitude, used in the rectangular survey system, serving as the main east-west reference line, with a principal meridian for a particular state or area.

Bench mark. Established point of reference used in calculating the height and location of other reference points.

Boston rod. Extension target rod used in surveying, made of two strips, one of which slides in a groove in the other, and provided with clamps to hold the two parts in any desired position.

Building line. Established location where the zoning code will allow the location of a structure.

Cairn. Pile of stones used as a marker.

Call. In surveyor's language, the statement or mention of a course and/or distance.

Cardinal points. Four major compass headings of north, east, south, and west.

Chain

1. Surveyor's chain measuring 66 ft in length.

2. Surveyor's measure. One link equals 7.92 in.; 100 links equal one chain.

Chord. Straight line connecting two points on a curve.

Contour line

1. Line joining points of equal elevation on a surface; representation of such a line on a drawing or map to scale.

2. Level line crossing a slope.

Contour map

1. Map that shows topography by means of contour lines. Each contour line connects points of equal elevation, and the elevation interval between lines is constant.

2. Map or drawing on which irregularities of land surface are shown by contour lines, the relative spacing of the lines indicating the relative slope of the surface.

Correction lines. East-west reference lines, used in the rectangular survey system, located at 24-mile intervals to the north and south of a baseline.

Course. Compass direction from one reference point to the next for each leg of a metes-and-bounds survey.

Crosshair. Hair mounted horizontally in a telescope so as to divide the field of view into halves.

Cross notches. Establishment of corners of properties by scratching crosses into pavements or sidewalks as permanent markers. Cross notches may be offset beyond the property corners by identifying the location of these crosses on the survey.

Cross section. Profile taken at right angles to the center-line of a project.

Cross-section paper. Paper ruled in squares for convenience in drawing and measuring.

Datum. Established reference altitude elevation, such as sea level.

Datum plane. Artificially established, well-surveyed horizontal plane against which elevations, depth, tides, etc., are measured (for example, mean sea level).

Degree. Unit of angular measure equal to the angle contained within the radii of a circle that describe an arc equal to $\frac{1}{360}$ part of the circumference of the circle. It is also used to define an arc equal to $\frac{1}{360}$ part of the circumference of a circle.

Degree of curve. Number of degrees at the center of a circle subtended by a chord of 100 ft at its rim. Occasionally in highway surveying it is defined as the central angle subtended by an arc of 100 ft.

Drainage system. Requirement of functional survey is to indicate on the survey plat the location of any existing drainage system or systems and the final disposition of the drainage.

Dumpy level. Instrument that has its vertical axis; the horizontal bar, and the supports of the telescope all in one piece, to which the spirit level is attached.

Easement. Right held by one person, group, or company to make use of a parcel, portion, or area of the land of another for a purpose; the purpose being limited in its use and also its length of duration by agreement with the party granting the easement.

Electric service. Requirement of a functional survey is to indicate on the survey plat the location of any existing underground or overhead electric service provided to the property line by the local utility company.

Elevation (surveying)
1. Height of a point above a plane of reference.
2. Vertical height of one point on the earth above a given datum plane, usually sea level.
3. Vertical distance of ground forms, usually measured in feet or meters, above mean sea level (plus elevation) or below mean sea level (minus elevation).
4. Altitude above a given established datum or bench mark.

Engineer's chain. Series of 100 wire links.

Equator. Parallel circling the middle of the earth, all points of which are equidistant from both the North and South Poles. The equator is designated as the starting line (0°) for measuring north or south latitude.

Flow gradient. Drainageway slope determined by the elevation and distance of the inlet and outlet, and by required volume and velocity.

Flying levels. In surveying a line of levels to determine approximate elevations in hundredths or tenths of a foot.

Forward direction. Direction of increasing stations.

Fractional section. "Adjusted" section of land generally containing less (sometimes more) than 1 square mile. The deficiency (or excess) may be the result of the convergence of meridians, the presence of bodies of water, or uncertainties in surveying.

Gas service. Requirement of a functional survey is to indicate on the survey plat the location of any existing underground gas line from the gas system provided by utility company to the property line.

Geodetic survey. Survey of large areas indicating established contour lines that can be used for controlling other surveys.

Grade. Usually the elevation of a real or planned surface or structure; also, surface slope.

Gradient. Slope along a specific route, as of a road surface, channel, or pipe.

Great circle. Line described on a sphere by a plane bisecting the sphere into equal parts. The equator is a great circle, as are pairs of opposing meridians.

Greenwich (prime) meridian. Meridian passing through the Royal Observatory at Greenwich, England, and designated as the starting line (0°) for measuring east and west longitude.

Grid. Set of surveyor's closely spaced reference lines laid out at right angles, with elevations taken at line intersections.

Grid lines. Lines drawn in parallel sequence in both directions at identical measured intervals. The elevations of the slope of land occurring at each intersection of the grid are identified and indicated on the survey.

Hachures. Shading used to indicate relief features consisting of lines drawn parallel to the slopes and varying in width with the degree of slope.

Height. The vertical difference in elevation between an object and its immediate surroundings.

Hydrant. Discharge pipe with valve and spouts at which water at established pressures may be drawn from the water mains. Hydrants should be specifically located and identified as to type and size on the survey plat.

Instrument. Telescopic level, such as a transit or a builder's level.

Invert. Inside bottom of a pipe, tunnel, catch basin, or manhole.

Latitude. Distance in degrees north or south of the equator.

Latitude location. Position of a point on the earth's surface north or south of the equator, stated as an angular measure (degrees, minutes, and seconds) of the meridian contained between that point and the equator.

Legal description. Written identification of the location and boundaries of a parcel of land. A legal description may be based on a metes-and bounds survey or the rectangular system of survey, or it may make reference to a recorded plat of survey.

Level. Instrument that can be used to indicate a horizontal line or plane.

Longitude. Position of a point on the earth's surface east or west of the Greenwich meridian, stated as an angular measure (degrees, minutes, and seconds) of the arc on the equator contained between a meridian passing through that point and the Greenwich meridian.

Meridian. Imaginary north-south line on the earth's surface described by a great circle arc from the North Pole to the South Pole. All points on a meridian arc of the same longitude.

Meter vault. Underground concrete enclosure for housing a water meter, as required in some municipalities. A survey plat should indicate its location.

Metes and bounds

1. System of land survey and description based on starting from a known reference point and tracing the boundary lines around an area.

2. Irregular system of metes and bounds is used entirely in Maine, New Hampshire, Vermont, Massachusetts, Rhode Island, Connecticut, New York, Pennsylvania, New Jersey, Maryland, Delaware, Virginia, North Carolina, South Carolina, Georgia, Tennessee, Kentucky, Texas, and parts of Ohio. Each parcel of land varies in size, is described independently, and is not tied in to any system of baselines.

Metes-and-bounds survey. Preparation of a legal document indicating the boundaries of a given tract of land, parcel, or piece of property drawn to scale, giving the dimension of each boundary or property line beginning at the starting point and circumventing the periphery of the properties and returning to the starting point.

Mil. Unit of angle equal to the angle subtended by an arc of $\frac{1}{6400}$ of a circumference.

Minus station. Stakes or points on the far side of the zero point from which a job was originally laid out.

Minute. Unit of angular measure equal to $\frac{1}{60}$ of a degree.

Monument

1. Permanent reference point for land surveying, whose location is recorded; either a man-made or a natural landmark.

2. Permanent object to mark a boundary, usually made of stone or concrete.

New York rod. Leveling rod marked with narrow lines, ruler fashion.

Pantograph. Drafting instrument consisting of wood and metal parts and metal parts and joined so as to form a parallelogram, and used for enlarging and reducing maps.

Parallel. Imaginary east-west line on the earth's surface, consisting of a circle on which all points are equidistant from one of the Poles. All points on a parallel are the same latitude.

Perched water table. Underground water lying over dry soil and sealed from it by an impervious layer.

Philadelphia rod. Leveling rod in which hundredths of feet or eighths of inches are marked by alternate bars of color the width of the measurement.

Photogrammetry. Art, science, or process of making maps and scale drawings from photographs, especially of maps from aerial photographs; process of making precise measurements by the use of photography.

Photomap. Mosaic map made from aerial photographs with physical and cultural features shown as on a planimetric map.

POB (place of beginning). Starting point of a metes-and-bounds survey or description.

Precision. Refinement of measurements or closeness with agreement among several measurements.

Principal meridian

1. North-south line projected through a prominent landmark established under the government survey system.

2. Meridian of specified longitude used in the rectangular survey system, serving as the main north-south reference line for a particular state or area. Guide meridians are north-south reference lines located at 24-mile intervals east and west of a principal meridian.

Properly line. Extreme boundary of a tract, parcel or area of land.

Protractor. Device for measuring angles on drawings.

Public land survey systems of the United States. There are two separate and distinct systems of land surveys in the United States: the system of metes and bounds in which each parcel of land is individually described and bounded and the system of rectangular surveys under which the land is divided basically into equal-sized townships, sections, and fractions thereof.

Range lines. North-south reference lines used in the rectangular survey system, located at 6-mile intervals between guide meridians.

Recorded plat. Plat that is recorded at an appropriate governmental office, usually the county recorder's office. A recorded plat, in addition to location, notes, and boundary line layout, may contain information such as restrictions, easements, approvals by zoning boards and planning commissions, and lot and block numbers for a subdivision.

Rectangular surveys. The system of rectangular surveys was inaugurated in 1784, and the laws governing its establishment have, with various modifications, been applied to all of the United States with exceptions. Under this system the lands are divided into "township" 6 miles square, which are related to baselines established by the federal government. Baseline running north and south are known as "principal meridians," while the east and west baselines are called simply "baselines." Township numbers east or west of the principal meridians are designated as ranges, where the numbers north and south of the baseline are tiers. Thus, the description of a township as "Township 16 North, Range 7 West" would mean that the township is situated 16 tiers north of the baseline for the principal meridian and 7 ranges west of that meridian. Guide meridians, at 24-mile intervals east and/or west of the principal meridian, are extended north and/or south from the baseline; standard parallels, at 24-mile intervals north and/or south of the baseline, are extended east and/or west from the principal meridian. A township is 6 miles square, and divided into 36 square-mile "sections" of 640 acres each, which may be divided and subdivided as desired.

Rectangular (government) survey system. Land survey system based on geographical coordinates of longitude and latitude originally established by acts of Congress to survey the lands of public domain and now used in 30 states.

Roads. Requirement of a functional survey is the indication on the survey plat of the location of existing public and private roads and curbs, including curb cuts and driveways to the property from the road.

Run levels. Survey an area or strip to determine elevations

Second. Unit of angular measure equal to one-sixtieth of a minute.

Section

1. Area of land used in the rectangular survey system approximately 1 mile square, bounded by section lines. The rectangular system provides for the further subdivision of sections into halves, quarters, and quarter-quarters.

2. Parcel of land containing 1 square mile or 640 acres.

Section lines. North-south reference lines used in the rectangular survey system parallel to the nearest range line to the east, and east-west lines parallel to the nearest township line to the south. These lines divide townships into 36 approximately equal squares called "sections."

Setback. Area confined by an established line governed by a zoning or deed restriction.

Sewage service. Requirement of a functional survey is the indication on the survey plat of the location of any existing system for the disposal of waste matter, the outlet being obtained immediately at the property line, in the case of a sewer line or the location of a sewage disposal system with septic tanks and disposal field.

Stadia. Measurement of distance by proportion to the space on a vertical rod seen between upper and lower instrument cross hairs. The usual proportion is 1 vertical to 100 horizontal.

Stakes. After corners of property have been established, the surveyor usually drives solid iron or pipe stakes as permanent markers.

Station. One of a series of stakes or points indicating distance from a point of beginning or reference.

Survey

1. Measure and marking of land accompanied by maps and held notes that describe the measures and marks made in the field.

2. Ascertain the exact area and boundaries of property.

Survey foot. A unit of length, used by the U.S. Coast and Geodetic Survey, equal to 1.000002 ft.

Surveying. Finding and recording elevations, locations, and directions by means of instruments.

Surveying instruments. Instruments with the minimum number of parts generally hold their accuracy longer. A transit should have standards high enough to permit a 360° vertical swing of the telescope with a prism eyepiece in place.

Surveyor. Registered professional who calculates areas and locates boundary lines in accordance with established standards.

Surveyor's chain. Series of wire links, each of which is 7.92 in. long. The total length of the chain is 4 rods or 60 ft. Ten square chains of land is 1 acre.

Survey plat

1. Legal document indicating the boundaries of a given tract of land, parcel, or piece of property, drawn to scale and giving the dimensions of each boundary or property line and the angular bearings in degrees, minutes, and seconds.

2. Map of surveyed land showing the location, boundaries, and dimensions of the parcel.

Tangent. Line that touches a circle and is perpendicular to its radius at the point of contact.

Target rod. Leveling rod.

Telephone service. Requirement of a functional survey is the indication on the survey plat on the location of any existing underground or overhead service provided by the local utility company.

Topograph map

1. Schematic drawing of prominent land forms indicated by conventionalized symbols, such as hachures or contours.

2. Map that presents the vertical position of features in measurable form, as well as their horizontal positions.

3. Map indicating surface elevation and slope.

Topographic plot. Representation by means of contour lines of the ground relief of an area shown in a stereoscopic model.

Topography

1. Shape of the earth's surface above and below sea level; set of land forms in a region; distribution of elevations.

2. Physical features, both natural and man-made, of the earth's surface. In terrain analysis the following categories of topographical features are considered; relief, drainage, surface materials, vegetation, special physical phenomena, and man-made (cultural) features.

Township. Area of land used in the rectangular survey system approximately 6 miles square, bounded by range lines and township lines.

Township lines. East-west reference lines used in the rectangular survey system, located at 6-mile intervals between correction lines.

Transit. Surveying instrument that can measure both vertical and horizontal angles.

Tripod. Three-legged support for a surveying instrument.

Turn angles. Measure the angle between directions with a surveying instrument.

Turning point (transfer point). Point whose elevation is taken from two or more instrument positions to determine their height in relation to each other.

United States Geological Survey maps. Topographic maps prepared by the USGS, are bounded on the north and south by parallels of latitude, and on the east and west by meridians of longitude.

Utility poles. Requirement of a functional survey is the indication on the survey plat the location of any existing wood or metal poles installed by the utility company for the installation of electric and telephone lines and the support of transformers on or off the property.

Walks. Requirement of a functional survey is the indication on the survey plat of the location of any existing public and private walks adjacent to the property lines or on the site.

Water service. Requirement of a functional survey is the indication on the survey flat of the location of any existing water service from any source to the property line system, including any lines that may exist on the site.

Well. Requirement of a functional survey is the indication on the survey plot of the location of any source of underground water brought to the surface by pumps, including the location of underground vaults or tanks.

Wye level. Instrument in which a spirit-level is attached to the telescope that rests in 2 "Y"-shaped supports, which in turn are fastened to a horizontal bar to which the vertical axis is attached.

Zoning. Continuous tract or land area that is distinguished for some purpose by an established set of rules or requirements.

B

Construction Categories

*The INFORM DIVISIONS presented here have been created by the author and are not in any way connected or related directly or indirectly with the Masterformat, published by the Construction Specifications Institute, Inc.

BUILDING TYPES AND USES

Animal clinic or animal hospital. A place where animals or pets are given medical or surgical treatment in emergency cases and are cared for during the time of such treatment. Use as a kennel is limited to a short-time boarding period, is only incidental to such clinic or hospital use and should be enclosed in a soundproof building or structure.

Animal shelter. A building, structure, or facility operated, owned, maintained, or used by a duly incorporated humane society, animal welfare society, or other not-for-profit organization whose purpose is to provide for and promote the welfare, protection, and humane treatment of animals, including animals impounded for rabies observation.

Arena or round theater. Any building, or portion of building, that is used for the presentation of performances on the floor level without elevated stage, scenery, curtains, apparatus, or stage properties.

Armory hall. Any building that, or a portion of which, is used as a drill hall for members of any military organization.

Assembly building. A building used, in whole or in part, for the gathering together of persons.

Assembly hall. Buildings used as an auditorium in which the public assembles, and whose primary and intended use is for the assembly of persons for the purpose of amusement, entertainment, instruction, workshop, transportation, sports, military drills, or similar purposes, with admission either public or restricted.

Auditorium. Building or portion thereof used or designed for use by an audience, including, but not limited to, public assembly, religious services, meetings, lectures, dances, entertainment, and similar uses. The largest room of a church, school, or college used for the assembly of persons is the auditorium when meeting the requirements of the local zoning ordinance.

Automobile filling station. A building, structure, premises, enclosure or other place used for the sale or offering for sale at retail of automobile fuels or oils, except hardware stores, painting and decorating shops, dyeing and cleaning shops, tailor shops, or drug stores, where such fuels or oils are not regularly dispensed to automobiles.

Automobile laundry. A building or portion thereof where automobiles are washed, using a conveyor, blower, steam-cleaning equipment, or other mechanical device of production-line nature.

Automobile parking structure. A structure used for the parking or storage of automobiles.

Aviary. Any lot, building, structure, enclosure or premise whereupon or wherein are kept ornamental or song birds, in any combination, whether such keeping is for pleasure, profit, breeding, or exhibiting but not including poultry or birds kept for production and sale of meat and/or eggs.

Bank. Building used primarily as an establishment for the custody, loan, exchange, or issue of money, for the extension of credit, and for facilitating the transmission of funds by drafts or bills of exchange.

Brewery. Building where brewing is carried on in the preparation of an alcoholic beverage.

Business building. A building occupied for the transaction of business, for the rendering of professional services, for the display or sale of goods, wares or merchandise, or for the performance of work or labor.

Business or commercial buildings. Buildings or parts of buildings, other than public or residential buildings, including among others, office buildings, stores, markets, restaurants, warehouses, freight depots, garages, factories, laboratories, and the like.

Carbarn. A structure for the storing and repairing of electric street cars and buses and other electrical conveyances.

Carport. Roofed structure providing space for the parking of motor vehicles and enclosed on not more than two sides. For the purposes of the ordinance a carport attached to a principal building is considered as part of the principal building and is subject to all local zoning ordinance yard requirements.

Church. Building, together with its accessory buildings and uses, where persons regularly assemble for religious worship, and which building, together with its accessory buildings and uses, is maintained and controlled by a religious body organized to sustain public worship.

Clinic. Building used by a group of doctors for the medical examination or treatment of persons on an outpatient or nonboarding basis only.

Club or lodge building. Building used by a fraternal, social, or similar organization for the private assembly of persons and not used by the general public as a place of assembly.

Cold storage plant. Building used for the storage of food or other material and for the preservation thereof by ice or refrigeration for a commercial purpose.

College. Building that, or a portion of which, is used for advanced educational purposes, or a particular building for a subject matter or course as part of a university.

Commercial building. Any building used in connection with direct trade with or service for the public.

Commercial stable. A stable where horses are let, hired, used, boarded, or sold on a commercial basis for remuneration.

Community garage. Group of private garages, detached or under one roof, arranged in a row or around a common means of access, and erected for use of residents in the immediate vicinity.

Community shopping center. Commercial establishment designed to provide the basic facilities found in a neighborhood center, with a wider range of commercial establishments. A community shopping center ranges from 30,000 to 300,000 sq ft. Subject to local zoning ordinances.

Covered mall buildings. A covered mall building is a single building enclosing a number of tenants and occupancies such as retail stores, restaurants, places of assemblage, recreation facilities, motion picture theaters, offices, banks, specialty shops, and anchor stores.

Creamery. Building where butter or cheese is made, or where milk or cream is processed.

Dental or medical clinic. Building in which a group of physicians, dentists, or physicians and allied professionals are associated for the purpose of carrying on their profession. The clinic may include a dental or medical laboratory. It does not include in-patient care or operating rooms for major surgery.

Department store. Building that is used for the housing and sale at retail of a variety of goods, which are organized in departments.

Detention facility. A local confinement institution for which the custodial authority is usually determined by the court; some persons can be confined in such facilities pending adjudication and for short-term sentences.

Distillery. Building where distilling of alcoholic liquors is carried on.

Dog kennel. Lot, building, structure, enclosure, or premises whereon or wherein a specified number of dogs over four months of age are kept and maintained for any purpose whatsoever.

Drive-in-theater. An open-air theater designed for viewing by the audience sitting in their motor vehicles.

Educational institutions. Colleges or universities supported wholly or in part by public funds and other colleges or universities giving general academic instruction, as prescribed by the state board of education.

Elementary and high schools. Institutions of learning that offer instructions in the several branches of learning and study required to be taught in the public schools by the education code of the state. High schools include junior and senior classifications.

Facility for the mentally retarded. Facility for the mentally retarded means any institution, place, building, or agency that maintains and operates facilities and care, treatment, or schooling for mentally retarded persons.

Factory. A building, the primary use of which is for the processing of materials or manufacture of products.

Fire station. Building or part of a building designed or used as a place for the housing of one or more pieces of fire fighting or salvaging equipment, together with sleeping quarters, locker rooms, toilet rooms, heating plant, and such other rooms or spaces as required by the fireman or the equipment.

Freight depot. Building used for storage incidental to transportation of freight or cargo.

Garage. A building or part of building or premises in which one or more motor vehicles, excluding motorcycles, are kept for storage, manufacture, repair, demonstration, sale, rental, painting, oiling, greasing, adjustment of equipment, or washing, including also a place of storage of motor vehicles for such work.

Gasoline service station. Building or structure designed or used for the retail sale or supply of fuels, lubricants, air, water, and other operating commodities for motor vehicles, and including the customary space and facilities for the installation of such commodities on or in such vehicles, but not including any operation specifically outlined in the local zoning ordinances.

Greenhouse. Building constructed mostly of glass, or similar material that is used for the protection or cultivation of plants.

Gymnasium. Building or part thereof used for athletic exercises or for athletic games.

Health facility. Includes buildings all or part of which accommodate patients or house services for patients such as laundry, power plant, laboratory, or kitchen. Health facility does not include separate buildings that are used exclusively to house personnel or provide activities not related to health facility patients.

Hospital. Institution in which patients are given medical or surgical care and that is licensed by the state to use the title "hospital" without a qualifying descriptive word.

House of correction. Building used as an institution to which offenders are committed for punishment, discipline, or reformation.

Industrial buildings. Buildings used in connection with production or process work or with storage or warehousing.

Institution. A building occupied by a nonprofit association, corporation, or other entity primarily for public or semipublic use.

Jail. A confinement facility, usually operated by a local law enforcement agency, that holds persons detained pending adjudication and/or persons committed after adjudication for short-term sentences; while intended for the confinement of adults, sometimes holds juveniles as well.

Kennel. Building or property where dogs and/or other small animals and/or pets are kept, sheltered, boarded, or medically treated either for or without compensation.

Library. Building or part thereof devoted to reading and to a collection of books, manuscripts, and similar reading matter that are for use, but not generally for sale.

Mall. A roofed or covered common pedestrian area within a covered mall building that serves as access for several tenants.

Mechanized parking garage. Parking garage in which motor vehicles are raised or lowered to tier levels by mechanical devices installed in the garage, and in which such vehicles are conveyed to stalls or areas by such mechanical devices or by employees, and in which the public is not admitted except in the drive-in and drive-away areas.

Medical or dental clinic. Organization of specializing physicians or dentists, or both, who have their offices in a common building. A clinic does not include in-patient care.

Mercantile building. Any building or a portion of which, used for the display and sale of merchandise.

Metal frame building. Building, metal frame unprotected construction if the enclosing walls are of unprotected metal or unprotected metal in combination with other noncombustible materials, and the other building elements are as set forth in the local building code unless otherwise exempted.

Monastery. Area containing buildings used for religious retirement or of seclusion for persons under religious vows, especially monks, representing any order.

Motion picture theater. Any building or part of a building regularly used for private or corporate profit as a place of assemblage for the witnessing of motion pictures, not having a stage capable of being used for theatricals, and not using movable scenery.

Museum. Building or part thereof in which are preserved and exhibited objects of permanent interest in one or more of the arts and sciences.

Neighborhood shopping center. Group of commercial establishments providing for the sale of convenience goods or personal services. Its size in square footage ranges from 5,000 to 30,000 sq ft. Subject to local zoning ordinances.

Nursery school. Any building used routinely for the daytime care or education of preschool age children, and including all accessory buildings and play areas.

Nursing home. Building used for the accommodation and care of persons with, or recuperating from, illness or incapacity, where nursing services are furnished.

Nursing or convalescent home. Building designed or used in whole or in part to provide, for compensation, the care of the ill, senile, or otherwise infirm persons resident on the premises, not in need of hospital care.

Office building. Building designed for or used as the offices of professional, commercial, industrial, religious, institutional, public, or semipublic persons or organizations, provided no goods, wares, or merchandise are prepared or sold on the premises except that a portion of an office building may be occupied and used as a drug-store, barber shop, cosmetologist's shop, cigar stand, or newsstand, when such uses are located entirely within the building with no entrance from the street nor visible from any sidewalk and having no sign or display visible from the outside of the building indicating the existence of such use.

Open parking structure. Unenclosed or partially enclosed structure for the parking of motor vehicles.

Open plan school. School building in which the school consists principally of individual teaching areas that are separated from each other only by informal dividers such as cabinets, bookcases, and partitions of less than ceiling height.

Parish house. Residence for a minister, priest, or rabbi in connection with the operation of a church, synagogue, or temple.

Parking garage. A garage for passenger motor vehicles involving only the parking or storing of automobiles and not including automobile repair or service work or the sale of gasoline or oil.

Parking structure. Unenclosed or partially enclosed multi-story structure for the parking of motor vehicles, with no provisions for the repairing or servicing of such vehicles.

Penthouse. An enclosed room or structure other than a "roof structure" located on the roof.

Petroleum bulk storage. Building or structure for the storage of lubricating oils with a flash point of 300°F or higher and enclosed storage space for not more than one motor vehicle.

Planned shopping center. Group of separate commercial retail or professional or personal service establishments developed as a unit on a site subject to local zoning ordinances.

Police station. Building or part of a building used by the police department for administrative offices and detention facilities and may include sleeping quarters, courtrooms, and such other rooms or spaces.

Power plant. Building used primarily to house equipment for the production, conversion, or distribution of energy.

Prison. Building for the safe custody or confinement of criminals or others committed by lawful authority.

Private stable. A detached accessory building for the keeping of horses, mules, or ponies owned by the occupants of the premises and not kept for remuneration, hire, or sale.

Public buildings. Buildings in which persons congregate for civic, political, educational, religious, social or recreational purposes including, but not limited to, courthouses, schools, colleges, libraries, museums, exhibition buildings, lecture halls, churches, assembly halls, lodge rooms, dance halls, theaters, taverns, bath houses, armories, recreation piers, stadiums, passenger stations, bowling alleys, skating rinks, gymnasiums, city halls, clubs, grandstands, motion picture theaters, auditoriums, restaurants, etc.

Public utility building. Includes telephone exchange buildings, transformer stations and substations, gas regulator stations, and similar structures.

Reformatory. Building used as an institution to which young offenders are committed for training or reformation.

Regional shopping center. Commercial establishment designed to provide a full scope of retail sales and services. It is designed to attract customers from an area of greater population than the county. Its size must be from 300,000 sq ft up. Subject to local zoning ordinances.

Repair garage. Main or accessory building used or designed for repairing motor vehicles; a service garage if accessory to an automobile salesroom.

Restaurant. Building in which food is prepared and sold for consumption within the building, as opposed to a drive-in restaurant establishment where food may be taken outside of the building for consumption either on or off the premises.

Roundhouse. A structure for the storing and repairing of locomotives using any fuel other than a volatile flammable liquid.

Sanitarium. Building used as an institution for the recuperation and treatment of persons having physical or mental disorders.

School. Any building or a portion of which used as a place of instruction. The school may be either public or private, or it may be operated in conjunction with a church or other organization.

Service station. Building or lot where gasoline, oil, greases, and accessories are supplied and dispensed to the motor vehicle trade; including battery, tire, and other similar services.

Shopping center. Group of commercial establishments planned, developed, owned, and managed as a unit, with on-site parking and of similar architectural characteristics.

Slaughterhouse. Building used for the slaughtering of animals and the scalding, dressing, butchering, and storage of carcasses for human consumption, but not including the rendering, smoking, curing, or other processing of meat, fat, bones, offal, blood, or other by-products of the permitted operations approved by the local zoning board.

Stable. A building occupied or used for the housing of quadrupeds.

Storage building. A building for the housing, except for purely display purposes, of airplanes, automobiles, railway cars, or other vehicles of transportation, for the sheltering of horses, livestock, or other animals, or exclusively for the storage of goods, wares, or merchandise, not excluding in any case, offices incidental to such uses.

Storage garage. A garage in which volatile inflammable oil, other than that contained in the fuel storage tanks of motor vehicles, is handled, stored, or kept.

Summer motion picture theater. Any building, or part of a building, used as a place of assemblage for witnessing motion pictures; does not have a stage capable of being used for theatricals nor movable scenery.

Telephone central office. Building or part of a building used for the transmission and exchange of telephone or radio telephone messages; in residence districts, such use does not include the transaction of business with the public, storage of materials, trucks, or repair facilities, or housing of repair crews.

Temporary building. A building not designed to be permanently located at its present site, or where it is intended to be temporarily erected, placed, or affixed.

Theater. Structure used for dramatic, operatic, motion pictures, or other performance, for admission to which entrance money is received and no audience participation or meal service is allowed.

Trade or industrial school. Establishment, public or private, for the purpose of training students in skills required for the practice of trades or industries.

Transportation station. Railroad station, bus station, airline terminal, or other public transportation building, used for the shelter of persons waiting for transportation facilities for which such building is an accessory.

Truck terminal. Commercial facility where truck freight is stored, handled, and dispatched between various locations by way of different major truck carriers and including facilities for the storage and repair of trucks and trailers while awaiting consignment.

Veterinary hospital or clinic (animal hospital). Institution for the treatment and care of illnesses and injuries of animals.

Warehouse. Any building used exclusively for the storage of merchandise or other products.

FIRE PROTECTION—MATERIALS AND CONSTRUCTION

Accessibility

1. Premises that are not readily accessible from public roads and that the fire department may be called upon to protect in case of fire should be provided with access roads or fire lanes so that all buildings on the premises are accessible to the fire department apparatus.

2. Access roads and fire lanes must not be obstructed.

Area of fire division. Maximum horizontal projected area of the division within the property lines including exterior walls, one or more of which may be party walls, and in the case of separation walls, within the property lines to the center of the separation wall.

Area of refuge. Room in a building or structure, another building or structure, or an open space outside of a building or structure affording escape and protection from fire, smoke, and gases.

Automatic. As applied to fire protection devices, a device or system providing an emergency function without the necessity for human intervention and activated as a result of a predetermined temperature rise, rate of rise of temperature, or combustion products; such as is incorporated in an automatic sprinkler system, automatic fire door, automatic fire shutter, or automatic fire vent.

Brick for fireproofing. Brick must be laid in Type M, S, N or O mortar. Solid clay and shale brick must conform to the "Standard Specifications for Facing Brick" ASTM C 216 or "Standard Specifications for Building Brick," ASTM C 62. Hollow clay and shale brick must conform to the "Standard Specification for Hollow Brick," ASTM C 652. Concrete brick must conform to the "Standard Specification for Concrete Building Brick," ASTM C 55. Sand-lime brick must conform to the "Standard Specification for Calcium Silicate Face Brick (Sand-Lime Brick," ASTM C 73).

Building. A structure wholly or partially enclosed within exterior walls, or within exterior and party walls, and a roof, affording shelter to persons, animals, or property.

Bureau or Bureau of Fire Prevention. The Bureau of Fire Prevention and Fire Safety Inspection in the fire department of the local area.

Ceiling protection. Fire protection membrane suspended beneath the floor or ceiling construction that, when included with the construction, develops the fire-resistive rating for the overall assembly.

Closure. Device for shutting off an opening through a construction assembly, such as a door or a shutter, and includes all components such as hardware, closing devices, and frames.

Combustible. Material or combination of materials that will ignite and support combustion when heated at any temperature up to 1382°F.

Combustible construction. Building or structure constructed in whole or part of combustible materials.

Combustible fibers. Includes readily ignitable and free burning fibers, such as cotton, sisal, henequen, ixtle, jute hemp, tow, cocoa fiber, oakum, baled waste paper, kapok, hay, straw, Spanish Moss, excelsior, certain synthetic fibers, and other like materials.

Combustible liquids. Liquids having a flash point at or above 100°F are known as Class II or III liquids.

Combustible materials. All materials not classified as noncombustible are considered combustible. The property of a material does not relate to its ability to structurally perform under fire exposure. The degree of combustibility is not defined by standard fire test procedures.

Combustible mixture. Means any liquid or solid mixture or substance or compound that does not emit a flammable vapor at a temperature below 500°F when tested in a Tagliabue open cup tester, but that may be ignited and cause burning.

Combustible plastic. A plastic material more than one twentieth ($\frac{1}{20}$) in. in thickness that burns at a rate of not more than two and one-half ($2\frac{1}{2}$) inches per minute when subjected to ASTM D 635. "Standard Method of Test for Flammability of Self-Supporting Plastics," listed in the local building code.

Combustion. Chemical process that involves oxidation sufficient to produce light or heat.

Conflagration hazard. The fire risk involved in the spread of fire by exterior exposure to and from adjoining buildings and structures.

Corridor. Passageway or hallway that provides a common way of travel to an exit or to another passageway leading to an exit.

Directly applied fire-resistive protection. Coating material applied directly to the structural element for the purpose of fire protection.

Draft curtain. Thin partition extending from the underside of the roof usually to the lower edge of the roof trusses, or a solid partition installed in attics extending from the underside of the roof to the ceiling of the top story. Draft curtains also are called "curtain boards" or "draft stops": they retard the spread of fire, hot gases, and smoke along the ceiling or through an attic or suspended ceiling areas.

Exit. That portion of the way of departure from the interior of a building or structure to the exterior at street, or grade level accessible to a street, consisting of: (a) corridors, stairways and lobbies enclosed in construction having a fire-resistance rating, including the door opening thereto from a habitable, public, or occupied space; or (b) an interior stairway; or (c) a horizontal exit; or (d) a door to the exterior at grade; or (e) an exterior stairway, or ramp.

Exit lighting and exit signs. Exits must be adequately lighted at all times when a building or structure is occupied; exit signs must be maintained in a clean and legible condition, unobstructed by decorations, furnishings, or equipment and illuminated at all times when the building or structure is occupied.

Exposed. Usually with reference to a part of a building and its fire separation or fire exposure, part that is in or on a side of the building that is not at right angle to, and faces toward the line, wall, or other object, to which the fire separation under consideration is measured.

Fire alarm system. An installation of equipment for sounding a fire alarm actuated by automatic protection devices.

Fire and smoke-detecting system. An installation of equipment that automatically actuates a fire alarm when the detecting element is exposed to fire, smoke, or abnormal rise in temperature.

Fire area. Adjacent buildings on the same premises constitute one fire area, unless the exterior walls of such buildings have proper fire-resistance ratings for outside exposure, and openings in such exterior walls are equipped with opening protectives. A portion of a building that is cut off from the other portions of a building by fire-resistive walls or partitions and floors and ceilings, having an approved fire-resistant rating established by local authorities, with communicating and exposed openings protected with approved opening protectives, so as to resist the spread of fire, smoke, heat, and noxious gases.

Fire barrier. A continuous membrane, either vertical or horizontal, such as a wall or floor assembly, that is designed and constructed with a specified fire-resistance rating to limit the spread of fire and that will also restrict the movement of smoke. Such barriers may have protected openings.

Fire compartment. Enclosed space in a building that is separated from all other parts of the building by enclosing construction providing a fire separation having a required fire-resistance rating.

Fire damper. A damper arranged to seal off air flow automatically through part of an air duct system, so as to restrict the passage of heat. The fire damper may also be used as a smoker damper if location lends itself to the dual purpose.

Fire damper operators. Designed to hold the damper in the open position under normal usage and release and automatically close the damper under fire conditions.

Fire dampers. Include single-blade, multiblade, or curtain types.

Fire department access. Every building more than one story in height that does not have windows opening directly upon a street in each story above the first, should be provided with suitable access for fire department use. Such access should be a window or door opening through the wall on each floor above the first story. The opening should be established sizes and locations according to fire department regulations. This requirement for access openings for fire department use does not apply where a building is equipped throughout with an automatic sprinkler system approved for fire protection purposes.

Fire division. The interior means of separation of one part of a floor area from another part together with fire-resistive floor construction to form a complete fire barrier between adjoining or superimposed floor areas in the same building or structure.

Fire drill. The organized procedure conducted with or without a private fire brigade for vacating the occupants of a building and for operating the first-aid fire appliances and equipment for the extinguishing of fire and safeguarding of life.

Fire exposure. Subjection of a material or construction to a high-heat flux from an external source, with or without flame impingement.

Fire grading of buildings. All buildings and structures are graded in accordance with the degree of fire hazard of their use in terms of hours and fractions of an hour and as regulated by the local fire department regulations.

Fire hazard. Anything or act that increases or may cause an increase of the hazard or menace of fire to a greater degree than that customarily recognized as normal by persons in the public service regularly engaged in preventing, suppressing, or extinguishing fire; or that may obstruct, delay, hinder, or interfere with the operations of the fire department or the egress of occupants in the event of fire.

Fire limits. Boundary line establishing an area in which there exists, or is likely to exist, a fire hazard requiring special fire protection.

Fire load. Combustible contents of a room or floor area expressed in terms of the average weight of combustible materials per square foot, from which the potential heat liberation may be calculated based on the calorific value of the materials, and includes the furnishings, finished floor, wall and ceiling finishes, trim, and contemporary and movable partitions.

Fire plan. Posted in areas or departments where personnel can readily see it and become familiar with it.

Fire prevention. Preventive measures that provide for fire-safe conduct and operations in buildings and includes the maintenance of fire-detection, fire-alarm, and fire-extinguishing equipment and systems, exit facilities, opening protectives, safety devices, good housekeeping practices, and fire drills.

Fireproof. Term that has been discontinued by many agencies as misleading, since no material is immune from intense heat or duration of fire. Term is used when referring to types of fire-resistive construction in certain building and fire protection codes.

Fireproof buildings. Buildings or structures composed entirely of materials that will resist the action of fire.

Fire protection. The provision of safeguards in construction and of exit facilities; and the installation of fire alarm, fire-detecting and fire-extinguishing service equipment to reduce fire risk and conflagration hazard.

Fire protection appliance. The apparatus, system, or equipment provided or installed close at hand for immediate use in the event of fire.

Fire protection equipment. Includes apparatus, assemblies, or systems either portable or fixed, for use to prevent, detect, control, or extinguish fire.

Fire protection system. A system including systems, devices, and equipment to detect a fire, actuate an alarm, or suppress fire, or any combination thereof.

Fire resistance. The property of a material or assembly to withstand fire or give protection from it; as applied to elements of buildings, it is characterized by the ability to confine a fire or to continue to perform a given structural function, or both.

Fire resistance and fire-resistive material. Materials having the properties to withstand fire or give protection from it. As applied to the elements of buildings or structures, it is characterized by the ability to confine a fire or to continue to perform a given structural design function.

Fire-resistance ratings. Time in hours that the material or construction will withstand the standard fire exposure as determined by a fire test made in conformity with the "Standard Methods of Fire Tests of Building Construction and Materials," ASTM E 119 or NFPA No. 251.

Fire-resisting ceiling. Ceiling that is part of a floor and ceiling, or roof and ceiling assembly having a minimum fire-resistance rating of one hour.

Fire resistive. In the absence of a specific ruling by the authority having jurisdiction, applies to materials for construction not combustible in the temperatures of ordinary fires and that will withstand such fires without serious impairment of their usefulness for a minimum of one hour.

Fire-resistive ceiling. Ceiling construction that is used as part of a floor assembly or roof system and has a minimum fire-resistance rating of one hour or more.

Fire-resistive materials. Materials that offer a degree of resistance to the passage or the effects of fire or heat sufficient to meet the minimum requirements of the code.

Fire-resistive protection. Thermal insulating materials applied directly, attached to, or suspended or sprayed to a structural assembly, to maintain the structural integrity of structural member or system for a specified time rating as established by the local fire code.

Fire-resistive separation. Every room containing a central heating plant in applicable occupancies as stated in the code should be separated from the balance of the building or structure by a minimum of one-hour fire-resistive separation with all openings in such separation protected by a fire assembly having a minimum of one-hour fire-resistive rating.

Fire retardant. Materials, buildings, or structures that are combustible in whole or in part, but have been treated or have surface applications to prevent or retard ignition or the spread of fire under the conditions for which they were designed to be used.

Fire-retardant ceiling. Ceiling used in a floor and ceiling assembly that has a minimum fire resistance rating of one hour.

Fire-retardant chemical. Chemical or preparation of chemicals or solutions used to reduce flammability or to retard spread of flame.

Fire-retardant lumber. Lumber treated by pressure impregnation to reduce combustibility.

Fire-retardant materials. Materials used in structures for which minimum fire retardant ratings have been developed.

Fire separation. A construction of specific fire resistance separating parts of a building.

Fire shutter. Shutter capable of resisting a fire as required by the fire code.

Firestop. Fire-resistive closure, placed so as to restrict the spread of smoke and fire in concealed spaces.

Firestopping. Effective prevention through the use of firestops for the retarding of horizontal or vertical spread of smoke, fire, and gases through hollow, concealed spaces in wall or floor assemblies, or attic spaces.

Fire treatment of materials. Surface-applied or pressure-type treatment for control of fire or flame spread, when specified, such as treatment by brushing, spraying, dipping, or by pressure application of chemical fire retardants to protect against combustion or flame spread.

Fire terrace. Level space or area at a setback of an exterior wall of a building and at approximately the same elevation as that of the curb or grade level of the higher street, to provide a safe termination for fire escapes from upper stories of the building.

Fire valve. An automatic self-opening noncombustible device designed to permit the passage of gases, smoke, or fire through an opening or vented area.

Fire watch. Uniformed members of the local fire department who will assist in providing adequate fire safety measures when conditions exist that may be hazardous to life and property.

Flameproof. Materials treated so that they will not propagate flame, such as decorations, draperies, curtains, scenery used on a stage, tents, woodwork, or other normally combustible materials of like use and purpose.

Flameproof material. Material that, because of its nature or as a result of chemical treatment, will not burst into flame at or below a temperature of 1200°F.

Flame resistance. Property of materials or combinations of component materials that restricts or resists the spread of flames, as determined by approved flame-resistance tests.

Flame resistant. That property of a material that is flame resistant by nature or has been made so by an accepted method.

Flame-resistant material. Material that is flame resistant by nature or has been made flame resistant in conformity with generally accepted standards.

Flame spread. The rate at which flames spread over surfaces of various materials as tested by ASTM E 84.

Flammable. Capable of igniting and continuing to burn within five seconds when exposed to flame.

Flash point. The minimum temperature in degrees Fahrenheit at which a flammable liquid will give off flammable vapor as determined by appropriate test procedure and apparatus as specified in the local fire code.

Fuel oil. Kerosene or any hydrocarbon oil conforming to nationally recognized standards and having a flash point not less than 100°F.

Horizontal compartment. Portion of a building bounded by exterior walls and the required fire rated and smoke barriers for floors and roof assembly.

Incombustible construction. Material that will not of and by itself ignite when its temperature and that of the surrounding air is 1200°F (649°C). Has been interpreted as having the same meaning as "noncombustible," and is also subject to misunderstanding due to the prefix. "Noncombustible" is preferred.

Incombustible. Synonymous with "noncombustible construction."

Incombustible material. Any material that will not ignite nor actively support combustion in a surrounding temperature of 1200°F during an exposure of five minutes and that will not melt when the temperature of the material is maintained at 900°F for a period of at least five minutes. Synonymous with "noncombustible material."

Labeled fire dampers. Only fire dampers that have been tested, listed, and labeled by Underwriters Laboratories, Inc., or an equivalent test and labeling by other accredited testing laboratories meet the requirements of the local fire code and for the recommended locations and use as listed in the local fire code.

Liquefied petroleum gas (LP gas). Any material that is composed predominantly of the following hydrocarbons, or mixtures of them: propane, propylene, butane (normal butane of isobutane) and butylenes.

Noncombustible. Materials or combination of material that, when tested in accordance with the provisions of ASTM E 136 exposed to a furnace temperature of 1382°F during an exposure of five minutes, will not ignite and support combustion.

Nonflammable. Not flammable; not combustible.

Opening protective. An assembly of materials and accessories, including any incidental frames, mullions, muntins, anchors, and hardware, that, when installed in an opening in a wall, partition, floor, or roof, prevents the passage of flame, heat, fumes, and smoke through the opening for a specified time.

Oxidizing materials. Includes substances that readily yield oxygen to stimulate combustion.

Protected construction. That in which all structural members are constructed, chemically treated, covered, or protected so that the individual unit or the combined assemblage of all such units has the required fire-resistance rating specified in the local building code for its particular use or application.

Protective assembly. Method of protecting an opening in a wall, partition, floor, or roof against the spread of fire, heat, and smoke.

Protective equipment. Any equipment, device, system, or apparatus, whether manual, mechanical, electrical, or otherwise, permitted or required by the bureau to be constructed or installed in any hotel or multiple dwelling for the protection of the occupants or intended occupants thereof, or of the general public.

Separation. System of walls, floors, or other construction serving to separate or cut off one unit of occupancy from another.

Shaft. A vertical opening or enclosed space extending through two or more floors of a building, or through a floor and roof.

Smoke and heat vents. For the purpose of localizing the heat released by a fire to the immediate fire area, minimizing sprinkler operation, lowering building temperatures, exhausting smoke, and improving accessibility for fire-fighting personnel to permit close approach and direct action against the seat of the fire.

Smoke barriers. Continuous from outside wall to outside wall, from a fire barrier to a fire barrier, from a floor to a floor, from a smoke barrier to a smoke barrier, or a combination thereof; including continuity through all concealed spaces such as those found above a ceiling, including interstitial spaces.

Smoke compartment. A space within a building enclosed by smoke barriers or fire barriers on all sides, including the top and bottom.

Smoke damper. Damper arranged to seal off air flow automatically through a part of an air duct system so as to restrict passage of smoke or gases.

Smoke detector. A device that senses visible or invisible particles of combustion.

Smokeproof enclosure. An enclosed stairway, with access from the floor area of a building either through outside balconies or ventilated vestibules, opening on a street or yard or open court, and with a separately enclosed direct exitway to the street at the grade floor.

Smokeproof tower. Stairway enclosure so designed that the movement into the smokeproof tower of products of combustion produced by a fire occurring in any part of the building will be limited.

Smoke stop. A partition in corridors, or between spaces, to retard the passage of smoke, with any opening in such partition protected by a door equipped with a self-closing device.

Smoke-stop door. A door or set of doors placed in a corridor to restrict the spread of smoke and to retard the spread of fire by reducing draft.

Sprinkler system. A system of piping and appurtenances designed and installed so that heat from a fire will automatically cause water to be discharged over the fire area to extinguish it or prevent its further spread.

Standpipe system. An installation of piping and appurtenances, whereby all parts of a building can be quickly reached with an effective stream of water.

Structure. An assembly of materials forming a construction framed of component structural parts for occupancy or use, including buildings.

Unprotected opening. Doorway, window, or opening other than one equipped with a closure having the required fire-protection rating, or any part of a wall forming part of the exposing building face that has a fire-resistance rating less than required for the exposing building face.

Vertical opening. Openings through floors, such as for stairways, elevators, ventilating shafts, etc., that if unprotected, may serve as channels for the spread of fire, smoke, and gases.

Vertical service space. Shaft oriented essentially vertically that is provided in a building to facilitate the installation of building services including mechanical, electrical, plumbing installations, and facilities such as elevators, refuse chutes, and linen chutes.

Volatile. Capable of emitting flammable vapors at a temperature below 75°F.

HAZARDOUS USE STRUCTURES—MATERIALS AND EQUIPMENT

Acid. A corrosive combustible material when used in conjunction with other ingredients.

Aerosol. Material that is dispensed from its container as a mist, spray, or foam by a propellant under pressure.

Aircraft service station. Portion of an airport where flammable liquids used as aircraft fuel are stored or dispensed from fixed equipment, including all facilities essential thereto.

Air-supported structure. Any structure constructed of lightweight fabric or film or any combination thereof that derives its sole support and stability from internal inflation pressure.

Ammunition. A metal or other shell containing a fulminate or black or smokeless powder for the purpose of propelling projectiles or shot; or black or smokeless powder packed for use as a propelling charge or for saluting purposes.

Atmospheric tank. Storage tank that has been designed to operate at pressures from atmospheric through 0.05 psig.

Automatic. Standpipe system so arranged through the use of approved devices as to admit water to the system automatically by opening a hose valve.

Automotive service station. Portion of property where flammable or combustible liquids used as motor fuels are stored and dispensed from fixed equipment into the fuel of motor vehicles.

Below-ground container. Storage installation in which the maximum liquid level in the container is below the surrounding grade or below a backfill berm and that is at least 10 ft wide at the top and then slopes away from the container at a natural angle of repose or is retained 10 ft from the container by a retaining wall constructed of earth, concrete, solid masonry, or suitable material designed to prevent the escape of liquid.

Black powder

1. Powder or gunpowder, general term for explosives including dynamite, but excluding caps.

2. Mixture consisting mostly of carbon, sodium or potassium nitrate, and sulfur, used as an explosive.

3. Gunpowder, that is, a mixture of carbon, sodium or potassium nitrate, and sulfur.

Blanket. Soil or broken rock left or placed over a blast to confine or direct throw of fragments.

Blast. Loosen or move rock or dirt by means of explosives or an explosion.

Blast area. Area in which explosives loading and blasting operations are being conducted.

Blaster. Person or persons authorized to use explosives for blasting purposes and meeting required qualifications.

Blast hole. Vertical drill hole 4 in. or more in diameter, used for a charge of explosives.

Blasting agent. Material or mixture consisting of a fuel and oxidizer, intended for blasting, not otherwise classified as an explosive, in which none of the ingredients are classified as an explosive, provided that the finished product, as mixed and packaged for use or shipment, cannot be detonated by means of a No. 8 test blasting cap when unconfined. Materials or mixtures classified as nitrocarbonitrates by Department of Transportation regulations are included in this definition.

Blasting cap. Metallic tube closed at one end, containing a charge of one or more detonating compounds, and designed for and capable of detonation from the sparks or flame from a safety fuse inserted and crimped into the open end.

Blasting gelatin

1. High explosive made by dissolving nitrocotton in nitroglycerin; strongest and highest velocity commercial explosive.

2. Jellylike high explosive made by dissolving nitrocotton in nitroglycerin.

Blasting machine (battery). Hand-operated generator used to supply firing current to blasting circuits.

Blasting mat. Steel blanket composed of woven cable or interlocked rings.

Blockholing. Blasting boulders by means of drilled holes.

Boiling point. Includes the boiling point of a liquid at a pressure of 14.7 psia (760 mm). Where an accurate boiling point is unavailable for the material in question or for mixtures that do not have a constant boiling point, the 10% point of a distillation performed in accordance with the Standard Method of Test for Distillation of Petroleum Products, ASTM D 86, may be used as the boiling point of the liquid.

Boilover. Expulsion of crude oil (or certain other liquids) from a burning tank in which the light fractions of the crude oil burn off, producing a heat wave in the residue, which on reaching a water stratum may result in the expulsion of a portion of the contents of the tank in the form of a froth.

Bulk oxygen system. Assembly of equipment, such as oxygen storage containers, pressure regulators, safety devices, vaporizers, manifolds, and interconnecting piping. The bulk oxygen system terminates at the point where oxygen at service pressure first enters the supply line. Oxygen may be stored as a liquid or gas in either stationary or portable containers.

Bulk plant. That portion of a property where refined flammable or combustible liquids are received by tank vessel, pipeline, tank car, or tank vehicle and are stored or blended in bulk for the purpose of distributing such liquids in tank vessel, pipeline, tank car, tank vehicle, or container.

Bullet resistant. Materials and construction methods capable of preventing penetration of a 180-grain, 30-caliber, soft-nose, hunting-type bullet, when it is propelled at a maximum velocity of 2700 ft/second.

Bus wire. Expendable wire, used in parallel or in series in parallel circuits, to which are connected the leg wires of electric blasting caps.

Cargo tank. Container having a liquid capacity in excess of 120 gal, used for the carrying of flammable or combustible liquids, LP gas, or hazardous chemicals, and mounted permanently or otherwise on a tank vehicle. The term "cargo tank" does not apply to a container used solely for the purpose of supplying fuel for propulsion of the vehicle on which it is mounted.

Cap. Detonator set off by electric current or a burning fuse.

Cartridge. Wrapped stick of dynamite or other explosive.

Cellulose nitrate plastic (pyroxylin). Any plastic substance, material, or compound, other than cellulose nitrate film, guncotton, or other explosive having cellulose nitrate as a base, by whatever name known when in the form of blocks, slabs, sheets, tubes, or fabricated shapes.

Chemical plant. Large integrated plant or the portion of such a plant, other than a refinery or distillery, where flammable or combustible liquids are produced by chemical reactions or used in chemical reactions.

Chip blasting. Shallow blasting of ledge rock.

Class A explosives. Explosives such as dynamite, nitroglycerin, picric acid, lead azide, fulminate of mercury, black powder, blasting caps, and detonating primers, possessing detonating or other maximum hazard.

Class A fires. Fires in ordinary combustible materials, such as wood, cloth, paper, and rubber.

Class B explosives. Explosives possessing flammable hazard, such as propellant explosives (including some smoke-less propellants), photographic flash powders, and some special fireworks.

Class B fires. Fires in flammable liquids, gases, and greases.

Class C explosives. Class that includes certain types of manufactured articles that contain Class A or Class B explosives, or both, as components but in restricted quantities.

Class C fires. Fires that involve energized electrical equipment where the electrical nonconductivity of the extinguishing mediums is of importance.

Class D fires. Fires in combustible metals, such as magnesium, titanium, zirconium, sodium, and potassium.

Class I flammable liquid. Liquid having a flash point at or below 20°F.

Class II flammable liquid. Liquid having a flash point above 20°F and below 70°F.

Class III flammable liquid. Liquid having a flash point above 70°F.

Closed container. Container so sealed by means of a lid or other device that neither liquid nor vapor will escape from it at ordinary temperatures.

Combustible dust. Fine particles of matter liable to spontaneous ignition or explosion or constituting a dust hazard, such as lint, shavings, sawdust, flour, starch, sulfur, metal powders, and powdered plastics, except when handled, stored, or confined to eliminate the hazard.

Combustible fibers. Readily ignitable and free-burning fibers, such as cotton, sisal, henequen, ixtle, jutes, hemp, tow, cocoa fibers, oakum, rags, waste, cloth, wastepaper, kapok, hay, straw, Spanish moss, excelsior, and other like materials.

Combustible fiber storage bins. Metal or metal lined containers with a capacity not exceeding 100 cu ft and equipped with a self-closing cover.

Combustible fiber storage rooms. Rooms with a capacity not exceeding 500 cu ft separated from the remainder of the building by a minimum one-hour occupancy separation and constructed as specified in the local building code.

Combustible liquid. Liquid having a flash point at or above 140°F and below 200°F and known as a "Class III liquid."

Combustible waste matter. Magazines; books; trimmings from lawns, trees, and flower gardens; pasteboard boxes; rags; paper; straw; sawdust; packing material; shavings; boxes; and all rubbish and refuse that will ignite through contact with flames of ordinary temperatures.

Compressed gas. Mixture or material having in the container either an absolute pressure exceeding 40 psi at 70°F or an absolute pressure exceeding 104 psi at 130°F or both; liquid flammable material having a vapor pressure exceeding 40 psi at 100°F.

Concussion. Shock or sharp air waves caused by an explosion or heavy blow.

Connecting wire. Insulated expendable wire used between electric blasting caps and the leading wires or between the bus wire and the leading wires.

Container. Can, bucket, barrel, drum, portable tank (except stationary tanks), tank vehicle, and tank car.

Corrosive acid storage building. Building or part of a building designed, intended, or used for no other purpose than the storage of hydrochloric acid, nitric acid, sulfuric acid, hydrofluoric acid, or any other corrosive acid.

Corrosive liquids. Hydrochloric acid, nitric acid, sulfuric acid, hydrofluoric acid, or any other corrosive acid or corrosive liquid that when in contact with living tissue will immediately cause severe damage of such tissue by chemical action, or in case of leakage will materially damage or destroy other containers or hazardous commodities by chemical action and cause the release of their contents, or are liable to cause a fire when in contact with organic matter or with certain chemicals.

Corrosive liquid storage building. Building or part of a building designed, intended, or used for no other purpose than the storage of corrosive liquids.

Crude petroleum. Hydrocarbon mixture that has a flash point below 150°F and that has not been processed in a refinery.

Cryogenic container. Cryogenic vessel used either for transportation or storage.

Cryogenic fluids. Fluids that have a normal boiling point below 200°F.

Cryogenic in-ground container. Container in which the maximum liquid level is below the normal surrounding grade. It is constructed essentially of natural materials, such as earth and rock, and is dependent on the freezing of water-saturated earth materials for its tightness or impervious nature.

Cryogenic vessel. Pressure vessel, low-pressure tank, or atmospheric tank on which venting, insulation, refrigeration, or a combination of these, are used in order to maintain the operating pressure within the design pressure, and the contents in a liquid phase.

Dangerous chemical. Any substance that is dangerous to life, limb, or property while being processed, stored or transported, when so designated in the local building or fire codes.

Decking. Separating charges of explosives by inert materials, which prevents the passing of concussion, and placing a primer in each charge.

Deck load. Charges of dynamite spaced well apart in a bore hole and fired by separate primers or by detonating cords.

Decorative materials. Materials such as curtains, draperies, streamers, and surface coverings applied over the building interior finish for decorative, acoustical, or other effect; also, cloth, cotton batting, straw, vines, leaves, trees, and moss used for decorative effect, but not including floor coverings, ordinary window shades, or materials $\frac{1}{40}$ of an inch or less in thickness applied directly to and adhering tightly to a base.

Deflagration. To burn with sudden and startling combustion. Describes explosion of black powder, in contrast with the more rapid detonation of dynamite.

Delay cap. Electric blasting cap that explodes at a set interval after current is passed through it.

Detonating cord. Flexible cord containing a center core of high explosives. When detonated, it has sufficient strength to detonate other cap-sensitive explosives with which it is in contact.

Detonating fuse. Stringlike core of PETN, a high explosive, contained within a waterproof reinforced sheath.

Detonation. Instantaneous decomposition or combustion of an unstable compound, with tremendous increase in volume.

Detonator

1. Blasting cap, electric blasting cap, delay electric blasting cap, and nonelectric delay blasting cap.

2. Device to start an explosion, as a fuse or cap.

Dip tank. Tank, vat, or container of flammable or combustible liquid in which articles or materials are immersed for the purpose of coating, finishing, treating, or similar processes.

Distillery. Plant or that portion of a plant where flammable or combustible liquids produced by fermentation are concentrated, and where the concentrated products may also be mixed, stored, or packaged.

Dry cleaning. Process of removing dirt, grease, paints, and other stains from wearing apparel, textiles, fabrics, rugs, or other material, by the use of nonaqueous liquids (solvents), including the process of dyeing clothes or other fabrics or textiles in a solution of dye colors and nonaqueous liquid solvents.

Dry cleaning building. Building or part of a building designed, intended, or used for no other purpose than dry cleaning or spotting.

Dry cleaning room. Room for the purpose of carrying on the dry cleaning process in a dry cleaning building.

Dry standpipe. Standpipe having no permanent water supply.

Dust. Pulverized particles that if mixed with air in the proper proportions become explosive and may be ignited by a flame or spark or other source of ignition.

Dynamite. Mixture of an explosive(s) with relatively inert material.

Electric blasting cap. Blasting cap designed for and capable of detonation by means of an electric current.

Electric delay blasting caps. Caps designed to detonate at a predetermined period of time after energy is applied to the ignition system.

Explosion hazard gases. Includes acetylene gas, ammonia gas, ether gas, ethyl chloride gas, ethylene gas, liquefied hydrocarbon gases, liquefied petroleum gases, hydrogen gas, illuminating gas, methyl chloride gas, and any other gas that is susceptible to explosion under any condition regardless of whether or not it is a poisonous, irritant, or corrosive gas.

Explosive

1. Chemical compound, mixture, or device the primary or common purpose of which is to function by explosion with substantially instantaneous release of gas and heat, unless such compound, mixture, or device is otherwise specifically classified by the U.S. Department of Transportation. The term "explosive" includes, but is not limited to, dynamite, black powder, pellet powder, initiating explosive, blasting cap, electric blasting cap, safety fuse, fuse lighter, fuse igniter, squib, cordeau detonant fuse, instantaneous fuse, igniter cord, igniter, small arms ammunition, small arms ammunition primer, smokeless propellant, cartridge for propellant-actuated power device, and cartridge for industrial gun.

2. Chemical compound that can decompose quickly and violently.

3. Substance or combination of substances that is commonly used for the purpose of detonation and that on

exposure to any external force or condition is capable of a relatively instantaneous release of gas and heat.

4. Chemical compound or mechanical mixture that is commonly used or intended for the purpose of producing an explosion or that contains oxidizing and combustible units or other ingredients in such proportions, quantities, or packing that an ignition by fire, friction, concussion, percussion, or detonator of any part of the compound or mixture may cause such a sudden generation of highly heated gases that the resultant gaseous pressures will be capable of producing destructive effects on contiguous objects or destroying life or limb.

5. Chemical compound, mixture, or device the primary or common purpose of which is to function by explosion, that is, with substantially instantaneous release of gas and heat, unless such compound, mixture, or device is otherwise specifically classified by the U.S. Department of Transportation.

Explosive gases. Acetylene, ether, ethyl chloride, ethylene, hydrogen illumination gas, petroleum, gases, methyl chloride gas, and oxygen.

Explosive material. Any dangerous chemical classified as an explosive material in the local fire or building codes.

Explosive power load. Substance, in any form, capable of producing a propellant force.

Extra hazard. Extra hazard exists where that amount of combustibles or flammable liquids present is such that fires of severe magnitude may be expected. Extra hazard conditions may be present in such occupations as woodworking, auto repair, and aircraft servicing; warehouses with high-pile (14 ft or higher) combustibles; and in processes such as flammable liquid handling, painting, or dipping.

Fast powder. Dynamites or other explosives having a high-speed detonation.

Fire area. Area of a building separated from the remainder of the building by construction having a fire-resistance rating of a minimum of one hour and having all communicating openings properly protected by an assembly having a fire-resistance rating of a minimum of one hour.

Fire hazard

1. Potential ease of ignition and potential degree of fire severity existing in the use of a building and classified as high, moderate, or low.

2. Thing or act that increases or may cause an increase of the hazard or menace of fire to a degree greater than that customarily recognized as normal by persons in the public service regularly engaged in preventing, suppressing, or extinguishing fire; also, thing or act that may obstruct, delay, hinder or interfere with the operations of the fire department or the egress of occupants in the event of fire.

Fireworks. Combustible or explosive composition, substance or combination of substances, or device prepared for the purpose of producing a visible or audible effect by combustion, explosion, deflagration, or detonation, including blank cartridges, toy pistols, toy cannons, toy canes, or toy guns in which explosives are used, and firecrackers, torpedoes, skyrockets, Roman candles, Daygo bombs, sparklers, or other devices of like construction, devices containing an explosive or flammable compound, or tablet or other device containing an explosive substance, but excluding auto flares, paper caps containing not in excess

of an average of twenty-five hundredths of a grain of explosive content per cap, and toy pistols, toy canes, toy guns, or other devices for the use of such caps.

First-class magazine. Magazine that contains more than 100 lb of explosives or fireworks and has a content of not more than 2500 lb of explosives or fireworks.

Flammable. Capable of igniting and continuing to burn within five seconds when exposed to flame.

Flammable aerosol. Aerosol that is required to be labeled "Flammable" under the Federal Hazardous Substances Labeling Act.

Flammable and combustible solids. Pyroxylin products, nitrocellulose, asphalt, coal tar, pitch, waste paper and rags, feathers, straw, hemp, excelsior, kapok, and greases and fats under 300°F flash point.

Flammable anesthetic. Compressed gas that is flammable and administered as an anesthetic, including among others, cyclopropane, divinyl ether, ethyl chloride, ethyl ether, and ethylene.

Flammable cryogenic fluids. Cryogenic fluids that are flammable in their vapor state.

Flammable dust. Any solid material sufficiently comminuted for suspension in still air that, when so suspended, is capable of self-sustained combustion.

Flammable fiber. Any free-burning material in a fibrous or shredded form such as cotton, sisal, rayon, henequin, ixtle, jute, hemp, tow, cocoa fiber, oakum, kapok, Spanish moss, excelsior, shredded paper, and other materials of a similar nature.

Flammable gas. Any gas having a flammability range with air greater than 1% by volume.

Flammable liquid. A liquid having a flash point below 200°F and having a vapor pressure of 40 psia or less at 100°F.

Flammable liquid container. Can, bucket, barrel, tank or other vessel in which flammable liquids are stored or kept for sale.

Flammable material. Material that will readily ignite from common sources of heat; material that will ignite at a temperature of 600°F or less.

Flammable solid. Solid substance, other than one classified as an explosive, that is liable to cause fires through friction, absorption of moisture, or spontaneous chemical changes or as a result of retained heat from the manufacturing or processing.

Flash point

1. Minimum temperature in degrees Fahrenheit at which a flammable liquid will give off flammable vapor, as determined by appropriate test procedure and apparatus.

2. Minimum temperature at which a liquid gives off vapor within a test vessel in sufficient concentration to form an ignitable mixture with air near the surface of the liquid.

Forbidden or not acceptable explosives. Explosives that are forbidden or not acceptable for transportation by common carriers, rail freight, rail express, highway, or water in accordance with the regulations of the U.S. Department of Transportation.

Fuel oil. Kerosene or hydrocarbon oil conforming to nationally recognized standards and having a flash point not less than 100°F.

Fume hazard gases. Ammonia gas, chlorine gas, phosgene gas, sulfur dioxide gas, and any other gas that is determined a poisonous irritant or corrosive gas and is not susceptible to fire or explosion.

Fume or explosion hazard building. Building, or part of a building, designed, intended, or used for manufacturing, compressing, or storing fume hazard or explosion hazard gas at a pressure of more than 15 psi in a quantity of more than 2500 cu ft.

Fume or explosion hazard room. Room designed, intended, or used that is located in a building other than a fume or explosion hazard building.

Fumigant. Substance that by itself or in combination with another substance emits or liberates a gas, fume, or vapor used for the destruction or control of insects, fungi, vermin, germs, rodents, or other pests and is distinguished from insecticides and disinfectants.

Fumigation. Use of a substance that emits or liberates a gas, fume, or vapor for the destruction or control of insects, fungi, vermin, germs, rodents, or other pests and is distinguished from insecticides and disinfectants.

Fuse. Thin core of black powder surrounded by wrappings that when lit at one end will burn to the other at a fixed speed.

Fuse lighters. Special devices for igniting safety fuses.

Grain elevator, malt house, and similar building. Building designed, intended, or used for receiving, storing, delivering, working with, or treating grain in bulk.

Grinding or dust-producing room. Room containing a machine or device for grinding, pulverizing, buffing, or polishing, which produces dust, lint, shavings, or other fine particles of matter liable to spontaneous ignition or explosion, or a room approved by the local fire and building codes, in which it is proposed to install such a machine or device.

Guncotton. Nitrocellulose chemically known as hexanitrocellulose, and generally used alone or in combination with other substances as a blasting explosive or as a propelling charge, including also all cellulose nitrates of a higher degree of nitration.

Gunpowder. Any of various powders used in firearms and small arms ammunition as propelling charges.

Hazardous areas. Areas of structures or buildings, or parts thereof, used for purposes that involve highly combustible, flammable, or explosive products or materials that are likely to burn with extreme rapidity or that may produce poisonous fumes or gases, including highly toxic, or noxious alkalies, acids, or other liquids or chemicals that involve flame, fume, explosive, poisonous, or irritant hazards; also, areas devoted to uses that cause division of material into fine particles or dust subject to explosion or spontaneous combustion and that constitute a high fire hazard because of the form, character, or volume of the material used.

Hazardous chemical. Chemical such as aluminum powder, calcium carbide, calcium, phosphide, metallic potassium, metallic sodium, phosphorus, sodium peroxide, or any other chemical or material that, as determined by testing, will in fact create an equally or more serious flame, explosion, flame and explosion hazard when coming in contact with water or moisture, or any solid substance other than one classified as an explosive that is liable to cause fires through friction, through spontaneous chemical changes, or as a result of retained heat from manufacturing or processing.

Hazardous chemical room

1. Room designed, intended, or used for storing or using aluminum powder, calcium carbide, calcium phosphide, metallic potassium, metallic sodium, phosphorus, sodium peroxide, or any other chemical or material that, as determined by tests will in fact create an equally or more serious flame, explosion, or flame and explosion hazard when coming in contact with water or moisture.

2. Room designed, intended, or used for storing or using hazardous chemicals or hazardous materials of a similar nature approved by the local fire department.

Hazardous gases. Includes, but is not limited to, such gases as ammonia, chlorine, phosgene, carbon bisulphide and other toxic irritant, corrosive or fume hazard gases such as acetylene, ether, ethyl chloride, ethylene, liquified hydrocarbons, and methyl chloride gas.

Hazardous location. Premises or buildings, or parts thereof, in which there exists the hazard of fire or explosion due to the fact that highly flammable gases, volatile liquids, mixtures, or other substances are manufactured, used, or stored in other than original containers; combustible dust or flyings are likely to be present in quantities sufficient to produce an explosive or combustible mixture; it is impracticable to prevent such dust or flyings from collecting in or on motors or other electrical equipment in such quantities as to produce overheating through the prevention of normal radiation or from being deposited on incandescent lamps; easily ignitible fibers or materials producing combustible flyings are manufactured, handled, or used in a free open state; or easily ignitible fibers or materials producing combustible flyings are stored in bales or containers but are not manufactured or handled in a free open state.

Hazardous material. Any material included under the definitions of "flammable dust," "flammable fiber," "combustible liquid," "dangerous chemical," "flammable gas," "liquified flammable gas," and "flammable liquid."

Hazardous piping. Service piping conveying oxygen, flammable liquids, flammable gases, or toxic gases.

Hazardous roofing materials. Readily ignitible and hazardous roofing materials. Roofing materials approved for use usually carry classification markings from the Underwriters Laboratories, Inc.

Hazardous use general units. Hazardous use unit other than a hazardous use storage unit or industrial unit.

Hazardous use industrial unit. Hazardous use unit designed, intended, or used for industrial purposes, including any operation or process incident to the producing, fabricating, assembling, developing, molding, pressing, preparing or adapting for use, repairing, or refinishing of any high-hazard material, product, article, or substance or high-hazard parts for appliances of any product or article.

Hazardous use storage unit. Hazardous use unit designed, intended, or used for the storage of high-hazard materials and products, and all other high-hazard storage uses.

High explosive. Material that detonates and explodes almost instantaneously.

High fire hazard. Usage that involves the storage, sale, handling, manufacture, or processing of volatile, flammable, or explosive, substances or products that burn with extreme rapidity or, when burning or subjected to heat, produce large volumes of smoke, poisonous fumes, or gases in quantities and under conditions dangerous to the health or safety of any person.

High-hazard industrial building. Includes industrial buildings whose use or occupancy involves substances or products that are volatile, flammable, or explosive, or that are hazardous when exposed to moisture, or that burn with extreme rapidity, or that when burning or subjected to heat produce toxic fumes or gases in quantities and under conditions dangerous to the safety or health of any person. Such substances and products include those having the fire-hazard characteristics of any of those listed in the code.

High hazardous material, product, article, or substance. Any material, product, article, or substance that is liable to burn with rapidity, or while burning to emit poisonous or noxious fumes or to cause explosions.

High-hazard occupancy

1. Occupancy or use of a building or structure, or any portion thereof, that involves highly combustible, flammable, or explosive material, or that has inherent characteristics that constitute a special fire hazard.

2. Occupancy of all buildings and parts of buildings used for purposes, processes, or storage involving highly combustible, flammable, or explosive products or materials, or products or materials that constitute a special hazard to life or limb because of the form, character, or volume of the materials used or that are hazardous because of special conditions incident to the processes or conditions of use or storage.

Highly flammable material storage building. Building or part of a building designed, intended, or used for storing hay, straw, broom, corn, hemp, jute, sisal, moss, sawdust, other wood dust, shavings, excelsior, fiber, hair, or any other similar material that involves an equally or more serious flame hazard.

Highly hazardous material, product, article, or substance. Material, product, article, or substance that is liable to burn with rapidity or while burning is likely to emit poisonous or noxious fumes or cause explosions.

Highly inflammable liquids. Liquids such as carbon bisulfide, naphta, ether, benzol, styrene, butadiene, collodion, ethyl, acetate, amyl acetate, acetone, amyl alcohol, herosene, turpentine, petroleum paint, varnish, dryer, gasoline, alcohol and oil in bulk quantities, and inflammable liquids used and stored in paint mixing and spraying rooms.

Highly toxic materials. Materials so toxic to man as to afford an unusual hazard to life and health during firefighting operations. Such materials as parathion, TEEP (tetraethyl phosphate), HETP (hexaethyl tetra phosphate), and similar insecticides or pesticides, as well as any material classed as poison A or poison B by the U.S. Department of Transportation is considered highly toxic.

Inflammable fibrous materials. Includes such materials as hay, straw, broomcorn, hemp, tow, jute, sisal, kapok, hair, excelsior, oakum, and the like. ("Inflammable" and "flammable" are identical in meaning.)

Leading wire. Insulated wire used between an electric power source and electric blasting cap circuit.

Lead wires. In blasting, the heavy wires that connect the firing current source or switch with the connecting or cap wires.

Light hazard. Light hazard exists where the amount of combustibles or flammable liquids present is such that fires of small size may be expected. These may include offices, schoolrooms, churches, assembly halls, telephone exchanges, etc.

Liquefied flammable gas. Any liquid or gas that is a liquid while under pressure and having a vapor pressure in excess of 27 psia at a temperature of 100°F and a flammability range with air greater than 1% by volume.

Liquefied petroleum gas (LP gas). Material composed predominantly of the following hydrocarbons or mixtures thereof: propane, propylene, butane (normal butane or isobutane), and butylenes.

Liquefied petroleum gas equipment. All containers, apparatus, piping (not including utility distribution piping systems), and equipment pertinent to the storage and handling of liquefied petroleum gas. Gas-consuming appliances are not considered liquefied petroleum gas equipment.

Load. Placing explosives in a hole.

Low fire hazard. Usage that involves the storage, sale, handling, manufacture, or processing of materials that do not ordinarily burn rapidly, produce excessive smoke or poisonous fumes in quantities dangerous to the health of any person, or explode in the event of fire.

Low-hazard industrial buildings. Includes industrial buildings in which the use and occupancy involve materials or substances that create not more than a "low fire hazard" as defined in the local building code.

Low-pressure tank. Storage tank that has been designed to operate at pressures above 0.5 but not more than 15 psig.

Magazine

1. Building or part of a building designed, intended, or used for the storage of an explosive or fireworks.

2. Any building or structure, other than an explosives manufacturing building, used for the storage of explosives.

Magnesium. Pure metal and alloys of which the major part is magnesium.

Marine service station. Portions of properties where flammable or combustible liquids and liquefied petroleum gases used as fuel for floating craft are stored and dispensed from fixed equipment on shore, piers, wharves, floats, or barges into the fuel tanks of floating craft, including all facilities used in connection therewith and intended for servicing small craft.

Mass shooting. Simultaneous exploding of charges in all of a large number of holes, in contrast to firing in sequence with delay caps.

Millisecond delay. Type of delay cap with a definite, but extremely short interval between the passing of current and explosion.

Millisecond delay cap. Detonating cap that fires from 20 to 500 thousandths of a second after the firing current passes through it.

Misfire. Failure of all or part of an explosive charge to go off.

Moderate fire hazard. A moderate fire hazard involves the storage, sale, handling, manufacture, or processing of materials that are likely to burn with moderate rapidity or produce a considerable volume of smoke, but that do not produce either poisonous fumes in quantities dangerous to the health of persons or explosion in the event of fire.

Moderate-hazard industrial buildings. Includes industrial buildings that are not low-hazard buildings, but in which the use and occupancy involve materials or substances that create not more than a "moderate fire hazard" as defined in the local building code.

Moisture hazard substances. Substances having serious flame or explosion hazard when coming in contact with water or moisture such as aluminum powder, magnesium powder, barium peroxide, calcium carbide, metallic potassium, metallic sodium, metallic calcium, phosphorous pentassulfide, sodium peroxide, strontium peroxide, potassium peroxide, sulfuric acid, zinc powder, cyanides, and similar chemicals.

Mudcapping. Blasting boulders or other rock by means of explosive laid on the surface and covered with mud.

Nitrocellulose. Substance, materials, or compound composed in whole or in part of soluble cotton or similar tetranitrate or of a higher nitrate of cellulose, including pyroxylin, plastic, celluloid, fibroid, viscoloid, pyralin, and all similar substances, materials, and compounds, including both new and reclaimed substances.

Nitrocellulose building. Building or part of a building designed, intended, or used for no other purpose than the nitration, manufacture, or storage of nitrocellulose; building in which nitrocellulose is nitrated; building in which discarded scraps of nitrocellulose are reclaimed.

Nitrocellulose product. Article or product either in the process of manufacture or fabrication or in the finished or completed state that is composed wholly or in part of nitrocellulose, including positive or negative nitrocellulose motion picture film, nitrocellulose photographic film, and nitrocellulose x-ray film, as well as pens, pencils, toilet articles, novelties, or other articles or products composed either wholly or partly of nitrocellulose, but excepting flammable liquids otherwise classified.

Nitrocellulose products building. Building or part of a building designed, intended, or used for no other purpose than the manufacture, fabrication, assembly, completion, receiving, shipping, distributing, or storing of either finished or unfinished nitrocellulose products or parts of such products.

Nitroglycerin. Powerful liquid explosive that is dangerously unstable unless combined with other materials.

Nonelectric delay blasting cap. Blasting cap with an integral delay element in conjunction with and capable of being detonated by a detonation impulse or signal from a miniaturized detonating cord.

Nonflammable medical gas. Compressed gas that is nonflammable and used for therapeutic purposes, including oxygen and nitrous oxide, among others.

Ordinary hazard. An ordinary hazard exists where the amount of combustibles or flammable liquids present is such that fires of moderate size may be expected. Areas of ordinary hazard may include mercantile storage and display, auto showrooms, parking garages, light manufacturing premises, and warehouses, not classified as extra hazard are schools, shops, etc.

Organic coating. Liquid mixture of binders, such as alkyds, nitrocellulose, acrylics, or oils, and flammable and combustible solvents, such as hydrocarbons, esters, ketones, or alcohols, that when spread in a thin film convert to a durable protective and decorative finish.

Oxidizing materials

1. Substances that readily yield oxygen to stimulate combustion.

2. Chlorates, permanganates, inorganic peroxides or nitrates, and other substances that readily yield free oxygen when heated.

Paint mixing or spraying room. Room designed, intended, or used for the purpose of mixing or spraying more than 10 gal of paint, varnish, lacquer, enamel, or other such volatile or flammable liquid or substance in solution or suspension during any 24-hour period.

Parallel. Arrangement of electric blasting caps in which the firing current passes through all of them at the same time.

Parallel series. Two or more series of electric blasting caps arranged in parallel.

Pellet powder. Black powder made up into hollow cartridges.

Permanent blasting wire. Permanently mounted insulated wire used between the electric power source and the electric blasting cap circuit.

Piped distribution systems. Central supply system with controlling equipment and a system of piping extending to one or more points where liquids or gases are used and a suitable station outlet valve located at each use point.

Plastic limit. Minimum amount of water in terms of the percentage oven-dry weight of soil that will make the soil plastic.

Poisonous, corrosive, or fume-hazard substance. Hydrochloric, nitric, sulfuric, hydrofluoric, perchloric, and other corrosive acids, corrosive, toxic, or noxious alkalies, cyanides, ammonia, chlorine, phosgene, sulfur dioxide, and similar substances providing like hazards.

Poisonous-irritant. Corrosive or fume-hazard gases such as ammonia, chlorine, phosgene, sulfur dioxide, and similar poisonous, irritant, corrosive, or fume-hazard gases.

Portable tank. Closed container having a liquid capacity of over 60 U.S. gal and not intended for fixed installation.

Potentially explosive chemicals

1. Chemical substance, other than one classified as an explosive, that can be exploded by heat or shock when it is unconfined and unmixed with air or other materials.

2. Chemical substance, other than one defined as an explosive or blasting agent, that has a tendency to be unstable and that can be exploded by heat or shock, or a combination thereof. Potentially explosive chemicals include organic peroxides, nitromethane, ammonium nitrate, and others that have an equal or greater potential explosive hazard.

Pressure delivery system or remote pumping system. Method of transferring flammable or combustible liquids from underground storage tanks to the fuel tanks of motor vehicles whenever the pump is located elsewhere than in the dispenser.

Pressure vessel. Storage tank or vessel that has been designed to operate at pressures above 15 psig.

Primacord. Trademark for a detonating fuse.

Primary blasting. Blasting operation by which the original rock formation is dislodged from its natural location.

Prime. Provide the means of starting a process, for example, in blasting, to pace a detonator in a cartridge or charge of explosive.

Primer

1. Cartridge or container of explosives into which a detonator or detonating cord is inserted or attached.

2. Usually the combination of a dynamite cartridge and a detonating cap.

Processing plant. Portion of a property in which flammable of combustible liquids or materials are mixed, heated, separated, or otherwise processed as principal business.

Propagation. Spread of an explosion through separated charges by concussion waves in water or mud.

Protected combustible fiber storage vault. Room with a capacity exceeding 1000 cu ft, separated from the remainder of the building by a minimum of two-hour occupancy separation, constructed as required in the local building code, and provided with an approved automatic extinguishing system.

Protection for exposure. Fire protection for structures on property in the vicinity of tanks or vessels housing or containing hazardous chemicals, solutions, or materials. When acceptable to the authority having jurisdiction, such structures, located within the jurisdiction of any public fire department or within or adjacent to plants having private fire brigades is to be considered as having adequate protection for exposures.

Pyrotechnics. Combustible or explosive compositions or manufactured articles designed and prepared for producing audible or visible effects that are commonly referred to as "fireworks."

Radioactive materials. Material or combination of materials that spontaneously emits ionizing radiation.

Refinery. Plant in which flammable or combustible liquids are produced on a commercial scale from crude petroleum, natural gasoline, or other hydrocarbon sources.

Remote control. Standpipe system arranged to admit water to the system through manual operation of approved remote control devices located at each hose station.

Rotation firing. Crushing a small piece of rock with a first explosion and timing other holes to throw their burdens toward the space made by that and other preceding explosions; row shooting.

Round. Blast including a succession of delay shots.

Row shooting. In a large blast, setting off the row of holes nearest the face first, and other rows behind it in succession.

Safety can. Approved container of not more than 5 gal capacity, having a spring-closing lid and spout cover, and so designed that it will safely relieve internal pressure when subjected to fire exposure.

Safety clearance. Space whose entire area is open to the sky and continuous on all sides of the building or structure; between such building and any other building, or property dividing lot line, or the opposite side of every adjoining and adjacent public way or public park, or any main line of a steam, electric, and elevated railway or any other railway right of way.

Safety factor. Ratio of the design burst pressure to the maximum working pressure, and a minimum of four.

Safety fuse. Flexible cord containing an internal burning medium by which fire is conveyed at a continuous and uniform rate for the purpose of firing blasting caps.

Sealed source. Quantity of radiation so enclosed as to prevent the escape of any radioactive materials but having control or permitting the release of radiation for special uses.

Secondary blasting. Reduction of oversize material to the dimension required for handling by the use of explosives, including mudcapping and blockholing.

Second-class magazine. Magazine of a size that will not contain more than 100 lb of explosives or fireworks.

Series. Arrangement of electric blasting caps in which all of the firing current passes through each of them in a single circuit.

Short-period delay. Electric blasting cap that explodes $\frac{1}{50}$ to $\frac{1}{2}$ second after passage of an electric current.

Shunt. Connection between the two wires of a blasting cap that prevents the build-up of opposed electric potential in the wires.

Slick hole. Hole column loaded with explosive, without springing.

Slick sheets. Thin steel plate spread on a tunnel floor before a blast to make hand mucking easier.

Slow powder. Black powder, often called gunpowder, and some of the slow-acting dynamites.

Small arms ammunition. Shotgun, rifle, pistol, or revolver cartridges and cartridges for propellant-actuated power devices and industrial guns.

Small arms ammunition primers. Small percussion-sensitive explosive charges encased in a cup and used to ignite propellant powder.

Smoke house. Building or part of a building, designed, intended, or used for no other purpose than that of smoking meats or fish.

Smokeless powder. A propellant for small arms or cannon, having for its explosive base nitrocellulose in varying proportions in the combustion of which smoke is largely eliminated.

Smokeless propellants. Solid propellants, commonly called "smokeless powders" in the trade, used in small arms ammunition, cannons, rockets, propellant-actuated power devices, etc.

Smoke room. Enclosure designed, intended, or used for the smoking of meats or fish and located within a building other than a smoke house.

Smoking. The carrying or use of lighted pipe, cigar, cigarette, or tobacco in any form.

Snake hole. Hole driven into a toe for blasting, with or without vertical holes.

Snakeholing. Drilling under a rock or face in order to blast it.

Solid loading. Filling a drill hole with all the explosive that can be crammed into it, except for the stemming space at the top.

Soluble cotton. Pyroxylin or nitrocellulose, including all cellulose nitrates below that chemically known as hexanitrocellulose, and soluble in a volatile flammable liquid.

Spaced loading. Loading in such a way that cartridges or groups of cartridges are separated by open spacers, which do not prevent the concussion from one charge from reaching the next.

Special industrial explosive devices

1. Explosive power pack containing an explosive charge in the form of a cartridge or construction device. The term includes, but is not limited to, explosive rivets, explosive bolts, explosive charges for driving pins or studs, cartridges for explosive-actuated power tools, and charges of explosives used in jet tapping of open-hearth furnaces and jet perforation of oil well casings.

2. Explosive- and propellant-actuated power devices.

Special industrial high-explosive materials

1. Shaped materials, sheet forms, and various other extrusions, pellets, and packages of high explosives, which include dynamite, trinitrotoluene, pentaerythritol tetranitrate, and other similar compounds used for high-energy-rate forming, expanding, and shaping in metal fabrication and for the dismemberment and quick reduction of scrap metal.

2. Sheets, extrusions, pellets, and packages of high explosives containing dynamite, trinitrotoluol, pentaerythritol tetranitrate, cyclotrimethylene-trinitramine, or other similar compounds used for high-energy-rate forming, expanding, and shaping in metal fabrication and for the dismemberment and quick reduction of scrap metal.

Spray booth. Power-ventilated structure of varying dimensions and construction provided to enclose or accommodate a spraying operation, and to confine and limit the escape of spray vapor and residue and to exhaust it safely.

Spraying area. Area in which dangerous quantities of flammable vapors or combustible residues, dusts, or deposits are present due to the operation of spraying processes; area in the direct path of spray; area containing dangerous quantities of air-suspended powder, combustible residue, dust, deposits, vapor, or mist as a result of spraying operations.

Spraying rooms. Enclosed and protected rooms designed to accommodate spraying operations, using mechanically operated ventilating equipment.

Springing. Enlarging the bottom of a drill hole by exploding a small charge in it.

Sprinkler alarm unit. Assembly of apparatus approved for the service and so constructed and installed that any flow of water from a sprinkler system equal to or greater than that from a single automatic sprinkler will result in an audible alarm signal on the premises.

Sprinkler system. Integrated system of underground and overhead piping designed in accordance with fire protection engineering standards. The system includes a suitable water supply, such as a gravity tank, fire pump, reservoir, or pressure tank, and/or connection by underground piping to a city main. The portion of the sprinkler system aboveground is a network of specially sized or hydraulically designed piping installed in a building, structure, or area, generally overhead, and to which sprinklers are connected in a systematic pattern. The system includes a controlling valve and a device for actuating an alarm when the system is in operation. The system is usually activated by heat from a fire and discharges water over the fire area.

Squib. Detonator consisting of a firing device and a chemical that will burn with a flash that will ignite black powder.

Standard drying room. Room designed, intended, or used for drying clothing, textiles, starch, candy, plaster work, or other articles or materials that do not give off explosive or flammable vapors during the drying process.

Straight dynamite. Dynamite in which nitroglycerin is the principal or only explosive.

Strength. In an explosive, the energy content in relation to its weight.

System. Assembly of equipment consisting of the container or containers, appurtenances, pumps, compressors, and connecting piping.

Tank. Vessel containing more than 60 gal.

Tank vehicle. Vehicle other than a railroad tank, car, or boat, with a cargo tank mounted thereon or built as an integral part thereof, used for the transportation of flammable or combustible liquids, LP gas, or hazardous chemicals. Tank vehicles include self-propelled vehicles, and full trailers and semitrailers, with or without motive power, that carry part or all of the load.

Tent. Structure, enclosure, or shelter constructed of canvas or pliable material and supported in any manner except by air or the contents it protects. Many local fire codes require portable fire fighting equipment.

Test blasting cap No. 8. Cap containing 2 g of a mixture of 80% mercury fulminate and 20% potassium chlorate, or a cap of equivalent strength.

Thermal insecticidal fogging. The use of insecticidal liquids that are passed through thermal-fog-generating units where they are, by means of heat, pressure, and turbulence, transformed and discharged in the form of fog or mist that is blown into the area to be treated.

Throw. Scattering of blast fragments.

Tight. Blasts or blast holes around which rock cannot break away freely.

Tracer. Bullet or projectile incorporating a feature that marks or traces the flight of the bullet or projectile by flame, smoke, or any other means that results in fire or heat.

Tracer charge. Bullet or projectile incorporating a feature designed to create a visible or audible effect by means that result in fire or heat; incendiary bullet or projectile.

Unprotected combustible fiber storage vault. Room with a capacity not exceeding 1000 cu ft separated from the remainder of the building by a minimum of a two-hour occupancy separation, constructed as required in the local building code, and provided with approved safety vents to the outside.

Unstable (reactive) liquid. Liquid that in the pure state or as commercially produced or transported will vigorously polymerize, decompose, condense, or become self-reactive under conditions of shock, pressure, or temperature.

Use group A—high-hazard buildings. All buildings and structures, or parts thereof, classified in the high-hazard use group, used for the storage, manufacture, or processing of highly combustible or explosive products or materials that are likely to burn with extreme rapidity or may produce poisonous fumes or explosions; for storage or manufacturing that involves highly corrosive, toxic, or noxious alkalies, acids, or other liquids or chemicals producing flame, fume, or explosive, poisonous, irritant, or corrosive gases; and for the storage or processing of materials producing explosive mixtures of dust or that result in the division of matter into fine particles subject to spontaneous ignition. Any occupancy or use that has an occupancy fire load of 36 or more fire units constitutes prima facie evidence of being high hazard.

Vapor area. Area containing quantities of flammable vapors. The fire department may determine the extent of the vapor area, taking into consideration the characteristics of the liquid, degree of sustained ventilation, and nature of the operations.

Vapor pressure. Pressure, measured in pounds per square inch (absolute), exerted by a volatile fluid, as determined in accordance with ASTM D 323, Standard Method of Test for Vapor Pressure of Petroleum Products (Reid Method).

Wadding. Paper or cloth placed over explosive in a hole.

Water gels or slurry explosives. Water gels or slurry explosives comprise a wide variety of materials used for blasting, and all contain substantial proportions of water and high proportions of ammonium nitrate, some of which is in solution in the water. Two broad classes of water gels are those that are sensitized by a material classed as an explosive, such as TNT or smokeless powder, and those that contain no ingredient classified as an explosive; these are sensitized with metals such as aluminum or with other fuels. Water gels may be premixed at an explosives plant or mixed at the site immediately before delivery into the bore hole.

Wet standpipe system. System having a supply valve open and water pressure maintained at all times.

HOUSING AND HABITATION

(Definitions will vary according to local zoning ordinances and building codes)

Accessory living quarters. Living quarters within an accessory building, for the sole use of persons employed on the premises, such quarters having no kitchen facilities and not rented or otherwise used as a separate dwelling.

Accessory room. Any room or enclosed floor space used for eating, cooking, bathrooms, water closet compartments, laundries, pantries, foyers, hallways, and similar floor spaces. Rooms designated as recreation, study, den, family room, office, etc., in addition to habitable rooms are considered accessory rooms.

Accessory sleeping quarters for nontransients. Space or building that is an accessory to another occupancy and that is used for sleeping quarters for a minimal amount of nontransients.

Apartment. A room or a suite of rooms within an apartment house, arranged, intended, or designed to be used as a home or residence of one family with kitchen facilities for the exclusive use of the one family.

Apartment building. A building used or intended to be used as a home or residence as defined in the local zoning ordinance for families living in separate apartments, in which the yard areas, hallways, stairways, balconies, and other common areas and facilities are shared by families living in the apartment units.

Apartment hotel. Any building that contains dwelling units and also satisfies the definition of a hotel, as defined in the local zoning ordinance.

Apartment house. A building defined in the local zoning ordinance made up of three apartment units so arranged that each unit has direct access, without common corridors, to a means of egress from the building, and which may or may not maintain an inner lobby for its tenants.

Apartment house or apartment hotel. A building designed, constructed or altered accommodations as provided in the local zoning ordinance for families living independently of each other, each unit of which is designed for cooking upon the premises.

Apartments for the elderly. An apartment building specifically designed for housing elderly individuals who are capable of self-preservation.

Attached dwelling. A single-family dwelling joined to another single-family dwelling by a wall or walls and roof.

Bachelor apartment. One or more rooms in an apartment house or dwelling occupied or intended or designed for occupancy by one person or family for sleeping or living purposes and containing one kitchen and utility room, one sleeping room, one bathroom and incidental closet space.

Bachelor dwelling unit. Dwelling unit for one or two adults with or without one bedroom.

Basement. Portion of a building between the floor and ceiling that is wholly or partly below grade and having more than one-half of its height below grade.

Bathroom. Enclosed space containing one bathtub or shower, and that may also contain a water closet, lavatory, or fixtures serving, similar purposes.

Bedroom. Any room within a residential dwelling unit that is designed to be used for sleeping purposes and contains a closet of sufficient size to hold clothing.

Boarding home for the aged. Any building, however named, operated in accordance with the zoning ordinance for the express or implied purpose of providing service or domiciliary care for elderly people who are not ill or in need of nursing care, and in which there is no agreement that such service shall include personal care or special attention.

Boarding house. Any building, not a hotel, inn, or tavern, in which persons are lodged or received for a consideration for a single day or night or longer period.

Bungalow court. Group of one-story detached buildings used or intended for use as dwellings grouped about a plot or parcel of land and a portion of which is used or intended to be used in common by the inhabitants of such dwellings, each building thereof having its own separate toilet and bath.

Cluster housing. Housing perceived as a complex of closely related structures.

Coach house. A structure separated from the principal building on a lot and containing a dwelling unit. Constructed in conjunction with, though not necessarily above, storage space that could be utilized by automobiles.

Combined rooms. Two or more adjacent habitable spaces that by their relationship, planning, and openness permit their common use.

Communal kitchen. A kitchen within a dwelling building used by the occupants of more than one dwelling unit, or shared by any persons residing in a rooming occupancy.

Condominium. An estate in real property consisting of an undivided interest in common in a portion of a parcel of real property together with a separate interest in airspace.

Condominium dwelling. A privately owned dwelling unit in a multiple dwelling structure, consisting of several rooms to be used primarily by a single family.

Convalescent home. Building defined in the local zoning ordinance and premise in which sick, injured, or infirm persons are housed or intended to be housed for compensation.

Culinary or cooking facilities. Space in a dwelling arranged, intended, designed, or used for the preparation of food for a family. Facilities may include a sink, stove, cabinets, and refrigerator, or any combination of these arranged in such space. A refrigerator alone does not constitute culinary or cooking facilities under this definition.

Density. The number of dwelling units per unit of land measure.

Detached dwelling. A dwelling entirely surrounded by open space on the same lot.

Detached house. One that has yard areas on all four sides.

Detached single-family dwelling. Single-family dwelling surrounded by yards or other open spaces.

Discriminating and discriminate. To segregate, separate, exclude, or treat any person unequally only because of race, color, religion, ancestry, or national origin.

Dormitory. A space in a unit where group sleeping accommodations are provided, with or without meals, for persons not members of the same family group, in one room, or in a series of closely associated rooms under joint occupancy and single management, as in college dormitories, fraternity houses, military barracks, and ski lodges.

Duplex. Building designed and/or used exclusively for residential purposes and containing two dwelling units separated by a common party wall or otherwise structurally attached.

Duplex dwelling. Single dwelling, providing housekeeping units for not more than two families with no interconnection between the two units, except that it may have a single entrance; all other exterior characteristics are that of a one family dwelling. Two single housekeeping units connected by a breezeway or corridor are not classified as a duplex.

Dwelling. House, apartment, or building, including dormitories, fraternities, and sororities, used primarily for human habitation. The word "dwelling" does not include hotels, motels, tourist courts, or other buildings for transients.

Dwelling bathroom. A bathroom must be planned to accommodate three fixtures that include a bathtub or shower, lavatory or wash basin, and a water closet.

Dwelling building. A building used or designed or intended to be used, all or in part, for residential purposes.

Dwelling building or structure. A building or structure used or designed or intended to be used, all or in part, for residential purposes.

Dwelling duplex. A residence building designed for, or used as the separate homes or residences of two separate and distinct families, but having the appearance of a single-family dwelling house. Each individual unit in the duplex shall comply with the requirements of the local zoning ordinance for a one-family dwelling.

Dwelling group. Group of two or more detached single dwellings located on a parcel of land in one ownership having a yard or court in common.

Dwelling house. Detached house designed for and occupied exclusively as the residence of not more than two families each living as an independent housekeeping unit.

Dwelling living room. Room designed for living use. When it is used only as a living room and space is provided elsewhere for cooking and eating, this room must have the minimum floor area prescribed by the zoning ordinance.

Dwelling room requirements. Every dwelling structure must have not less than a living room, bedroom, kitchen, and dining space and one bathroom.

Dwelling unit. Consists of one or more rooms arranged, designed, or used as living quarters for one family only. Individual bathrooms and complete kitchen facilities are permanently installed.

Efficiency. Accommodations that include kitchen facilities.

Efficiency apartment. A dwelling unit that has only one combined living and sleeping room; this dwelling unit, however, may also have a separate room containing only kitchen facilities and also a separate room containing only sanitary facilities.

Efficiency dwelling unit. A room located within an apartment house or apartment hotel used or intended to be used for residential purposes that combines a kitchen and living and sleeping quarters therein, and that complies with the requirements of the local zoning ordinance.

Efficiency living unit. Any room having cooking facilities and used for combined living, dining, and sleeping purposes.

Efficiency unit. A dwelling unit consisting of one principal room exclusive of bathroom, kitchen, hallway, closets, or dining alcove directly off the principal room.

Elderly housing. Planned multiple shelter of dwelling units as prescribed by the local zoning ordinance under one or more roofs, but upon one premises, designed and erected to be used exclusively for the housing of persons 60 years or more of age, and in which the normal operations of housekeeping and preparation of meals is customarily performed by the tenants or occupants, and provision is made therefore in each dwelling unit, excepting laundry and heating equipment; provided further that one dwelling unit may be occupied without restriction of age by a custodian or manager.

End-row dwelling. Semidetached dwelling.

Factory-built housing. Housing that is partly or totally built in a factory and then transported in sections or as a complete unit to a site where it is erected or stationed and provided with the necessary services to make it a habitable unit that, when occupied, is void of transport features such as wheels, tires, axles, brakes, or lamps.

Family. Consists of one person living individually or a group of persons living as a single household unit, using common housekeeping facilities, not to include, however, more than three persons unrelated by blood, marriage, or adoption.

Family unit. Room or group of rooms designed or used as a housekeeping unit for a family or for a group of not more than 10 persons other than a family and providing living, sleeping, eating, cooking, and sanitation facilities complying with the requirements of the local zoning ordinance.

Floor area. The area included within the surrounding exterior walls of a building or portion thereof, exclusive of vent shafts and courts. The floor area of a building, or portion thereof, not provided with surrounding exterior walls is the usable area under the horizontal projection of the roof or floor above.

Four-family dwelling. Building designed for or occupied exclusively by four families.

Fraternity or sorority building. Building, rented, occupied, or owned as a place of residence by a general or local chapter of some regularly organized college fraternity or sorority, or by or on its behalf by a building corporation or association composed of members of the local chapter of such fraternity or sorority.

Fraternity or sorority house. Building containing no more than one dwelling unit and more than two rooming units or guest rooms. Such rooming units or guest rooms are for residential purposes only.

Garage apartment. Accessory building, not a part of or attached to the main building, a portion of which contains living quarters and space for at least one automobile.

Garden apartment. Group of two or more multiple dwellings not over two stories in height, located on the same lot, that offer each dwelling unit direct access to an open yard area.

Garden-type apartments. Group of dwelling units on one lot in one or more buildings, connected or detached from one another, no one of which exceeds two stories and an attic in height, each unit having separate front and rear exterior entrances opening directly into the yard area.

Group dwelling. Two or more one-family, two-family, or multiple dwellings, apartment houses, or boarding or rooming houses, located on the same lot.

Group house. A group of attached single-family dwellings, separated from each other by vertical walls, but designed as a single structure not more than two rooms deep.

Guest. In connection with single-family and two-family occupancies means a person sharing single-family accommodations without profit on these accommodations.

Guest house. Living quarters within an accessory building for the sole use of persons employed on the premises or for temporary use by guests of the occupants of premises, which living quarters having no kitchen facilities and are not rented or otherwise used as a separate dwelling.

Guest house or cottage. A separate dwelling structure located on a lot with one or more main dwelling structures and used for the housing of guests or servants of the occupant of the premises; such building cannot have a kitchen and cannot be rented, leased, or sold separately from the rental, lease, or sale of the main dwelling.

Guest or servant's house. A detached accessory building located on the same zoning lot as the principal building and containing living quarters for temporary guests or servants.

Guest room. Any habitable room except a kitchen, designed or used for occupancy by one or more persons and not in a dwelling unit.

Habitable attic. A habitable attic is an attic that has a stairway as a means of access and egress and in which the ceiling area at a height of $7\frac{1}{3}$ ft above the attic floor is not more than $\frac{1}{3}$ the area of the floor next below.

Habitable room. Room or enclosed floor space in a basement, first, or upper story arranged for living, eating, or sleeping purposes, including bath or toilet rooms, but not including laundries, pantries, foyers, or communicating corridors or cellar recreation rooms, providing means of light and ventilation complying with the local zoning ordinance.

Habitable space. Space occupied by one or more persons for living, sleeping, eating, or cooking. Kitchenettes are not classified as habitable spaces.

Half-bath toilet room. Enclosed space, containing one water closet and a lavatory.

Half story. Story finished as living accommodations located wholly or partly within the roof frame and having a floor area at least half as large as the story below. Space with not less than 5 ft clear headroom at its lowest height considered floor area.

High-rise apartments. Residential structure four or more stories in height containing four or more separate dwelling units each with its own access to a common stairwell or elevator well.

Home. Place of human habitation.

Hotel. A building or part thereof in which rooms are available for hire as lodging to the general public, which rooms are occupied transiently and singly, rather than as dwelling units.

House. Building used for living, dining and sleeping quarters and the usual kitchen accessories for the occupants, such as refrigerator, stove, sink, and cabinets.

Household unit. Room or suite of rooms equipped with, or containing one or more facilities for cooking food and which room or suite of rooms is occupied as a residence of a single family, individual, or group of individuals.

Housekeeping residential use. All residential uses except those used for transient purposes.

Housekeeping unit. Room or combination of rooms containing living, sleeping, and kitchen facilities for one family.

House or household. Includes a dwelling house, lodging house, and also a students' residence, fraternity house, or other buildings in which any persons in attendance as students, pupils, or teachers, or employees in any capacity in or about a university, college, school, or other institution of learning resides or is lodged.

Housing. Any improved property used or occupied or intended to be used or occupied as a home or residence.

Housing for the elderly. New construction of multiple-family units or remodeling of existing multiple-family dwelling units, the occupation of which shall be limited to persons 62 years of age or more; provided that if two or more persons occupy a unit, at least one shall be 62 years of age or more, with a maximum annual income of such amount as may be determined by the senior citizen board by resolution.

Housing space. That portion of a multiple dwelling rented or offered for rent for living or dwelling purposes in which cooking equipment is supplied; includes all privileges, services, furnishings, furniture, equipment, facilities, and improvements connected with the use or occupancy of such portion of the property. The term does not include public housing or dwelling space in any hotel, motel, or established guest house, commonly regarded as a hotel, motel or established guest house, as the case may be in the local zoning ordinance.

Human occupancy. Use of a space or spaces in a building or structure in which any human or humans lives, works, or remains for continuous periods of time exceeding two or more hours.

Kitchen. A room used or adapted for cooking and containing a stove, range, hot-plate, or other cooking appliances, which burns coal, oil, gas, or other fuel or is heated by electricity.

Kitchenette. A space having a minimal floor area as defined in the local zoning ordinances, used for the preparation and cooking of food.

Landscaping. The provision of plantings and related improvements for the purpose of beautifying and enhancing a property and for the control of erosion and the reduction of glare, dust, and noise.

Let. To give another person the right to occupy any portion of a dwelling, family unit, or rooming unit. The act of "letting" is a continuing act for so long as the person given the right to occupy premises continues to do so. A further "letting" by any occupant of a dwelling, family unit, or rooming unit is, for the purpose of the local zoning ordinance, also a "letting" by the owner or operator of the dwelling.

Livability space. Portion of open space not devoted to motor vehicle parking or circulation that is landscaped, or improved as outdoor living space or recreation space for occupants of the premises. Public street and alley rights-of-way included in the computation of land area are not included as livability space. Beneficial open space included in the computation of land area may be included as livability space subject to the above limitations and provided that such space is directly accessible to occupants of the adjacent premises and is available for their leisure-time use.

Livability space ratio. For the purpose of site planning the minimum square foot of nonvehicular outdoor area that must be provided for each square foot of total floor area. The LSR is found by dividing the open space minus the open car space minus one-half the ground area of any open carports by the total floor area.

Livable room. Any room used for normal living purposes in a residence structure; does not include kitchens, laundry rooms, bathrooms, or storerooms.

Living room. Principal room designed for general living purposes in a dwelling unit. Each dwelling unit must have a living room.

Living unit. Residential unit providing complete, independent living facilities for one family including permanent provisions for living, sleeping, eating, cooking, and sanitation.

Lot coverage. The area of land covered by a building on a particular site. Lot coverage is the percentage of net lot area covered by the gross floor area of the first floor.

Main building. Any building having the predominant land use that is not an accessory building.

Member of a household. Person residing, boarding, or lodging in a house.

Migrant labor housing. One or more buildings or structures, trailers, or vehicles, together with the land appertaining thereto, established, operated, or used as living quarters for seasonal, temporary, or migrant workers, provided they are public lodging establishments, are not located within the farm, and the laborer pays rent. This definition does not apply to forestry or tobacco farm operation.

Minimum habitable room height. Clear height from finished floor to finished ceiling of not less than $7\frac{1}{2}$ ft, except that in attics and top half-stories, the height must be not less than $7\frac{1}{3}$ ft over not less than $\frac{1}{3}$ the floor area when used for sleeping, study, or similar activity.

Minimum habitable room size. No room may have a dimension less than 8 ft and must have a minimum total area as prescribed in the local zoning ordinance and building code between enclosing walls or partitions, exclusive of closet and storage spaces.

Minimum use requirement. In no case may a room, suite, or group of rooms comprising a family dwelling unit, in any dwelling, be so occupied as to provide less than 800 cu ft of airspace per occupant, exclusive of cubic airspace of bathroom, toilet rooms, closets, stairways, attics, utility rooms, and basements. No bedroom or room used as a bedroom, in any dwelling, may be so occupied as to provide less than 300 cu ft of airspace per occupant, exclusive of the cubic airspace of bathrooms, toilet rooms, and closets.

Model home. A one-family dwelling having all of the following characteristics: Dwelling is constructed upon a proposed lot previously designated as a model home site by the advisory agency in a subdivision for which the advisory agency has approved or conditionally approved a tentative map but for which a final map has not yet been recorded, and the proposed lot upon which the model home is constructed is recognized as a legal building site for the duration of the model home, constructed in accordance with the local zoning ordinance, and building code.

Multifamily apartment house. Any building or portion thereof used as a multiple dwelling for the purpose of providing three or more separate dwelling units with shared means of egress and other essential facilities.

Multifamily dwelling. Building or structure designed, erected, occupied, or intended to be occupied as living quarters for more than one family, and having cooking facilities for each occupancy.

Multifamily house. A building occupied as the home or residence of individuals, families, or households living independently of each other, of which four or more are doing cooking within their apartments, including a tenement house, apartment house, or flat. A row of four or more single-family houses not separated by fire walls is considered a multifamily house.

Multifamily residence. Building designed and/or used exclusively for residential purposes and containing more than one dwelling unit.

Multiple dwelling. Building or portion thereof designed for and occupied by more than two families including tenement houses, row houses, apartment houses, and apartment hotels.

Multiple family dwelling. Building, or portion thereof, other than a commercial or residential hotel, that contains three or more dwelling units regularly used or available for permanent occupancy only by households or families. A row of town houses is construed as a multiple family dwelling.

Multiple family housing development or project. Three or more single-family buildings, or more than one two-family buildings or more than one multiple family building on a building site, or any combination thereof.

Multiple unit dwelling. Building so constructed, altered, or used as to provide accommodation for more than one family to dwell in separately.

Net floor area. For the purpose of determining the number of persons for whom exitways are to be provided, net floor area is the actual occupied area, not including accessory unoccupied areas or thickness of walls.

Nonhabitable space. Space used as a kitchenette, pantry, laundry, closet, bath, toilet, dressing, locker, storage, utility, heater, or boiler room; and other spaces used only for service and maintenance of the building; and those spaces used only for access and vertical travel between stories.

Nonresidential usable floor area. Measurement of usable floor area for nonresidential uses is the sum of the area of the first floor, as measured to the exterior face of the exterior walls, plus that area, similarly measured, of all other stories that are accessible by a fixed stairway, ramp, escalator, or elevator that may be made fit for occupancy; the measurement includes the floor area of all accessory buildings measured similarly.

Nontransient. Person resident in the same building for a period of 30 days or more.

Occupant. A person over one year of age, living, sleeping, cooking, or eating in, or having actual possession of, a dwelling unit or a room used for rooming occupancy.

Occupiable room. A room or space, other than a habitable room designed for human occupancy or use, in which persons may remain for a period of time for rest, amusement, treatment, education, dining, shopping, or similar purposes or in which occupants are engaged at work.

Occupied area. Any place or area in which any person is, shall be, or may be required to inhabit, abide, or sojourn.

One-bedroom unit. A dwelling unit containing a minimum floor area as prescribed in the local zoning ordinance, consisting of not more than two rooms in addition to kitchen, dining, and necessary sanitary facilities, and for the purposes of computing density is considered a two room unit.

One-family detached dwelling. A one-family dwelling having a side yard on each side thereof.

One-family dwelling. Detached building arranged, intended, or designed to be occupied, or which is occupied, by not more than one family and that has not more than one kitchen.

One-family house. Building rented, leased, let or hired out to be occupied or is occupied or is intended, arranged, or designed to be occupied as the home or residence of not more than one family. All such buildings whether built singly, or in conjunction with others as double houses or terraces or attached or semidetached rows, are considered one-family houses when each such house complies with the definition of "building."

One-room apartment. Apartment containing one livable room.

Owner. Any person who has legal title to any dwelling, with or without accompanying actual possession thereof; or, who has equitable title and is either in actual possession or collects rents therefrom. Or, who as executor, executrix, trustee, guardian, or receiver of the estate of the owner, or as mortgagee or as vendee in possession either by virtue of a court order or by agreement or voluntary surrender of the premises by the person holding the legal title, or as collector of rent, has charge, care or control of any dwelling or rooming house, or any person who is a lessee or assignee subletting or assigning any part or all of any dwelling may have joint responsibility over the portion of the premises sublet or assigned.

Parties in interest. All individuals, associations, and corporations who have interests of record in a multiple dwelling, and who are in actual possession thereof; and any person authorized to receive rents payable for housing space in a multiple dwelling.

Patio. An open area or an accessory outdoor living structure not exceeding 14 ft in height and open on at least one side.

Patio home. An attached or detached single-family dwelling constructed with no side yard on one or more sides of the lot.

Permanent guest. A person who occupies or has the right to occupy a hotel accommodation, boarding house, or lodging house as his or her domicile and place of permanent residence.

Permanent occupancy. Rental of housing accommodations or rooms on a week-to-week, month-to-month, or year-to-year basis with a fixed rent for each period of occupancy.

Place of abode. Building or part of a building, such as apartment building, row house, rooming house, hotel, or dormitory.

Plumbing fixtures. Within each living unit the following plumbing fixture must be provided: A kitchen sink and a stove properly located to facilitate food preparation and dishwashing; a water closet, a built-in bathtub and shower, and a lavatory located in the bathroom. Each of the plumbing fixtures must be permanently installed and connected to an approved plumbing system.

Premises. Land, improvements thereon, or any part thereof.

Private. As applied to an exit way, toilet room, or other part of a building, means that such exit way or other part of the building is an adjunct to not more than one room or one suite of rooms.

Private dwelling. A dwelling occupied by one family and so designed and arranged as to provide cooking and kitchen accommodations for one family only.

Private residence. Separate dwelling or a separate apartment in a multiple dwelling occupied only by the members of a single family unit.

Public. As applied to an exit way, toilet room, or other part of a building, means that such exitway or other part of the building is subject to common use by those who occupy or enter the building and includes those parts of the building that are not included within the means of "private."

Public hallway. A public corridor or space separately enclosed that provides common access to all the exitways of the building in any story.

Public space. A legal open space on the premises, accessible to a public way or street, that abuts the premises, such as yards, courts, or open spaces permanently devoted to public use.

Public view. Any premises, or any part thereof, or any building or any part thereof, that may be lawfully viewed by the public, or any member thereof, from a sidewalk, street, alleyway, licensed open-air parking lot, or from any adjoining or neighboring premises.

Rental room. Room rented for permanent, not transient residence, as evidenced by a rental charge on a weekly or monthly basis, but not permitting preparation or cooking of food.

Residence. A detached part or whole of a building occupied exclusively as a family dwelling place and the usual accessory occupancies.

Residence building. Building in which sleeping accommodations are provided.

Residence hall. Dormitory, as applied to a building.

Residential accommodations. Any building or part of a building used or intended to be used for sleeping accommodations by a person or group of persons. Other housekeeping facilities may be provided.

Residential building. A building, or parts thereof, designed or used for one or more family units or designed or used for sleeping accommodations.

Residential dwelling unit. Building or portion of a building planned, designed, or usable as a residence for one family only, living independently of other families included in said unit. (For example, a one-family dwelling and four-family dwelling, each apartment in a four-family dwelling, each apartment in an apartment house, and each unit of a condominium or townhouse.)

Residential floor areas. Sum of the gross horizontal areas of the several floors of the dwelling, exclusive of garages, cellars, and open or roofed porches, measured from the exterior faces of the exterior walls of a dwelling.

Residential occupancy. Any building or part of a building in which a person or group of persons are provided with sleeping accommodations. Other housekeeping accommodations may also be provided.

Residential purposes. Any building used for residential purposes; one- or two-family residences, apartment houses, and multiple dwellings.

Residential structure. Any building or part of a building constructed with or as sleeping accommodations for a person or group of persons. Other housekeeping accommodations also may be provided.

Residential unit. Room or suite of rooms, with or without cooking facilities, occupied as the residence of a single individual, family, or group of individuals.

Residential usable floor area. Measurement of "usable floor area" for residential uses is in the sum of the area of the first floor, as measured to the exterior face of the exterior walls, plus that area, similarly measured, of all other stories, having more than 90 in. of headroom, that are accessible by affixed stairway and that may be made usable for human habitation; but excluding the floor area of uninhabitable basements, cellars, garages, accessory buildings, attics, breezeways, and unenclosed porches.

Residential use. Includes single and multiple dwellings, hotels, motels, dormitories, and mobile homes.

Room. Space within a building completely enclosed with walls, partitions, floor, and ceiling, except, when used for habitation, to be provided with the necessary openings for light and ventilation.

Room capacity. For determining the exits required, the minimum number of persons, or the occupant content of any floor area may in no case be taken less than specified in the local building code.

Room—habitable. An enclosing subdivision in a residential building commonly used for living purposes, but not including any lobby, hall, closet, storage space, water closet, bath, toilet, janitor's closet or general utility room or service porch.

Room height. Minimum clear height of a habitable room must be 96 in., including clear height under beams or other projections from the ceiling. In rooms with sloping ceilings, the minimum clear height of 84 in. is required for only 40% of the room area, and where side wall height is not less than 60 in., the area beyond that height cannot be made part of a habitable room.

Rooming house. Any dwelling or part thereof in which space is let by the owner, operator or occupant of one or more rooming units to three or more persons who are not husband or wife, son or daughter, mother or father, grandparents, grandchildren, sister or brother or niece or nephew of the owner or operator or tenant or his spouse of any of these, but any child lawfully under the care of any of the above members of a family is not considered a roomer.

Rooming occupancy. The term describes a use wherein a room or group of rooms in a dwelling building are used or intended to be used for sleeping, but not for cooking or eating purposes, provided, however, that where one or two roomers are so accommodated within a dwelling unit occupied by a family, the accommodations so provided are not considered a rooming occupancy.

Rooming unit. Any room, or group of rooms, forming a single habitable unit used for living and sleeping but that does not contain cooking or eating facilities.

Row dwelling. Dwelling, the walls on two sides of which are party or lot-line walls.

Row house. Attached house in a row or group, each house containing not more than two dwelling units and each house separated from adjoining houses in the same row or group by fire walls or fire separations.

Sanitation facilities. Water closet, lavatory or sink, and shower or bathtub.

Semidetached building. Building that has only one party wall in common with an adjacent building.

Semidetached dwelling. Dwelling, one side wall of which is a party or side lot-line wall.

Semidetached house. Surrounded on three sides by yard area and so constructed that one wall is on a side lot line, and abuts the neighboring house.

Senior citizens housing. Dwellings, including multiple dwellings, owned and operated by an educational, religious, or philanthropic organization, no part of the earnings of which inures to the benefit of any private shareholder, contributor, or individual, having accommodations for and occupied exclusively by persons who are 62 years of age or over.

Servants quarters. Accessory building located on the same lot or grounds with the main building, used as living quarters for servants employed on the premises and not rented or otherwise used as a separate domicile.

Service space. Space provided in a building to facilitate or conceal the installation of building service facilities, such as chutes, ducts, pipes, shafts, or wires.

Shaft. An enclosed space within a building, whether for air, light, ventilation, elevator-dumbwaiter or any other purpose connecting a series of two or more openings in successive floors and covered either by a skylight or by the roof. A vent shaft is used solely to ventilate or light waterclosets, compartments, or bathrooms.

Single-family attached dwelling. Building containing a single dwelling unit and having one or more exterior walls in common with, or contiguous to, another such dwelling. An attached single-family dwelling is capable of being located on a separate lot.

Single-family detached dwelling. Building designed for and occupied exclusively as a residence of only one family and having no party wall in common with an adjacent building.

Single-family dwelling. Building regularly used or available solely for permanent occupancy by one family or household.

Single-family semidetached dwelling. Building designed for and occupied exclusively as a residence for only one family and having a party wall in common with an adjacent one family dwelling.

Single-unit dwelling. Single detached structure having but one dwelling unit with a single kitchen and housing any number of persons immediately related by blood, marriage, or adoption, living together as a single nonprofit housekeeping unit, plus domestic servants employed for services on the premises.

Sleeping accommodations. Room, space, or portion thereof, used primarily for sleeping purposes.

Sleeping room. Room rented for sleeping or living quarters, but without cooking facilities and with or without an individual bathroom. In a suite of rooms not part of a dwelling unit, each room that provides sleeping accommodations is construed as one sleeping room.

Spacing of buildings. The required minimum horizontal distance between any wall of two or more buildings facing or overlapping each other in any manner either parallel or oblique. Such distance is measured at any given point and any given level by projecting or prolonging vertically and horizontally the perimeter lines of each wall from the lowest habitable floor to the ceiling of the highest habitable floor.

Split-level dwelling. A one- or two-family dwelling having part floors at staggered levels, so arranged to have a habitable room or rooms both above and below the first floor level.

Student cooperative housing. Facility used for housing students who therein largely perform their own household maintenance and meal preparation and who have a vote in the management of their household affairs. Such housing is associated with and recognized by the university and supervised by it in relation to membership, resident supervision, and operations, in the same fashion as fraternities and sororities are supervised.

Suite. A group of habitable rooms designed as a unit, and occupied by only one family, but not including a kitchen or other facilities for the preparation of food, with entrances and exits that are common to all rooms comprising the suite.

Suite of rooms. Two or more rooms that are arranged to be used as a unit and each room of which is accessible from within the unit without the use of a public exit way.

Tenement house. Any house or building or portion thereof that is rented, leased, let or hired out to be occupied or is occupied as the home or residence of three or more families, living independently of each other and doing their cooking upon the premises.

Terrace family dwelling. Building containing three or more dwelling units arranged side by side, separated from each other by a fire wall and having separate means of egress and ingress from the outside.

Terraces. A group of two or more, but not exceeding five single-family dwellings separated by walls without openings, not more than two rooms deep.

Three-family dwelling. Building in which three living units exist.

Three or more bedroom units. A dwelling unit wherein for each room in addition to the three rooms permitted in a two-bedroom unit, there is provided an additional area of 200 sq ft to the minimum floor area of 700 sq ft. For the purpose of computing density, said three bedroom unit is considered a four-room unit and each increase in a bedroom over three is an increase in the room count by one over the four.

Toilet room. Room containing one watercloset; one lavatory, urinal, and other plumbing fixtures.

Townhouse. A single-family dwelling unit constructed in a series or group of attached units with property lines separating such units.

Townhouse dwelling. Single family dwelling forming one of a group of two or more attached single-family dwellings separated from one another by party walls without doors, windows, or other provisions for human passage or visibility through such walls from basement or cellar.

Two-bedroom unit. Dwelling unit containing a minimum floor area of at least 700 sq ft per unit, consisting of not more than three rooms in addition to kitchen, dining, and necessary sanitary facilities, and for the purposes of computing density considered as a three-room unit.

Two-family detached dwelling. Building designed for and occupied exclusively as a residence for two families with one family living wholly or partly over the other and having no party wall in common with an adjacent building.

Two-family dwelling. Separate, detached building designed for or occupied exclusively as a residence by two families.

Two-family dwelling duplex. Detached building designed for or occupied exclusively by two families independently of each other.

Two-family dwelling or duplex. Detached building under one roof, arranged, intended or designed to be occupied by not more than two families, each apartment extending from basement to roof of the structure with a separate entrance to each dwelling though there may be but one front door and vestibule or hall.

Two-family house. A building rented, leased, let or hired out to be occupied or is occupied or is intended, arranged, or designed to be occupied as the home or residence of not more than two families. All such buildings whether built singly, or in conjunction with others as double houses or terraces or attached or semidetached rows, are two family houses when each such house complies with the definition of "building."

Unit.   One or more rooms arranged for the use of one or more individuals living together as a single housekeeping unit, with cooking, living, sanitary, and sleeping facilities.

PHYSICALLY HANDICAPPED, DISABLED, AND AGED

Access.   The appropriate spaces between walls, partitions, and objects facilitating capacity for the aged and the disabled to approach and enter.

Access aisle.   A pedestrian space between elements such as parking spaces, seating, and desks.

Accessible route.   A continuous unobstructed path connecting accessible elements and spaces in a building or facility.

Accommodate.   To provide the necessary requirements to make suitable for use or fit.

Aging.   Those manifestations of the aging processes that significantly reduce mobility, flexibility, coordination, and perceptiveness.

Ambulatory person.   Person who has the capacity and the ability to walk without difficulty or assistance.

Americans with Disabilities Act.   Purpose of the ADA is to extend to people with disabilities the same civil rights as stated in the Civil Rights Act of 1964, which is available to all citizens, regardless of race, color, sex, national origin, or religion.

Appropriate number.   Number of a specific item that would be necessary, in accordance with the purpose and function of a building or facility, to accommodate individuals with specific disabilities in proportion to the anticipated number of individuals with disabilities who would use a particular building or facility.

Automatic door.   A door for human passage equipped with a power-operated mechanism and controls that open and close the door upon receipt of a momentary actuating signal.

Barrier-free.   Spaces, buildings and facilities fully accessible to and usable by all people, including individuals with impairments to aspects of physical performance.

Building or facility.   All or any portions of buildings, structures, equipment, roads, walks, parking lots, parks, sites, or other real property, or interest in such facility or property.

Common areas.   Those interior and exterior spaces available for use by all occupants and users of a building or facility, exclusively of any spaces that are made available for the use of a restricted group of people or the use of which is restricted to particular functions.

Controls and switches.   Switches and controls for light, heat, ventilation, windows, draperies, fire alarms, and all similar controls of frequent or essential use, placed within the established standards for the reach of individuals in wheelchairs.

Corridors.   Corridors should be not less than 4 ft wide. The end of a corridor should have an area of not less than 5 sq ft to permit a wheelchair to be turned around.

Cross slope.   The incline that is perpendicular to the direction of travel.

Cue.   A signal warning of hazardous conditions, indicated by tactile, aural, or visual means.

Curb ramp.   A short ramp cutting through a curb or built up to it.

Detectable.   Item or object that can be identified without lengthy explanations.

Developmentally disabled.   Person having autism, cerebral palsey, epilepsy, or mental retardation.

Disability.   Any physiological disorder or condition, cosmetic disfigurement, or anatomical loss affecting one or more of the following bodily systems: neurological, musculloskeletal, special sense organs, respiratory, (including speech organs), cardiovascular, reproductive, digestive, genito-urinary, hemic and lymphatic, skin, and endocrine.

Disabilities of incoordination.   Faulty coordination or palsy from brain, spine, or peripheral nerve injury.

Door hardware.   Doors that are intended for normal use and for the handicapped should have the door handles or door knobs knurled.

Doors.   Doors should have a clear width of not less than 2 ft, 8 in. when open. In the case of double doors without a mullion, one of the pair should meet the 2-ft, 8-in. width requirement. The floor should be level for a distance of 5 ft from the door in the direction of the door swing and extend 1 ft to the side of the latch jamb of the door. The threshold should have a minimum rise and an incline to facilitate wheelchair travel.

Doors and doorways.   Doors should have a clear opening of no less than 32 in. when open and be operable by a single effort.

Drinking fountains.   An appropriate number of drinking fountains or other water-dispensing means should be accessible to, and usable by physically handicapped persons. A wall-mounted or semirecessed (a fully recessed fountain is not acceptable) drinking fountain or cooler should have a spout and control at the front of the unit, with the basin located not more than 3 ft above the floor. A floor model water cooler should have a side-mounted fountain 2 ft, 6 in. above the floor, accessible to a person in a wheelchair.

Egress or means of egress.   A continuous and unobstructed way of exit travel from any point in a building or facility to an exterior walk or out of a fire zone. It includes all intervening rooms, spaces, or elements.

Element.   An architectural or mechanical component of a building, facility, space, or site, e.g., telephone, curb ramp, door, drinking fountain, seating, and water closet.

Elevators

1. In multiple-story buildings elevators are essential to the successful functioning of physically disabled individuals.

2. Elevators should be accessible to and usable by the physically disabled on the level that they use to enter the building and at all levels normally used by the general public.

3. At least one elevator should be provided in a multi-story building to serve the floor containing the entrance and all other floors accessible to the public. The elevator cab should have a clear area of not less than 25 sq ft. However, an elevator cab with a clear area of not less than 23 sq ft and a clear length or width of not less than 3 ft, 8 in. may be used in a building that is not more than three stories in height

and would not normally have an elevator. The elevator door should have a minimum clear width of 2 ft, 8 in. when open. Braille plates (metal, plastic, or other suitable material) should be provided for floor selection, and other controls, such as open door, stop car, and emergency signal, should be adjacent to standard floor and control buttons within the cab. The maximum height of floor and control buttons should be 60 in. above the floor with a banked button arrangement where necessary. Braille plates (metal, plastic, or other suitable material should be provided for floor designation on each floor, 5 ft above the floor, on the fixed jamb at the open side of the elevator door.

4. All provisions should be provided for the deaf and hard of hearing by the installation of light signals, flashing or otherwise.

Entrances. At least one primary entrance to each building should be usable by individuals in wheelchairs, avoiding abrupt changes in levels and with thresholds flush with the floor. At least one entrance usable by individuals in wheelchairs should be on a level that makes the elevators accessible.

Essential features. Those elements and spaces that make a building or facility usable by, or serve the needs of, its occupants or users. Essential features include, but are not limited to, entrances, toilet rooms, and accessible routes. Essential features do not include those spaces that house the major activities for which the building or facility is intended, such as classrooms and offices.

Exterior accessible routes. May include, but are not limited to, parking access aisles, curb ramps, walks, ramps, and lifts.

Facilitate. To provide the conveniences for the handicapped, disabled, and aged, to make it comfortable to get about.

Facility for handicapped persons. The facility should include any ramp, handrail, elevator, or door with a specially treated surface and similar design. It should be convenient and should facilitate the health, safety, comfort, social and recreational needs and desires of the physically handicapped person.

Fixed turning radius—front structure to rear structure. Turning radius of a wheelchair, left front foot platform to right rear wheel, or right front foot platform to left rear wheel, when pivoting on a spot. Layouts provided to architects by the U.S. agency.

Fixed turning radius—wheel-to-wheel. Tracking of the caster wheels and large wheels of a wheelchair when pivoting on a spot. Layouts provided to architects by the U.S. agency.

Floor levels. Areas on the same floor should have a common level or be connected by a ramp.

Floors. Floors should have a nonslip surface.

Governmental discrimination. "No qualified individual with a disability shall, by reason of such disability, be excluded from the participation in, be denied the benefits of, or be subjected to discrimination by a department, agency, special purpose district, or other instrumentality of a state or local government."

Grading. Grading of ground, even contrary to existing topography, so that it attains a level with a normal entrance, making a facility accessible to individuals with physical disabilities.

Handicapped. Persons with reduced mobility, flexibility, coordination, or perceptiveness due to age or physical conditions.

Handicapped persons' use of public buildings. All buildings regulated by the local building code and that are owned by the state or any political subdivision thereof, which are open to and used by the public, must facilitate the reasonable access and use by all handicapped persons as required under the local building code. Lodging facilities are in compliance with these regulations if 10% of the units are accessible to handicapped persons. If all required units for the handicapped are located on one floor, other stories of the building need not comply. In these required units, switches and controls for lights, heating, and ventilating equipment, windows, draperies, fire alarms, and all similar controls of frequent or essential use required by the public, should be placed within reach of persons in wheelchairs and of a type easily operated by handicapped persons. Utility receptacles should be located at least 18 in. above the floor. Toilet rooms in these units should have space in special stalls to allow in and out movement of persons in wheelchairs.

Hearing disabilities. Deafness or hearing handicaps that might make an individual insecure in public areas because he is unable to communicate or hear warning signals.

Hemiplegia. Full or partial paralysis of one side of the body caused by brain damage, most often due to disease, stroke, or trauma.

Identification for the blind. Appropriate identification of specific facilities within a building used by the public is particularly essential to the blind. Raised letters or numbers should be used to identify rooms or offices.

Individual functioning on crutches. On the average, individuals 5 ft 6 in. tall require an average of 31 in. between crutch tips in the normally accepted gaits. On the average, individuals 6 ft 0 in. tall require an average of 32.5 in. between crutch tips in the normally accepted gaits.

Interior accessible routes. May include, but are not limited to, corridors, floors, ramps, elevators, lifts, and clear floor space at fixtures.

Involvement. Portion(s) of the human anatomy or physiology, or both, that have a loss or impairment of normal function as a result of genesis, trauma, disease, inflammation, or degeneration.

Lavatory. The lavatory should be wall mounted with the space under the unit open and unobstructed, have a narrow apron with the bowl near the front to permit use by a person in a wheelchair.

Light fixtures, protruding signs, and door closers. Light fixtures, protruding signs, door closers, and other similar hanging objects or signs and fixtures should be installed not less than 6 ft, 6 in. above the floor.

Light switches. Light switches should be located not more than 4 ft above the floor.

Mental disorder. Term used to indicate disturbances of thinking, feeling, and behaving that may be due to physical or psychological factors.

Mental retardation. Condition causing a person to have significantly below-average general intellectual functioning.

Nonambulatory disabilities. Impairments that, regardless of cause or manifestation, for all practical purposes individuals are confined to wheelchairs.

Operable part. A part of equipment or an appliance used to insert or withdraw objects, to activate or deactivate equipment, or to adjust the equipment.

Paraplegia. Paralysis of the lower half of the body involving the partial or total loss of function of both legs.

Parking lots and building approaches. A parking lot servicing an entrance should have an appropriate number of parking spaces adjacent to a walkway and be identified by signs as reserved for physically handicapped persons. Each reserved parking space should be suitably surfaced for wheelchair travel and not less than 12 ft wide, unless paralleling a walk. Where a curb exists between a parking lot surface and sidewalk surface, an inclined curb approach or a curb cut with a gradient of not more than 1 ft in 3 ft ($33\frac{1}{3}\%$) and a width of not less than 4 ft should be provided for wheelchair access.

Parking space. A parking space should be open on one side, allowing room for individuals in wheelchairs or on braces and crutches to get in and out of an automobile onto a level surface suitable for wheeling and walking.

Perceptual disability. Disorder in one of the one or more of the psychological processes involved in understanding or in using spoken or written language, which may effect one's ability to listen, read, speak, write, spell, or do any type of mathematics.

Physical handicap. Impairment that confines an individual to a wheelchair, causes an individual to walk with difficulty or insecurity, affects the sight or hearing to the extent that an individual functioning in public areas is insecure or exposed to danger, causes faulty coordination, or reduces mobility, flexibility, coordination, and perceptiveness to the extent that special facilities are needed to provide for the safety of that individual.

Public accommodations. "No individual shall be discriminated against on the basis of disability in the full and equal enjoyment of the goods, services, facilities, privileges, advantages, and accommodations of any place of public accommodations" (ADA).

Places of public accommodation. Places of public accommodation include public places of entertainment, amusement, or recreation; all public places where food or beverages are sold for consumption on the premises; all public places that are conducted for the lodging of transients or for the benefit, use, or accommodation of those seeking health or recreation; and all establishments that cater or offer their services, facilities, or goods to or solicit patronage from the members of the general public. Any residential house, residence in which fewer than five rooms are rented, private club, or place that is in its nature distinctly private is not a place of public accommodation.

Power-assisted door. A door used for human passage with a mechanism that helps to open the door or relieve the opening resistance of a door, upon activation of a switch or a continued force applied to the door itself.

Primary entrance. One primary entrance to the building should be accessible from the parking lot or the nearest street by way of a walk uninterrupted by steps or abrupt changes in level and have a width of not less than 5 ft and a gradient of not more than 1 ft in 20 ft.

Public telephones. An appropriate number of public telephones should be accessible to and usable by the physically disabled, and should not exceed 48 in. above the floor.

Public services. Services and activities of state and local governments, including actions applicable to public transportation provided by public entities.

Quadriplegia. Paralysis of the body involving partial or total loss of function in both arms and legs.

Ramp

1. Smooth, hard, prepared, nonslip sloped surface joining two different levels.
2. The outside or inside ramp should have a gradient of not more than 1 ft in 20 ft with a level platform 5 ft long at the top and bottom, at turns, and at an interval of not more than 30 ft on long runs having a clear width (inside handrails) of not less than 3 ft. The ramp, including platforms, should have handrails 2 ft, 8 in. high on both sides, with the handrail extended 1 ft beyond the top and bottom of the ramp on at least one side. Ramps, including platforms, should have nonslip surfaces.

Ramps with gradients. Ramps with gradients (or ramps with slopes) deviate from what would otherwise be considered the normal level. An exterior ramp, as distinguished from a "walk," is considered an appendage to a building leading to a level above or below existing ground level. As such, the ramp has requirements similar to those imposed upon stairs.

Reaching disability. Limitations in the extension or movement of both arms or in bending and stretching.

Room identification plates. Plate made of metal, plastic, or other suitable material having raised letters or numbers affixed to the wall surface approximately 5 ft above the floor in a horizontal line, adjacent to the latch side of a door.

Running slope. The slope that is parallel to the direction of travel.

Semiambulatory disabilities. Impairments that cause individuals to walk with difficulty or insecurity. Individuals using braces or crutches, amputees, arthritics, spastics, and those with pulmonary and cardiac ills may be semiambulatory.

Sight disabilities. Total blindness or impairments affecting sight to the extent that the individual functioning in public areas is insecure or exposed to danger.

Signage. The display of written, symbolic, tactile, or pictorial information.

Space. A definable area, such as a toilet room, hall, assembly area, parking area, entrance, storage room, alcove, courtyard, or lobby.

Speech impairment. Limited or difficult speech patterns. The presence of a speech impairment does not mean that there is a problem in hearing or mental ability.

Stairs. Outside or inside stairs should have risers not more than $7\frac{1}{2}$ in. and nosing with a radius of not less than $\frac{1}{8}$ in. Stairs should have handrails 2 ft, 8 in. above the stair nosing on both sides. The handrail on the wall side of main landings should extend 1 ft, 6 in. beyond the top and bottom step and the handrails at intermediate landings should be continuous on both sides. Stairs and landings should have nonslip surfaces (verify with OSHA).

Tactile. Perception through the sense of touch stimulated by surface textures and finishes.

Tactile warning. A surface texture applied to or built into walking surfaces or other elements to warn visually impaired persons of hazards in the path of travel.

Telecommunications. "Shall ensure that interstates and intrastate telecommunications relay services are available...to hearing-impaired and speech-impaired individuals in the United States" (ADA).

Telephones. An appropriate number of telephones should be accessible to and usable by physically handicapped persons. Such telephones should be located outside a conventional phone booth and have the dial and handset not more than 4 ft above the floor. An appropriate number of telephones should be equipped to assist persons with a hearing disability and be so designated with instruction for use.

Toilet rooms. An appropriate number of toilet rooms, in accordance with the nature and use of a specific building or facility, should be made accessible to and usable by the physically handicapped.

Urinals. Urinals should have the basin opening not more than 1 ft, 7 in. above the floor and should not be the type with extended shields.

Viewing height. The eye level of the average person in a wheelchair is between 45 and 48 in.

Walk

1. Smooth, hard prepared surface of concrete, bituminous concrete brick, asphalt, stabilized gravel, or other suitable material with the ground immediately adjacent thereto at the same level.

2. Predetermined prepared surface exterior pathway leading to or from a building or facility, or from one exterior area to another, placed on the existing ground level and not deviating from the level of the existing ground immediately adjacent.

Walking aid. Personal device used by the handicapped to assist in walking, such as a cane, crutch, walker, or brace.

Warning signals. Audible warning signals should be accompanied by simultaneous visual signals for the benefit of those with hearing disabilities. Visual signals should be accompanied by simultaneous audible signals for the benefit of the blind.

Water closet. At least one toilet room and the units within such room should be accessible to and usable by physically disabled and handicapped persons. The toilet room should have a clear space, excluding door swing, of not less than 5 by 4 ft to permit a wheelchair to be turned around. There should be a clear width of not less than 4 ft between the face of a water closet stall and a wall. The water closet stall should be not less than 3 ft wide and 4 ft, 8 in. deep and have an outswinging door, or an opening 2 ft, 8 in. wide. Handrails should be provided on both sides of a water closet (on stall partition) and should be not less than 3 ft 6 in. long and mounted 2 ft, 9 in. (lower for children) above and parallel to the floor. The water closet should have a narrow understructure that recedes sharply from the front (the trap should not extend in front of or be flush with the lip of the bowl).

Water fountains. An appropriate number of water fountains or other water-dispensing means should be accessible to and usable by the physically disabled.

C

Technical, Scientific, and Related Data

*The INFORM DIVISIONS presented here have been created by the author and are not in any way connected or related directly or indirectly with the Masterformat, published by the Construction Specifications Institute, Inc.

MATHEMATICS

Abstract number. Number that does not refer to any particular object.

Acute angle. Angle less than a right angle.

Acute triangle. Triangle that has three acute angles.

Algebra. Method of solving mathematical problems by the use of symbols that are usually letters for unknown quantities. Also the manipulation of equations involving numbers and the alphabet.

Altitude of a parallelogram or trapezoid. Perpendicular distance between its parallel sides.

Altitude of a prism. Perpendicular distance between its bases.

Altitude of a pyramid or cone. Perpendicular distance from its vertex to the plane of its base.

Altitude of a triangle. Line from any vertex perpendicular to the opposite side, prolonged if necessary. Every triangle has three altitudes.

Analysis. Process of investigating principles and solving problems independently of set rules.

Angle. Difference in direction of two lines proceeding from the same point called the "vertex."

Arc

1. Any part of a circle.

2. A continuous piece of the circumference of a circle.

Area. Surface included within the lines that bound a plane figure.

Arithmetic. Science of numbers and the art of computation.

Axiom. Statement about quantities in general that is accepted as true without proof.

Axis of symmetry. Line drawn through the center of a balanced equal design.

Base of a triangle. Side on which the triangle may be supposed to stand.

Bisect. To divide into two parts of equal size.

Bisector. A line or plane that bisects an angle or another line.

Calculus. Highly systematic method of treating problems by a special system of algebraic notation.

Chord. Line segment whose end points are on a circle.

Circle. Plane figure bounded by a curved line, called the "circumference," every point of which is equally distant from a point within, called the "center."

Complex fraction. Fraction whose numerator or denominator is a fraction.

Complementary angles. Two angles whose sum is a right angle.

Composite number. Number that can be divided by other integers besides itself and one.

Compound fraction. Fraction of a fraction.

Concentric circles. Circles having the same center.

Concrete number. Number used to designate objects or quantities.

Cone. Body having a circular base and whose convex surface tapers uniformly to the vertex.

Congruence. The property of geometric figures that can be made to coincide by a rigid transformation.

Congruent numbers. Two numbers having the same remainder when divided by a given quantity called the "modulus."

Cube. Parallelopipedon whose faces are all equal squares.

Cubic measure. Measure of volume involving three dimensions: length, breadth, and thickness.

Cylinder. Body bounded by a uniformly curved surface, its ends being equal and parallel circles.

Decimal scale. Scale in which the order of progression is uniformly 10.

Degree. Unit of measure used in measuring angles.

Demonstration. Process of reasoning by which a truth or principle is established.

Denomination. Name of the unit of a concrete number.

Denominator. Term of a fraction that shows the number of equal parts into which the unit is divided.

Diagonal (of a plane figure). Straight line joining the vertices of two angles not adjacent.

Diameter of a circle. Line passing through the center of a circle and terminated at both ends by the circumference.

Diameter of a sphere. Straight line passing through the center of the sphere and terminated at both ends by its surface.

Equal. Like or alike in quantity, value, or degree.

Equilateral triangle. Triangle that has all its sides and angles equal.

Even number. Number that can be exactly divided by two.

Exact divisor of a number. Whole number that will divide a number without a remainder.

Factors. One of two or more quantities that when multiplied together produce a given quantity.

Factors of a number. Numbers that when multiplied together make that number.

Fraction. Number that expresses equal parts of a whole thing or quantity.

Frustum of a pyramid or cone. Part that remains after cutting off the top by a plane parallel to the base.

Geometry. Branch of mathematics that deals with the relations and properties of solids, surfaces, lines, and angles.

Greatest common divisor. Greatest number that will exactly divide two or more numbers.

Hypotenuse of a right triangle. Side opposite the right angle.

Improper fraction. Fraction whose numerator equals or exceeds its denominator.

Integer. Number that represents whole things.

Involution. Multiplication of a quantity by itself any number of times; raising a number to a given power.

Isosceles triangle. Triangle that has two of its sides equal.

Least common multiple. Least number that is exactly divisible by two or more numbers.

Like numbers. Same kind of unit expressing the same kind of quantity.

Logarithm. Mathematical device consisting of the power to which a base number has been raised. For example the logarithm to the base of 10 of 10^2 (10×10) is 2. Thus, the logarithm of 100 is 2, of 1000 is 3, of 10,000 is 4, etc.

Mathematics. Science of quantity.

Measure. That by which extent, quantity, capacity, volume, or dimensions in general are ascertained by some fixed standard.

Median of a triangle. Line drawn from any vertex to the midpoint of the opposite side.

Mensuration. Process of measuring.

Metric system. System built on a foundation of six base units with multiples and submultiples expressed in a decimal system. The base units are given for length in meters (m), time in seconds (s), mass in kilograms (kg), temperature in kelvins (K), electric current in amperes (A), and luminous intensity in candelas (cd).

Multiple of a number. Any number exactly divisible by that number.

Notation. Writing down of figures to express a number.

Number. Unit or collection of units.

Numeration. Reading of numbers or a collection of figures already written.

Obtuse angle. Angle greater than a right angle and less than a straight angle.

Obtuse triangle. Triangle that has one obtuse angle.

Odd number. Number that cannot be divided into two whole numbers.

Parallelogram. Quadrilateral that has its opposite sides parallel.

Parallelopipedon. Prism bounded by six parallelograms, the opposite ones being parallel and equal.

Percentage. Rate per hundred.

Perimeter of a polygon. Sum of its sides.

Perpendicular of a right triangle. Side that forms a right angle with the base; intersection of two lines so that the adjacent angles formed are equal.

Plane figure. Plane surface.

Polygon. Plane figure bounded by straight lines.

Postulate. Any of the assumptions on which a mathematical theory is based.

Power. Product arising from multiplying a number.

Prime factor. Prime number used as a factor.

Prime number. Number exactly divisible by some number other than 1 or itself.

Prism. Solid whose ends are equal and parallel polygons, and whose sides are parallelograms.

Problem. Question requiring a solution.

Proper fraction. Fraction whose numerator is less than its denominator.

Proposition. Statement of a geometric truth.

Protractor. Instrument used in measuring degrees.

Pyramid. Body having for its base a square and for its other sides or facets four equal triangles that terminate in a common point called the "vertex."

Quadrilateral. Plane figure bounded by four straight lines (having four sides) and having four angles.

Quantity. That which can be increased, diminished, or measured.

Radius of a circle. Line extending from the center of a circle to any point on the circumference. It is one-half the diameter.

Radius of a sphere. Straight line drawn from the center of the sphere to any point on the surface.

Rectangle. Parallelogram with all its angles right angles.

Reflex angle. Angle greater than a straight angle and less than two straight angles.

Rhomboid. Parallelogram whose opposite sides only are equal, but whose angles are not right angles.

Rhombus. Parallelogram whose sides are all equal, but whose angles are not right angles.

Right angle. Each of the angles formed by perpendicular lines is called a "right angle."

Root. Factor repeated to produce a power.

Rule. Prescribed method of performing an operation.

Scale. Order of progression on which any system of notation is founded.

Scalene triangle. Triangle that has all of its sides unequal.

Simple fraction. Fraction whose numerator and denominator are whole numbers.

Simple number. Either an abstract number or a concrete number of but one denomination.

Slant height of a pyramid. Perpendicular distance from its vertex to one of the sides of the base.

Sphere. Body bounded by a uniformly curved surface, all the points of which are equally distant from a point within called the "center."

Square. Rectangle whose sides are equal.

Straight angle. A straight angle is an angle whose sides extend in opposite directions from the vertex and from a straight line.

Straight line. A straight line has length but no width or thickness.

Supplementary angles. Two angles whose sum is a straight line.

Theorem. Proposition to be proved.

Trapezium. Quadrilateral having no two sides parallel.

Trapezoid. Quadrilateral, two of whose sides are parallel and two oblique.

Triangle. Plane figure bounded by three sides and having three angles.

Uniform scale. Scale in which the order of progression is the same throughout the entire succession of units.

Unit. Single thing or definite quantity.

Unity. Unit of an abstract number.

Unlike numbers. Different kinds of units, used to express different kinds of quantity.

Varying scale. Scale in which the order of progression is not the same throughout the entire succession of units.

Vertical angles. When two straight lines intersect, a pair of nonadjacent angles are formed.

NATURAL GROWTH WOODS

Aggregate ray. Composite structure consisting of a number of small rays, fibers, and sometimes vessels also, which to the unaided eye or at low magnification appear as a single broad ray.

Alternate pitting. Type of pitting in which bordered pits are arranged in diagonal rows across the cell; when crowded, pits become polygonal in surface view.

Anistropic

1. Not isotropic, that is, not having the same properties in all directions. In general, fibrous materials such as wood are anisotropic.

2. Exhibiting different properties when tested along axes in different directions.

Annual growth. Layer of wood laid down during a given year. It is the same as an annual or seasonal increment.

Annual growth ring

1. Layer of wood consisting of many cells, formed around a tree stem during a single growing season. It is seen most easily in cross section.

2. Growth layer put on in a single growth year, including springwood and summerwood.

3. When exposed by conventional methods of sawing, growth rings provide the grain or characteristic pattern of the wood. The distinguishing features of the various species are thereby enhanced by the differences in growth ring formation.

4. Most species grown in temperate climates produce well-defined annual growth rings, which are the result of the difference in density and color between wood formed early and wood formed late in the growing season. The inner part of the growth ring formed first is called "spring wood," and the outer part formed later in the growing season is called "summer wood."

5. Growth layer as viewed in a cross section of a stem, branch, or root.

6. Layer of wood growth put on a tree during a single growing season. In the temperate zone the annual growth rings of many species (such as oaks and pines) are readily distinguished because of differences in the cells formed during the early and late parts of the season. In some temperate zone species (black gum and sweetgum) and many tropical species, annual growth rings are not easily recognized.

Annual ring. Annual increment of wood as it appears on a transverse surface or in a transverse section.

Annular. Ring shaped. The term is often used to describe annual growth rings of a tree.

Bark

1. Outer layer of a tree comprising the inner bark or thin, inner living part (phloem) and the outer bark, or corky layer, composed of dry, dead tissue.

2. Tissues in the cylindrical axis of a tree outside of the cambium. Bark is composed of inner living bark and outer dead brown bark.

Bark pocket

1. Opening between annual growth rings that contains bark. Bark pockets appear as dark streaks on radial surfaces and as rounded areas on tangential surfaces.

2. Small patches of bark embedded in wood.

Bird peck. Small hole or patch of distorted grain resulting from birds' pecking through the growing cells in the tree. In shape, bird peck usually resembles a carpet tack with the point toward the bark. Bird peck is usually accompanied by discoloration extending for considerable distance along the grain and to a much lesser extent across the grain.

Bird's-eye

1. Small central spot with the wood fibers arranged around it in the form of an ellipse so as to give the appearance of an eye.

2. Small localized areas in wood with the fibers indented and otherwise contorted to form few to many small circular or elliptical figures remotely resembling a bird's-eye, on the tangential surface. Common in sugar maple and is used for decorative purposes. It is rare in other hardwood species; not a defect if sound.

Bird's-eye figure. Figure found on the plain-sawed and rotary-cut surfaces of wood exhibiting numerous rounded areas resembling a bird's-eye, caused by local fiber distortions. It is most common in hard maple.

Black check. Bark pocket containing a certain amount of resin. It is common in western hemlock.

Black streaks. Black streaks in western hemlock caused by the maggots of a fly.

Blister. Elevation of the surface of an adherend, somewhat resembling in shape a blister on the human skin. Its boundaries may be indefinitely outlined and it may burst or become flattened.

Blister figure. Figure found on smooth plain-sawed and rotary-cut surfaces that appears to consist of small, more or less widely spaced, elevated or depressed areas of rounded contour.

Bole. Main stem of a tree of substantial diameter, roughly, capable of yielding sawtimber, veneer logs, or large poles. Seedlings, saplings, and small-diameter trees have stems, not boles.

Bolt. Short section of a tree trunk.

Boreholes. Minute openings in cell walls through which the hyphae of the fungus pass.

Boxed heart. Term used when the pith falls entirely within the four faces of a piece of wood anywhere in its length.

Brittle heart. Defect in hardwoods resulting from the presence of fibers with localized wrinkles that cause reduction in the strength of the wood.

Brown rot. Type of wood-destroying fungi that decompose cellulose and the associated pentosans, leaving the lignin in a more or less unaltered state. The resultant mass of decayed wood is of a powdery consistency, of varying shades of brown.

Burl

1. Hard, woody outgrowth on a tree, more or less rounded in form, usually resulting from the entwined growth of a cluster of adventitious buds. Such burls are the source of the highly figured burl veneers used for purely ornamental purposes. In lumber or veneer, localized severe distortion of the grain generally rounded in outline, usually resulting from overgrowth of dead branch stubs, varying from $\frac{1}{2}$ in. to several inches in diameter. A burl frequently includes one or more clusters of several small contiguous conical protuberances, each usually having a core or pith but no appreciable amount of end grain (in tangential view) surrounding it.

2. Large wartlike excrescence on a tree trunk containing the dark piths of a large number of buds which rarely develop. Formation of a burl apparently results from an injury to the tree.

3. Bulge or excrescence that forms on the trunk and branches of a tree; cambial fusiform initial; cambial

initial that through repeated division gives rise to a radially directed row of longitudinal elements of xylem and phloem. Such a cell is fusiform in shape. Longitudinal cambial initial.

Cambium

1. Layer of tissue just beneath the bark from which the new wood and bark cells of each year's growth develop.

2. Layer of actively dividing cells of a tree that produces wood cells on the side toward the pith and bark cells on the side away from the pith.

3. Layer of tissue, one cell wide in thickness, between the bark and wood that subdivides to form the new wood and bark cells of each year's growth.

4. Growing layer between the xylem and phloem; lateral meristem responsible for formation of xylem and phloem.

Capillaries. Thin-walled tubes or vessels found in wood.

Cell

1. Minute units of wood structure, including wood fibers, vessels members, and other elements of diverse structure and function.

2. Structural units of plant tissue.

Cell wall. Wall that encloses the cell contents. In a mature cell it is compound insofar as it consists of several layers.

Cellulose

1. Carbohydrate that is the principal constituent of wood and forms the framework of the wood cells.

2. Principal one of the substances that form the framework or walls of wood cells; also, an organic substance obtained from the cotton plant and used as raw material in the manufacture of paints and other materials.

3. Principal chemical constituent of the cell walls of the higher plants; complex carbohydrate occurring in the form of polymer chains.

Check

1. Lengthwise separation of the wood that usually extends across the rings of annual growth and commonly results from stresses set up in wood during seasoning.

2. Ruptures along the grain that develop during seasoning either because of a difference in radial and tangential shrinkage or because of uneven shrinkage of the tissue in adjacent portions of the wood.

Close-grained wood. Wood with narrow, inconspicuous annual rings. The term is sometimes used to designate wood having small and closely spaced pores, but in this sense the term "fine textured" is more often used.

Coarse-grained wood. Wood with wide conspicuous annual rings in which there is considerable difference between springwood and summerwood. The term is sometimes used to designate wood with large pores, such as oak, ash, chestnut, and walnut, but in this sense the term "coarse textured" is more often used.

Collapse

1. Flattening of single cells or rows of cells in heartwood during the drying or pressure treatment of wood, characterized by a caved-in or corrugated appearance of the wood surface.

2. Defect that sometimes develops above the fiber saturation point when very wet heartwood of certain species is dried. It is evidenced by abnormal and irregular shrinkage.

Compression failures. Localized buckling of fibers and other longitudinal elements produced by the compression of wood along the grain beyond its proportional limit. Compression failures sometimes develop in standing trees.

Compression wood

1. Wood formed on the lower side of branches and inclined trunks of softwood trees. Compression wood is identified by its relatively wide annual rings, usually eccentric; by a relatively large amount of summerwood, sometimes more than 50% of the width of annual rings in which it occurs; and by a lack of demarcation between springwood and summerwood in the same annual ring. Compression wood shrinks excessively lengthwise as compared with normal wood.

2. Wood formed on the lower side of branches and the curved stems of conifers. The tissue has abnormally high longitudinal shrinkage and physical properties that differ from normal wood.

Coniferous wood. Wood produced by coniferous trees. It is the same as softwood or nonporous wood.

Cross field. Term of convenience for the rectangle formed by the walls of a ray cell and a longitudinal tracheid, as seen in the radial section. Its principal application is to conifers.

Cross grain

1. Grain not parallel with the longitudinal axis of a piece as a result of sawing. The grain may be either diagonal or spiral, or a combination of both.

2. Pattern in which the fibers and other longitudinal elements deviate from a line parallel to the sides of the piece. Term applies to either diagonal or spiral grain, or a combination of the two.

3. In standing trees, grain in which the fiber alignment deviates from the axis of the stem; in wooden members, grain in which the fiber alignment deviates from the direction parallel to the long axis of the piece.

Cross-grained wood. Wood in which the fibers deviate from a line parallel to the sides of the piece. Cross grain may be either diagonal or spiral, or a combination of the two.

Cross section. In the examination of wood, the surface exposed when a tree stem is cut horizontally and the majority of the cells are cut transversely; in general, a cutting or section, or a piece of something cut off at right angles to an axis.

Crystallites. Regions in the cell wall of plants in which the cellulose is arranged in a highly ordered, crystal lattice of parallel chains. These regions are of limited size and are separated by regions of little crystalline order.

Curly grain. Grain that results from more or less abrupt and repeated right and left deviations from the vertical in fiber alignment. The radial split faces of such wood are corrugated, and the split tangential faces smooth or corrugated.

Curly-grained wood. Wood in which the fibers are distorted so that they have a curled appearance, as in bird's-eye wood. Areas showing curly grain may vary up to several inches in diameter.

Decay. Disintegration of wood or other substance through the action of fungi.

Deciduous trees. Trees where leaves fall at the end of the growing period.

Deliquescent growth. Growth in which the trunk divides rather abruptly into limbs.

Density

1. Weight of a body per unit volume. When expressed in the cgs (centimeter-gram-second) system, it is numerically equal to the specific gravity of the same substance.

2. Mass of wood per unit of volume.

Diagonal grain

1. Grain in which the longitudinal elements form an angle with the axis of the piece as a result of sawing at an angle with the bark of the tree or log; form of cross grain.

2. Type of cross grain resulting from failure to saw parallel to the growth increments.

Diagonal-grain wood. Wood in which the annual rings are at an angle with the axis of a piece as a result of sawing at an angle with the bark of the tree or log; form of cross-grained wood.

Diffuse porous woods

1. Woods of hardwood species in which the vessel diameter remains approximately constant throughout the annual ring.

2. Certain hardwoods in which the pores tend to be uniform in size and distribution throughout each annual ring or to decrease in size slightly and gradually toward the outer border of the ring.

3. Porous woods in which the pores exhibit little or no variation in size indicative of seasonal growth.

Discontinuous growth rings. Growth rings that are formed on only one side of the stem.

Double ring. Growth ring that appears to consist of two rings, one of which is a false ring.

Dry rot

1. Term loosely applied to any crumbly decay of wood, but especially to that which, when in an advanced stage, allows the wood to be crushed easily to a dry powder. The term does not accurately describe decay, since the fungi that cause the rot require considerable moisture for growth.

2. Special type of brown rot causing widespread damage in buildings. In the United States the causal organism is Poria Incrassata.

Early wood

1. Less dense part of a tree's growth ring made up of cells having thinner walls, a greater radial diameter, and shorter length than those formed later in the year.

2. Portion of the annual growth ring that is formed during the early part of the growing season. It is usually less dense and weaker mechanically than late wood.

3. Portion of a growth increment that is produced at the beginning of the growing season; springwood.

Edge grain. Figure in lumber that has been sawed so that the face of the board is the radial plane of the log. Commercially, lumber is considered edge-grained when the angle between the surface and the annual rings lies between 45 and 90° with the wide surface of the piece. The term is synonymous with "vertical grain," "rift grain," and "quarter-sawed."

Encased knot

1. Knot whose rings of annual growth are not intergrown with those of the surrounding wood.

2. Portion of a branch that becomes embedded in the hole of a tree after the branch dies.

Epidermis. Outermost, generally uniseriate, layer of primary tissue that is continuous over the younger portions of the aerial part of a plant, except where interrupted by stomatal openings. In woody plants the epidermis ceases to function after a peridem forms beneath it and is subsequently cast.

Excurrent growth. Growth in which the axis is prolonged, forming an undivided main trunk, as in pine.

Extractives. Substances in wood that are not integral parts of the cellular structure and can be removed by solution in hot or cold water, either by benzene or other solvents that do not react chemically with wood components.

False heartwood. Pathological heartwood formed in species that do not possess normal heartwood (judged on the basis of color).

Feather crotch. Figure with a design resembling a cluster of feathers, found in crotch veneer.

Fiber

1. Particle of organic or inorganic matter whose length is at least 100 times its diameter. Short fibers are called "staple fibers." Also, a general term used for any long, narrow cell of wood or bark, other than a vessel.

2. Wood fiber is a comparatively long, $\frac{1}{25}$ in. or less to $\frac{1}{3}$ in. narrow, tapering cell closed at both ends.

3. Elongated cell with pointed ends and a thick or, not infrequently, thin wall, including fiber tracheids with bordered pits and libriform fibers with simple pits.

Fibril. Threadlike component of cell walls, visible under a light microscope.

Figure

1. Pattern produced in a wood surface by annual growth rings, rays, knots, deviations from regular grain, such as interlocked and wavy grain, and irregular coloration.

2. In a broad sense, a design or distinctive markings on the longitudinal surfaces of wood; in a restricted sense, such decorative designs in wood as are prized in the furniture and cabinet-making industries.

Flat grain. Figure in lumber that has been sawed so that the wide face of the board is approximately perpendicular (tangent) to the radius of the log. Commercially, lumber is considered to be flat-grained if the wide surface of the board is less than 45° from a tangent to the annual rings. The term is synonymous with "slash grain."

Flat-grained lumber. Lumber that has been sawed so the wide surfaces extend approximately parallel to the annual growth rings. Lumber is considered flat-grained when the annual growth rings make an angle of less than 45° with the surface of the piece.

Flat-sawed. Wood so sawed that the tangential face of the wood is exposed on the surfaces of boards. The term is synonymous with "plain-sawed."

Frost ring. Brownish line extending circumferentially within a growth ring, consisting of collapsed cells and abnormal zones of parenchyma cells. Frost ring is traceable to injury of the cambium or young, unlignified wood cells by either early or late frost.

Grade. Designation of the quality of a log or of a processed or manufactured piece of lumber.

Grain

1. Quality, arrangement, direction, size, and appearance of the fibers in a piece of wood.

2. Vertical elements of wood as they occur in the living tree. Grain is perhaps most easily delineated in certain woods by the presence of annual layers of more densely aggregated cells, or in groups of prominent vessels that form the well-known growth rings. When severed, they may become quite pronounced, and the effect is referred to as "grain."

Grain of wood. Arrangement and direction of alignment of wood elements when considered en masse.

Green lumber

1. Freshly sawed lumber or lumber that has received no intentional drying, containing a moisture content in excess of 30%. It is also called "unseasoned" or "wet."

2. Freshly sawed lumber or lumber that has not received intentional drying; unseasoned. The term does not apply to lumber that may have become completely wet through waterlogging.

Growth ring. Ring of wood on a transverse surface or in a transverse section, resulting from periodic growing. If only one growth ring is formed during a year it is called an "annual ring."

Grub holes. Oval, circular, or irregular holes in wood, $\frac{3}{8}$ to 1 in. in diameter, caused by larvae and adult insects.

Gum. Comprehensive term for nonvolatile viscous plant exudates, which either dissolve or swell up in contact with water. Many substances referred to as gums, such as pine and spruce gum, are actually oleoresins.

Hardwoods

1. Wood produced by broad-leaved trees such as oak, elm, ash, or birch; same as porous wood.

2. Botanical group of trees that are broad leaved. The term has no reference to the actual hardness of the wood. Angiosperms is the botanical name for hardwoods.

3. Hardwoods that have broad leaves and softwoods that have needlelike or scalelike leaves. The botanical classification is sometimes confusing because there is no direct correlation between it and the hardness or softness of the wood. Generally, hardwoods are more dense than softwoods, but some hardwoods are softer than many softwoods.

Heart rot. Rot characteristically confined to the heartwood. It generally originates in the living tree.

Heart shake. Separation of wood across the ring, generally following the rays. It is also called "heart check" and "rift crack."

Heartwood

1. Inner part of the woody stem that in the growing tree no longer contains living cells. It is generally darker than sapwood, though the boundary is not always distinct.

2. Wood extending from the pith to the sapwood, the cells of which no longer participate in the life processes of the tree. Heartwood may contain phenolic compounds, gums, resins, and other materials that usually make it darker and more decay resistant than sapwood.

3. Heartwood consists of inactive cells formed by changes in the living cells of the inner sapwood rings, presumably after their use for sap conduction and other life processes of the tree have largely ceased. Cell cavities of heartwood may also contain deposits of various

materials that frequently provide a much darker color. All heartwood, however, is not darker. Infiltrations of material deposited in the cells of heartwood usually make lumber cut therefrom more durable when exposed to weather. All wood, with the possible exception of the heartwood of redwood and western red cedar should be preservative treated when used for exterior applications.

Honeycombing

1. Checks, often not visible at the surface, that occur in the interior of a piece of wood, usually along the wood rays.

2. Internal splitting in wood that develops in drying caused by internal stresses or by the closing of surface checks.

Incipient decay. Early stage of decay that has not proceeded far enough to soften or otherwise perceptibly impair the hardness of the wood. Incipient decay is usually accompanied by a slight discoloration or bleaching of the wood.

Included sapwood. Streaks or irregularly shaped areas of light-colored wood with the general appearance of normal sapwood, found embedded in the darker colored heartwood, common in western red-cedar.

Intergrown knot

1. Knot in which the annual growth layers are completely intergrown with those of the surrounding wood.

2. Portion of a branch that is embedded in the tree trunk while the branch is alive. It is also called a "tight knot."

Interior stain. Form of sap stain that is confined to the interior of an infected piece and is not evident on the surface. It is caused by the same fungi as the ordinary sap stain.

Interlocked grain

1. Grain in which the fibers incline in one direction in a number of rings of annual growth, then gradually reverse and incline in an opposite direction in succeeding rings, and then reverse again.

2. Condition produced in wood by the alternate orientation of fibers in successive layers of growth increments. The quarter-sawed face of such wood produces a ribbon figure.

Knot

1. Branch base that is embedded in the wood of a tree trunk or larger branch.

2. Portion of a branch or limb that has been surrounded by subsequent growth of the wood of the trunk or other portion of the tree. As a knot appears on the sawed surface, it is merely a section of the entire knot, its shape depending on the direction of the cut.

3. Branch or limb embedded in the tree and cut through in the process of lumber manufacture. It is classified according to size, quality, and occurrence. The size of the knot is determined as the average diameter on the surface of the piece.

4. Natural characteristic of wood that occurs where a branch base is embedded in the trunk of a tree. Generally the size of a knot is distinguishable by a difference in the color of limb wood and surrounding trunk wood, an abrupt change in growth ring width between the knot and bordering trunk wood, and a

diameter of circular or oval shape described by points where checks on the face of a knot that extend radially from its center to its side experience an abrupt change in direction.

Knotholes. Voids produced by the dropping of knots from the wood in which they are originally embedded.

Lamella. Thin sheet of microfibrils; sublayer in the cellulose aggregation of the secondary cell wall.

Late wood

1. Denser part of the growth ring. Late wood is made up of wood cells having thicker walls, smaller radial diameters, and generally greater length than those formed earlier in the growing season. It is also called "summerwood."

2. Portion of the annual growth ring that is formed after the early wood formation has ceased. It is usually denser and stronger mechanically than early wood.

3. Portion of an annual increment that is produced during the latter part of the growing season (during the summer); summerwood.

Lignin

1. Amorphous substance that infiltrates and surrounds the cellulose strands in wood, binding them together to give a strong mechanical structure. Lignin comprises 15 to 30% by weight of the wood substance.

2. Second most abundant constituent of wood (approximately 12 to 28%). It encrusts the cell walls and cements the cells together.

3. Second most abundant constituent of wood, located principally in the secondary wall and the middle lamella, which is the thin cementing layer between wood cells. Chemically it is an irregular polymer of substituted propylphenol groups, and thus no simple chemical formula can be written for it.

4. One of the principal constituents of woody cell walls whose exact chemical composition is still unknown; residue after treatment of solvent-extracted wood with strong mineral acids.

Longitudinal. Direction along the length of the grain of wood.

Longitudinal resin canal. Canal extending with the grain, appearing as an opening or fleck on the transverse surface to the naked eye or through a hand lens.

Loosened grain. Loosened small portions of wood on the flat-grained surfaces of boards, usually of the tops and edges of the growth increments.

Loose knot

1. Knot that is not held firmly in place by growth or position and that cannot be relied on to remain in place.

2. Portion of a branch that is incorporated into the bole of a tree after the death of that branch.

Lumen. Cavity of a cell.

Marine borers. Mollusks and crustaceans that attack submerged wood in salt and brackish water.

Medullary rays. Rays that extend radially from the pith of the log toward the circumference. The rays serve primarily to store food and transport it horizontally. They vary in height from a few cells in some species to 4 or more inches in the oaks, and produce the flake effect common to the quarter-sawed lumber in these species.

Microfibril. Bundle of cellulose polymer chains and associated polysaccharides of other types that are united at some regions in highly ordered crystalline lattices known as "crystallites" and are less highly ordered in the zones between the crystallites (amorphous regions); smallest natural unit of cell wall structure that can be distinguished with an electron microscope.

Mineral stain. Olive and greenish black streaks believed to designate areas of abnormal concentration of mineral matter. Mineral stain is common in hard maple, hickory, and basswood. It is also called "mineral streak."

Mineral streak. Olive to greenish black or brown discoloration of undetermined cause in hardwoods, particularly hard maples, commonly associated with bird pecks and other injuries, occurring in streaks usually containing accumulations of mineral matter.

Modified wood. Wood processed to impart properties quite different from those of the original wood by chemical, resin, compressive, heat, or radiation treatments.

Moisture content of wood. Amount of water contained in the wood, usually expressed as a percentage of the weight of the oven-dry wood.

Moisture in wood. Moisture in wood is found in two forms, hygroscopic and free. Hygroscopic moisture is present inside the hollow wood fibers; free moisture is found outside the cell walls, in a fiber bundle. When wood dries, the free moisture evaporates first. When it is gone, but the cells still contain their hygroscopic moisture, the wood is said to be at the "fiber saturation point." This occurs for most species when the wood has reached a moisture content of about 25% of the oven-dry weight. The moisture content of green wood may vary from about 35% for some species to nearly 200% in the sapwood of white pine and hemlock.

Moonshine crotch. Swirling figure found in crotch veneer.

Mottled figure. Broken stripe figure interrupted by irregular, horizontal waves in the grain.

Multiple ring. Growth ring that contains within its boundaries several false rings.

Multiseriate ray. Ray consisting of several to many rows of cells, as viewed in the tangential section.

Naval stores. Term applied to the oils, resins, tars, and pitches derived from oleoresin contained in, exuded by, or extracted from trees, chiefly species of pines (genus Pinus).

Old growth. Timber growing in or harvested from a mature, naturally established forest. When the trees have grown most or all of their individual lives in active competition with their companions for sunlight and moisture, the timber is usually straight and relatively free of knots.

Oleoresin. Solution of resin in an essential oil that occurs in or exudes from many plants, especially softwoods. The oleoresin from pine in a solution of pine resin (rosin) in turpentine.

Open grain. Common classification for woods with large pores such as oak, ash, chestnut, and walnut. It is also known as "coarse texture."

Open-grained wood. Common classification by painters for woods with large pores, such as oak, ash, chestnut, and walnut. It is also known as "coarse-textured wood."

Opposite pitting. Type of pitting in which bordered pits are arranged in transverse rows extending across the cell.

When crowded, the outlines of the pits become rectangular in surface view.

Orthotropic. Having unique and independent properties in three mutually orthogonal (perpendicular) planes of symmetry; special case of anisotropy.

Oxidative stain. Nonpathological stain in sapwood caused by chemical changes in the materials contained in the cells of the wood.

Parenchyma

1. Tissue composed of wood cells that are usually shaped like tiny bricks, have simple pits, and frequently only a primary wall. Parenchyma functions mainly in the storage and distribution of food material.

2. Parenchyma consists of short cells having simple pits and functioning primarily in the metabolism and storage of plant food materials. They remain alive longer than the tracheids, fibers, and vessel segments, sometimes for many years. Two kinds of parenchyma cells are recognized—those in vertical strands, known more specifically as "axial parenchyma," and those in horizontal series in the rays, known as "ray parenchyma."

3. Tissue consisting of short, relatively thin-walled cells, generally with simple pits. It is concerned primarily with the storage and distribution of carbohydrates and may be visible with a hand lens on the transverse surface of wood as dots, sheaths about pores, or broken or continuous lines or bands. The term is used specifically as a synonym for "axial parenchyma," which occurs in strands along the grain.

Peck. Pockets or areas of disintegrated wood caused by advanced stages of localized decay in the living tree. Peck is usually associated with cypress and incense cedar. There is no further development of peck once the lumber is seasoned.

Pecky dry rot. Rot characterized by finger-sized pockets of decay in the living trees of incense cedar caused by Polyporus Amarus, and baldycypress, caused by Stereum. It is also known as "peckiness" and "pocket dry rot."

Phellem. Outermost layer of periderm, composed of cork cells formed to the outside by the phellogen.

Phloem. Inner bark; principal tissue concerned with the distribution of elaborated foodstuffs, characterized by the presence of sieve tubes.

Pigment figure. Figure in wood occasioned by irregular infiltration, resulting in dark lines, bands, zones streaks, etc.

Pinholes. Small, round holes ($\frac{1}{100}$ to $\frac{1}{4}$ in. in diameter) in wood that result from the mining of ambrosia and similar beetles.

Pin knot. Knot that is not more than $\frac{1}{2}$ in. in diameter.

Pinoid pit. Term of convenience for the smaller types of early wood cross-field pits found in several species of Pinus. Characteristically these are simple or with narrow borders and often variable in size and shape. Excluding the windowlike, fenestriform pits in the ray parenchyma cross-fields of soft pines and Pinus resinosa and Pinus sylvestris.

Pit. Recess in the secondary wall of a cell, together with its external closing membranes. A pit is open internally to the lumen.

Pitch and bark pockets. A pitch pocket is an opening extending parallel to the annual growth rings containing, or having contained pitch, either solid or liquid. A bark pocket is an opening between annual growth rings that contains bark.

Pitch pocket. Lens-shaped opening in the grain at the common boundary of two growth increments, sometimes within a growth increment, that is empty or contains solid or liquid resin. It is found in certain coniferous woods.

Pitch streaks. Localized accumulation of resin, that permeates the cells forming resin-soaked patches or streaks in coniferous woods.

Pith

1. Central core of a woody stem, consisting mainly of parenchyma or soft tissue.
2. Small, soft core occurring near the center of a tree trunk, branch, twig, or log.
3. Small, soft core occurring in the structural center of a log.
4. Primary tissue in the form of a central parenchymatous cylinder found in stems and sometimes in roots.

Pith fleck

1. Narrow streak, resembling pith, on the surface of a piece. It is usually brownish and up to several inches in length. A pitch fleck results from the burrowing of larvae in the growing tissues of the tree.
2. Small area of wound tissue, darker or lighter than the surrounding tissue, produced in wood through injury to the cambium by the larvae of flies of the genus Agromyza and subsequent occlusion of the resulting tunnels with parenchymatous cells.

Plain-sawed lumber. Term for flat-grained lumber.

Pocket rot. Advanced decay that appears in the form of a hole or pocket, usually surrounded by apparently sound wood.

Porous woods

1. Another name for hardwoods, which frequently have vessels or pores large enough to be seen readily without magnification.
2. Woods containing pores (vessels). The term is another name for hardwoods, that is, woods produced by broad-leaved trees.

Powder-post damage. Small holes, $\frac{1}{16}$ to $\frac{1}{2}$ in. in diameter, filled with dry, pulverized wood, resulting from the work of beetles, largely Lyctus, in seasoned and unseasoned wood.

Quarter-sawed. Term applied to the wide face of the board or the radial face of the log; same as edge grained.

Quarter-sawed lumber. Edge-grained lumber.

Quarter section. Section cut along the grain parallel to the wood rays.

Quarter surface. Surface that is exposed when a log is cut along the grain in a radial direction (parallel to the wood rays).

Radial. Coincident with a radius from the axis of the tree or log to the circumference. A radial section is a lengthwise section in a plane that extends from pith to bark.

Radial face. Wood surface exposed when a stem is cut along a radius from pith to bark and the cut is parallel to the long axis of the majority of the cells.

Radial section. Section cut along the grain parallel to the wood rays and usually at right angles to the growth rings.

Rate of growth. Rate at which a tree has laid on wood, measured radially in the trunk or in lumber cut from the trunk. The measure in use is the number of annual growth rings per inch.

Ray. Ribbon-shaped strand of tissue extending in a radial direction across the grain, so oriented that the face of the ribbon is exposed as a fleck on the quarter surface.

Reaction wood. Wood with distinctive anatomical and physical characteristics, formed in parts of leaning or crooked stems and in branches. In dicotyledons reaction wood is known as "tension wood", and in gymosperms as "compression wood."

Resin. Inflammable, water-soluble, vegetable substances secreted by certain plants or trees and characterizing the wood of many coniferous species. The term is also applied to synthetic organic products related to the natural resins.

Resin canal. Tubular, intercellular space sheathed by secreting cells (epithelium), bearing resin in the sapwood.

Resin passages (or ducts). Intercellular passages that contain and transmit resinous materials. On a cut surface, they are usually inconspicuous. They may extend vertically parallel to the axis of the tree or at right angles to the axis and parallel to the rays.

Ribbon figure. Figure consisting of changeable (with light) darker and lighter bands, obtained by quarter-sawing or slicing interlocked grain wood.

Ring failure. Separation of the wood during seasoning, occurring along the grain and parallel to the growth rings.

Ring-porous woods

1. Wood of hardwood species in which the vessel diameter in the growth ring is considerably larger in the early wood than in the late wood.
2. Group of hardwoods in which the pores are comparatively large at the beginning of each annual ring and decrease in size more or less abruptly toward the outer portion of the ring, thus forming a distinct inner zone of pores known as the "springwood" (early wood) and an outer zone with smaller pores known as the "summerwood" (late wood).
3. Porous woods in which the pores formed at the beginning of the growing season (in the early wood) are much larger than those farther out in the ring, particularly if the transition from one to the other type is more or less abrupt.

Ring shake. Rupture in wood that occurs between increments or less frequently within an annual growth layer. It is sometimes called "wind shake."

Ripple marks. Striations across the grain on the tangential surface of a wood, occasioned by storied rays and/or other storied elements.

Round knot. Knot that is cut so that the exposed section is oval or circular.

Sap

1. The vital circulating fluid of a woody plant, including the special juice, extracted from some plants.
2. All the fluids in a tree except special secretions and excretions, such as oleoresin.

Sap stains. Stains in the sapwood caused by wood-staining fungi or the oxidation of compounds present in the lumina of living cells.

Sapwood

1. Outer (younger) portion of a woody stem (or log), usually distinguishable from the core (heartwood) by its lighter color.

2. Light-colored wood substances occurring in the outer portion of the tree.

3. Wood immediately inside the cambium of the living tree, containing living cells and reserve materials and in which most of the upward water movement takes place.

4. Living wood of pale color near the outside of the log. Under most conditions the sapwood is more susceptible to decay than heartwood.

5. Layers of wood next to the bark, usually lighter in color than the heartwood, $\frac{1}{2}$ in. to three or more inches wide, that are actively involved in the life processes of the tree. Under most conditions sapwood is more susceptible to decay than heartwood, and as a rule, it is more permeable to liquids that heartwood. Sapwood is not essentially weaker or stronger than heartwood of the same species.

6. Sapwood contains living cells and performs an active role in the life processes of the tree. It is located next to the cambium and functions in sap conduction and storage of food. Sapwood commonly ranges from $1\frac{1}{2}$ to 2 in. in thickness. Maples, hickories, ashes, and some of the southern yellow pines and ponderosa pines may have sapwood 3 to 6 in. in thickness, especially in second-growth trees.

Seasonal increment. Layer of wood laid down during a given year.

Seasoning. Removing moisture from green wood in order to improve its serviceability.

Secondary cell wall. Portion of the cell wall formed after the cell enlargement has been completed.

Second growth

1. Timber that has grown after one removal, whether by cutting, fire, wind, or other agency, of all or a large part of the previous stand.

2. Growth traceable to the activities of a lateral cambium. It is also called "secondary thickening."

Shake

1. Rupture of cells or between cells resulting in the formation of an opening in the grain of the wood. The opening may develop at the common boundary of two rings or within a growth ring.

2. Separation along the grain, the greater part of which occurs between the rings of annual growth. It is usually considered to have occurred in the standing tree or during felling.

Sieve-tube. Composite structure found in the phloem of all vascular plants, composed of a longitudinal series of sieve-tube elements whose protoplasts are connected by strands of protoplasm extending through small openings in sieve plates.

Simple pit. Pit in which the cavity becomes wider, remains of constant width, or only gradually narrows toward the cell lumen during the growth in thickness of the secondary wall.

Small knot. Knot over $\frac{1}{2}$ in. but not more than $\frac{3}{4}$ in. in diameter.

Soft rot

1. Special type of decay developing under very wet conditions in the outer wood layers, caused by cellulose-destroying microfungi that attack the secondary cell walls and not the intercellular layer.

2. Decay caused by the Ascomycetes and Fungi Imperfecti. The surface of the affected wood is typically softened.

Softwood

1. Botanical group of trees that have needles or scalelike leaves and are evergreen for the most part (cypress, larch and tamarack being exceptions). The term has no reference to the actual hardness of the wood.

2. Botanical group of trees that bear cones and in most cases have needle- or scalelike leaves; also, wood produced by such trees. The term has no reference to the actual hardness of the wood.

3. Wood produced by coniferous trees; nonporous wood.

Sound knot. Knot that is solid across its face and at least as hard as the surrounding wood, and shows no indication of decay.

Spike knot. Knot cut approximately parallel to its long axis so that the exposed section is definitely elongated.

Spiral grain

1. Form of cross grain in which the fibers extend spirally about instead of vertically along the bole of a tree or axis of a timber.

2. Grain in which the fibers are aligned in a helical orientation around the axis of the stem.

Spiral-grained wood. Wood in which the fibers take a spiral course about the trunk of a tree instead of the normal vertical course. The spiral may extend in a right-handed or left-handed direction around the tree trunk. Spiral grain is a form of cross grain.

Split. Lengthwise separation of the wood due to the tearing apart of the wood cells.

Springwood

1. Portion of the annual growth ring that is formed during the early part of the season's growth. In most softwoods and in ring-porous hardwoods, it is less dense and weaker mechanically than summerwood.

2. Wood characterized by cells having relatively large cavities and thin walls.

Stain. Discoloration in wood that may be caused by such diverse agencies as microorganisms, metals, or chemicals. The term also applies to materials used to impart color to wood.

Star shake. Heart shake that radiates from the pith.

Sticker markings. Special oxidative stain that occurs during air-seasoning and kiln-drying on and beneath the surface of the board, where stickers come in contact with it.

Straight grain. Term describing wood in which the direction of the fiber alignment is straight. In standing trees it also refers to the grain orientation that is parallel to the stem axis.

Straight-grained wood. Wood in which the fibers run parallel to the axis of a piece.

Strand tracheids. Tracheids in coniferous wood that arise from the further division of a cell that otherwise would have developed into the longitudinal tracheid. Strand tracheids differ from the latter in being shorter and having one or both end walls at right angles to the longitudinal walls.

Stump wood. Bell-shaped base of the tree just above the roots.

Substitute fiber. Fibrous parenchymatous cell in wood; fibrous cell in wood that remains living while it is a part of the sapwood.

Summerwood

1. Portion of the annual growth ring that is formed after the springwood formation has ceased. In most softwoods and in ring-porous hardwoods, summerwood is denser and stronger mechanically than springwood.

2. In summerwood cells have smaller cavities and thicker walls, and consequently are more dense than springwood.

Surface checks. Seasoning checks that develop on the surface and extend into the wood for varying distances.

Swirl crotch. Figure obtained from the section of the tree where the typical crotch figure faces into that of normal stem wood.

Tangential. Strictly, coincident with a tangent at the circumference of a tree or log, or parallel to such a tangent. In practice, however, it often means roughly coincident with a growth ring. A tangential section is a longitudinal section through a tree or limb perpendicular to a radius. Flat-grained lumber is sawed tangentially.

Tangential face. Surface exposed when a cut is made at right angles to the rays and parallel to the long axis of the majority of cells.

Tangential section. Section cut along the grain at right angles to the wood rays.

Taxodioid pit. Cross-field pit in early wood of conifers with a large, ovoid-to-circular included aperture that is wider than the lateral space on either side between the aperture and the border, as in sequoia, taxodia, and abies.

Tension wood

1. Form of wood found in leaning trees of some hardwood species, characterized by the presence of gelatinous fibers and excessive longitudinal shrinkage. Tension wood fibers hold together tenaciously, so that sawed surfaces usually have projecting fibers, and planed surfaces often are torn or have raised grain. Tension may cause warping.

2. Reaction wood formed typically on the upper sides of branches and the upper, usually concave side of leaning or crooked stems of hardwoods, characterized anatomically by lack of cell wall lignification and often by the presence of a gelatinous layer in the fibers.

Texture. Term often used interchangeably with "grain." It is sometimes used to combine the concepts of density and degree of contrast between springwood and summerwood.

Texture of wood. Expression that refers to the size and the proportional amounts of woody elements. In coniferous woods the average tangential diameter of the tracheids is the best indicator of texture; in the hardwoods tangential diameters and the number of vessels and rays are the best indicators.

Tiger grain. Curly grain in hard maple.

Tight knot. Portion of a branch that is embedded in the tree trunk while the branch is alive. It is also called an "intergrown knot."

Toughness. Ability to absorb energy without separation of the material as failure progresses; the opposite of brashness.

Tracheid

1. Long, pointed, supporting cell in wood. A tracheid is tubelike and conducts water.

2. Elongated cells that constitute the greater part of the structure of the softwoods (frequently referred to as fibers). They are also present in some hardwoods.

3. Fibrous lignified cell with bordered pits and imperforate ends. In coniferous wood the tracheids are very long (up to 7 + mm) and equipped with large prominent bordered pits on their radial walls. Tracheids in hardwoods are shorter fibrous cells (seldom over 1.5 mm) as long as the vessel elements with which they are associated and possess small bordered pits.

Transverse resin canals. Resin canals extending across the grain included in fusiform wood rays.

Tree branches. Most branches originate at the pith, and their bases are intergrown with the wood of the trunk as long as they are alive. Living branch bases constitute intergrown or tight knots. After the branches die, their bases continue to be surrounded by the wood of the growing trunk and thus loose or encased knots are formed. After the dead branches fall off, the stubs become overgrown, and subsequently clear wood is formed.

Tree growth. All growth in thickness takes place in the cambium layer by cell division. No growth in either diameter or length takes place in wood already formed; new growth is purely the addition of new cells, not the further development of existing cells.

Tree trunk. The cross section of a tree trunk shows the following well-defined features in succession from the outside to the center: bark and cambium layer, wood, which in most species is clearly differentiated into sapwood and heartwood; and pith, the small central core. The pith and bark are excluded from finished lumber.

Tyloses. Masses of cells appearing somewhat like froth in the pores of some hardwoods, notably white oak and black locust. In hardwoods tyloses are formed when walls of living cells surrounding vessels extend into the vessels. They are sometimes formed in softwoods in a similar manner by the extension of cell walls into resin passage cavities.

Typical or advanced decay. Stage of decay in which the disintegration is readily recognized because the wood has become punky, soft and spongy, stringy, pitted, or crumbly.

Uniseriate ray. Ray consisting of one row of cells, as viewed in the tangential section.

Upright ray cell. Short, high cell (at least twice the height of an ordinary ray cell), occurring on the margins and frequently, in addition, on the flanks and in the body of a heterocellular ray.

Vascular tracheids. Specialized cells in certain hardwoods, similar in shape, size, and arrangement to the small vessel elements, but differing from them in being imperforate at the ends.

Vasicentric tracheids. Short, irregularly shaped fibrous cells with conspicuous bordered pits. Vasicentric tracheids abound in the proximity of the large early wood vessels of certain ring-porous hardwoods; they differ from vascular tracheids not only in shape, but in arrangement (they are not arranged in definite longitudinal rows like vascular tracheids).

Vertical-grained lumber. Term for edge-grained lumber.

Vessel. Articulated, tubelike structure of indeterminate length in porous woods, formed through the fusion of the cells in a longitudinal row and the perforation of common walls in one of a number of ways.

Vessels

1. Series of cells extending longitudinally in the stem, the ends of the cells having fused together to form a long tube.

2. Wood cells of comparatively large diameter that have open ends and are set one above the other to form continuous tubes. The openings of the vessels on the surface of a piece of wood are usually referred to as "pores."

Virgin growth. Original growth of mature trees.

Wane. Bark or the lack of wood from any cause on the corner of a piece.

Warty layer. Isotropic layer of material deposited on the inner surface of the secondary wall of many kinds of wood. The layer frequently contains encysted globules of a dissimilar material. Inclusions produce the warts which lend the name to the layer.

Wavy grain

1. Grain in which the fibers and other longitudinal elements collectively take the form of waves or undulations.

2. Grain due to undulations in the direction of fiber alignment. When a wavy grained wood is split radially, the exposed surfaces are wavy.

Wavy-grained wood. Wood in which the fibers collectively take the form of waves or undulations.

Wetwood. Heartwood, and sometimes the inner sapwood, with higher moisture content than the adjacent sapwood. Excessive wetness is associated with bacterial action, resulting in fermentation odors. Wetwood is subject to excessive checking and collapse and causes difficulties in gluing.

White rot

1. Any decay or rot attacking both the cellulose and the lignin, producing a generally whitish residue that may be spongy or stringy rot or occur as pocket rot.

2. Type of wood-destroying fungi that attack both cellulose and lignin. The resultant mass is spongy or stringy, usually white in color, but may assume various shades of yellow, tan, and light brown.

Wood characteristics. Distinguishing features that by their extent and number determine the quality of a piece of wood.

Wood fiber. Comparatively long ($\frac{1}{25}$ or less to $\frac{1}{3}$ in.), narrow, tapering wood cell closed at both ends.

Wood fungi. Microscopic plants that live in damp wood and cause mold, stain, and decay.

Wood irregularities. Natural characteristics in or on wood that may lower its durability, strength, or utility.

Wood rays

1. Ribbonlike group of cells, usually parenchyma, extending radially in the wood and bark of a tree.

2. Strips of cells extending radially within a tree and varying in height from a few cells in some species to 4 or more inches in oak. The rays serve primarily to store food and transport it horizontally in the tree. On quarter-sawed oak, the rays form a conspicuous figure, sometimes referred to as "flecks."

Wood substance. Solid material of which wood is composed. The term usually, but not always, refers to the extractive-free solid substance of which the cell walls are composed. There is no wide variation in chemical composition or specific gravity between the wood substance of various species, the characteristic differences of species being largely due to differences in extractives and variations in relative amounts of cell walls and cell cavities.

Wound heartwood. Patches of dead sapwood that develop in the vicinity of wounds, such as sapwood in which the parenchyma is longer living; wound heartwood is similar to normal heartwood except in location.

PHYSICS AND CHEMISTRY

Absolute humidity. Mass of water vapor present in the atmosphere measured as grams per cubic meter. Absolute humidity may also be expressed in terms of the actual pressure of the water vapor present.

Absolute temperature. Temperature reckoned from the absolute zero.

Absolute units. System of units based on the smallest possible number of independent units; specifically, units of force, work, energy, and power not derived from or dependent on gravitation.

Absolute zero

1. Temperature at which the volume of an ideal gas would be reduced to zero. All molecular motion (except that due to quantum effects) ceases at absolute zero, which is $-273°C$.

2. Temperature at which a gas would show no pressure if the general law for gases should hold for all temperatures.

Absorptive power or absorptivity. For any body, absorptive power or absorptivity is measured by the fraction of the radiant energy falling on the body that is absorbed or transformed into heat. This ratio varies with the character of the surface and the wavelength of the incident energy. It is the ratio of the radiation absorbed by any substance to that absorbed under the same conditions by a blackbody.

Acceleration

1. Rate of change of velocity. An object is accelerated if its speed or its direction of motion changes.

2. Time rate or change of velocity in either speed or direction, measured by the change in unit time.

Acceleration due to gravity

1. Acceleration experienced by an object that falls (without friction).

2. Acceleration of a body freely falling in a vacuum.

Acids. Substances whose molecules ionize in water solution to give the hydrogen ion from their constituent elements. The strength of an acid is proportional to the concentration of hydrogen ions present.

Action. Measured by the product of work by time.

Active mass of a substance. Number of gram molecular weights per liter in solution or in gaseous form.

Addition polymerization. Polymerization in which monomers are linked together without the splitting off of water or other simple molecules.

Adiabatic. A body is said to undergo an adiabatic change when its condition is altered without gain or loss of heat. The line on the pressure volume diagram representing an adiabatic change is called an "adiabatic line."

Adsorption. Condensation of gases, liquids, or dissolved substances on the surfaces of solids.

Allotropy

1. Property shown by certain elements of being capable of existence in more than one form, due to differences in the arrangement of atoms or molecules.

2. The assumption by an element or other substance of 2 or more different forms or structures which are most frequently stable in different temperature ranges.

Alpha particle

1. Nucleus of a helium atom that consists of two protons and two neutrons bound together. Alpha particles are spontaneously emitted by certain radioactive nuclei.

2. Helium nucleus, that is, a helium atom that which has lost two electrons and has therefore a double positive charge.

Alpha rays. Strongly ionizing and weakly penetrating radiations deflected by a magnetic and electric field as positively charged particles. The particles are doubly charged helium atoms (ions) and are called "alpha particles."

Amorphous. Devoid of crystallinity, definite form, order, or pattern.

Amorphous polymers. Polymer chains that are too irregular in shape to permit the formation of repetitive arrangements of chain segments within the bulk shape. There is no specific structure or pattern in the relationship between the various chains; thus they resemble a liquid. These polymers do not solidify, but they form supercooled liquids or gases that appear to be solids because their viscosity is so high.

Amplitude. Maximum value of the displacement in an oscillatory motion.

Amplitude-modulated wave. Wave on which information has been impressed by varying the amplitude, with the frequency remaining the same.

Amplitude of a wave or a vibration. Maximum amount of displacement of the medium from its normal condition.

Angle. Ratio between the arc and the radius of the arc. The units of an angle include the radian, the angle subtended by an arc equal to the radius; the degree, $\frac{1}{36}$th part of a circumference.

Angular acceleration. Time rate of change of angular velocity either in angular speed or in direction of the axis of rotation (precession).

Angular harmonic motion or harmonic motion of rotation. Periodic, oscillatory angular motion in which the restoring torque is proportional to the angular displacement; torsional vibration.

Angular momentum of an object. Measure of the object's rotation around a particular axis. As long as no torque is exerted on an object, its angular momentum will remain constant.

Angular momentum or moment of momentum. Quantity of angular motion measured by the product of the angular velocity and the moment of inertia.

Angular velocity. Time rate of angular motion about an axis.

Anhydride (of acid or base). Oxide that when combined with water gives an acid or base.

Anion. Negatively charged ion.

Apogee of an earth satellite. Point in its elliptical orbit at which the satellite is farthest from the earth.

Archimedes' principle

1. The buoyant force exerted by a fluid on an object is equal to the weight of the fluid displaced by the object.

2. A body wholly or partly immersed in a fluid is buoyed up by a force equal to the weight of the fluid displaced.

Artificial gravity. Artificial gravity can be produced in a spacecraft by rotating the spacecraft around an axis. The centripetal reaction experienced by an object or person within the spacecraft will be the equivalent of gravity.

Astronomical unit. Unit of length used in many astronomical distance measurements.

Atmospheric pressure on a surface. Force per unit area exerted on a surface by the weight of the column of air above it.

Atom

1. Smallest bit of matter that can be identified as a particular chemical element.

2. Smallest particle of matter that has a distinct chemical identity. Each element is represented by a specific type of atom.

3. Smallest unit of mass of elements that participates in ordinary chemical changes. The atoms of a given element are unvarying in average mass but are different in such mass from atoms of all other elements.

Atomic mass unit. Unit used for expressing the masses of atoms and molecules.

Atomic number

1. Number of protons in the nucleus of an atom. Each element has a characteristic atomic number.

2. Number of excess positive charges on the atomic nucleus. This charge of the nucleus is the essential feature that distinguishes one element from another and determines the position of the element in the periodic table.

Atomic number of an element. The atomic number of an element is equal to the number of electrons in a normal electrically neutral atom of the element or the number of protons in the nucleus of the element.

Atomic theory. All elementary forms of matter are composed of very small unit quantities called "atoms." Atoms of a given element all have the same size and weight. Atoms of different elements have different sizes and weights. Atoms of the same or different elements unite with each other to form very small unit quantities of compound substances called "molecules."

Atomic weight

1. Sum of the weights of the protons and neutrons in the nucleus of an atom when each proton and neutron is assigned an arbitrary weight of 1.

2. Relative weight of the atom based on oxygen as 16. If these weights are expressed in grams, they are called "gram atomic weights."

Average chain length. Average number of mers in polymer chain. The average length of a polymer chain is evaluated by tests to determine certain characteristics of the polymer. Plastics with longer polymer chains have greater molecular weights.

Average molecular weight of a polymer. Production methods in polymer manufacture produce a range of molecular weights. The measure of the average molecular weight and the distribution of molecular weights are determined by laboratory tests yielding weight averages and/or number of averages for the molecular weight. These values are interdependent but not equivalent.

Avogadro's hypothesis. Equal volumes of all gases (at the same temperature and pressure) contain equal numbers of molecules.

Avogadro's law. Equal volumes of different gases at the same pressure and temperature contain the same number of molecules.

Avogadro's number

1. Number of molecules of a substance in 1 mole.

2. Number of molecules in a mole or in a mass in grams of substance equal numerically to its molecular weight.

Avogadro's principle (or theory). The number of molecules present in equal volumes of gases at the same temperature and pressure are equal.

Balanced or reversible action. Action that can be caused to proceed in either direction by suitable variation in the conditions of temperature, volume, or pressure or of the quantities of reacting substances.

Bases. Substances that ionize in water to give the hydroxyl ion from their constituent elements. The strength of a base is proportional to the concentration of hydroxyl ions.

Bernoulli's theorem. At any point in a tube through which a liquid is flowing, the sum of the pressure energy, potential energy, and kinetic energy is a constant.

Beta particle. Negatively charged particle that at rest has a mass about $\frac{1}{1845}$th that of a hydrogen atom; an electron.

Beta rays. Radiation more penetrating but less ionizing than x-rays. The rays are deflected by electric and magnetic fields as negatively charged particles. The particles consist of high-speed electrons.

Binding energy (or ionization energy) of an atom. Minimum energy required to free an electron from the atom. The binding energy of a nucleus is the energy required to separate the nucleus into free protons and neutrons.

Blackbody. If for all values of the wavelength of the incident radiant energy all of the energy is absorbed, the body is called a blackbody.

Black hole. Astronomical object with a gravitational field so intense that neither mass nor light can escape.

Bohr model of the hydrogen atom. The electron is considered to move around the nuclear proton in a planelike orbit. By allowing only discrete values for the angular momentum of the electron, Bohr succeeded in calculating the energies of hydrogen spectral lines that agreed with experiment. The Bohr model of atoms later gave way to quantum theory as a precise description of atomic matter.

Boiling point of a substance. Temperature at which the substance changes from the liquid to the gaseous state. The boiling point is usually stated for normal atmospheric pressure.

Boyle's law

1. the volume of a gas at constant temperature is inversely proportional to the pressure.

2. The volume occupied by a given mass of a gas at constant temperature varies, within moderate ranges of pressure, inversely with the pressure to which it is subjected.

3. At a constant temperature the volume of a given quantity of a gas varies inversely with the pressure to which the gas is subjected.

Branched polymer. Result of the growth of large polymer chain segments from active sites along a linear polymer. The mechanical properties vary greatly from a linear polymer though the chemical composition may be the same.

Breeder reactor. Nuclear reactor that generates power and also produces more nuclear fuel than it consumes.

Bulk modulus. Modulus of volume elasticity.

Buoyancy. Tendency of a fluid to exert an upward force on an immersed object owing to the weight of the displaced fluid.

Calorie. Unit of heat; 1 calorie is the energy required to raise the temperature of 1 kg of water by 1°C.

Carcel unit. Horizontal intensity of the carcel lamp, burning 42 g of colza oil per hour. For a consumption between 38 and 46 g/hour the intensity may be considered proportional to the consumption.

Catalytic agent. Substance that by its mere presence alters the velocity of a reaction. At the end of the reactions it may be recovered unaltered in nature or amount.

Cation. Positively charged ion.

Cellulose. Complex carbohydrate composed of long, unbranched molecules and making up 40 to 55% by weight of the cell wall substance. It is soluble in acid, but not in alkali.

Center of mass of an object or system. Point at which all of the mass may be considered to be concentrated when making dynamical calculations.

Centrifugal force. Outward directed reaction to centripetal force.

Centripetal acceleration. Inward acceleration experienced by an object moving in a circle.

Centripetal force

1. Inward directed force that is required to maintain an object moving in a circular path.

2. Force required to keep a moving mass in a circular path. "Centrifugal force" is the name given to the reaction against centripetal force.

Ceramic. Combination of one or more metals, such as aluminum (or semimetals, such as silicon), with a nonmetallic element, usually oxygen.

cgs. Centimeter-gram-second.

Charles's law and Gay-Lussac's law

1. The volume of a gas at constant pressure is directly proportional to the absolute temperature.

2. The volumes assumed by a given mass of a gas at different temperatures, the pressure remaining constant, are, within moderate ranges of temperature, directly proportional to the corresponding absolute temperatures.

3. At a constant pressure, the volume of a given quantity of a gas increases about $\frac{1}{273}$th of its volume at 0°C for each rise of 1°C and at constant volume the pressure of a given quantity of gas increases about $\frac{1}{273}$th of the pressure at 0°C for each rise of 1°C in temperature.

Chemical change. Interaction of two or more atoms to alter their electronic structure to a more stable condition, thus producing new material(s).

Chemical compound. Substance composed of the molecules of two or more different elements. The smallest bit of matter that retains the properties of a compound is a molecule.

Chemical equilibrium. When a chemical reaction and its reverse reaction take place at equal velocities, the concentrations of reacting substances remain constant.

Coefficient of friction. Between two surfaces, the ratio of the force required to move one over the other to the total force pressing the two together.

Coefficient of restitution. For two bodies on impact, the ratio of the difference in velocity before impact to the difference after impact.

Colligative property. Property numerically the same for a group of substances, independent of their chemical nature.

Colloid. Phase dispersed to such a degree that the surface forces become an important factor in determining its properties.

Combining volumes. Under comparable conditions of pressure and temperature the volume ratios of gases involved in chemical reactions are simple whole numbers.

Combining weight. The combining weight of an element or radical is its atomic weight divided by its valence.

Compounds. Substances containing more than one constituent element and having properties, on the whole, different from those that their constituents have as elementary substances. The composition of a given pure compound is perfectly definite and is always the same no matter how that compound may have been formed.

Compressibility. Reciprocal of the bulk modulus.

Concentration. Amount of a substance in weight, moles, or equivalents contained in unit volume.

Condensation polymerization. Reaction of two or more different organic molecules to form a polymer. The reaction is usually accompanied by the production of a small molecule, such as water, as a by-product.

Configuration. Spatial arrangement of atoms within a molecular unit of a polymer chain that is fixed by bonding and does not change with temperature.

Conformation. General shape of a polymer molecule resulting from the spatial arrangement of the atoms and molecular groups within the polymer chain.

Conservation of energy

1. In a chemical change there is no loss or gain but merely a transformation of energy from one form to another.

2. In every modification of a material system not affected by forces foreign to the system, the sum of its potential and kinetic energies remains constant.

Conservation of mass. In all ordinary chemical changes the total mass of the reactants is always equal to the total mass of the products.

Constitutive property. Property that depends on the constitution or structure of the molecule.

Constructive interference. Two (or more) waves exhibit constructive interference if they combine in phase. The interference is destructive if the waves are out of phase.

Convection. Flow that takes place within a fluid due to density differences arising from temperature differences.

Copolymer. Polymer molecule formed from two or more types of monomers or repeating structural units.

Copolymerization. Bonding of two or more different organic molecules into a single polymer chain without loss of

their individual chemical identities. Mechanical properties can vary with the specific arrangement of the chain.

Cosmic rays. High-speed particles of various types that are produced in violent events in stars, travel through space, and enter the earth's atmosphere.

Couple

1. Pair of forces that have the same magnitudes but opposite directions and that are applied off-center to an object, thereby producing a torque.

2. Two equal and oppositely directed parallel but not collinear forces acting on a body form a couple. The moment of the couple or torque is given by the product of one of the forces by the perpendicular distance between them.

Covalent bond. Interatomic bonding based on the sharing of valence electrons. Covalent bonds are found primarily in organic and polymeric materials.

Covalent bonding. Joining together of atoms to form a molecule by sharing electrons in their outermost shells.

Critical temperature. Temperature above which a gas cannot be liquefied by pressure alone. The pressure under which a substance may exist as a gas in equilibrium with the liquid at the critical temperature is the critical pressure.

Cross-linking. Chemical bonding of polymer chains to form a network structure.

Cryohydrate. Solid that separates when a saturated solution freezes. It contains the solvent and the solute in the same proportions as they were in the saturated solution.

Crystal

1. Solid composed of atoms that are arranged in a regular geometrical pattern.

2. Physically homogeneous solid formed by atoms arranged in a repetitive pattern.

3. Homogeneous portion of a substance bounded by plane surfaces making definite angles with each other, giving a regular geometrical form.

Cure. Chemically cross-link polymer chains by heating and/or adding a chemical agent.

Curie. Unit that is used to specify the amount of radioactivity in a sample.

Dalton's law. The pressure exerted by each component in a gaseous mixture is independent of other gases in the mixture, and the total pressure of the mixture of gases is equal to the sum of the pressures of the separate components.

Dalton's law of partial pressures. The pressure exerted by a mixture of gases is equal to the sum of the separate pressures that each gas would exert if it alone occupied the whole volume.

Damping capacity. Ability of a material to absorb vibrational energy.

De Broglie wavelength of a particle. The wavelength of a particle is inversely proportional to its momentum. A particle with a wavelength will exhibit wavelike properties that are the same as those of electromagnetic radiation with a wavelength.

Decibel. Unit of sound intensity. An increase of 10 dB corresponds to an increase of sound intensity by a factor of 10; an increase of 20 dB corresponds to an intensity increase by a factor of 100; and so forth.

Declination. Angle between the vertical plane containing the direction of the earth's field at any point and a plane containing the geographic north and south meridian.

Decomposition. Chemical separation of a substance into two or more substances, which may differ from each other and from the original substances.

Degree of crystallinity. Proportion of the structure of a bulk polymer that is crystalline. The degree of crystallinity can vary from 40 to 95% depending on the conformation of the polymer and the processing techniques.

Degree of freedom. Number of variables determining the state of a system (usually pressure, temperature, and concentrations of the components) to which arbitrary values can be assigned.

Degree of polymerization. Number of units in a polymer, determined as an average value from a sample of many individual polymer chains. The average molecular weight of the chains in a sample is divided by the weight of 1 mer of the sample.

Definite proportions law. In every chemical compound there is always a definite proportion by mass of each constituent element.

Density. Concentration of matter measured by the mass per unit volume, expressed as grams per cubic centimeter.

Density of a substance. Mass per unit volume.

Deuterium. Name given to hydrogen if the nucleus consists of one neutron in addition to the one proton of ordinary hydrogen. Deuterium is an isotope of hydrogen.

Dew point. Temperature at which the condensation of water vapor in the air takes place.

Dimensional formula. If mass, length, and time are considered fundamental quantities, the relation of other physical quantities and their units to these three may be expressed by a formula with appropriate exponents.

Dip. Angle measured in a vertical plane between the direction of the earth's magnetic field and the horizontal.

Dislocation. Imperfection in a crystal structure that can cause plastic deformation.

Displacement. Reaction in which an elementary substance displaces and sets free a constituent element from a compound.

Displacement or elongation. At any instant, the distance of a vibrating or oscillating particle from its position of equilibrium.

Distribution law

1. A substance distributes itself between two immiscible solvents so that the ratio of its concentrations in the two solvents is approximately a constant (and equal to the ratio of the solubilities of the substance in each solvent).

2. The law states that if a substance is dissolved in 2 immiscible liquids, the ratio of its concentration in each is constant.

Double decomposition. Simple exchange of the parts of two substances to form two new substances.

Dulong and Petit's law of thermal capacity. For simple substances the atoms all have approximately the same thermal capacity. The product of the specific heat by the atomic weight is a constant, about 6.38 calories (264 joules) per degree celsius.

Earth's magnetic field. Earth's magnetic field is similar to that of a giant bar magnet and is believed to be due to intense electrical currents circulating in the molten iron core.

Earth's radiation belts. Earth's radiation belts are due to the trapping of electrons and protons in the earth's magnetic field.

Efficiency. Ratio of the work done by a machine to the work done upon it.

Efficiency of a device or process. Ratio of the amount of work or energy delivered to the input amount of work or energy. Efficiency = (work done)/(energy used).

Elasticity. Property by virtue of which a body resists and recovers from deformation produced by force.

Elastic limit. Smallest value of the stress producing permanent alteration.

Elastomer. Material that can be extended to at least twice its original length at room temperature and immediately return to its original length when the load is removed. Natural rubber represents the structure of essentially unconnected polymer chains that exhibit this type of behavior.

Electrolysis. Process by which free elements are liberated from an ionic solution as the result of an electrical current passing through the solution.

Electrolytic dissociation or ionization theory. When an acid, base, or salt is dissolved in water or any other dissociating solvent, a part or all of the molecules of the dissolved substance are broken up into parts called "ions," some of which are charged with positive electricity and called "cations," and an equivalent number of which are charged with negative electricity and called "anions."

Electrolytic solution tension theory. (Helmholtz double layer theory). When a metal or any other substance capable of existing in solution as ions is placed in water or any other dissociating solvent, a part passes into solution in the form of ions, thus leaving the remainder charged with an equivalent amount of electricity of opposite sign from that carried by the ions. This establishes a difference in potential between the substance and the solvent in which it is immersed.

Electromotive series. List of metals arranged in decreasing order of their tendencies to pass into ionic form by losing electrons.

Electron

1. Very small negatively charged particle. Electrons appear to be uniform in mass and charge and to be one of the basic elements of which atoms are made.

2. Ultimate indivisible negative charge. The electron and the proton or positive charge constitute the basic elements in the formation of atoms. The electronic ratio is the ratio of the charge of an electron to its mass.

3. Elementary particles that constitute the outer portions of the atoms.

4. Basic carriers of negative electrical charge.

5. Negatively charged subatomic particles found in shells moving about the atomic nucleus. Electrons participate in chemical bonding, electrical conduction, and heat conduction.

Electron cloud. In metallic bonding, the common pool or cloud to which atoms donate their valence electrons. The cloud of electrons serves to bind the metallic ions together.

Electron shells. Energy levels in which electrons are located as they move about the nucleus of an atom.

Electron theory of matter. An atom is believed to consist of a nucleus bearing a positive charge, different for each sort of atom, surrounded by electrons or negative charges equal in total charge to the positive charge of the nucleus. The nucleus may consist of a certain number of protons or elementary positive charges and part of the electrons. The remaining electrons revolve as satellites around the nucleus. The electron and proton have equal negative and positive charges; hence a neutral atom will contain as many electrons as protons. Protons contain practically all of the mass of the atom, the number of protons determining the atomic weight. The number of satellite electrons determines the chemical properties of the atom. According to recent views the nucleus consists only of heavy particles, neutrons, and protons.

Electron volt. Unit of energy.

Electroplating. Deposition of one metal on another at the cathode of an electrolytic solution when an electrical current is passed through the solution.

Elementary particle. Particle that cannot be broken down into smaller components. Over 100 elementary particles are known, most of which have extremely short half-lives. Common elementary particles are electrons, protons, neutrons, and photons.

Elements

1. Substances that cannot be decomposed or transformed into one another by chemical means. Just over 100 chemical elements are known.

2. Substances that cannot be decomposed by ordinary types of chemical change or made by chemical union.

Emission spectrum (or bright-line spectrum). Series of lines of definite wavelength produced when light from a source is passed through a prism. The lines in an emission spectrum are characteristic of the source of the light.

Emissive power or emissivity. Emissive power or emissivity is measured by the energy radiated from unit area of a surface in unit time for unit difference of temperature between the surface in question and surrounding bodies.

Emissivity of a material. The emissivity of a material determines how much electromagnetic energy will be radiated by the material at a particular temperature. A good emitter is also a good absorber.

Enantiotropic. Term describing crystal forms capable of existing in reversible equilibrium with each other.

Energy

1. Quality possessed by an object that enables it to do work. Energy due to motion is kinetic energy, and the energy due to position in a field of force is potential energy. The metric unit of energy is the joule (J). Energy is often measured in kilowatt hours.

2. Capability of doing work.

Entropy. Quantity depending on the quantity of heat in a body and its temperature, which when multiplied by any lower temperature (minimum available), gives the unavailable energy or unavoidable waste when mechanical work is derived from the heat energy of the body.

Equilibrium. An object or system is said to be in equilibrium if there is no net force acting. Often, a condition of equilibrium is one of rest, but it can be one of uniform motion.

Equilibrium constant. Product of the concentrations (or activities) of the substances produced at equilibrium in a chemical reaction divided by the product of the concentrations of the reacting substances, each concentration raised to the power that is the coefficient of the substance in the chemical equation.

Equivalence principle. Effects due to gravity cannot be distinguished from effects due to an accelerated reference frame.

Equivalent or combining weight. The equivalent or combining weight of an element or ion is its atomic or formula weight divided by its valence. Elements entering into combination always do so in quantities proportional to their equivalent weights.

Eutectic

1. Term applied to the mixture of two or more substances that has the lowest melting point.

2. An alloy or solution that has the lowest possible constant melting point.

Evaporation. Process by which molecules escape from the surface of a liquid, thereby gradually converting the liquid into a gaseous vapor.

Exclusion principle. No two electrons in an atom can have exactly the same four quantum numbers. The exclusion principle accounts for the occurrence of electron shells in atoms.

Expansion of gases (Charles's law or Gay-Lussac's law). The volume of a gas at constant pressure increases proportionately to the absolute temperature.

Faraday's laws. In the process of electrolytic changes equal quantities of electricity charge or discharge equivalent quantities of ions at each electrode. One gram of equivalent weight of matter is chemically altered at each electrode for 96,500 C, or 1 faraday of electricity passed through the electrolyte.

Filament. Fiber of indefinite length, such as a silk or nylon fiber.

Fission. Splitting of a nucleus into two more or less equal fragments with the release of a substantial amount of energy.

Fluidity. Reciprocal of viscosity.

Force

1. Push or a pull that can alter the state of motion of an object (produce an acceleration). The unit of force is the newton.

2. That which changes the state of rest or motion in matter, measured by the rate of change of momentum. The absolute unit is 1 cm/second/second, the dyne, the force that will produce an acceleration of 1 cm/second/second in a gram mass. The gram weight or weight of a gram mass is the cgs gravitational unit. The poundal is the force that will give an acceleration of 1 ft/second/second to a pound mass.

Formula. Combination of symbols with their subscripts representing the constituents of a substance and their proportions by weight.

Fossil fuels. Natural fuels that are derived from previously living matter: coal, oil, and natural gas.

Foucault's pendulum. A swinging weight supported by a long wire, so that the wire's upper support restrains the wire only in the vertical direction, and the weight is set swinging with no lateral or circular motion; the plane of the pendulum gradually changes, demonstrating the rotation of the earth on its axis.

Frequency. In uniform circular motion or in any periodic motion frequency is the number of revolutions or cycles completed in unit time.

Frequency-modulated wave. Wave on which information has been impressed by varying the frequency, with the amplitude remaining the same.

Friction. Resistance to motion that an object experiences because it moves through, moves over, or rests on another object or substance. Moving (or kinetic) friction and static friction are two distinct types of friction.

Fuel cell. Device that produces electrical power through the flameless oxidation of a fuel.

Fundamental temperature scale. Absolute, thermodynamic or Kelvin scale in which the temperature measure is based on the average kinetic energy per molecule of a perfect gas.

Fusion. Combining of two light nuclei into a more massive nucleus with the release of a substantial amount of energy.

Gamma ray

1. Bundle (or photon) of very high frequency electromagnetic radiation. Gamma rays are often emitted by nuclei following radioactive decay.

2. Highly penetrating radiations from radioactive substances, undeflected by electric or magnetic fields, representing a high-frequency electromagnetic radiation. Gamma rays have the same nature as x-rays but are of higher frequency.

Gas. State of matter in which the molecules are practically unrestricted by cohesive forces. Gas has neither definite shape nor volume.

Gauss. Metric unit of magnetic field strength.

Gay-Lussac's law of combining volumes. If gases interact and form a gaseous product, the volumes of the reacting gases and the volumes of the gaseous products are to each other in very simple proportions, which can be expressed by small whole numbers.

Genetic radiation damage. Damage to human genes caused by exposure to radiation. The effects of such damage can be manifest in the progeny of the exposed person.

Geomagnetic poles. North and south poles of the earth's magnetic field. The north pole is located in the Northern Hemisphere, about 800 miles from the geographic North Pole.

Geothermal energy. Thermal energy within the earth's crust due to heating caused by the decay of radioactive materials. Geothermal energy can be tapped by drilling wells to release steam or hot water.

Gibbs' phase rule. The number of degrees of freedom of a system is the number of variable factors (temperature, pressure, and concentration) of the components, which must be arbitrarily fixed in order that the condition of the system be perfectly defined.

Glass. Amorphous, noncrystalline solid made by fusing silica with a basic oxide.

Glass transition. Reversible change in an amorphous polymer or in amorphous regions of a partially cystalline polymer from (or to) a viscous or rubbery condition to (or from) a hard and relatively brittle one.

Glass transition range. Temperature range through which liquid amorphous polymers become supercooled amorphous polymers, and vice versa. As the liquid plastic cools, the motion of the polymer chains slows down. Below the glass temperature chain motion is so limited that the structure appears to be a solid.

Graham's law. The relative rates of diffusion of gases under the same conditions are inversely proportional to the square roots of the densities of those gases.

Gram atom or gram atomic weight. Mass in grams numerically equal to the atomic weight.

Gram equivalent of a substance. Weight of a substance displacing or otherwise reacting with 1.008 g of hydrogen or combining with one-half of a gram atomic weight (8.00 g) of oxygen.

Gram molecular weight or gram molecule. Mass in grams of a substance numerically equal to its molecular weight; gram mole.

Gram mole, gram formula weight, gram equivalent. Mass in grams numerically equal to the molecular weight, formula weight, or chemical equivalent, respectively.

Gravitation. Universal attraction existing between all material bodies.

Gravitational field. Condition in space set up by a mass to which any other mass will react.

Gravitational force

1. One of the four basic forces in nature. Newton's law of universal gravitation states that the gravitational force between two objects is directly proportional to the product of their masses and inversely proportional to the square of the distance separating their centers.

2. Gravitational force is always attractive.

Half-life of a radioactive substance. Time required for one-half of the atoms in any sample of the substance to undergo decay.

Hardness. Property of substances determined by their ability to abrade or indent one another.

Heat. Thermal energy in transit. If a hot object is placed in contact with a cold object; heat will flow from the hot to the cold object; some of the molecular motion of the hot object will be transferred to the cold object. The unit of heat is the calorie.

Heat equivalent or latent heat of fusion. Quantity of heat necessary to change 1 g of a solid to a liquid with no temperature change.

Heat of combustion (of a substance). Amount of heat evolved by the combustion of 1 gram molecular weight of the substance.

Heat of fusion (of a substance). Amount of energy required to convert a substance from the solid state to the liquid state at the freezing temperature.

Heat of vaporization (of a substance). Amount of energy required to convert a substance from the liquid state to the gaseous state at the boiling temperature.

Heat quantity. Heat quantity is measured by the change of temperature produced.

Hefner unit. Horizontal intensity of the Hefner lamp burning amyl acetate with a flame 4 cm high.

Henry's law. The height of a slightly soluble gas that dissolves in a definite weight of a liquid at a given temperature is very nearly directly proportional to the partial pressure of that gas. This holds for gases that do not unite chemically with the solvent.

Hess's law of constant heat summation. The amount of heat generated by a chemical reaction is the same whether the reaction takes place in one step or in several steps; all chemical reactions that start with the same original substances and end with the same final substances liberate the same amounts of heat, irrespective of the process by which the final state is reached.

Holography. Process by which three-dimensional optical images are produced by laser beams.

Homopolymer. Polymer molecule formed from a single type of monomer or repeating structural unit.

Hooke's law

1. The extension of an elastic object is directly proportional to the stretching force that is applied.

2. Law of mechanics that states that within the elastic zone of a material, the stresses are directly proportional to strains. This proportionality is constant and is known as the "modulus of elasticity."

3. Within the elastic limit of any body, the ratio of the stress to the strain produced is constant.

Hybrid bonding. Carbon atoms enter into compounds by using their S and P electrons in an equivalent way.

Hydrogen bonds. Strong electrical bonds between polar molecules that are due to the exposed nuclear proton in molecules containing hydrogen.

Hydrogen equivalent (of a substance). Number of replaceable hydrogen atoms in one molecule or the number of atoms of hydrogen with which one molecule would react.

Hydrogen ion concentration or pH value. Logarithm of the reciprocal of the gram ionic hydrogen equivalents per liter.

Hydrolysis. Double decomposition reaction involving the splitting of water into its ions and the formation of a weak acid or base, or both.

Hydronium ion. A hydronium ion is formed when a hydrogen ion (H^+) attaches itself to a water molecule in solution. The hydronium ion is the principal hydrogen-containing positive ion in water solutions.

Hysteresis. The magnetization of a sample of iron or steel due to a magnetic field which, made to vary through a cycle of values, lags behind the field. This phenomenon is called "hysteresis."

Ideal gas law. Law that combines the laws of Boyle and of Charles and Gay-Lussac; that is, the pressure, volume, and absolute temperature of a gas are related according to PV/Tk = constant.

Illumination. Density of the luminous flux on a surface. It is the quotient of the flux by the area of the surface when the latter is uniformly illuminated.

Index of refraction of a transparent medium. Measure of the ability of the medium to refract a light wave. n is equal to the ratio of the speed of light in vacuum to the speed of light in the medium.

Indicators. Substances that change from one color to another when the hydrogen ion concentration reaches a certain value, which is different for each indicator.

Inert gases. Elements in the eighth row of the periodic table that have completed valence shells. These elements do not react chemically, and they exist as gases at room temperature.

Inert gases (or noble gases). Gases that have completely filled outer shells and therefore are chemically inactive.

Inertia. Resistance offered by a body to a change of its state of rest or motion. Inertia is a fundamental property of matter.

Inertia of an object. Quality of an object that resists a change in its state of motion. The measure of an object's inertia is its mass.

Inertial reference frame. Unaccelerated frame; frame in which Newton's laws are valid. A frame that moves with constant velocity with respect to an inertial frame is also an "inertial frame."

Inhibitor. Substance that prevents a chemical reaction.

Inorganic. Chemical compound of unliving matter.

Intensity of magnetization. The intensity of magnetization is given by the quotient of the magnetic moment of a magnet by its volume. The unit intensity of magnetization is the intensity of a magnet that has unit magnetic moment per cubic centimeter.

Intensity of radiation. Radiant energy emitted in a specified direction per unit time, per unit area of surface, per unit solid angle.

Ion

1. Atom that has lost or gained one or more electrons than is possessed by the normal, electrically neutral atom.

2. Atom or group of atoms exhibiting a net electronic charge by acquiring or losing valence electrons.

3. Acids, bases, and salts (electrolytes) when dissolved in certain solvents are more or less dissociated into electrically charged units, or parts of the molecules, called "ions." Some electrolytes dissociate into ions when fused. Ions carry charges of electricity and consequently have different properties from the uncharged radicals. Positive ions are atoms or group of atoms that have lost valence electrons; negative ions are those to which additional electrons have been added.

Ionic binding. Joining together of atoms in the form of positive and negative ions to produce a chemical compound.

Ionic bond. Atoms are bonded when valence electrons are interchanged to produce completed valence shells. Inorganic refractories are commonly bonded ionically.

Ionization energy of an atom or molecule. Minimum energy required to remove an electron and produce an ion.

Ionization potential. Potential required to transfer an electron from its normal quantum level to infinity.

Ionizing radiations. Particles or rays that produce ionization in matter. Usually, this term includes all charged-particle high-energy electromagnetic radiations.

Isomer. Organic molecule that has a fixed chemical composition but exhibits two or more structural arrangements of the component atoms.

Isomerism. Existence of molecules having the same number and kinds of atoms but in different configurations.

Isotactic. Pertaining to a type of polymeric molecular structure containing a sequence of regularly spaced asymmetric atoms arranged in like configuration in a polymer chain.

Isothermal. When a gas passes through a series of pressure and volume variations without change of temperature, the changes are called "isothermal." The line on a pressure-volume diagram representing these changes is called an "isothermal line."

Isotopes. Elements occupying the same place in the periodic system, having the same nuclear charge, but differing somewhat in atomic weight. Most of the ordinary inactive elements have been shown to consist of a mixture of isotopes.

Isotopes of a particular chemical element. All isotopes of a particular chemical element have nuclei with the same number of protons but with a different number of neutrons. Such isotopes have identical chemical properties but different nuclear properties.

Joule. Metric unit of work or energy. The work is done by a force of one newton acting through a distance of 1 meter. The joule is 10^7 ergs.

Kepler's laws

1. Empirical laws formulated to describe the motions of planets. The most significant of these laws states that planets move around the sun in elliptical orbits.

2. Planets move about the sun in ellipses, at one focus of which the sun is situated. The radius vector joining each planet with the sun describes equal areas in equal times. The cubes of the mean distances of the planets from the sun are proportional to the squares of their times of revolution about the sun.

Kilowatt-hour. Common unit of energy or work. The equivalent energy supplied by a power of 1000 watts for 1 hour, (KWh).

Kinetic energy. Energy due to motion.

Kinetic energy of an object. Energy possessed by the object because of its motion. The metric unit of kinetic energy is the joule.

Kinetic theory of gases. Theory that relates the bulk properties of a gas to the microscopic motions of the constituent molecules.

Kirchoff's laws. The algebraic sum of the currents that meet at any point is zero. In any closed circuit the algebraic sum of the products of the current and the resistance in each conductor in the circuit is equal to the electromotive force in the circuit.

Laser. Device that emits a narrow beam of light with a pure frequency and with all of the photons in phase.

Latent heat of vaporization. Quantity of heat necessary to change 1 g of liquid to vapor without change of temperature. Both the above quantities are measured as calories per gram.

Lattice energy. Energy required to separate the ions of a crystal to an infinite distance from each other.

Lattice parameter. Distance between atoms or ions in a crystal.

Lattice structure. Regular, repetitive arrangement of atoms or ions.

Law of combining weights. If the weights of elements that combine with each other are called their "combining weights," then elements always combine either in the ratio of their combining weights or of simple multiples of these weights.

Law of component substances. Every material consists of one substance or is a mixture of two or more substances, each of which exhibits a specific set of properties, independent of the other substances.

Law of definite proportions. In every sample of each compound substance, the proportions by weight of the constituent elements are always the same.

Law of mass action. At a constant temperature the product of the active masses on one side of a chemical equation when divided by the product of the active masses on the other side of the chemical equation is a constant, regardless of the amounts of each substance present at the beginning of the action.

Law of multiple proportions. If two elements form more than one compound, the weights of the first element that combine with a fixed weight of the second element are in the ratio of integers to each other.

Laws of thermodynamics. When mechanical work is transformed into heat or heat into work, the amount of work is always equivalent to the quantity of heat. It is impossible by any continuous self-sustaining process for heat to be transferred from a colder to a hotter body.

LeChatelier's principle. The principle establishes that when an external force is applied to a system at equilibrium, the system adjusts so as to minimize the effect of the applied force.

Length contraction. Length contraction takes place when two observers are in relative motion. An observer in motion with respect to an object will see that object with its length contracted compared to the length seen by an observer at rest with respect to the object. This effect is usually important only at relativistic speeds.

Lenz's law

1. If any electromagnetic change causes an effect, then that effect will always induce a reaction that tends to oppose the original change.

2. When an electromotive force is induced in a conductor by any change in the relation between the conductor and the magnetic field, the direction of the electromotive force is such as to produce a current whose magnetic field will oppose the change.

Light year. Distance that light will travel in 1 year.

Linear momentum of an object. Product of the object's mass and its velocity. The linear momentum of an object can be changed only by the application of a force.

Linear polymer. Chainlike polymer composed of repeating molecules with very slight branching.

Liquid. State of matter in which the molecules are relatively free to change their positions with respect to each other but are restricted by cohesive forces so as to maintain a relatively fixed volume.

Loschmidt's number

1. Equivalent to Avogadro's number.

2. Number of molecules per unit volume of an ideal gas at 0°C and normal atmospheric pressure.

Macerate. Dissolve out the bonding material between plant cells to obtain separate, entire cells for examination.

Macromolecule. Molecule composed of many smaller molecules. Polymers are macromolecules.

Macroscopic matter. Large-scale matter; objects that are visible to the unaided eye, ranging in size to the largest astronomical objects.

Macrostructure. Structure of material that can be observed by the unaided eye or low magnification.

Magnet. Every magnet has a north (N) and south (S) pole. Like poles repel and unlike poles attract. Permanent magnets retain their magnetism; electromagnets are magnetic only as long as the exciting electrical current flows in the windings.

Magnetic domain. Tiny crystal that uses permanent magnetic properties. If the domains of a piece of iron are aligned, a net magnetism results; if the domains are oriented at random, the sample as a whole is not magnetic.

Magnetic field. Condition in space set up by a magnet to which other magnets or moving charged particles react.

Magnetic field intensity or magnetizing force. Magnetic field intensity is measured by the force acting on a unit pole. Unit field intensity, the gauss, is the field that exerts a force of 1 dyne on a unit magnetic pole. The field intensity is also specified by the number of lines of force intersecting a unit area normal to the field, equal numerically to the field strength in gauss. The magnetizing force is measured by the space rate of variation of magnetic potential and as such its unit may be the gilbert per centimeter.

Magnetic field strength. Magnetic field strength is given in terms of the magnetic force on a charged particle moving in the field. The unit of magnetic field strength is the tesla or the gauss.

Magnetic flux. The magnetic flux through an area perpendicular to a magnetic field is measured as the product of the area by the field strength. The unit of magnetic flux, the maxwell, is the flux through a square centimeter normal to a field of 1 g. The line is also a unit of flux. It is equivalent to the maxwell.

Magnetic induction. Magnetic induction results when a substance is subjected to a magnetic field. It is measured as the magnetic flux per unit area taken perpendicular to the direction of the flux. The unit is the maxwell per square centimeter or its equivalent, the gauss.

Magnetic moment of a magnet. The magnetic moment of a magnet is measured by the torque experienced when it is at right angles to a uniform field of unit intensity. The value of the magnetic moment is given by the product of the magnetic pole strength by the distance between the poles. The unit magnetic moment is that possessed by a magnet formed by two poles of opposite sign and of unit strength 1 cm apart.

Magnetic permeability. Property of materials modifying the action of magnetic poles placed therein and modifying the magnetic induction resulting when the material is subjected to a magnetic field or magnetizing force. The permeability of a substance may be defined as the ratio of the magnetic induction in the substance to the magnetizing field to which it is subjected. The permeability of a vacuum is unity.

Magnetic pole or quantity of magnetism. Two unit quantities of magnetism concentrated at points a unit distance apart in a vacuum repel each other with unit force. If the distance involved is 1 cm and the force 1 dyne, the quantity of magnetism at each point is 1 cgs unit of magnetism.

Magnetic poles of a magnet. The magnetic poles of a magnet are labeled N and S. The N pole of a compass magnet is north-seeking and points toward the earth's geomagnetic pole in the Northern Hemisphere (which is actually an S pole).

Magnetic potential or magnetomotive force. The magnetic potential or magnetomotive force at a point is measured by the work required to bring a unit positive pole from an infinite distance (zero potential) to the point. The unit is the Gilbert, the magnetic potential against which an erg of work is done when a unit magnetic pole is transferred.

Mass. Quantity of matter.

Mass energy. Energy associated with a quantity of matter according to the Einstein equation.

Mass number of a nucleus. The mass number of a nucleus is equal to the sum of the number of protons and neutrons in the nucleus.

Mass of an object. Measure of the object's inertia (its resistance to change in state of motion). Mass may be considered the quantity of matter in an object.

Mass resistivity. Longitudinal resistance per unit length of a uniform bar of the substance of such a sectional area that it contains one unit of mass per unit of length.

Mechanical advantage. Ratio of the resistance overcome to the force applied.

Mechanical equivalent of heat

1. Conversion factor connecting mechanical work and heat energy: 1 calorie = 4186 J.

2. Quantity of energy that when transformed into heat is equivalent to a unit quantity of heat 4.18×10^7 ergs = 1 calorie (20°C).

Melting point of a substance. Temperature at which the substance changes from the solid state to the liquid state.

Mer. Basic repeating structural unit of a polymer chain.

Metallic elements. In general, metallic elements are distinguished from the nonmetallic elements by their luster, malleability, conductivity, and usual ability to form positive ions.

Micelle. Crystalline portion of a bulk polymer structure produced when segments of a single chain are folded together or when neighboring chains are parallel so that the structural units of the chains adopt a repetitive pattern.

Microhardness. Hardness of microscopic parts (usually, the individual crystal or constituents) of material.

Microscopic matter. Matter too small to be seen by the unaided eye, ranging in size down to atomic and subatomic particles.

Mixtures. Mixtures consist of two or more substances intermingled with no constant percentage composition and with each component retaining its essential original properties.

Modulus of elasticity. Stress required to produce unit strain. It may be a change of length (Young's modulus), a twist or shear (modulus of rigidity or modulus of torsion), or a change of volume (bulk modulus), expressed in dynes per square centimeter.

Molal solution. Solution that contains 1 mole/1000 g of solvent.

Molar solution. Solution that contains 1 mole or gram molecular weight of the solute in 1 liter of solution.

Mole. Mass numerically equal to the molecular weight.

Molecular architecture. Arrangement of the constituent atoms of polymers and, by extension, the intentional alteration or redesigning of macromolecules.

Molecular or atomic rotatory power. Product of the specific rotatory power by the molecular or atomic weight.

Molecular refractivity. Product of specific refractivity by the molecular weight.

Mole (one) of a substance. An amount with a mass in grams equal to the molecular mass of the substance expressed in atomic mass units.

Molecular volume. Volume occupied by 1 mole. It is numerically equal to the molecular weight divided by the density.

Molecular weight. Sum of the atomic weights of all the atoms in a molecule.

Molecule

1. Smallest unit of a particular substance that exists in nature. Some elements exist naturally as molecules, and all compounds exist as molecules.

2. Unit of matter consisting of two or more chemically bonded atoms. The sum of the atomic weights is the molecular weight.

3. Smallest unit quantity of matter that can exist by itself and retain all the properties of the original substance.

Mol volume. The volume occupied by a mol or a gram molecular weight of any gas measured at standard conditions is 22.414 liters.

Moment of force or torque. Effectiveness of a force to produce rotation about an axis, measured by the product of the force and the perpendicular distance from the line of action of the force to the axis.

Moment of inertia. The sum of the products formed by multiplying the mass (or sometimes the area) of each element of a figure by the square of its distance from a specified line.

Momentum. Quantity of motion measured by the product of mass and velocity.

Monochromatic emissive power. Ratio of the energy of certain defined wavelengths radiated at definite temperatures to the energy of the same wavelengths radiated by a blackbody at the same temperature and under the same conditions.

Monomer

1. Organic molecule capable of being converted into a polymer by chemical reaction with similar molecules or with other organic molecules.

2. Simple compound that can react to form a polymer.

Monotropic. Crystal forms one of which is always metastable with respect to the other.

Muons. Short-lived elementary particles that result from the decay of positively and negatively charged pions. Muons decay into electrons and neutrinos.

Network polymers. Structure of thermosets. The individual polymer chains are bound together by chemical crosslinking and/or mechanical entanglements.

Neutralization. Reaction in which the hydrogen ion of an acid and the hydroxyl ion of a base unite to form water, the other product being a salt.

Neutrinos. Elementary particles that are produced and emitted in certain radioactive decay processes and in the decay of muons. Neutrinos have no mass and carry no electrical charge. There are four distinct types of neutrinos.

Neutron

1. Subatomic particle located in the nucleus of an atom. It has an arbitrary weight of 1 and is electrically neutral.

2. Elementary particle with approximately the mass of a hydrogen atom but without any electric charge. The neutron is one of the constituents of the atomic nucleus.

Newton (N). Metric unit of force.

Newton's laws of motion

1. An object will maintain its state of rest or motion unless acted upon by a force. An object will accelerate in the direction of a force applied to it. For every force there is an equal and opposite reaction force.

2. Every body continues in its state of rest or of uniform motion in a straight line except in so far as it may be compelled to change that state by the action of some outside force. A change of motion is proportional to force applied and takes place in the direction of the line of action of the force. To every action there is always an equal and opposite reaction.

Node. Position of no vibration in a standing wave.

Nonmetallic elements. Nonmetallic elements are not malleable, have low conductivity, and never form positive ions.

Normal salt. Ionic compound containing neither replaceable hydrogen nor hydroxyl ions.

Normal solution. Contains 1 gram molecular weight of the dissolved substance divided by the hydrogen equivalent of the substance (that is, 1 gram equivalent) per liter of solution.

Nuclear force. Strongest of the four basic forces in nature. Nuclear force is responsible for binding together protons and neutrons in nuclei.

Nucleus. Heavy central particle of an atom in which most of the mass and the total positive electric charge is concentrated. The charge of the nucleus, an integral multiple (Z) of the electronic charge, is the essential factor that distinguishes one element from another. Z is called the "atomic number."

Nucleus of an atom. Tiny central core that carries most of the mass and all of the positive charge of the atom. All nuclei consist of protons and neutrons.

Ohm. Unit of electrical resistance.

Ohm's law. The current flow through a conductor is directly proportional to the potential difference across the conductor.

Oligomer. Polymer consisting of only a few monomer units such as a dimer, trimer, or tetramer, or their mixtures.

Organic matter. Either natural or synthetic matter that is derived from living matter.

Oxidation. Process that increases the proportion of oxygen or acid-forming elements or radicals in a compound.

Parabolic motion. An object that is projected with some horizontal velocity component will undergo parabolic motion.

Parallel laminate. Laminate in which all the layers of a material or process are oriented approximately parallel with respect to the grain or strongest direction in tension.

Paramagnetic bodies. Bodies that tend to set the longest dimension parallel to the magnetic field. The permeability of a paramagnetic substance is greater than unity.

Pascal's law. The pressure exerted at any point on a confined liquid is transmitted undiminished in all directions.

Pascal's principle. The pressure applied to one part of a fluid is transmitted undiminished to every other part of the fluid.

Perigee of an earth satellite. Point in its elliptical orbit at which the satellite is closest to the earth. For the motion of a planet around the sun, the corresponding term is "perihelion."

Periodic law. Elements when arranged in the order of their atomic weights or atomic numbers show regular variations in most of their physical and chemical properties.

Period in uniform circular motion. Time of one complete revolution. In any oscillatory motion it is the time of a complete oscillation.

Period of a motion or a wave. Time required to complete one revolution or cycle and to return to the initial condition.

Periodic table of the chemical elements

1. Tabular arrangement of the elements by increasing atomic number which shows similarities in physical and chemical properties that can be explained by the electronic structure of the elements.

2. Way of ordering the elements to show the periodicity of similar chemical properties.

Permeance. Reciprocal of reluctance. Unit permeance is the permeance of a cylinder with 1-sq cm cross section and 1-cm length taken in a vacuum.

Phase of oscillatory motion. Fraction of a whole period that has elapsed since the moving particle last passed through its middle position in a positive direction.

Phosphorescence. Delayed reradiation of electromagnetic energy by excited electrons.

Photoconductivity. Ability of covalent compounds to conduct electricity when their valence electrons are excited by incident light on other electromagnetic radiations.

Photoemission. Excitation of electrons to leave a material by the application of electromagnetic radiation of the proper wavelength.

Photons. Bundles of electromagnetic wave energy. The energy of a photon is proportional to its frequency.

Piezoelectric effect. Reversible conversion of mechanical energy into electrical energy by complex metal oxide ceramic crystals.

Pions. Short-lived elementary particles that are produced in many types of high-speed collisions between nuclear particles. There are three types of pions, identified by their electrical charge.

Planck's constant. Proportionality factor between the energy and frequency of a photon.

Plasma. Matter at such a high temperature that all the atoms are ionized; intimate mixture of nuclei and electrons.

Poisson's ratio. Ratio of the transverse contraction per unit dimension of a bar of uniform cross section to its elongation per unit length, when the bar is subjected to a tensile stress.

Polar molecule. Molecule in which there is a preponderance of negative charge at one end and positive charge at the other end. Polar molecules usually contain hydrogen and can be joined together by hydrogen bonds.

Polarization. Separation of the charge in an object so that one part bears a positive charge and another part bears an equal negative charge.

Polarized light. Light in which the electric field vector points preferentially in one direction.

Polymer. Very high molecular weight compound formed from smaller molecules that contain covalent bonds that permit them to join together. Polymers can be made from a single type of molecule or from several types of molecule. The properties of polymers are based on their high molec-ular weight, large molecular size, and the bonding of individual polymer chains into a bulk shape.

Polymerization

1. Process by which small organic molecules react together to form a long chain.

2. Process by which long-chain molecules are constructed from small molecules.

Polymorphism. Property of a chemical substance crystallizing into 2 or more forms having different structures, such as diamond and graphite.

Polysaccharides. Carbohydrates, such as starch or cellulose, decomposable into two or more molecules of monosaccharides (simple sugars like glucose) or their derivatives.

Positron

1. Elementary particle that is the same as an ordinary electron except that it carries a positive charge.

2. Particle with a mass equal to the electron, but possessing a positive charge.

Potential energy. Energy due to the position of one body with respect to another or to the relative parts of the same body.

Potential energy of an object. Energy possessed by an object by virtue of its position in a field of force. If an object is at a height above the surface of the earth, an amount of potential energy equal to mg/hour can be released if the object falls to the earth. The metric unit of potential energy is the joule.

Pot life. Period of time during which a reacting thermosetting plastic or rubber composition remains suitable for its intended processing after mixing with reaction-initiating agents.

Power

1. Rate at which work is done or energy is expended.

2. Time rate at which work is done.

Pressure

1. Force per unit area exerted on an object.

2. Force applied to or distributed over a surface, measured as force per unit area.

Proton

1. Subatomic particle located in the nucleus of the atom and exhibiting a positive charge. It has an arbitrary weight of 1.

2. Elementary particle having a positive charge equivalent to the negative charge of the electron but possessing a mass approximately 1845 times as great. The proton is in effect the positive nucleus of the hydrogen atom.

3. Protons are positively charged elementary particles found in the nuclei of all atoms. The magnitude of the charge carried by a proton is exactly equal to that carried by an electron.

Proton-proton chain. Sequence of nuclear reactions that converts hydrogen into helium in stars.

Pyrometric cone. Small, tapered ceramic slab that slumps when heated and is used as a measure of temperature when firing ceramics.

Quantity of electricity or charge. Electrostatic unit of charge; quantity that when concentrated at a point and placed at unit distance from an equal and similarly concentrated quantity, is repelled with unit force. If the distance is 1 cm, the force of repulsion 1 dyne, and the surrounding medium a vacuum, we have the electrostatic unit of quantity. The electromagnetic unit of quantity may be defined as the quantity transferred by unit current in unit time. The quantity transferred by 1 A in 1 second is the coulomb (C), the practical unit. The faraday is the electrical charge carried by 1 gram equivalent.

Quantum

1. A discrete portion of energy of a definite amount.
2. An entity resulting from a quantization of a field or wave.
3. The photon is the quantum of an electromagnetic field.

Quantum number. One of the quantities, usually discrete with integer or half-integer values, needed to characterize a quantum state of a physical system.

Quantum numbers. Quantum numbers are necessary to specify completely the state of an electron in an atom.

Quantum theory. Mathematical theory of the behavior of microscopic matter. It is completely a wave theory and does not invoke any mechanical models.

Rad. Unit that specifies the amount of radiation energy absorbed by an object.

Radiation damage. Damage done to living biological matter by the action of ionizing radiations.

Radioactive substances. Substances that continuously undergo a process of atomic disintegration in which energy is liberated.

Radioactivity. A property exhibited by certain elements, the atomic nuclei of which spontaneously disintegrate and gradually transmute the original element into stable isotopes of that element or into another element with different chemical properties. The process is accompanied by the emission of alpha particles, beta particles, gamma rays, positrons, or similar radiations.

Radius of gyration. Distance from the axis of rotation at which the total mass of a body might be concentrated without a change in its moment of inertia. The product of total mass and the square of the radius of gyration will give the moment of inertia.

Ranking scale of temperature. Absolute Fahrenheit scale.

Raoult's law. The molar weights of nonvolatile nonelectrolytes when dissolved in a definite weight of a given solvent under the same conditions lower the solvent's freezing point, elevate its boiling point, and reduce its vapor pressure equally for all such solutes.

Real image. The image formed by an optical system is a real image if that image can be projected onto a screen.

Reduction. Process that increases the proportion of hydrogen, base-forming elements, or radicals in a compound.

Reflection. Turning back of a wave when it is incident on a surface.

Refractive index. Ratio of the speed of light in a material to the speed of light in a vacuum. This ratio indicates the ability of a material to transmit light.

Relative biological effectiveness of an ionizing radiation. Measure of the amount of biological damage produced by radiation compared to the damage produced by electrons of x-rays of the same energy.

Relative humidity. Ratio of the quantity of water vapor present in the atmosphere to the quantity that would saturate at the existing temperature. It is also the ratio of the pressure of water vapor present to the pressure of saturated water vapor at the same temperature.

Relative humidity of a mass of air. Amount of water vapor in the air expressed as a percentage of the amount that the air is capable of holding at the particular temperature. A relative humidity of 100% means that the air is saturated.

Relative visibility. The relative visibility factor for a particular wavelength is the ratio of the visibility factor for that wavelength to the maximum visibility factor.

Reluctance. Property of a magnetic circuit that determines the total magnetic flux in the circuit when a given magnetomotive force is applied. The unit is the reluctance of 1-cm length and 1-sq cm cross section of space taken in a vacuum.

Reluctivity or specific reluctance. Reciprocal of magnetic permeability. The reluctivity of empty space is taken as unity.

Rem. Unit that is used to measure the radiation dosage in living tissue.

Replacement series. Arrangement of the metals in order of the values of their oxidation potentials.

Rest mass of an object. Mass as measured by an observer at rest with respect to the object.

Reversible reactions. Reactions in which the products of the reaction may in turn react upon each other to form the original reacting substances.

Right-hand rule. The right-hand rule for determining the direction of the magnetic field lines surrounding a current-carrying wire is as follows: Grasp the wire with the right hand, thumb pointing in the direction of current flow; the fingers will then encircle the wire in the direction of the magnetic field lines. There are similar right-hand rules for determining the direction of the angular momentum vector and the direction of the magnetic force on a moving charged particle.

Rotation. Movement of atomic and molecular segments of a polymer caused by thermal energy. The presence of bulky side groups can hinder rotation.

Salt. Substance that yields ions, other than hydrogen or hydroxyl ions. Salt is obtained by displacing the hydrogen of an acid by a metal.

Satellite. Object that orbits around an astronomical body. The planets are satellites of the sun; the moon is a satellite of the earth; and artificial satellites have been placed into orbit around the earth.

Shock wave. Concentration of wave motion along a surface due to the motion of the source through the medium at a speed greater than the speed of the wave in the medium. A sonic boom is an example of a shock wave.

Silicate unit. The basic structural pattern of most ceramic materials is that of a silicon atom surrounded by four tetrahedrally arranged oxygen atoms.

Snell's law of optics. Snell's law relates the angle of refraction of a light ray to the angle of incidence and to the indexes of refraction of the two materials.

Solar wind. The solar wind consists of rapidly moving charged particles ejected from the sun. These particles influence the magnetic field of the earth in space.

Solid. State of matter in which the relative motion of the molecules is restricted. The molecules tend to retain a definite fixed position relative to each other, giving rise to a crystal structure. A solid may be said to have a definite shape and volume.

Solid angle. Angle measured by the ratio of the surface of the portion of a sphere enclosed by the conical surface forming the angle to the square of the radius of the sphere. The unit of a solid angle is the steradian, the solid angle that encloses a surface on the sphere equivalent to the square of the radius.

Solubility of a gas. Ratio of the concentration of gas in the solution to the concentration of gas above the solution.

Solubility of one liquid or solid in another. When the mass of a substance contained in a solution is in equilibrium with an excess of the substance, the solution is said to be saturated.

Solubility product or precipitation value. Product of the concentrations of the ions of a substance in a saturated solution of the substance.

Solute. Constituent of a solution that is considered to be dissolved in the other, the solvent. The solvent is usually present in larger amount than the solute.

Solvent. Constituent of a solution that is present in larger amount; constituent of a solution that is liquid in the pure state, in the case of solutions of solids or gases in liquids.

Somatic effects of radiation. Effects that appear in an individual exposed to a large dose of radiation.

Specific gravity. Ratio of the mass of a body to the mass of an equal volume of water at 4°C or other specified temperature.

Specific heat of a substance

1. Amount of heat required to change the temperature of 1 kg of the substance 1°C.

2. Ratio of the thermal capacity of a substance to that of water at 15°C.

Specific volume. Reciprocal of density.

Speed

1. Rate at which an object moves regardless of the direction of motion.

2. Time rate of motion measured by the distance moved over in unit time.

Spherical candlepower. The spherical candlepower of a lamp is the average candlepower of the lamp in all directions in space. It is equal to the total luminous flux of the lamp in lumens divided by 4π.

Spin. Common name for the intrinsic angular momentum of an elementary particle.

Standard conditions for gases. Measured volumes of gases are quite generally recalculated to 0°C temperature and 760-mm pressure, which have been arbitrarily chosen as standard conditions.

Standing wave. Periodic disturbance set up between two boundaries such that reflections cause a regular pattern of reinforcements and cancellations.

Stimulated emission by an atom of a photon. Stimulated omission with a certain frequency occurs when another photon with the same frequency passes close to the excited atom. Stimulated emission is the basic process that takes place in lasers.

Stoichiometric. Pertaining to weight relations in chemical reactions.

Stoke's law. Stoke's law gives the rate of fall of a small sphere in a viscous fluid. When a small sphere falls under the action of gravity through a viscous medium, it ultimately acquires a constant velocity.

Strain. Deformation resulting from a stress, measured by the ratio of the change to the total value of the dimension in which the change occurred.

Stress. Force producing or tending to produce deformation in the body, measured by the force applied per unit area.

Subatomic particles. Bits of matter with no specific chemical identity that combine to form atoms. The common subatomic particles are protons, neutrons, and electrons.

Surface tension. Two fluids in contact exhibit phenomena, due to molecular attractions, which appear to arise from a tension in the surface of separation.

Symbol. The symbol for an element is not only an abbreviation of the name, but represents 1 atom and 1 gram atomic weight of that element.

Synchronous satellite. Satellite that revolves in its orbit around the earth at the same rate that the earth rotates on its axis and therefore maintains a fixed position relative to the earth.

Temperature

1. Measure of the internal motion of an object's constituent molecules. The greater the motion, the greater the internal energy and the higher the temperature.

2. Condition of a body that determines the transfer of heat to or from other bodies. The customary unit of temperature is the degree Celsius, $\frac{1}{100}$ of the difference between the temperature of melting ice and that of boiling water at standard atmospheric pressure. The degree Fahrenheit is $\frac{1}{180}$, and the degree Reaumur $\frac{1}{80}$ the same temperature difference.

Tesla. Metric unit of magnetic field strength.

Theory of relativity. The theory of relativity improves upon Newtonian theory, especially in the description of phenomena that takes place at high speeds. The special theory relates to situations in which two reference frames move relative to one another at constant velocity. The general theory treats cases of acceleration and gravitation.

Thermal capacity of a substance. Quantity of heat necessary to produce a unit change of temperature in unit mass. It is ordinarily expressed as calories per gram per degree Celsius and is numerically equivalent to specific heat.

Thermal capacity or water equivalent. Total quantity of heat necessary to raise a body or system unit temperature, measured as calories per degree Celsius in the cgs system.

Thermal conductivity

1. Time rate of transfer of heat by conduction, through unit thickness, across unit area for unit difference of temperature, measured as calories per second per square centimeter for a thickness of 1 cm and a difference of temperature of (1°C).

2. Ability of materials to transmit heat energy. Thermal conductivity occurs at the atomic or molecular level and, in metals, at the electronic level.

Thermal energy. Internal energy of an object due to the motion of the constituent molecules. The greater the thermal energy of an object, the higher is its temperature.

Thermal expansion. The coefficient of linear expansion or expansivity is the ratio of the change in length per degree to the length at 0°C. The coefficient of surface expansion is twice the linear coefficient. The coefficient of volume expansion (for solids) is three times the linear coefficient. The coefficient of volume expansion for liquids is the ratio of the change in volume per degree to the volume at 0°C. The value of the coefficient varies with temperature. The coefficient of volume expansion for a gas under constant pressure is nearly the same for all gases and temperatures and is equal to 0.00367 for 1°C.

Thermal pollution. Result of waste heat exhausted by power plants into bodies of water or into the air.

Thermionic emission. Excitation of electrons to leave a material by the application of thermal energy.

Thermodynamics. The branch of physics that seeks to derive, from a few basic postulates, relationships between properties of matter, especially those that are affected by changes in temperature, and a description of the conversion of energy from one form to another.

Thermonuclear reaction. Fusion reaction that will proceed only if the reactants are at an extremely high temperature.

Time dilation. Time dilation takes place when two observers are in relative motion. An observer in motion with respect to a clock will see that clock run more slowly than will an observer who is at rest with respect to the clock.

Torque. An object experiences a torque when a force is applied in such a way that the object tends to rotate.

Triangle or polygon of forces. If three or more forces acting on the same point are in equilibrium, the vectors representing them form, when added, a closed figure.

Tritium. Name given to the radioactive form of hydrogen, in which the nucleus consists of two neutrons in addition to the one proton of ordinary hydrogen. Tritium is an isotope of hydrogen.

True solution. Liquid, solid or gaseous mixture, in which the components are uniformly distributed throughout. The proportion of the constituents may be varied within certain limits.

Uncertainty principle. It is not possible to measure simultaneously the position and the momentum of a particle or photon with unlimited precision. The uncertainty principle is a key ingredient of quantum theory.

Unit of area. Square centimeter; area of a square whose sides are 1 cm in length. Other units of area are similarly derived.

Valence electrons of the atom. Electrons that are gained, lost, or shared in chemical reactions.

Valence of an atom of an element. Property that is measured by the number of atoms of hydrogen (or its equivalent) one atom of that element can hold in combination if negative, or can displace in a reaction if positive.

Valence shell. Outermost electron shell, containing the high-energy electrons that take part in chemical reactions. Atoms or ions with a filled valence shell are in a stable, low-energy configuration.

Van der Waals forces. Long-range interatomic forces that can bond gas atoms and molecules. The forces result because the motions of valence electrons in nearby atoms influence each other.

Van't Hoff's principle. If the temperature of interacting substances in equilibrium is raised, the equilibrium concentrations of the reaction are changed so that the products of that reaction which absorb heat are increased in quantity, or if the temperature for such an equilibrium is lowered, the products that evolve heat in their formation are increased in amounts.

Vapor pressure of a substance at a given temperature. Pressure of the vapor in a confined space above the substance. At the boiling point of an unconfined liquid, the vapor pressure is equal to the atmospheric pressure.

Vector. Quantity that requires both magnitude and direction for its complete specification.

Velocity

1. Combination of the rate at which an object moves (the speed) with the direction of motion. Velocity is a vector quantity.

2. Time rate of motion in a fixed direction.

Velocity ratio. Ratio of the distance through which force is applied to the distance through which resistance is overcome.

Virtual image. An optical system forms a virtual image if that image can only be perceived by the eye but cannot be projected onto a screen.

Viscosity

1. Resistance exhibited by a liquid to flow under an applied load or pressure. Viscosity is an important property of molten polymers and glassy solids, among other materials.

2. All fluids possess a definite resistance to change of form and many solids show a gradual yielding to forces tending to change their form. This property, a sort of internal friction, is called "viscosity."

3. Property of resistance of flow exhibited within the body of a material.

Viscosity coefficient. Shearing stress necessary to induce a unit velocity flow gradient in a material.

Weak force. One of the four basic forces in nature. The weak force is responsible for radioactive decay and for various processes involving elementary particles.

Weight. Force with which a body is attracted toward the earth.

Weight of an object. Gravitational force acting on the object. Weight is proportional to mass.

Wien's displacement law. When the temperature of a radiating blackbody increases, the wavelength corresponding to maximum energy decreases in such a way that the product of the absolute temperature and wavelength is constant.

Work

1. Product of force and the distance through which the force acts. The metric unit of work is the joule.

2. When a force acts against resistance to produce motion in a body, the force is said to do work. Work is measured by the product of the force acting and the distance moved through against the resistance.

Work function of a material. Minimum energy required to release an electron from the surface of the material by the photoelectric effect.

X-ray electromagnetic radiation. Radiations emitted by atoms when transitions occur in the inner electron shells.

ROCKS AND MINERALS

Acid. Term describing ingeneous rock containing a high percentage of silica.

Aggregate. Natural or man-made material with rocklike particles ranging from $\frac{1}{4}$ to $2\frac{1}{2}$ in. in diameter, with or without sand and artificial binder; used as a subgrade, base, or surface for a road. Term usually refers to gravel or crushed rock.

Alkali feldspars. Feldspars containing potassium or sodium occur typically in igneous rocks such as granites and rhyolites, whereas those of higher calcium content are found in igneous rocks of lower silica content, such as diorite, gabbro, andesite, and basalt.

Alum shales. Shales bearing alum, formed by the decomposition of pyrite.

Amphibole. Any of a group of related minerals that are essentially silicates of aluminum, magnesium, calcium, and iron.

Amphiboles and pyroxenes. Various types of igneous and metamorphic rocks containing characteristic dark green to black minerals, generally silicates of iron or magnesium, or both. Minerals of the amphibole and pyroxene groups are frequently present as prisms. The most common amphibole is hornblende; the most common pyroxene is augite. Amphiboles, pyroxenes, and micas are also found in marbles.

Amphibolite

1. Medium-to-coarse-grained rock composed mainly of hornblende and plagioclase feldspar. Schistosity, due to parallel alignment of hornblende grains, is commonly less obvious than in typical schists.

2. Metamorphic rock containing mostly amphibole and plagioclase feldspar.

Andesite. Eruptive rock whose mineral composition is plagioclase and hornblende or angite.

Ankerite. Objectionable calcium, magnesium, and iron carbonate sometimes found in calcareous rocks.

Anorthite. Calcium feldspar often present in diabase.

Apatite. Accessory mineral sometimes occurring in granites. It is a calcium phosphate.

Aphanitic texture. In igneous rocks, a grain size that is so uniformly small that crystals are invisible to the naked eye.

Aphte. Fine-grained granite consisting of quartz and feldspar. Muscovite may be sparingly present. Outcrop usually occurs as a dike.

Aragonite. Mineral having the same composition as calcite but crystallizing in the orthohombic system.

Arenaceous limestones. Carbonate rocks containing 10 to 50% sand are arenaceous (or sandy) limestones (or dolomites).

Argillaceous. Term applied to limestones and sandstones containing clayey matter.

Argillaceous limestones. Limestones containing 10 to 50% clay are argillaceous (or clayey or shaly) limestones (or dolomites).

Argillites

1. Massive, firmly indurated, fine-grained argillaceous rocks consisting of quartz, feldspar, and micaceous minerals. Argillites do not slake in water as some shales do. To distinguish these fine-grained sediments from fine-grained, foliated metamorphic rocks such as slates and phyllites, it should be noted that cleavage surfaces of shales are generally dull and earthy, while those of slates are more lustrous. Phyllite has a higher glossy luster resembling a silky sheen.

2. Argillaceous schist or clay slate that breaks readily into thin slabs.

Arkose

1. Coarse-grained sandstone derived from granite, containing conspicuous amounts of feldspar.

2. Variety of sandstone containing abundant feldspar and quartz, frequently in angular, poorly sorted grains.

3. Variety of sandstone containing an appreciable quantity of feldspar.

Augen gneiss. Gneiss containing phenocrysts or porphyroblasts that have been deformed into eye-shaped grains or grain clusters.

Augite

1. Most common pyroxene, a ferromagnesian mineral.

2. Aluminous pyroxene, most common species of the pyroxene group.

Aureole. Area surrounding an intrusion that has been affected by contact metamorphism.

Banded iron ore. Sediment consisting of layers of chert alternating with bands of ferric iron oxides (hematite and limonite) in valuable concentrations.

Barite. Common barium mineral and the major barium ore. Occurs in veins transecting many kinds of rocks and is concentrated in sedimentary rocks. In many of its occurrences it is accompanied by clay or a calcium sulfate mineral (gypsum or anhydrite), or both. Barite has a hardness of 3 to $3\frac{1}{2}$ and a specific gravity of 4.50 in the pure mineral; the color ranges from colorless to white to many usually pale colors.

Basalt

1. Fine-grained extrusive equivalent of gabro and diabase. When basalt contains natural glass, the glass is generally lower in silica content than that of the lighter colored extrusive rocks and hence is not deleteriously reactive with the alkalies in portland cement paste.

2. Fine-grained, dark, mafic igneous rock composed largely of plagioclase feldspar and pyroxene.

3. Basic igneous rock consisting of plagioclase and angite. Olivine is a common constituent.

Basement. Oldest rocks recognized in a given area; complex of metamorphic and igneous rocks that underlies all the sedimentary formations. Basement is usually Precambrian or Paleozoic in age.

Basic. Term applied to igneous rocks high in bases and low in their silica content.

Basic rock. Igneous rock containing mafic minerals rich in iron and magnesium but no quartz and little sodium-rich plagioclase feldspar.

Batholith. Irregular mass of coarse-grained igneous rock with an exposed surface of more than 100 km^2 which has either intruded the country rock or been derived from it through metamorphism.

Bauxite. Rock composed primarily of hydrous aluminum oxides and formed by weathering in tropical areas with good drainage; major ore of aluminum.

Bedding. Characteristic of sedimentary rocks in which parallel planar surfaces separating different grain sizes or compositions indicate successive depositional surfaces that existed at the time of sedimentation.

Biotite

1. Dark mica. Biotite may also be considered a ferromagnesian mineral. It readily forms cleaved flakes or plates.

2. Mineral of the mica family. Essentially, silicate of aluminum, magnesium, and iron. Often called "black mica," "iron mica," or "magnesium mica."

Blueschist facies. High-pressure form of metamorphic rock containing the blue amphibole glaucophane.

Borocalcite. Turkish borate ores, which have been referred to as "borocalcite," but are probably ulexite or colemanite, or mixtures of the two, used in shielding concrete.

Boron frit glasses. Clear, colorless, synthetic glasses produced by fusion and quenching, used in making ceramic glazes. May be obtained in many compositions, but those most useful in shielding concrete contain calcium, relatively high amounts of silica and alumina, and low amounts of alkalies. Increased silica and alumina decrease the solubility of the frits and thus diminish their retarding effect in shielding concrete.

Boron minerals. Commercially important sources of boron are principally sodium, calcium, and magnesium borate precipitates from waters in arid volcanic regions, or alteration products of such precipitates.

Breccia

1. Clastic rock composed principally of large angular fragments. Usually the clasts are all derived from the same parent formation.

2. Rock made up of angular fragments produced by crushing and then recemented by infiltrating mineral matter.

Calcareous. Containing calcium carbonate.

Calcite. Mineral consisting of calcium carbonate but crystalizing in the hexagonal system. Its cleavage is rhomohedral and perfect.

Calcite and dolomite. Most common carbonate mineral is calcite (calcium carbonate). Mineral dolomite consists of calcium carbonate and magnesium carbonate in equivalent molecular amounts. The hardness of calcite is 3 and that of dolomite $3\frac{1}{2}$ to 4 on the Mohs scale. They have rhombohedral cleavage, which results in their breaking into fragments with smooth parallelogram-shaped sides.

Calcite dolomite. If 50 to 90% of a rock is mineral dolomite, the rock is called calcitic dolomite. Most carbonate rocks contain some noncarbonate impurities such as quartz, chert, clay minerals, organic matter, gypsum, and sulfides.

California colemanite deposits. These deposits contain ulexite, but apparently are not regularly worked. However, colemanite ores have been obtained from them for use in shielding concrete.

Carbonate rocks

1. Rock composed of carbonate minerals, especially limestone and dolomite.

2. Most carbonate rock contains some noncarbonate impurities such as quartz, chert, clay minerals, organic matter, gypsum, and sulfides.

3. Known as chalk or "lime rock."

Carbonate rocks and shales. These rocks and shales may contain sulfates as impurities. Most abundant sulfate mineral is gypsum (hydrous calcium sulfate); anhydrite (anhydrous calcium sulfate) is less common. Gypsum is usually white or colorless and characterized by a perfect cleavage along one plane and its softness, representing a hardness of 2 on the Mohs scale.

Chalcedony. Chalcedony is considered both a distinct mineral and a variety of quartz and is frequently composed of a mixture of microscopic fibers of quartz with a large number of submicroscopic pores filled with water and air. The properties of chalcedony are intermediate between those of opal and quartz, from which it can sometimes be distinguished only by laboratory tests. It frequently occurs as a constituent of chert and is reactive with the alkalies in portland cement paste.

Chemical weathering. Total set of all chemical reactions that act on rock exposed to water and atmosphere and change its minerals to more stable forms.

Chert

1. Very fine-grained siliceous rock characterized by hardness and conchoidal (shell-like) fracture in dense varieties, the fracture becoming splintery in porous varieties. Dense varieties are very tough; they are usually gray to black or white to brown, and less frequently green, red, or blue, with a waxy to greasy luster.

2. Sedimentary form of amorphous or extremely fine-grained silica, partially hydrous, found in concretions and beds.

3. Cryptocrystalline variety of quartz. Applied to hornstone and any impure flinty rock.

Chlorites

1. The green micaceous minerals often found in schists are usually chlorites. Chlorites may be distinguished from the micas because they form comparatively nonelastic flakes.

2. Group name for the greenish colored micaceous minerals. They are silicates of aluminum with magnesium and iron. Secondary mineral derived from the alteration of pyroxene, amphibole, biotite, and vesuvianite.

Clastic rock. Sedimentary rock formed from mineral particles (clasts) that were mechanically transported.

Clay minerals

1. "Clay" refers to natural material composed of particles in a specific size range, generally less than 0.002 mm. Principal clay mineral groups found in and associated with natural mineral aggregates are clay-micas (illites), the kaolin group (very finely divided chlorites), and the swelling clays (montmorillonites).

2. Hydrous aluminum, magnesium, and iron silicates that may contain calcium, magnesium, potassium or sodium and other ions, formed by alteration and weathering of other silicates and volcanic glass. Clay minerals are major constituents of clays and shales. They are found disseminated in carbonate rocks and altered and weathered igneous and metamorphic rocks.

Claystones or siltstones. Fine-grained rocks largely composed of or derived by erosion of sedimentary silts and clays or any type of rock that contained clay. When soft they are known as "clay-stones" or "siltstones," depending on the size of the majority of the particles of which they are composed. Siltstones consist predominantly of silt-sized particles (0.0625 to 0.002 mm in diameter) and are intermediate rocks between claystones and sandstones. When the claystones are harder and platy, or fissile, they are known as "shales." Claystones and shales may be gray, black, reddish, or green, and may contain some carbonate minerals (calcareous shales).

Cobblestone. Rounded or partially rounded rock or mineral fragment between 3 and 10 in. in diameter.

Columnar jointing. Division of an igneous rock body into prismatic columns by cracks produced by thermal contraction on cooling.

Common iron oxide minerals. There are two common iron oxide minerals, black, magnetic, magnetite and red or reddish (when powdered) hematite, and one common hydrous oxide mineral, brown, or yellowish goethite. Another common iron-bearing mineral is black, weakly magnetic, ilmenite. Magnetite and ilmenite are important accessory minerals in many dark igneous rocks and are common detrital minerals in sediments. Very minor amounts of iron minerals color many rocks, such as ferruginous sandstones, shales, clay ironstones, and granites. Magnetite, ilmenite, and hematite ores are used as heavy aggregates.

Compensation (gravity). Mechanism by which segments of the earth's crust rise or sink to equilibrium positions, depending on the mass and density of the rocks above and below a certain depth, called the "compensation depth."

Competence (rock). Ability of a stratum to withstand deformation without fracturing or changing thickness.

Concordant contact. Planar contact of an intrusion that follows the bedding of the country rock.

Concretion. Rounded body of mineral matter such as chert in limestone or calcium carbonate or iron carbonate in certain clays and shales.

Conformable succession. Sequence of sedimentary rocks that indicates continuous deposition with no erosion over a geologically long period.

Conglomerate

1. Rock mass composed chiefly of rounded fragments. Such rocks are sometimes called "pudding stone." Conglomerate often includes the breccias.

2. Sedimentary rock, a significant fraction of which is composed of rounded pebbles and boulders; the lithified equivalent of gravel.

Conglomerates—sandstones and quartzites. Rocks that consist of particles of sand or gravel, or both, with or without interstitial and cementing material. If the particles include a considerable proportion of gravel, the rock is a conglomerate. If the particles are in the sand sizes, that is, less than 2 but more than 0.06 mm in major diameter, the rock is sandstone or quartzite; if the rock breaks around the sand grains, it is sandstone; if the grains are largely quartz and the rock breaks through the grains, it is quartzite.

Contact metamorphism. Mineralogical and textural changes and deformation of rock resulting from the heat and pressure of an igneous intrusion in the near vicinity.

Coquina. Limestone composed of loosely cohering shell fragments cemented together by an infiltration of carbonate of lime.

Country rock. Rock into which an igneous rock intrudes or a mineral deposit is emplaced.

Cross-bedding. Inclined beds of depositional origin in a sedimentary rock. Formed by currents of wind or water in the direction toward which the bed slopes downward.

Cryptocrystalline. Term applied to the varieties of quartz that are finely crystalline. Term is also applied to some igneous rocks.

Crystalline rocks. Metamorphic rocks composed of crystalline mineral grains.

Curie point. Temperature above which a given mineral cannot retain any permanent magnetization.

Daughter element. Element that occurs in a rock as the end product of the radioactive decay of another element.

Depositional remanent magnetization. Weak magnetization created in sedimentary rocks by the rotation of magnetic crystals into line with the ambient field during settling.

Diabase. Basic igneous rock consisting essentially of plagioclase, augite, and magnetite. Olivine may be present. It includes most trap rocks.

Diabase or dolerite. Rock of composition similar to gabbro and basalt, but intermediate in mode of origin, usually occurring in smaller intrusions than gabbro and having a medium-to-fine grained texture. Term "trap" or "trap rock" is a collective term for dark-colored, medium-to-fine-grained igneous rocks, especially diabase and basalt.

Diallage. Thin, foliated variety of pyroxene present in many gabbros.

Diatomite. Siliceous chertlike sediment formed from the hard parts of diatoms.

Dike. Roughly planar body of intrusive igneous rock that has discordant contacts with the surrounding rock.

Dike swarm. Group of dikes emanating from a common magma chamber.

Dilorite

1. Medium-to-coarse-grained rock composed essentially of plagioclase feldspar and one or more ferromagnesian minerals, such as hornblende, biotite, or pyroxene. Plagioclase is intermediate in composition, usually of the variety andesine, and is more abundant than the ferromagnesian minerals. Diorite usually is darker in color than granite or syenite and lighter than gabbro. If quartz is present, the rock is called "quartz diorite."

2. Basic igneous rock consisting of plagioclase, usually andesine, and hornblende; often porphyritic.

Diorite porphyry. Basic igneous porphyritic rock of the same mineral composition as diorite.

Disseminated deposit. Deposit of ore in which the metal is distributed in small amounts throughout the rock and not concentrated in veins.

Dolerite. Coarsely crystalline variety of basalt.

Dolomite

1. Mineral consisting of the double carbonates of calcium and magnesium; rock made up chiefly of the mineral dolomite.

2. Reaction of the dolomite in certain carbonate rocks with alkalies in portland cement paste has been found to be associated with deleterious expansion of concrete containing such rocks as coarse aggregate. Carbonate rocks capable of such reaction possess a characteristic texture, in which large crystals of dolomite are scattered in a fine-grained matrix of calcite and clay. Except in certain areas, such rocks are of relatively infrequent occurrence and seldom make up a significant proportion of the material present in a deposit of rock being considered for use in making aggregate for concrete.

Ductile rocks. Rock that can withstand 5 to 10% strain without fracturing.

Eclogite. Extremely high-pressure metamorphic rock containing garnet and pyroxene.

Epidote. Yellowish green mineral sometimes occurring in granites. In composition it is a silicate of calcium, aluminum, and iron.

Eruptive. Term describing igneous rocks that have been extruded.

Esker. Relatively long, narrow, winding ridges of mixed sand and gravel considered to have been deposited by streams of meltwater flowing through crevasses and tunnels in stagnant ice sheets.

Evaporite. Chemical sedimentary rock consisting of minerals, precipitated by evaporating waters, especially salt and gypsum.

Extrusive. Term describing igneous rocks that have cooled after reaching the surface.

Extrusive and intrusive rocks. Extrusive rocks commonly are so fine grained that the individual mineral grains are usually not visible to the naked eye. Porphyritic textures are common, and the rocks may be partially or wholly glassy. The glassy portion of a partially glassy rock usually has a higher silica content than the crystalline portion. Some intrusive rocks that originated in shallow depths may not be distinguishable in texture and structure from extrusive rocks.

Facies. Set of all the characteristics of a sedimentary rock that indicate its particular environment of deposition and distinguish it from other facies in the same rock unit.

Feldspars

1. Minerals of the feldspar group are the most abundant rock-forming minerals in the earth's crust. They are important constituents of all three major rock groups, igneous, sedimentary, and metamorphic. Since all feldspars have good cleavages in two directions, particles of feldspar usually show several smooth surfaces. Frequently, the smooth cleavage surfaces show fine parallel lines. All feldspars are slightly less hard than quartz. Various members of the group are differentiated by chemical composition and crystallographic properties.

2. Group of minerals possessing several characteristics in common. They are silicates of aluminum with potassium or sodium, or both, or with calcium, or sodium and calcium. Iron and manganese are absent. Their cleavage is perfect.

Felsic. Light-colored igneous rock poor in iron and magnesium content, abundant in feldspars and quartz.

Felsite

1. Light-colored, very fine-gained igneous rocks are collectively known as "felsites." The felsite group includes rhyolite, dacite, andesite, and trachyte, which are the equivalents of granite, quartz diorite, diorite, and syenite, respectively. Rocks are usually light colored, but they may be gray, green, dark red, or black. When they are dark, they may incorrectly be classed as "trap." When they are microcrystalline or contain natural glass, rhyolites, dacites, and andesites, they are reactive with the alkalies in portland cement concrete.

2. Cryptocrystalline mixture of quartz and feldspar. Ground mass of the quartz porphyries.

Felspathoid. Group of minerals that replace the feldspars in the formation of igneous rocks. Leucite, nephelite, and sodalite are the most common.

Ferromagnesian. Dark-colored silicates that contain both iron and magnesium.

Ferromagnesian minerals. Various types of igneous and metamorphic rocks containing characteristic dark green to black minerals, generally silicates of iron or magnesium, or both. They include the minerals of the amphibole and pyroxene groups.

Ferrophosphorus. Material produced in the production of phosphorus, consisting of a mixture of iron phosphides and used as coarse and fine aggregate in radiation-shielding concrete. Published specific gravities range from 5.72 to 6.50 for coarse aggregate. Coarse aggregate is reported to degrade easily and has been associated with extreme retardation of set in concrete. Ferrophosphorus in concrete releases flammable and possibly toxic gases that can develop high pressures if confined. Several iron phosphides are known, including silver gray to blue gray, with specific gravity of 6.50. Ferrophosphorus aggregates are silver gray but develop some rusty staining on exposure.

Ferruginous. Containing either the anhydrous or hydrous oxides of iron.

Fine-grained and glassy extrusive igneous rocks. Fine-grained equivalents of the coarse- and medium-grained igneous rocks have similar chemical compositions and may contain the same minerals. Extrusive rocks commonly are fine grained; and the individual mineral grains are usually not visible to the naked eye. The glassy portion of a partially glassy rock usually has a higher silica content than the crystalline portion. Some intrusive rocks that originated in shallow depths may not be distinguishable in texture and structure from extrusive rocks.

Fluorite. Mineral consisting of calcium fluoride.

Gabbro

1. Medium-to-coarse-grained dark-colored rock consisting essentially of ferromagnesian minerals and plagioclase feldspar. Ferromagnesian minerals may be pyroxenes or amphiboles, or both. Plagioclase is one of the calcium-rich varieties, namely, labradorite, bytownite, or anorthite. Ferromagnesian minerals are usually more abundant than feldspar.

2. Black, coarse-grained, intrusive igneous rock, composed of calcic feldspars and pyroxene, intrusive equivalent of basalt.

3. Basic igneous rock consisting of plagioclase, usually labradorite, and a pyroxene, usually augite or diallage. Magnetite is often present, and some varieties bear olivine.

Garnet. Silicate of aluminum, calcium, iron, or magnesium, which crystallizes in the isometric system. Common garnet in metamorphic limestones is grossularite, that in mica schists is almandine.

Gneiss

1. Coarse-grained regional metamorphic rock that shows compositional banding and parallel alignment of minerals.

2. Metamorphosed igneous rock having its ferromagnesian minerals arranged in more or less massive bands or layers; metamorphosed highly feldspathic sedimentary rock.

3. One of the most common metamorphic rocks usually formed from igneous or sedimentary rocks by a higher degree of metamorphism than the schists. Characterized by a layered or foliated structure resulting from

approximately parallel lenses and bands of platy minerals, usually micas, or prisms, usually amphiboles, and granular minerals, usually quartz and feldspars.

Gneissoid granite. Structure somewhat resembling a gneiss.

Goethite. Mineral having the same chemical composition as lepidocrocite but crystallizing differently. Color varies with the form, from crystals that are blackish brown with imperfect admantine metallic luster to dull or silky luster in fibrous varieties. Massive goethite is yellowish brown to reddish brown; clayey material is brownish yellow to ocher yellow, and streak is brownish yellow to ocher yellow.

Goethite ores. Ores that range from hard, tough, massive rocks to soft crumbling earths; these alternations frequently occur within fractions of an inch.

Granite

1. Medium-to-coarse-grained light-colored rock characterized by the presence of potassium and plagioclase feldspars and quartz. Characteristic potassium feldspars are orthoclase or microcline, or both; the common plagioclase feldspars are albite and oligoclase. Feldspars are more abundant than quartz. Dark-colored mica (biotite) is usually present, and light-colored mica (muscovite) is frequently present. Other dark-colored ferromagnesian minerals, especially hornblende, may be present in amounts smaller than those of the light-colored constituents.

2. Coarse-grained intrusive igneous rock composed of quartz, orthoclase feldspar, sodic plagioclase feldspar, and micas; sometimes a metamorphic product.

Granite porphyry. Igneous rock with the same mineral composition as granite but with a porphyritic texture.

Granitization. Formation of metamorphic granite from other rocks by recrystallization with or without complete melting.

Granitoid. Having a texture like that of granite.

Granodiorite. Diorite bearing an appreciable amount of quartz.

Granulite. Metamorphic rock with coarse interlocking grains and little or no foliation.

Gravel

1. Coarsest of alluvial sediments, containing mostly particles larger than 2 mm in size and including cobbles and boulders.

2. Small stones or fragments of stones.

3. Loose or unconsolidated coarse granular material, larger than sand grains, resulting from reduction of rock by natural or artificial means. Sizes range from $\frac{3}{16}$ in. (No. 4 sieve) to 3 in. in diameter. Coarse gravel ranges from 3 to $\frac{3}{4}$ in., while fine gravel ranges from $\frac{3}{4}$ to $\frac{3}{16}$ in.

Gravel, sand, silt, and clay. Gravel, sand, silt, and clay form the group of unconsolidated sediments. Although the distinction between these four members is made on the basis of their size, a general trend in the composition exists. Gravel and, to a lesser degree, coarse sands usually consist of rock fragments; fine sands and silt consist predominantly of mineral grains; and clay consists exclusively of mineral grains, largely of the group of clay minerals. All types of rocks and minerals may be represented in unconsolidated sediments.

Graywacke. Compact sandstone consisting of quartz, feldspars, and argillaceous matter.

Graywackes and subgraywackes. Gray to greenish-gray sandstones containing angular quartz and feldspar grains, and sand-sized rock fragments in an abundant matrix resembling claystone, shale, argillite, or slate. Graywackes grade into subgraywackes, the most common sandstones of the geologic column.

Greenschist. Metamorphic schist containing chlorite and epidote (which are green) and formed by low-temperature, low-pressure metamorphism.

Greenschist (mica schist). Green schistose rock whose color is due to an abundance of one or more of the green minerals, chlorite or amphibole, and commonly derived from altered volcanics.

Greenstone. Basic igneous rocks of greenish color due to the presence of chlorite.

Ground moraine. Heterogeneous accumulation of earth, sand, gravel, and boulders, deposited by a glacier, ordinarily thin compared to its areal extent, and usually with irregular topographic expression and unstratified. Ground moraine is thought to have accumulated largely by lodgment beneath the glacier ice, but also partly from being let down from the upper surface as the ice melted or evaporated.

Gypsum

1. Carbonate rocks and shales may contain sulfates as impurities. The most abundant sulfate mineral is gypsum (hydrous calcium sulfate). Anhydrite (anhydrous calcium sulfate) is less common. Gypsum is usually white or colorless, characterized by a perfect cleavage along one plane and by its softness, representing a hardness of 2 on the Mohs scale. Gypsum may form a whitish pulverulent or crystalline coating on sand and gravel. It is slightly soluble in water. Gypsum and anhydrite occurring in aggregates offer risks of sulfate attack on concrete and mortar.

2. Hydrous calcium sulfate, which is 2 on the scale of hardness.

Hematite

1. Frequently found as an accessory mineral in reddish rocks.

2. Mineral with a hardness of 5 to 6 on the Mohs scale and a specific gravity of 5.26 in the pure mineral. The color varies from bright red to dull red to steel gray, and the luster from metallic to submetallic to dull; the streak is cherry red or reddish brown. Hematite is nonmagnetic.

3. Anhydrous oxide of iron whose fine powder is cherry red or blood red in color.

Hematite ores. Rocks of which hematite is the major constituent vary in specific gravity, toughness, compactness, amount of impurities, degree of weathering, and suitability for use as concrete aggregate from one deposit to another and within the deposit. Hematite appears to be the iron ore mineral most exploited as a source of iron. Ores of the Lake Superior region are banded sedimentary ores consisting of layers rich in hematite and sometimes goethite, iron silicates, such as stipnomelane, minesotaite, greenalite, grunerite, and iron carbonate, alternating with silica-rich layers of chert or fine-grained quartz, or a mixture. Birmingham, Ala. ores are oolitic with hematite replacements of oolites and fossils in a matrix that ranges from fine-grained earthy hematite, with or without calcite, to crystalline calcite. Hematite ores dust in handling, the dust ranging in color from moderately red to dusky red to moderately reddish brown.

Hornblende. The most common amphibole is hornblende, a ferromagnesian mineral.

Hornfels. Equigranular, massive, and usually tough rock produced by complete recrystallization of sedimentary, igneous, or metamorphic rocks through thermal metamorphism, sometimes with the addition of components of molten rock. Mineral compositions vary widely.

Hypersthene. Mineral of the pyroxene group present in norite.

Igneous rock

1. Rock formed from molten rock matter either above or below the earth's surface.

2. Rock formed by congealing rapidly or slowly from a molten state.

3. Rock formed by the action of heat with sufficient intensity to effect fusion.

4. Rock formed by cooling from a molten rock mass. May be divided into two classes: coarse-grained, with grain size over 5 mm in diameter (intrusive, deep-seated), and fine-grained, with grain size less than 1 mm in diameter; frequently, partially or completely glassy (shallow intrusive, extrusive, surface, volcanic).

5. Igneous rocks are usually classified and named on the basis of their texture, internal structure, and mineral composition, which, in turn, depends to a large extent on their chemical composition. Rocks in the intrusive class generally have chemical equivalents in the extrusive class.

6. Fine-grained equivalents of the coarse- and medium-grained igneous rocks have similar chemical compositions and may contain the same minerals.

Ignimbrite. Igneous rock formed by the lithification of volcanic ash and volcanic breccia.

Illites (clay micas). Principal clay minerals groups found in and associated with natural mineral aggregates.

Illmenite. Mineral with a hardness of 5 to 6. Its color is iron black with metallic to submetallic luster, and the streak is black. Ilmenite is feebly magnetic.

Illmenite ores. Ores consisting of crystalline ilmenite with either magnetite or hematite and constituents of the associated gabbroic or anorthositic rocks. Massive ilmenite ores can form coarsely crystalline massive tough rocks but vary in specific gravity, composition, hardness, and suitability for use as concrete aggregate from deposit to deposit and within a deposit. Many ilmenite ores consist of ilmenite disseminated in rock-forming mineral. Ilmenite concentrated from beach sands is usually altered to a variable degree, and its mechanical properties probably differ from those of unaltered ilmenite. Ilmenite ore is one of the most widely used types of heavy aggregates.

Intrusion. Igneous rock body that has forced its way in a molten state into surrounding country rock.

Intrusive rock. Interpreted as a former intrusion from its cross-cutting contacts, chilled margins, or other field relations.

Iron sulfide minerals. Sulfides of iron, pyrite, marcasite, and pyrrhotite are frequently found in natural aggregates.

Irruptive. Term describing igneous rocks that have worked their way upward from the zone of flowage through other rocks, but have not flowed out over the surface.

Itacolamyte. Sandstone that is friable and flexible, especially when in thin slabs. Flexibility is due to the interlocking of the quartz grains.

Jasper. Cryptocrystalline variety of quartz, usually red or brown in color; term also applied to the marbles of Vermont, which in color closely resemble the mineral jasper.

Kaolin. The kaolin group constitutes very finely divided chlorites in the clay mineral group associated with natural mineral aggregates.

Kaolinite. Mineral that is a hydrous silicate of aluminum resulting from the decomposition of feldspars.

Kimberlite. Periodotite containing garnet and olivine and found in volcanic pipes, through which it may come from the upper mantle.

Labradorite. Mineral of the feldspar group, usually gray in color, crystallizing in the triclinic system, and with cleavage planes finely striated.

Lacolith. Sill-like igneous intrusion that forces apart two strata and forms a round lens-shaped body many times wider than it is thick.

Lepidocrocite. Mineral with a hardness of 5. Its color varies from ruby red to reddish brown, and the streak is dull orange. Lepidocrocite and goethite occur together, and lepidocrocite may be a constituent of goethite and limonite ores.

Limonite

1. General name for hydrous iron oxides of unknown composition, frequently cryptocrystalline goethite with adsorbed and capillary water, and probably mixtures of such goethite with similar lepidocrocite or hematite, or both, with adsorbed and capillary water. The color ranges from brownish black through browns to yellows. Limonite deposits range from recognizably crystalline goethite to dull, massive material of indefinite composition and therefore, properly limonite. Limonites of high iron content are also called brown iron ores. Frequently they contain sand, colloidal silica, clays, and other impurities.

2. Brown or yellow mineral that chemically is a hydrous oxide of iron.

3. Brown weathering product of iron-bearing minerals; field name for a variety of hydrous iron oxide minerals including goethite. It frequently contains adsorbed water and various impurities, such as colloidal or crystalline silica, clay minerals, and organic matter.

Lineation. Linear arrangement of features found in a rock.

Lithification. Processes that convert a sediment into a sedimentary rock.

Loess. Uniform aeolian deposit of silty material having an open structure and relatively high cohesion due to cementation of clay or calcareous material at grain contacts. A characteristic of loess deposits is their ability to stand with nearly vertical slopes.

Mafic mineral. Dark-colored mineral rich in iron and magnesium, especially a pyroxene, amphibole, or olivine.

Magma. Molten material from which the igneous rocks are formed.

Magnesite. White to brown magnesium carbonate crystallizing in the hexagonal system, often found massive.

Magnetite

1. Mineral with a hardness of $5\frac{1}{2}$ to $6\frac{1}{2}$ and strongly magnetic. The color is black with metallic to semimetallic luster; the streak is black.

2. Black iron oxide that is strongly magnetic both before and after hearing. Crystallizes in regular octahedrons.

Magnetic ores. Dense, tough, usually coarse-grained rocks with few impurities; magnetite ores are associated with metamorphic, igneous, or sedimentary rocks, and therefore, the impurities associated with magnetite ores may include a wide variety of rock-forming and accessory minerals. Magnetite occurs in association with hematite and ilmenite; magnetite ores are widely distributed, but many are not suitable for use as heavy aggregate because the magnetic occurs disseminated through rock rather than as a major rock-forming mineral. Magnetite is one of the most widely used types of heavy aggregates.

Mantlerock. Loose fragmental material that results from the disintegration of both igneous and sedimentary rocks.

Marcasite

1. Mineral that readily oxidizes with the liberation of sulfuric acid to form iron oxides and hydroxides, and, to a much smaller extent, sulfates; pyrite and pyrrhotite do so less readily. Marcasite and certain forms of concrete, producing a brown stain accompanied by a volume increase, which has been reported as one source of pop-outs in concrete. Reactive forms of iron sulfides may be recognized by immersion in saturated limewater (calcium hydroxide solution); the reactive varieties produce a brown precipitate within a few minutes.

2. White iron pyrite that crystallizes in the orthorhombic system. More readily decomposed than the other sulfides of iron.

Marl. Clayey limestone, fine-strained and commonly soft.

Metamorphic rocks

1. Rocks that have changed in response to different physical and chemical conditions below the earth's surface from their original texture, structure, and mineral composition, or with respect to one or two of these. Rocks are dense and may be massive, but are more frequently foliated and tend to break into platy particles.

2. Rocks formed from preexisting rocks by the action of heat, pressure, or shearing forces in the earth's crust. Not only igneous but also sedimentary and metamorphic rocks may be weathered and eroded to form new sedimentary rocks. Similarly, metamorphic rocks may again be metamorphosed. Most of the metamorphic rocks may derive either from igneous or sedimentary rocks, but a few, such as marbles and slates, originated only from sediments.

Metaphyre. Any igneous porphyry with a dark ground mass.

Metaquartzite. Granular rock consisting essentially of recrystallized quartz. Strength and resistance to weathering are derived from the interlocking of the quartz grains.

Mica. Group of minerals with eminent basal cleavage. Essentially hydrous silicates of aluminum with potassium, magnesium, and iron. Sodium and lithium micas are well known.

Micaceous minerals. Minerals that characteristically have a perfect cleavage in one direction and can therefore usually be split into extremely thin flakes. Micas are usually colorless or light green (muscovite), or black to dark brown or dark green (biovite), and form elastic flakes. Micas are abundant and occur in all three major rock groups.

Micaceous sandstone. Sandstone containing many scales of mica.

Microline. Potassium aluminum silicate of the feldspar group, crystallizing in the triclinic system. Frequently occurs in granite.

Migmatite. Rock with both igneous and metamorphic characteristics that shows large crystals and laminar flow structures. Probably formed metamorphically in the presence of water and without melting.

Mineral. Naturally occurring element or compound with a precise chemical formula and a regular internal lattice structure. Organic products are usually not included.

Montmorillonites. The swelling clays. Some clay minerals are made up of alternating layers of micas or chlorites with montmorillonites; such compositions resemble the montmorillonites in their fairly large volume changes upon wetting and drying. When such rocks are present in hardened concrete, the concrete will manifest increased volume change on wetting and drying.

Monzonite. Rock that is intermediate in mineral composition between a syenite and a diorite.

Muscovite. Potassium member of the mica group, often called the "white mica" or "potassium mica."

Nephelite. Silicate of aluminum, sodium, and potassium that crystallizes in the hexagonal system and occurs as small crystals or grains in the intermediate igneous rocks. Variety eleolite is distinguished by its greasy luster.

Obsidian

1. Dense, dark natural glass of high silica content.

2. Dark volcanic glass of felsic composition.

3. Common name for volcanic glass.

Oiigoclase. White sodium, calcium and aluminum silicate of the feldspar group. Common plagioclase mineral in granites.

Olivine

1. Common mineral, usually olive green, glassy in luster, and granular, found in dark igneous rocks of relatively low silica content. Does not occur in quartz-bearing igneous rocks, is uncommon in metamorphic rocks, and is not found in sediments except those deposited close to the olivine-bearing source rocks.

2. Olive green silicate of magnesium and iron crystallizing in the orthohombric system. Occurs in the basic and ultra basic igneous rocks.

Oolite. Granular limestone made up of concentric coats of the carbonate of calcium deposited around minute nuclei.

Opal

1. Hydrous form of silica, which occurs without characteristic external form or internal crystalline arrangement as determined by ordinary visible light methods. When x-ray diffraction methods are used, opal may show some evidences of internal crystalline arrangement. Opal has a variable water content, generally ranging from 3 to 9%. Specific gravity and hardness are always less than those of quartz. The color is variable, and the luster is resinous to glassy.

2. Noncrystalline hydrous variety of quartz.

Ophimagnesite. Rock consisting of crystallized magnesite and disseminated serpentine.

Ophitic. Consisting of interlacing lath-shaped crystals of feldspars whose interspaces are chiefly filled with pyroxenes of later growth.

Orthoclase, sanidine, and microcline. The feldspars orthoclase, sanidine, and microline are potassium aluminum silicates and are frequently referred to as the "potash" or potassium feldspars.

Paigeite. Mineral with a hardness of 5, coal black or greenish black in color, insoluble in water, and tough; high-temperature mineral occurring with magnetite in contact with metamorphic deposits. Paigeite has been used as a heavy boron-containing aggregate in Japan.

Pebble. Small stone worn smooth by the action of water, ice, sand, etc.; stone between 4 and 64 mm (about 0.16 to 2.5 in.) in diameter.

Pegmatite

1. Extremely coarse-grained varieties of igneous rocks are known as "pegmatites." They are usually light colored and most frequently equivalent to granite or syenite in mineral composition.

2. Igneous rock with extremely large grains, more than a centimeter in diameter. Pegmatite may be of any composition but is most frequently granitic.

3. Very coarse-grained phase of the granite rocks that usually occurs in dikes or lenses intruded in granites and metamorphic rocks.

Peridotite

1. Rock composed of olivine and pyroxene. Rocks composed almost entirely of pyroxene are known as "pyroxenites," and those composed of olivine as "dunites." Rocks of these types are relatively rare, but their metamorphosed equivalent, serpentinite, is more common.

2. Coarse-grained, mafic igneous rock composed of olivine with accessory amounts or pyroxene and amphibole, but little or no feldspar.

3. Ultrabasic igneous rock consisting essentially of olivine.

Perlite. High-silica glassy lava with an onionlike structure and a pearly luster, containing 2 to 5% water.

Phaneritic texture. Rock texture in which individual crystals are visible to the unaided eye.

Phenocryst. Prominent crystals in a rock of porphyritic texture.

Pholerite. Hydrous silicate of aluminum derived from the decomposition of orthoclase.

Phyllite

1. Fine-grained rock. Minerals, such as micas and chlorite, are noticeable and impart a silky sheen to the surface of schistosity. Phyllites are intermediate between slates and schists in grain size and mineral composition and are derived from agrillaceous sedimentary rocks or fine-grained extrusive igneous rocks, such as felsites.

2. Fine-grained metamorphic rock intermediate between a slate and schist.

Pisolite. Concretionary limestone with the globules about the size of small peas.

Pitchstone. Glass with up to 10% water and a dull resinous luster. Glassy rocks, particularly the more siliceous ones, are reactive with the alkalies of portland cement paste.

Plagioclase. Triclinic feldspars other than microline.

Plagioclase feldspars. Included in this group are sodium aluminum silicates and calcium aluminum silicates. This group, frequently referred to as the "soda-lime" group, includes a continuous series of varying chemical composition and optical properties, from albite, the sodium aluminum feldspar, to anorthite, the calcium aluminum feldspar, with intermediate members of the series designated oligoclase, andesine, labradorite, and bytownite. Alkali feldspars containing potassium or sodium occur typically in igneous rocks such as granites and rhyolites, whereas those of higher calcium content are found in igneous rocks of lower silica content, such as diorite, gabbro, andesite, and basalt.

Plutonic. Term characterizing granitoid igneous rock that has cooled a considerable distance below the surface of the earth.

Porphyries. Rock characterized by the presence of large mineral grains in a relatively finer-grained or glassy ground mass. Texture is the result of a sharp change in the rate of cooling or other physiochemical conditions during the solidification of the igneous rock.

Porphyritic. Term characterizing igneous rocks that contain phenocrysts of some mineral in a finer grained ground mass.

Porphyroblast. Large crystal in a finer-grained matrix of metamorphic rock, resembling a phenocryst in an igneous formation.

Porphyry. Igneous rock containing abundant phenocrysts of one mineral, but very fine grains of the other minerals.

Potassium feldspars. The feldspars orthoclase, sanidine, and microline are potassium aluminum silicates and frequently referred to as the "potash" or potassium feldspars.

Pudding stone. Conglomerate rock containing numerous rounded pebbles.

Pumice

1. Light-colored, finely vesicular, glassy froth filled with elongated, tubular bubbles. When heated quickly to the softening temperature, perlite puffs to become an artificial pumice. Pumices are usually silica-rich (corresponding to rhyolites or dacites), whereas scorias usually are more basic (corresponding to basalts).

2. Form of volcanic glass, usually of silicic composition, so filled with vesicles that it resembles a sponge and is very light.

Pyrite

1. Found in igneous, sedimentary, and metamorphic rocks; marcasite is much less common and found mainly in sedimentary rocks.

2. Common, yellow, metallic sulphide of iron which crystallizes in the isometric system.

3. Brass yellow, pyrrhotite bronze brown, both have a metallic luster; marcasite is also metallic, but lighter in color. Pyrite is often found in cubic crystals.

Pyroxene

1. Rocks composed almost entirely of pyroxene are known as "pyroxenites," and those composed of olivine as "dunites." Rocks of these types are relatively rare, but their metamorphosed equivalent, serpentinite, is more common.

2. Group of bisilicate minerals. Augite is the most important member of the group.

Pyroxene granulite. Coarse-grained contact metamorphic rock containing pyroxene, formed at high temperatures and low pressures.

Pyroxenite. Basic igneous rock consisting essentially of pyroxenes.

Pyrrhotite. May be found in many types of igneous and metamorphic rocks.

Quartz

1. Very common hard mineral composed of silica. When pure it is colorless with a glassy (vitreous) luster and a shell-like (conchoidal) fracture. It lacks a visible cleavage (the ability to break in definite directions along even planes), and when present in massive rocks such as granite, it usually has no characteristic shape. It is resistant to weathering and therefore an important constituent of many sand and gravel deposits and many sandstones. Quartz is abundant in many light-colored igneous and metamorphic rocks.

2. Form of silica occurring in hexagonal crystals or in cryptocrystalline massive forms.

3. Crystalline silica. In its most common form it is colorless and transparent, but it takes a large variety of forms of varying degrees of opaqueness and color. It is the most common solid mineral.

Quartz arenite. Sandstone containing very little except pure quartz grains and cement.

Quartz diorite. If quartz is present in diorite, the rock is called "quartz diorite."

Quartzite

1. Metamorphic sandstone whose cement is silica.

2. Very hard, clean, white metamorphic rock formed from a quartz arenite sandstone.

Quartz monzonite. Igneous rock of granitic texture containing quartz, with the plagioclase minerals equal or in excess of the orthoclase.

Quartz monozonite and granodiorite. Rocks similar to granite but containing more plagioclase feldspar.

Quartzose sandstone. Clean quartz sandstone, less pure than a quartz arenite, that may contain a moderate amount of other detrital minerals and/or calcite cement.

Rhodonite. Silicate of manganese that is susceptible of a high polish and suitable for decorative interior work.

Rhyolite. Fine-grained volcanic or extrusive equivalent of granite, light brown to gray and compact.

Rhyolitic. Term applied to igneous rock with the same mineral composition as granite, but usually with a porphyritic texture.

Rock cycle. Geologic cycle, with emphasis on the rocks produced; sedimentary rocks are metamorphosed to metamorphic rocks or melted to create igneous rocks, and all rocks may be uplifted and eroded to make sediments, which lithify to sedimentary rocks.

Rock flour. Glacial sediment of extremely fine (silt- and clay-sized) ground rock formed by abrasion of rocks at the base of the glacier.

Rock glacier. Glacierlike mass of rock fragments or talus with interstitial ice that moves downhill under the force of gravity.

Rocks and minerals. Materials found as constituents of natural mineral aggregates are rocks and minerals. Minerals are naturally occurring inorganic substances of more or less definite chemical composition and usually of a specific crystalline structure. Most rocks are composed of several minerals, but some are composed of only one mineral.

Certain examples of the rock quartzite are composed exclusively of the mineral quartz, and certain limestones are composed exclusively of the mineral calcite. Individual sand grains frequently are composed of particles of rock, but they may be composed of a single mineral, particularly in the finer sizes. Rocks are classified according to origin into three major divisions: igneous, sedimentary, and metamorphic.

Rounding. Degree to which the edges and corners of a particle become worn and rounded as a result of abrasion during transportation. Expressed as angular, subrounded, well rounded, etc.

Rubble. Unconsolidated accumulation of angular, rough rock fragments coarser than sand, broken from larger masses either by natural forces or artificially by quarrying or blasting.

Rubble land. Land areas with 90% or more of the surface covered with stones and boulders.

Sand. Finely pulverized fragments and water-worn particles of rocks.

Sandstone. Sedimentary rock consisting of grains of sand held together by some cementing material.

Schist

1. Strongly schistose rock in which the grain is coarse enough to permit identification of the principal minerals. Schists are subdivided into varieties on the basis of the most prominent mineral present in addition to quartz or quartz and feldspars.

2. Metamorphic rock characterized by strong foliation or schistosity.

3. Metamorphic rock that has a parallel or foliated structure secondarily developed by shearing. Schists frequently consist of grains of quartz and scales of mica arranged in more or less parallel layers. Feldspathic particles may be present.

Schistose. Having the structure of a schist.

Schistosity. Parallel arrangement of sheety or prismatic minerals like micas and amphiboles resulting from nonhydrostatic stress in metamorphism.

Scoria. Dark-colored, coarsely vesicular types of volcanic glass containing more or less spherical bubbles.

Sedimentary rock

1. Stratified rocks laid down for the most part under water, although wind and glacial action occasionally are important. Sediments may be composed of particles of preexisting rocks derived by mechanical agencies, or they may be of chemical or organic origin. Sediments are usually indurated by cementation or compaction during geologic time, although the degree of consolidation may vary widely.

2. Rocks formed at the earth's surface by the accumulation and consolidation of the products of weathering and erosion of existing rocks.

3. Rocks that have been deposited after being more or less sorted by running water.

Sedimentary structure. Structure of a sedimentary or weakly metamorphosed rock that was formed at the time of deposition; includes bedding, cross bedding, graded bedding, ripples, scour marks, and mudcracks.

Selenite. Variety of gypsum that is transparent and usually occurs in plates.

Series. Set of rocks formed in one area during a geologic epoch.

Serpentine. Metamorphic rock consisting essentially of the mineral serpentine. The coarser massive varieties are used as structural stone, the more highly colored varieties for interior decoration.

Serpentinite. Relatively soft, light to dark green to almost black rock, formed usually from silica-poor igneous rocks, such as pyroxenites, peridotites, and dunites. It may contain some of the original pyroxene or olivine but is largely composed of softer hydrous ferromagnesian minerals of the serpentine group. Very soft talclike material is often present in serpentinite.

Shale

1. Very fine-grained detrital sedimentary rock, composed of silt and clay, which tends to part along bedding planes.

2. Fine-grained, laminated, argillaceous rock, usually with friable structure.

Shales. When the claystones are hard and platy or fissile, they are known as shales. Aggregates containing abundant shale are detrimental to concrete because they can produce high shrinkage, but not all shales are harmful. Some argillites are alkali-silica reactive.

Shield volcano. Large, broad volcanic cone with very gentle slopes built up by nonviscous basalt lavas.

Shingle. Small, rounded, water-worn stones. Shingle is similar to gravel, but with the average size of stones generally larger.

Silicate. Any of a vast class of minerals containing silicon and oxygen and constructed from the tetrahedral group. The bulk of the earth's crust is composed of silicate minerals.

Silicic rock. Igneous rock containing more than two-thirds silicon-oxygen tetrahedra by weight, usually as quartz and feldspar (for example, granite).

Silicous. Containing an appreciable amount of silica as an impurity.

Siltstones. Rocks consisting predominantly of silt-sized particles (0.0625 to 0.002 mm in diameter), intermediate between claystones and sandstones.

Slaty cleavage. Foliation consisting of the parallel arrangement of sheety metamorphic minerals; at an angle to bedding planes; related to deformational structures.

Solidus. Curve on a pressure-versus-temperature graph representing the beginning of melting of a rock or mineral.

Sorting. Measure of the homogeneity of the sizes of particles in a sediment or sedimentary rock.

Spall. Chip or fragment of rock broken off by hammering or by natural agencies. The thin, curved pieces split off by exfoliation are examples.

Spalling. Chipping or fragmenting of a surface or surface coating caused, for example, by differential thermal expansion or contraction.

Stone. Rock fragment greater than 10 in. in diameter if rounded, and greater than 15 in. along the greater axis if flat.

Stratification. Structure of sedimentary rocks that have recognizable parallel beds of considerable lateral extent.

Syenite

1. Medium-to-coarse-grained, light-colored rock composed essentially of alkali feldspars, namely, microcline, orthoclase, or albite. Quartz is generally absent. Dark ferromagnesian minerals, such as hornblende, biotite, or pyroxene, are usually present.

2. Intermediate igneous rock whose mineral composition is essentially orthoclase and hornblende.

Syenite gneiss. Sheared syenite, syenite with its ferromagnesian minerals arranged in parallel layers.

Syenite porphyry. Rock with porphyritic texture, but of the same mineral composition as a syenite.

Talc schist. Schistose rock consisting essentially of talc and quartz.

Talus

1. Deposit of large angular fragments of physically weathered bedrock, usually at the base of a cliff or steep slope.

2. Collection of earth and broken rock at the foot of a cliff or steep slope. Material when angular is sometimes cemented together as a talus breccia.

Texture (rock). Rock characteristics of grain or crystal size, size variability, rounding or angularity, and preferred orientation.

Thermoremanent magnetization. Permanent magnetization, acquired by igneous rocks in the presence of the earth's magnetic field as they cool through the Curie point.

Tourmaline. Mineral with a hardness of 7. Tourmaline ranges widely in color, but common varieties are brown or black. It is characteristically a mineral of granites, pegmatites, and pneumatolytic veins, but persists as a detrital mineral in sediments. Concrete in which the coarse aggregate was serpentine and the fine aggregate a tourmaline sand concentrate has been described as having effective neutron-shielding characteristics.

Trachyte. Intermediate eruptive rock consisting essentially of orthoclase and one or more ferromagnesian minerals.

Trap. Any dark, fine-grained igneous rock.

Tremolite. White variety of amphibole, which sometimes occurs as an objectionable constituent in marbles.

Tridymite and cristobalite. Crystalline forms of silica sometimes found in volcanic igneous rocks. They are metastable at ordinary temperatures and pressures. Tridymite and cristobalite are rare minerals in aggregates except in areas where volcanic rocks are abundant. A type of cristobalite is a common constituent of opal. Tridymite and cristobalite are reactive with the alkalies in portland cement paste.

Tuff. Consolidated rock composed of pyroclastic fragments and fine ash. If particles are melted slightly together from their own heat, it is a "welded tuff."

Turbidite. Sedimentary deposit of a turbidity current, typically showing graded bedding and sedimentary structures on the undersides of the sandstones.

Ultramafic rock. Igneous rock consisting primarily of mafic minerals and containing less than 10% feldspar. Includes dunite, peridotite, amphibolite, and pyroxenite.

Unakite. Granite whose mineral composition is orthoclase, quartz, and epidote.

Varve. Thin layer of sediment grading upward from coarse to fine and light to dark, found in a lake bed and representing one year's deposition of glacial outwash.

Vein. Deposit of foreign minerals within a rock fracture or joint.

Ventifact. Rock that exhibits the effects of sandblasting or "snowblasting" on its surfaces, which become flat with sharp edges in between.

Verdeantique. Metamorphic rock consisting essentially of serpentine traversed by veinlets of talc, calcite, or dolomite. It's susceptible of a polish and is a marble only in a commercial sense.

Vermiculite. Dark micalike mineral formed by alteration of other micas.

Vesicle. Cavity in an igneous rock that was formerly occupied by a bubble of escaping gas.

Volcanic block. Pyroclastic rock fragment ranging from about the size of a fist to that of a car.

Volcanic breccia. Pyroclastic rock in which all fragments are more than 2 mm in diameter.

Volcanic glass. Igneous rocks composed wholly of glass are named on the basis of their texture and internal structure.

Weathering. Set of all processes that decay and break up bedrock by a combination of physical fracturing or chemical decomposition.

Witherite. Mineral with a hardness of 3 to $3\frac{1}{2}$. Its color ranges from colorless to white to grayish or many pale colors. Like calcite and aragonite, witherite is decomposed with effervescence by dilute hydrochloric acid. Witherite, the second most common barium mineral, occurs with barite and galena.

Zeolite

1. Class of silicates found in cavities within the crystal structure. Formed by alternation at low temperature and the pressure of other silicates, often volcanic glass.

2. Group of closely related minerals that are essentially hydrous silicates of aluminum, with potassium, sodium, calcium, and, more rarely, barium. They are made of secondary origin.

Zeolite minerals. Large group of hydrated aluminum silicates of the alkalies and alkaline earth elements that are soft and usually white or light colored. They are formed as a secondary filling in cavities or fissures in igneous rocks, or within the rock itself as a product of hydrothermal alteration of original minerals, especially feldspars. Some zeolites, particularly heulandite, natrolite, and laumontite, reportedly produce deleterious effects in concrete, the first two having been reported to augment the alkali content in concrete by releasing alkalies through cation exchange and thus increasing alkali reactivity when certain siliceous aggregates are present. Laumontite and its partially dehydrated variety leonhardite are notable for their substantial volume change with wetting and drying.

SHOP TOOLS AND EQUIPMENT

Abrasive paper disks. Abrasive disks are made from a selected even-sized abrasive grit bonded to a paper backing and then mounted on a backing pad; also for use with a grinder.

Adjustable box wrench. Tool to work nuts and bolts of a limited size, that automatically fits the bolt or nut and tightens up on either when force is applied. The wrench is made of tempered steel and is approximately 9 in. long.

Adjustable wrenches. Wrenches that are usually open-ended with one movable jaw. There are several types with heads set at 15, 45, and 90 degrees. Depending on the type, the movable jaw may be operated from an adjusting screw located in several positions on the shaft.

Adjusting spokeshaves (draw knife). Tool with winglike handles of wood or metal, on either side of the stock with provisions to secure and adjust the cutter on the stock.

Adze. Cutting tool used by the carpenter to trim and shape large sections of timber or logs.

Aligning punch. Steel tapered shaft about 12 in. in length, used to line up holes in two sheets of materials to prepare it for receiving a fastening. The tool is not actually a punch, but is inserted in the holes for the fastener, which could be a pin or a small bolt.

Alloy steel. Steel alloyed with other elements to modify its mechanical properties.

Anvil. Heavy iron or steel shaped blocks, used for hammering or shaping metal parts, either cold or in heated form.

Arc welder. Welding equipment designed to operate from 30 to 250 amperes, using welding rods or electrodes depending on the material being welded. The unit is portable and compact, containing the transformer and amperage selector, electrode holder and ground cable.

Army folding knife. Knife equipped with blades capable of being used as a screwdriver, can opener, bottle cap lifter, corkscrew, scissors, saw, and file. The tool is the most convenient portable tool for emergency jobs in the shop.

Auger. Steel shaft with twisted flutes, used to drill holes in wood. The shaft is approximately 24 in. long with a short collar to take a removable handle of wood or steel.

Auger bit. Commonly known as a "twist bit," made of tempered steel from 12 to 18 in. in length, with two types of twist patterns: single helical twist and the double helical twist, each with sufficient shank lead and lead screw starting at the beginning of the twist.

Automatic backknife lathe. Lathe that differs from the hand-operated wood lathe in that the cutters are mounted on a shaft at the rear of the machine. When the machine is in operation, the cutters revolve at a moderate speed while the work is moved slowly into the cutters and revolved a complete turn.

Automatic hand drill. Commonly known as a drill used for drilling small holes to receive nails or screws. The drills are fluted and without twist.

Aviation shears. Commonly known as compound action shears designed primarily for the aviation industry, for use in providing the required leverage at a short distance when making straight and curved cuts in sheet metal.

Awls. Tempered steel tool with sharp point used to start holes in lumber for screws or nails, and to pierce thick materials like leather.

Axe heads. Head of a tool in combination with a wood handle used for cutting or splitting wood. The head is fabricated in many patterns and designs depending on its use. It is usually made of forged steel, with the main body or bit several inches wide and deep, beveled at one end and sharpened like a knife, the shoulder being at least $\frac{1}{4}$ in. thick or more, with the eye prepared to receive the axe handle.

Back or tenon saw. Saw that has a wide parallel blade with saw points spaced 11 to 20/in. with the blade held by a wood or plastic handle. Handles are prepared with turnbolts to remove the blade. The back of the saw is usually stiffened with metal edge seam.

Ball peen hammer. Hammer used primarily for shopwork in clinching rivets or riveting bolt ends to prevent the nut from being unscrewed.

Band saw. Electrically operated piece of equipment used to cut large sections of lumber to required sizes and to cut curves.

Bar clamp. Malleable iron steel bar from 24 to 60 in. in length, drilled with evenly spaced holes for adjusting the tail slide. The upper portion of the bar is fitted with an adjustable clamp, whereas the tail slide is pinned to the bar.

Basin wrench. Long-stemmed bar or rod with a Stilson type head at one end, used where an ordinary wrench will not reach a fitting.

Bayonet awl blades. Blades used without handles for working with thick materials such as leather.

Beam compass. Compass that consists of a light metal beam or wood, with sliding trammel heads operated to stay in any position by a set screw. The heads have fine steel points for transcribing on wood or metal. One of the points is removable and can be replaced with a pencil for use on paper.

Belt sander. Electrically operated portable sander having an attached dust collecting bag, using continuous abrasive belts for the sanding action.

Bench grinder. Electrically operated grinder with positions for two wheels. When the grinding or polishing wheels are in position the operator is protected with plastic eye shields mounted on the grinder. Most grinders handle wheels from 5 to 10 in. in diameter.

Bench jigsaw. Electrically operated scroll saw used for cutting fine scroll work.

Bench rebate plane. Plane in which the blade extends across the full width of the sole. Made in 9 to 13 in. lengths and widths of 2 in. and over. It is used to cut wide rabbets.

Bench rule. Straight wood rule usually 36. in. in length, with U.S. standard and metric graduations.

Bit extension. Device used to extend the reach of an auger bit. Bit extensions are 15 to 18 in. or 24 in. long.

Bit gauge. A bit gauge serves as a stop when boring holes to a definite depth.

Bit stock drill. Drill used for wood or metal, available in sizes from $\frac{1}{16}$ to $\frac{3}{4}$ in. by thirty-seconds. Sizes are stamped on the shank.

Blade of a chisel. The blade of a chisel may vary in width by eighths of an inch up to 1 in. and by quarters of an inch from 1 to 2 in. Length and sturdiness of the blade determines its purpose.

Bleeder spray gun. Gun which delivers the paint and the air to the work surface in the required proportions and also can be adjusted to control the shape of the spray cone.

Block plane. Plane made in a variety of sizes and styles. The cutter is placed at a low angle to obtain smooth cuts on the end grain of wood. Its chief uses are for fitting short surfaces, such as miters, and planing end grain.

Blow torch. Torch used for heating metals for the preparation of brazing and soldering. Made in various size fuel capacities and of several types of metals.

Bolt cutters. Tool made from high strength steel, used for cutting small rods and bolts. The tool is available in 14- to 42-in. lengths.

Boring machines. Electrically operated boring machines are of two general types: horizontal and vertical. A vertical boring machine (drill press) is designed to do a variety of work on flat surfaces. The horizontal boring machine is primarily designed to bore holes for dowels in the edges and ends of face work, such as stiles and rails. A vertical boring machine with one spindle is designed for detail work, while the multiple-spindle horizontal borer is designed for quantity production. Specially designed, multiple-spindle borers are made for the mass production of doors and furniture.

Boring tools. The most commonly used boring tools are the auger bit, the twist drill, and various patented bits for special uses.

Bow calipers. Spring joint calipers used to transfer measurements from a pattern to the work piece. Knurled nut on a small stem adjusts the opening of the caliper legs.

Bow saw. Saw with two handles attached to vertical cheeks, stiffened with a center stretcher rail approximately 24 in. in length, with the saw blade at the bottom side and an adjustable wire cable at the top side to stiffen the tool. The saw is used to rip, cross cut, and cut curves in lumber.

Box wrench. Commonly known as a "ring spanner wrench," has a completely enclosed head and is used to loosen or tighten nuts or bolts. The wrench is made with 6 or 12 points inside of the head.

Brace. Hand operated drilling tool, consisting of a metal U-shaped frame, with a holding head at one end and chuck for drill bits at the other end. It is used primarily for accurate boring of holes in wood. Can be used with screwdriver bit and countersink bit.

Bradawl. Awl type tool with wood handle and removable tips. Tips are used to prevent splitting in soft wood grains.

Brad awl. While not a boring tool, a brad awl is used to start a hole for brads or screws.

Brad driver. Tool used to drive small nails instead of using a hammer. The barrel stem is spring loaded and holds the pin or nail magnetically in the barrel.

Brad pusher. A brad pusher is used to drive brads in difficult corners.

Breast drill. Hand-operated drill, having a vertical shaft, with a plate at one end to apply the necessary pressure required, a set of gear wheels attached to drive crank, a side handle for further control, and the chuck end of the drill for attaching the drill bits. The gear wheels drive the pinion that revolves the chuck and the bit.

Bull nose plane. Small cutting plane having a cast-iron body with a steel cutting blade, which is adjustable and removable. Several types are made and available.

Burnisher. A burnisher is used to turn the edge of scraper blades after sharpening the edges.

Butt chisel

1. A butt chisel is used in fitting hinges and other hardware. It has a short blade, about 3 in. long, making the chisel well balanced.

2. Chisel used primarily for light woodworking, made from alloy steel, with hardwood handle.

Butt gauge. A butt gauge is used for laying out the width and depth of a hinge seat, such as the butt hinge.

Cabinet rasp (file). The rasp is half rounded in shape with a flat bottom, having a cutting surface that is used for removing wood quickly, leaving a very rough surface.

Cabinet scraper. Scraper fitted with a beveled-edge scraper blade, which when properly sharpened will smooth any ridges or torn grain left by the smoothing plane. Primarily used on large surfaces.

Caliper rule and gauge. Instrument made of polished steel, with one jaw attached to the rule blade that moves in a slotted piece also having a jaw which can be fixed at any point by a clamp nut. The rule is graduated in U.S. customary or metric scale.

Calipers. Calipers are principally used for measuring outside and inside diameters of turnings or in places where a measurement cannot be established with an ordinary rule.

Carbide tipped blade. Circular blade with hardened tipped teeth used for cutting particle boards and other dense hardboards.

Carbide tipped blades. For circular saws, blades that stay sharp 50 to 60 times longer than ordinary steel blades, depending on the material being cut; blades used in woodworking shops to prepare special hardware and reinforcements.

Carpenter's hammers. Manufactured in many types and shapes. Common types are the one-piece solid steel head and handle, with a nylon-vinyl cushion grip covering on the handle and the combination steel head and wood handle. The head is with either the curved or straight claw.

Carpenter's mallet. Mallet made completely of wood and used instead of the ordinary iron head hammer to prevent damaging the work.

Carpenter's pincers. Tool used by the carpenter to remove nails from wood.

Carpenter's steel square. The typical square used by the carpenter to square up assembly work. The blade portion of the square is usually longer and wider than the tongue blade of the square.

Casting. Form or mold prepared to receive a liquid solution of a material that hardens when cooled.

Catapunch. A completely metal device that has a point activated by a spring to mark centers on metal or wood without the use of a hammer.

C clamp. Widely used device, made in many sizes and types, of aluminum, malleable iron, and pressed steel.

Center bit. Bit used with hand drills or braces, to drill holes in wood, having a screw point at the end of the bit to help start the bits into the wood.

Center punch. Heat treated steel tool used to mark centers or enlarge punch marks to help guide the point of the drill.

Chain wheel hoist. Hoist used in shops to lift heavy weights, usually up to 10 tons if properly supported.

Chamfer. Flat surface formed at the 90° corner of a material. Similar to a bevel.

Chisel. Tools used for cutting mortises, fitting hardware, and other irregular shaping that cannot be done with a plane or saw. There are two types of chisels, socket or firmer chisel and tang chisel. In the socket type the handle fits into a socket that is actually a part of the blade. In the

tang type the tang of the chisel runs into the handle. Socket-type chisels withstand pounding better than the tang type.

Chuck. The device that holds and secures the bit used for drilling or other purposes. The chuck is an integral part of a drill, lathe, or brace.

Circle plane. A circle plane operates on the same principle as the ordinary plane and is used for planing convex and concave surfaces. It has a flexible steel bottom that can be adjusted to fit any size curve within a minimum radius.

Circular saw blades. Saw blades made from temper steel and available chrome plated for crosscut and ripwork with all sizes and shapes of teeth, depending on the type of materials to be cut. They are used in woodworking shops to cut all types of woods.

Clamp-on-vise. Small portable bench or table vise used for light work.

Clamps. Clamps in general use by woodworking shops are of four types: the handscrew is used for gluing stock face to face or for clamping small pieces; the bar clamp is used for edge-clamping larger pieces; the C clamp is used for light work when neither a handscrew nor bar clamp can be used; and the column clamp is indispensable for gluing stave columns.

Claw hammer. Most common of all the tools used by the carpenter, having a split peen that forms the claw for the removal of nails.

Club hammer. Hammer having a double-faced head weighing from two to four lbs, and used in combination when using cold chisels, and any other heavy-duty work where weight is a factor.

Cold chisel. Chisel made of heated treated steel, used for cutting and chipping metal, in combination with a hammer as the driving force.

Combination blade. Circular blade designed for cutting lumber in any direction.

Combination metal square. The square is a combination of a 12-in. rule that is clamped in a head piece which has the control unit, and contains the spring loaded pin that locks the rule in place at any desired position. The head also includes a level vial.

Combination square. Square used to measure, mark, and check right angles, 45° angles, or, by use of the protractor feature, any degree angle desired.

Combination wrench. Wrench that has one open end and one closed end.

Compasses and dividers. Tools used for striking circles or segments of circles and for stepping off equal spaces.

Compass saw. The compass saw has a narrow tapered blade and is used to cut keyholes, lock holes, and letterbox openings. It is used primarily where a saw is needed without a frame. The hardwood handle is fitted with a slot to remove the blade.

Coping saw. Saw blade that is fitted in a bow frame. The blade is very narrow and has very fine teeth, averaging at 14/in. It is held under tension by the spring of the bow frame. The wood handle that holds the frame is made of hardwood.

Countersink bit

1. Bit used to recess the head of the flat-head and oval-head wood screws flush with the wood.

2. Bit used with hand drill or brace to cut a recessed hole to allow the countersunk head screw to fit flush with

the surface of the material. Countersink bits are also made with straight shafts for use with electric drills.

Crank and handle. Name of a tool in the trade for a hand drill that has a chuck that grips the straight round shank of the twist drill. It is used in boring holes larger than is possible with the automatic hand drill.

Crescent wrench. The most commonly used adjustable type wrench, having one fixed jaw and one movable jaw.

Cross cut and panel saws. Saws used primarily for cutting lumber across the grain.

Cross cut blade. Circular blade designed to cut across the grain of solid lumber.

Crosscut saw and ripsaw. The hand crosscut saw is used for cutting across the grain of wood and the ripsaw for cutting with the grain.

Crow bar. Steel bar with one chisel end and one end with a claw for removing nails. The bar is used primarily for leverage work in moving heavy objects.

Curved claw hammer. A claw hammer or nail hammer is selected on the basis of balance and quality of steel. The claw slot should be sharp enough to enable it to grip the smallest brad or nail, even if the head of the nail is missing.

Curved or straightback handsaws. Handsaws are available with either a curved (skew) back or a straight back. Straightback saw can be used as a straight edge in drawing lines. Skewback saws are preferred by some workers because of balance and design.

Cutoff saw. Electrically operated cutoff saw to cut rough stock to approximate length with an allowance for trimming after it is milled. Two basic types are in use: the stationary type, where the material moves into the saw, and the movable type, where the saw moves into the material.

Cutting gauge. All wood tool consisting of a 10 in. beam with a sliding stock block that can be set by a thumb screw after the distance has been established. One end of the beam accommodates a steel marking pin held with a steel wedge. It is used to mark a line parallel to an edge across the grain of the lumber.

Depth gauge. Metal tool used to measure the depth of holes and mortices. Consists of a graduated rule in U.S. standard inches or metric sale, with a sliding head that can be held with a clamp nut.

Diagonal cutting nippers. Nippers or pliers used to cut metal wires close to a surface. Also used for stripping insulation off of electrical wires.

Diamond point chisel. Chisel used to cut V grooves and square shoulders. Usually about 15 in. in length. Used primarily for wood but is the same name for a chisel used in metal work which also used to cut V grooves and to clean out burrs and shape up metal corners.

Die nut. The nut is used as a device for rethreading or recutting threaded bolts that have been rusted or where the thread has been damaged by a wrong start.

Dies. Made in many types and gauges for various threads. Primary use is to cut an external screw thread on a plain rod or bolt.

Disk sander. Metal disk with abrasive paper glued to the disk for use with electric drills.

Disk sanding tool. Tool using a rubber disk mounting pad for the abrasive sheet, and is electrically operated.

Double-planer surface. Electrically operated surface that has a top and bottom cutter head and in one operation, surface rough stock on two sides to finished thickness.

Dovetail saw. Saw used to cut very fine joints. The saw is a small back saw with very small teeth.

Dowel bit. Bit used to drill holes in the side and end grain of wood. The bit has a center point that keeps it from wandering off course and following the grain.

Dowel locator. Cast aluminum tool used as a device for aligning holes for dowel joints and to guide the drill bit true and square.

Drill and counter bore bits. The function of the bits are to drill a pilot hole, shank clearance hole, and counterbore in one operation.

Drill bit sharpener. Electrically operated device used to sharpen twist drill bits from $\frac{1}{8}$ to $\frac{3}{8}$ in.

Drill press. Electrically operated drill equipment that has a feed level to deliver the drill bit to the work. Primarily used in metalwork.

Electric belt sander. An electrical belt sander is manufactured in many sizes shapes, styles, and speeds. Sandpaper belts vary in widths and lengths depending on the unit.

Electrician's pliers. Steel pliers with insulating sleeves on the handles, used to grip, bend, and cut electrical conductors and wires.

Engineer's square. An engineer's square is a try square used for accurate squaring in metalwork. Commonly used in the shop having an 8-in. blade with graduations in U.S. standard inches or in metric.

End cutting nippers. Nippers or pliers used to cut nail heads and wires close to a surface.

Expansive bit. Bit made to fit both the hand drill or the electric drill, and used to cut holes in doors to install hardware due to its adjustable cutter, which can be set from $\frac{1}{2}$ to 3 in. All such bits have a lead screw starter at the end of the bit.

Extension rule. Wood or metal rule made from 6 ft in length to 8 ft, and is folded zig-zag style. The rule has a metal slide that fits into the first section and is used as a gauge and is used to extend the necessary inches to complete a dimension longer than the extended length of the rule.

Feeler metal gauge. Gauge used to measure very fine gaps. The gauge has several thin metal blades of various thicknesses that fan out from a metal case.

Field wind-up tape. Measuring device used primarily in measuring long distances in the range of 100 ft. Tape is made of flexible steel or reinforced plasticized linen. The tape is housed in a metal case and is recoiled after use by a winding handle which fits into the case after use.

Files. Files are used to smooth out metal and wood, remove burrs, to enlarge holes and slots and sharpen cutting tools. While there are a number of types and sizes of files, the most commonly used are the mill file, auger bit file, and various kinds of saw files. Files are equipped with hardwood handles.

Filing handsaws. It is important to select the right file. The choice depends on the filer's preference.

Firmer chisel. General purpose wood cutting tool used to trim and cut wood.

Flaring tool. Tool used to flare the end of a tube to fit a pipe fitting.

Flat file. Tapered steel file used to file flat surfaces.

Flat nosed pliers. Pliers with serrated jaws, used for light-weight metalwork.

Flat wood files. Files used for fine work. Files are equipped with handles.

Flat wood rasp. Tapered steel rasp having the same shape as a flat file, but is rasp cut for general rasping.

Flexible drill shaft. Steel shaft in a flexible casing, having a chuck at one end for securing drill bits, and the extension of the drive shaft cable that is inserted in an electric drill.

Flexible saw. Continuous string of hinged toothed sections with looped ends for finger control and use.

Flexible steel tape. Measuring device, from 6 to 8 ft in length, rolled up and housed in a small metal case usually 2×2 sq. in. and used as part of the measuring tape. The tape reacts to a spring that recoils it after it is used. The tape is available in U.S. standard and metric graduations on steel or on reinforced plasticized linen.

Flooring blade. Circular blade made for electric saws and used to cut through floors and the nails if encountered.

Flooring saw. Saw used primarily for cutting through floor boards. The saw blade is curved and has teeth on the bottom edge and a portion of the front top edge.

Folding rule. Single folding wood rule made from two pieces, which overlap each other and are jointed at one end to swing apart. Convenient rule used in confined spaces.

Forstner bit. Bit that has no feed screw and relies on the sharp circular rim for centering. It may be used in place of the auger bit and is particularly adapted to boring holes part way in thin wood where the feed screw from the auger bit would go entirely through the board. It produces a flat-bottomed hole suited for such jobs as housing a stair stringer.

Frame clamp. Device used to clamp a four-sided frame while it is being glued or nailed.

Friction blade. Circular blade designed for cutting through thin sheets of metal.

Gas pipe pliers. Pliers designed having jaws that are shaped to fit a pipe and serrated to hold it from slipping. The front end of the jaws have a small V notch to grip wire when the jaws are in the closed position. The ends of the handles are finished with one side having a pipe reamer and the other providing a screw driver end.

Gimlet bit or drill. The gimlet is used to bore shallow holes in wood, in preparation to install a screw in hardwood.

Glass cutter. Cutter used to scribe glass for snapping the sections apart. The glass cutter consists of a hardened steel wheel or in some instances an industrial diamond. The cutter does not actually cut the glass.

Glass drill. High strength steel bit with a tungsten carbide tip for drilling into glass with a hand or electric drill.

Glazier's pliers. Pliers used to snap off strips of glass after the glass has been scribed with the glass cutter.

Glue gun. Electrically operated device in which a dry stick of glue is loaded into the gun and when heated by the electrical current provides a thin stream of liquid glue at the squeeze of the trigger.

Gouges and carving tools. Gouges come in various diameters and may be sharpened with an inside or outside bevel, depending on the use. The two most commonly used types are the straight-shank and the bent-shank gouge, the latter being used when working or gouging out short, deep curves. Carving tools are made in numerous shapes for carving various shapes and designs in wood.

Grabber wrench. Wrench that automatically tightens on the bolt or nut as pressure is applied.

Gullet. Term that describes the notch between the teeth of a saw produced by a file.

Hack saw. Saw used primarily for cutting metal. The steel blade is removable, and can be fitted to be held in tubular metal frames or adjustable flat frames. The saw handle can be pistol grip type or straight type of hardwood or plastic.

Half-round cabinet files and cabinet rasps. Tools used for smoothing and shaping wood. Rasps are used where a large amount of stock is to be removed. Files equipped with handles.

Half-round file. File used for many purposes having one side half rounded and the other side absolutely flat. It is used for filing both wood and metal.

Half-round wood rasp. Rasp used like the ordinary half-round file, but is made in several rasping styles such as bastard, second cut, and smooth.

Hand drill. Drill that combines the use of the hand to drive the gear wheel in turn drives the pinions attached to the shaft and the chuck. It is used for close accurate work in drilling holes in wood and metal.

Hand file. File used for general purposes, and is slightly different from other files because it is actually flat with the sides running the same thickness from the butt end to the tip.

Hand saws. The hand saws generally have long tapered unsupported blades fitted with wood or plastic handles.

Hand scraper. Tool having a piece of steel about 0.035 in. thick, 2 or 3 in. wide, and 4 to 6 in. long. It has a square edge, which when properly sharpened is turned to form a fine cutting surface. A properly sharpened scraper will actually remove a very fine shaving. It is preferred by some shops for all uses and is especially adaptable in close quarters where the grain meets at right angles, as in the joining of a stile and rail.

Handscrew clamps. Hardwood clamps made from two horizontal blocks adjusted and manipulated by two wood rods that have been threaded and terminate in a handle. The clamp is made with threaded metal spindles and capped with wood handles. The purpose of the clamp is to be used for any shape desired to hold an object while it is being worked or to hold it in position for gluing or nailing.

Hand stone. Stone made from silicon carbide and used for sharpening axe and knife blades.

Hinge-butt template. Template used to establish the location of hinge butts. It will do the mortise work accurately.

Hole cutter. The cutter or saw is a combination of devices that consist of a backing plate that has a center drill bit which fits into the chuck of an ordinary electric drill, with the bit part acting as the starter for the hole to be cut or sawed. The backing plate has circular slots to fit at least six different circular sizes of cutting saw blades, any one of which can be locked in to saw or cut the size hole required.

Hole saws. High-speed hole saws are high-production tools for sawing circular holes in all machinable materials.

They are used primarily for installing hardware in doors and panels.

Horse rasp. Term describing the coarsest rasp, made for use on very rough work.

Jack plane. General all purpose plane that has many uses. Most popular sizes are 14 in. long, with a 2-in. cutter, and 15 in. long, with a $2\frac{2}{8}$-in. cutter.

Jaws. The securing and holding devices of a tool or machine. Jaws may be serrated to provide greater holding power.

Jigsaw or scroll saw. Electrically operated saw designed for sawing irregular curves in scroll or fretwork. The saw blade has a vertical reciprocal action. It is removable so that it may be inserted through a hole bored for that purpose when doing inside sawing. Various mechanical features reduce vibration caused by the reciprocal action.

Jointer. Electrically operated equipment designed for straightened wood by planing the surfaces. The operation of straightening the face of a board is called facing. The operation of straightening the edge is called jointing or edging. Jointing usually implies that the edge is to be jointed at right angles to the face side. Other operations that may be performed on the jointer include beveling, chamfering, tapering, and rabbeting. Hollow glue joints can also be made. A standard jointer is fed by hand.

Jointer plane. Plane similar to a jack plane, except that the length of the plane lends itself to square long edges of lumber for jointing.

Junior hack saw. Saw used for fine metal work. Made with a steel rod frame which includes the handle or flat steel frame with a wood or plastic handle. The blade in either case is removable.

Length of handsaw. The length of a handsaw is measured from the heel to the toe. The standard length is 26 in. The number of points per inch is stamped on the blade near the heel.

Level. Tool used for testing the vertical and horizontal planes of a material or an installation. The main body is made from wood, plastic, or metal and includes the closed vial that contains the clear nonfreeze liquid and the air bubble that accounts for the use of the level in one or more places on the body. Levels are made in many sizes and styles.

Lightweight fret saw. Saw with a flat steel U frame, wood handle and a very thin blade held at both ends of the frame by thumb screws. The blade is cut with 80 teeth per inch.

Line level. The line level consists of a rubber tubing that is terminated at each end with a short glass or plastic tube. The entire tubing is filled with water and each end of the level is held apart at two points to establish a given height, which can accurately be determined since water finds its own level.

Lock mortiser. The lock mortiser is fastened to the door edge with two self-centering clamps, and height rods inserted to ensure the location of the mortiser. A ratchet arrangement automatically feeds the machine into the cut.

Long nosed pliers. Pliers having long serrated jaws and used for working on small screws, nuts and bolts in confined spaces. Very lightweight, easy to handle for a variety of electrical jobs.

Mallet. The mallet was the original hammer, but now is understood to mean a tool acting as a hammer having a rubber, rawhide, or wood head.

Marking gauge. A marking gauge is used principally for laying out lines parallel to the side of stock, locating the center for dowels, laying out mortise and tenon joints, or gaining in hinges.

Marking knife. Knife used to mark lumber in preparation for cutting. The sharp edge of the blade is ground on one side only.

Masonry drill bit. Bit made from heat-treated high-strength steel, with the tip of the bit of hard tungsten carbide electronically brazed to the shank.

Matchers. Heavy-duty surfacers with four cutter heads, designed to surface in one operation dimension stock and timbers on all four sides to net thickness and width. Matchers were primarily designed for production of matched (tongue-and-groove) flooring or ceiling, hence the name.

Metal-cutting blades. Blades for circular saws. A course-toothed blade is used for such nonferrous metals as aluminum, magnesium, or lead. A fine-toothed blade is used for such nonferrous metals as copper, brass, thin aluminum sheets, and magnesium less than $\frac{3}{8}$ in. in thickness. The blades are used in woodworking shops to prepare special hardware and reinforcements.

Micrometer caliper. Instrument used to produce very close measurements. The graduations are read on the barrel and is actuated as the spindle touches the object between the anvil and the object.

Mill file. File, usually about 10 in. long, used for fine work and for sharpening tools such as circular saws, lawn mower blades, axes, shears, and knives.

Miter box. Box made usually of beech wood to act as a guide for the blade of a small saw, to produce the accurate 45° miters. The open-ended box also includes an opening or slot for the 90° angle cut.

Miter clamp. Tool used to clamp mitered joints and to hold the material in position for gluing or nailing.

Molder (sticker). Equipment designed to shape moldings with irregular faces. Molders are built to meet the requirements of various types of work. Examples are high-speed molders for production, molders specially adapted for quick change to produce short runs of custom work, and molders designed to handle short stock, such as furniture parts. The sash sticker is specially designed for the production of milled sash stock and has attachments for boring and slotting for sash cord. Matchers, built on the same principle as molders, are designed for mass production of flooring and other production items.

Molding plane. A molding plane is used when a shape or sticker is not available. Moldings can be shaped with this tool.

Monkey wrench. Most common type wrench used for all purposes, since it is adjustable and can fit most small bolt heads and nuts, and for working on pipes having screw on fittings.

Mortise chisel. Chisel that has a very thick stiff blade and a long beveled sharp edge and is used for making mortises by hand.

Mortise gauge. Gauge provided with two pins for marking the width of a mortise or tenon with one stroke and for keeping the shoulder offset parallel.

Mortiser. Electrically operated mortiser used except in the manufacture of sash, has been replaced by the horizontal boring machine for mass-production work. However, the mortiser continues in favor for certain types of detail work. The hollow chisel mortiser is made in a variety of horizontal and vertical designs and with either or both power feed and foot feed.

Multiplier. Plier that includes a combination of other tools such as an adjustable wrench on one handle and a medium sized butt tip for use as a screwdriver on the other. The jaw portion of the plier is also adjusted on a slip joint to the increase the opening of the jaws. The jaws are serrated for greater holding power.

Nail sets. Nail sets are available in sizes to fit any size finish nail head. Nail sets are used to drive a nail below the surface of the wood. The hole is then filled by the finisher.

Needle files. Small files used for precision filing. Files are marketed in sets and are available in lengths from 3 to 12 inches.

Numbered drills. A numbered drill is primarily the tool of a metal worker. However, the cabinetmaker sometimes finds it necessary, when fitting hardware, to bore holes of a size not found in the fractional sized drills. Numbered drills are made in sizes from No. 80 to No. 1 to correspond with standard wire gauge numbers. The largest is No. 1, which has a diameter of 0.2280 in., while the smallest is No. 80, with a diameter of 0.0135 in.

Nut socket driver. Tool with interchangable socket heads for various size nuts or heads of bolts. The use of that similar to a screwdriver, having a long shank and wood handle.

Offset screwdriver. Screwdriver shaped from a small sized bar having turned out legs at 90° angles and resembles a Z. The tool is used where it is impossible to use a straight screwdriver.

Offset wrench. Wrenches made with offset shanks to allow better use of the tool in obstructed positions.

Oil can. Can for applying oil to tools, equipment, and machinery. Made in all sizes, types and styles, and in various materials.

Oil stones. Oil stones are used for whetting sharp-edged tools. Stones may be purchased in coarse, medium, or fine grits, depending on the results desired. Some stones are combination stones having a coarse grit on one side, and a fine grit on the other. Stones are made from silicon carbide, aluminum oxide, or from natural stone.

Open-ended wrench. Wrench made from alloy steel in many sizes and in sets, having an open-ended jaw.

Overhead router. Electrically operated router designed for dadoing and shaping recessed moldings into flat surfaces, for example, housing a stair stringer or shaping a recessed panel in the end of a church pew. Two types of machines are built. An arm router provides for the bit to move into the work, guided by the arm. A spindle router is provided with a stationary spindle, and the work is guided into the cutter by means of a template.

Panel saw. Saw for use in lighter work, such as fine panels. It is available in 20-, 22-, and 24-in. lengths and carries the same specifications as the 26-in. handsaw.

Paring chisel. Chisel used to pare long housings such as grooves in stair construction.

Paring gouge. Gouge with long blade used to shape design work. Available in sizes from $\frac{1}{4}$ to $1\frac{1}{2}$ in.

Pawl. Device attached to tool or equipment to allow rotation in one direction only.

Phillips screw driver. A screw driver made for Philips screws must be selected to fit the cross-slot in the screw head. Four sizes, 1, 2, 3, and 4, will fit the entire range of Philips screws. Size 1 is for screws up to and including No. 4, size 2 for screws No. 5 to 9 inclusive, size 3 for screws No. 10 to 16 inclusive, and size 4 for screws No. 18 and over.

Pillar file. File used to file narrow openings such as slots and keyways.

Pin punch. Punch used primarily to drive out a pin from an assembly of materials.

Pipe burring reamer. Tool used to remove burrs from the end of cut pipe.

Pipe clamp. Clamp designed to be used with a piece of pipe with the sliding jaw spaced to accommodate a large sized panel or a door.

Pipe cutter. Tool having alloy steel cutters with an adjustable feature to accommodate the size of pipe being cut and to hold it in position until the pipe is cut by a back and forth action of the tool and the tightening of the adjusting screw.

Pipe die. Die used to cut a thread on a pipe. The die fits a diestock which has sockets for handles.

Pipe wrenches. Wrenches used in securing fittings on piping and also for the removable of rusted or calcined fittings from piping by the use of 2 wrenches in opposite directions.

Planar. Electrically operated planar is designed to surface lumber to a specific thickness. The cabinet planer is designed for precision work and is used only for stock that has been faced on the jointer.

Planer blade. Circular blade used to produce a fine finish on lumber and finished boards.

Planes. Tools having a metal or wood body which houses a cutting blade and its bottom surface. The position of the blade is adjustable. Planes are made in various sizes, types and styles.

Pliers. Combination slip-joint plier, made from selected steel with serrated jaws.

Plumb bob

1. A plumb bob is used to establish a true vertical position between two points when a straight edge of sufficient length is not readily obtainable. A stair builder may use a plumb bob to establish a vertical position when measuring a building for a flight of steps.

2. Device consisting of a weighted tapered bob (resembling a spinning top) suspended from a nylon, silk, or linen cord, and hung vertically to establish the position of two points over each other.

Pocket chisel. A pocket chisel has a blade about $4\frac{1}{2}$ in. long and is used to mortise hardwoods or other rugged work.

Points and teeth in handsaws. The number of points per inch of a saw (there is always one more point to the inch than there are teeth) determines to some extent the use of the saw. Fine work can be done with a saw having finer teeth (more points to the inch), while coarser saws are better for green lumber and faster work. For average work a crosscut should have 8 points/in., a ripsaw $5\frac{1}{2}$ or 6. For fine joinery work a 10- or 11-point saw is used.

Portable circular saw. Electrically operated tool designed to use circular blades to cut lumber and some metals.

Portable electric router. Router used to cut grooves, rabbets, mouldings, dados, and for trimming. The bits used by the router determine the shape and the design of the finish work.

Power drill. Electrically operated drill used for many other purposes and functions besides drilling. The accessories and equipment available provide this tool with varied uses such as, sanding, buffing, cutting, sharpening tools, and other bench-mounted work.

Power screwdriver bits. Bits made to fit the electric power drill and drive screws using various type heads.

Power tools. Power tools are used for the production of a variety of work in custom woodworking plants to speed up production time.

Propane torch. Torch used for providing a heat source for brazing and soldering. The torch has a burner controlled by a gas flow valve that sets the flame size.

Push drill. Drill that operates on the torque principle in which pressure is applied to the handle which revolves the shaft and the chuck which holds the bit.

Rabbet planes. Rabbet planes are made in a variety of styles for special purposes. Cabinetmakers prefer the low-angle plane for all-round work in smoothing rabbets.

Rasp plane. Plane that embodies the principle of the rasp file, except that the rough, sharp teeth, have openings that permit the waste to penetrate and prevent the teeth from clogging. The rasp blade is made from a steel sheet in which the teeth were pressed out.

Rasps. Rasp teeth are formed individually to slice off slivers of wood quickly. The rasp is used on wood, but must be used with care not to remove more wood than is necessary. It is also used on soft metals.

Ratchet bit brace. Brace used for holding boring tools with rectangularly shaped bit tang. The jaws of the chuck are made to center the bit, and the tapered square part of the bit tang keeps the bit from turning in the chuck. A brace with a 10-in. sweep is best suited for ordinary work. Most braces have a ratchet arrangement to permit boring holes in corners and other tight places.

Ratchet screwdriver. Screwdriver with adjustment that changes the position of travel for the shaft, from clockwise to counter clockwise by the operation of a thumb slide on the barrel below the handle.

Reciprocating electric saw. Saw in which the blade moves backward and forward from 1600 to 3000 strokes per minute. Blades are available to cut both wood and metal.

Ripping hammer. A ripping hammer, so-called because of its claw, is used in ripping work apart or in splitting boards.

Ripsaw. Electrically operated ripsaw used to rip rough stock to approximate width with an allowance made for jointing and sizing. The ripsaw is ordinarily not designed for fine accuracy.

Rip saw. Hand saw designed to be used to cut lumber with the grain.

Router plane. A router plane is used for leveling the bottom of dadoes, or rabbets if a uniform depth is desired.

Rules. The rules for measuring distances in woodwork are based on the standard English yard measure and are divided into feed, inches, and fractions of an inch—usually to sixteenths.

Saber saw. Electrically operated saw, in which the blade operates in an up-and-down motion at a rate of 2700 to 3300 strokes per minute. Blades have been designed to cut many types of materials.

Sander. Many types of electrically operated sanders are designed to accommodate different kinds of work. The drum sander is built on the principle of the planer in that it has a power feed and is set for thickness of the material to be sanded. Power-feed drums sanders are usually made with several drums. Each drum is covered with a different grade of abrasive so that the material, in being sanded, passes the coarser drums first. Final sanding is done by a drum covered with a fine abrasive.

Saw clamp. Clamp used to hold a handsaw while it is being filed.

Saw files. Files used primarily for sharpening handsaws, are single cut and come in various lengths from 4 to 10 in. Saw files are three-cornered and tapered. There are four sizes available: regular taper, slim taper, extra-slim taper, and double-extra-slim taper. The number of points per inch of the saw to be filed determines the length and degree of slimness of the file. Files are equipped with handles.

Saws. Saws are used primarily for hand-cutting wood stock into various widths and lengths. There are several types of saws, each designed for a particular job.

Saw set. Tool used to set saw teeth to the correct angles, after the saw has been sharpened.

Scrapers. Tool used to smooth a surface after it has been planed. There are several types of scrapers used in shop-work.

Scratch awl or scriber. A sharpened hard steel tool used for scratching marking lines on metal surfaces. Some scribers have attached or removable tungsten carbide tips.

Screwdriver. Tool that consists of a metal shank with many tip sizes designed to drive or remove screws from wood or metal. The metal shank terminates in a handle that may be made of wood, plastic, or hard rubber.

Screw holder. Device attached to a screwdriver for holding the screws in place when the screw cannot be reached or held by hand while starting it into the material.

Screws. Metal fastening devices made from steel, aluminum, stainless steel, specially coated metals, copper, brass, and bronze. The heads of screws have been designed by manufacturers for a specific use or a specific assembler of materials and products. The following screw heads are available in limited sizes: Phillips, Pozidriv, Reed and Prince, clutch, Robertson, and Torx. The parallel cut head is the most popular and available screwhead.

Scribers. Scribers are used to mark the edges of cabinets to make them fit irregular surfaces, such as plastered walls. They are also useful when coping one molding to another.

Scribing. Method used to transfer the shape of object on to a blank piece of material to reproduce the exact shape.

Self-centering punch. Punch used to start holes for wood screws. It will center the punch mark for a screw hole, for example, for a hole on a hinge.

Setscrew wrenches. Wrenches made from hexagonal bar material and shaped like an L are used for the setting of setscrews that hold wheels on turning shafts. The wrenches usually are marketed in a Set of 10 or more. Also widely known as "Allen keys."

Shank. The straight or vertical section of a tool to which the handle and the operating or action end of the tool usually are fastened.

Sheet metal punch. High-strength steel tool used to produce the selected holes required for self-tapping screws and screw nails.

Single-spindle drum sander. Electrically operated drum sander made to sand flat-faced curves that are fed by hand. The belt sander is used in combination with the drum sander to smooth flat surfaces ready for finishing.

Spade bit. Bit used with an electric drill to drill holes in various types of materials. The bit is available from $\frac{3}{8}$ to $1\frac{1}{2}$ inches. It has lead point to start the penetration of the material and the cutter edges engaged the material to cut the hole.

Spindle shapers. Spindle shapers are designed for shaping moldings on curved stock. They are versatile machines that may be adapted to numerous operations, including the shaping of joints, fluting, reeding, grooving, rabbeting, slotting, and tenoning.

Soldering iron. Electric or pot-heated tool used to heat metals and soft solder for brazing or joining.

Spirit level. Level used when work is installed permanently on the job, for example, when a cabinet or fixture is being set in a true horizontal or vertical position.

Spokeshave. Tool used to shape irregularly curve surfaces. Some spokeshaves have a straight bottom; others a convex or concave bottom.

Spring dividers. Plated steel instrument instrument used to scribe circles on wood or metal. A threaded bar and nut are used for adjusting the size of the opening desired.

Standard shears. Shears with straight jaw blades and long handles to provide leverage for shearing sheet metal.

Stapling hammers and tackers. Tool used with preformed staples for upholstering furniture, light crate assembly, installation of carpet padding, and generally light tacking instead of nailing.

Star drill. Tool used to drill holes in masonry when fastening wood to the masonry wall by means of an expansion sleeve and lag bolt or by other types of fasteners.

Steel chisel for wood-cutting. An all-steel chisel, having a beveled edge blade and a hexagonal handle.

Stilson wrench. A heavy duty wrench used for pipework and for gripping large nuts or bolt heads. The leverage on the long handle against the loose hook jaw places the grip on the head jaw as the tool is pulled downward.

Stock. Usually the body of a device, tool, or instrument to which is attached the functioning or working parts.

Straightedge. A straightedge is used for marking and checking straight lines when doing layout work and in checking the straightness of surfaces such as counter tops or paneled frames.

Straightedge rule. Finished heavy steel strip with one beveled edge, usually used in the shop or drafting room in a 36 in. length with U.S. customary or metric graduations.

Stub screwdriver. Short screwdriver, about 3 in. long and used in restricted spaces.

Taps. A tap is made from heat-treated high-strength steel and fluted to form cutting edges. The tap is used to cut internal screw threads into a pipe or hole which has been drilled in steel stock. There are various types of taps made and they are marketed in sets.

Tap wrench. Wrench and short shank and a chuck for clamping the tap.

T bevel. Often called a "bevel square," the T bevel is used for laying out any desired angle.

Tenoner. Electrically operated equipment designed primarily to make tenons. However, with the addition of coping heads and trim saw, together with the development of the double-end tenoner, it is now a highly versatile and important production machine used for tenoning, coping, shaping, trimming, and the milling of joints.

Tenon or back saw. Saw for 8 to 14 in. in length, having a beech or plastic handle and a steel blade. The blade is removable and the points vary from 11 to 20 per inch. It is the type of saw that can be used in a miter box.

Tinner's snips. Tool used for cutting sheet metal, made of solid steel, tempered jaws, polished blades, and nonslip enamel handles.

Trammel. Trammel points attached to a straight rod are used to strike circles with diameters too large for the ordinary compass.

Triangular saw files. Saw file made for filing all types of hand saws with 60° teeth. Edges are set and cut for filing the gullet between the saw teeth.

Trim saw. Electrically operated trim saw used as a bench saw, table saw, combination saw, tilting arbor saw, and variety saw. The term "trim saw" is comprehensive, as it accurately designates the type of work done on it. If properly adjusted, the trim saw will do accurate work in sizing jointed stock to width and trimming sized stock to length, and it will perform many other milling operations, such as beveling, chamfering, mitering, grooving, dadoing, rabbeting, slotting, and tenoning.

Try square. Square used mainly for checking the squareness of surfaces, laying out lines square with an edge, and testing 90° angles. The blade is marked in eighths of an inch along the top edge.

Tubular box wrench. Wrench made from heavy stock metal tubing with the edges crimped in to a hexagonal-shaped opening and used as a wrench for nuts and bolts.

Turning lathe. Electrically operated lathe with a powered spindle which revolves the wood at various rpm while the cutting tools are manipulated by hand to shape the required design. The lathe may be adapted for boring or performing semiautomatic work.

Wing divider-compass. A spring-controlled instrument that operates on a sliding wing with provisions for fine adjustments for the two pointers, which can be used as dividers for transfer of information or as a compass for inscribing.

Wire brushes. Brushes made from brass or steel fibers and mounted on a spindle to be used with an electric drill tool. The brushes are used for cleaning and scoring metal surfaces.

Wire gauge. Gauge in the form of a circular plate having slots for the gauging of wire sizes and sheet metal.

Wood chisel. The chisel is an all purpose general use chisel for wood-cutting, with a blade approximately 4 in. long, held in combination with a hardwood handle.

Wood file. File used to smooth up wood surfaces after a rasp is used. The file does not leave a smooth finish surface, but does remove the ridges caused by the rasp.

Wood mallets. Wood mallets come in many shapes and sizes. Mallets are used primarily to provide a more resilient impact on the chisel handle and to prevent the head of the tool handle from mushrooming. When used for assembling work, the surface of the material should be protected by a scrap piece of wood.

Wood turning tools. Tools consisting of chisels and gouges for cutting of shapes and patterns on wood placed in a lathe.

Woodworking vise. An all metal device designed to be attached to a bench top. The primary purpose is to hold lumber in its grip between the two jaws of the vise.

Wrenches. Wrenches are used by woodworking shops when parts are assembled with bolts. The adjustable wrench is a convenient tool. The wrench is available in lengths of from 4 to 20 in. depending on the type.

SOIL MECHANICS

AA type lava. Basaltic lava flows typified by a rough, jagged, spinose, clinkery surface.

ABC soil. Soil with a distinctly developed profile, including A, B, and C horizons.

AB horizon. Transitional horizon between the A and B horizons, having features of the A horizon on its upper part and features of the B horizon in its lower part, but without a clearly defined point to indicate where these features separate.

Acid. A substance that dissolves in water to release hydrogen ions.

Acid soil

1. Soil with a preponderance of hydrogen and aluminum ions in proportion to hydroxyl ions.

2. Soil with a pH value less than 7.0; for most practical purposes a soil with a pH value less than 6.6. The pH values obtained vary greatly with the method used and, consequently, there is no unanimous agreement as to what constitutes an acid soil. The term is usually applied to the surface layer or to the root zone unless specified otherwise.

3. Soil deficient in available bases, particularly calcium, and which gives an acid reaction when tested by standard methods.

AC soil. Soil having a profile containing only A and C horizons with no clearly developed B horizon.

Actinomycetes. Group of organisms intermediate between the bacteria and the true fungi that usually produce a characteristic branched mycelium; organism belonging to the order of Actinomycetales.

Active acidity. Activity of the hydrogen ion in the aqueous phase of a soil; measured and expressed as a pH value.

Activity of clay. Ratio of the plasticity index to percent by weight of the total sample, which is smaller than 0.002 mm in grain size. The property is correlated with the type of clay material.

Adhesion. Molecular attraction that holds the surfaces of two substances in contact, for example, water and sand particles.

Adsorption

1. Phenomenon that causes all solids to adsorb or condense on their surface any gas or vapors with which they are in contact.

2. Attraction of ions or compounds to the surface of a solid. Soil colloids adsorb large amounts of ions and water.

3. Increased concentration of molecules or ions at a surface, including exchangeable cations and anions on soil particles.

Adsorption complex. Group of substances in soil capable of adsorbing other materials. Colloidal particles account for most of this adsorption.

Aerate. Impregnate with a gas, usually air.

Aeration. Amount of interstitial space in a soil that contains air and gases; process of increasing this volume.

Aerobic. Having molecular oxygen as a part of the environment; growing only in the presence of molecular oxygen, as aerobic organisms; occurring only in the presence of molecular oxygen, for example, aerobic decomposition.

Aggregate (soil)

1. Many soil particles held in a single mass or cluster, such as a clod, crumb, block, or prism.

2. Group of soil particles cohering so as to behave mechanically as a unit.

Aggregation. Act or process of forming aggregates; state of being aggregated.

Agric horizon. Horizon, immediately below the plow layer of cultivated soils containing accumulated clay and humus to the extent of at least 15% of the horizon volume.

A horizon

1. Surface horizon of a mineral soil having maximum organic matter accumulation, maximum biological activity, and/or eluviation of materials such as iron and aluminum oxides and silicate clays.

2. Uppermost layer of the soil profile from which inorganic colloids and other soluble materials have been leached. It usually contains remnants of organic life.

3. Horizon formed at or near the surface, but within the mineral soil, having properties that reflect the influence of accumulating organic matter or eluviation, alone or in combination.

4. Uppermost layer of a soil, containing organic material and leached minerals.

Air-dry

1. State of dryness of a soil at equilibrium with the moisture content in the surrounding atmosphere. Actual moisture content will depend on the relative humidity and the temperature of the surrounding atmosphere.

2. State of dryness of a soil at equilibrium with the moisture contained in the surrounding atmosphere.

3. Allow to reach equilibrium in moisture content with the surrounding atmosphere.

Air porosity. Proportion of the bulk volume of soil that is filled with air at any given time or under a given condition, such as a specified moisture tension. Usually the large pores; that is, those drained by a tension of less than approximately 100 cm of water.

Airy isotatic compensation. Variation in thickness of a constant-density crust that serves to balance out the excess or deficient weight of topographic features; hence mountains might be underlaid by thick crustal "roots."

Albic horizon. Light-colored surface or lower horizon from which clay and free iron oxides have been removed or so segregated as to permit the color to be determined primarily by the primary sand and silt particles.

Alfisols

1. Mineral soils that have no mollic epipedon, oxic or spodic horizon, but do have an argillic or natric horizon that is at least 35% base saturated. Most soils are classified as gray-brown podzolic, noncalcic brown, and gray wooded.

2. Order of soils having a B2 horizon high in crystalline clay and a moderately high level of exchangeable bases. These soils usually occur in the climatic range associated with scrub to well-developed deciduous forests.

Alkali metal. Strongly basic metal like potassium or sodium.

Alkaline soil

1. Soil containing an excessive amount of the alkaline salts.

2. Soil that has a pH greater than 7.0.

3. Surface layer or root zone used to characterize a horizon or a sample thereof.

Alkali soil. Soil with a high degree of alkalinity or a high exchangeable sodium content, or both; soil that contains sufficient alkali (sodium) to interfere with the growth of most crop plants.

Alkalization. Process whereby the exchangeable sodium content of a soil is increased.

Alluvial soil

1. Soil developing from recently deposited alluvium and exhibiting essentially no horizon development or modification of the recently deposited materials.

2. Great soil group of the azonal order.

Alpine meadow soil. Soil group of the intrazonal order, comprised of dark soils of grassy meadows at altitudes above the timberline.

Aluminosilicates. Compounds containing aluminum, silicon, and oxygen as main constituents.

Amino acids. Nitrogen-containing organic acids that couple together to form proteins. Each acid molecule contains one or more amino groups and at least one carboxyl group. In addition, some amino acids contain sulfur.

Ammonification. Biochemical process whereby ammoniacal nitrogen is released from nitrogen-containing organic compounds.

Ammonium fixation. Adsorption or absorption of ammonium ions by the mineral or organic fractions of the soil in such a manner that they are relatively insoluble in water and relatively unexchangeable by the usual methods of cation exchange.

Amygdule. Vesicle or gas bubble in an igneous rock that has been filled with another mineral after the solidification of the lava.

Anaerobe. An organism that does not require air or free oxygen to maintain its life processes.

Anaerobic

1. The molecular oxygen. Growing in the absence of molecular oxygen like anaerobic bacteria.

2. The absence of molecular oxygen. Living or functioning in the absence of air or free oxygen.

Ando soils. Zonal group of dark-colored soils high in organic matter, developed in volcanic ash deposits.

Angular momentum. Product of a body's angular velocity or rotation rate and its moment of inertia.

Angular unconformity. Unconformity in which the bedding planes of the rocks above and below are not parallel.

Anion. Negatively charged ion; opposite of a cation.

Anion-exchange capacity. Sum total of exchangeable anions that a soil can adsorb, expressed as milliequivalents per 100 g of soil.

Anisotropic soil. Soil mass having different properties in different directions at any given point, primarily with reference to stress-strain or permeability characteristics.

Antibiotic. Substance produced by one species of organism that in low concentrations will kill or inhibit the growth of certain other organisms.

Anticline. Fold, usually from 100 m to 300 km in width, that is convex upward with the oldest stratum at the center.

Apatite. Naturally occurring complex calcium phosphate, which is the original source of most of the phosphate fertilizers.

Apron. Extensive alluvial deposit with generally low outward slope; essentially a mature coalesced series of alluvial fans or a well-developed bajada.

Argillaceous rock. Sedimentary rock in which clay, minerals, or low-grade micas (chlorite, sericite) predominate. Examples are shale, slate, argilite, or claystone.

Argillic horizon. Diagnostic illuvial subsurface horizon characterized by an accumulation of silicate clays.

Arid soils

1. Order of soils at apparent dynamic equilibrium with the climate of dry regions. These soils show limited profile development because of a low climatic intensity, the A horizon containing less than 1% organic matter and a high level of exchangeable bases.

2. Soils characteristic of dry places, including soils such as desert, red desert, sierozems, and solochak.

Arroyo. Deep, usually dry gully or channel in an arid area. Typically, it is cut in unconsolidated material and has steep vertical walls at least several feet high.

Asymmetrical ripple. Ripple whose cross section is asymmetric, with a gentle slope on the upcurrent side and a steeper face on the downcurrent side.

Atmosphere (unit). Unit of pressure equal to 101,325 N/m^2 or about 14.7 psi.

Atterberg limits

1. Water contents that correspond to the boundaries between the states of consistency of a remolded, cohesive soil.

2. Limits measured for soil material passing the No. 40 sieve.

Autotrophic. Capable of utilizing carbon dioxide or carbonates as the sole source of carbon and obtaining energy for life processes from the oxidation of inorganic elements or compounds such as iron, sulfur, hydrogen, ammonium, and nitrites, or from radiant energy. Contrast with heterotrophic.

Azonal soils. Soils without distinct genetic horizons.

Badland. Miscellaneous land type void of vegetation and occurring on severely eroded, soft geologic material in arid or semiarid regions.

Bar. Unit of pressure equal to 1,000,000 dynes/sq cm, which is nearly equal to the standard atmosphere.

Barchan. Crescent-shaped sand dune moving across a clean surface with its convex face upwind and its concave slip face downwind.

Barranca. Dry wash or ravine having steep sides and a narrow bed less than 2 m wide.

Barrens. Relatively desolate area where vegetation is lacking (as in an icecap or desert) or is scanty and restricted to a few species, as compared with adjacent areas, because of adverse soil, wind, or other environmental factors.

Basalt. Any one of a group of fine-grained, dark, heavy, widely distributed volcanic rocks. No strict definition of basalt as a mineralogic type has been agreed upon.

Base exchange. Physicochemical process in which one species of ions on the lattice of a clay particle are replaced by another species, thereby altering the plasticity and physical properties of the clay.

Base-saturation percentage. Extent to which the adsorption complex of a soil is saturated with exchangeable cations other than hydrogen and aluminum, expressed as a percentage of the total cation-exchange capacity.

Batholith. Great irregular mass of coarse-grained igneous rock with an exposed surface of more than 100 km², which has either intruded the country rock or been derived from it through metamorphism.

Bauxite. Rock composed primarily of hydrous aluminum oxides and formed by weathering in tropical areas with good drainage; major ore of aluminum.

BC soil. Soil profile with B and C horizons, but with little or no A horizon. Most BC soils have lost their A horizons by erosion.

Bed. Layer of rock differing from layers above and below or set off by more or less well-marked divisional planes; layer in a series of stratified (sedimentary) rocks.

Bedding. Characteristic of sedimentary rocks in which parallel planar surfaces separating different grain sizes or compositions indicate successive depositional surfaces that existed at the time of sedimentation.

Bedding cleavage. Cleavage that is parallel to the bedding.

Bedrock

1. Solid rock underlying soils and the regolith at depths ranging from zero where exposed by erosion to several hundred feet.

2. More or less solid, undisturbed rock in place either at the surface or beneath superficial deposits of gravel, sand, or soil. According to local conditions and usages, bedrock may be soft or hard, consolidated or unconsolidated.

Bench. Strip of relatively level earth or rock, raised and narrow, small terrace or comparatively level platform breaking the continuity of a declivity.

Bench terraces. Embankment constructed across sloping soils or fields with a steep drop on the downslope side.

Betonite. Type of mineral deposit consisting principally of montmorillonite clay.

B horizon

1. Layer of a soil profile in which material leached from the overlying A horizon is accumulated.

2. Horizon immediately beneath the A horizon, characterized by a higher colloid (clay) or (humus) content or by a darker or brighter color than the soil immediately above or below, the color usually being associated with the colloidal materials. Colloids may be of illuvial origin, as clay or humus; they may have been formed in place (e.g., clays, including sesquioxides); or they

may have been derived from a texturally layered parent material.

Binder (soil binder). Portion of soil passing No. 40 U.S. standard sieve.

Biomass. Amount of living matter in a given area.

Black earth. Term used by some as synonymous with "chernozem" and by others (in Australia) to describe self-mulching black clays, such as vertisols.

Black soils. Term used in Canada to describe soils with dark surface horizons of the black (chernozem) zone, including black earth or chernozem, wiesenboden, solenetz, etc.

Bleicherde. Light-colored, leached A2 horizon of podzol soils.

Blocky (or blocklike) structure. Soil aggregates have a blocky shape, irregularly six-faced, and with the three dimensions nearly equal. The size of these aggregates ranges from a fraction of an inch to 3 or 4 in. in thickness. This structure is found in the B horizon of many soils. When the edges of the cube are sharp and rectangular faces are distinct, the type is identified as a blocky or angular blocky. If subrounding is apparent, the aggregates are identified as nutlike, nuciform, or subangular blocky.

Blowout. Shallow circular or elliptical depression in sand or dry soil formed by wind erosion.

Blowout dune. Term applied to the accumulations of sand derived from blowout troughs or basins, particularly where the accumulation is of large size and rises to considerably height above the source area.

Bluff. Cliff with a broad face; relatively long strip of land rising abruptly above the surrounding land or a body of water.

Bog iron ore. Impure ferruginous deposits developed in bogs or swamps by the chemical or biochemical oxidation of iron carried in solution.

Bog soil. Great soil group of the intrazonal order and hydromorphic suborder. The group includes muck and peat.

Bolson. In arid regions, a basin filled with alluvium and intermittent playa lakes and having no outlet.

Bottomset bed. Flat-lying bed of fine sediment deposited in front of a delta and then buried by continued delta growth.

Boulder. Piece of rock, separated from bedrock, more than 256 mm in maximum dimension. The term is sometimes considered to apply only to rounded stones of this size.

Breccia

1. Rock composed of coarse angular fragments cemented together.

2. Clastic rock composed principally of large angular fragments. Usually the clasts are all derived from the same parent formation.

Broad base terrace. Low embankment with such gentle slopes that it can be farmed, constructed across sloping fields to reduce erosion and runoff.

Brown earths. Soils with a mull horizon, but having no horizon accumulation of clay or sesquioxides. The term is generally used as a synonym for "brown forest soils," but sometimes for similar soils acid in reaction.

Brown forest soils. Soil group of the intrazonal order and calcimorphic suborder, formed on calcium-rich parent materials under deciduous forest and possessing a high base status, but lacking a pronounced illuvial horizon.

Brown podzolic soils. Zonal great soil group similar to podzols, but lacking the distinct A horizon characteristic of the podzol group.

Brown soils. Great soil group of the temperate to cool arid regions, composed of soils with a brown surface and light-colored transitional subsurface horizon over calcium carbonate accumulation. They develop under short grasses.

Buffer. Substance that when added to a solution, causes a resistance to any change in pH.

Buffering. Chemically, the ability to resist change in pH.

Buffer solution. Solution to which large amounts of acid or base may be added with only a small resultant change in the hydrogen-ion concentration.

Bulking. Increase in volume of a material due to manipulation. Rock bulks upon being excavated; damp sand bulks if loosely deposited, as by dumping, because the apparent cohesion prevents movement of the soil particles to form a reduced volume.

Buried soil. Soil covered by an alluvial, loessal, or other deposit, usually to a depth greater than the thickness of the solum.

Butte. Steep-sided and flat-topped hill formed by erosion of flat-lying strata where remnants of a resistant layer protect softer rocks underneath.

Calcareous soil. Soil containing sufficient calcium carbonate, often with magnesium carbonate, to effervesce visibly when treated with cold hydrochloric acid.

Calcimorphic soils. Suborder of intrazonal soils consisting of two great soils groups whose properties reflect a predominating influence of highly calcareous parent material on soil development.

Caliche. Layer near the surface, weakly to strongly cemented by secondary carbonates of calcium or magnesium precipitated from the soil solution. Caliche may occur as a soft, thin, soil horizon; a hard, thick bed just beneath the solum; or a surface layer exposed by erosion. It is not a geologic deposit.

Cambic horizon. Horizon that has been altered or changed by soil-forming processes. It usually occurs below a diagnostic surface horizon (epipedon).

Capillary attraction. Liquid's movement over or retention by a solid surface due to the interaction of adhesive and cohesive forces.

Carbonation. Chemical weathering in which minerals are altered to carbonates by carbonic acid.

Carbon cycle. Sequence of transformations whereby carbon dioxide is fixed in living organisms by photosynthesis or chemosynthesis, liberated by respiration and by the death and decomposition of the fixing organism, used by heterotrophic species, and ultimately returned to its original state.

Carbon-nitrogen ratio. Ratio of the weight of organic carbon to the weight of total nitrogen in a soil or an organic material. It is obtained by dividing the percentage of organic carbon by the percentage of total nitrogen.

Cataclastic rock. Breccia or powdered rock formed by crushing and shearing during tectonic movements.

Cat clays. Wet clay soils high in reduced forms of sulfur which upon being drained become extremely acidic owing to the oxidation of the sulfur compounds.

Catena

1. Sequence of soils of about the same age, derived from similar parent material, and occurring under similar climatic conditions but having different characteristics owing to variation in relief and drainage.

2. Group of soil series within any one soil zone developed from similar parent material, but with contrasting characteristics of the solum due to difference in relief or drainage.

Cation. Positive ion; ion that travels to the cathode in electrolysis.

Cation exchange. Interchange between a cation in solution and another cation on the surface of any surface-active material, such as clay, organic matter, or organic colloids.

Cation exchange capacity. Sum total of exchangeable cations that a soil can adsorb, sometimes called "total-exchange capacity," "Base-exchange capacity," or "cation-adsorption capacity." It is expressed in milliequivalents per 100 g of soil or other adsorbing material, such as clay.

Cementation. Weakly cemented, strongly cemented, and indurated.

Cemented. Indurated; having a hard, brittle consistency because the particles are held together by cementing substances such as humus, calcium carbonate, or the oxides of silicon, iron, and aluminum. Hardness and brittleness persist even when wet.

Channery. Thin, flat fragments of limestone, sandstone, or schist up to 6 in. in major diameter.

Chelate. Type of chemical compound in which a metallic atom is firmly combined with a molecule by means of multiple chemical bonds. The term refers to the claw of a crab and is illustrative of the way in which the atom is held.

Chemical weathering. Weathering in which a change in composition occurs.

Chernozem

1. Dark brown or black soil rich in humus and lime.

2. Zonal great soil group consisting of soils with a thick, nearly black or black, organic matter; rich A horizon high in exchangeable calcium, underlaid by a lighter colored transitional horizon above a zone of calcium carbonate accumulation. Chernozem occurs in a cool subhumid climate under a vegetation of tall and midgrass prairie.

Chert. Structureless form of silica closely related to flint which breaks into angular fragments.

Chestnut soil. Zonal great soil groups consisting of soils with a moderately thick, dark brown A horizon over a lighter colored horizon that is above a zone of calcium carbonate accumulation. This type of soil develops under mixed tall and short grasses in a temperate to cool and subhumid to semiarid climate.

Chiseling. Breaking or shattering of compact soil or subsoil layers by use of a chisel.

Choker aggregate. Screenings or fines that fill the interstitial voids of coarse aggregates.

C horizon

1. Undisturbed parent material from which the overlying soil profile has been developed.

2. Horizon generally beneath the solum that is relatively little affected by biological activity and pedogenesis

and is lacking properties diagnostic of an A or B horizon. It may or may not be like the material from which the A and B have formed.

3. Horizon that normally lies beneath the B horizon, but may lie beneath the A horizon where the only significant change caused by soil development is an increase in organic matter, which produces an A horizon. In concept, the C horizon is unaltered or slightly altered parent material.

Cirque

1. Deep, steep-walled recess in a mountain, caused by glacial erosion.

2. Steep-walled basin at the head of a glacial valley.

Clastics. Deposits that are made up of fragments of preexisting rocks or of the solid products formed during the chemical weathering of such older rocks. Familiar examples of sediments belonging to this group are gravel, sand, mud, clay, and their consolidated equivalents—conglomerate, sandstone, and shale.

Clay

1. In the Unified Soil Classification System, clay is defined as a soil that contains more than 50% particles with diameters less than 0.047 mm and has a plasticity in relation to its liquid limit. Types are identified as lean clay, heavy clay, organic clay or low plasticity, and organic clay of high plasticity. Also a soil separate consisting of particles less than 0.002 mm in equivalent diameter. Clay is the U.S. Department of Agriculture textural class name for soil that contains 40% or more clay, less than 45% sand, and less than 40% of silt.

2. Mineral soil particles less than 0.002 mm in diameter; as a soil textural class, soil material that contains 40% or more clay, 45% sand, and less than 40% of silt.

3. Any of a number of hydrous aluminosilicate minerals formed by weathering and hydration of other silicates; also, mineral fragment smaller than 1/256 mm.

Clay loam. Soil material that contains 27 to 40% clay and 20 to 45% sand.

Clay mineral. Naturally occurring inorganic crystalline or amorphous material found in soils and other earthy deposits, the particles being predominantly greater than 0.002 mm in diameter. It is largely of secondary origin.

Claypan

1. Compact, slowly permeable soil horizon rich in clay and separated more or less abruptly from the overlying soil. Claypans are commonly hard when dry and plastic or still when wet.

2. Dense, compact layer in the subsoil, having a much higher clay content than the overlying material, from which it is separated by a sharply defined boundary. It is formed by downward movement of clay or by synthesis of clay in place during soil formation. Claypans are usually hard when dry, and plastic and sticky when wet. Also, they usually impede the movement of water and air and the growth of plant roots.

Clay size fraction. Portion of the soil that is finer than 0.002 mm. It is not a positive measure of the plasticity of the material or its characteristics as a clay.

Cleavage. Tendency to break in the same direction, thus yielding fragments of predictable shape.

Clod. Artificially compact, coherent mass of soil, produced by plowing or digging, ranging in size from 8 to 10 in.

Coarse fragments. Rock or mineral particles greater than 2.0 mm in diameter.

Coarse texture. Texture exhibited by sands, loamy sands, and sandy loams, except very fine sandy loam.

Cobblestone. Rounded or partially rounded rock or mineral fragments between 3 and 10 in. in diameter.

Cobbly. Term applied to soils containing a significant amount of cobbles, which are rounded, coarse mineral fragments from 3 to 10 in. in diameter.

Coefficient of permeability. Rate of discharge of water under laminar flow conditions through a unit cross-sectional area of a porous medium under a unit hydraulic gradient and standard temperature conditions (usually 20°C).

Coefficient of thermal expansion. Measure of the increase in volume of a material relative to the original volume with increasing temperature.

Cohesion. Capacity of sticking or adhering together. In effect, the cohesion of soil is the part of its shear strength that does not depend on interparticle friction. True cohesion is attributed to the shearing strength of the cement or the absorbed water films that separate the individual grains at their areas of contact. Apparent cohesion of moist soils is due to surface tension in capillary openings and disappears completely on immersion.

Colloid

1. In soils, mineral or organic particle of submicroscopic size.

2. Particle of such size that it will exhibit Brownian movement when suspended in water.

Colluvium. Deposit of rock fragments and soil material accumulated at the base of steep slopes as a result of gravitational action.

Columnar structure

1. Structure with the vertical axis of aggregates longer than the horizontal and with rounded tops. When the tops are level and clean cut, the structure is identified as prismatic. Columnar structure is found in the B horizon when present.

2. Soil structure type with a vertical axis much longer than the horizontal axis and a distinctly rounded upper surface.

Compaction. Changing of loose sediments to hard, firm rocks. Process by which soil grains are rearranged to decrease void space and bring them into closer contact with one another, thereby increasing the weight of solid material per cubic foot. Noncohesive soils are most effectively compacted by vibration; moderately cohesive soils are compacted by sheepsfoot or other types of rollers.

Composite cone. Volcanic cone of a stratovolcano composed of both cinders and lava flows.

Concretion. Hardened local concentrations of certain chemical compounds, such as calcium carbonate and iron and manganese oxides, that form indurated grains or nodules of various sizes, shapes, and colors.

Conglomerate. Hard rock formed by the natural cementing together of rounded pebbles (gravel). A similar rock formed of larger fragments may be called a "cobble conglomerate" or a "boulder conglomerate," as the case may be.

Consistence

1. Combination of properties of soil material that determine its resistance to crushing and its ability to be

molded or changed in shape. Such terms as "loose," "friable," "firm," "soft," "plastic," and "sticky" describe soil consistence.

2. Resistance of a material to deformation or rupture; degree of cohesion or adhesion of soil mass.

Consolidation. Gradual reduction in volume of a soil mass resulting from an increase in compressive stress.

Continental glacier. Continuous, thick glacier covering more than $50,000$ km^2 and moving independently of minor topographic features.

Continental rise. Broad and gently sloping ramp that rises from an abyssal plain to the continental slope at a rate of less than $1-40$.

Contour. Imaginary line connecting points of equal elevation on the surface of the soil. A contour terrace is laid out on a sloping soil at right angles to the direction of the slope and level throughout its course. In contour plowing, the plowman keeps to a level line at right angles to the direction of the slope, which usually results in a curving furrow.

Coquina. Soft porous limestone composed of broken shells, with or without corals and other organic debris.

Coral. Hard, calcareous skeleton of various small sea animals (polyps) or the stony solidified mass of a number of such skeletons. In warm waters colonial coral forms extensive limestone reefs. In cool or cold water coral usually appears in the form of isolated solitary individuals. Occasionally large reefs formed in cold waters by calcareous algae have been referred to as "coral."

Coral reef. Ridge or mass of limestone built up of detrital material deposited around a framework of the skeletal remains of mollusks, colonial coral, and massive calcareous algae. coral may constitute less than half of the reef material.

Cordillera. Continuous mountain system extending from Alaska to extreme South America and ranging up to 1500 km in width.

Craton. Portion of a continent that has not been subjected to major deformation for a prolonged time, typically since Precambrian or early Paleozoic time.

Cross section. Drawing showing the features that would be exposed by a vertical cut through a structure.

Crotovina. Former animal burrow in one soil horizon that has been filled with organic matter or material from another horizon (also, "krotivina").

Crumb. soft, porous, more or less rounded natural unit of structure from 1 to 5 mm in diameter.

Crumb structure. Small, soft, porous aggregates irregular in shape and rarely larger than $\frac{1}{3}$ in. in size. If the aggregates are relatively nonporous, they are identified as granular. Both types are found in surface soils, especially those high in organic matter.

Crust

1. Thin, brittle layer of hard soil that forms on the surface of many soils when they are dry; exposed hard layer of materials cemented by calcium carbonate, gypsum, or other binding agents. Most desert crusts are formed by the exposure of such layers through removal of the upper soil by wind or running water and their subsequent hardening.

2. Surface layer on soils, ranging in thickness from a few millimeters to perhaps as much as an inch, that is much more compact, hard, and brittle when dry than the material immediately beneath it.

Crystal. Homogeneous inorganic substance of definite chemical composition bounded by plane surfaces that form definite angles with each other, thus giving the substance a regular geometrical form.

Crystalline rock. Rock consisting of various minerals that have crystallized in place from magma.

Cuesta. Ridge with one steep and one gentle face formed by the outcrop and slower erosion of a resistant, gently dipping bed.

Debris avalanche. Fast downhill mass movement of soil and rock.

Decomposition. Chemical breakdown of mineral or organic matter.

Deep soil. Soil deeper than 40 in. to rock or other strongly contrasting materials. Also, a soil with a deep black surface layer; soil deeper than about 40 in. to parent material or to other unconsolidated rock material not modified by soil-forming processes; soil in which the total depth of unconsolidated material, whether true soil or not, is 40 in. or more.

Deflation. Removal of clay and dust from dry soil by strong winds.

Deflocculate

1. Separate the individual components of compound particles by chemical and/or physical means; to cause the particles of the disperse phase of a colloidal system to become suspended in the dispersion medium.

2. Separate or to break up soil aggregates into individual particles; to disperse the particles of a granulated clay to form a clay that runs together or puddles.

Delta kame. Deposit having the form of a steep, flat-topped hill, left at the front of a retreating continental glacier.

Densification of soils. Loose soils are grouted for densification by cement, often mixed with fly ash or bentonite to provide more mobility, into the pores, or with silt intrusion or chemical or asphalt emulsion. These procedures replace the void content, which is partly or wholly filled with moisture, with more solid filler.

Density

1. Mass per volume of a substance, commonly expressed in grams per cubic centimeter.

2. Usually dry unit weight or unit dry weight. Although it is recognized that "density" is defined as mass per unit volume, in the field of soil mechanics the term is frequently used in place of unit weight.

3. Unit weight of a soil in pounds per cubic foot. The type of density, such as natural, in place, wet, dry, remolded natural, or relative, should be specified. It is the equivalent of specific gravity or the ratio, at atmospheric pressure, of the weight of a given volume of sea water to that of an equal volume of distilled water at $4°C$.

Dentrification. Biochemical reduction of nitrate or nitrite to gaseous molecular nitrogen or an oxide of nitrogen.

Deposition. General term for the accumulation of sediments by either physical or chemical sedimentation.

Depth creep. Continuous gradual movement of a slope occurring at a substantial depth under the influence of gravity forces. Depth creep is distinguished from seasonal creep, which occurs at the ground surface owing to thermal expansion and contraction, swelling, shrinkage, freezing and thawing, and other seasonal processes.

Desert. Region where precipitation is insufficient to support any plant life except xerophilous vegetation; region of extreme aridity.

Desert crust. Hard layer containing calcium carbonate, gypsum, or other binding material, exposed at the surface in desert regions.

Desert flats. Essentially flat surface extending from the edges of playas to the alluvial fans or bajadas.

Desert pavement

1. Layer of gravel or stones left on the land surface in desert regions after the removal of the fine materials by wind erosion.

2. Residual deposit produced by continued deflation, which removes the fine grains of a soil and leaves a surface covered with close-packed cobbles.

3. Mosaic of closely packed pebbles and broken rock fragments usually coated with a stain or crust of manganese or iron oxide and caused by wind removal of sand, silt, and clay particles.

Desert soils

1. Zonal group of soils that have light-colored surface soils and are usually underlaid by calcareous material and frequently by hard layers. Such soils have developed under extremely scanty scrub vegetation in warm to cool arid climates.

2. Zonal great soil group consisting of soils with a very thin, light-colored surface horizon, which may be vesicular and is ordinarily underlaid by calcareous material. Desert soils are formed in arid regions under sparse shrub vegetation.

Desert varnish. Glossy sheen or coating on stones and gravel in arid regions.

Desorption. Removal of sorbed material from surfaces.

Detritus. Accumulation of the fragments resulting from the distintegration of rocks.

Diatomaceous earth. Geologic deposit of fine, grayish, siliceous material composed chiefly or wholly of the remains of diatoms. It may occur as a powder or as a porous, rigid material.

Diatoms. Algae having siliceous cell walls that persist as a skeleton after death; any of the microscopic unicellular or colonial algae constituting the class Bacillariaceae. Diatoms occur abundantly in fresh and salt waters, and their remains are widely distributed in soils.

Diffusion

1. Transport of matter as a result of the movement of the constituent particles. The intermingling of two gases or liquids in contact with each other takes place by diffusion.

2. Independent or random movement of ions or molecules that tends to bring about their uniform distribution within a continuous system.

Dike. Bank of earth, stones, etc., constructed to prevent low-lying land from being inundated by the sea, a river, etc.

Dip. Angle by which a stratum or other planar feature deviates from the horizontal. The angle is measured in a plane perpendicular to the strike.

Discordant contact. Contact that cuts across bedding or foliation planes, such as the contact between a dike and the surrounding rocks.

Disintegration

1. Breakdown of rock and mineral particles into smaller particles by physical forces such as frost action.

2. Physical weathering of rocks.

Disperse. Break up compound particles, such as aggregates, into individual component particles; to distribute or suspend fine particles, such as clay, in or throughout a dispersion medium, such as water.

Degraded chernozem. Zonal great soil group consisting of soils with a very dark brown or black A1 horizon underlaid by a dark gray, weakly expressed A2 horizon and a possible brown B horizon. It is formed in the forest-prairie transition of cool climates.

Draw. Shallow dry wash or arroyo characterized by low banks that are gentler than those of the wash or arroyo.

Drift

1. Material of any sort deposited by geological processes in one place after having been removed from another. Glacial drift includes material moved by the glaciers and by the streams and lakes associated with them.

2. Collective term for all the rock, sand, and clay that are transported and deposited by a glacier either as till or as outwash.

Drumlin. Long, smooth, cigar-shaped low hills of glacial till, with their long axes parallel to the direction of ice movement.

Dry soil consistency. Loose, soft, slightly hard, hard, very hard, and extremely hard.

Duff. Matted, partly decomposed organic surface layer of forest soils.

Dune

1. Heap of sand or other material accumulated by wind. Its external form may be that of a hill or ridge. Sand wave of approximately triangular cross section (in a vertical plane in the direction of flow) with gentle upstream slope and steep downstream slope, which travels downstream by the movement of the sediment up the upstream slope and its deposition on the downstream slope.

2. Mount or ridge of loose sand piled up by the wind. Occasionally during periods of extreme drought, granulated soil material of fine texture may be piled into low dunes, sometimes called "clay dunes."

3. Elongated mound of sand formed by wind or water.

Duripan (hardpan)

1. Indurated horizon cemented by materials such as aluminum silicate, silica, and iron.

2. A horizon in a soil characterized by cementation by silica or other materials.

Earth. Solid matter of the globe, as contrasted with water and air; loose or softer material composing part of the surface of the globe, as distinguished from firm rock.

Earth flow. Detachment of soil and broken rock and its subsequent downslope movement at slow or moderate rates in a stream or tonguelike form.

Ecology

1. Science that deals with the interrelations of organisms and their environments.

2. Science of the life cycles, populations, and interactions of various biological species as controlled by their

physical environment, including also the effect of life forms on the environment.

Economic geology. Study of and location of ores, fossil fuels, and other useful materials that occur in sufficient concentration to be marketable in a particular economy.

Edaphic. Of or pertaining to the soil; resulting from or influenced by factors inherent in the soil or other substrate, rather than by climatic factors.

Electron microprobe. Instrument that bombards a very minute sample with electrons to determine its chemical composition from the resulting x-radiation.

Eluvial horizon. Soil horizon that has been formed by the process of eluviation.

Entisol. Order of soils in which profile development is minimal. Characteristics are largely inherited from the parent material.

Epipedon. Diagnostic surface horizon that includes the upper part of the soil that is darkened by organic matter or the upper eluvial horizons, or both.

Epoch. One subdivision of a geologic period, often chosen to correspond to a stratigraphic series. The term is also used for a division of time corresponding to a paleomagnetic interval.

Equivalent diameter. In sedimentation analysis, the diameter assigned to a nonspherical particle. It is numerically equal to the diameter of a spherical particle of the same density and velocity of fall.

Era. Time period that includes several periods but is smaller than an eon. Commonly recognized eras are Precambrian, Paleozoic, Mesozoic, and Cenozoic.

Erratic. Rock fragment, usually of large size, that has been transported from a distant source, especially by the action of glacial ice.

Exchange acidity. Titratable hydrogen and aluminum that can be replaced from the adsorption complex by a neutral salt solution. It is usually expressed as milliequivalents per 100 g of soil.

Exchange capacity. Total ionic charge of the adsorption complex active in the adsorption of ions.

Exit gradient. Hydraulic gradient (difference in piezometric levels at two points divided by the distance between them) near an exposed surface through which seepage is moving.

Exfoliation. Scaling off a surface in flakes or layers; breaking or spalling off of thin concentric shells, scales, or lamellae from rock surfaces. The action is due to changes in temperature, to the action of frost, and, in the opinion of some observers, to obscure chemical effects.

Fine-grained soil. Soil of which more than 50% of the grains, by weight, will pass a No. 200 sieve (with mesh smaller than 0.074 mm in diameter).

Firm. Consistence of a moist soil that offers distinctly noticeable resistance to crushing but can be crushed with moderate pressure between the thumb and forefinger.

Flocculate. Aggregate or clump together individual tiny soil particles, especially fine clay, into small groups or granules.

Flocculent structure. Arrangement composed of flocs of soil particles instead of individual soil particles.

Flow slide. Shear failure in which a soil mass moves over a relatively long distance in a fluidlike manner. A flow slide occurs rapidly on flat slopes in loose, saturated, uniform sands or in highly sensitive clays.

Fluorapatite. Member of the apatite group of minerals rich in fluorine. It is the most common mineral in rock phosphate.

Foothill. One of the lower subsidiary hills at the foot of a mountain or of higher hills.

Forest land. Land bearing a stand of trees of any age or stature, including seedlings, and of species attaining a minimum of 6 ft average height at maturity; land from which such a stand has been removed, but for which no other use has been substituted. The term is commonly limited to land not on farms. Forests on farms are commonly called "woodlands" or "farm forests."

Fracture. Random breakage that results in fragments of unpredictable shape.

Fragipan

1. Dense and brittle pan or layer in soils that owes its hardness mainly to extreme density or compactness rather than high clay content or cementation. Removed fragments are friable but the material in place is so dense that roots cannot penetrate and water moves through it very slowly.

2. Natural subsurface horizon with high bulk density relative to the solum above, seemingly cemented when dry, but when moist showing a moderate to weak brittleness. The layer is low in organic matter, mottled, and slowly or very slowly permeable to water, and it shows occasional or frequent bleached cracks forming polygons. It may be found in profiles of either cultivated or virgin soils but not in calcareous material.

Friable. Consistency term pertaining to the ease of crumbling of soils.

Frost. Feathery deposit of minute ice crystals or grains on a surface or object, formed directly from vapor in the air; process by which such ice crystals are formed; temperature at which frost forms. Frost often forms when the close-lying air is 32°F especially in calm, clear weather when radiation or evaporation reduces a surface temperature to a point of freezing or below.

Fulvic acid. Term that describes a mixture of organic substances remaining in solution upon acidification of a dilute alkali extract from the soil.

Fungi. Simple plant organisms lacking chlorophyll, which enables higher plants and algae to manufacture their own food. Because of this inability to synthesize food, fungi are wholly dependent on organic materials for their nutrient supply. This dependence renders fungi capable of degrading almost all materials of an organic origin.

Genetic. Resulting from soil genesis or developments, that is, produced by soil-forming processes as, for example, a genetic soil profile or a genetic horizon.

G horizon. Formerly defined as a horizon of intense reduction, with ferrous iron and gray to brown mottling. Now denotes gleyzation as a soil-forming process.

Gilgai. Term used to describe the microrelief of soils produced by expansion and contraction with changes in moisture. Condition is found in soils that contain large amounts of clay, which swells and shrinks considerably with wetting and drying. Usually a succession of microbasins and microknolls in nearly level areas or of microvalleys and microridges parallel to the direction of the slope.

Gilsonite. Form of natural asphalt, hard and brittle, occurring in rock crevices or veins.

Gley soil. Soil developed under conditions of poor drainage, resulting in the reduction of iron and other elements and in gray colors and mottles.

Gorge. Canyon; rugged and deep ravine or gulch.

Granular structure

1. Soil structure in which the individual grains are grouped into spherical aggregates with indistinct sides. Highly porous granules are commonly called "crumbs." Well-granulated soil has the best structure for most ordinary crop plants.

2. Small, soft, relatively nonporous aggregates of soil particles, irregular in shape and rarely larger than $\frac{1}{3}$ in. in size. This type of structure is found in surface soils, especially those high in organic matter.

Granule. Aggregate similar in size to a crumb, but more dense and, therefore, less porous than a crumb.

Gray brown podzolic soils. Zonal great soil group consisting of soils with a thin, moderately dark A1 horizon and with a grayish brown A2 horizon underlaid by a B horizon containing a high percentage of exchangeable bases and an appreciable quantity of illuviated silicate clay. These soils are formed on relatively young land surfaces, mostly glacial deposits, from material relatively rich in calcium, under deciduous forests in humid temperate regions.

Gray desert soil. Term used in the United States; synonymous with "desert soil."

Great soil group

1. Any one of several broad groups of soils with fundamental characteristics in common. Examples are chernozems, gray-brown podzolic, and podzol.

2. Category in the system of soil classification used in the United States consisting of classes produced by subdividing suborders and, therefore, representing a range in conditions of soil formation that is narrower than the range allowed for suborders.

Grus

1. Deposits resulting from the weathering of the various minerals forming igneous rock. Deposits consist of the accumulation of countless discrete particles on the surface of the rock, sometimes to a depth of over 1 m.

2. An accumulation of angular, coarse-grained fragments resulting from the granular disintegration of crystalline rocks, generally in an arid or semiarid region.

Hammada. Desert surface that is either bedrock or bedrock covered only by a very thin veneer of sand or pebbles. The term was originally applied in the Sahara where it referred to a desert plateau of stones, but it is now used for similar desert surfaces in other parts of the world.

Hardpan

1. Hardened or cement soil layer. Soil material may be sandy or clayey and may be cemented by iron oxide, silica, calcium carbonate, or other substances.

2. Hardened soil layer in the lower A or B horizon caused by cementation of soil particles with organic matter or with materials such as silica, sesquioxides, or calcium carbonate. The hardness does not change appreciably with changes in moisture content, and pieces of the hard layer do not slake in water.

Heavy soil. Soil with a high content of the fine separates, particularly clay; soil with a high drawbar pull and hence difficult to cultivate.

Heterotrophic. Capable of deriving energy for life processes only from the decomposition of organic compounds and incapable of using inorganic compounds as sole sources of energy or for organic synthesis. Contrast with autotrophic.

Histic epipedon. Horizon at or near the surface, saturated with water at some seasons and containing a minimum of 20% organic matter if no clay is present and at least 30% organic matter if it has 50% clay or more.

Histisols

1. Soils characterized by their high organic matter content. Bog soils and half-bog soils are included in this soil order.

2. Soil order consisting of organic soils such as peat or muck.

Honeycomb structure. Arrangement of soil particles having a comparatively loose, stable structure resembling a honeycomb.

Horizon (soil horizon). One of the layers of the soil profile, distinguished principally by its texture, color, structure, and chemical content.

Horst. Long, narrow block of the earth's crust that has been relatively uplifted between faults along the sides. Although a horst may appear as an elevated ridge, the term applies to fundamental structure and not topographic expression.

Humic acid. Any of various complex organic acids obtained from humus.

Hydrous mica. Silicate clay with 2 to 1 lattice structure, but of indefinite chemical composition, since usually part of the silicon in the silica tetrahedral layer has been replaced by aluminum. It contains a considerable amount of potassium, which serves as an additional bonding between the crystal units, resulting in particles larger than normal in montmorillonite and, consequently, in a lower cation-exchange capacity. It is sometimes referred to as "illite."

Hydroxyapatite. Member of the apatite group of minerals rich in hydroxyl groups. It is a nearly insoluble calcium phosphate.

Hygroscopic coefficient. Amount of moisture in a dry soil when it is in equilibrium with some standard relative humidity near a saturated atmosphere (about 98%), expressed in terms of percentage on the basis of oven-dry soil.

Igneous rocks. Rocks formed by solidification of hot mobile rock material (magma), including those formed and cooled at great depths (plutonic rocks), which are crystalline throughout, and those that have poured out on the earth's surface in the liquid state or have been blown as fragments into the air (volcanic rocks).

Illite

1. Hydrous mica.

2. Layered clay in which a negative charge caused by ion substitution is neutralized predominantly by potassium ions.

Illuvial horizon

1. Soil layer or horizon in which material carried from an overlying layer has been precipitated from solution or deposited from suspension; layer of accumulation.

2. Soil layer or horizon produced by illuviation.

Illuviation

1. Deposition of colloidal soil material, removed from one horizon to another in the soil—usually from an upper to lower horizon in the soil profile.

2. Accumulation of material in a soil horizon through the deposition either mechanically or chemically of suspended mineral and organic matter originating from horizons above. Since at least part of the fine clay in the B horizons (or subsoils) of many soils have moved into them from the A horizons above, these are called illuvial horizons.

Immature soil. Soil lacking clear individual horizons because of the relatively short time for soil-building forces to act on the parent material since its deposition or exposure.

Inceptisols

1. Soils with one or more diagnostic horizons that are thought to form rather quickly and that do not represent significant illuviation, eluviation, or extreme weathering. Soils classified as brown forest, subarctic brown forest, ando, sols bruns acides, and associated humic gley and low-humic gley soils are included in this order.

2. Soil order consisting of soils showing a moderate degree of profile development, but not enough to be considered at equilibrium with the soil-forming environment.

Inherited soil characteristic. Characteristic of a soil that is due directly to the nature of the material from which it formed, in contrast to characteristics that are wholly or partly the result of soil-forming processes acting on parent material. For example, some soils are red because the parent material was red, although the color of most red soils is due to soil-forming processes.

Inorganic. Not organic; substances occurring as minerals in nature or obtainable from them by chemical means; all matter except the compounds of carbon, but including carbonates.

In place. Rock may decay and break down into small particles where it is first exposed in the land surface. It is then said to have weathered "in place" or "in situ".

Intrazonal soil. Any one of the great groups of soils having more or less well-developed soil characteristics that reflect a dominating influence of some local factor of relief or of parent material over the normal influences of the climate and the vegetation on the soil-forming processes. Such groups of soils may be geographically associated with two or more of the zonal groups of soils having characteristics dominated by the influence of climate and vegetation.

Intergrade. Soil that possesses moderately well-developed distinguishing characteristics of two or more genetically related great soil groups.

Intrusion. Process of forcible implacement of one body of mobile rock material into or between other rocks. The term generally refers to the invasion of older rocks at depth by molten rock or magma, but it is also used to describe the plastic injection of salt domes into overlying rocks.

Intrusive rock. Igneous rock that has ascended in a hot mobile state from the depths of the earth, but that has been arrested and cooled before reaching the surface. Forms usually assumed by intrusive rocks are tabular or sheetlike dikes and sills, cylindrical necks, and larger masses with steep walls and no apparent floor, such as stocks and batholiths.

Ions. Atoms that are positively charged (cations) because of the loss of one or more electrons, or that are negatively charged (anions) because of a gain in electrons.

Iron-pan. Indurated soil horizon in which iron oxide is the principal cementing agent.

Isomorphous substitution

1. In a crystal lattice, the replacement of one atom by another of similar size without disrupting or changing the crystal structure of the mineral.

2. Replacement of an ion considered normal to a mineral structure by another during the formation of a mineral.

Kame. Fluvioglacial deposit occurring as a mound, knob, or hillock, in which one or more sides were in contact with the glacier ice. Kames are diverse in size, shape, and composition and generally, but not universally, consist of poorly sorted, poorly stratified material.

Kame terrace. Stratified drift deposited by meltwater between glacier ice and adjacent higher ground and left as a constructional terrace after disappearance of the ice.

Kaolin. Aluminosilicate mineral of the crystal lattice group, that is, a mineral consisting of one silicon tetrahedral layer and one aluminum oxide-hydroxide octahedral layer.

Karst topography. Type of landform developed in a region of easily soluble limestone bedrock. It is characterized by vast numbers of depressions of all sizes, sometimes by great outcrops of fluted limestone ledges, sinks, and other solution passages; an almost total lack of surface streams; and large springs in the deeper valleys.

Laminated structure. Platy structure with the plates or very thin layers lying horizontal or parallel to the surface.

Land. Total natural and cultural environment within which production takes place. "Land" is a broader term than "soil." In addition to soil, its attributes include other physical conditions such as mineral deposits and water supply; location in relation to centers of commerce, populations, and other land; the size of the individual tracts or holdings; and existing plant cover, works of improvement, and the like.

Land classification. Arrangement of land units into various categories based on the properties of the land or its suitability for some particular purpose.

Landform. Physical expression of the land surface.

Land-use planning. Development of plans for the uses of land that over long periods will best serve the general welfare, together with the formulation of ways and means for achieving such uses.

Lateritic soil. Suborder of zonal soils formed in warm, temperate, and topical regions and including the following great soil groups: yellow podzolic, red podzolic, yellowish brown lateritic, and lateritic.

Latosol. Suborder of zonal soils that includes soils formed under forested, tropical, humid conditions and characterized by low silicasesquioxide ratios of the clay fractions, low base exchange capacity, low activity of the clay, low content of most primary materials and soluble constituents, a high degree of aggregate stability, and, usually, a red color.

Lattice structure. Orderly arrangement of atoms in a crystalline material.

Lava. Fluid rock, such as that which issues from a volcano or a fissure in the earth's surface; also, the same material solidified by cooling.

Lava flow. Stream of molten lava that has flowed over a part of the earth's surface. If it spreads out over a large area it is called a "lava bed" or a "lava field." Lava flows or lava beds may be nearly smooth and level or they may be extremely rough.

Lime pan. Playa whose surface layer is cemented by calcium carbonate precipitated as the water evaporates.

Limestone

1. Sedimentary rock composed primarily of calcite. If dolomite is present in appreciable quantities, it is called a "dolomitic limestone."

2. Bedded sedimentary deposit consisting chiefly of calcium carbonate which yields lime when burned. In a broader sense, the term has been used for combinations or mixtures with magnesium carbonate in which the proportion of calcium carbonate is less than half.

Liquefaction

1. Sudden large decrease of the shearing resistance of a cohesionless soil, caused by shock or other type of strain. It is associated with a sudden but temporary increase of the porefluid pressure and involves a temporary transformation of the material into a fluid mass.

2. Sudden large decrease of shearing resistance of a cohesionless soil caused by collapse of the soil structure due to shock or small shear strains and associated with sudden but temporary increase of pore water pressures.

Lithosols. Great soil group of azonal soils characterized by an incomplete solum or no clearly expressed soil morphology and consisting of freshly and imperfectly weathered rock or rock fragments.

Lithosphere. Solid outer portion of the earth; crust of the earth. The term is usually used in contexts wherein the lithosphere is said to make contact with the atmosphere and hydrosphere.

Loess

1. Material transported and deposited by wind and consisting of predominantly silt-sized particles.

2. Uniform aeolian deposit of silty material having an open structure and relatively high cohesion due to cementation of clay or calcareous material at grain contacts. A characteristic of loess deposits is that they can stand with nearly vertical slopes.

Luxury consumption. Intake by a plant of an essential nutrient in amounts exceeding what it needs. Thus if potassium is abundant in the soil, alfalfa may take in more than is required.

Lysimeter. Device for measuring percolation and leaching losses from a column of soil under controlled conditions.

Macrostructure. Structural features of a soil that are visible to the naked eye.

Magma. Hot mobile rock material generated within the earth, from which igneous rock results by cooling and crystallization. It is usually conceived of as a pasty or liquid material, or a mush of crystals together with a noteworthy amount of liquid phase having the composition of silicate melt.

Mantle rock. Unconsolidated material lying upon and in most cases derived from the bedrock; regolith.

Mapping unit. Soil or combination of soils delineated on a map and, where possible, named to show the taxonomic unit or units included. Principally, mapping units on maps of soils depict soil types, phases, associations, or complexes.

Marl. Soft and unconsolidated calcium carbonate, usually mixed with varying amounts of clay or other impurities.

Mass flow. Unidirectional flow of a liquid or gas along with any suspended or dissolved components that they contain.

Massive. Without definite crystalline structure.

Massive state. Nonstructural state in soils that contain cohesive particles occurring in a continuous mass having no well-defined cleavage pattern. When crushed, massive soil breaks into irregular fragments of unpredictable size and shape.

Massive structure. Large uniform masses of cohesive soil, structureless.

Mature soil. A soil with well-developed soil horizons, having characteristics produced by the natural processes of soil formation and in near equilibrium with its present environment.

Maximum density (maximum unit weight). Dry unit weight defined by the peak of a compaction curve.

Mellow soil. Very soft, very friable, porous soil without any tendency toward hardness or harshness.

Mesa. Flat-topped mountain bounded on at least one side by a steep cliff; plateau terminating on one or more sides in a steep cliff; tableland.

Metamorphic rocks

1. Rock that has been greatly altered from its previous condition through the combined action of heat and pressure. For example, marble is a metamorphic rock produced from limestone; gneiss is one produced from granite; and slate is produced from shale.

2. Rock derived from preexisting rocks but differing from them in physical, chemical, and mineralogical properties as a result of natural geological processes manifested principally by heat and pressure originating within the earth. The preexisting rocks may have been igneous, sedimentary, or another form of metamorphic rock.

3. One of the three great groups of rocks. Metamorphic rocks are formed from original igneous or sedimentary rocks through alterations produced by pressure, heat, or the infiltration of other materials at depths below the surface zones of weathering and cementation.

Micas. Primary aluminosilicate minerals in which two silica layers alternate with one alumina layer. They separate readily into thin sheets or flakes.

Microfauna. Part of the animal population that consists of individual organisms too small to be clearly distinguished without the use of a microscope. The classification includes protozoa and nematodes.

Microflora. Part of the plant population that consists of individual plants too small to be clearly distinguished without the use of microscope. The classification includes actinomycetes, algae, bacteria, and fungi.

Microrelief. Small-scale, local differences in topography, including mounds, swales, or pits that are only a few feet in diameter and with elevation differences of up to 6 ft.

Mineral. Substance occurring in inorganic nature, though not necessarily of inorganic origin, that has definite chemical composition or, more commonly, a characteristic range of chemical composition and distinctive physical properties or molecular structure.

Mineral deposit. An occurrence of one or more minerals in such concentrations as to make removal and processing a possibility.

Mineralization. Process of inducing or the result of an induced change in a body of rock. Conversion of an element from an organic form to an inorganic state as a result of microbial decomposition.

Modified loess. Loess that has lost its typical characteristics by secondary processes, including immersion, erosion, and subsequent deposition; chemical changes involving the destruction of the bond between the particles; chemical decomposition of the more perishable constituents, such as feldspar.

Mollic epipedon. Thick, dark surface layer, more than 50% base saturated (dominantly with bivalent cations), having a narrow C/N ratio (less with cultivated soils), a strong soil structure, and a relatively soft consistency when dry.

Mollisols

1. Soils characterized by a thick, dark mineral surface horizon that is dominantly saturated with bivalent cations and has moderate to strong structure. In the old classification system, such soils are included as chernozem, prairie, and chestnut.

2. Soil order consisting of soils having a thick A horizon with more than 1% organic matter and a base saturation percentage above 50%. Normally, they are formed under grass vegetation. They are distinguished from vertisols in that they are not self-inverting.

Montmorillonite. Aluminosilicate clay mineral with a 2 : 1 expanding crystal lattice, that is, with two silicon tetrahedral layers enclosing an aluminum octahedral layer. Considerable expansion may be caused along the C axis by water moving between silica layers of contiguous units.

Mottled. Soil horizons irregularly marked with spots of color. A common cause of mottling is imperfect or impeded drainage, although there are other causes, such as soil development from an unevenly weathered rock. Different kinds of minerals may cause mottling.

Mycorrhiza. Association, usually symbiotic, of fungi with the roots of seed plants.

Native boulders. Rock or stone used as landscape decorative material in combination with plant growth.

Natric horizon. Special argillic horizon that has prismatic or columnar structure and is more than 15% saturated with exchangeable sodium. This is common in solonetz and solodized solonetz soils.

Nitrate reduction. Biochemical reduction of nitrate.

Nitrification. Biochemical oxidation of ammonium to nitrate.

Nitrogen assimilation. Incorporation of nitrogen into organic cell substances by living organisms.

Nitrogen fixation. Conversion of elemental nitrogen to organic combinations or to forms readily utilizable in biological processes.

Nucleic acids. Complex compounds found in the nuclei of plant and animal cells and usually combined with proteins as nucleoproteins.

Nutrient. Chemical element essential for the growth and development of an organism.

Ochric epipedon. Light-colored surface horizon generally low in organic matter, which includes the eluvial layers near the surface. It is often hard and massive when dry. Although characteristics of aridisols, it is found in several of the other soil orders.

O horizon

1. Organic horizon of mineral soils.

2. Natural layer of horizon of fresh or partly decomposed plant litter on the surface of the mineral soil.

Order

1. Highest category in soil classification; the three orders are zonal soils, intrazonal soils, and azonal soils. Ten orders are recognized in the new comprehensive classification system: entisol, vertisol, inceptisol, aridisol, mollisol, spodisol, alifisol, ultisol, oxisol, and histisol.

2. Highest category in soil classification; the three orders are zonal soils, intrazonal soils, and azonal soils. Category in the classification of plants and animals, ranking below the class and above the family.

Ortstein. Indurated layer in the B horizon of podzols in which the cementing material consists of illuviated sesquioxides (mostly iron) and organic matter.

Osmotic. Type of pressure exerted in living bodies as a result of unequal concentration of salts on both sides of a cell wall or membrane. Water will move from the area having the least salt concentration through the membrane into the area having the highest salt concentration and, therefore exerts additional pressure on this side of the membrane.

Osmotic pressure. In concept, the force per unit area required to equal the attractive (hydration) force of water exerted by ions dissolved in a solution.

Outcrop. Part of a body of rock that appears, bare and exposed, at the surface of the ground. In a more general sense the term applies also to areas where the rock formation occurs just beneath the soil, even though it is not exposed.

Oven-dry soil. Soil that has been dried at 150°C until it reaches constant weight.

Oxic horizon. Highly weathered diagnostic subsurface horizon from which most of the combined silica has been removed leaving a mixture dominated by hydrous oxide clays with some silicate minerals and quartz present.

Oxisols

1. Soils of tropical and subtropical regions characterized by the presence of a horizon (oxic) from which most of the combined silica has been removed by weathering leaving oxides of iron and aluminum and some quartz. Included are soils referred to as latosols and some called "groundwater laterites."

2. Soil order consisting of soils containing a B horizon in which clays are largely kaolinite and sesquioxides and in which sands are essentially void of easily weathered minerals. Oxisols occur typically under the forested conditions of the humid tropics. They are usually red in color and have a high degree of aggregate stability, a low cation-exchange capacity, and a moderate to low degree of base saturation.

Pahoehoe. Solidified lava that is characterized by a smooth, billowy, or ropy surface having a skin of glass a fraction of an inch to several inches thick. Pahoehoe (Hawaiian term) is distinguished from the AA type lava by its smooth surface.

Pans. In soils, horizons, or layers that are strongly compacted, indurated, or very high in clay content.

Parent material

1. Unconsolidated and more or less chemically weathered mineral or organic matter from which the solum of soils is developed by pedogenic processes.

2. Relatively unaltered, unconsolidated mass of material from which a soil profile develops.

Parent rock. Rock from which a parent material is derived.

Particle density. Mass per unit volume of soil particles. In technical work, particle density is usually expressed as grams per cubic centimeter.

Particle size. Effective diameter of a particle measured by sedimentation, sieving, or micrometric methods.

Particle size analysis. Determination of the various amounts of the different separates in a soil sample, usually by sedimentation, sieving, micrometry, or combinations of these methods.

Particle size distribution. Amounts of the various soil separates in a soil sample, usually expressed as weight percentages.

Parts per million (ppm). Weight units of any given substance per one million equivalent weight units of oven-dry soil; or in the case of soil solution or other solution, the weight units of solute per million weight units of solution.

Patterned ground. Collective term for ground surface patterns resulting from frost action; surface expression of soil structures. Patterned ground is common but not exclusive to permafrost regions. The term includes a large group of soil forms that have distinct linear or polygonal elements in ground plan, such as stone rings, stone nets, and stone garlands.

Ped. Unit of soil structure, such as an aggregate, crumb, prism, block, or granule, formed by natural processes (in contrast with a clod, which is formed artificially).

Pedalfer

1. Subdivision of a soil order comprising a large group of soils in which the sesquioxides increased relative to silica during soil formation.

2. Common soil type in humid regions, characterized by an abundance of iron oxides and clay minerals deposited in the B horizon by leaching.

3. Soil in which there has been a shifting of alumina and iron oxide downward in the soil profile, but with no horizon of carbonate accumulation.

Pedestal rock. Residual or erosional rock supported by a relatively slender column or pedestal.

Pediment

1. Planar, sloping rock surface forming a ramp up to the front of a mountain range in an arid region. It may be covered locally by a thin alluvium.

2. A pediment is gently inclined, erosion surfaced, of low relief and typically developed in arid or semiarid regions at the foot of a receding mountain slope. A sediment may be bare or mantled by a thin layer of alluvium which is in transit to the adjoining basin.

Some pediments resemble alluvial fans in outward form.

3. Gently inclined planate erosion surfaces carved in bedrock and generally veneered with fluvial gravels.

Pedocal

1. Soil with a well-developed horizon of accumulated calcium carbonate or, more rarely, some other carbonate.

2. Subdivision of a soil order comprising a large group of soils in which calcium accumulated during soil formation.

3. Common soil type of arid regions, characterized by accumulation of calcium carbonate in the A horizon.

Pedology. Science dealing with the soil as a natural body.

Pedon. Smallest volume of a soil body that displays the normal range of variation in soil properties. Where properties such as horizon thickness vary little along a lateral dimension, the pedon may occupy an area of a square yard or less. Where such a property varies substantially along a lateral dimension, a large pedon several square yards may be required to show the full range in variation.

Peneplain. Once high, rugged area that has been reduced by erosion to a low, gently rolling surface resembling a plain.

Penetrability. Ease with which a probe can be pushed into the soil. It may be expressed in units of distance, speed, force, or work, depending on the type of penetrometer used.

Percent compaction. Ratio, expressed as a percentage, of dry unit weight of a soil to maximum unit weight obtained in a laboratory compaction test.

Permafrost. Permanently frozen ground; soil thickness, superficial deposit, or bedrock of variable depth beneath the surface of the earth in which below-freezing temperature has existed continuously for a long time.

Permanent charge. Net negative (or positive) charge of clay particles inherent in the crystal lattice of the particle; charge not affected by changes in pH or by ion exchange reactions.

Petrology. Study of the composition, structure, and origin of rocks.

pH

1. Numerical designation of relative acidity and alkalinity in soils and other biological systems. Technically, pH is the common logarithm of the reciprocal of the hydrogen ion concentration of a solution. A pH of 7.00 indicates precise neutrality; higher values indicate increasing alkalinity; and lower values indicate increasing acidity.

2. Negative logarithm of the hydrogen ion activity of a soil; degree of acidity (or alkalinity) of a soil as determined by means of a glass, quinhydrone, or other suitable electrode or indicator at a specified moisture content or soil-water rating, expressed in terms of the pH scale.

3. Dependent charge; portion of the total charge of the soil particles; which is affected by and varies with changes in pH.

Physical properties of soils. Characteristics, processes, or reactions of a soil that are caused by physical forces and can be described by or expressed in physical terms or equations. Examples of physical properties are bulk density, water-holding capacity, hydraulic conductivity, porosity, and pore-size distribution.

Plaggen epipedon. Man-made surface layer more than 20 in. thick, produced by long continued manuring.

Plain. Comparatively flat, smooth, and level tract without noticeable hills, mountains, or valleys.

Planosol

1. Great soil group of the intrazonal order and hydromorphic suborder consisting of soils with eluviated surface horizons underlaid by B horizons more strongly eluviated, cemented, or compacted than associated normal soil.

2. Intrazonal group of soils with eluviated surface horizons underlaid by claypans or fragipans, developed on nearly flat or gently sloping uplands in humid or subhumid climates.

Plasticity. Property of a soil that allows it to be deformed beyond the point of recovery without cracking or appreciable volume change.

Plastic soil. Soil capable of being molded or deformed continuously and permanently by relatively moderate pressure into various shapes.

Plateau

1. Level on which little or no change takes place; this can be represented on a graph. Tableland or flat-topped area of considerable extent elevated above the surrounding country on at least one side. A plateau is larger and more extensive than a mesa, but the two cannot be strictly separated. The surface may be fairly smooth, but not necessarily so. Large mountain masses may rise above it, and deep canyons may be cut into it.

2. Extensive upland region at high elevation with respect to its surroundings.

Platelike (platy) structure. Flat aggregates of soil particles with the vertical dimensions much less than the horizontal dimensions. Although this structure is found most often in surface horizons, it may be found in the subsoil, as it is often inherited from the parent materials.

Platy. Consisting of soil aggregates that are developed predominantly along the horizontal axes; laminated; flaky.

Platy soil structure. Soil aggregates with thin vertical axes and long horizontal axes. Flat, tabular, three-dimensional object that has one dimension much smaller than the other two.

Plinthite. Highly weathered mixture of sesquioxides of iron and aluminum with quartz and other diluents. It occurs as red mottles and changes irreversibly to hardpan upon alternate wetting and drying.

Podzol

1. Great soil group of the zonal order consisting of soils formed in cool-temperate to temperate, humid climates, under coniferous or mixed coniferous and deciduous forest, and characterized particularly by a highly leached, whitish gray A2 horizon.

2. Zonal group of soils having surface or organic mats and thin, organic-mineral horizons above gray leached horizons that rest on illuvial dark-brown horizons developed under coniferous or mixed forests or underneath vegetation in a cool-temperate, moist climate.

Plucking. Process by which rock fragments are detached from bedrock by the movement of glacial ice.

Podzolization

1. Soil formation resulting in the genesis of podzols and podzolic soils.

2. Process by which soils are depleted of bases, become more acid, and develop leached surface layers from which clay has been removed.

Polarity epoch. In paleomagnetism, a segment of geologic time in which the earth's magnetic field was either predominantly in the present direction or predominantly reversed.

Polygonal soil. More or less regular-sided ground surface patterns created by frost action, thawing, ground ice wedges, or a combination thereof. Widespread phenomenon over the permafrost area indicating poor drainage.

Polynya. Enclosed sea water area in pack ice, other than a lead, not large enough to be called open water. In summer it may be referred to as a "lake;" in winter with a covering of relatively thin ice it may be called an "ice skylite."

Pore-size distribution. Volume of the various sizes of pores in a soil; expressed as percentages of the bulk volume (soil plus pore space).

Porosity

1. Volume percentage of the total bulk not occupied by solid particles.

2. Total volume of pore space, usually expressed as a percentage of the total soil volume.

3. Property of a rock containing interstices, without regard to size, interconnection, or arrangement of openings; expressed as percentage of total volume occupied by interstices.

Potassium fixation. Process of converting exchangeable or water-soluble potassium to a form not easily exchanged with a cation of a neutral salt solution.

Potential acidity. Amount of exchangeable hydrogen ion in a soil that can be rendered free or active in the soil solution by cation exchange.

Prairie soils. Zonal great soil group consisting of soils formed in temperature to cool-temperature, humid regions under tall grass vegetation.

Pratt isostatic compensation. Mechanism in which variations in crustal density act to counterbalance the varying weight of topographic features. The crust is here assumed to be of approximately uniform thickness; thus a mountain range would be underlaid by lighter rocks.

Primary mineral. Mineral that has not been altered chemically since deposition and crystallization from molten lava.

Primary structure. Arrangement of the particles in a soil.

Prismatic soil structure

1. Soil structure type with prismlike aggregates that have the vertical axes much longer than the horizontal axes.

2. Elongated column structure with level and clean-cut tops. If the tops are rounded, the structure is identified as columnar. Prismatic structure is found in the B horizon when present.

3. Soil structure type with a vertical axis much longer than the horizontal axis and a flat or indistinct upper surface.

Protein. Any of a group of nitrogen-containing compounds that yield amino acids on hydrolysis and have higher molecular weights. They are essential parts of living matter and are one of the essential food substances of animals.

Pumice. Highly siliceous igneous glasses that are so extremely light and frothy that they will float on water. Open spaces are minute vesicles formed originally by the expulsion of water vapor or other gas from highly heated lava.

Radiolarian ooze. Siliceous deep-sea sediment composed largely of the skeletons of radiolaria.

Recrystallization. Growth of new mineral grains in a rock at the expense of old grains, which supply the material.

Red desert soils. Zonal great soil group consisting of soils formed in warm-temperature to hot, dry regions under desert-type vegetation, mostly shrubs.

Red earth. Highly leached, red clayey soils of the humid tropics, usually with very deep profiles that are low in silica and high in sesquioxides.

Reduction. Decrease in positive valence or increase in negative valence, caused by a gain in electrons by an ion or atom.

Red yellow podzolic soils. Combination of the zonal great soil groups red podzolic and yellow podzolic, consisting of soils formed in warm-temperature to tropical, humid climates, under deciduous or coniferous forest vegetation, and usually, except for a few members of the yellow podzolic group, under conditions of good drainage.

Regolith. Unconsolidated mantle of weathered rock and soil material on the earth's surface; loose earth materials above solid rock. It is approximately equivalent to the term "soil" as used by many engineers.

Regosol

1. Soil of the azonal order without definite genetic horizons, developing from or on deep, unconsolidated, soft mineral deposits such as sands, loess, or glacial drift.

2. Any one of an azonal group of soils that have undergone little or no pedological development, lack clearcut soil morphology, and consist mainly of unconsolidated material such as sand or silt.

Regur. Intrazonal group of dark calcareous soils high in clay, mainly montmorillonitic, and formed mainly from rocks low in quartz.

Relative density. Ratio of the difference between the void ratio of a cohesionless soil in the loosest state and any given void ratio to the difference between its void ratio in the loosest and in the densest states.

Relief. Irregularities of the land surface.

Remolded soil. Soil that has had its natural structure modified by manipulation.

Rendzina. Great soil group of the intrazonal order and calcimorphic suborder consisting of soils with brown or black friable surface horizons underlaid by light gray to pale yellow calcereous material. Rendzina is developed from soft, highly calcereous parent material under grass vegetation or mixed grasses and forest in humid and semiarid climates.

Required crushing strength. Force required to crush a mass of dry soil or, conversely, the resistance of the dry soil mass to crushing, expressed in units of force per unit area (pressure).

Residual material

1. Unconsolidated and partly weathered parent material from soils presumed to have developed from the same kind of rock as that on which it lies.

2. Unconsolidated and partly weathered mineral material accumulated by disintegration of consolidated rock in place.

Residual soil. Soil derived in place by weathering of the underlying material.

Reticulate mottling. Network of streaks of different colors most commonly found in the deeper profiles of lateritic soils.

Reversible reaction. Chemical reaction that can proceed in either direction, depending on the concentration of reacting materials.

Rheidity. Ability of a substance to yield to viscous flow under large strains; one thousand times the time required for a substance to stop changing shape when stress is no longer applied.

Rhizobia. Bacteria capable of living symbiotically with higher plants, usually legumes, from which they receive their energy, and capable of using atmospheric, nitrogen—hence the term "symbiotic nitrogen-fixing bacteria" (derived from the generic name rhizobium).

R horizon. Consolidated bedrock underlying the C horizon; degree of acidity or alkalinity of a soil, usually expressed as a pH value.

Ridge. Long, narrow, and usually sharply crested land form that may be more or less independent or a part of a large mountain or hill.

Ring dike. Dike in the form of a segment of a cone or cylinder, having an accurate outcrop.

Rippability. Characteristic of dense and rocky soils, which can be excavated without blasting after ripping with a rock rake or ripper.

Rock

1. Material that forms the essential part of the earth's solid crust, including loose incoherent masses such as sand and gravel, as well as solid masses of granite and limestone.

2. Natural solid mineral matter occurring in large masses or fragments.

3. Firm and coherent or consolidated earth material that cannot normally be excavated by manual methods alone.

Rockland. Areas containing frequent rock outcrops and shallow soils. Rock outcrops usually occupy from 25% to 90% of the area.

Rockslide. Landslide involving mainly large blocks of detached bedrock with little or no soil or sand.

Rolling ground. Undulating land surface; succession of low hills giving a wave effect to the surface; land surface much varied by small hills and valleys.

Rough broken land. Land with very steep topography and numerous intermittent drainage channels, but usually covered with vegetation.

Sag and swell topography. Undulating topography characteristic of sheets of till. Till is usually thick enough to obliterate completely all traces of former topography, and the postglacial drainage is then controlled by the surface configuration of the till.

Salic horizon. Horizon at least 6 in. thick with secondary enrichment of salts more soluble in cold water than gypsum.

Salinzation. Accumulation of salts in soil.

Sand

1. Soil separate consisting of particles between 0.05 and 2.0 mm in diameter; a soil textural class.

2. In soils, individual rock or mineral fragments having diameters ranging from 0.5 to 2.0 mm. Usually sand grains consist chiefly of quartz, but they may be of any mineral composition. Sand is the textural class name of any soil that contains 85% or more sand and not more than 10% clay.

3. Particles of rock that will pass the No. 4 sieve and be retained on the No. 200 U.S. Standard sieve.

Sandblasting. Physical weathering process in which rock is eroded by the impact of sand grains carried by the wind, frequently leading to ventifact formation of pebbles and cobbles.

Sand dune. Mount, ridge, or hill of sand piled up by the wind on the shore or in a desert.

Sandy clay. The soil of this textural class contains 35% or more clay and 45% or more sand.

Sandy clay loam. The soil of this textural class contains 20 to 35% clay, less than 28% silt, and 45% or more sand.

Sandy loam. The soil of the sand loam textural class has 50% and and less than 20% clay.

Schist. Crystalline metamorphic rock that has closely spaced foliation and tends to split readily into thin flakes or slabs. There is complete gradation between slates and schists on the one hand and schists and gneisses on the other.

Scoria. Rough, cinderlike, more or less vesicular lava thrown out by an explosive eruption or appearing on a lava stream. Expansion and escape of enclosed gases produce the typical structure. The term is usually restricted to basaltic or closely allied lavas.

Scoria land. Areas of slaglike clinkers, burned shale, and fine-grained sandstone; characteristic of burned-out coal beds. Such areas commonly support a sparse cover of grasses, but are of low agricultural value.

Secondary structure. Structure that develops after a soil is deposited. The structure is often produced by shrinkage caused by drying of cohesive soils. Cracks form which separate the soil into irregular or more or less regular blocks which are secondary particles. The cracks may later fill with some other soil to form a monolithic but nonhomogeneous mass. Faulting, brought about by landslides, also may produce a secondary structure.

Sedimentary rock

1. Rock formed from materials deposited from suspension or precipitated from solution and usually more or less consolidated. Principal sedimentary rocks are sandstones, shales, limestones, and conglomerates.

2. Rock composed of sediment, mechanical, chemical, or organic, formed through the agency of water, wind, glacial ice, or organisms, and deposited at the surface of the earth at ordinary temperatures.

Sedimentations. Deposition of mineral grains or precipitates in beds or other accumulations.

Seif dune. Longitudinal dune that shows the sculpturing effect of cross-winds not parallel to its axis.

Sensitivity. Effect of remodeling on the consistency of a clay, regardless of the physical nature of the causes of the change. The degree of sensitivity is different for different clays, and it may also be different for the same clay at different water contents.

Series (soil series). Group of soils developed from the same parent material, having similar soil horizons and essentially the same characteristics throughout the profile except for the texture of the A or surface horizon.

Shrinkage limit. Moisture content, expressed as a percentage of the weight of the oven-dried soil, at which a reduction in the amount of water will not cause a decrease in the volume of the soil mass, but at which an increase in the amount of water will cause an increase in the volume of the soil mass.

Sierozem. Zonal great soil group consisting of soils with pale grayish A horizons grading into calcareous material at a depth of 1 ft or less, and formed in temperature to cool, arid climates under a vegetation of desert plants, short grass, and scattered brush.

Silica-alumina ratio. Molecules of silicon dioxide per molecule of aluminum oxide in clay minerals or in soils.

Silica-sesquioxide ratio. Molecules of silicon dioxide per molecule of aluminum oxide plus ferric oxide in clay minerals or in soils.

Siliceous. Adjective describing a substance that is of or like silica.

Silt

1. Soil separate consisting of particles between 0.05 and 0.002 mm in equivalent diameter soil textural class.

2. Individual mineral particles of soil that range in diameter between the upper size of clay, 0.002 mm., and the lower size of very fine sand, 0.05 mm; soil of the textural class silt containing 80% or more silt and less than 12% clay; sediments deposited from water in which the individual grains are approximately the size of silt, although the term is sometimes applied loosely to sediments containing considerable sand and clay.

Silviculture. Art of producing and caring for a forest.

Single-grained state. Nonstructural state normally observed in soils containing a preponderance of large particles such as sand. Because of a lack of cohesion, the sand grains tend not be assemble in aggregate form.

Single-grained structure

1. Structure having no aggregation of the particles, such as in dune sand.

2. Arrangement composed of individual soil particles; characteristic structure of coarse-grained soils.

Sinkhole. Small, steep depression caused in karst topography by the dissolution and collapse of subterranean caverns in carbonate formations.

Site. Area considered in terms of its ecological factors, with reference to its capacity to produce forests or other vegetation; combination of biotic, climatic, and soil conditions of an area.

Site index. Quantitative evaluation of the productivity of a soil for forest growth under the existing or specified environment; height in feet of the dominant forest vegetation taken at or calculated to an index age, usually 50 to 100 years.

Slickenside. Secondary structural feature of some soils that is produced by movements along the walls or joints.

Soil is slickensided if it has inclined planes of weakness that are slick and glossy in appearance.

Slickensides. Surfaces within a soil mass that have been smoothed and striated by shear movements on these surfaces.

Slick spots. Small areas in a field that are slick when wet due to a high content of alkali or exchangeable sodium.

Slope

1. Incline of the surface of a soil; usually expressed in percentage of slope, which equals the number of feet of fall per 100 ft of horizontal distance.

2. Inclined surface of a hill, mountain, plateau, plain, or any part of the surface of the earth; angle at which such surfaces deviate from the horizontal. A slope of 45° is a slope of 1 on 1 or 100%.

Sodic soil. Soil that contains sufficient exchangeable sodium to interfere with the growth of most crop plants; soil whose exchangeable-sodium percentage is 15% or more.

Soil

1. Sediments or other unconsolidated accumulations of solid particles produced by the physical and chemical disintegration of rocks, which may or may not contain organic matter.

2. Unconsolidated mineral material on the immediate surface of the earth that serves as a natural medium for the growth of land plants; unconsolidated mineral matter on the surface of the earth that has been subjected to and influenced by genetic and environmental factors of parent material, climate (including moisture and temperature), macro- and microorganisms, and topography, all acting over a period of time and producing a product soil that differs from the material from which it is derived in many physical, chemical, biological, and morphological properties.

3. Surface accumulation of sand, clay, and humus that compose the regolith, but excluding the larger fragments of unweathered rock.

Soil aeration

1. Process by which air in the soil is replaced by air from the atmosphere. The rate of aeration depends largely on the size and continuity of empty pores within the soil.

2. In a well-aerated soil, the soil air is very similar in composition to the atmosphere above the soil. Poorly aerated soils usually contain a much higher percentage of carbon diode and a correspondingly lower percentage of oxygen than the atmosphere above the soil. The rate of aeration depends largely on the volume and continuity of pores within the soil.

Soil air. Soil atmosphere; gaseous phase of the soil which is the volume not occupied by solid or liquid.

Soil association. Group of defined and named taxonomic soil units occurring together in an individual and characteristic pattern over a region, comparable to plant associations in many ways; mapping unit used on general soil maps, in which two or more defined taxonomic units occurring together in a characteristic pattern are combined because the scale of the map or the purpose for which it is being made does not require delineation of the individual soils.

Soil buffer compounds. Clay, organic matter, and compounds such as carbonates and phosphates which enable the soil to resist appreciable change in pH.

Soil bulk density. Mass of dry soil per unit bulk volume including the air space. The bulk volume is determined before drying to constant weight at 105°C.

Soil category. Any one of the ranks of a system of soil classification in which soils are grouped on the basis of their characteristics.

Soil class. Group of soils having a definite range with respect to a particular property, such as acidity, degree of slope, texture, structure, land-use capability, degree of erosion, or drainage.

Soil classification. Systematic arrangement of soils into groups of categories on the basis of their characterization. Broad groupings are made on the basis of general characteristics and subdivisions on the basis of more detailed differences in specific properties. The three higher categories, which are broadly defined, are orders, suborders, and great groups. The lowest category is the soil series, with each series consisting of many individual occurrences or bodies of soil that are very similar in most respects.

Soil colloid. Organic or inorganic matter of very small particle size and very large surface area per unit of mass. Inorganic colloidal matter consists almost entirely of clay minerals of various kinds. Not all clay particles are colloid; usually only particles smaller than 0.00024 mm are so designated.

Soil complex. Mapping unit used in detailed soil surveys where two or more defined taxonomic units are so intimately intermixed geographically that it is undesirable or impractical, because of the scale being used, to separate them; more intimate mixing of smaller areas of individual taxonomic units than that described under soil association.

Soil creep. Imperceptibly slow downward movement of a slope forming soil or rock under shear stress less than that required to produce shear failure. Seasonal creep in response to freezing and thawing, thermal expansion, or changes in water content is confined to the upper layers of ground.

Soil extract. The solution separated from a soil or soil suspension.

Soil family. In soil classification, one of the categories intermediate between the great soil group and the soil series.

Soil formation factors. Variable, usually interrelated natural agencies that are active in and responsible for the formation of soil. Factors are usually grouped into five major categories as follows: parent rock, climate, organisms, topography, and time.

Soil-forming factors. Factors, such as parent material, climate, vegetation, topography, organisms, and time, involved in the transformation of an original geologic deposit into a soil profile.

Soil genesis

1. Mode of origin of the soil, with special reference to the processes responsible for the development of the solum or true soil from the unconsolidated parent material.

2. Formation of soils; creation of new characteristics by soil development processes.

Soil geography. Subspecialization of physical geography concerned with the areal distributions of soil types.

Soil horizon

1. Layer of soil, approximately parallel to the soil surface, with distinct characteristics produced by soil-forming processes.

2. Layer of soil or soil material approximately parallel to the land surface and differing from adjacent genetically related layers in physical, chemical, and biological properties or characteristics such as color, structure, texture, consistence, or pH.

Soil management. Sum total of all tillage operations, cropping practices, and fertilizer, lime, and other treatments conducted on or applied to a soil for the production of plants.

Soil map. Map showing the distribution of soil types or other soil mapping units in relation to the prominent physical and cultural features of the earth's surface.

Soil mechanics and engineering

1. Application of the laws and principles of mechanics and hydraulics to engineering problems dealing with soil as an engineering material.

2. Subspecialization of soil science concerned with the effect of forces on the soil and the application of engineering principles to problems involving the soil.

Soil microbiology. Subspecialization of soil science that deals with soil-inhabiting microorganisms and their relation to agriculture, including both plant and animal growth.

Soil mineralogy. In practical use, the kinds and proportions of minerals present in a soil.

Soil monolith. Vertical section of a soil profile removed from the soil and mounted for display or study.

Soil morphology

1. Constitution of the soil, including the texture, structure, consistence, color, and other physical, chemical, and biological properties of the various soil horizons that make up the soil profile.

2. Physical constitution, particularly the structural properties, of a soil profile as exhibited by the kinds, thickness, and arrangement of the horizons in the profile and by the texture, structure, consistency, and porosity of each horizon.

Soil order. Highest category in the system of soil classification used in the United States. Orders consist of broad groupings of soils organized principally to show which soil formation factor has had the greatest influence in determining the properties of a soil.

Soil permeability

1. Quality of a soil horizon that enables water or air to move through it. Permeability can be measured quantitatively in terms of rate, temperature, and hydraulic conditions. Values for saturated soils usually are called "hydraulic conductivity." The permeability of a soil may be limited by the presence of one nearly impermeable horizon even though the others are permeable.

2. Ease with which gases, liquids, or plant roots penetrate or pass through a bulk mass of soil or a layer of soil.

Soil pH. Degree of acidity or alkalinity expressed by the negative logarithm of the hydrogen ion activity of a soil.

Soil phase

1. Subdivision of a soil type or other unit of classification having characteristics that affect the use and management of the soil but do not vary sufficiently to differentiate it as a separate type; variation in a property or characteristic, such as degree of slope, degree of erosion, and content of stones.

2. Subdivision of a soil type or other classificational soil unit having variations in characteristics not significant to the classification of the soil in its natural landscape, but significant to the use and management of the soil. Examples of the variations recognized by phases of soil types include differences in slope, stoniness, and thickness because of accelerated erosion.

Soil physics. Organized body of knowledge concerned with the physical characteristics of soil and with the methods employed in their determinations.

Soil profile

1. Vertical section of a soil showing the nature and sequence of the various layers as developed by deposition or weathering, or both.

2. Vertical section of the soil through all its horizons, extending into the parent material.

Soil reaction. Degree of acidity or alkalinity of a soil, usually expressed as a pH value.

Soil salinity. Amount of soluble salts in a soil, expressed in terms of percentage parts per million or other convenient ratios.

Soil science. Science dealing with soils as a natural resource in the surface of the earth, including soil formation, classification, and mapping; the physical, chemical, biological, and fertility properties of soils per se; and these properties in relation to their management for crop production.

Soil separates

1. One of the individual size groups of mineral soil particles and sand, silt, or clay.

2. Groups of mineral particles separated on the basis of a range in size. Principal separates are sand, silt, and clay.

Soil series. The basic unit of soil classification is a subdivision of a family and consists of soils that are essentially alike in all major profile characteristics except the texture of the A horizon.

Soil solution. Aqueous liquid phase of the soil and its solutes, consisting of ions dissociated from the surfaces of the soil particles and of other soluble materials.

Soil—stabilized. Soil hardened by the addition of a binder such as cement.

Soil structure

1. Combination or arrangement of primary soil particles into secondary particles, units, or peds. These secondary units may be, but usually are not, arranged in the profile in such a manner as to give a distinctive characteristic pattern. Secondary units are characterized and classified on the basis of size, shape, and degree of distinctness into classes, types, and grades, respectively.

2. Aggregation of soil particles into clusters of particles, which are separated from adjoining aggregates by surfaces of weakness. Structure is judged by observation or by the breaking or by dropping the clods. If the clods are easily broken with cleavage planes visible, the soil is structured. If there is difficulty in breaking the clods and an irregular surface results, then the soil is not structured.

3. Arrangement and state of aggregation of soil particles in a soil mass.

4. Arrangement of primary soil particles into compound particles or clusters that are separated from adjoining aggregates and have properties unlike those of an equal mass of unaggregated primary soil particles.

Soil suborder. Subdivision of the soil order; the second highest category of soil classification. Classes within suborders are based on soil properties that reflect a narrower range of variation in soil-forming factors than is allowed for soil orders.

Soil survey. Systematic examination, description, classification, and mapping of soils in an area.

Soil textural class. Soils grouped on the basis of a specified range in texture. In the United States 12 textural classes are recognized.

Soil texture

1. Relative proportions of the various soil separates in a soil.

2. Relative proportions of the various size groups of individual soil grains in a mass of soil.

Soil type

1. In mapping soils, a subdivision of a soil series based on differences in the texture of the A horizon.

2. Lowest unit in the natural system of soil classification; subdivision of a soil series and consisting of or describing soils that are alike in all characteristics including the texture of the A horizon.

Soil variant. Soil whose properties are believed to be sufficiently different from other known soils to justify a new series name, but comprising such a limited geographic area that creation of a new series is not justified.

Solidized soil. Soil that has been subjected to the processes responsible for the development of a soloth and having at least some of the characteristics of a soloth.

Solid solution series. Series of minerals of identical structure that can contain a mixture of two elements over a range of ratios (for example, plagioclase feldspars).

Solonchak soils

1. Intrazonal group of soils with high concentrations of soluble salt in relation to those in other soils, usually light-colored, without characteristic structural form, developed under salt-loving plants, and occurring mostly in a subhumid or semiarid climate.

2. Great soil group of the intrazonal order and halomorphic suborder, consisting of soils with a gray, thin, salty crust on the surface and with a fine granular mulch immediately below underlaid with grayish, friable, salty soil. These soils are formed in subhumid to arid, hot or cool climate under conditions of poor drainage and under a sparse growth of salt-tolerant grasses and shrubs and some trees.

Solonetz soils

1. Intrazonal group of soils having surface horizons of varying degrees of friability, underlaid by dark, hard soil, ordinarily with columnar structure (prismatic structure with rounded tops). This hard layer is usually highly alkaline. Such soils are developed under grass or shrub vegetation, mostly in subhumid or semiarid climates.

2. Great soil group of the intrazonal order and halomorphic suborder, consisting of soils with a very thin friable surface soil underlaid by a dark, hard columnar layer, usually highly alkaline. These soils are formed in subhumid to arid, hot to cool climates under better drainage than solonchaks, and under a native vegetation of salt-tolerant plants.

Soloth. Great soil group of the intrazonal order and halomorphic suborder having a gray, leached surface horizon that rests on a fine-textured brown or dark-brown horizon, developed under grass or shrub vegetation, mostly in a subhumid or semiarid climate.

Soluble sodium percentage. Proportion of sodium ions in solution in relation to the total cation concentration.

Solubility. Mass of a substance that can be dissolved in a certain amount of solvent if chemical equilibrium is attained.

Solum

1. Upper and most weathered part of the soil profile; A and B horizons. (Pl. sola.)

2. Part of the soil profile, above the parent material, in which the processes of soil formation are taking place. In mature soils this includes the A and B horizons, and the character of the material may be greatly unlike that of the parent material.

Spodic horizon. Subsurface diagnostic horizon containing an illuvial accumulation of free sesquioxides of iron and aluminum and of organic matter.

Spodosols

1. Soils characterized by the presence of a spodic horizon, an eluvial horizon in which active organic matter and amorphous oxides of aluminum and iron have precipitated. These soils include most podzols, brown podzolics, and ground water podzols of the old classification system.

2. Soil order consisting of soils having a B2 horizon containing illuvial humus or clay, alone or in combination, with the clay being principally of noncrystalline types. The undisturbed profile also has distinct 0 and A2 horizons, occurring in that order, over the B2 horizon. Typically, spodosols are formed in cool to cold moist climates under coniferous forest vegetation and are moderately to strongly acid throughout their profile.

Star dune. Eolian deposit normally composed of a large peak from which four or more ridges radiate. The cross-sectional shapes may include combinations of rolling and crested deposits.

Stones. Rock fragments that are greater than 10 in. in diameter if rounded and greater than 15 in. along the greater axis if flat.

Stony. Term applied to soils that contain a significant amount of stones.

Stratification. Characteristic structural feature of sedimentary rocks produced by the deposition of sediments in beds, layers, strata, laminae, lenses, wedges, and other essentially tabular units.

Stratified

1. Composed of or arranged in layers. The term is applied to such geological materials as stratified alluvium. Layers in soils that are produced by the soil-forming processes are called "horizons," while those inherited from the parent material are called "strata."

2. Composed of or arranged in layers. Stratification is typical of soils deposited under water. If the individual

layers are not thicker than about 1 in. and are of roughly equal thickness, the soil is called "laminated."

3. Arranged in or composed of strata or layers.

Stratigraphic sequence. Set of deposited beds that reflects the geologic history of a region.

Stratigraphy. Science of the description, correlation, and classification of strata in sedimentary rocks, including the interpretation of the depositional environments of those strata.

Strike. Direction of a line formed by the intersection of a bedding lane, vein, fault, slaty cleavage, schistosity, or similar geologic structure with a horizontal plane. It is at right angles to the dip.

Stromatolite. Fossil form representing the growth habit or an algal mat; concentric spherules, stacked hemispheres, or flat sheets of calcium carbonate and trapped silt encountered in limestones.

Subsidence. Settling of surface soils, particularly unconsolidated materials, either by the introduction of moisture into upper layers and resulting lubrication, or by removal of moisture (either by pumping or lowering the water table) from lower strata, leaving voids filled by the weight of the overburden.

Subsoil

1. Soil below a subtrate or fill and that part of a soil profile occurring below the A horizon.

2. Part of the soil below the plow layer.

3. Soil that lies beneath another soil and is unlike the upper soil in some distinctive way. For example, the natural or undisturbed soil on which a fill is placed is often called the "subsoil." In pedology, subsoil refers to the B horizon of soils with distinct profiles. In soils with weak profiles, it is the soil below the surface soil. For pedologic uses, it is considered an undesirable term.

4. Soil below the plowed soil (or its equivalent of surface soil), in which roots normally grow.

Subsoiling. Breaking of compact subsoils, without inverting them, with a special knifelike instrument (chisel) which is pulled through the soil at depths usually of 12 to 24 in. and at spacings usually of 2 to 5 ft.

Summit. Highest point of undulating land, for example, highest point of a rolling plain; mountain; apex, top, or highest point of a land form.

Surface soil

1. Soil ordinarily moved in tillage or its equivalent in uncultivated soil, about 5 to 8 in. in thickness.

2. Uppermost part of the soil ordinarily moved in tillage or its equivalent in uncultivated soils, ranging in depth from 3 to 4 in. to 8 to 10 in. Frequently designated as the "plow layer."

Swale. Slight depression in generally level land, usually covered with a rank growth of grass and often marshy.

Symbiosis. Living together in intimate association of two dissimilar organisms, the cohabitation being mutually beneficial.

Tableland. Flat or undulating elevated area; plateau or mesa.

Talus

1. Fragments of rock and other soil material accumulated by gravity at the foot of cliffs or steep slopes.

2. Collection of fallen disintegrated material that has formed a slope at the foot of a steeper declivity.

Tar sand. Sandstone containing the densest asphaltic components of petroleum; end-product of evaporation of volatile components or of some thickening process.

Taxonomic unit. Arbitrarily defined unit consisting either of an individual or a combination of individuals established for the purpose of classification.

Terrace

1. Embankment or ridge constructed across sloping soils on the contour or at a slight angle to the contour. A terrace intercepts surplus runoff in order to retard it for infiltration into the soil and so that any excess may flow slowly to a prepared outlet without harm.

2. Level, usually narrow, plain bordering a river, lake, or the sea. Rivers are sometimes bordered by terraces at different levels. A terrace is a raised more or less level or horizontal strip of earth, usually constructed on or nearly on a contour and designed to make the land suitable for tillage and to prevent accelerated erosion.

3. Berm or discontinuous segments of a berm in a valley at some height above the flood plain, representing a former abandoned flood plain of the stream; relatively narrow plain or bench on the side of a slope terminating in a short declivity; plain, natural or artificial, from which the surface descends on one side and ascends on the other.

Textural class. Classification of soil material according to the proportions of sand, silt, and clay.

Thermal analysis (differential thermal analysis). Method of analyzing a soil sample for constituents, based on a differential rate of heating of the unknown and standard samples when a uniform source of heat is applied.

Thermokarst. Karstlike topographic features produced by the melting of ground ice and the subsequent settling or caving of ground, characterized by an uneven topography with short ravines, sinkholes, tunnels, and caverns similar to those produced in a limestone terrain by the solvent action of water.

Thermophilic organisms. Organisms that grow readily at temperatures above 45°C.

Tight soil. Compact, relatively impervious and tenacious soil (or subsoil) which may or may not be plastic.

Till

1. Unstratified glacial drift deposited directly by the ice, consisting of clay, sand, gravel, and boulders intermingled in any proportion.

2. Plow and prepare for seeding; to seed or cultivate the soil.

3. Unsorted and unstratified rock fragments ranging in size from clay to boulders, deposited directly by a glacier. It is sometimes called "boulder clay" or "glacial till."

Till plain. Relatively level area of ground moraine consisting of till.

Tilth. Physical condition of soil with relation to its ease of tillage, fitness as a seedbed, and impedance to seedling emergence and root penetration.

Time scale. Division of geologic history into eras, periods, and epochs, accomplished through stratigraphy and paleontology.

Topsoil

1. Term used in at least four different senses: (1) presumed fertile soil or soil material, usually rich in organic matter, used to topdress roadbanks, lawns, and gardens; (2) surface plow layer of a soil and thus a synonym for surface soil; (3) original or present dark-colored upper soil, which ranges from a mere fraction of an inch to 2 or 3 ft on different kinds of soil; (4) applied to soils in the field, the term has no precise meaning unless defined as to depth or productivity in relation to a specific kind of soil.

2. Layer of soil moved in cultivation; A horizon, A1 horizon; presumably fertile soil material used to topdress roadbanks, gardens, and lawns.

Topography. Shape of the physical features of the exposed ground surface.

Toposaic. Photomap on which topographic or terrain form lines are shown, as on standard topographic quadrangles.

Toposequence. Sequence of related soils that differ, one from the other, primarily because of topography as a soil formation factor.

Toxic. Poisonous or injurious to animals or plants through contact or systematic action.

Transitional soil. Soil with properties intermediate between those of two different soils and genetically related to them.

Transverse dune. Ridge of sand oriented at a right angle to the direction of the prevailing wind. Its cross-sectional shape is generally asymmetric and crested, with some areas rolling.

Truncated. Having lost all or part of the upper soil horizon or horizons.

Tuff. Volcanic ash usually more or less stratified and in various states of consolidation.

Tundra

1. Flat or gently rolling area having a muck-to-rock surface over permafrost above or north of the timberline.

2. Level or undulating treeless plain characteristic of arctic regions.

Trundra soils

1. One of a series of a zonal group of soils having a tough fibrous peaty mat underlaid by a dark-colored humus-rich stratum, which grades into lighter colored gray or mottled soil beneath.

2. Soils characteristic of tundra regions; zonal great soil group consisting of soils with dark-brown peaty layers over grayish horizons mottled with rust and having continually frozen substrata; formed in frigid, humid climates, with poor drainage and a native vegetation of lichens, moss, flowering plants, and shrubs.

Ultisols. Soils of humid areas characterized by the presence of either an argillic horizon or a fragipan, each of which is less than 35% saturated with bases. They have no spodic horizon and no oxic or natric horizons. They include soils formerly classified as red-yellow podzolics, reddish brown laterites, and rubrozems.

Umbric epipedon. Thick, dark surface layer that is less than 50% saturated with bases. Ando soils have umbric epipedrons.

Unconfined compressive strength. Load per unit area at which an unconfined prismatic or cylindrical specimen of soil will fail in a simple compression test.

Unconsolidated material. Nonlithified sediment that has no mineral cement or matrix binding its grains.

Undifferentiated soil groups. Soil mapping units in which two or more similar taxonomic soil units occur, but not in a regular geographic association. For example, the steep phases of two or more similar soils might be shown as a unit on a map because topography dominates the properties.

Undisturbed. Geologic structures in which the strata lie essentially horizontally or, as in a coastal plain, with gentle seaward dip.

Undue compaction. Subsurface layers in soil that have been so compacted by the application of weight (applied, for example, by machines or tractors) that the penetration of water and roots is interfered with. Because the traffic of machines is not the only cause of these pans, they are sometimes called "pressure pans" or "traffic pans."

Valley. Depression in the land surface, generally elongated and usually containing a stream; low land bounded by hills or mountains.

Varve

1. Distinct band representing the annual deposit of sedimentary materials, regardless of origin, and usually consisting of two layers, one a thick, light-colored layer of silt and fine sand and the other a thin, dark-colored layer of clay.

2. Pair of sediment layers ideally representing a year's record of melt water deposition in a glacier-fed lake or bay. The laminated soil structure of varves is similar in appearance to annual growth rings in trees. Sediments are known as "varved clays" or "varved sediments."

Varved clay. Alternating thin layers of silt (or fine sand) and clay formed by variations in sedimentation during the various seasons of the year, often exhibiting contrasting colors when partially dried.

Vermiculite. Layered clay similar to illite except that the negative charge caused by ion substitution is neutralized by hydrated cations, some of them potassium, rather than predominantly by unhydrated potassium, as is characteristic of illite.

Vertisols

1. Soils high in swelling clays which crack widely upon drying, resulting in shrinking, shearing, and soil mass movement.

2. Soil order consisting of soils that have formed from parent materials high in expanding clay, usually montmorillonite. Development is strongly influenced by the formation of deep, wide cracks into which surface soil material is sloughed or washed so that the surface soil undergoes continual mixing or inversion. Natural mixing results in an A horizon that is relatively uniform to the depth of crack formation.

Vesicular. Containing numerous small air pockets or spaces, like a sponge.

Vesicular structure. Soil structure containing many small cavities or pores, smooth on the inside as though formed by gas bubbles.

Viscosity. Measure of resistance to flow in a liquid.

Void. Space in a soil mass not occupied by solid mineral matter. This space may be occupied by air, water, or other gaseous or liquid material.

Void ratio. Ratio of the volume of void space to the volume of solid particles in a given soil mass.

Volcanic ash. Unconsolidated fine-grained material thrown out in volcanic eruptions. It consists of minute fragments of glass and other rock material, which in color and general appearance may resemble organic ashes.

Volcanic breccia. More or less indurated volcanic rocks consisting chiefly of angular ejecta 32 mm or more in diameter.

Volcanic cinders. Uncemented volcanic fragments that range from 4 to 32 mm in diameter. Such fragments are usually glassy or vesicular.

Wasteland. Land not suitable for or capable of producing materials or services of value.

Water table. Surface of underground gravity-controlled water.

Weathering. All physical and chemical changes produced in rocks or near the earth's surface by atmospheric agents.

Whaleback. Tremendous sand ridge built by movement of dunes over the same path for long periods of time. Ridges are elongated in plan and exhibit gentle, rounded crests, although one or more longitudinal dunes may be superimposed.

Wilting point. Moisture content of soil, on an oven-dry basis, at which plants (specifically sunflower plants) wilt and fail to recover their turgidity when placed in a dark humid atmosphere.

Windbreak. Planting of trees, shrubs, or other vegetation, usually perpendicular or nearly so to the principal wind direction, to protect soil, crops, homesteads, roads, etc., against the effects of winds, such as wind erosion and the drifting of soil and snow.

Xerophytes. Plants that grow in or on extremely dry soils or soil materials.

Yellow podzolic soils. Classification formerly used for a zonal group of soils having thin organic and organic mineral layers over grayish yellow leached horizons that rest on yellow B horizons and developed under coniferous or mixed coniferous and deciduous forests in a warm temperate to warm moist climate. These soils are now combined into the red yellow pedzolic group.

Zonal soil. Soil that is characteristically of a large zonal area; one of the three primary subdivisions (orders) in soil classification formerly used in the United States. Subdivisions of zonal soils were based on properties reflecting differing influences of climate and vegetation on soil development.

Zoned crystal. Single crystal of one mineral that has a different chemical composition in its inner and outer parts. It is formed from minerals belonging to a solid-solution series and caused by the changing concentration of elements in a cooling magma that results from crystals settling out.

TECHNICAL AND SCIENTIFIC UNITS OF VALUE

Absolute humidity. Mass of water vapor present in the atmosphere measured as grams per cubic meter.

Absolute temperature. Temperature reckoned from the absolute zero. Temperature measured in theory on the thermodynamic temperature scale.

Absolute units. System of units based on the smallest possible number of independent units, specifically, units of force, work, energy, and power not derived from or dependent on gravitation.

Absolute zero. The lowest temperature theoretically possible, $-273°C$.

Acceleration. Rate of change of velocity, measured as change in velocity (feet per second) in unit time (per second). It is expressed as feet per second per second (ft/sec^2).

Action. Force that is measured by the product of work by time. Cgs units of action are the erg-second and the joule-second.

Adiabatic. A body is said to undergo an adiabatic change when its condition is altered without gain or loss of heat.

Aerodynamics

1. Field of dynamics concerned with the motion of air and other gaseous fluids, or of the forces acting on bodies in motion relative to such fluids.

2. Branch of dynamics that deals with the motion of air and other gases, and the forces acting upon bodies passing through them.

Albedo. Fraction of the power from the sun incident on the earth that is reflected back into space unchanged. Averaged over the earth's surface, the albedo is 0.29.

Allowable stress. Permissible maximum stress used in the design of a structure or component that takes into account efficiency in the use of material and uncertainties in expected conditions of service, in properties of the material, and in stress analysis.

Alpha particles. Nuclei of the dominant isotope of helium consisting of two protons and two neutrons. Alpha particles are often emitted in the radioactive decay of heavy elements such as radium, thorium, uranium; they have a mass of 4.00260 (atomic mass units).

Ambient vibration. All-encompassing vibration associated with a given environment, usually a composite of vibration from many sources near and far.

Ampere. Electric current of 1 C (coulomb)/second.

Amplitude. The maximum value of the displacement in an oscillatory motion.

Angstrom. Unit of length, equal to one hundred-millionth of a centimeter, 10^{-8} cm.

Archimedes' principle. An object immersed in a fluid is pushed up with a force that is equal to the weight of the displaced fluid.

Atom. Smallest part of an element that can take part in a chemical change. Every atom consists of a positively charged core or nucleus orbited by negatively charged electrons. The nucleus contains positively charged protons and neutral neutrons of almost equal mass.

Atomic energy. Energy associated with the nucleus of an atom. It is released when the nucleus is split, or is derived from mass that is lost when the nucleus is fused together.

Atomic mass unit. An arbitrarily defined unit in terms of which the masses of individual atoms are expressed; the standard used is the unit of mass equal to $\frac{1}{12}$ the mass of the carbon atom, having as nucleus the isotope with mass number 12.

Atomic number. Number of protons in a nucleus, which equals the number of electrons in a neutral atom.

Atterberg limits. Measures of soil consistency for differentiation between materials of appreciable plasticity (clays) and slightly plastic or nonplastic materials (silts). The measures include the liquid limit, the plastic limit, and the plasticity index.

Audible sound. Sound containing frequency components lying between about 15 and 20,000 Hz with sufficient sound pressure to be heard.

Avogadro's law. A law that states that equal volumes of different gases at the same pressure and temperature contain the same number of molecules.

Bar. Unit of pressure equal to 10^5 pascals, or 10^5 newtons per square meter, or 10^6 dynes per square centimeter.

Barrel. Unit of volume used in petroleum production. One barrel of oil equals 42 U.S. gal or about 135 kg.

Bel. Unit expressing the relation between amounts of signal power and differences in sound-sensation levels. The number of bels is equal to the common logarithm of the ratio of the two powers or sound levels involved. Two powers or levels differ by 1 bel when their actual ratio is $10:1$.

Beta particle. Positive or negative electron emitted by the nucleus during radioactive transformation; also called "beta ray." By extension, electrons accelerated to extremely high speeds (kinetic energies about 1 MeV or more) are called "beta particles."

Blackbody. A solid that radiates or absorbs energy with no internal reflection of the energy at any wavelength.

Boiling point. Temperature at which a substance changes its state from liquid to gas. The boiling point of water is $100°C$ or $212°F$.

Boyle's law for gases. At a constant temperature the volume of a given quantity of any gas varies inversely as the pressure to which the gas is subjected.

British thermal unit (Btu)

1. $1/8$ Btu = 0.252 calorie = 0.000293 kWh.

2. Quantity of heat needed to raise 1 lb of water through $1°F$.

Brittleness. Quality of a material that leads to crack propagation without appreciable plastic deformation.

Bulk modulus. The number that expresses a material's resistance to elastic changes in volume. For example the number of lbs/sq in. necessary to cause a specified change in volume. The modulus of volume elasticity.

Calorie

1. Measure of heat energy: 1 calorie is the amount of heat necessary to raise or lower 1 kg of pure water through a temperature difference of $1°C$. When converted into mechanical energy units, 1 calorie = 4184 J = 0.001163 kWh. Btu is defined as that amount of heat needed to change the temperature of 1 lb of pure water by $1°F$. Since 1 kg is the mass of 2.204 lb, 1 calorie = $(9/5)(2.204) = 3.97$ Btu.

2. Quantity of heat needed to raise 1 g of water from $14.5°C$ to $15.5°C$. A kilogram-calorie is equal to 1000 calories; this is the calorie that dieticians use to measure the energy value of foods.

3. Amount of heat necessary to raise the temperature of 1 g of water $1°C$.

Candlepower. Luminous intensity expressed in candles.

Capacity. Maximum weight of water vapor a given quantity of air can hold at a given temperature.

Celsius scale. Temperature scale that takes the melting point of ice as 0 and the boiling point of water as 100. It is used by scientists throughout the world and usually where the metric system is used. To convert celsius to Fahrenheit, multiply the reading by 9, divide by 5, and add 32.

Centrifugal force. The force which provides the reaction against centripetal force. The force outward exerted by a body moving in a curved path.

Centripetal force. The force required to keep a moving mass in a circular path.

Chemical energy. Energy released or absorbed when atoms form compounds; generally becomes available when atoms have lost or gained electrons, and often appears in the form of heat.

Circular mil. Area of a circle diameter of which is 1 mil or $\frac{1}{1000}$ inch; 7.854×10^7 square inch; 0.78540 square mil.

CLO. Unit of measurement used in evaluating the insulative quality of clothing.

Compatibility. Particular quality or characteristic of a component, item of equipment, or system that permits it to function in harmony with other equipment, environments, or systems with a minimum amount of adapters, extensions, transformers, or other equalizer units.

Compound. Combination of the atoms or ions of different elements. The mechanism by which they are combined is called a "bond."

Conductors. A class of bodies that are incapable of supporting electric strain. A charge given to a conductor spreads to all parts of the body.

Cooling load. Amount of heat that must be removed from a building to maintain a comfortable temperature, measured in British thermal units per hour or tons of air conditioning (1 ton = 12,000 Btu/hour).

Corona. Set of one or more prismatically colored rings of small radii, concentrically surrounding the disk of the sun, moon, or other luminary when veiled by a thin cloud. A corona is due to diffraction by numerous water drops.

Coulomb

1. Amount of electric charge found on 6.24×10^{18} electrons.

2. Unit of quantity of electricity, roughly 6×10^{18} electrons.

Current (electric). The rate of transfer of electricity. The practical unit of current is the ampere.

Dalton. $\frac{1}{16}$ the mass of an atom of oxygen; 1.650×10^{-24} gram.

Decomposition. Chemical separation of a substance into two or more substances, which may differ from each other and from the original substances.

Decompression. Act or process of lowering the air pressure within a cabin, chamber, etc., or of subjecting to or undergoing a decrease in air or atmospheric pressure.

Decontamination. Process of making person, object, or area safe by absorbing, destroying, neutralizing, making harmless, or removing chemical or biological agents or by removing radioactive material clinging to or around it.

Deflection. Amount of downward vertical movement of a surface due to the application of a load to the surface.

Degradation. Deterioration, usually in the sense of a physical or chemical process rather than a mechanical one.

There may be a specific amount of degradation permitted as a result of performance of environmental testing.

Degrees of freedom. In a mechanical system, the minimum number of independent coordinates required to define completely the positions of all parts of the system at any instant of time. In general, a degree of freedom is equal to the number of independent displacements that are possible.

Deliquescence. Change undergone by certain substances that become damp and finally liquefy when exposed to the air, owing to the very low vapor pressure of their saturated solutions.

Density. Ratio of the mass of an object to its volume. Although the mass and volume depend on the specific object, the density is characteristic of the material making up the object. Thus, the density of water is 1 kg/liter (1 g/cu cm), which is equivalent to 0.946 kg/U.S. qt.

Deterioration. Loss in the value of a material or a decrease in the ability of a product to fulfill the function for which it was intended; process of transition from a higher to a lower energy level.

Dew point

1. Temperature at which air becomes saturated.

2. The temperature at which condensation of water vapor in the air takes place.

Displacement. Vector quantity that specifies the change of position of a body or particle and is usually measured from the mean position or position of rest. In general, displacement can be represented as a rotation vector or a translation vector, or both.

Doppler effect. An apparent change in the wavelength of a radiation where there is relative motion between the source of radiation and the receiver.

Drag. Frictional impedance offered by a fluid to the motion of bodies passing through it; more precisely, the component of aerodynamic or hydrodynamic force parallel to the direction of mean flow.

Dry-adiabatic lapse rate. The rate—$5\frac{1}{2}°$F per 1000 feet—at which rising air cools by expansion.

Dynamic loading. Loads introduced into a machine or its components by forces in motion.

Dynamic pressure. Difference between static and Pitot pressure due to relative motion of a fluid when compressibility of the fluid is not considered. Pitot pressure is the pressure at the open end of a Pitot tube.

Dynamic stress. Stress induced in an elastic element by the dynamic deflection applied to it.

Dyne

1. Unit of force that will produce an acceleration of 1 cm/second if applied to a body with a mass of 1 g.

2. Unit of force; force that will give a mass of 1 g an acceleration of 1 cm/second^2.

Efficiency. Ratio of the energy "out" to the energy "in." For a mechanical system, the energy in is the work done in producing the initial configuration; the energy out is the "useful" work that can be done by the system.

Elastic energy. The energy stored within a solid during elastic deformation, and released during elastic rebound.

Elasticity. Property of a material by virtue of which it tends to automatically recover its original size and shape after deformation.

Elastic limit. The maximum stress that produces only elastic deformation.

Electric resistance heat. Conversion of electrical energy to heat energy by means of an electrical resistor. May be applied to space heating or water heating (1 kWh = 3413 Btu).

Electromotive force. Measure of the intensity of electrical energy needed to produce a current in a circuit. The practical unit is the volt.

Electrons

1. Universal, negatively charged constituents of atoms. The mass of the electron is negligible compared to the mass of the entire atom. The electrons normally circulate around the positive nucleus of a normal, electrically neutral atom as planets do around the sun.

2. Elementary particles that orbit the atomic nucleus. The movement of electrons constitutes an electrical current.

Electron volt (eV)

1. Unit of energy often used in physics. It is defined as the energy gained when a charge equal to that of the electron falls through an electrical potential difference of 1 V.

2. Unit of energy used in nuclear physics; the energy used to raise an electron through a potential of 1 V.

Element. In chemistry, a substance consisting entirely of atoms of the same atomic number.

Emissive power (emissivity). Power measured by the energy radiated from unit area of a surface in unit time for unit difference of temperature between the surface in question and surrounding bodies.

Energy

1. Ability to do work. One joule of energy can do work of 1 J or 1 N-m.

2. Capacity to do work. Potential energy arises by virtue of the position or configuration of matter. Kinetic energy is energy of motion.

Engineer's chain. Chain used for measuring in the field, 100 feet in length, 100 links of 1 foot each, 30.4801 meters.

Entropy. A quantity depending on the quantity of heat in a body and on its temperature, which, when multiplied by any lower temperature, gives the unavailable energy, or unavoidable waste when mechanical work is derived from heat energy of the body.

Erg

1. An erg is a unit of energy; the capacity for doing work; the energy expended when a force of 1 dyne acts through a distance of 1 centimeter.

2. Unit of energy or work in the centimeter-gram-second (cgs) system.

Fahrenheit scale. Temperature scale in which the boiling point of water is 212 degrees F and its freezing point 32 degrees F. To convert degree Fahrenheit to degree Celsius subtract 32, multiply by 5 and divide the product by 9.

Failure. Operational or performance degradation or irreversible material or structural change, when examined in accordance with specific failure criteria.

Fatigue. Reaction that takes place in material under repeated cyclic stressing, resulting in a tendency for that material to fail below its static ultimate strength.

Fatigue life. Number cycles of stress or stress reversals that can be sustained prior to failure for a stated test condition.

Fatigue limit. Maximum stress below which a material can presumably endure an infinite number of stress cycles. If the stress is not completely reversed, the value of the mean stress, the minimum stress, or the stress ratio should be stated. The fatigue limit of a material is frequently referred to as its "endurance limit."

Footcandle. Unit of measurement of the intensity of light on a surface that is everywhere 1 ft from a uniform point source of light from a standard (sperm whale oil) candle and equal to 1 lumen/sq ft.

Foot per minute. 0.005080 meter per second, 0.011364 mile per hour, 0.016667 foot per second, 0.3048 meter per minute.

Foot per second. 0.011364 mile per minute, 0.6818 mile per hour, 18.29 meters per minute.

Foot-pound. Unit of measurement equivalent to the work of raising 1 lb vertically a distance of 1 ft.

Force

1. Push or pull that alters the motion of a moving body or moves a stationary body. The unit of force is the dynel or the poundal.

2. Originally, the perception of the muscular exertion involved in pushing, pulling, throwing, or lifting some object. Force is characterized by a magnitude (a "large" force, a "weak" force), as well as by a direction (a vertical force-lift, a horizontal force). Once the mass of an object is determined, force can be given a quantitative definition (independent of our feeling of muscular exertion) by use of Newton's second law.

Footlambert. Unit of photometric brightness (luminance) equal to 1π candela/sq ft. Theoretically perfectly diffusing surface emitting or reflecting flux at the rate of 1 lumen.

Freezing point. Temperature at which a substance changes its state from liquid to solid.

Fresnel. Unit of frequency equal to 10^{12} Hz.

Friction

1. Force that resists the relative motion of two surfaces in contact with each other, proportional to the force (usually gravity) that holds the surfaces together.

2. Rubbing together of two substances or bodies in contact with each other, of a body in contact with a gas or fluid, etc.; the resistance to relative motion caused by this contact.

Fusion. Phase transition of a substance passing from the solid to the liquid state; melting. Additional heat at the melting point is required to fuse any substance. This quantity of heat is called the "latent heat" of fusion. In nuclear technology, fusion is the transformation of nuclei by combining two light nuclei to form a heavier nucleus.

Gamma rays

1. Electromagnetic waves emitted during nuclear transformations and having wavelengths less than 3×10^{-11} m.

2. Electromagnetic radiation similar to x-rays. The names merely identify the radiation source; x-rays come from electrical machines, gamma rays from nuclear fission.

Gravity

1. Universal attraction of every bit of matter in the universe for every other bit. The attractive force is proportional to the product of the two masses involved and inversely proportional to the square of the distance between the two masses.

2. Gravitational force, as modified by centrifugal force due to rotation, exerted by the earth on bodies at or near its surface, resulting in their having weight; unit of acceleration equal to the acceleration resulting from the average force of gravity at the earth's surface. By international agreement this unit is equal to 9.80665 m/second/second.

Hardness. Combination of properties of a substance determined by arbitrary tests, such as the resistance of the substance to indenting or scratching by specific objects.

Harmonic. Sinusoidal quantity having a frequency that is an integral multiple of the frequency of a periodic quantity to which it is related; wave of vibration having a frequency that is an integral multiple of the fundamental (lowest) or other reference frequency of vibration of a physical system.

Heat. Form of energy that flows from body to body as temperature differences occur. The natural flow is always from the warmer to the cooler body; flow ceases when there is no temperature difference. It is measured in calories or British thermal units and is interpreted as the part of the molecular energy due to the random motion of the molecules of a body.

Heat energy. Special manifestation of kinetic energy in atoms. The temperature of a substance depends on the average kinetic energy of its component particles. When heat is applied to a substance, the average kinetic energy increases.

Hertz (Hz)

1. Unit of frequency. 1 Hz means 1 cycle or vibration per second.

2. Unit of frequency or a periodic process equal to 1 cycle/second.

Horsepower

1. Unit of power or the capacity of a mechanism to do work. It is the equivalent to raising 33,000 lb 1 ft in 1 minute or 550 lb ft in 1 second. One horsepower equals 746 W.

2. Unit of power; power needed to raise 550 lb through 1 ft in 1 second.

Hypersonic. Of or pertaining to speeds equal to or in excess of five times the speed of sound.

Impact. Single collision of a mass in motion with a second mass that may be either in motion or at rest.

Impact pressure. Pressure exerted when one object strikes another. It consists of pressure derived from both static and dynamic pressure.

Impulse. Product of a force and the time during which the force is applied.

Induced stress. Stress that is a component of the man-made phase of the environment, such as acceleration, shock, and vibration.

Inert. Destitute of power to move or actively to resist motion impressed; not having active properties; powerless for a desired effect.

Inertia

1. Property of matter by which it resists any change in its state of rest or uniform motion in a straight line.

2. Tendency of a body to resist force.

Intensity of radiation. The radiant energy emitted in a specified direction per unit time, per unit area of surface, per unit solid angle.

Ion. An ion is an electrically charged particle. An electrically unbalanced form of an atom or group of atoms, produced by the gain or loss of electrons.

Ionization. Breaking up of electrically neutral atoms or molecules into positively and negatively charged parts (ions).

Isotopes

1. Atoms having the same number of protons in their nuclei but differing in the number of neutrons.

2. Atoms of the same element that have the same atomic numbers, but different mass numbers. A radioisotope is a radioactive isotope made by exposing nonradioactive elements to radiation in a nuclear reactor.

Joule (J). Unit of work or energy equal to 1 N-m. One joule is identical to 1 W-second or 0.278×10^{-6} kWh. When converted to heat energy units, $1\ J = 2.39 \times 10^{-4}$ calories.

Kilogram (kg). Unit of mass in the metric system; mass of an arbitrary piece of metal stored in a vault that is approximately the same as the mass of 1000 cu cm (1 liter) of water (1 kg = 1000 g).

Kilowatt (kW). Unit of power equal to 1000 W.

Kilowatt hour (kWh). Amount of energy resulting from a power of 1 kW (1000 W) flowing for 1 hours. Since there are 3600 seconds in 1 hour, $1\ kWh = 3.6 \times 10^6\ J = 860$ calories = 3414 Btu.

Kinetic energy. Ability of a body to do work through moving. The kinetic energy of a body is proportional to the mass of the body and the square of its velocity.

Knot. Unit of speed in the nautical system; 1 nautical mile/hour, equal to 1.1508 statute miles/hour or 0.5144 m/second.

Langley. Measure of irradiation in terms of Langleys per minute, where 1 langley equals 1 calorie/1 sq cm.

Lapse rate. Rate at which atmospheric temperature changes with altitude.

Latent heat of fusion. The number of calories per unit volume that must be added to a material at the melting point to complete the process of melting.

Lever. Rigid bar turning on a fixed point of support called the "fulcrum," generally used for raising weights.

Light year. Distance light travels in one year—about 6 million miles.

Load. Force available at the business end of a lever. Energy tapped from any power source.

Lumen. Unit measure of the light output of a lamp, where 1 lumen provides an intensity of 1 footcandle at a distance of 1 ft from the light source.

Lumen-hour. Unit of the quantity of light delivered in 1 hour by a flux of 1 lumen.

Lux. Measure of light intensity on a plane, denoting lumens per square meter.

Mach number. Speed expressed as a multiple of the speed of sound. Near the earth's surface "mach 1" means about 750 miles/hour.

Magnetic field. Field of force that exists around electrons. It travels along a conductor or orbits the atomic nucleus.

Magnetic variation or declination. Angle by which the compass needle varies from true north.

Mass

1. A number that measures the quantity of matter. It is obtained on the earth's surface by dividing the weight of a body by the acceleration due to gravity.

2. There are two ways of determining the mass of a body. One is to weigh it and apply Newton's law. The other is to apply a force to it and to calculate the mass from the resulting acceleration.

3. In shock and vibration terminology, a rigid body whose acceleration, according to Newton's second law, is proportional to the resultant of all forces acting upon it.

4. Quantity measuring the resistance of a body to changes in motion. The more massive a body is, the more difficult it is (that is, the more force is required) to change its state of motion, that is, bring it to rest if it is moving or start it up if it is at rest. When two bodies exert forces on each other, the ratio of the resultant acceleration is inverse to the ratio of their masses.

Mass number. Number of protons and neutrons in the nucleus of an atom.

Mass unit. A unit that is $\frac{1}{16}$ the mass of the oxygen atom. Approximately the mass of the hydrogen atom.

Mechanical energy. Sum of the kinetic and potential energies of a system. The mechanical energy of an isolated physical system does not change with time, but there may be a transfer between kinetic and potential energies.

Mechanics. Study of how matter behaves under the influence of force.

Meter (m). Conversion to the English system of customary units is 1 m = 39.4 in. A meter is usually divided into 100 equal smaller units each called the centimeter (cm): 1 in. = 2.54 cm. Similarly, the centimeter is divided into 10 smaller units, the millimeter (mm). For longer distances, the kilometer (km) is used: 1 km = 1000 m = 0.621 miles, so that 1 mile = 1/0.621 km = 1.61 km = 1610 m.

Metric system. Decimal system of weights and measures.

Micro. Prefix meaning one one-millionth of (10^{-6}).

Microbar. Unit of pressure commonly used in acoustics. One microbar is equal to one dyne per square centimeter.

Micron. Unit of length equal to 10^{-6} m.

Microwaves. Electromagnetic waves with wavelengths, in a vacuum, in the millimeter and centimeter range. Used for information (radar) and power (microwave ovens).

Mile per hour. This measurement equals the following: 0.016667 mile per minute, 1.4667 feet per second, and 3.281 feet per minute.

Mile per minute. 60 miles per hour, 88 feet per second, and 196.8 feet per minute.

Millibar (mbar). Unit of pressure equal to 1000 dynes/sq cm or $\frac{1}{1000}$ of a bar. The millibar is used as a unit of measure of atmospheric pressure, the standard of pressure, a standard atmosphere, being equal to 1013.25 mbars or 22.92 in. of mercury.

Mirage. Refraction phenomenon wherein an image of some objects is made to appear displaced from its true position.

Modulus of elasticity. Slope of the elastic portion of the stress-strain curve in mechanical testing.

Molecule. Smallest part of a substance that can exist separately while still having the chemical properties of the original substance.

Moment of inertia. A measure of the effectiveness of mass in rotation, the product of total mass, and the square of the radius of gyration.

Momentum. Measure of the motion of a body, determined by multiplying its mass by its velocity.

Nadir. Point on the celestial sphere direction beneath the observer and directly opposite the zenith.

Nephelometry. Measurement of concentration or other property of a suspension by means of its light transmission or dispersion.

Neutrons

1. Fundamental nuclear particle with no electric charge. This neutral body has about the same mass as that of a proton (neutron mass). Except for the absence of charge, all of its physical properties are similar to those of the proton. Hence the neutron and the proton are known as nucleons. Together they constitute all atomic nuclei. In reactors, fast neutrons have energies greater than 1 MeV; slow neutrons are those with energies less than 1 eV.

2. Elementary particle found in almost all nuclei.

Newton (N). Unit of force equal (by Newton's second law) to 1 kg-m/second². The force of 1 N is the weight on an object of $3\frac{1}{2}$ oz. A 1-lb weight has a gravitational force of 4.54 N acting on it.

Newton's law of gravitation. Every particle in the universe attracts every other particle; the attraction between any two particles is proportional to their masses and inversely proportional to the square of the distance between them.

Octave. Interval between two oscillations having a basic frequency ratio of 2.

Ohm. Unit that defines the resistance of a conductor to electric current.

Ohm's law. The current in amperes through a wire is equal to the potential drop in volts along the wire divided by the resistance of the wire in ohms.

Period

1. Time interval needed to complete a cycle; smallest increment of the independent variable for which the function repeats itself.

2. The period of a wave or other periodic disturbance is the time interval between successive repeats of the same configuration.

pH. Numerical designation of relatively weak acidity and alkalinity, as in soils and other biological systems. Technically, pH is the common logarithm of the reciprocal of the hydrogen ion concentration of a solution. A pH of 7.0 indicates precise neutrality, higher values indicate increasing alkalinity, and lower values indicate increasing acidity.

Phase. Measure of the stage of progress in the cycle of a periodic motion, usually expressed as a phase angle, one complete cycle representing a phase angle of 360° for example, a phase angle of 90° indicates that the cycle is one-quarter completed.

Phon. Unit of loudness level of a sound, defined as numerically equal to the median sound pressure level, in decibels, relative to 0.0002 μbar of a free progressive wave of a frequency of 1000 Hz, presented to listeners facing the source, which in a number of trials is judged by the listeners to be equally loud.

Photon. Indivisible package of electromagnetic energy. The energy in a photon of light is proportional to the frequency of the light.

Plasticity. Property of a material that enables it to undergo permanent deformation without elastic rebound and without rupture.

Plastic solid. Solid that undergoes change of shape continuously and indefinitely after the stress applied to it passes a critical point.

Potential energy. Ability of a body to do work because of its position. An object of mass (m) held at a height (h) above the ground can do an amount of work equal to mgh as it falls to the ground after being released; it thus has a potential energy of mgh.

Poundal. Unit of force; force needed to give a mass of 1 lb an acceleration of 1 ft/second.

Power

1. Rate at which work is done or energy expended. If work is being done at the rate of 1 J/second, the power is said to be 1 W. Thus 1 W-second is 1 J (power multiplied by time is energy).

2. Measure of the rate of doing work.

Power factor. Ratio of actual power to apparent power.

Pressure

1. Force per unit area exerted by a gas or liquid on the walls of its container. If the end wall of a container has an area of 5 m² and if the fluid in the container exerts a pressure of 25 N/m², then the force exerted by the fluid on the end wall is 125 N. If the wall is not strong enough to resist this force, the container will burst. In general, the pressure exerted by a fluid increases with the temperature of the fluid.

2. Measure of force per unit area, expressed in pounds per square inch (psi) or grams per square centimer (g/sq cm).

Protons

1. Positively charged nucleus of the hydrogen atom. It has a mass 1840 times as great as that of the electron and a charge equal in magnitude though opposite in sign to that of the electron. All atomic nuclei are composed of protons and neutrons.

2. Elementary particle found in all nuclei.

Pulse. Variation of a quantity whose value is normally constant (often zero, the variation being characterized by a rise and a decay. A common example is a very short burst of electromagnetic energy.

Rad. Unit of absorbed dose of radiation. A rad represents the absorption of 100 ergs of nuclear (or ionizing) radiation per gram of the absorbing material or tissue.

Radar. Locating, navigating, or guiding system that employs microwaves. The word "radar" is an acronym of the phrase "radio detection and ranging." Radar works independently of weather and visibility conditions because it depends on the ability of microwaves to bounce back from solid objects.

Radiac. Term that designates various types of radiological measuring instruments or equipment. The term is derived from the words "radioactivity detection, indication, and computation and is normally used as an adjective.

Radius of gyration. The distance from the axis of rotation at which the total mass of a body might be concentrated without changing its moment of inertia.

Random motion. Motion of molecules, atoms, or ions (or other extremely small particles) resulting from incessant random collisions within a fluid.

Rayleigh wave. Two-dimensional barotropic disturbance in a fluid having one or more discontinuities in the vorticity profile; wave propagated along the surface of a semiinfinite elastic solid, bearing certain analogies to a surface gravity wave in a fluid.

Reciprocity. Method of calibration wherein two measuring devices, such as accelerometers, are compared while being used in opposing positions. The devices are then reversed in these positions and the comparison is repeated.

Reflection. Return or change in the direction of travel of particles, radiant energy, sound waves, or other longitudinal waves that impinge on a surface but do not enter the substance providing the reflecting surface.

Refraction. Process in which the direction of energy propagation is changed as the result of a change in density with the propagating medium, or as the energy passes through the interface representing a density discontinuity between two mediums.

Relative humidity. Ratio of the amount of water vapor present in the atmosphere at any given temperature to the amount needed for saturation at that temperature, expressed as a percentage.

Rem. Measure of the radiation dose received from a radioactive source. A dose of 1 rem to a body means that nuclear particles have deposited between 10^{-2} and 10^{-3} (depending on the particles) J of energy for each kilogram of body mass.

Resistance. Except for certain materials at low temperatures, all conductors of electricity resist the flow of current, turning some of it into heat.

Roentgen. As gamma rays travel through matter, they may disrupt the atoms, breaking them into positive and negative particles (ions). One roentgen is the amount of radiation that will produce 2×10^9 pairs of ions in each cubic centimeter of air traversed (under standard conditions).

Root mean square (rms). Square root of the mean of the sum of the squares, often used as a shorthand for the rms of the distance in microinches, above or below a mean reference line of corresponding points on a surface; measured of surface roughness.

Rupture. The act of breaking or bursting in the testing of materials, by applying stresses or loads beyond the design capacities.

Sabin

1. Unit used in measuring or expressing the capability of a surface to absorb sound, equivalent to 1 sq ft of a perfectly absorptive surface.

2. The unit of measure of sound absorption; the amount of sound absorbed by a theoretically perfect absorbtive surface of 1 sq ft area. Named for Wallace C. W. Sabin, American physicist.

Saturation. Condition or state of something holding something else to its fullest capacity; condition in which the partial pressure of an atmospheric constituent (usually water vapor) is equal to the maximum possible under the existing environmental conditions; impregnation of one substance with another until no more can be received; combination of two substances until they neutralize each other; degree of purity of a color as measured by the absence of an admixture of white light.

Shear stress. Action or stress resulting from applied forces, which causes or tends to cause two contiguous parts of a body to slide relative to each other in a direction parallel to their plane of contact.

Shock motion. Motion causing or resulting from a shock excitation.

Sigma. Ratio of the maximum peak acceleration of a random acceleration time history to its root-mean-square acceleration.

Solidus. On a temperature composition diagram, the curve joining the points at which the last of liquid phase solidifies at each composition; referring to a given composition, a point on the solidus curve.

Sone. Unit of loudness. A simple tone of a frequency of 1000 Hz, 40 dB above a listener's threshold, produces a loudness of 1 sone.

Sorption. Generalized term including absorption and absorption when the nature or mechanism of the phenomenon is unknown.

Sound level. Weighted sound pressure level, obtained by the use of metering characteristics and the weightings A, B, or C specified in American National Standards Institute, "Sound Level Meters for Measurement of Noise and Other Sounds."

Sound power level. The ratio, expressed on the decibel scale, of the sound power under consideration to the standard reference power of one picowatt (1pW) (ASTM C 634).

Specific conductivity. Quantity of electricity transferred across unit area per unit potential gradient per unit time. In the metric system, $K = amp/sq\ cm$ divided by volts per centimeter.

Specific gravity

1. The ratio of the weight of a piece of wood to the weight of an equal volume of water. Since wood shrinks and swells depending on the moisture content, the moisture content of the wood must be specified to determine its volume.

2. The ratio of the weight of a body to the weight of an equal volume of water at 4°C or the specified temperature.

3. As applied to wood, the ratio of the oven-dry weight of a sample to the weight of a volume of water equal to the volume of the sample at a specified moisture content (green, air-dry, or oven-dry).

4. Ratio of the weight of a substance to that of an equal volume of some other substance at the same or a standard temperature. The usual standard for liquids and solids is chemically pure water at 4°C.

5. Density of a substance relative to some standard substance taken as unity (usually water = 1).

Specific heat of a substance

1. Quantity of heat (in calories) needed to raise 1 g of the substance through 1°C.

2. Heat capacity of unit mass of a substance.

Spectrum

1. Description of a resolution into components, each of different frequency and (usually) different amplitude and phase; continuous range of components, usually wide in extent, within which waves have some specified common characteristics, for example, the audiofrequency spectrum; relative energy, power, or flux density per unit frequency (or wavelength) interval as a fraction of frequency (or wavelength); graphical representation of a distribution function.

2. The band of colors formed when light is split into its separate wavelengths.

Speed. Ratio of distance covered to the time taken to cover it by a moving body.

Standard atmospheric pressure. SAP equals 29.92 in. of mercury of 1013.2 millibars at sea level.

Standardize. Reduce to or compare with a standard; to calibrate. In the environmental field, the verification process by whatever steps are necessary to ensure that all investigators in a given field are performing in a similar manner and will produce similar results.

Static electricity. Negative or positive charge of electricity that an object accumulates. The charge creates a spark when the object comes near another object to which it may transmit its charge or from which it may receive a charge.

Static stress. Stress induced in an elastic element by the static deflection applied to it.

Stiffness. Ratio of change of force (or torque) to the corresponding change in translational or rotational deflection of an elastic element.

Strain. Deformation caused by stress. Technically, strain is the elongation or shortening per unit of original length of a body under tension or compression, or the distortion in angle between two planes in a body under shear stress.

Stress. Resultant condition of applied force; condition existent in a body when its internal structure or surfaces resist a force that produces or tends to produce deformation in the body; molecular resistance to change in shape or size.

Stress-corrosion cracking. Cracking resulting from the combined effect of corrosion and stress.

Stress endurance limit. Value of alternating stress repetitively applied to an elastic element, below which the element is not expected to experience fatigue failure after an infinite (or extremely large) number of stress reversals.

Supersonic. Of or pertaining to speed in excess of the speed of sound.

Surveyor's chain. Chain used in the field, 66 feet in length, 22 yards, 4 rods, 100 links, 798 inches, 20.117 meters, and 2011.7 centimeters.

Telemetry. Science concerned with measuring a quantity or quantities, transmitting the results to a distant station, and interpreting, indicating, or recording the quantities measured.

Tensile strength. Force per unit area required to break a material in tension.

Therm. Unit of heat; 100,000 Btu.

Thermal conductivity. Time rate of transfer of heat by conduction, through unit thickness, and across unit area for unit difference of temperature. It is measured as calories per second/sq cm for a thickness of 1 cm and a difference of temperature of 1°C.

Thermodynamics. General study of energy processes, particularly those that involve heat.

Torque. A motion that produces or tends to produce torsion or rotation.

Torr. Unit for measuring a state of vacuum, defined as $\frac{1}{760}$ of a standard atmosphere. A torr is very nearly equal to a millimeter of mercury.

Torsion. The twisting of a body by two equal and opposite torques.

Toughness. Ability of material to absorb energy during plastic deformation.

Ultrasonic. Of or pertaining to frequencies above those that affect the human ear, that is, above 20,000 Hz.

Ultraviolet rays. Rays of shorter wavelength than visible light but longer than x rays.

Umbra. Darkest part of a shadow in which light is completely cut off by an intervening object. The lighter part surrounding the umbra, in which the light is only partly cut off, is called the "penumbra."

Unit. A specific magnitude of a quantity, set apart by appropriate definition, which serves as a basis of comparison or measurement for other quantities of the same nature.

Unstable wave. Wave motion whose amplitude increases with time or whose total energy increases at the expense of its environment.

Velocity. Rate of motion in a given direction.

Vibration. Periodic or random motion of the particles in an elastic body in alternately opposite directions; variation with time of the magnitude of a quantity, which is descriptive of the motion or position of a mechanical system, when the magnitude is alternately greater and smaller than some average value.

Viscosity. Molecular property of a fluid that enables it to support tangential stresses for a finite time and thus to resist deformation; property of a fluid that resists internal flow by releasing counteracting forces.

Volt (V)

1. Measure of potential (electrical potential energy difference per unit charge). Two points have a potential difference of 1 V when a charge of 1 C gains or loses 1 J of energy in being transported between the two points.

2. Unit of electromotive force or of potential difference. A difference of 1 V potential across the ends of a conductor whose resistance is 1 ohm causes a current of 1 A to flow.

3. Voltage that will produce a current of 1 A through a resistance of 1 ohm.

Volume. Amount of space taken up by an object. For a rectangular structure, the volume is the product of the length, width, and depth.

Watt (W)

1. Unit of power defined as 1 J/second. Also commonly used are kilowatts (kW = 1000 W) and megawatts (MW = 1,000,000 W); 1 kW = 1.341 horsepower.

2. Unit of power; the capacity to do work at the rate of 10 ergs/second.

3. Energy rate of 1 J/second or the power of an electric current of 1 A with an intensity of 1 V.

Wave. Undulation or ridge on the surface of a fluid. Wind waves are generated by friction between wind and the fluid surface. Ocean waves are produced principally in this way. A wave is a disturbance propagated in such a manner that it may progress from point to point. An electromagnetic wave is produced by oscillation of an electric charge. Also, marked variation from normal weather, as a heat wave or a cold wave.

Wavelength. Distance between two successive comparable points in a traveling wave, such as the distance between two neighboring crests in a water wave. The velocity of the traveling wave is the product of the wavelength and the frequency.

Weight. The force with which a body is attracted toward the earth.

Work. Force that acts against resistance to produce motion in a body.

X-rays. Electromagnetic radiation of the same kind as light, but with much shorter wavelength. X-rays are produced when beams of high-speed electrons strike matter, or when high-energy orbital electrons lose energy and fall to a lower orbit.

WATER RESOURCES AND SOIL EROSION

Ablation zone. Lower part of a glacier where the annual water loss exceeds snow accumulation.

Absorption. Process by which one substance is taken into and included within another substance, as the absorption of water by soil or nutrients by plants.

Abyssal hill. Low, rounded, submarine hill with a relief of 100 to 200 m, common in deep ocean basins.

Abyssal plain. Flat, sediment-covered province of the sea floor that slopes at less than 1 to 1000.

Accelerated erosion. Erosion much more rapid than normal, natural, geological erosion, primarily a result of the influence of the activities of man or in some cases of animals.

Algal mat. Layered communal growth of algae observed in fossils and in present-day tidal zones associated with carbonate sedimentation.

Alkali flat. Bed of a dried-up saline lake that is heavily impregnated with alkaline salts.

Alluvial fan

1. Deposit of a stream where it emerges from a steep mountain valley on open level land.

2. Low, cone-shaped deposit of terrestrial sediment formed where a stream undergoes an abrupt reduction in slope.

Alluvium

1. Soil whose constituents have been transported in suspension by flowing water and subsequently deposited by sedimentation.

2. Sediment deposited on land from streams.

3. Unconsolidated terrestrial sediment composed of sorted or unsorted sand, gravel, and clay deposited by water.

4. All detrital material in transit or deposited permanently by streams, including gravel, sand, silt, and clay and all variations and mixtures of these; usually, the deposits of streams in their channels and over their flood plains and deltas. Unless otherwise noted, alluvium is unconsolidated.

Alpine glacier. Glacier confined to a stream valley.

Anchor ice. Very solid film of ice that develops on rocks and other obstructions on the bottom of a stream, lake, or shallow sea, irrespective of its nature of formation.

Angle of repose. Steepest slope angle in which a particular sediment will lie without cascading down.

Annual flood. Highest peak discharge in a water year.

Antecedent stream. Stream that maintains the same course after uplift that it originally followed prior to uplift.

Antiroot. Accumulation of higher density material in the suboceanic crust that compensates for the low density of seawater.

Apogean tidal current. Tidal current of decreased speed occurring at the time of apogean tide.

Aquiclude. Impermeable stratum that acts as a barrier to the flow of groundwater.

Aquifer

1. Permeable formation that stores and transmits groundwater in sufficient quantity to supply wells.

2. Water-bearing bed or stratum of earth, gravel, or porous stone.

Arroyo. Steep-sided and flat-bottomed gulley in an arid region that is occupied by a stream only intermittently, after rains.

Artesian water. Groundwater that is under sufficient pressure to rise above the level at which it is encountered by a well, but that does not necessarily rise to or above the surface of the ground.

Artesian well

1. Well that penetrates an aquiclude to reach an aquifer containing water under pressure. Thus water in the well rises above the surrounding water table.

2. Well in which the water comes from an aquifer below an impervious layer.

Asymmetrical ripple. Ripple whose cross section is asymmetric, with a gentle slope on the upcurrent side and a steeper face on the downcurrent side.

Atoll

1. Continuous or broken circle of coral reef and low coral islands surrounding a central lagoon.

2. Ring-shaped organic reef that is surrounded by the open sea and encloses a lagoon in which there is no land above sea level.

Atterberg limits. Measures of soil consistency for differentiation between materials of appreciable plasticity (clays) and slightly plastic or nonplastic materials (silts). Measures include the liquid limit, plastic limit, and plasticity index.

Available water. Portion of water in a soil that can be readily absorbed by plant roots; water held in the soil against a pressure of up to approximately 15 bars.

Average discharge. In the annual series of the Geological Survey's reports on surface-water supply, the arithmetic average of all complete water years of record whether or not they are consecutive. Average discharge is not published for fewer than five years of record. The term "average" is generally reserved for average of record and "mean" is used for averages of shorter periods, namely, daily mean discharge.

Awash. Tossed about or bathed by waves or tide. A rock exposed or just bare at any stage of the tide between the datum of mean high water and the sounding.

Back swamp. Marshy area of a flood plain at some distance from, and lower than the banks of a river confined by natural levees.

Backwash. Return flow of water down a beach after a wave has broken.

Bank. Elevation of the sea floor located on a continental (or island) shelf and over which the depth of water is relatively shallow but sufficient for safe navigation. A bank supports shoals or bars on its surface which are dangerous to navigation. Shallow area consisting of shifting forms of silt, sand, mud, gravel, etc. In this case the term is only used with a qualifying word, for example, "sandbank," or "gravelbank." Ridge of such material as earth, rock, or snow or anything resembling such a ridge, as a fog bank or cloud bank; border or shore of a river; margin of a channel.

Bankfull stage. In a stream, the height of water that just corresponds to the level of the surrounding floodplain.

Bar (stream). Accumulation of sediment, usually sandy, that forms at the borders or in the channels of streams or offshore from a beach.

Bar finger sand. Elongated lens of sand deposited during the growth of a distributary in a delta. The bar at the distributary mouth is the growing segment of the bar finger.

Barrier island. Long, narrow island parallel to the shore, composed of sand and built by wave action.

Barrier reef. Coral reef parallel to and separated from the coast by a lagoon that is too deep for coral growth. Generally, barrier reefs follow the coasts for long distances and are cut through at irregular intervals by channels or passes.

Base level

1. Level of a body of water into which a stream flows.

2. Level below which a stream cannot erode, usually sea level but sometimes locally the level of a lake or resistant formation.

Basin

1. In sedimentology, the site of accumulation of a large thickness of sediments.

2. Depression of the sea floor more or less equidimensional in form and of variable extent; area in a tidal region in which water can be kept at a desired level by means of a gate. It is also called a "tidal basin." Relatively small cavity in the bottom or shore, usually created or enlarged by excavation, large enough to receive one or more vessels for a specific purpose; area of land that drains into a particular lake or sea through a river and its tributaries; drainage or catchment area of a stream or lake.

Bathymeter. Instrument primarily designed for measuring the depth of water. Bathymetric surveys, previously done by lead line, are now performed by using an echo sounder and precision depth recorder.

Bathymetry. Study and mapping of sea-floor topography.

Beach. Area extending from the shore line inland to a marked change in physiographic form or material or to the line of permanent vegetation (coastline); gently sloping area of wave-deposited unconsolidated material bordering a sea or lake; also, unconsolidated material making up such a beach.

Bed load. Sediment that a stream moves along the bottom of its channel by rolling and bouncing.

Berm. Barely horizontal portion of a beach or bankshore having an abrupt fall and formed by deposition of material by wave action. It marks the limit of ordinary high tides.

Biochemical precipitate. Sediment, especially of limestone or iron, formed from elements extracted from seawater by living organisms.

Blown-out land. Areas from which all or almost all of the soil and soil material has been removed by wind erosion; usually barren, shallow depressions with a flat or irregular floor consisting of a more resistant layer and/or an accumulation of pebbles or a wet zone immediately above a water table. It is usually unfit for crop production.

Bog. Quagmire or morass; area of wet, peaty, spongy ground, usually lacking in mineral nutrients and often interspersed with pools of open water, where any dense body is likely to sink; vegetation of saturated, peaty land in open or forest areas—hence moss bog, juncus bog, carex or sedge bog, sphagnum bog, birch bog, tamarack bog, black spruce bog.

Border strip. Water is applied at the upper end of a strip with earth borders to confine the water to the strip.

Bore. Restricted tidal current of considerable force and size; also called an "cagre."

Bottom. Ground covered by water. "Bed" refers more specifically to the whole submerged basin, and "floor" is the essentially horizontal surface of the ground beneath the water.

Bottomset bed. Flat-lying bed of fine sediment deposited in front of a delta and then buried by continued delta growth.

Brackish water

1. Water containing salt to a moderate degree, such as seawater that has been diluted by freshwater near the mouth of a river. Brackish water has salinity values ranging from approximately 0.50 to 17.00 parts per thousands.

2. Water that contains concentrations of minerals considerably lower than in seawater, but too high for human consumption.

Braided stream

1. Stream so choked with sediment that it divides and recombines numerous times, forming many small and meandering channels.

2. Overloaded stream that winds in and out among sandbars.

3. Stream flowing in several channels that divide and reunite in a pattern resembling the strands of a braid.

Breaker

1. Ocean wave that becomes unstably steep on encountering shallow water and collapses turbulently.

2. Wave breaking on the shore, over a reef, etc. Breakers may be roughly classified into three kinds, although the categories may overlap: Spilling breakers break gradually over a considerable distance; plunging breakers tend to curl over and break with a crash; and

surging breakers peak up, but then instead of spilling or plunging they surge up on the beach face.

Breaks. Area in rolling land eroded by small ravines and gullies; also, sudden change in topography, as from a plain to hilly country.

Breakup. Spring melting of snow, ice, and frozen ground; specifically, the destruction of the ice cover on rivers during the spring thaw. The term is also applied to the time when the solid sheet of ice on rivers breaks into pieces that move with the current.

Brine

1. Seawater containing a higher concentration of dissolved salt than that of the ordinary ocean. Brine is produced by the evaporation or freezing of seawater. In the latter case, the sea ice formed is much less saline than the initial liquid, leaving the adjacent unfrozen water with increased salinity. The liquid remaining after seawater has been concentrated by evaporation until the salt has crystallized is called "bittern."

2. Seawater whose salinity has been increased by evaporation; groundwater with an unusual concentration of salts.

Calcium carbonate compensation depth. Ocean depth below which the solution rate becomes so great that no carbonate organisms are preserved on the sea floor.

Canyon. Very large, deep valley with precipitous walls formed principally by stream downcutting.

Capacity (stream). Measure of the amount of sediment and detritus a stream can transport past any point in a given time.

Capillary fringe. Zone just above the water table that is maintained in an essentially saturated state by capillary forces of lift.

Capillary porosity. Small pores (or the bulk volume of small pores) that hold water in soils against a tension usually greater than 60 cm of water.

Capillary stresses. Pore water pressures less than atmospheric values produced by surface tension of pore water acting on the miniscus formed in the void spaces between soil particles.

Capillary water. Water held in the "capillary" or small pores of a soil, usually with a tension greater than 60 cm of water.

Carbonate platform. Submarine or intertidal shelf whose elevation is maintained by active shallow-water carbonate deposition.

Cavitation. A process of erosion in a stream channel caused by a sudden collapse of vapor bubbles against the channel wall.

Check-basin. Water is applied rapidly to relatively level plots surrounded by levees. The basin is a small check.

Chemical sediment. Sediment that is formed at or near its place of deposition by chemical precipitation, usually from seawater.

Chemical weathering. Total set of all chemical reactions that act on rock exposed to water and atmosphere and so change its mineral to more stable forms.

Chlorinity. Measure of the chloride content, by mass, of seawater (grams per kilogram of seawater or per mille). Originally chlorinity was defined as the weight of chlorine in grams per kilogram of seawater after the bromides and iodides had been replaced by chlorides.

Choppy. As applied to the sea, having short, abrupt, breaking waves dashing against each other; chopping. As applied to the wind, variable, unstable, changeable; chopping.

Cirque. Head of a glacial valley, usually formed like one-half of an inverted cone. The upper edges have the steepest slopes, approaching vertical, and the base may be flat or hollowed out and occupied by a small lake or pond.

Coastal plain

1. Low plain of little relief adjacent to the ocean and covered with gently dipping sediments.

2. Any plain that has its margin on the shore of a large body of water, particularly the sea.

Compaction. Reduction in pore space between individual grains of soil, preventing soil absorption of water, as a result of pressure of overlying sediments or pressures resulting from earth movement.

Competence (stream). Measure of the largest particle a stream is able to transport, but not the total volume.

Condensation. Process by which water vapor changes into liquid water or solid ice crystals.

Confined water reservoir. Body of groundwater surrounded by impermeable strata.

Consumptive use. Water used per unit of time in plant growth by the combined processes of evaporation, transpiration, and retention in the plant. It is approximated by evapotranspiration.

Continental divide. Imaginary line connecting high points across a continent and dividing regions whose streams drain into one ocean from regions that drain into another.

Continental shelf

1. Gentle sloping submerged edge of a continent, commonly extending to a depth of about 200 m or the edge of the continental slope.

2. Zone adjacent to a continent or around an island, extending from the low water line to the depth at which there is usually a marked increase of slope to greater depth.

Continental slope

1. Region of steep slopes between the continental shelf and continental rise.

2. Declivity seaward from a continental shelf edge into greater depth.

Connate water. The water trapped in sedimentary rocks at the time they were formed.

Consequent stream. Stream that follows a course that is a direct consequence of the original slope of the surface on which it developed.

Corrugation. Water is applied to small, closely spaced furrows, frequently in grain and forage crops, to confine the flow of irrigation water to one direction.

Coulee. Steep-walled, trenchlike valley of considerable size through which water flows intermittently; more specifically, any of a number of steep-walled, trenchlike valleys cut into the Columbia Plateau lava sheets in the state of Washington and formerly occupied by glacial meltwater rivers.

Creep. Slow mass movement of soil and soil material down relatively steep slopes primarily under the influence of gravity, but facilitated by saturation with water and alternate freezing and thawing.

Cycle of erosion. A qualitative description of river valleys and regions passing through the stages of youth, maturity, and old age with respect to the amount of erosion that has been effected.

Debacle. Rush of water or ice in a stream immediately following the breakup; violent rush or flood of water.

Deflation. Removal of fine soil particles from soil by wind.

Degree of saturation (percent saturation). Ratio, expressed as a percentage, of the volume of water in a given soil mass to the total volume of intergranular space (voids).

Delta

1. Body of sediment deposited in an ocean or lake at the mouth of a stream.

2. Deposit of alluvium at the mouth(s) of a river. The term refers particularly to that part of the deposit forming a tract of land above water, usually roughly triangular in shape, resembling the Greek letter.

Dendritic drainage

1. Stream system that branches irregularly and resembles a branching tree in plan.

2. Drainage pattern in which streams run in many directions and branch irregularly like a tree. The positions of the streams are not influenced by differences in rock structure or hardness.

Density current. Subaqueous current that flows on the bottom of a sea or lake because the entering water is more dense due to temperature or suspended sediments.

Depression. Low place of any size on a plain surface, with drainage underground or by evaporation; hollow completely surrounded by higher ground and having no natural outlet for surface drainage.

Desiccant. Drying or dehydrating agent that absorbs water vapor by physical or chemical means.

Desiccation. Process of shrinkage or consolidation of the fine-grained soil produced by increase of effective stresses in the grain skeleton accompanying the development of capillary stresses in the pore water.

Detrital sediment. Sediment deposited by a physical process.

Dew. Water condensed onto grass and other objects near the ground, the temperatures of which have fallen below the dew point of the surface air owing to radiational cooling during the night, but are still above freezing. If the temperature falls below freezing after dew has formed, the frozen dew is known as "white dew."

Diagenesis. Physical and chemical changes undergone by a sediment during lithification and compaction, excluding erosion and metamorphism.

Diatom. One-celled plant that has a siliceous framework and grows in oceans and lakes.

Diatom ooze. Fine muddy sediment consisting of the hard parts of diatoms.

Direction of current. Direction toward which a current is flowing, called the "set" of the current.

Direct runoff. Runoff entering stream channels promptly after rainfall or snowmelt. Superimposed on the base runoff, it forms the bulk of the hydrograph of a flood.

Discharge. Rate of water movement through a stream, measured in volume units per unit time.

Distributary. Smaller branch of a large stream that receives water from the main channel; opposite of a tributary.

Diversion dam. Structure or barrier built to divert part or all of the water of a stream to a different course.

Divide

1. Ridge of high ground separating two drainage basins emptied by different streams.

2. Line of separation between drainage systems; the summit of an interfluve; highest summit of a pass or gap.

Drainage area. Area, measured in a horizontal plane, enclosed by a drainage divide.

Drainage basin

1. Region of land surrounded by divides and crossed by streams that eventually converge to one river or lake.

2. Part of the surface of the earth that is occupied by a drainage system, which consists of a surface stream or body of impounded surface water, together with all the tributary surface streams and bodies of the impounded surface water.

Drift. Any material laid down directly or deposited in lakes, streams, oceans, or seas as a result of glacial activity.

Drowned valley. Valley whose lower part has been inundated by the sea as a result of submergence of the land margin.

Dry wash

1. Broad, dry bed of a stream; dry stream channel.

2. Intermittent stream bed in an arroyo or canyon that carries water only briefly after a rain.

Earthflow. Detachment of soil and broken rock and its subsequent downslope movement at slow or moderate rates in a tonguelike stream.

Ebb current. Tidal current associated with the decrease in the height of a tide. Ebb currents generally set seaward or in an opposite direction to the tide progression.

Ebb tide. Part of the tide cycle during which the water level falls.

Echo sounder. Oceanographic instrument that emits sound pulses into the water and measures its depth by the time elapsed before they return.

Emergence. Process by which part of the sea or lake floor becomes dry land.

Equilibrium water. Water content of a solid that will remain unchanged by further exposure to air of a given humidity, temperature, and pressure.

Equivalent fluid pressure. Horizontal pressures of soil or soil and water in combination that increase linearly with depth and are equivalent to those that would be produced by a heavy fluid of a selected unit weight.

Erode. Wear away or remove the land surface by wind, water, or other agents.

Erosion

1. Set of all processes by which soil and rock are loosened and moved downhill or downwind.

2. Removal of soil or soil material by the action of wind or water.

3. Wearing away of the land surface by running water, wind, ice, or other geological agents, as well as such processes as gravitational creep; detachment and movement of soil or rock by water, wind, ice, or gravity.

4. Wearing away of the land surface by detachment and transport of soil and rock materials through the action of moving water, wind, or other geological agents.

Erosional flood plain. Flood plain that has been created by the lateral erosion and the gradual retreat of the valley walls.

Esker. Glacial deposit in the form of a continuous winding ridge, formed from the deposits of a stream flowing beneath the ice.

Esturary. Widened channel at the mouth of a river in which the influences of the tides is felt. The water in an estuary is normally brackish.

Eugeosyncline. Seaward part of a geosyncline, characterized by clastic sediments and volcanism.

Eustatic changes. Sea-level changes that affect the whole earth.

Eutrophication. Superabundance of algal life in a body of water, caused by an unusual influx of nitrate, phosphate, or other nutrients.

Evapotranspiration. Combined processes by which water is transferred from the earth's surface to the atmosphere; evaporation of liquid or solid water plus transpiration from plants.

Falling tide. Portion of the tide cycle between high water and the following low water.

Fast ice. Any type of sea, river, or lake ice attached to the shore (ice foot, ice shelf), beached (short ice), stranded in shallow water, or frozen to the bottom of shallow waters.

Fetch

1. Area in which ocean waves are generated by the wind. It is generally delineated by coastlines, fronts, or areas of wind curvature or divergence. The length of the fetch area is measured in the direction of the wind.

2. Extent of water over which the wind blows with nearly constant direction and speed.

Field moisture capacity. Percentage of water remaining in a soil two or three days after having been saturated and after free drainage has practically ceased. The percentage may be expressed on the basis of weight or volume.

Field moisture deficiency. Quantity of water that would be required to restore the soil moisture to field moisture capacity.

Field moisture equivalent. Minimum water content, expressed as a percentage of the weight of the oven-dried soil, at which a drop of water placed on a smoothed surface of the soil will not immediately be absorbed by the soil but will spread out over the surface and give it a shiny appearance.

Fiord

1. Former glacial valley with steep walls and a U-shaped profile now occupied by the sea.

2. Deep, narrow, and steep-walled inlet of the sea formed in most instances by intense glacial erosion of a valley. Some fiords may have been invaded by ocean water after glaciation owing to general subsidence of the land, but most are the direct result of erosion by tongues of ice that actually entered the ocean and moved along the bottom.

First bottom. Normal flood plain of a stream subject to frequent or occasional flooding.

Flash flood. Flood that rises and falls quite rapidly with little or no advance warning, usually as the result of intense rainfall over a relatively small area. Other possible causes are ice jams, dam failure, etc.

Flash floods. Floods occurring suddenly in narrow valleys as a result of heavy downpours or cloudbursts.

Flood. Overflow or inundation that comes from a river or other body of water and causes or threatens damage; relatively high stream flow overtopping the natural or artificial banks in a reach of a stream; relatively high flow as measured by either gauge height or discharge quantity.

Flooding. Water is released from field ditches and allowed to flood over the land.

Flood plain

1. Level plain of stratified alluvium on either side of a stream. It is submerged during floods and built up by silt and sand carried out of the main channel.

2. Any plain that borders a stream and is covered by its waters in time of flood.

3. Strip of relatively smooth land bordering a stream, built of sediment carried by the stream and dropped in the slack water beyond the influence of the swiftest current. It is called a "living flood plain" if it is overflowed in times of high water, but a "fossil flood plain" if it is beyond the reach of the highest flood. Low land that borders a river, usually dry, but subject to flooding; land outside of a stream channel described by the perimeter of the maximum probable flood.

Floor plain of aggradation. Flood plain formed by the building up of the valley floor by sedimentation.

Flood stage. Gauge height of the lowest bank of the reach in which the gauge is situated. "Lowest bank" is, however, not to be taken to mean an unusually low place or break in the natural bank through which the water inundates an unimportant and small area. Stage at which overflow of the natural banks of a stream begins to cause damage in the reach in which the elevation is measured.

Flood tide. Part of the tide cycle during which the water rises or levels off at high water.

Flood wave. Distinct rise in stage culminating in a crest and followed by recession to lower stages.

Flood zone. Land bordering a stream that is subject to floods of about equal frequency, for example, a strip of the flood plain subject to flooding more often than once, but not as frequently as twice in a century.

Flume. Laboratory model of stream flow and sedimentation, consisting of a rectangular channel filled with sediment and running water.

Foraminifer. Class of oceanic protozoa most of which have shells composed of calcite.

Foraminiferal ooze. Calcareous sediment composed of the shells of dead Foraminifera.

Ford. Place where a stream or other water is commonly passed by man or beast by wading; place where a road or trail crosses a body of water without a bridge or ferry.

Foreset bed. One of the inclined beds found in cross-bedding; also an inclined bed deposited on the outer front of a delta.

Fringing reef

1. Coral reef that is directly attached to a land mass not made of coral.

2. Coral reef that is closely attached to the shore. There is no lagoon or open water between the reef and the land to which it is attached. It is usually built on a shallow platform extending outward from the shore line and may be laid bare at very low tide.

Frost action

1. Freezing and thawing of moisture in materials and the resultant effects on these materials and on structures of which they are a part or with which they are in contact.

2. Freezing and thawing cycles of the water contained in natural or man-made materials and, especially, the disruptive effects of this action. In geology there are two basic types of fraction, the shattering or splitting of rock material, and congeliturbation, the churning, heaving, and thrusting of soil material.

Frost heave. Upward or sideway movement of surface soils, rocks, and vegetation through the expansion of freezing subsurface soil and gravel.

Frost line. Maximum depth of frozen ground during the winter. The term may refer to an individual winter, to the average over a number of years, or to the greatest depth recorded since observations began. The frost line varies with the nature of the soil and the protection afforded by vegetal ground cover and snow cover, as well as with the amount of seasonal cooling.

Frost weathering. Mechanical disintegration of earth materials brought about by frost action.

Gauging station. Particular site on a stream, canal, lake, or reservoir where systematic observations of gauge height or discharge are obtained.

Geomorphic cycle. Idealized model of erosion wherein a plain is uplifted epeirogenically, then dissected by rapid streams (youth), rounded by downslope movement into a landscape of steep hills (maturity), and finally reduced to a new peneplain at sea level.

Geyser

1. Hot spring that throws hot water and steam into the air. The heat is thought to result from the contact of groundwater with magma bodies.

2. Special type of hot spring that throws forth intermittent jets of hot water and steam. The action results from the contact of groundwater and rock or vapor hot enough to generate steam under conditions that prevent continuous circulation.

Glacial drift

1. All rock material in transport by glacier ice, all deposits made by glacier ice, and all deposits predominantly of glacial origin made in the sea or in bodies of glacial meltwater, whether rafted in icebergs or transported in the water itself. Included are till, stratified drift, and scattered rock fragments.

2. Rock debris that has been transported by glaciers and deposited, either directly from the ice or from the meltwater. The debris may or may not be heterogeneous.

Glaciofluvial. Term applied to stream(s) formed by the meltwater of a glacier(s) or deposits made by such streams.

Glaciofluvial deposits. Material moved by glaciers and subsequently sorted and deposited by streams flowing from the melting ice. The deposits are stratified and may occur in the form of outwash plains, deltas, kames, eskers, and kame terraces.

Gley soil. Soil horizon in which waterlogging and lack of oxygen have caused the material to be a neutral gray in color. The term "gleyed" is applied, as in "moderately gleyed soil," to soil horizons with yellow and gray mottling caused by intermittent waterlogging.

Gleyzation. Process affecting soil development under strongly reducing conditions, waterlogging, for example. Iron may be extensively solubilized and partly reprecipitated as rust-colored mottles or stains.

Graded bedding. Bed in which the coarsest particles are concentrated at the bottom and grade gradually upward into fine silt, the whole bed having been deposited by a waning current.

Graded stream. Stream whose smooth profile is unbroken by resistant ledges, lakes, or waterfalls and that maintains exactly the velocity required to carry the sediment provided to it.

Gradient (stream). Slope of a stream bed usually expressed in feet per mile.

Gravitational water. Water that moves into, through, or out of the soil under the influence of gravity.

Groundwater

1. Water that fills all the unblocked pores of underlying material below the water table, which is the upper limit of saturation.

2. Water that is free to move through a soil mass under the influence of gravity.

3. Portion of the total precipitation that at any particular time is either passing through a standing in the soil and the underlying strata and is free to move under the influence of gravity.

4. Mass of water in the ground below the phreatic zone, occupying the total pore space in the rock and moving slowly downhill where permeability allows.

5. Water in the ground that is in the zone of saturation, from which wells, springs, and groundwater runoff are supplied.

Groundwater laterite soil. Great soil group of the intrazonal order and hydromorphic suborder, consisting of soils characterized by hardpans or concretional horizons rich in iron and aluminum (and sometimes manganese) that have formed immediately above the water table.

Groundwater outflow. Part of the discharge from a drainage basin that occurs through the groundwater. The term "underflow" is often used to describe the groundwater outflow that takes place in valley alluvium instead of the surface channel and thus is not measured at a gauging station.

Groundwater podzol soil. Great soil group of the intrazonal order and hydromorphic suborder, consisting of soils with an organic mat on the surface over a very thin layer of acid humus material underlaid by a whitish gray leached layer, which may be as much as 2 or 3 ft in thickness, underlaid by a brown or very dark-brown cemented hardpan layer. This soil is formed under various types of forest vegetation in cool to tropical humid climates under conditions of poor drainage.

Groundwater runoff. Part of the runoff that has passed into the ground, become groundwater, and been discharged into a stream channel as spring or seepage water.

Gully

1. Small steep-sided valley or erosional channel.

2. Small ravine or miniature valley, especially one cut by running water, through which water flows only after a rain.

Gully erosion. Erosion process whereby water accumulates in narrow channels and, over short periods, removes the soil from this narrow area to considerable depths.

Guyot. Flat-topped submerged mountain or sea mount found in the ocean.

Gyre. Circular rotation of the waters of each major sea, driven by prevailing winds and the Coriolis effect.

Half-bog soils. Great soil group of the intrazonal order and hydromorphic suborder, consisting of soil with dark-brown or black peaty material over grayish and rust mottled mineral soil. These soils are formed under conditions of poor drainage under forest, sedge, or grass vegetation in cool to tropical humid climates.

Halomorphic soil. Suborder of the intrazonal soil order, consisting of saline and alkali soils formed under imperfect drainage in arid regions and including the great soil groups of solonchak or saline soils, solonetz soils, and soloth soils.

Hard water

1. Water that contains sufficient dissolved calcium and magnesium to cause a carbonate scale to form when the water is boiled or to prevent the sudsing of soap.

2. Water characterized by the presence of dissolved mineral salts, especially those of magnesium and calcium.

Head. Difference in elevation of water-producing discharge.

Homogeneous earth dam. Earth dam whose embankment is formed of one soil type without a systematic zoning of fill materials.

Hot spring. Spring whose water is above both human body and soil temperature as a result of plutonism at depth.

Humic gley soils. Soils of the intrazonal order and hydromorphic suborder, including weisenboden and related soils, such as half-bog soils, which have a thin muck of peat 02 horizon and an A1 horizon. They develop in wet meadows and forested swamps.

Hydration. Physical binding of water molecules to ions, molecules, particles, or other matter.

Hydraulic conductivity

1. Expression of the readiness with which a liquid such as water flows through a soil in response to a given potential gradient.

2. Ability of the soil to transmit water in liquid form through pores.

Hydraulic pressure. Pressure within water produced by a combination of forces, such as capillary and gravitational action on the water.

Hydrograph. Graphical representation of a stage or discharge at a point on a stream as a function of time. The most common type, the observed hydrograph, represents river gauge readings plotted at the time of observation.

Hydrographic chart. Nautical chart showing the depths of the water, the nature of the bottom, the contours of the bottom and coastline, and the tides and currents in a given sea or sea and land area.

Hydrographic datum. Plane of reference of soundings, depth curves, and elevations of foreshore and offshore features.

Hydrologic cycle

1. Cyclical movement of water from the ocean to the atmosphere through rain to the surface, runoff and groundwater to streams, and the return to the sea.

2. Circulation of water from the sea, through the atmosphere, to the land, and thence, with many delays, back to the sea by overland and subterranean routes, and in part by way of the atmosphere; also, the many short circuits of the water that is returned to the atmosphere without reaching the sea.

Hydrologic geometry feature. Channel, stream, pond, lake, or other depression that exhibits a water depth of 25 cm or greater for a total period of at least 1 week of the year.

Hydrology. Science of the hydrologic cycle between rain and the return of water to the sea; study of water on and within the land.

Hydrolysis. Chemical reaction of a compound with water, whereupon the anion from the compound combines with the hydrogen and the cation from the compound combines with the hydroxl from the water to form an acid and a base.

Hydromorphic soils. Suborder of iontrazonal soils, consisting of seven great soil groups, all formed under conditions of poor drainage in marshes, swamps, seepage areas, or flats.

Hydrophillic

1. That which absorbs, holds, or attracts water.

2. Having an attraction for water; readily wet by water.

Hydrophobic. Having little or no affinity for water; water repellent; not wet by water.

Hydrostatic excess pore pressures. Increments of pore water pressures greater that hydrostatic values, produced by consolidation stresses in compressible materials or by shear strain.

Hydrostatic pore pressures. Pore water pressures or groundwater pressures exerted under conditions of no flow where the magnitude of pore pressures increases linearly with depth below the ground surface.

Hydrothermal activity. Activity involving high-temperature groundwaters, especially the alteration and emplacement of minerals and the formation of hot springs and geysers.

Hydrous. Containing water.

Hygroscopic water. Water that approximates the film water held in air-dry soil.

Infiltration. Movement of groundwater or hydrothermal water into rock or soil through joints and pores.

Infiltration capacity. Maximum rate at which in a given condition the soil can absorb falling rain or melting snow.

Infiltration index. Average rate of infiltration, in inches per hour, equal to the average rate of rainfall such that the volume of rainfall at greater rates equals the total direct runoff.

Infiltration rate. Soil characteristic determining or describing the maximum rate at which water can enter the soil under specified conditions, including the presence of an excess of water.

Inlet. Small, narrow bay or creek; small body of water leading into a larger; narrow strip of water running into the land or between islands.

Interflow. Water, derived from precipitation, that infiltrates the soil surface and then moves laterally through the upper layers of soil above the water table until it reaches a stream channel or returns to the surface at some point downslope from its point of infiltration.

Interfluve. Area between two rivers; especially, a more or less undissected upland between two adjacent streams flowing in more or less parallel courses; surface area of alluvial fans between dry washes.

Interior drainage. System of streams that converge in a closed basin and evaporate without reaching the sea.

Intermediate belt. Subdivision of zone of aeration. The belt that lies between the belt of soil moisture and the capillary fringe.

Intermittent stream. Stream that flows only part of the time. It may be either an ephemeral stream, which flows for a few hours or days after rain, or a seasonal stream, which flows for several months of the year, as during the rainy season or during the season of snow melt.

Jet flow. Type of flow, related to turbulent flow, occurring when a stream reaches high velocity along a sharply inclined stretch, or over a waterfall, and the water moves in plunging jet-like surges.

Jetty. Structure, such as a wharf or pier, so located as to influence current or protect the entrance to a harbor or river. A jetty extending into the sea to protect the coast from erosion is called a "groin". A jetty that breaks the force of the sea at any place is called a "breakwater." A jetty, wall, or bank, often submerged, built to direct or confine the flow of a river or tidal current is called a "training wall." A wall or embankment along a waterfront, built to resist the encroachment of the sea, is called a "sea wall."

Juvenile water. Water derived from or existing in molten igneous rock or magma, and brought to the surface or added to underground supplies.

Lacustrine deposit. Material deposited in lake water and later exposed either by the lowering of the water level or the elevation of the land.

Lacustrine terraces. Benches or flats that mark the shore lines of ancient lakes or earlier high-water stages of existing lakes. Nearly horizontal surfaces with relatively steep slopes facing the central portion of the lake characterize this land form.

Lagoon. Shallow sound, channel, pond, or lake, especially one near or communicating with the sea.

Lahar. Mudflow of unconsolidated volcanic ash, dust, breccia, and boulders mixed with rain or the water of a lake displaced by a lava flow.

Lake. Standing body of inland water, generally of considerable size.

Laminar flow. Flow in which the water moves slowly along a smooth channel, with fluid particles following straight-line paths parallel to the channel or walls.

Leaching. Removal of elements from a soil by dissolution in water moving downward in the ground.

Levee

1. Low ridge along a stream bank, formed by deposits left when floodwater decelerates on leaving the channel; also, an artificial barrier to floods built in the same form.

2. Embankment along the shore of a river or arm of the sea to prevent overflow. A natural levee is one built by a river in times of flood by deposition of material on the banks. Natural levees are relatively low and wide.

Liquid limit. Water content corresponding to the arbitrary limit between the liquid and plastic states of consistency of a soil; water content at which a pat of soil, cut by a groove of standard dimensions, will flow together for a distance of $\frac{1}{2}$ in. under the impact of 25 blows in a standard liquid limit apparatus.

Longshore current

1. Current that moves parallel to a shore and is formed from the momentum of breaking waves that approach the shore obliquely.

2. Resultant current produced by waves being deflected at an angle by the shore. The current runs roughly parallel to the shore line.

Longshore drift. Movement of sediment along a beach by the swash and backwash of waves that approach the shore obliquely.

Lunar tide. Portion of a tide that is due to the tide-producing force of the moon.

Magmatic water. Water that is dissolved in a magma or that is derived from such water.

Manganese nodule. Small, rounded concretion found on the deep ocean floor that may contain as much as 20% manganese and smaller amounts of iron, copper, and nickel oxides and hydroxides.

Marsh. Area of continuously saturated or spongy ground having poor drainage; hence the term is synonymous with "swamp." An open, treeless, meadowlike or tussocky salt- or freshwater tract of wet or spongy land, usually with occasional open shallow pools of water and a vegetation of more or less dense, erect, aquatic or amphibious plants, including cattails, grasses, sedges, reeds, rushes, or other succulent herbs.

Maximum water-holding capacity. Average moisture content of a disturbed sample of soil 1 cm high, which is at equilibrium with a water table at its lower surface.

Meander

1. Broad, semicircular curves in a stream that develop as the stream erodes the outer bank of a curve and deposits sediment against the inner bank.

2. One of a series of somewhat regular winding or looping bends in a stream.

Meander belt. The zone along a valley floor that encloses a meandering stream.

Meteoric water

1. Rainwater, snow, hail, and sleet.

2. Ground water derived primarily from precipitation.

Mire. Small muddy marsh or bog; wet spongy earth; soft deep mud.

Miscellaneous land types. Mapping units for land areas that have little or no natural soils, that are too nearly inaccessible for orderly examination, or whose soil for some reason it is not feasible to classify. Examples are badland, riverwash, rough broken land, swamp, and wasteland.

Moist soil consistence. Loose, very friable, friable, firm, very firm, and extremely firm.

Moisture content

1. Ratio, expressed as a percentage, of the weight of water in a given soil mass to the weight of solid particles.

2. Water content of a soil expressed as a percentage of the dry weight of the soil. The weight of the water is determined by differential weightings before and after oven-drying a sample.

Moisture equivalent. Weight percentage of water retained by a previously saturated sample of soil 1 cm thick after it has been subjected to a centrifugal force of 1000 times gravity for 30 minutes.

Moisture tension (or pressure). Equivalent negative pressure in the soil water. It is equal to the equivalent pressure that must be applied to the soil water to bring it to hydraulic equilibrium, through a porous permeable wall or membrane, with a pool of water of the same composition.

Morass. Swamp, marsh, or bog having rank vegetation and a muddy or offensive appearance.

Mouth. Place of discharge of a stream into the ocean; entrance to a bay from the ocean.

Mud. Slimy, sticky mixture of water and finely divided particles of a solid, such as dirt, having little or no plasticity.

Mudflow

1. Mass movement of material finer than sand, lubricated with large amounts of water.

2. Well-mixed mass of water and alluvium that because of its high viscosity and low fluidity as compared with water moves at a much slower rate, usually piling up and spreading over the fan like a sheet of wet mortar or concrete.

Narrow. Contracted part of a stream, lake, or sea; strait connecting two bodies of water.

Natural erosion. Wearing away of the earth's surface by water, ice, or other natural agents under natural environmental conditions of climate, vegetation, etc., undisturbed by man.

Neap tide. Tide cycle of unusually small amplitude, which occurs twice monthly when the lunar and solar tides are opposed—that is, when the gravitational pull of the sun is at right angles to that of the moon.

Neck. Narrow strip of land connecting two larger areas; narrow body or channel of water between two larger bodies of water; narrow band of water flowing swiftly seaward through the surf.

Neck cutoff. The breakthrough of a stream across the narrow neck separating two meanders, where downstream migration of one has been slowed and the other meander upstream has overtaken it.

Neutral stress (pore pressure). Stress transmitted through the pore water; water filling the voids of the soil.

Nivation. Erosion behind and peripheral to a snowbank, caused by frost action, mass movement, transport by meltwater, or other related processes. Nivation is most noticeable behind summer snowbanks when nightly freezing alternates with daytime melting.

Normal erosion. Gradual erosion of land used by man that does not greatly exceed natural erosion.

Oasis. Fertile area or spot in the desert, resulting from a spring or stream.

Ocean current. Movement of ocean water characterized by regularity either of a cyclic nature or more commonly as a continuous flowing along a definable path.

Offshore bar. Narrow ridge of sand deposited parallel to the shore in shallow water. Its accumulation is due to the deposition of sand in the breaker zone.

Old age. Stage in the geomorphic cycle characterized by the formation of a peneplain near sea level.

Oolite. Sedimentary carbonate particle composed of spherical grains precipitated from warm ocean water on carbonate platforms; rock composed of such particles.

Ophiolite suite. Assemblage of mafic and ultramafic igneous rocks with deep-sea sediments supposedly associated with divergence zones and the sea-floor environment.

Outlet. Lower end of a lake or pond; point at which a lake or pond discharges into the stream that drains it.

Outwash. Glaciofluvial sediment that is deposited by melt-water streams emanating from a glacier.

Outwash plain. Broad deposit of detrital material sloping gently away from a glacial terminus or terminal moraine. The outwash plain may merge downstream with the valley train.

Overland flow. Flow of rainwater or snowmelt over the land surface toward stream channels. After it enters a stream, it becomes runoff.

Overloaded stream. Stream so heavily loaded with sediment that it deposits material all along its course.

Overtopping. Condition that occurs when the headwater passes over a retaining structure.

Oxbow. River bend shaped like an oxbow with only a neck of land left between two parts of the stream; crescent-shaped lake remaining in the abandoned channel after a river has formed a cutoff and the ends of the original bend have been filled.

Oxbow lake. Long, broad, crescent-shaped lake formed when a stream abandons a meander and takes a new course.

Peat. Marsh or swamp deposit of water-soaked plant remains containing more than 50% carbon.

Pediment. Broad, smooth erosional surface developed at the expense of a highland mass in an arid climate. Underlaid by beveled rock, which is covered by a veneer of gravel and rock debris. It is the final stage of a cycle of erosion in a dry climate.

Pelagic sediment. Deep-sea sediments composed of fine-grained detritus that slowly settles from surface waters. Common constituents are clay, radiolarian ooze, and foraminiferal ooze.

Peneplain

1. Hypothetical extensive area of low elevation and relief reduced to near sea level by a long period of erosion and representing the end product of the ideal geomorphic cycle.

2. An extensive, nearly flat surface developed by subaerial erosion, and close to base level, toward which the streams of the region reduce it.

Perched groundwater. Isolated body of groundwater that is perched above and separated from the main water table by an aquiclude.

Perched water table

1. The top of a zone of saturation that bottoms on an impermeable horizon above the level of the ground water table in the area. It is generally near the surface and frequently supplies a hillside spring.

2. Water table of a discontinuous saturated zone in a soil.

Percolation

1. Movement of gravitational water through soil.

2. Movement, under hydrostatic pressure, of water through the interstices of a rock or soil. Percolation does not include the movement through large openings such as caves.

Permafrost. Permanently frozen material underlying the solum; perennially frozen soil horizon.

Permeability

1. Quality of the soil that enables it to transmit water or air, measured in terms of rate of flow through a unit cross section of saturated soil in unit time.

2. Ability of a formation to transmit groundwater or other fluids through pores and cracks.

Permeability (coefficient of permeability). Rate of discharge of water under laminar flow conditions through a unit cross-sectional area of a porous medium under a unit hydraulic gradient and standard temperature conditions (usually 20°C).

Phreatic zone. Zone of soil and rock in which pores are completely filled with groundwater. It is also called the "saturated zone."

Physical sedimentation. Deposition of classic particles derived by erosion.

Pillow lava. Type of lava flow formed underwater, in which many small pillow-shaped tongues break through the chilled surface and quickly solidify, leading to a rock formation resembling a pile of sandbags.

Piping. Movement of soil particles as the result of unbalanced seepage forces produced by percolating water, leading to the development of boils or erosion channels.

Piracy (stream). Headward growth of one stream until it intersects the channel of another, whereupon all the water is diverted from the latter stream below that point.

Plasticity index

1. Numerical difference between the liquid limit and the plastic limit.

2. Difference in the water content of a soil at its liquid limit and at its plastic limit. Clayey soils, for example, have higher plasticity indexes than nonclay soils because they remain plastic over a wider range of water content.

Plastic limit

1. Water content corresponding to an arbitrary limit between the plastic and the semisolid states of consistency of a soil; water content at which a soil will just begin to crumble when rolled into a thread approximately $\frac{1}{4}$ in. in diameter.

2. One of the measures of soil consistency; lowest moisture content, expressed as a percentage of the weight of the oven-dried soil, at which the soil can be rolled into threads $\frac{1}{4}$ in. in diameter without the thread breaking into pieces.

Playa

1. Flat floor of a closed basin in an arid region. It may be occupied by an intermittent lake.

2. Flat or nearly flat low part of an enclosed basin or temporary lake without outlet. It is also known as a "dry lake." Included are dry, moist, crystal body, compound, lime pan, and artificial playas.

Playa lake. Lake occupying a playa. Such lakes are usually very shallow and temporary in nature.

Point bar. Deposit of sediment on the inner bank of a meander that forms because the stream velocity is lower against the inner bank.

Polder. Land reclaimed from the sea or other body of water by the construction of an embankment to restrain the water.

Pond. Relatively small body of water, usually surrounded on all sides by land. A larger body of water is called a "lake."

Pondage. Small-scale storage at a water power plant to equalize daily or weekly fluctuations in river flow or to permit irregular hourly use of the water for power generation that accords with fluctuations in load.

Pool. Small body of water, usually smaller than a pond, especially one that is quite deep. A pool left by an ebb tide is called a "tide pool." Small and comparatively still, deep part of a larger body of water such as a river or harbor.

Positive cutoff. Provision of a line of tight sheeting or a barrier of impervious material extending downward to an essentially impervious lower boundary to intercept completely the path of subsurface seepage.

Potable water. Water that is agreeable to the taste and not dangerous to the health.

Potential evapotranspiration. Water loss that will occur if at no time there is a deficiency of water in the soil for the use of vegetation.

Potential natural water loss. Water loss during years when the annual precipitation greatly exceeds the average water loss. It represents the approximate upper limit to water loss under the type and density of vegetation native to a basin, actual conditions of moisture supply, and other basin characteristics, whereas potential evapotranspiration represents the hypothetical condition of no deficiency of water in the soil at any time for use of the type and density of vegetation that would develop.

Pothole. Semispherical hole in the bedrock of a stream bed, formed by abrasion of small pebbles and cobbles in a strong current.

Quagmire. Saturated area with a surface of soft mud, or, at best, a surface providing a shaky and precarious footing.

Quaking bog. Peat deposit so wet and unconsolidated that the surface oscillates with the impact of a person walking on it; late vegetational stage in the filling of a lake or pond by encroaching floating mats of plants and plant debris.

Quick condition (quicksand). Condition in which water flows upward with sufficient velocity to reduce significantly the bearing capacity of the soil through a decrease in intergranular pressure.

Radial drainage. System of streams running in a radial pattern away from the center of a circular elevation such as a volcano or dome.

Radiolarian. Class of one-celled marine animals with siliceous skeletons that existed in the ocean throughout the Phanerozoic eon.

Rainfall excess. Volume of rainfall available for direct runoff. It is equal to the total rainfall minus interception, depression storage, and absorption.

Rainfall frequency. Number of times during a specified period of years that precipitation of a certain magnitude or greater occurs or will occur at a specific location.

Rain forest. Forest that grows in a region of heavy annual precipitation.

Rapid. Portion of a stream in swift, disturbed motion, but without cascade or waterfall (usually plural).

Ravine. Depression worn out by running water, larger than a gully and smaller than a valley; small gorge or canyon, the sides of which have comparatively uniform steep slopes.

Reach. Arm of the sea extending into the land; straight section of restricted waterway of considerable extent. A reach may be similar to a narrows, except much longer in extent.

Recharge. In hydrology, the replenishment of groundwater by infiltration of meteoric water through the soil.

Rectangular drainage. System of streams in which each straight segment of each stream takes one of two characteristic perpendicular directions, with right-angle bends between. Streams usually follow two perpendicular sets of joints.

Reef. Rocky or coral elevation at or near enough to the surface of the sea to be a danger to surface vessels. A barrier reef roughly parallels land but is some distance offshore with deeper water intervening. A fringing reef is closely attached to a shore.

Refraction of water waves. Process by which the direction of a wave moving in shallow water at an angle to the contours is changed. That part of the wave advancing in shallower water moves more slowly than the other part still advancing in deeper water, causing the wave crest to bend toward alignment with the underwater contours. Bending of wave crests by currents.

Regimen of a stream. System or order characteristic of a stream, that is, its habits with respect to velocity and volume, form of and changes in channel, capacity to transport sediment, and amount of material supplied for transportation.

Regression. Drop in sea level that causes an area of the earth to be uncovered by seawater, ending marine deposition.

Rejuvenated stream. Stream that has become active in erosion because of an uplift in the land.

Rejuvenation. Change in conditions of erosion that causes a stream to begin more active erosion and a new cycle.

Reversing current. Tital current that flows alternately in approximately opposite directions, with periods of slack water at each reversal.

Revetment. Facing of erosion-resistive materials to sustain or protect an embankment.

Ria. Long narrow arm of the ocean penetrating the coastline more or less at a right angle and narrowing gradually inland.

Ria shore line. Shore line characterized by numerous rias. Such shore lines may form by the submergence of a land dissected by normal river valleys.

Ridge (midocean). Major linear elevated land form of the ocean floor, from 200 to 20,000 km in extent. It is not a single ridge but resembles a mountain range and may have a central rift valley.

Riparian. Pertaining to the banks of a stream, or any other body of water.

Rip current. Current that flows strongly away from the seashore through gaps in the surf zone at intervals along the shore line.

Ripple. Very small dune of sand or silt whose long dimension is formed at right angles to the current.

Rip-rap. Protective work on the sloping bank of a river or canal, generally consisting of stones, either arranged or just dumped.

River basin. Total area drained by a river and its tributaries.

River wash. Barren alluvial land, usually coarse textured, exposed along streams at low water and subject to shifting during normal high water.

Runoff

1. Amount of rainwater directly leaving an area in surface drainage, as opposed to the amount that seeps out as groundwater.

2. Surface flow of water from an area; total volume of surface flow during a specified time.

3. Portion of the precipitation on an area is discharged from the area through stream channels. That which is lost without entering the soil is called "surface runoff" and that which enters the soil before reaching the stream is called "groundwater runoff" or "seepage flow from groundwater."

Runoff cycle. Part of the hydrologic cycle undergone by water between the time it reaches the land as precipitation and its subsequent evapotranspiration or discharge through stream channels.

Saline water. Water with a salt content of 17,000 parts per million or more.

Salinity. Measure of the quantity of dissolved salts in seawater. Salinity is normally defined as the total amount of dissolved solids in seawater in parts per thousand by weight when all the carbonate has been converted to oxide, all the bromide and iodide has been converted to chloride, and all organic matter is completely oxidized. Salinity is not determined directly but is computed from chlorinity, electrical conductivity, refractive index, or some other property whose relationship to salinity is well established.

Saltation

1. Movement of soil and mineral particles by intermittent leaps from the ground when the particles are being moved by wind or water.

2. Movement of sand or fine sediment by short jumps above the stream bed under the influence of a current too weak to keep it permanently suspended.

Salt marsh. Flat, poorly drained coastal swamps that are flooded by most high tides.

Salt pan. Pool used for obtaining salt by the natural evaporation of seawater, undrained natural depression such as an extinct crater or tectonic basin, in which water gathers and leaves a deposit of salt in evaporation.

Sand bar. Ridge of sand built up to the surface or near the surface of a river or along a beach.

Saturate. Fill all the voids between soil particles with a liquid; to form the most concentrated solution possible under a given set of physical conditions in the presence of an excess of the solute.

Saturated unit weight. Wet unit weight of a soil mass when saturated.

Sea-floor spreading. Mechanism by which new sea floor crust is created at ridges in divergence zones and adjacent

plates are moved apart to make room. The process may continue at 0.5 to 10 cm/year through many geologic periods.

Seamount. Isolated tall mountain on the sea floor that may extend more than 1 km from base to peak.

Sediment. Material carried in suspension by water; deposits of water-borne materials.

Seepage. Water that has seeped through porous material; amounts of such material expressed in terms of volume.

Seepage (percolation). Slow movement of ground water through the soil.

Semidiurnal tide. Type of tide having two high waters and two low waters each tidal day, with small inequalities between successive high- and successive low-water heights and durations. It is the most common type of tide throughout the world.

Settling velocity. Rate at which a sedimentary particle of a given size falls through water or air.

Sheet erosion. Removal of a fairly uniform layer of soil from the land surface by runoff water.

Sheetwash. Flow of rainwater that covers the entire ground surface with a thin film and is not concentrated into streams.

Shoal. Submerged ridge, bank, or bar consisting of or covered by unconsolidated sediments (mud, sand, gravel), at or near enough to the water surface. If composed of rock or coral, it is called a "reef."

Shooting flow. Very fast form of water flow with a high velocity induced by steep slopes in the flow bed, typically developed in rapids.

Shore. Land bordering a body of water.

Shoreline. Boundary line between a body of water and the land. The shoreline shown in charts generally approximates the mean high-water line. Instantaneous line marking the junction of water and land.

Shoreline of emergence. Shoreline formed by the emergence of a lake or sea floor.

Shrinkage limit. Maximum water content at which a reduction in water content will not cause a decrease in the volume of the soil mass. This defines the arbitrary limit between the solid and semisolid states.

Sill (topographic). Small submarine ridge that nearly separates or restricts water flow between two adjacent bodies of water.

Silting. Deposition of waterborne sediments in stream channels, lakes, and reservoirs or on flood plains usually resulting from a decrease in the velocity of the water.

Silt size. Portion of the soil finer than 0.02 and coarser than 0.002 mm (0.05 and 0.005 mm in some cases).

Sinking current. Downward movement of lake or sea water that has become denser through cooling or increased salinity or that sinks because an onshore wind piles up water against the shore.

Sinuosity. Length of the channel of a stream divided by the straight-line distance between its ends.

Skimming. Diversion of water from a stream or conduit by a shallow overflow, used to avoid diversion of sand, silt, or other debris carried as bottom load.

Slack water. Short period of time in which there is no movement of the water between changing tides.

Slope wash. Motion of water and sediment down a slope by the mechanism of sheet wash.

Slush. Snow or ice on the ground that has been reduced to a soft watery mixture by rain, warm temperature, or chemical treatment.

Snowfield. Area of permanent snow that is not moving or whose movement is not visible.

Spray. Ensemble of water droplets torn by the wind from the surface of an extensive body of water, generally from the crests of waves, and carried up a short distance into the air.

Soft water. Water that is free of calcium and magnesium carbonates and other dissolved materials of hard water.

Soil creep

1. Downward mass movement of sloping soil. Movement is usually slow and irregular and occurs most commonly when the lower soil is nearly saturated with water.

2. Sliding or rolling of particles over the soil surface, caused by wind.

3. Imperceptible downhill flow of soil under the force of gravity; shear flow with velocity decreasing downward, occurring even on gentle slopes.

Soil drainage. Rapidity and extent to the removal of water from a soil in relation to additions, especially by surface runoff and flow through the soil.

Soil moisture. Moisture or water contained in a belt identified as the upper subdivision of the zone of aeration. This zone contains plant roots and water available for plant growth.

Soil moisture tension. Equivalent negative pressure of suction of water in soil.

Soil porosity. Degree to which the soil mass is permeated with pores or cavities. Porosity can be generally expressed as a percentage of the whole volume of a soil horizon that is unoccupied by solid particles.

Soil suction. Measure of the force of water retention in unsaturated soil. Suction is equal to a force per unit area that must be exceeded by an externally applied suction to initiate water flow from the soil. Suction is expressed in standard pressure terms.

Soil water. Term emphasizing the physical rather than the chemical properties and behavior of the soil solution.

Soil water percolation. Downward movement of water through soil; especially, the downward flow of water in saturated or nearly saturated soil at hydraulic gradients of the order of 1.0 or less.

Soil water potential. Potential energy of a unit quantity of water produced by the interaction of the water with such forces as capillary (matric), ion hydration (osmotic), and gravity, expressed relative to an arbitrarily selected reference potential. In practical application, potentials are used to predict the direction and rate of water flow through soils or between the soil and some other system, such as plants or the outer atmosphere. Flow occurs spontaneously between points of differing water potential, the direction of flow being toward the site of lower potential.

Soil water tension. Expression, in positive terms, of the negative hydraulic pressure of soil water.

Solifluction

1. Slow downslope flow or creep of saturated soil, often initiated and augmented by frost action in high latitudes and at high elevations.

2. Soil creep of material saturated with water and/or ice, most common in polar regions.

Spillover. Part of orographic precipitation that is carried along by the wind so that it reaches the ground in the nominal rain shadow on the lee side of the barrier.

Spit

1. Long ridge of sand deposited by longshore current and drift where the coast takes an abrupt inward turn. It is attached to land at the upstream end.

2. Small point of land or narrow shoal projecting into a body of water from the shore.

Splash erosion. Spattering of small soil particles caused by the impact of raindrops on very wet soils. Loosened and separated particles may or may not be subsequently removed by surface runoff.

Splay deposit. Small delta deposited on a flood plain when the stream breaches a levee during a flood.

Spring. Place where water issues naturally from the rock or soil upon the land or into a body of surface water.

Spring tide. Tide cycle of unusually large amplitude that occurs twice monthly when the lunar and solar tides are in phase.

Stack. Lofty, isolated mass of rock left standing in the sea by the quarrying and erosive action of the waves.

Stage

1. Height of a water surface above an established datum plane.

2. Elevation of the water level of a stream measured against some constant reference.

Storage. Water artificially impounded in surface or underground reservoirs for future use; water naturally detained in a drainage basin, such as groundwater channel storage and depression storage.

Strait. Relatively narrow body of water connecting two larger bodies.

Stream. Course of water flowing between approximately parallel banks, such as a river.

Stream channel. Bed where a natural stream of water runs; the trench or depression washed in the surface of the earth by running water, a wash, arroyo, or coulee.

Streamflow. Discharge that occurs in a natural channel. Although the term "discharge" can be applied to the flow of a canal, the word "streamflow" uniquely describes the discharge in a surface stream course. Streamflow may be applied to discharge whether or not it is affected by diversion or regulation.

Stream order

1. Method of numbering streams as part of a drainage basin network. The smallest unbranched mapped tributary is called "first order," the stream receiving the tributary is called "second order," and so on.

2. Hierarchical number of a stream segment in dendritic drainage; the smallest tributary streams have order one, and at each junction of streams of equal order the order of the subsequent segment is one higher.

Subglacial stream. Stream that flows in a naturally created tunnel beneath a glacier.

Subirrigation. Water is applied in open ditches or tile lines until the water table is raised sufficiently to wet the soil.

Submarine canyon. Underwater canyon in the continental shelf.

Submerged unit weight. Weight of the solids in air minus the weight of water displaced by the solids per unit of volume of soil mass; the saturated unit weight minus the unit weight of water.

Subsurface runoff. Part of precipitation that infiltrates the surface soil and moves toward the streams as ephemeral, shallow, perched groundwater above the main groundwater level.

Subsurface water. Water that exists below the surface of the lithosphere.

Subterranean stream. Body of flowing water that passes through a very large interstice, such as a cave or cavern, or a group of large communicating interstices.

Superposed stream. Stream that flows through resistant formations because its course was established at a higher level on uniform rocks before downcutting began.

Surf. Breaking or tumbling forward of water waves as they approach the shore.

Surface runoff. Part of the runoff that travels over the soil surface to the nearest stream channel; also, part of the run off of a drainage basin that has not passed beneath the surface since precipitation.

Surface storage. Part of precipitation retained temporarily at the ground surface as interception or depression storage so that it does not appear as infiltration or surface runoff either during the rainfall or shortly thereafter.

Surface water. Water that occurs on the surface of the earth as opposed to water that occurs within the voids of a soil mass.

Surf zone. Offshore belt along which the waves collapse into breakers as they approach the shore.

Suspended load. Fine sediment kept suspended in a stream because the settling velocity is lower than the upward velocity of eddies.

Swale. Slight depression in generally level land, usually covered with a rank growth of grass and often marshy.

Swamp. Area of continuously saturated or spongy ground having poor drainage and therefore the same as a marsh; area of continuously saturated ground support]ing large aquatic plants having submerged or floating leafy shoots, often dominated by shrubs and trees.

Swash. Landward rush of water from a breaking wave up the slope of the beach.

Swell

1. Ocean waves that have traveled out of their generating area. A swell characteristically exhibits a more regular and longer period and has flatter crests than waves within their fetch. Long, broad elevation that rises gently and generally smoothly from the sea floor.

2. Oceanic water wave with a wavelength on the order of 30 m or more and a height of perhaps 2 m or less that may travel great distances from its source.

Tarn. Small mountain lake or pool.

Terrestrial sediment. Deposit of sediment that accumulated above sea level in lakes, alluvial fans, flood plains, moraines, etc., regardless of its present elevation.

Terrigenous sediments. The ocean sediments derived from land areas.

Thalweg. Sinuous imaginary line following the deepest part of a stream.

Thermal spring. Spring that brings warm or hot water to the surface.

Thermocline. Transitional layer between warm surface waters and cold bottom waters in lakes and seas.

Tidal current

1. Horizontal displacement of ocean water under the gravitational influence of the sun and moon, causing the water to pile up against the coast at high tide and move outward at low tide.

2. Alternating horizontal movement of water associated with the rise and fall of the tide, caused by the astronomical tide-producing forces. In relatively open locations the direction of tidal currents rotates continuously through 360° diurnally or semidiurnally. In coastal regions the nature of tidal currents is determined by local topography as well.

Tidal flat

1. Broad, flat region of muddy or sandy sediment covered and uncovered in each tidal cycle.

2. Marsh or sandy or muddy coastal flatland that is covered and uncovered by the rise and fall of the tide.

Tidal inlet. Waterway from open water into a lagoon.

Tidal marsh. Marsh or flatland whose surface is wetted by a tidal flow.

Tide. Periodic rising and falling of the earth's oceans and atmosphere resulting from the tide-producing forces of the moon and sun acting upon the rotating earth. A disturbance actually propagates as a wave through the atmosphere and through the surface layer of the oceans.

Tombolo. Sandbar built by the sea, tying an island to the mainland.

Training walls. Hydraulic structures built to cause flow to perform in a specified manner.

Transported soil. Soil transported from the place of its origin by wind, water, or ice and redeposited.

Travertine. Ground water deposit composed of calcite.

Trellis drainage. System of streams in which tributaries tend to lie in parallel valleys formed in steeply dipping beds in folded belts.

Trench. Long, narrow, deep trough in the sea floor. It is interpreted as marking the line along which a plate bends down into a subduction zone.

Tropical rain forest. Type of forest that exists in tropical regions where precipitation is heavy (generally more than 100 in. year). It consists mainly of a wide variety of lofty broadleaf evergreen trees which carry a profusion of parasitic or climbing plants.

Tsunami. Gigantic waves that result from earthquakes and landslides on the sea floor.

Turbidity current. Mass of mixed water and sediment that flows downhill along the bottom of an ocean or lake because it is denser than the surrounding water. It may reach high speeds and erode rapidly.

Turbulent flow. High-velocity flow in which stream lines are neither parallel nor straight, but curled into small tight eddies.

Undercurrent. Water current flowing beneath a surface current at a different speed or in a different direction.

Underflow. Downstream flow of water through the permeable deposits that underlie a stream and that are more or less limited by rocks of low permeability.

Undertow. Seaward flow near the bottom of a sloping beach; subsurface return by gravity flow of the water carried up on shore by waves or breakers.

Unit hydrograph. Hydrograph of a direct runoff from a storm uniformly distributed over the drainage basin during a specified unit of time. The hydrograph is reduced in vertical scale to correspond to a volume of runoff of 1 in. from the drainage basin.

Unsaturated flow. Movement of water in a soil that is not filled to capacity with water.

Upland

1. Land lying above the flood plain.

2. Highland; ground elevated above the lowlands along rivers or between hills.

Uplift. Broad and gentle epeirogenic increase in the elevation of a region without a eustatic change of sea level.

Upwelling. Process by which water rises from a lower to a higher depth, usually as a result of divergence and offshore currents. Upwelling is most prominent where a persistent wind blows parallel to a coastline so that the resultant wind-driven current sets away from the coast.

Upwelling current. Upward movement of cold bottom water in the sea which occurs when wind or currents displace the lighter surface water.

Vadose zone. Region in the ground between the surface and the water table in which pores are not filled with water. It is also called the "unsaturated zone."

Valley glacier. Glacier that flows down a valley.

Valley train. Outwash deposit extending downstream from a glacier's terminus or terminal moraine, usually confined by valley sides. A valley train may or may not merge with the outwash plain.

Velocity of a stream. Rate of motion of a stream measured in terms of distance its water travels in a unit time, usually in feet per second.

Virga. Wisps or streaks of water or ice particles falling out of a cloud but evaporating before reaching the earth's surface as precipitation.

V-shaped valley. Valley whose walls have a more or less uniform slope from top to bottom, usually formed by stream erosion.

Wadi. Steep-sided valley containing an intermittent stream in an arid region.

Wash. Dry wash alluvium collected, carried, and deposited by the action of water.

Watercourse. Stream of water; natural channel through which water may or does run.

Water equivalent of snow. Amount of water that would be obtained if the snow were to melt completely.

Water gap. Pass in a mountain ridge through which a stream flows.

Waterlogged. Saturated with water.

Water loss. Difference between the average precipitation over a drainage basin and the water yield from the basin for a given period.

Water mass. Mass of water that fills part of an ocean or lake and is distinguished by its uniform physical and chemical properties, such as temperature and salinity.

Watershed. The entire area that is drained by a stream and its tributaries.

Water-stable aggregate. Soil aggregate stable to the action of water, to falling drops, for example, or agitation as in wet-sieving analysis.

Water table

1. Level in saturated soil where the hydraulic pressure is zero.
2. Upper limit of the part of the soil or underlying rock material that is wholly saturated with water. In some places an upper or perched water table may be separated from a lower one by a dry zone.
3. Upper surface of a zone of saturation in the soil or parent material.
4. Surface of underground gravity-controlled water.
5. Gently curved surface below the ground at which the vadose zone ends and the phreatic zone begins; level to which a well would fill with water.

Water year. With reference to surface water supply, the 12-month period, October 1 through September 30. A water year is designated by the calendar year in which it ends and which includes 9 of the 12 months.

Water yield. Runoff from the drainage basin, including groundwater outflow that appears in the stream plus groundwater outflow that bypasses the gauging station and leaves the basin underground. Water yield is the precipitation minus the evapotranspiration.

Wave-built terrace. Coarser upper portion of a beach where the waves have thrown the pebbles up in low ridges parallel to the shore line and a few feet above mean high-water level.

Wave-cut terrace

1. Level surface formed by wave erosion of coastal bedrock to the bottom of the turbulent breaker zone. It may appear above sea level if uplifted.
2. Flat or gently sloping surface, usually covered by shallow water, which waves have cut by removal of bedrock or unconsolidated material.

Wet soil consistency. Nonsticky, slightly sticky, sticky, very sticky, nonplastic, slightly plastic, plastic, and very plastic.

Wet snow. In the International Snow Classification, snow that is saturated or almost saturated with water. If free water entirely fills the airspaces in the snow, it is classified as "very wet snow."

Wet unit weight. Weight (solids plus water) per unit of total volume of soil mass, irrespective of the degree of saturation.

Wild-flooding. Water is released at high points in the field, and distribution is uncontrolled.

Wind erosion. Process of wearing away a surface by wind action and the abrasion of windborne materials.

Yazoo stream. Tributary stream that flows parallel to the main stream for some distance through its flood plain.

Youth (geomorphology). Stage in the geomorphic cycle in which a landscape has just been uplifted and is beginning to be dissected by canyons cut by young streams.

Zoned earth dam. Earth dam embankment zoned by the systematic distribution of soil types according to their strength and permeability characteristics, usually with a central impervious core and shells of coarser materials

Zone of aeration. Region of soil or rock above a water table where the pore spaces contain air as well as water.

Zone of maximum precipitation. In a mountain region, the belt of elevation at which the annual precipitation is greatest.

Zone of saturation

1. Soil or rock beneath the water table. Pore spaces in the zone of saturation are filled with water, in contrast to the pore spaces above the water table which may contain considerable air.
2. The area beneath the earth's surface that is saturated with water.

Reference Data Sources

The following professional, technical, reference standards and scientific organizations, contractors, fabricators, and manufacturers and their respective associations, bureaus, committees, councils, federations, foundations, and societies, including code writing organizations and United States Government departments and agencies are listed as available reference data sources. The addresses indicated have been obtained from 1991 publications.

Acoustical Society of America
500 Sunnyside Blvd., Woodbury, NY 11797

Adhesive and Sealant Council, Inc.
1627 K St. NW, Washington, DC 20006

Adhesives Manufacturers Association of America
111 E. Wacker Dr., Chicago, IL 60601

Air and Waste Management Assoc.
PO Box 2861, Pittsburgh, PA 15230

Air-Conditioning and Refrigeration Institute
1501 Wilson Blvd., Arlington, VA 22209

Allied Stone Industries
Box 288145, Chicago, IL 60628

Aluminum Association, Inc.
818 Connecticut Ave. NW, Washington, DC 20006

Aluminum Extruders Council
1000 N. Rand Rd., Wauconda, IL 60084

American Arbitration Association
140 W. 51st St., New York, NY 10020

American Architectural Association
2700 River Rd., Des Plaines, IL 60018

American Ceramic Society, Inc.
757 Brooksedge Plaza Dr.
Westerville, Ohio 43081

American Chemical Manufacturers Association
2700 River Rd., Des Plaines, IL 60018

American Chemical Society
1155 16th St. NW, Washington, DC 20036

American Concrete Institute
PO Box 19150, Redford Sta., Detroit, MI 48219

American Concrete Paving Assoc.
3800 N. Wilke Rd., Arlington Heights, IL 60004

American Concrete Pipe Association
8300 Boone Blvd., Vienna, VA 22182

American Consulting Engineers Council
1015 15th St. NW, Washington, DC 20005

American Forest Council
1250 Connecticut Ave. NW, Washington, DC 20036

American Forestry Association
1516 P St., NW, Washington, DC 20005

American Galvanizers Association
PO Box 80, Clarendon Hills, IL 60514

American Gas Association
1515 Wilson Blvd., Arlington, VA 22209

American Hardboard Association
520 N. Hicks Rd., Palatine, IL 60067

American Hardware Manufacturers Association
931 N. Plum Grove Rd., Schaumburg, IL 60173

American Hotel and Motel Assoc.
1201 New York Ave. NW, Washington, DC 20005

American Institute of Architects
1735 New York Ave. NW, Washington, DC 20006

American Institute of Chemical Engineers
345 E. 47th St., New York, NY 10017

American Institute of Steel Construction, Inc.
One E. Wacker Dr., Chicago, IL 60601

American Institute of Timber Construction
11818 SE Mill Plane Blvd., Vancouver, WA 98684

American Insurance Association
1130 Connecticut Ave. NW, Washington, DC 20036

American Iron and Steel Institute
1133 15th St. NW, Washington, DC 20005

American Lumber Standards Committee
PO Box 210, Germantown, MD 20875

American Mobilehome Association, Inc.
12929 West 26th Ave., Golden, CO 80401

American National Standards Institute
1430 Broadway, New York, NY 10018

American Nuclear Society, Inc.
555 N. Kensington Ave., La Grange Park, IL 60525

American Petroleum Institute
1220 L St. NW, Washington, DC 20005

American Pipe Fitting Association
6203 Old Keene Mill Ct., Springfield, VA 22151

American Planning Association
1776 Massachusetts Ave. NW, Washington, DC 20036

American Plywood Association
PO Box 11700, Tacoma, WA 98411

American Powder Metallurgy Institute
105 College Rd. E. Princeton, NJ 08540

American Public Gas Association
PO Box 1426, Vienna, VA 22183

American Public Works Association
1313 E. 60th St., Chicago, IL 60637

American Railway Car Institute
Bldg. 5 19900 Grovernors Dr.,
Olympia Fields, IL 60461

American Society for Concrete Construction
3330 Dundee Rd., Northbrook, IL 60062

American Society for Metals, Intl.
9639 Kinsman, Materials Park, OH 44073

American Society for Testing and Materials
1916 Race St., Philadelphia, PA 19103

American Society of Civil Engineers
345 E. 47th St., New York, NY 10017

American Society of Consulting Engineers
1750 Old Meadow Rd., McLean, VA 22101

American Society of Heating, Refrigeration and Air
  Conditioning Engineers, Inc.
1791 Tullie Circle NE, Atlanta, GA 30329

American Society of Interior Designers
608 Massachusetts Ave. NE, Washington, DC 20002

American Society of Irrigation Consultants
425 Oak St., Brentwood, CA 94513

American Society of Landscape Architects
4401 Connecticut Ave. NW, Washington, DC 20008

American Society of Mechanical Engineers
345 E. 47th St., New York, NY 10017

American Society of Plumbing Engineers
3617 Thousand Oaks Blvd., Westlake, CA 91362

American Society of Professional Appraisers
100 Galeria Parkway, Atlanta, GA 30339

American Society of Professional Estimators
11141 Georgia Ave., Wheaton, MD 20902

American Society of Safety Engineers
1800 E. Oakton St., Des Plaines, IL 60016

American Society of Sanitary Engineering
PO Box 40362, Bay Village, OH 44140

American Subcontractors Association
1004 Duke St., Alexandria, VA 22314

American Welding Society
550 LeJeune Rd. NW, Miami, FL 33125

American Wood Preservers Assoc.
PO Box 849, Stevensville, MD 21666

American Wood Preservers Bureau
PO Box 5283, Springfield, VA 22150

American Wood Preservers Institute
1945 Old Gallows Rd., Vienna, VA 22182

American Woods Council
1250 Connecticut Ave. NW, Washington, DC 20036

Appalachian Hardwood Manufacturers Association, Inc.
PO Box 427, High Point, NC 27261

Architectural Precast Association
825 E 64th St., Indianapolis, IN 46220

Architectural Woodwork Institute
2310 S. Walter Reed Dr., Arlington, VA 22206

Asphalt Roofing Manufacturers Association
6288 Montrose Rd., Rockville, MD 20852

Associated General Contractors of America
1957 E. St. NW, Washington, DC 20006

Associated Landscape Contractors of America
405 N. Washington St., Falls Church, VA 22046

Associated Specialty Contractors, Inc.
7315 Wisconsin Ave., Bethesda, MD 20814

Association of American Railroads
American Railroads Bldg.
50 F St. NW, Washington, DC 20001

Association of Iron and Steel Engineers
3 Gateway Center, Pittsburgh, PA 15222

Brick Institute of America
11490 Commerce Park Dr., Reston, VA 22091

Builders Hardware Manufacturers Association
355 Lexington Ave., New York, NY 10017

Building Officials and Code Administrators,
  International, Inc.
405 Flossmor Rd., Country Club Hill, IL 60477-5795

Building Research Board
2101 Constitution Ave. NW, Washington, DC 20418

Building Stone Institute
420 Lexington Ave., New York, NY 10017

California Redwood Association
405 Enfrente Dr., Novato, CA 94949

Carpet and Rug Institute
PO Box 2048, Dalton, GA 30722

Cast Iron Soil Pipe Institute
5959 Shallow Rd., Chattanooga, TN 37421

Cedar Shake and Shingle Bureau
515 116th Ave. NE, Bellview, WA 98004

Ceilings and Interior Systems Contractors Association
104 Wilmot Rd., Deerfield, IL 60015-5195

Ceramic Tile Institute of America
700 N. Virgil Ave., Los Angeles, CA 90029

Community Planning and Development Department
451 7th St. SW, Washington, DC 20410

Compressed Gas Association, Inc.
1235 Jefferson Davis Highway, Arlington, VA 22202

Concrete Pipe Association, Inc.
8300 Boon Blvd., Vienna, VA 22182

Concrete Reinforcing Steel Institute
933 N. Plum Grove Rd., Schaumburg, IL 60173

Construction Specifications Institute
601 Madison St., Alexandria, VA 22314

Copper Development Association, Inc.
Two Greenwich Office Park, Box 1840
Greenwich, CT 06836

Cultured Marble Institute
435 N. Michigan Ave., Chicago, IL 60611

Deep Foundation Institute
PO Box 281, Sparta, NJ 07871

Door and Hardware Institute
7711 Old Springhouse Rd., McLean, VA 22102

Door Operator and Remote Controls
  Manufacturers Association
655 West Irving Park Dr., Chicago, IL 60613

Ductile Iron Society
28938 Lorrain Rd., North Olmstead, OH 44070

Electronic Industries Association
2001 Pennsylvania Ave. NW, Washington, DC 20006

Engineering Foundation
345 E. 47th St., New York, NY 10017

Expanded Shale Clay and Slate Institute
2225 E. Murray Holladay Rd., Salt Lake City, UT 84117

Fabricators and Manufacturers Association
5411 E. State St., Rockford, IL 61108

Facing Tile Institute
Box 8880, Canton, OH 44711

Factory Insurance Association
85 Woodland St., Hartford, CN 06102

Factory Mutual System
1151 Boston-Providence Turnpike, Norwood, MA 02062

Federal Commission Communications
199 M St. NW, Washington, DC 20554

Fine Hardwoods Veneer Association
5603 W. Raymond St., Indianapolis, IN 46241

Fire Equipment Manufacturers Association
1230 Keith Blvd., Cleveland, OH 44115

Flat Glass Marketing Association
3310 Harrison St., Topeka, KS 66611

Flexicore Manufacturers Assoc.
PO Box 825, Dayton, OH 45401

Forest Products Research Society
2801 Marshall Ct., Madison, WI 53705

Forging Industries Association
25 Prospect Ave. W, Cleveland, OH 44115

General Services Administration
18th and F St. NW, Washington, DC 20405

Glass Tempering Association
3310 Harrison St., Topeka, KS 66611

Gypsum Association
810 1st St. NE, Washington, DC 20002

Hardwood Manufacturers Association
2831 Airways Blvd., Memphis, TN 38132

Hardwood Plywood Manufacturers Association
1825 Michael Faraday Dr., PO Box 2789
Reston, VA 22090

Hardwood Research Council
PO Box 34518, Memphis, TN 38184

Heat Exchange Institute
1230 Keith Bldg. 1621 Euclid Ave.,
Cleveland, OH 44115

Hoist Manufacturers Institute
8720 Red Oak Blvd., Charlotte, NC 28217

Home Ventilating Institute
30 West University Dr., Arlington Heights, IL 60004

Hydraulic Tool Manufacturers Association
Box 1337, Milwaukee, WI 53201

IEEE Professional Communications Society
345 E. 47th St., New York, NY 10017

Illuminating Engineering Society
345 E. 47 St., New York, NY 10017

Independent Computer Consultants Association
933 Gardenview Office Parkway, St. Louis, MO 63144

Indiana Limestone Institute of America
Stone City National Bank Bldg., Bedford, IN 47421

Industrial Designers Society of America
1142 E. Walker Rd., Great Falls, VA 22066

Industrial Heating Equipment Association
1901 N. Moore St., Arlington, VA 22209

Institute of Electrical and Electronics Engineers
345 E. 47th St., New York, NY 10017

Institute of Gas Technology
3424 S. State St., Chicago, IL 60616

Insurance Information Institute
100 N. Interregional, Austin, TX 78701

Interior Design Society
PO Box 2396, High Point, NC 27261

International Conference of Building Officials
5360 S. Workman Mill Rd., Whittier, CA 90601

International Masonry Institute
823 15th St. NW, Washington, DC 20005

Interstate Commerce Commission
12th St. and Constitution Ave. NW
Washington, DC 20423

Jute Carpet Backing Council, Inc.
30 Rockefeller Plaza, 24th Floor, New York, NY 10112

Lead Industries Association, Inc.
295 Madison Ave., New York, NY 10017

Lighting Research Institute
345 E. 47th St., New York, NY 10017

Lightning Protection Institute
PO Box 1039, Woodstock, IL 60098

Manufactured Housing Institute
1745 Jefferson Davis Highway, Arlington, VA 22202

Maple Flooring Manufacturers Association, Inc.
60 Revere Dr., Northbrook, IL 60062

Marble Institute of America
33505 State St., Farmington, MI 48335

Masons Contractors Association of America
17W601 14th St., Oak Brook Terrace, IL 60181

Mechanical Contractors Association of America
1385 Piccard Dr., Rockville, MD 20832

Metal Building Manufacturers Association
1230 Keith Bldg., Cleveland, OH 44115

Metal Fabricating Institute
PO Box 1178, Rockford, IL 61105

Metal Framing Association
404 N. Michigan Ave., Chicago, IL 60611

Metal Lath Steel Framing Association
600 S. Federal St., Chicago, IL 60605

Metal Powder Industries Federation
105 College Rd., Princeton, NJ 08540

Metal Treating Institute, Inc.
302 Third St., Neptune Beach, FL 32233

Modular Building Systems Council
15th and M St. NW, Washington, DC 20005

Mortgage Bankers Association of America
1125 15th St. NW, Washington, DC 20005

Mortgage Insurance Companies of America
805 15th St. NW, Washington, DC 20005

National Asphalt Pavement Association
5100 Forbes Blvd., Lanhan, MD 20706

National Association of Architectural
Metal Manufacturers
600 S. Federal St., Chicago, IL 60605

National Association of Floor Covering Distributors
85 West Algonquin Rd., Arlington, IL 60005

National Association of Garage Door
Manufacturers
1230 Keith Bldg., Cleveland, OH 44115

National Association of Metal Finishers
One Illinois Center, 111 E. Wacker Dr.
Chicago, IL 60601

National Association of Minority Contractors
1333 F St. NW, Washington, DC 20004

National Association of Mirror Manufacturers
9005 Congressional Ct., Potomac, MD 20854

National Association of Mutual Insurance Companies
3707 Woodview Terrace, Box 68700
Indianapolis, IN 46268

National Association of Plastic Distributors
6333 Long St., Shawnee Mission, KS 66216

National Association of Plumbing, Heating,
and Cooling Contractors
PO Box 6808, 180 S. Washington St.
Falls Church, VA 22040

National Building Granite Quarries Association
PO Box 482, Barre, VT 05641

National Bureau of Standards
US Dept. of Commerce, Building Research Division
Washington, DC 20234

National Certified Pipe Welding Bureau
1385 Piccard Dr., Rockville, MD 20850

National Clay Pipe Institute
PO Box 759, Lake Geneva, WI 53147

National Concrete Masonry Association
PO Box 781, Hemden, VA 22070

National Conference of States on Building Codes
& Standards
505 Huntmar Park Dr., Herndon, VA 22070

National Constructors Association
1730 M St. NW, Washington, DC 20036

National Electrical Manufacturers Association
2101 L St. NW, Washington, DC 20037

National Elevator Industry, Inc.
185 Bridge Plaza N., Fort Lee, NJ 07024

National Environmental Development Association
1440 New York Ave. NW, Washington, DC 20005

National Environmental Health Association
720 S. Colorado Blvd., Denver, CO 80222

National Fire Protection Association
One Batterymarch Park, PO Box 9101, Quincy, MA 02269

National Fire Sprinkler Association
Box 1000, Patterson, NJ 12563

National Forest Products Association
1250 Connecticut Ave. NW, Washington, DC 20036

National Glass Association
8200 Greensboro Dr., McLean, VA 22102

National Greenhouse Manufacturers Association
PO Box 567, Pana, IL 62557

National Hardwood Lumber Association
PO Box 34518, Memphis, TN 38184

National Institute of Standards and Technology
Route I-270 and Quince Orchard Rd.
Gaithersburg, MD 20899

National Insulators Association
3557 Nicklaus Dr., Titusville, FL 32780

National Labor Relations Board
1717 Pennsylvania Ave. NW, Washington, DC 20570

National Landscape Association, Inc.
1250 I St. NW, Washington, DC 20005

National Lime Association
3601 N. Fairfax Dr., Arlington, VA 22201

National Limestone Institute
1300 L St. NW, Washington, DC 20005

National Oak Flooring Manufacturing Association
PO Box 3009, Memphis, TN 38173

National Ornamental and Miscellaneous
Metals Association
804-10 Main St., Forest Park, GA 30050

National Paint and Coatings Association
1500 Rhode Island Ave. NW, Washington, DC 20005

National Paint Distributors
701 Lee St., Des Plaines, IL 60016

National Particleboard Association
18928 Premiere Ct., Gaithersburg, MD 20879

National Precast Concrete Association
825 E. 64th St., Indianapolis, IN 46220

National Propane Gas Association
1600 Eisenhower Lane, Lisle, IL 60532

National Ready Mixed Concrete Association
900 Spring St., Silver Spring, MD 20910

National Roof Deck Contractors Association
600 S. Federal St., Chicago, IL 60605

National Roofing Contractors Association
10255 W. Higgins Rd., Rosemont, IL 60018

National Slag Association
300 S. Washington St., Alexandria, VA 22314

National Society of Professional Engineers
1420 King St., Alexandria, VA 22314

National Society of Professional Sanitarians
1224 Hoffman Dr., Jefferson, City, MO 65101

National Society of Professional Surveyors
5410 Grosvenor Lane, Bethesda, MD 20814

National Swimming Pool Foundation
10803 Golfdale, San Antonio, TX 78216

National Terrazzo and Mosaic Association, Inc.
3166 Des Plaines Ave., Des Plaines, IL 60018

National Water Resource Association
3800 N. Fairfax Dr., Arlington, VA 22203

National Water Well Association
6375 Riverside Dr., Dublin, OH 43017

National Wood Flooring Association
11046 Manchester Rd., St. Louis, MO 63122

National Wood Window and Door Association
1400 East Touhy Ave., Des Plaines, IL 60018

Northeastern Lumber Manufacturers Association
272 Tuttle Rd., Box 87A, Cumberland Center, ME 04021

Northern Textile Association
230 Congress St., Boston, MA 02110

Occupational Safety and Health Review Commission
1825 K St. NW, Washington, DC 20006

Painting and Decorating Contractors of America
3913 Old Lee Highway, Fairfax, VA 22030

Perlite Institute
88 New Dorp Plaza, Staten Island, NY 10306

Pipe Fabrication Institute
PO Box 173, Springdale, PA 15144

Pipe Line Contractors Association
1700 Pacific Ave., Dallas, TX 75201

Plastics Pipe Institute
Wayne Interchange Plaza II
155 Route 46 W., Wayne, NJ 07470

Plumbing and Drainage Institute
1106 West 77th St. S. Dr., Indianapolis, IN 46260

Plumbing Manufacturers Institute
800 Roosevelt Rd., Glen Ellyn, IL 60137

Plywood Research Foundation
PO Box 11700, Tacoma, WA 98411

Porcelain Enamel Institute
1101 Connecticut Ave. NW, Washington, DC 20006

Portland Cement Association
5420 Old Orchard Rd., Skokie, IL 60077

Power Tool Institute
PO Box 818, Yachats, OR 97498

Prestressed Concrete Institute
175 West Jackson Blvd., Chicago, IL 60604

Radiation Research Society
1819 Preston White Dr., Weston, VA 22091

Resilient Floor Covering Institute
966 Hungerford Dr., Rockville, MD 20805

Rubber Manufacturers Association
1400 K St. NW, Washington, DC 20005

Scaffolding, Shoring and Forming Institute
2130 Keith Bldg., Cleveland, OH 44115

Sealant Waterproofing and Restoration Institute
3101 Broadway, Kansas City, MO 64111

Sealed Insulating Glass Manufacturing Association
111 East Wacker Dr., Chicago, IL 60601

Seismological Society of America
El Cerrito Professional Building Suite 201
El Cerrito, CA 94530

Sheet Metal and Air Conditioning Contractors'
  National Association
PO Box 70, Merrifield, VA 22116

Sheet Metal Workers International Association
1750 New York Ave. NW, Washington, DC 20006

Society of the Plastics Industry
1275 K St. NW, Washington, DC 20005

Society of Wood Science and Technology
One Gifford Pinchot Dr., Madison, WI 53705

Soil and Water Conservation Society
7515 NE Ankeny Rd., Ankeny, IA 50021

Soil Science Society of America
677 S. Segoe Rd., Madison, WI 53711

Southern Building Code Congress International
900 Montclair Rd., Birmingham, AL 35213

Southern Cypress Manufacturers Association
2831 Airways Blvd., Memphis, TN 38132

Southern Forest Products Association
PO Box 52468, New Orleans, LA 70152

Southern Pine Inspection Bureau
4709 Scenic Highway, Pensacola, FL 32504

Southern Woodwork Association
PO Box 148, Macon, GA 31202

Specialty Steel Industry of the United States
3050 K St. NW, Washington, DC 20007

Specialty Tools and Fasteners Distributors Association
500 Elm Grove Rd., Box 44, Elm Grove, WI 53122

Stained Glass Association of America
4050 Broadway, Kansas City, MO 64111

Standards Committee TI—Telecommunications
5430 Grosvenor Lane, Bethesda, MD 20814

Standards Engineering Society
11 West Monument Ave., Dayton, OH 45401

Steel Deck Institute
PO Box 9506, Canton, OH 44711

Steel Door Institute
30200 Detroit Rd., Cleveland, OH 44145

Steel Joist Institute
1205 48th Ave., Myrtle Beach, SC 29577

Steel Plate Fabricators Association
2400 S. Downing Ave., Westchester, IL 60154

Steel Structures Painting Council
4400 Fifth Ave., Pittsburgh, PA 15213

Steel Tank Institute
570 Oakwood Rd., Lake Zurich, IL 60047

Steel Tube Institute of North America
522 Westgate Tower, Cleveland, OH 44116

Steel Window Institute
1230 Keith Bldg. Cleveland, OH 44115

Surety Association of America
100 Wood Avenue S., Iselin, NJ 08830

Technology Administration
14th St. & Constitution Ave. NW, Washington, DC 20230

Telecommunications Industry Association
2001 Pennsylvania Ave. NW, Washington, DC 20006

Test Boring Association, Inc.
PO Box 5126, Old Bridge, NJ 08857

Tile Contractors Association of America
112 N. Alfred St., Alexandria, VA 22314

Tile Council of America
PO Box 326, Princeton, NJ 08542

Timber Products Manufacturers
951 E. 3rd Ave., Spokane, WA 99202

Tin Research Institute, Inc.
1353 Perry St., Columbus, OH 43201

Truck Trailer Manufacturers Association
1020 Princess St., Alexandria, VA 22314

Tube and Pipe Fabricators Association—International
5411 E. State St., Rockford, IL 61108

Underwriters Laboratories, Inc.
333 Pfingsten Rd., Northbrook, IL 60062

US Air Force Department
The Pentagon, Washington, DC 20330

US Army Department
The Pentagon, Washington, DC 20310

US Consumer Product Safety Commission
1111 18th St. NW, Washington, DC 20207

US Department of Agriculture
14th St. & Independence Ave. SW
Washington, DC 20250

US Department of Commerce Commodity Standards
  Division
14th and Constitution Ave. NW, Washington, DC 20230

US Department of Energy
1000 Independence Ave. SW, Washington, DC 20585

US Department of Labor
200 Constitution Ave. NW, Washington, DC 20503

US Government Printing Office
Superintendent of Documents
Washington, DC 20402

US Navy Department
The Pentagon, Washington, DC 20350

Vermiculite Association
600 S. Federal St., Chicago, IL 60605

Wall Covering Manufacturers Association, Inc.
355 Lexington Ave., New York, NY 10017

Water Pollution Control Federation
601 Wythe St., Alexandria, VA 22314

Welding Research Council
345 E. 47th St., New York, NY 10017

West Coast Lumber Inspection Bureau
PO Box 23145, Portland, OR 97223

Western Forest Industries Association
1500 SW Taylor St., Portland, OR 97205

Western Hardware Association
133 Second Ave. SW, Portland, OR 97204

Western Red Cedar Lumber Association
522 SW Fifth Ave., Portland, OR 97204

Western Wood Products Association
522 SW Fifth Ave., Portland, OR 97204

Window Coverings Association of America
1050 N. Lindbergh Blvd., St. Louis, MO 63132

Wire Reinforcement Institute
1760 Reston Parkway, Reston, VA 22090

Wood and Synthetic Flooring Institute
4415 W. Harrison St., Hillside, IL 60612

Wood Truss Council of America
111 E. Wacker Dr., Chicago, IL 60601

Woven Wire Products Association
2515 N. Nordica Ave., Chicago, IL 60635

1

Abbreviations for Scientific, Engineering, and Construction Terms

abrasive hardness	Ha	candlepower	cp
absolute	abs	capacitance	C
absolute temperature	T	carbon dioxide	CO_2
acre	spell out	Celsius (degree)	°C
acre-foot	acre-ft	center to center	c to c
air horsepower	air hp	centi	c
alternating current	ac	centigrade/Celsius	C
American Standard wire gauge	ASWG	centigram	cg
American wire gauge	AWG	centiliter	cl
ampere	A, amp	centimeter	cm
ampere-hour	Ah, amp-hr	centimeter-gram-second (system)	cgs
ampere per meter	A/m	chemical	chem
amplitude, an elliptic function	am	chemically pure	cp
angstrom	Å	circular	cir
antilogarithm	antilog	circular mil	cmil, cir mil
atmosphere	atm or spell out	coefficient	coef
atomic mass unit	amu	cologarithm	colog
atomic weight	at. wt	concentrate	conc
atto	a	conductance	C
average	avg	conductivity	cond
avoirdupois	avdp	constant	const
azimuth	az, α	continental horsepower	cont hp
		cord	cd
bar	spell out	cosecant	csc
barometer	bar.	cosine	cos
barrel	bbl	cotangent	cot
billion electron volts	BeV, GeV	coulomb	C or spell out
board feet (feet board measure)	fbm	counter electromotive force	cemf
boiler pressure	spell out	cubic	^{3}, cu
boiling point	bp	cubic centimeter	cm^3, cu cm
brake horsepower	bhp	cubic foot	ft^3, cu ft
brake horsepower-hour	bhp-hr	cubic feet per minute	cfm
Brinell hardness number	Bhn	cubic feet per second	cfs
British thermal unit	Btu	cubic inch	$in.^3$, cu in.
British thermal unit per hour	Btu/hr, Btuh	cubic meter	m^3, cu m
Brown and Sharp Gauge	B & SG	cubic micron	μ^3, cu μ
bushel	bu	cubic millimeter	mm^3, cu mm
		cubic yard	yd^3, cu yd
calorie	cal	current density	spell out
candela	cd	cycles per second	Hz, cps
candle	c or spell out	cylinder	cyl
candle-hour	c-hr		

day	spell out	hectare	ha
deca	da	hecto	h
deci	d	henry	H
decibel	dB	high frequency	hf
degree	deg, °	high-pressure	h-p
degree Celsius	°C	high tension	ht
degree centigrade	°C	high voltage	hv
degree Fahrenheit	°F	horsepower	hp or spell out
degree Kelvin	°K	horsepower-hour	hp-hr
delta amplitude, an elliptic function	dn	hour	hr, h, or spell out
		hundred	C
depth or least dimension	d	hundredweight (112 lb)	cwt
diameter	diam or spell out	hyperbolic cosine	cosh
direct current	dc	hyperbolic sine	sinh
dram	dr	hyperbolic tangent	tanh
dyne	dyn		
		impact noise rating	INR
efficiency	eff	inch	in.
electric	elec	inch-pound	in.-lb
electromotive force	emf	inches per second	ips
electron volt	eV	indicated horsepower	ihp
elevation	el	indicated horsepower-hour	ihp-hr
equation	eq	infrared	ir
equivalent direct radiation	EDR	inside diameter	ID
erg	spell out	intermediate-pressure	i-p
external	ext	internal	int
		iron pipe size	I.P.S.
Fahrenheit	F		
farad	F or spell out	joule	J
fathom	spell out		
feet or foot	ft	kelvin	K
feet board measure (board feet)	fbm	kilo	k
feet per minute	fpm	kilocalorie	kcal
feet per second	fps	kilocycles per second	kc
femto	f	kilogram	kg
film conductance	F	kilogram-calorie	kg-cal
fire standpipe	fsp	kilogram-meter	kg-m
fluid	fl	kilograms per cubic meter	kg/m^3
fluid ounce	fl oz	kilograms per second	kgps
foot	ft	kilohertz	kHz
foot-candle	ft-c	kilojoule	kJ
foot-lambert	ft-L	kilometer	km
foot-pound	ft-lb	kilometer per second	kmps
foot-pound-second (system)	fps	kilonewton	kN
foot-second, *see* cubic foot per second		kilopascal	kPa
		kilovolt	kV
force of gravity	g	kilovolt-ampere	kVa
for example	e.g.	kilowatt	kW
freezing point	fp	kilowatt-hour	kWh
frequency	spell out	knott	spell out
frequency modulation	FM	Krebs unit	KU
fusion point	fnp		
		lambert	L or spell out
gallon	gal	latitude	lat, ϕ
gallons per hour	gph	least common multiple	lcm
gallons per minute	gpm	length	l
gallons per second	gps	licence	lic
gauss	G	linear foot	lin ft
giga	G	liquid	liq
gram	g or spell out	liter	spell out
gram-calorie	g-cal	logarithm, common	log
grams per cubic centimeter	g/cm^3	logarithm, natural	log$_e$, ln
grams per square foot	g/ft^2	longitude	long, λ

low-pressure	l-p	ohm-centimeter	Ω-cm, ohm-cm
low tension	lt	on center	o.c.
lumen	lm or spell out	ounce	oz
lumen-hour	lm-hr	ounce-foot	oz-ft
lumen per watt	lm/W, lpw	ounce-inch	oz-in.
lux	lx or spell out	outside diameter	OD
magnetomotive force	mmf	parts per million	ppm
mass	M or spell out	pascal	Pa
mathematics, -al	math	peck	pk
maximum	max	penny, pence	d
maxwell	Mx	pennyweight	dwt
mean effective pressure	mep	pico	p
mega	M	picofarad	pF
megacycle	Mc or spell out	pint	pt
megagram	Mg	poise	P or spell out
megohm	spell out	polyethylene	PE
melting point	mp	polyvinyl acetate	PVA
meter	m	polyvinyl chloride	PVC
meter-kilogram	m-kg	potential	spell out
mho	spell out	potential difference	spell out
micro	μ	pound	lb
microampere	μA, μamp	pound-foot	lb-ft
microbar	μbar	pound-inch	lb-in.
microfarad	μF	pounds per brake	
microinch	μin.	horsepower-hour	lb/bhp-hr
micrometer	μm	pounds per cubic foot	lb/ft^3, lb/cu ft
micron	μ	pounds per square foot	psf
microvolt	μV	pounds per square inch	psi
microwatt	mW	pounds per square inch absolute	psia
mile	spell out	pounds per square inch gauge	psig
miles per hour	mph, miles/hr	power factor	pf or spell out
miles per hour per second	mphps	pyrometric cone equivalent	P.C.E.
milli	m		
milliampere	mA, ma	quart	qt
millibar	mbar		
milligram	mg	radian	spell out
millihenry	mH	reactive kilovolt-ampere	kvar
millilambert	mL	reactive volt-ampere	var
milliliter	ml	revolutions per minute	rpm
millimeter	mm	revolutions per second	rps
millimicron	mμ	roentgen	R
million	spell out	root mean square	rms
million electron volts	MeV		
million gallons per day	mgd	secant	sec
millivolt	mV	second	spell out
minimum	min	as angular measure	$''$
minute	spell out	second-foot, *see* cubic feet per	
as angular measure	$'$	second	
mole	spell out	shaft horsepower	shp
molecular weight	mol wt	siemens	S
		sine of the amplitude, an elliptic	sn
nano	n	function	
nanometer	nm	sound transmission class	STC
nautical mile	spell out	specific gravity	sp gr
neutron	n	specific heat	sp ht
newton	N	spherical candlepower	scp
noise factor	nf	square	^{2}, sq
number	no.	square centimeter	cm^2, sq cm
		square foot	ft^2, sq ft
oersted	Oe $\rightarrow$	square inch	in.^{2}, sq in.
ohm	Ω or spell out	square hectometer	hm^2, sq hm

square kilometer	km^2, sq km	ton	spell out
square meter	m^2, sq m	ton-mile	spell out
square micron	μ^2, sq μ		
square millimeter	mm^2, sq mm	ultraviolet	uv
standard	std	ultrahigh frequency	uhf
standard temperature and			
pressure	STP	versed sine	vers
steam working pressure	swp	volt	V
steradian	sr	volt-ampere	VA
stere	s	volt-coulomb	VC or spell out
		volume units	vu
Tagliabue	Tag		
tangent	tan or spell out	water vapor transmission	WVT
temperature	temp	water working pressure	wwp
tensile strength	ts	watt	W
tesla	T	watt-hour	Wh
tera	T	watts per candle	wpc
thousand	M	weber	Wb
thousand foot-pounds	kip-ft	weight	wt
thousand pounds	kip	yard	yd

2

Weights and Measures

U.S. SYSTEM

LIQUID MEASURE

4	gills	= 1 pint (pt)
2	pints	= 1 quart (qt)
4	quarts	= 1 gallon (gal)
31.5	gallons	= 1 barrel (bbl)
2	barrels	= 1 hogshead
60	minims	= 1 fluid dram (fl dr)
8	fluid drams	= 1 fluid ounce (fl oz)
16	fluid ounces	= 1 pint

SQUARE MEASURE

144	square inches (sq in)	= 1 square foot (sq ft)
9	square feet	= 1 square yard (sq yd)
30.25	square yards	= 1 square rod (sq rd)
160	square rods	= 1 acre (A)
640	bcres	= 1 square mile (sq mi)

AVOIRDUPOIS WEIGHT

27.34	grains (gr)	= 1̄ dram (dr av)
16	drams	= 1 ounce (oz av)
16	ounces	= 1 pound (lb av)
200	pounds	= 1 short ton (sh tn)
2240	pounds	= 1 long ton (1 tn)

CUBIC MEASURE

144 cubic inches (cu in)	= 1 board foot (bd ft)
1728 cubic inches	= 1 cubic foot (cu ft)
27 cubic feet	= 1 cubic yard (cu yd)
128 cubic feet	= 1 cord (cd)

LINEAR MEASURE

1	mil	= 0.001 inch (in)
12	inches	= 1 foot (ft)
3	feet	= 1 yard (yd)
6	feet	= 1 fathom
5.5	yards	= 1 rod (rd)
40	rods	= 1 furlong
5280	feet	= 1 mile (mi)
1760	yards	= 1 mile

APOTHECARIES' WEIGHT

20 grains (gr.)	= 1 scruple
3 scruples	= 1 dram (dr)
8 drams	= 1 ounce (oz)
12 ounces	= 1 pound (lb)

TROY WEIGHT

24 grains	= 1 pennyweight (dwt)
20 pennyweight	= 1 ounce (oz t)
12 ounces	= 1 pound (lb t)

DRY MEASURE

2	pints	= 1 quart (qt)
8	quarts	= 1 peck (pk)
4	pecks	= 1 bushel (bu)
3.28	bushels	= 1 barrel (bbl)

SYMBOLS

lb, #	pounds(s)
'	foot, feet
"	inch(es)

APOTHECARIES' MEASURE

lb	pounds(s)
℥	dram(s)
℥	ounce(s)
℈	scruple(s)
M,ɱ,M	minim
i, j	one: used as a coefficient

METRIC SYSTEM

LENGTH

Unit		Metric Equivalent		U.S. Equivalent	
millimeter	(mm)	0.001	meter	0.03937	inch
centimeter	(cm)	0.01	meter	0.3937	inch
decimeter	(dm)	0.1	meter	3.937	inches
METER	(m)	1.0	meter	39.37	inches
dekameter	(dkm)	10.0	meters	10.93	yards
hectometer	(hm)	100.0	meters	328.08	feet
kilometer	(km)	1000.0	meters	0.6214	mile

WEIGHT OR MASS

Unit		Metric Equivalent		U.S. Equivalent	
milligram	k(mg)	0.001	gram	0.0154	grain
centigram	(cg)	0.01	gram	0.1543	grain
decigram	k(dg)	0.1	gram	1.543	grains
GRAM	(g)	1.0	gram	15.43	grains
dekagram	(dkg)	10.0	grams	0.3527	ounce avoirdupois
hectogram	(hg)	100.0	grams	3.527	ounces avoirdupois
kilogram	(kg)	1000.0	grams	2.2	pounds avoirdupois

CAPACITY

Unit		Metric Equivalent		U.S. Equivalent	
milliliter	(ml)	0.001	liter	0.034	fluid ounce
centiliter	(cl)	0.01	liter	0.338	fluid ounce
deciliter	(dl)	0.1	liter	0.38	fluid ounces
LITER	(l)	1.0	liter	1.05	liquid quarts
dekaliter	(dkl)	10.0	liters	0.284	bushel
hectoliter	(hl)	100.0	liters	2.837	bushels
kiloliter	(kl)	1000.0	liters	264.18	gallons

AREA

Unit		Metric Equivalent		U.S. Equivalent	
square millimeter	(mm^2)	0.000001	centare	0.00155	square inch
square centimeter	(cm^2)	0.001	centare	0.155	square inch
square decimeter	(dm^2)	0.01	centare	15.5	square inches
CENTARE *also*	(ca)	1.0	centare	10.76	square feet
square meter	(m^2)				
are *also*	(a)	100.0	centares	0.0247	acre
square dekameter	(dkm^2)				
hectare *also*	(ha)	10,000.0	centares	2.47	acres
square hectometer	(hm^2)				
square kilometer	(km^2)	1,000,000.0	centares	0.386	square mile

VOLUME

Unit		Metric Equivalent		U.S. Equivalent	
cubic millimeter	(mm^3)	0.001	cubic centimeter	0.016	minim
cubic centimeter	(cc, cm^3)	0.001	cubic decimeter	0.061	cubic inch
cubic decimeter	(dm^3)	0.001	cubic meter	61.023	cubic inches
STERE *also*	(s)	1.0	cubic meter	1.308	cubic yards
cubic meter	(m^3)				
cubic dekameter	(dkm^3)	1000.0	cubic meters	1307.943	cubic yards
cubic hectometer	(hm^3)	1,000,000.0	cubic meters	1,307,942.8	cubic yards
cubic kilometer	(km^3)	1,000,000,000.0	cubic meters	0.25	cubic mile

SYMBOLS

g	10^9 times (a unit); giga-
m	10^6 times (a unit); mega-
k	10^3 times (a unit); kilo-
h	10^2 times (a unit); hecto-
dk	10 times (a unit); decka-
d	10^{-1} times (a unit); deci-
c	10^{-2} times (a unit); centi-
m	10^{-3} times (a unit); milli-
μ	10^{-4} times (a unit); micro-
n	10^{-9} times (a unit); nano-
$\mu\mu$	10^{-12} times (a unit); micromicro-
Å, λ	Angstrom unit
$\mu\mu$	micromicron
μ	micron

Carpentry Abbreviations

air-dried	A.D	kiln-dried	K.D.
all lengths	a.l.	kilo-pound (1,000 pounds)	kip
		knocked-down	k.d.
beam and stringer	B&S		
better	btr.	less carload lots	L.C.L.
board foot	bd ft	lumber	lbr.
board (foot) measure	bm	post and timber	P&T
center matched	CM	random lengths	r.l.
clear	clr.	resawed	res.
common	com.	ripped	rip.
		sapwood	Sap.
dimension	dim.	softwood	Sftwd.
dressed and headed—that is,		square edge	Sq.E.
dressed one or two sides and		standard matched	SM
worked to tongue and groove		surfaced and matched—that is,	
joints on both the edge and		surfaced on one or two sides	
the ends	D&H	and tongued and grooved on	
dressed and matched—that is,		the edges—with the match	
dressed one or two sides and		either center or standard	S&M
tongued and grooved on the		surfaced four sides	S4S
edges; the match may be		surfaced four sides with a calking	
center or standard	D&M	seam on each edge	S4SCS
dressed one or two sides and		surfaced one edge	S1E
standard matched	D&SM	surfaced one or two sides and	
dressed two sides and center		center matched	S&CM
matched	D2S&CM	surfaced one or two sides and	
dressed two sides and center or		standard matched	S&SM
standard matched	D2S&M	surfaced one side	S1S
dressed two sides and standard		surfaced one side and one edge	S1S1E
matched	D2S&SM	surfaced one side and two edges	S1S2E
		surfaced two edges	S2E
edge grain	E.G.	surfaced two sides	S2S
ends center matched	ECM	surfaced two sides and center	
ends matched, either center or		matched	S2S&CM
standard	EM	surfaced two sides and (center or	
ends standard matched	ESM	standard) matched	S2S&M
		surfaced two sides and one edge	S2S1E
flat grain	F.G.	surfaced two sides and standard	
		matched	S2S&SM
hardwood	Hdwd.	thousand (feet) board measure	Mbm
heartwood	Hrtwd.	timbers	Tbrs.
		tongued and grooved	T&G
joist and plank	J&P	vertical grain	V.G.

INDEX

Boldface numbers in index indicate the Division number.